AF889472

24 Years

2025-2013 JEE Main & 2012-2002 AIEEE

JEE Main

Chapterwise Topicwise Solved Papers

PHYSICS

Preparation Strategy
& Analysis Videos

24 Years

2025-2013 JEE Main & 2012-2002 AIEEE

JEE Main

Chapterwise Topicwise Solved Papers

PHYSICS

Authors
Dharmendra Singh
Sumit Shukla

arihant
SUCCESS AND NOTHING LESS

SUCCESS AND NOTHING LESS

ARIHANT PRAKASHAN (Series), MEERUT

Administrative & Production Offices

Regd. Office

'Ramchhaya' 4577/15, Agarwal Road, Darya Ganj, New Delhi -110002
Tele: 011- 47630600, 43518550

Head Office

Kalindi, TP Nagar, Meerut (UP) - 250002
Tel: 0121-7156203, 7156204

Sales & Support Offices

Agra, Ahmedabad, Bengaluru, Bareilly, Chennai, Delhi, Guwahati, Hyderabad, Jaipur, Jhansi, Kolkata, Lucknow, Nagpur & Pune.

PRICE ₹810.00

PO No : TXT-XX-XXXXXXX-X-XX

Published by Arihant Publications (India) Ltd.

For further information about the books published by Arihant, log on to www.arihantbooks.com or e-mail at info@arihantbooks.com

Follow us on

Preface

JEE Main is a gateway examination for candidates expecting to seek admission in Bachelor in Engineering (BE), Bachelor of Technology (B.Tech) and Bachelor of Architecture (B.Arch) at Indian Institutes of Information Technology (IIITs), National Institutes of Technology (NITs), Delhi Technological University and other Centrally Funded Technical Institutes (CFTIs).

JEE Main is also an examination which is like screening examination for JEE Advanced (The gateway examination to India's most reputed Technical Institutes, Indian Institutes of Technology— IITs).

To make the students well-versed with the pattern as well as the level of the questions asked in the exam, this book contains **Chapterwise Topicwise Solutions of Questions asked in Last 24 Years' Examinations of JEE Main** (formerly known as AIEEE).

The Unique and exclusive features of this book are:

- Questions are further sorted according to their difficulty level in each topic.
- Questions on frequently asked concepts are marked with concept name to get edge over others.

Solutions to all the questions have been kept very detailed and accurate with ingredient of special features like Key idea, Concept Enhancer & Alternate/Time saver techniques.

We hope this book would be highly beneficial for the students. We would be grateful if any discrepancy or mistake in the questions or answers is brought to our notice so that these could be rectified in subsequent editions.

Publisher

Contents

Class 11

Class 12

SYLLABUS

Physics

UNIT 1 Physics and Measurement

Units of measurements, System of Units, S I Units, fundamental and derived units, least count, significant figures, Errors in measurements, Dimensions of Physics quantities, dimensional analysis, and its applications.

UNIT 2 Kinematics

The frame of reference, motion in a straight line, Position- time graph, speed and velocity; Uniform and non-uniform motion, average speed and instantaneous velocity, uniformly accelerated motion, velocity-time, position-time graph, relations for uniformly accelerated motion, Scalars and Vectors, Vector. Addition and subtraction, scalar and vector products, Unit Vector, Resolution of a Vector. Relative Velocity, Motion in a plane, Projectile Motion, Uniform Circular Motion.

UNIT 3 Laws of Motion

Force and inertia, Newton's First law of motion; Momentum, Newton's Second law of motion, Impulses; Newton's Third law of motion. Law of conservation of linear momentum and its applications. Equilibrium of concurrent forces.
Static and Kinetic friction, laws of friction, rolling friction.
Dynamics of uniform circular motion: centripetal force and its applications: vehicle on a level circular road, vehicle on a banked road.

UNIT 4 Work, Energy and Power

Work done by a constant force and a variable force; kinetic and potential energies, work-energy theorem, power.
The potential energy of spring conservation of mechanical energy, conservative and non-conservative forces;
motion in a vertical circle: Elastic and inelastic collisions in one and two dimensions.

UNIT 5 Rotational Motion

Centre of the mass of a two-particles system, Centre of the mass of a rigid body; Basic concepts of rotational motion; moment of a force; torque, angular momentum, conservation of angular momentum and its applications;
The moment of inertia, the radius of gyration, values of moments of inertia for simple geometrical objects, parallel and perpendicular axes theorems and their applications.
Equilibrium of rigid bodies, rigid body rotation and equations of rotational motion, comparison of linear and rotational motions.

UNIT 6 Gravitation

The universal law of gravitation. Acceleration due to gravity and its variation with altitude and depth. Kepler's law of planetary motion. Gravitational potential energy; gravitational potential. Escape velocity, Motion of a satellite, orbital velocity, time period and energy of satellite.

UNIT 7 Properties of Solids and Liquids

Elastic behaviour, Stress-strain relationship, Hooke's Law. Young's modulus, bulk modulus, and modulus of rigidity. Pressure due to a fluid column; Pascal's law and its applications. Effect of gravity on fluid pressure.

Viscosity, Stokes' law, terminal velocity, streamline, and turbulent flow, critical velocity. Bernoulli's principle and its applications. Surface energy and surface tension, angle of contact, excess of pressure across a curved surface,

application of surface tension - drops, bubbles, and capillary rise. Heat, temperature, thermal expansion; specific heat capacity, calorimetry; change of state, latent heat. Heat transfer conduction, convection and radiation.

UNIT 8 Thermodynamics

Thermal equilibrium, zeroth law of thermodynamics, the concept of temperature. Heat, work and internal energy. The first law of thermodynamics, isothermal and adiabatic processes. The second law of thermodynamics: reversible and irreversible processes.

UNIT 9 Kinetic Theory of Gases

Equation of state of a perfect gas, work done on compressing a gas, Kinetic theory of gases -assumptions, the concept of pressure. Kinetic interpretation of temperature: RMS speed of gas molecules: Degrees of freedom. Law of equipartition of energy and applications to specific heat capacities of gases; Mean free path. Avogadro's number.

UNIT 10 Oscillations and Waves

Oscillations and periodic motion – time period, frequency, displacement as a function of time. Periodic functions. Simple harmonic motion (SHM) and its equation; phase: oscillations of a spring -restoring force and force constant: energy in SHM - Kinetic and potential energies; Simple pendulum - derivation of expression for its time period:

Wave motion. Longitudinal and transverse waves, speed of the travelling wave. Displacement relation for a progressive wave. Principle of superposition of waves, reflection of waves. Standing waves in strings and organ pipes, fundamental mode and harmonics. Beats.

UNIT 11 Electrostatics

Electric charges: Conservation of charge. Coulomb's law forces between two point charges, forces between multiple charges: superposition principle and continuous charge distribution.

Electric field: Electric field due to a point charge, Electric field lines. Electric dipole, Electric field due to a dipole. Torque on a dipole in a uniform electric field

Electric flux. Gauss's law and its applications to find field due to infinitely long uniformly charged straight wire uniformly charged infinite plane sheet and uniformly charged thin spherical shell. Electric potential and its calculation for a point charge, electric dipole and system of charges; potential difference, Equipotential surfaces, Electrical potential energy of a system of two point charges and of electric dipole in an electrostatic field.

Conductors and insulators: Dielectrics and electric polarization, capacitors and capacitances, the combination of capacitors in series and parallel, and capacitance of a parallel plate capacitor with and without dielectric medium between the plates. Energy stored in a capacitor.

UNIT 12 Current Electricity

Electric current. Drift velocity, mobility, and their relation with electric current. Ohm's law, Electrical resistance. V-I characteristics of Ohmic and non-ohmic conductors. Electrical energy and power. Electrical resistivity and conductivity. Series and parallel combinations of resistors; Temperature dependence of resistance.

Internal resistance, potential difference, and emf of a cell, a combination of cells in series and parallel. Kirchhoff's laws and their applications. Wheatstone bridge. Metre bridge.

UNIT 13 Magnetic Effects of Current and Magnetism

Biot - Savart' law and its application to the current-carrying circular loop. Ampere's law and its applications to infinitely long current carrying straight wire and solenoid. Force on a moving charge in uniform magnetic and electric fields.

Force on a current-carrying conductor in a uniform magnetic field. The force between two parallel currents carrying conductors-definition of ampere. Torque experienced by a current loop in a uniform magnetic field: Moving coil galvanometer, its sensitivity, and conversion to ammeter and voltmeter.

Current loop as a magnetic dipole and its magnetic dipole moment. Bar magnet as an equivalent solenoid, magnetic field lines; Magnetic field due to a magnetic dipole (bar magnet) along its axis and perpendicular to its axis. Torque on a magnetic dipole in a uniform magnetic field. Para, dia and ferromagnetic substances with examples, the effect of temperature on magnetic properties.

UNIT 14 Electromagnetic Induction and Alternating Currents

Electromagnetic induction: Faraday's law. Induced emf and current: Lenz's Law, Eddy currents. Self and mutual inductance. Alternating currents, peak and RMS value of alternating current/ voltage: reactance and impedance: LCR series circuit, resonance: power in AC circuits, wattless current. AC generator and transformer.

UNIT 15 Electromagnetic Waves

Displacement current. Electromagnetic waves and their characteristics, Transverse nature of electromagnetic waves, Electromagnetic spectrum (radio waves, microwaves, infrared, visible, ultraviolet, X-rays, Gamma rays), Applications of EM waves.

UNIT 16 Optics

Reflection of light, spherical mirrors, mirror formula. Refraction of light at plane and spherical surfaces, thin lens formula, and lens maker formula. Total internal reflection and its applications. Magnification. Power of a Lens. Combination of thin lenses in contact. Refraction of light through a prism. Microscope and Astronomical Telescope (reflecting and refracting) and their magnifying powers.

Wave optics: wavefront and Huygens' principle. Laws of reflection and refraction using Huygens principle. Interference, Young's double-slit experiment, and expression for fringe width, coherent sources, and sustained interference of light. Diffraction due to a single slit, width of central maximum. Polarization, plane-polarized light: Brewster's law, uses of plane-polarized light and Polaroid.

UNIT 17 Dual Nature of Matter and Radiation

Dual nature of radiation. Photoelectric effect. Hertz and Lenard's observations; Einstein's photoelectric equation: particle nature of light. Matter waves-wave nature of particle, de Broglie relation.

UNIT 18 Atoms and Nuclei

Alpha-particle scattering experiment; Rutherford's model of atom; Bohr model, energy levels, hydrogen spectrum. Composition and size of nucleus, atomic masses, Mass-energy relation, mass defect; binding energy per nucleon and its variation with mass number, nuclear fission, and fusion.

UNIT 19 Electronic Devices

Semiconductors; semiconductor diode: I-V characteristics in forward and reverse bias; diode as a rectifier; I-V characteristics of LED, the photodiode, solar cell, and Zener diode; Zener diode as a voltage regulator. Logic gates (OR, AND, NOT, NAND and NOR).

UNIT 20 Experimental Skills

Familiarity with the basic approach and observations of the experiments and activities:

1. Vernier calipers -its use to measure the internal and external diameter and depth of a vessel.
2. Screw gauge - its use to determine the thickness/ diameter of thin sheet/wire.
3. Simple Pendulum-dissipation of energy by plotting a graph between the square of amplitude and time.
4. Metre Scale - the mass of a given object by the principle of moments.
5. Young's modulus of elasticity of the material of a metallic wire.
6. Surface tension of water by capillary rise and effect of detergents.
7. Co-efficient of Viscosity of a given viscous liquid by measuring the terminal velocity of a given spherical body.
8. Speed of sound in air at room temperature using a resonance tube.
9. Specific heat capacity of a given (i) solid and (ii) liquid by method of mixtures.
10. The resistivity of the material of a given wire using a metre bridge.
11. The resistance of a given wire using Ohm's law.
12. Resistance and figure of merit of a galvanometer by half deflection method.
13. The focal length of;
    (i) Convex mirror
    (ii) Concave mirror
    (iii) Convex lens, using the parallax method.
14. The plot of the angle of deviation vs angle of incidence for a triangular prism.
15. The refractive index of a glass slab using a travelling microscope.
16. Characteristic curves of a p-n junction diode in forward and reverse bias.
17. Characteristic curves of a Zener diode and finding reverse breakdown voltage.
18. Identification of Diode. LED, Resistor. A capacitor from a mixed collection of such items.

C-01

JEE Main Chapterwise Topicwise Physics

In JEE Main there is usually 0-1 question per paper is visible from this chapter. In past 3 years, 66 conceptual and 3 numerical questions were asked of easy to moderate level from this chapter.
For detailed analysis scan QR code

UNITS AND MEASUREMENT

Topic 1 Physical Quantities and Their Units

Level 1

01 The unit of $\sqrt{\frac{2I}{\varepsilon_0 c}}$ is (I = intensity of an electromagnetic wave, c = speed of light) *[2025, 7 April Shift-II]*

(a) NC (b) Nm
(c) NC^{-1} (d) Vm

02 In an expression $a \times 10^b$ *[2024, 8 April Shift-I]*

(a) b is order of magnitude for $a \geq 5$
(b) b is order of magnitude for $a \leq 5$
(c) a is order of magnitude for $b \leq 5$
(d) b is order of magnitude for $5 < a \leq 10$

03 Given below are two statements

Statement I Astronomical unit (AU), Parsec (pc) and Light year (ly) are units for measuring astronomical distances.

Statement II AU < Parsec (pc) < ly

In the light of above statements, choose the most appropriate answer from the options given below. *[2023, 11 April Shift-I]*

(a) Both Statement I and Statements II are correct.
(b) Statement I is correct but Statement II is incorrect.
(c) Statement I is incorrect but Statement II is correct.
(d) Both Statement I and Statement II are incorrect.

04 Match List I with List II.

List I	List II
A. Surface tension	I. $kgm^{-1}s^{-1}$
B. Pressure	II. $kg\,ms^{-1}$
C. Viscosity	III. $kgm^{-1}s^{-2}$
D. Impulse	IV. kgs^{-2}

Choose the correct answer from the options given below. *[2023, 25 Jan Shift-I]*

(a) A-III, B-IV, C-I, D-II
(b) A-IV, B-III, C-I, D-II
(c) A-II, B-I, C-III, D-IV
(d) A-IV, B-III, C-II, D-I

05 If E and H represent the intensity of electric field and magnetising field respectively, then the unit of E/H will be *[2021, 27 Aug Shift-I]*

(a) ohm (b) mho (c) joule (d) newton

06 A metal sample carrying a current along X-axis with density J_x is subjected to a magnetic field B_z (along Z-axis). The electric field E_y developed along Y-axis is directly proportional to J_x as well as B_z. The constant of proportionality has SI unit *[JEE Main 2013]*

(a) $\frac{m^2}{A}$ (b) $\frac{m^3}{As}$ (c) $\frac{m^2}{As}$ (d) $\frac{As}{m^3}$

07 The rad is the correct unit used to report the measurement of *[AIEEE 2006]*

(a) the ability of a beam of gamma ray photons to produce ions in a target
(b) the energy delivered by radiation to a target
(c) the biological effect of radiation
(d) the rate of decay of a radioactive source

Level 2

08 Match List I with List II. *[2025, 2 April Shift-II]*

List I	List II
A. Heat capacity of body	I. $J\,kg^{-1}$
B. Specific heat capacity of body	II. $J\,K^{-1}$
C. Latent heat	III. $J\,kg^{-1}\,K^{-1}$
D. Thermal conductivity	IV. $J\,m^{-1}\,K^{-1}\,s^{-1}$

Choose the correct answer from the options given below.
(a) (A)-(III), (B)-(IV), (C)-(I), (D)-(II)
(b) (A)-(II), (B)-(III), (C)-(I), (D)-(IV)
(c) (A)-(III), (B)-(I), (C)-(II), (D)-(IV)
(d) (A)-(IV), (B)-(III), (C)-(II), (D)-(I)

09 Match List-I with List-II.

List-I		List-II	
(A)	Magnetic induction	(I)	Ampere metre^2
(B)	Magnetic intensity	(II)	Weber
(C)	Magnetic flux	(III)	Gauss
(D)	Magnetic moment	(IV)	Ampere metre

Choose the correct answer from the options given below. **[2025, 29 Jan Shift-I]**
(a) (A)-(I), (B)-(II), (C)-(III), (D)-(IV)
(b) (A)-(III), (B)-(IV), (C)-(I), (D)-(II)
(c) (A)-(III), (B)-(II), (C)-(I), (D)-(IV)
(d) (A)-(III), (B)-(IV), (C)-(II), (D)-(I)

Imp. Derived Units

10 Match List I with List II.

List-I	List-II
A. Torque	I. kg $m^{-1}s^{-2}$
B. Energy density	II. kg ms^{-1}
C. Pressure gradient	III. kg $m^{-2}s^{-2}$
D. Impulse	IV. kg $m^2 s^{-2}$

Choose the correct answer from the options given below. **[2023, 30 Jan Shift-II]**
(a) A-IV, B-I, C-II, D-III
(b) A-IV, B-I, C-III, D-II
(c) A-IV, B-III, C-I, D-II
(d) A-I, B-IV, C-III, D-II

11 Electric field in a certain region is given by $E = \left(\frac{A}{x^2}\hat{i} + \frac{B}{y^3}\hat{j}\right)$. The SI unit of A and B are **[2023, 30 Jan Shift-I]**
(a) $Nm^2 C^{-1}, Nm^3 C^{-1}$ (b) $Nm^2 C, Nm^3 C$
(c) $Nm^3 C^{-1}, Nm^2 C^{-1}$ (d) $Nm C, Nm^2 C$

12 Match List I with List II.

List I	List II
A. Torque	I. Nms^{-1}
B. Stress	II. J kg^{-1}
C. Latent Heat	III. Nm
D. Power	IV. Nm^{-2}

Choose the correct answer from the options given below. **[2022, 29 July Shift-II]**
(a) A-III, B-II,C-I, D-IV (b) A-III, B-IV, C-II, D-I
(c) A-IV, B-I, C-III, D-II (d) A-II, B-III, C-I, D-IV

Level 3

13 Velocity v and acceleration a in two systems of units 1 and 2 are related as $v_2 = \frac{n}{m^2} v_1$ and $a_2 = \frac{a_1}{mn}$, respectively. Here, m and n are constants. The relations for distance and time in two systems respectively are: **[2022, 28 June Shift-II]**
(a) $\frac{n^3}{m^3} L_1 = L_2$ and $\frac{n^2}{m} T_1 = T_2$
(b) $L_1 = \frac{n^4}{m^2} L_2$ and $T_1 = \frac{n^2}{m} T_2$
(c) $L_1 = \frac{n^2}{m} L_2$ and $T_1 = \frac{n^4}{m^2} = T_2$
(d) $\frac{n^2}{m} L_1 = L_2$ and $\frac{n^4}{m^2} T_1 = T_2$

14 Match List I with List II.

List I	List II
A. R_H (Rydberg constant)	1. kg $m^{-1}s^{-1}$
B. h (Planck's constant)	2. kg m^2s^{-1}
C. μ_B (Magnetic field energy density)	3. m^{-1}
D. η (Coefficient of viscosity)	4. kg $m^{-1}s^{-2}$

Choose the most appropriate answer from the options given below. **[2021, 27 Aug Shift-II]**

	A	B	C	D		A	B	C	D
(a)	2	3	4	1	(b)	3	2	4	1
(c)	4	2	1	3	(d)	3	2	1	4

15 The density of a material in SI unit is 128 kg m^{-3}. In certain units in which the unit of length is 25 cm and the unit of mass is 50 g, the numerical value of density of the material is **[2019, 10 Jan Shift-I]**
(a) 40 (b) 16 (c) 640 (d) 410

Topic 2 Errors in Measurement and Significant Figures

Level 1

01 Match List I with List II.

	List I (Number)		List II (Significant figure)
A.	1001	I.	3
B.	010.1	II.	4
C.	100.100	III.	5
D.	0.0010010	IV.	6

Choose the correct answer from the options given below. **[2024, 1 Feb Shift-II]**
(a) A-II, B-I, C-IV, D-III
(b) A-III, B-IV, C-II, D-I
(c) A-IV, B-III, C-I, D-II
(d) A-I, B-II, C-III, D-IV

02 If the percentage errors in measuring the length and the diameter of a wire are 0.1% each. The percentage error in measuring its resistance will be *[2024, 31 Jan Shift-I]*

(a) 0.1% (b) 0.144%
(c) 0.2% (d) 0.3%

03 Given below are two statements, one is labelled as Assertion (A) and the other is labelled as Reason (R).

Assertion (A) A spherical body of radius (5 ± 0.1)mm having a particular density is falling through a liquid of constant density. The percentage error in the calculation of its terminal velocity is 4%.

Reason (R) The terminal velocity of the spherical body falling through the liquid is inversely proportional to its radius.

In the light of the above statements, choose the most appropriate answer from the options given below. *[2023, 13 April Shift-II]*

(a) Both A and R are true but R is not the correct explanation of A.
(b) A is true but R is false.
(c) A is false but R is true.
(d) Both A and R are true and R is the correct explanation of A.

Imp. Percentage Error

04 A cylindrical wire of mass (0.4 ± 0.01) g has length (8 ± 0.04) cm and radius (6 ± 0.03) mm. The maximum error in its density will be *[2023, 8 April Shift-I]*

(a) 5% (b) 3.5%
(c) 4% (d) 1%

05 If $Z = \dfrac{A^2B^3}{C^4}$, then the relative error in Z will be *[2022, 25 June Shift-I]*

(a) $\dfrac{\Delta A}{A} + \dfrac{\Delta B}{B} + \dfrac{\Delta C}{C}$ (b) $\dfrac{2\Delta A}{A} + \dfrac{3\Delta B}{B} - \dfrac{4\Delta C}{C}$

(c) $\dfrac{2\Delta A}{A} + \dfrac{3\Delta B}{B} + \dfrac{4\Delta C}{C}$ (d) $\dfrac{\Delta A}{A} + \dfrac{\Delta B}{B} - \dfrac{\Delta C}{C}$

06 The density of a material in the shape of a cube is determined by measuring three sides of the cube and its mass. If the relative errors in measuring the mass and length are respectively 1.5% and 1% , the maximum error in determining the density is *[JEE Main 2018]*

(a) 2.5% (b) 3.5%
(c) 4.5% (d) 6%

07 Resistance of a given wire is obtained by measuring the current flowing in it and the voltage difference applied across it. If the percentage errors in the measurement of the current and the voltage difference are 3% each, then error in the value of resistance of the wire is *[AIEEE 2012]*

(a) 6% (b) zero
(c) 1% (d) 3%

08 The respective number of significant figures for the numbers 23.023, 0.0003 and 2.1×10^{-3} are *[AIEEE 2010]*

(a) 5, 1, 2 (b) 5, 1, 5 (c) 5, 5, 2 (d) 4, 4, 2

Level 2

09. A quantity Q is formulated as $X^{-2}Y^{+\frac{3}{2}}Z^{-\frac{2}{5}}$. X, Y and Z are independent parameters which have fractional errors of 0.1, 0.2 and 0.5, respectively in measurement. The maximum fractional error of Q is *[2025, 8 April Shift-II]*

(a) 0.7 (b) 0.8 (c) 0.1 (d) 0.6

10. For an experimental expression $y = \dfrac{32.3 \times 1125}{27.4}$, where all the digits are significant. Then, to report the value of y we should write *[2025, 24 Jan Shift-I]*

(a) $y = 1330$ (b) $y = 1326.19$
(c) $y = 1326.186$ (d) $y = 1326.2$

11 The radius r, length l and resistance R of a metal wire was measured in the laboratory as $r = (0.35 \pm 0.05)$ cm, $R = (100 \pm 10)$ ohm and $l = (15 \pm 0.2)$ cm
The percentage error in resistivity of the material of the wire is *[2024, 1 Feb Shift-I]*

(a) 39.9% (b) 25.6% (c) 37.3% (d) 35.6%

12 A physical quantity Q is found to depend on quantities a, b, c by the relation $Q = \dfrac{a^4b^3}{c^2}$.
The percentage error in a, b and c are 3%, 4% and 5% respectively. Then, the percentage error in Q is *[2024, 29 Jan Shift-II]*

(a) 14% (b) 43% (c) 34% (d) 66%

13 A physical quantity P is given as $P = \dfrac{a^2b^3}{c\sqrt{d}}$.
The percentage error in the measurement of a, b, c and d are 1%, 2%, 3% and 4% respectively. The percentage error in the measurement of quantity P will be *[2023, 10 April Shift-I]*

(a) 14% (b) 13% (c) 16% (d) 12%

14 The maximum error in the measurement of resistance, current and time for which current flows in an electric circuit are 1%, 2% and 3%, respectively. The maximum percentage error in the detection of the dissipated heat will be *[2022, 25 July Shift-II]*

(a) 2 (b) 4 (c) 6 (d) 8

15 A silver wire has mass (0.6 ± 0.006) g, radius (0.5 ± 0.005) mm and length (4 ± 0.04) cm. The maximum percentage error in the measurement of its density will be *[2022, 27 June Shift-I]*

(a) 4% (b) 3% (c) 6% (d) 7%

16 A student determined Young's modulus of elasticity using the formula $Y = \frac{MgL^3}{4bd^3\delta}$. The value of g is taken to be 9.8 m/s^2, without any significant error, his observations are as following.

Physical quantity	Least count of the equipment used for measurement	Observed value
Mass (M)	1 g	2 kg
Length of bar (L)	1 mm	1 m
Breadth of bar (b)	0.1 mm	4 cm
Thickness of bar (d)	0.01 mm	0.4 cm
Depression (δ)	0.01 mm	5 mm

Then, the fractional error in the measurement of Y is *[2021, 1 Sep Shift-II]*

(a) 0.0083 (b) 0.0155 (c) 0.155 (d) 0.083

17 If the length of the pendulum in pendulum clock increases by 0.1%, then the error in time per day is *[2021, 26 Aug Shift-II]*

(a) 86.4 s (b) 4.32 s (c) 43.2 s (d) 8.64 s

18 A physical quantity z depends on four observables a, b, c and d, as $z = \frac{a^2 b^{2/3}}{\sqrt{c} d^3}$. The percentages of error in the measurement of a, b, c and d are 2%, 1.5%, 4% and 2.5% respectively. The percentage of error in z is *[2020, 5 Sep Shift-I]*

(a) 13.5% (b) 16.5% (c) 14.5% (d) 12.25%

Imp. Relative Error

19 The diameter and height of a cylinder are measured by a meter scale to be 12.6 ± 0.1 cm and 34.2 ± 0.1 cm, respectively. What will be the value of its volume in appropriate significant figures? *[2019, 10 Jan Shift-II]*

(a) $4300 \pm 80 \text{ cm}^3$ (b) $4260 \pm 80 \text{ cm}^3$

(c) $4264.4 \pm 81.0 \text{ cm}^3$ (d) $4264 \pm 81 \text{ cm}^3$

20 The current voltage relation of diode is given by $I = (e^{1000V/T} - 1)$ mA, where the applied voltage V is in volt and the temperature T is in kelvin. If a student makes an error measuring ± 0.01V while measuring the current of 5 mA at 300K, what will be the error in the value of current in mA? *[JEE Main 2013]*

(a) 0.2 mA (b) 0.02 mA (c) 0.5 mA (d) 0.05 mA

Level 3

21 A person measures mass of 3 different particles as 435.42 g. 226.3 g and 0.125 g. According to the rules for arithmetic operations with significant figures, the addition of the masses of 3 particles will be *[2025, 3 April Shift-I]*

(a) 662 g (b) 661.84 g

(c) 661.8 g (d) 661.845 g

22 The maximum percentage error in the measurement of density of a wire is

[Given, mass of wire = (0.60 ± 0.003) g

radius of wire = (0.50 ± 0.01) cm

length of wire = (10.00 ± 0.05) cm] *[2025, 22 Jan Shift-II]*

(a) 5 (b) 8 (c) 4 (d) 7

23 The energy of a system is given as $E(t) = \alpha^3 e^{-\beta t}$, where t is the time and $\beta = 0.3 \text{ s}^{-1}$. The errors in the measurement of α and t are 1.2% and 1.6%, respectively. At $t = 5$ s, maximum percentage error in the energy is *[2025, 23 Jan Shift-II]*

(a) 8.4% (b) 6% (c) 11.6% (d) 4%

24 The measured value of the length of a simple pendulum is 20 cm with 2 mm accuracy. The time for 50 oscillations was measured to be 40 s with 1 s resolution. From these measurements, the accuracy in the measurement of acceleration due to gravity is N%. The value of N is *[2024, 31 Jan Shift-II]*

(a) 4 (b) 6 (c) 8 (d) 5

25 Two resistances are given as $R_1 = (10 \pm 0.5)\,\Omega$ and $R_2 = (15 \pm 0.5)\,\Omega$. The percentage error in the measurement of equivalent resistance when they are connected in parallel is *[2023, 6 April Shift-I]*

(a) 6.33% (b) 5.33% (c) 2.33% (d) 4.33%

Numerical Value Type Questions

26 A physical quantity C is related to four other quantities p, q, r and s as follows

$$C = \frac{pq^2}{r^3 \sqrt{s}}$$

The percentage errors in the measurement of p, q, r and s are 1%, 2%, 3% and 2%, respectively. The percentage error in the measurement of C will be .... %. *[2025, 3 April Shift-II]*

27 A physical quantity Q is related to four observables a, b, c, d as follows.

$Q = ab^4 / cd$, where $a = (60 \pm 3)$ Pa, $b = (20 \pm 0.1)$ m, $c = (40 \pm 0.2)$ Nsm^{-2} and $d = (50 \pm 0.1)$m, then the percentage error in Q is $\frac{x}{1000}$, where x =........ *[2025, 29 Jan Shift-II]*

28 The resistance $R = \frac{V}{I}$, where $V = (50 \pm 2)$ V and $I = (20 \pm 0.2)$ A. The percentage error in R is x %. The value of x to the nearest integer is......... . *[2021, 16 March Shift-I]*

29 The density of a solid metal sphere is determined by measuring its mass and its diameter. The maximum error in the density of the sphere is $(x/100)$ %. If the relative errors in measuring the mass and the diameter are 6.0% and 1.5% respectively, the value of x is .... *[2020, 6 Sep Shift-I]*

Topic 3 Dimensions

Level 1

01 If μ_0 and ε_0 are the permeability and permittivity of free space, respectively, then the dimensions of $\left(\frac{1}{\mu_0\varepsilon_0}\right)$ is [2025, 2 April Shift-II]

(a) $\left[\frac{L^2}{T^2}\right]$ (b) $\left[\frac{L}{T^2}\right]$ (c) $\left[\frac{T^2}{L^2}\right]$ (d) $\left[\frac{T^2}{L}\right]$

02 Match the List-I with List-II. [2025, 3 April Shift-I]

	List-I	List-II
A.	Gravitational constant	I. $[LT^{-2}]$
B.	Gravitational potential energy	II. $[L^2T^{-2}]$
C.	Gravitational potential	III. $[ML^2T^{-2}]$
D.	Acceleration due to gravity	IV. $[M^{-1}L^3T^{-2}]$

Choose the correct answer from the options given below.

(a) A-II, B-IV, C-III, D-I (b) A-I, B-III, C-IV, D-II
(c) A-III, B-II, C-I, D-IV (d) A-IV, B-III, C-II, D-I

03 The dimension of $\sqrt{\frac{\mu_0}{\varepsilon_0}}$ is equal to that of

(μ_0 = vacuum permeability and ε_0 = vacuum permittivity) [2025, 7 April Shift-II]

(a) Capacitance (b) Inductance
(c) Voltage (d) Resistance

04 If B magnetic field and μ_0 is permeability of free space, then the dimensions of $\left(\frac{B}{\mu_0}\right)$ is [2025, 22 Jan Shift-I]

(a) $[ML^2T^{-2}A^{-1}]$ (b) $[L^{-1}A]$
(c) $[MT^{-2}A^{-1}]$ (d) $[LT^{-2}A^{-1}]$

05 Which one of the following is the correct dimensional formula for the capacitance in F? M, L, T and C stand for unit of mass, length, time and charge. [2025, 22 Jan Shift-II]

(a) $[F] = [C^2M^{-1}L^{-2}T^2]$ (b) $[F] = [CM^{-2}L^{-2}T^{-2}]$
(c) $[F] = [CM^{-1}L^{-2}T^2]$ (d) $[F] = [C^2M^{-2}L^2T^2]$

06 The dimensional formula of latent heat is [2024, 9 April Shift-I]

(a) $[MLT^{-2}]$ (b) $[M^0LT^{-2}]$
(c) $[ML^2T^{-2}]$ (d) $[M^0L^2T^{-2}]$

07 If ε_0 is the permittivity of free space and E is the electric field, then ε_0E^2 has the dimensions [2024, 8 April Shift-II]

(a) $[M^0L^{-2}TA]$ (b) $[ML^2T^{-2}]$
(c) $[M^{-1}L^{-3}T^4A^2]$ (d) $[ML^{-1}T^{-2}]$

08 The dimensional formula of angular impulse is [2024, 1 Feb Shift-I]

(a) $[MLT^{-1}]$ (b) $[ML^{-2}T^{-1}]$
(c) $[ML^2T^{-2}]$ (d) $[ML^2T^{-1}]$

Imp. Dimensions

09 Match List I with List II.

	List I		List II
(A)	Coefficient of viscosity	(I)	$[ML^2T^{-2}]$
(B)	Surface tension	(II)	$[ML^2T^{-1}]$
(C)	Angular momentum	(III)	$[ML^{-1}T^{-1}]$
(D)	Rotational kinetic energy	(IV)	$[ML^0T^{-2}]$

Choose the correct answer from the options given below. [2024, 30 Jan Shift-I]

(a) (A)-(I), (B)-(II), (C)-(III), (D)-(IV)
(b) (A)-(III), (B)-(IV), (C)-(II), (D)-(I)
(c) (A)-(II), (B)-(I), (C)-(IV), (D)-(III)
(d) (A)-(IV), (B)-(III), (C)-(II), (D)-(I)

10 If force F, velocity v and time t are considered as fundamental physical quantity, then dimensional formula of density will be [2023, 11 April Shift-II]

(a) $[Fv^{-4}t^{-2}]$ (b) $[Fv^4t^{-6}]$
(c) $[F^2v^{-2}t^6]$ (d) $[Fv^{-2}t^2]$

11 Dimension of $\frac{1}{\mu_0\varepsilon_0}$ should be equal to [2023, 8 April Shift-I]

(a) $\left[\frac{L^2}{T^2}\right]$ (b) $\left[\frac{T^2}{L^2}\right]$
(c) $\left[\frac{T}{L}\right]$ (d) $\left[\frac{L}{T}\right]$

12 The speed of a wave produced in water is given by $v = \lambda^a g^b \rho^c$. Where λ, g and ρ are wavelength of wave, acceleration due to gravity and density of water respectively. The values of a, b and c respectively, are [2023, 15 April Shift-I]

(a) $\frac{1}{2}, \frac{1}{2}, 0$ (b) $\frac{1}{2}, 0, \frac{1}{2}$
(c) $1, -1, 0$ (d) $1, 1, 0$

13 Match List I with List II.

List-I	List-II
A. Planck's constant (h)	I. $[M^1L^2T^{-2}]$
B. Stopping potential (V_s)	II. $[M^1L^1T^{-1}]$
C. Work function (ϕ)	III. $[M^1L^2T^{-1}]$
D. Momentum (p)	IV. $[M^1L^2T^{-3}A^{-1}]$

Choose the correct answer from the options given below. [2023, 24 Jan Shift-I]

(a) A-I, B-III, C-IV, D-II
(b) A-II, B-IV, C-III, D-I
(c) A-III, B-IV, C-I, D-II
(d) A-III, B-I, C-II, D-IV

14 Match List I with List II.

List I (Physical Quantity)	List II (Dimensional Formula)
A. Pressure gradient	I. $[M^0L^2T^{-2}]$
B. Energy density	II. $[M^1L^{-1}T^{-2}]$
C. Electric Field	III. $[M^1L^{-2}T^{-2}]$
D. Latent heat	IV. $[M^1L^1T^{-3}A^{-1}]$

Choose the correct answer from the options given below. *[2023, 29 Jan Shift-I]*

(a) A-II, B-III, C-I, D-IV
(b) A-III, B-II, C-IV, D-I
(c) A-II, B-III, C-IV, D-I
(d) A-III, B-II, C-I, D-IV

15 Match List I with List II.

List I	List II
A. Angular momentum	I. $[ML^2T^{-2}]$
B. Torque	II. $[ML^{-2}T^{-2}]$
C. Stress	III. $[ML^2T^{-1}]$
D. Pressure gradient	IV. $[ML^{-1}T^{-2}]$

Choose the correct answer from the options given below. *[2023, 31 Jan Shift-II]*

(a) A-II, B-III, C-IV, D-I
(b) A-III, B-I, C-IV, D-II
(c) A-I, B-IV, C-III, D-II
(d) A-IV, B-II, C-I, D-III

16 The SI unit of a physical quantity is pascal-second. The dimensional formula of this quantity will be *[2022, 27 June Shift-II]*

(a) $[ML^{-1}T^{-1}]$ (b) $[ML^{-1}T^{-2}]$
(c) $[ML^2T^{-1}]$ (d) $[M^{-1}L^3T^0]$

17 Match List I with List II. *[2021, 31 Aug Shift-I]*

List I	List II
(A) Torque	(i) MLT^{-1}
(B) Impulse	(ii) MT^{-2}
(C) Tension	(iii) ML^2T^{-2}
(D) Surface tension	(iv) MLT^{-2}

Choose the most appropriate answer from the options given below.

(a) (A)-(iii), (B)-(i), (C)-(iv), (D)-(ii)
(b) (A)-(ii), (B)-(i), (C)-(iv), (D)-(iii)
(c) (A)-(i), (B)-(iii), (C)-(iv), (D)-(ii)
(d) (A)-(iii), (B)-(iv), (C)-(i), (D)-(ii)

Imp.

18 Match List I with List II. *[2021, 31 Aug Shift-I]*

List I	List II
(A) Capacitance, C	(i) $[M^1L^1T^{-3}A^{-1}]$
(B) Permittivity of free space, ε_0	(ii) $[M^{-1}L^{-3}T^4A^2]$
(C) Permiability of free space, μ_0	(iii) $[M^{-1}L^{-2}T^4A^2]$
(D) Electric field, E	(iv) $[M^1L^1T^{-2}A^{-2}]$

Choose the most appropriate answer from the options given below.

(a) (A)-(iii), (B)-(ii), (C)-(iv), (D)-(i)
(b) (A)-(iii), (B)-(iv), (C)-(ii), (D)-(i)
(c) (A)-(iv), (B)-(ii), (C)-(iii), (D)-(i)
(d) (A)-(iv), (B)-(iiii), (C)-(ii), (D)-(i)

19 The dimension of $\frac{B^2}{2\mu_0}$, where B is magnetic field and μ_0 is the magnetic permeability of vacuum, is *[2020, 7 Jan Shift-II]*

(a) $[ML^{-1}T^{-2}]$ (b) $[MLT^{-2}]$
(c) $[ML^2T^{-1}]$ (d) $[ML^2T^{-2}]$

20 Let L, R, C, and V represent inductance, resistance, capacitance and voltage, respectively. The dimension of L / RCV in SI units will be *[2019, 12 Jan Shift-II]*

(a) $[LT^2]$ (b) $[LTA]$
(c) $[A^{-1}]$ (d) $[LA^{-2}]$

21 In the following I refers to current and other symbols have their usual meaning. Choose the option that corresponds to the dimensions of electrical conductivity *[JEE Main 2016]*

(a) $[M^{-1}L^{-3}T^3I]$ (b) $[M^{-1}L^{-3}T^3I^2]$
(c) $[M^{-1}L^3T^3I]$ (d) $[ML^{-3}T^{-3}I^2]$

22 If the time period t of the oscillation of a drop of liquid of density d, radius r, vibrating under surface tension s is given by the formula $t = \sqrt{r^{2b} s^c d^{a/2}}$. It is observed that the time period is directly proportional to $\sqrt{d/s}$. The value of b should be *[JEE Main 2013]*

(a) $\frac{3}{4}$ (b) $\sqrt{3}$ (c) $\frac{3}{2}$ (d) $\frac{2}{3}$

23 The dimensions of magnetic field in M, L, T and C (coulomb) is given as *[AIEEE 2008]*

(a) $[MLT^{-1}C^{-1}]$ (b) $[MT^2C^{-2}]$
(c) $[MT^{-1}C^{-1}]$ (d) $[MT^2C^{-1}]$

24 Out of the following pairs, which one does not have identical dimensions? *[AIEEE 2005]*

(a) Angular momentum and Planck's constant
(b) Impulse and momentum
(c) Moment of inertia and moment of a force
(d) Work and torque

25 Which one of the following represents the correct dimensions of the coefficient of viscosity ? *[AIEEE 2004]*
(a) $[ML^{-1}T^{-2}]$ (b) $[MLT^{-1}]$
(c) $[ML^{-1}T^{-1}]$ (d) $[ML^{-2}T^{-2}]$

26 The physical quantities not having same dimensions are *[AIEEE 2003]*
(a) torque and work
(b) momentum and Planck's constant
(c) stress and Young's modulus
(d) speed and $(\mu_0\varepsilon_0)^{-1/2}$

Level 2

27 Match List - I with List - II. *[2025, 2 April Shift-I]*

List-I		List-II	
A.	Coefficient of viscosity	I.	$[ML^0T^{-3}]$
B.	Intensity of wave	II.	$[ML^{-2}T^{-2}]$
C.	Pressure gradient	III.	$[M^{-1}LT^2]$
D.	Compressibility	IV.	$[ML^{-1}T^{-1}]$

Choose the correct answer from the options given below.
(a) (A)-(II), (B)-(III), (C)-(IV), (D)-(I)
(b) (A)-(IV), (B)-(I), (C)-(II), (D)-(III)
(c) (A)-(I), (B)-(IV), (C)-(III), (D)-(II)
(d) (A)-(IV), (B)-(II), (C)-(1), (D)-(III)

28 Given a charge q, current I and permeability of vacuum μ_0. Which of the following quantity has the dimension of momentum? *[2025, 2 April Shift-II]*
(a) $\frac{q\mu_0}{I}$ (b) $q^2\mu_0 I$ (c) $q\mu_0 I$ (d) $\frac{qI}{\mu_0}$

29 Match List I with List - II. *[2025, 7 April Shift-II]*

	List - I		List - II
(A)	Mass density	(I)	$[ML^2T^{-3}]$
(B)	Impulse	(II)	$[MLT^{-1}]$
(C)	Power	(III)	$[ML^2T^0]$
(D)	Moment of inertia	(IV)	$[ML^{-3}T^0]$

Choose the correct answer from the options given below.
(a) (A)-(IV), (B)-(II), (C)-(I), (D)-(III)
(b) (A)-(I), (B)-(III), (C)-(IV), (D)-(II)
(c) (A)-(II), (B)-(III), (C)-(IV), (D)-(I)
(d) (A)-(IV), (B)-(II), (C)-(III), (D)-(I)

30 In an electromagnetic system, the quantity representing the ratio of electric flux and magnetic flux has dimension of $M^PL^QT^RA^S$, where value of Q and R are *[2025, 4 April Shift-I]*
(a) (– 2, 1) (b) (3, – 5)
(c) (– 2, 2) (d) (1, – 1)

31 In an electromagnetic system, a quantity defined as the ratio of electric dipole moment and magnetic dipole moment has dimension of $[M^PL^QT^RA^S]$. The value of P and Q are *[2025, 4 April Shift-II]*
(a) 0, –1 (b) –1, 1
(c) 1, –1 (d) –1, 0

32 The expression given below shows the variation of velocity (v) with time (t),
$v = At^2 + \frac{Bt}{C+t}$. The dimension of ABC is *[2025, 29 Jan Shift-I]*
(a) $[M^0L^1T^{-3}]$ (b) $[M^0L^2T^{-2}]$
(c) $[M^0L^2T^{-3}]$ (d) $[M^0L^1T^{-2}]$

33 The pair of physical quantities not having same dimensions is *[2025, 29 Jan Shift-I]*
(a) surface tension and impulse
(b) angular momentum and Planck's constant
(c) pressure and Young's modulus
(d) torque and energy

34 The energy E and momentum p of a moving body of mass m are related by some equation. Given that c represents the speed of light, identify the correct equation *[2025, 24 Jan Shift-II]*
(a) $E^2 = pc^2 + m^2c^4$ (b) $E^2 = p^2c^2 + m^2c^4$
(c) $E^2 = pc^2 + m^2c^2$ (d) $E^2 = p^2c^2 + m^2c^2$

35 The electric flux is $\phi = \alpha\sigma + \beta\lambda$,
where λ and σ are linear and surface charge density, respectively, $\left(\frac{\alpha}{\beta}\right)$ represents *[2025, 23 Jan Shift-I]*
(a) electric field (b) area
(c) charge (d) displacement

36 Given below are two statements
Statement I Dimensions of specific heat is $[L^2T^{-2}K^{-1}]$.
Statement II Dimensions of gas constant is $[ML^2T^{-1}K^{-1}]$.
In the light of the above statements, choose the most appropriate answer from the options given below. *[2024, 6 April Shift-II]*
(a) Both Statement I and Statement II are incorrect.
(b) Statement I is correct but Statement II is incorrect.
(c) Both Statement I and Statement II are correct.
(d) Statement I is incorrect but Statement II is correct.

Imp. Principle of Homogeneous

37 What is the dimensional formula of ab^{-1} in the equation $\left(p + \frac{a}{V^2}\right)(V - b) = RT$, where letters have their usual meaning. *[2024, 5 April Shift-II]*
(a) $[M^0L^3T^{-2}]$ (b) $[ML^2T^{-2}]$
(c) $[M^{-1}L^5T^3]$ (d) $[M^6L^7T^4]$

38 Applying the principle of homogeneity of dimensions, determine which one is correct ?
(Where T is time period, G is gravitational constant, M is mass and r is radius of orbit.) *[2024, 4 April Shift-II]*

(a) $T^2 = \frac{4\pi^2 r^3}{GM}$ (b) $T^2 = 4\pi^2 r^3$

(c) $T^2 = \frac{4\pi^2 r^2}{GM}$ (d) $T^2 = \frac{4\pi^2 r}{GM^2}$

39 The equation of stationary wave is

$$y = 2a\sin\left(\frac{2\pi nt}{\lambda}\right)\cos\left(\frac{2\pi x}{\lambda}\right).$$

Which of the following is not correct? *[2024, 4 April Shift-I]*

(a) The dimensions of nt is [L]
(b) The dimensions of x is [L]
(c) The dimensions of $\frac{n}{\lambda}$ is [T]
(d) The dimensions of n is $[LT^{-1}]$

40 A force is represented by $F = ax^2 + bt^{1/2}$, where x = distance and t = time. The dimensions of b^2/a are *[2024, 31 Jan Shift-I]*

(a) $[ML^{-1}T^{-1}]$ (b) $[ML^3T^{-3}]$
(c) $[ML^2T^{-3}]$ (d) $[MLT^{-2}]$

41 If mass is written as $m = k c^P G^{-\frac{1}{2}} h^{\frac{1}{2}}$, then the value of P will be (Constants have their usual meaning with k a dimensionless constant) *[2024, 30 Jan Shift-II]*

(a) $\frac{-1}{3}$ (b) 2 (c) $\frac{1}{2}$ (d) $\frac{1}{3}$

42 The equation of state of a real gas is given by $\left(p + \frac{a}{V^2}\right)(V - b) = RT$, where p, V and T are pressure, volume and temperature, respectively and R is the universal gas constant. The dimensions of a/b^2 is similar to that of *[2024, 27 Jan Shift-II]*

(a) RT (b) p (c) pV (d) R

43 Match List I with List II.

List I	List II
A. Spring constant	I. $[T^{-1}]$
B. Angular speed	II. $[ML^{-2}]$
C. Angular momentum	III. $[ML^2]$
D. Moment of inertia	IV. $[ML^2T^{-1}]$

Choose the correct answer from the options given below. *[2023, 12 April Shift-I]*

(a) A-II, B-I, C-IV, D-III
(b) A-I, B-III, C-II, D-IV
(c) A-II, B-III, C-I, D-IV
(d) A-IV, B-I, C-III, D-II

44 Match List I with List II.

List I	List II
A. Torque	I. $[ML^{-2}T^{-2}]$
B. Stress	II. $[ML^2T^{-2}]$
C. Pressure gradient	III. $[ML^{-1}T^{-1}]$
D. Coefficient of viscosity	IV. $[ML^{-1}T^{-2}]$

Choose the correct answer from the options given below. *[2023, 8 April Shift-I]*

(a) A-IV, B-II, C-III, D-I (b) A-II, B-I, C-IV, D-III
(c) A-II, B-IV, C-I, D-III (d) A-III, B-IV, C-I, D-II

45 Match List I with List II.

List I	List II
A. Young's modulus (Y)	I. $[ML^{-1}T^{-1}]$
B. Coefficient of viscosity (η)	II. $[ML^2T^{-1}]$
C. Planck's constant (h)	III. $[ML^{-1}T^{-2}]$
D. Work function (ϕ)	IV. $[ML^2T^{-2}]$

Choose the correct answer from the options given below. *[2023, 25 Jan Shift-II]*

(a) A-III, B-I, C-II, D-IV (b) A-II, B-III, C-IV, D-I
(c) A -I, B-II, C-III, D-IV (d) A-I, B-III, C-IV, D-II

46 The equation of a circle is given by $x^2 + y^2 = a^2$, where a is the radius. If the equation is modified to change the origin other than (0, 0) , then find out the correct dimensions of A and B in a new equation $(x - At)^2 + \left(y - \frac{t}{B}\right)^2 = a^2$. The dimensions of t is given as $[T^{-1}]$ *[2023, 29 Jan Shift -II]*

(a) $A = [L^{-1}T^{-1}]$, $B = [LT]$ (b) $A = [LT]$, $B = [L^{-1}T^{-1}]$
(c) $A = [L^{-1}T]$, $B = [LT^{-1}]$ (d) $A = [L^{-1}T^{-1}]$, $B = [LT^{-1}]$

47 Identify the pair of physical quantities which have different dimensions. *[2022, 24 June Shift-I]*

(a) Wave number and Rydberg's constant
(b) Stress and coefficient of elasticity
(c) Coercivity and Magnetisation
(d) Specific heat capacity and latent heat

48 Given below are two statements. One is labelled as Assertion (A) and other is labelled as Reason (R).

Assertion (A) Time period of oscillation of a liquid drop depends on surface tension (S), if density of the liquid is ρ and radius of the drop is r, then $T = k\sqrt{\frac{\rho r^3}{S^{3/2}}}$ is dimensionally correct, where k is dimensionless.

Reason (R) Using dimensional analysis we get RHS having different dimension than that of time period.

In the light of above statements, choose the correct answer from the options given below.
[2022, 29 July Shift-I]

(a) Both A and R are true and R is the correct explanation of A.
(b) Both A and R are true but (R) is not the correct explanation of A.
(c) A is true but R is false.
(d) A is false but R is true.

Imp. Application of Dimensions

49 If momentum p, area A and time t are taken as fundamental quantities, then the dimensional formula for coefficient of viscosity is *[2022, 25 July Shift-I]*
(a) $[pA^{-1}t^0]$ (b) $[pAt^{-1}]$
(c) $[pA^{-1}t]$ (d) $[pA^{-1}t^{-1}]$

50 Which of the following physical quantities have the same dimensions? *[2022, 25 July Shift-I]*
(a) Electric displacement D and surface charge density.
(b) Displacement current and electric field.
(c) Current density and surface charge density.
(d) Electric potential and energy.

51 If velocity $[v]$, time $[t]$ and force $[F]$ are choosen as the base quantities, the dimensions of the mass will be *[2021, 31 Aug Shift-II]*
(a) $[Ft^{-1}v^{-1}]$ (b) $[Ftv^{-1}]$ (c) $[Ft^2v]$ (d) $[Fvt^{-1}]$

52 Which of the following is not a dimensionless quantity ? *[2021, 27 Aug Shift-I]*
(a) Relative magnetic permeability (μ_r)
(b) Power factor
(c) Permeability of free space (μ_0)
(d) Quality factor

53 Dimensional formula for thermal conductivity is (here, K denotes the temperature) *[2020, 4 Sep Shift-I]*
(a) $[MLT^{-2}K]$ (b) $[MLT^{-2}K^{-2}]$
(c) $[MLT^{-3}K^{-1}]$ (d) $[MLT^{-3}K]$

54 In form of G (universal gravitational constant), h (Planck's constant) and c (speed of light), the time period will be proportional to *[2019, 9 Jan Shift-II]*
(a) $\sqrt{\frac{Gh}{c^5}}$ (b) $\sqrt{\frac{hc^5}{G}}$ (c) $\sqrt{\frac{c^3}{Gh}}$ (d) $\sqrt{\frac{Gh}{c^3}}$

55 The force of interaction between two atoms is given by $F = \alpha\beta \exp\left(-\frac{x^2}{\alpha kT}\right)$, where x is the distance, k is the Boltzmann constant and T is temperature and α and β are two constants. The dimension of β is *[2019, 11 Jan Shift-I]*
(a) $[MLT^{-2}]$ (b) $[M^0L^2T^{-4}]$
(c) $[M^2LT^{-4}]$ (d) $[M^2L^2T^{-2}]$

56 If electronic charge e, electron mass m, speed of light in vacuum c and Planck's constant h are taken as fundamental quantities, the permeability of vacuum μ_0 can be expresed in units of *[JEE Main 2014]*
(a) $\left(\frac{h}{me^2}\right)$ (b) $\left(\frac{hc}{me^2}\right)$ (c) $\left(\frac{h}{ce^2}\right)$ (d) $\left(\frac{mc^2}{he^2}\right)$

57 From the following combinations of physical constants (expressed through their usual symbols) the only combination, that would have the same value in different systems of units, is *[JEE Main 2014]*
(m_e = mass of electron)
(a) $\frac{ch}{2\pi\varepsilon_0^2}$ (b) $\frac{e^2}{2\pi\varepsilon_0 Gm_e^2}$
(c) $\frac{\mu_0\varepsilon_0}{c^2}\frac{G}{he^2}$ (d) $\frac{2\pi\sqrt{\mu_0\varepsilon_0}}{ce^2}\frac{h}{G}$

58 The dimensions of angular momentum, latent heat and capacitance are, respectively. *[JEE Main 2013]*
(a) $[ML^2T^1A^2]$, $[L^2T^{-2}]$, $[M^{-1}L^{-2}T^2]$
(b) $[ML^2T^{-2}]$, $[L^2T^2]$, $[M^{-1}L^{-2}T^4A^2]$
(c) $[ML^2T^{-1}]$, $[L^2T^{-2}]$, $[ML^2TA^2]$
(d) $[ML^2T^{-1}]$, $[L^2T^{-2}]$, $[M^{-1}L^{-2}T^4A^2]$

59 Let $[\varepsilon_0]$ denotes the dimensional formula of the permittivity of vacuum. If M = mass, L = length, T = time and A = electric current, then *[JEE Main 2013]*
(a) $[\varepsilon_0] = [M^{-1}L^{-3}T^2A]$
(b) $[\varepsilon_0] = [M^{-1}L^{-3}T^4A^2]$
(c) $[\varepsilon_0] = [M^{-2}L^2T^{-1}A^{-2}]$
(d) $[\varepsilon_0] = [M^{-1}L^2T^{-1}A^2]$

60 Which of the following units denotes the dimensions $[ML^2/Q^2]$, where Q denotes the electric charge? *[AIEEE 2006]*
(a) Wb/m^2 (b) Henry (H)
(c) H/m^2 (d) Weber (Wb)

Level 3

61 Match the List I with List II. *[2025, 3 April Shift-II]*

	List I		List II
A.	Boltzmann constant	I.	$[ML^2T^{-1}]$
B.	Coefficient of viscosity	II.	$[MLT^{-3}K^{-1}]$
C.	Planck's constant	III.	$[ML^2T^{-2}K^{-1}]$
D.	Thermal conductivity	IV.	$[ML^{-1}T^{-1}]$

Choose the correct answer from the options given below.
(a) A-III, B-IV, C-I, D-II (b) A-III, B-IV, C-II, D-I
(c) A-III, B-II, C-I, D-IV (d) A-II, B-III, C-IV, D-I

62 The position of a particle moving on X-axis is given by $x(t) = A\sin t + B\cos^2 t + Ct^2 + D$, where t is time. The dimension of $\frac{ABC}{D}$ is *[2025, 23 Jan Shift-I]*
(a) $[L^3T^{-2}]$ (b) $[L^2]$ (c) $[L^2T^{-2}]$ (d) $[L]$

63 Match List-I with List-II. **[2025, 29 Jan Shift-II]**

List-I		List-II	
(A)	Young's modulus	(I)	$[ML^{-1}T^{-1}]$
(B)	Torque	(II)	$[ML^{-1}T^{-2}]$
(C)	Coefficient of viscosity	(III)	$[M^{-1}L^{3}T^{-2}]$
(D)	Gravitational constant	(IV)	$[ML^{2}T^{-2}]$

Choose the correct answer from the options given below.

(a) (A)-(I), (B)-(III), (C)-(II), (D)-(IV)
(b) (A)-(IV), (B)-(II), (C)-(III), (D)-(I)
(c) (A)-(II), (B)-(IV), (C)-(I), (D)-(III)
(d) (A)-(II), (B)-(I), (C)-(IV), (D)-(III)

64 Match List-I with List-II. **[2025, 28 Jan Shift-II]**

List-I		List-II	
(A)	Angular impulse	(I)	$[M^{0}L^{2}T^{-2}]$
(B)	Latent heat	(II)	$[ML^{2}T^{-3}A^{-1}]$
(C)	Electrical resistivity	(III)	$[ML^{2}T^{-1}]$
(D)	Electromotive force	(IV)	$[ML^{3}T^{-3}A^{-2}]$

Choose the correct answer from the options given below.

(a) (A)-(I), (B)-(III), (C)-(IV), (D)-(II)
(b) (A)-(III), (B)-(I), (C)-(II), (D)-(IV)
(c) (A)-(III), (B)-(I), (C)-(IV), (D)-(II)
(d) (A)-(II), (B)-(I), (C)-(IV), (D)-(III)

65 Match List-I with List-II. **[2025, 23 Jan Shift-II]**

List-I		List-II	
(A)	Permeability of free space	(I)	$[ML^{2}T^{-2}]$
(B)	Magnetic field	(II)	$[MT^{-2}A^{-1}]$
(C)	Magnetic moment	(III)	$[MLT^{-2}A^{-2}]$
(D)	Torsional constant	(IV)	$[L^{2}A]$

Choose the correct answer from the options given below.

(a) (A)-(III), (B)-(II), (C)-(IV), (D)-(I)
(b) (A)-(I), (B)-(IV), (C)-(II), (D)-(III)
(c) (A)-(II), (B)-(I), (C)-(III), (D)-(IV)
(d) (A)-(IV), (B)-(III), (C)-(I), (D)-(II)

66 Match List I with List II.

List I		List II	
A.	Torque	I.	$[M^{1}L^{1}T^{-2}A^{-2}]$
B.	Magnetic field	II.	$[L^{2}A^{1}]$
C.	Magnetic moment	III.	$[M^{1}T^{-2}A^{-1}]$
D.	Permeability of free space	IV.	$[M^{1}L^{2}T^{-2}]$

Choose the correct answer from the options given below. **[2024, 6 April Shift-I]**

(a) A-IV, B-III, C-II, D-I
(b) A-I, B-III, C-II, D-IV
(c) A-IV, B-II, C-III, D-I
(d) A-III, B-I, C-II, D-IV

67 Consider two physical quantities A and B related to each other as $E = \frac{B - x^2}{At}$ where E, x and t have dimensions of energy, length and time respectively. The dimension of AB is **[2024, 31 Jan Shift-II]**

(a) $[L^{-2}M^{1}T^{0}]$
(b) $[L^{2}M^{-1}T^{1}]$
(c) $[L^{0}M^{-1}T^{1}]$
(d) $[L^{-2}M^{-1}T^{1}]$

68 $\left(p + \frac{a}{V^2}\right)(V - b) = RT$ represents the equation of state of some gases, where p is the pressure, V is the volume, T is the temperature and a, b, R are the constants. The physical quantity, which has dimensional formula as that of b^2/a, will be **[2023, 1 Feb Shift-I]**

(a) bulk modulus
(b) compressibility
(c) modulus of rigidity
(d) energy density

69 If the velocity of light c, universal gravitational constant G and Planck's constant h are choosen as fundamental quantities. The dimensions of mass in the new system is **[2023, 1 Feb Shift-II]**

(a) $[h^{-1/2}c^{1/2}G^{1/2}]$
(b) $[h^{1}c^{1}G^{-1}]$
(c) $[h^{1/2}c^{1/2}G^{-1/2}]$
(d) $[h^{1/2}c^{-1/2}G^{1}]$

70 An expression of energy density is given by $u = \frac{\alpha}{\beta}\sin\left(\frac{\alpha x}{k_B T}\right)$, where α, β are constants, x is displacement, k_B is Boltzmann constant and T is the temperature. The dimensions of β will be **[2022, 27 July Shift-II]**

(a) $[ML^{-2}T^{-1}q^{-1}]$
(b) $[M^{0}L^{2}T^{-2}]$
(c) $[M^{0}L^{0}T^{0}]$
(d) $[M^{0}L^{2}T^{0}]$

71 Consider the efficiency of Carnot's engine is given by $\eta = \frac{\alpha\beta}{\sin\theta}\log_e \frac{\beta x}{k_B T}$, where α and β are constants. If T is temperature, k_B is Boltzmann constant, θ is angular displacement and x has the dimension of length. Then, choose the incorrect option. **[2022, 28 July Shift-II]**

(a) Dimensions of β is same as that of force.
(b) Dimensions of $\alpha^{-1}x$ is same as that of energy.
(c) Dimensions of $\eta^{-1}\sin\theta$ is same as that of $\alpha\beta$.
(d) Dimensions of α is same as that of β.

72 Identify the pair of physical quantities that have same dimensions. *[2022, 24 June Shift-II]*
(a) Velocity gradient and decay constant
(b) Wien's constant and Stefan constant
(c) Angular frequency and angular momentum
(d) Wave number and Avogadro number

73 The force is given in terms of time t and displacement x by the equation $F = A\cos Bx + C\sin Dt$. The dimensional formula of AD/B is *[2021, 25 July Shift-II]*
(a) $[M^0LT^{-1}]$ (b) $[ML^2T^{-3}]$
(c) $[M^1L^1T^{-2}]$ (d) $[M^2L^2T^{-3}]$

74 In a typical combustion engine, the work done by a gas molecule is given $W = \alpha^2\beta e^{\frac{-\beta x^2}{k_BT}}$, where x is the displacement, k_B is the Boltzmann constant and T is the temperature. If α and β are constants, dimensions of α will be *[2021, 26 Feb Shift-I]*
(a) $[MLT^{-2}]$ (b) $[M^0LT^0]$
(c) $[M^2LT^{-2}]$ (d) $[MLT^{-1}]$

75 A quantity f is given by $f = \sqrt{hc^5/G}$, where c is speed of light, G universal gravitational constant and h is the Planck's constant. Dimension of f is that of *[2020, 9 Jan Shift-I]*
(a) area (b) volume
(c) momentum (d) energy

76 Which of the following combinations has the dimension of electrical resistance (ε_0 is the permittivity of vacuum and μ_0 is the permeability of vacuum)? *[2019, 12 April Shift-I]*
(a) $\sqrt{\frac{\mu_0}{\varepsilon_0}}$ (b) $\frac{\mu_0}{\varepsilon_0}$
(c) $\sqrt{\frac{\varepsilon_0}{\mu_0}}$ (d) $\frac{\varepsilon_0}{\mu_0}$

Numerical Value Type Questions

77 In a measurement, it is asked to find modulus of elasticity per unit torque applied on the system. The measured quantity has dimension of $[M^aL^bT^c]$. If $b = 3$, the value of c is .... *[2025, 28 Jan Shift-I]*

Key Index

Topic 1

1 (c)	2 (c)	3 (b)	4 (b)	5 (a)	6 (b)	7 (c)	8 (b)	9 (*)	10 (b)
11 (a)	12 (b)	13 (a)	14 (b)	15 (a)					

Topic 2

1 (a)	2 (d)	3 (b)	4 (c)	5 (c)	6 (c)	7 (a)	8 (a)	9 (a)	10 (a)
11 (a)	12 (c)	13 (b)	14 (d)	15 (a)	16 (b)	17 (c)	18 (c)	19 (b)	20 (a)
21 (c)	22 (a)	23 (b)	24 (b)	25 (d)	26 (15)	27 (77)	28 (5)	29 (1050)	

Topic 3

1 (a)	2 (d)	3 (d)	4 (b)	5 (a)	6 (d)	7 (d)	8 (d)	9 (b)	10 (a)
11 (a)	12 (a)	13 (c)	14 (b)	15 (b)	16 (a)	17 (a)	18 (a)	19 (a)	20 (c)
21 (b)	22 (c)	23 (c)	24 (c)	25 (c)	26 (b)	27 (b)	28 (c)	29 (a)	30 (d)
31 (a)	32 (c)	33 (a)	34 (b)	35 (d)	36 (b)	37 (b)	38 (a)	39 (c)	40 (b)
41 (c)	42 (b)	43 (a)	44 (c)	45 (a)	46 (b)	47 (d)	48 (d)	49 (a)	50 (a)
51 (b)	52 (c)	53 (c)	54 (a)	55 (c)	56 (c)	57 (b)	58 (d)	59 (b)	60 (b)
61 (a)	62 (c)	63 (c)	64 (c)	65 (a)	66 (a)	67 (b)	68 (b)	69 (c)	70 (d)
71 (d)	72 (a)	73 (a)	74 (b)	75 (b)	76 (d)	77 (0)			

Detailed **Explanations**

Topic 1

01. *(c)* We know that

Intensity, $I = \frac{1}{2}\varepsilon_0 E_0^2 c$

$\Rightarrow \quad E_0^2 = \frac{2I}{\varepsilon_0 c}$

$\Rightarrow \quad E_0 = \sqrt{\frac{2I}{\varepsilon_0 c}}$

Here, E_0 is the electric field.

Thus its unit is NC^{-1}.

02. *(c)* In scientific rotation of a physical quanity is given as $a \times 10^b$ and $a \leq 5$, then, b is order of magnitude of the quantity.

03. *(b)* Astronomical unit (AU), Parsec and Light year are the unit of distance.

Light year $\rightarrow$ distance travelled by light in one year

1 Light year $= 9.5 \times 10^{15}$ m

Parsec $= 3.262$ light year

Astronomical unit (AU)

$= 1.58 \times 10^{-5}$ light year

AU < ly < Parsec

Statement I correct and Statement II incorrect.

04. *(b)* A $\rightarrow$ IV

Surface tension

$= \frac{\text{Force}}{\text{Length}} = \frac{\text{kg ms}^{-2}}{\text{m}} = \text{kg s}^{-2}$

B $\rightarrow$ III

$\text{Pressure} = \frac{\text{Force}}{\text{Area}} = \frac{\text{kg ms}^{-2}}{\text{m}^2} = \text{kgm}^{-1}\text{s}^{-2}$

C $\rightarrow$ I

We know that coefficient of viscosity is given by

$\eta = \frac{F}{A(dv/dx)} = \frac{\text{kg ms}^{-2}}{\text{m}^2\text{s}^{-1}} = \text{kg m}^{-1}\text{s}^{-1}$

D $\rightarrow$ II

Impulse = Force × Time interval

$= \text{kg ms}^{-2} \times \text{s} = \text{kg ms}^{-1}$

05. *(a)* Unit of intensity of electric field E is Vm^{-1}. Unit of intensity of magnetising field H is Am^{-1}.

Unit of E/H can be calculated as

$\frac{\text{Unit of } E}{\text{Unit of } H} = \frac{\text{Vm}^{-1}}{\text{Am}^{-1}} = \frac{\text{V}}{\text{A}} = \text{ohm}.$

Thus, the unit of E/H will be ohm.

06. *(b)* As given, $E_y \propto J_x B_z$

$\therefore$ Constant of proportionality

$$K = \frac{E_y}{B_z J_x} = \frac{c}{J_x} = \frac{\text{m}^3}{\text{As}}$$

> • Concept **Enhancer**
>
> In EM-waves, $\frac{E}{B} = c$ (speed of light) and
>
> $J = \frac{I}{\text{Area}}$

07. *(c)* Rad is used to measure biological effect of radiation.

08. *(b)* Heat capacity $= \frac{\Delta Q}{\Delta T}$ (JK^{-1})

Specific heat capacity

$C = \frac{Q}{m\Delta T}$ $(\text{J kg}^{-1}\text{ K}^{-1})$

Latent heat, $L = \frac{Q}{m}$ (J/kg)

Thermal conductivity,

$K = \frac{Ql}{tA\Delta T}$ $(\text{Jm}^{-1}\text{ K}^{-1}\text{ s}^{-1})$

09. *(*)* A. Magnetic induction has unit Tesla or Gauss.

B. Magnetic intensity

$H = \frac{m}{\text{Volume}} = \frac{\text{Ampere metre}^2}{\text{metre}^3}$

= Ampere/metre

C. Magnetic flux has unit Weber.

D. Magnetic moment has unit

$m = IA = \text{Ampere metre}^2$

10. *(b)* A $\rightarrow$ IV

Torque = Force × perpendicular distance

= Mass × acceleration

× perpendicular distance

$= \text{kg} \times \text{ms}^{-2} \times \text{m} = \text{kg m}^2\text{s}^{-2}$

B $\rightarrow$ I Energy density $= \frac{\text{Energy}}{\text{Volume}}$

$= \frac{\text{kg m}^2\text{s}^{-2}}{\text{m}^3} = \text{kg m}^{-1}\text{s}^{-2}$

C $\rightarrow$ III Pressure gradient

$p = \frac{\text{Pressure}}{\text{distance}} = \frac{\text{Force}}{\text{Area} \times \text{distance}}$

$= \frac{\text{kg ms}^{-2}}{\text{m}^2 \times \text{m}} = \text{kg m}^{-2}\text{s}^{-2}$

D $\rightarrow$ II, Impulse = Force × time

$= \text{kg ms}^{-2} \times \text{s} = \text{kg ms}^{-1}$

11. *(a)* Electric field in a region is given,

$$\text{E} = \left(\frac{A}{x^2}\hat{\text{i}} + \frac{B}{y^3}\hat{\text{j}}\right)$$

By principle of homogeneity of physical equation, every independent term in expression of field must have same dimensions and units.

As $\text{E} = \text{NC}^{-1}$ and $x^2 \rightarrow \text{m}^2$

also, $y^3 \rightarrow \text{m}^3$

So, units of A/x^2 and B/y^3 are must be NC^{-1}

$\Rightarrow \quad A = \text{Nm}^{+2}\text{C}^{-1}$

and $\quad B = \text{Nm}^{+3}\text{C}^{-1}$

12. *(b)* As we know,

$\text{Torque} = F(\text{N}) \times r_\perp(\text{m}) \rightarrow \text{Nm}$

$\text{Stress} = \frac{\text{Force (N)}}{\text{Area (m}^2)} \rightarrow \text{N/m}^2$

$\text{Latent heat} = \frac{\text{Energy (J)}}{\text{Mass (kg)}} \rightarrow \text{J kg}^{-1}$

$\text{Power} = \frac{\text{Work (Nm)}}{\text{Time (s)}} \rightarrow \text{N ms}^{-1}$

13. *(a)* Given, $v_2 = \frac{n}{m^2} v_1$

$\because$ Velocity, $v = \frac{\text{Displacement}}{\text{Time}}$

$\therefore \quad \frac{L_2}{T_2} = \frac{n}{m^2} \cdot \frac{L_1}{T_1} \quad \ldots\text{(i)}$

and $\quad a_2 = \frac{a_1}{mn}$

$\therefore \quad \frac{L_2}{T_2^2} = \frac{L_1}{T_1^2} \cdot \frac{1}{mn} \quad \ldots\text{(ii)}$

On dividing Eq. (ii) by Eq. (i), we get

$$\frac{L_2/T_2^2}{L_2/T_2} = \frac{(L_1/T_1^2) \times (1/mn)}{(n/m^2) \times (L_1/T_1)}$$

$\Rightarrow \quad T_2 = \frac{n^2}{m} T_1$

as, $\quad L_2 = \frac{T_2}{T_1} \cdot L_1 \frac{n}{m^2} = \frac{n^2}{m} L_1 \frac{n}{m^2}$

$= \frac{n^3}{m^3} L_1$

14. *(b)* As we know that, SI unit of following terms are

A. R_H (Rydberg constant) $= m^{-1}$

B. h (Planck's constant)

$$= Js = kg\ m^2\ s^{-1}$$

C. Magnetic field energy density

$$(\mu_B) = \frac{\text{Energy}}{\text{Volume}} = \frac{kg\ m^2 s^{-2}}{m^3} = kgm^{-1}s^{-2}$$

D. η (coefficient of viscosity)

$$\because \quad F = \eta A \frac{dv}{dx}$$

$$\therefore \quad \eta = \frac{Fdx}{A\,dv} = \frac{kg\ ms^{-2} \cdot m}{m^2 . ms^{-1}}$$

$$= kgm^{-1}s^{-1}$$

So, the correct match is A-3, B-2, C-4 and D-1.

15. *(a)* Since, we know,

$$N_1 u_1 = N_2 u_2$$

$$128 \frac{kg}{m^3} = N_2 \frac{50g}{(25\ cm)^3}$$

$$\Rightarrow \frac{128 \times 1000\ g}{100 \times 100 \times 100\ cm^3} = \frac{N_2 \times 50\ g}{25 \times 25 \times 25\ cm^3}$$

$$\Rightarrow \quad N_2 = \frac{128 \times 1000 \times 25 \times 25 \times 25}{50 \times 100 \times 100 \times 100}$$

$$= 40$$

• Concept Enhancer

Physical quantity (Q) = Numerical value $(n) \times$ Unit (u)

If n_1 and n_2 are numerical values quantity (Q) of the same physical quantity corresponding to the unit u_1 and u_2, then

$$Q = n_1 u_1 = n_2 u_2$$

It is based on the fact that the magnitude of a physical quantity remains the same, whatever may be the system of units.

$$\Rightarrow \quad n \propto \frac{1}{u}$$

Thus, the numerical value (n) is inversaly proportional to the size of the unit (u). If a physical quantity A has dimensional formula, $[M^{\alpha}L^{\beta}T^{\gamma}]$ and if the unit of that physical quantity in two systems are $[M_1^{\alpha}L_1^{\beta}T_1^{\gamma}]$ and $[M_2^{\alpha}L_2^{\beta}T_2^{\gamma}]$, respectively and n_1 and n_2 be the numerical values in the two systems, respectively. Then,

$$n_1[u_1] = n_2[u_2]$$

$$\Rightarrow \quad n_1[M_1^{\alpha}L_1^{\beta}T_1^{\gamma}] = n_2[M_2^{\alpha}L_2^{\beta}T_2^{\gamma}]$$

$$\therefore \quad n_2 = \left[\frac{M_1}{M_2}\right]^{\alpha}\left[\frac{L_1}{L_2}\right]^{\beta}\left[\frac{T_1}{T_2}\right]^{\gamma} n_1$$

Topic 2

01. *(a)* A-II, B-I, C-IV, D-III

(A) 1001 = 4 significant figures

(B) 010.1 = 3 significant figures

(C) 100.100 = 6 significant figures

(D) 0.0010010 = 5 significant figures

02. *(d)* Resistance of wire is,

$$R = \rho \frac{l}{A} = \frac{4\rho l}{\pi D^2}$$

So, percentage error in resistance is

$$\frac{\Delta R}{R} \times 100 = \frac{\Delta l}{l} \times 100 + \frac{2\Delta D}{D} \times 100$$

$$= 0.1\% + 2 \times 0.1\% = 0.3\%$$

03. *(b)* A is true but R is false

$$v_T = \frac{2}{9\eta}(\sigma - \rho) r^2 g$$

$$v_T \propto r^2$$

$$\text{Percentage error in } v_T = 2 \times \left(\frac{0.1}{5} \times 100\%\right)$$

$$= 4\%$$

04. *(c)* Density of wire ρ

$$= \frac{\text{Mass}}{\text{Volume}} \text{ or } \rho = \frac{m}{\pi r^2 l}$$

Hence, maximum % error in density is

$$\Rightarrow \quad \frac{\Delta\rho}{\rho} \times 100\% = \frac{\Delta m}{m} \times 100\% + \frac{2\Delta r}{r} \times 100\% + \frac{\Delta l}{l} \times 100\%$$

$$\text{So, } \frac{\Delta\rho}{\rho} \times 100\% = \frac{0.01}{0.4} \times 100\% + \frac{2 \times 0.03}{6} \times 100\% + \frac{0.04}{8} \times 100\%$$

$$= 2.5 + 1 + 0.5 = 4\%$$

05. *(c)*

> **key idea** Error in measurement always adds up. Therefore, some options can be eliminated easily.

$$Z = \frac{A^2 B^3}{C^4}$$

Relative error in Z will be

$$\frac{\Delta Z}{Z} = 2\frac{\Delta A}{A} + 3\frac{\Delta B}{B} + 4\frac{\Delta C}{C}$$

06. *(c)* $\because$ Density, $\rho = \frac{\text{Mass}}{\text{Volume}} = \frac{M}{L^3}$

$\Rightarrow$ Error in density

$$\frac{\Delta\rho}{\rho} = \frac{\Delta M}{M} + \frac{3\Delta L}{L}$$

So, maximum % error in measurement of ρ will be

$$\frac{\Delta\rho}{\rho} \times 100 = \frac{\Delta M}{M} \times 100 + \frac{3\Delta L}{L} \times 100$$

or % error in density $= 1.5 + 3 \times 1$

% error = 4.5%

07. *(a)* From Ohm's law,

$$R = \frac{V}{I}$$

Fractional error is resistance will be given by $\frac{\Delta R}{R} = \frac{\Delta V}{V} + \frac{\Delta I}{I} = 3\% + 3\% = 6\%$

08. *(a)* The reliable digit plus the first uncertain digit is known as significant figures.

For the number 23.023, all the non-zero digits and zeroes, between non-zero digits are significant, hence 5.

For the number 0.0003, number is less than 1, the zero(s) on the right of decimal point but to the left of the first non-zero digit are not significant, hence 1.

For the number 2.1×10^{-3}, significant figures are 2.

09. *(a)* $Q = X^{-2}Y^{3/2}Z^{-2/5}$

$\therefore$ Maximum fractional error in Q,

$$\frac{\Delta Q}{Q} = 2\frac{\Delta X}{X} + \frac{3}{2}\frac{\Delta Y}{Y} + \frac{2}{5}\frac{\Delta Z}{Z}$$

$$= 2(0.1) + \frac{3}{2}(0.2) + \frac{2}{5}(0.5)$$

$$= 0.2 + 0.3 + 0.2 = 0.7$$

10. *(a)* Given,

$$y = \frac{32 \cdot 3 \times 1125}{27 \cdot 4} = 1326 \cdot 186$$

Least significant digits are 3 in operands, so results should be rounded off to 3 digits.

$$\therefore \quad y = 1330$$

11. *(a)* Resistivity of a wire is given by,

$$\rho = \frac{R \times A}{l} \Rightarrow \rho = \frac{R \times \pi r^2}{l}$$

So, percentage error in resistivity is,

$$\frac{\Delta\rho}{\rho} \times 100 = \frac{\Delta R}{R} \times 100 + 2\frac{\Delta r}{r} \times 100 + \frac{\Delta l}{l} \times 100$$

Substituting given values we get, percentage error in resistivity is,

$$= \frac{10}{100} \times 100 + 2 \times \frac{0.05}{0.35} \times 100 + \frac{0.2}{15} \times 100$$

$$= (0.1 + 0.286 + 0.013) \times 100$$

$$= 0.399 \times 100 = 39.9\%$$

12. *(c)* $Q = \dfrac{a^4 b^3}{c^2}$

$\therefore \dfrac{\Delta Q}{Q} = 4\dfrac{\Delta a}{a} + 3\dfrac{\Delta b}{b} + 2\dfrac{\Delta c}{c}$

$\Rightarrow \dfrac{\Delta Q}{Q} \times 100$

$= 4 \times \dfrac{\Delta a}{a} \times 100 + 3 \times \dfrac{\Delta b}{b} \times 100 + 2 \times \dfrac{\Delta c}{c} \times 100$

$= 4 \times 3\% + 3 \times 4\% + 2 \times 5\%$

$= 34\%$

13. *(b)* Given, $P = \dfrac{a^2 b^3}{c\sqrt{d}}$

So, maximum % error in measurement of P is,

$\dfrac{\Delta P}{P} \times 100\% = 2\left(\dfrac{\Delta a}{a} \times 100\%\right) + 3\left(\dfrac{\Delta b}{b} \times 100\%\right) + \left(\dfrac{\Delta c}{c} \times 100\%\right) + \dfrac{1}{2}\left(\dfrac{\Delta d}{d} \times 100\%\right)$

$= \left(2 \times 1 + 3 \times 2 + 3 + \dfrac{1}{2} \times 4\right)\%$

$= 2 + 6 + 3 + 2$

$= 13\%$

14. *(d)* Given, $\dfrac{\Delta R}{R} \times 100 = 1\%$

$\dfrac{\Delta I}{I} \times 100 = 2\%$

and $\dfrac{\Delta t}{t} \times 100 = 3\%$

$\therefore$ Heat produced in R due to I,

$H = I^2 Rt$

$\therefore \dfrac{\Delta H}{H} \times 100 = 2\left(\dfrac{\Delta I}{I} \times 100\right) + \dfrac{\Delta R}{R} \times 100 + \dfrac{\Delta t}{t} \times 100$

$= 2 \times 2\% + 1\% + 3\% = 8\%$

15. *(a)* We know,

$\rho = \dfrac{m}{V} = \dfrac{m}{\pi r^2 h}$

$\Rightarrow \dfrac{\Delta \rho}{\rho} = \dfrac{\Delta m}{m} + \dfrac{2\Delta r}{r} + \dfrac{\Delta h}{h}$

$= \dfrac{0.006}{0.6} + \dfrac{2 \times 0.005}{0.5} + \dfrac{0.04}{4}$

$= 0.01 + 2 \times 0.01 + 0.01$

$= 0.01 + 0.02 + 0.01 = 0.04$

So, $\dfrac{\Delta \rho}{\rho} \times 100 = 4\%$

16. *(b)* The formula of Young's modulus of elasticity is given by,

$Y = \dfrac{MgL^3}{4bd^3 \delta}$

where, symbols have their usual meaning

There is no error in the value of the g.

The fractional error in Y,

$\dfrac{\Delta Y}{Y} = \dfrac{\Delta M}{M} + 3\dfrac{\Delta L}{L} + \dfrac{\Delta b}{b} + 3\dfrac{\Delta d}{d} + \dfrac{\Delta \delta}{\delta}$

$\therefore \dfrac{\Delta Y}{Y} = \dfrac{10^{-3}}{2} + 3\dfrac{(1 \times 10^{-3})}{1} + \dfrac{0.1 \times 10^{-3}}{4 \times 10^{-2}} + 3\dfrac{(0.01 \times 10^{-3})}{0.4 \times 10^{-2}} + \dfrac{(0.01 \times 10^{-3})}{5 \times 10^{-3}}$

$\dfrac{\Delta Y}{Y} = 0.0155$

The fractional error in is 0.0155.

17. *(c)* Given, $\dfrac{\Delta L}{L} \times 100 = 0.1$

$\therefore \quad T = 2\pi\sqrt{\dfrac{l}{g}}$

$\therefore$ The % change in T,

$\dfrac{\Delta T}{T} \times 100 = \dfrac{1}{2}\dfrac{\Delta L}{L} \times 100$

$\Rightarrow \dfrac{\Delta T}{T} \times 100 = \dfrac{1}{2} \times 0.1$

$\Rightarrow \quad \Delta T = 0.05 \times \dfrac{T}{100}$

In one single day,

$T = 24 \times 60 \times 60 = 86400\text{s}$

$\therefore \quad \Delta T = 0.05 \times \dfrac{86400}{100} = 43.2 \text{ s}$

Thus, the error in time per day is 43.2 s.

18. *(c)* Given, $z = \dfrac{a^2 b^{2/3}}{\sqrt{c}\, d^3}$

According to question,

% error in $z = (2)\%$ error in $a + \left(\dfrac{2}{3}\right)\%$ error in $b + \left(\dfrac{1}{2}\right)\%$ error in $c + (3)\%$ error in d

$\dfrac{\Delta z}{z} \times 100$

$= \left(2\dfrac{\Delta a}{a} + \dfrac{2}{3}\dfrac{\Delta b}{b} + \dfrac{1}{2}\dfrac{\Delta c}{c} + 3\dfrac{\Delta d}{d}\right) \times 100$

$= 2 \times 2\% + \dfrac{2}{3} \times 1.5\% + \dfrac{1}{2} \times 4\% + 3 \times 2.5\% = 14.5\%$

19. *(b)* Volume of a cylinder of radius r and height h is given by

$V = \pi r^2 h = \dfrac{\pi D^2 h}{4}$

Here, $D = 12.6$ cm

and $h = 34.2$ cm

$\Rightarrow \quad V = \dfrac{\pi}{4} \times (12.6)^2 \times (34.2)$

$V = 4262.22 \text{ cm}^3$

$V = 4260$ (in three significant numbers)

Now, error calculation can be done as

$\dfrac{\Delta V}{V} = 2\left(\dfrac{\Delta D}{D}\right) + \dfrac{\Delta h}{h}$

$= \dfrac{2 \times 0.1}{12.6} + \dfrac{0.1}{34.2}$

$\Rightarrow \dfrac{\Delta V}{V} = 0.0158 + 0.0029$

$\Rightarrow \Delta V = (0.01879) \times (4262.22)$

$\Rightarrow \Delta V = 80.01 = 80 \text{ cm}^3$

(In one significant digit)

$\therefore$ For proper significant numbers, volume reading will be

$V = 4260 \pm 80 \text{ cm}^3$

20. *(a)* Given, $I = (e^{1000V/T} - 1)$ mA

$dV = \pm 0.01$ V, $T = 300$ K

$I = 5$ mA

$\Rightarrow \quad I = e^{1000V/T} - 1$

$I + 1 = e^{1000V/T}$

Taking log on both sides, we get

$\log(I + 1) = \dfrac{1000V}{T}$

$\Rightarrow \dfrac{d(I+1)}{I+1} = \dfrac{1000}{T} dV$

$\Rightarrow \dfrac{dI}{I+1} = \dfrac{1000}{T} dV$

$\Rightarrow \quad dI = \dfrac{1000}{T} \times (I + 1)\, dV$

$dI = \dfrac{1000}{300} \times (5 + 1) \times 0.01$

$= 0.2$ mA

So, error in the value of current is 0.2 mA.

21. *(c)*

$m_1 + m_2 + m_3 = 435.42 + 226.3 + 0.125$

$= 661.845$

According to least significant digits,

$m_1 + m_2 + m_3 = 661.8$ g

22. *(a)* Density $= \dfrac{\text{Mass}}{\text{Volume}}$

$\rho = \dfrac{M}{\pi R^2 l} = \dfrac{MR^{-2} l^{-1}}{\pi} = \dfrac{1}{\pi} MR^{-2} l^{-1}$

$$\ln \rho = \ln\left(\frac{1}{\pi}\right) + \ln M + \ln R^{-2} + \ln l^{-1}$$

After differentiation -

$$\frac{d\rho}{\rho} = 0 + \frac{dM}{M} + \frac{2dR}{R} + \frac{dl}{l}$$

$$\frac{d\rho}{\rho} \times 100 = \frac{dM}{M} \times 100 + \frac{2dR}{R} \times 100 + \frac{dl}{l} \times 100$$

$$\frac{d\rho}{\rho} \times 100 = \left(\frac{dM}{M} + \frac{2dR}{R} + \frac{dl}{l}\right) \times 100$$

$$= \left(\frac{0.003}{0.6} + \frac{2 \times 0.01}{0.5} + \frac{0.05}{10}\right) \times 100$$

$$\frac{d\rho}{\rho} \times 100 = 5\%$$

23. *(b)* $E = \alpha^3 e^{-\beta t}$

$$\ln E = \ln(\alpha^3 e^{-\beta t}) = \ln \alpha^3 + \ln e^{-\beta t}$$

$$= 3 \ln \alpha - \beta t \ln e$$

$$\ln E = 3 \ln \alpha - \beta t \quad \text{...(i) } [\ln e = 1]$$

Differentiate Eq. (i)

$$\frac{dE}{E} \times 100 = 3\frac{d\alpha}{\alpha} \times 100 - \beta(100)dt \quad \text{...(ii)}$$

$$\frac{d\alpha}{\alpha} \times 100 = 1.2\%, \beta = 0.3 \text{ s}^{-1}$$

$$\frac{dt}{t} \times 100 = 1.6\%$$

Put the value in Eq. (ii)

$$\frac{dE}{E} \times 100 = 3(1.2) + 0.3 \times 1.6 \times 5$$

$$= 3.6 + 1.5 \times 1.6$$

$$= 3.6 + 2.4 = 6\%$$

24. *(b)* Time period of simple pendulum is,

$$T = 2\pi\sqrt{\frac{l}{g}} \Rightarrow g = \frac{4\pi^2 l}{T^2}$$

$$\therefore \quad \frac{\Delta g}{g} \times 100 = \frac{\Delta l}{l} \times 100 + 2\frac{\Delta T}{T} \times 100$$

⇒ Percentage accuracy in the measurement of acceleration due to gravity is,

Percentage accuracy

$$= \frac{\Delta l}{l} \times 100 + 2\frac{\Delta T}{T} \times 100$$

Here, $\Delta l = 2$ mm, $l = 20$ cm

$\Delta T = 1$ s, $T = 40$ s

So, percentage accuracy

$$= \frac{2 \times 10^{-1}}{20} \times 100 + \frac{2 \times 1}{40} \times 100$$

$$= \left(\frac{2}{200} + \frac{2}{40}\right) \times 100 = \frac{12}{2} = 6\%$$

25. *(d)* Given, $R_1 = (10 \pm 0.5)\,\Omega$

$R_2 = (15 \pm 0.5)\,\Omega$

$$\frac{1}{R_{eq}} = \frac{1}{R_1} + \frac{1}{R_2} \quad \text{(Parallel combination)}$$

$$= \frac{1}{10} + \frac{1}{15}$$

$$\frac{1}{R_{eq}} = \frac{5}{30} = \frac{1}{6} \Rightarrow R_{eq} = 6$$

Now for error calculation.

$$\frac{dR_{eq}}{R_{eq}^2} = \frac{dR_1}{R_1^2} + \frac{dR_2}{R_2^2}$$

$$\frac{dR_{eq}}{36} = \frac{dR_1}{100} + \frac{dR_2}{225}$$

$$\frac{dR_{eq}}{36} = \frac{0.5}{100} + \frac{0.5}{225}$$

$$dR_{eq} = 36 \times 0.5 \times \frac{325}{225 \times 100}$$

$$= 36 \times 0.5 \times \frac{13}{900} = 18 \times \frac{13}{900}$$

$$= \frac{26}{100} = 0.26$$

$$\text{Now, } \frac{dR_{eq}}{R_{eq}} \times 100 = \frac{0.26}{6} \times 100 = \frac{26}{6}$$

$$= 4.33\%$$

26. *(15)* $C = \dfrac{pq^2}{r^3\sqrt{s}}$

% error in $p = 1$

$q = 2$

$x = 3$

$s = \frac{1}{2}$

$$\%\frac{\Delta C}{C} = \left[1\left(\frac{\Delta p}{p}\right) + 2\left(\frac{\Delta q}{q}\right) + 3\left(\frac{\Delta r}{r}\right) + \frac{1}{2}\left(\frac{\Delta s}{s}\right)\right]\%$$

$$\Rightarrow \quad \left[1(1) + 2(2) + 3(3) + \frac{1}{2}(2)\right]\%$$

$$\Rightarrow \quad (1 + 4 + 9 + 1)\%$$

$$\Rightarrow \quad 15\%$$

27. *(77)* $a = 60 \pm 3$

$$\frac{\Delta a}{a} \times 100 = \frac{3}{60} \times 100 = 5\%$$

$$b = 20 \pm 0.1$$

$$\frac{\Delta b}{b} \times 100 = \frac{0.1}{20} \times 100 = 0.5\%$$

$$c = 40 \pm 0.2$$

$$\frac{\Delta c}{c} \times 100 = \frac{0.2}{40} \times 100 = 0.5\%$$

$$d = 50 \pm 0.1$$

$$\frac{\Delta d}{d} \times 100 = \frac{0.1}{50} \times 100 = 0.2\%$$

Now, $Q = \dfrac{ab^4}{cd}$

Percentage error in Q is given by

$$\frac{\Delta Q}{Q} \times 100 = \frac{\Delta a}{a} \times 100 + 4\frac{\Delta b}{b} \times 100 + \frac{\Delta c}{c} \times 100 + \frac{\Delta d}{d} \times 100$$

$$= 5\% + 4 \times 0.5\% + 0.5\% + 0.2\%$$

$$= 7.7\%$$

Now, $\dfrac{7.7}{100} = \dfrac{x}{1000}$

$\Rightarrow \quad x = 77$

28. *(5)* Given, $R = V / I$ ...(i)

where, $V = (50 \pm 2)$ V

$I = (20 \pm 0.2)$ A

From Eq. (i)

$$\frac{\Delta R}{R} \times 100\% = \frac{\Delta V}{V} \times 100\% + \frac{\Delta I}{I} \times 100\%$$

$$\% \text{ error in } R = \left[\frac{2}{50} \times 100 + \frac{0.2}{20} \times 100\right]\%$$

$$= [2 \times 2 + 0.2 \times 5]\% = 5\%$$

Thus, the value of $x = 5$.

29. *(1050)* Given, percentage error in mass,

$$\frac{\Delta m}{m} \times 100 = 6\%$$

Percentage error in diameter,

$$\frac{\Delta d}{d} \times 100 = 1.5\%$$

Density of sphere,

$$\rho = \frac{\text{Mass}}{\text{Volume}} = \frac{M}{\frac{4}{3}\pi r^3} = \frac{M}{\frac{4}{3}\pi\left(\frac{d}{2}\right)^3}$$

$$\rho = \frac{6}{\pi} Md^{-3} \text{ or } \rho \propto Md^{-3}$$

For maximum error in density,

$$\frac{\Delta \rho}{\rho} \times 100\% = \frac{\Delta M}{M} \times 100\% + 3 \times \frac{\Delta d}{d} \times 100\%$$

$$= 6\% + 3 \times 1.5\%$$

$$\frac{\Delta \rho}{\rho} \times 100 = 10.5\%$$

$$\frac{1050}{100}\% = \frac{x}{100}\% \quad \text{(given)}$$

$$\therefore \quad x = 1050$$

Topic 3

01. (a) We know that,

$$c = \frac{1}{\sqrt{\mu_0 \varepsilon_0}}$$

$$\Rightarrow c^2 = \frac{1}{\mu_0 \varepsilon_0} \Rightarrow \left[\frac{1}{\mu_0 \epsilon_0}\right] = [c^2]$$

$$= [(LT^{-1})^2] = [L^2T^{-2}] = \left[\frac{L^2}{T^2}\right]$$

02. (d) Since, $F = \frac{Gm_1m_2}{r^2}$

$$\Rightarrow \quad [G] = \left[\frac{Fr^2}{m_1m_2}\right]$$

$$= \left[\frac{MLT^{-2}L^2}{M^2}\right] = [M^{-1}L^3T^{-2}]$$

Gravitational potential energy,

$$[U] = [ML^2T^{-2}]$$

Gravitational potential,

$$V = \frac{W}{m}$$

$$[V] = \frac{[ML^2T^{-2}]}{[M]} = [L^2T^{-2}]$$

Acceleration due to gravity,

$$[g] = [LT^{-2}]$$

03. (d) We know that

$$L = \frac{\mu_0 N^2 A}{l}$$

and $C = \frac{\varepsilon_0 A}{l}$

Thus, $\frac{L}{C} = \frac{\mu_0 N^2}{\varepsilon_0} \Rightarrow \left[\frac{L}{C}\right] = \left[\frac{\mu_0}{\varepsilon_0}\right]$

Since, N is dimensionless

$$\therefore \sqrt{\left[\frac{\mu_0}{\varepsilon_0}\right]} = \sqrt{\frac{L}{C}} = \sqrt{\frac{\tau R}{\tau / R}} = \sqrt{R^2} = R$$

Thus, dimension of $\sqrt{\left[\frac{\mu_0}{\varepsilon_0}\right]}$

= Dimension of $[R]$

04. (b) The relation between field and permeability is

$$B = \mu_0 ni$$

$$\frac{B}{\mu_0} = ni$$

For $n = \frac{N}{L}$, for $i = A$

$$\frac{B}{\mu_0} = [L^{-1}A]$$

05. (a) $C = \frac{q}{V} = \frac{q \cdot q}{V \cdot q} = \frac{q^2}{W} \quad \left[V = \frac{W}{q}\right]$

$$= \frac{[C^2]}{[ML^2T^{-2}]} = [M^{-1}L^{-2}T^2C^2]$$

06. (d) Latent heat is specific heat

$$Q = ML$$

$$L = \frac{Q}{M} = \frac{[ML^2T^{-2}]}{[M]}$$

$$L = [M^0L^2T^{-2}]$$

07. (d) Energy density of electric field is

$$u_E = \frac{1}{2}\varepsilon_0 E^2$$

$\therefore$ Dimensions of $\varepsilon_0 E^2$ are dimension of energy per unit volume

$$= \frac{[ML^2T^{-2}]}{[L^3]} = [ML^{-1}T^{-2}]$$

08. (d) Angular Impulse = Torque × Time
= Force × Distance × Time
$= [MLT^{-2}] \times [L] \times [T] = [ML^2T^{-1}]$

09. (b)

Term	SI unit	Dimension
Coefficient of viscosity	$Ns\,m^{-2}$	$ML^{-1}T^{-1}$
Surface tension	Nm^{-1}	ML^0T^{-2}
Angular momentum	$kg\,m^2s^{-1}$	ML^2T^{-1}
Rotational kinetic energy	J or $kg\,m^2s^{-2}$	ML^2T^{-2}

Therefore, the correct match is
A-III, B-IV, C-II, D-I

10. (a) $\rho = KF^a v^b t^c$

K is dimensionless constant

$$[ML^{-3}] = [MLT^{-2}]^a [LT^{-1}]^b [T]^c$$

$$[ML^{-3}] = [M^a L^{a+b} T^{-2a-b+c}]$$

$$a = 1, \; a + b = -3, \; 1 + b = -3$$

$$b = -4, \; -2a - b + c = 0$$

$$-2 \times 1 - (-4) + c = 0$$

$$-2 + 4 + c = 0$$

$$2 + c = 0, \; c = -2$$

$$\rho = KF^1 v^{-4} t^{-2}$$

$$= [Fv^{-4}t^{-2}]$$

11. (a) Speed of light in vacuum is given by $c = \frac{1}{\sqrt{\mu_0 \varepsilon_0}} \Rightarrow c^2 = \frac{1}{\mu_0 \varepsilon_0}$

Hence dimensions of $\frac{1}{\mu_0 \varepsilon_0}$ are dimensions of

$$c^2 = [LT^{-1}]^2 = [L^2T^{-2}]$$

12. (a) Speed of wave produced in water,

$$v = \lambda^a g^b \rho^c \quad \text{...(i)}$$

where, λ = wavelength, ρ = density,

$\therefore$ Writing the dimensions of each term of Eq. (i), we get

$$[LT^{-1}] = [L]^a [LT^{-2}]^b [ML^{-3}]^c$$

$$\Rightarrow [LT^{-1}] = [M^c L^{a+b-3c} T^{-2b}]$$

Comparing the powers, we get

$$c = 0, \; -2b = -1 \Rightarrow b = \frac{1}{2}$$

$$a + b - 3c = 1$$

$$\Rightarrow a + b - 0 = 1 \Rightarrow a + \frac{1}{2} = 1$$

$$\Rightarrow \quad a = \frac{1}{2}$$

13. (c) Planck's constant, $h = \frac{E}{f}$

$$= \frac{[ML^2T^{-2}]}{[T^{-1}]} = [ML^2T^{-1}]$$

So, A → III.

Stopping potential,

$$V = \frac{U}{q} = \frac{[ML^2T^{-2}]}{[A] \cdot [T]}$$

$$= [ML^2T^{-3}A^{-1}]$$

So, B → IV

Work function = Minimum energy
$= [ML^2T^{-2}]$

So, C → I

Momentum = Mass × Velocity
$= [MLT^{-1}]$

So, D → II

14. (b) We have,

$$\text{Pressure gradient} = \frac{dp}{dx}$$

$$= \frac{[ML^{-1}T^{-2}]}{[L]} = [ML^{-2}T^{-2}]$$

(A → III)

$$\text{Energy density} = \frac{\text{Energy}}{\text{Volume}} = \frac{[ML^2T^{-2}]}{[L^3]}$$

$$= [ML^{-1}T^{-2}]$$

(B → II)

$$\text{Electric field} = \frac{\text{Force}}{\text{Charge}} = \frac{[MLT^{-2}]}{[AT]}$$

$$= [MLT^{-3}A^{-1}]$$

(C → IV)

$$\text{Latent heat} = \text{Energy/mass} = \frac{[ML^2T^{-2}]}{[M]}$$

$$= [M^0L^2T^{-2}]$$

(D → I)

15. *(b)* Angular momentum, $L = mvr$

$\Rightarrow$ $[L] = [M][LT^{-1}][L] = [ML^2T^{-1}]$

Torque, $\tau = rF$

$\Rightarrow$ $\tau = [L][MLT^{-2}] = [ML^2T^{-2}]$

Stress, $\sigma = \dfrac{F}{A}$

$\Rightarrow$ $\sigma = \dfrac{[MLT^{-2}]}{[L^2]} = [ML^{-1}T^{-2}]$

Pressure gradient $= \Delta p / \Delta x$

$\Rightarrow$ $\dfrac{\Delta p}{\Delta x} = \dfrac{[ML^{-1}T^{-2}]}{[L]} = [ML^{-2}T^{-2}]$

Correct match code is A-III, B-I, C-IV, D-II.

16. *(a)* Given, SI unit of a quantity is pascal-second.

Pressure × Time $= [ML^{-1}T^{-2}][T]$

$= [ML^{-1}T^{-1}]$

17. *(a)* Since we know,

Torque τ = force × perpendicular distance

$= [MLT^{-2}][L] = [ML^2T^{-2}]$

Impulse I = force × time

$= [MLT^{-2}][T] = [MLT^{-1}]$

Tension force $= MLT^{-2}$

Surface tension, $S = \dfrac{F}{L} = MT^{-2}$

18. *(a)* Charge, $q = CV$ and $V = \dfrac{W}{q}$

$\Rightarrow$ $C = \dfrac{q}{V} = \dfrac{(A \times T)^2}{ML^2T^{-2}}$

$= [M^{-1}L^{-2}T^4A^2]$

Electric Field, $E = \dfrac{F}{q} = \dfrac{[MLT^{-2}]}{[AT]}$

$= [MLT^{-3}A^{-1}]$

Force, $F = \dfrac{q_1 q_2}{4\pi\varepsilon_0 r^2}$

$\Rightarrow$ $\varepsilon_0 = [M^{-1}L^{-3}T^4A^2]$

Speed of light,

$c = \dfrac{1}{\sqrt{\mu_0\varepsilon_0}} \Rightarrow \mu_0 = \dfrac{1}{\varepsilon_0 c^2}$

$= \dfrac{1}{[M^{-1}L^{-3}T^4A^2][LT^{-1}]^2}$

$= [M^1L^1T^{-2}A^{-2}]$

19. *(a)* As, $\dfrac{B^2}{2\mu_0}$ = Energy density of magnetic field $= \dfrac{\text{Energy}}{\text{Volume}}$

So, $\left[\dfrac{B^2}{2\mu_0}\right]$ = [Energy /Volume]

$= \dfrac{[ML^2T^{-2}]}{[L^3]} = [ML^{-1}T^{-2}]$

20. *(c)* L/R is a time constant,

So, $\dfrac{L}{R} = [T]$

$CV = Q$

So, $\left[\dfrac{L}{R \times CV}\right] = \left[\dfrac{T}{Q}\right] = [A^{-1}]$

21. *(b)* We know that,

$\sigma = \dfrac{ne^2\tau}{m}$

$[\sigma] = \dfrac{[L^{-3}I^2T^2T]}{[M]} = [M^{-1}L^{-3}I^2T^3]$

22. *(c)* According to question,

$[t] = [r]^b [s]^{c/2} [d]^{a/4}$

$[T] = [L]^b [MT^{-2}]^{c/2} [ML^{-3}]^{a/4}$

So, $\dfrac{c}{2} + \dfrac{a}{4} = 0$, $c = -1$ and $b - \dfrac{3a}{4} = 0$

On solving above, we get

$b = 3/2$

23. *(c)* From the relation

$F = qvB$

$\Rightarrow$ $[MLT^{-2}] = [C][LT^{-1}][B]$

$\Rightarrow$ $[B] = [MC^{-1}T^{-1}]$

24. *(c)* Moment of inertia, $I = mr^2$

$\therefore [I] = [ML^2]$

and τ = Moment of force $= \mathbf{r} \times \mathbf{F}$

$\therefore$ $[\tau] = [L][MLT^{-2}] = [ML^2T^{-2}]$

Therefore, moment of inertia and moment of force do not have identical dimensions.

25. *(c)* Coefficient of viscosity is given as

$\eta = \dfrac{F}{A(\Delta v_x / \Delta x)}$

$\therefore$ Dimensions of η

$= \dfrac{\text{Dimensions of force}}{\text{Dimensions of area} \times \text{Dimensions of velocity gradient}}$

$= \dfrac{[MLT^{-2}]}{[L^2][T^{-1}]} = [ML^{-1}T^{-1}]$

26. *(b)* Planck's constant (in terms of unit)

h = J-s $= [ML^2T^{-2}][T]$

$= [ML^2T^{-1}]$

Momentum (p) = kg - ms^{-1}

$= [M][L][T^{-1}] = [MLT^{-1}]$

Hence, momentum (p) and Planck's constant do not have same dimensions.

27. *(b)* Coefficient of viscosity

$F = 6\pi\eta r v_T$

$\eta = \dfrac{F}{6\pi r v_T} = \dfrac{[MLT^{-2}]}{[L][LT^{-1}]}$

$[\eta] = [ML^{-1}T^{-1}]$

Intensity of wave, $I = \dfrac{P}{A} = \dfrac{P}{4\pi R^2}$

$P \to$ Power

$A \to$ Area

$[P] = [ML^2T^{-3}]$

$I = \dfrac{[ML^2T^{-3}]}{[L^2]} = [ML^0T^{-3}]$

$[I] = [ML^0T^{-3}]$

Pressure gradient $= \dfrac{\text{Pressure difference}}{\text{Distance}}$

$= \dfrac{\text{Pascal}}{\text{Metre}} = \dfrac{[ML^{-1}T^{-2}]}{[L]}$

$= [ML^{-2}T^{-2}]$

Compressibility $= \dfrac{1}{\text{Bulk Modulus}}$

$= \dfrac{1}{\dfrac{-p}{\dfrac{\Delta V}{V}}} = [M^{-1}L^1T^{-2}]$

A-(IV), B-(I), C-(II), D-(III)

28. *(c)* As we know,

$[\mu_0] = [MLT^{-2}I^{-2}]$

$[q] = [IT]$

$\therefore [q\mu_0 I] = [q][\mu_0][I]$

$= [IT][MLT^{-2}I^{-2}][I] = [MLT^{-1}]$

= Dimension of momentum.

29. *(a)* [Mass density] $= \left[\dfrac{M}{V}\right] = [ML^{-3}]$

[Impulse] $= [F][\Delta t] = [MLT^{-2}][T]$

Power, $[P] = \left[\dfrac{W}{t}\right] = \dfrac{[ML^2T^{-2}]}{[T]}$

$= [ML^2T^{-3}]$

Moment of inertia,

$[I] = [M][r^2] = [M][L^2] = [ML^2]$

30. *(d)* $\dfrac{\phi_E}{\phi_M} = \dfrac{EA}{BA} = \dfrac{E}{B} = c = [LT^{-1}]$

$B = \dfrac{[MLT^{-2}]}{[ATLT^{-1}]}$

$F = qvB$

$B = \dfrac{F}{qv}$

$[q] = [AT]$

$B = [MT^{-2}A^{-1}]$

$[F] = [MLT^{-2}]$

$[v] = [LT^{-1}]$

$[E] = \frac{F}{q} = \frac{[MLT^{-2}]}{[AT]} = [MLT^{-3}A^{-1}]$

So $\left[\frac{E}{B}\right] = \frac{ML^{-3}A^{-1}}{MT^{-2}A^{-1}} = [LT^{-1}]$

or $c = \frac{E}{B}$, where c is speed of light

$[c] = [LT^{-1}]$

After compare

$Q = 1, R = -1$

31. *(a)* Electric dipole moment, $\mathbf{p} = q \times 2l$

Magnetic dipole moment, $\mathbf{M} = IA$

$\left[\frac{p}{M}\right] = \frac{q \times 2l}{I \times A} = \frac{It \times 2l}{I \times A}$

$= \frac{t \times 2l}{A} = \frac{t \times 2l}{l \times x} = [L^{-1}T^{1}]$

$= [M^0L^{-1}T^1A^0]$

After comparing $[M^PL^QT^RA^S]$

P and Q are 0, −1.

32. *(c)* $v = At^2 + \frac{Bt}{C+t}$

By principle of homogeneity

1. $C = [T]$

2. $v = At^2$

$A = \frac{v}{t^2} = \frac{[LT^{-1}]}{[T^2]} = [LT^{-3}]$

3. $v = \frac{Bt}{t}$

$B = [LT^{-1}]$

Now $ABC = [LT^{-3}][LT^{-1}][T]$

$= [L^2T^{-3}] = [M^0L^2T^{-3}]$

33. *(a)* a. Surface tension $= [MT^{-2}]$

Impulse $= [MLT^{-1}]$

b. Angular momentum $= [ML^2T^{-1}]$

Planck's constant $= [ML^2T^{-1}]$

c. Pressure $= [ML^{-1}T^{-2}]$

Young's modulus $= [ML^{-1}T^{-2}]$

d. Torque $= [ML^2T^{-2}]$

Energy $= [ML^2T^{-2}]$

34. *(b)* $[E] = [M^1L^2T^{-2}]$

$[pc] = [MLT^{-1}L^1T^{-1}]$

$p \to$ momentum, $c \to$ speed of light

$= [ML^2T^{-2}]$

$[mc^2] = [ML^2T^{-2}]$

$[E^2] = [M^2L^4T^{-4}]$

$E^2 = p^2c^2 + m^2c^4$ is correct

Alternate Solution

We know that energy in terms of momentum in relativistic case, $E = pc$

and energy in rest mass, $E = mc^2$

$\therefore \quad E^2 = p^2c^2 + m^2c^4$

35. *(d)* $\phi = \alpha\sigma + \beta\lambda$

$[\phi] = [\alpha\sigma] = [\beta\lambda]$

$[\alpha] = \frac{[\phi]}{[\sigma]}, [\beta] = \frac{[\phi]}{[\lambda]}$

$\frac{[\alpha]}{[\beta]} = \frac{\frac{[\phi]}{[\sigma]}}{\frac{[\phi]}{[\lambda]}} = \frac{[\lambda]}{[\sigma]} = \frac{\left[\frac{Q}{L}\right]}{\left[\frac{Q}{A}\right]} = [L]$

$\frac{[\alpha]}{[\beta]} = [L] =$ Displacement

36. *(b)* We know that,

$\Delta Q = ms\Delta T$

$\Rightarrow \quad s = \frac{\Delta Q}{m\Delta T} \Rightarrow [s] = \frac{[Q]}{[m][\Delta T]}$

$= \frac{[ML^2T^{-2}]}{[M][K]} = [L^2T^{-2}K^{-1}]$

Thus, Statement I is correct.

Again, using ideal gas equation

$pV = nRT \Rightarrow R = \frac{pV}{nT}$

$\Rightarrow \quad [R] = \frac{[p][V]}{[n][T]} = \frac{[ML^{-1}T^{-2}][L^3]}{[mol][K]}$

$= [ML^2T^{-2} \text{ mol}^{-1} K^{-1}]$

Thus, Statement II is incorrect.

37. *(b)* $\left(p + \frac{a}{V^2}\right)(V - b) = RT$

Dimensional formula of $b = [V]$

$\Rightarrow \quad [b] = [L^3]$

$\therefore \quad [b^{-1}] = [L^{-3}]$

$\left[\frac{a}{V^2}\right] = [p] = [ML^{-1}T^{-2}]$

$\therefore \quad [a] = [ML^{-1}T^{-2}][L^6]$

$= [ML^5T^{-2}]$

$\therefore \quad [ab^{-1}] = [ML^5T^{-2}][L^{-3}]$

$= [ML^2T^{-2}]$

38. *(a)* We know that,

$F = \frac{Gm_1m_2}{r^2}$

$\therefore \quad G = \frac{[F][r^2]}{[m_1][m_2]} = \frac{[MLT^{-2}][L^2]}{[M^2]}$

$= [M^{-1}L^3T^{-2}]$

$\therefore$ For option (a),

$[T^2] = \frac{[r^3]}{[G][M]}$

$\Rightarrow [T^2] = \frac{[L^3]}{[M^{-1}L^3T^{-2}][M]} = [T^2]$

Thus, option (a) is correct.

39. *(c)* Compare the given equation with standard equation of standing wave.

$y = A \sin \omega t \cos kx$

Thus, $\omega = \frac{2\pi n}{\lambda}, k = \frac{2\pi}{\lambda}$

2π is dimensionless

$[\omega] = T^{-1} = \left[\frac{n}{\lambda}\right]$

$[nT] = [\lambda] = [L]$

$[n] = \left[\frac{\lambda}{T}\right] = [\lambda\omega]$

$[n] = [LT^{-1}]$

$\Rightarrow \quad \left[\frac{n}{\lambda}\right] = [T^{-1}]$

Hence, $\frac{n}{\lambda}$ has dimension of T^{-1} not T.

40. *(b)* Given force, $F = ax^2 + bt^{1/2}$

$x =$ distance and $t =$ time

Clearly, dimensions of ax^2 and $bt^{1/2}$ must be that of force.

So, $\quad [ax^2] = [MLT^{-2}]$

$\Rightarrow \quad [a][L^2] = [MLT^{-2}]$

or $\quad [a] = [ML^{-1}T^{-2}]$

Also, $\quad [bt^{1/2}] = [MLT^{-2}]$

$[b][T^{1/2}] = [MLT^{-2}]$

$[b] = [MLT^{-5/2}]$

So, $\quad \left[\frac{b^2}{a}\right] = \left[\frac{M^2L^2T^{-5}}{ML^{-1}T^{-2}}\right]$

$= [ML^3T^{-3}]$

41. *(c)* Given,

$m = k\, c^P \cdot G^{-1/2} \cdot h^{1/2}$

Substituting dimensions of different quantities on both sides we get,

$[M] = [L^PT^{-P}][M^{-1}L^3T^{-2}]^{\frac{-1}{2}}$

$[ML^2T^{-1}]^{1/2}$

$$\Rightarrow \quad [M] = [M]\left[L^{p-\frac{1}{2}}\right]\cdot\left[T^{\frac{1}{2}-p}\right]$$

Equating dimensions of length on both sides, we get

$$p - \frac{1}{2} = 0 \text{ or } p = \frac{1}{2}$$

42. *(b)* $\left(p + \frac{a}{V^2}\right)(V - b) = RT$

As $[V] = [b]$

$\therefore \quad [b] = [L^3]$ and $[p] = [a]/[V^2]$

$\Rightarrow \quad [a] = [V^2][p] = [ML^{-1}T^{-2}][L^6]$

$= [ML^5T^{-2}]$

$\therefore$ Dimension of $\left[\frac{a}{b^2}\right]$

$= \frac{[ML^5T^{-2}]}{[L^3]^2} = \frac{[ML^5T^{-2}]}{[L^6]}$

$= [ML^5T^{-2}][L^{-6}]$

$= [ML^{-1}T^{-2}] = [p]$

• Concept Enhancer

More than one physical quantity may have the same dimensions thus a physical quantity can not be uniquely defined from its dimension.

43. *(a)* Spring constant.

$$k = \frac{[F]}{[x]} = \frac{[MLT^{-2}]}{[L]} = [MT^{-2}]$$

Angular speed, $\omega = \frac{[\theta]}{[t]} = \frac{1}{[T]} = [T^{-1}]$

Angular momentum, $L = mvr$

$= [M]\,[LT^{-1}] \times [L] = [ML^2T^{-1}]$

Moment of inertia, $I = [Mr^2]$

$= [M]\,[L^2] = [ML^2]$

Hence the correct match is

A → II, B → I, C → IV, D → III

44. *(c)* Torque = Force × Distance

$= [MLT^{-2}] \times [L] = [ML^2T^{-2}]$

Stress $= \frac{\text{Force}}{\text{Area}} = \frac{[MLT^{-2}]}{[L^2]} = [ML^{-1}T^{-2}]$

Pressure gradient $= \frac{\text{Pressure}}{\text{Length}} = \frac{F}{A \times L}$

$= \frac{[MLT^{-2}]}{[L^3]} = [ML^{-2}T^{-2}]$

Coefficient of viscosity $= \frac{\text{Force} \times \text{Length}}{\text{Area} \times \text{Velocity}}$

$= \frac{[MLT^{-2} \times L]}{[L^2] \times [LT^{-1}]} = [ML^{-1}T^{-1}]$

45. *(a)* A → III

Young's modulus, $Y = \frac{\text{Stress}}{\text{Strain}}$

$[Y] = \frac{[\text{Stress}]}{[\text{Strain}]} = \frac{[F/A]}{[\text{Strain}]}$

$= \frac{\frac{[MLT^{-2}]}{[L^2]}}{[M^0L^0T^0]}$

$= [ML^{-1}T^{-2}]$

B → I We know that coefficient of viscosity,

$$\eta = \frac{F}{A\left(\frac{dv}{dx}\right)}$$

$\therefore \quad [\eta] = \frac{[F]}{[A]\left[\frac{dv}{dx}\right]} = \frac{[MLT^{-2}]}{[L^2]\,[T^{-1}]}$

$= [ML^{-1}T^{-1}]$

C → II. Since, $E = h\nu \Rightarrow h = \frac{E}{\nu}$

$\Rightarrow \quad [h] = \frac{[E]}{[\nu]} = \frac{[ML^2T^{-2}]}{[T^{-1}]}$

$= [ML^2T^{-1}]$

D → IV. Work function, $\phi = h\nu_0$

$= [ML^2T^{-1}]\,[T^{-1}]$

$= [ML^2T^{-2}]$

46. *(b)*

key idea To find the dimensions of A and B, you can use the principle of homogeneity.

Given equation is,

$$(x - At)^2 + \left(y - \frac{t}{B}\right)^2 = a^2$$

As a is radius, its dimension, $[a] = [L]$

So, both At and $\frac{t}{B}$ must have dimensions of length.

$\Rightarrow \quad [A][t] = [L]$

$\therefore \quad [t] = [T^{-1}]$

$\Rightarrow \quad [A][T^{-1}] = [L]$

$\Rightarrow \quad [A] = [LT]$

Also, $\frac{[t]}{[B]} = [L]$ or $[B] = \frac{[t]}{[L]}$

$\Rightarrow \quad [B] = [L^{-1}T^{-1}]$

47. *(d)* Since we know, wave number and Rydberg constant have same unit m^{-1}. Stress and coefficient of elasticity have same unit Pascal (Pa).

Coercivity and magnetisation have same unit A/m. Whereas, specific heat capacity, $S = \frac{\Delta Q}{m\Delta T}$ has unit $\frac{J}{kg\,K}$ and latent heat, $L = \frac{\Delta Q}{m}$ has unit $\frac{J}{kg}$.

48. *(d)* We know,

$[S] = [MT^{-2}]$

$\Rightarrow \quad [\rho] = [ML^{-3}]$

$[r] = [L]$

$\therefore$ Dimension of RHS $= \frac{\left[M^{\frac{1}{2}}L^{-\frac{3}{2}}\right]\left[L^{\frac{3}{2}}\right]}{[MT^{-2}]^{\frac{3}{4}}}$

$= \left[M^{\left(\frac{1}{2}-\frac{3}{4}\right)}\,L^{\left(-\frac{3}{2}+\frac{3}{2}\right)}\,T^{\frac{6}{4}}\right]$

$= M^{-\frac{1}{4}}L^0T^{\frac{3}{2}}$

Dimension of LHS = [T]

$\therefore$ Dimension of LHS ≠ Dimension of RHS.

49. *(a)* Fundamental quantities taken are momentum (p), area (A) and time (t). Let coefficient of viscosity (η) depends on p, A and t are

$\eta = k \cdot p^a \cdot A^b \cdot t^c$, where k is a constant

Then writing dimensions of all quantities and comparing, we get

$[ML^{-1}T^{-1}] = [MLT^{-1}]^a[L^2]^b[T]^c$

$\Rightarrow \quad M = M^a \Rightarrow a = 1$

$L^{-1} = L^{a+2b}$

$\Rightarrow \quad a + 2b = -1 \quad \text{or} \quad 2b = -2$

$\Rightarrow \quad b = -1 \text{ and } T^{-1} = T^{-a+c}$

$\Rightarrow \quad -a + c = -1 \quad \text{or} \quad -1 + c = -1$

$c = 0$

So, coefficient of viscosity has dimension

$[\eta] = [p^1A^{-1}t^0] = [pA^{-1}t^0]$

50. *(a)* Surface charge density,

$\sigma = \frac{Q}{A} = \frac{I.t}{A} \Rightarrow \sigma = \frac{[A]\,[T]}{[L^2]}$

$\sigma = [L^{-2}TA]$

Electric displacement, $D = \varepsilon_0 E$

$\Rightarrow \quad D = \frac{Q^2}{F \cdot r^2} \cdot \frac{F}{Q} = \frac{Q}{r^2}$

$\because \quad \varepsilon_0 = \frac{Q^2}{F \cdot r^2}$ and $E = \frac{F}{Q}$

$\Rightarrow \quad D = \frac{I \cdot t}{r^2} \Rightarrow [D] = \frac{[I]\,[t]}{[r^2]} = \frac{[A][T]}{[L^2]}$

$\Rightarrow [D] = [L^{-2}TA]$ ...(ii)

From Eqs. (i) and (ii), we can observe that dimensions of surface charge density and electric displacement are same.

51. *(b)* As per question

$$m \propto v^x t^y F^z \quad \text{...(i)}$$

Thus, we have,

$$[ML^0T^0] = [LT^{-1}]^x [T]^y [MLT^{-2}]^z$$

$$[M^1L^0T^0] = [M^zL^{x+z}T^{-x+y-2z}]$$

On Comparing we get, $z = 1$

$x + z = 0$ and $-x + y - 2z = 0$

$\Rightarrow \quad x + 1 = 0 \Rightarrow x = -1$

and $-(-1) + y - 2(1) = 0$

$\Rightarrow \quad 1 + y - 2 = 0 \Rightarrow y = 1$

Substituting these values in Eq. (i), we get

$$m \propto v^{-1}t^1F^1$$

$$\Rightarrow \quad m = [Ftv^{-1}]$$

52. *(c)* Relative magnetic permeability is given by $\mu_r = \dfrac{\mu_m}{\mu_0}$

As, it is a ratio of permeabilities. So, it is unitless and dimensionless quantity.

Power factor $= \cos\phi$, so it is also a unitless and dimensionless.

Permeability of free space,

$$\mu_0 = \frac{B}{H} \Rightarrow [\mu_0] = \frac{[B]}{[H]} \quad \text{...(i)}$$

$$[H] = [n][I] = [L^{-1}][A] = [AL^{-1}]$$

$\because n =$ number of terms per unit length

$B =$ magnetic field, $[MT^{-2}A^{-1}]$

$$[\mu_0] = \frac{[MT^{-2}A^{-1}]}{[AL^{-1}]} = [MLT^{-2}A^{-2}]$$

Hence, permeability of free space is not a dimensionless quantity.

Quality factor

$$= \frac{\text{Energy stored per cycle}}{\text{Energy dissipated per cycle}}$$

So, it will be unitless and dimensionless quantity.

53. *(c)* For conduction of heat,

$$\frac{dQ}{dt} = KA\frac{dT}{dx}$$

$$K = \frac{\left(\dfrac{dQ}{dt}\right)}{A\left(\dfrac{dT}{dx}\right)} = \frac{dQ \times dx}{A \times dt \times dT}$$

$$= \frac{\text{joule} \times \text{metre}}{(\text{metre})^2 \times \text{second} \times \text{kelvin}}$$

$$= \frac{\dfrac{\text{kilogram} \times (\text{metre})^2}{(\text{second})^2} \times \text{metre}}{(\text{metre})^2 \times \text{second} \times \text{kelvin}}$$

$$= \frac{\text{kg}\cdot\text{m}^2\cdot\text{m}}{\text{s}^2\cdot\text{m}^2\cdot\text{s}\cdot\text{K}} = \frac{\text{kg}\cdot\text{m}}{\text{s}^3\cdot\text{K}}$$

$$[K] = \frac{[M^1][L^1]}{[T^3][K^1]}$$

$$\Rightarrow [K] = [M^1L^1T^{-3}K^{-1}]$$

54. *(a)* For the given question, the time t can be written as

$$t = kG^ah^bc^c \quad \text{...(i)}$$

For calculating the values of a, b and c, compare the dimensional formula for both side.

LHS t = time $= [M^0L^0T^1]$

RHS $G = [M^{-1}L^3T^{-2}]^a$

$h = [M^1L^2T^{-1}]^b$

$c = [M^0L^1T^{-1}]^c$

Comparing both side for powers of M, L and T,

$M \Rightarrow 0 = -a + b$ ...(i)

$L \Rightarrow 0 = 3a + 2b + c$ ...(ii)

$T \Rightarrow 1 = -2a - b - c$ ...(iii)

On solving Eqs. (i), (ii) and (iii), we get

$$a = \frac{1}{2}, \quad b = \frac{1}{2} \text{ and } c = \frac{-5}{2}$$

So, put these values in Eq. (i), we get

$$t = k\,G^{1/2}\,h^{1/2}\,c^{-5/2}$$

$$t = k\sqrt{\frac{Gh}{c^5}}$$

55. *(c)* Force of interaction between two atoms is given as

$$F = \alpha\beta \exp(-x^2/\alpha kT)$$

As we know, exponential terms are always dimensionless, so

$$\text{dimensions of}\left(\frac{-x^2}{\alpha\,kT}\right) = [M^0L^0T^0]$$

$\Rightarrow$ Dimensions of $\alpha =$ Dimension of (x^2/kT)

Now, substituting the dimensions of individual term in the given equation,

$$\text{we get} = \frac{[M^0L^2T^0]}{[M^1L^2T^{-2}]}$$

$$= [M^{-1}L^0T^2] \quad \text{...(i)}$$

{$\because$ Dimensions of kT equivalent to the dimensions of energy i.e. $[M^1L^2T^{-2}]$}

Now from given equation, we have dimensions of

$F =$ dimensions of $\alpha \times$ dimensions of β

$\Rightarrow$ Dimensions of $\beta =$ Dimensions of (F/α)

$$= \frac{[M^1L^1T^{-2}]}{[M^{-1}L^0T^2]} \quad [\because \text{using Eq. (i)}]$$

$$= [M^2L^1T^{-4}]$$

56. *(c)* $\mu_0 = ke^am^bc^ch^d$

$$[\mu_0] = [e]^a[m]^b[c]^c[h]^d$$

$$[MLT^{-2}A^{-2}] = [AT]^a[M]^b[LT^{-1}]^c[ML^2T^{-1}]^d$$

$$[MLT^{-2}A^{-2}] = [M^{b+d}L^{c+2d}T^{a-c-d}A^a]$$

Comparing LHS and RHS

$b + d = 1$

$c + 2d = 1$

$a - c - d = -2$

$\Rightarrow \quad a = -2$

Solving for a, b and c

$a = -2,\ d = 1,\ c = -1,\ b = 0$

So, $[\mu_0] = [e^{-2}m^0c^{-1}h^1] = \left[\dfrac{h}{e^2c}\right]$

57. *(b)* We know,

$e = [M^0L^0T^1A^1]$

$\varepsilon_0 = [M^{-1}L^{-3}T^4A^2]$

$G = [M^{-1}L^3T^{-2}]$ and $m_e = [M^1L^0T^0]$

According to question, $\dfrac{e^2}{2\pi\varepsilon_0Gm_e^2}$

$$= \frac{[M^0L^0T^1A^1]^2}{2\pi[M^{-1}L^{-3}T^4A^2][M^{-1}L^3T^{-2}][M^1L^0T^0]^2}$$

$$= \frac{1}{2\pi}$$

$\because \dfrac{1}{2\pi}$ is dimensionless thus the combination $\dfrac{e^2}{2\pi\varepsilon_0Gm_e^2}$ would have the same value in different systems of units.

58. *(d)* As we know, angular momentum

$$= m \times v \times r = [ML^2T^{-1}]$$

Latent heat

$$L = \frac{Q}{m} = \frac{[ML^2T^{-2}]}{[M]} = [L^2T^{-2}]$$

Capacitance

$$C = \frac{\text{Charge}}{\text{Potential difference}}$$

$$C = \frac{\text{Charge}}{\dfrac{\text{Work}}{\text{Charge}}}$$

$$C = \frac{\text{Charge}^2}{\text{Work}} = \frac{q^2}{W} = \frac{[AT]^2}{[ML^2T^{-2}]}$$

$$= M^{-1}L^{-2}T^4A^2$$

59. *(b)* Electrostatic force between two charges,

$$F = \frac{1}{4\pi\varepsilon_0}\frac{q_1 q_2}{R^2} \Rightarrow \varepsilon_0 = \frac{q_1 q_2}{4\pi F R^2}$$

Substituting the units, we get

$$\varepsilon_0 = \frac{C^2}{N\text{-}m^2} = \frac{[AT]^2}{[MLT^{-2}][L^2]}$$

$$= [M^{-1}L^{-3}T^4A^2]$$

60. *(b)* Magnetic energy

$$= \frac{1}{2}LI^2 = \frac{Lq^2}{2t^2} \quad \left[\text{as } I = \frac{q}{t}\right]$$

where L = inductance, I = current

Energy has the dimensions

$$= [ML^2T^{-2}]$$

Equate the dimensions, we have

$$[ML^2T^{-2}] = [\text{Henry}] \times \frac{[Q^2]}{[T^2]}$$

$$\Rightarrow [\text{Henry}] = \frac{[ML^2]}{[Q^2]}$$

61. *(a)* Boltzmann constant,

$$k_B = \frac{R}{N}$$

$$= \frac{pV}{TN} \quad \left[\because pV = RT,\ R = \frac{PV}{T}\right]$$

$$= \frac{[ML^2T^{-2}]}{[K]}$$

$$= [ML^2T^{-2}K^{-1}]$$

Coefficient of viscosity,

$$\eta = \frac{F}{6\pi rv}$$

$$[\eta] = \frac{[F]}{[r][v]}$$

$$= \frac{[MLT^{-2}]}{[L][LT^{-1}]}$$

$$= [ML^{-1}T^{-1}]$$

Planck's constant,

$$[h] = \left[\frac{E}{\nu}\right] = \frac{[ML^2T^{-2}]}{[T^{-1}]}$$

$$= [ML^2T^{-1}]$$

Thermal conductivity,

$$K = \frac{\Delta Q}{\Delta t}\cdot\frac{dx}{dT}\cdot\frac{1}{A}$$

$$[K] = \left[\frac{ML^2T^{-2}}{T}\cdot\frac{L}{K}\cdot\frac{1}{L^2}\right]$$

$$= [MLT^{-3}K^{-1}]$$

62. *(c)* $x(t) = A\sin t + B\cos^2 t + Ct^2 + D$

$[x(t)] = [L]$

$[A] = [L]$

(sin t and $\cos^2 t$ both are dimensionless)

$[B] = [L]$

$x = Ct^2$

$$C = \frac{x}{t^2} = [LT^{-2}]$$

$[C] = [LT^{-2}]$

$[D] = [L]$

$$\left[\frac{ABC}{D}\right] = \left[\frac{L \times L \times LT^{-2}}{L}\right]$$

$$= [L^2T^{-2}]$$

63. *(c)* A. Young's modulus = Pressure

$= [ML^{-1}T^{-2}]$

B. Torque = Work = $[ML^2T^{-2}]$

C. Coefficient of viscosity (η)

$F = 6\pi r\eta v$

$$\Rightarrow \quad \eta = \frac{F}{rv} = \frac{[MLT^{-2}]}{[L][LT^{-1}]}$$

$\therefore \quad \eta = [ML^{-1}T^{-1}]$

D. Gravitational constant (G)

$$F = \frac{Gm_1m_2}{r^2}$$

$$\Rightarrow \quad G = \frac{Fr^2}{m_1m_2} = \frac{[MLT^{-2}][L^2]}{[M][M]}$$

$\therefore \quad G = [M^{-1}L^3T^{-2}]$

64. *(c)* Angular impulse = $[ML^2T^{-1}]$

Latent heat = $[M^0L^2T^{-2}]$

Electrical resistivity = $[ML^3T^{-3}A^{-2}]$

Electromotive force = $[ML^2T^{-3}A^{-1}]$

A $\rightarrow$ III, B $\rightarrow$ I, C $\rightarrow$ IV, D $\rightarrow$ II

65. *(a)* $\frac{B^2}{2\mu_0}$ – Energy density

$$\mu_0 = \frac{B^2}{2(\text{Energy density})}$$

$$= \frac{[M^2T^{-4}A^{-2}]}{\frac{[ML^2T^{-2}]}{[L^3]}}$$

$$= [MT^{-2}LA^{-2}]$$

$$= [MLT^{-2}A^{-2}]$$

Magnetic moment, $= NIA = [L^2A]$

$$\text{Torsional constant} = \frac{\text{Torque}}{\theta}$$

$\therefore \quad F = qvB$

$$B = \frac{[F]}{[q][v]}$$

$$= \frac{[MLT^{-2}]}{[AT][LT^{-1}]}$$

$$= [MT^{-2}A^{-1}]$$

66. *(a)* Torque, $\tau = F \times r$

$[\tau] = [MLT^{-2}][L]$

$= [ML^2T^{-2}]$

$\therefore$ A $\rightarrow$ IV

Since, $F = Bqv$

$$\Rightarrow \quad [B] = \frac{[F]}{[q][v]}$$

$$= \frac{[MLT^{-2}]}{[AT][LT^{-1}]}$$

$$= [MA^{-1}T^{-2}] = [MT^{-2}A^{-1}]$$

$\therefore$ B $\rightarrow$ III

Magnetic moment, $M = I.A$

$[M] = [I][A] = [A][L^2]$

$= [AL^2]$

$\therefore$ C $\rightarrow$ II

$$\text{Since, } B = \frac{\mu_0}{4\pi}\cdot\frac{Idl\sin\theta}{r^2}$$

$$\therefore \quad [\mu_0] = \frac{[B][r^2]}{[I][dl]}$$

$$= \frac{[MT^{-2}A^{-1}][L^2]}{[A][L]}$$

$$= [MLT^{-2}A^{-2}]$$

$\therefore$ D $\rightarrow$ I

The correct match is

A-IV, B-III, C-II, D-I

67. *(b)* Given, $E = \dfrac{B - x^2}{At}$

E = Energy, x = length

t = time

Dimensions of B and x^2 must be same.

Also dimensions of EAt must be same as that of x^2.

$\therefore \quad EAt = B - x^2$

So, $\quad [B] = [x^2] = [L^2]$

Also, $\quad [E][A][t] = [x^2]$

$$[ML^2T^{-2}]\,[A]\,[T]=[L^2]$$

$$\therefore \quad [A]=\frac{[L^2]}{[ML^2T^{-1}]}$$

$$\Rightarrow \quad [A]=[M^{-1}T^1]$$

So, dimension of AB is,

$$[A][B]=[M^{-1}T^1]\,[L^2]$$
$$=[M^{-1}L^2T^1]$$

68. *(b)*

key idea For this question, you can use the fact that a physical equation will be dimensionally correct only, if quantity of having same dimensional formula is added or subtracted.

Given equation is as follows,

$$\left(p+\frac{a}{V^2}\right)(V-b)=RT$$

According to principle of homogeneity,

$$[p]=\frac{[a]}{[V^2]} \text{ or } [a]=[p][V^2]$$

$$[a]=[ML^{-1}T^{-2}]\,[L^3]^2$$
$$[a]=[ML^5T^{-2}]$$

and $\quad [V]=[b]=[L^3]$

$$\text{Now, } \frac{[b^2]}{[a]}=\frac{[L^3]^2}{[ML^5T^{-2}]}=[M^{-1}LT^2]$$

The above dimensional formula is same dimensional formula of compressibility.

Also, Bulk modulus $B=[M^1L^{-1}T^{-2}]$

Compressibility $\frac{1}{B}=k=[M^{-1}LT^2]$

Modulus of rigidity $=[ML^{-1}T^{-2}]$

Energy density $=[M^1L^{-1}T^{-2}]$

69. *(c)* According to given question,

$m=kG^x h^y c^z$

where k is dimensionless constant

$$[M]=[M^{-1}L^3T^{-2}]^x\,[ML^2T^{-1}]^y\,[LT^{-1}]^z$$
$$=[M^{-x+y}L^{3x+2y+z}T^{-2x-y-z}]$$

On comparing, we get

$$-x+y=1$$
$$\Rightarrow \quad y=1+x \quad \ldots(i)$$
$$3x+2y+z=0 \quad \ldots(ii)$$
$$-2x-y-z=0 \quad \ldots(iii)$$

On adding Eqs. (ii) and (iii), we get

$$x+y=0$$
$$x+1+x=0$$
$$2x=-1 \Rightarrow x=\frac{-1}{2}$$

$$\therefore \quad y=1+\left(-\frac{1}{2}\right)=\frac{1}{2} \text{ and } z=\frac{1}{2}$$

$$\text{So, } [M]=[G]^{-\frac{1}{2}}[h]^{\frac{1}{2}}[c]^{\frac{1}{2}}$$

$$[M]=(h^{\frac{1}{2}}c^{\frac{1}{2}}G^{-\frac{1}{2}})$$

70. *(d)* Energy density, $u=\frac{\alpha}{\beta}\sin\left(\frac{\alpha x}{kt}\right)$

where, $\alpha, \beta=$ constants

$x=$ Displacement

$k=$ Boltzmann constant

$t=$ Temperature

We know that, energy density $=\frac{\text{energy}}{\text{volume}}$

and $\alpha x/kt=$ dimensionless

$$\alpha=\frac{kt}{x}=\frac{[ML^2T^{-2}K^{-1}][K]}{[L]}=[MLT^{-2}] \quad \text{[Eq. (i)]}$$

$$\text{Now, } \frac{\alpha}{\beta}=\frac{\text{energy}}{\text{volume}}$$

$$\Rightarrow \quad \beta=\frac{\alpha\times\text{volume}}{\text{energy}}=\frac{\alpha\,[L^3]}{[ML^2T^{-2}]}$$

$$\Rightarrow \quad \beta=\frac{[MLT^{-2}][L^3]}{[ML^2T^{-2}]} \quad \text{[From Eq. (i)]}$$

$$\Rightarrow \quad \beta=[L^2]$$
$$\Rightarrow \quad \beta=[M^0L^2T^0]$$

• Concept Enhancer

All trignometric functions are dimensionless.

71. *(d)* Since, ratios of trigonometric function and logarithmic function are dimensionless quantities.

$$\therefore \quad [\eta]=[M^0L^0T^0]$$

Also, dimensions of temperature,

$$[T]=[M^0L^0T^0K]$$

Dimensions of Boltzmann constant,

$$[k_B]=[ML^2T^{-2}K^{-1}]$$

Dimension of $x=[M^0LT^0]$

The logarithmic term is dimensionless

$$\therefore \quad \left[\frac{\beta x}{kT}\right]=[M^0L^0T^0]$$

$$\Rightarrow \quad [\beta]=\frac{[k_BT]}{[x]}=\frac{[ML^2T^{-2}K^{-1}][K]}{[L]}$$
$$=[MLT^{-2}]=[\text{Force}]$$

Hence, dimensions of β is same as that of force.

Also, $\quad [\eta]=[\alpha\beta]$

$$[\alpha\beta]=[M^0L^0T^0]$$

$$\Rightarrow \quad [\alpha]=\frac{1}{[\beta]}=\frac{1}{[MLT^{-2}]}$$

$$\text{or} \quad [\alpha^{-1}]=[\beta]=[MLT^{-2}]$$

$$\therefore \quad [\alpha^{-1}x]=[MLT^{-2}][L]=[ML^2T^{-2}]$$
$$=[\text{Energy}]$$

i.e. Dimensions of $\alpha^{-1}x$ is same as that of energy.

Since, the quantity $\alpha\beta$ is dimensionless, so dimensions of α cannot be same as that of β.

72. *(a)* Velocity gradient $=\frac{\Delta v}{\Delta x}$

$$[\text{Velocity gradient}]=\frac{[\Delta v]}{[\Delta x]}=\frac{[LT^{-1}]}{[L]}$$
$$=[T^{-1}]$$

Decay constant, $\lambda=\frac{1}{\tau}$

where, τ is average life time

$$\therefore \quad [\lambda]=\frac{1}{[\tau]}=\frac{1}{[T]}=[T^{-1}]$$

Therefore, velocity gradient and decay constant have same dimensional formula.

According to Stefan's Boltzmann law,

$E/At=\sigma T^4$,

where σ is Stefan's constant.

$$\Rightarrow \quad \sigma=\frac{E}{At\,T^4}\Rightarrow[\sigma]=\frac{[ML^2T^{-2}]}{[L^2][T][K^4]}$$

$$[\sigma]=[MT^{-3}K^{-4}]$$

Wien's constant, $b=\lambda_m T$

$$\therefore \quad [b]=[LK]$$

Therefore, Wien's constant and Stefan's constant have different dimensions.

Angular frequency, $\omega=2\pi/T$

$$\therefore \quad [\omega]=[T^{-1}]$$

Angular momentum,

$$L=mvr$$
$$[L]=[M]\,[LT^{-1}]\,[L]=[ML^2T^{-1}]$$

Thus, angular frequency and angular momentum have different dimensions.

Wave number, $\bar{\nu}=\frac{1}{\lambda}$

$$[\bar{\nu}]=\frac{1}{[\lambda]}=\frac{1}{[L]}=[L^{-1}]$$

Avogadro number is a dimensionless constant,

Hence, wave number and Avogadro number also have different dimensions.

73. *(a)* According to question,

$$\left[\frac{\pi p a^4}{8\eta L}\right]=\frac{[ML^{-1}T^{-2}][L^4]}{[ML^{-1}T^{-1}][L^1]}$$

and $[v] = L^3$

$= [M^0 L^3 T^{-1}]$ ... (ii)

Thus, option (a) is dimensionally incorrect.

Now, since formula of capillary rise in tube, $h = \frac{2s\cos\theta}{\rho g r}$

Thus, $= \frac{[s]}{[\rho][g][r]} = \frac{[MT^{-2}]}{[ML^{-3}][LT^{-2}][L]}$

$= [L]$

Hence, $h = \frac{2s\cos\theta}{\rho r g}$ is dimensionally correct.

So, option (b) will also be dimensionally correct.

In option (c),

$$J = \varepsilon \frac{dE}{\partial t} \quad \ldots\text{(iii)}$$

$$\Rightarrow \quad J = \varepsilon \frac{E}{t}$$

Since, $J = \frac{I}{A}$

$$\therefore \quad [J] = \frac{[I]}{[A]} = \frac{[A]}{[L^2]}$$

$$\Rightarrow \quad [J] = [AL^{-2}] \quad \ldots\text{(iv)}$$

Again, we know that

$$E = \frac{1}{4\pi\varepsilon}\cdot\frac{q}{r^2}$$

$$\Rightarrow \quad \varepsilon E = \frac{1}{4\pi}\cdot\frac{q}{r^2} \Rightarrow \frac{\varepsilon E}{t} = \frac{1}{4\pi}\cdot\frac{q}{tr^2}$$

$$\Rightarrow \quad \left[\frac{\varepsilon E}{t}\right] = \frac{[q]}{[t][r^2]} = \frac{[AT]}{[T][L^2]}$$

$$\Rightarrow \quad \left[\frac{\varepsilon E}{t}\right] = [AL^{-2}] \quad \ldots\text{(v)}$$

Form Eqs. (iv) and (v), we see that Eq. (iii) is dimensionally correct.

In option (d) $W = \tau\theta$ and $\tau = r \times F$

So, dimensional formula of

$$[\tau][\theta] = [r][F] = [L][MLT^{-2}]$$

$$= [ML^2T^{-2}] = [\text{Work}]$$

74. *(b)* Given,

Equation of force,

$$F = A\cos Bx + C\sin Dt$$

where, x is displacement and t is time.

Now, trigonometric ratios are dimensionless,

$$\therefore \quad [Bx] = [M^0L^0T^0]$$

$$\Rightarrow [B][L^1] = [M^0L^0T^0]$$

$$\Rightarrow \quad [B] = [M^0 L^{-1}T^0] \quad \ldots\text{(i)}$$

Similarly,

$$[Dt] = [M^0L^0T^0]$$

$$\Rightarrow \quad [D] = \frac{[M^0 L^0T^0]}{[t]}$$

$$= [M^0 L^0T^{-1}] \quad \ldots\text{(ii)}$$

and $[A] = [F] = [MLT^{-2}]$ ...(iii)

From Eqs. (i), (ii) and (iii), we get

$$\frac{[AD]}{[B]} = \frac{[MLT^{-2}][M^0L^0T^{-1}]}{[M^0L^{-1}T^0]}$$

$$= \frac{[MLT^{-3}]}{[L^{-1}]} = [ML^2T^{-3}]$$

75. *(b)*

> **key idea** Deducing (Establising) relation among different physical quantities by using the principle of homogenity of dimensions we can derive an expression for a physical quantity.
>
> If a physical quantity Q depends on three physical quantities X, Y and Z, then $Q \propto X^aY^bZ^c \Rightarrow Q = K X^aY^bZ^c$
>
> K = Dimensionless constant.

Given, work done by gas molecule,

$$W = \alpha^2\beta e^{-\beta x^2/kT}$$

Here, x is displacement, k is Boltzmann constant, α and β are constants and T is temperature.

Dimensional formula of

$$[W] = [ML^2T^{-2}]$$

$\therefore$ Dimensions of $[\alpha^2\beta] = [ML^2T^{-2}]$

$$\Rightarrow \quad \alpha = \left[\frac{ML^2T^{-2}}{\beta}\right]^{1/2} \quad \ldots\text{(i)}$$

The term $\frac{\beta x^2}{kT}$ should be dimensionless, i.e. $[M^0L^0T^0]$.

$$\Rightarrow \left[\frac{\beta x^2}{kT}\right] = [M^0L^0T^0]$$

$$\Rightarrow \quad [\beta] = \frac{[k][T]}{[x^2]} \quad \ldots\text{(ii)}$$

Energy of gaseous molecule

$$E = \frac{7}{2}k_BT$$

$$\Rightarrow \quad [k] = [E]/[T] = [ML^2T^{-2}K^{-1}]$$

Substituting the value of k_B in Eq. (ii), we get

$$[\beta] = \frac{[ML^2T^{-2}K^{-1}][K]}{[L^2]}$$

$$= [MT^{-2}]$$

Substituting the value of β in Eq. (i), we get

$$[\alpha] = \left\{\frac{[ML^2T^{-2}]}{[MT^{-2}]}\right\}^{1/2} = [M^0L^1T^0]$$

76. *(d)* Dimensions of quantity f are

$$[f] = \frac{[h]^{1/2}[c]^{5/2}}{[G]^{1/2}} \quad \ldots\text{(i)}$$

As, $h = \frac{E}{\nu}$,

$$[h] = [ML^2T^{-2}][T] = [ML^2T^{-1}]$$

$$c = [LT^{-1}]$$

and $G = \frac{F\cdot r^2}{m^2}$

$$\Rightarrow [G] = \frac{[MLT^{-2}][L^2]}{[M^2]} = [M^{-1}L^3T^{-2}]$$

So, dimensions of f using Eq. (i),

$$[f] = \frac{[ML^2T^{-1}]^{\frac{1}{2}}[LT^{-1}]^{\frac{5}{2}}}{[M^{-1}L^3T^{-2}]^{\frac{1}{2}}}$$

$$= \left[M^{\frac{1}{2}+\frac{1}{2}}L^{\frac{5}{2}-\frac{3}{2}+1}T^{-\frac{1}{2}-\frac{5}{2}+\frac{2}{2}}\right]$$

$$= [ML^2T^{-2}]$$

Thus, it has the dimensions of energy.

[**Note** To produce dimensions of different constants, just remember/recall nearest formulae containing their constants.]

77. *(0)* Modulus of elasticity = $\frac{\text{Force}}{\text{Area}}$

$$= \frac{[MLT^{-2}]}{[L^2]} = [ML^{-1}T^{-2}] \quad \ldots\text{(i)}$$

Torque = Force × perpendicular distance

$$= [MLT^{-2}][L^1]$$

$$= [ML^2T^{-2}] \quad \ldots\text{(ii)}$$

$$\frac{\text{Modulus of elasticity}}{\text{Torque}} = \frac{[ML^{-1}T^{-2}]}{[ML^2T^{-2}]}$$

$$= [M^0L^{-3}T^0]$$

Measured quantity

$$= [M^aL^bT^c] = [M^0L^{-3}T^0]$$

For $b = -3, c = 0$

If $b = 3$, then also $c = 0$

C-02

JEE Main Chapterwise
Topicwise Physics

In JEE Main there is usually 0-1 question per paper are visible from this chapter. In past 3 years, 29 conceptual and 8 numerical questions were asked of moderate to difficult level from this chapter.

For detailed analysis scan QR code

MOTION IN A STRAIGHT LINE

Topic 1 Distance, Displacement, Speed, Velocity and Acceleration

Level 1

01 The position of a particle related to time is given by $x = (5t^2 - 4t + 5)$ m. The magnitude of velocity of the particle at $t = 2$ s will be *[2023, 15 April Shift-I]*

(a) 06 ms^{-1} (b) 10 ms^{-1}
(c) 14 ms^{-1} (d) 16 ms^{-1}

02 The distance travelled by an object in time t is given by $s = (2.5)t^2$. The instantaneous speed of the object at $t = 5$ s will be *[2023, 13 April Shift-II]*

(a) 25 ms^{-1} (b) 5 ms^{-1}
(c) 62.5 ms^{-1} (d) 12.5 ms^{-1}

03 A vehicle travels 4 km with speed of 3 km/h and another 4 km with speed of 5 km/h, then its average speed is *[2023, 30 Jan Shift-II]*

(a) 3.50 km/h (b) 3.75 km/h
(c) 4.00 km/h (d) 4.25 km/h

04 The distance travelled by a particle is related to time t as $x = 4t^2$. The velocity of the particle at $t = 5$s is *[2023, 25 Jan Shift-II]*

(a) 8 ms^{-1} (b) 40 ms^{-1} (c) 20 ms^{-1} (d) 25 ms^{-1}

05 A person travels x distance with velocity v_1 and then x distance with velocity v_2 in the same direction. The average velocity of the person is v, then the relation between v, v_1 and v_2 will be *[2023, 10 April Shift-II, 25 Jan Shift-I]*

(a) $v = \frac{v_1 + v_2}{2}$ (b) $\frac{2}{v} = \frac{1}{v_1} + \frac{1}{v_2}$
(c) $v = v_1 + v_2$ (d) $\frac{1}{v} = \frac{1}{v_1} + \frac{1}{v_2}$

06 The relation between time t and distance x for a moving body is given as $t = mx^2 + nx$, where m and n are constants. The retardation of the motion is (where, v stands for the velocity) *[2021, 25 July Shift-II]*

(a) $2nv^3$ (b) $2mnv^3$ (c) $2mv^3$ (d) $2n^2v^3$

07 An object moving with a speed of 6.25 m/s, is decelerated at a rate given by $dv/dt = -2.5\sqrt{v}$, where v is the instantaneous speed. The time taken by the object to come to rest, would be *[AIEEE 2011]*

(a) 2 s (b) 4 s (c) 8 s (d) 1 s

Level 2

08 A particle moving in a straight line covers half the distance with speed 6 m/s. The other half is covered in two equal time intervals with speeds 9 m/s and 15 m/s respectively. The average speed of the particle during the motion is *[2024, 9 April Shift-I]*

(a) 8.8 m/s (b) 8 m/s
(c) 10 m/s (d) 9.2 m/s

09 A train starting from rest first accelerates uniformly up to a speed of 80 km/h for time t, then it moves with a constant speed for time $3t$. The average speed of the train for this duration of journey will be (in km/h) *[2024, 6 April Shift-I]*

(a) 70 (b) 40 (c) 80 (d) 30

10 A body travels 102.5 m in nth second and 115.0 m in $(n+2)$th second. The acceleration is *[2024, 4 April Shift-I]*

(a) 5 m/s^2 (b) 9 m/s^2 (c) 6.25 m/s^2 (d) 12.5 m/s^2

11 Train A is moving along two parallel rail tracks towards north with speed 72 km/h and train B is moving towards south with speed 108 km/h. Velocity of train B with respect to A and velocity of ground with respect to B are (in ms^{-1}). *[2024, 1 Feb Shift-II]*

(a) –30 and 50 (b) –50 and –30
(c) 50 and –30 (d) –50 and 30

12 The relation between time t and distance x is $t=\alpha x^2+\beta x$, where α and β are constants. The relation between acceleration a and velocity v is **[2024, 31 Jan Shift-I]**

(a) $a=-5\alpha v^5$ (b) $a=-2\alpha v^3$
(c) $a=-3\alpha v^2$ (d) $a=-4\alpha v^4$

13 A particle is moving in a straight line. The variation of position x as a function of time t is given as $x=(t^3-6t^2+20t+15)$ m. The velocity of the body when its acceleration becomes zero is **[2024, 29 Jan Shift-II]**

(a) 8 m/s (b) 10 m/s (c) 4 m/s (d) 6 m/s

14 A passenger sitting in a train A moving at 90 km/h observes another train B moving in the opposite direction for 8s. If the velocity of the train B is 54 km/h, then length of train B is **[2023, 13 April Shift-II]**

(a) 200 m (b) 320 m (c) 120 m (d) 80 m

Imp. Average Speed

15 An object moves with speed v_1, v_2 and v_3 along a line segment AB, BC and CD, respectively as shown in figure. where $AB=BC$ and $AD=3AB$, then average speed of the object will be **[2023, 1 Feb Shift-I]**

(a) $\dfrac{v_1v_2v_3}{3(v_1v_2+v_2v_3+v_3v_1)}$ (b) $\dfrac{3v_1v_2v_3}{(v_1v_2+v_2v_3+v_3v_1)}$

(c) $\dfrac{(v_1+v_2+v_3)}{3v_1v_2v_3}$ (d) $\dfrac{(v_1+v_2+v_3)}{3}$

16 The instantaneous velocity of a particle moving in a straight line is given as $v=\alpha t+\beta t^2$, where α and β are constants. The distance travelled by the particle between 1s and 2s is **[2021, 25 July Shift-II]**

(a) $3\alpha+7\beta$ (b) $\dfrac{3}{2}\alpha+\dfrac{7}{3}\beta$

(c) $\dfrac{\alpha}{2}+\dfrac{\beta}{3}$ (d) $\dfrac{3}{2}\alpha+\dfrac{7}{2}\beta$

17 The position of a particle as a function of time t, is given by $x(t)=at+bt^2-ct^3$ where a, b and c are constants. When the particle attains zero acceleration, then its velocity will be **[2019, 9 April Shift-II]**

(a) $a+\dfrac{b^2}{2c}$ (b) $a+\dfrac{b^2}{4c}$ (c) $a+\dfrac{b^2}{3c}$ (d) $a+\dfrac{b^2}{c}$

18 A particle located at $x=0$ at time $t=0$, starts moving along the positive x-direction with a velocity v that varies as $v=\alpha\sqrt{x}$. The displacement of the particle varies with time as **[AIEEE 2006]**

(a) t^2 (b) t (c) $t^{1/2}$ (d) t^3

Level 3

19 A clock has 75 cm, 60 cm long second hand and minute hand, respectively. In 30 min duration the tip of second hand will travel x distance more than the tip of minute hand. The value of x in metre is nearly (Take, $\pi=3.14$) **[2024, 8 April Shift-I]**

(a) 139.4 (b) 118.9
(c) 140.5 (d) 220.0

20 Two trains A and B of length l and $4l$ are travelling into a tunnel of length L in parallel tracks from opposite directions with velocities 108 km/h and 72 km/h, respectively. If train A takes 35 s less time than train B to cross the tunnel then, length L of tunnel is (Given, $L=60l$) **[2023, 13 April Shift-I]**

(a) 1200 m (b) 1800 m (c) 2700 m (d) 900 m

21 Two buses P and Q start from a point at the same time and move in a straight line and their positions are represented by $X_P(t)=\alpha t+\beta t^2$ and $X_Q(t)=ft-t^2$. At what time, both the buses have same velocity? **[2022, 25 June Shift-II]**

(a) $\dfrac{\alpha-f}{1+\beta}$ (b) $\dfrac{\alpha+f}{2(\beta-1)}$ (c) $\dfrac{\alpha+f}{2(1+\beta)}$ (d) $\dfrac{f-\alpha}{2(1+\beta)}$

22 A boy reaches the airport and finds that the escalator is not working. He walks up the stationary escalator in time t_1. If he remains stationary on a moving escalator, then the escalator takes him up in time t_2. The time taken by him to walk up on the moving escalator will be **[2021, 20 July Shift-II]**

(a) $\dfrac{t_1t_2}{t_2-t_1}$ (b) $\dfrac{t_1+t_2}{2}$ (c) $\dfrac{t_1t_2}{t_2+t_1}$ (d) t_2-t_1

23 A balloon is moving up in air vertically above a point A on the ground. When it is at a height h_1, a girl standing at a distance d (point B) from A (see figure) sees it at an angle $45°$ with respect to the vertical. When the balloon climbs up a further height h_2, it is seen at an angle $60°$ with respect to the vertical if the girl moves further by a distance $2.464\,d$ (point C). Then, the height h_2 is (Given, $\tan 30°=0.5774$) **[2020, 5 Sep Shift-I]**

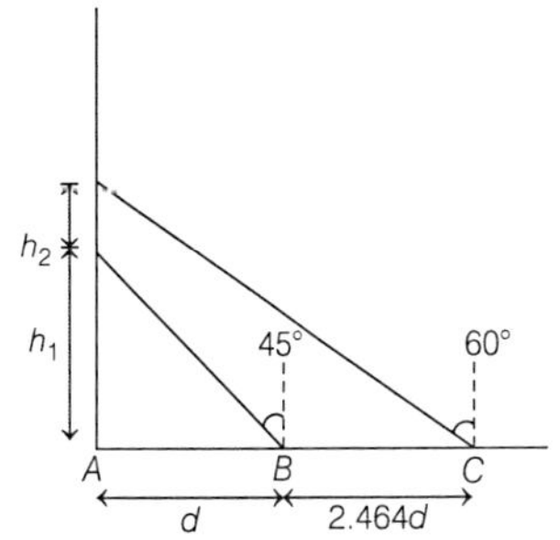

(a) $1.464\,d$ (b) d (c) $0.464\,d$ (d) $0.732\,d$

Numerical Value Type Questions

24 A person travelling on a straight line moves with a uniform velocity v_1 for a distance x and with a uniform velocity v_2 for the next $\frac{3}{2}x$ distance. The average velocity in this motion is $\frac{50}{7}$ m/s. If v_1 is 5 m/s, then $v_2 =$ .... m/s *[2025, 2 April Shift-I]*

25 Two cars P and Q are moving on a road in the same direction. Acceleration of car P increases linearly with time, whereas car Q moves with a constant acceleration. Both cars cross each other at time $t = 0$, for the first time. The maximum possible number of crossing(s) (including the crossing at $t = 0$) is......... *[2025, 29 Jan Shift-I]*

26 A particle moves in a straight line so that its displacement x at any time t is given by $x^2 = 1 + t^2$. Its acceleration at any time t is x^{-n}, where $n =$.......... *[2024, 6 April Shift-II]*

Imp. Instantaneous Acceleration

27 A particle initially at rest starts moving from reference point $x = 0$ along X-axis, with velocity v that varies as, $v = 4\sqrt{x}$ m/s. The acceleration of the particle is .... ms^{-2}. *[2024, 1 Feb Shift-II]*

28 A particle is moving in one dimension (along X-axis) under the action of a variable force. It's initial position was 16 m right of origin. The variation of its position x with time t is given as
$x = -3t^3 + 18t^2 + 16t$, where x is in m and t is in s.
The velocity of the particle when its acceleration becomes zero is .... m/s. *[2024, 1 Feb Shift-I]*

29 A car covers AB distance with first one–third at velocity v_1 ms^{-1}, second one–third at v_2 ms^{-1} and last one–third at v_3 ms^{-1}. If $v_3 = 3v_1, v_2 = 2v_1$ and $v_1 = 11\ ms^{-1}$ then the average velocity of the car is .... ms^{-1}. *[2022, 28 June Shift-II]*

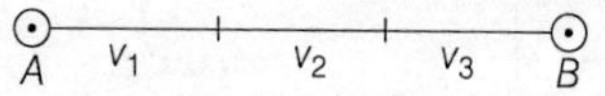

30 If the velocity of a body related to displacement x is given by $v = \sqrt{5000 + 24x}$ m/s, then the acceleration of the body is ...... m/s^2. *[2021, 27 Aug Shift-I]*

Topic 2 Kinematics Equation of Uniformly Accelerated Motion

Level 1

01 A ball is thrown vertically upward with an initial velocity of 150 m/s. The ratio of velocity after 3s and 5s is $\frac{x+1}{x}$. The value of x is [Take, $g = 10\ m/s^2$] *[2023, 12 April Shift-I]*

(a) 10 (b) 5 (c) 6 (d) −5

Imp. Uniformly Accelerated Motion

02 A juggler throws balls vertically upwards with same initial velocity in air. When the first ball reaches its highest position, he throws the next ball. Assuming the juggler throws n balls per second, the maximum height the balls can reach is *[2022, 29 July Shift-II]*

(a) $g/2n$ (b) g/n (c) $2gn$ (d) $g/2n^2$

03 When a ball is dropped into a lake from a height 4.9 m above the water level, it hits the water with a velocity v and then sinks to the bottom with the constant velocity v. If reaches the bottom of the lake 4.0 s after it is dropped. The approximate depth of the lake is *[2022, 27 June Shift-II]*

(a) 19.6 m (b) 29.4 m (c) 39.2 m (d) 73.5 m

04 A scooter accelerates from rest for time t_1 at constant rate a_1 and then retards at constant rate a_2 for time t_2 and comes to rest. The correct value of t_1/t_2 will be *[2021, 26 Feb Shift-II]*

(a) $\frac{a_1 + a_2}{a_2}$ (b) $\frac{a_2}{a_1}$

(c) $\frac{a_1}{a_2}$ (d) $\frac{a_1 + a_2}{a_1}$

05 A car moving with a speed of 50 km/h, can be stopped by brakes after atleast 6 m. If the same car is moving at a speed of 100 km/h, the minimum stopping distance is *[AIEEE 2003]*

(a) 12 m (b) 18 m (c) 24 m (d) 6 m

06 From a building, two balls A and B are thrown such that A is thrown upwards and B downwards (both vertically). If v_A and v_B are their respective velocities on reaching the ground, then *[AIEEE 2002]*

(a) $v_B > v_A$
(b) $v_A = v_B$
(c) $v_A > v_B$
(d) their velocities depend on their masses

Level 2

07 A particle moves along the X-axis and has its displacement x varying with time t according to the equation $x = c_0(t^2 - 2) + c(t-2)^2$

where c_0 and c are constants of appropriate dimensions. Then, which of the following statements is correct? *[2025, 3 April Shift-II]*

(a) The acceleration of the particle is $2c$.
(b) The acceleration of the particle is $2(c + c_0)$.
(c) The initial velocity of the particle is $4c$.
(d) The acceleration of the particle is $2c_0$.

08 Two cars are travelling towards each other at speed of $20\ ms^{-1}$ each. When the cars are 300 m apart, both the drivers apply brakes and the cars retard at the rate of $2\ ms^{-2}$. The distance between them when they come to rest is *[2024, 9 April Shift-II]*

(a) 25 m (b) 200 m (c) 100 m (d) 50 m

09 A body starts moving from rest with constant acceleration covers displacement s_1 in first $(p-1)$ seconds and s_2 in first p seconds. The displacement $s_1 + s_2$ will be made in time *[2024, 29 Jan Shift-I]*

(a) $(2p^2 - 2p + 1)$ s (b) $(2p+1)$ s
(c) $(2p-1)$ s (d) $\sqrt{(2p^2 - 2p + 1)}$ s

10 A NCC parade is going at a uniform speed of 9 km/h under a mango tree on which a monkey is sitting at a height of 19.6 m. At any particular instant, the monkey drops a mango. A cadet will receive the mango whose distance from the tree at time of drop is (Given, $g = 9.8\ m/s^2$) *[2022, 28 July Shift-I]*

(a) 5 m (b) 10 m (c) 19.8 m (d) 24.5 m

11 A small toy starts moving from the position of rest under a constant acceleration. If it travels a distance of 10m in t s. the distance travelled by the toy in the next t s will be *[2022, 29 June Shift-II]*

(a) 10 m (b) 20 m (c) 30 m (d) 40 m

12 An engine of a train moving with uniform acceleration, passes the signal-post with velocity u and the last compartment with velocity v. The velocity with which middle point of the train passes the signal post is *[2021, 25 Feb Shift-I]*

(a) $\sqrt{\frac{v^2 + u^2}{2}}$ (b) $\frac{v-u}{2}$
(c) $\frac{u+v}{2}$ (d) $\sqrt{\frac{v^2 - u^2}{2}}$

13 A helicopter rises from rest on the ground vertically upwards with a constant acceleration g. A food packet is dropped from the helicopter when it is at a height h. The time taken by the packet to reach the ground is close to (Here, g is the acceleration due to gravity). *[2020, 5 Sep Shift-I]*

(a) $t = \frac{2}{3}\sqrt{\frac{h}{g}}$ (b) $t = 1.8\sqrt{\frac{h}{g}}$
(c) $t = \sqrt{\frac{2h}{3g}}$ (d) $t = 3.4\sqrt{\frac{h}{g}}$

14 From a tower of height H, a particle is thrown vertically upwards with a speed u. The time taken by the particle to hit the ground is n times that taken by it to reach the highest point of its path. The relation between H, u and n is *[JEE Main 2014]*

(a) $2gH = n^2 u^2$ (b) $gH = (n-2)^2 u^2$
(c) $2gH = nu^2 (n-2)$ (d) $gH = (n-2)^2 u^2$

15 A parachutist after bailing out falls 50 m without friction. When parachute opens, it decelerates at $2\ ms^{-2}$. He reaches the ground with a speed of $3\ ms^{-1}$. At what height, did he bail out? *[AIEEE 2005]*

(a) 91 m (b) 182 m
(c) 293 m (d) 111 m

16 A ball is released from the top of a tower of height h metre. It takes T second to reach the ground. What is the position of the ball in $T/3$ second? *[AIEEE 2004]*

(a) $h/9$ m from the ground
(b) $7h/9$ m from the ground
(c) $8h/9$ m from the ground
(d) $17h/18$ m from the ground

17 If a body losses half of its velocity on penetrating 3 cm in a wooden block, then how much will it penetrate more before coming to rest ? *[AIEEE 2002]*

(a) 1 cm (b) 2 cm
(c) 3 cm (d) 4 cm

Level 3

Imp. Uniformly Accelerated Motion

18 A ball is released from a height h. If t_1 and t_2 be the time required to complete first half and second half of the distance, respectively. Then, choose the correct relation between t_1 and t_2. *[2022, 29 July Shift-II]*

(a) $t_1 = (\sqrt{2})t_2$ (b) $t_1 = (\sqrt{2}-1)t_2$
(c) $t_2 = (\sqrt{2}+1)t_1$ (d) $t_2 = (\sqrt{2}-1)t_1$

19 A ball is thrown up vertically with a certain velocity so that, it reaches a maximum height h. Find the ratio of the times in which it is at height $h/3$ while going up and coming down, respectively. *[2022, 29 July Shift-I]*

(a) $\frac{\sqrt{2}-1}{\sqrt{2}+1}$ (b) $\frac{\sqrt{3}-\sqrt{2}}{\sqrt{3}+\sqrt{2}}$ (c) $\frac{\sqrt{3}-1}{\sqrt{3}+1}$ (d) $\frac{1}{3}$

20 Water drops are falling from a nozzle of a shower onto the floor, from a height of 9.8 m. The drops fall at a regular interval of time. When the first drop strikes the floor, at that instant, the third drop begins to fall.

Locate the position of second drop from the floor when the first drop strikes the floor.

[2021, 27 Aug Shift-II]

(a) 4.18 m (b) 2.94 m (c) 2.45 m (d) 7.35 m

21 A balloon was moving upwards with a uniform velocity of 10 m/s. An object of finite mass is dropped from the balloon when it was at a height of 75 m from the ground level. The height of the balloon from the ground when object strikes the ground was around, is
(Take, the value of $g = 10\ m/s^2$)

[2021, 25 July Shift-II]

(a) 300 m (b) 200 m (c) 125 m (d) 250 m

22 Water droplets are coming from an open tap at a particular rate. The spacing between a droplet observed at 4th second after its fall to the next droplet is 34.3 m. At what rate, the droplets are coming from the tap? (Take, $g = 9.8\ m/s^2$)

[2021, 25 July Shift-I]

(a) 3 drops / 2 s (b) 2 drops / s
(c) 1 drop / s (d) 1 drop / 7 s

23 A rubber ball is released from a height of 5 m above the floor. It bounces back repeatedly always rising to $\frac{81}{100}$ of the height through which it falls. Find the average speed of the ball. (Take, $g = 10\ m\ s^{-2}$)

[2021, 17 March Shift-II]

(a) $3.0\ ms^{-1}$ (b) $2.50\ ms^{-1}$ (c) $2.0\ ms^{-1}$ (d) $3.50\ ms^{-1}$

Numerical Value Type Questions

24 A bus moving along a straight highway with speed of 72 km/h is brought to halt within 4 s after applying the brakes. The distance travelled by the bus during this time (assume the retardation is uniform) is ..... m.

[2024, 4 April Shift-II]

25 The displacement and the increase in velocity of a moving particle in the time interval of t to $(t + 1)$ s are 125 m and 50 m/s, respectively. The distance travelled by the particle in $(t + 2)$th s is .... m.

[2024, 30 Jan Shift-I]

26 A tennis ball is dropped on to the floor from a height of 9.8 m. It rebounds to a height 5.0 m. Ball comes in contact with the floor for 0.2 s. The average acceleration during contact is .... ms^{-2}
(Given, $g = 10\ ms^{-2}$)

[2023, 29 Jan Shift-I]

27 A ball is thrown vertically upwards with a velocity of $19.6\ ms^{-1}$ from the top of a tower. The ball strikes the ground after 6 s. The height from the ground up to which the ball can rise will be $\left(\frac{k}{5}\right)$m.
The value of k is .... (Use, $g = 9.8\ m/s^2$)

[2022, 28 July Shift-II]

28 A particle is moving in a straight line such that its velocity is increasing at $5\ ms^{-1}$. The acceleration of the particle is ...... ms^{-2} at a point where its velocity is $20\ ms^{-1}$.

[2022, 25 July Shift-II]

29 A ball is projected vertically upward with an initial velocity of $50\ ms^{-1}$ at $t = 0$ s. At $t = 2$ s, another ball is projected vertically upward with same velocity. At t = ... s, second ball will meet the first ball (Take, $g = 10\ ms^{-2}$).

[2022, 26 June Shift-II]

30 Two spherical balls having equal masses with radius of 5 cm each are thrown upwards along the same vertical direction at an interval of 3s with the same initial velocity of 35 m/s, then these balls collide at a height of .... m. (Take, $g = 10\ m/s^2$)

[2021, 26 Aug Shift-I]

31 A ball is dropped from the top of a 100 m high tower on a planet. In the last $\frac{1}{2}$ s before hitting the ground, it covers a distance of 19 m. Acceleration due to gravity (in ms^{-2}) near the surface on that planet is ....

[2020, 8 Jan Shift-II]

Topic 3 Graphs in Motion

Level 1

01 The velocity-time graph of an object moving along a straight line is shown in figure. What is the distance covered by the object between $t = 0$ to $t = 4$ s?

[2025, 28 Jan Shift-II]

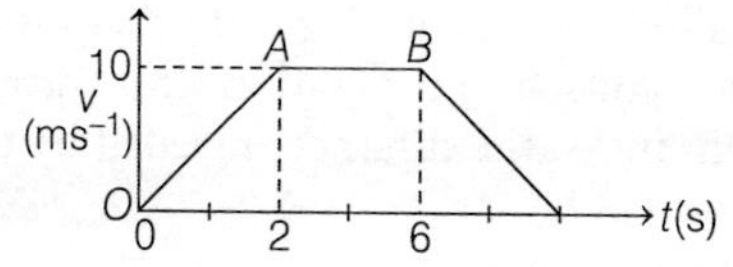

(a) 30 m (b) 13 m (c) 11 m (d) 10 m

02 The position-time graphs for two students A and B returning from the school to their homes are shown in figure.

A. A lives closer to the school
B. B lives closer to the school
C. A takes lesser time to reach home
D. A travels faster than B
E. B travels faster than A

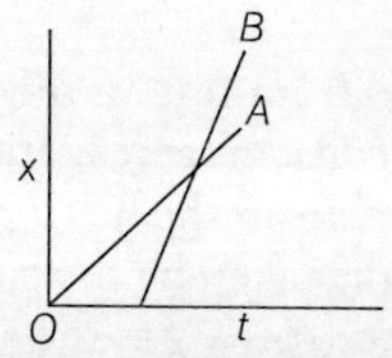

Choose the correct answer from the options given below **[2023, 10 April Shift-I]**

(a) A, C and E only (b) A, C and D only
(c) A and E only (d) B and E only

03 Given below are two statements.

Statement I Area under velocity-time graph gives the distance travelled by the body in a given time.

Statement II Area under acceleration- time graph is equal to the change in velocity in the given time.

In the light of given statements, choose the correct answer from the options given below. **[2023, 8 April Shift-II]**

(a) Statement I is incorrect but Statement II is correct.
(b) Statement I is correct but Statement II is incorrect.
(c) Both Statement I and Statement II are correct.
(d) Both Statement I and Statement II are incorrect.

04 The velocity-time graph of a body moving in a straight line is shown in figure.

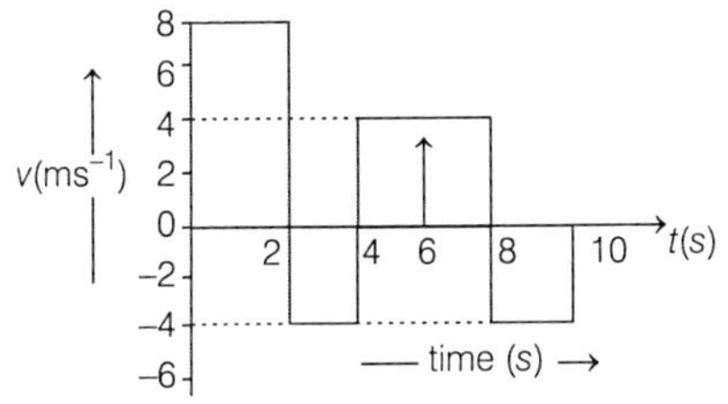

The ratio of displacement to distance travelled by the body in time 0 to 10s is **[2023, 24 Jan Shift -II]**

(a) 1 : 2 (b) 1 : 3 (c) 1 : 1 (d) 1 : 4

Imp. Velocity Time Graph

05 If the velocity-time graph has the shape *AMB*, what would be the shape of the corresponding acceleration time graph? **[2021, 24 Feb Shift-I]**

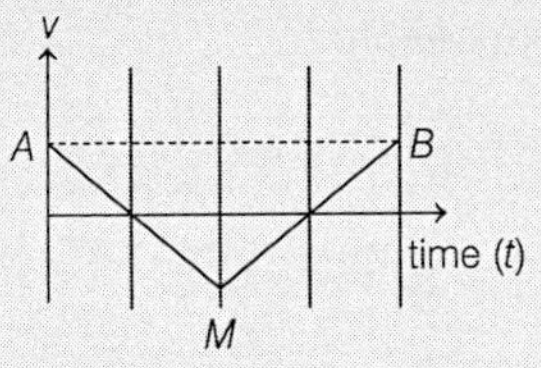

(a) 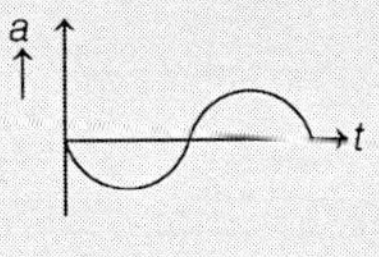
(b) 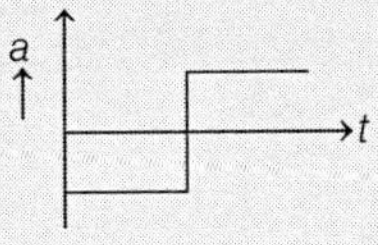
(c) 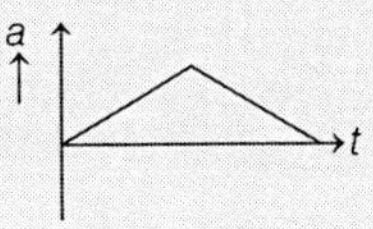
(d) 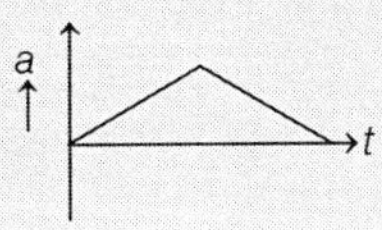

06 A body is thrown vertically upwards. Which one of the following graphs correctly represent the velocity *versus* time ? **[JEE Main 2017]**

(a) 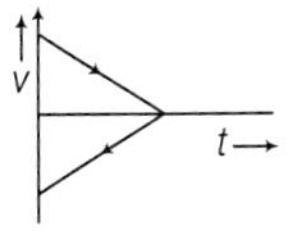
(b) 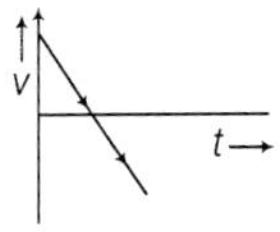
(c) 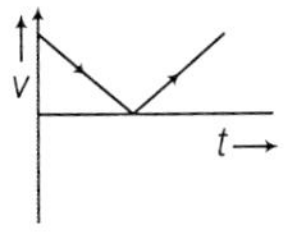
(d) 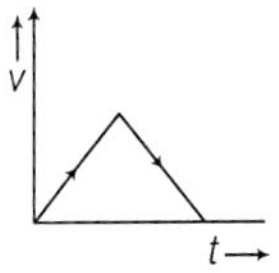

Level 2

07 The displacement (x) *versus* time (t) graph is shown below. **[2025, 4 April Shift-II]**

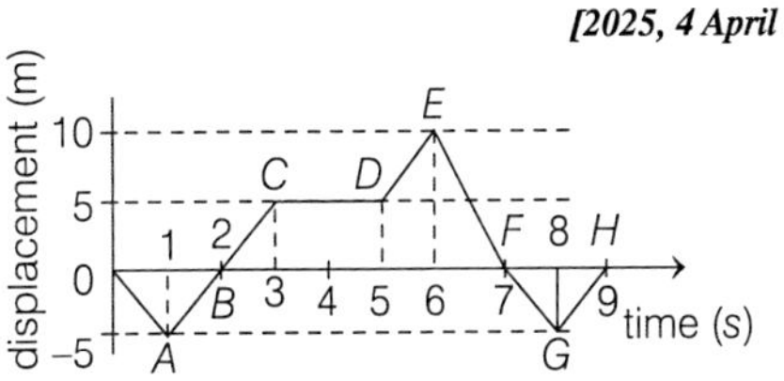

(A) The average velocity during 0 to 3 s is 10 m/s.
(B) The average velocity during 3 to 5 s is 0 m/s.
(C) The instantaneous velocity at $t = 2$ s is 5 m/s.
(D) The average velocity during 5 to 7 s and instantaneous velocity at $t = 6.5$ s are equal.
(E) The average velocity from $t = 0$ to $t = 9$ s is zero.

Choose the correct answer from the options given below.

(a) (B), (D), (E) only (b) (A), (D), (E) only
(c) (B), (C), (E) only (d) (B), (C), (D) only

08 The motion of an airplane is represented by velocity-time graph as shown below. The distance covered by airplane in the first 30.5 second is ...... km. **[2025, 23 Jan Shift-I]**

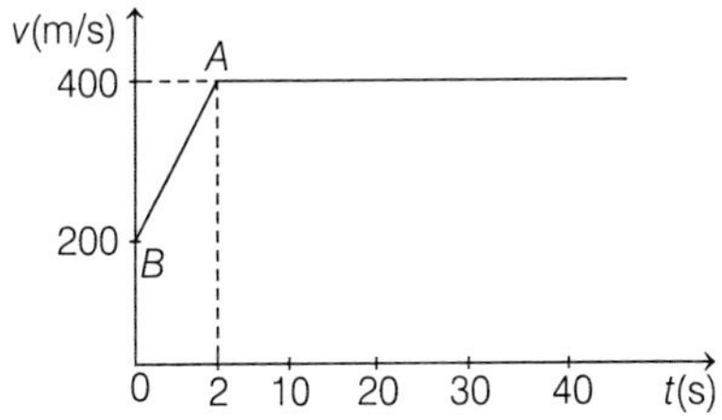

(a) 3 (b) 6
(c) 9 (d) 12

09 The coordinates of a particle moving in XY-plane are given by

$$x = 2 + 4t,\ y = 3t + 8t^2.$$

The motion of the particle is **[2024, 4 April Shift-I]**

(a) non-uniformly accelerated.
(b) uniformly accelerated having motion along a straight line.
(c) uniform motion along a straight line.
(d) uniformly accelerated having motion along a parabolic path.

10 From the v-t graph shown, the ratio of distance to displacement in 25 s of motion is *[2023, 11 April Shift-I]*

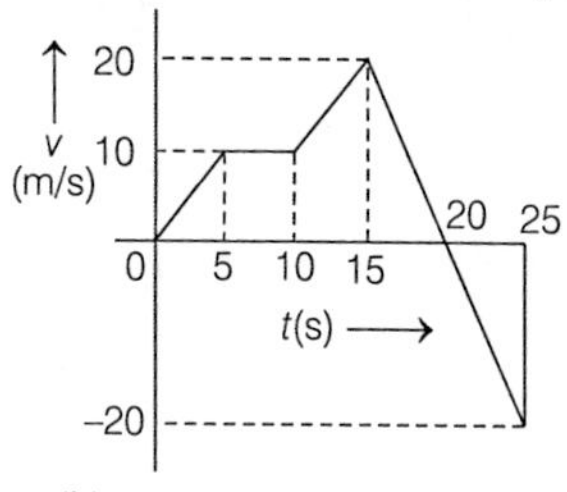

(a) 5/3 (b) 3/5 (c) 1/2 (d) 1

11 A particle is moving with constant acceleration a. Following graph shows v^2 versus x (displacement) plot. The acceleration of the particle is .... m/s^2 *[2021, 31 Aug Shift-II]*

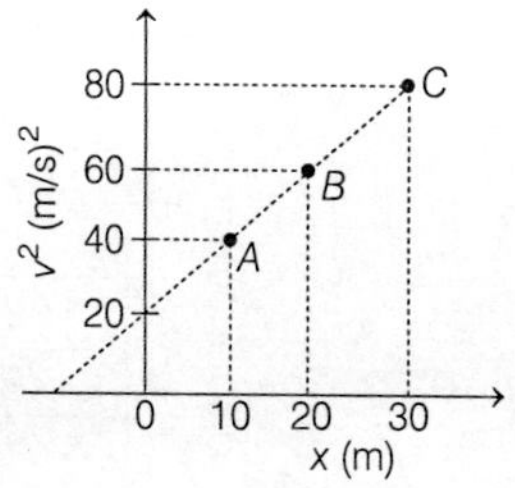

(a) 1 (b) 2
(c) 3 (d) 4

12 The velocity-time graphs of a car and a scooter are shown in the figure. (i) The difference between the distance travelled by the car and the scooter in 15 s and (ii) the time at which the car will catch up with the scooter are, respectively. *[JEE Main 2018]*

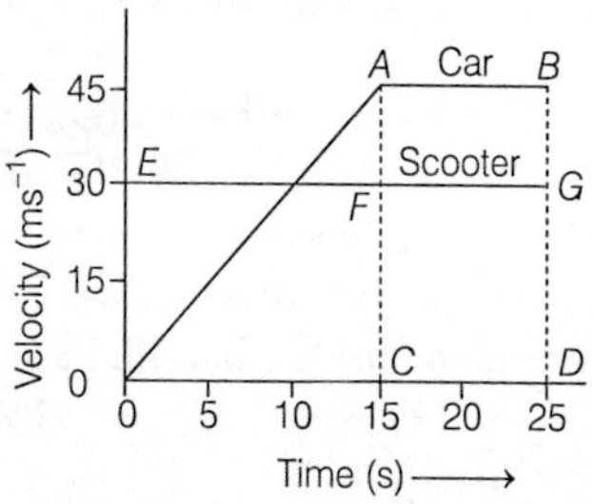

(a) 112.5 m and 15 s (b) 337.5 m and 25 s
(c) 225.5 m and 10 s (d) 112.5 m and 22.5 s

Level 3

13 Which of the following curves possibly represent one-dimensional motion of a particle? *[2025, 3 April Shift-I]*

A. 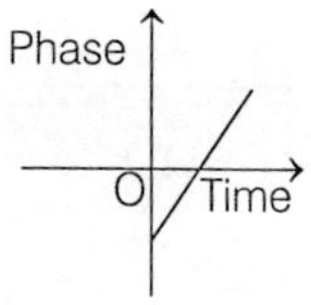

B. 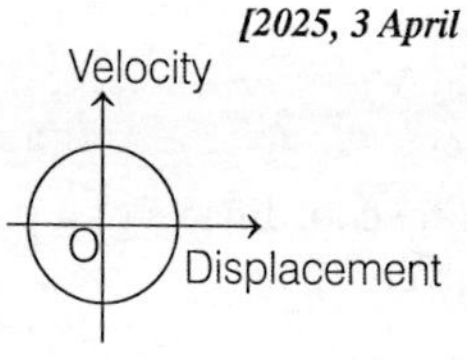

C. 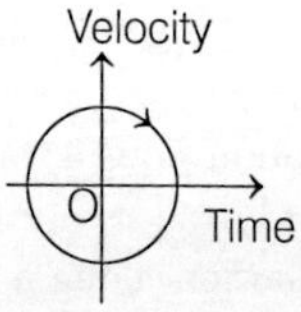

D. 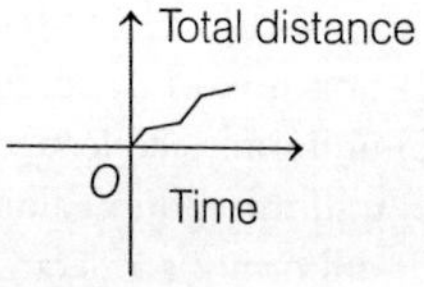

Choose the correct answer from the options given below.

(a) A and B only (b) A, C and D only
(c) A, B and D only (d) A, B and C only

14 Match List I with List II.

	List I		List II
A.	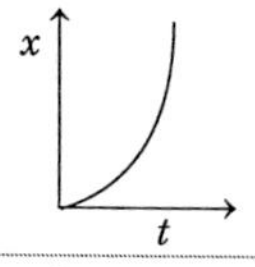	I.	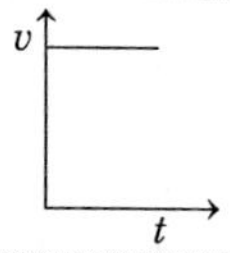
B.	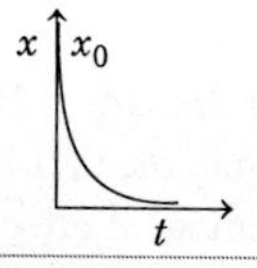	II.	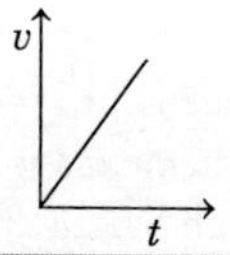
C.	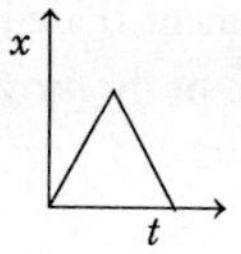	III.	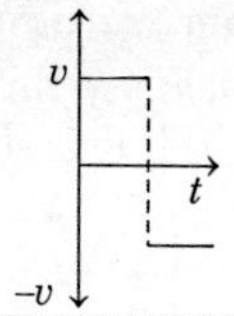
D.	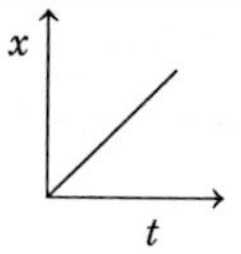	IV.	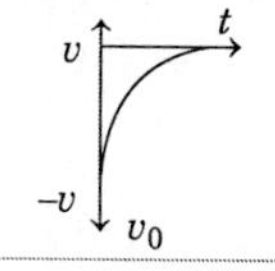

Choose the correct answer from the options given below *[2023 Jan Shift-I]*

(a) A-I, B-II, C-III, D-IV
(b) A-II, B-IV, C-III, D-I
(c) A-I, B-III, C-IV, D-II
(d) A-II, B-III, C-IV, D-I

15 The velocity- displacement graph of a particle is shown in the figure.
The acceleration-displacement graph of the same particle is represented by *[2021, 18 March Shift-II]*

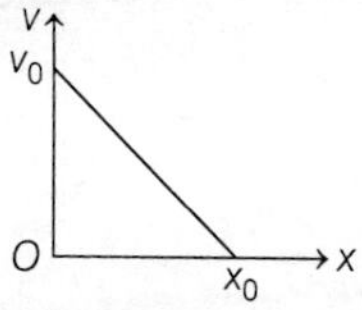

(a) 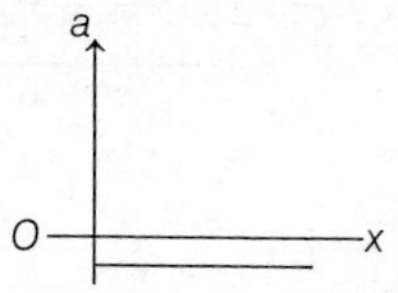

(b) 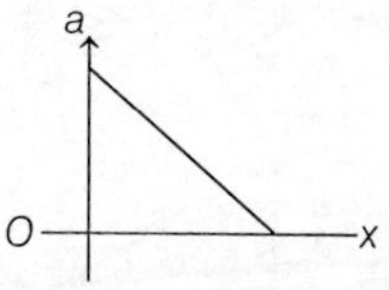

(c) 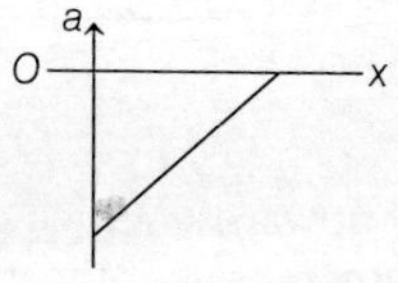

(d) 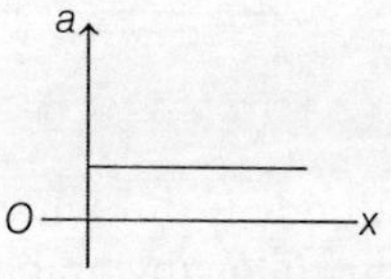

16 The position, velocity and acceleration of a particle moving with a constant acceleration can be represented by **[2021, 18 March Shift-I]**

(a)

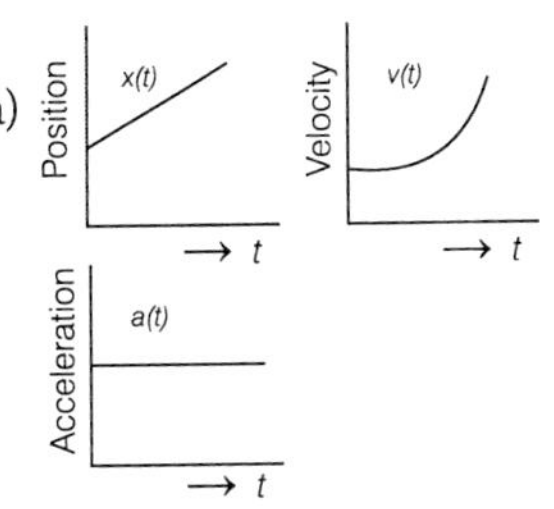

(b)

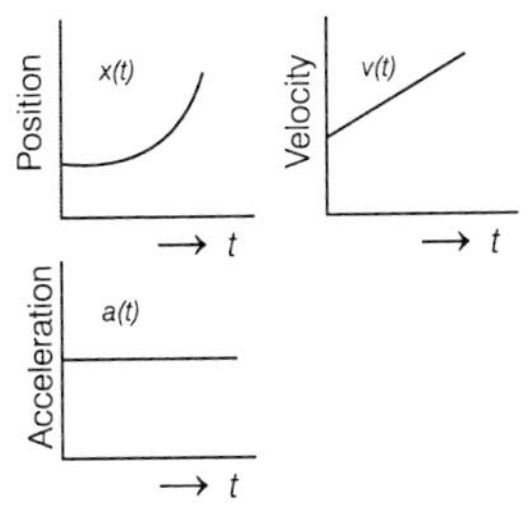

(c)

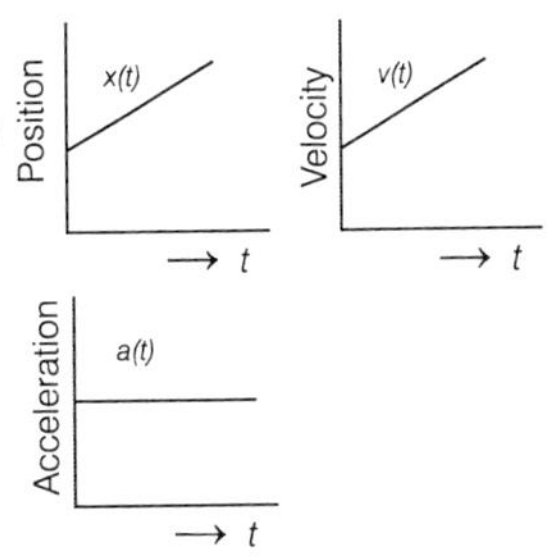

(d)

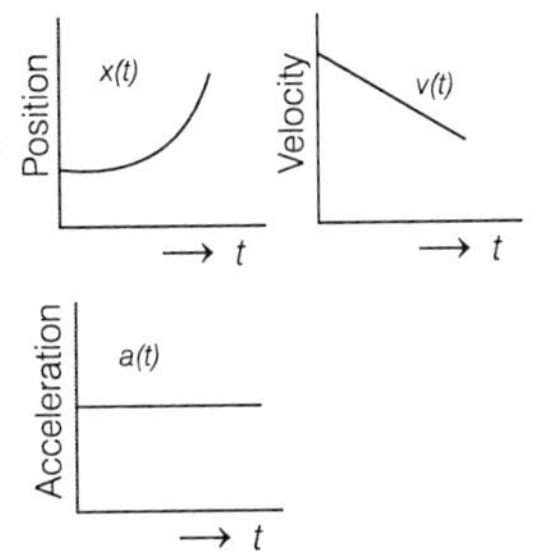

17 The velocity-displacement graph describing the motion of a bicycle is shown in the following figure.

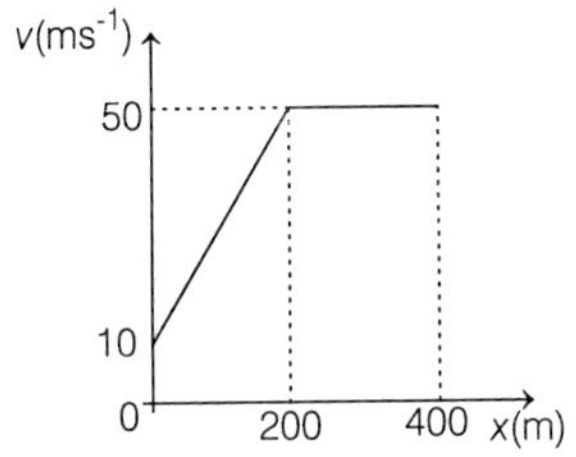

The acceleration-displacement graph of the bicycle's motion is best described by **[2021, 16 March Shift-I]**

(a)

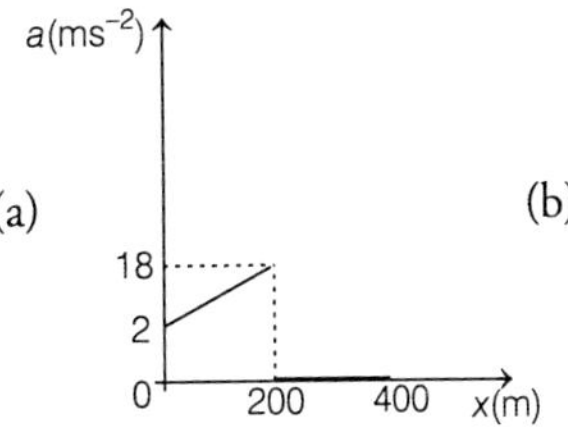

(b)

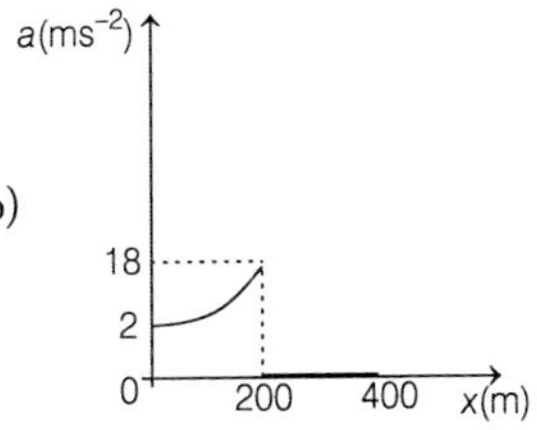

(c)

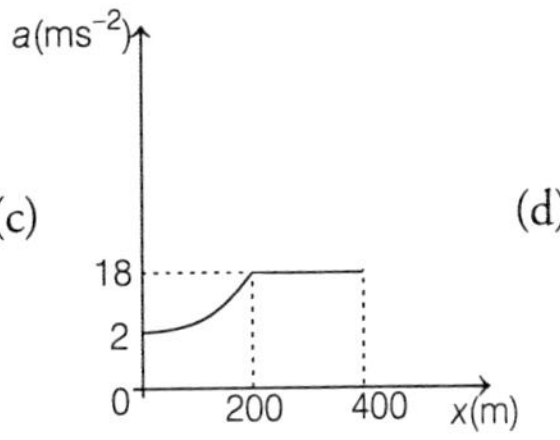

(d)

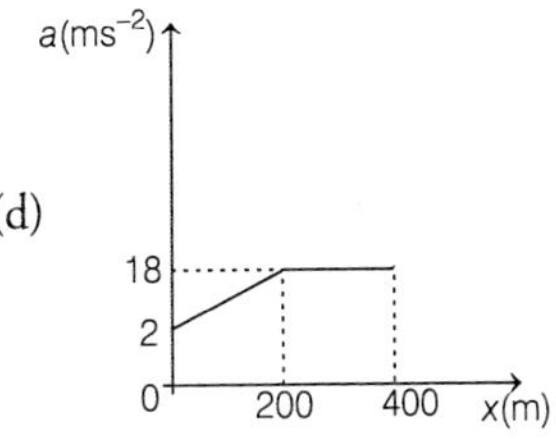

Numerical Value Type Questions

Imp. Slope of Graph

18 The speed *versus* time graph for a particle is shown in the figure. The distance travelled (in metre) by the particle during the time interval $t = 0$ s to $t = 5$ s will be ....

[2020, 4 Sep Shift-II]

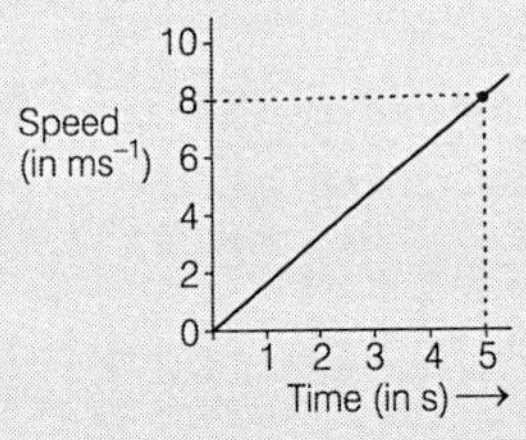

• Key **Index**

Topic 1

1 (d)	**2** (a)	**3** (b)	**4** (b)	**5** (b)	**6** (c)	**7** (a)	**8** (b)	**9** (a)	**10** (c)
11 (d)	**12** (b)	**13** (a)	**14** (b)	**15** (b)	**16** (b)	**17** (c)	**18** (a)	**19** (a)	**20** (b)
21 (d)	**22** (c)	**23** (b)	**24** (10)	**25** (3)	**26** (3)	**27** (8)	**28** (52)	**29** (18)	**30** (12)

Topic 2

1 (b)	**2** (d)	**3** (b)	**4** (b)	**5** (c)	**6** (b)	**7** (c)	**8** (c)	**9** (d)	**10** (a)
11 (c)	**12** (a)	**13** (d)	**14** (c)	**15** (c)	**16** (c)	**17** (a)	**18** (d)	**19** (b)	**20** (d)
21 (c)	**22** (c)	**23** (b)	**24** (40)	**25** (175)	**26** (120)	**27** (392)	**28** (100)	**29** (6)	**30** (50)
31 (8)									

Topic 3

1 (a)	**2** (c)	**3** (a)	**4** (b)	**5** (b)	**6** (b)	**7** (c)	**8** (d)	**9** (d)	**10** (a)
11 (a)	**12** (d)	**13** (c)	**14** (b)	**15** (c)	**16** (b)	**17** (a)	**18.** (20)		

Detailed **Explanations**

Topic 1

01. *(d)* Given, $x = 5t^2 - 4t + 5$

$\therefore$ Velocity, $v = \frac{dx}{dt} = \frac{d}{dt}(5t^2 - 4t + 5)$

$\Rightarrow \quad v = 10t - 4$

at $\quad t = 2$ s, $v = 10 \times 2 - 4 = 16$ m/s

02. *(a)* Given, $s = 2.5t^2$

$$\frac{ds}{dt} = 2.5 \times 2t$$

$$v = 5t,\ t = 5 \text{ s}$$

$$v = 5 \times 5 = 25 \text{ m/s}$$

03. *(b)* Time taken by vehicle to travel first 4 km with speed of 3 km/h.

$$t_1 = \frac{4}{3} \text{ h}$$

Time taken by the vehicle to travel second 4 km with speed of 5 km/h,

$$t_2 = \frac{4}{5} \text{ h}$$

$\therefore$ Average speed

$$= \frac{\text{Total distance travelled}}{\text{Total time taken}}$$

$$= \frac{4+4}{t_1 + t_2} = \frac{8}{\frac{4}{3} + \frac{4}{5}} = \frac{8 \times 15}{32}$$

$$= 3.75 \text{ km/h}$$

04. *(b)*

> **key idea** If distance travelled by a particle is given in terms of time (t) or as a function of time, then you can calculate its velocity by using following formula, $v = \frac{dx}{dt}$

Distance travelled $x = 4t^2$

$\therefore$ Velocity of the particle is given as

$$v = \frac{dx}{dt} = \frac{d}{dt}(4t^2) = 8t$$

At $t = 5$ s, $v = 8 \times 5$

$$= 40 \text{ ms}^{-1}$$

05. *(b)* Average velocity is given by formula as

Average velocity,

$$v = \frac{x_1 + x_2}{\frac{x_1}{v_1} + \frac{x_2}{v_2}} = \frac{2x}{x\left(\frac{1}{v_1} + \frac{1}{v_2}\right)}$$

$$(\because x_1 = x_2 = x)$$

$$\Rightarrow \quad \frac{2}{v} = \frac{1}{v_1} + \frac{1}{v_2}$$

06. *(c)* Given, $t = mx^2 + nx$,

differentiating w.r.t. t

$$\frac{d}{dt}(t) = m.2x\frac{dx}{dt} + n\frac{dx}{dt}$$

So, $\quad 1 = m.\,2xv + nv \qquad [\therefore dx/dt = v]$

$\Rightarrow \quad 1/v = (2mx + n)$

differentiating again, we have

$$-\frac{1}{v^2}\frac{dv}{dt} = 2m.\frac{dx}{dt} \text{ or } \frac{dv}{dt} = -2mv^3$$

$\Rightarrow \quad a = -2mv^3$

07. *(a)* Given, $dv/dt = -2.5\sqrt{v}$

$$\Rightarrow \quad \frac{dv}{\sqrt{v}} = -2.5\,dt$$

$$\Rightarrow \quad \int_{6.25}^{0} v^{-1/2}\,dv = -2.5\int_0^t dt$$

$$\Rightarrow \quad -2.5\,[t]_0^t = [2v^{1/2}]_{6.25}^{0}$$

$$\Rightarrow \quad t = 2 \text{ s}$$

08. *(b)*

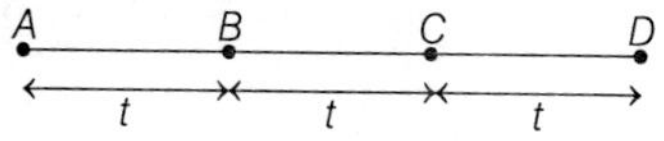

$BD = s = 9t + 15t = 24t$

$AB = s = 6t_1 = 24t \Rightarrow t_1 = 4t$

Distance $= AB + BD \Rightarrow 24t + 6t_1$

$= 24t + 24t = 48t$

$$\text{Speed} = \frac{\text{distance}}{\text{time}} = \frac{48t}{2t + t_1} = \frac{48t}{2t + 4t}$$

$$= \frac{48t}{6t} = 8 \text{ m/s}$$

09. *(a)* Total time $= t + 3t = 4t$

For first part of motion, distance travelled,

$$s_1 = \frac{1}{2}at^2 = \frac{1}{2}(at) \cdot t = \frac{1}{2}vt$$

$$= \frac{1}{2} \times 80 \times t$$

At constant speed, $s_2 = v \times 3t = 80 \times 3t$

$$\text{Average speed} = \frac{s_1 + s_2}{4t}$$

$$= \frac{\frac{1}{2} \times 80 \times t + 80 \times 3t}{4t}$$

$$= 70 \text{ km/h}$$

10. *(c)*

> **key idea** Use the formula of distance travelled in nth second to solve this question.

Given,

Distance travel in nth second $= 102.5$ m

Distance travel in $(n+2)$th second $= 115$ m

Acceleration, $a = ?$

Using the formula of nth second distance travel,

$$d = u + \frac{a}{2}(2n - 1)$$

$$102.5 = u + \frac{a}{2}(2n - 1) \qquad \text{...(i)}$$

$$115 = u + \frac{a}{2}[2(n+2) - 1]$$

$$115 = u + \frac{a}{2}[2n + 3] \qquad \text{...(ii)}$$

Now, Eqs. (ii) – (i), we get

$12.5 = 2a \quad \Rightarrow \quad a = 6.25 \text{ m/s}^2$

11. *(d)*

Given, speed of $A = 72$ kmph

$$= 72 \times \frac{5}{18} \text{ m/s} = 20 \text{ m/s}$$

$$\text{Speed of } B = 108 \times \frac{5}{18} = 30 \text{ m/s}$$

Taking north $(+x)$ as positive direction, for A the B is moving towards south with a speed,

$$v_{BA} = v_B - v_A$$

$$= -30 - 20 = -50 \text{ m/s}$$

Also, with respect to B, ground appear to move towards north with same speed.

$\therefore$ Speed of ground w.r.t. B is towards north with 30 m/s.

$$v_{GB} = v_G - v_B$$

$$v_{GB} = 0 - (-30)$$

$$v_{GB} = 30 \text{ m/s}$$

12. *(b)* Given, time t and distance x are related as,

$$t = \alpha x^2 + \beta x$$

$$\Rightarrow \quad \frac{dt}{dt} = 2\alpha xv + \beta v \qquad \left(\because \frac{dx}{dt} = v\right)$$

$$\Rightarrow \quad 1 = 2\alpha xv + \beta v$$

or $v = \frac{1}{2\alpha x + \beta}$

Now, differentiating with respect to x,

$$\frac{dv}{dx} = \frac{-2\alpha}{(2\alpha x + \beta)^2} = -2\alpha v^2$$

Acceleration is,

$$a = v\frac{dv}{dx} = -2\alpha v^3$$

13. *(a)* Given,

$$x = (t^3 - 6t^2 + 20t + 15)\ \text{m}$$

Velocity, $v = \frac{dx}{dt}$

$$= \frac{d}{dt}(t^3 - 6t^2 + 20t + 15)$$

$$v = 3t^2 - 12t + 20 \qquad \ldots\text{(i)}$$

$\therefore$ Acceleration, $a = \frac{dv}{dt}$

$$= 6t - 12$$

When, $a = 0$

$6t - 12 = 0$

$t = 2\ \text{s}$

$\therefore$ From Eq. (i), $v = 3 \times 2^2 - 12 \times 2 + 20$

$= 12 - 24 + 20$

$= 8\ \text{m/s}$

14. *(b)* $v_A = 90\ \text{km/h} = \frac{90 \times 5}{18} = 25\ \text{m/s}$

$v_B = 54\ \text{km/h} = \frac{54 \times 5}{18} = 15\ \text{m/s}$

$v_{BA} = v_B - v_A = 15 - (-25)$

(Move opposite to each other)

$= 15 + 25$

$= 40\ \text{m/s}$

Time of crossing $= \frac{\text{Length of train}}{\text{Relative velocity}}$

$8 = \frac{l}{40}$

$l = 8 \times 40 = 320$

$l = 320\ \text{m}$

15. *(b)* From given figure,

$AB = BC$ and $AD = 3AB$

$(AB + BC + CD) = 3AB$

$CD = AB \quad [\because AB = BC]$

Average speed,

$$v_{\text{avg}} = \frac{\text{Total distance}}{\text{Total time}}$$

$$= \frac{AB + BC + CD}{t_1 + t_2 + t_3} = \frac{x + x + x}{\frac{x}{v_1} + \frac{x}{v_2} + \frac{x}{v_3}}$$

$$= \frac{3x}{x\left(\frac{1}{v_1} + \frac{1}{v_2} + \frac{1}{v_3}\right)}$$

$$v_{\text{avg}} = \frac{3v_1v_2v_3}{v_1v_2 + v_2v_3 + v_1v_3}$$

16. *(b)* Given, instantaneous velocity

$$v = \alpha t + \beta t^2$$

Here, $t_1 = 1\text{s}$ and $t_2 = 2\text{s}$

As, $v = \frac{ds}{dt}$

$\Rightarrow \quad ds = v\,dt \Rightarrow ds = (\alpha t + \beta t^2)$

Integrating both sides, we get

$$s = \left|\left(\alpha\frac{t^2}{2} + \frac{\beta t^3}{3}\right)\right|_1^2$$

$$= \frac{\alpha}{2}[2^2 - 1^2] + \frac{\beta}{3}[2^3 - 1^3]$$

$$= \frac{\alpha}{2}[4 - 1] + \frac{\beta}{3}[8 - 1]$$

$$= \frac{3\alpha}{2} + \frac{7\beta}{3}$$

17. *(c)*

> **key idea** If position (x) of a moving body is given as a function of time (t), then you can find the velocity (v) and acceleration (a) of the particle with the help of following formula
>
> $$v = \frac{dx}{dt} \text{ and } a = \frac{dv}{dt} = \frac{d^2x}{dt^2}$$

Position of particle is,

$$x(t) = at + bt^2 - ct^3$$

$\therefore \quad v = \frac{dx}{dt} = a + 2bt - 3ct^2$

and $\quad a = \frac{dv}{dt} = 2b - 6ct$

Acceleration is zero, then

$2b - 6ct = 0$

$\Rightarrow \quad t = \frac{2b}{6c} = \frac{b}{3c}$

Substituting this t in expression of velocity, we get

$$v = a + 2b\left(\frac{b}{3c}\right) - 3c\left(\frac{b}{3c}\right)^2$$

$$= a + \frac{2b^2}{3c} - \frac{b^2}{3c} = a + \frac{b^2}{3c}$$

18. *(a)* Given, $v = \alpha\sqrt{x}$

or $\quad \frac{dx}{dt} = \alpha\sqrt{x} \qquad \left[\because v = \frac{dx}{dt}\right]$

or $\quad \frac{dx}{\sqrt{x}} = \alpha\,dt$

On integrating, we get

$$\int_0^x \frac{dx}{\sqrt{x}} = \int_0^t \alpha\,dt$$

[$\because$ at $t = 0$, $x = 0$ and let at any time t, particle be at x]

$$\Rightarrow \left[\frac{x^{1/2}}{1/2}\right]_0^x = \alpha t \text{ or } x^{1/2} = \frac{\alpha}{2}t$$

or $\quad x = \frac{\alpha^2}{4} \times t^2$

or $\quad x \propto t^2$

19. *(a)* In 1min, a second hand travels a distance $= 2\pi R_1$

R_1 = length of second hand

So, total distance travelled by a second hand in 30 min

$= 30 \times 2\pi R_1$

$\therefore \quad R_1 = 75\ \text{cm}$

$= 30 \times 2\pi \times 75\ \text{cm}$

In 30 min, minute hand travels by half round

$= \frac{1}{2} \times 2\pi R_2 = \pi \times 60\ \text{cm}$

R_2 = length of minute hand

$= 60\ \text{cm}$

So, difference in distance travelled by second and minute hand

$= 30 \times 2\pi \times 75 - \pi \times 60$

$= 13940\ \text{cm}$

$= 139.4\ \text{m}$

20. *(b)* Velocity of train A,

$v_A = 108\ \text{km/h}$

$= 180 \times \frac{5}{18}\ \text{m/s} = 30\ \text{m/s}$

Velocity of train B,

$v_B = 72\ \text{km/h}$

$= 72 \times \frac{5}{18} = 20\ \text{m/s}$

According to given situation, time taken to cross the tunnel by train B – time taken to cross the tunnel by train $A = 35$ s

$$\Rightarrow \frac{L + 4l}{20} - \frac{L + l}{30} = 35$$

$$\Rightarrow \frac{60l + 4l}{20} - \frac{60l + l}{30} = 35$$

$(\because L = 60l)$

$$\Rightarrow \frac{180l + 12l - 120l - 2l}{60} = 35$$

$\Rightarrow l = 30\ \text{m}$

$\therefore$ Length of the tunnel,

$L = 60l = 60 \times 30 = 1800\ \text{m}$

21. *(d)* Given, position of bus P,

$$X_P(t) = \alpha t + \beta t^2$$

Its velocity, $v_P = \dfrac{d}{dt} X_P(t)$

$$= \frac{d}{dt}(\alpha t + \beta t^2)$$

$\Rightarrow \quad v_P = \alpha + 2\beta t$

Position of bus Q,

$$X_Q(t) = ft - t^2$$

Its velocity, $v_Q = \dfrac{d}{dt} X_Q(t)$

$$= \frac{d}{dt}(ft - t^2) = f - 2t$$

Since, $\quad v_P = v_Q$

$\alpha + 2\beta t = f - 2t$

$\Rightarrow \quad 2\beta t + 2t = f - \alpha$

$\Rightarrow \quad t(2\beta + 2) = f - \alpha$

$\Rightarrow \quad t = \dfrac{f-\alpha}{2\beta + 2} = \dfrac{f - \alpha}{2(\beta + 1)}$

22. *(c)* $\therefore$ Velocity of boy with respect to stationary escalator

$$v_1 = \frac{L}{t_1} \quad \ldots(i)$$

When escalator is moving and boy is at rest velocity of escalator

$$v_2 = \frac{L}{t_2} \quad \ldots(ii)$$

So, the resultant velocity can be given as when both escalator and boy are moving,

$$v = v_1 + v_2 \quad \ldots(iii)$$

From Eqs. (i), (ii) and (iii), we get

$$v = \frac{L}{t_1} + \frac{L}{t_2} \Rightarrow \frac{v}{L} = \frac{1}{t_1} + \frac{1}{t_2}$$

$\Rightarrow \quad \dfrac{v}{L} = \dfrac{t_1 + t_2}{t_1 t_2} \Rightarrow \dfrac{L}{v} = \dfrac{t_1 t_2}{t_1 + t_2}$

$\Rightarrow \quad t = \dfrac{t_1 t_2}{t_1 + t_2} \quad [\because t = L/v]$

23. *(b)* The given figure is shown below,

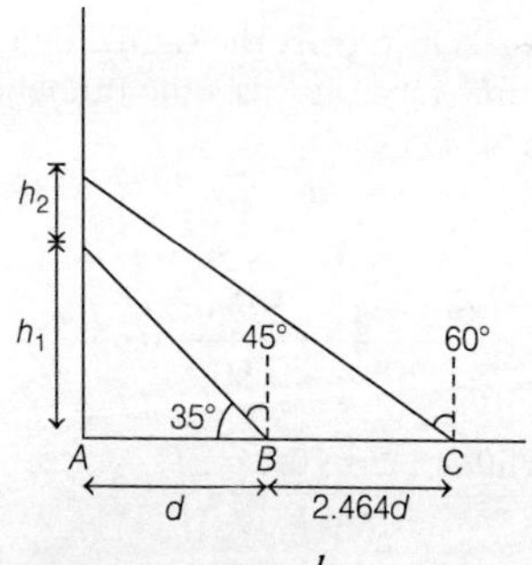

In ΔABD, $\tan 45° = \dfrac{h_1}{d}$

$\because \quad \tan 45° = 1 \Rightarrow 1 = h_1/d$

$\Rightarrow \quad h_1 = d \quad \ldots(i)$

Now, in ΔACE,

$$\tan 30° = \frac{AD + DE}{AB + BC}$$

$\Rightarrow \quad \dfrac{1}{\sqrt{3}} = \dfrac{h_1 + h_2}{d + 2.464d}$

$\Rightarrow \quad 0.5774 = \dfrac{h_1 + h_2}{3.464d}$

$\Rightarrow \quad h_1 + h_2 = (0.5774) \times (3.464d)$

$h_1 + h_2 = 2d$

$d + h_2 = 2d \quad$ [Using Eq. (i)]

$h_2 = 2d - d$

$\Rightarrow \quad h_2 = d$

24. *(10)* $v_{avg} = \dfrac{x_1 + x_2}{t_1 + t_2}$

$$\frac{50}{7} = \frac{x + \frac{3x}{2}}{\frac{x}{5} + \frac{3x}{2v_2}}$$

$$\frac{50}{7} = \frac{\frac{5}{2}}{\frac{1}{5} + \frac{3}{2v_2}} \Rightarrow \frac{1}{5} + \frac{3}{2v_2} = \frac{7}{20}$$

$$\frac{3}{2v_2} = \frac{7}{20} - \frac{1}{5} = \frac{7-4}{20} \Rightarrow \frac{3}{20}$$

$$\frac{3}{2v_2} = \frac{3}{20} \Rightarrow 2v_2 = 20$$

$$v_2 = 10 \text{ m/s}$$

25. *(3)* $a_p = kt$

$\therefore \quad v_p = \int a_p dt$

$$v_p = \frac{1}{2}\frac{kt^2}{2} + C$$

and $x_p = \int v_p dt = \int \left(\dfrac{1}{2}kt^2 + C\right) dt$

$$x_p = \frac{1}{6}kt^3 + Ct + D$$

$a_q = a$

$v_q = \int a_p dt = at + C_1$

$x_p = \int v_q dt = \int (at + C_1) dt$

$x_p = \dfrac{1}{2}at^2 + C_1 t + D_1$

$x_p = x_q$

$$\frac{1}{6}kt^3 + Ct + D = \frac{1}{2}at^2 + C_1 t + D_1$$

$$\frac{1}{6}kt^3 - \frac{1}{2}at^2 + (C - C_1)t + (D - D_1) = 0$$

Equation of order 3

$\therefore$ 3 crossing

26. *(3)*

> **key idea** To get acceleration of the particle at any instant, differentiate the displacement-time relation, two times with respect to time.

Distance-time relation is given by

$$x^2 = 1 + t^2$$

Differentiating w.r.t x, we get

$$2x\frac{dx}{dt} = 0 + 2t$$

$\Rightarrow \quad xv = t \quad \ldots(i)$

Differentiating again with respect to t, we get

$$x\frac{dv}{dt} + v\frac{dx}{dt} = 1 \Rightarrow xa + v^2 = 1$$

$$a = \frac{1 - v^2}{x}$$

Substitute the value of v from Eq. (i)

$$= \frac{1 - \left(\frac{t}{x}\right)^2}{x} = \frac{1 - \frac{t^2}{x^2}}{x} = \frac{x^2 - t^2}{x^3}$$

$\Rightarrow \quad a = \dfrac{x^2 - t^2}{x^3}$

Substitute the value of x^2

$$= \frac{1 + t^2 - t^2}{x^3} = \frac{1}{x^3} = x^{-3} = x^{-n}$$

(given)

$\therefore \quad n = 3$

27. *(8)* Velocity of particle is given,

$$v = 4\sqrt{x} \text{ m/s}$$

$\Rightarrow \quad v = \dfrac{dx}{dt} = 4\sqrt{x}$

$\Rightarrow \quad \dfrac{dx}{\sqrt{x}} = 4dt$

Integrating we have,

$$\int_0^x \frac{dx}{\sqrt{x}} = 4\int_0^t dt$$

$$2\sqrt{x} = 4t \Rightarrow 4x = 16t^2$$

So, position of the particle is

$\Rightarrow \quad x = 4t^2$

$\Rightarrow$ Velocity, $v = \dfrac{dx}{dt} = 4 \times 2t$

Acceleration, $a = \dfrac{dv}{dt} = 8 \text{ m/s}^2$

28. *(52)* Position of particle is given by,

$$x = -3t^3 + 18t^2 + 16t$$

Velocity of particle

$$v = \frac{dx}{dt} = -3 \times 3t^2 + 18 \times 2t + 16$$

$$= -9t^2 + 36t + 16$$

Acceleration of particle is,

$$a = \frac{dv}{dt} = -18t + 36$$

Acceleration is zero when, $-18t + 36 = 0$

or $t = 2\,s$

Velocity at $t = 2\,s$ is,

$$v = -9 \times 2^2 + 36 \times 2 + 16$$
$$= -36 + 72 + 16$$
$$= 52 \text{ m/s}$$

29. *(18)* Given situation is

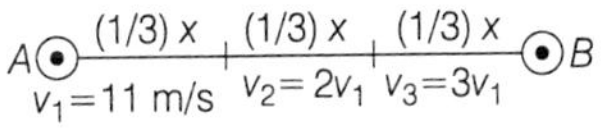

Now, let total displacement $= x$

Then, time to cover first $\frac{x}{3}$

Time,

$$t_1 = \frac{\text{displacement}}{\text{velocity}}$$

$$\Rightarrow \quad t_1 = \frac{x}{3 \times 11}$$

Similarly, $t_2 = \frac{x}{3 \times 22}$ and $t_3 = \frac{x}{3 \times 33}$

Now, $v_{\text{average}} = \frac{\text{Total displacement}}{\text{Total time}}$

$$= \frac{x}{\frac{x}{33} + \frac{x}{66} + \frac{x}{99}}$$

$$= \frac{1}{\frac{1}{33}\left(1 + \frac{1}{2} + \frac{1}{3}\right)} = \frac{33}{\left(\frac{6+3+2}{6}\right)}$$

$$= \frac{33 \times 6}{11} = 18 \text{ m/s}$$

> ***Alternate Solution***
>
> If a car travels one-third of total distance with speed v_1, next one-third of total distance with speed v_2 and last one third of distance with speed v_3, then average velocity of car is given as,
>
> $$v_{\text{avg}} = \frac{3v_1v_2v_3}{v_1v_2 + v_2v_3 + v_3v_1} \quad \text{...(i)}$$
>
> Here, $v_1 = 11$ m/s
>
> $v_2 = 2v_1 = 2 \times 11 = 22$ m/s
>
> $v_3 = 3v_1 = 3 \times 11 = 33$ m/s
>
> $\therefore$ From Eq. (i), we get
>
> $$\therefore \ v_{\text{avg}} = \frac{3 \times 11 \times 22 \times 33}{11 \times 22 + 22 \times 33 + 33 \times 11} = 18 \text{ m/s}$$

30. *(12)* Velocity of body is

$$v = \sqrt{5000 + 24x} \text{ m/s}$$

Acceleration of body can be calculated as

$$a = \frac{dv}{dt} = \frac{dv}{dx} \times \frac{dx}{dt}$$

$$\Rightarrow \quad a = v\frac{dv}{dx} \quad \text{...(i)}$$

$$\frac{dv}{dx} = \frac{d}{dx}(\sqrt{5000 + 24x})$$

$$\Rightarrow \quad \frac{dv}{dx} = \frac{1}{2(\sqrt{5000 + 24x})} \times 24$$

$$= \frac{12}{\sqrt{5000 + 24x}}$$

Substituting the value of $\frac{dv}{dx}$ in Eq. (i), we get

$$a = \sqrt{5000 + 24x} \times \frac{12}{\sqrt{5000 + 24x}}$$

$$= 12 \text{ m/s}^2$$

Topic 2

01. *(b)* Given, initial velocity, $v = 150$ m/s

From $v = u + at$, we have

$$v = 150 - 10t$$

($\because a = -g$ for vertically thrown object)

$$v_{(3)} = 150 - 30 = 120 \text{ m/s}$$
$$v_{(5)} = 150 - 50 = 100 \text{ m/s}$$

According to question,

$$\frac{x+1}{x} = \frac{120}{100} = \frac{6}{5} \Rightarrow x = 5$$

02. *(d)* Let initial speed of the ball is u

Time taken by the ball to reach maximum height is given as,

$$0 = u - gt \Rightarrow t = \frac{u}{g} \quad \text{... (i)}$$

Time interval between 2 balls $= t$

$$\text{Frequency} = \frac{1}{t}$$

According to question,

$$\text{Frquency} = n = \frac{1}{t}$$

$$\text{So,} \quad t = \frac{1}{n} \quad \text{... (ii)}$$

From Eqs. (i) and (ii), $\frac{u}{g} = \frac{1}{n} \Rightarrow u = \frac{g}{n}$

$\therefore$ Maximum height,

$$h = \frac{u^2}{2g} = \frac{g^2}{2gn^2} = \frac{g}{2n^2}$$

03. *(b)* The time taken by the ball to reach at the surface of water is given by

$$t_1 = \sqrt{\frac{2h}{g}}$$

$$\therefore \quad t_1 = \sqrt{\frac{2 \times 4.9}{9.8}} = 1.0 \text{ s}$$

Now, the velocity acquired by the ball to reach water surface,

$$v = gt_1 = 9.8 \times 1.0 = 9.8 \text{ m s}^{-1}$$

Now, time taken by the ball to reach the bottom of the lake from the water surface is $t_2 = 4.0 - 1.0 = 3.0$ s

Given that, velocity of the ball in water is constant, therefore depth of the lake will be,

$$d = vt_2 = 9.8 \times 3.0$$
$$= 29.4 \text{ m}$$

04. *(b)* By using first equation of motion,

$$v_1 = u + a_1t_1$$

$$\Rightarrow \quad v_1 = a_1t_1 \quad \text{... (i)}$$

and for second case,

$$v_2 = v_1 - a_2t_2$$

$$\Rightarrow \quad 0 = v_1 - a_2t_2$$

$$\Rightarrow \quad v_1 = a_2t_2 \quad \text{...(ii)}$$

On dividing Eq. (i) by Eq. (ii), we get

$$v_1 / v_1 = 1 = \frac{a_1t_1}{a_2t_2}$$

$$\Rightarrow a_2t_2 = a_1t_1 \Rightarrow t_1 / t_2 = a_2 / a_1$$

05. *(c)*

> **key idea** To solve such type of problem, you can use the following equation of motion,
>
> $$v^2 = u^2 + 2as$$

As the third equation of motion,

$$v^2 = u^2 + 2as$$

a is –ve because retardation,

so, $r = -a$

$$0 = u^2 - 2rs$$

$$s = u^2 / 2r \Rightarrow s \propto u^2$$

$$\frac{s_2}{s_1} = \left(\frac{u_2}{u_1}\right)^2 \Rightarrow s_2 = s_1\left(\frac{100}{50}\right)^2$$

$$s_2 = 6 \times 4 = 24 \text{ m}$$

06. *(b)* The given situation is shown in figure

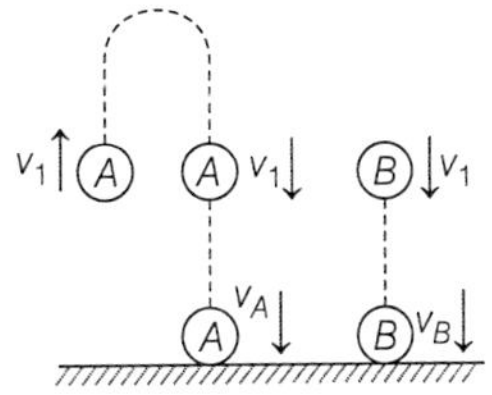

From the figure, it is clear that ball A attains same downward velocity at point of projection as ball B.

Thus, $v_A = v_B$ because both move under the effect of same value of gravitational acceleration g.

07. *(c)* $x = c_0(t^2 - 2) + c(t-2)^2$

$$\therefore \quad v = \frac{dx}{dt} = 2tc_0 + 2c(t-2)$$

$$\Rightarrow v = 2tc_0 + 2ct - 4c$$

$\therefore$ Acceleration

$$a = \frac{dv}{dt} = 2c_0 + 2c = 2(c_0 + c)$$

08. *(c)* Given, relative speed between cars,

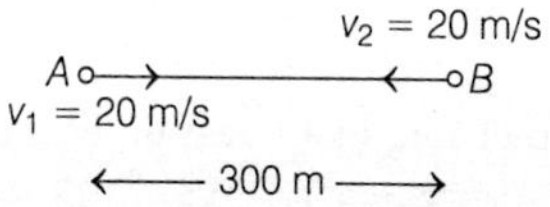

$$v_r = 40 \text{ ms}^{-1}$$

Relative retardation, $a_r = 4 \text{ ms}^{-2}$

We know that, $v^2 - u^2 = 2as$

$$0 - (40)^2 = 2 \times (-4) \times s$$

$$\Rightarrow \quad -1600 = -8\,s \Rightarrow s = 200 \text{ m}$$

Distance moved = 200 m

Distance between them

$$= 300 - 200 = 100 \text{ m}$$

09. *(d)* s_1 in first $(p-1)$s

$$s_1 = \frac{1}{2}a(p-1)^2,\ 2s_1 = a(p-1)^2$$

s_2 in first p s

$$s_2 = \frac{1}{2}ap^2,\ 2s_2 = ap^2$$

Let $s_1 + s_2$ displacement is in time t s then,

$$s_1 + s_2 = \frac{1}{2}at^2$$

$$2(s_1 + s_2) = at^2$$

$$2s_1 + 2s_2 = at^2$$

$$a(p-1)^2 + ap^2 = at^2$$

$$(p-1)^2 + p^2 = t^2$$

$$t = \sqrt{(p-1)^2 + p^2}$$

$$= \sqrt{p^2 + 1 - 2p + p^2}$$

$$= \sqrt{2p^2 - 2p + 1}$$

10. *(a)* Speed of cadet performing NCC parade, $v = 9 \text{ km/h} = 9 \times \frac{5}{18} = 2.5 \text{ m/s}$

Height of monkey, $H = 19.6$ m

$$g = 9.8 \text{ m/s}^2$$

If t be the time taken to catch mango after dropping by monkey, then

$$H = \frac{1}{2}gt^2$$

$$\Rightarrow \quad t = \sqrt{\frac{2H}{g}} = \sqrt{\frac{2 \times 19.6}{9.8}} = 2 \text{ s}$$

Hence, if x be the distance of cadet from the tree at the time of dropping, then

$$x = v \times t = 2.5 \times 2 = 5 \text{ m}$$

11. *(c)* The given situation is shown below

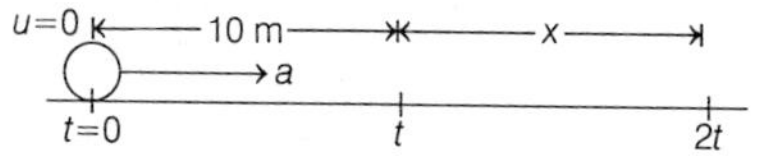

For first t seconds,

$$u = 0,\ s = 10 \text{ m},\ a = \text{constant}$$

Using $s = ut + \frac{1}{2}at^2$, we get

$$10 = \frac{1}{2}at^2 \qquad \ldots\text{(i)}$$

For first $2t$ seconds,

$$u = 0,\ s = 10 + x,\ a = \text{constant}$$

$$\therefore 10 + x = \frac{1}{2}a(2t)^2$$

$$\Rightarrow 10 + x = \frac{1}{2}at^2 \times 4$$

$$\Rightarrow 10 + x = 10 \times 4$$

$$\left[\because \frac{1}{2}at^2 = 10 \text{ from Eq. (i)}\right]$$

$$\Rightarrow \quad x = 30 \text{ m}$$

12. *(a)* Let the length of train be l and the speed of mid-point of train be v'.

Using third equation of motion,

$$\Rightarrow \quad v'^2 = u^2 + 2al/2$$

$$\Rightarrow \quad v'^2 - u^2 = al \qquad \ldots\text{(i)}$$

Also, $v^2 - u^2 = 2al$

$$\Rightarrow \quad a = \frac{v^2 - u^2}{2l}$$

From Eq. (i), we get

$$v'^2 - u^2 = \frac{v^2 - u^2}{2l} \cdot l = \frac{v^2 - u^2}{2}$$

$$\Rightarrow \quad v' = \sqrt{\frac{v^2 - u^2}{2} + u^2} = \sqrt{\frac{v^2 + u^2}{2}}$$

13. *(d)* Let t be the time taken by the packet to reach the ground. As, the helicopter rises from rest in upward direction, its final velocity is

$$v = \sqrt{0^2 + 2gh} \Rightarrow v = \sqrt{2gh}$$

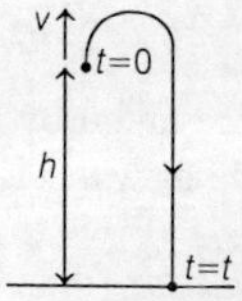

From second equation of motion,

$$s = ut + \frac{1}{2}at^2$$

Here, $\quad s = -h$

$$u \text{ or } v = \sqrt{2gh} \Rightarrow a = -g$$

Substituting all these values in above equation, we get

$$-h = \sqrt{2gh}\,t + \frac{1}{2}(-g)t^2$$

$$\Rightarrow \frac{1}{2}gt^2 - \sqrt{2gh}\,t - h = 0$$

This is a quadratic equation in t.

$$\therefore t = \frac{\sqrt{2gh} \pm \sqrt{(\sqrt{2gh})^2 - 4 \times \frac{g}{2}(-h)}}{2 \times \frac{g}{2}}$$

$$= \frac{\sqrt{2gh} \pm \sqrt{2gh + 2gh}}{g} = \frac{\sqrt{2gh}}{g}(1 + \sqrt{2})$$

$$= \sqrt{\frac{2h}{g}}(1 + \sqrt{2})$$

$$= (2 + \sqrt{2})\sqrt{\frac{h}{g}} = 3.4\sqrt{\frac{h}{g}}$$

14. *(c)* Time taken to reach the maximum height

$$t_1 = u/g$$

If t_2 is the time taken to hit ground, then

$$-H = ut_2 - \frac{1}{2}gt_2^2$$

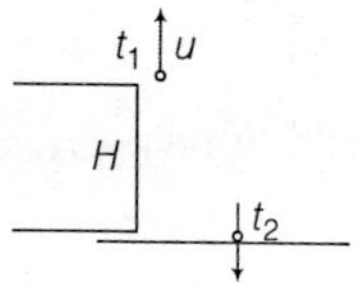

But $\quad t_2 = nt_1 = \frac{nu}{g}$ [given]

So, $\quad -H = u\frac{nu}{g} - \frac{1}{2}g\frac{n^2u^2}{g^2}$

$$\Rightarrow \quad -H = \frac{nu^2}{g} - \frac{1}{2}\frac{n^2u^2}{g}$$

$$\Rightarrow \quad 2gH = nu^2(n-2)$$

15. *(c)* Parachute bails out at height H from ground velocity at A

$$v = \sqrt{2gh}$$

$$= \sqrt{2 \times 9.8 \times 50} = \sqrt{980} \text{ ms}^{-1}$$

Velocity at ground, $v_1 = 3 \text{ ms}^{-1}$ [Given]

Acceleration $= -2 \text{ ms}^{-2}$ [Given]

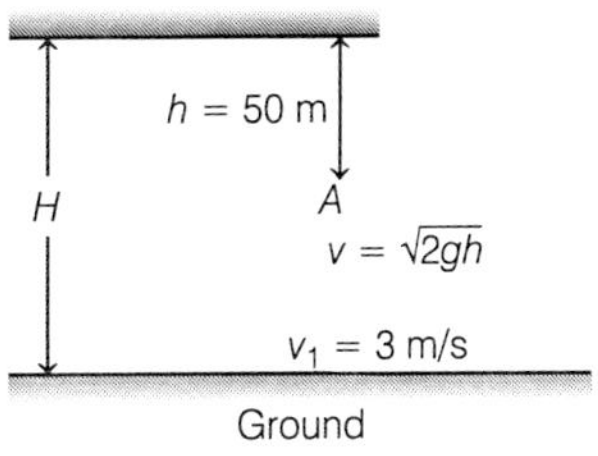

$$\therefore\ H - h = \frac{v^2 - v_1^2}{2 \times 2} = \frac{980 - 9}{4}$$

$$= \frac{971}{4} = 242.75$$

$$\Rightarrow \quad H = 242.75 + h$$

$$= 242.75 + 50 \approx 293 \text{ m}$$

16. *(c)* Second law of motion gives

$$s = ut + \frac{1}{2} gT^2$$

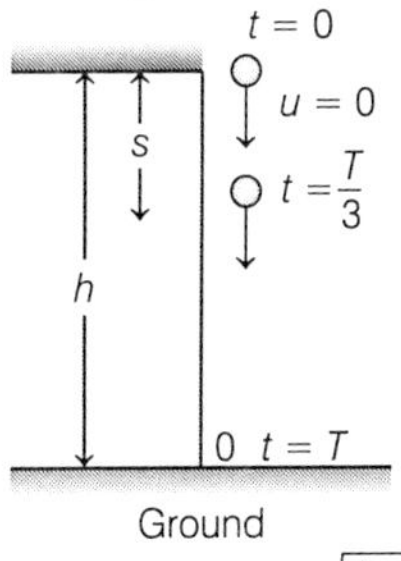

or $h = 0 + \frac{1}{2} gT^2 \Rightarrow T = \sqrt{\left(\frac{2h}{g}\right)}$ $[\because u = 0]$

At $t = \frac{T}{3}$ s, $s = 0 + \frac{1}{2} g \left(\frac{T}{3}\right)^2$

or $s = \frac{1}{2} g \cdot \frac{T^2}{9} \Rightarrow s = \frac{g}{18} \times \frac{2h}{g}$

or $s = \frac{h}{9}$ m $\quad [\because T = \sqrt{2h/g}]$

Hence, the position of ball from the ground $= h - \frac{h}{9} = \frac{8h}{9}$ m

17. *(a)* Let initial velocity of body at point A be v, AB is 3 cm.

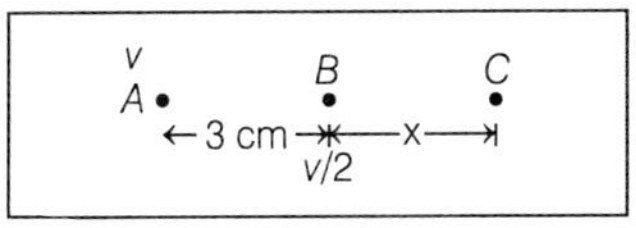

From $\quad v^2 = u^2 + 2as$

$$\Rightarrow \quad \left(\frac{v}{2}\right)^2 = v^2 + 2a \times 3 \text{ or } a = \frac{-v^2}{8}$$

Let after penetrating 3 cm in a wooden block, the body moves x distance from B to C.

So, for B to C, $u = v/2$, $\quad v = 0$,

$$s = x,\ a = \frac{v^2}{8} \qquad \text{[deceleration]}$$

$$\therefore (0)^2 = \left(\frac{v}{2}\right)^2 - 2 \cdot \frac{-v^2}{8} \cdot x \text{ or } x = 1 \text{ cm}$$

18. *(d)* The given situation is shown below.

For the first-half of distance travelled,

$$h_1 = \frac{1}{2} gt_1^2$$

$$\Rightarrow \quad \frac{h}{2} = \frac{1}{2} gt_1^2$$

$$\Rightarrow \quad h = gt_1^2 \qquad \text{...(i)}$$

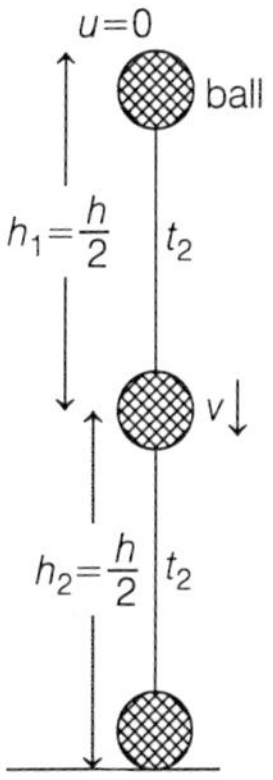

If v be the velocity of ball after travelling distance h_1, then

$$v = 0 + gt_1$$

$$\Rightarrow \quad v = gt_1 \qquad \text{...(ii)}$$

Distance travelled by the ball in second-half,

$$h_2 = vt_2 + \frac{1}{2} gt_2^2$$

$$\Rightarrow \quad \frac{h}{2} = gt_1t_2 + \frac{1}{2} gt_2^2 \qquad \text{[From Eq. (ii)]}$$

$$\Rightarrow \quad \frac{gt_1^2}{2} = gt_1t_2 + \frac{1}{2} gt_2^2 \qquad \text{[From Eq. (i)]}$$

$$\Rightarrow \quad t_1^2 = 2t_1t_2 + t_2^2$$

$$\Rightarrow t_2^2 + 2t_1t_2 - t_1^2 = 0$$

$$\Rightarrow \quad t_2 = \frac{-2t_1 \pm \sqrt{4t_1^2 + 4t_1^2}}{2}$$

$$= \frac{-2t_1 \pm 2\sqrt{t_1^2 + t_1^2}}{2}$$

$$\Rightarrow \quad t_2 = -t_1 \pm t_1\sqrt{2}$$

Taking + ve sign only.

$$t_2 = -t_1 + t_1\sqrt{2}$$

$$= t_1(-1 + \sqrt{2})$$

$$\Rightarrow \quad t_2 = t_1(\sqrt{2} - 1)$$

19. *(b)* According to the question,

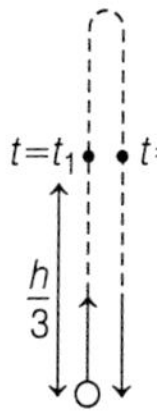

Let the two times be t_1 and t_2, respectively while going up and coming down when they are at $h/3$ height above ground.

Now, as given

$$H_{max} = h$$

$$\therefore \quad h = \frac{u^2}{2g}$$

(using $v^2 = u^2 - 2gh$ and $v = 0$ at topmost point)

$$\Rightarrow \quad u = \sqrt{2gh}$$

$$\therefore \quad \frac{h}{3} = \sqrt{2gh}\, t - \frac{1}{2} gt^2$$

$$\Rightarrow \quad \frac{1}{2} gt^2 - \sqrt{2gh}\, t + \frac{h}{3} = 0$$

$$\Rightarrow 3gt^2 - 6\sqrt{2gh}\, t + 2h = 0$$

Now, using quadratic formula

$$t_1 = \frac{6\sqrt{2gh} - \sqrt{72gh - 24gh}}{6g}$$

$$= \frac{6\sqrt{2gh} - \sqrt{24}\sqrt{2gh}}{6g}$$

$$= \frac{\sqrt{2gh}(6 - \sqrt{24})}{6g} \qquad \text{...(i)}$$

Similarly,

$$t_2 = \frac{\sqrt{2gh}(6 + \sqrt{24})}{6g} \qquad \text{...(ii)}$$

$$\therefore \quad \frac{t_1}{t_2} = \frac{6 - \sqrt{24}}{6 + \sqrt{24}}$$

$$= \frac{\sqrt{6}\sqrt{6} - \sqrt{6}\sqrt{4}}{\sqrt{6}\sqrt{6} + \sqrt{6}\sqrt{4}}$$

$$\Rightarrow \quad \frac{t_1}{t_2} = \frac{\sqrt{3} - \sqrt{2}}{\sqrt{3} + \sqrt{2}}$$

20. *(d)* Given, height of nozzle from floor, $H = 9.8$ m

Let, t_1 = time taken by first drop to reach the ground and t_2 = time spent by 2nd drop in air

$$t_2 = \frac{t_1}{2}$$

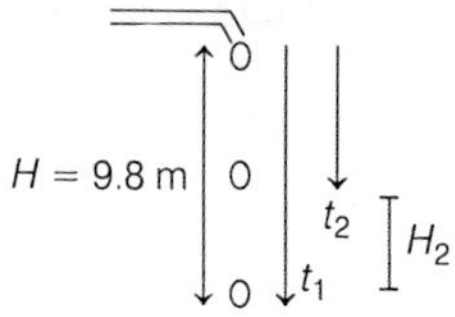

g = acceleration due to gravity ($9.8\ \text{ms}^{-2}$)

$$u = \text{initial speed} = 0$$

Since, $H = ut + \frac{1}{2} g t_1^2$ ...(i)

$\therefore \quad t_1 = \sqrt{\frac{2H}{g}}$

$\Rightarrow \quad t_1 = \sqrt{\frac{2 \times 9.8}{9.8}} = \sqrt{2}\ \text{s}$

$\therefore \quad t_2 = \frac{\sqrt{2}}{2} = \frac{1}{\sqrt{2}}\ \text{s}$

Again from Eq. (i)

Distance covered by 2nd drop,

$$x = \frac{1}{2} g t_2^2$$

$\therefore \quad x = \frac{1}{2} \times 9.8 \times \left(\frac{1}{\sqrt{2}}\right)^2 = 2.45\text{m}$

and position of second drop from ground,

$$H_2 = H - x = 9.8 - 2.45 = 7.35\ \text{m}$$

21. *(c)* Given,

$$u_b = 10\ \text{ms},\ h = 75\ \text{m}$$

Initial speed of object,

$$u_a = -10\ \text{ms}^{-1}$$

Let the time taken by object to reach ground be t.

As, $\quad h = ut + \frac{1}{2} g t^2$

For object a,

$\Rightarrow \quad 75 = -10t + \frac{1}{2} \times 10 \times t^2$

$= -10t + 5t^2$

$\Rightarrow \quad t^2 - 2t - 15 = 0$

$\Rightarrow t^2 - 5t + 3t - 15 = 0$

$\Rightarrow t(t-5) + 3(t-5) = 0$

$\Rightarrow \quad t = 5, -3$

$\therefore \quad t = 5\ \text{s} \quad (\because t \neq -\text{ve})$

$\therefore$ Final height, of balloon at $t = 5$ s is,

$h_f = h + s$

$= h + u_b \times t$

$= 75 + 10 \times 5$

$= 75 + 50 = 125\ \text{m}$

22. *(c)* Given, distance travelled in 4th second,

$$s_2 = 34.3\ \text{m},$$

Acceleration due to gravity,

$$g = 9.8\ \text{ms}^{-2}$$

Let distance covered by 1st drop in time t (= 4 s) be s_1.

As we know that,

$$s = ut + \frac{1}{2} g t^2 \quad \text{...(i)}$$

where, u = initial speed of drop

$= 0\ \text{ms}^{-1}$

$\Rightarrow \quad s_1 = \frac{1}{2} g t^2 = \frac{1}{2} \times 9.8 \times 4^2$

$= \frac{1}{2} \times 9.8 \times 16 = 9.8 \times 8 = 78.4\ \text{m}$

Now, distance covered by 2nd drop will be s.

$s = s_1 - s_2$

$= 78.4 - 34.3 = 44.1\ \text{m}$

Again, By using Eq. (i)

$$s = \frac{1}{2} 9.8\, t'^2$$

where, t' is the time taken by 2nd drop to cover 44.1 m.

$\therefore \quad 44.1 = 4.9 t'^2$

$\Rightarrow \quad t'^2 = \frac{44.1}{4.9}$

$\Rightarrow \quad t' = \sqrt{\frac{441}{49}} = \sqrt{9} = 3\ \text{s} \quad \text{...(ii)}$

Since, time taken by 1st drop is 4 s, and time taken by 2nd drop is 3 s, so one drop is falling per second.

23. *(b)* Ball is released from height, $h = 5$ m Total distance travelled by ball

$$H = 5 + 2 \times \left[5 \times \frac{81}{100} + 5 \times \left(\frac{81}{100}\right)^2 + \ldots\right]$$

$$= 5 + 2\left[\frac{5 \times \frac{81}{100}}{1 - \frac{81}{100}}\right] \quad \left[\text{using, } S_\infty = \frac{a}{1-r}\right]$$

$$= \frac{905}{19}\ \text{m}$$

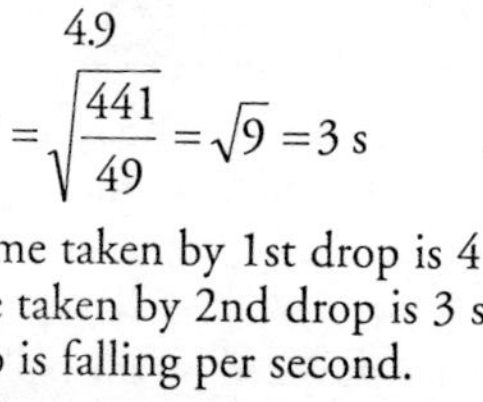

Using second equation of motion,

$$h = ut + \frac{1}{2} g t^2$$

$u = 0$ (Ball is released)

$h = \frac{1}{2} g t^2$ or $t = \sqrt{\frac{2h}{g}}$

So total time,

$$T = \sqrt{\frac{2h}{g}} + 2\left[\sqrt{\frac{2h}{g} \times \frac{81}{100}} + \sqrt{\frac{2h}{g} \times \left(\frac{81}{100}\right)^2} \ldots\right]$$

$$= \sqrt{\frac{2 \times 5}{10}} + 2\left[\sqrt{\frac{2 \times 5}{10} \times \frac{81}{100}} + \sqrt{\frac{2 \times 5}{10} \times \left(\frac{81}{100}\right)^2}\right]$$

$$= 1 + 2\left[\frac{9}{10} + \left(\frac{9}{10}\right)^2 + \ldots\right]$$

$$= 1 + \frac{2 \times \frac{9}{10}}{1 - \frac{9}{10}} = 19\ \text{s} \quad \left[\because S_\infty = \frac{a}{1-r}\right]$$

Average speed of ball, v_{av}

$$= \frac{\text{Total distance travelled}}{\text{Total time}} = \frac{H}{T}$$

or $v_{av} = \frac{905}{19 \times 19} = \frac{905}{361} = 2.5\ \text{m/s}$

24. *(40)* Given, initial speed of bus,

$$u = 72\ \text{km/h} = 72 \times \frac{5}{18} = 20\ \text{m/s}$$

$$t = 4\ \text{s},\ v = 0$$

Using, $v = u + at$

$$0 = 20 + a \times 4 \Rightarrow a = -5\ \text{m/s}^2$$

If the distance travelled by the bus is s, then

$$v^2 = u^2 + 2as$$

$\Rightarrow \quad 0 = 20^2 + 2(-5)s \Rightarrow s = 40\ \text{m}$

25. *(175)* Let u is initial speed

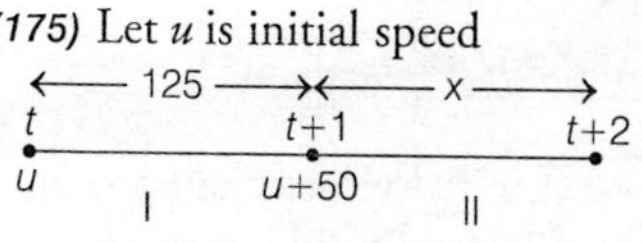

For region I,

$$s = ut + \frac{1}{2} a t^2 \quad (\text{Time} = 1\ \text{s})$$

$$125 = u \times 1 + \frac{1}{2} \times 50 \times 1$$

($a = 50\ \text{m/s}^2$ because time is 1s)

$$u = 100\ \text{m/s}$$

Now, for region II,

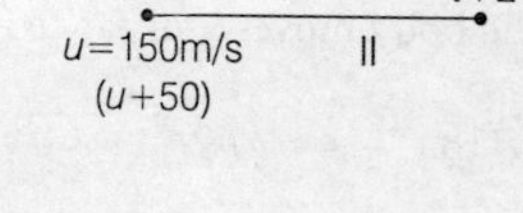

Using equation of motion

$$s = ut + \frac{1}{2}at^2$$

$$x = 150 \times 1 + \frac{1}{2} \times 50 \times 1$$

$$x = 175 \text{ m}$$

26. *(120)* Velocity with which ball strikes from v_1

$$= \sqrt{2gh} = \sqrt{2 \times 10 \times 9.8} \simeq 14 \text{ m/s}$$

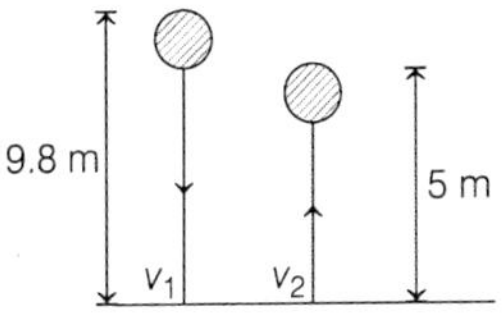

$$\Delta t = 0.2 \text{ s}$$

Velocity with which ball rebounces,

$$v_2 = \sqrt{2gh} = \sqrt{2 \times 10 \times 5} = 10 \text{ m/s}$$

$$\text{Acceleration of ball} = \frac{10 - (-14)}{0.2}$$

$$= 120 \text{ m/s}^2$$

27. *(392)* Given, $v = 19.6 \text{ ms}^{-1}$

$$t = 6 \text{ s}, \; g = 9.8 \text{ ms}^{-2}$$

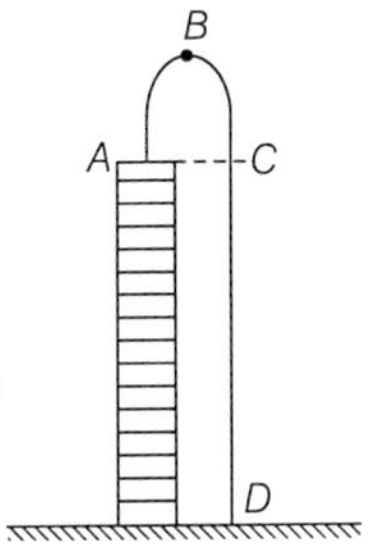

Let t' be the time taken by the ball to reach B from position A, then using equation of motion,

$$v = u + gt'$$

$$\Rightarrow \quad 0 = 19.6 + (-9.8)t'$$

$$\Rightarrow \quad t' = \frac{19.6}{9.8} = 2 \text{ s}$$

So, time taken from B to D,

$$t'' = (6 - 2) \text{ s} = 4 \text{ s}$$

If h be the height from the ground to the top, then

$$h = ut + \frac{1}{2}gt^2$$

$$\Rightarrow \quad h = 0 + \frac{1}{2}(gt'')^2 = \frac{1}{2} \times 9.8 \times (4)^2$$

$$= 78.4 \text{ m}$$

Given, $h = \dfrac{k}{5}$

$$\Rightarrow \quad k = h \times 5 = 78.4 \times 5 = 392$$

28. *(100)* Given, increase of velocity of particle with respect to distance

$$\frac{dv}{dx} = 5 \text{ ms}^{-1}/\text{m}$$

$\therefore$ Acceleration of particle,

$$a = \frac{dv}{dt} = \frac{dv}{dx} \cdot \frac{dx}{dt} = \frac{dv}{dx} \times v \quad \left(\because v = \frac{dx}{dt}\right)$$

$$= 5 \times 20 \qquad (\because v = 20 \text{ m/s})$$

$$= 100 \text{ m/s}^2$$

29. *(6)* Initial velocity of ball,

$$u = 50 \text{ m/s}$$

Let both balls meet together after t seconds.

For the first ball, height attained by it in t seconds,

$$h_1 = ut - \frac{1}{2}gt^2$$

$$\Rightarrow \quad h_1 = 50t - \frac{1}{2}gt^2 \qquad \ldots\text{(i)}$$

After 2 s, second ball is projected vertically upward with $u = 50$ m/s,

i.e. $\quad t' = t - 2$

$\therefore$ Height attained by second ball,

$$h_2 = u(t-2) - \frac{1}{2}g(t-2)^2$$

$$\Rightarrow \quad h_2 = 50(t-2) - \frac{1}{2}g(t-2)^2 \quad \ldots\text{(ii)}$$

Both balls will meet, if

$$h_1 = h_2$$

$$50t - \frac{1}{2}gt^2 = 50(t-2) - \frac{1}{2}g(t-2)^2$$

$$\Rightarrow 50t - 50(t-2) = \frac{1}{2}gt^2 - \frac{1}{2}g(t-2)^2$$

$$\Rightarrow \quad t = 6 \text{ s}$$

30. *(50)* Given,

$$m_1 = m_2 = m$$

$$r_1 = r_2 = r = 5 \text{ cm} = 5 \times 10^{-2} \text{ m}$$

$$u_1 = u_2 = 35 \text{ m/s}$$

$$t_1 = 0 \text{ and } t_2 = 3 \text{ s}$$

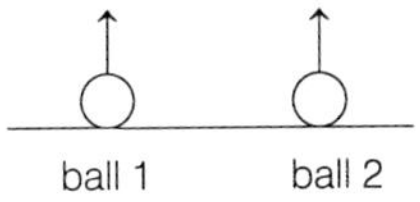

When both ball will collide,

$$(s_1)_{\text{vertical}} = (s_2)_{\text{vertical}} \qquad \ldots\text{(i)}$$

Let after time t both ball will collide.

$$(s_1)_{\text{vertical}} = u_1 t - \frac{1}{2}gt^2$$

$$(s_2)_{\text{vertical}} = u_2(t-3) - \frac{1}{2}g(t-3)^2$$

From Eq. (i), we have

$$\Rightarrow u_1 t - \frac{1}{2}gt^2 = u_2(t-3) - \frac{1}{2}g(t-3)^2$$

$$\Rightarrow \; 35t = \frac{1}{2} \times 10 \times (2t - 3)$$

$$\Rightarrow \quad t = 5 \text{ s}$$

So, the height at which ball will collide is

$$(s_1)_{\text{vertical}} = u_1 t - \frac{1}{2}gt^2$$

$$= 35 \times 5 - \frac{1}{2} \times 10 \times (5)^2$$

$$= 175 - 125 = 50 \text{ m}$$

31. *(8)* Let ball covers 81 m from top in time t_1

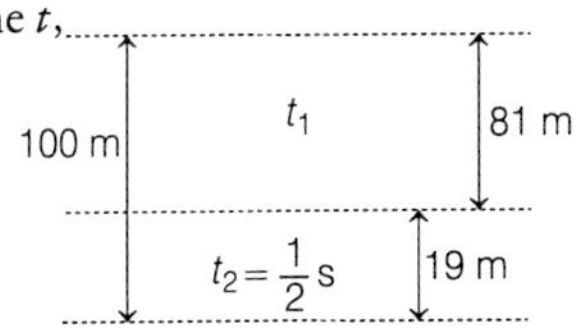

Then, $\quad s = ut + \dfrac{1}{2}at^2$

$$\Rightarrow \quad 81 = \frac{1}{2}at_1^2 \qquad \ldots\text{(i)}$$

As ball covers 100 m in time $t_1 + \dfrac{1}{2}$ seconds, we have

$$100 = \frac{1}{2}a\left(t_1 + \frac{1}{2}\right)^2 \qquad \ldots\text{(ii)}$$

From Eqs. (i) and (ii), we get

Substituting $t_1 = \dfrac{9}{2}$ in Eq. (i), we get

$$a = 8 \text{ ms}^{-2}$$

Topic 3

01. *(a)* $t = 0$ to $t = 4$s

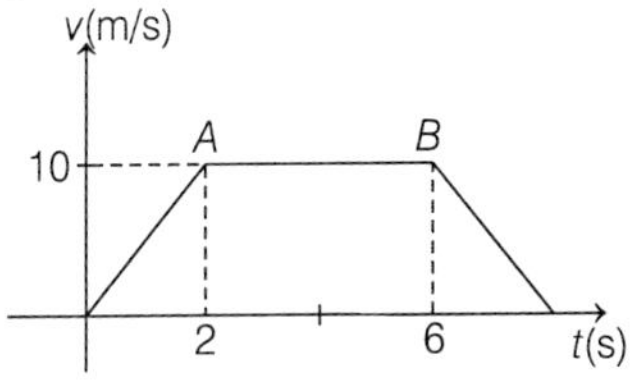

Distance = Area under v-t graph

$$= \frac{1}{2} \times 2 \times 10 + 2 \times 10$$

$$= 10 + 20 = 30 \text{ m}$$

02. *(c)*

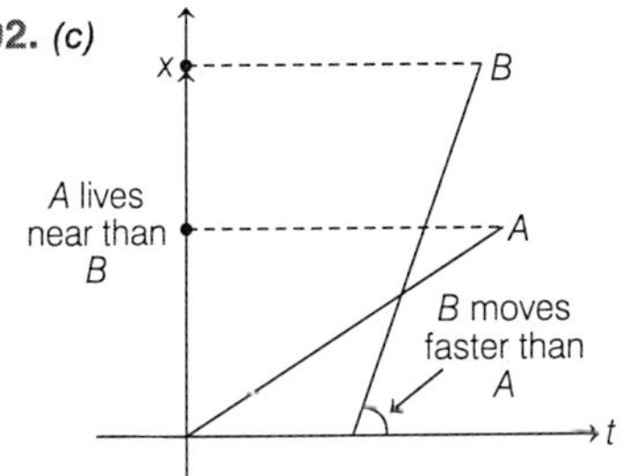

Given, slope of $B >$ slope of A. So, B is fast.

Projection of A on Y-axis (position) than projection of B. So, A lives near to school.

Therefore, option (c) is correct.

03. *(a)*

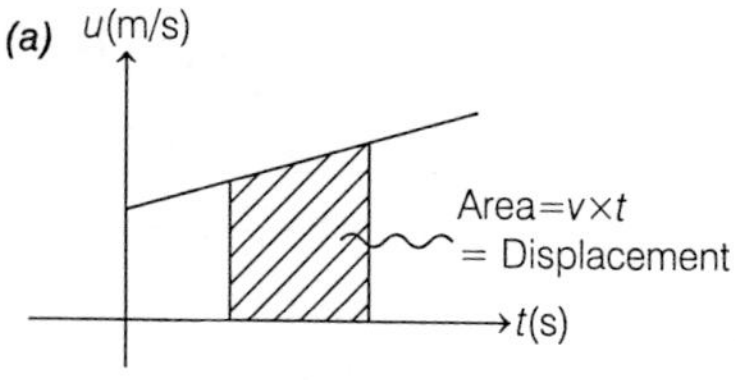

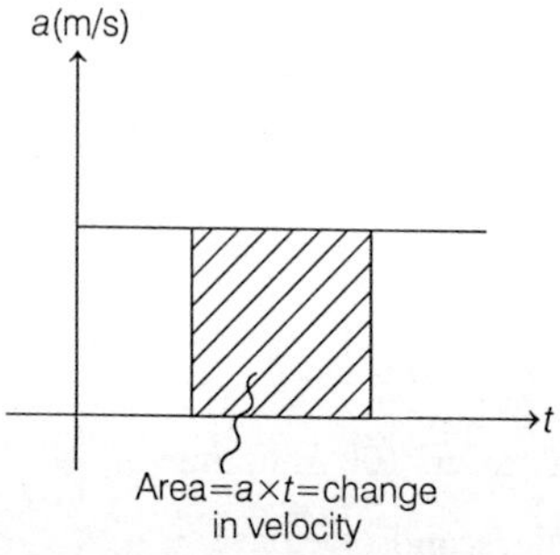

Area $= a \times t =$ change in velocity.
So, Statement I is incorrect and Statement II is correct.

04. *(b)* Displacement = Net area of velocity-time graph

$= 8 \times 2 + 4 \times 4 - 4 \times 2 - 4 \times 2$

$= 16 + 16 - 8 - 8 = 16$ m

Distance = Magnitude of total area under velocity-time graph

$= 16 + 16 + 8 + 8 = 48$ m

Hence, ratio $= \dfrac{\text{Displacement}}{\text{Distance}} = \dfrac{16}{48} = \dfrac{1}{3}$

05. *(b)* The slope of velocity-time graph gives the value of acceleration.

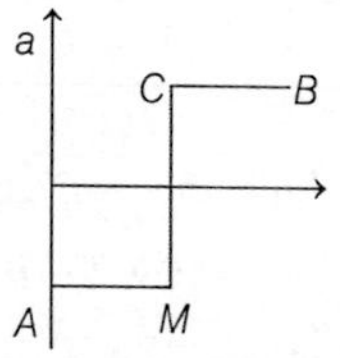

In part AM, acceleration is constant and negative. In part CB acceleration is constant and positive.

06. *(b)* Initially velocity keeps on decreasing at a constant rate, then it increases in negative direction with same rate.

07. *(c)* $< v > = \dfrac{\Delta s}{\Delta t} = \dfrac{s_f - s_i}{t_f - t_i}$

$\mathbf{v} = \dfrac{ds}{dt} =$ Slope

(A) 0 to 3 s, $< \mathbf{v} > = \dfrac{5-0}{3} = \dfrac{5}{3}$ m/s

(B) 3 to 5 s, $< \mathbf{v} > = \dfrac{5-5}{5-3} = 0$

(C) $t = 2$ s, $< \mathbf{v} > = 5$ m/s

(D) $t = 5$ to 7 s, $< \mathbf{v} > = \dfrac{0-5}{2}$

$= -2.5$ m/s

At $t = 6.5$ s, $\mathbf{v} = 10$

(E) $t = 0$ to $t = 9$, $< \mathbf{v} > = 0$

Options (B), (C), (E) only

08. *(d)* Area under v-t graph represents distance covered.

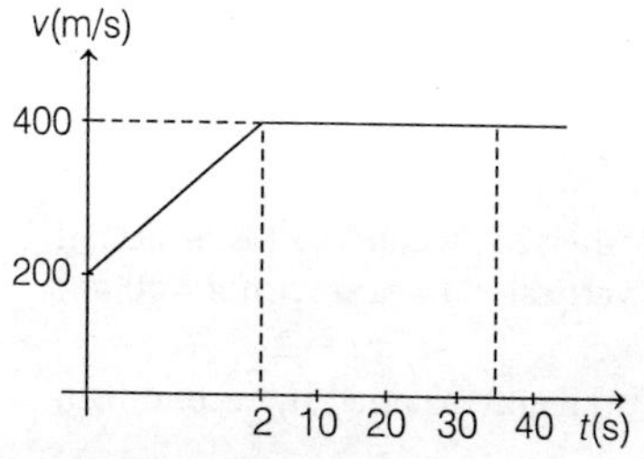

Total area under curve

$= \dfrac{1}{2} \times (400 + 200) \times 2 + 28.5 \times 400$

$= 600 + 11400 = 12000$ m $= 12$ km

09. *(d)* Coordinate of moving particle,

$x = 2 + 4t,\ y = 3t + 8t^2$

Velocity $dx/dt = 4$ unit,

$\dfrac{dy}{dt} = (3 + 16t)$ unit

Acceleration $\dfrac{d^2x}{dt^2} = 0,\ \dfrac{d^2y}{dt^2} = 16$ units

So, its motion is uniformly accelerated and follows a parabolic path because velocity and acceleration are not parallel to each other.

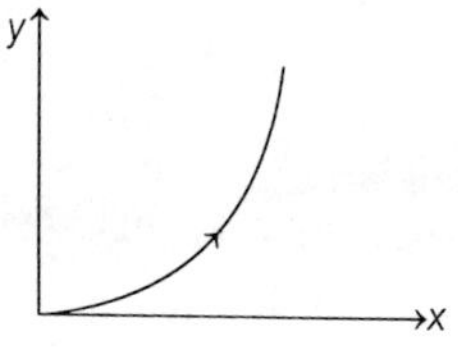

> **• Concept Enhancer**
> Always remember, if **v** and **a** are parallel to each other, then motion is along a straight line and if **v** and **a** are perpendicular to each other then motion is along a circular path.

10. *(a)*

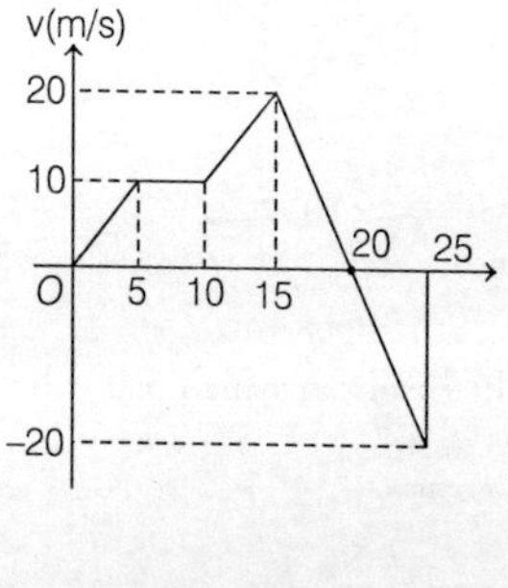

Displacement = Area of graph

∴ Displacement

$= \left(\dfrac{1}{2} \times 10 \times 5\right) + (10 \times 5) + \left(\dfrac{1}{2} \times 5 \times 30\right)$

$+ \left(\dfrac{1}{2} \times 5 \times 20\right) - \dfrac{1}{2}(5)(20)$

$= 25 + 50 + 75 + 50 - 50 = 150$ m

Distance → Area of graph with the value

Distance $= 25 + 50 + 75 + 50 + 50 = 250$

$\dfrac{\text{Distance}}{\text{Displacement}} = \dfrac{250}{150} = \dfrac{5}{3}$

11. *(a)* v^2 *versus* x graph is shown below.

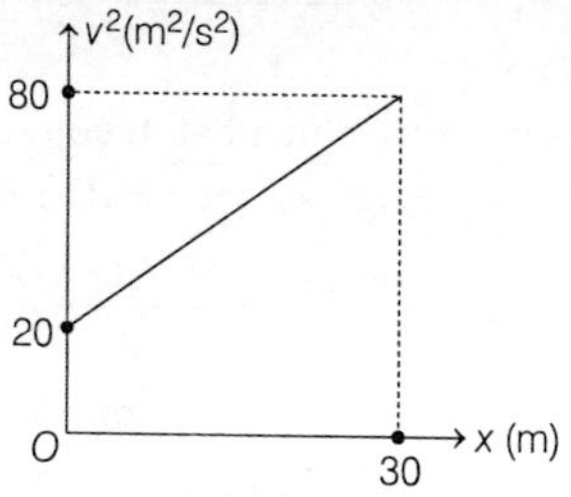

Using equation of motion,

$v^2 = u^2 + 2ax$ ...(i)

We know that, the equation of straight line.

$y = mx + c$ ...(ii)

Comparing Eqs. (i) and (ii),

Slope, $m = 2a = \dfrac{80 - 20}{30 - 0} = \dfrac{60}{30} = 2$

Now, the acceleration,

$a = 2/2 = 1$ m/s^2

12. *(d)* Distance travelled by car in 15 s = Area of $\Delta AOC = \dfrac{1}{2} \times AC \times OC$

$= \dfrac{1}{2}(45)(15) = \dfrac{675}{2}$ m

Distance travelled by scooter in 15

$s = v \times t$

$= 30 \times 15 = 450$ m

Difference in distance

$= 450 - \dfrac{675}{2}$

$= \dfrac{225}{2} = 112.5$ m

Let car catches scooter in time t

then, $\dfrac{675}{2} + 45(t - 15) = 30t$

$337.5 + 45t - 675 = 30t$

$\Rightarrow \quad 15t = 337.5$

$\Rightarrow \quad t = 22.5$ s

13. *(c)* From figure A, $\phi = kt + C$

which represents one-dimensional motion.

From figure B, $v^2 + x^2 = \text{constant}$ which is one-dimensional motion

For graph C, time is negative, which is not possible.

Graph D is also possible for one-dimensional motion.

14. *(b)*

> **key idea** Area under the velocity-time curve represents displacement.

(x-t graphs)

For

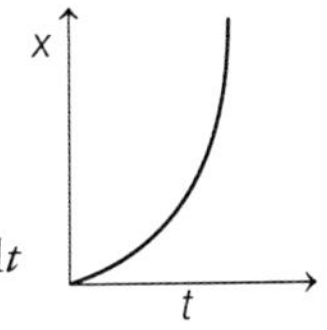

Here, $x = At^2$

So, $v = \dfrac{d}{dt} x = 2At$

$\Rightarrow \quad v \propto t$

$\therefore$ For A $\rightarrow$ II

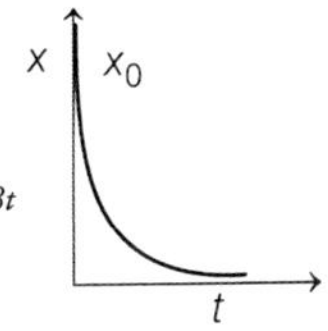

Here, $x = x_0 e^{-Bt}$

$\Rightarrow \quad v = \dfrac{dx}{dt} = -x_0 Be^{-Bt}$

$\Rightarrow \quad v = -v_0 e^{-Bt}$

$\therefore$ For B $\rightarrow$ IV

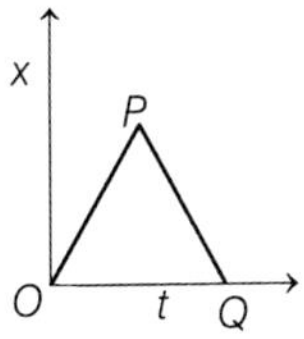

Here, for region *OP* velocity is positive and for region *PQ*, velocity is negative

For C $\rightarrow$ III

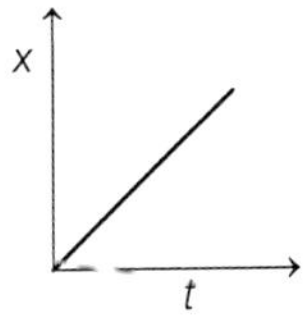

Here, $x = At$

$\Rightarrow \quad v = \dfrac{dx}{dt} = A = \text{constant}$

$\therefore \quad$ D $\rightarrow$ I

15. *(c)* For the given velocity-displacement graph,

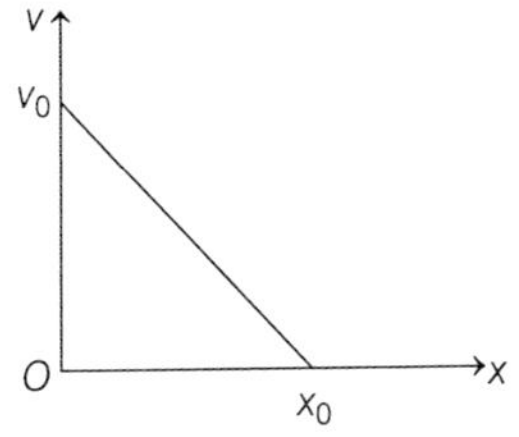

Intercept, $\quad c = v_0$

Slope, $\quad m = -\dfrac{v_0}{x_0}$

$\therefore$ Slope, $\quad m = \dfrac{dv}{dx} = -\dfrac{v_0}{x_0} \quad \ldots$(i)

For given diagram, the equation of line,

$$y = mx + c$$

$$v = -\frac{v_0}{x_0} x + v_0 \quad \ldots\text{(ii)}$$

Acceleration, $a = \dfrac{dv}{dx} \times \dfrac{dx}{dt}$

$\Rightarrow \quad a = v \dfrac{dv}{dx}$

$\Rightarrow \quad a = \left(-\dfrac{v_0}{x_0} x + v_0 \right)\left(-\dfrac{v_0}{x_0} \right)$

[From Eqs. (i) and (ii)]

$\Rightarrow \quad a = \left(\dfrac{v_0}{x_0} \right)^2 x - \dfrac{v_0^2}{x_0}$

Hence, the intercept is negative and the slope is positive, so the correct graph is given in option (c).

16. *(b)* Acceleration = Constant, (the acceleration-time graph is parallel to the time axis, and acceleration is positive)

$$a = \frac{dv}{dt}$$

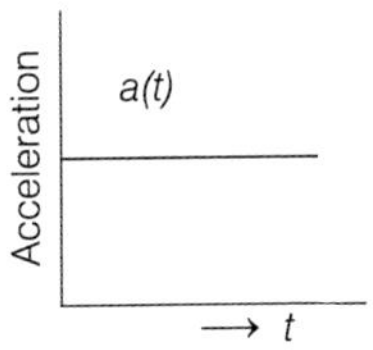

As we know,

$$\frac{dv}{dt} = C \Rightarrow \int dv = C \int dt$$

$\Rightarrow \quad v = Ct + C_0$

The velocity-time graph is straight line with positive slope

$$v = dx / dt$$

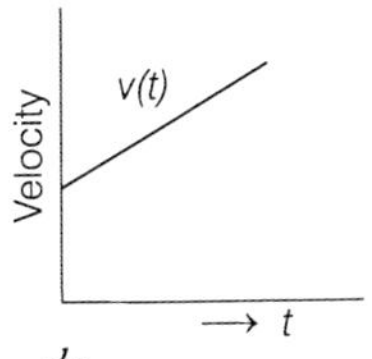

As we know, $\dfrac{dx}{dt} = Ct + C_0$

$\Rightarrow \quad \int dx = \int (Ct + C_0)\, dt$

$\Rightarrow \quad x = \dfrac{Ct^2}{2} + C_0 t + C_1$

This equation shows that the distance-time graph is upward concave parabola.

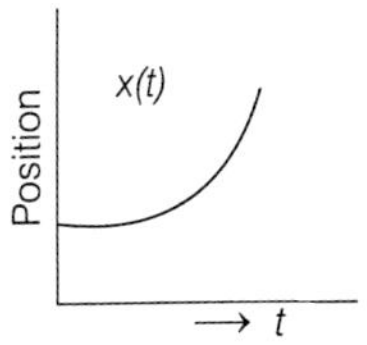

17. *(a)* From the given velocity-displacement graph, for $0 \le x \le 200$, the graph is a straight line i.e., varying linearly.

As, the equation of straight line is

$$y = mx + c$$

Here, it can be given as $v = mx + c$

where, $\quad m = \text{slope}$

$$= \frac{(50-10)}{200-0} = \frac{1}{5} \text{ and } c = 10$$

$\Rightarrow \quad v = \dfrac{x}{5} + 10 \quad \ldots$(i)

As, acceleration,

$$a = \frac{dv}{dt} = \frac{vdv}{dx}$$

$$a = \left(\frac{x}{5} + 10 \right) \frac{d}{dx} \left(\frac{x}{5} + 10 \right)$$

[using Eq. (i)]

$$= \left(\frac{x}{5} + 10 \right)\left(\frac{1}{5} \right) = 2 + \frac{x}{25}$$

For $x = 0$, $a = 2 + \dfrac{0}{25} = 2 \text{ ms}^{-2}$

and for $x = 200$,

$$a = 2 + \frac{200}{25} = 10 \text{ ms}^{-2}$$

Therefore, the graph will have straight line till $x = 200$ and for $x > 200$;

$v = \text{constant}$ and $a = 0$

18. *(20)* Distance travelled = Area under speed-time graph

Distance travelled

(from $t = 0$ s to $t = 5$ s)

$= \text{Area of } \Delta\, OAB = \dfrac{1}{2} \times \text{Base} \times \text{Height}$

$= \dfrac{1}{2} \times (5 \text{ s}) \times (8 \text{ m/s}) = 20 \text{ m}$

> **• Concept Enhancer**
>
> Area under speed-time graph gives the distancce travelled by the moving object. Slope of speed time graph gives acceleration of moving object.

C-03

JEE Main Chapterwise Topicwise Physics

In JEE Main there is usually 0-1 question per paper are visible from this chapter. In past 3 years, 23 conceptual and 9 numerical questions were asked of moderate to difficult level from this chapter.
For detailed analysis scan QR code

MOTION IN A PLANE

Topic 1 Vectors

Level 1

01 If two vectors **A** and **B** having equal magnitude R are inclined at an angle θ, then *[2024, 31 Jan Shift-II]*

(a) $|\mathbf{A}+\mathbf{B}| = 2R\sin\left(\frac{\theta}{2}\right)$ (b) $|\mathbf{A}-\mathbf{B}| = 2R\cos\left(\frac{\theta}{2}\right)$

(c) $|\mathbf{A}-\mathbf{B}| = \sqrt{2}R\sin\left(\frac{\theta}{2}\right)$ (d) $|\mathbf{A}+\mathbf{B}| = 2R\cos\left(\frac{\theta}{2}\right)$

02 If two vectors $\mathbf{P} = \hat{\mathbf{i}} + 2m\hat{\mathbf{j}} + m\hat{\mathbf{k}}$ and $\mathbf{Q} = 4\hat{\mathbf{i}} - 2\hat{\mathbf{j}} + m\hat{\mathbf{k}}$ are perpendicular to each other then the value of m will be *[2023, 24 Jan Shift-II]*

(a) 3 (b) -1 (c) 2 (d) 1

03 A is a vector quantity such that $|\mathbf{A}|$ = non-zero constant. Which of the following expressions is true for A? *[2022, 25 June Shift-I]*

(a) $\mathbf{A}\cdot\mathbf{A} = 0$ (b) $\mathbf{A}\times\mathbf{A} < 0$ (c) $\mathbf{A}\times\mathbf{A} = 0$ (d) $\mathbf{A}\times\mathbf{A} > 0$

04 Match List I with List II. *[2021, 25 July Shift-1]*

List I		List II
(A) $\mathbf{C}-\mathbf{A}-\mathbf{B}=0$	(i)	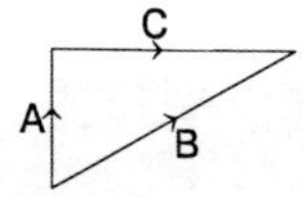
(B) $\mathbf{A}-\mathbf{C}-\mathbf{B}=0$	(ii)	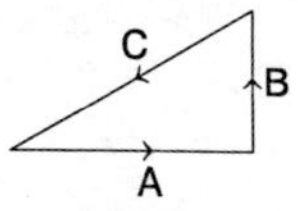
(C) $\mathbf{B}-\mathbf{A}-\mathbf{C}=0$	(iii)	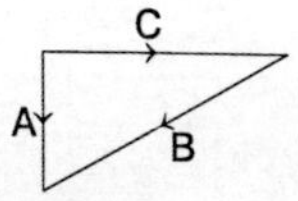
(D) $\mathbf{A}+\mathbf{B}=-\mathbf{C}$	(iv)	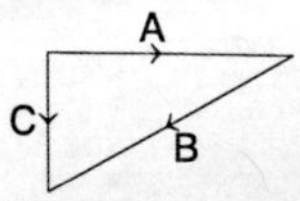

Choose the correct answer from the options given below.

(a) (A) → (iv), (B) → (i), (C) → (iii), (D) → (ii)
(b) (A) → (iv), (B) → (iii), (C) → (i), (D) → (ii)
(c) (A) → (iii), (B) → (ii), (C) → (iv), (D) → (i)
(d) (A) → (i), (B) → (iv), (C) → (ii), (D) → (iii)

05 What will be the projection of vector $\mathbf{A} = \hat{\mathbf{i}} + \hat{\mathbf{j}} + \hat{\mathbf{k}}$ on vector $\mathbf{B} = \hat{\mathbf{i}} + \hat{\mathbf{j}}$? *[2021, 22 July Shift-II]*

(a) $\sqrt{2}(\hat{\mathbf{i}} + \hat{\mathbf{j}} + \hat{\mathbf{k}})$ (b) $2(\hat{\mathbf{i}} + \hat{\mathbf{j}} + \hat{\mathbf{k}})$

(c) $\sqrt{2}(\hat{\mathbf{i}} + \hat{\mathbf{j}})$ (d) $(\hat{\mathbf{i}} + \hat{\mathbf{j}})$

06 If A and B are two vectors satisfying the relation $\mathbf{A}\cdot\mathbf{B} = |\mathbf{A}\times\mathbf{B}|$. Then, the value of $|\mathbf{A}-\mathbf{B}|$ will be *[2021, 20 July Shift-I]*

(a) $\sqrt{A^2+B^2}$ (b) $\sqrt{A^2+B^2+\sqrt{2}AB}$

(c) $\sqrt{A^2+B^2+2AB}$ (d) $\sqrt{A^2+B^2-\sqrt{2}AB}$

07 Let $|\mathbf{A}_1| = 3$, $|\mathbf{A}_2| = 5$ and $|\mathbf{A}_1 + \mathbf{A}_2| = 5$. The value of $(2\mathbf{A}_1 + 3\mathbf{A}_2)\cdot(3\mathbf{A}_1 - 2\mathbf{A}_2)$ is *[2019, 8 April Shift-II]*

(a) -106.5 (b) -112.5 (c) -99.5 (d) -118.5

08 If $\mathbf{A}\times\mathbf{B} = \mathbf{B}\times\mathbf{A}$, then the angle between A and B is *[AIEEE 2004]*

(a) π (b) $\pi/3$ (c) $\pi/2$ (d) $\pi/4$

Level 2

09 The angle between vector **Q** and the resultant of $(2\mathbf{Q}+2\mathbf{P})$ and $(2\mathbf{Q}-2\mathbf{P})$ is *[2024, 5 April Shift-I]*

(a) 0° (b) $\tan^{-1}\left(\frac{2Q}{P}\right)$

(c) $\tan^{-1}\left(\frac{P}{Q}\right)$ (d) $\tan^{-1}\frac{(2\mathbf{Q}-2\mathbf{P})}{2\mathbf{Q}+2\mathbf{P}}$

10 Two vectors **A** and **B** have equal magnitudes. If magnitude of $\mathbf{A}+\mathbf{B}$ is equal to two times the magnitude of $\mathbf{A}-\mathbf{B}$, then the angle between **A** and **B** will be *[2022, 29 June Shift-I]*

(a) $\sin^{-1}\left(\frac{3}{5}\right)$ (b) $\sin^{-1}\left(\frac{1}{3}\right)$ (c) $\cos^{-1}\left(\frac{3}{5}\right)$ (d) $\cos^{-1}\left(\frac{1}{3}\right)$

11 Which of the following relations is true for two unit vectors $\hat{A}$ and $\hat{B}$ making an angle θ with each other? *[2022, 25 June Shift-I]*

(a) $|\hat{A}+\hat{B}|=|\hat{A}-\hat{B}|\tan\frac{\theta}{2}$

(b) $|\hat{A}-\hat{B}|=|\hat{A}+\hat{B}|\tan\frac{\theta}{2}$

(c) $|\hat{A}+\hat{B}|=|\hat{A}-\hat{B}|\cos\frac{\theta}{2}$

(d) $|\hat{A}-\hat{B}|=|\hat{A}+\hat{B}|\cos\frac{\theta}{2}$

Imp. Resolution of Vectors

12 The resultant of these forces OP, OQ, OR, OS and OT is approximately ...... N.
[Take, $\sqrt{3}=1.7$, $\sqrt{2}=1.4$ and given $\hat{i}$ and $\hat{j}$ unit vectors along X, Y-axes] *[2021, 27 Aug Shift-I]*

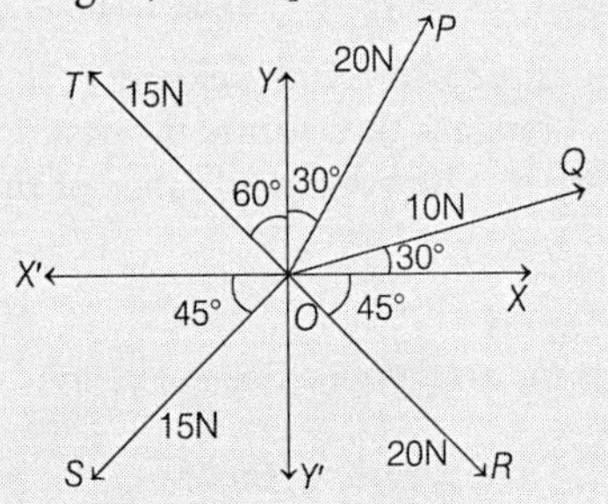

(a) $9.25\hat{i}+5\hat{j}$
(b) $3\hat{i}+15\hat{j}$
(c) $2.5\hat{i}-14.5\hat{j}$
(d) $-1.5\hat{i}-15.5\hat{j}$

13 The angle between vector **A** and (**A** − **B**) is *[2021, 26 Aug Shift-II]*

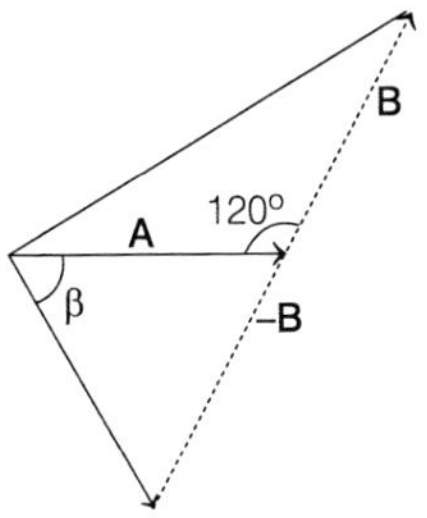

(a) $\tan^{-1}\left(\dfrac{-\frac{B}{2}}{A-B\frac{\sqrt{3}}{2}}\right)$
(b) $\tan^{-1}\left(\dfrac{A}{0.7B}\right)$
(c) $\tan^{-1}\left(\dfrac{\sqrt{3}B}{2A-B}\right)$
(d) $\tan^{-1}\left(\dfrac{B\cos\theta}{A-B\sin\theta}\right)$

14 Assertion (A) If A, B, C, D are four points on a semi-circular arc with centre at O such that $|\mathbf{AB}|=|\mathbf{BC}|=|\mathbf{CD}|$, then $\mathbf{AB}+\mathbf{AC}+\mathbf{AD}=4\mathbf{AO}+\mathbf{OB}+\mathbf{OC}$

Reason (R) Polygon law of vector addition yields $\mathbf{AB}+\mathbf{BC}+\mathbf{CD}+\mathbf{AD}=2\mathbf{AO}$ *[2021, 27 July Shift-I]*

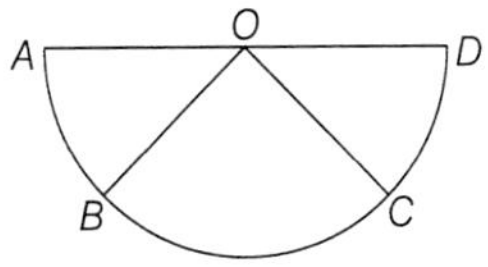

In the light of the above statements, choose the most appropriate answer from the options given below.
(a) Both A and R are correct and R is the correct explanation of A.
(b) Both A and R are correct but R is not the correct explanation of A.
(c) A is correct but R is not correct.
(d) A is not correct but R is correct.

15 In the cube of side a shown in the figure, the vector from the central point of the face $ABOD$ to the central point of the face $BEFO$ will be *[2019, 10 Jan Shift-I]*

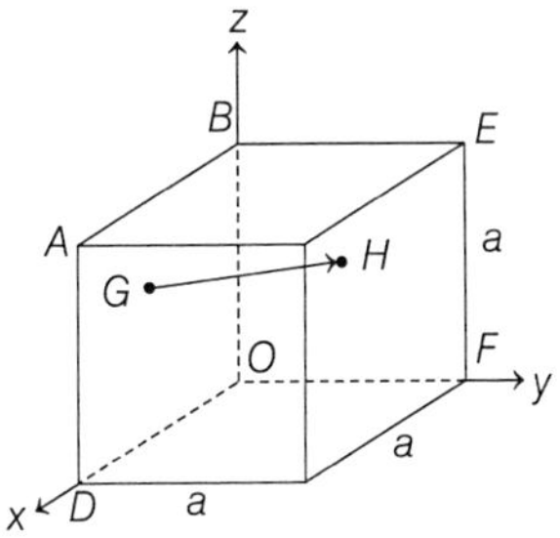

(a) $\frac{1}{2}a(\hat{i}-\hat{k})$
(b) $\frac{1}{2}a(\hat{j}-\hat{i})$
(c) $\frac{1}{2}a(\hat{j}-\hat{k})$
(d) $\frac{1}{2}a(\hat{k}-\hat{i})$

Level 3

16 Statement I Two forces $(\mathbf{P}+\mathbf{Q})$ and $(\mathbf{P}-\mathbf{Q})$ where $\mathbf{P}\perp\mathbf{Q}$, when act at an angle θ_1 to each other, the magnitude of their resultant is $\sqrt{3(P^2+Q^2)}$, when they act at an angle θ_2, the magnitude of their resultant becomes $\sqrt{2(P^2+Q^2)}$. This is possible only when $\theta_1<\theta_2$.

Statement II In the situation given above. $\theta_1=60°$ and $\theta_2=90°$.

In the light of the above statements, choose the most appropriate answer from the options given below. *[2021, 31 Aug Shift-II]*

(a) Statement I is false but Statement II is true.
(b) Both statement I and Statement II are true.
(c) Statement I is true but Statement II is false.
(d) Both statement I and Statement II are false.

17 The magnitude of vectors **OA**, **OB**, and **OC** in the given figure are equal. The direction of **OA** + **AB** − **OC** with X-axis will be *[2021, 26 Aug Shift-I]*

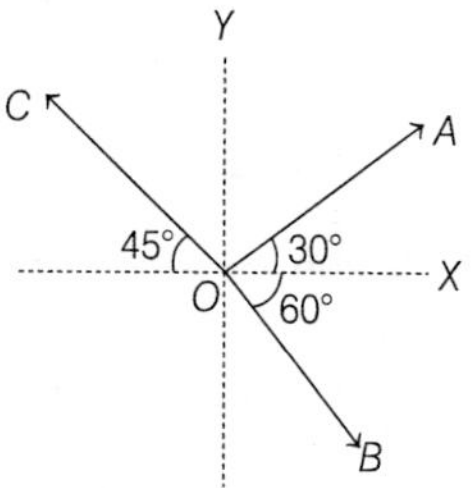

(a) $\tan^{-1}\dfrac{(1-\sqrt{3}-\sqrt{2})}{(1+\sqrt{3}+\sqrt{2})}$ (b) $\tan^{-1}\dfrac{(\sqrt{3}-1+\sqrt{2})}{(1+\sqrt{3}-\sqrt{2})}$

(c) $\tan^{-1}\dfrac{(\sqrt{3}-1+\sqrt{2})}{(1-\sqrt{3}+\sqrt{2})}$ (d) $\tan^{-1}\dfrac{(1+\sqrt{3}-\sqrt{2})}{(1-\sqrt{3}-\sqrt{2})}$

Numerical Value Type Questions

18 Two particles are located at equal distance from origin. The position vectors of those are represented by $\mathbf{A} = 2\hat{i} + 3n\hat{j} + 2\hat{k}$ and $\mathbf{B} = 2\hat{i} - 2\hat{j} + 4p\hat{k}$, respectively. If both the vectors are at right angle to each other, the value of n^{-1} is .... *[2025, 23 Jan Shift-I]*

19 The resultant of two vectors **A** and **B** is perpendicular to **A** and its magnitude is half that of **B**. The angle (in degree) between vectors **A** and **B** is .... *[2024, 9 April Shift-II]*

20 If **a** and **b** makes an angle $\cos^{-1}\left(\dfrac{5}{9}\right)$ with each other, then $|\mathbf{a}+\mathbf{b}| = \sqrt{2}\,|\mathbf{a}-\mathbf{b}|$ for $|\mathbf{a}| = n|\mathbf{b}|$. The integer value of n is .... *[2024, 9 April Shift-I]*

21 A vector has magnitude same as that of $\mathbf{A} = 3\hat{i} + 4\hat{j}$ and is parallel to $\mathbf{B} = 4\hat{i} + 3\hat{j}$. The x and y components of this vector in first quadrant are x and 3, respectively where $x =$ .... *[2024, 30 Jan Shift-II]*

22 If $\mathbf{P} = 3\hat{i} + \sqrt{3}\hat{j} + 2\hat{k}$ and $\mathbf{Q} = 4\hat{i} + \sqrt{3}\hat{j} + 2.5\hat{k}$, then the unit vector in the direction of $\mathbf{P} \times \mathbf{Q}$ is $\dfrac{1}{x}(\sqrt{3}\hat{i} + \hat{j} - 2\sqrt{3}\hat{k})$. The value of x is .... *[2023, 25 Jan Shift-II]*

23 Vectors $a\hat{i} + b\hat{j} + \hat{k}$ and $2\hat{i} - 3\hat{j} + 4\hat{k}$ are perpendicular to each other when $3a + 2b = 7$, the ratio of a to b is $x/2$. The value of x is ..... . *[2023, 24 Jan Shift-I]*

24 If $\mathbf{A} = (2\hat{i} + 3\hat{j} - \hat{k})$ m and $\mathbf{B} = (\hat{i} + 2\hat{j} + 2\hat{k})$ m. The magnitude of component of vector **A** along vector **B** will be .... m. *[2022, 26 July Shift-II]*

25 If $\mathbf{P} \times \mathbf{Q} = \mathbf{Q} \times \mathbf{P}$, the angle between **P** and **Q** is θ $(0° < \theta < 360°)$. The value of θ (in degree) will be .... *[2021, 25 Feb Shift-II]*

26 The sum of two forces **P** and **Q** is **R** such that $|\mathbf{R}| = |\mathbf{P}|$. The angle θ (in degrees) that the resultant of 2**P** and **Q** will make with **Q** is, ... *[2020, 7 Jan Shift-II]*

Topic 2 Motion in a Plane and Projectile Motion

Level 1

01 Two balls with same mass and initial velocity, are projected at different angles in such a way that maximum height reached by first ball is 8 times higher than that of the second ball. T_1 and T_2 are the total flying times of first and second ball, respectively, then the ratio of T_1 and T_2 is *[2025, 8 April Shift-II]*

(a) 4 : 1 (b) 2 : 1
(c) $\sqrt{2}:1$ (d) $2\sqrt{2}:1$

02 A particle is projected with velocity u, so that its horizontal range is three times the maximum height attained by it. The horizontal range of the projectile is given as $\dfrac{nu^2}{25g}$, where value of n is
(Given, g is the acceleration due to gravity) *[2025, 3 April Shift-II]*

(a) 6 (b) 24 (c) 18 (d) 12

03 A helicopter flying horizontally with a speed of 360 km/h at an altitude of 2 km, drops an object at an instant. The object hits the ground at a point O, 20 s after it is dropped. Displacement of O from the position of helicopter where the object was released is (Use acceleration due to gravity $g = 10$ m/s^2 and neglect air resistance) *[2025, 7 April Shift-II]*

(a) $2\sqrt{2}$ km (b) 4 km (c) $2\sqrt{5}$ km (d) 7.2 km

04 The position vector of a moving body at any instant of time is given as $\mathbf{r} = (5t^2\hat{i} - 5t\hat{j})$ m. The magnitude and direction of velocity at $t = 2$ s is, *[2025, 24 Jan Shift-II]*

(a) $5\sqrt{17}$ m/s, making an angle of $\tan^{-1}(4)$ with + ve X-axis.
(b) $5\sqrt{15}$ m/s, making an angle of $\tan^{-1}(4)$ with − ve Y-axis.
(c) $5\sqrt{15}$ m/s, making an angle of $\tan^{-1}(4)$ with + ve X-axis.
(d) $5\sqrt{17}$ m/s, making an angle of $\tan^{-1}(4)$ with − ve Y-axis.

05 The angle of projection for a projectile to have same horizontal range and maximum height is *[2024, 8 April Shift-II]*

(a) $\tan^{-1}\left(\dfrac{1}{2}\right)$ (b) $\tan^{-1}\left(\dfrac{1}{4}\right)$
(c) $\tan^{-1}(2)$ (d) $\tan^{-1}(4)$

06 A projectile is projected at 30° from horizontal with initial velocity 40 ms^{-1}. The velocity of the projectile at $t = 2$ s from the start will be

(Given, $g = 10$ m/s^2) *[2023, 11 April Shift-II]*

(a) $40\sqrt{3}$ ms^{-1} (b) 20 ms^{-1}

(c) $20\sqrt{3}$ ms^{-1} (d) zero

Imp. Maximum Height

07 Two projectiles are projected at 30° and 60° with the horizontal with the same speed. The ratio of the maximum height attained by the two projectiles respectively is *[2023, 10 April Shift-II]*

(a) $1:\sqrt{3}$ (b) $1:3$

(c) $2:\sqrt{3}$ (d) $\sqrt{3}:1$

08 The range of the projectile projected at an angle of 15° with horizontal is 50 m. If the projectile is projected with same velocity at an angle of 45° with horizontal, then its range will be *[2023, 10 April Shift-I]*

(a) $100\sqrt{2}$ m (b) 50 m

(c) $50\sqrt{2}$ m (d) 100 m

09 The trajectory of projectile, projected from the ground is given by $y = x - \frac{x^2}{20}$. Where x and y are measured in metre. The maximum height attained by the projectile will be *[2023, 8 April Shift-II]*

(a) 200 m (b) 5 m

(c) 10 m (d) $10\sqrt{2}$ m

Imp. Projection Through a Height

10 A child stands on the edge of the cliff 10 m above the ground and throws a stone horizontally with an initial speed of 5 ms^{-1}. Neglecting the air resistance, the speed with which the stone hits the ground will be .... ms^{-1}. (Given, $g = 10$ ms^{-2})

[2023, 1 Feb Shift-I]

(a) 25 (b) 30 (c) 15 (d) 20

11 The maximum vertical height to which a man can throw a ball is 136 m. The maximum horizontal distance upto which he can throw the same ball is *[2023, 24 Jan Shift-I]*

(a) 136 m (b) 192 m

(c) 68 m (d) 272 m

12 A ball is projected from the ground with a speed 15 ms^{-1} at an angle θ with horizontal so that its range and maximum height are equal, then $\tan\theta$ will be equal to *[2022, 25 July Shift-II]*

(a) 1/4 (b) 1/2

(c) 2 (d) 4

13 A bomb is dropped by fighter plane flying horizontally. To an observer sitting in the plane, the trajectory of the bomb is a *[2021, 26 Aug Shift-II]*

(a) hyperbola

(b) parabola in the direction of motion of plane

(c) straight line vertically down to the plane

(d) parabola in a direction opposite to the motion of plane

14 The position vector of particle changes with time according to the relation $\mathbf{r}(t) = 15t^2\hat{\mathbf{i}} + (4 - 20t^2)\hat{\mathbf{j}}$. What is the magnitude of the acceleration (in ms^{-2}) at $t = 1$? *[2019, 9 April Shift-II]*

(a) 50 (b) 100 (c) 25 (d) 40

15 A projectile is given an initial velocity of $(\mathbf{i} + 2\mathbf{j})$ m/s, where $\mathbf{i}$ is along the ground and $\mathbf{j}$ is along the vertical. If $g = 10$ m/s^2, the equation of its trajectory is *[JEE Main 2013]*

(a) $y = x - 5x^2$ (b) $y = 2x - 5x^2$

(c) $4y = 2x - 5x^2$ (d) $4y = 2x - 25x^2$

16 A boy can throw a stone upto a maximum height of 10 m. The maximum horizontal distance that the boy can throw the same stone upto will be *[AIEEE 2012]*

(a) $20\sqrt{2}$ m (b) 10 m (c) $10\sqrt{2}$ m (d) 20 m

17 A particle has an initial velocity $3\,\mathbf{i} + 4\,\mathbf{j}$ and an acceleration of $0.4\,\mathbf{i} + 0.3\,\mathbf{j}$. Its speed after 10 s is *[AIEEE 2009]*

(a) 10 units (b) $7\sqrt{2}$ units (c) 7 units (d) 8.5 units

18 A ball is thrown from a point with a speed v_0 at an angle of projection θ. From the same point and at the same instant, a person starts running with a constant speed $v_0/2$ to catch the ball. Will the person be able to catch the ball? If yes, what should be the angle of projection? *[AIEEE 2004]*

(a) Yes, 60° (b) Yes, 30° (c) No (d) Yes, 45°

19 A boy playing on the roof of a 10 m high building throws a ball with a speed of 10 m/s at an angle of 30° with the horizontal. How far from the throwing point will the ball be at the height of 10 m from the ground? [$g = 10$ m/s^2, $\sin 30° = 1/2$, $\cos 30° = \sqrt{3}/2$] *[AIEEE 2003]*

(a) 5.20 m (b) 4.33 m (c) 2.60 m (d) 8.66 m

Level 2

20 Two projectiles are fired from ground with same initial speeds from same point at angles $(45° + \alpha)$ and $(45° - \alpha)$ with horizontal direction. The ratio of their times of flights is *[2025, 7 April Shift-I]*

(a) $\frac{1 + \sin 2\alpha}{1 - \sin 2\alpha}$ (b) $\frac{1 - \tan\alpha}{1 + \tan\alpha}$

(c) 1 (d) $\frac{1 + \tan\alpha}{1 - \tan\alpha}$

21 The angle of projection of a particle is measured from the vertical axis as ϕ and the maximum height reached by the particle is h_m. Here h_m as function of ϕ can be presented as *[2025, 3 April Shift-I]*

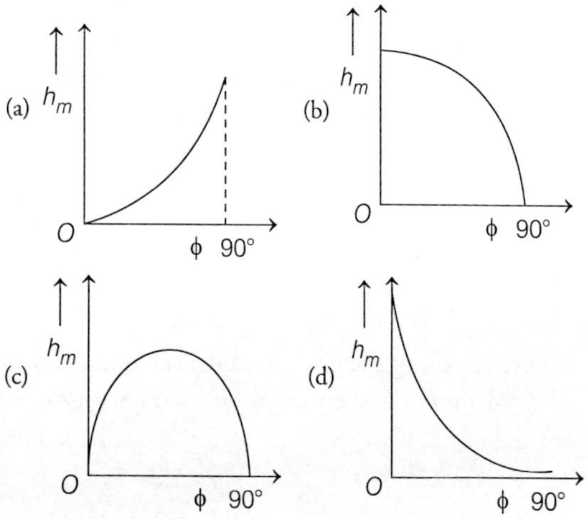

22 Projectiles A and B are thrown at angles of 45° and 60° with vertical, respectively from top of a 400 m high tower. If their ranges and times of flight are same, the ratio of their speeds of projection $v_A : v_B$ is [Take, $g = 10\ ms^{-2}$] *[2024, 30 Jan Shift-II]*

(a) $\sqrt{2}:1$ (b) $1:2$ (c) $1:\sqrt{3}$ (d) $1:\sqrt{2}$

23 The initial speed of a projectile fired from ground is u. At the highest point during its motion, the speed of projectile is $\sqrt{3}/2\,u$. The time of flight of the projectile is *[2023, 31 Jan Shift-I]*

(a) $\frac{2u}{g}$ (b) $\frac{u}{2g}$ (c) $\frac{u}{g}$ (d) $\frac{\sqrt{3}u}{g}$

Imp. Time of Flight and Range

24 A projectile is projected with velocity of 25 m/s at an angle θ with the horizontal. After t seconds its inclination with horizontal becomes zero. If R represents horizontal range of the projectile, the value of θ will be [use, $g = 10\ m/s^2$] *[2022, 24 June Shift-I]*

(a) $\frac{1}{2}\sin^{-1}\left(\frac{5t^2}{4R}\right)$ (b) $\frac{1}{2}\sin^{-1}\left(\frac{4R}{5t^2}\right)$

(c) $\tan^{-1}\left(\frac{4t^2}{5R}\right)$ (d) $\cot^{-1}\left(\frac{R}{20t^2}\right)$

25 The ranges and heights for two projectiles projected with the same initial velocity at angles 42° and 48° with the horizontal are R_1, R_2 and H_1, H_2, respectively. Choose the correct option. *[2021, 1 Sep Shift-II]*

(a) $R_1 > R_2$ and $H_1 = H_2$ (b) $R_1 = R_2$ and $H_1 < H_2$
(c) $R_1 < R_2$ and $H_1 < H_2$ (d) $R_1 = R_2$ and $H_1 = H_2$

26 A helicopter is flying horizontally with a speed v at an altitude h has to drop a food packet for a man on the ground. What is the distance of helicopter from the man when the food packet is dropped? *[2021, 31 Aug Shift-I]*

(a) $\sqrt{\frac{2ghv^2+1}{h^2}}$ (b) $\sqrt{2ghv^2+h^2}$

(c) $\sqrt{\frac{2v^2h}{g}+h^2}$ (d) $\sqrt{\frac{2gh}{v^2}+h^2}$

27 The trajectory of a projectile in a vertical plane is $y = \alpha x - \beta x^2$, where α and β are constants and x and y are respectively the horizontal and vertical distances of the projectile from the point of projection. The angle of projection θ and the maximum height attained H are respectively given by *[2021, 26 Feb Shift-II]*

(a) $\tan^{-1}\alpha, \frac{\alpha^2}{4\beta}$ (b) $\tan^{-1}\beta, \frac{\alpha^2}{2\beta}$

(c) $\tan^{-1}\alpha, \frac{4\alpha^2}{\beta}$ (d) $\tan^{-1}\left(\frac{\beta}{\alpha}\right), \frac{\alpha^2}{\beta}$

28 Starting from the origin at time $t = 0$, with initial velocity $5\hat{j}\ ms^{-1}$, a particle moves in the xy-plane with a constant acceleration of $(10\hat{i} + 4\hat{j})\ ms^{-2}$. At time t, its coordinates are (20 m, y_0 m). The values of t and y_0 respectively, are *[2020, 4 Sep Shift-I]*

(a) 2 s and 18 m (b) 5 s and 25 m
(c) 2 s and 24 m (d) 4 s and 52 m

29 A particle is moving with velocity $\mathbf{v} = k(y\hat{i} + x\hat{j})$, where k is a constant. The general equation for its path is *[AIEEE 2010]*

(a) $y = x^2$ + constant (b) $y^2 = x$ + constant
(c) xy = constant (d) $y^2 = x^2$ + constant

Level 3

Imp. Uniformly Accelerated Motion

30 Two projectiles fired with same initial speed from same point on ground at angles of $(45° - \alpha)$ and $(45° + \alpha)$, respectively, with the horizontal direction. The ratio of their maximum heights attained is *[2025, 29 Jan Shift-I]*

(a) $\frac{1+\sin\alpha}{1-\sin\alpha}$ (b) $\frac{1-\tan\alpha}{1+\tan\alpha}$

(c) $\frac{1+\sin 2\alpha}{1-\sin 2\alpha}$ (d) $\frac{1-\sin 2\alpha}{1+\sin 2\alpha}$

31 A particle moving in a circle of radius R with uniform speed, takes time T to complete one revolution. If this particle is projected with the same speed at an angle θ to the horizontal, the maximum height attained by it is equal to $4R$. The angle of projection θ is then given by *[2024, 1 Feb Shift-I]*

(a) $\sin^{-1}\left[\frac{2gT^2}{\pi^2 R}\right]^{\frac{1}{2}}$ (b) $\sin^{-1}\left[\frac{\pi^2 R}{2gT^2}\right]^{\frac{1}{2}}$

(c) $\cos^{-1}\left[\frac{2gT^2}{\pi^2 R}\right]^{\frac{1}{2}}$ (d) $\cos^{-1}\left[\frac{\pi R}{2gT^2}\right]^{\frac{1}{2}}$

32 A body of mass 10 kg is projected at an angle of 45° with the horizontal. The trajectory of the body is observed to pass through a point (20, 10). If T is the time of flight, then its momentum vector, at time $t = \frac{T}{\sqrt{2}}$, is .... [Take, $g = 10\ \text{m/s}^2$] *[2022, 27 July Shift-II]*

(a) $100\hat{i} + (100\sqrt{2} - 200)\hat{j}$ (b) $100\sqrt{2}\,\hat{i} + (100 - 200\sqrt{2})\hat{j}$

(c) $100\hat{i} + (100 - 200\sqrt{2})\hat{j}$ (d) $100\sqrt{2}\,\hat{i} + (100\sqrt{2} - 200)\hat{j}$

33 A projectile is launched at an angle α with the horizontal with a velocity 20 ms^{-1} . After 10 s, its inclination with horizontal is β. The value of $\tan\beta$ will be ($g = 10\ \text{ms}^{-2}$) *[2022, 27 June Shift-I]*

(a) $\tan\alpha + 5\sec\alpha$ (b) $\tan\alpha - 5\sec\alpha$

(c) $2\tan\alpha - 5\sec\alpha$ (d) $2\tan\alpha + 5\sec\alpha$

34 A shell is fired from a fixed artillery gun with an initial speed u such that it hits the target on the ground at a distance R from it. If t_1 and t_2 are the values of the time taken by it to hit the target in two possible ways, the product $t_1 t_2$ is *[2019, 12 April Shift-I]*

(a) $\frac{R}{4g}$ (b) $\frac{R}{g}$ (c) $\frac{R}{2g}$ (d) $\frac{2R}{g}$

35 Two particles are projected from the same point with the same speed u such that they have the same range R, but different maximum heights h_1 and h_2. Which of the following is correct? *[2019, 12 April Shift-II]*

(a) $R^2 = 4h_1h_2$ (b) $R^2 = 16h_1h_2$

(c) $R^2 = 2h_1h_2$ (d) $R^2 = h_1h_2$

Numerical Value Type Questions

36 A particle is projected at an angle of 30° from horizontal at a speed of 60 m/s. The height traversed by the particle in the first second is h_0 and height traversed in the last second, before it reaches the maximum height, is h_1. The ratio $h_0 : h_1$ is .... [Take, $g = 10\ \text{m/s}^2$] *[2025, 22 Jan Shift-I]*

37 The maximum speed of a boat in still water is 27 km/h. Now, this boat is moving downstream in a river flowing at 9 km/h. A man in the boat throws a ball vertically upwards with speed of 10 m/s. Range of the ball as observed by an observer at rest on the river bank, is ...... cm. (Take, $g = 10\ \text{m/s}^2$) *[2025, 29 Jan Shift-I]*

38 A body of mass M thrown horizontally with velocity v from the top of the tower of height H touches the ground at a distance of 100 m from the foot of the tower. A body of mass $2M$ thrown at a velocity $\frac{v}{2}$ from the top of the tower of height $4H$ will touch the ground at a distance of ...... m. *[2024, 8 April Shift-II]*

39 The maximum height reached by a projectile is 64 m. If the initial velocity is halved, the new maximum height of the projectile is .... m. *[2024, 5 April Shift-II]*

40 A body starts falling freely from height H hits an inclined plane in its path at height h. As a result of this perfectly elastic impact, the direction of the velocity of the body becomes horizontal. The value of $\frac{H}{h}$ for which the body will take the maximum time to reach the ground is .... *[2024, 31 Jan Shift-I]*

41 A ball rolls off the top of a stairway with horizontal velocity u. The steps are 0.1 m high and 0.1 m wide. The minimum velocity u with which that ball just hits the step 5 of the stairway will be $\sqrt{x}\ \text{ms}^{-1}$ where $x =$ .... [use, $g = 10\ \text{m/s}^2$]. *[2024, 29 Jan Shift-I]*

42 Two bodies are projected from ground with same speeds 40 ms^{-1} at two different angles with respect to horizontal. The bodies were found to have same range. If one of the body was projected at an angle of 60°, with horizontal then sum of the maximum heights, attained by the two projectiles, is .... m. (Given, $g = 10\ \text{ms}^{-2}$) *[2023, 31 Jan Shift-II]*

43 A projectile fired at 30° to the ground is observed to be at same height at time 3 s and 5 s after projection, during its flight. The speed of projection of the projectile is .... ms^{-1}. (Given, $g = 10\ \text{ms}^{-2}$) *[2023, 11 April Shift-I]*

44 A body is projected from the ground at an angle of 45° with the horizontal. Its velocity after 2s is 20 ms^{-1}. The maximum height reached by the body during its motion is ...... m. (Use, $g = 10\ \text{ms}^{-2}$) *[2022, 24 June Shift-II]*

Topic 3 Relative Velocity

Level 1

01 A particle is moving Eastwards with a velocity of 5 ms^{-1}. In 10 s, the velocity changes to 5 ms^{-1} Northwards. The average acceleration in this time is *[AIEEE 2005]*

(a) $\frac{1}{\sqrt{2}}\ \text{ms}^{-2}$ towards North-East

(b) $\frac{1}{2}\ \text{ms}^{-2}$ towards North

(c) zero

(d) $\frac{1}{\sqrt{2}}\ \text{ms}^{-2}$ towards North-West

Level 2

02 A butterfly is flying with a velocity $4\sqrt{2}$ m/s in North-East direction. Wind is slowly blowing at 1 m/s from North to South. The resultant displacement of the butterfly in 3 s is *[2021, 20 July Shift-I]*

(a) 3 m (b) 20 m
(c) $12\sqrt{2}$ m (d) 15 m

Level 3

03 A river is flowing from west to east direction with speed of 9 km h^{-1}. If a boat capable of moving at a maximum speed of 27 km h^{-1} in still water, crosses the river in half a minute, while moving with maximum speed at an angle of 150° to direction of river flow, then the width of the river is *[2025, 2 April Shift-I]*

(a) 112.5 m (b) 300 m
(c) $112.5 \times \sqrt{3}$ m (d) 75 m

04 When a car is at rest, its driver sees rain drops falling on it vertically. When driving the car with speed v, he sees that rain drops are coming at an angle 60° from the horizontal. On further increasing the speed of the car to $(1+\beta)v$, this angle changes to 45°. The value of β is close to *[2020, 6 Sep Shift-II]*

(a) 0.50 (b) 0.41
(c) 0.37 (d) 0.73

Numerical Value Type Questions

Imp. River Boat Problem

05 The speed of a swimmer is 4 km h^{-1} in still water. If the swimmer makes his strokes normal to the flow of river of width 1 km, he reaches a point 750 m down the stream on the opposite bank.
The speed of the river water is .... km h^{-1}. *[2023, 31 Jan Shift-I]*

06 A swimmer wants to cross a river from point A to point B. Line AB makes an angle of 30° with the flow of river. Magnitude of velocity of the swimmer is same as that of the river. The angle θ with the line AB should be ......°, so that the swimmer reaches point B *[2021, 27 July Shift-II]*

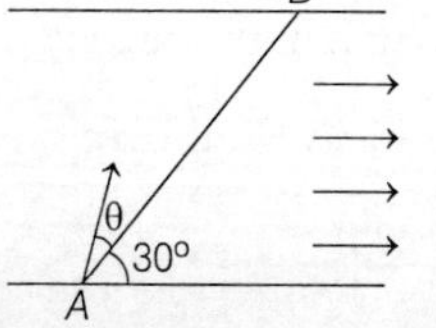

07 A person is swimming with a speed of 10 m/s at an angle of 120° with the flow and reaches to a point directly opposite on the other side of the river. The speed of the flow is x m/s. The value of x to the nearest integer is .... *[2021, 18 March Shift-I]*

08 A swimmer can swim with velocity of 12 km/h in still water. Water flowing in a river has velocity 6 km/h. The direction with respect to the direction of flow of river water he should swim in order to reach the point on the other bank just opposite to his starting point is ......° (in degree). (Round off to the nearest integer) *[2021, 16 March Shift-II]*

09 A particle is moving along the X-axis with its coordinate with the t given by $x(t) = 10 + 8t - 3t^2$. Another particle is moving along the Y-axis with its coordinate as a function of time given by $y(t) = 5 - 8t^3$. At $t = 1$ s, the speed of the second particle as measured in the frame of the first particle is given as $\sqrt{v}$. Then, v (in m/s) is .... *[2020, 8 Jan Shift-I]*

Topic 4 Uniform Circular Motion

Level 1

01 For particle P revolving round the centre O with radius of circular path r and angular velocity ω, as shown in below figure, the projection of OP on the X-axis at time t is *[2023, 8 April Shift-II]*

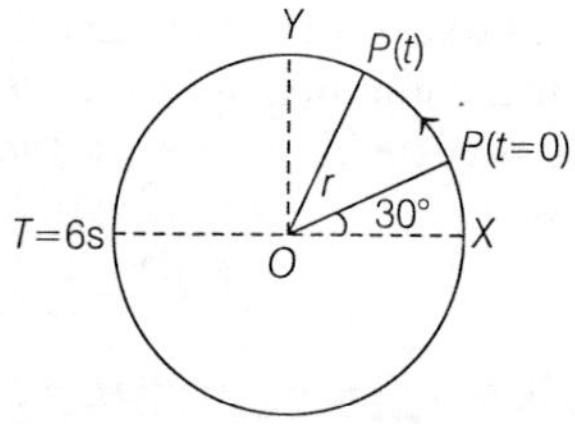

(a) $x(t) = r\cos\left(\omega t + \frac{\pi}{6}\right)$ (b) $x(t) = r\cos(\omega t)$
(c) $x(t) = r\sin\left(\omega t + \frac{\pi}{6}\right)$ (d) $x(t) = r\cos\left(\omega t - \frac{\pi}{6}\omega\right)$

02 A clock has a continuously moving second's hand of 0.1 m length. The average acceleration of the tip of the hand (in ms^{-2}) is of the order of *[2020, 6 Sep Shift-I]*

(a) 10^{-3} (b) 10^{-4}
(c) 10^{-2} (d) 10^{-1}

03 Two cars of masses m_1 and m_2 are moving in circles of radii r_1 and r_2, respectively. Their speeds are such that they make complete circles in the same time t.
The ratio of their centripetal acceleration is *[AIEEE 2012]*

(a) $m_1 r_1 : m_2 r_2$ (b) $m_1 : m_2$
(c) $r_1 : r_2$ (d) $1 : 1$

04 Which of the following statements is false for a particle moving in a circle with a constant angular speed ?

(a) The velocity vector is tangent to the circle. *[AIEEE 2004]*
(b) The acceleration vector is tangent to the circle.
(c) The acceleration vector points to the centre of the circle.
(d) The velocity and acceleration vectors are perpendicular to each other.

Level 2

Imp. Average Velocity in Uniform Circular Motion

05 As shown in the figure, a particle is moving with constant speed π m/s. Considering its motion from A to B, the magnitude of the average velocity is *[2023, 6 April Shift-II]*

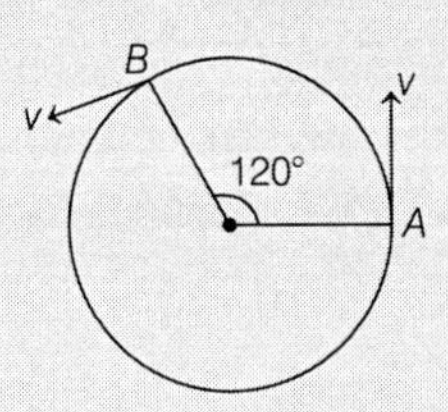

(a) π m/s (b) $\sqrt{3}$ m/s
(c) $1.5\sqrt{3}$ m/s (d) $2\sqrt{3}$ m/s

06 A person moved from A to B on a circular path as shown in figure. If the distance travelled by him is 60 m, then the magnitude of displacement would be (Given, $\cos 135° = -0.7$) *[2022, 25 July Shift-I]*

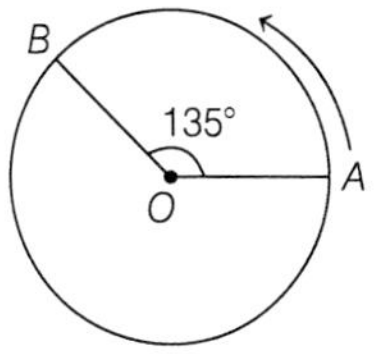

(a) 42 m (b) 47 m
(c) 19 m (d) 40 m

Level 3

07 A particle is moving with constant speed in a circular path. When the particle turns by an angle 90°, the ratio of instantaneous velocity to its average velocity is $\pi : x\sqrt{2}$. The value of x will be *[2023, 6 April Shift-I]*

(a) 1 (b) 2
(c) 7 (d) 5

08 A huge circular arc of length 4.4 ly subtends an angle 4s at the centre of the circle. How long it would take for a body to complete 4 revolution if its speed is 8 AU per sec ?
[Given, 1 ly $= 9.46 \times 10^{15}$ m, 1 AU $= 1.5 \times 10^{11}$ m] *[2021, 27 Aug Shift-I]*

(a) 4.1×10^8 s (b) 4.5×10^{10} s
(c) 3.5×10^6 s (d) 7.2×10^8 s

• Key **Index**

Topic 1

1 (d)	2 (c)	3 (c)	4 (b)	5 (d)	6 (d)	7 (d)	8 (a)	9 (a)	10 (c)
11 (b)	12 (a)	13 (c)	14 (c)	15 (b)	16 (b)	17 (a)	18 (3)	19 (150)	20 (3)
21 (4)	22 (4)	23 (1)	24 (2)	25 (180)	26 (90)				

Topic 2

1 (d)	2 (b)	3 (a)	4 (d)	5 (d)	6 (c)	7 (b)	8 (d)	9 (b)	10 (c)
11 (d)	12 (d)	13 (c)	14 (a)	15 (b)	16 (d)	17 (b)	18 (a)	19 (d)	20 (d)
21 (b)	22 (d)	23 (c)	24 (d)	25 (b)	26 (c)	27 (a)	28 (a)	29 (d)	30 (d)
31 (a)	32 (d)	33 (b)	34 (d)	35 (b)	36 (5)	37 (2000)	38 (100)	39 (16)	40 (2)
41 (2)	42 (80)	43 (80)	44 (20)						

Topic 3

1 (d)	2 (d)	3 (a)	4 (d)	5 (3)	6 (30)	7 (5)	8 (120)	9 (580)

Topic 4

1 (a)	2 (a)	3 (c)	4 (b)	5 (c)	6 (b)	7 (b)	8 (b)

Detailed **Explanations**

Topic 1

01. *(d)* We have,

$$|A+B|^2 = (A+B)\cdot(A+B)$$
$$= |A|^2 + |B|^2 + 2|A||B|\cos\theta$$

Given, $|A| = |B| = R$

So, $|A+B|^2 = 2R^2(1+\cos\theta)$

$\Rightarrow |A+B|^2 = 2R^2 \cdot 2\cos^2(\theta/2)$

or, $|A+B| = 2R\cos(\theta/2)$

02. *(c)*

> **key idea** When two vectors (**P** and **Q**) are perpendicular to each other, then you can apply the following condition as, **P** · **Q** = 0

Given vectors, $P = \hat{i} + 2m\hat{j} + m\hat{k}$

and $Q = 4\hat{i} - 2\hat{j} + m\hat{k}$

As, P and Q are given perpendicular vectors, so $P\cdot Q = 0$

$\Rightarrow (\hat{i} + 2m\hat{j} + m\hat{k})\cdot(4\hat{i} - 2\hat{j} + m\hat{k}) = 0$

$\Rightarrow \quad 4 - 4m + m^2 = 0$

$\Rightarrow \quad m = 2$

03. *(c)* $A \times A = |A||A|\sin 0°\,\hat{n}$

$= |A|^2 \times 0 \times \hat{n} \quad [\because |A| \neq 0]$

$= 0$

04. *(b)* Using triangular law of vector addition for each case,

(A) $C - A - B = 0$

$\Rightarrow C = A + B$ i.e., C is resultant vector.

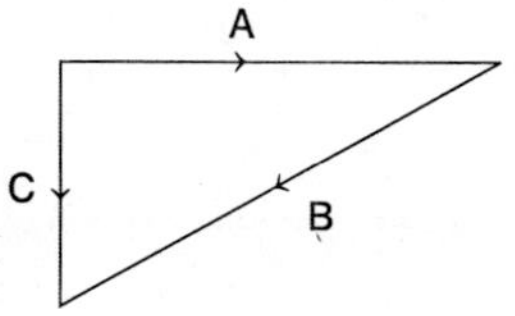

(B) $A - C - B = 0$

$\Rightarrow A = C + B$ i.e., A is resultant vector.

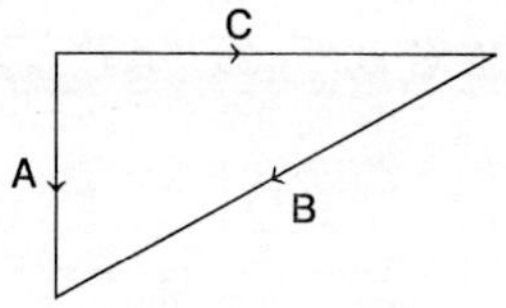

(C) $B - A - C = 0$

$\Rightarrow B = A + C$ i.e., B is resultant vector.

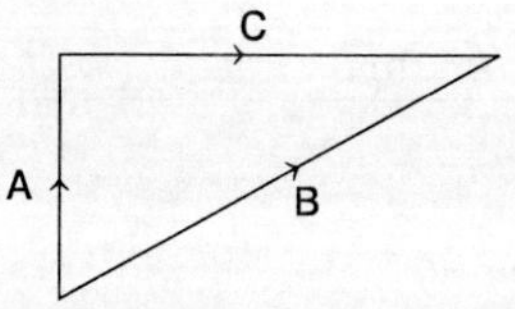

(D) $A + B = -C$

$A + B + C = 0$ i.e., these are cyclic vectors.

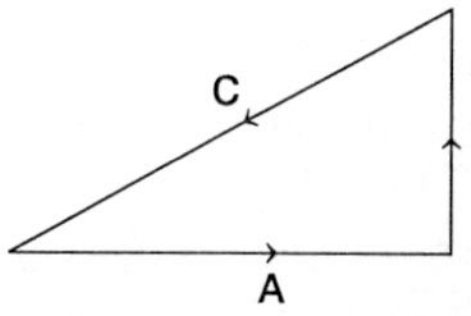

05. *(d)* Given, $A = \hat{i} + \hat{j} + \hat{k}$ and $B = \hat{i} + \hat{j}$

As we know that, vector projection of A on B is $\frac{A\cdot B}{|B|}\cdot\hat{B}$

Therefore, projection of A on B will be

$$\frac{(\hat{i}+\hat{j}+\hat{k})\cdot(\hat{i}+\hat{j})}{\sqrt{1^2+1^2}} \cdot \frac{(\hat{i}+\hat{j})}{\sqrt{1^2+1^2}}$$
$$= \frac{1+1+0}{\sqrt{2}} \cdot \frac{(\hat{i}+\hat{j})}{\sqrt{2}}$$
$$= \frac{2}{2}(\hat{i}+\hat{j}) = (\hat{i}+\hat{j})$$

06. *(d)* Given, $|A\cdot B| = |A \times B|$ ...(i)

Also, we know that

$A\cdot B = AB\cos\theta$ ...(ii)

and $A \times B = AB\sin\theta$ ...(iii)

From Eqs. (i), (ii) and (iii), we get

$AB\cos\theta = AB\sin\theta$

$\Rightarrow \cos\theta = \sin\theta$

$\Rightarrow \frac{\sin\theta}{\cos\theta} = 1$

$\Rightarrow \tan\theta = 1 \Rightarrow \theta = 45°$

$$\therefore |A - B| = \sqrt{A^2 + B^2 - 2AB\cos\theta}$$
$$= \sqrt{A^2 + B^2 - 2AB\cos(45°)}$$
$$= \sqrt{A^2 + B^2 - \sqrt{2}AB}$$

> • Concept **Enhancer**
>
> Parallelogram law of vector addition is used to find resultant of two vectors.
>
> 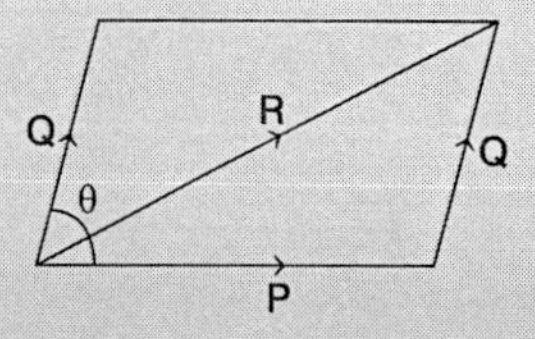
>
>
> According to this law,
>
> $R = \sqrt{P^2 + Q^2 + 2PQ\cos\theta}$

07. *(d)* For vector $A_1 + A_2$, we have

$|A_1 + A_2|^2 = (A_1 + A_2)\cdot(A_1 + A_2)$

$[\because x\cdot x = |x|^2]$

$\Rightarrow |A_1 + A_2|^2 = |A_1|^2 + |A_2|^2 + 2A_1\cdot A_2$

Given, $|A_1| = 3, |A_2| = 5$

and $|A_1 + A_2| = 5$

So, we have,

$(5)^2 = 9 + 25 + 2A_1\cdot A_2$

$\Rightarrow \quad A_1\cdot A_2 = -\frac{9}{2}$

Now, $(2A_1 + 3A_2)\cdot(3A_1 - 2A_2)$

$= 6|A_1|^2 - 4A_1\cdot A_2 + 9A_1\cdot A_2 - 6|A_2|^2$

$= 6|A_1|^2 - 6|A_2|^2 + 5A_1\cdot A_2$

$= 6\,(9) - 6(25) + 5\left(-\frac{9}{2}\right)$

$= -118.5$

> ***Alternate solution***
>
> As we know, $|A_1 + A_2|$ can also be written as
>
> $|A_1 + A_2| = \sqrt{|A_1|^2 + |A_2|^2 + 2|A_1||A_2|\cos\theta}$
>
> Substituting the given values, we get
>
> $5 = \sqrt{(3)^2 + (5)^2 + 2\times3\times5\cos\theta}$
>
> or $\cos\theta = -\frac{9}{2\times3\times5} = -\frac{3}{10}$
>
> So, $(2A_1 + 3A_2)\cdot(3A_1 - 2A_2)$
>
> $= 6|A_1|^2 - 6|A_2|^2 + 5A_1\cdot A_2$
>
> $= 6|A_1|^2 - 6|A_2|^2 + 5|A_1||A_2|\cos\theta$
>
> $= 6\times9 - 6\times25 + 5\times3\times5\times\left(\frac{-3}{10}\right) = -118.5$

08. *(a)* As $(A \times B) = (B \times A)$ [given]

$\Rightarrow (A\times B) - (B \times A) = 0$

or $(A \times B) + (A \times B) = 0$

$[\because (B\times A) = -(A\times B)]$

or $2(A\times B) = 0$

$\Rightarrow \quad 2AB\sin\theta = 0$ or $\sin\theta = 0$

$[\because |A| = A \neq 0, |B| = B \neq 0]$

$\Rightarrow \quad \theta = 0$ or π

09. *(a)* Let, vector $A = 2Q + 2P$,

Vector $B = 2Q - 2P$

Resultant $= R = A + B$

$= 2Q + 2P + 2Q - 2P$

$R = 4Q$ ...(i)

Now angle between R and Q,

$$\cos\theta = \frac{R\cdot Q}{|R||Q|}$$

$$=\frac{4Q\cdot Q}{4Q^2}=1$$

$$\theta=0^\circ$$

Hence, angle is 0 degree.

10. *(c)* Given that,

$$|A+B|=2|A-B| \quad ...(i)$$

$$\therefore |A+B|=\sqrt{A^2+B^2+2AB\cos\theta}$$

$$=\sqrt{A^2+A^2+2A^2\cos\theta}$$

$$=\sqrt{2A^2(1+\cos\theta)} \quad (\because A=B)$$

Similarly,

$$|A-B|=\sqrt{A^2+B^2-2AB\cos\theta}$$

$$=\sqrt{A^2+A^2-2A^2\cos\theta}$$

$$=\sqrt{2A^2(1-\cos\theta)} \quad (\because A=B)$$

∴From Eq. (i),

$$\sqrt{2A^2(1+\cos\theta)}=2\sqrt{2A^2(1-\cos\theta)}$$

$$\Rightarrow 2A^2(1+\cos\theta)=4[2A^2(1-\cos\theta)]$$

$$\Rightarrow 2A^2(1+\cos\theta)=8A^2(1-\cos\theta);$$

$$\Rightarrow 1+\cos\theta=4(1-\cos\theta)$$

$$\Rightarrow \cos\theta=\frac{3}{5} \text{ or } \theta=\cos^{-1}\left(\frac{3}{5}\right)$$

11. *(b)* Given, $\hat{A}$ and $\hat{B}$ are unit vectors.

$$\therefore \quad |\hat{A}|=1 \text{ and } |\hat{B}|=1$$

We know that,

$$|\hat{A}+\hat{B}|=\sqrt{|\hat{A}|^2+|\hat{B}|^2+2|\hat{A}||\hat{B}|\cos\theta}$$

$$=\sqrt{1+1+2\cdot1\cdot1\cos\theta}$$

$$=\sqrt{2+2\cos\theta}$$

$$=\sqrt{2(1+\cos\theta)}$$

$$=\sqrt{2\cdot2\cos^2\theta/2}$$

$$[\because 1+\cos\theta=2\cos^2\theta/2]$$

$$|\hat{A}+\hat{B}|=2\cos(\theta/2) \quad ...(i)$$

and

$$|\hat{A}-\hat{B}|=\sqrt{|\hat{A}|^2+|\hat{B}|^2-2|\hat{A}||\hat{B}|\cos\theta}$$

$$=\sqrt{1+1-2\cdot1\cdot1\cdot\cos\theta}$$

$$=\sqrt{2(1-\cos\theta)}$$

$$=\sqrt{2\cdot2\sin^2(\theta/2)}$$

$$[\because 1-\cos\theta=2\sin^2\theta/2]$$

$$|\hat{A}-\hat{B}|=2\sin(\theta/2) \quad ...(ii)$$

Dividing Eq. (ii) by Eq. (i), we get,

$$\frac{|\hat{A}-\hat{B}|}{|\hat{A}+\hat{B}|}=\frac{2\sin\theta/2}{2\cos\theta/2}$$

$$=\tan(\theta/2)$$

$$\Rightarrow |\hat{A}-\hat{B}|=|\hat{A}+\hat{B}|\tan(\theta/2)$$

12. *(a)*

> **key idea** If various vectors are acting at a point, then to find resultant vector, resolve all vectors along *X*-axis and *Y*-axis.

From given figure, we can write all the forces in vector notation as

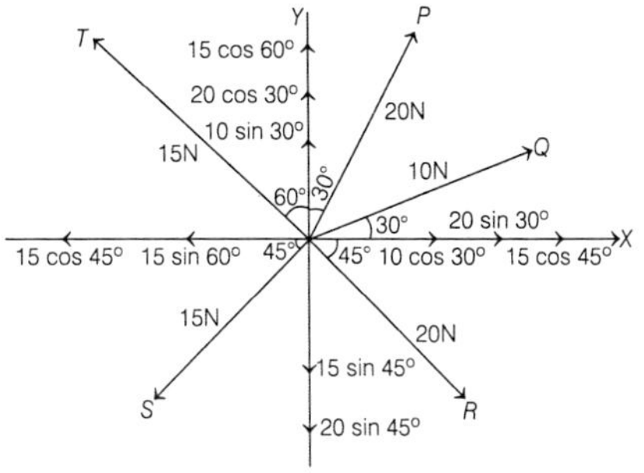

$$OP=(20\sin30°\hat{i}+20\cos30°\hat{j})\text{ N}$$

$$=(10\hat{i}+17\hat{j})\text{ N}$$

$$OQ=(10\cos30°\hat{i}+10\sin30°\hat{j})\text{ N}$$

$$=(8.5\hat{i}+5\hat{j})\text{ N}$$

$$OR=(20\cos45°\hat{i}-20\sin45°\hat{j})\text{ N}$$

$$=(14\hat{i}-14\hat{j})\text{ N}$$

$$OS=(-15\cos45°\hat{i}-15\sin45°\hat{j})\text{ N}$$

$$=(-10.5\hat{i}-10.5\hat{j})\text{ N}$$

$$OT=(-15\sin60°\hat{i}+15\cos60°\hat{j})\text{ N}$$

$$=(-12.75\hat{i}+7.5\hat{j})\text{ N}$$

The resultant of these forces is given as

$$F=OP+OQ+OR+OS+OT$$

$$F=[(10\hat{i}+17\hat{j})+(8.5\hat{i}+5\hat{j})+(14\hat{i}-14\hat{j})+(-10.5\hat{i}-10.5\hat{j})+(-12.75\hat{i}+7.5\hat{j})]$$

$$F=9.25\hat{i}+5\hat{j}\text{ N}$$

Thus, the resultant force is

$$F=9.25\hat{i}+5\hat{j}\text{ N}$$

13. *(c)* The angle between **A** and − **B**,

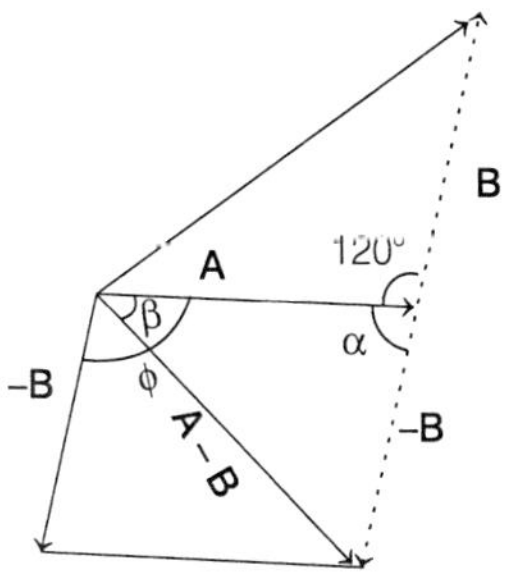

$$\alpha=180°-120°=60°$$

$$\Rightarrow \tan\beta=\frac{B\sin60°}{A+(-B\cos60°)}$$

$$=\frac{B\frac{\sqrt{3}}{2}}{A-\frac{B}{2}}=\frac{\sqrt{3}B}{2A-B}$$

$$\Rightarrow \beta=\tan^{-1}\left(\frac{\sqrt{3}B}{2A-B}\right)$$

14. *(c)* Applying triangular law of vector addition in the given figure in question, we can write

$$OA+AB=OB \quad ...(i)$$

$$OB+BC=OC \quad ...(ii)$$

$$OC+CD=OD \quad ...(iii)$$

Also,

$$AB=OB-OA \quad ...(iv)$$

$$AC=OC-OA \quad ...(v)$$

$$AD=OD-OA$$

$$=-OA-OA \quad ...(vi)$$

On adding Eqs. (iv), (v) and (vi), we get,

$$AB+AC+AD=OB+OC-4OA$$

$$=4AO+OB+OC \quad ...(vii)$$

According to polygon law of vector addition,

$$AB+BC+CD=AD$$

$$\Rightarrow AB+BC+CD+AD=2AD=4AO$$

So, Reason is incorrect.

15. *(b)* In the given cube, coordinates of point *G*(centre of face *ABOD*) are

$$x_1=a/2,\ y_1=0,\ z_1=a/2$$

where, a = side of cube

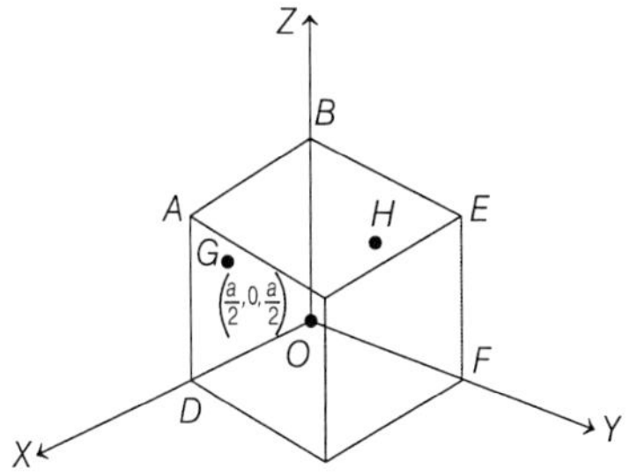

and coordinates of point *H* are

$$x_2=0,\ y_2=\frac{a}{2},\ z_2=\frac{a}{2}$$

So, vector *GH* is

$$GH=(x_2-x_1)\hat{i}+(y_2-y_1)\hat{j}+(z_2-z_1)\hat{k}$$

$$=-\frac{a}{2}\hat{i}+\frac{a}{2}\hat{j}=\frac{a}{2}(\hat{j}-\hat{i})$$

16. *(b)* Given, force vectors $P \perp Q$,

i.e., $\theta=90°$

Let resultant of **P** + **Q** = **x**

and resultant of **P** − **Q** = **y**

$$\therefore \quad x^2=P^2+Q^2+2PQ\cos90°$$

$$=P^2+Q^2 \quad ...(i)$$

$$y^2 = P^2 + Q^2 - 2PQ\cos 90°$$
$$= P^2 + Q^2 \quad \text{...(ii)}$$

When θ_1 is the angle between $(\mathbf{P}+\mathbf{Q})$ and $(\mathbf{P}-\mathbf{Q})$, then, their resultant is given by

$$|\mathbf{x}+\mathbf{y}| = \sqrt{3(P^2+Q^2)}$$
$$\therefore \sqrt{x^2+y^2+2xy\cos\theta_1}$$
$$= \sqrt{3P^2+3Q^2} \quad \text{...(iii)}$$

Substituting the values of x^2 and y^2 in Eq. (iii), we get

$$P^2+Q^2+P^2+Q^2$$
$$+2\sqrt{P^2+Q^2}\sqrt{P^2+Q^2}\cos\theta_1$$
$$= 3P^2+3Q^2$$
$$2P^2+2Q^2+(2P^2+2Q^2)\cos\theta_1$$
$$= 3P^2+3Q^2$$
$$\Rightarrow \cos\theta_1 = \frac{P^2+Q^2}{2(P^2+Q^2)} = \frac{1}{2}$$
$$\therefore \quad \theta_1 = 60°$$

When θ_2 is the angle between $(\mathbf{P}+\mathbf{Q})$ and $(\mathbf{P}-\mathbf{Q})$, then the magnitude of their resultant is given by

$$|\mathbf{x}+\mathbf{y}| = \sqrt{2(P^2+Q^2)}$$
$$x^2+y^2+2xy\cos\theta_2 = 2P^2+2Q^2 \quad \text{...(iv)}$$

Substituting the values, we get

$$P^2+Q^2+P^2+Q^2+2(P^2+Q^2)\cos\theta_2$$
$$= 2P^2+2Q^2$$
$$\cos\theta_2 = 0 \Rightarrow \theta_2 = 90°$$

So, $\theta_1 = 60°$ and $\theta_2 = 90°$

and $\theta_1 < \theta_2$

17. *(a)* In the given figure, the magnitude of vectors **OA**, **OB** and **OC** are equal. Let it be m. From figure, we can write

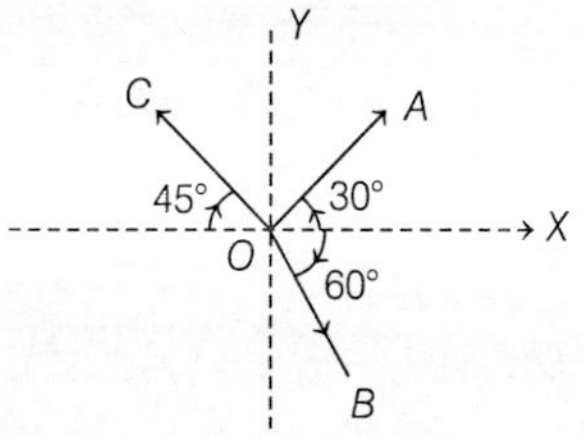

$$\mathbf{OA} = m(\cos 30°\,\hat{\mathbf{i}} + \sin 30°\,\hat{\mathbf{j}})$$
$$= m\left(\frac{\sqrt{3}}{2}\hat{\mathbf{i}} + \frac{1}{2}\hat{\mathbf{j}}\right)$$
$$\mathbf{OB} = m(\cos 60°\,\hat{\mathbf{i}} - \sin 60°\,\hat{\mathbf{j}})$$
$$= m\left(\frac{1}{2}\hat{\mathbf{i}} - \frac{\sqrt{3}}{2}\hat{\mathbf{j}}\right)$$
$$\mathbf{OC} = m(\cos 45°(-\hat{\mathbf{i}}) + \sin 45°\,\hat{\mathbf{j}})$$
$$= m\left\{\frac{-1}{\sqrt{2}}\hat{\mathbf{i}} + \frac{1}{\sqrt{2}}\hat{\mathbf{j}}\right\}$$

According to the question,

$$\mathbf{OA}+\mathbf{OB}-\mathbf{OC} = m\left(\frac{\sqrt{3}}{2}\hat{\mathbf{i}} + \frac{1}{2}\hat{\mathbf{j}}\right)$$
$$+ m\left(\frac{1}{2}\hat{\mathbf{i}} - \frac{\sqrt{3}}{2}\hat{\mathbf{j}}\right) - m\left(-\frac{1}{\sqrt{2}}\hat{\mathbf{i}} + \frac{1}{\sqrt{2}}\hat{\mathbf{j}}\right)$$
$$= m\hat{\ }\left(\frac{\sqrt{3}}{2} + \frac{1}{2} + \frac{1}{\sqrt{2}}\right)$$
$$+ m\hat{\mathbf{j}}\left(\frac{1}{2} - \frac{\sqrt{3}}{2} - \frac{1}{\sqrt{2}}\right)$$

$\therefore$ Angle with X-axis is given by

$$\tan\theta = \frac{y}{x} = \frac{m\left(\frac{1}{2} - \frac{\sqrt{3}}{2} - \frac{1}{\sqrt{2}}\right)}{m\left(\frac{\sqrt{3}}{2} + \frac{1}{2} + \frac{1}{\sqrt{2}}\right)}$$
$$\Rightarrow \tan\theta = \frac{\frac{1}{2} - \frac{\sqrt{3}}{2} - \frac{\sqrt{2}}{2}}{\frac{\sqrt{3}}{2} + \frac{1}{2} + \frac{\sqrt{2}}{2}}$$
$$\Rightarrow \tan\theta = \frac{1-\sqrt{3}-\sqrt{2}}{\sqrt{3}+1+\sqrt{2}}$$
$$\Rightarrow \quad \theta = \tan^{-1}\left(\frac{1-\sqrt{3}-\sqrt{2}}{1+\sqrt{3}+\sqrt{2}}\right)$$

• Concept Enhancer

Direction of resultant vector is found by the following formula.

$$\beta = \tan^{-1}\frac{Q\sin\theta}{P+Q\cos\theta}$$

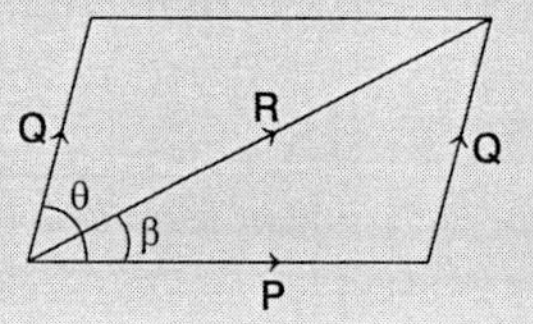

18. *(3)* $\mathbf{A}\cdot\mathbf{B} = 0$

$$(2\hat{\mathbf{i}} + 3n\hat{\mathbf{j}} + 2\hat{\mathbf{k}})\cdot(2\hat{\mathbf{i}} - 2\hat{\mathbf{j}} + 4p\hat{\mathbf{k}}) = 0$$
$$4 - 6n + 8p = 0 \quad \text{...(i)}$$
$$|\mathbf{A}| = |\mathbf{B}|$$
$$4 + 9n^2 + 4 = 4 + 4 + 16p^2$$
$$9n^2 = 16p^2$$

So, $p = +\frac{3}{4}n,\ p = -\frac{3n}{4}$

$$4 - 6n + 8\left(\frac{3}{4}n\right) = 0$$
$$4 - 6n + 8\left(-\frac{3}{4}n\right) = 0$$
$$4 - 6n + 6n = 0$$
$$4 - 6n - 6n = 0$$
$$4 = 12n,\ n = \frac{1}{3} = 3^{-1}$$
$$\Rightarrow \quad n^{-1} = 3$$

19. *(150)* Given

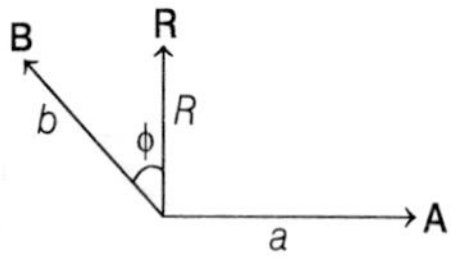

$$|\mathbf{R}| = \frac{|\mathbf{B}|}{2}$$
$$b\cos\phi = R$$
$$\Rightarrow b\cos\phi = b/2$$
$$\Rightarrow \cos\phi = \frac{1}{2} \Rightarrow \phi = 60°$$

Thus, angle between **A** and **B**

$$= 90° + 60° = 150°$$

Time Saver Tech

From the figure

$$B\sin\theta = R = \frac{B}{2}$$

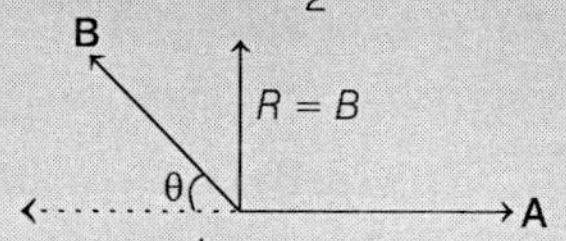

$$\Rightarrow \quad \sin\theta = \frac{1}{2}$$
$$\therefore \quad \theta = 30°$$

Angle between the vector

$$= 180° - \theta = 180° - 30° = 150°$$

20. *(3)* Given, $\cos\theta = \frac{5}{9}, \frac{\mathbf{a}\cdot\mathbf{b}}{ab} = \frac{5}{9}$ ...(i)

$$\mathbf{a}\cdot\mathbf{b} = ab\cos\theta$$
$$\cos\theta = \frac{\mathbf{a}\cdot\mathbf{b}}{ab}$$
$$|\mathbf{a}+\mathbf{b}| = \sqrt{2}|\mathbf{a}-\mathbf{b}|$$
$$a^2 + b^2 + 2\mathbf{a}\cdot\mathbf{b} = 2a^2 + 2b^2 - 4\mathbf{a}\cdot\mathbf{b}$$
$$6\mathbf{a}\cdot\mathbf{b} = a^2 + b^2$$
$$6\times\frac{5}{9}ab = a^2 + b^2$$
$$\frac{10}{3}ab = a^2 + b^2 \text{ and } a = nb$$
$$\frac{10}{3}nb^2 = n^2b^2 + b^2$$
$$\Rightarrow 3n^2 - 10n + 3 = 0$$
$$n = 1/3 \text{ and } n = 3$$

$\therefore$ Integer value of n is 3.

21. *(4)* Magnitude of vector A

$$= \sqrt{3^2 + 4^2} = 5$$

$$\text{Direction of } \mathbf{B} = \hat{\mathbf{B}} = \frac{\mathbf{B}}{|\mathbf{B}|} = \frac{4\hat{\mathbf{i}} + 3\hat{\mathbf{j}}}{\sqrt{4^2+3^2}}$$
$$= \frac{4\hat{\mathbf{i}} + 3\hat{\mathbf{j}}}{5}$$

Required vector has direction of B and magnitude of A. So, required vector is,
vector = magnitude × direction

$$=5\times\left(\frac{4\hat{i}+3\hat{j}}{5}\right)=4\hat{i}+3\hat{j}$$

22. *(4)* Given, $\mathbf{P}=3\hat{i}+\sqrt{3}\hat{j}+2\hat{k}$

$$\mathbf{Q}=4\hat{i}+\sqrt{3}\hat{j}+2.5\hat{k}$$

$$\therefore\quad \mathbf{P}\times\mathbf{Q}=\begin{vmatrix}\hat{i} & \hat{j} & \hat{k}\\ 3 & \sqrt{3} & 2\\ 4 & \sqrt{3} & 2.5\end{vmatrix}$$

$$=\hat{i}(2.5\sqrt{3}-2\sqrt{3})-\hat{j}(7.5-8)+\hat{k}(3\sqrt{3}-4\sqrt{3})$$

$$\Rightarrow\quad \mathbf{P}\times\mathbf{Q}=\frac{\sqrt{3}}{2}\hat{i}+\frac{1}{2}\hat{j}-\sqrt{3}\hat{k}$$

$$\therefore\quad |\mathbf{P}\times\mathbf{Q}|=\sqrt{\left(\frac{\sqrt{3}}{2}\right)^2+\left(\frac{1}{2}\right)^2+(-\sqrt{3})^2}$$

$$=\sqrt{\frac{3}{4}+\frac{1}{4}+3}=\sqrt{4}=2$$

∴ Unit vector in the direction of $\mathbf{P}\times\mathbf{Q}$

$$\hat{n}=\frac{\mathbf{P}\times\mathbf{Q}}{|\mathbf{P}\times\mathbf{Q}|}$$

$$=\frac{1}{2}\left(\frac{\sqrt{3}}{2}\hat{i}+\frac{1}{2}\hat{j}-\sqrt{3}\hat{k}\right)$$

$$=\frac{1}{4}(\sqrt{3}\hat{i}+\hat{j}-2\sqrt{3}\hat{k})$$

$$=\frac{1}{x}(\sqrt{3}\hat{i}+\hat{j}-2\sqrt{3}\hat{k})\text{ (given)}$$

$$\therefore\quad x=4$$

23. *(1)*

> **key idea** You can use the fact that when two vetors (**a** and **b**) are perpendcular to each other, then **a** · **b** = 0

$$\mathbf{a}=a\hat{i}+b\hat{j}+\hat{k}$$

$$\mathbf{b}=2\hat{i}-3\hat{j}+4\hat{k}$$

$\mathbf{a}\perp\mathbf{b}$, when $3a+2b=7$

As vectors **a** and **b** are given perpendicular,

$$\mathbf{a}\cdot\mathbf{b}=0$$

$$(a\hat{i}+b\hat{j}+\hat{k})\cdot(2\hat{i}-3\hat{j}+4\hat{k})=0$$

$$\Rightarrow\quad 2a-3b+4=0$$

$$\Rightarrow\quad 2a-3b=-4$$

Also, $\quad 4a-6b=-8 \quad \ldots(\text{i})$

Since, $\quad 3a+2b=7$

$$\Rightarrow\quad 9a+6b=21 \quad \ldots(\text{ii})$$

Adding Eqs. (i) and (ii), we get

$$13a=13$$

$$\Rightarrow\quad a=1$$

Substituting a in Eq. (i), we get

$$\Rightarrow\quad 4\times1-6b=-8$$

$$\Rightarrow\quad b=2$$

$$a=1\text{ and }b=2$$

$$\frac{a}{b}=\frac{1}{2}=\frac{x}{2}\quad\text{(given)}$$

$$\therefore\quad x=1$$

24. *(2)* Given, $\mathbf{A}=(2\hat{i}+3\hat{j}-\hat{k})$ m

$$\mathbf{B}=(\hat{i}+2\hat{j}+2\hat{k})\text{ m}$$

We know that,

$$\mathbf{A}\cdot\mathbf{B}=AB\cos\theta\quad\ldots(\text{i})$$

The magnitude of component of **A** along **B**

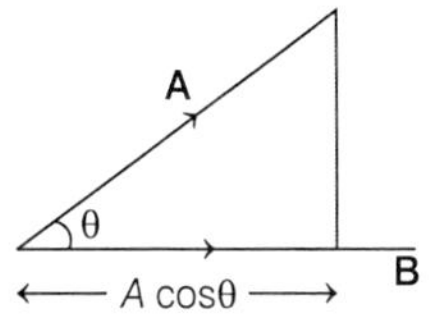

$$A\cos\theta=\frac{\mathbf{A}\cdot\mathbf{B}}{B}\quad\text{[From Eq. (i)]}$$

$$=\frac{2.1+3.2+(-1).2}{\sqrt{1^2+2^2+2^2}}$$

$$=\frac{2+6-2}{\sqrt{9}}$$

$$=\frac{6}{3}=2$$

25. *(180)* Given, $\mathbf{P}\times\mathbf{Q}=\mathbf{Q}\times\mathbf{P}$

$$\Rightarrow\quad \mathbf{P}\times\mathbf{Q}=-\mathbf{P}\times\mathbf{Q}$$

$$\Rightarrow\quad 2(\mathbf{P}\times\mathbf{Q})=0\text{ or }\mathbf{P}\times\mathbf{Q}=0$$

It is possible only, if $\mathbf{P}=0$ or $\mathbf{Q}=0$, angle between them is 0° or 180°.

$$\because\quad 0°<\theta<360°$$

$$\therefore\quad \theta=180°$$

26. *(90)* Then, resultant of **P** and **Q**,

$$|\mathbf{R}|=\sqrt{|\mathbf{P}|^2+|\mathbf{Q}|^2+2|\mathbf{P}||\mathbf{Q}|\cos\beta}$$

As, $|\mathbf{R}|=|\mathbf{P}|$ (given)

So,

$$|\mathbf{P}|^2=|\mathbf{P}|^2+|\mathbf{Q}|^2+2|\mathbf{P}||\mathbf{Q}|\cos\beta$$

$$\text{or } |\mathbf{P}|\cos\beta=-\frac{|Q|}{2}\quad\ldots(\text{i})$$

If resultant of 2**P** and **Q** makes angle θ with **Q**, then angle θ is given by

$$\tan\theta=\frac{|2\mathbf{P}|\sin\beta}{|\mathbf{Q}|+|2\mathbf{P}|\cos\beta}$$

Substituting the value of $|\mathbf{P}|\cos\beta$ from Eq. (i) in above equation, we get

$$\tan\theta=\infty$$

$$\Rightarrow\quad \theta=\frac{\pi}{2}=90°$$

Topic 2

01. *(d)* Here, $(H_{max})_1=8(H_{max})_2$

$$\Rightarrow\quad \frac{u^2\sin^2\theta_1}{2g}=8\times\frac{u^2\sin^2\theta_2}{2g}$$

$$\Rightarrow\quad \sin\theta_1=2\sqrt{2}\sin\theta_2$$

$$\Rightarrow\quad \frac{\sin\theta_1}{\sin\theta_2}=2\sqrt{2}\quad\ldots(\text{i})$$

$$\therefore\quad \frac{T_1}{T_2}=\frac{\frac{2u\sin\theta_1}{g}}{\frac{2u\sin\theta_2}{g}}=\frac{\sin\theta_1}{\sin\theta_2}$$

$$=2\sqrt{2}\quad\text{[from Eq. (i)]}$$

$$\therefore T_1:T_2=2\sqrt{2}:1$$

02. *(b)* Given, $R=3H_{max}$

$$\frac{u^2\sin2\theta}{g}=3\cdot\frac{u^2\sin^2\theta}{2g}$$

$$2\sin2\theta=3\sin^2\theta$$

$$\Rightarrow\quad 2\times2\sin\theta\cos\theta=3\sin^2\theta$$

$$\Rightarrow\quad \tan\theta=\frac{4}{3}$$

$$\therefore\quad \sin\theta=\frac{4}{5}\text{ and }\cos\theta=\frac{3}{5}$$

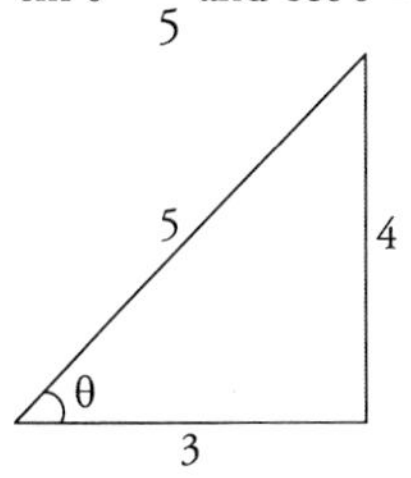

$$\text{Thus, } R=\frac{u^2\sin2\theta}{g}=\frac{u^2\times2\sin\theta\cos\theta}{g}$$

$$=\frac{u^2\times2\times\frac{4}{5}\times\frac{3}{5}}{g}=\frac{24u^2}{25g}$$

$$=\frac{nu^2}{25g}\quad\text{(given)}$$

$$\therefore\quad n=24$$

03. *(a)* Initial speed of object $u=360$ km/h

$$=360\times\frac{5}{18}\text{ m/s}=100\text{ m/s}$$

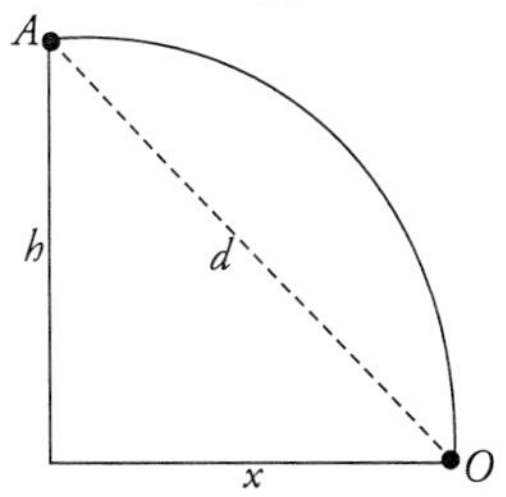

$\therefore x = u \times t = 100 \times 20 = 2000 \text{ m} = 2 \text{ km}$

$\therefore h = 0 + \frac{1}{2}gt^2 = \frac{1}{2}gt^2 = \frac{1}{2} \times 10 \times 20^2$

$= 2000 \text{ m} = 2 \text{ km}$

$\therefore d = \sqrt{h^2 + x^2} = \sqrt{2^2 + 2^2} = 2\sqrt{2} \text{ km}$

04. *(d)* $\mathbf{r} = 5t^2\hat{\mathbf{i}} - 5t\hat{\mathbf{j}}$

at $t = 2$s

$\mathbf{v} = \frac{d\mathbf{r}}{dt} = 10t\hat{\mathbf{i}} - 5\hat{\mathbf{j}} = 10 \times 2\hat{\mathbf{i}} - 5\hat{\mathbf{j}}$ $(t = 2\text{s})$

$\mathbf{v} = 20\hat{\mathbf{i}} - 5\hat{\mathbf{j}}$

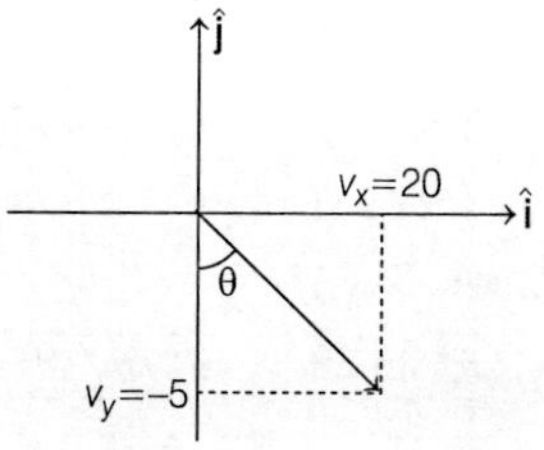

$|\mathbf{v}| = \sqrt{(20)^2 + (5)^2} = \sqrt{400 + 25}$

$= \sqrt{425} = 5\sqrt{17}$ m/s

$\tan\theta = \frac{20}{5} = 4 \Rightarrow \theta = \tan^{-1}(4)$

From −ve Y-axis.

05. *(d)* Given, projectile range = Maximum height

$\Rightarrow \frac{u^2 \sin 2\theta}{g} = \frac{u^2 \sin^2\theta}{2g} \Rightarrow \tan\theta = 4$

$\Rightarrow \theta = \tan^{-1}(4)$

06. *(c)* At $t = 2$ s, $g = 10 \text{ m/s}^2$

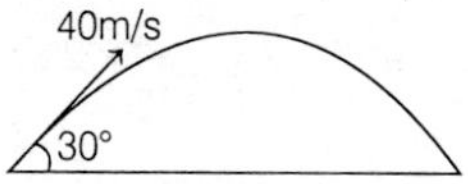

$u_x = 40\cos 30° = 20\sqrt{3}$

$u_y = 40\sin 30° = 20$

At $v_y = u_y + a_y t = 20 - 10 \times 2 = 0$

$\Rightarrow v = v_x\hat{\mathbf{i}} + v_y\hat{\mathbf{j}} = 20\sqrt{3}$ m/s

07. *(b)* Maximum height of projectile is given by,

$$H_{max} = \frac{u^2 \sin^2\theta}{2g}$$

Therefore, $\frac{H_{max_1}}{H_{max_2}} = \frac{(u^2 \sin^2 30°)/2g}{(u^2 \sin^2 60°)/2g}$

$$= \frac{\sin^2 30°}{\sin^2 60°} = \frac{\left(\frac{1}{2}\right)^2}{\left(\frac{\sqrt{3}}{2}\right)^2} = \frac{1}{3}$$

08. *(d)*

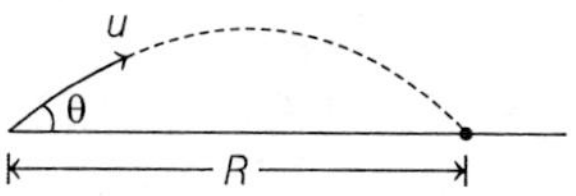

Range of a projectile, $R = \frac{u^2 \sin 2\theta}{g}$

$\Rightarrow$ Ratio of ranges of a projectile thrown at two different angles θ_2 and θ_1 is;

$$\frac{R_2}{R_1} = \frac{\sin(2\theta_2)}{\sin(2\theta_1)}$$

Here $R_1 = 50$ m, $\theta_1 = 15°$

and $\theta_2 = 45°$

So, $R_2 = R_1 \cdot \frac{\sin(2\theta_2)}{\sin(2\theta_1)} = 50 \times \frac{\sin 2(45°)}{\sin 2(15°)}$

$= \frac{50 \times \sin 90°}{\sin 30°} = 100$ m

09. *(b)*

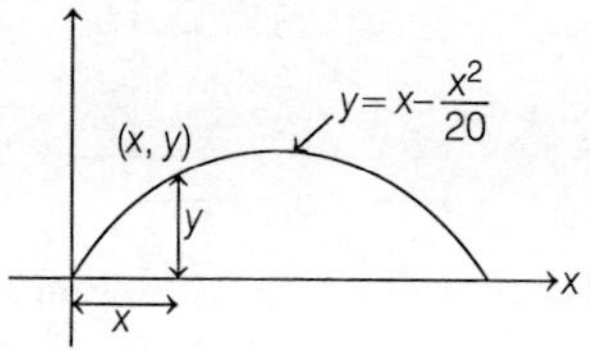

Height above ground is $y = x - \frac{x^2}{20}$

For maximum height,

$$\frac{dy}{dx} = 0 \Rightarrow \frac{d}{dx}\left(x - \frac{x^2}{20}\right) = 0$$

$\Rightarrow 1 - \frac{x}{10} = 0 \Rightarrow x = 10$ m

So, y_{max} (at $x = 10$ m)

$= 10 - \frac{10^2}{20} = 10 - 5 = 5$ m

10. *(c)* The given situation in question is as follows

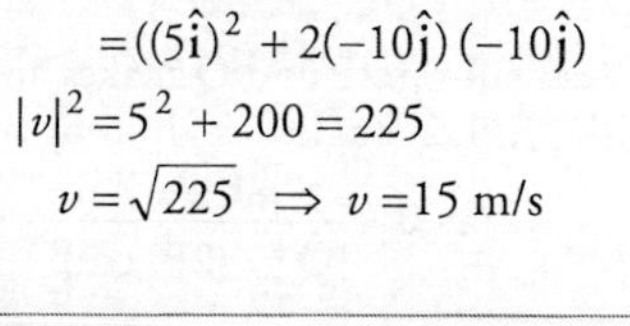

Let, the striking speed be v.

Using third equation of motion,

$v^2 = u^2 + 2as$

$= ((5\hat{\mathbf{i}})^2 + 2(-10\hat{\mathbf{j}})(-10\hat{\mathbf{j}})$

$|v|^2 = 5^2 + 200 = 225$

$\Rightarrow v = \sqrt{225} \Rightarrow v = 15$ m/s

11. *(d)*

> **key idea** You can directly calculate the maximum horizontal distance (R_{max}) for the known value of maximum vertical height (H_{max}) with following relation.
>
> $R_{max} = 2 \times H_{max}$

Maximum horizontal distance

= Maximum range

$= \left.\frac{u^2 \sin 2\theta}{g}\right|_{(\theta = 45°)} = \frac{u^2}{g}$

Maximum vertical height (from $v^2 - u^2 = 2gh$) is

$H_{max} = \frac{u^2}{2g} = 136$ m (given)

Hence, maximum horizontal distance

$= 2 \times H_{max} = 2 \times 136$

$= 272$ m

12. *(d)* Projection speed of ball,

$u = 15 \text{ ms}^{-1}$

Given that, $R = H$

$\Rightarrow \frac{u^2 \sin 2\theta}{g} = \frac{u^2 \sin^2\theta}{2g}$

$\Rightarrow \frac{\sin 2\theta}{1} = \frac{\sin^2\theta}{2}$

$\Rightarrow 2\sin\theta . \cos\theta = \frac{\sin^2\theta}{2}$

$[\because \sin 2\theta = 2\sin\theta \cdot \cos\theta]$

$\Rightarrow 4 = \frac{\sin\theta}{\cos\theta} \Rightarrow 4 = \tan\theta$

$\Rightarrow \tan\theta = 4$

> • Concept **Enhancer**
>
> If we throw a projectile at an angle of projection, $\theta = \tan^{-1}(4) = 76°$, then it covers horizontal range (R) equal to maximum height (H).

13. *(c)* The bomb is dropped from a fighter plane flying in horizontal direction only.

Let us assume the velocity of fighter plane, $\mathbf{v}_a = v_x\hat{\mathbf{i}}$. As fighter plane is moving in horizontal direction only. Hence, it has only one component.

When fighter plane drops the bomb, the bomb will free fall from the fighter plane and have a horizontal velocity in the direction of plane and vertical velocity due to gravity.

Velocity of bomb can be given as, $\mathbf{v}_b = v_x\hat{\mathbf{i}} - v_y\hat{\mathbf{j}}$. Velocity in vertical direction will be negative as bomb is moving downwards.

Now, for the person sitting inside the plane, the relative velocity of bomb with respect to plane can be calculated as

$\mathbf{v}_{ba} = \mathbf{v}_b - \mathbf{v}_a = v_x\hat{\mathbf{i}} - v_y\hat{\mathbf{j}} - v_x\hat{\mathbf{i}}$

$= -v_y\hat{\mathbf{j}}$

The velocity of bomb with respect to person sitting in plane is only in vertical direction. Thus, the trajectory of bomb with respect to the person will be a straight line vertically down to the plane.

14. *(a)* Position vector of particle is given as

$$\mathbf{r} = 15t^2\hat{\mathbf{i}} + (4 - 20t^2)\hat{\mathbf{j}}$$

Velocity of particle is

$$\mathbf{v} = \frac{d\mathbf{r}}{dt} = \frac{d}{dt}\,[15t^2\hat{\mathbf{i}} + (4 - 20t^2)\hat{\mathbf{j}}]$$

$$= 30t\,\hat{\mathbf{i}} - 40\,t\,\hat{\mathbf{j}}$$

Acceleration of particle is

$$\mathbf{a} = \frac{d}{dt}(\mathbf{v}) = \frac{d}{dt}(30t\hat{\mathbf{i}} - 40t\hat{\mathbf{j}})$$

$$= 30\hat{\mathbf{i}} - 40\hat{\mathbf{j}}$$

So, magnitude of acceleration at $t = 1$ s is

$$|\mathbf{a}|_{t=1\,\text{s}} = \sqrt{a_x^2 + a_y^2}$$

$$= \sqrt{30^2 + 40^2}$$

$$= 50\ \text{ms}^{-2}$$

15. *(b)* As, initial velocity $\mathbf{v} = (\mathbf{i} + 2\,\mathbf{j})$ m/s

For projectile

Motion along horizontal direction,

$$x = u_x t = t \qquad \ldots\text{(i)}$$

Motion along vertical direction,

$$y = u_y t - \frac{1}{2}gt^2$$

$$y = 2t - \frac{1}{2}10t^2 \qquad \ldots\text{(ii)}$$

$$= 2t - 5t^2$$

From Eqs. (i) and (ii), we get

$$y = 2x - 5x^2$$

16. *(d)* Given, $H_{\text{max}} = 10$ m

$$\therefore \quad R_{\text{max}} = 2\,H_{\text{max}}$$

$$= 2 \times 10 = 20\ \text{m}$$

17. *(b)* Given, $\mathbf{u} = 3\,\mathbf{i} + 4\,\mathbf{j}$,

$$\mathbf{a} = 0.4\,\mathbf{i} + 0.3\,\mathbf{j}$$

$$\mathbf{v} = \mathbf{u} + \mathbf{a}\,t$$

$$= 3\,\mathbf{i} + 4\,\mathbf{j} + (0.4\,\mathbf{i} + 0.3\,\mathbf{j})10$$

$$= 3\,\mathbf{i} + 4\mathbf{j} + 4\,\mathbf{i} + 3\,\mathbf{i} = 7\,\mathbf{i} + 7\,\mathbf{j}$$

So, speed is equal to magnitude of velocity $= \sqrt{7^2 + 7^2} = 7\sqrt{2}$ units

18. *(a)* Man will catch the ball, if the horizontal component of velocity becomes equal to the constant speed of man, i.e.,

$$v_0 \cos\theta = v_0/2 \ \text{ or } \ \cos\theta = 1/2$$

$$\Rightarrow \quad \cos\theta = \cos 60°$$

or $\quad \theta = 60°$

19. *(d)* The given situation is shown below

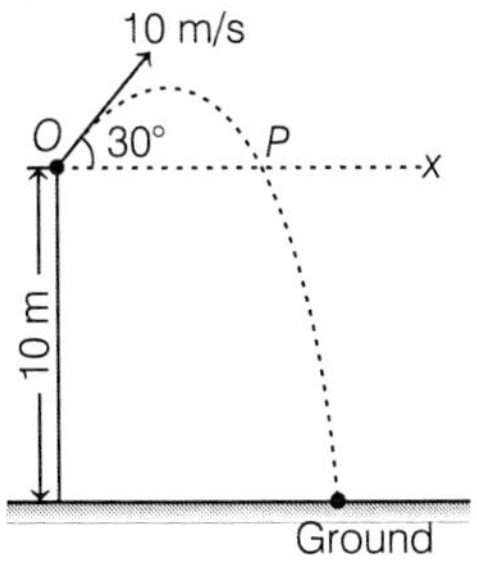

$$OP = R = \frac{u^2 \sin 2\theta}{g}$$

$$= \frac{10^2 \times \sin(2 \times 30°)}{10}$$

$$= \frac{10\sqrt{3}}{2} = 5\sqrt{3} = 8.66\ \text{m}$$

20. *(d)* $\theta_1 = 45° + \alpha,\ \theta_2 = 45° - \alpha$

Time of flight, $T = \dfrac{2u\sin\theta}{g}$

$$\frac{T_1}{T_2} = \frac{\sin(45° + \alpha)}{\sin(45° - \alpha)}$$

$$\frac{T_1}{T_2} = \frac{\sin 45° \cdot \cos\alpha + \cos 45° \cdot \sin\alpha}{\sin 45° \cdot \cos\alpha - \cos 45° \cdot \sin\alpha}$$

$$= \frac{\frac{1}{\sqrt{2}}\cos\alpha + \frac{1}{\sqrt{2}}\sin\alpha}{\frac{1}{\sqrt{2}}\cos\alpha - \frac{\sin\alpha}{\sqrt{2}}}$$

$$\frac{T_1}{T_2} = \frac{\cos\alpha + \sin\alpha}{\cos\alpha - \sin\alpha} = \frac{1 + \tan\alpha}{1 - \tan\alpha}$$

21. *(b)* According to given condition,

$$h_m = H_{\text{max}} = \frac{u^2 \sin^2(90° - \phi)}{2g}$$

$$\Rightarrow h_m = \frac{u^2\cos^2\phi}{2g}$$

At $\phi = 90°$, $h_m = 0'$

and at $\phi = 0°$, $h_m = \dfrac{u^2}{2g}$ (maximum)

22. *(d)*

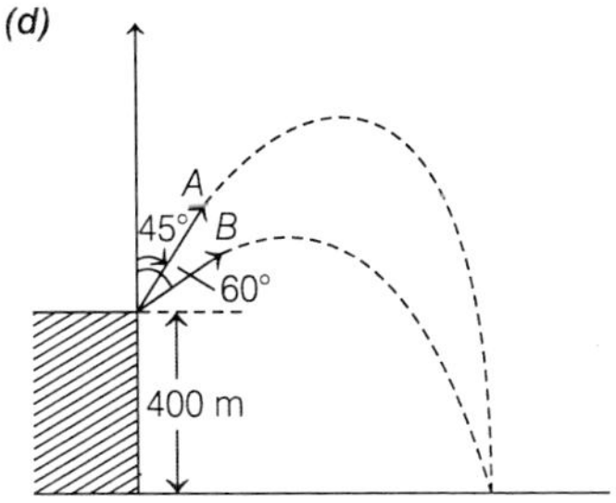

In vertical direction,

For projectile A,

$$u_{yA} = v_A \cdot \cos 45° = \frac{v_A}{\sqrt{2}}$$

$$s_{yA} = -400\ \text{m},\ a_{yA} = -10\ \text{m/s}^2$$

Using $s = ut + \dfrac{1}{2}at^2$ we have,

$$-400 = \frac{v_A}{\sqrt{2}} \times t + \frac{1}{2}(-10)t^2 \qquad \ldots\text{(i)}$$

Similarly for projectile B,

$$-400 = \frac{v_B}{2} \times t + \frac{1}{2}(-10)t^2 \qquad \ldots\text{(ii)}$$

From Eqs. (i) and (ii),

$$\frac{v_A}{\sqrt{2}}t = \frac{v_B}{2}t$$

or $\quad v_A : v_B = 1 : \sqrt{2}$

Alternate Solution

Since, $\quad R_A = R_B$

$(v_A \cos 45°)T = (v_B \cos 60°)T$

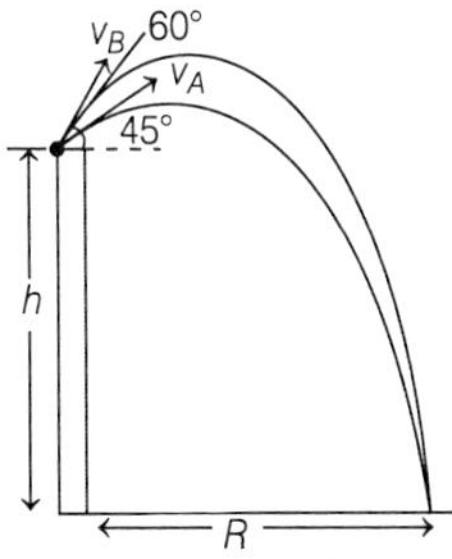

$$\Rightarrow \quad \frac{v_A}{\sqrt{2}} = \frac{v_B}{2} \Rightarrow \frac{v_A}{v_B} = \frac{\sqrt{2}}{2} = \frac{1}{\sqrt{2}}$$

23. *(c)* During the projectile motion, horizontal component of velocity does not change.

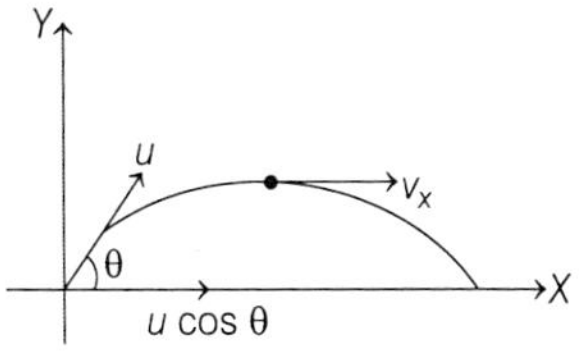

Given that, $v_x = \dfrac{\sqrt{3}}{2}u$

$$u\cos\theta = \frac{\sqrt{3}}{2}u \Rightarrow \cos\theta = \frac{\sqrt{3}}{2}$$

$$\theta = 30°$$

Time of flight,

$$T = \frac{2u\sin\theta}{g} = \frac{2u\sin 30°}{g}$$

$$= \frac{2u}{g} \times \frac{1}{2} \Rightarrow T = \frac{u}{g}$$

24. *(d)* Since, after t seconds, its inclination becomes 0° i.e., $\theta = 0°$, hence it concludes that projectile reaches at maximum height after t seconds.

$$\therefore \quad t = \frac{\text{time of flight}}{2} = \frac{2u\sin\theta}{2g}$$

$$t = \frac{u\sin\theta}{g} \Rightarrow u = \frac{tg}{\sin\theta} \quad \ldots(i)$$

Again we know that, horizontal range,

$$R = \frac{u^2 \sin 2\theta}{g}$$

$$= \frac{\frac{t^2 g^2}{\sin^2\theta} \times 2\sin\theta\cos\theta}{g}$$

[From Eq. (i)]

$$\Rightarrow \quad R = 2t^2 g \cot\theta$$

$$\Rightarrow \quad \frac{1}{\cot\theta} = \frac{2t^2 g}{R} = \frac{2t^2 \times 10}{R} = \frac{20t^2}{R}$$

$$\Rightarrow \cot\theta = \frac{R}{20t^2}$$

$$\Rightarrow \quad \theta = \cot^{-1}\left(\frac{R}{20t^2}\right)$$

25. *(b)* We know that,

$$\theta_1 + \theta_2 = 90°$$

Complementary angles

Therefore,

$$R_1 = R_2$$

Now, as we know that

$$H = \frac{u^2 \sin^2\theta}{2g}$$

$$\theta_1 < \theta_2$$

$$\sin\theta_1 < \sin\theta_2$$

Therefore, $H_1 < H_2$.

26. *(c)* The given condition can be drawn in the figure given below

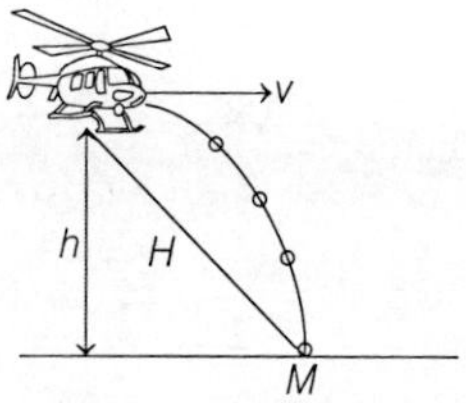

Initial velocity of food packet along Y-axis, $u_y = 0$

Since, $h = u_y t + \frac{1}{2} g t^2$

$$\therefore \quad t = \sqrt{2h/g}$$

and distance travelled by helicopter in time t along horizontal direction

$$\Rightarrow \quad x = u_x t = v\sqrt{\frac{2h}{g}}$$

Hence, distance of helicopter from man,

$$H = \sqrt{h^2 + x^2} = \sqrt{h^2 + \left(v\sqrt{\frac{2h}{g}}\right)^2}$$

$$= \sqrt{h^2 + \frac{v^2 2h}{g}}$$

27. *(a)* Given, equation of trajectory of projectile,

$$y = \alpha x - \beta x^2 \quad \ldots (i)$$

The standard equation of trajectory of projectile,

$$y = x\tan\theta - \frac{gx^2}{2u^2\cos^2\theta} \quad \ldots (ii)$$

Now, from Eqs. (i) and (ii), we get

$$\alpha = \tan\theta$$

and $$\beta = \frac{g}{2u^2\cos^2\theta} \Rightarrow \alpha^2 = \frac{\sin^2\theta}{\cos^2\theta}$$

$$\therefore \quad \frac{\beta}{\alpha^2} = \frac{g}{2u^2\cos^2\theta \times \frac{\sin^2\theta}{\cos^2\theta}}$$

$$= \frac{g}{2u^2\sin^2\theta} \times \frac{2}{2}$$

$$\Rightarrow \quad \frac{4\beta}{\alpha^2} = \frac{2g}{u^2\sin^2\theta} = \frac{1}{H}$$

$$\Rightarrow \quad H = \alpha^2/4\beta$$

$$\therefore \quad \theta = \tan^{-1}\alpha$$

and $$H = \alpha^2/4\beta$$

28. *(a)* $\mathbf{u} = (5\hat{j})$ m/s $= (0\hat{i} + 5\hat{j})$ m/s

$$\mathbf{a} = (10\hat{i} + 4\hat{j}) \text{ m/s}^2$$

$$\mathbf{s} = [(20 - 0)\hat{i} + (y_0 - 0)\hat{j}] \text{ m}$$

$$= (20\hat{i} + y_0\hat{j}) \text{ m}$$

Using second equation of motion in x-direction,

$$s_x = u_x t + \frac{1}{2} a_x t^2$$

$$20 = (0)t + \frac{1}{2}(10)(t)^2$$

$$20 = 0 + 5t^2 \Rightarrow t^2 = 4$$

$$\Rightarrow \quad t = 2 \text{ s}$$

Using second equation of motion in y-direction,

$$s_y = u_y t + \frac{1}{2} a_y t^2$$

$$= 5 \times 2 + \frac{1}{2} \times 4 \times (2)^2$$

$$y_0 = 10 + 8 = 18 \text{ m}$$

29. *(d)* Given, $\mathbf{v} = ky\,\hat{i} + kx\,\hat{j}$

$$\Rightarrow \frac{dx}{dt}\hat{i} + \frac{dy}{dt}\hat{j} = ky\hat{i} + kx\hat{j}$$

$$\Rightarrow \quad \frac{dx}{dt} = ky, \frac{dy}{dt} = kx$$

$$\frac{dy}{dx} = \frac{dy}{dt} \times \frac{dt}{dx} = \frac{kx}{ky}$$

$$\Rightarrow \quad y\,dy = x\,dx$$

Integrating and using

$$\int x^n\,dx = \frac{x^{n+1}}{n+1}, \text{ we have}$$

$$y^2 = x^2 + c$$

30. *(d)* Maximum height $= \frac{u^2\sin^2\theta}{2g}$

For $\theta = 45° - \alpha$

$$H_1 = \frac{u^2\sin^2(45° - \alpha)}{2g}$$

for $\theta = 45° + \alpha$

$$H_2 = \frac{u^2\sin(45° + \alpha)}{2g}$$

$$\frac{H_1}{H_2} = \frac{\sin^2(45° - \alpha)}{\sin^2(45° + \alpha)}$$

$$= \frac{1 - \cos(90° - 2\alpha)}{1 - \cos(90° + 2\alpha)} = \frac{1 - \sin 2\alpha}{1 + \sin 2\alpha}$$

31. *(a)* In T time particle covers a distance $2\pi R$.

$\therefore$ Speed of particle is, $u = \frac{2\pi R}{T}$

Now, when particle is projected with speed u at angle θ, then it's maximum height is $4R$.

Then, $$H_{max} = 4R$$

$$\frac{u^2\sin^2\theta}{2g} = 4R$$

$$\Rightarrow \quad \frac{(2\pi R)^2\sin^2\theta}{T^2 \times 2g} = 4R$$

$$\Rightarrow \quad \sin^2\theta = \frac{4R \times T^2 \times 2g}{4\pi^2 R^2}$$

$$\Rightarrow \quad \theta = \sin^{-1}\sqrt{\frac{2gT^2}{\pi^2 R}}$$

32. *(d)* Given situation is shown below,

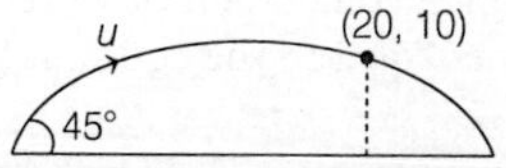

Equation of trajectory of a projectile is given as,

$$y = x\tan\theta - \frac{1}{2}\frac{gx^2}{u^2\cos^2\theta}$$

$$\Rightarrow 10 = 20 \tan 45° - \frac{1 \times 10 \times (20)^2}{2u^2 (\cos 45°)^2}$$

$\Rightarrow \quad u = 20$ m/s

Time of flight,

$$T = \frac{2u \sin \theta}{g}$$

$$= \frac{2 \times 20 \times \sin 45°}{10} = 2\sqrt{2}$$

Momentum at $t = \frac{T}{\sqrt{2}} = 2$ s is

$$\mathbf{p} = m\,[u \cos \theta \hat{\mathbf{i}} + (u \sin \theta - gt)\,\hat{\mathbf{j}}]$$

$$= 10\,[20 \cos 45° \hat{\mathbf{i}} + (20 \sin 45° - 10 \times 2)\hat{\mathbf{j}}]$$

$$= 100\sqrt{2}\,\hat{\mathbf{i}} + (100\sqrt{2} - 200)\hat{\mathbf{j}}$$

33. *(b)* At $t = 0$, the motion of projectile is given as

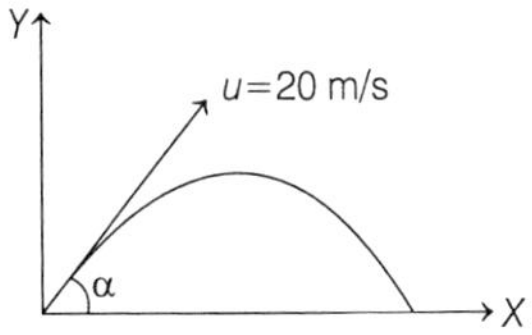

$$\tan \alpha = \frac{u_y}{u_x} \quad \ldots \text{(i)}$$

At $t = 10$ s, the motion of projectile is given as

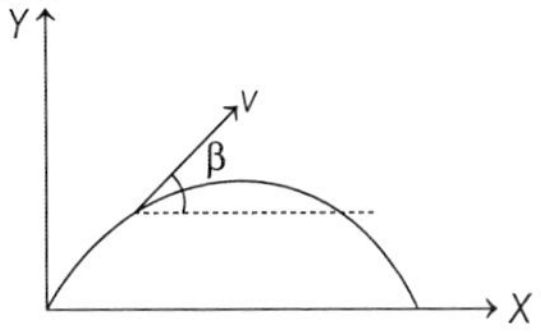

$$\tan \beta = \frac{v_y}{v_x} \quad \ldots \text{(ii)}$$

where, v_y is the vertical component of final velocity after $t = 10$ s and v_x is the horizontal component of final velocity.

From Eqs. (i) and (ii), we get

$$\frac{\tan \beta}{\tan \alpha} = \frac{v_y}{u_y} \quad (\text{as } v_x = u_x) \quad \ldots \text{(iii)}$$

Applying second equation of motion in vertical direction, we get

$$v_y = u_y - gt$$

$$\therefore \quad v_y = u_y - 100 \quad \ldots \text{(iv)}$$

Using Eq, (iii)

$$\frac{\tan \beta}{\tan \alpha} = \frac{u_y - 100}{u_y}$$

$$\frac{\tan \beta}{\tan \alpha} = 1 - \frac{100}{u_y} = 1 - \frac{100}{20 \sin \alpha}$$

$$(\because u_y = 20 \sin \alpha)$$

$$= 1 - \frac{5}{\sin \alpha}$$

$$\Rightarrow \quad \tan \beta = \tan \alpha \left(1 - \frac{5}{\sin \alpha}\right)$$

$$= \tan \alpha - 5 \sec \alpha$$

34. *(d)*

> **key idea** Range of a projectile motion is given by $R = \frac{u^2 \sin 2\theta}{g}$ and time of flight is given by $t = \frac{2u \sin \theta}{g}$

Let θ_1 and θ_2 are the two angles at which shell is fired. As, range in both cases is same, i.e. $R_1 = R_2 = R$

Here,

$$R_1 = \frac{u^2 \sin 2\theta_1}{g} \text{ and } R_2 = \frac{u^2 \sin 2\theta_2}{g}$$

$$\Rightarrow \quad R = \frac{u^2 \sin 2\theta_1}{g} = \frac{u^2 \sin 2\theta_2}{g} \quad \ldots \text{(i)}$$

$$\Rightarrow \quad \sin 2\theta_1 = \sin 2\theta_2$$

$$\Rightarrow \quad \sin 2\theta_1 = \sin(180° - 2\theta_2)$$

$$[\because \sin(180° - \theta) = \sin \theta]$$

$$\Rightarrow 2(\theta_1 + \theta_2) = 180° \text{ or } \theta_1 + \theta_2 = 90°$$

$$\Rightarrow \quad \theta_2 = 90° - \theta_1 \quad \ldots \text{(ii)}$$

So, time of flight in first case,

$$t_1 = \frac{2u \sin \theta_1}{g} \quad \ldots \text{(iii)}$$

and time of flight in second case,

$$t_2 = \frac{2u \sin \theta_2}{g}$$

$$= \frac{2u \sin(90° - \theta_1)}{g}$$

$$= \frac{2u \cos \theta_1}{g} \quad \ldots \text{(iv)}$$

From Eqs. (iii) and (iv), we get

$$t_1 t_2 = \frac{2u \sin \theta_1}{g} \times \frac{2u \cos \theta_1}{g}$$

$$\Rightarrow \quad t_1 t_2 = \frac{4u^2 \sin \theta_1 \cos \theta_1}{g^2}$$

$$\Rightarrow \quad t_1 t_2 = \frac{2u^2 \sin 2\theta_1}{g^2}$$

$$(\because \sin 2\theta = 2 \sin \theta \cos \theta) \quad \ldots \text{(v)}$$

From Eq. (i), we get

$$R = \frac{u^2 \sin 2\theta_1}{g}$$

$$\therefore \quad t_1 t_2 = \frac{2R}{g}$$

35. *(b)*

> **key idea** For two particles thrown with same initial speed when they are projected at complementary angle as.
>
> $\therefore \quad \theta_1 + \theta_2 = 90°$
>
> (where, θ_1 and θ_2 are angles of projection.)

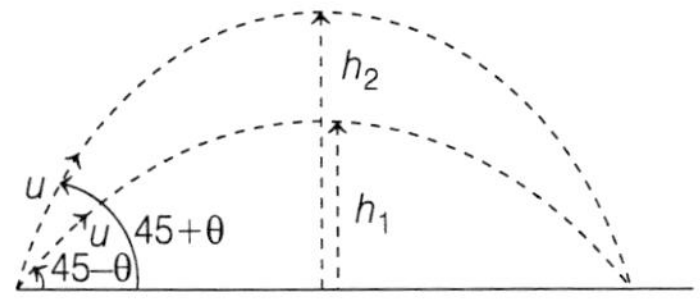

As maximum range occurs at $\theta = 45°$ for a given initial projection speed, we take angles of projection of two particles as

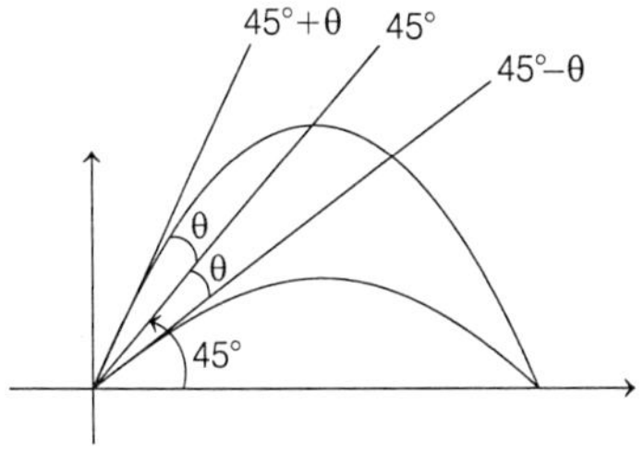

$$\theta_1 = 45° + \theta, \quad \theta_2 = 45° - \theta$$

where, θ is angle of projectiles with 45° line.

So, range of projectiles will be

$$R = R_1 = R_2 = \frac{u^2 \sin 2(\theta_1)}{g}$$

$$\Rightarrow \quad R = \frac{u^2 \sin 2(45° + \theta)}{g}$$

$$\Rightarrow \quad R = \frac{u^2 \sin(90° + 2\theta)}{g}$$

$$\Rightarrow \quad R = \frac{u^2 \cos 2\theta}{g}$$

$$\Rightarrow \quad R^2 = \frac{u^4 \cos^2 2\theta}{g^2} \quad \ldots \text{(i)}$$

Maximum heights achieved in two cases are

$$h_1 = \frac{u^2 \sin^2(45° + \theta)}{2g}$$

and

$$h_2 = \frac{u^2 \sin^2(45° - \theta)}{2g}$$

$$\text{So, } h_1 h_2 = \frac{u^4 \sin^2(45° + \theta) \sin^2(45° - \theta)}{4g^2}$$

Using $2 \sin A \cdot \sin B = \cos(A - B) - \cos(A + B)$,

we have

$$\sin(45° + \theta)\sin(45° - \theta) = \frac{1}{2}(\cos 2\theta - \cos 90°)$$

$$\Rightarrow \sin(45° + \theta)\sin(45° - \theta) = \frac{\cos 2\theta}{2}$$

$[\because \cos 90° = 0]$

So, we have

$$h_1 h_2 = \frac{u^4\left(\frac{\cos 2\theta}{2}\right)^2}{4g^2}$$

$$\Rightarrow \quad h_1 h_2 = \frac{u^4 \cos^2 2\theta}{16 g^2} \quad \ldots(ii)$$

From Eqs. (i) and (ii), we get

$$\Rightarrow \quad h_1 h_2 = \frac{R^2}{16}$$

$$\Rightarrow R^2 = 16\, h_1 h_2$$

• Concept Enhancer

For same range, 2 angles are complementary

$\theta_1 = \theta$ and $\theta_2 = (90 - \theta)$

$$R = R_1 = \frac{u^2 \sin 2\theta}{g},$$

$$R_2 = R = \frac{u^2 \sin 2(90° - \theta)}{g} = \frac{u^2 \sin 2\theta}{g}$$

$$R^2 = \frac{u^4 \sin 2\theta \sin 2\theta}{g^2} \quad \ldots(i)$$

$$h_1 = \frac{u^2 \sin^2\theta}{2g};\ h_2 = \frac{u^2 \sin^2(90 - \theta)}{2g} = \frac{u^2 \cos^2\theta}{2g}$$

$$h_1 h_2 = \frac{u^4}{4g^2}\sin^2\theta \cos^2\theta \quad \ldots(ii)$$

Divide Eq. (i) by Eq. (ii)

$$\frac{R^2}{h_1 h_2} = \frac{u^4 \sin 2\theta \sin 2\theta}{g^2 \times \frac{u^4 \sin^2\theta \cos^2\theta}{4g^2}}$$

$$= \frac{4 \times 2\sin\theta\cos\theta \times 2\sin\theta\cos\theta}{\sin^2\theta\cos^2\theta}$$

$$R^2 = 16 h_1 h_2$$

36. *(5)*

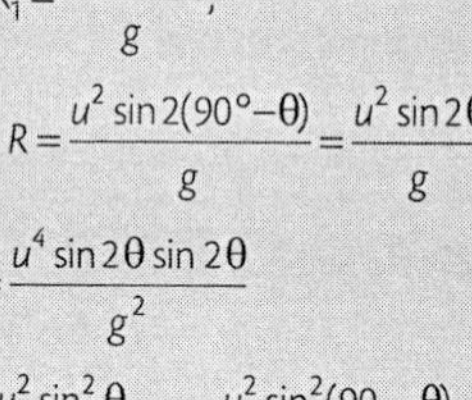

Using 2nd equation of motion for first second,

$$s = ut + \frac{1}{2}at^2$$

$$h_0 = 30 \times 1 - \frac{1}{2} \times 10 \times 1 = 25$$

Time to reach maximum height

$$t = \frac{T}{2}$$

$$= \frac{u \sin\theta}{g} = 3\text{s}$$

For last second

$$h_1 = 30 + \left(\frac{-10}{2}\right) \times (2 \times 3 - 1) = 5$$

$$\frac{h_0}{h_1} = \frac{25}{5} = 5$$

37. *(2000)* $v_r = 9 \text{ km/h} = \frac{9 \times 5}{18} = 2.5 \text{ m/s}$

v_b in still water = 27 km/h

$$= 27 \times \frac{5}{18} = 7.5 \text{ m/s}$$

Velocity of boat in moving river as observed from ground

$$= 2.5 + 7.5 = 10 \text{ m/s}$$

This becomes horizontal speed of ball i.e. 10 m/s

Vertical speed of ball is 10 m/s

Net speed of ball

$$= \sqrt{10^2 + 10^2} = 10\sqrt{2} \text{ m/s}$$

Angle of throw as observed from ground, $\tan\theta = \frac{u_y}{u_x} = \frac{10}{10} = 1$

$$\theta = 45°$$

$$\text{Range} = \frac{u^2 \sin 2\theta}{g}$$

$$= \frac{(10\sqrt{2})^2 \sin 2 \times 45°}{10} = \frac{200}{10} = 20 \text{ m}$$

$= 2000$ cm

38. *(100)* For ball of mass M, time to reach ground

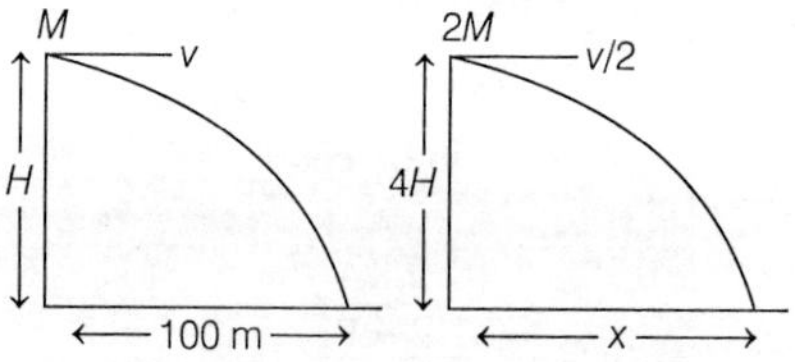

Case 1 Case 2

$$t_1 = \sqrt{\frac{2H}{g}} \quad \left(\text{using } H = \frac{1}{2}gt^2\right)$$

So, horizontal distance covered

$$= 100 = v_x \times t_1$$

$$\Rightarrow \quad 100 = v \times \sqrt{\frac{2H}{g}} \quad \ldots(i)$$

For ball of mass $2M$,

Time to reach ground

$$t_2 = \sqrt{\frac{2(4H)}{g}} = 2\sqrt{\frac{2H}{g}}$$

Horizontal distance covers by 2nd ball

$$= v_x \times t_2 = \frac{v}{2} \times 2\sqrt{\frac{2H}{g}}$$

$$= v\sqrt{\frac{2H}{g}} = 100 \text{ m [from Eq, (i)]}$$

39. *(16)* Maximum height reached by the projectile,

$$H_{max} = \frac{u^2 \sin^2\theta}{2g}\bigg|_{\theta = 90°} = \frac{u^2}{2g}$$

$$\therefore \frac{(H_{max})_1}{(H_{max})_2} = \left(\frac{u_1}{u_2}\right)^2 = \left(\frac{u}{u/2}\right)^2 = 4$$

$$\therefore (H_{max})_2 = \frac{(H_{max})_1}{4} = \frac{64}{4} = 16 \text{ m}$$

40. *(2)*

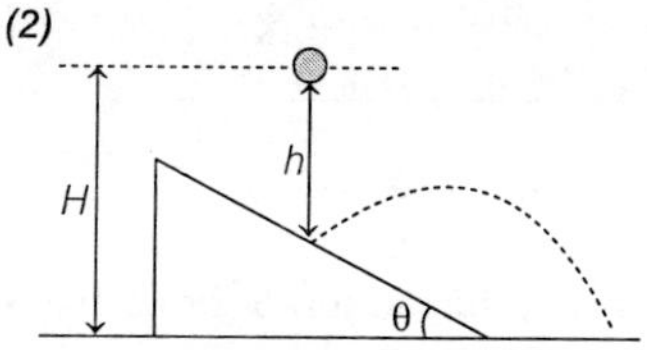

Time taken by ball to fall to ground is, t = time to drop by $(H - h)$ + time to drop by h

$$\Rightarrow \quad t = \sqrt{\frac{2(H-h)}{g}} + \sqrt{\frac{2h}{g}}$$

$$\left[\text{Using } s = \frac{1}{2}at^2\right]$$

Now, $\frac{dt}{dh} = 0$ for maximum t.

$$\frac{dt}{dh} = \frac{\sqrt{2}}{\sqrt{g}}\left(\frac{1}{2\sqrt{h}} - \frac{1}{2\sqrt{H-h}}\right) = 0$$

or $\quad \sqrt{h} = \sqrt{H - h}$

$$\Rightarrow \quad 2h = H \quad \text{or} \quad \frac{H}{h} = 2$$

41. *(2)*

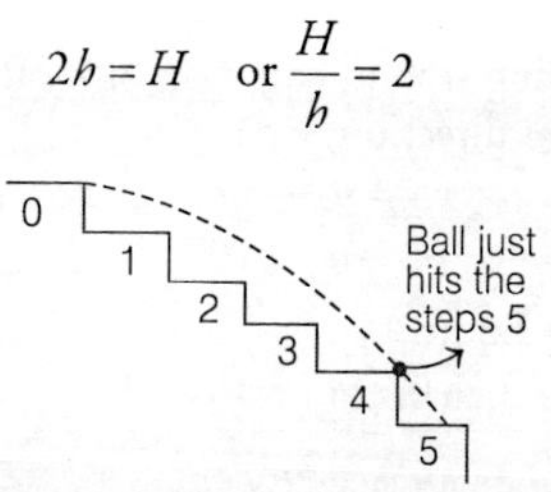

The ball needs to just cross 4 steps to just hit 5th step.

Therefore, horizontal range, $R = 0.4$ m

$$R = ut,\ t = R/u$$

Similarly, in vertical direction

$$h=\frac{1}{2}gt^2 \Rightarrow 0.4=\frac{1}{2}gt^2$$

$$0.4=\frac{1}{2}g\left(\frac{R}{u}\right)^2$$

$$0.4=\frac{1}{2}\times\frac{0.4\times0.4}{u^2}\times10$$

$$u^2=0.2\times10=2$$

$$u=\sqrt{2}\text{ m/s}$$

Therefore, $x=2$

42. *(80)*

> **key idea** If two bodies are projected with same speed and cover equal horizontal range, then their angle of projection must be complementary, i.e., if $\theta_1=\theta$, then $\theta_2=90°-\theta$.

For same range if projection speed for two projectiles are same, then angles are complementary.

For projectile-1,

$\theta_1=60°$, $u=40$ m/s

Maximum height,

$$H_1=\frac{u^2\sin^2\theta}{2g}$$

$$=\frac{(40)^2\times\sin^2 60°}{2\times10}$$

$$H_1=60\text{ m}$$

For projectile-2,

$\theta_2=90°-\theta_1=90°-60°=30°$, $u=40$ m/s

Maximum height,

$$H_2=\frac{u^2\sin^2 30°}{2g}=\frac{(40)^2}{2\times10}\times\frac{1}{4}=20\text{ m}$$

Sum of heights,

$H_1+H_2=60+20=80$ m

43. *(80)* Time of flight $=5+3=8$ s

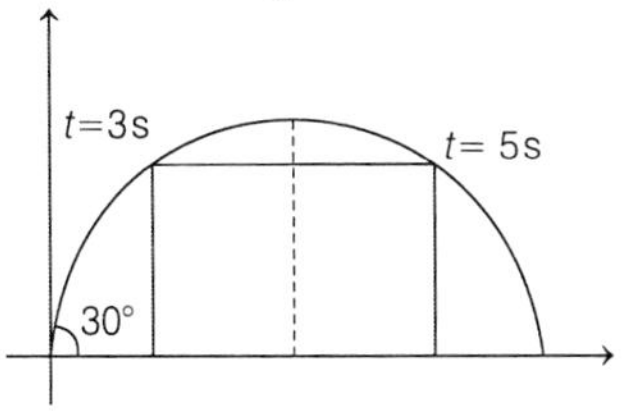

$$T=\frac{2u\sin\theta}{g}$$

$$\Rightarrow\quad 8=\frac{2u\sin30°}{10}\Rightarrow u\sin30°=40$$

$$\Rightarrow\quad u\times\frac{1}{2}=40$$

$$\Rightarrow\quad u=80\text{ m/s}$$

44. *(20)* Given, angle of projection, $\theta=45°$

Velocity after $t=2$s, $v'=20$ m/s

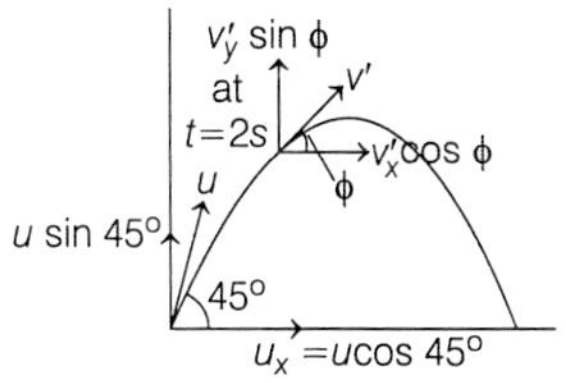

The resultant velocity at time t after the projection,

$$v'=\sqrt{u^2+g^2t^2-2ugt\sin45°}$$

$$20=\sqrt{u^2+10^2\times2^2-2u\times10\times2\times\frac{1}{\sqrt{2}}}$$

$$\Rightarrow\quad 400=u^2+400-20\sqrt{2}u$$

$$\Rightarrow u^2-20\sqrt{2}u=0\Rightarrow u(u-20\sqrt{2})=0$$

But $u\neq0$, $\therefore u-20\sqrt{2}=0$

$$\Rightarrow\quad u=20\sqrt{2}\text{ m/s}$$

$\therefore$ Maximum height,

$$H_{max}=\frac{u^2\sin^2\theta}{2g}$$

$$=\frac{(20\sqrt{2})^2\sin^2 45°}{2\times10}=\frac{800\times\frac{1}{2}}{20}=20\text{ m}$$

Topic 3

01. *(d)* As, velocity of eastward $\mathbf{v}_1=+5\,\mathbf{i}$

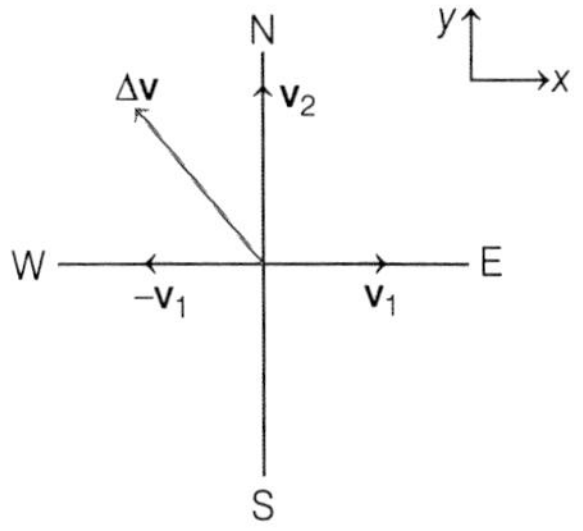

and velocity of Northward

$$\mathbf{v}_2=+5\,\mathbf{j}$$

$$\Delta\mathbf{v}=\mathbf{v}_2-\mathbf{v}_1=5\,\mathbf{j}-5\,\mathbf{i}$$

$$|\Delta\mathbf{v}|=\sqrt{5^2+5^2}=5\sqrt{2}$$

$$\therefore\quad a=\frac{|\Delta\mathbf{v}|}{t}=\frac{5\sqrt{2}}{10}=\frac{1}{\sqrt{2}}\text{ ms}^{-2}$$

For direction,

$$\tan\alpha=-\frac{5}{5}=-1\Rightarrow\alpha=135°$$

Average acceleration is $\frac{1}{\sqrt{2}}$ ms^{-2} towards North-West.

02. *(d)* The given situation can be represented as

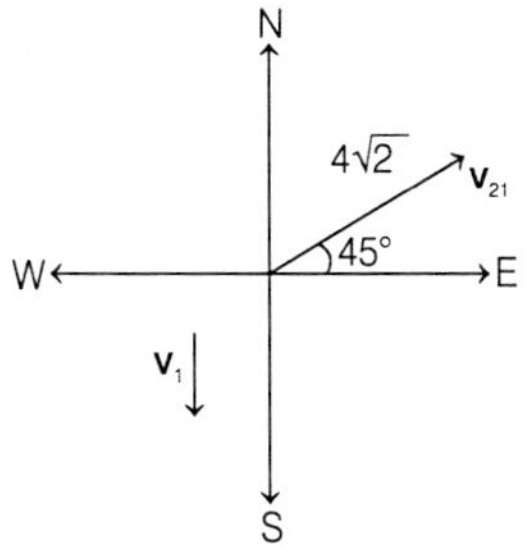

In the above figure, $\mathbf{v}_1$ is the speed of wind and $\mathbf{v}_{21}$ is the speed of butterfly with respect to wind.

So, $\mathbf{v}_{21}$ can be given as

$$\mathbf{v}_{21}=4\sqrt{2}\cos45°\hat{\mathbf{i}}+4\sqrt{2}\sin45°\hat{\mathbf{j}}$$

$$=4\sqrt{2}\times\frac{1}{\sqrt{2}}\hat{\mathbf{i}}+4\sqrt{2}\times\frac{1}{\sqrt{2}}\hat{\mathbf{j}}$$

$$=4\hat{\mathbf{i}}+4\hat{\mathbf{j}}$$

and $\mathbf{v}_1$ can be given as $\mathbf{v}_1=-\hat{\mathbf{j}}$

$\therefore$ Velocity of butterfly can be given as

$$\mathbf{v}_2=\mathbf{v}_1+\mathbf{v}_{21}=4\hat{\mathbf{i}}+4\hat{\mathbf{j}}-\hat{\mathbf{j}}$$

$$=4\hat{\mathbf{i}}+3\hat{\mathbf{j}}$$

$\therefore$ Displacement of butterfly,

$$D=\mathbf{v}_2\times t=(4\hat{\mathbf{i}}+3\hat{\mathbf{j}})\times3=12\hat{\mathbf{i}}+9\hat{\mathbf{j}}$$

$\therefore$ Magnitude of displacement,

$$|\mathbf{D}|=\sqrt{12^2+9^2}=15\text{ m}$$

03. *(a)*

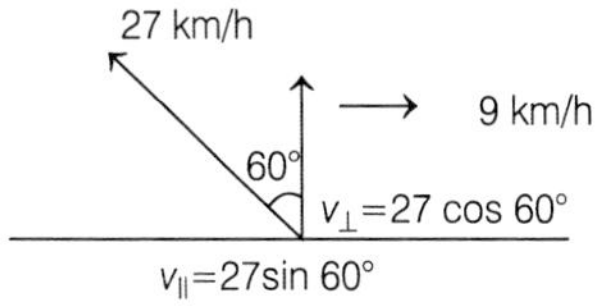

$$v_1=27\cos60°=27\times\frac{1}{2}=\frac{27}{2}\text{ km/h}$$

Time taken $=30$ s

$$s=vt=\frac{27}{2}\times\frac{5}{18}\times30=112.5\text{ m}$$

04. *(d)* Given, rain drop is falling vertically downwards (w.r.t ground).

So, velocity of rain $\mathbf{v}_r=v_r(-\hat{\mathbf{j}})$,

velocity of man $\mathbf{v}_m=v\hat{\mathbf{i}}$.

Velocity of rain w.r.t to man

$$\mathbf{v}_{r/m}=\mathbf{v}_r-\mathbf{v}_m$$

Drawing velocity vectors

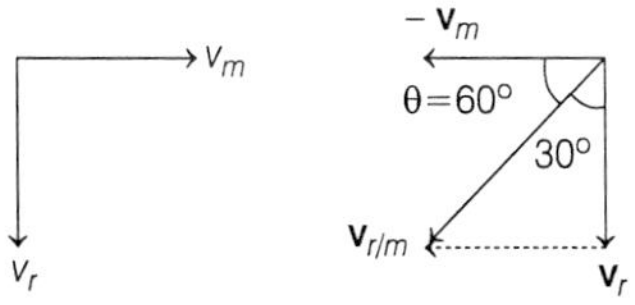

From diagram,

$$\tan 30° = \frac{v_m}{v_r}$$

$$\Rightarrow \frac{1}{\sqrt{3}} = \frac{v_m}{v_r} \text{ or } v_r = \sqrt{3}v$$

Now, man increases velocity of car to $(1+\beta)v$ in same direction, then

$$\mathbf{v}'_m = (1+\beta)v\hat{\mathbf{i}}$$

Drawing velocity vectors

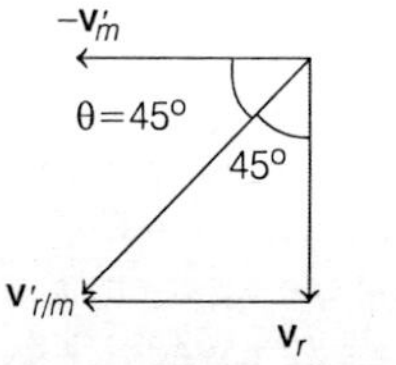

$$\tan 45° = \frac{v'_m}{v_r} \Rightarrow 1 = \frac{(1+\beta)v}{\sqrt{3}v}$$

$$\Rightarrow 1+\beta = \sqrt{3} \text{ or } \beta = \sqrt{3}-1 = 0.73$$

05. *(3)* Given that, Speed of swimmer with respect to ground.

$$v_{MR} = 4\text{km/h}$$

Width of river,

$$w = 1 \text{ km}$$

Drift in downstream,

$$x = 750 \text{ m or } 0.75 \text{ km}$$

Swimmer makes his strokes normal to the flow of river and due to addition of river velocity to swimmer, man has drift.

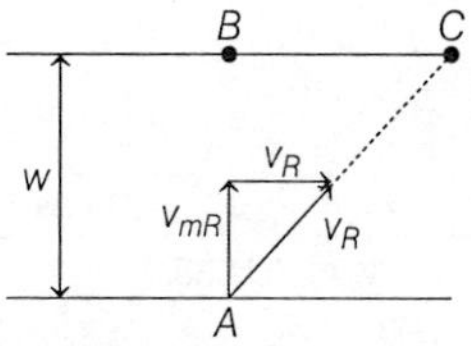

Time of crossing,

$$t = \frac{w}{v_{MR}} \Rightarrow t = \frac{1}{4} \text{ h}$$

Drift, $x = v_R \times t$

$$\text{or} \quad v_R = \frac{x}{t} = \frac{0.75}{1/4} = 3 \text{ km/h}$$

06. *(30)* Let v_M the velocity of swimmer with respect to ground, v_{MR} be the velocity of swimmer with respect to river and v_R be the velocity of river

$$\therefore \quad \mathbf{v}_M = \mathbf{v}_{MR} + \mathbf{v}_R$$

$\mathbf{v}_M$ should be along line AB.

As per question,

Magnitude of velocity of swimmer = Magnitude of velocity of river

$$\Rightarrow \quad |\mathbf{v}_{MR}| = |\mathbf{v}_R|$$

If we represent the given scenario as follows

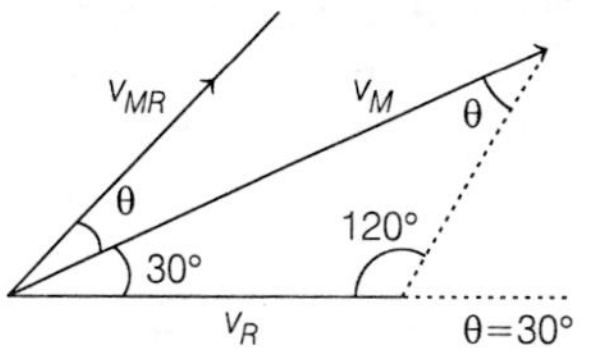

$\therefore v_M$ should be along angle of bisector of angle between $\mathbf{v}_{MR}$ and $\mathbf{v}_R$.

$$\therefore \quad \theta = 30°$$

07. *(5)* Given,

A person's swimming speed with respect to the river, $v_{mr} = 10$ m/s

As shown in the figure,

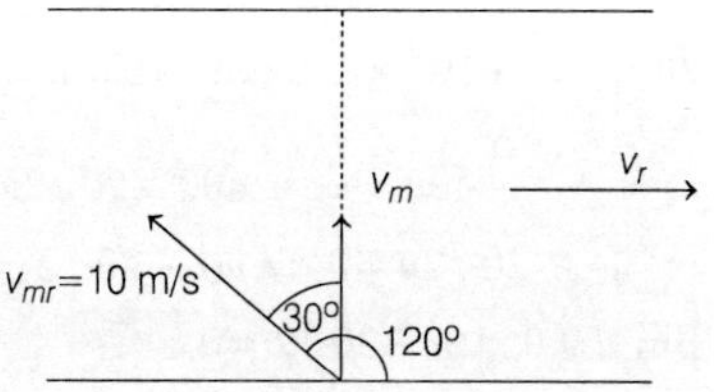

Let's determine the speed of the river flow, v_r

In x-direction,

$$v_r = v_{mr} \sin 30°$$
$$v_r = 10 \times \sin 30°$$
$$v_r = 5 \text{ m/s}$$

Hence, the speed of the river flow is 5 m/s. So, the value of the x to the nearest integer is 5.

08. *(120)* v_{MR} = velocity of man

$$= 12 \text{ km/h}$$

and v_R = velocity of water flow in river

$$= 6 \text{ km/h}$$

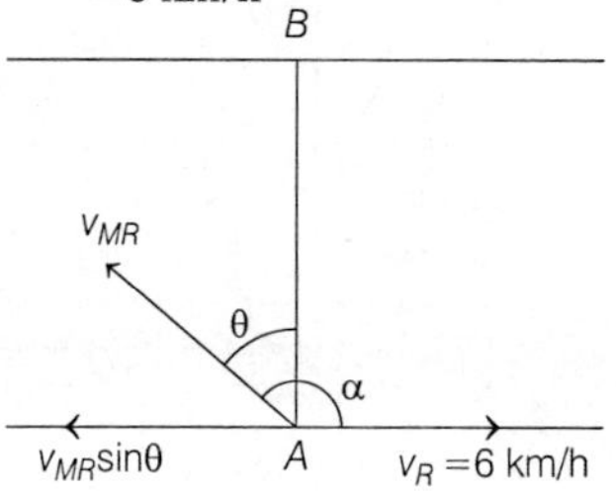

$$\Rightarrow v_R - v_{MR} \sin\theta = 0$$

$$\Rightarrow 6 - 12\sin\theta = 0$$

$$\Rightarrow \quad \sin\theta = \frac{6}{12} \Rightarrow \sin\theta = \frac{1}{2}$$

$$\Rightarrow \quad \theta = \sin^{-1}\left(\frac{1}{2}\right) = \sin^{-1}(\sin 30°)$$

$$\left[\because \sin 30° = \frac{1}{2}\right]$$

$$\Rightarrow \quad \theta = 30°$$

$$\therefore \quad \alpha = 90° + \theta = 90° + 30° = 120°$$

$$\Rightarrow \quad \alpha = 120°$$

09. *(580)* Let particle A moves along X-axis such that

$$x = 10 + 8t - 3t^2$$

and let particle B moves over Y-axis such that

$$y = 5 - 8t^3$$

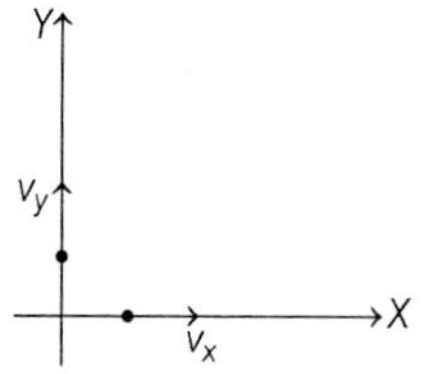

Velocities of particles A and B are

$$v_A = v_x = \frac{dx}{dt} = \frac{d}{dt}(10 + 8t - 3t^2)$$
$$= 8 - 6t$$

or v_A at $t = 1$ s, $v_A = 8 - 6 \times 1 = 2 \text{ ms}^{-1}$

$$\Rightarrow \quad \mathbf{v}_A = +2\hat{\mathbf{i}} \text{ ms}^{-1}$$

Velocity of particle B,

$$v_B = v_y = \frac{d}{dt}(5 - 8t^3) = -8 \times 3t^2$$

At $t = 1$ s,

$$v_B = -8 \times 3 \times 1^2 = -24 \text{ ms}^{-1}$$

$$\Rightarrow \mathbf{v}_B = -24\hat{\mathbf{j}} \text{ ms}^{-1}$$

Speed of particle B w.r.t. particle A,

$$\mathbf{v}_{BA} = \text{velocity of particle } B \text{ w.r.t. } A$$
$$= \mathbf{v}_B - \mathbf{v}_A$$
$$= -24\hat{\mathbf{j}} - (2\hat{\mathbf{i}}) = -2\hat{\mathbf{i}} - 24\hat{\mathbf{j}}$$

$\Rightarrow$ Magnitude of velocity,

$$|\mathbf{v}_{BA}| = \sqrt{(2)^2 + (24)^2}$$

But given $|\mathbf{v}_{BA}| = \sqrt{v}$

So, $\sqrt{v} = \sqrt{24^2 + 2^2}$ or $v = 580 \text{ ms}^{-1}$

Topic 4

01. *(a)* Projection of particle P on X-axis is

$$x = r\cos\theta$$

$$\text{or} \quad x = r\cos(\omega t + 30°)$$

$$= r\cos\left(\omega t + \frac{\pi}{6}\right)$$

02. *(a)* Given that, radius,

$$R = 0.1 \text{ m}$$

Angular frequency,

$$\omega = \frac{2\pi}{T} = \frac{2\pi}{60} = 0.105 \text{ rad/s}$$

$\therefore$ Average acceleration,

$$a = \omega^2 R$$
$$= (0.105)^2(0.1)$$
$$= 1.102 \times 10^{-3} \text{ m/s}^2$$

and it is of the order of 10^{-3}.

03. *(c)* As their period of revolution is same, so their angular speed is also same.

Centripetal acceleration in circular path, $a = \omega^2 r$.

Thus, $\dfrac{a_1}{a_2} = \dfrac{\omega^2 r_1}{\omega^2 r_2} = \dfrac{r_1}{r_2}$

04. *(b)* For a particle moving in a circle with constant angular speed, velocity vector is always along the tangent to the circle and the acceleration vector always points towards the centre of circle or is always along radius of the circle.

Since, tangential vector is perpendicular to radial vector, therefore velocity vector will be perpendicular to the acceleration vector.

But in no case, acceleration vector is tangent to the circle.

05. *(c)* Given,

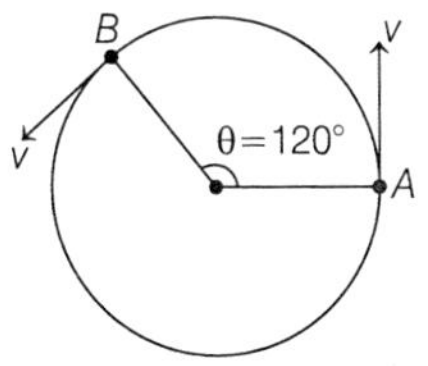

Given, $v = \pi$ m/s or $R\omega = \pi$

$$\omega = \frac{\pi}{R} \text{ rad/s}$$

Angular displacement, $\theta = 120° = \dfrac{2\pi}{3}$

Using $\theta = \omega t$

$$t = \frac{\theta}{\omega} = \frac{2\pi/3}{\pi/R} = \frac{2R}{3}$$

linear displacement,

$$d = 2R\sin(\theta/2)$$

$$d = 2R\sin\left(\frac{120°}{2}\right) = 2R\sin 60°$$

$$= 2R\frac{\sqrt{3}}{2} = R\sqrt{3}$$

Average velocity $= \dfrac{d}{t} = \dfrac{R\sqrt{3}}{2R/3} = \dfrac{3\sqrt{3}}{2}$

$$= 1.5\sqrt{3} \text{ m/s}$$

06. *(b)* The given situation is shown below

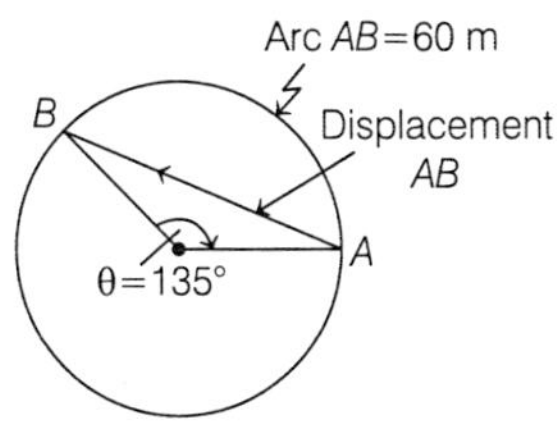

Displacement AB = (chord AB)

$$= 2 \times R \times \sin(\theta/2)$$

Also, arc $AB = 2\pi R \times \dfrac{\theta}{360°}$

$$\Rightarrow \quad R = \frac{AB \times 360°}{\theta \times 2\pi}$$

So, displacement AB

= chord AB

$$= 2 \times \frac{AB \times 360°}{\theta \times 2\pi} \times \sin\left(\frac{\theta}{2}\right)$$

$$= 2 \times \frac{AB \times 360°}{\theta \times 2\pi} \times \left\{\left(\frac{1-\cos\theta}{2}\right)\right\}^{1/2}$$

$$\left[\because 2\sin^2\frac{\theta}{2} = 1 - \cos\theta \quad \Rightarrow \sin\frac{\theta}{2} = \left(\frac{1-\cos\theta}{2}\right)^{1/2}\right]$$

$$= \frac{2 \times 60 \times 360°}{135° \times 2\pi} \times \left(\frac{1-(-0.7)}{2}\right)^{1/2}$$

$$= 47 \text{ m}$$

07. *(b)* Given,

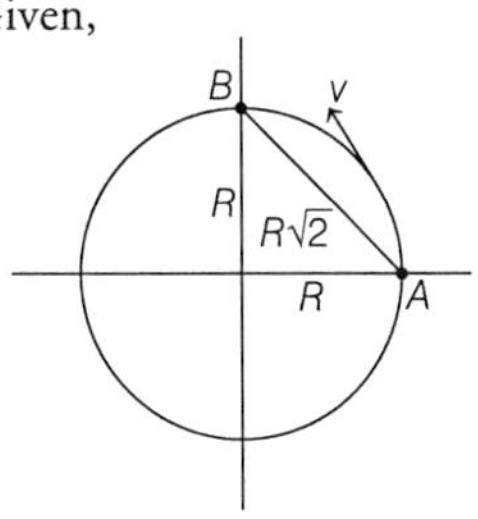

$$v_A = v\hat{\mathbf{j}} \Rightarrow v_B = -v\hat{\mathbf{i}}$$

Time to reach from A to B

$$= \frac{2\pi R}{4} \times \frac{1}{v} = \frac{\pi R}{2v}$$

Displacement from A to $B = R\sqrt{2}$

Now, average velocity from A to B

$$= \frac{\text{displacement}}{\text{time}}$$

$$= \frac{R\sqrt{2}}{\pi R/2v} = \frac{R\sqrt{2} \times 2v}{\pi R} = \frac{2\sqrt{2}v}{\pi}$$

Instantaneous velocity at B is $-v\hat{\mathbf{i}}$

According to question,

$$\frac{\text{Allign at centre}}{\text{Average velocity}} = \frac{v}{\dfrac{2\sqrt{2}v}{\pi}}$$

$$= \frac{\pi}{2\sqrt{2}} = \frac{\pi}{x\sqrt{2}} \quad \text{(given)}$$

Thus, $x = 2$

08. *(b)* Length of arc,

$$l = 4.4 \text{ ly} = 4.4 \times 9.46 \times 10^{15}$$

$$= 4.1624 \times 10^{16} \text{ m}$$

$$[\because 1 \text{ ly} = 9.46 \times 10^{15} \text{ m}]$$

Angle subtended by arc,

$$\theta = 4 \text{ s}$$

Number of revolution,

$$n = 4$$

Speed of body,

$$v = 8 \text{ AUs}^{-1} = 1.5 \times 10^{11} \times 8$$

$$= 1.2 \times 10^{12} \text{ ms}^{-1}$$

$$[\because 1 \text{ AU} = 1.5 \times 10^{11} \text{ m}]$$

We know that, length of arc subtended by an angle,

$$l = R\theta$$

Here, R = radius of arc

Substituting the values in above expression, we get

$$4.1624 \times 10^{16} = R \times 4 \times \left(\frac{1°}{3600}\right) \times \left(\frac{\pi}{180°}\right)$$

$$\left(\because 1 \text{ s} = \frac{1}{3600} \text{ degree}\right)$$

$$R = 2.146 \times 10^{21} \text{ m}$$

We know that, time = distance/speed

$$t = \frac{n \times 2\pi R}{v}$$

($\because$ Distance $= n \times$ length of circular path)

Substituting the values in above expression, we get

$$t = \frac{4 \times 2\pi \times 2.146 \times 10^{21}}{1.2 \times 10^{12}}$$

$$= 4.5 \times 10^{10} \text{ s}$$

C-04

JEE Main Chapterwise Topicwise Physics

In JEE Main there is usually 1-2 question per paper are visible from this chapter. In past 3 years, 5 conceptual and 5 numerical questions were asked of easy to moderate level from this chapter.
For detailed analysis scan QR code

LAWS OF MOTION

Topic 1 Newton's Laws of Motion and Conservation of Momentum

Level 1

01 An object with mass 500 g moves along X-axis with speed $v = 4\sqrt{x}$ m/s. The force acting on the object is *[2025, 7 April Shift-II]*

(a) 6 N (b) 5 N
(c) 8 N (d) 4 N

02 A body of mass 1000 kg is moving horizontally with a velocity 6 m/s. If 200 kg extra mass is added, then the final velocity (in m/s) is *[2024, 27 Jan Shift-I]*

(a) 3 (b) 2 (c) 5 (d) 6

03 A bullet of 10 g leaves the barrel of gun with a velocity of 600 m/s. If the barrel of gun is 50 cm long and mass of gun is 3 kg, then value of impulse supplied to the gun will be *[2023, 13 April Shift-I]*

(a) 36 N-s (b) 6 N-s (c) 3 N-s (d) 12 N-s

04 An average force of 125 N is applied on a machine gun firing bullets each of mass 10 g at the speed of 250 m/s to keep it in position. The number of bullets fired per second by the machine gun is *[2023, 11 April Shift-I]*

(a) 25 (b) 5 (c) 50 (d) 100

05 A machine gun of mass 10 kg fires 20 g bullets at the rate of 180 bullets per minute with a speed of 100 ms^{-1} each. The recoil velocity of the gun is *[2023, 30 Jan Shift-II]*

(a) 0.6 m/s (b) 0.02 m/s (c) 1.5 m/s (d) 2.5 m/s

06 A balloon has mass of 10 g in air. The air escapes from the balloon at a uniform rate with velocity 4.5 cm/s. If the balloon shrinks in 5 s completely, the average force acting on that balloon will be (in dyne). *[2022, 28 July Shift-I]*

(a) 3 (b) 9 (c) 12 (d) 18

07 A block of metal weighing 2 kg is resting on a frictionless plane (as shown in figure). It is struck by a jet releasing water at a rate of 1 kg s^{-1} and at a speed of 10 m s^{-1}. Then, the initial acceleration of the block in m s^{-2} will be *[2022, 29 June Shift-I]*

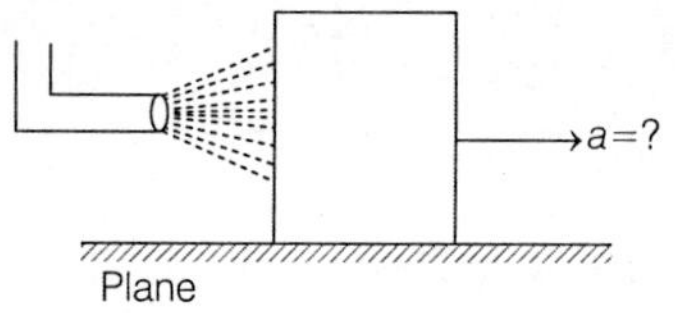

(a) 3 (b) 6 (c) 5 (d) 4

08 An object is thrown vertically upwards. At its maximum height, which of the following quantity becomes zero ? *[2022, 26 June Shift-I]*

(a) Momentum (b) Potential energy
(c) Acceleration (d) Force

Imp. Impulse

09 Two billiard balls of equal mass 30 g strike a rigid wall with same speed of 108 kmph (as shown) but at different angles. If the balls get reflected with the same speed then the ratio of the magnitude of impulses imparted to ball a and ball b by the wall along x-direction is *[2021, 25 July Shift-I]*

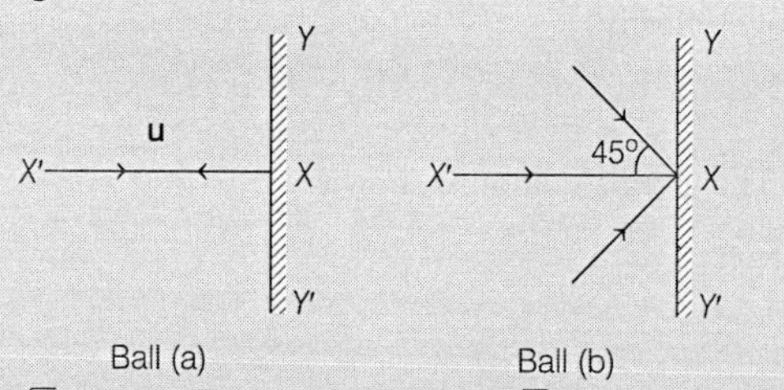

(a) $\sqrt{2}:1$ (b) $1:\sqrt{2}$
(c) 1 : 1 (d) 2 : 1

10 A particle of mass m is moving in a straight line with momentum p. Starting at time $t = 0$, a force $F = kt$ acts in the same direction on the moving particle during time interval T so that its momentum changes from p to $3p$. Here, k is a constant. The value of T is *[2019, 11 Jan Shift-II]*

(a) $2\sqrt{\frac{p}{k}}$ (b) $\sqrt{\frac{2k}{p}}$ (c) $\sqrt{\frac{2p}{k}}$ (d) $2\sqrt{\frac{k}{p}}$

11 A particle of mass m is acted upon by a force F given by the empirical law $F = \frac{R}{t^2} v(t)$. If this law is to be tested experimentally by observing the motion starting from rest, the best way is to plot *[JEE Main 2016]*

(a) $\log v(t)$ against $1/t$ (b) $v(t)$ against t^2
(c) $\log v(t)$ against $1/t^2$ (d) $\log v(t)$ against t

12 A body of mass $m = 3.513$ kg is moving along the X-axis with a speed of 5.00 m s^{-1}. The magnitude of its momentum is recorded as *[AIEEE 2008]*

(a) 17.57 kg m s^{-1} (b) 17.6 kg m s^{-1}
(c) 17.565 kg m s^{-1} (d) 17.56 kg m s^{-1}

13 A player caught a cricket ball of mass 150 g moving at a rate of 20 m/s. If the catching process is completed in 0.1 s, the force of the blow exerted by the ball on the hand of the player is equal to *[AIEEE 2006]*

(a) 150 N (b) 3 N
(c) 30 N (d) 300 N

14 A particle of mass 0.3 kg is subjected to a force $F = -kx$ with $k = 15$ Nm^{-1}. What will be its initial acceleration, if it is released from a point 20 cm away from the origin? *[AIEEE 2005]*

(a) 3 ms^{-2} (b) 15 ms^{-2}
(c) 5 ms^{-2} (d) 10 ms^{-2}

Level 2

15 A body of mass 2 kg moving with velocity of $\mathbf{v}_{in} = 3\hat{i} + 4\hat{j}$ ms^{-1} enters into a constant force field of 6N directed along positive Z-axis. If the body remains in the field for a period of $\frac{5}{3}$ s, then velocity of the body when it emerges from force field is. *[2025, 8 April Shift-II]*

(a) $3\hat{i} + 4\hat{j} + \sqrt{5}\hat{k}$ (b) $3\hat{i} + 4\hat{j} - 5\hat{k}$
(c) $4\hat{i} + 3\hat{j} + 5\hat{k}$ (d) $3\hat{i} + 4\hat{j} + 5\hat{k}$

16 A balloon and its content having mass M is moving up with an acceleration a. The mass that must be released from the content, so that the balloon starts moving up with an acceleration $3a$ will be (Take, g as acceleration due to gravity) *[2025, 28 Jan Shift-II]*

(a) $\frac{3Ma}{2a+g}$ (b) $\frac{3Ma}{2a-g}$
(c) $\frac{2Ma}{3a+g}$ (d) $\frac{2Ma}{3a-g}$

17 A wooden block, initially at rest on the ground, is pushed by a force which increases linearly with time t. Which of the following curve best describes acceleration of the block with time ? *[2024, 4 April Shift-I]*

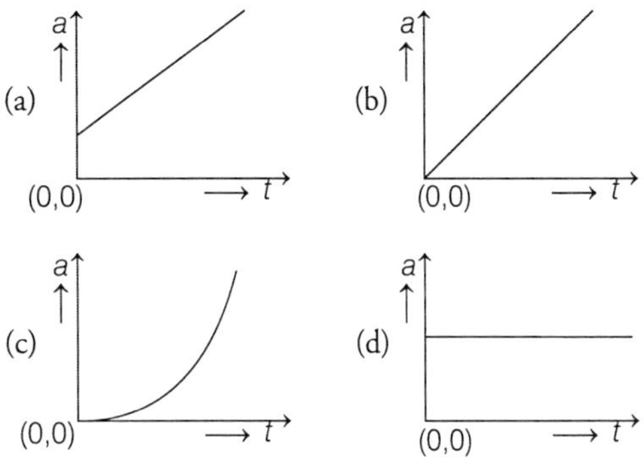

18 The position vector of a particle related to time t is given by $\mathbf{r} = (10t\hat{i} + 15t^2\hat{j} + 7\hat{k})$ m. The direction of net force experienced by the particle is *[2023, 15 April Shift-I]*

(a) positive X-axis (b) positive Z-axis
(c) positive Y-axis (d) In XY-plane

Imp. Newton's Second Law of Motion

19 100 balls each of mass m moving with speed v simultaneously strike a wall normally and reflected back with same speed, in time t s. The total force exerted by the balls on the wall is *[2023, 31 Jan Shift-I]*

(a) $\frac{100\,mv}{t}$ (b) $\frac{mv}{100\,t}$ (c) $\frac{200\,mv}{t}$ (d) $200\,mvt$

20 The figure represents the momentum-time (p-t) curve for a particle moving along an axis under the influence of a force. Identify the regions on the graph where the magnitude of the force is maximum and minimum respectively? *[2023, 30 Jan Shift-I]*

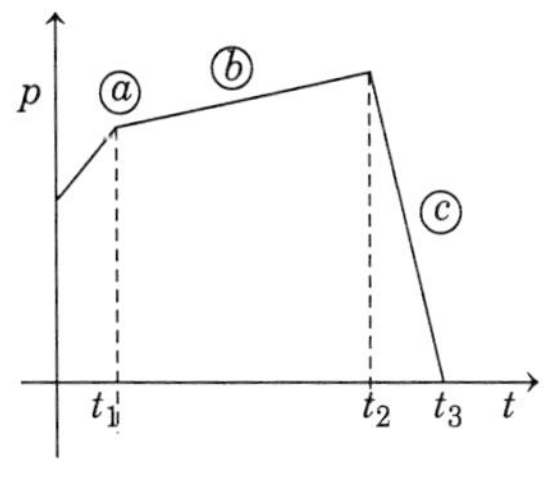

(a) a and b (b) c and b
(c) c and a (d) b and c

21 In two different experiments, an object of mass 5 kg moving with a speed of 25 ms^{-1} hits two different walls and comes to rest within (i) 3 s, (ii) 5 s, respectively.

Choose the correct option out of the following. *[2022, 28 July Shift-I]*

(a) Impulse and average force acting on the object will be same for both the cases.

(b) Impulse will be same for both the cases but the average force will be different.

(c) Average force will be same for both the cases but the impulse will be different.

(d) Average force and impulse will be different for both the cases.

22 A body of mass M at rest explodes into three pieces, in the ratio of masses 1:1:2. Two smaller pieces fly off perpendicular to each other with velocities of 30 m s^{-1} and 40 m s^{-1} respectively. The velocity of the third piece will be *[2022, 29 June Shift-I]*

(a) 15 m s^{-1} (b) 25 m s^{-1}

(c) 35 m s^{-1} (d) 50 m s^{-1}

23 The initial mass of a rocket is 1000 kg. Calculate at what rate the fuel should be burnt, so that the rocket is given an acceleration of 20 ms^{-2}. The gases come out at a relative speed of 500 ms^{-1} with respect to the rocket [Use, $g = 10$ m/s^2] *[2021, 26 Aug Shift-I]*

(a) 6.0×10^2 kg s^{-1} (b) 500 kg s^{-1}

(c) 10 kg s^{-1} (d) 60 kg s^{-1}

24 A bullet of 4 g mass is fired from a gun of mass 4 kg. If the bullet moves with the muzzle speed of 50 ms^{-1}, the impulse imparted to the gun and velocity of recoil of gun are *[2021, 22 July Shift-II]*

(a) 0.4 kg-ms^{-1}, 0.1 ms^{-1}

(b) 0.2 kg-ms^{-1}, 0.05 ms^{-1}

(c) 0.2 kg-ms^{-1}, 0.1 ms^{-1}

(d) 0.4 kg-ms^{-1}, 0.05 ms^{-1}

Imp. Conservation of Linear Momentum

25 A man (mass = 50 kg) and his son (mass = 20 kg) are standing on a frictionless surface facing each other. The man pushes his son, so that he starts moving at a speed of 0.70 ms^{-1} with respect to the man. The speed of the man with respect to the surface is *[2019, 12 April Shift-I]*

(a) 0.28 ms^{-1} (b) 0.20 ms^{-1}

(c) 0.47 ms^{-1} (d) 0.14 ms^{-1}

26 A bullet of mass 20 g has an initial speed of 1 ms^{-1}, just before it starts penetrating a mud wall of thickness 20 cm. If the wall offers a mean resistance of 2.5×10^{-2} N, the speed of the bullet after emerging from the other side of the wall is close to *[2019, 10 April Shift-II]*

(a) 0.3 ms^{-1} (b) 0.4 ms^{-1}

(c) 0.1 ms^{-1} (d) 0.7 ms^{-1}

27 A particle of mass m is at rest at the origin at time $t = 0$. It is subjected to a force $F(t) = F_0 e^{-bt}$ in the x-direction. Its speed $v(t)$ is depicted by which of the following curves? *[AIEEE 2012]*

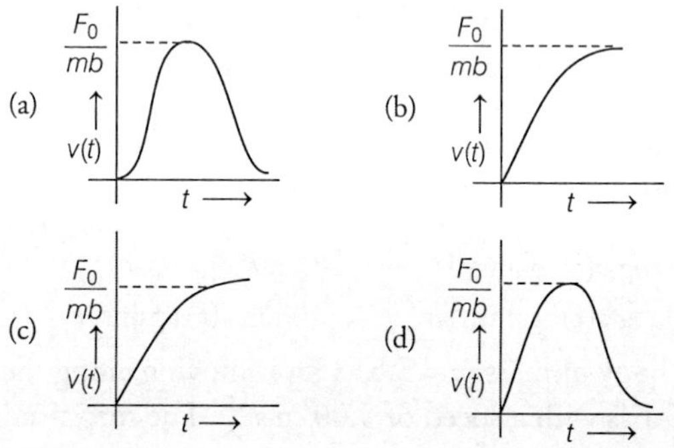

28 The figure shows the position-time (x-t) graph of one-dimensional motion of a body of mass 0.4 kg. The magnitude of each impulse is *[AIEEE 2010]*

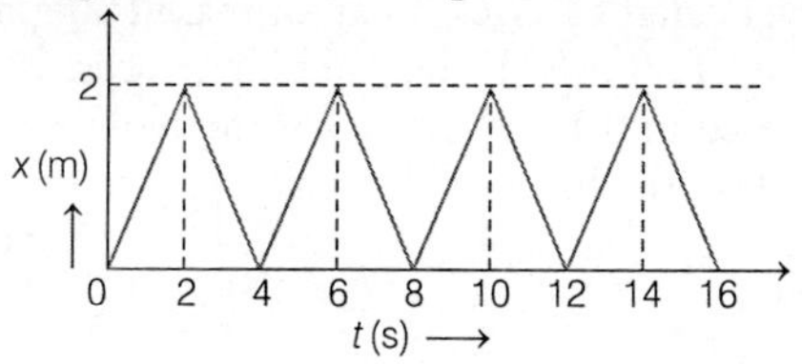

(a) 0.4 N- s (b) 0.8 N-s

(c) 1.6 N-s (d) 0.2 N-s

29 A ball of mass 0.2 kg is thrown vertically upwards by applying a force by hand. If the hand moves 0.2 m while applying the force and the ball goes upto 2 m height further, find the magnitude of the force. (Consider, $g = 10$ m/s^2). *[AIEEE 2006]*

(a) 4 N (b) 16 N

(c) 20 N (d) 22 N

Level 3

30 A block of mass 1 kg, moving along x with speed $v_i = 10$ m/s enters a rough region ranging from $x =$ 0.1 m to $x = 1.9$ m. The retarding force acting on the block in this range is $F_r = -kx$ N, with $k = 10$ N/m. Then the final speed of the block as it crosses rough region is *[2025, 3April Shift-II]*

(a) 10 m/s (b) 4 m/s

(c) 6 m/s (d) 8 m/s

31 A body of mass 500 g moves along X-axis such that it's velocity varies with displacement x according to the relation $v = 10\sqrt{x}$ m/s. The force acting on the body is *[2023, 11 April Shift-II]*

(a) 5 N (b) 166 N (c) 25 N (d) 125 N

32 At any instant the velocity of a particle of mass 500 g is $(2t\hat{i}+3t^2\hat{j})$ ms^{-1}. If the force acting on the particle at $t=1$ s is $(\hat{i}+x\hat{j})$ N. Then, the value of x will be *[2023, 8 April Shift-I]*

(a) 3 (b) 4 (c) 6 (d) 2

33 A ball of mass 200 g rests on a vertical post of height 20 m. A bullet of mass 10g, travelling in horizontal direction, hits the centre of the ball. After collision both travel independently. The ball hits the ground at a distance of 30m and the bullet at a distance of 120m from the foot of the post. The value of initial velocity of the bullet will be (if $g = 10$ m/s^2) *[2023, 30 Jan Shift-I]*

(a) 60 m/s (b) 400 m/s
(c) 360 m/s (d) 120 m/s

34 A force acts for 20 s on a body of mass 20 kg, starting from rest, after which the force ceases and then body describes 50 m in the next 10 s. The value of force will be *[2023, 29 Jan Shift- II]*

(a) 20 N (b) 40 N
(c) 10 N (d) 5 N

Imp. Wooden-Bullet Problem

35 The velocity of a bullet becomes one-third after it penetrates 4 cm in a wooden block. Assuming that bullet is facing a constant resistance during its motion in the block. The bullet stops completely after travelling at $(4+x)$ cm inside the block. The value of x is *[2022, 27 July Shift-II]*

(a) 2.0 (b) 1.0
(c) 0.5 (d) 1.5

36 An object of mass 5 kg is thrown vertically upwards from the ground. The air resistance produces a constant retarding force of 10 N throughout the motion. The ratio of time of ascent to the time of descent will be equal to [Use, $g = 10$ ms^{-2}] *[2022, 24 June Shift-II]*

(a) 1 : 1 (b) $\sqrt{2}:\sqrt{3}$ (c) $\sqrt{3}:\sqrt{2}$ (d) 2 : 3

37 A particle of mass M originally at rest is subjected to a force whose direction is constant but magnitude varies with time according to the relation

$$F = F_0\left[1-\left(\frac{t-T}{T}\right)^2\right].$$

Where, F_0 and T are constants. The force acts only for the time interval $2T$. The velocity v of the particle after time $2T$ is *[2021, 27 July Shift-II]*

(a) $\frac{2F_0T}{M}$ (b) $\frac{F_0T}{2M}$
(c) $\frac{4F_0T}{3M}$ (d) $\frac{F_0T}{3M}$

38 A spaceship in space sweeps stationary interplanetary dust. As a result, its mass increases at a rate $\frac{dM(t)}{dt} = bv^2(t)$, where $v(t)$ is its instantaneous velocity. The instantaneous acceleration of the satellite is *[2020, 5 Sep Shift-II]*

(a) $-bv^3(t)$ (b) $-\frac{bv^3}{M(t)}$ (c) $-\frac{2bv^3}{M(t)}$ (d) $-\frac{bv^3}{2M(t)}$

39 A ball is thrown vertically up (taken as $+Z$-axis) from the ground. The correct momentum-height (p-h) diagram is *[2019, 9 April Shift-I]*

(a) 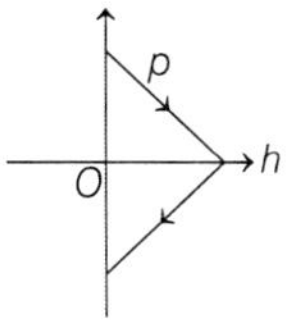

(b) 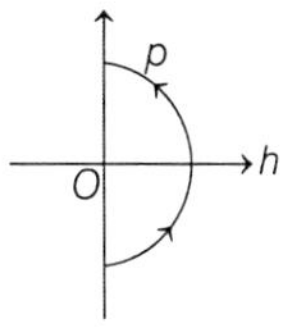

(c) 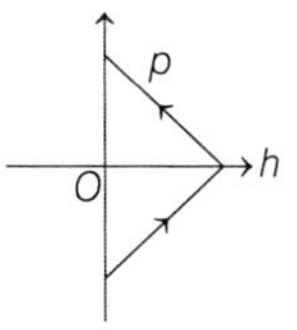

(d) 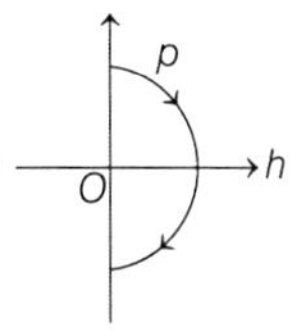

Numerical Value Type Questions

40 A man of 60 kg is running on the road and suddenly jumps into a stationary trolley car of mass 120 kg. Then, the trolley car starts moving with velocity 2 ms^{-1}. The velocity of the running man was ...... ms^{-1}, when he jumps into the car. *[2022, 28 June Shift-I]*

41 A batsman hits back a ball of mass 0.4 kg straight in the direction of the bowler without changing its initial speed of 15 ms^{-1}. The impulse imparted to the ball is ...... N-s. *[2022, 26 June Shift-II]*

42 The projectile motion of a particle of mass 5 g is shown in the figure.

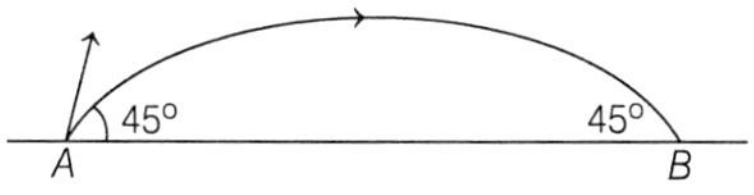

The initial velocity of the particle is $5\sqrt{2}$ m/s and the air resistance is assumed to be negligible. The magnitude of the change in momentum between the points A and B is $x \times 10^{-2}$ kg m s^{-1}. The value of x, to the nearest integer, is .... *[2021, 18 March Shift-II]*

43 A bullet of mass 0.1 kg is fired on a wooden block to pierce through it, but it stops after moving a distance of 50 cm into it. If the velocity of bullet before hitting the wood is 10 m/s and it slows down with uniform deceleration, then the magnitude of effective retarding force on the bullet is x N. The value of x to the nearest integer is .... *[2021, 18 March Shift-I]*

Topic 2 Equilibrium of a Particle and Common Forces in Mechanics

Level 1

01 A body of mass m is suspended by two strings making angles θ_1 and θ_2 with the horizontal ceiling with tensions T_1 and T_2 simultaneously. T_1 and T_2 are related by $T_1 = \sqrt{3}T_2$, the angles θ_1 and θ_2 are [2025, 4 April Shift-I]

(a) $\theta_1 = 30°, \theta_2 = 60°$ with $T_2 = \frac{3mg}{4}$

(b) $\theta_1 = 30°, \theta_2 = 60°$ with $T_2 = \frac{4mg}{5}$

(c) $\theta_1 = 60°, \theta_2 = 30°$ with $T_2 = \frac{mg}{2}$

(d) $\theta_1 = 45°, \theta_2 = 45°$ with $T_2 = \frac{3mg}{4}$

02 A body of weight 200 N is suspended from a tree branch through a chain of mass 10 kg. The branch pulls the chain by a force equal to (if $g = 10 \text{ m/s}^2$) [2024, 6 April Shift-II]

(a) 100 N (b) 150 N
(c) 200 N (d) 300 N

03 A cricket player catches a ball of mass 120 g moving with 25 m/s speed. If the catching process is completed in 0.1 s, then the magnitude of force exerted by the ball on the hand of player will be (in SI unit). [2024, 1 Feb Shift-II]

(a) 30 (b) 12
(c) 24 (d) 25

04 As shown in figure, a 70 kg garden roller is pushed with a force of F = 200 N at an angle of 30° with horizontal. The normal reaction on the roller is (Given, $g = 10 \text{ m s}^{-2}$) [2023, 31 Jan Shift-I]

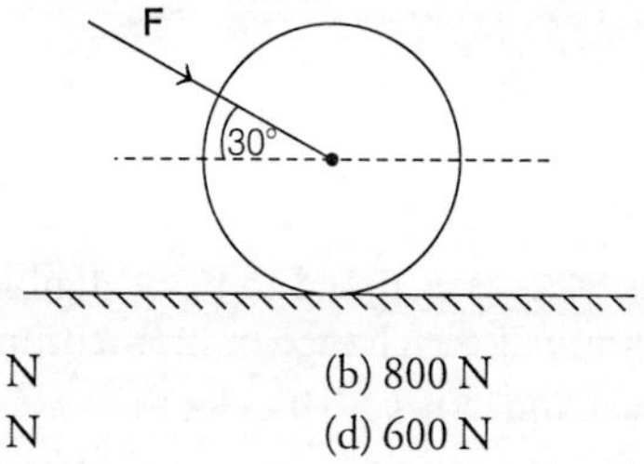

(a) $800\sqrt{2}$ N (b) 800 N
(c) $200\sqrt{3}$ N (d) 600 N

05 A block of mass M placed inside a box descends vertically with acceleration a. The block exerts a force equal to one-fourth of its weight on the floor of the box. The value of a will be [2022, 29 June Shift-I]

(a) $\frac{g}{4}$ (b) $\frac{g}{2}$
(c) $\frac{3g}{4}$ (d) g

06 In the arrangement shown in figure a_1, a_2, a_3 and a_4 are the accelerations of masses m_1, m_2, m_3 and m_4, respectively. Which of the following relation is true for this arrangement? [2022, 26 June Shift-II]

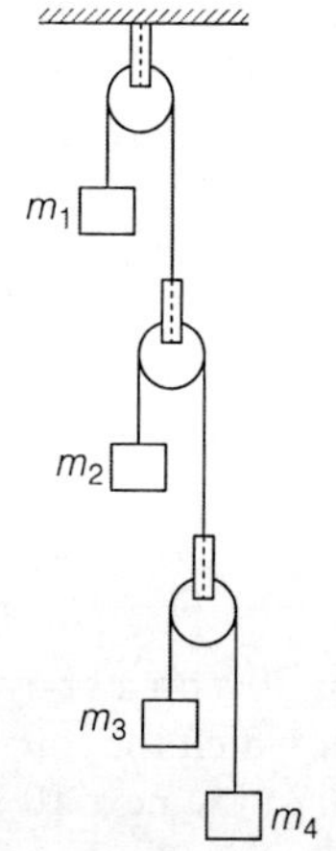

(a) $4a_1 + 2a_2 + a_3 + a_4 = 0$
(b) $a_1 + 4a_2 + 3a_3 + a_4 = 0$
(c) $a_1 + 4a_2 + 3a_3 + 2a_4 = 0$
(d) $2a_1 + 2a_2 + 3a_3 + a_4 = 0$

07 A block of mass m is connected to another block of mass M by a spring (massless) of spring constant k. The blocks are kept on a smooth horizontal plane. Initially the blocks are at rest and the spring is unstretched. Then, a constant force F starts acting on the block of mass M to pull it. Find the force on the block of mass m. [AIEEE 2007]

(a) $\frac{mF}{M}$ (b) $\frac{(M+m)F}{m}$
(c) $\frac{mF}{(m+M)}$ (d) $\frac{MF}{(m+M)}$

08 A block is kept on a frictionless inclined surface with angle of inclination α. The incline is given an acceleration a to keep the block stationary. Then, a is equal to [AIEEE 2005]

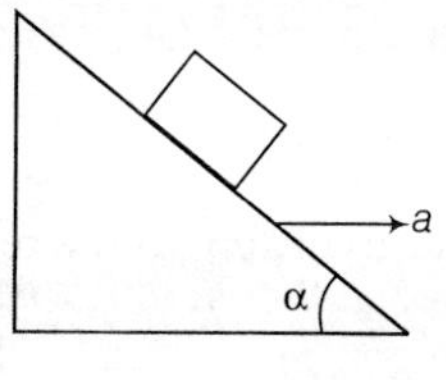

(a) $\frac{g}{\tan\alpha}$ (b) g cosec
(c) g (d) $g \tan\alpha$

09 A block of mass M is pulled along a horizontal frictionless surface by a rope of mass m. If a force P is applied at the free end of the rope, the force exerted by the rope on the block is [AIEEE 2003]

(a) $\frac{Pm}{M+m}$ (b) $\frac{Pm}{M-m}$ (c) P (d) $\frac{PM}{M+m}$

10 A light spring balance hangs from the hook of the other light spring balance and a block of mass M kg hangs from the former one. Then, the true statement about the scale reading is **[AIEEE 2003]**

(a) Both the scales read M kg each
(b) The scale of the lower one reads M kg and of the upper one zero
(c) The reading of the two scales can be anything but the sum of the readings will be M kg
(d) Both the scales read $M/2$ kg

11 A lift is moving down with acceleration a. A man in the lift drops a ball inside the lift. The acceleration of the ball as observed by the man in the lift and a man standing stationary on the ground are respectively **[AIEEE 2002]**

(a) g, g (b) $g-a, g-a$
(c) $g-a, g$ (d) a, g

Level 2

12

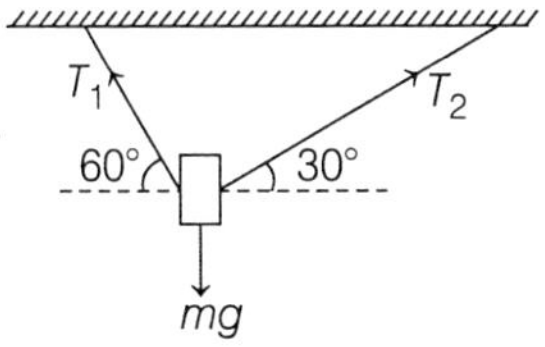

A body of mass 1 kg is suspended with the help of two strings making angles as shown in figure. Magnitudes of tensions T_1 and T_2, respectively, are (in N) (Take acceleration due to gravity 10 m/s^2) **[2025, 2 April Shift-II]**

(a) $5\sqrt{3}, 5$ (b) $5, 5\sqrt{3}$
(c) 5, 5 (d) $5\sqrt{3}, 5\sqrt{3}$

13 A square lamina $OABC$ of length 10 cm is pivoted at O. Forces act at lamina as shown in figure. If lamina remains stationary, then the magnitude of F is **[2025, 2 April Shift-I]**

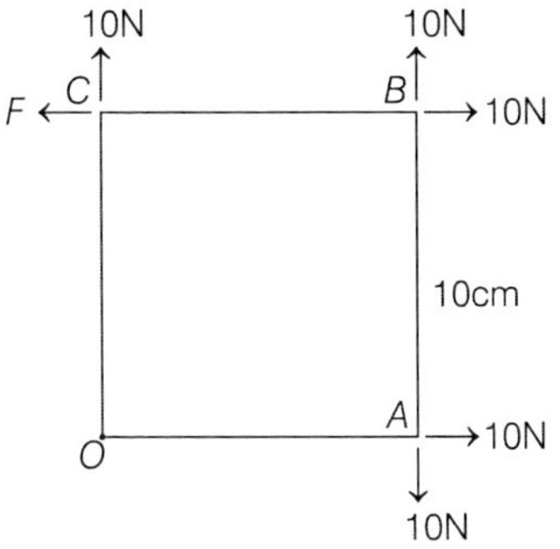

(a) 10 N (b) 0 (zero)
(c) $10\sqrt{2}$ N (d) 20 N

14 A massless spring gets elongated by amount x_1 under a tension of 5 N. Its elongation is x_2 under the tension of 7 N. For the elongation of $(5x_1 - 2x_2)$, then tension in the spring will be **[2025, 23 Jan Shift-II]**

(a) 11 N (b) 39 N
(c) 20 N (d) 15 N

Imp. Equilibrium of Forces

15 A 1 kg mass is suspended from the ceiling by a rope of length 4m. A horizontal force F is applied at the mid-point of the rope, so that the rope makes an angle of 45° with respect to the vertical axis as shown in figure.

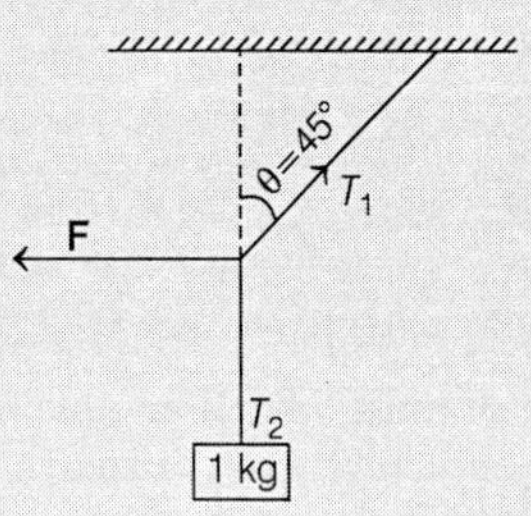

The magnitude of F is (assume that the system is in equilibrium and $g = 10$ m / s^2) **[2024, 9 April Shift-II]**

(a) $\frac{1}{10\times\sqrt{2}}$ N (b) 1 N (c) 10 N (d) $\frac{10}{\sqrt{2}}$ N

16 A light string passing over a smooth light fixed pulley connects two blocks of masses m_1 and m_2. If the acceleration of the system is $g/8$, then the ratio of masses is **[2024, 9 April Shift-I, 31 Jan Shift-II]**

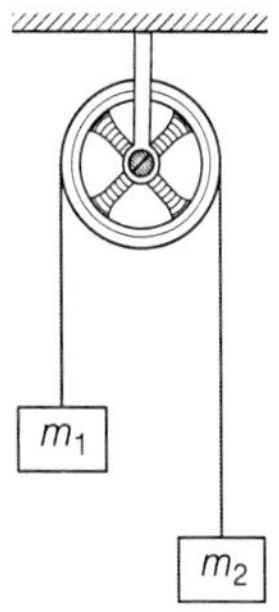

(a) $\frac{4}{3}$ (b) $\frac{8}{1}$ (c) $\frac{5}{3}$ (d) $\frac{9}{7}$

17 A player caught a cricket ball of mass 150 g moving at a speed of 20 m/s. If the catching process is completed in 0.1 s. The magnitude of force exerted by the ball on the hand of the player is **[2024, 8 April Shift-I]**

(a) 150 N (b) 30 N (c) 3 N (d) 300 N

18 A light string passing over a smooth light pulley connects two blocks of masses m_1 and m_2 (where $m_2 > m_1$). If the acceleration of the system is $g/\sqrt{2}$, then the ratio of the masses m_1/m_2 is **[2024, 8 April Shift-I]**

(a) $\frac{\sqrt{2}-1}{\sqrt{2}+1}$ (b) $\frac{1+\sqrt{5}}{\sqrt{2}-1}$ (c) $\frac{1+\sqrt{5}}{\sqrt{5}-1}$ (d) $\frac{\sqrt{3}+1}{\sqrt{2}-1}$

19 A body of mass 4 kg experiences two forces $\mathbf{F}_1 = 5\hat{\mathbf{i}} + 8\hat{\mathbf{j}} + 7\hat{\mathbf{k}}$ and $\mathbf{F}_2 = 3\hat{\mathbf{i}} - 4\hat{\mathbf{j}} - 3\hat{\mathbf{k}}$. The acceleration acting on the body is *[2024, 1 Feb Shift-II]*

(a) $-2\hat{\mathbf{i}} - \hat{\mathbf{j}} - \hat{\mathbf{k}}$ (b) $4\hat{\mathbf{i}} + 2\hat{\mathbf{j}} + 2\hat{\mathbf{k}}$
(c) $2\hat{\mathbf{i}} + 3\hat{\mathbf{j}} + 3\hat{\mathbf{k}}$ (d) $2\hat{\mathbf{i}} + \hat{\mathbf{j}} + \hat{\mathbf{k}}$

20 A spherical body of mass 100 g is dropped from a height of 10 m from the ground. After hitting the ground, the body rebounds to a height of 5 m. The impulse of force imparted by the ground to the body is given by (Given, $g = 9.8\ \text{m/s}^2$) *[2024, 30 Jan Shift-I]*

(a) 43.2 kg ms^{-1} (b) 4.32 kg ms^{-1}
(c) 23.9 kg ms^{-1} (d) 2.39 kg ms^{-1}

21 Three forces $F_1 = 10$ N, $F_2 = 8$ N, $F_3 = 6$ N are acting on a particle of mass 5 kg. The forces F_2 and F_3 are applied perpendicularly so that particle remains at rest. If the force F_1 is removed, then the acceleration of the particle is *[2023, 12 April Shift-I]*

(a) 0.5 ms^{-2} (b) 4.8 ms^{-2} (c) 2 ms^{-2} (d) 7 ms^{-2}

22 A block A takes 2 s to slide down a frictionless incline of 30° and length l kept inside a lift going up with uniform velocity v. If the incline is changed to 45°, the time taken by the block, to slide down the incline, will be approximately. *[2022, 27 July Shift-II]*

(a) 2.66 s (b) 0.83 s (c) 1.68 s (d) 0.70 s

23 For a free body diagram shown in the figure, the four forces are applied in the x and y-directions. What additional force must be applied and at what angle with positive X-axis, so that net acceleration of body is zero? *[2022, 25 July Shift-II]*

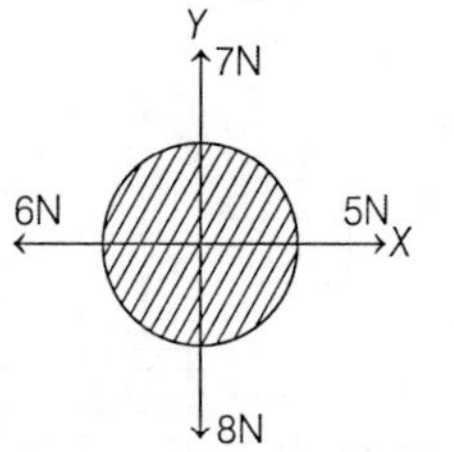

(a) $\sqrt{2}$ N, 45° (c) $\sqrt{2}$ N, 135°
(c) $\frac{2}{\sqrt{3}}$ N, 30° (d) 2 N, 45°

24 An object of mass m is being moved with a constant velocity under the action of an applied force of 2 N along a frictionless surface with following surface profile. *[2021, 1 Sep Shift-II]*

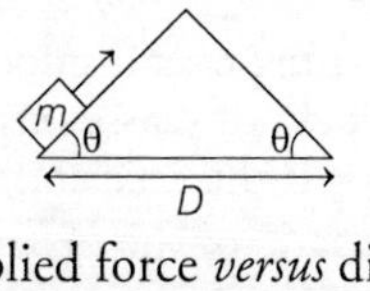

The correct applied force *versus* distance graph will be

(a)
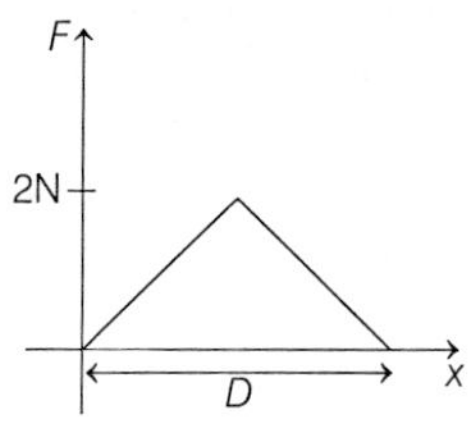

(b)
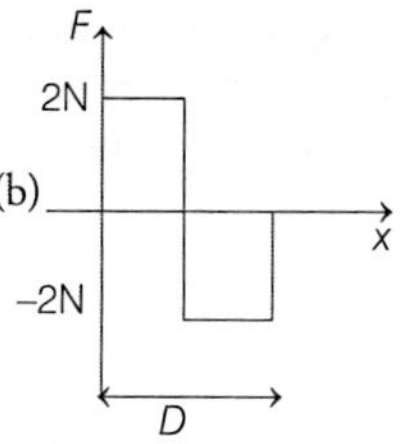

(c)
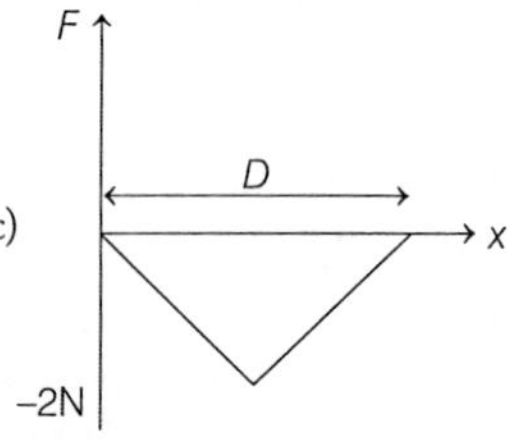

(d)
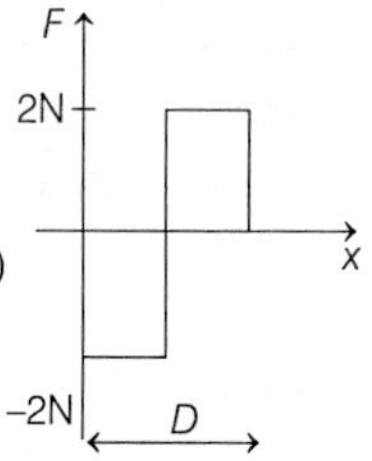

25 **Statement I** If three forces $\mathbf{F}_1$, $\mathbf{F}_2$ and $\mathbf{F}_3$ are represented by three sides of a triangle and $\mathbf{F}_1 + \mathbf{F}_2 = -\mathbf{F}_3$, then these three forces are concurrent forces and satisfy the condition for equilibrium.
Statement II A triangle made up of three forces $\mathbf{F}_1$, $\mathbf{F}_2$ and $\mathbf{F}_3$ as its sides taken in the same order, satisfy the condition for translatory equilibrium.
In the light of the above statements, choose the most appropriate answer from the options given below. *[2021, 31 Aug Shift-II]*

(a) Statement I is false but Statement II is true.
(b) Statement I is true but Statement II is false.
(c) Both Statement I and Statement II are false.
(d) Both Statement I and Statement II are true.

26 A mass of 10 kg is suspended vertically by a rope from the roof. When a horizontal force is applied on the mass, the rope deviated at an angle of 45° at the roof point. If the suspended mass is at equilibrium, the magnitude of the force applied is (Take, $g = 10\ \text{ms}^{-2}$) *[2019, 9 Jan Shift-II]*

(a) 70 N (b) 200 N (c) 100 N (d) 140 N

27 Two fixed frictionless inclined plane making the angles 30° and 60° with the vertical are shown in the figure. Two blocks A and B are placed on the two planes. What is the relative vertical acceleration of A with respect to B? *[AIEEE 2010]*

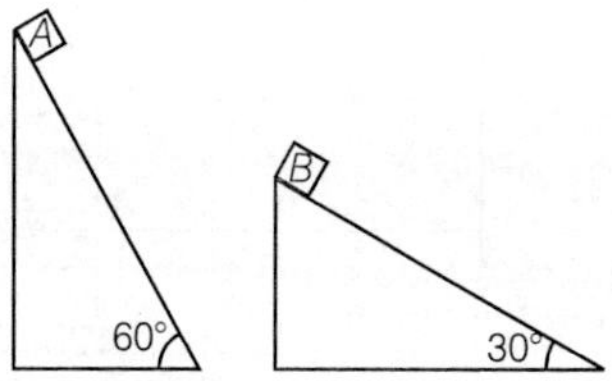

(a) 4.9 ms^{-2} in horizontal direction
(b) 9.8 ms^{-2} in vertical direction
(c) Zero
(d) 4.9 ms^{-2} in vertical direction

Level 3

28 A wooden block of mass 5 kg rests on a soft horizontal floor. When an iron cylinder of mass 25 kg is placed on the top of the block, the floor yields and the block and the cylinder together go down with an acceleration of $0.1\ ms^{-2}$. The action force of the system on the floor is equal to *[2024, 5 April Shift-I]*

(a) 294 N (b) 297 N (c) 196 N (d) 291 N

29 Consider a block and trolley system as shown in figure. If the coefficient of kinetic friction between the trolley and the surface is 0.04, the acceleration of the system in ms^{-2} is (consider that the string is massless and unstretchable and the pulley is also massless and frictionless) *[2024, 1 Feb Shift-I]*

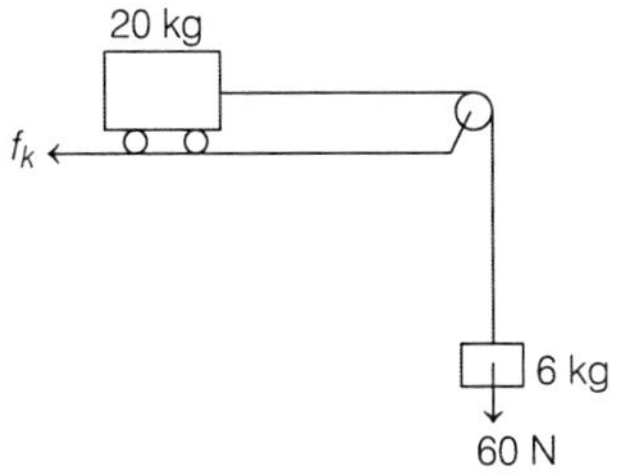

(a) 2 (b) 3 (c) 4 (d) 1.2

30 In the given arrangement of a doubly inclined plane two blocks of masses M and m are placed. The blocks are connected by a light string passing over an ideal pulley as shown. The coefficient of friction between the surface of the plane and the blocks is 0.25. The value of m, for which $M = 10$ kg will move down with an acceleration of 2 m/s^2 is (Take, $g = 10\ m/s^2$ and $\tan 37° = 3/4$) *[2024, 31 Jan Shift-I]*

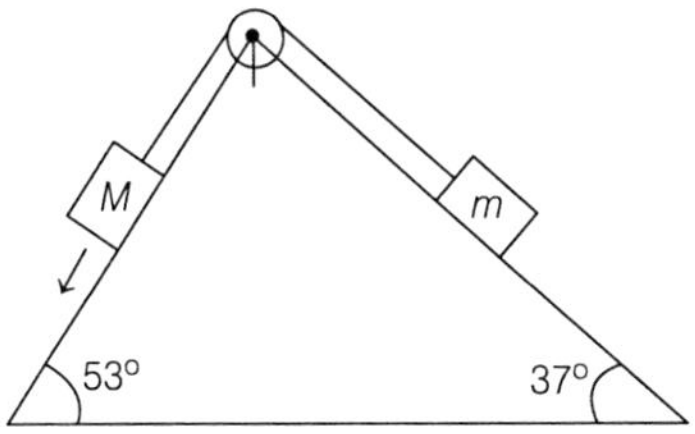

(a) 6.5 kg (b) 4.5 kg (c) 9 kg (d) 2.25 kg

31 Three blocks A, B and C are pulled on a horizontal smooth surface by a force of 80 N as shown in figure. *[2024, 30 Jan Shift-II]*

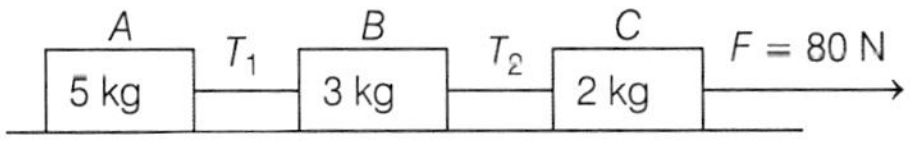

The tensions T_1 and T_2 in the string are respectively

(a) 60 N, 80 N (b) 88 N, 96 N
(c) 40 N, 64 N (d) 80 N, 100 N

32 All surfaces shown in figure are assumed to be frictionless and the pulleys and the string are light. The acceleration of the block of mass 2 kg is *[2024, 30 Jan Shift-I]*

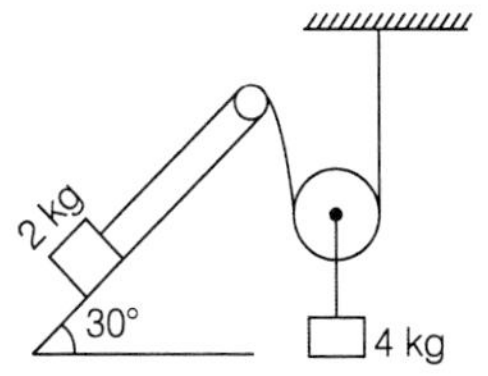

(a) $g/3$ (b) g (c) $g/2$ (d) $g/4$

Imp. Newton's Second Law of Motion

33 As per given figure, a weightless pulley P is attached on double inclined frictionless surfaces. The tension in the string (massless) will be (if $g = 10\ m/s^2$) *[2023, 24 Jan Shift-I]*

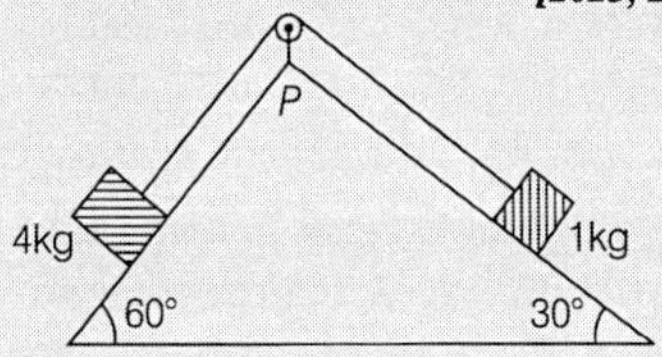

(a) $4(\sqrt{3}+1)$ N (b) $(4\sqrt{3}+1)$N
(c) $(4\sqrt{3}-1)$N (d) $4(\sqrt{3}-1)$N

34 Two bodies of masses $m_1 = 5$ kg and $m_2 = 3$ kg are connected by a light string going over a smooth light pulley on a smooth inclined plane as shown in the figure. The system is at rest. The force exerted by the inclined plane on the body of mass m_1 will be [Take, $g = 10\ m\ s^{-2}$] *[2022, 29 July Shift-II]*

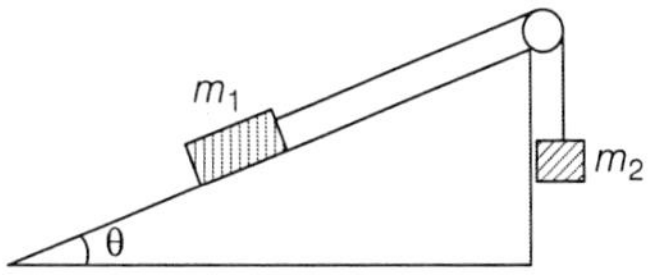

(a) 30 N (b) 40 N (c) 50 N (d) 60 N

35 A uniform metal chain of mass m and length L passes over a massless and frictionless pulley. It is released from rest with a part of its length l is hanging on one side and rest of its length $L - l$ is hanging on the other side of the pulley. At a certain point of time, when $l = L/x$, the acceleration of the chain is $g/2$. The value of x is .... *[2022, 28 July Shift-II]*

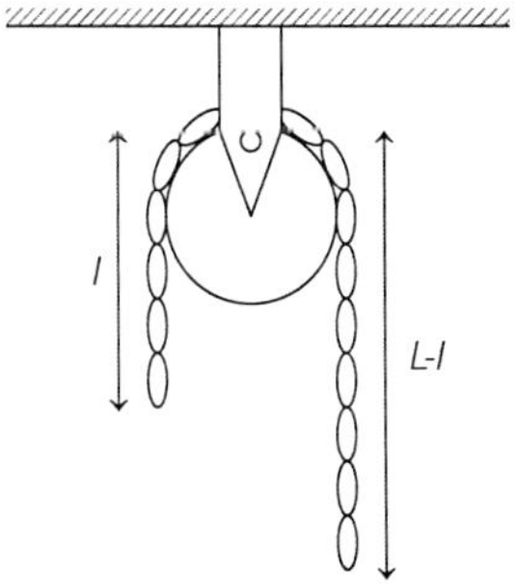

(a) 6 (b) 2 (c) 1.5 (d) 4

36 Three masses $M = 100$ kg, $m_1 = 10$ kg and $m_2 = 20$ kg are arranged in a system as shown in figure. All the surfaces are frictionless and strings are inextensible and weightless. The pulleys are also weightless and frictionless. A force F is applied on the system so that the mass m_2 moves upward with an acceleration of 2 ms^2. The value of F is (Take, $g = 10\ ms^{-2}$)

[2022, 26 July Shift-I]

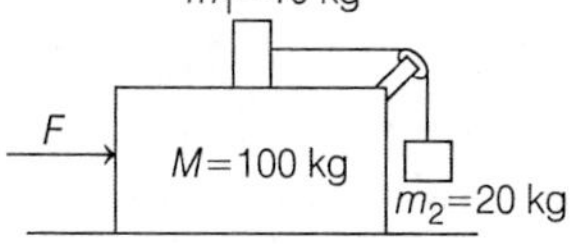

(a) 3360 N (b) 3380 N
(c) 3120 N (d) 3240 N

37 Two masses M_1 and M_2 are tied together at the two ends of a light inextensible string that passes over a frictionless pulley. When the mass M_2 is twice that of M_1, the acceleration of the system is a_1. When the mass M_2 is thrice that of M_1, the acceleration of the system is a_2. The ratio a_1/a_2 will be

[2022, 26 July Shift-I]

(a) 1/3 (b) 2/3
(c) 3/2 (d) 1/2

38 A block of mass m slides on the wooden wedge, which in turn slides backward on the horizontal surface. The acceleration of the block with respect to the wedge is [Given, $m = 8$ kg, $M = 16$ kg]
Assume all the surfaces shown in the figure to be frictionless.

[2021, 1 Sep Shift-II]

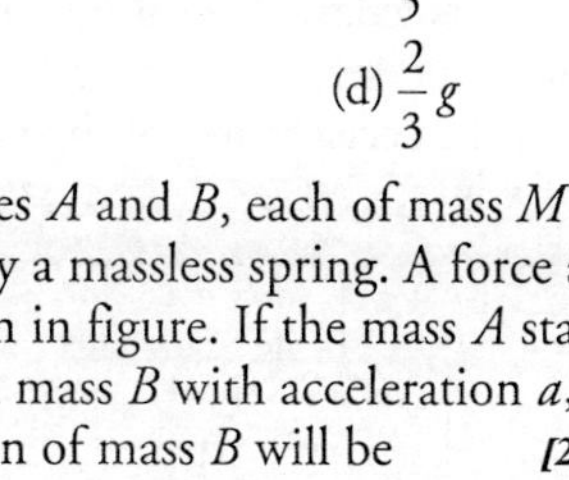

(a) $\frac{4}{3}g$ (b) $\frac{6}{5}g$
(c) $\frac{3}{5}g$ (d) $\frac{2}{3}g$

39 Two masses A and B, each of mass M are fixed together by a massless spring. A force acts on the mass B as shown in figure. If the mass A starts moving away from mass B with acceleration a, then the acceleration of mass B will be

[2021, 26 Feb Shift-II]

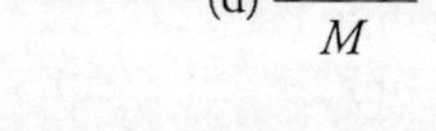

(a) $\frac{Ma - F}{M}$ (b) $\frac{MF}{F + Ma}$
(c) $\frac{F + Ma}{M}$ (d) $\frac{F - Ma}{M}$

40 The boxes of masses 2 kg and 8 kg are connected by a massless string passing over smooth pulleys. Calculate the time taken by box of mass 8 kg to strike the ground starting from rest. (Use, $g = 10\ m/s^2$)

[2021, 27 Aug Shift-II]

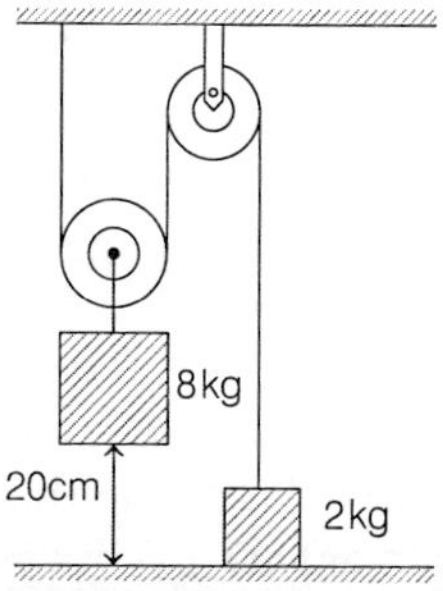

(a) 0.34 s (b) 0.2 s
(c) 0.25 s (d) 0.4 s

Numerical Value Type Questions

41 Three blocks m_1, m_2, m_3 having masses 4 kg, 6 kg and 10 kg, respectively are hanging from a smooth pully using rope 1, 2 and 3 as shown in figure. The tension in the rope 1, T_1 when they are moving upward with acceleration of 2 ms^{-2} is .... N. (If, $g = 10\ m/s^2$).

[2024, 5 April Shift-I]

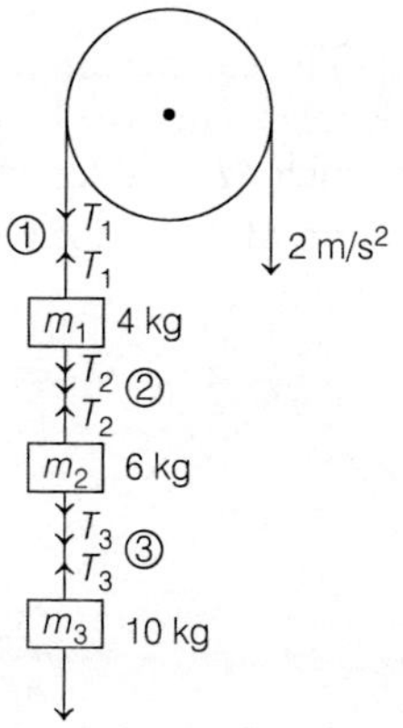

42 An elastic spring under tension of 3 N has a length a. Its length is b under tension 2 N. For its length $(3a - 2b)$, the value of tension will be......... N.

[2024, 4 April Shift-I]

43 Two inclined planes are placed as shown in figure. A block is projected from the point A of inclined plane AB along its surface with a velocity just sufficient to carry it to the top point B at a height 10 m. After reaching point B the block slides down on inclined plane BC. Time taken to reach to point C from point A is $t(\sqrt{2} + 1)$ s. The value of t is ....
(Use, $g = 10\ m/s^2$)

[2022, 27 July Shift-II]

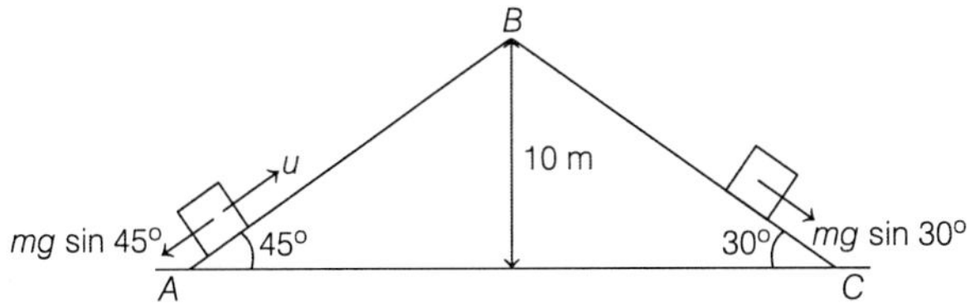

44 Four forces are acting at a point P in the equilibrium as shown in figure. The ratio of force F_1 to F_2 is $1:x$, where x is *[2022, 25July Shift-I]*

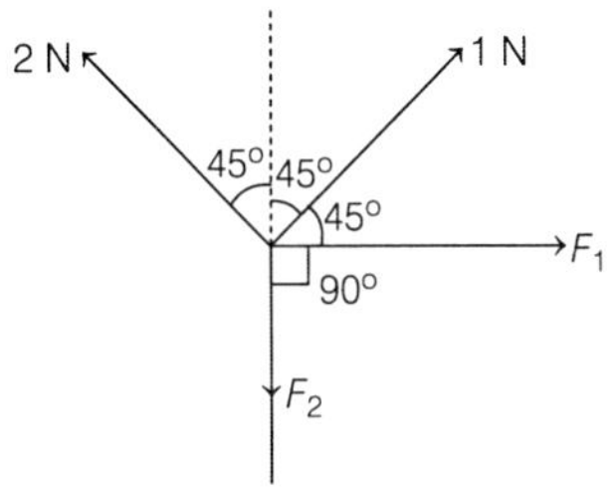

45 A block of mass 200 g is kept stationary on a smooth inclined plane by applying a minimum horizontal force $F = \sqrt{x}$ N as shown in figure. The value of $x =$ .... *[2022, 25 June Shift-II]*

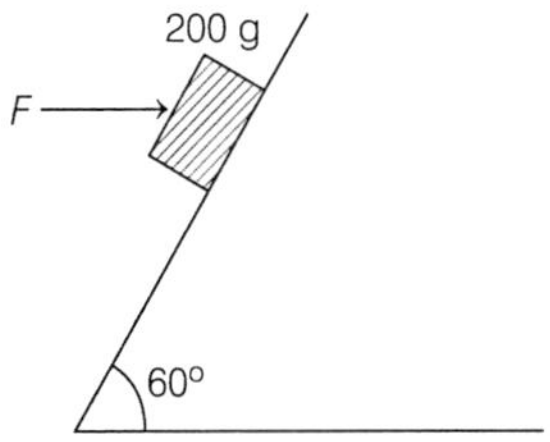

46 A car is moving on a plane inclined at 30° to the horizontal with an acceleration of 10 ms^{-2} parallel to the plane upward. A bob is suspended by a string from the roof of the car. The angle in degrees which the string makes with the vertical is ....
(Take, g = 10 ms^{-2}) *[2021, 31 Aug Shift-I]*

47 Consider a frame that is made up of two thin massless rods AB and AC as shown in the figure. A vertical force **P** of magnitude 100 N is applied at point A of the frame. Suppose, the force is P resolved parallel to the arms AB and AC of the frame. The magnitude of the resolved component along the arm AC is x N. The value of x, to the nearest integer, is ....
[Given, $\sin(35°) = 0.573$, $\cos(35°) = 0.819, \sin(110°) = 0.939$, $\cos(110°) = -0.342$]

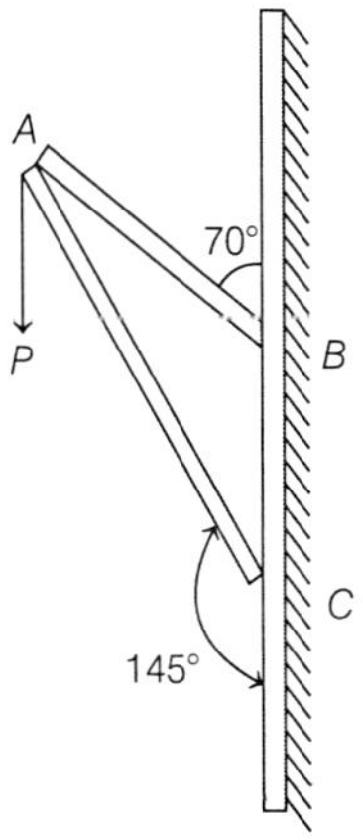

[2021, 16 March Shift-I]

Topic 3 Friction

Level 1

01 A cubic block of mass m is sliding down on an inclined plane at 60° with an acceleration of $\frac{g}{2}$, the value of coefficient of kinetic friction is *[2025, 7 April Shift-I]*

(a) $\frac{\sqrt{2}}{3}$ (b) $\sqrt{3} - 1$
(c) $1 - \frac{\sqrt{3}}{2}$ (d) $\frac{\sqrt{3}}{2}$

02 Given below are two statements.
Statement I The limiting force of static friction depends on the area of contact and independent of materials.
Statement II The limiting force of kinetic friction is independent of the area of contact and depends on materials.
In the light of the above statements, choose the most appropriate answer from the options given below. *[2024, 27 April Shift-II]*

(a) Statement I is incorrect but Statement II is correct.
(b) Statement I is correct but Statement II is incorrect.
(c) Both Statement I and Statement II are incorrect.
(d) Both Statement I and Statement II are correct.

03 A heavy box of mass 50 kg is moving on a horizontal surface. If coefficient of kinetic friction between the box and horizontal surface is 0.3, then force of kinetic friction is *[2024, 5 April Shift-II]*

(a) 1.47 N (b) 14.7 N (c) 147 N (d) 1470 N

04 The vehicles carrying inflammable fluids usually have metallic chains touching the ground. *[2024, 5 April Shift-II]*

(a) It is a custom
(b) To alert other vehicles
(c) To conduct excess charge due to air friction to ground and prevent sparking
(d) To protect tyres from catching dirt from ground

05 A body of mass 10 kg is moving with an initial speed of 20 m/s. The body stops after 5 s due to friction between body and the floor. The value of the coefficient of friction is
(Take, acceleration due to gravity $g = 10$ ms^{-2}) *[2023, 31 Jan Shift-II]*

(a) 0.3 (b) 0.5 (c) 0.4 (d) 0.2

Imp. Kinetic Friction

06 A block of mass m slides down the plane inclined at angle 30° with an acceleration $g/4$. The value of coefficient of kinetic friction will be *[2023, 29 Jan Shift-I]*

(a) $\frac{2\sqrt{3}+1}{2}$ (b) $\frac{\sqrt{3}}{2}$ (c) $\frac{2\sqrt{3}-1}{2}$ (d) $\frac{1}{2\sqrt{3}}$

07 A block of mass M slides down on a rough inclined plane with constant velocity. The angle made by the incline plane with horizontal is θ. The magnitude of the contact force will be *[2022, 27 July Shift-II]*

(a) Mg (b) $Mg \cos\theta$

(c) $\sqrt{Mg \sin\theta + Mg \cos\theta}$ (d) $Mg \sin\theta \sqrt{1+\mu}$

08 A bag is gently dropped on a conveyor belt moving at a speed of 2 m/s. The coefficient of friction between the conveyor belt and bag is 0.4. Initially the bag slips on the belt before it stops due to friction. The distance travelled by the bag on the belt during slipping motion is [Take, $g = 10\ m/s^2$] *[2022, 27 July Shift-I]*

(a) 2 m (b) 0.5 m (c) 3.2 m (d) 0.8 m

09 Given in the figure are two blocks A and B of weight 20 N and 100 N respectively. These are being pressed against a wall by a force F as shown in figure. If the coefficient of friction between the blocks is 0.1 and between block B and the wall is 0.15, the frictional force applied by the wall in block B is *[JEE Main 2015]*

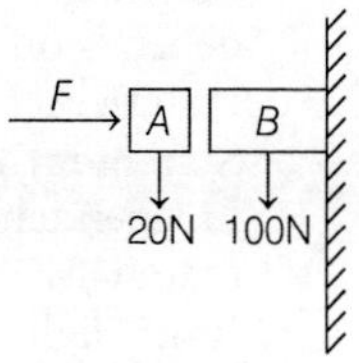

(a) 100 N (b) 80 N (c) 120 N (d) 150 N

10 Consider a car moving on a straight road with a speed of $100\ ms^{-1}$. The distance at which car can be stopped, is [$\mu_k = 0.5$] *[AIEEE 2005]*

(a) 800 m (b) 1000 m

(c) 100 m (d) 400 m

11 A horizontal force of 10 N is necessary to just hold a block stationary against a wall. The coefficient of friction between the block and the wall is 0.2. The weight of the block is *[AIEEE 2003]*

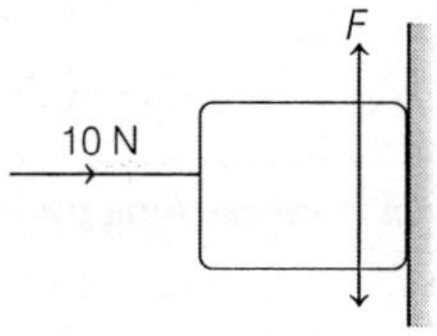

(a) 20 N (b) 50 N (c) 100 N (d) 2 N

Level 2

12 A 2 kg brick begins to slide over a surface which is inclined at an angle of $45°$ with respect to horizontal axis. The coefficient of static friction between their surfaces is *[2024, 4 April Shift-II]*

(a) 1.7 (b) 1 (c) 0.5 (d) $\frac{1}{\sqrt{3}}$

13 A block of mass 5 kg is placed at rest on a table of rough surface. Now, if a force of 30N is applied in the direction parallel to surface of the table, the block slides through a distance of 50 m in an interval of time 10s. Coefficient of kinetic friction is (Given, $g = 10\ ms^{-2}$) *[2023, 1 Feb Shift-I]*

(a) 0.50 (b) 0.75 (c) 0.25 (d) 0.60

14 A block of mass 40 kg slides over a surface, when a mass of 4 kg is suspended through an inextensible massless string passing over frictionless pulley as shown below. the coefficient of kinetic friction between the surface and block is 0.02. The acceleration of block is (Given, $g = 10\ ms^{-2}$). *[2022, 29 June Shift-II]*

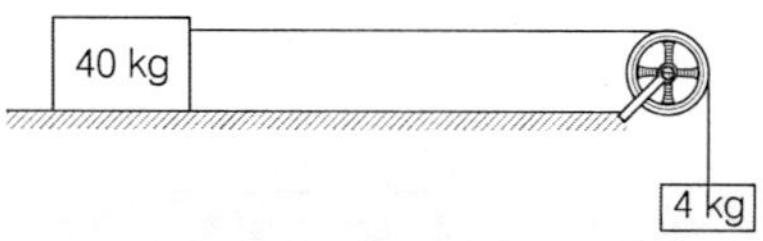

(a) $1\ ms^{-1}$ (b) $1/5\ ms^{-2}$ (c) $4/5\ ms^{-2}$ (d) $8/11\ ms^{2}$

15 A system of two blocks of masses $m = 2$ kg and $M = 8$ kg is placed on a smooth table as shown in figure. The coefficient of static friction between two blocks is 0.5. The maximum horizontal force F that can be applied to the block of mass M, so that the blocks move together will be *[2022, 27 June Shift-I]*

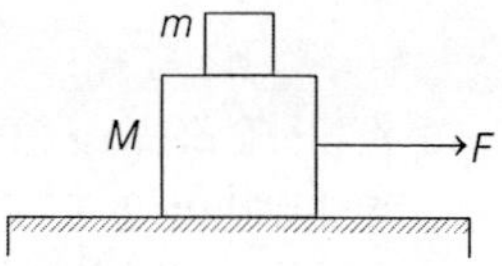

(a) 9.8 N (b) 39.2 N (c) 49 N (d) 78.4 N

16 A steel block of 10 kg rests on a horizontal floor as shown. When three iron cylinders are placed on it as shown, the block and cylinders go down with an acceleration $0.2\ m/s^2$. The normal reaction R by the floor, if mass of the iron cylinders are equal and of 20 kg each, is ...... N. [Take, $g = 10\ m/s^2$ and $\mu_s = 0.2$] *[2021, 20 July Shift-I]*

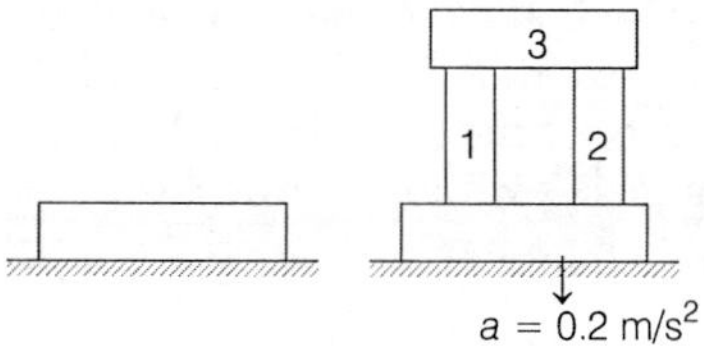

(a) 716 (b) 686 (c) 714 (d) 684

17 A block of mass m slides along a floor, while a force of magnitude F is applied to it at an angle θ as shown in figure. The coefficient of kinetic friction is μ_k. Then, the block's acceleration a is given by (g is acceleration due to gravity) *[2021, 16 March Shift-I]*

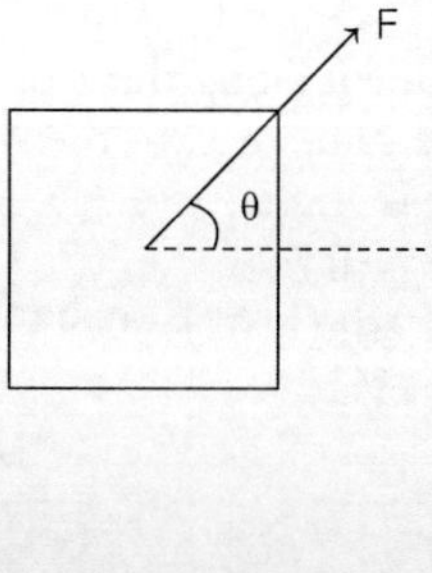

(a) $-\frac{F}{m}\cos\theta-\mu_k\left(g-\frac{F}{m}\sin\theta\right)$

(b) $\frac{F}{m}\cos\theta-\mu_k\left(g-\frac{F}{m}\sin\theta\right)$

(c) $\frac{F}{m}\cos\theta-\mu_k\left(g+\frac{F}{m}\sin\theta\right)$

(d) $\frac{F}{m}\cos\theta+\mu_k\left(g-\frac{F}{m}\sin\theta\right)$

18 Two masses $m_1=5$ kg and $m_2=10$ kg connected by an inextensible string over a frictionless pulley, are moving as shown in the figure. The coefficient of friction of horizontal surface is 0.15. The minimum weight m that should be put on top of m_2 to stop the motion is *[JEE Main 2018]*

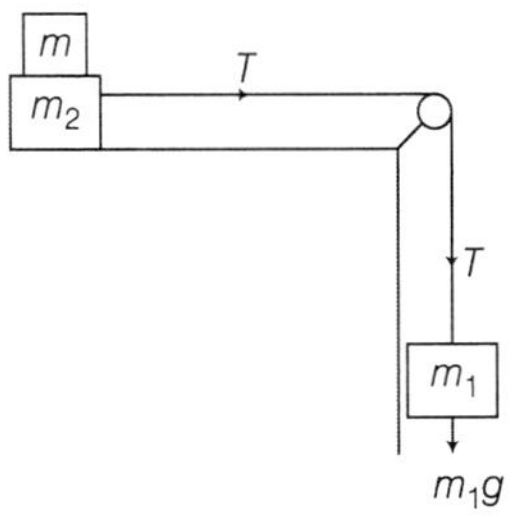

(a) 18.3 kg (b) 27.3 kg (c) 43.3 kg (d) 10.3 kg

19 The minimum force required to start pushing a body up a rough (frictional coefficient μ) inclined plane is F_1 while the minimum force needed to prevent it from sliding down is F_2. If the inclined plane makes an angle θ from the horizontal such that $\tan\theta=2\mu$, then the ratio F_1/F_2 is *[AIEEE 2011]*

(a) 4 (b) 1 (c) 2 (d) 3

Level 3

20 A given object takes n times the time to slide down 45° rough inclined plane as it takes the time to slide down an identical perfectly smooth 45° inclined plane. The coefficient of kinetic friction between the object and the surface of inclined plane is *[2024, 8 April Shift-II 2023, 29 Jan Shift-II]*

(a) $1-n^2$ (b) $1-\frac{1}{n^2}$ (c) $\sqrt{1-n^2}$ (d) $\sqrt{1-\frac{1}{n^2}}$

21

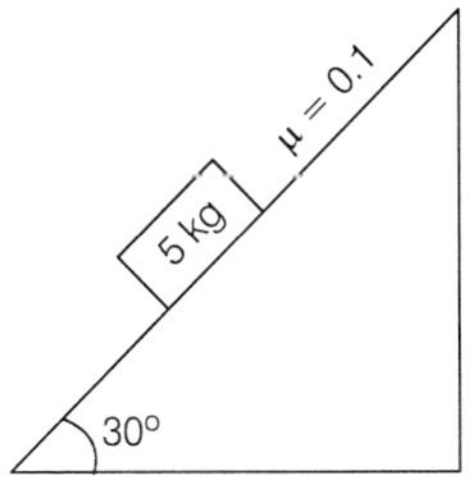

A block of mass 5 kg is placed on a rough inclined surface as shown in the figure. If F_1 is the force required to just move the block up the inclined plane and F_2 is the force required to just prevent the block from sliding down, then the value of $|F_1|-|F_2|$ is [Use, $g=10\ m/s^2$] *[2024, 31 Jan Shift-II]*

(a) 10 N (b) $50\sqrt{3}$ N

(c) $\frac{5\sqrt{3}}{2}$ N (d) $25\sqrt{3}$ N

22 A block of mass m is placed on a surface having vertical cross-section given by $y=x^2/4$. If coefficient of friction is 0.5, the maximum height above the ground at which block can be placed without slipping is *[2024, 30 Jan Shift-II]*

(a) $\frac{1}{6}$ m (b) $\frac{1}{4}$ m (c) $\frac{1}{3}$ m (d) $\frac{1}{2}$ m

23 As shown in the figure, a block of mass 10 kg lying on a horizontal surface is pulled by a force F acting at an angle 30° with horizontal. For $\mu_s=0.25$, the block will just start to move for the value of F.
(Given, $g=10\ ms^{-2}$) *[2023, 1 Feb Shift-II]*

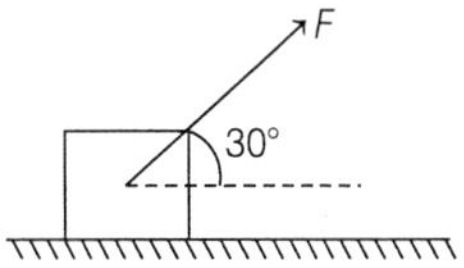

(a) 25.2 N (b) 33.3 N
(c) 20 N (d) 35.7 N

24 Consider a block kept on an inclined plane (inclined at 45°) as shown in the figure.
If the force required to just push it up the incline is 2 times the force required to just prevent it from sliding down. The coefficient of friction between the block and inclined plane (μ) is equal to *[2023, 25 Jan Shift-II]*

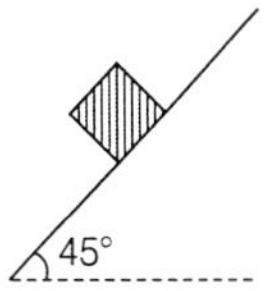

(a) 0.50 (b) 0.25 (c) 0.33 (d) 0.60

25 A block kept on a rough inclined plane, as shown in the figure, remains at rest upto a maximum force 2 N down the inclined plane. The maximum external force up the inclined plane that does not move the block is 10 N. The coefficient of static friction between the block and the plane is
(Take, $g=10\ m/s^2$) *[2019, 12 Jan Shift-II]*

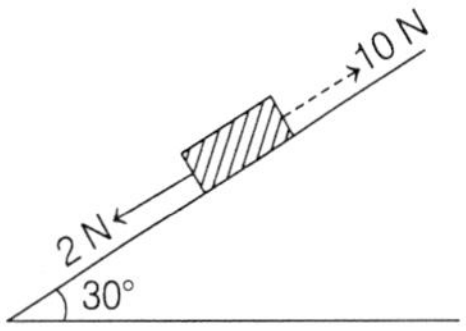

(a) 2/3 (b) $\sqrt{3}/2$ (c) $\sqrt{3}/4$ (d) 1/2

26 A block of mass m is placed on a surface with a vertical cross-section given by $y = x^3 / 6$. If the coefficient of friction is 0.5, the maximum height above the ground at which the block can be placed without slipping is *[JEE Main 2014]*

(a) $\frac{1}{6}$ m (b) $\frac{2}{3}$ m (c) $\frac{1}{3}$ m (d) $\frac{1}{2}$ m

Numerical Value Type Questions

27 A uniform chain of 6 m length is placed on a table such that a part of its length is hanging over the edge of the table. The system is at rest. The co-efficient of static friction between the chain and surface of the table is 0.5, the maximum length of the chain hanging from the table is .... m. *[2022, 25 June Shift-I]*

28 A body of mass m is launched up on a rough inclined plane making an angle of 30° with the horizontal. The coefficient of friction between the body and plane is $\frac{\sqrt{x}}{5}$ if the time of ascent is half of the time of descent. The value of x is .... *[2021, 20 July Shift-II]*

29 A body of mass 1 kg rests on a horizontal floor with which it has a coefficient of static friction $1/\sqrt{3}$. It is desired to make the body move by applying the minimum possible force F N. The value of F will be .... (Round off to the nearest integer).

[Take, $g = 10 \text{ m s}^{-2}$] *[2021, 17 March Shift-II]*

30 When a body slides down from rest along a smooth inclined plane making an angle of 30° with the horizontal, it takes time T. When the same body slides down from the rest along a rough inclined plane making the same angle and through the same distance, it takes time αT, where α is a constant greater than 1. The coefficient of friction between the body and the rough plane is $\frac{1}{\sqrt{x}}\left(\frac{\alpha^2 - 1}{\alpha^2}\right)$, where x is .... *[2021, 1 Sep Shift-II]*

31 A boy of mass 4 kg is standing on a piece of wood having mass 5 kg . If the coefficient of friction between the wood and the floor is 0.5, the maximum force that the boy can exert on the rope, so that the piece of wood does not move from its place is .... N. (Round off to the nearest integer)

(Take, $g = 10 \text{ ms}^{-2}$) *[2021, 17 March Shift-II]*

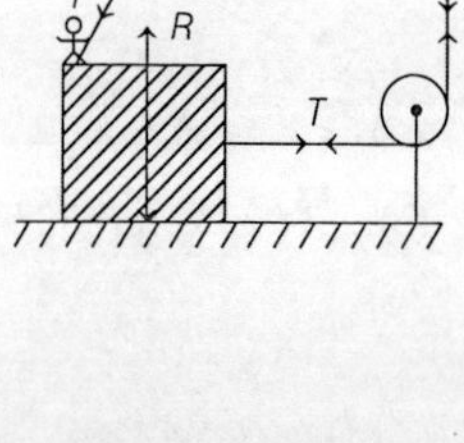

Imp. Static Friction

32 Two blocks ($m = 0.5$ kg and $M = 4.5$ kg) are arranged on a horizontal frictionless table as shown in figure. The coefficient of static friction between the two blocks is 3/7. Then, the maximum horizontal force that can be applied on the larger block so that the blocks move together is ...... N. (Round off to the nearest integer.
Take, $g = 9.8 \text{ ms}^{-2}$) *[2021, 17 March Shift-I]*

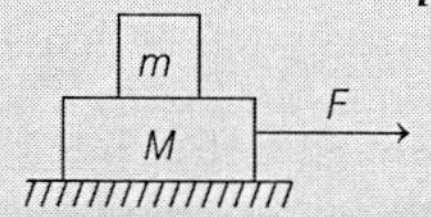

33 As shown in the figure, a block of mass $\sqrt{3}$ kg is kept on a horizontal rough surface of coefficient of friction $1/3\sqrt{3}$. The critical force to be applied on the vertical surface as shown at an angle 60° with horizontal such that it does not move, will be $3x$. The value of x will be ....

[$g = 10 \text{ ms}^{-2}$ $\sin 60° = \frac{\sqrt{3}}{2}$, $\cos 60° = \frac{1}{2}$] *[2021, 26 Feb Shift-I]*

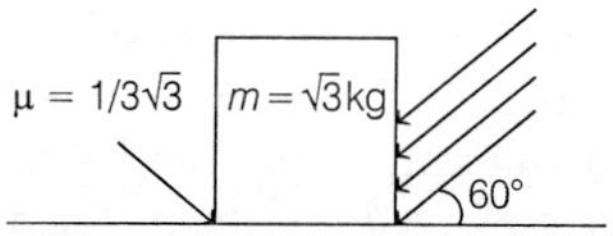

34 An inclined plane is bent in such a way that the vertical cross-section is given by $y = \frac{x^2}{4}$ where, y is in vertical and x in horizontal direction. If the upper surface of this curved plane is rough with coefficient of friction $\mu = 0.5$, the maximum height in cm at which a stationary block will not slip downward is .... cm. *[2021, 24 Feb Shift-I]*

35 The coefficient of static friction between a wooden block of mass 0.5 kg and a vertical rough wall is 0.2. The magnitude of horizontal force that should be applied on the block to keep it adhere to the wall will be .... N. [Take, $g = 10 \text{ ms}^{-2}$] *[2021, 24 Feb Shift-I]*

Topic 4 Dynamics of Circular Motion

01 A vehicle of mass 200 kg is moving along a levelled curved road of radius 70 m with angular velocity of 0.2 rad/s. The centripetal force acting on the vehicle is *[2023, 13 April Shift-II]*

(a) 560 N (b) 2240 N (c) 14 N (d) 2800 N

02 A child of mass 5 kg is going round a merry-go-round that makes 1 rotation in 3.14 s. The radius of the merry-go-round is 2 m. The centrifugal force on the child will be *[2023, 6 April Shift-II]*

(a) 40 N (b) 100 N (c) 80 N (d) 50 N

Imp. Motion on Curved Road

03 A car is moving on a horizontal curved road with radius 50 m. The approximate maximum speed of car will be, if friction between tyres and road is 0.34. [Take, $g = 10\ \text{ms}^{-2}$]. **[2023, 29 Jan Shift-I]**

(a) $17\ \text{ms}^{-1}$ (b) $3.4\ \text{ms}^{-1}$
(c) $13\ \text{ms}^{-1}$ (d) $22.4\ \text{ms}^{-1}$

04 A car is moving with a constant speed of 20 m/s in a circular horizontal track of radius 40 m. A bob is suspended from the roof of the car by a massless string. The angle made by the string with the vertical will be (Take, $g = 10\ \text{m/s}^2$) **[2023, 25 Jan Shift-I]**

(a) $\frac{\pi}{3}$ (b) $\frac{\pi}{2}$ (c) $\frac{\pi}{4}$ (d) $\frac{\pi}{6}$

05 A disc with a flat small bottom beaker placed on it at a distance R from its centre is revolving about an axis passing through the center and perpendicular to its plane with an angular velocity ω. The coefficient of static friction between the bottom of the beaker and the surface of the disc is μ. The beaker will revolve with the disc if **[2022, 25 June Shift-II]**

(a) $R \le \frac{\mu g}{2\omega^2}$ (b) $R \le \frac{\mu g}{\omega^2}$ (c) $R \ge \frac{\mu g}{2\omega^2}$ (d) $R \ge \frac{\mu g}{\omega^2}$

06 A smooth circular groove has a smooth vertical wall as shown in figure. A block of mass m moves against the wall with a speed v. Which of the following curve represents the correct relation between the normal reaction on the block by the wall (N) and speed of the block (v) ? **[2022, 29 July Shift-I]**

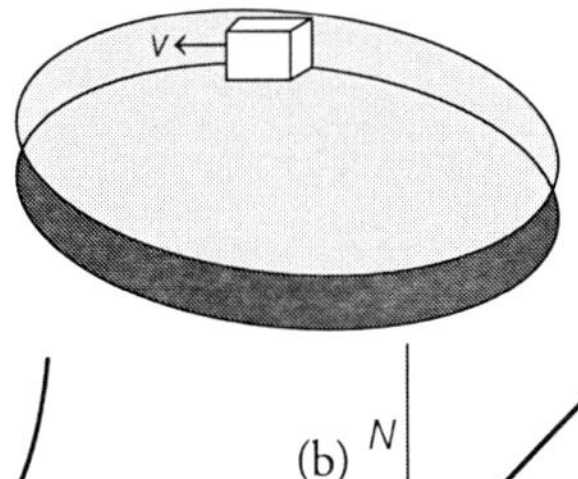

(a) 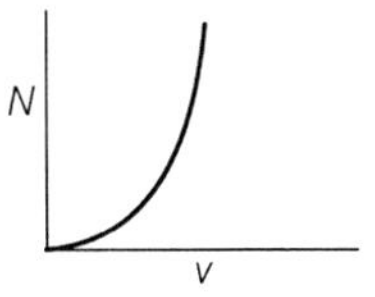

(b) 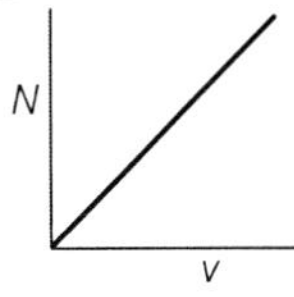

(c) 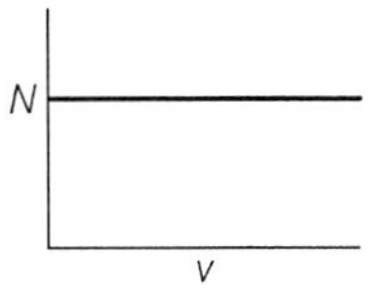

(d) 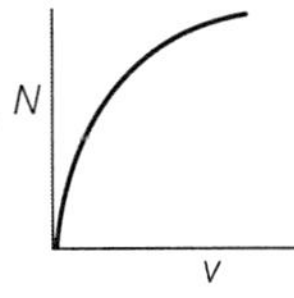

07 A stone of mass m, tied to a string is being whirled in a vertical circle with a uniform speed. The tension in the string is **[2022, 24 June Shift-II]**

(a) the same throughout the motion
(b) minimum at the highest position of the circular path
(c) minimum at the lowest position of the circular path
(d) minimum when the rope is the horizontal position

08 A particle of mass m is fixed to one end of a light spring having force constant k and unstretched length l. The other end is fixed. The system is given an angular speed ω about the fixed end of the spring such that it rotates in a circle in gravity free space. Then, the stretch in the spring is **[2020, 8 Jan Shift-I]**

(a) $\frac{ml\omega^2}{k - \omega m}$ (b) $\frac{ml\omega^2}{k - m\omega^2}$ (c) $\frac{ml\omega^2}{k + m\omega^2}$ (d) $\frac{ml\omega^2}{k + \omega m}$

09 A spring mass system (mass m, spring constant k and natural length l) rests in equilibrium on a horizontal disc. The free end of the spring is fixed at the centre of the disc. If the disc together with spring mass system, rotates about its axis with an angular velocity ω, ($k >> m\omega^2$) the relative change in the length of the spring is best given by the option **[2020, 9 Jan Shift-II]**

(a) $\frac{2m\omega^2}{k}$ (b) $\frac{m\omega^2}{3k}$
(c) $\frac{m\omega^2}{k}$ (d) $\sqrt{\frac{2}{3}}\left(\frac{m\omega^2}{k}\right)$

10 Concrete mixture is made by mixing cement, stone and sand in a rotating cylindrical drum. If the drum rotates too fast, the ingredients remain stuck to the wall of the drum and proper mixing of ingredients does not take place. The maximum rotational speed of the drum in revolutions per minute (rpm) to ensure proper mixing is close to (Take the radius of the drum to be 1.25 m and its axis to the horizontal) **[JEE main 2016]**

(a) 27.0 (b) 0.4 (c) 1.3 (d) 8.0

Level 2

Imp. Banking of Road

11 A car of mass m moves on a banked road having radius r and banking angle θ. To avoid slipping from banked road, the maximum permissible speed of the car is v_0. The coefficient of friction μ between the wheels of the car and the banked road is **[2025, 24 Jan Shift-I]**

(a) $\mu = \frac{v_0^2 - rg\tan\theta}{rg - v_0^2\tan\theta}$ (b) $\mu = \frac{v_0^2 - rg\tan\theta}{rg + v_0^2\tan\theta}$
(c) $\mu = \frac{v_0^2 + rg\tan\theta}{rg + v_0^2\tan\theta}$ (d) $\mu = \frac{v_0^2 + rg\tan\theta}{rg - v_0^2\tan\theta}$

12 A coin is placed on a disc. The coefficient of friction between the coin and the disc is μ. If the distance of the coin from the centre of the disc is r, the maximum angular velocity which can be given to the disc, so that the coin does not slip away, is **[2024, 31 Jan Shift-I]**

(a) $\sqrt{r/\mu g}$ (b) $\mu/\sqrt{rg}$ (c) $\mu g/r$ (d) $\sqrt{\mu g/r}$

13 If the radius of curvature of the path of two particles of same mass are in the ratio 3 : 4, then in order to have constant centripetal force, their velocities will be in the ratio of *[2024, 29 Jan Shift-I]*

(a) $2:\sqrt{3}$ (b) $\sqrt{3}:2$ (c) $1:\sqrt{3}$ (d) $\sqrt{3}:1$

14 A train is moving with a speed of 12 m/s on rails which are 1.5 m apart. To negotiate a curve of radius 400 m, the height by which the outer rail should be raised with respect to the inner rail is (Given, $g = 10\ \text{m/s}^2$) *[2024, 27 Jan Shift-I]*

(a) 4.8 cm (b) 5.4 cm (c) 6.0 cm (d) 4.2 cm

15 A stone of mass 1 kg is tied to end of a massless string of length 1 m. If the breaking tension of the string is 400 N, then maximum linear velocity the stone can have without breaking the string while rotating in horizontal plane is *[2023, 31 Jan Shift-II]*

(a) $10\ \text{ms}^{-1}$ (b) $40\ \text{ms}^{-1}$ (c) $400\ \text{ms}^{-1}$ (d) $20\ \text{ms}^{-1}$

16 One end of a massless spring of spring constant k and natural length l_0 is fixed while the other end is connected to a small objects of mass m lying on a frictionless table. The spring remains horizontal on the table. If the object is made to rotate at an angular velocity ω about an axis passing through fixed end, then the elongation of the spring will be *[2022, 27 June Shift-II]*

(a) $\dfrac{k - m\omega^2 l_0}{m\omega^2}$ (b) $\dfrac{m\omega^2 l_0}{k + m\omega^2}$

(c) $\dfrac{m\omega^2 l_0}{k - m\omega^2}$ (d) $\dfrac{k + m\omega^2 l_0}{m\omega^2}$

17 For a particle in uniform circular motion, the acceleration **a** at any point $P(R, \theta)$ on the circular path of radius R is (when θ is measured from the positive X-axis and v is uniform speed) *[2022, 25 June Shift-II]*

(a) $-\dfrac{v^2}{R}\sin\theta\,\hat{\mathrm{i}} + \dfrac{v^2}{R}\cos\theta\,\hat{\mathrm{j}}$ (b) $-\dfrac{v^2}{R}\cos\theta\,\hat{\mathrm{i}} + \dfrac{v^2}{R}\sin\theta\,\hat{\mathrm{j}}$

(c) $-\dfrac{v^2}{R}\cos\theta\,\hat{\mathrm{i}} - \dfrac{v^2}{R}\sin\theta\,\hat{\mathrm{j}}$ (d) $-\dfrac{v^2}{R}\hat{\mathrm{i}} + \dfrac{v^2}{R}\hat{\mathrm{j}}$

18 A block of 200 g mass moves with a uniform speed in a horizontal circular groove, with vertical side walls of radius 20 cm. If the block takes 40 s to complete one round, the normal force by the side walls of the groove is *[2021, 16 March Shift-I]*

(a) 0.0314 N (b) 9.859×10^{-2} N

(c) 6.28×10^{-3} N (d) 9.859×10^{-4} N

19 A uniform rod length l is being rotated in a horizontal plane with a constant angular speed about an axis passing through one of its ends. If the tension generated in the rod due to rotation is $T(x)$ at a distance x from the axis, then which of the following graphs depicts it most closely ? *[2019, 12 April Shift-I]*

(a)
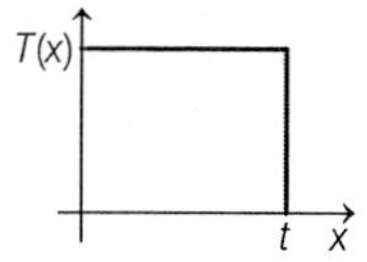

(b)
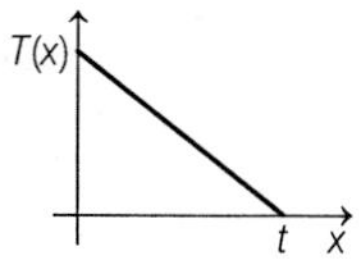

(c)
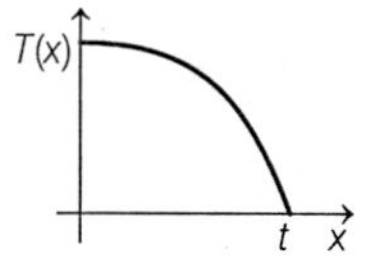

(d)
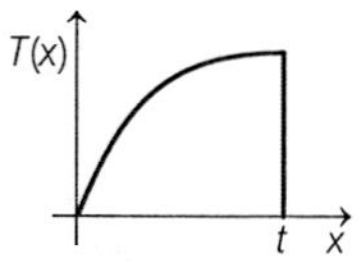

20 A disc rotates about its axis of symmetry in a horizontal plane at a steady rate of 3.5 revolutions per second. A coin placed at a distance of 1.25 cm from the axis of rotation remains at rest on the disc. The coefficient of friction between the coin and the disc is $[g = 10\ \text{m/s}^2]$ *[JEE main 2018]*

(a) 0.7 (b) 0.5 (c) 0.3 (d) 0.6

21 A particle is moving with a uniform speed in a circular orbit of radius R in a central force inversely proportional to the nth power of R. If the period of rotation of the particle is T, then *[JEE Main 2018]*

(a) $T \propto R^{3/2}$ for any n (b) $T \propto R^{\frac{n}{2}+1}$

(c) $T \propto R^{(n+1)/2}$ (d) $T \propto R^{n/2}$

22 A conical pendulum of length 1m makes an angle $\theta = 45°$ w.r.t Z-axis and moves in a circle in the XY plane. The radius of the circle is 0.4 m and its centre is vertically below O. The speed of the pendulum, in its circular path, will be (Take, $g = 10\ \text{m s}^{-2}$) *[JEE Main 2017]*

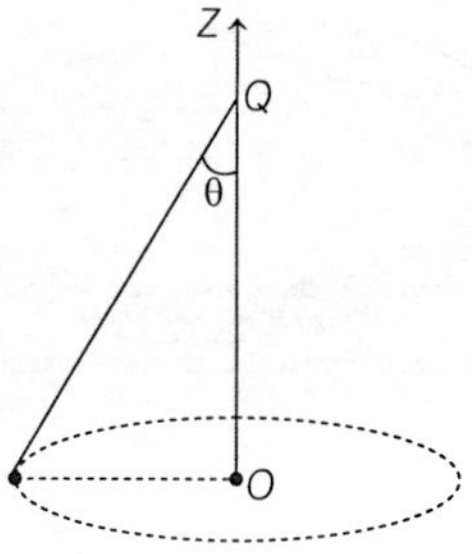

(a) 0.4 m/s (b) 2 m/s
(c) 0.2 m/s (d) 4 m/s

Level 3

23 A body of mass 200 g is tied to a spring of spring constant 12.5 N/m, while the other end of spring is fixed at point O. If the body moves about O in a circular path on a smooth horizontal surface with constant angular speed 5 rad/s, then the ratio of extension in the spring to its natural length will be *[2023, 24 Jan Shift-II]*

(a) $2:5$ (b) $1:2$ (c) $2:3$ (d) $1:1$

24 A ball is released from rest from point P of a smooth semi-spherical vessel as shown in figure. The ratio of the centripetal force and normal reaction on the ball at point Q is A while angular position of point Q is α with respect to point P. Which of the following graphs represent the correct relation between A and α when ball goes from Q to R? ***[2022, 26 June Shift-I]***

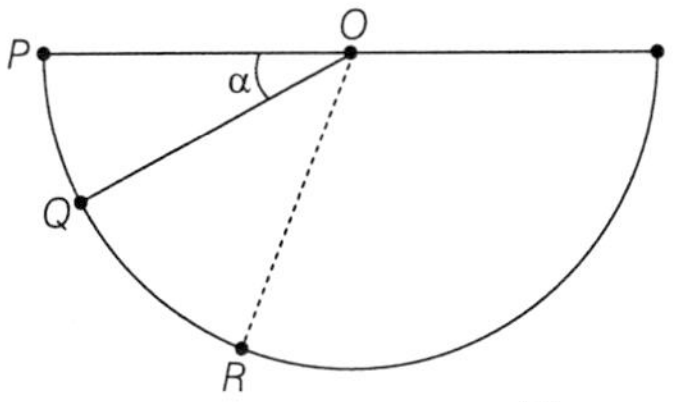

(a)
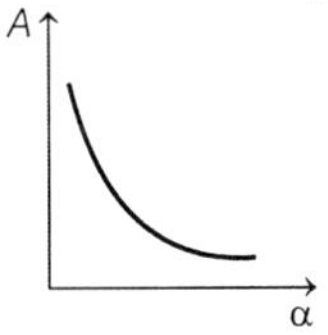

(b)
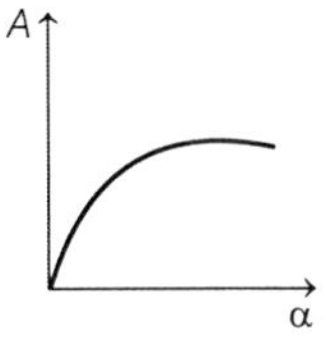

(c)
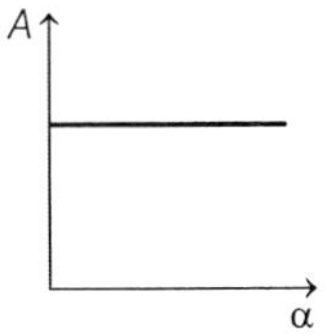

(d)
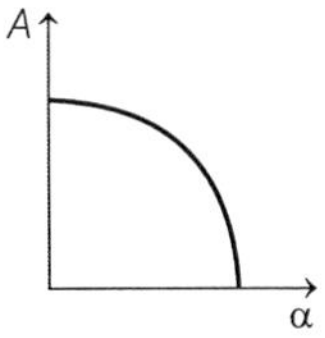

25 A boy ties a stone of mass 100 g to the end of a 2m long string and whirls it around in a horizontal plane. The string can withstand the maximum tension of 80 N. If the maximum speed with which the stone can revolve is $\frac{K}{\pi}$ rev/min. The value of K is (Assume the string is massless and unstretchable) ***[2022, 24 June Shift-I]***

(a) 400 (b) 300 (c) 600 (d) 800

26 A particle of mass m is suspended from a ceiling through a string of length L. The particle moves in a horizontal circle of radius r such that $r = \frac{L}{\sqrt{2}}$. The speed of particle will be ***[2021, 26 Aug Shift-II]***

(a) $\sqrt{rg}$ (b) $\sqrt{2rg}$ (c) $2\sqrt{rg}$ (d) $\sqrt{\frac{rg}{2}}$

27 The normal reaction N for a vehicle of 800 kg mass, negotiating a turn on a 30° banked road at maximum possible speed without skidding is ... $\times 10^3$ kg-m/s^2. [Take, $\mu_s = 0.2$] ***[2021, 20 July Shift-I]***

(a) 10.2 (b) 7.2 (c) 12.4 (d) 6.96

28 A modern grand-prix racing car of mass m is travelling on a flat track in a circular arc of radius R with a speed v. If the coefficient of static friction between the tyres and the track is μ_s, then the magnitude of negative lift f_L acting downwards on the car is

(Assume forces on the four tyres are identical and g = acceleration due to gravity) ***[2021, 17 March Shift-I]***

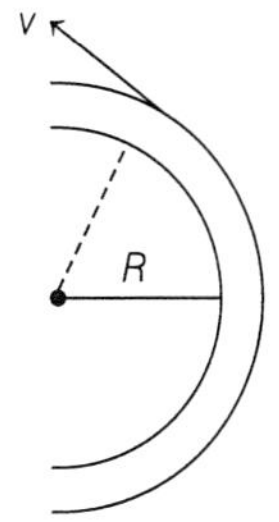

(a) $m\left(\frac{v^2}{\mu_s R} + g\right)$ (b) $m\left(\frac{v^2}{\mu_s R} - g\right)$

(c) $m\left(g - \frac{v^2}{\mu_s R}\right)$ (d) $-m\left(g + \frac{v^2}{\mu_s R}\right)$

29 **Statement I** A cyclist is moving on an unbanked road with a speed of 7 kmh^{-1} and takes a sharp circular turn along a path of radius of 2 m without reducing the speed The static friction coefficient is 0.2. The cyclist will not slip and pass the curve ($g = 9.8$ m / s^2)

Statement II If the road is banked at an angle of 45°, cyclist can cross the curve of 2 m radius with the speed of 18.5 kmh^{-1} without slipping.

In the light of the above statements, choose the correct answer from the options given below. ***[2021, 16 March Shift-II]***

(a) Statement I is false and Statement II is true.
(b) Statement I is true and Statement II is false.
(c) Both statement I and Statement II are false.
(d) Both statement I and Statement II are true.

Numerical Value Type Questions

30

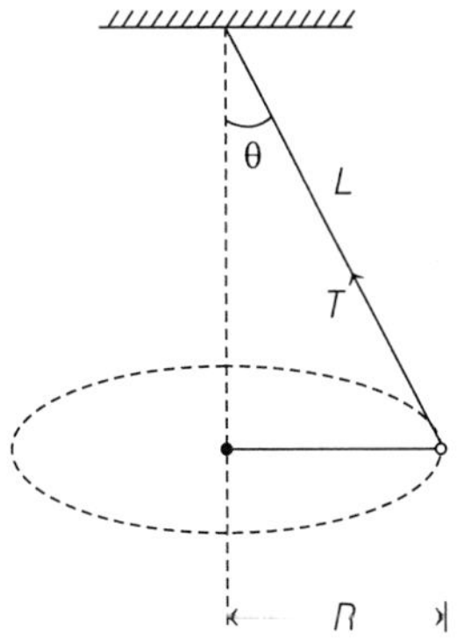

A string of length L is fixed at one end and carries a mass of M at the other end. The mass makes $3/\pi$ rotations per second about the vertical axis passing through end of the string as shown. The tension in the string is .... ML. ***[2025, 24 Jan Shift-II]***

31 A tube of length 1 m is filled completely with an ideal liquid of mass $2M$ and closed at both ends. The tube is rotated uniformly in horizontal plane about one of its ends. If the force exerted by the liquid at the other end is F, then angular velocity of the tube is $\sqrt{F/\alpha M}$ in SI unit. The value of α is ....

[2025, 22 Jan Shift-II]

32 A car is moving on a circular path of radius 600 m such that the magnitudes of the tangential acceleration and centripetal acceleration are equal. The time taken by the car to complete first quarter of revolution, if it is moving with an initial speed of 54 km/h is $t(1-e^{-\pi/2})$ s. The value of t is ....

[2023, 29 Jan Shift-II]

33 A curved in a level road has a radius 75 m. The maximum speed of a car turning this curved road can be 30 m/s without skidding. If radius of curved road is changed to 48 m and the coefficient of friction between the tyres and the road remains same, then maximum allowed speed would be .... m/s.

[2022, 25 June Shift-II]

34 A small bob tied at one end of a thin string of length 1 m is describing a vertical circle so that the maximum and minimum tension in the string are in the ratio 5 : 1. The velocity of the bob at the highest position is .... m/s. (Take, $g = 10\ m/s^2$)

[2021, 25 Feb Shift-I]

• Key Index

Topic 1

1 (d)	2 (c)	3 (b)	4 (c)	5 (a)	6 (b)	7 (c)	8 (a)	9 (a)	10 (a)
11 (a)	12 (b)	13 (c)	14 (d)	15 (d)	16 (c)	17 (b)	18 (c)	19 (c)	20 (b)
21 (b)	22 (b)	23 (d)	24 (b)	25 (b)	26 (d)	27 (b)	28 (b)	29 (d)	30 (d)
31 (c)	32 (a)	33 (c)	34 (d)	35 (c)	36 (b)	37 (c)	38 (b)	39 (d)	40 (6)
41 (12)	42 (5)	43 (10)							

Topic 2

1 (c)	2 (d)	3 (a)	4 (b)	5 (c)	6 (a)	7 (c)	8 (d)	9 (d)	10 (a)
11 (c)	12 (a)	13 (a)	14 (a)	15 (c)	16 (d)	17 (b)	18 (a)	19 (d)	20 (d)
21 (c)	22 (c)	23 (a)	24 (b)	25 (d)	26 (c)	27 (d)	28 (d)	29 (a)	30 (b)
31 (c)	32 (a)	33 (a)	34 (b)	35 (d)	36 (a)	37 (b)	38 (d)	39 (d)	40 (d)
41 (240)	42 (5)	43 (2)	44 (3)	45 (12)	46 (30)	47 (82)			

Topic 3

1 (b)	2 (a)	3 (c)	4 (c)	5 (c)	6 (d)	7 (a)	8 (b)	9 (c)	10 (b)
11 (d)	12 (b)	13 (a)	14 (d)	15 (c)	16 (b)	17 (b)	18 (b)	19 (d)	20 (b)
21 (*)	22 (b)	23 (a)	24 (c)	25 (b)	26 (a)	27 (2)	28 (3)	29 (5)	30 (3)
31 (30)	32 (21)	33 (3.33)	34 (25)	35 (25)					

Topic 4

1 (a)	2 (a)	3 (c)	4 (c)	5 (b)	6 (a)	7 (b)	8 (b)	9 (c)	10 (a)
11 (b)	12 (d)	13 (b)	14 (b)	15 (d)	16 (c)	17 (c)	18 (d)	19 (c)	20 (d)
21 (c)	22 (b)	23 (c)	24 (c)	25 (c)	26 (a)	27 (a)	28 (b)	29 (d)	30 (36)
31 (1)	32 (40)	33 (24)	34 (5)						

Detailed **Explanations**

Topic 1

01. *(d)* Speed of object

$v = 4\sqrt{x}$ m/s

∴ Acceleration

$$a = \frac{dv}{dt} = \frac{dv}{dx}\cdot\frac{dx}{dt} = \frac{dv}{dx}\cdot v$$

$$= v\frac{d}{dx}(4\sqrt{x}) = 4\sqrt{x}\cdot 4\cdot\frac{1}{2\sqrt{x}}$$

$$= 8 \text{ m/s}^2$$

Thus, $F = m\times a = 0.5\times 8 = 4$ N

02. *(c)* Linear momentum is conserved,

$$p_i = p_f$$

$$m_1v_1 = m_2v_2$$

$$1000\times 6 = 1200\times v_2$$

$$v_2 = 5 \text{ m/s}$$

03. *(b)* Mass of the bullet,

$$m_b = 10 \text{ g} = 0.01 \text{ kg}$$

Speed of the bullet, $v_b = 600$ m/s

$$\mathbf{I}_b = \mathbf{p}_f - \mathbf{p}_i$$

$$= 0.01\times 600 - 0 = 6 \text{ N-s}$$

$$|\mathbf{I}_g| = |\mathbf{I}_b| = 6 \text{ N-s}$$

04. *(c)* Given, $F = 125$ N

$$F = \frac{dp}{dt}$$

$$p = nmv$$

$$F = \frac{d}{dt}(nmv) = mv\frac{dn}{dt}$$

$$125 = \frac{10}{1000}\times 250\times\frac{dn}{dt}$$

$$\frac{dn}{dt} = \frac{125\times 100}{250}$$

$$\Rightarrow \frac{dn}{dt} = 50$$

05. *(a)* Given, Mass of machine gun,

$$m_1 = 10 \text{ kg}$$

Mass of bullet,

$$m_2 = 20 \text{ g} = 0.02 \text{ kg}$$

Number of bullets per second $= \frac{180}{60} = 3$

Force required to hold the gun,

F = Change in momentum per second

$= 3\times 20\times 10^{-3}\times 100 = 6$ N

Let v' be the recoil velocity of the gun, then

$$\frac{m_1v'}{t} = F \Rightarrow \frac{10\times v'}{1} = 6$$

$$\Rightarrow v' = 0.6 \text{ m/s}$$

> ***Alternate Solution***
>
> By momentum of conservation
>
> Momentum of gun = Momentum of bullets (per second)
>
> $$m_gv_g = 3m_bv_b$$
>
> $$v_g = \frac{3m_bv_b}{m_g} = \frac{3\times 0.02\times 100}{10} = 0.6 \text{ m/s}$$

06. *(b)* Given,

$$m = 10 \text{ g}$$

Air escaping rate from balloon,

$$v = 4.5 \text{ cm/s}, t = 5 \text{ s}$$

$$F_{avg} = \frac{dp}{dt}$$

$$= m\cdot\frac{dv}{dt} + v\frac{dm}{dt}$$

For constant velocity $\frac{dv}{dt} = 0$

$$= 4.5\times\frac{10}{5}$$

$$= 9 \text{ dyne}$$

07. *(c)* Given, $m = 2$ kg

Rate of water striking the block,

$$\frac{dm}{dt} = 1 \text{ kg s}^{-1}$$

Water released by the jet striking the block with a rate of 1 kg s^{-1} at a speed of 10 m/s will exert a force on the block given by, $F = v\frac{dm}{dt} = 10\times 1 = 10$ N

According to Newton's second law of motion, $a = \frac{F}{m} = \frac{10}{2} = 5 \text{ m s}^{-2}$

08. *(a)* At maximum height, the velocity of the body is zero.

$$v = 0$$

We have momentum, $p = mv$

Since, the velocity is zero, its momentum at the highest point is also zero.

09. *(a)* Mass of each ball, $m = 30$ g

Speed, $u = 108$ km/h = 30 m/s

For ball (a) Impulse = Change in momentum

$$I_a = m[u-(-u)] = 2mu$$

For ball (b)

$$I_b = m[\{u\cos\theta - (-u\cos\theta)\}]$$

$$= 2mu\cos\theta$$

So, $\frac{I_a}{I_b} = \frac{2mu}{2mu\cos 45°}$

$$= \sqrt{2}:1$$

10. *(a)* As $F = \frac{dp}{dt}$ or $kt\ dt = dp$

Integrating both sides,

$$k\int_0^T t\ dt = \int_p^{3p} dp$$

$$\frac{kT^2}{2} = [3p - p] = 2p$$

$$\therefore \quad T = 2\sqrt{\frac{p}{k}}$$

11. *(a)* Here, $F = \frac{R}{t^2}v(t)$

$$\Rightarrow m\frac{dv}{dt} = \frac{R}{t^2}v(t)$$

$$\Rightarrow \frac{dv}{v(t)} = \frac{R}{m}\frac{dt}{t^2}$$

On integrating both sides, we get

$$\int\frac{dv}{v(t)} = \frac{R}{m}\int\frac{dt}{t^2}$$

$$\Rightarrow \ln v = -\left(\frac{R}{m}\right)\left(\frac{1}{t}\right) + C$$

Graph between $\ln v$ and $1/t$ is a straight line.

12. *(b)* Given, $m = 3.513$ kg (4 significant digits)

$v = 5.00$ m/s (3 significant digit)

$p = mv = 3.513\times 5.00 = 17.565$ kg m/s

(p can have 3 significant digit)

Hence, $p = 17.6$ kg m/s

13. *(c)* Using impulse- momentum theorem.

$$|F\cdot\Delta t| = |\text{Change in momentum}|$$

$$\Rightarrow F\times 0.1 = |p_f - p_i|$$

As the ball will stop after catching,

$$p_i = mv_i = 0.15\times 20 = 3, p_f = 0$$

$$\Rightarrow F\times 0.1 = 3 \Rightarrow F = 30 \text{ N}$$

14. *(d)* Negative means that a is opposite to displacement

Given, $m = 0.3$ kg, $x = 20$ cm

and $k = 15$ N / m

$\because \quad F = -kx \quad$ ...(i)

But, $\quad F = ma \quad$ ...(ii)

$$\therefore \quad ma = -kx$$

$$\Rightarrow a = -\frac{15}{0.3}\times 20\times 10^{-2}$$

$$\Rightarrow a = -\frac{15}{3}\times 2 = -10 \text{ ms}^{-2}$$

∴ Initial acceleration, $a = 10 \text{ ms}^{-2}$

15. *(d)* Here, $t = \frac{5}{3}$ s

$$\mathbf{F} = 6\hat{\mathbf{k}}$$

$\therefore$ Acceleration, $a = \frac{F}{m} = \frac{6\hat{\mathbf{k}}}{2} = 3\,\hat{\mathbf{k}}$

Since, $\mathbf{v}_{in} = (3\hat{\mathbf{i}} + 4\hat{\mathbf{j}})$ m/s

$\therefore$ Velocity of the body when it emerges from force field, is given by

$$\mathbf{v}' = \mathbf{v}_{in} + a t$$
$$= 3\hat{\mathbf{i}} + 4\hat{\mathbf{j}} + 3\hat{\mathbf{k}} \times \frac{5}{3}$$
$$= 3\hat{\mathbf{i}} + 4\hat{\mathbf{j}} + 5\hat{\mathbf{k}}$$

16. *(c)*

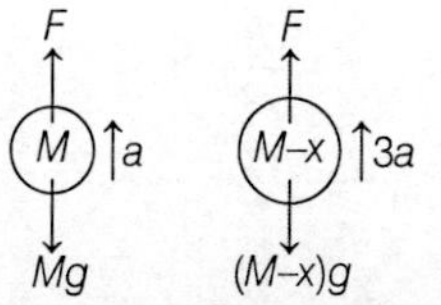

$$F - Mg = Ma$$
$$F = Mg + Ma \quad \ldots\text{(i)}$$
$$F - (M - x)g = (M - x)3a \quad \ldots\text{(ii)}$$

Put Eq. (i) in Eq. (ii)

$$Mg + Ma - Mg + xg = 3Ma - 3xa$$
$$3xa + xg = 3Ma - Ma$$
$$x(3a + g) = 2Ma$$
$$x = \frac{2Ma}{3a + g}$$

17. *(b)* Given,

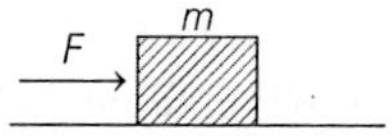

From second law of motion,

$$F = ma$$
$$a = \frac{F}{m}$$

Force applied is linearly dependent on time

$F \propto t \Rightarrow F = kt$ (k = constant)

$$a = \frac{kt}{m}$$

or $a \propto t$

We plot *a versus t* it will be linear from origin.

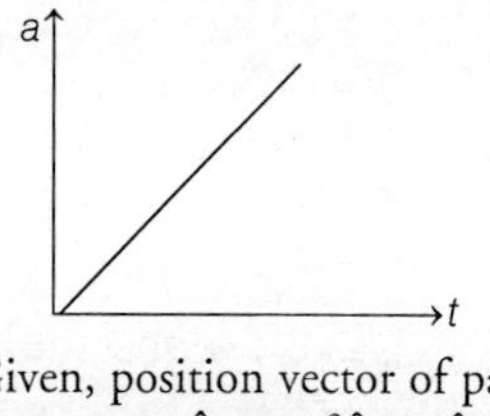

18. *(c)* Given, position vector of particle

$$\mathbf{r} = (10t\hat{\mathbf{i}} + 15t^2\hat{\mathbf{j}} + 7\hat{\mathbf{k}})\text{ m}$$

$\therefore$ Velocity of particle,

$$\mathbf{v} = \frac{d\mathbf{r}}{dt} = 10\hat{\mathbf{i}} + 30t\hat{\mathbf{j}}$$

Acceleration of particle,

$$\mathbf{a} = \frac{d\mathbf{v}}{dt} = 30\hat{\mathbf{j}}$$

$\therefore$ Net force applied on the particle,

$$\mathbf{F} = m\mathbf{a} = 30m\hat{\mathbf{j}}$$

Clearly, net force is applied along Y-axis.

19. *(c)* Let N number of balls strike normally to a wall and be reflected back with the same speed.

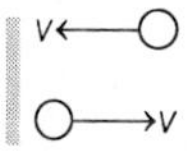

Change in momentum of a ball,

$$\Delta \mathbf{p} = \mathbf{p}_f - \mathbf{p}_i$$
$$= mv\hat{\mathbf{i}} - mv(-\hat{\mathbf{i}}) = 2mv\hat{\mathbf{i}}$$

For N balls $\Delta p_1 = N \times 2mv\hat{\mathbf{i}}$

Average force,

$$(\mathbf{F}_{avg})_{balls} = \frac{\Delta p_1}{\Delta t} = \frac{N \times 2mv\hat{\mathbf{i}}}{t}$$

Given, $N = 100$, $\mathbf{F}_{avg} = \frac{200\,mv\hat{\mathbf{i}}}{t}$

or $F_{avg} = \frac{200\,mv}{t}$

$$|\mathbf{F}_{wall}| = |\mathbf{F}_{balls}|$$
$$= \frac{200\,mv}{t}$$

20. *(b)* Force $= \frac{d}{dt}(p) =$ Slope of p-t graph.

From the graph, it is clear that slope of line c is maximum while slope of line b is minimum, therefore, in region c, magnitude of force is maximum whereas in region b, magnitude of force is minimum.

21. *(b)* Given, $m = 5$ kg, $v = 25$ ms^{-1}

$$\Delta t_1 = 3\text{ s}, \Delta t_2 = 5\text{ s}$$

We know,

Impulse, $I = \Delta p = \mathbf{F} \cdot \Delta t$

$$\therefore \quad F = \frac{\Delta p}{\Delta t} \quad \ldots\text{(i)}$$

$\because$ Δp is equal in both cases and

$$\Delta t_1 < \Delta t_2$$

Thus, $F_1 > F_2$ [From Eq. (i)]

22. *(b)* Given that ratio of three masses are 1 : 1 : 2.

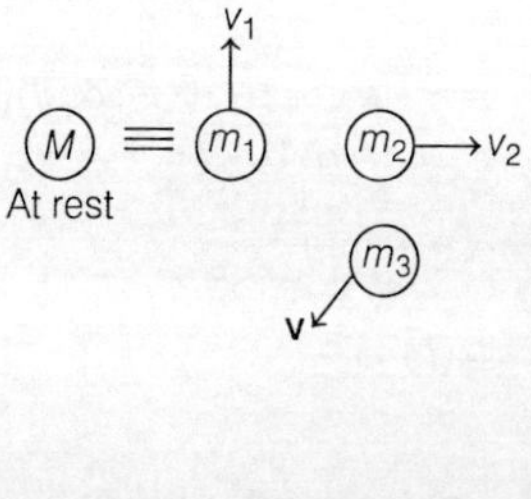

$$\therefore \quad m_1 = m_2 = \frac{M}{4} \text{ and } m_3 = \frac{M}{2}$$

As given two smaller pieces fly off perpendicularly with velocities $v_1 = 30$ m s^{-1} and $v_2 = 40$ m s^{-1}.

Let the velocity of the third pieces be v.

Since, the body of mass M is at rest, thus initial momentum is zero.

On applying law of conservation of momentum, we get

$$m_2 v_2 \hat{\mathbf{i}} + m_1 v_1 \hat{\mathbf{j}} + m_3 \mathbf{v} = 0$$
$$\therefore \quad \mathbf{v} = -15\hat{\mathbf{i}} - 20\hat{\mathbf{j}}$$
$$\Rightarrow \quad |\mathbf{v}| = \sqrt{(-15)^2 + (-20)^2} = 25\text{ m/s}$$

23. *(d)* Given, $M = 1000$ kg, $a = 20$ m/s^2

$v_{relative} = 500$ m/s, $g = 10$ m/s^2

The given situation is shown below

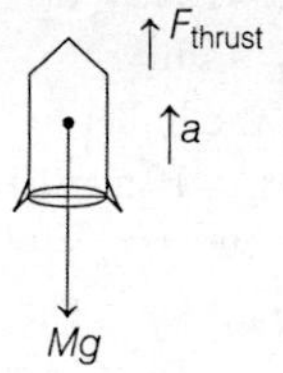

$$\therefore \quad F_{thrust} = v_{relative}\left(\frac{dM}{dt}\right)$$

By Newton's second law of motion,

$$\Rightarrow F_{thrust} - Mg = Ma$$
$$\Rightarrow 500\left(\frac{dM}{dt}\right) - 1000 \times 10 = 1000 \times 20$$
$$\Rightarrow 500\left(\frac{dM}{dt}\right) = 1000\,(20 + 10)$$
$$\Rightarrow \quad \frac{dM}{dt} = \frac{1000 \times 30}{500} = 60\text{ kg/s}$$

24. *(b)* Given,

$$m_B = 4\text{g} = 4 \times 10^{-3}\text{ kg}$$
$$m_G = 4\text{ kg}$$
$$v_B = 50\text{ ms}^{-1}$$

Let recoiling velocity of gun be v_G.

By using law of conservation of momentum

$$m_B v_B = -m_G v_G$$
$$\Rightarrow \quad v_G = -\frac{m_B v_B}{m_G}$$
$$= -\frac{4 \times 10^{-3} \times 50}{4}$$
$$= -50 \times 10^{-3} = -0.05\text{ ms}^{-1}$$

Impulse (I) = Change in momentum of gun

$$I = m_G(0 - v_G) = 4 \times 0.05$$
$$= 0.2\text{ kg ms}^{-1}$$

25. *(b)* Using momentum conservation law,

$$(\text{Total momentum})_{\text{before collision}} = (\text{Total momentum})_{\text{after collision}}$$

$$(m_1 \times 0) + (m_2 \times 0) = m_1 \mathbf{v}_1 + m_2 \mathbf{v}_2$$

$$0 = m_1(-v_1)\hat{\mathbf{i}} + m_2 v_2 \hat{\mathbf{i}}$$

$$\Rightarrow \quad m_1 v_1 = m_2 v_2$$

$$\Rightarrow \quad 50 v_1 = 20 v_2$$

$$\Rightarrow \quad v_2 = 2.5 v_1 \quad \ldots \text{(i)}$$

$\because$ Relative velocity $= 0.70$ m/s

$$\therefore \quad v_1 + v_2 = 0.7 \quad \ldots \text{(ii)}$$

From Eqs. (i) and (ii), we get

$$v_1 + 2.5 v_1 = 0.7$$

$$\Rightarrow \quad v_1(3.5) = 0.7$$

$$v_1 = \frac{0.7}{3.5} = 0.20 \text{ m/s}$$

26. *(d)* Given, resistive force

$$= F = -2.5 \times 10^{-2} \text{ N}$$

$$\therefore \quad a = \frac{F}{m} = \frac{-2.5 \times 10^{-2}}{20 \times 10^{-3}} = -\frac{5}{4} \text{ ms}^{-2}$$

$$(\because m = 20 \text{ g} = 20 \times 10^{-3} \text{ kg})$$

$$\because v^2 - u^2 = 2as$$

$$\therefore \quad v^2 = 1 + 2\left(-\frac{5}{4}\right)(20 \times 10^{-2})$$

$$(\because u = 1 \text{ ms}^{-1} \text{ and } s = 20 \text{ cm} = 20 \times 10^{-2} \text{ m})$$

$$\Rightarrow \quad v^2 = \frac{1}{2}$$

$$\therefore \quad v = \frac{1}{\sqrt{2}} \approx 0.7 \text{ ms}^{-1}$$

27. *(b)* Given, force $F = F_0 e^{-bt}$

$$\therefore \quad a = \frac{F}{m} = \frac{F_0}{m} e^{-bt} = \frac{dv}{dt}$$

$$\Rightarrow \quad \int_0^v dv = \int_0^t \frac{F_0}{m} e^{-bt} dt$$

$$\Rightarrow v = \frac{F_0}{m}\left[\left(\frac{1}{-b}\right) e^{-bt}\right]_0^t = \frac{F_0}{mb}[e^{-bt}]_t^0$$

$$= \frac{F_0}{mb}(e^0 - e^{-bt})$$

$$= \frac{F_0}{mb}(1 - e^{-bt})$$

The graph from the relation $(1 - e^{-bt})$ is converging in nature towards $\frac{F_0}{mb}$

Hence, option (b) is corret $v_{\max} = \frac{F_0}{mb}$

28. *(b)* From the given graph, due to impulse the direction of velocity changes from the slope of the graph.

Initial velocity, $v_1 = \frac{2}{2} = 1 \text{ ms}^{-1}$

Final velocity, $v_2 = -\frac{2}{2} = -1 \text{ ms}^{-1}$

$$\mathbf{p}_i = mv_1 = 0.4 \text{ N-s}$$

$$\mathbf{p}_f = mv_2 = -0.4 \text{ N-s}$$

$$\mathbf{I} = \mathbf{p}_f - \mathbf{p}_i = -0.4 - 0.4 = -0.8 \text{ N-s}$$

[I = impulse]

$$\therefore |\mathbf{I}| = 0.8 \text{ N-s}$$

29. *(d)* As per question, let the acceleration is a (for P to A).

$$v^2 = 0^2 + 2a \times 0.2 \quad \ldots \text{(i)}$$

For AB, $\quad 0 = v^2 - 2 \times g \times 2$

$$\Rightarrow \quad v^2 = 2g \times 2 \quad \ldots \text{(ii)}$$

From Eqs. (i) and (ii), we get

$$a = 10g = 100 \text{ ms}^{-2}$$

Then, for PA, FBD of ball is

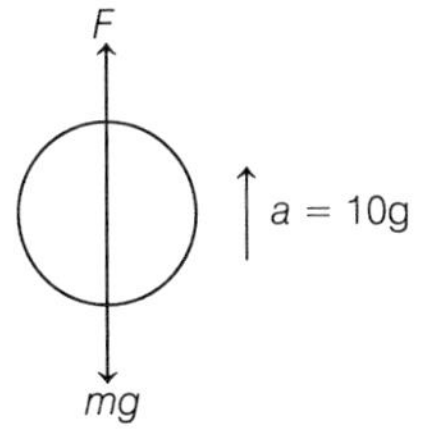

$$F - mg = ma$$

[F is the force exerted by hand on ball]

$$\Rightarrow \quad F = m(g + a)$$

$$= 0.2(11g) = 22 \text{ N}$$

30. *(d)* $F = -kx$

Work done $= \int_{x_1}^{x_2} F\, dx$

$$W = -\int_{x_1}^{x_2} kx\, dx = -k\left[\frac{x^2}{2}\right]_{0.1}^{1.9}$$

$$= -10\left[\frac{(1.9)^2}{2} - \frac{(0.1)^2}{2}\right] = -18 \text{ J}$$

Initial kinetic energy $= \frac{1}{2} m v_i^2$

$$K_i = \frac{1}{2} \times 1 \times (10)^2 \Rightarrow 50 \text{ J}$$

Final kinetic energy can be evaluated using work energy theorem

$$K_f = K_i + W$$

$$K_f = 50 - 18 = 32 \text{ J}$$

$$K_f = \frac{1}{2} m v_f^2$$

$$32 = \frac{1}{2} \times 1 \times (v_f^2)$$

$$64 = (v_f)^2$$

$$8 \text{ m/s} = v_f$$

$$v_f = 8 \text{ m/s}$$

31. *(c)* Given question is

$$v = 10\sqrt{x},\ m = 500 \text{ g} = \frac{500}{1000} \text{ kg}$$

$$F = ma \Rightarrow a = v\frac{dv}{dx}$$

$$v = 10\sqrt{x}$$

$$\frac{dv}{dx} = 10 \times \frac{1}{2} x^{-1/2} \Rightarrow \frac{dv}{dx} = \frac{5}{\sqrt{x}}$$

$$a = v\frac{dv}{dx} = 10\sqrt{x} \cdot \frac{5}{\sqrt{x}}$$

$$a = 50 \text{ m/s}^2$$

$$F = ma$$

$$F = \frac{500}{1000} \times 50$$

$$F = 25 \text{ N}$$

32. *(a)* Mass of particle $= 500$ g

$$= 500 \times 10^{-3} \text{ kg}$$

$$= 0.5 \text{ kg}$$

Velocity of the particle is,

$$v = 2t\hat{\mathbf{i}} + 3t^2\hat{\mathbf{j}} \text{ (m/s)}$$

Acceleration of particle is,

$$a = \frac{dv}{dt} = 2\hat{\mathbf{i}} + 6t\hat{\mathbf{j}} \text{ m/s}^2$$

So, acceleration of the particle at $t = 1$s is,

$$a(t = 1) = 2\hat{\mathbf{i}} + 6\hat{\mathbf{j}} \text{ m/s}^2$$

$\therefore$ Force on the particle is

$$F = ma \text{ (at } t = 1\text{s)}$$

$$\Rightarrow \quad F = 0.5(2\hat{\mathbf{i}} + 6\hat{\mathbf{j}})$$

$$= \hat{\mathbf{i}} + 3\hat{\mathbf{j}} \text{ N}$$

Hence, $x = 3$

33. *(c)* For both ball and bullet, time to reach the ground is equal to time of flight which is given as

$$T = \sqrt{\frac{2H}{g}} = \sqrt{\frac{2 \times 20}{10}} = 2 \text{ s}$$

Let, $\quad m_1 =$ mass of bullet

$\quad\quad m_2 =$ mass of ball

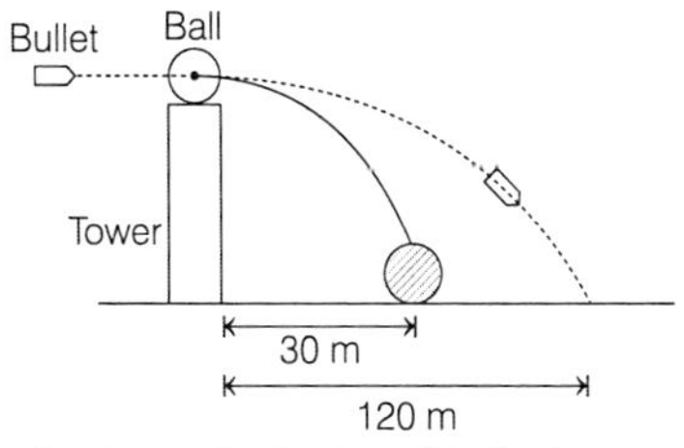

So, horizontal velocity of ball after collision,

$$v_2 = \frac{30}{2} = 15 \text{ m/s}$$

and for bullet $v_1 = \frac{120}{2} = 60$ m/s

Now, from conservation of linear momentum.

We have,

$$\Rightarrow m_1u_1 + m_2u_2 = m_2v_2 + m_1v_1$$

$$\Rightarrow 0.01u + 0.2(0) = 0.2(15) + 0.01(60)$$

$$u_1 = u \Rightarrow u = 360 \text{ m/s}$$

34. *(d)*

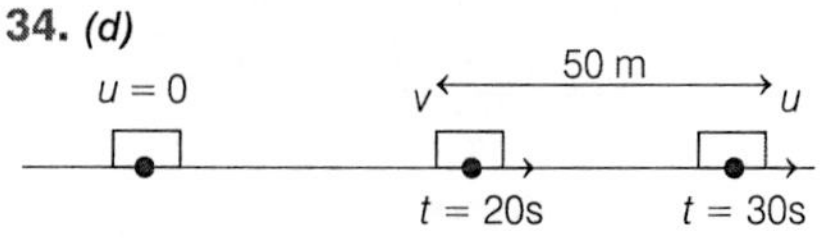

As particle covers 50m in 10s while moving with a constant velocity ($F = 0$, $a = 0$).

Velocity of particle is,

$$v = \frac{d}{t} = \frac{50}{10} = 5 \text{ m/s}$$

Now, for first 20 s of journey,

$$u = 0, v = 5 \text{ m/s}, t = 20 \text{ s}$$

Acceleration, $a = \frac{v-u}{t} = \frac{5}{20}$ m/s^2

Force magnitude is

$$F = m \times a = 20 \times \frac{5}{20} = 5 \text{ N}$$

35. *(c)* According to the situation given in the question,

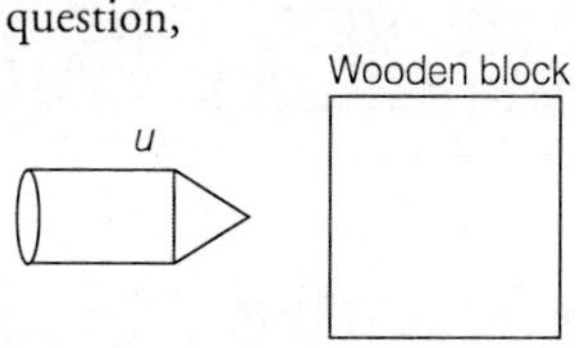

after penetrating 4 cm in the wooden block, its velocity becomes $(u/3)$ due to constant resistance.

Therefore, $v^2 = u^2 - 2as$

$$\Rightarrow \left(\frac{u}{3}\right)^2 = u^2 - 2 \times a \times \frac{4}{100}$$

$$\Rightarrow a = \frac{8u^2}{9} \times \frac{100}{8} = \frac{100}{9}u^2$$

Now, since the bullet stops completely after travelling a further x cm.

$$\therefore \quad v^2 = u^2 - 2ax$$

$$0 = u^2 - 2 \times \frac{-100}{9}u^2 \frac{(4+x)}{100}$$

$$\Rightarrow \frac{9}{2} = 4 + x$$

$$\therefore \quad x = 0.5 \text{ m}$$

Alternate Solution

According to work energy theorem,

$\Delta \text{KE} = W_{\text{net}}$

In Ist case $\frac{1}{2}mv_2^2 - \frac{1}{2}mv_1^2 = -F.4$

$$\frac{1}{2}m\frac{v_1^2}{9} - \frac{1}{2}mv_1^2 = -F.4$$

$$\Rightarrow \frac{1}{2}m\frac{8v_1^2}{9} = F.4 \Rightarrow \frac{4mv_1^2}{9} = F.4 \quad \ldots\text{(i)}$$

In IInd case $0 - \frac{1}{2}mv_1^2 = -F(4+x)$

$$\Rightarrow \frac{mv_1^2}{2} = F(4+x) \quad \ldots\text{(ii)}$$

From Eqs. (i) and (ii)

$$\Rightarrow \frac{4m_1^2/9}{mv_1^2/2} = \frac{4}{4+x}$$

$$\Rightarrow \frac{8}{9} = \frac{4}{4+x} \Rightarrow \frac{2}{9} = \frac{1}{4+x}$$

$$\Rightarrow x = \frac{1}{2} = 0.5 \text{ cm}$$

36. *(b)* Given $m = 5$ kg

For upward motion, acceleration,

$$a_1 = \frac{mg+10}{m} = g + \frac{10}{5}$$

$$= 10 + 2 = 12 \text{ m/s}^2$$

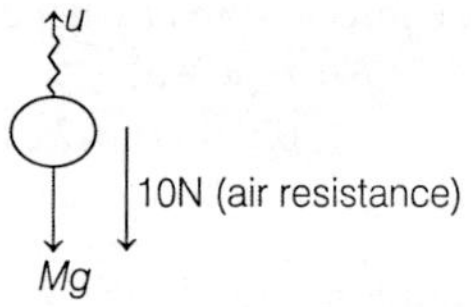

Using, $v = u - at$

Time of ascent, $t_1 = \frac{u}{a_1}$ ...(i)

($\because v = 0$ at maximum height)

and from $v^2 = u^2 - 2as$

Height, $h = \frac{u^2}{2a_1}$ ...(ii)

For downward motion,

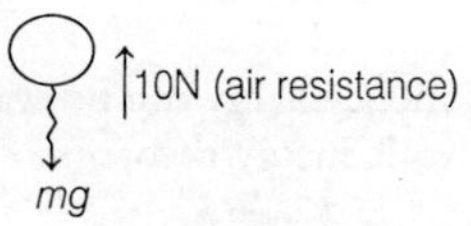

Acceleration, $a_2 = \frac{mg-10}{m} = g - \frac{10}{5}$

$$= 10 - 2 = 8 \text{ m/s}^2$$

Using, $h = ut + \frac{1}{2}at^2$

$\therefore$ Time of desent,

$$t_2 = \sqrt{\frac{2h}{a_2}} \quad (\because u = 0) \ldots\text{(iii)}$$

From Eqs. (i) and (iii), using (ii)

$$\frac{t_1}{t_2} = \frac{\frac{u}{a_1}}{\sqrt{\frac{2h}{a_2}}} = \frac{\frac{u}{a_1}}{\sqrt{\frac{2 \cdot u^2}{2a_2a_1}}} = \frac{\frac{u}{a_1}}{\frac{u}{\sqrt{a_1a_2}}}$$

$$= \sqrt{\frac{a_2}{a_1}} = \sqrt{\frac{8}{12}} = \frac{\sqrt{2}}{\sqrt{3}}$$

37. *(c)* Given, force on particle,

$$F = F_0\left[1 - \left(\frac{t-T}{T}\right)^2\right]$$

$\because$ Acceleration of ball,

$$a = \frac{F_0}{M}\left[1 - \left(\frac{t-T}{T}\right)^2\right]$$

$$\Rightarrow \frac{dv}{dt} = \frac{F_0}{M}\left[1 - \left(\frac{t-T}{T}\right)^2\right]$$

$$\Rightarrow dv = \frac{F_0}{M}\left[1 - \left(\frac{t-T}{T}\right)^2\right]dt$$

$$\Rightarrow \int dv = \frac{F_0}{M}\int_0^{2T}\left[1 - \left(\frac{t-T}{T}\right)^2\right]dt$$

$$\Rightarrow v = \frac{F_0}{M}\left[t - \frac{1}{3T^2}(t-T)^3\right]_0^{2T}$$

$$\Rightarrow v = \frac{F_0}{M}\left\{\left[2T - \frac{1}{3T^2}(2T-T)^3\right] - \left[0 - \frac{(-T)^3}{3T^2}\right]\right\}$$

$$\Rightarrow v = \frac{4F_0T}{3M}$$

38. *(b)* Given that mass increases at the rate,

$$\frac{dM(t)}{dt} = bv^2(t)$$

Now, from Newton's second law,

i.e., $F = \frac{dp}{dt} = \frac{d}{dt}(Mv)$

$$= M\frac{dv}{dt} + v\frac{dM}{dt}$$

$$= M\frac{dv}{dt} + bv^3(t)$$

$\because$ External force on the spaceship is zero,

$$F_{ext} = 0$$

$$\Rightarrow \quad M\frac{dv}{dt} = -bv^3 \Rightarrow \quad \frac{dv}{dt} = -\frac{bv^3}{M(t)}$$

$\therefore$ Acceleration of the satellite,

$$a = -\frac{bv^3}{M(t)}$$

Hence, correct option is (b)

39. *(d)* For upward motion,

$$v = \sqrt{u^2 - 2gh}$$

$$\because \quad p = mv = m\sqrt{u^2 - 2gh}$$

$$p^2 = m^2(u^2 - 2gh)$$

For downward motion,

$$v = \sqrt{2gh} \qquad (\text{at } h_{max}, u = 0)$$

$$\therefore \quad p = mv = m\sqrt{2gh}$$

$$p^2 = 2m^2 gh$$

For upward motion, **v** and **p** will be +ve.

For downward motion, **v** and **p** will be –ve.

Therefore, momentum vs height graph will be represented by option (d).

40. *(6)* Given, $m_1 = 60$ kg, $v_1 = ?$

$m_2 = 120$ kg, $v_2 = 2$ m/s

Total momentum of (man + trolley) system is always conserved

So,

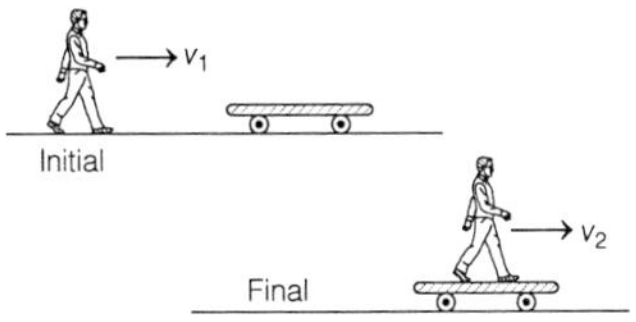

$$\Rightarrow \quad m_1v_1 + 0 = (m_1 + m_2)v_2$$

$$\Rightarrow \quad 60 \times v_1 = (60 + 120) \times 2$$

$$\Rightarrow \quad v_1 = \frac{(60 + 120) \times 2}{60}$$

$$= 6 \text{ m/s}$$

41. *(12)* Given, $m = 0.4$ kg, $u = 15$ m/s

Final speed of the ball,

$$v = -u = -15 \text{ m/s}$$

$\therefore$ Impulse imparted to the ball,

$$I = \text{Change in momentum}$$

$$= m(v - u)$$

$$= 0.4(-15 - 15) = -0.4 \times 30$$

$$= -12 \text{ N-s}$$

The impulse imparted to the ball is 12 Ns.

42. *(5)* As per question, change in momentum between the points A and B,

$$\Delta \mathbf{p} = \mathbf{p}_B - \mathbf{p}_A$$

Here, $\mathbf{p}_A = m(u\cos 45°\hat{\mathbf{i}} + u\sin 45°\hat{\mathbf{j}})$

and $\mathbf{p}_B = m(u\cos 45°\hat{\mathbf{i}} - u\sin 45°\hat{\mathbf{j}})$

$$\therefore \quad \Delta\mathbf{p} = -2mu\sin 45°\hat{\mathbf{j}}$$

or $|\Delta \mathbf{p}| = 2 \times 5 \times 10^{-3} \times 5\sqrt{2} \times \frac{1}{\sqrt{2}}$

$$= 5 \times 10^{-2} \text{ kg m s}^{-1}$$

43. *(10)* Given,

The mass of the bullet, $m = 0.1$ kg

The initial velocity of the bullet $u = 10$ m/s

The final velocity of the bullet

$$v = 0$$

$$v^2 - u^2 = 2as$$

$$\Rightarrow 0^2 - (10)^2 = 2a(0.50)$$

$$\Rightarrow \quad a = -100 \text{ m/s}^2$$

The retarding force on the bullet,

$$F = ma$$

$$F = 0.1(-100)$$

$$F = -10 \text{ N}$$

Hence, the value of x to the nearest integer is 10.

Topic 2

01. *(c)*

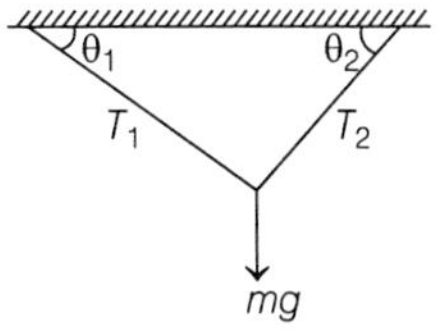

$$T_1\sin\theta_1 + T_2\sin\theta_2 = mg$$

$$T_1 = \sqrt{3}T_2$$

$$T_2[\sqrt{3}\sin\theta_1 + \sin\theta_2] = mg$$

For $\theta_1 = 60°$ and $\theta_2 = 30°$

$$T_2 = \frac{mg}{2}$$

02. *(d)* The branch pulls the chain by a force given by

F = weight of body + weight of chain

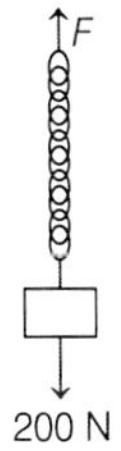

$$= 200 + mg$$

$$= 200 + (10 \times 10)$$

$$= 300 \text{ N}$$

03. *(a)* Force $= \frac{\text{Change in momentum}}{\text{Time}}$

$$= \frac{p_i - p_f}{\Delta t} = \frac{m(u - v)}{\Delta t}$$

$$= \frac{120 \times 10^{-3}(25 - 0)}{0.1}$$

$$= 3000 \times 10^{-2}$$

$$= 30 \text{ kg-m/s}^2$$

$$= 30 \text{ N}$$

> ***Alternate Solution***
>
> $m = 120$ g $= 0.12$ kg, $u = 25$ m/s, $v = 0$
>
> $\therefore$ Retardation applied on the ball
>
> $$a = \frac{v - u}{t} = \frac{0 - 25}{0.1} = -250 \text{ m/s}^2$$
>
> $\therefore \quad |a| = 250 \text{ m/s}^2$
>
> $\therefore \quad F = ma = 0.12 \times 250$
>
> $= 30.00$ N $= 30$ N

04. *(b)* FBD of the sphere

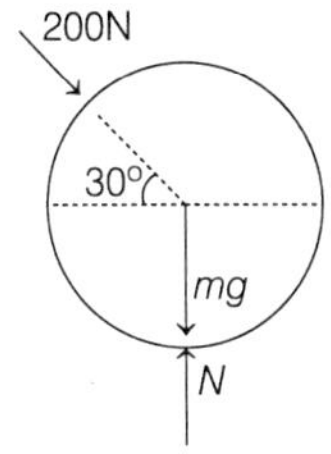

Therefore,

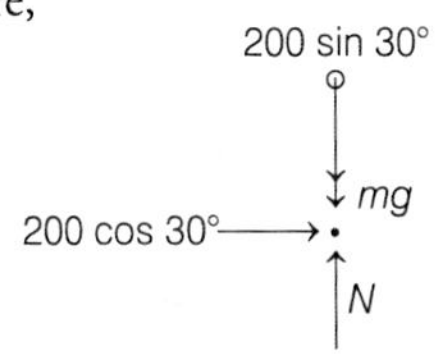

Since, $a_y = 0$

Thus, $N = mg + 200\sin 30°$

$$N = 70 \times 10 + 200 \times \sin 30° = 800 \text{ N}$$

05. *(c)* As per question, $N = Mg/4$

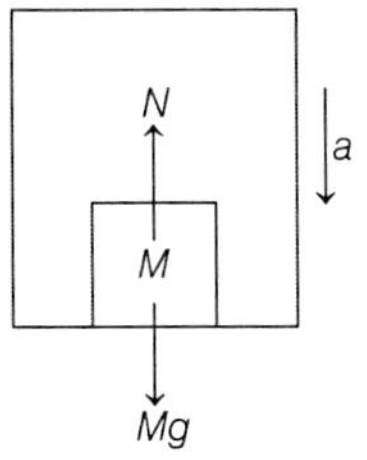

By using Newton's second law,

$$Mg - N = Ma$$

$$Mg - \frac{Mg}{4} = Ma, \; a = \frac{3g}{4}$$

06. *(a)* By the method of virtual work,

$$\Rightarrow \quad T \cdot a = 0,$$

So, $4Ta_1 + 2Ta_2 + Ta_3 + Ta_4 = 0$

Hence, $4a_1 + 2a_2 + a_3 + a_4 = 0$

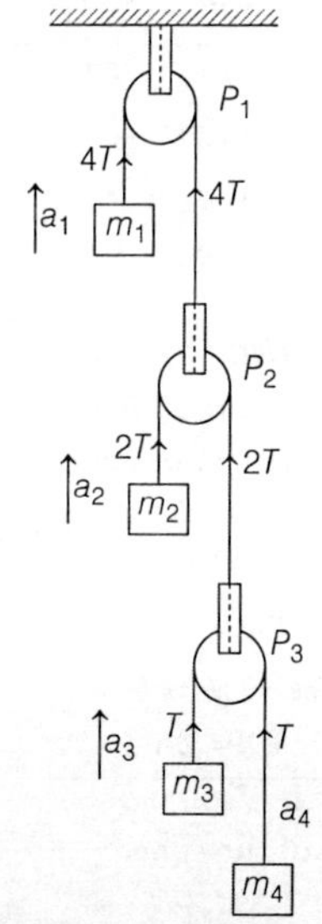

07. *(c)* Acceleration of system,

$$a = \frac{F}{m+M}$$

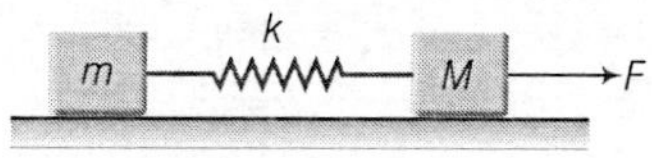

So, force acting on mass,

$$F = ma = \frac{mF}{m+M}$$

08. *(d)* In the frame of wedge, the FBD of block is shown in figure.

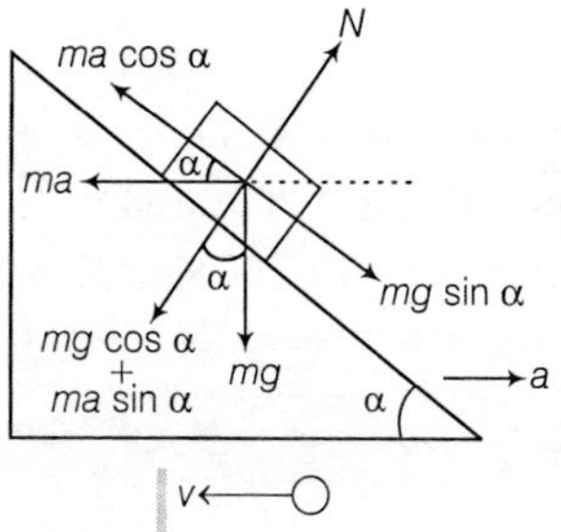

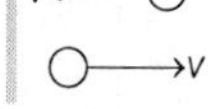

For block to remain stationary,

$ma \cos \alpha = mg \sin \alpha$

or $\quad a = g \tan \alpha$

09. *(d)* Acceleration of the system

(rope + block) $a = \dfrac{P}{M+m}$

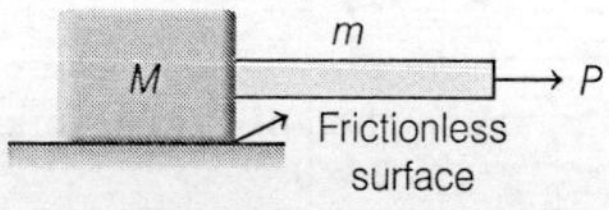

The FBD of block and rope as shown in figure.

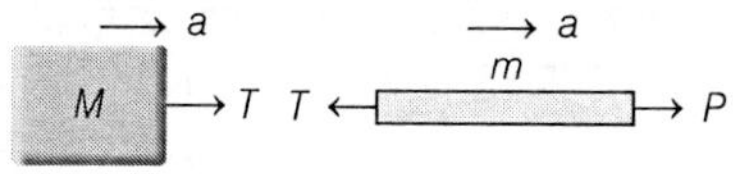

Where, T is the required parameter.

For block, $T = Ma \Rightarrow T = \dfrac{MP}{M+m}$

10. *(a)* Because both springs are light.

Hence, $T_1 = T_2 = Mg$

So, both scales read M kg.

11. *(c)* Apparent weight of ball,

$w' = w - R$

[where, R=normal reaction]

$R = ma$ [acting upward]

$w' = mg - ma = m(g-a)$

Hence, apparent acceleration in the lift is $g - a$. Now, if the man is standing stationary on the ground, then the apparent acceleration of the falling ball is g.

12. *(a)* Drawing free body diagram as

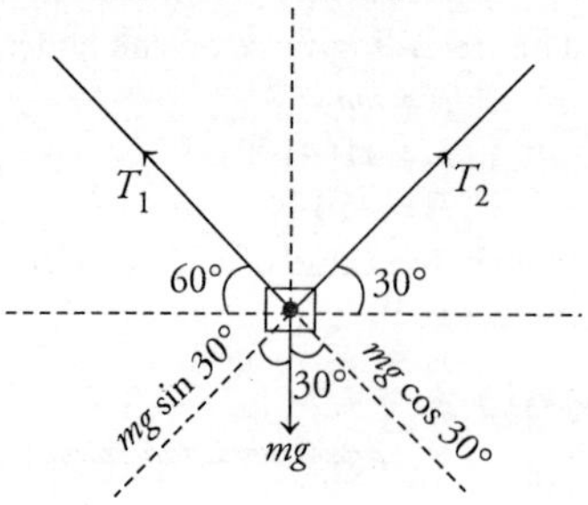

Here, $m = 1$ kg

$T_1 = mg \cos 30°$

$= 1 \times 10 \times \dfrac{\sqrt{3}}{2} = 5\sqrt{3}$ N

$T_2 = mg \sin 30° = 1 \times 10 \times \dfrac{1}{2} = 5$ N

13. *(a)*

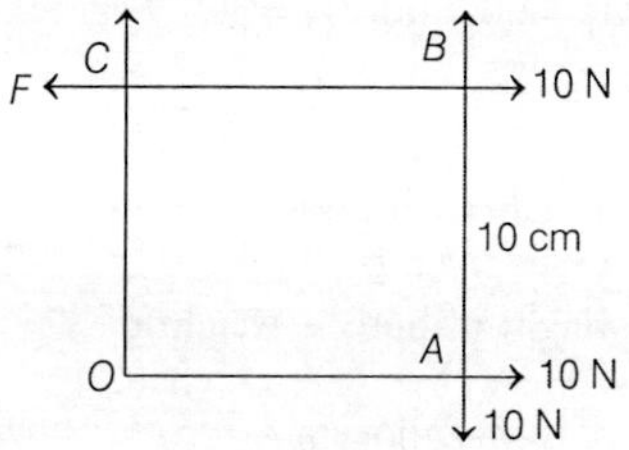

Since the lamina is equilibrium

$F_{net} = 0$ and $\tau_{net} = 0$

Torque about $0 = 10l - Fl$

$0 = 10l - Fl$

$Fl = 10l$

$F = 10$ N

14. *(a)*

$F = kx$

$5 = kx_1$ ...(i)

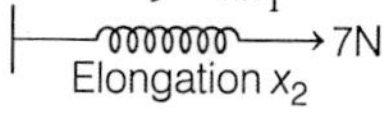

$7 = kx_2$ ...(ii)

Multiply by 5 in Eq. (i)

$5 \times 5 = k5x_1$ ...(iii)

Multiply by 2 in Eq. (ii)

$7 \times 2 = k \times 2x_2$ ...(iv)

From Eq. (iii) and Eq. (iv)

$k \cdot 5x_1 - k \cdot 2x_2 = 25 - 14$

$k(5x_1 - 2x_2) = 25 - 14$

$k(5x_1 - 2x_2) = 11$

$\Rightarrow$ Tension in the string,

$T = k(5x_1 - 2x_2) = 11$ N

15. *(c)* Here, $m = 1$ kg

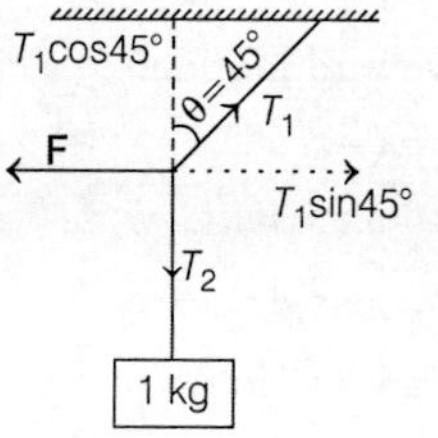

$T_1 \sin 45° = F$

$T_1 \cos 45° = T_2 = 1 \times g$

$\therefore \quad \tan 45° = \dfrac{F}{g}$ $\quad [g = 10 \text{ m/s}^2]$

$\therefore \quad F = 10$ N

16. *(d)* FBD for m_1 and m_2 are

(Taking, $m_1 > m_2$)

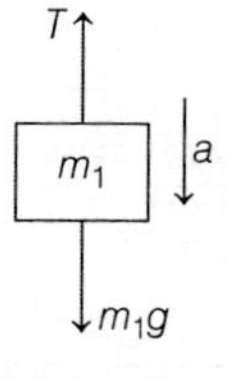

$\Rightarrow \quad m_1 a = m_1 g - T$ ...(i)

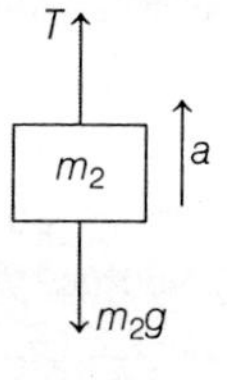

$m_2 a = T - m_2 g$ ...(ii)

Adding Eqs. (i) and (ii) we have,

$(m_1 + m_2)\, a = (m_1 - m_2)\, g$

Now, given $a = g/8$

So, $\quad (m_1 + m_2)\, g/8 = (m_1 - m_2)\, g$

$\Rightarrow \quad 7m_1 = 9m_2$

$\Rightarrow$ Ratio, $\dfrac{m_1}{m_2} = \dfrac{9}{7}$

> ***Alternate Solution***
>
> We know that, common acceleration
>
> $$a = \left(\frac{m_1 - m_2}{m_1 + m_2}\right) g \Rightarrow \frac{g}{8} = \left(\frac{m_1 - m_2}{m_1 + m_2}\right) g$$
>
> $$\Rightarrow \quad m_1 + m_2 = 8m_1 - 8m_2$$
>
> $$\Rightarrow \quad 9m_2 = 7m_1 \Rightarrow \frac{m_1}{m_2} = \frac{9}{7}$$

17. *(b)* Given, mass of ball $= 150$ g

$= 0.15$ kg

Initial speed of ball $= 20$ m/s

Final speed of ball $= 0$ m/s

Time for process of catching $= 0.1$ s

Force exerted by ball on the hands of the player

$$= \text{Rate of change of momentum}$$

$$= \frac{m(v-u)}{t}$$

$$= \frac{0.15 \times (20-0)}{0.1} = 30 \text{ N}$$

18. *(a)* Given,

Common acceleration $a = g/\sqrt{2}$

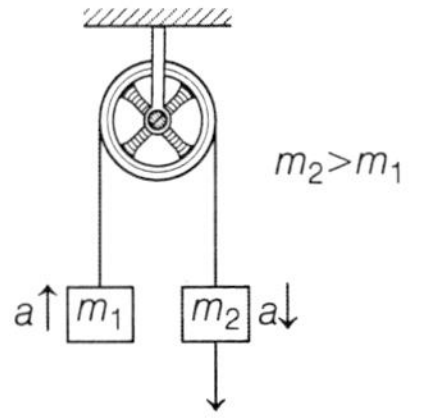

$$\Rightarrow \quad \left(\frac{m_2 - m_1}{m_1 + m_2}\right) g = \frac{g}{\sqrt{2}}$$

$$\Rightarrow \quad \frac{m_2 - m_1}{m_1 + m_2} = \frac{1}{\sqrt{2}}$$

$$\Rightarrow \sqrt{2}m_2 - \sqrt{2}m_1 = m_1 + m_2$$

$$\Rightarrow \quad (\sqrt{2} - 1)m_2 = (\sqrt{2} + 1)m_1$$

$$\Rightarrow \quad \frac{m_1}{m_2} = \frac{\sqrt{2} - 1}{\sqrt{2} + 1}$$

19. *(d)* Net force on the object,

$$F_{net} = \mathbf{F}_1 + \mathbf{F}_2$$

$$= 8\hat{i} + 4\hat{j} + 4\hat{k}$$

Acceleration of object, $\mathbf{a} = \frac{\mathbf{F}_{net}}{m}$

$$= \frac{8\hat{i} + 4\hat{j} + 4\hat{k}}{4} = 2\hat{i} + \hat{j} + \hat{k} \text{ m/s}^2$$

20. *(d)* Given,

Mass of body, $m = 100$ g $= 0.1$ kg

Height from where it dropped,

$h_1 = 10$ m

Rebound height, $h_2 = 5$ m

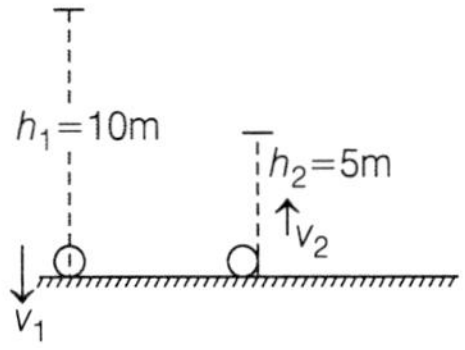

Impulse is given by

$$J = \Delta p$$

$$J = mv_2 - (-mv_1) = m(v_2 + v_1)$$

$$= m(\sqrt{2gh_1} + \sqrt{2gh_2})$$

$$= 0.1(\sqrt{2 \times 9.8 \times 10} + \sqrt{2 \times 9.8 \times 5})$$

$$= 2.39 \text{ kg-m/s}$$

21. *(c)* Given,

$F_1 = 10$ N, $F_2 = 8$ N,

$F_3 = 6$ N

Mass, $m = 5$ kg

Since, particle remains stationary under forces F_1, F_2 and F_3, where F_2 and F_3 are mutually perpendicular. Thus, for equilibrium

$$F_1 = \sqrt{F_2^2 + F_3^2}$$

In the absence of force F_1, acceleration,

$$a = \frac{\sqrt{F_2^2 + F_3^2}}{m} = \frac{F_1}{m} = \frac{10}{5} = 2 \text{ m/s}^2$$

22. *(c)* For the block's motion,

$$l = 0 + \frac{1}{2}(g \sin 30°)(2)^2 \quad \ldots \text{(i)}$$

$$l = 0 + \frac{1}{2}(g \sin 45°)(t)^2 \quad \ldots \text{(ii)}$$

From Eqs. (i) and (ii), we get

$$\Rightarrow 0 + \frac{1}{2}(g \sin 30°)(2)^2$$

$$= 0 + \frac{1}{2}(g \sin 45°)(t)^2$$

By solving, $t = (2)^{3/4} = 1.68$ s

23. *(a)* The given situation is shown below.

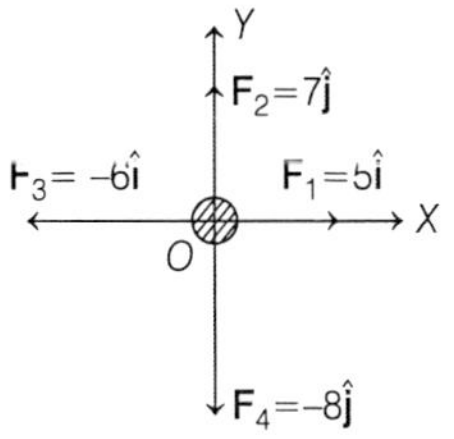

$\therefore$ Resultant force

$$\mathbf{F} = \mathbf{F}_1 + \mathbf{F}_2 + \mathbf{F}_3 + \mathbf{F}_4$$

$$= 5\hat{i} + 7\hat{j} - 6\hat{i} - 8\hat{j}$$

$$\Rightarrow \quad \mathbf{F} = -\hat{i} - \hat{j}$$

Since, the net force is zero

$$\mathbf{F} + \mathbf{F}' = 0$$

$$\Rightarrow \quad \mathbf{F} = -\mathbf{F} = -(-\hat{i} - \hat{j})$$

$$= \hat{i} + \hat{j}$$

$$\therefore \quad |\mathbf{F}'| = \sqrt{1^2 + 1^2}$$

$$= \sqrt{2} \text{ N}$$

If θ be the angle made by force $\mathbf{F}'$ from X-axis, then

$$\theta = \tan^{-1}\left(\frac{1}{1}\right) = \tan^{-1}(1) = 45°$$

24. *(b)* According to the question,

$F = 2$N and velocity is constant for A to B and B to C.

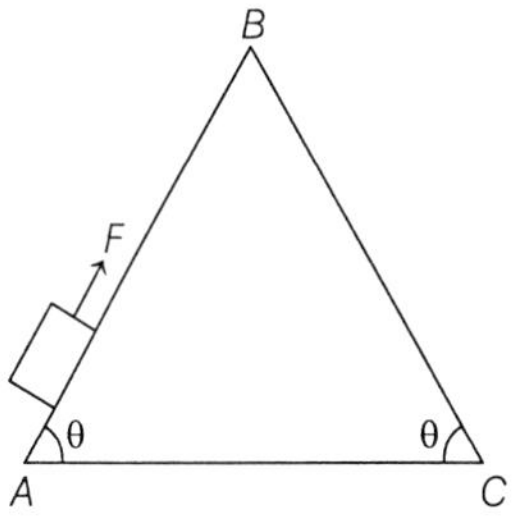

For upward motion (A to B) force and velocity are in same direction, both will be positive.

For downward motion (B to C) force and velocity are opposite to each other. In this case force will be negative (because it is opposite to the velocity).

25. *(d)* Three forces $\mathbf{F}_1$, $\mathbf{F}_2$ and $\mathbf{F}_3$ are acting on a body and if this body is in equilibrium, then resultant of these three forces must be zero i.e.

$$\mathbf{F}_{net} = \mathbf{F}_1 + \mathbf{F}_2 + \mathbf{F}_3 = 0$$

$$\Rightarrow \quad \mathbf{F}_1 + \mathbf{F}_2 = -\mathbf{F}_3$$

This situation can be shown graphically by three concurrent forces at 120° with each others.

or, by three forces in the same order along three sides of a triangle.

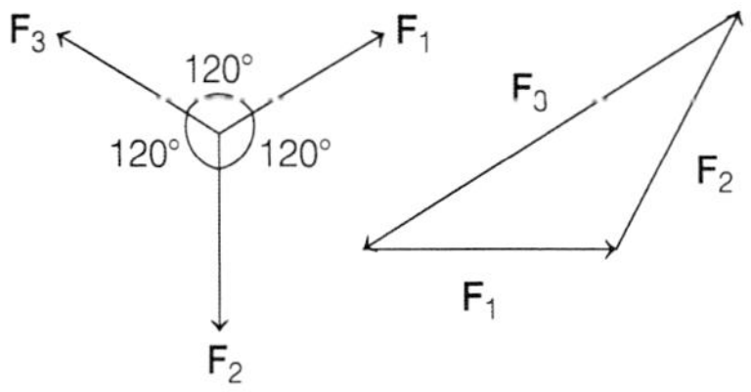

Hence, both Statement I and Statement II are true.

26. *(c)* FBD of the given system is follow

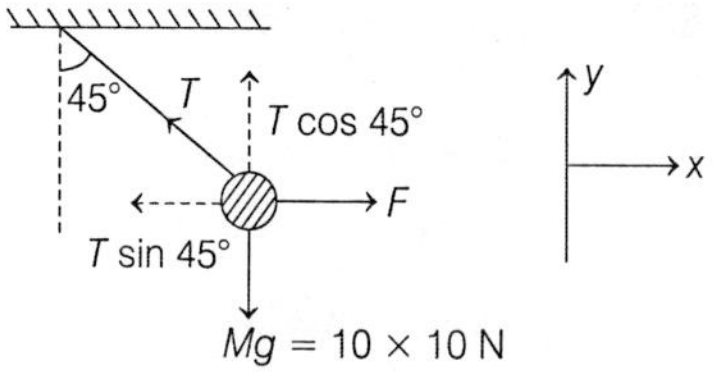

$$F = T \sin 45° \quad \ldots(i)$$

$$Mg = T \cos 45° \quad \ldots(ii)$$

Using Eqs. (i) and (ii), we get

$$\Rightarrow \frac{F}{Mg} = \frac{T \sin 45°}{T \cos 45°}$$

$$\Rightarrow F = Mg \tan 45°$$

$$= 10 \times 10 = 100 \text{ N}$$

27. *(d)* Force on the block

$$F = mg \sin\theta \text{ or } mg \sin\theta = ma$$

$$\therefore \quad a = g \sin\theta$$

where, a is along the inclined plane.

$\therefore$ Vertical component of acceleration is $g \sin^2 \theta$.

$\therefore$ Relative vertical acceleration of A with respect to B is

$$g(\sin^2 60° - \sin^2 30°) = \frac{g}{2} = 4.9 \text{ ms}^{-2}$$

[in vertical direction]

28. *(d)* Given, Mass of block, $m_1 = 5$ kg

Iron cylinder mass, $m_2 = 25$ kg

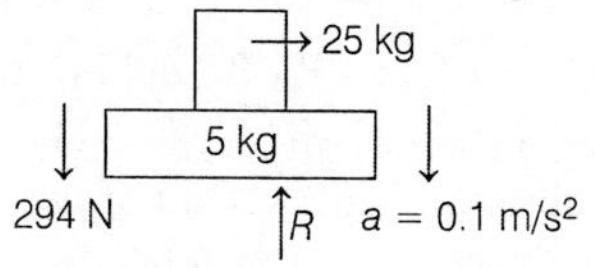

Free body diagram

Acceleration $= 0.1 \text{ m/s}^2$

Total mass $= m_1 + m_2 = 30$ kg

Total downward force,

$$F_d = mg = 30 \times 9.8 = 294 \text{ N} \quad \ldots(i)$$

Net force,

$$294 - R = ma \quad \ldots(ii)$$

$$294 - R = 30 \times 0.1 = R = 291 \text{ N}$$

Hence, action force of system on the floor is 291 N.

29. *(a)*

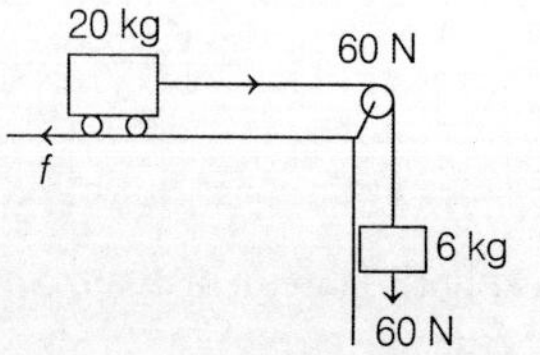

Net accelerating force on trolley is,

$$F_{net} = T - f$$

$$= 60\text{ N} - \mu(20) \times 10$$

$$= 60 - 0.04 \times 200 = 52 \text{ N}$$

Acceleration of system, $a = \dfrac{F}{m + M}$

$$= \frac{52}{20 + 6} = \frac{52}{26} = 2 \text{ m/s}^2$$

Alternate method

FBD of trolley

$20a = T - f = T - \mu N$

$= T - 20 \times 0.04 \times 10$

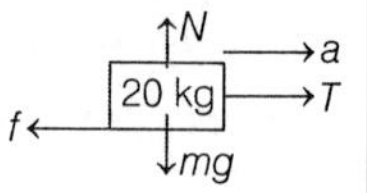

$20a = T - 8 \quad \ldots(i)$

FBD of block

$6a = 60 - T \quad \ldots(i)$

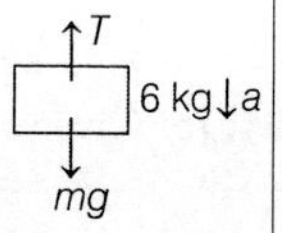

Solving Eqs. (i) and (ii)

$26a = 52$

$\Rightarrow a = 2 \text{ m/s}^2$

30. *(b)*

key idea Use following stepwise approach to solve this problem.

Step I. Draw the free body diagram for left block and write the equation of motion.

Step II. Again, draw the free body diagram for the right block and write the equation of motion.

Now, solve the two equations obtained in step I and step II and get required result.

Remember as the given blocks are not fixed at any point hence you can draw FBD for any of the block initially.

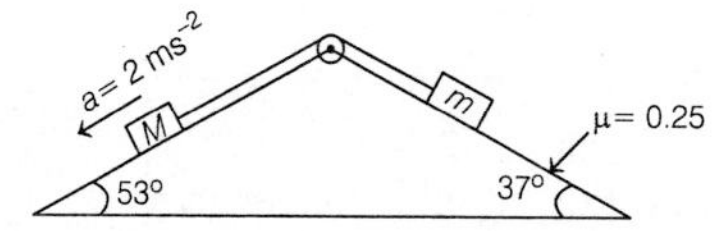

FBD for mass M,

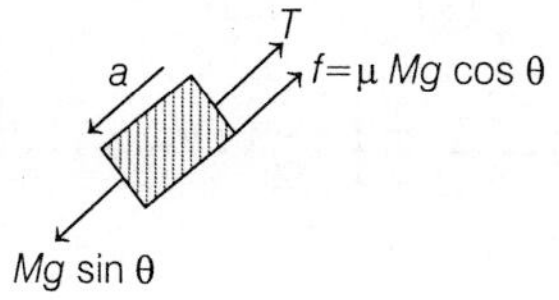

For mass M,

$$F_{net} = Ma$$

$$\Rightarrow Mg \sin 53° - \mu Mg \cos 53° - T = Ma$$

or, $\quad Mg(\sin 53° - \mu \cos 53°) - T = 2M \quad \ldots(i)$

FBD for mass m,

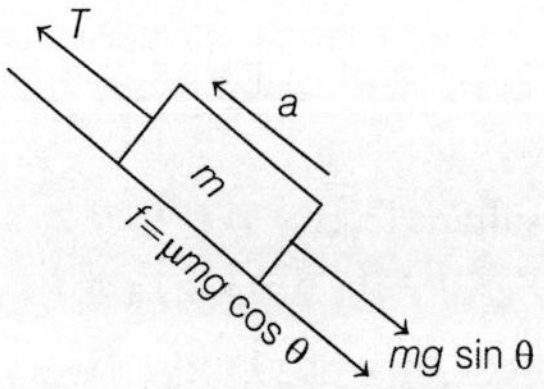

$$T - mg(\sin 37° + \mu \cos 37°)$$

$$= 2m \quad \ldots(ii)$$

Adding Eq. (i) and Eq. (ii) and substituting values, we get

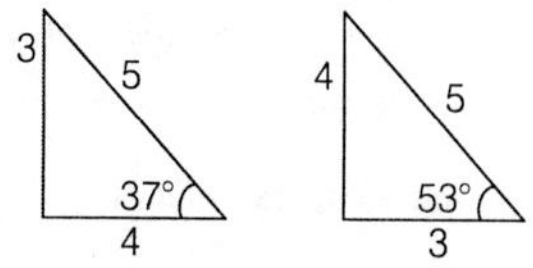

$$100\left(\frac{4}{5} - \frac{1}{4} \times \frac{3}{5}\right) - m \times 10\left(\frac{3}{5} + \frac{1}{4} \times \frac{4}{5}\right)$$

$$= 20 + 2m \Rightarrow m = \frac{45}{10} = 4.5 \text{ kg}$$

31. *(c)*

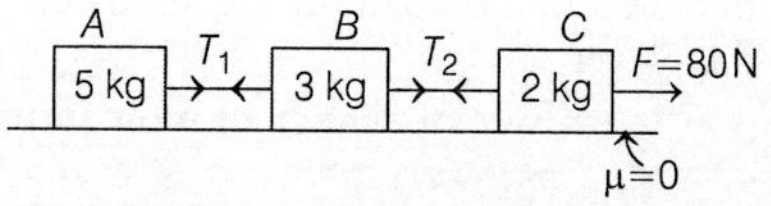

Acceleration of system $= \dfrac{\text{Net force}}{\text{Total mass}}$

$$= \frac{80 \text{ N}}{10 \text{ kg}} = 8 \text{ m/s}^2$$

Force on block A,

$$T_1 = m_A \times a$$

$$= 5 \times 8 = 40 \text{ N}$$

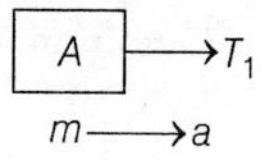

Force on block B,

$$T_2 - T_1 = m_B a$$

$$T_2 - 40 = 3 \times 8$$

$$T_2 = 64 \text{ N}$$

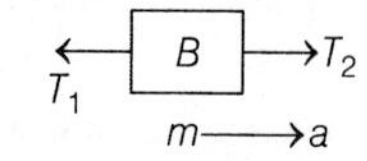

32. *(a)*

key idea Use following stepwise approach to solve this problem.

Step I. Draw the free body diagram for 2 kg block and write the equation of motion.

Step II. Again, draw the free body diagram for 4 kg block and write the equation of motion.

Now, solve the equations of motion obtained in Step I and Step II, and get required result.

Remember, 4 kg block and pulley associated with it must be considered as a single system in the given problem.

Given,

Slope of inclined, $\theta = 30°$

Mass on inclined plane, $m = 2$ kg

Mass hanging $= 4$ kg

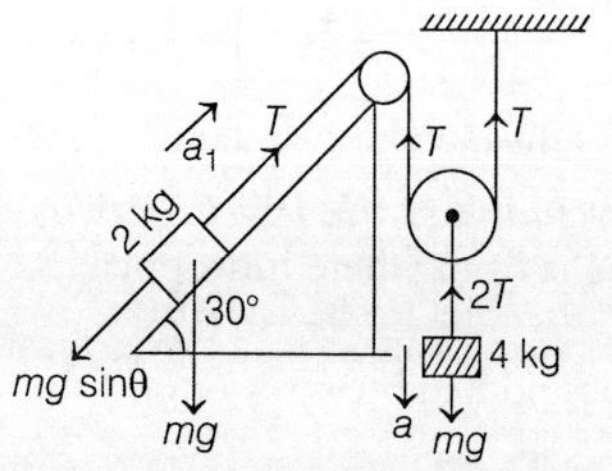

Using the relation

$$\Sigma T \cdot a = 0$$

$$Ta_1 - 2Ta = 0$$

$$a_1 = 2a$$

Net force on 2 kg mass

$$T - mg \sin 30° = 4a$$

$$T - g = 4a \quad \ldots\text{(i)}$$

Net force on 4 kg mass

$$4g - 2T = 4a \quad \ldots\text{(ii)}$$

From Eq. (i) and (ii), we get

$$4g - 2g = 12a$$

$$a = \frac{g}{6}$$

Now, $a_1 = 2a = 2 \times \frac{g}{6} = \frac{g}{3}$

33. *(a)* Let, a be acceleration of system FBD of the system

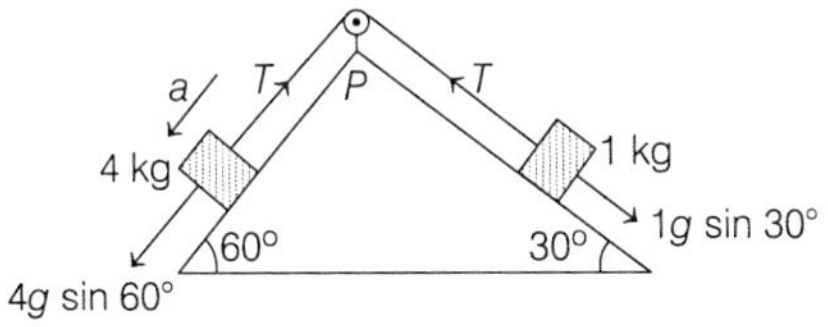

On applying Newton's second law, we get

$$4 \times a = 4 \times g \sin 60° - T$$

Since, $4g \sin 60° > 1g \sin 30°$

and $\quad 1 \times a = T - 1 \times g \sin 30°$

or, $\quad 4a = 20\sqrt{3} - T \quad \ldots\text{(i)}$

and $\quad a = T - 5 \quad \ldots\text{(ii)}$

From Eqs. (i) and (ii), we get

$$5a = 20\sqrt{3} - 5 \text{ or } a = 4\sqrt{3} - 1$$

From Eq. (ii),

$$4\sqrt{3} - 1 = T - 5$$

or $\quad T = 4(\sqrt{3} + 1)$ units

34. *(b)* Consider forces acting on m_1 and m_2 as shown in figures.

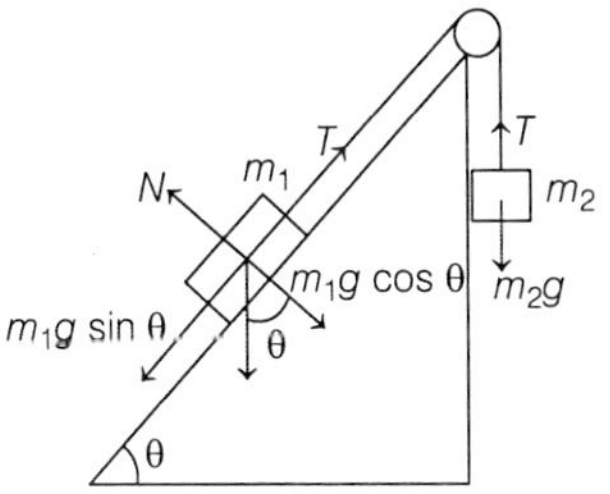

The force exerted by the inclined plane on m_1 is

$$N = m_1 g \cos \theta$$

$$N = 5 \times 10 \times \cos \theta \quad \ldots\text{(i)}$$

Also, $\quad T = m_1 g \sin \theta = m_2 g$

$$\Rightarrow \quad \sin \theta = \frac{m_2}{m_1} = \frac{3}{5}$$

$$\Rightarrow \quad \cos \theta = \sqrt{1 - \frac{9}{25}} = \frac{4}{5} \quad \ldots\text{(ii)}$$

From (i) and (ii),

$$N = 5 \times 10 \times \frac{4}{5} = 40 \text{ N}$$

35. *(d)* Mass of complete chain $= m$

Length of chain $= L$

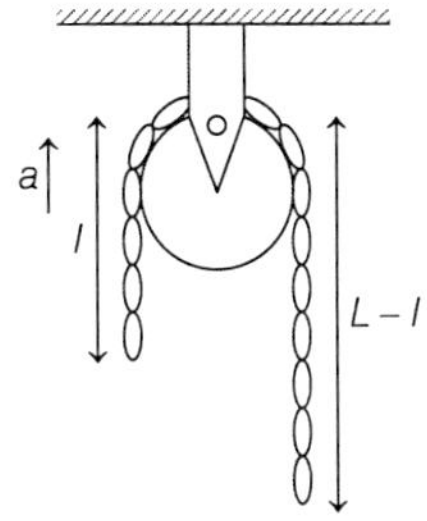

Mass of $(L - l)$ length chain $= \frac{m}{L}(L - l)$

Force on $(L - l)$ length chain

$$= \frac{m}{L}(L - l)\,g$$

Mass of l length chain $= ml / L$

Force on l length chain $= \frac{ml}{L}$ g

Now, $\frac{m(L - l)}{L} g - \frac{ml}{L} g = ma$

$$a = \frac{(L - 2l)}{L} g \Rightarrow \frac{g}{2} = \frac{L - 2l}{L} g$$

$$L = 4l \Rightarrow l = \frac{L}{4} = \frac{1}{x}$$

$$x = 4$$

36. *(a)* Given, $M = 100$ kg,

$m_1 = 10$ kg,

$m_2 = 20$ kg

The given situation is shown below.

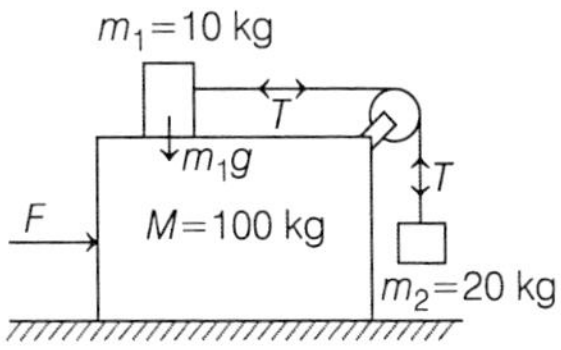

Let m_1 moves with acceleration a.

If m_2 block moves with $a = 2 \text{ m/s}^2$ up wards m_2 will move with acceleration of 2 m/s^2 in leftwards direction.

Hence, for m_1,

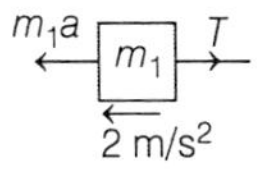

($m_1 a$ is pseudo force on m_1)

$$m_1 a - T = 2m_1$$

$$\Rightarrow \quad 10a - T = 2 \times 10$$

$$\Rightarrow \quad 10a - T = 20 \quad \ldots\text{(i)}$$

For m_2

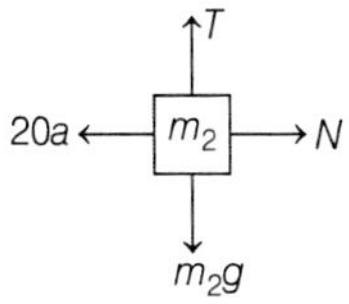

$$T - m_2 g = 2m_2 \Rightarrow T - 20g = 2 \times 20$$

$$\Rightarrow \quad T - 20g = 40 \quad \ldots\text{(ii)}$$

On adding Eqs. (i) and (ii), we get

$$10a - 20g = 60$$

$$a = \frac{60 + 20g}{10}$$

$$= \frac{60 + 20 \times 10}{10}$$

$$\Rightarrow \quad a = 26 \text{ m/s}^2$$

$$\Rightarrow \quad T = 240 \text{ N}$$

$$N = 520$$

FBD of 100 kg block

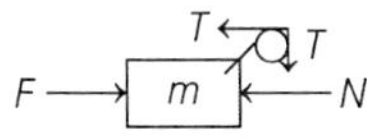

$$F - T - N = 100a$$

$$F = 100 \times 26 + 240 + 520 = 3360 \text{ N}$$

37. *(b)* Let the acceleration is a and tension in the string be T.

For mass M_1,

$$M_1 a = T - M_1 g \quad \ldots\text{(i)}$$

For mass M_2,

$$M_2 a = M_2 g - T \quad \ldots\text{(ii)}$$

From Eqs. (i) and (ii), we get

$$a = \frac{(M_2 - M_1)g}{M_1 + M_2}$$

When, $M_2 = 2M_1$, then acceleration of the system is,

$$a_1 = \frac{(2M_1 - M_1)g}{M_1 + 2M_1} = \frac{g}{3}$$

When, $M_2 = 3M_1$, then acceleration of the system is,

$$a_2 = \frac{(3M_1 - M_1)g}{M_1 + 3M_1} = \frac{g}{2}$$

$$\therefore \quad \frac{a_1}{a_2} = \frac{g}{3} \times \frac{2}{g} = \frac{2}{3}$$

38. *(d)* Here, both block and wedge are moving.

Consider the acceleration of the block with respect to the wedge is a_1 and the acceleration of the wedge is a_2.

Given, $\quad M = 16$ kg

$m = 8$ kg

FBD of the wedge

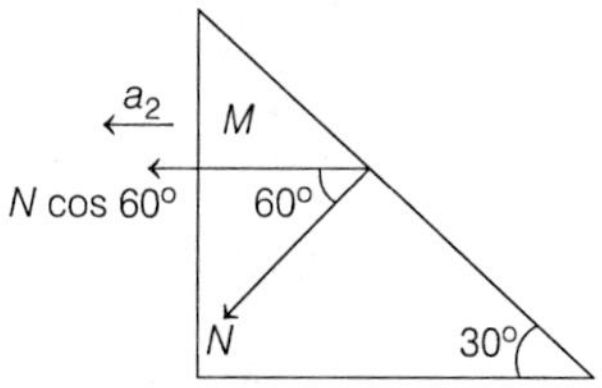

In the x-directions,

$N\cos 60° = Ma_2$

$N(0.5) = 16 \times a_2 \Rightarrow N = 32a_2$

Now, FBD of the block with respect to the wedge.

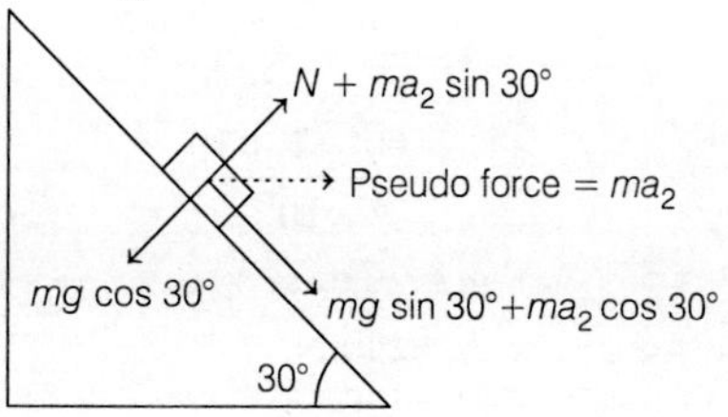

Along the perpendicular to the inclined plane,

$N = 8g\cos 30° - 8a_2 \sin 30°$

$32a_2 = 4\sqrt{3}g - 4a_2$

$36a_2 = 4\sqrt{3}g \Rightarrow a_2 = \frac{\sqrt{3}}{9}g$

Along the inclined plane,

$mg\sin 30° + ma_2\cos 30° = ma_1$

$8g \times \frac{1}{2} + 8 \times \frac{\sqrt{3}}{9}g \times \frac{\sqrt{3}}{2} = 8a_1$

$\Rightarrow \quad a_1 = 2g/3$

39. *(d)* Let spring is stretched by x

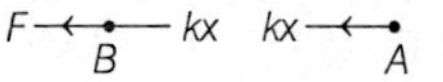

$F - kx - Ma_B \quad \ldots(\text{i})$

$kx = Ma_A = Ma \quad \ldots(\text{ii})$

$\frac{F - Ma}{M} = a_B$

40. *(d)* According to the given figure free body diagram

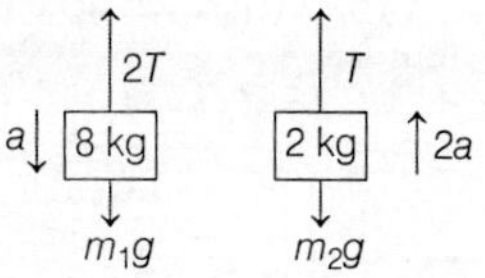

Thus equation of motion will be

$(m_1 g - 2T) = m_1 a \quad \ldots(\text{i})$

$T - m_2 g = m_2(2a)$

$\Rightarrow 2T - 2m_2 g = 4m_2 a \quad \ldots(\text{ii})$

From Eq. (i) and (ii), we get

$m_1 g - 2m_2 g = (m_1 + 4m_2)a$

$\Rightarrow \quad a = \frac{8-4}{8+8}g = \frac{g}{4}$

As we know that, $s = ut + \frac{1}{2}at^2$

$0.2 = 0 + \frac{g}{8}t^2 \quad (\because s = 0.2, \text{ given})$

$\Rightarrow \quad t = 0.4$ s

41. *(240)* Given,

$m_1 = 4$ kg, $m_2 = 6$ kg, $m_3 = 10$ kg

Acceleration, $a = 2$ m/s^2

We have to calculate the tension T_1 in rope 1.

Drawing the free body diagram,

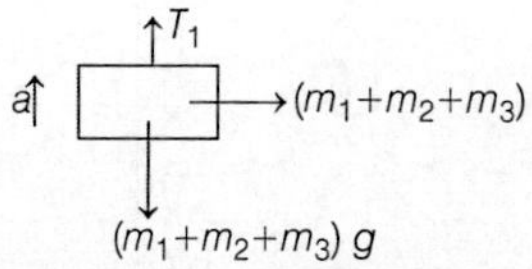

$T_1 - (m_1 + m_2 + m_3)g$
$= (m_1 + m_2 + m_3)a \quad \ldots(\text{i})$

$T_1 - (20)10 = 20 \times 2$

$T_1 - 200 = 40$

$T_1 = 240$ N

Hence, tension is 240 N.

42. *(5)* Let natural length of spring $= l_0$

$3 = k(a - l_0) \quad \ldots(\text{i})$

$2 = k(b - l_0) \quad \ldots(\text{ii})$

Now, Eqs. (i) ÷ (ii), We get

$\frac{3}{2} = \frac{a - l_0}{b - l_0}$

$2a - 2l_0 = 3b - 3l_0$

$l_0 = 3b - 2a \quad \ldots(\text{iii})$

Eq. (i) – Eq. (ii)

$k(a - b) = +1 \quad \ldots(\text{iv})$

Now, for new length,

$k(3a - 2b - l_0) = T$

$k(3a - 2b - 3b + 2a) = T$

$5k(a - b) = T$

[Put value of $k(a - b)$]

$5 = T$

$\Rightarrow \quad T = 5$ N

43. *(2)* According to the question,

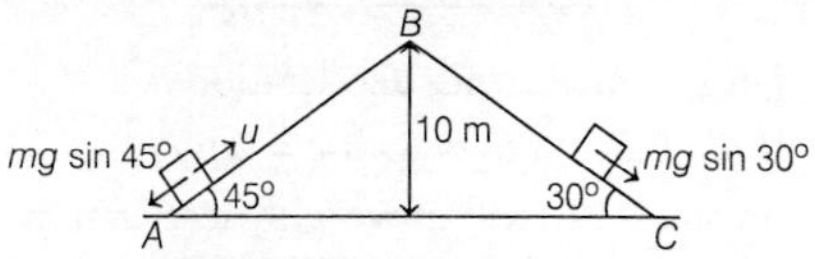

In upward motion,

$\frac{1}{2}g\sin 45° t_1^2 = 10\sqrt{2}$

$\Rightarrow \quad \frac{1}{2}\frac{10}{\sqrt{2}} \times t_1^2 = 10\sqrt{2}$

$t_1^2 = 4$ or $t_1 = 2$ s

In downward motion,

$\frac{1}{2}g\sin 30° t_2^2 = \frac{10}{1/2} = 20$

$\Rightarrow \frac{1}{2} \times \frac{10}{2} \times t_2^2 = 20$

$\Rightarrow t_2 = 2\sqrt{2}$ s

Total time $= t_1 + t_2 = 2 + 2\sqrt{2}$
$= 2(1 + \sqrt{2})$

On comparing the given value with $t(\sqrt{2} + 1)$ s, we get, $t = 2$

44. *(3)* Resolving the forces components along X-axis

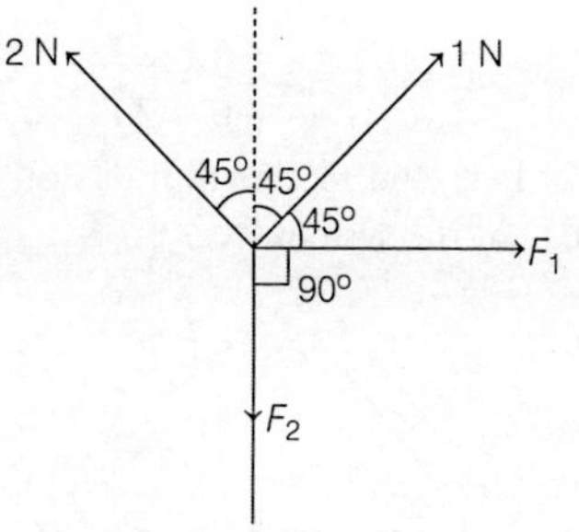

$F_x = F_1 + 1\cos 45° - 2\cos 45°$

$F_x = 0$ for equilibrium

$F_1 = +\cos 45° \quad \ldots(\text{i})$

$F_y = 1\sin 45° + 2\sin 45° - F_2$

$F_y = 0$

$F_2 = 3\sin 45° \quad \ldots(\text{ii})$

From Eqs. (i) and (ii), we get

$\frac{F_1}{F_2} = +\frac{\cos 45°}{3\sin 45°} = +\frac{1}{3}$

So, $\quad x = 3$

45. *(12)* From free body diagram,

$F\cos 60° = mg\sin 60°$

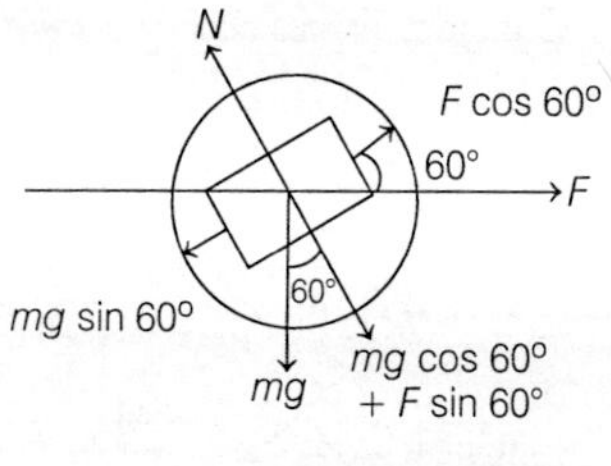

$F = mg\tan 60°$

$= 0.2 \times 10 \times \sqrt{3} = 2\sqrt{3}$

$F = \sqrt{12}$ N

So, $\quad x = 12 \quad (\because F = \sqrt{x}$ N$)$

46. *(30)* Given, Angle of inclination,

$\theta = 30°$

Acceleration, $a = 10$ ms^{-2}, $g = 10$ ms^{-2}

According to the question ,

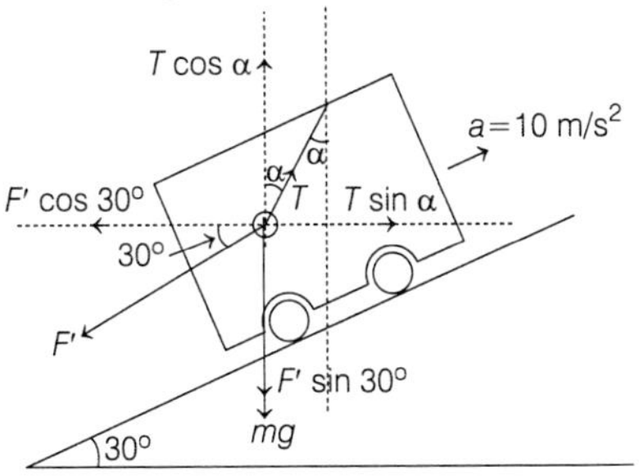

Here, F' is the pseudo force on the bob car's frame and T is the tension in the string.

In equilibrium, $\Sigma F_x = 0$ and $\Sigma F_y = 0$

$\Rightarrow \quad F' \cos 30° = T \sin \alpha$

$$\frac{ma \cos 30°}{\sin \alpha} = T \quad \ldots \text{(i)}$$

where, m is the mass of the bob.

$$F' \sin 30° + mg = T \cos \alpha$$

$$\Rightarrow \quad ma \sin 30° + mg = \frac{ma \cos 30°}{\sin \alpha} (\cos \alpha)$$

[∵ Using Eq. (i)]

$$\Rightarrow \quad a \sin 30° + g = \frac{a \cos 30°}{\sin \alpha} (\cos \alpha)$$

$$10 \times \frac{1}{2} + 10 = \frac{10 \times \sqrt{3}}{2} \cot \alpha$$

$$\Rightarrow \quad \frac{1}{2} + 1 = \frac{\sqrt{3}}{2} \cot \alpha$$

$$\Rightarrow \quad \frac{3}{2} = \frac{\sqrt{3}}{2} \cot \alpha \text{ or } \cot \alpha = \sqrt{3}$$

$$\Rightarrow \quad \alpha = 30°$$

47. *(82)* If the force **P** of magnitude 100 N is resolved parallel to the arms AB and AC of the frame, the figure will be as follows

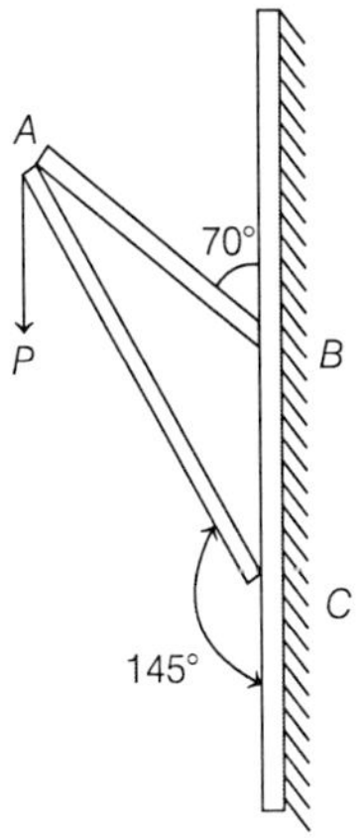

Component of force along

$$AC = 100 \cos 35° \text{ N}$$
$$= 100 \times 0.819 \text{ N}$$
$$= 81.9 \text{ N} \approx 82 \text{ N}$$

This is magnitude of the component along the arm AC.

So, $x = 82$.

Topic 3

01. *(b)*

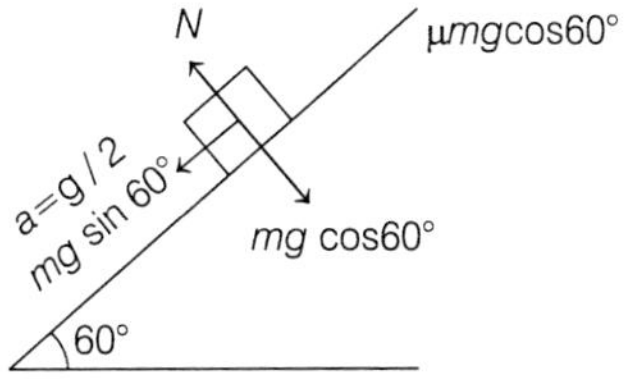

$$mg \sin 60° - \mu\, mg \cos 60° = ma$$

$$g \sin 60° - \mu g \cos 60° = \frac{g}{2}$$

$$\frac{\sqrt{3}}{2} - \frac{\mu}{2} = \frac{1}{2}$$

$$\mu = \sqrt{3} - 1$$

02. *(a)* Limiting force of static friction and kinetic friction does not depend on the area of contact but they depends on materials.

03. *(c)* Mass of box, $m = 50$ kg

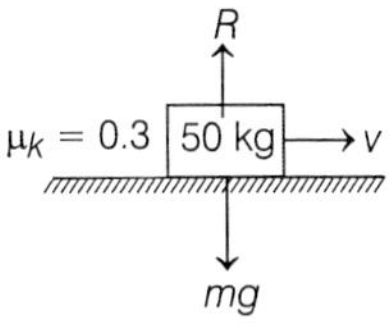

Normal force, $R = mg$

$$= 50 \times 9.8 = 490 \text{ N}$$

∴ Force of kinetic friction $= \mu_k R$

$$= 0.3 \times 490 = 147 \text{ N}$$

04. *(c)* Moving vehicle gets charged due to friction. The inflammable material may catch fire due to spark produced by charged vehicle.

When metallic chains touches the ground, the charge developed on the vehicle is transferred to the ground which prevents sparking.

05. *(c)* Given,

$$u = 20 \text{ m/s}, t = 5 \text{ s}$$

Let, μ be the coefficient of friction, then R = retardation,

$$R = \mu g = -a$$

Using, $v = u + at$

$$\Rightarrow \quad 0 = 20 - \mu g \times 5$$

$$\mu = \frac{20}{10 \times 5} = 0.4$$

06. *(d)* According to the question,

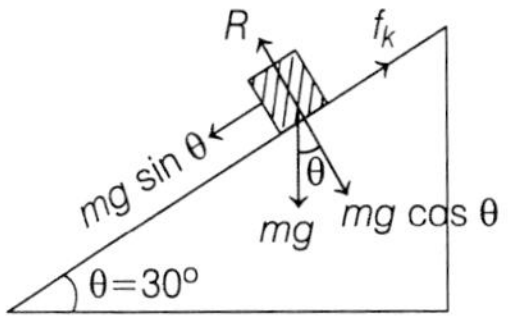

Acceleration of block

$$= \frac{mg \sin \theta - \mu_k mg \cos \theta}{m}$$

$$= g(\sin \theta - \mu_k \cos \theta) = \frac{g}{4} \text{ [Given]}$$

So, $\sin \theta - \mu_k \cos \theta = \dfrac{1}{4}$

As $\theta = 30°$; we have $\dfrac{1}{2} - \dfrac{\sqrt{3}}{2} \mu_k = \dfrac{1}{4}$

or $\quad \mu_k = \dfrac{2 \times 1}{4\sqrt{3}} = \dfrac{1}{2\sqrt{3}}$

07. *(a)* Since, the block is moving with constant velocity, thus acceleration of the block is zero.

Therefore, net force on the block is zero.

$$F_{\text{plane}} + Mg = 0 \Rightarrow F_{\text{plane}} = -Mg$$

Alternate Solution

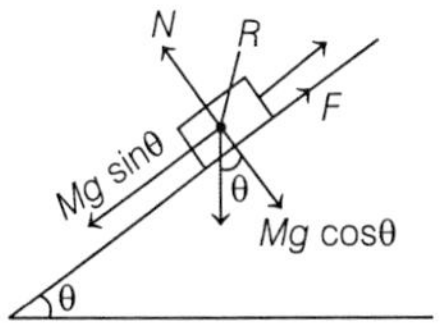

From the figure, $N = Mg \cos \theta, f = Mg \sin \theta$

$$R = \sqrt{N^2 + f^2} = Mg$$

08. *(b)* Since the bag is slipping on the conveyor belt, retardation due to friction will be

$$a = \frac{-\mu Mg}{M} = -\mu g$$

or $\quad a = -0.4 \times 10 = -4 \text{ m s}^{-2}$

In the frame of conveyor, $v = 0$ and $u = 2$ m/s

Using , $v^2 - u^2 = 2as$

$$\therefore \quad (0)^2 - (2)^2 = 2 \times -4 \times s$$

or $\quad -4 = -8s$ or $s = \dfrac{1}{2} = 0.5$ m

09. *(c)*

key idea In vertical direction, weights are balanced by frictional forces.

09. *(c)*

> **key idea** In vertical direction, weights are balanced by frictional forces.

Consider FBD of block A and B as shown in diagram below.

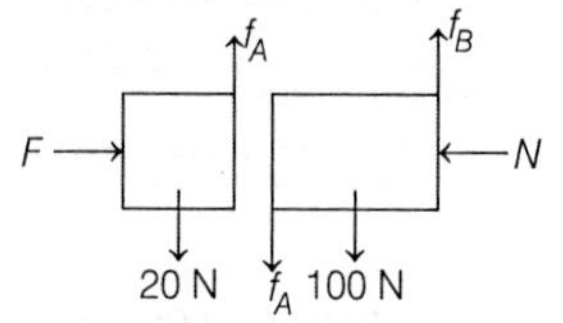

As the blocks are in equilibrium, balance forces are in vertical direction.

For the system of blocks (**A** + **B**)

$$F = N$$

For block A, $f_A = 20$ N and for block B, $f_B = f_A + 100 = 120$ N

10. *(b)* From Newton's equations, we have

$$v^2 = u^2 - 2as$$

Given, $v = 0$ [car is stopped]

As friction provide the retardation

$$F = ma = mN = \mu mg$$

$\therefore$ $a = \mu g,\ u = 100\ \text{ms}^{-1}$

$\therefore$ $(100)^2 = 2\,\mu gs$

$\Rightarrow$ $s = \dfrac{100 \times 100}{2 \times 0.5 \times 10}$

$= \dfrac{100 \times 100}{5 \times 2}$

$= 1000$ m

11. *(d)*

> **key idea** To solve these kind of problems, draw the free body diagram first. The frictional force between the block and the wall will balance the weight of the block.

$\therefore f_L = w \Rightarrow \mu R = Mg$ ($\because f_L = \mu R$ and $w = Mg$)

Let R be the normal contact force by wall on the block.

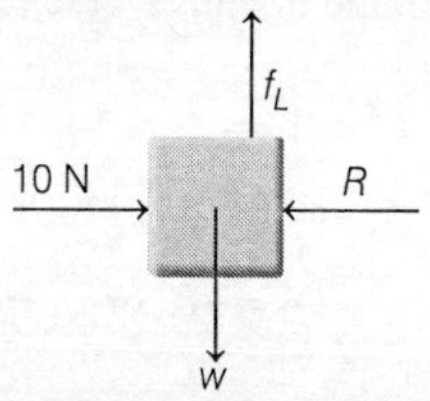

$R = 10$ N

$f_L = w$ and $f_L = \mu R$

$\therefore$ $\mu R = w$

or $w = 0.2 \times 10 = 2$ N

12. *(b)* According to given situation

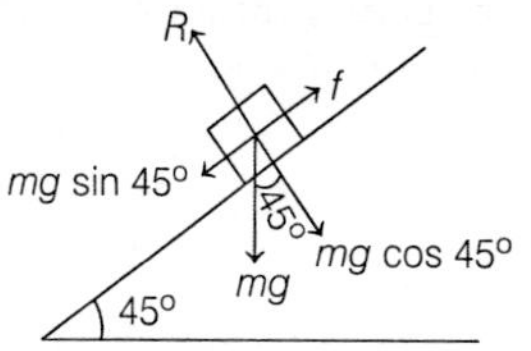

Friction = Component of weight along the plane

i.e. $\mu R = mg \sin 45°$

$\Rightarrow \mu mg \cos 45° = mg \sin 45°$

$\Rightarrow$ $\mu = 1$

13. *(a)* The given situation in question is as follows.

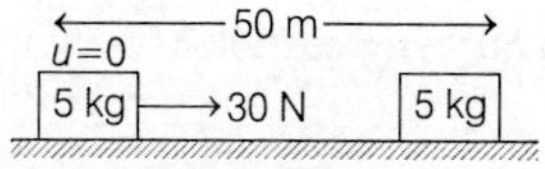

Let, the coefficient of kinetic friction be μ

Acceleration, $a = \dfrac{F - f}{m} = \dfrac{F - \mu mg}{m}$

$$a = \frac{30 - \mu \times 5 \times 10}{5} = 6 - 10\mu$$

Using second equation of motion,

$$s = ut + \frac{1}{2}at^2$$

$\Rightarrow$ $50 = 0 + \dfrac{1}{2}(6 - 10\mu) \times 10^2$

$\Rightarrow 6 - 10\mu = 1.0 \Rightarrow \mu = 0.5$

14. *(d)* The given situation is shown below,

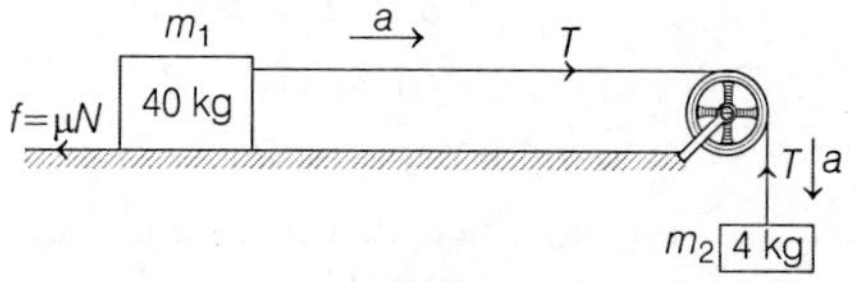

For, $m_1 = 40$ kg

$f = \mu N = \mu m_1 g = 40\,\mu g$

$= 40 \times 0.02\,g = 0.8\,g$

and $T - f = m_1 a$

$T - 0.8\,g = 40a$ ...(i)

For $m_2 g - T = m_2 a$

$\Rightarrow$ $4g - T = 4a$ ...(ii)

On adding Eqs. (i) and (ii), we get

$$3.2g = 44a$$

$\Rightarrow$ $a = \dfrac{3.2 \times 10}{44} = \dfrac{32}{44} = \dfrac{8}{11}$ m/s^2

15. *(c)*

> **key idea** The required force for the upper block is provided by the frictional force between the blocks
>
> $ma = f_L = \mu N$

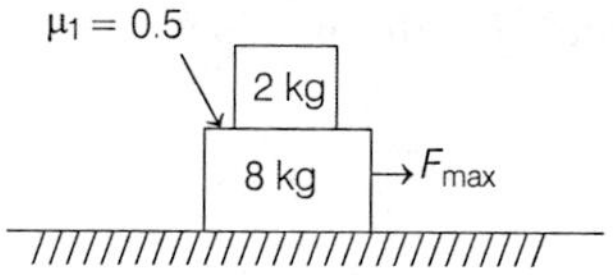

If both blocks move together, then

$$a = \mu_1 g = 0.5 \times 9.8 = 4.9 \text{ m/s}^2$$

Now taking 2 kg and 8 kg as combined (single) mass, $m = 10$ kg

$$F_{\max} = ma = 49 \text{ N}$$

16. *(b)* The force equation in vertical direction is $Mg - R = Ma$

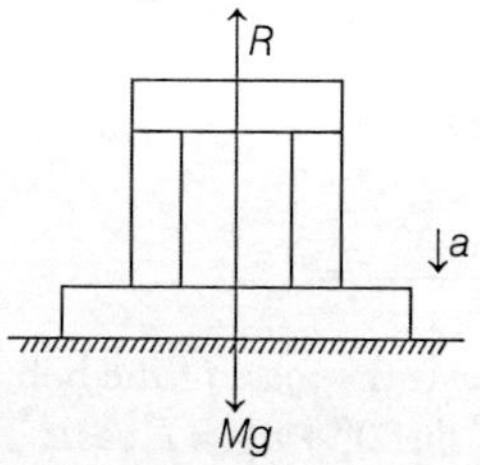

where, M = collective mass of block and iron cylinders

$= 10 + 3 \times 20 = 70$ kg

a = acceleration of block $= 0.2$ ms^{-2}

$g = 10$ ms^{-2} and $\mu_s = 0.2$

and R = normal reaction

Force along verticle axis $Mg - R = Ma$

$\therefore$ $70g - R = 70 \times 0.2$

$\Rightarrow$ $R = 70 \times 10 - 14$

$= 700 - 14 = 686$ N

17. *(b)* The FBD of the block is shown below

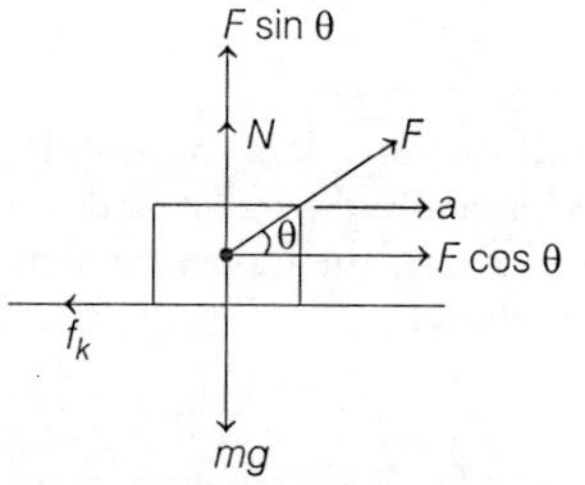

We have, $N = mg - F \sin\theta$ ...(i)

and $F\cos\theta - f_k = ma$

$\Rightarrow$ $F\cos\theta - \mu_k N = ma$ ...(ii)

From Eq. (i) and Eq. (ii), we get

$$F\cos\theta - \mu_k(mg - F\sin\theta) = ma$$

$\Rightarrow$ $a = \dfrac{F}{m}\cos\theta - \mu_k\left(g - \dfrac{F}{m}\sin\theta\right)$

18. *(b)* Motion stops when pull due to $m_1 \le$ force of friction between m and m_2 and surface.

$\Rightarrow$ $m_1 g \le \mu(m_2 + m)g$

$\Rightarrow$ $5 \times 10 \le 0.15(10 + m) \times 10$

$\Rightarrow$ $m \ge 23.33$ kg

Here, nearest value is 27.3 kg

So, $m_{\min} = 27.3$ kg

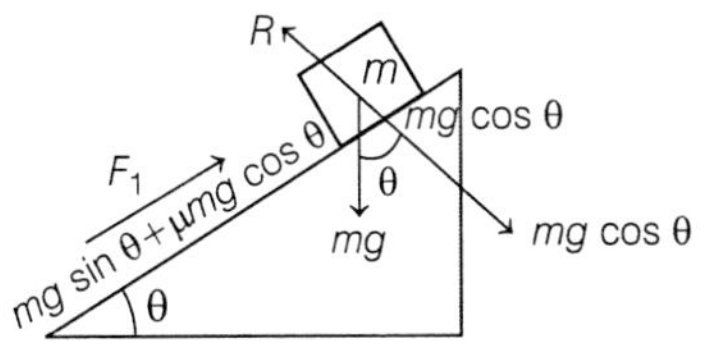

$$\therefore \quad \frac{F_1}{F_2} = \frac{\sin\theta + \mu\cos\theta}{\sin\theta - \mu\cos\theta}$$

$$= \frac{\tan\theta + \mu}{\tan\theta - \mu} = \frac{2\mu + \mu}{2\mu - \mu} = 3$$

20. *(b)* Over a smooth plane, (frictionless)

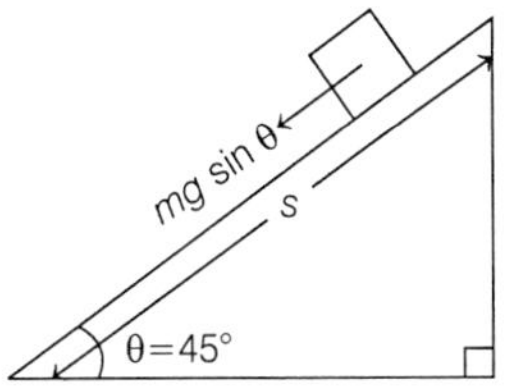

acceleration along the plane,

$$a = g\sin\theta$$

We have $u = 0$, $s =$ length of incline.

$\therefore\ s = ut + \frac{1}{2}at^2$,

Time of sliding down,

$$t_1 = \sqrt{\frac{2s}{g\sin\theta}} = \sqrt{\frac{2s}{g/\sqrt{2}}}$$

When friction is present, acceleration

$$a = g(\sin\theta - \mu\cos\theta)$$

$\mu =$ coefficient of friction

$$\Rightarrow \quad a = \frac{g}{\sqrt{2}}(1-\mu)$$

$\therefore$ Time of sliding down

$$t_2 = \sqrt{\frac{2s}{\frac{g}{\sqrt{2}}(1-\mu)}}$$

Given that,

$$t_2 = n \times t_1$$

$$\Rightarrow \quad \sqrt{\frac{2s}{\frac{g}{\sqrt{2}}(1-\mu)}} = n \cdot \sqrt{\frac{2s}{(g/\sqrt{2})}}$$

$$\Rightarrow \quad \frac{1}{(1-\mu)} = n^2 \Rightarrow \frac{1}{n^2} = 1 - \mu$$

or $$\mu = 1 - \frac{1}{n^2}$$

21. *(*)* **Case 1** When block is pushed up, the inclined force are

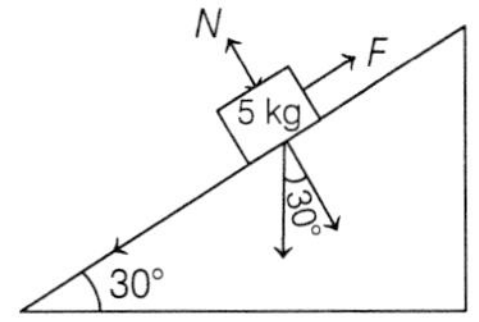

$$f_{max} = \mu N = \mu mg\cos 30°$$

$$= 0.1 \times 5 \times 10 \times \frac{\sqrt{3}}{2} = \frac{5\sqrt{3}}{2}$$

$$F_1 = mg\sin\theta + f_{max}$$

$$= 5 \times 10\sin 30° + \frac{5\sqrt{3}}{2} = 25 + \frac{5\sqrt{3}}{2}$$

Case 2

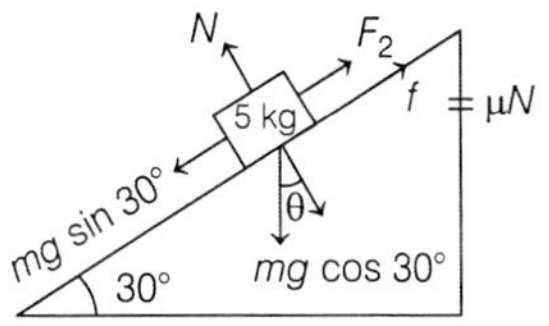

$$f_{max} = \mu N = 0.1 \times mg\cos 30° = \frac{5\sqrt{3}}{2}$$

$$F_2 + f_{max} = mg\sin 30°$$

$$F_2 = mg\sin 30° - f_{max}$$

$$= 5 \times 10 \times \frac{1}{2} - \frac{5\sqrt{3}}{2}$$

$$= 25 - \frac{5\sqrt{3}}{2}$$

$$\text{So, } |F_1| - |F_2| = \left|25 + \frac{5\sqrt{3}}{2}\right| - \left|25 - \frac{5\sqrt{3}}{2}\right|$$

$$= \frac{10\sqrt{3}}{2} = 5\sqrt{3}\text{ N}$$

22. *(b)* For no slipping,

$$\mu\, mg\cos\theta \ge mg\sin\theta$$

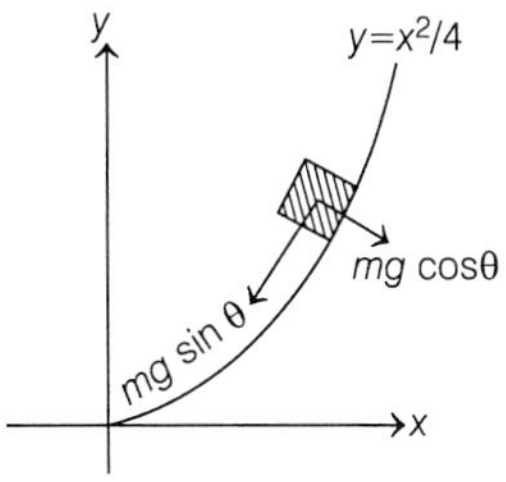

or $$\mu \ge \tan\theta$$

$$\Rightarrow \quad \mu \ge \frac{dy}{dx} \quad \ldots\text{(i)}$$

Here, $\mu = 0.5$

and $y = \frac{x^2}{4} \Rightarrow \frac{dy}{dx} = \frac{x}{2}$

So from Eq. (i),

$$0.5 \ge \frac{x}{2} \Leftrightarrow x \le 1$$

As, $x \le 1$ or $x^2 \le 1$ and $y = \frac{x^2}{4}$

So, $x^2 = 4y$

$$\Rightarrow \quad 4y \le 1 \Rightarrow y \le \frac{1}{4}$$

$$\therefore \quad y_{max} = \frac{1}{4}\text{ m}$$

23. *(a)* Drawing FBD of block,

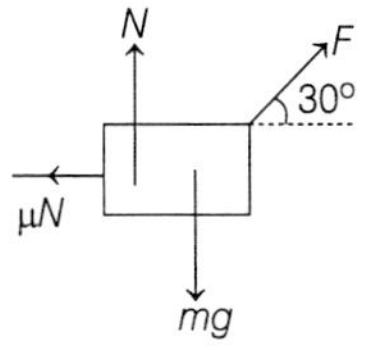

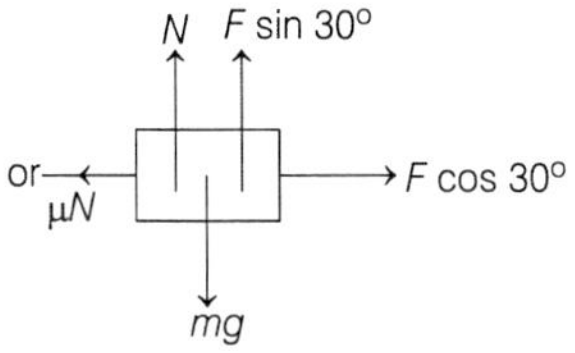

In y-direction,

$$\Sigma F_y = ma_y$$

$$N + F\sin 30° = mg \quad (\because a_y = 0)$$

or $$N = mg - F\sin 30° \quad \ldots\text{(i)}$$

In x-direction,

$$\Sigma F_x = ma_x$$

$$F\cos 30° - \mu N = ma_x$$

As block is on the verge of slipping then

$$a_x = 0$$

$$F\cos 30° = \mu N \quad \ldots\text{(ii)}$$

On substituting value of N from Eq. (i) to Eq. (ii), we get

$$F\cos 30° = \mu(mg - F\sin 30°)$$

$$F\cos 30° = \mu mg - \mu F\sin 30°$$

$$F(\cos 30° + \mu\sin 30°) = \mu mg$$

$$F = \frac{\mu mg}{\cos 30° + \mu\sin 30°}$$

Given that

$\mu_s = 0.25$, $m = 10$ kg,

$g = 10\text{ m/s}^2$

$$F = \frac{0.25 \times 10 \times 10}{\frac{\sqrt{3}}{2} + 0.25 \times \frac{1}{2}}$$

$$F = 25.22\text{ N}$$

24. *(c)* When pushing upward, friction force would be in downward direction.

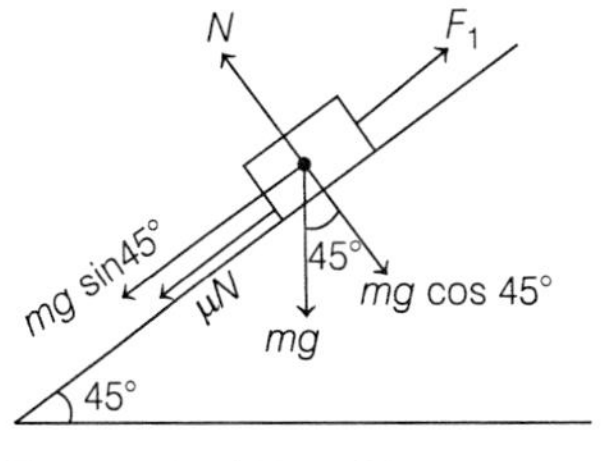

$$F_1 = mg\sin 45° + \mu N$$

$= mg \sin 45° + \mu\, mg \cos 45°$

$$F_1 = \frac{mg}{\sqrt{2}} + \frac{\mu mg}{\sqrt{2}}$$

When preventing it from sliding, friction force would be in upward direction.

$$F_2 + \mu N = mg \sin 45°$$

$$F_2 = mg \sin 45° - \mu mg \cos 45°$$

$$= \frac{mg}{\sqrt{2}} - \frac{\mu mg}{\sqrt{2}}$$

According to given situation,

$$F_1 = 2F_2$$

$$\frac{mg}{\sqrt{2}} + \frac{\mu mg}{\sqrt{2}} = 2\left(\frac{mg}{\sqrt{2}} - \frac{\mu mg}{\sqrt{2}}\right)$$

$$3\,\mu mg = mg$$

$$\mu = \frac{1}{3} = 0.33$$

25. *(b)* Block does not move upto a maximum applied force of 2N down the inclined plane.

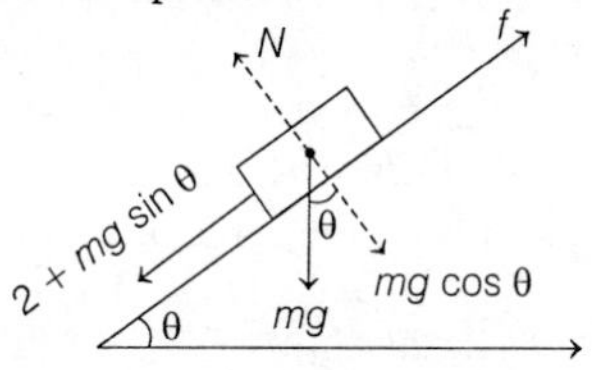

So, equating forces, we get

$$2 + mg \sin \theta = f$$

or $\quad 2 + mg \sin \theta = \mu\, mg \cos\theta \quad \ldots(i)$

Similarly, block also does not move upto a maximum applied force of 10 N up the plane.

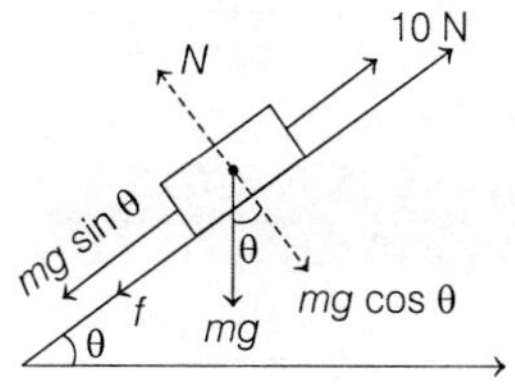

Now, equating forces, we have

$$mg \sin \theta + f = 10 \text{ N}$$

or $mg \sin \theta + \mu mg \cos\theta = 10 \quad \ldots(ii)$

Now, solving Eqs. (i) and (ii), we get

$$mg \sin \theta = 4 \quad \ldots(iii)$$

and $\quad \mu\, mg \cos\theta = 6 \quad \ldots(iv)$

On dividing, Eqs. (iii) and (iv), we get

$$\mu \cot \theta = \frac{3}{2}$$

$$\Rightarrow \quad \mu = \frac{3 \tan \theta}{2} = \frac{3 \tan 30°}{2}$$

$$\Rightarrow \quad \mu = \frac{\sqrt{3}}{2}$$

26. *(a)* A block of mass m is placed on a surface with a vertical cross-section, then

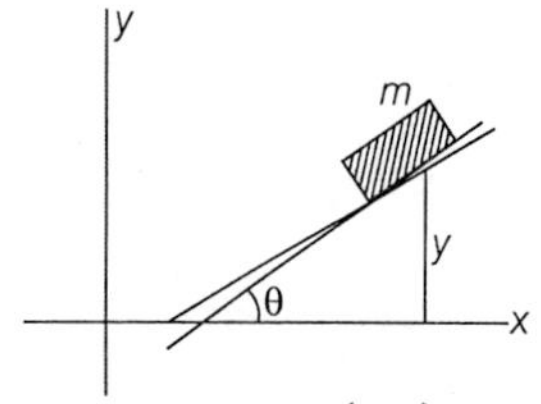

$$\tan \theta = \frac{dy}{dx} = \frac{d\left(\frac{x^3}{6}\right)}{dx} = \frac{x^2}{2}$$

At limiting equilibrium, we get

$$\mu = \tan \theta, \; 0.5 = x^2/2$$

$$\Rightarrow \quad x^2 = 1 \Rightarrow x = \pm 1$$

Now, putting the value of x in $y = x^3/6$, we get

When $x = 1$	When $x = -1$
$\therefore \quad y = \frac{(1)^3}{6} = \frac{1}{6}$	$y = \frac{(-1)^3}{6} = \frac{-1}{6}$

So, the maximum height above the ground at which the block can be placed without slipping is $\frac{1}{6}$ m.

27. *(2)* Given, $\mu_s = 0.5$

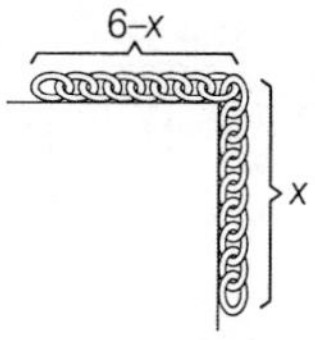

Let x length is hanging, then $6 - x$ on the table

$$\text{Hanging mass} = \frac{m}{6} \cdot x$$

Mass on the table

$$= (6 - x)\frac{m}{6}$$

For not to slide

$$\left(\frac{m}{6}x\right) g \le \mu_s \left((6 - x)\frac{m}{6}\right) g$$

$$x \le 2, \; x_{\text{max}} = 2 \text{ m}$$

28. *(3)* Given, $\mu = \frac{\sqrt{x}}{5}$

$$\text{Time of ascent} = \frac{1}{2}\text{ time of descent}$$

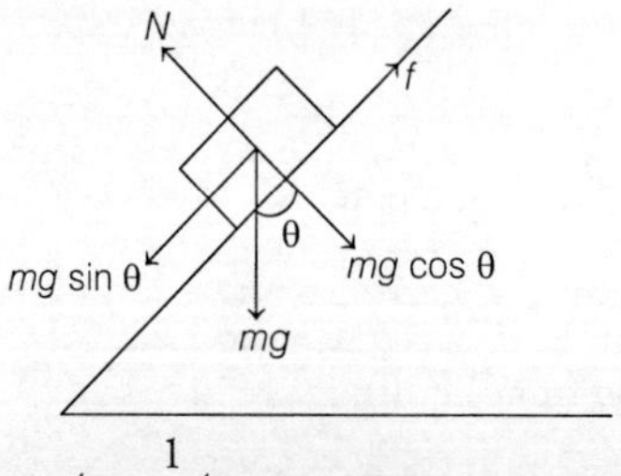

$$t_a = \frac{1}{2} t_d$$

Let L is the length of the incline, a_a is acceleration of ascent and a_d is acceleration of descent.

So, $\sqrt{\frac{2L}{a_a}} = \frac{1}{2}\sqrt{\frac{2L}{a_d}}; \; \frac{2L}{a_a} = \frac{1}{4} \times \frac{2L}{a_d}$

$$4a_d = a_a \quad \ldots(i)$$

Since, we know,

Acceleration of ascent,

$$a_a = g\,(\sin \theta + \mu \cos \theta)$$

$$= g(\sin 30° + \mu \cos 30°)$$

$$= g\left(\frac{1}{2} + \frac{\mu\sqrt{3}}{2}\right)$$

Acceleration of descent,

$$a_d = g\,(\sin \theta - \mu \cos \theta)$$

From Eq. (i), $4 \cdot g\,(\sin \theta - \mu \cos \theta)$

$$= g\,(\sin \theta + \mu \cos \theta)$$

$$= g\,[\sin 30° + \mu \cos 30°]$$

$$= g\left[\frac{1}{2} + \frac{\mu\sqrt{3}}{2}\right]$$

$$\Rightarrow 4\left[\frac{1}{2} - \mu\frac{\sqrt{3}}{2}\right] = \left[\frac{1}{2} + \mu\frac{\sqrt{3}}{2}\right]$$

$$\Rightarrow \quad 4 - 4\mu\sqrt{3} = 1 + \mu\sqrt{3}$$

$$3 = 5\mu\sqrt{3} \Rightarrow \mu = \frac{\sqrt{3}}{5}$$

Thus, $x = 3$

Alternate Solution

Let T_A be the time of ascent and T_D be the time of descent

Using, $s = ut + \frac{1}{2}at^2$

In time t_A, the distance covered by the body,

$$s = \frac{1}{2}a_A t_A^2 \quad \ldots(i)$$

and in time t_B the distance covered by the body,

$$s = \frac{1}{2}a_D t_D^2 \quad \ldots(ii)$$

$\therefore$ From Eqs. (i) and (ii), we get

$$\frac{1}{2}a_A t_A^2 = \frac{1}{2}a_D t_D^2$$

$$\Rightarrow \quad a_A t_A^2 = a_D t_D^2 \Rightarrow \frac{t_A^2}{t_D^2} = \frac{a_D}{a_A}$$

$$\Rightarrow \quad \frac{t_A^2}{t_D^2} = \frac{g \sin\theta - \mu g \cos\theta}{g \sin\theta + \mu g \cos\theta}$$

$$\Rightarrow \quad \frac{t_A}{t_0} = \sqrt{\frac{g \sin 30° - \mu g \cos 30°}{g \sin 30° + \mu g \cos 30°}}$$

$$\Rightarrow \quad \frac{1}{2} = \sqrt{\frac{1 - \sqrt{3}\mu}{1 + \sqrt{3}\mu}} \quad \left\{\because \frac{t_A}{t_D} = \frac{1}{2} \text{(given)}\right\}$$

$$\Rightarrow 1 + \sqrt{3}\mu = 4 - 4\sqrt{3}\mu$$

$$\Rightarrow \mu(5\sqrt{3}) = 3$$

$$\Rightarrow \quad \mu = \frac{\sqrt{3}}{5} \quad \ldots(iii)$$

Thus, $\quad x = 3$

29. *(5)* Given, Mass of body, $m = 1$ kg
Coefficient of static friction

$$\mu = \frac{1}{\sqrt{3}}$$

Minimum force,

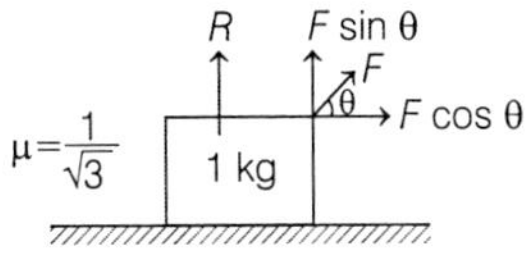

$$F\cos\theta = \mu R \quad \ldots\text{(i)}$$
$$F\sin\theta + R = mg$$
$$\Rightarrow \quad F\sin\theta = mg - R = mg - \frac{F\cos\theta}{\mu}$$
[From Eq. (i)]
$$\Rightarrow \mu F\sin\theta + F\cos\theta = \mu mg$$
$$\Rightarrow \quad F = \frac{\mu mg}{\cos\theta + \mu\sin\theta}$$
$$\Rightarrow F_{min} = \frac{\mu mg}{\sqrt{1+u^2}}$$
$$F = \frac{\mu mg}{\sqrt{1+\mu^2}} = \frac{\frac{1}{\sqrt{3}}\times 1\times 10}{\sqrt{1+\frac{1}{3}}} = 5\text{ N}$$

30. *(3)* Let's draw the free body diagram when body slides down on smooth surface

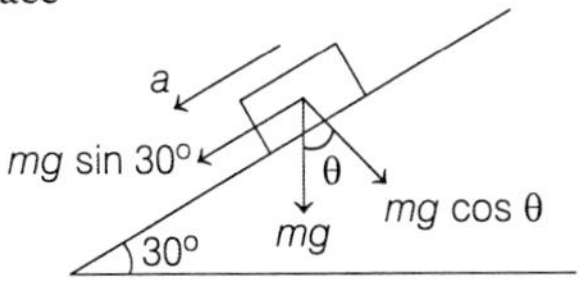

$$a = g\sin 30° = \frac{g}{2}$$

Using, $\quad s = ut + \frac{1}{2}at^2$

$\because \quad u = 0$

$$\Rightarrow \quad s = \left(\frac{g}{4}\right)T^2 \quad \ldots\text{(i)}$$

FBD for the

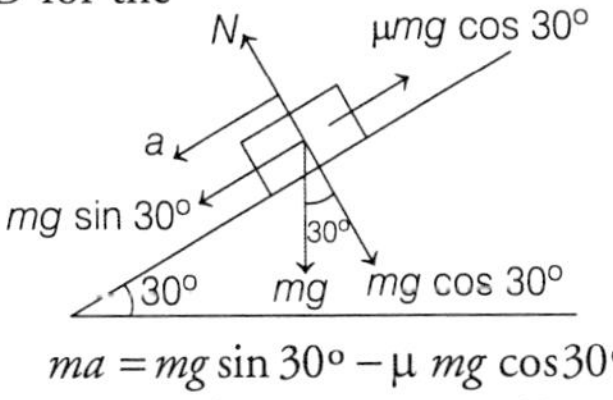

$$ma = mg\sin 30° - \mu\, mg\cos 30°$$
$$\Rightarrow \quad a = g\sin 30° - \mu\, g\cos 30°$$

Distance covered on the rough surface in time αT,

$$s = 0 + \frac{1}{2}(g\sin 30° - \mu\, g\sin 30°)\,t^2$$
$$s = \frac{g}{4}(1 - \sqrt{3}\,\mu)(\alpha T)^2 \quad \ldots\text{(ii)}$$

Distance covered by the block is same for both the case,

$$\Rightarrow \frac{g}{4}(1-\sqrt{3}\,\mu)(\alpha T)^2 = \frac{g}{4}T^2$$
[From Eq. (i) and Eq. (ii)]
$$\Rightarrow 1 - \sqrt{3}\mu = \frac{1}{\alpha^2}$$
$$\Rightarrow \quad \mu = \left(\frac{\alpha^2 - 1}{\alpha^2}\right)\frac{1}{\sqrt{3}}$$

On comparing with $\mu = \left(\frac{\alpha^2-1}{\alpha^2}\right)\frac{1}{\sqrt{x}}$

The value of the $x = 3$.

31. *(30)* The free body diagram of the block

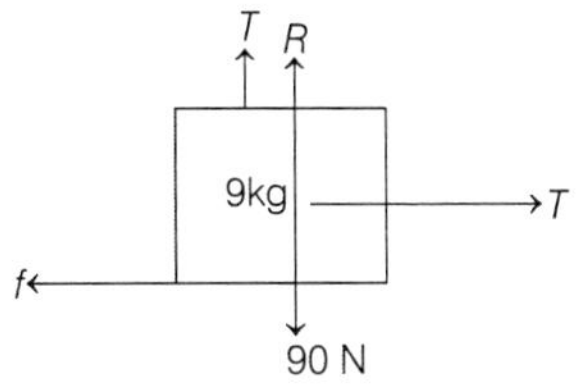

In the x-direction,

$$T = f$$
$$T = 0.5\,R \quad \ldots\text{(i)}$$

In the y-direction,

$$T + R = 90$$
$$\Rightarrow \quad R = 60\text{ N}$$

Hence, the normal force on the block is 60 N.

Using the Eq. (i),

$$T = 0.5(60) = 30\text{ N}$$

Hence, the maximum value of the tension in the rope, so that wooden block will not move is 30N.

32. *(21)* Since, both blocks move togther,

$$a = \frac{F}{m+M} \quad \ldots\text{(i)}$$

Force on mass, m

$$f = ma \quad \ldots\text{(ii)}$$

From Eqs. (i) and (ii), we get

$$f = m\left(\frac{F}{m+M}\right)$$

For no slipping, $f \le \mu mg$
[$\because$ μ being the coefficient of static friction]

$$\Rightarrow m\left(\frac{F}{m+M}\right) \le \mu mg$$
$$\Rightarrow \quad F \le \mu(m+M)g$$
$$\therefore \quad F_{max} = \frac{3}{7}(0.5 + 4.5)\times 9.8$$
$$\Rightarrow \quad F_{max} = 21\text{ N}$$

33. *(3.33)* Given, $m = \sqrt{3}$ kg, $\mu = 1/3\sqrt{3}$

According to diagram,
Let F be the force applied on the body,
For no movement of body along X-axis, net force along X-axis should be zero.

$$\therefore \quad N = F\sin 60° + mg$$
$$\Rightarrow \quad N = \frac{\sqrt{3}}{2}F + 10\sqrt{3} \quad \ldots\text{(i)}$$

Similarly, $F_x = F\cos 60° - \mu N = 0$

From Eq. (i), we get

$$\Rightarrow \frac{F}{2} - \frac{1}{3\sqrt{3}}\left(\frac{\sqrt{3}}{2}F + 10\sqrt{3}\right) = 0$$
$$\Rightarrow \quad \frac{F}{2} = \frac{F}{6} + \frac{10}{3} \Rightarrow \frac{F}{2} - \frac{F}{6} = \frac{10}{3}$$
$$\Rightarrow \quad F = 10\text{ N}$$

Given, $\quad F = 3x$

$$\Rightarrow \quad x = \frac{10}{3} = 3.33$$

34. *(25)* The graph for given equation is shown below

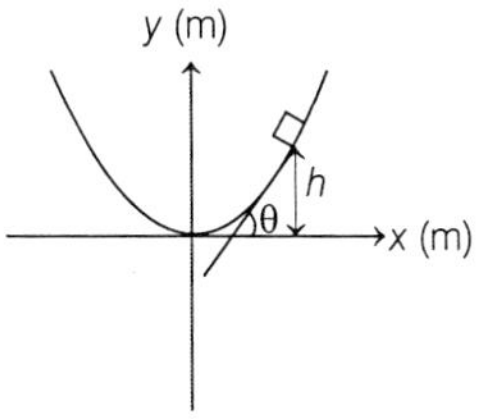

At maximum height, the slope of tangent drawn,

$$\tan\theta = \frac{dy}{dx} = \frac{2x}{4} = \frac{x}{2} \quad \left[\because y = \frac{x^2}{4}\right]$$
$$\Rightarrow \quad 0.5 = \frac{x}{2} \quad [\because \mu = \tan\theta]$$
$$\Rightarrow \quad x = 1\text{ m}$$
$$\therefore \quad y = \frac{x^2}{4} = \frac{1}{4} = 0.25\text{ m} = 25\text{ cm}$$

35. *(25)* Given, coefficient of static friction, $\mu_s = 0.2$

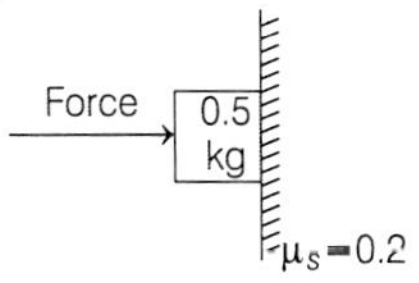

Various forces acting on block are shown below

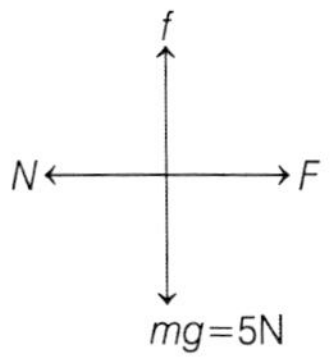

Frictional force $\leq mg$

$\Rightarrow \quad N \times 0.2 \leq 5$

$\Rightarrow \quad N \leq 25$

$\therefore$ Magnitude of horizontal force,

$$F = N = 25 \text{ N}$$

Topic 4

01. *(a)* Given,

$$m = 200 \text{ kg}, r = 70 \text{ m}$$
$$\omega = 0.2 \text{ rad/s}$$
$$F_c = m\omega^2 r = 200 \times (0.2)^2 \times 70$$
$$= 200 \times 0.04 \times 70$$
$$= 2 \times 4 \times 70 = 560 \text{ N}$$

02. *(a)* Given, $m = 5$ kg, $R = 2$ m

Time t for 1 rev $= 3.14$ s or π s

θ for 1 rev $= 2\pi$ rad

Therefore, $\omega = \frac{\theta}{t} = \frac{2\pi}{\pi} = 2$ rad/s

Centrifugal force, $F = mR\omega^2$

$$F = 5 \times 2 \times 2^2$$
$$F = 40 \text{ N}$$

03. *(c)* Maximum safe speed of car over a horizontal turn is $v_{\max} = \sqrt{\mu rg}$

Here, $r = 50$ m, $\mu = 0.34$, $g = 10$ m/s^2

So, $v_{\max} = \sqrt{0.34 \times 50 \times 10}$

$= \sqrt{170} \approx 13$ m/s

> **• Concept Enhancer**
> When a car is turning on a horizontal curved road of radius r then, maximum speed of the car on the turn to avoid slipping is given as $v_{\max} = \sqrt{\mu rg}$
>
> where, μ is the coefficient of friction between tyres and road.

04. *(c)* A bob suspended from the roof of the car by a massless string, then angle θ between the string and the vertical,

$$\theta = \tan^{-1}\left(\frac{v^2}{rg}\right) = \tan^{-1}\left(\frac{20^2}{40 \times 10}\right)$$
$$= \tan^{-1}(1) = \tan^{-1}\left(\tan\frac{\pi}{4}\right) = \frac{\pi}{4}$$

05. *(b)*

> **key idea** From the free body diagram and from the diagram conclude the required centripetal force as provided by the frictional force between the disc and the beaker.
>
> $f_L \geq mR\omega^2 \Rightarrow \mu N \geq mR\omega^2$

Centripetal force, $F_c = mR\omega^2$

The force $\mu N \geq mR\omega^2$

$$\mu mg \geq mR\omega^2 \quad (\because N = mg)$$

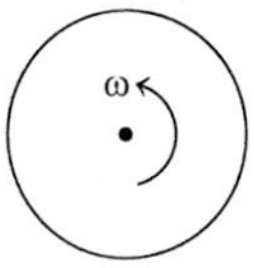

$$R \leq \frac{\mu g}{\omega^2}$$

06. *(a)* Normal reaction will provide centripetal acceleration

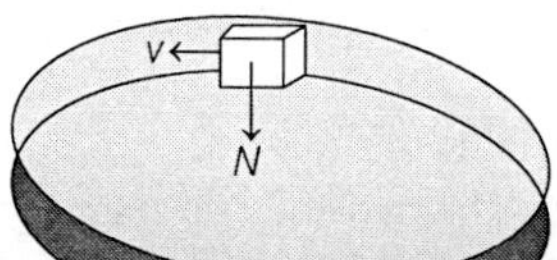

$$N = \frac{mv^2}{r}, N \propto v^2$$

This is the equation of parabola, symmetric about N

$\therefore$Graph (a) is correct.

07. *(b)*

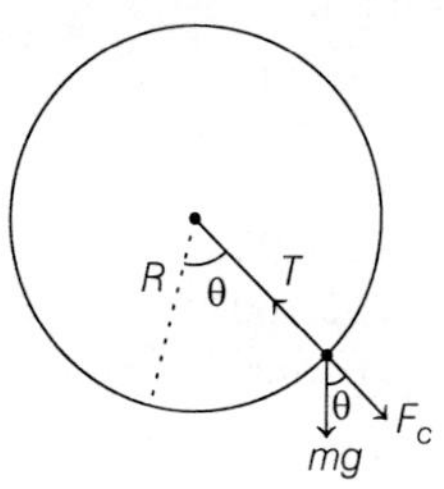

F_c is centrifugal force $= \frac{mv^2}{R}$

$$T - mg\cos\theta = F_c$$
$$T - mg\cos\theta = \frac{mv^2}{R}$$

Since, v is constant.

$\Rightarrow T$ is minimum when $\cos\theta$ is minimum which is when $\theta = 180°$

$\therefore$Tension is minimum at highest point position of the circular path.

08. *(b)*

> **key idea** The restoring force of the spring will provide the required centripetal force.
>
> $\therefore \quad kx = mR\omega^2$

Let the extension of spring is Δl.

So, $k\Delta l = m\omega^2(l + \Delta l) = m\omega^2 l + m\omega^2 \Delta l$

$$\Delta l = \frac{m\omega^2 l}{k - m\omega^2}$$

09. *(c)* Let x be the extension of the string.

So, $kx = m\omega^2(x + l)$

or $\frac{x}{l} = \frac{m\omega^2}{k - m\omega^2}$ or $\frac{x}{l} = \frac{m\omega^2}{k}$

[Using $k >> m\omega^2$]

10. *(a)* Given, $R = 1.25$ m

For just one complete rotation, speed of the drum at top position, $v = \sqrt{Rg}$

Angular velocity of the drum,

$$\omega = \frac{v}{R} = \sqrt{\frac{g}{R}} \Rightarrow \omega = \sqrt{\frac{10}{1.25}} \text{ rad s}^{-1}$$
$$= \frac{60}{2\pi}\sqrt{\frac{10}{1.25}} \text{ rpm} = 27 \text{ rpm}$$

11. *(b)* Drawing the free body diagram,

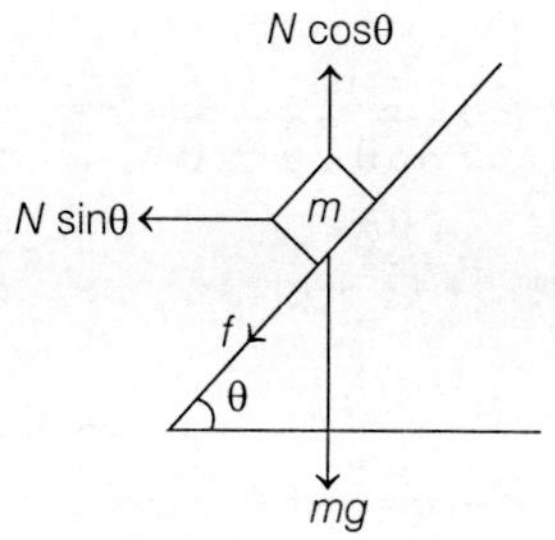

$$N\sin\theta + f\cos\theta = \frac{mv_0^2}{r}$$
$$N\cos\theta - f\sin\theta = mg$$
$$\frac{\sin\theta + \mu\cos\theta}{\cos\theta - \mu\sin\theta} = \frac{v_0^2}{rg}$$
$$rg\tan\theta + \mu rg = v_0^2 - v_0^2\mu\tan\theta$$
$$\mu = \frac{v_0^2 - rg\tan\theta}{rg + v_0^2\tan\theta}$$

12. *(d)*

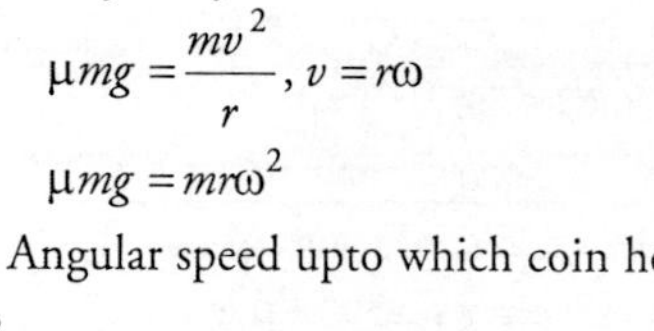

Forces on the coin is centripetal force which is provided by the friction.

So, $f = F_c$

$$\mu mg = \frac{mv^2}{r}, v = r\omega$$
$$\mu mg = mr\omega^2$$

$\therefore$ Angular speed upto which coin holds is,

$$\omega = \sqrt{\mu g / r}$$

13. *(b)* Given, $m_1 = m_2$ and $\frac{r_1}{r_2} = \frac{3}{4}$

as centripetal force, $F = \frac{mv^2}{r}$

In order to have constant (same in this question centripetal force)

$$F_1 = F_2$$

$$\frac{m_1 v_1^2}{r_1} = \frac{m_2 v_2^2}{r_2} \Rightarrow \frac{v_1^2}{v_2^2} = \frac{m_2}{m_1}\frac{r_1}{r_2}$$

$$\frac{v_1}{v_2} = \sqrt{\frac{m_2}{m_1}\cdot\frac{r_1}{r_2}} \Rightarrow \frac{v_1}{v_2} = \sqrt{\frac{m_1}{m_2}\cdot\frac{3}{4}} = \frac{\sqrt{3}}{2}$$

14. *(b)* Resolving normal,

$$N\cos\theta = mg \quad \text{...(i)}$$

$$N\sin\theta = \frac{mv^2}{R} \quad \text{...(ii)}$$

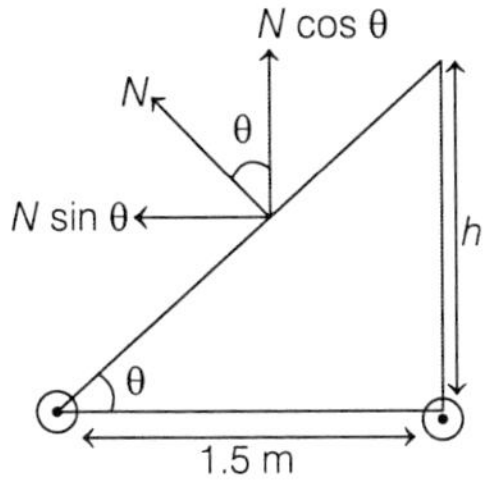

Dividing, $\tan\theta = \frac{v^2}{Rg}$

$$\frac{h}{1.5} = \frac{12\times 12}{400\times 10} \Rightarrow h = \frac{1.5\times 12\times 12}{400\times 10}$$

$$= \frac{54}{1000} = 0.054 \text{ m} = 5.4 \text{ cm}$$

15. *(d)* Given, $m = 1$ kg, $l = 1$ m

Breaking tension, $T_{max} = 400$ N (breaking tension)

Since, stone rotates horizontally,

Thus, $T_{max} = \frac{mv_{max}^2}{l}$

or $v_{max}^2 = \frac{T_{max}\, l}{m}$

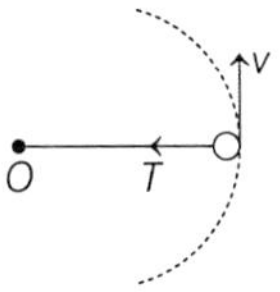

or $v_{max} = \sqrt{\frac{T_{max}}{m}\, l} = \sqrt{\frac{400\times 1}{1}}$

or $v_{max} = 20$ m/s

16. *(c)* Given, the natural length of the spring $= l_0$

Let x be extension in the spring.

Angular speed of the mass m is ω.

The net length of the spring is $(l_0 + x)$

Thus, $kx = m(l_0 + x)\omega^2$

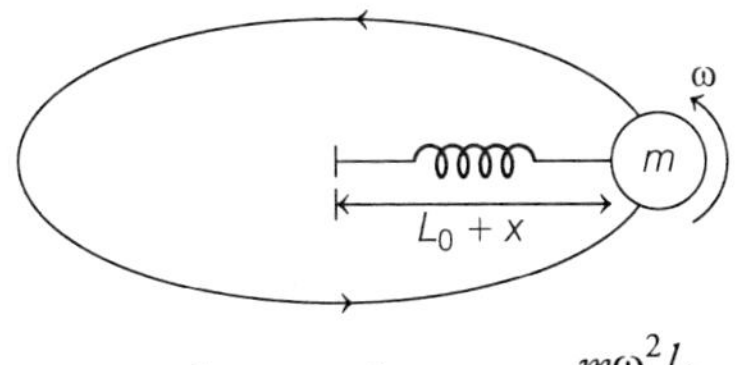

or $(k - m\omega^2)x = m\omega^2 l_0$ or $x = \frac{m\omega^2 l_0}{k - m\omega^2}$

17. *(c)* The given situation is shown below

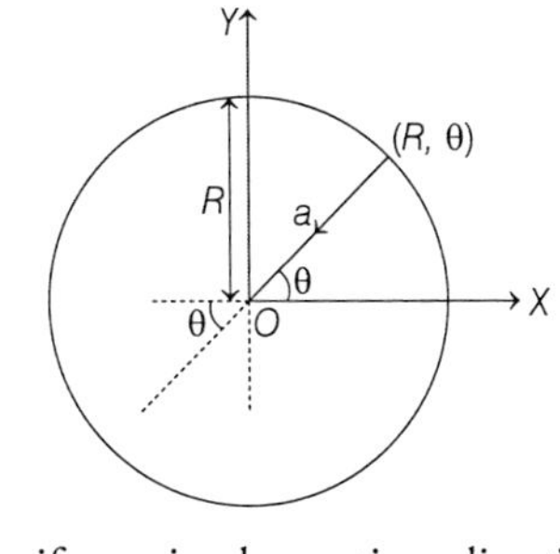

In uniform circular motion, direction of centripetal acceleration a is towards centre O, which is given as,

$$\mathbf{a} = -a\cos\theta\hat{\mathbf{i}} - a\sin\theta\hat{\mathbf{j}}$$

Since, $a = |\mathbf{a}| = v^2/R$

$$\therefore \quad \mathbf{a} = -\frac{v^2}{R}\cos\theta\hat{\mathbf{i}} - \frac{v^2}{R}\sin\theta\hat{\mathbf{j}}$$

18. *(d)* The normal force by the side walls of the groove will be equal to the centripetal force acting on it.

i.e. $N = mv^2/r$ ...(i)

where, $r = 20$ cm $= 0.2$ m

$m = 200$ g $= 200\times 10^{-3}$ kg

and $v = r\omega = \frac{2\pi r}{T} = \frac{2\pi\times 0.2}{40}$ m/s

On substituting the given values in Eq. (i), we get

$$N = \frac{(200\times 10^{-3})\times\left(\frac{2\pi\times 0.2}{40}\right)^2}{0.2}$$

$$\approx 9.859\times 10^{-4}\text{ N}$$

19. *(c)* Tension in the rod will be due to centrifugal force acting on different parts along the length of the rod.

$$T = \int_x^l dm\,\omega^2 x = \int_x^l \frac{m}{l}\omega^2 x\,dx$$

$$\Rightarrow \quad T = \frac{m}{l}\omega^2\frac{(l^2 - x^2)}{2}$$

The graph between x and T will be parabolic.

At $x = 0$, T will be maximum.

20. *(d)* The centripetal force is provided by the friction force between the coin and disc.

$$f = m\omega^2 R \text{ or } \mu mg = m\omega^2 r$$

$$\text{or} \quad \mu = \frac{\omega^2 r}{g} = \frac{(2\pi\nu)^2 r}{g}$$

$$\mu = \frac{4\pi^2(3.5)^2\times 1.25\times 10^{-2}}{10}$$

$$= 604\times 10^{-3} \approx 0.6$$

21. *(c)* Force = Mass × Acceleration

$$= m\omega^2 R$$

and given, $F \propto \frac{1}{R^n} \Rightarrow F = \frac{k}{R^n}$

So, we have $\frac{k}{R^n} = m\left(\frac{2\pi}{T}\right)^2 R$

$$\Rightarrow \quad T^2 = \frac{4\pi^2 m}{k}\cdot R^{n+1}$$

$$\Rightarrow \quad T \propto R^{\frac{n+1}{2}}$$

22. *(b)*

> **key idea** The horizontal component of the tension in the string provides the required centripetal force.
>
> $$T_x = \frac{mv^2}{r}$$

FBD of the pendulum is shown in the figure.

$$T\sin\theta = \frac{mv^2}{r},$$

$$T\cos\theta = mg,$$

$$\tan\theta = \frac{v^2}{rg}$$

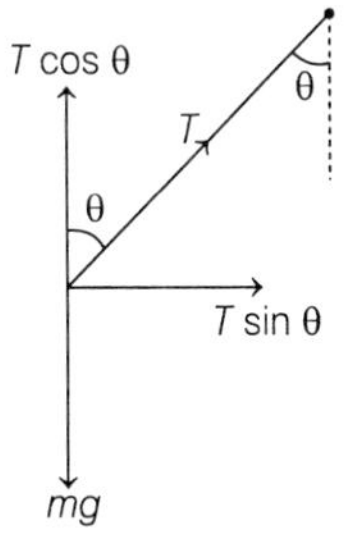

$\theta = 45°$, $r = 0.4$ m,

$\therefore \quad v^2 = rg \Rightarrow v = \sqrt{rg}$

$= \sqrt{0.4\times 10} = 2$ m/s

23. *(c)*

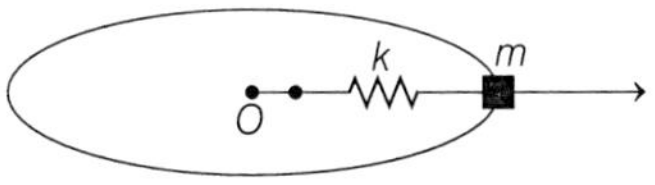

Extension in spring due to centripetal force, $F = mv^2/r = mr\omega^2$

Given, $m = 200$ g $= 0.2$ kg

$r = x_0 + x$

= natural length + extension.

$\because \quad F = kx$

where, $k = 12.5$ N/m

Thus, $kx = m(x_0 + x)\omega^2$

$$12.5x = 0.2(x_0 + x)5^2$$

$$12.5x = 5x_0 + 5x \Rightarrow 7.5x = 5x_0$$

$$\Rightarrow \quad \frac{x}{x_0} = \frac{5}{7.5} = \frac{1}{1.5} = \frac{2}{3}$$

24. *(c)* According to the question,

$QC = R\cos\alpha$

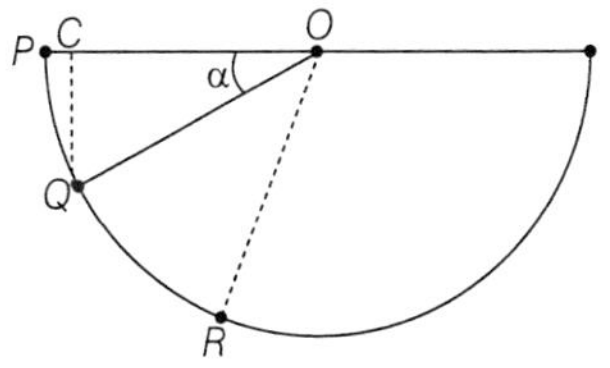

Velocity, $v = \sqrt{2g\,QC}$, $v = \sqrt{2gR\cos\alpha}$

Centripetal force,

$$F_c = \frac{mv^2}{R} = \frac{m \times 2gR \cos\alpha}{R} = 2mg \cos\alpha$$

Now, normal reaction,

$$N = mg \cos\alpha + \frac{mv^2}{R}$$

$$= mg \cos\alpha + 2mg \cos\alpha$$

$$N = 3mg \cos\alpha$$

So, $\frac{F_c}{N} = \frac{2mg \cos\alpha}{3mg \cos\alpha} = \frac{2}{3} = \text{constant}$

So, the graph is straight line.

25. *(c)* Given, mass of stone,

$$m = 100 \text{ g} = 0.1 \text{ kg}$$

length of string, $l = 2$ m

maximum tension, $T = 80$ N

Tension in the string provides required centripetal force to the stone to revolve in circular orbit.

i.e $T = m\omega^2 l \Rightarrow 80 = 0.1 \times \omega^2 \times 2$

$\Rightarrow \omega^2 = 400 \Rightarrow \omega = 20$

$\Rightarrow 2\pi f = 20$

$\Rightarrow f = \frac{20}{2\pi} = \frac{10}{\pi}$ rev/s

$= \frac{10 \times 60}{\pi}$ rev/min

$= \frac{600}{\pi}$ rev/min $= K/\pi$ rev/min (given)

$\therefore K = 600$

26. *(a)* Given, radius of horizontal circle, $r = L/\sqrt{2}$

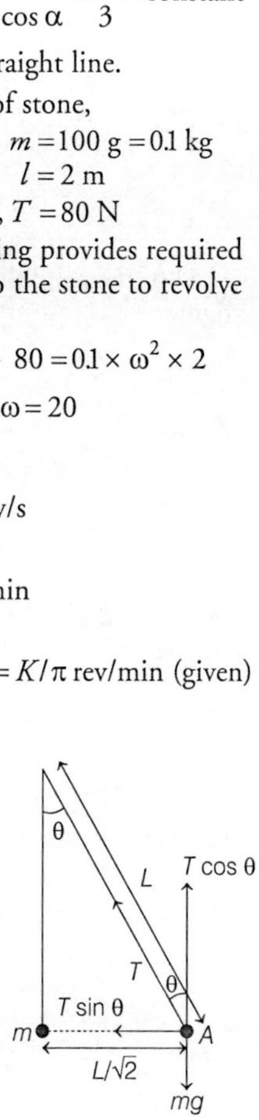

According to the question,

From the diagram

$\tan\theta = v^2 / rg$

$\Rightarrow v = \sqrt{rg \tan\theta}$ ...(i)

Now, from figure, we can write

$$\sin\theta = \frac{L/\sqrt{2}}{L} = \frac{1}{\sqrt{2}}$$

$\Rightarrow \theta = 45°$

On substituting the value of θ in Eq. (i), we get

$$v = \sqrt{rg \tan 45°} = \sqrt{rg}$$

Thus, the value of speed of particle is $v = \sqrt{rg}$

27. *(a)* The given situation can be represented as

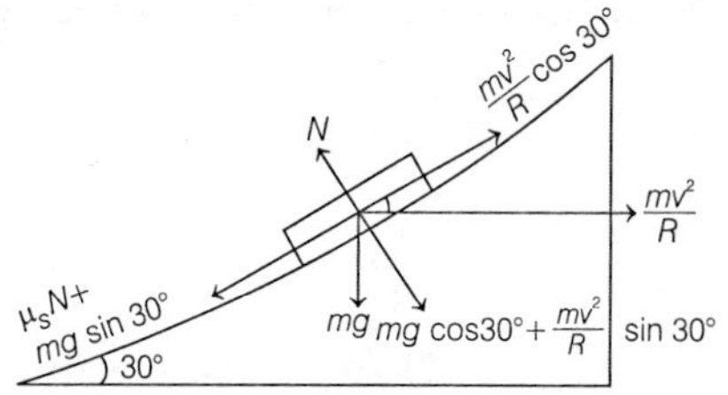

Equating forces perpendicular to the inclined plane,

$$N = mg \cos 30° + \frac{mv^2}{R} \sin 30°$$

$$\Rightarrow N - mg \cos 30° = \frac{mv^2}{R} \sin 30° \quad \text{...(i)}$$

Equating forces along the inclined plane,

$$mg \sin 30° + \mu_s N = \frac{mv^2}{R} \cos 30° \quad \text{...(ii)}$$

On dividing Eq. (i) by Eq. (ii), we get

$$\frac{N - mg \cos 30°}{mg \sin 30° + \mu_s N} = \tan 30°$$

[$\because \cos 30° = \sqrt{3}/2$ and $\sin 30° = 1/2$]

$$\Rightarrow \frac{N - mg(\sqrt{3}/2)}{mg(1/2) + (0.2)N} = \frac{1}{\sqrt{3}}$$

$$N\sqrt{3} - mg\frac{\sqrt{3}}{2} \cdot \sqrt{3} = \frac{mg}{2} + 0.2N$$

$$\Rightarrow (\sqrt{3} - 0.2)N = mg\frac{(1+3)}{2} = 2mg$$

$$\Rightarrow N = \frac{2\,mg}{\sqrt{3} - 0.2} = \frac{2 \times 800 \times 10}{1.532}$$

$$= 10.44 \times 10^3 \text{ N}$$

Therefore, $N = 10.2 \times 10^3$ kg-m/s^2

Alternate Solution

At maximum possible speed, friction will be limitng in nature,

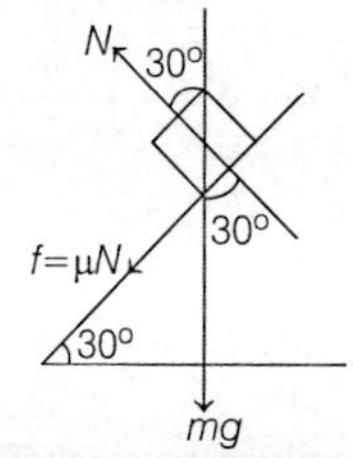

$\therefore$ Balancing the forces in vertical direction, we get

$N \cos 30° - mg - \mu N \cos 60° = 0$

$\Rightarrow N(\cos 30° - \mu \cos 60°) = mg$

$$\Rightarrow N = \frac{mg}{\cos 30° - \mu \cos 60°}$$

$$\Rightarrow N = \frac{800 \times 9.8}{(3.87 - 0.2 \times 0.5)} \simeq 10.2 \times 10^3 \text{ kg m/s}^2$$

28. *(b)* We know that, static friction force, $f_s = \mu_s N$

Since, the car is travelling on a circular track, so centripetal force is also acting on it.

$\Rightarrow F_c = mv^2/R$

In limiting condition,

$$f_s = F_c \Rightarrow \mu_s N = mv^2/R$$

$$\Rightarrow N = mv^2/\mu_s R \quad \text{...(i)}$$

The magnitude of negative lift f_L acting downwards on the car is given by

$$f_L = mg - N = mg - \frac{mv^2}{\mu_s R}$$

$$\Rightarrow f_L = m\left(g - \frac{v^2}{\mu_s R}\right)$$

$$\Rightarrow f_L = -m\left(\frac{v^2}{\mu_s R} - g\right)$$

$$\Rightarrow |f_L| = m\left(\frac{v^2}{\mu_s R} - g\right)$$

29. *(d)* The maximum speed of cyclist on turn of unbanked road without slipping is given as

$$v_{\max} = \sqrt{\mu g R} = \sqrt{0.2 \times 10 \times 2} = 2 \text{ ms}^{-1}$$

[$\because \mu = 0.2$ (given)]

Given, speed = 7 km/h

$$= \frac{7000}{3600} \text{ ms}^{-1} = \frac{70}{36} = 1.94 \text{ ms}^{-1}$$

As given speed is lesser than $v_{\max}$, so the cyclist will not slip. Therefore, Statement I is true. As per Statement II, angle of banking, $\theta = 45°$

We know that for banked road,

$$v_{\max} = \sqrt{\frac{gR(\mu + \tan\theta)}{(1 - \mu\tan\theta)}}$$

and $$v_{\min} = \sqrt{\frac{gR(\tan\theta - \mu)}{1 + \mu\tan\theta}}$$

$$\Rightarrow v_{\max} = \sqrt{\frac{10 \times 2(0.2 + \tan 45°)}{1 - 0.2\tan 45°}}$$

and $$v_{\min} = \sqrt{\frac{10 \times 2(1 - 0.2)}{1 + 0.2}}$$

$\Rightarrow v_{\max} = 5.47 \text{ ms}^{-1}$

and $v_{\min} = 3.65 \text{ ms}^{-1}$

$\because v = 18.5$ km/h

$$= \frac{18.5 \times 1000}{5600} = 5.13 \text{ ms}^{-1}$$

[$\because$ given speed = 18.5 kmh^{-1}]

As, $v_{\min} < v < v_{\max}$, so the cyclist will not slip.

$\therefore$ Statement II is also true.

30. *(36)*

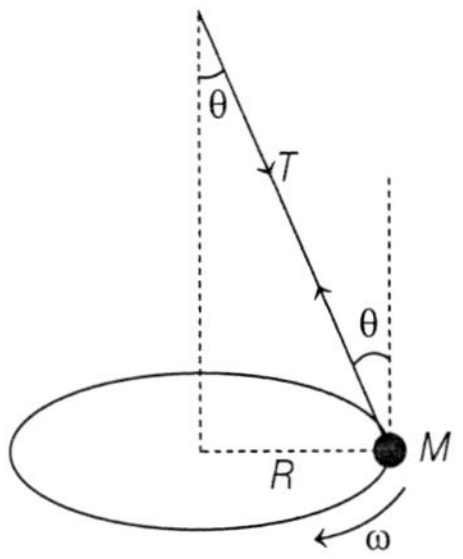

$$T\cos\theta = Mg \quad \ldots\text{(i)}$$

$$T\sin\theta = M\omega^2 R \quad \ldots\text{(ii)}$$

Using Eqs. (i) and (ii)

$$T\sin\theta = M\omega^2(L\sin\theta) \quad \left[\sin\theta = \frac{R}{L}\right]$$

$$T = M\omega^2 L$$

$$= M\left(\frac{3}{\pi}\times 2\pi\right)^2 \times L \quad \left(\omega = 2\pi\times\frac{3}{\pi}\right)$$

$$T = 36\,ML$$

31. *(1)*

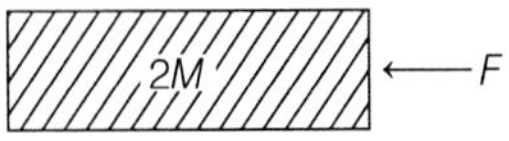

$$F = 2M\omega^2\frac{l}{2} = M\omega^2 l \Rightarrow \omega^2 = \frac{F}{Ml}$$

$$\omega = \sqrt{\frac{F}{Ml}} \quad [l = 1\text{ m}]$$

$$\omega = \sqrt{\frac{F}{M(1)}} \Rightarrow \alpha = 1$$

32. *(40)*

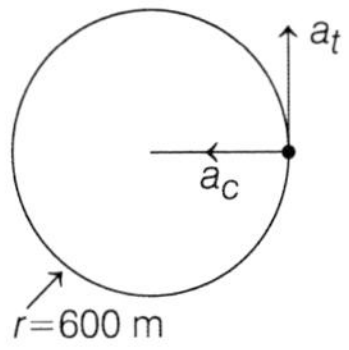

Given, initial speed of car,

$$u = 54\text{ km/h} = 54\times\frac{5}{18} = 15\text{ m/s}$$

Given, magnitude of tangential acceleration = Magnitude of centripetal acceleration.

$$\Rightarrow \frac{dv}{dt} = \frac{v^2}{r} \Rightarrow \frac{dv}{ds}\cdot\frac{ds}{dt} = \frac{v^2}{r}$$

$$\Rightarrow \frac{dv}{ds}\cdot v = \frac{v^2}{r} \Rightarrow \frac{dv}{v} = \frac{ds}{r}$$

$$\Rightarrow \int_{u=15}^{v}\frac{dv}{v} = \frac{1}{r}\int_0^s ds \Rightarrow [\ln v]_{15}^{v} = \frac{s}{r}$$

$$v = 15e^{s/r}$$

$$\Rightarrow ds/dt = 15e^{s/r}$$

$$\Rightarrow \int_0^{t_0} dt = \int_0^{\pi r/2}\frac{1}{15}e^{-s/r}\,ds$$

$$\Rightarrow t_0 = 40\,(1 - e^{-\pi/2})$$

$$\Rightarrow t = 40$$

33. *(24)* Given, $r = 75$ m, $v = 30$ m/s

$$r_1 = 48\text{ m}$$

Let the new speed is v_1 for curved path,

$\tan\theta = v^2/rg$

As coefficient of friction is same.

So, $\frac{v_1^2}{v^2} = \frac{r_1}{r}, \frac{v_1^2}{30^2} = \frac{48}{75}, v_1 = 24$ m/s

34. *(5)* Let the speed of bob at L be v_1 and at H be v_2.

Maximum tension is at L position and minimum tension is at H.

Now,

$$T_{\max} - mg = \frac{mv_1^2}{L}$$

$$\Rightarrow T_{\max} = mg + \frac{mv_1^2}{L} \quad \ldots\text{(i)}$$

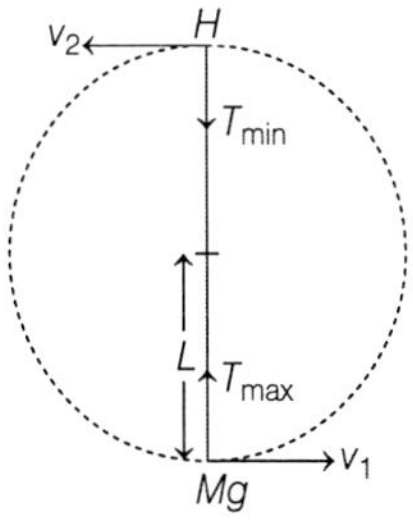

$$T_{\min} + mg = \frac{mv_2^2}{L}$$

$$\Rightarrow T_{\min} = \frac{mv_2^2}{L} - mg \quad \ldots\text{(ii)}$$

From Eqs. (i) and (ii), we get

$$\frac{T_{\max}}{T_{\min}} = \frac{mg + \frac{mv_1^2}{L}}{\frac{mv_2^2}{L} - mg} \quad \ldots\text{(iii)}$$

Also, $$\frac{T_{\max}}{T_{\min}} = \frac{5}{1} \quad \ldots\text{(iv)}$$

From Eqs. (iii) and (iv), we get

$$\frac{mg + \frac{mv_1^2}{L}}{\frac{mv_2^2}{L} - mg} = \frac{5}{1}$$

$$\Rightarrow mg + \frac{mv_1^2}{L} = \left[\frac{mv_2^2}{L} - mg\right]\times 5$$

$$\Rightarrow mg + \frac{m}{L}[v_2^2 + 4gL] = \frac{5mv_2^2}{L} - 5mg$$

$$\Rightarrow 10mg = \frac{4mv_2^2}{L}$$

$$\Rightarrow v_2^2 = \frac{10\times 10\times 1}{4} \quad (\because L = 1\text{ m})$$

$$\Rightarrow v_2 = 5\text{ m/s}$$

C-05

JEE Main Chapterwise
Topicwise Physics

In JEE Main there is usually 1-2 question per paper is visible from this chapter. In past 3 years, 46 conceptual and 22 numerical questions were asked of moderate to difficult level from this chapter.
For detailed analysis scan QR code

WORK, ENERGY AND POWER

Topic 1 Work and Energy

Level 1

01 Which one of the following forces cannot be expressed in terms of potential energy?
[2025, 7 April Shift-II]
(a) Restoring force (b) Coulomb's force
(c) Gravitational force (d) Frictional force

02 A force $\mathbf{F} = 2\hat{\mathbf{i}} + b\hat{\mathbf{j}} + \hat{\mathbf{k}}$ is applied on a particle and it undergoes a displacement $\hat{\mathbf{i}} - 2\hat{\mathbf{j}} - \hat{\mathbf{k}}$. What will be the value of b, if work done on the particle is zero?
[2025, 22 Jan Shift-II]
(a) 2 (b) 0 (c) 1/3 (d) 1/2

03 Four particles A, B, C, D of mass $\frac{m}{2}$, m, $2m$, $4m$, have same momentum, respectively. The particle with maximum kinetic energy is *[2024, 6 April Shift-I]*
(a) A (b) C (c) D (d) B

04 A body of mass 50 kg is lifted to a height of 20 m from the ground in the two different ways as shown in the figures. The ratio of work done against the gravity in both the respective cases will be
[2024, 5 April Shift-I]

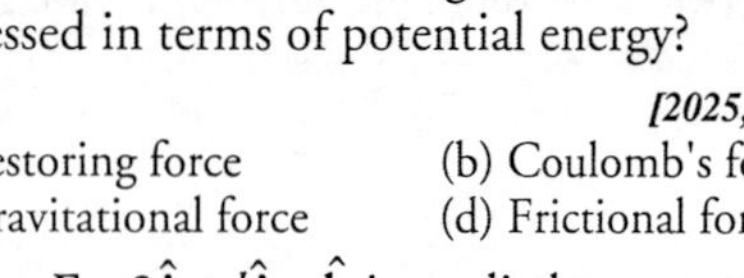

Case-1→ Pulled straight line

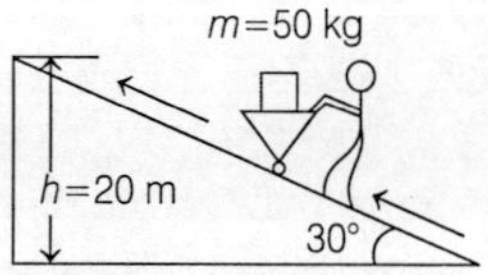

Case-2→ Along the ramp

(a) 1 : 1 (b) $\sqrt{3} : 2$
(c) 1 : 2 (d) 2 : 1

05 Two bodies are having kinetic energies in the ratio 16 : 9. If they have same linear momentum, the ratio of their masses respectively is
[2023, 13 April Shift-I]
(a) 3 : 4 (b) 4 : 3
(c) 9 : 16 (d) 16 : 9

06 For a body projected at an angle with the horizontal from the ground. Choose the correct statement.
[2023, 1 Feb Shift-II]
(a) The horizontal component of velocity is zero at the highest point.
(b) The vertical component of momentum is maximum at the highest point.
(c) Gravitational potential energy is maximum at the highest point.
(d) The kinetic energy (KE) is zero at the highest point of projectile motion.

07 Identify the correct statements from the following
A. Work done by a man in lifting a bucket out of a well by means of a rope tied to the bucket is negative.
B. Work done by gravitational force in lifting a bucket out of a well by a rope tied to the bucket is negative.
C. Work done by friction on a body sliding down an inclined plane is positive.
D. Work done by an applied force on a body moving on a rough horizontal plane with uniform velocity is zero.
E. Work done by the air resistance on an oscillating pendulum is negative.

Choose the correct answer from the options given below *[2023, 29 Jan Shift-II]*
(a) B, D and E only
(b) B and D only
(c) B and E only
(d) A and C only

08 A stone is projected at angle 30° to the horizontal. The ratio of kinetic energy of the stone at point of projection to its kinetic energy at the highest point of flight will be **[2023, 29 Jan Shift-I]**

(a) 1 : 2 (b) 4 : 1 (c) 1 : 4 (d) 4 : 3

Imp. Kinetic Energy-Momentum Relation

09 A body of mass 8 kg and another of mass 2 kg are moving with equal kinetic energy. The ratio of their respective momenta will be **[2022, 26 July Shift-II]**

(a) 1 : 1 (b) 2 : 1 (c) 1 : 4 (d) 4 : 1

10 In the given figure, the block of mass m is dropped from the point A. The expression for kinetic energy of block when it reaches point B is **[2022, 29 June Shift-II]**

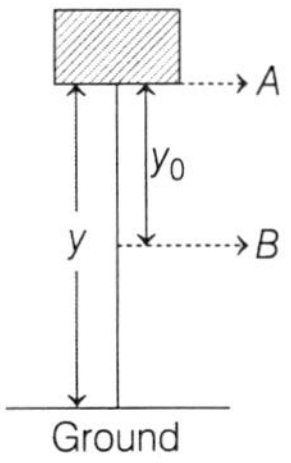

(a) $\frac{1}{2} mgy_0^2$ (b) $\frac{1}{2} mgy^2$

(c) $mg(y - y_0)$ (d) mgy_0

11 Arrange the four graphs in descending order of total work done, where W_1, W_2, W_3 and W_4 are the work done corresponding to figures A, B, C and D, respectively **[2022, 26 June Shift-II]**

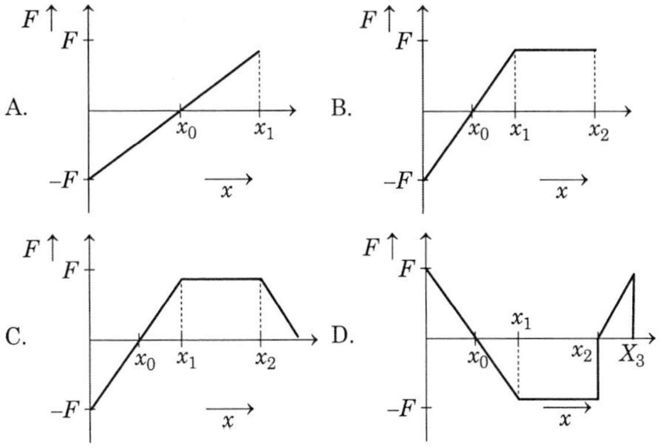

(a) $W_3 > W_2 > W_1 > W_4$ (b) $W_3 > W_2 > W_4 > W_1$

(c) $W_2 > W_3 > W_4 > W_1$ (d) $W_2 > W_3 > W_1 > W_4$

12 Potential energy as a function of r is given by $U = \frac{A}{r^{10}} - \frac{B}{r^5}$, where r is the interatomic distance, A and B are positive constants. The equilibrium distance between the two atoms will be **[2022, 24 June Shift-II]**

(a) $\left(\frac{A}{B}\right)^{1/5}$ (b) $\left(\frac{B}{A}\right)^{1/5}$ (c) $\left(\frac{2A}{B}\right)^{1/5}$ (d) $\left(\frac{B}{2A}\right)^{1/5}$

13 A porter lifts a heavy suitcase of mass 80 kg and at the destination lowers it down by a distance of 80 cm with a constant velocity. Calculate the work done by the porter in lowering the suitcase.
[Take, $g = 9.8 \text{ ms}^{-2}$] **[2021, 22 July Shift-II]**

(a) −62720.0 J (b) −627.2 J
(c) +627.2 J (d) 784.0 J

14 Consider a force $\mathbf{F} = -x\hat{\mathbf{i}} + y\hat{\mathbf{j}}$. The work done by this force in moving a particle from point $A(1, 0)$ to $B(0, 1)$ along the line segment is (all quantities are in SI units) **[2020, 09 Jan Shift-I]**

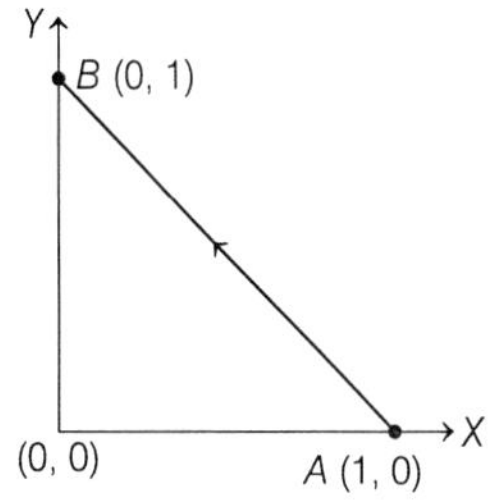

(a) 3/2 (b) 2 (c) 1 (d) 1/2

15 An athlete in the olympic games covers a distance of 100 m in 10 s. His kinetic energy can be estimated to be in the range (assume $m = 60$ kg) **[AIEEE 2008]**

(a) 200 J − 500 J (b) 2×10^5 J − 3×10^5 J
(c) 20000 J − 50000 J (d) 2000 J − 5000 J

16 A force $\mathbf{F} = (5\hat{\mathbf{i}} + 3\hat{\mathbf{j}} + 2\hat{\mathbf{k}})$ N is applied over a particle which displaces it from its origin to the point $\mathbf{r} = (2\hat{\mathbf{i}} - \hat{\mathbf{j}})$ m. The work done on the particle in joule is **[AIEEE 2004]**

(a) −7 (b) +7 (c) +10 (d) +13

17 Consider the following two statements.

I. Linear momentum of a system of particles is zero.
II. Kinetic energy of a system of particles is zero. Then,

(a) I does not imply II and II does not imply I. **[AIEEE 2003]**
(b) I implies II but II does not imply I.
(c) I does not imply II but II implies I.
(d) I implies II and II implies I.

Level 2

Imp. Kinetic-Energy During Projectile Motion

18 A ball having kinetic energy KE, is projected at an angle of 60° from the horizontal. What will be the kinetic energy of ball at the highest point of its flight? **[2025, 23 Jan Shift-II]**

(a) $\frac{(KE)}{4}$ (b) $\frac{(KE)}{16}$ (c) $\frac{(KE)}{2}$ (d) $\frac{(KE)}{8}$

Imp. Work-Energy Theorem

19 A ball of mass 100 g is projected with velocity 20 m/s at 60° with horizontal. The decrease in kinetic energy of the ball during the motion from point of projection to highest point is *[2025, 22 Jan Shift-II]*

(a) 20 J (b) 5 J (c) zero (d) 15 J

20 Given below are two statements. One is labelled as Assertion (A) and the other is labelled as Reason (R).

Assertion (A) In a central force field, the work done is independent of the path chosen.

Reason (R) Every force encountered in mechanics does not have an associated potential energy.

In the light of the above statements, choose the most appropriate answer from the option given below. *[2025, 28 Jan Shift-I]*

(a) A is true but R is false.
(b) Both A and R are true and R is the correct explanation of A.
(c) A is false but R is true.
(d) Both A and R are true but R is not the correct explanation of A.

21 A force $F = \alpha + \beta x^2$ acts on an object in the x-direction. The work done by the force is 5 J, when the object is displaced by 1 m. If the constant $\alpha = 1$ N, then β will be *[2025, 24 Jan Shift-I]*

(a) 12 N/m^2 (b) 10 N/m^2
(c) 15 N/m^2 (d) 8 N/m^2

22 A particle of mass m moves on a straight line with its velocity increasing with distance according to the equation $v = \alpha\sqrt{x}$, where α is a constant. The total work done by all the forces applied on the particle during its displacement from $x = 0$ to $x = d$, will be *[2024, 9 April Shift-I]*

(a) $\frac{m\alpha^2 d}{2}$ (b) $2m\alpha^2 d$
(c) $\frac{m}{2\alpha^2 d}$ (d) $\frac{md}{2\alpha^2}$

23 A bullet of mass 50 g is fired with a speed 100 m/s on a plywood and emerges with 40 m/s. The percentage loss of kinetic energy is *[2024, 6 April Shift-I]*

(a) 44% (b) 84% (c) 32% (d) 16%

24 An artillery piece of mass M_1 fires a shell of mass M_2 horizontally. Instantaneously after the firing, the ratio of kinetic energy of the artillery and that of the shell is *[2024, 31 Jan Shift-I]*

(a) $M_2/(M_1 + M_2)$ (b) $\frac{M_2}{M_1}$
(c) $\frac{M_1}{(M_1 + M_2)}$ (d) $\frac{M_1}{M_2}$

25 A block of mass 1 kg is pushed up a surface inclined to horizontal at an angle of 60° by a force of 10 N parallel to the inclined surface as shown in figure. When the block is pushed up by 10 m along inclined surface, the work done against frictional force is $[g = 10 \text{ m/s}^2]$ *[2024, 30 Jan Shift-II]*

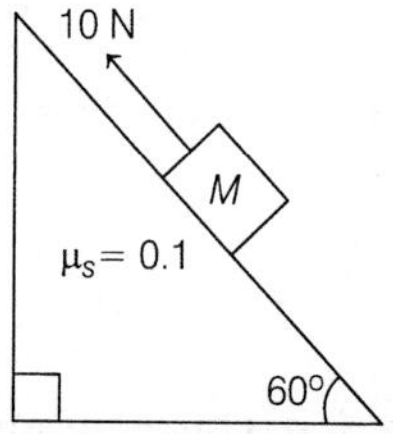

(a) 5×10^3 J (b) 5 J (c) 10 J (d) $5\sqrt{3}$ J

26 The bob of a pendulum was released from a horizontal position. The length of the pendulum is 10 m. If it dissipates 10% of its initial energy against air resistance, the speed with which the bob arrives at the lowest point is [Use, $g = 10 \text{ ms}^{-2}$] *[2024, 29 Jan Shift-II]*

(a) $6\sqrt{5}$ ms^{-1} (b) $5\sqrt{6}$ ms^{-1}
(c) $2\sqrt{5}$ ms^{-1} (d) $5\sqrt{5}$ ms^{-1}

27 A bob of mass m is suspended by a light string of length L. It is imparted a minimum horizontal velocity at the lowest point A such that it just completes half circle reaching the top most position B. The ratio of kinetic energies $\frac{(\text{KE})_A}{(\text{KE})_B}$ is *[2024, 29 Jan Shift-II]*

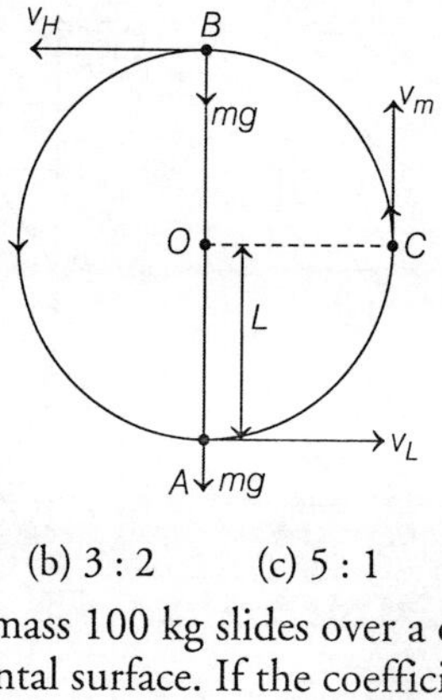

(a) 2 : 5 (b) 3 : 2 (c) 5 : 1 (d) 1 : 5

28 A block of mass 100 kg slides over a distance of 10 m on a horizontal surface. If the coefficient of friction between the surfaces is 0.4, then the work done against friction (in J) is *[2024, 29 Jan Shift-I]*

(a) 4200 (b) 4500 (c) 4000 (d) 3900

29 Given below are two statements.

Statement I A truck and a car moving with the same kinetic energy are brought to rest by applying breaks which provide equal retarding forces. Both come to rest in equal distance.

Statement II A car moving towards east takes a turn and moves towards north, the speed remains unchanged. The acceleration of the car is zero.
In the light of above statements, choose the most appropriate answer from the options given below. *[2023, 12 April Shift-I]*

(a) Statement I is incorrect but Statement II is correct.
(b) Statement I is correct but Statement II is incorrect.
(c) Both Statement I and Statement II are correct.
(d) Both Statement I and Statement II are incorrect.

30 If momentum of a body is increased by 20%, then its kinetic energy increases by *[2022, 29 July Shift-II]*
(a) 36% (b) 40% (c) 44% (d) 48%

31 A ball is projected with kinetic energy E, at an angle of 60° to the horizontal. The kinetic energy of this ball at the highest point of its flight will become *[2022, 29 July Shift-I]*
(a) zero (b) $E/2$ (c) $E/4$ (d) E

32 As per the given figure, two blocks each of mass 250 g are connected to a spring of spring constant 2 Nm^{-1}. If both are given velocity v in opposite directions, then maximum elongation of the spring is *[2022, 26 July Shift-I]*

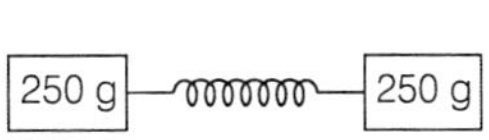

(a) $\frac{v}{2\sqrt{2}}$ (b) $\frac{v}{2}$ (c) $\frac{v}{4}$ (d) $\frac{v}{\sqrt{2}}$

33 Two identical blocks A and B each of mass m resting on the smooth horizontal floor are connected by a light spring of natural length L and spring constant k. A third block C of mass m moving with a speed v along the line joining A and B collides with A. The maximum compression in the spring is *[2021, 17 March Shift-II]*

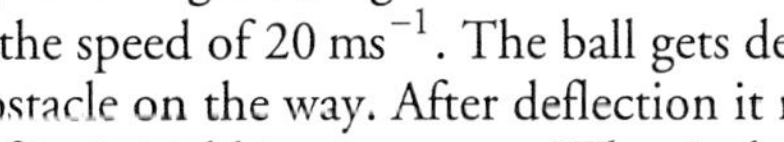

(a) $v\sqrt{\frac{m}{2k}}$ (b) $\sqrt{\frac{mv}{2k}}$
(c) $\sqrt{\frac{mv}{k}}$ (d) $\sqrt{\frac{m}{2k}}$

34 A boy is rolling a 0.5 kg ball on the frictionless floor with the speed of 20 ms^{-1}. The ball gets deflected by an obstacle on the way. After deflection it moves with 5% of its initial kinetic energy. What is the speed of the ball now ? *[2021, 17 March Shift-I]*
(a) 19.0 ms^{-1} (b) 4.47 ms^{-1}
(c) 14.41 ms^{-1} (d) 1.00 ms^{-1}

35 If the kinetic energy of a moving body becomes four times of its initial kinetic energy, then the percentage change in its momentum will be *[2021, 20 July Shift-II]*
(a) 100% (b) 200%
(c) 300% (d) 400%

36 A particle of mass m moves in a circular orbit under the central potential field, $U(r) = \frac{-C}{r}$, where C is a positive constant. The correct radius-velocity graph of the particle's motion is *[2021, 18 March Shift-II]*

(a) 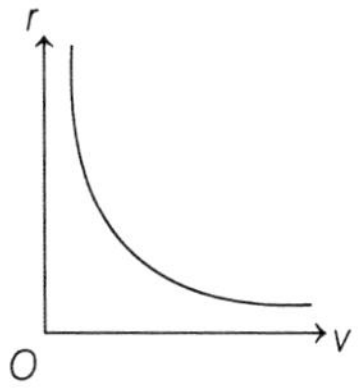

(b) 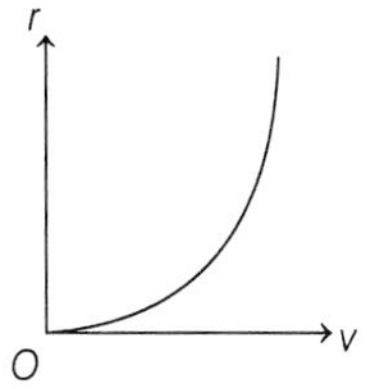

(c) 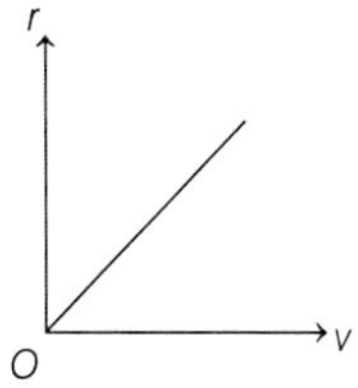

(d) 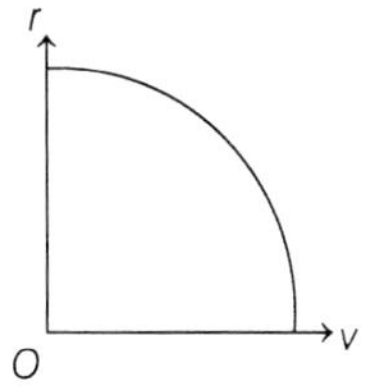

37 A particle is moving in a circular path of radius a under the action of an attractive potential $U = -\frac{k}{2r^2}$. Its total energy is *[JEE Main 2018]*
(a) $-\frac{k}{4a^2}$ (b) $\frac{k}{2a^2}$
(c) zero (d) $-\frac{3}{2}\frac{k}{a^2}$

38 A person trying to lose weight by burning fat, lifts a mass of 10 kg upto a height of 1 m, 1000 times. Assume that the potential energy lost each time he lowers the mass is dissipated. How much fat will he use up, considering the work done only when the weight is lifted up? Fat supplies 3.8×10^7 J of energy per kg which is converted into mechanical energy with a 20% of efficiency rate. (Take, $g = 9.8$ ms^{-2}) *[JEE Main 2016]*
(a) 2.45×10^{-3} kg
(b) 6.45×10^{-3} kg
(c) 9.89×10^{-3} kg
(d) 12.89×10^{-3} kg

39 A point particle of mass m, moves along the uniformly rough track PQR as shown in the figure. The coefficient of friction between the particle and the rough track equals μ.
The particle is released, from rest, from the point P and it comes to rest at a point R. The energies, lost by the ball, over the parts, PQ and QR, of the track, are equal to each other, and no energy is lost when particle changes direction from PQ to QR. The values of the coefficient of friction μ and the distance $x(= QR)$, are respectively close to *[JEE Main 2016]*

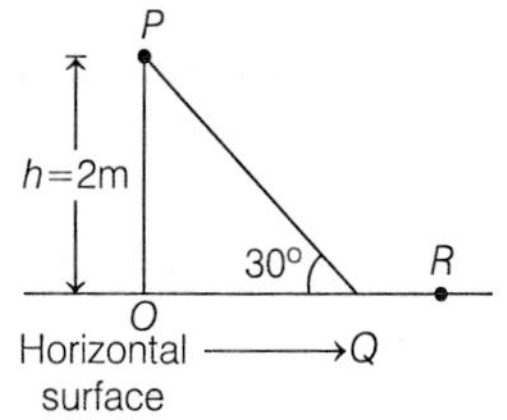

(a) 0.2 and 6.5 m (b) 0.2 and 3.5 m
(c) 0.29 and 3.5 m (d) 0.29 and 6.5 m

40 When a rubber band is stretched by a distance x, it exerts a restoring force of magnitude $F = ax + bx^2$, where a and b are constants. The work done in stretching the unstretched rubber band by L is *[JEE Main 2014]*

(a) $aL^2 + bL^3$ (b) $\frac{1}{2}(aL^2 + bL^3)$
(c) $\frac{aL^2}{2} + \frac{bL^3}{3}$ (d) $\frac{1}{2}\left(\frac{aL^2}{2} + \frac{bL^3}{3}\right)$

41 This question has Statement I and Statement II. Of the four choices given after the statements, choose the one that best describes the two statements.
If two springs S_1 and S_2 of force constants k_1 and k_2 respectively are stretched by the same force, it is found that more work is done on spring S_1 than on spring S_2.
Statement I If stretched by the same amount, work done on S_1 will be more than that on S_2.
Statement II $k_1 < k_2$ *[AIEEE 2012]*

(a) Statement I is false, Statement II is true.
(b) Statement I is true, Statement II is false.
(c) Statement I is true, Statement II is true; Statement II is the correct explanation of Statement I.
(d) Statement I is true, Statement II is true; Statement II is not the correct explanation of Statement I.

42 The potential energy function for the force between two atoms in a diatomic molecule is approximately given by $U(x) = \frac{a}{x^{12}} - \frac{b}{x^6}$, where a and b are constants and x is the distance between the atoms. If the dissociation energy of the molecule is $D = [U(x = \infty) - U_{\text{at equilibrium}}]$, D is *[AIEEE 2010]*

(a) $\frac{b^2}{2a}$ (b) $\frac{b^2}{12a}$
(c) $\frac{b^2}{4a}$ (d) $\frac{b^2}{6a}$

43 A 2 kg block slides on a horizontal floor with a speed of 4 m/s. It strikes a uncompressed spring and compresses it till the block is motionless. The kinetic friction force is 15 N and spring constant is, 10000 N/m. The spring compresses by *[AIEEE 2007]*

(a) 5.5 cm (b) 2.5 cm
(c) 11.0 cm (d) 8.5 cm

44 The potential energy of a 1 kg particle free to move along the X-axis is given by

$$V(x) = \left(\frac{x^4}{4} - \frac{x^2}{2}\right) \text{ J}$$

The total mechanical energy of the particle is 2 J. Then, the maximum speed (in ms^{-1}) is *[AIEEE 2006]*

(a) $3/\sqrt{2}$ (b) $\sqrt{2}$
(c) $1/\sqrt{2}$ (d) 2

45 The block of mass M moving on the frictionless horizontal surface collides with the spring of spring constant k and compresses it by length L. The maximum momentum of the block after collision is *[AIEEE 2005]*

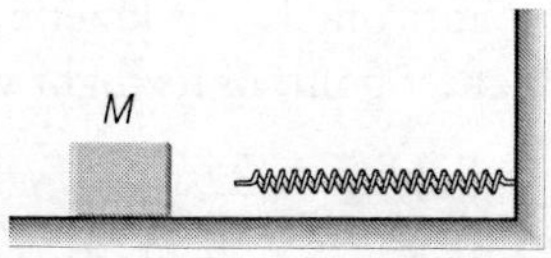

(a) $L\sqrt{Mk}$ (b) $\frac{kL^2}{2M}$
(c) zero (d) $\frac{ML^2}{k}$

46 A uniform chain of length 2 m is kept on a table such that a length of 60 cm hangs freely from the edge of the table. The total mass of the chain is 4 kg. What is the work done in pulling the entire chain on the table? *[AIEEE 2004]*

(a) 7.2 J (b) 3.6 J
(c) 120 J (d) 1200 J

47 A spring of spring constant 5×10^3 N/m is stretched initially by 5 cm from the unstretched position. Then, the work required to stretch it further by another 5 cm is *[AIEEE 2003]*

(a) 12.50 N-m (b) 18.75 N-m
(c) 25.00 N-m (d) 6.25 N-m

48 A ball whose kinetic energy is E, is projected at an angle of 45° to the horizontal. The kinetic energy of the ball at the highest point of its flight will be *[AIEEE 2002]*

(a) E (b) $E/\sqrt{2}$
(c) $E/2$ (d) zero

Level 3

49 A block of mass 25 kg is pulled along a horizontal surface by a force at an angle 45° with the horizontal. The friction coefficient between the block and the surface is 0.25. The block travels at a uniform velocity. The work done by the applied force during a displacement of 5 m of the block is *[2025, 4 April Shift-II]*

(a) 970 J (b) 490 J
(c) 245 J (d) 735 J

50 A block is simply released from the top of an inclined plane as shown in the figure below. The maximum compression in the spring when the block hits the spring is

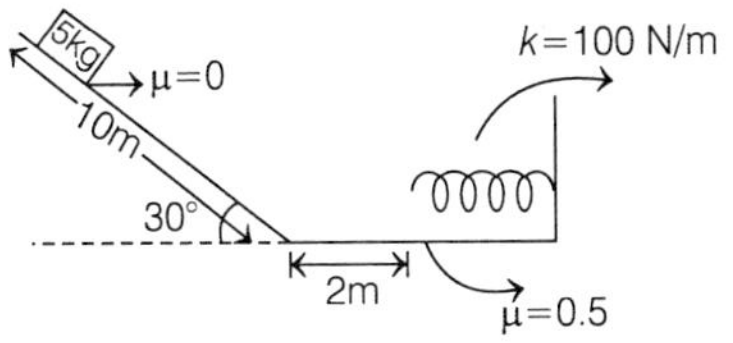

[2024, 8 April Shift-II]

(a) 1 m (b) $\sqrt{5}$ m
(c) $\sqrt{6}$ m (d) 2 m

51 Three bodies A, B and C have equal kinetic energies and their masses are 400 g, 1.2 kg and 1.6 kg respectively. The ratio of their linear momenta is [2024, 8 April Shift-I]

(a) $1:\sqrt{3}:\sqrt{2}$ (b) $\sqrt{2}:\sqrt{3}:1$
(c) $1:\sqrt{3}:2$ (d) $\sqrt{3}:\sqrt{2}:1$

52 A stationary particle breaks into two parts of masses m_A and m_B which move with velocities v_A and v_B respectively. The ratio of their kinetic energies $(K_B:K_A)$ is [2024, 8 April Shift-I]

(a) 1 : 1 (b) $m_Bv_B : m_Av_A$
(c) $m_B : m_A$ (d) $v_B : v_A$

53 A simple pendulum of length 1 m has a wooden bob of mass 1 kg. It is struck by a bullet of mass 10^{-2} kg moving with a speed of 2×10^2 ms^{-1}. The bullet gets embedded into the bob. The height to which the bob rises before swinging back is (use, $g=10$ m/s^2) [2024, 1 Feb Shift-I]

(a) 0.30 m (b) 0.35 m
(c) 0.20 m (d) 0.40 m

54 A bullet of mass 200 g having initial kinetic energy 90 J is shot inside a long swimming pool as shown in the figure. If it's kinetic energy reduces to 40 J within 1s, the minimum length of the pool, the bullet has to travel, so that it completely comes to rest is [2022, 28 July Shift-II]

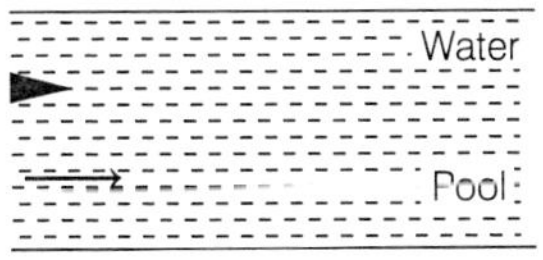

(a) 45 m (b) 90 m
(c) 125 m (d) 25 m

55 A block moving horizontally on a smooth surface with a speed of 40 m/s splits into two parts with masses in the ratio of 1:2. If the smaller part moves at 60 m/s in the same direction, then the fractional change in kinetic energy is [2021, 31 Aug Shift-II]

(a) $\frac{1}{3}$ (b) $\frac{2}{3}$
(c) $\frac{1}{8}$ (d) $\frac{1}{4}$

56 A force acts on a 2 kg object, so that its position is given as a function of time as $x=3t^2+5$. What is the work done by this force in first 5 s ? [2019, 09 Jan Shift-II]

(a) 850 J (b) 900 J (c) 950 J (d) 875 J

57 A body of mass $m=10^{-2}$kg is moving in a medium and experiences a frictional force $F=-kv^2$. Its initial speed is $v_0=10$ ms^{-1}. If, after 10 s, its energy is $1/8\,mv_0^2$, the value of k will be [JEE Main 2017]

(a) 10^{-3} kgs^{-1} (b) 10^{-4} kgm^{-1}
(c) 10^{-1} kgm^{-1}s^{-1} (d) 10^{-3} kgm^{-1}

Numerical Value Type Questions

58 A force $F=x^2y\hat{i}+y^2\hat{j}$ acts on a particle in a plane $x+y=10$. The work done by this force during a displacement from (0, 0) to (4 m, 2 m) is .... J (round off to the nearest integer) [2025, 23 Jan Shift-I]

59 A force $(3x^2+2x-5)$ N displaces a body from $x=2$ to $x=4$ m. Work done by this force is ....J . [2024, 9 April Shift-II]

60 A block of mass 10 kg is moving along X-axis under the action of force $F=5x$ N. The work done by the force in moving the block from $x=2$ m to 4 m will be .... J. [2023, 15 April Shift-I]

61 A car accelerates from rest to u m/s. The energy spent in this process is E J. The energy required to accelerate the car from u m/s to $2u$ m/s is nE J. The value of n is .... [2023, 13 April Shift-II]

62 To maintain a speed of 80 km/h by a bus of mass 500 kg on a plane rough road for 4 km distance, the work done by the engine of the bus will be ...... kJ. [The coefficient of friction between tyre of bus and road is 0.04.] [2023, 12 April Shift-I]

63 A force $\mathbf{F}=(2+3x)\hat{i}$ acts on a particle in the x-direction where F is in Newton and x is in metre. The work done by this force during a displacement from $x=0$ to $x=4$ m, is .... J. [2023, 11 April Shift-I]

64 A body of mass 5 kg is moving with a momentum of 10 kg ms^{-1}. Now, a force of 2 N acts on the body in the direction of its motion for 5 s. The increase in the kinetic energy of the body is .... J. [2023, 8 April Shift-II]

65 The momentum of a body is increased by 50%. The percentage increase in the kinetic energy of the body is .... %. [2023, 8 April Shift-I]

66 A particle of mass 10 g moves in a straight line with retardation $2x$, where x is the displacement in SI units. Its loss of kinetic energy for above displacement is $\left(\frac{10}{x}\right)^{-n}$ J. The value of n will be .... *[2023, 6 April Shift-I]*

67 A force $\mathbf{F} = (5 + 3y^2)\hat{\mathbf{j}}$ acts on a particle in the y-direction, where F is in Newton and y is in metre. The work done by the force during a displacement from $y = 2$ m to $y = 5$ m is .... J. *[2023, 01 Feb Shift-II]*

68 A small particle moves to position $5\hat{\mathbf{i}} - 2\hat{\mathbf{j}} + \hat{\mathbf{k}}$ from its initial position $2\hat{\mathbf{i}} + 3\hat{\mathbf{j}} - 4\hat{\mathbf{k}}$ under the action of force $5\hat{\mathbf{i}} + 2\hat{\mathbf{j}} + 7\hat{\mathbf{k}}$ N. The value of work done will be .... J. *[2023, 01 Feb Shift-I]*

69 A lift of mass $m = 500$ kg is descending with speed of 2 ms^{-1}. Its supporting cable begins to slip thus allowing it to fall with a constant acceleration of 2 ms^{-2}. The kinetic energy of the lift at the end of fall through to a distance of 6 m will be .... kJ. *[2023, 31 Jan Shift-I]*

70 A 0.4 kg mass takes 8s to reach ground when dropped from a certain height P above surface of earth. The loss of potential energy in the last second of fall is .... J.
(Take, $g = 10 \text{ m/s}^2$) *[2023, 29 Jan Shift-I]*

71 A spherical body of mass 2 kg starting from rest acquires a kinetic energy of 10000 J at the end of 5th seconds. The force acted on the body is .... N. *[2023, 24 Jan Shift-I]*

72 An object of mass m initially at rest on a smooth horizontal plane starts moving under the action of force F = 2 N. In the process of its linear motion, the angle θ (as shown in figure) between the direction of force and horizontal varies as $\theta = kx$, where k is a constant and x is the distance covered by the object from its initial position. The expression of kinetic energy of the object will be $E = \frac{x}{k} \sin\theta$, The value of x is ....

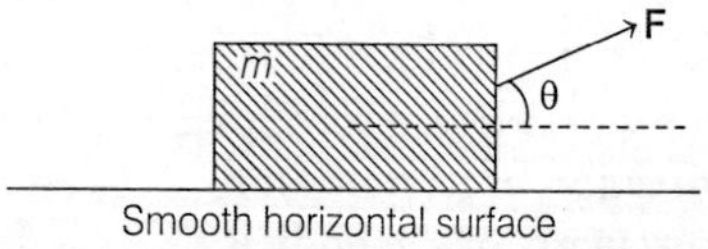

[2023, 25 Jan Shift-I]

73 A block of mass m (as shown in figure) moving with kinetic energy E compresses a spring through a distance 25 cm when its speed is halved. The value of spring constant of used spring will be nE Nm^{-1} for $n =$ ....

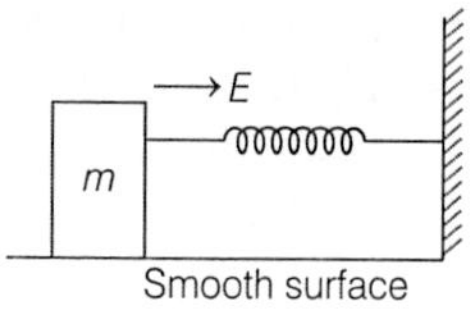

[2022, 28 July Shift-I]

74 A 0.5 kg block moving at a speed of 12 ms^{-1} compresses a spring through a distance 30 cm when its speed is halved. The spring constant of the spring will be ... Nm^{-1}. *[2022, 25 June Shift-I]*

75 A block moving horizontally on a smooth surface with a speed of 40 ms^{-1} splits into two equal parts. If one of the parts moves at 60 ms^{-1} in the same direction, then the fractional change in the kinetic energy will be $x : 4$, where x is .... *[2021, 31 Aug Shift-I]*

76 Two persons A and B perform same amount of work in moving a body through a certain distance d with application of forces acting at angle 45° and 60° with the direction of displacement, respectively. The ratio of force applied by person A to the force applied by person B is $1/\sqrt{x}$. The value of x is .... *[2021, 27 Aug Shift-I]*

77 A uniform chain of length 3 m and mass 3 kg overhangs a smooth table with 2 m laying on the table. If K is the kinetic energy of the chain in joule as it completely slips off the table, then the value of K is .... (Take, $g = 10 \text{ m/s}^2$) *[2021, 26 Aug Shift-I]*

78 A force of $F = (5y + 20)\hat{\mathbf{j}}$ N acts on a particle. The work done by this force when the particle is moved from $y = 0$ m to $y = 10$ m is ...... J. *[2021, 25 July Shift-II]*

79 In a spring gun having spring constant 100 N/m a small ball B of mass 100 g is put in its barrel (as shown in figure) by compressing the spring through 0.05 m. There should be a box placed at a distance d on the ground, so that the ball falls in it. If the ball leaves the gun horizontally at a height of 2 m above the ground. The value of d is .... m.
(Take, $g = 10 \text{ m/s}^2$)

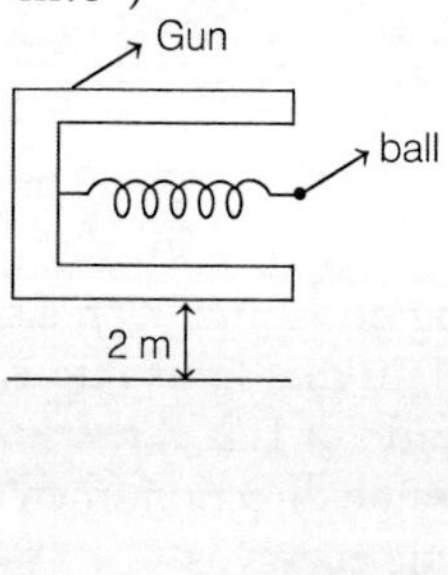

[2021, 20 July Shift-I]

Imp. Conservation of Energy

80 As shown in the figure, a particle of mass 10 kg is placed at a point A. When the particle is slightly displaced to its right, it starts moving and reaches the point B. The speed of the particle at B is x m/s. (Take, $g = 10\ \text{m/s}^2$)

The value of x to the nearest integer is .... *[2021, 18 March Shift-I]*

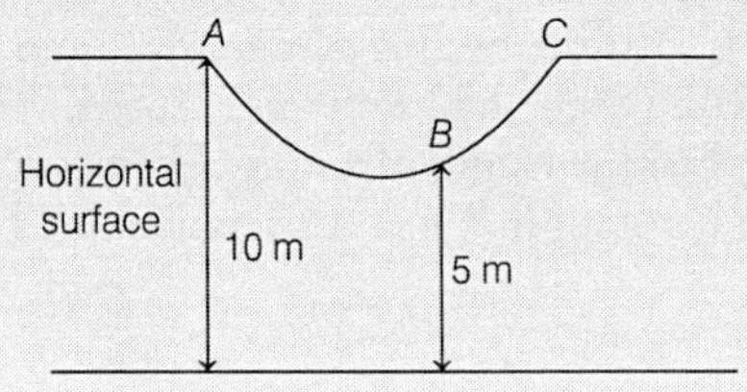

81 Two particles having masses 4 g and 16 g, respectively are moving with equal kinetic energies. The ratio of the magnitudes of their linear momentum is $n : 2$. The value of n will be .... *[2021, 25 Feb Shift-II]*

82 Two solids A and B of mass 1 kg and 2 kg, respectively are moving with equal linear momentum. The ratio of their kinetic energies $(KE)_A : (KE)_B$ will be $\frac{A}{1}$, so the value of A will be .... *[2021, 25 Feb Shift-II]*

Topic 2 Work-Energy Theorem and Vertical Circle

Level 1

01 A body of mass 0.5 kg travel on straight line path with velocity $v = (3x^2 + 4)$ m/s. The net work done by the force during its displacement from $x = 0$ to $x = 2$ m is *[2022, 25 July Shift-I]*
(a) 64 J (b) 60 J (c) 120 J (d) 128 J

02 A particle experiences a variable force $\mathbf{F} = (4x\hat{\mathbf{i}} + 3y^2\hat{\mathbf{j}})$ in a horizontal XY-plane. Assume, distance in metre and force in Newton. If the particle moves from point (1, 2) to point (2, 3) in the XY-plane, then kinetic energy changes by *[2022, 24 June Shift-I]*
(a) 50.0 J (b) 12.5 J (c) 25.0 J (d) 0 J

03 A particle is acted upon by a force of constant magnitude which is always perpendicular to the velocity of the particle. The motion of the particle takes place in a plane, it follows that *[AIEEE 2004]*
(a) its velocity is constant
(b) its acceleration is constant
(c) its kinetic energy is constant
(d) it moves in a straight line

Level 2

04 A block of mass 2 kg is attached to one end of a massless spring whose other end is fixed at a wall. The spring-mass system moves on a frictionless horizontal table. The spring's natural length is 2 m and spring constant is 200 N/m. The block is pushed such that the length of the spring becomes 1 m and then released. At distance x m $(x < 2)$ from the wall, the speed of the block will be *[2025, 8 April Shift-II]*
(a) $10[1-(2-x)^2]$ m/s (b) $10[1-(2-x)]^{\frac{3}{2}}$ m/s
(c) $10[1-(2-x)^2]^2$ m/s (d) $10[1-(2-x)^2]^{\frac{1}{2}}$ m/s

05 A particle is released from height s above the surface of the earth. At certain height its kinetic energy is three times its potential energy. The height from the surface of the earth and the speed of the particle at that instant are respectively. *[2025, 3 April Shift-I]*
(a) $\frac{s}{2}, \frac{3gs}{2}$ (b) $\frac{s}{2}, \sqrt{\frac{3gs}{2}}$ (c) $\frac{s}{4}, \frac{3gs}{2}$ (d) $\frac{s}{4}, \sqrt{\frac{3gs}{2}}$

06 A stone tied to a string of length L is whirled in a vertical circle with the other end of the string at the centre. At a certain instant of time, the stone is at its lowest position and has a speed u. The magnitude of change in its velocity, as it reaches a position where the string is horizontal, is $\sqrt{x(u^2 - gL)}$. The value of x is *[2022, 27 June Shift-II]*
(a) 3 (b) 2 (c) 1 (d) 5

07 A body of mass m dropped from a height h reaches the ground with a speed of $0.8\sqrt{gh}$. The value of workdone by the air-friction is *[2021, 01 Sep Shift-II]*
(a) $-0.68\ mgh$ (b) mgh
(c) $1.64\ mgh$ (d) $0.64\ mgh$

08 A particle moves in one dimension from rest under the influence of a force that varies with the distance travelled by the particle as shown in the figure. The kinetic energy of the particle after it has travelled 3 m is *[2019, 08 April Shift-I]*

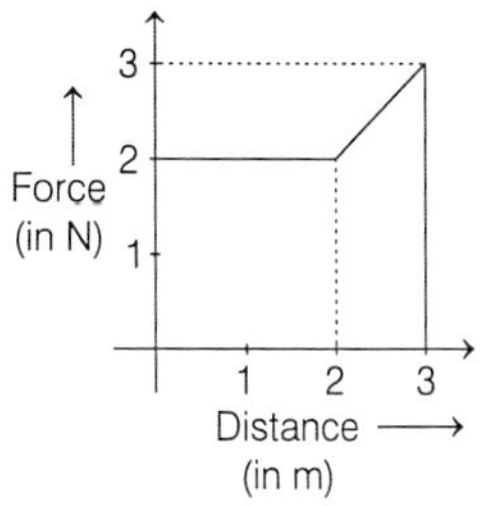

(a) 4 J (b) 2.5 J (c) 6.5 J (d) 5 J

09 A block of mass m lying on a smooth horizontal surface is attached to a spring (of negligible mass) of spring constant k. The other end of the spring is fixed as shown in the figure. The block is initially at rest in its equilibrium position. If now the block is pulled with a constant force F, the maximum speed of the block is *[2019, 09 Jan Shift-I]*

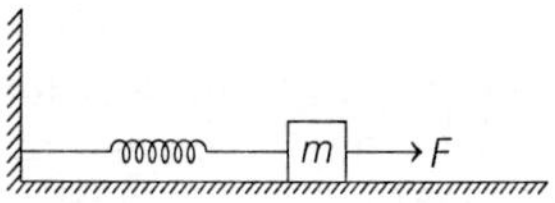

(a) $\frac{\pi F}{\sqrt{mk}}$ (b) $\frac{F}{\sqrt{mk}}$ (c) $\frac{2F}{\sqrt{mk}}$ (d) $\frac{F}{\pi\sqrt{mk}}$

10 A time dependent force $F = 6t$ acts on a particle of mass 1 kg. If the particle starts from rest, the work done by the force during the first 1s will be *[JEE Main 2017]*

(a) 22 J (b) 9 J (c) 18 J (d) 4.5 J

Level 3

Imp. Motion in Verticle Circle

11 A bob of mass m is suspended at a point O by a light string of length l and left to perform vertical motion (circular) as shown in figure. Initially, by applying horizontal velocity v_0 at the point A, the string becomes slack when, the bob reaches at the point D. The ratio of the kinetic energy of the bob at the points B and C is .... *[2025, 22 Jan Shift-I]*

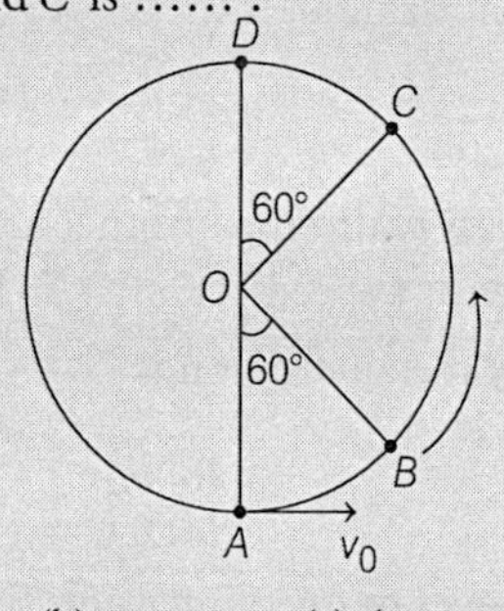

(a) 3 (b) 1 (c) 4 (d) 2

12 A body of mass 100 g is moving in circular path of radius 2 m on vertical plane as shown in figure. The velocity of the body at point A is 10 m/s. The ratio of its kinetic energies at point B and C is

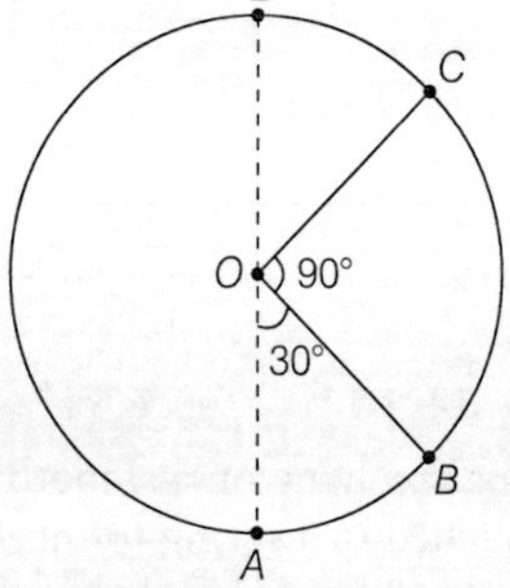

(Take acceleration due to gravity as 10 m/s^2) *[2025, 22 Jan Shift-II]*

(a) $\frac{2+\sqrt{2}}{3}$ (b) $\frac{3+\sqrt{3}}{2}$ (c) $\frac{2+\sqrt{3}}{3}$ (d) $\frac{3-\sqrt{2}}{2}$

13 A bead of mass m slides without friction on the wall of a vertical circular hoop of radius R as shown in figure. The bead moves under the combined action of gravity and a massless spring (k) attached to the bottom of the hoop. The equilibrium length of the spring is R. If the bead is released from top of the hoop with (negligible) zero initial speed, velocity of bead, when the length of spring becomes R, would be (spring constant is k, g is acceleration due to gravity) *[2025, 28 Jan Shift-I]*

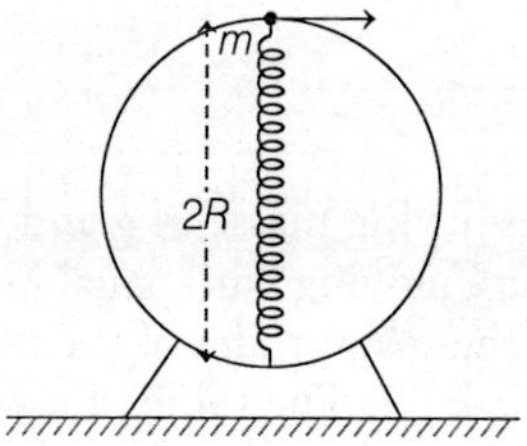

(a) $\sqrt{3Rg + \frac{kR^2}{m}}$ (b) $\sqrt{2Rg + \frac{4kR^2}{m}}$

(c) $2\sqrt{gR + \frac{kR^2}{m}}$ (d) $\sqrt{2Rg + \frac{kR^2}{m}}$

14 A body of mass m connected to a massless and unstretchable string goes in verticle circle of radius R under gravity g. The other end of the string is fixed at the centre of circle. If velocity at top of circular path is $n\sqrt{gR}$, where, $n \geq 1$, then ratio of kinetic energy of the body at bottom to that at top of the circle is *[2025, 29 Jan Shift-I]*

(a) $\frac{n+4}{n}$ (b) $\frac{n^2}{n^2+4}$ (c) $\frac{n}{n+4}$ (d) $\frac{n^2+4}{n^2}$

15 A body of m kg slides from rest along the curve of vertical circle from point A to B in frictionless path. The velocity of the body at B is

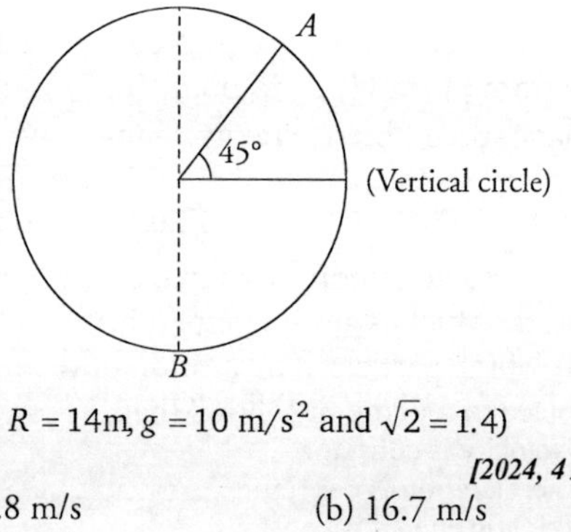

(given, $R = 14$m, $g = 10$ m/s^2 and $\sqrt{2} = 1.4$) *[2024, 4 April Shift-II]*

(a) 19.8 m/s (b) 16.7 m/s

(c) 21.9 m/s (d) 10.6 m/s

16 A particle is placed at the point A of a frictionless track ABC as shown in figure. It is gently pushed towards right. The speed of the particle, when it reaches the point B is (Take, $g = 10 \text{ m/s}^2$)

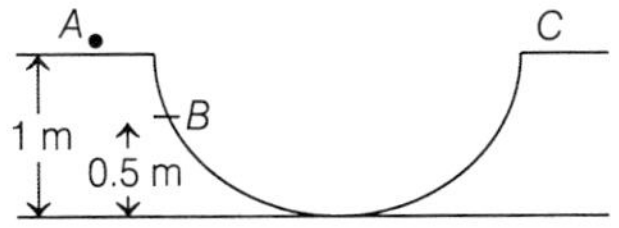

[2024, 30 Jan Shift-I]

(a) $\sqrt{10}$ m/s (b) 20 m/s (c) $2\sqrt{10}$ m/s (d) 10 m/s

Imp. Work-Energy Theorem

17 A ball suspended by a thread swings in a vertical plane, so that its magnitude of acceleration in the extreme position and lowest position are equal. The angle θ of thread deflection in the extreme position will be *[2024, 27 Jan Shift-II]*

(a) $\tan^{-1}(\sqrt{2})$ (b) $\tan^{-1}\left(\frac{1}{2}\right)$

(c) $2\tan^{-1}\left(\frac{1}{\sqrt{5}}\right)$ (d) $2\tan^{-1}\left(\frac{1}{2}\right)$

18 A mass of M kg is suspended by a weightless string. The horizontal force that is required to displace it until the string makes an angle of 45° with the initial vertical direction is *[AIEEE 2006]*

(a) $Mg(\sqrt{2}+1)$ (b) $Mg\sqrt{2}$

(c) $\frac{Mg}{\sqrt{2}}$ (d) $Mg(\sqrt{2}-1)$

19 A bullet fired into a fixed target loses half of its velocity after penetrating 3 cm. How much further it will penetrate before coming to rest, assuming that it faces constant resistance to motion? *[AIEEE 2005]*

(a) 3.0 cm (b) 2.0 cm
(c) 1.5 cm (d) 1.0 cm

Numerical Value Type Questions

20 A closed circular tube of average radius 15 cm, whose inner walls are rough, is kept in vertical plane. A block of mass 1 kg just fit inside the tube. The speed of block is 22 m/s, when it is introduced at the top of tube. After completing five oscillations, the block stops at the bottom region of tube. The work done by the tube on the block is .... J. (Given, $g = 10$ m/s). *[2023, 10 April Shift-I]*

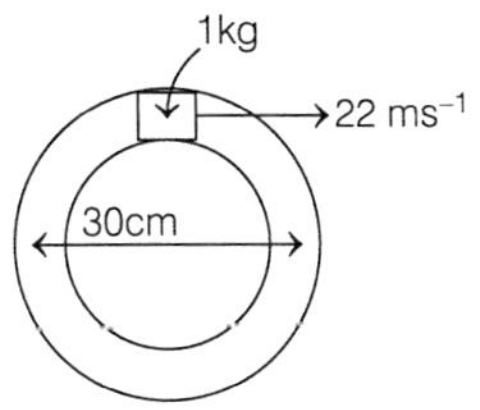

21 A pendulum of length 2 m consists of a wooden bob of mass 50 g. A bullet of mass 75 g is fired towards the stationary bob with a speed v. The bullet emerges out of the bob with a speed $v/3$ and the bob just completes the vertical circle. The value of v is ...... ms^{-1}. (If $g = 10 \text{ m/s}^2$) *[2022, 27 June Shift-I]*

22 A ball of mass 100 g is dropped from a height, h = 10 cm on a platform fixed at the top of vertical spring (as shown in figure). The ball stays on the platform and the platform is depressed by a distance $h/2$. The spring constant is .... Nm^{-1}. (Use, $g = 10 \text{ ms}^{-2}$)

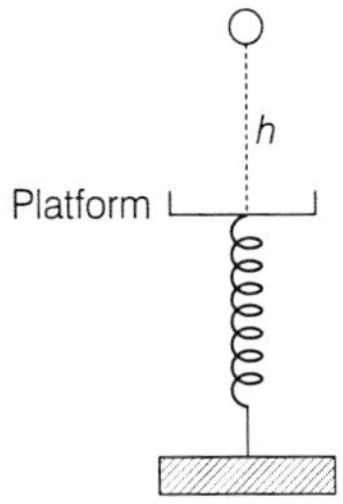

[2022, 24 June Shift-I]

23 An engine is attached to a wagon through a shock absorber of length 1.5 m. The system with a total mass of 40000 kg is moving with a speed of 72 kmh^{-1}, when the brakes are applied to bring it to rest. In the process of the system being brought to rest, the spring of the shock absorber gets compressed by 1.0 m. If 90% of energy of the wagon is lost due to friction, the spring constant is .... $\times 10^5$ N/m. *[2021, 01 Sep Shift-II]*

24 A small block slides down from the top of hemisphere of radius $R = 3$ m as shown in the figure. The height h at which the block will lose contact with the surface of the sphere is .... m. (Assume there is no friction between the block and the hemisphere) *[2021, 27 July Shift-II]*

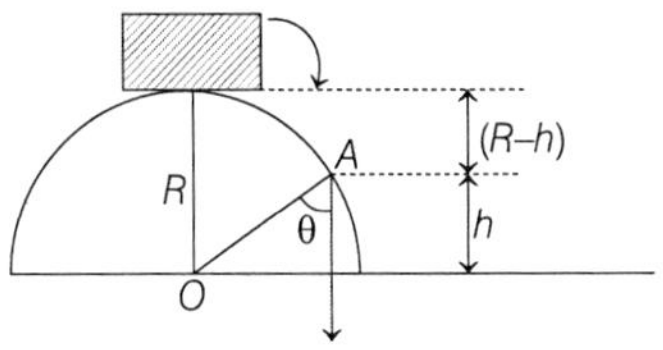

25 A pendulum bob has a speed of 3 m/s at its lowest position. The pendulum is 50 cm long. The speed of bob when the length makes an angle of 60° to the vertical will be .... m/s. (Take, $g = 10 \text{ m/s}^2$) *[2021, 25 July Shift-I]*

26 A block starts moving up an inclined plane of inclination 30° with an initial velocity of v_0. It comes back to its initial position with velocity $v_0/2$. The value of the coefficient of kinetic friction between the block and the inclined plane is close to $I/1000$, the nearest integert to I is .... *[2020, 3 Sep Shift-II]*

Topic 3 Power

Level 1

01 An automobile of mass m accelerates starting from origin and initially at rest, while the engine supplies constant power P. The position is given as a function of time by **[2021, 27 July Shift-II]**

(a) $\left(\frac{9P}{8m}\right)^{\frac{1}{2}} \cdot t^{\frac{3}{2}}$ (b) $\left(\frac{8P}{9m}\right)^{\frac{1}{2}} \cdot t^{\frac{2}{3}}$

(c) $\left(\frac{9m}{8P}\right)^{\frac{1}{2}} \cdot t^{\frac{3}{2}}$ (d) $\left(\frac{8P}{9m}\right)^{\frac{1}{2}} \cdot t^{\frac{3}{2}}$

02 A body at rest is moved along a horizontal straight line by a machine delivering a constant power. The distance moved by the body in time t is proportional to **[2021, 20 July Shift-II]**

(a) $t^{\frac{3}{2}}$ (b) $t^{\frac{1}{2}}$ (c) $t^{\frac{1}{4}}$ (d) $t^{\frac{3}{4}}$

03 A constant power delivering machine has towed a box, which was initially at rest, along a horizontal straight line. The distance moved by the box in time t is proportional to **[2021, 18 March Shift-I]**

(a) $t^{2/3}$ (b) $t^{3/2}$ (c) t (d) $t^{1/2}$

04 A body of mass m is accelerated uniformly from rest to a speed v in a time T. The instantaneous power delivered to the body as a function of time, is given by **[AIEEE 2005, 04]**

(a) $\frac{mv^2}{T^2}t$ (b) $\frac{mv^2}{T^2}t^2$ (c) $\frac{1}{2}\frac{mv^2}{T^2}t$ (d) $\frac{1}{2}\frac{mv^2}{T^2}t^2$

Level 2

05 An object of mass 1000 g experiences a time dependent force $\mathbf{F} = (2t\hat{i} + 3t^2\hat{j})$N. The power generated by the force at time t is **[2025, 7 April Shift-I]**

(a) $(2t^2 + 18t^3)$ W (b) $(2t^3 + 3t^5)$ W

(c) $(2t^2 + 3t^3)$ W (d) $(3t^3 + 5t^5)$ W

06 A body is moving unidirectionally under the influence of a constant power source. Its displacement in time t is proportional to **[2024, 5 April Shift-II]**

(a) t (b) $t^{2/3}$ (c) t^2 (d) $t^{3/2}$

Imp. Instantaneous Power

07 A body of mass 2 kg begins to move under the action of a time dependent force given by $\mathbf{F} = (6t\hat{i} + 6t^2\hat{j})$ N. The power developed by the force at the time t is given by **[2024, 31 Jan Shift-II]**

(a) $(9t^5 + 6t^3)$ W (b) $(6t^4 + 9t^5)$ W

(c) $(9t^3 + 6t^5)$ W (d) $(3t^3 + 6t^5)$ W

08 Sand is being dropped from a stationary dropper at a rate of 0.5 kg-s^{-1} on a conveyor belt moving with a velocity of 5 ms^{-1}. The power needed to keep the belt moving with the same velocity will be **[2022, 27 July Shift-I]**

(a) 1.25 W (b) 2.5 W (c) 6.25 W (d) 12.5 W

09 A particle of mass m is moving in a circular path of constant radius r, such that its centripetal acceleration a is varying with time t as $a = k^2rt^2$, where k is a constant. The power delivered to the particle by the force acting on it is given as **[2022, 28 June Shift-I]**

(a) zero (b) $mk^2r^2t^2$ (c) mk^2r^2t (d) mk^2rt

Level 3

10 A sand dropper drops sand of mass $m(t)$ on a conveyer belt at a rate proportional to the square root of speed (v) of the belt, i.e $\frac{dm}{dt} \propto \sqrt{v}$. If P is the power delivered to run the belt at constant speed, then which of the following relationship is true? **[2025, 29 Jan Shift-II]**

(a) $P \propto \sqrt{v}$ (b) $P \propto v$ (c) $P^2 \propto v^5$ (d) $P^2 \propto v^3$

11 A body of mass 4 kg is placed on a plane at a point P having coordinate (3,4) m. Under the action of force $\mathbf{F} = (2\hat{i} + 3\hat{j})$ N, it moves to a new point Q having coordinates (6,10) m in 4 s. The average power and instantaneous power at the end of 4 s are in the ratio of **[2025, 28 Jan Shift-II]**

(a) 4 : 3 (b) 6 : 13 (c) 13 : 6 (d) 1 : 2

12 The ratio of powers of two motors is $\frac{3\sqrt{x}}{\sqrt{x}+1}$, that are capable of raising 300 kg water in 5 min and 50 kg water in 2 min respectively from a well of 100 m deep. The value of x will be **[2023, 13 April Shift-I]**

(a) 4 (b) 2.4 (c) 16 (d) 2

13 An elevator in a building can carry a maximum of 10 persons with the average mass of each person being 68 kg. The mass of the elevator itself is 920 kg and it moves with a constant speed of 3 m/s. The frictional force opposing the motion is 6000 N.
If the elevator is moving up with its full capacity, the power delivered by the motor to the elevator ($g = 10$ m/s^2) must be at least **[2020, 07 Jan Shift-II]**

(a) 62360 W (b) 48000 W

(c) 56300 W (d) 66000 W

Numerical Value Type Questions

14 A block of mass 5kg starting from rest pulled up on a smooth incline plane making an angle of 30° with horizontal with an effective acceleration of 1 ms^{-2}.

The power delivered by the pulling force at $t = 10$ s from the start is .... W. [Use, $g = 10 \text{ ms}^{-2}$]
(Calculate the nearest integer value)
[2023, 11 April Shift-II]

15 If the maximum load carried by an elevator is 1400 kg (600 kg - Passengers + 800 kg - elevator), which is moving up with a uniform speed of 3 m s^{-1} and the frictional force acting on it is 2000 N, then the maximum power used by the motor is .... kW. ($g = 10 \text{ m/s}^2$) [2023, 10 April Shift-II]

16 A body of mass 2 kg is initially at rest. It starts moving unidirectionally under the influence of a source of constant power P. Its displacement in 4 s is $\frac{1}{3}\alpha^2\sqrt{P}$ m. The value of α will be .... [2023, 30 Jan Shift-II]

17 A body of mass 1 kg begins to move under the action of a time dependent force $\mathbf{F} = (t\,\hat{\mathbf{i}} + 3t^2\,\hat{\mathbf{j}})$ N, where $\hat{\mathbf{i}}$ and $\hat{\mathbf{j}}$ are the unit vectors along X and Y-axis. The power developed by above force, at the time $t = 2$s, will be .... W. [2023, 24 Jan Shift-II]

Topic 4 Collision

Level 1

01 A particle of mass m moving with velocity v collides with a stationary particle of mass 2m. After collision, they stick together and continue to move together with velocity [2023, 10 April Shift-I]
(a) $v/2$ (b) v (c) $v/4$ (d) $v/3$

Imp. Elastic Collision

02 As per the given figure, a small ball P slides down the quadrant of a circle and hits the other ball Q of equal mass which is initially at rest. Neglecting the effect of friction and assume the collision to be elastic, the velocity of ball Q after collision will be ($g = 10 \text{ m/s}^2$)

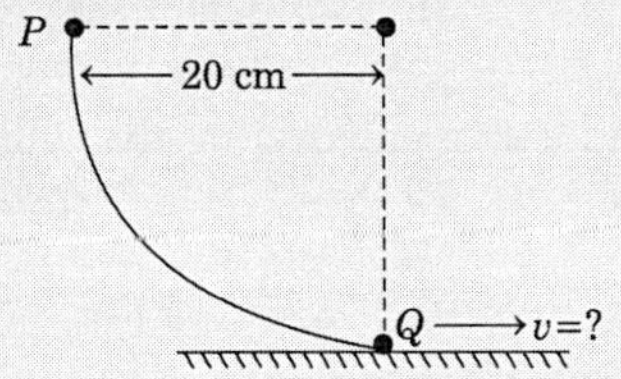

[2023, 30 Jan Shift-I]
(a) 4 m/s (b) 0 (c) 2 m/s (d) 0.25 m/s

03 Given below are two statements. One is labelled as Assertion A and the other is labelled as Reason R.
Assertion (A) Body P having mass M moving with speed u has head-on collision elastically with another body Q having mass m initially at rest. If $m << M$, body Q will have a maximum speed equal to $2u$ after collision.
Reason (R) During elastic collision, the momentum and kinetic energy are both conserved.
In the light of the above statements, choose the most appropriate answer from the options given below.
[2021, 26 Feb Shift-I]
(a) A is not correct but R is correct.
(b) Both A and R are correct but R is not the correct explanation of A.
(c) Both A and R are correct and R is the correct explanation of A.
(d) A is correct but R is not correct.

04 A particle of mass m with an initial velocity $u\hat{\mathbf{i}}$ collides perfectly elastically with a mass $3m$ at rest.
It moves with a velocity $v\,\hat{\mathbf{j}}$ after collision, then v is given by [2020, 02 Sep Shift-I]
(a) $v = \sqrt{\frac{2}{3}}u$ (b) $v = \frac{1}{\sqrt{6}}u$
(c) $v = \frac{u}{\sqrt{2}}$ (d) $v = \frac{u}{\sqrt{3}}$

05 A particle of mass m is moving with speed $2v$ and collides with a mass $2m$ moving with speed v in the same direction. After collision, the first mass is stopped completely while the second one splits into two particles each of mass m, which move at angle 45° with respect to the original direction.
The speed of each of the moving particle will be
[2019, 09 April Shift-II]
(a) $\sqrt{2}\,v$ (b) $\frac{v}{\sqrt{2}}$
(c) $\frac{v}{(2\sqrt{2})}$ (d) $2\sqrt{2}\,v$

06 **Statement I** Two particles moving in the same direction do not lose all their energy in a completely inelastic collision.
Statement II Principle of conservation of momentum holds true for all kinds of collisions.
[AIEEE 2010]
(a) Statement I is true, Statement II is true; Statement II is the correct explanation of Statement I.
(b) Statement I is true, Statement II is true; Statement II is not the correct explanation of Statement I.
(c) Statement I is true, Statement II is true.
(d) Statement I is true, Statement II is false.

07 A bomb of mass 16 kg at rest explodes into two pieces of masses 4 kg and 12 kg. The velocity of the 12 kg mass is 4 ms^{-1}. The kinetic energy of the other mass is [AIEEE 2006]
(a) 144 J (b) 288 J
(c) 192 J (d) 96 J

Level 2

08 Given below are two statements. One is labelled as Assertion (A) and the other is labelled as Reason (R).

Assertion (A)

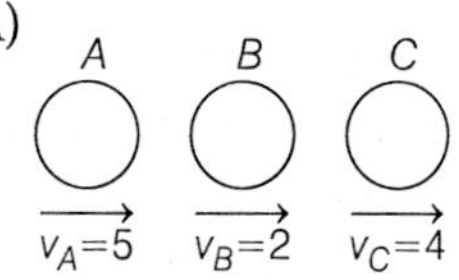

Three identical spheres of same mass undergo one dimensional motion as shown in figure with initial velocities $v_A = 5$ m/s, $v_B = 2$ m/s, $v_C = 4$ m/s. If we wait sufficiently long for elastic collision to happen, then $v_A = 4$ m/s, $v_B = 2$ m/s, $v_C = 5$ m/s will be the final velocities.

Reason (R) In an elastic collision between identical masses, two objects exchange their velocities.

In the light of the above statements, choose the correct answer from the options given below. *[2025, 29 Jan Shift-II]*

(a) A is false but R is true.
(b) Both A and R are true and R is not the correct explanation of A.
(c) Both A and R are true and R is the correct explanation of A.
(d) A is true but R is false.

09 If a rubber ball falls from a height h and rebounds upto the height of $\frac{h}{2}$. The percentage loss of total energy of the initial system as well as velocity of ball before it strikes the ground, respectively, are *[2024, 4 April Shift-I]*

(a) 50%, $\sqrt{gh/2}$ (b) 40%, $\sqrt{2gh}$
(c) 50%, $\sqrt{2gh}$ (d) 50%, $\sqrt{gh}$

10 A bullet of mass 0.1 kg moving horizontally with speed 400 ms^{-1} hits a wooden block of mass 3.9 kg kept on a horizontal rough surface. The bullet gets embedded into the block and moves 20 m before coming to rest. The coefficient of friction between the block and the surface is (Given, $g = 10$ m/s^2) *[2023, 8 April Shift-II]*

(a) 0.50 (b) 0.90 (c) 0.25 (d) 0.65

11 Two billiard balls of mass 0.05 kg each moving in opposite directions with 10 ms^{-1} collide and rebound with the same speed. If the time duration of contact is $t = 0.005$ s, then what is the force exerted on the ball due to each other ? *[2022, 25 July Shift-II]*

(a) 100 N (b) 200 N (c) 300 N (d) 400 N

12 Three objects A, B and C are kept in a straight line on a frictionless horizontal surface. The masses of A, B and C are m, $2m$ and $2m$, respectively. A moves towards B with a speed of 9 m/s and makes an elastic collision with it. There after B makes a completely inelastic collision with C. All motions occur along same straight line. The final speed of C is *[2021, 27 July Shift-I]*

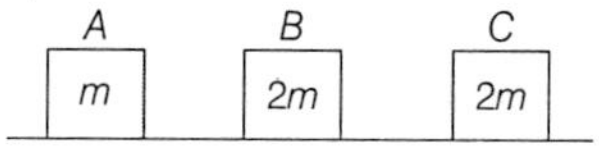

(a) 6 m/s (b) 9 m/s (c) 4 m/s (d) 3 m/s

13 An object of mass m_1 collides with another object of mass m_2, which is at rest. After the collision, the objects move with equal speeds in opposite direction. The ratio of the masses $m_2 : m_1$ is *[2021, 18 March Shift-II]*

(a) 3 : 1 (b) 2 : 1 (c) 1 : 2 (d) 1 : 1

Imp. Perfectly Inelastic Collision

14 A large block of wood of mass $M = 5.99$ kg is hanging from two long massless cords. A bullet of mass $m = 10$ g is fired into the block and gets embedded in it. The system (block + bullet) then swing upwards, their centre of mass rising a vertical distance $h = 9.8$ cm before the (block + bullet) pendulum comes momentarily to rest at the end of its arc. The speed of the bullet just before collision is (Take, $g = 9.8$ ms^{-2}) *[2021, 16 March Shift-II]*

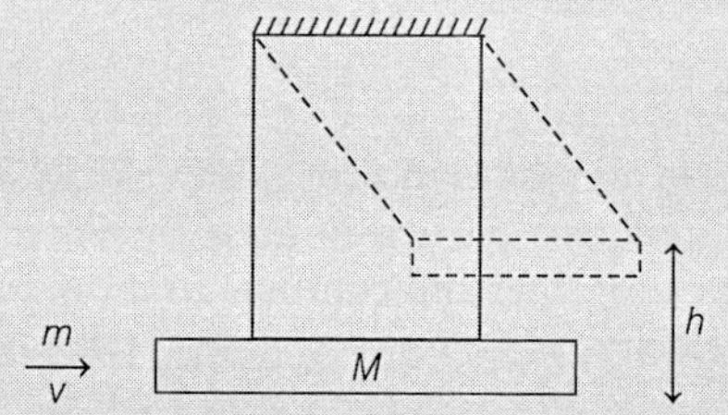

(a) 841.4 m/s (b) 811.4 m/s
(c) 831.4 m/s (d) 821.4 m/s

15 A block of mass 1.9 kg is at rest at the edge of a table of height 1 m. A bullet of mass 0.1 kg collides with the block and sticks to it. If the velocity of the bullet is 20 m/s in the horizontal direction just before the collision, then the kinetic energy just before the combined system strikes the floor, is

(Take, $g = 10$ m/s^2 and assume there is no rotational motion and loss of energy after the collision is negligible) *[2020, 03 Sep Shift-II]*

(a) 20 J (b) 19 J
(c) 21 J (d) 23 J

16 A body of mass 2 kg makes an elastic collision with a second body at rest and continues to move in the original direction but with one-fourth of its original speed. What is the mass of the second body? *[2019, 09 April Shift-I]*

(a) 1.5 kg (b) 1.2 kg
(c) 1.8 kg (d) 1.0 kg

17 Three blocks A, B and C are lying on a smooth horizontal surface as shown in the figure. A and B have equal masses m while C has mass M. Block A is given an initial speed v towards B due to which it collides with B perfectly inelastically. The combined mass collides with C, also perfectly inelastically 5/6th of the initial kinetic energy is lost in whole process. What is value of M/m? ***[2019, 09 Jan Shift-I]***

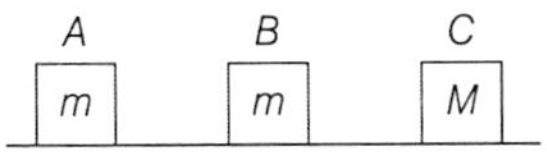

(a) 4 (b) 2
(c) 3 (d) 5

18 In a collinear collision, a particle with an initial speed v_0 strikes a stationary particle of the same mass. If the final total kinetic energy is 50% greater than the original kinetic energy, the magnitude of the relative velocity between the two particles after collision, is ***[JEE Main 2018]***

(a) $\frac{v_0}{4}$ (b) $\sqrt{2}\, v_0$ (c) $\frac{v_0}{2}$ (d) $\frac{v_0}{\sqrt{2}}$

19 A proton of mass m collides elastically with a particle of unknown mass at rest. After the collision, the proton and the unknown particle are seen moving at an angle of 90° with respect to each other. The mass of unknown particle is ***[JEE Main 2018]***

(a) m (b) $2m$ (c) $m/\sqrt{3}$ (d) $m/2$

20 A particle of mass m moving in the x-direction with speed $2v$ is hit by another particle of mass $2m$ moving in the y-direction with speed v. If the collision is perfectly inelastic, the percentage loss in the energy during the collision is close to ***[JEE Main 2015]***

(a) 44% (b) 50% (c) 56% (d) 62%

21 This question has Statement I and Statement II. Of the four choices given after the statements, choose the one that best describes the two statements.

Statement I A point particle of mass m moving with speed v collides with stationary point particle of mass M. If the maximum energy loss possible is given as

$$f\left(\frac{1}{2}mv^2\right), \text{ then } f = \left(\frac{m}{M+m}\right).$$

Statement II Maximum energy loss occurs when the particles get stuck together as a result of the collision. ***[JEE Main 2013]***

(a) Statement I is true, Statement II is true; Statement II is the correct explanation of Statement I.
(b) Statement I is true, Statement II is true; Statement II is not the correct explanation of Statement I.
(c) Statement I is true, Statement II is false.
(d) Statement I is false, Statement II is true.

22 A block of mass 0.50 kg is moving with a speed of 2.00 ms^{-1} on a smooth surface. It strikes another mass of 1.00 kg and then they move together as a single body. The energy loss during the collision is ***[AIEEE 2008]***

(a) 0.16 J (b) 1.00 J (c) 0.67 J (d) 0.34 J

23 A mass m moves with a velocity v and collides inelastically with another identical mass. After collision, the 1st mass moves with velocity $v/\sqrt{3}$ in a direction perpendicular to the initial direction of motion. Find the speed of the second mass after collision. ***[AIEEE 2005]***

(a) v (b) $\sqrt{3}\,v$
(c) $\frac{2}{\sqrt{3}}v$ (d) $\frac{v}{\sqrt{3}}$

Level 3

24

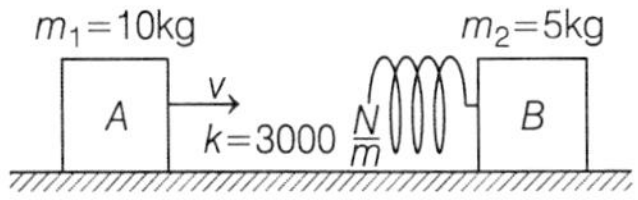

Consider two blocks A and B of masses $m_1 = 10$ kg and $m_2 = 5$ kg that are placed on a frictionless table. The block A moves with a constant speed $v = 3$ m/s towards the block B kept at rest. A spring with spring constant $k = 3000$ N/m is attached with the block B as shown in the figure. After the collision, suppose that the blocks A and B, along with the spring in constant compression state, move together, then the compression in the spring is, (Neglect the mass of the spring) ***[2025, 3 April Shift-II]***

(a) 0.4 m (b) 0.2 m (c) 0.1 m (d) 0.3 m

25 As shown below, bob A of a pendulum having massless string of length R is released from 60° to the vertical. It hits another bob B of half the mass that is at rest on a frictionless table in the centre. Assuming elastic collision, the magnitude of the velocity of bob A after the collision will be (take g as acceleration due to gravity) ***[2025, 29 Jan Shift-I]***

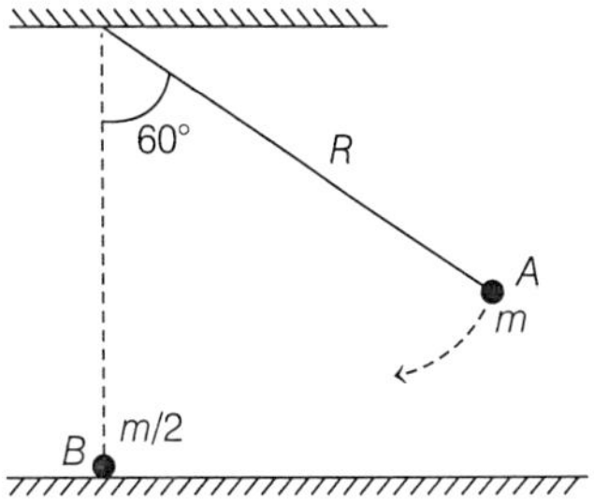

(a) $\sqrt{Rg}$ (b) $\frac{1}{3}\sqrt{Rg}$
(c) $\frac{2}{3}\sqrt{Rg}$ (d) $\frac{4}{3}\sqrt{Rg}$

26 What percentage of kinetic energy of a moving particle is transferred to a stationary particle when it strikes the stationary particle of 5 times its mass? (Assume the collision to be head-on elastic collision) *[2022, 27 June Shift-I]*

(a) 50.0% (b) 66.6%
(c) 55.5% (d) 33.3%

Imp. Oblique Collision

27 A body of mass M moving at speed v_0 collides elastically with a mass m at rest. After the collision, the two masses move at angles θ_1 and θ_2 with respect to the initial direction of motion of the body of mass M. The largest possible value of the ratio M/m, for which the angles θ_1 and θ_2 will be equal, is *[2021, 31 Aug Shift-I]*

(a) 4 (b) 1
(c) 3 (d) 2

28 A rubber ball is released from a height of 5 m above the floor. It bounces back repeatedly, always rising to 81 / 100 of the height through which it falls. Find the average speed of the ball.
(Take, $g = 10\ ms^{-2}$) *[2021, 17 March Shift-II]*

(a) $3.0\ ms^{-1}$ (b) $3.5\ ms^{-1}$
(c) $2.0\ ms^{-1}$ (d) $2.5\ ms^{-1}$

29 A wedge of mass $M = 4$ m lies on a frictionless plane. A particle of mass m approaches the wedge with speed v. There is no friction between the particle and the plane or between the particle and the wedge. The maximum height climbed by the particle on the wedge is given by *[2019, 9 April Shift-II]*

(a) $\frac{2v^2}{7g}$ (b) $\frac{v^2}{g}$
(c) $\frac{2v^2}{5g}$ (d) $\frac{v^2}{2g}$

30 Two particles A and B of equal mass M are moving with the same speed v as shown in the figure. They collide completely inelastically and move as a single particle C. The angle θ that the path of C makes with the X-axis is given by *[JEE Main 2017]*

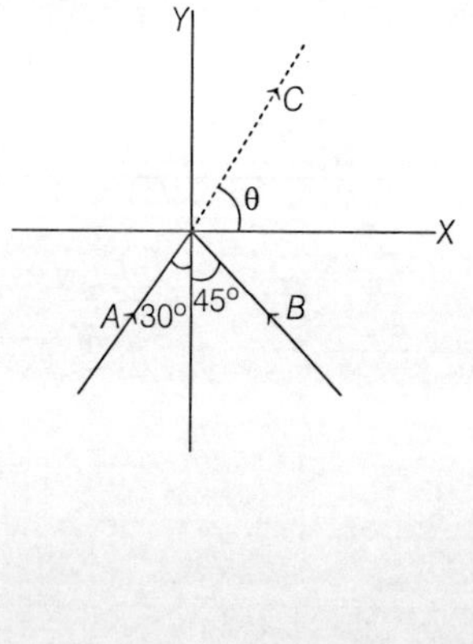

(a) $\tan\theta = \frac{\sqrt{3}+\sqrt{2}}{1-\sqrt{2}}$

(b) $\tan\theta = \frac{1-\sqrt{3}}{1+\sqrt{2}}$

(c) $\tan\theta = \frac{\sqrt{3}-\sqrt{2}}{1-\sqrt{2}}$

(d) $\tan\theta = \frac{1-\sqrt{2}}{\sqrt{2}(1+\sqrt{3})}$

Numerical Value Type Questions

31 A body is dropped on ground from a height h_1 and after hitting the ground, it rebounds to a height h_2. If the ratio of velocities of the body just before and after hitting ground is 4, then percentage loss in kinetic energy of the body is $x/4$. The value of x is .... *[2023, 6 April Shift-II]*

32 A body of mass 1 kg collides head on elastically with a stationary body of mass 3 kg. After collision, the smaller body reverses its direction of motion and moves with a speed of 2 m/s. The initial speed of the smaller body before collision is .... ms^{-1}. *[2023, 25 Jan Shift-II]*

33 A ball of mass 10 kg moving with a velocity $10\sqrt{3}$ m/s along the X-axis, hits another ball of mass 20 kg which is at rest. After the collision, first ball comes to rest while the second ball disintegrates into two equal pieces. One piece starts moving along Y-axis with a speed of 10 m/s.

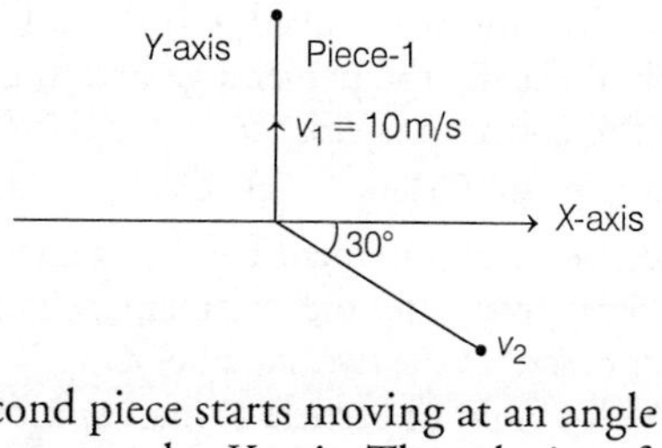

The second piece starts moving at an angle of 30° with respect to the X-axis. The velocity of the ball moving at 30° with X-axis is x m/s. The configuration of pieces after collision is shown in the figure below. The value of x to the nearest integer is .... *[2021, 18 March Shift-I]*

34 A ball of mass 10 kg moving with a velocity $10\sqrt{3}\ ms^{-1}$ along X-axis, hits another ball of mass 20 kg, which is at rest. After collision, the first ball comes to rest and the second one disintegrates into two equal pieces. One of the pieces starts moving along Y-axis at a speed of 10 m/s. The second piece starts moving at a speed of 20 m/s at an angle θ (degree) with respect to the X-axis.

The configuration of pieces after collision is shown in the figure. The value of θ to the nearest integer is …… . *[2021, 16 March Shift-I]*

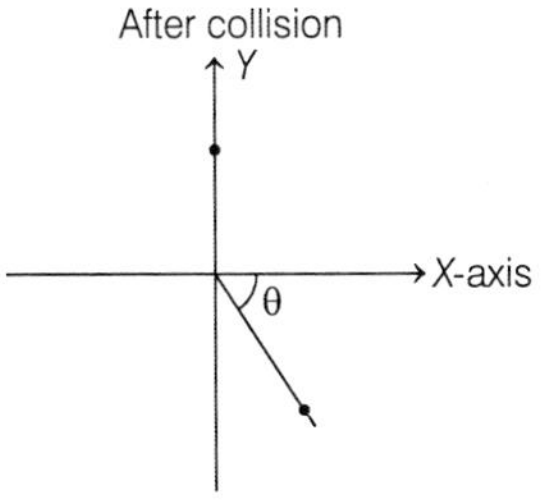

35 A particle of mass m is moving along the X-axis with initial velocity $u\hat{i}$. It collides elastically with a particle of mass 10 m at rest and then moves with half its initial kinetic energy (see figure). If $\sin\theta_1 = \sqrt{n}\sin\theta_2$, then value of n is …….. . *[2020, 02 Sep Shift-II]*

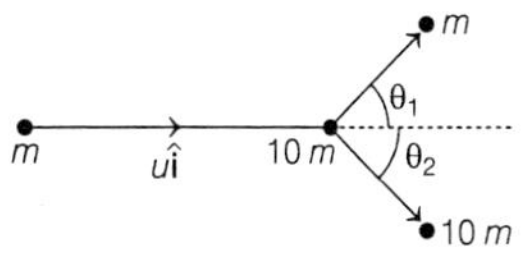

Key Index

Topic 1

1 (d)	2 (d)	3 (a)	4 (a)	5 (c)	6 (c)	7 (c)	8 (d)	9 (b)	10 (d)
11 (a)	12 (c)	13 (b)	14 (c)	15 (d)	16 (b)	17 (c)	18 (a)	19 (d)	20 (d)
21 (a)	22 (a)	23 (b)	24 (b)	25 (b)	26 (a)	27 (c)	28 (c)	29 (b)	30 (c)
31 (c)	32 (b)	33 (a)	34 (b)	35 (a)	36 (a)	37 (c)	38 (d)	39 (c)	40 (c)
41 (a)	42 (c)	43 (a)	44 (a)	45 (a)	46 (b)	47 (b)	48 (c)	49 (c)	50 (d)
51 (c)	52 (d)	53 (c)	54 (a)	55 (c)	56 (b)	57 (b)	58 (152)	59 (58)	60 (30)
61 (3)	62 (784)	63 (32)	64 (30)	65 (125)	66 (2)	67 (132)	68 (40)	69 (7)	70 (300)
71 (40)	72 (2)	73 (24)	74 (600)	75 (5)	76 (2)	77 (40)	78 (450)	79 (1)	80 (10)
81 (1)	82 (2)								

Topic 2

1 (b)	2 (c)	3 (c)	4 (d)	5 (d)	6 (b)	7 (a)	8 (c)	9 (b)	10 (d)
11 (d)	12 (b)	13 (a)	14 (b)	15 (c)	16 (a)	17 (d)	18 (d)	19 (d)	20 (245)
21 (10)	22 (120)	23 (16)	24 (2)	25 (2)	26 (346)				

Topic 3

1 (d)	2 (a)	3 (b)	4 (a)	5 (b)	6 (d)	7 (c)	8 (d)	9 (c)	10 (c)
11 (b)	12 (c)	13 (d)	14 (300)	15 (48)	16 (4)	17 (100)			

Topic 4

1 (d)	2 (c)	3 (c)	4 (c)	5 (d)	6 (a)	7 (b)	8 (a)	9 (c)	10 (c)
11 (b)	12 (d)	13 (a)	14 (c)	15 (c)	16 (b)	17 (a)	18 (b)	19 (a)	20 (c)
21 (d)	22 (c)	23 (c)	24 (c)	25 (b)	26 (c)	27 (c)	28 (d)	29 (c)	30 (a)
31 (375)	32 (4)	33 (20)	34 (30)	35 (10)					

Detailed **Explanations**

Topic 1

01. *(d)* Potential energy does not defined for non-conservative force i.e. for frictional force.

02. *(d)* $\mathbf{F} = 2\hat{i} + b\hat{j} + \hat{k}$, $W = 0$ (given)

$$\mathbf{s} = \hat{i} - 2\hat{j} - \hat{k}$$

$$W = \mathbf{F} \cdot \mathbf{s} = (2\hat{i} + b\hat{j} + \hat{k}) \cdot (\hat{i} - 2\hat{j} - \hat{k})$$

$$0 = 2 - 2b - 1$$

$$2b = 1 \Rightarrow b = 1/2$$

03. *(a)* Kinetic energy K and momentum p are related as

$$K = \frac{p^2}{2m}$$

For same mometnum,

$$K \propto \frac{1}{m}$$

Particle A has least mass.

$\therefore$ Particle A has maximum kinetic energy.

04. *(a)* Work done by gravity is independent of path taken as it is a conservative force.

It only depends on vertical displacement in both cases.

As height in both cases are same, so the work done will be also same.

So, ratio will be 1 : 1.

05. *(c)* We know that

$$\text{KE} = \frac{p^2}{2m}, \; K \propto \frac{p^2}{m}$$

$$K \propto \frac{1}{m} \text{ if } p_1 = p_2$$

$$\frac{K_1}{K_2} = \frac{m_2}{m_1}$$

$$\Rightarrow \frac{m_2}{m_1} = \frac{16}{9} \Rightarrow \frac{m_1}{m_2} = \frac{9}{16}$$

$$\therefore \; m_1 : m_2 = 9 : 16$$

06. *(c)* At highest point during projectile motion, the velocity of projectile in vertical direction becomes zero but it has horizontal component of velocity. Because of this, kinetic energy will not be equal to zero.

Gravitational potential energy is maximum at highest point and equal to

$$mgH = mg\left(\frac{u^2 \sin^2 \theta}{2g}\right)$$

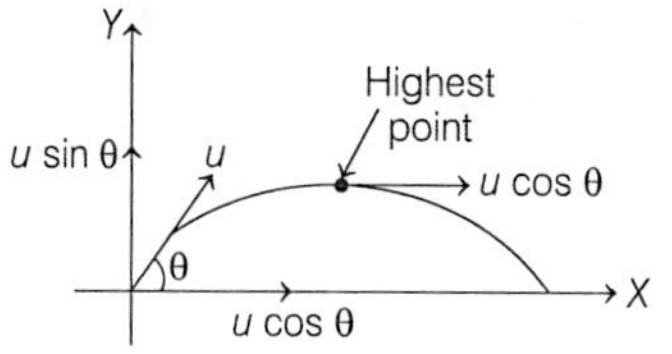

07. *(c)* Work done by a force is negative when force is acting in opposite direction of displacement. So, work of gravitational force while lifting a bucket and due to air resistance on oscillating pendulum are negative.

08. *(d)*

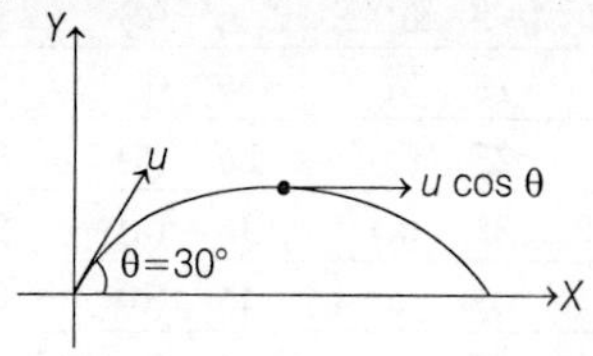

Kinetic energy of particle at point of projection,

$$K_1 = \frac{1}{2} mu^2$$

Kinetic energy at highest point,

$$K_2 = \frac{1}{2} mu^2 \cos^2 \theta$$

$$= \left(\frac{1}{2} mu^2\right) \times \left(\frac{\sqrt{3}}{2}\right)^2 = \frac{3}{4} K_1$$

$$\Rightarrow \frac{K_1}{K_2} = \frac{4}{3} \qquad [\because \theta = 30°]$$

09. *(b)*

> **key idea** To find the ratio of momenta of two bodies having different masses and same kinetic energy, you can use following formula:
>
> $$\frac{p_1}{p_2} = \sqrt{\frac{m_1}{m_2}}$$

We know that kinetic energy (K) of a moving body is related as its momentum (p) as given below

$$p = \sqrt{2mK} \Rightarrow \frac{p_1}{p_2} = \sqrt{\frac{m_1}{m_2} \cdot \frac{K_1}{K_2}}$$

But $K_1 = K_2$

$$\frac{p_1}{p_2} = \sqrt{\frac{m_1}{m_2}} = \sqrt{\frac{8}{2}} = \frac{2}{1}$$

$$p_1 : p_2 = 2 : 1$$

10. *(d)*

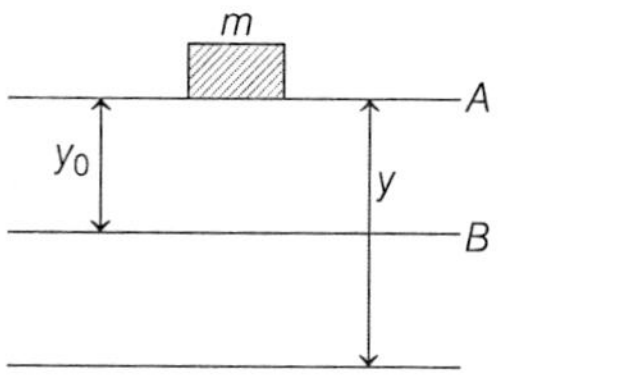

Kinetic energy at B = Potential energy at A − Potential energy at B

$$\Rightarrow mgy - mg(y - y_0) = mgy_0$$

11. *(a)* Work done from force-displacement (F-x) graph is equal to area enclosed under F-x curve.

The area enclosed below the X-axis is negative and area above the X-axis is taken as positive.

From the given F-x graph in Fig. (a), Fig. (b), Fig. (c) and Fig. (d), area under F-x curve are arranged as

$$(\text{Area})_{\text{Fig. (c)}} > (\text{Area})_{\text{Fig. (b)}} > (\text{Area})_{\text{Fig. (a)}} > (\text{Area})_{\text{Fig. (d)}}$$

$$\therefore \quad W_3 > W_2 > W_1 > W_4$$

12. *(c)* Potential energy, $U = \dfrac{A}{r^{10}} - \dfrac{B}{r^5}$

At the equilibrium distance between the two atoms, force between the atoms is zero.

$$F = -\frac{dU}{dr} = 0$$

Hence, $$\frac{dU}{dr} = 0$$

$$\Rightarrow \frac{d}{dr}\left[\frac{A}{r^{10}} - \frac{B}{r^5}\right] = 0$$

$$\Rightarrow \frac{-10A}{r^{11}} + \frac{5B}{r^6} = 0$$

$$\Rightarrow \frac{5}{r^6}\left[\frac{-2A}{r^5} + B\right] = 0$$

$$\Rightarrow \frac{-2A}{r^5} + B = 0$$

$$\Rightarrow B = \frac{2A}{r^5} \Rightarrow r^5 = \frac{2A}{B}$$

$$\Rightarrow r = \left(\frac{2A}{B}\right)^{1/5}$$

13. *(b)* Given, mass of suitcase,

$$m = 80 \text{ kg}$$

Let height lowered by porter,

$$h = 80 \text{ cm} = 80 \times 10^{-2} \text{ m}$$

As we know that,
Work $(W) = -mgh$

$\therefore \quad W = -80 \times 9.8 \times 80 \times 10^{-2}$

$= -627.2$ J

14. *(c)* Work done by a variable force on the particle,

$$W = \int \mathbf{F} \cdot d\mathbf{r} = \int \mathbf{F} \cdot (dx\hat{\mathbf{i}} + dy\hat{\mathbf{j}})$$

$\therefore$ In two dimension, $d\mathbf{r} = dx\hat{\mathbf{i}} + dy\hat{\mathbf{j}}$

and it is given $\mathbf{F} = -x\hat{\mathbf{i}} + y\hat{\mathbf{j}}$.

$$\therefore \quad W = \int (-x\hat{\mathbf{i}} + y\hat{\mathbf{j}}) \cdot (dx\hat{\mathbf{i}} + dy\hat{\mathbf{j}})$$

$$= \int -x\,dx + y\,dy$$

$$= \int -x\,dx + \int y\,dy$$

As particle is displaced from $A(1, 0)$ to $B(0, 1)$, so x varies from 1 to 0 and y varies from 0 to 1.

So, with limits, work will be

$$W = \int_1^0 -x\,dx + \int_0^1 y\,dy$$

$$= \left[\frac{-x^2}{2}\right]_1^0 + \left[\frac{y^2}{2}\right]_0^1$$

$$= -\frac{1}{2}[0 - (1)^2] + \frac{1}{2}[(1)^2 - 0] = 1 \text{ J}$$

15. *(d)* The average velocity (v)

$$v = \frac{\text{Distance}}{\text{Time}}$$

Given that the distance is 100 m and the time is 10 s

$$v = \frac{100 \text{ m}}{10 \text{ s}} = 10 \text{ m/s}$$

Kinetic energy $\text{KE} = \frac{1}{2}mv^2$

where $m = 60$ kg

$v = 10$ m/s

$$\text{KE} = \frac{1}{2} \times 60 \text{ kg} \times (10 \text{ m/s})^2$$

$$\text{KE} = \frac{1}{2} \times 60 \times 100$$

$$\text{KE} = 30 \times 100 = 3000 \text{ J}$$

16. *(b)* Work done in displacing the particle

$$W = \mathbf{F} \cdot \mathbf{r} = (5\hat{\mathbf{i}} + 3\hat{\mathbf{j}} + 2\hat{\mathbf{k}}) \cdot (2\hat{\mathbf{i}} - \hat{\mathbf{j}})$$

$$= 5 \times 2 + 3 \times (-1) + 2 \times 0$$

$$= 10 - 3 = 7 \text{ J}$$

17. *(c)* Here, Statement II is implying Statement I but Statement I is not implying Statement II as kinetic energy of a system of particles is zero means speed of each and every particle is zero which says that momentum of every particle is zero.

But Statement I means linear momentum of a system of particles is zero, which may be true even, if particles have equal and opposite momentums and hence having non-zero kinetic energy.

18. *(a)*

> **key idea** At highest point of it's hight a projectile has only horizontal velocity component ($v = u\cos\theta$). It's vertical velocity component vanishes at highest point.

$$\text{KE} = \frac{1}{2}mv^2$$

$$\text{KE}_H = \frac{1}{2}m(v\cos 60°)^2$$

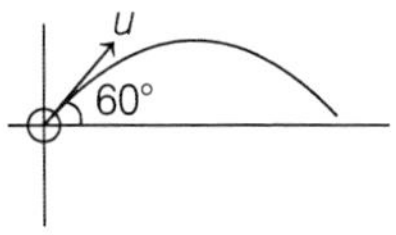

$$\text{KE}_H = \frac{1}{2}mv^2 \times \left(\frac{1}{2}\right)^2$$

$$\text{KE}_H = \frac{mv^2}{8} = \frac{1}{2}\frac{mv^2}{4} = \frac{1}{4}\text{KE}$$

19. *(d)* Given, $m = 100 \text{ g} = 0.1$ kg

$u = 20$ m/s

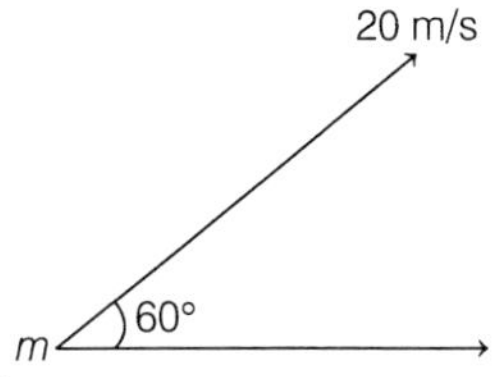

$$\text{KE}_i = \frac{1}{2}mv^2$$

$$\text{KE}_f = \frac{1}{2}m(v\cos 60°)^2$$

$$= \frac{1}{2}mv^2\left(\frac{1}{2}\right)^2$$

$$= \frac{1}{2}mv^2 \times \frac{1}{4} = \frac{1}{8}mv^2$$

$$\Delta K = K_i - K_f = \frac{1}{2}mv^2 - \frac{1}{8}mv^2$$

$$= \frac{3}{8}mv^2 = \frac{3}{8} \times 0.1 \times (20)^2 = 15 \text{ J}$$

20. *(d)* In a central force filed, the force is conservative. This means the work done by the force is independent of path chosen. In mechanics, all conservative forces have an associated potential energy. Non-conservative forces do not have patential energy. Reason does not correctly explain the Assertion.

21. *(a)* Given force, $F = \alpha + \beta x^2$

Work done is given by, $W = \int F \cdot dx$

$$5 = \int (\alpha + \beta x^2)dx$$

$$5 = \alpha x + \left.\frac{\beta x^3}{3}\right|_0^1 = \alpha + \frac{\beta}{3} \quad [\alpha = 1]$$

$$4 = \frac{\beta}{3} \Rightarrow \beta = 12 \text{ N/m}^2$$

22. *(a)* Given, $v = \alpha\sqrt{x}$, at $x = 0, v = 0$

at $\quad x = d, v = \alpha\sqrt{d}$

$$\text{WD} = K_f - K_i = \frac{1}{2}mv_f^2 - \frac{1}{2}mv_i^2$$

$$\Rightarrow \frac{1}{2}m(v_f^2 - v_i^2)$$

$$= \frac{1}{2}m[(\alpha\sqrt{d})^2 - \frac{1}{2}m(0)^2]$$

$$= \frac{1}{2}m\alpha^2 d$$

23. *(b)* According to given situation,

$$K_i = \frac{1}{2}mv_i^2$$

$$= \frac{1}{2}m(100)^2$$

and $\quad K_f = \frac{1}{2}mv_f^2 = \frac{1}{2}m(40)^2$

$\therefore$ Percentage loss (in KE)

$$= \left(\frac{K_i - K_f}{K_i}\right) \times 100$$

$$= \frac{\frac{1}{2}m(100)^2 - \frac{1}{2}m(40)^2}{\frac{1}{2}m(100)^2} \times 100$$

$$= \frac{10000 - 1600}{10000} \times 100 = 84\%$$

24. *(b)*

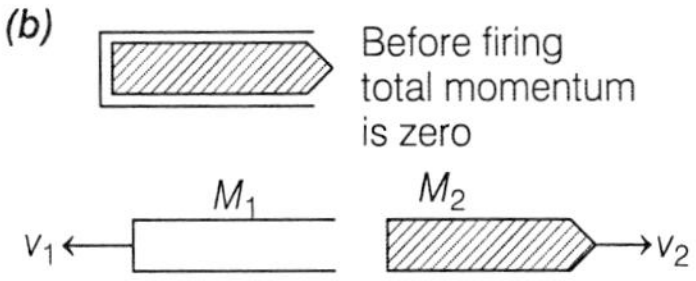

After firing let speeds are v_1 and v_2. Then, conservation of momentum gives,

$$M_1 v_1 = M_2 v_2 \quad \text{...(i)}$$

Ratio of kinetic energies of artillery and shell is

$$\frac{E_1}{E_2} = \frac{\frac{1}{2}M_1 v_1^2}{\frac{1}{2}M_2 v_2^2} = \frac{M_1 v_1^2}{M_2 v_2^2}$$

$$= \frac{M_1}{M_2} \cdot \frac{(M_2)^2}{(M_1)^2} = \frac{M_2}{M_1}$$

25. *(b)*

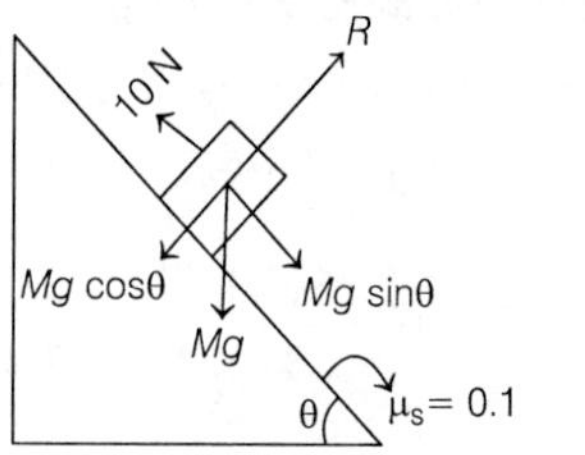

Work done against friction = Force of friction × displacement

$$= \mu R \times s = (\mu \, Mg \cos\theta) \times s$$

Here $\mu = 0.1$, $M = 1$ kg

$g = 10 \text{ m/s}^2$, $\theta = 60°$, $s = 10$ m

So, Work done

$$= 0.1 \times 1 \times 10 \times \cos 60° \times 10 = 5\text{J}$$

26. *(a)* According to given condition,

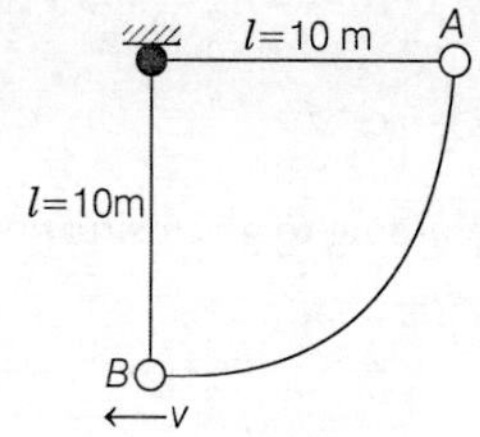

$$\frac{1}{2}mv^2 = 90\% \text{ of } mgl$$

$$\Rightarrow \quad \frac{v^2}{2} = 0.9gl$$

$$\Rightarrow \quad v^2 = 1.8gl$$

$$= 1.8 \times 10 \times 10 = 180$$

$$\therefore \quad v = \sqrt{180}$$

$$= 6\sqrt{5} \text{ m/s}$$

27. *(c)* According to given condition, velocity of bob at point A, $v_A = \sqrt{5gL}$

and at point B, $v_B = \sqrt{gL}$

$$\therefore \quad \frac{(\text{KE})_A}{(\text{KE})_B} = \frac{\frac{1}{2}mv_A^2}{\frac{1}{2}mv_B^2}$$

$$= \frac{v_A^2}{v_B^2} = \frac{5gL}{gL} = \frac{5}{1}$$

$$\therefore \quad (\text{KE})_A : (\text{KE})_B = 5:1$$

28. *(c)* $m = 100$ kg

$s = 10$ m, $\mu = 0.4$

Force of friction, $f = \mu \, mg$

$$= 0.4 \times 100 \times 10 = 400 \text{ N}$$

Work done against friction,

$$W = F \cdot s$$

$$= 400 \times 10 = 4000$$

$$W = 4000 \text{ J}$$

29. *(b)* **For Statement I**

Work done = ΔKE

$$\Rightarrow \quad -Fs = 0 - \text{KE}$$

$$\Rightarrow \quad \text{KE} = sF \Rightarrow s = \frac{\text{KE}}{F}$$

Hence, both car and truck will cover same distance.

For Statement II

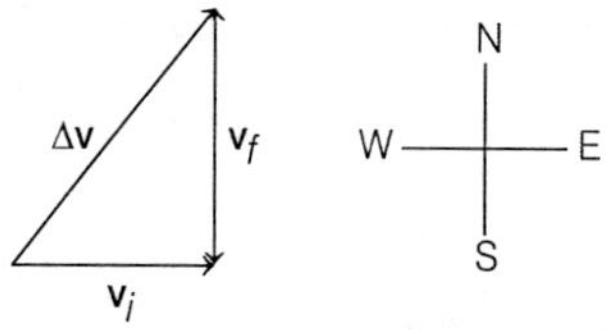

From the diagram, $\Delta\mathbf{v} = \mathbf{v}_f - \mathbf{v}_i$

$|\Delta\mathbf{v}| = \sqrt{2}\, v$

Thus, the velocity is changing, therefore $\mathbf{a} \neq 0$.

Hence, Statement II is incorrect.

30. *(c)* Initial momentum of body,

$$p_1 = p$$

Final momentum of body,

$$p_2 = p_1 + 20\% \text{ of } p_1$$

$$= p_1 + \frac{20}{100} p_1 = \frac{6p_1}{5}$$

$$p_2 = 6p/5$$

we know that, $E_K = \dfrac{p^2}{2m}$

$\therefore$ Initial kinetic energy, $E_{K_1} = \dfrac{p_1^2}{2m} = \dfrac{p^2}{2m}$

Final kinetic energy,

$$E_{K_2} = \frac{p_2^2}{2m} = \frac{(6p/5)^2}{2m} = \frac{36p^2}{50m}$$

$\therefore$ % increase in kinetic energy

$$= \frac{E_{K_2} - E_{K_1}}{E_{K_1}} \times 100\%$$

$$= \frac{\frac{36p^2}{50m} - \frac{p^2}{2m}}{\frac{p^2}{2m}} \times 100\%$$

$$= \frac{11}{25} \times 100 = 44\%$$

31. *(c)* According to the question.

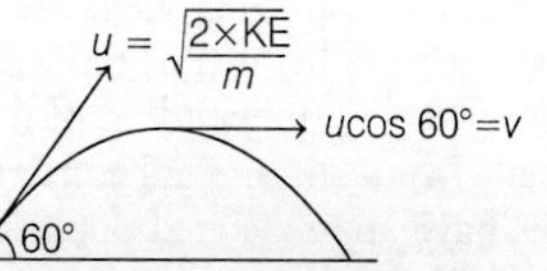

$$u = \sqrt{\frac{2KE}{m}} \qquad \left(\because \text{KE} = \frac{1}{2}mu^2\right)$$

$$\Rightarrow \quad u = \sqrt{\frac{2 \times E}{m}} \qquad (\because \text{Given KE} = E)$$

At topmost point,

$$v = \text{speed} = u\cos 60° = \sqrt{\frac{2E}{m}} \times \frac{1}{2}$$

$\therefore$ Kinetic energy at topmost point

$$= \frac{1}{2}mv^2 = \frac{1}{2} \times m \times \frac{2E}{m} \times \frac{1}{4} = \frac{E}{4}$$

32. *(b)* Given, spring constant,

$$k = 2 \text{ Nm}^{-1}$$

If x be the maximum elongation in spring

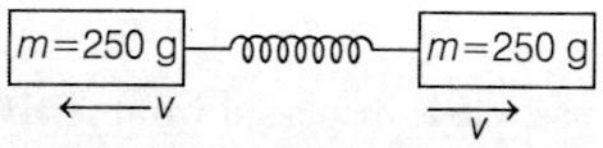

According to conservation of energy,

Loss in KE = Gain in PE of spring

$$\frac{1}{2}mv^2 + \frac{1}{2}mv^2 = \frac{1}{2}kx^2$$

$$\Rightarrow \quad 2 \times \frac{1}{2}mv^2 = \frac{1}{2}kx^2$$

$$\Rightarrow \quad x^2 = \frac{2mv^2}{k} = \frac{2 \times 0.250}{2}v^2$$

$$(\because m = 250 \text{ g} = 0.250 \text{ kg})$$

$$= 0.25v^2$$

$$\Rightarrow \quad x^2 = (0.5\,v)^2$$

$$\Rightarrow \quad x = 0.5v = \frac{v}{2}$$

33. *(a)* Let v is the speed of the third block C.

The velocity of centre of mass of A and B is

$$v_{\text{CM}} = \frac{v}{2}$$

The spring is compressed maximum by x distance.

Using the law of conservation of energy,

$$\frac{1}{2}mv^2 = \frac{1}{2}m\left(\frac{v}{2}\right)^2 + \frac{1}{2}m\left(\frac{v}{2}\right)^2 + \frac{1}{2}kx^2$$

$$\Rightarrow \frac{1}{4}mv^2 = \frac{1}{2}kx^2 \Rightarrow x = \sqrt{\frac{mv^2}{2k}}$$

$$\Rightarrow \quad x = v\sqrt{\frac{m}{2k}}$$

34. *(b)* Before deflection,

Initial kinetic energy,

$$\text{KE}_i = \frac{1}{2}mu^2 = \frac{1}{2} \times 0.5 \times (20)^2$$

$$= \frac{1}{2} \times 0.5 \times 400 = \frac{1}{2} \times \frac{1}{2} \times 400 = 100 \text{ J}$$

It is given in the question that after deflection the ball moves with 5% of its initial kinetic energy

$$KE_f = 5\% \text{ of } KE_i$$

$$\Rightarrow \quad KE_f = \frac{5}{100} \times 100 = 5 \text{ J}$$

If the final speed of the ball is v ms^{-1}, then

$$KE_f = \frac{1}{2}mv^2 \Rightarrow 5 = \frac{1}{2} \times 0.5 \times v^2$$

$$\Rightarrow \quad v = 4.47 \text{ ms}^{-1}$$

35. *(a)* We know that,

Kinetic energy, $K = \dfrac{p^2}{2m}$

$$p \propto \sqrt{K} \Rightarrow \frac{p_2}{p_1} = \frac{\sqrt{K_2}}{\sqrt{K_1}}$$

$$\frac{p_2}{p_1} = \frac{\sqrt{4K_1}}{\sqrt{K_1}} \Rightarrow \frac{p_2}{p_1} = 2$$

$$\Rightarrow \quad p_2 = 2p_1$$

$\therefore$ Percentage change in momentum

$$= \frac{p_2 - p_1}{p_1} \times 100 = \frac{2p_1 - p_1}{p_1} \times 100$$

$$= \frac{p_1}{p_1} \times 100 = 100\%$$

36. *(a)* The central potential field when particle moves in circular orbit,

$$U(r) = -\frac{C}{r}$$

We know that,

$$F = -\frac{dU}{dr} \Rightarrow F = -\frac{d}{dr}\left(-\frac{C}{r}\right)$$

$$\Rightarrow \quad |F| = +\frac{C}{r^2}$$

$$\Rightarrow \quad \frac{mv^2}{r} = +\frac{C}{r^2} \Rightarrow v^2 \propto \frac{1}{r}$$

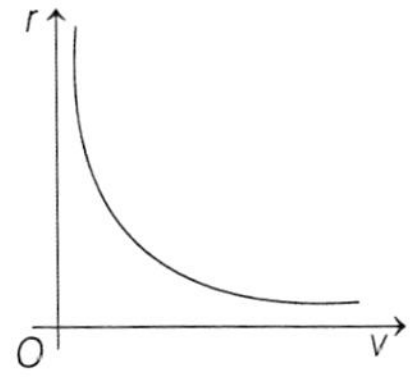

The graph between velocity and radius is hyperbolic.

37. *(c)* $\therefore$ Force $= -\dfrac{dU}{dr}$

$$\Rightarrow \quad F = -\frac{d}{dr}\left(\frac{-k}{2r^2}\right) = -\frac{k}{r^3}$$

As particle is on circular path, this force must be centripetal force.

$$\Rightarrow \quad |F| = \frac{mv^2}{r}$$

So, $\quad \dfrac{k}{r^3} = \dfrac{mv^2}{r}$

$$\Rightarrow \quad \frac{1}{2}mv^2 = \frac{k}{2r^2}$$

$\therefore$ Total energy of particle = KE + PE

$$= \frac{k}{2r^2} - \frac{k}{2r^2} = 0$$

Total energy $= 0$

38. *(d)* Given, potential energy burnt by lifting weight

$$= mgh = 10 \times 9.8 \times 1 \times 1000$$
$$= 9.8 \times 10^4$$

If mass lost by a person be m, then

energy dissipated $= m \times \dfrac{2}{10} \times 3.8 \times 10^7$ J

$$\Rightarrow 9.8 \times 10^4 = m \times \frac{1}{5} \times 3.8 \times 10^7$$

$$\Rightarrow \quad m = \frac{5}{3.8} \times 10^{-3} \times 9.8$$
$$= 12.89 \times 10^{-3} \text{ kg}$$

39. *(c)* Energy lost over path PQ

$$= \mu\, mg \cos\theta \times 4$$

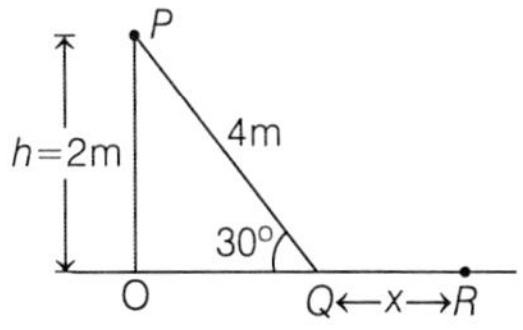

Energy lost over path $QR = \mu mgx$

i.e. $\mu\, mg \cos 30° \times 4 = \mu mgx$ $\quad (\because \theta = 30°)$

$$x = 2\sqrt{3} = 3.45 \text{ m}$$

From Q to R energy loss is half of the total energy loss.

i.e. $\quad \mu mgx = \dfrac{1}{2} \times mgh \Rightarrow \mu = 0.29$

The values of the coefficient of friction μ and the distance $x (= QR)$ are 0.29 and 3.5.

40. *(c)*

> **key idea** As, we know that change in potential energy of a system corresponding to a conservative internal force as
>
> $$U_f - U_i = -W = -\int_i^f F.dr$$

Given, $F = ax + bx^2$

According to work-energy theorem, we know that work done in stretching the rubber band by L is

$dW = Fdx$ (W_R = Work done by restoring force)

$$-W_R = \int_0^L (ax + bx^2)\, dx$$

$$= \left[\frac{ax^2}{2}\right]_0^L + \left[\frac{bx^3}{3}\right]_0^L$$

$$= \left[\frac{aL^2}{2} - \frac{a \times (0)^2}{2}\right] + \left[\frac{b \times L^3}{3} - \frac{b \times (0)^3}{3}\right]$$

$$\therefore -W_R = \frac{aL^2}{2} + \frac{bL^3}{3}$$

$$\Rightarrow W_{ext} = -W_R = \frac{aL^2}{2} + \frac{bL^3}{3}$$

41. *(a)* As no relation between k_1 and k_2 is given in the question, that is why, nothing can be predicted about Statement I. But as in Statement II, $k_1 < k_2$.

Then, for same force,

$$W = F \cdot x = F \cdot \frac{F}{k} = \frac{F^2}{k}$$

$$\Rightarrow \quad W \propto \frac{1}{k}$$

Since, $W_1 > W_2 \Rightarrow k_1 < k_2$

But for same displacement,

$$W = F \cdot x = \frac{1}{2}kx \cdot x = \frac{1}{2}kx^2$$

$$\Rightarrow \quad W \propto k \text{ i.e., } W_1 < W_2$$

42. *(c)* Given, potential energy function,

$$U(x) = \frac{a}{x^{12}} - \frac{b}{x^6}$$

$$U(x = \infty) = 0$$

As, $\quad F = -\dfrac{dU}{dx} = \left[-\dfrac{12a}{x^{13}} + \dfrac{6b}{x^7}\right]$

At equilibrium,

$$F = 0$$

$\therefore \quad x^6 = 2a/b$

$$\therefore U_{\text{at equilibrium}} = \frac{a}{\left(\frac{2a}{b}\right)^2} - \frac{b}{\left(\frac{2a}{b}\right)} = \frac{-b^2}{4a}$$

$$\therefore D = [U(x = \infty) - U_{\text{at equilibrium}}] = \frac{b^2}{4a}$$

43. *(a)* Loss in kinetic energy = Work done against friction + Potential energy of spring

$$\frac{1}{2}mv^2 = Fx + \frac{1}{2}kx^2$$

$$\Rightarrow \frac{1}{2} \times 2(4)^2 = 15x + \frac{1}{2} \times 10000\, x^2$$

$$\Rightarrow \quad 5000\,x^2 + 15x - 16 = 0$$

$$\therefore \quad x = 0.055 \text{ m} = 5.5 \text{ cm}$$

44. *(a)* Given, $V(x) = \left(\frac{x^4}{4} - \frac{x^2}{2}\right)$

For minimum value of V, $\frac{dV}{dx} = 0$

$$\Rightarrow \quad \frac{4x^3}{4} - \frac{2x}{2} = 0 \Rightarrow x = 0, x = \pm 1$$

So, $V_{\min}\,(x = \pm 1) = \frac{1}{4} - \frac{1}{2} = \frac{-1}{4}$ J

Now, $K_{\max} + V_{\min}$ = Total mechanical energy

$$\Rightarrow \quad K_{\max} = \left(\frac{1}{4}\right) + 2$$

or $\quad K_{\max} = \frac{9}{4}$

or $\quad \frac{mv^2}{2} = \frac{9}{4}$ or $v = \frac{3}{\sqrt{2}}$ ms^{-1}

45. *(a)* Momentum would be maximum, when KE would be maximum and this is the case when total elastic PE is converted into KE.

According to conservation of energy,

$$\frac{1}{2}kL^2 = \frac{1}{2}Mv^2 \Rightarrow kL^2 = \frac{(Mv)^2}{M}$$

or $\quad MkL^2 = p^2 \qquad [\because p = Mv]$

$$\Rightarrow \quad p = L\sqrt{Mk}$$

46. *(b)* Mass per unit length $= \frac{M}{L}$

$$= \frac{4}{2} = 2 \text{ kg/m}$$

The mass of 0.6 m of chain

$$= 0.6 \times 2 = 1.2 \text{ kg}$$

The height of centre of mass of hanging part

$$h = \frac{0.6 + 0}{2} = 0.3 \text{ m}$$

Hence, work done in pulling the chain on the table = Work done against gravity force

i.e, $W = mgh = 1.2 \times 10 \times 0.3 = 3.6$ J

47. *(b)* Work done to stretch the spring by 5 cm from mean position

$$= W_1 = \frac{1}{2}kx_1^2$$

$$= \frac{1}{2} \times 5 \times 10^3 \times (5 \times 10^{-2})^2$$

$$= 6.25 \text{ J}$$

Work done to stretch the spring by 10 cm from mean position,

$$W_2 = \frac{1}{2}k(x_1 + x_2)^2$$

$$= \frac{1}{2} \times 5 \times 10^3\,(5 \times 10^{-2} + 5 \times 10^{-2})^2$$

$$= 25 \text{ J}$$

Net work done to stretch the spring from 5 cm to 10 cm $= W_2 - W_1$

$$= 25 - 6.25 = 18.75 \text{ J} = 18.75 \text{ N-m}$$

48. *(c)* At the highest point of its flight, vertical component of velocity is zero and only horizontal component is left which is $u_x = u\cos\theta$

Given, $\theta = 45°$

$$\therefore \quad u_x = u\cos 45° = \frac{u}{\sqrt{2}}$$

Hence, at the highest point, kinetic energy is

$$E' = \frac{1}{2}mu_x^2$$

$$= \frac{1}{2}m\left(\frac{u}{\sqrt{2}}\right)^2 = \frac{1}{2}m\left(\frac{u^2}{2}\right)$$

$$= \frac{E}{2} \qquad \left[\because \frac{1}{2}mu^2 = E\right]$$

49. *(c)*

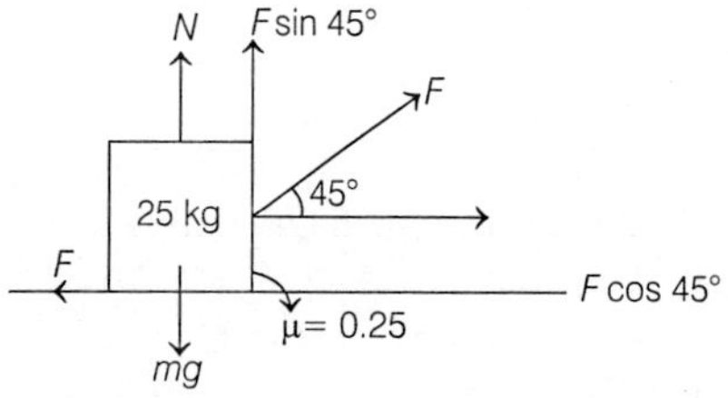

We have,

$$N + \frac{F}{\sqrt{2}} = mg$$

$$N = mg - \frac{F}{\sqrt{2}}$$

Block travel with uniform velocity so, $a = 0$

$F\cos 45°$ = friction (μN)

$$\Rightarrow \quad \frac{F}{\sqrt{2}} = \mu\left(mg - \frac{F}{\sqrt{2}}\right)$$

$$\Rightarrow \quad \frac{F}{\sqrt{2}} = 0.25\left(25 \times 9.8 - \frac{F}{\sqrt{2}}\right)$$

$$\Rightarrow \quad 1.25\frac{F}{\sqrt{2}} = 61.25$$

$$\Rightarrow \quad F = \frac{61.25\sqrt{2}}{1.25} = 49\sqrt{2}$$

$$W_{\text{ext}} = Fs\cos 45° = 49\sqrt{2} \times 5 \times \frac{1}{\sqrt{2}}$$

$$= 245 \text{ J}$$

50. *(d)* Initially,

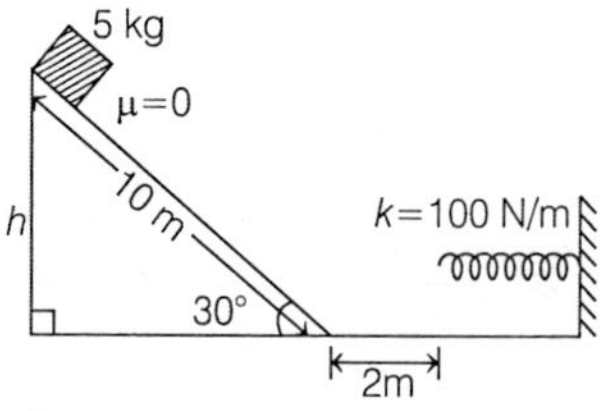

Finally,

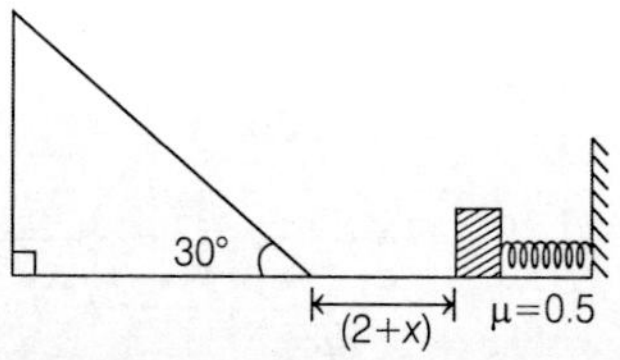

Initial PE of block is used in doing work against friction and is stored in form of PE of spring.

So, by energy conservation,

$$mgh = \mu mg\,s + \frac{1}{2}kx^2$$

$$\Rightarrow \quad 5 \times 10 \times 5 = 0.5 \times 5 \times 10 \times 2 + \frac{1}{2} \times 100 \times x^2$$

$$\Rightarrow \quad 250 = 50 + 50x^2 \Rightarrow 50x^2 = 200$$

$$\Rightarrow \quad x^2 = 4 \Rightarrow x = 2 \text{ m}$$

(Note friction is not present after 2 m.)

51. *(c)*

> **key idea** To solve such question, first of all find the relation between kinetic energy and momentum and after that you can easily get the required result.

Given,

Kinetic energies of A, B and C are equal

$$\Rightarrow \quad K_A = K_B = K_C$$

$$\Rightarrow \quad \frac{1}{2}m_A v_A^2 = \frac{1}{2}m_B v_B^2 = \frac{1}{2}m_C v_C^2$$

$$\Rightarrow \quad \frac{m_A^2 v_A^2}{m_A} = \frac{m_B^2 v_B^2}{m_B} = \frac{m_C^2 v_C^2}{m_C}$$

$$\Rightarrow \quad \frac{p_A^2}{m_A} = \frac{p_B^2}{m_B} = \frac{p_C^2}{m_C}$$

$$\Rightarrow \quad \frac{p_A}{\sqrt{m_A}} = \frac{p_B}{\sqrt{m_B}} = \frac{p_C}{\sqrt{m_C}}$$

$$\Rightarrow p_A : p_B : p_C = \sqrt{m_A} : \sqrt{m_B} : \sqrt{m_C}$$

$$= \sqrt{400} : \sqrt{1200} : \sqrt{1600}$$

$$= 2 : 2\sqrt{3} : 4$$

$$= 1 : \sqrt{3} : 2$$

52. *(d)* By momentum conservation,

$$M \times 0 = m_A v_A + m_B v_B$$

or $$m_A v_A = -m_B v_B$$

or $$\frac{v_A}{v_B} = -\left(\frac{m_B}{m_A}\right)$$

Ratio of kinetic energies is,

$$\frac{K_A}{K_B} = \frac{\frac{1}{2} m_A v_A^2}{\frac{1}{2} m_B v_B^2}$$

$$= \frac{m_A}{m_B} \times \left(\frac{v_A}{v_B}\right)^2$$

$$= \frac{m_A}{m_B} \times \left(-\frac{m_B}{m_A}\right)^2$$

$$= \frac{m_B}{m_A}$$

$$\because \quad \frac{K_B}{K_A} = \frac{m_A}{m_B} = \frac{v_B}{v_A}$$

53. *(c)*

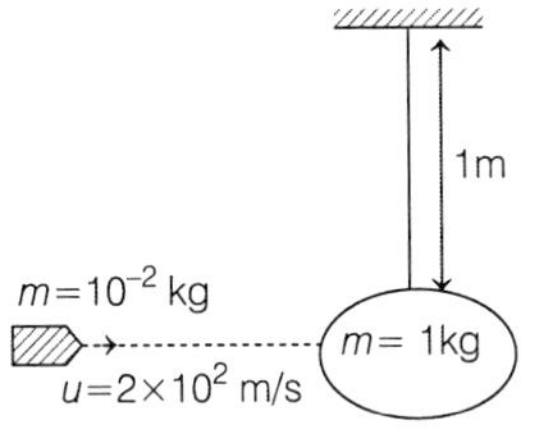

Let pendulum and bullet embedded moves with speed v after collision, then by conservation of linear momentum,

$$mu = (m + M)\, v,$$

or $$v = \frac{mu}{m + M}$$

$$\Rightarrow \quad v = \frac{10^{-2} \times 2 \times 10^2}{(1 + 10^{-2})}$$

$$= \frac{2}{1.01} \text{ m/s}$$

$$\approx 2 \text{ m/s}$$

The KE of (bullet + bob) is converted into PE

So, $$\frac{1}{2}(m + M)\, v^2 = (m + M)\, gh$$

or $$\frac{v^2}{2g} = h$$

$$\Rightarrow \quad h = \frac{2 \times 2}{2 \times 10} = 0.2 \text{ m}$$

54. *(a)*

> **key idea** To solve this question, you can follow these steps :
>
> (1) Find the initial velocity and final velocity using kinetic energy formula
>
> i.e. $KE = \frac{1}{2} mv^2$
>
> (2) Find the retardation produced in bullet using first equation of motion i.e. $v_f = v_i + at$
>
> $\Rightarrow \quad a = \frac{v_f - v_i}{t}$
>
> (3) Now, using third equation of motion and get distance travelled by the bullet before coming to rest
>
> i.e. $v_f^2 - v_i^2 = 2as \Rightarrow s = \frac{v_f^2 - v_i^2}{2a}$

Given, mass of bullet, $m = 200 \text{ g} = 0.2 \text{ kg}$

Initial kinetic energy $KE_i = 90 \text{ J}$

As, we know, $KE = \frac{1}{2} mv_i^2$

$$\Rightarrow \quad 90 = \frac{1}{2} \times 0.2 \times v_i^2$$

$$\Rightarrow \quad v_i^2 = 900$$

or $$v_i = 30 \text{ ms}^{-1}$$

Since, KE reduces to 40 J in 1s,

$$\Rightarrow \quad \frac{1}{2} mv_f^2 = 40$$

$$\Rightarrow \quad v_f^2 = \frac{40 \times 2}{0.2} = 400$$

or $$v_f = 20 \text{ ms}^{-1}$$

Retardation produced in bullet

$$v_f = v_i + at$$

$$\Rightarrow \quad 20 = 30 + a(1) \Rightarrow a = -10 \text{ ms}^{-2}$$

If l be the length of pool travelled by the bullet before coming to rest. Then, using third equation of motion.

$$v^2 - u^2 = 2as \text{ or } v_f^2 - v_i^2 = 2al$$

$$\Rightarrow 0 - (30)^2 = 2 \times (-10) l \qquad [\because v_f = 0]$$

$$\Rightarrow \quad l = \frac{-900}{-20} = 45 \text{ m}$$

55. *(c)* Let a block of mass M splits into two masses m_1 and m_2 in ratio 1 : 2. Then, mass of smaller part,

$$m_1 = \frac{M}{1 + 2} = \frac{M}{3}$$

Mass of bigger part,

$$m_2 = 2M/3$$

Given, speed of mass M,

$$v = 40 \text{ m/s}$$

Speed of mass m_1,

$$v_1 = 60 \text{ m/s}$$

Let speed of mass m_2,

$$v_2 = v$$

Using conservation of linear momentum,

$$\mathbf{p}_i = \mathbf{p}_f$$

$$\Rightarrow \quad M\mathbf{v} = m_1 \mathbf{v}_1 + m_2 \mathbf{v}_2$$

$$M \times 40 = \frac{M}{3} \times 60 + \frac{2M}{3} \times v$$

$$\Rightarrow \quad v = 30 \text{ m/s}$$

$\therefore$ The fractional change in kinetic energy,

$$\frac{\Delta K}{K} = \frac{K_f - K_i}{K_i} = \frac{K_f}{K_i} - 1$$

$$= \frac{\frac{1}{2} m_1 v_1^2 + \frac{1}{2} m_2 v_2^2}{\frac{1}{2} Mv^2} - 1$$

$$= \frac{\frac{1}{2}\left[\frac{M}{3} \times (60)^2 + \frac{2M}{3} \times (30)^2\right]}{\frac{1}{2} M \times (40)^2} - 1$$

$$= \frac{9}{8} - 1 = \frac{1}{8}$$

56. *(b)* Here, the displacement of an object is given by

$$x = (3t^2 + 5) \text{ m}$$

Therefore, velocity

$$v = \frac{dx}{dt} = \frac{d(3t^2 + 5)}{dt}$$

or $$v = 6t \text{ m/s} \qquad \text{...(i)}$$

The work done in moving the object from $t = 0$ to $t = 5$ s

$$W = \int_{x_0}^{x_5} F \cdot dx \qquad \text{...(ii)}$$

The force acting on this object is given by

$$F = ma = m \times \frac{dv}{dt}$$

$$= m \times \frac{d(6t)}{dt} \qquad \text{[using Eq. (i)]}$$

$$F = m \times 6 = 6 \text{ m} = 12 \text{ N}$$

Also, $x_0 = 3t^2 + 5 = 3\times(0)^2 + 5 = 5$ m

and at $t = 5$ s,

$x_5 = 3\times(5)^2 + 5 = 80$ m

Put the values in Eq. (ii), we get

$$W = 12 \times \int_{x_0}^{x_5} dx = 12\,[80 - 5]$$

$$W = 12 \times 75 = 900 \text{ J}$$

Alternate Solution

Using work – kinetic energy theorem,

$$W = \Delta KE = \frac{1}{2}m(v_f^2 - v_i^2)$$

$$= \frac{1}{2}m\times(30^2 - 0^2) = \frac{1}{2}\times 2\times 900 = 900 \text{ J}$$

57. *(b)* Given, force, $F = -kv^2$

$\therefore$ Acceleration, $a = \frac{-k}{m}v^2$

or $\frac{dv}{dt} = \frac{-k}{m}v^2 \Rightarrow \frac{dv}{v^2} = -\frac{k}{m}.\,dt$

Now, with limits, we get

$$\int_{10}^{v}\frac{dv}{v^2} = -\frac{k}{m}\int_0^t dt$$

$$\Rightarrow \left(-\frac{1}{v}\right)_{10}^{v} = -\frac{k}{m}t \Rightarrow \frac{1}{v} = 0.1 + \frac{kt}{m}$$

$$\Rightarrow v = \frac{1}{0.1 + \frac{kt}{m}} = \frac{1}{0.1 + 1000k}$$

$$\Rightarrow \frac{1}{2}\times m\times v^2 = \frac{1}{8}mv_0^2$$

$$\Rightarrow v = \frac{v_0}{2} = 5 \Rightarrow \frac{1}{0.1 + 1000\,k} = 5$$

$$\Rightarrow 1 = 0.5 + 5000\,k \Rightarrow k = \frac{0.5}{5000}$$

$$\Rightarrow k = 10^{-4} \text{ kg/m}$$

58. *(152)* $F = x^2 y\hat{i} + y^2\hat{j}$

$x + y = 10,\ y = 10 - x$

$$\int dW = \int F_x dx + \int F_y dy$$

$$W = \int_0^4 x^2(10 - x)dx + \int_0^2 y^2 dy$$

$$= \left[\frac{10x^3}{3} - \frac{x^4}{4}\right]_0^4 + \left[\frac{y^3}{3}\right]_0^2$$

$$= \frac{640}{3} - 64 + \frac{8}{3} = 152$$

59. *(58)* Given, force, $F = 3x^2 + 2x - 5$ N

Work done, $W = \int_2^4 F\cdot dx$

$$= \int_2^4 (3x^2 + 2x - 5)\cdot dx$$

$$= \left[\frac{3x^3}{3} + \frac{2x^2}{2} - 5x\right]_2^4$$

$$= [x^3 + x^2 - 5x]_2^4$$

$$= [4^3 + 4^2 - 5\times 4 - 2^3 - 2^2 + 5\times 2]$$

$$= [64 + 16 - 20 - 8 - 4 + 10]$$

$$= 58 \text{ J}$$

60. *(30)* Work done by variable force,

$$W = \int F dx = \int_2^4 5x\,dx = \left[5\cdot\frac{x^2}{2}\right]_2^4$$

$$= \frac{5}{2}[4^2 - 2^2]$$

$$= \frac{5}{2}[12] = 30 \text{ J}$$

61. *(3)* Given, $E_1 = \frac{1}{2}mu^2 - 0$

$$E_1 = \frac{1}{2}mu^2 = E$$

$$E_2 = \frac{1}{2}m(2u)^2 - \frac{1}{2}mu^2$$

$$= 2mu^2 - \frac{1}{2}mu^2$$

$$E_2 = \frac{3}{2}mu^2 = 3\left(\frac{1}{2}mu^2\right) = 3E \text{ J}$$

62. *(784)* Speed of bus, $v = 80$ km/h

mass of bus, $m = 500$ kg

distance, $x = 4$ km $= 4\times 10^3$ m

For a constant speed,

Work done by engine + Work done by friction = 0

Thus,

$$WD_{\text{engine}} = -WD_{\text{friction}} = -(-\mu\, mg\, x)$$

$$= 0{\cdot}04\times 500\times 9.8\times 4\times 10^3 = 784 \text{ kJ}$$

63. *(32)* Given,

$$F = (2 + 3x)\hat{\mathbf{i}}$$

$$W = \int_0^4 F\cdot dx = \int_0^4 (2 + 3x)\,dx$$

$$W = \left[2x + \frac{3x^2}{2}\right]_0^4$$

$$= [8 + 24]$$

$$W = 32 \text{ J}$$

64. *(30)* Acceleration of body is;

$$a = \frac{F}{m} = \frac{2}{5} \text{ m/s}^2$$

Initial velocity of body $= \frac{p}{m} = \frac{10}{5} = 2$ m/s

Final velocity,

$$v = u + at \Rightarrow v = 2 + \frac{2}{5}\times 5 = 4 \text{ m/s}$$

So, change in kinetic energy is,

$$\Delta K = K_f - K_i$$

$$= \frac{1}{2}m(v_f^2 - v_i^2) = \frac{1}{2}\times(4^2 - 2^2)\times 5$$

$$= \frac{1}{2}\times 5\times 12 = 30 \text{ J}$$

65. *(125)* Percentage increase in kinetic energy $= \frac{K_f - K_i}{K_i}\times 100$

$$= \left(\frac{p_f^2}{2m} - \frac{p_i^2}{2m}\right)\Big/\frac{p_i^2}{2m}\times 100$$

$$= \frac{\frac{9}{4}\left(\frac{p_i^2}{2m}\right) - \frac{p_i^2}{2m}}{\left(\frac{p_i^2}{2m}\right)}\times 100$$

[Since, $p_f = p_i + 50\%$ of $p_i = 3/2\,p_i$]

$$= \left(\frac{9}{4} - 1\right)\times 100 = \frac{5}{4}\times 100 = 125\%$$

66. *(2)* Given, $a = -2x$

$$v\frac{dv}{dx} = -2x$$

$$\Rightarrow vdx = -2xdx$$

$$\int_{v_1}^{v_2} vdv = -2\int_0^x xdx$$

$$\Rightarrow \frac{v_2^2}{2} - \frac{v_1^2}{2} = -\frac{2x^2}{2}$$

$$\Rightarrow \frac{mv_2^2}{2} - \frac{mv_1^2}{2} = -mx^2$$

$$\Rightarrow \frac{mv_1^2}{2} - \frac{mv_2^2}{2} = mx^2$$

$\because$ Loss in KE $= mx^2 = \frac{10}{1000}x^2$

$$= 10^{-2}x^2$$

$$= \left(\frac{10}{x}\right)^{-2}$$

$$n = 2$$

67. *(132)* Given that, $\mathbf{F} = (5 + 3y^2)\hat{\mathbf{j}}$ N

Displacement is also along Y-axis.

So, work done, $W = \int \mathbf{F}\cdot dy$

$$W = \int (5 + 3y^2)\hat{\mathbf{j}}\cdot dy\hat{\mathbf{j}} = \int_2^5 (5 + 3y^2)\,dy$$

$$= \left[5y + \frac{3y^3}{3}\right]_2^5 = [5y + y^3]_2^5$$

$$= [5 \times 5 + (5)^3] - [5 \times 2 + (2)^3]$$
$$= [25 + 125] - [10 + 8] = 150 - 18$$
$$\Rightarrow W = 132 \text{ J}$$

68. *(40)* Force, $\mathbf{F} = (5\hat{i} + 2\hat{j} + 7\hat{k})$ N

Displacement vector,

$$\Delta \mathbf{r} = \mathbf{r}_f - \mathbf{r}_i$$
$$= (5\hat{i} - 2\hat{j} + \hat{k}) - (2\hat{i} + 3\hat{j} - 4\hat{k})$$
$$\Delta \mathbf{r} = (3\hat{i} - 5\hat{j} + 5\hat{k}) \text{ m}$$

Work done, $W = \mathbf{F} \cdot \Delta \mathbf{r}$

$$= (5\hat{i} + 2\hat{j} + 7\hat{k}) \cdot (3\hat{i} - 5\hat{j} + 5\hat{k})$$
$$= (5)(3) + (2)(-5) + (7)(5)$$
$$= 15 - 10 + 35 = 40 \text{ J}$$

> ***Alternate Solution***
>
> $$\mathbf{F} = 5\hat{i} + 2\hat{j} + 7\hat{k}$$
> $$F = |\mathbf{F}| = \sqrt{5^2 + 2^2 + 7^2} = \sqrt{78}$$
> $$\Delta \mathbf{r} = r_f - r_i = 3\hat{i} - 5\hat{j} + 5\hat{k}$$
> $$\Delta \mathbf{r} = \sqrt{3^2 + (-5)^2 + 5^2} = \sqrt{59}$$
>
> Angle between **F** and Δ **r** is given as,
>
> $$\cos\theta = \frac{\mathbf{F} \cdot \Delta \mathbf{r}}{|F||\Delta r|} = \frac{5 \times 3 + 2(-5) + 7 \times 5}{\sqrt{78}\sqrt{59}}$$
> $$= \frac{40}{\sqrt{78}\sqrt{59}}$$
>
> $\therefore$ Work done, $W = |\mathbf{F}| \times |\Delta \mathbf{r}| \cos\theta$
>
> $$= \sqrt{78} \times \sqrt{59} \times \frac{40}{\sqrt{78} \times \sqrt{59}} = 40 \text{ J}$$

69. *(7)* From third equation of motion,

$$v^2 = u^2 + 2as = 2^2 + 2 \times 2 \times 6$$
$$\Rightarrow v^2 = 28 \text{ m}^2/\text{s}^2$$

Kinetic energy of lift, $K = \frac{1}{2}mv^2$

$$K = \frac{1}{2} \times 500 \times 28$$
$$K = 7 \times 10^3 \text{ J} \Rightarrow K = 7 \text{ kJ}$$

70. *(300)* Distance covered by mass in 8s.

Using, $h = ut + \frac{1}{2}at^2$

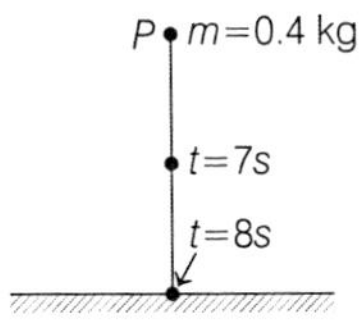

$$h_1 = 0 \times 8 + \frac{1}{2}(10) \times 8^2 = 320 \text{ m}$$

Distance covered by mass in 7s is

$$h_2 = \frac{1}{2} \times 10 \times 7^2 = 245 \text{ m}$$

Distance covered by mass in last 1s

$$= 320 - 245 = 75 \text{ m}$$

So, loss of PE in last second

$$= mgh = 0.4 \times 10 \times 75 = 300 \text{ J}$$

> ***Alternate Solution***
>
> Distance covered by the body in last second (8th second)
>
> $$h_8 = 0 + \frac{1}{2}g(2t - 1)$$
> $$= \frac{1}{2} \times 10(2 \times 8 - 1) = 75 \text{ m}$$
>
> $\therefore$ Loss of PE in last second $= mgh_8$
>
> $$= 0.4 \times 10 \times 75 = 300 \text{ J}$$

71. *(40)* Given, KE = 10000 J

As, $\quad \text{KE} = \frac{1}{2}mv^2$

$$10000 = \frac{1}{2} \times 2 \times v^2 \Rightarrow v = 100 \text{ m/s}$$

Now using, $v = u + at$

we have, $\quad v = at$ [Given, $u = 0$]

$$v = a \times 5$$
$$a = \frac{100}{5} = 20 \text{ ms}^{-2}$$

Force on the body,

$$F = ma = 2 \times 20 = 40 \text{ N}$$

72. *(2)* Given that, $F = 2$ N

If a be the acceleration produced in object, then

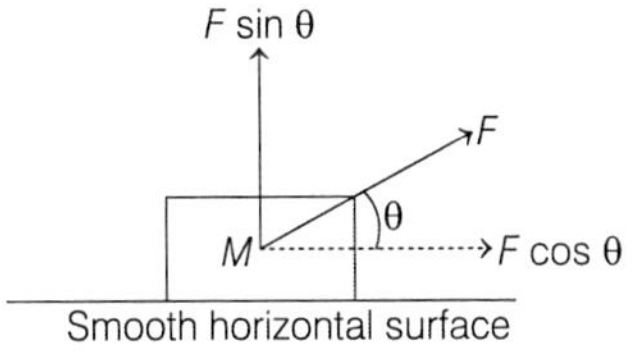

$$F\cos\theta = ma$$
$$\Rightarrow F\cos kx = ma \quad [\because \theta = kx]$$
$$\Rightarrow F\cos kx = m\frac{dv}{dt} \quad \left[\because a = \frac{dv}{dt}\right]$$
$$\Rightarrow 2\cos kx = m\frac{dv}{dt} \quad [\because F = 2\text{N}]$$
$$\Rightarrow 2\cos kx = m \cdot \frac{dv}{dx} \cdot \frac{dx}{dt}$$
$$\Rightarrow 2\cos kx = mv\frac{dv}{dx}$$
$$\Rightarrow 2\cos kx \, dx = mv \, dv$$

On integrating both sides, we get

$$\int 2\cos kx \, dx = \int mv \, dv$$
$$\Rightarrow \frac{2\sin kx}{k} = \frac{mv^2}{2}$$
$$\Rightarrow \frac{2\sin\theta}{k} = \text{KE} \quad [\because \theta = kx]$$
$$\Rightarrow \text{KE} = \frac{2\sin\theta}{k} = \frac{x}{k}\sin\theta \quad \text{[Given]}$$
$$\therefore \quad x = 2$$

73. *(24)* Given, compression in the spring,

$$x = 25 \text{ cm} = 25 \times 10^{-2} \text{ m}$$
$$v_1 = v \text{ and } v_2 = v/2$$

According to given situation.

Change in kinetic energy = Gain in potential energy

$$\Rightarrow \frac{1}{2}mv^2 - \frac{1}{2}m\left(\frac{v}{2}\right)^2 = \frac{1}{2}kx^2$$
$$\Rightarrow \frac{1}{2}mv^2 - \frac{1}{4} \times \frac{1}{2}mv^2$$
$$= \frac{1}{2}k(25 \times 10^{-2})^2$$
$$\Rightarrow E_K - \frac{1}{4}E_K = \frac{625\,k \times 10^{-4}}{2}$$
$$\Rightarrow \frac{3}{4}E_K = \frac{1}{2} \times \frac{k}{16}$$
$$\Rightarrow k = 24E_K = 24E \quad [\because E_K = E]$$
$$= nE \quad \text{(given)}$$
$$\therefore \quad n = 24$$

74. *(600)* Initial speed of block,

$$v_i = 12 \text{ m/s}$$

Mass of block, $m = 0.5$ kg

$$x = 30 \text{ cm} = 0.3 \text{ m}$$

Final speed of block,

$$v_f = \frac{v_i}{2} = \frac{12}{2} = 6 \text{ m/s}$$

According to conservation of energy,

$$U_i + K_i = U_f + K_f$$
$$\Rightarrow 0 + \frac{1}{2}mv_i^2 = \frac{1}{2}kx^2 + \frac{1}{2}mv_f^2$$
$$\Rightarrow \frac{1}{2} \times 0.5 \times 12^2 = \frac{1}{2} \times k \times (0.3)^2 + \frac{1}{2} \times 0.5 \times 6^2$$
$$\Rightarrow k = 600 \text{ N/m}$$

75. *(5)* As we know that, kinetic energy,

$$\text{KE} = 1/2\, mv^2$$

$\therefore$ Initial kinetic energy $\text{KE}_i = \frac{1}{2}mu^2$

$$= \frac{1}{2}m(40)^2 = 800\,m \quad \ldots\text{(i)}$$

By using law of conservation of momentum,

$$mu = m_1v_1 + m_2v_2$$
$$\therefore \quad v_2 = \frac{m \times 40 - m/2 \times 60}{m/2}$$
$$= \frac{10m}{m/2} = 20 \text{ ms}^{-1}$$

So, final kinetic energy

$$\text{KE}_f = \frac{1}{2}m_1v_1^2 + \frac{1}{2}m_2v_2^2$$

$$=\frac{1}{2}\frac{m}{2}(v_1^2+v_2^2)=\frac{m}{4}(60^2+20^2)$$

$$=m/4\,(3600+400)=1000\,m \quad \ldots\text{(ii)}$$

$\therefore$ On dividing Eq (ii) by Eq. (i), we get

$$\frac{\text{KE}_f}{\text{KE}_i}=\frac{1000\,m}{800m}=\frac{5}{4}=\frac{x}{4} \quad \text{[given]}$$

So, $x=5$

76. *(2)* Given, work done by both person is same, $W_A=W_B$

Direction of first force with displacement, $\theta_1=45°$

Direction of second force with displacement, $\theta_2=60°$

Ratio of force applied by person A to force applied by person B is

$F_A:F_B=1:\sqrt{x}$.

Work done by both person is same,

$W_A=W_B \Rightarrow F_A d\cos\theta_1=F_B d\cos\theta_2$

($\because$ distance is same for both $=d$)

$$\frac{F_A}{F_B}=\frac{\cos\theta_2}{\cos\theta_1}=\frac{\cos 60°}{\cos 45°}$$

$$\Rightarrow \frac{F_A}{F_B}=\frac{\frac{1}{2}}{\frac{1}{\sqrt{2}}}=\frac{1}{\sqrt{2}}$$

$$\Rightarrow \frac{F_A}{F_B}=\frac{1}{\sqrt{x}}=\frac{1}{\sqrt{2}}$$

$\Rightarrow x=2$

Thus, the value of x is 2.

77. *(40)* The given situation is shown below as

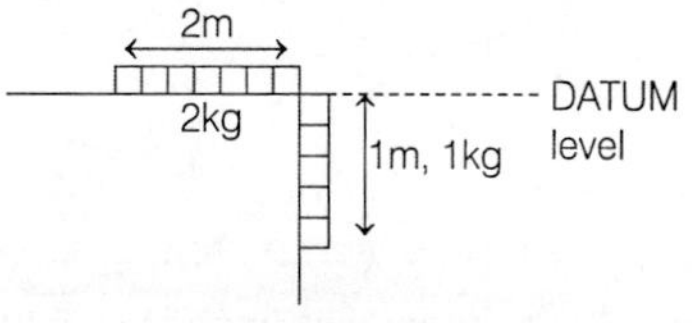

Initial condition

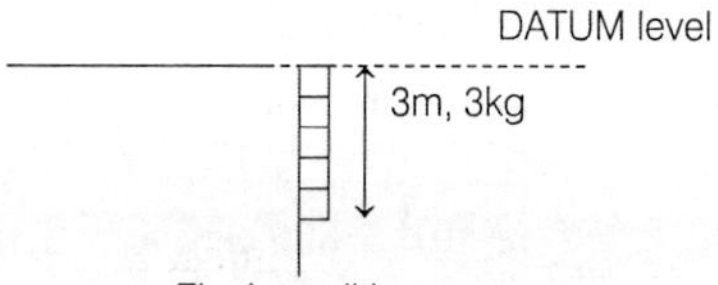

Final condition

On applying law of conservation of energy, we get

$$K_{\text{initial}}+U_{\text{initial}}=K_{\text{final}}+U_{\text{final}}$$

$$K_{\text{initial}}=0$$

$$\Rightarrow U_{\text{initial}}=-1\times 10\times\frac{1}{2}$$

(1 m lie below DATUM level, so negative)

$$K_{\text{final}}=?$$

$$K_{\text{final}}=-3\times 10\times\frac{3}{2}$$

(3 m lie below DATUM level, so negative)

$$0-1\times 10\times\frac{1}{2}=K_{\text{final}}-3\times 10\times\frac{3}{2}$$

$$\Rightarrow \quad -5=K_{\text{final}}-45$$

$$\Rightarrow \quad K_{\text{final}}=45-5=40\text{ J}$$

78. *(450)* Given, force, $\mathbf{F}=(5y+20)\hat{\mathbf{j}}$ N

Since, work done $W=\int_0^{10}F\cdot dy$

$$\therefore W=\int_0^{10}(5y+20)dy$$

$$=\frac{5y^2}{2}\Big|_0^{10}+20y\Big|_0^{10}$$

$$=\frac{5}{2}(10^2-0^2)+20(10-0)$$

$$=450\text{ J}$$

79. *(1)* Given, $k=100$ N/m

$m=100$ g $=0.1$ kg

$x=0.05$ m and $H=2$m

By conservation of energy

$$\frac{1}{2}kx^2=\frac{1}{2}mv^2$$

$$\Rightarrow v=x\sqrt{\frac{k}{m}}=0.05\times\sqrt{\frac{100}{0.1}}$$

$$=0.5\sqrt{10}\text{ ms}^{-1} \quad \ldots\text{(i)}$$

Time of flight of a ball, $t=\sqrt{\dfrac{2H}{g}}$

$$\Rightarrow t=\sqrt{\frac{2\times 2}{10}}=\frac{2}{\sqrt{10}}\text{ s} \quad \ldots\text{(ii)}$$

$\therefore$Range of a ball, $d=vt$

$$=0.5\sqrt{10}\times\left(\frac{2}{\sqrt{10}}\right)$$

[From Eqs. (i) and (ii)]

$$=0.5\times 2=1\text{ m}$$

80. *(10)* Given,

Mass of the particle, $m=10$ kg

Speed of the particle at point A,

$$v_A=0\text{ m/s}$$

The elevation of point A from point B,

$h_A=5+h_B$

Let's consider the speed of the particle at point $B=v_B$

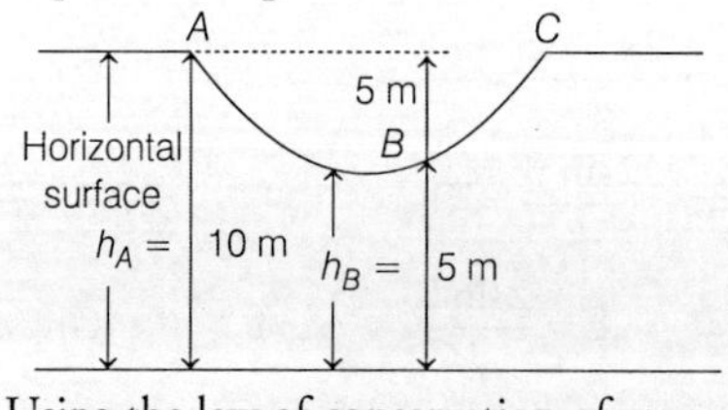

Using the law of conservation of energy.

Energy at point A = Energy at point B

$$\frac{1}{2}mv_A^2+mgh_A=\frac{1}{2}mv_B^2+mgh_B$$

On substituting the values in the above equations, we get

$$\frac{1}{2}(10)(0)^2+10\times 10\times(5+h_B)$$

$$=\frac{1}{2}(10)v_B^2+10\times 10\times h_B$$

$$v_B=10\text{ m/s}=x\text{ m/s}$$

$\therefore \quad x=10$

81. *(1)* Given, mass of particle A,

$$m_A=4\text{ g}$$

Mass of particle B, $m_B=16$ g

Kinetic energy of A and B is same

i.e. $\quad \text{KE}_A=\text{KE}_B$

As, kinetic energy (KE) $=p^2/2m$

$$\therefore \quad \frac{(p_A)^2}{m_A}=\frac{(p_B^2)}{m_B}$$

$$\Rightarrow \quad \frac{p_A^2}{4}=\frac{p_B^2}{16}$$

$$\Rightarrow \quad \frac{p_A}{p_B}=\frac{1}{2}=\frac{n}{2} \quad \text{[given]}$$

$\therefore \quad n=1$

82. *(2)* Given, $m_A=1$ kg, $m_B=2$ kg,

$(\text{KE})_A:(\text{KE})_B=A:1$

Linear momentum of A and B are equal.

$$\Rightarrow \quad p_A=p_B$$

$\because$ Kinetic energy (KE) $=p^2/2m$

$$\therefore \quad \frac{\text{KE}_A}{\text{KE}_B}=\frac{m_B}{m_A}=\frac{2}{1}=\frac{A}{1}$$

$$\Rightarrow \quad A=2$$

Topic 2

01. *(b)* Net work done is given as,

According to work-energy theorem,

$$W=K_f-K_i=\frac{1}{2}m(v_f^2-v_i^2)$$

$$W_{\text{net}}=\frac{1}{2}\times 0.5[(3(2)^2+4)^2-(3(0)^2+4)^2]$$

$$=60\text{ J}$$

02. *(c)* Given, $\mathbf{F}=4x\hat{\mathbf{i}}+3y^2\hat{\mathbf{j}}$

Particle moves from point (1, 2) to point (2, 3)

According to work-energy theorem,

Work done = Change in KE

i.e. $W=\Delta\text{KE}$

$$\Rightarrow \Delta\text{KE}=W=\int\int \mathbf{F}\cdot(dx\hat{\mathbf{i}}+dy\hat{\mathbf{j}})$$

$$=\int\int(4x\hat{\mathbf{i}}+3y^2\hat{\mathbf{j}})\cdot(dx\hat{\mathbf{i}}+dy\hat{\mathbf{j}})$$

$$= \int_1^2 4x\,dx + \int_2^3 3y^2\,dy$$
$$= [2x^2]_1^2 + [y^3]_2^3$$
$$= (8-2) + (27-8)$$
$$= 6 + 19 = 25\text{ J}$$

03. *(c)* When a force of constant magnitude acts on velocity of particle perpendicularly, then there is no change in the kinetic energy of particle. Hence, kinetic energy remains constant.

04. *(d)* Natural length of the spring, $l_0 = 2$ m.

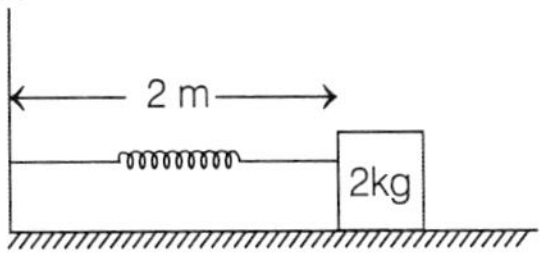

Initial compression, $l_i = 1$ m

Final compression in the spring,

$l_f = (l_0 - x)$ m

$= (2 - x)$ m

Using law of conservation of energy.

$$K_i + U_i = K_f + U_f$$
$$\Rightarrow 0 + \frac{1}{2}kl_i^2 = \frac{1}{2}mv^2 + \frac{1}{2}kl_f^2$$
$$\Rightarrow \frac{1}{2}mv^2 = \frac{1}{2}k(l_i^2 - l_f^2)$$
$$\Rightarrow \frac{1}{2} \times 2 \times v^2 = \frac{1}{2} \times 200\,[1^2 - (2-x)^2]$$
$$\Rightarrow v^2 = 100[1-(2-x)^2]$$
$$\Rightarrow v = 10[1-(2-x)^2]^{\frac{1}{2}}\text{ m/s}$$

05. *(d)* At point B,

KE = 3 PE

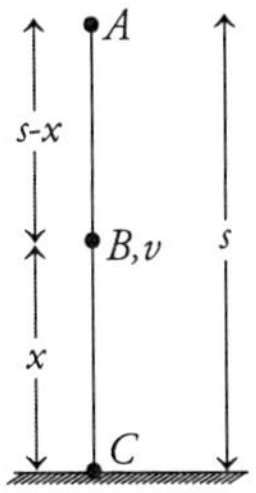

$$\Rightarrow \quad \frac{1}{2}mv^2 = 3mgx$$
$$v = \sqrt{6gx} \quad \ldots\text{(i)}$$

But $v^2 = 0^2 + 2g(s-x)$

$$\Rightarrow \quad v = \sqrt{2g(s-x)} \quad \ldots\text{(ii)}$$

Now from Eqs. (i) and (ii),

$$\sqrt{6gx} = \sqrt{2g(s-x)}$$
$$\Rightarrow \quad 6gx = 2gs - 2gx$$
$$\Rightarrow \quad 8gx = 2gs$$
$$\Rightarrow \quad x = \frac{s}{4}$$

$\therefore$ From Eq. (i),

$$v = \sqrt{6g \times \frac{s}{4}} = \sqrt{\frac{3gs}{2}}$$

06. *(b)* A stone has speed u at lowest position and magnitude of change in velocity when string is horizontal, is $\sqrt{x(u^2 - gL)}$.

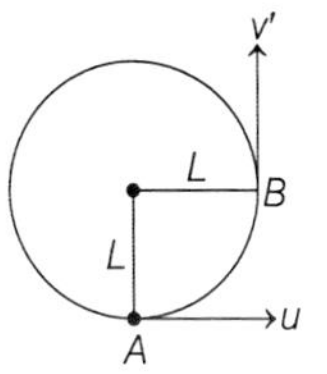

To find x,

When string is horizontal, lets assume speed of stone is v'.

By conservation of energy,

$$\text{TE}_B = \text{TE}_A$$
$$\text{PE}_B + \text{KE}_B = \text{PE}_A + \text{KE}_A$$
$$mgL + \frac{1}{2}mv'^2 = 0 + \frac{1}{2}mu^2$$
$$\frac{1}{2}m(v'^2 - u^2) = -mgL$$
$$v'^2 = u^2 - 2gL$$
$$\Rightarrow \quad v' = \sqrt{u^2 - 2gL}$$

Since, v' is along Y-axis

$$\therefore \quad \mathbf{v}' = (\sqrt{u^2 - 2gL})\hat{\mathbf{j}}$$

Initial velocity u is along X-axis.

So, $\mathbf{u}' = u\hat{\mathbf{i}}$

$\therefore$ Change in velocity vector,

$$|\Delta \mathbf{v}| = \sqrt{|\mathbf{v}'^2| + |\mathbf{u}^2|}$$
$$= \sqrt{(u^2 - 2gL) + u^2}$$
$$= \sqrt{2u^2 - 2gL}$$
$$|\Delta \mathbf{v}| = \sqrt{2(u^2 - gL)}$$

Equating this value with given value in question, we have

$$\sqrt{x(u^2 - gL)} = \sqrt{2(u^2 - gL)}$$

Thus, $x = 2$

07. *(a)* The speed of the body when reached the ground,

$$v_f = 0.8\sqrt{gh}$$

Initial velocity of the body, $v = 0$ m/s

Using the work-energy theorem,

Work done by gravity + Work done by air-friction = Final kinetic energy − Initial kinetic energy

$$W_{mg} + W_{\text{air-friction}} = \frac{1}{2}mv_f^2 - \frac{1}{2}mv_i^2$$

Here, work done by gravity $= mgh$

$$\Rightarrow mgh + W_{\text{air-friction}} = \frac{1}{2}m(0.8\sqrt{gh})^2 - \frac{1}{2}m(0)^2$$
$$\Rightarrow \quad W_{\text{air-friction}} = \frac{0.64\,mgh}{2} - mgh$$
$$= 0.32mgh - mgh$$
$$= -0.68\,mgh$$

08. *(c)*

> **key idea** Area under force-displacement graph gives the amount of work done.

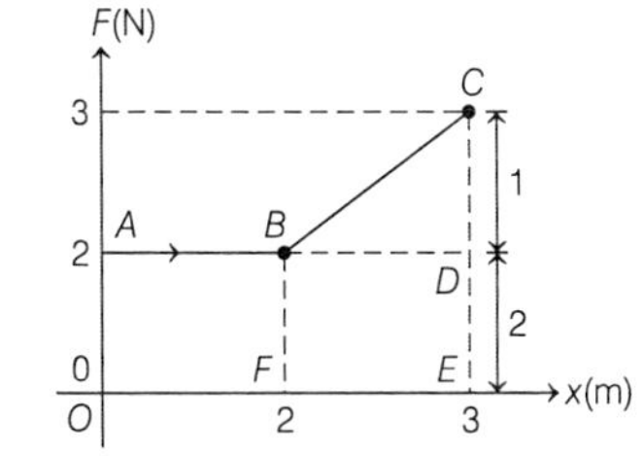

$\therefore$ Work done on the particle

= Area under the curve ABC

W = Area of square $ABFO$ + Area of ΔBCD + Area of rectangle $BDEF$

$$= 2 \times 2 + \frac{1}{2} \times 1 \times 1 + 2 \times 1$$
$$= 6.5\text{ J}$$

Now, from work-energy theorem,

$$\Delta W = K_f - K_i$$
$$\Rightarrow \quad K_f = \Delta W = 6.5\text{ J} \quad [\because K_i = 0]$$

09. *(b)* In a spring-block system, when a block is pulled with a constant force F, then its speed is maximum at the mean position. Also, it's acceleration will be zero.

In that case, force on the system is given as,

$$F = kx \quad \ldots\text{(i)}$$

Where, x is the extension produced in the spring.

or $\quad x = F/k$

Now we know that, for a system vibrating at its mean position, its maximum velocity is given as, $v_{max} = A\omega$

Where, A is the amplitude and ω is the angular velocity.
Since, the block is at its mean position.

So, $\quad A = x = \dfrac{F}{k}$

$$v_{\max} = \frac{F}{k}\sqrt{\frac{k}{m}} \qquad \left[\because \omega = \sqrt{\frac{k}{m}}\right]$$

$$= \frac{F}{\sqrt{km}}$$

Alternate Solution

Here, net work done = work done due to external force (W_{ext}) + work done due to the spring (W_{spr}).

As, $W_{ext} = F \cdot x$ and $W_{spr} = \dfrac{-1}{2}kx^2$

$\Rightarrow \Delta KE = F \cdot x + \left(-\dfrac{1}{2}kx^2\right)$

$\Rightarrow KE_f - KE_i = F \cdot x - \dfrac{1}{2}kx^2$

$\Rightarrow \dfrac{1}{2}mv_{\max}^2 - \dfrac{1}{2}m(0)^2 = F \cdot \left(\dfrac{F}{k}\right) - \dfrac{1}{2}k\left(\dfrac{F}{k}\right)^2$

[using Eq. (i)]

$\Rightarrow \dfrac{1}{2}mv_{\max}^2 = \dfrac{F^2}{k} - \dfrac{F^2}{2k} = \dfrac{F^2}{2k}$ or $v_{\max}^2 = \dfrac{F^2}{km}$

$\Rightarrow \quad v_{\max} = F/\sqrt{km}$

10. *(d)* From Newton's second law,

$$\frac{dp}{dt} = F$$

$\Rightarrow \quad dp = F dt$

$\therefore \quad p = \int dp = \int_0^1 F\,dt$

$\Rightarrow \quad p = \int_0^1 6t\,dt = 3\ \text{kg}\left(\dfrac{\text{m}}{\text{s}}\right)$

Also, change in kinetic energy

$$\Delta k = \frac{\Delta p^2}{2m} = \frac{3^2}{2 \times 1} = 4.5$$

From work-energy theorem,
work done = change in kinetic energy.
So, work done $= \Delta k = 4.5$ J

Alternate Solution

$F = 6t$

$\dfrac{mdv}{dt} = 6t$

$\int_0^t dv = \int_0^t 6t\,dt$

$v = 3t^2$

$ds = 3t^2 dt$

$W = \int \mathbf{F} \cdot d\mathbf{s}$

$W = \int_0^1 6t \times 3t^2 dt = \dfrac{18t^4}{4} = 4.5\ \text{J}$

11. *(d)*

key idea As particle rises in the vertical plane, it's kinetic energy is converted into potential energy. Also as no loss of energy is given, so total initial energy (at A) and total final energy at some point (B or C) remains same.

For slacking at Point D

$$v_A = \sqrt{5gl}$$
$$v_D = \sqrt{gl}$$

Applying conservation of energy at point A,

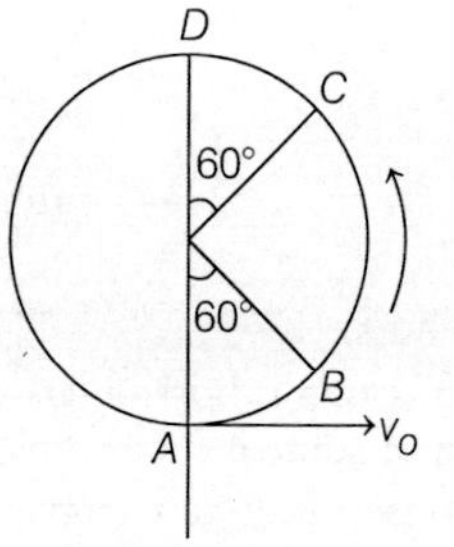

$$\frac{1}{2}mv_A^2 = \frac{1}{2}mv_B^2 + mgh$$

As string becomes slack at top,
At lowest point $v_A = \sqrt{5gl}$

$\therefore \quad \dfrac{1}{2}m(5gl) = \dfrac{1}{2}mv_B^2 + mg\dfrac{l}{2}$

$\Rightarrow \quad \dfrac{1}{2}mv_B^2 = \dfrac{5mgl}{2} - \dfrac{mgl}{2}$

$\Rightarrow \quad (\text{KE})_B = 2mgl \qquad \ldots(\text{i})$

Now at point C

$\dfrac{1}{2}mv_C^2 = \dfrac{1}{2}mv_D^2 + mg\dfrac{l}{2}$

$(\text{KE})_C = \dfrac{1}{2}mgl + mg\dfrac{l}{2} = mgl \qquad \ldots(\text{ii})$

$$\frac{(\text{KE})_B}{(\text{KE})_C} = \frac{2}{1}$$

12. *(b)*

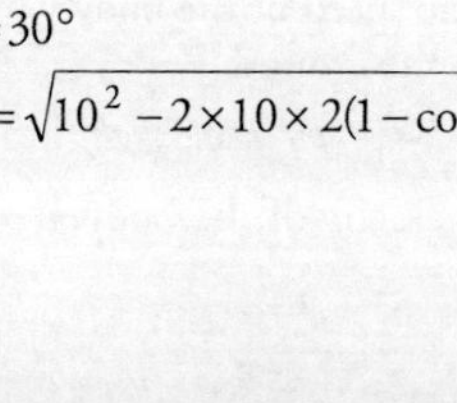

$$v = \sqrt{v_0^2 - 2gr(1 - \cos\theta)}$$

For $\theta = 30°$

$$v_B = \sqrt{10^2 - 2 \times 10 \times 2(1 - \cos 30°)}$$

$$v_B = \sqrt{100 - 40\left(1 - \frac{\sqrt{3}}{2}\right)}$$

For $\theta = 120°$

$$v_C = \sqrt{10^2 - 2 \times 10 \times 2(1 - \cos 120°)}$$
$$= \sqrt{100 - \frac{40 \times 3}{2}}$$

As per question

$$\frac{\text{KB}_B}{\text{KE}_C} = \frac{\frac{1}{2}mv_B^2}{\frac{1}{2}mv_C^2}$$

$$= \frac{100 - 40\left(1 - \frac{\sqrt{3}}{2}\right)}{40}$$

$$= \frac{60 + 20\sqrt{3}}{40} = \frac{3 + \sqrt{3}}{2}$$

13. *(a)*

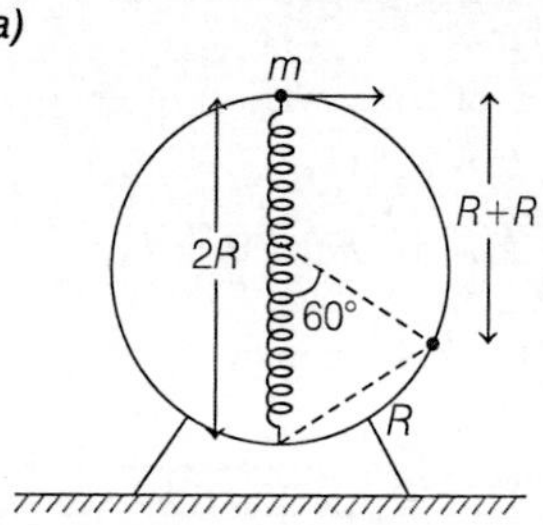

Using work - energy theorem

$$mg(R + R\cos 60°) + \frac{1}{2}k(R^2 - 0^2) = \frac{1}{2}mv^2$$

$$mg\frac{3R}{2} + \frac{kR^2}{2} = \frac{1}{2}mv^2$$

$$v = \sqrt{3gR + \frac{kR^2}{m}}$$

14. *(b)* Given,

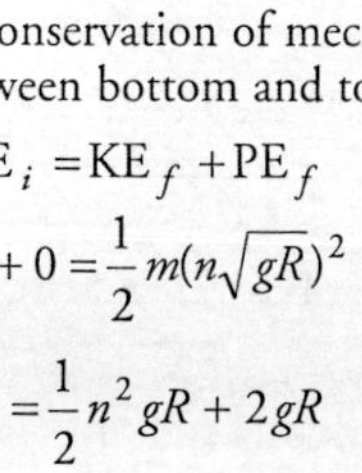

$$v_t = n\sqrt{gR}$$

Applying conservation of mechanical energy between bottom and top points

$$\text{KE}_i + \text{PE}_i = \text{KE}_f + \text{PE}_f$$

$$\frac{1}{2}mv_b^2 + 0 = \frac{1}{2}m(n\sqrt{gR})^2 + mg \times 2R$$

$$\frac{1}{2}v_b^2 = \frac{1}{2}n^2 gR + 2gR$$

$$v_b^2 = n^2 gR + 4gR$$

$$v_b^2 = gR(n^2+4)$$

$$v_b = \sqrt{gR(n^2+4)}$$

Now, $\dfrac{\text{KE at top}}{\text{KE at bottom}} = \dfrac{\frac{1}{2}mv_t^2}{\frac{1}{2}mv_b^2} = \left(\dfrac{v_t}{v_b}\right)^2$

$$= \left(\frac{n\sqrt{gR}}{\sqrt{gR(n^2+4)}}\right)^2$$

$$= \frac{n^2 gR}{gR(n^2+4)} = \frac{n^2}{n^2+4}$$

Alternate Solution

If v_b is the speed at the bottom, then speed at any other point is given by

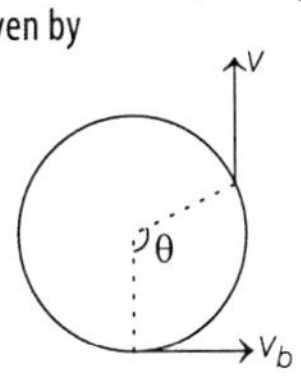

$$v = \sqrt{v_0^2 - 2gR(1-\cos\theta)}$$

So, according to question

$$v = n\sqrt{gR},\ v_0 = v_b,\ \theta = 180°$$

Using the formula

$$v = \sqrt{v_0^2 - 2gR(1-\cos\theta)}$$

$$n\sqrt{gR} = \sqrt{v_b^2 - 2gR[1-\cos 180°]}$$

$$\Rightarrow v_b = \sqrt{gR(n^2+4)}$$

Now, $\dfrac{\text{KE at top}}{\text{KE at bottom}} = \dfrac{\frac{1}{2}mv_t^2}{\frac{1}{2}mv_b^2} = \dfrac{n^2}{n^2+4}$

15. *(c)* Let v_B be the velocity of body at point B, then according to principle of conservation of energy.

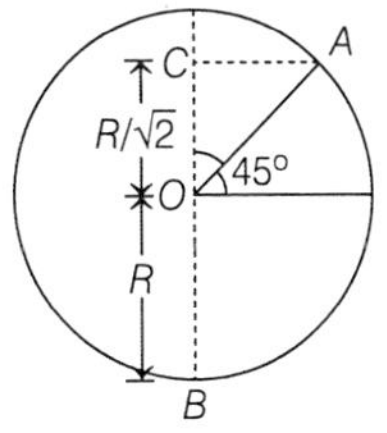

$$(\text{KE}+\text{PE})_{\text{at } A} = (\text{KE}+\text{PE})_{\text{at } B}$$

$$\Rightarrow 0 + mg(OC+OB) = \frac{1}{2}mv_B^2 + 0$$

$$\Rightarrow mg(R\sin 45° + R) = \frac{1}{2}mv_B^2$$

$$\Rightarrow \quad 2g\left(\frac{R}{\sqrt{2}} + R\right) = v_B^2$$

$$\Rightarrow \quad v_B^2 = \sqrt{2}g(\sqrt{2}+1)R$$

$$= g(2+\sqrt{2})R$$

$$= 10(2+1.4)\times 14 = 476$$

$$\therefore \quad v_B = \sqrt{476} = 21.8 \text{ m/s}$$

$$\simeq 21.9 \text{ m/s}$$

16. *(a)* Given,

Height of point, $A = 1$ m

Height of point, $B = 0.5$ m

Now using conservation of mechanical energy,

$(\text{Total Energy})_A = (\text{Total Energy})_B$

$(\text{PE}+\text{KE})_A = (\text{PE}+\text{KE})_B$

At point A, particle is stationary ($v_A = 0$)

$$mgh_A = mgh_B + \frac{1}{2}mv_B^2$$

$$gh_A = gh_B + \frac{1}{2}v_B^2$$

$$g = 0.5\,g + \frac{1}{2}v_B^2$$

$$v_B^2 = g$$

$$v_B = \sqrt{g}$$

$$= \sqrt{10} \text{ m/s}$$

17. *(d)*

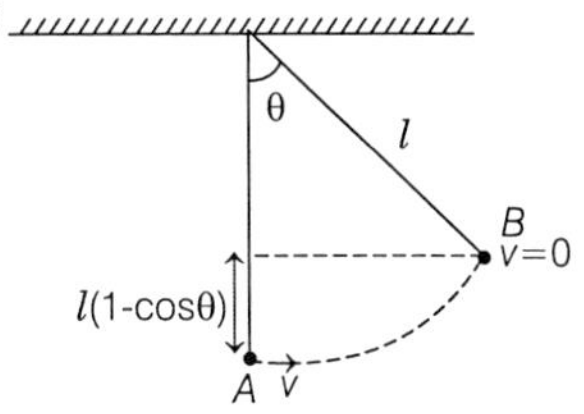

By energy conservation,

$$\frac{1}{2}mv^2 + 0 = 0 + mgl(1-\cos\theta)$$

$$v^2 = 2gl(1-\cos\theta)$$

Given, $\quad a_A = a_B$

$$\therefore \quad \frac{v^2}{l} = g\sin\theta$$

$$\Rightarrow \quad 2g(1-\cos\theta) = g\sin\theta$$

$$\Rightarrow \quad 2\left[2\sin^2\frac{\theta}{2}\right] = 2\sin\frac{\theta}{2}\cos\frac{\theta}{2}$$

$$\Rightarrow \quad \tan\frac{\theta}{2} = \frac{1}{2}$$

$$\Rightarrow \quad \theta = 2\tan^{-1}\frac{1}{2}$$

Alternate Solution

Apply work energy theorem,

$$mgl(1-\cos\theta) = \frac{1}{2}mv_0^2$$

$$\frac{v_0^2}{l} = 4g\sin^2\left(\frac{\theta}{2}\right) \quad \text{...(i)}$$

$$g\sin\theta = \frac{v_0^2}{l} \quad \text{...(ii)}$$

From Eqs. (i) and (ii), we get

$$\tan\left(\frac{\theta}{2}\right) = \frac{1}{2} \Rightarrow \theta = 2\tan^{-1}\left(\frac{1}{2}\right)$$

18. *(d)* Here, the constant horizontal force required to take the body from position 1 to position 2 can be calculated by using work-energy theorem. Let us assume that body be taken slowly, so that its speed does not change, then

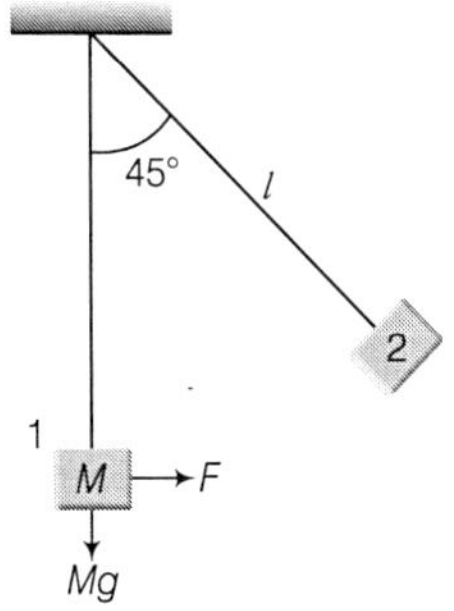

$$\Delta K = 0 = W_F + W_{Mg} + W_{\text{tension}}$$

[symbols have their usual meanings]

$$W_F = F\times l\sin 45° = \frac{Fl}{\sqrt{2}}$$

$$W_{Mg} = Mg(l - l\cos 45°),$$

$$W_{\text{tension}} = 0$$

$$\therefore \quad F = Mg(\sqrt{2}-1)$$

19. *(d)* According to work-energy theorem,

$$W = \Delta K$$

Case I $\quad -F\times 3 = \dfrac{1}{2}m\left(\dfrac{v_0}{2}\right)^2 - \dfrac{1}{2}mv_0^2$

where, F is resistive force and v_0 is initial speed.

Case II Let the further distance travelled by the bullet before coming to rest be s.

$$\therefore -F(3+s) = K_f - K_i = -\frac{1}{2}mv_0^2$$

$$\Rightarrow \quad -\frac{1}{8}mv_0^2(3+s) = -\frac{1}{2}mv_0^2$$

or $\quad \dfrac{1}{4}(3+s) = 1$ or $\dfrac{3}{4} + \dfrac{s}{4} = 1$

or $\quad s = 1$ cm

20. *(245)* By work energy theorem

$$W_{\text{friction}} + W_{\text{gravity}} = \Delta K$$

$$\Rightarrow W_f + (10 \times 0.3) = 0 - \frac{1}{2}(1)(22)^2$$

$$\Rightarrow \quad W_f = -245 \text{ J}$$

21. *(10)* From the conservation of momentum, we have

$$75 \times 10^{-3} \times v = 50 \times 10^{-3} \times v' + 75 \times 10^{-3} \times \frac{v}{3}$$

$$75 \times 10^{-3} v = 50 \times 10^{-3} \sqrt{5rg} \times \left(75 \times 10^{-3} \times \frac{v}{3}\right)$$

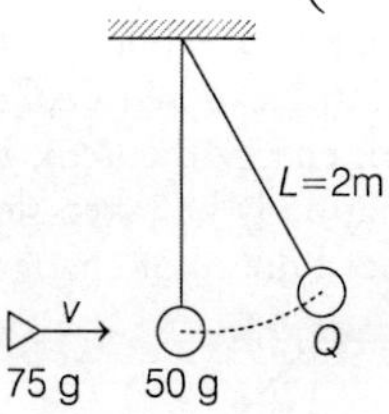

According to question, the bob completes a vertical circle of 2m radius, therefore

$$r = 2 \text{ m}, \; g = 10 \text{ms}^{-2}$$

$$75 \times 10^{-3} v = 50 \times 10^{-3} \times \sqrt{5 \times 2 \times 10} + 75 \times 10^{-3} \times \frac{v}{3}$$

$$75 \times 10^{-3}\left(\frac{2v}{3}\right) = 50 \times 10^{-3} \times 10$$

$$150 \times 10^{-3} \times v = (150) \times 10^{-2}$$

$$v = 10 \text{ m/s}$$

22. *(120)* The given situation is shown below

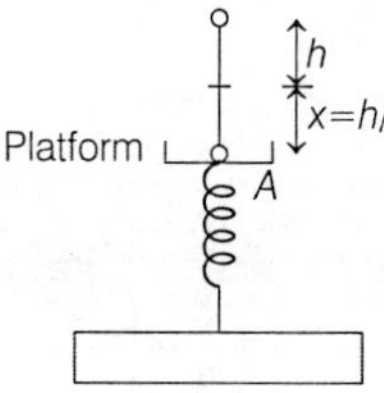

Mass of a ball $= 100 \text{ g} = 0.1 \text{ kg}$

$h = 10 \text{ cm} = 0.1 \text{ m}$

Displacement in the spring,

$$x = \frac{h}{2} = \frac{0.1}{2} = 0.05 \text{ m}$$

By conservation of energy at position A, we get

Total change in potential energy of the ball = Total potential energy of spring

i.e. $$mg(h + x) = \frac{1}{2} kx^2$$

$$\Rightarrow 0.1 \times 10(0.1 + 0.05) = \frac{1}{2} k(0.05)^2$$

$$\Rightarrow \quad k = 120 \text{ N/m}$$

23. *(16)* Given, $l = 1.5$ m

The total mass of the system,

$$M = 40000 \text{ kg}$$

The speed of the wagon, $v = 72$ km/h

When brakes are applied, the final velocity, $v_f = 0$

The compressed spring of the shock absorber, $x = 1$ m

On applying the work-energy theorem,

Work done by the system = Change in kinetic energy

$$W = \Delta \text{KE}$$

$$W_{\text{friction}} + W_{\text{spring}} = \frac{1}{2} mv_f^2 + \frac{1}{2} mv_i^2$$

$$-\frac{90}{100}\left(\frac{1}{2} mv^2\right) + W_{\text{spring}} = 0 - \frac{1}{2} mv_i^2$$

($\because$ 90% energy lost due to friction)

$$W_{\text{spring}} = -\frac{10}{100} \times \frac{1}{2} mv^2$$

$$-\frac{1}{2} kx^2 = \frac{1}{20} mv^2 \Rightarrow k = \frac{mv^2}{10 \times x^2}$$

$$k = \frac{40000 \times \left(72 \times \frac{5}{18}\right)^2}{10(1)^2}$$

$$= 16 \times 10^5 \text{ N/m}$$

$$= x \times 10^5 \text{ N/m} \quad \text{[Given]}$$

$$\therefore \quad x = 16$$

24. *(2)* The given figure can also be represented as

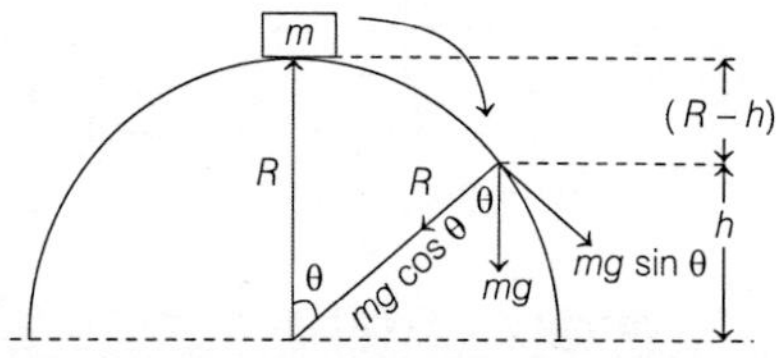

From work - energy theorem, we have

Work done = Change in kinetic energy

$$\Rightarrow \quad W = \Delta \text{KE}$$

$$\Rightarrow \quad mg(R - R\cos\theta) = \frac{1}{2} mv^2$$

$$\Rightarrow \quad v = \sqrt{2gR(1 - \cos\theta)} \quad \ldots \text{(i)}$$

To loose contact,

$$\frac{mv^2}{R} = mg\cos\theta \quad \ldots \text{(ii)}$$

From Eqs. (i) and (ii), we get

$$\frac{m[2gR(1 - \cos\theta)]}{R} = mg\cos\theta$$

$$\Rightarrow 2(1 - \cos\theta) = \cos\theta$$

$$\Rightarrow 2 - 2\cos\theta = \cos\theta \Rightarrow \cos\theta = \frac{2}{3}$$

$$\Rightarrow \quad \cos\theta = \frac{h}{R} \quad \text{(using the figure)}$$

$$\Rightarrow \quad \cos\theta = \frac{2}{3} = \frac{h}{R} = \frac{h}{3} \Rightarrow h = 2 \text{ m}$$

25. *(2)* Given, speed of bob at lowest point,

$$v_1 = 3 \text{ ms}^{-1}$$

Length of string, $l = 50 \text{ cm} = 50 \times 10^{-2}$ m

Angle, $\theta = 60°$

Let v_2 be the speed of bob at 60° from mean position.

By using law of conservation of energy

$$\text{KE}_i + \text{PE}_i = \text{KE}_f + \text{PE}_f$$

$$\Rightarrow \frac{1}{2} mv_1^2 + 0 = \frac{1}{2} mv_2^2 + mgl(1 - \cos\theta)$$

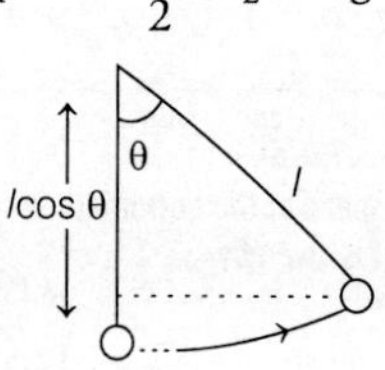

$$\Rightarrow \quad \frac{1}{2} mv_2^2 = \frac{1}{2} mv_1^2 - mgl(1 - \cos\theta)$$

$$\Rightarrow \quad v_2^2 = v_1^2 - 2gl(1 - \cos\theta)$$

$$\Rightarrow \quad v_2 = \sqrt{3^2 - 2 \times 10 \times 50 \times 10^{-2}(1 - \cos 60°)}$$

$$= \sqrt{9 - 10\left(1 - \frac{1}{2}\right)} = \sqrt{9 - 5}$$

$$= \sqrt{4} = 2 \text{ ms}^{-1}$$

26. *(346)* Let μ be the coefficient of kinetic friction

According to the question,

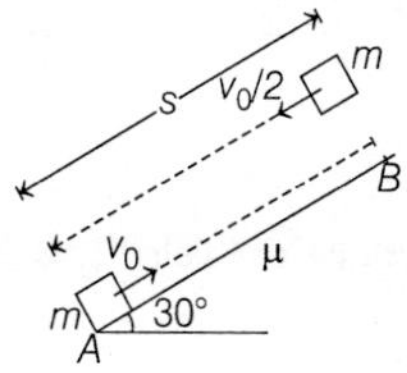

While going from A to B, the acceleration of the block is

$$a_1 = g\sin 30° + \mu g\cos 30°$$

$$\Rightarrow \quad a_1 = \frac{g}{2} + \frac{\mu g\sqrt{3}}{2}$$

$$\Rightarrow \quad a_1 = 5 + 5\sqrt{3}\mu \ldots \text{(i)} \; (g = 10 \text{ ms}^{-2})$$

Let s be the distance between A and B

$$v_0^2 - 0 = 2a_1 s$$

$$\Rightarrow \quad s = \frac{v_0^2}{2a_1} = \frac{v_0^2}{2(5 + 5\sqrt{3}\mu)} \quad \ldots \text{(ii)}$$

[Using Eq. (i)]

When the block comes back to its initial position (i.e., from B to A), its velocity is $v_0/2$.

So, the acceleration of the block while coming from B to A is

$$a_2 = g\sin 30° - \mu g\cos 30°$$

$$\Rightarrow \quad a_2 = \frac{g}{2} - \frac{\mu g\sqrt{3}}{2}$$

$$\Rightarrow \quad a_2 = 5 - 5\sqrt{3}\mu \quad \ldots\text{(iii)}$$

Again, using third equation of motion, we get

$$\left(\frac{v_0}{2}\right)^2 - 0 = 2a_2 s$$

$$\Rightarrow \quad \frac{v_0^2}{4} = 2a_2 s \Rightarrow s = \frac{v_0^2}{8a_2}$$

$$\Rightarrow \quad s = \frac{v_0^2}{8(5-5\sqrt{3}\mu)} \quad \ldots\text{(iv)}$$

[Using Eq. (iii)]

On equating Eqs. (ii) and (iv), we get

$$\frac{v_0^2}{2(5+5\sqrt{3}\mu)} = \frac{v_0^2}{8(5-5\sqrt{3}\mu)}$$

$$\Rightarrow \quad 4(5-5\sqrt{3}\mu) = 5+5\sqrt{3}\mu$$

$$\Rightarrow \quad 20-20\sqrt{3}\mu = 5+5\sqrt{3}\mu$$

$$\Rightarrow \quad 25\sqrt{3}\mu = 15$$

$$\Rightarrow \quad \mu = \frac{15}{25\sqrt{3}} = \frac{3}{5\sqrt{3}}$$

$$\Rightarrow \quad \mu = \sqrt{3}/5$$

So, $\quad \frac{I}{1000} = \frac{\sqrt{3}}{5} = 0.346$

$$\Rightarrow \quad I = 0.346 \times 1000 = 346$$

Alternate Solution

Apply work energy theorem on the block we get

$W_g + W_f + W_N = (\text{KE})_f - (\text{KE})_i$

W_g = work done by gravity

W_f = Work done due to friction

W_N = Work done due to normal force

$$\Rightarrow \frac{1}{2}mv_0^2 = \left(mg\frac{s}{2} + \mu mg\frac{\sqrt{3}}{2}s\right) \quad \ldots\text{(i)}$$

(For upward motion)

$$\text{and } \frac{I}{2}m\frac{v_0^2}{4} = \left(mg\frac{s}{2} - \mu mg\frac{\sqrt{3}}{2}s\right) \quad \ldots\text{(ii)}$$

(For downward motion)

$$\Rightarrow 4 = \frac{1+\mu\sqrt{3}}{1-\mu\sqrt{3}} \Rightarrow 4 - 4\mu\sqrt{3} = 1+\mu\sqrt{3}$$

$$\Rightarrow 3 = 5\mu\sqrt{3}$$

$$\therefore \mu = \frac{\sqrt{3}}{5} = \frac{I}{1000} \Rightarrow I = 0.346 \times 1000 = 346$$

Topic 3

01. *(d)* According to given situation energy supplied in delivering the constant power P is equal to the kinetic energy of the automobile.

$\therefore$ Energy supplied = Kinetic energy

$$\Rightarrow \quad Pt = \frac{1}{2}mv^2 \Rightarrow v = \left(\frac{2Pt}{m}\right)^{1/2}$$

or $\quad \frac{ds}{dt} = \left(\frac{2Pt}{m}\right)^{1/2} \quad \left[\because v = \frac{ds}{dt}\right]$

Integrating both sides of above equation, we get

$$\int_0^s ds = \left(\frac{2P}{m}\right)^{1/2}\int_0^t t^{1/2}\cdot dt$$

$$\Rightarrow \quad s = \left(\frac{2P}{m}\right)^{1/2}\cdot\frac{2}{3}t^{3/2}$$

$$\Rightarrow \quad s = \frac{2}{3}\left(\frac{2P}{m}\right)^{1/2}\cdot t^{3/2}$$

$$\Rightarrow \quad s = \left(\frac{8P}{9m}\right)^{1/2}\cdot t^{3/2}$$

02. *(a)* $\therefore$ Energy supply $= Pt$

where, P = power

and $\quad t$ = time taken.

$$\Rightarrow \quad \frac{1}{2}mv^2 = Pt \Rightarrow v^2 \propto t \Rightarrow v \propto \sqrt{t}$$

or $\quad ds/dt = c\sqrt{t} \quad \ldots\text{(i)}$

where, c is any constant.

Now, integrating Eq. (i), we get

$$\int_0^S ds = \int_0^t c\sqrt{t}\,dt \Rightarrow s = c\int_0^t t^{\frac{1}{2}}\cdot dt$$

$$\Rightarrow \quad s = \frac{2c}{3}t^{\frac{3}{2}} \Rightarrow s \propto t^{3/2}$$

03. *(b)* We know that,

Power = force × velocity

[$\because$ Given, power = constant]

$\therefore$ Force × velocity = constant

$$\Rightarrow \quad ma \times v = \text{constant}$$

$$\Rightarrow \quad a \times v = \text{constant}$$

$$\Rightarrow \quad \left(v\frac{dv}{dx}\right)\left(\frac{dx}{dt}\right) = C \Rightarrow \int v\,dv = \int C\,dt$$

$$\Rightarrow \quad \frac{v^2}{2} = Ct$$

$$\Rightarrow \quad v = \sqrt{2Ct}$$

$$\Rightarrow \quad \frac{dx}{dt} = \sqrt{2Ct} \quad \left[\because v = \frac{dx}{dt}\right]$$

$$\Rightarrow \quad \int dx = \int\sqrt{2Ct}\,dt$$

$$\Rightarrow \quad x = \sqrt{2C}\left(\frac{t^{3/2}}{\frac{3}{2}}\right)$$

$$\Rightarrow \quad x \propto t^{3/2}$$

04. *(a)* A body of mass m with uniform acceleration, then force

$$F = ma = \frac{m(v-0)}{T} = \frac{mv}{T}$$

Instantaneous power $= Fv$

$$= \frac{mv}{T}\cdot at \quad [\because v = at]$$

$$= \frac{mv}{T}\cdot\frac{v}{T}\cdot t = \frac{mv^2}{T^2}t$$

05. *(b)* $\mathbf{F} = 2t\hat{\mathbf{i}} + 3t^2\hat{\mathbf{j}}$, $m = 1000\text{ g} = 1\text{ kg}$

$$\mathbf{F} = m\mathbf{a}$$

$$\mathbf{a} = \frac{F}{m} = \frac{2t\hat{\mathbf{i}} + 3t^2\hat{\mathbf{j}}}{1} = 2t\hat{\mathbf{i}} + 3t^2\hat{\mathbf{j}}$$

$$\frac{dv}{dt} = 2t\hat{\mathbf{i}} + 3t^2\hat{\mathbf{j}}$$

$$\int dv = \int(2t\hat{\mathbf{i}} + 3t^2\hat{\mathbf{j}})dt \Rightarrow v = t^2\hat{\mathbf{i}} + t^3\hat{\mathbf{j}}$$

$$\text{Power } P = \mathbf{F}\cdot\mathbf{v} = (2t\hat{\mathbf{i}} + 3t^2\hat{\mathbf{j}})\cdot(t^2\hat{\mathbf{i}} + t^3\hat{\mathbf{j}})$$

$$P = (2t^3 + 3t^5)\text{W}$$

06. *(d)* Since, $P = Fv = mav \quad [\because F = ma]$

$$= ma\cdot at$$

$$P = ma^2t \quad [\because v = at]$$

$$\Rightarrow \quad a = \sqrt{\frac{P}{mt}}$$

We know that, $s = ut + \frac{1}{2}at^2$

$$= \frac{1}{2}at^2 = \frac{1}{2}\sqrt{\frac{P}{mt}}\cdot t^2 \quad [\because u = 0]$$

$$s = \frac{1}{2}\sqrt{\frac{P}{m}}\cdot t^{3/2} \Rightarrow s \propto t^{3/2}$$

Time Saver Tech

When a body is moving undirectionally under the influence of constant power, then its displacement in time t is proportional to $t^{3/2}$.

07. *(c)* Given force, $\mathbf{F} = 6t\,\hat{\mathbf{i}} + 6t^2\hat{\mathbf{j}}$

and mass, $m = 2$ kg

$$\text{Acceleration } \mathbf{a} = \frac{\mathbf{F}}{m} = 3t\,\hat{\mathbf{i}} + 3t^2\hat{\mathbf{j}}$$

So, velocity at time t is,

$$\mathbf{v} = \int\mathbf{a}\cdot dt = \int(3t\hat{\mathbf{i}} + 3t^2\hat{\mathbf{j}})\,dt$$

$$= \frac{3}{2} t^2 \hat{i} + t^3 \hat{j}$$

Power $= \mathbf{F} \cdot \mathbf{v}$

$$= (6t\hat{i} + 6t^2\hat{j}) \cdot \left(\frac{3}{2} t^2 \hat{i} + t^3 \hat{j}\right)$$

$$= (9t^3 + 6t^5) \text{ W}$$

08. *(d)* Given, rate of dropped sand,

$$\frac{dm}{dt} = 0.5 \text{ kgs}^{-1}$$

Velocity of conveyor, $v = 5$ m/s

Force applied on the belt, $F = \frac{dp}{dt}$

$$= m\frac{dv}{dt} + v\frac{dm}{dt} = 5 \times 0.5 = 2.5 \text{ N}$$

$$\left(\because \frac{dv}{dt} = 0\right)$$

Power needed to keep the belt moving with velocity, $v = 5$ m/s, is given as

$$P = F \times v$$

$$= 2.5 \times 5 = 12.5 \text{ W}.$$

09. *(c)* Given, centripetal acceleration of particle is

$$a_c = k^2 r t^2 \text{ or } \frac{v^2}{r} = k^2 r t^2$$

$$\Rightarrow \quad v = krt$$

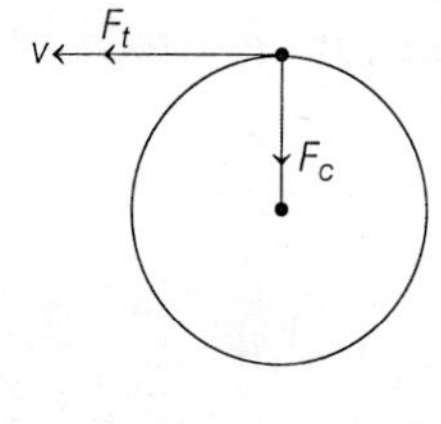

Tangential acceleration of particle,

$$a_t = \frac{dv}{dt} = \frac{d}{dt} krt = kr$$

Power delivered, $P = \mathbf{F}_t \cdot \mathbf{v} = m(\mathbf{a}_t \cdot \mathbf{v})$

$$= m(kr)(krt)\cos 0° = mk^2 r^2 t$$

> • Concept **Enhancer**
>
> Centripetal force F_c is perpendicular to velocity and does not deliver any power to particle as $\mathbf{F}_c \cdot \mathbf{v} = 0$ at every instant.

10. *(c)* Given, $\frac{dm}{dt} \propto \sqrt{v}$

$$\Rightarrow \quad \frac{dm}{dt} = K\sqrt{v}$$

By Newton's 2nd law

$$\mathbf{F}_{\text{ext}} = \frac{m d\mathbf{v}}{dt} + \mathbf{v}\frac{dm}{dt}$$

As speed is constant

$$\Rightarrow \quad \frac{dv}{dt} = 0$$

$$\Rightarrow \quad \mathbf{F}_{\text{ext}} = \mathbf{v}\frac{dm}{dt} = v \times K\sqrt{v} = Kv^{3/2}$$

Now power $= Fv = Kv^{3/2} \times v$

$$P = Kv^{5/2}$$

$$P^2 = K^2 v^5$$

$$\therefore \quad P^2 \propto v^5$$

11. *(b)*

$$m = 4 \text{ kg}$$

$$\mathbf{F} = 2\hat{i} + 3\hat{j}$$

$$m\mathbf{a} = 2\hat{i} + 3\hat{j}$$

$$(4)\mathbf{a} = 2\hat{i} + 3\hat{j}$$

$$\mathbf{a} = \frac{2\hat{i}}{4} + \frac{3\hat{j}}{4} = \frac{1}{2}\hat{i} + \frac{3}{4}\hat{j}$$

$$\frac{dv}{dt} = \frac{1}{2}\hat{i} + \frac{3}{4}\hat{j}$$

$$\mathbf{v} = \left(\frac{1}{2}\hat{i} + \frac{3}{4}\hat{j}\right)t$$

at $t = 4$ s

$$\mathbf{v} = \left(\frac{1}{2}\hat{i} + \frac{3}{4}\hat{j}\right)4 = 2\hat{i} + 3\hat{j}$$

$$P_{\text{inst}} = \mathbf{F} \cdot \mathbf{v} = (2\hat{i} + 3\hat{j}) \cdot (2\hat{i} + 3\hat{j})$$

$$= 4 + 9 = 13$$

$$s = ut + \frac{1}{2}at^2$$

$$s = 0 + \frac{1}{2}\frac{(2\hat{i} + 3\hat{j})}{4}(4)^2$$

$$= 4\hat{i} + 6\hat{j}$$

If $\quad r_i = 3\hat{i} + 4\hat{j}$

and $r_f = 6\hat{i} + 10\hat{j}$, then

Final position will be

$$\Delta s = r_f - r_i$$

$$= (6\hat{i} + 10\hat{j}) - (3\hat{i} + 4\hat{j})$$

$$\Delta s = (3\hat{i} + 6\hat{j})$$

$$P_{\text{avg}} = \frac{\Delta W}{\Delta t} = \frac{F\Delta s}{\Delta t}$$

$$P_{\text{avg}} = \frac{(2\hat{i} + 3\hat{j})(3\hat{i} + 6\hat{j})}{4}$$

$$= \frac{6 + 18}{4} = \frac{24}{4} = 6$$

$$P_{\text{avg}} = 6$$

$$\frac{P_{\text{avg}}}{P_{\text{inst}}} = \frac{6}{13} = 6:13$$

12. *(c)* Let's two motors are A and B

For motor A

$$p_A = F_1 v_1 \quad \text{...(i)}$$

given $m = 300$ kg

$$\therefore \quad F_1 = mg = 3000 \text{ N}$$

$$d = 100 \text{ m}; \; t = 5 \text{ min}$$

$$t = 5 \times 60 = 300 \text{ s}$$

$$\therefore \quad v_1 = \frac{d}{t} = \frac{100}{300} = \frac{1}{3} \text{ m/s}$$

$\therefore$ From Eqs. (i),

$$p_A = 3000 \times \frac{1}{3} = 1000 \text{ W}$$

For motor B

$$m = 50 \text{ kg}, F_2 = mg \Rightarrow 50 \times 10 = 500 \text{ N}$$

$$v_2 = \frac{100}{2 \times 60} \Rightarrow v_2 = \frac{5}{6} \text{ m/s}$$

$$\therefore P_B = F_2 v_2 = 500 \times \frac{5}{6} = \frac{2500}{6}$$

$$\therefore \quad \frac{P_A}{P_B} = \frac{1000}{\frac{2500}{6}} \Rightarrow \frac{P_A}{P_B} = \frac{12}{5} \quad \text{...(ii)}$$

But, $\frac{P_A}{P_B} = \frac{3\sqrt{x}}{\sqrt{x} + 1}$ ... Put these values from Eq. (ii)

$$\Rightarrow \quad \frac{3\sqrt{x}}{\sqrt{x} + 1} = \frac{12}{5} \Rightarrow 15\sqrt{x} = 12\sqrt{x} + 12$$

$$\Rightarrow \quad 3\sqrt{x} = 12 \Rightarrow \sqrt{x} = 4 \Rightarrow x = 16$$

13. *(d)* Given, mass of elevator,

$$M = 920 \text{ kg}$$

Mass of all 10 passengers carried by elevator $= 10 \times m = 10 \times 68 = 680$ kg

Total weight of elevator and passengers

$$= (M + 10m)g$$

$$= (920 + 680) \times 10 = 16000 \text{ N}$$

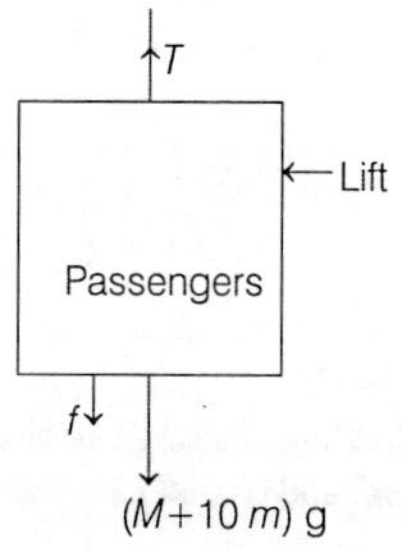

Force of friction $= 6000$ N

Total force (T) applied by the motor of elevator

$$= 16000 + 6000 = 22000 \text{ N}$$

Power delivered by elevator's motor,

$$P = F \cdot v = 22000 \times 3 \quad [\because v = 3 \text{ ms}^{-1}]$$

$$= 66000 \text{ W}$$

14. *(300)*

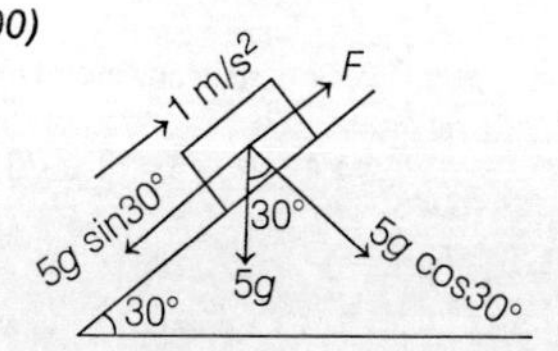

F at $t = 10$ s

$$F - 5g\sin 30° = 5a$$

$$F - 5\times10\times\frac{1}{2} = 5\times1$$

$$F = 30\text{ N},$$

$$\therefore \quad v = u + at$$

$$v = 0 + 1(10),\ v = 10\text{ m/s}$$

$$P = Fv = 30\times10 = 300\text{ W}$$

15. *(48)* Given, maximum load = 1400 kg

Speed of elevator $(v) = 3\text{ ms}^{-1}$

Friction force = 2000 N

Force due to load $= Ma = mg$

$= 1400\times10 = 14000$ N

Maximum force = Force on elevator + Friction

$= 14000 + 2000 = 16000$ N

$$P_{max} = F_{max}\times v$$

$$= 16000\times3 = 48000\text{ W} = 48\text{ kW}$$

16. *(4)*

> **key idea** If a body is moving with velocity *v* under the effect of force *F*, then you can calculate power applied on the body directly by the formula,
>
> $P = F\times v$

Given, $m = 2$ kg

We know that, $P = F\times v$

$$P = m\times a\times v$$

$$\Rightarrow \quad P = m\frac{dv}{dt}\cdot v \Rightarrow Pdt = mvdv$$

$$\Rightarrow \int_0^t Pdt = \int_0^v mv\,dv$$

$$Pt = \frac{mv^2}{2}$$

$$\Rightarrow \quad v = \sqrt{\frac{2Pt}{m}} \Rightarrow \frac{ds}{dt} = \sqrt{\frac{2Pt}{m}}$$

$$\Rightarrow ds = \sqrt{\frac{2Pt}{m}}\cdot dt \Rightarrow \int_0^s ds = \int_0^4 \sqrt{\frac{2Pt}{m}}\,dt$$

$$\Rightarrow \quad s = \sqrt{\frac{2P}{m}}\cdot\frac{t^{3/2}}{3/2} = \frac{2}{3}\sqrt{\frac{2P}{m}}\cdot[t^{3/2}]_0^4$$

$$= \frac{2}{3}\sqrt{\frac{2P}{2}}(4)^{3/2}$$

$$(\because t = 4\text{ s and } m = 2\text{ kg})$$

$$= \frac{16}{3}\sqrt{P} = \frac{4^2}{3}\sqrt{P} = \frac{\alpha^2\sqrt{P}}{3}\quad\text{(given)}$$

$$\therefore\ \alpha = 4$$

17. *(100)* Given, force acting on body,

$$\mathbf{F} = t\hat{i} + 3t^2\hat{j}$$

Acceleration of body,

$$\mathbf{a} = \frac{\mathbf{F}}{m} = (t\hat{i} + 3t^2\hat{j})\text{ m/s}^2$$

(given, mass = 1 kg)

$$a = t\hat{i} + 3t^2\hat{j}$$

$$\int_0^v dv = \int_0^t (t\hat{i} + 3t^2\hat{j})\,dt$$

$$v = \frac{t^2}{2}\hat{i} + t^3\hat{j}$$

$$P = \mathbf{F}\cdot\mathbf{v} = (t\hat{i} + 3t^2\hat{j})\cdot\left(\frac{t^2}{2}\hat{i} + t^3\hat{j}\right)$$

$$= \frac{t^3}{2} + 3t^5$$

at $t = 2$

$$= 4 + 96 = 100\text{ W}$$

Topic 4

01. *(d)* By momentum conservation,

$$mv = (m + 2m)v'$$

$$\Rightarrow \quad v' = \frac{v}{3}$$

02. *(c)* Velocity of a ball *P* when it strikes ball *Q* is given by,

$$v = \sqrt{2gh} = \sqrt{2\times10\times0.2} = 2\text{ ms}^{-1}$$

As balls are identical, they exchange velocities in an elastic collision. So, velocity of $Q = 2$ m/s.

> • Concept **Enhancer**
>
> When two bodies of equal masses suffers one dimensional elastic collision, their velocities will get exchanged.

03. *(c)* Let v_1 and v_2 are the speed of *P* and *Q* after collision.

By using law of conservation of mementum,

$$m_1u_1 + m_2u_2 = m_1v_1 + m_2v_2$$

$$\Rightarrow \quad Mu + m\cdot0 = Mv_1 + mv_2$$

$$\Rightarrow \quad \frac{M(u - v_1)}{m} = v_2 \quad \ldots\text{(i)}$$

and by using law of conservation of energy,

$$\frac{1}{2}m_1u_1^2 + \frac{1}{2}m_2u_2^2 = \frac{1}{2}m_1v_1^2 + \frac{1}{2}m_2v_2^2$$

$$\Rightarrow \quad Mu^2 + 0 = Mv_1^2 + mv_2^2$$

$$\Rightarrow \quad M(u^2 - v_1^2) = mv_2^2$$

$$\Rightarrow M(u - v_1)(u + v_1)/m = v_2^2 \quad \ldots\text{(ii)}$$

On substituting the value of $M\frac{(u - v_1)}{m}$ from Eq. (i) in Eq. (ii), we get

$$v_2(u + v_1) = v_2^2$$

$$\Rightarrow \quad u + v_1 = v_2$$

$$\because \quad M >> m$$

$$\therefore \quad v_1 = u$$

and $\quad v_2 = 2u$

04. *(c)* As collision is elastic as shown below,

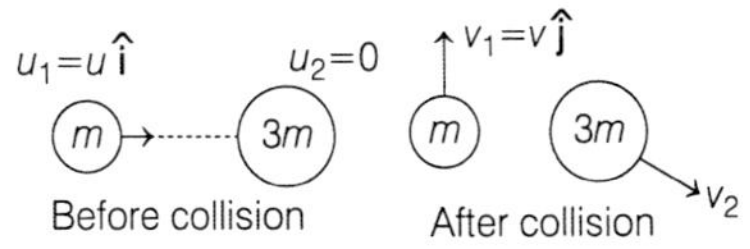

Momentum conservation gives,

$$mu\hat{i} = mv\hat{j} + 3m\mathbf{v}_2$$

$$\Rightarrow \mathbf{v}_2 = \frac{1}{3}(u\hat{i} - v\hat{j}) \Rightarrow |\mathbf{v}_2| = \sqrt{\frac{u^2 + v^2}{9}}$$

or $v_2^2 = (u^2 + v^2)/9 \quad \ldots\text{(i)}$

Kinetic energy conservation gives,

$$\Rightarrow \quad \frac{1}{2}mu^2 = \frac{1}{2}mv^2 + \frac{1}{2}3mv_2^2$$

$$\Rightarrow \quad u^2 = v^2 + 3v_2^2 \quad \ldots\text{(ii)}$$

On substituting value of v_2 from Eq (i) into Eq (ii), we get

$$u^2 = v^2 + 3\left(\frac{u^2 + v^2}{9}\right)$$

$$\Rightarrow \quad \frac{2}{3}u^2 = \frac{4}{3}v^2$$

$$\Rightarrow \quad v = \frac{u}{\sqrt{2}}$$

05. *(d)* According to the questions,

Initial condition,

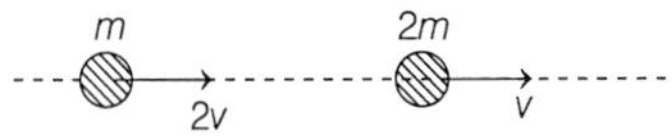

Final condition,

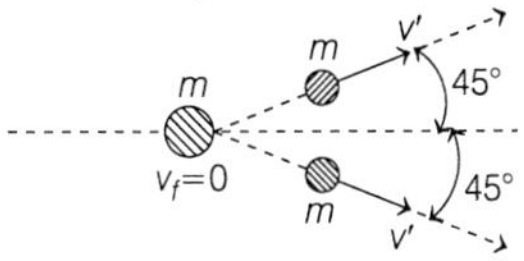

As we know that, in collision, linear momentum is conserved in both *x* and *y*-directions separately.

So, $\quad (p_x)_{\text{initial}} = (p_x)_{\text{final}}$

$$m(2v) + 2m(v) = 0 + mv'\cos45° + mv'\cos45°$$

$$\Rightarrow \quad 4mv = \frac{2m}{\sqrt{2}}v'$$

$$\Rightarrow \quad v' = 2\sqrt{2}v$$

06. *(a)* If it is a completely inelastic collision, then

$$m_1v_1 + m_2v_2 = m_1v + m_2v$$

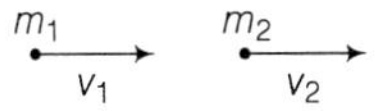

$$v = \frac{m_1v_1 + m_2v_2}{m_1 + m_2},$$

$$\text{KE} = \frac{p_1^2}{2m_1} + \frac{p_2^2}{2m_2}$$

As p_1 and p_2 both simultaneously cannot be zero. Therefore, total KE cannot be lost.

07. *(b)* Here, momentum of the system is remaining conserved as no external force is acting on the bomb (system).

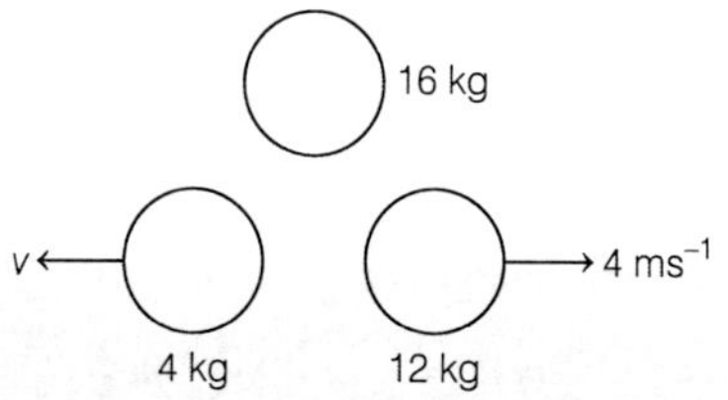

Initial momentum (before explosion) = Final momentum (after explosion)

Let velocity of 4 kg mass be v ms^{-1}.

From momentum conservation, we can say that its direction is opposite to velocity of 12 kg mass.

From $p_i = p_f$

$\Rightarrow \quad 0 = 12 \times 4 - 4 \times v$ or $v = 12$ m/s

$\therefore$ KE of 4 kg mass $= \dfrac{4 \times (12)^2}{2} = 288$ J

08. *(a)*

> **key idea** In an elastic collision both linear momentum and total kinetic energy of system remains conserved.

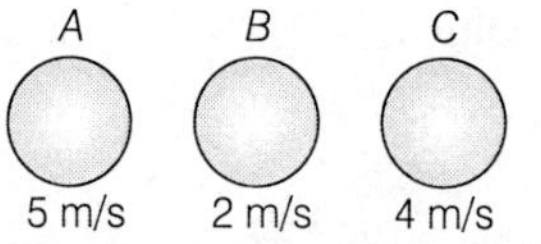

Reason is true as an elastic collision between identical masses, two objects exchange their velocities. Now, in assertion, first A and B collide and exchange their velocities.

v_A becomes 2 m/s.

v_B becomes 5 m/s.

After that balls B and C collide and exchange their velocities.

v_B becomes 4 m/s.

v_C becomes 5 m/s.

So, final velocities are

$v_A = 2$ m/s, $v_B = 4$ m/s, $v_C = 5$ m/s

Thus, assertion is wrong.

09. *(c)* Let, m = mass of ball

Velocity of ball just before collision

$$v^2 = u^2 + 2gh$$

$$v_1 = \sqrt{2gh} \qquad (u = 0) \ldots\text{(i)}$$

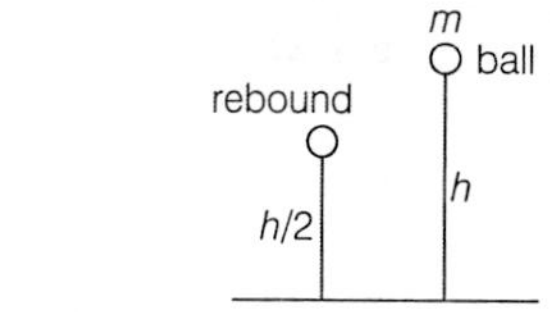

Velocity just after collision,

$$v_2 = \sqrt{2g\frac{h}{2}} \qquad \ldots\text{(ii)}$$

Change in kinetic energy,

$$\Delta\text{KE} = \frac{1}{2}m(2gh) - \frac{1}{2}mgh \qquad \ldots\text{(iii)}$$

Percentgage loss in kinetic energy

$$= \frac{1}{2}mgh$$

$$\frac{\Delta\text{KE}}{\text{KE}_i} \times 100 = \frac{\frac{1}{2}mgh}{\frac{1}{2}mg2h} \times 100 = 50\%$$

10. *(c)* Before collision

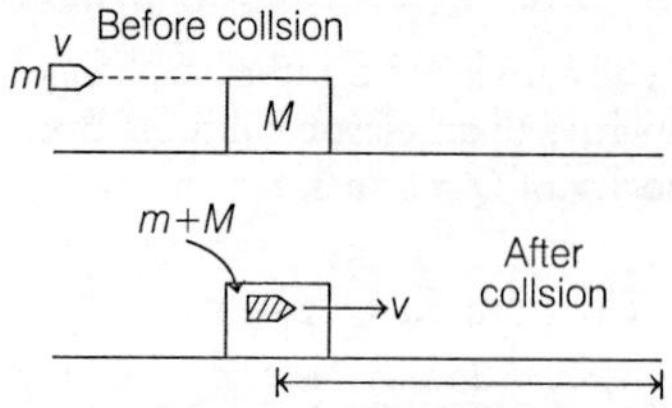

By conservation of momentum,

$$mu = (m + M)v$$

$\Rightarrow \quad 0.1 \times 400 = (0.1 + 3.9)v$

$\Rightarrow \quad v = \dfrac{0.1 \times 400}{4} = 10$ m/s

KE of block and bullet system = work done against friction

$\Rightarrow \quad \dfrac{1}{2}(m + M)v^2 = \mu(m + M)g \times s$

$\Rightarrow \quad \mu = \dfrac{v^2}{2gs} = \dfrac{10^2}{2 \times 10 \times 20} = \dfrac{1}{4} = 0.25$

11. *(b)* Masses of each billiard ball,

$$m_1 = m_2 = m = 0.05 \text{ kg}$$

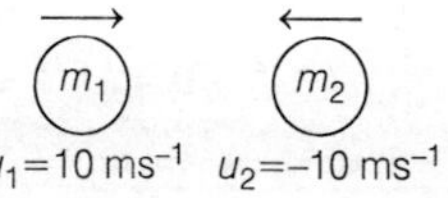

Situation of balls after collision is given below.

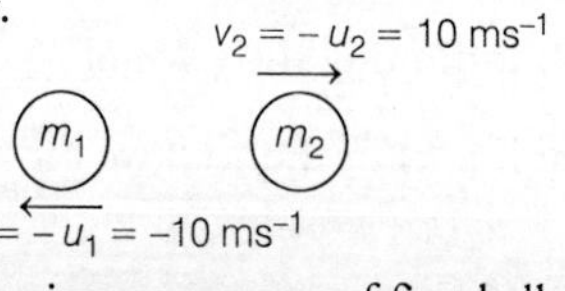

Change in momentum of first ball,

$$\Delta p = p_i - p_f = m_1u_1 - m_1v_1$$

$$= m_1(u_1 - v_1) = 0.05[10 - (-10)]$$

$$= 0.05 \times 20 = 1 \text{ kg-ms}^{-1}$$

Average force exerted on the ball due to each other,

$$F_{\text{avg}} = \frac{\Delta p}{\Delta t} = \frac{1}{0.005} = 200 \text{ N}$$

12. *(d)* If v'_A and v'_B be the velocity of body A and B after first collision, then by conservation of linear momentum

$$mv_A = mv'_A + 2m\,v'_B$$

$$v_A = v'_A + 2v'_B \qquad \ldots\text{(i)}$$

Again, $e = \dfrac{v'_A - v'_B}{0 - v_A} \Rightarrow 1 = \dfrac{v'_A - v'_B}{-v_A}$

[For elastic collision, $e = 1$]

$\Rightarrow \quad -v_A = v'_A - v'_B \qquad \ldots$(ii)

On adding Eqs. (i) and (ii), we get

$$0 = 2v'_A + v'_B$$

$\Rightarrow \quad v'_B = -2v'_A \qquad \ldots$(iii)

From Eqs. (ii) and (iii), we get

$$-v_A = v'_A - (-2v'_A)$$

$\Rightarrow \quad -v_A = v'_A + 2v'_A = 3v'_A$

$\Rightarrow \quad v'_A = -\dfrac{v_A}{3} = -\dfrac{9}{3} = -3$ m/s

$\therefore$ From Eq. (iii), $v'_B = -2(-3) = 6$ m/s

Again, after 2nd collision,

$$2mv'_B = (2m + 2m)v_c = 4mv_c$$

$\Rightarrow \quad v_c = \dfrac{v'_B}{2} = \dfrac{6}{2} = 3$ m/s

13. *(a)*

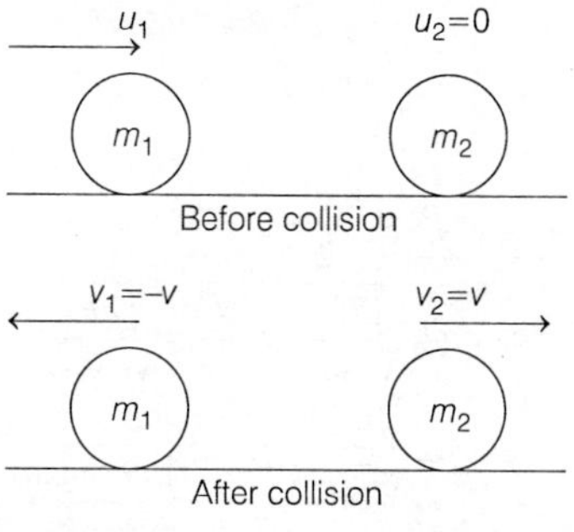

Using the law of conservation of linear momentum,

$$m_1u_1 + m_2u_2 = m_1v_1 + m_2v_2$$

$\Rightarrow \quad m_1u_1 + m_2(0) = m_1(-v) + m_2v$

$\Rightarrow \quad m_1u_1 = (-m_1 + m_2)v \qquad \ldots$(i)

Let the collision is elastic coefficient of restitution, $e = 1$

$\therefore \quad e = \dfrac{v_2 - v_1}{u_1 - u_2} \Rightarrow 1 = \dfrac{v - (-v)}{u_1 - 0}$

$$u_1 = 2v \qquad \ldots\text{(ii)}$$

Putting the above value in Eq. (i), we get

$$m_1(2v) = (-m_1 + m_2)v$$

$\Rightarrow \quad 3m_1 = m_2 \Rightarrow \dfrac{m_2}{m_1} = \dfrac{3}{1}$

$\Rightarrow \quad m_2 : m_1 = 3 : 1$

14. *(c)* From the law of conservation of energy,

Energy of the system when bullet gets embedded = Energy of the system till it momentarily comes to rest.

$$\Rightarrow \quad \frac{1}{2}(M+m)v_1^2 = (M+m)gh$$

where, v_1 = velocity of bullet + block system

$$\Rightarrow \quad v_1 = \sqrt{2gh} \quad \ldots \text{(i)}$$

According to law of conservation of momentum,

Momentum before collision = Momentum after collision.

$$\Rightarrow \quad mv = (M+m)v_1$$

[where, v = velocity of bullet before collision]

$$\Rightarrow mv = (M+m)\sqrt{2gh} \quad \text{[using Eq. (i)]}$$

$$\Rightarrow v = \left(\frac{M+m}{m}\right)\sqrt{2gh}$$

$$\Rightarrow v = \frac{(5.99+0.01)}{10\times10^{-3}} \times \sqrt{2\times9.8\times9.8\times10^{-2}}$$

$$\Rightarrow v = 831.55 \text{ ms}^{-1} \approx 831.4 \text{ ms}^{-1}$$

15. *(c)* When bullet + block system falls from height h, then its total energy (kinetic + potential) becomes kinetic energy, so kinetic energy of bullet + block system at bottom just before collision is equal to total energy just after collision.

Now, by law of conservation of momentum, we have

$$mu = (m+M)v$$

$$\Rightarrow \quad v = \frac{mu}{m+M} = \frac{0.1\times 20}{(0.1+1.9)} = 1 \text{ ms}^{-1}$$

Total energy of bullet and block just after collision

$$= \text{KE} + \text{PE} = \frac{1}{2}(m+M)v^2 + (m+M)gh$$

$$= \frac{1}{2}\times 2\times 1^2 + 2\times 10\times 1 = 1 + 20 = 21 \text{ J}$$

16. *(b)* Given, mass of small body,

$$m = 2 \text{ kg}$$

Given situation is as shown

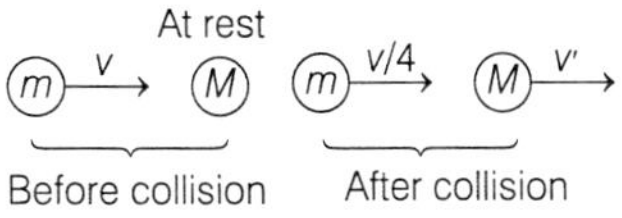

Using momentum conservation law for the given system,

$$(\text{Total momentum})_{\text{before collision}} = (\text{Total momentum})_{\text{after collision}}$$

$$\Rightarrow m(v) + M(0) = m\left(\frac{v}{4}\right) + M(v') \quad \ldots \text{(i)}$$

$\because \quad e = 1$ and we know that,

$$e = -\frac{v_2 - v_1}{u_2 - u_1}$$

$$\Rightarrow \quad 1 = -\frac{v' - v/4}{0 - v} \Rightarrow v = v' - v/4$$

$$\text{or} \quad v' = 5v/4 \quad \ldots \text{(ii)}$$

Using value from Eq. (ii) into Eq. (i), we get

$$mv = \frac{mv}{4} + M\left(\frac{5v}{4}\right)$$

$$\Rightarrow \quad m\left(v - \frac{v}{4}\right) = M\left(\frac{5v}{4}\right)$$

$$\Rightarrow \quad \frac{3}{4}mv = \frac{5}{4}Mv$$

$$M = \frac{3}{5}m = \frac{3}{5}\times 2 = 1.2 \text{ kg}$$

17. *(a)* Initially, block A is moving with velocity v as shown in the figure below,

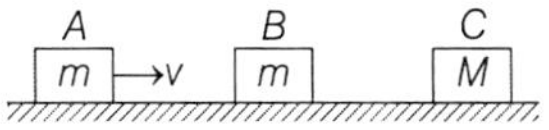

Now, A collides with B such that they collide inelastically. Thus, the combined mass (say) move with the velocity v' as shown below,

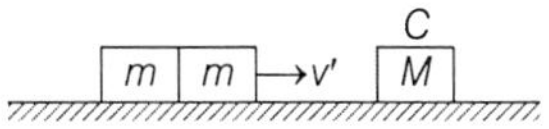

Then, if this combined system collided inelastically again with the block C velocity of system be v'' as shown below.

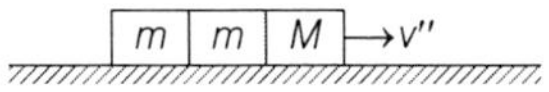

Thus, according to the principle of conservation of momentum,

initial momentum of the system = final momentum of the system

$$\Rightarrow \quad mv = (2m+M)v''$$

$$\text{or} \quad v'' = \left(\frac{mv}{2m+M}\right) \quad \ldots \text{(i)}$$

Initial kinetic energy of the system,

$$(\text{KE})_i = \frac{1}{2}mv^2 \quad \ldots \text{(ii)}$$

Final kinetic energy of the system,

$$(\text{KE})_f = \frac{1}{2}(2m+M)(v'')^2$$

$$= \frac{1}{2}(2m+M)\left(\frac{mv}{2m+M}\right)^2 \quad [\because \text{using Eq. (i)}]$$

$$= \frac{1}{2}\cdot\frac{v^2m^2}{(2m+M)} \quad \ldots \text{(iii)}$$

On dividing Eq. (iii) and Eq. (ii), we get

$$\frac{(\text{KE})_f}{(\text{KE})_i} = \frac{\dfrac{\frac{1}{2}m^2v^2}{(2m+M)}}{\frac{1}{2}mv^2}$$

$$= \frac{m}{2m+M} \quad \ldots \text{(iv)}$$

It is given that 5/6 of (KE) is lost in this process.

$$\Rightarrow \quad (\text{KE})_f = \frac{1}{6}(\text{KE})_i$$

$$\Rightarrow \quad \frac{(\text{KE})_f}{(\text{KE})_i} = \frac{1}{6} \quad \ldots \text{(v)}$$

On comparing Eq. (iv) and Eq. (v), we get

$$\frac{m}{2m+M} = \frac{1}{6}$$

$$\Rightarrow \quad 6m = 2m + M$$

$$4m = M \Rightarrow \frac{M}{m} = 4$$

• Concept Enhancer

For a perfectly inelastic collision, the momentum of the system remains conserved but there is some of loss of kinetic energy. Also, after collision the objects of the system are stuck to each other and move as a combined system.

18. *(b)* Final kinetic energy is 50% more than initial kinetic energy

$$\Rightarrow \quad \frac{1}{2}mv_2^2 + \frac{1}{2}mv_1^2 = \frac{150}{100}\times\frac{1}{2}mv_0^2 \quad \ldots \text{(i)}$$

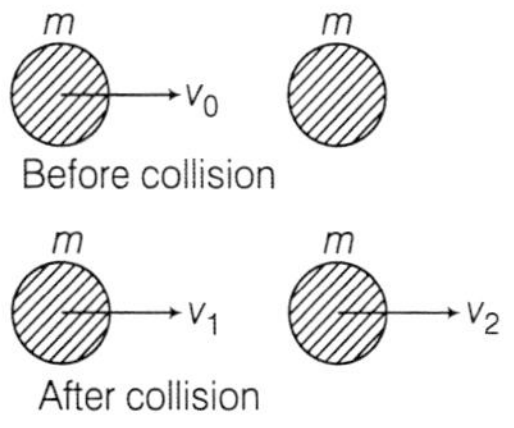

Conservation of momentum gives,

$$mv_0 = mv_1 + mv_2$$

$$v_0 = v_2 + v_1 \quad \ldots \text{(ii)}$$

From Eqs. (i) and (ii), we have

$$v_1^2 + v_2^2 + 2v_1v_2 = v_0^2$$

$$\Rightarrow \quad 2v_1v_2 = \frac{-v_0^2}{2}$$

$$\therefore \quad (v_1 - v_2)^2 = (v_1 + v_2)^2 - 4v_1v_2$$

$$= 2v_0^2$$

or $\quad v_{rel} = \sqrt{2} v_0$

> • Concept **Enhancer**
> Momentum is conserved in all type of collisions.

19. *(a)* According to conservation of momentum

$$mv = mv_1 \cos 45° + Mv_2 \cos 45°$$

$$\Rightarrow mu = (mv_1 + Mv_2)\frac{1}{\sqrt{2}} \quad \ldots (i)$$

and similarly along *y*-direction

$$0 = (mv_1 - Mv_2)\frac{1}{\sqrt{2}} \quad \ldots (ii)$$

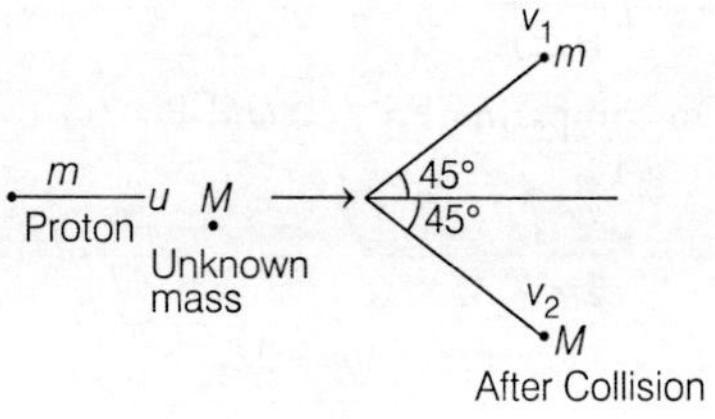

$$e = \frac{v_2 - v_1 \cos 90°}{u \cos 45°} = 1$$

$$\Rightarrow \quad u = \sqrt{2} v_2 \quad \ldots (iii)$$

Solving Eqs. (i), (ii) and (iii) we are get,

$$M = m$$

20. *(c)* Consider the movement of two particles as shown below.

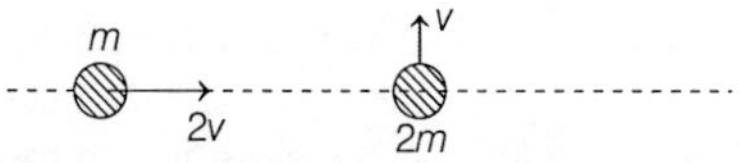

Conserving linear momentum in *x*-direction

$$(p_i)x = (p_f)x$$

or $\quad 2mv = (2m + m) v_x$ or $v_x = \frac{2}{3} v$

Conserving linear momentum in *y*-direction

$$(p_i)y = (p_f)y$$

or $\quad 2mv = (2m + m) v_y$ or $v_y = \frac{2}{3} v$

Initial kinetic energy of the two particles system is

$$E_i = \frac{1}{2} m (2v)^2 + \frac{1}{2}(2m)(v)^2$$

$$= \frac{1}{2} \times 4mv^2 + \frac{1}{2} \times 2mv^2$$

$$= 2mv^2 + mv^2 = 3mv^2$$

Final energy of the combined two particles system is

$$E_f = \frac{1}{2}(3m)(v_x^2 + v_y^2)$$

$$= \frac{1}{2}(3m)\left[\frac{4v^2}{9} + \frac{4v^2}{9}\right]$$

$$= \frac{3m}{2}\left[\frac{8v^2}{9}\right] = \frac{4mv^2}{3}$$

Loss in the energy $\Delta E = E_i - E_f$

$$= mv^2\left[3 - \frac{4}{3}\right] = \frac{5}{3} mv^2$$

Percentage loss in the energy during the collision

$$\frac{\Delta E}{E_i} \times 100 = \frac{(5/3)\, mv^2}{3mv^2} \times 100 = \frac{5}{9} \times 100$$

$$\simeq 56\%$$

21. *(d)* Energy, $E = \frac{p^2}{2m}$, where p is momentum, m is the mass moving of the particle. Maximum energy loss occurs when the particles get stuck together as a result of the collision.

Maximum energy loss (ΔE)

$$= \frac{p^2}{2m} - \frac{p^2}{2(m + M)}$$

where, $(m + M)$ is the resultant mass when the particles get stuck.

$$\Delta E = \frac{p^2}{2m}\left[1 - \frac{m}{m + M}\right]$$

$$= \frac{p^2}{2m}\left[\frac{M}{m + M}\right]$$

Also, $\quad p = mv$

$$\therefore \quad \Delta E = \frac{m^2 v^2}{2m}\left[\frac{M}{m + M}\right]$$

$$= \frac{mv^2}{2}\left[\frac{M}{m + M}\right]$$

Comparing the expression with

$$\Delta E = f\left(\frac{1}{2} mv^2\right),$$

$$f = \frac{M}{m + M}$$

22. *(c)* From law of conservation of momentum, we have

$$m_1 v_1 + m_2 v_2 = (m_1 + m_2) v$$

Given, $\quad m_1 = 0.50$ kg, $v_1 = 2$ ms^{-1},

$m_2 = 1$ kg, $v_2 = 0 \quad$ [at rest]

$$0.5 \times 2 + 1 \times 0 = 1.5 \times v$$

[assumed that 2nd body is at rest]

$$\Rightarrow \quad v = \frac{2}{3}$$

$$\therefore \quad \Delta K = K_f - K_i$$

$$= \frac{1.5 \times \left(\frac{2}{3}\right)^2}{2} - (0.5) \times \frac{2^2}{2}$$

$$= -\frac{2}{3} \text{ J} = -0.67 \text{ J}$$

23. *(c)* In *x*-direction, apply conservation of momentum, we get

$$mu_1 + 0 = mv_x$$

$$\Rightarrow \quad mv = mv_x \Rightarrow v_x = v$$

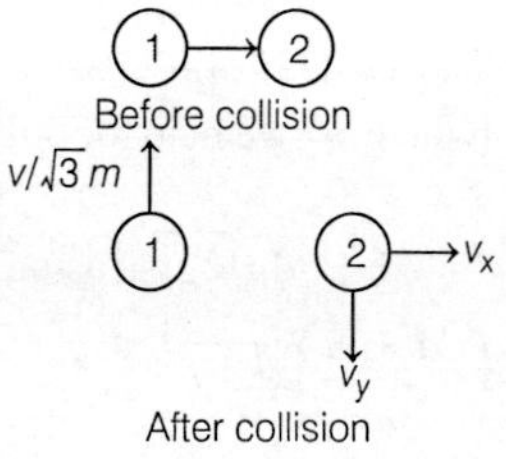

In *y*-direction, apply conservation of momentum, we get

$$0 + 0 = m\left(\frac{v}{\sqrt{3}}\right) - mv_y$$

$$\Rightarrow \quad v_y = \frac{v}{\sqrt{3}}$$

Velocity of second mass after collision

$$v' = \sqrt{\left(\frac{v}{\sqrt{3}}\right)^2 + v^2} = \sqrt{\frac{4}{3} v^2}$$

or $\quad v' = \frac{2}{\sqrt{3}} v$

24. *(c)* According to conservation of linear momentum,

$$m_1 v_1 + m_2 v_2 = (m_1 + m_2) v_{cm}$$

$$\Rightarrow \quad 10 \times 3 + 5 \times 0 = (10 + 5) v_{cm}$$

$$\Rightarrow \quad v_{cm} = \frac{30}{19} = 2 \text{ m/s}$$

According to conservation of energy

$$\frac{1}{2} kx^2 = \frac{1}{2} m_1 v^2 - \frac{1}{2}(m_1 + m_2) v_{cm}^2$$

$$\frac{1}{2} \times 3000 x^2 = \frac{1}{2} \times 10 \times 3^2 - \frac{1}{2}(10 + 5) \times 2^2$$

$$\Rightarrow \quad 1500 x^2 = 45 - 30$$

$$\Rightarrow \quad 1500 x^2 = 15$$

$$\Rightarrow \quad x^2 = \frac{1}{100}$$

$$\Rightarrow \quad x = \frac{1}{10} \text{ m}$$

$$= 0.1 \text{ m}$$

25. *(b)* To calculate the speed of bob

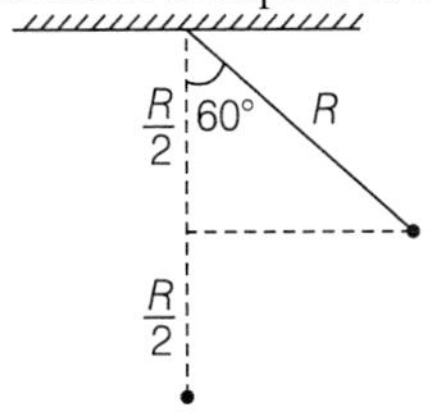

A just before the collision, apply conservation of mechanical energy.

$$\text{KE}_i + \text{PE}_i = \text{KE}_f + \text{PE}_f$$

$$0 + \frac{mgR}{2} = \frac{1}{2}mv_1^2 + 0 \Rightarrow v_1 = \sqrt{gR}$$

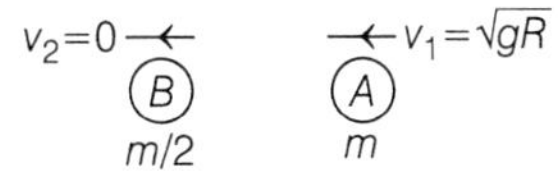

Let final velocities of A and B after collisions be v_1' and v_2'

By applying conservation of momentum

$$mv_1 + \frac{mv_2}{2} = mv_1' + \frac{mv_2'}{2}$$

$$\sqrt{gR} + 0 = v_1' + \frac{v_2'}{2} \quad \ldots\text{(i)}$$

By using coefficient of restitution

$$e = \frac{v_2' - v_1'}{v_1 - v_2} = 1 \Rightarrow v_2' - v_1' = v_1 - v_2$$

$$v_2' - v_1' = v \quad [v_1 = v]$$

$$v_2' = v + v_1' \quad \ldots\text{(ii)}$$

Put value of v_2' in Eq. (i)

$$\sqrt{gR} = v_1' + \frac{v}{2} + \frac{v_1'}{2}$$

$$\Rightarrow \sqrt{gR} - \frac{v}{2} = \frac{3v_1'}{2}$$

$$\sqrt{gR} - \frac{\sqrt{gR}}{2} = \frac{3v_1'}{2}$$

$$\frac{\sqrt{gR}}{2} = \frac{3v_1'}{2} \Rightarrow v_1' = \frac{\sqrt{gR}}{3}$$

26. *(c)* Let m be the mass of lighter body travelling with speed u.

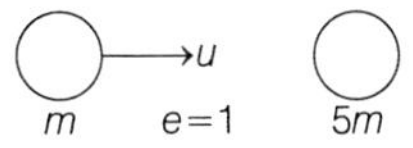

By conservation of momentum,

$$\Rightarrow m_1u_1 + m_2u_2 = m_1v_1 + m_2v_2$$

where m_1 is the lighter mass and m_2 is the heavier stationary mass

$$\Rightarrow m \times u + (5m \times 0) = mv_1 + 5mv_2 \quad \ldots\text{(i)}$$

As, $e = \dfrac{v_2 - v_1}{u_1 - u_2} = 1$, for elastic collision

$$\Rightarrow 1 = \frac{v_2 - v_1}{u - 0} \Rightarrow v_2 - v_1 = u$$

or $v_2 = u + v_1 \quad \ldots\text{(ii)}$

On putting value from Eq. (ii) in Eq. (i), we get

$$mu = mv_1 + 5m(u + v_1)$$

$$mu = mv_1 + 5mu + 5mv_1$$

$$-4mu = 6mv_1$$

$$\Rightarrow v_1 = \frac{-4u}{6} = \frac{-2u}{3}$$

Now $\text{KE}_i = \dfrac{1}{2}mu^2 \quad \ldots\text{(iii)}$

$$\text{KE}_f = \frac{1}{2}mv_1^2 = \frac{1}{2}m\left(-\frac{2u}{3}\right)^2$$

$$= \frac{1}{2} \times \frac{4}{9}mu^2$$

From Eq. (iii), we get

$$\text{KE}_f = \frac{4}{9}\text{KE}_i$$

So, the percentage of kinetic energy transferred is

$$\frac{\Delta\text{KE}}{\text{KE}_i} \times 100\% = \left(\frac{\text{KE}_i - \text{KE}_f}{\text{KE}_i}\right) \times 100\%$$

$$= \left(1 - \frac{4}{9}\right) \times 100\% = 55.5\%$$

27. *(c)* Given, mass of body 1 $= M$

Mass of body 2 $= m$

Initial speed of body 1, $u_1 = v_0$

Initial speed of body 2, $u_2 = 0$

Final speed of body 1 and 2 $= v_1$ and v_2

Angle made by body 1 and 2 after collision with respect to initial direction $= \theta_1, \theta_2$

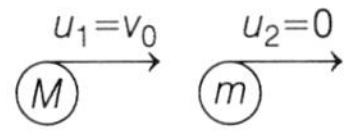

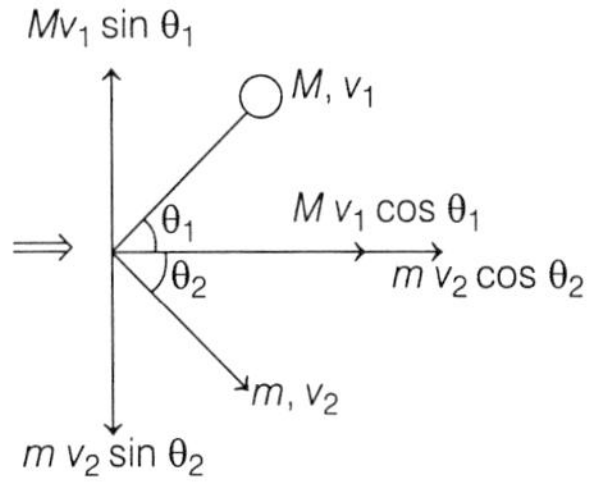

By using law of conservation of momentum,

Along X-axis,

$$Mv_0 + m \cdot 0 = Mv_1 \cos\theta_1 + mv_2 \cos\theta_2$$

If $\theta_1 = \theta_2 = \theta$

$$\therefore \; Mv_0 = Mv_1 \cos\theta + mv_2 \cos\theta \quad \ldots\text{(i)}$$

Along Y-axis,

$$Mv_1 \sin\theta_1 = mv_2 \sin\theta_2$$

[since, $\theta_1 = \theta_2 = \theta$]

$$v_2 = \frac{Mv_1}{m} \quad \ldots\text{(ii)}$$

On substituting the value of v_2 in Eq. (i), we get

$$Mv_0 = Mv_1 \cos\theta + m\left(\frac{Mv_1}{m}\right)\cos\theta$$

$$= 2Mv_1 \cos\theta$$

$$v_1 = \frac{v_0}{2\cos\theta} \quad \ldots\text{(iii)}$$

By using law of conservation of energy along X-axis

$$\frac{1}{2}Mv_0^2 = \frac{1}{2}Mv_1^2 + \frac{1}{2}mv_2^2$$

$$\Rightarrow Mv_0^2 = Mv_1^2 + m\left(\frac{M}{m}v_1\right)^2$$

[from Eq. (ii)]

$$\Rightarrow Mv_0^2 = Mv_1^2 + \frac{M^2v_1^2}{m}$$

$$= \frac{Mv_1^2}{m}(m + M)$$

$$\Rightarrow v_0^2 = \left(\frac{v_0}{2\cos\theta}\right)^2\left(\frac{m + M}{m}\right)$$

[From Eq. (iii)]

$$\Rightarrow v_0^2 = \frac{v_0^2}{4\cos^2\theta}\left(1 + \frac{M}{m}\right)$$

$$\Rightarrow 4\cos^2\theta = 1 + \frac{M}{m}$$

$$\Rightarrow \frac{M}{m} = 4\cos^2\theta - 1$$

$$\left[\text{For largest possible value of } \frac{M}{m}, \theta = 0\right]$$

$$= 4\cos^2 0° - 1$$

$$= 4 - 1 = 3$$

28. *(d)* Velocity of rubber ball when it strikes to the ground, $v_0 = \sqrt{2gh_0}$

Using the formula of coefficient of restitution,

$$e = \frac{\text{velocity after collision}}{\text{velocity before collision}}$$

$$\Rightarrow e = \frac{v_1 - 0}{v_0 - 0} \Rightarrow e = \frac{v_1}{\sqrt{2gh_0}}$$

$$\Rightarrow v_1 = e\sqrt{2gh_0} \quad \ldots\text{(i)}$$

As, initial height of the ball $= h_0$.

$\therefore$ The first height of the rebound,

$$h_1 = \frac{v_1^2}{2g}$$

$$\Rightarrow h_1 = e^2h_0 \quad [\text{Using Eq. (i)}]$$

The nth height of the ball to the rebound,

$$h_n = \frac{v_n^2}{2g} \Rightarrow h_n = e^{2n}h_0$$

The velocity of the ball after nth rebound, $v_n = e^n v_0$

Now, the total distance travelled by the ball after nth rebound is

$$H = h_0 + 2h_1 + 2h_2 + 2h_3 \ldots$$

$$H = h_0 + 2e^2 h_0 + 2e^4 h_0 + 2e^6 h_0 \ldots$$

$$H = h_0 [1 + 2e^2 (1 + e^2 + e^4 + e^6 \ldots)]$$

Using the formula,

$$1 + e^2 + e^4 + \ldots = \frac{1}{1 - e^2}$$

$$\Rightarrow \quad H = h_0 \left[1 + 2e^2 \left(\frac{1}{1 - e^2}\right)\right]$$

$$\Rightarrow \quad H = h_0 \left(\frac{1 + e^2}{1 - e^2}\right)$$

$$\Rightarrow \quad H = 5 \left[\frac{1 + (0.81)}{1 - (0.81)}\right]$$

$$\left(\because e^2 = \frac{h_1}{h_0} = \frac{81}{100}\right)$$

$$H = 47.6 \text{ m}$$

Now, the total time taken by the ball to come to rest

$$T = t_0 + 2t_1 + 2t_2 + \ldots$$

$$T = \sqrt{\frac{2h_0}{g}} + 2\sqrt{\frac{2h_1}{g}} + 2\sqrt{\frac{2h_2}{g}} + \ldots$$

$$T = \sqrt{\frac{2h_0}{g}} [1 + 2e + 2e^2 + \ldots]$$

$$T = \sqrt{\frac{2h_0}{g}} [1 + 2e(1 + e + e^2 + \ldots)]$$

$$T = \sqrt{\frac{2h_0}{g}} \left(\frac{1 + e}{1 - e}\right)$$

$$\Rightarrow \quad T = \sqrt{\frac{2(5)}{10}} \left(\frac{1 + \sqrt{0.81}}{1 - \sqrt{0.81}}\right) = 19 \text{ s}$$

The average velocity,

$$v_{avg} = \frac{H}{T} = \frac{47.6}{19} = 2.5 \text{ m/s}$$

29. *(c)* Initial condition can be shown in the figure below.

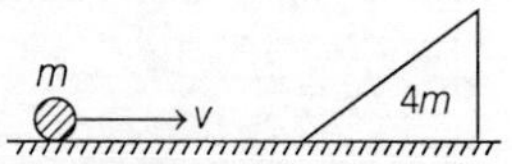

As mass m collides with wedge, let both wedge and mass move with speed v'. Where it reaches to maximum height.

Then,

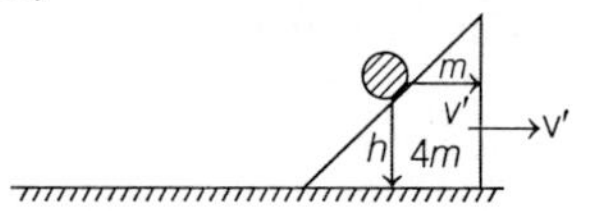

By applying linear momentum conservation, we have initial momentum of the system.

= Final momentum of the system

$$mv + 0 = (m + 4m) v'$$

$$\Rightarrow \quad v' = v/5 \quad \ldots \text{(i)}$$

Now, if m rises upto height h over wedge, then by applying conservation of mechanical energy, we have initial energy of the system

= Final energy of the system

$$\frac{1}{2} mv^2 + 0 = \frac{1}{2} mv'^2 + mgh + \frac{1}{2}(4m)v'^2$$

$$mv^2 = (m + 4m)v'^2 + 2mgh$$

$$\Rightarrow \quad v^2 = 5v'^2 + 2gh$$

$$\Rightarrow \quad v^2 = \frac{1}{5} v^2 + 2gh \quad \text{[using Eq. (i)]}$$

$$\Rightarrow \quad \frac{4}{5} v^2 = 2gh \Rightarrow h = \frac{2v^2}{5g}$$

30. *(a)* During completely inelastic collision both particles A and B stick together, thus situation is shown below.

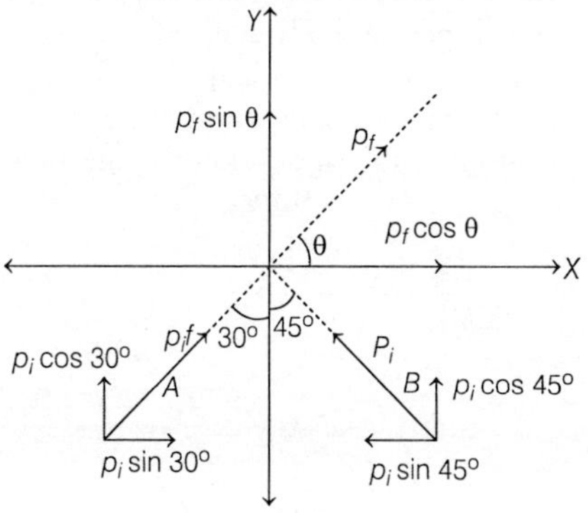

Using conservation of linear momentum, we get

Along X-axis,

$$p_f \cos \theta = p_i \sin 30° - p_i \sin 45° \quad \ldots \text{(i)}$$

Along Y-axis,

$$p_f \sin \theta = p_i \cos 30° + p_i \cos 45° \ldots \text{(ii)}$$

Dividing Eqs. (ii) by Eq. (i), we get

$$\frac{\sin \theta}{\cos \theta} = \frac{\cos 30° + \cos 45°}{\sin 30° - \sin 45°} = \frac{\frac{\sqrt{3}}{2} + \frac{1}{\sqrt{2}}}{\frac{1}{2} - \frac{1}{\sqrt{2}}}$$

$$\Rightarrow \quad \tan \theta = \frac{\sqrt{3} + \sqrt{2}}{1 - \sqrt{2}}$$

31. *(375)* Let u and v be speeds, just before and after body strikes the ground.

Given, $\frac{u}{v} = \frac{4}{1}$

$$\text{loss in KE} \Rightarrow \Delta K = \frac{\frac{1}{2} mu^2 - \frac{1}{2} mv^2}{\frac{1}{2} mu^2}$$

$$\Delta\text{KE} = 1 - \left(\frac{1}{4}\right)^2 = 1 - \frac{1}{16} = \frac{15}{16}$$

$$\text{Percentage loss} = \frac{15}{16} \times 100 = \frac{x}{4} \Rightarrow x = 375$$

32. *(4)* Before collision,

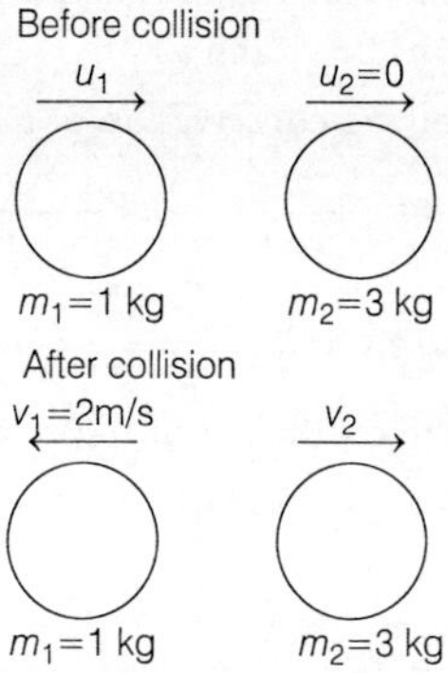

By the conservation of momentum,

$$p_i = p_f$$

$$\Rightarrow \quad m_1 u_1 + m_2 u_2 = m_1 v_1 + m_2 v_2$$

$$\Rightarrow \quad 1.u_1 + 3 \times 0 = 1 \times (-2) + 3v_2$$

$$\Rightarrow \quad u_1 = -2 + 3v_2$$

$$\Rightarrow \quad v_2 = \frac{u_1 + 2}{3} \quad \ldots \text{(i)}$$

In elastic collision, kinetic energy is conserved, hence.

$$\Rightarrow \frac{1}{2} m_1 u_1^2 + \frac{1}{2} m_2 u_2^2 = \frac{1}{2} m_1 v_1^2 + \frac{1}{2} m_2 v_2^2$$

$$\Rightarrow \frac{1}{2} \times 1 \times u_1^2 + \frac{1}{2} \times 3 \times 0$$

$$= \frac{1}{2} \times 1 \times (-2)^2 + \frac{1}{2} \times 3 \times v_2^2$$

$$\Rightarrow \frac{u_1^2}{2} = 2 + \frac{3}{2} v_2^2 = 2 + \frac{3}{2}\left(\frac{u_1 + 2}{3}\right)^2$$

$$\frac{u_1^2}{2} = 2 + \frac{u_1^2 + 4 + 4u_1}{6}$$

$$\Rightarrow \quad u_1^2 - 2u_1 - 8 = 0$$

$$\Rightarrow \quad u_1^2 - 4u_1 + 2u_1 - 8 = 0$$

$\Rightarrow\ u_1(u_1-4)+2(u_1-4)=0$

$\Rightarrow\ (u_1-4)(u_1+2)=0$

$\Rightarrow\ u_1=4\text{ m/s}$

33. *(20)* Let's consider the final velocity of the second piece of the second ball $= v_3$

As shown in the figure,

The net external force on the system is to be zero. Hence, we can use the law of conservation of linear momentum in both directions.

In x-direction the linear momentum conserved,

$$m_1u_{1x}+m_2u_{2x}=m_1v_{1x}+m_2v_{2x}+m_3v_{3x}$$

$$\Rightarrow\quad 10(10\sqrt{3})+0=10(0)+0+10v_3\cos 30^\circ$$

$$\Rightarrow\quad v_3=20\text{ m/s}$$

Hence, the velocity of the ball moving at 30° with respect to the X-axis is 20 m/s. So, the value of x to the nearest integer is 20.

34. *(30)* We can represent the given situation in figure as

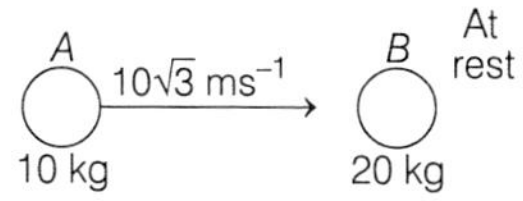

Before collision

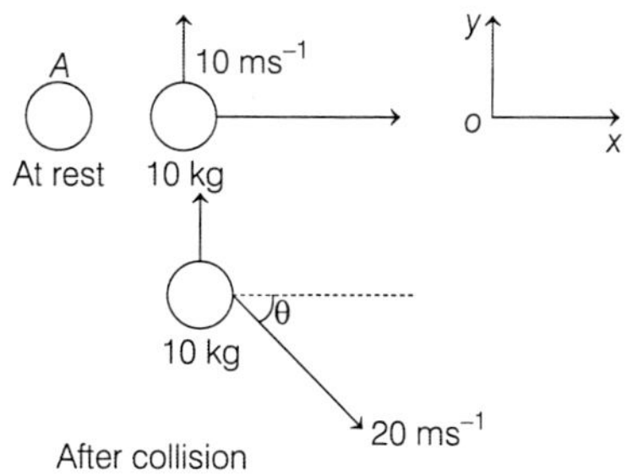

After collision

It means linear momentum is conserved along X-axis.

According to the law of conservation of linear momentum, $\mathbf{p}_i=\mathbf{p}_f$

$$m_Au_A+m_Bu_B=m_Av_A+(m_Bv_B\cos\theta)$$

$$=m_Av_A+m_{B_1}v_{B_1}\cos 90^\circ+m_{B_2}v_{B_2}\cos\theta$$

$$\Rightarrow 10\times 10\sqrt{3}+20\times 0=10\times 0+10\times 10\times 0+10\times 20\cos\theta$$

$$\Rightarrow\quad 10\times 10\sqrt{3}=200\cos\theta$$

where, $\cos\theta$ being the horizontal component i.e., along X-axis

$$\Rightarrow\quad \cos\theta=\sqrt{3}/2$$

$$\Rightarrow\quad \theta=30^\circ\qquad [\because \cos 30^\circ=\sqrt{3}/2]$$

35. *(10)* Given, For particle 1, final KE is equals to half of its initial value,

$$\Rightarrow\quad KE_f=\frac{1}{2}KE_i$$

$$\Rightarrow\quad \frac{1}{2}m_1v_1^2=\left(\frac{1}{2}m_1u^2\right)\times\frac{1}{2}$$

$\Rightarrow$ Final velocity of m_1 will be,

$$v_1=\frac{u}{\sqrt{2}}$$

Momentum in y-direction is conserved, i.e. $10mv_2\sin\theta_2=mv_1\sin\theta_1$

Here, $v_1=u/\sqrt{2}$

and $\sin\theta_1=\sqrt{n}\sin\theta_2$

So, we have

$$10mv_2\cdot\sin\theta_2=\frac{mu}{\sqrt{2}}\cdot\sqrt{n}\sin\theta_2$$

$$\Rightarrow\quad v_2=\frac{u\sqrt{n}}{10\sqrt{2}}\qquad\ldots\text{(i)}$$

Also collision is elastic, so KE is conserved.

$$\Rightarrow\quad \frac{1}{2}mu^2=\frac{1}{2}mv_1^2+\frac{1}{2}(10m)v_2^2$$

On substituting values of v_1 and v_2, we get

$$u^2=\left(\frac{u}{\sqrt{2}}\right)^2+10\left(\frac{u^2n}{100\times 2}\right)$$

$$\Rightarrow\quad \frac{u^2}{2}=\frac{u^2n}{10\times 2}$$

$$\Rightarrow\quad n=10$$

C-06

JEE Main Chapterwise Topicwise Physics

In JEE Main there is usually 1-2 question per paper are visible from this chapter. In past 3 years, 31 conceptual and 44 numerical questions were asked of easy to moderate level from this chapter.

For detailed analysis scan QR code

SYSTEM OF PARTICLES AND ROTATIONAL MOTION

Topic 1 Centre of Mass, Torque and Angular Momentum

Level 1

01 A rod of length 5 L is bent right angle keeping one side length as 2 L.

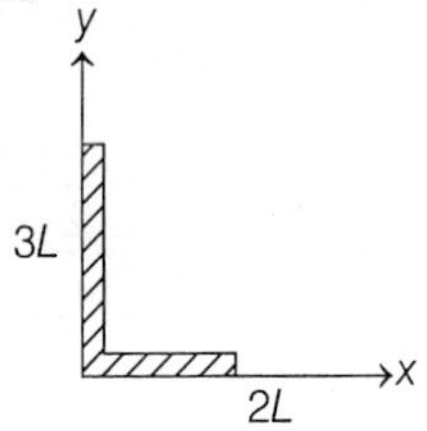

The position of the centre of mass of the system. (Consider $L = 10$, cm) *[2025, 07 April Shift-I]*

(a) $5\hat{i} + 8\hat{j}$ (b) $4\hat{i} + 9\hat{j}$ (c) $3\hat{i} + 7\hat{j}$ (d) $2\hat{i} + 3\hat{j}$

02 A uniform rod of mass 250 g having length 100 cm is balanced on a sharp edge at 40 cm mark. A mass of 400 g is suspended at 10 cm mark. To maintain the balance of the rod, the mass to be suspended at 90 cm mark is *[2025, 28 Jan Shift-II]*

(a) 290 g (b) 190 g (c) 300 g (d) 200 g

Imp. Torque

03 The torque due to the force $(2\hat{i} + \hat{j} + 2\hat{k})$ about the origin, acting on a particle whose position vector is $(\hat{i} + \hat{j} + \hat{k})$, would be *[2025, 22 Jan Shift-II]*

(a) $\hat{i} + \hat{k}$ (b) $\hat{i} - \hat{j} + \hat{k}$ (c) $\hat{j} + \hat{k}$ (d) $\hat{i} - \hat{k}$

04 The torque of a force $5\hat{i}+3\hat{j}-7\hat{k}$. about the origin is τ. If the force acts on a particle whose position vector is $2\hat{i}+2\hat{j}+\hat{k}$, then the value of τ will be *[2022, 29 July Shift-II]*

(a) $11\hat{i}+19\hat{j}-4\hat{k}$ (b) $-11\hat{i}+9\hat{j}-16\hat{k}$

(c) $-17\hat{i}+19\hat{j}-4\hat{k}$ (d) $17\hat{i}+9\hat{j}+16\hat{k}$

05 Angular momentum of a single particle moving with constant speed along circular path *[2021, 31 Aug Shift-1]*

(a) changes in magnitude but remains same in the direction

(b) remains same in magnitude and direction

(c) remains same in magnitude but changes in the direction

(d) is zero

06 Four point masses, each of mass m are fixed at the corners of a square of side l. The square is rotating with angular frequency ω, about an axis passing through one of the corners of the square and parallel to its diagonal, as shown in the figure.

The angular momentum of the square about this axis is *[2020, 06 Sep Shift-I]*

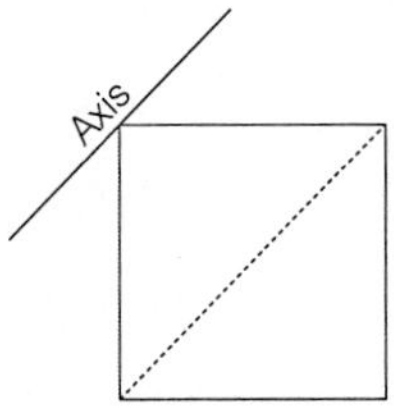

(a) $ml^2\omega$ (b) $4\,ml^2\omega$ (c) $3\,ml^2\omega$ (d) $2\,ml^2\omega$

07 Three point particles of masses 1.0 kg, 1.5 kg and 2.5 kg are placed at three corners of a right angle triangle of sides 4.0 cm, 3.0 cm and 5.0 cm as shown in the figure. The centre of mass of the system is at a point *[2020, 7 Jan Shift-I]*

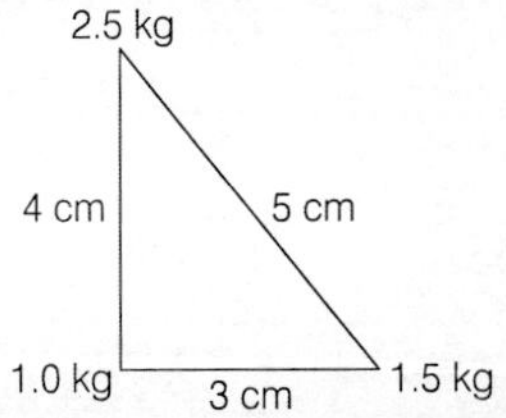

(a) 2.0 cm right and 0.9 cm above 1 kg mass

(b) 0.6 cm right and 2.0 cm above 1 kg mass

(c) 1.5 cm right and 1.2 cm above 1 kg mass

(d) 0.9 cm right and 2.0 cm above 1 kg mass

Imp. Principle of Torque

08 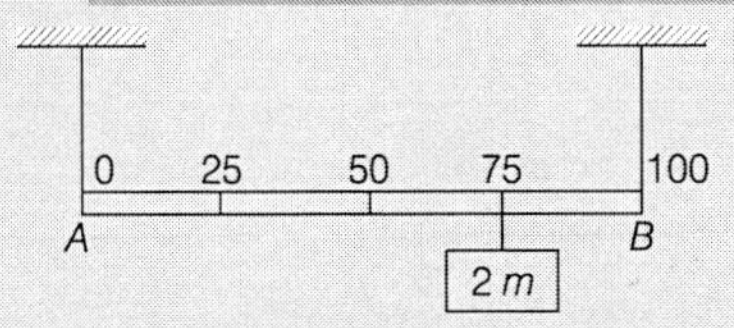

Shown in the above figure, a rigid and uniform 1 m long rod AB held in horizontal position by two strings tied to its ends and attached to the ceiling. The rod is of mass m and has another weight of mass 2 m hung at a distance of 75 cm from A. The tension in the string at A is *[2020, 2 Sep Shift-I]*

(a) 1 mg (b) 2 mg
(c) 0.75 mg (d) 0.5 mg

09 Four particles A, B, C and D with masses $m_A = m$, $m_B = 2m$, $m_C = 3m$ and $m_D = 4m$ are at the corners of a square. They have accelerations of equal magnitude with directions as shown. The acceleration of the centre of mass of the particles (in ms^{-2}) is *[2019, 8 April Shift-I]*

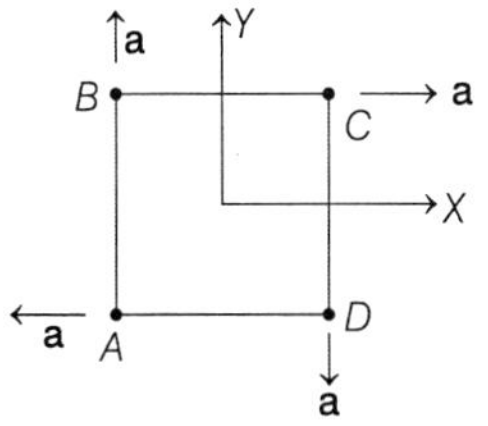

(a) $\frac{a}{5}(\hat{i}-\hat{j})$ (b) $a(\hat{i}-\hat{j})$
(c) zero (d) $\frac{a}{5}(\hat{i}+\hat{j})$

Imp. Centre of Mass

10 Three particles of masses 50 g, 100 g and 150 g are placed at the vertices of an equilateral triangle of side 1 m (as shown in the figure). The (x, y) coordinates of the centre of mass will be *[2019, 12 April Shift-II]*

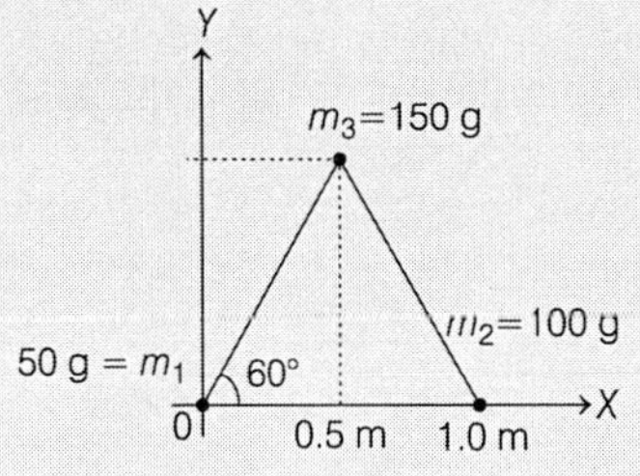

(a) $\left(\frac{\sqrt{3}}{4}\text{ m}, \frac{5}{12}\text{ m}\right)$ (b) $\left(\frac{7}{12}\text{ m}, \frac{\sqrt{3}}{8}\text{ m}\right)$
(c) $\left(\frac{7}{12}\text{ m}, \frac{\sqrt{3}}{4}\text{ m}\right)$ (d) $\left(\frac{\sqrt{3}}{8}\text{ m}, \frac{7}{12}\text{ m}\right)$

11 The magnitude of torque on a particle of mass 1 kg is 2.5 N-m about the origin. If the force acting on it is 1 N and the distance of the particle from the origin is 5 m, then the angle between the force and the position vector is (in radian) *[2019, 11 Jan Shift-II]*

(a) $\frac{\pi}{8}$ (b) $\frac{\pi}{4}$ (c) $\frac{\pi}{3}$ (d) $\frac{\pi}{6}$

12 A uniform rod AB is suspended from a point X, at a variable distance x from A, as shown. To make the rod horizontal, a mass m is suspended from its end A. A set of (m, x) values is recorded. The appropriate variables that give a straight line, when plotted, are *[JEE main 2018]*

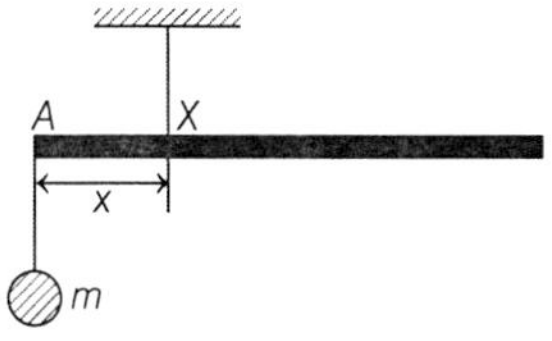

(a) $m, 1/x^2$ (b) m, x^2 (c) m, x (d) $m, 1/x$

13 Distance of the centre of mass of a solid uniform cone from its vertex is z_0. If the radius of its base is R and its height is h, then z_0 is equal to *[JEE Main 2015]*

(a) $\frac{h^2}{4R}$ (b) $\frac{3h}{4}$ (c) $\frac{5h}{8}$ (d) $\frac{3h^2}{8R}$

14 A bob of mass m attached to an inextensible string of length l is suspended from a vertical support. The bob rotates in a horizontal circle with an angular speed ω rad/s about the vertical support. About the point of suspension, *[JEE Main 2014]*

(a) angular momentum is conserved
(b) angular momentum changes in magnitude but not in direction
(c) angular momentum changes in direction but not in magnitude
(d) angular momentum changes both in direction and magnitude

15 A small particle of mass m is projected at an angle θ with the x-axis with an intial velocity v_0 in the xy plane as shown in the figure. At a time $t < \frac{v_0 \sin\theta}{g}$, the angular momentum of the particle is *[AIEEE 2010]*

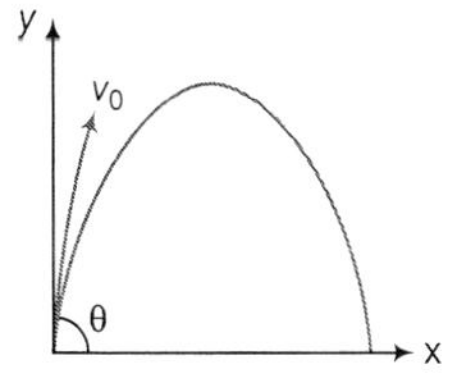

(a) $-mgv_0t^2\cos\theta\,\hat{j}$ (b) $mgv_0t\cos\theta\,\hat{k}$
(c) $-\frac{1}{2}mgv_0t^2\cos\theta\,\hat{k}$ (d) $\frac{1}{2}mgv_0t^2\cos\theta\,\hat{i}$

16 Angular momentum of the particle rotating with a central force is constant due to *[AIEEE 2007]*

(a) constant force
(b) constant linear momentum
(c) zero torque
(d) constant torque

17 Consider a two particle system with particles having masses m_1 and m_2. If the first particle is pushed towards the centre of mass through a distance d, by what distance should the second particle be moved, so as to keep the centre of mass at the same position ? *[AIEEE 2006]*

(a) $\frac{m_2}{m_1}d$ (b) $\frac{m_1}{m_1+m_2}d$ (c) $\frac{m_1}{m_2}d$ (d) d

Imp. Torque

18 A force of $-F_k$ acts on O, the origin of the coordinate system. The torque about the point $(1, -1)$ is *[AIEEE 2006]*

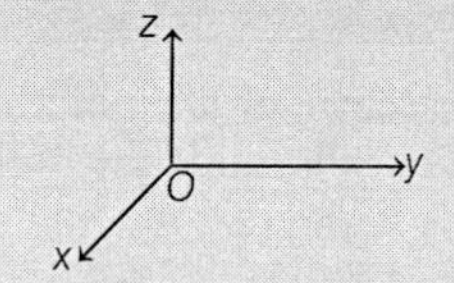

(a) $F(\hat{i}-\hat{j})$ (b) $-F(\hat{i}-\hat{j})$ (c) $F(\hat{i}+\hat{j})$ (d) $-F(\hat{i}+\hat{j})$

19 A thin circular ring of mass m and radius R is rotating about its axis with a constant angular velocity ω. Two objects each of mass M are attached gently to the opposite ends of a diameter of the ring. The ring now rotates with an angular velocity $\omega' =$ *[AIEEE 2006]*

(a) $\frac{\omega(m+2M)}{m}$ (b) $\frac{\omega(m-2M)}{(m+2M)}$
(c) $\frac{\omega m}{(m+M)}$ (d) $\frac{\omega m}{(m+2M)}$

20 A body A of mass M while falling vertically downwards under gravity breaks into two parts; a body B of mass $\frac{1}{3}M$ and a body C of mass $\frac{2}{3}m$. The centre of mass of bodies B and C taken together shifts compared to that of body A towards *[AIEEE 2005]*

(a) depends on height of breaking
(b) does not shift
(c) body C
(d) body B

21 A T shaped object with dimensions shown in the figure, is lying on a smooth floor. A force **F** is applied at the point P parallel to AB, such that the object has only the translational motion without rotation. Find the location of P with respect to C. *[AIEEE 2005]*

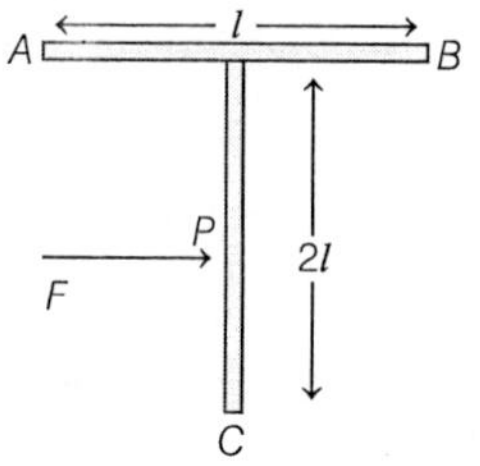

(a) $\frac{2}{3}l$ (b) $\frac{3}{2}l$
(c) $\frac{4}{3}l$ (d) l

22 A solid sphere is rotating in free space. If the radius of the sphere is increased keeping mass same, which one of the following will not be affected? *[AIEEE 2004]*

(a) Moment of inertia (b) Angular momentum
(c) Angular velocity (d) Rotational kinetic energy

23 A particle performing uniform circular motion has angular momentum L. If its angular frequency is doubled and its kinetic energy halved, then the new angular momentum is *[AIEEE 2003]*

(a) $\frac{L}{4}$ (b) $2L$
(c) $4L$ (d) $L/2$

24 Let **F** be the force acting on a particle having position vector **r** and τ be the torque of this force about the origin. Then, *[AIEEE 2003]*

(a) $\mathbf{r}\cdot\tau=0$ and $\mathbf{F}\cdot\tau\neq0$ (b) $\mathbf{r}\cdot\tau\neq0$ and $\mathbf{F}\cdot\tau=0$
(c) $\mathbf{r}\cdot\tau\neq0$ and $\mathbf{F}\cdot\tau\neq0$ (d) $\mathbf{r}\cdot\tau=0$ and $\mathbf{F}\cdot\tau=0$

25 Initial angular velocity of a circular disc of mass M is ω_1. Then, two small spheres of mass m are attached gently to two diametrically opposite points on the edge of the disc. What is the final angular velocity of the disc? *[AIEEE 2002]*

(a) $\left(\frac{M+m}{M}\right)\omega_1$ (b) $\left(\frac{M+m}{m}\right)\omega_1$
(c) $\left(\frac{M}{M+4m}\right)\omega_1$ (d) $\left(\frac{M}{M+2m}\right)\omega_1$

26 A particle of mass m moves along line PC with velocity v as shown. What is the angular momentum of the particle about O? *[AIEEE 2002]*

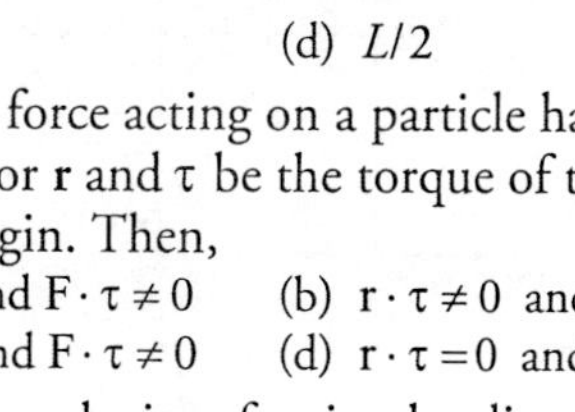

(a) mvL (b) mvl
(c) mvr (d) Zero

Imp. Motion of Centre of Mass

27 Two identical particles move towards each other with velocity $2v$ and v, respectively. The velocity of centre of mass is [AIEEE 2002]

(a) v (b) $\frac{v}{3}$ (c) $\frac{v}{2}$ (d) zero

Level 2

28 If $\mathbf{L}$ and $\mathbf{p}$ represent the angular momentum and linear momentum respectively of a particle of mass m having position vector as $\mathbf{r} = a(\hat{\mathbf{i}} \cos \omega t + \hat{\mathbf{j}} \sin \omega t)$. The direction of force is [2025, 4 April Shift-I]

(a) opposite to the direction of $\mathbf{r}$
(b) opposite to the direction of $\mathbf{p}$
(c) opposite to the direction of $\mathbf{L}$
(d) opposite to the direction of $\mathbf{L} \times \mathbf{p}$

29 Which of the following are correct expression for torque acting on a body? [2025, 4 April Shift-I]

A. $\tau = \mathbf{r} \times \mathbf{L}$ B. $\tau = \frac{d}{dt}(\mathbf{r} \times \mathbf{p})$

C. $\tau = \mathbf{r} \times \frac{d\mathbf{p}}{dt}$ D. $\tau = I\alpha$

E. $\tau = \mathbf{r} \times \mathbf{F}$

($\mathbf{r}$ = position vector; $\mathbf{p}$ = linear momentum; $\mathbf{L}$ = angular momentum; α = angular acceleration; I = moment of inertia; $\mathbf{F}$ = force; t = time)

Choose the correct answer from the options given below.

(a) B, C, D and E Only (b) C and D Only
(c) A, B, D and E Only (d) B, D and E Only

Imp. Center of Mass

30 Consider a circular disc of radius 20 cm with centre located at the origin. A circular hole of radius 5 cm is cut from this disc in such a way that the edge of the hole touches the edge of the disc. The distance of centre of residual or remaining disc from the origin will be [2025, 23 Jan Shift-I]

(a) 0.5 cm (b) 2.0 cm (c) 1.0 cm (d) 1.5 cm

31

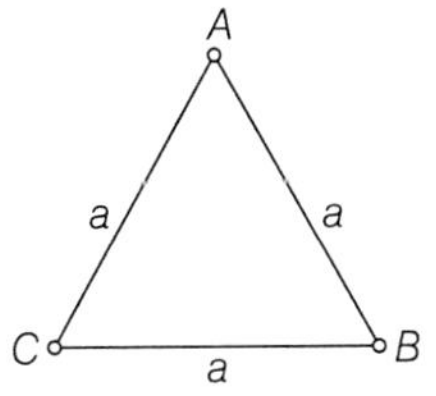

Three equal masses m are kept at vertices (A, B, C) of an equilateral triangle of side a in free space. At $t = 0$, they are given an initial velocity $\mathbf{v}_A = v_0\mathbf{AC}$, $\mathbf{v}_B = v_0\mathbf{BA}$ and $\mathbf{v}_C = v_0\mathbf{CB}$. Here, $\mathbf{AC}$, $\mathbf{CB}$ and $\mathbf{BA}$ are unit vectors along the edges of the triangle. If the three masses interact gravitationally, then the magnitude of the net angular momentum of the system at the point of collision is [2025, 29 Jan Shift-II]

(a) $3\, a\, mv_0$ (b) $\frac{3}{2}\, a\, mv_0$ (c) $\frac{\sqrt{3}}{2}\, a\, mv_0$ (d) $\frac{1}{2}\, a\, mv_0$

32 A heavy iron bar of weight w is having its one end on the ground and the other on the shoulder of a person. The bar makes an angle θ with the horizontal. The weight experienced by the person is [2024, 9 April Shift-I]

(a) w (b) $w \cos \theta$ (c) $w \sin \theta$ (d) $w/2$

33 A particle of mass m is projected with a velocity u making an angle of 30° with the horizontal. The magnitude of angular momentum of the projectile about the point of projection, when the particle is at its maximum height h is [2024, 30 Jan Shift-I]

(a) zero (b) $\frac{\sqrt{3}}{2}\frac{mu^2}{g}$ (c) $\frac{mu^3}{\sqrt{2}\,g}$ (d) $\frac{\sqrt{3}}{16}\frac{mu^3}{g}$

34 A heavy iron bar of weight 12 kg is having its one end on the ground and the other on the shoulder of a man. The rod makes an angle 60° with the horizontal, the weight experienced by the man is [2024, 27 Jan Shift-II]

(a) 6 kg (b) $6\sqrt{3}$ kg (c) 3 kg (d) 12 kg

35 Two bodies of mass 1 kg and 3 kg have position vectors $\hat{\mathbf{i}} + 2\hat{\mathbf{j}} + \hat{\mathbf{k}}$ and $-3\hat{\mathbf{i}} - 2\hat{\mathbf{j}} + \hat{\mathbf{k}}$, respectively. The position vector of centre of mass of this system will be similar to the vector [2022, 29 July Shift-I]

(a) $\hat{\mathbf{i}} - 2\hat{\mathbf{j}} + \hat{\mathbf{k}}$ (b) $-3\hat{\mathbf{i}} - 2\hat{\mathbf{j}} + \hat{\mathbf{k}}$
(c) $-2\hat{\mathbf{i}} + 2\hat{\mathbf{k}}$ (d) $-2\hat{\mathbf{i}} - \hat{\mathbf{j}} + \hat{\mathbf{k}}$

36 A mass M hangs on a massless rod of length l which rotates at a constant angular frequency. The mass M moves with steady speed in a circular path of constant radius. Assume that the system is in steady circular motion with constant angular velocity ω. The angular momentum of M about point A is L_A which lies in the positive z-direction and the angular momentum of M about B is L_B. The correct statement for this system is [2021, 17 March Shift-I]

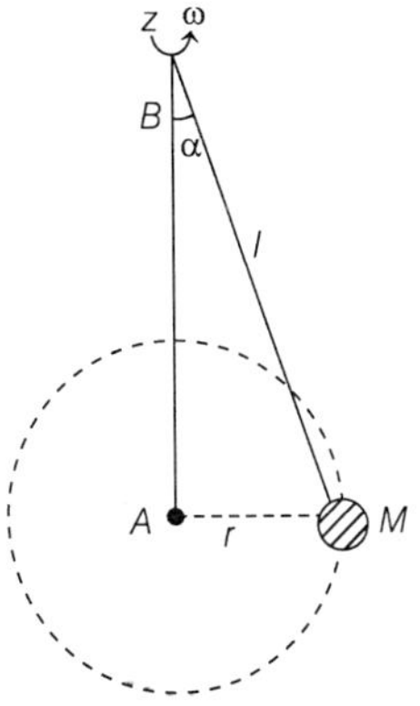

(a) L_A and L_B are both constant in magnitude and direction
(b) L_B is constant in direction with varying magnitude
(c) L_B is constant, both in magnitude and direction
(d) L_A is constant, both in magnitude and direction

Imp. Centre of Mass of Removed Part

37 A circular hole of radius $\left(\frac{a}{2}\right)$ is cut out of a circular disc of radius a as shown in figure. The centroid of the remaining circular portion with respect to point O will be *[2021, 24 Feb Shift-II]*

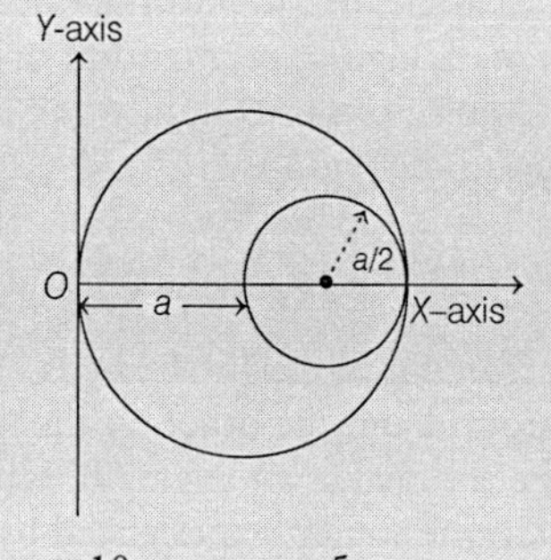

(a) $\frac{1}{6}a$ (b) $\frac{10}{11}a$ (c) $\frac{5}{6}a$ (d) $\frac{2}{3}a$

38 A uniform cylinder of mass M and radius R is to be pulled over a step of height a $(a < R)$ by applying a force F at its centre O perpendicular to the plane through the axes of the cylinder on the edge of the step (see figure). The minimum value of F required is *[2020, 02 Sep Shift-I]*

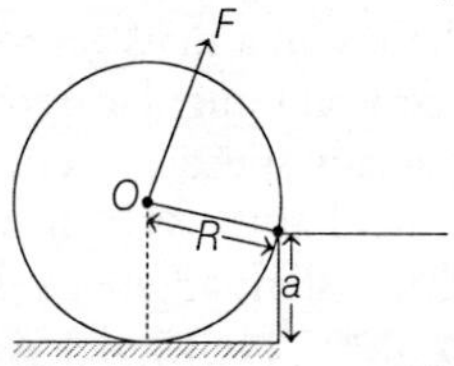

(a) $Mg\frac{a}{R}$ (b) $Mg\sqrt{1-\frac{a^2}{R^2}}$

(c) $Mg\sqrt{\left(\frac{R}{R-a}\right)^2-1}$ (d) $Mg\sqrt{1-\left(\frac{R-a}{R}\right)^2}$

39 A rod of length L has non-uniform linear mass density given by $\rho(x) = a + b(x/L)^2$,

where a and b are constants and $0 \le x \le L$. The value of x for the centre of mass of the rod is at *[2020, 9 Jan Shift-II]*

(a) $\frac{3}{4}\left(\frac{2a+b}{3a+b}\right)L$ (b) $\frac{4}{3}\left(\frac{a+b}{2a+3b}\right)L$

(c) $\frac{3}{2}\left(\frac{a+b}{2a+b}\right)L$ (d) $\frac{3}{2}\left(\frac{2a+b}{3a+b}\right)L$

40 The time dependence of the position of a particle of mass $m = 2$ is given by $\mathbf{r}(t) = 2t\hat{\mathbf{i}} - 3t^2\hat{\mathbf{j}}$. Its angular momentum, with respect to the origin, at time $t = 2$ is *[2019, 10 April Shift-II]*

(a) $36\,\hat{\mathbf{k}}$ (b) $-34(\hat{\mathbf{k}} - \hat{\mathbf{i}})$ (c) $-48\,\hat{\mathbf{k}}$ (d) $48(\hat{\mathbf{i}} + \hat{\mathbf{j}})$

41 A particle of mass m is moving along a trajectory given by $x = x_0 + a\cos\omega_1 t$ and $y = y_0 + b\sin\omega_2 t$. The torque acting on the particle about the origin at $t = 0$ is *[2019, 10 April Shift-I]*

(a) zero (b) $m(-x_0 b + y_0 a)\,\omega_1^2\,\hat{\mathbf{k}}$

(c) $-m(x_0 b\omega_2^2 - y_0 a\omega_1^2)\hat{\mathbf{k}}$ (d) $+my_0 a\,\omega_1^2\,\hat{\mathbf{k}}$

42 An L-shaped object made of thin rods of uniform mass density is suspended with a string as shown in figure. If $AB = BC$ and the angle is made by AB with downward vertical is θ, then *[2019, 9 Jan Shift-I]*

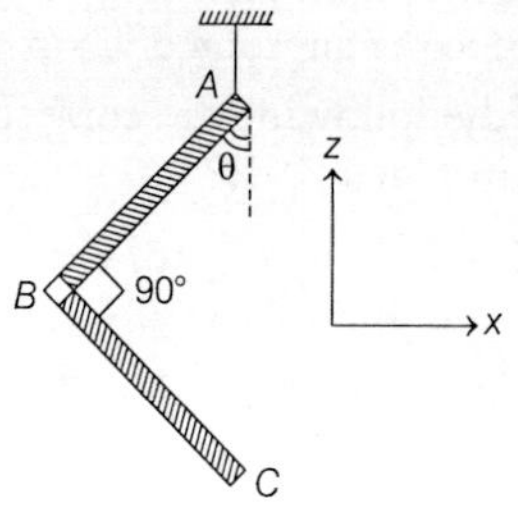

(a) $\tan\theta = \frac{2}{\sqrt{3}}$ (b) $\tan\theta = \frac{1}{2\sqrt{3}}$

(c) $\tan\theta = \frac{1}{2}$ (d) $\tan\theta = \frac{1}{3}$

43 A slab is subjected to two forces $\mathbf{F}_1$ and $\mathbf{F}_2$ of same magnitude F as shown in the figure. Force $\mathbf{F}_2$ is in xy-plane while force $\mathbf{F}_1$ acts along Z-axis at the point $(2\hat{\mathbf{i}} + 3\hat{\mathbf{j}})$. The moment of these forces about point O will be *[2019, 11 Jan Shift-I]*

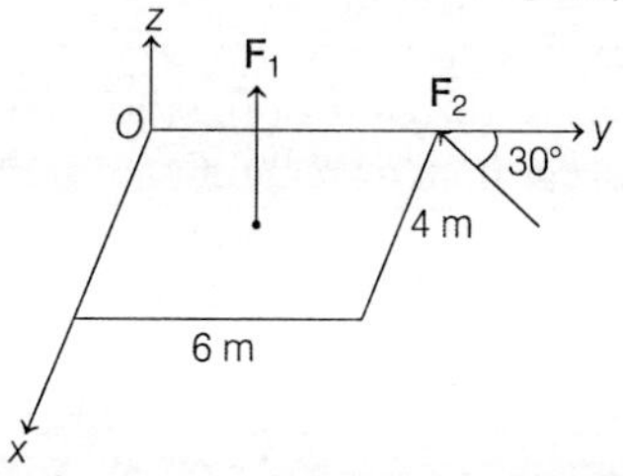

(a) $(3\hat{\mathbf{i}} + 2\hat{\mathbf{j}} - 3\hat{\mathbf{k}})F$ (b) $(3\hat{\mathbf{i}} - 2\hat{\mathbf{j}} + 3\hat{\mathbf{k}})F$

(c) $(3\hat{\mathbf{i}} - 2\hat{\mathbf{j}} - 3\hat{\mathbf{k}})F$ (d) $(3\hat{\mathbf{i}} + 2\hat{\mathbf{j}} + 3\hat{\mathbf{k}})F$

44 The position vector of the centre of mass $\mathbf{r}_{cm}$ of an asymmetric uniform bar of negligible area of cross-section as shown in figure is *[2019, 12 Jan Shift-I]*

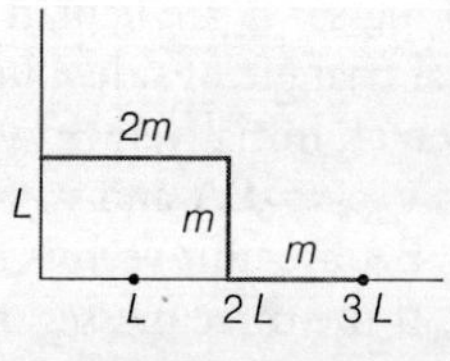

(a) $\mathbf{r} = \frac{13}{8} L\hat{\mathbf{x}} + \frac{5}{8} L\hat{\mathbf{y}}$ (b) $\mathbf{r} = \frac{11}{8} L\hat{\mathbf{x}} + \frac{3}{8} L\hat{\mathbf{y}}$

(c) $\mathbf{r} = \frac{3}{8} L\hat{\mathbf{x}} + \frac{11}{8} L\hat{\mathbf{y}}$ (d) $\mathbf{r} = \frac{5}{8} L\hat{\mathbf{x}} + \frac{13}{8} L\hat{\mathbf{y}}$

45 A particle of mass 20 g is released with an initial velocity 5 m/s along the curve from the point *A*, as shown in the figure. The point *A* is at height *h* from point *B*. The particle slides along the frictionless surface. When the particle reaches point *B*, its angular momentum about *O* will be
(Take, $g = 10 \text{ m/s}^2$) ***[2019, 12 Jan Shift-II]***

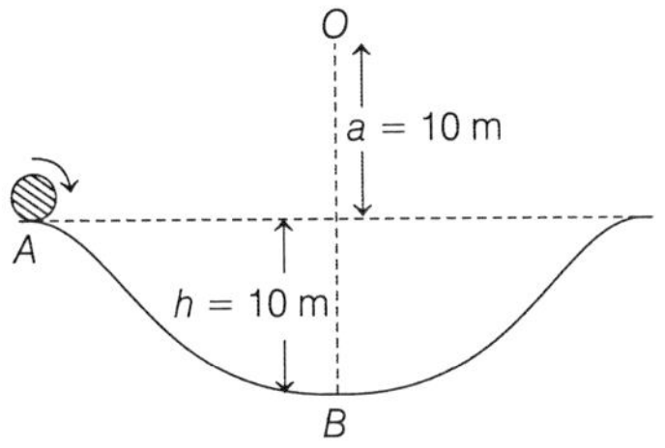

(a) 8 kg-m^2/s (b) 3 kg-m^2/s (c) 2 kg-m^2/s (d) 6 kg-m^2/s

46 To mop-clean a floor, a cleaning machine presses a circular mop of radius *R* vertically down with a total force *F* and rotates it with a constant angular speed about its axis. If the force *F* is distributed uniformly over the mop and if coefficient of friction between the mop and the floor is μ, the torque applied by the machine on the mop in (N-m) is ***[2019, 10 Jan Shift-I]***

(a) $\frac{2}{3}\mu FR$ (b) $\frac{\mu FR}{6}$ (c) $\frac{\mu FR}{3}$ (d) $\frac{\mu FR}{2}$

47 In the figure shown *ABC* is a uniform wire. If centre of mass of wire lies vertically below point *A*, then *BC* / *AB* is close to ***[JEE Main 2016]***

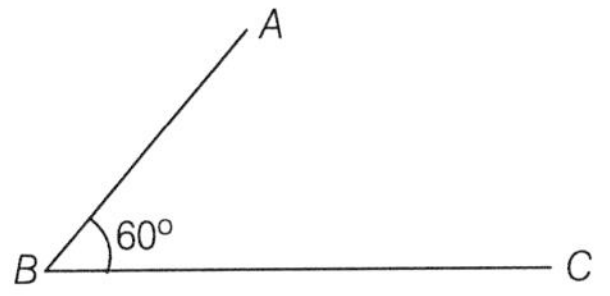

(a) 1.85 (b) 1.5 (c) 1.37 (d) 3

48 A circular disc of radius *R* is removed from a bigger circular disc of radius 2 *R*, such that the circumference of the discs coincide. The centre of mass of thc ncw disc is α / *R* from the centre of the bigger disc. The value of α is ***[AIEEE 2007]***

(a) 1/3 (b) 1/2 (c) 1/6 (d) 1/4

Level 3

49 The centre of mass of a thin rectangular plate with sides of length *a* and *b*, whose mass per unit area (σ) varies as $\sigma = \frac{\sigma_0 x}{ab}$ (where σ_0 is a constant), would be …… ***[2025, 28 Jan Shift-I]***

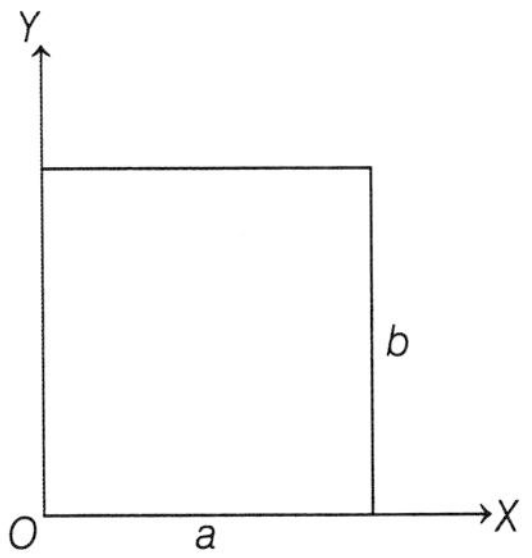

(a) $\left(\frac{1}{3}a, \frac{b}{2}\right)$ (b) $\left(\frac{2}{3}a, \frac{b}{2}\right)$

(c) $\left(\frac{a}{2}, \frac{b}{2}\right)$ (d) $\left(\frac{2}{3}a, \frac{2}{3}b\right)$

50 An object of mass *m* is projected from origin in a vertical *XY*-plane at an angle 45° with the *X*-axis with an initial velocity v_0. The magnitude and direction of the angular momentum of the object with respect to origin, when maximum height, will be

(*g* is acceleration due to gravity) ***[2025, 24 Jan Shift-I]***

(a) $\frac{mv_0^3}{2\sqrt{2}g}$ along positive *Z*-axis

(b) $\frac{mv_0^3}{4\sqrt{2}g}$ along positive *Z*-axis

(c) $\frac{mv_0^3}{2\sqrt{2}g}$ along negative *Z*-axis

(d) $\frac{mv_0^3}{4\sqrt{2}g}$ along negative *Z*-axis

51 Two blocks of masses 10 kg and 30 kg are placed on the same straight line with coordinates (0, 0) cm and (*x*, 0) cm, respectively. The block of 10 kg is moved on the same line through a distance of 6 cm towards the other block. The distance through which the block of 30 kg must be moved to keep the position of centre of mass of the system unchanged is

(a) 4 cm towards the 10 kg block ***[2022, 27 June Shift-I]***
(b) 2 cm away from the 10 kg block
(c) 2 cm towards the 10 kg block
(d) 4 cm away from the 10 kg block

52

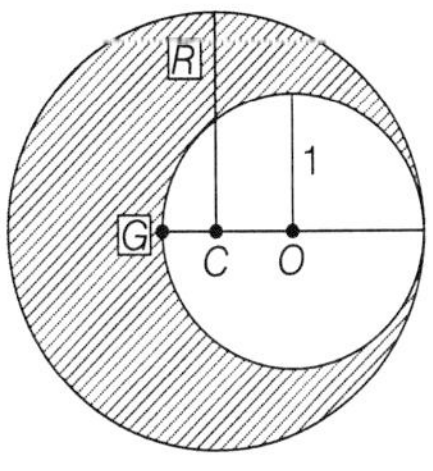

As shown in figure, when a spherical cavity (centred at *O*) of radius 1 is cut out of a uniform sphere of radius R (centred at C), the centre of mass of remaining (shaded) part of sphere is at G, i.e. on the surface of the cavity. R can be determined by the equation *[2020, 8 Jan Shift-II]*

(a) $(R^2 + R + 1)(2 - R) = 1$ (b) $(R^2 + R - 1)(2 - R) = 1$
(c) $(R^2 - R - 1)(2 - R) = 1$ (d) $(R^2 - R + 1)(2 - R) = 1$

Imp. Centre of Mass of Remaining Part

53 A uniform rectangular thin sheet $ABCD$ of mass M has length a and breadth b, as shown in the figure. If the shaded portion $HBGO$ is cut-off, the coordinates of the centre of mass of the remaining portion will be *[2019, 8 April Shift-II]*

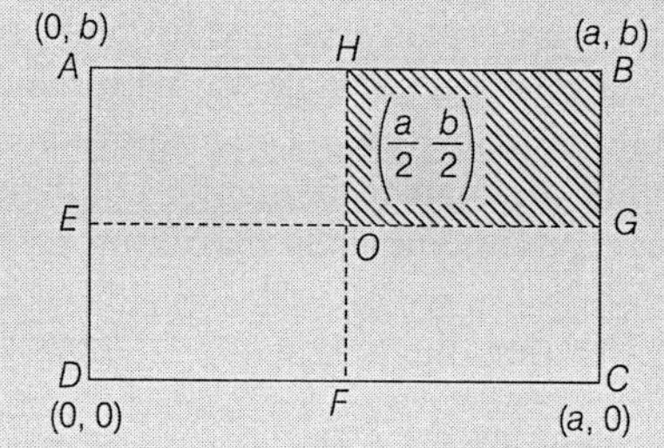

(a) $\left(\frac{2a}{3}, \frac{2b}{3}\right)$ (b) $\left(\frac{5a}{12}, \frac{5b}{12}\right)$
(c) $\left(\frac{3a}{4}, \frac{3b}{4}\right)$ (d) $\left(\frac{5a}{3}, \frac{5b}{3}\right)$

Numerical Value Type Questions

54 The position vectors of two 1 kg particles, A and B, are given by
$\mathbf{r}_A = (\alpha_1 t^2 \hat{\mathbf{i}} + \alpha_2 t\, \hat{\mathbf{j}} + \alpha_3 t\, \hat{\mathbf{k}})$m and
$\mathbf{r}_B = (\beta_1 t\, \hat{\mathbf{i}} + \beta_2 t^2\, \hat{\mathbf{j}} + \beta_3 t\, \hat{\mathbf{k}})$m, respectively;
($\alpha_1 = 1\text{m}/\text{s}^2, \alpha_2 = 3n$ m/s, $\alpha_3 = 2$ m/s, $\beta_1 = 2$ m/s, $\beta_2 = -1\text{m}/\text{s}^2, \beta_3 = 4p$ m/s), where t is time, n and p are constants. At $t = 1$ s, $|\mathbf{v}_A| = |\mathbf{v}_B|$ and velocities $\mathbf{v}_A$ and $\mathbf{v}_B$ of the particles are orthogonal to each other. At $t = 1$ s, the magnitude of angular momentum of particle A with respect to the position of particle B is $\sqrt{L}$ kgm^{2}s^{-1}. The value of L is .... *[2025, 22 Jan Shift-I]*

55 The coordinates of a particle with respect to origin in a given reference frame is (1, 1, 1) m. If a force of $\mathbf{F} = \hat{\mathbf{i}} - \hat{\mathbf{j}} + \hat{\mathbf{k}}$ acts on the particle, then the magnitude of torque (with respect to origin) in z-direction is .... *[2025, 29 Jan Shift-I]*

56 A uniform thin metal plate of mass 10 kg with dimensions is shown. The ratio of x and y-coordinates of centre of mass of plate in $n/9$. The value of n is .... *[2024, 8 April Shift-I]*

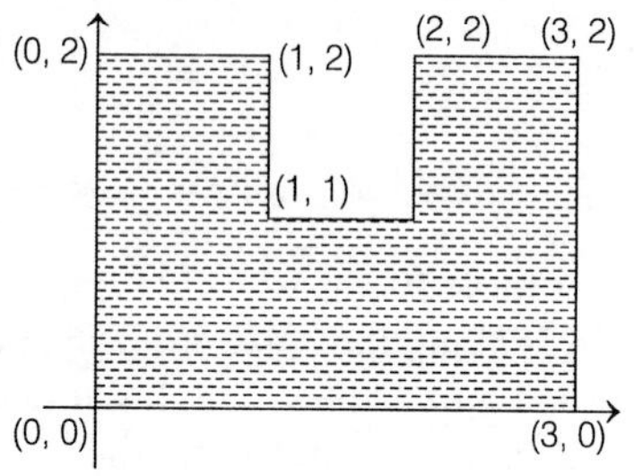

57 In a system two particles of masses $m_1 = 3$ kg and $m_2 = 2$ kg are placed at certain distance from each other. The particle of mass m_1 is moved towards the centre of mass of the system through a distance 2 cm. In order to keep the centre of mass of the system at the original position, the particle of mass m_2 should move towards the centre of mass by the distance......... cm. *[2024, 4 April Shift-II]*

58 The identical spheres each of mass $2m$ are placed at the corners of a right angled triangle with mutually perpendicular sides equal to 4 m each. Taking point of intersection of these two sides as origin, the magnitude of position vector of the centre of mass of the system is $\frac{4\sqrt{2}}{x}$, where the value of x is .... *[2024, 1 Feb Shift-I]*

59 A body of mass m is projected with a speed u making an angle of 45° with the ground. The angular momentum of the body about the point of projection, at the highest point is expressed as $\sqrt{2}mu^3/Xg$. The value of X is .... *[2024, 31 Jan Shift-II]*

60 A body of mass 5 kg moving with a uniform speed $3\sqrt{2}$ ms^{-1} in XY-plane along the line $y = x - 4$. The angular momentum of the particle about the origin will be .... kg m^{2}s^{-1}. *[2024, 29 Jan Shift-II]*

61 A force of $-P\hat{\mathbf{k}}$ acts on the origin of the coordinate system. The torque about the point (2, − 3) is $P(a\hat{\mathbf{i}} + b\hat{\mathbf{j}})$, The ratio of $\frac{a}{b}$ is $\frac{x}{2}$. The value of x is *[2023, 10 April Shift-II]*

62 A particle of mass 100 g is projected at time $t = 0$ with a speed 20 ms^{-1} at an angle 45° to the horizontal as given in the figure. The magnitude of the angular momentum of the particle about the starting point at time $t = 2$ s is found to be $\sqrt{K}$ kg-m^2/s. The value of K is .... (Take, $g = 10$ ms^{-2}) *[2023, 29 Jan Shift-II]*

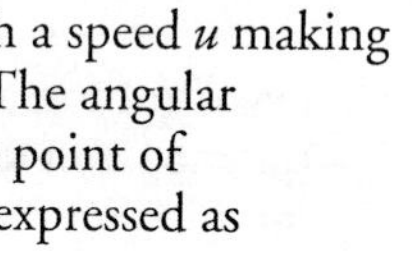

63 A tube of length 50 cm is filled completely with an incompressible liquid of mass 250 g and closed at both ends. The tube is then rotated in horizontal plane about one of its ends with a uniform angular velocity $x\sqrt{F}$ rad s^{-1}. If F be the force exerted by the liquid at the other end, then the value of x will be .... [2022, 29 July Shift-II]

64 The distance of centre of mass from end A of a one dimensional rod AB having mass density $\rho = \rho_0\left(1-\frac{x^2}{L^2}\right)$ kg/m and length L (in metre) is $\frac{3L}{\alpha}$. The value of α is .... (where, x is the distance from end A) [2022, 28 July Shift-II]

65 Three identical spheres each of mass M are placed at the corners of a right angled triangle with mutually perpendicular sides equal to 3 m each. Taking point of intersection of mutually perpendicular sides as origin, the magnitude of position vector of centre of mass of the system will be $\sqrt{x}$ m. The value of x is .... [2022, 25 July Shift-II]

66 The position vector of 1 kg object is $\mathbf{r} = (3\hat{i} - \hat{j})$ m and its velocity, $\mathbf{v} = (3\hat{j} + \hat{k})$ ms^{-1}. The magnitude of its angular momentum is $\sqrt{x}$ N-m, where x is . [2022, 28 June Shift-I]

67 A metre scale is balanced on a knife edge at its centre. When two coins, each of mass 10 g are put one on the top of the other at the 10.0 cm mark the scale is found to be balanced at 40.0 cm mark. The mass of the metre scale is found to be $x \times 10^{-2}$ kg. The value of x is .... [2022, 24 June Shift-I]

68 A body of mass 2 kg moving with a speed of 4 m/s makes an elastic collision with another body at rest and continues to move in the original direction but with one-fourth of its initial speed. The speed of the two body centre of mass is x / 10 m/s. Then, the value of x is .... [2021, 25 July Shift-I]

69 A particle of mass m is moving in time t on a trajectory given by $\mathbf{r} = 10\alpha t^2\hat{i} + 5\beta(t-5)\hat{j}$ where, α and β are dimensional constants. The angular momentum of the particle becomes the same as it was for $t = 0$ at time t is ...... s. [2021, 25 July Shift-I]

70 The position of the centre of mass of a uniform semi-circular wire of radius R placed in XY-plane with its centre at the origin and the line joining its ends as X-axis is given by $\left(0, \frac{xR}{\pi}\right)$. Then, the value of $|x|$ is .... [2021, 22 July Shift-II]

71 The disc of mass M with uniform surface mass density σ is shown in the figure. The centre of mass of the quarter disc (the shaded area) is at the position $\frac{x}{3}\frac{a}{\pi}, \frac{x}{3}\frac{a}{\pi}$, where x is .... (Round off to the nearest integer) (a is an area as shown in the figure) [2021, 17 March Shift-II]

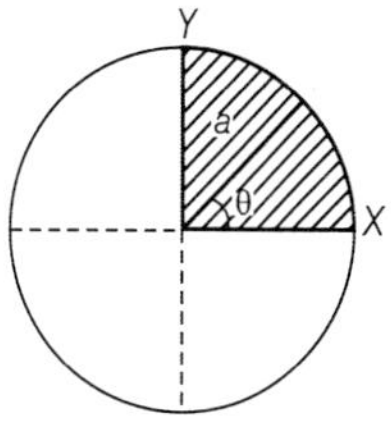

72 A force $\mathbf{F} = 4\hat{i} + 3\hat{j} + 4\hat{k}$ is applied on an intersection point of $x = 2$ plane and X-axis. The magnitude of torque of this force about a point (2, 3, 4) is .... (Round off to the nearest integer) [2021, 16 March Shift-II]

73 A person of 80 kg mass is standing on the rim of a circular platform of mass 200 kg rotating about its axis at 5 rpm. The person now starts moving towards the centre of the platform. What will be the rotational speed (in rpm) of the platform when the person reaches its centre? [2020, 03 Sep Shift-I]

74 The centre of mass of a solid hemisphere of radius 8 cm is x cm from the centre of the flat surface. Then, value of x is .... [2020, 06 Sep Shift-II]

Topic 2 Moment of Inertia

Level 1

01 The moment of inertia of a circular ring of mass M and diameter r about a tangential axis lying in the plane of the ring is [2025, 2 April Shift-II]
(a) $\frac{3}{8}Mr^2$ (b) $\frac{3}{2}Mr^2$ (c) $2Mr^2$ (d) $\frac{1}{2}Mr^2$

02 A rod of linear mass density λ and length L is bent to form a ring of radius R. Moment of inertia of ring about any of its diameter is [2025, 8 April Shift-I]
(a) $\frac{\lambda L^3}{8\pi^2}$ (b) $\frac{\lambda L^3}{4\pi^2}$ (c) $\frac{\lambda L^3}{16\pi^2}$ (d) $\frac{\lambda L^3}{12}$

03 A circular disk of radius R metre and mass M kg is rotating around the axis perpendicular to the disk. An external torque is applied to the disk such that $\theta(t) = 5t^2 - 8t$, where $\theta(t)$ is the angular position of the rotating disc as a function of time t.
How much power is delivered by the applied torque, when $t = 2$ s? [2025, 23 Jan Shift-II]
(a) $108MR^2$ (b) $8MR^2$ (c) $60MR^2$ (d) $72MR^2$

04 A system consists of two identical spheres each of mass 1.5 kg and radius 50 cm at the end of light rod. The distance between the centres of the two spheres is 5 m. What will be the moment of inertia of the system about an axis perpendicular to the rod passing through its mid-point ? *[2021, 31 Aug Shift-II]*

(a) 18.75 kg-m^2 (b) 1.905×10^5 kg-m^2
(c) 19.05 kg-m^2 (d) 1.875×10^5 kg-m^2

05 Consider a uniform wire of mass M and length L. It is bent into a semicircle. Its moment of inertia about a line perpendicular to the plane of the wire passing through the centre is *[2021, 18 March Shift-II]*

(a) $\frac{1}{4}\frac{ML^2}{\pi^2}$ (b) $\frac{2}{5}\frac{ML^2}{\pi^2}$
(c) $\frac{ML^2}{\pi^2}$ (d) $\frac{1}{2}\frac{ML^2}{\pi^2}$

06 Moment of inertia (MI) of four bodies, having same mass and radius, are reported as
I_1 = MI of thin circular ring about its diameter,
I_2 = MI of circular disc about an axis perpendicular to the disc and going through the centre,
I_3 = MI of solid cylinder about its axis
and I_4 = MI of solid sphere about its diameter.
Then, *[2021, 24 Feb Shift-I]*

(a) $I_1 + I_2 = I_3 + \frac{5}{2} I_4$ (b) $I_1 + I_3 < I_2 + I_4$
(c) $I_1 = I_2 = I_3 < I_4$ (d) $I_1 = I_2 = I_3 > I_4$

07 Consider a uniform square plate of side a and mass m. The moment of inertia of this plate about an axis perpendicular to its plane and passing through one of its corners is *[AIEEE 2008]*

(a) $\frac{5}{6} ma^2$ (b) $\frac{1}{12} ma^2$
(c) $\frac{7}{12} ma^2$ (d) $\frac{2}{3} ma^2$

08 The moment of inertia of uniform semi-circular disc of mass M and radius r about a line perpendicular to the plane of the disc through the centre is *[AIEEE 2005]*

(a) $\frac{1}{4} Mr^2$ (b) $\frac{2}{5} Mr^2$
(c) Mr^2 (d) $\frac{1}{2} Mr^2$

09 One solid sphere A and another hollow sphere B are of same mass and same outer radii. Their moment of inertia about their diameters are respectively I_A and I_B such that *[AIEEE 2004]*

(a) $I_A = I_B$ (b) $I_A > I_B$
(c) $I_A < I_B$ (d) $\frac{I_A}{I_B} = \frac{d_A}{d_B}$

10 Moment of inertia of a circular wire of mass M and radius R about its diameter is *[AIEEE 2002]*

(a) $MR^2/2$ (b) MR^2
(c) $2MR^2$ (d) $MR^2/4$

Level 2

11 Moment of inertia of a rod of mass M and length L about an axis passing through its center and normal to its length is α. Now the rod is cut into two equal parts and these parts are joined symmetrically to form a cross shape. Moment of inertia of cross about an axis passing through its center and normal to plane containing cross is *[2025, 2 April Shift-I]*

(a) α (b) $\alpha/4$ (c) $\alpha/2$ (d) $\alpha/8$

12 A uniform circular disc of radius R and mass M is rotating about an axis perpendicular to its plane and passing through its centre. A small circular part of radius $\frac{R}{2}$ is removed from the original disc as shown in the figure. Find the moment of inertia of the remaining part of the original disc about the axis as given above. *[2025, 22 Jan Shift-I]*

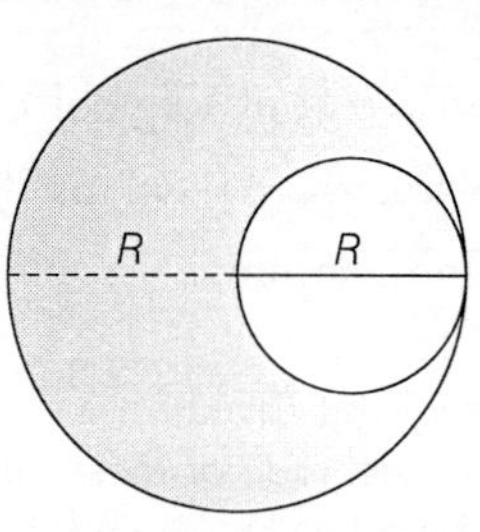

(a) $\frac{13}{32} MR^2$ (b) $\frac{7}{32} MR^2$ (c) $\frac{9}{32} MR^2$ (d) $\frac{17}{32} MR^2$

13 Ratio of radius of gyration of a hollow sphere to that of a solid cylinder of equal mass, for moment of inertia about their diameter axis AB as shown in figure is $\sqrt{8/x}$. The value of x is *[2024, 5 April Shift-I]*

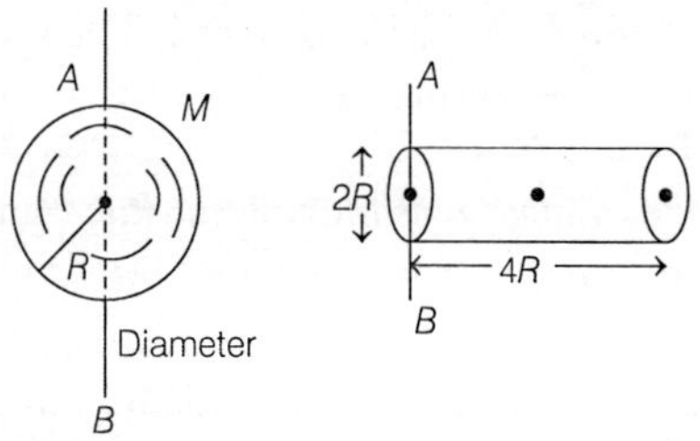

(a) 51 (b) 34 (c) 67 (d) 17

14 Match List I with List II.

	List I		List II
(A)	Moment of inertia of solid sphere of radius R about any tangent.	(i)	$\frac{5}{3} MR^2$
(B)	Moment of inertia of hollow sphere of radius (R) about any tangent.	(ii)	$\frac{7}{5} MR^2$
(C)	Moment of inertia of circular ring of radius (R) about its diameter.	(iii)	$\frac{1}{4} MR^2$
(D)	Moment of inertia of circular disc of radius (R) about any diameter.	(iv)	$\frac{1}{2} MR^2$

Choose the correct answer from the options given below. *[2022, 28 June Shift-I]*

(a) (A)-(ii), (B)-(i), (C)-(iv), (D)-(iii)
(b) (A)-(i), (B)-(ii), (C)-(iv), (D)-(iii)
(c) (A)-(ii), (B)-(i), (C)-(iii), (D)-(iv)
(d) (A)-(i), (B)-(ii), (C)-(iii), (D)-(iv)

15 Moment of inertia of a square plate of side l about the axis passing through one of the corner and perpendicular to the plane of square plate is given by *[2021, 27 Aug Shift-I]*

(a) $Ml^2/6$ (b) Ml^2 (c) $Ml^2/12$ (d) $\frac{2}{3}Ml^2$

16 The solid cylinder of length 80 cm and mass M has a radius of 20 cm. Calculate the density of the material used, if the moment of inertia of the cylinder about an axis CD parallel to AB as shown in figure is 2.7 kg m^2. *[2021, 26 Aug Shift-II]*

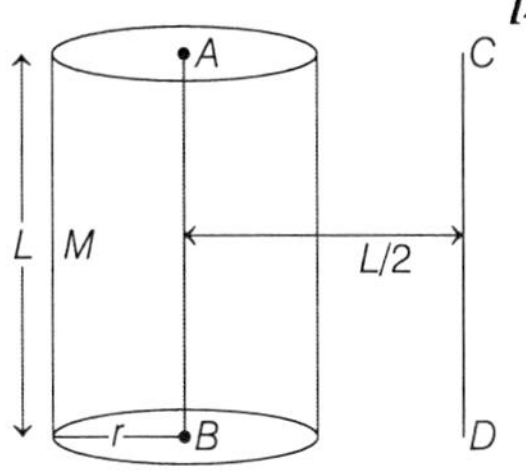

(a) 14.9 kg/m^3 (b) $7.5\times 10^1 \text{ kg/m}^3$
(c) $7.5\times 10^2 \text{ kg/m}^3$ (d) $1.49\times 10^2 \text{ kg/m}^3$

17 The figure shows two solid discs with radius R and r, respectively. If mass per unit area is same for both, what is the ratio of MI of bigger disc around axis AB (which is perpendicular to the plane of the disc and passing through its centre) of MI of smaller disc around one of its diameters lying on its plane? Given, M is the mass of the larger disc. (MI stands for moment of inertia) *[2021, 27 July Shift-I]*

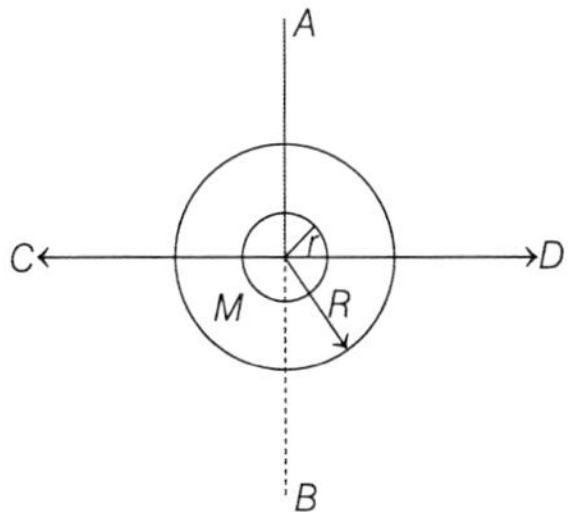

(a) $R^2 : r^2$ (b) $2r^4 : R^4$
(c) $2R^2 : r^2$ (d) $2R^4 : r^4$

18 **Assertion** (A) Moment of inertia of a circular disc of mass M and radius R about X, Y-axis (passing through its plane) and Z-axis which is perpendicular to its plane were found to be I_x, I_y and I_z respectively. The respective radii of gyration about all the three axis will be the same.

Reason (R) A rigid body making rotational motion has fixed mass and shape.

In the light of the above statements, choose the most appropriate answer from the options given below. *[2021, 25 July Shift-1]*

(a) Both A and R are correct and R is the correct explanation of A.
(b) Both A and R are correct but R is not the correct explanation of A.
(c) A is correct but R is not correct.
(d) A is not correct but R is correct.

19 Four equal masses, m each are placed at the corners of a square of length (l) as shown in the figure. The moment of inertia of the system about an axis passing through A and parallel to DB would be *[2021, 16 March Shift-I]*

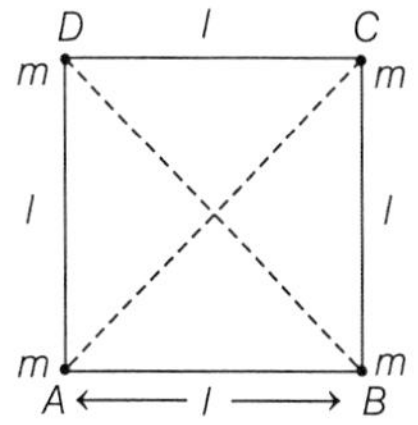

(a) $1ml^2$ (b) $2ml^2$
(c) $3ml^2$ (d) $\sqrt{3}ml^2$

20 Four identical solid spheres each of mass m and radius a are placed with their centres on the four corners of a square of side b. The moment of inertia of the system about one side of square, where the axis of rotation is parallel to the plane of the square is *[2021, 26 Feb Shift-I]*

(a) $\frac{4}{5}ma^2 + 2mb^2$ (b) $\frac{8}{5}ma^2 + mb^2$
(c) $\frac{8}{5}ma^2 + 2mb^2$ (d) $\frac{4}{5}ma^2$

21 The radius of gyration of a uniform rod of length l, about an axis passing through a point $\frac{l}{4}$ away from the centre of the rod and perpendicular to it, is *[2020, 7 Jan Shift-I]*

(a) $\frac{1}{8}l$ (b) $\sqrt{\frac{3}{8}}l$
(c) $\sqrt{\frac{7}{48}}l$ (d) $\frac{1}{4}l$

22 Let the moment of inertia of a hollow cylinder of length 30 cm (inner radius 10 cm and outer radius 20 cm) about its axis be I. The radius of a thin cylinder of the same mass such that its moment of inertia about its axis is also I, is *[2019, 12 Jan Shift-I]*

(a) 16 cm (b) 14 cm
(c) 12 cm (d) 18 cm

23 The moment of inertia of a solid sphere, about an axis parallel to its diameter and at a distance of x from it, is $I(x)$. Which one of the graphs represents the variation of $I(x)$ with x correctly? *[2019, 12 Jan Shift-II]*

(a) 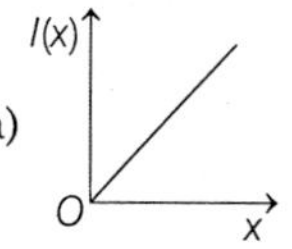

(b) 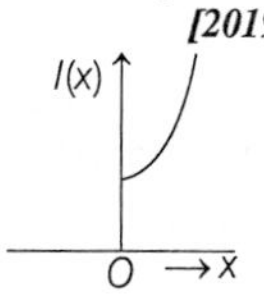

(c) 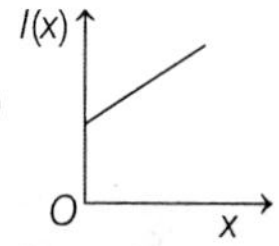

(d) 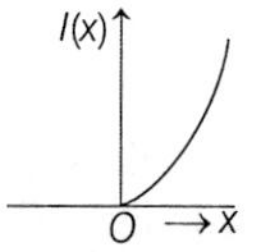

24 From a uniform circular disc of radius R and mass $9\,M$, a small disc of radius $R/3$ is removed as shown in the figure. The moment of inertia of the remaining disc about an axis perpendicular to the plane of the disc and passing through centre of disc is *[JEE Main 2018]*

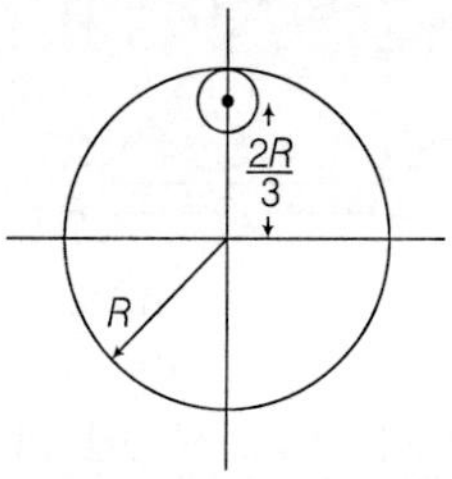

(a) $4MR^2$ (b) $\frac{40}{9}MR^2$ (c) $10MR^2$ (d) $\frac{37}{9}MR^2$

25 A circular hole of radius $R/4$ is made in a thin uniform disc having mass M and radius R, as shown in figure. The moment of inertia of the remaining portion of the disc about an axis passing through the point O and perpendicular to the plane of the disc is *[JEE Main 2017]*

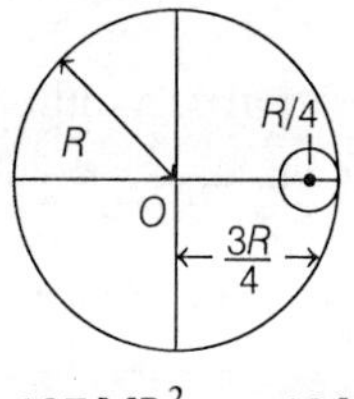

(a) $\frac{219MR^2}{256}$ (b) $\frac{197MR^2}{256}$ (c) $\frac{19MR^2}{512}$ (d) $\frac{237MR^2}{512}$

26 For the given uniform square lamina $ABCD$, whose centre is O *[AIEEE 2007]*

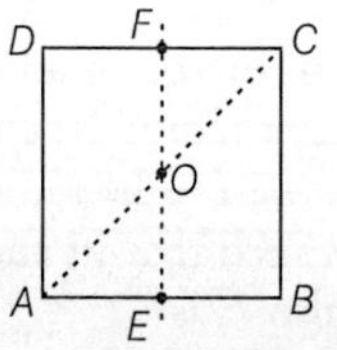

(a) $\sqrt{2}\,I_{AC} = I_{EF}$ (b) $I_{AD} = 3\,I_{EF}$

(c) $I_{AD} = 4I_{EF}$ (d) $I_{AC} = \sqrt{2}\,I_{EF}$

Level 3

27

List I	List II
A. Moment of inertia of the rod (length L, mass M, about an axis perpendicular to the rod passing through the mid-point)	I. $\frac{8\,ML^2}{3}$
B. Moment of inertia of the rod (length L, mass $2M$, about an axis perpendicular to the rod passing through one of its end)	II. $\frac{ML^2}{3}$
C. Moment of inertia of the rod (length $2L$, mass M, about an axis perpendicular to the rod passing through its midpoint)	III. $\frac{ML^2}{12}$
D. Moment of inertia of the rod (length $2L$, mass $2M$, about an axis perpendicular to the rod passing through one of its end)	IV. $\frac{2\,ML^2}{3}$

Choose the correct answer from the options given below. *[2021, 27 July Shift-I]*

	A	B	C	D
(a)	(II)	(III)	(I)	(IV)
(b)	(II)	(I)	(III)	(IV)
(c)	(III)	(IV)	(II)	(I)
(d)	(III)	(IV)	(I)	(II)

28 Moment of inertia of a cylinder of mass M, length L and radius R about an axis passing through its centre and perpendicular to the axis of the cylinder is $I = M\left(\frac{R^2}{4} + \frac{L^2}{12}\right)$. If such a cylinder is to be made for a given mass of a material. To have minimum possible moment of inertia, the ratio L/R for cylinder is *[2020, 03 Sep Shift-I]*

(a) $\frac{2}{3}$ (b) $\frac{3}{2}$

(c) $\sqrt{\frac{3}{2}}$ (d) $\sqrt{\frac{2}{3}}$

29 Consider two uniform discs of same thickness and different radii $R_1 = R$ and $R_2 = \alpha R$ made of the same material. If the ratio of their moments of inertia I_1 and I_2 respectively, about their axis is $I_1 : I_2 = 1:16$, then the value of α is *[2020, 04 Sep Shift-II]*

(a) 2 (b) $2\sqrt{2}$

(c) 4 (d) $\sqrt{2}$

30 The linear mass density of a thin rod AB of length L varies from A to B as $\lambda(x) = \lambda_0\left(1 + \frac{x}{L}\right)$, where x is the distance from A. If M is mass of the rod, then its moment of inertia about an axis passing through A and perpendicular to the rod is *[2020, 06 Sep Shift-II]*

(a) $\frac{5}{12}ML^2$ (b) $\frac{7}{18}ML^2$

(c) $\frac{2}{5}ML^2$ (d) $\frac{3}{7}ML^2$

31 Mass per unit area of a circular disc of radius a depends on the distance r from its centre as $\sigma(r) = A + Br$. The moment of inertia of the disc about the axis, perpendicular to the plane and passing through its centre is *[2020, 7 Jan Shift-II]*

(a) $2\pi a^4 \left(\frac{A}{4} + \frac{B}{5}\right)$ (b) $2\pi a^4 \left(\frac{aA}{4} + \frac{B}{5}\right)$

(c) $\pi a^4 \left(\frac{A}{4} + \frac{aB}{5}\right)$ (d) $2\pi a^4 \left(\frac{A}{4} + \frac{aB}{5}\right)$

32. A thin circular plate of mass M and radius R has its density varying as $\rho(r) = \rho_0 r$ with ρ_0 as constant and r is the distance from its centre. The moment of inertia of the circular plate about an axis perpendicular to the plate and passing through its edge is $I = aMR^2$. The value of the coefficient a is *[2019, 8 April Shift-I]*

(a) 1/2 (b) 3/5 (c) 8/5 (d) 3/2

33 A thin disc of mass M and radius R has mass per unit area $\sigma(r) = kr^2$, where r is the distance from its centre. Its moment of inertia about an axis going through its centre of mass and perpendicular to its plane is *[2019, 10 April Shift-I]*

(a) $\frac{MR^2}{2}$ (b) $\frac{MR^2}{6}$ (c) $\frac{MR^2}{3}$ (d) $\frac{2MR^2}{3}$

34 A solid sphere of mass M and radius R is divided into two unequal parts. The first part has a mass of $7M/8$ and is converted into a uniform disc of radius $2R$. The second part is converted into a uniform solid sphere. Let I_1 be the moment of inertia of the disc about its axis and I_2 be the moment of inertia of the new sphere about its axis. The ratio I_1/I_2 is given by *[2019, 10 April Shift-II]*

(a) 285 (b) 185
(c) 65 (d) 140

35 A circular disc of radius b has a hole of radius a at its centre (see figure). If the mass per unit area of the disc varies as $\left(\frac{\sigma_0}{r}\right)$, then the radius of gyration of the disc about its axis passing through the centre is *[2019, 12 April Shift-I]*

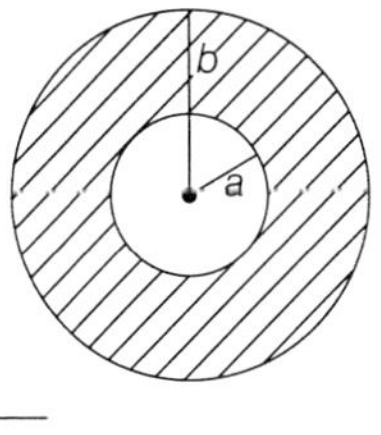

(a) $\sqrt{\frac{a^2 + b^2 + ab}{2}}$ (b) $\frac{a+b}{2}$

(c) $\sqrt{\frac{a^2 + b^2 + ab}{3}}$ (d) $\frac{a+b}{3}$

Imp. Moment of Inertia

36 Two identical spherical balls of mass M and radius R each are stuck on two ends of a rod of length $2R$ and mass M (see figure).

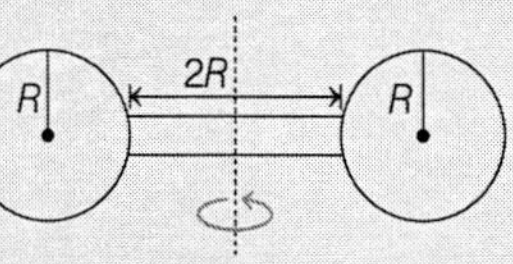

The moment of inertia of the system about the axis passing perpendicularly through the centre of the rod is *[2019, 10 Jan Shif-II]*

(a) $\frac{137}{15} MR^2$ (b) $\frac{209}{15} MR^2$ (c) $\frac{17}{15} MR^2$ (d) $\frac{152}{15} MR^2$

37 An equilateral triangle ABC is cut from a thin solid sheet of wood. (see figure) D, E and F are the mid points of its sides as shown and G is the centre of the triangle. The moment of inertia of the triangle about an axis passing through G and perpendicular to the plane of the triangle is I_0. If the smaller triangle DEF is removed from ABC, the moment of inertia of the remaining figure about the same axis is I. Then *[2019, 11 Jan Shift-I]*

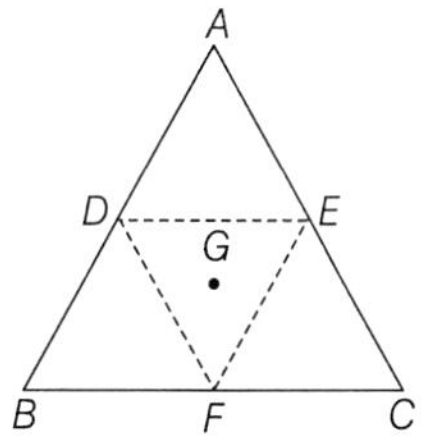

(a) $I' = \frac{3}{4} I_0$ (b) $I' = \frac{15}{16} I_0$

(c) $I' = \frac{I_0}{4}$ (d) $I' = \frac{9}{16} I_0$

Imp. Theorem of Parallel Axes

38 A circular disc D_1 of mass M and radius R has two identical discs D_2 and D_3 of the same mass M and radius R attached rigidly at its opposite ends (see figure). The moment of inertia of the system about the axis OO' passing through the centre of D_1, as shown in the figure will be *[2019, 11 Jan Shift-II]*

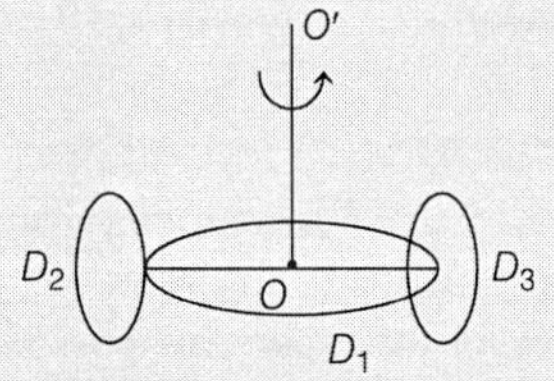

(a) $\frac{2}{3} MR^2$ (b) $\frac{4}{5} MR^2$ (c) $3MR^2$ (d) MR^2

39 Seven identical circular planar discs, each of mass M and radius R are welded symmetrically as shown in the figure. The moment of inertia of the arrangement about the axis normal to the plane and passing through the point P is *[JEE Main 2018]*

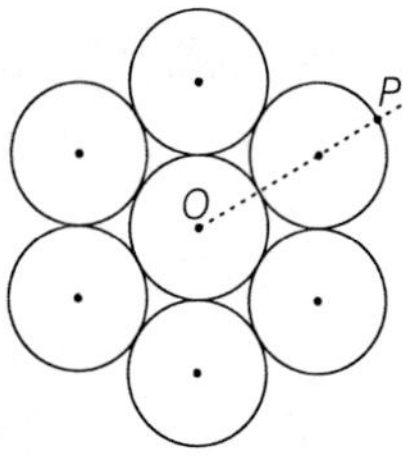

(a) $\frac{19}{2}MR^2$ (b) $\frac{55}{2}MR^2$ (c) $\frac{73}{2}MR^2$ (d) $\frac{181}{2}MR^2$

40 The moment of inertia of a uniform cylinder of length l and radius R about its perpendicular bisector is I. What is the ratio l/R such that the moment of inertia is minimum? *[JEE Main 2017]*

(a) $\sqrt{3}/2$ (b) 1 (c) $3/\sqrt{2}$ (d) $\sqrt{3/2}$

41 From a solid sphere of mass M and radius R, a cube of maximum possible volume is cut. Moment of inertia of cube about an axis passing through its centre and perpendicular to one of its faces is *[JEE Main 2015]*

(a) $\frac{MR^2}{32\sqrt{2}\pi}$ (b) $\frac{MR^2}{16\sqrt{2}\pi}$ (c) $\frac{4MR^2}{9\sqrt{3}\pi}$ (d) $\frac{4MR^2}{3\sqrt{3}\pi}$

Numerical Value Type Questions

42

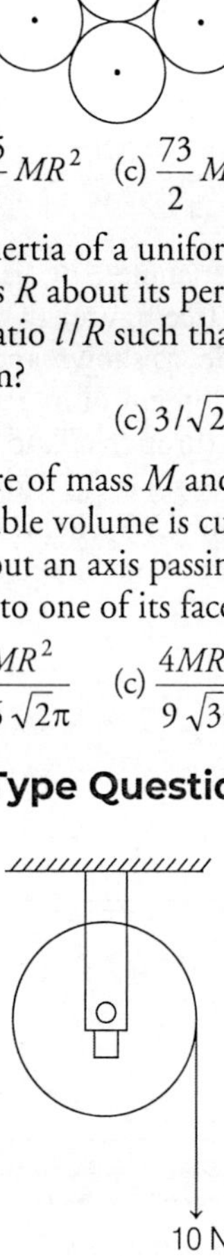

A wheel of radius 0.2 m rotates freely about its center when a string that is wrapped over its rim is pulled by force of 10 N as shown in figure. The established torque produces an angular acceleration of 2 rad/s^2. Moment of intertia of the wheel is .... kg m^2. (Acceleration due to gravity $= 10$ m/s^2) *[2025, 2 April Shift-II]*

43 M and R be the mass and radius of a disc. A small disc of radius $R/3$ is removed from the bigger disc as shown in figure. The moment of inertia of remaining part of bigger disc about an axis AB passing through the centre O and perpendicular to the plane of disc is $\frac{4}{x}MR^2$. The value of x is .... *[2025, 7 April Shift-II]*

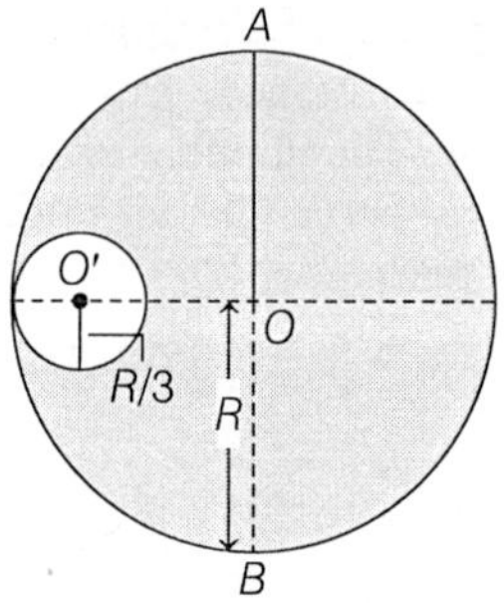

44 A, B and C are disc, solid sphere and spherical shell respectively with same radii and masses. These masses are placed as shown in figure. *[2025, 7 April Shift-I]*

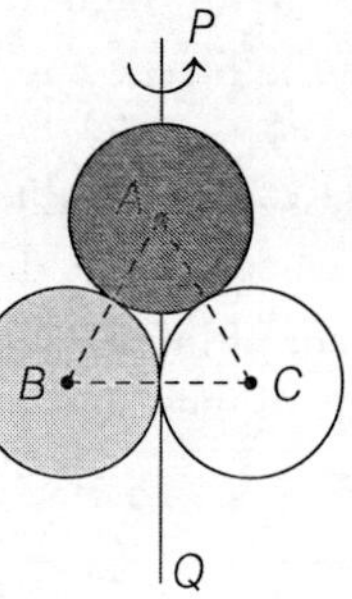

The moment of inertia of the given system about PQ axis is $\frac{x}{15}I$, where I is the moment of inertia of the disc about its diameter. The value of x is ....

45 Two iron solid discs of negligible thickness have radii R_1 and R_2 and moment of inertia I_1 and I_2, respectively. For $R_2 = 2R_1$, the ratio of I_1 and I_2 would be $\frac{1}{x}$. Where, $x =$........ *[2025, 28 Jan Shift-I]*

Imp. Moment of Inertia

46 The moment of inertia of a solid disc rotating along its diameter is 2.5 times higher than the moment of inertia of a ring rotating in similar way. The moment of inertia of a solid sphere which has same radius as the disc and rotating in similar way, is n times higher than the moment of inertia of the given ring. Here, $n =$ ....

Consider all the bodies have equal masses. *[2025, 28 Jan Shift-I]*

47 There balls of masses 2 kg, 4 kg and 6 kg respectively are arranged at centre of the edges of an equilateral triangle of side 2 m. The moment of intertia of the system about an axis through the centroid and perpendicular to the plane of triangle, will be .... kg m^2. *[2024, 4 April Shift-II]*

48 Two identical spheres each of mass 2 kg and radius 50 cm are fixed at the ends of a light rod, so that the separation between the centers is 150 cm. Then, moment of inertia of the system about an axis perpendicular to the rod and passing through its middle point is $\frac{x}{20}$ kg m^2, where the value of x is .... *[2024, 31 Jan Shift-II]*

49 Four particles each of the mass 1 kg are placed at four corners of a square of side 2m. Moment of inertia of system about an axis perpendicular to its plane and passing through one of its vertex is .... kg m^2. *[2024, 27 Jan Shift-I]*

50 Two identical solid spheres each of mass 2 kg and radii 10 cm are fixed at the ends of a light rod. The separation between the centres of the spheres is 40 cm. The moment of inertia of the system about an axis perpendicular to the rod passing through its middle point is ....$\times 10^{-3}$ kg-m^2. *[2023, 6 April Shift-I]*

51 The moment of inertia of a semicircular ring about an axis, passing through the centre and perpendicular to the plane of ring, is $\frac{1}{x}MR^2$ where R is the radius and M is the mass of the semicircular ring. The value of x will be .... *[2023, 8 April Shift-I]*

52 A ring and a solid sphere rotating about an axis passing through their centre have same radii of gyration. The axis of rotation is perpendicular to plane of ring. The ratio of radius of ring to that of sphere is $\sqrt{2/x}$. The value of x is .... *[2023, 6 April Shift-II]*

53 A solid sphere of mass 500 g and radius 5 cm is rotated about one of its diameter with angular speed of 10 rad s^{-1}. If the moment of inertia of the sphere about its tangent is $x \times 10^{-2}$ times its angular momentum about the diameter. Then the value of x will be .... *[2023, 11 April Shift-I]*

54 Two discs of same mass and different radii are made of different materials such that their thicknesses are 1 cm and 0.5 cm respectively. The densities of materials are in the ratio 3 : 5. The moment of inertia of these discs respectively about their diameters will be in the ratio of $x/6$. The value of x is .... *[2023, 31 Jan Shift-II]*

55 Moment of inertia of a disc of mass M and radius R about any of its diameter is $\frac{MR^2}{4}$. The moment of inertia of this disc about an axis normal to the disc and passing through a point on its edge will be, $\frac{x}{2}MR^2$. The value of x is .... *[2023, 1 Feb Shift-II]*

Imp. Theorem of Parallel and Perpendicular Axes

56 I_{CM} is the moment of inertia of a circular disc about an axis (CM) passing through its centre and perpendicular to the plane of disc. I_{AB} is it's moment of inertia about an axis AB perpendicular to plane and parallel to axis CM at a distance $2/3\,R$ from centre. Where R is the radius of the disc. The ratio of I_{AB} and I_{CM} is $x:9$. The value of x is .... *[2023, 25 Jan Shift-I]*

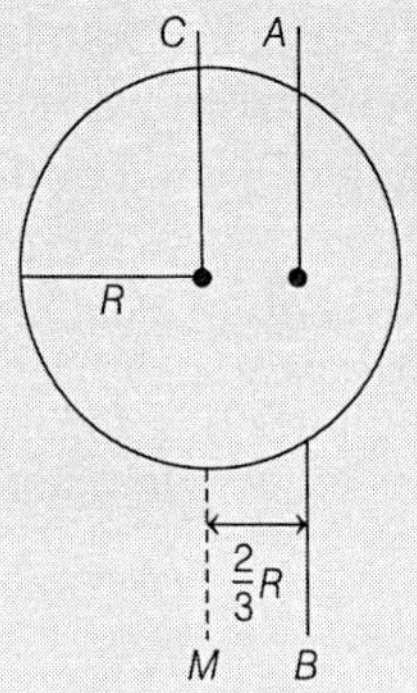

57 If a solid sphere of mass 5 kg and a disc of mass 4 kg have the same radius. Then, the ratio of moment of inertia of the disc about a tangent in its plane to the moment of inertia of the sphere about its tangent will be $x/7$. The value of x is .... *[2023, 25 Jan Shift-II]*

58 Solid spheres A is rotating about an axis PQ. If the radius of the sphere is 5 cm then its radius of gyration about PQ with will be $\sqrt{x}$ cm. The value of x is .... *[2023, 24 Jan Shift-I]*

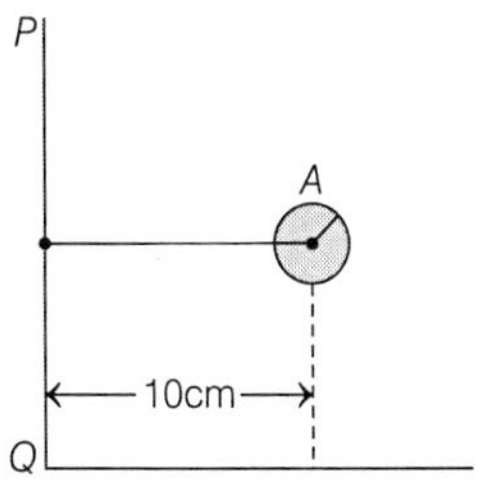

59 A uniform solid cylinder with radius R and length L has moment of inertia I_1, about the axis of the cylinder. A concentric solid cylinder of radius $R' = \frac{R}{2}$ and length $L' = \frac{L}{2}$ is carved out of the original cylinder. If I_2 is the moment of inertia of the carved out portion of the cylinder then $\frac{I_1}{I_2} =$ ....

(Both I_1 and I_2 are about the axis of the cylinder) *[2023, 24 Jan Shift-II]*

60 Four identical discs each of mass M and diameter a are arranged in a small plane as shown in figure. If the moment of inertia of the system about OO' is $\frac{x}{4}Ma^2$. Then, the value of x will be ....

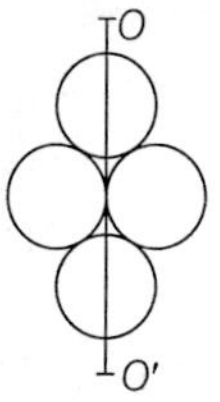

[2022, 28 July Shift-I]

61 The radius of gyration of a cylindrical rod about an axis of rotation perpendicular to its length and passing through the center will be .... m. (Given, the length of the rod is $10\sqrt{3}$ m)

[2022, 26 July Shift-II]

62 The moment of inertia of a uniform thin rod about a perpendicular axis passing through one end is I_1. The same rod is bent into a ring and its moment of inertia about a diameter is I_2. If $\frac{I_1}{I_2}$ is $\frac{x\pi^2}{3}$, then the value of x will be......... . [2022, 29 June Shift-II]

63 A uniform thin bar of mass 6 kg and length 2.4 m is bent to make an equilateral hexagon. The moment of inertia about an axis passing through the centre of mass and perpendicular to the plane of hexagon is .... $\times 10^{-1}$ kg-m^2. [2021, 24 Feb Shift-II]

64 A massless equilateral triangle EFG of side a (as shown in figure) has three particles of mass m situated at its vertices. The moment of inertia of the system about the line EX perpendicular to EG in the plane of EFG is $\frac{N}{20}ma^2$, where N is an integer. The value of N is .... [2020, 03 Sep Shift-II]

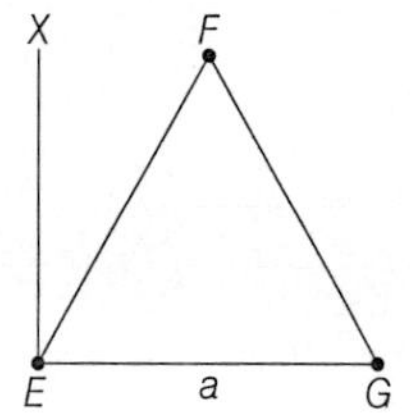

Topic 3 Kinematics and Dynamics of Rotational Motion

Level 1

01 A wheel is rolling on a plane surface. The speed of a particle on the highest point of the rim is 8 m/s. The speed of the particle on the rim of the wheel at the same level as the centre of wheel, will be

[2025, 4 April Shift-II]

(a) 4 m/s
(b) $8\sqrt{2}$ m/s
(c) 8 m/s
(d) $4\sqrt{2}$ m/s

02 A uniform solid cylinder of mass m and radius r rolls along an inclined rough plane of inclination 45°. If it starts to roll from rest from the top of the plane then, the linear acceleration of the cylinder's axis will be

[2025, 24 Jan Shift-I]

(a) $\frac{1}{3\sqrt{2}}g$ (b) $\frac{\sqrt{2}g}{3}$ (c) $\frac{1}{\sqrt{2}}g$ (d) $\sqrt{2}g$

03 A disc is rolling without slipping on a surface. The radius of the disc is R. At $t = 0$, the topmost point on the disc is A as shown in figure. When the disc completes half of its rotation, the displacement of point A from its initial position is [2023, 13 April Shift-I]

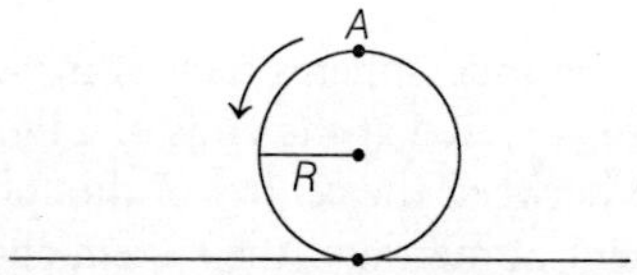

(a) $2R\sqrt{(1+4\pi^2)}$ (b) $R\sqrt{(\pi^2+4)}$
(c) $2R$ (d) $R\sqrt{(\pi^2+1)}$

04 A body rolls down an inclined plane without slipping. The kinetic energy of rotation is 50% of its translational kinetic energy. The body is

[2021, 20 July Shift-II]

(a) solid sphere (b) solid cylinder
(c) hollow cylinder (d) ring

05 A cord is wound round the circumference of wheel of radius r. The axis of the wheel is horizontal and the moment of inertia about it is I. A weight mg is attached to the cord at the end. The weight falls from rest. After falling through a distance h, the square of angular velocity of wheel will be [2021, 26 Feb Shift-II]

(a) $\frac{2mgh}{I+2mr^2}$ (b) $\frac{2mgh}{I+mr^2}$
(c) $2gh$ (d) $\frac{2gh}{I+mr^2}$

06 A rectangular solid box of length 0.3 m is held horizontally, with one of its sides on the edge of a platform of height 5 m. When released, it slips off the table in a very short time $\tau = 0.01$ s, remaining essentially horizontal. The angle by which it would rotate when it hits the ground will be (in radians) close to [2019, 8 April Shift-II]

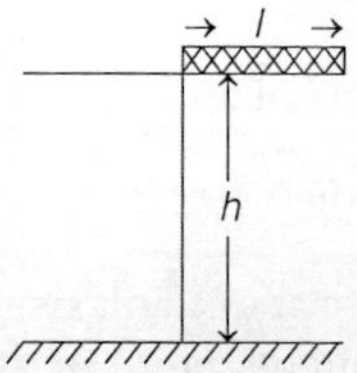

(a) 0.02 (b) 0.3
(c) 0.5 (d) 0.28

07 A slender uniform rod of mass M and length l is pivoted at one end so that it can rotate in a vertical plane (see the figure). There is negligible friction at the pivot. The free end is held vertically above the pivot and then released. The angular acceleration of the rod when it makes an angle θ with the vertical, is *[JEE Main 2017]*

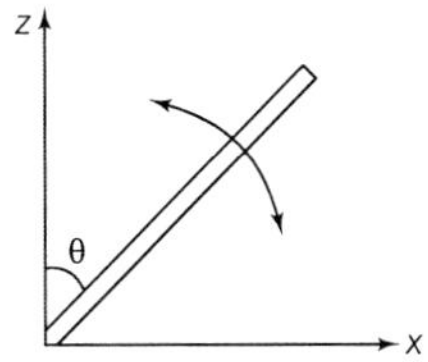

(a) $\frac{2g}{3l}\sin\theta$ (b) $\frac{3g}{2l}\cos\theta$ (c) $\frac{2g}{3l}\cos\theta$ (d) $\frac{3g}{2l}\sin\theta$

08 A roller is made by joining together two corners at their vertices O. It is kept on two rails AB and CD which are placed a symmetrically (see the figure), with its axis perpendicular to CD and its centre O at the centre of line joining AB and CD (see the figure). It is given a light push, so that it starts rolling with its centre O moving parallel to CD in the direction shown. As it moves, the roller will tend to *[JEE Main 2016]*

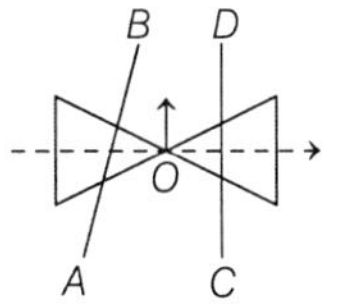

(a) turn left
(b) turn right
(c) go straight
(d) turn left and right alternately

Imp. Dynamics of Rotational Motion

09 A mass m supported by a massless string wound around a uniform hollow cylinder of mass m and radius R. If the string does not slip on the cylinder, with what acceleraton will the mass fall on release? *[JEE Main 2014]*

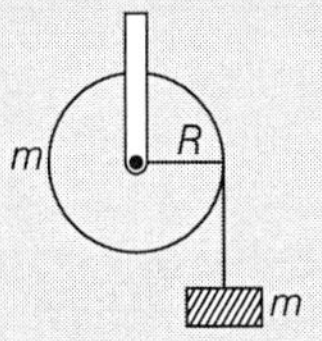

(a) $\frac{2g}{3}$ (b) $\frac{g}{2}$ (c) $\frac{5g}{6}$ (d) g

10 A thin horizontal circular disc is rotating about a vertical axis passing through its centre. An insect is at rest at a point near the rim of the disc. The insect now moves along a diameter of the disc to reach its other end. During the journey of the insect, the angular speed of the disc *[AIEEE 2011]*

(a) continuously decreases
(b) continuously increases
(c) first increases and then decreases
(d) remains unchanged

11 A thin uniform rod of length l and mass m is swinging freely about a horizontal axis passing through its end. Its maximum angular speed is ω. Its centre of mass rises to a maximum height of *[AIEEE 2009]*

(a) $\frac{1}{3}\frac{l^2\omega^2}{g}$ (b) $\frac{1}{6}\frac{l\omega}{g}$ (c) $\frac{1}{2}\frac{l^2\omega^2}{g}$ (d) $\frac{1}{6}\frac{l^2\omega^2}{g}$

12 A solid sphere, a hollow sphere and a ring are released from top of an inclined plane (frictionless), so that they slide down the plane. Then, maximum acceleration down the plane is for (no rolling) *[AIEEE 2002]*

(a) solid sphere (b) hollow sphere
(c) ring (d) All same

Level 2

13 A force of 49 N acts tangentially at the highest point of a sphere (solid) of mass 20 kg, kept on a rough horizontal plane. If the sphere rolls without slipping, then the acceleration of the center of the sphere is *[2025, 3 April Shift-I]*

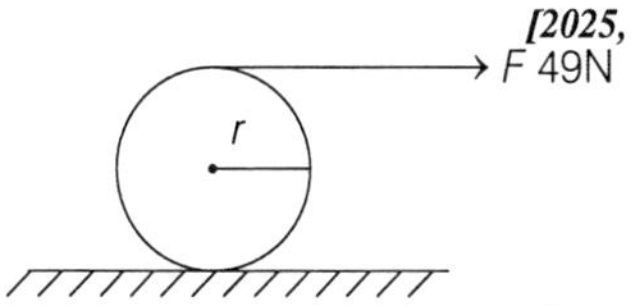

(a) 2.5 m/s^2 (b) 0.35 m/s^2
(c) 0.25 m/s^2 (d) 3.5 m/s^2

14 A cord of negligible mass is wound around the rim of a wheel supported by spokes with negligible mass. The mass of wheel is 10 kg and radius is 10 cm and it can freely rotate without any friction. Initially the wheel is at rest. If a steady pull of 20 N is applied on the cord, the angular velocity of the wheel, after the cord is unwound by 1 m, would be *[2025, 2 April Shift-I]*

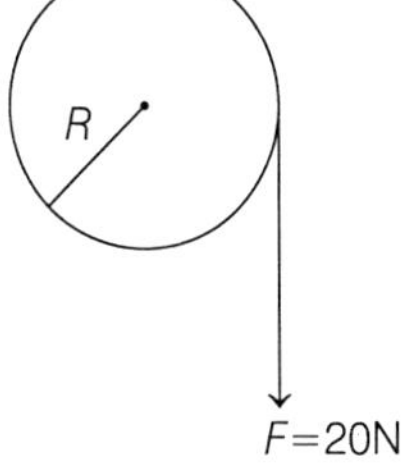

(a) 0 rad/s (b) 10 rad/s (c) 30 rad/s (d) 20 rad/s

15 A solid sphere is rolling without slipping on a horizontal plane. The ratio of the linear kinetic energy of the centre of mass of the sphere and rotational kinetic energy is *[2025, 24 Jan Shift-II]*

(a) 2/5 (b) 4/3 (c) 3/4 (d) 5/2

16 A thin circular disc of mas M and radius R is rotating in a horizontal plane about an axis passing through its centre and perpendicular to its plane with angular velocity ω. If another disc of same dimensions but of mass $M/2$ is placed gently on the first disc coaxially, then the new angular velocity of the system is *[2024, 8 April Shift-II]*

(a) $\frac{4}{5}\omega$ (b) $\frac{3}{2}\omega$

(c) $\frac{2}{3}\omega$ (d) $\frac{5}{4}\omega$

17 A disc of radius R and mass M is rolling horizontally without slipping with speed v. It then moves up an inclined smooth surface as shown in figure. The maximum height that the disc can go up the incline is

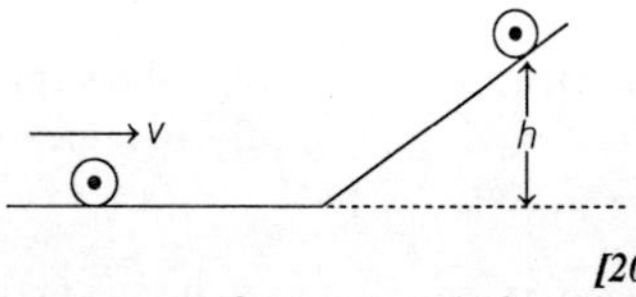

[2024, 1 Feb Shift-II]

(a) $\frac{2}{3}\frac{v^2}{g}$ (b) $\frac{3}{4}\frac{v^2}{g}$ (c) $\frac{1}{2}\frac{v^2}{g}$ (d) $\frac{v^2}{g}$

18 A ball of mass 0.5 kg is attached to a string of length 50 cm. The ball is rotated on a horizontal circular path about its vertical axis. The maximum tension that the string can bear is 400 N. The maximum possible value of angular velocity of the ball in rad/s is *[2024, 1 Feb Shift-I]*

(a) 1600 (b) 20 (c) 40 (d) 1000

19 A stone of mass 900 g is tied to a string and moved in a vertical circle of radius 1 m making 10 rpm. The tension in the string, when the stone is at the lowest point is (If, $\pi^2 = 9.8$ and $g = 9.8\ \text{m/s}^2$) *[2024, 29 Jan Shift-II]*

(a) 9.8 N (b) 8.82 N
(c) 17.8 N (d) 97 N

20 A coin placed on a rotating table just slips when it is placed at a distance of 1 cm from the centre. If the angular velocity of the table is halved, it will just slip when placed at a distance of ...... from the centre *[2023, 11 April Shift-I]*

(a) 2 cm (b) 4 cm (c) 8 cm (d) 1 cm

21 A spherical shell of 1 kg mass and radius R is rolling with angular speed ω on horizontal plane as shown in figure. The magnitude of angular momentum of the shell about the origin O is $\frac{a}{3}R^2\omega$. The value of a will be *[2022, 29 June Shift-I]*

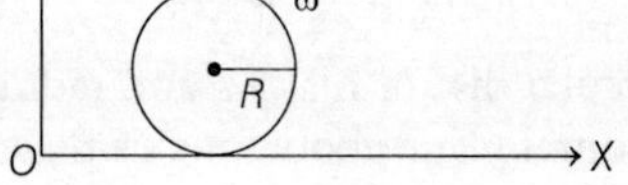

(a) 2 (b) 3 (c) 5 (d) 4

22 A fly wheel is accelerated uniformly from rest and rotates through 5 rad in the first second. The angle rotated by the fly wheel in the next second, will be *[2022, 24 June Shift-II]*

(a) 7.5 rad (b) 15 rad (c) 20 rad (d) 30 rad

23 As shown in the figure, a bob of mass m is tied by a massless string whose other end portion is wound on a flywheel (disc) of radius r and mass m. When released from rest the bob starts falling vertically. When it has covered a distance of h, the angular speed of the wheel will be *[2020, 7 Jan Shift-I]*

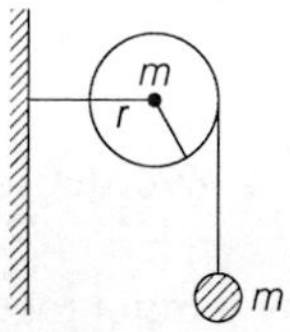

(a) $\frac{1}{r}\sqrt{\frac{2gh}{3}}$ (b) $r\sqrt{\frac{3}{4gh}}$ (c) $\frac{1}{r}\sqrt{\frac{4gh}{3}}$ (d) $r\sqrt{\frac{3}{2gh}}$

24 A uniform sphere of mass 500 g rolls without slipping on a plane horizontal surface with its centre moving at a speed of 5.00 cm/s. Its kinetic energy is *[2020, 8 Jan Shift-II]*

(a) 6.25×10^{-4} J (b) 1.13×10^{-3} J
(c) 8.75×10^{-4} J (d) 8.75×10^{-3} J

25 A rod of length 50 cm is pivoted at one end. It is raised such that it makes an angle of 30° from the horizontal as shown and released from rest. Its angular speed when it passes through the horizontal (in rad s^{-1}) will be (Take, $g = 10\ \text{ms}^{-2}$) *[2019, 9 Jan Shift-II]*

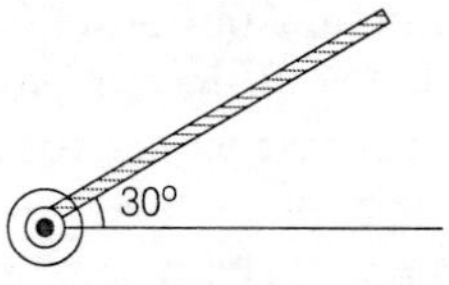

(a) $\frac{\sqrt{30}}{2}$ (b) $\sqrt{30}$ (c) $\frac{\sqrt{20}}{3}$ (d) $\sqrt{\frac{30}{2}}$

26 A stationary horizontal disc is free to rotate about its axis. When a torque is applied on it, its kinetic energy as a function of θ, where θ is the angle by which it has rotated, is given as $k\theta^2$. If its moment of inertia is I, then the angular acceleration of the disc is *[2019, 9 April Shift-I]*

(a) $\frac{k}{2I}\theta$ (b) $\frac{k}{I}\theta$ (c) $\frac{k}{4I}\theta$ (d) $\frac{2k}{I}\theta$

27 Moment of inertia of a body about a given axis is 1.5 kg m^2. Initially, the body is at rest. In order to produce a rotational kinetic energy of 1200 J, the angular acceleration of 20 rad/s^2 must be applied about the axis for a duration of *[2019, 9 April Shift-II]*

(a) 5 s (b) 2 s (c) 3 s (d) 2.5 s

28 A uniform solid cylindrical roller of mass m is being pulled on a horizontal surface with force F parallel to the surface and applied at its centre. If the acceleration of the cylinder is a and it is rolling without slipping then the value of F is *[JEE Main 2015]*

(a) ma (b) $2ma$ (c) $\left(\frac{3}{2}\right)ma$ (d) $\left(\frac{5}{3}\right)ma$

29 A string is wound around a hollow cylinder of mass 5 kg and radius 0.5 m. If the string is now pulled with a horizontal force of 40 N and the cylinder is rolling without slipping on a horizontal surface (see figure), then the angular acceleration of the cylinder will be (Neglect the mass and thickness of the string) *[2019, 11 Jan Shift-II]*

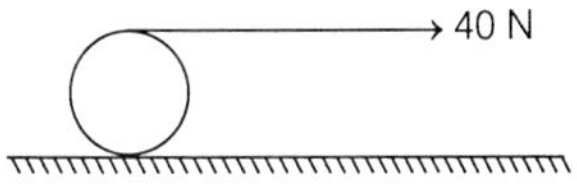

(a) 10 rad/s^2 (b) 16 rad/s^2
(c) 20 rad/s^2 (d) 12 rad/s^2

Level 3

30 A solid sphere and a hollow sphere of the same mass and of same radius are rolled on an inclined plane. Let the time taken to reach the bottom by the solid sphere and the hollow sphere be t_1 and t_2, respectively, then *[2025, 24 Jan Shift-II]*

(a) $t_1 < t_2$ (b) $t_1 > t_2$ (c) $t_1 = t_2$ (d) $t_1 = 2t_2$

31 A solid sphere of mass m and radius r is allowed to roll without slipping from the highest point of an inclined plane of length L and makes an angle 30° with the horizontal. The speed of the particle at the bottom of the plane is v_1. If the angle of inclination is increased to 45° while keeping L constant. Then the new speed of the sphere at the bottom of the plane is v_2. The ratio $v_1^2 : v_2^2$ is *[2025, 23 Jan Shift-I]*

(a) 1 : 2 (b) $1:\sqrt{3}$
(c) 1 : 3 (d) $1:\sqrt{2}$

32 A solid cylinder and a solid sphere, having same mass M and radius R, roll down the same inclined plane from top without slipping. They start from rest. The ratio of velocity of the solid cylinder to that of the solid sphere, with which they reach the ground, will be *[2022, 25 July Shift I]*

(a) $\sqrt{\frac{5}{3}}$ (b) $\sqrt{\frac{4}{5}}$ (c) $\sqrt{\frac{3}{5}}$ (d) $\sqrt{\frac{14}{15}}$

33 A ball spins with angular acceleration $\alpha = 6t^2 - 2t$ where t is in second and α is in rad-s^{-2}. At $t = 0$, the ball has angular velocity of 10 rad-s^{-1} and angular position of 4 rad. The most appropriate expression for the angular position of the ball is *[2022, 28 June Shift-II]*

(a) $\frac{3}{4}t^4 - t^2 + 10t$ (b) $\frac{t^4}{2} - \frac{t^3}{3} + 10t + 4$
(c) $\frac{3t^4}{3} - \frac{t^3}{6} + 10t + 12$ (d) $2t^4 - \frac{t^3}{2} + 5t + 4$

34 A solid spherical ball is rolling on a frictionless horizontal plane surface about its axis of symmetry. The ratio of rotational kinetic energy of the ball to its total kinetic energy is *[2022, 26 June Shift-II]*

(a) $\frac{2}{5}$ (b) $\frac{2}{7}$
(c) $\frac{1}{5}$ (d) $\frac{7}{10}$

35 Two discs have moments of inertia I_1 and I_2 about their respective axes perpendicular to the plane and passing through the centre. They are rotating with angular speeds, ω_1 and ω_2, respectively and are brought into contact face to face with their axes of rotation co-axial. The loss in kinetic energy of the system in the process is given by *[2021, 27 Aug Shift-II]*

(a) $\frac{I_1 I_2}{(I_1 + I_2)}(\omega_1 - \omega_2)^2$ (b) $\frac{(I_1 - I_2)^2 \omega_1 \omega_2}{2(I_1 + I_2)}$
(c) $\frac{I_1 I_2}{2(I_1 + I_2)}(\omega_1 - \omega_2)^2$ (d) $\frac{(\omega_1 - \omega_2)^2}{2(I_1 + I_2)}$

36 Consider a situation in which a ring, a solid cylinder and a solid sphere roll down on the same inclined plane without slipping. Assume that they start rolling from rest and having identical diameter.The correct statement for this situation. *[2021, 22 July Shift-II]*

(a) The sphere has the greatest and the ring has the least velocity of the centre of mass at the bottom of the inclined plane.
(b) The ring has the greatest and the cylinder has the least velocity of the centre of mass at the bottom of the inclined plane.
(c) All of them will have same velocity.
(d) The cylinder has the greatest and the sphere has the least velocity of the centre of mass at the bottom of the inclined plane.

37 A sphere of mass 2 kg and radius 0.5 m is rolling with an initial speed of 1 ms^{-1} goes up an inclined plane which makes an angle of 30° with the horizontal plane, without slipping. How long will the sphere take to return to the starting point A ? *[2021, 17 March Shift-II]*

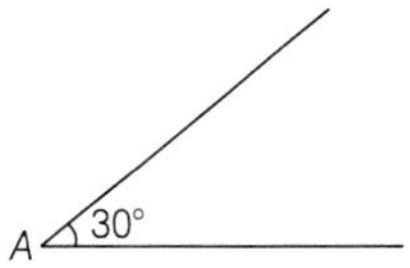

(a) 0.60 s (b) 0.52 s
(c) 0.57 s (d) 0.80 s

38 A sphere of radius a and mass m rolls along a horizontal plane with constant speed v_0. It encounters an inclined plane at angle θ and climbs upwards. Assuming that it rolls without slipping, how far up the sphere will travel? *[2021, 25 Feb Shift-II]*

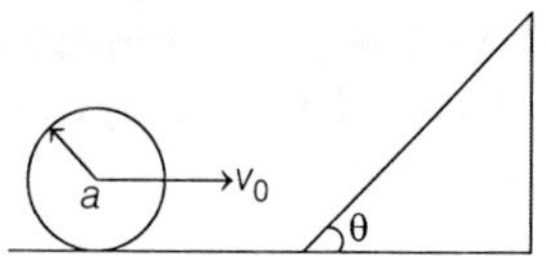

(a) $\frac{v_0^2}{2g\sin\theta}$ (b) $\frac{v_0^2}{5g\sin\theta}$

(c) $\frac{7v_0^2}{10g\sin\theta}$ (d) $\frac{2}{5}\frac{v_0^2}{g\sin\theta}$

Imp. Conservation of Angular Momentum

39 Two uniform circular discs are rotating independently in the same direction around their common axis passing through their centres. The moment of inertia and angular velocity of the first disc are 0.1 kg-m^2 and 10 rad s^{-1}, respectively while those for the second one are 0.2 kg-m^2 and 5 rad s^{-1}, respectively. At some instant they get stuck together and start rotating as a single system about their common axis with some angular speed. The kinetic energy of the combined system is *[2020, 02 Sep Shift-II]*

(a) $\frac{10}{3}$ J (b) $\frac{20}{3}$ J (c) $\frac{5}{3}$ J (d) $\frac{2}{3}$ J

40 A wheel is rotating freely with an angular speed ω on a shaft. The moment of inertia of the wheel is I and the moment of inertia of the shaft is negligible. Another wheel of moment of inertia $3I$ initially at rest is suddenly coupled to the same shaft. The resultant fractional loss in the kinetic energy of the system is *[2020, 05 Sep Shift-I]*

(a) 3/4 (b) 5/6 (c) 0 (d) 1/4

41 A uniformly thick wheel with moment of inertia I and radius R is free to rotate about its centre of mass (see figure). A massless string is wrapped over its rim and two blocks of masses m_1 and $m_2 (m_1 > m_2)$ are attached to the ends of the string. The system is released from rest. The angular speed of the wheel when m_1 descents by a distance h is *[2020, 9 Jan Shift-II]*

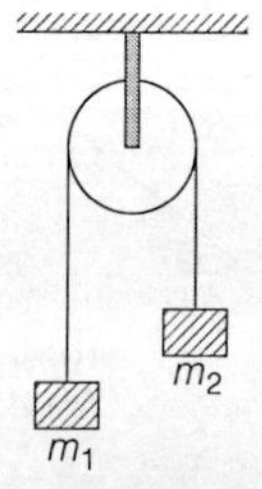

(a) $\left[\frac{2(m_1+m_2)gh}{(m_1+m_2)R^2+I}\right]^{1/2}$

(b) $\left[\frac{m_1+m_2}{(m_1+m_2)R^2+I}\right]^{1/2} gh$

(c) $\left[\frac{m_1-m_2}{(m_1+m_2)R^2+I}\right]^{1/2} gh$

(d) $\left[\frac{2(m_1-m_2)gh}{(m_1+m_2)R^2+I}\right]^{1/2}$

42 A rigid massless rod of length $3l$ has two masses attached at each end as shown in the figure. The rod is pivoted at point P on the horizontal axis (see figure). When released from initial horizontal position, its instantaneous angular acceleration will be *[2019, 10 Jan Shift-II]*

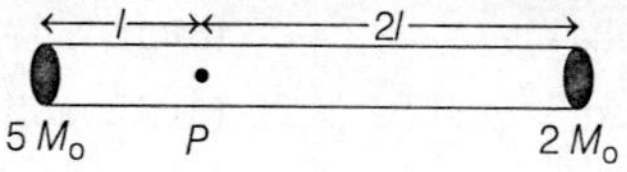

(a) $g/13l$ (b) $g/2l$
(c) $7g/3l$ (d) $g/3l$

43 A solid sphere and solid cylinder of identical radii approach an incline with the same linear velocity (see figure). Both roll without slipping all throughout. The two climb maximum heights h_{sph} and h_{cyl} on the incline. The ratio h_{sph}/h_{cyl} is given by *[2019, 8 April Shift-II]*

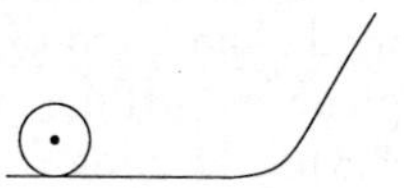

(a) $\frac{2}{\sqrt{5}}$ (b) $\frac{14}{15}$

(c) 1 (d) $\frac{4}{5}$

44 Two coaxial discs having moments of inertia I_1 and $I_1/2$ are rotating with respective angular velocities ω_1 and $\frac{\omega_1}{2}$, about their common axis. They are brought in contact with each other and thereafter they rotate with a common angular velocity. If E_f and E_i are the final and initial total energies, then $(E_f - E_i)$ is *[2019, 10 April Shift-I]*

(a) $-\frac{I_1\omega_1^2}{24}$ (b) $-\frac{I_1\omega_1^2}{12}$

(c) $\frac{3}{8}I_1\omega_1^2$ (d) $\frac{I_1\omega_1^2}{6}$

45 A homogeneous solid cylindrical roller of radius R and mass m is pulled on a cricket pitch by a horizontal force. Assuming rolling without slipping, angular acceleration of the cylinder is *[2019, 10 Jan Shift-I]*

(a) $\frac{F}{2mR}$ (b) $\frac{2F}{3mR}$ (c) $\frac{3F}{2mR}$ (d) $\frac{F}{3mR}$

46 A pulley of radius 2 m is rotated about its axis by a force $F=(20t-5t^2)$ N (where, t is measured in seconds) applied tangentially. If the moment of inertia of the pulley about its axis of rotation is 10 kg-m^2, then the number of rotations made by the pulley before its direction of motion is reversed, is *[AIEEE 2011]*

(a) more than 3 but less than 6
(b) more than 6 but less than 9
(c) more than 9
(d) less than 3

47 A round uniform body of radius R, mass M and moment of inertia I, rolls down (without slipping) an inclined plane making an angle θ with the horizontal. Then, its acceleration is *[AIEEE 2007]*

(a) $\frac{g\sin\theta}{1+I/MR^2}$ (b) $\frac{g\sin\theta}{1+MR^2/I}$ (c) $\frac{g\sin\theta}{1-I/MR^2}$ (d) $\frac{g\sin\theta}{1-MR^2/I}$

Numerical Value Type Questions

48 A circular ring and a solid sphere having same radius roll down on an inclined plane from rest without slipping. The ratio of their velocities when reached at the bottom of the plane is $\sqrt{\frac{x}{5}}$, where $x=$ .... *[2025, 4 April Shift-I]*

49 A thin solid disk of 1 kg is rotating along its diameter axis at the speed of 1800 rpm. By applying an external torque of 25π Nm for 40s, the speed increases to 2100 rpm. The diameter of the disk is .... m. *[2025, 8 April Shift-II]*

50 A solid sphere with uniform density and radius R is rotating initially with constant angular velocity (ω_1) about its diameter. After some time during the rotation its starts loosing mass at a uniform rate, with no change in its shape. The angular velocity of the sphere when its radius become $\frac{R}{2}$ is $x\omega_1$. The value of x is .... *[2025, 4 April Shift-II]*

51 If the radius of earth is reduced to three-fourth of its present value without change in its mass, then value of duration of the day of earth will be .... h 30 min. *[2024, 6 April Shift-I]*

52 A uniform rod AB of mass 2 kg and length 30 cm at rest on a smooth horizontal surface. An impulse of force 0.2 N-s is applied to end B. The time taken by the rod to turn through at right angles will be $\frac{\pi}{x}$ s, where $x=$ ... . *[2024, 1 Feb Shift-II]*

53 A solid circular disc of mass 50 kg rolls along a horizontal floor, so that its center of mass has a speed of 0.4 m/s. The absolute value of work done on the disc to stop it is .... J. *[2024, 31 Jan Shift-I]*

54

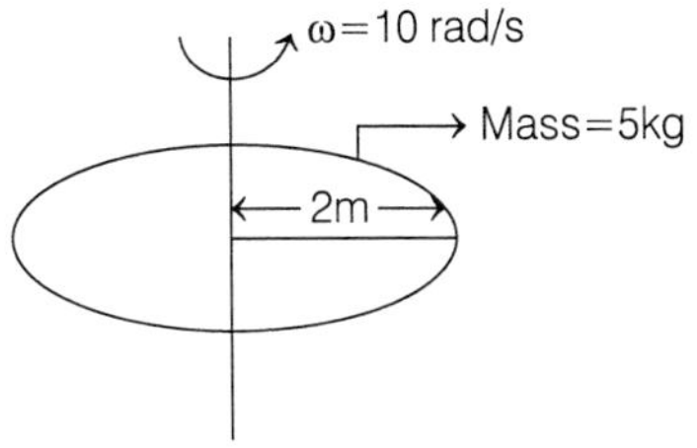

Consider a disc of mass 5 kg, radius 2 m, rotating with angular velocity of 10 rad/s about an axis perpendicular to the plane of rotation. An identical disc is kept gently over the rotating disc along the same axis. The energy dissipated, so that both the discs continue to rotate together without slipping is .... J. *[2024, 30 Jan Shift-I]*

55 Two discs of moment of inertia $I_1=4$ kg m^2 and $I_2=2$ kg m^2, about their central axes and normal to their planes, rotating with angular speeds 10 rad/s and 4 rad/s, respectively are brought into contact face to face with their axes of rotation coincident. The loss in kinetic energy of the system in the process is .... J. *[2024, 30 Jan Shift-II]*

56 A light rope is wound around a hollow cylinder of mass 5 kg and radius 70 cm. The rope is pulled with a force of 52.5 N. The angular acceleration of the cylinder will be .... rad s^{-2}. *[2023, 13 April Shift-II]*

57 For a rolling spherical shell, the ratio of rotational kinetic energy and total kinetic energy is $x/5$. The value of x is .... *[2023, 12 April Shift-I]*

58 A solid sphere is rolling on a horizontal plane without slipping. If the ratio of angular momentum about axis of rotation of the sphere to the total energy of moving sphere is $\pi:22$, then the value of its angular speed will be .... rad/s. *[2023, 13 April Shift-I]*

59 A hollow spherical ball of uniform density rolls up a curved surface with an initial velocity 3 m/s (as shown in figure). Maximum height with respect to the initial position covered by it will be .... cm. (take, $g=10$ m/s^2) *[2023, 8 April Shift-II]*

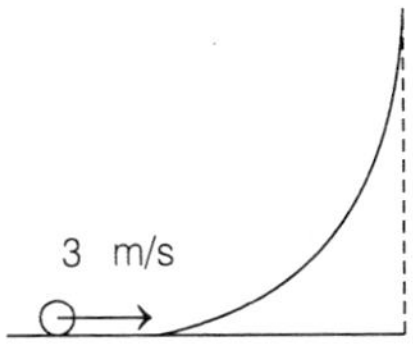

60 A circular plate is rotating in a horizontal plane about an axis passing through its centre and perpendicular to the plate with an angular velocity ω. A person sits at the centre having two dumbbells in his hands, the moment of inertia of the system become triple. If E be the initial kinetic energy of the system, then final kinetic energy will be E / x. The value of x is .... *[2023, 11 April Shift-II]*

61 A uniform disc of mass 0.5 kg and radius r is projected with velocity 18 m/s at $t = 0$ s on a rough horizontal surface. It starts off with a purely sliding motion at $t = 0$ s. After 2 s it acquires a purely rolling motion (see figure).
The total kinetic energy of the disc after 2 s will be .... J (given, coefficient of friction is 0.3 and $g = 10 \text{ m/s}^2$). *[2023, 30 Jan Shift-II]*

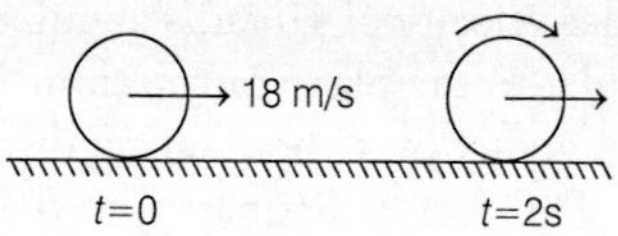

62 A solid sphere of mass 2 kg is making pure rolling on a horizontal surface with kinetic energy 2240 J. The velocity of centre of mass of the sphere will be .... ms^{-1}. *[2023, 29 Jan Shift-I]*

63 A solid cylinder is released from rest from the top of an inclined plane of inclination 30° and length 60 cm. If the cylinder rolls without slipping, its speed upon reaching the bottom of the inclined plane is .... ms^{-1}. (Given, $g = 10 \text{ ms}^{-2}$) *[2023, 1Feb Shift-I]*

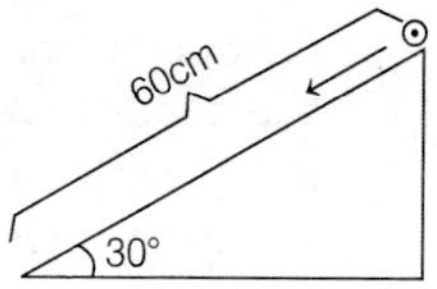

64 A solid sphere of mass 1 kg rolls without slipping on a plane surface. Its kinetic energy is 7×10^{-3} J. The speed of the centre of mass of the sphere is .... cm s^{-1}. *[2023, 31 Jan Shift-I]*

65 A uniform disc with mass $M = 4$ kg and radius $R = 10$ cm is mounted on a fixed horizontal axle as shown in figure. A block with mass $m = 2$ kg hangs from a massless cord that is wrapped around the rim of the disc. During the fall of the block, the cord does not slip and there is no friction at the axle. The tension in the cord is .... N.
(Take, $g = 10 \text{ m s}^{-2}$) *[2022, 28 June Shift-II]*

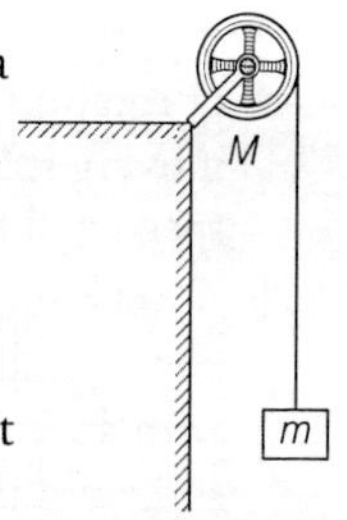

66 A solid cylinder length is suspended symmetrically through two massless strings, as shown in the figure. The distance from the initial rest position, the cylinder should be unbinding the strings to achieve a speed of 4 m/s, is .... cm.
(Take, $g = 10 \text{ m/s}^2$) *[2022, 27 July Shift-II]*

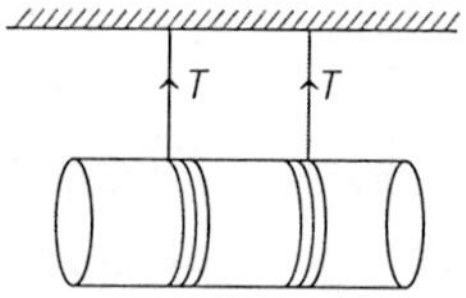

67 A pulley of radius 1.5 m is rotated about its axis by a force $F = (12t - 3t^2)$ N applied tangentially (while t is measured in seconds). If moment of inertia of the pulley about its axis of rotation is 4.5 kg - m^2, the number of rotations made by the pulley before its direction of motion is reversed, will be K/π. The value of K is .... *[2022, 27 July Shift-I]*

68 A rolling wheel of 12 kg is on an inclined plane at position P and connected to a mass of 3 kg through a string of fixed length and pulley as shown in figure. Consider PR as friction free surface. The velocity of centre of mass of the wheel when it reaches at the bottom Q of the inclined plane PQ will be $\frac{1}{2}\sqrt{xgh}$ m/s. The value of x is . *[2022, 27 June Shift-II]*

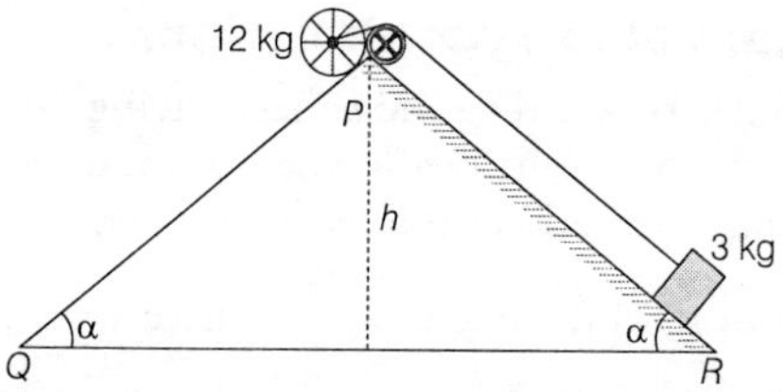

69 In the given figure, two wheels P and Q are connected by a belt B. The radius of P is three times as that of Q. In case of same rotational kinetic energy, the ratio of rotational inertias $\left(\frac{I_1}{I_2}\right)$ will be $x : 1$.
The value of x will be .... *[2021, 27 July Shift-II]*

70 A rod of mass M and length L is lying on a horizontal frictionless surface. A particle of mass m travelling along the surface hits at one end of the rod with a velocity u in a direction perpendicular to the rod. The collision is completely elastic. After collision, particle comes to rest. The ratio of masses $\left(\frac{m}{M}\right)$ is $\frac{1}{x}$.
The value of x will be .... *[2021, 20 July Shift-I]*

71 The centre of a wheel rolling on a plane surface moves with a speed v_0. A particle on the rim of the wheel at the same level as the centre will be moving at a speed $\sqrt{x}\, v_0$. Then, the value of x is .... *[2021, 22 July Shift-II]*

72 A circular disc reaches from top to bottom of an inclined plane of length L. When it slips down the plane, it takes time t_1. When it rolls down the plane, it takes time t_2. The value of $\frac{t_2}{t_1}$ is $\sqrt{\frac{3}{x}}$. The value of x will be .... *[2021, 20 July Shift-I]*

73 Two bodies, a ring and a solid cylinder of same material are rolling down without slipping an inclined plane. The radii of the bodies are same. The ratio of velocity of the centre of mass at the bottom of plane of the ring to that of the cylinder is $\sqrt{x}/2$. Then, the value of x is .... *[2021, 20 July Shift-II]*

74 A body rotating with an angular speed of 600 rpm is uniformly accelerated to 1800 rpm in 10 s. The number of rotations made in the process is .... *[2021, 20 July Shift-II]*

75 The following bodies,

1. a ring
2. a disc
3. a solid cylinder
4. a solid sphere

of same mass m and radius R are allowed to roll down without slipping simultaneously from the top of the inclined plane. The body which will reach first at the bottom of the inclined plane is .... (Mark the body as per their respective numbering given in the question) *[2021, 17 March Shift-I]*

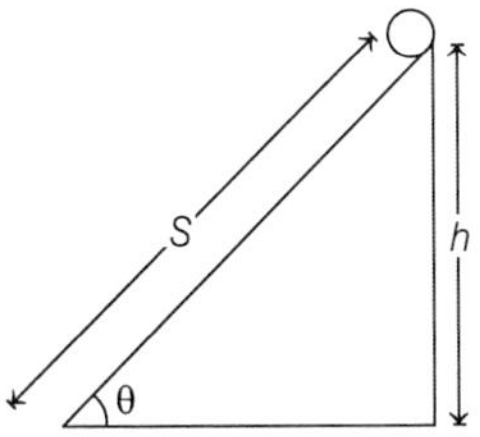

76 The angular speed of truck wheel is increased from 900 rpm to 2460 rpm in 26 s. The number of revolutions by the truck engine during this time is .... (Assuming the acceleration to be uniform). *[2021, 17 March Shift-I]*

77 A solid disc of radius a and mass m rolls down without slipping on an inclined plane making an angle θ with the horizontal. The acceleration of the disc will be $\frac{2}{b} g \sin\theta$, where b is .... (Round off to the nearest integer)

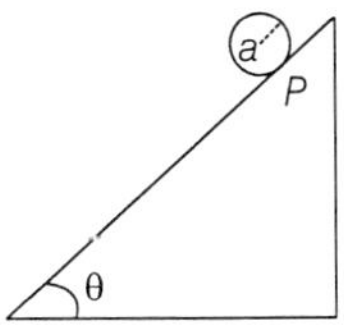

(g = acceleration due to gravity)
(θ = angle as shown in figure) *[2021, 16 March Shift-II]*

78 Consider a 20 kg uniform circular disc of radius 0.2 m. It is pin supported at its centre and is at rest initially. The disc is acted upon by a constant force $F = 20$ N through a massless string wrapped around its periphery as shown in the figure.

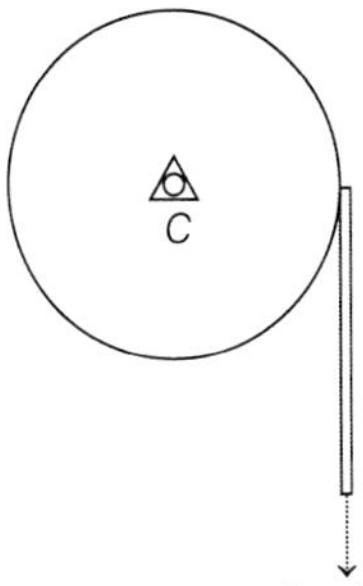

Suppose the disc makes n number of revolutions to attain an angular speed of 50 rad s^{-1}. The value of n to the nearest integer, is .... (Given, in one complete revolution, the disc rotates by 6.28 rad.) *[2021, 16 March Shift-I]*

79 A thin rod of mass 0.9 kg and length 1 m is suspended at rest from one end, so that it can freely oscillate in the vertical plane. A particle of mass 0.1 kg moving in a straight line with velocity 80 m/s hits the rod at its bottom-most point and sticks to it (see figure).
The angular speed (in rad/s) of the rod immediately after the collision will be .... *[2020, 05 Sep Shift-II]*

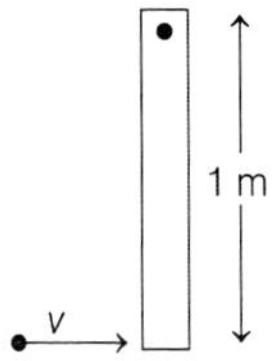

80 A circular disc of mass M and radius R is rotating about its axis with angular speed ω_1. If another stationary disc having radius $R/2$ and same mass M is dropped co-axially on to the rotating disc. Gradually, both discs attain constant angular speed ω_2. The energy lost in the process is p% of the initial energy. Value of p is .... *[2020, 04 Sep Shift-I]*

Key Index

Topic 1

1 (b)	**2** (b)	**3** (d)	**4** (c)	**5** (b)	**6** (c)	**7** (d)	**8** (a)	**9** (a)	**10** (c)
11 (d)	**12** (d)	**13** (b)	**14** (c)	**15** (c)	**16** (c)	**17** (c)	**18** (c)	**19** (d)	**20** (b)
21 (c)	**22** (b)	**23** (a)	**24** (d)	**25** (c)	**26** (b)	**27** (c)	**28** (a)	**29** (a)	**30** (c)
31 (c)	**32** (d)	**33** (d)	**34** (c)	**35** (d)	**36** (d)	**37** (c)	**38** (d)	**39** (a)	**40** (c)
41 (d)	**42** (d)	**43** (b)	**44** (a)	**45** (d)	**46** (a)	**47** (c)	**48** (a)	**49** (b)	**50** (d)
51 (c)	**52** (a)	**53** (b)	**54** (90)	**55** (2)	**56** (15)	**57** (3)	**58** (3)	**59** (8)	**60** (60)
61 (3)	**62** (800)	**63** (4)	**64** (8)	**65** (2)	**66** (91)	**67** (6)	**68** (25)	**69** (10)	**70** (2)
71 (4)	**72** (20)	**73** (9)	**74** (3)						

Topic 2

1 (a)	**2** (a)	**3** (c)	**4** (c)	**5** (c)	**6** (d)	**7** (d)	**8** (d)	**9** (c)	**10** (a)
11 (b)	**12** (a)	**13** (c)	**14** (a)	**15** (d)	**16** (d)	**17** (d)	**18** (d)	**19** (c)	**20** (c)
21 (c)	**22** (a)	**23** (b)	**24** (a)	**25** (d)	**26** (c)	**27** (c)	**28** (c)	**29** (a)	**30** (b)
31 (d)	**32** (c)	**33** (d)	**34** (d)	**35** (c)	**36** (a)	**37** (b)	**38** (c)	**39** (d)	**40** (d)
41 (c)	**42** (1)	**43** (9)	**44** (199)	**45** (16)	**46** (4)	**47** (4)	**48** (53)	**49** (16)	**50** (176)
51 (1)	**52** (5)	**53** (35)	**54** (5)	**55** (3)	**56** (17)	**57** (5)	**58** (110)	**59** (32)	**60** (3)
61 (5)	**62** (8)	**63** (8)	**64** (25)						

Topic 3

1 (d)	**2** (b)	**3** (b)	**4** (b)	**5** (b)	**6** (c)	**7** (d)	**8** (a)	**9** (b)	**10** (c)
11 (d)	**12** (d)	**13** (d)	**14** (d)	**15** (d)	**16** (c)	**17** (b)	**18** (c)	**19** (a)	**20** (b)
21 (c)	**22** (b)	**23** (c)	**24** (c)	**25** (b)	**26** (d)	**27** (b)	**28** (c)	**29** (b)	**30** (a)
31 (d)	**32** (d)	**33** (b)	**34** (b)	**35** (c)	**36** (a)	**37** (c)	**38** (c)	**39** (b)	**40** (a)
41 (d)	**42** (a)	**43** (b)	**44** (a)	**45** (b)	**46** (a)	**47** (a)	**48** (4)	**49** (40)	**50** (32)
51 (13)	**52** (4)	**53** (6)	**54** (250)	**55** (24)	**56** (15)	**57** (2)	**58** (4)	**59** (75)	**60** (3)
61 (54)	**62** (40)	**63** (2)	**64** (10)	**65** (10)	**66** (120)	**67** (18)	**68** (3)	**69** (9)	**70** (4)
71 (2)	**72** (2)	**73** (3)	**74** (200)	**75** (4)	**76** (728)	**77** (3)	**78** (20)	**79** (20)	**80** (20)

Detailed **Explanations**

Topic 1

01. *(b)* $x_{com} = \dfrac{2m(10) + 3m(0)}{2m + 3m}$

$= \dfrac{20m}{5m} = 4$ cm

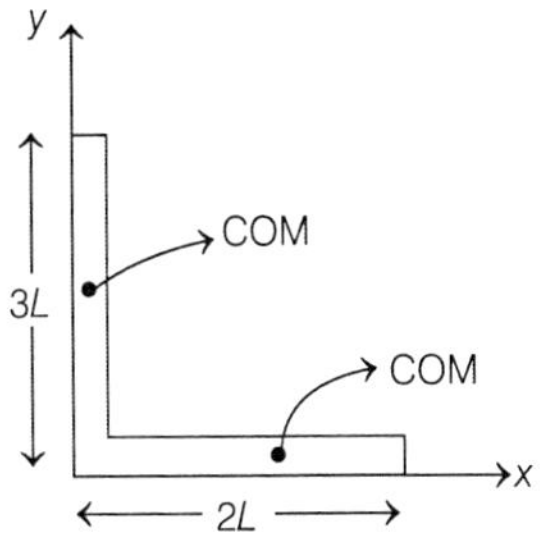

$y_{com} = \dfrac{2m(0) + 3m(15)}{5m} = 9$ cm

$r_{com} = 4\hat{i} + 9\hat{j}$

02. *(b)* $\tau_{net} = 0$

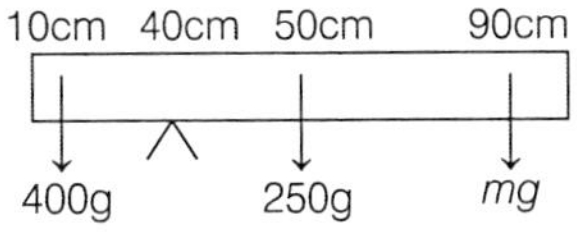

$\tau_{net} = 0$

$\Rightarrow \quad (400g \times 30) = (250g \times 10) + mg \times 50$

$m = 190$ g

03. *(d)* $\mathbf{r} = \hat{i} + \hat{j} + \hat{k}$

$\mathbf{F} = 2\hat{i} + \hat{j} + 2\hat{k}$

$$\tau = \mathbf{r} \times \mathbf{F} = \begin{vmatrix} \hat{i} & \hat{j} & \hat{k} \\ 1 & 1 & 1 \\ 2 & 1 & 2 \end{vmatrix}$$

$= \hat{i} - 0\hat{j} - \hat{k} = \hat{i} - \hat{k}$

04. *(c)* $\therefore$Torque applied on the body,

$\tau = \mathbf{r} \times \mathbf{F}$

$= (2\hat{i} + 2\hat{j} + \hat{k}) \times (5\hat{i} + 3\hat{j} - 7\hat{k})$

$$= \begin{vmatrix} \hat{i} & \hat{j} & \hat{k} \\ 2 & 2 & 1 \\ 5 & 3 & -7 \end{vmatrix}$$

$= \hat{i}(-14 - 3) - \hat{j}(-14 - 5) + \hat{k}(6 - 10)$

$= -17\hat{i} + 19\hat{j} - 4\hat{k}$

05. *(b)* As we know that,

Angular momentum,

$L = r \times p = mvr \sin\theta$

$\theta =$ angle between v and $r = 90°$ (for circular motion) Since, m, v, r are constant.

$\therefore L$ will always remain same in magnitude and direction.

06. *(c)* The angular momentum of the square about the given axis will be the sum of angular momentum due to each point masses.

i.e., $L = L_1 + L_2 + L_3 + L_4$

$= I_1\omega_1 + I_2\omega_2 + I_3\omega_3 + I_4\omega_4$

Here, angular frequency will be same.

i.e., $\omega_1 = \omega_2 = \omega_3 = \omega_4 = \omega$ (say)

$\therefore \quad L = (I_1 + I_2 + I_3 + I_4)\omega$

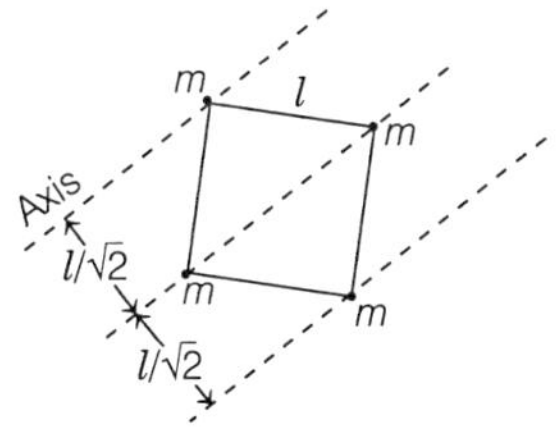

$$L = \left\{ m(0)^2 + 2 \times m\left(\frac{l}{\sqrt{2}}\right)^2 + m(\sqrt{2}l)^2 \right\}\omega$$

$= 3ml^2\omega$

07. *(d)* We choose origin as shown in the figure.

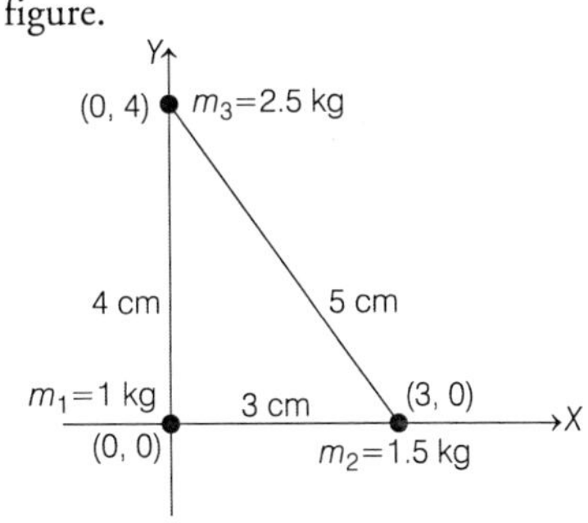

Using $x_{CM} = \dfrac{m_1x_1 + m_2x_2 + m_3x_3}{m_1 + m_2 + m_3}$,

we have

$x_{CM} = \dfrac{1 \times 0 + 1.5 \times 3 + 2.5 \times 0}{1 + 1.5 + 2.5} = 0.9$ cm

Similarly,

$y_{CM} = \dfrac{1 \times 0 + 1.5 \times 0 + 2.5 \times 4}{1 + 1.5 + 2.5} = 2.0$ cm

So, centre of mass (CM) is 0.9 cm right and 2.0 cm above 1 kg mass.

08. *(a)* Given, arrangement is shown below,

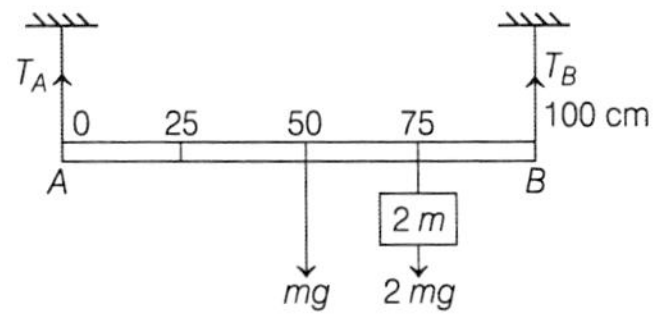

As rod is in equilibrium, net torque about point B is zero.

$\Rightarrow T_A \times 100 - mg \times 50 - 2mg \times 25 = 0$

$\Rightarrow \quad T_A = 1\,mg$

09. *(a)* For a system of discrete masses, acceleration of centre of mass (CM) is given by

$$\mathbf{a}_{CM} = \frac{m_A\mathbf{a}_A + m_B\mathbf{a}_B + m_C\mathbf{a}_C + m_D\mathbf{a}_D}{m_A + m_B + m_C + m_D}$$

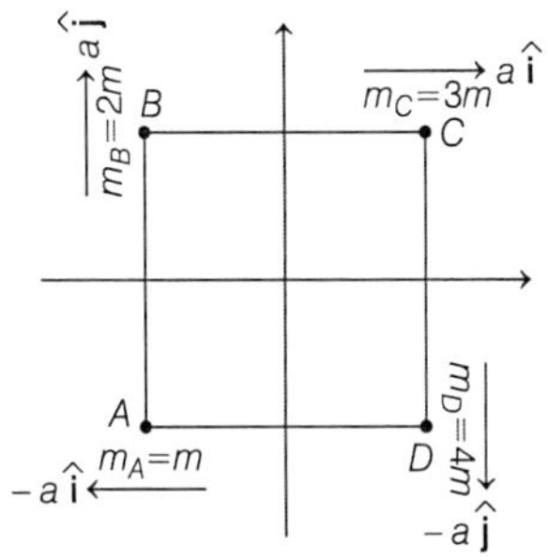

$$a_{CM} = \frac{-ma\hat{i} + 2ma\hat{j} + 3ma\hat{i} - 4ma\hat{j}}{m + 2m + 3m + 4m}$$

$$= \frac{2a\hat{i} - 2a\hat{j}}{10} = \frac{a}{5} \cdot (\hat{i} - \hat{j}) \text{ ms}^{-2}$$

10. *(c)* The height of equilateral Δ is

$h = y_3 = \sqrt{(1)^2 - (0.5)^2} = \sqrt{3}/2$ m

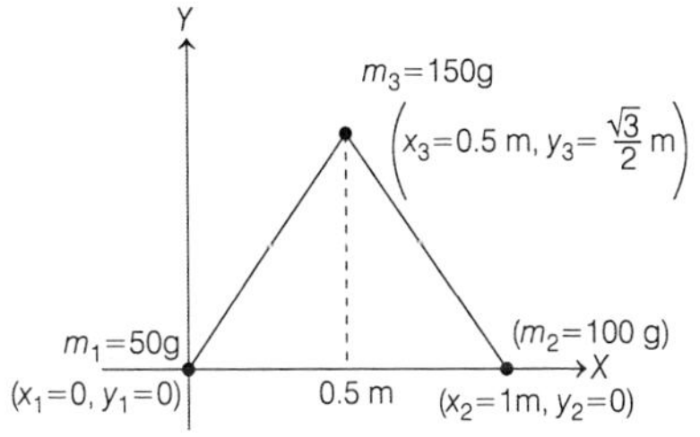

Using,

$$X_{CM} = \frac{m_1x_1 + m_2x_2 + m_3x_3}{m_1 + m_2 + m_3},$$

we have,

$$X_{CM} = \frac{50 \times 0 + 100 \times 1 + 150 \times 0.5}{50 + 100 + 150}$$

$$= \frac{175}{300} = \frac{7}{12} \text{ m}$$

Similarly, $Y_{CM} = \dfrac{m_1 y_1 + m_2 y_2 + m_3 y_3}{m_1 + m_2 + m_3}$

$$= \frac{50 \times 0 + 100 \times 0 + 150 \times \frac{\sqrt{3}}{2}}{50 + 100 + 150}$$

$$= \frac{\sqrt{3}}{4} \text{ m}$$

11. *(d)* Given, $m = 1$ kg

$|\tau| = 2.5$ N-m,

$F = 1$ N and $r = 5$ m

We know that,

torque $|\tau| = rF \sin\theta$

$\Rightarrow \quad 2.5 = 5 \times 1 \times \sin\theta$

$\Rightarrow \quad \sin\theta = 1/2$ or $\theta = \pi/6$ rad

12. *(d)* The given situation is shown below.

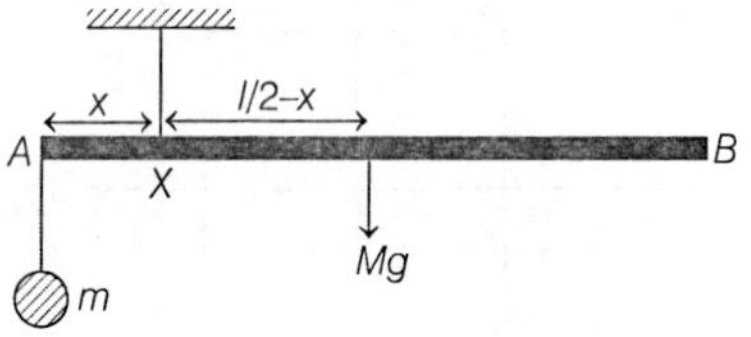

Balancing torque about point X,

$$mgx = Mg\left(\frac{l}{2} - x\right) \Rightarrow mx = M\frac{l}{2} - Mx$$

$$\Rightarrow \quad m = \left(M\frac{l}{2}\right)\frac{1}{x} - M$$

This is equation of straight line with variables m and $1/x$.

13. *(b)* We know that centre of mass of a uniform solid cone of height h is at height $h/4$ from base, therefore

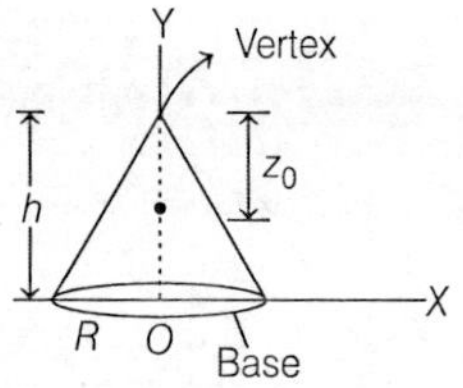

$$h - z_0 = \frac{h}{4}$$

or $\quad z_0 = h - \dfrac{h}{4} = \dfrac{3h}{4}$

14. *(c)* Angular momentum of the pendulum about the suspension point O is

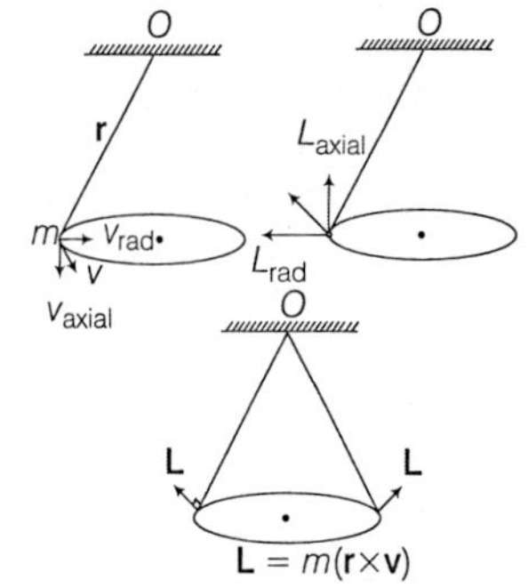

Then, v can be resolved into two components, radial component v_{rad} and axial component v_{axial}. Due to v_{rad}, will be axial and due to v_{axial}, L will be radially outwards as shown.

So, net angular momentum will be as shown in figure whose value will be constant ($|L| = mv$). But its direction will change as shown in the figure.

$$L = m(r \times v)$$

15. *(c)* Angular momentum of the particle is,

$$\mathbf{L} = m(\mathbf{r} \times \mathbf{v})$$

$$\mathbf{L} = m\left[v_0 \cos\theta t\,\hat{\mathbf{i}} + \left(v_0 \sin\theta t - \frac{1}{2}gt^2\right)\hat{\mathbf{j}}\right] \times [v_0 \cos\theta\,\hat{\mathbf{i}} + (v_0 \sin\theta - gt)\,\hat{\mathbf{j}}]$$

$$= mv_0 \cos\theta t\left(-\frac{1}{2}gt\right)\hat{\mathbf{k}}$$

$$= -\frac{1}{2}mgv_0 t^2 \cos\theta\,\hat{\mathbf{k}}$$

16. *(c)* According to the principle of conservation of angular momentum, in the absence of external torque, the total angular momentum of the system is constant.

17. *(c)*

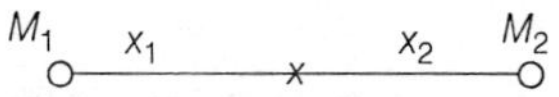

$$x_{CM} = \frac{m_1 x_1 + m_2 x_2}{m_1 + m_2} \quad \text{...(i)}$$

After moving m_1 through a distance d towards CM and to keep the position of CM unchanged, let d' be the shift of m_2.

$$x_{CM} = \frac{m_1(x_1 - d) + m_2(x_2 + d')}{m_1 + m_2} \quad \text{...(ii)}$$

From Eqs. (i) and (ii), we get

$$\frac{m_1 x_1 + m_2 x_2}{m_1 + m_2} = \frac{m_1(x_1 - d) + m_2(x_2 + d)}{m_1 + m_2}$$

$$\Rightarrow \quad -m_1 d + m_2 d' = 0$$

$$\therefore \quad d' = \frac{m_1}{m_2} d$$

18. *(c)* The torque about the given position, $\tau = \mathbf{r} \times \mathbf{F}$

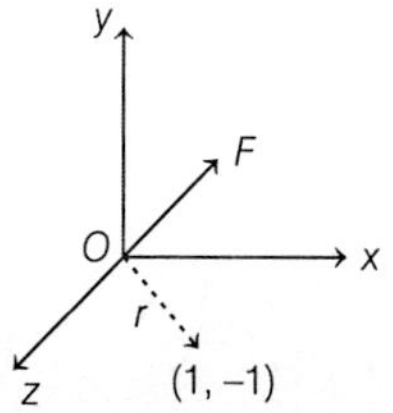

Here, $\mathbf{r} = \hat{\mathbf{i}} - \hat{\mathbf{j}}$ and $\mathbf{F} = -F\hat{\mathbf{k}}$

$$\therefore \quad \tau = (\hat{\mathbf{i}} - \hat{\mathbf{j}}) \times (-F\hat{\mathbf{k}})$$

$$= F[(-\hat{\mathbf{i}} \times \hat{\mathbf{k}}) + (\hat{\mathbf{j}} \times \hat{\mathbf{k}})]$$

$$= F(\hat{\mathbf{j}} + \hat{\mathbf{i}}) = F(\hat{\mathbf{i}} + \hat{\mathbf{j}})$$

19. *(d)* As no external torque is acting on the system, angular momentum of system remains conserved.

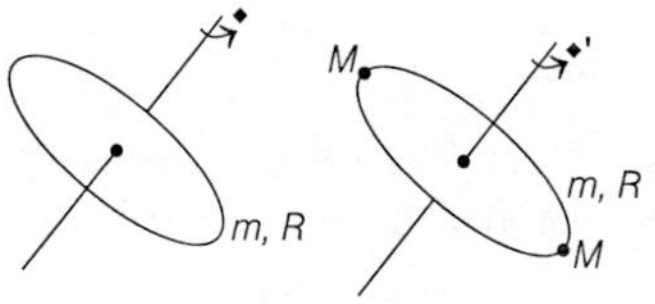

So, by conservation of angular momentum, *i.e.*, $I_1\omega = I_2\omega'$

$$\Rightarrow mR^2\omega = (mR^2 + 2MR^2)\omega'$$

$$\Rightarrow \quad \omega' = \left(\frac{m}{m + 2M}\right)\omega$$

20. *(b)* Since, the acceleration of centre of mass in both the cases is same equal to g, so the centre of mass of the bodies B and C taken together does not shift compared to that of body A.

21. *(c)* For pure translatory motion, net torque about centre of mass should be zero. Thus, F is applied at centre of mass of system,

$$OP = \frac{m_1 \times 0 + m_2 \times l}{m_1 + m_2}$$

where, m_1 and m_2 are masses of horizontal and vertical section of the object. Assuming object is uniform,

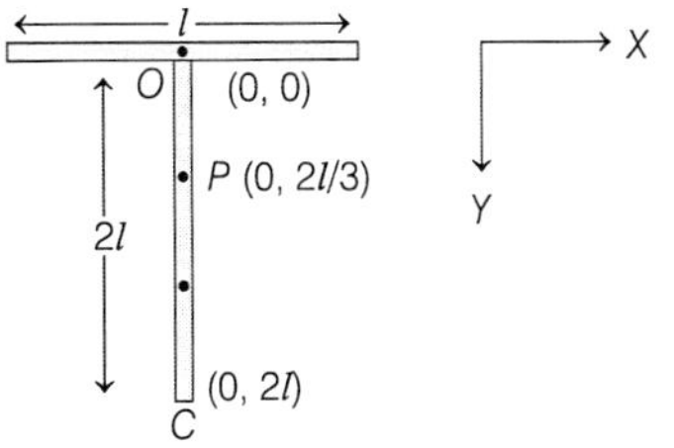

$$m_2 = 2m_1 \Rightarrow OP = \frac{2l}{3}$$

$$PC = \left(l - \frac{2l}{3} + l\right) = \left(2l - \frac{2l}{3}\right) = \frac{4l}{3}$$

22. *(b)* In free space, neither acceleration due to gravity nor external torque act on the rotating solid sphere. Therefore, taking the same mass of sphere, if radius is increased, then moment of inertia, rotational kinetic energy and angular velocity will change but according to law of conservation of momentum, angular momentum will not change.

23. *(a)* Angular momentum,

$$L = I\omega \quad \ldots(\text{i})$$

Kinetic energy,

$$K = \frac{1}{2}I\omega^2 = \frac{1}{2}L\omega \quad \text{[from Eq. (i)]}$$

$$\therefore \quad L = 2\frac{K}{\omega}$$

Now, $L' = \dfrac{2\left(\dfrac{K}{2}\right)}{2\omega} \Rightarrow L' = \dfrac{L}{4}$

24. *(d)* As $\tau = \mathbf{r} \times \mathbf{F}$

τ is perpendicular to both $\mathbf{r}$ and $\mathbf{F}$, so $\mathbf{r} \cdot \tau$ as well as $\mathbf{F} \cdot \tau$ has to be zero.

25. *(c)* Conservation of angular momentum gives

$$\frac{1}{2}MR^2\omega_1 = \left(\frac{1}{2}MR^2 + 2mR^2\right)\omega_2$$

$$\Rightarrow \frac{1}{2}MR^2\omega_1 = \frac{1}{2}R^2(M + 4m)\omega_2$$

or $\quad \omega_2 = \left(\dfrac{M}{M + 4m}\right)\omega_1$

26. *(b)* Angular momentum of particle about O,

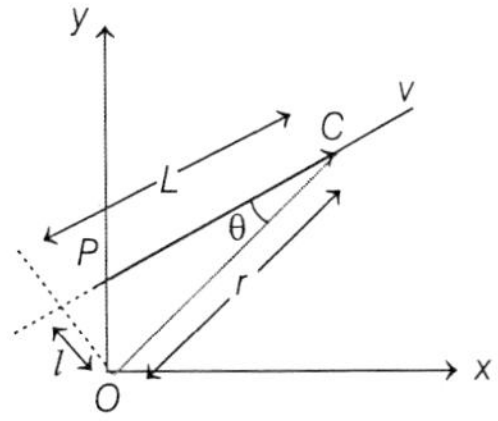

$$\mathbf{L} = m(\mathbf{r} \times \mathbf{v})$$

$$|\mathbf{v}| = mrv \sin\theta = mv(r\sin\theta) = mvl$$

27. *(c)* Let mass of each body be m. Their motion is represented as shown in figure.

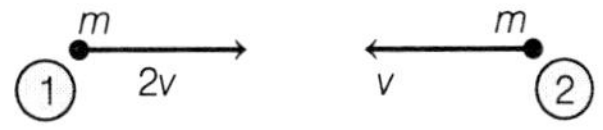

From $\mathbf{v}_{CM} = \dfrac{m_1\mathbf{v}_1 + m_2\mathbf{v}_2}{m_1 + m_2}$

$$\mathbf{v}_{CM} = \frac{m \times 2v - mv}{m + m} = \frac{v}{2}$$

[the direction of motion of first particle is taken as positive]

So, velocity of centre of mass of the system is $\frac{v}{2}$ in the direction of motion of particle having larger speed.

28. *(a)* As we know

$$\mathbf{a} = -\omega^2\mathbf{r}$$

$\therefore$ F opposite to $\mathbf{r}$

$\mathbf{r} = a(\hat{\mathbf{i}}\cos\omega t + \hat{\mathbf{j}}\sin\omega t)$

So, the direction of force is opposite to $\mathbf{r}$.

29. *(a)* $\tau = \mathbf{r} \times \mathbf{F}$

$$\mathbf{F} = \frac{d\mathbf{p}}{dt}$$

$$\tau = \frac{d}{dt}(\mathbf{r} \times \mathbf{p})$$

$$\tau = \mathbf{r} \times \frac{d\mathbf{p}}{dt}$$

$$\tau = I\alpha$$

So option B, C, D, and E only correct.

30. *(c)*

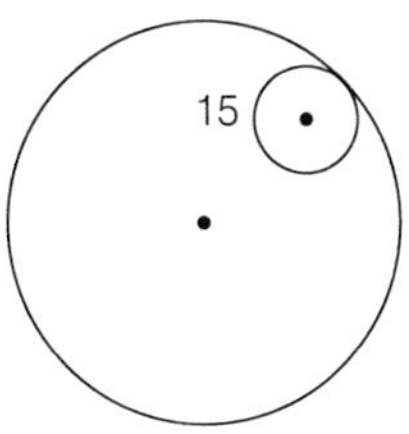

Mass of disc $= m$

Mass of cut part $= \dfrac{m}{16}$

$$x_{com} = \frac{m \times 0 - \frac{m}{16} \times 15}{m - \frac{m}{16}}$$

$$x_{com} = 1 \text{ cm}$$

31. *(c)* Since only internal forces are acting, hence, net external torque is zero on the system. Thus angular momentum is conserved.

$$\Rightarrow \quad \mathbf{L}_i = \mathbf{L}_f$$

By symmetry, the masses collide at the centroid of the triangle. So, total angular momentum at point C.

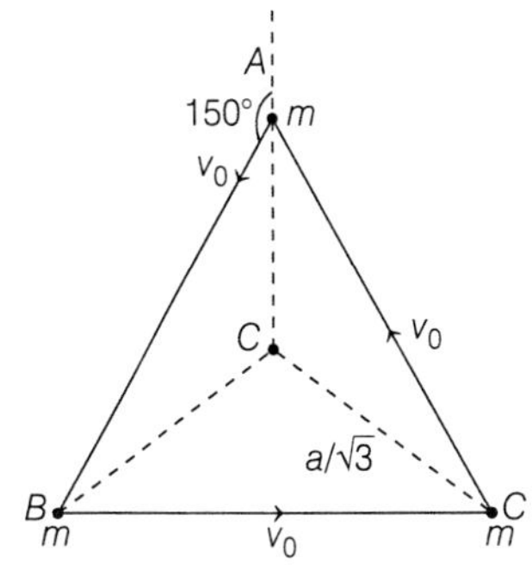

$$\mathbf{L}_C = 3 \times mv_0 \times \frac{a}{\sqrt{3}}\sin 150°$$

$$= \frac{\sqrt{3}}{2}mv_0 a$$

32. *(d)*

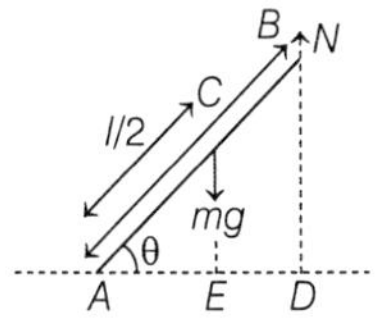

Taking torque around point A.

Using principle of moments of force.

$$N(AD) - w(AE) = 0$$

$$AD = l\cos\theta$$

$$AE = \frac{l}{2}\cos\theta$$

$$mg \times \frac{l}{2}\cos\theta = N \times l\cos\theta$$

$$\Rightarrow \quad N = \frac{mg}{2} = \frac{w}{2}$$

33. *(d)* Given,

Mass of particle $= m$

Initial velocity $= u$

Angle of projection $= 30°$

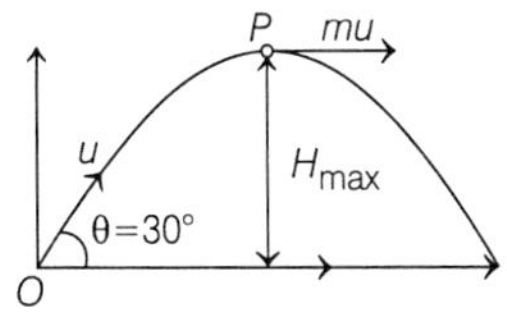

At point P, velocity will only has horizontal component,

$$u_x = u\cos\theta \quad \ldots(\text{i})$$

Angular momentum at point P,

$$L = mu_x H_{max} \quad \ldots(\text{ii})$$

$$= \frac{mu\cos\theta u^2 \sin^2\theta}{2g}$$

$$= \frac{mu\cos 30° u^2 \sin^2 30°}{2g}$$

$$= \frac{\sqrt{3}mu^3}{16g}$$

34. *(c)*

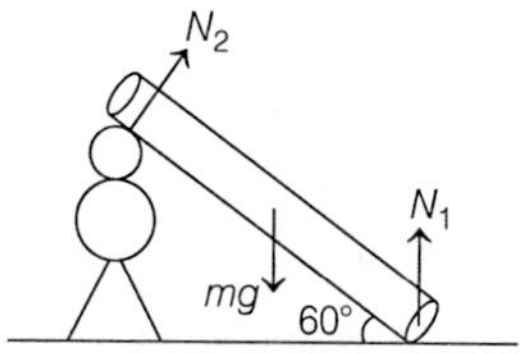

Taking torque about N_1,
and using principle of moments of force

$$mg\frac{l}{2}\cos 60° = N_2 l$$

$$12\times 10\times\frac{l}{2}\times\frac{1}{2} = N_2\times l$$

$$N_2 = \frac{120}{4} = 30\text{ N}$$

$$N_2 = 30\text{ N}$$

$$\therefore \text{Mass} = \frac{N_2}{g} = \frac{30}{10} = 3\text{ kg}$$

35. *(d)* $m_1 = 1$ kg

$$\mathbf{r}_1 = \hat{\mathbf{i}} + 2\hat{\mathbf{j}} + \hat{\mathbf{k}}$$

$m_2 = 3$ kg

$$\mathbf{r}_2 = -3\hat{\mathbf{i}} - 2\hat{\mathbf{j}} + \hat{\mathbf{k}}$$

We know,

$$\mathbf{r}_{CM} = \frac{m_1\mathbf{r}_1 + m_2\mathbf{r}_2}{m_1 + m_2}$$

$$\Rightarrow \mathbf{r}_{CM} = \frac{1(\hat{\mathbf{i}} + 2\hat{\mathbf{j}} + \hat{\mathbf{k}}) + 3(-3\hat{\mathbf{i}} - 2\hat{\mathbf{j}} + \hat{\mathbf{k}})}{4}$$

$$\Rightarrow \mathbf{r}_{CM} = \frac{\hat{\mathbf{i}} + 2\hat{\mathbf{j}} + \hat{\mathbf{k}} - 9\hat{\mathbf{i}} - 6\hat{\mathbf{j}} + 3\hat{\mathbf{k}}}{4}$$

$$\Rightarrow \mathbf{r}_{CM} = \frac{-8\hat{\mathbf{i}} - 4\hat{\mathbf{j}} + 4\hat{\mathbf{k}}}{4} = -2\hat{\mathbf{i}} - \hat{\mathbf{j}} + \hat{\mathbf{k}}$$

36. *(d)* We know that, $\mathbf{L} = m(\mathbf{r}\times\mathbf{v})$

Since, L_A is the angular momentum of M about point A which lies in positive z-direction. Therefore, with respect to point A, we will get the direction of **L** along positive Z-axis and of constant magnitude of mvr.

Since, L_B is the angular momentum of M about point B, so with respect to point B, we will get the constant magnitude of **L** but its direction will be continuously changing. Hence, option (d) is correct.

37. *(c)* Given, radius of hole, $r = a/2$
and radius of disc, $R = a$

Let x_{CM} be the centre of mass of remaining portion,
m_1, x_1 be the mass and centre of mass of disc
m_2, x_2 be the mass and centre of mass of circular hole.

$$\therefore \quad m_2 = \frac{m_1}{\pi R^2}\pi r^2$$

$$\Rightarrow \quad m_2 = \frac{m_1 r^2}{R^2} = \frac{m_1(a/2)^2}{a^2} = \frac{m_1}{4}$$

and
$$x_{CM} = \frac{m_1x_1 + m_2x_2}{m_1 + m_2}$$

$$\therefore \quad x_{CM} = \frac{m_1 a - (m_1/4)(3a/2)}{m_1 - (m_1/4)}$$

$$\Rightarrow \quad x_{CM} = \frac{m_1 a(1 - 3/8)}{3m_1/4}$$

$$\Rightarrow \quad x_{CM} = 5a/6$$

38. *(d)* Force F is applied to cylinder as shown,

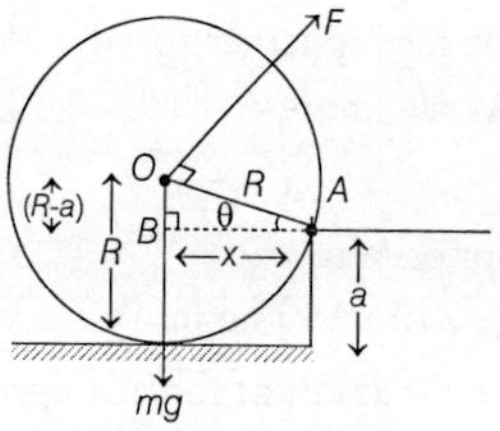

Force F must be large enough to create a torque about point A, so that it is able to counteract the torque due to weight about point A.

$\therefore$Torque due to force about point A
= Torque due to weight about point A

$$\Rightarrow \quad F\cdot R = Mgx$$

From ΔOAB, $x = \sqrt{R^2 - (R-a)^2}$

$$\Rightarrow \quad F\cdot R = Mg\sqrt{R^2 - (R-a)^2}$$

$$\Rightarrow \quad F = \frac{Mg}{R}\sqrt{R^2 - (R-a)^2}$$

$$\Rightarrow \quad F = Mg\sqrt{1 - \left(\frac{R-a}{R}\right)^2}$$

39. *(a)* For a continuous mass distribution,

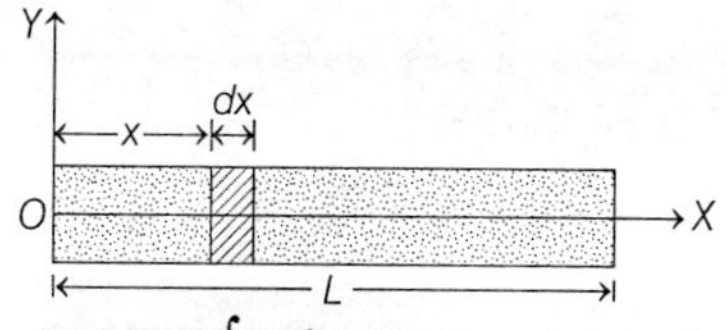

$$x_{CM} = \frac{\int x\,dm}{\int dm}$$

Here, mass of an elemental length dx of rod,

$$dm = \rho dx = \left(a + b\left(\frac{x}{L}\right)^2\right)dx$$

$$\text{So, } x_{CM} = \frac{\int_0^L x\left(a + \frac{bx^2}{L^2}\right)dx}{\int_0^L \left(a + \frac{bx^2}{L^2}\right)dx}$$

$$= \frac{\int_0^L \left(ax + \frac{bx^3}{L^2}\right)dx}{\int_0^L \left(a + \frac{bx^2}{L^2}\right)dx}$$

$$= \frac{\left[\left(\frac{ax^2}{2} + \frac{b}{L^2}\cdot\frac{x^4}{4}\right)\right]_0^L}{\left[\left(ax + \frac{bx^3}{3L^2}\right)\right]_0^L}$$

$$= \frac{\frac{aL^2}{2} + \frac{bL^2}{4}}{aL + \frac{bL}{3}} = \left(\frac{\frac{a}{2} + \frac{b}{4}}{a + \frac{b}{3}}\right)L$$

$$\text{So, } x_{CM} = \frac{3}{4}\left(\frac{2a + b}{3a + b}\right)L$$

40. *(c)*

> **key idea** First we found moment of inertia (MI) of system using parallel axis theorem about centre of mass, then we use it to find moment of inertia about given axis.

Velocity of particle is

$$\mathbf{v} = d\mathbf{r}/dt = 2\hat{\mathbf{i}} - 6t\hat{\mathbf{j}}$$

Now, angular momentum of particle with respect to origin is given by

$$\mathbf{L} = m(\mathbf{r}\times\mathbf{v})$$
$$= m\{(2t\hat{\mathbf{i}} - 3t^2\hat{\mathbf{j}})\times(2\hat{\mathbf{i}} - 6t\hat{\mathbf{j}})\}$$
$$= m(-12t^2(\hat{\mathbf{i}}\times\hat{\mathbf{j}}) - 6t^2(\hat{\mathbf{j}}\times\hat{\mathbf{i}}))$$

As, $\hat{\mathbf{i}}\times\hat{\mathbf{i}} = \hat{\mathbf{j}}\times\hat{\mathbf{j}} = 0$

$$\Rightarrow \quad \mathbf{L} = m(-12t^2\hat{\mathbf{k}} + 6t^2\hat{\mathbf{k}})$$

As, $\hat{\mathbf{i}}\times\hat{\mathbf{j}} = \hat{\mathbf{k}}$ and $\hat{\mathbf{j}}\times\hat{\mathbf{i}} = -\hat{\mathbf{k}}$

$$\Rightarrow \quad \mathbf{L} = -m(6t^2)\hat{\mathbf{k}}$$

So, angular momentum of particle of mass 2 kg at time $t = 2$ s is

$$\mathbf{L} = (-2\times 6\times 2^2)\hat{\mathbf{k}} = -48\,\hat{\mathbf{k}}$$

41. *(d)* Given equations of trajectory are

$$x = x_0 + a\cos\omega_1 t$$
$$y = y_0 + b\sin\omega_2 t$$

$$\text{Now, } v_x = \frac{dx}{dt} = -a\omega_1\sin\omega_1 t$$

$$\text{and} \quad v_y = \frac{dy}{dt} = b\omega_2\cos\omega_2 t$$

$$\text{Similarly, } a_x = \frac{dv_x}{dt} = -a\omega_1^2\cos(\omega_1 t)$$

$$\text{Similarly, } a_y = \frac{dv_y}{dt} = -b\omega_2^2\sin(\omega_2 t)$$

Now, at $t = 0$; $x = x_0 + a$; $y = y_0$

and $a_x = -a\omega_1^2$; $a_y = 0$ ...(i)

Torque acting on the particle is given as

$$\tau = \mathbf{r} \times \mathbf{F} = m(\mathbf{r} \times \mathbf{a})$$

Here, $\mathbf{r} = (x_0 + a)\hat{\mathbf{i}} + y_0\hat{\mathbf{j}} + 0\hat{\mathbf{k}}$

and $\mathbf{a} = -a\omega_1^2\hat{\mathbf{i}} + 0\hat{\mathbf{j}} + 0\hat{\mathbf{k}}$

So, torque at $t = 0$,

$$\tau = m(-a\omega_1^2) \times y_0\,(-\hat{\mathbf{k}}) \text{ [using Eq. (i)]}$$

$$\Rightarrow \tau = +my_0 a\omega_1^2\hat{\mathbf{k}}$$

> ***Alternate Solution***
>
> We can calculate total force from Eq. (i),
>
> $\mathbf{F} = -m(a\omega_1^2 \cos\omega_1 t\,\hat{\mathbf{i}} + b\omega_2^2 \sin\omega_2 t\,\hat{\mathbf{j}})$
>
> and $\mathbf{r} = x\hat{\mathbf{i}} + y\hat{\mathbf{j}} = (x_0 + a\cos\omega_1 t)\hat{\mathbf{i}} + (y_0 + b\sin\omega_2 t)\hat{}$
>
> Now, torque, $\tau = \mathbf{r} \times \mathbf{F}$
>
> $= [(x_0 + a\cos\omega_1 t)\hat{} + (y_0 + b\sin\omega_2 t)\hat{}] \times [-m(a\omega_1^2 \cos\omega_1 t\hat{} + b\omega_2^2 \sin\omega_2 t\hat{})]$
>
> $= -m\begin{vmatrix} \hat{} & \hat{} & \hat{} \\ x_0 + a\cos\omega_1 t & y_0 + b\sin\omega_2 t & 0 \\ a\omega_1^2 \cos\omega_1 t & b\omega_2^2 \sin\omega_2 t & 0 \end{vmatrix}$
>
> $= -m[x_0 b\omega_2^2 \sin\omega_2 t + ab\omega^2 \sin\omega_2 t \cos\omega_1 t - y_0 a\omega_1^2 \cos\omega_1 t - ab\omega_1^2 \sin\omega_2 t \cos\omega_1 t]\hat{}$
>
> On solving it at $t = 0$, we get $\tau = ma\omega_1^2 y_0 \hat{\mathbf{k}}$

42. *(d)*

> **key idea** The centre of mass of a thin rod of uniform density lies at its centre.

The given system of rods can be drawn using geometry as,

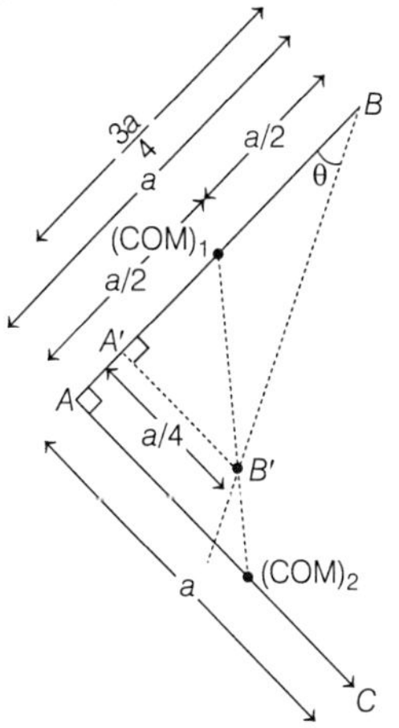

where, $(COM)_1$ and $(COM)_2$ are the centre of mass of both rods AB and AC, respectively.

So, in $\Delta A'BB'$,

$$\tan\theta = \frac{A'B'}{A'B} = \frac{\frac{a}{4}}{\frac{3a}{4}} = \frac{1}{3}$$

or $\tan\theta = \dfrac{1}{3}$

43. *(b)* According to the question as shown in the figure below

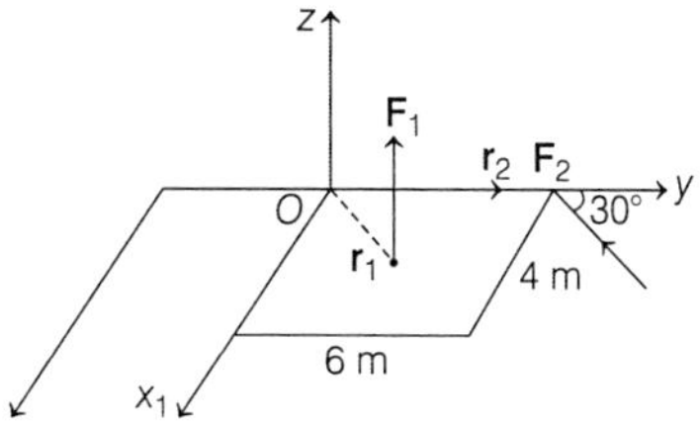

$$\mathbf{r}_1 = 2\hat{\mathbf{i}} + 3\hat{\mathbf{j}} \text{ and } \mathbf{r}_2 = 6\hat{\mathbf{j}}$$

$$\mathbf{F}_1 = F\hat{\mathbf{k}}$$

and $\mathbf{F}_2 = (-\sin 30°\,\hat{\mathbf{i}} - \cos 30°\,\hat{\mathbf{j}})\,F$

Moment of force is given as,

$$\tau = \mathbf{r} \times \mathbf{F}$$

$\therefore$ Moment due to $\mathbf{F}_1$

$$\tau_1 = (2\hat{\mathbf{i}} + 3\hat{\mathbf{j}}) \times (F\hat{\mathbf{k}})$$

$$= -2F\hat{\mathbf{j}} + 3F\hat{\mathbf{i}} \quad \text{...(i)}$$

Moment due to $\mathbf{F}_2$

$$\tau_2 = (6\hat{\mathbf{j}}) \times (-\sin 30°\,\hat{\mathbf{i}} - \cos 30°\,\hat{\mathbf{j}})F$$

$$= 6\sin 30°\,F\,\hat{\mathbf{k}} = 3F\,\hat{\mathbf{k}} \quad \text{...(ii)}$$

$\therefore$ Resultant torque,

$$\tau = \tau_1 + \tau_2 = 3F\hat{\mathbf{i}} - 2F\hat{\mathbf{j}} + 3F\hat{\mathbf{k}}$$

$$= (3\hat{\mathbf{i}} - 2\hat{\mathbf{j}} + 3\hat{\mathbf{k}})F.$$

44. *(a)* Coordinates of centre of mass (COM) are given by

$$X_{COM} = \frac{m_1x_1 + m_2x_2 + m_3x_3}{m_1 + m_2 + m_3}$$

and $$Y_{COM} = \frac{m_1y_1 + m_2y_2 + m_3y_3}{m_1 + m_2 + m_3}$$

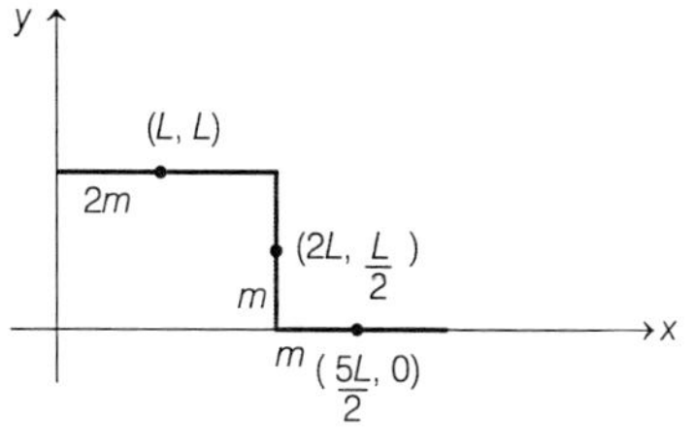

$$\text{So, } X_{COM} = \left(\frac{2mL + m2L + m\frac{5L}{2}}{4m}\right)$$

$$= \frac{13}{8}L$$

$$\text{and } Y_{COM} = \frac{2mL + m \times \frac{L}{2} + m \times 0}{4m} = \frac{5L}{8}$$

So, position vector of COM is

$$\mathbf{r}_{COM} = X_{COM}\hat{\mathbf{x}} + Y_{COM}\hat{\mathbf{y}}$$

$$= \frac{13}{8}L\hat{\mathbf{x}} + \frac{5}{8}L\hat{\mathbf{y}}$$

45. *(d)* The given figure is shown below as

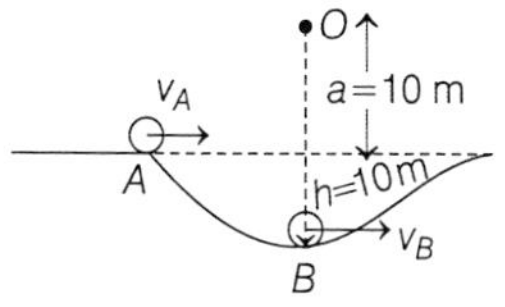

As friction is absent, energy at A = energy at B

$$\Rightarrow \frac{1}{2}mv_A^2 + mgh = \frac{1}{2}mv_B^2$$

$$\Rightarrow v_A^2 + 2gh = v_B^2$$

or $$v_B^2 = (5)^2 + 2 \times 10 \times 10 = 225$$

$$\Rightarrow v_B = 15 \text{ ms}^{-1}$$

Angular momentum about point O,

$$= mv_Br_B = 20 \times 10^{-3} \times 15 \times 20$$

$$= 6 \text{ kg-m}^2\text{s}^{-1}$$

46. *(a)* Let a small strip of mop has width dx and radius x, as shown below,

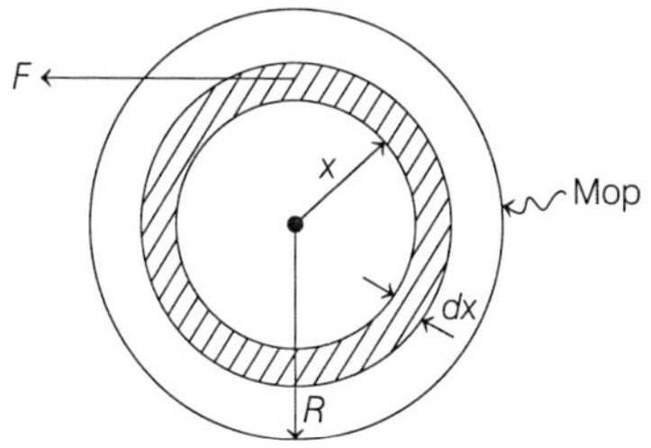

Torque applied to move this strip is

$d\tau$ = Force on strip × Perpendicular distance from the axis

$\Rightarrow d\tau$ = Force per unit area × Area of strip × Perpendicular distance from the axis.

$$= \frac{\mu F}{\pi R^2} \cdot 2\pi x dx \cdot x$$

$$\Rightarrow d\tau = \frac{2\mu F x^2}{R^2} \cdot dx$$

So, total torque to be applied on the mop is

$$\tau = \int_{x=0}^{x=R} d\tau = \int_0^R \frac{2\mu F x^2}{R^2} \cdot dx$$

$$= \frac{2\mu F}{R^2} \times \frac{R^3}{3} = \frac{2}{3}\mu FR \text{ (N-m)}$$

47. *(c)* Let $AB = p$, $BC = q$, $\lambda =$ linear mass density of the rod according to given situation centre of mass of the rod lies vertically below point A.

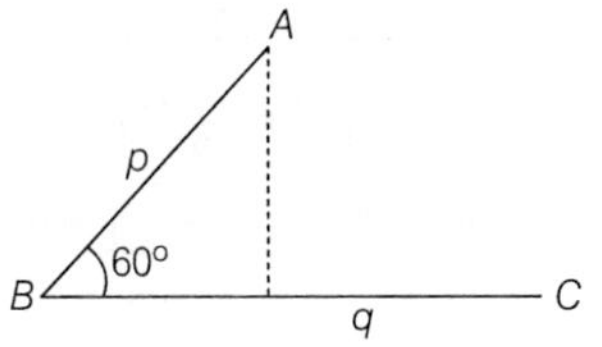

$$\therefore X_{CM} = p\cos 60°$$

$$= \frac{(\lambda q)\left(\frac{q}{2}\right) + (\lambda p)\left(\frac{p}{2}\right)\cos 60°}{\lambda(p+q)}$$

$$\Rightarrow \frac{p}{2} = \frac{\frac{q^2}{2} + \frac{p^2}{4}}{(p+q)} \Rightarrow p^2 + pq = q^2 + \frac{p^2}{2}$$

$$\Rightarrow 1 + \frac{q}{p} = \frac{q^2}{p^2} + \frac{1}{2} \Rightarrow \left(\frac{q}{p}\right)^2 - \frac{q}{p} - \frac{1}{2} = 0$$

$$\frac{q}{p} = \frac{-(-1) \pm \sqrt{(-1)^2 - 4(1)\left(-\frac{1}{2}\right)}}{2\times 1} = \frac{1 \pm \sqrt{3}}{2}$$

$$\therefore \text{Possible value of } \frac{q}{p} = \frac{1+\sqrt{3}}{2}$$

$$= 1.366 \approx 1.37$$

48. *(a)* In this question, distance of centre of mass of new disc from the centre of mass of remaining disc is αR.

Mass of original disc = M

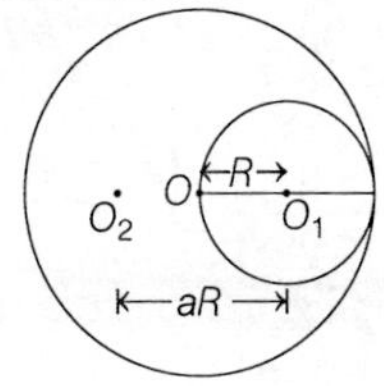

Mass of disc removed

$$= \frac{M}{\pi(2R)^2} \times \pi R^2 = \frac{M}{4}$$

$$\text{Mass of remaining disc} = M - \frac{M}{4} = \frac{3M}{4}$$

$$\therefore -\frac{3M}{4}\alpha R + \frac{M}{4}R = 0 \Rightarrow \alpha = \frac{1}{3}$$

49. *(b)* As there is no dependence of mass density in y-direction

So, $y_{cm} = \dfrac{b}{2}$

For x-direction

$$dm = \left(\frac{\sigma_0 x}{ab}\right) b \cdot dx$$

For centre of mass in x-direction

$$x_{cm} = \frac{\int_0^a x\, dm}{\int_0^a dm} = \frac{\frac{\sigma_0}{a}\int_0^a x^2\, dx}{\frac{\sigma_0}{a}\int_0^a x\, dx}$$

$$x_{cm} = \frac{a^3 \times 2}{3 \times a^2} = \frac{2}{3}a$$

So, coordinates of $r_{cm} = \left(\dfrac{2}{3}a, \dfrac{b}{2}\right)$

50. *(d)* Angle of projection, $\theta = 45°$, initial velocity v_0.

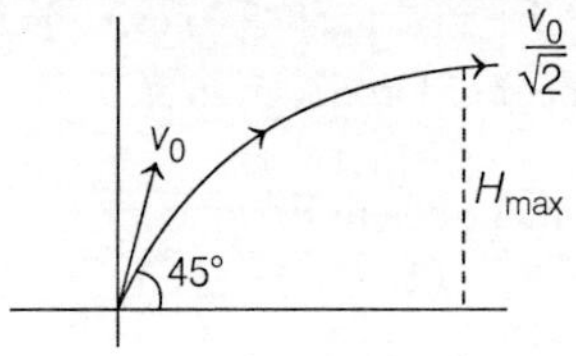

At maximum height

$$H = \frac{\left(\frac{v_0}{\sqrt{2}}\right)^2}{2g} = \frac{v_0^2}{4g}$$

Angular momentum is given by

$$L = mvh \qquad (h = H)$$

$$= \frac{mv_0 v_0^2}{\sqrt{2}\,4g} = \frac{mv_0^3}{4\sqrt{2}g}$$

along negative Z-axis

51. *(c)* For the given system,

$$x_{CM} = \frac{m_1 x_1 + m_2 x_2}{m_1 + m_2}$$

$$\text{Also, } dx_{CM} = \frac{m_1 dx_1 + m_2 dx_2}{m_1 + m_2} \qquad \ldots\text{(i)}$$

$dx_{CM} = 0$, since, centre of mass is at same position.

Therefore, Eq. (i) can be written as

$$0 = m_1(dx_1) + m_2(dx_2) \qquad \ldots\text{(ii)}$$

Here, $m_1 = 10$ kg, $m_2 = 30$ kg,

$dx_1 = 6$ cm, $dx_2 = ?$

$$\Rightarrow 0 = 10(6) + 30(dx_2)$$

$$\Rightarrow dx_2 = \frac{-60}{30} = -2 \text{ cm}$$

So, mass m_2 must be moved 2 cm towards 10 kg block.

52. *(a)* If we combine the remaining and removed spherical parts at their initial relative positions, we get complete sphere as shown in the figure.

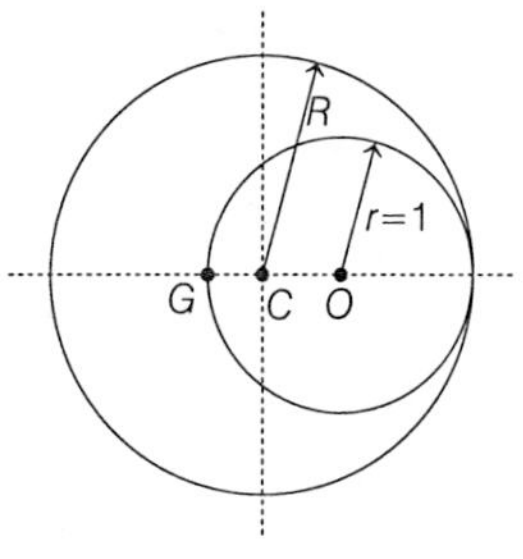

So, for complete sphere,

$$X_{CM} = \frac{m_{\text{remaining}} \times x_{\text{remaining}} + m_{\text{removed}} \times x_{\text{removed}}}{m_{\text{remaining}} + m_{\text{removed}}}$$

As, X_{CM} is at C which is taken at origin of our chosen reference axis, so

$$X_{CM} = 0$$

$$\Rightarrow m_{\text{remaining}} \times x_{\text{remaining}} + m_{\text{removed}} \times x_{\text{removed}} = 0 \quad \ldots\text{(i)}$$

Here,

$$m_{\text{remaining}} = \left(\frac{4}{3}\pi R^3 - \frac{4}{3}\pi 1^3\right)\rho$$

where, $\rho =$ density of sphere,

$x_{\text{remaining}} = R - 2$,

$$m_{\text{removed}} = +\left(\frac{4}{3}\pi 1^3\right)\rho$$

and $x_{\text{removed}} = R - 1$.

So, from Eq. (i),

we get

$$\left(\frac{4}{3}\pi R^3 - \frac{4}{3}\pi\right)\rho \times (R-2) = \left(\frac{4}{3}\pi\right)\rho \times (R-1)$$

$$\Rightarrow (R^3 - 1)(2 - R) = R - 1$$

$$\Rightarrow (R^2 + R + 1)(2 - R) = 1$$

53. *(b)* The given rectangular thin sheet $ABCD$ can be drawn as shown in the figure below,

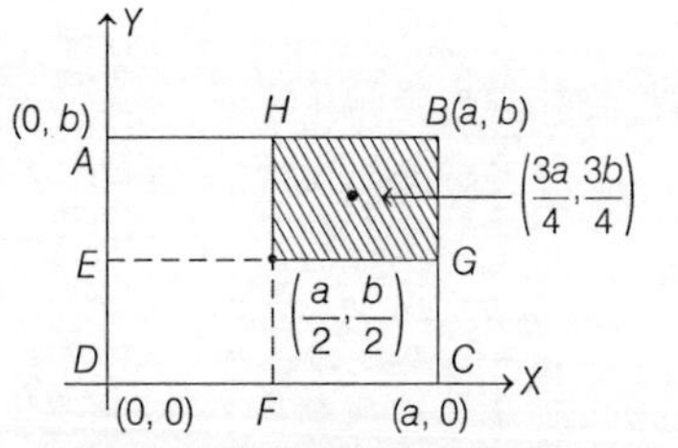

Here, Area of complete lamina, $A_1 = ab$

Area of shaded part of lamina

$$= \frac{a}{2} \times \frac{b}{2} = \frac{ab}{4}$$

(x_1, y_1) = coordinates of centre of mass of complete lamina $= \left(\frac{a}{2}, \frac{a}{2}\right)$

(x_2, y_2) = coordinates of centre of mass of shaded part of lamina $= \left(\frac{3a}{4}, \frac{3b}{4}\right)$

∴ Using formula for centre of mass, we have

$$X_{CM} = \frac{A_1 x_1 - A_2 x_2}{A_1 - A_2}$$

$$= \frac{ab\left(\frac{a}{2}\right) - \frac{ab}{4}\left(\frac{3a}{4}\right)}{ab - \frac{ab}{4}}$$

$$= \frac{\frac{8a^2b - 3a^2b}{16}}{\frac{3ab}{4}} = \frac{5a}{12}$$

Similarly, $Y_{CM} = \frac{A_1 y_1 - A_2 y_2}{A_1 - A_2}$

$$= \frac{ab\left(\frac{b}{2}\right) - \frac{ab}{4}\left(\frac{3b}{4}\right)}{ab - \frac{ab}{4}} = \frac{5b}{12}$$

Alternate Solution

Let m be the mass of entire rectangular lamina.

So, the mass of the shaded portion of lamina $= \frac{m}{4}$

Using the relation, $X_{CM} = \frac{m_1 x_1 - m_2 x_2}{m_1 - m_2}$, we get

$$X_{CM} = \frac{m\left(\frac{a}{2}\right) - \frac{m}{4}\left(\frac{3a}{4}\right)}{m - \frac{m}{4}} = \frac{\frac{a}{2} - \frac{3a}{16}}{\frac{3}{4}} = \frac{5a}{12}$$

Similarly, $Y_{CM} = \frac{m_1 y_1 - m_2 y_2}{m_1 - m_2}$, we get

$$Y_{CM} = \frac{m\left(\frac{b}{2}\right) - \frac{m}{4}\left(\frac{3b}{4}\right)}{m - \frac{m}{4}} = \frac{\frac{b}{2} - \frac{3b}{16}}{\frac{3}{4}} = \frac{5b}{12}$$

∴ The coordinates of the centre of mass of remaining portion will be $\left(\frac{5a}{12}, \frac{5b}{12}\right)$.

54. *(90)* Vector $\mathbf{v}_A = (2t\hat{i} + 3n\hat{j} + 2\hat{k})$

$$\mathbf{v}_B = (2\hat{i} - 2t\hat{j} + 4p\hat{k})$$

$$\mathbf{v}_A \cdot \mathbf{v}_B = 0$$

$$4 - 6n + 8p = 0$$

$$2 - 3n + 4p = 0$$

$$3n = 2 + 4p$$

$$|\mathbf{v}_A| = |\mathbf{v}_B|$$

$$4 + 9n^2 + 4 = 4 + 4 + 16p^2$$

$$p = \frac{-1}{4} \Rightarrow n = \frac{1}{3}$$

$$\mathbf{L} = m_{A/B}(\mathbf{r}_{A/B} \times \mathbf{v}_{A/B})$$

$$\mathbf{r}_{A/B} = (\alpha_1 - \beta_1)\hat{i} + (\alpha_2 - \beta_2)\hat{j} + (\alpha_3 - \beta_3)\hat{k}$$

$$= (1-2)\hat{i} + (1+1)\hat{j} + 3\hat{k}$$

$$\therefore \mathbf{L} = \begin{vmatrix} \hat{i} & \hat{j} & \hat{k} \\ -1 & 2 & 3 \\ 2 & 1 & 2 \end{vmatrix} = \hat{i} + 8\hat{j} - 5\hat{k}$$

$$|\mathbf{L}| = \sqrt{1 + 64 + 25} = \sqrt{90}$$

55. *(2)* $\tau = \mathbf{r} \times \mathbf{F}$

Point of force (1, 1, 1)

Point about which torque needed to calculate (0, 0, 0)

Thus $\mathbf{r} = (1-0)\hat{i} + (1-0)\hat{j} + (1-0)\hat{k}$

$$= \hat{i} + \hat{j} + \hat{k}$$

$$\mathbf{F} = \hat{i} - \hat{j} + \hat{k}$$

Now $\tau = \mathbf{r} \times \mathbf{F}$

$$= (\hat{i} + \hat{j} + \hat{k}) \times (\hat{i} - \hat{j} + \hat{k})$$

$$\begin{vmatrix} \hat{i} & \hat{j} & \hat{k} \\ 1 & 1 & 1 \\ 1 & -1 & 1 \end{vmatrix}$$

$$\tau = 2\hat{i} + 0\hat{j} - 2\hat{k}$$

Magnitude of torque in z-direction is 2.

56. *(15)*

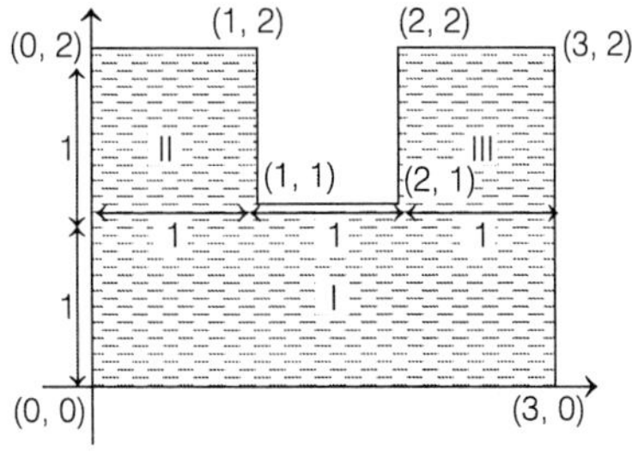

As metal plate is uniform, mass ∝ area

By symmetry along X-axis

$$x_{cm} = 3/2 = 1.5 \text{ units}$$

$$y_{cm} = \frac{3 \times \frac{1}{2} + 1 \times \frac{3}{2} + 1 \times \frac{3}{2}}{3 + 1 + 1}$$

$$= \frac{9}{5 \times 2} = \frac{9}{10} \text{ units}$$

Ratio $\frac{x_{cm}}{y_{cm}} = \frac{15}{9}$

So, $n = 15$

57. *(3)* If x_1 and x_2 are the positions of masses m_1 and m_2, the position of the centre of mass is given by

$$x_{cm} = \frac{m_1 x_1 + m_2 x_2}{m_1 + m_2}$$

If x_1 change by Δx_1 and x_2 change by Δx_2, the change in x_{cm} will be

$$\Delta x_{cm} = \frac{m_1 \Delta x_1 + m_2 \Delta x_2}{m_1 + m_2} \quad \text{...(i)}$$

Given, $\Delta x_{cm} = 0$ and $\Delta x_1 = d = 2$ cm

Using these values in Eq. (i), we get

$$m_1 d + m_2 \Delta x_2 = 0$$

$$\Rightarrow \quad \Delta x_2 = -\frac{m_1}{m_2} \cdot d$$

∴ Distance moved by second mass m_2,

$$\Delta x_2 = \frac{m_1 d}{m_2}$$

Here, $m_1 = 3$ kg, $m_2 = 2$ kg, $d = 2$ cm

$$\Delta x_2 = \frac{3}{2} \times 2$$

$$= 3 \text{ cm}$$

58. *(3)* Given system of masses is,

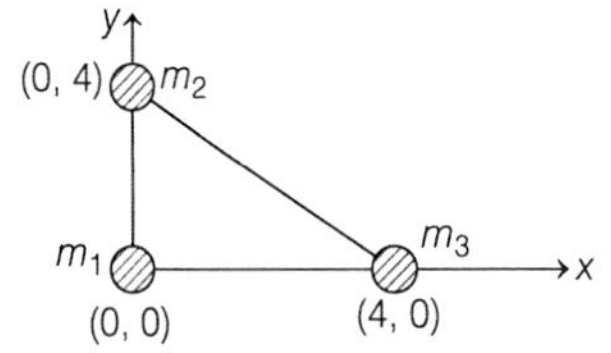

x-coordinate of COM,

$$x = \frac{m_1 x_1 + m_2 x_2 + m_3 x_3}{m_1 + m_2 + m_3}$$

$$= \frac{2m \times 0 + 2m \times 0 + 2m \times 4}{2m + 2m + 2m}$$

$$= \frac{8}{6} = \frac{4}{3}$$

y-coordinate of COM,

$$y = \frac{m_1 y_1 + m_2 y_2 + m_3 y_3}{m_1 + m_2 + m_3}$$

$$= \frac{2m \times 0 + 2m \times 4 + 2m \times 0}{2m + 2m + 2m} = \frac{4}{3}$$

Position vector of COM is,

$$\mathbf{r} = x\hat{i} + y\hat{j} = \frac{4}{3}\hat{i} + \frac{4}{3}\hat{j}$$

Magnitude of position vector,

$$|\mathbf{r}| = \sqrt{\left(\frac{4}{3}\right)^2 + \left(\frac{4}{3}\right)^2}$$

$$= \frac{4\sqrt{2}}{3}$$

59. *(8)* Angular momentum of projectile about the point of projection, when projectile is at highest point is,

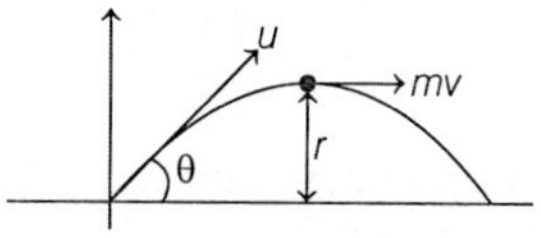

$$L = mvr = mu\cos\theta \cdot \frac{u^2\sin^2\theta}{2g}$$

$$= \frac{mu^3\cos\theta\sin^2\theta}{2g}$$

with $\theta = 45°$, we have

$$L = \frac{mu^3}{2g}\cdot\frac{1}{\sqrt{2}}\cdot\frac{1}{2}$$

$$= \frac{1}{4\sqrt{2}}\frac{mu^3}{g} = \frac{\sqrt{2}}{8}\frac{mu^3}{g}$$

$$= \frac{\sqrt{2}mu^3}{xg}$$

$\therefore \quad x = 8$

60. *(60)* Given, $m = 5$ kg

$$v = 3\sqrt{2}\ \text{ms}^{-1}$$

$$y = x + 4$$

$$x - y + 4 = 0$$

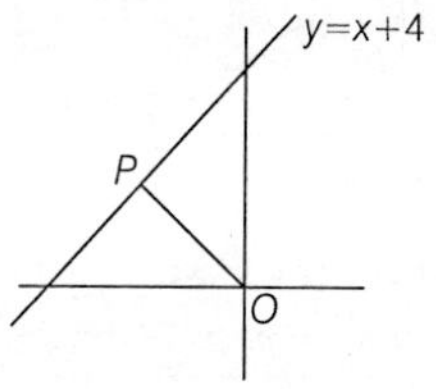

Perpendicular difference on the line from origin,

$$OP = \frac{4}{\sqrt{1^2+1^2}} = 2\sqrt{2}$$

∴Angular momentum

$$L = mv \times OP$$

$$= 5\times 3\sqrt{2}\times 2\sqrt{2}$$

$$= 60\ \text{kg m}^2\text{s}^{-1}$$

61. *(3)* We know that, $\tau = \mathbf{r}\times\mathbf{F}$

Where, $\mathbf{r} = 2\hat{i} - 3\hat{j}$ and $\mathbf{F} = -P\hat{k}$

$$\Rightarrow \quad \tau = (+2\hat{i} - 3\hat{j})\times(-P)\hat{k}$$

$$= P(2\hat{j} + 3\hat{i}) = P(+3\hat{i} + 2\hat{j})$$

$$= P(a\hat{i} + b\hat{j}) \quad \text{(Given)}$$

So, $\quad a = +3$ and $b = +2$

Thus, $\quad \dfrac{a}{b} = \dfrac{3}{2} = \dfrac{x}{2}$

$\therefore \quad x = 3$

62. *(800)*

key idea First, get the position vector **r** and velocity **v** of the particle after 2 s. After that you can calculate the value of angular momentum using the following formula.

$$\mathbf{L} = m(\mathbf{r}\times\mathbf{v})$$

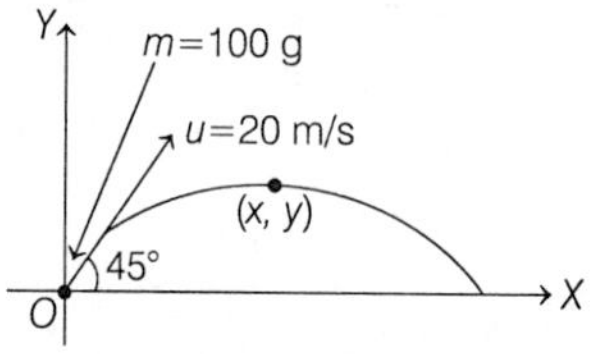

At time $t = 2$ s, let particle is at point (x, y)

Then, $x = v_x t = x\cos\theta t$

$$= 20\times\cos 45°\times 2 = \frac{40}{\sqrt{2}}$$

$$= 20\sqrt{2}\ \text{m}$$

and $\quad y = u\sin\theta\cdot t - \dfrac{1}{2}gt^2$

$$= 20\times\frac{1}{\sqrt{2}}\times 2 - \frac{1}{2}\times 10\times 2^2$$

$$= 20\sqrt{2} - 20 = 20(\sqrt{2}-1)\ \text{m}$$

So, position vector of particle,

$$\mathbf{r} = x\hat{i} + y\hat{j}$$

$$= 20\sqrt{2}\hat{i} + 20(\sqrt{2}-1)\hat{j}$$

Also, velocity of particle at $t = 2$ s is

$$\mathbf{v} = \mathbf{u} + \mathbf{a}t$$

$$\Rightarrow \quad \mathbf{v} = (u\cos\theta\hat{i} + u\sin\theta\hat{j}) - (g\hat{j}t)$$

$$= \frac{20}{\sqrt{2}}\hat{i} + \frac{20}{\sqrt{2}}\hat{j} - 20\hat{j}$$

$$= 10\sqrt{2}\hat{i} + 10\sqrt{2}\hat{j} - 20\hat{j}$$

$$= 10\sqrt{2}\hat{i} + 10(\sqrt{2}-2)\hat{j}$$

So, Angular momentum at $t = 2$s is given by,

$$\mathbf{L} = m(\mathbf{r}\times\mathbf{v})$$

$$= 0.1\begin{vmatrix}\hat{i} & \hat{j} & \hat{k}\\ 20\sqrt{2} & 20(\sqrt{2}-1) & 0\\ 10\sqrt{2} & 10(\sqrt{2}-2) & 0\end{vmatrix}$$

$$= 0.1(-200\sqrt{2})\,\hat{k} = -20\sqrt{2}\,\hat{k}$$

So, magnitude of angular momentum is

$$|\mathbf{L}| = |-20\sqrt{2}\,\hat{k}| = 20\sqrt{2} = \sqrt{800}$$

$$= \sqrt{K} \quad \text{(Given)}$$

$\therefore \quad K = 800$

63. *(4)* Given, length of tube,

$$l = 50\ \text{cm} = 0.5\ \text{m}$$

Mass of incompressible liquid,

$$m = 250\ \text{g} = 0.25\ \text{kg}$$

The given situation is shown below.

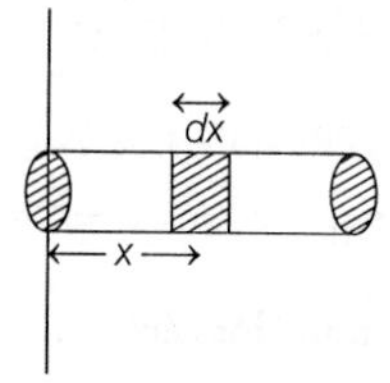

Force exerted by the liquid at the other end due to rotation,

$$F = \int (dm)\omega^2 x$$

where, $dm = \dfrac{m}{L}\cdot dx$

$$\therefore \quad F = \int_0^L \frac{m}{L}dx\cdot\omega^2 x = \frac{m}{L}\cdot\omega^2\left[\frac{x^2}{2}\right]_0^L$$

$$F = \frac{m\omega^2 L}{2}$$

$$\Rightarrow \quad \omega = \sqrt{\frac{2F}{mL}} = \sqrt{\frac{2F}{0.25\times 0.5}}$$

$$= \sqrt{16F} = 4\sqrt{F}$$

$$= x\sqrt{F} \quad \text{(given)}$$

$\therefore \quad x = 4$

64. *(8)* Given, mass density,

$$\rho = \rho_0\left(1 - \frac{x^2}{L^2}\right)\text{kg/m}$$

Mass of element,

$$dm = \rho dx = \rho_0\left(1 - \frac{x^2}{L^2}\right)dx$$

The distance of centre of mass is

$$x_{CM} = \frac{\int_0^L x\,dm}{\int_0^L dm} = \frac{\int_0^L \rho_0\left(1-\frac{x^2}{L^2}\right)x\,dx}{\int_0^L \rho_0\left(1-\frac{x^2}{L^2}\right)dx}$$

$$= \frac{\left[\frac{x^2}{2} - \frac{x^4}{4L^2}\right]_0^L}{\left[x - \frac{x^3}{3L^2}\right]_0^L}$$

$$= \frac{\left(\frac{L^2}{2} - \frac{L^4}{4L^2} - 0\right)}{\left(L - \frac{L^3}{3L^2} - 0\right)} = \frac{\left(\frac{L^2}{2} - \frac{L^2}{4}\right)}{\left(L - \frac{L}{3}\right)}$$

$$= \frac{L^2/4}{2L/3} = \frac{3L}{8} = \frac{3L}{\alpha} \quad \text{(Given)}$$

$\therefore \quad \alpha = 8$

65. *(2)* The given situation is shown below.

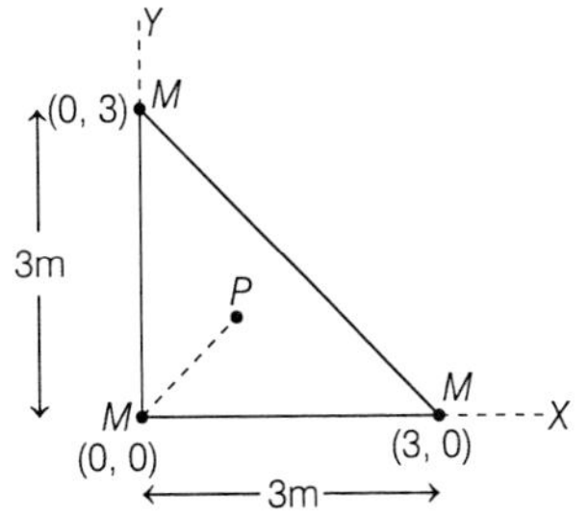

If (X_{CM}, Y_{CM}) be the positions of centre of mass, then

$$X_{CM} = \frac{\mathbf{x}_1 m_1 + \mathbf{x}_2 m_2 + \mathbf{x}_3 m_3}{m_1 + m_2 + m_3}$$

$$= \frac{0 \times m + 3\hat{\mathbf{i}} \times m + 0 \times m}{m + m + m} = \hat{\mathbf{i}}$$

Similarly,

$$Y_{CM} = \frac{\mathbf{y}_1 m_1 + \mathbf{y}_2 m_2 + \mathbf{y}_3 m_3}{m_1 + m_2 + m_3}$$

$$= \frac{0 \times m + 0 \times m + 3\hat{\mathbf{j}} \times m}{3} = \hat{\mathbf{j}}$$

$\therefore$ Position vector of centre of mass,

$$\mathbf{OP} = \mathbf{X}_{CM} + \mathbf{Y}_{CM}; \quad \mathbf{OP} = \hat{\mathbf{i}} + \hat{\mathbf{j}}$$

$\therefore$Magnitude of position of centre of mass,

$$|\mathbf{OP}| = \sqrt{1^2 + 1^2} = \sqrt{2} = \sqrt{x} \quad \text{(given)}$$

$\therefore \quad x = 2$

66. *(91)* Angular momentum of a body,

$$\mathbf{L} = \mathbf{r} \times \mathbf{p} = (\mathbf{r} \times m\mathbf{v})$$

Here, $\mathbf{r} = (3\hat{\mathbf{i}} - \hat{\mathbf{j}})$ m

$\mathbf{v} = (3\hat{\mathbf{j}} + \hat{\mathbf{k}})$ m/s and $m = 1$ kg

$$\text{So,} \quad \mathbf{L} = \begin{vmatrix} \hat{\mathbf{i}} & \hat{\mathbf{j}} & \hat{\mathbf{k}} \\ 3 & -1 & 0 \\ 0 & 3 & 1 \end{vmatrix}$$

Or $\mathbf{L} = -\hat{\mathbf{i}} - 3\hat{\mathbf{j}} + 9\hat{\mathbf{k}}$

Magnitude of angular momentum will be

$$|\mathbf{L}| = \sqrt{1^2 + 3^2 + 9^2} = \sqrt{91} \text{ N-ms}$$

67. *(6)* The given situation is shown below

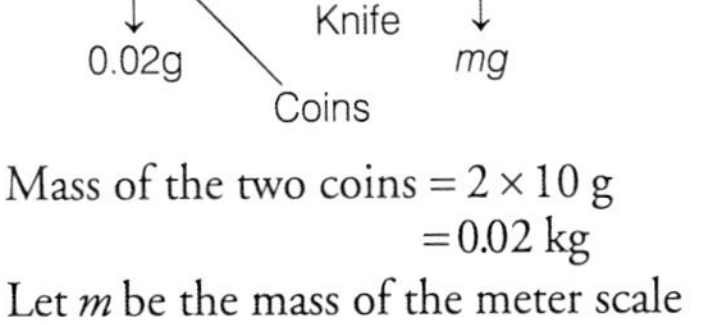

Mass of the two coins $= 2 \times 10$ g $= 0.02$ kg

Let m be the mass of the meter scale which is balanced at point O.

By the principle of moments about knife edge.

$$(0.02\,g) \times (30 \times 10^{-2}) = mg \times 10 \times 10^{-2}$$

$$\Rightarrow \quad m = 6 \times 10^{-2} \text{ kg}$$

Thus, $\quad x = 6$

68. *(25)* Given, mass of 1st body,

$m_1 = 2$ kg

Initial speed of 1st body, $u_1 = 4$ ms^{-1}

Initial speed of 2nd body, $u_2 = 0$ ms^{-1}

Let v_1 be the speed of 1st body after collision. So, as per question,

$$v_1 = \frac{u_1}{4} = \frac{4}{4} = 1 \text{ ms}^{-1}$$

and v_2 be the speed of 2nd body after collision.

By using law of conservation of momentum,

Initial momentum of system p_i = Final momentum of system p_f

$$\Rightarrow \quad m_1 u_1 + m_2 u_2 = m_1 v_1 + m_2 v_2$$

$$\Rightarrow \quad 2 \times 4 + m_2 \times 0 = 2 \times 1 + m_2 v_2$$

$$\Rightarrow \quad m_2 v_2 = 8 - 2 = 6 \text{ kg-ms}^{-1} \quad \ldots\text{(i)}$$

Since it is an elastic collision.

Coefficient of restitution

$$e = \frac{v_2 - v_1}{u_1 - u_2} = 1$$

$$\Rightarrow \quad v_2 - 1 = 4 - 0 \quad \Rightarrow \quad v_2 = 5 \text{ ms}^{-1}$$

On substituting in Eq. (i), we get

$$m_2 \times 5 = 6 \text{ kg-ms}^{-1}$$

$$\Rightarrow \quad m_2 = 6/5 \text{ kg}$$

As we know that

Speed of centre of mass,

$$v_{CM} = \frac{m_1 v_1 + m_2 v_2}{m_1 + m_2} = \frac{2 \times 1 + \frac{6}{5} \times 5}{2 + \frac{6}{5}}$$

$$= \frac{8 \times 5}{16} = \frac{5}{2} = 2.5 \text{ ms}^{-1}$$

$$= \frac{25}{10} \text{ ms}^{-1}$$

$\therefore x = 25$

69. *(10)* Given,

$$\mathbf{r} = 10\alpha t^2 \hat{\mathbf{i}} + 5\beta(t - 5)\hat{\mathbf{j}}$$

Since, velocity $\mathbf{v} = \dfrac{d\mathbf{r}}{dt}$

$$\Rightarrow \quad \mathbf{v} = \frac{d}{dt}[10\alpha t^2 \hat{\mathbf{i}} + 5\beta(t - 5)\hat{\mathbf{j}}]$$

$$= 20\alpha t\hat{\mathbf{i}} + 5\beta\hat{\mathbf{j}}$$

Now $\mathbf{v}(t = 0) = 5\beta\hat{\mathbf{j}}$

and angular momentum (L) at $t = 0$ will be

$$L = mvr$$

and $r(t = 0) = 5\beta(0 - 5)\hat{\mathbf{j}} = -25\beta\hat{\mathbf{j}}$

$$\mathbf{L}(t = 0) = m(\mathbf{r} \times \mathbf{v}) = 0$$

[$\because \sin\theta$ (angle between v and r) is 0].

Now, $\mathbf{r}(t) \times \mathbf{v}(t) = \mathbf{v}(t) \times \mathbf{r}(t) = 0$

$$\Rightarrow \quad \mathbf{v} \times \mathbf{r} = \begin{vmatrix} \hat{\mathbf{i}} & \hat{\mathbf{j}} & \hat{\mathbf{k}} \\ 20\alpha t & 5\beta & 0 \\ 10\alpha t^2 & 5\beta(t - 5) & 0 \end{vmatrix}$$

$$\Rightarrow \quad \hat{\mathbf{i}}(0) - \hat{\mathbf{j}}(0) + \hat{\mathbf{k}}$$

$$[20\alpha t \times 5\beta(t - 5) - 5\beta \times 10\alpha t^2] = 0$$

$$\Rightarrow \hat{\mathbf{k}}[100\alpha\beta t(t - 5) - 50\alpha\beta t^2] = 0$$

$$\Rightarrow 100\alpha\beta t(t - 5) = 50\alpha\beta t^2$$

$$\Rightarrow \quad 2(t - 5) = t \Rightarrow 2t - 10 = t$$

$$\Rightarrow \quad t = 10 \text{ s}$$

70. *(2)* Let R be the radius of semicircle, dl be the elemental arc length for $d\theta$ angle

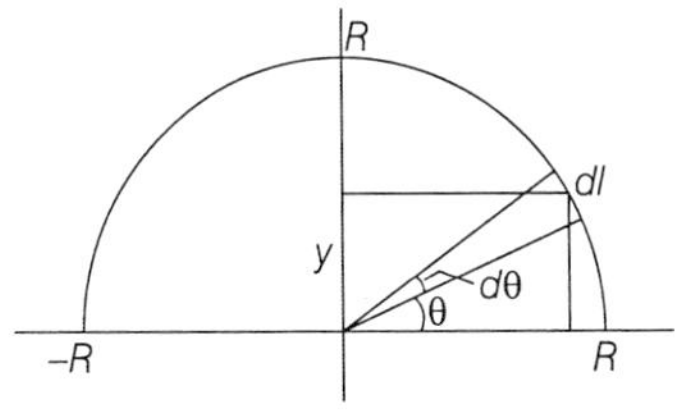

$\therefore \quad dl = Rd\theta$

For πR length, mass be m

$\therefore$Mass per unit length $= m/\pi R$

$$\therefore dm = \frac{m}{\pi R} dl \Rightarrow dm = \frac{m}{\pi R} Rd\theta = \frac{m}{\pi} d\theta$$

$$\sin\theta = \frac{y}{R} \Rightarrow y = R\sin\theta$$

and as we know that

$X_{CM} = 0$ (by symmetry)

$$\text{and } Y_{CM} = \frac{1}{m}\int y\,dm = \frac{1}{m}\int_0^{\pi} R\sin\theta \frac{m}{\pi} d\theta$$

$$= -\frac{R}{\pi}[\cos\theta]_0^{\pi} = \frac{2R}{\pi} = \frac{xR}{\pi}$$

(given)

Hence, $x = 2$

71. *(4)* As we know the centre of mass of the quarter disc, $= \left(\dfrac{4R}{3\pi}, \dfrac{4R}{3\pi}\right)$

Since, $R = a$

$\therefore$The centre of mass of the quarter disc is $\left(\dfrac{4a}{3\pi}, \dfrac{4a}{3\pi}\right)$.

Hence, the value of x is 4.

- **Concept Enhancer**

It is not necessary that centre of mass would lie inside the body, i.e., it may be outside the body.

72. *(20)* Given, Force, $\mathbf{F} = 4\hat{i} + 3\hat{j} + 4\hat{k}$

We know that

Torque, $\tau = \mathbf{r} \times \mathbf{F}$

Where, **r** is the perpendicular distance.

$$\mathbf{r} = (2\hat{i}) - (2\hat{i} + 3\hat{j} + 4\hat{k}) = -3\hat{j} - 4\hat{k}$$

$$\Rightarrow \tau = \mathbf{r} \times \mathbf{F} = \begin{vmatrix} \hat{i} & \hat{j} & \hat{k} \\ 0 & -3 & -4 \\ 4 & 3 & 4 \end{vmatrix}$$

$$\Rightarrow \tau = \hat{i}(-12+12) - \hat{j}(0+16) + \hat{k}(0+12)$$

$$\Rightarrow \tau = -16\hat{j} + 12\hat{k}$$

$\therefore$ Magnitude of torque

$$|\tau| = \sqrt{(16)^2 + (12)^2}$$

$$= \sqrt{256 + 144} = \sqrt{400}$$

$$\Rightarrow \quad |\tau| = 20$$

73. *(9)*

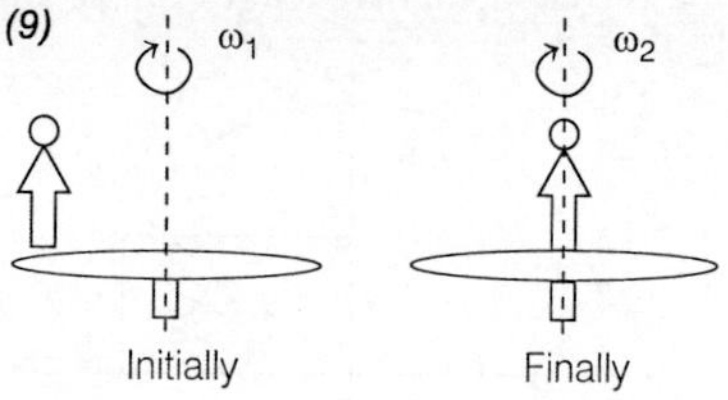

According to conservation of angular momentum,

$$I_1 \omega_1 = I_2 \omega_2$$

$$\Rightarrow \left(80R^2 + \frac{200R^2}{2}\right)\omega_1 = \left(\frac{200R^2}{2}\right)\omega_2$$

$$\Rightarrow \quad \frac{360R^2}{2} \times 5 = \frac{200R^2}{2} \times \omega_2$$

$$\omega_2 = 9 \text{ rpm}$$

74. *(3)* The distance of centre of mass of a solid hemisphere from the centre of the flate surface is given by

$$d = \frac{3R}{8}$$

Here, $R = 8$ cm

$$\therefore \quad d = \frac{3 \times 8}{8} = 3 \text{ cm}$$

Given, $d = x$ cm

$$\therefore \quad x = 3$$

Topic 2

01. (a) Diameter of ring $= r$

$\therefore$ radius, $a = \dfrac{r}{2}$

$$I_{\text{tangent}} = \frac{Ma^2}{2} + Ma^2$$

$$= \frac{M}{2}\left(\frac{r}{2}\right)^2 + M\left(\frac{r}{2}\right)^2$$

$$= \frac{Mr^2}{8} + \frac{Mr^2}{4} = \frac{3}{8}Mr^2$$

02. *(a)* Here, $L = 2\pi R$

$$\Rightarrow \quad R = \frac{L}{2\pi}$$

$\therefore$ Mass, $M = \lambda L$

$\therefore$ Moment of inertia of ring about any of its diameter.

$$I = \frac{MR^2}{2} = \frac{\lambda L}{2}\left(\frac{L}{2\pi}\right)^2 = \frac{\lambda L^3}{8\pi^2}$$

03. *(c)* $\theta = 5t^2 - 8t$

$$\omega = \frac{d\theta}{dt} = 10t - 8$$

$$\alpha = \frac{d\omega}{dt} = 10$$

$$\therefore \quad P = \tau\omega = I\,\alpha\,\omega \qquad [\tau = I\alpha]$$

$$= \left(\frac{MR^2}{2}\right)\alpha\,\omega \qquad \left[I = \frac{MR^2}{2}\right]$$

$$= \frac{MR^2}{2} \times (10)(10t - 8)$$

At $t = 2$s, $P = 60\,MR^2$

04. *(c)* Given, mass of each sphere,

$$M = 1.5 \text{ kg}$$

Radius of each sphere, $R = 50$ cm

$$= 50 \times 10^{-2} \text{ m}$$

Distance between centre of spheres,

$$l = 5 \text{ m}$$

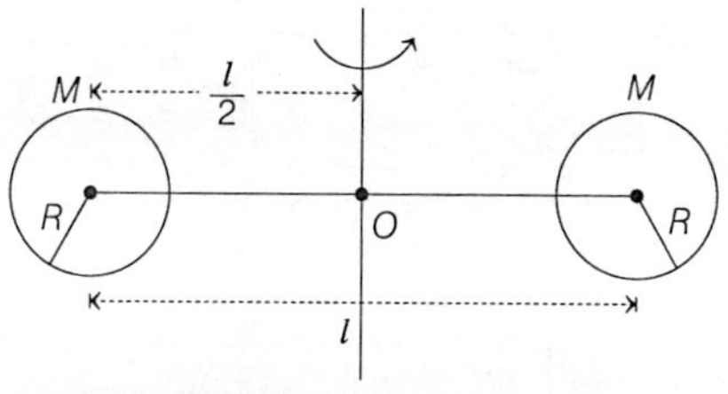

By using parallel axis theorem, moment of inertia of system is

$$I = 2\left[\frac{2}{5}MR^2 + \frac{Ml_2}{4}\right]$$

$$\Rightarrow \quad I = 2M\left[\frac{2}{5}R^2 + \frac{l_2}{4}\right]$$

On substituting the values, we get

$$I = 2 \times 1.5\left[\frac{2}{5} \times (50 \times 10^{-2})^2 + \frac{(5)^2}{4}\right]$$

$$= 19.05 \text{ kg-m}^2$$

05. *(c)* The uniform wire whose mass is M and length is L and bent into a semicircle of radius r.

The length of the semicircle wire,

$$L = \pi r$$

$$\Rightarrow \quad r = \frac{L}{\pi}$$

The moment of inertia,

$$I = Mr^2$$

$$\Rightarrow \quad I = M\left(\frac{L}{\pi}\right)^2$$

$$\Rightarrow \quad I = \frac{ML^2}{\pi^2}$$

06. *(d)* Let M and R be the mass and radius of four bodies. Then, as per question, their moment of inertia are

$$I_1 = \frac{MR^2}{2}, I_2 = \frac{MR^2}{2}, I_3 = \frac{MR^2}{2},$$

$$I_4 = \frac{2}{5}MR^2$$

Clearly,

$$I_1 = I_2 = I_3 > I_4$$

07. *(d)* Moment of inertia of square plate about xx' is $\dfrac{ma^2}{6}$, therefore moment of inertia about zz' can be computed using parallel axis theorem,

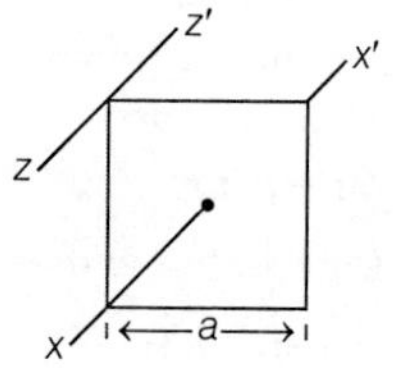

$$I_{zz'} = I_{xx'} + m\left(\frac{a}{\sqrt{2}}\right)^2$$

$$= \frac{ma^2}{6} + \frac{ma^2}{2} = \frac{2\,ma^2}{3}$$

08. *(d)* The mass of complete (circular) disc is

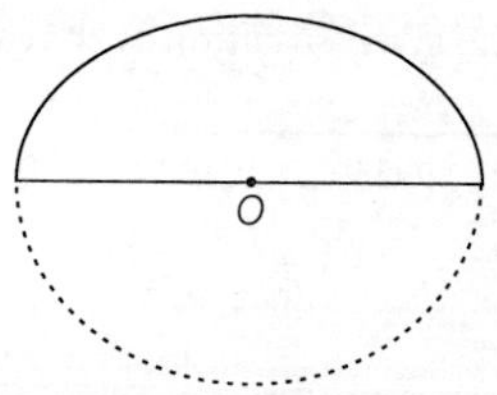

The moment of inertia of disc about the given axis is

$$I = \frac{2\,Mr^2}{2} = Mr^2$$

Let the moment of inertia of semi-circular disc be I_1.
The disc may be assumed as combination of two semi-circular parts.
Thus, $I_1 = I - I_1$

or $\quad I_1 = \frac{I}{2} = \frac{Mr^2}{2}$

09. *(c)* Let same mass and same outer radii of solid sphere and hollow sphere be M and R, respectively.
The moment of inertia of solid sphere A about its diameter,

$$I_A = \frac{2}{5}MR^2 \qquad \ldots(i)$$

Similarly, the moment of inertia of hollow sphere (spherical shell) B about its diameter,

$$I_B = \frac{2}{3}MR^2 \qquad \ldots(ii)$$

It is clear from Eqs. (i) and (ii),

$$I_A < I_B$$

10. *(a)* MI of circular wire about its diameter $= \frac{MR^2}{2}$

11. *(b)*

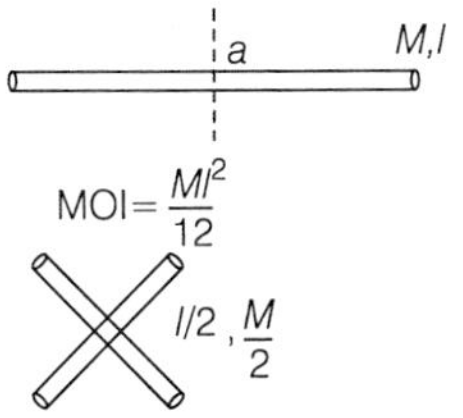

$$\alpha' = \left[\frac{\left(\frac{M}{2}\right)\left(\frac{l}{2}\right)^2}{12}\right] \times 2$$

$$\alpha' = \frac{Ml^2}{48}$$

$$= \frac{Ml^2}{12} \times \frac{1}{4}$$

$$\Rightarrow \quad \alpha' = \frac{\alpha}{4}$$

12. *(a)* Moment of inertia, when the section is taken out is

$$I = \frac{MR^2}{2} - \left[\frac{\frac{M}{4}\left(\frac{R}{2}\right)^2}{2} + \frac{M}{4}\left(\frac{R}{2}\right)^2\right]$$

$$I = \frac{13}{32}MR^2$$

Hence, option (a) is correct.

13. *(c)* For hollow sphere, moment of inertia,

$$I_1 = \frac{2}{3}mR^2 \text{ or } \left(\sqrt{\frac{2}{3}}R\right)^2 m$$

$$\therefore \quad K_1 = \sqrt{\frac{2}{3}}R \qquad \ldots(i)$$

[As $I = mK^2$]

For solid cylinder, moment of inertia,

$$I_2 = \frac{1}{4}mR^2 + \frac{m}{3}(4R)^2 = \left(\sqrt{\frac{67}{12}}R\right)^2 m$$

$$\therefore K_2 = \sqrt{\frac{67}{12}}R \qquad \ldots(ii)$$

Taking the ratio of Eqs. (i) and (ii),

$$\frac{K_1}{K_2} = \sqrt{\frac{2}{3}}R \times \sqrt{\frac{12}{67}} \times \frac{1}{R} = \sqrt{\frac{8}{67}}$$

Matching with the given equation, the value of x is 67.

14. *(a)* Moment of inertia of solid sphere about a tangential axis, $I = \frac{7}{5}MR^2$

Moment of inertia of hollow sphere about tangential axis, $I = \frac{5}{3}MR^2$

Moment of inertia of circular ring about its diametre,

$$I = \frac{1}{2}MR^2$$

Moment of inertia of circular disc of mass M about its diametre, $I = \frac{1}{4}MR^2$

Hence, (A)→(ii), (B) → (i), (C) → (iv), (D) → (iii)

15. *(d)*

> **key idea** Use parallel axis theorem to solve such type of problems

Let the length of side of square plate is l.
The diagram of square plate having an axis passing from centre as $X - X'$ and an axis passing through one of the corner and perpendicular to plane of square plate $Z - Z'$ is shown below.

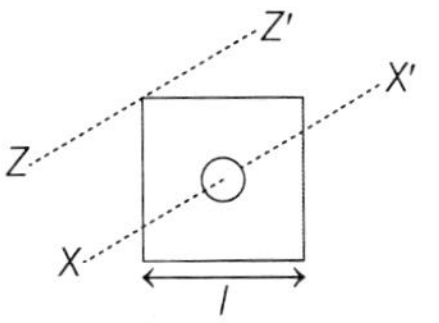

The distance between the centre of square plate and the corner of square plate is

$$d = \sqrt{\left(\frac{l}{2}\right)^2 + \left(\frac{l}{2}\right)^2} = \frac{l}{\sqrt{2}}$$

$$I_{XX'} = Ml^2/6$$

and $I_{ZZ'} = I_{XX'} + M\left(\frac{l}{\sqrt{2}}\right)^2$

$$I_{ZZ'} = \frac{Ml^2}{6} + \frac{Ml^2}{2} = Ml^2\left[\frac{1}{6} + \frac{1}{2}\right]$$

$$\Rightarrow I_{ZZ'} = \frac{2Ml^2}{3}$$

16. *(d)* Given, $L = 80 \text{ cm} = 0.8 \text{ m}$

$r = 20 \text{ cm} = 0.2 \text{ m}$

$I_{CD} = 2.7 \text{ kg-m}^2$

Now, $I_{AB} = \frac{1}{2}Mr^2$

By parallel axis theorem,

$$I_{CD} = I_{AB} + M\left(\frac{L}{2}\right)^2$$

$$\Rightarrow \quad 2.7 = \frac{Mr^2}{2} + \frac{ML^2}{4}$$

$$\Rightarrow \quad 2.7 = M\left[\frac{(0.2)^2}{2} + \frac{(0.8)^2}{4}\right]$$

$$\Rightarrow \quad M = 15 \text{ kg}$$

Density of cylinder is given as

$$\rho = \frac{M}{V} = \frac{M}{\pi r^2 L} = \frac{15}{\pi(0.2)^2(0.8)}$$

$$= 149.2 = 1.49 \times 10^2 \text{ kg m}^{-3}$$

17. *(d)* Let mass per unit area be σ.
$\therefore$ Mass of smaller disc, $M' = \pi r^2 \sigma$
Mass of larger disc, $M = \pi R^2 \sigma$
$\therefore$Moment of inertia of smaller disc around one of its diameter lying on its plane, $I' = \frac{1}{4}M'r^2$

$$\Rightarrow \quad I' = \frac{\pi\sigma r^4}{4}$$

and moment of inertia of bigger disc around axis AB (perpendicular to the plane) of disc and passing through its centre,

$$I = \frac{1}{2}MR^2 = \frac{\pi\sigma R^4}{2}$$

$$\therefore \quad \frac{I}{I'} = \frac{\frac{\pi\sigma R^4}{2}}{\frac{\pi\sigma r^4}{4}} = \frac{2R^4}{r^4} \Rightarrow \frac{I}{I'} = \frac{2R^4}{r^4}$$

18. *(d)* Given, I_X, I_Y and I_Z be the moment of inertia of disc about X, Y and Z-axis.

By using perpendicular axis theorem,

$$I_Z = I_X + I_Y$$

where, I_Z is the moment of inertia about its centre of mass $= \frac{MR^2}{2}$

and in symmetrical body, $I_X = I_Y$ (Disc in given case)

$\therefore \quad I_Z = 2I_X$

$\Rightarrow \quad I_X = I_y = \frac{I_Z}{2} = \frac{MR^2}{4}$

Now, $I_X = I_Y \neq I_Z$

As, radius of gyration,

$$k = \sqrt{I/M}$$

$\Rightarrow \quad k \propto I$

So, radii of gyration about all axis will not be the same.

A rigid body in rotational motion maintains a fixed mass and shape, meaning its particles retain their relative position. No deformation occurs and all points move in circular paths about a fixed axis.

19. *(c)* The given situation can be represented as follows

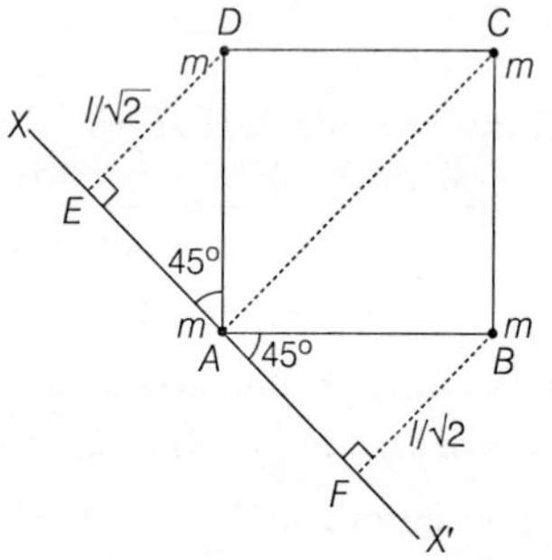

In the above figure, XX' be the axis which passes through A and is parallel to DB.

$\therefore$Moment of inertia of the system

= Mass × (Perpendicular distance from axis)^2

$$I = m(0)^2 m(AC)^2 + m(ED)^2 + m(FB)^2$$

$$= m(0)^2 + m(l\sqrt{2})^2 + m\left(\frac{l}{\sqrt{2}}\right)^2 + m\left(\frac{l}{\sqrt{2}}\right)^2 = 3ml^2$$

20. *(c)* Given, mass of solid sphere $= m$

Radius of solid sphere $= a$

Side of square $= b$

Moment of inertia of sphere, $I_s = \frac{2}{5}ma^2$

Axis of rotation is BC.

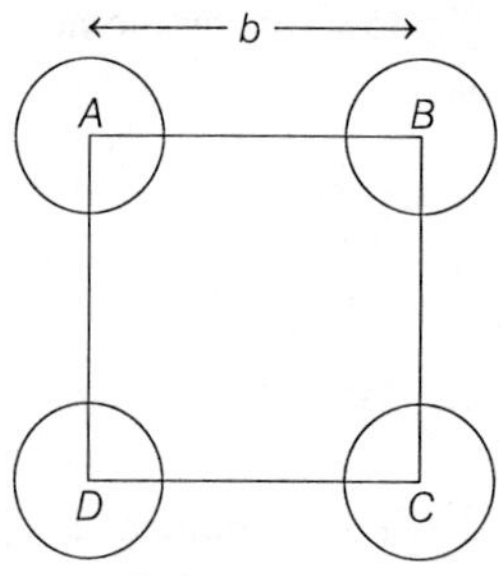

$\therefore$Moment of inertia of any body at distance,

$$d = md^2 = mb^2 \qquad (\because d = b)$$

$\therefore \quad I_{\text{net}} = 4\left(\frac{2}{5}ma^2\right) + 2(mb^2)$

$\Rightarrow \quad I_{\text{net}} = \frac{8}{5}ma^2 + 2mb^2$

21. *(c)* To calculate radius of gyration, first we calculate moment of inertia of rod about given axis.

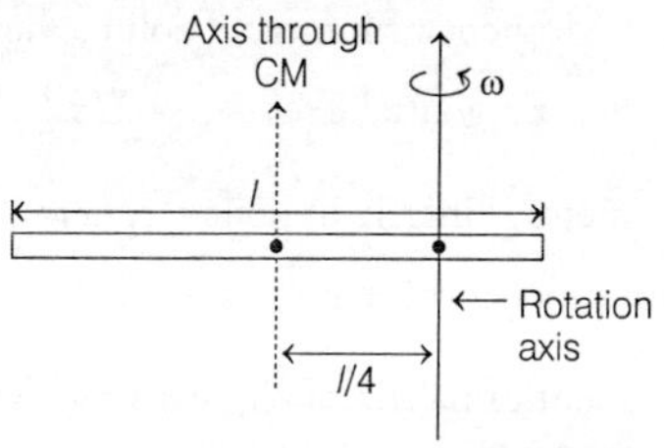

Moment of inertia of rod about an axis through centre of mass, $I_c = \frac{ml^2}{12}$

Moment of inertia of rod about given axis of rotation by parallel axis theorem will be

$$I = I_c + mh^2 = m\frac{l^2}{12} + m\left(\frac{l}{4}\right)^2$$

$$= \frac{7}{48}ml^2$$

If K = radius of gyration of rod, then moment of inertia of rod, $I = mK^2$

$\Rightarrow mK^2 = \frac{7}{48}ml^2$

$\therefore$Radius of gyration of rod, $K = \sqrt{\frac{7}{48}}l$

22. *(a)* Moment of inertia of hollow cylinder about its axis is

$$I_1 = \frac{M}{2}(R_1^2 + R_2^2)$$

where, R_1 = inner radius and R_2 = outer radius. Moment of inertia of thin hollow cylinder of radius R about its axis is.

$$I_2 = MR^2$$

Given, $I_1 = I_2$ and both cylinders have same mass (M). So, we have

$$\frac{M}{2}(R_1^2 + R_2^2) = MR^2$$

$$(10^2 + 20^2)/2 = R^2$$

$\Rightarrow \quad R^2 = 250 = 15.8$

$$R \approx 16 \text{ cm}$$

23. *(b)* Moment of inertia of a solid sphere about an axis through its centre of mass is $I_C = \frac{2}{5}MR^2$

Moment of inertia about a parallel axis at a distance x from axis through its COM is

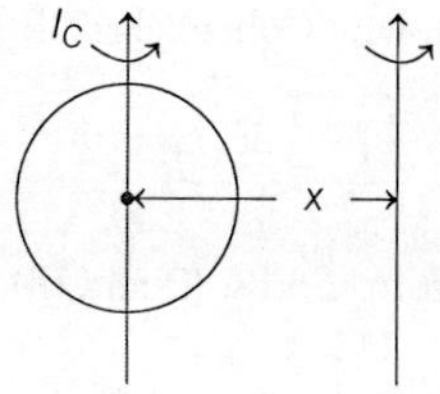

$I = I_C + Mx^2$ (by parallel axis theorem)

So graph of I *versus* x is parabolic are shown

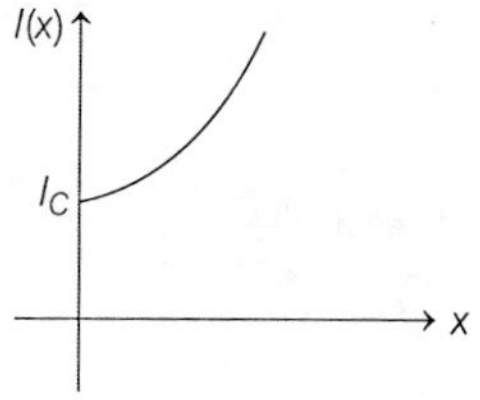

24. *(a)* Moment of inertia of remaining solid

= Moment of inertia of complete solid

− Moment of inertia of removed portion

$$\therefore I = \frac{9MR^2}{2} - \left[\frac{M(R/3)^2}{2} + M\left(\frac{2R}{3}\right)^2\right]$$

$\Rightarrow I = 4MR^2$

25. *(d)* Mass of removed portion

$$= \frac{M}{\pi R^2} \times \pi\left(\frac{R}{4}\right)^2 = \frac{M}{16}$$

Moment of inertia of removed portion about the given axis (Using parallel axis theorem)

$$I_R = \frac{1}{2}\frac{M}{16}\frac{R^2}{16} + \frac{M}{16} \times \frac{9R^2}{16} = \frac{19MR^2}{512}$$

Required moment of inertia,

$$I = I_D - I_R = \frac{1}{2}MR^2 - \frac{19MR^2}{512}$$

$$= \frac{237}{512}MR^2$$

26. *(c)* Let the each side of square lamina be d.

So, $I_{EF} = I_{GH}$ [due to symmetry]

and $I_{AC} = I_{BD}$ [due to symmetry]

Now, according to theorem of perpendicular axes,

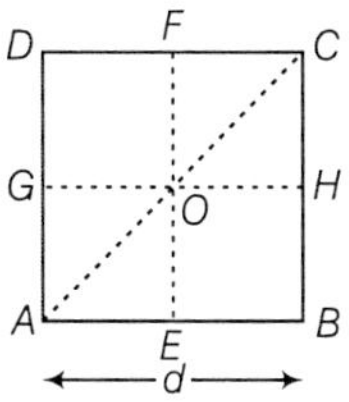

$$I_{AC} + I_{BD} = I_0$$

$$\Rightarrow \quad 2I_{AC} = I_0 \quad \ldots(\text{i})$$

and $I_{EF} + I_{GH} = I_0$

$$\Rightarrow \quad 2I_{EF} = I_0 \quad \ldots(\text{ii})$$

From Eqs. (i) and (ii), we get

$$I_{AC} = I_{EF}$$

$$\therefore \quad I_{AD} = I_{EF} + \frac{md^2}{4}$$

$$= \frac{md^2}{12} + \frac{md^2}{4} \quad \left[\text{as } I_{EF} = \frac{md^2}{12}\right]$$

So, $I_{AD} = \dfrac{md^2}{3} = 4I_{EF}$

27. *(c)* **Case I**

Moment of inertia of the rod (length L, mass M, about an axis perpendicular to the rod passing through the mid point)

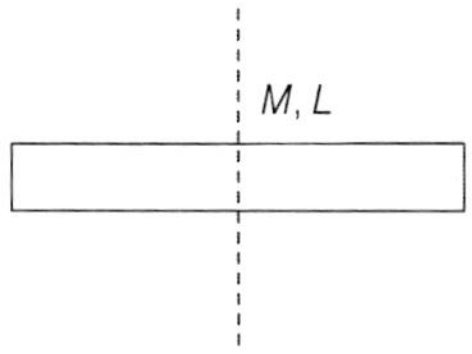

$$= \frac{ML^2}{12}$$

Case II

Moment of inertia of rod (length L, mass $2M$ about an axis perpendicular to the rod passing through one of its end)

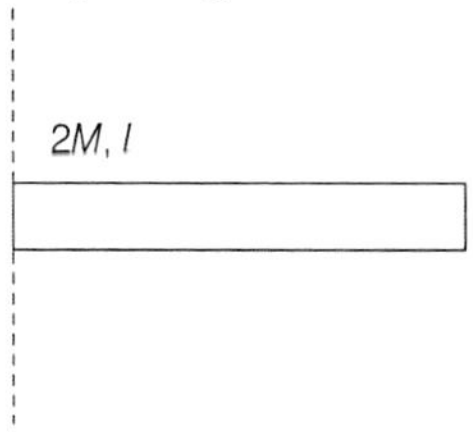

$$= \frac{(2M)(L^2)}{3} = \frac{2ML^2}{3}$$

Case III Moment of inertia of rod (length $2L$, mass M, about an axis perpendicular to the rod passing through its mid point)

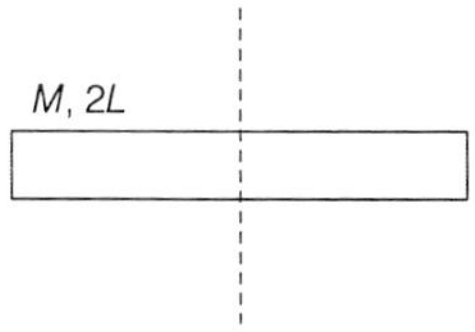

$$= \frac{M(2L)^2}{12} = \frac{ML^2}{3}$$

Case IV Moment of inertia of rod (length $2L$, mass $2M$ about an axis perpendicular to the rod passing through one of its end)

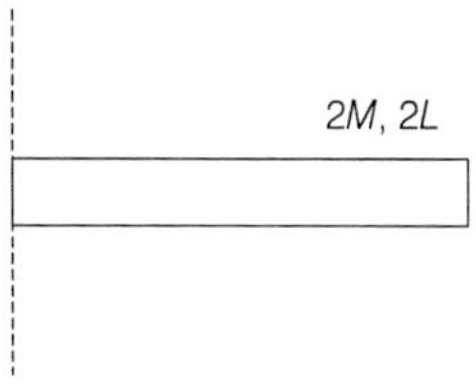

$$= \frac{2M(2L)^2}{3} = \frac{8}{3}ML^2$$

28. *(c)*

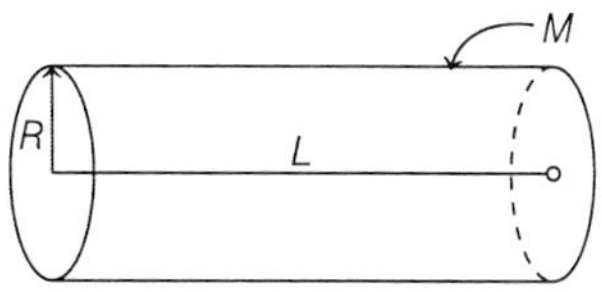

Mass of cylinder, $M = \rho\pi R^2 L$ ...(i)

(where, ρ = density of material)

$$\Rightarrow \quad L = \frac{M}{\rho\pi R^2} \quad \ldots(\text{ii})$$

Moment of inertia of a cylinder,

$$I = M\left(\frac{R^2}{4} + \frac{L^2}{12}\right)$$

$$\Rightarrow \quad I = M\left(\frac{R^2}{4} + \frac{M^2}{12\rho^2\pi^2R^4}\right)$$

[From Eq. (ii)]

For moment of inertia to be minimum,

$$\frac{dI}{dR} = 0$$

$$\Rightarrow \frac{d}{dR}\left[M\left(\frac{R^2}{4} + \frac{M^2}{12\rho^2\pi^2R^4}\right)\right] = 0$$

$$\Rightarrow \quad \frac{2R}{4} + \frac{M^2}{12\rho^2\pi^2}\cdot(-4R^{-5}) = 0$$

[$\because M$ = constant]

$$\Rightarrow \quad \frac{2R}{4} = \frac{4M^2}{12\rho^2\pi^2R^5}$$

$$\Rightarrow \quad R^6 = \frac{2}{3}\frac{M^2}{\rho^2\pi^2}$$

$$\Rightarrow \quad R^6 = \frac{2}{3}\cdot\frac{\rho^2\pi^2R^4L^2}{\rho^2\pi^2} \quad \text{[from Eq. (i)]}$$

$$\Rightarrow \quad \left(\frac{R}{L}\right)^2 = \frac{2}{3} \Rightarrow \frac{L}{R} = \sqrt{\frac{3}{2}}$$

29. *(a)* Both the discs are made up of same material.

Let the density of material used for both discs be ρ and their thickness be t.

Now, moment of inertia of first disc,

$$I_1 = \frac{M_1R_1^2}{2} = \frac{\rho A_1 t R_1^2}{2}$$

$$= \frac{\rho(\pi R_1^2)tR_1^2}{2} = \frac{\pi\rho R_1^4 t}{2}$$

$\because \quad R_1 = R$ (given)

$$\therefore \quad I_1 = \frac{\pi\rho R^4 t}{2} \quad \ldots(\text{i})$$

And, moment of inertia of second disc,

$$I_2 = \frac{M_2R_2^2}{2} = \frac{\rho A_2 t R_2^2}{2}$$

$$= \frac{\rho(\pi R_2^2)tR_2^2}{2} = \frac{\pi\rho R_2^4 t}{2}$$

$\because \quad R_2 = \alpha R$ (given)

$$\therefore \; I_2 = \frac{\pi\rho(\alpha R)^4 t}{2} = \frac{\alpha^4\pi\rho R^4 t}{2} \quad \ldots(\text{ii})$$

Dividing Eq. (i) by Eq. (ii), we get

$$\frac{I_1}{I_2} = \frac{\left(\dfrac{\pi\rho R^4 t}{2}\right)}{\left(\dfrac{\alpha^4\pi\rho R^4 t}{2}\right)}$$

$$\Rightarrow \quad \frac{I_1}{I_2} = \frac{1}{\alpha^4} \quad \ldots(\text{iii})$$

It is given that, $\dfrac{I_1}{I_2} = \dfrac{1}{16}$ ...(iv)

On comparing the Eqs. (iii) and (iv), we get

$$\frac{1}{\alpha^4} = \frac{1}{16}$$

$$\Rightarrow \quad \alpha = (16)^{1/4} = (2^4)^{1/4} = 2$$

30. *(b)* According to question,

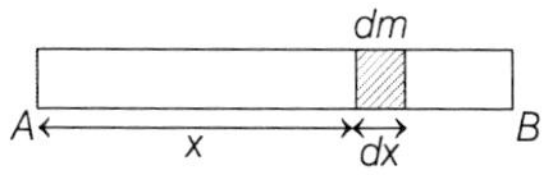

Linear mass density, $\lambda(x)=\lambda_0\left(1+\frac{x}{L}\right)$

Moment of inertia of small element about given axis,

$$dI=(dm)x^2=(\lambda dx)x^2$$

$$\Rightarrow dI=\lambda_0\left(1+\frac{x}{L}\right)x^2dx$$

Total moment of inertia about same axis due to the complete rod,

$$I=\int dI=\int_0^L\lambda_0\left(1+\frac{x}{L}\right)x^2dx$$

$$=\lambda_0\left[\frac{x^3}{3}+\frac{x^4}{4L}\right]_0^L=\frac{7L^3\lambda_0}{12} \quad \ldots\text{(i)}$$

Now, mass of complete rod,

$$M=\int dm=\int\lambda dx$$

$$=\int_0^L\lambda_0\left(1+\frac{x}{L}\right)dx=\lambda_0\left[x+\frac{x^2}{2L}\right]_0^L$$

$$=\lambda_0\left[L+\frac{L^2}{2L}-0\right]=\frac{3\lambda_0L}{2}$$

$$\Rightarrow \lambda_0=\frac{2M}{3L}$$

Putting the value of λ_0 in Eq. (i), we get

$$I=\frac{7}{18}ML^2$$

31. *(d)* Consider an elemental ring of thickness d at a distance from the centre.

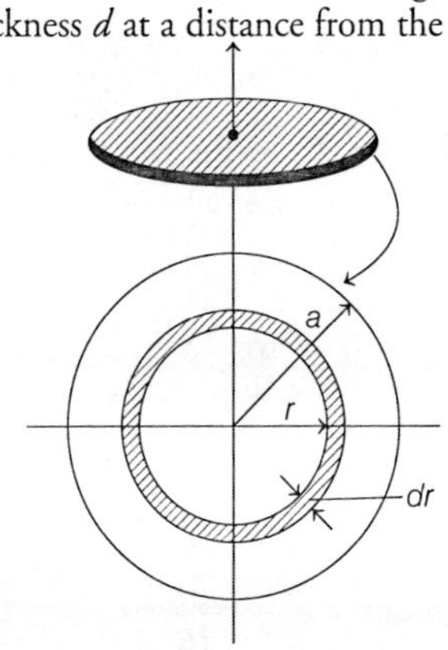

Area of this elemental ring,

$$dA=2\pi rdr$$

Given, mass per unit area,

$$\sigma(r)=A+Br$$

Mass of elemental ring,

$$dm=\sigma(r)\cdot dA=(A+Br)2\pi r\,dr$$

Moment of inertia of this elemental ring about axis, $dI=dm\cdot r^2$

$$=2\pi r^3\,(A+Br)\cdot dr$$

Moment of inertia of complete disc of radius a,

$$I=\int_0^a dI=\int_0^a 2\pi r^3\,(A+Br)\,dr$$

$$=\left[2\pi A\frac{r^4}{4}+2\pi B\frac{r^5}{5}\right]_0^a$$

$$=\frac{2\pi Aa^4}{4}+\frac{2\pi Ba^5}{5}$$

$$=2\pi a^4\left(\frac{A}{4}+\frac{aB}{5}\right)$$

32. *(c)*

$$M=\int_0^R\rho_0 r(2\pi rdr)=\frac{\rho_0\times 2\pi\times R^3}{3}$$

$$\int dI=\int_0^R\rho_0 r(2\pi rdr)r^2=\frac{\rho_0\times 2\pi R^5}{5}$$

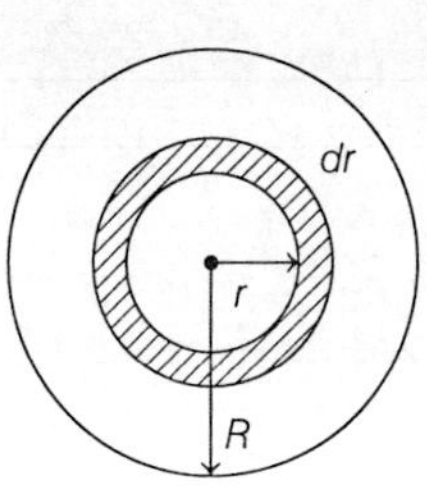

By parallel axis theorem,

$$I=I_0+MR^2$$

$$=\frac{\rho_0\times 2\pi R^5}{5}+\frac{\rho_0\times 2\pi R^3\times R^2}{3}$$

$$=\frac{8}{5}MR^2=aMR^2$$

$$\Rightarrow \quad a=\frac{8}{5}$$

33. *(d)* Given, surface mass density,

$$\sigma=kr^2$$

So, mass of the disc can be calculated by considering small element of area $2\pi rdr$ on it and then integrating it for complete disc, i.e.

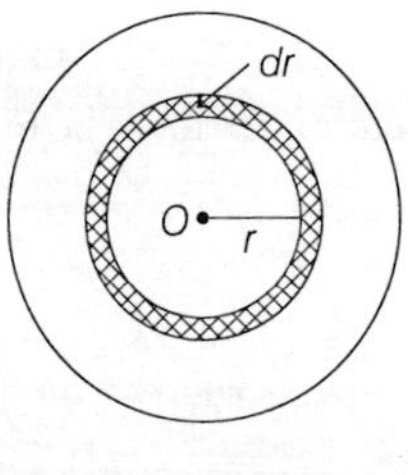

$$dm=\sigma\,dA=\sigma\times 2\pi rdr$$

$$\int dm=M=\int_0^R(kr^2)\,2\pi rdr$$

$$\Rightarrow M=2\pi k\frac{R^4}{4}=\frac{1}{2}\pi kR^4 \quad \ldots\text{(i)}$$

Moment of inertia about the axis of the disc,

$$I=\int dI=\int dmr^2=\int\sigma dAr^2$$

$$=\int_0^R kr^2\,(2\pi rdr)\,r^2$$

$$\Rightarrow I=2\pi k\int_0^R r^5\,dr=\frac{2\pi kR^6}{6}=\frac{\pi kR^6}{3} \quad \ldots\text{(ii)}$$

From Eqs. (i) and (ii), we get

$$I=2/3\,MR^2$$

34. *(d)* The given situation is shown in the figure given below

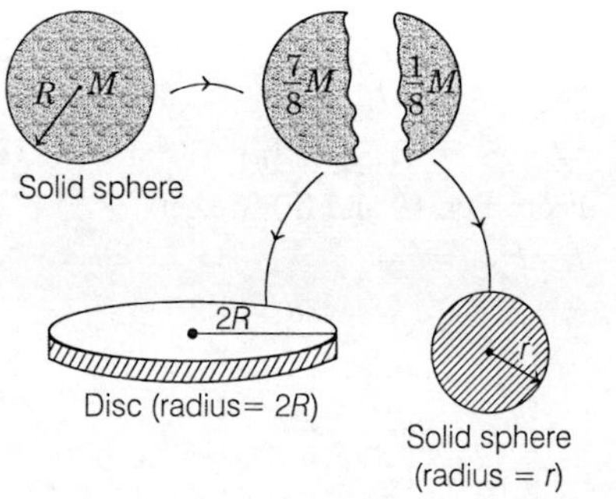

Density of given sphere of radius R is

$$\rho=\frac{\text{Mass}}{\text{Volume}}=\frac{M}{\frac{4}{3}\pi R^3}$$

Let radius of sphere formed from second part is r, then mass of second part

$$=\text{volume}\times\text{density}$$

$$\frac{1}{8}M=\frac{4}{3}\pi r^3\times\frac{M}{\frac{4}{3}\pi R^3}$$

$$\therefore \quad r^3=R^3/8 \Rightarrow \quad r=R/2$$

Now, I_1 = moment of inertia of disc (radius $2R$ and mass $7/8M$) about its axis

$$=\frac{\text{Mass}\times(\text{Radius})^2}{2}=\frac{\frac{7}{8}M\times(2R)^2}{2}$$

$$=\frac{7}{4}MR^2$$

and I_2 = moment of inertia of sphere (radius $\frac{R}{2}$ and mass $\frac{1}{8}M$) about its axis

$$=\frac{2}{5}\times\text{Mass}\times(\text{Radius})^2$$

$$=\frac{2}{5}\times\frac{1}{8}M\times\left(\frac{R}{2}\right)^2=\frac{MR^2}{80}$$

$$\therefore \text{Ratio } \frac{I_1}{I_2}=\frac{\frac{7}{4}MR^2}{\frac{1}{80}MR^2}=140$$

35. *(c)*

key idea Radius of gyration K of any structure is given by

$$I = MK^2 \text{ or } K = \sqrt{\frac{I}{M}}$$

To find K, we need to find both moment of inertia I and mass M of the given structure.

Given, variation in mass per unit area (surface mass density),

$$\sigma = \frac{\sigma_0}{r} \quad \ldots(i)$$

Calculation of Mass of Disc

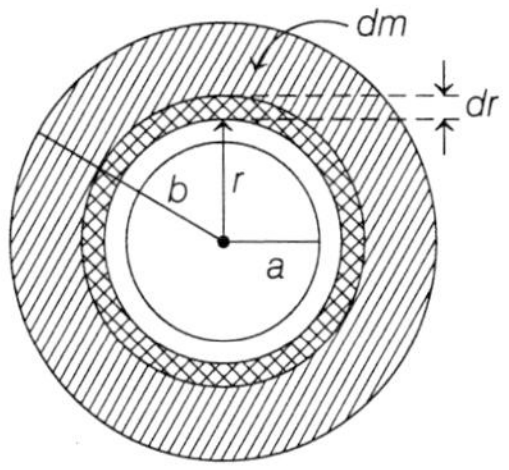

Let us divide whole disc in small area elements, one of them shown at r distance from the centre of the disc with its width as dr.

Mass of this element is

$$dm = \sigma \cdot dA$$

$$\Rightarrow \quad dm = \frac{\sigma_0}{r} \times 2\pi r dr \quad \text{[from Eq. (i)]} \quad \ldots(ii)$$

Mass of the disc can be calculated by integrating it over the given limits of r,

$$\int_0^M dm = \int_a^b \sigma_0 \times 2\pi \times dr$$

$$\Rightarrow \quad M = \sigma_0 2\pi(b-a) \quad \ldots(iii)$$

Calculation of Moment of Inertia

$$I = \int_0^M r^2 dm = \int_a^b r^2 \cdot \frac{\sigma_0}{r} \times 2\pi r dr$$

[from Eq. (i)]

$$= \sigma_0 2\pi \int_a^b r^2 dr = \sigma_0 2\pi \left[\frac{r^3}{3}\right]_a^b$$

$$\Rightarrow \quad I = \frac{1}{3}\sigma_0 2\pi [b^3 - a^3] \quad \ldots(iv)$$

Now, radius of gyration,

$$K = \sqrt{\frac{I}{M}} = \sqrt{\frac{\frac{2\pi\sigma_0}{3}(b^3 - a^3)}{2\pi\sigma_0 (b-a)}}$$

$$\Rightarrow \quad K = \sqrt{\frac{1}{3}\frac{(b^3 - a^3)}{b-a}}$$

As we know,

$$b^3 - a^3 = (b-a)(b^2 + a^2 + ab)$$

$$\therefore \quad K = \sqrt{\frac{1}{3}(b^2 + a^2 + ab)}$$

$$\text{or} \quad K = \sqrt{\frac{a^2 + b^2 + ab}{3}}$$

36. *(a)*

key idea This problem will be solved by applying parallel axis theorem, which states that moment of inertia of a rigid body about any axis is equals to its moment of inertia about a parallel axis through its centre of mass plus the product of the mass of the body and the square of the perpendicular distance between the axis.

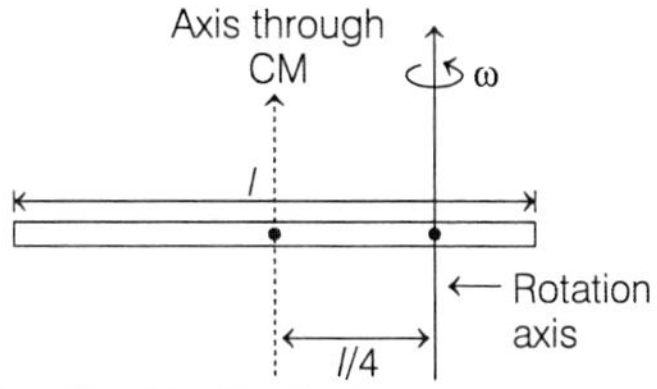

We know that moment of inertia (MI) about the principle axis of the sphere is given by

$$I_{\text{sphere}} = \frac{2}{5}MR^2 \quad \ldots(i)$$

Using parallel axis theorem, moment of inertia about the given axis in the figure below will be

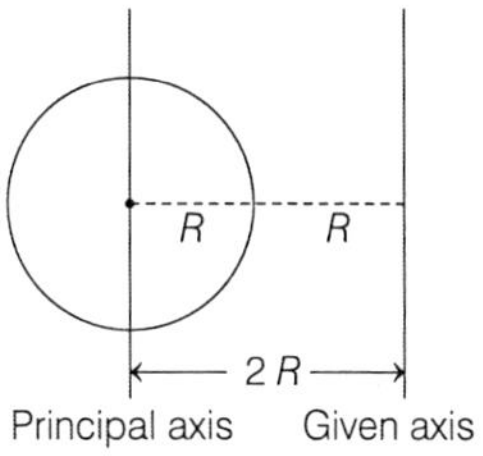

$$I_1 = \frac{2}{5}MR^2 + M(2R)^2$$

$$I_1 = \frac{22}{5}MR^2 \quad \ldots(ii)$$

Considering both spheres at equal distance from the axis, moment of inertia due to both spheres about this axis will be

$$2I_1 = 2 \times 22/5\, MR^2$$

Now, moment of inertia of rod about its perpendicular bisector axis is given by

$$I_2 = \frac{1}{12}ML^2$$

Here, given that

$$L = 2R$$

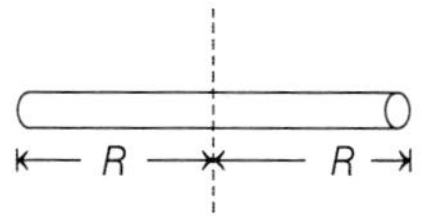

$$\therefore \quad I_2 = \frac{1}{12}M(2R)^2 = \frac{1}{3}MR^2 \ldots \text{(iii)}$$

So, total moment of inertia of the system is

$$I = 2I_1 + I_2$$

$$= 2 \times \frac{22}{5}MR^2 + \frac{1}{3}MR^2$$

$$\Rightarrow \quad I = \left(\frac{44}{5} + \frac{1}{3}\right)MR^2 = \frac{137}{15}MR^2$$

37. *(b)*

key idea When a rod is pivoted at any point, its angular acceleration is given by $\tau_{net} = I\alpha$.

Suppose the mass of the ΔABC be M and length of the side be l.

When the ΔDEF is being removed from it, then the mass of the removed Δ will be $M/4$ and length of its side will be $l/2$ as shown below

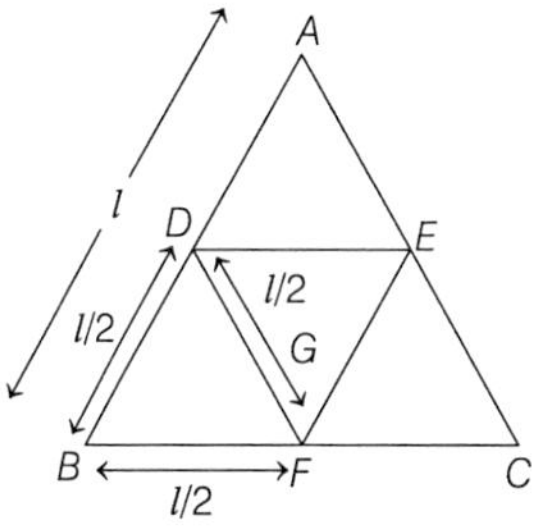

Since we know, moment of inertia of the triangle about the axis, passing through its centre of gravity is, $I = kml^2$, where k is a constant.

Then for ΔDEF, moment of inertia of the triangle about the axis,

$$I = k\left(\frac{M}{4}\right)\left(\frac{l}{2}\right)^2 = \frac{kMl^2}{16} \quad \ldots(i)$$

Moment of inertia of ΔABC is

$$I_0 = kMl^2 \quad \ldots(ii)$$

The moment of inertia of the remaining part will be

$$I' = I_0 - I = kMl^2 - \frac{kMl^2}{16}$$

[$\because$ using Eqs. (i) and (ii)]

$$= \frac{15kMl^2}{16}$$

$$\text{or} \quad I' = \frac{15}{16}I_0$$

38. *(c)* For disc D_1, moment of inertia across axis OO' will be

$$I_1 = \frac{1}{2}MR^2 \quad \ldots\text{(i)}$$

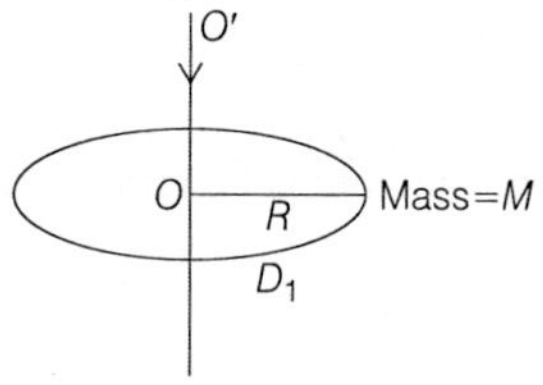

For discs D_2 and D_3, OO' is an axis parallel to the diameter of disc. Using parallel axis theorem,

$$I_2 = I_3 = I_{\text{diameter}} + Md^2 \quad \ldots\text{(ii)}$$

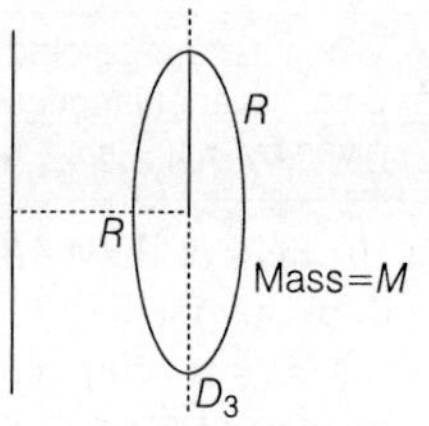

Here, $I_{\text{diameter}} = \frac{1}{4}MR^2$ and $d = R$

$$\therefore\ I_2 = I_3 = \frac{1}{4}MR^2 + MR^2 = \frac{5}{4}MR^2$$

Now, total MI of the system

$$I = I_1 + I_2 + I_3 = \frac{1}{2}MR^2 + 2 \times \frac{5}{4}MR^2$$

$$= 3MR^2$$

39. *(d)* Moment of inertia of an outer disc about the axis through centre is

$$= \frac{MR^2}{2} + M(2R)^2$$

$$= MR^2\left(4 + \frac{1}{2}\right) = \frac{9}{2}MR^2$$

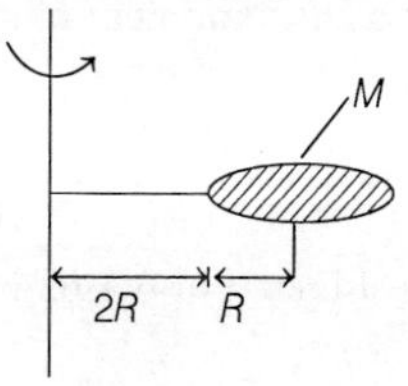

For 6 such discs,
Moment of inertia

$$= 6 \times \frac{9}{2}MR^2 = 27MR^2$$

So, moment of inertia of system

$$= \frac{MR^2}{2} + 27MR^2 = \frac{55}{2}MR^2$$

Hence, $I_P = \frac{55}{2}MR^2 + (7M \times 9R^2)$

$$\Rightarrow \quad I_P = \frac{181}{2}MR^2$$

$$I_{\text{system}} = \frac{181}{2}MR^2$$

40. *(d)* MI of a solid cylinder about its perpendicular bisector of length is

$$I = m\left(\frac{l^2}{12} + \frac{R^2}{4}\right)$$

$$\Rightarrow \quad I = \frac{mR^2}{4} + \frac{ml^2}{12} = \frac{m^2}{4\pi\rho l} + \frac{ml^2}{12}$$

$$[\because \rho\pi R^2 l = m]$$

For I to be maximum,

$$\frac{dI}{dl} = -\frac{m^2}{4\pi\rho}\left(\frac{1}{l^2}\right) + \frac{ml}{6} = 0$$

$$\Rightarrow \frac{m^2}{4\pi\rho} = \frac{ml^3}{6}$$

Now, putting $m = \rho\pi R^2 l$

$$\therefore \quad l^3 = \frac{3}{2\pi\rho} \cdot \rho\pi R^2 l \Rightarrow \frac{l^2}{R^2} = \frac{3}{2}$$

$$\therefore \quad \frac{l}{R} = \sqrt{\frac{3}{2}}$$

41. *(c)*

> **key idea** Use geometry of the figure to calculate mass and side length of the cube in terms of M and R respectively.

Consider the cross-sectional view of a diametric plane as shown in the adjacent diagram.

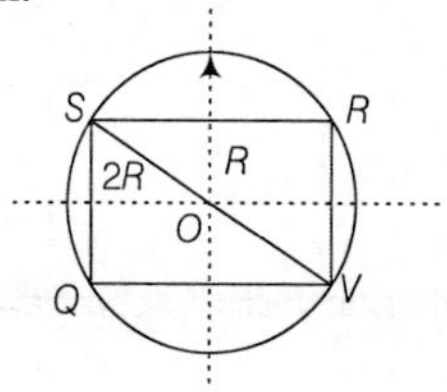

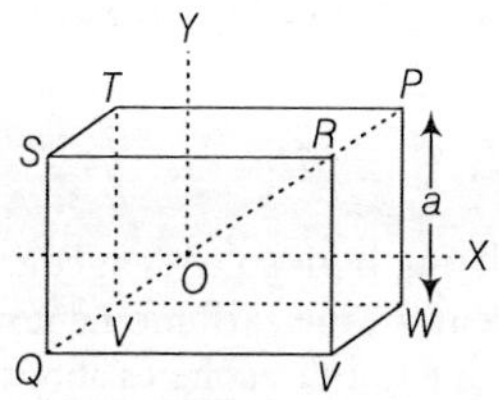

Using geometry of the cube

$$PQ = 2R = (\sqrt{3})\,a \text{ or } a = \frac{2R}{\sqrt{3}}$$

Volume density of the solid sphere

$$\rho = \frac{M}{(4/3)\,\pi R^3} = \frac{3}{4\pi}\left(\frac{M}{R^3}\right)$$

Mass of cube $(m) = (\rho)\,(a)^3$

$$= \left(\frac{3}{4\pi} \times \frac{M}{R^3}\right)\left[\frac{2R}{\sqrt{3}}\right]^3$$

$$= \frac{3M}{4\pi R^3} \times \frac{8R^3}{3\sqrt{3}} = \frac{2M}{\sqrt{3}\pi}$$

Moment of inertia of the cube about given axis is

$$I_Y = \frac{m}{12}(a^2 + a^2) = \frac{ma^2}{6}$$

$$\Rightarrow I_Y = \frac{ma^2}{6} = \frac{2M}{\sqrt{3}\pi} \times \frac{1}{6} \times \frac{4R^2}{3} = \frac{4MR^2}{9\sqrt{3}\pi}$$

42. *(1)* Angular acceleration, $\alpha = 2\ \text{rad/s}^2$

$R = 0.2$ m and $F = 10$ N

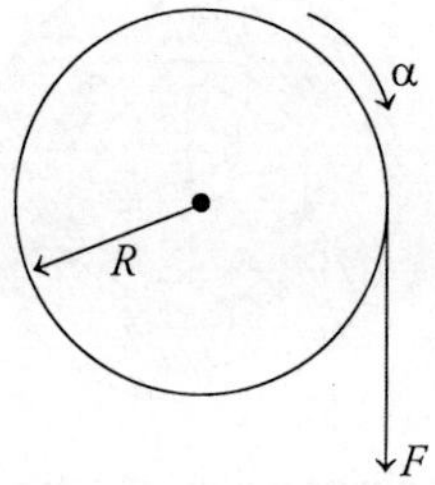

According to given situation,

$$\tau = I\alpha$$

$$\Rightarrow \quad FR = I\alpha$$

$$\Rightarrow \quad I = \frac{FR}{\alpha} = \frac{10 \times 0.2}{2} = 1\ \text{kg-m}^2$$

43. *(9)* Moment of inertia without cavity,

$$I_1 = \frac{MR^2}{2}$$

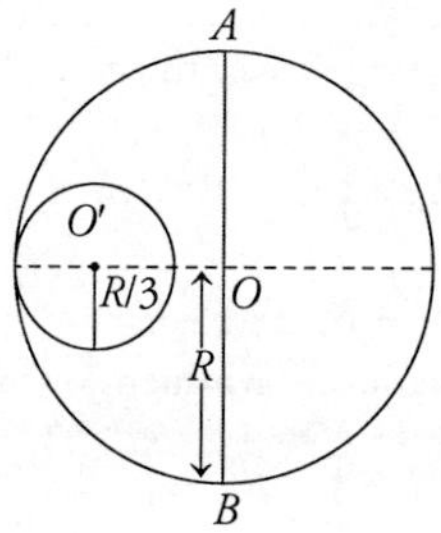

Mass of removed disc $= \frac{M}{\pi R^2} \times \left(\frac{R}{3}\right)^2 \pi$

$$= \left(\frac{M}{9}\right)$$

MI of removed disc

$$I_2 = \frac{\frac{M}{9}\left(\frac{R}{3}\right)^2}{2} + \frac{M}{9} \times \left(\frac{2R}{3}\right)^2 = \frac{MR^2}{18}$$

$$I = I_1 - I_2 = \frac{MR^2}{2} - \frac{MR^2}{18} = \frac{4MR^2}{9}$$

$$= \frac{4}{x}MR^2 \quad \text{(given)}$$

$$\therefore \quad x = 9$$

44. *(199)*

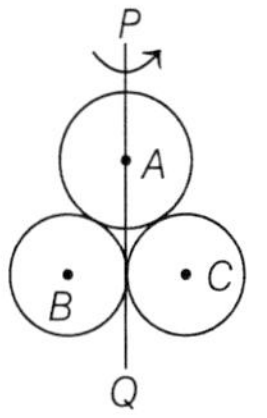

$I = \frac{MR^2}{4}$

All bodies have same mass and same radius

$A \to$ disc

$B \to$ solid sphere

$C \to$ spherical shell

$$I_{\text{disc}} = \frac{MR^2}{4}, I_{\text{solid}} = \frac{2}{5}MR^2$$

$$I_{\text{shell}} = \frac{2}{3}MR^2$$

$$I_{PQ} = \frac{MR^2}{4} + \left(\frac{2}{5}MR^2 + MR^2\right)$$

$+\left(\frac{2}{3}MR^2 + MR^2\right) \to$ from parallel axis theorem

$$= \frac{189}{15}\left(\frac{MR^2}{4}\right) = \frac{199}{15}(I)$$

$x = 199$

45. *(16)* Given, $R_2 = 2R_1$

$$M_1 = \sigma \times \pi R_1^2 = M_0$$

$$M_2 = \sigma \times \pi R_2^2 = \sigma \times \pi[2R_1]^2$$

$$= \sigma \times 4\pi R_1^2 = 4M_0$$

$$\frac{I_1}{I_2} = \frac{\frac{M_1R_1^2}{2}}{\frac{M_2R_2^2}{2}} = \frac{M_1R_1^2}{M_2R_2^2} = \frac{1}{4} \times \frac{1}{4} = \frac{1}{16}$$

$x = 16$

46. *(4)* $I_{\text{disc}} = \frac{MR_1^2}{4}, I_{\text{ring}} = \frac{MR_2^2}{2},$

$$I_{\text{sphere}} = \frac{2MR_1^2}{5}$$

According to the question,

$$\frac{I_{\text{disc}}}{I_{\text{ring}}} = 2.5 \Rightarrow \frac{\frac{MR_1^2}{4}}{\frac{MR_2^2}{2}} = \frac{5}{2}$$

$$\Rightarrow \frac{R_1^2}{R_2^2} = 5 \quad \ldots\text{(i)}$$

Also, we are given that

$$\frac{I_{\text{sphere}}}{I_{\text{ring}}} = n \Rightarrow \frac{\frac{2MR_1^2}{5}}{\frac{MR_2^2}{2}} = n$$

$$\Rightarrow \frac{4R_1^2}{5R_2^2} = n \quad \ldots\text{(ii)}$$

From Eqs. (i) and (ii)

$$n = 4$$

47. *(4)* According to question

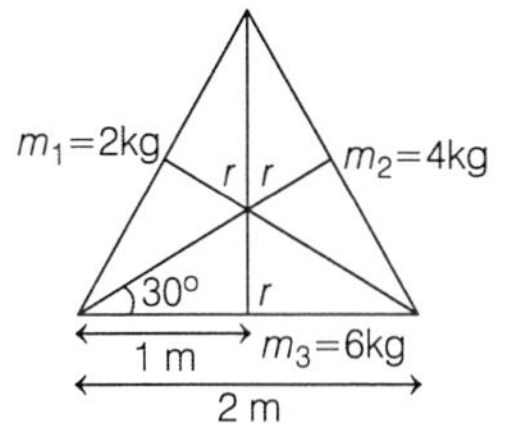

$$\frac{r}{1} = \tan 30° \therefore r = \frac{1}{\sqrt{3}} \text{ m}$$

$$\therefore \quad I = m_1r_1^2 + m_2r_2^2 + m_3r_3^2$$

$$= m_1r^2 + m_2r^2 + m_3r^2$$

$$= (m_1 + m_2 + m_3)r^2$$

$$= (2 + 4 + 6)\left(\frac{1}{\sqrt{3}}\right)^2$$

$$= \frac{12}{3} = 4 \text{ kgm}^2$$

48. *(53)*

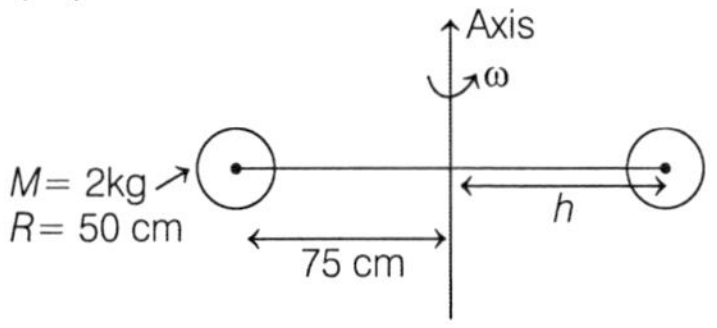

MI of system

$I = 2 \times$ MI of each sphere about axis shown

$$I = 2\left[\frac{2}{5}MR^2 + Mh^2\right]$$

$h =$ distance of centre of sphere and axis of rotation

$$\Rightarrow \quad I = 2\left[\frac{2}{5} \times 2 \times (50 \times 10^{-2})^2 + 2 \times (75 \times 10^{-2})^2\right]$$

$$= 2\left[\frac{2 \times 2 \times 50 \times 50}{5 \times 10^4} + \frac{2 \times 75 \times 75}{10^4}\right]$$

$$\Rightarrow I = \frac{2}{10^4}\left[\frac{10000}{5} + 11250\right]$$

$$= \frac{53}{20} \text{ kg-m}^2$$

$\therefore x = 53$

49. *(16)*

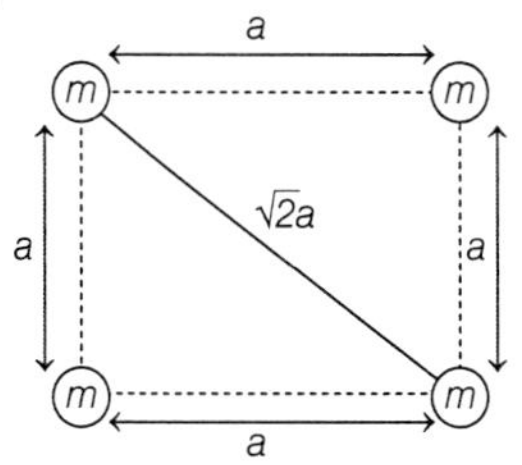

Moment of inertia of system is,

$$I = \Sigma mr^2 = ma^2 + ma^2 + m(\sqrt{2}a)^2$$

$$= ma^2 + ma^2 + 2ma^2 = 4ma^2$$

$$I = 4 \times 1 \times (2)^2$$

$$= 4 \times 1 \times 4 = 16 \text{ kg/m}^2$$

50. *(176)* Using parallel axis theorem

$$I_{\text{sys}} = \left(\frac{2}{5}mr^2 + md^2\right) \times 2$$

Here, $r = 10$ cm $= 0.1$ m

$$d = \frac{40}{2} \text{ cm} = 0.2 \text{ m}$$

$$I_{\text{sys}} = \left(\frac{2}{5} \times 2 \times 0.01 + 2 \times 0.04 \times 2\right)$$

$$= (0.008 + 0.08) \times 2$$

$$= 0.088 \times 2$$

$$= 176 \times 10^{-3} \text{ kg-m}^2$$

51. *(1)* Moment of inertia of a semicirular ring about an axis through centre and perpendicular to plane of ring is;

$$I = MR^2$$

And given $I = \frac{1}{x}MR^2$

$$\therefore \quad x = 1$$

52. *(5)* Given radius of gyration is same for ring and solid sphere

$$K_R = K_{SS} \Rightarrow R_R = \sqrt{\frac{2}{5}}R_{SS}$$

$$\Rightarrow \frac{R_R}{R_{SS}} = \sqrt{\frac{2}{5}} = \sqrt{\frac{2}{x}}$$

$x = 5$

53. *(35)* Moment of inertia of sphere about its diameter

$$I_1 = \frac{2}{5}mR^2$$

Moment of inertia of the sphere about its tangent,

$$I_2 = \frac{2}{5}MR^2 + MR^2 = \frac{7}{5}MR^2$$

Angular momentum about diameter,

$$L = I_1\omega = \frac{2}{5}MR^2\omega$$

$$\therefore \quad \frac{I_2}{L} = \frac{7}{2\omega} = \frac{7}{2 \times 10} = \frac{7}{20}$$

$$= \frac{7}{20} \times 100 \times 10^{-2}$$

$$= 35 \times 10^{-2} = x \times 10^{-2} \text{ (given)}$$

$\therefore \quad x = 35$

54. *(5)* Given that, both discs have same mass.

$$M_1 = M_2$$

$$\rho_1 V_1 = \rho_2 V_2$$

$$\rho_1 \pi R_1^2 t_1 = \rho_2 \pi R_2^2 t_2$$

or $$\frac{R_1^2}{R_2^2} = \left(\frac{\rho_2}{\rho_1}\right)\left(\frac{t_2}{t_1}\right) \quad \ldots \text{(i)}$$

Moment of inertia of a disc about diameter is given as

$$I = \frac{MR^2}{4}$$

or $$\frac{I_1}{I_2} = \frac{M_1}{M_2} \cdot \frac{R_1^2}{R_2^2}$$

$$\frac{I_1}{I_2} = \frac{M_1}{M_2} \cdot \frac{\rho_2}{\rho_1} \cdot \frac{t_2}{t_1} \text{ [from Eq. (i)]}$$

Given that,

$$\frac{M_1}{M_2} = 1, \frac{\rho_2}{\rho_1} = \frac{5}{3}, \frac{t_2}{t_1} = \frac{1}{0.5}$$

$$\frac{I_1}{I_2} = (1)\left(\frac{5}{3}\right)\left(\frac{0.5}{1}\right) = \frac{5}{6}$$

$$= \frac{x}{6} \quad \text{(given)}$$

$\therefore \quad x = 5$

55. *(3)* For disc,

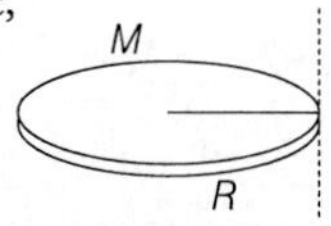

By using parallel axis theorem,

$$I = I_{CM} + Md^2$$

$$= \frac{MR^2}{2} + MR^2 \Rightarrow I = \frac{3}{2} MR^2$$

Given, $I = \frac{x}{2} MR^2$

On comparing, we get, $x = 3$

56. *(17)* $I_{CM} = \frac{1}{2} MR^2$

Using parallel axis theorem,

$$I_{AB} = I_{CM} + M\left(\frac{2}{3}R\right)^2$$

$$= \frac{1}{2} MR^2 + \frac{4}{9} MR^2 = \frac{17}{18} MR^2$$

$$\therefore \quad \frac{I_{AB}}{I_{CM}} = \frac{\frac{17}{18} MR^2}{\frac{1}{2} MR^2}$$

$$= \frac{17}{9} = \frac{x}{9} \text{ (given)}$$

$\therefore \quad x = 17$

57. *(5)* Mass of solid sphere, $M_s = 5$ kg

Mass of disc, $M_d = 4$ kg

$$R_s = R_d = R$$

Moment of inertia of disc about its diameter

$$I_{\text{dia}} = \frac{M_d R^2}{4}$$

Moment of inertia of disc about its tangent in its plane.

$$I' = I_{\text{dia}} + M_d \cdot R^2$$

According to parallel axis theorem

$$I' = \frac{M_d R^2}{4} + M_d R^2$$

$$I' = \frac{5}{4} M_d R^2 \quad \ldots \text{(i)}$$

Moment of inertia of solid sphere about its tangent,

$$I'' = \frac{2}{5} M_s R^2 + M_s R^2$$

$$= \frac{7}{5} M_s R^2 \quad \ldots \text{(ii)}$$

$\therefore$From Eqs. (i) and (ii),

$$\frac{I'}{I''} = \frac{\frac{5}{4} M_d R^2}{\frac{7}{5} M_s R^2}$$

$$= \frac{25}{28} \times \frac{M_d}{M_s} = \frac{25}{28} \times \frac{4}{5} = \frac{5}{7} = \frac{x}{7} \quad \text{(given)}$$

$\therefore \quad x = 5$

58. *(110)*

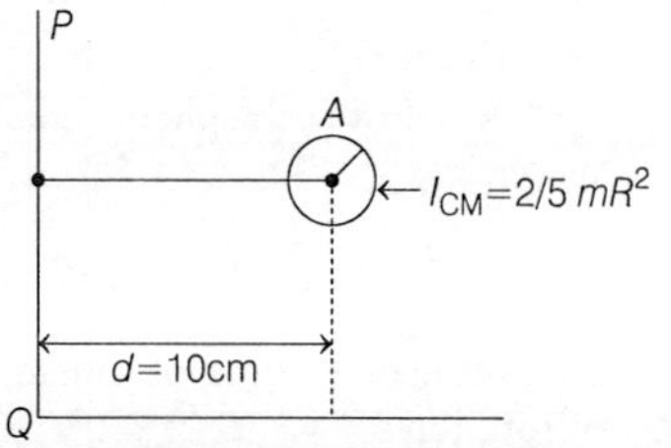

By parallel axis theorem, moment of inertia of sphere about axes PQ,

$$I_{PQ} = I_{CM} + M \times d^2$$

or $$I_{PQ} = \frac{2}{5} MR^2 + Md^2$$

$$= \frac{2}{5} M \times 5^2 + M \times 10^2$$

$$= M(10 + 100) = M \times 110$$

If radius of gyration about PQ is K, then $M \cdot K^2 = I_{PQ}$

$$\Rightarrow \quad MK^2 = M \times 110$$

$$\Rightarrow \quad K = \sqrt{110} = \sqrt{x} \quad \text{(given)}$$

$\therefore \quad x = 110$

59. *(32)*

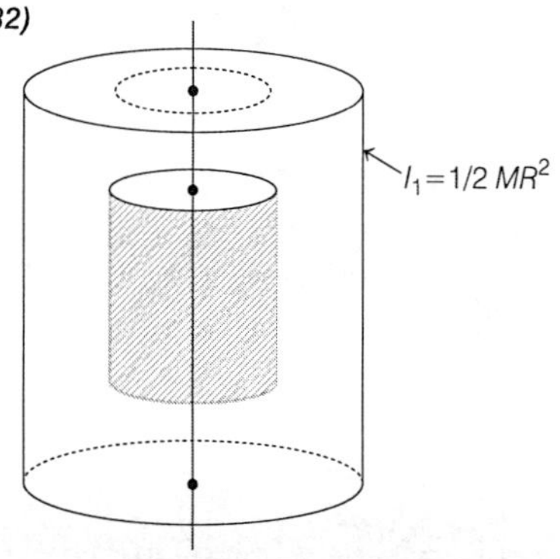

Moment of inertia of a solid cylinder about its axis, $I_1 = \frac{1}{2} MR^2$

$$= \frac{1}{2} (\pi R^2)(L)(\rho) R^2$$

$$= \frac{\pi\rho}{2} \cdot R^4 L$$

Here, M = Volume × Density

Moment of inertia of removed portion

$$I_2 = \frac{\pi\rho}{2} \cdot \left(\frac{R}{2}\right)^4 \cdot \left(\frac{L}{2}\right)$$

$$= \frac{1}{32} \cdot \left(\frac{\pi\rho}{2} R^4 L\right) = \frac{1}{32} \cdot I_1$$

So, $\frac{I_1}{I_2} = 32$

60. *(3)* The given situation is shown below

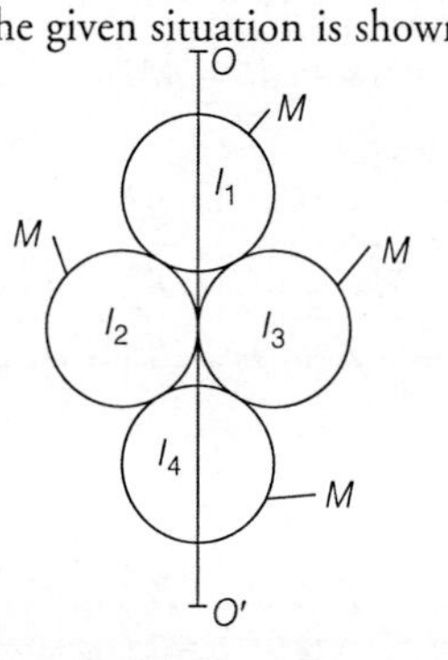

Diameter of each disc = a

Radius of each disc, $r = \frac{a}{2}$

Moment of inertia of the given system of arrangement of disc,

$$I = I_1 + I_2 + I_3 + I_4$$

But $I_1 = I_4$

and $I_3 = I_2$

$$\therefore \quad I = 2I_1 + 2I_2 = 2(I_1 + I_2)$$

$$= 2\left[\frac{Mr^2}{4} + \left(\frac{Mr^2}{4} + Mr^2\right)\right]$$

$$= 2\left[\frac{Mr^2}{4} + \frac{5Mr^2}{4}\right]$$

$$= 2 \times \frac{6Mr^2}{4} = 3Mr^2$$

$$= 3M\left(\frac{a}{2}\right)^2 = \frac{3}{4}Ma^2 = \frac{x}{4}Ma^2 \quad \text{(given)}$$

$\therefore \quad x = 3$

61. *(5)* Given, $l = 10\sqrt{3}$ m

$$I = Ml^2/12 \quad \ldots\text{(i)}$$

If K be the radius of gyration, then

$$I = MK^2 \quad \ldots\text{(ii)}$$

From Eqs. (i), and (ii), we get

$$MK^2 = Ml^2/12$$

$$\Rightarrow \quad K = \frac{l}{\sqrt{12}} = \frac{l}{2\sqrt{3}}$$

$$= \frac{10\sqrt{3}}{2\sqrt{3}} = 5 \text{ m}$$

62. *(8)* Moment of inertia of a rod about a perpendicular axis passing through one end $= I_1 = \dfrac{ML^2}{3}$

Now, same rod is bent into a ring.

Mass of ring $= M$

Circumference of ring $= 2\pi R = L$

or $\quad R = \dfrac{L}{2\pi}$

Moment of inertia of ring $= I_2 = \dfrac{MR^2}{2}$

$$= \frac{ML^2/4\pi^2}{2} = \frac{ML^2}{8\pi^2}$$

So, $\quad \dfrac{I_1}{I_2} = \dfrac{ML^2}{3} \times \dfrac{8\pi^2}{ML^2} = \dfrac{8}{3}\pi^2$

$$= \frac{x\pi^2}{3} \quad \text{(given)}$$

So, $\quad x = 8$

63. *(8)* Given, mass of uniform bar, $m = 6$ kg

Length of bar, $l = 2.4$ m

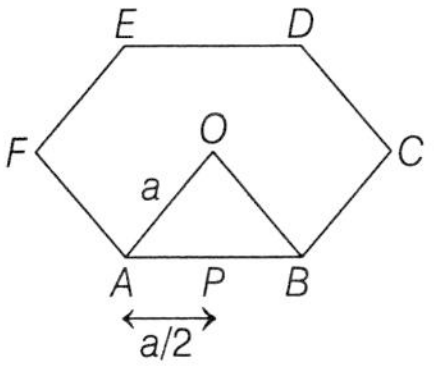

Side of hexagon, $a = \dfrac{2.4}{6} = 0.4$ m

Mass of each side,

$$m' = 6/6 = 1 \text{ kg}$$

and $\quad OP = \sqrt{a^2 - a^2/4} = a\sqrt{3}/2$

$$= 0.2\sqrt{3}\text{m}$$

Now, by using parallel axis theorem,

$$I_{OP} = \frac{m'a^2}{12} + m'\left(\frac{\sqrt{3}\,a}{2}\right)^2$$

$$= \frac{a^2}{12} + \frac{3a^2}{4}$$

$$= \frac{10a^2}{12} = \frac{5a^2}{6}$$

$$\Rightarrow \quad I_{OP} = \frac{5}{6} \times 0.4 \times 0.4$$

and I_{net} (net moment of inertia at O)

$$= 6 \times \frac{5}{6} \times 0.4 \times 0.4$$

$$= 8 \times 10^{-1} \text{ kg-m}^2$$

64. *(25)*

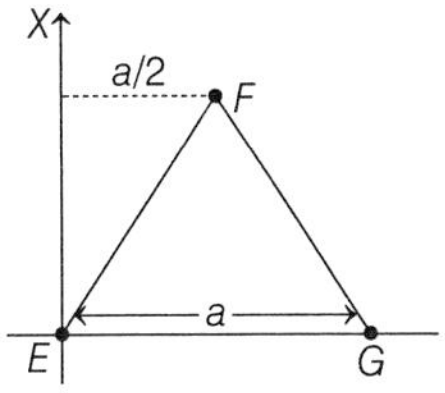

Moment of inertia of system

$$= \sum_{i=1}^{3} m_i r_i^2$$

$$= m_1 r_1^2 + m_2 r_2^2 + m_3 r_3^2$$

$$= m \times 0 + m \times a^2 + m \times \left(\frac{a}{2}\right)^2$$

$$= \frac{5}{4}ma^2 = \frac{25}{20}ma^2 = \frac{N}{20}ma^2$$

$\therefore \quad N = 25$

Topic 3

01. *(d)*

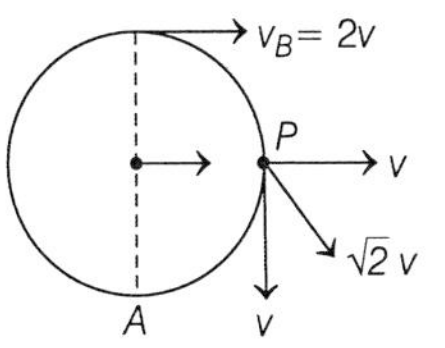

If $v_B = 2v$

Point A is instantaneous centre of rotation,

Given, $v_B = 8$ m/s $\Rightarrow 2v = 8$

$v = 4$ m/s

$v = 4$ m/s

$v_p = \sqrt{2}v = 4\sqrt{2}$ m/s

02. *(b)* Linear acceleration is,

$$a = \frac{g\sin\theta}{1 + \dfrac{I}{mR^2}}$$

$$a = \frac{\dfrac{g}{\sqrt{2}}}{1 + \dfrac{1}{2}} = \frac{2 \times \dfrac{g}{\sqrt{2}}}{3} = \frac{\sqrt{2}g}{3}$$

03. *(b)* The given situation is shown below.

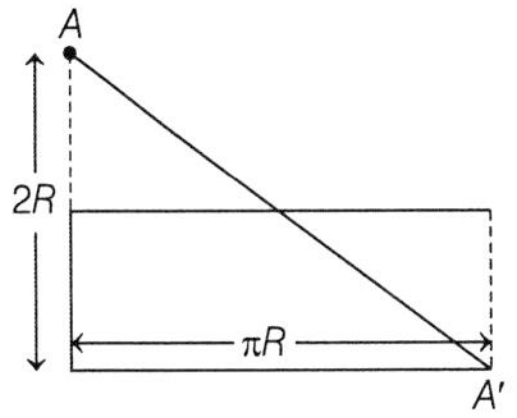

A is the initial top most position of disc.

A' is the final position of disc after completing half rotation.

$\therefore$ Required displacement,

$$AA' = \sqrt{(\pi R)^2 + (2R)^2}$$

$$= R\sqrt{\pi^2 + 4}$$

04. *(b)* It is given that $K_{\text{rotational}} = 50\%$ of $K_{\text{translational}}$

$$\Rightarrow \quad \frac{1}{2}I\omega^2 = \frac{1}{2}\left(\frac{1}{2}mv^2\right) \quad \ldots\text{(i)}$$

$\because$ This case is of pure rolling, therefore

$$v = \omega R \quad \ldots\text{(iii)}$$

From Eqs. (i) and (ii), we get

$$\frac{1}{2}I\omega^2 = \frac{1}{4}mR^2\omega^2$$

$$\Rightarrow \quad I = \frac{1}{2}mR^2$$

As this the value of moment of inertia is for solid cylinder. Therefore, the body is solid cylinder.

05. *(b)* Moment of inertia of wheel, $I = mR^2$

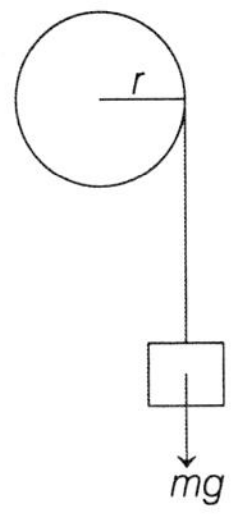

Let v be the velocity and

$v = r\omega \Rightarrow \omega = v/r$

By using law of conservation of energy,

$$1/2(I + mr^2)\omega^2 = mgh$$

$$\Rightarrow \quad \omega^2 = \frac{2mgh}{I + mr^2}$$

06. *(c)*

> **key idea** The rectangular box rotates due to torque of weight about its centre of mass.

Now, angular impulse of weight

= Change in angular momentum.

$$\therefore mg\frac{l}{2}\times\tau=\frac{ml^2}{3}\omega \Rightarrow \omega=\frac{3g\times\tau}{2\times l}$$

Substituting the given values, we get

$$=\frac{3\times 10\times 0.01}{2\times 0.3}=0.5 \text{ rad s}^{-1}$$

Time of fall of box,

$$t=\sqrt{\frac{2h}{g}}=\sqrt{\frac{2\times 5}{10}}\approx 1 \text{ s}$$

So, angle turned by box in reaching ground is $\theta=\omega t=0.5\times 1=0.5$ rad

07. *(d)* As the rod rotates in vertical plane so a torque is acting on it, which is due to the vertical component of weight of rod.

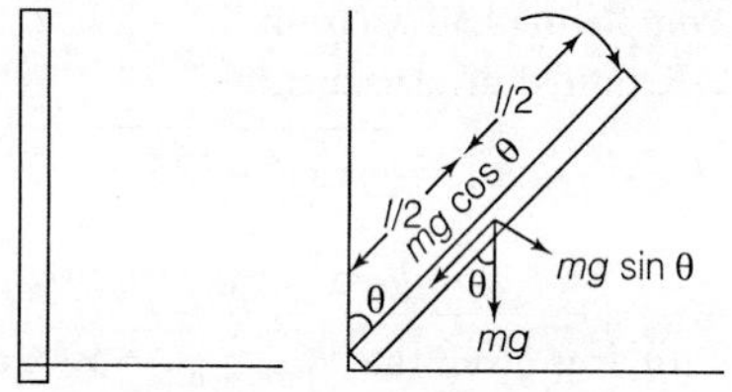

Initial condition   At any time t

Now, torque τ = force × perpendicular distance of line of action of force from axis of rotation $= mg\sin\theta\times\frac{l}{2}$

Again, torque $\tau=I\alpha$

where, I = moment of inertia $=ml^2/3$

α = angular acceleration

$$\therefore mg\sin\theta\times\frac{l}{2}=\frac{ml^2}{3}\alpha$$

$$\therefore \quad \alpha=\frac{3g\sin\theta}{2l}$$

08. *(a)* As, the wheel rolls forward the radius of the wheel decreases along AB, hence for the same number of rotations it moves less distance along AB, hence it turns left.

09. *(b)* For the mass m, $mg-T=ma$

As we know, $a=R\alpha$

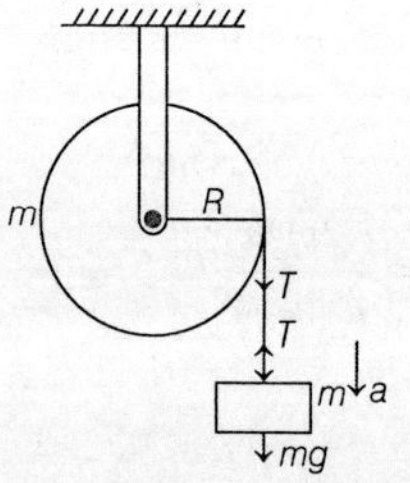

So, $mg-T=mR\alpha$ ...(i)

As torque $=T\times R=mR^2\alpha$ ...(ii)

From Eqs. (i) and (ii), we get

$$a=g/2$$

10. *(c)* Moment of inertia, I

$$=\frac{1}{2}MR^2+mx^2$$

where, x = distance of insect from centre. Clearly, as the insect moves along the diameter of the disc, Moment of inertia first decreases, then increases. By conservation of angular momentum, angular speed first increases, then decreases.

11. *(d)* By conservation of energy,

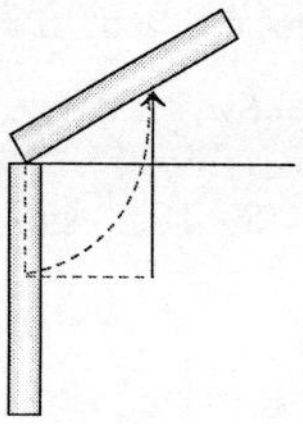

Initial total energy = Final total energy

$$TE_i=TE_f$$

$$\frac{1}{2}I\omega^2=mgh$$

$$\frac{1}{2}\times\frac{1}{3}ml^2\omega^2=mgh$$

$$\Rightarrow \quad h=\frac{1}{6}\frac{l^2\omega^2}{g}$$

12. *(d)* Since, the inclined plane is frictionless, then there will be no rolling and the mass will only slide down. Hence, acceleration is same for all the given bodies.

13. *(d)* Since, sphere rolls without slipping,

Thus, $F\times 2r=I\alpha$

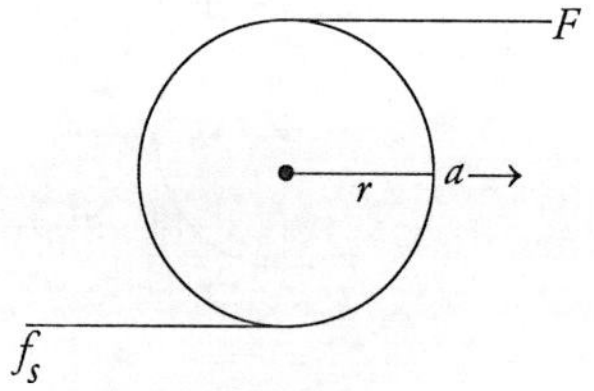

$$\Rightarrow \quad 49\times 2r=\frac{7}{5}mr^2\alpha$$

$$\Rightarrow \quad 49\times 2r=\frac{7}{5}\times 20\times r^2\alpha$$

$$\Rightarrow \quad r\alpha=\frac{7}{2}$$

But $a=r\alpha=\frac{7}{2}=3.5 \text{ m/s}^2$

14. *(d)*

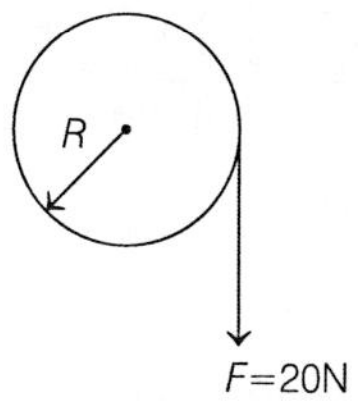

$$I=MR^2=10\times(0.1)^2$$

$$=0.1 \text{ kg m}^2$$

$$\Delta K=20 \text{ J}=\frac{1}{2}I\omega^2$$

$$20=\frac{1}{2}\times 0.1\times\omega^2$$

$$\omega^2=\frac{40}{0.1}=400$$

$$\Rightarrow \omega=20 \text{ rad/s}$$

15. *(d)* $\frac{\text{Linear KE}}{\text{Rotational KE}}=\frac{\frac{1}{2}mv_{cm}^2}{\frac{1}{2}I\omega^2}$

$$=\frac{\frac{1}{2}mv_{cm}^2}{\frac{1}{2}\times\frac{2}{5}mR^2\left(\frac{v}{R}\right)^2} \quad (v=R\omega)$$

$$=\frac{mv_{cm}^2}{\frac{2}{5}mR^2\frac{v^2}{R^2}}$$

$$=\frac{1}{\frac{2}{5}}=\frac{5}{2}$$

16. *(c)*

> **key idea** Use conservation of angular momentum to solve such type of problem.

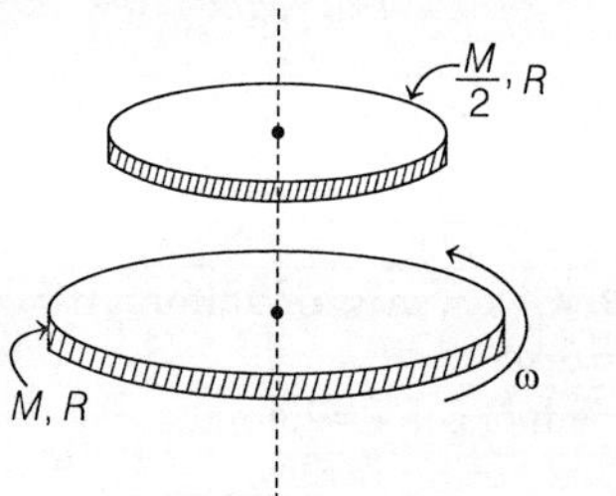

As there is no external torque, angular momentum of system is conserved.

$$\Rightarrow \quad I_1\omega_1=I_2\omega_2 \quad ...(i)$$

Here I_1 = Moment of inertia of disc of mass M and radius $R=MR^2/2$

ω_1 = Angular speed of disc 1 = ω

I_2 = Moment of inertia of both discs combined.

$$= \frac{1}{2}\left(M + \frac{M}{2}\right) \times R^2 = \frac{3}{4} MR^2$$

So, from Eq. (i)

$$\frac{MR^2}{2} \times \omega = \frac{3}{4} MR^2 \times \omega_2$$

or $$\omega_2 = \frac{2}{3}\omega$$

17. *(b)*

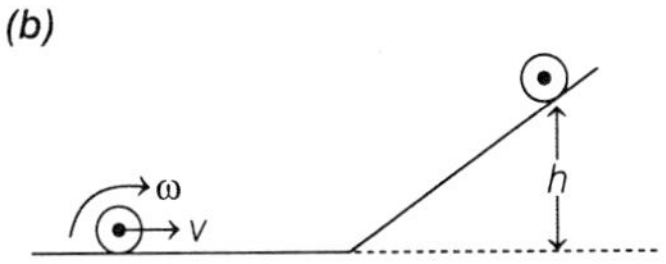

Energy conservation gives

$$\frac{1}{2}mv^2 + \frac{1}{2}I\omega^2 = mgh.$$

For disc, $I = \frac{mr^2}{2}$ and $v = r\omega$, hence disc is not slipping.

So, we have

$$\frac{1}{2}mv^2 + \frac{1}{2}\frac{mr^2}{2} \times \frac{v^2}{r^2} = mgh$$

$$\Rightarrow \left(\frac{1}{2} + \frac{1}{4}\right)mv^2 = mgh$$

or $$h = \frac{3v^2}{4g}$$

18. *(c)*

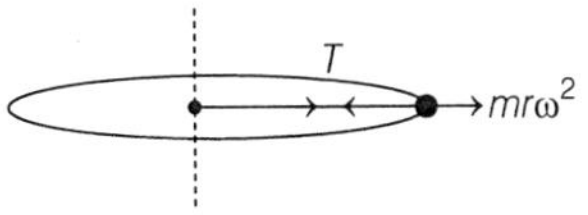

Tension provides necessary centripetal pull for rotation.

$$\therefore \quad T = mr\omega^2$$

or $$\omega^2 = \frac{T}{mr}$$

$$= \frac{400}{0.5 \times 50 \times 10^{-2}}$$

$$\Rightarrow \quad \omega^2 = 1600$$

$$\Rightarrow \quad \omega = 40 \text{ rad/s}$$

19. *(a)* Given, Mass of stone,

$$m = 900 \text{ g} = 0.9 \text{ kg}$$

$$r = 1 \text{ m}$$

$$f = 10\text{rpm} = \frac{10}{60}\text{rps} = \frac{1}{6}\text{rps}$$

Tension at the lowest point,

$$T = mg + mr\omega^2$$

$$= mg + mr(2\pi f)^2$$

$$= mg + mr(4\pi^2 f^2)$$

$$= mg + 4\pi^2 mrf^2$$

$$= 0.9 \times 9.8 + 4 \times 9.8 \times 0.9 \times 1 \times \frac{1}{36}$$

$$= 8.82 + 0.98 = 9.8 \text{ N}$$

20. *(b)* Given,

$$f_r = m\omega^2 r$$

$$\Rightarrow \mu mg = m\omega^2 r = \text{constant}$$

$$\omega^2 r = \text{constant}$$

$$\omega_1^2 r_1 = \omega_2^2 r_2$$

given, $r_1 = 1$, $\omega_2 = \frac{\omega}{2}$ and $\omega_1 = \omega$

$$\therefore \quad \omega^2(1) = \left(\frac{\omega}{2}\right)^2 \cdot r_2$$

$$\Rightarrow \quad r_2 = 4 \text{ cm}$$

21. *(c)* Angular momentum about origin = Angular momentum due to translation + Angular momentum due to rotation

$$\Rightarrow \quad L = L_T + L_R = MvR + I\omega$$

Now, as $v = R\omega$ and $I = \frac{2}{3}MR^2$

So, $$L = MR^2\omega + \frac{2}{3}MR^2\omega$$

$$= \frac{5}{3}MR^2\omega$$

As $$L = \frac{a}{3}MR^2\omega \qquad \text{(given)}$$

So, $$a = 5$$

22. *(b)* When, $t = 1$ s, then $\theta = 5$ rad

$$\omega_0 = 0 \text{ rad/s},$$

By the equation of rotational motion,

$$\theta = \omega_0 t + \frac{1}{2}\alpha t^2$$

$$\Rightarrow \quad 5 = 0 \times 1 + \frac{1}{2}\alpha \times 1^2$$

$$\Rightarrow \quad 5 = \alpha / 2$$

$$\Rightarrow \quad \alpha = 10 \text{ rad/s}^2$$

If θ' be angle rotated by wheel in $t' = 2$s, then

$$\theta' = \omega_0 t' + \frac{1}{2}\alpha(t')^2$$

$$= 0 \times 2 + \frac{1}{2} \times 10 \times (2)^2$$

$$\theta' = 20 \text{ rad}$$

$\therefore$ Angle rotated by wheel is $\Delta\theta = \theta' - \theta$

$$= 20 - 5 = 15 \text{ rad}$$

23. *(c)* Decrease in potential energy of bob = kinetic energy of bob + kinetic energy of disc

$$\Rightarrow \quad mgh = \frac{1}{2}mv^2 + \frac{1}{2}I\omega^2 \qquad \text{...(i)}$$

If there is no slip, then $v = r\omega$

Also, moment of inertia of disc,

$$I = \frac{mr^2}{2}$$

Substituting these in Eq. (i), we get

$$mgh = \frac{1}{2}m(r\omega)^2 + \frac{1}{2} \times \frac{mr^2}{2} \times \omega^2$$

$$\Rightarrow \quad mgh = \frac{3}{4}mr^2\omega^2$$

$$\Rightarrow \quad \omega^2 = \frac{4gh}{3r^2}$$

or $$\omega = \frac{1}{r}\sqrt{\frac{4gh}{3}}$$

24. *(c)* The given situation is as shown in the figure below.

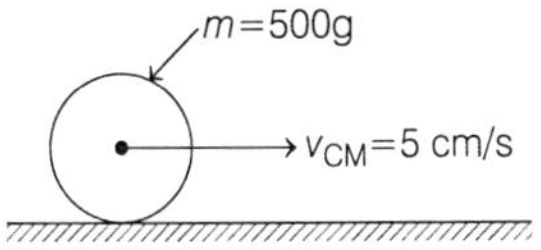

In pure rolling object possesses both translational kinetic energy and rotational kinetic energy.

So, $$\text{KE}_{\text{sphere}} = \frac{1}{2}mv_{\text{CM}}^2 + \frac{1}{2}I\omega^2$$

Here, $$I = \frac{2}{5}mr^2 \text{ and } \omega = \frac{v}{r}$$

$$[\because v_{\text{CM}} = v]$$

So, $$\text{KE}_{\text{sphere}} = \frac{1}{2}mv^2 + \frac{1}{2} \times \frac{2}{5}mr^2 \times \frac{v^2}{r^2}$$

$$= \left(\frac{1}{2} + \frac{1}{5}\right)mv^2$$

$$= \frac{7}{10} \times 500 \times 10^{-3} \times 25 \times 10^{-4}$$

$$= 8.75 \times 10^{-4} \text{ J}$$

25. *(b)* Loss of potential energy of rod = Gain of kinetic energy

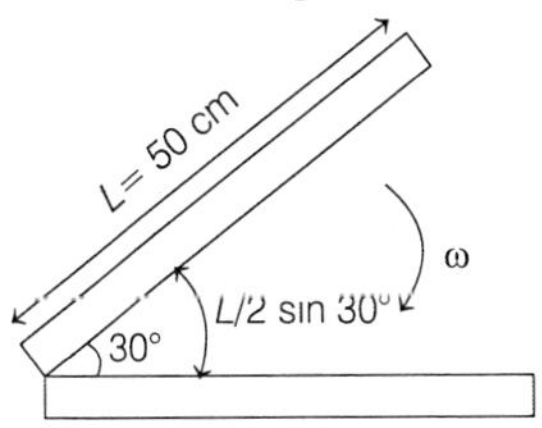

$$\therefore \quad \Delta\text{PE} = \frac{1}{2}I\omega^2$$

(where, I = moment of inertia of rod and

ω = angular frequency of rod)

$$\Rightarrow Mg \times \frac{L}{2}\sin 30° = \frac{1}{2} \times I \times \omega^2$$

$$\Rightarrow \quad Mg\frac{L}{2}\times\frac{1}{2}=\frac{1}{2}\times I\times\omega^2$$

$$\Rightarrow \quad \frac{Mg\,L\times 2}{4\times I}=\omega^2 \Rightarrow \sqrt{\frac{MgL}{4\times\frac{ML^2}{3}}\times\frac{2}{1}}=\omega$$

$$\left[\because I=\frac{ML^2}{3} \text{ for rod}\right]$$

$$\omega=\sqrt{30}\text{ rad/s}$$

26. *(d)* Given, kinetic energy $= k\theta^2$

We know that, kinetic energy of a rotating body about its axis $=\frac{1}{2}I\omega^2$

where, I is moment of inertia and ω is angular velocity.

So, $\frac{1}{2}I\omega^2=k\theta^2$ or $\omega^2=\frac{2k\theta^2}{I}$

$$\Rightarrow \quad \omega=\sqrt{\frac{2k}{I}}\,\theta \quad \ldots\text{(i)}$$

Differentiating the above equation w.r.t. time on both sides, we get

$$\frac{d\omega}{dt}=\sqrt{\frac{2k}{I}}\cdot\frac{d\theta}{dt}=\sqrt{\frac{2k}{I}}\cdot\omega$$

$$\left[\because \omega=\frac{d\theta}{dt}\right]$$

$\therefore$ Angular acceleration,

$$\alpha=\frac{d\omega}{dt}=\sqrt{\frac{2k}{I}}\cdot\omega=\sqrt{\frac{2k}{I}}\cdot\sqrt{\frac{2k}{I}}\,\theta$$

[using Eq. (i)]

or $\quad \alpha=\frac{2k}{I}\theta$

Alternate Solution

As, $\omega^2=\frac{2k\theta^2}{I} \Rightarrow 2\omega\frac{d\omega}{dt}=\frac{2k}{I}\cdot 2\theta\frac{d\theta}{dt}$

or $\omega\,d\omega=\frac{2k}{I}\theta\,d\theta \Rightarrow \omega\frac{d\omega}{d\theta}(=\alpha)=\frac{2k}{I}\cdot\theta$

or $\quad \alpha=\frac{2k}{I}\theta$

27. *(b)* Rotational kinetic energy of a body is given by

$$\text{KE}_{\text{rotational}}=\frac{1}{2}I\omega^2$$

where $\quad \omega=\omega_0+\alpha t$

So, $\text{KE}_{\text{rotational}}=\frac{1}{2}I(\omega_0+\alpha t)^2 \quad \ldots$ (i)

Here, $\quad I=1.5\text{ kgm}^2$,

KE $= 1200$ J and

$\alpha=20\text{ rad/s}^2$ and $\omega_0=0$

Substituting these values in Eq. (i), we get

$$1200=\frac{1}{2}(1.5)(20\times t)^2$$

$$\Rightarrow \quad t^2=\frac{2\times 1200}{1.5\times 400}=4$$

$\therefore \quad t=2$ s

28. *(c)* Equation of motion for solid cylinder is given as,

$$F-f=ma \quad \ldots\text{(i)}$$

but, $\quad \tau=fR=I\alpha$

For pure rolling $a=\alpha R$;

$$\therefore \quad fR=\frac{mR^2}{2}\cdot\frac{a}{R}$$

$$f=\frac{ma}{2} \quad \ldots\text{(ii)}$$

From Eqs. (i) and (ii), we get

$$F-\frac{ma}{2}=ma \Rightarrow F=\frac{3}{2}ma$$

29. *(b)* As, cylinder is rolling without slipping. Hence, torque is producing rotation about centre O.

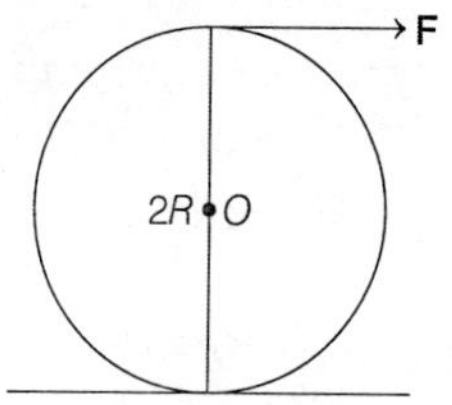

So, $\quad \tau=\mathbf{r}\times\mathbf{F}$ (Here, $r=R$)

$\theta=90°$

So, $\quad \tau=\mathbf{r}\times\mathbf{F}=RF$

or $\quad \tau=0.5\times 40=20$ N-m $\quad \ldots$(i)

If α is acceleration of centre of mass 'O' then torque is, $\tau=I\alpha$

where, $\quad I=MR^2$

$$\therefore \quad \tau=MR^2\alpha \quad \ldots\text{(ii)}$$

Comparing Eq. (i) with Eq. (ii),

$$MR^2\alpha=20$$

$$\Rightarrow \quad \alpha=\frac{20}{5\times(0.5)^2} \text{ or } \alpha=16\text{ rad/s}^2$$

30. *(a)*

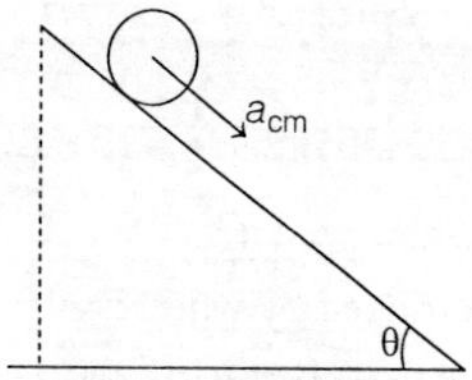

$$t=\sqrt{\frac{2l}{a_{cm}}}$$

$$a_{cm}=\frac{g\sin\theta}{1+\frac{I_{cm}}{MR^2}}$$

$$a_1=a_{cm_1}=\frac{5g\sin\theta}{7} \to \text{Solid}$$

$$\left[I=\frac{2}{5}MR^2\right]$$

$$a_2=a_{cm_2}=\frac{3g\sin\theta}{5} \to$$

$$\text{Hollow}\left[I=\frac{2}{3}MR^2\right]$$

$$a_1>a_2$$

$$\therefore \quad t_1<t_2$$

31. *(d)*

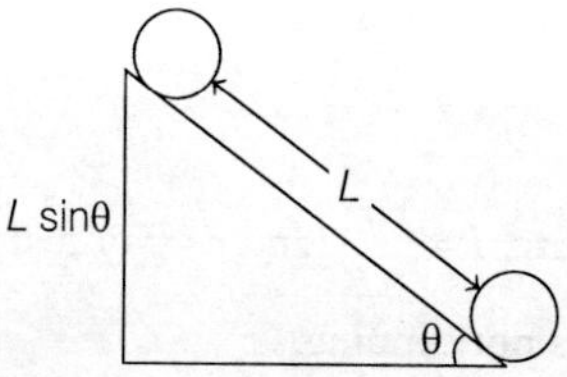

Using work-energy theorem

$W_g=K_f-K_i$

$mg\,L\sin\theta=K_f-K_i$

KE in pure rolling $=\frac{1}{2}mv_{cm}^2+\frac{1}{2}I_{cm}\omega^2$

$$=\frac{1}{2}mv^2+\frac{1}{2}\times\frac{2}{5}mR^2\frac{v^2}{R^2}=\frac{7}{10}mv^2$$

$$mgl\sin\theta=\frac{7}{10}mv_f^2-0$$

$$v_f^2\propto\sin\theta$$

$$\left(\frac{v_1}{v_2}\right)^2=\frac{\sin\theta_1}{\sin\theta_2}=\frac{\sin 30°}{\sin 45°}=\frac{1}{\sqrt{2}}$$

$$\frac{v_1^2}{v_2^2}=\frac{1}{\sqrt{2}}$$

Alternate Solution

Speed of the body at the bottom is

$$v=\sqrt{\frac{2gL\sin\theta}{1+\left(\frac{K}{r}\right)^2}}$$

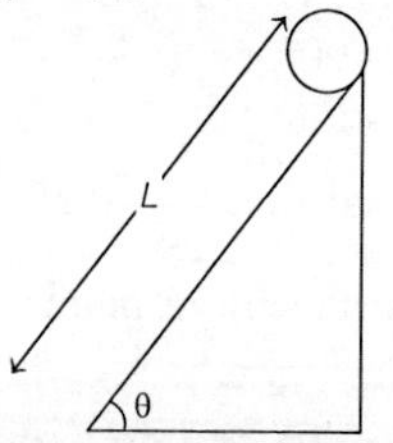

$\Rightarrow v^2\propto\sin\theta$

$$\Rightarrow \left(\frac{v_1}{v_2}\right)^2=\frac{\sin\theta_1}{\sin\theta_2}=\frac{\sin 30°}{\sin 45°}=\frac{1}{\sqrt{2}}$$

32. *(d)* When a body rolls down an inclined plane, then

$$\Rightarrow \quad Mgh = \frac{1}{2}I\omega^2 + \frac{1}{2}Mv^2$$

$$\Rightarrow \quad Mgh = \frac{1}{2}MK^2\left(\frac{v^2}{r^2}\right) + \frac{1}{2}Mv^2 \quad \left[\because \omega = \frac{v}{r}\right]$$

$$\Rightarrow \quad Mgh = \frac{1}{2}Mv^2\left(1 + \frac{K^2}{R^2}\right) \quad \ldots(i)$$

For solid cylinder, $\frac{K^2}{R^2} = \frac{1}{2}$

For solid sphere, $\frac{K^2}{R^2} = \frac{2}{5}$

Now, if we take ratio of Eq. (i) for solid cylinder and solid sphere (for both mgh term will be same), we have

$$\frac{(mgh)_{cyl}}{(mgh)_{sphere}} = \frac{\frac{1}{2}Mv_{cyl}^2\left(1 + \frac{K^2}{R^2}\right)_{cyl}}{\frac{1}{2}Mv_{sphere}^2\left(1 + \frac{K^2}{R^2}\right)_{sphere}}$$

(For both bodies, mass and height are same)

$$\Rightarrow \quad 1 = \frac{v_{cyl}^2}{v_{sphere}^2} \cdot \frac{\left(1 + \frac{1}{2}\right)}{\left(1 + \frac{2}{5}\right)}$$

$$\Rightarrow \quad \frac{v_{cyl}^2}{v_{sphere}^2} = \frac{1 + \frac{2}{5}}{1 + \frac{1}{2}} = \frac{7}{5} \times \frac{2}{3}$$

$$\Rightarrow \quad \frac{v_{cyl}}{v_{sphere}} = \sqrt{\frac{14}{15}}$$

33. *(b)* As angular acceleration is given by,

$$\alpha = \frac{d\omega}{dt}$$

$$\Rightarrow \quad d\omega = \alpha \cdot dt$$

Integrating

$$\omega = \int d\omega = \int \alpha \cdot dt$$

Here, $\alpha = 6t^2 - 2t$

So, $\omega = \int (6t^2 - 2t)dt$

$$\Rightarrow \quad \omega = \frac{6t^3}{3} - \frac{2t^2}{2} + C_1 \quad \ldots(i)$$

Now, given at $t = 0$, $\omega = 10$ rad/s

Substituting it in Eq. (i), we get

$$\Rightarrow \quad C_1 = 10$$

$\therefore$ Angular velocity,

$$\omega = \frac{6t^3}{3} - \frac{2t^2}{2} + 10 = 2t^3 - t^2 + 10$$

Also, $\omega = \frac{d\theta}{dt}$, where θ = angular position

$$\Rightarrow \quad d\theta = \omega dt$$

$$\Rightarrow \quad \theta = \int_0 d\theta = \int (2t^3 - t^2 + 10)dt$$

$$\Rightarrow \quad \theta = \frac{2t^4}{4} - \frac{t^3}{3} + 10\,t + C_2 \quad \ldots(ii)$$

Here, given at $t = 0$, $\theta = 4$ rad

Substituting in Eq. (ii), we get

$$C_2 = 4$$

So, angular position is given by

$$\theta = \frac{t^4}{2} - \frac{t^3}{3} + 10t + 4$$

34. *(b)* Rotational kinetic energy of spherical ball,

$$K_{rot} = \frac{1}{2}I\omega^2 = \frac{1}{2}\left[\frac{2}{5}MR^2\right]\left(\frac{v_{CM}}{R}\right)^2$$

$$\left(\because I_{solid\ sphere} = \frac{2}{5}MR^2 \text{ and } \omega = \frac{v_{CM}}{R}\right)$$

$$K_{rot} = \frac{1}{5}Mv_{CM}^2 \quad \ldots(i)$$

Total kinetic energy of spherical ball,

$$K_{tot} = K_{rot} + K_{trans}$$

$$= \frac{1}{5}Mv_{CM}^2 + \frac{1}{2}Mv_{CM}^2$$

$$K_{tot} = \frac{7}{10}Mv_{CM}^2 \quad \ldots(ii)$$

Dividing Eq. (i) by Eq. (ii), we get,

$$\frac{K_{rot}}{K_{rot}} = \frac{\frac{1}{5}Mv_{CM}^2}{\frac{7}{10}Mv_{CM}^2} = \frac{10}{7} \times \frac{1}{5} = \frac{2}{7}$$

35. *(c)* Given, moments of inertia of two discs are I_1, I_2.

Their initial angular velocities is ω_1 and ω_2.

Let final angular velocity is ω

By using law of conservation of angular momentum,

$$I_1\omega_1 + I_2\omega_2 = (I_1 + I_2)\omega$$

$$\Rightarrow \quad \omega = \frac{I_1\omega_1 + I_2\omega_2}{I_1 + I_2}$$

$\therefore$ Change in kinetic energy ΔKE

$$= \frac{1}{2}I_1\omega_1^2 + \frac{1}{2}I_2\omega_2^2 - \frac{1}{2}(I_1 + I_2)\omega^2$$

$$= \frac{1}{2}I_1\omega_1^2 + \frac{1}{2}I_2\omega_2^2 - \frac{1}{2}(I_1 + I_2)\left(\frac{I_1\omega_1 + I_2\omega_2}{I_1 + I_2}\right)^2$$

$$= \frac{1}{2}I_1\omega_1^2 + \frac{1}{2}I_2\omega_2^2 - \frac{1}{2}\frac{(I_1\omega_1 + I_2\omega_2)^2}{I_1 + I_2}$$

$$= \frac{1}{2}I_1\omega_1^2 + \frac{1}{2}I_2\omega_2^2 - \frac{1}{2}\frac{(I_1^2\omega_1^2 + I_2^2\omega_2^2 + 2I_1I_2\omega_1\omega_2)}{I_1 + I_2}$$

$$= \frac{\left[\frac{1}{2}I_1^2\omega_1^2 + \frac{1}{2}I_1I_2\omega_1^2 + \frac{1}{2}I_2^2\omega_2^2 + \frac{1}{2}I_1I_2\omega_2^2 - \frac{1}{2}I_1^2\omega_1^2 - \frac{1}{2}I_2^2\omega_2^2 - \frac{1}{2}2I_1I_2\omega_1\omega_2\right]}{I_1 + I_2}$$

$$= \frac{1}{2}\frac{I_1I_2(\omega_1^2 + \omega_2^2 - 2\omega_1\omega_2)}{I_1 + I_2}$$

$$= \frac{1}{2}I_1I_2\frac{(\omega_1 - \omega_2)^2}{I_1 + I_2}$$

36. *(a)* Moment of inertia of ring, solid cylinder and solid sphere about its centre of mass be

$$I_{ring} = MR^2 \quad \ldots(i)$$

$$I_{solid\ cylinder} = \frac{MR^2}{2} \quad \ldots(ii)$$

$$I_{solid\ sphere} = \frac{2}{5}MR^2 \quad \ldots(iii)$$

and speed of body to roll on inclined plane is

$$v = \sqrt{\frac{2gh}{1 + \frac{K^2}{R^2}}} \quad \ldots(iv)$$

where, K is the of gyration.

Now from Eqs. (i), (ii) and (iii), we get

$$K_{ring} = R$$

$$K_{solid\ cylinder} = \frac{R}{\sqrt{2}}$$

and $K_{solid\ sphere} = R\sqrt{\frac{2}{5}}$

Substituting these values sequently in Eq. (iv), we get

$$v_{ring} = \sqrt{\frac{2gh}{1 + \frac{R^2}{R^2}}} = \sqrt{gh}$$

$$v_{\text{solid cylinder}} = \sqrt{\frac{2gh}{1+\frac{R^2/2}{R^2}}} = \sqrt{\frac{4gh}{3}}$$

$$= \sqrt{1.33\,gh}$$

$$v_{\text{solid sphere}} = \sqrt{\frac{2gh}{1+\frac{\frac{2}{5}R^2}{R^2}}} = \sqrt{\frac{10}{7}gh}$$

$$= \sqrt{1.43\,gh}$$

$$v_{\text{solid sphere}} > v_{\text{solid cylinder}} > v_{\text{ring}}$$

37. *(c)* As we know the moment of inertia of the solid sphere, $I = \frac{2}{5}mR^2$

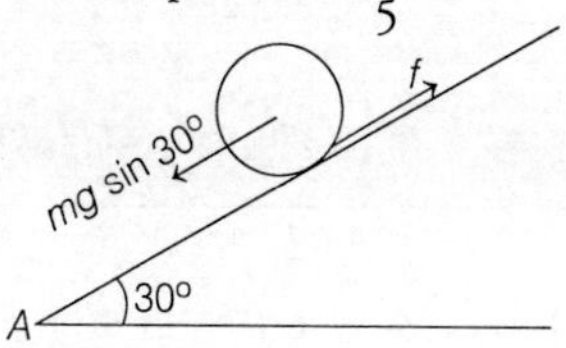

Acceleration of sphere on inclined plane,

$$a = \frac{g\sin\theta}{1+\frac{I}{mR^2}}$$

Substituting the values in the above equation, we get

$$a = \frac{g\sin 30°}{1+\frac{\frac{2}{5}mR^2}{mR^2}} \Rightarrow a = 3.5\ \text{m/s}^2$$

At the top of incline,

$$v = u + at \Rightarrow 0 = 1 - 3.5t$$

$$t = 1/3.5\ \text{s}$$

∴Total time = time of ascent + time of decent

As, time of ascent = time of decent

⇒ Total time = 2 (time of ascent)

Total time = 2(1/3.5) s

Total time = 0.57 s

38. *(c)* Given, radius of sphere is a, mass of sphere is m, horizontal speed of sphere is v_0.

Moment of inertia of solid sphere,

$$I = 2/5\,ma^2$$

Let, h be the height of inclined plane and l be the length of inclined plane.

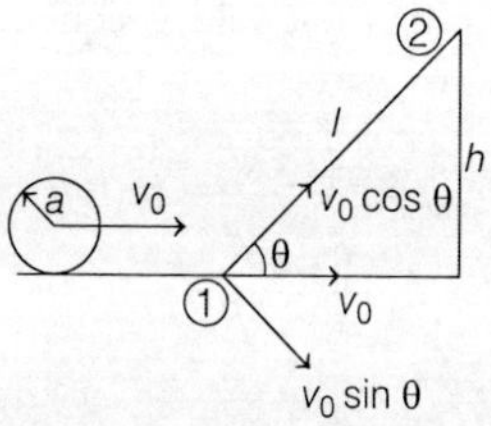

By using law of conservation of energy,

Total energy at 1 = Total energy at 2

$$\therefore \quad PE_1 + KE_1 = PE_2 + KE_2$$

$$\Rightarrow 0 + 1/2\,mv_0^2 + 1/2\,I\omega^2 = mgh \quad \ldots\text{(i)}$$

As, $v_0 = a\omega \Rightarrow \omega = \frac{v_0}{a}$

$$\therefore \frac{1}{2}mv_0^2 + \frac{1}{2}\left(\frac{2}{5}ma^2\right)\frac{v_0^2}{a^2} = mgh$$

$$\Rightarrow \quad \frac{1}{2}mv_0^2(1+2/5) = mgh$$

$$\Rightarrow \quad \frac{7v_0^2}{10} = gh \quad \ldots\text{(ii)}$$

As $\quad \sin\theta = h/l \Rightarrow h = l\sin\theta$

Put this value in Eq. (ii), we get

$$\Rightarrow \frac{7}{10}v_0^2 = gl\sin\theta \Rightarrow l = \frac{7v_0^2}{10\,g\sin\theta}$$

39. *(b)* The given situation is shown in figure.

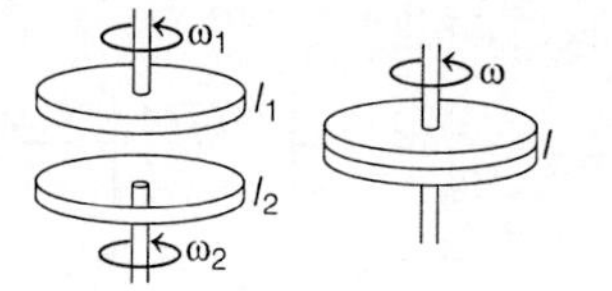

As there is no external torque, angular momentum of system is conserved

$$L_1 + L_2 = L_{\text{system}}$$

$$\Rightarrow I_1\omega_1 + I_2\omega_2 = (I_1 + I_2)\omega$$

$$\Rightarrow \omega = \frac{I_1\omega_1 + I_2\omega_2}{I_1 + I_2}$$

$$\omega = \frac{0.1\times 10 + 0.2\times 5}{(0.1+0.2)} \text{ or } \omega = \frac{20}{3}\ \text{rad/s}$$

Kinetic energy of combined system is

$$K = \frac{1}{2}I_{\text{system}}\omega^2$$

$$= \frac{1}{2}(0.1+0.2)\times\left(\frac{20}{3}\right)^2 = \frac{20}{3}\ \text{J}$$

40. *(a)* Let ω_c be the final angular speed.

According to law of conservation of angular momentum,

Initial angular momentum = Final angular momentum

$$I\omega + 0 = I\omega_c + 3I\omega_c$$

$$I\omega = 4I\omega_c \Rightarrow \omega_c = \omega/4$$

Loss of rotational kinetic energy,

$$(\Delta KE)_{\text{loss}} = \text{Initial KE} - \text{Final KE}$$

$$\Rightarrow (\Delta KE)_{\text{loss}} = \frac{1}{2}I\omega^2 - \frac{1}{2}(I+3I)\omega_c^2$$

$$= \frac{1}{2}I\omega^2 - \frac{1}{2}(I+3I)\left(\frac{\omega}{4}\right)^2$$

$$= \frac{1}{2}I\omega^2 - \frac{1}{2}(4I)\left(\frac{\omega}{4}\right)^2$$

$$= \frac{1}{2}I\omega^2\left(1-\frac{1}{4}\right) = \frac{3}{8}I\omega^2$$

The resultant fractional loss in KE of the system

$$= \frac{(\Delta KE)_{\text{loss}}}{(KE)_{\text{initial}}} = \frac{3/8\,I\omega^2}{1/2\,I\omega^2}$$

$$\left(\because (KE)_{\text{initial}} = \frac{1}{2}I\omega^2\right)$$

$$= 3/4$$

41. *(d)* Loss of gravitational potential energy of system appears in form of kinetic energy of system.

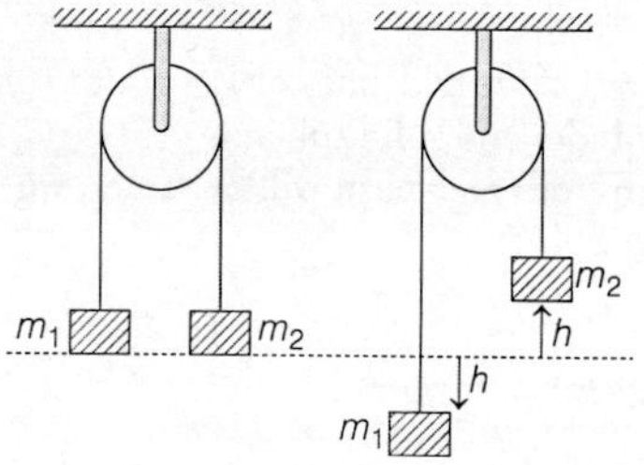

Loss of potential energy of masses = Kinetic energy of masses + Kinetic energy of rotation of wheel

$$\Rightarrow m_1gh - m_2gh = \frac{1}{2}m_1v^2 + \frac{1}{2}m_2v^2 + \frac{1}{2}I\omega^2$$

Assuming no slip, we have $v = \omega R$

$$\therefore (m_1 - m_2)gh = \frac{1}{2}(m_1 + m_2)(\omega^2R^2) + \frac{1}{2}I\omega^2$$

$$\Rightarrow \quad \omega = \sqrt{\frac{2(m_1 - m_2)gh}{\left(m_1 + m_2 + \frac{I}{R^2}\right)R^2}}$$

$$\text{or} \quad \omega = \sqrt{\frac{2(m_1 - m_2)gh}{(m_1 + m_2)R^2 + I}}$$

$$\text{or} \quad \omega = \left[\frac{2(m_1 - m_2)gh}{(m_1 + m_2)R^2 + I}\right]^{1/2}$$

42. *(a)* The given condition can be drawn in the figure below

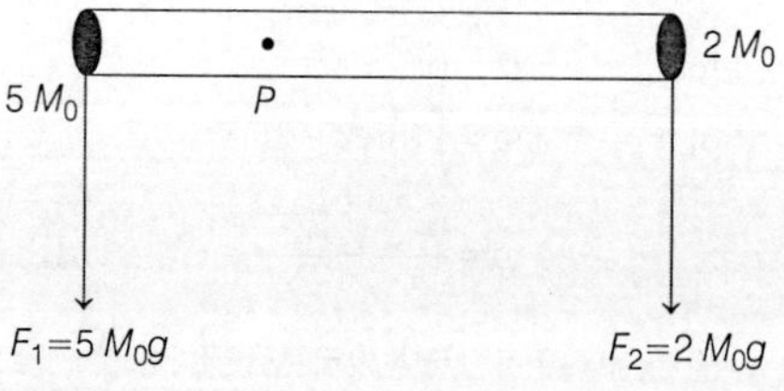

Torque (τ) about

$$P = r_1 \times F_1 + r_2 \times F_2 \quad \text{... (i)}$$

$\Rightarrow \quad \tau = l \times 5M_0 g$ (outwards)

$\quad - 2l \times 2M_0 g$ (outwards)

$\Rightarrow \quad \tau = 5M_0 gl - 4M_0 gl$ (outwards)

$\Rightarrow \quad \tau = M_0 gl$ (outwards)

or $\quad \tau = M_0 gl \quad \text{...(ii)}$

Now we know that, torque is also given by

$$\tau = I\alpha \quad \text{...(iii)}$$

Here, I = moment of inertia (w.r.t. point P) of rod and α = angular acceleration.

For point P,

$$I = (5M_0) \times l^2 + (2M_0)(2l)^2 \quad [\because I = MR^2]$$

$$\Rightarrow \quad I = 13M_0 l^2 \quad \text{...(iv)}$$

Putting value of I from Eq. (iv) in Eq. (iii), we get

$$\tau = (13M_0 l^2)\,\alpha \quad \text{...(v)}$$

From Eqs. (ii) and (v), we get

$$M_0 gl = 13M_0 l^2 \alpha \Rightarrow \alpha = g/13l$$

43. *(b)* $K_{\text{total}} = K_{\text{translation}} + K_{\text{rotation}}$

$$= \frac{1}{2}mv^2 + \frac{1}{2}I\omega^2$$

$$= \frac{1}{2}mv^2 + \frac{1}{2}I \cdot \frac{v^2}{r^2} \quad \left[\because \omega = \frac{v}{r}\right]$$

Let v be the linear velocity and R be the radius for both solid sphere and solid cylinder.

$\therefore$ Kinetic energy of the given solid sphere will be

$$K_{\text{sph}} = \frac{1}{2}mv^2 + \frac{1}{2}I_{\text{sph}}\frac{v^2}{R^2}$$

$$= \frac{1}{2}mv^2 + \frac{1}{2} \times \frac{2}{5}mR^2 \times \frac{v^2}{R^2}$$

$$= \frac{7}{10}mv^2 \quad \text{...(i)}$$

Similarly, kinetic energy of the given solid cylinder will be

$$K_{\text{cyl}} = \frac{1}{2}mv^2 + \frac{1}{2}I_{\text{cyl}}\frac{v^2}{R}$$

$$= \frac{1}{2}mv^2 + \frac{1}{2} \times \frac{mR^2}{2} \times \frac{v^2}{R^2}$$

$$= \frac{3}{4}mv^2 \quad \text{...(ii)}$$

Now, from the conservation of mechanical energy, $mgh = K_{\text{total}}$

$\therefore$ For solid sphere,

$$mgh_{\text{sph}} = \frac{7}{10}mv^2 \quad \text{[using Eq. (i)]...(iii)}$$

Similarly, for solid cylinder,

$$mgh_{\text{cyl}} = \frac{3}{4}mv^2 \quad \text{...(iv)}$$

[using Eq. (ii)]

Taking the ratio of Eqs. (iii) and (iv), we get

$$\frac{mgh_{\text{sph}}}{mgh_{\text{cyl}}} = \frac{\frac{7}{10}mv^2}{\frac{3}{4}mv^2} \Rightarrow \frac{h_{\text{sph}}}{h_{\text{cyl}}} = \frac{7}{10} \times \frac{4}{3} = \frac{14}{15}$$

Alternate Solution

Total kinetic energy for a rolling body without slipping can also be given as

$$K_{\text{total}} = \frac{1}{2}mv^2\left[1 + \frac{K^2}{R^2}\right]$$

where, K is the radius of gyration.

$\therefore$ From law of conservation,

$$mgh = \frac{1}{2}mv^2\left[1 + \frac{K^2}{R^2}\right] \text{ or } h \propto \left(1 + \frac{K^2}{R^2}\right)$$

As we know that, for solid sphere,

$$K = \sqrt{\frac{2}{5}}R \Rightarrow \frac{K^2}{R^2} = \frac{2}{5}$$

Similarly, for solid cylinder, $K = \frac{R}{\sqrt{2}} \Rightarrow \frac{K^2}{R^2} = \frac{1}{2}$

So, $$\frac{h_{\text{sph}}}{h_{\text{cyl}}} = \frac{1 + \frac{2}{5}}{1 + \frac{1}{2}} = \frac{\frac{7}{5}}{\frac{3}{2}} = \frac{7}{5} \times \frac{2}{3} = \frac{14}{15}$$

44. *(a)* Initial kinetic energy of the given system,

$$KE_i = \frac{1}{2}I_1\omega_1^2 + \frac{1}{2}\left(\frac{I_1}{2}\right)\left(\frac{\omega_1}{2}\right)^2$$

$$= \left(\frac{1}{2} + \frac{1}{16}\right)I_1\omega_1^2 = \frac{9}{16}I_1\omega_1^2 \quad \text{... (i)}$$

Now, using angular momentum conservation law (assuming angular velocity after contact is ω)

Initial angular momentum = Final angular momentum

$$I_1\omega_1 + \left(\frac{I_1}{2}\right)\left(\frac{\omega_1}{2}\right) = I_1\omega' + \frac{I_1}{2}\omega'$$

$$\Rightarrow \quad \frac{5}{4}\omega_1 = \frac{3}{2}\omega' \quad \text{or} \quad \omega' = \frac{5}{6}\omega_1 \quad \text{...(ii)}$$

Now, final kinetic energy (after contact) is

$$KE_f = \frac{1}{2}I_1\omega'^2 + \frac{1}{2}\left(\frac{I_1}{2}\right)\omega'^2$$

$$= \frac{1}{2}I_1\left(\frac{5}{6}\omega_1\right)^2 + \frac{1}{4}I_1\left(\frac{5}{6}\omega_1\right)^2$$

[using Eq. (ii)]

$$= \left(\frac{25}{72} + \frac{25}{144}\right)I_1\omega_1^2 = \frac{25}{48}I_1\omega_1^2 \quad \text{...(iii)}$$

Hence, change in KE,

$$\Delta KE = KE_f - KE_i$$

$$= \frac{25}{48}I_1\omega_1^2 - \frac{9}{16}I_1\omega_1^2$$

[using Eqs. (i)]

$$\Delta KE = -\frac{1}{24}I_1\omega_1^2$$

45. *(b)* When force F is applied at the centre of roller of mass m as shown in the figure below

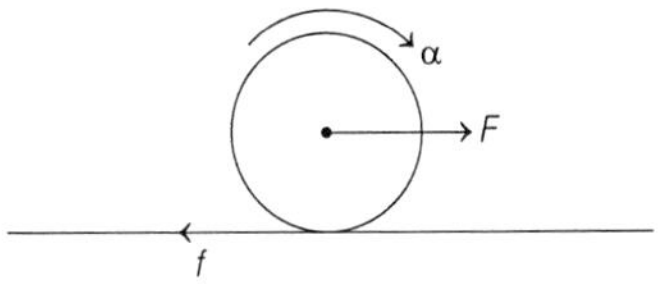

Its acceleration is given by

$$\frac{(F - f)}{m} = a \quad \text{... (i)}$$

Since, $\quad \tau = fR = I\alpha \quad \text{...(ii)}$

where, I = moment of inertia of solid cylindrical roller.

$= mR^2/2$ and α = angular acceleration of cylinder $= a/R$.

Hence, $\quad \tau = \frac{mR^2}{2} \cdot \frac{a}{R} = \frac{maR}{2}$

From eq. (ii), ($\tau = fR$)

$$f = \frac{ma}{2} \quad \text{...(iii)}$$

From Eqs. (i) and (iii), we get

$$F = \frac{3}{2}ma \Rightarrow a = \frac{2F}{3m}$$

So, $$\alpha = \frac{2F}{3mR} \quad \left[\because \alpha = \frac{a}{R}\right]$$

46. *(a)* Given force,

$$F = 20t - 5t^2$$

$$\alpha = \frac{FR}{I} = \frac{(20t - 5t^2)2}{10} = 4t - t^2$$

$$\Rightarrow \frac{d\omega}{dt} = 4t - t^2$$

$$\Rightarrow \int_0^\omega d\omega = \int_0^t (4t - t^2)\,dt$$

$$\Rightarrow \quad \omega = 2t^2 - \frac{t^3}{3}$$

When direction is reversed,

$\omega = 0$, i.e., $t = 0$, 6s

Now, $\quad d\theta = \omega\,dt$

$$\Rightarrow \quad \int_0^\theta d\theta = \int_0^6 \left(2t^2 - \frac{t^3}{3}\right)dt$$

$$\Rightarrow \quad \theta = \left[\frac{2t^3}{3} - \frac{t^4}{12}\right]_0^6$$

$$\Rightarrow \quad \theta = 144 - 108 = 36 \text{ rad}$$

$\therefore$ Number of rotations, $n = \frac{\theta}{2\pi} = \frac{36}{2\pi} < 6$

47. *(a)*

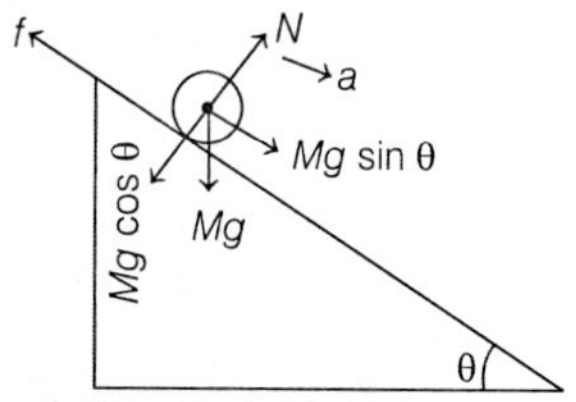

$Mg \sin\theta - f = Ma$ ...(i)

As torque $= I\alpha = I\frac{a}{R}$ [as $a = r\alpha$]

Torque $\tau = f \cdot R$

So, $f \cdot R = \frac{Ia}{R} \Rightarrow f = \frac{Ia}{R^2}$

From Eq. (i),

$$Mg\sin\theta - \frac{Ia}{R^2} = Ma$$

$$\Rightarrow a\left[M + \frac{I}{R^2}\right] = Mg\sin\theta$$

$$a = \frac{Mg\sin\theta}{M + \frac{I}{R^2}} \quad \therefore \quad a = \frac{g\sin\theta}{1 + \frac{I}{MR^2}}$$

48. *(4)* Applying of mechanical energy conservation-

$$K_i + U_i = K_f + U_f$$

$$0 + mgh = \frac{1}{2}mv^2 + \frac{1}{2}mv^2\frac{K^2}{R^2} + 0$$

$$mgh = \frac{1}{2}mv^2\left(1 + \frac{K^2}{R^2}\right) + 0$$

$$v = \sqrt{\frac{2gh}{1 + \frac{K^2}{R^2}}}$$

So ratio of velocities

$$\frac{v_{\text{Ring}}}{v_{\text{Solid sphere}}} = \sqrt{\frac{1 + \frac{2}{5}}{1 + 1}} = \sqrt{\frac{7}{5 \times 2}}$$

$x = 3.5$ Rounding off we get $x = 4$

49. *(40)* Here, $m = 1$ kg

$$\omega_i = 1800 \text{ rpm} = 1800 \times \frac{2\pi}{60} \text{ rad/s}$$

$$= 60\pi \text{ rad/s}$$

$$\omega_f = 2100 \text{ rpm}$$

$$= 2100 \times \frac{2\pi}{60} = 70\pi \text{ rad/s}$$

$\therefore$ Using equation of rotational motion,

For ring, $k^2 = R^2$ and for sphere,

$$k^2 = \frac{2}{5}R^2 \Rightarrow \omega_f = \omega_i + \alpha t$$

$$70\pi = 60\pi + \alpha \times 40$$

$$\Rightarrow \quad \alpha = \frac{\pi}{4} \text{ rad/s}^2$$

Now, $\tau = I\alpha$

$$\tau = \frac{MR^2}{4}\alpha \Rightarrow 25\pi = \frac{1 \times R^2}{4} \times \frac{\pi}{4}$$

$$\Rightarrow R^2 = 25 \times 16 \Rightarrow R = 5 \times 4 = 20 \text{ m}$$

Diameter $= 2R$

Diameter $= 2 \times 20 = 40$ m

50. *(32)* When sphere is of radius R, its mass is M.

When radius is reduced to $\frac{R}{2}$, mass will reduced to $\frac{M}{8}$.

$$\text{As } M' = M\left(\frac{R'}{R}\right)^3 \Rightarrow M' = M\left(\frac{1}{2}\right)^3$$

Now by conservation of angular momentum

$$\tau_{\text{ext}} = 0$$

$$L_1 = L_2$$

$$I_1\omega_1 = I_2\omega_2$$

$$\left(\frac{2}{5}MR^2\right)\omega_1 = \left[\frac{2}{5}\left(\frac{M}{8}\right)\left(\frac{R}{2}\right)^2\right]\omega_2$$

$$\omega_2 = 32\omega_1$$

51. *(13)* Using conservation of angular momentum.

$$I_1\omega_1 = I_2\omega_2$$

$$\Rightarrow \quad \frac{2}{5}MR^2 \cdot \left(\frac{2\pi}{T_1}\right) = \frac{2}{5}M\left(\frac{3}{4}R\right)^2 \cdot \frac{2\pi}{T_2}$$

$$\Rightarrow \quad \frac{1}{T_1} = \frac{9}{16T_2} \Rightarrow T_2 = \frac{9}{16}T_1$$

$$= \frac{9}{16} \times 24 = 13.5 \text{ h}$$

$$= 13 \text{ h } 30 \text{ min}$$

52. *(4)* Given,

$m = 2$kg, $L = 0.3$ m, $J = 0.2$ N-s

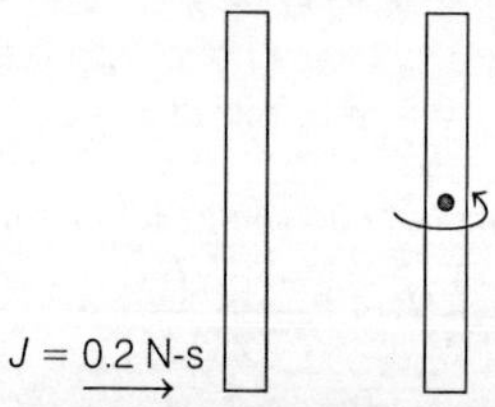

Angular impulse about centre of mass is given as,

Angular impulse, $\mathbf{I} = \mathbf{r} \times \mathbf{J} = \mathbf{L}_f - \mathbf{L}_i$

$$\Rightarrow \frac{L}{2} \times 0.2 \times \sin 90° = I_{\text{cm}}\,\omega - 0$$

$$\Rightarrow \frac{L}{2} \times 0.2 \times 1 = \frac{ML^2}{12}\omega \Rightarrow \omega = \frac{0.2 \times 6}{ML}$$

$$= \frac{1.2}{2 \times 0.3} = 2$$

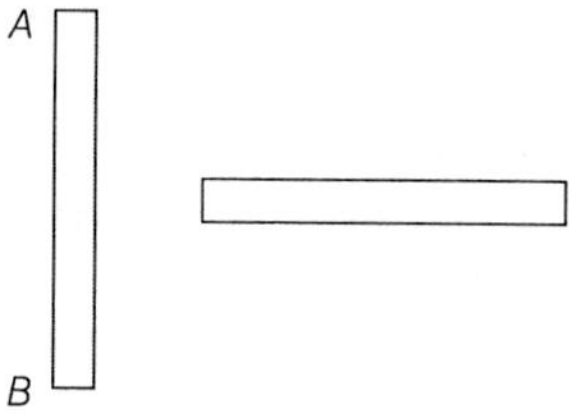

$$\theta = \frac{\pi}{2} \Rightarrow \theta = \omega t$$

$$t = \frac{\theta}{\omega} = \frac{\pi}{2 \times 2} = \frac{\pi}{4} = \frac{\pi}{x} \quad \text{(given)}$$

$\therefore \quad x = 4$

53. *(6)* To stop a rolling disc,

Work done = Total energy of the disc

$\therefore$ Work done = KE of COM + Rotational KE of body about its COM

$$= \frac{1}{2}mv^2 + \frac{1}{2}I\omega^2 = \frac{1}{2}mv^2 + \frac{1}{2}\frac{mr^2}{2} \cdot \omega^2$$

$$= \frac{1}{2}mv^2 + \frac{1}{4}mv^2 \quad \left[\because \omega = \frac{v}{r}\right]$$

$$= \frac{3}{4}mv^2 = \frac{3}{4} \times 50 \times (0.4)^2 = 6 \text{ J}$$

54. *(250)* Given,

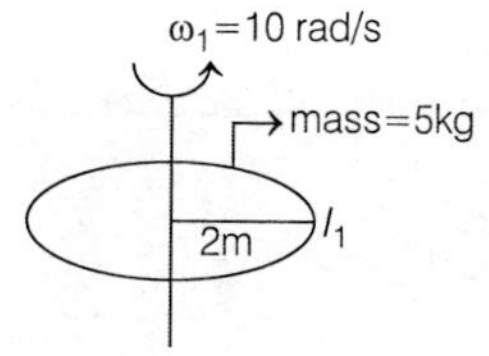

Case I Only one disc

Angular velocity, $\omega = 10$ rad/s

Mass, $m = 5$ kg

Radius = 2 m

Rotational Energy is given by

$$E_1 = \frac{1}{2}I_1\omega_1^2$$

Moment of inertia of disc about centre

$$I_1 = \frac{MR^2}{2}$$

$$I_1 = I_2 = 10 \text{ kg m}^2$$

Case II Identical disc added to same axis

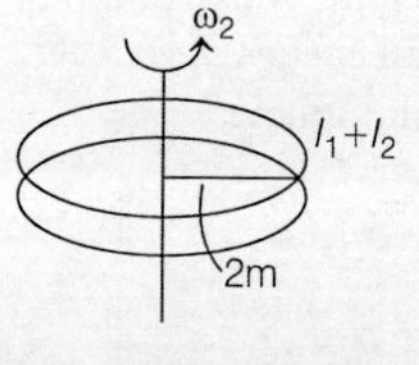

Now applying conservation of angular momentum.

$$I_1\omega_1 = (I_1 + I_2)\,\omega_2$$

$$\omega_2 = \frac{I_1\omega_1}{I_1 + I_2} \quad \ldots\text{(i)}$$

Rotational energy in second case

$$E_2 = \frac{1}{2}(I_1 + I_2)\,\omega_2^2 \quad \ldots\text{(ii)}$$

Substitute the value of ω_2 in Eq. (ii)

$$E_2 = \frac{1}{2}(I_1 + I_2)\frac{(I_1\omega_1)^2}{(I_1 + I_2)^2}$$

$$\Rightarrow \quad E_2 = \frac{1}{2}\frac{(I_1\omega_1)^2}{(I_1 + I_2)}$$

Energy dissipated $= E_1 - E_2$

$$= \frac{1}{2}I_1\omega_1^2 - \frac{1}{2}\frac{(I_1\omega_1)^2}{(I_1 + I_2)}$$

$$= \frac{1}{2}\times 10 \times 100 - \frac{1}{2}\frac{(100)(100)}{20}$$

$$= 500 - 250 = 250 \text{ J}$$

55. *(24)*

key idea To solve this problem you can use following step.

Step 1 Find the angular speed of combination.

Step 2 Now to find loss of energy, subtract final energy of combination from total initial energy.

When discs are brought together, angular speed of combination is ω then,

$I_1\omega_1 + I_2\omega_2 = (I_1 + I_2)\,\omega$

(Angular momentum conservation)

$$\Rightarrow \quad \omega = \frac{I_1\omega_1 + I_2\omega_2}{I_1 + I_2}$$

$$= \frac{4\times 10 + 2\times 4}{4+2} = 8 \text{ rad s}^{-1}$$

$$\text{Loss of energy} = \frac{1}{2}I_1\omega_1^2 + \frac{1}{2}I_2\omega_2^2 - \frac{1}{2}(I_1 + I_2)\,\omega^2$$

$$= \frac{1}{2}\times 4\times(10)^2 + \frac{1}{2}\times 2\times(4)^2 - \frac{1}{2}(4+2)(8)^2$$

$$= 24 \text{ J}$$

56. *(15)*

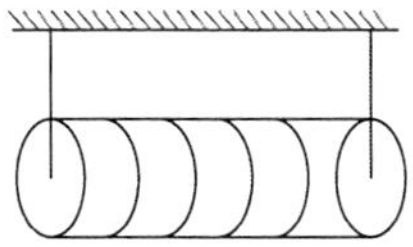

We know that

$$\tau = I\alpha$$

$$\Rightarrow \quad FR = mR^2\alpha \Rightarrow \alpha = \frac{F}{mR} = \frac{52.5}{5\times 0.7}$$

$$\Rightarrow \quad \alpha = 15 \text{ rad/s}^2$$

57. *(2)*

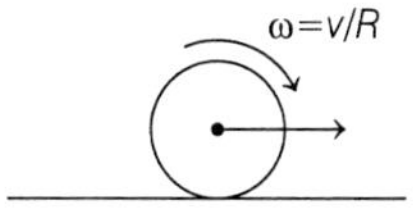

$$\text{Rotational energy} = \frac{1}{2}I\omega^2$$

$$= \frac{1}{2}\left(\frac{2}{3}mR^2\right)\left(\frac{v}{R}\right)^2$$

Total energy = Translational KE + Rotational energy

$$= \frac{1}{2}mv^2 + \frac{1}{2}\times\left(\frac{2}{3}mR^2\right)\left(\frac{v}{R}\right)^2$$

According to question,

$$\frac{K_{\text{rot}}}{K_{\text{Total}}} = \frac{\frac{1}{2}\left(\frac{2}{3}\right)mR^2 \times \frac{v^2}{R^2}}{\frac{1}{2}mv^2 + \frac{1}{2}\times\frac{2}{3}mR^2\times\frac{v^2}{R^2}}$$

$$= \frac{2}{5} = \frac{x}{5} \quad \text{(given)}$$

Thus, $x = 2$

58. *(4)* For solid sphere Angular momentum, $L = I\omega = \frac{2}{5}MR^2 \times \frac{v}{R}$

$$L = \frac{2MRv}{5} \quad \ldots\text{(i)}$$

Total energy of sphere,

$$E = \frac{1}{2}I\omega^2 + \frac{1}{2}Mv^2$$

$$= \frac{1}{2}\left(\frac{2}{5}MR^2\right)\frac{v^2}{R^2} + \frac{1}{2}Mv^2$$

$$\Rightarrow \quad E = \frac{7}{10}Mv^2 \quad \ldots\text{(ii)}$$

From Eqs. (i) and (ii), we get

$$\frac{L}{E} = \frac{2MRv/5}{7Mv^2/10}$$

$$\Rightarrow \quad \frac{\pi}{22} = \frac{4R}{7v} \Rightarrow \frac{v}{R} = \frac{4}{7}\times\frac{22}{\pi}$$

$$\Rightarrow \quad \omega = 4 \text{ rad/s}$$

59. *(75)* By energy conservation,

$$\frac{1}{2}mv^2 + \frac{1}{2}I\omega^2 = mgh$$

$$\frac{1}{2}mv^2 + \frac{1}{2}\times\frac{2}{3}mv^2 = mgh$$

$$\frac{5mv^2}{6} = mgh \Rightarrow h = 75 \text{ cm}$$

60. *(3)* We know that,

Rotational kinetic energy

$$E = \frac{1}{2}I\omega^2 = \frac{(I\omega)^2}{2I} = \frac{L^2}{2I} \quad (\because L = I\omega)$$

$$\therefore \quad E \propto \frac{1}{I} \Rightarrow \frac{E_f}{E_i} = \frac{I_i}{I_f} = \frac{I}{3I} = \frac{1}{3}$$

$$\Rightarrow \quad E_f = \frac{E_i}{3} = \frac{E}{3} = \frac{E}{x} \quad \text{(given)}$$

$$\therefore \quad x = 3$$

61. *(54)* Given, $m = 0.5$ kg, $u = 18$ m/s, $\mu = 0.3$

For disc, $\quad I = \frac{1}{2}mR^2$

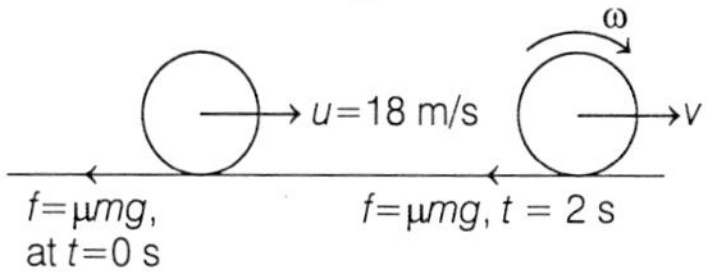

At time, $t = 0$ (Pure sliding)

$$f = ma \Rightarrow \mu mg = ma$$

$$\therefore \quad a = \mu g = 0.3\times 10 = 3 \text{ m/s}^2$$

At time, $t = 2$ s (Pure rolling)

$$v = u - at$$

$$= 18 - 3\times 2 = 12 \text{ m/s}$$

$$\therefore \quad v = 12 \text{ m/s}$$

Also, $\quad v = \omega r \Rightarrow \omega = v/r = 12/r$

$$\therefore \ (\text{KE})_{\text{Total}} = \frac{1}{2}mv^2 + \frac{1}{2}I\omega^2$$

$$= \frac{1}{2}\times 0.5\times 12^2 + \frac{1}{2}\times\frac{0.5\times r^2}{2}\times\frac{12^2}{r^2}$$

$$= 54 \text{ J}$$

62. *(40)*

key idea In pure rolling motion of sphere on a horizontal surface you can calculate total kinetic energy of sphere by adding its rotational kinetic energy and translational kinetic energy.

For a rolling sphere, KE $= \text{KE}_{\text{rotational}} + \text{KE}_{\text{translational}}$

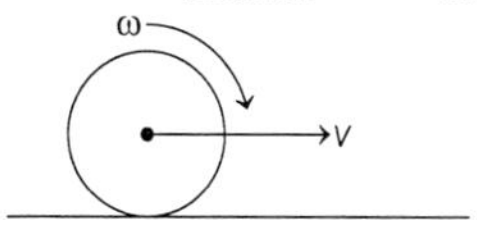

$$\text{KE} = \frac{1}{2}I\omega^2 + \frac{1}{2}mv^2$$

$$= \frac{1}{2}\times\frac{2}{5}mr^2\omega^2 + \frac{1}{2}mv^2 = \left(\frac{1}{5} + \frac{1}{2}\right)mv^2$$

$$\therefore \ \left[I = \frac{2}{5}mr^2 \text{ and } v = r\omega\right]$$

$$= \frac{7}{10}mv^2 = \frac{14}{10}v^2 \quad (\because m = 2 \text{ kg})$$

Given, KE = 2240 J

So, $\frac{14}{10}v^2 = 2240 \Rightarrow v^2 = 1600$

$$\Rightarrow \quad v = 40 \text{ m/s}$$

63. *(2)* Given that, $\theta = 30°$, $l = 60$ cm

For solid cylinder, $K^2 / R^2 = 1/2$

Speed of a body in pure rolling motion is given as

$$v = \sqrt{\frac{2gs \sin\theta}{1 + \frac{K^2}{R^2}}} = \sqrt{\frac{2 \times 10 \times 0.6 \times \sin 30°}{1 + \frac{1}{2}}}$$

$v = 2$ m/s

64. *(10)*

> **key idea** For pure rolling motion, you can calculate the kinetic energy of solid sphere with the help of following formula.
>
> $$K_T = \frac{1}{2} m v_{cm}^2 \left(1 + \frac{k^2}{R^2}\right)$$
>
> $$\left(\text{for solid sphere } \frac{k^2}{R^2} = \frac{2}{5}\right)$$

Given that, mass, $m = 1$ kg

Total kinetic energy, $K_T = 7 \times 10^{-3}$ J

In pure rolling motion,

$$K_T = \frac{1}{2} m v_{CM}^2 \left(1 + \frac{K^2}{R^2}\right)$$

For solid sphere, $\frac{K^2}{R^2} = \frac{2}{5}$

$$\therefore K_T = \frac{1}{2} m v_{CM}^2 \left(1 + \frac{2}{5}\right)$$

$$\Rightarrow K_T = \frac{7}{10} m v_{CM}^2 \text{ or } v_{CM}^2 = \frac{10 K_T}{7m}$$

$$\text{or } v_{CM} = \sqrt{\frac{10 K_T}{7m}} = \sqrt{\frac{10 \times 7 \times 10^{-3}}{7 \times 1}}$$

$v_{CM} = 10^{-1}$ m/s $\Rightarrow v_{CM} = 10$ cm/s

65. *(10)* The given situation is shown in the figure.

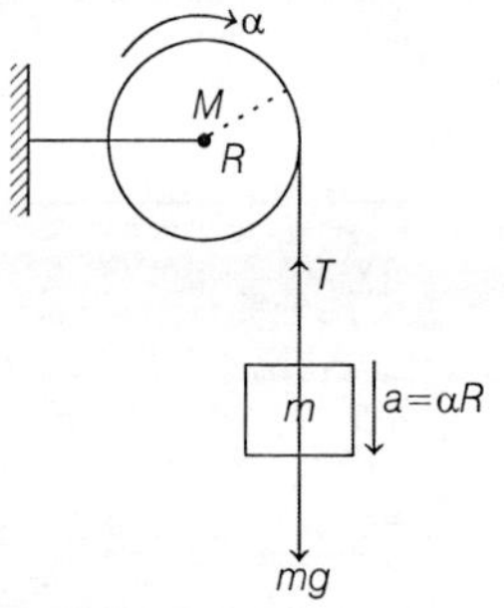

Fro figure, we have

$mg - T = ma;\ 20 - T = 2a$ ... (i)

We known that, moment of inertia of uniform disc, $I = MR^2 / 2$

Acceleration of the disc, $\alpha = a / R$, Since, torque, $\tau = I\alpha$

$$\therefore R \times T = I\alpha \text{ or } R \times T = \frac{MR^2}{2} \times \frac{a}{R} \text{ or}$$

$T = Ma / 2$ ...(ii)

From Eqs. (i) and (ii),

$$20 - T = 2 \times \frac{2T}{M}$$

$$\therefore \quad T = \frac{20 \times 4}{8} = 10 \text{ N}$$

66. *(120)* Let it unbinds and falls h height to attain speed of 4 m/s

By energy conservation,

$$\therefore \quad mgh = \frac{1}{2} mv^2 + \frac{1}{2} I\omega^2$$

($\because$ Decrease in gravitational potential energy = increase in translational kinetic energy + increase in rotational kinetic energy)

$$\Rightarrow \quad mgh = \frac{1}{2} mv^2 + \frac{1}{2} \frac{mR^2}{2} \omega^2$$

$$\Rightarrow \quad gh = \frac{1}{2} v^2 + \frac{1}{4} R^2 \omega^2$$

$$\Rightarrow \quad 10h = \frac{4^2}{2} + \frac{1}{4}(v^2) \quad (\because v = R\omega)$$

$$\Rightarrow \quad 10h = \frac{16}{2} + \frac{4^2}{4}$$

$$\Rightarrow \quad 10h = \frac{16}{2} + \frac{16}{4} \Rightarrow 10h = 12$$

$$\Rightarrow \quad h = \frac{12}{10} \text{ m} = 1.2 \text{ m}$$

$= 1.2 \times 100$ cm $= 120$ cm

67. *(18)* Radius of pulley, $R = 1.5$ m

Tangential applied force,

$F = (12t - 3t^2)$ N

Moment of inertia, $I = 4.5$ kg-m^2

Torque, $\tau = I \times \alpha$ ...(i)

Also, torque, $\tau = F \times R$ ...(ii)

From Eqs. (i) and (ii), we have

$FR = I\alpha$

$$\Rightarrow \alpha = \frac{FR}{I} = \frac{(12t - 3t^2) \times 1.5}{4.5} = 4t - t^2$$

$\therefore$ Angular velocity,

$$\omega = \int \alpha dt = \int (4t - t^2) dt = 4\frac{t^2}{2} - \frac{t^3}{3}$$

$$= 2t^2 - \frac{t^3}{3}$$

During the instant when the direction of rotation is reversed,

$\omega = 0$

$$2t^2 - \frac{t^3}{3} = 0 \Rightarrow t^2\left(2 - \frac{t}{3}\right) = 0$$

$t \neq 0, \quad \therefore \quad t = 6$ s

Total angular displacement,

$$\theta = \int_0^6 \omega dt = \int_0^6 \left(2t^2 - \frac{t^3}{3}\right) dt$$

$$= \left[2 . \frac{t^3}{3} - \frac{t^4}{4 \times 3}\right]_0^6 = 2 \times \frac{6^3}{3} - \frac{6^4}{4 \times 3} = 36$$

$\therefore$ Number of rotations made by the pulley,

$$n = \frac{\theta}{2\pi} = \frac{36}{2\pi} = \frac{18}{\pi} = \frac{K}{\pi} \quad \text{(given)}$$

$K = 18$

68. *(3)* Given,

M = mass of rolling wheel = 12 kg

$m = 3$ kg

To find value of x.

Velocity of centre of mass of wheel when it reaches at bottom of inclined plane,

$$PQ = \frac{1}{2} \sqrt{xgh} \text{ m/s} \quad \text{...(i)}$$

Free body diagram,

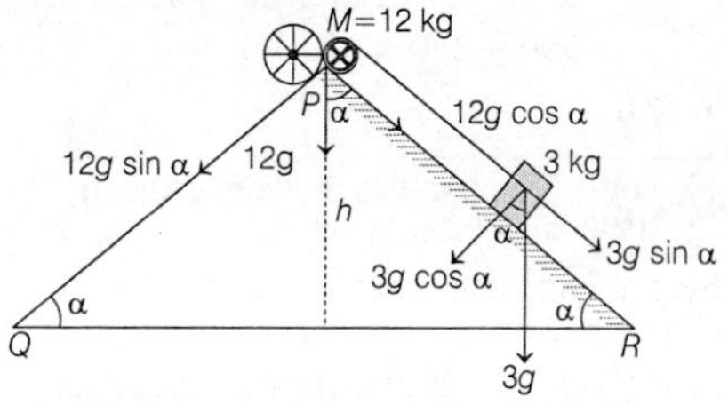

For rolling wheel, $\tau = I\alpha$

$(12\, g \sin\alpha - 3\, g \sin\alpha) \times R$

$$= (2 \times 12R^2 + 3R^2) \times \frac{a}{R}$$

$$27R^2 \times \frac{a}{R} = 9\, g \sin\alpha \times R$$

$$a = \frac{g \sin\alpha}{3}$$

Since, $v = \sqrt{2as}$

$$\Rightarrow \quad v = \sqrt{\frac{2 \times g \sin\alpha}{3} \times \frac{h}{\sin\alpha}} = \sqrt{\frac{2gh}{3}}$$

$$v = \frac{1}{2} \times \sqrt{\frac{8gh}{3}} \quad \text{...(ii)}$$

Comparing Eqs. (i) and (ii), we get

$$x = \frac{8}{3} = 2.67 \simeq 3$$

69. *(9)* From the given situation,

$KE_P = KE_Q$

$$\Rightarrow \quad \frac{1}{2} I_1 (\omega_1)^2 = \frac{1}{2} I_2 (\omega_2)^2$$

$$\Rightarrow \frac{1}{2} I_1 \left(\frac{v}{3R}\right)^2 = \frac{1}{2} I_2 \left(\frac{v}{R}\right)^2$$

[$\because$ Radius of P is 3 times to that of Q]

$$\Rightarrow \quad I_1 \times \frac{v^2}{9R^2} = I_2 \times \frac{v^2}{R^2}$$

$$\Rightarrow \quad \frac{I_1}{9} = I_2$$

$$\Rightarrow \quad \frac{I_1}{I_2} = \frac{9}{1} = \frac{x}{1} \quad \text{(given)}$$

70. *(4)* The given situation can be shown as

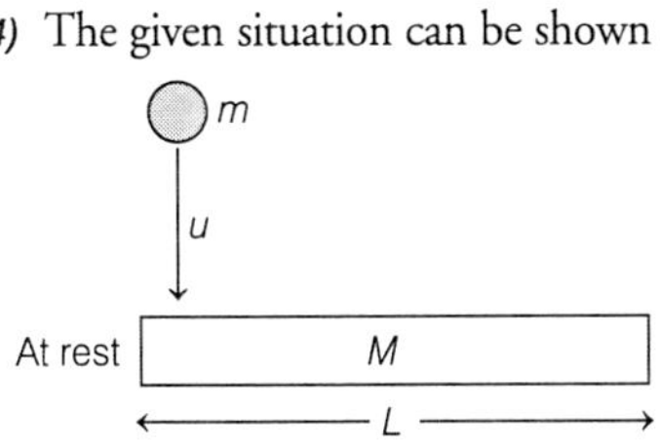

Before collision,

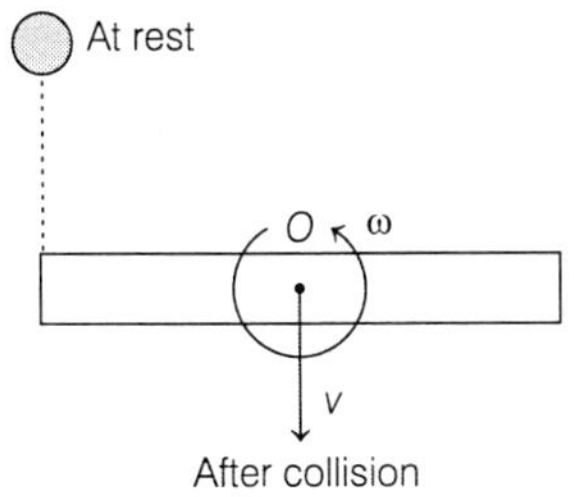

As the collision is perfectly elastic, therefore momentum is conserved, i.e.

$$p_{\text{initially}} = p_{\text{finally}}$$

$$\Rightarrow \quad mu = Mv \quad \text{...(i)}$$

Angular momentum will also be conserved about point O.

$$\Rightarrow \quad mu \cdot \frac{L}{2} = \frac{ML^2}{12}\omega$$

$$\Rightarrow \quad \omega = \frac{6mu}{ML} \quad \text{...(ii)}$$

$\because$ Coefficient of restitution,

$$e = \frac{\text{Relative velocity after collision}}{\text{Relative velocity before collision}}$$

$$\Rightarrow 1 = \frac{v + \frac{\omega L}{2}}{u} \Rightarrow v + \frac{\omega L}{2} = u \quad \text{...(iii)}$$

From Eqs. (ii) and (iii), we get

$$v + \frac{3mu}{M} = u$$

$$\Rightarrow \frac{mu}{M} + \frac{3mu}{M} = u \quad \text{[using Eq.(i)]}$$

$$\Rightarrow \quad \frac{4mu}{M} = u$$

$$\Rightarrow \quad \frac{m}{M} = \frac{1}{4} \quad \text{...(iv)}$$

$$= 1/x \quad \text{(given)}$$

$$x = 4$$

71. *(2)*

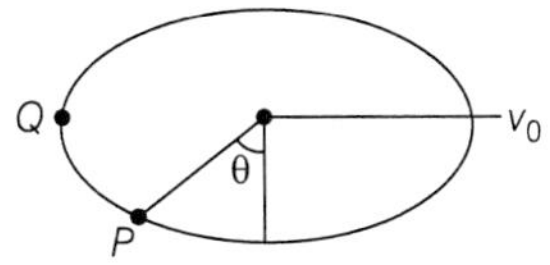

$$v_P = 2v_0 \sin\frac{\theta}{2}$$

For *pt* Q, $\theta = 90°$

$$v_Q = 2v_0 \sin 45 = \sqrt{2}v_0 = v_0\sqrt{x}$$

$$x = 2$$

72. *(2)* According to question, a circular disc reaches from top to bottom of an inclined plane of length L. This can be shown as

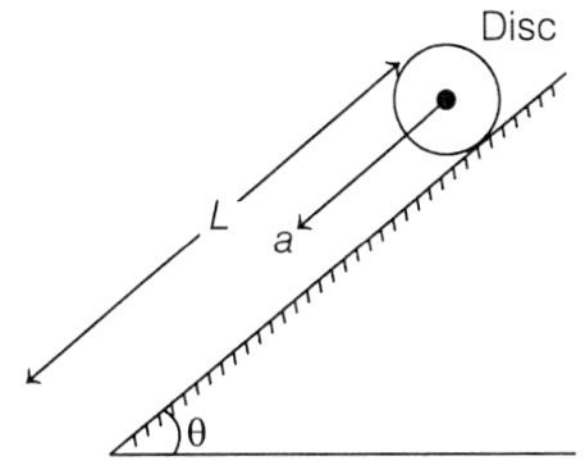

When the disc slips down the inclined plane, it takes time t_1.
Therefore, in this case its acceleration, $a_1 = g\sin\theta$

$$\because \quad s = ut_1 + \frac{1}{2}a_1 t_1^2$$

$$\Rightarrow \quad L = \frac{1}{2}g\sin\theta t_1^2 \quad \text{...(i)}$$

And when the disc rolls down the inclined plane, it takes time t_2.
Therefore in this case, its acceleration,

$$a_2 = \frac{g\sin\theta}{1 + \frac{K^2}{R^2}} = \frac{g\sin\theta}{1+\frac{1}{2}} = \frac{2}{3}g\sin\theta$$

$$\left[\because \text{for disc, } \frac{K^2}{R^2} = \frac{1}{2}\right]$$

$$\because \quad s = ut_2 + \frac{1}{2}a_2 t_2^2$$

$$L = \frac{1}{2}\cdot\frac{2}{3}g\sin\theta t_2^2 \quad \text{...(ii)}$$

On dividing Eq. (i) by Eq. (ii), we get

$$\frac{t_2}{t_1} = \sqrt{\frac{3}{2}} = \sqrt{\frac{3}{x}} \quad \text{(given)}$$

$$\therefore \quad x = 2$$

73. *(3)* Rotational kinetic energy of ring,

$$K = \frac{1}{2}I\omega^2$$

$\because$ Moment of inertia for ring about the point of contact,

$$I_{\text{ring}} = 2mR^2 \therefore K = \frac{1}{2}(2mR^2)\omega^2$$

$$\therefore \quad K = \frac{1}{2}(2mR^2)\frac{v_R^2}{R^2}\left\{\because \omega = \frac{v}{R}\right\}$$

As, this is a case of pure rolling and the ring is rolling down, then by conservation of energy

$$\therefore \quad PE = mgh = \frac{1}{2}(2mR^2)\frac{v_R^2}{R^2}$$

$$\Rightarrow \quad v_R = \sqrt{gh} \quad \text{...(i)}$$

Now, consider the case of solid cylinder,
$\because$ Moment of inertia in this case as well as about the point of contact is given as

$$I_{\text{cylinder}} = \frac{3}{2}mR^2 \Rightarrow mgh = \frac{1}{2}I_{\text{cylinder}}\omega^2$$

$$\Rightarrow \quad mgh = \frac{1}{2}\left(\frac{3}{2}mR^2\right)\cdot\frac{v_c^2}{R^2}\left\{\because \omega = \frac{v}{R}\right\}$$

$$\Rightarrow \quad v_c = \sqrt{\frac{4gh}{3}} \quad \text{...(ii)}$$

On dividing Eq. (i) by Eq. (ii), we get

$$\frac{v_R}{v_c} = \frac{\sqrt{3}}{2} = \frac{\sqrt{x}}{2} \quad \text{(given)}$$

74. *(200)* Given,

$$\omega_0 = \frac{600 \times 2\pi}{60}\text{ rad/s} = 20\pi \text{ rad/s}$$

$$\omega = \frac{1800 \times 2\pi}{60}\text{ rad/s} = 60\,\pi \text{ rad/s}$$

we know that,

$$\omega = \omega_0 + \alpha t \Rightarrow 60\pi = 20\pi + 10\alpha$$

$$\alpha = 4\pi \text{ rad/s}^2$$

$$\therefore \theta = \omega_0 t + \frac{1}{2}\alpha t^2$$

$$= 20\pi \times 10 + \frac{1}{2} \times 4\pi \times 10^2 = 400\pi \text{ rad}$$

$$\therefore \text{Number of rotations} = \frac{\theta}{2\pi} = \frac{400\pi}{2\pi}$$

$$= 200$$

75. *(4)* For rolling without slipping on an inclined plane, we can write

$$Rmg\sin\theta = (mK^2 + mR^2)\alpha$$

$$\Rightarrow Rmg\sin\theta = m(K^2 + R^2)\alpha$$

$$\Rightarrow \quad \alpha = \frac{Rg\sin\theta}{K^2 + R^2} \Rightarrow R\alpha = \frac{g\sin\theta}{1 + \frac{K^2}{R^2}}$$

$$\Rightarrow \quad a = \frac{g\sin\theta}{1 + \frac{K^2}{R^2}} \quad [\because R\alpha = a] \text{...(i)}$$

Time period, $t = \sqrt{\frac{2s}{a}}$ ...(ii)

From Eqs. (i) and (ii), we get

$$t = \sqrt{\frac{2s}{g\sin\theta}\left(1 + \frac{K^2}{R^2}\right)}$$

For least time acceleration a should be maximum and K should be minimum and we know that K is least for solid sphere.

So, time will be least for sphere.

It means the body which will reach first at the bottom of the inclined plane is 4, i.e., solid sphere.

76. *(728)* Initial angular speed,

$$\omega_0 = 900 \text{ rpm} = 900 \times \frac{2\pi}{60} = 30\pi \text{ rad/s}$$

Final angular speed, $\omega = 2460$ rpm

$$= 2460 \times \frac{2\pi}{60} = 41 \times 2\pi = 82\pi \text{ rad/s}$$

We know that,

$$\omega = \omega_0 + \alpha t$$

$\Rightarrow$ $\omega - \omega_0 = \alpha t$

$\Rightarrow$ $82\pi - 30\pi = \alpha \times 26$

$\Rightarrow$ $52\pi = 26\alpha$

$\Rightarrow$ $\alpha = 2\pi \text{ rad/s}^2$

Also we know that, $\theta = \omega_0 t + \frac{1}{2}\alpha t^2$

$\Rightarrow$ $2\pi n = 30\pi \times 26 + \frac{1}{2} \times 2\pi \times (26)^2$

$\Rightarrow$ $2n = 30 \times 26 + (26)^2$

$\Rightarrow$ $n = \frac{30 \times 26 + (26)^2}{2}$

$$= \frac{26(30 + 26)}{2} = 13 \times 56 = 728$$

77. *(3)* Acceleration, $a = \frac{g\sin\theta}{1 + \frac{I}{mR^2}}$

$\Rightarrow$ $a = \frac{g\sin\theta}{1 + \frac{1}{2}}$

$\left[\because \text{For disc, } I = \frac{mR^2}{2}\right]$

$\Rightarrow$ $a = \frac{2}{3}g\sin\theta$ ... (i)

As per question, acceleration of the disc will be $\frac{2}{b}g\sin\theta$.

Comparing it with Eq. (i), we get

$$b = 3$$

> **• Concept Enhancer**
>
> When a solid disc of radius R and mass m rolls down without slipping on an inclined plane making θ angle with horizontal then acceleration of disc will be $a = \frac{g\sin\theta}{1 + I/mR^2}$

78. *(20)* Given, $M = 20$ kg, $R = 0.2$ m,

$F = 20$ N, $\omega = 50 \text{ rad s}^{-1}$,

$\theta = 6.28$ rad (for one revolution).

We know that,

Torque = Moment of inertia × Angular acceleration

$\Rightarrow$ $\tau = I \times \alpha$

$\Rightarrow$ $\alpha = \frac{\tau}{I} = \frac{F \times R}{\frac{MR^2}{2}} = \frac{2F}{MR}$

$\left[\because \tau = F \times R \text{ and } I_{disc} = \frac{1}{2}MR^2\right]$

$\Rightarrow$ $\alpha = \frac{2 \times 20}{20 \times (0.2)} = 10 \text{ rad/s}^2$

In angular terms, third equation of motion,

$$\omega^2 = \omega_0^2 + 2\alpha\Delta\theta$$

$\Rightarrow$ $(50)^2 = (0)^2 + 2 \times 10 \times \Delta\theta$

$\Rightarrow$ $\Delta\theta = \frac{2500}{20} \Rightarrow \Delta\theta = 125$ rad

$\therefore$ Number of revolution

$$= \frac{\Delta\theta}{\theta} = \frac{125}{6.28} \approx 20$$

79. *(20)* The given situation is shown in the following fig.,

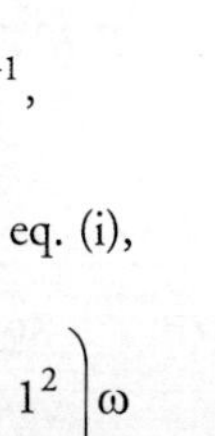

Applying law of conservation of angular momentum about pivotal point,

$$L_i = L_f$$

$\Rightarrow$ $mvl = I\omega$

$\Rightarrow$ $mvl = \left(\frac{1}{3}Ml^2 + ml^2\right)\omega$ ...(i)

Given, $m = 0.1$ kg, $v = 80 \text{ ms}^{-1}$,

$M = 0.9$ kg, $l = 1$ m

On substituting these values in eq. (i), we get

$$0.1 \times 80 \times 1 = \left(\frac{0.9 \times 1^2}{3} + 0.1 \times 1^2\right)\omega$$

$$8 = \left(\frac{3}{10} + \frac{1}{10}\right)\omega$$

$\Rightarrow$ $8 = \frac{4}{10}\omega \Rightarrow \omega = 20$ rad/s

80. *(20)* From law of conservation of angular momentum,

$$(L_{sys})_i = (L_{sys})_f$$

$\Rightarrow$ $I_1\omega_1 + 0 = (I_1 + I_2)\omega_2$

$\Rightarrow$ $\frac{MR^2}{2} \times \omega_1 = \left[\frac{MR^2}{2} + \frac{M(R/2)^2}{2}\right]\omega_2$

$\Rightarrow$ $\frac{MR^2}{2} \times \omega_1 = \left[\frac{MR^2}{2} + \frac{MR^2}{8}\right]\omega_2$

$\Rightarrow$ $\frac{MR^2}{2} \times \omega_1 = \frac{5MR^2}{8} \times \omega_2$

$\Rightarrow$ $\omega_1 = \frac{5}{4}\omega_2 \Rightarrow \omega_2 = \frac{4}{5}\omega_1$ ...(i)

Now, the loss in rotational kinetic energy,

$$(\Delta K_{sys})_{loss} = (K_{sys})_i - (K_{sys})_f$$

$$= \left[\frac{1}{2}I_1\omega_1^2 + 0\right] - \left[\frac{1}{2}(I_1 + I_2)\omega_2^2\right]$$

$$= \left[\frac{1}{2} \times \frac{MR^2}{2} \times \omega_1^2 + 0\right] - \left[\frac{1}{2} \times \left(\frac{MR^2}{2} + \frac{MR^2}{8}\right) \times \left(\frac{4}{5}\omega_1\right)^2\right]$$

$$= \frac{MR^2\omega_1^2}{4} - \frac{1}{2} \times \frac{5MR^2}{8} \times \frac{16\omega_1^2}{25}$$

$$= \frac{MR^2\omega_1^2}{4} - \frac{MR^2\omega_1^2}{5} = \frac{MR^2\omega_1^2}{20}$$

Now, the % loss in rotational kinetic energy,

$$\%(\Delta K_{sys})_{loss} = \frac{(\Delta K_{sys})_{loss}}{(K_{sys})_i} \times 100\%$$

$$= \frac{\left(\frac{MR^2\omega_1^2}{20}\right)}{\left(\frac{MR^2\omega_1^2}{4}\right)} \times 100\% = \frac{4}{20} \times 100 = 20\%$$

and given that $\%(\Delta K_{sys})_{loss} = p\%$

On comparing, we get

$$p = 20$$

> **• Concept Enhancer**
>
> Concept enhancer: If no external torque is applied on a rotating body, then its angular momentum remains conserved.

C-07

JEE Main Chapterwise Topicwise Physics

In JEE Main there is usually 1-2 questions per paper are visible from this chapter. In past 3 years, 56 conceptual and 5 numerical questions were asked of moderate to difficult level from this chapter.

For detailed analysis scan QR code

GRAVITATION

Topic 1 Kepler's Law and Universal Law of Gravitation

Level 1

01 A satellite is launched into a circular orbit of radius R around the earth. A second satellite is launched into an orbit of radius 1.03 R. The time period of revolution of the second satellite is larger than the first one approximately by *[2025, 24 Jan Shift-I]*

(a) 2.5% (b) 9%
(c) 4.5% (d) 3%

02 A satellite revolving around a planet in stationary orbit has time period 6 h. The mass of planet is one-fourth the mass of earth. The radius of orbit planet is
(Given, radius of geostationary orbit for earth is 4.2×10^4 km) *[2024, 5 April Shift-II]*

(a) 8.4×10^4 km (b) 1.4×10^4 km
(c) 1.68×10^5 km (d) 1.05×10^4 km

03 A metal wire of uniform mass density having length L and mass M is bent to form a semicircular arc and a particle of mass m is placed at the centre of the arc. The gravitational force on the particle by the wire is *[2024, 4 April Shift-I]*

(a) $\frac{2GmM\pi}{L^2}$ (b) $\frac{GmM\pi}{2L^2}$
(c) $\frac{GmM\pi^2}{L^2}$ (d) 0

04 A planet takes 200 days to complete one revolution around the sun. If the distance of the planet from sun is reduced to one-fourth of the original distance, how many days will it take to complete one revolution? *[2024, 29 Jan Shift-II]*

(a) 25 (b) 20
(c) 100 (d) 50

05 Given below are two statements. One is labelled as Assertion (A) and other is labelled as Reason (R).
Assertion (A) The angular speed of the moon in its orbit about the earth is more than the angular speed of the earth in its orbit about the sun.
Reason (R) The moon takes less time to move around the earth than the time taken by the earth to move around the sun.
In the light of the above statements, choose the most appropriate answer from the options given below *[2024, 27 Jan Shift-II]*

(a) Both A and R are correct but R is not the correct explanation of A.
(b) Both A and R are correct and R is the correct explanation of A.
(c) A is correct but R is not correct.
(d) A is not correct but R is correct.

06 Two identical particles each of mass m go round a circle of radius a under the action of their mutual gravitational attraction. The angular speed of each particle will be *[2023, 15 April Shift-I]*

(a) $\sqrt{\frac{Gm}{4a^3}}$ (b) $\sqrt{\frac{Gm}{2a^3}}$ (c) $\sqrt{\frac{Gm}{a^3}}$ (d) $\sqrt{\frac{Gm}{8a^3}}$

07 The time period of a satellite of earth is 24 h. If the separation between the earth and the satellite is decreased to one-fourth of the previous value, then its new time period will become *[2023, 29 Jan Shift-II]*

(a) 1 h (b) 6 h (c) 3 h (d) 4 h

08 Two particles of equal mass m move in a circle of radius r under the action of their mutual gravitational attraction. The speed of each particle will be *[2023, 29 Jan Shift-I]*

(a) $\sqrt{\frac{Gm}{4r}}$ (b) $\sqrt{\frac{4Gm}{r}}$
(c) $\sqrt{\frac{Gm}{2r}}$ (d) $\sqrt{\frac{Gm}{r}}$

Imp. Keplar's Law : Law of Periods

09 If the distance of the earth from the sun is 1.5×10^6 km, then the distance of an imaginary planet from the sun, if its period of revolution is 2.83 years is *[2023, 24 Jan Shift-II]*

(a) 3×10^7 km (b) 6×10^6 km
(c) 3×10^6 km (d) 6×10^7 km

10 The time period of a satellite revolving around earth in a given orbit is 7 h. If the radius of orbit is increased to three times its previous value, then approximate new time period of the satellite will be *[2022, 29 June Shift-II]*

(a) 40 h (b) 36 h (c) 30 h (d) 25 h

11 Two planets A and B of equal mass are having their period of revolutions T_A and T_B, such that $T_A = 2T_B$. These planets are revolving in the circular orbits of radii r_A and r_B, respectively. Which of the following would be the correct relationship for their orbits? *[2022, 28 June Shift-I]*

(a) $2r_A^2 = r_B^3$ (b) $r_A^3 = 2r_B^3$
(c) $r_A^3 = 4r_B^3$ (d) $T_A^2 - T_B^2 = \frac{\pi^2}{GM}(r_B^3 - 4r_A^3)$

12 The distance between sun and earth is R. The duration of year if the distance between sun and earth becomes 3R will be *[2022, 24 June Shift-II]*

(a) $\sqrt{3}$ years (b) 3 years
(c) 9 years (d) $3\sqrt{3}$ years

13 A planet revolving in elliptical orbit has

I. a constant velocity of revolution
II. the least velocity when it is nearest to the Sun
III. areal velocity is directly proportional to its velocity
IV. areal velocity is inversely proportional to its velocity.
V. to follow a trajectory such that the areal velocity is constant.

Choose the correct answer from the options given below. *[2021, 26 Feb Shift-I]*

(a) Only I (b) Only IV (c) Only III (d) Only V

14 The time period of a satellite in a circular orbit of radius R is T. The period of another satellite in a circular orbit of radius $9R$ is *[2021, 18 March Shift-I]*

(a) $9\,T$ (b) $27\,T$ (c) $12\,T$ (d) $3\,T$

15 The maximum and minimum distance of a comet from the sun are 1.6×10^{12} m and 8.0×10^{10} m, respectively. If the speed of the comet at the nearest point is $6 \times 10^4\ \text{ms}^{-1}$, then the speed at the farthest point is *[2021, 16 March Shift-I]*

(a) 1.5×10^3 m/s (b) 6.0×10^3 m/s
(c) 3.0×10^3 m/s (d) 4.5×10^3 m/s

16 Two particles of equal mass m go around a circle of radius R under the action of their mutual gravitational attraction. The speed of each particle with respect to their centre of mass is *[AIEEE 2011]*

(a) $\sqrt{\frac{Gm}{R}}$ (b) $\sqrt{\frac{Gm}{4R}}$
(c) $\sqrt{\frac{Gm}{3R}}$ (d) $\sqrt{\frac{Gm}{2R}}$

Level 2

17 Given below are two statements. One is labelled as Assertion (A) and the other is labelled as Reason (R).

Assertion (A) The radius vector from the Sun to a planet sweeps out equal areas in equal intervals of time and thus areal velocity of planet is constant.

Reason (R) For a central force field the angular momentum is a constant.

In the light of the above statements, choose the most appropriate answer from the options given below. *[2025, 7 April Shift-II]*

(a) A is not correct but R is correct.
(b) Both A and R are correct and R is the correct explanation of A .
(c) Both A and R are correct but R is not the correct explanation of A.
(d) A is correct but R is not correct.

18 If a satellite orbiting the Earth is 9 times closer to the Earth than the Moon, what is the time period of rotation of the satellite? Given rotational time period of Moon = 27 days and gravitational attraction between the satellite and the Moon is neglected. *[2025, 23 Jan Shift-II]*

(a) 27 days (b) 3 days (c) 81 days (d) 1 day

19 If G be the gravitational constant and u be the energy density then which of the following quantity have the dimensions as that of that $\sqrt{uG}$? *[2024, 5 April Shift-I]*

(a) Energy per unit mass
(b) Gravitational potential
(c) Pressure gradient per unit mass
(d) Force per unit mass

Imp. Newton's Law of Gravitation

20 Four identical particles of mass m are kept at the four corners of a square. If the gravitational force exerted on one of the masses by the other masses is $\left(\frac{2\sqrt{2}+1}{32}\right)\frac{Gm^2}{L^2}$, the length of the sides of the square is *[2024, 31 Jan Shift-I]*

(a) $3L$ (b) $2L$ (c) $4L$ (d) $\frac{L}{2}$

21 Every planet revolves around the sun in an elliptical orbit.

A. The force acting on a planet is inversely proportional to square of distance from sun.
B. Force acting on planet is inversely proportional to product of the masses of the planet and the sun.
C. The centripetal force acting on the planet is directed away from the sun.
D. The square of time period of revolution of planet around sun is directly proportional to cube of semi-major axis of elliptical orbit.

Choose the correct answer from the options given below. *[2023, 25 Jan Shift-II]*

(a) C and D only (b) A and D only
(c) A and C only (d) B and C only

22 Three identical particles A, B and C of mass 100 kg each are placed in a straight line with $AB = BC = 13$ m. The gravitational force on a fourth particle P of the same mass is F, when placed at a distance 13 m from the particle B on the perpendicular bisector of the line AC. The value of F will be approximately *[2022, 25 July Shift-I]*

(a) 21 G (b) 100 G (c) 59 G (d) 42 G

23 Two objects of equal masses placed at certain distance from each other attracts each other with a force of F. If one-third mass of one object is transferred to the other object, then the new force will be *[2022, 28 June Shift-II]*

(a) $\frac{2}{9}F$ (b) $\frac{16}{9}F$
(c) $\frac{8}{9}F$ (d) F

24 Consider a binary star system of star A and star B with masses m_A and m_B revolving in a circular orbit of radii r_A and r_B, respectively. If T_A and T_B are the time period of star A and star B, respectively, then *[2021, 20 July Shift-I]*

(a) $\frac{T_A}{T_B} = \left(\frac{r_A}{r_B}\right)^{3/2}$ (b) $T_A > T_B$ (if $r_A > r_B$)
(c) $T_A = T_B$ (d) $T_A > T_B$ (if $m_A > m_B$)

25 Find the gravitational force of attraction between the ring and sphere as shown in the diagram, where the plane of the ring is perpendicular to the line joining the centres. If $\sqrt{8}\,R$ is the distance between the centres of a ring (of mass m) and a sphere (mass M) where both have equal radius R. *[2021, 25 Feb Shift-II]*

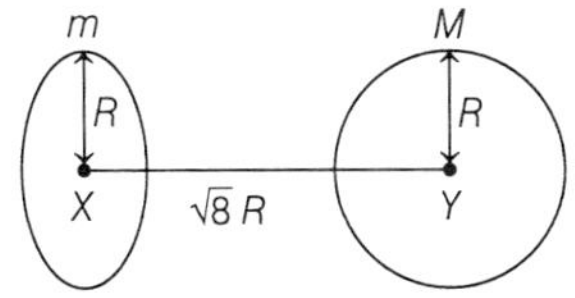

(a) $\frac{1}{3\sqrt{8}}\cdot\frac{GMm}{R^2}$ (b) $\frac{2\sqrt{2}}{3}\cdot\frac{GMm}{R^2}$
(c) $\frac{\sqrt{8}}{27}\cdot\frac{GMm}{R^2}$ (d) $\frac{\sqrt{8}}{5}\cdot\frac{GMm}{R^2}$

26 A straight rod of length L extends from $x = a$ to $x = L + a$. The gravitational force it exerts on a point mass m at $x = 0$, if the mass per unit length of the rod is $A + Bx^2$, is given by *[2019, 12 Jan Shift-I]*

(a) $Gm\left[A\left(\frac{1}{a+L} - \frac{1}{a}\right) - BL\right]$
(b) $Gm\left[A\left(\frac{1}{a+L} - \frac{1}{a}\right) + BL\right]$
(c) $Gm\left[A\left(\frac{1}{a} - \frac{1}{a+L}\right) + BL\right]$
(d) $Gm\left[A\left(\frac{1}{a} - \frac{1}{a+L}\right) - BL\right]$

27 Take the mean distance of the moon and the sun from the earth to be 0.4×10^6 km and 150×10^6 km respectively. Their masses are 8×10^{22} kg and 2×10^{30} kg respectively. The radius of the earth is 6400 km. Let ΔF_1 be the difference in the forces exerted by the moon at the nearest and farthest points on the earth and ΔF_2 be the difference in the force exerted by the sun at the nearest and farthest points on the earth. Then, the number closest to $\frac{\Delta F_1}{\Delta F_2}$ is *[JEE Main 2018]*

(a) 2 (b) 0.6 (c) 6 (d) 10^{-2}

Level 3

28 A small point of mass m is placed at a distance $2R$ from the centre O of a big uniform solid sphere of mass M and radius R. The gravitational force on m due to M is F_1. A spherical part of radius $\frac{R}{3}$ is removed from the big sphere as shown in the figure and the gravitational force on m due to remaining part of M is found to be F_2. The value of ratio $F_1 : F_2$ is *[2025, 22 Jan Shift-I]*

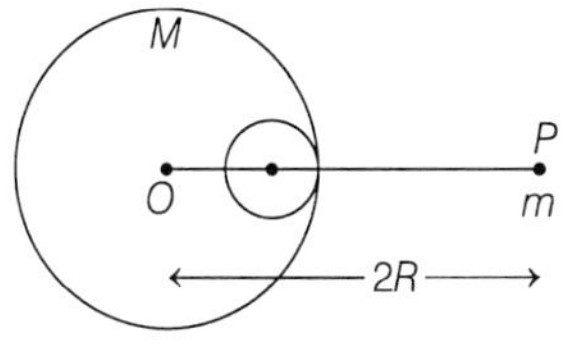

(a) 16 : 9 (b) 12 : 11 (c) 11 : 10 (d) 12 : 9

29 Two planets A and B having masses m_1 and m_2 move around the sun in circular orbits of r_1 and r_2 radii respectively. If angular momentum of A is L and that of B is $3L$, the ratio of time period $\left(\frac{T_A}{T_B}\right)$ is *[2024, 8 April Shift-I]*

(a) $\left(\frac{r_1}{r_2}\right)^3$ (b) $\left(\frac{r_2}{r_1}\right)^{3/2}$ (c) $27\left(\frac{m_1}{m_2}\right)^3$ (d) $\frac{1}{27}\left(\frac{m_2}{m_1}\right)^3$

30 A light planet is revolving around a massive star in a circular orbit of radius R with a period of revolution T. If the force of attraction between planet and star is proportional to $R^{-3/2}$, then choose the correct option. *[2024, 1 Feb Shift-II]*

(a) $T^2 \propto R^3$ (b) $T^2 \propto R^{7/2}$
(c) $T^2 \propto R^{5/2}$ (d) $T^2 \propto R^{3/2}$

31 Two identical particles of mass 1 kg each go round a circle of radius R, under the action of their mutual gravitational attraction. The angular speed of each particle is *[2021, 22 July Shift-II]*

(a) $\sqrt{\frac{G}{2R^3}}$ (b) $\frac{1}{2}\sqrt{\frac{G}{R^3}}$ (c) $\frac{1}{2R}\sqrt{\frac{1}{G}}$ (d) $\sqrt{\frac{2G}{R^3}}$

32 The angular momentum of a planet of mass M moving around the sun in an elliptical orbit is L. The magnitude of the areal velocity of the planet is *[2021, 18 March Shift-II]*

(a) $\frac{4L}{M}$ (b) $\frac{L}{M}$ (c) $\frac{2L}{M}$ (d) $\frac{L}{2M}$

33 A solid sphere of radius R gravitationally attracts a particle placed at $3R$ from its centre with a force F_1. Now, a spherical cavity of radius $R/2$ is made in the sphere (as shown in figure) and the force becomes F_2. The value of $F_1 : F_2$ is *[2021, 25 Feb Shift-I]*

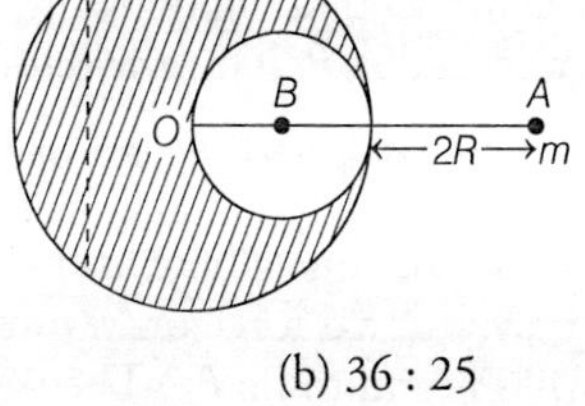

(a) 25 : 36 (b) 36 : 25
(c) 50 : 41 (d) 41 : 50

34 Four identical particles of equal masses 1 kg made to move along the circumference of a circle of radius 1 m under the action of their own mutual gravitational attraction. The speed of each particle will be *[2021, 24 Feb Shift-I]*

(a) $\frac{1}{2}\sqrt{\frac{(1+2\sqrt{2})G}{1}}$ (b) $\sqrt{\frac{G}{2}(1+2\sqrt{2})}$
(c) $\sqrt{G(1+2\sqrt{2})}$ (d) $\sqrt{\frac{G}{2}(2\sqrt{2}-1)}$

35 The mass density of a spherical galaxy varies as K/r over a large distance r from its centre. In that region, a small star is in a circular orbit of radius R. Then, the period of revolution T depends on R as *[2020, 2 Sep Shift-I]*

(a) $T \propto R$ (b) $T^2 \propto R^3$
(c) $T^2 \propto R$ (d) $T^2 \propto \frac{1}{R^3}$

Imp. Application of Gravitational Force

36 Four identical particles of mass M are located at the corners of a square of side a. What should be their speed, if each of them revolves under the influence of other's gravitational field in a circular orbit circumscribing the square? *[2019, 8 April Shift-I]*

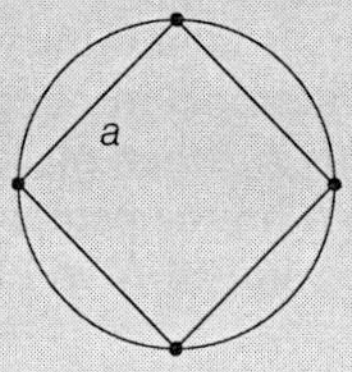

(a) $1.35\sqrt{\frac{GM}{a}}$ (b) $1.16\sqrt{\frac{GM}{a}}$
(c) $1.21\sqrt{\frac{GM}{a}}$ (d) $1.41\sqrt{\frac{GM}{a}}$

37 A test particle is moving in a circular orbit in the gravitational field produced by mass density $\rho(r) = \frac{K}{r^2}$. Identify the correct relation between the radius R of the particle's orbit and its period T *[2019, 9 April Shift-II]*

(a) T^2/R^3 is a constant (b) T/R^2 is a constant
(c) TR is a constant (d) T/R is a constant

38 Suppose the gravitational force varies inversely as the nth power of distance. Then, the time period of a planet in circular orbit of radius R around the sun will be proportional to *[AIEEE 2004]*

(a) $R^{\left(\frac{n+1}{2}\right)}$ (b) $R^{\left(\frac{n-1}{2}\right)}$
(c) R^n (d) $R^{\left(\frac{n-2}{2}\right)}$

Numerical Value Type Question

39. Three identical spheres of mass m, are placed at the vertices of an equilateral triangle of length a. When released, they interact only through gravitational force and collide after a time $T = 4$ s. If the sides of the triangle are increased to length $2a$ and also the masses of the spheres are made $2m$, then they will collide after .... seconds. *[2025, 3 April Shift-I]*

40 Two satellites revolve around a planet in coplanar circular orbits in anti-clockwise direction. Their period of revolutions are 1 h and 8 h, respectively. The radius of the orbit of nearer satellite is 2×10^3 km. The angular speed of the farther satellite as observed from the nearer satellite at the instant when both the satellites are closest is π/x rad h^{-1}, where x is .... *[2021, 1 Sep. Shift-II]*

Topic 2 Acceleration due to Gravity and its Variation

Level 1

01 The acceleration due to gravity on the surface of earth is g. If the diameter of earth reduces to half of its original value and mass remains constant, then acceleration due to gravity on the surface of earth would be *[2024, 27 Jan Shift-I]*

(a) $g/2$ (b) $4g$ (c) $g/4$ (d) $2g$

02 Given below are two statements

Statement I Rotation of the Earth shows effect on the value of acceleration due to gravity g.

Statement II The effect of rotation of the Earth on the value of g at the equator is minimum and that at the pole is maximum.

In the light of the above statements, choose the correct answer from the options given below. *[2023, 10 April Shift-II]*

(a) Both Statement I and Statement II are true.
(b) Statement I is true but Statement II is false.
(c) Both Statement I and Statement II are false.
(d) Statement I is false but Statement II is true.

03 The acceleration due to gravity at height h above the earth if $h << R$ (radius of earth) is given by *[2023, 8 April Shift-II]*

(a) $g' = g\left(1-\frac{h^2}{2R^2}\right)$ (b) $g' = g\left(1-\frac{2h^2}{R^2}\right)$

(c) $g' = g\left(1-\frac{2h}{R}\right)$ (d) $g' = g\left(1-\frac{h}{2R}\right)$

Imp. Variation in g on Going Below the Earth's Surface

04 The weight of a body on the earth is 400 N. Then weight of the body when taken to a depth half of the radius of the earth will be *[2023, 8 April Shift-I]*

(a) 300 N (b) zero
(c) 100 N (d) 200 N

05 Given below are two statements.

Statement I Acceleration due to earth's gravity decreases as go up or down from earth's surface.

Statement II Acceleration due to earth's gravity is same at a height h and depth d from earth's surface, if $h = d$.

In the light of the above Statements, choose the most appropriate answer from the options given below. *[2023, 24 Jan Shift-II]*

(a) Both Statement I and Statement II are correct.
(b) Statement I is correct but Statement II is incorrect.
(c) Both Statement I and Statement II are incorrect.
(d) Statement I is incorrect but Statement II is correct.

06 The percentage decrease in the weight of a rocket, when taken to a height of 32 km above the surface of Earth will be (radius of earth = 6400 km) *[2022, 26 July Shift-I]*

(a) 1% (b) 3% (c) 4% (d) 0.5%

07 Given below are two statements.

Statement I The law of gravitation holds good for any pair of bodies in the universe.

Statement II The weight of any person becomes zero when the person is at the centre of the earth.

In the light of the above statements, choose the correct answer from the options given below. *[2022, 27 June Shift-I]*

(a) Both Statement I and Statement II are true.
(b) Both Statement I and Statement II are false.
(c) Statement I is true but Statement II is false.
(d) Statement I is false but Statement II is true.

08 The variation of acceleration due to gravity g with distance r from the centre of the earth is correctly represented by (Given, R = radius of earth) *[2022, 26 June Shift-I]*

(a) 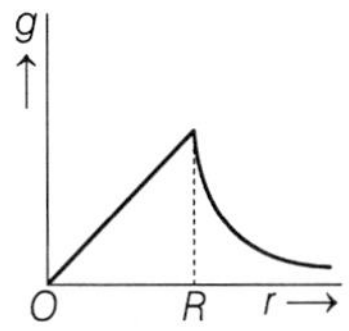

(b) 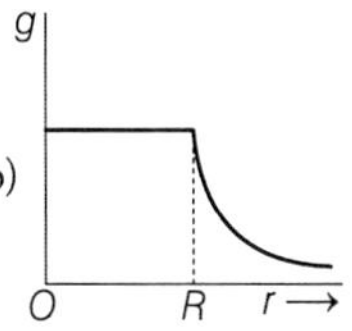

(c) 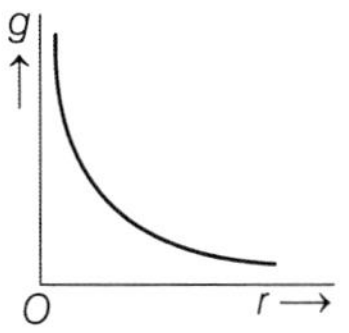

(d) 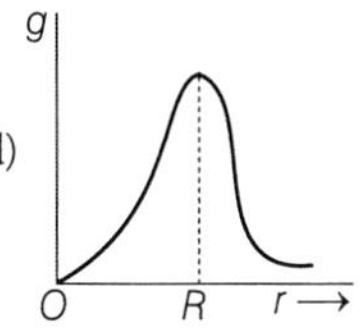

09 The height of any point P above the surface of earth is equal to diameter of earth. The value of acceleration due to gravity at point P will be (Given, g = acceleration due to gravity at the surface of earth) *[2022, 25 June Shift-I]*

(a) $g/2$ (b) $g/4$ (c) $g/3$ (d) $g/9$

10 Suppose that the angular velocity of rotation of earth is increased. Then, as a consequence *[Online 2018]*

(a) there will be no change in weight anywhere on the earth
(b) weight of the object, everywhere on the earth, will increase
(c) except at poles, weight of the object on the earth will decrease
(d) weight of the object, everywhere on the earth, will decrease.

Level 2

11 Two planets A and B of radii R and $1.5R$ have densities ρ and $\rho/2$ respectively. The ratio of acceleration due to gravity at the surface of B to A is *[2023, 13 April Shift-II]*

(a) 3 : 4 (b) 4 : 3 (c) 2 : 3 (d) 2 : 1

12 Assuming the earth to be a sphere of uniform mass density, the weight of a body at a depth $d = \frac{R}{2}$ from the surface of earth, if its weight on the surface of the earth is 200 N, will be
(Given, R = radius of earth) *[2023, 10 April Shift-I]*

(a) 500 N (b) 300 N (c) 400 N (d) 100 N

13 Given below are two statements.

Statement I Acceleration due to gravity is different at different places on the surfaces of earth.

Statement II Acceleration due to gravity increases as we go down below the earth's surface.

In the light of the above statements, choose the correct answer from the options given below. *[2023, 1 Feb Shift-I]*

(a) Both Statement I and Statement II are true.
(b) Both Statement I and Statement II are false.
(c) Statement I is false but Statement II is true.
(d) Statement I is true but Statement II is false.

Imp. Variation in g on Going Above the Earth's Surface

14 A body weight w, is projected vertically upwards from earth's surface to reach a height above the earth which is equal to nine times the radius of earth. The weight of the body at that height will be *[2023, 31 Jan Shift-II]*

(a) $\frac{w}{100}$ (b) $\frac{w}{3}$ (c) $\frac{w}{9}$ (d) $\frac{w}{91}$

15 At a certain depth d below surface of earth, value of acceleration due to gravity becomes four times that of its value at a height $3R$ above earth surface, where R is Radius of earth (Take, $R = 6400$ km). The depth d is equal to *[2023, 31 Jan Shift-I]*

(a) 640 km (b) 4800 km
(c) 5260 km (d) 2560 km

16 If the radius of earth shrinks by 2% while its mass remains same. The acceleration due to gravity on the earth's surface will approximately *[2022, 28 July Shift-I]*

(a) decrease by 2% (b) decrease by 4%
(c) increase by 2% (d) increase by 4%

17 An object is taken to a height above the surface of earth at a distance $\frac{5}{4}R$ from the centre of the earth. Where radius of earth, $R = 6400$ km. The percentage decrease in the weight of the object will be *[2022, 25 July Shift-II]*

(a) 36% (b) 50% (c) 64% (d) 25%

18 The approximate height from the surface of earth at which the weight of the body becomes 1/3 of its weight on the surface of earth is [Radius of earth, $R = 6400$ km and $\sqrt{3} = 1.732$] *[2022, 27 June Shift I]*

(a) 3840 km (b) 4685 km (c) 2133 km (d) 4267 km

19 Given below are two statements. One is labelled as Assertion A and the other is labelled as Reason R.

Assertion (A) If we move from poles to equator, the direction of acceleration due to gravity of earth always points towards the centre of earth without any variation in its magnitude.

Reason (R) At equator, the direction of acceleration due to the gravity is towards the centre of earth.
In the light of above statements, choose the correct answer from the options given below. *[2022, 26 June Shift-II]*

(a) Both A and R are true and R is the correct explanation of A.
(b) Both A and R are true but R is not the correct explanation of A.
(c) A is true but R is false.
(d) A is false but R is true.

20 A body weights 49 N on a spring balance at the north pole. What will be its weight recorded on the same weighing machine, if it is shifted to the equator?
(Use, $g = \frac{GM}{R^2} = 9.8\ \text{ms}^{-2}$ and radius of earth, $R = 6400$ km) *[2021, 24 Feb Shift-II]*

(a) 49 N (b) 48.83 N (c) 49.83 N (d) 49.17 N

21 The value of the acceleration due to gravity is g_1 at a height $h = R/2$ (where, R = radius of the earth) from the surface of the earth. It is again equal to g_1 at a depth d below the surface of the earth. The ratio d/R equals *[2020, 5 Sep Shift-I]*

(a) 7/9 (b) 1/3 (c) 4/9 (d) 5/9

22 Consider two solid spheres of radii $R_1 = 1$ m, $R_2 = 2$ m and masses M_1 and M_2, respectively. The gravitational field due to sphere 1 and 2 are shown. The value of $\frac{M_1}{M_2}$ is *[2020, 8 Jan Shift-I]*

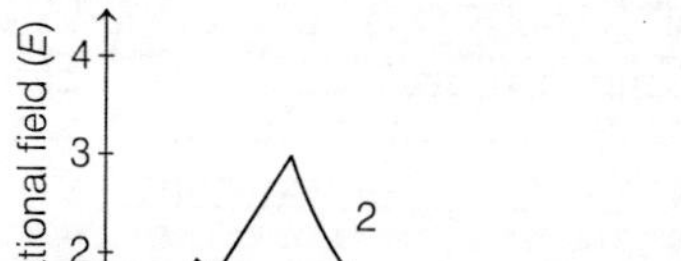

(a) 1/3 (b) 1/2 (c) 1/6 (d) 2/3

23 The value of acceleration due to gravity at earth's surface is 9.8 ms^{-2}. The altitude above its surface at which the acceleration due to gravity decreases to 4.9 ms^{-2}, is close to (Take, radius of earth $= 6.4 \times 10^6$ m) *[2019, 10 April Shift-I]*
(a) 9.0×10^6 m (b) 2.6×10^6 m
(c) 6.4×10^6 m (d) 1.6×10^6 m

Imp. Variation in g Due to Rotation of Earth

24 If the earth has no rotational motion, the weight of a person on the equator is w. Determine the speed with which the earth would have to rotate about its axis so that the person at the equator will weigh 3/4 w. (Radius of the earth is 6400 km and $g = 10$ m/s^2). *[JEE Main 2017]*
(a) 0.63×10^{-3} rad/s (b) 0.83×10^{-3} rad/s
(c) 0.28×10^{-3} rad/s (d) 1.1×10^{-3} rad/s

Level 3

25 Assuming the earth to be a sphere of uniform mass density, a body weighed 300 N on the surface of earth. How much it would weight at $R/4$ depth under surface of earth ? *[2024, 6 April Shift-II]*
(a) 225 N (b) 375 N
(c) 300 N (d) 75 N

26 At what distance above and below the surface of the earth a body will have same weight. (Take, radius of earth as R) *[2024, 29 Jan Shift-I]*
(a) $\sqrt{5}R - R$ (b) $\frac{\sqrt{5}R - R}{2}$
(c) $\frac{R}{2}$ (d) $\frac{\sqrt{3}R - R}{2}$

27 The weight of a body at the surface of earth is 18 N. The weight of the body at an altitude of 3200 km above the earth's surface is (Given, radius of earth, $R_e = 6400$ km). *[2023, 24 Jan Shift-I]*
(a) 9.8 N (b) 19.6 N (c) 8 N (d) 4.9 N

28 If R_E be the radius of earth, then the ratio between the acceleration due to gravity at a depth r below and a height r above the earth surface is (Given, $r < R_E$) *[2021, 31 Aug Shift-II]*
(a) $1 + \frac{r}{R_E} - \frac{r^2}{R_E^2} - \frac{r^3}{R_E^3}$
(b) $1 + \frac{r}{R_E} + \frac{r^2}{R_E^2} + \frac{r^3}{R_E^3}$
(c) $1 - \frac{r}{R_E} - \frac{r^2}{R_E^2} - \frac{r^3}{R_E^3}$
(d) $1 + \frac{r}{R_E} - \frac{r^2}{R_E^2} + \frac{r^3}{R_E^3}$

29 Consider a planet in some solar system which has a mass double the mass of earth and density equal to the average density of earth. If the weight of an object on earth is w, the weight of the same object on that planet will be *[2021, 25 July Shift-II]*
(a) $2w$ (b) w (c) $2^{1/3}w$ (d) $\sqrt{2}w$

30 A person whose mass is 100 kg travels from earth to mars in a spaceship. Neglect all other objects in sky and take acceleration due to gravity on the surface of the earth and mars as 10 m/s^2 and 4 m/s^2, respectively.
Identify from the below figures, the curve that fits best for the weight of the passenger as a function of time. *[2021, 20 July Shift-I]*

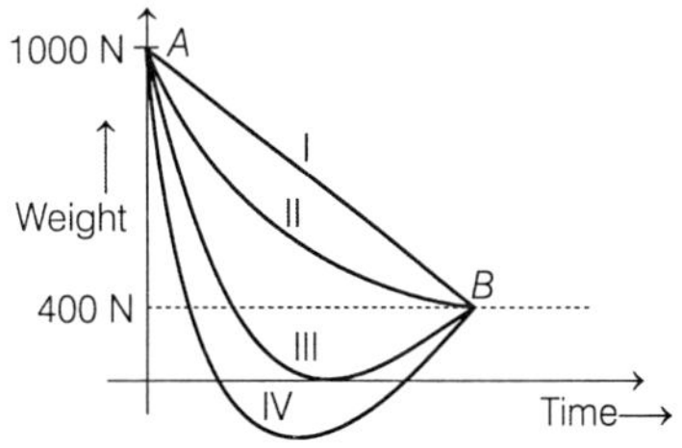

(a) III (b) I (c) IV (d) II

31 If the angular velocity of earth's spin is increased such that the bodies at the equator start floating, the duration of the day would be approximately
(Take, $g = 10$ ms^{-2}, the radius of earth, $R = 6400 \times 10^3$ m, take $\pi = 3.14$) *[2021, 18 March Shift-II]*
(a) 60 min (b) does not change
(c) 1200 min (d) 84 min

Numerical Value Type Questions

32 Acceleration due to gravity on the surface of earth is g. If the diameter of earth is reduced to one-third of its original value and mass remains unchanged, then the acceleration due to gravity on the surface of the earth is .... g. *[2025, 24 Jan Shift-II]*

33 If the acceleration due to gravity experienced by a point mass at a height h above the surface of earth is same as that of the acceleration due to gravity at a depth αh ($h << R_e$) from the earth's surface. The value of α will be .... (Use, $R_e = 6400$ km) *[2022, 29 July Shift-I]*

34 In the reported figure of earth, the value of acceleration due to gravity is same at point A and C but it is smaller than that of its value at point B (surface of the earth). The value of $OA : AB$ will be $x : y$. The value of x is ....

[2021, 26 Feb Shift-II]

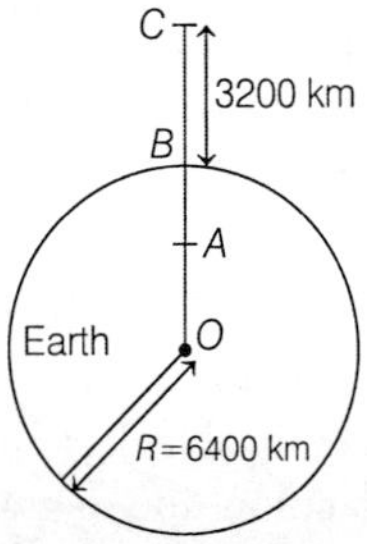

Topic 3 Gravitational Potential and Gravitational Potential Energy

Level 1

01 To project a body of mass m from earth's surface to infinity, the required kinetic energy is (assume, the radius of earth is R_E, g = acceleration due to gravity on the surface of earth) *[2024, 6 April Shift-I]*

(a) $\frac{1}{2} mgR_E$ (b) mgR_E (c) $4mgR_E$ (d) $2mgR_E$

02 Two bodies of masses m and $4\,m$ are placed at a distance r. The gravitational potential at a point on the line joining them, where the gravitational field is zero, is *[AIEEE 2011]*

(a) $-\frac{4\,Gm}{r}$ (b) $-\frac{6\,Gm}{r}$

(c) $-\frac{9\,Gm}{r}$ (d) zero

Level 2

03 Given below are two statements. One is labelled as Assertion (A) and the other is labelled as Reason (R).

Assertion (A) The kinetic energy needed to project a body of mass m from Earth surface to infinity is $\frac{1}{2} mgR$, where R is the radius of Earth.

Reason (R) The maximum potential energy of a body is zero when it is projected to infinity from Earth surface.

In the light of the above statements, choose the correct answer from the options given below. *[2025, 4 April Shift-I]*

(a) Both A and R are true and R is the correct explanation of A.
(b) A is true but R is false.
(c) A is false but R is true.
(d) Both A and R are true but R is not the correct explanation of A.

04 An astronaut takes a ball of mass m from earth to space. He throws the ball into a circular orbit about earth at an altitude of 318.5 km. From earth's surface to the orbit, the change in total mechanical energy of the ball is $x\frac{GM_e m}{21R_e}$. The value of x is

(Take, $R_e = 6370$ km) *[2024, 9 April Shift-I]*

(a) 9 (b) 11
(c) 10 (d) 12

05 Match List I with List II.

	List I		List II
A.	Kinetic energy of planet	I.	$-GMm/a$
B.	Gravitational potential energy of Sun-planet system	II.	$GMm/2a$
C.	Total mechanical energy of planet	III.	$\frac{Gm}{r}$
D.	Escape energy at the surface of planet for unit mass object	IV.	$-GMm/2a$

(where a = radius of planet orbit, r = radius of planet, M = mass of Sun, m = mass of planet)

Choose the correct answer from the options given below. *[2024, 5 April Shift-I]*

(a) A-I, B-II, C-III, D-IV (b) A-II, B-I, C-IV, D-III
(c) A-I, B-IV, C-II, D-III (d) A-III, B-IV, C-I, D-II

06 If V is the gravitational potential due to sphere of uniform density on it's surface, then it's value at the centre of sphere will be *[2023, 11 April Shift-II]*

(a) $\frac{V}{2}$ (b) V (c) $\frac{3V}{2}$ (d) $\frac{4}{3}V$

07 If the gravitational field in the space is given as $\left(-\frac{K}{r^2}\right)$. Taking the reference point to be at $r = 2$ cm with gravitational potential $V = 10$ J/kg. Find the gravitational potential at $r = 3$ cm in SI unit (Given, $K = 6$ J-cm/kg) *[2023, 30 Jan Shift-I]*

(a) 12 (b) 9
(c) 10 (d) 11

08 Four spheres each of mass m form a square of side d (as shown in figure). A fifth sphere of mass M is situated at the centre of square. The total gravitational potential energy of the system is *[2022, 27 June Shift-II]*

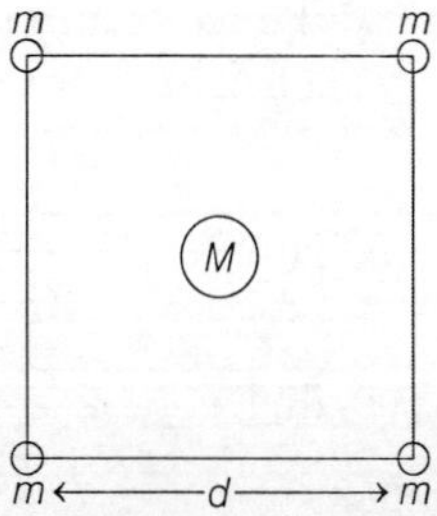

(a) $-\frac{Gm}{d}[(4+\sqrt{2})m + 4\sqrt{2}\,M]$

(b) $-\frac{Gm}{d}[(4+\sqrt{2})M + 4\sqrt{2}\,m]$

(c) $-\frac{Gm}{d}[3m^2+4\sqrt{2}M]$

(d) $-\frac{Gm}{d}[6m^2+4\sqrt{2}M]$

09 A particle of mass 10 g is kept on the surface of a uniform sphere of mass 100 kg and radius 10 cm. Find the work to be done against the gravitational force between them, to take the particle far away from the sphere, (you may take $G = 6.67\times 10^{-11}\ \text{Nm}^2/\text{kg}^2$) *[AIEEE 2005]*

(a) 13.34×10^{-10} J (b) 3.33×10^{-10} J
(c) 6.67×10^{-9} J (d) 6.67×10^{-10} J

Level 3

10 The gravitational potential at a point above the surface of earth is -5.12×10^7 J/kg and the acceleration due to gravity at that point is 6.4 m/s^2. Assume that the mean radius of earth to be 6400 km. The height of this point above the earth's surface is *[2024, 30 Jan Shift-I]*

(a) 1200 km (b) 540 km
(c) 1000 km (d) 1600 km

11 An object of mass 1 kg is taken to a height from the surface of earth which is equal to three times the radius of earth. The gain in potential energy of the object will be
(If, $g=10\ \text{ms}^{-2}$ and radius of earth $=6400$ km) *[2022, 29 July Shift-II]*

(a) 48 MJ (b) 24 MJ
(c) 36 MJ (d) 12 MJ

12 A body is projected vertically upwards from the surface of Earth with a velocity sufficient enough to carry it to infinity. The time taken by it to reach height h is .... s. *[2021, 22 July Shift-II]*

(a) $\sqrt{\frac{R_e}{2g}}\left[\left(1+\frac{h}{R_e}\right)^{3/2}-1\right]$ (b) $\sqrt{\frac{2R_e}{g}}\left[\left(1+\frac{h}{R_e}\right)^{3/2}-1\right]$

(c) $\frac{1}{3}\sqrt{\frac{R_e}{2g}}\left[\left(1+\frac{h}{R_e}\right)^{3/2}-1\right]$ (d) $\frac{1}{3}\sqrt{\frac{2R_e}{g}}\left[\left(1+\frac{h}{R_e}\right)^{3/2}-1\right]$

13 Two planets have masses M and $16M$ and their radii are a and $2a$, respectively. The separation between the centres of the planets is $10a$. A body of mass m is fired from the surface of the larger planet towards the smaller planet along the line joining their centres. For the body to be able to reach at the surface of smaller planet, the minimum firing speed needed is *[2020, 6 Sep Shift-II]*

(a) $2\sqrt{\frac{GM}{a}}$ (b) $4\sqrt{\frac{GM}{a}}$ (c) $\sqrt{\frac{GM^2}{ma}}$ (d) $\frac{3}{2}\sqrt{\frac{5GM}{a}}$

Numerical Value Type Questions

Imp. Gravitational Potential Energy

14 A body of mass $2M$ splits into four masses $\{m, M-m, m, M-m\}$, which are rearranged to form a square as shown in the figure. The ratio of M/m for which, the gravitational potential energy of the system becomes maximum is $x:1$. The value of x is......... . *[2021, 27 Aug Shift-I]*

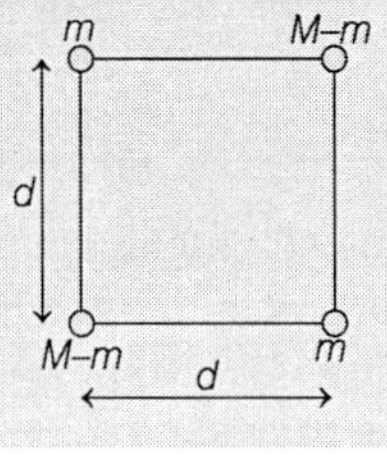

15 Suppose two planets (spherical in shape) of radii R and $2R$, but mass M and $9M$ respectively have a centre to centre separation $8R$ as shown in the figure. A satellite of mass m is projected from the surface of the planet of mass M directly towards the centre of the second planet. The minimum speed v required for the satellite to reach the surface of the second planet is $\sqrt{\frac{a}{7}\frac{GM}{R}}$, then the value of a is ....
[Take, the two planets are fixed in their position] *[2021, 27 July Shift-I]*

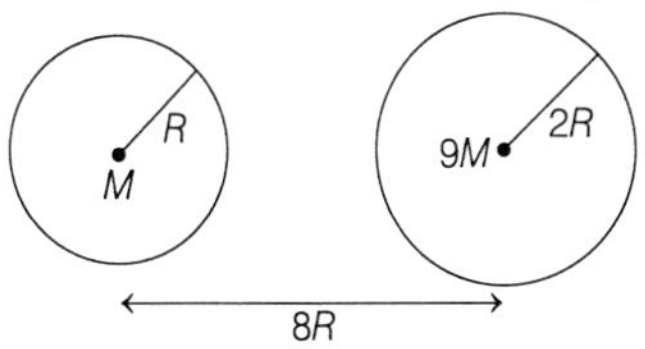

16 If one wants to remove all the mass of the earth to infinity in order to break it up completely. The amount of energy that need to be supplied will be $\frac{x}{5}\frac{GM^2}{R}$, where x is ....
(Round off to the nearest integer) (M is the mass of earth, R is the radius of earth and G is the gravitational constant.) *[2021, 16 March Shift-II]*

Topic 4 Escape Speed and Motion of Satellites

Level 1

01 Earth has mass 8 times and radius 2 times that of a planet. If the escape velocity from the earth is 11.2 km/s the escape velocity in km/s from the planet will be *[2025, 28 Jan Shift-II]*

(a) 2.8 (b) 5.6 (c) 8.4 (d) 11.2

02 Two satellite A and B go round a planet in circular orbits having radii $4R$ and R respectively. If the speed of A is $3v$, the speed of B will be *[2024, 8 April Shift-II]*

(a) $12v$ (b) $3v$ (c) $6v$ (d) $\frac{4}{3}v$

03 The mass of the moon is 1/144 times the mass of a planet and its diameter is 1/16 times the diameter of a planet. If the escape velocity on the planet is v, then the escape velocity on the moon will be *[2024, 31 Jan Shift-II]*

(a) $v/4$ (b) $v/3$
(c) $v/6$ (d) $v/12$

04 Given below are two statements.

Statement I For a planet, if the ratio of mass of the planet to its radius increases, the escape velocity from the planet also increases.

Statement II Escape velocity is independent of the radius of the planet.

In the light of the above statements, choose the correct answer from the options given below. *[2023, 13 April Shift-II]*

(a) Statement I is incorrect but Statement II is correct.
(b) Both Statement I and Statement II are correct.
(c) Statement I is correct but Statement II is incorrect.
(d) Both Statement I and Statement II are incorrect.

05 Two satellites of masses m and $3m$ revolve around the earth in circular orbits of radii r and $3r$ respectively. The ratio of orbital speeds of the satellites respectively is *[2023, 10 April Shift-I]*

(a) 9 : 1 (b) $\sqrt{3}$: 1 (c) 1 : 1 (d) 3 : 1

06 Choose the incorrect statement from the following. *[2023, 6 April Shift-II]*

(a) The linear speed of a planet revolving around the sun remains constant.
(b) When a body falls towards the earth, the displacement of earth towards the body is negligible.
(c) The speed of satellite in a given circular orbit remains constant.
(d) For a planet revolving around the sun in an elliptical orbit, the total energy of the planet remains constant.

07 Given below are two statements. One is labelled as Assertion (A) and the other is labelled as Reason (R).

Assertion (A) Earth has atmosphere whereas moon doesn't have any atmosphere.

Reason (R) The escape velocity on moon is very small as compared to that on earth.

In the light of the above statements, choose the correct answer from the options given below.

(a) Both A and R are correct but R is not the correct explanation of A. *[2023, 6 April Shift-I]*
(b) A is true but R is false.
(c) A is false but R is true.
(d) Both A and R are correct and R is the correct explanation of A.

Imp. Escape Velocity

08 If earth has a mass nine times and radius twice to that of a planet P, then $\frac{v_e}{3}\sqrt{x}\ \text{ms}^{-1}$ will be the minimum velocity required by a rocket to pull out of gravitational force of P, where v_e is escape velocity on earth. The value of x is *[2023, 1 Feb Shift I]*

(a) 18 (b) 1 (c) 2 (d) 3

09 A body is moving in a low circular orbit about a planet of mass M and radius R. The radius of the orbit can be taken to be R itself. Then, the ratio of the speed of this body in the orbit to the escape velocity from the planet is *[2020, 4 Sep Shift-II]*

(a) 1 (b) $1/\sqrt{2}$ (c) $\sqrt{2}$ (d) 2

10 Planet A has mass M and radius R. Planet B has half the mass and half the radius of planet A. If the escape velocities from the planets A and B are v_A and v_B respectively, then $\frac{v_A}{v_B}=\frac{n}{4}$. The value of n is *[2020, 9 Jan Shift-II]*

(a) 1 (b) 2 (c) 3 (d) 4

Imp. Energy of Satellite

11 Two satellites A and B have masses m and $2m$ respectively. A is in a circular orbit of radius R and B is in a circular orbit of radius $2R$ around the earth. The ratio of their kinetic energies, T_A/T_B is *[2019, 12 Jan Shift-II]*

(a) 1/2 (b) 2 (c) $\sqrt{1/2}$ (d) 1

12 A satellite of mass M is in a circular orbit of radius R about the centre of the earth. A meteorite of the same mass falling towards the earth collides with the satellite completely inelastically. The speeds of the satellite and the meteorite are the same just before the collision. The subsequent motion of the combined body will be *[2019, 12 Jan Shift-I]*

(a) in the same circular orbit of radius R
(b) in an elliptical orbit
(c) such that it escapes to infinity
(d) in a circular orbit of a different radius

Level 2

13 An object is kept at rest at a distance of $3R$ above the earth's surface, where R is earth's radius. The minimum speed with which it must be projected so that it does not return to earth is

(Assume M = mass of earth, G = Universal gravitational constant) *[2025, 4 April Shift-II]*

(a) $\sqrt{\frac{GM}{R}}$ (b) $\sqrt{\frac{2GM}{R}}$
(c) $\sqrt{\frac{3GM}{R}}$ (d) $\sqrt{\frac{GM}{2R}}$

14 Correct formula for height of a satellite from earth's surface is *[2024, 4 April Shift-II]*

(a) $\left(\frac{T^2R^2}{4\pi^2 g}\right)^{\frac{1}{3}} - R$ (b) $\left(\frac{T^2R^2 g}{4\pi^2}\right)^{\frac{1}{3}} - R$

(c) $\left(\frac{T^2R^2 g}{4\pi^2}\right)^{\frac{-1}{3}} + R$ (d) $\left(\frac{T^2R^2 g}{4\pi}\right)^{\frac{1}{2}} - R$

15 Escape velocity of a body from earth is 11.2 km/s. If the radius of a planet be one-third the radius of earth and mass be one-sixth that of earth, the escape velocity from the planet is *[2024, 30 Jan Shift-II]*

(a) 8.4 km/s (b) 11.2 km/s (c) 4.2 km/s (d) 7.9 km/s

16 A planet having mass 9 M_e and radius $4R_e$, where M_e and R_e are mass and radius of earth respectively, has escape velocity in km/s given by

(Given, escape velocity on earth $v_e = 11.2 \times 10^3$ m/s) *[2023, 13 April Shift-I]*

(a) 11.2 (b) 33.6 (c) 67.2 (d) 16.8

17 The ratio of escape velocity of a planet to the escape velocity of the earth will be

(Given : Mass of the planet is 16 times mass of earth and radius of the planet is 4 times the radius of the earth.) *[2023, 12 April Shift-I]*

(a) 1 : 4 (b) $1:\sqrt{2}$ (c) 2 : 1 (d) 4 : 1

18 Two satellites A and B move round the earth in the same orbit. The mass of A is twice the mass of B. The quantity which is same for the two satellites will be *[2023, 12 April Shift-I]*

(a) total energy (b) kinetic energy
(c) speed (d) potential energy

19 A space ship of mass 2×10^4 kg is launched into a circular orbit close to the earth's surface. The additional velocity to be imparted to the space ship in the orbit to overcome the gravitational pull will be (if $g = 10$ m/s^2 and radius of earth = 6400 km) *[2023, 11 April Shift-II]*

(a) $11.2(\sqrt{2}-1)$ km/s (b) $7.9(\sqrt{2}-1)$ km/s
(c) $8(\sqrt{2}-1)$ km/s (d) $7.4(\sqrt{2}-1)$ km/s

20 The time period of a satellite, revolving above earth's surface at a height equal to R will be

(Given, $g = \pi^2$ m/s^2 and R = radius of earth) *[2023, 10 April Shift-II]*

(a) $\sqrt{2R}$ (b) $\sqrt{8R}$
(c) $\sqrt{4R}$ (d) $\sqrt{32R}$

21 The escape velocities of two planets A and B are in the ratio 1 : 2. If the ratio of their radii respectively is 1 : 3, then the ratio of acceleration due to gravity of planet A to the acceleration of gravity of planet B will be *[2023, 1 Feb Shift II]*

(a) 4/3 (b) 3/4 (c) 3/2 (d) 2/3

22 Two satellites A and B, having masses in the ratio 4 : 3, are revolving in circular orbits of radii $3r$ and $4r$, respectively around the earth. The ratio of total mechanical energy of A to B is *[2022, 27 July Shift-I]*

(a) 9 : 16 (b) 16 : 9 (c) 1 : 1 (d) 4 : 3

23 The escape velocity of a body on a planet A is 12 kms^{-1}. The escape velocity of the body on another planet B, whose density is four times and radius is half of the planet A is *[2022, 29 June Shift-I]*

(a) 12 kms^{-1} (b) 24 kms^{-1}
(c) 36 kms^{-1} (d) 6 kms^{-1}

24 The masses and radii of the earth and Moon are (m_1, R_1) and (m_2, R_2), respectively. Their centres are at a distance r apart. Find the minimum escape velocity for a particle of mass m to be projected from the middle of these two masses. *[2021, 31 August Shift-I]*

(a) $v = \frac{1}{2}\sqrt{\frac{4G(m_1+m_2)}{r}}$ (b) $v = \sqrt{\frac{4G(m_1+m_2)}{r}}$

(c) $v = \frac{1}{2}\sqrt{\frac{2G(m_1+m_2)}{r}}$ (d) $v = \frac{\sqrt{2G}\,(m_1+m_2)}{r}$

25 The planet mars has two moons, if one of them has a period 7 h 30 min and an orbital radius of 9.0×10^3 km, then the mass of mars is *[2021, 27 July Shift-II]*

$$\left[\text{Take, } \frac{4\pi^2}{G} = 6 \times 10^{11}\ \text{N}^{-1}\text{m}^{-2}\text{kg}^2\right]$$

(a) 5.96×10^{19} kg (b) 3.25×10^{21} kg
(c) 7.02×10^{25} kg (d) 6.00×10^{23} kg

26 A satellite is launched into a circular orbit of radius R around earth, while a second satellite is launched into a circular orbit of radius $1.02\,R$. The percentage difference in the time periods of the two satellites is *[2021, 20 July Shift-II]*

(a) 1.5 (b) 3.0 (c) 0.7 (d) 2.0

27 Given below are two statements. One is labelled as Assertion (A) and the other is labelled as Reason (R). *[2021, 25 Feb Shift-I]*

Assertion (A) The escape velocities of planet A and B are same. But A and B are of unequal mass.

Reason (R) The product of their mass and radius must be same, $M_1R_1 = M_2R_2$.

In the light of the above statements, choose the most appropriate answer from the options given below.

(a) Both A and R are correct but R is not the correct explanation of A.
(b) A is correct but R is not correct.
(c) Both A and R are correct and R is the correct explanation of A.
(d) A is not correct but R is correct.

28 Consider two satellites S_1 and S_2 with periods of revolution 1 h and 8 h respectively, revolving around a planet in circular orbits. The ratio of angular velocity of satellite S_1 to the angular velocity of satellite S_2 is [2021, 24 Feb Shift-I]

(a) 8 : 1 (b) 1 : 8 (c) 2 : 1 (d) 1 : 4

Imp. Motion of Satellite

29 A satellite is revolving in a circular orbit at a height h from the earth surface such that $h << R$, where R is the radius of the earth. Assuming that the effect of earth's atmosphere can be neglected, so the minimum increase in the speed required so that the satellite could escape from the gravitational field of earth is [2019, 11 Jan Shift-I, JEE Main 2017]

(a) $\sqrt{gR/2}$ (b) $\sqrt{gR}$ (c) $\sqrt{2gR}$ (d) $\sqrt{gR}(\sqrt{2}-1)$

30 A satellite is moving with a constant speed v in circular orbit around the earth. An object of mass 'm' is ejected from the satellite such that it just escapes from the gravitational pull of the earth. At the time of ejection, the kinetic energy of the object is [2019, 10 Jan Shift-I]

(a) $\frac{3}{2}mv^2$ (b) $2mv^2$ (c) mv^2 (d) $\frac{1}{2}mv^2$

31 What is the minimum energy required to launch a satellite of mass m from the surface of a planet of mass M and radius R in a circular orbit at an altitude of $2R$? [JEE Main 2013]

(a) $\frac{5GmM}{6R}$ (b) $\frac{2GmM}{3R}$ (c) $\frac{GmM}{2R}$ (d) $\frac{GmM}{3R}$

32 A planet in a distant solar system is 10 times more massive than the earth and its radius is 10 times smaller. Given that the escape velocity from the earth is 11 kms^{-1}, the escape velocity from the surface of the planet would be [AIEEE 2008]

(a) 1.1 kms^{-1} (b) 11 kms^{-1}
(c) 110 kms^{-1} (d) 0.11 kms^{-1}

Level 3

33 A satellite of 10^3 kg mass is revolving in circular orbit of radius $2R$. If $\frac{10^4 R}{6}$ J energy is supplied to the satellite, it would revolve in a new circular orbit of radius. [2024, 9 April Shift-II]
(Use, $g = 10 \text{ m/s}^2$, R = radius of earth)

(a) $3R$ (b) $2.5R$
(c) $6R$ (d) $4R$

34 A body of mass m is projected with velocity λv_e in vertically upward direction from the surface of the earth into space. It is given that v_e is escape velocity and $\lambda < 1$. If air resistance is considered to be negligible, then the maximum height from the centre of earth, to which the body can go, will be (R is radius of earth) [2022, 27 July Shift-II]

(a) $\frac{R}{1+\lambda^2}$ (b) $\frac{R}{1-\lambda^2}$ (c) $\frac{R}{1-\lambda}$ (d) $\frac{\lambda^2 R}{1-\lambda^2}$

35 A body is projected vertically upwards from the surface of earth with a velocity equal to one-third of escape velocity. The maximum height attained by the body will be (Take, radius of earth = 6400 km and $g = 10 \text{ ms}^{-2}$) [2022, 26 July Shift-II]

(a) 800 km (b) 1600 km (c) 2133 km (d) 4800 km

36 Two satellites A and B of masses 200 kg and 400 kg are revolving around the Earth at height of 600 km and 1600 km, respectively. If T_A and T_B are the time periods of A and B respectively, then the value of $T_B - T_A$ is [2021, 25 Feb Shift-I]

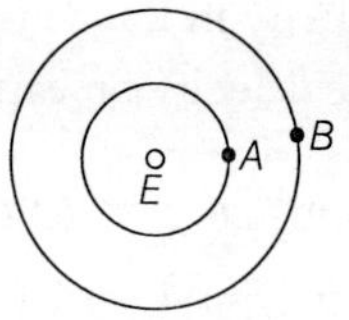

(Given, radius of earth = 6400 km, mass of earth $= 6 \times 10^{24}$ kg)

(a) 1.30×10^3 s (b) 3.33×10^2 s
(c) 4.24×10^3 s (d) 4.24×10^2 s

37 A satellite is moving in a low nearly circular orbit around the earth. Its radius is roughly equal to that of the earth's radius R_e. By firing rockets attached to it, its speed is instantaneously increased in the direction of its motion, so that it becomes $\sqrt{3/2}$ times larger.

Due to this, the farthest distance from the centre of the earth that the satellite reaches is R. Value of R is [2020, 3 Sep Shift-I]

(a) $4R_e$ (b) $2.5R_e$ (c) $3R_e$ (d) $2R_e$

38 A rocket has to be launched from earth in such a way that it never returns. If E is the minimum energy delivered by the rocket launcher, what should be the minimum energy that the launcher should have, if the same rocket is to be launched from the surface of the moon? Assume that the density of the earth and the moon are equal and that the earth's volume is 64 times the volume of the moon. [2019, 8 April Shift-II]

(a) $\frac{E}{64}$ (b) $\frac{E}{16}$ (c) $\frac{E}{32}$ (d) $\frac{E}{4}$

39 The energy required to take a satellite to a height h above earth surface (where, radius of earth $= 6.4 \times 10^3$ km) is E_1 and kinetic energy required for the satellite to be in a circular orbit at this height is E_2. The value of h for which E_1 and E_2 are equal is [2019, 9 Jan Shift-II]

(a) 3.2×10^3 km (b) 1.28×10^4 km
(c) 6.4×10^3 km (d) 1.6×10^3 km

Numerical Value Type Questions

40 A satellite of mass 1000 kg is launched to revolve around the earth in an orbit at a height of 270 km from the earth's surface. Kinetic energy of the satellite in this orbit is .... $\times 10^{10}$ J.

(Mass of earth $= 6 \times 10^{24}$ kg, Radius of earth $= 6.4 \times 10^{6}$ m, Gravitational constant $= 6.67 \times 10^{-11}$ Nm^2 kg^{-2}) *[2025, 2 April Shift-II]*

41 Two planets, A and B are orbiting a common star in circular orbits of radii R_A and R_B, respectively, with $R_B = 2R_A$. The planet B is $4\sqrt{2}$ times more massive the planet A. The ratio (L_B / L_A) of angular momentum L_B of planet B to that of planet $A(L_A)$ is closest to integer........... *[2025, 29 Jan Shift-II]*

42 A satellite of mass $M/2$ is revolving around Earth in a circular orbit at a height of $R/3$ from Earth surface. The angular momentum of the satellite is $M\sqrt{\frac{GMR}{x}}$. The value of x is.........m, where M and R are the mass and radius of Earth, respectively. (G is the gravitational constant) *[2025, 23 Jan Shift-II]*

43 Two satellites S_1 and S_2 are revolving in circular orbits around a planet with radius , $R_1 = 3200$ km and $R_2 = 800$ km, respectively. The ratio of speed of satellite S_1 to the speed of satellite S_2 in their respective orbits would be $1/x$, where $x =$ .... *[2022, 25 June Shift-II]*

44 An asteroid is moving directly towards the centre of the earth. When at a distance of 10 R (R is the radius of the earth) from the earth's centre, it has a speed of 12 km/s. Neglecting the effect of earth's atmosphere, what will be the speed of the asteroid when it hits the surface of the earth (escape velocity from the earth is 11.2 km/s)? Give your answer to the nearest integer in km/s .... *[2020, 8 Jan Shift-II]*

• Key **Index**

Topic 1

1 (c)	2 (d)	3 (a)	4 (a)	5 (b)	6 (a)	7 (c)	8 (a)	9 (c)	10 (b)
11 (c)	12 (d)	13 (d)	14 (b)	15 (c)	16 (b)	17 (b)	18 (d)	19 (d)	20 (c)
21 (b)	22 (b)	23 (c)	24 (c)	25 (c)	26 (c)	27 (a)	28 (b)	29 (d)	30 (c)
31 (b)	32 (d)	33 (c)	34 (a)	35 (c)	36 (b)	37 (d)	38 (a)	39 (8)	40 (3)

Topic 2

1 (b)	2 (b)	3 (c)	4 (d)	5 (b)	6 (a)	7 (a)	8 (a)	9 (d)	10 (c)
11 (a)	12 (d)	13 (d)	14 (a)	15 (b)	16 (d)	17 (a)	18 (b)	19 (d)	20 (b)
21 (d)	22 (c)	23 (b)	24 (a)	25 (a)	26 (b)	27 (c)	28 (a)	29 (c)	30 (a)
31 (d)	32 (9)	33 (2)	34 (4)						

Topic 3

1 (b)	2 (c)	3 (c)	4 (b)	5 (b)	6 (c)	7 (d)	8 (a)	9 (d)	10 (d)
11 (a)	12 (d)	13 (d)	14 (2)	15 (4)	16 (3)				

Topic 4

1 (b)	2 (c)	3 (b)	4 (c)	5 (b)	6 (a)	7 (a)	8 (c)	9 (b)	10 (d)
11 (d)	12 (b)	13 (d)	14 (b)	15 (d)	16 (d)	17 (c)	18 (c)	19 (c)	20 (d)
21 (b)	22 (b)	23 (a)	24 (b)	25 (d)	26 (b)	27 (b)	28 (a)	29 (d)	30 (c)
31 (a)	32 (c)	33 (c)	34 (b)	35 (a)	36 (a)	37 (c)	38 (b)	39 (a)	40 (3)
41 (8)	42 (0.75)	43 (2)	44 (16)						

Detailed **Explanations**

Topic 1

01. *(c)* According to Kepler's third law,

$$T^2 = KR^3$$

$$\frac{2\Delta T}{T} = \frac{3\Delta R}{R}$$

$$\frac{2\Delta T}{T} = 3 \times \frac{0.03R}{R}$$

$$\frac{\Delta T}{T} = \frac{3 \times 0.03}{2} \times 100 = 4.5\%$$

02. *(d)* We know that,

Time period of revolving satellite,

$$T = \frac{2\pi r^{3/2}}{\sqrt{GM}}$$

$$\therefore \quad \frac{T_1}{T_2} = \left(\frac{r_1}{r_2}\right)^{3/2} \cdot \left(\frac{M_2}{M_1}\right)^{1/2}$$

$$\Rightarrow \quad \frac{6}{24} = \frac{r_1^{3/2}}{(4.2 \times 10^4)^{3/2}} \times \left(\frac{M}{M/4}\right)^{1/2}$$

$$\Rightarrow \quad r_1 = 1.05 \times 10^4 \text{ km}$$

03. *(a)* Given,

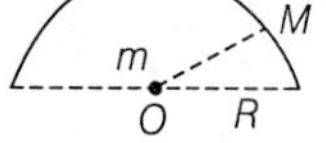

Let R be the radius of arc,

M = mass of metal wire,

L = length of metal wire,

m = mass placed at centre

$$R = \frac{L}{\pi}$$

Gravitational field at the centre due to arc,

$$E = \frac{2GM\pi}{L^2}$$

Net force on mass, m

$$F = mE$$

$$F = \frac{2GMm\pi}{L^2}$$

04. *(a)* For the planet,

$$T_1 = 200 \text{ days}$$

$$R_1 = R \Rightarrow R_2 = \frac{R}{4}$$

$$\therefore \quad T_2 = ?$$

According to Kepler's law

$$T^2 \propto R^3$$

$$\therefore \quad \frac{T_2}{T_1} = \left(\frac{R_2}{R_1}\right)^{3/2}$$

$$= \left(\frac{R/4}{R}\right)^{3/2} = \left(\frac{1}{4}\right)^{3/2}$$

$$= \frac{1}{8}$$

$$\therefore \quad T_2 = \frac{T_1}{8} = \frac{200}{8}$$

$$= 25 \text{ days}$$

05. *(b)* $T_{\text{earth}} = 365$ days

$T_{\text{moon}} = 27$ days

As, $T = \frac{2\pi}{\omega}$, then $\omega_{\text{moon}} > \omega_{\text{earth}}$

Both Assertion and Reason are correct and Reason is the correct explanation of Assertion.

06. *(a)* Due to mutual gravitational attraction, the particles will be at the position shown below.

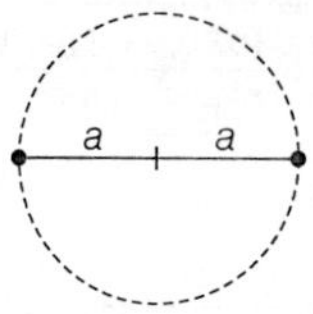

Here, centripetal force

= gravitational force

$$\Rightarrow \quad ma\omega^2 = \frac{Gm^2}{(2a)^2}$$

$$\Rightarrow \quad \omega = \sqrt{\frac{Gm}{4a^3}}$$

07. *(c)* According to Kepler's third law,

$$T^2 \propto r^3$$

We get, $\frac{T_2^2}{T_1^2} = \frac{r_2^3}{r_1^3}$

or $\quad T_2^2 = T_1^2 \left(\frac{r_2}{r_1}\right)^3$

Here, $\quad T_1 = 24 \text{ h}, \frac{r_2}{r_1} = \frac{1}{4}$

So, $\quad T_2^2 = (24)^2 \times \frac{1}{4^3}$

or $\quad T_2 = 3 \text{ h}$

08. *(a)* Particles moves along their COM,

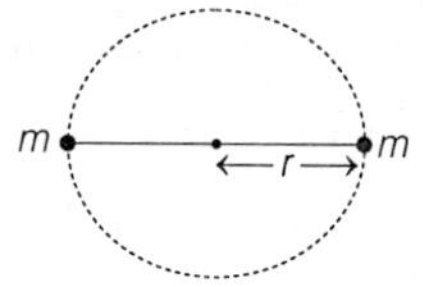

$$\text{Force on each particle} = \frac{Gmm}{(2r)^2} = \frac{Gm^2}{4r^2}$$

This force acts like centripetal pull so,

$$\frac{mv^2}{r} = \frac{Gm^2}{4r^2} \Rightarrow v = \sqrt{\frac{Gm}{4r}}$$

09. *(c)* Using $T^2 \propto r^3$ we get,

$$\frac{T_2^2}{T_1^2} = \frac{(r_2)^3}{(r_1)^3}$$

Here, $T_1 = 1$ year, $r_1 = 1.5 \times 10^6$ km,

$T_2 = 2.83$ years, $r_2 = ?$

So, $\quad r_2^3 = \frac{(2.83)^2}{1^2} \times (1.5 \times 10^6)^3$

$$\Rightarrow \quad r_2^3 = 8.0089 \times (1.5 \times 10^6)^3$$

or $\quad r_2 \approx [8 \times (1.5 \times 10^6)^3]^{1/3}$

$$\Rightarrow \quad r_2 = 2 \times 1.5 \times 10^6 \approx 3 \times 10^6 \text{ km}$$

> • Concept **Enhancer**
>
> If period of revolutions of two planets around the sun are T_1 and T_2 and their distances from the sun are R_1 and R_2, then
>
> $$\frac{R_1}{R_2} = \left(\frac{T_1}{T_2}\right)^{3/2} \Rightarrow R_2 = \left(\frac{T_2}{T_1}\right)^{2/3} R_1$$

10. *(b)* For a satellite, time period of revolution is dependent on its orbit as,

$$T^2 \propto R^3$$

Therefore, $\quad \frac{T_2}{T_1} = \left(\frac{R_2}{R_1}\right)^{3/2}$

Here, $\quad T_1 = 7 \text{ h}, \frac{R_2}{R_1} = 3$

So, $\quad T_2 = 7 \times (3)^{3/2} = 7 \times 5.19$

$$= 36.37 \text{ h} \approx 36 \text{ h}$$

11. *(c)* For a planet in a circular orbit,

$T^2 \propto R^3$ (Kepler's law)

For planets A and B, we have,

$$\Rightarrow \quad \frac{T_A^2}{T_B^2} = \frac{r_A^3}{r_B^3}$$

Given, $T_A = 2T_B$

so, $$\frac{(4T_B^2)}{T_B^2} = 4 = \frac{r_A^3}{r_B^3} \Rightarrow r_A^3 = 4r_B^3$$

12. *(d)* Distance between sun and earth,

$$r_1 = R$$

Time taken by earth to complete one revolution around sun, $T_1 = 1$ year

According to Kepler's law,

$$\frac{T_2}{T_1} = \left(\frac{r_2}{r_1}\right)^{3/2} = \left(\frac{3R}{R}\right)^{3/2} = (3)^{3/2} = 3\sqrt{3}$$

$$\Rightarrow \quad T_2 = 3\sqrt{3}\,T_1 = 3\sqrt{3} \text{ years}$$

13. *(d)* According to Kepler's second law of planetary motion, areal velocity of every planet moving around the sun should remain constant in a closed orbit.

14. *(b)* According to Kepler's third law,

$$T^2 \propto R^3$$

$$\Rightarrow \quad \frac{T_2}{T_1} = \left(\frac{R_2}{R_1}\right)^{3/2}$$

$$\Rightarrow \quad \frac{T_2}{T} = \left(\frac{9R}{R}\right)^{3/2} \quad \left[\because R_1 = R, \; R_2 = 9R\right]$$

$$\Rightarrow \quad T_2 = 27T$$

15. *(c)* Given, maximum distance of comet from the sun,

$$r_1 = 1.6 \times 10^{12} \text{ m}$$

Minimum distance of comet from the sun,

$$r_2 = 8.0 \times 10^{10} \text{m}$$

Speed of the comet at nearest point,

$$v_2 = 6 \times 10^4 \text{ ms}^{-1}$$

Applying law of conservation of angular momentum,

$$mv_1 r_1 = mv_2 r_2$$

($\because$ mass of comet will remain same)

$$\Rightarrow \quad v_1 = \frac{v_2 r_2}{r_1} = \frac{6 \times 10^4 \times 8 \times 10^{10}}{1.6 \times 10^{12}}$$

$$= \frac{48 \times 10^{14}}{1.6 \times 10^{12}} = 3 \times 10^3 \text{ m/s}$$

16. *(b)* As gravitational force provides necessary centripetal force.

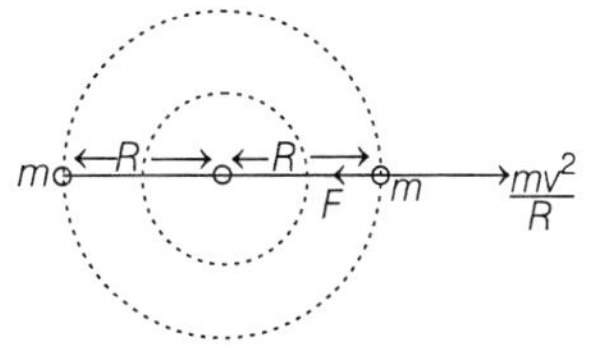

i.e., $$F = \frac{Gm^2}{(2R)^2} = \frac{mv^2}{R} \Rightarrow v = \sqrt{\frac{Gm}{4R}}$$

17. *(b)* According to Kepler's second law, a planet sweeps out equal areas in equal interval of time. This implies that areal velocity is constant.

In central force, the force acts along radius vector. Therefore, there is no torque about the center and thus angular momentum is conserved.

Conservation of angular momentum directly leads to constant areal velocity, which is why (R) is the correct explanation of (A).

18. *(d)* Given in question, $9R_s = R_m$,

$T_m = 27$ days

According to Kepler's law

$T^2 \propto R^3$

So, $$\frac{T_m^2}{T_s^2} = \frac{R_m^3}{(R_s)^3} \quad \text{...(i)}$$

Given, $9R_s = R_m$

$$R_s = \frac{R_m}{9}$$

Put the value of R_s in Eq. (i)

$$\frac{T_m^2}{T_s^2} = \frac{R_m^3}{\left(\frac{R_m}{9}\right)^3}$$

$$\frac{T_m^2}{T_s^2} = \frac{R_m^2}{\frac{R_m^3}{(9)^3}} = (9)^3$$

$$\left(\frac{T_m}{T_s}\right)^2 = 9^3$$

$$\Rightarrow \quad \frac{T_m}{T_s} = 9^{\frac{3}{2}}$$

$$T_s = \frac{T_m}{9^{\frac{3}{2}}} = \frac{27}{27} = 1 \text{ day}$$

19. *(d)* Given,

Gravitational constant $= G$,

Energy density $= u$

Dimensional formula of gravitational constant,

$$[G] = [\text{M}^{-1}\text{L}^3\text{T}^{-2}] \quad \text{...(i)}$$

As, $$F = \frac{GM_1 M_2}{R^2}$$

From here, $$G = \frac{FR^2}{M_1 M_2}$$

Dimensional formula of $u = \dfrac{\text{Energy}}{\text{Volume}}$

$$[u] = \frac{[\text{ML}^2\text{T}^{-2}]}{[\text{L}^3]} \quad \text{...(ii)}$$

Substitute the value of $[u]$ and $[G]$ in $\sqrt{uG}$.

$$[\sqrt{uG}] = \sqrt{\frac{[\text{ML}^2\text{T}^{-2}][\text{M}^{-1}\text{L}^3\text{T}^{-2}]}{[\text{L}^3]}}$$

$$= \sqrt{\text{L}^2\text{T}^{-4}} = [\text{LT}^{-2}] = \frac{\text{Force}}{\text{Mass}}$$

It has dimension of force per unit mass.

20. *(c)* Let side of square be x.

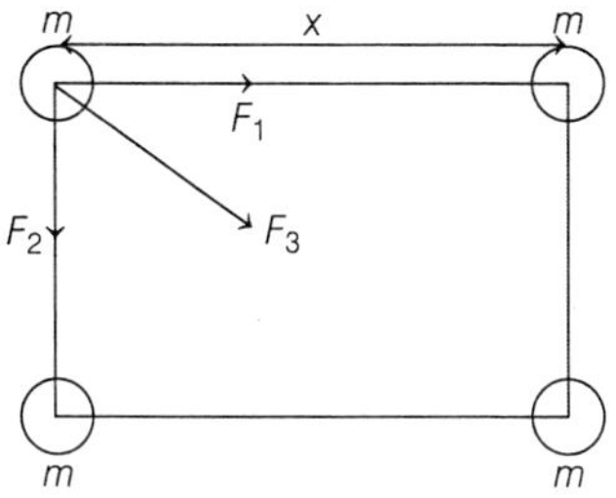

Gravitational force on any of mass due to other three masses is,

$$F = \text{F}_1 + \text{F}_2 + \text{F}_3$$

$$= \frac{\sqrt{2} \cdot Gm^2}{x^2} + \frac{Gm^2}{(\sqrt{2}x)^2}$$

$$= \frac{Gm^2}{x^2}\left(\sqrt{2} + \frac{1}{2}\right)$$

$$= \frac{Gm^2}{x^2}\left(\frac{2\sqrt{2}+1}{2}\right)$$

This force is given

$$F = \left(\frac{2\sqrt{2}+1}{32}\right)\frac{Gm^2}{L^2}$$

So, $$\frac{1}{x^2}\left(\frac{2\sqrt{2}+1}{2}\right) = \frac{2\sqrt{2}+1}{(32)\,L^2} \Rightarrow x = 4\,L$$

21. *(b)* Every planet revolve around the sun in elliptical path and gravitational force acting on the planet is directly proportional to the product of mass of the planet m and mass of sun M and inversely proportional to the square of distance r between sun and planet.

i.e. $$F \propto \frac{Mm}{r^2} \Rightarrow F = \frac{GMm}{r^2}$$

When a planet revolve around the sun, then centripetal force acting on the planet is directed towards the sun. Thus, statement C is incorrect.

According to Kepler's third law.

$$T^2 \propto a^3$$

Thus, statements A and D are correct. Hence, option (b) is correct.

22. *(b)* Given,

$$m_A = m_B = m_C = m_P = 100 \text{ kg}$$

The given situation is shown below.

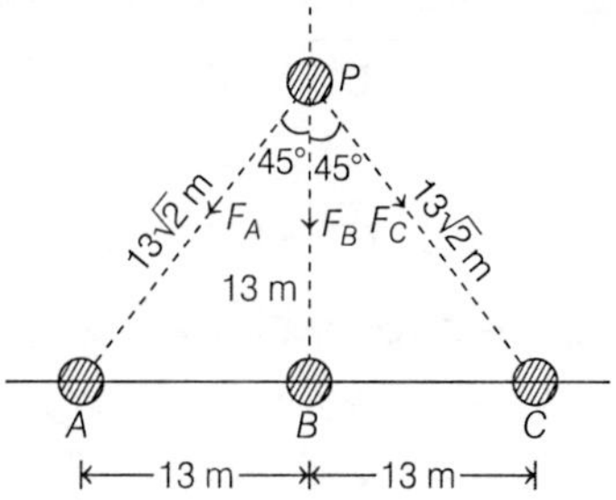

Force F on particle at P is resultant of F_A, F_C and F_B.

$\therefore \quad F = \sqrt{F_A^2 + F_C^2} + F_B$

As $\quad F_A = F_C$, we have

$$F = \sqrt{2} \cdot F_A + F_B$$
$$= \sqrt{2} \times \frac{Gm^2}{13^2 \times 2} + \frac{Gm^2}{13^2}$$
$$= \frac{Gm^2}{13^2} \times \left(\frac{1}{\sqrt{2}} + 1\right)$$
$$\approx \frac{G \times 100^2}{13^2} \times \left(\frac{2.4}{1.4}\right) \approx 100\,G$$

23. *(c)* Initially, force between two masses

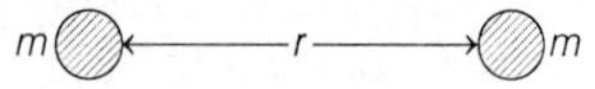

$$F = \frac{Gm \times m}{r^2} = \frac{Gm^2}{r^2}$$

When $\frac{1}{3}$rd of first mass is transferred to other, masses will be $\frac{4}{3}m$ and $\frac{2}{3}m$.

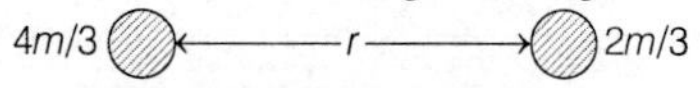

Now, force of attraction will be

$$F' = \frac{G \frac{4m}{3} \times \frac{2m}{3}}{r^2}$$

$$\Rightarrow \quad F' = \frac{8}{9}\frac{Gm^2}{r^2} \Rightarrow F' = \frac{8}{9}F$$

24. *(c)* Mass of star $A = m_A$, mass of star $B = m_B$

Radius of orbit for star $A = r_A$

Radius of orbit for star $B = r_B$

According to given condition,

$$m_A r_A = m_B r_A \qquad \text{...(i)}$$

Force acting on both stars i.e., gravitational force and centripetal force are balanced.

$$\therefore \quad \frac{Gm_A m_B}{r^2} = \frac{m_A v_A^2}{r_A} = \frac{m_B v_B^2}{r_B} \qquad \text{...(ii)}$$

Also, $v_A = \frac{2\pi r_A}{T_A}$ and $v_B = \frac{2\pi r_B}{T_B}$

Putting v_A and v_B in Eq. (ii), we get

$$\frac{m_A r_A}{(T_A)^2} = \frac{m_B r_B}{(T_B)^2} \qquad \text{...(iii)}$$

From Eqs. (i) and (iii), we have

$T_A = T_B$

25. *(c)* Gravitational field due to ring at a distance x is given as,

$$E = G \cdot \frac{mx}{(R^2 + x^2)^{3/2}}$$

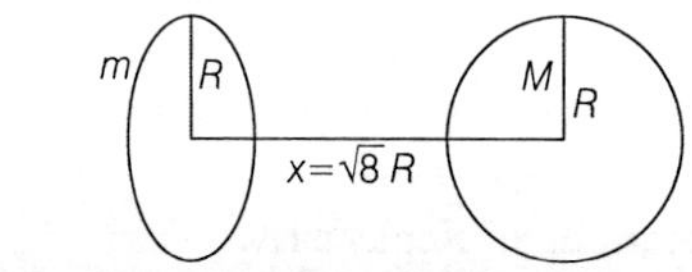

Gravitational force on sphere,

$$F = ME$$
$$F = \frac{MGmx}{(R^2 + x^2)^{3/2}}$$
$$\Rightarrow \quad F = \frac{GMm \cdot \sqrt{8}R}{(R^2 + 8R^2)^{3/2}}$$
$$\Rightarrow \quad F = \frac{GMm \cdot \sqrt{8}R}{9^{3/2} \cdot R^3} = \frac{GMm\sqrt{8}}{3^3 \cdot R^2}$$
$$\Rightarrow \quad F = \frac{\sqrt{8}}{27} \cdot \frac{GMm}{R^2}$$

26. *(c)* Given, situation is,

Force of attraction between mass m and an elemental mass dm of rod is

$$dF = \frac{Gmdm}{x^2} = \frac{Gm(A + Bx^2)\,dx}{x^2}$$

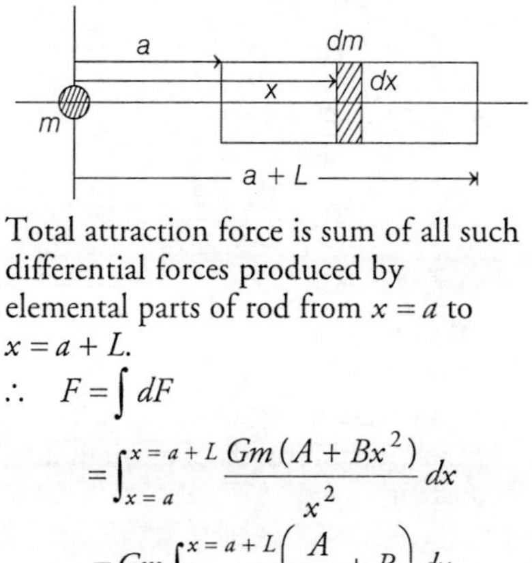

Total attraction force is sum of all such differential forces produced by elemental parts of rod from $x = a$ to $x = a + L$.

$$\therefore \quad F = \int dF$$
$$= \int_{x=a}^{x=a+L} \frac{Gm(A + Bx^2)}{x^2}\,dx$$
$$= Gm \int_{x=a}^{x=a+L} \left(\frac{A}{x^2} + B\right) dx$$
$$= Gm \left[-\frac{A}{x} + Bx\right]_{x=a}^{x=a+L}$$
$$= Gm \left(\frac{-A}{a+L} + B(a+L) + \frac{A}{a} - Ba\right)$$
$$= Gm \left[A\left(\frac{1}{a} - \frac{1}{a+L}\right) + BL\right]$$

27. *(a)* Gravitational forces are given as,

$$F_1 = \frac{GM_e M_m}{r_1^2}$$
$$F_2 = \frac{GM_e M_s}{r_2^2}$$

According to given situation,

$$\Delta F_1 = \frac{2GM_e M_m}{r_1^3} \Delta r_1,$$
$$\Delta F_2 = -\frac{2GM_e M_s}{r_2^3} \Delta r_2$$
$$\frac{\Delta F_1}{\Delta F_2} = \frac{M_m \Delta r_1}{r_1^3} \frac{r_2^3}{M_s \Delta r_2}$$
$$= \left(\frac{M_m}{M_s}\right)\left(\frac{r_2^3}{r_1^3}\right)\left(\frac{\Delta r_1}{\Delta r_2}\right) \qquad \text{...(i)}$$

Here, $\quad \Delta r_1 = \Delta r_2 = 2R_e$

$M_m = 8 \times 10^{22}$ kg,

$M_s = 2 \times 10^{30}$ kg

$r_1 = 0.4 \times 10^6$ km,

$r_2 = 150 \times 10^6$ km

So, on putting these values in Eq. (i) and solving,

$\therefore$ We get $\frac{\Delta F_1}{\Delta F_2} \simeq 2$

28. *(b)*

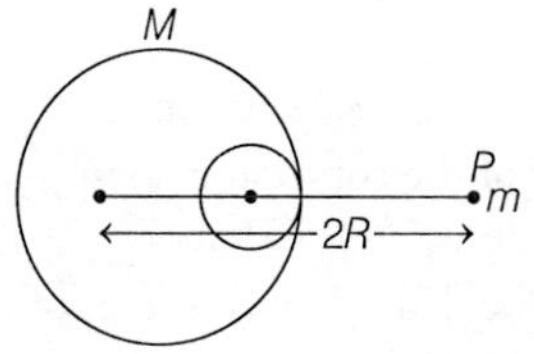

According to Newton's gravitational law,

When whole mass is accounted.

$$F_1 = \frac{GMm}{(2R)^2} \qquad \text{...(i)}$$

When section of mass (of radius $\frac{R}{3}$) is removed;

$$F_2 = \frac{GMm}{(2R)^2} - \left[\frac{G\left(\frac{M}{27}\right)m}{\left(\frac{4R}{3}\right)^2}\right]$$

$$F_2 = \frac{11}{48}\frac{GMm}{R^2} \qquad \text{...(ii)}$$

From Eqs. (i) and (ii)

$F_1 : F_2 = 12 : 11$

29. *(d)* Given for planets A and B, time periods are T_A and T_B, radii are r_1 and r_2 and masses m_1 and m_2 respectively. Now, by Kepler's law

$$T^2 \propto r^3$$

$$\Rightarrow \quad \frac{T_A^2}{T_B^2} = \left(\frac{r_1}{r_2}\right)^3$$

$$\text{or} \quad \left(\frac{T_A}{T_B}\right)^{2/3} = \frac{r_1}{r_2} \quad \ldots\text{(i)}$$

Also angular momentum of planet is,

$$L = I\omega = I \cdot \frac{2\pi}{T} = \frac{mr^2 \cdot 2\pi}{T}$$

$$\Rightarrow \quad T = \frac{2\pi mr^2}{L}$$

$$\therefore \quad \frac{T_A}{T_B} = \frac{m_1 r_1^2}{L_A} \times \frac{L_B}{m_2 r_2^2}$$

$$\frac{T_A}{T_B} = \frac{m_1}{m_2} \cdot \left(\frac{r_1}{r_2}\right)^2 \cdot \frac{L_B}{L_A}$$

As from Eq. (i), $\frac{r_1}{r_2} = \left(\frac{T_A}{T_B}\right)^{2/3}$

and given, $L_B / L_A = 3$

$$\text{Hence,} \quad \frac{T_A}{T_B} = \frac{m_1}{m_2}\left(\frac{T_A}{T_B}\right)^{4/3} \cdot 3$$

$$\Rightarrow \quad \frac{1}{3}\left(\frac{m_2}{m_1}\right) = \left(\frac{T_A}{T_B}\right)^{\frac{4}{3}-1}$$

$$\Rightarrow \quad \left(\frac{T_A}{T_B}\right)^{1/3} = \frac{1}{3}\left(\frac{m_2}{m_1}\right)$$

$$\text{or} \quad T_A / T_B = \frac{1}{27} \cdot \left(\frac{m_2}{m_1}\right)^3$$

30. *(c)* Force $\propto R^{-3/2}$

$$\Rightarrow \quad F = kR^{-3/2}, \text{ where } k = \text{constant}$$

This force is acting like centripetal force so,

$$\frac{mv^2}{R} = kR^{-3/2}$$

$$\Rightarrow \quad m \cdot \frac{\omega^2 R^2}{R} = kR^{-3/2}$$

$$\text{or} \quad \omega^2 = \frac{k}{m} \cdot R^{-5/2} \Rightarrow \omega = \sqrt{\frac{k}{m}} \cdot R^{-5/4}$$

Time period of rotation,

$$T = \frac{2\pi}{\omega} = \frac{2\pi}{\sqrt{\left(\frac{k}{m}\right)}} \cdot R^{+5/4}$$

$$\Rightarrow \quad T^2 = \frac{4\pi^2 m}{k} \cdot R^{+5/2}$$

$$\text{or} \quad T^2 \propto R^{5/2}$$

31. *(b)*

> **key idea** Equate the gravitational force to the centripetal force and determine the angular speed of each particle.

$\because$ Gravitational force,

$$F = \frac{Gm^2}{R^2} \quad \ldots\text{(i)}$$

$\because$ Both are tracing a circular path, centripetal force will be acting on them.

$$F = \frac{Gm^2}{(2R)^2}$$

$$\Rightarrow \quad F_c = \frac{m(\omega R)^2}{R} \quad [\because v = \omega R]$$

$$\Rightarrow \quad F_c = m\omega^2 R \quad \ldots\text{(ii)}$$

During this circular motion,
Gravitational force = centripetal force

$$\Rightarrow \quad \frac{Gm^2}{4R^2} = m\omega^2 R$$

$$\Rightarrow \quad \omega = \frac{1}{2}\sqrt{\frac{Gm}{R^3}} = \frac{1}{2}\sqrt{\frac{Gm}{R^3}}$$

$$(\because m = 1\text{ kg})$$

$$\omega = \frac{1}{2}\sqrt{\frac{G \times 1}{R^3}} = \frac{1}{2}\sqrt{\frac{G}{R^3}}$$

32. *(d)* $dA = \frac{1}{2} \times r \times v\, dt \sin(\pi - \theta)$

$$dA = \frac{1}{2} r v\, dt \sin\theta$$

$$\frac{dA}{dt} = \frac{1}{2}(\mathbf{r} \times \mathbf{v})\frac{M}{M} = \frac{L}{2M} = \text{constant}$$

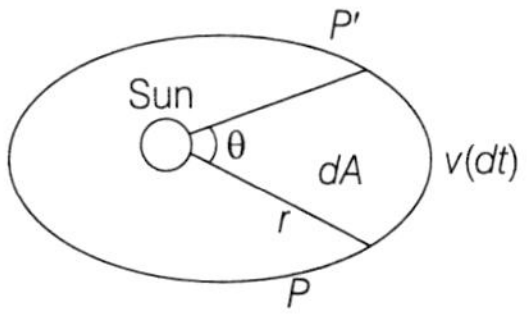

33. *(c)* Distance of particle from centre of Earth $= 3R$
Force between sphere and particle $= F_1$
Radius of cavity $= \frac{R}{2}$
Let mass of sphere $= M$
Mass of sphere with cavity $= M'$
Mass of particle $= m$

$$\text{Now,} \quad M' = \rho \cdot \frac{4}{3}\pi\left(\frac{R}{2}\right)^3 = \frac{M}{8}$$

By using concept of gravitational force,

$$F_1 = \frac{GMm}{(3R)^2} = \frac{GMm}{9R^2} \quad \ldots\text{(i)}$$

$$\text{and } F_{\text{cavity}} = \frac{GM'm}{(AB)^2} = \frac{G(M/8)m}{(2R + R/2)^2}$$

$$= \frac{4GMm}{25 \times 8\, R^2} \quad \ldots\text{(ii)}$$

$$\therefore \quad F_2 = F_1 - F_{\text{cavity}}$$

$$= \frac{GMm}{9R^2} - \frac{4}{25 \times 8}\frac{GMm}{R^2}$$

$$= \frac{GMm}{R^2}\left(\frac{1}{9} - \frac{1}{50}\right)$$

$$= \frac{41}{50 \times 9}\frac{GMm}{R^2}$$

$$\because \quad \frac{F_1}{F_2} = \frac{50}{41}$$

34. *(a)*

> **key idea** Calculate the net force on one particle and then determine the required speed of each particle.

Given, $m = 1$ kg, $R = 1$ m

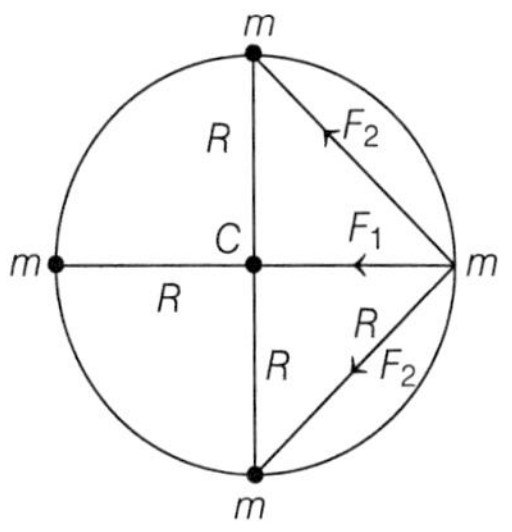

We know that,

$$F = \frac{Gm_1 m_2}{r^2}$$

$$\because \quad F_1 = \frac{Gmm}{(2R)^2} = \frac{Gm^2}{4R^2}$$

$$\text{and} \quad F_2 = \frac{Gmm}{(\sqrt{2}R)^2} = \frac{Gm^2}{2R^2}$$

Net force on one particle,

$$F_{\text{net}} = F_1 + F_2 \cos 45° + F_2 \cos 45°$$

$$= F_1 + 2F_2 \cos 45°$$

$$= \frac{Gm^2}{4R^2} + 2\left(\frac{Gm^2}{2R^2}\right) \cdot \frac{1}{\sqrt{2}}$$

$$= \frac{Gm^2}{4R^2} + \frac{Gm^2}{\sqrt{2}R^2}$$

$$=\frac{Gm^2}{R^2}\left[\frac{1}{4}+\frac{1}{\sqrt{2}}\right]$$

As the gravitational force provides the necessary centripetal force, so

$$F_{net}=F_C=\frac{mv^2}{R}$$

Here, F_C = centripetal force.

$$\Rightarrow \quad \frac{Gm^2}{R^2}\left[\frac{1}{4}+\frac{1}{\sqrt{2}}\right]=\frac{mv^2}{R}$$

$$\Rightarrow \quad v=\frac{1}{2}\sqrt{\frac{Gm}{R}(1+2\sqrt{2})}$$

$$\Rightarrow \quad v=\frac{1}{2}\sqrt{G(1+2\sqrt{2})}$$

35. *(c)* Force on star is provided by the mass enclosed inside region of radius R as shown below.

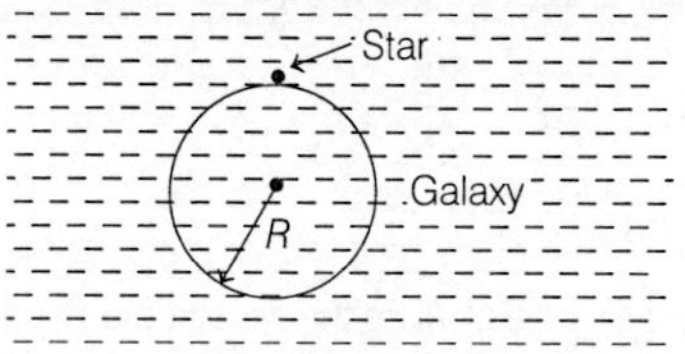

where, R = orbital radius of star. Now, consider a spherical region of radius r and thickness dr as shown below.

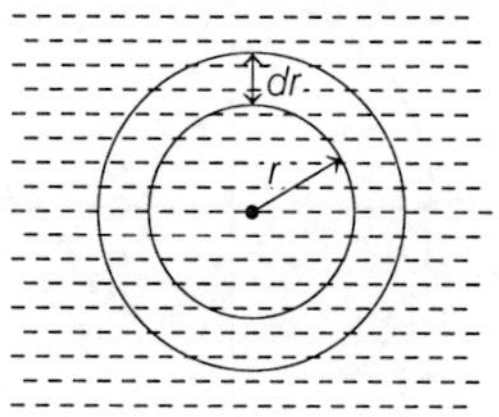

Mass enclosed in this region is given by

dm = density × volume

$$=\frac{k}{r}\cdot 4\pi r^2\cdot dr$$

So, total mass enclosed in a region of radius R is given by

$$M=\int_0^R dm=\int_0^R \frac{k}{r}\cdot 4\pi r^2 dr$$

$$M=4\pi k\left[\frac{r^2}{2}\right]_0^R=2k\pi R^2$$

Gravitational pull of this mass provides necessary centripetal force for rotation of star.

$$\text{So, } \frac{GMm}{R^2}=\frac{mv^2}{R}$$

$$\Rightarrow \quad v^2=\frac{GM}{R}=\frac{G2k\pi R^2}{R}$$

$$\Rightarrow \quad v^2=2Gk\pi R$$

$$\Rightarrow \quad v=\sqrt{2Gk\pi R}$$

Time period of rotation,

$$T=\frac{2\pi R}{v}=\frac{2\pi R}{\sqrt{2Gk\pi R}}$$

$$\Rightarrow \quad T\propto\sqrt{R} \text{ or } T^2\propto R$$

36. *(b)*

> **key idea** In given configuration of masses, net gravitational force provides the necessary centripetal force for rotation.

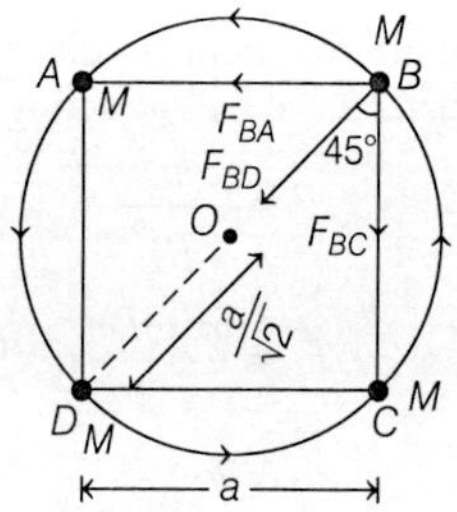

Net force on mass M at position B towards centre of circle is

$$F_{BO\,net}=F_{BD}+F_{BA}\sin 45°+F_{BC}\cos 45°$$

$$=\frac{GM^2}{(\sqrt{2}a)^2}+\frac{GM^2}{a^2}\left(\frac{1}{\sqrt{2}}\right)+\frac{GM^2}{a^2}\left(\frac{1}{\sqrt{2}}\right)$$

[where, diagonal length BD is $\sqrt{2}a$]

$$=\frac{GM^2}{2a^2}+\frac{GM^2}{a^2}\left(\frac{2}{\sqrt{2}}\right)$$

$$=\frac{GM^2}{a^2}\left(\frac{1}{2}+\sqrt{2}\right)$$

This force will act as centripetal force. Distance of particle from centre of circle is $\frac{a}{\sqrt{2}}$.

Here,

$$F_{centripetal}=\frac{Mv^2}{r}=\frac{Mv^2}{\frac{a}{\sqrt{2}}}=\frac{\sqrt{2}Mv^2}{a}$$

$$\left(\because r=\frac{a}{\sqrt{2}}\right)$$

So, for rotation about the centre,

$$F_{centripetal}=F_{BO(net)}$$

$$\Rightarrow \sqrt{2}\frac{Mv^2}{a}=\frac{GM^2}{a^2}\left(\frac{1}{2}+\sqrt{2}\right)$$

$$\Rightarrow \quad v^2=\frac{GM}{a}\left(1+\frac{1}{2\sqrt{2}}\right)$$

$$=\frac{GM}{a}(1.35)$$

$$\Rightarrow \quad v=1.16\sqrt{\frac{GM}{a}}$$

37. *(d)* Let the mass of the test particle be m and its orbital linear speed be v.

⇒ Gravitational force

(F_g) = Centripetal force (F_c)

Let us now assume an elementary ring at a distance r from the centre of the mass density, such that mass of this elementary ring

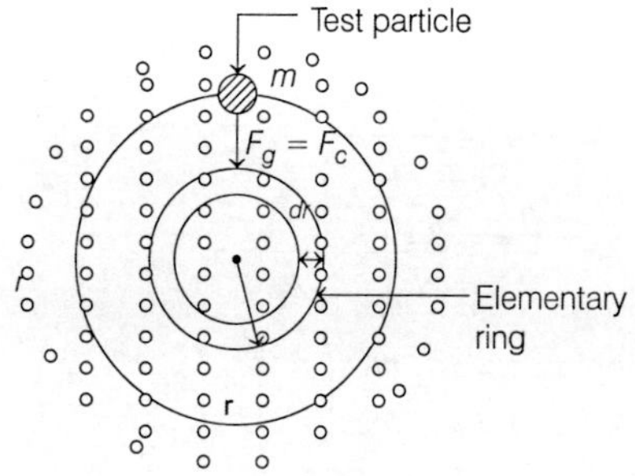

$$\Rightarrow \quad \frac{GMm}{R^2}=\frac{mv^2}{R}$$

where, $M=\int_0^R \rho\cdot 4\pi r^2\cdot dr$

$$\Rightarrow \quad \frac{G\left(\int_0^R \frac{K}{r^2}\cdot 4\pi r^2\cdot dr\right)}{R^2}=\frac{v^2}{R}$$

$$\left[\because \text{given, } \rho(r)=\frac{K}{r^2}\right]$$

$$\Rightarrow \quad G\cdot 4\pi K=v^2$$

∴ Orbital speed of mass, m

$$v=\sqrt{G4\pi K} \quad \ldots\text{(i)}$$

Time period of rotation of the test particle is

$$T=\frac{2\pi R}{v}=\frac{2\pi R}{\sqrt{G\cdot 4\pi K}}$$

[∵ using Eq. (i)]

$$\text{Hence, } \frac{T}{R}=\sqrt{\frac{\pi}{GK}}=\text{constant}$$

38. *(a)* The necessary centripetal force required for a planet to move round the sun

= Gravitational force exerted on it

$$\text{i.e., } \frac{mv^2}{R}=\frac{GM_e m}{R^n}$$

$$\text{or} \quad v=\left(\frac{GM_e}{R^{n-1}}\right)^{1/2}$$

$$\text{Now, } T=\frac{2\pi R}{v}=2\pi R\times\left(\frac{R^{n-1}}{GM_e}\right)^{1/2}$$

$$= 2\pi\left(\frac{R^2 \times R^{n-1}}{GM_e}\right)^{1/2}$$

$$= 2\pi\left(\frac{R^{(n+1)/2}}{(GM_e)^{1/2}}\right)$$

or $T \propto R^{(n+1)/2}$

39. *(8)* Gravitational force between two masses is given by

$$F = \frac{Gm_1m_2}{r^2}$$

Here, $m_1 = m_2 = m$
and $r = a$ [length of side of triangle]

$$F = \frac{Gm^2}{a^2} \quad \ldots(i)$$

Also, $F = ma'$

$$a' = \frac{F}{m} \quad \ldots(ii)$$

Combining Eqs. (i) and (ii)

$$a' = \frac{F}{m} = \frac{1}{m}\left[\frac{Gm^2}{a^2}\right] = \frac{Gm}{a^2}$$

Using, $s = ut + \frac{1}{2}at^2$

$$a = 0 + \frac{1}{2}\left[\frac{Gm}{a^2}\right]t^2$$

$$a = \frac{Gm}{2a^2}t^2$$

$$\sqrt{\frac{2a^3}{Gm}} = t$$

$$t \propto \sqrt{\frac{a^3}{Gm}}$$

Therefore, time $= T_1 \propto \sqrt{\frac{\alpha^3}{m}}$

Now, length is increased to $2a$
and mass is increased to $2m$

$$\therefore \text{New time} = T_2 \propto \sqrt{\frac{(2a)^3}{2m}} = 2\sqrt{\frac{a^3}{m}}$$

$$\frac{T_2}{T_1} \Rightarrow \frac{2\sqrt{a^3/m}}{\sqrt{a^3/m}} = 2$$

$$T_2 = 2T_1$$
$$T_2 = 2 \times 4$$
$$T_2 = 8 \text{ s}$$

40. *(3)* Given, the time period of the first satellite,

$$T_1 = 1 \text{ h}$$

The time period of the second satellite,

$$T_2 = 8 \text{ h}$$

The radius of the orbit of nearer satellite, $R_1 = 2000$ km

Applying the Kepler's third law,

$$T^2 \propto R^3$$

$$\left(\frac{T_1}{T_2}\right)^2 = \left(\frac{R_1}{R_2}\right)^3 \Rightarrow \frac{2000}{R_2} = \left(\frac{1}{8}\right)^{2/3}$$

$$\Rightarrow \quad R_2 = 8000 \text{ km}$$

As we know that $\omega = \frac{\theta}{T}$

So, the angular speed of the nearer satellite to the orbit, $\omega_1 = \frac{2\pi}{1}$ rad/h

and the angular speed of the farther satellite to the orbit,

$$\omega_2 = \frac{2\pi}{8} = \frac{\pi}{4} \text{ rad/h}$$

The speed of the nearer satellite to orbit,

$$v_1 = \omega_1 R_1$$
$$= (2\pi) \times 2 \times 10^3 \text{ km/h}$$

The speed of the farther satellite to the orbit $v_2 = \omega_2 R_2 = \left(\frac{\pi}{4}\right) \times 8000$

$$= \pi \times 2 \times 10^3 \text{ km/h}$$

Thus, the relative angular speed of the nearer satellite to the farther satellite,

$$\omega = \frac{v_1 - v_2}{R_1 - R_2}$$

$$= \frac{2\pi \times 2 \times 10^3 - \pi \times 2 \times 10^3}{-8000 + 2000}$$

$$= \frac{2\pi \times 10^3}{-6000} = -\frac{\pi}{3} \text{ rad/h}$$

$$\Rightarrow \quad |\omega| = \frac{\pi}{3}$$

Comparing with, $\omega = \frac{\pi}{x}$

The value of $x = 3$

Topic 2

01. *(b)* Initially, $g = \frac{GM}{R^2}$

Under given condition

$$g' = \frac{GM}{(R/2)^2} = \frac{4GM}{R^2} = 4g$$

02. *(b)* Statement I is true as rotation causes variation in value of g.

i.e. $g' = g - r\omega^2 \cos^2 \lambda$

At pole, $g' = g$ (no effect)

At equator, $g' = g - r\omega^2 \cos^2 \lambda$

$$= g - r\omega^2$$

($\because \cos^2 \lambda_{max}$ at $\lambda = 0°$, is at equator)

Hence, the effect of rotation of the earth on the value of g is maximum at equator. Therefore, Statement II is false.

03. *(c)* Acceleration due to gravity at surface of earth, $g = \frac{GM}{R^2}$

At height h, acceleration due to gravity will be,

$$g' = \frac{GM}{(R+h)^2}$$

$$\Rightarrow \quad \frac{g'}{g} = \frac{R^2}{(R+h)^2} = \left(1 + \frac{h}{R}\right)^{-2}$$

$$= 1 - \frac{2h}{R} \text{ (when } h << R)$$

$$\Rightarrow \quad g' = g\left(1 - \frac{2h}{R}\right)$$

04. *(d)* Acceleration due to gravity on the surface of the Earth is,

$$g = \frac{GM}{R^2} = \frac{4}{3}\pi G\rho R$$

And acceleration due to gravity at depth d is,

$$g' = \frac{4}{3}\pi G \cdot \rho(R - d)$$

Here $d = \frac{R}{2}$

So, $g' = \frac{4}{3}\pi G\rho \cdot \frac{R}{2} = \frac{g}{2}$

Hence, weight of a body at a depth $d = \frac{R}{2}$ is,

$$mg' = \frac{mg}{2} = \frac{400}{2} = 200 \text{ N}$$

05. *(b)* As, we go up or down from earth's surface, acceleration due to gravity decreases. So, Statement I is correct.

Now, $g_h = g_d$ gives,

Now, for small $h (h << R)$ we have,

$$g\left(1 - \frac{2h}{R}\right) = g\left(1 - \frac{d}{R}\right)$$

$$2h = d \text{ or } h = d/2$$

So, Statement II is incorrect.

06. *(a)* Weight of rocket, $w = mg$

$$w = m\left(\frac{Gm}{R^2}\right) \Rightarrow w \propto \frac{1}{R^2}$$

$$\Rightarrow \quad \frac{\Delta w}{w} = \frac{2\Delta R}{R} = 2 \times \frac{32}{6400}$$

($\because R = 6400$ km, $\Delta R = 32$ km)

$$= \frac{1}{100}$$

∴ Percentage decrease in weight

$$=\frac{\Delta w}{w}\times 100\% = \frac{1}{100}\times 100\% = 1\,\%$$

07. *(a)* Since, Newton's law of gravitation is universal.

Hence, it holds good for any pair of bodies in the universe.

Variation of g with depth from the surface of earth is given by

$$g' = g\left(1-\frac{d}{R}\right)$$

At the centre of earth,

$$d = R$$

$$\therefore \quad g' = g\left(1-\frac{R}{R}\right) = 0$$

Thus, apparent weight of any person at the centre of earth $= mg' = 0$

08. *(a)* With the depth d, the acceleration due to gravity changes as,

$$g' = g\left(1-\frac{d}{R}\right)$$

So, as depth increases, acceleration due to gravity decreases and reaches minimum at centre.

So, g' decreases as r increases. Therefore, the graph of g *versus* r is plotted as

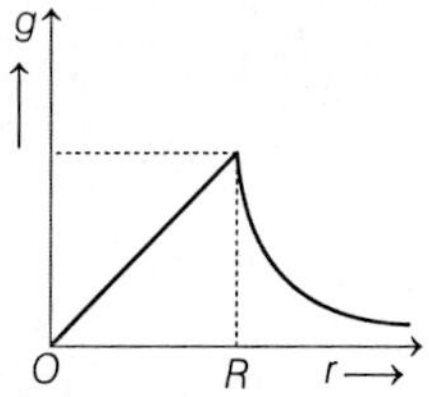

09. *(d)* Acceleration due to gravity at height h from the surface of earth is given as

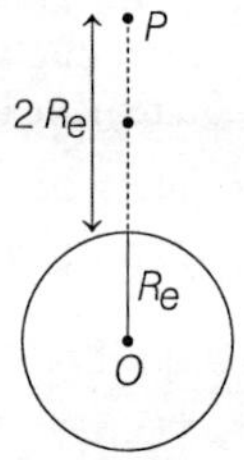

$$g_h = \frac{g}{\left(1+\frac{h}{R_e}\right)^2}$$

$$= \frac{g}{\left(1+\frac{2R_e}{R_e}\right)^2} = \frac{g}{(3)^2} = \frac{g}{9}$$

10. *(c)* Effect of rotation of earth on acceleration due to gravity is given by $g' = g - \omega^2 R\cos^2\phi$, where ϕ is latitude. There will be no change in gravity at poles as $\phi = 90°$ at all other points as ω increases, g' will decrease.

11. *(a)*

	Planet-A	Planet-B
Radius	R	$1.5\,R$
Density	ρ	$\frac{\rho}{2}$

$$g = \frac{4}{3}\pi G\rho R$$

$$\Rightarrow \frac{g_B}{g_A} = \frac{\frac{4}{3}\pi G\rho_B R_B}{\frac{4}{3}\pi G\rho_A\cdot R_A}$$

$$\frac{g_B}{g_A} = \frac{\rho_B R_B}{\rho_A\cdot R_A} = \frac{\left(\frac{\rho}{2}\right)\times(1.5)R}{\rho R}$$

$$= \frac{1.5}{2} = \frac{3}{2\times 2} = \frac{3}{4}$$

12. *(d)* As we know,

$$g_d = g\left(\frac{R-d}{R}\right)$$

g_d = Acceleration at depth d

g = Acceleration at surface

R = Radius of earth

$$\Rightarrow mg_d = mg\left(\frac{R-d}{R}\right)$$

$$\Rightarrow \text{Weight at depth } d = mg\left(\frac{R-d}{R}\right)$$

Here, $mg = 200$ N and $d = \frac{R}{2}$

So, weight at depth d is,

$$w' = 200\left(\frac{R-\frac{R}{2}}{R}\right) = \frac{200}{2} = 100\text{ N}$$

13. *(d)* The acceleration due to gravity at the surface of the earth is given as,

$$g = \frac{GM}{R^2}$$

where, R is the distance from centre at that place.

Since, the radius of poles is slightly less than that of equator, hence the value of g at poles is greater than at the equator. As we move from equator to poles, the value of g increases.

Thus, it is clear that the value of g is different at different places on the surface of the earth.

The expression of acceleration due to gravity at depth d from surface of the earth is given as

$$g' = g\left(1-\frac{d}{R}\right)$$

If d is increased, the g' decreases.

So, Statement I given in question is true and Statement II is false.

14. *(a)* Mass of body,

$$m = \frac{\text{Weight of body}}{\text{Acceleration due to gravity at that point}}$$

$$m = \frac{w}{g}$$

At height h, acceleration due to gravity is given as,

$$g_h = \frac{g}{\left(1+\frac{h}{R}\right)^2}$$

Here, $h = 9R$, $g_h = \frac{g}{\left(1+\frac{9R}{R}\right)^2}$

$$= \frac{g}{(1+9)^2} = \frac{g}{(10)^2} = \frac{g}{100}$$

Weight of body at height of $9R$,

$$w_h = mg_h = \left(\frac{w}{g}\right)\times\left(\frac{g}{100}\right)$$

$$\Rightarrow w_h = \frac{w}{100}$$

> **Concept Enhancer**
>
> Weight of the body decreases on going above the surface of earth due to decrease of the value of g. Weight of the body at height is calculated by
>
> $$w_h = \frac{mg}{\left(1+\frac{h}{r}\right)^2} = \frac{w}{\left(1+\frac{h}{R}\right)^2}$$

15. *(b)* $g_d = g\left(1-\frac{d}{R}\right)$

At height h from surface of Earth,

$$g_h = \frac{g}{\left(1+\frac{h}{R}\right)^2}$$

According to question,

$$g_d = 4g_h$$

$$g\left(1-\frac{d}{R}\right) = \frac{4g}{\left(1+\frac{h}{R}\right)^2}$$

$$\Rightarrow \left(1-\frac{d}{R}\right) = \frac{4}{\left(1+\frac{3R}{R}\right)^2} = \frac{4}{(1+3)^2}$$

$$\Rightarrow 1-\frac{d}{R} = \frac{1}{4} \Rightarrow \frac{d}{R} = 1-\frac{1}{4} = \frac{3}{4}$$

or $d = \frac{3R}{4} = \frac{3}{4} \times 6400 = 4800$ km

• Concept Enhancer

On going below the earth's surface at depth '*d*', acceleration due gravity decreases with following relation

$$g_d = g\left(1 - \frac{d}{R_e}\right)$$

and on going above the earth's surface at height '*h*', acceleration due to gravity decreases with following relation.

$$g_h = \frac{g}{\left(1 + \frac{h}{R_e}\right)^2}$$

where, *g* is acceleration due to gravity on the surface of earth.

16. *(d)* Given, radius of earth shrinks by 2% i.e. $\frac{\Delta R}{R} \times 100 = -2\%$

We know that, acceleration due to gravity,

$$g = \frac{GM}{R^2} \Rightarrow g \propto \frac{1}{R^2}$$

$$\Rightarrow \quad \frac{\Delta g}{g} = -2\frac{\Delta R}{R}$$

$$= -2(-2\%) = 4\ \%$$

Hence, acceleration due to gravity is increased by 4%.

17. *(a)* Weight of object on the surface of earth,

$$w = mg \quad \ldots(\text{i})$$

At height, $h = \frac{5R}{4} - R = \frac{R}{4}$, the value of acceleration due to gravity,

$$g' = \frac{g}{\left(1 + \frac{h}{R}\right)^2} = \frac{g}{\left(1 + \frac{\frac{R}{4}}{R}\right)^2}$$

$$= \frac{g}{\left(\frac{5}{4}\right)^2} = \frac{16g}{25} \Rightarrow g' = \frac{16g}{25}$$

Weight of object at this height,

$$w' = mg' = m \times \frac{16g}{25} \Rightarrow w' = \frac{16mg}{25}$$

∴ percentage decrease in weight

$$= \frac{w - w'}{w} \times 100\%$$

$$= \frac{mg - \frac{16mg}{25}}{mg} \times 100\%$$

$$= \frac{9}{25} \times 100\% = 36\%$$

18. *(b)* Let *h* be required height, i.e. weight of body at height *h*

$$= \frac{\text{weight of body on the surface of earth}}{3}$$

$$\Rightarrow \quad mg_h = \frac{mg}{3} \Rightarrow g_h = \frac{g}{3}$$

$$\Rightarrow \quad \frac{g}{\left(1 + \frac{h}{R_e}\right)^2} = \frac{g}{3}$$

$$\Rightarrow \quad \left(1 + \frac{h}{R_e}\right)^2 = 3$$

$$1 + \frac{h}{R_e} = \sqrt{3}$$

$$\Rightarrow \quad h = (\sqrt{3} - 1)R_e$$

$$= (1.732 - 1)6400$$

$$= 0.732 \times 6400$$

$$= 4684.8 \text{ km}$$

$$\simeq 4685 \text{ km}$$

19. *(d)* **Controversial Question** Direction of acceleration also changes if we consider the effect of rotation. If is not always towards the centre.

At every location, the direction of acceleration due to gravity is always towards the centre of earth, but its magnitude changes at different locations, i.e. if we move from poles to equator, the direction of acceleration due to gravity of the earth always points towards the centre of earth but its magnitude decreases and becomes minimum on reaching at equator.

Hence, A is false but R is true.

• Concept Enhancer

Acceleration due to gravity is different at different locations (on moving from pole to equator)

20. *(b)* Given, weight of body at north pole,

$$w_p = mg = 49 \text{ N}$$

Radius of earth,

$$R = 6400 \text{ km}$$

Let weight of body at equator be w_e.

At equator,

$$g_e = g - R\omega^2$$

$$\therefore \quad w_e = mg_e = m(g - R\omega^2)$$

Since, $w_p > w_e \Rightarrow w_e < 49$ N

21. *(d)* Given that, acceleration due to gravity at height *h* from the surface of earth = acceleration due to gravity at depth *d* below the surface of earth $= g_1$

$$\Rightarrow \quad g_1 = \frac{GM}{(R+h)^2} = \frac{GM}{R^3}(R - d)$$

$$\Rightarrow \quad \frac{GM}{\left(R + \frac{R}{2}\right)^2} = \frac{GM}{R^3}(R - d) \quad \left[\because h = \frac{R}{2}\right]$$

$$\Rightarrow \quad \frac{4GM}{9R^2} = \frac{GM}{R^2}\left(1 - \frac{d}{R}\right)$$

$$\Rightarrow \quad 1 - \frac{d}{R} = \frac{4}{9}$$

$$\Rightarrow \quad \frac{d}{R} = 1 - \frac{4}{9} = \frac{5}{9}$$

22. *(c)* Gravitational field is maximum at surface $(r = R)$

$$E = \frac{GM}{R^2}$$

From graph given in the question,

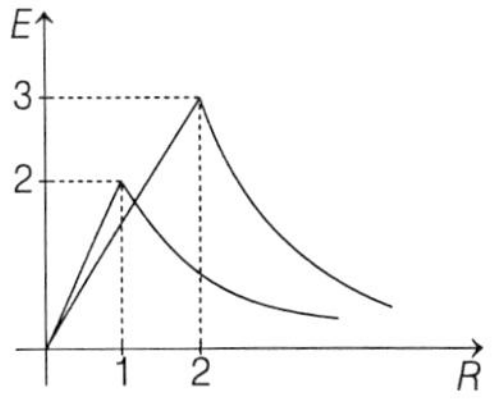

We can observe that

$$E_1 = \frac{GM_1}{(1)^2} = 2$$

and $$E_2 = \frac{GM_2}{(2)^2} = 3$$

or $$GM_1 = 2 \quad \ldots(\text{i})$$

and $$GM_2 = 12 \quad \ldots(\text{ii})$$

On dividing Eq. (i) by Eq. (ii), we get

$$\frac{M_1}{M_2} = \frac{2}{12} = \frac{1}{6}$$

23. *(b)*

key idea The relation of gravitational acceleration at an altitude *h* above the earth's surface is given as

$$g_h = g\left(1 + \frac{h}{R_e}\right)^{-2}$$

where, *g* is the acceleration due to gravity at earth's surface and R_e is the radius of the earth.

Given that at some height *h*, acceleration due to gravity,

$$g_h = 4.9 \text{ m/s}^2 \approx \frac{g}{2} \quad \ldots(\text{i})$$

∴ The ratio of acceleration due to gravity at earth's surface and at some altitude h is

$$\left(1+\frac{h}{R_e}\right)=\sqrt{\frac{g}{g_h}}=\sqrt{2}$$

[From Eq. (i)]

$$\therefore \quad \frac{h}{R_e}=\sqrt{2}-1 \text{ or } h=0.414\times R_e$$

$$h=0.414\times 6400 \text{ km}$$

$$(\because \text{Given, } R_e=6.4\times 10^6 \text{ m})$$

or $h=2649.6$ km $=2.6\times 10^6$ m

24. *(a)* If earth rotate about its axis, then

weight $=\dfrac{3w}{4}$

Radius of the earth $=6400$ km

The acceleration due to gravity at the equator,

$$g_e=g-\omega^2 R\cos^2\theta$$

or $$\frac{3}{4}g=g-\omega^2 R\cos^2 0°$$

or $$\omega^2 R=\frac{g}{4}$$

$$\omega=\sqrt{\frac{g}{4R}}=\sqrt{\frac{10}{4\times 6400\times 10^3}}$$

$$=0.63\times 10^{-3} \text{ rad/s}$$

25. *(a)* $g_d=g\left(1-\dfrac{d}{R}\right)$

At the depth, $d=R/4$

$$=g\left(1-\frac{R/4}{R}\right)=\frac{3g}{4}$$

Weight on the surface of earth,

$$w=mg=300 \text{ N}$$

Weight of body at depth,

$$d=\frac{R}{4}$$

$$w'=mg_d$$

$$=m\times\frac{3g}{4}=\frac{3}{4}mg$$

$$=\frac{3}{4}\times 300=225 \text{ N}$$

> **• Concept Enhancer**
> The weight of a body decreases more rapidly when it goes below the Earth's surface than when it goes above the surface.

26. *(b)* Given,

$$g_p=\frac{gR^2}{(R+h)^2}$$

$$g_q=g\left(1-\frac{h}{R}\right)$$

$$g_p=g_q$$

$$\frac{g}{\left(1+\frac{h}{R}\right)^2}=g\left(1-\frac{h}{R}\right)$$

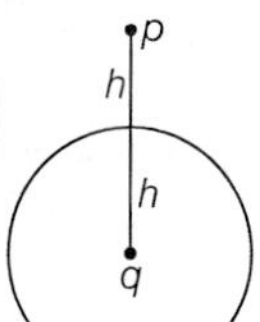

$$\Rightarrow$$

$$\left(1+\frac{h}{R}\right)\left(1+\frac{h}{R}\right)\left(1-\frac{h}{R}\right)=1$$

$$\left(1-\frac{h^2}{R^2}\right)\left(1+\frac{h}{R}\right)=1$$

Take, $\dfrac{h}{R}=x$

So, $$x^3-x+x^2=0$$

$$x=\frac{\sqrt{5}-1}{2} \Rightarrow h=\frac{R}{2}(\sqrt{5}-1)$$

option, $$h=\frac{\sqrt{5}R-R}{2}$$

> **• Concept Enhancer**
> Remember, the value of gravitational acceleration not only decreases as we go up or down from the Earth's surface but also decreases due to the rotation of earth about its own axis.
> If g' be the gravitational acceleration due to rotation of earth, then
> $$g'=g-\omega^2 R\cos^2\lambda$$
> where, $\lambda\rightarrow$ angle of latitude,
> $\omega\rightarrow$ angular speed of earth.

27. *(c)* Acceleration due to gravity on surface of earth,

$$g=\frac{GM}{R^2} \quad \ldots\text{(i)}$$

At height h acceleration is,

$$g'=\frac{GM}{(R+h)^2} \quad \ldots\text{(ii)}$$

Dividing Eq. (i) by Eq. (ii), we get

$$\frac{g}{g'}=\frac{(R+h)^2}{R^2}$$

$$\Rightarrow \quad g'=\frac{gR^2}{(R+h)^2}$$

If mass of body is m, then weight of body at height h is

$$mg'=\frac{(mg)\cdot R^2}{(R+h)^2}$$

Substituting given values, we get weight at height h,

$$w'=mg'=\frac{18\times(6400)^2}{(6400+3200)^2}$$

$$=\frac{18\times 6400\times 6400}{9600\times 9600}=8 \text{ N}$$

> ***Alternate Solution***
> Weight of the body on the surface of the earth,
> $$w=18 \text{ N}$$
> weight of the body at altitude h,
> $$w_h=\frac{w}{\left(1+\frac{h}{R_e}\right)^2}=\frac{18}{\left(1+\frac{3200}{6400}\right)^2}=\frac{18}{\left(\frac{3}{2}\right)^2}=8 \text{ N}$$

28. *(a)* The acceleration due to gravity at depth d,

$$g'=g\left(1-\frac{d}{R_E}\right) \quad \ldots\text{(i)}$$

The acceleration due to gravity at height r,

$$g''=g\left(\frac{R_E}{R_E+r}\right)^2 \quad \ldots\text{(ii)}$$

From Eqs. (i) and (ii), we get

$$\frac{g'}{g''}=\frac{g\left(1-\frac{d}{R_E}\right)}{g\left(\frac{R_E}{R_E+r}\right)^2}$$

$$=\frac{\left(1-\frac{r}{R_E}\right)(R_E+r)^2}{R_E^2} \quad [\because d=r]$$

$$\Rightarrow \quad \frac{g'}{g''}=\left(1-\frac{r}{R_E}\right)\frac{R_E^2}{R_E^2}\left(1+\frac{r}{R_E}\right)^2$$

$$\Rightarrow \quad \frac{g'}{g''}=\left(1-\frac{r}{R_E}\right)\left(1+\frac{r^2}{R_E^2}+\frac{2r}{R_E}\right)$$

$$\Rightarrow \quad \frac{g'}{g''}=1+\frac{r}{R_E}-\frac{r^2}{R_E^2}-\frac{r^3}{R_E^3}$$

29. *(c)*

> **key idea** Determine the weight of object on earth and on the planet, then establish the required relation between both the weights.

Let M be the mass of an object.

Given, $m_p=2m$

Density of planet (ρ_p)

= Density of earth (ρ)

As we know that,

Weight of object on earth is w

$$w=MG\rho\frac{4}{3}\pi R \quad \ldots\text{(i)}$$

As densities of planets and earth are same.

i.e., $\rho_p=\rho$

$$\Rightarrow \frac{m_p}{\frac{4}{3}\pi R_p^3} = \frac{m}{\frac{4}{3}\pi R^3}$$

$$\Rightarrow \quad R_p = \left(\frac{m_p}{m} R^3\right)^{1/3} = \left(\frac{2m}{m} R^3\right)^{1/3}$$

$$\Rightarrow \quad R_p = (2)^{1/3} R$$

Weight on planet,

$$w_p = \frac{4}{3}(2)^{1/3} \pi M G \rho R \quad \ldots\text{(ii)}$$

Dividing Eq. (ii) by Eq. (i) , we get

$$\frac{w_p}{w} = \frac{\frac{4}{3}(2)^{1/3} \pi M G \rho R}{M \rho G \frac{4}{3}\pi R}$$

$$\Rightarrow \quad w_p = 2^{1/3} w$$

30. *(a)* Given, Acceleration due to gravity on earth, $g_E = 10 \text{ m/s}^2$

and acceleration due to gravity on mars, $g_M = 4 \text{ m/s}^2$

We know that,

$$g'_E \text{ (at height } h) = \frac{g_E}{\left(1 + \frac{h}{R}\right)^2}$$

$\therefore$ Weight at earth,

$$m g_E = 100 \times 10 = 1000 \text{ N}$$

As the spaceship moves far away from earth, the value of g_E decreases to zero at a point where $g_E + g_M = 0$ and hence, weight will also be zero.

This point is called neutral point and is shown by graph (III) in the given figure.

Then, g_E increases 4 ms^{-2} at mars surface and weight becomes 400 N, which is also exhibited by graph (III).

31. *(d)* Condition of weightlessness,

$$g_{\text{equator}} = 0$$

$$g_{\text{equator}} = g - \omega^2 R$$

The radius of the earth,

$$R = 6400 \text{ km} = 6.4 \times 10^6 \text{ m}$$

$$\Rightarrow \quad \omega = \sqrt{\frac{g}{R}}$$

$$\Rightarrow \quad \omega = \sqrt{\frac{10}{6.4 \times 10^6 \text{ m}}}$$

$$\Rightarrow \quad \omega = \frac{1}{800} \text{ rad/s}$$

The duration of the day,

$$T = \frac{2\pi}{\omega}$$

$$\Rightarrow \quad T = \frac{2\pi}{\frac{1}{800}} \Rightarrow T = 5024 \text{ s}$$

$$\Rightarrow \quad T = \frac{5024}{60} \text{ min}$$

$$\Rightarrow \quad T \simeq 84 \text{ min}$$

32. *(9)* Acceleration due to gravity on surface is given by

$$g = \frac{GM}{R_e^2}$$

Now since diameter is reduced to 1/3rd, radius also reduces to 1/3rd, keeping mass constant.

New value of acceleration due to gravity on earth surface is

$$g' = \frac{GM}{\left(\frac{R_e}{3}\right)^2} = \frac{GM}{\frac{R_e^2}{9}}$$

$$g' = \frac{9\,GM}{R_e^2}$$

$$g' = 9g \qquad \left[g = \frac{GM}{R_e^2}\right]$$

33. *(2)* Since, given $h << R_e$

$\therefore$ We can use the approximation formula for gravity at a height, h

$$g_{\text{height}} = g_{\text{surface}}\left(1 - \frac{2h}{R_e}\right) \quad \ldots\text{(i)}$$

We also know,

Gravity at any depth d below the surface of earth

$$g_{\text{depth}} = g_{\text{surface}}\left(1 - \frac{d}{R_e}\right) \quad \ldots\text{(ii)}$$

In question, given that,

$$d = \alpha h$$

$\therefore$ Eq. (ii) becomes

$$g_{\text{depth}} = g_{\text{surface}}\left(1 - \frac{\alpha h}{R_e}\right)$$

According to the questions,

$$g_{\text{height}} = g_{\text{depth}}$$

$$\Rightarrow \quad g_{\text{surface}}\left(1 - \frac{2h}{R_e}\right) = g_{\text{surface}}\left(1 - \frac{\alpha h}{R_e}\right)$$

$$\Rightarrow \quad \frac{2h}{R_e} = \frac{\alpha h}{R_e}$$

$$\Rightarrow \quad \alpha = 2$$

34. *(4)* Given, $g_A = g_C$

Radius of earth, $R = 6400$ km

Height, $h = 3200 \text{ km} = \frac{R}{2}$

Since, $g = \frac{GM}{R^2}$

where, G is gravitational constant.

$$\therefore \quad g_A = \frac{GM(OA)}{R^3} \quad \ldots\text{(i)}$$

$$\text{and } g_C = \frac{GM}{(OC)^2} = \frac{GM}{(R+h)^2}$$

$$= \frac{GM}{(R + R/2)^2} = \frac{GM}{(3R/2)^2} \ldots\text{(ii)}$$

According to given information from Eqs. (i) and (ii), we get

$$\frac{GM(OA)}{R^3} = \frac{GM}{9R^2/4}$$

$$\Rightarrow \quad OA = \frac{4}{9} R = \frac{4}{9} \times 6400$$

$$\text{and} \quad AB = OB - OA$$

$$\Rightarrow \quad AB = 6400 - \frac{4}{9} \times 6400$$

$$= 6400\,(1 - 4/9)$$

$$= 6400 \times 5/9$$

$$\text{Now, } OA : AB = \frac{6400 \times 4/9}{5/9 \times 6400} = \frac{4}{5}$$

Hence, $x = 4$

Topic 3

01. *(b)* According to conservation of energy,

$$\frac{1}{2} m v_E^2 = \frac{G M_E m}{R_E}$$

$$\Rightarrow \quad K = \frac{G M_E m}{R_E}$$

$$= \frac{g R_E^2 \cdot m}{R_E} \qquad [\because G M_E = g R_E^2]$$

$$\Rightarrow \quad K = g m R_E$$

02. *(c)* Let gravitational field be zero at P as shown in figure.

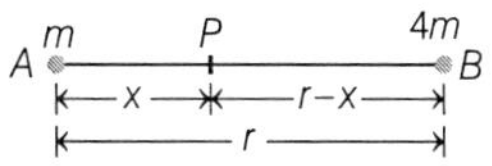

$$\therefore \quad \frac{Gm}{x^2} = \frac{G(4m)}{(r-x)^2} \Rightarrow 4x^2 = (r-x)^2$$

$$\Rightarrow \quad 2x = r - x \Rightarrow \quad x = \frac{r}{3}$$

$\therefore \quad V_p = -\dfrac{Gm}{x} - \dfrac{G(4m)}{r-x}$

$= -\dfrac{3Gm}{r} - \dfrac{6Gm}{r} = -\dfrac{9Gm}{r}$

03. *(c)* $KE = \dfrac{1}{2}mv^2$

$v^2 = \dfrac{2GM}{R}$

$KE = \dfrac{1}{2}m\dfrac{2GM}{R}$

$KE = m\dfrac{GM}{R}$

$g = \dfrac{GM}{R^2}$

$KE = m\dfrac{GM}{R^2}R$

$KE = mgR$

So Assertion wrong at ∞ $U = 0$ and Reason is correct.

04. *(b)* At earth surface, $E_1 = \dfrac{-GMm}{R_e}$

In the orbit, $E_2 = \dfrac{-GMm}{2r}$

$\Delta E = GMm\left[\dfrac{1}{R_e} - \dfrac{1}{2r}\right]$

$\therefore \quad r = \left(R_e + \dfrac{R_e}{20}\right)$

$r = \dfrac{21}{20}R_e$

$= GMm\left[\dfrac{1}{R_e} - \dfrac{1}{2.1R_e}\right] = \dfrac{11}{21}\dfrac{GMm}{R_e}$

$x = 11$

05. *(b)* The correct match is A-II, B-I, C-IV, D-III.

Kinetic energy of planet $= \dfrac{GMm}{2a}$

Gravitational potential energy of sun-planet system $= -\dfrac{GMm}{a}$

Total mechanical energy of planet $= -\dfrac{GMm}{2a}$

Escape energy $= \dfrac{Gm}{r}$

06. *(c)* $V = \dfrac{GM}{2R^3}(3R^2 - r^2)$

at $r = R \Rightarrow V = \dfrac{GM}{R}$

at $r = 0, V_0 = \dfrac{3GM}{2R}$

$\Rightarrow \quad V = \dfrac{GM}{R}$

$V_0 = \dfrac{3}{2}\cdot\dfrac{GM}{R}$

Put the value of V, $V_0 = \dfrac{3}{2}V$

07. *(d)* So, gravitational potential V is given by

$\dfrac{dV}{dr} = -E$

or $\quad dV = -Edr$

$\int_{V_i}^{V_f} dV = -\int_{2\text{ cm}}^{3\text{ cm}} Edr = -\int_2^3 \dfrac{-K}{r^2}dr$

$= \left[\dfrac{-K}{r}\right]_2^3 = -K\left[\dfrac{1}{3} - \dfrac{1}{2}\right]$

$= \dfrac{K}{6} = \dfrac{6}{6} = 1\text{ Jkg}^{-1}$

$\Rightarrow \quad V_f - V_i = 1$

$\Rightarrow \quad V_f - 10 = 1$

$\Rightarrow \quad V_f = 11\text{ J kg}^{-1}$

08. *(a)*

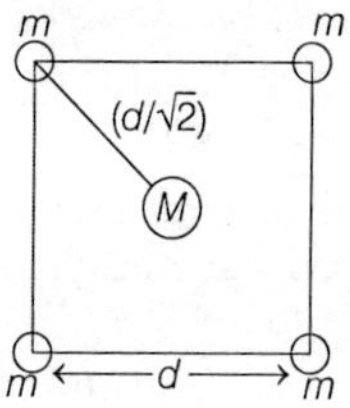

Since, gravitational PE of 2 particle system is $-\dfrac{GM_1M_2}{r_{12}}$, where r_{12} is distance between particle-1 and particle-2.

4 pairs of masses m and M separated by distance d and 2 pairs of masses m and m separated by distances $\sqrt{2}d$ (diagonally).

$U = -\dfrac{4GMm}{(d/\sqrt{2})} - \dfrac{4Gm^2}{d} - \dfrac{2Gm^2}{d\sqrt{2}}$

$= -\dfrac{4\sqrt{2}\,GMm}{d} - (4+\sqrt{2})\dfrac{Gm^2}{d}$

$= \dfrac{-Gm}{d}[4\sqrt{2}M + (4+\sqrt{2})m]$

09. *(d)* The initial potential energy of the particle,

$U_i = -\dfrac{GMm}{r}$

$U_i = -\dfrac{6.67\times10^{-11}\times100\times10^{-2}}{0.1}$

$U_i = -\dfrac{6.67\times10^{-11}}{0.1} = -6.67\times10^{-10}\text{ J}$

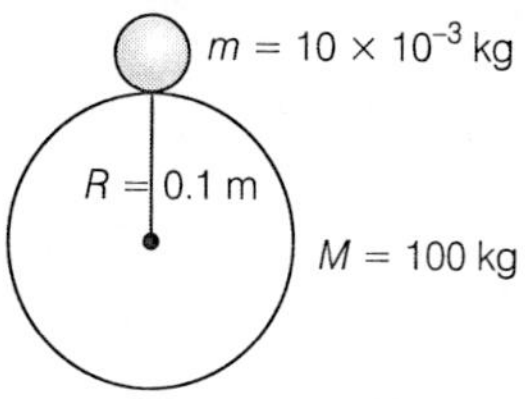

We know that, work done = Difference in potential energy

$\therefore \quad W = \Delta U = U_f - U_i$

$\Rightarrow \quad W = -U_i \qquad [\because U_f = 0]$

$= 6.67\times10^{-10}\text{ J}$

10. *(d)* Given,

Gravitational potential at point P,

$V_P = -5.12\times10^7$ J/kg

Acceleration due to gravity at point P,

$g_P = 6.4\text{ m/s}^2$

Radius of earth, $R = 6400$ km

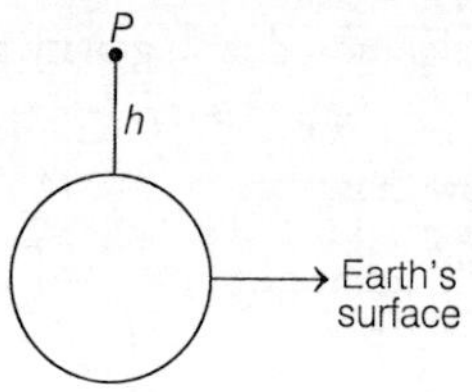

Gravitational potential at point P is given by

$V_P = -\dfrac{GM}{R+h} \quad \ldots(i)$

Acceleration due to gravity at point P above earth's surface is given by

$g_P = \dfrac{GM}{(R+h)^2} \quad \ldots(ii)$

On dividing Eq. (i) by Eq. (ii), we get

$\dfrac{V_P}{g_P} = -(R+h)$

$\Rightarrow \quad \dfrac{-5.12\times10^7}{6.4} = -(6.4\times10^6 + h)$

$\Rightarrow \quad 8\times10^6 = 6.4\times10^6 + h$

$h = 1.6\times10^6$ m

$h = 1600$ km

11. *(a)* On the surface of earth, gravitational potential energy,

$U_1 = -\dfrac{GMm}{R} = -\dfrac{GM\cdot1}{R}$

$U_1 = -\dfrac{GM}{R}$

At height $h = 3R$, gravitational potential energy,

$U_2 = -\dfrac{GMm}{r} = -\dfrac{GMm}{R+h}$

$$= -\frac{GM \times 1}{R+3R} = -\frac{GM}{4R}$$

$\therefore$ Gain in potential energy,

$$\Delta U = U_2 - U_1$$

$$= -\frac{GM}{4R} - \left(-\frac{GM}{R}\right) = \frac{3GM}{4R}$$

$$\Rightarrow \quad \Delta U = \frac{3gR^2}{4R} \qquad [\because GM = gR^2]$$

$$= \frac{3}{4} gR = \frac{3}{4} \times 10 \times 6400 \times 10^3$$

$$= 48 \times 10^6 \text{ J} = 48 \text{ MJ}$$

12. *(d)* Let v be the velocity and h be the height achieved in time t and U_1, KE and TE be the potential energy, kinetic energy and total energy, respectively. Now, by using law of conservation of energy,

$$\text{TE} = 0 = -\frac{GMm}{h} + \frac{1}{2}mv^2$$

$$\Rightarrow \quad \frac{1}{2}mv^2 = \frac{GMm}{h} \Rightarrow v = \sqrt{\frac{2GM}{h}}$$

Since, $\quad v = \frac{dh}{dt} \Rightarrow \frac{dh}{dt} = \sqrt{2\frac{GM}{h}}$

$$\Rightarrow \quad \int_{R_e}^{R_e+h} h^{1/2}\, dh = \int_0^t \sqrt{2GM}\, dt$$

$$\Rightarrow \quad \left[\frac{h^{1/2+1}}{3/2}\right]_{R_e}^{R_e+h} = \sqrt{2GM} \cdot t$$

$$\Rightarrow \quad \frac{2}{3}[h^{3/2}]_{R_e}^{R_e+h} = \sqrt{2GM}\, t$$

$$\Rightarrow \frac{2}{3}[(R_e+h)^{3/2} - R_e^{3/2}] = \sqrt{2GM} \cdot t$$

$$\Rightarrow \quad t = \frac{2}{3}\frac{[(R_e+h)^{3/2} - R_e^{3/2}]}{\sqrt{2GM}} \qquad \ldots\text{(i)}$$

$\because$ Acceleration due to gravity,

$$g = \frac{GM}{R_e^2}$$

$$\therefore \quad 2GM = 2gR_e^2$$

Substituting in Eq. (i), we get

$$t = \frac{2}{3}\left[\frac{(R_e+h)^{3/2} - R_e^{3/2}}{(2gR_e^2)^{1/2}}\right]$$

$$= \frac{1}{3}\left[\sqrt{\frac{4}{2gR_e^2}}\{(R_e+h)^{3/2} - R_e^{3/2}\}\right]$$

$$= \frac{1}{3}\sqrt{2\frac{R_e}{g}}\left[\left(1+\frac{h}{R_e}\right)^{3/2} - 1\right]$$

13. *(d)*

> **key idea** For the body to be able to reach at the surface of smaller planet, it just has to overcome the gravitational field of bigger planet and at that point, net force will be zero.

After crossing this P point, body will move towards smaller planet itself. So, the body need to be thrown with such a speed that it reaches at point P only. Let the distance of point P from bigger planet be x.

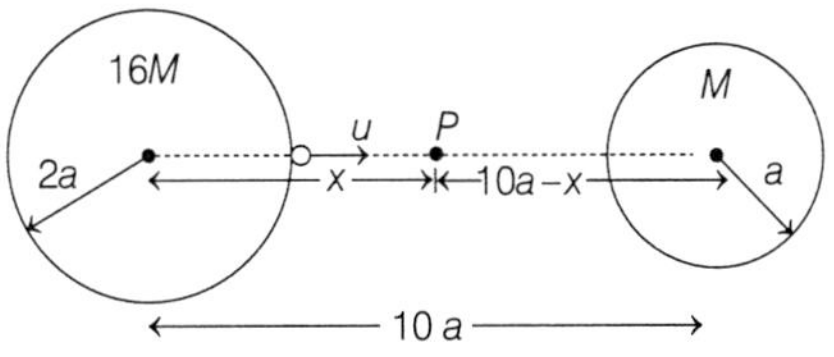

At point P, $F_{\text{net}} = 0$

$$\Rightarrow \quad F_{16M} = F_M$$

$$\Rightarrow \quad \frac{G(16M)m}{x^2} = \frac{G(M)(m)}{(10a-x)^2}$$

$$\Rightarrow \quad x = 8a \qquad \ldots\text{(i)}$$

Since, there is only conservative forces on the system, so we can apply conservation of mechanical energy from the surface of bigger planet to point P.

i.e., $\quad K_i + U_i = K_f + U_f$

$$\frac{1}{2}mu^2 + \left(-\frac{G(16M)m}{2a} - \frac{GMm}{8a}\right)$$

$$= 0 + \left(\frac{-G(16M)m}{8a} - \frac{GMm}{2a}\right)$$

$$\frac{1}{2}mu^2 - \frac{65}{8}\frac{GMm}{a} = -\frac{5}{2}\frac{GMm}{a}$$

$$\frac{1}{2}mu^2 = \frac{45}{8}\frac{GMm}{a}$$

$$\Rightarrow \quad u = \frac{3}{2}\sqrt{\frac{5GM}{a}}$$

14. *(2)* Given, total mass of body is $2M$. Potential energy is maximum at $M/m = x/1$

The arrangement of masses to form a square is shown in diagram.

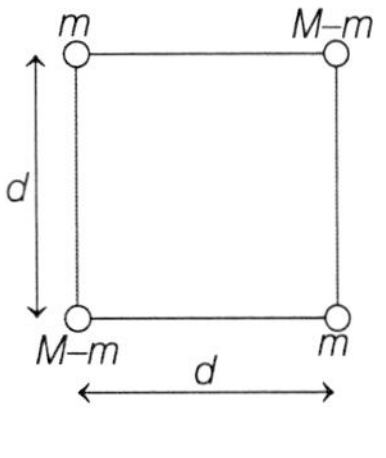

The gravitational potential energy of a body is given as

$$U = -\frac{GMm}{r}$$

For the given system, the potential energy will be

$$U_T = -\frac{Gm(M-m)}{d} - \frac{Gm(M-m)}{d} - \frac{Gm(M-m)}{d} - \frac{Gm(M-m)}{d} - \frac{Gm^2}{(\sqrt{2}d)} - \frac{G(M-m)^2}{(\sqrt{2}d)}$$

$$U_T = -\frac{4Gm(M-m)}{d} - \frac{Gm^2}{\sqrt{2}d} - \frac{G(M-m)^2}{\sqrt{2}d}$$

For maximum potential energy,

$$\frac{dU_T}{dm} = 0 \Rightarrow \frac{4G}{d}[M-2m] - \frac{G}{\sqrt{2}d}[2m]$$

$$-\frac{G}{\sqrt{2}d}[2(M-m) \times -1] = 0$$

$$\frac{M}{m} = \frac{2(4-\sqrt{2})}{4-\sqrt{2}} = 2$$

Comparing it with the given value, we get $x = 2$.

Thus, potential energy will be maximum when $x = 2$.

15. *(4)* Assume that at a distance x from the planet of mass M, the net gravitational field becomes zero.

$$\therefore \quad \frac{GM}{x^2} = \frac{G \times 9M}{(8R-x)^2} \Rightarrow \frac{1}{x^2} = \frac{9}{(8R-x)^2}$$

$$\Rightarrow \quad \left(\frac{1}{x}\right)^2 = \left(\frac{3}{8R-x}\right)^2 \Rightarrow \frac{1}{x} = \frac{3}{8R-x}$$

$$\Rightarrow \quad 3x = 8R - x \Rightarrow 4x = 8R$$

$$\Rightarrow \quad x = 2R \qquad \ldots\text{(i)}$$

By conservation of energy

$$\text{KE}_i + \text{PE}_i = \text{KE}_f + \text{PE}_f$$

$$\frac{1}{2}mv^2 - \frac{G9M \times m}{7R} - \frac{GMm}{R} = 0 - \frac{GMm}{2R} - \frac{G9M \times m}{6R}$$

$$v = \sqrt{\frac{4GM}{7R}} = \sqrt{\frac{aGM}{7R}}$$

$$\Rightarrow \quad a = 4$$

16. *(3)* We know that binding energy of earth, $\text{BE} = -\frac{3}{5}\frac{GM^2}{R}$

$\therefore$ Energy required to break the earth into pieces

$$= -\text{BE} = \frac{3}{5}\frac{GM^2}{R} \qquad \ldots\text{(i)}$$

According to question, the amount of energy that needs to be supplied is

$\frac{x}{5}\frac{GM^2}{R}$

Comparing it with value in Eq. (i), we get, $x = 3$

Topic 4

01. *(b)* $v_{\text{escape}} = \sqrt{\frac{2GM}{R}}$

$$\frac{(v_{\text{esp}})_{\text{planet}}}{(v_{\text{esp}})_{\text{earth}}} = \sqrt{\frac{M_P}{M_E} \times \frac{R_E}{R_P}}$$

$$(v_{\text{esp}})_{\text{planet}} = \frac{1}{2}(v_{\text{esp}})_{\text{earth}}$$

$$= \frac{1}{2} \times 11.2 = 5.6 \text{ km/s}$$

02. *(c)*

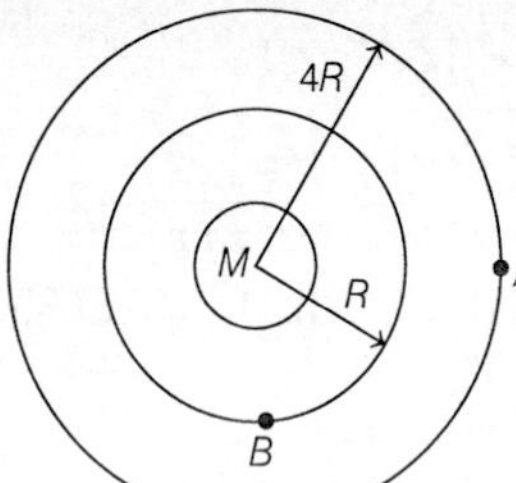

Orbital speed of a satellite is, $v = \sqrt{\frac{GM}{R}}$

For satellite A, $3v = \sqrt{\frac{GM}{4R}}$

For satellite B, $v_B = \sqrt{\frac{GM}{R}}$

So, $\frac{v_B}{3v} = \frac{\sqrt{(GM/R)}}{\sqrt{(GM/4R)}} = 2$

$\Rightarrow$ $v_B = 6v$

03. *(b)* Escape velocity on the surface of planet is, $v = \sqrt{\frac{2GM}{R}}$

Escape velocity on the surface of moon is,

$$v' = \sqrt{\frac{2GM'}{R'}}$$

$$= \sqrt{\left[\frac{2G\left(\frac{M}{144}\right)}{\left(\frac{R}{16}\right)}\right]} = \frac{1}{3}\sqrt{\frac{2GM}{R}} = \frac{v}{3}$$

04. *(c)* Statement I is correct but Statement II is incorrect.

$$v_e = \sqrt{\frac{2GM}{R}} \simeq 11.2 \text{ km/s}$$

$$v_e \propto \sqrt{\frac{M}{R}}$$

$\sqrt{\frac{M}{R}}$ increase then v_e increase

v_e increase $\Rightarrow \frac{M}{R}$ increases

$$v_e \propto \frac{1}{\sqrt{R}}$$

v_e depends on R

05. *(b)* Orbital speed of a satellite orbiting in a radius r is

$$v = \sqrt{\frac{GM}{r}}$$

So, ratio of orbital speeds for two satellites is,

$$\frac{v_1}{v_2} = \sqrt{\frac{r_2}{r_1}} \Rightarrow \frac{v_1}{v_2} = \sqrt{\frac{3r}{r}} = \frac{\sqrt{3}}{1}$$

(Given $r_2 = 3r$ and $r_1 = r$)

06. *(a)* Since planets revolve around the sun in an elliptical orbit, its linear speed is not constant, hence option (a) is not correct.

Other statement are correct as per theory.

07. *(a)* $v_{\text{escape}} = \sqrt{\frac{2GM}{R}} = \sqrt{2gR}$

Radius of moon is less than that of the Earth and acceleration due to gravity is also less on moon as compared to that on the earth.

Thus, v_{esp} of moon < v_{esp} of earth

This is also the reason behind escape of atmosphere from the moon.

08. *(c)* The escape velocity from surface of any planet is given as

$$v = \sqrt{\frac{2GM}{R}} \quad \text{...(i)}$$

For given planet, $M = \frac{M_e}{9}$, $R = \frac{R_e}{2}$

From Eq. (i),

$$v = \sqrt{\frac{2G\left(\frac{M_e}{9}\right)}{\frac{R_e}{2}}} = \sqrt{\frac{2}{9}}\sqrt{\frac{2GM_e}{R_e}}$$

$$= \sqrt{\frac{2}{9}}v_e = \frac{\sqrt{2}}{3}v_e \Rightarrow x = 2$$

09. *(b)* Orbital velocity of a body,

$$v_o = \sqrt{\frac{GM}{r}} \quad \text{...(i)}$$

Escape velocity, $v_e = \sqrt{\frac{2GM}{R}}$ ...(ii)

$$\frac{v_o}{v_e} = \frac{\sqrt{\frac{GM}{R}}}{\sqrt{\frac{2GM}{R}}} = \frac{1}{\sqrt{2}}$$

10. *(d)* Escape velocity is given by

$$v_e = \sqrt{\frac{2GM}{R}}$$

where, m = mass, R = radius of planet.

So, $v_A = \sqrt{\frac{2GM}{R}}$

and $v_B = \sqrt{\frac{2G\left(\frac{M}{2}\right)}{\left(\frac{R}{2}\right)}} = \sqrt{\frac{2GM}{R}}$

$\Rightarrow$ $\frac{v_A}{v_B} = 1$ or $\frac{n}{4} = 1 \Rightarrow n = 4$

11 *(d)* Orbital speed of a satellite in a circular orbit is

$$v_0 = \sqrt{\left(\frac{GM}{r_0}\right)}$$

where r_0 is the radius of the circular orbit.

So, kinetic energies of satellites A and B are

$$T_A = \frac{1}{2}m_A v_A^2 = \frac{GMm}{2R}$$

$$T_B = \frac{1}{2}m_B v_B^2$$

$$= \frac{GM(2m)}{2(2R)} = \frac{GMm}{2R}$$

So, ratio of their kinetic energies is

$$\frac{T_A}{T_B} = 1$$

12. *(b)* $v_S \to$ Speed of satellite

$v_m \to$ Speed of metorite

$v_R \to$ Resultant speed

$$mv\hat{\mathbf{i}} + mv(-\hat{\mathbf{j}}) = 2m\mathbf{v}_R$$

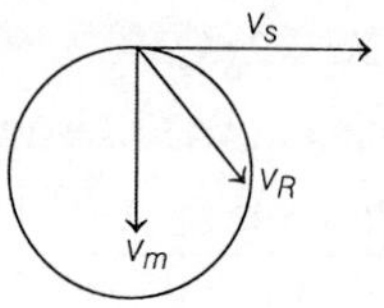

$$|\mathbf{v}_R| = \frac{v}{2}\hat{\mathbf{i}} - \frac{v}{2}\hat{\mathbf{j}} \Rightarrow |\mathbf{v}_R| = \frac{v}{\sqrt{2}}$$

$$|\mathbf{v}_R| = \frac{v}{\sqrt{2}} = \frac{1}{\sqrt{2}}\sqrt{\frac{GM}{R}}$$

$v_R < \sqrt{\frac{GM}{R}}$, hence orbit is elliptical

13. *(d)*

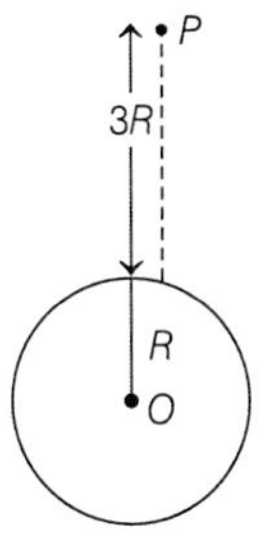

By conservation of mechanical energy,

$$P_P + K_P = P_o + K_o$$

$$-\frac{GMm}{4R} + \frac{1}{2}mv_P^2 = 0$$

$$\frac{1}{2}mv_P^2 = \frac{GMm}{4R}$$

$$v_P^2 = \frac{GM}{2R}$$

$$v_P = \sqrt{\frac{GM}{2R}}$$

14. *(b)* Time period of a satellite revolving at height h above the earth's surface,

$$T = \frac{2\pi(R+h)}{v} = \frac{2\pi(R+h)}{\sqrt{\frac{GM}{(R+h)}}}$$

$$T = \frac{2\pi(R+h)}{\sqrt{\frac{gR^2}{R+h}}} \quad [\because GM = gR^2]$$

$$\Rightarrow \quad T^2 = \frac{4\pi^2(R+h)^2}{gR^2/(R+h)}$$

$$\Rightarrow \quad gR^2T^2 = 4\pi^2(R+h)^3$$

$$\Rightarrow (R+h)^3 = \frac{gR^2T^2}{4\pi^2}$$

$$\Rightarrow \quad R+h = \left(\frac{gR^2T^2}{4\pi^2}\right)^{1/3}$$

$$\Rightarrow \quad h = \left(\frac{gR^2T^2}{4\pi^2}\right)^{1/3} - R$$

Time Saver Tech

Orbital radius of satellite having time period T can be directly calculated by the formula

$$r = \left(\frac{gR^2T^2}{4\pi^2}\right)^{1/3}$$

Here, $r = R + h$

15. *(d)* Escape speed from earth is

$$v_e = \sqrt{\frac{2GM}{R}} = 11.2 \text{ km/s}$$

Now for planet, $R_p = \frac{1}{3}R$

and $M_P = \frac{1}{6}M$

So, escape speed for planet is

$$v_p = \sqrt{\frac{2GM_p}{R_p}} = \sqrt{\frac{2G\left(\frac{M}{6}\right)}{\left(\frac{R}{3}\right)}}$$

$$= \sqrt{\frac{GM}{R}} = \frac{1}{\sqrt{2}}v_e$$

$$= 11.2 \times \frac{1}{\sqrt{2}} = 7.92 \frac{\text{km}}{\text{s}}$$

$$\approx 7.9 \text{ km/s}$$

16. *(d)* We know that, escape velocity on the surface of the Earth.

$$v_e = \sqrt{\frac{2GM_e}{R_e}}$$

and on the surface of planet

$$v_p = \sqrt{\frac{2GM_p}{R_p}} = \sqrt{2G\cdot\frac{9M_e}{4R_e}}$$

$$(\because M_p = 9M_e, R_p = 4R_e)$$

$$= \frac{3}{2}\sqrt{\frac{2GM_e}{R_e}} = \frac{3}{2}v_e = \frac{3}{2}\times 11.2$$

$$= 16.8 \text{ km/s}$$

17. *(c)* Given, mass of planet $M_P = 16M_e$

radius of planet, $R_P = 4R_e$

Escape velocity, $v_e = \sqrt{\frac{2GM}{R}}$

$\therefore v_e$ for planet

$$= \sqrt{\frac{2G(16M_e)}{(4R_e)}} = 2\sqrt{\frac{2GM_e}{R_e}}$$

$$= 2\,(v_e \text{ for earth})$$

i.e., $\frac{v_p \text{ for planet}}{v_e \text{ for earth}} = \frac{2}{1}$

18. *(c)* Given, mass of satellite A is twice the mass of B. i.e.,

$$M_A = 2M_B$$

For satellite, we know that,

$$\text{PE} = \frac{-GM_PM_A}{R} \text{ and KE} = \frac{+GM_PM_A}{2R}$$

$$\text{TE} = \frac{-GM_PM_A}{2R}$$

and Speed, $v = \sqrt{\frac{GM_p}{R}}$

Therefore, orbital speed of the satellite is independent of mass of the satellite.

19. *(c)* $v_{\text{circular}} = \sqrt{\frac{GM}{R}} = \sqrt{gR}$

$$v_{\text{escape}} = \sqrt{\frac{2GM}{R}} = \sqrt{2gR}$$

$$\Delta v = v_{\text{esp}} - v_{\text{circular}}$$

$$= \sqrt{2gR} - \sqrt{gR} = (\sqrt{2}-1)\sqrt{gR}$$

$$= (\sqrt{2}-1)\sqrt{\frac{10}{1000}\times 6400}$$

$$= (\sqrt{2}-1)8 \text{ km/s}$$

20. *(d)* Height of satellite from centre of earth $= 2R$

Therefore,

$$\frac{mv^2}{2R} = \frac{GMm}{(2R)^2} \Rightarrow v = \sqrt{\frac{GM}{2R}} = \sqrt{\frac{Rg}{2}}$$

If satellite takes time T to revolve around the Earth with velocity, v, then

$$T = \frac{2\pi(2R)}{v} = \frac{4\pi R\sqrt{2}}{\sqrt{Rg}}$$

$$\approx \sqrt{32\,R}$$

21. *(b)* Escape velocity of any planet from surface is given as

$$v_e = \sqrt{2gR}$$

For two planets A and B,

$$\frac{(v_e)_A}{(v_e)_B} = \sqrt{\frac{g_A}{g_B}}\sqrt{\frac{R_A}{R_B}} \quad \ldots\text{(i)}$$

Given that, $\frac{(v_e)_A}{(v_e)_B} = \frac{1}{2}, \frac{R_A}{R_B} = \frac{1}{3}$

From Eq. (i), $\frac{1}{2} = \sqrt{\frac{g_A}{g_B}}\sqrt{\frac{1}{3}}$

or $\sqrt{\frac{g_A}{g_B}} = \frac{\sqrt{3}}{2}$ or $\frac{g_A}{g_B} = \frac{3}{4}$

22. *(b)* Ratio of masses of two satellites A and B, $m_A : m_B = 4:3$

$$r_A = 3r \text{ and } r_B = 4r$$

$$\frac{r_A}{r_B} = \frac{3r}{4r} = \frac{3}{4}$$

We know that, total mechanical energy of satellite revolving around the earth,

$$E = -\frac{GM_em}{2r} \Rightarrow E \propto \frac{m}{r}$$

$$\Rightarrow \quad \frac{E_A}{E_B} = \frac{m_A}{m_B}\times\frac{r_B}{r_A} = \frac{4}{3}\times\frac{4}{3} = \frac{16}{9}$$

$$\therefore \quad E_A : E_B = 16:9$$

23. *(a)* Escape velocity of a mass over a planet,

$$v_{\text{esc}} = \sqrt{\frac{2GM}{R}}, M = \frac{4}{3}\pi R^3\cdot\rho$$

$\Rightarrow v_{esc} = \sqrt{\frac{8}{3} G\pi R^2 \cdot \rho}$ or $v_{esc} \propto R\sqrt{\rho}$

$\Rightarrow \frac{v_B}{v_A} = \frac{R_B}{R_A} \cdot \sqrt{\frac{\rho_B}{\rho_A}}$

Here, $v_A = 12$ km/s, $\rho_B = 4\rho_A$

and $R_B = \frac{1}{2} R_A$

Hence $\frac{v_B}{v_A} = \frac{1}{2} \times \sqrt{4}$

$\Rightarrow v_B = v_A$ or $v_B = 12$ km/s

24. *(b)* Given, masses and radii of earth and moon are (m_1, R_1), (m_2, R_2) and separation between their centre of mass $= r$

If v be the minimum escape velocity of particle of mass m, then by using law of conservation of energy,

Initial energy – Final energy = 0

$$\frac{1}{2} mv^2 - \frac{Gm_1 m}{r/2} - \frac{Gm_2 m}{r/2} = 0$$

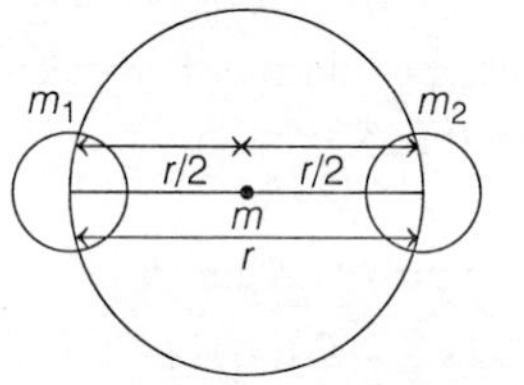

$$\Rightarrow \frac{1}{2} mv^2 = \frac{2Gm}{r}(m_1 + m_2)$$

$$\Rightarrow v = \sqrt{\frac{4G(m_1 + m_2)}{r}}$$

25. *(d)* Given,

$r = 9 \times 10^3 \text{ km} = 9 \times 10^6 \text{m}$

$T = 7h + 30$ min

$= 7 \times 60 \times 60 + 30 \times 60$

$= 27000$ s

$\frac{4\pi^2}{G} = 6 \times 10^{11} \text{ N}^{-1}\text{m}^{-2}\text{kg}^2$

Let m be the mass of moon and M be the mass of mars.

$\therefore$ Gravitational force acting between them = Centripetal force making both of them move in circular orbit

$\Rightarrow \frac{GMm}{r^2} = m\omega^2 r \Rightarrow \omega^2 = \frac{GM}{r^3}$

$\Rightarrow \frac{4\pi^2}{T^2} = \frac{GM}{r^3} \Rightarrow T^2 = \frac{4\pi^2 r^3}{GM}$

$\Rightarrow M = \frac{4\pi^2 r^3}{GT^2}$ ...(i)

$= \frac{6 \times 10^{11} \times (9 \times 10^6)^3}{(27000)^2}$

$M = 6.0 \times 10^{23}$ kg

• Concept Enhancer

Both mars and its one moon will have gravitational force of attraction between them and as both of them are revolving in a circular orbit, then centripetal force will be acting on them.

26. *(b)* According to question, a satellite is launched into a circular orbit of radius R around the earth, while a second satellite is launched into a circular orbit of radius 1.02 R.

As we know, according to Kepler's law of planetary motion,

$$T^2 \propto R^3 \Rightarrow T \propto R^{3/2}$$

Differentiating above equation, we get

$$\frac{dT}{T} = \frac{3}{2}\frac{dR}{R}$$

Here, $dR = 1.02R - R = 0.02R$

$$\Rightarrow \frac{dT}{T} = \frac{3}{2} \times \frac{0.02R}{R} = 0.03$$

$\therefore$ Percentage difference in time periods of two satellites is

$$\frac{dT}{T} \times 100 = 0.03 \times 100 = 3\%$$

27. *(b)* Let v_A, v_B, M_A, M_B and R_A, R_B be the escape velocities, masses and radii of planet A and B, respectively.

As we know that,

$$v = \sqrt{2gR} = \sqrt{\frac{2GM}{R^2} \cdot R} = \sqrt{\frac{2GM}{R}}$$

where, G is the gravitational constant.

$$\Rightarrow v \propto \sqrt{\frac{M}{R}}$$

Since, $v_A = v_B$

$\because \frac{M_A}{R_A} = \frac{M_B}{R_B}$

$\Rightarrow M_A R_B = M_B R_A$

or $M_1 R_2 = M_2 R_1$

28. *(a)* Given, period of revolution of first satellite,

$$T_1 = 1 \text{ h}$$

Period of revolution of second satellite,

$$T_2 = 8 \text{ h}$$

$\therefore \frac{T_1}{T_2} = \frac{1}{8}$

We know that,

$$\omega = \frac{2\pi}{T} \Rightarrow \omega \propto \frac{1}{T}$$

$\because \frac{\omega_1}{\omega_2} = \frac{T_2}{T_1} \Rightarrow \frac{\omega_1}{\omega_2} = \frac{8}{1}$

or $\omega_1 : \omega_2 = 8 : 1$

29. *(d)* Orbital velocity of the satellite is given as,

$$v_o = \sqrt{\frac{GM}{R+h}}$$

Since, $R >> h$

$\therefore v_o = \sqrt{\frac{GM}{R}} = \sqrt{gR} \left[\because g = \frac{GM}{R^2}\right]$

Escape velocity of the satellite,

$$v_e = \sqrt{\frac{2\,GM}{R+h}} = \sqrt{\frac{2\,GM}{R}} = \sqrt{2gR}$$

Since, we know that in order to escape. change in velocity,

$$\Delta v = v_e - v_o = \sqrt{gR}\,(\sqrt{2} - 1)$$

30. *(c)* Kinetic energy must be given to an object so that it just escapes to infinity as shown in the figure given below.

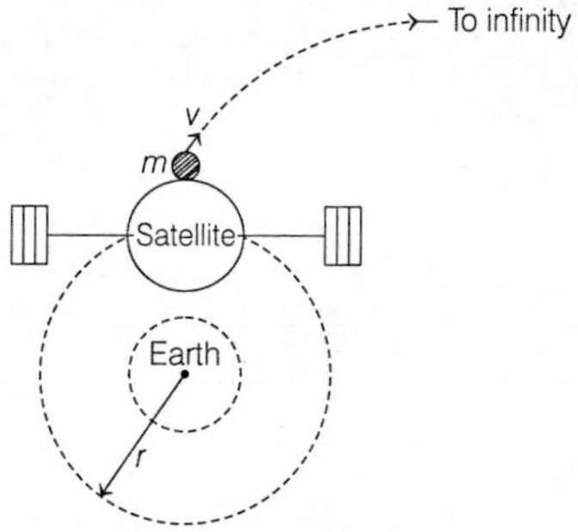

Applying energy-conservation for mass m, we get

Initial energy of m at satellite = Final energy of m at infinity.

$\Rightarrow U_i + K_i = U_f + K_f$ ...(i)

Here, U_i = initial PE of m

$= \frac{-G \cdot Mm}{r} = -(v^2)m$

As, $v = \sqrt{\frac{GM}{r}}$ = orbital speed of the satellite.

So, from Eq. (i), we have

$-v^2 m + K_i = 0 + 0$

or $K_i = mv^2$

31. *(a)* From conservation of energy,

Total energy at the planet

= Total energy at altitude

$-\frac{GMm}{R} + (\text{KE})_{\text{surface}}$

$= -\frac{GMm}{3R} + \frac{1}{2} mv_A^2$ ...(i)

In the orbital of planet, the necessary centripetal force is obtained by gravitational force.

$$\therefore \quad \frac{mv_A^2}{R+2R} = \frac{GMm}{(R+2R)^2}$$

$$\Rightarrow \quad v_A^2 = \frac{GM}{3R} \quad \ldots\text{(ii)}$$

From Eqs. (i) and (ii), we get

$$(\text{KE})_{\text{surface}} = \frac{5}{6}\frac{GMm}{R}$$

32. *(c)* Mass of planet, $M_p = 10 M_e$, where M_e is mass of earth.

Radius of planet, $R_p = R_e/10$, where R_e is radius of earth.

Escape velocity is given by,

$$v_e = \sqrt{\frac{2GM}{R}}$$

For planet,

$$v_p = \sqrt{\frac{2G \times M_p}{R_p}} = \sqrt{\frac{100 \times 2GM_e}{R_e}}$$

$$= 10 \times v_e = 10 \times 11 = 110 \text{ kms}^{-1}$$

33. *(c)* Given, mass of satellite, $m = 10^3$ kg

Radius of satellite, $r_1 = 2R$

Mass of earth, $m = 5.972 \times 10^{24}$ kg

Radius of earth, $R = 6.4 \times 10^6$ m

Energy required to jump from r_1 to r_2 orbits is given by,

$$\Delta E = \frac{GMm}{2}\left(\frac{1}{r_1} - \frac{1}{r_2}\right)$$

$$\Rightarrow \frac{10^4 R}{6} = \frac{6.67 \times 10^{-11} \times 5.972 \times 10^{24} \times 10^3}{2} \times \left(\frac{1}{r_1} - \frac{1}{r_2}\right)$$

$$\Rightarrow \frac{10^4 R}{6}\left(\frac{r_1 r_2}{r_2 - r_1}\right) = 19.91 \times 10^{16}$$

$$\Rightarrow R\left(\frac{r_1 r_2}{r_2 - r_1}\right) = 119.49 \times 10^{12}$$

$$\Rightarrow \frac{6.4 \times 10^6 \times 2 \times 6.4 \times 10^6 \times r_2}{r_2 - 2 \times 6.4 \times 10^6} = 119.49 \times 10^{12}$$

$$\Rightarrow 81.92 \times 10^{12} r_2 = 119.49 \times 10^{12} (r_2 - 2 \times 6.4 \times 10^6)$$

$$\Rightarrow 81.92 \times 10^{12} r_2 = 119.49 \times 10^{12} r_2 - 119.49 \times 2 \times 6.4 \times 10^6 \times 10^{12}$$

$$\Rightarrow (119.49 - 81.92) \times 10^{12} r_2 = 119.49 \times 2 \times 6.4 \times 10^6 \times 10^{12}$$

$$\Rightarrow 37.57 r_2 = 1529.472 \times 10^6$$

$$\Rightarrow \quad r_2 = 40.70 \times 10^6 = 6.3R \simeq 6R$$

34. *(b)* The given situation is shown below.

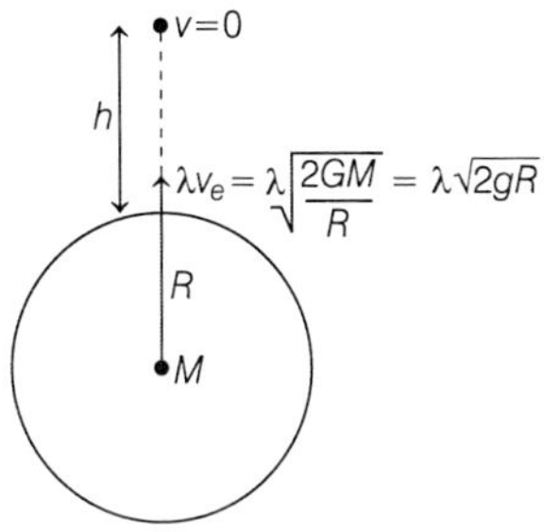

By energy conservation,

Total energy of system, when body is at the surface of earth = Total energy of the system when the body is at height h above the surface of earth

$$-\frac{GMm}{R} + \frac{1}{2}m(\lambda v_e)^2 = -\frac{GMm}{(R+h)} + \frac{1}{2}mv^2$$

$$\Rightarrow \quad -\frac{GM}{R} + \frac{1}{2}\lambda^2 v_e^2 = -\frac{GM}{(R+h)} + \frac{1}{2}v^2$$

Since at topmost point,

$$v = 0$$

$$\therefore \quad -\frac{GM}{R} + \frac{1}{2}\lambda^2 v_e^2 = \frac{-GM}{(R+h)} + 0$$

$$\Rightarrow \quad -\frac{GM}{R} + \frac{1}{2}\lambda^2 \times \frac{2GM}{R} = -\frac{GM}{(R+h)}$$

$$\left(\because v_e = \sqrt{\frac{2GM}{R}}\right)$$

$$\Rightarrow \quad -\frac{1}{R} + \frac{\lambda^2}{R} = -\frac{1}{R+h}$$

$$\Rightarrow \quad \frac{\lambda^2 - 1}{R} = -\frac{1}{R+h}$$

$$\Rightarrow \quad R + h = \frac{R}{1-\lambda^2}$$

= distance from centre of earth

35. *(a)* Given, radius of earth,

$$R_e = 6400 \text{ km} = 6.4 \times 10^6 \text{ m}$$

$$g = 10 \text{ m/s}^2$$

Escape velocity on the surface of earth,

$$v_e = \sqrt{\frac{2GM_e}{R_e}}$$

$\therefore$ Velocity of projection of body,

$$v = \frac{1}{3} \times v_e = \frac{1}{3}\sqrt{\frac{2GM_e}{R_e}}$$

If h be the maximum height attained by the projectile, then by applying the law of conservation of energy.

$$\Rightarrow \quad \frac{-GM_e m}{R_e} + \frac{1}{2}mv^2 = 0 + \left(-\frac{GM_e m}{R_e + h}\right)$$

$$\Rightarrow \quad \frac{-GM_e m}{R_e} + \frac{1}{2}m\frac{1}{9} \times \frac{2GM_e}{R_e} = \frac{-GM_e m}{R_e + h}$$

$$\Rightarrow -\frac{1}{R_e} + \frac{1}{9R_e} = -\frac{1}{R_e + h}$$

$$\Rightarrow \quad \frac{-8}{9R_e} = -\frac{1}{R_e + h}$$

$$\Rightarrow \quad h = \frac{R_e}{8} = \frac{6.4 \times 10^6}{8}$$

$$= 8 \times 10^5 \text{ m}$$

$$= 800 \times 10^3 \text{ m} = 800 \text{ km}$$

36. *(a)* Given,

$$M_A = 200 \text{ kg}, M_B = 400 \text{ kg},$$

$$H_A = 600 \text{ km}, H_B = 1600 \text{ km}$$

and

$$R_A = R_E + H_A = (6400 + 600) \text{ km}$$

$$= 7000 \text{ km} = 7000000 \text{ m}$$

$$R_B = R_E + H_B = (6400 + 1600) \text{ km}$$

$$= 8000 \text{ km} = 8000000 \text{ m}$$

Let T_A, T_B, ω_A, ω_B, R_A and R_B be the time period, angular frequencies and radii of satellites A and B, respectively.

Force on satellite A,

$$F_A = m_A \omega_A^2 R_A = \frac{GMm_A}{R_A^2}$$

$$\Rightarrow \quad \omega_A^2 = \frac{GM}{R_A^3} \quad \text{but,} \quad \omega_A = \frac{2\pi}{T_A}$$

$$\therefore \quad \left(\frac{2\pi}{T_A}\right)^2 = \frac{GM}{R_A^3} \Rightarrow T_A = \sqrt{\frac{4\pi^2 R_A^3}{GM}}$$

Similarly, $T_B = \sqrt{\frac{4\pi^2 R_B^3}{GM}}$

$$\therefore T_B - T_A = \sqrt{\frac{4\pi^2}{GM}}\left(\sqrt{R_B^3} - \sqrt{R_A^3}\right)$$

$$= \frac{2\pi}{\sqrt{GM}}\left[\sqrt{(8 \times 10^6)^3} - \sqrt{(7 \times 10^6)^3}\right]$$

$$= \frac{2\pi \times 10^9}{\sqrt{GM}}(8\sqrt{8} - 7\sqrt{7})$$

$$= \frac{2\pi \times 10^9}{\sqrt{6.67 \times 10^{-11} \times 6 \times 10^{24}}} (4.107)$$

$$= \frac{2\pi}{\sqrt{4 \times 10^{14}}} \times 10^9 (4.107)$$

$$= \pi \times 10^2 \times 4.107$$

$$= 12.9 \times 10^2 \text{ s}$$

$$\simeq 1.30 \times 10^3 \text{ s}$$

37. *(c)*

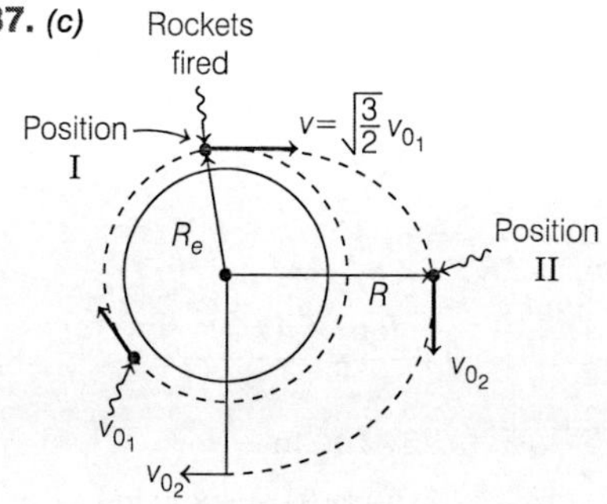

Using law of conservation of energy for position I and position II of satellite, we have

$$(\text{PE} + \text{KE})_{\text{Position I}} = (\text{PE} + \text{KE})_{\text{Position II}}$$

$$\Rightarrow \frac{-GMm}{R_e} + \frac{1}{2} m v_1^2 = \frac{-GMm}{R} + \frac{1}{2} m v_2^2$$

$$\Rightarrow \frac{-GMm}{R_e} + \frac{1}{2} m \left(\sqrt{\frac{3}{2}} v_{0_1} \right)^2 = \frac{-GMm}{R} + \frac{1}{2} m v_{0_2}^2$$

As, orbital speed is $v_{0_1} = \sqrt{\frac{GM}{R}}$

(where, R = orbital radius)

Therefore,

$$\frac{-GMm}{R_e} + \frac{1}{2} m \left(\frac{3}{2} \frac{GM}{R_e} \right) = -\frac{GMm}{R} + \frac{1}{2} m (v_{0_2}^2) \quad \ldots \text{(i)}$$

Conservation of angular momentum

$$M v_{0_1} R_e = M v_{0_2} R \quad \ldots \text{(ii)}$$

On solving Eq. (i) and (ii)

$$R = 3R_e$$

38. *(b)* Given, volume of earth (V_e) is 64 times of volume of moon (V_m), i.e.

$$\frac{V_e}{V_m} = 64 = \frac{\frac{4}{3}\pi R_e^3}{\frac{4}{3}\pi R_m^3}$$

Then, $\frac{R_e}{R_m} = 4$ ...(i)

Also, since the density of moon and earth are equal, i.e. $\rho_m = \rho_e$

$\Rightarrow \frac{M_e}{V_e} = \frac{M_m}{V_m}$, where M_e and M_m are the mass of the earth and moon, respectively.

$$\Rightarrow \frac{M_e}{M_m} = \frac{V_e}{V_m} = 64 \quad \ldots \text{(ii)}$$

The minimum energy or escape energy

$$E_e = \frac{GM_e m}{R_e} = E$$

Similarly, minimum energy that a launcher should have to escape or to never return, if rocket is launched from surface of the moon is

$$E_m = \frac{GM_m m}{R_m}$$

$$\therefore \frac{E_e}{E_m} = \frac{\left(\frac{GM_e m}{R_e} \right)}{\left(\frac{GM_m m}{R_m} \right)} = \frac{M_e}{M_m} \cdot \frac{R_m}{R_e}$$

$$= 64 \times \frac{1}{4} = 16$$

[using Eqs. (i) and (ii)]

$$\therefore E_m = \frac{E_e}{16} = \frac{E}{16}$$

39. *(a)* The energy required for taking a satellite upto a height h from earth's surface is the difference between the energy at h height and energy at surface, then

$$\Rightarrow E_1 = U_f - U_i$$

$$E_1 = -\frac{GM_e m}{R_e + h} + \frac{GM_e m}{R_e} \quad \ldots \text{(i)}$$

(where, U = potential energy)

$\therefore$ Orbital velocity of satellite,

$$v_o = \sqrt{\frac{GM_e}{(R_e + h)}}$$

(where, M_e = mass of earth)

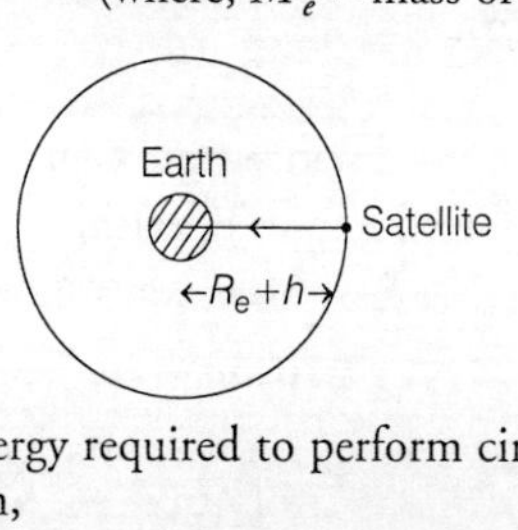

So, energy required to perform circular motion,

$$\Rightarrow E_2 = \frac{1}{2} m v_o^2 = \frac{GM_e m}{2(R_e + h)}$$

$$E_2 = \frac{GM_e m}{2(R_e + h)} \quad \ldots \text{(ii)}$$

According to the question,

$$E_1 = E_2$$

$$\therefore \frac{-GM_e m}{R_e + h} + \frac{GM_e m}{R_e} = \frac{GM_e m}{2(R_e + h)}$$

$$\Rightarrow 3R_e = 2R_e + 2h$$

$$h = \frac{R_e}{2}$$

As radius of earth, $R_e \approx 6.4 \times 10^3$ km

Hence, $h = \frac{6.4 \times 10^3}{2}$ km

or 3.2×10^3 km

40. *(3)* Kinetic energy of satellite

$$= \frac{GM_e m}{2(R_e + h)}$$

$$= \frac{6.67 \times 10^{11} \times 6 \times 10^{24} \times 6.4 \times 10^6}{2(6.4 \times 10^6 + 2.7 \times 10^5)}$$

$$= 3 \times 10^{10} \text{ J}$$

41. *(8)* Angular momentum, $L = mvr$

For planet A,

$$v_A = \sqrt{\frac{GM}{R_A}}$$

$$r_A = R_A$$

$$m_A = m$$

$$L_A = m_A v_A r_A = m\sqrt{\frac{GM}{R_A}} \times R_A$$

$$= m\sqrt{GMR_A} \quad \ldots \text{(i)}$$

For planet B

$$v_B = \sqrt{\frac{GM}{2R_A}}$$

$$r_B = 2R_A$$

$$m_B = 4\sqrt{2}\, m$$

$$L_B = m_B r_B v_B$$

$$= 4\sqrt{2} m \times 2R_A \times \sqrt{\frac{GM}{2R_A}}$$

$$= 8m\sqrt{GMR_A} \quad \ldots \text{(ii)}$$

Divide Eq. (ii) by Eq. (i)

$$\frac{L_B}{L_A} = \frac{8m\sqrt{GMR_A}}{m\sqrt{GMR_A}} = 8$$

42. *(0.75)*

Given, $L = M\sqrt{\frac{GMR}{x}}$

$h = R + \frac{R}{3} = \frac{4R}{3}$

$L = Mvr$

$$M\sqrt{\frac{GMR}{x}} = \frac{M}{2}\sqrt{\frac{GM}{\left(R+\frac{R}{3}\right)}}\cdot\left(R+\frac{R}{3}\right)$$

$$M\sqrt{\frac{GMR}{x}} = \frac{M}{2}\sqrt{\frac{3GM}{4R}}\cdot\frac{4R}{3}$$

$$\sqrt{\frac{GMR}{x}} = \frac{1}{4}\sqrt{\frac{3GM}{16R}}\cdot\frac{4R}{3}$$

$$\frac{GMR}{x} = \frac{3GM}{4R}\cdot\frac{16R^2}{9}$$

$$x = 3$$

43. *(2)* Given, $R_1 = 3200$ km

$R_2 = 800$ km

Speed of satellite revolving around the planet in circular orbit is given as,

$$v = \sqrt{\frac{GM}{R}}$$

$$\Rightarrow \quad v \propto \frac{1}{\sqrt{R}}$$

$$\Rightarrow \quad \frac{v_1}{v_2} = \sqrt{\frac{R_2}{R_1}} = \sqrt{\frac{800}{3200}} = \sqrt{\frac{1}{4}}$$

$$\Rightarrow \quad \frac{v_1}{v_2} = \frac{1}{2} = \frac{1}{x} \qquad \text{[Given]}$$

$\therefore \quad x = 2$

44. *(16)*

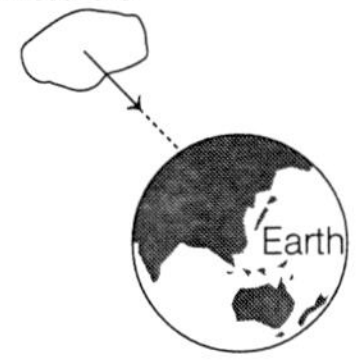

Applying energy conservation, we have Total energy of asteroid at 12 km = Total energy of asteroid at surface of earth

$$\Rightarrow \quad U_1 + K_1 = U_2 + K_2$$

$$\Rightarrow \quad \frac{-GM_e m}{10R_e} + \frac{1}{2}mv_0^2 = \frac{-GM_e m}{R_e} + \frac{1}{2}mv^2$$

$$\Rightarrow \quad \frac{9}{10}\frac{GM_e m}{R_e} + \frac{1}{2}mv_0^2 = \frac{1}{2}mv^2$$

$$\Rightarrow \quad \frac{9}{10}\times\frac{2GM_e}{R_e} + v_0^2 = v^2$$

$$\Rightarrow \quad \frac{9}{10}\times(\sqrt{v_e})^2 + v_0^2 = v^2$$

$$\Rightarrow \quad \frac{9}{10}\times(11.2)^2 + (12)^2 = v^2$$

$$\Rightarrow \quad v^2 = \sqrt{\frac{9}{10}\times(11.2)^2 + (12)^2}$$

$$\Rightarrow \quad v \approx 16\ \text{kms}^{-1}$$

C-08

JEE Main Chapterwise Topicwise Physics

In JEE Main there is usually 0-1 question per paper is visible from this chapter. In past 3 years, 18 conceptual and 17 numerical questions were asked of moderate of difficult level from this chapter.

For detailed analysis scan QR code

MECHANICAL PROPERTIES OF SOLIDS

Topic 1 Mechanical Properties of Solids

Level 1

01 A 3 m long wire of radius 3 mm shows an extension of 0.1 mm when loaded vertically by a mass of 50 kg in an experiment to determine Young's modulus. The value of Young's modulus of the wire as per this experiment is $P \times 10^{11}\ \text{Nm}^{-2}$, where the value of P is (Take $g = 3\pi\ \text{m/s}^2$) *[2025, 8 April Shift-II]*

(a) 25 (b) 2.5
(c) 10 (d) 5

02 A cylindrical rod of length 1 m and radius 4 cm is mounted vertically. It is subjected to a shear force of 10^5 N at the top. Considering infinitesimally small displacement in the upper edge, the angular displacement θ of the rod axis from its original position would be (shear moduli, $G = 10^{10}\ \text{N/m}^2$) *[2025, 4 April Shift-II]*

(a) $\frac{1}{2\pi}$ (b) $\frac{1}{40\pi}$
(c) $\frac{1}{4\pi}$ (d) $\frac{1}{160\pi}$

03 The fractional compression $\left(\frac{\Delta V}{V}\right)$ of water at the depth of 2.5 km below the sea level is .... %. Given, the Bulk modulus of water $= 2 \times 10^9\ \text{Nm}^{-2}$, density of water $= 10^3\ \text{kg m}^{-3}$ acceleration due to gravity $= 10\ \text{ms}^{-2}$. *[2025, 29 Jan Shift-I]*

(a) 1.5 (b) 1.0
(c) 1.75 (d) 1.25

04 Match List I with List II.

	List I		List II
A.	A force that restores an elastic body of unit area to its original state	I.	Bulk modulus
B.	Two equal and opposite forces parallel to opposite faces	II.	Young's modulus
C.	Forces perpendicular everywhere to the surface per unit area same everywhere	III.	Stress
D.	Two equal and opposite forces perpendicular to opposite faces	IV.	Shear modulus

Choose the correct answer from the options given bellow. *[2024, 5 April Shift-II]*

(a) A-II, B-IV, C-I, D-III (b) A-III, B-IV, C-I, D-II
(c) A-IV, B-II, C-III, D-I (d) A-III, B-I, C-II, D-IV

05 With rise in temperature, the Young's modulus of elasticity *[2024, 1 Feb Shift-I]*

(a) remains unchanged (b) changes erratically
(c) decreases (d) increases

06 Young's modulus of material of a wire of length L and cross-sectional area A is Y. If the length of the wire is doubled and cross-sectional area is halved, then Young's modulus will be *[2024, 30 Jan Shift-I]*

(a) $4Y$ (b) Y (c) $2Y$ (d) $Y/4$

07 Given below are two statements one is labelled as Assertion (A) and other is labelled as Reason (R).

Assertion (A) The property of body, by virtue of which it tends to regain its original shape when the external force is removed, is elasticity.

Reason (R) The restoring force depends upon the bonded interatomic and intermolecular force of solids.

In the light of the above statements, choose the correct answer from the options given below. *[2024, 27 Jan Shift-II]*

(a) Both A and R are true but R is not the correct explanation of A.
(b) A is false but R is true.
(c) Both A and R are true and R is the correct explanation of A.
(d) A is true but R is false.

Imp. Young's Modulus

08 Young's moduli of the material of wires A and B are in the ratio of 1 : 4, while its area of cross-sections are in the ratio of 1 : 3. If the same amount of load is applied to both the wires, the amount of elongation produced in the wires A and B will be in the ratio of (Assume length of wires A and B are same) *[2023, 10 April Shift-II]*
(a) 1 : 12 (b) 36 : 1
(c) 12 : 1 (d) 1 : 36

09 A 100 m long wire having cross-sectional area 6.25×10^{-4} m^2 and Young's modulus is 10^{10} Nm^{-2} is subjected to a load of 250 N, then the elongation in the wire will be *[2023, 24 Jan Shift - I]*
(a) 4×10^{-4} m (b) 4×10^{-3} m
(c) 6.25×10^{-3} m (d) 6.25×10^{-6} m

10 Under the same load, wire A having length 5.0 m and cross-section 2.5×10^{-5} m^2 stretches uniformly by the same amount as another wire B of length 6.0 m and a cross-section of 3.0×10^{-5} m^2 stretches. The ratio of the Young's modulus of wire A to that of wire B will be *[2023, 31 Jan Shift - I]*
(a) 1 : 2 (b) 1 : 4 (c) 1 : 1 (d) 1 : 10

11 If the length of a wire is made double and radius is halved of its respective values, then the Young's modulus of the material of the wire will *[2022, 29 July Shift-I]*
(a) remain same
(b) become 8 times its initial value
(c) become 1/4 of its initial value
(d) become 4 times of its initial value

12 The force required to stretch a wire of cross-section 1 cm^2 to double its length will be (Given, Young's modulus of the wire $= 2 \times 10^{11}$ N/m^2) *[2022, 28 July Shift-I]*
(a) 1×10^7 N (b) 1.5×10^7 N
(c) 2×10^7 N (d) 2.5×10^7 N

13 The area of cross-section of the rope used to lift a load by a crane is 2.5×10^{-4} m^2. The maximum lifting capacity of the crane is 10 metric tons. To increase the lifting capacity of the crane to 25 metric tons, the required area of cross-section of the rope should be (Take, $g = 10$ ms^{-2}) *[2022, 26 July Shift-II]*
(a) 6.25×10^{-4} m^2 (b) 10×10^{-4} m^2
(c) 1×10^{-4} m^2 (d) 1.67×10^{-4} m^2

14 A wire of length L is hanging from a fixed support. The length changes to L_1 and L_2 when masses 1 kg and 2 kg are suspended, respectively from its free end, then the value of L is equal to *[2022, 29 June Shift-I]*
(a) $\sqrt{L_1 L_2}$ (b) $\frac{L_1 + L_2}{2}$
(c) $2L_1 - L_2$ (d) $3L_1 - 2L_2$

15 The bulk modulus of a liquid is 3×10^{10} Nm^{-2}. The pressure required to reduce the volume of liquid by 2 % is *[2022, 24 June Shift-I]*
(a) 3×10^8 Nm^{-2} (b) 9×10^8 Nm^{-2}
(c) 6×10^8 Nm^{-2} (d) 12×10^8 Nm^{-2}

16 A steel wire having a radius of 2.0 mm, carrying a load of 4 kg, is hanging from a ceiling. Given that $g = 3.1\pi$ ms^{-2}, what will be the tensile stress that would be developed in the wire ? *[2019, 8 April Shift-I]*
(a) 6.2×10^6 Nm^{-2} (b) 5.2×10^6 Nm^{-2}
(c) 3.1×10^6 Nm^{-2} (d) 4.8×10^6 Nm^{-2}

17 A man grows into a giant such that his linear dimensions increase by a factor of 9. Assuming that his density remains same, the stress in the leg will change by a factor of *[JEE Main 2017]*
(a) 1/9 (b) 81
(c) 1/81 (d) 9

18 A thin 1 m long rod has a radius of 5 mm. A force of 50 πkN is applied at one end to determine its Young's modulus. Assume that the force is exactly known. If the least count in the measurement of all lengths is 0.01 mm, which of the following statements is false ? *[JEE Main 2016]*
(a) The maximum value of Y that can be determined is 2×10^{14} Nm^2.
(b) $\Delta Y / Y$ gets minimum contribution from the uncertainty in the length.
(c) $\Delta Y / Y$ gets its maximum contribution from the uncertainty in strain.
(d) The figure of merit is the largest for the length of the rod.

19 In materials like aluminium and copper, the correct order of magnitude of various elastic moduli is *[JEE Main 2014]*
(a) Young's modulus < shear modulus < bulk modulus.
(b) Bulk modulus < shear modulus < Young's modulus
(c) Shear modulus < Young's modulus < bulk modulus.
(d) Bulk modulus < Young's modulus < shear modulus.

20 Two blocks of masses m and M are connected by means of a metal wire of cross-sectional area A passing over a frictionless fixed pulley as shown in the figure. The system is then released. If $M = 2$ m, then the stress produced in the wire is *[JEE Main 2013]*

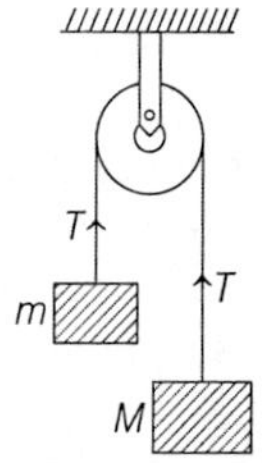

(a) $\frac{2mg}{3A}$ (b) $\frac{4mg}{3A}$ (c) $\frac{mg}{A}$ (d) $\frac{3mg}{4A}$

21 A steel wire can sustain 100 kg weight without breaking. If the wire is cut into two equal parts, then each part can sustain a weight of *[JEE Main 2013]*
(a) 50 kg (b) 400 kg (c) 100 kg (d) 200 kg

22 The load versus elongation graphs for four wires of same length and made of the same material are shown in the figure. The thinnest wire is represented by the line. *[AIEEE 2012]*

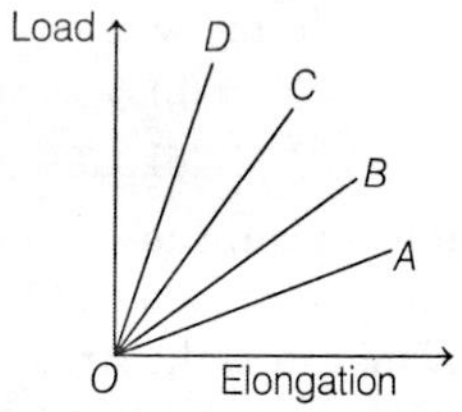

(a) OA (b) OC (c) OD (d) OB

23 Two wires are made of the same material and have the same volume. However, wire 1 has cross-sectional area A and wire 2 has cross-sectional area $3A$. If the length of wire 1 increases by Δx on applying force F, how much force is needed to stretch wire 2 by the same amount? *[AIEEE 2009]*
(a) F (b) $4F$ (c) $6F$ (d) $9F$

24 A wire elongates by l mm when a load w is hanged from it. If the wire goes over a pulley and two weights w each are hung at the two ends, the elongation of the wire will be (in mm) *[AIEEE 2006]*
(a) l (b) $2l$
(c) zero (d) $l/2$

Level 2

25 Two wires A and B are made of same material having ratio of lengths $\frac{L_A}{L_B} = \frac{1}{3}$ and their diameters ratio $\frac{d_A}{d_B} = 2$. If both the wires are stretched using same force, what would be the ratio of their respective elongations? *[2025, 7 April Shift-I]*
(a) 1 : 3 (b) 1 : 6
(c) 1 : 12 (d) 3 : 4

26 Given below are two statements. One is labelled as Assertion (A) and the other is labelled as Reason (R).
Assertion (A) A sound wave has higher speed in solids than gases.
Reason (R) Gases have higher value of Bulk modulus than solids.
In the light of the above statements, choose the correct answer from the options given below. *[2025, 28 Jan Shift-I]*
(a) A is true but R is false.
(b) Both A and R are true and R is the correct explanation of A.
(c) A is false but R is true.
(d) Both A and R are true but R is not the correct explanation of A.

27 Young's modulus is determined by the equation given by $Y = 49000 \frac{M}{l} \frac{\text{dyne}}{\text{m}^2}$, where M is the mass and l is the extension of wire used in the experiment. Now error in Young modulus Y is estimated by taking data from M-l plot is graph paper. The smallest scale division are 5 g and 0.02 cm along load axis and extension axis respectively. If the value of M and l are 500 g and 2 cm respectively, then percentage error of Y is *[2024, 8 April Shift-I]*
(a) 0.5% (b) 0.2% (c) 2% (d) 0.02%

28 A wire of length L and radius r is clamped at one end. If its other end is pulled by a force F, its length increases by l. If the radius of the wire and the applied force both are reduced to half of their original values keeping original length constant, the increase in length will become *[2024, 29 Jan Shift-II]*
(a) 3/2 times (b) 4 times (c) 2 times (d) 3 times

29 A wire of length L and radius r is clamped rigidly at one end. When the other end of the wire is pulled by a force f, its length increases by l. Another wire of same material of length $2L$ and radius $2r$ is pulled by a force $2f$. Then the increase in its length will be *[2023, 15 April Shift-I]*
(a) $2l$ (b) l (c) $4l$ (d) $l/2$

30 A force is applied to a steel wire A, rigidly clamped at one end. As a result, elongation in the wire is 0.2 mm. If same force is applied to another steel wire B of double the length and a diameter 2.4 times that of the wire A, the elongation in the wire B will be (wires having uniform circular cross-sections) *[2023, 30 Jan Shift - II]*
(a) 6.06×10^{-2} mm (b) 6.9×10^{-2} mm
(c) 2.77×10^{-2} mm (d) 3.0×10^{-2} mm

31 The Young's modulus of a steel wire of length 6 m and cross-sectional area 3 mm^2 is $2 \times 10^{11} \text{ N/m}^2$. The wire is suspended from its support on a given planet. A block of mass 4 kg is attached to the free end of the wire. The acceleration due to gravity on the planet is 1/4 of its value on the earth. The elongation of wire is
(Take, g on the earth $= 10 \text{ m/s}^2$) *[2023, 1 Feb Shift-II]*
(a) 0.1 mm (b) 1 mm (c) 1 cm (d) 0.1 cm

32 A steel wire of length 3.2 m ($Y_S = 2.0 \times 10^{11} \text{ Nm}^{-2}$) and a copper wire of length 4.4 m ($Y_C = 1.1 \times 10^{11} \text{ Nm}^{-2}$), both of radius 1.4 mm are connected end to end. When stretched by a load, the net elongation is found to be 1.4 mm. The load applied, in Newton, will be (Given, $\pi = 22/7$) *[2022, 27 July, Shift-II]*
(a) 360 (b) 180 (c) 1080 (d) 154

33 A uniform heavy rod of weight 10 kg, cross-sectional area 100 cm^2 and length 20 cm is hanging from a fixed support. Young's modulus of the material of the rod is $2 \times 10^{11} \text{ Nm}^{-2}$. Neglecting the lateral contraction, find the elongation of rod due to its own weight. *[2021, 31 Aug Shift-I]*
(a) 2×10^{-9} m (b) 5×10^{-8} m
(c) 4×10^{-8} m (d) 5×10^{-10} m

34 Two wires of same length and radius are joined end-to-end and loaded. The Young's moduli of the materials of the two wires are Y_1 and Y_2. The combination behaves as a single wire, then its Young's modulus is *[2021, 25 July Shift-I]*
(a) $Y = \dfrac{2Y_1Y_2}{3(Y_1 + Y_2)}$ (b) $Y = \dfrac{2Y_1Y_2}{Y_1 + Y_2}$
(c) $Y = \dfrac{Y_1Y_2}{2(Y_1 + Y_2)}$ (d) $Y = \dfrac{Y_1Y_2}{Y_1 + Y_2}$

35 The value of tension in a long thin metal wire has been changed from T_1 to T_2. The lengths of the metal wire at two different values of tension T_1 and T_2 are l_1 and l_2, respectively. The actual length of the metal wire is *[2021, 20 July Shift-I]*
(a) $\dfrac{T_1l_2 - T_2l_1}{T_1 - T_2}$ (b) $\dfrac{T_1l_1 - T_2l_2}{T_1 - T_2}$
(c) $\dfrac{l_1 + l_2}{2}$ (d) $\sqrt{T_1T_2l_1l_2}$

36 A cube of metal is subjected to a hydrostatic pressure of 4 GPa. The percentage change in the length of the side of the cube is close to
(Take, bulk modulus of metal, $B = 8 \times 10^{10}$ Pa) *[2020, 4 Sep Shift-II]*
(a) 1.67 (b) 0.6 (c) 20 (d) 5

37 The elastic limit of brass is 379 MPa. What should be the minimum diameter of a brass rod, if it is to support a 400 N load without exceeding its elastic limit? *[2019, 10 April Shift-II]*
(a) 0.90 mm (b) 1.00 mm (c) 1.16 mm (d) 1.36 mm

38 A solid sphere of radius r made of a soft material of bulk modulus K is surrounded by a liquid in a cylindrical container. A massless piston of area a floats on the surface of the liquid, covering entire cross-section of cylindrical container. When a mass m is placed on the surface of the piston to compress the liquid, the fractional decrement in the radius of the sphere, $\left(\dfrac{dr}{r}\right)$ is *[JEE Main 2018]*
(a) $\dfrac{Ka}{mg}$ (b) $\dfrac{Ka}{3mg}$ (c) $\dfrac{mg}{3Ka}$ (d) $\dfrac{mg}{Ka}$

39 A bottle has an opening of radius a and length b. A cork of length b and radius $(a + \Delta a)$, where ($\Delta a << a$) is compressed to fit into the opening completely (see figure). If the bulk modulus of cork is B and frictional coefficient between the bottle and cork is μ, then the force needed to push the cork into the bottle is *[JEE Main 2016]*

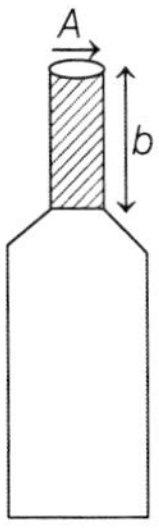

(a) $(\pi\mu B b)a$ (b) $(2\pi\mu B b)\Delta a$
(c) $(\pi\mu B b)\Delta a$ (d) $(4\pi\mu B b)\Delta a$

40 A copper wire of length 1.0 m and a steel wire of length 0.5 m having equal cross-sectional areas are joined end to end. The composite wire is stretched by a certain load which stretches the copper wire by 1 mm. If the Young's moduli of copper and steel are, respectively $1.0 \times 10^{11} \text{ Nm}^{-2}$ and $2.0 \times 10^{11} \text{ Nm}^{-2}$, the total extension of the composite wire is *[JEE Main 2013]*
(a) 1.75 mm (b) 2.0 mm
(c) 1.50 mm (d) 1.25 mm

41 If the ratio of lengths, radii and Young's moduli of steel and brass wires in the figure are a, b and c respectively, then the corresponding ratio of increase in their lengths is *[JEE Main 2013]*

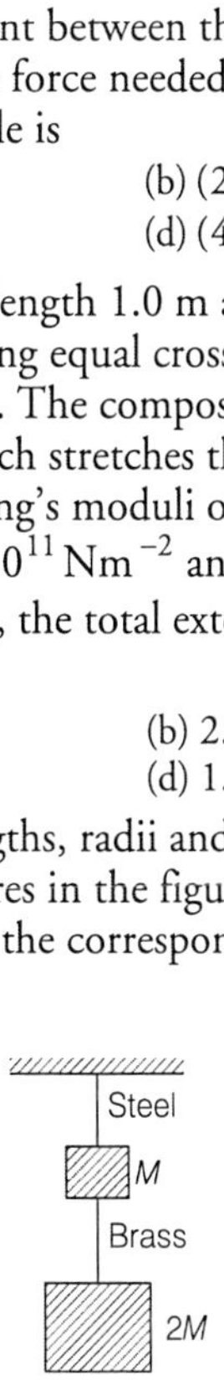

(a) $\dfrac{3c}{2ab^2}$ (b) $\dfrac{2a^2c}{b}$
(c) $\dfrac{3a}{2b^2c}$ (d) $\dfrac{2ac}{b^2}$

Level 3

42 Two blocks of masses 3 kg and 5 kg are connected by a metal wire going over a smooth pulley. The breaking stress of the metal is $(24/\pi) \times 10^2$ Nm^{-2}. What is the minimum radius of the wire ? (Take, $g = 10$ ms^{-2}) *[2021, 26 Aug Shift-II]*

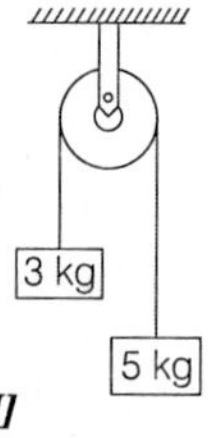

(a) 125 cm (b) 1250 cm (c) 12.5 cm (d) 1.25 cm

Numerical Value Type Questions

43. Two slabs with square cross-section of different materials (1, 2) with equal sides (l) and thickness d_1 and d_2 such that $d_2 = 2d_1$ and $l > d_2$. Considering lower edges of these slabs are fixed to the floor, we apply equal shearing force on the narrow faces. The angle of deformation is $\theta_2 = 2\theta_1$. If the shear moduli of material 1 is 4×10^9 N/m^2, then shear moduli of material 2 is $x \times 10^9$ N/m^2, where value of x is .... *[2025, 4 April Shift-I]*

44. A sample of a liquid is kept at 1 atm. It is compressed to 5 atm which leads to change of volume of 0.8 cm^3. If the bulk modulus of the liquid is 2 GPa, then the initial volume of the liquid was .... L. (Take 1 atm $= 10^5$ Pa) *[2025, 8 April Shift-II]*

45 The increase in pressure required to decrease the volume of a water sample by 0.2% is $p \times 10^5$ Nm^{-2}. Bulk modulus of water is 2.15×10^9 Nm^{-2}. The value of p is .... *[2025, 24 Jan Shift-II]*

Imp. Bulk Modulus

46 The volume contraction of a solid copper cube of edge length 10 cm, when subjected to a hydraulic pressure of 7×10^6, Pa, would be......mm^3. (Given, bulk modulus of copper $= 1.4 \times 10^{11}$ Nm^{-2}) *[2025, 28 Jan Shift-II]*

47. A steel wire of length 2 m and Young's modulus 2.0×10^{11} $N\,m^{-2}$ is stretched by a force. If Poisson ratio and transverse strain for the wire are 0.2 and 10^{-3} respectively, then the elastic potential energy density of the wire is .... $\times 10^5$ (in SI units). *[2025, 2 April Shift-I]*

48 Two persons pull a wire towards themselves. Each person exerts a force of 200 N on the wire. Young's modulus of the material of wire is 1×10^{11} Nm^{-2}. Original length of the wire is 2 m and the area of cross-section is 2 cm^2. The wire will extend in length by .... μm. *[2024, 9 April Shift-I]*

49 A wire of cross-sectional area A, modulus of elasticity 2×10^{11} Nm^{-2} and length 2 m is stretched between two vertical rigid supports. When a mass of 2 kg is suspended at the middle it sags lower from its original position making angle $\theta = \frac{1}{100}$ rad on the points of support. The value of A is .... $\times 10^{-4}$ m^2 (consider $x << L$)

Given, $g = 10$ m/s^2

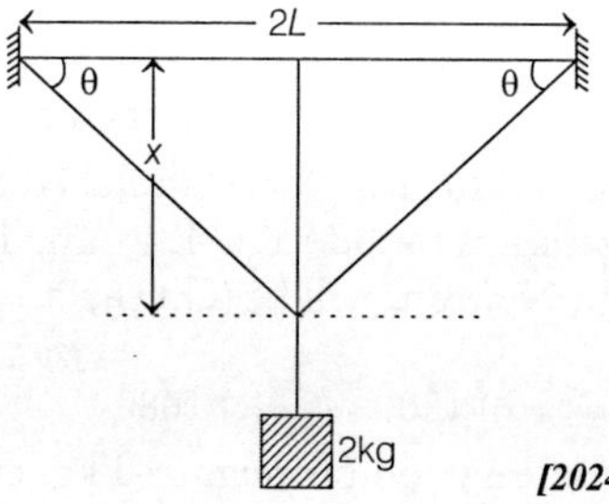

[2024, 6 April Shift-II]

50 The density and breaking stress of a wire are 6×10^4 kg/m^3 and 1.2×10^8 N/m^2 respectively. The wire is suspended from a rigid support on a planet where acceleration due to gravity is 1/3rd of the value on the surface of earth. The maximum length of the wire with breaking is .... m. (Take, $g = 10$ m/s^2) *[2024, 5 April Shift-I]*

51 One end of a metal wire is fixed to a ceiling and a load of 2 kg hangs from the other end. A similar wire is attached to the bottom of the load and another load of 1 kg hangs from this lower wire. Then, the ratio of longitudinal strain of upper wire to that of the lower wire will be .... [Area of cross-section of wire $= 0.005$ cm^2, $Y = 2 \times 10^{11}$ Nm^{-2} and $g = 10$ ms^{-2}] *[2024, 1 Feb Shift-II]*

52 Two blocks of mass 2 kg and 4 kg are connected by a metal wire going over a smooth pulley as shown in figure. The radius of wire is 4.0×10^{-5} m and Young's modulus of the metal is 2.0×10^{11} N/m^2. The longitudinal strain developed in the wire is $\frac{1}{\alpha\pi}$. The value of α is .... [Use, $g = 10$ m/s^2] *[2024, 31 Jan Shift-II]*

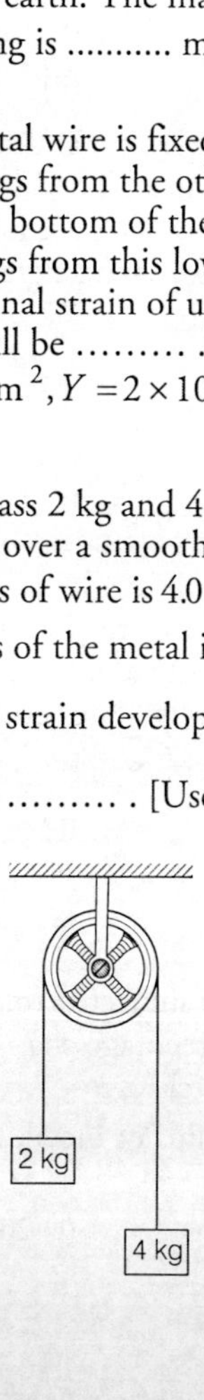

53 Each of three blocks P, Q and R shown in figure has a mass of 3 kg. Each of the wires A and B has cross-sectional area 0.005 cm^2 and Young's modulus 2×10^{11} N m^{-2}. Neglecting friction, the longitudinal strain on wire B is ...... $\times 10^{-4}$. (Take, $g = 10$ m/s^2) **[2024, 30 Jan Shift-I]**

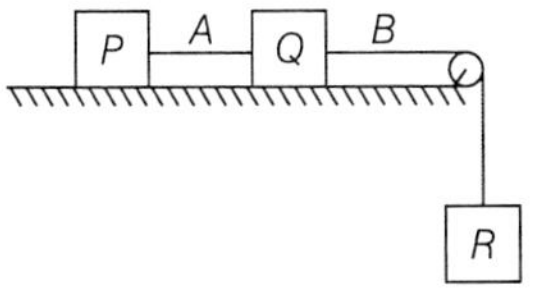

54 Two metallic wires P and Q have same volume and are made up of same material. If their area of cross-sections are in the ratio 4 : 1 and force F_1 is applied to P, an extension of Δl is produced. The force which is required to produce same extension in Q is F_2. The value of $\frac{F_1}{F_2}$ is .... **[2024, 29 Jan Shift-II]**

55 A steel rod has a radius of 20 mm and a length of 2.0 m. A force of 62.8 kN stretches it along its length. Young's modulus of steel is 2.0×10^{11} N/m^2. The longitudinal strain produced in the wire is .... $\times 10^{-5}$ **[2023, 6 April Shift-I]**

56 A metal block of mass m is suspended from a rigid support through a metal wire of diameter 14 mm. The tensile stress developed in the wire under equilibrium state is 7×10^5 Nm^{-2}. The value of mass m in kg. (Take, $g = 9.8$ ms^{-2} and $\pi = 22/7$) **[2023, 6 April Shift-II]**

57 Two wires each of radius 0.2 cm and negligible mass, one made of steel and the other made of brass are loaded as shown in the figure. The elongation of the steel wire is .... $\times 10^{-6}$ m.
(Young's modulus for steel $= 2 \times 10^{11}$ Nm^{-2} and $g = 10$ ms^{-2}) **[2023, 10 April Shift-I]**

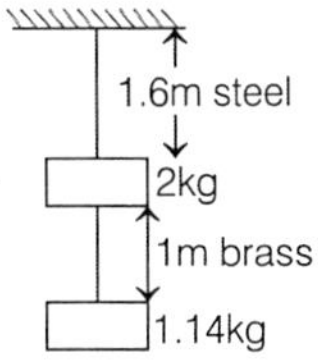

58 The length of a wire becomes l_1 and l_2 when 100 N and 120 N tensions are applied, respectively. If $10\, l_2 = 11 l_1$, the natural length of wire will be $\frac{1}{x} l_1$. Here the value of x is .... **[2023, 11 April Shift-I]**

Imp. Shear Modulus

59 A square aluminium (shear modulus is 25×10^9 Nm^{-2}) slab of side 60 cm and thickness 15 cm is subjected to a shearing force (on its narrow face) of 18.0×10^4 N. The lower edge is riveted to the floor. The displacement of the upper edge is .... μm. **[2022, 27 July Shift-I]**

60 A metal wire of length 0.5 m and cross-sectional area 10^{-4} m^2 has breaking stress 5×10^8 Nm^{-2}. A block of 10 kg is attached at one end of the string and is rotating in a horizontal circle. The maximum linear velocity of block will be ...... ms^{-1}. **[2022, 29 July Shift-II]**

61 The elongation of a wire on the surface of the earth is 10^{-4} m. The same wire of same dimensions is elongated by 6×10^{-5} m on another planet. The acceleration due to gravity on the planet will be .... ms^{-2}. (Take, acceleration due to gravity on the surface of earth = 10 m/s^2) **[2022, 26 June Shift-I]**

62 When a rubber ball is taken to a depth of ...... m in deep sea, its volume decreases by 0.5%. (The bulk modulus of rubber = 9.8×10^8 Nm^{-2}. Density of sea water = 10^3 kg m^{-3}, $g = 9.8$ m/s^2) **[2021, 31 Aug Shift-I]**

63 A uniform metallic wire is elongated by 0.04 m when subjected to a linear force F. The elongation, if its length and diameter is doubled and subjected to the same force will be .... cm. **[2021, 24 Feb Shift-II]**

64 Two separate wires A and B are stretched by 2 mm and 4 mm respectively, when they are subjected to a force of 2 N. Assume that both the wires are made up of same material and the radius of wire B is 4 times that of the radius of wire A. The length of the wires A and B are in the ratio of $a : b$. Then, a/b can be expressed as $1/x$, where x is .... **[2021, 18 March Shift-I]**

Topic 2 Stress-Strain Curve, Thermal Stress and Elastic Potential Energy

Level 1

01 An aluminium rod with Young's modulus $Y = 7.0 \times 10^{10}$ N/m^2 undergoes elastic strain of 0.04%. The energy per unit volume stored in the rod in SI unit is **[2023, 8 April Shift-I]**
(a) 2800 (b) 11200
(c) 8400 (d) 5600

02 A uniform cylindrical rod of length L and radius r, is made from a material whose Young's modulus of Elasticity equals Y. When this rod is heated by temperature T and simultaneously subjected to a net longitudinal compressional force F, its length remains unchanged. The coefficient of volume expansion, of the material of the rod, is (nearly) equal to **[JEE Main 2014]**
(a) $9F/(\pi r^2 YT)$ (b) $6F/(\pi r^2 YT)$
(c) $3F/(\pi r^2 YT)$ (d) $F/(\pi r^2 YT)$

Imp. Energy Stored in Stretched Wire

03 If S is stress and Y is Young's modulus of material of a wire, the energy stored in the wire per unit volume is *[AIEEE 2005]*

(a) $2S^2Y$ (b) $\frac{S^2}{2Y}$ (c) $\frac{2Y}{S^2}$ (d) $\frac{S}{2Y}$

Level 2

04 Two steel wires having same length are suspended from a ceiling under the same load. If the ratio of their energy stored per unit volume is 1 : 4, the ratio of their diameters is *[2020, 9 Jan Shift-II]*

(a) $\sqrt{2}:1$ (b) $1:\sqrt{2}$ (c) 2 : 1 (d) 1 : 2

Level 3

05 A boy's catapult is made of rubber cord which is 42 cm long, with 6 mm diameter of cross-section and of negligible mass. The boy keeps a stone weighing 0.02 kg on it and stretches the cord by 20 cm by applying a constant force. When released the stone flies off with a velocity of 20 ms^{-1}. Neglect, the change in the area of cross-section of the cord while stretched. The Young's modulus of rubber is closest to *[2019, 8 April Shift-I]*

(a) $10^6 Nm^{-2}$ (b) $10^4 Nm^{-2}$
(c) $10^8 Nm^{-2}$ (d) $10^3 Nm^{-2}$

06 The pressure that has to be applied to the ends of a steel wire of length 10 cm to keep its length constant when its temperature is raised by 100°C is (For steel, Young's modulus is 2×10^{11} Nm^{-2} and coefficient of thermal expansion is 1.1×10^{-5} K^{-1}) *[JEE Main 2014]*

(a) 2.2×10^8 Pa
(b) 2.2×10^9 Pa
(c) 2.2×10^7 Pa
(d) 2.2×10^6 Pa

Numerical Value Type Questions

Imp. Energy Stored in Stretched Wire

07 The elastic potential energy stored in a steel wire of length 20 m stretched through 2 cm is 80 J. The cross-sectional area of the wire is .... mm^2. (Given, $Y = 2.0\times 10^{11}$ Nm^{-2}). *[2023, 13 April Shift-I]*

08 The elastic behaviour of material for linear stress and linear strain, is shown in the figure. The energy density for a linear strain of 5×10^{-4} is .... kJ/m^3. (Assume that material is elastic upto the linear strain of 5×10^{-4}). *[2022, 26 June Shift-I]*

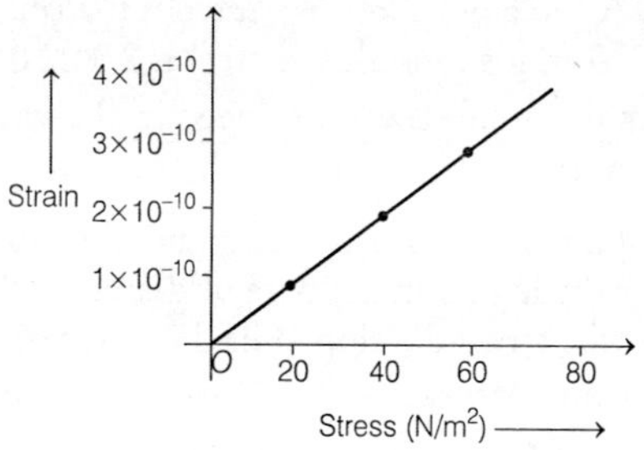

09 A stone of mass 20 g is projected from a rubber catapult of length 0.1 m and area of cross-section 10^{-6} m^2 stretched by an amount 0.04 m. The velocity of the projected stone is .... m/s. (Take, Young's modulus of rubber $= 0.5\times 10^9$ N/m^2) *[2021, 27 July Shift-I]*

Key Index

Topic 1

1 (d)	2 (d)	3 (d)	4 (b)	5 (c)	6 (b)	7 (c)	8 (c)	9 (b)	10 (c)
11 (a)	12 (c)	13 (a)	14 (c)	15 (c)	16 (c)	17 (d)	18 (a)	19 (c)	20 (b)
21 (c)	22 (a)	23 (d)	24 (a)	25 (c)	26 (a)	27 (c)	28 (c)	29 (b)	30 (b)
31 (a)	32 (d)	33 (d)	34 (b)	35 (a)	36 (a)	37 (c)	38 (c)	39 (d)	40 (b)
41 (c)	42 (c)	43 (43)	44 (50)	45 (1)	46 (4)	47 (25)	48 (20)	49 (1)	50 (600)
51 (3)	52 (12)	53 (2)	54 (16)	55 (25)	56 (11)	57 (20)	58 (2)	59 (48)	60 (50)
61 (6)	62 (500)	63 (2)	64 (32)						

Topic 2

1 (d)	2 (c)	3 (b)	4 (a)	5 (a)	6 (a)	7 (40)	8 (25)	9 (20)

Detailed **Explanations**

Topic 1

01. *(d)* We know that Young's modulus

$$Y = \frac{mgl}{\pi r^2 \Delta l} = \frac{50 \times 3\pi \times 3}{\pi \times (3 \times 10^{-3})^2 \times 0.1 \times 10^{-3}}$$
$$= 5 \times 10^{11} \text{ N/m}^2$$
$$\therefore \quad P = 5$$

02. *(d)*

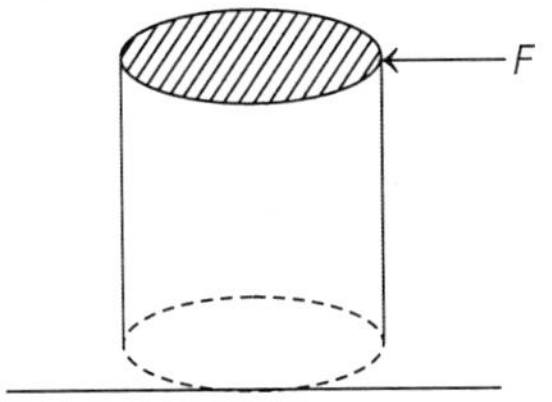

$$\sigma_{\text{shear}} = \frac{F}{A} = \frac{10^5}{\pi \times 16 \times 10^{-4}}$$

$$\text{Shear moduli} = \frac{\sigma_{\text{shear}}}{\theta}$$

$$10^{10} = \frac{10^5}{\pi \times 16 \times 10^{-4}} \times \frac{1}{\theta}$$

$$\theta = \frac{1}{160\pi} \text{ rad}$$

03. *(d)* $\dfrac{\Delta V}{V} = \dfrac{-\Delta p}{B} = -\dfrac{(p_0 + \rho g h)}{B}$

$$= -\left(\frac{10^5 + 1000 \times 10 \times 2.5 \times 10^3}{2 \times 10^9}\right)$$
$$= -\left(\frac{10^5 + 250 + 10^5}{2 \times 10^9}\right)$$
$$= -\frac{251 \times 10^5}{2 \times 10^9} = -125.5 \times 10^{-4}$$
$$\frac{\Delta V}{V} = -125.5 \times 10^{-4}$$
$$\frac{\Delta V}{V} \times 100 = -125.5 \times 10^{-2}\ \% = -1.25\%$$

04. *(b)* Since, Stress $= \dfrac{F_{\text{restoring}}}{\text{Area } (A)}$

If Area is unity, then

Stress $= F_{\text{restoring}}$

$\therefore$ A $\rightarrow$ III

Two equal and opposite forces parallel to opposite faces produce shear stress.

Thus, B $\rightarrow$ IV

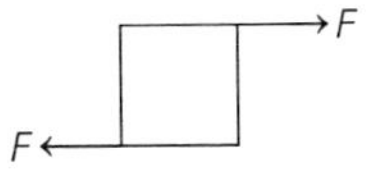

Forces perpendicular everywhere to the surface per unit area same everywhere is known as volumetric stress.

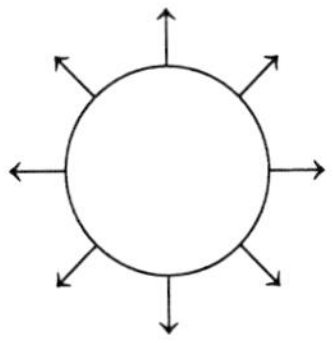

$\therefore$ C $\rightarrow$ I

Two equal and opposite forces perpendicular to opposite faces per unit area is equal to longitudinal stress.

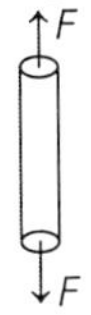

$\therefore$ D $\rightarrow$ II

Thus, the correct match is A-III, B-IV, C-I, D-II.

05. *(c)* On increasing temperature, change in length will be more for same load.

$\therefore$ Strain will be more so, Young's modulus will decreases.

$$\because \quad Y \propto \frac{1}{\Delta L}$$

06. *(b)* Young's modulus describe the relation between stress and strain.

$$Y = \frac{\text{Stress}}{\text{Strain}}$$

It is independent of geometrical factor. Hence, it will remain same as Y.

07. *(c)* Elasticity is properties of body by virtue of which it tends to regain their shape and size after removing external force and restoring force depends upon the bonded interatomic and intermolecular force of solid.

Both Assertion and Reason are true and Reason is correct explanation of Assertion.

08. *(c)* Young's modulus of elasticity is given by

$$Y = \frac{FL}{A\Delta L}$$

According to question,

$$\frac{Y_A}{Y_B} = \frac{1}{4} = \frac{\dfrac{FL}{A\Delta L_A}}{\dfrac{FL}{3A\Delta L_B}} \Rightarrow \frac{\Delta L_A}{\Delta L_B} = \frac{3A}{A} \times 4$$
$$= 12$$
$$\therefore \quad \frac{\Delta L_A}{\Delta L_B} = \frac{12}{1}$$

09. *(b)* Given, $A = 6.25 \times 10^{-4}$ m^2

$Y = 10^{10}$ Nm^{-2}

$l = 100$ m, $F = 250$ N

$$\because \quad \Delta l = \frac{Fl}{AY} = \frac{250 \times 100}{6.25 \times 10^{-4} \times 10^{10}}$$
$$= 4 \times 10^{-3} \text{ m}$$

10. *(c)*

> **key idea** Using Hooke's law, you can determine the required ratio of Young's modulus.

Given that $L_A = 5$ m,

$A_A = 2.5 \times 10^{-5}$ m^2

$L_B = 6$ m,

$A_B = 3 \times 10^{-5}$ m^2

From Hooke's law, $Y = \dfrac{FL}{A\Delta L}$

If F and ΔL are same, then

$$Y \propto \frac{L}{A}$$

For two wires A and B,

$$\frac{Y_A}{Y_B} = \frac{L_A}{L_B} \cdot \frac{A_B}{A_A}$$
$$= \frac{5}{6} \cdot \frac{3 \times 10^{-5}}{2.5 \times 10^{-5}}$$
$$\Rightarrow \quad \frac{Y_A}{Y_B} = \frac{1}{1}$$

11. *(a)* We Know, Young's modulus depends on the material of wire only. i.e, it is a characteristic property of a material.

$\therefore$There will be no effect on the value of Young's modulus on changing the radius and length of wire.

Thus, Young's modulus will remain unchanged.

12. *(c)* Given, Young's modulus of wire,

$Y = 2 \times 10^{11}$ N / m^2

$A = 1$ cm^2 $= 10^{-4}$ m^2

$l_1 = l$

$l_2 = 2l_1 = 2l$

$$\therefore \quad \Delta l = l_2 - l_1 = 2l - l = l$$
$$\therefore \quad Y = \frac{F/A}{\Delta l / l} = \frac{Fl}{A\Delta l} = \frac{Fl}{Al} \quad [\because \Delta l = l]$$

$$Y = \frac{F}{A} \Rightarrow F = YA$$
$$= 2\times10^{11}\times10^{-4}$$
$$= 2\times10^{7}\text{ N}$$

13. *(a)*

> **key idea** Apply the condition of elastic limit, then determine the required area of cross-setion.

Given, $A_1 = 2.5\times10^{-4}\text{ m}^2$

Load, $F_1 = 10$ metric tonne $\times g$
$$= 10\times1000\text{ kg}\times g = 10^4 g\text{N}$$

For second condition,
$$F_2 = 25\text{ metric tonne}\times g$$
$$= 25\times1000\times g = 2.5\times10^4\ g\text{ N}$$

For the condition of elastic limit,
$$\frac{F_1}{A_1} = \frac{F_2}{A_2} \Rightarrow A_2 = \frac{F_2}{F_1}A_1$$
$$= \frac{2.5\times10^4 g}{10^4 g}\times2.5\times10^{-4}$$
$$= 6.25\times10^{-4}\text{ m}^2$$

14. *(c)* Using, $\Delta L = \frac{Mg}{AY}\cdot L$

If with mass of 1 kg, length becomes L_1 and with mass of 2 kg length becomes L_2, then

So, $L_1 - L = \frac{1\cdot g}{AY}\cdot L$ ...(i)

and $L_2 - L = \frac{2\cdot g}{AY}\cdot L$ ...(ii)

Now, on multiplying Eq. (i) by 2 and then subtracting Eq. (ii) from it, we get
$$2(L_1 - L) - (L_2 - L) = \frac{2gL}{AY} - \frac{2gL}{AY}$$
or $2L_1 - 2L - L_2 + L = 0$

or $2L_1 - L_2 = L$

15. *(c)* Given, Bulk modulus of a liquid,
$$B = 3\times10^{10}\text{ Nm}^{-2}$$
$$-\frac{\Delta V}{V} = 2\% = -\frac{2}{100} = -0.02$$

We know that bulk modulus,
$$B = \frac{\Delta p}{-\frac{\Delta V}{V}} \Rightarrow \Delta p = -B\left(\frac{\Delta V}{V}\right)$$
$$= -3\times10^{10}(-0.02)$$
$$= 6\times10^{8}\text{ N/m}^2$$

16. *(c)* Given, radius of wire, $r = 2$ mm
$= 2\times10^{-3}$ m

Weight of load,

$m = 4$ kg, $g = 3.1\,\pi\text{ ms}^{-2}$

$\therefore$ Tensile stress $= \frac{\text{Force }(F)}{\text{Area }(A)} = \frac{mg}{\pi r^2}$
$$= \frac{4\times3.1\pi}{\pi\times(2\times10^{-3})^2} = 3.1\times10^{6}\text{ Nm}^{-2}$$

17. *(d)* $\because$ Stress $= \frac{\text{Weight}}{\text{Area}}$

Volume will become (9^3) times.

So, weight = volume × density × g will also become $(9)^3$ times.

Area of cross-section will become $(9)^2$ times.
$$= \frac{9^3\times W_0}{9^2\times A_0} = 9\left(\frac{W_0}{A_0}\right)$$

Hence, the stress increases by a factor of 9.

18. *(a)* As $Y = \frac{MgL}{\pi r^2 l} = \frac{FL}{\pi r^2 l}$

As r^2 and l are lesser than L, figure of merit is largest for length.
$$\therefore \frac{\Delta Y}{Y} = \frac{\Delta F}{F} + \frac{2\Delta r}{r} + \frac{\Delta L}{L} + \frac{\Delta l}{l}$$

As fouce is exactly known, $\Delta F = 0$

So, $\frac{\Delta Y}{Y}$ gets maximum contribution from the uncertainity in radius.

Hence, option (b), (c) and (d) are correct.

Also,
$$Y = \frac{FL}{\pi r^2 l} = \frac{50\pi\times10^3\times1}{\pi\times(5\times10^{-3})^2\times x\times10^{-3}}$$

(Here change in length is $x\times10^{-3}$ m)

So, $Y = \frac{2\times10^{12}}{x}$

$\therefore Y < 2\times10^{12}$ for any

Finite x option (a) is incorrect.

19. *(c)* For metals shear modulus η or G is less then Young's modulus

Emperially, $G \approx \frac{Y}{3}$.

As change in volume ΔV for solids is very less, they have a greater bulk modulus 'B' than Young's modulus.

For aluminium,
$$B = 72\times10^{9}\text{ N/m}^2,$$
$$G = 25\times10^{9}\text{ N/m}^2$$
and $Y = 70\times10^{9}\text{ N/m}^2$

20. *(b)* Tension in the wire,
$$T = \left(\frac{2mM}{m+M}\right)g$$
$$\text{Stress} = \frac{\text{Force (Tension)}}{\text{Area}} = \frac{2mM}{A(m+M)}g$$
$$= \frac{2(m\times2m)g}{A(m+2m)} \quad (M = 2\,m\text{, given})$$
$$= \frac{4m^2}{3mA}g = \frac{4mg}{3A}$$

21. *(c)* Breaking force $\propto$ area of cross-section of wire. Load hold by wire is independent of length of the wire. So, the breaking stress in each part of the wire will be same as the original because of the same material. Since area is same, each part can sustain a weight of 100 kg.

22. *(a)* Let m = slope of the given curve
$$F = \frac{\Delta l}{l}YA \Rightarrow F = m\Delta l \quad \left\{m = \frac{YA}{l}\right\}$$
$\Rightarrow \quad m \propto A$

As slope of OA is minimum, so its area is also minimum

23. *(d)* As volume is same.

$\therefore \quad A_1 l_1 = A_2 l_2$
$$\Rightarrow l_2 = \frac{A_1 l_1}{A_2} = \frac{A\times l_1}{3A} = \frac{l_1}{3} \Rightarrow \frac{l_1}{l_2} = 3$$

$\because \quad \Delta x_1 = \frac{F_1}{A\gamma}\times l_1$ ...(i)

and $\quad \Delta x_2 = \frac{F_2}{3A\gamma}l_2$ ...(ii)

Here, $\quad \Delta x_1 = \Delta x_2 \Rightarrow \frac{F_2}{3A\gamma}l_2 = \frac{F_1}{A\gamma}l_1$

$\Rightarrow \quad F_2 = 3F_1\times\frac{l_1}{l_2}$
$$= 3F_1\times3 = 9F$$

24. *(a)* Let us consider the length of wire as L and cross-sectional area A, the material of wire has Young's modulus as Y.

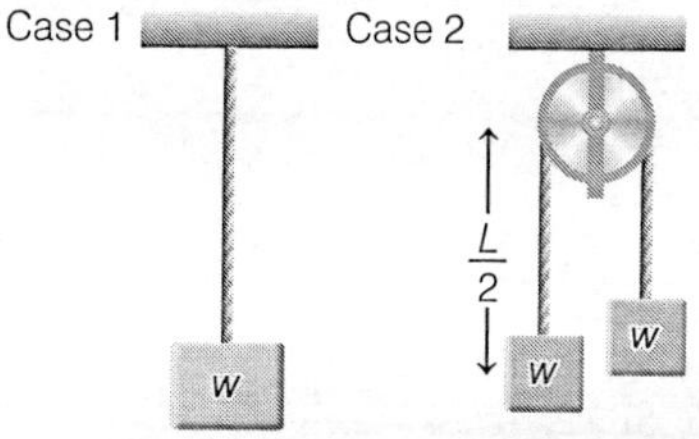

Then, for 1st case, $Y = \frac{w/A}{l/L}$

For 2nd case, $Y = \frac{w/A}{2l'/L} \Rightarrow l' = \frac{l}{2}$

So, total elongation of both sides
$$= 2l' = l$$

25. *(c)* $\frac{L_A}{L_B} = \frac{1}{3}$ and $\frac{d_A}{d_B} = 2$
$$\Delta L_A = \frac{F_A L_A}{Y_A A_A},\ \Delta L_B = \frac{F_B L_B}{A_B Y_B}$$

Given $F_A = F_B$ and $Y_A = Y_B$

$$\frac{\Delta L_A}{\Delta L_B} = \frac{\dfrac{F_A L_A}{A_A Y_A}}{\dfrac{F_B L_B}{A_B Y_B}} = \frac{L_A}{L_B} \cdot \frac{A_B}{A_A}$$

$$\frac{\Delta L_A}{\Delta L_B} = \left(\frac{L_A}{L_B}\right)\left(\frac{\frac{\pi}{4} d_B^2}{\frac{\pi}{4} d_A^2}\right) = \frac{L_A}{L_B} \times \left(\frac{d_B}{d_A}\right)^2$$

$$\frac{\Delta L_A}{\Delta L_B} = \left(\frac{1}{3}\right)\left(\frac{1}{2}\right)^2 = \frac{1}{3} \times \frac{1}{4} = \frac{1}{12}$$

$$\frac{\Delta L_A}{\Delta L_B} = \frac{1}{12}$$

26. *(a)* Speed of sound in medium depends on elastic and inertial property of medium.

For gas, $v = \sqrt{\dfrac{B}{\rho}}$

For solids, $v = \sqrt{\dfrac{Y}{\rho}}$

The elastic property of solids happens to be many fold greater than that of elastic property of gases. Bulk modulus of gas depends on process, $B = -\dfrac{\Delta p}{\Delta V / V}$ varies between 0 to ∞. Therefore, Reason is incorrect.

27. *(c)* Given,

$$Y = 49000 \times \frac{M}{l} \frac{\text{dyne}}{\text{m}^2}$$

$\therefore$ Percentage error in Y is,

$$\frac{\Delta Y}{Y} \times 100 = \frac{\Delta M}{M} \times 100 + \frac{\Delta l}{l} \times 100$$

$\Rightarrow$ Percentage error in Y

$$= \frac{\Delta M}{M} \times 100 + \frac{\Delta l}{l} \times 100$$

$$= \frac{5}{500} \times 100 + \frac{0.02}{2} \times 100$$

$$= 1 + 1 = 2\%$$

28. *(c)* We know that,

$$Y = \frac{F/A}{\Delta l / l} = \frac{Fl}{A \Delta l}$$

$$Y = \frac{Fl}{\pi r^2 \Delta l}$$

$$\Rightarrow \quad \Delta l = \frac{Fl}{\pi r^2 Y}$$

$$\Rightarrow \quad \Delta l \propto \frac{F}{r^2}$$

$$\therefore \quad \frac{\Delta l_2}{\Delta l_1} = \frac{F_2}{F_1} \times \left(\frac{r_1}{r_2}\right)^2$$

$$= \frac{F_1/2}{F_1} \times \left(\frac{r_1}{r_1/2}\right)^2$$

$$= \frac{1}{2} \times 4 = 2$$

$$\therefore \quad \Delta l_2 = 2 \Delta l_1$$

29. *(b)* We know that,

$$Y = \frac{fL}{\pi r^2 l} \Rightarrow l = \frac{fL}{\pi r^2 Y}$$

$$\Rightarrow \quad l \propto \frac{fL}{r^2}$$

$$\Rightarrow \quad \frac{l_2}{l_1} = \frac{f_2}{f_1} \times \frac{L_2}{L_1} \times \left(\frac{r_1}{r_2}\right)^2$$

Here, $f_2 = 2f$, $f_1 = f$

$L_2 = 2L$, $L_1 = L$

$r_2 = 2r$, $r_1 = r$

$$\therefore \quad \frac{l_2}{l_1} = \frac{2f}{f} \times \frac{2L}{L} \times \left(\frac{r}{2r}\right)^2 = 1$$

$$\Rightarrow \quad l_2 = l_1 = l$$

30. *(b)* For steel wire A, $Y_1 = \dfrac{F_1 l_1}{A_1 \Delta l_1}$

For steel wire B, $Y_2 = \dfrac{F_2 l_2}{A_2 \Delta l_2}$

Since, both wire are made of same material as steel, hence

$$Y_1 = Y_2$$

$$\Rightarrow \quad \frac{F_1 l_1}{A_1 \Delta l_1} = \frac{F_2 l_2}{A_2 \Delta l_2} \quad \ldots \text{(i)}$$

Here, $F_1 = F_2$, $l_2 = 2l_1$ $d_2 = 2.4 d_1$

$\Rightarrow \quad r_2 = 2.4 r_1$

$\therefore \quad A_2 = \pi r_2^2 = \pi (2.4 r_1)^2$

$\therefore$ From Eq. (i), we get

$$\frac{F_1 l_1}{\pi r_1^2 \cdot \Delta l_1} = \frac{F_1 (2 l_1)}{\pi (2.4 r_1)^2 \Delta l_2}$$

$$\Rightarrow \quad \Delta l_2 = \frac{2 \times \Delta l_1}{(2.4)^2} = \frac{2 \times 0.2}{(2.4)^2}$$

$$= 0.069 = 6.9 \times 10^{-2} \text{ mm}$$

31. *(a)* Given that length, $l = 6$ m

Cross-sectional area,

$$A = 3 \text{ mm}^2 \text{ or } 3 \times 10^{-6} \text{ m}^2$$

Young's modulus, $Y = 2 \times 10^{11}$ N/m^2

Mass of block, $m = 4$ kg

Acceleration due to gravity,

$$g_p = \frac{1}{4} \times g_e = \frac{1}{4} \times 10 = 2.5 \text{ m/s}^2$$

Elongation in wire,

$$\Delta l = \frac{Fl}{AY} \text{ or } \frac{m g_p l}{AY}$$

$$\Rightarrow \quad \Delta l = \frac{4 \times 2.5 \times 6}{3 \times 10^{-6} \times 2 \times 10^{11}}$$

$$\Rightarrow \quad \Delta l = 10^{-4} \text{ m} \Rightarrow \quad \Delta l = 0.1 \text{ mm}$$

32. *(d)* Young's modulus,

$$Y = \frac{F/A}{\dfrac{\Delta l}{l}} = \frac{Fl}{A \Delta l} \Rightarrow \Delta l = \frac{Fl}{AY}$$

Net elongation, $\Delta l = \Delta l_1 + \Delta l_2$

$$\Rightarrow \quad \Delta l = \frac{F l_1}{A_1 Y_s} + \frac{F l_2}{A_2 Y_c}$$

$$\Rightarrow \quad F = \frac{\Delta l}{\dfrac{l_1}{A_1 Y_S} + \dfrac{l_2}{A_2 Y_C}}$$

$$= \frac{\Delta l A}{\dfrac{l_1}{Y_S} + \dfrac{l_2}{Y_C}} \quad (\because A_1 = A_2 = A)$$

$$= \frac{1.4 \times 10^{-3} \times \pi \times (1.4 \times 10^{-3})^2}{\dfrac{3.2}{2 \times 10^{11}} + \dfrac{4.4}{1.1 \times 10^{11}}}$$

$$= 1.54 \times 10^2 = 154 \text{ N}$$

33. *(d)* $w = 10$ kg

$A = 100 \times 10^{-4}$ m^2

$l = 20 \times 10^{-2}$ m

Let extension is dy in length dx

$$Y = \frac{\text{Stress}}{\text{Strain}} = \frac{T/A}{dy/dx} = \frac{T}{A} \frac{dx}{dy}$$

$$dy = \frac{T dx}{AY}$$

Tension at a x distance from lower end $= \dfrac{mgx}{l}$

$$\int_0^{\Delta l} dy = \int_0^l \frac{mg}{l} \frac{x dx}{AY}$$

$$\Delta l = \frac{mgl}{2AY} = \frac{wl}{2AY} = \frac{10 \times 20 \times 10^{-2}}{2 \times 10^{-2} \times 2 \times 10^{11}}$$

$$= 5 \times 10^{-10} \text{ m}$$

34. *(b)* Given, length of two wires,

$$l_1 = l_2 = l$$

Radius of two wires, $r_1 = r_2 = r$

Since, Young's moduli are different.

$\therefore$ Change in length of wire 1 and 2 will be Δl_1 and Δl_2.

As we know that,

$$Y = \frac{Fl}{A \Delta l} \Rightarrow \Delta l = \frac{Fl}{YA}$$

As the wires are in series,

$\therefore$ Net change in length,

$$\Delta l = \Delta l_1 + \Delta l_2$$

$$\frac{F2l}{Y\pi r^2}=\frac{Fl}{Y_1\pi r^2}+\frac{Fl}{Y_2\pi r^2}$$

$$\Rightarrow \frac{2}{Y}=\frac{1}{Y_1}+\frac{1}{Y_2}\Rightarrow \frac{2}{Y}=\frac{Y_2+Y_1}{Y_1Y_2}$$

$$\Rightarrow Y=\frac{2Y_1Y_2}{Y_1+Y_2}$$

This is the Young's modulus for the combination of a single wire.

> • Concept **Enhancer**
>
> The combination of wires is in series, therefore, use the formula for net change in length accordingly.

35. *(a)* Suppose, l_0 be the actual length of metal wire and Y be its Young's modulus.

From Hooke's law, $Y=\dfrac{Tl_0}{A\Delta l}$

where, $\Delta l = l-l_0$

$$\Rightarrow Y=\frac{Tl_0}{A(l-l_0)} \text{ or } l-l_0=\frac{Tl_0}{AY}$$

$$\therefore \frac{l_1-l_0}{l_2-l_0}=\frac{T_1l_0}{AY}\times\frac{AY}{T_2l_0}=\frac{T_1}{T_2}$$

$$\Rightarrow l_1T_2-l_0T_2=l_2T_1-l_0T_1$$

$$\Rightarrow l_0=\frac{l_1T_2-l_2T_1}{T_2-T_1}=\frac{T_1l_2-T_2l_1}{(T_1-T_2)}$$

36. *(a)* Bulk modulus, $B=\dfrac{\Delta p}{\left(-\dfrac{\Delta V}{V}\right)}$

$$\Rightarrow \frac{\Delta V}{V}=-\frac{\Delta p}{B}$$

$$\frac{\Delta V}{V}\times 100\%=-\frac{\Delta p}{B}\times 100\%$$

$$\Rightarrow \frac{\Delta V}{V}=\frac{-4\times10^9}{8\times10^{10}}\times100\%=-\frac{1}{2}\times10\%$$

$$\Rightarrow \frac{\Delta V}{V}=-0.5\times10\%=-5\%$$

Now, $V=l^3$ (for cube)

$$\Rightarrow l=(V)^{1/3}$$

$$\Rightarrow \frac{\Delta l}{l}\times100\%=\frac{1}{3}\left(\frac{\Delta V}{V}\times100\%\right)$$

$$=\frac{1}{3}(-5\%)=-1.67\ \%$$

So, length of cube will be decreased by 1.67%.

37. *(c)* Let d_{min} = minimum diameter of brass.

Then, stress in brass rod is given by

$$\sigma=\frac{F}{A}=\frac{4F}{\pi d_{min}^2}\quad\left[\because A=\frac{\pi d^2}{4}\right]$$

For stress not to exceed elastic limit, we have

$$\sigma\le 379\text{ MPa}$$

$$\Rightarrow \frac{4F}{\pi d_{min}^2}\le 379\times10^6$$

Here, $F=400$ N

$$\therefore d_{min}^2=\frac{1600}{\pi\times379\times10^6}$$

$$\Rightarrow d_{min}=1.16\times10^{-3}\text{ m}$$

$$=1.16\text{ mm}$$

38. *(c)* Bulk modulus,

$$K=\frac{\text{Volumetric stress}}{\text{Volumetric strain}}=\frac{\Delta p}{-\dfrac{\Delta V}{V}}$$

$$\Rightarrow K=-\frac{mg}{a\left(\dfrac{3\Delta r}{r}\right)}$$

$$\left[\because V=\frac{4}{3}\pi r^3,\text{ so }\frac{\Delta V}{V}=\frac{3\Delta r}{r}\right]$$

$$\Rightarrow \frac{\Delta r}{r}=\frac{-mg}{3aK}$$

Decrease is $\dfrac{mg}{3ak}$

39. *(d)* Stress $=\dfrac{\text{Normal force}}{\text{Area}}$

$$=\frac{N}{A}=\frac{N}{(2\pi a)b}$$

$$V=\pi a^2b$$

$$\Delta V=2\pi ab\,\Delta a$$

Stress $= B\times$ strain

$$\frac{N}{(2\pi a)b}=B\frac{2\pi a\Delta a\times b}{\pi a^2b}$$

$$\left[\because \text{Strain}=\frac{\Delta V}{V}\right]$$

$$\Rightarrow N=B\frac{(2\pi a)^2\Delta ab^2}{\pi a^2b}$$

Force needed to push the cork

$$F=\mu N=\frac{\mu\times B\times4\pi^2a^2b^2\Delta a}{\pi a^2b}$$

$$=(4\pi\mu Bb)\Delta a$$

40. *(d)* $Y_c\times(\Delta L_c/L_c)=Y_s\times(\Delta L_s/L_s)$

$$1\times10^{11}\times\left(\frac{1\times10^{-3}}{1}\right)=2\times10^{11}\times\left(\frac{\Delta L_s}{0.5}\right)$$

$$\therefore \Delta L_s=\frac{0.5\times10^{-3}}{2}=0.25\text{ mm}$$

Therefore, total extension of the composite wire

$$=\Delta L_c+\Delta L_s=1\text{ mm}+0.25\text{mm}$$

$$=1.25\text{ mm}$$

41. *(c)* According to questions,

$$\frac{l_s}{l_b}=a,\ \frac{r_s}{r_b}=b,\ \frac{y_s}{y_b}=c,\ \frac{\Delta l_s}{\Delta l_b}=?$$

As, $y=\dfrac{Fl}{A\Delta l}$

$$\Rightarrow \Delta\ell=\frac{Fl}{Ay}$$

$$\Delta l_s=\frac{3Mgl_s}{\pi r_s^2y_s}\quad[\because F_s=(M+2M)g]$$

$$\Delta l_b=\frac{2Mgl_b}{\pi r_b^2y_b}\quad[\therefore F_b=2Mg]$$

$$\therefore \frac{\Delta l_s}{\Delta l_b}=\frac{\dfrac{3Mgl_s}{\pi r_s^2y_s}}{\dfrac{2Mgl_b}{\pi r_b^2y_b}}=\frac{3a}{2b^2c}$$

42. *(c)* Given, breaking stress of wire,

$$\sigma=\frac{24}{\pi}\times10^2\text{ Nm}^{-2}$$

Free body diagram of 5 kg block is given as

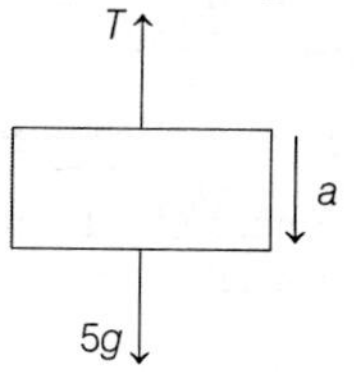

where, a is common acceleration.

From free body diagram of block of mass 5 kg,

$$5g-T=5a\Rightarrow 5\times10-T=5a$$

$$\Rightarrow 50-T=5a\qquad\ldots\text{(i)}$$

Free body diagram of 3 kg block is given as

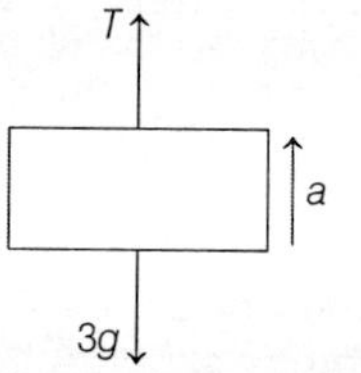

From free body diagram of block of mass 3 kg, $T-3g=3a$

$$\Rightarrow T-3\times10=3a$$

$$\Rightarrow T-30=3a\qquad\ldots\text{(ii)}$$

Add Eqs. (i) and (ii), we get

$$50-T+T-30=5a+3a$$

$$\Rightarrow 20=8a\Rightarrow a=2.5\text{ ms}^{-2}$$

Substituting the value of a in Eq. (i), we get

$$50-T=5\times2.5\Rightarrow T=37.5\text{ N}$$

Let us assume the minimum radius of wire be r. The breaking stress is expressed as

$$\sigma = \frac{T}{\pi r^2} \Rightarrow \frac{24}{\pi} \times 10^2 = \frac{37.5}{\pi r^2}$$

$$\Rightarrow \quad r^2 = \frac{37.5}{24 \times 10^2} = \frac{1}{64}$$

$$\Rightarrow \quad r = \frac{1}{8}\text{ m} = \frac{100}{8}\text{ cm} = 12.5\text{ cm}$$

Thus, the minimum radius of wire should be 12.5 cm.

43. *(43)* Bulk modulus is given as

$$B = \frac{-\Delta p}{\left(\frac{\Delta V}{V}\right)}$$

$$2.15 \times 10^9 = -\frac{\Delta p}{-\left(\frac{0.2}{100}\right)}$$

$$\Delta p = 2.15 \times 10^9 \times 2 \times 10^{-3} = 4.3 \times 10^6$$

$$\Delta p = 43 \times 10^5\text{ N/m}^2$$

44. *(50)*

$$B = 1.4 \times 10^{11}\text{ N/m}^2,\ p = 7 \times 10^6\text{ Pa}$$

$$V = (10 \times 10^{-2})^3$$

$$B = \frac{\Delta p}{\frac{\Delta V}{V}}$$

$$B = \frac{\Delta p \cdot V}{\Delta V}$$

$$\Delta V = \frac{\Delta p \cdot V}{B}$$

$$= \frac{7 \times 10^6}{1.4 \times 10^{11}} \times (10 \times 10^{-2})^3$$

$$\Delta V = 50\text{mm}^3$$

45. *(1)* $n_1 = 4 \times 10^9$ (given)

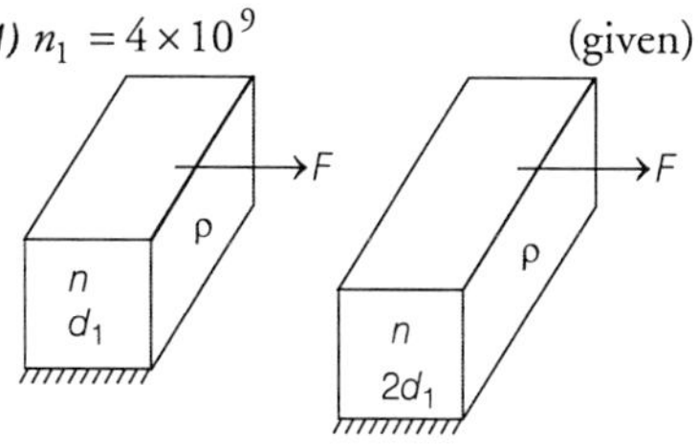

Deformation angle $2\theta_1 = \theta_2$

$$2\frac{\sigma_1}{n_1} = \frac{\sigma_2}{n_2}$$

$$2\frac{F}{ld_1 n_1} = \frac{F}{ld_2 n_2}$$

$$n_2 = \frac{n_1}{4} = \frac{4 \times 10^9}{4}$$

$$n_2 = 1 \times 10^9\text{ N/m}^2$$

$$x = 1$$

46. *(4)* $p_i = 1$ atm

$p_f = 5$ atm

$$\Delta p = p_f - p_i = 5\text{ atm} - 1\text{ atm} = 4\text{ atm} = 4 \times 10^5\text{ Pa}$$

$$\Delta V = -0.8\text{ cm}^3 = -0.8 \times 10^{-6}\text{ m}^3$$

We know that,

$$B = \frac{\Delta p}{-\frac{\Delta V}{V}}$$

$$2 \times 10^9 = \frac{4 \times 10^5}{-\frac{(-0.8 \times 10^{-6})}{V}}$$

$$2 \times 10^9 = V \times 5 \times 10^{11}$$

$$\Rightarrow V = 4 \times 10^{-3}\text{m}^3 = 4\text{ L}$$

47. *(25)* $l = 2\text{m}, Y = 2 \times 10^{11}\text{ N/m}^2$

$$\mu = \frac{-\frac{\Delta r}{r}}{\frac{\Delta l}{l}} \Rightarrow \frac{\Delta l}{l} = \frac{1}{\mu}\left(\frac{\Delta r}{r}\right)$$

$$\frac{\Delta l}{l} = \frac{1}{0.2} \times 10^{-3} = 5 \times 10^{-3}$$

$$U = \frac{1}{2}Y\left(\frac{\Delta l}{l}\right)^2$$

$$= \frac{1}{2} \times 2 \times 10^{11} \times (5 \times 10^{-3})^2$$

$$= 25 \times 10^5$$

48. *(20)* According to question,

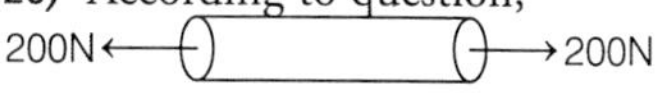

$$\frac{F}{A} = Y\frac{\Delta l}{l}$$

$$\Delta l = \frac{Fl}{YA} = \frac{200 \times 2}{2 \times 10^{-4} \times 10^{11}}$$

$$= 2 \times 10^{-5} = 20\ \mu\text{m}$$

49. *(1)* The given situation is shown below.

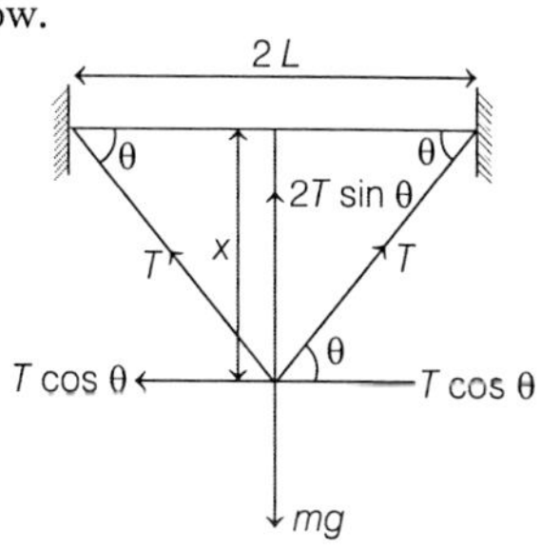

In vertical direction

$$2T\sin\theta = mg$$

$$2T\sin\theta = 2 \times 10$$

$$\Rightarrow \quad 2T\sin\theta = 20$$

$$T = \frac{10}{\sin\theta} = \frac{10}{\theta} \quad [\because \sin\theta \simeq \theta]$$

$$= \frac{10}{\frac{1}{100}}$$

$$T = 1000\text{ N}$$

$\therefore$ Change in length,

$$\Delta L = 2\sqrt{x^2 + L^2} - 2L$$

$$= 2L\left[1 + \frac{x^2}{2L^2} - 1\right] = \frac{x^2}{L}$$

$\therefore$ Modulus of elasticity, $E = \dfrac{\text{Stress}}{\text{Strain}}$

$$2 \times 10^{11} = \frac{10^3 \times 2L}{A \times \frac{x^2}{L}} \Rightarrow 10^{11} = \frac{10^3 \times L^2}{A \times x^2}$$

$$\Rightarrow \quad A = 1 \times 10^{-4}\text{ m}^2$$

50. *(600)* Given,

Density, $\rho = 6 \times 10^4\text{ kg/m}^3$

Breaking stress, $\sigma = 1.2 \times 10^8\text{ N/m}^2$

$$g_{\text{planet}} = \frac{g_{\text{earth}}}{3}$$

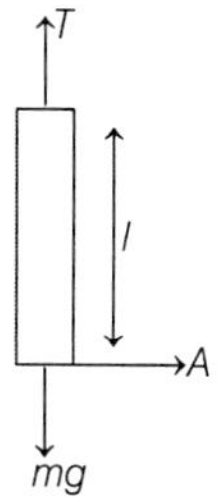

Now, $T = mg_{\text{planet}}$ ...(i)

$$\text{Stress} = \frac{T}{A} = \frac{mg_{\text{planet}}}{A}$$

$$\sigma = \frac{(\rho A l)}{A}g_{\text{planet}}$$

$$l = \frac{\sigma}{\rho g_{\text{planet}}} \quad \text{...(ii)}$$

$$= \frac{1.2 \times 10^8 \times 3}{6 \times 10^4 \times 10} = 600\text{ m}$$

$$l = 600\text{ m}$$

51. *(3)* Wires are similar

$$\Rightarrow \quad Y_{\text{I}} = Y_{\text{II}}$$

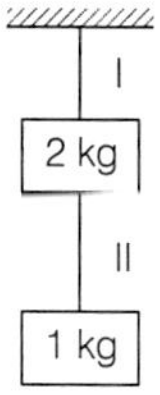

Since, wires are similar, then area of both wires

Let, $A_1 = A_2 = A$

$$\frac{\frac{F_{\text{I}}}{A}}{E_{\text{I}}} = \frac{\frac{F_{\text{II}}}{A}}{E_{\text{II}}} \Rightarrow \frac{F_{\text{I}}}{F_{\text{II}}} = \frac{E_{\text{I}}}{E_{\text{II}}}$$

$\Rightarrow$ Ratio of strains in wire is

$$\frac{E_I}{E_{\text{II}}} = \frac{F_{\text{I}}}{F_{\text{II}}} = \frac{3}{1}$$

52. *(12)*

> **key idea** Use following stepwise approach to solve this problem.
>
> **Step I.** Find tension in wire using pulley-mass system.
> **Step II.** Use this value of T as a force in Young's modulus formula and get the required result.

In given pulley mass system,

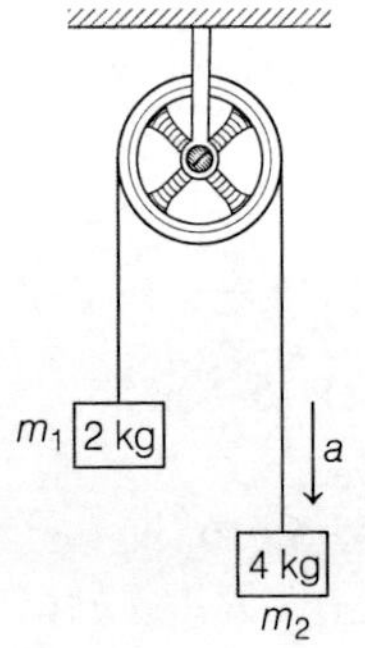

We have,

$$T - m_1 g = m_1 a \quad \dots\text{(i)}$$

and $$m_2 g - T = m_2 a \quad \dots\text{(ii)}$$

So, acceleration of system

$$a = g\left(\frac{m_2 - m_1}{m_2 + m_1}\right)$$

$$\Rightarrow \quad a = 10\left(\frac{4-2}{4+2}\right) = \frac{10}{3}\ \text{m/s}^2$$

So, from Eq. (i), we get

$$T - 20 = \frac{20}{3} \Rightarrow T = \frac{80}{3}\ \text{N}$$

This tension causes extension in metal wire,

$$Y = \frac{Fl}{A\Delta l} = \frac{T}{\pi R^2 \varepsilon}$$

$$\varepsilon = \frac{\Delta l}{\Delta} = \text{strain}$$

So strain is,

$$\Rightarrow \quad \varepsilon = \frac{T}{\pi R^2 \times Y}$$

$$= \frac{\frac{80}{3}}{\pi \times 16 \times 10^{-10} \times 2 \times 10^{11}}$$

$$= \frac{5}{60\pi} = \frac{1}{12\pi} = \frac{1}{\alpha\pi} \quad \text{(given)}$$

$\therefore \quad \alpha = 12$

53. *(2)* Given, each block mass = 3 kg

Cross-section area of wire A and B is

$$= 0.005\ \text{cm}^2$$

Young's modulus of wires

$$Y = 2 \times 10^{11}\ \text{N/m}^2$$

First we have to calculate tension in wire,

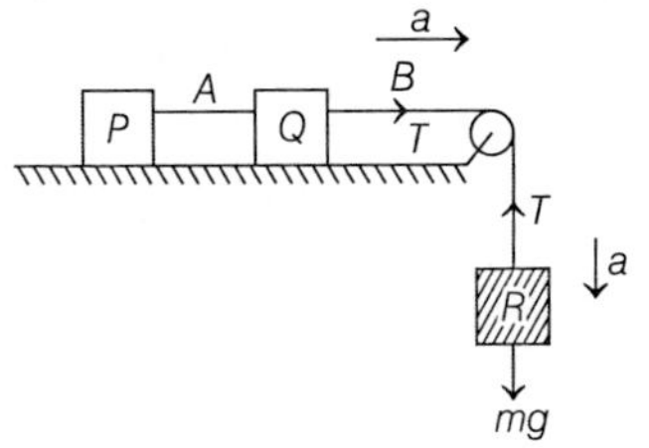

Let a is the acceleration,

Net force on R block,

$$mg - T = ma$$

$$3 \times 10 - T = 3a \quad \dots\text{(i)}$$

On horizontal block,

$$T = (m_1 + m_2)a$$

$$T = 6a \quad \dots\text{(ii)}$$

From Eq. (i) and Eq.(ii),

$$a = \frac{30}{9}\ \text{m/s}^2$$

$$T = \frac{6 \times 30}{9} \quad \text{[From Eq. (ii)]}$$

$$\Rightarrow \quad T = 20\ \text{N}$$

Strain is given by formula,

$$\text{Strain} = \frac{\text{Tension}}{\text{Area of cross-section} \times \text{Young's modulus}}$$

$$= \frac{20}{5 \times 10^{-7} \times 2 \times 10^{11}} = 2 \times 10^{-4}$$

54. *(16)* Given,

$$V_1 = V_2 \Rightarrow A_1 l_1 = A_2 l_2$$

$$\Rightarrow \quad \frac{A_1}{A_2} = \frac{l_2}{l_1} \quad \dots\text{(i)}$$

We know that,

$$Y = \frac{F/A}{\Delta l/l} = \frac{Fl}{A\Delta l} \Rightarrow F = \frac{YA\Delta l}{l}$$

$$\Rightarrow \quad F \propto \frac{A\Delta l}{l} \Rightarrow \frac{F_1}{F_2} = \frac{A_1}{A_2} \cdot \frac{l_2}{l_1} \cdot \frac{\Delta l_1}{\Delta l_2}$$

$$= \frac{A_1}{A_2} \cdot \frac{A_1}{A_2}$$

[$\because \Delta l_1 = \Delta l_2$ and using Eq. (i)]

$$= \left(\frac{A_1}{A_2}\right)^2 = \left(\frac{4}{1}\right)^2 = 16$$

55. *(25)* We know, $Y = \dfrac{\text{Stress}}{\text{Strain}}$

$$\therefore \text{Strain} = \frac{\text{Stress}}{Y}, \ \text{Stress} = \frac{F}{A}$$

$$\text{Strain} = \frac{F}{A \cdot Y}$$

Putting the given value, we get longitudinal strain

$$= \frac{62.8 \times 1000}{\pi r^2 \times 2 \times 10^{11}}$$

$$= \frac{62.8 \times 1000}{3.14 \times 400 \times 10^{-6} \times 2 \times 10^{11}}$$

$$= \frac{200}{8} \times 10^{-5} = 25 \times 10^{-5}$$

$$= x \times 10^{-5} \quad \text{(Given)}$$

So, $x = 25$

56. *(11)* We know, Stress $= \dfrac{\text{Force}}{\text{Area}} = \dfrac{mg}{A}$

$$\therefore \quad m = \frac{S \times A}{g} = \frac{7 \times 10^5 \times \pi R^2}{g}$$

$$= \frac{7 \times 10^5 \times \frac{22}{7} \times (7 \times 10^{-3})^2}{9.8}$$

$$= 11\ \text{kg}$$

57. *(20)* Steel wire is elongated due to a load of mass

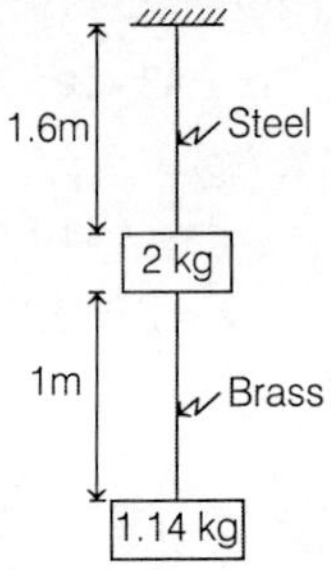

$$m = 2 + 1.14$$

$$= 3.14\ \text{kg}$$

$$= \pi\ \text{kg}$$

Now using,

$$Y = \frac{Fl}{A\Delta l}$$

$$\therefore \quad \Delta l = \frac{F \cdot l}{Y \cdot A}$$

$$= \frac{\pi(1.6)g}{2 \times 10^{11} \times \pi \times (0.2 \times 10^{-2})^2}$$

$$= \frac{\pi \times 1.6 g}{2 \times 10^{11} \times \pi \times (0.2)^2 \times 10^{-4}}$$

$$= \frac{\pi \times 16 \times 10^{-1} g}{2 \times 10^{11} \times \pi \times 2^2 \times 10^{-2} \times 10^{-4}}$$

$$= 20 \times 10^{-6}\ \text{m}$$

58. *(2)* We know that,

$$F = kx$$

$$\therefore \quad F = \frac{YA}{l_0} \cdot x$$

When $F = 100$ N

$$100 = k(l_1 - l_0) \quad \dots\text{(i)}$$

When $F = 120$ N

$$120 = k(l_2 - l_0) \quad \dots\text{(ii)}$$

So, $$\frac{120}{100} = \frac{k(l_2 - l_0)}{k(l_1 - l_0)}$$

Given, $10 l_2 = 11 l_1,\ l_2 = 1 \cdot 1 l_1$

$$\frac{120}{100} = \frac{(1 \cdot 1 l_1 - l_0)}{l_1 - l_0}$$

$1\cdot 2l_1 - 1\cdot 2l_0 = 1\cdot 1l_1 - l_0$

$0\cdot 1l_1 = 0\cdot 2l_0 \Rightarrow l_0 = \frac{l_1}{2}$

So, $x = 2$

59. *(48)* $\eta = \frac{F}{A\tan\theta} = \frac{F}{A\left(\frac{x}{\ell}\right)}$

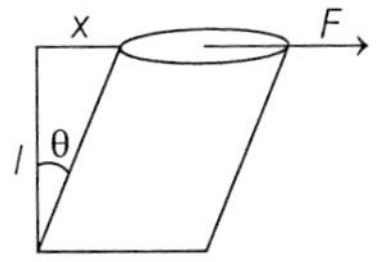

$\frac{F}{A} = \eta\frac{x}{l} \Rightarrow \frac{Fl}{A\eta} = x$

$\Rightarrow x = \frac{18\times 10^4 \times 60\times 10^{-2}}{60\times 10^{-2}\times 15\times 10^{-2}\times 25\times 10^9}$

$= 48\times 10^{-6}\ \text{m} = 48\ \mu\text{m}$

60. *(50)* Given, $l = 0.5$ m

$A = 10^{-4}\ \text{m}^2$

Breaking stress $= 5\times 10^8\ \text{Nm}^{-2}$

Mass of block, $m = 10$ kg

Tension in string,

$T = \frac{mv^2}{l}$

i.e. $v = v_{\text{max}}, T = T_{\text{max}}$

$\therefore \quad T_{\text{max}} = \frac{mv^2_{\text{max}}}{l}$ ...(i)

But T_{max} = Breaking stress × Area

$= 5\times 10^8 \times 10^{-4}$

$= 5\times 10^4$ N

Hence, from Eq. (i), we have

$\frac{mv^2_{\text{max}}}{l} = 5\times 10^4$

$\Rightarrow \frac{10}{0.5}\times v^2_{\text{max}} = 5\times 10^4$

$\Rightarrow \quad v_{\text{max}} = 50$ m/s

61. *(6)* Given, elongation in the wire on earth, $\Delta l = 10^{-4}$ m

Elongation in the wire on another planet, $\Delta l' = 6\times 10^{-5}$ m

The Young's modulus remains the same,

$\Rightarrow \quad Y = \frac{F/A}{\Delta l/l} = \frac{Fl}{A\Delta l} = \frac{mgl}{A\Delta l}$

$\Rightarrow \quad Y' = \frac{mg'l}{A\Delta l'}$

$\therefore \quad \frac{mgl}{A\Delta l} = \frac{mg'l}{A\Delta l'} \Rightarrow g' = g\times\frac{\Delta l'}{\Delta l}$

$g' = 10\times\frac{6\times 10^{-5}}{10^{-4}} = 6\ \text{m/s}^2$

62. *(500)* Given, decrease in volume

$\Delta V/V = -\frac{0.5}{100}$

$B = 9.8\times 10^8\ \text{Nm}^{-2}$

$\rho = 10^3\ \text{kgm}^{-3}$

$g = 9.8\ \text{ms}^{-2}$

Let h be the depth at which ball is dipped.

Since, $B = \frac{\Delta p}{-\Delta V/V}$

where, Δp is change in pressure $= \rho gh$

$\therefore \rho gh = -B\frac{\Delta V}{V} \Rightarrow h = -\frac{1}{\rho g}B\frac{\Delta V}{V}$

$\Rightarrow h = -\frac{1}{10^3\times 9.8}\times 9.8\times 10^8\times\left(-\frac{0.5}{100}\right)$

$= 0.5\times 10^{8-3-2}$

$= 0.5\times 10^3 = 500$ m

63. *(2)* Let initial length and diameter be l_1 and d_1, whereas final length and diameter be l_2 and d_2.

Given, $l_2 = 2l_1$, $d_2 = 2d_1$, $\Delta l_1 = 0.04$ m

By using formula of Young's modulus of elasticity,

$Y = \frac{F\cdot l}{A\Delta l}$

$\because \quad Y_1 = Y_2$

$\Rightarrow \quad \frac{Fl_1}{A_1\times\Delta l_1} = \frac{Fl_2}{A_2\times\Delta l_2}$

$\Rightarrow \quad \frac{Fl_1}{\pi(d_1/2)^2\times 0.04} = \frac{F2l_1}{\pi(2d_1/2)^2\Delta l_2}$

$\Rightarrow \quad \frac{1}{1/4\times 0.04} = \frac{2}{\Delta l_2}$

$\Rightarrow \quad \Delta l_2 = 0.02\ \text{m} = 2\ \text{cm}$

64. *(32)* Given,

Wire A, $\Delta L_A = 2\ \text{mm} = 0.002$ m

Wire B, $\Delta L_B = 4\ \text{mm} = 0.004$ m

The force $F = 2$ N

$\because \quad \frac{r_B}{r_A} = \frac{4}{1}$

Since, $Y_A = Y_B$ ...(i)

Using Hooke's law, Stress $= Y$ (Strain)

$\frac{F}{A} = Y\left(\frac{\Delta L}{L}\right) \Rightarrow L = \frac{Y\Delta LA}{F}$

$\Rightarrow \quad \frac{L_A}{L_B} = \frac{Y_A}{Y_B}\times\frac{\Delta L_A}{\Delta L_B}\times\frac{A_A}{A_B}\times\frac{F_B}{F_A}$

$\Rightarrow \quad \frac{L_A}{L_B} = \frac{Y_A}{Y_B}\times\frac{0.002}{0.004}\times\frac{\pi r_A^2}{\pi r_B^2}\times\frac{2}{2}$

$\Rightarrow \quad \frac{L_A}{L_B} = \frac{0.002}{0.004}\times\frac{r_A^2}{16r_A^2}$ [Using (i)]

$\Rightarrow \quad \frac{L_A}{L_B} = \frac{a}{b} = \frac{1}{32}$

Comparing this equation with $1/x$, we get the value of the x is 32.

Topic 2

01. *(d)* Strain energy per unit volume in a rod is given by

$u = \frac{U}{V} = \frac{1}{2}\times\text{Stress}\times\text{Strain} = \frac{1}{2}\varepsilon^2 Y$

ε = strain

Given, $\varepsilon = 0.04\% = \frac{0.04}{100} = 4\times 10^{-4}$

and $Y = 7\times 10^{10}\ \frac{\text{N}}{\text{m}^2}$

Hence, strain energy per unit volume is,

$u_E = \frac{1}{2}\varepsilon^2 Y = \frac{1}{2}\times(4\times 10^{-4})^2\times 7\times 10^{10}$

$= \frac{1}{2}\times 16\times 10^{-8}\times 7\times 10^{10}$

$= 5600\ \text{J/m}^3$

02. *(c)* As we know, $\Delta_{\text{temp}} = \Delta_{\text{force}}$

or $L\alpha(\Delta T) = \frac{FL}{AY}$

$\therefore \quad \alpha = \frac{F}{AYT} = \frac{F}{\pi r^2 YT}$

Coefficient of volume expansion

$\gamma = 3\alpha = \frac{3F}{\pi r^2 YT}$

03. *(b)* Energy stored in wire

$= \frac{1}{2}\times$ Stress× Strain× Volume

and Young's modulus $= \frac{\text{Stress}}{\text{Strain}}$

$\Rightarrow$ Strain $= S/Y$

$\therefore \frac{\text{Energy stored in wire}}{\text{Volume}} = \frac{1}{2}\times\text{Stress}\times\text{Strain}$

$= \frac{1}{2}\times S\times\frac{S}{Y} = \frac{S^2}{2Y}$

04. *(a)* Elastic potential energy stored in a loaded wire,

$U = \frac{1}{2}$ (Stress × Strain × Volume)

$\therefore$ Energy stored per unit volume,

$u = \frac{U}{\text{Volume}} = \frac{1}{2}\times\text{Stress}\times\text{Strain}$

$= \frac{1}{2}\left(\frac{F}{A}\right)^2\times\frac{1}{Y}$

$$\therefore \quad \frac{u_A}{u_B} = \frac{\frac{1}{2Y} \cdot \frac{F^2}{A_A^2}}{\frac{1}{2Y} \cdot \frac{F^2}{A_B^2}} = \frac{A_B^2}{A_A^2}$$

$$\Rightarrow \quad \frac{u_A}{u_B} = \frac{d_B^4}{d_A^4} \qquad \left(\because A = \pi \frac{d^2}{4}\right)$$

Here, $\frac{u_A}{u_B} = \frac{1}{4}$

So, $\frac{d_B^4}{d_A^4} = \frac{1}{4}$ or $\frac{d_B}{d_A} = \frac{1}{\sqrt{2}}$

$$\Rightarrow \quad \frac{d_A}{d_B} = \sqrt{2} : 1$$

05. *(a)*

key idea In the given situation, the potential energy of stretched cord is equal to the kinetic energy of stone.

When rubber cord is stretched, it stores potential energy and when released, this potential energy is given to the stone as kinetic energy.

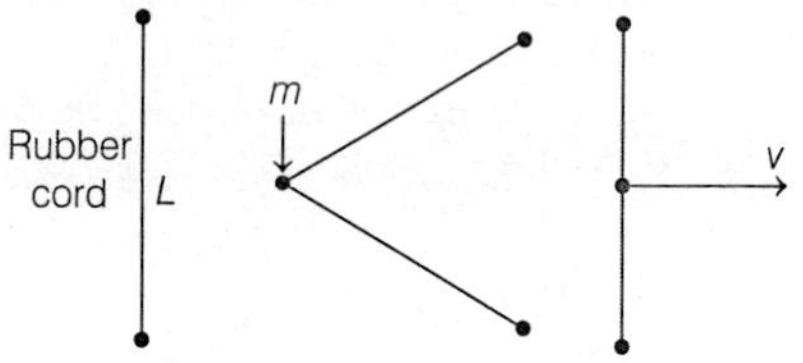

So, potential energy of stretched cord = kinetic energy of stone

$$\Rightarrow \frac{1}{2} Y \left(\frac{\Delta L}{L}\right)^2 A \cdot L = \frac{1}{2} mv^2$$

Here, $\Delta L = 20 \text{ cm} = 0.2 \text{ m}$,

$L = 42 \text{ cm} = 0.42 \text{ m}$,

$v = 20 \text{ ms}^{-1}$, $m = 0.02 \text{ kg}$,

$d = 6 \text{ mm} = 6 \times 10^{-3} \text{ m}$

$$\therefore \quad A = \pi r^2 = \pi \left(\frac{d}{2}\right)^2 = \pi \left(\frac{6 \times 10^{-3}}{2}\right)^2$$

$$= \pi (3 \times 10^{-3})^2 = 9\pi \times 10^{-6} \text{ m}^2$$

On substituting values, we get

$$Y = \frac{mv^2 L}{A(\Delta L)^2} = \frac{0.02 \times (20)^2 \times 0.42}{9\pi \times 10^{-6} \times (0.2)^2}$$

$$\approx 3.0 \times 10^6 \text{ Nm}^{-2}$$

So, the closest value of Young's modulus is 10^6 Nm^{-2}.

06. *(a)* Given length of a steel wire, $L = 10 \text{ cm}$

Temperture change, $\Delta\theta = 100°C$

As length is constant,

$$\therefore \quad \text{Strain} = \frac{\Delta L}{L} = \alpha \Delta\theta$$

$$[\text{as } \Delta L = (L \alpha \Delta\theta)]$$

Now, pressure = stress = $Y \times$ strain

[given, $Y = 2 \times 10^{11} \text{ N/m}^2$ and $\alpha = 1.1 \times 10^{-5} \text{ K}^{-1}$]

$$= 2 \times 10^{11} \times 1.1 \times 10^{-5} \times 100$$

$$= 2.2 \times 10^8 \text{ Pa}$$

07. *(40)* Length of steel wire, $L = 20$ m

Change in length,

$$\Delta L = 2 \text{ cm} = 2 \times 10^{-2} \text{ m}$$

$$Y = 2 \times 10^{11} \text{ Nm}^{-2}$$

We know that,

Elastic potential energy,

$$U = \frac{1}{2} \times \text{Stress} \times \text{Strain} \times \text{Volume}$$

$$U = \frac{1}{2} \times Y \times (\text{Strain})^2 \times (A \times L)$$

$$\Rightarrow A = \frac{2U}{Y \times (\text{Strain})^2 \times L}$$

$$= \frac{2U}{Y \times \left(\frac{\Delta L}{L}\right)^2 \times L}$$

$$= \frac{2UL}{Y \times (\Delta L)^2}$$

$$= \frac{2 \times 80 \times 20}{2 \times 10^{11} \times (2 \times 10^{-2})^2}$$

$$= \frac{400}{10^7} = 4 \times 10^{-5} \text{ m}^2$$

$$= 40 \times 10^{-6} \text{ m}^2 = 40 \text{ mm}^2$$

08. *(25)* From the graph, the Young's modulus can be calculated as

$$Y = \frac{\text{Stress}}{\text{Strain}} = \frac{20}{1 \times 10^{-10}} = 20 \times 10^{10} \text{ N/m}^2$$

Energy density, $E = \frac{1}{2} \times \text{Stress} \times \text{Strain}$

$$= \frac{1}{2} \times Y \times \text{Strain} \times \text{Strain}$$

$$= \frac{1}{2} \times Y \times (\text{Strain})^2$$

Given, strain $= 5 \times 10^{-4}$

$$\therefore \; E = \frac{1}{2} \times 20 \times 10^{10} \times (5 \times 10^{-4})^2$$

$$= 25 \times 10^3 \text{ J/m}^3 = 25 \text{ kJ/m}^3$$

09. *(20)* Given, mass of stone,

$$m = 20 \text{ g} = 0.020 \text{ kg}$$

$$l = 0.1 \text{ m} \Rightarrow A = 10^{-6} \text{ m}^2$$

$$Y = 0.5 \times 10^9 \text{ N/m}^2$$

During the projection of stone from catapult,

Work done in stretching a catapult = Kinetic energy of stone after releasing from catapult

$$\Rightarrow \frac{1}{2} \times Y \times (\text{Strain})^2 \times \text{Volume} = \frac{1}{2} mv^2$$

$$\Rightarrow \frac{1}{2} \times 0.5 \times 10^9 \times 16 \times 10^{-2} \times 10^{-7}$$

$$= \frac{1}{2} \times 20 \times 10^{-3} \times v^2$$

$$\Rightarrow \quad v^2 = 400 \Rightarrow v = 20 \text{ ms}^{-1}$$

C-09
JEE Main Chapterwise
Topicwise Physics

In JEE Main there is usually 0-1 question per paper is visible from this chapter. In past 3 years, 40 conceptual and 23 numerical questions were asked of moderate to difficult level from this chapter.
For detailed analysis scan QR code

MECHANICAL PROPERTIES OF FLUIDS

Topic 1 Pressure, Density, Pascal's Law and Archimede's Principle

Level 1

01 Given below are two statements.

Statement I Pressure in a reservoir of water is same at all points at the same level of water.

Statement II The pressure applied to enclosed water is transmitted in all directions equally.

In the light of the above statements, choose the correct answer from the options given below *[2023, 10 April Shift-I]*

(a) Both Statement I and Statement II are false.
(b) Statement I is false but Statement II is true.
(c) Both Statement I and Statement II are true.
(d) Statement I is true but Statement II is false.

02 Given below are two statements. One is labelled as Assertion (A) and the other is labelled as Reason (R).

Assertion (A) When you squeeze one end of a tube to get toothpaste out from the other end, Pascal's principle is observed.

Reason (R) A change in the pressure applied to enclosed incompressible fluid is transmitted undiminished to every portion of the fluid and to the walls of its container.

In the light of the above statements, choose the most appropriate answer from the options given below. *[2023, 6 April Shift-II]*

(a) Both A and R are correct and R is the correct explanation of A.
(b) A is correct but R is not correct.
(c) Both A and R are correct but R is not the correct explanation of A.
(d) A is not correct but R is correct.

03 A pressure-pump has a horizontal tube of cross-sectional area $10\ cm^2$ for the outflow of water at a speed of 20 m/s. The force exerted on the vertical wall just infront of the tube which stops water horizontally flowing out of the tube, is (given, density of water $= 1000\ kg/m^3$) *[2022, 28 July Shift-II]*

(a) 300 N (b) 500 N (c) 250 N (d) 400 N

04 A submarine experiences a pressure of 5.05×10^6 Pa at a depth of d_1 in a sea. When it goes further to a depth of d_2, it experiences a pressure of 8.08×10^6 Pa, then $d_2 - d_1$ is approximately (density of water $= 10^3\ kg/m^3$ and acceleration due to gravity $= 10\ ms^{-2}$) *[2019, 10 April Shift-II]*

(a) 500 m (b) 400 m (c) 600 m (d) 300 m

05 A ball is made of a material of density ρ, where $\rho_{oil} < \rho < \rho_{water}$ with ρ_{oil} and ρ_{water} representing the densities of oil and water, respectively. The oil and water are immiscible. If the above ball is in equilibrium in a mixture of this oil and water, which of the following pictures represents its equilibrium position? *[AIEEE 2010]*

(a)
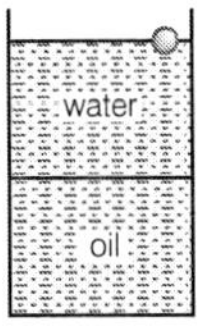

(b)
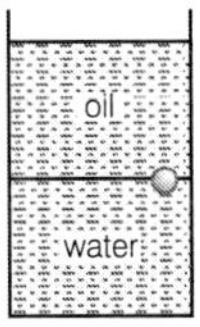

(c)
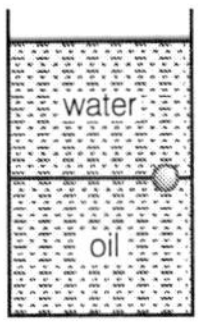

(d)
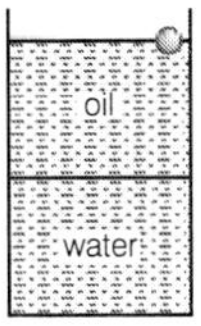

06 A jar is filled with two non-mixing liquids 1 and 2 having densities ρ_1 and ρ_2, respectively. A solid ball, made of a material of density ρ_3, is dropped in the jar. It comes to equilibrium in the position shown in the figure. Which of the following is true for ρ_1, ρ_2 and ρ_3? *[AIEEE 2008]*

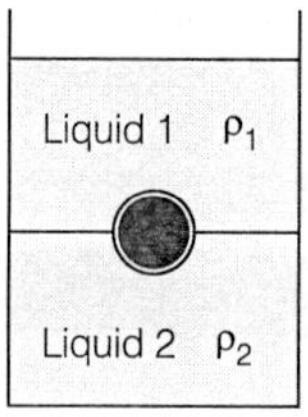

(a) $\rho_3 < \rho_1 < \rho_2$ (b) $\rho_1 > \rho_3 > \rho_2$
(c) $\rho_1 > \rho_2 > \rho_3$ (d) $\rho_1 < \rho_3 < \rho_2$

Level 2

07 The pressure acting on a submarine is 3×10^5 Pa at a certain depth. If the depth is doubled, the percentage increase in the pressure acting on the submarine would be (Assume that atmospheric pressure is 1×10^5 Pa, density of water is 10^3 kg m^{-3}, $g = 10$ ms^{-2}) *[2021, 16 March Shift-I]*

(a) $\frac{200}{3}\%$ (b) $\frac{200}{5}\%$ (c) $\frac{5}{200}\%$ (d) $\frac{3}{200}\%$

08 An air bubble of radius 1 cm in water has an upward acceleration 9.8 cm s^{-2}. The density of water is 1 g/cm^3 and water offers negligible drag force on the bubble. The mass of the bubble is (Take, $g = 980$ cm/s^2) *[2020, 4 Sep Shift-I]*

(a) 1.52 g (b) 4.51 g
(c) 4.15 g (d) 3.15 g

09 Two liquids of densities ρ_1 and $\rho_2 (\rho_2 = 2\rho_1)$ are filled up behind a square wall of side 10 m as shown in figure. Each liquid has a height of 5 m. The ratio of the forces due to these liquids exerted on upper part MN to that at the lower part NO is (assume that the liquids are not mixing) *[2020, 8 Jan Shift-II]*

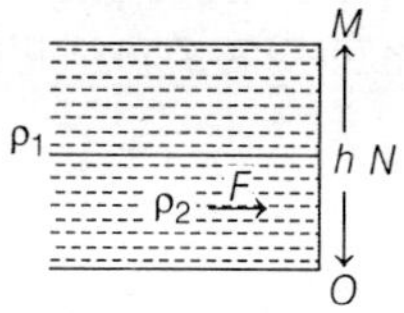

(a) 1/2 (b) 2/3
(c) 1/4 (d) 1/3

Imp. Relative Density

10 A wooden block floating in a bucket of water has 4/5 of its volume submerged. When certain amount of an oil is poured into the bucket, it is found that the block is just under the oil surface with half of its volume under water and half in oil. The density of oil relative to that of water is *[2019, 9 April Shift-II]*

(a) 0.6 (b) 0.8
(c) 0.7 (d) 0.5

11 A liquid of density ρ is coming out of a hose pipe of radius a with horizontal speed v and hits a mesh. 50% of the liquid passes through the mesh unaffected, 25% losses all of its momentum and, 25% comes back with the same speed. The resultant pressure on the mesh will be *[2019, 11 Jan Shift-I]*

(a) ρv^2 (b) $\frac{1}{2}\rho v^2$
(c) $\frac{1}{4}\rho v^2$ (d) $\frac{3}{4}\rho v^2$

12 A thin uniform tube is bent into a circle of radius r in the vertical plane. Equal volumes of two immiscible liquids, whose densities are ρ_1 and ρ_2 $(\rho_1 > \rho_2)$, fill half the circle. The angle θ between the radius vector passing through the common interface and the vetrical is *[JEE Main 2018]*

(a) $\theta = \tan^{-1} \frac{\pi}{2}\left(\frac{\rho_1 + \rho_2}{\rho_1 - \rho_2}\right)$ (b) $\theta = \tan^{-1}\left[\frac{\pi}{2}\left(\frac{\rho_1 - \rho_2}{\rho_1 + \rho_2}\right)\right]$

(c) $\theta = \tan^{-1} \pi\left(\frac{\rho_1}{\rho_1}\right)$ (d) None of these

13 There is a circular tube in a vertical plane. Two liquids which do not mix and of densities d_1 and d_2 are filled in the tube. Each liquid subtends 90°angle at centre. Radius joining their interface makes an angle α with vertical. Ratio d_1 / d_2 is *[JEE Main 2014]*

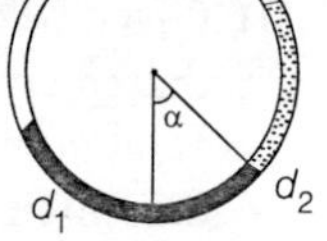

(a) $\frac{1 + \sin\alpha}{1 - \sin\alpha}$ (b) $\frac{1 + \cos\alpha}{1 - \cos\alpha}$

(c) $\frac{1 + \tan\alpha}{1 - \tan\alpha}$ (d) $\frac{1 + \sin\alpha}{1 - \cos\alpha}$

14 A uniform cylinder of length L and mass M having cross-sectional area A is suspended with its length vertical from a fixed point by a massless spring such that it is half submerged in a liquid of density σ at equilibrium position. The extension x_0 of the spring when it is in equilibrium, is *[JEE Main 2013]*

(a) $\frac{Mg}{k}$ (b) $\frac{Mg}{k}\left(1 - \frac{LA\sigma}{M}\right)$

(c) $\frac{Mg}{k}\left(1 - \frac{LA\sigma}{2M}\right)$ (d) $\frac{Mg}{k}\left(1 + \frac{LA\sigma}{M}\right)$

Level 3

15 Two cylindrical vessels of equal cross-sectional area of 2 m^2 contain water upto heights 10 m and 6 m, respectively. If the vessels are connected at their bottom then the work done by the force of gravity is (Density of water is 10^3 kg/m^3 and $g = 10$ m/s^2) *[2025, 3 April Shift-II]*

(a) 1×10^5 J (b) 6×10^4 J
(c) 8×10^4 J (d) 4×10^4 J

16 A 400 g solid cube having an edge of length 10 cm floats in water. How much volume of the cube is outside the water ?
(Given, density of water $= 1000\ \text{kg m}^{-3}$)
[2025, 28 Jan Shift-II]

(a) $4000\ \text{cm}^3$ (b) $1400\ \text{cm}^3$ (c) $600\ \text{cm}^3$ (d) $400\ \text{cm}^3$

Imp. Buoyant Force

17 A sphere of relative density σ and diameter D has concentric cavity of diameter d. The ratio of D/d, if it just floats on water in a tank is *[2024, 9 April Shift-I]*

(a) $\left(\frac{\sigma+1}{\sigma-1}\right)^{1/3}$ (b) $\left(\frac{\sigma-1}{\sigma}\right)^{1/3}$

(c) $\left(\frac{\sigma}{\sigma-1}\right)^{1/3}$ (d) $\left(\frac{\sigma-2}{\sigma+2}\right)^{1/3}$

18 A cube of ice floats partly in water and partly in kerosene oil. The ratio of volume of ice immersed in water to that in kerosene oil (specific gravity of kerosene oil = 0.8, specific gravity of ice = 0.9)
[2024, 8 April Shift-II]

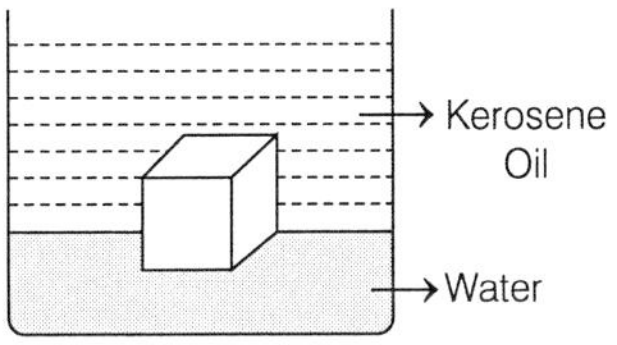

(a) 9 : 10 (b) 8 : 9 (c) 1 : 1 (d) 5 : 4

19 Two cylindrical vessels of equal cross-sectional area $16\ \text{cm}^2$ contain water upto heights 100 cm and 150 cm, respectively. The vessels are interconnected so that the water levels in them become equal. The work done by the force of gravity during the process, is (Take, density of water $= 10^3\ \text{kg/m}^3$ and $g = 10\ \text{ms}^{-2}$)
[2022, 27 July Shift-I]

(a) 0.25 J (b) 1 J (c) 8 J (d) 12 J

20 A soap bubble, blown by a mechanical pump at the mouth of a tube, increases in volume, with time, at a constant rate. The graph that correctly depicts the time dependence of pressure inside the bubble is given by *[2019, 12 Jan Shift-II]*

(a)
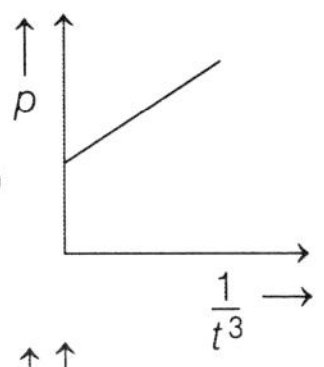

(b)
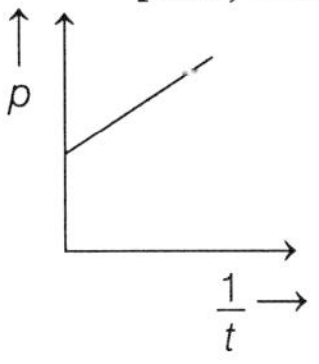

(c)
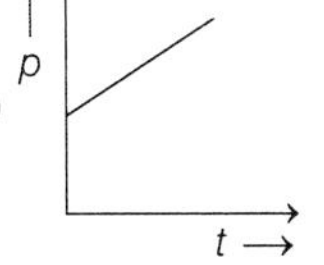

(d)
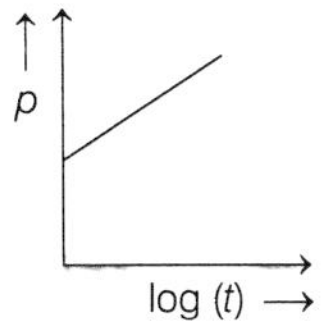

21 A long cylindrical vessel is half-filled with a liquid. When the vessel is rotated about its own vertical axis, the liquid rises up near the wall. If the radius of the vessel is 5 cm and its rotational speed is 2 rotations per second, then the difference in the heights between the centre and the sides (in cm) will be
[2019, 12 Jan Shift-II]

(a) 0.1 (b) 1.2
(c) 0.4 (d) 2.0

Numerical Value Type Questions

22 In a hydraulic lift, the surface area of the input piston is $6\ \text{cm}^2$ and that of the output piston is $1500\ \text{cm}^2$. If 100 N force is applied to the input piston to raise the output piston by 20 cm, then the work done is .... kJ. *[2025, 29 Jan Shift-I]*

23 A cube having a side of 10 cm with unknown mass and 200 g mass were hung at two ends of a uniform rigid rod of 27 cm long. The rod along with masses was placed on a wedge keeping the distance between wedge point and 200 g weight as 25 cm. Initially the masses were not at balance. A beaker is placed beneath the unknown mass and water is added slowly to it. At given point the masses were in balance and half volume of the unknown mass was inside the water. (Take the density of unknown mass is more than that of the water, the mass did not absorb water and water density is $1\ \text{gm/cm}^3$.) The unknown mass is .... kg *[2025, 8 April Shift-II]*

24 A vessel with square cross-section and height of 6 m is vertically partitioned. A small window of $100\ \text{cm}^2$ with hinged door is fitted at a depth of 3 m in the partition wall. One part of the vessel is filled completely with water and the other side is filled with the liquid having density $1.5 \times 10^3\ \text{kg/m}^3$. What force one needs to apply on the hinged door so that it does not get opened?
(Acceleration due to gravity $= 10\ \text{m/s}^2$)
[2025, 2 April Shift-I]

25

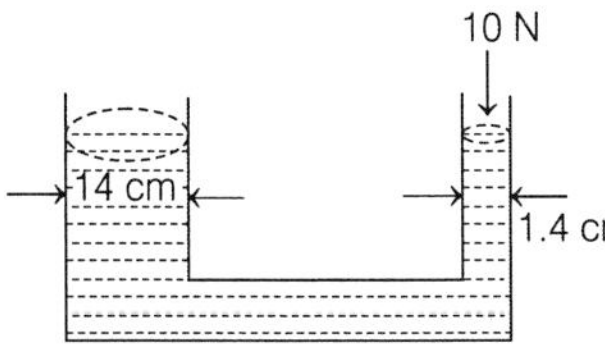

A hydraulic press containing water has two arms with diameters as mentioned in the figure. A force of 10 N is applied on the surface of water in the thinner arm. The force required to be applied on the surface of water in the thicker arm to maintain equilibrium of water is .... N. *[2024, 5 April Shift-II]*

26 Mercury is filled in a tube of radius 2 cm up to a height of 30 cm. The force exerted by mercury on the bottom of the tube is...... N.

(Given, atmospheric pressure $= 10^5 \text{ Nm}^{-2}$, density of mercury $= 1.36 \times 10^4 \text{ kg m}^{-3}$, $g = 10 \text{ ms}^{-2}$, $\pi = \frac{22}{7}$)

[2024, 4 April Shift-II]

27 The velocity of a small ball of mass 0.3 g and density 8 g/cc when dropped in a container filled with glycerine becomes constant after some time. If the density of glycerine is 1.3 g/cc, then the value of viscous force acting on the ball will be $x \times 10^{-4}$ N. The value of x is .... [use, $g = 10 \text{ m/s}^2$]

[2022, 29 July Shift-II]

28 A hydraulic press can lift 100 kg when a mass m is placed on the smaller piston. It can lift ...... kg when the diameter of the larger piston is increased by 4 times and that of the smaller piston is decreased by 4 times keeping the same mass m on the smaller piston.

[2021, 24 Feb Shift-I]

Topic 2 Fluid Flow, Bernoulli's Principle and Viscosity

Level 1

01 A solid steel ball of diameter 3.6 mm acquired terminal velocity 2.45×10^{-2} m/s while falling under gravity through an oil of density 925 kg m^{-3}. Take density of steel as 7825 kg m^{-3} and g as 9.8 m/s^2. The viscosity of the oil in SI unit is *[2025, 3 April Shift-II]*

(a) 2.38 (b) 1.99
(c) 1.68 (d) 2.18

02

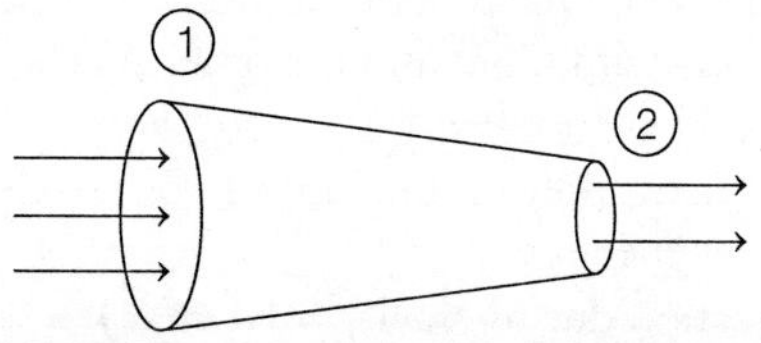

A tube of length L is shown in the figure. The radius of cross-section at the point 1 is 2 cm and at the point 2 is 1 cm, respectively. If the velocity of water entering at point 1 is 2 m/s, then velocity of water leaving the point 2 will be *[2025, 22 Jan Shift-II]*

(a) 4 m/s (b) 8 m/s
(c) 6 m/s (d) 2 m/s

03 In the experiment for measurement of viscosity η of given liquid with a ball having radius R, consider following statements. *[2025, 28 Jan Shift-I]*

A. Graph between terminal velocity v and R will be a parabola.
B. The terminal velocities of different diameter balls are constant for a given liquid.
C. Measurement of terminal velocity is dependent on the temperature.
D. This experiment can be utilised to assess the density of a given liquid.
E. If balls are dropped with some initial speed, the value of η will change.

Choose the correct answer from the options given below.

(a) C, D and E only (b) B, D and E only
(c) A, C and D only (d) A, B and E only

04 Water flows in a horizontal pipe whose one end is closed with a valve. The reading of the pressure gauge attached to the pipe is p_1. The reading of the pressure gauge falls to p_2 when the valve is opened. The speed of water flowing in the pipe is proportional to

[2025, 23 Jan Shift-II]

(a) $(p_1 - p_2)^2$ (b) $p_1 - p_2$ (c) $(p_1 - p_2)^4$ (d) $\sqrt{p_1 - p_2}$

05 A small rigid spherical ball of mass M is dropped in a long vertical tube containing glycerine. The velocity of the ball becomes constant after some time. If the density of glycerine is half of the density of the ball, then the viscous force acting on the ball will be (consider g as acceleration due to gravity)

[2025, 22 Jan Shift-I]

(a) $\frac{Mg}{2}$ (b) $\frac{3}{2} Mg$ (c) $2 Mg$ (d) Mg

06 Correct Bernoulli's equation is (symbols have their usual meaning) *[2024, 8 April Shift-I]*

(a) $p + \frac{1}{2}\rho g h + \frac{1}{2}\rho v^2 = \text{constant}$

(b) $p + \rho g h + \rho v^2 = \text{constant}$

(c) $p + \rho g h + \frac{1}{2}\rho v^2 = \text{constant}$

(d) $p + mgh + \frac{1}{2} m v^2 = \text{constant}$

07 Given below are two statements.

Statement I Viscosity of gases is greater than that of liquids.

Statement II Surface tension of a liquid decreases due to the presence of insoluble impurities.

In the light of the above statements, choose the most appropriate answer from the options given below.

[2024, 27 Jan Shift-I]

(a) Both Statement I and Statement II are correct.
(b) Statement I is incorrect but Statement II is correct.
(c) Both Statement I and Statement II are incorrect.
(d) Statement I is correct but Statement II is incorrect.

08 Eight equal drops of water are falling through air with a steady speed of 10 cm/s. If the drops coalesce, the new velocity is *[2023, 11 April Shift-II]*

(a) 16 cm/s (b) 40 cm/s (c) 10 cm/s (d) 5 cm/s

09 A hydraulic automobile lift is designed to lift vehicles of mass 5000 kg. The area of cross-section of the cylinder carrying the load is 250 cm^2. The maximum pressure the smaller piston would have to bear is (Assume $g = 10\ m/s^2$) *[2023, 8 April Shift-II]*

(a) $2 \times 10^{+5}$ Pa (b) $200 \times 10^{+6}$ Pa
(c) $20 \times 10^{+6}$ Pa (d) $2 \times 10^{+6}$ Pa

10 A small ball of mass M and density ρ is dropped in a viscous liquid of density ρ_0. After some time, the ball falls with a constant velocity. What is the viscous force on the ball ? *[2023, 6 April Shift-I]*

(a) $F = Mg\left(1 + \frac{\rho}{\rho_0}\right)$ (b) $F = Mg\left(1 + \frac{\rho_0}{\rho}\right)$
(c) $F = Mg\left(1 - \frac{\rho_0}{\rho}\right)$ (d) $F = Mg\ (1 \pm \rho\rho_0)$

11 A water drop of radius 1 μm falls in a situation where the effect of buoyant force is negligible. Coefficient of viscosity of air is 1.8×10^{-5} N-s-m^{-2} and its density is negligible as compared to that of water (10^6 gm^{-3}). Terminal velocity of the water drop is (Take, acceleration due to gravity = 10 ms^{-2}) *[2022, 28 June Shift-II]*

(a) 145.4×10^{-6} ms^{-1} (b) 118.0×10^{-6} ms^{-1}
(c) 132.6×10^{-6} ms^{-1} (d) 123.4×10^{-6} ms^{-1}

12 The velocity of a small ball of mass m and density d_1, when dropped in a container filled with glycerine, becomes constant after sometime. If the density of glycerine is d_2, then the viscous force acting on the ball, will be *[2022, 27 June Shift-I]*

(a) $mg\left(1 - \frac{d_1}{d_2}\right)$ (b) $mg\left(1 - \frac{d_2}{d_1}\right)$
(c) $mg\left(\frac{d_1}{d_2} - 1\right)$ (d) $mg\left(\frac{d_2}{d_1} - 1\right)$

13 The terminal velocity v_t of the spherical rain drop depends on the radius r of the spherical rain drop as *[2022, 25 June Shift-I]*

(a) $r^{1/2}$ (b) r (c) r^2 (d) r^3

14 A raindrop with radius $R = 0.2$ mm falls from a cloud at a height $h = 2000$ m above the ground. Assume that, the drop is spherical throughout its fall and the force of buoyancy may be neglected, then the terminal speed attained by the raindrop is (Take, density of water, $\rho_w = 1000$ kg m^{-3} and density of air, $\rho_a = 1.2$ kg m^{-3}, $g = 10$ m/s, coefficient of viscosity of air, $\eta = 1.8 \times 10^{-5}$ N-s m^{-2}) *[2021, 27 July Shift-II]*

(a) 250.6 ms^{-1} (b) 43.56 ms^{-1}
(c) 4.94 ms^{-1} (d) 14.4 ms^{-1}

15 A light cylindrical vessel is kept on a horizontal surface. Area of base is A. A hole of cross-sectional area a is made just at its bottom side. The minimum coefficient of friction necessary to prevent sliding the vessel due to the impact force of the emerging liquid is ($a << A$). *[2021, 27 July Shift-I]*

(a) $\frac{A}{2a}$ (b) None (c) $\frac{2a}{A}$ (d) $\frac{a}{A}$

Imp. Equation of Continuity

16 An ideal fluid flows (laminar flow) through a pipe of non-uniform diameter. The maximum and minimum diameters of the pipes are 6.4 cm and 4.8 cm, respectively. The ratio of the minimum and the maximum velocities of fluid in this pipe is *[2020, 7 Jan Shift-II]*

(a) 9/16 (b) 81/256 (c) $\sqrt{3}/2$ (d) 3/4

17 A solid sphere of radius R acquires a terminal velocity v_1 when falling (due to gravity) through a viscous fluid having a coefficient of viscosity η. The sphere is broken into 27 identical solid spheres. If each of these spheres acquires a terminal velocity, v_2 when falling through the same fluid, the ratio (v_1 / v_2) equals *[2019, 12 April Shift-II]*

(a) 9 (b) 1/27 (c) 1/9 (d) 27

18 Water flows into a large tank with flat bottom at the rate of 10^{-4} $m^3 s^{-1}$. Water is also leaking out of a hole of area 1 cm^2 at its bottom. If the height of the water in the tank remains steady, then this height is *[2019, 10 Jan Shift-I]*

(a) 4 cm (b) 2.9 cm
(c) 5.1 cm (d) 1.7 cm

19 If a ball of steel (density, $\rho = 7.8$ g cm^{-3}) attains a terminal velocity of 10 cms^{-1} when falling in a tank of water (coefficient of viscosity $\eta_{water} = 8.5 \times 10^{-4}$ Pa-s), then its terminal velocity in glycerine (ρ = 1.2 g cm^{-3}, $\eta = 13.2$ Pa-s) would be nearly *[AIEEE 2011]*

(a) 1.6×10^{-5} cms^{-1} (b) 6.25×10^{-4} cms^{-1}
(c) 6.45×10^{-4} cms^{-1} (d) 1.5×10^{-5} cms^{-1}

20 A spherical solid ball of volume V is made of a material of density ρ_1. It is falling through a liquid of density $\rho_2 (\rho_2 < \rho_1)$. [Assume that the liquid applies a viscous force on the ball that is proportional to the square of its speed v, i.e., $F_{viscous} = -kv^2 (k > 0)$]. The terminal speed of the ball is *[AIEEE 2008]*

(a) $\sqrt{\frac{Vg(\rho_1 - \rho_2)}{k}}$ (b) $\frac{Vg\rho_1}{k}$
(c) $\sqrt{\frac{Vg\rho_1}{k}}$ (d) $\frac{Vg(\rho_1 - \rho_2)}{k}$

Level 2

21 Consider a completely full cylindrical water tank of height 1.6 m and of cross-sectional area $0.5\ m^2$. It has a small hole in its side at a height 90 cm from the bottom. Assume, the cross-sectional area of the hole to be negligibly small as compared to that of the water tank. If a load 50 kg is applied at the top surface of the water in the tank, then the velocity of the water coming out at the instant when the hole is opened is *[2025, 3 April Shift-I]*

(a) 2 m/s (b) 4 m/s (c) 5 m/s (d) 4.3 m/s

22 A small ball of mass m and density ρ is dropped in a viscous liquid of density ρ_0. After sometime the ball falls with constant velocity. The viscous force on the ball is *[2024, 6 April Shift-I]*

(a) $mg\left(\frac{\rho_0}{\rho}-1\right)$ (b) $mg\left(1-\frac{\rho_0}{\rho}\right)$

(c) $mg(1-\rho\rho_0)$ (d) $mg\left(1+\frac{\rho}{\rho_0}\right)$

23 A small steel ball is dropped into a long cylinder containing glycerine. Which one of the following is the correct representation of the velocity-time graph for the transit of the ball ? *[2024, 31 Jan Shift-I]*

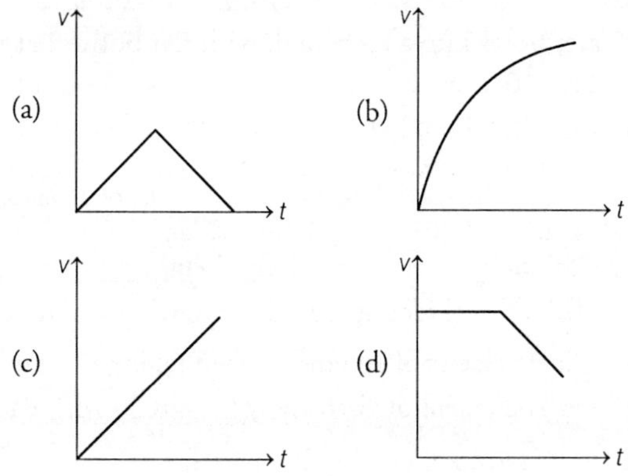

24 The figure shows a liquid of given density flowing steadily in horizontal tube of the section. Cross-sectional areas at A is $1.5\ cm^2$ and B is $25\ mm^2$, if speed of liquid at B is 60 cm/s, then $p_A - p_B$ is

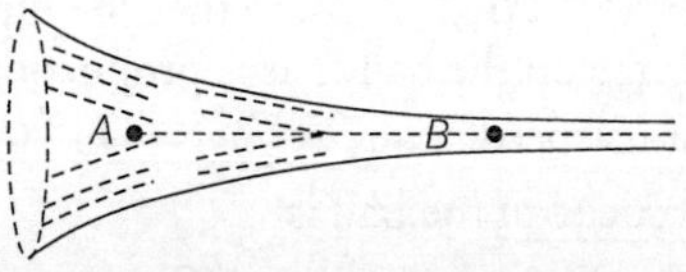

(Given, p_A and p_B are liquid pressure at A and B points. density, $\rho = 1000\ kg\ m^{-3}$

A and B are on the axis of tube *[2023, 13 April Shift-I]*

(a) 36 Pa (b) 135 Pa (c) 175 Pa (d) 27 Pa

25 When a ball is dropped into a lake from a height 4.9 m above the water level, it hits the water with a velocity v and then sinks to the bottom with the constant velocity v. It reaches the bottom of the lake 4.0 s after it is dropped. The approximate depth of the lake is *[2022, 27 June Shift-II]*

(a) 19.6 m (b) 29.4 m

(c) 39.2 m (d) 73.5 m

Imp. Viscous Force

26 In Millikan's oil drop experiment, what is viscous force acting on an uncharged drop of radius 2.0×10^{-5} m and density $1.2\times10^3\ kgm^{-3}$? Take viscosity of liquid $=1.8\times10^{-5}\ Nsm^{-2}$. (Neglect buoyancy due to air). *[2021, 27 Aug Shift-I]*

(a) 3.8×10^{-11} N (b) 3.9×10^{-10} N

(c) 1.8×10^{-10} N (d) 5.8×10^{-10} N

27 Two tubes of radii r_1 and r_2, and lengths l_1 and l_2, respectively, are connected in series and a liquid flows through each of them in streamline conditions. p_1 and p_2 are pressure differences across the two tubes. If p_2 is $4p_1$ and l_2 is $\frac{l_1}{4}$, then the radius r_2 will be equal to *[JEE Main 2017]*

(a) $2r_1$ (b) $\frac{r_1}{2}$

(c) $4r_1$ (d) r_1

28 Consider a water jar of radius R that has water filled up to height H and is kept on a stand of height h (see figure). Through a hole of radius $r(r<<R)$ at its bottom, the water leaks out and the stream of water coming down towards the ground has a shape like a funnel as shown in the figure.

If the radius of the cross-section of water stream when it hits the ground is x. Then *[JEE Main 2016]*

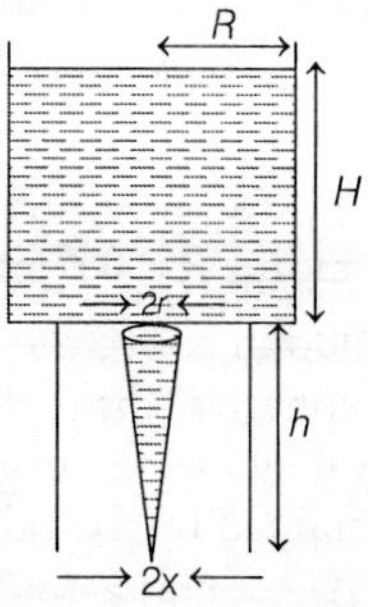

(a) $x=r\left(\frac{H}{H+h}\right)^{1/4}$ (b) $x=r\left(\frac{H}{H+h}\right)$

(c) $x=r\left(\frac{H}{H+r}\right)^{2}$ (d) $x=r\left(\frac{H}{H+h}\right)^{1/2}$

29 On heating water, bubbles being formed at the bottom of the vessel detatch and rise. Take the bubbles to be spheres of radius R and making a circular contact of radius r with the bottom of the vessel. If $r << R$ and the surface tension of water is T, value of r just before bubbles detatch is (density of water is ρ) *[JEE Main 2014]*

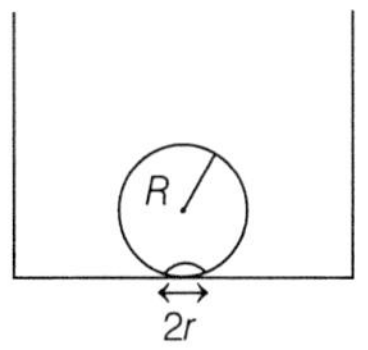

(a) $R^2\sqrt{\frac{2\rho_w g}{3T}}$ (b) $R^2\sqrt{\frac{\rho_w g}{6T}}$
(c) $R^2\sqrt{\frac{\rho_w g}{T}}$ (d) $R^2\sqrt{\frac{3\rho_w g}{T}}$

30 Water is flowing continuously from a tap having an internal diameter 8×10^{-3} m. The water velocity as it leaves the tap is $0.4\ \text{ms}^{-1}$. The diameter of the water stream at a distance 2×10^{-1} m below the tap is close to *[AIEEE 2011]*

(a) 7.5×10^{-3} m (b) 9.6×10^{-3} m
(c) 3.6×10^{-3} m (d) 5.0×10^{-3} m

Level 3

31 A spherical ball of radius 1×10^{-4} m and density $10^5\ \text{kg/m}^3$ falls freely under gravity through a distance h before entering a tank of water. If after entering in water the velocity of the ball does not change, then the value of h is approximately (The coefficient of viscosity of water is $9.8 \times 10^{-6}\ \text{N s/m}^2$) *[2024, 9 April Shift-II]*

(a) 2296 m (b) 2519 m
(c) 2396 m (d) 2249 m

Imp. Terminal Velocity

32 A small spherical ball of radius r, falling through a viscous medium of negligible density has terminal velocity v. Another ball of the same mass but of radius $2r$, falling through the same viscous medium will have terminal velocity *[2024, 31 Jan Shift-II]*

(a) $\frac{v}{4}$ (b) $\frac{v}{2}$ (c) $4v$ (d) $2v$

33 A fully loaded boeing aircraft has a mass of 5.4×10^5 kg. Its total wing area is $500\ \text{m}^2$. It is in level flight with a speed of 1080 km/h. If the density of air ρ is $1.2\ \text{kg m}^{-3}$, the fractional increase in the speed of the air on the upper surface of the wing relative to the lower surface in percentage will be ($g = 10\ \text{m/s}^2$) *[2023, 29 Jan Shift-II]*

(a) 10 (b) 8
(c) 16 (d) 6

34 Consider a cylindrical tank of radius 1 m is filled with water. The top surface of water is at 15 m from the bottom of the cylinder. There is a hole on the wall of cylinder at a height of 5 m from the bottom. A force of 5×10^5 N is applied an the top surface of water using a piston. The speed of efflux from the hole will be (given atmosphere pressure $p_A = 1.01 \times 10^5$ Pa, density of water $\rho_w = 1000\ \text{kg/m}^3$ and gravitational acceleration $g = 10\ \text{m/s}^2$) *[2022, 28 July Shift-II]*

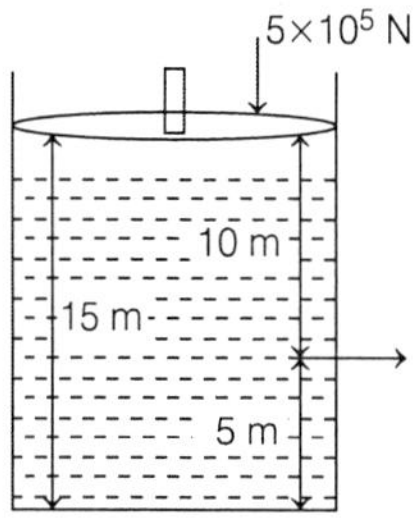

(a) 11.6 m/s (b) 10.8 m/s (c) 17.8 m/s (d) 14.4 m/s

35 Two identical cylindrical vessels are kept on the ground and each contains the same liquid of density d. The area of the base of both vessels is S but the height of liquid in one vessel is x_1 and in the other, x_2. When both cylinders are connected through a pipe of negligible volume very close to the bottom, the liquid flows from one vessel to the other until it comes to equilibrium at a new height. The change in energy of the system in the process is *[2020, 4 Sep Shift-II]*

(a) $gdS(x_2^2 + x_1^2)$ (b) $\frac{3}{4}gdS(x_2 - x_1)^2$
(c) $gdS(x_2 + x_1)^2$ (d) $\frac{1}{4}gdS(x_2 - x_1)^2$

36 A cylindrical vessel containing a liquid is rotated about its axis, so that the liquid rises at its sides as shown in the figure. The radius of vessel is 5 cm and the angular speed of rotation is ω rad s^{-1}. The difference in the height h (in cm) of liquid at the centre of vessel and at the side will be *[2020, 2 Sep Shift-I]*

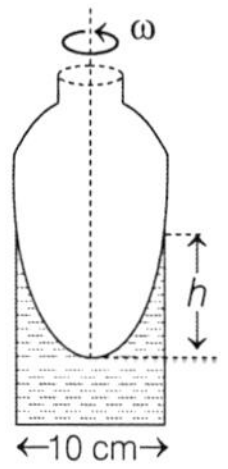

(a) $\frac{25\omega^2}{2g}$ (b) $\frac{5\omega^2}{2g}$ (c) $\frac{2\omega^2}{25g}$ (d) $\frac{2\omega^2}{5g}$

37 Water flows in a horizontal tube (see figure). The pressure of water changes by $700\ \text{Nm}^{-2}$ between A and B, where the area of cross- section are $40\ \text{cm}^2$ and $20\ \text{cm}^2$, respectively. Find the rate of flow of water through the tube. (Take, density of water $= 1000 \text{ kgm}^{-3}$) *[2020, 9 Jan Shift-I]*

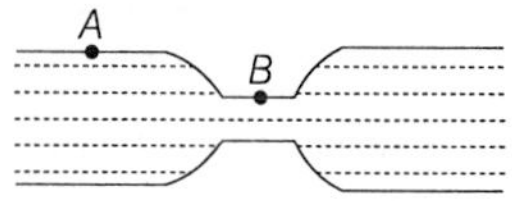

(a) $3020 \text{ cm}^3/\text{s}$
(b) $2420 \text{ cm}^3/\text{s}$
(c) $2720 \text{ cm}^3/\text{s}$
(d) $1810 \text{ cm}^3/\text{s}$

38 Which of the following options correctly describes the variation of the speed v and acceleration a of a point mass falling vertically in a viscous medium that applies a force $F = -kv$, where k is a constant, on the body? (Graphs are schematic and not drawn to scale) *[JEE Main 2016]*

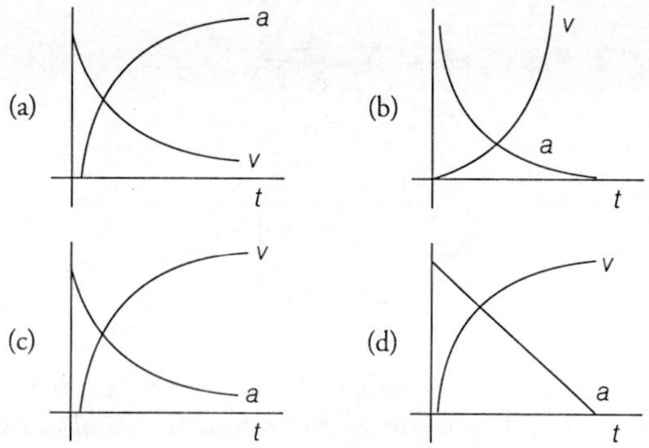

Numerical Value Type Questions

39 Small water droplets of radius 0.01 mm are formed in the upper atmosphere and falling with a terminal velocity of 10 cm/s. Due to condensation, if 8 such droplets are coalesced and formed a larger drop, the new terminal velocity will be .... cm/s. *[2024, 8 April Shift-II]*

40 In a test experiment on a model aeroplane in wind tunnel, the flow speeds on the upper and lower surfaces of the wings are 70 ms^{-1} and 65 m/s, respectively. If the wing area is 2 m^2, the lift of the wing is .... N.
(Given, density of air $= 1.2 \text{ kg m}^{-3}$) *[2024, 29 Jan Shift-I]*

Imp. Bernoulli's Principle

41 Glycerin of density $1.25 \times 10^3 \text{ kg m}^{-3}$ is flowing through the conical section of pipe. The area of cross-section of the pipe at its ends are 10 cm^2 and 5 cm^2 and pressure drop across its length is 3 Nm^{-2}. The rate of flow of glycerin through the pipe is $x \times 10^{-5} \text{ m}^3\text{s}^{-1}$. The value of x is .... *[2023, 12 April Shift-I]*

42 Figure below shows a liquid being pushed out of the tube by a piston having area of cross-section 2.0 cm^2. The area of cross-section at the outlet is 10 mm. If the piston is pushed at a speed of 4 cm s^{-1}, the speed of outgoing fluid in cm s^{-1} is .... *[2023, 10 April Shift-II]*

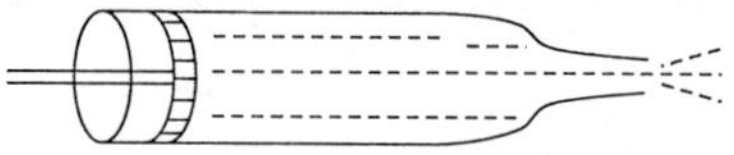

43 The surface of water in a water tank of cross-section area 750 cm^2 on the top of a house is h m above the top level. The speed of water coming out through the top of cross-section area 500 mm^2 is 30 cm/s. At that instant, dh/dt is $x \times 10^{-3}$ m/s. The value of x will be .... *[2023, 1 Feb Shift-II]*

44 A spherical ball of radius 1 mm and density 10.5 g/cc is dropped in glycerine of coefficient of viscosity 9.8 poise and density 1.5 g/cc. Viscous force on the ball by which it attains constant velocity is 3696×10^{-x} N. The value of x is ....
(Given, $g = 9.8 \text{ m/s}^2$ and $\pi = 22/7$) *[2023, 24 Jan Shift-II]*

45 The diameter of an air bubble which was initially 2 mm, rises steadily through a solution of density 1750 kg m^{-3} at the rate of 0.35 cms^{-1}. The coefficient of viscosity of the solution is .... poise (in nearest integer). (The density of air is negligible). *[2022, 28 July Shift-I]*

46 A small spherical ball of radius 0.1 mm and density 10^4 kg m^{-3} falls freely under gravity through a distance h before entering a tank of water. If after entering the water, the velocity of ball does not change and it continues to fall with same constant velocity inside water, then the value of h will be......m.
(Given $g = 10 \text{ ms}^{-2}$, viscosity of water $= 1.0 \times 10^{-5} \text{ N-sm}^{-2}$). *[2022, 29 June Shift-II]*

47 A liquid of density 750 kg m^{-3} flows smoothly through a horizontal pipe that tapers in cross-sectional area from $A_1 = 1.2 \times 10^2 \text{ m}^2$ to $A_2 = \frac{A_1}{2}$.
The pressure difference between the wide and narrow sections of the pipe is 4500 Pa. The rate of flow of liquid is .... $\times 10^{-3} \text{ m}^3\text{s}^{-1}$ *(2022, 28 June Shift-II)*

48 The area of cross-section of a large tank is 0.5 m^2. It has a narrow opening near the bottom having area of cross-section 1 cm^2. A load of 25 kg is applied on the water at the top in the tank.

Neglecting the speed of water in the tank, the velocity of the water, coming out of the opening at the time when the height of water level in the tank is 40 cm above the bottom, will be ...... cms^{-1}.
(Take, $g = 10\ ms^{-2}$) *[2022, 27 June Shift-I]*

49 An ideal fluid of density 800 kgm^{-3} flows smoothly through a bent pipe (as shown in figure) that tapers in cross-sectional area from a to $a/2$. The pressure difference between the wide and narrow sections of pipe is 4100 Pa. At wider section, the velocity of fluid is $\frac{\sqrt{x}}{6}\ ms^{-1}$ for $x =$ .... (Given, $g = 10\ m/s^2$) *[2022, 26 June Shift-I]*

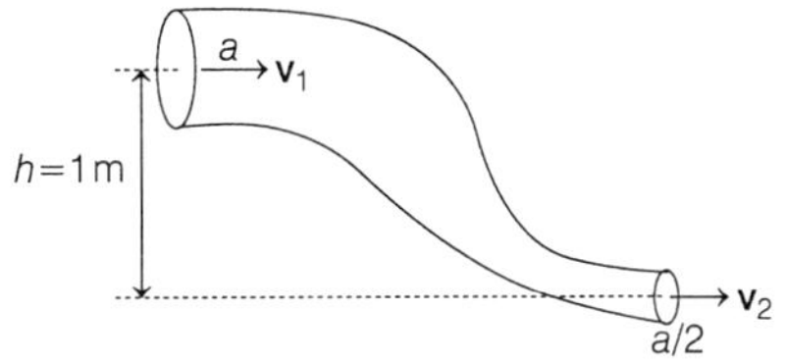

50 The velocity of upper layer of water in a river is 36 km h^{-1}. Shearing stress between horizontal layers of water is $10^3\ N\ m^{-2}$. Depth of the river is .... m. (Coefficient of visosity of water is 10^{-2} Pa-s) *[2022, 25 June Shift-I]*

51 The water is filled upto height of 12 m in a tank having vertical sidewalls. A hole is made in one of the walls at a depth h below the water level. The value of h for which the emerging stream of water strikes the ground at the maximum range is ...... m. *[2021, 27 July Shift-II]*

52 Consider a water tank as shown in the figure. It's cross-sectional area is 0.4 m^2. The tank has an opening B near the bottom whose cross-sectional area is 1 cm^2. A load of 24 kg is applied on the water at the top when the height of the water level is 40 cm above the bottom, the velocity of water coming out the opening B is $v\ ms^{-1}$. The value of v, to the nearest integer, is ....

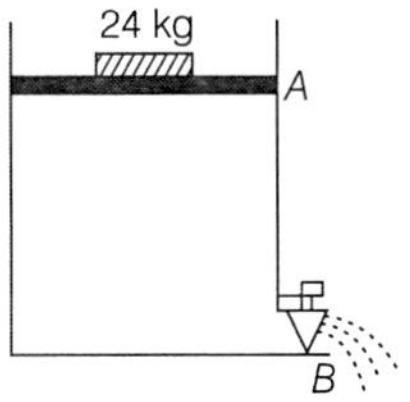

(Take, value of g to be 10 ms^{-2}) *[2021, 18 March Shift-II]*

Topic 3 Surface Tension, Excess Pressure and Capillarity

Level 1

01 Two liquids A and B have θ_A and θ_B as contact angles in a capillary tube. If $K = \cos\theta_A / \cos\theta_B$, then identify the correct statement. *[2025, 4 April Shift-I]*
(a) K is negative, then liquid A and liquid B have concave meniscus.
(b) K is zero, then liquid A has convex meniscus and liquid B has concave meniscus.
(c) K is negative, then liquid A and liquid B have convex meniscus.
(d) K is negative, then liquid A has concave meniscus and liquid B has convex meniscus.

02 Given below are two statements.
Statement I The hot water flows faster than cold water.
Statement II Soap water has higher surface tension as compared to fresh water.
In the light above statements, choose the correct answer from the options given below. *[2025, 23 Jan Shift-I]*
(a) Statement I is true but Statement II is false.
(b) Both Statement I and Statement II is false.
(c) Statement I is false but Statement II is true.
(d) Both Statement I and Statement II are true.

03 Consider following statements. *[2025, 28 Jan Shift-I]*
A. Surface tension arises due to extra energy of the molecules at the interior as compared to the molecules at the surface, of a liquid.
B. As the temperature of liquid rises, the coefficient of viscosity increases.
C. As the temperature of gas increases, the coefficient of viscosity increases.
D. The onset of turbulence is determined by Reynold's number.
E. In a steady flow two stream lines never intersect.
Choose the correct answer from the options given below.
(a) A, D, E only
(b) A, B, C only
(c) C, D E only
(d) B, C, D only

04 An air bubble of radius 0.1 cm lies at a depth of 20 cm below the free surface of a liquid of density 1000 kg/m^3. If the pressure inside the bubble is 2100 N/m^2 greater than the atmospheric pressure, then the surface tension of the liquid in SI unit is (use $g = 10m/s^2$) *[2025, 24 Jan Shift-I]*
(a) 0.02 (b) 0.25
(c) 0.05 (d) 0.1

05 Pressure inside a soap bubble is greater than the pressure outside by an amount (Given, R = Radius of bubble, S = Surface tension of bubble) *[2024, 6 April Shift-II]*
(a) $\frac{4R}{S}$ (b) $\frac{2S}{R}$ (c) $\frac{4S}{R}$ (d) $\frac{S}{R}$

06 Given below are two statements.
Statement I If a capillary tube is immersed first in cold water and then in hot water, the height of capillary rise will be smaller in hot water.

Statement II If a capillary tube is immersed first in cold water and then in hot water, the height of capillary rise will be smaller in cold water.

In the light of the above statements, choose the most appropriate from the options given below.

[2024, 29 Jan Shift-I]

(a) Both Statement I and Statement II are true.
(b) Both Statement I and Statement II are false.
(c) Statement I is false but Statement II is true.
(d) Statement I is true but Statement II is false.

07 If 1000 droplets of water of surface tension 0.07 N/m, having same radius 1 mm each, combine to form a single drop. In the process, the released surface energy is $\left(\text{Take}, \pi = \frac{22}{7}\right)$ *[2023, 31 Jan Shift-I]*

(a) 7.92×10^{-6} J (b) 8.8×10^{-5} J
(c) 9.68×10^{-4} J (d) 7.92×10^{-4} J

Imp. Angle of Contact

08 Given below are two statements. One is labelled as Assertion (A) and the other is labelled as Reason (R).

Assertion (A) Clothes containing oil or grease stains cannot be cleaned by water wash.

Reason (R) Because the angle of contact between the oil/ grease and water is obtuse.

In the light of the above statements, choose the correct answer from the option given below.

[2022, 29 July Shift-I]

(a) Both A and R are true and R is the correct explanation of (A).
(b) Both A and R are true but R is not the correct explanation of A.
(c) A is true but R is false.
(d) A is false but R is true.

09 If M is the mass of water that rises in a capillary tube of radius r, then mass of water which will rise in a capillary tube of radius $2r$ is *[2019, 9 April Shift-I]*

(a) $2M$ (b) $4M$ (c) $M/2$ (d) M

10 If two glass plates have water between them and are separated by very small distance (see figure), it is very difficult to pull them apart. It is because the water in between forms cylindrical surface on the side that gives rise to lower pressure in the water in comparison to atmosphere. If the radius of the cylindrical surface is R and surface tension of water is T, then the pressure in water between the plates is lower by *[Online 2015]*

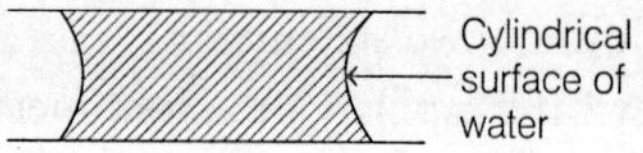

(a) $T/4R$ (b) $T/2R$ (c) $4T/R$ (d) $2T/R$

11 Assume that a drop of liquid evaporates by decrease in its surface energy, so that its temperature remains unchanged. What should be the minimum radius of the drop for this to be possible? The surface tension is T, density of liquid is ρ and L is its latent heat of vaporisation. *[JEE Main 2013]*

(a) $\rho L/T$ (b) $\sqrt{T/\rho L}$ (c) $T/\rho L$ (d) $2T/\rho L$

12 A thin liquid film formed between a U-shaped wire and a light slider supports a weight of 1.5×10^{-2} N (see figure). The length of the slider is 30 cm and its weight negligible. The surface tension of the liquid film is *[AIEEE 2012]*

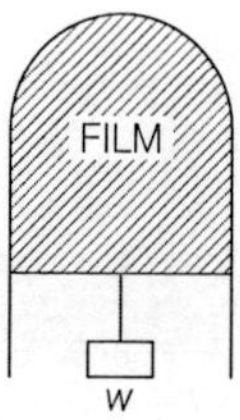

(a) 1.0 Nm^{-1} (b) 0.05 Nm^{-1}
(c) 0.025 Nm^{-1} (d) 0.0125 Nm^{-1}

13 Two mercury drops (each of radius r) merge to form a bigger drop. The surface energy of the bigger drop, if T is the surface tension, is *[AIEEE 2011]*

(a) $2^{5/3}\pi r^2 T$ (b) $4\pi r^2 T$
(c) $2\pi r^2 T$ (d) $2^{8/3}\pi r^2 T$

Level 2

14 Two water drops each of radius r coalesce to form a bigger drop. If T is the surface tension, the surface energy released in this process is *[2025, 2 April Shift-II]*

(a) $4\pi r^2 T\,[2 - 2^{\frac{2}{3}}]$ (b) $4\pi r^2 T\,[\sqrt{2} - 1]$
(c) $4\pi r^2 T\,[2 - 2^{\frac{1}{3}}]$ (d) $4\pi r^2 T\,[1 + \sqrt{2}]$

Imp. Surface Energy

15 The amount of work done to break a big water drop of radius R into 27 small drops of equal radius is 10 J. The work done required to break the same big drop into 64 small drops of equal radius will be *[2025, 24 Jan Shift-I]*

(a) 20 J (b) 15 J (c) 5 J (d) 10 J

16 The excess pressure inside a soap bubble is thrice the excess pressure inside a second soap bubble. The ratio between the volume of the first and the second bubble is *[2024, 9 April Shift-II]*

(a) 1 : 9 (b) 1 : 81 (c) 1 : 3 (d) 1 : 27

17 Given below are two statements .

Statement I When a capillary tube is dipped into a liquid, the liquid neither rises nor falls in the capillary. The contact angle may be 0°.

Statement II The contact angle between a solid and a liquid is a property of the material of the solid and liquid as well.

In the light of the above statement, choose the correct answer from the options given below.

[2024, 5 April Shift-I]

(a) Statement I is true and Statement II is false.
(b) Both Statement I and Statement II are true.
(c) Statement I is false but Statement II is true.
(d) Both Statement I and Statement II are false.

18 Given below are two statements.

Statement I The contact angle between a solid and a liquid is a property of the material of the solid and liquid as well.

Statement II The rise of a liquid in a capillary tube does not depend on the inner radius of the tube.

In the light of the above statements, choose the correct answer from the options given below.

[2024, 4 April Shift-II]

(a) Statement I is false but Statement II is true.
(b) Both Statement I and Statement II are false.
(c) Both Statement I and Statement II are true.
(d) Statement I is true but Statement II is false.

19 A small liquid drop of radius R is divided into 27 identical liquid drops.. If the surface tension is T, then the work done in the process will be

[2024, 29 Jan Shift-II]

(a) $3\pi R^2T$ (b) $4\pi R^2T$ (c) $\frac{1}{8}\pi R^2T$ (d) $8\pi R^2T$

20 A mercury drop of radius 10^{-3} m is broken into 125 equal size droplets. Surface tension of mercury is $0.45\ Nm^{-1}$. The gain in surface energy is

[2023, 1 Feb Shift-I]

(a) 5×10^{-5} J (b) 2.26×10^{-5} J
(c) 17.5×10^{-5} J (d) 28×10^{-5} J

21 The height of liquid column raised in a capillary tube of certain radius when dipped in liquid A vertically is 5 cm. If the tube is dipped in a similar manner in another liquid B of surface tension and density double the values of liquid A, the height of liquid column raised in liquid B would be .... m.

[2023, 30 Jan Shift-I]

(a) 0.10 (b) 0.05 (c) 0.20 (d) 0.5

22 A water drop of radius 1 cm is broken into 729 equal droplets. If surface tension of water is 75 dyne/cm, then the gain in surface energy upto first decimal place will be (Given, $\pi = 3.14$) *[2022, 26 July Shift-I]*

(a) 8.5×10^{-4} J (b) 8.2×10^{-4} J
(c) 7.5×10^{-4} J (d) 5.3×10^{-4} J

23 A drop of liquid of density ρ is floating half immersed in a liquid of density σ and surface tension $7.5\times10^{-4}\ N\ cm^{-1}$. The radius of drop in cm will be (Take, $g = 10\ ms^{-2}$) *[2022, 25 July Shift-II]*

(a) $\frac{15}{\sqrt{(2\rho-\sigma)}}$ (b) $\frac{15}{\sqrt{(\rho-\sigma)}}$
(c) $\frac{3}{2\sqrt{(\rho-\sigma)}}$ (d) $\frac{3}{20\sqrt{(2\rho-\sigma)}}$

24 Two spherical soap bubbles of radii r_1 and r_2 in vacuum combine under isothermal conditions. The resulting bubble has a radius equal to

[2021, 25 July Shift-II]

(a) $\frac{r_1r_2}{r_1+r_2}$ (b) $\sqrt{r_1r_2}$
(c) $\sqrt{r_1^2+r_2^2}$ (d) $\frac{r_1+r_2}{2}$

Imp. Excess Pressure

25 When two soap bubbles of radii a and $b(b > a)$ coalesce, the radius of curvature of common surface is *[2021, 17 March Shift-I]*

(a) $\frac{ab}{b-a}$ (b) $\frac{a+b}{ab}$ (c) $\frac{b-a}{ab}$ (d) $\frac{ab}{a+b}$

26 A small spherical droplet of density d is floating exactly half immersed in a liquid of density ρ and surface tension T. The radius of the droplet is (take note that the surface tension applies an upward force on the droplet) *[2020, 9 Jan Shift-II]*

(a) $r=\sqrt{\frac{3T}{(2d-\rho)g}}$ (b) $r=\sqrt{\frac{T}{(d-\rho)g}}$
(c) $r=\sqrt{\frac{T}{(d+\rho)g}}$ (d) $r=\sqrt{\frac{2T}{3(d+\rho)g}}$

27 A small soap buble of radius 4 cm is trapped inside another bubble of radius 6 cm without any contact. Let p_2 be the pressure inside the inner bubble and p_0, the pressure outside the outer bubble. Radius of another bubble with pressure difference $p_2 - p_0$ between its inside and outside would be *[JEE Main 2018]*

(a) 12 cm (b) 4.8 cm
(c) 2.4 cm (d) 6 cm

Level 3

28 A capillary tube of radius 0.1 mm is partly dipped in water (surface tension 70 dyne/cm and glass water contact angle $\approx 0°$) with 30° inclined with the vertical. The length of water risen in the capillary is .... cm. (Take $g = 9.8\ m/s^2$)

[2025, 7 April Shift-II]

(a) $\frac{68}{5}$ (b) $\frac{82}{5}$
(c) $\frac{71}{5}$ (d) $\frac{57}{2}$

29 A big drop is formed by coalescing 1000 small droplets of water. The surface energy will become *[2024, 1 Feb Shift-II]*

(a) $\frac{1}{100}$ th (b) 100 times
(c) $\frac{1}{10}$ th (d) 10 times

30 A water drop of diameter 2 cm is broken into 64 equal droplets. The surface tension of water is 0.075 N/m. In this process, the gain in surface energy will be *[2022, 28 June Shift-I]*

(a) 2.8×10^{-4} J (b) 1.5×10^{-3} J
(c) 1.9×10^{-4} J (d) 9.4×10^{-5} J

31 Two narrow bores of diameter 5.0 mm and 8.0 mm are joined together to form a U-shaped tube open at both ends. If this U-tube contains water, what is the difference in the level of two limbs of the tube?
[Take surface tension of water $T = 7.3\times 10^{-2}$ Nm^{-1}, angle of contact $=0$, $g = 10$ ms^{-2} and density of water $= 1.0\times10^3$ kg m^{-3}] *[2021, 26 Aug Shift-I]*

(a) 3.62 mm (b) 2.19 mm
(c) 5.34 mm (d) 4.97 mm

32 Two small drops of mercury each of radius R coalesce to form a single large drop. The ratio of total surface energy before and after the change is *[2021, 20 July Shift-II]*

(a) $2^{1/3}:1$ (b) $1:2^{1/3}$
(c) $2:1$ (d) $1:2$

33 A large number of water drops, each of radius r, combine to have a drop of radius R. If the surface tension is T and mechanical equivalent of heat is J, the rise in heat energy per unit volume will be *[2021, 26 Feb Shift-I]*

(a) $\frac{2T}{J}\left(\frac{1}{r}-\frac{1}{R}\right)$ (b) $\frac{2T}{rJ}$
(c) $\frac{3T}{rJ}$ (d) $\frac{3T}{J}\left(\frac{1}{r}-\frac{1}{R}\right)$

Imp. Capillary Rise

34 A capillary tube made of glass of radius 0.15 mm is dipped vertically in a beaker filled with methylene iodide, which rises to height h in the tube. It is observed that the two tangents drawn from liquid-glass interfaces (from opposite sides of the capillary) make an angle of 60° with one another. Then, h is close to
(Given, surface tension $=0.05$ Nm^{-1}, density $=667$ kg m^{-3} and $g = 10$ ms^{-2}) *[2020, 2 Sep Shift-II]*

(a) 0.049 m (b) 0.087 m
(c) 0.137 m (d) 0.172 m

Numerical Value Type Questions

35 Two soap bubbles of radius 2 cm and 4 cm, respectively, are in contact with each other. The radius of curvature of the common surface, in cm, is ...... *[2025, 22 Jan Shift-I]*

36 The excess pressure inside a soap bubble A in air is half the excess pressure inside another soap bubble B in air. If the volume of the bubble A is n times the volume of the bubble B, then, the value of n is .... *[2025, 3 April Shift-II]*

37 An air bubble of radius 1.0 mm is observed at a depth of 20 cm below the free surface of a liquid having surface tension 0.095 J/m^2 and density 10^3 kg/m^3. The difference between pressure inside the bubble and atmospheric pressure is ....N/m^2.
(Take, $g = 10$ m/s^2) *[2025, 23 Jan Shift-II]*

38 A liquid column of height 0.04 cm balances excess pressure of a soap bubble of certain radius. If density of liquid is 8×10^3 kg m^{-3} and surface tension of soap solution is 0.28 Nm^{-1}. Then, diameter of the soap bubble is .... cm. (If $g = 10$ m s^{-2}) *[2024, 8 April Shift-I]*

39 A big drop is formed by coalescing 1000 small droplets of water. The ratio of surface energy of 1000 droplets to that of energy of big drop is $\frac{10}{x}$. The value of x is .... *[2024, 6 April Shift-I]*

40 A soap bubble is blown to a diameter of 7 cm. If 36960 erg of work is done in blowing it further. If surface tension of soap solution is 40 dyne/cm, then the new radius is...... cm. $\left(\text{Take, } \pi = \frac{22}{7}\right)$. *[2024, 4 April Shift-I]*

41 A plane is in level flight at constant speed and each of its two wings has an area of 40 m^2. If the speed of the air is 180 km/h over the lower wing surface and 252 km/h over the upper wing surface, the mass of the plane is .... kg. (Take, air density to be 1 kg m^{-3} and $g = 10$ ms^{-2}) *[2024, 1 Feb Shift-I]*

42 A big drop is formed by coalescing 1000 small identical drops of water. If E_1 be the total surface energy of 1000 small drops of water and E_2 be the surface energy of single big drop of water, then $E_1 : E_2$ is $x:1$ where $x =$ .... *[2024, 30 Jan Shift-II]*

43 The reading of pressure metre attached with a closed pipe is 4.5×10^4 N/m^2. On opening the valve, water starts flowing and the reading of pressure metre falls to 2.0×10^4 N/m^2. The velocity of water is found to be $\sqrt{v}$ m/s. The value of v is .... *[2024, 27 Jan Shift-II]*

44 There is an air bubble of radius 1.0 mm in a liquid of surface tension 0.075 Nm^{-1} and density 1000 kg m^{-3} at a depth of 10 cm below the free surface. The amount by which the pressure inside the bubble is greater than the atmospheric pressure is .... Pa ($g = 10\, m^{-2}$) *[2023, 15 April Shift-I]*

45 The surface tension of soap solution is $3.5 \times 10^{-2} Nm^{-1}$. The amount of work done required to increase the radius of soap bubble from 10 cm to 20 cm is .... $\times 10^{-4}$ J. (Take, $\pi = 22/7$) *[2023, 11 April Shift-II]*

46 A spherical drop of liquid splits into 1000 identical spherical drops. If U_i is the surface energy of the original drop and U_f is the total surface energy of the resulting drops, the (ignoring evaporation), $\frac{U_f}{U_i} = \left(\frac{10}{x}\right)$. Then, value of x is .... *[2023, 25 Jan Shift-II]*

47 A spherical soap bubble of radius 3 cm is formed inside another spherical soap bubble of radius 6 cm. If the internal pressure of the smaller bubble of radius 3 cm in the above system is equal to the internal pressure of the another single soap bubble of radius r cm. The value of r is .... *[2022, 27 July Shift-II]*

48 A soap bubble of radius 3 cm is formed inside the another soap bubble of radius 6 cm. The radius of an equivalent soap bubble which has the same excess pressure as inside the smaller bubble with respect to the atmospheric pressure is .... cm. *[2021, 26 Aug Shift-I]*

49 Suppose you have taken a dilute solution of oleic acid in such a way that its concentration becomes 0.01 cm^3 of oleic acid per cm^3 of the solution. Then you make a thin film of this solution (monomolecular thickness) of area 4 cm^2 by considering 100 spherical drops of radius $(3/40\pi)^{1/3} \times 10^{-3}$ cm. Then the thickness of oietic acid layer will be $x \times 10^{-14}$ m where x is .... *[2021, 17 March Shift-II]*

50 When a long glass capillary tube of radius 0.015 cm is dipped in a liquid, the liquid rises to a height of 15 cm within it.If the contact angle between the liquid and glass is close to 0°, the surface tension of the liquid, (in millinewton m^{-1}) to the nearest integer is (Take, $\rho_{(liquid)} = 900$ kgm^{-3}, $g = 10\, ms^{-2}$) .... *[2020, 3 Sep Shift-I]*

• Key **Index**

Topic 1

1 (c)	2 (a)	3 (d)	4 (d)	5 (b)	6 (d)	7 (a)	8 (c)	9 (c)	10 (a)
11 (d)	12 (d)	13 (c)	14 (c)	15 (c)	16 (c)	17 (c)	18 (c)	19 (b)	20 (a)
21 (d)	22 (5)	23 (3)	24 (150)	25 (1000)	26 (177)	27 (25)	28 (25600)		

Topic 2

1 (b)	2 (b)	3 (c)	4 (d)	5 (a)	6 (c)	7 (b)	8 (b)	9 (d)	10 (c)
11 (d)	12 (b)	13 (c)	14 (c)	15 (c)	16 (a)	17 (a)	18 (c)	19 (b)	20 (a)
21 (b)	22 (b)	23 (b)	24 (c)	25 (b)	26 (b)	27 (b)	28 (a)	29 (a)	30 (c)
31 (b)	32 (b)	33 (a)	34 (c)	35 (d)	36 (a)	37 (c)	38 (c)	39 (40)	40 (810)
41 (4)	42 (80)	43 (2)	44 (7)	45 (11)	46 (20)	47 (24)	48 (300)	49 (363)	50 (100)
51 (6)	52 (3)								

Topic 3

1 (d)	2 (a)	3 (c)	4 (c)	5 (c)	6 (d)	7 (d)	8 (a)	9 (a)	10 (d)
11 (d)	12 (c)	13 (d)	14 (a)	15 (b)	16 (d)	17 (c)	18 (d)	19 (d)	20 (b)
21 (b)	22 (c)	23 (a)	24 (c)	25 (a)	26 (a)	27 (c)	28 (b)	29 (c)	30 (a)
31 (b)	32 (a)	33 (d)	34 (b)	35 (4)	36 (8)	37 (2190)	38 (7)	39 (1)	40 (7)
41 (9600)	42 (10)	43 (50)	44 (1130)	45 (264)	46 (1)	47 (2)	48 (2)	49 (25)	50 (101)

Detailed **Explanations**

Topic 1

01. *(c)* For a fluid mass, pressure at all points on same level is equal.

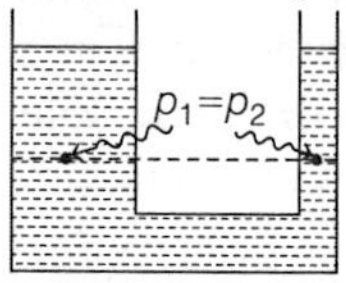

Also by pascal's law, pressure applied anywhere in a fluid is equally transmitted in all directions.

02. *(a)* As per Pascal's law, when we apply pressure to an ideal liquid, it is equally distributed in the entire liquid and to the walls as well since due to applied pressure, every morning, the tooth paste does not get compressed and we can safely consider it an incompressible liquid.

Therefore, both statement are true and R is correct explanation of A.

Option (a) is correct.

03. *(d)*

> **key idea** The force exerted by water flow on the wall,
>
> F_w = momentum loss of water per second
>
> = mass of hitting the wall per second × speed of water

Given, $A = 10\ \text{cm}^2$

$= 10\times10^{-4}\ \text{m}^2$

Speed of outflow of water, $v = 20\ \text{ms}^{-1}$

Density of water, $\rho_w = 1000\ \text{kgm}^{-3}$

$\Rightarrow F_w = (\rho_w A v_w)\times v_w = \rho_w A v_w^2$

$= 1000\times10\times10^{-4}\times(20)^2 = 400\ \text{N}$

04. *(d)* Pressure inside a fluid volume open to atmosphere is

$$p = p_0 + h\rho g$$

when $h = d_1$, $p_1 = 5.05\times10^6$ Pa

$h = d_2$, $p_2 = 8.08\times10^6$ Pa

Thus, $p_1 = p_0 + d_1\rho g = 5.05\times10^6$

and $p_2 = p_0 + d_2\rho g = 8.08\times10^6$

$\Rightarrow p_2 - p_1 = (d_2 - d_1)\rho g = 3.03\times10^6$

$$\Rightarrow d_2 - d_1 = \frac{3.03\times10^6}{\rho g}$$

Given, $\rho = 10^3\ \text{kgm}^{-3}$ and $g = 10\ \text{ms}^{-2}$

$$\therefore d_2 - d_1 = \frac{3.03\times10^6}{10^3\times10}$$

$= 303\ \text{m} \approx 300\ \text{m}$

05. *(b)* $\rho_{\text{oil}} < \rho < \rho_{\text{water}}$

Oil is least dense of them, so it should settle at the top with water at the base. Now, the ball is denser than oil but less denser than water. So, it will sink through oil but will not sink in water. So, it will stay at the oil-water interface.

06. *(d)* $\rho_1 < \rho_2$ as denser liquid acquires lowest position of vessel. $\rho_3 > \rho_1$ as ball sinks in liquid 1 and $\rho_3 < \rho_2$ as ball does not sink in liquid 2, so

$\rho_1 < \rho_3 < \rho_2$.

07. *(a)* Given, pressure on submarine at a certain depth,

$$p_i = 3\times10^5\ \text{Pa}$$

At a certain depth, pressure is given by

$p_i = p_0 + \rho g h$

$\Rightarrow 3\times10^5 = p_0 + \rho g h$

$\Rightarrow 3\times10^5 = 1\times10^5 + \rho g h$

$(\because p_0 = 1\times10^5\ \text{Pa})$

$\Rightarrow 3\times10^5 - 1\times10^5 = \rho g h$

$\Rightarrow 2\times10^5 = \rho g h$ ...(i)

At $h_1 = 2h$,

$p_f = p_0 + \rho g 2h$

$= 1\times10^5 + 2(\rho g h)$

$= 1\times10^5 + 2(2\times10^5)$ [using Eq. (i)]

$= 1\times10^5 + 4\times10^5$

$\Rightarrow p_f = 5\times10^5\ \text{Pa}$

$\therefore$ Percentage increase in pressure

$$= \frac{p_f - p_i}{p_i}\times100$$

$$= \frac{(5-3)\times10^5}{3\times10^5}\times100 = \frac{200}{3}\%$$

08. *(c)* From free body diagram,

$\rho_w V g - mg = ma$

$\rho_w V g = mg + ma$

$\rho_w V g = m(g + a)$

$$\Rightarrow m = \frac{\rho_w V g}{g + a} \quad \text{...(i)}$$

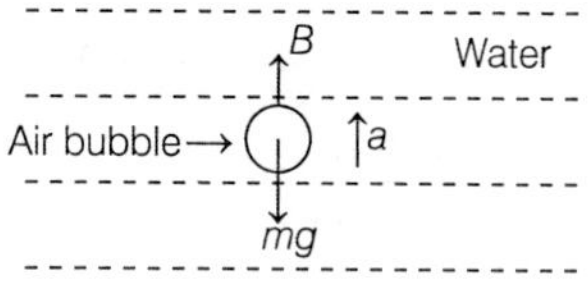

Given, $\rho_w = 1\ \text{g/cm}^3$, $g = 980\ \text{cm/s}^2$,

$a = 9.8\ \text{cm/s}^2$ and $r = 1$ cm

$$\therefore m = \frac{(1)\times\frac{4}{3}\pi r^3\times980}{980 + 9.8}$$

$$= \frac{4\times22\times980}{3\times7\times989.8} = 4.148$$

$m \simeq 4.15$ g

Hence, correct option is (c).

09. *(c)* Force on a vertical surface of area A and dipped upto height h in a fluid of density ρ is

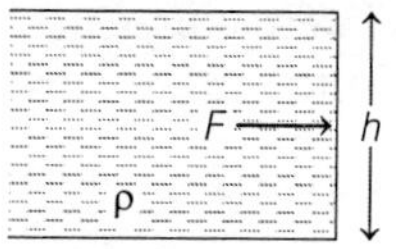

Force, $F = \frac{1}{2}(\rho g h\times A)$

So, in given case,

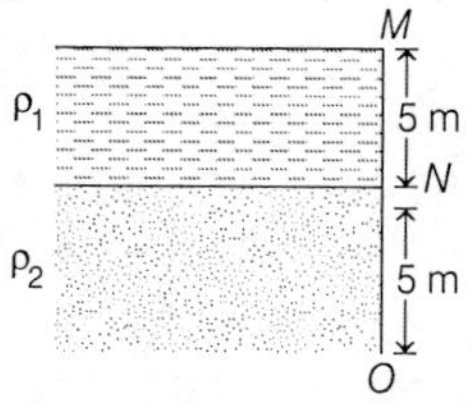

Force on portion $MN = \frac{\rho_1 g h}{2}\times A$

Force on portion NO

$= \rho_1 g h A + \frac{1}{2}(\rho_2 g h A)$

$= \rho_1 g h A + \frac{1}{2}\times 2\rho_1\times g h A$

$(\because \rho_2 = 2\rho_1)$

$= 2\rho_1 g h A$

$$\text{Required ratio, } \frac{F_{MN}}{F_{NO}} = \frac{\frac{1}{2}(\rho_1 g h A)}{2\rho_1 g h A} = \frac{1}{4}$$

10. *(a)* In first situation, $V_b\rho_b g = V_s\rho_w g$

$$\Rightarrow V\rho_b g = \frac{4}{5}V\rho_w g$$

where, ρ_b = density of block

and ρ_w = density of water

$$\Rightarrow \quad \frac{\rho_b}{\rho_w} = \frac{4}{5} \quad \dots \text{(i)}$$

In second situation, when in equilibrium

$$\Rightarrow \quad V\rho_b g = \frac{V}{2}\rho_o g + \frac{V}{2}\rho_w g$$

where, ρ_o = density of oil.

$$\Rightarrow \quad 2\rho_b = \rho_0 + \rho_w \quad \text{[using Eq. (i)]}$$

$$\Rightarrow \quad \frac{\rho_0}{\rho_w} = \frac{3}{5} = 0.6$$

11. *(d)* Mass per unit time of a liquid flow is given by

$$\frac{dm}{dt} = \rho A v$$

where, ρ is density of liquid, A is area through which it is flowing and v is velocity.

∴Rate of change in momentum of the 25% of liquid which loses all momentum is

$$\frac{dp_1}{dt} = \frac{1}{4}\left(\frac{dm}{dt}\right)v = \frac{1}{4}\rho A v^2 \quad \dots \text{(i)}$$

and the rate of change in momentum of the 25% of the liquid which comes back with same speed.

$$\frac{dp_2}{dt} = \frac{1}{4}\left(\frac{dm}{dt}\right) \times 2v = \frac{1}{2}\rho A v^2 \quad \dots \text{(ii)}$$

[∵ Net change in velocity is = $2v$]

∴ Net pressure on the mesh is

$$p = \frac{F_{net}}{A} = \frac{(dp_1/dt + dp_2/dt)}{A} \quad \dots \text{(iii)}$$

$$\left[\because F = \frac{dp}{dt}\right]$$

∴From Eqs. (i) and (ii), we get

$$p = \frac{3}{4}\rho v^2 A / A = \frac{3}{4}\rho v^2$$

12. *(d)*

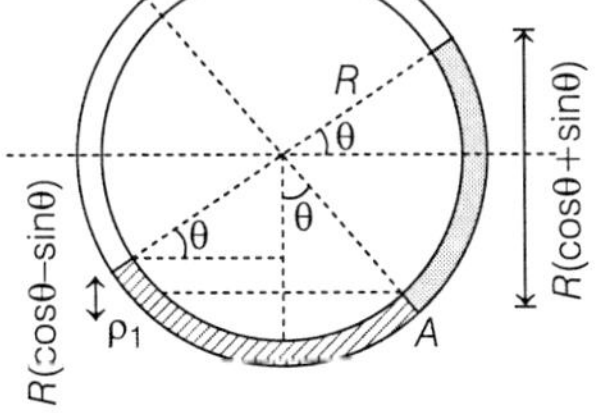

Balancing pressure at point A

$$\rho_1 gR(\cos\theta - \sin\theta) = \rho_2 gR(\sin\theta + \cos\theta)$$

$$\frac{\rho_1}{\rho_2} = \frac{\sin\theta + \cos\theta}{\cos\theta - \sin\theta} = \frac{\tan\theta + 1}{1 - \tan\theta}$$

$$\rho_1 - \rho_1 \tan\theta = \rho_2 + \rho_2$$

$$\tan\theta(\rho_1 + \rho_2) = \rho_1 - \rho_2$$

$$\theta = \tan^{-1}\left(\frac{\rho_1 - \rho_2}{\rho_1 + \rho_2}\right)$$

None of the given options is correct.

13. *(c)* Equating pressure at A, we get

$$R\sin\alpha\, d_2 + R\cos\alpha\, d_2 + R(1-\cos\alpha)\, d_1 = R(1-\sin\alpha)\, d_1$$

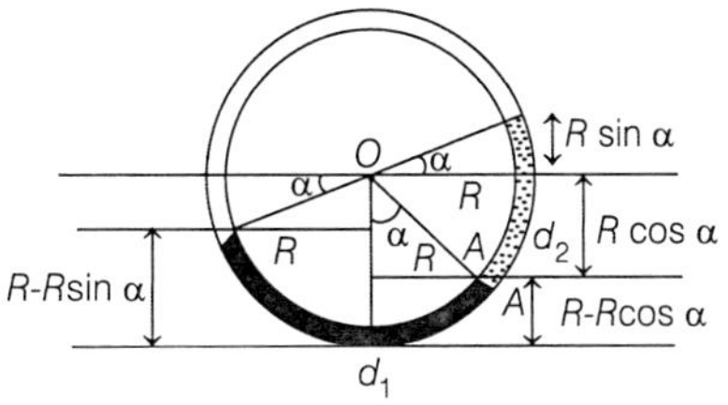

$$(\sin\alpha + \cos\alpha)\, d_2 = d_1(\cos\alpha - \sin\alpha)$$

$$\Rightarrow \quad \frac{d_1}{d_2} = \frac{1+\tan\alpha}{1-\tan\alpha}$$

14. *(c)* The various forces acting on the cylinder are shown in the figure.

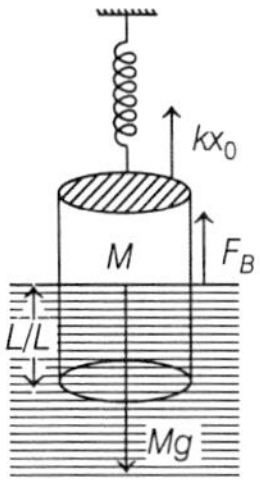

Let x_0 be the extension in spring. At equilibrium,

Upward forces = Downward forces

$$kx_0 + F_B = mg$$

where, k is spring constant, F_B is force due to buoyancy which is equal to weight of liquid displaced.

$$F_B = \sigma \cdot \frac{L}{2} \cdot A \cdot g$$

$$\therefore \; kx_0 + \sigma\frac{L}{2}Ag = Mg$$

$$\Rightarrow x_0 = \frac{Mg - \dfrac{\sigma L A g}{2}}{k} = \frac{Mg}{k}\left(1 - \frac{\sigma L A}{2M}\right)$$

15. *(c)* As the cross-sectional areas are equal, water will redistribute evenly between two cylinders.

$$\text{Average height of water} = \frac{10+6}{2} = 8 \text{ m}$$

Gravitatioinal potential energy of a water column at height h is

$$U = \frac{\rho g A h^2}{2}$$

∴Initial potential energy

$$= \frac{1}{2}\rho g A[h_1^2 + h_2^2]$$

$$U_i = \frac{1}{2} \times 1000 \times 10 \times 2[(10)^2 + (6)^2]$$

$$= 10^4[136] = 1.36 \times 10^6 \text{ J}$$

Final potential energy

$$= \frac{1}{2}\rho g A[(h_1')^2 + (h_2')^2]$$

$$U_f = \frac{1}{2} \times 1000 \times 10 \times 2[(8)^2 + (8)^2]$$

$$U_f = 10^4[128] = 1.28 \times 10^6 \text{ J}$$

Work done = Change in potential energy

$$W = U_i - U_f = (1.36 \times 10^6) - (1.28 \times 10^6)$$

$$= 0.08 \times 10^6 \text{ J} = 8 \times 10^4 \text{ J}$$

16. *(c)* $Mg = F_B$

$$M = 400 \text{ g}, \; l = 10 \text{ cm}, \; \rho = 10^3 \text{ kg/m}^3$$

$$M = 400 \times 10^{-3} \text{ kg}$$

$$Mg = V_d \rho g \qquad \left[\rho = \frac{M}{V}, \; M = \rho V\right]$$

$$400 \times 10^{-3} = V_d \times 10^3$$

$$V_d = 400 \times 10^{-6} = 400 \times 10^{-6} \text{ m}^3$$

$$(\text{Volume})_{\text{out}} = (10 \times 10^{-2})^3 - 400 \times 10^{-6}$$

$$= 1000 \times 10^{-6} - 400 \times 10^{-6}$$

$$= 600 \times 10^{-6} = 600 \text{ cm}^3$$

Time Saver Tech

Volume of cube outside the water is given by

$$V_{\text{outside}} = h^3\left[1 - \frac{\rho_b}{\rho_f}\right]$$

where h is edge length

According to question

$$\rho_b = \frac{\text{mass}}{h^3} = \frac{400 \times 10^{-3}}{(10 \times 10^{-2})^3}$$

$$= \frac{4 \times 10^{-1}}{10^{-3}} = 4 \times 10^2$$

$$V_{\text{outside}} = h^3\left[1 - \frac{\rho_b}{\rho_f}\right]$$

$$= (10 \times 10^{-2})^3\left[1 - \frac{400}{1000}\right]$$

$$= \frac{10^{-3} \times 600}{1000} = 600 \text{cm}^2$$

17. *(c)* Weight, $W = \frac{4}{3}\pi\left(\frac{D^3 - d^3}{8}\right)\sigma g$

$$\text{Buoyant force} = 1 \times \frac{4}{3}\pi\frac{D^3}{8} \cdot g$$

For just float ⇒ $W = F_1$

$$(D^3 - d^3)\sigma = D^3$$

$$1 - \frac{d^3}{D^3} = \frac{1}{\sigma} \Rightarrow \frac{-d^3}{D^3} = \frac{1}{\sigma} - 1$$

$$\frac{d^3}{D^3} = 1 - \frac{1}{\sigma} \Rightarrow \frac{d^3}{D^3} = \frac{\sigma - 1}{\sigma}$$

$$\frac{D^3}{d^3}=\frac{\sigma}{\sigma-1}$$

$$\Rightarrow \quad \frac{D}{d}=\left(\frac{\sigma}{\sigma-1}\right)^{\frac{1}{3}}$$

18. *(c)*

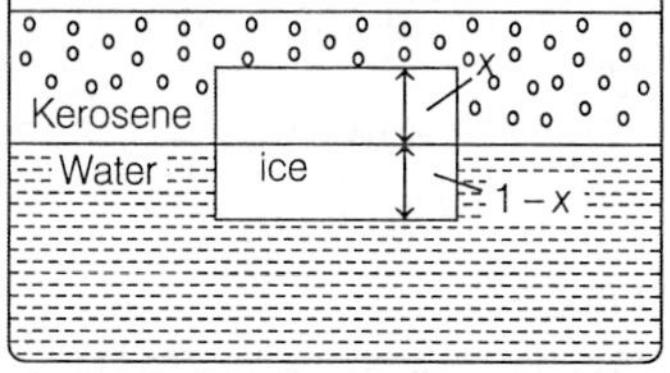

Let x part of ice is in kerosene and $1-x$ part is in water.

Then, weight of ice cube

= upthrust by kerosene

+ upthrust by water on ice

$$\Rightarrow V\cdot\rho_{\text{ice}}\cdot g = xV\cdot\rho_{\text{kerosene}}\cdot g + (1-x)V\cdot\rho_{\text{water}}\cdot g$$

$$\Rightarrow \quad \rho_{\text{ice}} = x\rho_{\text{kerosene}} + (1-x)\,\rho_{\text{water}}$$

$$\Rightarrow \quad \frac{\rho_{\text{ice}}}{\rho_{\text{water}}} = x\left(\frac{\rho_{\text{kerosene}}}{\rho_{\text{water}}}\right) + (1-x)$$

Substituting values,

$$0.9 = x0.8 + 1 - x$$

$$\Rightarrow \quad 0.2x = 0.1 \Rightarrow x = 1/2$$

So, remaining 1/2 is in water

$\therefore$ Ratio of volume immersed $=\frac{1}{2}:\frac{1}{2}$

or 1 : 1

19. *(b)* The given situation is shown below,

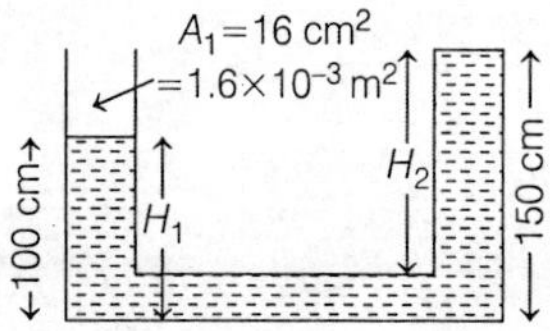

Here, $H_1 = 100$ cm $= 1$ m,

$H_2 = 150$ cm $= 1.5$ m

Density of water, $\rho = 10^3$ kg/m^3,

Total initial energy of vessels before interconnection,

$$E_i = \rho A H_1\, g\frac{H_1}{2} + \rho A H_2 g\frac{H_2}{2}$$

$$= \frac{\rho A g}{2}(H_1^2 + H_2^2) = \frac{\rho A g}{2}(1^2 + 1.5^2)$$

$$\Rightarrow E_i = \frac{3.25}{2}\rho A g \qquad \ldots\text{(i)}$$

Similarly, total final energy of the vessels after interconnecting

$$E_f = \frac{\rho A g}{2}(H_1^2 + H_2^2)$$

$$(\because H_1 = H_2 = H)$$

$$= \frac{\rho A g}{2}(2H^2)$$

$$\left(H = \frac{H_1 + H_2}{2} = \frac{1 + 1.5}{2} = 1.25\text{m}\right)$$

$$= \frac{\rho A g 2(1.25)^2}{2} = \frac{3.125\,\rho A g}{2}$$

Work done by the force of gravity,

$$W = E_i - E_f = \frac{\rho A g}{2}(3.25 - 3.125)$$

$$= \frac{10^3 \times 1.6 \times 10^{-3} \times 10 \times 0.125}{2} = 1\text{ J}$$

20. *(a)* $V \neq$ constant as moles are changing

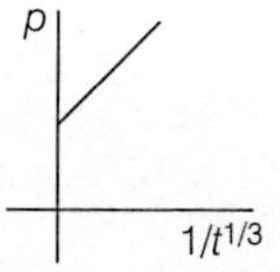

$$V = ct$$

$$\frac{4}{3}\pi r^3 = ct$$

$$r = kt^{1/3}$$

$$p = p_0 + \frac{4T}{kt^{1/3}}$$

21. *(d)*

> **key idea** When liquid filled vessel is rotated, the liquid profile becomes a paraboloid due to centripetal force.

The given situation is shown in the figure.

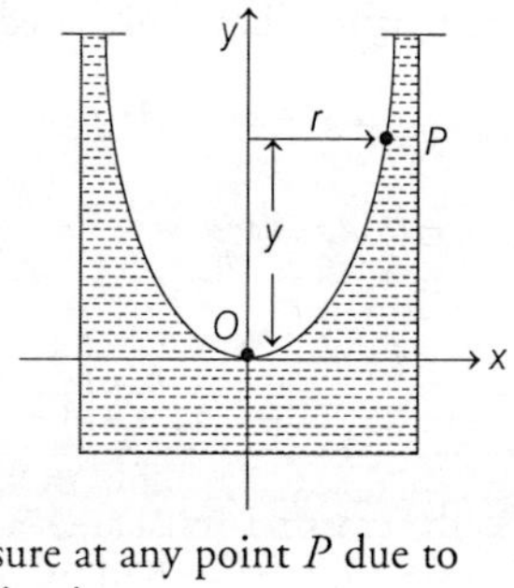

Pressure at any point P due to rotation is

For any point on surface of rotating fluid,

$$p = p_0$$

Hence, for any surface point,

$$\frac{1}{2}\rho r^2\omega^2 = \rho g y \text{ or } y = \frac{r^2\omega^2}{2g} \qquad \ldots\text{(i)}$$

In the given case,

Angular speed, $\omega = 2$rps

$= 2 \times 2\pi = 4\pi$ rad s^{-1}

Radius of vessel, $r = 5$ cm $= 0.05$ m

and $\qquad g = 10$ ms^{-2}

Hence, substituting these values in Eq. (i), we get

$$y = \frac{\omega^2 r^2}{2g} = \frac{(4\pi)^2(0.05)^2}{2\times 10}$$

$= 0.02$ m $= 2$ cm

22. *(5)* $F_2 = \frac{F_1 A_2}{A_1} = \frac{100 \times 1500}{6}$

$$= \frac{15}{6}\times 10^4 \text{ N}$$

Work done = Force × Displacement

$$= \frac{15}{6}\times 10^4 \times 0.2$$

$$= \frac{3}{6}\times 10^4 = 5000\text{ J} = 5\text{ kJ}$$

23. *(3)*

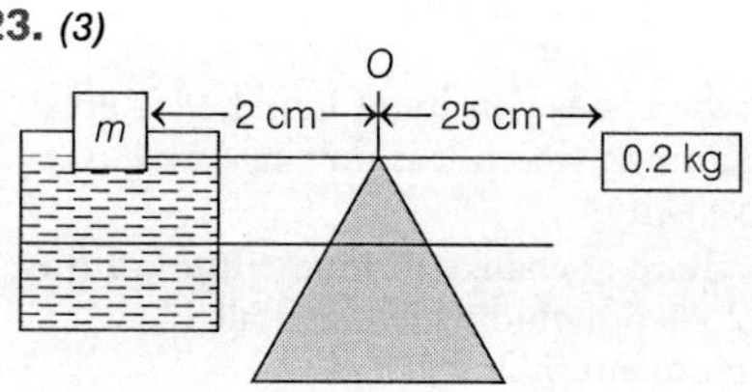

Given, volume of block $= (10 \times 10^{-2})^3$

$= 10^{-3}$ m^3

Let density of block $= \rho$ kg/m^3

mass of block $= \rho \times 10^{-3}$ kg

Buoyant Force $(F_B) = 1000 \times \frac{10^{-3}}{2} \times 10 = 5$ N

FBD of blocks

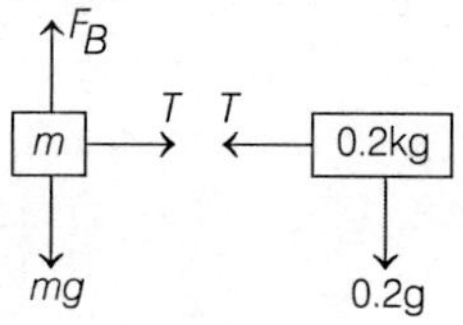

Balancing torque about point O, we get

$$mg(2\times 10^{-2}) - F_B(2\times 10^{-2}) = 0.2\,g\,(25 \times 10^{-2})$$

$$\rho \times 10^{-3} \times 10 \times 2 - 10 = 50$$

$$\rho = 3000\text{ kg/m}^3$$

Hence, mass of block $= \rho \times 10^{-3}$

$= 3000 \times 10^{-3} = 3$ kg

24. *(150)*

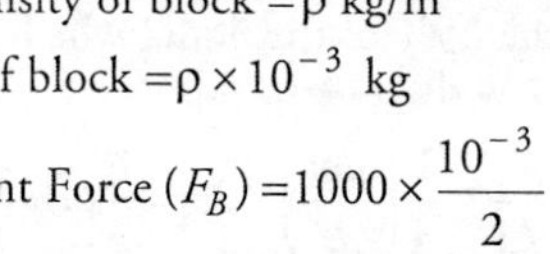

In equilibrium,

$$F_{\text{ext}} + F_w = F_l$$

$$F_{\text{ext}} = F_l - F_w$$

$$= (p_0 + \rho_l gh)A - (p_0 + \rho_w gh)A$$

$$= (\rho_l - \rho_w)\,ghA$$

$$= (1500 - 1000) \times 10 \times 3 \times 100 \times 10^{-4}$$

$$= 150\text{ N}$$

25. *(1000)* If F_1 be the force required to applied on the thicker arm, then by using Pascal's law

$$\frac{F_1}{A_1}=\frac{F_2}{A_2}$$

$$\Rightarrow F_2=\frac{F_1A_2}{A_1}=\frac{10\times\pi r_2^2}{\pi r_1^2}$$

$$=10\cdot\left(\frac{r_2}{r_1}\right)^2=10\times\left(\frac{7}{0.7}\right)^2$$

$$=1000\text{ N}$$

26. *(177)* Atmospheric pressure,

$$p_0=10^5\text{ N/m}^2$$

The pressure at the bottom of the tube

$$p=p_0+\rho gh$$
$$=10^5+1.36\times10^4\times10\times0.3$$
$$[\because h=30\text{ cm}=0.3\text{ m}]$$
$$=1.408\times10^5\text{ N/m}^2$$

Area of cross-section of tube,

$$A=\pi r^2$$
$$=\frac{22}{7}\times(2\times10^{-2})^2$$
$$=\frac{22}{7}\times4\times10^{-4}\text{ m}^2$$
$$=\frac{88}{7}\times10^{-4}\text{ m}^2$$

$\therefore$ Force exerted by the mercury on the bottom of the tube,

$$F=pA$$
$$=1.408\times10^5\times\frac{88}{7}\times10^{-4}$$
$$=177.001\simeq177\text{ N}$$

27. *(25)* Given, $m=0.3$ g,
density, $\rho=8$ g/cc, $\sigma=1.3$ g/cc

$$g=10\text{ m/s}^2$$

Thus, Buoyant force,

$$F=mg-f$$

where f is viscous force

$$f=mg-\text{Buoyant force},$$
$$f=mg-V\sigma g\Rightarrow f=mg-\frac{m}{\rho}\sigma g$$
$$f=0.3\times10^{-3}\times10\left[1-\frac{1.3}{8}\right]$$

Thus, $f=25.125\times10^{-4}$ N, $x=25$

28. *(25600)* According to Pascal's law,

$$\frac{F_1}{A_1}=\frac{F_2}{A_2}$$

Initially, $$\frac{100g}{A_1}=\frac{mg}{A_2}\qquad\ldots\text{(i)}$$

Finally $$\frac{Mg}{16A_1}=\frac{mg}{\left(\frac{A_2}{16}\right)}\qquad\ldots\text{(ii)}$$

On dividing Eqs. (i) by (ii), we get

$$\frac{100\times16}{M}=\frac{1}{16}$$

$$\therefore\qquad M=25600\text{ kg}$$

Topic 2

01. *(b)* Terminal velocity

$$v=\frac{2(\rho_s-\rho_0)r^2g}{9\eta}$$
$$\Rightarrow\eta=\frac{2(\rho_s-\rho_0)r^2g}{9v}$$
$$=\frac{2(7825-925)(1.8\times10^{-3})^2\times9.8}{9\times2.45\times10^{-2}}$$
$$\simeq1.99$$

02. *(b)*

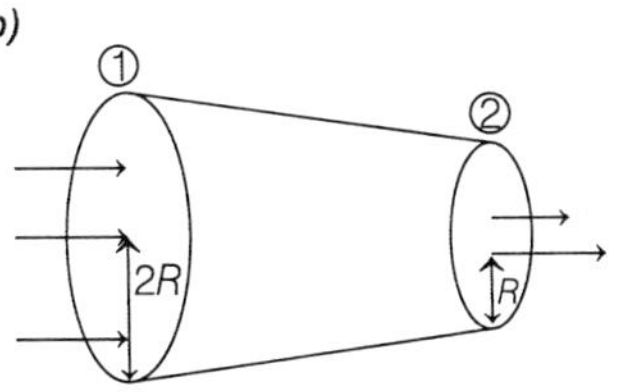

$$A_1v_1=A_2v_2\Rightarrow2\pi(2R)^2=v_2\pi R^2$$
$$8\pi R^2=v_2\pi R^2\Rightarrow v_2=8\text{ m/s}$$

03. *(c)*

> **key idea** Viscosity of fluid mainly depends upon the temperature.
>
> For a liquid, viscosity decreases with increase of temperature.
>
> For a gas, viscosity increases with increase of temperature.

We know,

$$v_T=\frac{2}{9}R^2\frac{g}{\eta}(d-\rho)$$

Statements A, C and D are correct.

04. *(d)* By Bernoulli equation

$$p_1+\frac{1}{2}\rho\times0^2=p_2+\frac{1}{2}\rho v^2$$
$$v=\sqrt{\frac{2(p_1-p_2)}{\rho}}$$
$$v\propto\sqrt{(p_1-p_2)}$$

05. *(a)*

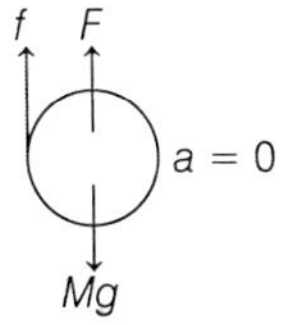

$$Mg-F_B=f\Rightarrow Mg-\frac{Mg}{2}=f$$
$$f=\frac{2Mg-Mg}{2}$$
$$f=\frac{Mg}{2}$$

06. *(c)* By Bernoulli's theorem, total energy per unit volume of a flowing fluid is constant throughout the flow.

$$\Rightarrow p+\rho gh+\frac{1}{2}\rho v^2=\text{Constant}$$

p = pressure energy per unit volume.

ρgh = Potential energy per unit volume.

$\frac{1}{2}\rho v^2$ = Kinetic energy per unit volume.

> **Concept Enhancer**
>
> - Bernoulli's principle is a fundamental principle of fluid dynamics based on the law of conservation of energy.
> - When fluid moves at a constant depth, then Bernoulli's equation is given is
>
> $$p+\frac{1}{2}\rho v^2=\text{constant}$$
>
> Thus, for the streamline flow of an ideal liquid flowing horizontally, the pressure decreases when velocity increases and *vice-versa.*

07. *(b)* Of all the fluids, gases have the lowest viscosities. Highly soluble impurities increases the value of surface tension of liquid whereas insoluble impurities decreases the surface tension of liquid.

08. *(b)* Given, $v_s=10$ cm/s, $R=n^{1/3}r$

$$v_T\propto(\text{Radius})^2$$
$$\frac{v_f}{v_s}=\frac{R^2}{r^2}\qquad[\therefore R=8^{1/3}r]$$
$$\Rightarrow\frac{v_f}{v_s}=\frac{(8^{1/3}r)^2}{r^2}\Rightarrow\frac{v_f}{v_s}=\frac{8^{2/3}r^2}{r^2}$$
$$\Rightarrow\frac{v_f}{v_s}=[(8)^{1/3}]^2\Rightarrow\frac{v_f}{v_s}=(2)^2$$
$$\Rightarrow v_f=4v_s$$
$$\Rightarrow v_f=4\times10=40\text{ cm/s}$$

09. *(d)* By Pascal's law, pressure at larger limb = pressure at smaller limb.

So,

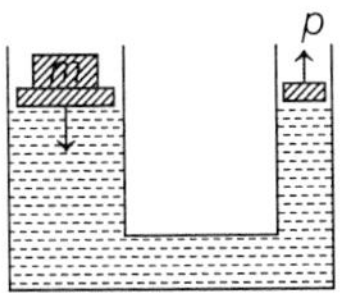

$$p=\frac{Mg}{A}=\frac{5000\times10}{250\times10^{-4}}$$
$$=2\times10^6\frac{\text{N}}{\text{m}^2}$$
$$=2\times10^6\text{ Pa}$$

10. *(c)* At terminal velocity, net force on the ball is zero.

$$Mg = F + B$$

$$Mg = F + V_{\text{ball}}\rho_0 g \quad \text{...(i)}$$

Volume of ball $= M/\rho$

From Eq. (i),

$$Mg = F + \frac{M}{\rho}\cdot\rho_0 g$$

$$\Rightarrow F = Mg - \frac{M}{\rho}\rho_0 g$$

$$\Rightarrow F = Mg\left(1 - \frac{\rho_0}{\rho}\right)$$

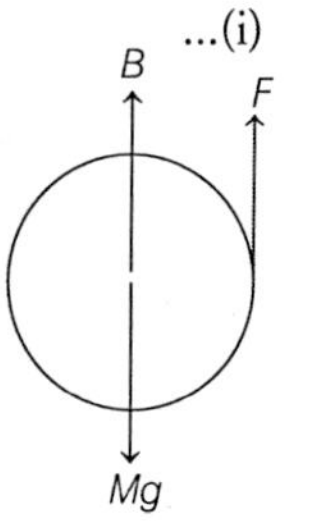

11. *(d)* Let v be the terminal velocity, then force in equilibrium is

$$6\pi\eta rv = mg$$

$$\Rightarrow 6\pi\eta rv = \frac{4}{3}\pi r^3\cdot\rho\cdot g$$

or terminal velocity, $v = \dfrac{2}{9}\dfrac{\rho r^2 g}{\eta}$

Substituting, $r = 1\times10^{-6}$ m,

$\rho = 10^3$ kg m^{-3}

$g = 10$ ms^{-2}, $\eta = 1.8\times10^{-5}$ Ns m^{-2}

$$\Rightarrow v = \frac{2}{9}\times\frac{10^3\times(10^{-6})^2\times10}{1.8\times10^{-5}}$$

$$= 123.4\times10^{-6}\text{ m/s}$$

12. *(b)* Let F_B be buoyant force and F_v be viscous force. When terminal velocity is achieved, equation of motion is written as

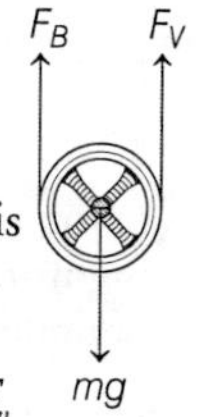

$$F_{\text{net}} = 0$$

$$\Rightarrow \Sigma F_{\text{down}} = \Sigma F_{\text{up}} \Rightarrow mg = F_B + F_v$$

$$\Rightarrow mg = \left(\frac{m}{d_1}\right)d_2 g + F_v$$

$$\Rightarrow F_v = mg\left(1 - \frac{d_2}{d_1}\right)$$

13. *(c)* We know that, terminal velocity of spherical rain drop is given as

$$v_t = \frac{2gr^2(\rho-\sigma)}{9\eta} \Rightarrow v_t \propto r^2$$

where, ρ = density of rain drop,

σ = density of air

and η = coefficient of viscosity of air.

14. *(c)*

> **key idea** When spherical rain drop falls with terminal velocity v_T, then
>
> Viscous force = Weight of rain drop

Given, $R = 0.2$ mm $= 2\times10^{-4}$ m,

$h = 2000$ m, $\rho_w = 1000$ kgm^{-3},

$\rho_a = 1.2$ kgm^{-3}

and $\eta = 1.8\times10^{-5}$ N-sm^{-2}

According to the question,

$$6\pi\eta Rv_T = mg$$

$$\Rightarrow 6\pi\eta Rv_T = \left(\frac{4}{3}\pi R^3\right)\rho_w\times g$$

$$\Rightarrow v_T = \frac{2R^2\rho_w g}{9\times\eta}$$

$$= \frac{2\times(2\times10^{-4})^2\times1000\times10}{9\times1.8\times10^{-5}}$$

$$= 4.94\text{ ms}^{-1}$$

15. *(c)* $\because$ Force exerted by leaving water on the container, $F = \rho av^2$

Also, by conservation of energy,

$$v = \sqrt{2gh} \Rightarrow F = \rho a(\sqrt{2gh})^2$$

$$\Rightarrow F = \rho a(2gh)$$

$\because$ For no sliding of vessel,

$$\mu mg \geq F$$

$$\Rightarrow \mu mg \geq \rho a(2gh)$$

$$\Rightarrow \mu\rho(Ah)g \geq \rho a\, 2gh$$

$$\Rightarrow \mu_{\min} = \frac{2a}{A}$$

16. *(a)* From equation of continuity, we have

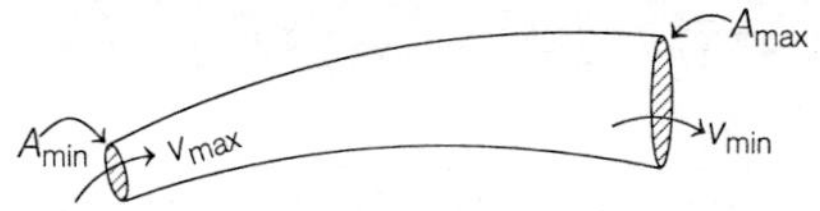

$$\frac{\pi}{4}d^2_{\min}\cdot v_{\max} = \frac{\pi}{4}d^2_{\max}\cdot v_{\min}$$

$$\Rightarrow \frac{v_{\min}}{v_{\max}} = \left(\frac{d_{\min}}{d_{\max}}\right)^2$$

Here, $d_{\max} = 6.4$ cm

and $d_{\min} = 4.8$ cm

So, $$\frac{v_{\min}}{v_{\max}} = \left(\frac{4.8}{6.4}\right)^2 = \left(\frac{3}{4}\right)^2 = \frac{9}{16}$$

> **• Concept Enhancer**
>
> According to equation of continuity, we have
>
> $A_{\min}v_{\max} = A_{\max}v_{\min}$

17. *(a)*

> **key idea** Terminal speed of a sphere falling in a viscous fluid,
>
> $$v_T = \frac{2}{9}\frac{r^2}{\eta}(\rho_0 - \rho_f)g$$
>
> where, η = coefficient of viscosity of fluid,
>
> ρ_0 = density of falling sphere and
>
> ρ_f = density of fluid.

As we know, if other parameters remains constant, terminal velocity is proportional to square of radius of falling sphere.

i.e. $$v_T \propto r^2 \quad \text{...(i)}$$

Now, when sphere of radius R is broken into 27 identical solid sphere of radius r, then

Volume of sphere of radius $R = 27\times$ Volume of sphere of radius r

$$\Rightarrow \frac{4}{3}\pi R^3 = 27\times\frac{4}{3}\pi r^3$$

$$\Rightarrow R = 3r$$

$$\Rightarrow r = \frac{R}{3}$$

So, from Eq. (i), we have

$$\frac{v_1}{v_2} = \frac{R^2}{\left(\frac{R}{3}\right)^2} = 9$$

18. *(c)* As, level of water in tank remains constant with time, so (water inflow rate) = (outflow rate of water)

$\Rightarrow 10^{-4}$ m^{3}s^{-1} = Area of orifice × Velocity of outflow

$\Rightarrow 10^{-4}$ m^{3}s^{-1} $= 10^{-4}\times\sqrt{2gh}$

where, h = Height of water above the orifice or hole.

$$\Rightarrow \sqrt{2gh} = 1 \text{ or } 2\times9.8\times h = 1$$

$$\Rightarrow h = \frac{1}{19.6}\text{ m} = \frac{100}{19.6}\text{ cm}$$

or $h = 5.1$ cm

19 *(b)* As terminal velocity,

$$u = \frac{2}{9}r^2\frac{(\rho-\rho_0)}{\eta} \Rightarrow v \propto \frac{\rho-\rho_0}{\eta}$$

$$\therefore \frac{v_2}{v_1} = \frac{\rho-\rho_{02}}{\rho-\rho_{01}}\times\frac{\eta_1}{\eta_2}$$

$$= \frac{7.8-1.2}{7.8-1}\times\frac{8.5\times10^{-4}\times10}{13.2}$$

$$= 6.25\times10^{-4}\text{ cms}^{-1}$$

20. *(a)* The forces acting on the ball are gravity force, buoyancy force and viscous force. When ball acquires terminal speed, it is in dynamic equilibrium, let terminal speed of ball be v_T.

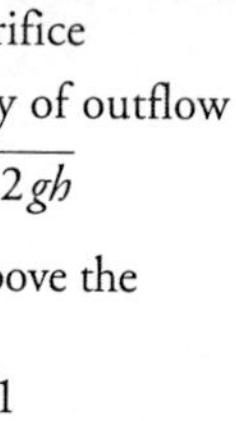

So, $V\rho_2 g + kv_T^2 = V\rho_1 g$

or $$v_T = \sqrt{\frac{V(\rho_1-\rho_2)g}{k}}$$

21. *(b)* The given situation is shown below.

Using Bernoulli's principle,

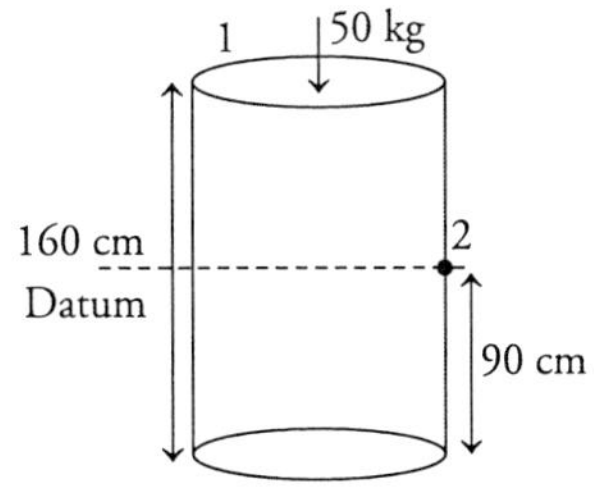

$$p_1 + \frac{1}{2}\rho v_1^2 + \rho g h = p_2 + \frac{1}{2}\rho v_2^2 + 0$$

$$\Rightarrow \quad p_0 + \frac{mg}{A} + \rho g \times \frac{70}{100} = p_0 + \frac{1}{2}\rho v_2^2$$

$$\Rightarrow \quad \frac{mg}{A} + \rho g \times \frac{70}{100} = \frac{1}{2}\rho v_2^2$$

$$\Rightarrow \quad \frac{50 \times 10}{0.5} + 10^3 \times 10 \times \frac{70}{100} = \frac{1}{2} \times 10^3 \times v_2^2$$

$$\Rightarrow \quad v_2^2 = 16$$

$$\Rightarrow \quad v_2 = 4 \text{ m/s}$$

22. *(b)* For the given situation,

$mg - F_{\text{upthrust}} - F_{\text{viscous}} = ma$

For constant velocity, $a = 0$

$$\therefore \quad mg = F_{\text{upthrust}} + F_{\text{viscous}}$$

$$F_{\text{viscous}} = mg - F_{\text{upthrust}}$$

$$= mg - \frac{m}{\rho} \cdot \rho_0 g$$

$$= mg\left(1 - \frac{\rho_0}{\rho}\right)$$

23. *(b)* Initially velocity of ball increases and that results in increase of opposing viscous force.

$$F_{\text{viscous}} = 6\pi\eta r v$$

When viscous force and buoyant force are equals to gravitational force, net force on ball is zero.

After that ball falls with constant terminal speed.

So, correct graph is

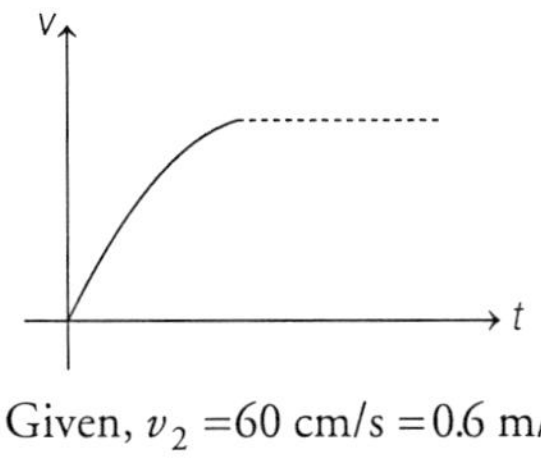

24. *(c)* Given, $v_2 = 60 \text{ cm/s} = 0.6 \text{ m/s}$

$$A_1 = 1.5 \text{ cm}^2 = 1.5 \times 10^{-4} \text{ m}^{-2}$$

$$A_2 = 25 \text{ mm}^2 = 25 \times 10^{-6} \text{ m}^2$$

According to equation of continuity,

$$A_1 v_1 = A_2 v_2$$

$$\Rightarrow \quad v_1 = \frac{A_2 v_2}{A_1} = \frac{25 \times 10^{-6} \times 0.6}{1.5 \times 10^{-4}}$$

$$= 0.1 \text{ m/s}$$

$\therefore$ According to Bernoulli's principle,

$$p_A - p_B = \frac{1}{2}\rho(v_2^2 - v_1^2)$$

$$= \frac{1}{2} \times 1000[(0.6)^2 - (0.1)^2]$$

$$= 500[0.35] = 175.00$$

$$= 175 \text{ Pa}$$

25. *(b)* Given, $h = 4.9$ m

Throughout the motion, the velocity is v.

Velocity after falling through a distance (height) of 4.9 m can be calculated as,

$$v^2 - u^2 = 2as$$

$$v^2 - 0^2 = 2 \times 9.8 \times 4.9$$

$$v = \sqrt{(9.8)^2}$$

$$v = 9.8 \text{ m/s} \quad \ldots\text{(i)}$$

Time taken to fall through 4.9 m will be calculated as, $v = u + gt$

$$9.8 = 0 + 9.8(t) \ \therefore \ t = 1 \text{ s}$$

Remaining 3 s, it went inside water with constant speed 9.8 m/s.

$\therefore$ Depth of water = Speed × Time

$$= v \times t = 9.8 \times 3 = 29.4 \text{ m}$$

26. *(b)* Given, $r = 2.0 \times 10^{-5}$ m

$$\rho = 1.2 \times 10^3 \text{ kgm}^{-3}$$

$$\eta = 1.8 \times 10^{-5} \text{ Nsm}^{-2}$$

To neglect buoyancy, consider the density of air, $\rho_a = 0$

The terminal velocity of drop can be given as

$$v = \frac{2r^2(\rho - \rho_a)g}{9\eta}$$

$$\Rightarrow v = \frac{2 \times (2.0 \times 10^{-5})^2 (1.2 \times 10^3 - 0) \times 9.8}{9 \times 1.8 \times 10^{-5}}$$

$$= 5.807 \times 10^{-2} \text{ ms}^{-1}$$

Viscous force, $F = 6\pi\eta r v$ ...(i)

Here, v is terminal velocity.

Putting values

$$F = 6\pi \times 1.8 \times 10^{-5} \times 2.0 \times 10^{-5} \times 5.807 \times 10^{-2}$$

$$= 3.94 \times 10^{-10} \text{ N} \simeq 3.9 \times 10^{-10} \text{ N}$$

27 *(b)* Rate of flow of liquid through the tube, $\frac{dv}{dt} = \frac{\pi p r^4}{8\eta l}$

Since both the given tubes are connected in series, so rate of flow of liquid is same.

$$\therefore \quad \frac{\pi p_1 r_1^4}{8\eta l_1} = \frac{\pi p_2 r_2^4}{8\eta l_2}$$

$$\Rightarrow \quad r_2^4 = \left(\frac{p_1}{p_2}\right)\left(\frac{l_2}{l_1}\right) r_1^4$$

Here, $p_2 = 4p_1, l_2 = l_1/4$

$$\text{So,} \quad r_2^4 = \left(\frac{p_1}{4p_1}\right)\left(\frac{l_1}{4l_1}\right) r_1^4$$

$$\Rightarrow \quad r_2^4 = \frac{r_1^4}{16} \quad \Rightarrow \quad r_2 = \frac{r_1}{2}$$

28. *(a)* Let v_1 and v_2 be the velocities of water when it leaks out through the hole and when it hits the ground respectively.

Then, as per Bernoulli's principle,

$$v_1^2 + 2gh = v_2^2$$

Now, according to Torricelli's law

$$v_1 = \sqrt{2gH} \quad \ldots\text{(i)}$$

$$\therefore \quad 2gH + 2gh = v_2^2 \quad \ldots\text{(ii)}$$

According to continuity equation,

$$a_1 v_1 = a_2 v_2$$

$$\text{or } \pi r^2 \cdot \sqrt{2gH} = \pi x^2 \cdot \sqrt{2g(H+h)}$$

[Using Eqs. (i) and (ii)]

$$x^2 = r^2 \sqrt{\frac{H}{H+h}} \text{ or } x = r\left(\frac{H}{H+h}\right)^{1/4}$$

29. *(a)*

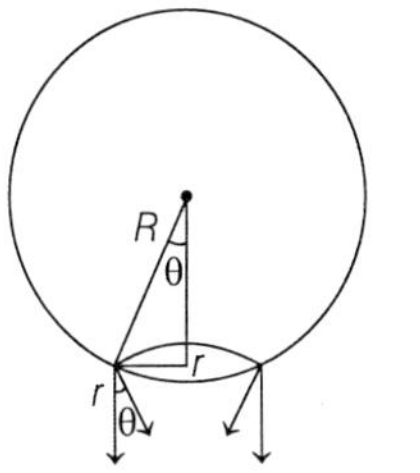

$T \times dl = T(2\pi r)\sin\theta$

$$\text{i.e.,} \ \frac{4}{3}\pi R^3 \rho_w g \geq \int T \times dl \sin\theta$$

$$(\rho_w)\left(\frac{4}{3}\pi R^3\right) g \geq (T)(2\pi r)\sin\theta$$

$$\sin\theta = \frac{r}{R}$$

$$\text{Solving,} \quad r = \sqrt{\frac{2\rho_w R^4 g}{3T}} = R^2\sqrt{\frac{2\rho_w g}{3T}}$$

30. *(c)* From Bernoulli's theorem,

$$\rho g h = \frac{1}{2}\rho(v_2^2 - v_1^2)$$

$$\Rightarrow \quad gh = \frac{1}{2}v_1^2\left[\left(\frac{v_2}{v_1}\right)^2 - 1\right]$$

$$\Rightarrow \quad gh = \frac{1}{2}v_1^2\left[\left(\frac{A_1}{A_2}\right)^2 - 1\right]$$

$$[\because A_1v_1 = A_2v_2]$$

$$\Rightarrow \left(\frac{A_1}{A_2}\right)^2 = 1 + \frac{2hg}{v_1^2}$$

$$\Rightarrow \left(\frac{D_1}{D_2}\right)^4 = 1 + \frac{2gh}{v_1^2}$$

$$\Rightarrow \quad D_2 = \frac{D_1}{\left(1 + \frac{2gh}{v_1^2}\right)^{1/4}}$$

$$= \frac{8 \times 10^{-3}}{\left(1 + \frac{2 \times 10 \times 0.2}{(0.4)^2}\right)^{1/4}}$$

$$\approx 3.6 \times 10^{-3} \text{ m}$$

31. *(b)* Given, radius of the ball,

$$r = 10^{-4} \text{ m}$$

Density of ball, $\rho = 10^5 \text{ kg/m}^3$

Density of water, $\sigma = 10^3 \text{ kg/m}^3$

Coefficient of viscosity of water,

$$\eta = 9.8 \times 10^{-6} \text{ Ns/m}^2$$

According to question,

Velocity at surface = Terminal velocity

$$\sqrt{2gh} = \frac{2}{9}r^2 g\left(\frac{\rho - \sigma}{\eta}\right)$$

$$= \frac{2}{9} \times (10^{-4})^2 \times 10 \times \left(\frac{10^5 - 10^3}{9.8 \times 10^{-6}}\right)$$

$$= \frac{2}{9} \times 10 \times 10^{-8} \times 10 \times \frac{99000}{9.8 \times 10^{-6}}$$

$$\Rightarrow \sqrt{2gh} = 224.489$$

$$\Rightarrow h = \frac{224.48 \times 224.48}{2 \times 10} \simeq 2519 \text{ m}$$

32. *(b)* Terminal speed of a ball of radius r is given by,

$$v = \frac{2}{9}\frac{r^2(\rho - \sigma)g}{\eta}$$

ρ = density of ball

σ = density of fluid

η = viscosity of fluid

If fluid has negligible density then,

$$v = \frac{2}{9}\frac{r^2 \rho g}{\eta}$$

Another ball has same mass as that of first and has radius $2r$.

$$\therefore \frac{4}{3}\pi r^3 \cdot \rho = \frac{4}{3}\pi(2r)^3 \cdot \rho'$$

$\Rightarrow$ For second ball, density is, $\rho' = \frac{\rho}{8}$

Terminal speed of second ball is,

$$v' = \frac{2}{9}\frac{(r')^2 \cdot \rho' g}{\eta} = \frac{2}{9}\frac{(2r)^2\left(\frac{\rho}{8}\right)g}{\eta}$$

$$= \frac{1}{2} \cdot \frac{2}{9}\left(\frac{r^2 \rho g}{\eta}\right) = \frac{1}{2} \cdot v$$

33. *(a)* For balancing the weight of aircrafts, $Mg = (p_1 - p_2)A$

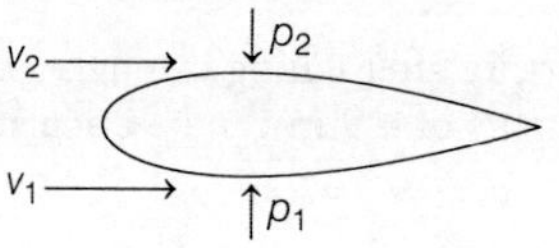

$$\Rightarrow p_1 - p_2 = \frac{Mg}{A} = \frac{5 \cdot 4 \times 10^5 \times 10}{500}$$

$$= 10 \cdot 8 \times 10^3 \text{ units}$$

Now, by Bernoulli's theorem,

$$p_1 + \frac{1}{2}\rho v_1^2 = p_2 + \frac{1}{2}\rho v_2^2$$

(Assuming height of plane small)

$$\Rightarrow p_1 - p_2 = \frac{1}{2}\rho(v_2^2 - v_1^2)$$

$$\Rightarrow \quad (v_2 - v_1)\left(\frac{v_2 + v_1}{2}\right) = \frac{p_1 - p_2}{\rho}$$

Here, $\frac{v_2 + v_1}{2} = 1080 \frac{\text{km}}{\text{h}} = 300 \frac{\text{m}}{\text{s}}$ and

$$\rho = 1 \cdot 2 \text{ kg/m}^3$$

$$\text{So, } v_2 - v_1 = \frac{p_1 - p_2}{\rho\left(\frac{v_1 + v_2}{2}\right)}$$

$$= \frac{10.8 \times 10^3}{1.2 \times 300} = 30$$

Also, $v_1 + v_2 = 600$

So, $v_2 = 315$ m/s and $v_1 = 285$ m/s

Percentage change in velocity

$$= \frac{v_2 - v_1}{v_1} \times 100$$

$$= \frac{315 - 285}{285} \times 100$$

$$= 10.52\% \simeq 10\%$$

34. *(c)* Given, radius of tank, $r = 1$ m

Force on top surface,

$$F = 5 \times 10^5 \text{ N}$$

$$p_A = 1.01 \times 10^5 \text{ Pa}$$

$$\rho_w = 1000 \text{ kg m}^{-3}$$

$$g = 10 \text{ ms}^{-2}$$

The height of hole from top surface, $h = 15 - 5 = 10$ m

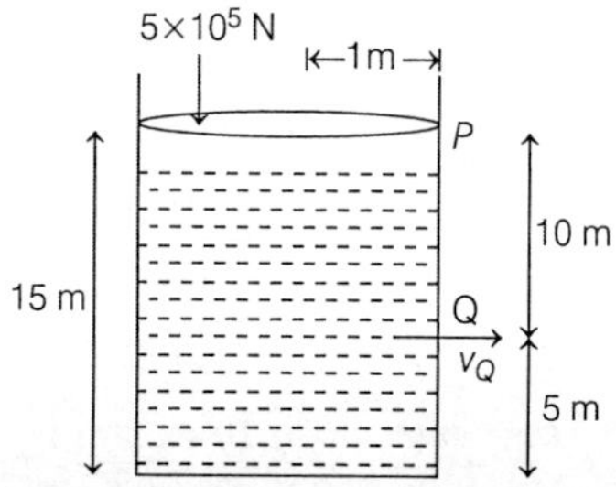

Using Bernoulli's theorem, at points P and Q,

$$p_1 + \frac{1}{2}\rho v_P^2 + \rho g h_1 = p_2 + \frac{1}{2}\rho v_Q^2 + \rho g h_2$$

$$\Rightarrow \frac{F}{A_A} + 0 + \rho g(10) = p_A + \frac{1}{2}\rho v_Q^2 + 0$$

$$\Rightarrow \quad \frac{5 \times 10^5}{\pi(1)^2} + 1000 \times 10 \times 10$$

$$= 1.01 \times 10^5 + \frac{1}{2} \times 1000 \times v_Q^2$$

$$\Rightarrow v_Q^2 = (1.592 \times 10^5 + 10^5 - 1.01 \times 10^5) \times \frac{2}{1000}$$

$$= 3.164 \times 10^2$$

$$\text{or } v_Q = 17.8 \text{ ms}^{-1}$$

35. *(d)*

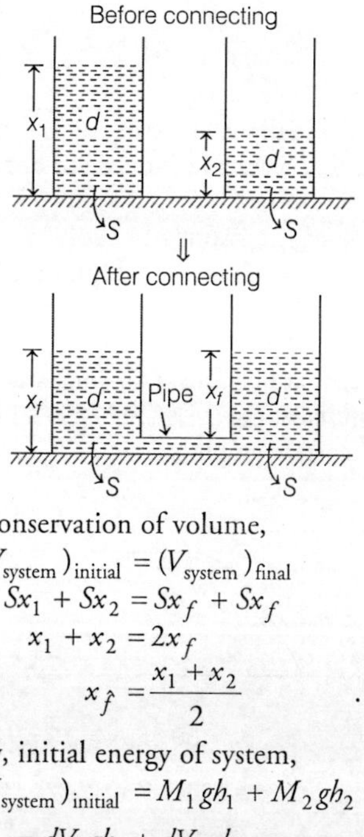

By conservation of volume,

$$(V_{\text{system}})_{\text{initial}} = (V_{\text{system}})_{\text{final}}$$

$$Sx_1 + Sx_2 = Sx_f + Sx_f$$

$$x_1 + x_2 = 2x_f$$

$$x_f = \frac{x_1 + x_2}{2} \quad \ldots(i)$$

Now, initial energy of system,

$$(U_{\text{system}})_{\text{initial}} = M_1 g h_1 + M_2 g h_2$$

$$= dV_1 g h_1 + dV_2 g h_2$$

$$= dSx_1 g\left(\frac{x_1}{2}\right) + dSx_2 g\left(\frac{x_2}{2}\right)$$

$$= \frac{1}{2} dSg(x_1^2 + x_2^2) \quad \ldots\text{(ii)}$$

and final energy of system,

$$(U_{\text{system}})_{\text{final}} = M' gh' + M' gh'$$

$$= dV' gh' + dV' gh'$$

$$= dSx_f g\left(\frac{x_f}{2}\right) + dSx_f g\left(\frac{x_f}{2}\right)$$

$$= dSgx_f^2 = dSg\left(\frac{x_1 + x_2}{2}\right)^2 \quad \ldots\text{(iii)}$$

[Using Eq. (i)]

Therefore, the change in energy of system,

$$\Delta U_{\text{system}} = (U_{\text{system}})_{\text{final}} - (U_{\text{system}})_{\text{initial}}$$

$$= dSg\left(\frac{x_1 + x_2}{2}\right)^2 - \frac{1}{2} dSg\,(x_1^2 + x_2^2)$$

$$= \frac{dSg}{4}(x_1^2 + x_2^2 + 2x_1x_2 - 2x_1^2 - 2x_2^2)$$

$$= \frac{dSg}{4}(-x_1^2 - x_2^2 + 2x_1x_2)$$

$$= -\frac{1}{4} gdS(x_2 - x_1)^2$$

So, energy of the system will be decreased by $\frac{1}{4} gdS(x_2 - x_1)^2$.

Hence, option (d) is correct.

36. *(a)* Let tangent at any point $P(x, y)$ makes angle θ with X-axis as shown below.

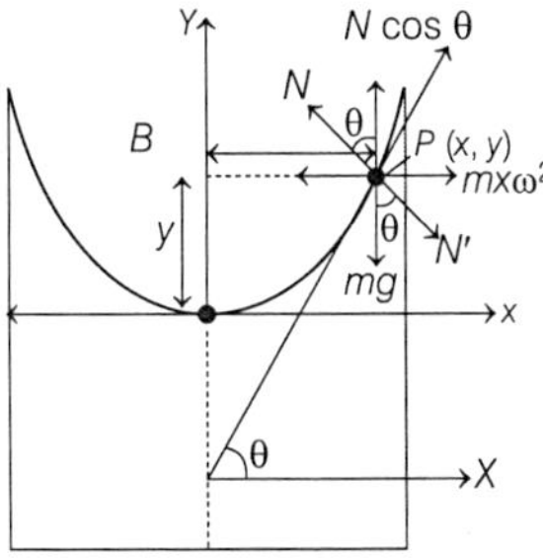

In rotating frame of vessel, let N be normal reaction. Then,

$$N \cos\theta = mg$$

and $N \sin\theta = mx\omega^2 \Rightarrow \tan\theta = \frac{\omega^2 x}{g}$

But $\tan\theta = \frac{dy}{dx}$ = slope of tangent

$$\therefore \quad \frac{dy}{dx} = \frac{\omega^2 x}{g} \Rightarrow y = \int dy = \int \frac{\omega^2 x}{g} \cdot dx$$

$$\Rightarrow \quad y = \frac{\omega^2}{g} \cdot \frac{x^2}{2}$$

At topmost point of profile of rotating fluid body,

$$y = h \text{ and } x = R \Rightarrow h = \frac{\omega^2 R^2}{2g}$$

Given, $R = 5$ cm

$$\Rightarrow \quad h = \frac{\omega^2 (5)^2}{2g} = \frac{25\,\omega^2}{2g} \text{ cm}$$

37. *(c)* By equation of continuity for sections A and B, we have

$$A_A v_A = A_B v_B$$

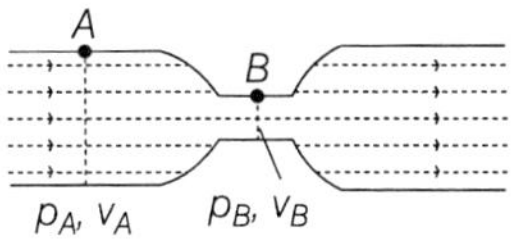

$$\Rightarrow \quad 40\, v_A = 20\, v_B$$

$$\Rightarrow \quad 2\, v_A = v_B \quad \ldots\text{(i)}$$

Now, using Bernoulli's equation (for horizontal tube), we have

$$p_A + \frac{1}{2}\rho v_A^2 = p_B + \frac{1}{2}\rho v_B^2$$

$$\Rightarrow \; p_A - p_B = \frac{1}{2}\rho(v_B^2 - v_A^2)$$

Here, $p_A - p_B = 700 \text{ Nm}^{-2}$

and $\rho = 1000 \text{ kg m}^{-3}$

$$\Rightarrow \quad 700 = \frac{1}{2} \times 1000\,(v_B^2 - v_A^2)$$

$$\Rightarrow \; v_B^2 - v_A^2 = 1.4 \quad \ldots\text{(ii)}$$

From Eqs. (i) and (ii), we have

$$3\,v_A^2 = 1.4$$

$$\Rightarrow \quad v_A = \sqrt{0.467} = 0.68 \text{ ms}^{-1}$$

$$= 68 \text{ cm s}^{-1}$$

So, volume flow rate of water

$$= A_A \cdot v_A$$

$$= 40 \times 68 = 2720 \text{ cm}^3/\text{s}$$

38. *(c)* From the equation of motion of a point mass,

$$ma = mg - kv \quad \ldots\text{(i)}$$

or $\frac{dv}{dt} = \frac{mg - kv}{m} \Rightarrow \frac{dv}{mg - kv} = \frac{dt}{m}$

Integrating $\int_0^v \frac{dv}{mg - kv} = \frac{1}{m} = \int_0^v dt$

$$\Rightarrow -\frac{1}{k}[\ln(mg - kv)]_0^v = \frac{t}{m}$$

$$\Rightarrow \quad \ln\left(\frac{mg - kv}{mg}\right) = -\frac{k}{m}t$$

$$\Rightarrow 1 - \frac{kv}{mg} = e^{\frac{-kt}{m}} \Rightarrow \frac{kv}{mg} = 1 - e^{\frac{-kt}{m}}$$

$$\Rightarrow \quad v = \frac{mg}{k}\left(1 - e^{\frac{-kt}{m}}\right) \quad \ldots\text{(ii)}$$

Putting Eq. (ii) in Eq. (i), we get

$$ma = mg - k \times \frac{mg}{k}\left(1 - e^{\frac{-kt}{m}}\right) \text{ or } a = ge^{\frac{-kt}{m}}$$

Hence, option (c) represents the correct variation.

39. *(40)* Terminal velocity of a ball is

$$v_T = \frac{2}{9} \cdot r^2 \cdot \left(\frac{\rho - \sigma}{\eta}\right) \cdot g$$

r = radius, ρ = density of ball
σ = density of medium
η = viscosity of medium.

It is given 8 balls of radius 0.01 mm combined to form a large drop.

As volume of water remains same.

$$\therefore 8 \times V_{\text{small drop}} = V_{\text{large drop}}$$

$$\Rightarrow \quad 8 \times \frac{4}{3}\pi r^3 = \frac{4}{3}\pi R^3$$

or $\quad 8r^3 = R^3$

$$\Rightarrow \quad 2r = R$$

As terminal velocity,

$v_T \propto r^2$ (other things are same)

$$\Rightarrow \frac{v_{T_2}}{v_{T_1}} = \left(\frac{R}{r}\right)^2 = (2)^2 \Rightarrow v_{T_2} = 4 \times v_{T_1}$$

$$= 4 \times 10 = 40 \frac{\text{cm}}{\text{s}}$$

40. *(810)* $F = \frac{1}{2}\rho\,(v_1^2 - v_2^2)\,A$,

$$\rho = 1.2 \text{ kg/m}^3$$

$$F = \frac{1}{2} \times 1.2 \times (70^2 - 65^2) \times 2$$

$$= 1.2 \times (4900 - 4225) = 810 \text{ N}$$

41. *(4)* Given, density of glycerine,

$\rho = 1.25 \times 10^2 \text{ kgm}^{-3}$

Area of cross-section, $A_1 = 10 \text{ cm}^2$

Area of cross-section, $A_2 = 5 \text{ cm}^2$

$A_1 = 10 \text{ cm}^2$

Pressure difference, $\Delta p = 3 \text{ Nm}^{-2}$

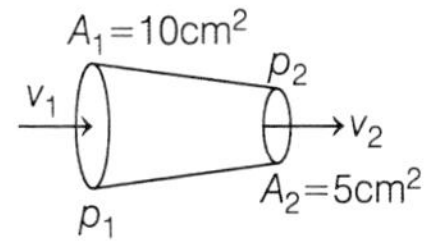

By equation of continuity, we have

$$A_1 v_1 = A_2 v_2$$

$\therefore \quad v_1 = \frac{A_2}{A_1} v_2$ ...(i)

By Bernoull's equation

$$p_1 + \frac{1}{2}\rho v_1^2 = p_2 + \frac{1}{2}\rho v_2^2$$

$$\Rightarrow p_1 - p_2 = \frac{1}{2}\rho(v_2^2 - v_1^2)$$

$$\Rightarrow \quad \Delta p = \frac{1}{2}\rho\left(v_2^2 - \frac{A_2^2}{A_1^2} v_2^2\right)$$

[using Eq. (i)]

$$\Rightarrow \quad \Delta p = \frac{1}{2}\rho\left[1 - \left(\frac{A_2}{A_1}\right)^2\right] v_2^2$$

$$\Rightarrow \quad 3 = \frac{1}{2} \times 1.25 \times 10^3 \left[1 - \left(\frac{5}{10}\right)^2\right] v_2^2$$

$$= \frac{1}{2} \times 1.25 \times 10^3 \times \frac{3}{4} v_2^2$$

$\therefore \quad v_2 = 8 \times 10^{-2}$ m/s

So, discharge rate $= A_2 v_2$

$= 5 \times 10^{-4} \times 8 \times 10^{-2}$

$= 4 \times 10^{-5} \text{m}^3/\text{s}$

42. *(80)* From equation of continuity,

$A_1 v_1 = A_2 v_2$

Here, $A_1 = 2 \text{ cm}^2$, $A_2 = 10 \text{ mm}^2$

$= 10 \times 10^{-2} \text{ cm}^2$

$v_1 = 4 \text{ cm s}^{-1}$

Thus, $v_2 = \frac{A_1 v_1}{A_2} = \frac{2 \times 4}{10 \times 10^{-2}}$

$= 80$ cm/s

43. *(2)* By using equation of continuity,

$A_1 v_1 = A_2 v_2$

Given that,

$A_1 = 750 \text{ cm}^2$,

$A_2 = 500 \text{ mm}^2$

$= 500 \times 10^{-2} \text{ cm}^2$

$v_2 = 30$ cm/s

$v_1 = \frac{A_2 v_2}{A_1} = \frac{500 \times 10^{-2} \times 30}{750}$

$v_1 = 2 \times 10^{-1}$ cm/s

$v_1 = 2 \times 10^{-3}$ m/s

Given that $\frac{dx}{dt} = v = x \times 10^{-3}$ m/s

Therefore, $\quad x = 2$

44. *(7)* Ball attains constant velocity when net force on it is zero.

or $F_v + F_b = F_g$

(F_v = viscous force,

F_b = force of buoyancy, F_g = weight)

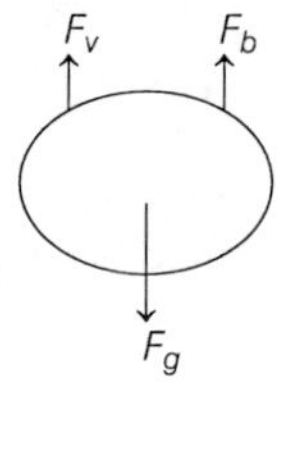

$\Rightarrow \quad F_v = F_g - F_b$

$= (mg) - (m_f \cdot g)$

$= V \cdot \rho_b \cdot g - V \cdot \rho_f \cdot g$

$= Vg(\rho_b - \rho_f)$

Here, ρ_b = density of ball

ρ_f = density of glycerine

V = volume of ball

= volume of glycerine displaced.

$= \frac{4}{3}\pi r^3$

$= \frac{4}{3}\pi \times (1 \times 10^{-3})^3$

$= 4.2 \times 10^{-9} \text{ m}^3$

So, viscous force is

$$F_v = 4.2 \times 10^{-9} \times 9.8\,(10.5 - 1.5) \times \frac{10^{-3}}{10^{-6}}$$

$= 4.2 \times 9.8 \times 9 \times 10^3 \times 10^{-9}$

$= 370.44 \times 10^{-6}$ N

$\approx 3700 \times 10^{-7}$ N

But $F_v = 3696 \times 10^{-x}$ N (Given)

So, $\quad x = 7$

45. *(11)* Diameter of air bubble

$d = 2 \text{ mm} = 2 \times 10^{-3}$ m

$\therefore$Radius, $r = \frac{2}{2} \times 10^{-3} \text{ m} = 10^{-3}$ m

Density of solution, $\rho = 1750 \text{ kgm}^{-3}$

Speed of bubble, $v = 0.35 \text{ cms}^{-1}$

$= 3.5 \times 10^{-3} \text{ ms}^{-1}$

According to given situation, bubble moves with terminal velocity v which is given as,

$$v = \frac{2r^2 g(\rho - \sigma)}{9\eta}$$

Here, σ = density of air ≈ 0

$$v = \frac{2r^2 g\rho}{9\eta} \Rightarrow \eta = \frac{2r^2 g\rho}{9v}$$

$$= \frac{2 \times (10^{-3})^2 \times 10 \times 1750}{9 \times 3.5 \times 10^{-3}}$$

$= 1.1 \text{ N-s/m}^2 = 11$ Poise

[$\because 1 \text{ N-sm}^{-2} = 10$ Poise]

46. *(20)* Given, radius of ball

$= 0.1 \text{ mm} = 10^{-4}$ m

$\eta = 10^{-5} \text{ N-sm}^{-2}$

$\rho = 10^4 \text{ kg m}^{-3}$

Velocity of ball does not change as it falls through water with terminal velocity.

So, $\quad v = v_T$

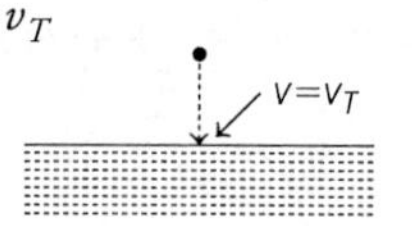

$$\Rightarrow \sqrt{2gh} = \frac{2r^2(\rho - \rho_0)g}{9\eta}$$

$$\Rightarrow \sqrt{2 \times 10 \times h} = \frac{2 \times (10^{-4})^2 (10^4 - 10^3) \times 10}{9 \times 1 \times 10^{-5}}$$

$\Rightarrow 20h = (20)^2 \Rightarrow h = 20$ m

47. *(24)* Given that, $A_1 = 1.2 \times 10^{-2} \text{ m}^2$

and $A_2 = \frac{A_1}{2} = 0.6 \times 10^{-2} \text{ m}^2$

Density of liquid, $\rho = 750 \text{ kg m}^{-3}$

Pressure difference, $p_1 - p_2 = 4500$ Pa

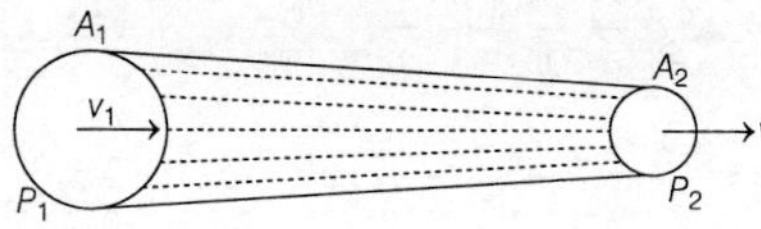

Form equation of continuity,

$A_1 v_1 = A_2 v_2$

or $\quad \frac{v_1}{v_2} = \frac{A_2}{A_1} = \frac{0.6 \times 10^{-2}}{1.2 \times 10^{-2}} = \frac{1}{2}$

or $\quad v_2 = 2v_1$ ...(i)

From Bernoulli's theorem for a horizontal pipe,

$$p_1 + \frac{1}{2}\rho v_1^2 = p_2 + \frac{1}{2}\rho v_2^2$$

or $p_1 - p_2 = \frac{1}{2}\rho(v_2^2 - v_1^2)$

or $\quad 4500 = \frac{1}{2} \times 750$

$[(2v_1)^2 - v_1^2]$ (Using Eq. (i))

or $\quad v_1^2 = 4$ or $v_1 = 2 \text{ ms}^{-1}$

Now, rate of flow $= A_1 v_1$

$= 1.2 \times 10^{-2} \times 2 = 24 \times 10^{-3} \text{ m}^3\text{s}^{-1}$

48. *(300)* The pressure at the surface of the water,

$\therefore \quad p_A = p_0 + \frac{25 \times 10}{0.5}$

$= p_0 + 500 \text{ N/m}^2$

Applying Bernoulli's equation, at top of water tank and its bottom,

$$\Rightarrow p_A + \rho gh + \frac{1}{2}\rho v_A^2 = p_B + \frac{1}{2}\rho v_B^2$$

$$\Rightarrow p_0 + 500 + (1000 \times 10 \times 0.4) + 0 = p_0 + \left(\frac{1}{2} \times 1000\right) v_B^2$$

$500 + 4000 = 500\, v_B^2$

$$\frac{4500}{500} = v_B^2 \Rightarrow v_B^2 = 9$$

$\Rightarrow \quad v_B = 3 \text{ ms}^{-1} = 300 \text{ cms}^{-1}$

49. *(363)* Given, density, $\rho = 800 \text{ kg/m}^3$, height, $h_1 = 1$ m

$$a_2 = \frac{a}{2} \Rightarrow a_1 = a$$

Pressure difference,

$$p_1 - p_2 = 4100 \text{ Pa}$$

From Bernoulli's theorem,

$$\frac{p_1}{\rho} + gh_1 + \frac{1}{2}v_1^2 = \frac{p_2}{\rho} + gh_2 + \frac{1}{2}v_2^2$$

$$\frac{p_1 - p_2}{\rho} = g(h_2 - h_1) + \frac{1}{2}(v_2^2 - v_1^2)$$

$$\frac{4100}{800} = 10 \times (0 - 1) + \frac{1}{2}(v_2^2 - v_1^2)$$

$$v_2^2 - v_1^2 = 30.25 = \frac{121}{4} \quad \ldots\text{(i)}$$

From equation of continuity;

$$a_1 v_1 = a_2 v_2$$

$$av_1 = \frac{a}{2}.v_2 \Rightarrow v_2 = 2v_1$$

$\therefore$ Eq. (i) becomes,

$$(2v_1)^2 - v_1^2 = 121/4 \Rightarrow 3v_1^2 = 121/4$$

$$v_1 = \frac{\sqrt{363}}{6} = \frac{\sqrt{x}}{6}$$

$\therefore \quad x = 363$

50. *(100)* Given : $\Delta v = 36$ km/h $= 10$ m/s

Shearing stress $= 10^{-3}$ N/m^2,

$\eta = 10^{-2}$ Pa-s

Let the depth be Δx.

$$= -\eta A \frac{\Delta v}{\Delta x}, \frac{F}{A} = \eta \frac{\Delta v}{\Delta x}$$

$$10^{-3} = 10^{-2} \times \frac{10}{\Delta x}, \Delta x = \frac{10^{-1}}{10^{-3}},$$

$$\Delta x = 100 \text{ m}$$

51. *(6)* As per question, the water is filled up to a height of 12 m in a tank having vertical side walls. A hole is made in one of the walls at a depth h below the water level.

Horizontal range will be

$$R - \sqrt{2gh} \times \sqrt{\frac{(12 - h) \times 2}{g}}$$

$$\Rightarrow \quad R = \sqrt{4h(12 - h)}$$

For range to be maximum, $dR/dh = 0$

$$\Rightarrow \quad \frac{d}{dh}(\sqrt{4h(12 - h)}) = 0$$

On solving, we will get $h = 6$ m

52. *(3)* Given, cross-sectional area of the tank, $A = 0.4 \text{ m}^2$

Cross-sectional area of the opening at B,

$$A = 1 \text{ cm}^2 = 1 \times 10^{-4} \text{ m}^2$$

Load applied on the top of the tank,

$$m = 24 \text{ kg}$$

Height of the water level, $h = 40$ cm

From the figure,

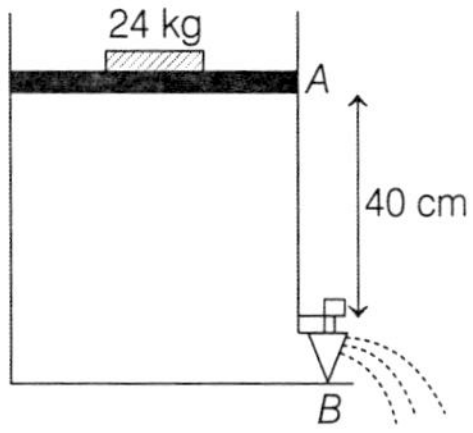

Using the Bernoulli's principle equation,

$$p_0 + \frac{mg}{A} + \rho gh + \frac{\rho v_A^2}{2} = p_0 + \frac{\rho v_B^2}{2}$$

$$\Rightarrow \quad 0 + \frac{(24)10}{0.4} + 1000 \times 10 \times 0.4 + \frac{\rho(0)^2}{2} = 0 + \frac{1000 \times v_B^2}{2}$$

$$v_B = 3 \text{ m/s}$$

Hence, the value of v to the nearest integer is 3.

Topic 3

01. *(d)* $K = \frac{\cos\theta_A}{\cos\theta_B}$ (given)

It is negative when $\cos\theta_A$ and $\cos\theta_B$ are opposite sign.

Liquid A has concave meniscus.

Liquid B has convex meniscus.

02. *(a)* **Statement I** Hot water is less viscous than cold water, hence hot water flows faster than cold water, so Statement I is true.

Statement II Soap water has a lower surface tension as compared to fresh water because soap molecules disrupt the hydrogen bond between water molecules at the surface, effectively weaking the cohesive forces that create surface tension, so Statement II is false.

03. *(c)* A. Surface tension is due to less stability or unbalanced forces on the molecules at the surface.

B. For liquid, coefficient of viscosity decreases with temperature.

Statements C, D and E are correct.

04. *(c)* Let T is the surface tension.

Excess pressure p_e in the air bubble

$$= \rho gh + \frac{2T}{R}$$

$$p_{\text{in}} - p_0 = \rho gh + \frac{2T}{R} = 2100$$

$$\frac{2T}{R} = 2100 - \rho gh$$

$$T = \frac{R}{2}(2100 - 10^3 \times 10 \times 0.2)$$

$$= 0.05 \qquad [\text{where } R = 0.1 \text{ cm}]$$

05 *(c)* Excess pressure inside soap bubble,

$$p = \frac{4S}{R}$$

06. *(d)* Surface tension will be less as temperature increases,

$$h = \frac{2T\cos\theta}{\rho gr}$$

Height of capillary rise will be smaller in hot water and large in cold water.

07. *(d)* Energy released in process when n small identical drops of radius r are combined to make a big drop of radius R is given as

$$\Delta U = 4\pi r^2 \times T(n - n^{2/3})$$

where, T = Surface tension

Here, $T = 0.07$ N/m or 7×10^{-2}N/m

$$r = 1 \text{ mm} = 10^{-3} \text{ m}$$

$$\Delta U = 4 \times \frac{22}{7} \times (10^{-3})^2 \times 7 \times 10^{-2} \times (1000 - 100)$$

$$\Rightarrow \quad \Delta U = 7.92 \times 10^{-4} \text{ J}$$

> **Alternate Solution**
>
> If r and R be the radius of small and big droplet, then
>
> $$1000 \times \frac{4}{3}\pi r^3 = \frac{4}{3}\pi R^3 \Rightarrow R = 10r$$
>
> Change in surface area,
>
> $$\Delta A = 1000 \times 4\pi r^2 - 4\pi R^2$$
> $$= 1000 \times 4\pi r^2 - 4\pi(10r)^2$$
> $$= 100 \times 4\pi r^2(10 - 1) = 3600\pi(1 \times 10^{-3})^2$$
> $$= 0.0036 \times 3.14 = 0.0113 \text{ m}^2$$
>
> $\therefore$ Released energy,
>
> $$\Delta U = T \times \Delta A = 0.07 \times 0.0113$$
> $$= 0.000792 = 7.92 \times 10^{-4} \text{ J}$$

08. *(a)* The water-oil contact angle is as shown below.

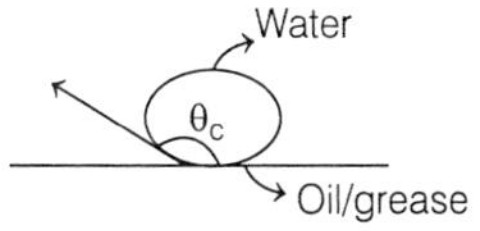

where; θ_c = Contact angle

Clearly $\theta_c > 90°$ (obtuse contact angle)

$\therefore$ Cohesive force > adhesive force
Hence, clothes containing oil or grease cannot be cleaned by water.

09. *(a)* Height of liquid rise in capillary tube,

$$h = \frac{2T\cos\theta_c}{\rho r g} \Rightarrow h \propto \frac{1}{r}$$

So, when radius is doubled, height becomes half.

$$\therefore \quad h' = \frac{h}{2}$$

Now, density $(\rho) = \frac{\text{mass}(M)}{\text{volume}(V)}$

$$\Rightarrow \quad M = \rho \times \pi r^2 h$$

$$\therefore \quad M' = \rho \pi r'^2 h'$$

$$\therefore \quad \frac{M'}{M} = \frac{r'^2 h'}{r^2 h} = \frac{(2r)^2\left(\frac{h}{2}\right)}{r^2 h} = 2$$

$$\Rightarrow \quad M' = 2M$$

> ***Alternate Solution***
> According to the given figure, force inside the capillary tube is
>
> 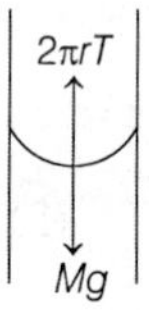
>
>
> $2\pi rT = Mg \Rightarrow M \propto r$
> When $r' = 2r$, then $M' = 2M$

10. *(d)* For cylindrical shape, excess pressure is given by $\Delta p = 2T/R$

11. *(d)* If a layer of thickness dr is evaporated, then
Change in surface energy
$= \text{Change in surface area} \times T$
$= [4\pi(r + dr)^2 - 4\pi r^2]T$
$= [4\pi r^2 + 4\pi\, dr^2 + 8\pi r\, dr - 4\pi r^2]$
As dr is very small, so dr^2 neglected.
So, change in surface energy $= 8\pi r dr\, T$
Energy required to evaporate layer of thickness

$$dr = (4\pi r^2 dr)\rho L$$

The process of evaporation just starts if change in surface energy is just sufficient to evaporate the water layer.

$$\therefore (4\pi r^2 dr)\rho L = (8\pi r dr)\,T$$

$$\Rightarrow \quad r = 2T/\rho L$$

12. *(c)* The force due to the surface tension will balance the weight.

$$F = w$$

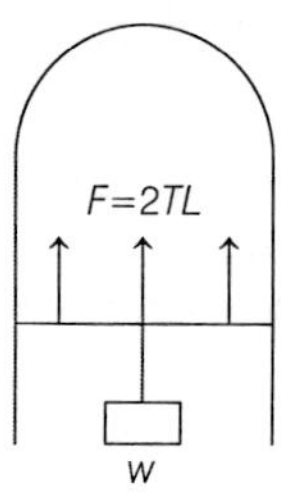

$$2TL = w \Rightarrow T = \frac{w}{2L}$$

$$T = \frac{1.5\times10^{-2}\ \text{N}}{2\times30\times10^{-2}\ \text{m}} = 0.025\ \text{Nm}^{-1}$$

13. *(d)* Let R be the radius of the bigger drop, then
Volume of bigger drop $= 2 \times$ Volume of small drop

$$\frac{4}{3}\pi R^3 = 2\times\frac{4}{3}\pi r^3 \Rightarrow R = 2^{1/3} r$$

Surface energy of bigger drop,

$$E = 4\pi R^2 T = 4\times 2^{2/3}\,\pi r^2 T$$
$$= 2^{8/3}\,\pi r^2 T$$

14. *(a)* If R be the radius of bigger drop, then

$$2\times\frac{4}{3}\pi r^3 = \frac{4}{3}\pi R^3 \Rightarrow R = 2^{1/3} r$$

$$U_i = 2\times 4\pi r^2 T$$

$$U_f = 4\pi R^2 T = 4\pi\times 2^{\frac{2}{3}} r^2 T$$

$\therefore$ Released surface energy

$$= U_i - U_f = 2\times 4\pi r^2 T - 4\pi\times 2^{\frac{2}{3}} r^2 T$$

$$= 4\pi r^2 T\left(2 - 2^{\frac{2}{3}}\right)$$

15. *(b)* Work done, W = Surface energy

$$W = \Delta U = S\Delta A$$

where, S is surface tension
One drop change to n drop

$$\frac{4}{3}\pi R^3 = n\frac{4}{3}\pi r^3$$

$$r = \frac{R}{n^{1/3}}$$

(Small r is radius of small drop)

$$W = S[n4\pi r^2 - 4\pi R^2]$$

$$= S4\pi R^2[n^{\frac{1}{3}} - 1]$$

For one drop to 27 drops

$$W = S4\pi R^2[27^{1/3} - 1] = 10 \quad \text{...(i)}$$

For one drop to 64 drops

$$W' = S4\pi R^2(64^{1/3} - 1) \quad \text{...(ii)}$$

Dividing Eq. (ii) by Eq. (i)

$$\frac{W'}{W} = \frac{4-1}{3-1} = \frac{3}{2}$$

$$W' = \frac{3}{2}W = 15\ \text{J}$$

> ***Alternate Solution***
> Work required to break a drop into n smaller drops is given by
>
> $W \propto (n^{1/3} - 1)$
>
> So, as per question
> For $n_1 = 27$, $W_1 = 10\ \text{J}$
> For $n_2 = 64$, $W_2 = ?$
> Applying the formula
>
> $\frac{W_2}{W_1} = \frac{(n_2^{1/3} - 1)}{(n_1^{1/3} - 1)} = \frac{64^{1/3} - 1}{27^{1/3} - 1}$
>
> $\frac{W_2}{10} = \frac{3}{2}$
>
> $W_2 = 15\ \text{J}$

16. *(d)* Given, $p_A = 3p_B$
We know that excess pressure inside soap bubble, $\Delta p = \frac{4T}{r}$

$$\Rightarrow \quad \frac{p_A}{p_B} = \frac{r_B}{r_A} \Rightarrow \frac{3}{1} = \frac{r_B}{r_A}$$

$$\Rightarrow \quad \frac{r_A}{r_B} = \frac{1}{3}$$

Hence, $\frac{V_A}{V_B} = \frac{1}{27}$

17. *(c)* Statement I is false, but Statement II is true.
The correct form of Statement I is, the capillary rise in a capillary tube is given by

$$h = \frac{2T\cos\theta}{\rho g r} \text{ [where } T \text{ is surface tension]}$$

θ = angle of contact
If $\theta = 0°$, then rise is non-zero as

$$h = \frac{2T}{\rho g r}$$

Hence, option (c) is correct.

18. *(d)* Contact angle between a solid and liquid depends on the adhesive and cohesive forces between the two media. These forces are the properties of material of both the solid and the liquid.
Rise of liquid in capillary tube

$$h = \frac{2T\cos\theta}{r\rho g} \Rightarrow h \propto \frac{1}{r}$$

where r is inner radius of the tube,
Hence, Statement I is true but Statement II is false.

19. *(d)* According to given condition,

$$\frac{4}{3}\pi R^3 = 27 \times \frac{4}{3}\pi r^3$$

where, r is radius of small liquid drop

$$\therefore \quad R = 3r$$

$$\Rightarrow \quad r = \frac{R}{3} \quad \ldots(i)$$

$$\therefore \text{Work done} = T \times \Delta A$$

$$= T[27 \times 4\pi r^2 - 4\pi R^2]$$

$$= 4T\left[27\pi \times \left(\frac{R}{3}\right)^2 - \pi R^2\right] \quad \text{[From Eq. (i)]}$$

$$= T[8\pi R^2]$$

$$= 8\pi TR^2$$

20. *(b)* Let the radius of bigger drop and smaller drops of mercury be R and r respectively.

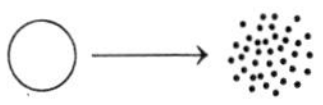

$$V = \text{constant}$$

$$\frac{4}{3}\pi R^3 = n \times \frac{4}{3}\pi r^3$$

$$\Rightarrow \quad R = n^{1/3} r$$

Gain in surface energy,

$$\Delta U = U_f - U_i = T \cdot \Delta A$$

$$= T[n \times 4\pi r^2 - 4\pi R^2]$$

$$= T\left[n \times 4\pi\left(\frac{R}{n^{1/3}}\right)^2 - 4\pi R^2\right]$$

$$= T[n^{1/3} \times 4\pi R^2 - 4\pi R^2]$$

$$\Delta U = T \times 4\pi R^2 [n^{1/2} - 1]$$

Given that, $T = 0.45\ \text{Nm}^{-1}$

$$R = 10^{-3}\ \text{m} \Rightarrow n = 125$$

$$\therefore \quad \Delta U = 0.45 \times 4\pi \times (10^{-3})^2 [(125)^{1/3} - 1]$$

$$\Delta U = 2.26 \times 10^{-5}\ \text{J}$$

> **Alternate Solution**
>
> When a mercury drop of radius R is broken into n equal size droplets, then gain in surface energy is given as, $\Delta E = T[n^{1/3} - 1] \cdot 4\pi R^2$
>
> Here, $T = 0.45\ \text{Nm}^{-1}$, $R = 10^{-3}$ m
>
> $n = 125$
>
> $\therefore \ \Delta E = 0.45[125^{1/3} - 1] \times 4 \times 3.14 \times (10^{-3})^2$
>
> $= 0.45 \times 4 \times 4 \times 3.14 \times 10^{-6}$
>
> $= 2.26 \times 10^{-5}$ J

21. *(b)* Height of a liquid column in a capillary is given by,

$$H = \frac{2S\cos\theta}{\rho R g}$$

In liquid A, $H = 5$ cm,

$$5 = \frac{2S_A \cos\theta}{\rho_A R g} \quad \ldots(i)$$

In liquid B, $h = \dfrac{2S_B \cos\theta}{\rho_B R g} \quad \ldots(ii)$

Form Eqs. (i) and (ii), we get

$$\frac{h}{5} = \frac{S_B}{S_A} \cdot \frac{\rho_A}{\rho_B} = \frac{2S_A}{S_A} \cdot \frac{\rho_A}{2\rho_A}$$

$$(\because s_B = 2s_A \text{ and } \rho_B = 2\rho_A)$$

$$= 1$$

$$\Rightarrow \quad h = 5\ \text{cm} = 0.05\ \text{m}$$

22. *(c)* Let R and r be the radius of big water drop and small water droplet, then

Volume of 1 big water drop $= 729 \times$ Volume of one small droplet

$$\Rightarrow \quad \frac{4}{3}\pi R^3 = 729 \times \frac{4}{3}\pi r^3$$

$$\Rightarrow \quad R^3 = (9r)^3$$

$$\Rightarrow \quad R = 9r$$

$$\Rightarrow \quad r = \frac{R}{9} = \frac{1}{9}\ \text{cm} \quad (\because R = 1\ \text{cm})$$

Gain in surface energy, $\Delta E = T.\Delta A$

$$\Rightarrow \Delta E = T \times (729 \times 4\pi r^2 - 4\pi R^2)$$

$$= 75\left(729 \times 4\pi \times \frac{1}{9^2} - 4\pi(1)^2\right)$$

$$= 75(36\pi - 4\pi) = 32\pi \times 75$$

$$= 7.54 \times 10^3\ \text{dyne-cm}$$

$$= 7.54 \times 10^3 \times 10^{-7}\ \text{N-m}$$

$$[\because 1\ \text{dyne-cm} = 10^{-7}\ \text{N-m}]$$

$$= 7.54 \times 10^{-4}\ \text{Nm}$$

23. *(a)* Given, density of drop of liquid, $\rho_d = \rho$, density of liquid, $\rho_l = \sigma$ and surface tension of liquid,

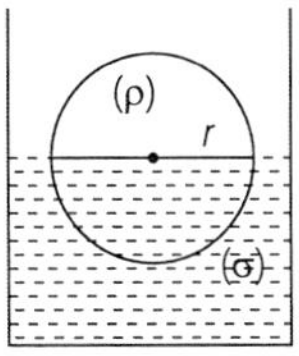

$$T = 7.5 \times 10^{-4}\ \text{N/cm}$$

$$= 7.5 \times 10^{-2}\ \text{N/m}$$

Since, drop of liquid is floating half immersed in the liquid as shown in figure, hence at balanced situation.

Buoyant force at surface tension $= mg$

$$\Rightarrow \quad \sigma\frac{V}{2}g + 2\pi RT = \rho v g$$

$$\Rightarrow 2\pi RT = \frac{(2\rho - \sigma)}{2} \times \frac{4}{3}\pi R^3 g$$

$$\Rightarrow R = \sqrt{\frac{3T}{(2\rho - \sigma)g}}$$

$$= \sqrt{\frac{3T}{g(2\rho - \sigma)}}$$

$$= \sqrt{\frac{3 \times 7.5 \times 10^{-2}}{10(2\rho - \sigma)}} = \frac{15}{\sqrt{(2\rho - \sigma)}}\ \text{cm}$$

24. *(c)* Given, radius of bubbles is r_1 and r_2, respectively.

Let, the radius of combined new bubble be r.

Excess pressure inside a soap bubble is

$$p = \frac{4T}{r}$$

where, T is surface tension.

By using law of conservation of energy

$$pV = p_1 V_1 + p_2 V_2$$

$$\frac{4T}{r} \times \frac{4}{3}\pi r^3 = \frac{4T}{r_1} \times \frac{4}{3}\pi r_1^3 + \frac{4T}{r_2} \times \frac{4}{3}\pi r_2^3$$

$$\Rightarrow \quad r^2 = r_1^2 + r_2^2 \Rightarrow r = \sqrt{r_1^2 + r_2^2}$$

25. *(a)* When two soap bubbles of radii a and b $(b > a)$ coalesce, it can be shown as follows

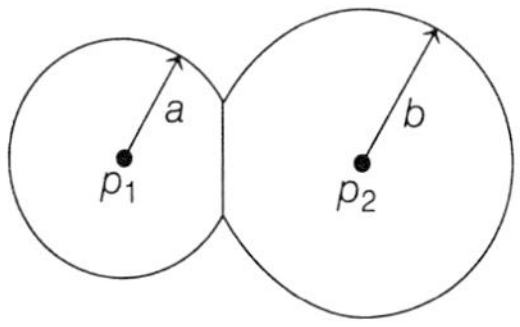

Pressure inside the bubble 1,

$$p_1 = p_0 + \frac{4T}{a}$$

where, p_0 is the initial pressure, T is the surface tension and a is the radius of the bubble 1.

Similarly, pressure inside the bubble 2,

$$p_2 = p_0 + \frac{4T}{b}$$

where, b is the radius of the bubble 2.

Excess pressure at common surface is given by

$$p_{ex} = p_1 - p_2$$

Let r be the radius of common surface, then

$$\Rightarrow \quad \frac{4T}{r} = \left(p_0 + \frac{4T}{a}\right) - \left(p_0 + \frac{4T}{b}\right)$$

$$\Rightarrow \quad \frac{4T}{r} = p_0 + \frac{4T}{a} - p_0 - \frac{4T}{b}$$

$$\Rightarrow \quad \frac{1}{r} = \frac{1}{a} - \frac{1}{b} \Rightarrow \frac{1}{r} = \frac{b - a}{ab}$$

$$\Rightarrow \quad r = \frac{ab}{b - a}$$

which is the required expression for radius of curvature at common surface.

26. *(a)*

> **key idea** Weight of the drop is balanced by upthrust or buoyant force (F_B) and surface tension (T).

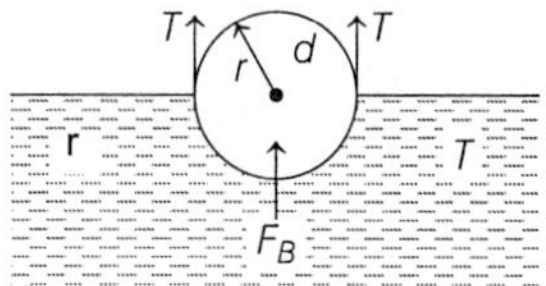

So, weight, $w = F_B + T \times 2\pi r$

$$\Rightarrow \quad d \cdot V \cdot g = \rho \cdot \frac{V}{2} g + T \times 2\pi r$$

$$[\because w = dVg \text{ and } F_B = \rho \frac{V}{2} g]$$

$$\Rightarrow d \cdot \frac{4}{3}\pi r^3 \cdot g = \rho \cdot \frac{2}{3}\pi r^3 \cdot g + T \times 2\pi r$$

$$\Rightarrow \quad r = \sqrt{\frac{3T}{(2d - \rho)g}}$$

27. *(c)* For inner bubble, excess pressure for outer bubble,

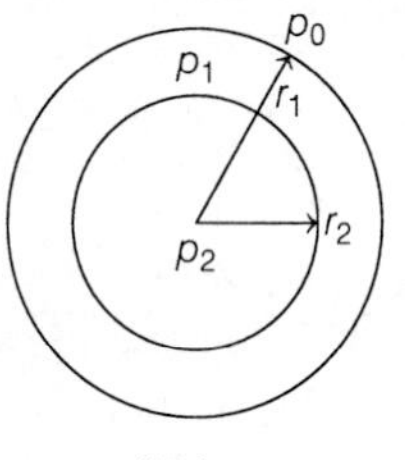

$$p_2 - p_1 = 4T / r_2 \quad \ldots(\text{i})$$

Excess pressure,

$$p_1 - p_0 = \frac{4T}{r_1} \quad \ldots(\text{ii})$$

From Eq. (i) and Eq. (ii), $p_2 - p_0$

$$= 4T\left(\frac{1}{r_2} + \frac{1}{r_1}\right) = \frac{4T}{r}$$

Here, r is required radius of a soap bubble,

$$\therefore \quad r = \frac{r_2 r_1}{r_1 + r_2}$$

$$= \frac{4 \times 6}{4 + 6} = \frac{24}{10} = 2.4 \text{ cm}$$

28. *(b)* Radius of capillary tube, $r = 0.1$ mm $= 0.01$ cm

Surface tension $= T = 70$ dyn/cm

Contact angle $= \theta = 0°$

$\cos\theta = 1$

Inclination $= \alpha = 30°$

Density of water $= 1$ g/cm^3

Acceleration due to gravity $= 9.8$ ms^{-2} $= 980$ cms^{-2}

Capillary rise formula for vertical tube is

$$h = \frac{2T\cos\theta}{r\rho g} = \frac{2 \times 70 \times 1}{0.01 \times 1 \times 980} = 14.285 \text{ cm}$$

≈ 14.29 cm

As the tube is inclined at an angle $\alpha = 30°$

The actual length of water column $L°$ is

$$L = \frac{h}{\cos\alpha} = \frac{14.29}{\sqrt{3}/2} = 16.50$$

$$L = \frac{82}{5}$$

29. *(c)* A big drop is formed by combining 1000 small drops.

Volume of big drop $= 1000 \times$ volume of 1 small drop

$$\Rightarrow \quad \frac{4}{3}\pi R^3 = 1000 \times \frac{4}{3}\pi r^3$$

$$\Rightarrow \quad R = 10r$$

Now, Surface energy = Area × Surface tension

Surface energy of 1000 small drops

$$E_1 = 1000 \times \text{ST} \times \text{Area of 1 drop}$$

$$E_1 = 1000 \times S \times 4\pi r^2$$

Surface energy of big drop,

$$E_2 = S \times 4\pi R^2$$

$$= S \times 4\pi \times (10r)^2$$

$$= 100\, S \times 4\pi r^2 = \frac{1}{10} E_1$$

30. *(a)* According to question,

Volume of 64 small drops = Volume of 1 large drop

$$\Rightarrow \quad \frac{4}{3}\pi r^3 \times 64 = \frac{4}{3}\pi \times R^3$$

$$\Rightarrow \quad (4r)^3 = R^3 \text{ or } r = \frac{R}{4} \quad \ldots(\text{i})$$

Now, increase in surface area of liquid drops

= surface area of 64 drops − surface area of initial large drop

$$= 64 \times 4\pi r^2 - 4\pi R^2$$

$$= 64 \times 4\pi\left(\frac{R}{4}\right)^2 - 4\pi R^2 \quad \text{[using Eq. (i)]}$$

$$= 16\pi R^2 - 4\pi R^2 = 4\pi R^2(4 - 1)$$

$$= 12\pi R^2$$

Now, increase in surface energy,

ΔE = Surface tension × increase in surface area

$$= T \times 12\pi R^2$$

Here, $T = 0.075$ N / m,

$R = 1$ cm $= 10^{-2}$ m

$$\Rightarrow \quad \Delta E = 0.075 \times 12 \times 3.14 \times (10^{-2})^2$$

$$= 2.8 \times 10^{-4} \text{ J}$$

31. *(b)* The given situation can be shown as below.

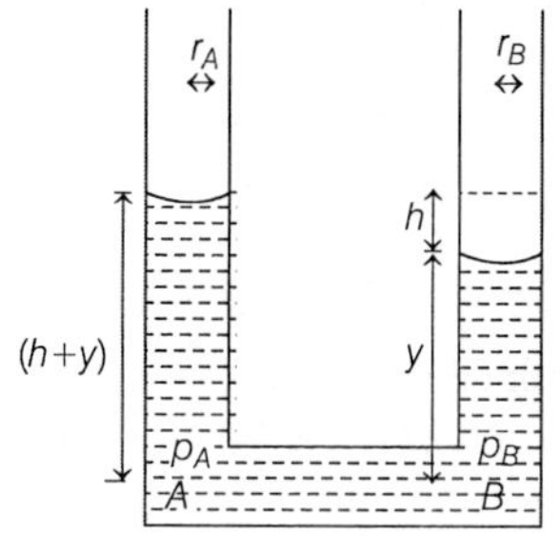

Points A and B are at same horizontal level.

So, the pressure at A and B will be equal

i.e. $\quad p_A = p_B \quad \ldots(\text{i})$

$$\Rightarrow p_0 - \frac{2T}{r_A} + (h + y)\rho g = p_0 - \frac{2T}{r_B} + \rho g y$$

or $\quad \rho g h = 2T\left(\frac{1}{r_A} - \frac{1}{r_B}\right) \quad \ldots(\text{ii})$

Here, $r_A = 2.5$ mm $= 2.5 \times 10^{-3}$ m,

and $T = 7.3 \times 10^{-2}$ N / m,

$r_B = 4$ mm $= 4 \times 10^{-3}$ m,

$g = 10$ m/s^2

and $\rho = 1 \times 10^3$ kg m^{-3}

Substituting the given value in Eq. (ii), we get

$$h = \frac{2 \times 7.3 \times 10^{-2}}{1 \times 10^3 \times 10}\left(\frac{1}{2.5 \times 10^{-3}} - \frac{1}{4 \times 10^{-3}}\right)$$

$$h = 2.19 \times 10^{-3} \text{ m} = 2.19 \text{ mm}$$

32. *(a)* According to question,

Volume of 1st small drop + Volume of 2nd small drop = Volume of a big drop

$$\frac{4}{3}\pi R^3 + \frac{4}{3}\pi R^3 = \frac{4}{3}\pi R'^3$$

$$\Rightarrow \quad R^3 + R^3 = R'^3$$

$$\Rightarrow \quad R' = 2^{1/3}(R) \quad \ldots(\text{i})$$

$\because$ So, considering the given scenario, we can write

$$\text{Area}_{\text{initially}} = 4\pi R^2 + 4\pi R^2 = 2(4\pi R^2)$$

and $\quad \text{Area}_{\text{finally}} = 4\pi R'^2$

$$\therefore \frac{\text{Surface energy initially}}{\text{Surface energy finally}}, \frac{U_i}{U_f} = \frac{A_i}{A_f}$$

$$\Rightarrow \frac{U_i}{U_f} = \frac{2(4\pi R^2)}{4\pi R'^2} \Rightarrow \frac{U_i}{U_f} = \frac{2R^2}{R'^2} \quad \ldots(\text{ii})$$

$\therefore$ From Eqs. (i) and (ii), we get

$$\frac{U_i}{U_f}=\frac{2R^2}{(2^{1/3})^2\cdot(R)^2}=\frac{2R^2}{2^{2/3}\cdot R^2}$$

$$\Rightarrow \quad \frac{U_i}{U_f}=2^{1-\frac{2}{3}} \Rightarrow \frac{U_i}{U_f}=2^{1/3}$$

$$\Rightarrow \quad U_i : U_f = 2^{1/3} : 1$$

33. *(d)* Given, radius of small drop $= r$
Radius of big drop $= R$
Surface tension $= T$
and mechanical equivalent of heat $= J$
As, small drops combine to form big drop.
$\therefore$ Volume of big drop $(V_B) = n \times$ Volume of small drop (V_S)

$$\Rightarrow \quad \frac{4}{3}\pi R^3 = n.\frac{4}{3}\pi r^3 \Rightarrow nr^3 = R^3$$

$$\Rightarrow \quad r=\frac{R}{n^{1/3}} \quad \ldots \text{(i)}$$

Surface energy of small drop $(E_S) =$ Surface tension $(T) \times$ Area (A)

$$\Rightarrow \quad E_S = n\times 4\pi r^2 T \text{ and } E_B = 4\pi R^2 T$$

Now, change in energy will be

$$\Delta E = E_s - E_B = 4\pi T(nr^2 - R^2)$$

$\therefore$ Heat energy per unit volume

$$=\frac{\Delta E}{V}=\frac{4\pi T(nr^2-R^2)}{J\times\frac{4}{3}\pi R^3}$$

$$=\frac{3T}{J}\left(\frac{nr^2}{R^3}-\frac{1}{R}\right)=\frac{3T}{J}\left(n\frac{R^2}{n^{2/3}R^3}-\frac{1}{R}\right)$$

[using Eq. (i)]

$$=\frac{3T}{J}\left[\frac{n^{1/3}}{R}-\frac{1}{R}\right]=\frac{3T}{J}\left[\frac{1}{r}-\frac{1}{R}\right]$$

[From Eq. (i)]

34. *(b)* It is given that tangents drawn from point of contact of meniscus makes 60° angle with each other. We have following situation, as shown in figure.

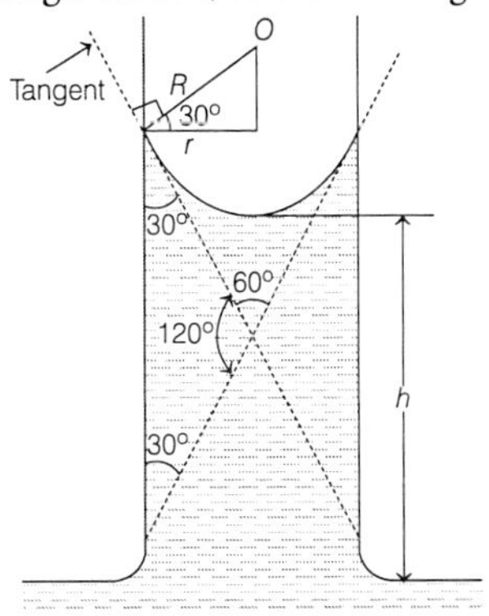

A normal drawn on the tangent, will pass through the centre O of meniscus. Let r be the radius of capillary tube and R be the radius of meniscus.
From geometry of figure,

$$\frac{\text{radius of capillary}}{\text{radius of meniscus}}=\frac{r}{R}=\cos 30°$$

$$\Rightarrow \quad \frac{r}{R}=\frac{\sqrt{3}}{2} \text{ or } R=\frac{2r}{\sqrt{3}}=\frac{2\times 0.15\times 10^{-3}}{\sqrt{3}}$$

$$\Rightarrow \quad R=\sqrt{3}\times 10^{-4} \text{ m}$$

Now, by ascent formula,

$$h=\frac{2T}{\rho g R}=\frac{2\times 0.05}{667\times 10\times\sqrt{3}\times 10^{-4}}$$

$$=0.087 \text{ m}$$

> **Alternate Solution**
>
> Angle of contact with walls of capillary tube, $\theta = 30°$
>
> $\therefore$ Height of vertical column, $h=\frac{2T\cos\theta}{r\rho g}$
>
> $$=\frac{2\times 0.05\times\cos 30°}{0.15\times 10^{-3}\times 667\times 10}=0.087 \text{ m}$$
>
> Hence, correct option is (b).

35. *(4)* Radius, $r_1 = 2$ cm
Radius, $r_2 = 4$ cm
The radius of curvature of common surface.

$$r=\frac{r_1\cdot r_2}{r_1-r_2}$$

$$=\frac{2\times 4}{4-2}=4 \text{ cm}$$

36. *(8)* Excess preffure

$$P=\frac{4T}{R}$$

$$\therefore \quad \frac{P_A}{P_B}=\frac{R_B}{R_A}\Rightarrow\frac{P_{B/2}}{P_B}=\frac{R_B}{R_A}$$

$$\Rightarrow \quad \frac{1}{2}=\frac{R_B}{R_A}\Rightarrow\frac{R_A}{R_B}=2$$

$$\text{Thus,} \quad \frac{v_A}{v_B}=\left(\frac{R_A}{R_B}\right)^3=2^3=8$$

37. *(2190)* Given, $S = 0.095 \text{ J/m}^2$

$\rho = 10^3 \text{ kg/m}^3$

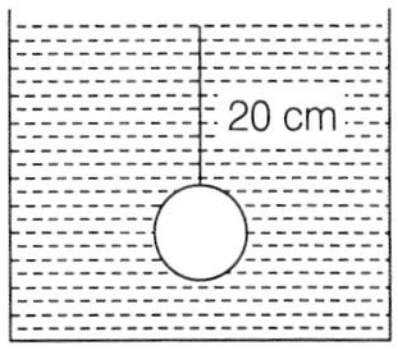

Pressure just outside the bubble,

$$p_A = p_0 + \rho g h$$

Pressure of air inside the bubble,

$$p_{\text{in}} = p_A + \frac{2S}{R} = p_0 + \rho g h + \frac{2S}{R}$$

$$p_{\text{in}} - p_0 = \rho g h + \frac{2S}{R}$$

$$=10^3\times 10\times\frac{20}{100}+\frac{2\times 0.095}{1\times 10^{-3}}$$

$$p_{\text{in}} - p_0 = 2000 + 190 = 2190 \text{ N/m}^2$$

38. *(7)* Excess pressure inside a soap bubble is

$$\Delta p = 4T/R$$

If this pressure is balanced by h height of liquid column then,

$$\Delta p = \rho g h = \frac{4T}{R}$$

Then, radius of soap bubble is

$$R=\frac{4T}{\rho g h}$$

Substituting given values,

$$R=\frac{4\times 0.28}{8\times 10^3\times 10\times 0.04\times 10^{-2}}$$

$$=3.5\times 10^{-2} \text{m}$$

So, diameter of bubble = 7 cm

39. *(1)* According to given condition,

$$1000\times\frac{4}{3}\pi r^3=\frac{4}{3}\pi R^3$$

$$\Rightarrow \quad R = 10r$$

[R = radius of larger drop
r = radiusof smaller drop]

$$\frac{\text{Surface energy of 1000 droplets}}{\text{Surface energy of big drop}}$$

$$=\frac{1000\times 4\pi r^2\times T}{4\pi R^2\times T}$$

$$=\frac{1000 r^2}{R^2}=\frac{1000\, r^2}{(10r)^2}=\frac{10}{1}$$

$$=\frac{10}{x} \quad \text{(given)}$$

$$\therefore \quad x = 1$$

40. *(7)* Given,
Work done, $W = 36960$ erg
Initial radius, $R_1 = 7$ cm,
Final radius $= R_2$
Surface tension,

$$T = 40 \text{ dyne/cm}$$

Work done is given by,

$$\Delta W = 8\pi(R_2^2 - R_1^2)T$$

$$36960 = 8\times\frac{22}{7}\times 40\left(R_2^2-\frac{49}{4}\right)$$

$$R_2 = 7 \text{ cm}$$

41. *(9600)*

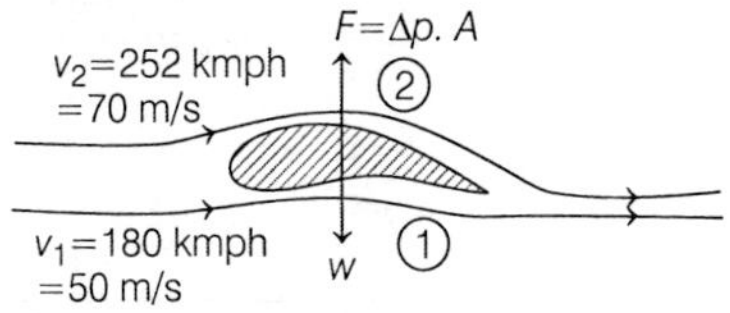

1 kmph = 5/18 m/s

Applying Bernoulli's equation to upper and lower wing surface and taking $h_1 \approx h_2$, we have

$$p_1 + \frac{1}{2}\rho v_1^2 = p_2 + \frac{1}{2}\rho v_2^2$$

$\Rightarrow$ Pressure difference,

$$p_1 - p_2 = \frac{1}{2}\rho (v_2^2 - v_1^2)$$

$$\Rightarrow \Delta p = \frac{1}{2} \times 1 \times (70^2 - 50^2) = 1200 \frac{\text{N}}{\text{m}^2}$$

Lifting force on two wings of plane (Area = $2 \times 40 = 80\ \text{m}^2$) due to pressure difference must be equal to weight of plane in a level flight.

So, $\Delta p \cdot 2A = mg$

$1200 \times 80 = m \times 10$

So, mass of plane $m = 9600$ kg

42. *(10)* Surface energy = Surface tension × Area

$\Rightarrow \quad E_1 = S \times 4\pi r^2 \times 1000$

and $\quad E_2 = S \times 4\pi R^2$

where, $1000 \times \frac{4}{3}\pi r^3 = \frac{4}{3}\pi R^3$

So, $\frac{r}{R} = \frac{1}{10}$

Hence, $\frac{E_1}{E_2} = 1000 \times \left(\frac{r}{R}\right)^2 = \frac{1000}{100} = 10$

43. *(50)* In given question,

$p_1 = 4.5 \times 10^4\ \text{N/m}^2$

$p_2 = 2.0 \times 10^4\ \text{N/m}^2$

Applying Bernoulli's principle,

$$p_1 = p_2 + \frac{1}{2}\rho v^2$$

$$4.5 \times 10^4 = 2.0 \times 10^4 + \frac{1}{2}\rho v^2$$

$$4.5 \times 10^{-4} - 2.0 \times 10^4 = \frac{1}{2}\rho v^2$$

$$2.5 \times 10^4 = \frac{1}{2}\rho v^2$$

$$\Rightarrow 5.0 \times 10^4 = \rho v^2$$

$$\frac{5.0 \times 10^4}{\rho} = v^2 \Rightarrow \frac{5.0 \times 10^4}{10^3} = v^2$$

$$5.0 \times 10^4 \times 10^{-3} = v^2$$

$$\Rightarrow \quad v = \sqrt{50}\ \text{m/s}$$

44. *(1130)* Given, depth,

$h = 10\ \text{cm} = 0.1$ m

Radius of bubble, r = 1 mm = 10^{-3} m

Surface, tension, $T = 0.075$ N/m

$\therefore$ Excess pressure inside the air bubble

$$\Delta p = \rho g h + \frac{2T}{r}$$

$$= 10^3 \times 9.8 \times 0.1 + \frac{2 \times 0.075}{10^{-3}}$$

$$= 1130\ \text{N/m}^2$$

45. *(264)* Given,

S.T = 3.5×10^{-2} N/m

$r_2 = 20$ cm

$= 20 \times 10^{-2}\ \text{m} = 0.2$ m

$r_1 = 0.1$ m

Workdone = ST × (Change in Area) × 2

$$= 3.5 \times 10^{-2} \times 2 \times (4\pi r_2^2 - 4\pi r_1^2)$$

$$= 3.5 \times 10^{-2} \times 2 \times 4\pi \times (r_2^2 - r_1^2)$$

$$= 3.5 \times 10^{-2} \times 8\pi(4-1) \times 10^{-4}$$

$$= 3.5 \times 10^{-2} \times 8\pi \times 3 \times 10^{-4}$$

$$= 3.5 \times 10^{-2} \times 8 \times \frac{22}{7} \times 3 \times 10^{-4}$$

$$= 264 \times 10^{-4}\ \text{J}$$

46. *(1)* If R be the radius of big (original) spherical drop and r be the radius of small drops, then

$$\frac{4}{3}\pi R^3 = 1000 \times \frac{4}{3}\pi r^3$$

$\Rightarrow \quad R = 10r \Rightarrow r = R/10 \quad \ldots$(i)

Initial surface area of spherical drop,

$$A = 4\pi R^2$$

$\therefore$Initial surface energy of original drop,

$$U_i = T \cdot A = T \cdot 4\pi R^2$$

$$U_i = 4\pi R^2 T \quad \ldots\text{(ii)}$$

When original drop splits into 1000 drops, then final energy,

$$U_f = 1000 \times 4\pi r^2 \cdot T$$

$$= 1000 \times 4\pi \times \left(\frac{R}{10}\right)^2 T$$

[From Eq. (i)]

$$= 40\pi R^2 T$$

$$\therefore \frac{U_f}{U_i} = \frac{40\pi R^2 T}{4\pi R^2 T} = \frac{10}{1}$$

$$= \left(\frac{10}{x}\right) \quad \text{[given]}$$

$\therefore \quad x = 1$

47. *(2)* According to the given situation,

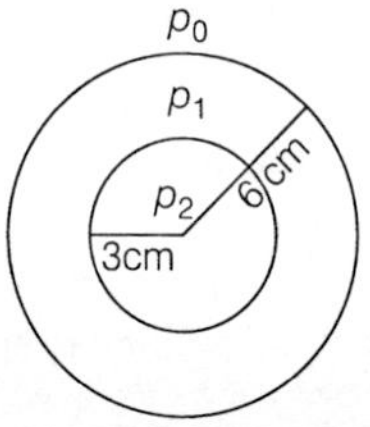

We know, excess pressure inside a soap bubble = $4T/r$

where, T = surface tension of liquid soap solution

$\therefore p_1 - p_0 = \frac{4T}{6}$ and $p_2 - p_1 = \frac{4T}{3}$

Adding both we get,

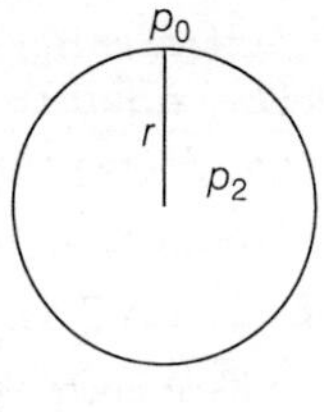

$$p_2 - p_0 = \frac{4T}{6} + \frac{4T}{3}$$

$$= \frac{4T + 8T}{6}$$

$$= \frac{12T}{6}$$

$\therefore p_2 - p_0 = 2T \quad \ldots$(i)

Also, according to the given situation,

$$p_2 - p_0 = \frac{4T}{r} \quad \ldots\text{(ii)}$$

$\therefore$From Eqs. (i) and (ii), we get

$$2T = \frac{4T}{r} \Rightarrow r = 2\ \text{cm}$$

48. *(2)* The given situation is shown below.

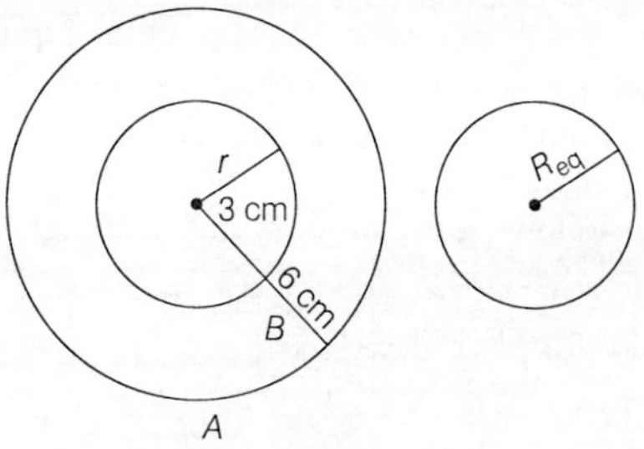

$r_A = 6$ cm, $r_B = 3$ cm

Excess pressure inside the soap bubble is given by

$$(\Delta p)_{\text{excess}} = \frac{4S}{r_A} + \frac{4S}{r_B}$$

According to the given question,

$$\frac{4S}{r_A}+\frac{4S}{r_B}=\frac{4S}{R_{eq}}$$

$$\Rightarrow \quad \frac{1}{r_A}+\frac{1}{r_B}=\frac{1}{R_{eq}}$$

$$\Rightarrow \quad \frac{1}{6}+\frac{1}{3}=\frac{1}{R_{eq}}$$

$$\Rightarrow \quad \frac{1+2}{6}=\frac{1}{R_{eq}}$$

$$\Rightarrow \quad \frac{1}{R_{eq}}=\frac{1}{2}$$

$$\Rightarrow \quad R_{eq}=2 \text{ cm}$$

49. *(25)* Since we know,
Volume of film

= area of film × thickness

Also, we know, volume of film

= volume of 100 spherical drops

$$4t_T=100\times\frac{4}{3}\pi r^3$$

$$\Rightarrow \quad =100\times\frac{4}{3}\pi\left[\left(\frac{3}{40\pi}\right)^{1/3}\times 10^{-3}\right]^3$$

$$=100\times\frac{4\pi}{3}\times\left(\frac{3}{40\pi}\right)\times 10^{-9}$$

$$=10^{-8} \text{ cm}^3$$

Hence, thickness of film,

$$t_T=\frac{10^{-8}}{4} \text{ cm}$$

$$=25\times 10^{-12} \text{ m}$$

Thickness of oleic acid layer

$$=0.01\, t_T$$

$$=0.01\times 25\times 10^{-12} \text{ m}$$

$$=25\times 10^{-14} \text{ m}$$

On comparing with

$x\times 10^{-14}$ m, $x=25$

50. *(101)* Height of liquid in capillary tube,

$$h=\frac{2S\cos\theta}{\rho g r}$$

$$\Rightarrow \quad S=\frac{\rho g r h}{2\cos\theta}$$

Here, $\rho=900 \text{ kgm}^{-3}$, $g=10 \text{ ms}^{-2}$,

$r=0.015 \text{ cm}=15\times 10^{-5}$ m,

$h=15 \text{ cm}=15\times 10^{-2}$ m

and $\theta\approx 0°$

$\Rightarrow \quad \cos\theta=1$

Hence,

$$S=\frac{900\times 10\times 15\times 10^{-5}\times 15\times 10^{-2}}{2\times 1}$$

$$=101.25\times 10^{-3} \text{ N/m}$$

$$\approx 101 \text{ mN/m}$$

C-10

JEE Main Chapterwise Topicwise Physics

In JEE Main there is usually 0-1 question per paper are visible from this chapter. In past 3 years, 16 conceptual and 6 numerical questions were asked of moderate to difficult level from this chapter.

For detailed analysis scan QR code

THERMAL PROPERTIES OF MATTER

Topic 1 Thermometry and Thermal Expansion

Level 1

01 Which of the following figure represents the relation between Celsius and Fahrenheit temperatures? *[2025, 24 Jan Shift-II]*

(a)
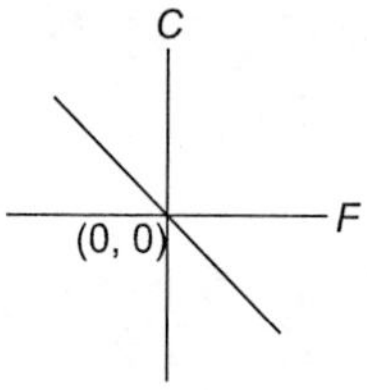

(b)
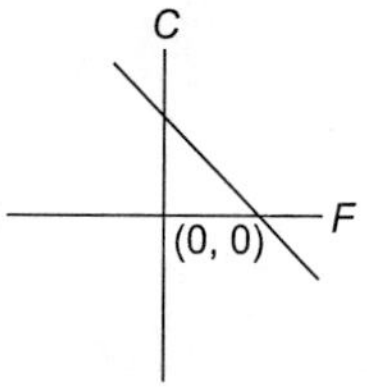

(c)
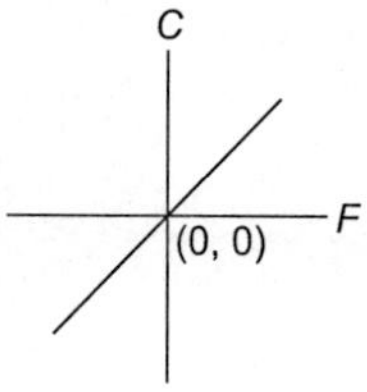

(d)
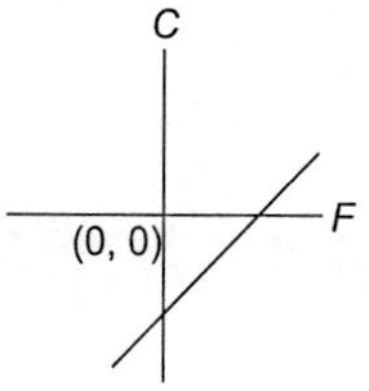

02 On celsius scale, the temperature of body increases by 40°C. The increase in temperature on fahrenheit scale is *[2024, 4 April Shift-I]*

(a) 72°F (b) 75°F (c) 70°F (d) 68°F

03 On a temperature scale X, the boiling point of water is 65°X and the freezing point is $-15°$X. Assume that the X scale is linear. The equivalent temperature corresponding to $-95°$X on the Farenheit scale would be *[2023, 11 April Shift-I]*

(a) -148°F (b) -112°F (c) -48°F (d) -63°F

04 A bimetallic strip consists of metals A and B. It is mounted rigidly as shown. The metal A has higher coefficient of expansion compared to that of metal B. When the bimetallic strip is placed in a cold bath, it will *[2021, 16 March Shift-II]*

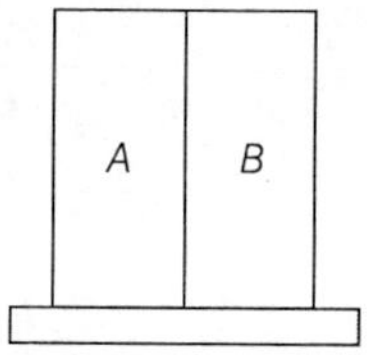

(a) bend towards the right (b) not bend but shrink
(c) Neither bend nor shrink (d) bend towards the left

05 Given below are two statements: one is labelled as Assertion (A) and the other is labelled as Reason (R).

Assertion (A) When a rod lying freely is heated, no thermal stress is developed in it.

Reason (R) On heating, the length of the rod increases.

In the light of the above statements, choose the correct answer from the options given below. *[2021, 25 Feb Shift-I]*

(a) A is true but R is false.
(b) Both A and R are true and R is the correct explanation of A.
(c) Both A and R are true but R is not the correct explanation of A.
(d) A is false but R is true.

06 Each side of a box made of metal sheet in cubic shape is a at room temperature T, the coefficient of linear expansion of the metal sheet is α. The metal sheet is heated uniformly, by a small temperature ΔT, so that its new temperature is $T + \Delta T$. Calculate the increase in the volume of the metal box. *[2021, 24 Feb Shift-I]*

(a) $4\pi a^3 \alpha \Delta T$ (b) $4a^3 \alpha \Delta T$
(c) $\frac{4}{3}\pi a^3 \alpha \Delta T$ (d) $3a^3 \alpha \Delta T$

Imp. Linear Expansion

07 Two different wires having lengths L_1 and L_2 and respective temperature coefficients of linear expansion α_1 and α_2, are joined end-to-end. Then the effective temperature coefficient of linear expansion is *[2020, 5 Sep Shift-II]*

(a) $\frac{\alpha_1 L_1 + \alpha_2 L_2}{L_1 + L_2}$ (b) $2\sqrt{\alpha_1 \alpha_2}$

(c) $\frac{\alpha_1 + \alpha_2}{2}$ (d) $4\frac{\alpha_1 \alpha_2}{\alpha_1 + \alpha_2}\frac{L_2 L_1}{(L_2 + L_1)^2}$

08 A uniform cylindrical rod of length L and radius r, is made from a material whose Young's modulus of elasticity equals to Y. When this rod is heated by temperature T and simultaneously subjected to a net longitudinal compressional force F, its length remains unchanged. The coefficient of volume expansion, of the material of the rod, is (nearly) equal to *[2019, 12 April Shift-II]*

(a) $\frac{9F}{(\pi r^2 YT)}$ (b) $\frac{F}{(3\pi r^2 YT)}$ (c) $\frac{6F}{(\pi r^2 YT)}$ (d) $\frac{3F}{(\pi r^2 YT)}$

09 A thermometer graduated according to a linear scale reads a value x_0, when in contact with boiling water and $x_0/3$, when in contact with ice. What is the temperature of an object in °C, if this thermometer in the contact with the object reads $x_0/2$? *[2019, 11 Jan Shift-II]*

(a) 35 (b) 60 (c) 40 (d) 25

10 A rod of length L at room temperature and uniform area of cross-section A, is made of a metal having coefficient of linear expansion α/°C. It is observed that an external compressive force F, is applied on each of its ends, prevents any change in the length of the rod, when its temperature rises by ΔT K. Young's modulus Y, for this metal is *[2019, 9 Jan Shift-I]*

(a) $\frac{2F}{A\alpha\Delta T}$ (b) $\frac{F}{A\alpha\Delta T}$

(c) $\frac{F}{2A\alpha\Delta T}$ (d) $\frac{F}{A\alpha(\Delta T - 273)}$

11 An external pressure p is applied on a cube at 0°C so that it is equally compressed from all sides. K is the bulk modulus of the material of the cube and α is its coefficient of linear expansion. Suppose we want to bring the cube to its original size by heating. The temperature should be raised by *[JEE Main 2017]*

(a) $\frac{p}{\alpha K}$ (b) $\frac{3\alpha}{pK}$

(c) $3pK\alpha$ (d) $\frac{p}{3\alpha K}$

12 A steel rail of length 5 m and area of cross-section 40 cm^2 is prevented from expanding along its length while the temperature rises by 10°C. If coefficient of linear expansion and Young's modulus of steel are $1.2 \times 10^{-5}\ K^{-1}$ and $2 \times 10^{11}\ N\ m^{-2}$ respectively, then the forced developed in the rail is approximately equals to *[JEE Main 2017]*

(a) 2×10^9 N (b) 3×10^{-5} N

(c) 2×10^7 N (d) 1×10^5 N

13 The pressure that has to be applied to the ends of a steel wire of length 10 cm to keep its length constant when its temperature is raised by 100°C is (For steel Young's modulus is $2 \times 10^{11}\ Nm^{-2}$ and coefficient of thermal expansion is $1.1 \times 10^{-5}\ K^{-1}$) *[JEE Main 2014]*

(a) 2.2×10^6 Pa (b) 2.2×10^8 Pa

(c) 2.2×10^9 Pa (d) 2.2×10^7 Pa

Level 2

Imp. Thermometry

14 The graph between two temperature scales P and Q is shown in the figure. Between upper fixed point and lower fixed point, there are 150 equal divisions of scale P and 100 divisions on scale Q. The relationship for conversion between the two scales is given by *[2023, 25 Jan Shift-II]*

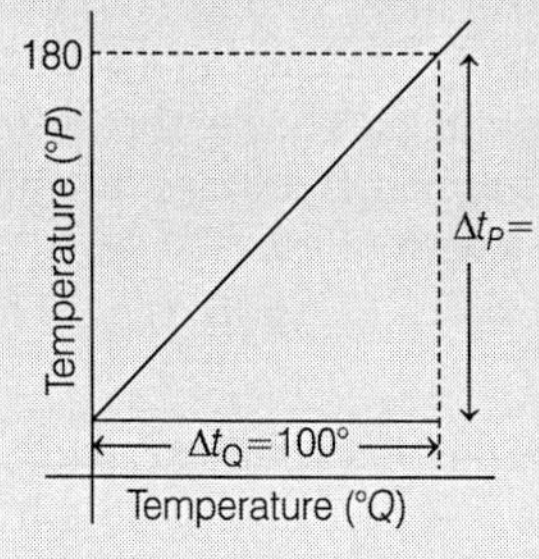

(a) $\frac{t_P}{180} - \frac{t_Q - 40}{100}$ (b) $\frac{t_Q}{100} = \frac{t_P - 30}{150}$

(c) $\frac{t_Q}{150} = \frac{t_P - 180}{100}$ (d) $\frac{t_P}{100} = \frac{t_Q - 180}{150}$

15 A solid metallic cube having total surface area 24 m^2 is uniformly heated. If its temperature is increased by 10°C, calculate the increase in volume of the cube (Given, $\alpha = 5.0 \times 10^{-4}\ °C^{-1}$) *[2022, 25 June Shift-II]*

(a) $2.4 \times 10^6\ cm^3$

(b) $1.2 \times 10^5\ cm^3$

(c) $6.0 \times 10^4\ cm^3$

(d) $4.8 \times 10^5\ cm^3$

16 At 40°C, a brass wire of 1 mm radius is hung from the ceiling. A small mass, M is hung from the free end of the wire. When the wire is cooled down from 40°C to 20°C, it regains its original length of 0.2 m. The value of M is close to (coefficient of linear expansion and Young's modulus of brass are $10^{-5}\,°C$ and $10^{11}\ N/m^2$ respectively, $g = 10\ m\ s^{-2}$)

[2019, 12 April Shift-I]

(a) 1.5 kg (b) 9 kg (c) 0.5 kg (d) 6.3 kg

Numerical Value Type Questions

17 A steel rod of length 1 m and cross-sectional area $1\ cm^2$ is heated from 0°C to 200°C without being allowed to extend or bend. The compressive tension produced in the rod is .... $\times 10^4$ N. (Given Young's modulus of steel $= 2 \times 10^{11}\ Nm^{-2}$, coefficient of linear expansion $= 10^{-5}\ K^{-1}$)

[2023, 8 April Shift-II]

18 A hole is drilled in a metal sheet. At 27°C, the diameter of hole is 5 cm. When the sheet is heated to 177°C, the change in the diameter of hole is $d \times 10^{-3}$ cm. The value of d will be .... if coefficient of linear expansion of the metal is 1.6×10^{-5}/°C.

[2023, 24 Jan Shift-I]

19 A unit scale is to be prepared whose length does not change with temperature and remains 20 cm, using a bimetallic strip made of brass and iron each of different length. The length of both components would change in such a way that difference between their lengths remains constant. If length of brass is 40 cm then length of iron will be .... cm
{$\alpha_{iron} = 1.2 \times 10^{-5} K^{-1}$ and $\alpha_{brass} = 1.8 \times 10^{-5}\ K^{-1}$)

[2022, 25 July Shift-I]

20 A bakelite beaker has volume capacity of 500 cc at 30°C. When it is partially filled with V_m volume (at 30°C) of mercury, it is found that the unfilled volume of the beaker remains constant as temperature is varied. If $\gamma_{beaker} = 6 \times 10^{-6}\ °C^{-1}$ and $\gamma_{mercury} = 1.5 \times 10^{-4}\ °C^{-1}$, where γ is the coefficient of volume expansion, then V_m (in cc) is close to ....

[2020, 3 Sep Shift-I]

Topic 2 Specific Heat Capacity, Calorimetry & Change of State

01 Water falls from a height of 200 m into a pool. Calculate the rise in temperature of the water assuming no heat dissipation from the water in the pool.
(Take $g = 10\ m/s^2$, specific heat of water $= 4200\ J/(kg\text{-}K)$)

[2025, 8 April Shift-II]

(a) 0.23 K (b) 0.36 K
(c) 0.14 K (d) 0.48 K

02 A gun fires a lead bullet of temperature 300 K into a wooden block. The bullet having melting temperature of 600 K penetrates into the block and melts down. If the heat required for the process is 625 J, then the mass of the bullet is .... grams.
(Latent heat of fusion of lead $= 2.5 \times 10^4\ Jkg^{-1}$ and specific heat capacity of lead $= 125\ Jkg^{-1}\ K^{-1}$)

[2025, 23 Jan Shift-I]

(a) 15 (b) 10
(c) 5 (d) 20

03 A block of ice at −10°C is slowly heated and converted to steam at 100°C. Which of the following curves represent the phenomenon qualitatively?

[2024, 30 Jan Shift-II]

(a)
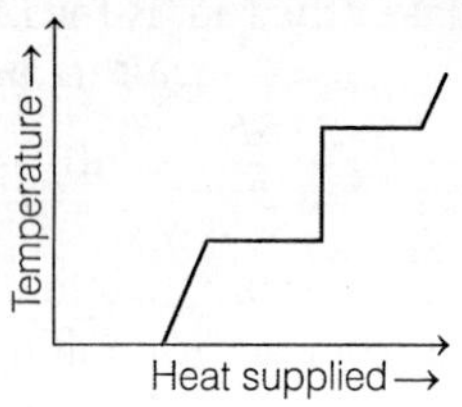

(b)
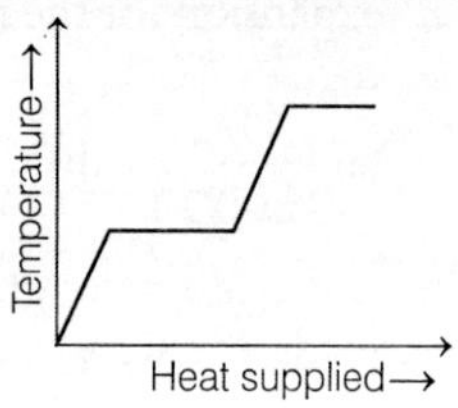

(c)
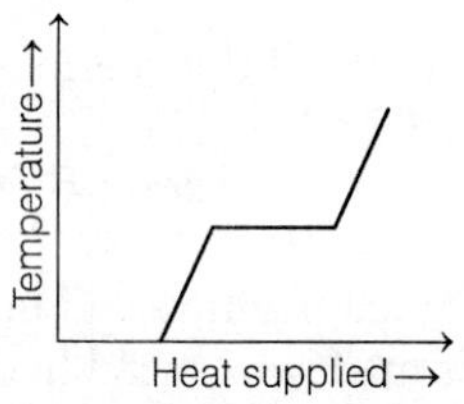

(d)
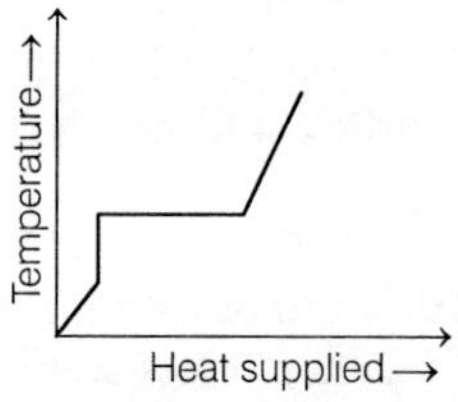

04 0.08 kg air is heated at constant volume through 5°C. The specific heat of air at constant volume is 0.17 kcal/kg°C and J = 4.18 J/cal. The change in its internal energy is approximately.

[2024, 27 Jan Shift-I]

(a) 318 J (b) 142 J
(c) 298 J (d) 284 J

05 The height of victoria falls is 63 m. What is the difference in temperature of water at the top and at the bottom of fall ?
[Given, 1 cal = 4.2 J and specific heat of water $= 1\ cal\ g^{-1}°C^{-1}$]

[2021, 27 Aug Shift-II]

(a) 0.147°C (b) 14.76°C
(c) 1.476°C (d) 0.014°C

06 The specific heat of water $= 4200\ J\ kg^{-1}K^{-1}$ and the latent heat of ice $= 3.4 \times 10^5\ J\ kg^{-1}$. 100 g of ice at 0°C is placed in 200 g of water at 25°C. The amount of ice that will melt as the temperature of water reaches 0°C is close to (in grams)

[2020, 4 Sep Shift-I]

(a) 63.8 (b) 69.3
(c) 64.6 (d) 61.7

07 A massless spring (k = 800 N/m), attached with a mass (500 g) is completely immersed to 1 kg of water. The spring is stretched by 2 cm and released, so that it starts vibrating. What would be the order of magnitude of the change in the temperature of water when the vibrations stop completely? (Assume that the water container and spring receive negligible heat and specific heat of mass = 400 J/kg K, specific heat of water = 4184 J/kg K) ***[2019, 9 April Shift-II]***

(a) 10^{-4} K (b) 10^{-3} K
(c) 10^{-1} K (d) 10^{-5} K

08 Assume that a drop of liquid evaporates by decrease in its surface energy and while this happens, its temperature remains unchanged. What should be the minimum radius of the drop for this to be possible? The surface tension is T, density of liquid is ρ and L is its latent heat of vaporisation. ***[JEE Main 2013]***

(a) $\frac{2T}{\rho L}$ (b) $\frac{\rho L}{T}$
(c) $\sqrt{\frac{T}{\rho L}}$ (d) $\frac{T}{\rho L}$

Level 2

09 There are two vessels filled with an ideal gas where volume of one is double the volume of other. The large vessel contains the gas at 8 kPa at 1000 K while the smaller vessel contains the gas at 7 kPa at 500 K. If the vessels are connected to each other by a thin tube allowing the gas to flow and the temperature of both vessels is maintained at 600 K, at steady state the pressure in the vessels will be (in kPa). ***[2025, 4 April Shift-II]***

(a) 4.4 (b) 24
(c) 6 (d) 18

10 Consider a rectangular sheet of solid material of length $l = 9$ cm and width $d = 4$ cm. The coefficient of linear expansion is $\alpha = 3.1 \times 10^{-5} K^{-1}$ at room temperature and one atmospheric pressure. The mass of sheet $m = 0.1$ kg and the specific capacity $C_V = 900$ J kg^{-1} K^{-1}. If the amount of heat to the material is 8.1×10^2 J then rectangular sheet is ***[2025, 4 April Shift-II]***

(a) 2.0×10^{-6} m^2 (b) 4.0×10^{-7} m^2
(c) 6.0×10^{-7} m^2 (d) 3.0×10^{-7} m^2

11 Heat energy of 184 kJ is given to ice of mass 600 g at − 12°C. Specific heat of ice is 2222.3 J kg^{-1} $°C^{-1}$ and latent heat of ice is 336 kJ kg^{-1}. ***[2023, 29 Jan Shift-II]***

A. Final temperature of system will be 0°C.
B. Final temperature of the system will be greater than 0°C
C. The final system will have a mixture of ice and water in the ratio of 5 : 1.
D. The final system will have a mixture of ice and water in the ratio of 1: 5.
E. The final system will have water only.

Choose the correct answer from the options given below.

(a) A and C Only (b) A and D Only
(c) B and D Only (d) A and E Only

12 A 100 g of iron nail is hit by a 1.5 kg hammer striking at a velocity of 60 ms^{-1}. What will be the rise in the temperature of the nail if one-fourth of energy of the hammer goes into heating the nail? [Specific heat capacity of iron = 0.42 Jg^{-1} $°C^{-1}$] ***[2022, 24 June Shift-II]***

(a) 675°C (b) 1600°C
(c) 16.07°C (d) 6.75°C

13 Two different metal bodies A and B of equal mass are heated at a uniform rate under similar conditions. The variation of temperature of the bodies is graphically represented as shown in the figure. The ratio of specific heat capacities is ***[2021, 25 July Shift-I]***

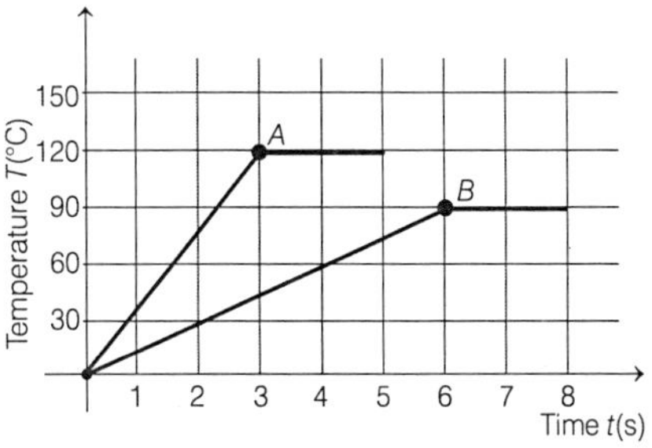

(a) $\frac{8}{3}$ (b) $\frac{3}{8}$ (c) $\frac{3}{4}$ (d) $\frac{4}{3}$

14 A bullet of mass 5 g, travelling with a speed of 210 m/s, strikes a fixed wooden target. One-half of its kinetic energy is converted into heat in the bullet while the other half is converted into heat in the wood. The rise of temperature of the bullet, if the specific heat of its material is 0.030 cal g^{-1} $°C^{-1}$, is close to ***[2020, 5 Sep Shift-I]***

(a) 87.5°C (b) 83.3°C
(c) 119.2°C (d) 38.4°C

Imp. Principle of Calorimetry

15 When M_1 g of ice at −10°C (specific heat = 0.5 cal g^{-1} $°C^{-1}$) is added to M_2 g of water at 50°C, finally no ice is left and the water is at 0°C. The value of latent heat of ice, in cal g^{-1} is ***[2019, 12 April Shift-I]***

(a) $\frac{50M_2}{M_1} - 5$ (b) $\frac{50M_1}{M_2} - 50$
(c) $\frac{50M_2}{M_1}$ (d) $\frac{5M_2}{M_1} - 5$

16 Ice at $-20°C$ is added to 50 g of water at 40°C. When the temperature of the mixture reaches 0°C, it is found that 20 g of ice is still unmelted. The amount of ice added to the water was close to (Take, specific heat of water $= 4.2$ J/g/°C, specific heat of ice $= 2.1$ J/g/°C and heat of fusion of water at 0°C $= 334$ J/g) *[2019, 11 Jan Shift-I]*
(a) 40 g (b) 50 g
(c) 60 g (d) 100 g

Level 3

Imp. Calorimetry

17 An amount of ice of mass 10^{-3} kg and temperature $-10°C$ is transformed to vapour of temperature 110°C by applying heat. The total amount of work required for this conversion is, (Take, specific heat of ice $= 2100$ J $kg^{-1}K^{-1}$, specific heat of water $= 4180$ J $kg^{-1}K^{-1}$, specific heat of steam $= 1920$ J $kg^{-1}K^{-1}$, latent heat of ice $= 3.35 \times 10^5$ Jkg^{-1} and latent heat of steam $= 2.25 \times 10^6$ Jkg^{-1}). *[2025, 22 Jan Shift-I]*
(a) 3043 J (b) 3003 J
(c) 3022 J (d) 3024 J

18 A lead bullet penetrates into a solid object and melts. Assuming that 40% of its kinetic energy is used to heat it, the initial speed of bullet is
(Given, initial temperature of the bullet $= 127°C$,
Melting point of the bullet $= 327°C$,
Latent heat of fusion of lead $= 2.5 \times 10^4$ J kg^{-1})
Specific heat capacity of lead $= 125$ J/kg K) *[2022, 27 June Shift-II]*
(a) 125 m s^{-1} (b) 500 m s^{-1}
(c) 250 m s^{-1} (d) 600 m s^{-1}

19 A copper block of mass 5.0 kg is heated to a temperature of 500°C and is placed on a large ice block. What is the maximum amount of ice that can melt? [Specific heat of copper, 0.39 $Jg^{-1}°C^{-1}$ and latent heat of fusion of water, 335 J g^{-1}] *[2022, 25 June Shift-II]*
(a) 1.5 kg (b) 5.8 kg
(c) 2.9 kg (d) 3.8 kg

20 The temperature of equal masses of three different liquids x, y and z are 10°C, 20°C and 30°C, respectively. The temperature of mixture when x is mixed with y is 16°C and that when y is mixed with z is 26°C. The temperature of mixture when x and z are mixed will be *[2021, 26 Aug Shift-II]*
(a) 28.32°C (b) 25.62°C
(c) 23.84°C (d) 20.28°C

21 When 100 g of a liquid A at 100°C is added to 50 g of a liquid B at temperature 75°C, the temperature of the mixture becomes 90°C. The temperature of the mixture, if 100 g of liquid A at 100°C is added to 50 g of liquid B at 50°C will be *[2019, 11 Jan Shift-II]*
(a) 60°C (b) 80°C
(c) 70°C (d) 85°C

22 A metal ball of mass 0.1 kg is heated upto 500°C and dropped into a vessel of heat capacity 800 JK^{-1} and containing 0.5 kg water. The initial temperature of water and vessel is 30°C. What is the approximate percentage increment in the temperature of the water? [Take, specific heat capacities of water and metal are respectively 4200 $Jkg^{-1}K^{-1}$ and 400 $Jkg^{-1}K^{-1}$] *[2019, 11 Jan Shift-II]*
(a) 25% (b) 15%
(c) 30% (d) 20%

23 An unknown metal of mass 192 g heated to a temperature of 100°C was immersed into a brass calorimeter of mass 128 g containing 240 g of water at a temperature of 8.4°C. Calculate the specific heat of the unknown metal, if water temperature stabilises at 21.5°C. (Take, specific heat of brass is 394 J $kg^{-1}K^{-1}$) *[2019, 10 Jan Shift-II]*
(a) 916 J kg^{-1} K^{-1} (b) 654 J kg^{-1} K^{-1}
(c) 1232 J kg^{-1} K^{-1} (d) 458 J kg^{-1} K^{-1}

Numerical Value Type Questions

24 A block of ice of mass 120 g at temperature 0°C is put in 300 g of water at 25°C. The x g of ice melts as the temperature of the water reaches 0°C. The value of x is ………
[Use specific heat capacity of water $= 4200$ J kg^{-1} K^{-1}, Latent heat of ice $= 3.5 \times 10^5$ J kg^{-1}] *[2022, 25 July Shift-II]*

25 A geyser heats water flowing at a rate of 2.0 kg per minute from 30°C to 70°C. If geyser operates on a gas burner, the rate of combustion of fuel will be ……… g min^{-1}.
[Heat of combustion $= 8 \times 10^3$ Jg^{-1}]
Specific heat of water $= 4.2$ Jg^{-1} $°C^{-1}$] *[2022, 26 June Shift-II]*

Imp. Concept of Latent Heat

26 M g of steam at 100°C is mixed with 200 g of ice at its melting point in a thermally insulated container. If it produces liquid water at 40°C [heat of vaporisation of water is 540 cal/g and heat of fusion of ice is 80 cal/g], the value of M is … . *[2020, 7 Jan Shift-II]*

Topic 3 Heat Transfer

Level 1

01 The difference of temperature in a material can convert heat energy into electrical energy. To harvest the heat energy, the material should have **[2025, 29 Jan Shift-II]**

(a) high thermal conductivity and high electrical conductivity
(b) low thermal conductivity and low electrical conductivity
(c) high thermal conductivity and low electrical conductivity
(d) low thermal conductivity and high electrical conductivity

02 Two spherical bodies of same materials having radii 0.2 m and 0.8 m are placed in same atmosphere. The temperature of the smaller body is 800 K and temperature of the bigger body is 400 K. If the energy radiated from the smaller body is E, the energy radiated from the bigger body is (assume, effect of the surrounding temperature to be negligible), **[2025, 22 Jan Shift-I]**

(a) 256 E (b) 16 E (c) 64 E (d) E

03 Two metallic blocks M_1 and M_2 of same area of cross-section are connected to each other (as shown in figure). If the thermal conductivity of M_2 is K, then the thermal conductivity of M_1 will be :[Assume steady state heat conduction] **[2022, 24 June Shift-I]**

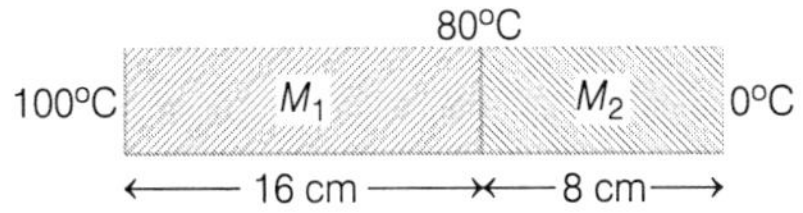

(a) 10 K (b) 8 K (c) 12.5 K (d) 2 K

04 The temperature θ at the junction of two insulating sheets, having thermal resistances R_1 and R_2 as well as top and bottom temperatures θ_1 and θ_2 (as shown in figure) is given by **[2021, 26 Feb Shift-I]**

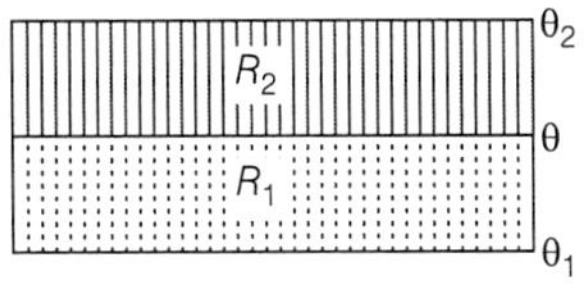

(a) $\dfrac{\theta_2 R_2 - \theta_1 R_1}{R_2 - R_1}$ (b) $\dfrac{\theta_1 R_2 - \theta_2 R_1}{R_2 - R_1}$

(c) $\dfrac{\theta_1 R_2 + \theta_2 R_1}{R_1 + R_2}$ (d) $\dfrac{\theta_1 R_1 + \theta_2 R_2}{R_1 + R_2}$

05 Two materials having coefficients of thermal conductivity $3K$ and K and thickness d and $3d$ respectively, are joined to form a slab. The temperatures of the outer surfaces are θ_2 and θ_1 respectively, $(\theta_2 > \theta_1)$. The temperature at the interface is **[2019, 9 April Shift-II]**

(a) $\dfrac{\theta_2 + \theta_1}{2}$ (b) $\dfrac{\theta_1}{3} + \dfrac{2\theta_2}{3}$

(c) $\dfrac{\theta_1}{6} + \dfrac{5\theta_2}{6}$ (d) $\dfrac{\theta_1}{10} + \dfrac{9\theta_2}{10}$

Imp. Thermal Conductivity

06 A cylinder of radius R is surrounded by a cylindrical shell of inner radius R and outer radius $2R$. The thermal conductivity of the material of the inner cylinder is K_1 and that of the outer cylinder is K_2. Assuming no loss of heat, the effective thermal conductivity of the system for heat flowing along the length of the cylinder is **[2019, 12 Jan Shift-I]**

(a) $\dfrac{K_1 + K_2}{2}$ (b) $\dfrac{K_1 + 3K_2}{4}$

(c) $\dfrac{2K_1 + 3K_2}{5}$ (d) $K_1 + K_2$

07 A long metallic bar is carrying heat from one of its ends to the other end under steady-state. The variation of temperature θ along the length x of the bar from its hot end is best described by which of the following figure. **[AIEEE 2009]**

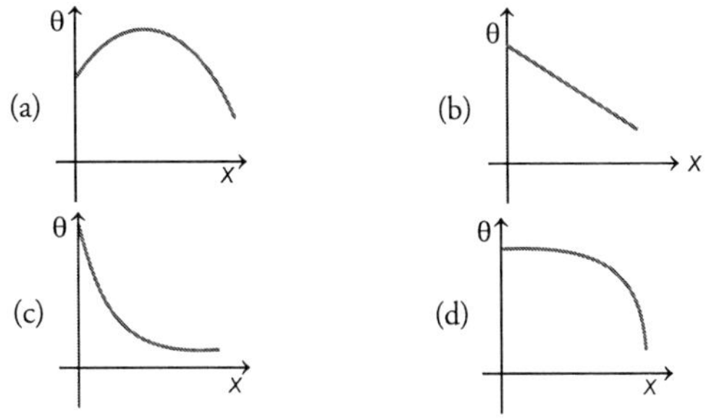

Level 2

08 The temperature of a body in air falls from 40°C to 24°C in 4 min. The temperature of the air is 16°C. The temperature of the body in the next 4 min will be **[2025, 29 Jan Shift-I]**

(a) $\dfrac{28}{3}$ °C (b) $\dfrac{42}{3}$ °C

(c) $\dfrac{14}{3}$ °C (d) $\dfrac{56}{3}$ °C

09 If K_1 and K_2 are the thermal conductivities, L_1 and L_2 are the lengths and A_1 and A_2 are the cross-sectional areas of steel and copper rods respectively, such that $\dfrac{K_2}{K_1} = 9$, $\dfrac{A_1}{A_2} = 2$, $L_1 / L_2 = 2$.

Then, for the arrangement as shown in the figure, the value of temperature T of the steel-copper junction in the steady state will be *[2022, 27 July Shift-I]*

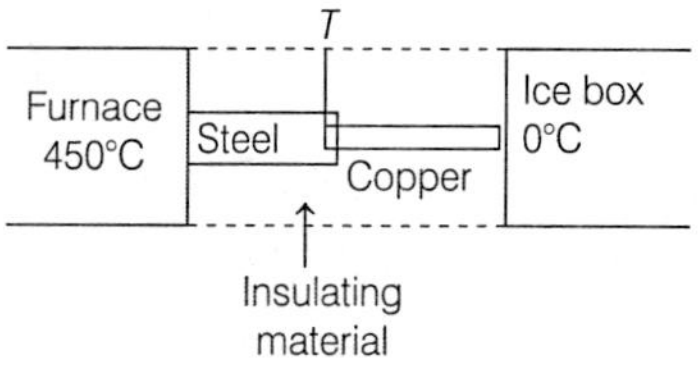

(a) 18°C (b) 14°C (c) 45°C (d) 150°C

10 An ice cube of dimensions 60 cm × 50 cm × 20 cm is placed in an insulation box of wall thickness 1 cm. The box keeping the ice cube at 0°C of temperature is brought to a room of temperature 40°C. The rate of melting of ice is approximately (Latent heat of fusion of ice is 3.4×10^5 J kg^{-1} and thermal conducting of insulation wall is 0.05 Wm^{-1}°C^{-1}) *[2022, 26 July Shift-II]*

(a) 61×10^{-3} kg s^{-1} (b) 61×10^{-5} kg s^{-1}
(c) 208 kg s^{-1} (d) 30×10^{-5} kg s^{-1}

11 Two identical metal wires of thermal conductivities K_1 and K_2, respectively are connected in series. The effective thermal conductivity of the combination is *[2021, 17 March Shift-I]*

(a) $\dfrac{2K_1 K_2}{K_1 + K_2}$ (b) $\dfrac{K_1 + K_2}{2K_1 K_2}$ (c) $\dfrac{K_1 + K_2}{K_1 K_2}$ (d) $\dfrac{K_1 K_2}{K_1 + K_2}$

12 Three rods of identical cross-section and lengths are made of three different materials of thermal conductivity K_1, K_2 and K_3, respectively. They are joined together at their ends to make a long rod (see figure). One end of the long rod is maintained at 100°C and the other at 0°C (see figure). If the joints of the rod are at 70°C and 20°C in steady state and there is no loss of energy from the surface of the rod, the correct relationship between K_1, K_2 and K_3 is *[2020, 6 Sep Shift-II]*

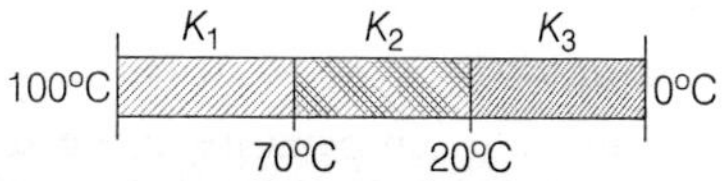

(a) $K_1 : K_3 = 2:3$; $K_2 : K_3 = 2:5$
(b) $K_1 < K_2 < K_3$
(c) $K_1 : K_2 = 5:2$; $K_1 : K_3 = 3:5$
(d) $K_1 > K_2 > K_3$

13 A heat source at $T = 10^3$ K is connected to another heat reservoir at $T = 10^2$ K by a copper slab which is 1 m thick. Given that the thermal conductivity of copper is 0.1 W K^{-1}m^{-1}, the energy flux through it in the steady state is *[2019, 10 Jan Shift-I]*

(a) 90 Wm^{-2} (b) 65 Wm^{-2}
(c) 120 Wm^{-2} (d) 200 Wm^{-2}

14 One end of a thermally insulated rod is kept at a temperature T_1 and the other at T_2. The rod is composed of two sections of lengths l_1 and l_2 and thermal conductivities K_1 and K_2, respectively. The temperature at the interface of the two sections is *[AIEEE 2007]*

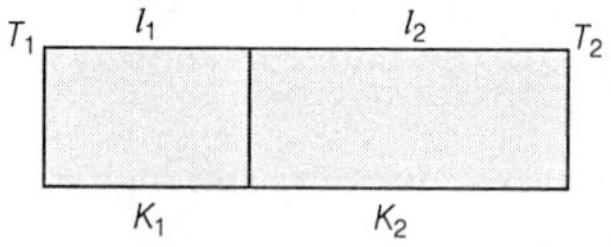

(a) $\dfrac{(K_2 l_2 T_1 + K_1 l_1 T_2)}{(K_1 l_1 + K_2 l_2)}$ (b) $\dfrac{(K_2 l_1 T_1 + K_1 l_2 T_2)}{(K_2 l_1 + K_1 l_2)}$
(c) $\dfrac{(K_1 l_2 T_1 + K_2 l_1 T_2)}{(K_1 l_2 + K_2 l_1)}$ (d) $\dfrac{(K_1 l_1 T_1 + K_2 l_2 T_2)}{(K_1 l_1 + K_2 l_2)}$

15 Assuming the sun to be a spherical body of radius R at a temperature of T K, evaluate the total radiant power, incident on earth, at a distance r from the sun. *[AIEEE 2006]*

(a) $\dfrac{4\pi r_0^2 R^2 \sigma T^4}{r^2}$ (b) $\dfrac{\pi r_0^2 R^2 \sigma T^4}{r^2}$
(c) $\dfrac{r_0^2 R^2 \sigma T^4}{4\pi r^2}$ (d) $\dfrac{R^2 \sigma T^4}{r^2}$

Where, r_0 is the radius of the earth and σ is Stefan's constant.

Imp. Stefan's Boltzmann Law

16 If the temperature of the sun were to increase from T to $2T$ and its radius from R to $2R$, then the ratio of the radiant energy received on earth to what it was previously, will be *[AIEEE 2004]*

(a) 4 (b) 16
(c) 32 (d) 64

Level 3

17 A cup of coffee cools from 90°C to 80°C in t minutes when the room temperature is 20°C. The time taken by the similar cup of coffee to cool from 80°C to 60°C at the same room temperature is *[2025, 29 Jan Shift-I]*

(a) $\dfrac{13}{10}t$ (b) $\dfrac{10}{13}t$ (c) $\dfrac{5}{13}t$ (d) $\dfrac{13}{5}t$

18 Two thin metallic spherical shells of radii r_1 and r_2 $(r_1 < r_2)$ are placed with their centres coinciding. A material of thermal conductivity K is filled in the space between the shells. The inner shell is maintained at temperature θ_1 and the outer shell at temperature $\theta_2 (\theta_1 < \theta_2)$. The rate at which heat flows radially through the material is *[2021, 31 Aug Shift-II]*

(a) $\dfrac{4\pi K r_1 r_2 (\theta_2 - \theta_1)}{r_2 - r_1}$ (b) $\dfrac{\pi r_1 r_2 (\theta_2 - \theta_1)}{r_2 - r_1}$
(c) $\dfrac{K(\theta_2 - \theta_1)}{r_2 - r_1}$ (d) $\dfrac{K(\theta_2 - \theta_1)(r_2 - r_1)}{4\pi r_1 r_2}$

19 Three rods of copper, brass and steel are welded together to form a Y-shaped structure. Area of cross-section of each rod = 4 cm^2. End of copper rod is maintained at 100°C whereas ends of brass and steel are kept at 0°C. Lengths of the copper, brass and steel rods are 46, 13 and 12 cm, respectively.
The rods are thermally insulated from surrounding except at ends. Thermal conductivities of copper, brass and steel are 0.92, 0.26 and 0.12 CGS units, respectively. Rate of heat flow through copper rod is *[JEE Main 2014]*

(a) 1.2 cal/s (b) 2.4 cal/s
(c) 4.8 cal/s (d) 6.0 cal/s

20 The figure shows a system of two concentric spheres of radii r_1 and r_2 are kept at temperatures T_1 and T_2, respectively. The radial rate of flow of heat in a substance between the two concentric spheres, is proportional to *[AIEEE 2005]*

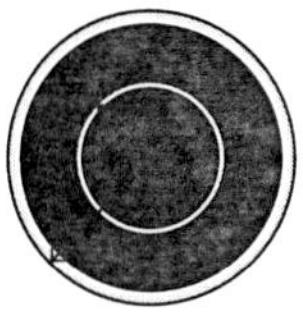

(a) $\frac{(r_2 - r_1)}{(r_1 r_2)}$ (b) $\ln\left(\frac{r_2}{r_1}\right)$ (c) $\frac{r_1 r_2}{(r_2 - r_1)}$ (d) $(r_2 - r_1)$

21 The temperature of the two outer surfaces of a composite slab, consisting of two materials having coefficients of thermal conductivity K and $2K$ and thickness x and $4x$, respectively are T_2 and $T_1 (T_2 > T_1)$. The rate of heat transfer through the slab, in a steady state is $\left(\frac{A(T_2 - T_1)K}{x}\right)f$, then f is equal to *[AIEEE 2004]*

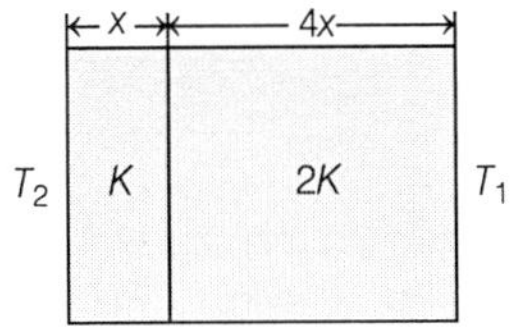

(a) 1 (b) 1/2 (c) 2/3 (d) 1/3

Numerical Value Type Questions

22 Two cylindrical rods A and B made of different materials, are joined in a straight line. The ratios of lengths, radii and thermal conductivites of these rods are: $\frac{L_A}{L_B} = \frac{1}{2}, \frac{r_A}{r_B} = 2$ and $\frac{K_A}{K_B} = \frac{1}{2}$. The free ends of rods A and B are maintained at 400 K, 200 K, respectively. The temperature of rods interface is .... K when equilibrium is established. *[2025, 7 April Shift-II]*

23 A wire of length 10 cm and diameter 0.5 mm is used in a bulb. The temperature of the wire is 1727°C and power radiated by the wire is 94.2 W. Its emissivity is $\frac{x}{8}$ where x = ...
(Given $\sigma = 6.0 \times 10^{-8}$ W m^{-2} K^{-4}, $\pi = 3.14$ and assume that the emissivity of wire material is same at all wavelength) *[2025, 7 April Shift-I]*

24 Three conductors of same length having thermal conductivity K_1, K_2 and K_3 are connected as shown in figure.

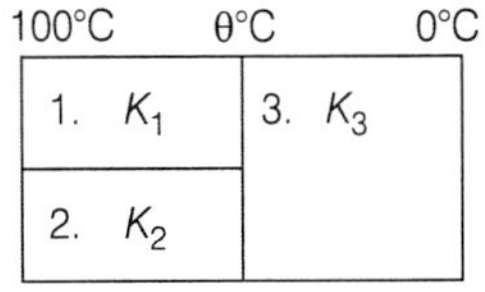

Area of cross-sections of 1st and 2nd conductor are same and for 3nd conductor it is double of the 1st conductor. The temperature are given in the figure. In steady state condition, the value of θ is ...... °C.
(Given, $K_1 = 60$ $Js^{-1}m^{-1}K^{-1}$, $K_2 = 120$ $Js^{-1}m^{-1}K^{-1}$, $K_3 = 135$ $Js^{-1}m^{-1}K^{-1}$) *[2025, 22 Jan Shift-I]*

25 Two plates A and B have thermal conductivities 84 $Wm^{-1}K^{-1}$ and 126 $Wm^{-1}K^{-1}$, respectively. They have same surface area and same thickness. They are placed in contact along their surfaces. If the temperatures of the outer surfaces of A and B are kept at 100°C and 0°C, respectively, then the temperature of the surface of contact in steady state is .... °C. *[2023, 13 April Shift-II]*

26 As per the given figure, two plates A and B of thermal conductivity K and $2K$ are joined together to form a compound plate. The thickness of plates are 4.0 cm and 2.5 cm, respectively and the area of cross-section is 120 cm^2 for each plate. The equivalent thermal conductivity of the compound plate is $\left(1 + \frac{5}{\alpha}\right)$K, then the value of α will be . *[2022, 29 June Shift-I]*

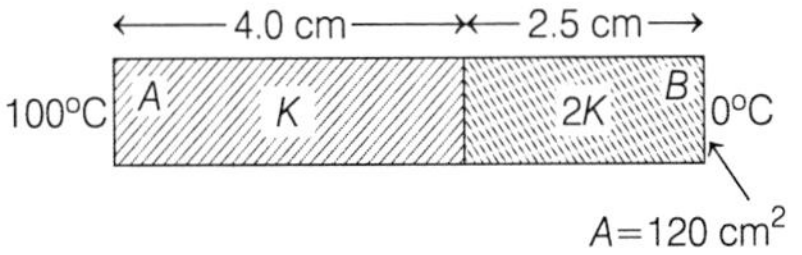

27 A rod CD of thermal resistance 10.0 kW^{-1} is joined at the middle of an identical rod AB as shown in figure. The end A, B and D are maintained at 200°C, 100°C and 125°C, respectively. The heat current in CD is P W. The value of P is .... *[2021, 27 Aug Shift-I]*

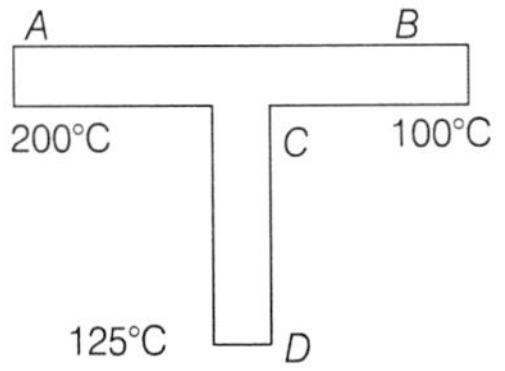

Key **Index**

Topic 1

1 (d)	**2** (a)	**3** (a)	**4** (d)	**5** (c)	**6** (d)	**7** (a)	**8** (d)	**9** (d)	**10** (b)
11 (d)	**12** (d)	**13** (b)	**14** (b)	**15** (b)	**16** (d)	**17** (4)	**18** (12)	**19** (60)	**20** (20)

Topic 2

1 (d)	**2** (b)	**3** (b)	**4** (d)	**5** (a)	**6** (d)	**7** (d)	**8** (a)	**9** (c)	**10** (a)
11 (b)	**12** (c)	**13** (b)	**14** (a)	**15** (a)	**16** (a)	**17** (a)	**18** (b)	**19** (c)	**20** (c)
21 (b)	**22** (d)	**23** (a)	**24** (90)	**25** (42)	**26** (40)				

Topic 3

1 (d)	**2** (d)	**3** (b)	**4** (c)	**5** (d)	**6** (b)	**7** (b)	**8** (d)	**9** (c)	**10** (b)
11 (a)	**12** (a)	**13** (a)	**14** (c)	**15** (b)	**16** (d)	**17** (d)	**18** (a)	**19** (c)	**20** (c)
21 (d)	**22** (360)	**23** (5)	**24** (40)	**25** (40)	**26** (21)	**27** (2)			

Detailed **Explanations**

Topic 1

01. *(d)* $C = \frac{5}{9}(F - 32)$

$$C = \frac{5F}{9} - \frac{32 \times 5}{9}$$

$$C = \frac{5F}{9} - \frac{160}{9}$$

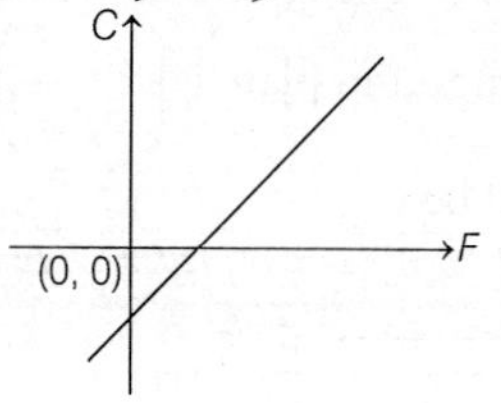

02. *(a)* The relation between fahrenheit and celsius scale is

$$\frac{F - 32}{9} = \frac{C}{5}$$

Change in celsius and fahrenheit is related as,

$$\Delta C = \frac{5}{9}\Delta F$$

Now, $\Delta C = 40°C$ (given)

$$40 = \frac{5}{9}\Delta F$$

$\Rightarrow$ $\Delta F = 72°F$

Hence, option (a) is correct.

03. *(a)* $\frac{X_T - X_L}{X_H - X_L} = \frac{T_F - 32}{212 - 32}$

$\Rightarrow$ $\frac{-95 - (-15°)}{65 - (-15°)} = \frac{T_F - 32}{180}$

$\Rightarrow$ $\frac{-80°}{80} = \frac{T_F - 32}{180}$

$\Rightarrow$ $-180 = T_F - 32$

$\Rightarrow$ $T_F = -180 + 32$

$\Rightarrow$ $T_F = -148°F$

04. *(d)* Given, $\alpha_A > \alpha_B$

$$\Delta l = l_0 \alpha \Delta T$$

Thus, $\Delta l_A > \Delta l_B$ (for equal, ΔT and l_0)

$\therefore$ The bimetallic strip will bend towards left.

It can be shown as below.

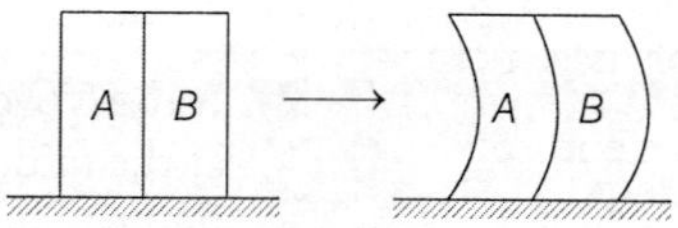

05. *(c)* If a rod is fixed between supports, it will have thermal stress. In the given situation there will be no thermal stress because the rod is free to expand due to increase in temperature.

06. *(d)* We know that $\gamma = 3\alpha$ ...(i)

and $\frac{\Delta V}{V} = \gamma \Delta T \Rightarrow \frac{\Delta V}{V} = 3\alpha \Delta T$

[From Eq. (i)]

$$\Delta V = 3a^3 \alpha \Delta T$$

[$\because$ volume of cube$= a^3$]

07. *(a)* Let α be the effective temperature coefficient of linear expansion.

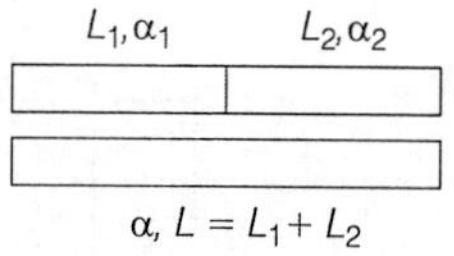

Therefore, $\Delta L = \Delta L_1 + \Delta L_2$

$\Rightarrow$ $L\alpha\Delta T = L_1\alpha_1\Delta T + L_2\alpha_2\Delta T$

($\because$ $L = L_1 + L_2$)

$\therefore$ $(L_1 + L_2)\,\alpha\,\Delta T = \Delta T\,(L_1\alpha_1 + L_2\alpha_2)$

$\Rightarrow$ $\alpha = \frac{L_1\alpha_1 + L_2\alpha_2}{L_1 + L_2}$

Hence, correct option is (a).

08. *(d)* For the given question,

$\therefore$ $(\text{Stress})_{\text{Compressive}} = (\text{Stress})_{\text{Thermal}}$

$$\frac{F}{A} = Y\,\alpha T \Rightarrow \frac{F}{\pi r^2} = Y\,\alpha\,T$$

where, symbols have their usual meaning

$\therefore$ $\alpha = \frac{F}{\pi r^2 YT}$

$\because$ Coefficient of volume expansion,

$$\gamma = 3\alpha$$

$\therefore$ $\gamma = \frac{3F}{\pi r^2 YT}$

09. *(d)* By principle of thermometry for any linear temperature scale,

$$\frac{T - T_{\text{LFP}}}{T_{\text{UFP}} - T_{\text{LFP}}} = a \quad \text{(constant)}$$

where, symbols have their usual meaning, we have

$$\frac{T - 0°\text{C}}{100°\text{C} - 0°\text{C}} = \frac{\frac{x_0}{2} - \frac{x_0}{3}}{x_0 - \frac{x_0}{3}} \Rightarrow \frac{T}{100} = \frac{1}{4}$$

or $\quad T = 25°\text{C}$

10. *(b)* As the temperature of rod rises, extension in the rod, $\Delta L = L\,\alpha\Delta T$

Young's modulus of the rod,

$$Y = \frac{F/A}{\Delta L/L} \Rightarrow Y = \frac{F \times L}{A\Delta L} = \frac{F}{A\alpha\Delta T}$$

11. *(d)* $K = \dfrac{p}{(-\Delta V/V)}$

$$\Rightarrow \quad -\frac{\Delta V}{V} = \frac{p}{K} \Rightarrow -\Delta V = \frac{pV}{K}$$

Change in volume, $\Delta V = \gamma V \Delta T$

where, γ = coefficient of volume expansion

$\therefore \quad \gamma = 3\alpha$

$\therefore \quad \Delta V = V(3\alpha)\,\Delta T$

$\therefore \quad \dfrac{pV}{K} = V(3\alpha)\,\Delta T$

$\therefore \quad \Delta T = \dfrac{p}{3\alpha K}$

12. *(d)* Given, $A = 40\text{ cm}^2 = 4 \times 10^{-3}\text{ m}^2$

$\Delta\theta = 10°\text{C}, Y = 2 \times 10^{11}\text{ Nm}^2,$

$\alpha = 1.2 \times 10^{-5}\text{ K}^{-1}, F = ?$

As $F = Y A \alpha \Delta\theta$

$= 2 \times 10^{11} \times 4 \times 10^{-3} \times 1.2 \times 10^{-5} \times 10$

$= 9.6 \times 10^4\text{ N} \approx 1 \times 10^5\text{ N}$

13. *(b)* Given, $\Delta T = 100°\text{C}$,

$Y = 2 \times 10^{11}\text{ N m}^{-2}$

$\alpha = 1.1 \times 10^{-5}\text{ K}^{-1}$

Thermal strain in the wire $= \alpha\Delta T$

Thermal stress in rod is the pressure due to the thermal strain.

Required pressure $= Y\,\alpha\Delta T$

$= 2 \times 10^{11} \times 1.1 \times 10^{-5} \times 100$

$= 2.2 \times 10^8\text{ Pa}$

14. *(b)* In the given graph, for scale P

Lower fixed point $= 30°$

Upper fixed point $= 180°$

and for the scale Q,

Lower fixed point (LFP) $= 0°$

Upper fixed point $= 100°$

$\because$ The relationship between the two scale P and Q is given by

$$\frac{t_P - (\text{LFP})_P}{\text{Number of divisions on scale } P} = \frac{t_Q - (\text{LFP})_Q}{\text{Number of divisions on scale } Q}$$

$$\Rightarrow \frac{t_P - 30}{150} = \frac{t_Q - 0}{100} \Rightarrow \frac{t_P - 30}{150} = \frac{t_Q}{100}$$

15. *(b)* Given, $\quad A = 24\text{ m}^2$

$\Delta T = 10\text{ °C}, \alpha = 5.0 \times 10^{-4}\text{ °C}^{-1}.$

We know, $\quad \gamma = 3\alpha = 3 \times 5 \times 10^{-4}$

$= 1.5 \times 10^{-3}\text{ °C}^{-1}$

Since, total surface area of cube,

$6a^2 = 24 \Rightarrow a = 2\text{ m}$

$\therefore$ Volume of cube,

$V = a^3 = 2^3 = 8\text{ m}^3$

$\because \quad \Delta V = \gamma V \Delta T = 1.5 \times 10^{-3} \times 8 \times 10$

$= 1.2 \times 10^5\text{ cm}^3$

16. *(d)* Extension due to the weight will be equal to contraction by cooling.

$$Y = \frac{Mg/A}{\Delta l/l} \Rightarrow \Delta l = \frac{Mgl}{AY}$$

Also, $\Delta l = l\,\alpha\Delta T \Rightarrow l\,\alpha\Delta T = \dfrac{Mgl}{AY}$

$$\Rightarrow \quad M = \frac{\alpha\Delta T A Y}{g}$$

$$= \frac{10^{-5} \times 10^{11} \times 20 \times \pi \times (10^{-3})^2}{10}$$

$\therefore \quad M = 2\pi\text{ kg} = 6.28\text{ kg} \simeq 6.3\text{ kg}$

17. *(4)* We know that,

$F = YA\alpha\Delta T$

$\Rightarrow \quad F = 2 \times 10^{11} \times 10^{-4} \times 10^{-5} \times 200$

$= 4 \times 10^4\text{ N}$

$\therefore \quad x = 4$

18. *(12)* Change in diameter is given by

$\Delta d = d_0\,\alpha\,\Delta T = d \times 10^{-3}$

Given, $d_0 = 5 \times 10^{-2}\text{ m}$

$\alpha = 1.6 \times 10^{-5}/°\text{C}$

$\Delta T = (177 - 27)°\text{C} = 150°\text{C}$

$\therefore \quad \Delta d = 5 \times 10^{-2} \times 1.6 \times 10^{-5} \times 150$

$= 1200 \times 10^{-7} = 12 \times 10^{-5}\text{ m}$

$= 12 \times 10^{-3}\text{ cm} = d \times 10^{-3}\text{ cm}$

So, $\quad d = 12$

19. *(60)* As given,

$\alpha_{\text{iron}} = 1.2 \times 10^{-5}\text{ K}^{-1},$

$\alpha_{\text{brass}} = 1.8 \times 10^{-5}\text{ K}^{-1}$

$l_{\text{iron}} - l_{\text{brass}} = K \quad \ldots\text{(i)}$

$l_{0(\text{iron})}(1 + \alpha_{\text{iron}}\Delta T) - l_{0(\text{brass})}(1 + \alpha_{\text{brass}}\Delta T) = K$

So, $l_{0(\text{iron})} - l_{0(\text{brass})} + l_{0\text{iron}}\,\alpha_{\text{iron}}\,\Delta T - l_{0\text{brass}}\,\alpha_{\text{brass}}\Delta T = K$

$K + (l_{0(\text{iron})} \times \alpha_{\text{iron}} - l_{0(\text{brass})}\,\alpha_{\text{brass}})\,\Delta T = K \quad$ (using (i))

$\alpha_{\text{iron}} \times l_{0(\text{iron})} = \alpha_{\text{brass}} \times l_{0(\text{brass})}$

$1.2 \times 10^{-5} \times l_{0(\text{iron})} = 1.8 \times 10^{-5} \times l_{0(\text{brass})}$

$2\,l_{0(\text{iron})} = 3 \times 40, \; l_{0(\text{iron})} = 60\text{ cm}$

20. *(20)* Volume of empty part

= Volume of beaker − Volume of mercury

Since given that there is no change in unfilled volume of the beaker with the varying temperature.

So, change in volume of beaker

= change in volume of mercury

$\Rightarrow V_b\gamma_b\Delta T = V_m\gamma_m\Delta T$

$\Rightarrow \quad V_b\gamma_b = V_m\gamma_m$

$$\Rightarrow \quad V_m = \frac{V_b\gamma_b}{\gamma_m} = \frac{500 \times 6 \times 10^{-6}}{1.5 \times 10^{-4}}$$

$= 20\text{ cm}^3 = 20\text{ cc}$

Topic 2

01. *(d)* According to given condition,

$ms\Delta T = mgh$

$$\Rightarrow \quad \Delta T = \frac{gh}{s} = \frac{10 \times 200}{4200}$$

$$= \frac{10}{21}\text{ K} = 0.48\text{ K}$$

02. *(b)* $\quad Q = ms\Delta T + mL$

$625 = m[125 \times 300 + 2.5 \times 10^4]$

$625 = m[37500 + 25000]$

$$m = \frac{625}{62500} = \frac{1}{100}\text{ kg}$$

$m = 10\text{ g}$

03. *(b)*

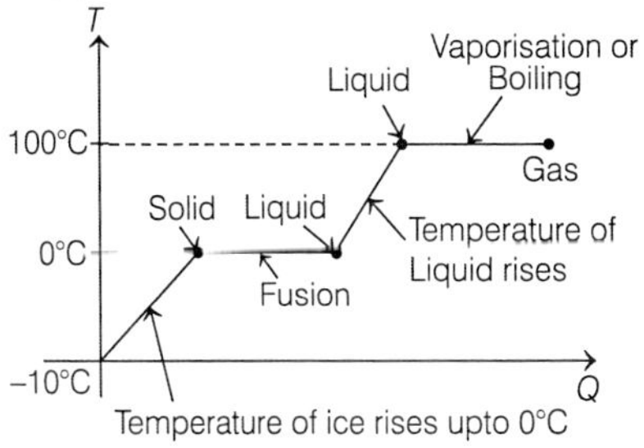

04. *(d)* $m = 0.08\text{ kg}$

$S_V = 0.17\text{ kcal/kg-°C}$

$\Delta T = 5°\text{C}$

Change in internal energy is,

$\Delta U = mS_V\Delta T$

$\Delta U = 0.08 \times 0.17 \times 10^3 \times 5$

$\Delta U = 68 \text{ cal} \Rightarrow 68 \times 4.184 \text{ J}$

$\Delta U = 284.24 \text{ J} \simeq 284 \text{ J}$

05. *(a)* Given, $H = 63$ m

$g = 9.8 \text{ ms}^{-2}$

Specific heat of water,

$s = 1 \text{ cal g}^{-1} \text{ C}^{-1}$

$= 4.2 \times 10^3 \text{ J kg}^{-1} \text{ C}^{-1}$

Since energy, $E = mgH = ms\Delta T$

$\therefore \quad \Delta T = \dfrac{gH}{s} = \dfrac{9.8 \times 63}{4.2 \times 10^3} = 0.147°\text{C}$

06. *(d)* Let the mass of ice which gets melted be m_i. Now,

Heat lost by water = Heat gained by ice

$\Rightarrow \quad m_w c_w \Delta T_w = m_i L_i$

$\Rightarrow \quad \dfrac{200}{1000} \times 4200 \times (25 - 0) = m_i \times 3.4 \times 10^5$

$\Rightarrow \quad 21000 = m_i \times 3.4 \times 10^5$

$\Rightarrow \quad m_i = \dfrac{21000}{3.4 \times 10^5} = 0.0617 \text{ kg}$

$= 61.7$ g

Hence, correct option is (d).

07. *(d)* Applying the principle of energy

$\dfrac{1}{2} kx^2 = (m_1 s_1 + m_2 s_2)\Delta T,$

$\Delta T = \dfrac{kx^2}{2(m_1 s_1 + m_2 s_2)}$

$= \dfrac{800 \times 4 \times 10^{-4}}{2(500 \times 10^{-3} \times 400 + 1 \times 4184)}$

$= 3.6 \times 10^{-5}$ K

08. *(a)*

> **key idea** Decrease in surface energy = Heat required in vaporisation

Decrease in surface energy

$= T \times 2(4\pi r dr)$

Heat required in vapourisation

$= ML = V.\rho L = 4\pi r^2 dr.\rho L$

Thus, $T.2(4\pi r dr) = 4\pi r^2 dr \, \rho L$

$\Rightarrow \quad r = 2T/\rho L$

09. *(c)*

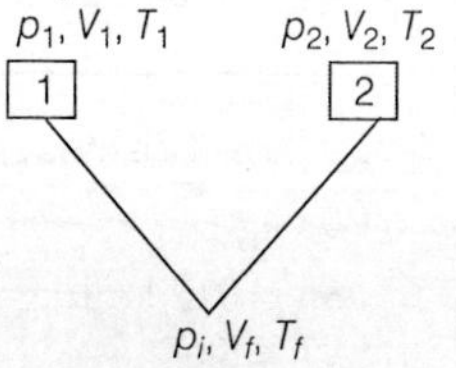

Number of masses will remain constant.

$n_1 + n_2 = n_f$

$\dfrac{p_1 V_1}{RT_1} + \dfrac{p_2 V_2}{RT_2} = \dfrac{p_f V_f}{RT_f}$

$\dfrac{8 \times 2V}{R \times 1000} + \dfrac{7 \times V}{R \times 500} = \dfrac{p_f (3V)}{R \times 600}$

$\dfrac{16}{1000} \times \dfrac{14}{1000} = \dfrac{p_f}{R \times 600}$

$\dfrac{30}{1000} = \dfrac{p_f}{200}$

$p_f = 6$ kPa

10. *(a)* $A_0 = 9 \times 4 \text{ cm}^2 = 36 \times 10^{-4} \text{ m}^2$

$\alpha = 3.1 \times 10^{-5} \text{ K}^{-1}$

As we know,

$\Delta Q = ms\Delta T$

$\Delta Q = 8.1 \times 10^2$ J

$m = 0.1$ kg

$s = 900 \text{ J kg}^{-1} \text{ K}^{-1}$

$\Delta T = ?$

$8.1 \times 10^2 = 0.1 \times 900 \times \Delta T$

$\therefore \quad \Delta T = \dfrac{8.1 \times 10^2}{0.1 \times 900} = 9$

$\Delta A = A_0 2\alpha\Delta T$

$= 36 \times 10^{-4} \times 2 \times 3.1 \times 10^{-5} \times 9$

$= 2008.8 \times 10^{-9}$

$= 2.0 \times 10^{-6} \text{ m}^2$

11. *(b)*

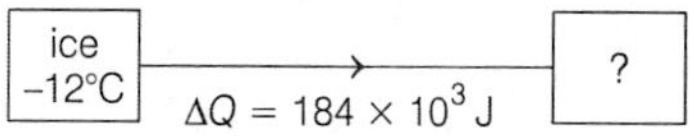

First ice absorbs heat and it's temperature rises to 0°C.

Now using, $\Delta Q = mc\Delta T$

We have,

$\Delta Q = (600 \times 10^{-3})(2222.3)(12)$

$= 16000.56 \text{ J} \approx 16 \text{ kJ}$

Now, ice at 0°C starts melting. Let m kg of ice melts due to remaining heat.

Heat available $= 184 - 16 = 168$ kJ.

Now using,

$\Delta Q = mL$, we have

$168 = m(336)$

$\Rightarrow \quad m = 0.5 \text{ kg} = 500 \text{ g}$

As out of 600 g of ice, 500 g is converted into water, final mixture contains 100 g ice and 500 g water.

12. *(c)* Mass of iron ball,

$m_1 = 100 \text{ g} = 0.1 \text{ kg}$

Mass of hammer, $m_2 = 1.5$ kg

$v = 60$ m/s, $s = 0.42 \text{ Jg}^{-1} \text{°C}^{-1}$

$= 420 \text{ J kg}^{-1} \text{°C}^{-1}$

According to question,

$\dfrac{\text{Total kinetic energy of hammer}}{4}$

= Heat energy gain by nail

$\Rightarrow \quad \dfrac{1}{4}\left[\dfrac{1}{2} m_2 v^2\right] = m_1 s\Delta T$

$\Rightarrow \dfrac{1}{4}\left[\dfrac{1}{2} \times 1.5 \times 60^2\right] = 0.1 \times 420 \times \Delta T$

$\Rightarrow \quad \Delta T = \dfrac{675}{42} = 16.07 \text{ °C}$

13. *(b)* Given, $m_A = m_B = m$ (Let)

As we know that,

$\Delta Q / \Delta t = ms \, \Delta T / \Delta t$

Since, $\left(\dfrac{\Delta Q}{\Delta t}\right)_A = \left(\dfrac{\Delta Q}{\Delta t}\right)_B$

$\Rightarrow \quad ms_A \left(\dfrac{\Delta T}{t}\right)_A = ms_B \left(\dfrac{\Delta T}{t}\right)_B$

$\Rightarrow \dfrac{s_A}{s_B} = \dfrac{\left(\dfrac{\Delta T}{t}\right)_B}{\left(\dfrac{\Delta T}{t}\right)_A} = \dfrac{\dfrac{90}{6}}{\dfrac{120}{3}} = \dfrac{3}{2 \times 4} = \dfrac{3}{8}$

14. *(a)* We know that,

for bullet $\Delta Q = (ms\Delta T)_{\text{bullet}}$ ...(i)

$\Delta Q = ms\Delta T$

Given $\quad \Delta Q = \dfrac{1}{2}\left(\dfrac{1}{2} mv^2\right)$ ...(ii)

From Eqs. (i) and (ii), we get

$(ms\Delta T)_{\text{bullet}} = \dfrac{1}{2}\left(\dfrac{1}{2} mv^2\right)$

$\Rightarrow \quad \Delta T = v^2 / 4s$

Since, $\quad s = 0.030$ cal/g° C

and $\quad v = 210$ m/s

On putting values,

$\Delta T = \dfrac{(210)^2}{4 \times 4.2 \times 0.03 \times 1000} = 87.5 \text{ °C}$

Hence, correct option is (a).

15. *(a)*

> **key idea** In such kind of heat transfer problems,
>
> Heat given by water = Heat gained by ice and
>
> $(\text{heat})_{\text{liquid}}$ = mass × specific heat × temperature
>
> $= \{m_l \, s_l \, \Delta T_l\}$
>
> $(\text{Heat})_{\text{solid}}$ = (mass × latent heat) + (mass × specific heat × temperature)
>
> $= (m_s \times L) + m_s \times s_s \times T_s$

Heat given by water is (specific heat of water is $1 \text{ cal g}^{-1} \text{°C}^{-1}$)

$(\Delta H)_{\text{water}} = M_2 \times 1 \times (50 - 0)$

$= 50 M_2$ ...(i)

Heat taken by ice is

$(\Delta H)_{ice} = M_1 \times 0.5 \times [0-(-10)] + M_1 \times L_{ice}$

$\Rightarrow (\Delta H)_{ice} = 5M_1 + M_1 L_{ice}$ ...(ii)

Comparing Eqs. (i) and (ii), we get

$(\Delta H)_{water} = (\Delta H)_{ice}$

$\therefore \quad 50M_2 = 5M_1 + M_1 L_{ice}$

$\Rightarrow \quad L_{ice} = \dfrac{50M_2 - 5M_1}{M_1}$

$\Rightarrow \quad L_{ice} = 50\dfrac{M_2}{M_1} - 5$

16. *(a)* Let amount of ice be x g.
According to the principle of calorimeter, heat lost by water = heat gained by ice

Here, heat lost by water,

$\Delta Q = ms_{water}\,\Delta T$

Substituting the given values, we get

$\Delta Q = 50 \times 4.2 \times 40$

Heat gained by ice,

$\Delta Q = x\, s_{ice}\,\Delta T + (x-20)\,L$

$= x \times 2.1 \times 20 + (x-20) \times 334$

$= 20\,x \times 2.1 + 334x - 6680$

$\therefore\ 20\,x \times 2.1 + 334x - 6680 = 50 \times 4.2 \times 40$

$42x + 334x - 6680 = 8400$

$\Rightarrow\ 376\,x = 15080$ or $x = 40.10$ g

$x \approx 40$ g

17. *(a)* Ice from $-10°C$ changes to vapour with temperature.
The different stage of heat energy required is

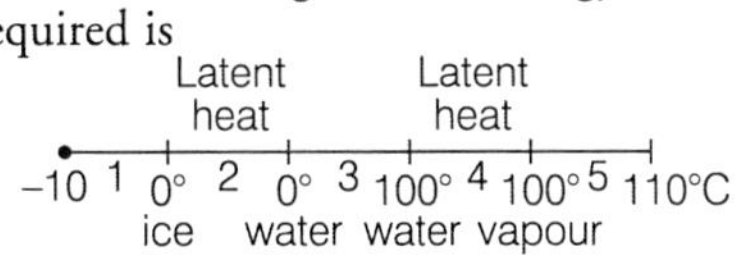

Heat required in each step is,

$\Delta Q_1 = ms_1 \Delta T = 10^{-3} \times 2100 \times 10$

$= 21\,J$

$\Delta Q_2 = mL_f = 10^{-3} \times 3.35 \times 10^5 = 335\,J$

$\Delta Q_3 = m \times s_w \times \Delta T$

$= 10^{-3} \times 4180 \times 100 = 418\,J$

$\Delta Q_4 = m \times L_v = 10^{-3} \times 2.25 \times 10^6$

$= 2250\,J$

$\Delta Q_5 = m \times s_v \times \Delta T = 10^{-3} \times 1920 \times 10$

$= 19.2\,J$

$\Delta Q_{total} = \Delta Q_1 + \Delta Q_2 + \Delta Q_3 + \Delta Q_4 + \Delta Q_5$

$\Delta Q_{total} = 3043.25\,J$

18. *(b)* Let the mass for the bullet be m.
Heat needed to increase the temperature of the bullet from 127°C to 327°C, $Q_1 = ms\Delta T$

$= m \times 125\ J/kg\ K \times (600-400)\ K$

$= m \times 2.5 \times 10^4\ J\ kg^{-1}$

Heat needed to melt the bullet,

$Q_2 = mL = m \times 2.5 \times 10^4\ J\ kg^{-1}$

The net heat needed to melt the bullet,

$Q = Q_1 + Q_2$

$= m \times (2.5 + 2.5) \times 10^4\ J\ kg^{-1}$

Let v be the initial speed of the bullet, then 40% of its kinetic energy will be equal to the heat Q.

$\therefore \dfrac{2}{5} \times \dfrac{1}{2} mv^2 = m \times 5 \times 10^4$

or $\quad v^2 = 25 \times 10^4$ or $v = 500\ ms^{-1}$

19. *(c)* Given, $m = 5$ kg

$T_1 = 500°C$ and

$T_2 = 0°C$ (Placed in the ice block)

$s_{Cu} = 0.39\ Jg^{-1}°C^{-1}$

$= 0.39 \times 10^3\ Jkg^{-1}°C^{-1}$

$\therefore$ Heat given by the block,

$\Delta Q = ms\Delta T$

$= 5 \times 0.39 \times 10^3 \times (500-0)$

$= 9.75 \times 10^5\ J$

Heat taken by ice to melt m kg of ice

$Q_2 = mL_f$

$Q_2 = m \times 335 \times 10^3$

$[\because L_f = 335\ Jg^{-1} = 335 \times 10^3\ Jkg^{-1}]$

Hence, according to principle of calorimetry,

$Q_2 = Q_1$

$m \times 335 \times 10^3 = 9.75 \times 10^5$

$m = \dfrac{9.75 \times 10^5}{335 \times 10^3} = 2.91$ kg

$\simeq 2.9$ kg

20. *(c)*

key idea Heat lost by a hot body = Heat gain by a cold body.

Given, $m_x = m_y = m_z = m$

Temperature of mixture of liquids x and y, $\quad T_1 = 16°\,C$

Temperature of mixture of liquids y and z, $\quad T_2 = 26°\,C$

While mixing the liquids x and y,

Heat gained by liquid x = Heat lost by liquid y

$ms_x\,(T_1 - 10) = ms_y\,(20 - T_1)$

$\Rightarrow \quad s_x\,(16-10) = s_y\,(20-16)$

$\Rightarrow \quad \dfrac{s_x}{s_y} = \dfrac{2}{3}$ ...(i)

Similarly, for liquid y and z,

$\therefore$ Heat gained by liquid y = Heat lost by liquid z

$ms_y\,(T_2 - 20) = ms_z\,(30 - T_2)$

$\Rightarrow \quad s_y\,(26-20) = s_z\,(30-26)$

$\Rightarrow \quad \dfrac{s_y}{s_z} = \dfrac{2}{3}$ ...(ii)

Multiply Eq. (i) by Eq. (ii), we get

$\dfrac{s_x}{s_z} = \dfrac{4}{9}$

Consider the mixing of liquids x and z, let the temperature of mixture be T_3.

$\therefore$ Heat gained by liquid x = Heat lost by liquid z

$ms_x\,(T_3 - 10) = ms_z\,(30 - T_3)$

$\Rightarrow \quad s_x\,(T_3 - 10) = s_z\,(30 - T_3)$

$\Rightarrow \quad \dfrac{s_x}{s_z} = \dfrac{30 - T_3}{T_3 - 10}$

$\Rightarrow \quad \dfrac{4}{9} = \dfrac{30 - T_3}{T_3 - 10}$

$\Rightarrow \quad 4T_3 - 40 = 270 - 9T_3$

$\Rightarrow \quad T_3 = 23.84°C$

21. *(b)* In first case, according to principle of calorimetry,

heat lost by liquid A = heat gained by liquid B

or $\quad m_A s_A \Delta T_A = m_B s_B \Delta T_B$

$\Rightarrow\ 100 \times s_A(100-90) = 50 \times s_B\,(90-75)$

$\Rightarrow \quad 1000 s_A = 50 \times 15 s_B$

or $\quad 4s_A = 3s_B$ ...(i)

Similarly, in second case,

$100 \times s_A(100 - T) = 50 \times s_B(T - 50)$

where, T = Final temperature of the mixture.

$\Rightarrow \quad 4s_A(100 - T) = 2s_B(T - 50)$

Using Eq. (i),

$3s_B(100 - T) = 2s_B(T - 50)$

or $\quad 300 - 3T = 2T - 100$

or $\quad 5T = 400$ or $T = 80°C$

22. *(d)* Using heat lost or gained without change in state is $\Delta Q = ms\Delta T$,

Let final temperature of ball be T.

Then heat lost by ball is,

$\Delta Q = 0.1 \times 400(500 - T)$ ...(i)

This lost heat by ball (ΔQ) is gained by water and vessel (ΔQ_2) and given as

$\therefore \quad \Delta Q_1 = 0.5 \times 4200(T - 30)$

and $\quad \Delta Q_2$ = heat capacity $\times \Delta T$

$= 800 \times (T - 30)$

According to principle of calorimetry, total heat lost = total heat gained

$\Rightarrow\ 0.1 \times 400(500 - T)$

$= 0.5 \times 4200(T - 30) + 800(T - 30)$

$\Rightarrow \quad (500 - T) = \dfrac{(2100+800)(T-30)}{40}$

$\Rightarrow \quad 500 - T = 72.5(T-30)$

$\Rightarrow \quad 500 + 2175 = 73.5T$

or $\quad T = 36.39$ K

So, percentage increment in temperature of water

$= \dfrac{36.39-30}{30} \times 100 \approx 20\%$

23. *(a)* Let specific heat of unknown metal is s and heat lost by this metal is ΔQ.

Heat lost

$\Delta Q = ms\Delta T \quad \ldots$ (i)

For unknown metal, $m = 192$ g and

$\Delta T = (100 - 21.5)$ °C

$\therefore \quad \Delta Q' = 192(100-21.5) \times s \quad \ldots$(ii)

Now, this heat is gained by the calorimeter and water inside it.

As, heat gained by calorimeter can be calculated by Eq. (i).

So, for brass specific heat,

$s = 394 \text{ J kg}^{-1} \text{ K}^{-1}$ (given)

$= 0.394 \text{ J g}^{-1} \text{ K}^{-1}$

Mass of calorimeter, $m = 128$ g

$\Delta T = (21.5 - 8.4)$ °C

So, using Eq. (i) for calorimeter, heat gained by brass

$\Delta Q_1 = 128 \times 0.394 \times (21.5 - 8.4) \quad \ldots$(iii)

Heat gained by water

$\Delta Q_2 = ms_w \Delta T$

$m = 240$g,

$s_w = 4.18 \text{ J g}^{-1}\text{K}^{-1}$,

$\Delta T = (21.5 - 8.4)$°C

$\Delta Q_2 = 240 \times 4.18 \times (21.5 - 8.4) \quad \ldots$(iv)

Now, according to the principle of calorimetry,

$\Delta Q' = \Delta Q_1 + \Delta Q_2$

Using Eqs. (ii), (iii) and (iv), we get

$192(100 - 21.5) \times s$

$= 128 \times 0.394 \times (21.5 - 8.4) + 240 \times 4.18 \times (21.5 - 8.4)$

$\Rightarrow 15072\, s = 660.65 + 13142$

$\Rightarrow \quad s = 0.916 \text{ J g}^{-1} \text{ K}^{-1}$

or $\quad s = 916 \text{ J kg}^{-1} \text{ K}^{-1}$

24. *(90)* Given,

$m_{\text{ice}} = 120 \text{ g} = 0.12$ kg, $T_{\text{ice}} = 0$°C

$m_{\text{water}} = 300 \text{ g} = 0.3$ kg, $T_w = 25$°C

$c_w = 4200$ J/kg/K,

$L = 3.5 \times 10^5$ J/kg

Heat lost by water

$H_1 = m_{\text{water}} \times c_w \times \Delta T$

$= 0.3 \times 4200 \times (25 - 0)$

$H_1 = 31500$ J

Heat gained by x gram ice, $H_2 = \dfrac{x \times L}{1000}$

$= \dfrac{x}{1000} \times 3.5 \times 10^5 = 350x$

According to calorimetry law,

$31500 = 350x$

$\therefore$Total ice melts, $x = 90$ g

25. *(42)* According to the question,

$\dfrac{\Delta Q}{\Delta t} = \dfrac{m}{\Delta t} \cdot s\Delta T$

$= m' \times$ Heat of combustion

$\dfrac{2}{60} \times 4.2 \times 10^3 \times (70 - 30)$

$= m' \times 8 \times 10^3 \times 10^3$

$m' = \dfrac{2 \times 4.2 \times 10^3 \times 40}{60 \times 8 \times 10^3 \times 10^3}$ kg/s

$m' = \dfrac{2 \times 4.2 \times 10^3 \times 40 \times 10^3 \times 60}{60 \times 8 \times 10^3 \times 10^3}$ g/min;

$m' = 42$ g/min

26. *(40)* M g of steam at 100° C is mixed with 200 g of ice at 0°C and mixture obtained is at 40°C.

Using the principle of energy conservation, we get

$\Rightarrow M_{\text{steam}} \times L_V + M_{\text{steam}} \times s_w \times \Delta T$

$= M_{\text{ice}} \times L_f + M_{\text{ice}} \times s_w \times \Delta T$

$\Rightarrow M \times 540 + M \times 1 \times (100 - 40)$

$= 200 \times 80 + 200 \times 1 \times (40 - 0)$

$\Rightarrow \quad M \times 600 = 24000$

$\Rightarrow \quad M = 40$ g

Topic 3

01. *(d)*

key idea This phenomenon is called the seeback effect, and materials with these properties are called thermoelectric materials.

Low thermal conductivity ensures that heat doesn't easily transfer through the material, allowing a significant temperature difference to be maintained between the hot and cold sides.

High electrical conductivity allows for efficient flow of electrical current generated by the temperature difference.

02. *(d)* According to Stefan law, rate of heat radiated is given by

$\dfrac{dQ}{dt} = \sigma e A T^4$

$\Rightarrow \quad P \propto AT^4$

$P \propto R^2 T^4$

$\dfrac{P_{\text{smaller}}}{P_{\text{larger}}} = \dfrac{(0.2)^2 \times (800)^4}{(0.8)^2 \times (400)^4} = 1$

$\therefore \quad P_{\text{larger}} = P_{\text{smaller}}$

03. *(b)* Since, both blocks M_1 and M_2 are connected in series, hence rate of heat flow through block M_1 = rate of heat flow through block M_2

$\Rightarrow \quad \dfrac{K_1 A(\Delta T_1)}{l_1} = \dfrac{K_2 A(\Delta T_2)}{l_2}$

$\Rightarrow \quad \dfrac{K_1 \Delta T_1}{l_1} = \dfrac{K_2 \Delta T_2}{l_2}$

$\Rightarrow \quad \dfrac{K_1(100-80)}{16 \times 10^{-2}} = \dfrac{K_2(80-0)}{8 \times 10^{-2}}$

$\Rightarrow \quad \dfrac{20K_1}{2} = 80K_2$

$\Rightarrow \quad K_1 = 8K_2 = 8K \quad [\because K_2 = K \text{ given}]$

04. *(c)* Since, we know, the heat currents,

$\therefore \quad Q = \dfrac{KA\Delta\theta}{l} = \dfrac{\Delta\theta}{R} \quad \left(\because R = \dfrac{l}{KA}\right)$

$\Rightarrow \quad \dfrac{\theta_2 - \theta}{R_2} = \dfrac{\theta - \theta_1}{R_1}$

$\Rightarrow R_1\theta_2 - R_1\theta = R_2\theta - R_2\theta_1$

$\Rightarrow \quad \theta = \dfrac{R_1\theta_2 + R_2\theta_1}{R_1 + R_2}$

05. *(d)* Let interface temperature in steady state conduction be θ, then assuming no heat loss through sides;

$\begin{pmatrix}\text{Rate of heat} \\ \text{flow through} \\ \text{first slab}\end{pmatrix} = \begin{pmatrix}\text{Rate of heat} \\ \text{flow through} \\ \text{second slab}\end{pmatrix}$

$\Rightarrow \quad \dfrac{(3K)\, A(\theta_2 - \theta)}{d} = \dfrac{KA(\theta - \theta_1)}{3d}$

$\Rightarrow 9(\theta_2 - \theta) = \theta - \theta_1$

$\Rightarrow 9\theta_2 + \theta_1 = 10\theta$

$\Rightarrow \quad \theta = \dfrac{9}{10}\theta_2 + \dfrac{1}{10}\theta_1$

06. *(b)* Both the given cylinders are in parallel as heat flow is given along length. In parallel, equivalent thermal conductivity of system is

$K_{\text{eq}} = \dfrac{K_1 A_1 + K_2 A_2}{A_1 + A_2}$

So, in given system

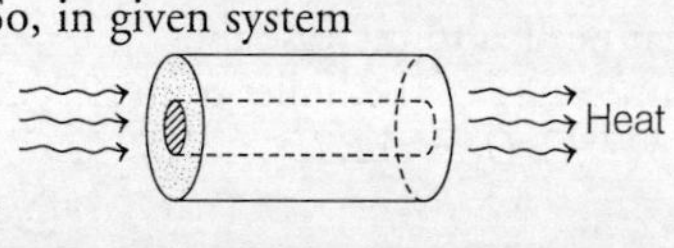

$$K_{eq} = \frac{K_1(\pi R^2) + K_2[\pi(2R)^2 - \pi R^2]}{(\pi R^2) + (4\pi R^2 - \pi R^2)}$$

or $$K_{eq} = \frac{K_1 + 3K_2}{4}$$

07. *(b)* We know that

Rate of flow of heat,

$$\frac{dQ}{dt} = KA\frac{d\theta}{dx}$$

In steady state, flow of heat,

$$d\theta = \frac{dQ}{dt} \cdot \frac{1}{KA} dx$$

$$\Rightarrow \quad \theta_H - \theta = K'x \Rightarrow \theta = \theta_H - K'x$$

Equation $\theta = \theta_H - K'x$ represents a straight line.

08. *(d)* $$\frac{T_2 - T_1}{t} = k[T_{avg} - T_s]$$

$T_1 = 24°C, T_2 = 40°\text{ C}, t = 4$ min,

$T_s = 16°C$ given

$$\frac{40-24}{4} = k[32-16]$$

$$k = \frac{4}{16} = \frac{1}{4}$$

Now, $$\frac{24-T}{4} = k\left[\frac{T+24}{2} - 16\right]$$

$$\frac{3T}{2} = 28$$

$$T = \frac{56}{3}°C$$

09. *(c)* Given, for steel and copper rods,

$$\frac{K_2}{K_1} = 9, \frac{A_1}{A_2} = 2, \frac{L_1}{L_2} = 2$$

From the diagram given in question, for steel rod

$$\frac{dQ}{dt} = K_1 A_1 \frac{(450 - T)}{L_1} \quad \ldots(i)$$

and for copper rod,

$$\frac{dQ}{dt} = K_2 A_2 \frac{(T - 0)}{L_2} \quad \ldots(ii)$$

Dividing Eq. (i) by Eq. (ii), we get

$$1 = \frac{K_1}{K_2} \frac{A_1}{A_2}\left(\frac{450 - T}{T}\right)\frac{L_2}{L_1}$$

$$\Rightarrow \quad \frac{450 - T}{T} = \frac{K_2}{K_1} \cdot \frac{A_2}{A_1} \cdot \frac{L_1}{L_2}$$

$$\Rightarrow \quad \frac{450}{T} - 1 = 9 \times \frac{1}{2} \times 2 \Rightarrow \frac{450}{T} = 9 + 1$$

$$\Rightarrow \quad T = \frac{450}{10} = 45°\text{ C}$$

10. *(b)* Thickness of wall of insulation box,

$$l = 1\text{ cm} = 1 \times 10^{-2}\text{ m}$$

$$T_1 = 0, T_2 = 40°\text{ C}$$

Latent heat of fusion of ice,

$$L_f = 3.4 \times 10^5\text{ Jkg}^{-1}$$

Thermal conductivity,

$$K = 0.05\text{ Wm}^{-1}\text{°C}^{-1}$$
$$= 5 \times 10^{-2}\text{ Wm}^{-1}\text{°C}^{-1}$$

Area,

$$A = 2[0.6 \times 0.5 + 0.5 \times 0.2 + 0.2 \times 0.6]$$
$$= 1.04\text{ m}^2$$

We know that rate of flow of heat

$$\frac{\Delta Q}{\Delta t} = \frac{KA(T_2 - T_1)}{l}$$

$$\Rightarrow \quad \frac{mL_f}{\Delta t} = \frac{KA(T_2 - T_1)}{l} \quad [\because Q = mL]$$

$$\Rightarrow \quad \frac{m}{\Delta t} = \frac{KA(T_2 - T_1)}{l L_f}$$

$$= \frac{5 \times 10^{-2} \times 1.04(40 - 0)}{1 \times 10^{-2} \times 3.4 \times 10^5}$$

$$= 61 \times 10^{-5}\text{ kg/s}$$

11. *(a)* Two identical metal wires of thermal conductivities K_1 and K_2, respectively connected in series are represented as follows

The above figure can also be represented as

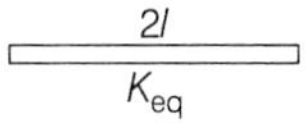

Therefore, the effective thermal conductivity of the combination will be given by

$$R_{eff} = \frac{l}{K_1 A} + \frac{l}{K_2 A} = \frac{2l}{K_{eq} A}$$

$$\Rightarrow \quad \frac{2l}{K_{eq} A} = \frac{l}{A}\left(\frac{1}{K_1} + \frac{1}{K_2}\right)$$

$$\Rightarrow \quad \frac{2l}{K_{eq} A} = \frac{l}{A}\left(\frac{K_1 + K_2}{K_1 K_2}\right)$$

$$\Rightarrow \quad \frac{2}{K_{eq}} = \frac{K_1 + K_2}{K_1 K_2}$$

$$\Rightarrow \quad K_{eq} = \frac{2K_1 K_2}{K_1 + K_2}$$

> • Concept **Enhancer**
>
> Form the correct combination of wires and then find the thermal conductivity of combination accordingly.

12. *(a)* Rate of heat transfer,

$$H = \frac{KA(\theta_H - \theta_L)}{L}$$

At steady state, the rate of heat transfer through each rod is same.

i.e., $\quad H_1 = H_2 = H_3$

For same cross-section and lengths,

$$K \times (\theta_H - \theta_L) = \text{constant}$$

$$K_1(100 - 70) = K_2(70 - 20)$$
$$= K_3(20 - 0)$$

$$K_1 \times 30 = K_2 \times 50 = K_3 \times 20$$

$$3K_1 = 2K_3$$

$$K_1 / K_3 = 2/3 \text{ or } K_1 : K_3 = 2:3$$

or $\quad 5K_2 = 2K_3$ or $K_2 / K_3 = 2/5$

or $K_2 : K_3 = 2:5$

13. *(a)* Energy flux is the rate of heat flow per unit area through the rod. Also, rate of flow of heat per unit time through a material of area of cross-section A and thermal conductivity K between the temperatures T_1 and T_2 $(T_1 > T_2)$ is given as,

$$\frac{\Delta Q}{\Delta t} = \frac{KA(T_1 - T_2)}{l} \quad \ldots(i)$$

Energy flux using Eq. (i), we get

$$\frac{1}{A} \cdot \frac{\Delta Q}{\Delta t} = \frac{K(T_1 - T_2)}{l}$$

Here, $K = 0.1\text{ W K}^{-1}\text{m}^{-1}, l = 1\text{ m}$

$$T_1 = 1000\text{ K and } T_2 = 100\text{ K}$$

$$\therefore \text{ Energy flux} = \frac{0.1(1000 - 100)}{1}$$

$$= 90\text{ Wm}^{-2}$$

14. *(c)* Let temperature at the interface be T.

For part AB,

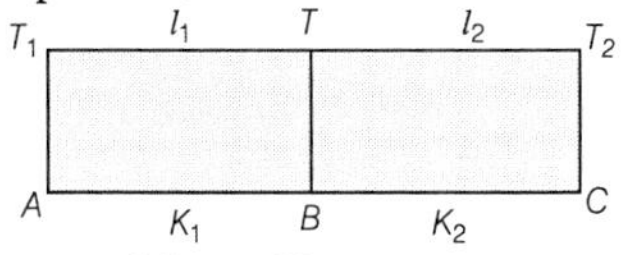

As rate of flow of heat,

$$\frac{Q_1}{t} \propto \frac{(T_1 - T)K_1}{l_1}$$

For part BC, $$\frac{Q_2}{t} \propto \frac{(T - T_2)K_2}{l_2}$$

At equilibrium, $$\frac{Q_1}{t} = \frac{Q_2}{t}$$

$$\Rightarrow \quad \frac{(T_1 - T)K_1}{l_1} = \frac{(T - T_2)K_2}{l_2}$$

or $$T = \frac{T_1 K_1 l_2 + T_2 K_2 l_1}{K_1 l_2 + K_2 l_1}$$

15. *(b)* From Stefan's law, the rate at which energy is radiated by sun at its surface is $P = \sigma \times 4\pi R^2 \times T^4$

[sun is a perfectly black body so, $e = 1$]

The intensity of this power at earth's surface (under the assumption $r >> r_0$) is

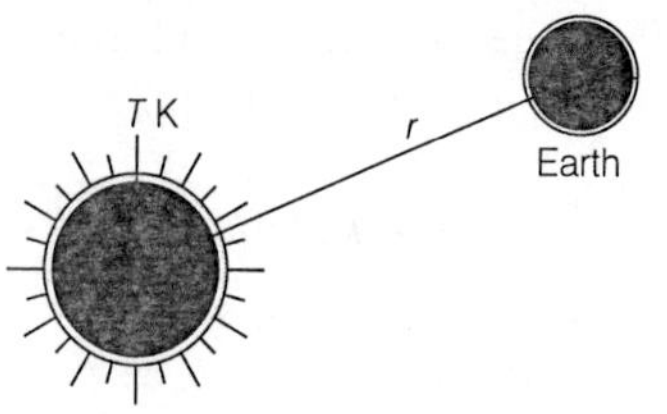

$$I = \frac{P}{4\pi r^2} = \frac{\sigma \times 4\pi R^2 T^4}{4\pi r^2} = \frac{\sigma R^2 T^4}{r^2}$$

The area of earth which receives this energy is only one half of total surface area of earth, whose projection would be πr_0^2.

$\therefore$ The total radiant power as received by earth

$$= \pi r_0^2 \times I = \frac{\pi r_0^2 \times \sigma R^2 T^4}{r^2} = \frac{\pi r_0^2 R^2 \sigma T^4}{r^2}$$

16. *(d)* From Stefan's law, the energy radiated by sun is given by $P = \sigma e A T^4$.

In 1st case, $P_1 = \sigma e \times 4\pi R^2 \times T^4$

In 2nd case,

$$P_2 = \sigma e \times 4\pi (2R)^2 \times (2T)^4$$
$$= \sigma e \times 4\pi R^2 \times T^4 \times 64 = 64 P_1$$

The rate at which energy received at earth is

$$E = \frac{P}{4\pi R_{SE}^2} \times A_E$$

where, A_E = area of earth

and R_{SE} = distance between sun and earth

So, in 1st case,

$$E_1 = \frac{P_1}{4\pi R_{SE}^2} \times A_E$$

$$\Rightarrow E_2 = \frac{P_2}{4\pi R_{SE}^2} \times A_E = 64 E_1$$

Hence, ratio is 64.

17. *(d)* According to Newton's law of cooling

$$-\frac{(T_f - T_i)}{t} = C\left[\frac{T_f + T_i}{2} - T_0\right]$$

For $T_i = 90°C$, $T_f = 80°C$, $T_0 = 20°C$, $t = t$

$$\Rightarrow \frac{-(80-90)}{t} = C\left[\frac{90+80}{2} - 20\right]$$

$$\Rightarrow \frac{10}{t} = 65\,C \quad \text{...(i)}$$

or $T_i = 80°\,C$, $T_f = 60°C$, $T_0 = 20°C$, $t = t_1$

$$\Rightarrow -\frac{60-80}{t_1} = C\left[\frac{80+60}{2} - 20\right]$$

$$\Rightarrow \frac{20}{t_1} = 50C \quad \text{...(ii)}$$

Divide Eq. (i) by Eq. (ii)

$$\frac{\frac{10}{t}}{\frac{20}{t_1}} = \frac{65C}{50C}$$

$$\Rightarrow \frac{t_1}{2t} = \frac{13}{10} \Rightarrow t_1 = \frac{13}{5} t$$

18. *(a)* Given that, r_1 and r_2 be the radii of inner and outer shells.

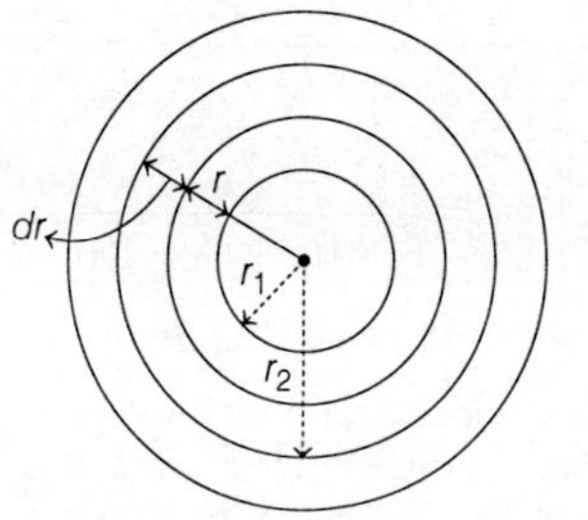

A material of thermal conductivity K is filled between region.

Consider an elementary sphere of radius r and thickness dr.

Now, thermal resistance of elementary sphere $dR = \frac{dr}{K 4\pi r^2}$

[$\because$ Inner surface area, $A = 4\pi r^2$]

So, total thermal resistance,

$$R = \int dR = \int_{r_1}^{r_2} \frac{dr}{4\pi K r^2} = -\frac{1}{4\pi K}\left[\frac{1}{r}\right]_{r_1}^{r_2}$$

$$R = \frac{r_2 - r_1}{4\pi K r_1 r_2}$$

Hence, heat current (or the rate of flow of heat)

$$\frac{dQ}{dt} = \frac{\Delta\theta}{R} = \frac{(\theta_2 - \theta_1)}{\frac{(r_2 - r_1)}{4\pi K r_1 r_2}}$$

$$= \frac{4\pi K r_1 r_2 (\theta_2 - \theta_1)}{r_2 - r_1}$$

19. *(c)* Then, $\frac{\Delta Q}{\Delta t} = \frac{KA\,(T_1 - T_2)}{x}$

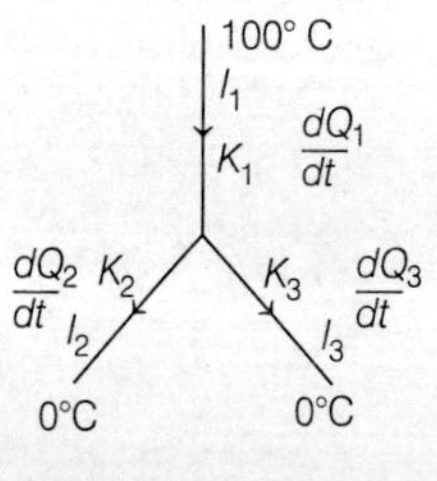

Since, there is no heat loss.

Thus,

$$\frac{dQ_1}{dt} = \frac{dQ_2}{dt} + \frac{dQ_3}{dt}$$

$$\frac{0.92\,(100 - T)}{46} = \frac{0.26\,(T - 0)}{13} + \frac{0.12\,(T - 0)}{12}$$

$$T = 40°C$$

$$\therefore \frac{dQ_1}{dt} = \frac{0.92 \times 4\,(100 - 40)}{46} = 4.8 \text{ cal/s}$$

20. *(c)*

> **key idea** To measure the radial rate of heat flow, we have to go for integration technique as here, the area of the surface through which heat will flow is not constant.

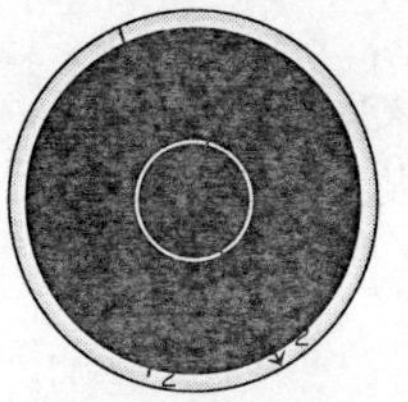

Let us consider an element (spherical shell) of thickness dx and radius x as shown in figure. Thermal resistance of shell,

$$dR = \frac{dx}{K \times 4\pi x^2}$$

$\left[\text{from } R = \frac{l}{KA} \text{ where, } K \text{ is thermal conductivity}\right]$

$$\Rightarrow \int dR = R = \int_{r_1}^{r_2} \frac{dx}{4\pi K x^2}$$

$$= \frac{1}{4\pi K}\left[\frac{1}{r_1} - \frac{1}{r_2}\right] = \frac{r_2 - r_1}{4\pi K (r_1 r_2)}$$

Rate of heat flow $= H = \frac{T_1 - T_2}{R}$

$$= \frac{T_1 - T_2}{r_2 - r_1} \times 4\pi K (r_1 r_2) \Rightarrow \propto \frac{r_1 r_2}{r_2 - r_1}$$

21. *(d)* Let the temperature of common interface be T°C.

Rate of heat flow

$$H = \frac{Q}{t} = \frac{KA\,\Delta T}{l}$$

$$\therefore H_1 = \left(\frac{Q}{t}\right)_1 = \frac{2KA(T - T_1)}{4x}$$

$$\text{and } H_2 = \left(\frac{Q}{t}\right)_2 = \frac{KA(T_2 - T)}{x}$$

In steady state, the rate of heat flow should be same in whole system i.e.,

$$H_1 = H_2$$

$$\Rightarrow \frac{2KA(T-T_1)}{4x} = \frac{KA(T_2-T)}{x}$$

or $$\frac{T-T_1}{2} = T_2 - T$$

or $$T = \frac{2T_2+T_1}{3} \quad \ldots(i)$$

Hence, heat flow from composite slab is

$$H = \frac{KA(T_2-T)}{x} = \frac{KA}{x}\left(T_2 - \frac{2T_2+T_1}{3}\right)$$

$$= \frac{KA}{3x}(T_2-T_1) \quad \ldots(ii)$$

Accordingly, $$H = \left[\frac{A(T_2-T_1)K}{x}\right] f \quad \ldots(iii)$$

By comparing Eqs. (ii) and (iii), we get

$f = 1/3$

22. *(360)*

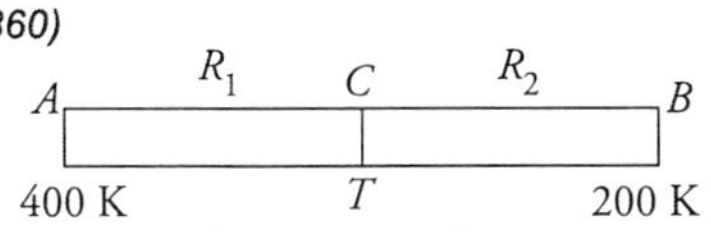

Here, $R_1 = \frac{l_1}{k_1 A_1}$, $R_2 = \frac{l_2}{k_2 A_2}$

We know that

$$\left(\frac{dQ}{dt}\right)_1 = \left(\frac{dQ}{dt}\right)_2$$

$$\frac{400-T}{R_1} = \frac{T-200}{R_2}$$

$$\Rightarrow \frac{400-T}{T-200} = \frac{R_1}{R_2}$$

$$= \left(\frac{l_1}{l_2}\right) \times \left(\frac{r_2}{r_1}\right)^2 \times \frac{k_2}{k_1}$$

$$= \frac{1}{2} \times \left(\frac{1}{2}\right)^2 \times 2 = \frac{1}{4}$$

$$\Rightarrow \frac{400-T}{T-200} = \frac{1}{4}$$

$$\Rightarrow 1600 - 4T = T - 200$$

$$\Rightarrow 5T = 1800$$

$$\Rightarrow T = 360 \text{ K}$$

23. *(5)* Given,

$L = 10$ cm, $d = 0.5$ mm, $T = 1727°C$

$T = 2000$ K

Power $\Rightarrow P = 94.2$ W

$P = \varepsilon \sigma A T^4$

$$94.2 = \varepsilon \times (6 \times 10^{-8})(\pi\, dl)(2000)^4$$

$$\varepsilon = \frac{94.2}{(6\times 10^{-8})(\pi dl)(2000)^4}$$

$$= \frac{94.2}{(6\times 10^{-8})(3.14)(0.5\times 10^{-3} \times 10 \times 10^{-2}) \times (2000)^4}$$

$$\varepsilon = \frac{94.2}{(94.2)(1.6)} = \frac{1}{1.6} = \frac{10}{16} = \frac{5}{8}$$

$x = 5$

24. *(40)* The diagram for the given data is

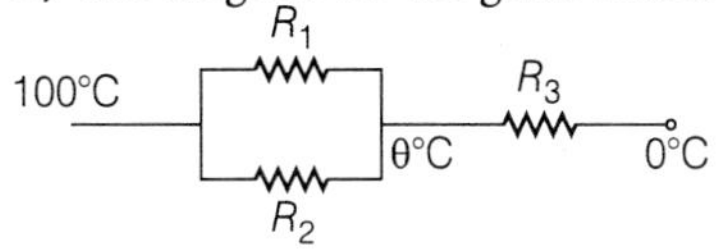

Resistance $R_1 = \frac{2L}{K_1 A}$

$$R_2 = \frac{2L}{K_2 A}$$

$$R_3 = \frac{L}{K_3 A}$$

$$\frac{\theta - 100}{\frac{R_1 R_2}{R_1 + R_2}} + \frac{\theta - 0}{R_3} = 0$$

$$\theta = 40°C$$

25. *(40)*

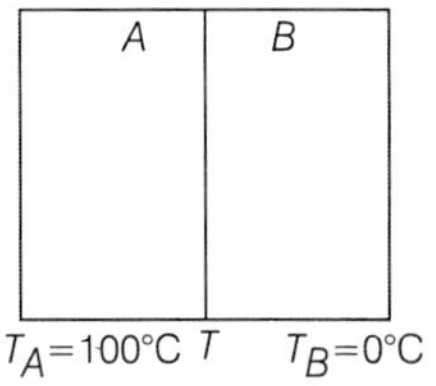

Let the temperature of contact surface is T then,

$$\frac{K_A A(T_A - T)}{L} = \frac{K_B A(T - T_B)}{L}$$

$\Rightarrow 84(100 - T) = 126(T - 0)$

$\Rightarrow 8400 - 84T = 126T \Rightarrow 8400 = 210T$

$\Rightarrow T = 8400/210 = 40°$ C

26. *(21)* The given figure for two plates A and B is given as

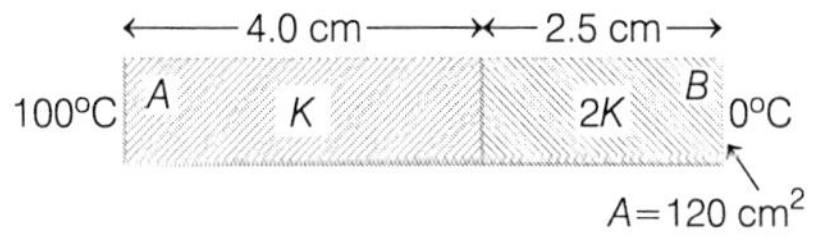

For series combination,

$$R_{\text{series}} = R_1 + R_2 + \ldots$$

$$\Rightarrow \frac{l_1 + l_2 + \ldots}{K_{eq} A} = \frac{l_1}{K_1 A} + \frac{l_2}{K_2 A} + \ldots$$

$\Rightarrow$ For two slabs of same area, we have

$$\frac{l_1 + l_2}{K_{eq}} = \frac{l_1}{K_1} + \frac{l_2}{K_2}$$

Here, $l_1 = 4$ cm, $K_1 = K$

$l_2 = 2.5$ cm and $K_2 = 2K$

So, $$\frac{4 + 2.5}{K_{eq}} = \frac{4}{K} + \frac{2.5}{2K} \Rightarrow \frac{6.5}{K_{eq}} = \frac{10.5}{2K}$$

or $$K_{eq} = \frac{13K}{10.5} = \frac{26}{21}K = \left(1 + \frac{5}{21}\right)K$$

$$= \left(1 + \frac{5}{\alpha}\right)K \quad \text{(given)}$$

$\alpha = 21$

27. *(2)* Given, thermal resistances

$R_{AB} = R_{CD} = 10.0 \text{ kW}^{-1}$

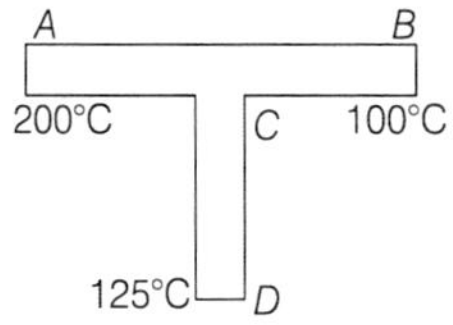

As, C is mid-point of AB, and as we know

$R \propto$ length

$R_{AC} = R_{BC} = 5 \text{ kW}^{-1}$

Now, we know that heat current is given as

$Q/t = \Delta T / R$

Let temperature of point C as T.

Consider at point C heat is coming in from A and going out at point B and D. Therefore, we can write,

$$\frac{T_A - T}{R_{AC}} = \frac{T - T_D}{R_{CD}} + \frac{T - T_B}{R_{BC}}$$

$$\Rightarrow \frac{200 - T}{5} = \frac{T - 125}{10} + \frac{T - 100}{5}$$

$$\Rightarrow 400 - 2T = T - 125 + 2T - 200$$

$5T = 725°C \Rightarrow T = 145°C$

The heat current in CD is

$$I_{CD} = \frac{T - T_D}{R_{CD}}$$

$$= \frac{145 - 125}{10} = \frac{20}{10} = 2 \text{ W}$$

Thus, the value of P is 2.

C-11

JEE Main Chapterwise Topicwise Physics

In JEE Main there is usually 0-1 question per paper is visible from this chapter. In past 3 years, 48 conceptual and 4 numerical questions were asked of moderate to difficult level from this chapter.

For detailed analysis scan QR code

THERMODYNAMICS

Topic 1 Zeroth and First Law of Thermodynamics

Level 1

01 Given are statements for certain thermodynamic variables, *[2025, 22 Jan Shift-II]*

(A) internal energy, volume (V) and mass (m) are extensive variables.

(B) pressure (p), temperature (T) and density (ρ) are intensive variables.

(C) volume (V), temperature (T) and density (ρ) are intensive variables.

(D) mass (m), temperature (T) and internal energy are extensive variables.

Choose the correct answer from the options given below.

(a) (C) and (D) Only (b) (B) and (C) Only

(c) (D) and (A) Only (d) (A) and (B) Only

02 A total 48 J heat is given to one mole of helium kept in a cylinder. The temperature of helium increases by 2°C. The work done by the gas is

(Given, $R = 8.3\ \text{J K}^{-1}\ \text{mol}^{-1}$) *[2024, 6 April Shift-II]*

(a) 24.9 J (b) 72.9 J

(c) 23.1 J (d) 48 J

03 A source supplies heat to a system at the rate of 1000 W. If the system performs work at a rate of 200 W. The rate at which internal energy of the system increases is *[2023, 6 April Shift-I]*

(a) 600 W (b) 500 W

(c) 800 W (d) 1200 W

04 Given below are two statements : One is labelled as **Assertion A** and the other is labelled as **Reason R**.

Assertion A If dQ and dW represent the heat supplied to the system and the work done on the system respectively. Then according to the first law of thermodynamics $dQ = dU - dW$.

Reason R First law of thermodynamics is based on the law of conservation of energy.

In the light of the above statements, choose the correct answer from the options given below

[2023, 29 Jan Shift-I]

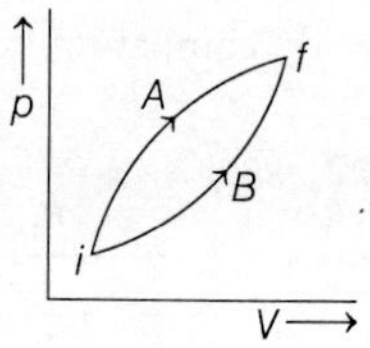

(a) Both A and R are correct and R is the correct explanation of A.

(b) Both A and R are correct but R is not the correct explanation of A.

(c) A is not correct but R is correct.

(d) A is correct but R is not correct.

05 In thermodynamics, heat and work are *[2021, 16 March Shift-I]*

(a) path functions

(b) intensive thermodynamic state variables

(c) extensive thermodynamic state variables

(d) point functions

06 Following figure shows two processes A and B for a gas. If ΔQ_A and ΔQ_B are the amount of heat absorbed by the system in two cases, and ΔU_A and ΔU_B are changes in internal energies respectively, then *[2019, 9 April Shift-I]*

(a) $\Delta Q_A > \Delta Q_B,\ \Delta U_A > \Delta U_B$

(b) $\Delta Q_A < \Delta Q_B,\ \Delta U_A < \Delta U_B$

(c) $\Delta Q_A > \Delta Q_B,\ \Delta U_A = \Delta U_B$

(d) $\Delta Q_A = \Delta Q_B;\ \Delta U_A = \Delta U_B$

07 100 g of water is heated from 30°C to 50°C. Ignoring the slight expansion of the water, the change in its internal energy is (specific heat of water is 4184 J/kg/K) *[AIEEE 2011]*

(a) 8.4 kJ (b) 84 kJ (c) 2.1 kJ (d) 4.2 kJ

Imp. First Law of Thermodynamics

08 When a system is taken from state i to state f along the path iaf, it is found that $Q = 50$ cal and $W = 20$ cal. Along the path ibf, $Q = 36$ cal. W along the path ibf is *[AIEEE 2007]*

(a) 6 cal (b) 16 cal (c) 66 cal (d) 14 cal

09 A system goes from A to B *via* two processes I and II as shown in figure. If ΔU_1 and ΔU_2 are the changes in internal energies in the processes I and II respectively, then **[AIEEE 2005]**

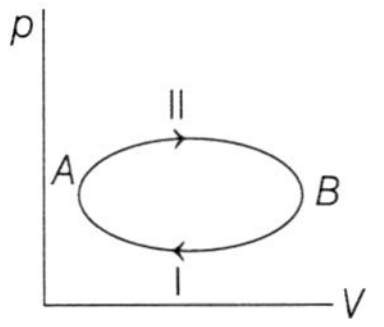

(a) $\Delta U_1 = \Delta U_2$
(b) relation between ΔU_1 and ΔU_2 cannot be determined
(c) $\Delta U_2 > \Delta U_1$
(d) $\Delta U_2 < \Delta U_1$

10 Which of the following parameters does not characterise the thermodynamic state of matter? **[AIEEE 2003]**
(a) Temperature (b) Pressure
(c) Work (d) Volume

Level 2

11 During the melting of a slab of ice at 273 K at atmospheric pressure. **[2025, 3 April Shift-I]**
(a) Internal energy of the ice-water system decreases.
(b) Positive work is done by the ice-water system on the atmosphere.
(c) Positive work is done on the ice-water system by the atmosphere.
(d) Internal energy of ice-water system remains unchanged.

Imp. Area Under p-V Curve

12 A thermodynamic system is taken from an original state A to an intermediate state B by a linear process as shown in the figure. It's volume is then reduced to the original value from B to C by an isobaric process. The total work done by the gas from A to B and B to C would be **[2024, 29 Jan Shift-I]**

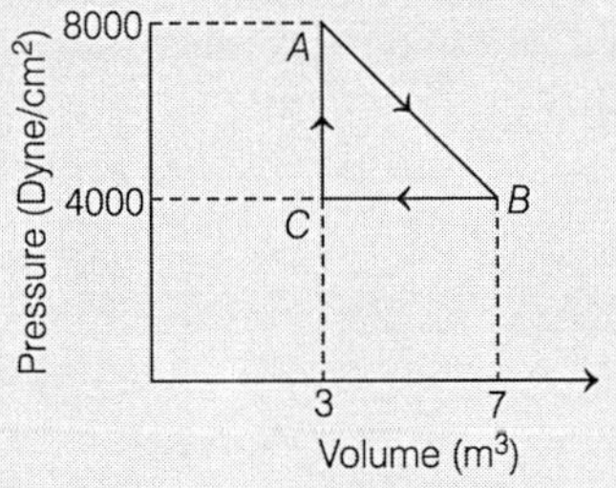

(a) 600 J (b) 2200 J (c) 33800 J (d) 1200 J

13 An electric appliance supplies 6000 J/min heat to the system. If the system delivers a power of 90W. How long it would take to increase the internal energy by 2.5×10^3 J? **[2021, 26 Aug Shift-I]**
(a) 2.5×10^2 s (b) 4.1×10^1 s
(c) 2.4×10^3 s (d) 2.5×10^1 s

14 A gas can be taken from A to B *via* two different processes ACB and ADB.

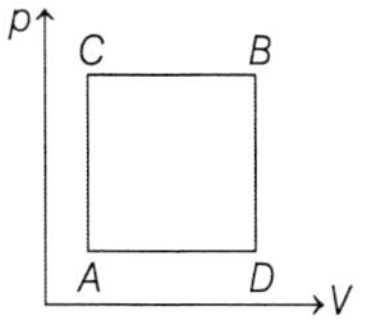

When path ACB is used 60 J of heat flows into the system and 30 J of work is done by the system. If path ADB is used work done by the system is 10 J the heat flow into the system in path ADB is **[2019, 9 Jan Shift-I]**
(a) 80 J (b) 40 J (c) 100 J (d) 20 J

15 Which of the following is incorrect regarding the first law of thermodynamics? **[AIEEE 2005]**
(A) It is not applicable to any cyclic process.
(B) It is a restatement of the principle of conservation of energy.
(C) It introduces the concept of the internal energy.
(D) It introduces the concept of the entropy.
Choose the most correct answer from the options given below.
(a) (B, C) (b) (A, C)
(c) (C, D) (d) (A, D)

Numerical Value Type Questions

16 The internal energy of air in 4 m × 4 m × 3 m sized room at 1 atmospheric pressure will be .... $\times 10^6$ J.
(Consider air as diatomic molecule) **[2025, 2 April Shift-II]**

Topic 2 Thermodynamics Process

Level 1

01 Identify the characteristics of an adiabatic process in a monoatomic gas.
(A) Internal energy is constant.
(B) Work done in the process is equal to the change in internal energy.
(C) The product of temperature and volume is a constant.
(D) The product of pressure and volume is a constant.
(E) The work done to change the temperature from T_1 to T_2 is proportional to $(T_2 - T_1)$.
Choose the correct answer from the options given below. **[2025, 2 April Shift-II]**
(a) (A), (C), (E) only (b) (B), (E) only
(c) (B), (D) only (d) (A), (C), (D) only

02 Pressure of an ideal gas, contained in a closed vessel, is increased by 0.4% when heated by 1°C. Its initial temperature must be **[2025, 3 April Shift-II]**
(a) 250 K (b) 25°C
(c) 2500 K (d) 250°C

03 A monoatomic gas having $\gamma = \frac{5}{3}$ is stored in a thermally insulated container and the gas is suddenly compressed to $\left(\frac{1}{8}\right)$th of its initial volume. The ratio of final pressure and initial pressure is
(γ is the ratio of specific heats of the gas at constant pressure and at constant volume) *[2025, 8 April Shift-II]*
(a) 40 (b) 16
(c) 32 (d) 28

04 The work done in an adiabatic change in an ideal gas depends upon only *[2025, 29 Jan Shift-I]*
(a) change in its pressure (b) change in its temperature
(c) change in its specific heat (d) change in its volume

05 Given below are two statements. One is labelled as Assertion (A) and the other is labelled as Reason (R).
Assertion (A) In an insulated container, a gas is adiabatically shrunk to half of its initial volume. The temperature of the gas decreases.
Reason (R) Free expansion of an ideal gas is an irreversible and an adiabatic process.
In the light of the above statements, choose the correct answer from the options given below. *[2025, 24 Jan Shift-II]*
(a) A is false but R is true.
(b) Both A and R are true but R is not the correct explanation of A.
(c) Both A and R are true and R is the correct explanation of A.
(d) A is true but R is false.

06 The volume of an ideal gas ($\gamma = 1.5$) is changed adiabatically from 5 L to 4 L. The ratio of initial pressure to final pressure is *[2024, 9 April Shift-I]*
(a) $\frac{8}{5\sqrt{5}}$ (b) $\frac{16}{25}$ (c) $\frac{4}{5}$ (d) $\frac{2}{\sqrt{5}}$

07 The temperature of a gas having 2.0×10^{25} molecules per cubic metre at 1.38 atm
(Given, $k = 1.38 \times 10^{-23}$ JK^{-1}) is *[2024, 29 Jan Shift-II]*
(a) 300 K (b) 200 K
(c) 500 K (d) 100 K

08 The initial pressure and volume of an ideal gas are p_0 and V_0. The final pressure of the gas when the gas is suddenly compressed to volume $V_0/4$ will be
(Given, γ = ratio of specific heats at constant pressure and at constant volume)
(a) $4p_0$ (b) p_0
(c) $p_0(4)^{1/\gamma}$ (d) $p_0(4)^{\gamma}$

09 A thermodynamic system is taken through cyclic process. The total work done in the process is *[2023, 13 April Shift-I]*

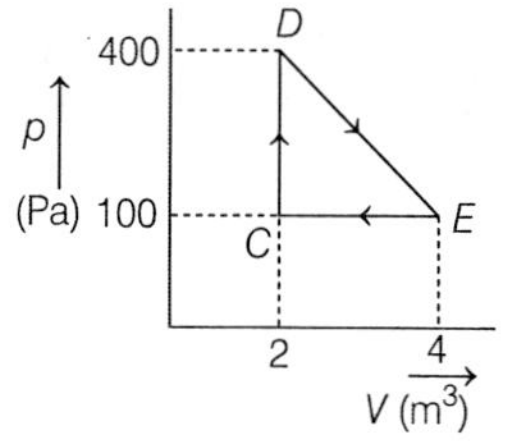

(a) 200 J (b) Zero
(c) 300 J (d) 100 J

10 A gas is compressed adiabatically, which one of the following statement is not true, *[2023, 10 April Shift-II]*
(a) There is no heat supplied to the system
(b) The temperature of the gas increases.
(c) There is no change in the internal energy
(d) The change in the internal energy is equal to the work done on the gas.

11 The thermodynamic process, in which internal energy of the system remains constant is *[2023, 10 April Shift-I]*
(a) isobaric (b) isochoric
(c) adiabatic (d) isothermal

12 Given below are two statements.
Statement I If heat is added to a system, its temperature must increase.
Statement II If positive work is done by a system in a thermodynamic process, its volume must increase.
In the light of the above statements, choose the correct answer from the options given below. *[2023, 8 April Shift-I]*
(a) Statement I is true but Statement II is false.
(b) Statement I is false but Statement II is true.
(c) Both Statement I and Statement II are false.
(d) Both Statement I and Statement II are true.

Imp. Isothermal Process

13 In an isothermal change, the change in pressure and volume of a gas can be represented for three different temperature $T_3 > T_2 > T_1$ as *[2023, 24 Jan Shift-II]*

(a)
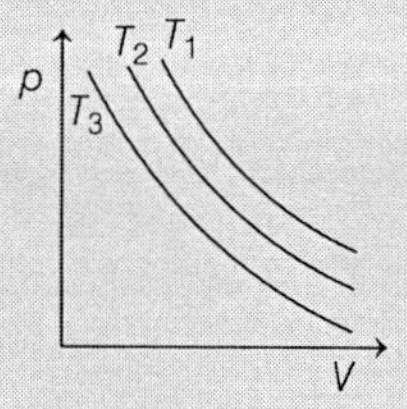

(b)
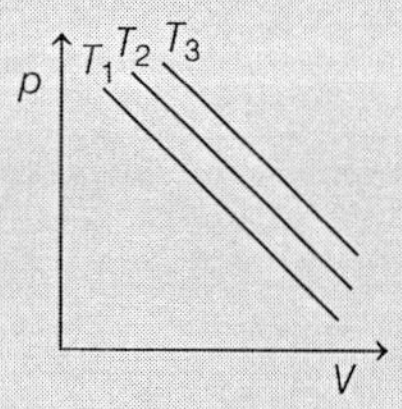

(c)
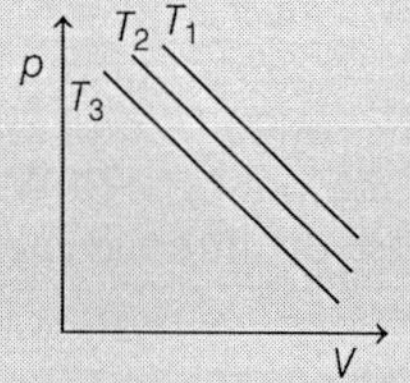

(d)
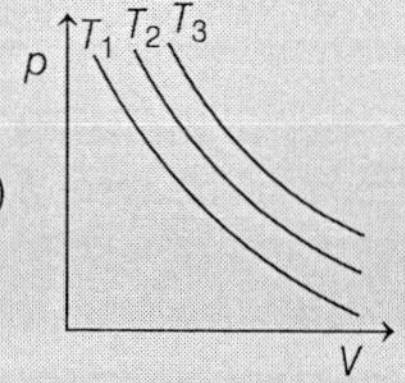

14 For an adiabatic expansion of an ideal gas, the fractional change in its pressure is equal to (where, γ is the ratio of specific heats) *[2021, 18 March Shift-II]*

(a) $-\gamma\dfrac{dV}{V}$ (b) $-\gamma\dfrac{V}{dV}$ (c) $-\dfrac{1}{\gamma}\dfrac{dV}{V}$ (d) $\dfrac{dV}{V}$

15 The p-V diagram of a diatomic ideal gas system going under cyclic process as shown in figure. The work done during an adiabatic process CD is (use, $\gamma = 1.4$) *[2021, 18 March Shift-I]*

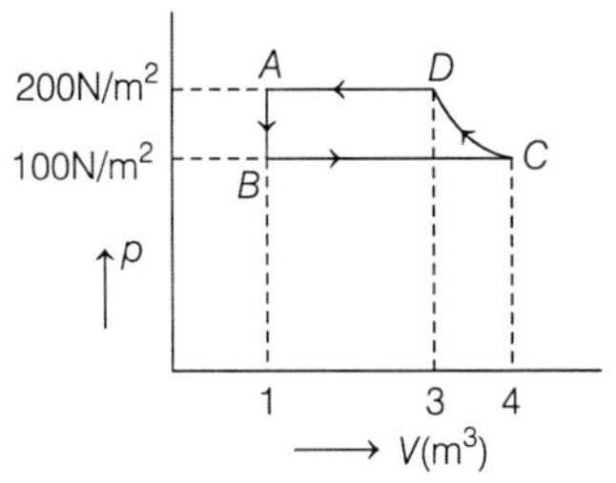

(a) −500 J (b) −400 J (c) 400 J (d) 200 J

Imp. Adiabatic and Isothermal Curves

16 Which one is the correct option for the two different thermodynamic processes? *[2021, 17 March Shift-II]*

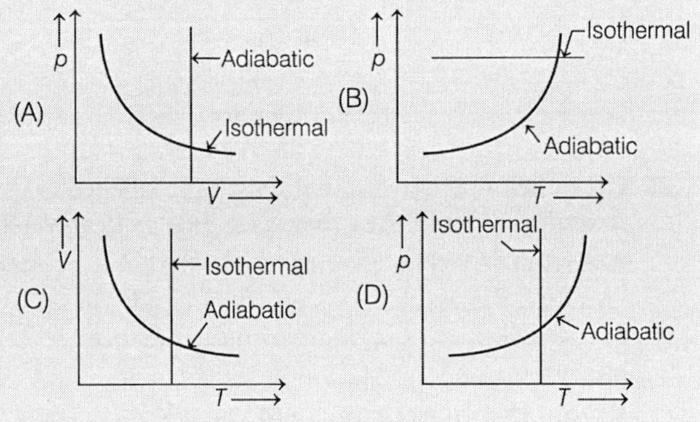

(a) C and A (b) C and D
(c) Only A (d) B and C

17 In an adiabatic process, the density of a diatomic gas becomes 32 times of its initial value. The final pressure of the gas is found to be n times the initial pressure. The value of n is *[2020, 5 Sep Shift-II]*

(a) 32 (b) 326
(c) 128 (d) 1/32

18 A balloon filled with helium (32°C and 1.7 atm) bursts. Immediately afterwards; the expansion of helium can be considered as *[2020, 3 Sep Shift-I]*

(a) irreversible isothermal
(b) irreversible adiabatic
(c) reversible adiabatic
(d) reversible isothermal

19 The given diagram shows four processes, i.e. isochoric, isobaric, isothermal and adiabatic. The correct assignment of the processes, in the same order is given by *[2019, 8 April Shift-II]*

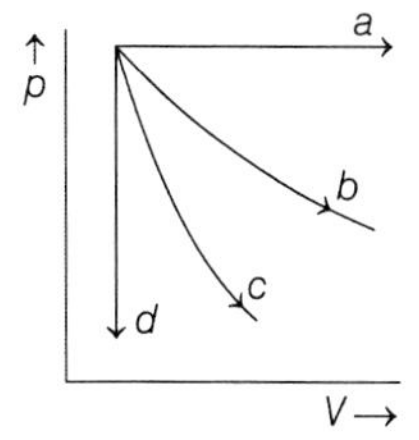

(a) $d\,a\,b\,c$ (b) $a\,d\,b\,c$ (c) $d\,a\,c\,b$ (d) $a\,d\,c\,b$

20 One mole of an ideal monoatomic gas is compressed isothermally in a rigid vessel to double its pressure at room temperature, 27°C. The work done on the gas will be *[JEE Main 2018]*

(a) $300\,R$ (b) $300\,R\ln 2$
(c) $300\,R\ln 6$ (d) $300\,R\ln 7$

21 For the p-V diagram given for an ideal gas,

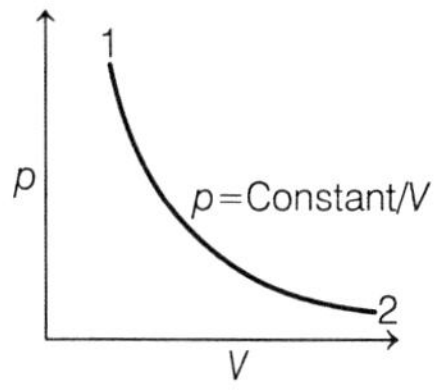

out of the following which one correctly represents the T-p diagram? *[JEE Main 2017]*

(a)
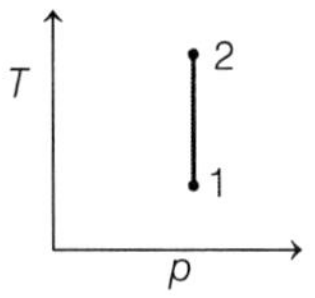

(b)
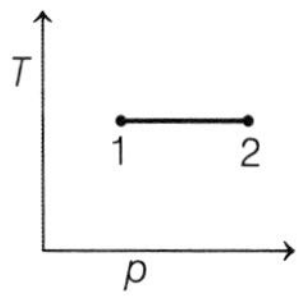

(c)
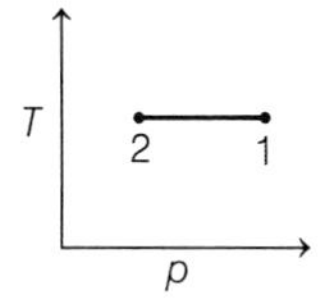

(d)
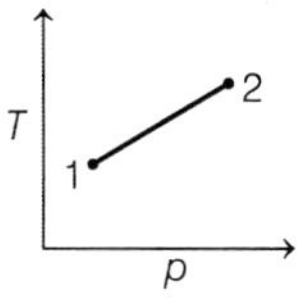

Level 2

22 Match List - I with List - II.

List-1		List-1I	
(A)	Isothermal	(I)	ΔW (work done) $= 0$
(B)	Adiabatic	(II)	ΔQ (supplied heat) $= 0$
(C)	Isobaric	(III)	ΔU (change in internal energy) $\neq 0$
(D)	Isochoric	(IV)	$\Delta U = 0$

Choose the correct answer from the options given below: *[2025, 7 April Shift-II]*

(a) (A)-(IV), (B)-(II), (C)-(III), (D)-(I)
(b) (A)-(II), (B)-(IV), (C)-(I), (D)-(III)
(c) (A)-(IV), (B)-(I), (C)-(III), (D)-(II)
(d) (A)-(III), (B)-(II), (C)-(I), (D)-(IV)

23 Match List - I with List II. *[2025, 4 April Shift-II]*

List - I		List - II	
(A)	Isobaric	(I)	$\Delta Q = \Delta W$
(B)	Isochoric	(II)	$\Delta Q = \Delta U$
(C)	Adiabatic	(III)	$\Delta Q =$ zero
(D)	Isothermal	(IV)	$\Delta Q = \Delta U + p\Delta V$

$\Delta Q =$ Heat supplied
$\Delta W =$ Work done by the system
$\Delta U =$ Change in internal energy
$p =$ Pressure of the system
$\Delta V =$ Change in volume of the system
Choose the correct answer from the options given below:
(a) (A)-(II), (B)-(IV), (C)-(III), (D)-(I)
(b) (A)-(IV), (B)-(III), (C)-(II), (D)-(I)
(c) (A)-(IV), (B)-(II), (C)-(III), (D)-(I)
(d) (A)-(IV), (B)-(I), (C)-(III), (D)-(II)

24 A gas is kept in a container having walls which are thermally non-conducting. Initially the gas has a volume of 800 cm^3 and temperature 27 °C. The change in temperature when the gas is adiabatically compressed to 200 cm^3 is
(Take $\gamma = 1.5$, γ is the ratio of specific heats at constant pressure and at constant volume) *[2025, 3 April Shift-I]*
(a) 522 K (b) 600 K
(c) 327 K (d) 300 K

25 In an adiabatic process, which of the following statements is true ? *[2025, 2 April Shift-I]*
(a) The molar heat capacity is infinite.
(b) Work done by the gas equals the increase in internal energy.
(c) The internal energy of the gas decreases as the temperature increases.
(d) The molar heat capacity is zero.

26 An ideal gas goes from an initial state to final state. During the process, the pressure of gas increases linearly with temperature. *[2025, 24 Jan Shift-I]*
A. The work done by gas during the process is zero.
B. The heat added to gas is different from change in its internal energy.
C. The volume of the gas is increased.
D. The internal energy of the gas is increased
E. The process is isochoric (constant volume process)
Choose the correct answer from the options given below.
(a) E only (b) A, C only
(c) A, B, C, D only (d) A, D, E only

27

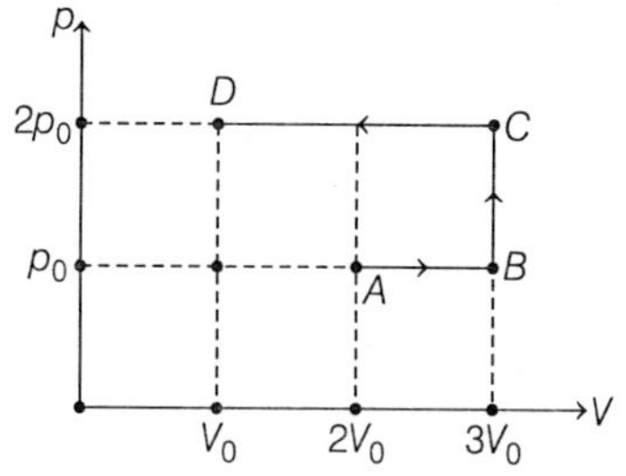

Using the given p-V diagram, the work done by an ideal gas along the path $ABCD$ is *[2025, 23 Jan Shift-II]*
(a) $-4p_0 V_0$ (b) $3p_0 V_0$
(c) $-3p_0 V_0$ (d) $4p_0 V_0$

28 The magnitude of heat exchanged by a system for the given cyclic process $ABCA$ (as shown in figure) is (in SI unit) *[2025, 24 Jan Shift-II]*

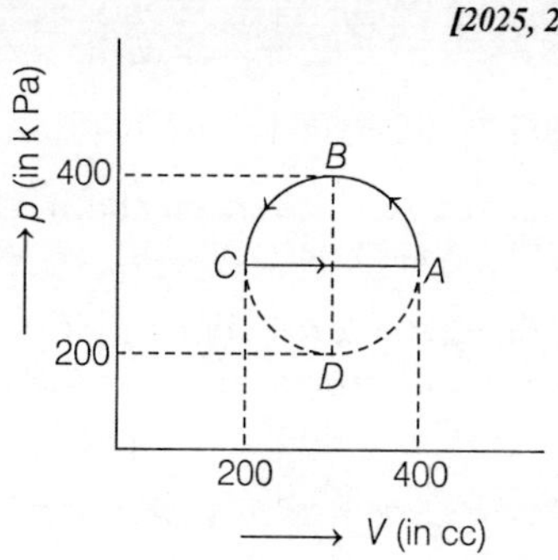

(a) 5π (b) zero
(c) 10π (d) 40π

29 Given below are two statements. One labelled as Assertion (A) and the other is labelled as Reason (R).
Assertion (A) With the increase in the pressure of an ideal gas, the volume falls off more rapidly in an isothermal process in comparison to adiabatic process.
Reason (R) In isothermal process, $pV =$ constant, while in adiabatic process $pV^{\gamma} =$ constant. Here γ is the ratio of specific heats, p is the pressure and V is the volume of the ideal gas.
In the light of the above statements, choose the correct answer from the options given below. *[2025, 29 Jan Shift-II]*
(a) Both A and R are true and R is the correct explanation of A.
(b) Both A and R are true and R is not the correct explanation of A.
(c) A is true but R is false.
(d) A is false but R is true.

30 A sample of 1 mole gas at temperature T is adiabatically expanded to double its volume. If adiabatic constant for the gas is $\gamma = \frac{3}{2}$, then the work done by the gas in the process is **[2024, 9 April Shift-I]**

(a) $\frac{R}{T}[2-\sqrt{2}]$ (b) $RT[2+\sqrt{2}]$

(c) $\frac{T}{R}[2+\sqrt{2}]$ (d) $RT[2-\sqrt{2}]$

Imp. Cyclic Process

31 The heat absorbed by a system in going through the given cyclic process is **[2024, 5 April Shift-I]**

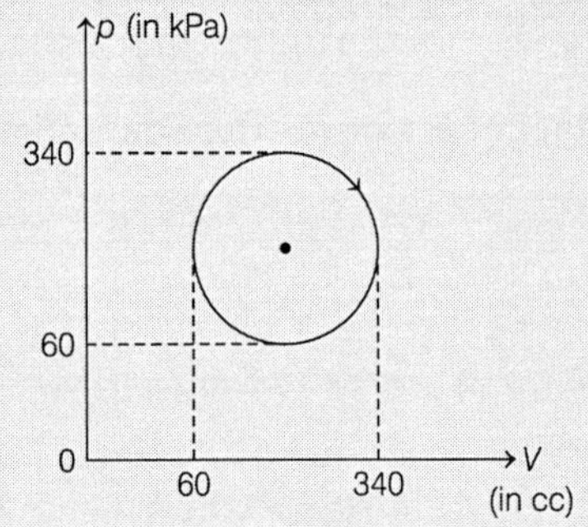

(a) 616 J (b) 431.2 J
(c) 19.6 J (d) 61.6 J

32 The given figure represents two isobaric processes for the same mass of an ideal gas, then **[2024, 31 Jan Shift-I]**

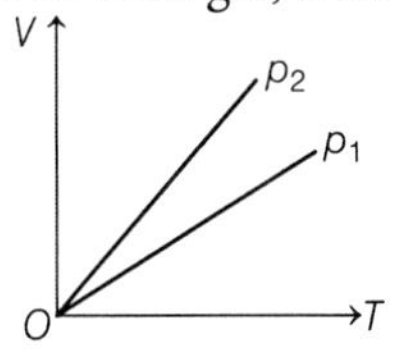

(a) $p_2 > p_1$ (b) $p_2 \geq p_1$ (c) $p_1 = p_2$ (d) $p_1 > p_2$

33 Under isothermal condition, the pressure of a gas is given by $p = aV^{-3}$, where a is a constant and V is the volume of the gas. The bulk modulus at constant temperature is equal to **[2023, 13 April Shift-I]**

(a) $3p$ (b) $\frac{p}{2}$ (c) $2p$ (d) p

34 Consider two containers A and B containing monoatomic gases at the same pressure p, volume V and Temperature T. The gas in A is compressed isothermally to $\frac{1}{8}$ of its original volume while the gas in B is compressed adiabatically to $1/8$ of its original volume. The ratio of final pressure of gas in B to that of gas in A is **[2023, 11 April Shift-II]**

(a) $\frac{1}{8}$ (b) 4 (c) $8^{\frac{3}{2}}$ (d) 8

35 An air bubble of volume 1 cm^{-3} rises from the bottom of a lake 40 m deep to the surface at a temperature of 12°C. The atmospheric pressure is 1×10^5 Pa, the density fo water is 1000 kg/m^3 and $g = 10\ m/s^2$. There is no difference of the temperature of water at the depth of 40 m and on the surface. The volume of air bubble when it reaches the surface will be **[2023, 8 April Shift-I]**

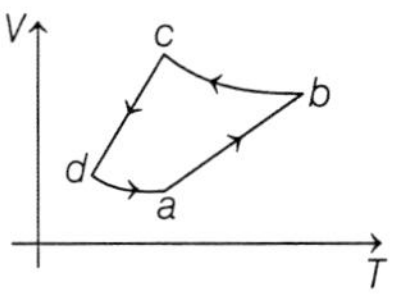

(a) 4 cm^3 (b) 3 cm^3 (c) 5 cm^3 (d) 2 cm^3

36 A hypothetical gas expands adiabatically such that its volume changes from 08 L to 27 L. If the ratio of final pressure of the gas to initial pressure of the gas is $\frac{16}{81}$. Then, the ratio of $\frac{C_p}{C_V}$ will be **[2023, 31 Jan Shift-II]**

(a) $\frac{3}{2}$ (b) $\frac{1}{2}$ (c) $\frac{4}{3}$ (d) $\frac{3}{1}$

37 Heat is given to an ideal gas in an isothermal process.
A. Internal energy of the gas will decrease.
B. Internal energy of the gas will increase.
C. Internal energy of the gas will not change.
D. The gas will do positive work.
E. The gas will do negative work.
Choose the correct answer from the options given below **[2023, 30 Jan Shift-I]**

(a) C and D only (b) A and E only
(c) B and D only (d) C and E only

38 1 g of a liquid is converted to vapour at 3×10^5 Pa pressure. If 10% of the heat supplied is used for increasing the volume by 1600 cm^3 during this phase change, then the increase in internal energy is the process will be **[2023, 24 Jan Shift-I]**

(a) 432×10^8 J (b) 432000 J
(c) 4320 J (d) 4800 J

39 A thermodynamic system is taken from an original state D to an intermediate state E by the linear process shown in the figure. Its volume is then reduced to the original volume from E to F by an isobaric process. The total work done by the gas from D to E to F will be **[2022, 29 July Shift-II]**

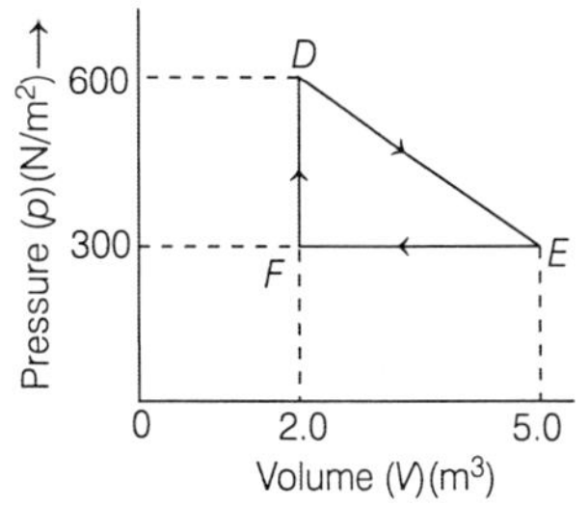

(a) −450 J (b) 450 J
(c) 900 J (d) 1350 J

40 A monoatomic gas at pressure p and volume V is suddenly compressed to one eighth of its original volume. The final pressure at constant-entropy will *[2022, 26 July Shift-I]*

(a) p (b) $8p$
(c) $32p$ (d) $64p$

41 Starting with the same initial conditions, an ideal gas expands from volume V_1 to V_2 in three different ways. The work done by the gas is W_1, if the process is purely isothermal. W_2, if the process is purely adiabatic and W_3, if the process is purely isobaric. Then, choose the correct option *[2022, 29 June Shift-II]*

(a) $W_1 < W_2 < W_3$ (b) $W_2 < W_3 < W_1$
(c) $W_3 < W_1 < W_2$ (d) $W_2 < W_1 < W_3$

42 Given below are two statements.

Statement I When μ amount of an ideal gas undergoes adiabatic change from state (p_1, V_1, T_1) to state (p_2, V_2, T_2), then work done is

$$W = \frac{\mu R(T_2 - T_1)}{1-\gamma},$$

where $\gamma = \frac{C_p}{C_V}$ and R = universal gas constant.

Statement II In the above case, when work is done on the gas, the temperature of the gas would rise.

Choose the correct answer from the options given below. *[2022, 28 June Shift-I]*

(a) Both Statement I and Statement II are true.
(b) Both Statement I and Statement II are false.
(c) Statement I is true but Statement II is false.
(d) Statement I is false but Statement II is true.

43 In the reported figure, there is a cyclic process $ABCDA$ on a sample of 1 mol of a diatomic gas. The temperature of the gas during the process $A \rightarrow B$ and $C \rightarrow D$ are T_1 and T_2 $(T_1 > T_2)$, respectively. *[2021, July 27 Shift-I]*

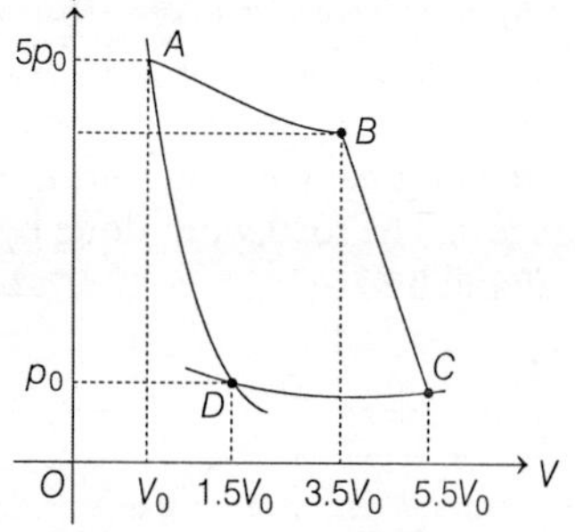

Choose the correct option out of the following for work done, if processes BC and DA are adiabatic.

(a) $W_{AB} = W_{DC}$ (b) $W_{AD} = W_{BC}$
(c) $W_{BC} + W_{DA} > 0$ (d) $W_{AB} < W_{CD}$

44 A monoatomic ideal gas, initially at temperature T_1 is enclosed in a cylinder fitted with a frictionless piston. The gas is allowed to expand adiabatically to a temperature T_2 by releasing the piston suddenly. If l_1 and l_2 are the lengths of the gas column, before and after the expansion respectively, then the value of T_1 / T_2 will be *[2021, 25 July Shift-I]*

(a) $\left(\frac{l_1}{l_2}\right)^{\frac{2}{3}}$ (b) $\left(\frac{l_2}{l_1}\right)^{\frac{2}{3}}$ (c) $\frac{l_2}{l_1}$ (d) $\frac{l_1}{l_2}$

45 Match List I with List II.

List-I	List-II
A. Isothermal	1. Pressure constant
B. Isochoric	2. Temperature constant
C. Adiabatic	3. Volume constant
D. Isobaric	4. Heat content is constant

Choose the correct answer from the options given below. *[2021, 24 Feb Shift-I]*

	A	B	C	D		A	B	C	D
(a)	1	3	2	4	(b)	3	2	1	4
(c)	2	4	3	1	(d)	2	3	4	1

46 A litre of dry air at STP expands adiabatically to a volume of 3 L.
If $\gamma = 1.40$, the work done by air is $(3^{1.4} = 4.6555)$
[Take, air to be an ideal gas] *[2020, 7 Jan Shift-I]*

(a) 100.8 J (b) 90.5 J (c) 48 J (d) 60.7 J

47 For the given cyclic process CAB as shown for a gas, the work done is *[2019, 12 Jan Shift-I]*

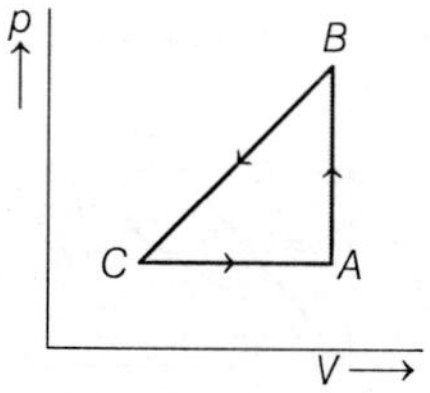

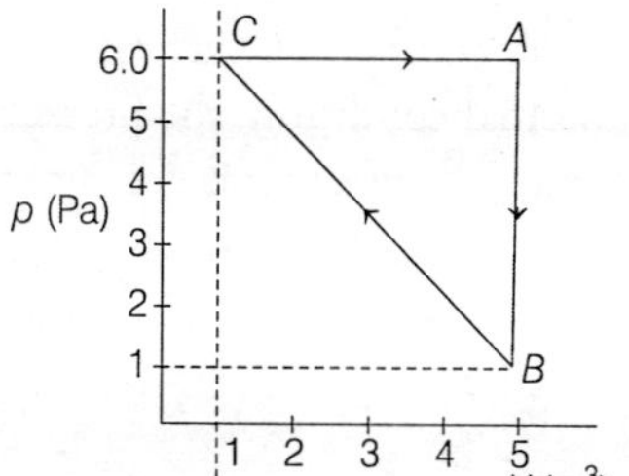

(a) 5 J (b) 10 J
(c) 1 J (d) 30 J

48 In a process, temperature and volume of one mole of an ideal monoatomic gas are varied according to the relation $VT = K$, where K is a constant. In this process, the temperature of the gas is increased by ΔT. The amount of heat absorbed by gas is (R is gas constant) *[2019, 11 Jan Shift-II]*

(a) $\frac{2K}{3}\Delta T$ (b) $\frac{1}{2}R\Delta T$ (c) $\frac{3}{2}R\Delta T$ (d) $\frac{1}{2}KR\Delta T$

49 Two moles of an ideal monoatomic gas occupies a volume V at 27°C. The gas expands adiabatically to a volume $2V$. *[JEE Main 2018]*

Calculate (i) the final temperature of the gas and
(ii) change in its internal energy.

(a) (i) 189 K (ii) 2.7 kJ (b) (i) 195 K (ii) −2.7 kJ
(c) (i) 189 K (ii) −2.7 kJ (d) (i) 195 K (ii) 2.7 kJ

50 The ratio of work done by an ideal monoatomic gas to the heat supplied to it in an isobaric process is *[JEE Main 2016)*

(a) 2/5 (b) 3/2 (c) 3/5 (d) 2/3

51 An ideal gas goes through a reversible cycle $a \to b \to c \to d$ has the V-T diagram shown below. Process $d \to a$ and $b \to c$ are adiabatic.

The corresponding p-V diagram for the process is (all figures are schematic and not drawn to scale) *[JEE Main 2015)*

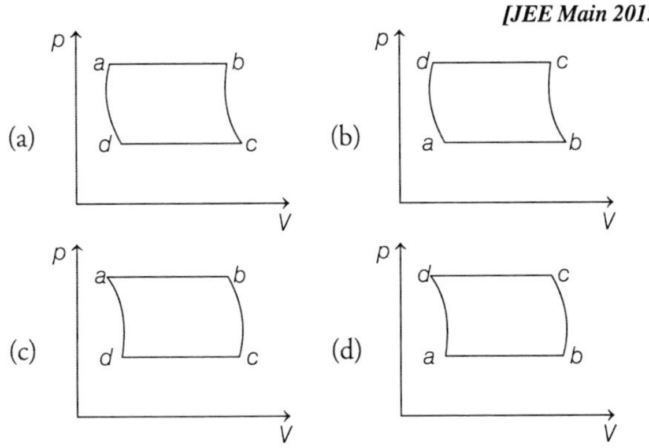

52 Consider a spherical shell of radius R at temperature T. The black body radiation inside it can be considered as an ideal gas of photons with internal energy per unit volume $u = \frac{U}{V} \propto T^4$ and pressure $p = \frac{1}{3}\left(\frac{U}{V}\right)$. If the shell now undergoes an adiabatic expansion, the relation between T and R is *[JEE Main 2015]*

(a) $T \propto e^{-R}$ (b) $T \propto e^{-3R}$
(c) $T \propto \frac{1}{R}$ (d) $T \propto \frac{1}{R^3}$

Level 3

53

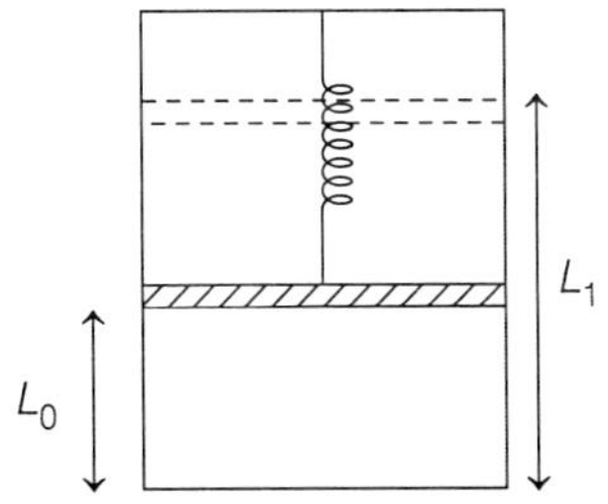

A piston of mass M is hung from a massless spring whose restoring force law goes as $F = -kx^3$, where k is the spring constant of appropriate dimension. The piston separates the vertical chamber into two parts, where the bottom part is filled with 'n' moles of an ideal gas. An external work is done on the gas isothermally (at a constant temperature T) with the help of a heating filament (with negligible volume) mounted in lower part of the chamber, so that the piston goes up from a height L_0 to L_1 the total energy delivered by the filament is (Assume spring to be in its natural length before heating) *[2025, 3 April Shift-I]*

(a) $nRT \ln\left(\frac{L_1}{L_0}\right) + Mg(L_1 - L_0) + \frac{3k}{4}(L_1^4 - L_0^4)$

(b) $nRT \ln\left(\frac{L_1^2}{L_0^2}\right) + \frac{Mg}{2}(L_1 - L_0) + \frac{k}{4}(L_1^4 - L_0^4)$

(c) $nRT \ln\left(\frac{L_1}{L_0}\right) + Mg(L_1 - L_0) + \frac{k}{4}(L_1^4 - L_0^4)$

(d) $3nRT \ln\left(\frac{L_1}{L_0}\right) + 2Mg(L_1 - L_0) + \frac{k}{3}(L_1^3 - L_0^3)$

54 An ideal gas exists in a state with pressure p_0, volume V_0. It is isothermally expanded to 4 times of its initial volume (V_0), then isobarically compressed to its original volume. Finally the system is heated isochorically to bring it to its initial state. The amount of heat exchanged in this process is *[2025, 3 April Shift-II]*

(a) $p_0V_0\,(2\ln 2 - 0.25)$ (b) $p_0V_0\,(\ln 2 - 0.25)$
(c) $p_0V_0\,(2\ln 2 - 0.75)$ (d) $p_0V_0\,(2\ln 2 - 0.75)$

55 Match the List-I with List-II. *[2025, 23 Jan Shift-I]*

List-I	List-II
A. Pressure varies inversely with volume of an ideal gas.	I. Adiabatic process
B. Heat absorbed goes partly to increase internal energy and partly to do work.	II. Isochoric process
C. Heat is neither absorbed nor released by a system	III. Isothermal process
D. No work is done on or by a gas.	IV. Isobaric process

Choose the correct answer from the options given below.

(a) A-I, B-IV, C-II, D-III (b) A-III, B-IV, C-I, D-II
(c) A-I, B-III, C-II, D-IV (d) A-III, B-I, C-IV, D-II

56

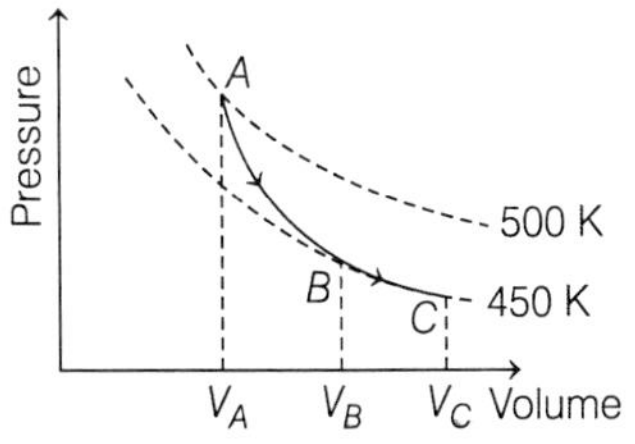

A polyatomic molecule ($C_V = 3R, C_p = 4R$, where R is gas constant) goes from phase space point A $(p_A = 10^5 \text{ Pa}, V_A = 4 \times 10^{-6} \text{m}^3)$ to point B $(p_B = 5 \times 10^4 \text{ Pa}, V_B = 6 \times 10^{-6} \text{m}^3)$ to point $C(p_C = 10^4 \text{ Pa}, V_C = 8 \times 10^{-6} \text{m}^3)$. A to B is an adiabatic path and B to C is an isothermal path. The net heat absorbed per unit mole by the system is **[2025, 29 Jan Shift-II]**

(a) $500R$ (In 3 + In 4) (b) $450R$ (In 4-In 3)
(c) $500R$ In2 (d) $400R$ In 4

57 A real gas within a closed chamber at 27°C undergoes the cyclic process as shown in figure. The gas obeys $pV^3 = RT$ equation for the path A to B. The net work done in the complete cycle is (assuming $R = 8$ J/mol K) **[2024, 9 April Shift-II]**

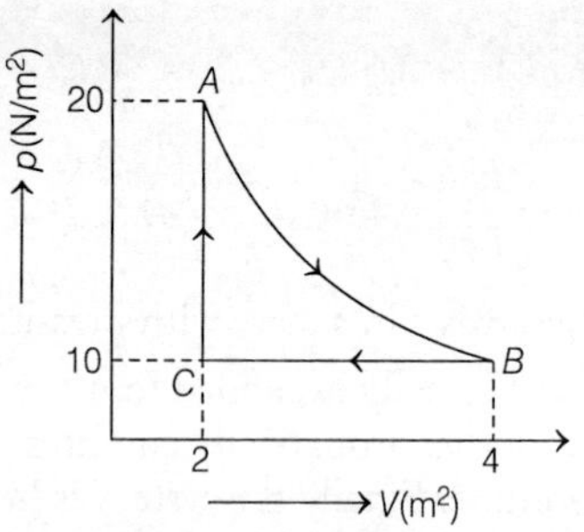

(a) – 20 J (b) 20 J
(c) 205 J (d) 225 J

58 The pressure and volume of an ideal gas are related as $pV^{3/2} = K$ (Constant). The work done when the gas is taken from state $A(p_1, V_1, T_1)$ to state $B(p_2, V_2, T_2)$ is **[2024, 1 Feb Shift-I]**

(a) $2(p_1V_1 - p_2V_2)$ (b) $2(p_2V_2 - p_1V_1)$
(c) $2(p_2\sqrt{V_2} - p_1\sqrt{V_1})$ (d) $2(\sqrt{p_1}V_1 - \sqrt{p_2}V_2)$

59 Choose the correct statement for processes A and B shown in figure **[2024, 30 Jan Shift-II]**

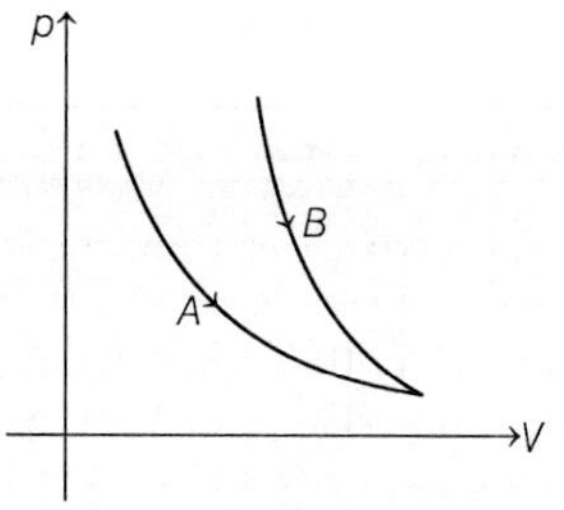

(a) $\dfrac{p^{\gamma-1}}{T^{\gamma}} = k$ for process B and $T = k$ for process A
(b) $pV = k$ for process B and A
(c) $\dfrac{T^{\gamma}}{p^{\gamma-1}} = k$ for process A and $pV = k$ for process B
(d) $pV^{\gamma} = k$ for process B and $pV = k$ for process A

60 Match List I with List II.

List I	List II
A. Isothermal process	I. Work done by the gas decreases internal energy
B. Adiabatic process	II. No change in internal energy
C. Isochoric process	III. The heat absorbed goes partly to increase internal energy and partly to do work
D. Isobaric process	IV. No work is done on or by the gas

Choose the correct answer from the options given below. **[2023, 25 Jan Shift-II]**

(a) A-II, B-I, C-IV, D-III (b) A-II, B-I, C-III, D-IV
(c) A-I, B-II, C-IV, D-III (d) A-I, B-II, C-III, D-IV

61 A certain amount of gas of volume V at 27°C temperature and pressure 2×10^7 Nm^{-2} expands isothermally until its volume gets doubled. Later it expands adiabatically until its volume gets redoubled. The final pressure of the gas will be (Use, $\gamma = 1.5$) **[2022, 25 July Shift-I]**

(a) 3.536×10^5 Pa (b) 3.536×10^6 Pa
(c) 1.25×10^6 Pa (d) 1.25×10^5 Pa

62 A sample of an ideal gas is taken through the cyclic process $ABCA$ as shown in figure. It absorbs, 40 J of heat during the part AB, no heat during BC and rejects 60 J of heat during CA. A work 50 J is done on the gas during the part BC. The internal energy of the gas at A is 1560 J. The work done by the gas during the part CA is **[2022, 28 June Shift-II]**

(a) 20 J (b) 30 J (c) – 30 J (d) – 60 J

63 Thermodynamic process is shown below on a p-V diagram for one mole of an ideal gas. If $V_2 = 2V_1$, then the ratio of temperature T_2/T_1 is **[2021, 25 Feb Shift-II]**

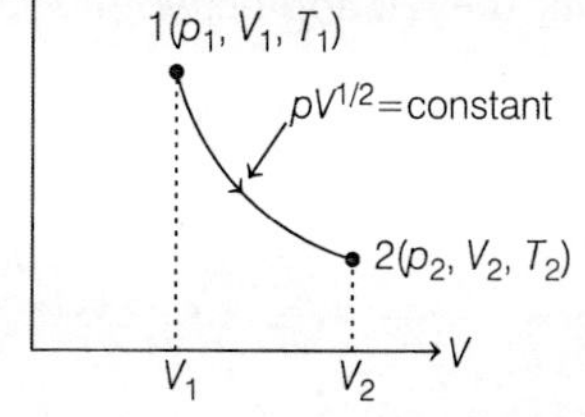

(a) $1/\sqrt{2}$ (b) $\sqrt{2}$
(c) 1/2 (d) 2

64 If one mole of an ideal gas at (p_1, V_1) is allowed to expand reversibly and isothermally (A to B), its pressure is reduced to one-half of the original pressure (see figure). This is followed by a constant volume cooling till its pressure is reduced to one-fourth of the initial value ($B \to C$). Then, it is restored to its initial state by a reversible adiabatic compression (C to A).

The net work done by the gas is equal to **[2021, 24 Feb Shift-II]**

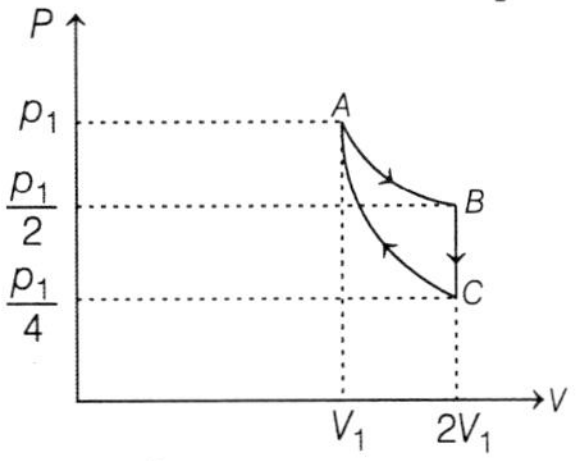

(a) $RT\left(\ln 2 - \frac{1}{2(\gamma - 1)}\right)$ (b) $-\frac{RT}{2(\gamma-1)}$

(c) 0 (d) $RT \ln 2$

65 n mole of a perfect gas undergoes a cyclic process $ABCA$ (see figure) consisting of the following processes.

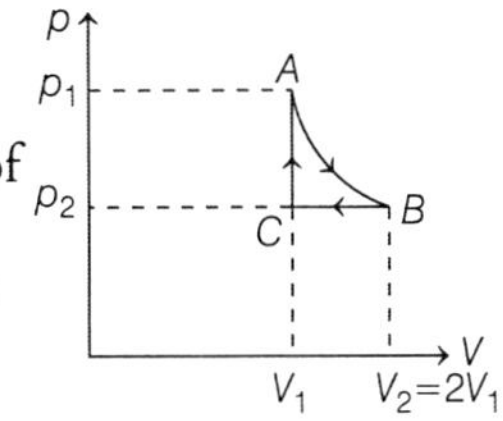

$A \to B$: Isothermal expansion at temperature T, so that the volume is doubled from V_1 to $V_2 = 2V_1$ and pressure changes from p_1 to p_2.

$B \to C$: Isobaric compression at pressure p_2 to initial volume V_1.

$C \to A$: Isochoric change leading to change of pressure from p_2 to p_1. Total work done in the complete cycle $ABCA$ is **[2021, 24 Feb Shift-I]**

(a) 0 (b) $nRT \ln 2$

(c) $nRT\left(\ln 2 + \frac{1}{2}\right)$ (d) $nRT\left(\ln 2 - \frac{1}{2}\right)$

Imp. *p-V* Diagram Under Cyclic Process

66 Which of the following is an equivalent cyclic process corresponding to the thermodynamic cyclic given in the figure? where, $1 \to 2$ is adiabatic.

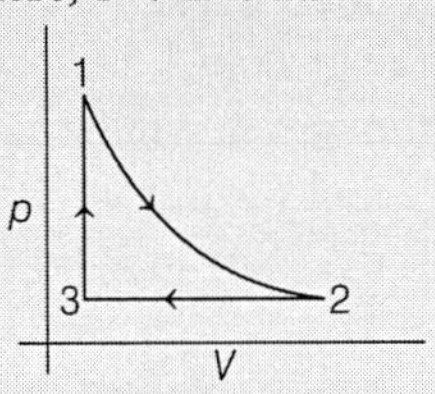

(Graphs are schematic and are not to scale) **[2020, 9 Jan Shift-I]**

(a)

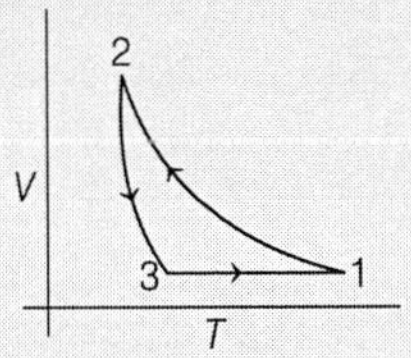

(b)

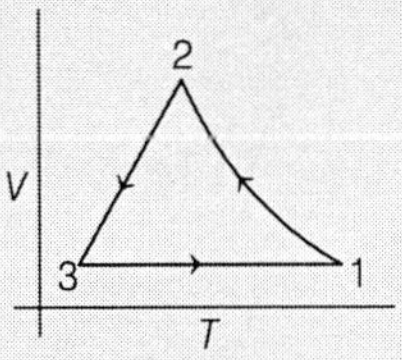

(c)

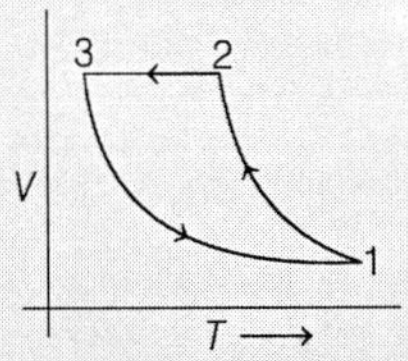

(d)

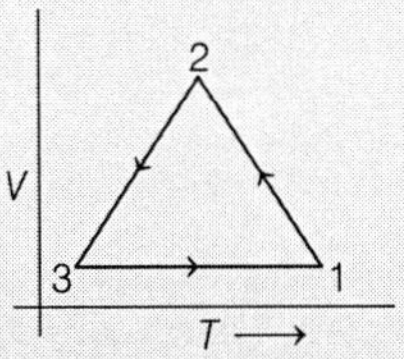

67 A sample of an ideal gas is taken through the cyclic process *abca* as shown in the figure. The change in the internal energy of the gas along the path *ca* is -180 J. The gas absorbs 250 J of heat along the path *ab* and 60 J along the path *bc*. The work done by the gas along the path *abc* is **[2019, 12 April Shift-I]**

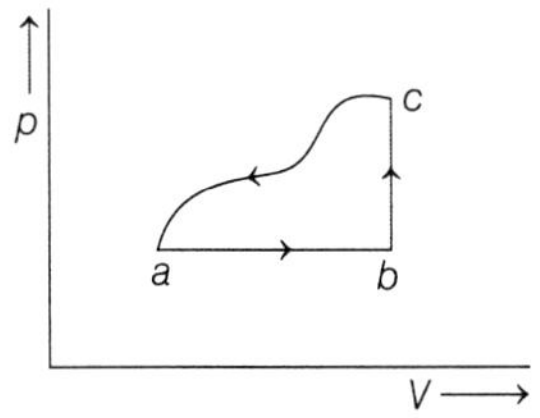

(a) 120 J (b) 130 J (c) 100 J (d) 140 J

68 One mole of an ideal gas passes through a process where pressure and volume obey the relation

$$p = p_0\left[1 - \frac{1}{2}\left(\frac{V_0}{V}\right)^2\right].$$ Here p_0 and V_0 are constants.

Calculate the change in the temperature of the gas if its volume changes from V_0 to $2V_0$. **[2019, 10 April Shift-II)**

(a) $\frac{1}{4}\frac{p_0V_0}{R}$ (b) $\frac{5}{4}\frac{p_0V_0}{R}$ (c) $\frac{1}{2}\frac{p_0V_0}{R}$ (d) $\frac{3}{4}\frac{p_0V_0}{R}$

69 One mole of an ideal monoatomiç gas is taken along the path $ABCA$ as shown in the p-V diagram. The maximum temperature attained by the gas along the path BC is given by **[JEE Main 2018)**

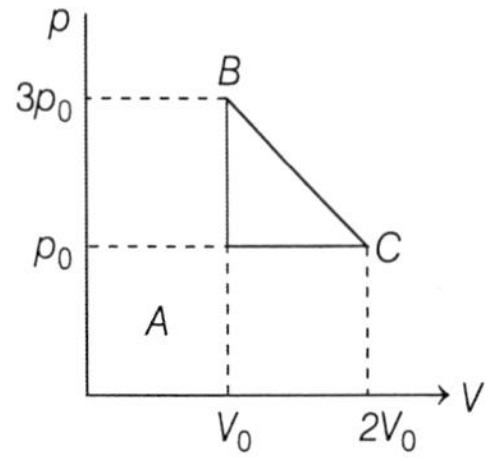

(a) $\frac{5}{8}\frac{p_0V_0}{R}$ (b) $\frac{25}{8}\frac{p_0V_0}{R}$ (c) $\frac{25}{4}\frac{p_0V_0}{R}$ (d) $\frac{25}{16}\frac{p_0V_0}{R}$

70 n moles of an ideal gas undergoes a process A and B as shown in the figure. The maximum temperature of the gas during the process will be **[JEE Main 2016]**

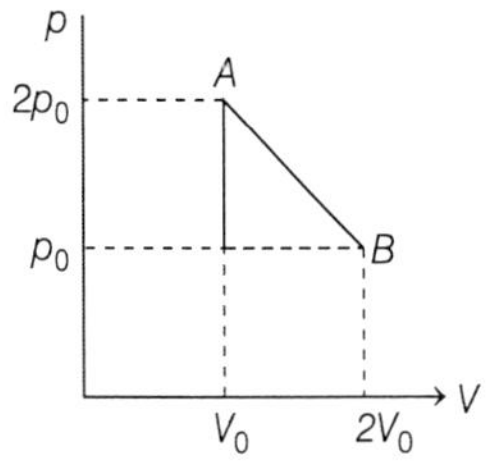

(a) $\frac{9}{4}\frac{p_0V_0}{nR}$ (b) $\frac{3}{2}\frac{p_0V_0}{nR}$ (c) $\frac{9}{2}\frac{p_0V_0}{nR}$ (d) $\frac{9p_0V_0}{nR}$

71 One mole of diatomic ideal gas undergoes a cyclic process ABC as shown in figure. The process BC is adiabatic. The temperatures at A, B and C are 400 K, 800 K and 600 K, respectively.

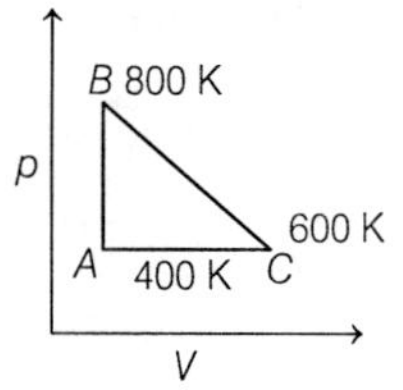

Choose the correct statement. *[JEE Main 2014]*

(a) The change in internal energy in whole cyclic process is 250 R.

(b) The change in internal energy in the process CA is 700 R.

(c) The change in internal energy in the process AB is -350 R.

(d) The change in internal energy in the process BC is $-500R$.

72 Helium gas goes through a cycle $ABCDA$ (consisting of two isochoric and isobaric lines) as shown in figure. Efficiency of this cycle is nearly (Assume the gas to be close to ideal gas) *[AIEEE 2012]*

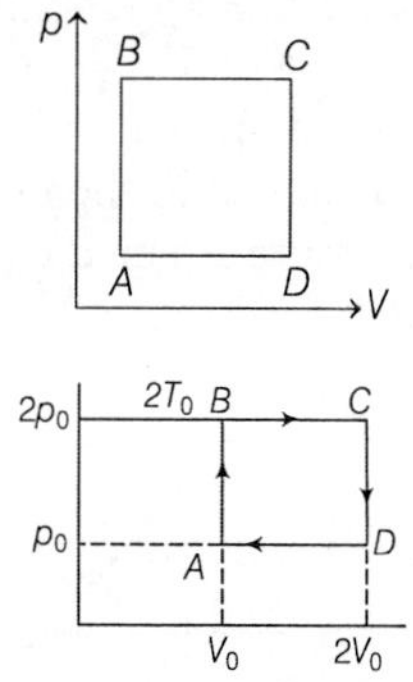

(a) 15.4% (b) 9.1% (c) 10.5% (d) 12.5%

Numerical Value Type Questions

73 An ideal gas has undergone through the cyclic process as shown in the figure. Work done by the gas in the entire cycle is .... $\times 10^{-1}$ J.

(Take $\pi = 3.14$) *[2025, 7 April Shift-I]*

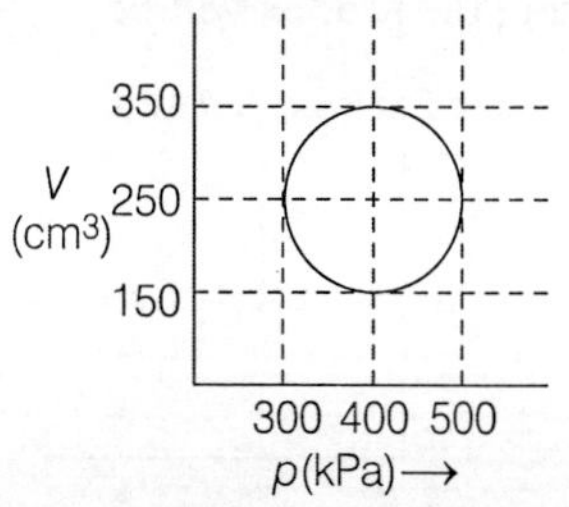

74 An ideal gas initially at 0°C temperature, is compressed suddenly to one-fourth of its volume. If the ratio of specific heat at constant pressure to that at constant volume is $\frac{3}{2}$, the change in temperature due to the thermodynamic process is .... K. *[2025, 23 Jan Shift-I]*

75 The temperature of 1 mole of an ideal monoatomic gas is increased by 50°C at constant pressure. The total heat added and change in internal energy are E_1 and E_2, respectively. If $\frac{E_1}{E_2} = \frac{x}{9}$, then the value of x is........... *[2025, 24 Jan Shift-I]*

76 The pressure p_1 and density d_1 of diatomic gas $\left(\gamma = \frac{7}{5}\right)$ changes suddenly to p_2 ($> p_1$) and d_2 respectively during an adiabatic process. The temperature of the gas increases and becomes ______ times of its initial temperature. $\left(\text{Given, } \frac{d_2}{d_1} = 32\right)$ *[2022, 29 July Shift-I]*

77 A diatomic gas ($\gamma = 1.4$) does 400 J of work when it is expanded isobarically. The heat given to the gas in the process is ______ J. *[2022, 27 June Shift-II]*

78 A sample of gas with $\gamma = 1.5$ is taken through an adiabatic process in which the volume is compressed from 1200 cm^3 to 300 cm^3. If the initial pressure is 200 kPa. The absolute value of the work done by the gas in the process is .... J. *[2021, 31 Aug Shift-II]*

79 One mole of an ideal gas at 27°C is taken from A to B as shown in the given p-V indicator diagram. The work done by the system will be ...... $\times 10^{-1}$ J. [Take, $R = 8.3$ J/ mol-K, $\ln 2 = 0.6931$] (Round off to the nearest integer) *[2021, 20 July Shift-II]*

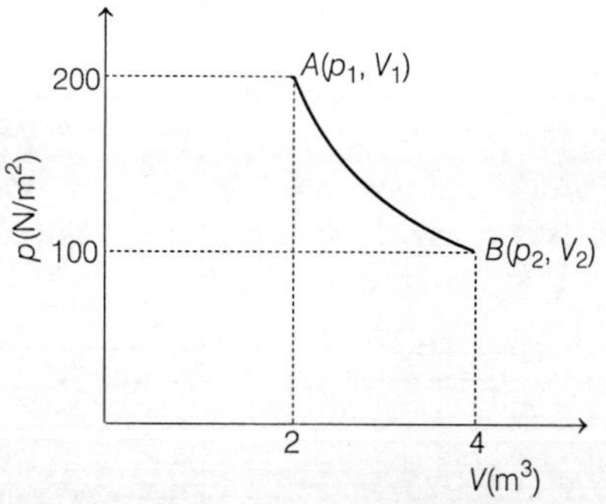

80 In the reported figure, heat energy absorbed by a system in going through a cyclic process is ...... πJ. *[2021, 20 July Shift-I]*

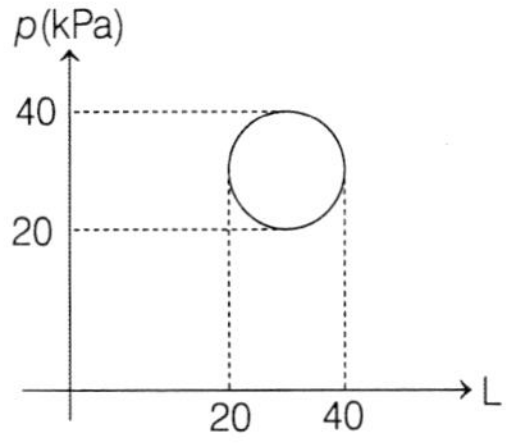

81 In a certain thermodynamical process, the pressure of a gas depends on its volume as kV^3. The work done when the temperature changes from 100°C to 300°C will be .... nR, where n denotes number of moles of a gas. *[2021, 25 Feb Shift-I]*

82 Starting at temperature 300 K, one mole of an ideal diatomic gas ($\gamma = 1.4$) is first compressed adiabatically from volume V_1 to $V_2 = \frac{V_1}{16}$. It is then allowed to expand isobarically to volume $2V_2$. If all the processes are the quasi-static, then the final temperature of the gas (in K) is (to the nearest integer) .... *[2020, 9 Jan Shift-II]*

Topic 3 Heat Engine and Second Law of Thermodynamics

Level 1

01 An engine operating between the boiling and freezing points of water will have

A. efficiency more than 27%.

B. efficiency less than the efficiency of a Carnot engine operating between the same two temperatures.

C. efficiency equal to 27%.

D. efficiency less than 27%.

Choose the correct answer from the options given below. *[2023, 12 April Shift-I]*

(a) B, C and D only (b) A and B only
(c) B and C only (d) B and D only

02 A solid body of constant heat capacity 1 J/°C is being heated by keeping it in contact with reservoirs in two ways *[JEE Main 2015]*

(i) sequentially keeping in contact with 2 reservoirs such that each reservoir supplies same amount of heat.

(ii) sequentially keeping in contact with 8 reservoirs such that each reservoir supplies same amount of heat.

In both the cases, body is brought from initial temperature 100°C to final temperature 200°C. Entropy change of the body in the two cases respectively, is

(a) ln 2, 4 ln 2 (b) ln 2, ln 2
(c) ln 2, 2 ln 2 (d) 2 ln 2, 8 ln 2

Level 2

03 Water of mass m gram is slowly heated to increase the temperature from T_1 to T_2. The change in entropy of the water, given specific heat of water is $1 \text{Jkg}^{-1}\text{K}^{-1}$, is *[2025, 23 Jan Shift-II]*

(a) zero (b) $m \ln\left(\frac{T_1}{T_2}\right)$
(c) $m(T_2 - T_1)$ (d) $m \ln\left(\frac{T_2}{T_1}\right)$

04 "Heat cannot be itself flow from a body at lower temperature to a body at higher temperature" is a statement or consequence of *[AIEEE 2003]*

(a) second law of thermodynamics
(b) conservation of momentum
(c) conservation of mass
(d) first law of thermodynamics

05 A diatomic gas ($\gamma = 1.4$) does 200 J of work when it is expanded isobarically. The heat given to the gas in the process is *[2024, 1 Feb Shift-II]*

(a) 800 J (b) 700 J
(c) 850 J (d) 600 J

06 The temperature-entropy diagram of a reversible engine cycle is given in the figure. Its efficiency is *[AIEEE 2005]*

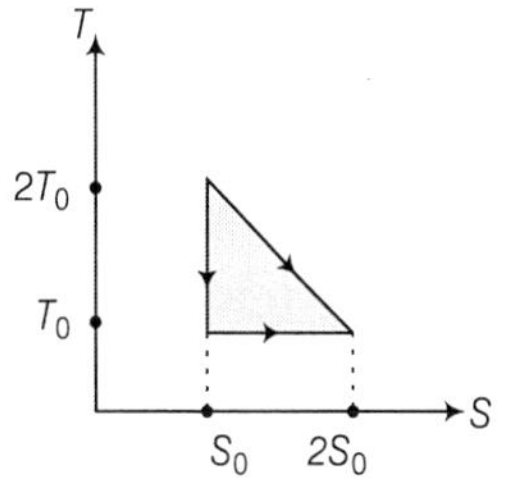

(a) $\frac{1}{2}$ (b) $\frac{1}{4}$ (c) $\frac{1}{3}$ (d) $\frac{2}{3}$

Level 3

07 A Carnot engine E is working between two temperatures 473 K and 273 K. In a new system, two engines engine E_1 works between 473 K to 373 K and engine E_2 works between 373 K to 273 K. If η_{12}, η_1 and η_2 are the efficiencies of the engines E, E_1 and E_2, respectively, then *[2025 28 Jan, Shift-I]*

(a) $\eta_{12} = \eta_1\eta_2$ (b) $\eta_{12} < \eta_1 + \eta_2$
(c) $\eta_{12} \geq \eta_1 + \eta_2$ (d) $\eta_{12} = \eta_1 + \eta_2$

08 Two different adiabatic paths for the same gas intersect two isothermal curves as shown in p-V diagram. The relation between the ratio $\frac{V_a}{V_d}$ and the ratio $\frac{V_b}{V_c}$ is *[2024 8 April, Shift-I]*

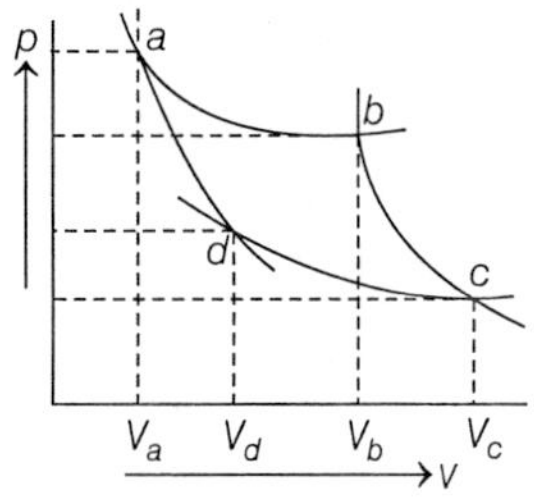

(a) $\frac{V_a}{V_d} = \left(\frac{V_b}{V_c}\right)^2$ (b) $\frac{V_a}{V_d} = \frac{V_b}{V_c}$

(c) $\frac{V_a}{V_d} \neq \frac{V_b}{V_c}$ (d) $\frac{V_a}{V_d} = \left(\frac{V_b}{V_c}\right)^{-1}$

Imp. Efficiency of Heat Engine

09 Let η_1 is the efficiency of an engine at $T_1 = 447°C$ and $T_2 = 147°C$ while η_2 is the efficiency at $T_1 = 947°C$ and $T_2 = 47°C$. The ratio η_1 / η_2 will be *[2022, 25 July Shift-II]*

(a) 0.41 (b) 0.56
(c) 0.73 (d) 0.70

10 A reversible engine has an efficiency of 1/4. If the temperature of the sink is reduced by 58°C, its efficiency becomes double. Calculate the temperature of the sink. *[2021, 31 Aug Shift-I]*

(a) 174°C (b) 280°C
(c) 180.4°C (d) 382°C

11 A heat engine has an efficiency of 1/6 . When the temperature of sink is reduced by 62°C, its efficiency get doubled. The temperature of the source is *[2021, 25 July Shift-II]*

(a) 124°C (b) 37°C (c) 62°C (d) 99°C

12 A heat engine is involved with exchange of heat of 1915 J, −40 J, +125 J and −QJ, during one cycle achieving an efficiency of 50.0%. The value of Q is *[2020, 2 Sep Shift-II]*

(a) 640 J (b) 40 J (c) 980 J (d) 400 J

13 An engine operates by taking n moles of an ideal gas through the cycle $ABCDA$ shown in figure. The thermal efficiency of the engine is (Take, $C_V = 1.5\,R$, where R is gas constant) *[JEE Main 2017]*

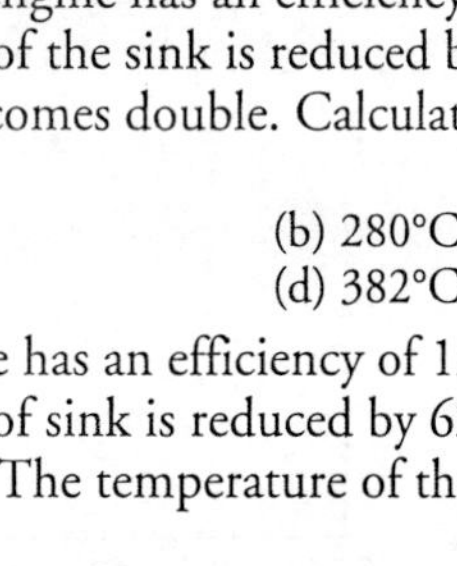

(a) 0.15 (b) 0.32 (c) 0.24 (d) 0.08

14 The specific heat capacity of a metal at low temperature T is given as

$$C_p(\text{kJ K}^{-1}\text{ kg}^{-1}) = 32\left(\frac{T}{400}\right)^3.$$

A 100 g vessel of this metal is to be cooled from 20 K to 4 K by a special refrigerator operating at room temperature (27°C). The amount of work required to cool the vessel is *[AIEEE 2011]*

(a) equal to 0.002 kJ
(b) greater than 0.148 kJ
(c) between 0.148 kJ and 0.028 kJ
(d) less than 0.028 kJ

Numerical Value Type Questions

Imp. Efficiency of Carnot's Engine

15 300 cal. of heat is given to a heat engine and it rejects 225 cal. of heat. If source temperature is 227°C, then the temperature of sink will be ....°C. *[2022, 29 June Shift-I]*

16 A heat engine operates with the cold reservoir at temperature 324 K. The minimum temperature of the hot reservoir, if the heat engine takes 300 J heat from the hot reservoir and delivers 180 J heat to the cold reservoir per cycle, is ...... K. *[2022, 26 June Shift-II]*

17 A steam engine intakes 50 g of steam at 100°C per minute and cools it down to 20°C. If latent heat of vaporization of steam is 540 cal g^{-1}, then the heat rejected by the steam engine per minute is .... $\times 10^3$ cal. (Given, specific heat capacity of water : 1 cal g^{-1}°C^{-1}) *[2022, 25 June Shift-I]*

18 A heat engine operates between a cold reservoir at temperature $T_2 = 400$ K and a hot reservoir at temperature T_1. It takes 300 J of heat from the hot reservoir and delivers 240 J of heat to the cold reservoir in a cycle. The minimum temperature of the hot reservoir has to be .... K. *[2021, 27 Aug Shift-II]*

19 For an ideal heat engine, the temperature of the source is 127°C. In order to have 60% efficiency the temperature of the sink should be ...... °C. (Round off to the nearest integer) *[2021, 16 March Shift-II]*

20 A reversible heat engine converts one-fourth of the heat input into work. When the temperature of the sink is reduced by 52 K, its efficiency is doubled. The temperature in kelvin of the source will be .... *[2021, 25 Feb Shift-II]*

21 If minimum possible work is done by a refrigerator in converting 100 g of water at 0°C to ice, how much heat (in cal) is released to the surroundings at temperature 27°C to the nearest integer ...... ? (Take, latent heat of ice = 80 cal/g) *[2020, 3 Sep Shift-II]*

• Key **Index**

Topic 1

1	(d)	**2**	(c)	**3**	(c)	**4**	(a)	**5**	(a)	**6**	(c)	**7**	(a)	**8**	(a)	**9**	(a)	**10**	(c)
11	(c)	**12**	(*)	**13**	(a)	**14**	(b)	**15**	(d)	**16**	(12)								

Topic 2

1	(b)	**2**	(a)	**3**	(c)	**4**	(b)	**5**	(a)	**6**	(a)	**7**	(c)	**8**	(d)	**9**	(c)	**10**	(c)
11	(d)	**12**	(b)	**13**	(d)	**14**	(a)	**15**	(a)	**16**	(b)	**17**	(c)	**18**	(b)	**19**	(a)	**20**	(b)
21	(c)	**22**	(a)	**23**	(c)	**24**	(d)	**25**	(d)	**26**	(d)	**27**	(c)	**28**	(a)	**29**	(a)	**30**	(d)
31	(d)	**32**	(d)	**33**	(a)	**34**	(b)	**35**	(c)	**36**	(c)	**37**	(a)	**38**	(c)	**39**	(b)	**40**	(c)
41	(d)	**42**	(a)	**43**	(b)	**44**	(b)	**45**	(d)	**46**	(b)	**47**	(b)	**48**	(b)	**49**	(c)	**50**	(a)
51	(a)	**52**	(c)	**53**	(c)	**54**	(d)	**55**	(b)	**56**	(b)	**57**	(c)	**58**	(a)	**59**	(a,d)	**60**	(a)
61	(b)	**62**	(b)	**63**	(b)	**64**	(a)	**65**	(d)	**66**	(b)	**67**	(b)	**68**	(b)	**69**	(b)	**70**	(a)
71	(d)	**72**	(a)	**73**	(314)	**74**	(273)	**75**	(15)	**76**	(4)	**77**	(1400)	**78**	(480)	**79**	(17258)	**80**	(100)
81	(50)	**82**	(1819)																

Topic 3

1	(d)	**2**	(b)	**3**	(d)	**4**	(a)	**5**	(b)	**6**	(c)	**7**	(b)	**8**	(b)	**9**	(b)	**10**	(a)
11	(d)	**12**	(c)	**13**	(a)	**14**	(c)	**15**	(102)	**16**	(540)	**17**	(31)	**18**	(500)	**19**	(–113)	**20**	(208)
21	(8791)																		

Detailed **Explanations**

Topic 1

01. *(d)* Extensive variable depends on size or mass of system

Example - Internal energy (ΔU)

Volume (V)

Mass (m)

Intensive variables are

Pressure (p), Temperature (T),

Density (ρ)

(A) and (B) are correct.

02. *(c)* Given, $\Delta Q = 48$ J, $\Delta T = 2°C$

According to first law of thermodynamics,

$$\Delta Q = \Delta U + W$$

$$\Rightarrow \Delta Q = nC_V \Delta T + W$$

$$\Rightarrow 48 = 1\left(\frac{3R}{2}\right)(2) + W$$

$$\Rightarrow W = 48 - 3R = 48 - 3 \times 8.3 = 23.1 \text{ J}$$

03. *(c)* From the first law of thermodynamics,

$$dQ = dU + dW$$

Also, we can write this as,

$$\frac{dQ}{dt} = \frac{dU}{dt} + \frac{dW}{dt}$$

$$1000 \text{ W} = \frac{dU}{dt} + 200 \text{ W}$$

$$\frac{dU}{dt} = 800 \text{ W}$$

04. *(a)* When a system is given heat dQ and the work done on system is dW

Then, $dQ + dW = dU$

or $\quad dQ = dU - dW$

Both, statements are correct and R explains A.

05. *(a)* In thermodynamic process ΔQ and ΔW depend on the path followed by the system between two states. Thus, they are path functions.

06. *(c)* According to the first law of thermodynamics,

$$\Delta Q = \Delta W + \Delta U$$

$\because \quad \Delta U_A = \Delta U_B$ (initial and final state are same for both A and B) and

$$\Delta W_A > \Delta W_B$$

As area below process A is more than that of process B.

Therefore, we can conclude that

$$\Delta Q_A > \Delta Q_B$$

07. *(a)* As work done $= 0$

[as $W = p\Delta V$ and $\Delta V = 0$, so $W = 0$]

$$\Delta U = mC\Delta T$$

$$= 100 \times 10^{-3} \times 4184 \times (50 - 30)$$

$$\approx 8.4 \text{ kJ}$$

08. *(a)* From first law of thermodynamics,

$$\Delta Q = \Delta U + W$$

For path iaf, $\quad 50 = \Delta U + 20$

$\therefore \quad \Delta U = U_f - U_i = 30$ cal

For path ibf, $\quad \Delta Q = \Delta U + W$

or $\quad W = Q - \Delta U$

$$= 36 - 30 = 6 \text{ cal}$$

09. *(a)* Internal energy is a state function i.e. it does not depend upon the path followed by the process. It depends only on initial and final states.

Hence, $\Delta U_1 = \Delta U_2$.

10. *(c)* Work does not characterise the thermodynamic state of matter, it is a path function gives only relationship between pressure and volume.

11. *(c)* During melting of ice, volume decreases, so positive work is done on ice water system by atmosphere. Heat absorbed by ice water, so ΔQ is positive.

Work done by ice water system is negative. Thus, by first law of thermodynamics,

$$\Delta Q = \Delta U + (-\Delta W)$$

$\Rightarrow \quad \Delta U = \Delta Q + \Delta W$

Thus, internal energy increases.

12. *(*)*

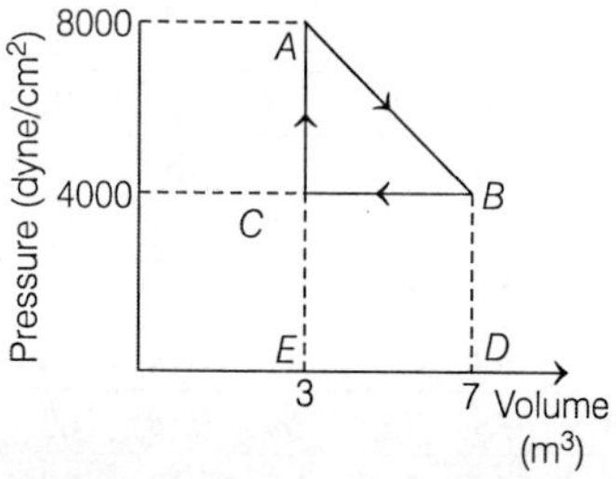

W_{AB} = Area of trapezium $ABDECA$

$$= \frac{1}{2} \times (8000 + 4000) \times 10^{-1} \times 4$$

$$= 2400 \text{ J}$$

W_{BC} = Area of rectangle $CBDEC$

$$= 4000 \times 10^{-1} \times 4 = 1600 \text{ J}$$

∴ Total work done,

$$W = W_{AB} - W_{BC} = 2400 - 1600$$

$$= 800 \text{ J}$$

13. *(a)* Given, heat supplied to the system,

$$\frac{\Delta Q}{\Delta t} = 6000 \text{ J/min} = \frac{6000}{60} \text{ J/s} = 100 \text{ J/s}$$

Power delivered, $P = \dfrac{\Delta W}{\Delta t} = 90$ W

Increase in internal energy,

$$\Delta U = 2.5 \times 10^3 \text{ J}$$

From first law of thermodynamics, we have $\Delta Q = \Delta U + \Delta W$

or $$\frac{\Delta Q}{\Delta t} = \frac{\Delta U}{\Delta t} + \frac{\Delta W}{\Delta t} \quad \ldots \text{(i)}$$

Putting values

$$100 = \frac{2.5 \times 10^3}{\Delta t} + 90$$

$$\Rightarrow \quad 10 = \frac{2.5 \times 10^3}{\Delta t}$$

$$\Rightarrow \quad \Delta t = \frac{2.5 \times 10^3}{10}$$

$$\Rightarrow \quad \Delta t = 2.5 \times 10^2 \text{ s}$$

14. *(b)* For the ACB as shown in the figure below,

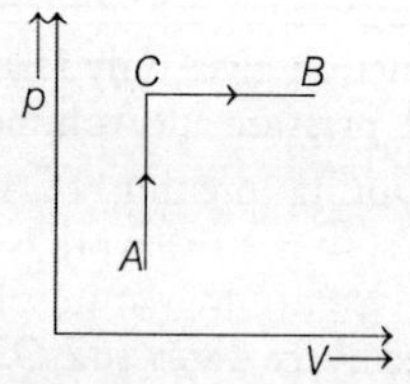

According to the first law of thermodynamics, heat supplied,

$$\Delta Q = \Delta W + \Delta U$$

$$\Rightarrow \quad \Delta Q_{ACB} = \Delta W_{ACB} + (U_B - U_A)$$

Substituting the given values,

$$U_B - U_A = 60 - 30 = 30 \text{ J} \quad \ldots \text{(i)}$$

Similarly for the ADB as shown in the figure below,

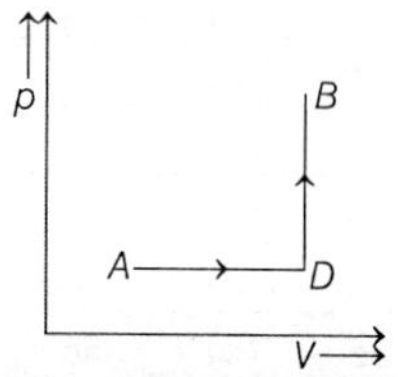

$$\Delta Q_{ADB} = \Delta W_{ADB} + (U_B - U_A)$$

$$\Rightarrow \quad \Delta Q_{ADB} = 10 + 30 \quad \text{[using Eq. (i)]}$$

$$= 40 \text{ J}$$

15. *(d)* Statements (A) and (D) are wrong. Concept of entropy is associated with second law of thermodynamics and first law of thermodynamics is applicable to cylic process.

16. *(12)* $V = 4 \times 4 \times 3 = 48 \text{ m}^3$

Internal energy,

$$U = nC_V T = n\left(\frac{5R}{2}\right)T$$

$$= \frac{5}{2} nRT = \frac{5}{2} pV \quad [\because pV = nRT]$$

$$= \frac{5}{2} \times 1 \times 10^5 \times 48 = 12 \times 10^6 \text{ J}$$

Topic 2

01. *(b)* We know that,

$$\Delta Q = \Delta U + W$$

For adiabatic process, $\Delta Q = 0$

$$\therefore \quad W = -\Delta U$$

$$\Rightarrow \quad W = -nC_V \Delta T \Rightarrow W \propto \Delta T$$

02. *(a)* For isochoric process,

$$p \propto T$$

$$\frac{\Delta p}{p} = \frac{\Delta T}{T}$$

$$\Rightarrow \quad \frac{0.4}{100} = \frac{1}{T} \Rightarrow T = 250 \text{ K}$$

03. *(c)* The given situation follow adiabatic process.

Thus, $p_i V_i^\gamma = p_f V_f^\gamma$

$$\Rightarrow \quad \frac{p_f}{p_i} = \left(\frac{V_i}{V_f}\right)^\gamma$$

$$= \left(\frac{V_i}{\frac{V_i}{8}}\right)^\gamma \quad \left[\because V_f = \frac{V_i}{8}\right]$$

$$= (8)^\gamma = (8)^{\frac{5}{3}}$$

$$= (2^3)^{\frac{5}{3}} = 2^5 = 32$$

04. *(b)* (Work done)$_a$

$$= \frac{p_1 V_1 - p_2 V_2}{\gamma - 1} = \frac{nR(T_1 - T_2)}{\gamma - 1}$$

So, depends only on change of temperature.

05. *(a)* (A) $T_1 V_1^{\gamma - 1} = T_2 V_2^{\gamma - 1}$

Temperature increases

(B) Free expansion is assumed fast, so adiabatic

(A) → false

(R) → true

06. *(a)* For adiabatic process,

$$p_i V_i^\gamma = p_f V_f^\gamma$$

$$p_i (5)^{1.5} = p_f (4)^{1.5}$$

$$\frac{p_i}{p_f} = \left(\frac{4}{5}\right)^{1.5}$$

$$\frac{p_i}{p_f} = \left(\frac{4}{5}\right)^{3/2} = \frac{4}{5}\left(\frac{4}{5}\right)^{1/2} = \frac{8}{5\sqrt{5}}$$

07. *(c)* Given,

$n = 2 \times 10^{25}$ molecules per cubic metre

$$p = 1.38 \text{ atm} = 1.38 \times 10^5 \text{ N/m}^2$$

$$K = 1.38 \times 10^{-23} \text{ J K}^{-1}$$

we know that

$$pV = nKT \Rightarrow T = \frac{pV}{nK}$$

$$= \frac{1.38 \times 10^5 \times 1}{2 \times 10^{25} \times 1.38 \times 10^{-23}} = 500 \text{ K}$$

08. *(d)* We know that,

$$PV^\gamma = \text{constant}$$

$$p_1 V_1^\gamma = p_2 V_2^\gamma \Rightarrow \frac{p_1}{p_2} = \left(\frac{V_2}{V_1}\right)^\gamma$$

Given, $p_1 = p_0$, $V_1 = V_0$

$$V_2 = \frac{V_0}{4}$$

$$\frac{p_0}{p_2} = \left(\frac{V_0/4}{V_0}\right)^\gamma \Rightarrow \frac{p_0}{p_2} = \left(\frac{V_0}{4V_0}\right)^\gamma$$

$$\frac{p_0}{p_2} = \left(\frac{1}{4}\right)^\gamma \Rightarrow \quad p_2 = p_0 (4)^\gamma$$

09. *(c)* Work done in the cyclic process,

W = area of closed curve

= Area of ΔDCE

$$= \frac{1}{2} \times CE \times CD$$

$$= \frac{1}{2} \times (4-2) \times (400-100)$$
$$= 300 \text{ J}$$

10. *(c)* Since, the given process is adiabatic, hence $\Delta Q = 0$
from first law of thermodynamics, we get
$$\Delta Q = \Delta U + \Delta W \Rightarrow \Delta U = -\Delta W$$
The process is adiabatic compression, therefore volume decreases.
Hence, $\Delta W = -\text{ve} \Rightarrow \Delta U = +\text{ve}$
Thus, internal energy of the system increases positively with change in temperature as gas is compressed adiabatically. Hence, option (c) is not true.

11. *(d)* $dU = nC_V \Delta T$
$$\Delta T = \text{Zero} \rightarrow \text{Isothermal process}$$

12. *(b)* When heat is added to a system isothermally its temperature remains constant. So, statement I is incorrect.
Expansion work of a gas is given by;
$$W = p(V_f - V_i)$$
As $V_f > V_i$ so, work done is positive
Hence, statement II is correct.

13. *(d)* For an isothermal change,
$$pV = \text{constant or } p \propto \frac{1}{V}.$$
Also pV value will be more at higher temperature.
$\therefore$ Correct isotherm graph will be,

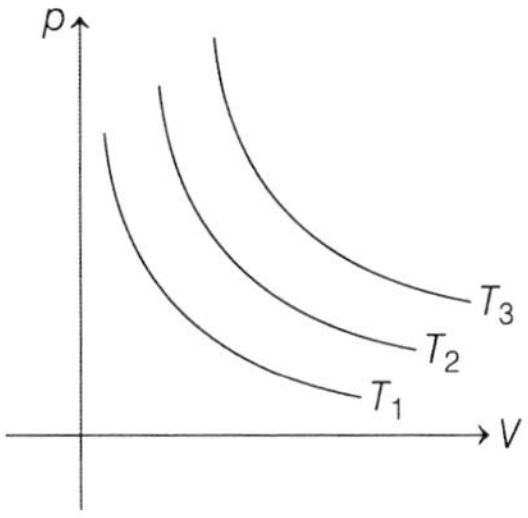

14. *(a)* The ideal gas equation for adiabatic expansion process,
$$pV^\gamma = C$$
$$\ln(p) + \gamma \ln(V) = C$$
where, C is constant.
On differentiating both sides, we get
$$\frac{dp}{p} + \gamma\left(\frac{dV}{V}\right) = 0 \Rightarrow \frac{dp}{p} = -\gamma\left(\frac{dV}{V}\right)$$

15. *(a)* Given,
The final pressure of the adiabatic process, $p_f = 200 \text{ N/m}^2$
The initial pressure of the adiabatic process, $p_i = 100 \text{ N/m}^2$
$$V_i = 4 \text{ m}^3 \Rightarrow V_f = 3 \text{ m}^3$$
We know that, for an adiabatic process
$$W = \frac{p_f V_f - p_i V_i}{1-\gamma}$$
$$\Rightarrow \quad W = \frac{200(3) - (100)(4)}{1-1.4}$$
$$\Rightarrow \quad W = -500 \text{ J}$$
Negative sign represents the work done on the system.

16. *(b)* In graph (A) vertical line is isochoric process, so the graph (A) is incorrect.
In graph (B) horizontal line is isobaric process, so the graph (B) is incorrect.
In graph (C) and (D) vertical line is isothermal process and the curve line is adiabatic process. So, the correct representation of thermodynamic process graph is (C) and (D).
So, the correct option is (b)

17. *(c)* Given that, density of the gas becomes 32 times and we know that density ρ is inversely proportional to volume V.
So, volume becomes 1/32 times of initial volume.
In adiabatic process,
$$pV^\gamma = \text{constant}$$
$$\Rightarrow \quad p_i V_i^\gamma = p_f V_f^\gamma$$
Now, $\gamma = \frac{7}{5}$ for diatomic gas
$$\Rightarrow \quad \frac{p_f}{p_i} = \left(\frac{V_i}{V_f}\right)^\gamma = (32)^{\frac{7}{5}} = 128$$
Hence, correct option is (c).

18. *(b)* Expansion after bursting of balloon is a fast process in which helium presses the atmosphere.
This expansion occurs at the expense of internal energy of helium molecules.
So, this process is irreversible and adiabatic.
Hence, option (b) is correct.

19. *(a)* Given processes are

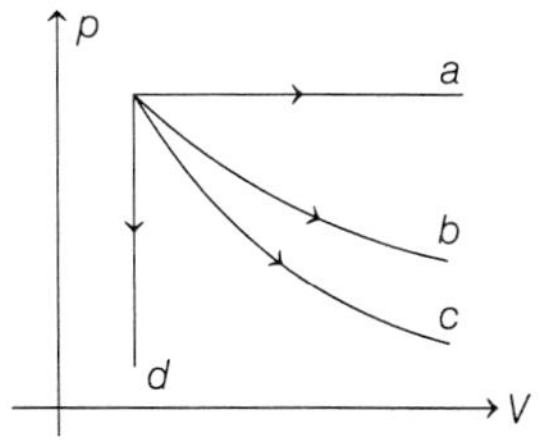

For process a, pressure is constant.
$\therefore a$ is isobaric.
For process d, volume is constant.
$\therefore d$ is isochoric.
Also, as we know that, slope of adiabatic curve in p-V diagram is more than that of isothermal curve.
$\therefore b$ is isothermal and c is adiabatic.

20. *(b)* Given, $n = 1, T = 300$ K
$$\text{So, } W = nRT \ln\left[\frac{p_1}{p_2}\right]$$
$$W = R(300)\ln(2) = 300R \ln 2$$

21. *(c)* Here, $pV = \text{constant}$ so given process is isothermal i.e., temperature is constant.
Pressure at point 1 is higher than that at point 2. So, correct option is (c).

22. *(a)* For isothermal process
$$\Delta T = 0$$
$$\Rightarrow \quad \Delta U = 0$$
For adiabatic process, $\Delta Q = 0$
For isobaric process, $\Delta p = 0$
but $\Delta U \neq 0$
For isochoric process,
$$\Delta V = 0$$
$$\therefore \quad \Delta W = p\Delta V = 0$$

23. *(c)* (A) Isobaric $\rightarrow$ Pressure = Constant
$$\Delta Q = \Delta U + p\Delta V$$
(B) Isochoric (V = constant)
$$\Delta Q = \Delta U$$
(C) Adiabatic: $\Delta Q = 0$
$$\Delta Q = 0$$
(D) Isothermal $\Delta U = 0$
$$\Delta Q = \Delta W$$
A-IV, B-II, C-III, D-I

24. *(d)* Here, $V_1 = 800 \text{ cm}^3$
$$= 8 \times 10^{-4} \text{ m}^3$$
$V_2 = 200 \text{ cm}^3 = 2 \times 10^{-4} \text{ m}^3$
$T_1 = 300$ K
We know that for adiabatic process,
$TV^{\gamma-1} = \text{constant}$
$$\Rightarrow \quad T_1 V_1^{\gamma-1} = T_2 V_2^{\gamma-1}$$
$$\Rightarrow \quad \frac{T_2}{T_1} = \left(\frac{V_1}{V_2}\right)^{\gamma-1} = \left(\frac{8 \times 10^{-4}}{2 \times 10^{-4}}\right)^{1.5-1}$$
$$= (4)^{0.5} = 2^{2 \times 0.5} = 2$$
$$\therefore \quad T_2 = 2T_1 = 2 \times 300 \text{ K} = 600 \text{ K}$$
$$\therefore \quad \Delta T = T_2 - T_1$$
$$= (600 - 300) \text{ K} = 300 \text{ K}$$

25. *(d)* For adiabatic process $d\theta = 0$
$\therefore$ Molar heat capacity $= 0$
$\because dQ = 0 \Rightarrow dU = -dW$

Also, $dU = \frac{f}{2} nRdT$

Molar heat capacity is zero.

26. *(d)* Among the given statements, A, D and E are correct, while B and C are incorrect.

The correct form of B and C are

$$p = kT$$

$$\frac{p}{T} = \text{constant}$$

$\therefore$ Volume is constant or isochoric process.

$\therefore \quad W = 0$

$\therefore \quad Q = \Delta U$

Hence, heat added is equal to change in internal energy.

27. *(c)* Work done = Area under curve

$$= -4p_0V_0 + p_0V_0$$

$$= -3p_0V_0$$

Alternate Solution

Work done in an isobaric process work A to B

$= p\Delta V$

$= p(V_f - V_i)$

$= p_0(3V_0 - 2V_0) = p_0V_0$

Work $BC = 0$ (process is isochoric)

Work $CD = 2p_0(V_0 - 3V_0)$

$= 2p_0(-2V_0) = -4p_0V_0$

Total work $A \to B \to C \to D$

$= p_0V_0 + 0 + (-4p_0V_0) = -3p_0V_0$

28. *(a)*

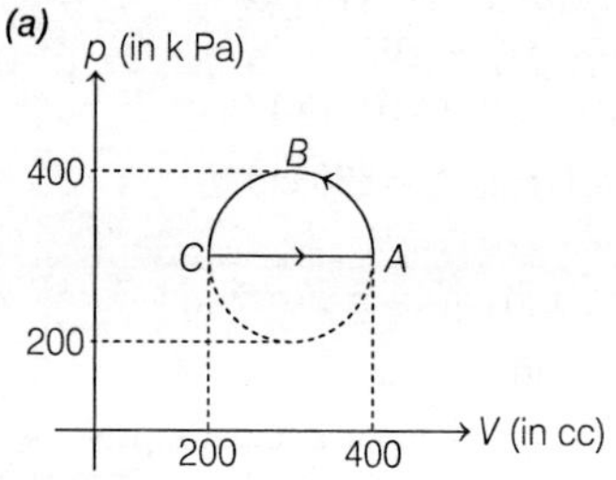

Work done = Area under ABC curve

$$= \frac{\pi R_1 R_2}{2}$$

$$= \frac{1}{2} \times \pi \times \left(\frac{200}{2} \times 10^3\right) \times \left(\frac{200}{2} \times 10^{-6}\right)$$

$$= \frac{10\pi}{2} = 5\pi \text{ J}$$

29. *(a)* For isothermal process, $pV = C$, which for adiabatic process $pV^\gamma = C$

The graphs are as shown.

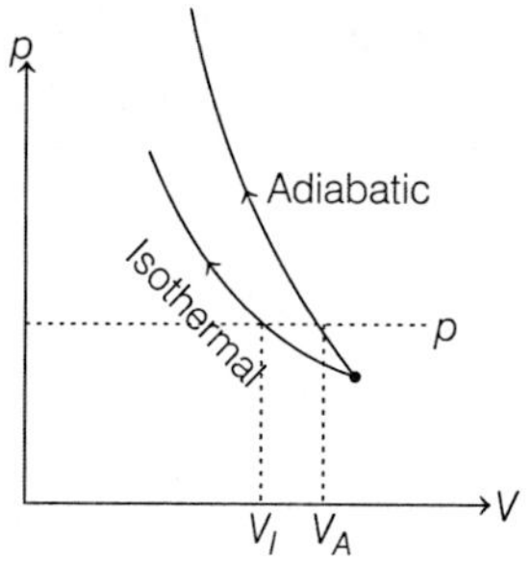

As seen from the graph, for the same pressure p,

$$V_I < V_A$$

i.e.

Volume falls more rapidly for isothermal process.

So, assertion is correct.

Reason is also true and explains the assertion.

30. *(d)* We know that, $TV^{\gamma - 1} = \text{Constant}$

$$T_i V^{\frac{3}{2}-1} = T_f (2V)^{\frac{3}{2}-1}$$

$$TV^{1/2} = T_f 2^{1/2} \cdot V^{1/2}$$

$$T_f = \frac{T}{\sqrt{2}}$$

Now, $W = \frac{nR\Delta T}{1-\gamma} = \frac{1R\left(\frac{T}{\sqrt{2}} - T\right)}{1 - \frac{3}{2}}$

$$= 2RT\left(1 - \frac{1}{\sqrt{2}}\right)$$

$$W = RT(2 - \sqrt{2})$$

• Concept Enhancer

In adiabatic expansion, work is done on the expense of internal enrgy, so temperature of the system decreases.

31. *(d)* From thermodynamic equation,

$$W = \Delta U + \Delta Q$$

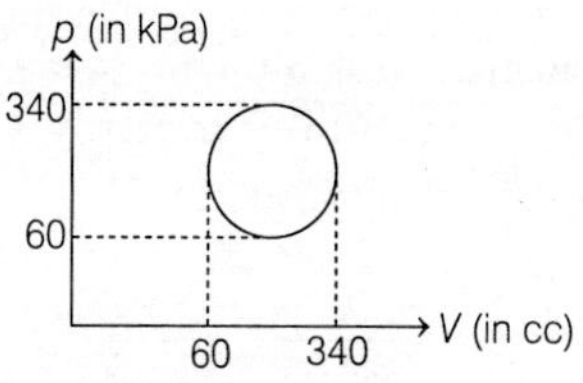

$\Delta U = 0$ for cyclic process

$\therefore \quad \Delta Q = W =$ Area of p-V curve

$= \pi \times (140 \times 10^3 \text{ Pa}) \times (140 \times 10^{-6} \text{ m}^3)$

$\Delta Q = 61.6$ J

The heat absorb is 61.6 J.

32. *(d)* For isobaric process, p = constant

So, from gas equation, we get

$$\frac{V}{T} = \frac{nR}{p} \text{ or } V = \frac{nR}{p} \cdot T$$

So, slope of V-T curve is proportional to $\frac{1}{p}$ value.

As slope of line 2 is more,

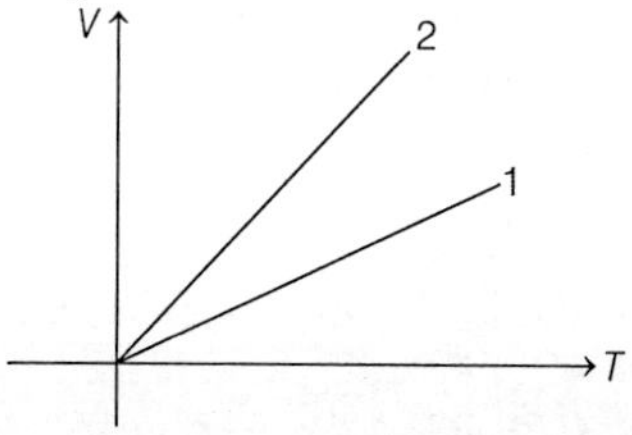

$$\frac{1}{p_2} > \frac{1}{p_1} \text{ or } p_2 < p_1$$

33. *(a)* Since, $p = aV^{-3}$

$$\frac{dp}{dV} = -3aV^{-4}$$

$$-V\frac{dp}{dV} = (-3aV^{-4})(-V)$$

$$-V\frac{dp}{dV} = +3aV^{-3}$$

$$-V\frac{dp}{dV} = 3p$$

Bulk modulus $= -V\frac{dp}{dV} = 3p$

34. *(b)* For container A, change is isothermal.

$$p, V, T \text{ to } p', \frac{V}{8}, T$$

Using Boyle's law, $p'\frac{V}{8} = pV$

or $\quad p' = 8p \quad$ ...(i)

For container B, change is adiabatic $\left(\gamma = \frac{5}{3}\right)$. (For monoatomic gas)

$$p, V, T \text{ to } p'', \frac{V}{8}, T'$$

Using $p \cdot V^\gamma = p''\left(\frac{V}{8}\right)^\gamma$

We have;

So, $\quad p'' = p \cdot (8)^\gamma = p(8)^{\frac{5}{3}} \quad$ ...(ii)

Ratio of final pressures of B and A samples is;

$$\frac{p''}{p'} = \frac{p \cdot 8^{\frac{5}{3}}}{p \cdot 8} = 8^{\frac{5}{3}-1} = 8^{\frac{2}{3}} = 2^2 = 4$$

35. *(c)*

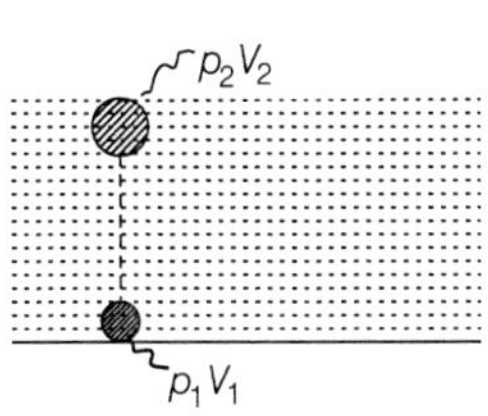

As bubble rises, volume of bubble increses due to decrease of pressure. It is also given temperature remians constant so.

$$p_1V_1 = p_2V_2 \quad \text{...(i)}$$

Here, $p_1 = p_{\text{atm}+p_{\text{water}}} = p_{\text{atm}} + h\rho g$

Now, $p_{\text{atm}} = 1\times 10^5$ Pa

and $h\rho g = 40\times 1000\times 10 = 4\times 10^5$ Pa

So, $p_1 = 1\times 10^5 + 4\times 10^5 = 5\times 10^5$ Pa

and $V_1 = 1 \text{ cm}^3$

Also p_2 = pressure at surface
= atmospheric pressure
$= 1\times 10^5$ Pa

So substituting value in Eq. (i) we get;

$$p_1V_1 = p_2V_2$$

$\Rightarrow (5\times 10^5 \text{ Pa})(1 \text{ cm}^3) = (1\times 10^5 \text{ Pa})(V_2)$

$\Rightarrow V_2 = 5 \text{ cm}^3$

36. *(c)* Given that,

$$V_i = 8 \text{ L}, V_f = 27 \text{ L}$$

$$\frac{p_f}{p_i} = \frac{16}{81}$$

From state equation of adiabatic process

$$pV^\gamma = \text{constant}$$

or $p_iV_i^\gamma = p_fV_f^\gamma$

or $\dfrac{p_f}{p_i} = \left(\dfrac{V_i}{V_f}\right)^\gamma$

$\Rightarrow \dfrac{16}{81} = \left(\dfrac{8}{27}\right)^\gamma$

or $\left(\dfrac{2}{3}\right)^4 = \left[\left(\dfrac{2}{3}\right)^3\right]^\gamma$

or $\left(\dfrac{2}{3}\right)^4 = \left(\dfrac{2}{3}\right)^{3\gamma}$

On comparing both sides, we get

$$3\gamma = 4$$

or $\gamma = \dfrac{4}{3}$ or $\dfrac{C_p}{C_V} = \dfrac{4}{3}$

37. *(a)* In an Isothermal process,

$$\Delta T = 0, \text{ So, } \Delta U = 0$$

By using First law of thermadynomics,

$$\Delta Q = \Delta U + \Delta W \quad \text{...(i)}$$

So, in isothermal process,

$$\Delta Q = \Delta W \quad \text{...(ii)}$$

Now, when $\Delta Q = +$ ve (Heat is added to system)

Then, $\Delta W = +$ ve [from Eq(ii)]

So, both C and D are correct.

38. *(c)* Total energy used in doing work

$$= p\Delta V$$
$$= 3\times 10^5 \times 1600\times 10^{-6} = 480 \text{ J}$$

If only 10% energy = 480 J, is used in doing work.

Then, total energy = 4800 J

Therefore, the rest 90% of the remaining energy is used in increasing internal energy.

So, increase in internal energy = 90% of 4800 J

$$= \frac{90}{100}\times 4800 \text{ J} = 4320 \text{ J}$$

39. *(b)* p-V diagram for a thermodynamic system is shown below.

Work done by the gas in the process DE,

W_{DE} = Area enclosed by the trapezium $DEABFD$

$$= \frac{1}{2}(DB + AE)\times EF$$
$$= \frac{1}{2}(600+300)\times 3 = 1350 \text{ J}$$

Work done on the gas in the process EF,

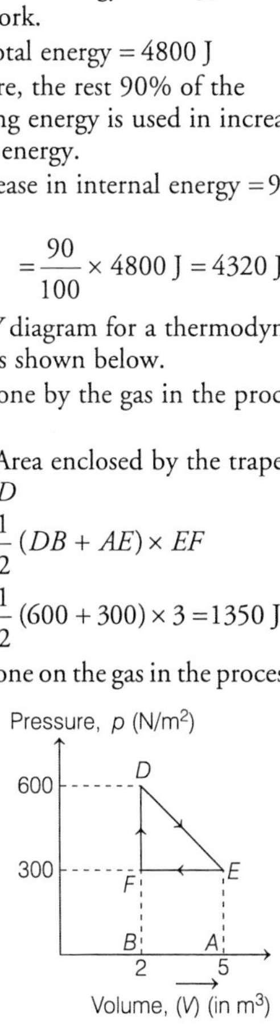

$W_{EF} = -$ Area enclosed by rectangle $FEAB$

$$= -(FE\times BF) = -3\times 300 = -900 \text{ J}$$

$\therefore$ Total work done by the gas from D to E to F

$$W_{DEF} = W_{DE} + W_{EF}$$
$$= 1350 - 900 = 450 \text{ J}$$

40. *(c)* Let the initial and final pressure of the gas be p_1 and p_2 respectively.

Let the initial and final volume of the gas be V_1 and V_2 respectively.

For an adiabatic process, ideal gas equation is given by, pV^γ = constant

or $p_1V_1^\gamma = p_2V_2^\gamma$

or $\dfrac{p_2}{p_1} = \left(\dfrac{V_1}{V_2}\right)^\gamma$

Given that, $V_2 = \dfrac{V_1}{8}$ and for monoatomic gas, $\gamma = \dfrac{5}{3}$

$\therefore \dfrac{p_2}{p_1} = \left(\dfrac{V_1}{\frac{V_1}{8}}\right) = 32$

or $p_2 = 32, p_1 = 32p$

41. *(d)* p-V graphs for given processes are as shown

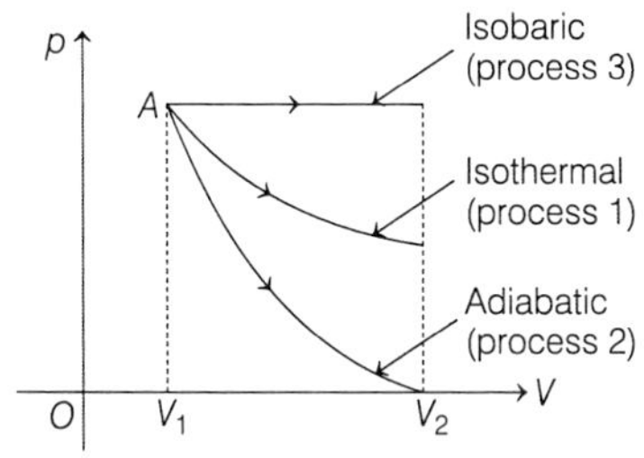

As work done in any process = Area under p-V graph, so we have

$$W_{\text{isobaric}} > W_{\text{isothermal}} > W_{\text{adiabatic}}$$

$\Rightarrow W_3 > W_1 > W_2$

42. *(a)* $\gamma = 1.4$ for an ideal gas and work done on the gas is negative.

$\Rightarrow \Delta W < 0$

$\Rightarrow \dfrac{\mu R(T_2 - T_1)}{1-\gamma} < 0$

$\Rightarrow T_2 - T_1 > 0$

$\Rightarrow T_2 > T_1$

Hence, temperature of gas rises when work is done on gas. So, both statement I and statement II are true.

> • Concept **Enhancer**
>
> Work done in an adiabatic process is
>
> $$\Delta W = \frac{\mu R(T_2 - T_1)}{1-\gamma}$$

43. *(b)* As per question, the process BC and DA are adiabatic.

$\because$ We know that work done in adiabatic process is given as

$$W = \frac{-nR}{\gamma - 1}(T_f - T_i)$$

$$\therefore \quad W_{AD} = \frac{-nR}{\gamma-1}(T_2 - T_1)$$

and $W_{BC} = \dfrac{-nR}{\gamma-1}(T_2 - T_1)$

$\Rightarrow \quad W_{AD} = W_{BC}$

44. *(b)* As we know that,

In adiabatic condition,

$$TV^{\gamma-1} = \text{constant}$$

$$T(Al)^{\gamma-1} = \text{constant}$$

where, A is constant.

$\Rightarrow \; Tl^{\gamma-1} = \text{constant} \qquad \ldots\text{(i)}$

For monoatomic gas, $\gamma = 5/3$

$$\Rightarrow T_1 l_1^{\frac{5}{3}-1} = T_2 l_2^{5/3-1}$$

$$\Rightarrow \quad \frac{T_1}{T_2} = \left(\frac{l_2}{l_1}\right)^{2/3}$$

45. *(d)* We know that, in isothermal process, $\Delta T = 0$

In isochoric process, $\Delta V = 0$

In adiabatic process, $\Delta Q = 0$

In isobaric process, $\Delta p = 0$

So, the correct match is,

A $\rightarrow$ 2, B $\rightarrow$ 3, C $\rightarrow$ 4, D $\rightarrow$ 1.

46. *(b)* The given expansion process is adiabatic in nature and we have the following data :

$$V_1 = 1\,\text{L} = 10^{-3}\ \text{m}^3$$

$$V_2 = 3\,\text{L} = 3\times 10^{-3}\ \text{m}^3$$

$$p_1 = 1\ \text{atm} = 1.01\times 10^5\ \text{Pa}$$

For adiabatic process

$$pV^{\gamma} = \text{constant}$$

$$p_1V_1^{\gamma} = p_2V_2^{\gamma}, \text{ we have}$$

$$1\times 1^{\gamma} = p_2\times(3)^{\gamma}$$

Here, $\gamma = 1.40$

and $(3)^{1.4} = 4.6555$

So, $p_2 = \dfrac{1}{4.6555} \approx 0.22$ atm

$$= 0.22\times 1.01\times 10^5\ \text{Pa}$$

Work done in an adiabatic expansion is given by

$$W = \frac{(p_1V_1 - p_2V_2)}{\gamma - 1}$$

$$= \frac{(1.01\times 10^5\times 10^{-3} - 0.22\times 1.01\times 10^5\times 3\times 10^{-3})}{(1.40-1)}$$

$$= 0.85\times 1.01\times 10^5\times 10^{-3} = 85.85\ \text{J}$$

Closest value of work done, $W \approx 90.5$ J

47. *(b)*

> **key idea** In a cyclic thermodynamic process work done = area under p-V diagram.

In clockwise cycle, work done is positive. In the given cyclic process, work done $= \oint pdV$ = area enclosed by the cycle

$$= \frac{1}{2}\times \text{base}\times \text{height of triangle } (CAB)$$

$$= \frac{1}{2}(5-1)(6-1) = 10\ \text{J}$$

48. *(b)* $VT = K, \; V\left(\dfrac{pV}{R}\right) = K \quad [pV = RT]$

or $pV^2 = $ constant is a polytropic process,

$\because \quad x = 2$

$$C_p = \frac{R}{1-x} + C_V$$

$$= \frac{R}{1-2} + \frac{3R}{2} = \frac{R}{2}$$

$$\therefore \; \Delta Q = nC_p\Delta T = \frac{1}{2}R\Delta T$$

49. *(c)* For adiabatic process relation of temperature and volume is,

$$T_2V_2^{\gamma-1} = T_1V_1^{\gamma-1}$$

$$\Rightarrow T_2(2V)^{2/3} = 300(V)^{2/3}$$

$$[\gamma = 5/3 \text{ for monoatomic gases}]$$

$$\Rightarrow \qquad T_2 = \frac{300}{2^{2/3}} \approx 189\ \text{K}$$

Also, in adiabatic process,

$$\Delta Q = 0,\ \Delta U = -\Delta W$$

or $\quad \Delta U = \dfrac{-nR(\Delta T)}{\gamma - 1}$

$$= -2\times\frac{3}{2}\times\frac{25}{3}(300-189)$$

$$\approx -2.7\ \text{kJ}$$

50. *(a)* For an ideal gas in an isobaric process,

Heat supplied,

$$Q = nC_p\Delta T,$$

Work done,

$$W = p\Delta V = nR\Delta T$$

$\therefore$ Required ratio

$$= \frac{W}{Q} = \frac{nR\Delta T}{nC_p\Delta T} = \frac{R}{\frac{5}{2}R} = \frac{2}{5}$$

(For monoatomic gas, $C_p = \frac{5}{2}R$)

51. *(a)* Process $a \rightarrow b$

a to b is a straight line.

$$pV = nRT$$

$$T = \frac{p}{nR}V$$

Compare $y = mx + c$

Slope $m_{ab} = \dfrac{p}{nR} = $ constant

Therefore

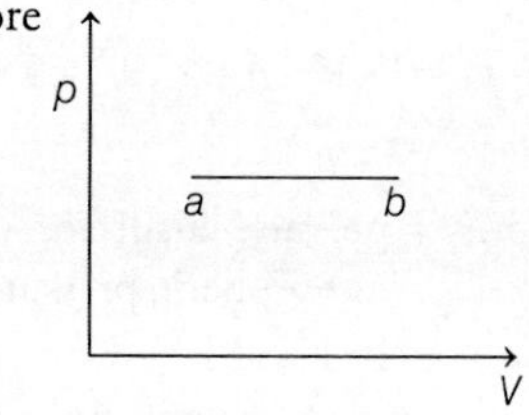

Process b to c

Adiabatic process (given)

$pV^{\gamma} = \text{constant} \Rightarrow p = c/V^{\gamma}$

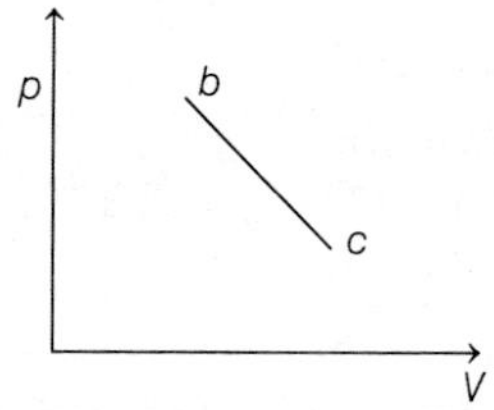

Similarly c to d is isobaric, and its graph will be similar to process a to b and process d to a is given as adiabatic.

So, its pV graph will be similar to b to c.

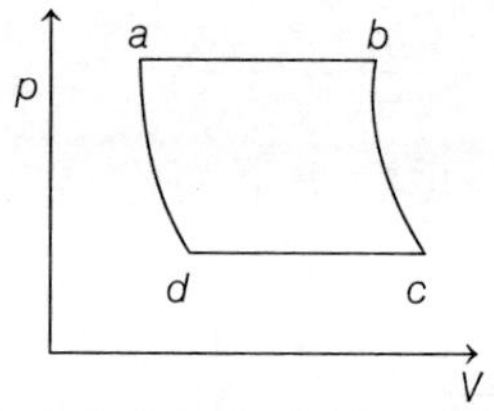

52. *(c)* According to question,

$$p = \frac{1}{3}\left(\frac{U}{V}\right)$$

$$\Rightarrow \quad \frac{nRT}{V} = \frac{1}{3}\left(\frac{U}{V}\right) \qquad [\because pV = nRT]$$

or $\quad \dfrac{nRT}{V} \propto \dfrac{1}{3}T^4$ or $VT^3 = $ constant

or $\dfrac{4}{3}\pi R^3T^3 = $ constant

or $TR = \text{constant} \Rightarrow \; T \propto \dfrac{1}{R}$

53. *(c)* A piston of mass m is hung from a massless spring whose restoring force is given by

$$F = -kx^3$$

Initial height of piston $= L_0$

Final height of piston $= L_1$

Work is done isothermally by gas using filament, therefore work done by gas can be evaluated by relation

$$W_{iso} = nRT \ln\left[\frac{V_1}{V_0}\right]$$

Here, volume of gas $\propto$ height of piston

$$V \propto L$$

$$\therefore \quad W_{iso} = nRT \ln\left[\frac{L_1}{L_0}\right] \quad \ldots(i)$$

Change in gravitational potential energy ΔU can be evaluated by

$$\Delta U_g = M \cdot g[L_1 - L_0] \quad \ldots(ii)$$

Change in potential energy of spring

$$\Delta U_{spring} = \int F\,dx$$

$$U_{spring} = \int_{L_0}^{L_1} kx^3\,dx$$

$$= k\left[\frac{x^4}{4}\right]_{L_0}^{L_1} = \frac{k}{4}[L_1^4 - L_0^4]$$

Total energy delivered by filament

$= W_{gas} + \Delta U_g + U_s$

$$W = nRT \ln\left[\frac{L_1}{L_0}\right] + Mg[L_1 - L_0] + \frac{k}{4}[L_1^4 - L_0^4]$$

54. *(d)* Work done in isothermal expansion $= nRT \ln\left[\frac{V_2}{V_1}\right]$

$$W_1 = nRT \ln\left[\frac{4V_0}{V_0}\right]$$

$$= nRT \ln 4 = p_0V_0 \ln 4$$

For isothermal process $p_1V_1 = p_2V_2$

$$\Rightarrow \quad p_0V_0 = p_1[4V_0]$$

$$\Rightarrow \quad p_1 = \frac{p_0}{4}$$

Work done in isobaric compression $= -pdV$

$$W_2 = -\frac{p_0}{4}[4V_0 - V_0]$$

$$W_2 = \frac{-3}{4}p_0V_0$$

Work done in isochoric process $= 0$

Amount of Heat exchanged = Change in internal energy + Net work done

$Q = \Delta U + W_{total}$

For cyclic process $\Delta U = 0$

$$Q = W_{Total}$$

$$Q = W_1 + W_2 + W_3$$

$$Q = p_0V_0 \ln 4 + \left[-\frac{3}{4}p_0V_0\right] + 0$$

$$= 2p_0V_0 \ln 2 - \frac{3}{4}p_0V_0$$

$$= p_0V_0[2\ln 2 - 0.75]$$

55. (b) A. $p \propto \frac{1}{V}$

$pV = $ constant

$pV = nRT$

$T = \frac{pV}{nR} = $ constant

$T = $ constant $\rightarrow$ Isothermal process

A $\rightarrow$ III

B. $W \neq 0, \Delta U \neq 0, \Delta Q \neq 0$

Only in isobaric process

B $\rightarrow$ IV

C. $\Delta Q = 0 \rightarrow$ Adiabatic process

C $\rightarrow$ I

D. $W = 0 \rightarrow$ Isochoric process

D $\rightarrow$ II

III, IV, I, II

56. *(b)*

> **key idea** Heat absorbed in an adiabatic process $\Delta Q = 0$. Heat absorbed in an isothermal process,
>
> $$\Delta Q_{isothermal} = W_{isothermal} = nRT \ln\left(\frac{V_f}{V_i}\right)$$
>
> $$\therefore \quad \Delta U_{isothermal} = 0$$

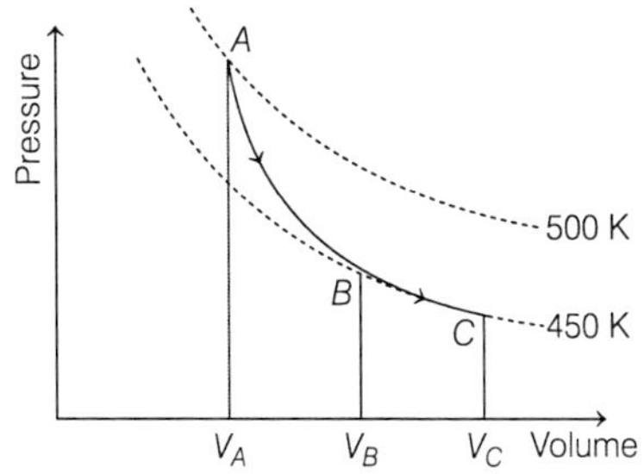

Since, process A to B is adiabatic, hence no heat get absorbed.

$$Q_{AB} = 0$$

or B to C, process is isothermal.

Temperature is 450 K.

$$\Delta Q = \Delta W = nRT \ln\frac{V_C}{V_B}$$

$$\frac{\Delta Q_{BC}}{n} = R \times 450 \times \ln\left(\frac{8\times10^{-6}}{6\times10^{-6}}\right)$$

$$= 450R \ln\left(\frac{4}{3}\right)$$

$$\frac{\Delta Q_{BC}}{n} = 450R(\ln 4 - \ln 3)$$

Thus, total heat from A to C per mole

$$\frac{Q_{AC}}{n} = \frac{Q_{AB}}{n} + \frac{Q_{BC}}{n}$$

$$= 0 + 450R(\ln 4 - \ln 3)$$

$$= 450R(\ln 4 - \ln 3)$$

57. *(c)* $W_{AB} = \int p\,dV$

(Assuming T to be constant)

$$= \int \frac{RT}{V^3}\,dV$$

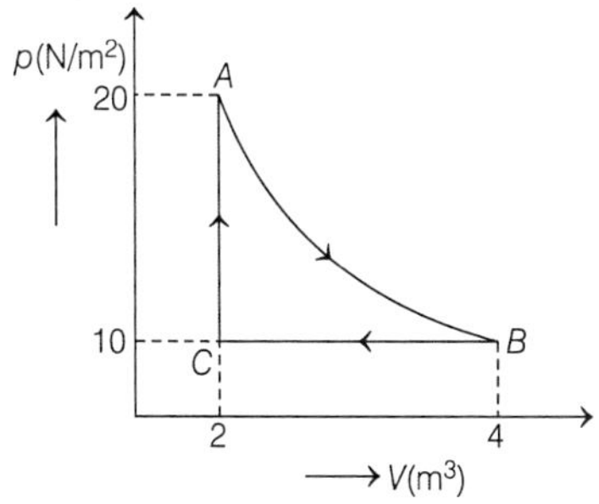

$$= RT\int_2^4 V^{-3}\,dV$$

$$= 8 \times 300 \times \left[-\frac{1}{2}\left(\frac{1}{4^2} - \frac{1}{2^2}\right)\right]$$

$$= 225\text{ J}$$

$$W_{BC} = p\int_4^2 dV = 10(2-4) = -20\text{ J}$$

$$W_{CA} = 0$$

$$\therefore W_{cycle} = 205\text{ J}$$

58. *(a)* Process eqaution given is

$$pV^{3/2} = K$$

Work done by gas, W

$$= \int_{V_1}^{V_2} p\,dV$$

$$= \int_{V_1}^{V_2} \frac{K}{V^{3/2}} \cdot dV$$

$$= K \cdot \int_{V_1}^{V_2} V^{-3/2} \cdot dV$$

$$= K\left[\frac{V^{-1/2}}{-1/2}\right]_{V_1}^{V_2}$$

$$= -2K\left\{\frac{1}{\sqrt{V_2}} - \frac{1}{\sqrt{V_1}}\right\}$$

$$= -2\left[\frac{K}{\sqrt{V_2}} - \frac{K}{\sqrt{V_1}}\right]$$

$$= -2\left[p_2V_2^{\frac{3}{2}-\frac{1}{2}} - p_1V_1^{\frac{3}{2}-\frac{1}{2}}\right]$$

$$= -2\left[p_2V_2 - p_1V_1\right]$$

$$= 2\left[p_1V_1 - p_2V_2\right]$$

59. *(a,d)* Process B is adiabatic (higher slope) and process A is isothermal (Lower slope).

So, $pV^{\gamma} = k$ for process B

and $pV = k$ for process A.

For process B,

$$pV^{\gamma} = k \quad \text{...(i)}$$

$$pV = RT$$

$\therefore$ Eq. (i) can be written as

$$\frac{pT^{\gamma}}{p^{\gamma}} = k$$

$$\frac{p^{1-\gamma}}{T^{-\gamma}} = k$$

Also, $\dfrac{p^{\gamma-1}}{T^{\gamma}} = k$

Similarly, for process B

$$pV = k$$

$$T = k \quad (\because pV/T = \text{constant})$$

Therefore, option (a) and (d) both are correct.

60. *(a)* A $\to$ II, For an isothermal process, the temperature remains constant. So, the change in internal energy is zero.

B $\to$ I, In adiabatic process, no heat transfer takes place.

Hence, by first law of thermodynamics, work done by the gas is equal to decrease in internal energy.

C $\to$ IV, In isochoric process, change in volume is zero.

i.e., $\Delta V = 0$

Hence, work done,

$$W = p\cdot\Delta V$$

$$= p \times 0 = 0$$

Thus, in isochoric process, no work is done on or by the gas.

D $\to$ III, In isobaric process, when heat is supplied, a part of it goes in the expansion of gas an remaining part is used to increase the temperature of the gas which is turn increase in the internal energy.

61. *(b)*

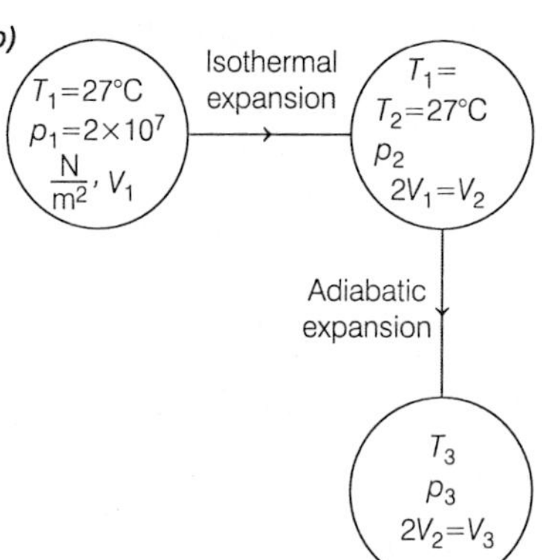

When gas is expanded isothermally,

$$p_1V_1 = p_2V_2$$

Given, $V_2 = 2V_1$

$$\Rightarrow \quad p_2 = p_1 \times \frac{V_1}{V_2}$$

or $$p_2 = p_1 \times \frac{V_1}{2V_1} = \frac{p_1}{2}$$

or $$p_2 = \frac{2\times10^7}{2}$$

$$= 1\times10^7\,\text{Nm}^{-2}$$

Now gas is expanded adiabatically,

$$\Rightarrow \quad p_2V_2^{\gamma} = p_3V_3^{\gamma}$$

Given, $V_3 = 2V_2$

So, $p_2\cdot V_2^{\gamma} = p_3(2V_2)^{\gamma}$

$$\Rightarrow \quad p_3 = \frac{p_2}{2^{\gamma}} = \frac{1\times10^7}{2^{1.5}}$$

$$\Rightarrow \quad p_3 = \frac{1\times10^7}{(2^{3/2})} = \frac{1\times10^7}{2\sqrt{2}}$$

$$= 0.3536\times10^7$$

$$= 3.536\times10^6\ \text{N/m}^2$$

62. *(b)* For the given cyclic process :

AB is isochoric process, hence work done is zero.

$$\Rightarrow \quad \Delta Q_{AB} = \Delta U_{AB} + \Delta W_{AB}$$

$$\Rightarrow \quad \Delta Q_{AB} = \Delta U_{AB}$$

$$\Rightarrow \quad \Delta U_{AB} = \Delta Q_{AB} = 40\ \text{J} \quad \text{(given)}$$

BC is adiabatic process and

$$\Delta Q_{BC} = 0 \quad \text{(given)}$$

$$\Rightarrow \Delta U_{BC} + \Delta W_{BC} = 0$$

$$\Rightarrow \quad \Delta U_{BC} = -\Delta W_{BC} = -(-50)$$

$$\Rightarrow \quad \Delta U_{BC} = 50\ \text{J}$$

(given, $\Delta W_{BC} = -50$ J)

Now, as process is cyclic,

$$\Delta U_{ABCA} = 0$$

$$\Rightarrow \Delta U_{AB} + \Delta U_{BC} + \Delta U_{CA} = 0$$

$$\Rightarrow \quad \Delta U_{CA} = -(\Delta U_{AB} + \Delta U_{BC})$$

$$= -(40 + 50) = -90\ \text{J}$$

Now, for process CA,

$$\Delta U_{CA} + \Delta W_{CA} = \Delta Q_{CA}$$

$$\Rightarrow -90 + \Delta W_{CA} = -60$$

Given, $\Delta Q_{CA} = -60$

So, $\Delta W_{CA} = 90 - 60 = 30$ J

> ***Alternate Solution***
>
> According to the question,
> for process A to B, using Ist law of thermodynamic $\Delta Q = \Delta W + \Delta U$
>
> $$\Delta Q_{AB} = \Delta W_{AB} + U_B - U_A \quad \text{...(i)}$$
>
> Similarly,
> for B to C, $\Delta Q_{BC} = \Delta W_{BC} + U_C - U_B$ ...(ii)
> and for C to A, $\Delta Q_{CA} = \Delta W_{CA} + U_A - U_C$ ...(iii)
> On adding Eqs. (i), (ii) and (iii)
>
> $$\Delta Q_{AB} + \Delta Q_{BC} + \Delta Q_{CA} = \Delta W_{AB} + \Delta W_{BC} + \Delta W_{CA} \quad \text{...(iv)}$$
>
> Given, $\Delta Q_{AB} = +40$ J, $\Delta Q_{BC} = 0$, $\Delta Q_{CA} = -60$ J,
> $\Delta W_{AB} = 0$ (Isochoric), $\Delta W_{BC} = -50$ J
> and $\Delta W_{CA} = ?$
> Putting values, in Eq. (iv)
>
> $$40 + 0 - 60 = 0 - 50 + \Delta W_{CA} \Rightarrow \Delta W_{CA} = 30\ \text{J}$$

63. *(b)* Here, p_1 and p_2, T_1 and T_2, V_1 and V_2 are initial and final pressures, temperatures and volumes, respectively.

Given,

$$V_2 = 2V_1$$

$$pV^{1/2} = \text{constant}$$

$$\frac{RT}{V}V^{1/2} = C$$

$$\Rightarrow \quad T = CV^{1/2}$$

$$\Rightarrow \frac{T_2}{T_1} = \left(\frac{V_2}{V_1}\right)^{1/2} = \left(\frac{2V_1}{V_1}\right)^{1/2} = \sqrt{2}$$

64. *(a)* Let p_i, p_f, V_i and V_f be the initial and final pressure and volume.

Given, AB is isothermal ($\Delta T = 0$),

BC is isochoric ($\Delta V = 0$)

and CA is adiabatic ($\Delta Q = 0$)

Since, isothermal work

$$(W_{AB}) = p_1V_1 \ln\frac{V_f}{V_i}$$

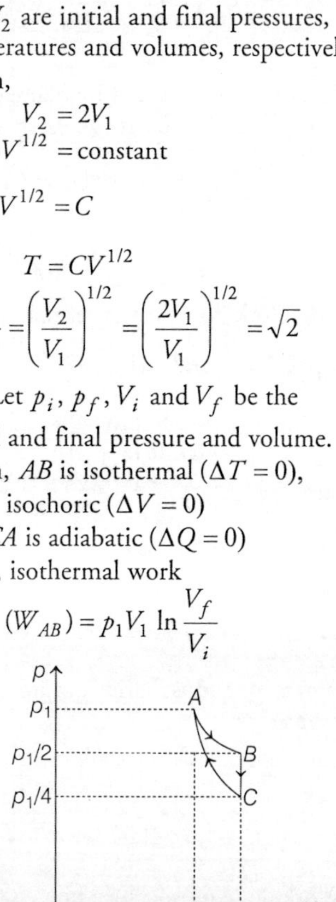

where, V_i and V_f are volume at A and B, respectively.

$\therefore \quad W_{AB} = p_1V_1 \ln \frac{2V_1}{V_1} = p_1V_1 \ln 2$

$\therefore \quad \Delta V = 0$

$\therefore \quad W_{BC} = 0$

Since, W_{CA} is an adiabatic work done, i.e.

$$W_{CA} = \frac{1}{1-\gamma}(p_fV_f - p_iV_i)$$

$$\Rightarrow \quad W_{CA} = \frac{1}{1-\gamma}\left(p_1V_1 - \frac{p_1}{4} \times 2V_1\right)$$

$$= \frac{1}{1-\gamma}\frac{p_1V_1}{2}$$

$\therefore$Net work done,

$$W_{net} = W_{AB} + W_{BC} + W_{CA}$$

$$= p_1V_1 \ln 2 + 0 + \frac{1}{1-\gamma}\frac{p_1V_1}{2}$$

$$= p_1V_1 [\ln 2 + 1/2(1-\gamma)]$$

From ideal gas law,

$$pV = nRT,\ p_1V_1 = RT$$

$$\therefore W_{net} = RT[\ln 2 - 1/2(\gamma - 1)] \quad (\because n = 1)$$

65. *(d)* As, AB is isothermal process, so work done in isothermal process is given by

$$W_{AB} = nRT \ln\left(\frac{V_2}{V_1}\right)$$

$$W_{AB} = nRT \ln\left(\frac{2V_1}{V_1}\right) = nRT \ln 2$$

As, BC is isobaric process, so work done in isobaric process is given by

$$W_{BC} = p\Delta V = p_2(V_1 - V_2)$$

$$= p_2\left(\frac{V_2}{2} - V_2\right)$$

$$= -\frac{p_2V_2}{2}$$

$$= \frac{-nRT}{2} \quad \left[\begin{array}{l}\because pV = nRT \\ \therefore p_2V_2 = nRT\end{array}\right]$$

As, CA is isochoric process, so work done in isochoric process is given by, $W_{CA} = 0 \quad [\because \Delta V = 0]$

Total work done in the cycle $ABCA$,

$$W_{ABCA} - W_{AB} + W_{BC} + W_{CA}$$

$$= nRT \ln 2 + \left(\frac{-nRT}{2}\right) + 0$$

$$= nRT\left(\ln 2 - \frac{1}{2}\right)$$

66. *(b)* In given p-V diagram,

1 to 2 is adiabatic expansion in which temperature reduces.

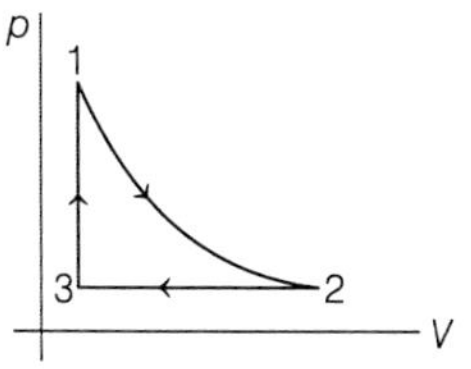

2 to 3 is isobaric compression, so volume decreases and temperature increases in this process.

3 to 1 is isochoric process in which pressure increases. So, it must be isochoric heating in which temperature increases. So, resultant V-T graph will be

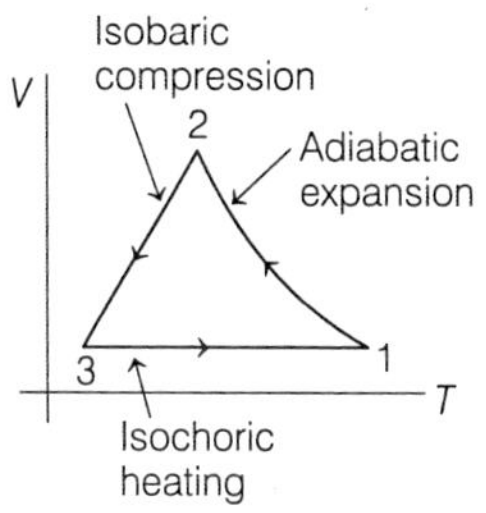

Thus, option (b) is correct.

67. *(b)*

> **key idea** In a p-V curve, the work done (dw), change in internal energy (ΔU) and heat transfor (ΔQ) are related by the first law of thermodynamics
>
> $\Delta Q = \Delta U + dW$
>
> For a complete thermodynamic cycle, the total change in internal energy is always zero. By applying this equation to different segments of the curve, we can solve the given problem effectively.

In Process $a \to b$

Given, $\Delta Q_{ab} = 250$ J

$\therefore \quad 250\text{J} = \Delta U_{ab} + dW_{ab} \quad \ldots\text{(ii)}$

In Process $b \to c$ (Isochoric process)

Given, $\Delta Q_{bc} = 60$ J

$\Rightarrow \quad dW_{bc} = 0$

$\Rightarrow \quad \Delta U_{bc} = 60\text{J} \quad \ldots\text{(iii)}$

In Process $c \to a$

Given,

$\Delta U_{ca} = -180\text{J} \quad \ldots\text{(iv)}$

Now, for complete cycle,

$\Delta U_{abca} = \Delta U_{ab} + \Delta U_{bc} + \Delta U_{ca} = 0 \quad \ldots\text{(v)}$

From Eqs. (iii), (iv) and (v), we get

$\Delta U_{ab} = -\Delta U_{bc} - \Delta U_{ca}$

$\Delta U_{ab} = -60 + 180 = 120 \text{ J} \quad \ldots\text{(vi)}$

From Eq. (ii), we get

$250\text{J} = 120\text{J} + dW_{ab}$

$\Rightarrow \quad dW_{ab} = 130\text{J} \quad \ldots\text{(vii)}$

From Eqs. (i) and (vii), we get

Work done by the gas along the path abc,

$$dW_{abc} = dW_{ab} + dW_{bc} = 130\text{J} + 0$$

$$\Rightarrow \quad dW_{abc} = 130 \text{ J}$$

68. *(b)* For one mole of ideal gas

$$pV = RT \quad \ldots\text{(i)}$$

Differentiating equation (i) w.r.t. V

$$\text{or } p + V\frac{dp}{dV} = R\frac{dT}{dV}$$

$$\text{or } p_0\left[1 - \frac{1}{2}\left(\frac{V_0}{V}\right)^2\right] + V\left(-\frac{p_0}{2}V_0^2\frac{(-2)}{V^3}\right) = R\frac{dT}{dV}$$

Total change in temperature by changing volume from V_0 to $2V_0$

$$\int_{T_1}^{T} dT = \frac{1}{R}\int_{V_0}^{2V_0}\left(p_0 + \frac{1}{2}p_0\frac{V_0^2}{V^2}\right)dV$$

$$= \frac{1}{R}\left(p_0V - \frac{1}{2}\frac{p_0V_0^2}{V}\right)\Bigg|_{V_0}^{2V_0}$$

$$R(T_2 - T_1) = p_0(2V_0 - V_0) - \frac{1}{2}p_0V_0^2\left(\frac{1}{2V_0} - \frac{1}{V_0}\right)$$

$$\Delta T = \frac{5}{4R}p_0V_0$$

69. *(b)* Equation of line BC is given by

$$p = p_0 - \frac{2p_0}{V_0}(V - 2V_0)$$

$$\text{or} \quad pV = p_0V - \frac{2p_0}{V_0}(V - 2V_0)V$$

$$T = \frac{p_0V - \frac{2p_0V^2}{V_0} + 4p_0V}{1 \times R}$$

$$(\because pV = nRT)$$

$$T = \frac{p_0}{R}\left(5V - \frac{2V^2}{V_0}\right)$$

For maximum value of T,

$$\frac{dT}{dV} = 0 \text{ or } 5 - \frac{4V}{V_0} = 0$$

$$\Rightarrow \quad V = \frac{5}{4}V_0$$

$$\therefore \quad T_{max} = \frac{P_0}{R}\left[5 \times \frac{5V_0}{4} - \frac{2}{V_0} \times \frac{25}{16}V_0^2\right]$$

$$= \frac{25}{8}\frac{p_0V_0}{R}$$

70. *(a)* As, T will be maximum temperature where product of pV is maximum.

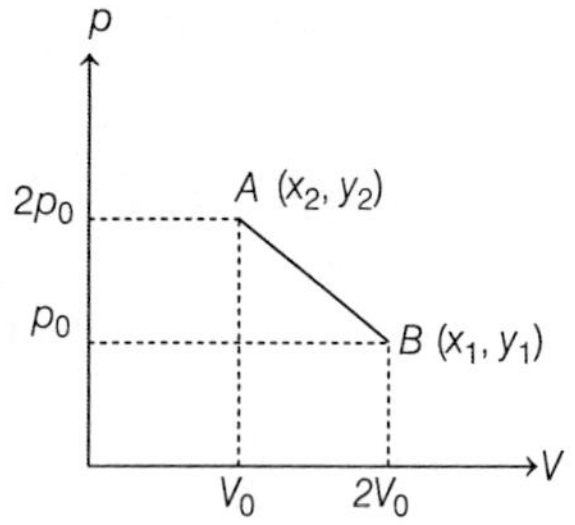

Equation of line AB, we have

$$y - y_1 = \frac{y_2 - y_1}{x_2 - x_1}(x - x_1)$$

$$\Rightarrow \quad p - p_0 = \frac{2p_0 - p_0}{V_0 - 2V_0}(V - 2V_0)$$

$$\Rightarrow \quad p = \frac{-p_0}{V_0}V + 3p_0$$

$$pV = \frac{-p_0}{V_0}V^2 + 3p_0V$$

$$nRT = \frac{-p_0}{V_0}V^2 + 3p_0V$$

$$T = \frac{1}{nR}\left(\frac{-p_0}{V_0}V^2 + 3p_0V\right)$$

For maximum temperature,

$$\frac{\partial T}{\partial V} = 0$$

$$\frac{-p_0}{V_0}(2V) + 3p_0 = 0$$

$$\frac{-p_0}{V_0}(2V) = -3p_0$$

$$\Rightarrow \quad V = \frac{3}{2}V_0$$

(Condition for maximum temperature)

Thus, the maximum temperature of the gas during the process will be

$$T_{\max} = \frac{1}{nR}\left(\frac{-p_0}{V_0} \times \frac{9}{4}V_0^2 + 3p_0 \times \frac{3}{2}V_0\right)$$

$$= \frac{1}{nR}\left(-\frac{9}{4}p_0V_0 + \frac{9}{2}p_0V_0\right)$$

$$= \frac{9}{4}\frac{p_0V_0}{nR}$$

Alternate Solution

Since, initial and final temperature are equal, hence maximum temperature is at the middle of line.

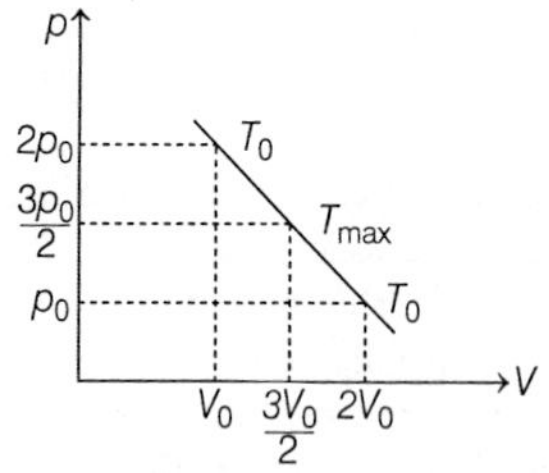

i.e. $\quad pV = nRT$

$$\Rightarrow \quad \frac{\left(\frac{3}{2}p_0\right)\left(\frac{3V_0}{2}\right)}{nR} = T_{\max} \Rightarrow \frac{9}{4}\frac{p_0V_0}{nR} = T_{\max}$$

71. *(d)* According to first law of thermodynamics, we get

(i) Change in internal energy from A to B

$$\Delta U_{AB} = nC_V(T_B - T_A)$$

$$= 1 \times \frac{5R}{2}(800 - 400) = 1000\,R$$

(ii) Change in internal energy from B to C

$$\Delta U_{BC} = nC_V(T_C - T_B)$$

$$= 1 \times \frac{5R}{2}(600 - 800) = -500\,R$$

(iii) $\Delta U_{\text{total}} = 0$ change in internal energy in cyclic process is zero.

(iv) Change in internal energy from C to A i.e., ΔU_{CA}

$$\Delta U_{CA} = nC_V(T_A - T_C)$$

$$= 1 \times \frac{5R}{2}(400 - 600) = -500\,R$$

• Concept Enhancer

For a cyclic process, work done is equal to the area under the curve and it is taken to be positive if the cycle is clockwise.

72. *(a)* Efficiency of a process is defined as the ratio of work done to energy supplied. Here,

$$\eta = \frac{\Delta W}{\Delta Q} = \frac{\text{Area under the } p\text{-}V \text{ diagram}}{\Delta Q_{AB} + \Delta Q_{BC}}$$

$$\therefore \eta = \frac{p_0V_0}{nC_V\Delta T_1 + nC_p\Delta T_2}$$

$$= \frac{p_0V_0}{\frac{3}{2}nR(T_B - T_A) + \frac{5}{2}nR(T_C - T_D)}$$

$$= \frac{p_0V_0}{\frac{3}{2}(2p_0V_0 - p_0V_0) + \frac{5}{4}(4p_0V_0 - 2p_0V_0)}$$

[as $pV = nRT$]

$$= \frac{p_0V_0}{\frac{3}{2}p_0V_0 + \frac{5}{2} \cdot 2p_0V_0} = \frac{1}{6.5} = 15.4\%$$

73. *(314)*

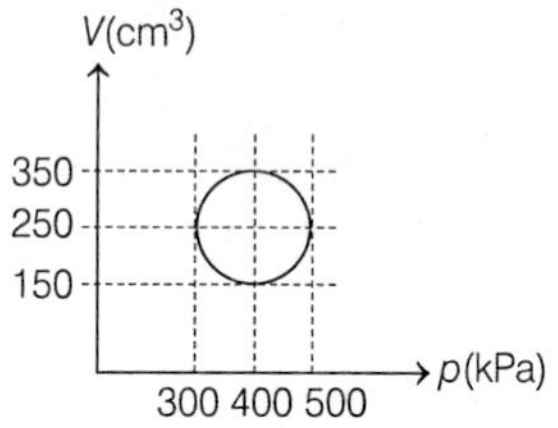

$$W = \text{Area of } p\text{-}V \text{ graph}$$

$$= \text{Area of circle}$$

$$= \frac{\pi}{4}d_1d_2$$

$$= \frac{\pi}{4}(500 - 300) \times 10^3 \times (350 - 150) \times 10^{-6}$$

$$= 31.4\text{ J}$$

$$W = 314 \times 10^{-1}\text{ J}$$

74. *(273)* $\gamma = \frac{3}{2}$

$$TV^{\gamma - 1} = C$$

$$273V_0^{1.5-1} = T\left(\frac{V_0}{4}\right)^{1.5-1}$$

$$273V_0^{0.5} = T\left(\frac{V_0}{4}\right)^{0.5}$$

$$273V_0^{\frac{1}{2}} = T\frac{V_0^{\frac{1}{2}}}{4^{\frac{1}{2}}}$$

$$T = 273 \times 2 = 546 = 546\text{ K}$$

$$T_f = 546\text{ K}$$

$$T_1 = 0°\text{C} = 273\text{ K}$$

$$\Delta T = T_f - T_i$$

$$= 546 - 273$$

$$\Delta T = 273\text{ K}$$

75. *(15)* Given that the process is isobaric

$\Delta T = 50°\text{C}$

Q in isobaric process $= nC_p\Delta T = E_1$

ΔU in isobaric process $= nC_V\Delta T = E_2$

$$\therefore \quad \frac{E_1}{E_2} = \frac{C_p}{C_V} = \gamma$$

As the gas is monoatomic

$$\gamma = 1 + \frac{2}{f} = 1 + \frac{2}{3} = \frac{5}{3}$$

As per question

$$\frac{5}{3} = \frac{x}{9} \Rightarrow x = 15$$

76. *(4)* Since we know, for an adiabatic process, $pV^{\gamma} = K$

$p_1 V_1^{\gamma} = p_2 V_2^{\gamma}$, m is constant

$$p_1 \left(\frac{m}{d_1}\right)^{\gamma} = p_2 \left(\frac{m}{d_2}\right)^{\gamma}, \frac{p_1}{p_2} = \left(\frac{d_1}{d_2}\right)^{\gamma}$$

$$\frac{p_1}{p_2} = \left(\frac{1}{32}\right)^{7/5}, \frac{p_1}{p_2} = \frac{1}{128}$$

As $\quad pV = nRT$

$$\frac{p_1 V_1}{p_2 V_2} = \frac{T_1}{T_2}, \frac{1}{128} \times \frac{d_2}{d_1} = \frac{T_1}{T_2};$$

$$\frac{1}{128} \times 32 = \frac{T_1}{T_2}, T_2 = 4T_1$$

77. *(1400)* Given, for diatomic gas, $\gamma = 1.4$

$W = 400$ J(isobaric expansion)

To find the heat given to the gas.

Work done, $W = nR\Delta T = 400$ J ... (i)

Heat given to gas,

$$\Delta Q = nC_p \,\Delta T$$

$$\Rightarrow \quad \Delta Q = \frac{n \times 7R}{2} \times \Delta T$$

Using Eq. (i), $\Delta Q = \frac{7}{2} \times 400 = 1400$ J

78. *(480)* Given, sample of an ideal gas with ($\gamma = 1.5$) is taken through adiabatic process, where

$$V_1 = 1200 \text{ cm}^3 = 12 \times 10^{-4} \text{ m}^3$$

$$V_2 = 300 \text{ cm}^3 = 3 \times 10^{-4} \text{ m}$$

$$p_1 = 200 \text{ kPa} = 2 \times 10^5 \text{ Pa}$$

Using equation of adiabatic process,

$$p_1 V_1^{\gamma} = p_2 V_2^{\gamma},$$

$$p_2 = p_1 \left(\frac{V_1}{V_2}\right)^{\gamma}$$

$$= 2 \times 10^5 \left(\frac{12 \times 10^{-4}}{3 \times 10^{-4}}\right)^{1.5}$$

$$= 16 \times 10^5 \text{ Pa}$$

Work done in adiabatic process,

$$W = \frac{p_1 V_1 - p_2 V_2}{\gamma - 1}$$

Substituting the values, we get

$$W = \frac{(2 \times 10^5 \times 12 \times 10^{-4} - 16 \times 10^5 \times 3 \times 10^{-4})}{1.5 - 1}$$

$$= -480 \text{ J}$$

Hence, work done in adiabatic compression is 480 J.

79. *(17258)* According to given p-V diagram in question, work done in isothermal process,

$$W = nRT \ln\left(\frac{V_f}{V_i}\right)$$

$$V_f = 2V_i \quad \text{(given)}$$

$$\Rightarrow \quad W = 8.3 \times 300 \times \ln(2)$$

[Since, $T = 273 + 27 = 300$ K]

$$W = 1725.819 \text{ J}$$

$$= 17258 \times 10^{-1} \text{ J}$$

This is the required value of the work done by the system.

80. *(100)* Consider the given diagram,

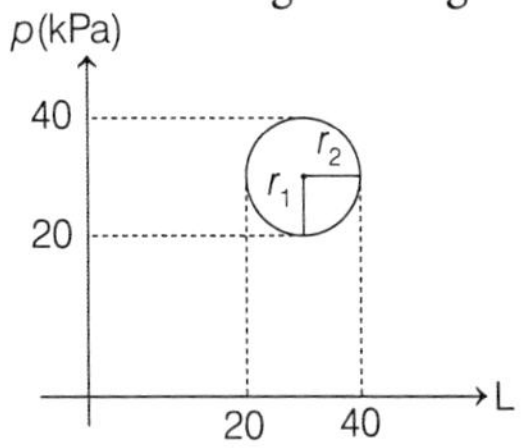

Here, $r_1 = 10 \times 10^3$ m

and $r_2 = 10 \times 10^{-3}$ m

We know that for a complete cyclic process, change in internal energy,

$(\Delta U) = 0$ ...(i)

According to 1st law of thermodynamics,

$\Delta Q = \Delta U + W$ ...(ii)

From Eqs. (i) and (ii), we get

$\Rightarrow \Delta Q = 0 + W$

$\Rightarrow \Delta Q = W$ ...(iii)

$\because \quad W = \text{Area} = \pi r_1 r_2$

$= \pi \times (10 \times 10^3) \times (10 \times 10^{-3})$

$W = 100\pi$ J ...(iv)

From Eqs. (iii) and (iv), we get

$\Delta Q = 100\,\pi$ J

81. *(50)* Given, pressure, $p = kV^3$

$$T_1 = 100°\text{C}, T_2 = 300°\text{C}$$

$$\Rightarrow \Delta T = T_2 - T_1 = 300 - 100 = 200°\text{C}$$

By using ideal gas equation,

$$pV = nRT$$

$$\Rightarrow kV^3 \cdot V = nRT \Rightarrow kV^4 = nRT$$

On differentiating both sides w.r.t temperature, we get

$$4kV^3 \frac{dV}{dT} = nR \Rightarrow 4kV^3\,dV = nRdT$$

$$\Rightarrow \quad kV^3\,dV = nRdT/4$$

$$\Rightarrow \quad pdV = nRdT/4$$

As, work done (W)

$$= pdV = nRdT/4$$

$$= \frac{nR}{4}\Delta T = \frac{nR}{4} \times 200$$

$$= 50\,nR$$

82. *(1819)* Initially gas is compressed adiabatically.

Initial temperature, $T_1 = 300$ K

Number of moles $= 1$, $\gamma = 1.4$

Initial volume $= V_1$

Final volume $= V_1/16$

Adiabatic process equation is,

$$T_1 V_1^{\gamma - 1} = T_2 V_2^{\gamma - 1}$$

$$\Rightarrow 300 V_1^{1.4-1} = T_2 \left(\frac{V_1}{16}\right)^{1.4-1}$$

$$\Rightarrow \quad T_2 = 300 \times 2^{8/5} \quad ...(i)$$

In next process, gas is expanded isobarically. Initial volume, $V_2 = V_1/16$

Final volume, $V_3 = 2\left(\frac{V_1}{16}\right) = \frac{V_1}{8}$

Now, for isobaric expansion,

$$\frac{V_2}{T_2} = \frac{V_3}{T_3} \Rightarrow \frac{V_2}{300 \times 2^{\frac{8}{5}}} = \frac{2V_2}{T_3}$$

$$\Rightarrow \quad T_3 = 2 \times 300 \times 2^{\frac{8}{5}}$$

$$= 300 \times 2^{\frac{13}{5}} = 1818.85$$

or $\quad T_3 = 1819$ K

Topic 3

01. *(d)* Given, temperature of reservoir (boiling point of water),

$$T_1 = 100°\text{C} = 373 \text{ K}$$

Temperature of sink (freezing point of water),

$$T_2 = 0°\text{C} = 273 \text{ K}$$

Efficiency of carnot engine,

$$\eta = 1 - \frac{T_2}{T_1} = \left(1 - \frac{273}{373}\right) \times 100\%$$

$$= 26.8\% = 27\%$$

No engine can be more efficient than carnot engine. Therefore, the efficiency of given engine must be less than 27%.

02. *(b)* Since, entropy is a state function, therefore change in entropy in both the processes must be same. Therefore, correct option should be (b).

03. *(d)*

> **key idea** Entropy (S) is a measure of randomness (disorder) in a system. It's SI unit is JK^{-1}. For a reversible process change in entropy dS due to a transfer of heat dQ at finite temperature T is,
>
> $$dS = \frac{dQ}{T}$$

$$d(\text{Entropy})_{sys} = \frac{dQ}{T}$$

$$\int d(\Delta E)_{sys} = \int_{T_1}^{T_2} \frac{dQ}{T}$$

$$(\Delta E)_{sys} = \int_{T_1}^{T_2} \frac{msdT}{T} = ms\int_{T_1}^{T_2} \frac{dT}{T}$$

$$= ms\,[\ln T]_{T_1}^{T_2} = ms\,[\ln T_2 - \ln T_1]$$

$$= ms\left[\ln\left(\frac{T_2}{T_1}\right)\right]$$

$$= ms \ln\frac{T_2}{T_1} = m(1)\ln\frac{T_2}{T_1} = m\ln\frac{T_2}{T_1}$$

04. *(a)* Heat cannot flow itself from a body at lower temperature to a body at higher temperature. This corresponds to second law of thermodynamics.

05. *(b)* Work done in an isobaric process

$$\Delta W = nR\Delta T$$

$$\Rightarrow \quad 200 = nR\Delta T \qquad \ldots\text{(i)}$$

For diatomic gas, specific heat at constant pressure $C_p = \frac{7}{2}R, C_V = \frac{5}{2}R$

and change in internal energy

$$\Delta U = nC_V\Delta T$$

$$= n\cdot\frac{5}{2}R\Delta T = \frac{5}{2}\cdot nR\Delta T$$

$$= \frac{5}{2}\times 200 \qquad \text{[From Eq. (i)]}$$

$$= 500\text{ J}$$

Now by first law of thermodynamics

$$\Delta Q = \Delta U + \Delta W$$

$$= 500 + 200 = 700\text{ J}$$

06. *(c)* According to the figure,

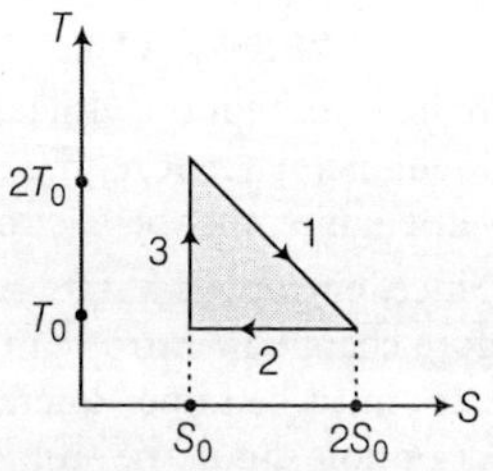

For process 1,

$$Q_1 = T_0S_0 + \frac{1}{2}T_0S_0 = \frac{3}{2}T_0S_0$$

For process 2,

$$Q_2 = T_0(2S_0 - S_0) = T_0S_0$$

For process 3, $Q_3 = 0$

$$\therefore \quad \eta = \frac{W}{Q_1} = \frac{Q_1 - Q_2}{Q_1}$$

$$= 1 - \frac{Q_2}{Q_1} = 1 - \frac{2}{3} = \frac{1}{3}$$

07. *(b)* Using efficiency of engine, we have

$$\eta = 1 - \frac{T_0}{T_1}, (T_2 < T_1)$$

For engine -E

$$\eta_{12} = 1 - \frac{273}{473} = \frac{200}{473}$$

For engine - E_1

$$\eta_1 = 1 - \frac{373}{473} = \frac{100}{473}$$

For engine - E_2

$$\eta_2 = 1 - \frac{273}{373} = \frac{100}{373}$$

So, $\eta_{12} - \eta_1 = \frac{100}{473} < \eta_2$

Which gives $\eta_{12} < \eta_1 + \eta_2$

08. *(b)* For isotherms ab and cd,

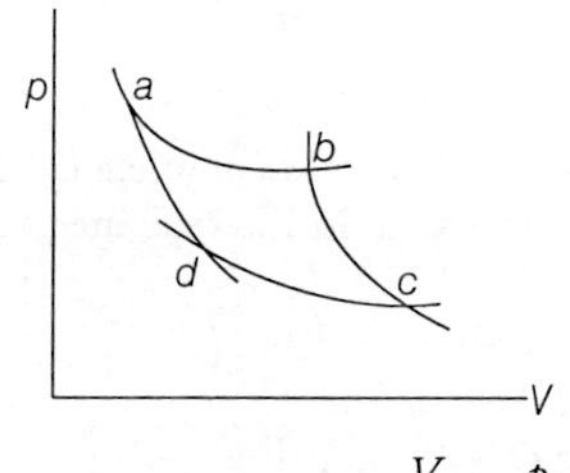

$$\Rightarrow \quad p_aV_a = p_bV_b \Rightarrow \frac{V_a}{V_b} = \frac{p_b}{p_a}$$

$$\text{and} \quad p_cV_c = p_dV_d \Rightarrow \frac{V_c}{V_d} = \frac{p_d}{p_c}$$

Now for adiabatic process bc and da,

$$p_bV_b^\gamma = p_cV_c^\gamma$$

$$\text{and} \quad p_dV_d^\gamma = p_aV_a^\gamma$$

$$\text{or,} \quad p_bV_b^\gamma = p_cV_c^\gamma$$

$$\text{and} \quad p_aV_a^\gamma = p_dV_d^\gamma$$

Dividing we have,

$$\frac{p_b}{p_a}\left(\frac{V_b}{V_a}\right)^\gamma = \frac{p_c}{p_d}\left(\frac{V_c}{V_d}\right)^\gamma$$

$$\frac{V_a}{V_b}\left(\frac{V_b}{V_a}\right)^\gamma = \frac{V_d}{V_c}\left(\frac{V_c}{V_d}\right)^\gamma$$

$$\Rightarrow \left(\frac{V_b}{V_a}\right)^{\gamma-1} = \left(\frac{V_c}{V_d}\right)^{\gamma-1}$$

$$\Rightarrow \quad \frac{V_b}{V_a} = \frac{V_c}{V_d} \Rightarrow \frac{V_a}{V_d} = \frac{V_b}{V_c}$$

09. *(b)* Case -1

Temperature of source,

$$T_1 = 447°\text{ C} = (447 + 273)\text{ K}$$
$$= 720\text{ K}$$

Temperature of sink,

$$T_2 = 147°\text{ C}$$
$$= (147 + 273)\text{ K} = 420\text{ K}$$

$\therefore$Efficiency,

$$\eta_1 = 1 - \frac{T_2}{T_1} = 1 - \frac{420}{720} = \frac{300}{720}$$

$$\eta_1 = \frac{5}{12}$$

Case-2 Temperature of source,

$$T_1' = 947°\text{C}$$
$$= (947 + 273)\text{ K} = 1220\text{ K}$$

Temperature of sink,

$$T_2' = 47°\text{C}$$
$$= (273 + 47)\text{ K} = 320\text{ K}$$

$\therefore$Efficiency,

$$\eta_2 = 1 - \frac{T_2'}{T_1'} = 1 - \frac{320}{1220} = \frac{900}{1220} = \frac{45}{61}$$

$$\therefore \quad \frac{\eta_1}{\eta_2} = \frac{\frac{5}{12}}{\frac{45}{61}} = \frac{5\times 61}{12\times 45} = 0.56$$

10. *(a)* Given, initial efficiency of engine,

$$\eta_1 = \frac{1}{4}$$

Let, initial temperature of sink $= T_2$

Final temperature of sink $= T_2 - 58°\text{C}$

Final efficiency of engine,

$$\eta_2 = 2\eta_1 = 2/4 = 1/2$$

Temperature of source $= T_1$

Since, $\eta = 1 - \frac{T_2}{T_1}$

$$\therefore \quad \eta_1 = \frac{1}{4} = 1 - \frac{T_2}{T_1}$$

$$\Rightarrow \quad \frac{T_2}{T_1} = 1 - \frac{1}{4} \Rightarrow T_2/T_1 = 3/4$$

$$\Rightarrow \quad T_1 = \frac{4}{3}T_2 \qquad \ldots\text{(i)}$$

$$\text{and} \quad \eta_2 = \frac{1}{2} = 1 - \frac{T_2 - 58}{T_1}$$

$$\Rightarrow \quad \frac{T_2 - 58}{T_1} = 1 - \frac{1}{2}$$

$$\Rightarrow \frac{T_2 - 58}{4/3\,T_2} = \frac{1}{2} \qquad \text{[From Eq. (i)]}$$

$$\Rightarrow 2T_2 - 116 = \frac{4}{3}T_2$$

$$\Rightarrow \frac{2T_2}{3} = 116$$

$$\Rightarrow T_2 = \frac{116 \times 3}{2} = 58 \times 3 = 174°C$$

11. *(d)* Given, efficiency of heat engine,

$$\eta_1 = \frac{1}{6}$$

Let T_1, T_2 be the initial temperature of source and sink, respectively and T_1', T_2' be the final temperature of source and sink, respectively.

Given, $T_2' = T_2 - 62$ and $T_1' = T_1$

Final efficiency, $\eta_2 = 2\eta_1 = \frac{2}{6} = \frac{1}{3}$

As we know that, $\eta = 1 - \frac{T_2}{T_1}$

For case 1

$$\therefore \quad \frac{1}{6} = 1 - \frac{T_2}{T_1}$$

$$\Rightarrow \frac{T_2}{T_1} = 1 - \frac{1}{6} = \frac{5}{6} \qquad \ldots\text{(i)}$$

For case 2 $\frac{1}{3} = 1 - \frac{T_2'}{T_1'} \Rightarrow \frac{T_2'}{T_1'} = 1 - \frac{1}{3}$

$$\Rightarrow \frac{T_2 - 62}{T_1} = \frac{2}{3} \Rightarrow \frac{T_2}{T_1} - \frac{62}{T_1} = \frac{2}{3}$$

$$\Rightarrow \frac{5}{6} - \frac{62}{T_1} = \frac{2}{3} \qquad \text{[From Eq. (i)]}$$

$$\Rightarrow \frac{62}{T_1} = \frac{5}{6} - \frac{2}{3} \Rightarrow \frac{62}{T_1} = \frac{1}{6}$$

$$\Rightarrow T_1 = 372\text{ K}$$

$$\therefore T_1 = 372 - 273°\text{C} = 99°\text{C}$$

12. *(c)* As processes involved in one cycle of a heat engine must be

Expansion ⟶ Compression
⟶ Expansion ⟶ Compression

So, Q must be negative.

Now, Work done in process
= Heat supplied − Heat rejected
$= (1915 + 125) - (40 + Q)$ J

And heat supplied to cycle
$= 1915 + 125$ J

Given, efficiency = 50%

$$\Rightarrow \eta = \frac{\text{Work done}}{\text{Heat supplied}}$$

$$\Rightarrow \frac{50}{100} = \frac{(1915+125)-(40+Q)}{1915+125}$$

$$\Rightarrow Q = 2000 - 1020 = 980\text{ J}$$

Hence, correct option is (c).

13. *(a)* Work done by engine = Area inclosed by the curve $= p_0V_0$

Heat given to the system,

$$Q = Q_{AB} + Q_{BC}$$
$$= nC_V\Delta T_{AB} + nC_p\Delta T_{BC}$$
$$= \frac{3}{2}(nRT_B - nRT_A) + (nRT_C - nRT_B)$$
$$= \frac{3}{2}(2p_0V_0 - p_0V_0) + \frac{5}{2}(4p_0V_0 - 2p_0V_0)$$
$$= \frac{13}{2}p_0V_0$$

Thermal efficiency,

$$\eta = \frac{W}{Q} = \frac{p_0V_0}{\frac{13}{2}p_0V_0} = \frac{2}{13} \approx 0.15$$

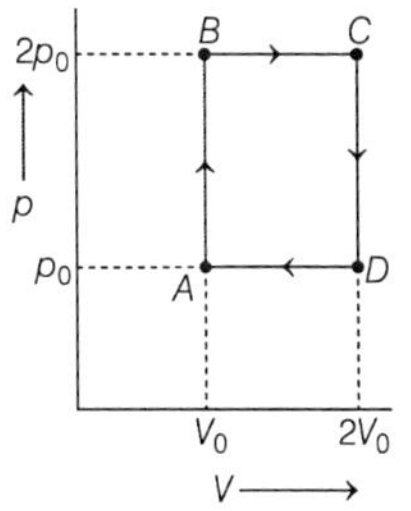

14. *(c)* Heat required to change the temperature of vessel by a small amount dT,

$$-dQ = mC_p dT$$

Total heat required

$$-Q = m\int_{20}^{4} 32\left(\frac{T}{400}\right)^3 dT$$

$$= \frac{100 \times 10^{-3} \times 32}{(400)^3}\left[\frac{T^4}{4}\right]_{20}^{4}$$

$$\Rightarrow Q = 0.001996\text{ kJ}$$

Work done required to maintain the temperature of sink to T_2,

$$W = Q_1 - Q_2 = \frac{Q_1 - Q_2}{Q_2}Q_2$$

$$= \left(\frac{T_1}{T_2} - 1\right)Q_2$$

$$\Rightarrow W = \left(\frac{T_1 - T_2}{T_2}\right)Q_2$$

For $T_2 = 20$ K,

$$W_1 = \frac{300 - 20}{20} \times 0.001996 = 0.028\text{ kJ}$$

For $T_2 = 4$ K,

$$W_2 = \frac{300 - 4}{4} \times 0.001996$$
$$= 0.148\text{ kJ}$$

As temperature is changing from 20 K to 4 K, work done required will be more than W_1 but less than W_2.

15. *(102)* Here, $T_1 = 227°C = 500$ K,

$T_2 = ?$, $Q_1 = 300$ cal

and $Q_2 = 225$ cal

Work done,

$$W = Q_1 - Q_2$$

$$\therefore \quad W = 300 - 225 = 75\text{ Cal}$$

Thus, we have $\frac{W}{Q_1} = 1 - \frac{T_2}{T_1}$

or $\frac{75}{300} = 1 - \frac{T_2}{500}$ or $\frac{1}{4} = 1 - \frac{T_2}{500}$

or $T_2 = \frac{3}{4} \times 500 = 375$ K or $T_2 = 102°$ C

> **• Concept Enhancer**
>
> Efficiency of a Carnot engine,
>
> $$\eta = 1 - \frac{T_2}{T_1} = \frac{W}{Q_1}$$
>
> where W is the work done, Q_1 is the energy input, T_1 source temperature and T_2 is temperature of the sink.

16. *(540)* Given, temperature of cold reservoir, $T_2 = 324$ K

$Q_1 = 300$ J, $Q_2 = 180$ J

Let T_1 be the minimum temperature of hot reservoir, then efficiency of heat engine is given as

$$\eta = \frac{W}{Q_1} = 1 - \frac{T_2}{T_1}$$

$$\Rightarrow \frac{W}{Q_1} = 1 - \frac{T_2}{T_1}$$

$$\Rightarrow \frac{Q_1 - Q_2}{Q_1} = 1 - \frac{T_2}{T_1}$$

$$\Rightarrow 1 - \frac{Q_2}{Q_1} = 1 - \frac{T_2}{T_1} \Rightarrow \frac{Q_2}{Q_1} = \frac{T_2}{T_1}$$

$$\Rightarrow T_1 = T_2 \cdot \frac{Q_1}{Q_2}$$

$$= 324 \times \frac{300}{180} = 540\text{ K}$$

17. *(31)* Given, mass of steam,

$m = 50$ g at $T = 100$ °C

Time, $t = 1$ min, $C_{\text{water}} = 1$ cal/g°C,

$T = 20$ °C, $L_V = 540$ cal/g

Heat rejected per minute is

$$\Delta Q = mL + mC\Delta T$$
$$\Delta Q = 50 \times 540 + 50 \times 1 \times (100 - 20);$$
$$\Delta Q = 540 \times 50 + 50 \times 80;$$
$$\Delta Q = 31000\text{ cal}, \Delta Q = 31 \times 10^3\text{ cal}$$

18. *(500)* Given, temperature of hot reservoir $= T_1$

Temperature of cold reservoir, $T_2 = 400$ K

Heat of hot reservoir, $Q_1 = 300$ J

Heat of cold reservoir, $Q_2 = 240$ J

Since, efficiency

$$\eta = 1 - \frac{T_2}{T_1} = \frac{Q_1 - Q_2}{Q_1}$$

$$1 - \frac{T_2}{T_1} = 1 - \frac{Q_2}{Q_1}$$

$$\Rightarrow 1 - \frac{400}{T_1} = 1 - \frac{240}{300} \Rightarrow \frac{400}{T_1} = \frac{4}{5}$$

$$\Rightarrow \quad T_1 = 500 \text{ K}$$

19. *(–113)* Given, temperature of source,

$$T_H = 127°\text{C} = 273 + 127 = 400 \text{ K}$$

Efficience, $\eta = 60\% = 0.6$

The efficiency of Carnot (ideal) heat engine is given by

$$\eta = \left(1 - \frac{T_L}{T_H}\right)$$

where, T_L = temperature of sink,

$$0.6 = \left(1 - \frac{T_L}{T_H}\right)$$

$$\Rightarrow \quad \frac{T_L}{T_H} = 1 - 0.6$$

$$\frac{T_L}{T_H} = 0.4$$

$$\Rightarrow \quad T_L = 0.4 \times T_H = 0.4 \times 400 = 160 \text{ K} = 160 - 273 = -113°\text{C}$$

20. *(208)* Given, initial efficiency

$$\eta_1 = \frac{1}{4}$$

$$\Rightarrow \quad \eta_1 = 1 - \frac{T_2}{T_1} = \frac{1}{4}$$

$$\Rightarrow \quad \frac{T_2}{T_1} = 1 - \frac{1}{4} = \frac{3}{4}$$

When temperature of sink is reduced by 52 K,

$$\eta_2 = 2\eta_1 = 1 - \frac{T_2 - 52}{T_1}$$

$$\Rightarrow \quad 2 \times \frac{1}{4} = 1 - \frac{T_2}{T_1} + \frac{52}{T_1}$$

$$\Rightarrow \quad \frac{1}{2} = 1 - \frac{3}{4} + \frac{52}{T_1}$$

$$\Rightarrow \quad \frac{1}{2} - \frac{1}{4} = \frac{52}{T_1} \Rightarrow \frac{1}{4} = \frac{52}{T_1}$$

$$\Rightarrow \quad T_1 = 208 \text{ K}$$

21. *(8791)* To convert 100 g water at 0°C to ice at 0°C, heat absorbed by refrigerator,

$$Q_1 = mL = 100 \times 80 = 8000 \text{ cal}$$

W = work done or energy supplied to refrigerator

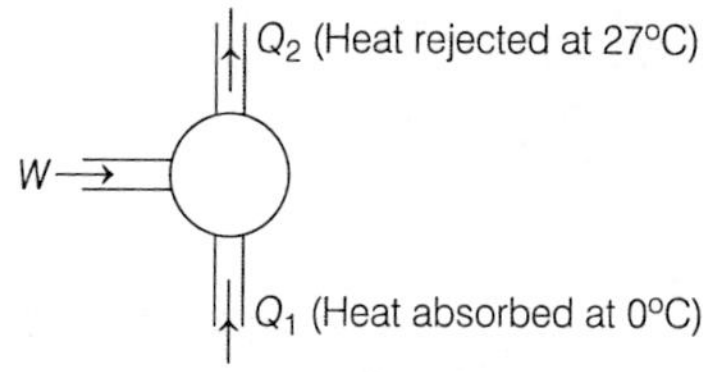

From figure, we have

$$W + Q_1 = Q_2$$

$$\Rightarrow \quad W = Q_2 - Q_1$$

Now, coefficient of performance of refrigerator is

$$\beta = \frac{Q_1}{W} = \frac{Q_1}{Q_2 - Q_1} = \frac{T_1}{T_2 - T_1}$$

Here, $T_1 = 0°\text{C} = 273$ K

$T_2 = 27°\text{C} = 300$ K

So, $$\frac{Q_1}{Q_2 - Q_1} = \frac{273}{300 - 273}$$

$$\Rightarrow \quad 27Q_1 = (Q_2 - Q_1)273$$

$$\Rightarrow \quad 300\, Q_1 = Q_2 \times 273$$

$$\Rightarrow \quad Q_2 = 300 \times \frac{8000}{273} = \frac{2400000}{273} = 8791.2 \text{ cal} \approx 8791 \text{ cal}$$

C-12

JEE Main Chapterwise Topicwise Physics

In JEE Main there is usually 0-1 question per paper is visible from this chapter. In past 3 years, 46 conceptual and 2 numerical questions were asked of moderate to difficult level from this chapter.

For detailed analysis scan QR code

KINETIC THEORY OF GASES

Topic 1 Kinetic Theory of Gases and Gas Laws

Level 1

01 The ratio of vapour densities of two gases at the same temperature is 4/25, then the ratio of rms velocities will be *[2025, 28 Jan Shift-II]*

(a) $\frac{5}{2}$ (b) $\frac{2}{5}$ (c) $\frac{25}{4}$ (d) $\frac{4}{25}$

02 A sample contains mixture of helium and oxygen gas. The ratio of root mean square speed of helium and oxygen in the sample is *[2024, 6 April Shift-I]*

(a) $\frac{1}{4}$ (b) $\frac{2\sqrt{2}}{1}$ (c) $\frac{1}{2\sqrt{2}}$ (d) $\frac{1}{32}$

Imp. RMS Velocity of Gas Molecules

03 If the root mean square velocity of hydrogen molecule at a given temperature and pressure is 2 km/s, the root mean square velocity of oxygen at the same condition in km/s is *[2024, 1 Feb Shift-II]*

(a) 0.5 (b) 2.0

(c) 1.5 (d) 1.0

04 The speed of sound in oxygen at STP will be approximately (Given, $R = 8.3\ \text{JK}^{-1}$, $\gamma = 1.4$) *[2024, 31 Jan Shift-II]*

(a) 341 m/s (b) 325 m/s

(c) 310 m/s (d) 333 m/s

05 The parameter that remains the same for molecules of all gases at a given temperature is *[2024, 31 Jan Shift-I]*

(a) speed (b) mass

(c) momentum (d) kinetic energy

06 A flask contains Hydrogen and Argon in the ratio 2 : 1 by mass. The temperature of the mixture is 30°C. The ratio of average kinetic energy per molecule of the two gases ($K_{\text{argon}}/K_{\text{hydrogen}}$) is

(Given, Atomic Weight of Ar = 39.9) *[2023, 15 April Shift-I]*

(a) 2 (b) 39.9

(c) 39.9 / 2 (d) 1

07 The temperature of an ideal gas is increased from 200 K to 800 K. If rms speed of gas at 200 K is v_0. Then, rms speed of the gas at 800 K will be *[2023, 6 April Shift-II]*

(a) $v_0/4$ (b) v_0 (c) $4v_0$ (d) $2v_0$

08 For three low density gases A, B, C, pressure *versus* temperature graphs are plotted while keeping them at constant volume, as shown in the figure.

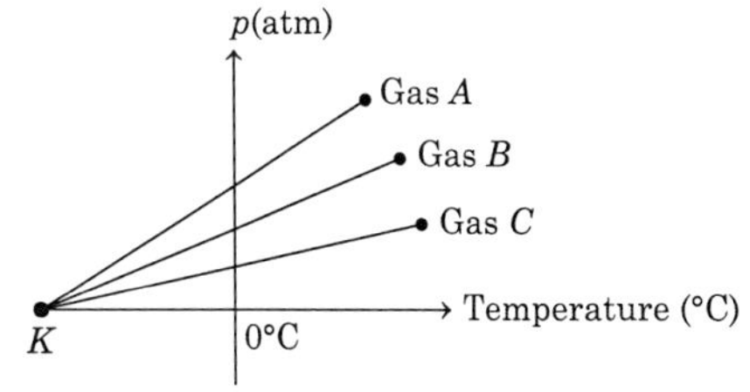

The temperature corresponding to the point K is *[2023, 1 Feb Shift-II]*

(a) − 40°C (b) −100°C

(c) − 273°C (d) −373°C

09 The correct relation between $\gamma = \frac{C_p}{C_V}$ and temperature T is *[2023, 31 Jan Shift-I]*

(a) $\gamma \propto T$ (b) $\gamma \propto \frac{1}{T}$ (c) $\gamma \propto T^0$ (d) $\gamma \propto \frac{1}{\sqrt{T}}$

10 The root mean square velocity of molecules of gas is *[2023, 25 Jan Shift-I]*

(a) inversely proportional to square root of temperature $(\sqrt{1/T})$

(b) proportional to square root of temperature $(\sqrt{T})$

(c) proportional to temperature (T)

(d) proportional to square of temperature (T^2)

11 Given below are two statements.

Statement I The average momentum of a molecule in a sample of an ideal gas depends on temperature.

Statement II The rms speed of oxygen molecules in a gas is v. If the temperature is doubled and the oxygen molecules dissociate into oxygen atoms, the rms speed will become $2v$.

In the light of the above statements, choose the correct answer from the options given below. *[2022, 28 July Shift-I]*

(a) Both Statement I and Statement II are true.
(b) Both Statement I and Statement II are false.
(c) Statement I is true but Statement II is false.
(d) Statement I is false but Statement II is true.

12 7 mole of a certain monoatomic ideal gas undergoes a temperature increase of 40 K at constant pressure. The increase in the internal energy of the gas in this process is
(Given, $R = 8.3\ \text{JK}^{-1}\ \text{mol}^{-1}$) *[2022, 26 July Shift-I]*

(a) 5810 J (b) 3486 J (c) 11620 J (d) 6972 J

13 Following statements are given.

(A) The average kinetic energy of a gas molecule decreases when the temperature is reduced.
(B) The average kinetic energy of a gas molecule increases with increase in pressure at constant temperature.
(C) The average kinetic energy of a gas molecule decreases with increase in volume.
(D) Pressure of a gas increases with increase in temperature at constant pressure.
(E) The volume of gas decreases with increase in temperature.

Choose the correct answer from the options given below. *[2022, 25 July Shift-I]*

(a) (A) and (D) only
(b) (A), (B) and (D) only
(c) (B) and (D) only
(d) (A), (B) and (E) only

14 A cylinder of fixed capacity of 44.8 L contains helium gas at standard temperature and pressure. The amount of heat needed to raise the temperature of gas in the cylinder by 20.0°C will be (Given, gas constant, $R = 8.3\ \text{JK}^{-1}\ \text{mol}^{-1}$) *[2022, 29 June Shift-I]*

(a) 249 J (b) 415 J
(c) 498 J (d) 830 J

15 What will be the effect on the root mean square velocity of oxygen molecules, if the temperature is doubled and oxygen molecule dissociates into atomic oxygen? *[2022, 28 June Shift-II]*

(a) The velocity of atomic oxygen remains same.
(b) The velocity of atomic oxygen doubles.
(c) The velocity of atomic oxygen becomes half.
(d) The velocity of atomic oxygen becomes four times.

16 A flask contains argon and oxygen in the ratio of 3 : 2 in mass and the mixture is kept at 27°C. The ratio of their average kinetic energy per molecule respectively, will be *[2022, 26 June Shift-II]*

(a) 3 : 2 (b) 9 : 4
(c) 2 : 3 (d) 1 : 1

17 The relation between root mean square speed v_{rms} and most probable speed v_p for the molar mass M of oxygen gas molecule at the temperature of 300 K will be *[2022, 25 June Shift-I]*

(a) $v_{\text{rms}} = \sqrt{\frac{2}{3}}\, v_p$ (b) $v_{\text{rms}} = \sqrt{\frac{3}{2}}\, v_p$

(c) $v_{\text{rms}} = v_p$ (d) $v_{\text{rms}} = \sqrt{\frac{1}{3}}\, v_p$

18 The pressure of the gas in a constant volume gas thermometer is 100 cm of mercury when placed in melting ice at 1 atm. When the bulb is placed in a liquid, the pressure becomes 180 cm of mercury. Temperature of the liquid is (Given, $0°\text{C} = 273\ \text{K}$) *[2022, 30 June Shift-II]*

(a) 300 K (b) 400 K (c) 600 K (d) 491 K

19 The root mean square speed of smoke particles of mass 5×10^{-17} kg in their Brownian motion in air at NTP is approximately.
[Given, $k_B = 1.38 \times 10^{-23}\ \text{J K}^{-1}$] *[2022, 29 July Shift-II]*

(a) 60 mm s^{-1} (b) 12 mm s^{-1}
(c) 15 mm s^{-1} (d) 36 mm s^{-1}

20 Which of the following graphs represent the behaviour of an ideal gas ? (Symbols have their usual meanings.) *[2021, 20 July Shift-II]*

(a)
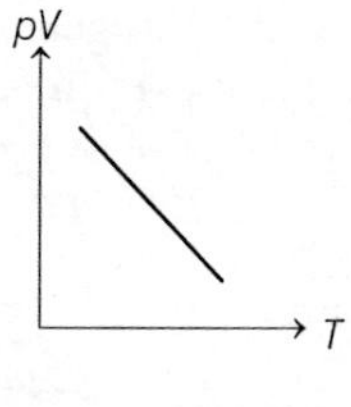

(b)
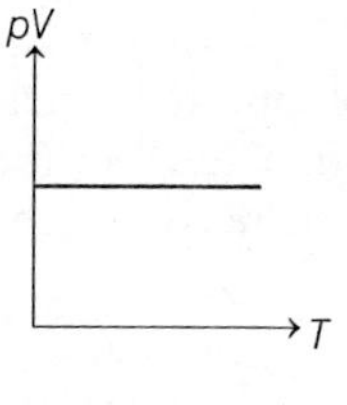

(c)
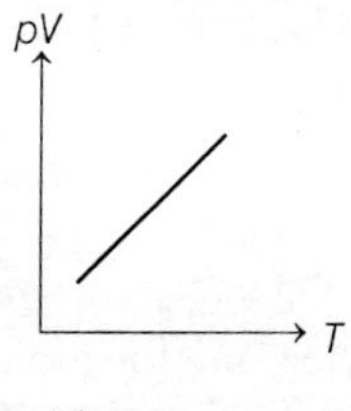

(d)
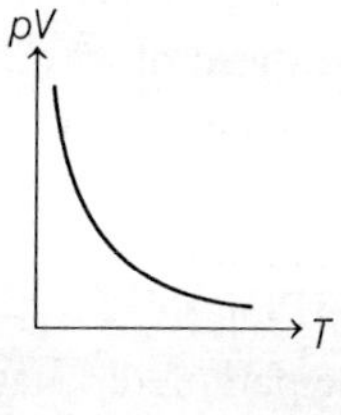

21 Consider a mixture of gas molecule of types A, B and C having masses $m_A < m_B < m_C$. The ratio of their root mean square speeds at normal temperature and pressure is *[2021, 20 July Shift-I]*

(a) $v_A = v_B = v_C = 0$ (b) $\frac{1}{v_A} > \frac{1}{v_B} > \frac{1}{v_C}$

(c) $v_A = v_B \neq v_C$ (d) $\frac{1}{v_A} < \frac{1}{v_B} < \frac{1}{v_C}$

22 Consider a sample of oxygen behaving like an ideal gas. At 300 K, the ratio of root mean square (rms) velocity to the average velocity of gas molecule would be (Molecular weight of oxygen is 32 g/mol, $R = 8.3\ \text{JK}^{-1}\ \text{mol}^{-1}$) *[2021, 18 March Shift-II]*

(a) $\sqrt{\frac{3}{3}}$ (b) $\sqrt{\frac{8}{3}}$

(c) $\sqrt{\frac{3\pi}{8}}$ (d) $\sqrt{\frac{8\pi}{3}}$

23 On the basis of kinetic theory of gases, the gas exerts pressure because its molecules *[2021, 24 Feb Shift-II]*

(a) continuously lose their energy till it reaches wall
(b) are attracted by the walls of container
(c) continuously stick to the walls of container
(d) suffer change in momentum when impinge on the walls of container

24 For a given gas at 1 atm pressure, rms speed of the molecules is 200 m/s at 127°C. At 2 atm pressure and at 227° C, the rms speed of the molecules will be *[2019, 9 April Shift-I]*

(a) $100\sqrt{5}$ m/s (b) 80 m/s

(c) 100 m/s (d) $80\sqrt{5}$ m/s

25 An ideal gas occupies a volume of $2\ \text{m}^3$ at a pressure of 3×10^6 Pa. The energy of the gas is *[2019, 12 Jan Shift-I]*

(a) 6×10^4 J (b) 10^8 J

(c) 9×10^6 J (d) 3×10^2 J

26 A mixture of 2 moles of helium gas (atomic mass $= 4u$) and 1 mole of argon gas (atomic mass $= 40u$) is kept at 300 K in a container. The ratio of their rms speeds $\left[\frac{v_{\text{rms(helium)}}}{v_{\text{rms(argon)}}}\right]$ is close to *[2019, 9 Jan Shift-I]*

(a) 0.32 (b) 2.24

(c) 3.16 (d) 0.45

27 The temperature of an open room of volume $30\ \text{m}^3$ increases from 17°C to 27°C due to the sunshine. The atmospheric pressure in the room remains 1×10^5 Pa. If n_i and n_f are the number of molecules in the room before and after heating, then $n_f - n_i$ will be *[JEE Main 2017]*

(a) 1.38×10^{23} (b) 2.5×10^{25}

(c) -2.5×10^{25} (d) -1.61×10^{23}

28 Cooking gas containers are kept in a lorry moving with uniform speed. The temperature of the gas molecules inside will *[AIEEE 2002]*

(a) increase
(b) decrease
(c) remain same
(d) decrease for some, while increase for others

Level 2

29 For a particular ideal gas which of the following graphs represents the variation of mean squared velocity of the gas molecules with temperature? *[2025, 28 Jan Shift-I]*

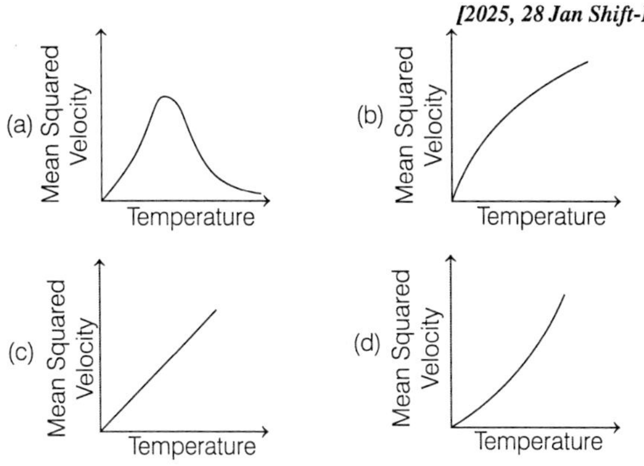

30 The kinetic energy of translation of the molecules in 50 g of CO_2 gas at 17°C is *[2025, 28 Jan Shift-II]*

(a) 4205.5 J (b) 3986.3 J (c) 4102.8 J (d) 3582.7 J

31 A diatomic gas ($\gamma = 1.4$) does 100 J of work in an isobaric expansion. The heat given to the gas is *[2024, 8 April Shift-II]*

(a) 150 J (b) 490 J (c) 350 J (d) 250 J

32 The specific heat at constant pressure of a real gas obeying $pV^2 = RT$ equation is *[2024, 6 April Shift-I]*

(a) R (b) $C_V + R$ (c) $\frac{R}{3} + C_V$ (d) $C_V + \frac{R}{2V}$

33 If the collision frequency of hydrogen molecules in a closed chamber at 27°C is Z, then the collision frequency of the same system at 127°C is *[2024, 5 April Shift-I]*

(a) $\frac{3}{4}Z$ (b) $\frac{2}{\sqrt{3}}Z$

(c) $\frac{4}{3}Z$ (d) $\frac{\sqrt{3}}{2}Z$

34 p-T diagram of an ideal gas having three different densities ρ_1, ρ_2, ρ_3 (in three different cases) is shown in the figure. Which of the following is correct ? *[2024, 4 April Shift-I]*

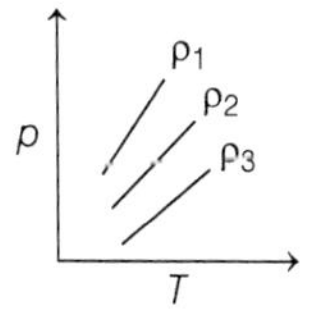

(a) $\rho_1 = \rho_2 = \rho_3$ (b) $\rho_1 < \rho_2$

(c) $\rho_2 < \rho_3$ (d) $\rho_1 > \rho_2$

35 At which temperature the rms velocity of a hydrogen molecule equal to that of an oxygen molecule at 47°C? *[2024, 30 Jan Shift-I]*

(a) 4 K (b) −73 K

(c) 20 K (d) 80 K

36 Two vessels A and B are of the same size and are at same temperature. A contains 1 g of hydrogen and B contains 1 g of oxygen. p_A and p_B are the pressures of the gases in A and B respectively, then $\frac{p_A}{p_B}$ is *[2024, 29 Jan Shift-I]*

(a) 4 (b) 8 (c) 32 (d) 16

37 The rms speed of oxygen molecule in a vessel at particular temperature is $\left(1+\frac{5}{x}\right)^{1/2} v$, where v is the average speed of the molecule. The value of x will be $\left(\text{Take}, \pi = \frac{22}{7}\right)$ *[2023, 13 April Shift-I]*

(a) 28 (b) 8 (c) 4 (d) 27

38 If the rms speed of chlorine molecule is 490 m/s at 27°C, the rms speed of argon molecules at the same temperature will be (Atomic mass of argon = 39.9 u, molecular mass of chlorine = 70.9 u) *[2023, 12 April Shift-I]*

(a) 751.7 m/s (b) 451.7 m/s
(c) 651.7 m/s (d) 551.7 m/s

39 The root mean square speed of molecules of nitrogen gas at 27°C is approximately. (Given, mass of a nitrogen molecule = 4.6×10^{-26} kg and take Boltzmann constant $k_B = 1.4 \times 10^{-23}\ \text{JK}^{-1}$) *[2023, 11 April Shift-II]*

(a) 91 m/s (b) 523 m/s
(c) 27.4 m/s (d) 1260 m/s

Imp. Kinetic Energy of Molecules

40 The temperature at which the kinetic energy of oxygen molecules becomes double than its value at 27°C is *[2023, 8 April Shift-II]*

(a) 927°C (b) 327°C (c) 627°C (d) 1227°C

41 At 300 K, the rms speed of oxygen molecules is $\sqrt{\frac{\alpha+5}{\alpha}}$ times to that of its average speed in the gas. Then, the value of α will be $\left(\text{use}, \pi = \frac{22}{7}\right)$ *[2023, 29 Jan Shift-II]*

(a) 28 (b) 24 (c) 32 (d) 27

42 Given below are two statements.

Statement I The temperature of a gas is −73°C. When the gas is heated to 527°C, the root mean square speed of the molecules is doubled.

Statement II The product of pressure and volume of an ideal gas will be equal to translational kinetic energy of the molecules. *[2023, 24 Jan Shift-I]*

In the light of the above statements, choose the correct answer from the options given below.

(a) Statement I is false but Statement II is true.
(b) Both Statement I and Statement II are true.
(c) Statement I is true but Statement II is false.
(d) Both Statement I and Statement II are false.

43 Same gas is filled in two vessels of the same volume at the same temperature. If the ratio of the number of molecules is 1 : 4, then

A. the rms velocity of gas molecules in two vessels will be the same.
B. the ratio of pressure in these vessels will be 1 : 4.
C. the ratio of pressure will be 1 : 1.
D. the rms velocity of gas molecules in two vessels will be in the ratio of 1 : 4.

Choose the correct answer from the options given below. *[2022, 27 July Shift-I]*

(a) A and C only (b) B and D only
(c) A and B only (d) C and D only

44 A vessel contains 16 g of hydrogen and 128 g of oxygen at standard temperature and pressure. The volume of the vessel in (cm^3) is *[2022, 29 June Shift-II]*

(a) 72×10^5 (b) 32×10^5
(c) 27×10^4 (d) 54×10^4

45 A mixture of hydrogen and oxygen has volume 2000 cm^3, temperature 300 K, pressure 100 kPa and mass 0.76 g. The ratio of number of moles of hydrogen to number of moles of oxygen in the mixture will be *[2022, 27 June Shift-I]*

(a) $\frac{1}{3}$ (b) $\frac{3}{1}$
(c) $\frac{1}{16}$ (d) $\frac{16}{1}$

46 For a perfect gas, two pressures p_1 and p_2 are shown in figure. The graph shows *[2022, 27 June Shift-II]*

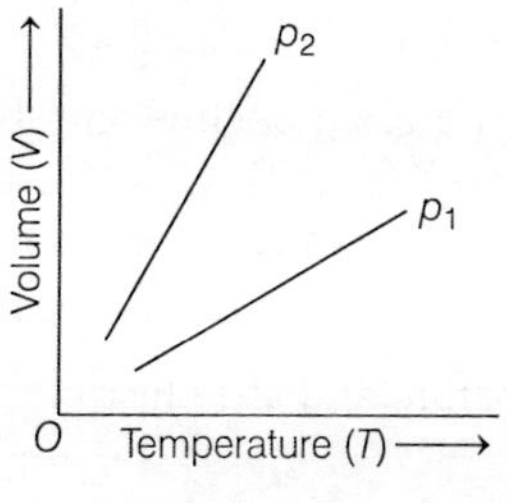

(a) $p_1 > p_2$ (b) $p_1 < p_2$ (c) $p_1 = p_2$
(d) insufficient data to draw any conclusion

47 If the rms speed of oxygen molecules at 0°C is 160 m/s, find the rms speed of hydrogen molecules at 0°C. *[2021, 27 Aug Shift-II]*

(a) 640 m/s (b) 40 m/s
(c) 80 m/s (d) 332 m/s

48 A cylindrical container of volume $4.0\times 10^{-3}\,m^3$ contains one mole of hydrogen and two moles of carbon dioxide. Assume the temperature of the mixture is 400 K. The pressure of the mixture of gases is [Take, gas constant $=8.3$ J $mol^{-1}K^{-1}$] *[2021, 26 Aug Shift-II]*

(a) 249×10^1 Pa (b) 24.9×10^3 Pa
(c) 24.9×10^5 Pa (d) 24.9 Pa

49 The rms speeds of the molecules of hydrogen, oxygen and carbon dioxide at the same temperature are v_H, v_O and v_{CO_2} respectively, then *[2021, 26 Aug Shift-I]*

(a) $v_H > v_O > v_{CO_2}$ (b) $v_{CO_2} > v_O > v_H$
(c) $v_H = v_O > v_{CO_2}$ (d) $v_H = v_O = v_{CO_2}$

50 The number of molecules in 1 L of an ideal gas at 300 K and 2 atm pressure with mean kinetic energy 2×10^{-9} J per molecules is *[2021, 27 July Shift-I]*

(a) 0.75×10^{11} (b) 3×10^{11}
(c) 1.5×10^{11} (d) 6×10^{11}

Imp. Mayer's Formula

51 For a gas $C_p - C_V = R$ in a state P and $C_p - C_V = 1.10\,R$ in a state Q. T_P and T_Q are the temperatures in two different states P and Q, respectively. Then, *[2021, 25 July Shift-I]*

(a) $T_P = T_Q$ (b) $T_P < T_Q$
(c) $T_P = 0.9\,T_Q$ (d) $T_P > T_Q$

52 The volume V of an enclosure contains a mixture of three gases, 16 g of oxygen, 28 g of nitrogen and 44 g of carbon dioxide at absolute temperature T. Consider R as universal gas constant. The pressure of the mixture of gases is *[2021, 16 March Shift-I]*

(a) $\frac{88\,RT}{V}$ (b) $\frac{3\,RT}{V}$
(c) $\frac{5}{2}\frac{RT}{V}$ (d) $\frac{4\,RT}{V}$

Imp. Mean Free Path

53 In a dilute gas at pressure p and temperature T, the mean time between successive collisions of a molecule varies with T as *[2020, 6 Sep Shift-II]*

(a) T (b) $\frac{1}{\sqrt{T}}$ (c) $\frac{1}{T}$ (d) $\sqrt{T}$

54 Number of molecules in a volume of 4 cm^3 of a perfect monoatomic gas at some temperature T and at a pressure of 2 cm of mercury is close to (Given, mean kinetic energy of a molecule at T is 4×10^{-14} erg, $g = 980$ cm/s^2, density of mercury $= 13.6$ g/cm^3) *[2020, 5 Sep Shift-I]*

(a) 5.8×10^{16} (b) 4.0×10^{16}
(c) 4.0×10^{18} (d) 5.8×10^{18}

55 An ideal gas in a closed container is slowly heated. As its temperature increases, which of the following statements are true? *[2020, 2 Sep Shift-II]*

A. The mean free path of the molecules decreases.
B. The mean collision time between the molecules decreases.
C. The mean free path remains unchanged.
D. The mean collision time remains unchanged.

(a) B and C (b) A and B
(c) C and D (d) A and D

56 Consider a mixture of n moles of helium gas and $2n$ moles of oxygen gas (molecules taken to be rigid) as an ideal gas. Its C_p / C_V value will be *[2020, 8 Jan Shift-II]*

(a) 40/27 (b) 23/15
(c) 19/13 (d) 67/45

57 A diatomic gas with rigid molecules does 10 J of work when expanded at constant pressure. What would be the heat energy absorbed by the gas, in this process? *[2019, 12 April Shift-II]*

(a) 25 J (b) 35 J (c) 30 J (d) 40 J

58 When heat Q is supplied to a diatomic gas of rigid molecules, at constant volume, its temperature increases by ΔT. The heat required to produce the same change in temperature, at a constant pressure is *[2019, 10 April Shift-II]*

(a) $\frac{2}{3}Q$ (b) $\frac{5}{3}Q$ (c) $\frac{3}{2}Q$ (d) $\frac{7}{5}Q$

59 The temperature, at which the root mean square velocity of hydrogen molecules equals their escape velocity from the earth, is closest to [Boltzmann constant $k_B = 1.38\times 10^{-23}$ J/K, Avogadro number $N_A = 6.02\times 10^{26}$/kg, Radius of earth $= 6.4\times 10^6$ m, Gravitational acceleration on earth $= 10$ ms^{-2}] *[2019, 8 April Shift-II]*

(a) 10^4 K (b) 650 K
(c) 3×10^5 K (d) 800 K

60 If 10^{22} gas molecules each of mass 10^{-26} kg collide with a surface (perpendicular to it) elastically per second over an area 1 m^2 with a speed 10^4 m/s, the pressure exerted by the gas molecules will be of the order of *[2019, 8 April Shift-I]*

(a) 10^4 N/m^2 (b) 10^8 N/m^2
(c) 10^3 N/m^2 (d) 10^{16} N/m^2

61 Half-mole of an ideal monoatomic gas is heated at constant pressure of 1 atm from 20°C to 90°C. Work done by gas is close to (Take, gas constant, $R = 8.31$ J/mol-K) *[2019, 10 Jan Shift-II]*

(a) 291 J (b) 581 J
(c) 146 J (d) 73 J

62 2 kg of a monoatomic gas is at a pressure of 4×10^4 N/m^2. The density of the gas is 8 kg/m^3. What is the order of energy of the gas due to its thermal motion ? *[2019, 10 Jan Shift-II]*
(a) 10^6 J (b) 10^3 J (c) 10^4 J (d) 10^5 J

63 An open glass tube is immersed in mercury in such a way that a length of 8 cm extends above the mercury level. The open end of the tube is then closed and sealed and the tube is raised vertically up by additional 46 cm. What will be length of the air column above mercury in the tube now? (Atmospheric pressure = 76 cm of Hg) *[JEE Main 2014]*
(a) 16 cm (b) 22 cm (c) 38 cm (d) 6 cm

64 Two rigid boxes 1/2 containing different ideal gases are placed on a table. Box A contains one mole of nitrogen at temperature T_0, while box B contains one mole of helium at temperature (7/3) T_0. The boxes are then put into thermal contact with each other and heat flows between them until the gases reach a common final temperature (ignore the heat capacity of boxes). Then, the final temperature of the gases, T_f, in terms of T_0 is *[AIEEE 2006]*
(a) $T_f = \frac{3}{7} T_0$ (b) $T_f = \frac{7}{3} T_0$
(c) $T_f = \frac{3}{2} T_0$ (d) $T_f = \frac{5}{2} T_0$

Level 3

65 A mixture of one mole of monoatomic gas and one mole of a diatomic gas (rigid) are kept at room temperature 27°C. The ratio of specific heat of gases at constant volume respectively is *[2024, 8 April Shift-I]*
(a) $\frac{3}{2}$ (b) $\frac{3}{5}$ (c) $\frac{5}{3}$ (d) $\frac{7}{5}$

66 The pressure p and temperature T relationship of an ideal gas obeys the equation pT^2 = constant. The volume expansion coefficient of the gas will be *[2023, 30 Jan Shift-I]*
(a) $3T^2$ (b) $3/T$ (c) $3/T^3$ (d) $3/T^2$

67 A vessel contains 14 g of nitrogen gas at a temperature of 27°C. The amount of heat to be transferred to the gas to double the rms speed of its molecules will be
(Take, $R = 8.32$ J mol^{-1}K^{-1}) *[2022, 28 July Shift-II]*
(a) 2229 J (b) 5616 J
(c) 9360 J (d) 13,104 J

68 Sound travels in a mixture of two moles of helium and n moles of hydrogen. If rms speed of gas molecules in the mixture is $\sqrt{2}$ times the speed of sound, then the value of n will be *[2022, 25 July Shift-II]*
(a) 1 (b) 2 (c) 3 (d) 4

69 A mixture of hydrogen and oxygen has volume 500 cm^3, temperature 300 K, pressure 400 kPa and mass 0.76 g. The ratio of masses of oxygen to hydrogen will be *[2021, 31 Aug Shift-II]*
(a) 3 : 8 (b) 3 : 16
(c) 16 : 3 (d) 8 : 3

70 For an ideal gas, the instantaneous change in pressure p with volume V is given by the equation $dp/dV = -ap$. If $p = p_0$ at $V = 0$ is the given boundary condition, then the maximum temperature one mole of gas can attain is (Here, R is the gas constant) *[2021, 31 Aug Shift-I]*
(a) $\frac{p_0}{aeR}$ (b) $\frac{ap_0}{cR}$
(c) infinity (d) 0°C

71 An ideal gas is expanding such that pT^3 = constant. The coefficient of volume expansion of the gas is *[2021, 27 Aug Shift-I]*
(a) $1/T$ (b) $2/T$ (c) $4/T$ (d) $3/T$

72 A balloon carries a total load of 185 kg at normal pressure and temperature of 27°C. What load will the balloon carry on rising to a height at which the barometric pressure is 45 cm of Hg and the temperature is −7°C ?
[Assuming, the volume constant.] *[2021, 27 Aug Shift-I]*
(a) 181.46 kg (b) 214.15 kg
(c) 219.07 kg (d) 123.54 kg

73 Two moles of helium gas is mixed with three moles of hydrogen molecules (taken to be rigid). What is the molar specific heat of mixture at constant volume? [Take, $R = 8.3$ J/mol-K] *[2019, 12 April Shift-I]*
(a) 19.7 J/mol-K (b) 15.7 J/mol-K
(c) 17.4 J/mol-K (d) 21.6 J/mol-K

74 A vertical closed cylinder is separated into two parts by a frictionless piston of mass m and of negligible thickness. The piston is free to move along the length of the cylinder. The length of the cylinder above the piston is l_1 and that below the piston is l_2, such that $l_1 > l_2$. Each part of the cylinder contains n moles of an ideal gas at equal temperature T. If the piston is stationary, its mass m, will be given by (where, R is universal gas constant and g is the acceleration due to gravity) *[2019, 12 Jan Shift-II]*
(a) $\frac{nRT}{g}\left[\frac{l_1 - l_2}{l_1 l_2}\right]$ (b) $\frac{nRT}{g}\left[\frac{1}{l_2} + \frac{1}{l_1}\right]$
(c) $\frac{RT}{g}\left[\frac{2l_1 + l_2}{l_1 l_2}\right]$ (d) $\frac{RT}{ng}\left[\frac{l_1 - 3l_2}{l_1 l_2}\right]$

75 *A* 15 g mass of nitrogen gas is enclosed in a vessel at a temperature 27°C. Amount of heat transferred to the gas, so that rms velocity of molecules is doubled is about (Take, R = 8.3 J / K - mole) *[2019, 9 Jan Shift-II]*
(a) 10 kJ (b) 0.9 kJ
(c) 14 kJ (d) 6 kJ

76 A gaseous mixture consists of 16 g of helium and 16 g of oxygen. The ratio C_p / C_V of the mixture is **[AIEEE 2005]**

(a) 1.59 (b) 1.62 (c) 1.4 (d) 1.54

77 One mole of ideal monoatomic gas ($\gamma = 5/3$) is mixed with one mole of diatomic gas ($\gamma = 7/5$). What is γ for the mixture ? γ denotes the ratio of specific heat at constant pressure, to that at constant volume. **[AIEEE 2004]**

(a) $\frac{3}{2}$ (b) $\frac{23}{15}$ (c) $\frac{35}{23}$ (d) $\frac{4}{3}$

Numerical Value Type Questions

78 A container of fixed volume contains a gas at 27°C. To double the pressure of the gas, the temperature of gas should be raised to .... °C. **[2025, 29 Jan Shift-I]**

79 A monoatomic gas performs a work of $Q/4$, where Q is the heat supplied to it. The molar heat capacity of the gas will be ...... R during this transformation. (where, R is the gas constant.) **[2022, 24 June Shift-II]**

80 The change in the magnitude of the volume of an ideal gas when a small additional pressure Δp is applied at a constant temperature, is the same as the change when the temperature is reduced by a small quantity ΔT at constant pressure. The initial temperature and pressure of the gas are 300 K and 2 atm, respectively. If $|\Delta T| = C|\Delta p|$, then value of C (in K/atm) is .... **[2020, 4 Sep Shift-II]**

81 Nitrogen gas is at 300°C temperature. The temperature (in K) at which the rms speed of a H_2 molecule would be equal to the rms speed of a N_2 molecule is ....

(Molar mass of N_2 gas is 28 g.) **[2020, 5 Sep Shift-II]**

82 Initially a gas of diatomic molecules is contained in a cylinder of volume V_1 at a pressure p_1 and temperature 250 K. Assuming that 25% of the molecules get dissociated causing a change in number of moles. The pressure of the resulting gas at temperature 2000 K, when contained in a volume $2V_1$ is given by p_2. The ratio p_2/p_1 is .... **[2020, 6 Sep Shift-I]**

Topic 2 Degree of Freedom and Law of Equipartition of Energy

Level 1

01 The mean free path and the average speed of oxygen molecules at 300 K and 1 atm are 3×10^{-7} m and 600 m/s, respectively. Find the frequency of its collisions. **[2025, 4 April Shift-I]**

(a) 5×10^{8}/s (b) 9×10^{5}/s
(c) 2×10^{10}/s (d) 2×10^{9}/s

02 The helium and argon are put in the flask at the same room temperature (300 K). The ratio of average kinetic energies (per molecule) of helium and argon is (Give: Molar mass of helium = 4 g/mol, Molar mass of argon = 40 g/mol) **[2025, 7 April Shift-II]**

(a) 10 : 1 (b) 1 : 10 (c) 1 : 1 (d) $1 : \sqrt{10}$

03 Energy of 10 non-rigid diatomic molecules at temperature T is **[2024, 6 April Shift-II]**

(a) $70\, k_B T$ (b) $\frac{7}{2} RT$ (c) $35 k_B T$ (d) $35 RT$

Imp. Degree of Freedom

04 The translational degree of freedom (f_t) and rotational degree of freedom (f_r) of CH_4 molecule are **[2024, 4 April Shift-II]**

(a) $f_t = 3$ and $f_r = 2$ (b) $f_t = 2$ and $f_r = 3$
(c) $f_t = 3$ and $f_r = 3$ (d) $f_t = 2$ and $f_r = 2$

05 A gas mixture consists of 8 moles of argon and 6 moles of oxygen at temperature T. Neglecting all vibrational modes, the total internal energy of the system is **[2024, 31 Jan Shift-II]**

(a) 21 RT (b) 29 RT (c) 27 RT (d) 20 RT

06 The average kinetic energy of a monoatomic molecule is 0.414 eV at temperature.

(Use $k_B = 1.38 \times 10^{-23}$ J/mol-K) **[2024, 27 Jan Shift-I]**

(a) 1600 K (b) 3000 K (c) 1500 K (d) 3200 K

07 Three vessels of equal volume contain gases at the same temperature and pressure. The first vessel contains neon (monoatomic), the second contains chlorine (diatomic) and third contains uranium hexafloride (polyatomic). Arrange these on the basis of their root mean square speed (v_{rms}) and choose the correct answer from the options given below **[2023, 11 April Shift-I]**

(a) v_{rms}(mono) = v_{rms} (dia) = v_{rms} (poly)
(b) v_{rms}(mono) < v_{rms} (dia) < v_{rms} (poly)
(c) v_{rms}(dia) < v_{rms} (poly) < v_{rms} (mono)
(d) v_{rms}(mono) > v_{rms} (dia) > v_{rms} (poly)

08 A gas mixture consists of 2 moles of oxygen and 4 moles of neon at temperature T. Neglecting all vibrational modes, the total internal energy of the system will be. **[2023, 10 April Shift-II]**

(a) $4RT$ (b) $11RT$ (c) $8RT$ (d) $16RT$

09 Match List I with List II.

List I	List II
A. 3 Translational degrees of freedom	I. Monoatomic gases
B. 3 Translational, 2 rotational degrees of freedoms	II. Polyatomic gases
C. 3 Translational, 2 rotational and 1 vibrational degrees of freedom	III. Rigid diatomic gases
D. 3 Translational, 2 rotational and more than one vibrational degrees of freedom	IV. Non-rigid diatomic gases

Choose the correct answer from the options given below. [2023, 10 April Shift-I]

(a) (A)-(IV), (B)-(III), (C)-(II), (D)-(I)
(b) (A)-(I), (B)-(III), (C)-(IV), (D)-(II)
(c) (A)-(IV). (B)-(II), (C)-(I), (D)-(III)
(d) (A)-(I), (B)-(IV), (C)-(III), (D)-(II)

10 According to law of equipartition of energy, the molar specific heat of a diatomic gas at constant volume where the molecule has one additional vibrational mode is [2023, 25 Jan Shift-II]

(a) $\frac{9}{2}R$ (b) $\frac{7}{2}R$ (c) $\frac{5}{2}R$ (d) $\frac{3}{2}R$

11 Let γ_1 be the ratio of molar specific heat at constant pressure and molar specific heat at constant volume of a monoatomic gas and γ_2 be the similar ratio of diatomic gas. Considering the diatomic gas molecules as a rigid rotator, the ratio, $\frac{\gamma_1}{\gamma_2}$ is [2023, 24 Jan Shift-II]

(a) 27/35 (b) 35/27 (c) 21/25 (d) 25/21

12 Which statements are correct about degrees of freedom?

A. A molecule with n degrees of freedom has n^2 different ways of storing energy.
B. Each degree of freedom is associated with $1/2RT$ average energy per mole.
C. A monoatomic gas molecule has 1 rotational degree of freedom whereas diatomic molecule has 2 rotational degrees of freedom.
D. CH_4 has a total of 6 degrees of freedom.

Choose the correct answer from the option given below. [2022, 27 July Shift-II]

(a) B and C only (b) B and D only
(c) A and B only (d) C and D only

13 According to kinetic theory of gases,

A. The motion of the gas molecules freezes at 0°C.
B. The mean free path of gas molecules decreases if the density of molecules is increased.
C. The mean free path of gas molecules increases if temperature is increased keeping pressure constant.
D. Average kinetic energy per molecule per degree of freedom is $\frac{3}{2}k_BT$ (for monoatomic gases).

Choose the most appropriate answer from the options given below. [2022, 27 June Shift-II]

(a) A and C only (b) B and C only
(c) A and B only (d) C and D only

Imp. Law of Equipartition of Energy

14 What will be the average value of energy for a monoatomic gas molecule in thermal equilibrium at temperature T? [2021, 22 July Shift-II]

(a) $\frac{2}{3}k_BT$ (b) k_BT (c) $\frac{3}{2}k_BT$ (d) $\frac{1}{2}k_BT$

15 The correct relation between the degrees of freedom f and the ratio of specific heat γ is [2021, 20 July Shift-II]

(a) $f=\frac{2}{\gamma-1}$ (b) $f=\frac{2}{\gamma+1}$ (c) $f=\frac{\gamma+1}{2}$ (d) $f=\frac{1}{\gamma+1}$

16 What will be the average value of energy along one degree of freedom for an ideal gas in thermal equilibrium at a temperature T? (k_B is Boltzmann constant) [2021, 18 March Shift-I]

(a) $\frac{1}{2}k_BT$ (b) $\frac{2}{3}k_BT$ (c) $\frac{3}{2}k_BT$ (d) k_BT

17 Given below are two statements.

Statement I In a diatomic molecule, the rotational energy at a given temperature obeys Maxwell's distribution.

Statement II In a diatomic molecule, the rotational energy at a given temperature equals the translational kinetic energy for each molecule.

In the light of the above statements, choose the correct answer from the options given below. [2021, 25 Feb Shift-II]

(a) Both Statement I and Statement II are true.
(b) Both Statement I and Statement II are false.
(c) Statement I is true but Statement II is false.
(d) Statement I is false but Statement II is true.

18 Three perfect gases at absolute temperatures T_1, T_2 and T_3 are mixed. The masses of molecules are m_1, m_2 and m_3 and the number of molecules are n_1, n_2 and n_3 respectively. Assuming no loss of energy, the final temperature of the mixture is [AIEEE 2011]

(a) $\frac{n_1T_1+n_2T_2+n_3T_3}{n_1+n_2+n_3}$ (b) $\frac{n_1T_1^2+n_2T_2^2+n_3T_3^2}{n_1T_1+n_2T_2+n_3T_3}$

(c) $\frac{n_1^2T_1^2+n_2^2T_2^2+n_3^2T_3^2}{n_1T_1+n_2T_2+n_3T_3}$ (d) $\frac{(T_1+T_2+T_3)}{3}$

Level 2

19 Match the List-I with List-II [2025, 7 April Shift-I]

	List - I		List -II
A.	Triatomic rigid gas	I.	$\frac{C_p}{C_V}=\frac{5}{3}$
B.	Diatomic non-rigid gas	II.	$\frac{C_p}{C_V}=\frac{7}{5}$
C.	Monoatomic gas	III.	$\frac{C_p}{C_V}=\frac{4}{3}$
D.	Diatomic rigid gas	IV.	$\frac{C_p}{C_V}=\frac{9}{7}$

Choose the correct answer from the options given below:

(a) A-III, B-IV, C-I, D-II (b) A-III, B-II, C-IV, D-I
(c) A-IV, B-II, C-III, D-I (d) A-II, B-IV, C-I, D-III

20 For a diatomic gas, if $\gamma_1 = \left(\frac{C_p}{C_V}\right)$ for rigid molecules and $\gamma_2 = \left(\frac{C_p}{C_V}\right)$ for another diatomic molecules, but also having vibrational modes. Then, which one of the following options is correct? (C_p and C_V are specific heats of the gas at constant pressure and volume) *[2025, 22 Jan Shift-II]*

(a) $\gamma_2 = \gamma_1$ (b) $2\gamma_2 = \gamma_1$ (c) $\gamma_2 < \gamma_1$ (d) $\gamma_2 > \gamma_1$

21 The temperature of a gas is $-78°C$ and the average translational kinetic energy of its molecules is K. The temperature at which the average translational kinetic energy of the molecules of the same gas becomes $2K$ is *[2024, 9 April Shift-II]*

(a) $-39°C$ (b) $127°C$ (c) $117°C$ (d) $-78°C$

22 Given below are two statements

Statement I The mean free path of gas molecules is inversely proportional to square of molecular diameter.

Statement II Average kinetic energy of gas molecules is directly proportional to absolute temperature of gas. *[2024, 8 April Shift-II]*

In the light of the above statements, choose the correct options given below.

(a) Both Statement I and Statement II are true.
(b) Statement I is true but Statement II is false.
(c) Statement I is false but Statement II is true.
(d) Both Statement I and Statement II are false.

23 Two moles a monoatomic gas is mixed with six moles of a diatomic gas. The molar specific heat of the mixture at constant volume is *[2024, 1 Feb Shift-I]*

(a) $\frac{3}{2}R$ (b) $\frac{7}{4}R$ (c) $\frac{9}{4}R$ (d) $\frac{5}{2}R$

24 The total kinetic energy of 1 mole of oxygen at 27°C is [Use universal gas constant $(R) = 8.31$ J/mole K] *[2024, 27 Jan Shift-II]*

(a) 6845.5 J (b) 6232.5 J
(c) 5670.5 J (d) 5942.0 J

25 Heat energy of 735 J is given to a diatomic gas allowing the gas to expand at constant pressure. Each gas molecule rotates around an internal axis but do not oscillate. The increase in the internal energy of the gas will be *[2023, 31 Jan Shift-II]*

(a) 572 J (b) 525 J
(c) 735 J (d) 441 J

26 A gas has n degrees of freedom. The ratio of specific heat of gas at constant volume to the specific heat of gas at constant pressure will be *[2022, 26 July Shift-II]*

(a) $\frac{n}{n+2}$ (b) $\frac{n+2}{n}$
(c) $\frac{n}{2n+2}$ (d) $\frac{n}{n-2}$

27 If one mole of the polyatomic gas is having two vibrational modes and β is the ratio of molar specific heats for polyatomic gas $\left(\beta = \frac{C_p}{C_V}\right)$, then the value of β is *[2021, 17 March Shift-II]*

(a) 1.02 (b) 1.2 (c) 1.25 (d) 1.35

28 A polyatomic ideal gas has 24 vibrational modes. What is the value of γ? *[2021, 17 March Shift-I]*

(a) 1.03 (b) 1.30 (c) 1.37 (d) 10.3

29 The internal energy U, pressure p and volume V of an ideal gas are related as $U = 3pV + 4$. The gas is *[2021, 26 Feb Shift-II]*

(a) diatomic only (b) polyatomic only
(c) either monoatomic (d) monoatomic only

30 Molecules of an ideal gas are known to have three translational degrees of freedom and two rotational degrees of freedom. The gas is maintained at a temperature of T. The total internal energy U of a mole of this gas, and the value of $\gamma (= C_p / C_V)$ are given respectively, by *[2020, 6 Sep Shift-I]*

(a) $U = \frac{5}{2}RT$ and $\gamma = \frac{6}{5}$ (b) $U = 5RT$ and $\gamma = \frac{7}{5}$
(c) $U = \frac{5}{2}RT$ and $\gamma = \frac{7}{5}$ (d) $U = 5RT$ and $\gamma = \frac{6}{5}$

31 Match the C_p/C_V ratio for ideal gases with different type of molecules. *[2020, 4 Sep Shift-I]*

Molecule type		C_p/C_V
(A) Monatomic molecules	I.	7/5
(B) Diatomic rigid molecules	II.	9/7
(C) Diatomic non-rigid molecules	III.	4/3
(D) Triatomic rigid molecules	IV.	5/3

	A	B	C	D		A	B	C	D
(a)	IV	I	II	III	(b)	III	IV	II	I
(c)	II	III	I	IV	(d)	IV	II	I	III

Imp. Degree of Freedom

32 A gas mixture consists of 3 moles of oxygen and 5 moles of argon at temperature T. Assuming the gases to be ideal and the oxygen bond to be rigid, the total internal energy (in units of RT) of the mixture is *[2020, 2 Sep Shift-I]*

(a) 15 (b) 13 (c) 11 (d) 20

33 Consider two ideal diatomic gases A and B at some temperature T. Molecules of the gas A are rigid and have a mass m. Molecules of the gas B have an additional vibrational mode and have a mass $m/4$. The ratio of the specific heats (C_V^A and C_V^B) of gas A and B respectively is *[2020, 9 Jan Shift-I]*

(a) 5 : 9 (b) 7 : 9 (c) 3 : 5 (d) 5 : 7

34 The specific heats, C_p and C_V of a gas of diatomic molecules, A are given (in units of J mol^{-1} K^{-1}) by 29 and 22, respectively. Another gas of diatomic molecules B, has the corresponding values 30 and 21. If they are treated as ideal gases, then *[2019, 9 April Shift-II]*

(a) A has a vibrational mode but B has none
(b) Both A and B have a vibrational mode each
(c) A has one vibrational mode and B has two
(d) A is rigid but B has a vibrational mode

35 An HCl molecule has rotational, translational and vibrational motions. If the rms velocity of HCl molecules in its gaseous phase is $\bar{v}$, m is its mass and k_B is Boltzmann constant, then its temperature will be *[2019, 9 April Shift-I]*

(a) $\frac{m\bar{v}^2}{3k_B}$ (b) $\frac{m\bar{v}^2}{7k_B}$ (c) $\frac{m\bar{v}^2}{5k_B}$ (d) $\frac{m\bar{v}^2}{6k_B}$

36 Two moles of an ideal monoatomic gas occupies a volume V at 27° C. The gas expands adiabatically to a volume 2V. Calculate (I) the final temperature of the gas and (II) change in its internal energy. *[JEE Main 2018]*

(a) (I) 189 K (II) 2.7 kJ (b) (I) 195 K (II) −2.7 kJ
(c) (I) 189 K (II) −2.7 kJ (d) (I) 195 K (II) 2.7 kJ

37 An ideal gas has molecules with 5 degrees of freedom. The ratio of specific heats at constant pressure C_p and at constant volume C_V is *[JEE Main 2017]*

(a) 6 (b) 7/2 (c) 5/2 (d) 7/5

38 One kg of a diatomic gas is at a pressure of 8×10^4 Nm^{-2}. The density of the gas is 4 kgm^{-3}. What is the energy of the gas due to its thermal motion? *[AIEEE 2009]*

(a) 3×10^4 J (b) 5×10^4 J (c) 6×10^4 J (d) 7×10^4 J

Level 3

39 If three moles of monoatomic gas ($\gamma = 5/3$) is mixed with two moles of a diatomic gas ($\gamma = 7/5$), the value of adiabatic exponent γ for the mixture is *[2024, 30 Jan Shift-II]*

(a) 1.75 (b) 1.52 (c) 1.40 (d) 1.35

40 N moles of a polyatomic gas ($f = 6$) must be mixed with two moles of a monoatomic gas, so that the mixture behaves as a diatomic gas. The value of N is *[2024, 29 Jan Shift-II]*

(a) 6 (b) 2 (c) 3 (d) 4

41 Two ideal polyatomic gases at temperatures T_1 and T_2 are mixed so that there is no loss of energy. If f_1 and f_2, m_1 and m_2, n_1 and n_2 be the degrees of freedom, masses, number of molecules of the first and second gas respectively, the temperature of mixture of these two gases is *[2021, 17 March Shift-I]*

(a) $\frac{n_1T_1 + n_2T_2}{n_1 + n_2}$ (b) $\frac{n_1 f_1 T_1 + n_2 f_2 T_2}{n_1 f_1 + n_2 f_2}$

(c) $\frac{n_1 f_1 T_1 + n_2 f_2 T_2}{f_1 + f_2}$ (d) $\frac{n_1 f_1 T_1 + n_2 f_2 T_2}{n_1 + n_2}$

Numerical Value Type Questions

42 γ_A is the specific heat ratio of monoatomic gas A having 3 translational degrees of freedom. γ_B is the specific heat ratio of polyatomic gas B having 3 translational, 3 rotational degrees of freedom and 1 vibrational mode. If $\frac{\gamma_A}{\gamma_B} = \left(1 + \frac{1}{n}\right)$, then the value of n is .... *[2025, 2 April Shift-I]*

43 The total internal energy of two mole monoatomic ideal gas at temperature, $T = 300$ K will be .... J. (Given, R = 8.31 J/mol- K) *[2022, 28 June Shift-I]*

44 The temperature of 3.00 mol of an ideal diatomic gas is increased by 40.0 °C without changing the pressure of the gas. The molecules in the gas rotate but do not oscillate. If the ratio of change in internal energy of the gas to the amount of workdone by the gas is $x/10$, then the value of x (round off to the nearest integer) is .... (Given, $R = 8.31$ J mol^{-1} K^{-1}) *[2021, 1 Sep Shift-II]*

45 One mole of rigid diatomic gas performs a work of $Q/5$ when heat Q is supplied to it. The molar heat capacity of the gas during this transformation is $\frac{xR}{8}$. The value of x is .... *[2021, 26 Feb Shift-II]*
[R = universla gas constant]

46 An engine takes in 5 moles of air at 20° C and 1 atm and compresses it adiabatically to 1/10th of the original volume. Assuming air to be a diatomic ideal gas made up of rigid molecules, the change in its internal energy during this process comes out to be XkJ. The value of X to the nearest integer is .... *[2020, 2 Sep Shift-I]*

• Key **Index**

Topic 1

[redacted]

Topic 2

[redacted]

Detailed **Explanations**

Topic 1

01. *(a)* Molecular mass of a gas is directly proportional to its vapour density

$$\Rightarrow \quad \frac{M_1}{M_2} = \frac{\rho_1}{\rho_2} = \frac{4}{25}$$

Now, $v_{\text{rms}} = \sqrt{\frac{3RT}{M}}$

As temperature is constant

$$\Rightarrow \quad \frac{v_{\text{rms}1}}{v_{\text{rms}2}} = \sqrt{\frac{M_2}{M_1}} = \sqrt{\frac{25}{4}} = \frac{5}{2}$$

02. *(b)* We know that, $v_{\text{rms}} = \sqrt{\frac{3RT}{M}}$

$$\therefore \quad \frac{(v_{\text{rms}})_{O_2}}{(v_{\text{rms}})_{\text{He}}} = \sqrt{\frac{M_{\text{He}}}{M_{O_2}}} = \sqrt{\frac{4}{32}} = \frac{1}{2\sqrt{2}}$$

$$\therefore \quad \frac{(v_{\text{rms}})_{\text{He}}}{(v_{\text{rms}})_{O_2}} = \frac{2\sqrt{2}}{1}$$

03. *(a)* $v_{\text{rms}} = \sqrt{\frac{3RT}{M}}$, M = molar mass

Now, for same temperature

$$\frac{v_{\text{rms}}(O_2)}{v_{\text{rms}}(H_2)} = \left(\frac{3RT}{M_{O_2}} \times \frac{M_{H_2}}{3RT}\right)^{\frac{1}{2}}$$

$$\Rightarrow \quad v_{\text{rms}}(O_2) = v_{\text{rms}}(H_2) \times \left(\frac{2}{32}\right)^{\frac{1}{2}}$$

$$= \frac{2}{4} = 0.5 \text{ km/s}$$

04. *(c)* Speed of sound in a gas at temperature T is, $v_s = \sqrt{\frac{\gamma RT}{M}}$

For oxygen molar mass, $M = 32$ g/mol

or $M = 32 \times 10^{-3}$ kg/mol, $\gamma = 1.4$

$$R = 8.3 \text{ JK}^{-1}\text{mol}^{-1} \text{ and } T = 273 \text{ K}$$

So, speed of sound is,

$$v_s = \sqrt{\frac{1.4 \times 8.3 \times 273}{32 \times 10^{-3}}}$$

$$= 314.85 \text{ m/s} \simeq 315 \text{ m/s}$$

Nearest answer is 310 m/s.

05. *(d)* At a given temperature, kinetic energy remains the same for the molecules of all gases.

06. *(d)* The average kinetic energy per molecule of any gas (hydrogen argon, O_2, N_2 etc) is always equal to $\frac{3}{2} k_B T$.

It depends only on temperature and is independent of the nature of the gas.

Since, hydrogen and argon both have the same temperature as 30° in the flask hence, the ratio of average kinetic energy (per molecule) of the two gases will be 1 : 1.

07. *(d)* Using $v_{\text{rms}} = \sqrt{\frac{3RT}{M}}$

$$v_0 = \sqrt{\frac{3R \times 200}{M}} \quad \ldots\text{(i)}$$

$$v' = \sqrt{\frac{3R \times 800}{M}} \quad \ldots\text{(ii)}$$

On dividing (ii) by (i), we get

$$\frac{v'}{v_0} = \sqrt{\frac{800}{200}} = \sqrt{4} = 2$$

$$v' = 2v_0$$

08. *(c)* From ideal gas equation,

$$pV = \mu RT$$

If volume is constant, $p \propto T$

From the given graph, it is clear that all the gases lines of graphs meet at same value or point.

On the horizontal axis (temperature axis), pressure is zero but temperature is negative and it will be equal to 0K or $-273°$C.

09. *(c)* Molar specific heat capacity C_p and C_V, both do not depend on temperature.
γ is ratio of C_p and C_V which is also independent of temperature.

$$\gamma \propto T^0$$

10. *(b)* The root mean square velocity of molecules of gas is given as

$$v_{rms} = \sqrt{\frac{3RT}{M}} \Rightarrow v_{rms} \propto \sqrt{T}$$

11. *(d)* Average momentum of a molecule in a sample of an ideal gas is always zero and it does not depend on temperature. Hence, statement I is false.

We know that, $v_{rms} = \sqrt{\frac{3RT}{M}}$

When temperature of gas is doubled to $2T$ and oxygen atoms are used, then atomic mass becomes $M/2$.

$$\therefore v'_{rms} = \sqrt{\frac{3R(2T)}{M/2}} = 2\sqrt{\frac{3RT}{M}}$$

$$= 2v_{rms} = 2v \; [(\because v_{rms} = v \text{ (given)}]$$

Hence, statement II is correct.

12. *(b)* Given, increase of temperature,

$$\Delta T = 40 \text{ K}$$

Number of moles, $n = 7$

$$R = 8.3 \text{ JK}^{-1} \text{ mol}^{-1}$$

Increase in internal energy of the monoatomic gas, $\Delta U = nC_V \Delta T$
For monoatomic gas,

$$C_V = 3R/2$$

$$\therefore \quad \Delta U = 7 \times \frac{3R}{2} \times 40$$

$$= 420\,R$$
$$= 420 \times 8.3$$
$$= 3486 \text{ J}$$

13. *(a)* According to kinetic theory of gases, KE decreases with decrease in temperature i.e., KE $\propto T$.
Pressure of a gas increases with increase in temperature at constant pressure.
Hence, (A) and (D) are correct statement.

14. *(c)* Number of moles,

$$n = \frac{\text{Gas Volume}}{\text{Molar Volume}} = \frac{44.8}{22.4} = 2$$

Now, amount of heat required,

$$\Delta Q = n\,C_V \Delta T = 2 \times \frac{3}{2} R \times \Delta T$$

$$(\text{He is monoatomic}, \therefore C_V = \frac{3}{2} R)$$

So, $\Delta Q = 2 \times \frac{3}{2} \times 8.3 \times 20 = 498$ J

15. *(b)* RMS speed of a gas molecule,

$$v_{rms} = \sqrt{\frac{3RT}{M_0}}$$

When temperature is doubled, $T' = 2T$ and when molecule is dissociated,

$$M' = \frac{M_0}{2}$$

Hence, $v'_{rms} = \sqrt{\frac{3R(2T)}{\left(\frac{M_0}{2}\right)}} = 2\sqrt{\frac{3RT}{M_0}}$

$$= 2v_{rms}$$

16. *(d)* The average translational kinetic energy per molecule of any gas (either monoatomic like argon or diatomic like O_2) is always equal to $\frac{3}{2} k_B T$, where T is temperature and k_B is Boltzmann constant. It depends only on temperature and is independent of the nature of the gas. Since, argon and oxygen both have the same temperature in the flask at 27°C, hence the ratio of average kinetic energy per molecule of the two gases will be 1 : 1.

17. *(b)* We know that, most probable speed v_p of gas,

$$v_p = \sqrt{\frac{2RT}{M}} \quad \ldots\text{(i)}$$

$\therefore$ Root mean square speed,

$$v_{rms} = \sqrt{\frac{3RT}{M}} = \sqrt{\frac{3}{2}}\sqrt{\frac{2RT}{M}}$$

$$= \sqrt{\frac{3}{2}} \cdot v_p \quad \text{[From Eq. (i)]}$$

18. *(d)* We know that at constant volume,

$$p \propto T$$

$$\text{so,} \quad T_2 = T_1 \times \frac{p_2}{p_1} = 273 \times \frac{180}{100}$$

$$= 273 \times \frac{9}{5} \approx 491 \text{ K}$$

19. *(c)* Given, Mass, $m = 5 \times 10^{-17}$ kg,

$$T = 273 \text{ K}$$

$$k_B = 1.38 \times 10^{-23} \text{ J K}^{-1}$$

Root mean square velocity of smoke particle is given as

$$v_{rms} = \sqrt{\frac{3k_B T}{m}} = \sqrt{\frac{3 \times 1.38 \times 10^{-23} \times 273}{5 \times 10^{-17}}}$$

$$\Rightarrow v_{rms} = 15 \text{ mm s}^{-1}$$

20. *(c)* According to ideal gas equation,

$$pV = nRT \quad \ldots\text{(i)}$$

and T = absolute temperature of gas.
Eq. (i) can be written as $pV = cT$
where, c is any constant.

$$\Rightarrow \quad pV \propto T$$

So, graph in option (c) is correct.

21. *(d)* Rms velocity of gas molecules,

$$v_{rms} = \sqrt{\frac{3RT}{m}} \quad \ldots\text{(i)}$$

According to question,

$$m_A < m_B < m_C$$

From Eq. (i), $v_{rms} \propto \frac{1}{\sqrt{m}}$

$\therefore$ We can write, $v_A > v_B > v_C$

$$\text{or} \quad \frac{1}{v_A} < \frac{1}{v_B} < \frac{1}{v_C}$$

22. *(c)* We know that, $v_{rms} = \sqrt{\frac{3RT}{M}}$

The expression of the average velocity of the gas molecules,

$$v_{av} = \sqrt{\frac{8RT}{\pi M}}$$

$$\frac{v_{rms}}{v_{av}} = \frac{\sqrt{\frac{3RT}{M}}}{\sqrt{\frac{8RT}{\pi M}}} = \sqrt{\frac{3\pi}{8}}$$

23. *(d)* On the basis of kinetic theory of gases, the gas exerts pressure because the molecules of the gas contain uniform speed, random motion and perform elastic collision with each other as well as with the walls of container. As a result of this, gaseous molecules suffer change in momentum when they impinge on the walls of the container.

24. *(a)*

> **key idea** You can find the rms speed of a gas molecule, with the help of following form.
>
> $$v_{rms} = \sqrt{\frac{3RT}{M}}$$

We know that, $v_{rms} \propto \sqrt{T}$

Let unknown rms speed be $v_{rms,2}$

at $\quad T_2 = 227°\text{C} \quad$ (or 500 K)

and at $\quad T_1 = 127°\text{C} \quad$ (or 400 K)

$$v_{rms,1} = 200 \text{ m/s}$$

$\therefore$ Using the relation $v_{rms} \propto \sqrt{T}$, we can write

$$\frac{v_{\text{rms},2}}{v_{\text{rms},1}} = \sqrt{\frac{T_2}{T_1}} \quad \ldots\text{(i)}$$

Substituting these given values in Eq. (i), we get

$$\therefore \; v_{\text{rms},2} = \sqrt{\frac{500}{400}} \times 200 \text{ m/s}$$

$$= \frac{1}{2}\sqrt{5} \times 200 = 100\sqrt{5} \text{ m/s}$$

25. *(c)* For an ideal gas, internal energy,

$$E = \frac{3}{2}nRT$$

$$= \frac{3}{2}p \cdot V \quad (\because \text{ from } pV = nRT)$$

Substituting the given values, we get

$$= \frac{3}{2} \times 3 \times 10^6 \times 2 = 9 \times 10^6 \text{ J}$$

26. *(c)* Root mean square (rms) velocity,

$$v_{\text{rms}} = \sqrt{\frac{3RT}{M}} \Rightarrow v_{\text{rms}} \propto \sqrt{\frac{1}{M}}$$

$$\therefore \; \frac{v_{\text{rms (helium)}}}{v_{\text{rms (argon)}}} = \sqrt{\frac{M_{\text{argon}}}{M_{\text{helium}}}} \quad \ldots\text{(i)}$$

Given, $M_{\text{argon}} = 40$ u

and $M_{\text{helium}} = 4$ u

$$\therefore \; \frac{v_{\text{rms (helium)}}}{v_{\text{rms (argon)}}} = \sqrt{\frac{40}{4}}$$

$$= \sqrt{10} = 3.16$$

27. *(c)* From $pV = nRT = \dfrac{N}{N_A}RT$

We have, $n_f - n_i = \dfrac{pVN_A}{RT_f} - \dfrac{pVN_A}{RT_i}$

$$\Rightarrow n_f - n_i = \frac{10^5 \times 30}{8.3} \times 6.02 \times 10^{23} \cdot \left(\frac{1}{300} - \frac{1}{290}\right)$$

$$= -2.5 \times 10^{25}$$

28. *(c)* Temperature of a gas is determined by the total translational kinetic energy measured with respect to the centre of mass of the gas. Therefore, the motion of centre of mass of the gas does not affect the temperature. Hence, the temperature of gas will remain same.

29. *(c)* We know that, relation between energy of gas molecules with temperature

$$\frac{1}{2}Mv_{\text{rms}}^2 = \frac{3}{2}RT$$

$$\Rightarrow \quad v_{\text{rms}}^2 \propto T$$

30. *(c)* $m = 50$ g at $T = 17°$C

$$\text{KE}_{\text{Translational}} = \left[\frac{3}{2}k_BT\right] \times \text{Number of molecule}$$

$$\text{Number of molecule} = \frac{50N_0}{12+32} = \frac{50N_0}{44}$$

$$\text{KE}_T = \frac{3}{2}k_BT \times \frac{50}{40}N_0$$

$$= \frac{3}{2}RT \times \frac{50}{40}$$

$$= \frac{3}{2} \times 8.3 \times (273+17) \times \frac{50}{44}$$

$$= 4102.84 \text{ J}$$

> ***Alternate Solution***
>
> $$\text{KE} = \frac{1}{2}mv^2 = \frac{1}{2} \times m \times \frac{3RT}{M}$$
>
> $$= \frac{1}{2} \times \frac{50 \times 3 \times 8.3 \times 290}{44}$$
>
> $$= 4102.84 \text{ J}$$

31. *(c)* In isobaric expansion given,

$\gamma = 1.4$ and $\Delta W = 100$ J

In isobaric process,

$$\frac{\Delta Q}{\Delta W} = \frac{nC_p\Delta T}{n\Delta T(C_p - C_V)}$$

$$= \frac{C_p}{C_p - C_V} = \frac{\gamma}{\gamma - 1}$$

So, $\Delta Q = \Delta W \cdot \left(\dfrac{\gamma}{\gamma-1}\right)$

Substituting values, we get

$$\Delta Q = 100 \times \left(\frac{1.4}{1.4-1}\right)$$

$$= \frac{100 \times 1.4}{0.4} = 350 \text{ J}$$

32. *(d)* From first law of thermodynamic,

$$dQ = dU + dW$$

$$\Rightarrow \quad CdT = C_VdT + pdV \quad \ldots\text{(i)}$$

Since, $pV^2 = RT$

$$\therefore \; p \cdot 2V\frac{dV}{dT} = R \quad [\because p \to \text{Constant}]$$

$$pdV = \frac{RdT}{2V} \quad \ldots\text{(ii)}$$

From Eqs. (i) and (ii), we get

$$CdT = C_VdT + \frac{RdT}{2V}$$

$$\Rightarrow \quad C = C_V + \frac{R}{2V}$$

33. *(b)* Relation between frequency (collision) and temperature is,

$$\nu = \sqrt{2}\pi d^2 \frac{N}{V}\sqrt{\frac{3RT}{m}}$$

or $\quad \nu \propto \sqrt{T} \quad \ldots$(i)

Now for $T_1 = 27°\text{C} = 300$ K

and $\quad T_2 = 127°\text{C} = 400$ K

$$\nu_1 \propto \sqrt{300}, \; \nu_2 \propto \sqrt{400}$$

$$\frac{\nu_1}{\nu_2} = \sqrt{\frac{300}{400}} \quad \ldots\text{(ii)}$$

$[\nu_1 = Z, \text{ given}]$

$$\nu_2 = \frac{2}{\sqrt{3}}Z$$

34. *(d)* From ideal gas equation,

$$pV = nRT = \frac{m}{M}RT$$

(m = mass of gas,

M = molecular mass of gas)

$$p = \left(\frac{m}{V}\right)\frac{RT}{M}$$

$$\Rightarrow \quad p = \frac{\rho RT}{M} \quad \ldots\text{(i)}$$

Now plotting the graph,

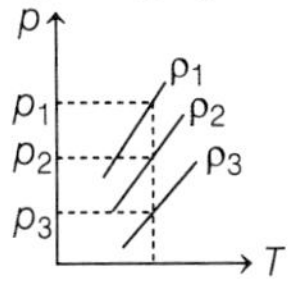

For same temperature,

$$p_1 > p_2 > p_3$$

So, $\quad \rho_1 > \rho_2 > \rho_3$

Hence, $\rho_1 > \rho_2$

35. *(c)* The rms velocity is given by

$v_{\text{rms}} = \sqrt{\dfrac{3RT}{M}}$ where, M is molar mass

According to question,

v_{rms} of H_2 = v_{rms} of O_2

$$\sqrt{\frac{3RT_{H_2}}{M_{H_2}}} = \sqrt{\frac{3RT_{O_2}}{M_{O_2}}}$$

$$\Rightarrow \quad \frac{T_{H_2}}{M_{H_2}} = \frac{T_{O_2}}{M_{O_2}}$$

$$T_{H_2} = \frac{T_{O_2}}{M_{O_2}} \times M_{H_2}$$

$$= \frac{(273+47) \times 2}{32}$$

[given $T_{O_2} = 47°$C]

$$T_{H_2} = 20 \text{ K}$$

36. *(d)* Ideal gas equation,

$$p_AV_A = n_ART_A \quad \ldots\text{(i)}$$

$$p_BV_B = n_BRT_B \quad \ldots\text{(ii)}$$

$$\frac{p_A V_A}{p_B V_B} = \frac{n_A RT_A}{n_B RT_B}$$

As, $V_A = V_B,\ T_A = T_B$

$$\frac{p_A V_B}{p_B V_B} = \frac{n_A RT_A}{n_B RT_A}$$

$$\frac{p_A V_B}{p_B V_B} = \frac{n_A T_A}{n_B T_A} \Rightarrow \frac{p_A}{p_B} = \frac{n_A}{n_B}$$

$$\frac{p_A}{p_B} = \frac{1/2}{1/32} = \frac{1}{2} \times \frac{32}{1}$$

$$\frac{p_A}{p_B} = \frac{16}{1}$$

37. *(a)* We know that,

$$v_{rms} = \sqrt{\frac{3RT}{M}} \quad \text{...(i)}$$

and $v_{avg} = \sqrt{\frac{8RT}{\pi M}} = v \Rightarrow v^2 = \frac{8RT}{\pi M}$

$$\Rightarrow \frac{\pi v^2}{8} = \frac{RT}{M} \quad \text{...(ii)}$$

From Eqs. (i) and (ii),we have

$$v_{rms} = \sqrt{\frac{3\pi v^2}{8}} = \sqrt{\frac{3\pi}{8}} v$$

$$= \sqrt{\frac{3 \times 22}{56}}\, v = \sqrt{\frac{33}{28}}\, v$$

$$= \left(1 + \frac{5}{28}\right)^{\frac{1}{2}} v = \left(1 + \frac{5}{x}\right)^{\frac{1}{2}} v \quad \text{[given]}$$

$\therefore \quad x = 28$

38. *(c)* Given, rms speed of chlorine molecule = 490 m/s

temperature $(T) = 27°C$
$= 273 + 27 = 300$ K

Atomic mass of argon = 39.9 u

Atomic mass of chlorine = 70.9 u

rms speed is given by,

$$v_{rms} = \sqrt{\frac{3RT}{M}} \Rightarrow \frac{v_{Ar}}{v_{Cl}} = \sqrt{\frac{M_{Cl}}{M_{Ar}}}$$

$$\Rightarrow v_{Ar} = 490\sqrt{\frac{70.9}{39.9}}$$

$$= 490 \times 1{\cdot}33$$

$$= 651{\cdot}7 \text{ m/s}$$

39. *(b)* $v_{rms} = \sqrt{\frac{3k_B T}{m}}$

$T = 27°$ C,

$T = (273 + 27) = 300$ K

$$= \sqrt{\frac{3 \times 1.4 \times 10^{-23} \times 300}{4.6 \times 10^{-26}}}$$

$$= \sqrt{\frac{1260 \times 10^{-23}}{4.6 \times 10^{-26}}}$$

$$= \sqrt{273.913 \times 10^3}$$

$$= \sqrt{273913} = 523 \text{ m/s}$$

40. *(b)* Kinetic energy K of a gas molecule at absolute temperature T is

$$K = \frac{3}{2} k_B T$$

where, k_B = Boltzmann constant
T = Absolute temperature

So, $\frac{K_2}{K_1} = \frac{T_2}{T_1}$ or $T_2 = \frac{K_2}{K_1} \times T_1$

Here $T_1 = 27°C = 273 + 27$
$= 300$ K

And $K_2 = 2K$,

$\Rightarrow T_2 = 2 \times T_1$
$= 2 \times 300 = 600$ K
$= (600 - 273)°C$
$= 327°C$

41. *(a)* RMS speed of gas molecules,

$$v_{rms} = \sqrt{\frac{3RT}{M}}$$

Average speed of gas molecules,

$$v_{avg} = \sqrt{\frac{8}{\pi}\frac{RT}{M}}$$

Given, $\frac{v_{rms}}{v_{avg}} = \sqrt{\frac{\alpha + 5}{\alpha}}$ at 300 K

$$\Rightarrow \frac{\sqrt{3\frac{RT}{M}}}{\sqrt{\frac{8}{\pi}\frac{RT}{M}}} = \sqrt{\frac{\alpha + 5}{\alpha}} \Rightarrow \frac{3\pi}{8} = \frac{\alpha + 5}{\alpha}$$

$$\Rightarrow \frac{3}{8} \times \frac{22}{7} \times \alpha = \alpha + 5$$

$$\alpha = 28$$

> **• Concept Enhancer**
>
> Rms speed (v_{rms}) and average speed of a gas molecule is calculated as
>
> $v_{rms} = \sqrt{\frac{3RT}{M}}$ and $v_{avg} = \sqrt{\frac{8RT}{\pi M}}$
>
> where, the symbols have their usual meanings.

42. *(c)* Statement I As, $v_{rms} = \sqrt{\frac{3k_B T}{m}}$

So, at $T = -73°C = 273 + (-73)$
$= 200$ K, we get

$$(v_{rms})_1 = \sqrt{\frac{3k_B(200)}{m}} = 10\sqrt{\frac{6k_B}{m}}$$

and at $T = 527°C = (527 + 273)$ K
$= 800$ K,

we get

$$(v_{rms})_2 = \sqrt{\frac{3k_B(800)}{m}} = 20\sqrt{\frac{6k_B}{m}}$$

So, Statement I is correct, as $(v_{rms})_2 = 2(v_{rms})_1$

Statement II

Translational kinetic energy $= \frac{3}{2} nRT$

As, $pV = nRT$, so, translational KE

$$= \frac{3}{2} pV$$

So, Statement II is false.

43. *(c)* We know that, rms velocity of gas molecule

$$v_{rms} = \sqrt{\frac{3RT}{M}}$$

Here, T and M are same and R is gas constant, hence the rms velocity of gas molecules in two vessels will be the same. So, Statement A is correct and Statement D is incorrect.

Again, using ideal gas equation,

$$pV = \mu RT$$

$$\Rightarrow \frac{p_1 V_1}{p_2 V_2} = \frac{\mu_1 T_1}{\mu_2 T_2}$$

For, $V_1 = V_2$ & $T_1 = T_2$

$$\Rightarrow \frac{p_1}{p_2} = \frac{\mu_1}{\mu_2} = \frac{1}{4}$$

Statement B is correct and statement C is incorrect.

44. *(c)* Number of moles of hydrogen present,

$$n_1 = \frac{m_1}{M_1} = \frac{16}{2} = 8 \text{ moles}$$

Number of moles of oxygen present,

$$n_2 = \frac{m_2}{M_2} = \frac{128}{32} = 4 \text{ moles}$$

Temperature of vessel = 27°C = 300 K

Pressure of vessel = 1atm
$= 1.01 \times 10^5$ Pa

Now using, $pV = nRT$, $(\because n = n_1 + n_2)$ we have,

$$V = \frac{(n_1 + n_2)RT}{p} = \frac{12 \times \frac{25}{3} \times 300}{1.01 \times 10^5}$$

$$= 29 \times 10^4 \text{ cm}^3$$

$$\approx 27 \times 10^4 \text{ cm}^3$$

45. *(b)* Let n_H and n_O be number of moles of hydrogen and oxygen.

Given, volume = 2000 cm^3

$= 2000 \times 10^{-6} m^3$

By Ideal gas equation,

$$pV = nRT$$

$$100 \times 10^3 \times 2000 \times 10^{-6} = n \times 8.3 \times 300$$

$$n = 0.08$$

$$n_H + n_O = 0.08 \quad ...(i)$$

Both hydrogen and oxygen are diatomic molecules. Therefore,

$$2n_H + 32n_O = 0.76 \quad ...(ii)$$

Solving Eqs. (i) and (ii), we get

$$n_H = 0.06$$

$$\Rightarrow \quad n_O = 0.02$$

$$\therefore \quad \frac{n_H}{n_O} = \frac{3}{1}$$

46. *(a)* According to ideal gas equation,

$$pV = nRT \Rightarrow V = \frac{nR}{p} T$$

$$\Rightarrow \quad V = mT, \text{ where } m = \text{slope of}$$

$$V\text{-}T_{curve} = \frac{nR}{p}$$

Now, as $m_2 > m_1$.

Therefore,

$$\frac{nR}{p_2} > \frac{nR}{p_1} \Rightarrow p_2 < p_1$$

47. *(a)* Given, rms speed of oxygen

$(v_{rms})_{O_2} = 160$ m/s

Let rms speed of hydrogen $= (v_{rms})_{H_2}$

$$\text{Since,} \quad v_{rms} = \sqrt{\frac{3RT}{M}}$$

$$\therefore \quad \frac{(v_{rms})_{O_2}}{(v_{rms})_{H_2}} = \sqrt{\frac{M_{H_2}}{M_{O_2}}}$$

$$\Rightarrow \quad \frac{160}{(v_{rms})_{H_2}} = \sqrt{\frac{1}{16}}$$

$$\Rightarrow \quad \frac{160}{(v_{rms})_{H_2}} = \frac{1}{4}$$

$$\Rightarrow \quad (v_{rms})_{H_2} = 160 \times 4$$

$$= 640 \text{ ms}^{-1}$$

48. *(c)* Given, volume of container,

$$V = 4.0 \times 10^{-3} \text{ m}^3$$

Gas constant, $R = 8.3$ J mol^{-1} K^{-1}

Number of moles of hydrogen,

$$n_1 = 1$$

Number of moles of carbon dioxide,

$$n_2 = 2$$

Temperature of mixture, $T = 400$ K

Using Dalton's law,

$$p = p_1 + p_2 = \frac{n_1 RT}{V} + \frac{n_2 RT}{V}$$

$$= \frac{RT}{V}(n_1 + n_2) = \frac{RT}{V}(1+2) = \frac{3RT}{V}$$

$$= \frac{3 \times 8.3 \times 400}{4.0 \times 10^{-3}} = 24.9 \times 10^5 \text{ Pa}$$

49. *(a)* The rms speed of the molecule of gas is given by

$$v_{rms} = \sqrt{\frac{3RT}{M}}$$

According to the, question T is constant and R is also a constant. So, we get

$$v_{rms} \propto \frac{1}{\sqrt{M}}$$

$$M_H = 1.00784 \text{ u} \approx 1 \text{ u}$$

$$M_O = 15.999 \text{ u} \approx 16 \text{ u}$$

$$M_{CO_2} = 44 \text{ u}$$

$$(v_{rms})_H > (v_{rms})_O > (v_{rms})_{CO_2}$$

$$\text{or} \quad v_H > v_O > v_{CO_2}$$

50. *(c)* Given, pressure of ideal gas,

$$p = 2 \text{ atm}$$

$$= 2 \times 1 \times 10^5 \text{ N/m}^2$$

$$= 2 \times 10^5 \text{ N/m}^2$$

Volume of gas, $V = 1 \text{ L} = 10^{-3} \text{ m}^3$

We know that,

$$KE = \frac{3}{2} k_B T = \frac{3}{2} \cdot \frac{RT}{N_A} \quad ...(i)$$

As per ideal gas equation,

$$pV = nRT \Rightarrow pV = \frac{N}{N_A} RT$$

$$\Rightarrow \quad pV = \frac{NRT}{N_A} \quad ...(ii)$$

From Eqs. (i) and (ii), we have

$$\therefore \quad N = \frac{pV}{\left(\frac{2}{3}\right) KE}$$

$$\Rightarrow \quad N = \frac{2 \times 10^5 \times 10^{-3}}{\frac{2}{3} \times 2 \times 10^{-9}} = 1.5 \times 10^{11}$$

51. *(d)* Given, $C_p - C_V = R$ for state P and $C_p - C_V = 1.1 R$ for state Q.

As we know that,

Ideal gas obeys the Mayer's equation, i.e. $\quad C_p - C_V = R$

$\therefore$ P will be the ideal gas, but Q will not be the ideal gas.

Also, a gas behave like an ideal gas at high temperature and low pressure.

$$\therefore \quad T_P > T_Q$$

52. *(c)* From ideal gas equation,

$$pV = (n_1 + n_2 + n_3)RT$$

where, n_1 = number of moles of oxygen

$$= \frac{16}{32} = \frac{1}{2}$$

n_2 = number of moles of nitrogen

$$= \frac{28}{28} = 1$$

n_3 = number of moles of carbon dioxide

$$= \frac{44}{44} = 1$$

$$\Rightarrow \quad pV = \left[\frac{1}{2} + 1 + 1\right] RT = \frac{5}{2} RT$$

$$\Rightarrow \quad p = \frac{5}{2} \frac{RT}{V}$$

53. *(b)* Averge time between to successive collision is

$$\tau = \frac{1}{n\pi v_{avg} \cdot d^2}$$

Here, v_{avg} = average speed $= \sqrt{\frac{8RT}{\pi m}}$

$$v_{avg} = \sqrt{T}$$

$$\therefore \quad \tau = \frac{1}{\sqrt{T}}$$

54. *(c)* According to gas equation,

$$pV = Nk_B T \Rightarrow N = \frac{pV}{k_B T}$$

For monoatomic gas,

$$\text{Energy, } U = \frac{3}{2} k_B T \Rightarrow k_B T = \frac{2}{3} U$$

$$\text{Now,} \quad N = \frac{3pV}{2U}$$

$$\text{So,} \quad N = \frac{3g\rho h V}{2U}$$

$(\because$ Pressure, $p = \rho g h)$

$$\therefore \quad N = \frac{3 \times 980 \times 13.6 \times 2 \times 4}{2 \times 4 \times 10^{-14}}$$

$$= 3.99 \times 10^{18} \approx 4 \times 10^{18}$$

55. *(a)* Mean free path,

$$\lambda = \frac{V}{\sqrt{2}\pi d^2 N}$$

Given that N and V both constants container is closed.

So, in given condition (closed container), mean free path remains constant or unchanged.

$\therefore$ Statement A is incorrect, but C is correct. Also, mean time between two collisions is

Average time, $\tau \propto \frac{1}{\sqrt{T}}$

As, the temperature is increasing, so mean collision time (τ) decreases.

$\therefore$ Statement B is correct.

56. *(c)* For a mixture of two gases,

$$\gamma_{\text{mixture}} = \frac{C_{p\text{mixture}}}{C_{V\text{mixture}}}$$

$$= \frac{n_1 C_{p_1} + n_2 C_{p_2}}{n_1 C_{V_1} + n_2 C_{V_2}} \quad \text{....(i)}$$

Here, gas 1 is helium which is monoatomic, so $n_1 = n$,

$$C_{p_1} = \frac{5}{2}R,\ C_{V_1} = \frac{3}{2}R$$

and gas 2 is oxygen which is diatomic, so $n_2 = 2n$,

$$C_{p_2} = \frac{7}{2}R,\ C_{V_2} = \frac{5}{2}R$$

Hence, from Eq. (i), we get

$$\gamma_{\text{mix}} = \frac{n \times \frac{5R}{2} + 2n \times \frac{7R}{2}}{n \times \frac{3R}{2} + 2n \times \frac{5R}{2}} = \frac{19}{13}$$

57. *(b)* Work done by a gas is

$W = p\,\Delta V = nR\Delta T = 10$ J (given) ...(i)

As the gas is diatomic,

C_p = specific heat of gas at constant pressure $= \frac{7}{2}R$

So, heat absorbed by gas at constant pressure is

$$\Delta Q = nC_p\,\Delta T = n\frac{7}{2}R\,\Delta T$$

$$= \frac{7}{2}(nR\,\Delta T) = \frac{7}{2} \times 10 \quad \text{[From Eq. (i)]}$$

$$= 35\text{ J}$$

58. *(d)* Given, heat supplied at constant volume is ΔQ.

$$\therefore \quad \Delta Q = nC_V\Delta T$$

For same change in temperature, if heat supplied at constant pressure is $\Delta Q'$, then

$$\Delta Q' = nC_p\Delta T$$

So, we have

$$\frac{\Delta Q'}{\Delta Q} = \frac{nC_p\Delta T}{nC_V\Delta T} \Rightarrow \frac{\Delta Q'}{\Delta Q} = \frac{C_p}{C_V}$$

$$\Rightarrow \quad \Delta Q' = \Delta Q\gamma \quad \left[\because \gamma = \frac{C_p}{C_V}\right]$$

As given gas is diatomic, so $\gamma = 7/5$

$$\therefore \quad \Delta Q' = \frac{7}{5}\Delta Q$$

59. *(a)* We know that, rms velocity,

$$v_{\text{rms}} = \sqrt{\frac{3k_BT}{m}}$$

Escape velocity, $v_e = \sqrt{2gR_e}$

Given, $v_{\text{rms}} = v_e$

or $\sqrt{\frac{3k_BT}{m}} = \sqrt{2gR_e} \Rightarrow T = \frac{2gR_e m}{3 \times k_B}$

Substituting the given values, we get

$$T = \frac{2 \times 10 \times 6.4 \times 10^6 \times 2}{3 \times 1.38 \times 10^{-23} \times 6.02 \times 10^{26}}$$

$$\approx 10^4\text{ K}$$

> **Alternate Solution**
>
> At rms speed, average thermal kinetic energy of a hydrogen gas molecule is $= \frac{3}{2}k_BT$
>
> and if v_e = escape velocity of the gas molecule from the earth, then its kinetic energy is
>
> $$\text{KE} = \frac{1}{2}mv_e^2$$
>
> Equating the above thermal and kinetic energies, we have
>
> $$\frac{3}{2}k_BT = \frac{1}{2}mv_e^2 \text{ or } T = \frac{mv_e^2}{3k_B} \quad \text{...(i)}$$
>
> Here, $k_B = 1.38 \times 10^{-23}$ JK^{-1},
>
> $v_e = 11.2 \times 10^3$ ms^{-1},
>
> $$m = \frac{2}{6.02 \times 10^{26}} = 0.332 \times 10^{-26}\text{ kg}$$
>
> Substituting these value in Eq. (i), we get
>
> $$T = \frac{0.332 \times 10^{-26} \times (11.2 \times 10^3)^2}{3 \times 1.38 \times 10^{-23}} \approx 10^4\text{ K}$$

60. *(*)* Momentum imparted to the surface in one collision,

$$\Delta p = (p_i - p_f) = mv - (-mv) = 2mv \quad \text{...(i)}$$

Force on the surface due to N collision per second, $F = \frac{N}{t}(\Delta p) = n\Delta p$

$$\left(\because \frac{N}{t} = n\right)$$

$$= 2\,mnv \quad \text{[from Eq. (i)]}$$

So, pressure on the surface,

$$p = \frac{F}{A} = \frac{2mnv}{A}$$

Here, $m = 10^{-26}$ kg, $n = 10^{22}$ s^{-1}, $v = 10^4$ ms^{-1}, $A = 1$ m^2

$$\therefore \text{ Pressure, } p = \frac{2 \times 10^{-26} \times 10^{22} \times 10^4}{1}$$

$$= 2\text{ N/m}^2$$

So, pressure exerted is of order of 10^0.

61. *(a)* Work done by gas during heat process at constant pressure is given by

$$\Delta W = p\Delta V$$

Using ideal gas equation,

$$pV = nRT$$

$$\Rightarrow \quad p\Delta V = nR\Delta T$$

So, $\quad \Delta W = nR\Delta T \quad \text{... (i)}$

Now, it is given that, $n = \frac{1}{2}$ and

$$\Delta T = 90°\text{C} - 20°\text{C}$$

$$= 363\text{ K} - 293\text{ K} = 70\text{ K}$$

Substituting these values in Eq. (i), we get

$$\Delta W = \frac{1}{2} \times 8.31 \times 70$$

$$= 290.85\text{ J}$$

$$\Delta W \simeq 291\text{ J}$$

62. *(c)* Volume of gas $= \frac{\text{Mass}}{\text{Density}}$

$$\Rightarrow \quad V = \frac{2}{8} = \frac{1}{4}\text{ m}^3 \quad \text{...(i)}$$

Using ideal gas equation,

$$pV = nRT$$

$$\Rightarrow \quad nRT = 4 \times 10^4 \times \frac{1}{4} = 10^4 \quad \text{...(ii)}$$

Internal energy of n moles of a monoatomic gas is given by,

$$U = \frac{3}{2}nRT$$

$$\Rightarrow \quad U = \frac{3}{2} \times 10^4\text{ J}$$

$$= 1.5 \times 10^4\text{ J}$$

i.e. in order of 10^4 J.

63. *(a)*

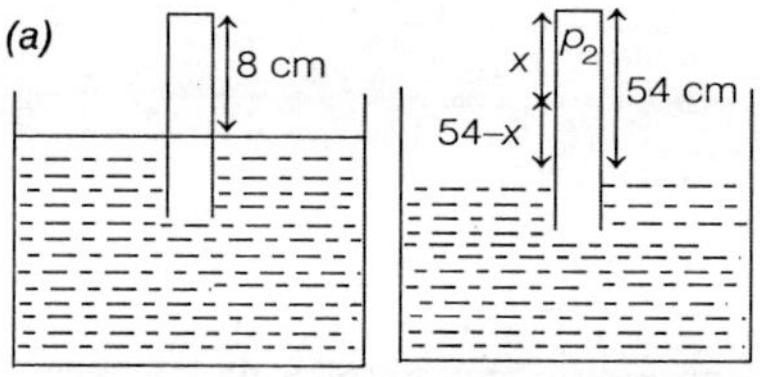

For air trapped in tube, $p_1V_1 = p_2V_2$

$$p_1 = p_{\text{atm}} = \rho g 76$$

$$V_1 = A \times 8$$

[$\because A$ = area of cross-section]

$$p_2 = p_{\text{atm}} - \rho g(54 - x) = \rho g(22 + x)$$

$$V_2 = A \cdot x$$

At constant temprature,

$$p_1V_1 = p_2V_2$$

$$\rho g 76 \cdot A8 = \rho g(22 + x)Ax$$

$$x^2 + 22x - 76 \times 8 = 0$$

$$\Rightarrow \quad x = 16\text{ cm}$$

64. *(c)* Heat flow from box B to A till both reaches the same temperature T_f,

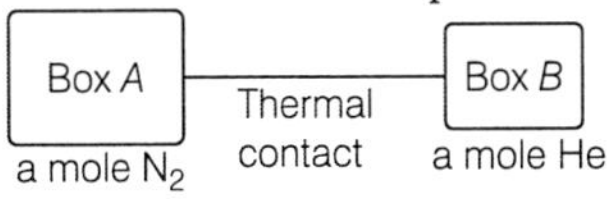

Change in internal energy in box A,

$$\Delta U_A = 1 \times \frac{5}{2} R (T_f - T_0)$$

Change in internal energy in box B

$$\Delta U_B = 1 \times \frac{3R}{2}\left(\frac{7T_0}{3} - T_f\right)$$

Now, $\Delta Q_A = \Delta Q_B \Rightarrow \Delta U_A = \Delta U_B$

As $\Delta W_A = \Delta W_B = 0$

$$\frac{5}{2}(T_f - T_0) = \frac{3}{2}\left(\frac{7T_0}{3} - T_f\right)$$

$$8T_f = 12T_0$$

$$T_f = \frac{3}{2} T_0$$

65. *(b)* Given,

Temperature, $T = 27°C = 300$ K

We know that $(C_V)_{\text{monoatomic}} = \frac{3}{2} R$

and $(C_V)_{\text{diatomic}} = \frac{5}{2} R$

$$\therefore \frac{(C_V)_{\text{monoatomic}}}{(C_V)_{\text{diatomic}}} = \frac{3}{5}$$

66. *(b)* Given for a process,

$$pT^2 = \text{constant} \quad \ldots\text{(i)}$$

From $pV = nRT$, $\Rightarrow p = \frac{nRT}{V}$

Substituting value of p in Eq (i), we get

$$\frac{nRT}{V} \cdot T^2 = \text{constant } (K)$$

$$\Rightarrow \frac{T^3}{V} = \frac{K}{nR} = \text{a constant}$$

$$\text{or} \quad T^3 = \frac{K}{nR} \cdot V \quad \ldots\text{(ii)}$$

Differentiating with respect to temperature, we get

$$3T^2 = \frac{K}{nR} \cdot \left(\frac{dV}{dT}\right)$$

$$\text{So, } \frac{1}{V} \cdot \frac{dV}{dT} = \frac{3T^2 \cdot nR}{V \cdot K} = \frac{3T^2 \cdot nR}{V \cdot \frac{nRT^3}{V}}$$

$$\Rightarrow \frac{1}{V} \cdot \left(\frac{dV}{dT}\right) = \frac{3}{T}$$

Trick : For an ideal gas expanding such that pT^n = constant. The coefficient of volume expansion is given as

$$\frac{1}{V}\frac{dV}{dT} = \frac{n+1}{T}$$

67. *(c)* Given, mass of N_2 gas, $m = 14$ g

Initial temperature,

$$T_1 = 27°C$$
$$= 27 + 273 = 300 \text{ K}$$
$$R = 8.32 \text{ J mol}^{-1} \text{ K}^{-1}$$

The rms speed of a gas is given by

$$v_{\text{rms}} = \sqrt{\frac{3k_B T}{m}}$$

$$\Rightarrow v_{\text{rms}} \propto \sqrt{T}$$

$$\frac{(v_{\text{rms}})_2}{(v_{\text{rms}})_1} = \sqrt{\frac{T_2}{T_1}}$$

Given, $(v_{\text{rms}})_2 = 2(v_{\text{rms}})_1 = 2v_{\text{rms}}$

$$\Rightarrow \frac{2v_{\text{rms}}}{v_{\text{rms}}} = \sqrt{\frac{T_2}{300}}$$

$$\Rightarrow T_2 = (2)^2 \times 300 = 1200 \text{ K}$$

As, heat transferred,

$$H = nC_V, \Delta T = n \times \frac{5}{2} R\Delta T$$

Also, $n = \frac{\text{mass } (m)}{\text{molar mass } (M)} = \frac{14}{28} = 0.5$

$$\therefore H = \frac{5}{2} \times 0.5 \times 8.32 \times 900 = 9360 \text{ J}$$

68. *(b)* Number of moles of He, $n_1 = 2$

Number of moles in H_2, $n_2 = n$

For diatomic gas of H_2,

$$C_{P_2} = \frac{7}{2} R, C_{V_2} = \frac{5}{2} R$$

For monoatomic gas of He,

$$C_{P_1} = \frac{5}{2} R, C_{V_1} = \frac{3}{2} R$$

Specific heat ratio,

$$\gamma = \frac{n_1 C_{P_1} + n_2 C_{P_2}}{n_1 C_{V_1} + n_2 C_{V_2}} = \frac{2 \times \frac{5}{2} R + n \times \frac{7}{2} R}{2 \times \frac{3}{2} R + n \times \frac{5}{2} R}$$

$$\gamma = \frac{10 + 7n}{6 + 5n} \quad \ldots\text{(i)}$$

It is given that $v_{\text{rms}} = \sqrt{2} \times v_{\text{sound}}$

$$\Rightarrow \sqrt{\frac{3RT}{M}} = \sqrt{2} \times \sqrt{\frac{\gamma RT}{M}}$$

$$\Rightarrow \sqrt{3} = \sqrt{2} \times \sqrt{\gamma}$$

$$\Rightarrow 3 = 2\gamma$$

$$\Rightarrow \gamma = 3/2$$

From Eq. (i), $\frac{3}{2} = \frac{10 + 7n}{6 + 5n}$

$$\Rightarrow 18 + 15n = 20 + 14n$$

$$\Rightarrow n = 2$$

69. *(c)* Given,

Volume, $V = 500 \text{ cm}^3$

$$= 5 \times 10^2 \times 10^{-6} \text{ m}^3$$
$$= 5 \times 10^{-4} \text{ m}^3$$

Pressure, $p = 400 \text{ kPa} = 4 \times 10^5$ Pa

Temperature, $T = 300$ K

Total mass, $m = 0.76$ g

We know that,

Molar mass of hydrogen, $M_1 = 2$ g

Molar mass of oxygen, $M_2 = 32$ g

Let n_1 and n_2 be the number of moles of hydrogen and oxygen, respectively.

Total number of moles of mixture,

$$n = n_1 + n_2$$

By using ideal gas equation,

$$pV = nRT$$

$$4 \times 10^5 \times 5 \times 10^{-4} = n \times 8.314 \times 300$$

$$n = 0.08 \text{ mol}$$

Then, $n_1 + n_2 = 0.08 \quad \ldots\text{(i)}$

Let mass of hydrogen in mixture $= m_1$ and mass of oxygen in oxygen $= m_2$

$$n_1 M_1 + n_2 M_2 = m$$

$$\left[\because n = \frac{m}{M} \Rightarrow m = nM\right]$$

$$2n_1 + 32n_2 = 0.76$$

$$n_1 + 16n_2 = 0.38 \quad \ldots\text{(ii)}$$

On solving Eqs. (i) and (ii), we get

$$n_1 = 0.06 \text{ and } n_2 = 0.02$$

$\therefore$ Ratio of masses,

$$\frac{m_2}{m_1} = \frac{n_2 M_2}{n_1 M_1} = \frac{0.02 \times 32}{0.06 \times 2} = \frac{16}{3}$$

i.e. $m_2 : m_1 = 16 : 3$

70. *(a)* Given, change in pressure with volume $\frac{dp}{dV} = -ap$

At $V = 0, p = p_0$

Number of moles, $n = 1$

Gas constant, $R = 8.314 \text{ J mol}^{-1}\text{K}^{-1}$

From given equation,

$$\frac{dp}{p} = -a dV$$

$$\Rightarrow \int_{p_0}^{p} \frac{dp}{p} = -a \int_0^V dV$$

$$\Rightarrow [\ln p]_{p_0}^{p} = -a[V]_0^V$$

$$\Rightarrow \ln(p/p_0) = -a(V - 0)$$

Taking anti log on both side,

$$\Rightarrow p/p_0 = e^{-aV}$$

$$\Rightarrow p = p_0 e^{-aV}$$

By using ideal gas law,

$$pV = nRT$$

$\Rightarrow \quad (p_0 e^{-aV})V = nRT$

$\Rightarrow \dfrac{1}{nR}(p_0 e^{-aV})V = T$

$\Rightarrow \dfrac{1}{R}(p_0\, e^{-aV})V = T \quad [\because n=1] \ldots$(i)

For maximum temperature ($T_{\max}$), differentiating both side with respect to V, we get

$$\Rightarrow \frac{1}{R}[p_0 \cdot e^{-aV} \cdot (-a)V + p_0 e^{-aV}] = \frac{dT}{dV} = 0$$

$\Rightarrow p_0 e^{-aV}(1 - aV) = 0$

$\Rightarrow \quad 1 = aV \Rightarrow V = 1/a$

Substituting the value in Eq. (i), we get

$$T = \frac{1}{R}(p_0 e^{-a \cdot 1/a})1/a$$

$$= \frac{p_0 e^{-1}}{R}(1/a) = \frac{p_0}{eRa}$$

71. *(c)* Given, pT^3 = constant ...(i)

From ideal gas equation, we have

$$pV = nRT \Rightarrow p = \frac{nRT}{V}$$

From Eq. (i), we get

$\therefore \quad \left(\dfrac{nRT}{V}\right)T^3$ = constant

$\Rightarrow \quad nRT^4 V^{-1}$ = constant

Differentiating above expression, we get

$$nR\left[4T^3V^{-1}dT - \frac{T^4}{V^2}dV\right] = 0$$

$\Rightarrow \qquad 4dT = \dfrac{T}{V}dV$ ...(ii)

We know that, $dV = \gamma V dT$

$\Rightarrow \qquad \gamma = \dfrac{dV}{VdT}$ ...(iii)

From Eq. (ii), $\quad \dfrac{4}{T} = \dfrac{dV}{VdT}$ ...(iv)

On comparing Eqs. (iii) and (iv),

we get $\qquad \gamma = \dfrac{4}{T}$

72. *(d)* Given, initial load,

$M_1 = 185$ kg

Initial pressure, $p_1 = 76$ cm of Hg

Initial temperature,

$T_1 = 27$ °C

$= (273 + 27)$ K $= 300$ K

Final pressure at height,

$p_2 = 45$ cm of Hg

Final temperature, $T_2 = -7$°C

$= [273 + (-7)]$K $= 266$ K

Volume of balloon is constant,

$V_1 = V_2 = V$

We know that for air, $p = \rho RT$

Now, for initial condition,

$p_1 = \rho_1 RT_1$ ...(i)

Similarly, for final condition,

$p_2 = \rho_2 RT_2$ ..(ii)

Dividing Eq. (i) by Eq. (ii), we get

$$\frac{p_1}{p_2} = \frac{\rho_1 T_1}{\rho_2 T_2}$$

$$\frac{76}{45} = \frac{\rho_1}{\rho_2}\left(\frac{300}{266}\right)$$

$$\Rightarrow \quad \frac{\rho_1}{\rho_2} = \frac{76 \times 266}{45 \times 300}$$

We know that density of balloon,

$$\rho = \frac{M}{V}$$

As, volume is constant, $\rho \propto M$

$$\therefore \quad \frac{\rho_1}{\rho_2} = \frac{M_1}{M_2} = \frac{76 \times 266}{45 \times 300}$$

$$\Rightarrow \quad \frac{M_1}{M_2} = \frac{20216}{13500}$$

Substituting the value of M_1 in above expression, we get

$$M_2 = 185 \times \frac{13500}{20216}$$

$= 123.54$ kg

73. *(c)* Let molar specific heat of the mixture is C_V.

Total number of molecules in the mixture $= 3 + 2 = 5$

$\therefore C_V$ can be determined using

$n\,C_V\,dT = n_1\,C_{V_1}\,dT + n_2\,C_{V_2}\,dT$

or $(n_1 + n_2)\,(C_V)_{\text{mix}} = n_1 C_{V_1} + n_2\,C_{V_2}$

[here, $n = n_1 + n_2$]

Here, $C_{V_1} = \dfrac{3R}{2}$ (for helium); $n_1 = 2$

$C_{V_2} = \dfrac{5R}{2}$ (for hydrogen); $n_2 = 3$

[For monoatomic gases, $C_V = \dfrac{3}{2}R$ and for diatomic gases, $C_V = \dfrac{5}{2}R$]

$$\therefore \quad 5 \times C_V = \left(2 \times \frac{3R}{2}\right) + \left(3 \times \frac{5R}{2}\right)$$

$$\Rightarrow 5C_V = \frac{21R}{2}$$

$$\text{or} \quad C_V = \frac{21R}{10} = \frac{21 \times 8.3}{10} = 17.4 \text{ mol-K}$$

74. *(a)*

key idea As piston is in equilibrium, so net force on piston is zero.

When the piston is stationary, i.e. on equilibrium as shown in the figure below,

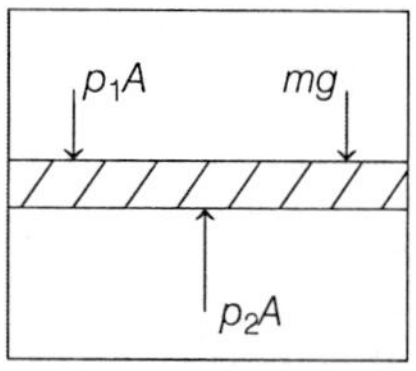

then $p_1 A + mg = p_2 A$

$\Rightarrow \qquad mg = p_2 A - p_1 A$

or $\qquad mg = \left(\dfrac{nRTA}{V_2} - \dfrac{nRTA}{V_1}\right)$

[$\because pV = nRT$ (ideal gas equation)]

$$mg = nRT\left(\frac{A}{Al_2} - \frac{A}{Al_1}\right)$$

$$mg = nRT\left(\frac{l_1 - l_2}{l_1 l_2}\right)$$

or $\qquad m = \dfrac{nRT}{g}\left(\dfrac{l_1 - l_2}{l_1 l_2}\right)$

75. *(a)* We know that

$$v_{\text{rms}} = \sqrt{\frac{3RT}{M}} \text{ or } v_{\text{rms}} \propto \sqrt{T}$$

Here, to double v_{rms}, the temperature must be 4 times that of the initial temperature.

$\therefore$ New temperature,

$T_2 = 4 \times T_1$

As, $\qquad T_1 = 27°$ C $= 300$ K

$\Rightarrow \qquad T_2 = 4 \times 300 = 1200$ K

As, gas is kept inside a closed vessel.

$\therefore \qquad \Delta Q = nC_V \cdot dT$

$$= \frac{15}{28} \times \frac{5R}{2}(1200 - 300)$$

$$\left[\text{As}, n = \frac{m}{M} = \frac{15}{28} \text{ for nitrogen}\right]$$

$$= \frac{15}{28} \times \frac{5}{2} \times 8.3 \times 900$$

[Given, $R = 8.3$ J/K-mol]

or $\qquad \Delta Q = 10004.4$ J ≈ 10 kJ

76. *(b)* For mixture, $C_V = \dfrac{n_1 C_{V_1} + n_2 C_{V_2}}{n_1 + n_2}$

For helium,

Number of moles, $n_1 = \dfrac{16}{4} = 4$

and $\qquad \gamma_1 = 5/3$

For oxygen,

Number of moles, $n_2 = \dfrac{16}{32} = \dfrac{1}{2}$

and $\qquad \gamma_2 = \dfrac{7}{5}$

Then, $C_{V_1} = \frac{R}{\gamma_1 - 1} = \frac{R}{\frac{5}{3}-1} = \frac{3}{2}R$

and $C_{V_2} = \frac{R}{\gamma_2 - 1} = \frac{R}{\frac{7}{5}-1} = \frac{5}{2}R$

∴ For mixture,

$$C_V = \frac{4\times\frac{3}{2}R + \frac{1}{2}\times\frac{5}{2}R}{4+\frac{1}{2}}$$

$$= \frac{6R + \frac{5}{4}R}{\frac{9}{2}} = \frac{29R\times 2}{9\times 4} = \frac{29R}{18}$$

Now, $C_V = \frac{R}{\gamma - 1} \Rightarrow \gamma - 1 = \frac{R}{C_V}$

or $\gamma = \frac{R}{C_V} + 1 = \frac{R}{\frac{29}{18}R} + 1$

$$\Rightarrow \frac{C_p}{C_V} = \frac{18}{29} + 1 = \frac{18+29}{29} = 1.62$$

77. *(a)* Mayer's formula is

$$C_p - C_V = R \quad \text{and} \quad \gamma = \frac{C_p}{C_V}$$

Therefore, using above two relations, we find

$$C_V = \frac{R}{\gamma - 1}$$

For a mole of monatomic gas,

$$\gamma = \frac{5}{3}$$

$$\therefore \quad C_V = \frac{R}{\left(\frac{5}{3}\right) - 1} = \frac{3}{2}R$$

For a mole of diatomic gas,

$$\gamma = \frac{7}{5}$$

$$\therefore \quad C_V = \frac{R}{\left(\frac{7}{5}\right) - 1} = \frac{5}{2}R$$

When these two moles are mixed, then heat required to raise the temperature to 1°C is

$$C_V = \frac{3}{2}R + \frac{5}{2}R = 4R$$

Hence, for one mole, heat required

$$= \frac{4R}{2} = 2R$$

$$\therefore \quad C_V = 2R$$

$$\Rightarrow \frac{R}{\gamma - 1} = 2R \text{ or } \gamma = \frac{3}{2}$$

> ***Alternate Solution***
>
> $$\frac{n_1 + n_2}{\gamma - 1} = \frac{n_1}{\gamma_1 - 1} + \frac{n_2}{\gamma_2 - 1}$$
>
> Here, $n_1 = 1, n_2 = 1, \gamma_1 = \frac{5}{3}, \gamma_2 = \frac{7}{5}$
>
> $$\therefore \frac{1+1}{\gamma - 1} = \frac{1}{\left(\frac{5}{3}\right)-1} + \frac{1}{\left(\frac{7}{5}\right)-1}$$
>
> $$\Rightarrow \frac{2}{\gamma - 1} = \frac{3}{2} + \frac{5}{2} \text{ or } \frac{2}{\gamma - 1} = \frac{8}{2}$$
>
> or $\frac{2}{\gamma - 1} = 4 \Rightarrow \gamma = \frac{2}{4} + 1$
>
> Hence, $\gamma = \frac{3}{2}$

78. *(327)* Since volume in fixed

$$\Rightarrow \frac{p_1}{T_1} = \frac{p_2}{T_2} \Rightarrow \frac{p}{300} = \frac{2p}{T_2}$$

$$T_2 = 600 \text{ K or } 327°\text{C}$$

79. *(2)* Work done, $W = \frac{Q}{4}$

According to first law of thermodynamics,

$$Q = \Delta U + W, \; Q = \Delta U + \frac{Q}{4}$$

$$\Rightarrow \quad \Delta U = \frac{3Q}{4} \quad \ldots\text{(i)}$$

But $\Delta U = nC_V\Delta T$

$$\Delta U = n\left(\frac{3R}{2}\right)\Delta T \quad \ldots\text{(ii)}$$

[For monoatomic gas, $C_V = \frac{3R}{2}$]

From Eqs. (i) and (ii), we get

$$n\left(\frac{3R}{2}\right)\Delta T = \frac{3Q}{4}$$

$$(n\,\Delta T)R = \frac{1}{2}Q = \frac{nC\Delta T}{2} \quad [\because Q = nC\Delta T]$$

$$\Rightarrow \quad R = \frac{C}{2} \Rightarrow C = 2R$$

80. *(150)* From ideal gas equation,

$$pV = nRT$$

At constant temperature, for small changes in p and V,

$$\Delta(pV) = \Delta(nRT)$$

$$\Rightarrow p\Delta V + V\Delta p = 0$$

$$\Rightarrow \quad V\Delta p = -p\Delta V$$

$$\Rightarrow \quad \Delta p = -\frac{p\Delta V}{V}$$

$$\Rightarrow \quad |\Delta p| = \left|-\frac{p\Delta V}{V}\right|$$

$$\Rightarrow \quad |\Delta p| = \frac{p\Delta V}{V} \quad \ldots\text{(i)}$$

Again, from ideal gas equation,

$$pV = nRT$$

At constant pressure, for small changes in V and T.

$$\Delta(pV) = \Delta(nRT)$$

$$\Rightarrow \quad p\,\Delta V = nR\Delta T$$

$$\Rightarrow \quad \Delta T = \frac{p\Delta V}{nR} \Rightarrow |\Delta T| = \left|\frac{p\Delta V}{nR}\right|$$

$$\Rightarrow \quad |\Delta T| = \frac{p\,\Delta V}{nR} \quad \ldots\text{(ii)}$$

If ΔV is same in both the cases (as given), then dividing eq. (ii) by eq. (i), we get

$$\frac{|\Delta T|}{|\Delta p|} = \frac{\left(\frac{p\Delta V}{nR}\right)}{\left(\frac{p\Delta V}{V}\right)} \Rightarrow \frac{|\Delta T|}{|\Delta p|} = \frac{V}{nR}$$

$$\Rightarrow \quad |\Delta T| = \frac{V}{nR}|\Delta p| \Rightarrow |\Delta T| = \frac{T}{p}|\Delta p|$$

$$(\because pV = nRT \Rightarrow \frac{V}{nR} = \frac{T}{p})$$

$$\Rightarrow \quad |\Delta T| = \frac{300}{2}|\Delta p|$$

(as $T = 300$ K and $p = 2$ atm)

$$\Rightarrow \quad |\Delta T| = 150\,|\Delta p|$$

Given that, $|\Delta T| = C\,|\Delta p|$

On comparing both equations, we get

$$C = 150$$

> ***Alternate Solution***
>
> For temperature To be constant
>
> $$pV = nRT \Rightarrow pdV = -Vdp \Rightarrow |\Delta V| = \left|\frac{Vdp}{p}\right|$$
>
> and at constant pressure
>
> $$pdV = nRdT \Rightarrow |\Delta V| = \left|\frac{NRdT}{p}\right|$$
>
> Therefore, $\frac{Vdp}{p} = \frac{NRdT}{p} \Rightarrow Vdp = NRCdp$
>
> where C is constant
>
> $$\therefore \quad V = NRC \Rightarrow C = \frac{V}{NR} = \frac{T}{p}$$
>
> $$\Rightarrow \quad C = \frac{300\text{K}}{2\,\text{atm}} = 150\left(\frac{\text{K}}{\text{atm}}\right)$$

81. *(41)* The rms speed is given by

$$v_{\text{rms}} = \sqrt{\frac{3RT}{M}}$$

For nitrogen molecule,

$$(v_{rms})_{N_2} = \sqrt{\frac{3RT_1}{M_1}}$$

and for hydrogen molecule,

$$(v_{rms})_{H_2} = \sqrt{\frac{3RT_2}{M_2}}$$

According to question,

$$(v_{rms})_{N_2} = (v_{rms})_{H_2}$$

$$\Rightarrow \sqrt{\frac{3RT_1}{M_1}} = \sqrt{\frac{3RT_2}{M_2}}$$

$$\Rightarrow \frac{T_1}{M_1} = \frac{T_2}{M_2} \Rightarrow \frac{T_1}{T_2} = \frac{M_1}{M_2}$$

$$\Rightarrow T_2 = \frac{T_1 M_2}{M_1}$$

Given, $T_1 = 300°C$

$$= 300 + 273 = 573\ K$$

$M_1 = 28$ g and $M_2 = 2$ g

$$T_2 = 573 \times \frac{2}{28} = 40.93\ K \approx 41\ K$$

82. *(5)*

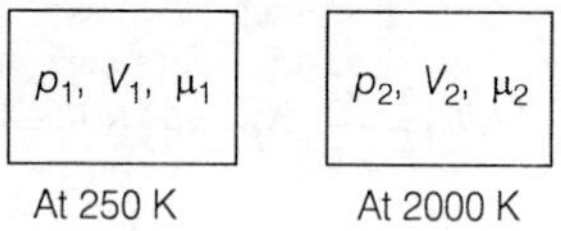

Let $\mu_1 = \mu_0, V_2 = 2V_1$ (given)

Total number of moles at 2000 K

= Number of moles of diatomic molecules + Number of moles of monoatomic

$$\mu_2 = 0.75\mu_0 + 2 \times 0.25\mu_0$$
$$= 1.25\mu_0$$

Ideal gas equation for situation 1,

$$p_1V_1 = \mu_0 RT_1$$

$$p_1V_1 = \mu_0 R(250) \quad \dots \text{(i)}$$

For situation 2,

$$p_2V_2 = (1.25\mu_0) R(2000)$$

$$p_2(2V_1) = (1.25\mu_0) R(2000) \quad \dots \text{(ii)}$$

Dividing Eq. (i) by Eq. (ii), we get

$$\frac{p_1V_1}{p_2(2V_1)} = \frac{\mu_0 R(250)}{(1.25\mu_0)R(2000)}$$

$$\Rightarrow \frac{p_1}{p_2} = \frac{1}{5} \text{ or } \frac{p_2}{p_1} = 5$$

Topic 2

01. *(d)* Given $\lambda = 3 \times 10^{-7}$ m

$$v_{avg} = 600 \text{ m/s}$$
$$T = 300\ K$$
$$p = 1 \text{ atm}$$

$$\text{Frequency} = \frac{1}{T} = \frac{v_{avg}}{\lambda}$$

$$= \frac{600}{3 \times 10^{-7}}$$
$$= 200 \times 10^7$$
$$= 2 \times 10^9/\text{s}$$

02. *(c)* Kinetic energy,

$$E_k = \frac{f}{2} k_B T$$

For He or Ar, $f = 3$

and $T = 300$ K

$$\therefore \frac{(E_k)_{He}}{(E_k)_{Ar}} = 1$$

$$\therefore (E_k)_{He} : (E_k)_{Ar} = 1:1$$

03. *(c)* Degree of freedom of non-rigid diatomic molecules,

$$= 5 + 2(3N - 5)$$
$$= 5 + 2(3 \times 2 - 5)$$
$$= 5 + 2 = 7$$
$$f = 7$$

Energy of one molecule $= \frac{f}{2} k_B T$

$$\therefore \text{Energy of 10 molecule} = 10\left(\frac{f}{2} k_B T\right)$$
$$= 10 \times \frac{7}{2} k_B T$$
$$= 35 k_B T$$

04. *(c)* For CH_4 molecule, there are 3 rotational degree of freedom and 3 translational degree of freedom.

i.e. $f_t = 3$ and $f_r = 3$

05. *(c)* Argon is monoatomic, its degree of freedom is 3.

Oxygen is diatomic with 5 degree of freedom.

Total internal energy of mixture

= Internal energy of Ar + Internal energy of O_2

$$= 8 \times \frac{3}{2} RT + 6 \times \frac{5}{2} RT$$
$$= (12 + 15)\ RT = 27\ RT$$

06. *(d)* Average KE = 0.414 eV,

for monoatomic $f = 3$

$$\frac{f}{2} k_B T = \frac{3}{2} k_B T$$
$$= 0.414 \times 1.6 \times 10^{-19}\ J$$
$$T = 3200\ K$$

07. *(d)* Given,

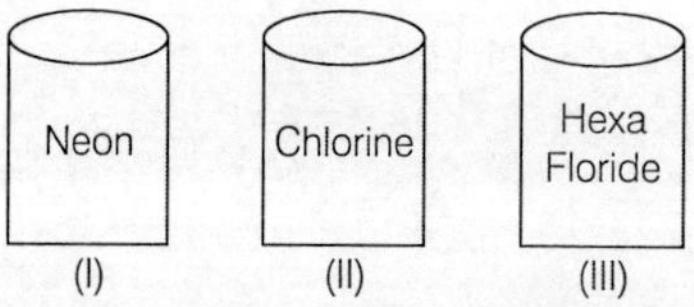

$$v_{rms} = \sqrt{\frac{\gamma RT}{m}} \quad \dots \text{(i)}$$

$$\gamma = 1 + \frac{2}{f}$$

f = degree of freedom

$f_{diatomic} = 5$, $f_{poly} \geq 6$

$f_{mono} = 3$

$\because \gamma_{monoatomic} > \gamma_{diatomic} > \gamma_{polyatomic}$

$$v_{rms} \propto \sqrt{\gamma / m}$$

and

$m_{Neon} < m_{Chlorine} < m_{Uranium\ hexafloride}$

So, $v_{mono} > v_{diatomic} > v_{poly}$

08. *(b)* Given, moles of oxygen = 2

Moles of neon = 4

Internal energy of mixture

$$= U_{O_2} + U_{Ne}$$
$$= n_{O_2} C_V RT + n_{Ne} C_V RT$$
$$= 2 \times \frac{5}{2} RT + 4 \times \frac{3}{2} RT$$

($\because C_r = \frac{5}{2}$ for diatomic gas and $C_V = \frac{3}{2}$ for monoatomic gas)

$$\Rightarrow U_{max} = 11RT$$

09. *(b)* A-I, B-III, C-IV, D-II

A monoatomic gas molecule has 3-translatory degree of freedom.

A rigid diatomic molecule has 3-translationl and 2-rotational degree of freedom.

A non-rigid diatomic gas molecule has 3-translationl, 2-rotational and 1-vibrational degree of freedom.

A polyatomic gas molecule has 3-translational, 2-rotational and more than one vibrational degree of freedom.

10. *(b)* Molecules of diatomic gas have 7 degree of freedom : 3 translational, 2 rotational and 2 vibrational.

i.e.,

$$f = 3 + 2 + 2 = 7$$

Thus, total internal energy of one mole of diatomic gas,

$$U = \frac{f}{2} RT = \frac{7}{2} RT$$

$\therefore$ Molar specific heat at constant volume,

$$C_V = \frac{dU}{dT}$$
$$= \frac{d}{dT}\left(\frac{7}{2} RT\right) = \frac{7}{2} R$$

11. *(d)* Ratio of molar specific heat at constant pressure and molar specific heat at constant volume is

$$\frac{C_p}{C_V} = 1 + \frac{2}{f}$$

where, f = degree of freedom of gas molecule.

As, $f = 3$ for monoatomic and $f = 5$ for diatomic gas, we have,

$$\gamma_1 = \frac{C_p}{C_V} = 1 + \frac{2}{3} = \frac{5}{3}$$ (monoatomic)

and $$\gamma_2 = 1 + \frac{2}{5} = \frac{7}{5}$$ (diatomic)

So, ratio $\frac{\gamma_1}{\gamma_2} = \frac{5}{3} \times \frac{5}{7} = \frac{25}{21}$

12. *(b)* Statement A is incorrect because a molecule with n degrees of freedom have n different ways of storing energy.

- Statement B is correct as each degree of freedom is associated with $1/2\ RT$ average energy per mole.
- Statement C is incorrect because a monoatomic gas molecule has no rotational degree of freedom and diatomic gas molecule has 2 rotational degrees of freedom.
- Statement D is correct as CH_4 molecule has 3 translational and 3 rotational, i.e. total 6 degrees of freedom.

13. *(b)* According to kinetic theory of gases,

A. The motion of gas molecules freezes at 0 K.

B. The mean free path of gas molecules decreases on increasing the number density of gas molecules as

$$\mu = \text{mean free path} = \frac{1}{\sqrt{2}\,\pi n d^2}$$

$$\Rightarrow \quad \mu \propto \frac{1}{n}$$

The mean free path increases with the increase of volume.

C. If pressure is constant and temperature is increased, then volume should increase, hence mean free path increases.

D. Average KE per molecule per degree of freedom is $\frac{1}{2} k_B T$.

Hence, statements (B) and (C) are correct.

14. *(c)* Given, nature of gas is monoatomic.

Let degree of freedom of monoatomic gas be f i.e. $f = 3$

As we know that,

Average energy, $$E_{av} = \frac{1}{2} f k_B T$$

For monoatomic gas, $$E_{av} = \frac{3}{2} k_B T$$

15. *(a)* Since, we know that,

$$\gamma = 1 + \frac{2}{f}$$

$$\Rightarrow \quad \frac{2}{f} = \gamma - 1 \Rightarrow f = \frac{2}{\gamma - 1}$$

16. *(a)* We know that, the average value of energy of an ideal gas in thermal equilibrium temperature T,

$$E = \frac{f}{2} k_B T$$

Given, degree of freedom, $f = 1$

Energy per degree of freedom,

$$E = \frac{1}{2} k_B T$$

17. *(c)* According to Statement 1 : In diatomic molecule, the rotational energy at a given temperature obeys Maxwell's distribution is correct.

But, according to Statement 2 : In diatomic molecule, the rotational energy at a given temperature equals translational energy for each molecule is false.

Statement II is incorrect, because kinetic energy (KE) of gaseous molecule is $\frac{f}{2} k_B T$.

where, f is the degree of freedom, and f(diatomic atom) $= 5$ [3 (translational) and 2(rotational)]

Therefore, translational KE of gas

$$= \frac{3}{2} k_B T \quad \ldots\text{(i)}$$

and rotational KE of gas

$$= 2/2\ k_B T$$
$$= k_B T \quad \ldots\text{(ii)}$$

$\because$ Eq. (i) is not equal to Eq. (ii)

18. *(a)* Total KE of n moleules $= n\left(\frac{1}{2} k_B T\right)$

Total KE of mixture

$$= n_1 \frac{1}{2} k_B T_1 + n_2 \frac{1}{2} k_B T_2 + n_3 \frac{1}{2} k_B T_3$$

As no loss of energy,

$$n\left(\frac{1}{2} k_B T\right) = \frac{1}{2} k_B (n_1 T_1 + n_2 T_2 + n_3 T_3)$$

$$\therefore T = \frac{n_1 T_1 + n_2 T_2 + n_3 T_3}{n_1 + n_2 + n_3}$$

19. *(a)* $\gamma = 1 + \frac{2}{f}$, $f \rightarrow$ degree of freedom

$f = 6$, Triatomic rigid gas
$f = 7$, Diatomic non-rigid gas
$f = 5$, Diatomic rigid gas
$f = 3$, Monoatomic rigid gas

$\gamma = 1 + \frac{2}{6} = \frac{4}{3}$ Triatomic

$\gamma = 1 + \frac{2}{7} = \frac{9}{7}$ (Diatomic, non-rigid)

$\gamma = 1 + \frac{2}{5} = \frac{7}{5}$ Diatomic rigid

$\gamma = 1 + \frac{2}{3} = \frac{5}{3}$ Monoatomic rigid

A-III, B-IV, C-I, D-II.

20. *(c)* $\gamma_1 = \frac{C_p}{C_V} \Rightarrow \gamma_2 = \frac{C_p}{C_V}$

$\gamma = \frac{2}{f} + 1$, where $f \rightarrow$ degree of freedom without vibration : $f = 5$ (for rigid molecules)

$$\gamma_1 = \frac{2}{5} + 1 = 0.4 + 1 = 1.40$$

For diatomic molecules with vibration $f = 7$

$$\gamma_2 = \frac{2}{f} + 1 = \frac{2}{7} + 1 = 1.28$$

$$\therefore \quad \gamma_1 > \gamma_2$$
$$\gamma_2 < \gamma_1$$

21. *(c)* Given, temperature of the gas,

$$T_1 = -78°\text{C}$$
$$= -78 + 273\ \text{K}$$
$$= 195\ \text{K}$$

Average translational kinetic energy at $T_1 = K$

We know that average translational kinetic energy, at given temperature

$$= \frac{3}{2} k_B T$$

$$K \propto T$$

According to question, $K \propto T_1$ and $2K \propto T_2$

Therefore, $$\frac{K}{2K} = \frac{T_1}{T_2}$$

$$\Rightarrow \quad \frac{1}{2} = \frac{195}{T_2}$$

$$\Rightarrow \quad T_2 = 195 \times 2 = 390\ \text{K}$$
$$= 117\ °\text{C}$$

22. *(a)* Mean free path, $\lambda = \frac{1}{\sqrt{2}\pi n d^2}$

n = Number of molecules per unit volume and d = diameter of gas molecule

$\therefore \quad \lambda \propto \frac{1}{d^2}$

Statement I is true.

Also, mean KE of a gas molecule

$E_K = \mu \cdot \frac{f}{2} \cdot RT$

μ = number of moles,
f = degree of freedom,
R = gas constant

$\therefore \quad E_K \propto T$

Statement II is true.

• Concept Enhancer

(i) Pressure and temperature also affect the density of the gas and thus affect the mean free path.

(ii) As the radius of the molecule of gas increases, the space between the molecules decreases causing the number of collisions to increase, thus decreasing the mean free path.

23. *(c)* Molar specific heat of a mixture is given by,

$$C_{V_{mix}} = \frac{n_1 C_{V_1} + n_2 C_{V_2}}{n_1 + n_2}$$

Here, $n_1 = 2, C_{V_1} = \frac{3}{2}R$

$n_2 = 6, C_{V_2} = \frac{5}{2}R$

So, $$C_{V_{mix}} = \frac{2 \times \frac{3}{2}R + 6 \times \frac{5}{2}R}{2+6} = \frac{9}{4}R$$

• Concept Enhancer

Molar specific heat at constant pressure is always more than molar specific heat at constant volume because when a gas is heated under constant pressure, work is done to overcome the pressure, and the volume expands, resulting in a rise in the system's internal energy.

24. *(b)* Given, $T = 27°C = 300$ K

$R = 8.31$ J/molK

The total kinetic energy of 1 mole of oxygen,

$$KE = \frac{1}{2} fRT$$

where f is degree of freedom f for O_2 molecule

$$f = 2 \times 3 - 1 = 5$$

$$\therefore \quad KE = \frac{1}{2} \times 5 \times 8.31 \times 300 = 6232.5 \text{ J}$$

25. *(b)* Heat given at constant pressure,

$Q_p = \mu C_p \Delta T = 735$ J

Gas molecule does not oscillate, hence vibrational degree of freedom is not considered.

So, Change in internal energy,

$$\Delta U = \mu C_V \Delta T$$

$$\because \quad \frac{C_p}{C_V} = \gamma \text{ or } C_V = \frac{C_p}{\gamma}$$

or $$\Delta U = \mu\left(\frac{C_p}{\gamma}\right)\Delta T = \frac{\mu C_p \Delta T}{\gamma}$$

For diatomic gas, $\gamma = \frac{7}{5}$

Thus, $\Delta U = \frac{735}{7/5} = 525$ J

26. *(a)* According to law of equipartition of energy, average kinetic energy of each molecule $= \frac{n}{2} k_B T$

where, n = Degree of freedom,
k_B = Boltzmann constant and
T = Temperature.

$\therefore$ Internal energy of one mole of the gas, $U = \frac{n}{2} k_B T \times N_A = \frac{n}{2} RT$

$$\therefore \quad C_V = \frac{dU}{dT} = \frac{d}{dT}\frac{n}{2}RT = \frac{nR}{2}$$

and $$C_p = C_V + R = \frac{nR}{2} + R = R\left(\frac{n}{2}+1\right)$$

$$\therefore \quad \frac{C_V}{C_p} = \frac{\frac{nR}{2}}{R\left(\frac{n}{2}+1\right)} = \frac{n}{n+2}$$

27. *(b)* For polyatomic gas, molecule has 3 rotational degrees of freedom, 3 translational degrees of freedom, and 2 vibrational modes.

So, number of vibrational degrees of freedom = 2(2) = 4

$\therefore$ Total number of degrees of freedom, $f = 3 + 3 + 4 = 10$

Here, ratio of molar specific heats is given as

$$\beta = \frac{C_p}{C_V}$$

In terms of degrees of freedom,

$$\beta = 1 + \frac{2}{f}$$

Substituting the value of f in above equation, we get

$$\beta = 1 + \frac{2}{10} \Rightarrow \beta = 1.2$$

28. *(a)* We know that,

$$\gamma = 1 + \frac{2}{f} \qquad \ldots(i)$$

As each vibrational mode has 2 degrees of freedom, hence total vibrational degrees of freedom = 2× 24 = 48

$\Rightarrow f = 3$ (rotational) +3 (translational) + 48 (vibrational)

$\Rightarrow \quad f = 3 + 3 + 48$

$\Rightarrow \quad f = 54$

Now, put the value of f in Eq. (i), we get

$$\gamma = 1 + \frac{2}{54} \Rightarrow \gamma = 1 + \frac{1}{27}$$

$$\Rightarrow \quad \gamma = \frac{28}{27} \text{ or } \gamma = 1.03$$

29. *(b)* $U = 3pV + 4$

$\Delta U = 3p\Delta V$

$n \cdot \frac{f}{2} R\Delta T = 3nR\Delta T$

$f = 6$

So, gas is polytomic.

30. *(c)* Given that, translational degree of freedom, $f_{trans} = 3$

Rotational degree of freedom, $f_{rot} = 2$

Total degree of freedom,

$f = f_{trans} + f_{rot} = 3 + 2 = 5$

At temperature T, total internal energy of an ideal gas is given by $U = \frac{f}{2}\mu RT$

$U = 5/2RT \qquad (\because \mu = 1)$

$$\gamma = \frac{C_p}{C_V} = 1 + \frac{2}{f} = 1 + \frac{2}{5} = \frac{7}{5}$$

31. *(a)* $\because \quad \gamma = \frac{C_p}{C_V} = 1 + \frac{2}{f}$

where, f = degree of freedom.

Now, (A) for monoatomic gas,

$f = 3$

so, $\gamma = 1 + \frac{2}{3} = \frac{5}{3}$

(B) for diatomic rigid gas,

$f = 5$

so, $\gamma = 1 + \frac{2}{5} = \frac{7}{5}$

(C) for diatomic non-rigid gas,

$f = 7$

so, $\gamma = 1 + \frac{2}{7} = \frac{9}{7}$

(D) for triatomic rigid gas,

$$f = 6$$

so, $\gamma = 1 + \frac{2}{6} = \frac{8}{6} = \frac{4}{3}$

Therefore, A-IV, B-I, C-II, D-III.

32. *(a)* Internal energy of a gas mixture = Sum of internal energies of its' components.

Also, internal energy of a gas is given by

$$U = \frac{f}{2}(nRT)$$

Here, gases are oxygen (diatomic, $f = 5$) and argon (monoatomic, $f = 3$).

So, total internal energy of the mixture is

$$U = \left(\frac{f}{2} nRT\right)_{O_2} + \left(\frac{f}{2} nRT\right)_{Ar}$$

$$= \frac{5}{2} \times 3 \times RT + \frac{3}{2} \times 5 \times RT$$

$$= 15RT$$

33. *(d)* For a gas, value of specific heat at constant volume, $C_V = (1/2) fR$

where, f = degrees of freedom

and R = gas constant

For a diatomic gas molecule, degrees of freedom is in general 5 but when vibration occurs, then degrees of freedom increases to 7.

So, for gas A, specific heat is

$$C_V^A = \frac{1}{2} 5R$$

and for gas B, specific heat is

$$C_V^B = \frac{1}{2} 7R$$

∴Ratio of specific heats of gas A and B is, $C_V^A / C_V^B = 5/7$

34. *(a)*

> **key idea** A diatomic gas molecule has 5 degrees of freedom, i.e. 3 translational and 2 rotational, at low temperature ranges (~ 250 K to 750 K). At temperatures above 750 K, molecular vibrations occurs and this causes two extra degrees of freedom.

Now, in given case,

For gas A, $C_p = 29$, $C_V = 22$

For gas B, $C_p = 30$, $C_V = 21$

By using $\gamma = \frac{C_p}{C_V} = 1 + \frac{2}{f}$

We have,

For gas A, $1 + \frac{2}{f} = \frac{29}{22} \approx 1.3$

$$\Rightarrow \quad f = 6.67 \approx 7$$

So, gas A has vibrational mode of degree of freedom.

For gas B, $1 + \frac{2}{f} = \frac{30}{21} \approx 1.4 \Rightarrow f = 5$

Hence, gas B does not have any vibrational mode of degree of freedom.

35. *(b)* Since, HCl is a diatomic molecule that has rotational, translational and vibrational motion.

So, $n = 7 \Rightarrow \frac{1}{2} mv_{rms}^2 = \frac{7}{2} k_B T$

Here, $v_{rms}^2 = \bar{v} \Rightarrow T = \frac{m\bar{v}^2}{7k_B}$

36. *(c)* We know that in adiabatic process,

$$TV^{\gamma-1} = \text{Constant}$$

or, $T_1 V_1^{\gamma-1} = T_2 V_2^{\gamma-1}$

For monoatomic gas, $\gamma = \frac{5}{3}$

$$(300)V^{2/3} = T_2(2V)^{2/3}$$

$$\Rightarrow \quad T_2 = \frac{300}{(2)^{2/3}}$$

$$T_2 = 189 \text{ K (final temperature)}$$

$$\therefore \quad \Delta U = n\frac{f}{2} R\Delta T$$

$$= 2\left(\frac{3}{2}\right)\left(\frac{25}{3}\right)(-111) = -2.7 \text{ kJ}$$

37. *(d)* We know that,

$$\frac{C_p}{C_V} = \gamma = \left(1 + \frac{2}{f}\right)$$

where, f is degree of freedom,

$$\frac{C_p}{C_V} = \left(1 + \frac{2}{5}\right) = \frac{7}{5}$$

38. *(b)* Volume $= \frac{\text{Mass}}{\text{Density}} = \frac{1}{4} \text{ m}^3$

Pressure $= 8 \times 10^4 \text{ Nm}^{-2}$

As internal energy $= \frac{f}{2} nRT = \frac{f}{2} pV$

Degree of freedom for diatomic = 5

∴ Internal energy $= \frac{5}{2} p \times V$

Putting the values of

$$p = 8 \times 10^4 \text{ Nm}^{-2}$$

and $V = \frac{1}{4} \text{ m}^3$, we get

Internal energy $= \frac{5}{2} \times 8 \times 10^4 \times \frac{1}{4}$

$$= 5 \times 10^4 \text{ J}$$

39. *(b)* For monoatomic gas, number of moles, $n_1 = 3$

and degree of freedom,

$$f_1 = 3$$

For diatomic gas, number of moles,

$$n_2 = 2$$

and degree of freedom,

$$f_2 = 5$$

Degree of freedom of mixed gas,

$$f_{mix} = \frac{n_1 f_1 + n_2 f_2}{n_1 + n_2}$$

$$= \frac{3 \times 3 + 2 \times 5}{3 + 2} = \frac{19}{5}$$

$$\therefore \gamma_{mixture} = 1 + \frac{2}{f_{mix}} = 1 + \frac{2}{(19/5)}$$

$$= 1 + \frac{10}{19} = \frac{29}{19} = 1.52$$

> ***Alternate Solution***
>
> For a mixture of gases
>
> $$\gamma_{mixture} = \left[\frac{\frac{n_1\gamma_1}{\gamma_1 - 1} + \frac{n_2\gamma_2}{\gamma_2 - 1}}{\frac{n_1}{\gamma_1 - 1} + \frac{n_2}{\gamma_2 - 1}}\right]$$
>
> Here, $n_1 = 3, \gamma_1 = \frac{5}{3}$ and $n_2 = 2, \gamma_2 = \frac{7}{5}$
>
> Substituting above values, we get,
>
> $$\gamma_{mixture} = 1.52$$

40. *(d)* Given,

$$\because \quad f = 6, \mu_1 = N, \mu_2 = 2$$

For diatomic gas

$$\gamma_1 = 1 + \frac{2}{f}$$

$$= 1 + \frac{2}{6}$$

$$= 1 + 0.33 = 1.33$$

For monoatomic gas

$$\gamma_2 = 1 + \frac{2}{3}$$

$$= 1.67$$

$$\therefore \gamma_{mix} = \frac{\frac{\mu_1\gamma_1}{\gamma_1 - 1} + \frac{\mu_2\gamma_2}{\gamma_2 - 1}}{\frac{\mu_1}{\gamma_1 - 1} + \frac{\mu_2}{\gamma_2 - 1}}$$

$$\because \quad 1.4 = \frac{\frac{N \times 1.33}{1.33 - 1} + \frac{2 \times 1.67}{1.67 - 1}}{\frac{N}{1.33 - 1} + \frac{2}{1.67 - 1}}$$

$$\Rightarrow \quad 1.4 = \frac{N \times 4 + 5}{3N + 3}$$

$$\Rightarrow 4.2N + 4.2 = 4N + 5$$

$$\Rightarrow \quad 0.2N = 0.8$$

$$N = 4$$

Alternate Solutions

$$\gamma_{mix} = \frac{n_1 f_1 + n_2 f_2}{n_1 + n_2} = \frac{N \times 6 + 2 \times 3}{N + 2}$$

Since, degree of freedom for diatomic gas is 5.

$$\therefore \quad 5 = \frac{6N + 6}{N + 2} \Rightarrow N = 4$$

41. *(b)* Let the final temperature of this mixture be T.

As per question, there is no loss of energy, it means

$$\Delta U = 0 \quad \dots \text{(i)}$$

As, we know,

$$\Delta U = \frac{f_1 n_1 R\Delta T}{2} + \frac{f_2 n_2 R\Delta T}{2} \quad \dots \text{(ii)}$$

From Eqs. (i) and (ii), we get

$$\frac{f_1 n_1 R\Delta T}{2} + \frac{f_2 n_2 R\Delta T}{2} = 0$$

$$\Rightarrow \quad f_1 n_1 R\Delta T + f_2 n_2 R\Delta T = 0$$

$$\Rightarrow f_1 n_1 R(T_1 - T) + f_2 n_2 R(T_2 - T) = 0$$

$$T = \frac{f_1 n_1 T_1 + f_2 n_2 T_2}{f_1 n_1 + f_2 n_2}$$

42. *(3)* $\frac{\gamma_A}{\gamma_B} = \frac{f_A + 2}{f_A} \times \frac{f_B}{f_B + 2}$

As we know, each vibrational mode contributes 2 degrees of freedom (1 potential +1 kinetic). Then,

$$= \frac{3+2}{3} \times \frac{6+2}{(6+2)+2}$$

$$\frac{\gamma_A}{\gamma_B} = \frac{5}{3} \times \frac{8}{10} = \frac{40}{30}$$

$$\frac{\gamma_A}{\gamma_B} = 1 + \frac{1}{n} \rightarrow \text{given}$$

$$\frac{40}{30} = 1 + \frac{1}{n}$$

$$\frac{40}{30} - 1 = \frac{1}{n}$$

$$\frac{10}{30} = \frac{1}{n}$$

$$\Rightarrow \quad n = 3$$

43. *(7479)* For a monoatomic gas, degree of freedom, $f = 3$

So, specific heat at constant volume

$$C_V = \frac{f}{2} R \Rightarrow C_V = \frac{3}{2} R$$

Hence, internal energy of 2 moles at 300 K is

$$U = n \cdot C_V T$$

$$= 2 \times \frac{3}{2} \times 8.31 \times 300 = 7479 \text{ J}$$

44. *(25)* Given, the number of diatomic moles, $n = 3$ mol

The increase in temperature of the diatomic mole,

$$\Delta T = 40^\circ\text{C}$$

Now, the degree of freedom,

f = linear + rotational + no oscillation

$$f = 3 + 2 + 0$$

$$\Rightarrow \quad f = 5$$

Change in internal energy,

$$\Delta U = nC_V \Delta T \quad \dots \text{(i)}$$

where, $C_V = \frac{f}{2} R = \frac{5}{2} R$

Substituting the value in Eq. (i), we get

$$\Delta U = \frac{5R}{2} n\Delta T$$

Now, work done by the gas for isobaric process,

$$W = p\Delta V = nR\Delta T$$

The ratio of the change in internal energy to the work done by the gas,

$$\frac{\Delta U}{W} = \frac{\frac{5}{2} nR\Delta T}{nR\Delta T}$$

$$\Rightarrow \quad \frac{\Delta U}{W} = \frac{5}{2}$$

Multiply and divide the above equation with 5, we get

$$\frac{\Delta U}{W} = \frac{5 \times 5}{2 \times 5} = \frac{25}{10}$$

Comparing with given equation,

$$\frac{\Delta U}{W} = \frac{x}{10}$$

The value of the $x = 25$.

45. *(25)* Given, number of mole, $n = 1$

Heat supplied $= Q$

Work done $= Q/5$

By using first law of thermodynamics,

$$\Delta Q = \Delta U + \Delta W$$

$$\Rightarrow \quad Q = \Delta U + Q/5$$

$$\Rightarrow \quad \Delta U = 4Q/5 = nC_V \Delta T \quad \dots \text{(i)}$$

$$Q = \frac{5}{4}\frac{f}{2} R\Delta T \quad \text{[from Eq. (i)]}$$

$$\Rightarrow \quad \frac{Q}{\Delta T} = C = \frac{5}{4} \times \frac{5}{2} R = \frac{25}{8} R$$

Hence, $x = 25$

46. *(46)* Given, air is diatomic gas

$\therefore$ Degrees of freedom,

$$f = 5$$

So, ratio of specific heats of gas,

$$\gamma = 1 + \frac{2}{f} = \frac{7}{5}$$

Also, given that $V_i = V$

and $\quad V_f = \frac{V}{10}$

$$T_i = 20^\circ\text{C}$$

$$= (273 + 20) \text{ K}$$

$$= 293 \text{ K}$$

As for adibatic change,

$$TV^{\gamma - 1} = \text{constant}$$

$$\Rightarrow \quad T_i V_i^{\gamma - 1} = T_f V_f^{\gamma - 1}$$

$$\Rightarrow 293 \cdot V^{\frac{7}{5} - 1} = T_f \cdot \left(\frac{V}{10}\right)^{\frac{7}{5} - 1}$$

$$\Rightarrow \quad T_f = 293 \times 10^{2/5} \text{K}$$

Change in internal energy is

$$\Delta U = \frac{n \cdot f \cdot R(T_f - T_i)}{2}$$

$$= \frac{5 \times 5 \times \frac{25}{3}(293 \times 10^{2/5} - 293)}{2}$$

$$= 4.195 \times 10^3 \text{ J}$$

$$= 46144 \text{ J} \approx 46 \text{ kJ}$$

C-13

JEE Main Chapterwise
Topicwise Physics

In JEE Main there is usually 0-1 question per paper is visible from this chapter. In past 3 years, 27 conceptual and 21 numerical questions were asked of moderate to difficult level from this chapter.

For detailed analysis scan QR code

OSCILLATIONS

Topic 1 Simple Harmonic Motion

Level 1

01 A particle is executing simple harmonic motion with time period 2 s and amplitude 1 cm. If D and d are the total distance and displacement covered by the particle in 12.5 s, then $\frac{D}{d}$ is

[2025, 24 Jan Shift-I]

(a) 15/4 (b) 16/5 (c) 10 (d) 25

02 In a linear Simple Harmonic Motion (SHM)

A. Restoring force is directly proportional to the displacement.
B. The acceleration and displacement are opposite in direction.
C. The velocity is maximum at mean position.
D. The acceleration is minimum at extreme points.

Choose the correct answer from the options given below. ***[2023, 15 April Shift-I]***

(a) A, B and C only
(b) A, B and D only
(c) A, C and D only
(d) C and D only

03 A particle executes SHM of amplitude A. The distance from the mean position when it's kinetic energy becomes equal to its potential energy is ***[2023, 13 April Shift-II]***

(a) $\frac{1}{\sqrt{2}}A$ (b) $\sqrt{2A}$
(c) $\frac{1}{2}A$ (d) $2A$

04 Which graph represents the difference between total energy and potential energy of a particle executing SHM *versus* it's distance from mean position? ***[2023, 13 April Shift-I]***

(a) 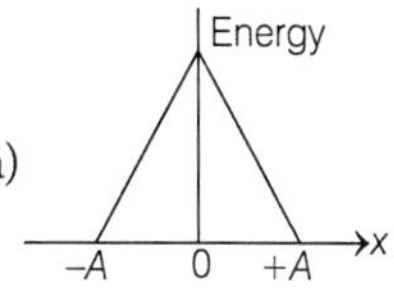

(b) 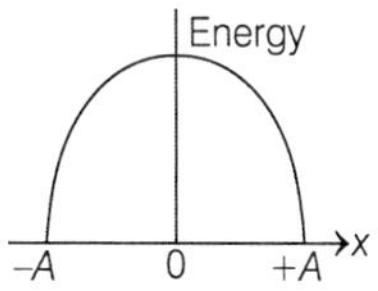

(c) 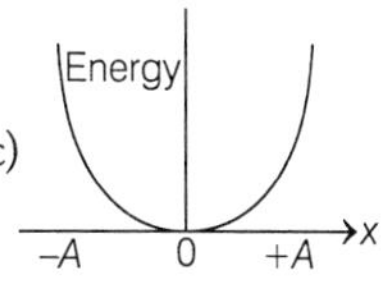

(d) 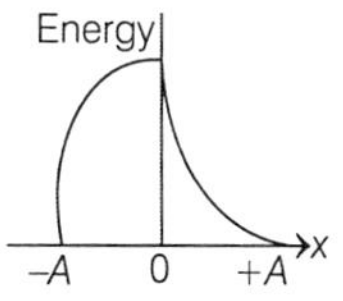

05 The variation of kinetic energy KE of a particle executing simple harmonic motion with the displacement x starting from mean position to extreme position A is given by ***[2023, 11 April Shift-I]***

(a) 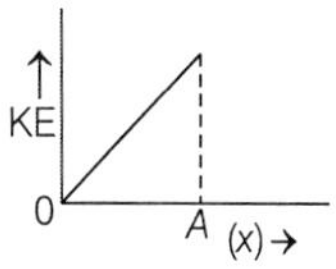

(b) 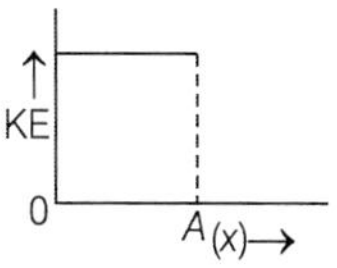

(c) 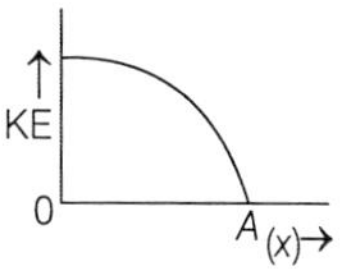

(d) 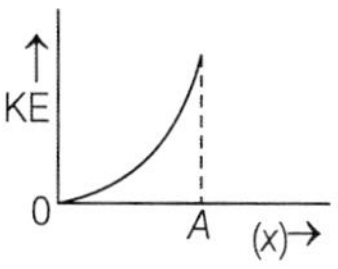

06 A particle executes SHM of amplitude A along X-axis. At $t = 0$, the position of the particle is $x = \frac{A}{2}$ and it moves along positive X-axis. The displacement of particle in time t is $x = A(\sin \omega t + \delta)$, then the value of phase difference will be ***[2023, 10 April Shift-I]***

(a) $\frac{\pi}{4}$ (b) $\frac{\pi}{6}$
(c) $\frac{\pi}{3}$ (d) $\frac{\pi}{2}$

Imp. Kinetic Energy of Block Performing SHM

07 The maximum potential energy of a block executing simple harmonic motion is 25 J. A is amplitude of oscillation. At $A/2$, the kinetic energy of the block is *[2023, 31 Jan Shift-I]*
(a) 18.75 J (b) 12.5 J (c) 37.5 J (d) 9.75 J

08 For a body executing SHM
A. potential energy is always equal to its kinetic energy.
B. average potential and kinetic energy over any given time interval are always equal.
C. sum of the kinetic and potential energy at any point of time is constant.
D. average kinetic energy in one time period is equal to average potential energy in one time period.
Choose the most appropriate option from the options given below. *[2021, 31 Aug Shift-II]*
(a) (C) and (D) (b) Only (C)
(c) (B) and (C) (d) Only (B)

09 The function of time representing a simple harmonic motion with a period of π/ω is *[2021, 18 March Shift-II]*
(a) $\sin(\omega t) + \cos(\omega t)$
(b) $\cos(\omega t) + \cos(2\omega t) + \cos(3\omega t)$
(c) $\sin^2(2\omega t)$
(d) $3\cos\left(\frac{\pi}{4} - 2\omega t\right)$

10 For what value of displacement the kinetic energy and potential energy of a simple harmonic oscillation become equal? *[2021, 17 March Shift-I]*
(a) $x = 0$ (b) $x = \pm A$ (c) $x = \pm\frac{A}{\sqrt{2}}$ (d) $x = \frac{A}{2}$

11 $Y = A\sin(\omega t + \phi_0)$ is the time-displacement equation of SHM. At $t = 0$, the displacement of the particle is $Y = \frac{A}{2}$ and it is moving along negative x-direction. Then, the initial phase angle ϕ_0 will be *[2021, 25 Feb Shift-II]*
(a) $\frac{\pi}{3}$ (b) $\frac{5\pi}{6}$ (c) $\frac{\pi}{6}$ (d) $\frac{2\pi}{3}$

12 When a particle executes SHM, the nature of graphical representation of velocity as a function of displacement is *[2021, 24 Feb Shift-II]*
(a) circular (b) elliptical
(c) parabolic (d) straight line

13 A particle executes simple harmonic motion with an amplitude of 5 cm. When the particle is at 4 cm from the mean position, the magnitude of its velocity in SI units is equal to that of its acceleration. Then, its periodic time (in seconds) is *[2019, 10 Jan Shift-II]*
(a) $\frac{4\pi}{3}$ (b) $\frac{8\pi}{3}$ (c) $\frac{7}{3}\pi$ (d) $\frac{3}{8}\pi$

14 A particle is executing simple harmonic motion with a time period T. At time $t = 0$, it is at its position of equilibrium. The kinetic energy-time graph of the particle will look, like *[JEE Main 2017]*

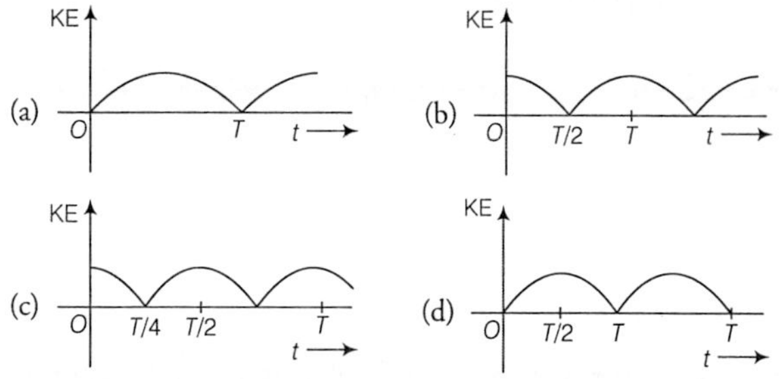

15 If x, v and a denote the displacement, the velocity and the acceleration of a particle executing simple harmonic motion of time period T, then which of the following does not change with time? *[AIEEE 2009]*
(a) $a^2T^2 + 4\pi^2v^2$ (b) $\frac{aT}{x}$
(c) $aT + 2\pi v$ (d) $\frac{aT}{v}$

16 A particle of mass m executes simple harmonic motion with amplitude a and frequency ν. The average kinetic energy during its motion from the position of equilibrium to the end is *[AIEEE 2007]*
(a) $\pi^2 ma^2\nu^2$ (b) $\frac{1}{4}ma^2\nu^2$
(c) $4\pi^2 ma^2\nu^2$ (d) $2\pi^2 ma^2\nu^2$

17 The maximum velocity of a particle, executing simple harmonic motion with an amplitude 7 mm, is 4.4 ms^{-1}. The period of oscillation is *[AIEEE 2006]*
(a) 0.01 s (b) 10 s
(c) 0.1 s (d) 100 s

18 If a simple harmonic motion is represented by $\frac{d^2x}{dt^2} + \alpha x = 0$, its time period is *[AIEEE 2005]*
(a) $\frac{2\pi}{\alpha}$ (b) $\frac{2\pi}{\sqrt{\alpha}}$
(c) $2\pi\alpha$ (d) $2\pi\sqrt{\alpha}$

19 The total energy of a particle, executing simple harmonic motion is *[AIEEE 2004]*
(a) $\propto x$ (b) $\propto x^2$
(c) independent of x (d) $\propto x^{1/2}$
where, x is the displacement from the mean position.

20 In a simple harmonic oscillator, at the mean position *[AIEEE 2002]*
(a) kinetic energy is minimum, potential energy is maximum
(b) both kinetic and potential energies are maximum
(c) kinetic energy is maximum, potential energy is minimum
(d) both kinetic and potential energies are minimum

Level 2

21 A particle is subjected to two simple harmonic motions as:

$x_1 = \sqrt{7} \sin 5t$ cm

and $x_2 = 2\sqrt{7} \sin\left(5t + \frac{\pi}{3}\right)$ cm

where x is displacement and t is time in seconds.

The maximum acceleration of the particle is $x \times 10^{-2}$ m/s^2. The value of x is ***[2025, 2 April Shift-I]***

(a) $5\sqrt{7}$ (b) 125
(c) 175 (d) $25\sqrt{7}$

22 Given below are two statements. One is labelled as Assertion (A) and the other is labelled as Reason (R).

Assertion(A) Knowing initial position x_0 and initial momentum p_0 is enough to determine the position and momentum at any time for a simple harmonic motion with a given angular frequency ω.

Reason (R) The amplitude and phase can be expressed in terms of x_0 and p_0.

In the light of the above statements, choose the correct answer from the options given below. ***[2025, 28 Jan Shift-II]***

(a) (A) is false but (R) is true.
(b) (A) is true but (R) is false.
(c) Both (A) and (R) are true and (R) is not the correct explanation of (A).
(d) Both (A) and (R) are true and (R) is the correct explanation of (A).

23 A particle is executing simple harmonic motion (SHM). The ratio of potential energy and kinetic energy of the particle when its displacement is half of its amplitude will be ***[2023, 12 April Shift-I]***

(a) 2 : 1 (b) 1 : 3 (c) 1 : 4 (d) 1 : 1

24 A particle executes simple harmonic motion between $x = -A$ and $x = +A$. If time taken by particle to go from $x = 0$ to $\frac{A}{2}$ is 2 s, then time taken by particle in going from $x = \frac{A}{2}$ to A is ***[2023, 25 Jan Shift-II]***

(a) 3 s (b) 1.5 s
(c) 4 s (d) 2 s

Imp. Velocity in SHM

25 The equation of a particle executing simple harmonic motion is given by $x = \sin \pi\left(t + \frac{1}{3}\right)$ m.

At $t = 1$s the speed of particle will be (Given, $\pi = 3.14$) ***[2022, 27 June Shift-II]***

(a) 0 cm s^{-1} (b) 157 cm s^{-1}
(c) 272 cm s^{-1} (d) 314 cm s^{-1}

26 The displacement of simple harmonic oscillator after 3 s starting from its mean position is equal to half of its amplitude. The time period of harmonic motion is ***[2022, 27 June Shift-I]***

(a) 6 s (b) 8 s
(c) 12 s (d) 36 s

27 The variation of displacement with time of a particle executing free simple harmonic motion is shown in the figure.

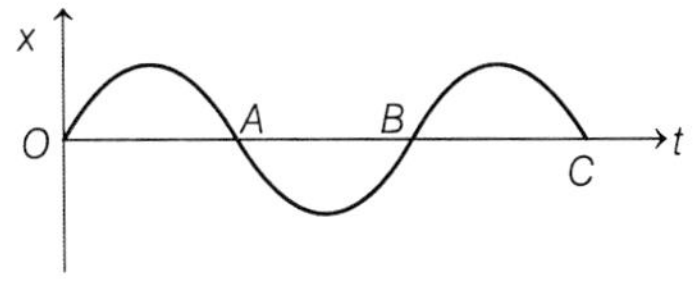

The potential energy $U(x)$ *versus* time (t) plot of the particle is correctly shown in figure ***[2021, 27 Aug Shift-I]***

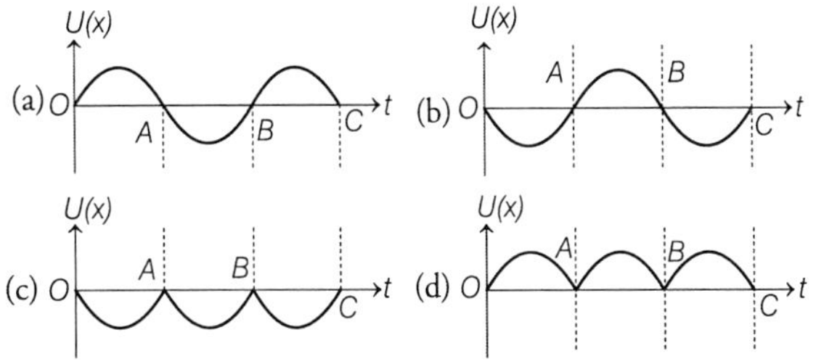

28 A particle starts executing simple harmonic motion (SHM) of amplitude a and total energy E. At any instant, its kinetic energy is $3E/4$, then its displacement y is given by ***[2021, 27 July Shift-I]***

(a) $y = a$ (b) $y = \frac{a}{\sqrt{2}}$
(c) $y = \frac{a\sqrt{3}}{2}$ (d) $y = \frac{a}{2}$

Imp. Mechanical Energy in SHM

29 In a simple harmonic oscillation, what fraction of total mechanical energy is in the form of kinetic energy, when the particle is mid-way between mean and extreme position. ***[2021, 25 July Shift-II]***

(a) $\frac{1}{2}$ (b) $\frac{3}{4}$ (c) $\frac{1}{3}$ (d) $\frac{1}{4}$

30 A particle is making simple harmonic motion along the X-axis. If at a distances x_1 and x_2 from the mean position, the velocities of the particle are v_1 and v_2 respectively, then the time period of its oscillation is given as ***[2021, 20 July Shift-II]***

(a) $T = 2\pi\sqrt{\frac{x_2^2 + x_1^2}{v_1^2 - v_2^2}}$ (b) $T = 2\pi\sqrt{\frac{x_2^2 + x_1^2}{v_1^2 + v_2^2}}$

(c) $T = 2\pi\sqrt{\frac{x_2^2 - x_1^2}{v_1^2 + v_2^2}}$ (d) $T = 2\pi\sqrt{\frac{x_2^2 - x_1^2}{v_1^2 - v_2^2}}$

31 A particle executes SHM, the graph of velocity as a function of displacement is *[2021, 26 Feb Shift-II]*
(a) a circle (b) a parabola (c) an ellipse (d) a helix

32 The displacement-time graph of a particle executing SHM is given in figure (sketch is schematic and not to scale) *[2020, 2 Sep Shift-II]*

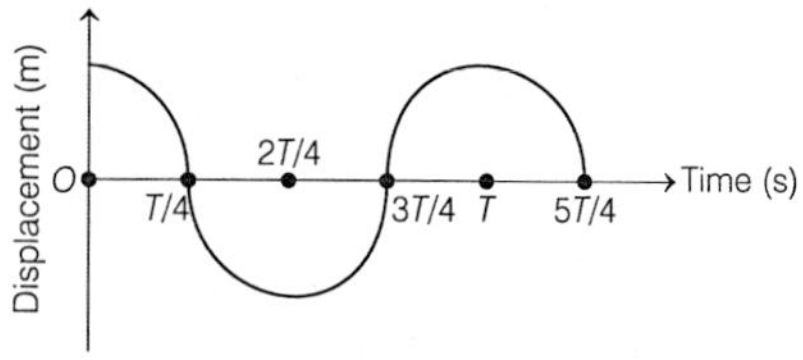

Which of the following statement(s) is/are true for this motion?

A. The force is zero at $t = \frac{3T}{4}$.
B. The acceleration is maximum at $t = T$.
C. The speed is maximum at $t = \frac{T}{4}$.
D. The potential energy is equal to kinetic energy of the oscillation at $t = \frac{T}{2}$.

(a) A, B and D (b) B, C and D
(c) A, B and C (d) A and D

33 A particle performs simple harmonic motion with amplitude A. Its speed is tripled at the instant that it is at a distance $2/3A$ from equilibrium position. The new amplitude of the motion is *[JEE Main 2016]*
(a) $\frac{A}{3}\sqrt{41}$ (b) $3A$
(c) $A\sqrt{3}$ (d) $\frac{7}{3}A$

34 A particle moves with simple harmonic motion in a straight line. In first t sec, after starting from rest, it travels a distance a and in next t s, it travels $2a$ in same direction, then *[JEE Main 2014]*
(a) amplitude of motion is $3a$
(b) time period of oscillations is 5π
(c) amplitude of motion is $4a$
(d) time period of oscillations is 6π

35 Two particles are executing simple harmonic motion of the same amplitude A and frequency ω along the x-axis. Their mean position is separated by distance X_0 $(X_0 > A)$. If the maximum separation between them is $(X_0 + A)$, the phase difference between their motion is *[AIEEE 2011]*
(a) $\frac{\pi}{3}$ (b) $\frac{\pi}{4}$
(c) $\frac{\pi}{6}$ (d) $\frac{\pi}{2}$

36 A point mass oscillates along the x-axis according to the law $x = x_0 \cos(\omega t - \pi/4)$. If the acceleration of the particle is written as $a = A\cos(\omega t + \delta)$, then *[AIEEE 2007]*
(a) $A = x_0, \delta = -\frac{\pi}{4}$ (b) $A = x_0\omega^2, \delta = \frac{\pi}{4}$
(c) $A = x_0\omega^2, \delta = -\frac{\pi}{4}$ (d) $A = x_0\omega^2, \delta = \frac{3\pi}{4}$

37 The function $\sin^2(\omega t)$ represents *[AIEEE 2005]*
(a) a periodic but not simple harmonic motion with a period $2\pi/\omega$
(b) a periodic but not simple harmonic motion with a period π/ω
(c) a simple harmonic motion with a period $2\pi/\omega$
(d) a simple harmonic motion with a period π/ω

38 The displacement of a particle varies according to the relation $x = 4(\cos \pi t + \sin \pi t)$. The amplitude of the particle is *[AIEEE 2003]*
(a) -4 (b) 4 (c) $4\sqrt{2}$ (d) 8

Level 3

39 The blocks of masses m and M, $(M > m)$ are placed on a frictionless table as shown in figure. A massless spring with spring constant k is attached with the lower block. If the system is slightly displaced and released, then (μ = coefficient of friction between the two blocks) *[2025, 3 April Shift-I]*

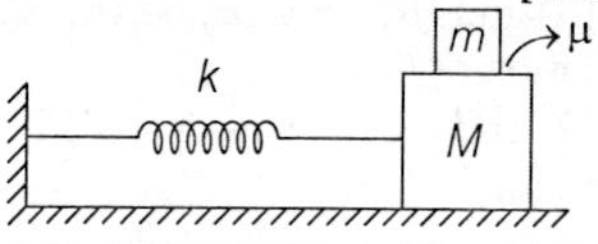

A. The time period of small oscillation of the two blocks is $T = 2\pi\sqrt{\frac{(m+M)}{k}}$
B. The acceleration of the blocks is $a = -\frac{kx}{M+m}$
(x = displacement of the blocks from the mean position)
C. The magnitude of the frictional force on the upper block is $\frac{m\mu|x|}{M+m}$.
D. The maximum amplitude of the upper block, if it does not slip, is $\frac{\mu(M+m)g}{k}$.
E. Maximum frictional force can be $\mu(M+m)g$.

Choose the correct answer from the options given below.
(a) A, B, D Only (b) A, B, C Only
(c) C, D, E Only (d) B, C, D Only

40 A particle oscillates along the X-axis according to the law, $x(t) = x_0 \sin^2\left(\frac{t}{2}\right)$, where $x_0 = 1$ m. The kinetic energy (K) of the particle as a function of x is correctly represented by the graph *[2025, 24 Jan Shift-I]*

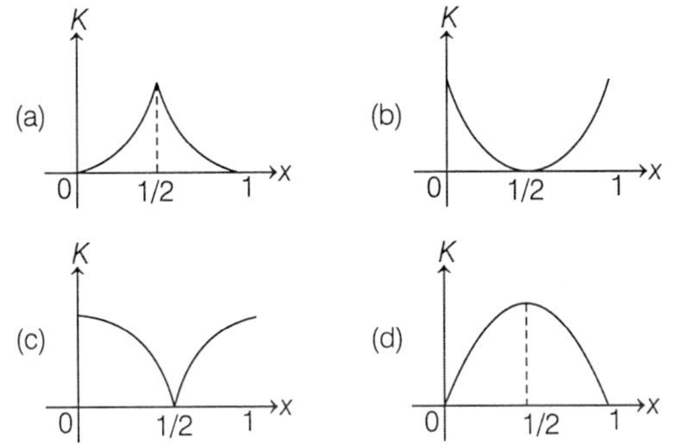

41 The point A moves with a uniform speed along the circumference of a circle of radius 0.36 m and covers 30° in 0.1 s. The perpendicular projection P from A on the diameter MN represents the simple harmonic motion of P. The restoration force per unit mass when P touches M will be *[2021, 25 Feb Shift-II]*

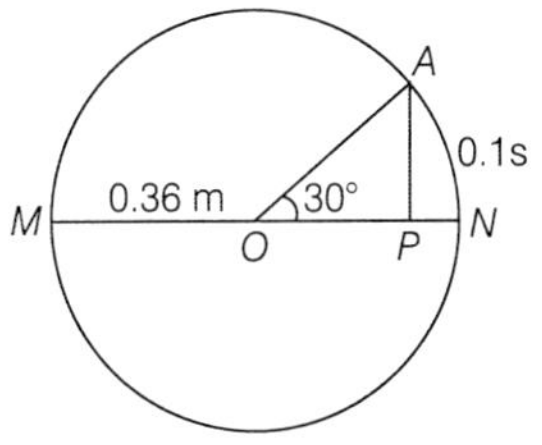

(a) 100 N (b) 9.87 N (c) 50 N (d) 0.49 N

42 A pendulum clock loses 12 s a day, if the temperature in 40°C and gains 4 s a day, if the temperature is 20°C. The temperature at which the clock will show correct time, and the coefficient of linear expansion respectively. *[JEE Main 2016]*

(a) $25°C, \alpha = 1.85 \times 10^{-5}/°C$

(b) $60°C, \alpha = 1.85 \times 10^{-4}/°C$

(c) $30°C, \alpha = 1.85 \times 10^{-3}/°C$

(d) $55°C, \alpha = 1.85 \times 10^{-2}/°C$

43 Starting from the origin, a body oscillates simple harmonically with a period of 2 s. After what time will its kinetic energy be 75% of the total energy? *[AIEEE 2006]*

(a) $\frac{1}{6}$ s (b) $\frac{1}{4}$ s (c) $\frac{1}{3}$ s (d) $\frac{1}{12}$ s

44 Two simple harmonic motions are represented by the equations $y_1 = 0.1 \sin\left(100\pi t + \frac{\pi}{3}\right)$ and $y_2 = 0.1 \cos \pi t$. The phase difference of the velocity of particle 1 with respect to the velocity of particle 2 is *[AIEEE 2005]*

(a) $\frac{-\pi}{6}$ (b) $\frac{\pi}{3}$

(c) $\frac{-\pi}{3}$ (d) $\frac{\pi}{6}$

Numerical Value Type Questions

45 A particle of mass 0.50 kg executes simple harmonic motion under force $F = -50\ (\text{Nm}^{-1})x$. The time period of oscillation is $\frac{x}{35}$ s. The value of x is ....

(Given, $\pi = \frac{22}{7}$) *[2024, 9 April Shift-II]*

46 The position, velocity and acceleration of a particle executing simple harmonic motion are found to have magnitudes of 4 m, 2 ms^{-1} and 16 ms^{-2} at a certain instant. The amplitude of the motion is $\sqrt{x}$, m where x is .... *[2024, 9 April Shift-I]*

47 An object of mass 0.2 kg executes simple harmonic motion along X-axis with frequency of $\left(\frac{25}{\pi}\right)$ Hz. At the position $x = 0.04$ m the object has kinetic energy 0.5 J and potential energy 0.4 J. The amplitude of oscillation is .... cm. *[2024, 8 April Shift-II]*

48 A particle is doing simple harmonic motion of amplitude 0.06 m and time period 3.14 s. The maximum velocity of the particle is ..... cm/s. *[2024, 6 April Shift-I]*

49 The displacement of a particle executing SHM is given by $x = 10 \sin\left(\omega t + \frac{\pi}{3}\right)$ m. The time period of motion is 3.14 s. The velocity of the particle at $t = 0$ is...... m/s. *[2024, 4 April Shift-II]*

50 A mass m is suspended from a spring of negligible mass and the system oscillates with a frequency f_1. The frequency of oscillations if a mass 9 m is suspended from the same spring is f_2. The value of f_1/f_2 is .... *[2024, 1 Feb Shift-II]*

51 A particle performs simple harmonic motion with amplitude A. Its speed is increased to three times at an instant when its displacement is $\frac{2A}{3}$. The new amplitude of motion is $nA/3$. The value of n is .... *[2024, 31 Jan Shift-I]*

52 A simple harmonic oscillator has an amplitude A and time period 6π s. Assuming the oscillation starts from its mean position, the time required by it to travel from $x = A$ to $x = \frac{\sqrt{3}}{2}$ A will be $\frac{\pi}{x}$ s, where $x =$ .... *[2024, 29 Jan Shift-II]*

53 When the displacement of a simple harmonic oscillator is one-third of its amplitude, the ratio of total energy to the kinetic energy is $x/8$, where $x =$ .... *[2024, 29 Jan Shift-I]*

54 A particle executes simple harmonic motion with an amplitude of 4 cm. At the mean position, velocity of the particle is 10 cm/s. The distance of the particle from the mean position, when its speed becomes 5 cm/s is $\sqrt{\alpha}$ cm, where $\alpha =$ .... *[2024, 27 Jan Shift-I]*

55 At a given point of time the value of displacement of a simple harmonic oscillator given as $y = A\cos(30°)$. If amplitude is 40 cm and kinetic energy at that time is 200 J, the value of force constant is $1.0 \times 10^{x}\ \text{Nm}^{-1}$. The value of x is .... *[2023, 13 April Shift-I]*

56 A simple pendulum with length 100 cm and bob of mass 250 g is executing SHM of amplitude 10 cm. The maximum tension in the string is found to be $x/40$ N. The value of x is .... *[2023, 6 April Shift-II]*

57 The amplitude of a particle executing SHM is 3 cm. The displacement at which its kinetic energy will be 25% more than the potential energy is .... cm. *[2023, 1 Feb Shift-I]*

58 The velocity of a particle executing SHM varies with displacement x as $4v^2 = 50 - x^2$. The time period of oscillations is $x/7$ s. The value of x is ....
$\left(\text{Take}, \pi = \dfrac{22}{7}\right)$ *[2023, 30 Jan Shift-II]*

59 A particle of mass 250 g executes a simple harmonic motion under a periodic force $F = (-25x)$ N. The particle attains a maximum speed of 4 m/s during its oscillation. The amplitude of the motion is .... cm. *[2023, 29 Jan Shift-II]*

60 A particle executes simple harmonic motion. Its amplitude is 8 cm and time period is 6s. The time it will take to travel from its position of maximum displacement to the point corresponding to half of its amplitude, is .... s. *[2022, 27 June Shift-II]*

61 Two simple harmonic motion, are represented by the equations $y_1 = 10\sin\left(3\pi t + \dfrac{\pi}{3}\right)$

and $y_2 = 5(\sin 3\pi t + \sqrt{3}\cos 3\pi t)$

Ratio of amplitude of y_1 to $y_2 = x : 1$. The value of x is .... *[2021, 27 Aug Shift-II]*

62 Two simple harmonic motions are represented by the equations

$x_1 = 5\sin\left(2\pi t + \dfrac{\pi}{4}\right)$ and $x_2 = 5\sqrt{2}(\sin 2\pi t + \cos 2\pi t)$

The amplitude of second motion is .... times the amplitude in first motion. *[2021, 26 Aug Shift-II]*

63 A particle executes simple harmonic motion represented by displacement function as

$$x(t) = A\sin(\omega t + \phi)$$

If the position and velocity of the particle at $t = 0$ s are 2 cm and $2\omega\ \text{cms}^{-1}$ respectively, then its amplitude is $x\sqrt{2}$ cm, where the value of x is .... *[2021, 27 July Shift-II]*

64 A particle performs simple harmonic motion with a period of 2 s. The time taken by the particle to cover a displacement equal to half of its amplitude from the mean position is $1/a$ s. The value of a to the nearest integer is .... *[2021, 18 March Shift-I]*

65 A particle executes SHM with amplitude a and time period T. The displacement of the particle when its speed is half of maximum speed is $\sqrt{x}a/2$. The value of x is .... *[2021, 26 Feb Shift-II]*

Topic 2 Some Systems Excecuting SHM

01 Two simple pendulums having lengths l_1 and l_2 with negligible string mass undergo angular displacements θ_1 and θ_2, from their mean positions, respectively. If the angular accelerations of both pendulums are same, then which expression is correct? *[2025, 4 April Shift-I]*

(a) $\theta_1 l_1 = \theta_2 l_2$
(b) $\theta_1 l_2 = \theta_2 l_1$
(c) $\theta_1 l_1^2 = \theta_2 l_2^2$
(d) $\theta_1 l_2^2 = \theta_2 l_1^2$

02 Given below are two statements. One is labelled as Assertion (A) and the other is labelled as Reason (R).
Assertion (A) Time period of a simple pendulum is longer at the top of a mountain than that at the base of the mountain.
Reason (R) Time period of a simple pendulum decreases with increasing value of acceleration due to gravity and *vice-versa*.
In the light of the above statements, choose the most appropriate answer from the options given below. *[2025, 29 Jan Shift-I]*

(a) Both A and R are true but R is not the correct explanation of A.
(b) A is false but R is true.
(c) Both A and R are true and R is the correct explanation of A.
(d) A is true but R is false.

03 In simple harmonic motion, the total mechanical energy of given system is E. If mass of oscillating particle P is doubled then the new energy of the system for same amplitude is

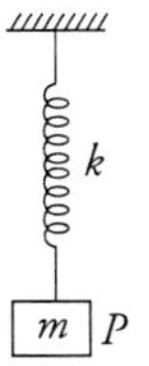

[2024, 4 April Shift-II]

(a) $\frac{E}{\sqrt{2}}$ (b) $E\sqrt{2}$ (c) $2E$ (d) E

Imp. Oscillation of a Body Connected Through Spring

04 A mass m is attached to two strings as shown in figure. The spring constants of two springs are k_1 and k_2. For the frictionless surface, the time period of oscillation of mass m is [2023, 6 April Shift-I]

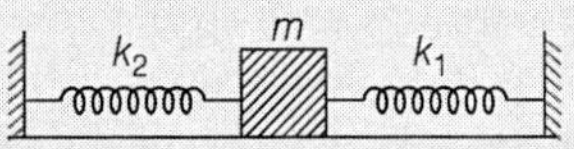

(a) $2\pi\sqrt{\frac{m}{k_1+k_2}}$ (b) $\frac{1}{2\pi}\sqrt{\frac{k_1+k_2}{m}}$

(c) $\frac{1}{2\pi}\sqrt{\frac{k_1-k_2}{m}}$ (d) $2\pi\sqrt{\frac{m}{k_1-k_2}}$

05 Choose the correct length L *versus* square of time period T^2 graph for a simple pendulum executing simple harmonic motion. [2023, 1 Feb Shift-II]

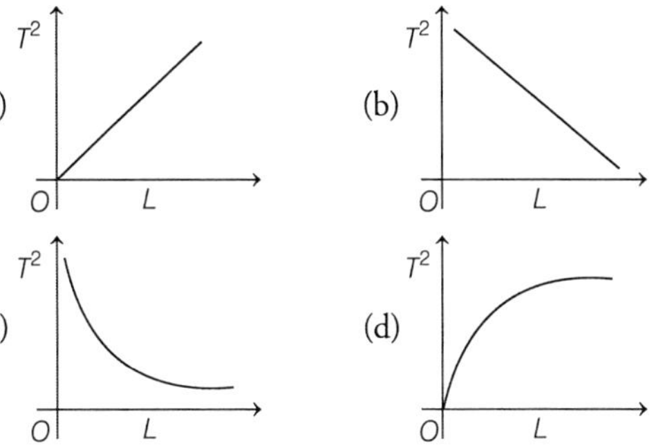

06 For a simple harmonic motion in a mass spring system shown, the surface is frictionless. When the mass of the block is 1 kg, the angular frequency is ω_1. When the mass block is 2 kg the angular frequency is ω_2. The ratio ω_2/ω_1 is [2023, 30 Jan Shift-II]

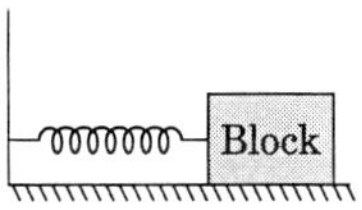

(a) $\frac{1}{\sqrt{2}}$ (b) $\frac{1}{2}$ (c) 2 (d) $\sqrt{2}$

07 Assume that the earth is a solid sphere of uniform density and a tunnel is dug along its diameter throughout the earth. It is found that when a particle is released in this tunnel. It executes a simple harmonic motion. The mass of the particle is 100 g. The time period of the motion of the particle will be (approximately)
(Take, $g = 10\ \text{ms}^{-2}$, radius of earth $= 6400$ km) [2023, 25 Jan Shift-I]

(a) 24 h (b) 1 h 40 min
(c) 1 h 24 min (d) 12 h

08 Given below are two statements. One is labelled as Assertion (A) and the other is labelled as Reason (R).
Assertion (A) A pendulum clock when taken to Mount Everest becomes fast.
Reason (R) The value of g (acceleration due to gravity) is less at Mount Everest than its value on the surface of earth.
In the light of the above statements, choose the most appropriate answer from the options given below. [2023, 24 Jan Shift-II]

(a) Both A and R are correct and R is the correct explanation of A.
(b) Both A and R are correct but R is not the correct explanation of A.
(c) A is not correct but R is correct.
(d) A is correct but R is not correct.

09 T_0 is the time period of a simple pendulum at a place. If the length of the pendulum is reduced to 1/16 times of its initial value, then the modified time period is [2021, 22 July Shift-II]

(a) T_0 (b) $8\pi T_0$ (c) $4T_0$ (d) $1/4\,T_0$

10 Two particles A and B of equal masses are suspended from two massless springs of spring constants k_1 and k_2, respectively. If the maximum velocities during oscillations are equal, the ratio of the amplitude of A and B is [2021, 17 March Shift-II, AIEEE 2003]

(a) $\frac{k_2}{k_1}$ (b) $\frac{k_1}{k_2}$ (c) $\sqrt{\frac{k_1}{k_2}}$ (d) $\sqrt{\frac{k_2}{k_1}}$

11 Given below are two statements.
Statement I A second's pendulum has a time period of 1 s.
Statement II It takes precisely one second to move between the two extreme positions.
In the light of the above statements, choose the correct answer from the options given below. [2021, 26 Feb Shift-II]

(a) Both Statement I and Statement II are false.
(b) Statement I is false but Statement II is true.
(c) Statement I is true but Statement II is false.
(d) Both Statement I and Statement II are true.

12 If two similar springs each of spring constant k_1 are joined in series, the new spring constant and time period would be changed by a factor [2021, 26 Feb Shift-I]

(a) $\frac{1}{2}, \sqrt{2}$ (b) $\frac{1}{4}, \sqrt{2}$ (c) $\frac{1}{4}, 2\sqrt{2}$ (d) $\frac{1}{2}, 2\sqrt{2}$

13 Two identical springs of spring constant $2k$ are attached to a block of mass m and to fixed support (see figure). When the mass is displaced from equilibrium position on either side, it executes simple harmonic motion. The time period of oscillations of this system is *[2021, 25 Feb Shift-II]*

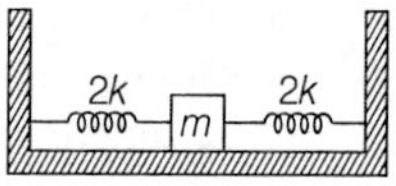

(a) $2\pi\sqrt{\frac{m}{2k}}$ (b) $2\pi\sqrt{\frac{m}{k}}$ (c) $\pi\sqrt{\frac{m}{k}}$ (d) $\pi\sqrt{\frac{m}{2k}}$

14 A child swinging on a swing in sitting position, stands up, then the time period of the swing will *[AIEEE 2002]*

(a) increase
(b) decrease
(c) remain same
(d) increase, if the child is long and decrease, if the child is short

Level 2

15 Given below are two statements. One is labelled as Assertion (A) and the other is labelled as Reason (R).
Assertion (A) A simple pendulum is taken to a planet of mass and radius, 4 times and 2 times, respectively, than the Earth. The time period of the pendulum remains same on Earth and the planet.
Reason (R) The mass of the pendulum remains unchanged at Earth and the other planet.
In the light of the above statements, choose the correct answer from the options given below. *[2025, 22 Jan Shift-II]*

(a) Both A and R are true but R is not the correct explanation of A.
(b) Both A and R are true and R is the correct explanation of A.
(c) A is false but R is true.
(d) A is true but R is false.

16 T is the time period of simple pendulum on the earth's surface. Its time period becomes xT when taken to a height R (equal to earth's radius) above the earth's surface. Then, the value of x will be *[2023, 25 Jan Shift-I]*

(a) 2 (b) 1/2 (c) 4 (d) 1/4

17 The time period of oscillation of a simple pendulum of length L suspended from the roof of a vehicle, which moves without friction down an inclined plane of inclination α, is given by *[2022, 29 July Shift-I]*

(a) $2\pi\sqrt{L/(g\cos\alpha)}$ (b) $2\pi\sqrt{L/(g\sin\alpha)}$
(c) $2\pi\sqrt{L/g}$ (d) $2\pi\sqrt{L/(g\tan\alpha)}$

18 One end of a massless spring of spring constant k and natural length l_0 is fixed, while the other end is connected to a small object of mass m lying on a frictionless table. The spring remains horizontal on the table. If the object is made to rotate at an angular velocity ω about an axis passing through fixed end, then the elongation of the spring will be *[2022, 27 June Shift-II]*

(a) $\frac{k-m\omega^2 l_0}{m\omega^2}$ (b) $\frac{m\omega^2 l_0}{k+m\omega^2}$
(c) $\frac{m\omega^2 l_0}{k-m\omega^2}$ (d) $\frac{k+m\omega^2 l_0}{m\omega^2}$

19 Time period of a simple pendulum in a stationary lift is T. If the lift accelerates with $g/6$ vertically upwards, then the time period will be (where, g = acceleration due to gravity) *[2022, 26 June Shift-I]*

(a) $\sqrt{\frac{6}{5}}T$ (b) $\sqrt{\frac{5}{6}}T$ (c) $\sqrt{\frac{6}{7}}T$ (d) $\sqrt{\frac{7}{6}}T$

20 Two massless springs with spring constants 2 k and 9 k, carry 50 g and 100 g masses at their free ends. These two masses oscillate vertically such that their maximum velocities are equal. Then, the ratio of their respective amplitudes will be *[2022, 24 June Shift-II]*

(a) 1 : 2 (b) 3 : 2 (c) 3 : 1 (d) 2 : 3

21 A mass of 5 kg is connected to a spring. The potential energy curve of the simple harmonic motion executed by the system is shown in the figure. A simple pendulum of length 4 m has the same period of oscillation as the spring system. What is the value of acceleration due to gravity on the planet where these experiments are performed? *[2021, 1 Sep Shift-II]*

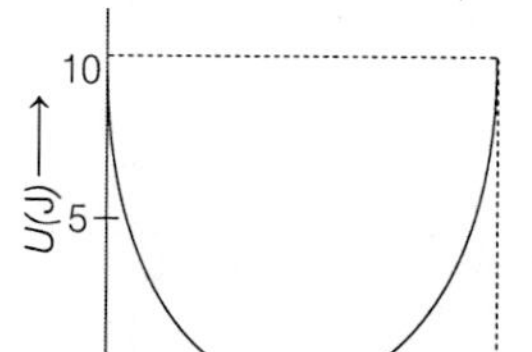

(a) $10\ m/s^2$ (b) $5\ m/s^2$
(c) $4\ m/s^2$ (d) $9.8\ m/s^2$

Imp. Oscillation of Simple Pendulum Inside a Lift

22 Time period of a simple pendulum is T inside a lift, when the lift is stationary. If the lift moves upwards with an acceleration $g/2$, then the time period of pendulum will be *[2021, 16 March Shift-I]*

(a) $\sqrt{3}T$ (b) $\frac{T}{\sqrt{3}}$ (c) $\sqrt{\frac{3}{2}}T$ (d) $\sqrt{\frac{2}{3}}T$

23 If the time period of a 2 m long simple pendulum is 2 s, the acceleration due to gravity at the place, where pendulum is executing SHM is *[2021, 25 Feb Shift-I]*

(a) $\pi^2 ms^{-2}$ (b) $9.8\ ms^{-2}$
(c) $2\pi^2\ ms^{-2}$ (d) $16\ ms^{-2}$

24 In the given figure, a body of mass M is held between two massless springs, on a smooth inclined plane. The free ends of the springs are attached to firm supports. If each spring has spring constant k, then the frequency of oscillation of given body is **[2021, 24 Feb Shift-II]**

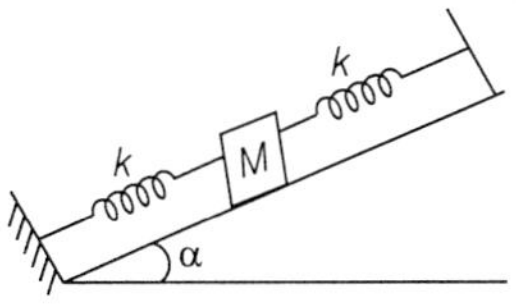

(a) $\frac{1}{2\pi}\sqrt{\frac{k}{2M}}$ (b) $\frac{1}{2\pi}\sqrt{\frac{2k}{Mg\sin\alpha}}$
(c) $\frac{1}{2\pi}\sqrt{\frac{2k}{M}}$ (d) $\frac{1}{2\pi}\sqrt{\frac{k}{Mg\sin\alpha}}$

25 A spring mass system (mass m, spring constant k and natural length l_0) rests in equilibrium on horizontal disc. The free end of the spring is fixed at the centre of the disc. If the disc together with spring mass system, rotates about it's axis with an angular velocity ω, ($k >> m\omega^2$), the relative change in the length of the spring is best given by the option **[2020, 9 Jan Shift-II]**
(a) $\frac{m\omega^2}{3k}$ (b) $\frac{2m\omega^2}{k}$ (c) $\sqrt{\frac{2}{3}}\left(\frac{m\omega^2}{k}\right)$ (d) $\frac{m\omega^2}{k}$

26 A particle of mass m is fixed to one end of a light spring having force constant k and unstretched length l. The other end is fixed. The system is given an angular speed ω about the fixed end of the spring such that it rotates in a circle in gravity free space. Then, the stretch in the spring is **[2020, 8 Jan Shift-I]**
(a) $\frac{ml\omega^2}{k+m\omega}$ (b) $\frac{ml\omega^2}{k+m\omega^2}$
(c) $\frac{ml\omega^2}{k-m\omega^2}$ (d) $\frac{ml\omega^2}{k-m\omega}$

27 A pendulum is executing simple harmonic motion and its maximum kinetic energy is K_1. If the length of the pendulum is doubled and it performs simple harmonic motion with the same amplitude as in the first case, its maximum kinetic energy is K_2. Then **[2019, 11 Jan Shift-II]**
(a) $K_2 = 2K_1$ (b) $K_2 = \frac{K_1}{2}$
(c) $K_2 = \frac{K_1}{4}$ (d) $K_2 = K_1$

28 A silver atom in a solid oscillates in simple harmonic motion in some direction with a frequency of 10^{12} per second. What is the force constant of the bonds connecting one atom with the other? (Take, molecular weight of silver = 108 and Avogadro number $= 6.02 \times 10^{23}$ g mol^{-1}) **[JEE Main 2018]**
(a) 6.4 N/m (b) 7.1 N/m
(c) 2.2 N/m (d) 5.5 N/m

29 A pendulum made of a uniform wire of cross-sectional area A has time period T. When an additional mass M is added to its bob, the time period changes T_M. If the Young's modulus of the material of the wire is Y, then $1/Y$ is equal to (g = gravitational acceleration) **[JEE Main 2015]**
(a) $\left[\left(\frac{T_M}{T}\right)^2 - 1\right]\frac{A}{Mg}$ (b) $\left[\left(\frac{T_M}{T}\right)^2 - 1\right]\frac{Mg}{A}$
(c) $\left[1-\left(\frac{T_M}{T}\right)^2\right]\frac{A}{Mg}$ (d) $\left[1-\left(\frac{T}{T_M}\right)^2\right]\frac{A}{Mg}$

30 A wooden cube (density of wood d) of side l floats in a liquid of density ρ with its upper and lower surfaces horizontal. If the cube is pushed slightly down and released, it performs simple harmonic motion of period, T. Then, T is equal to **[AIEEE 2011]**
(a) $2\pi\sqrt{\frac{l\rho}{(\rho-d)g}}$ (b) $2\pi\sqrt{\frac{ld}{\rho g}}$
(c) $2\pi\sqrt{\frac{l\rho}{dg}}$ (d) $2\pi\sqrt{\frac{ld}{(\rho-d)g}}$

31 A particle of mass m is attached to a spring (of spring constant k) and has a natural angular frequency ω_0. An external force $F(t)$ proportional to $\cos\omega t (\omega \neq \omega_0)$ is applied to the oscillator. The time displacement of the oscillator will be proportional to **[AIEEE 2004]**
(a) $\frac{m}{\omega_0^2-\omega^2}$ (b) $\frac{1}{m(\omega_0^2-\omega^2)}$
(c) $\frac{1}{m(\omega_0^2+\omega^2)}$ (d) $\frac{m}{\omega_0^2+\omega^2}$

32 A mass M is suspended from a spring of negligible mass. The spring is pulled a little and then released so that the mass executes SHM of time period T. If the mass is increased by m, the time period becomes $5T/3$, then the ratio of m/M is **[AIEEE 2003]**
(a) 3/5 (b) 25/9 (c) 16/9 (d) 5/3

Level 3

33 A light hollow cube of side length 10 cm and mass 10g, is floating in water. It is pushed down and released to execute simple harmonic oscillations. The time period of oscillations is $y\pi \times 10^{-2}$s, where the value of y is
(Acceleration due to gravity, $g = 10$ m/s^2, density of water $= 10^3$ kg/m^3) **[2025, 23 Jan Shift-I]**
(a) 2 (b) 4
(c) 6 (d) 1

34 Two bodies A and B of equal mass are suspended from two massless springs of spring constant k_1 and k_2, respectively. If the bodies oscillate vertically such that the their amplitudes are equal, the ratio of the maximum velocity of A to the maximum velocity of B is *[2025, 29 Jan Shift-II]*

(a) $\frac{k_1}{k_2}$ (b) $\sqrt{\frac{k_1}{k_2}}$ (c) $\sqrt{\frac{k_2}{k_1}}$ (d) $\frac{k_2}{k_1}$

35 If R is the radius of the earth and the acceleration due to gravity on the surface of earth is $g=\pi^2\ \text{m/s}^2$, then the length of the second's pendulum at a height $h=2R$ from the surface of earth will be *[2024, 1 Feb Shift-I]*

(a) $\frac{4}{9}$ m (b) $\frac{8}{9}$ m (c) $\frac{2}{9}$ m (d) $\frac{1}{9}$ m

36 Assume there are two identical simple pendulum clocks. Clock-1 is placed on the earth and Clock-2 is placed on a space station located at a height h above the earth surface. Clock-1 and Clock-2 operate at time periods 4 s and 6 s respectively. Then the value of h is (consider radius of earth $R_e=6400$ km and g on earth 10 m/s^2) *[2022, 28 July Shift-II]*

(a) 1200 km (b) 1600 km
(c) 3200 km (d) 4800 km

Imp. Second Pendulum

37 The length of a seconds pendulum at a height $h=2R$ from earth's surface will be (Given, $R=$ Radius of earth and acceleration due to gravity at the surface of earth, $g=\pi^2\ \text{ms}^{-2}$) *[2022, 25 July Shift-II]*

(a) $\frac{2}{9}$ m (b) $\frac{4}{9}$ m (c) $\frac{8}{9}$ m (d) $\frac{1}{9}$ m

38 A bob of mass m suspended by a thread of length l undergoes simple harmonic oscillations with time period T. If the bob is immersed in a liquid that has density 1/4 times that of the bob and the length of the thread is increased by 1/3rd of the original length, then the time period of the simple harmonic oscillations will be *[2021, 31 Aug Shift-II]*

(a) T (b) $\frac{3}{2}T$ (c) $\frac{3}{4}T$ (d) $\frac{4}{3}T$

39 Assume that a tunnel is dug along a chord of the earth, at a perpendicular distance $(R/2)$ from the earth's centre, where R is the radius of the earth. The wall of the tunnel is frictionless. If a particle is released in this tunnel, it will execute a simple harmonic motion with a time period? *[2021, 26 Feb Shift-I]*

(a) $\frac{2\pi R}{g}$ (b) $\frac{g}{2\pi R}$ (c) $\frac{1}{2\pi}\sqrt{\frac{g}{R}}$ (d) $2\pi\sqrt{\frac{R}{g}}$

40 In the given figure, a mass M is attached to a horizontal spring which is fixed on one side to a rigid support. The spring constant of the spring is k. The mass oscillates on a frictionless surface with time period T and amplitude A. When the mass is in equilibrium position as shown in the figure, another mass m is gently fixed upon it. The new amplitude of oscillation will be *[2021, 24 Feb Shift-I]*

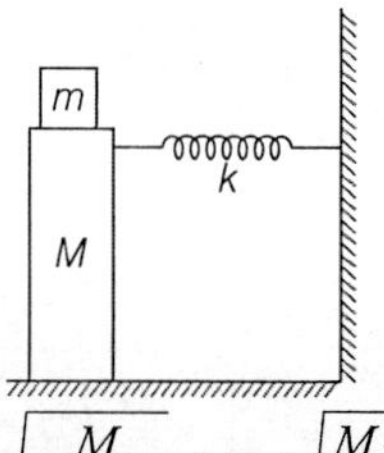

(a) $A\sqrt{\frac{M+m}{M}}$ (b) $A\sqrt{\frac{M}{M+m}}$ (c) $A\sqrt{\frac{M-m}{M}}$ (d) $A\sqrt{\frac{M}{M-m}}$

41 When a particle of mass m is attached to a vertical spring of spring constant k and released, its motion is described by $y(t)=y_0\sin^2\omega t$, where y is measured from the lower end of unstretched spring. Then, ω is *[2020, 6 Sep Shift-II]*

(a) $\frac{1}{2}\sqrt{\frac{g}{y_0}}$ (b) $\sqrt{\frac{g}{y_0}}$ (c) $\sqrt{\frac{g}{2y_0}}$ (d) $\sqrt{\frac{2g}{y_0}}$

42 Two light identical springs of spring constant k are attached horizontally at the two ends of an uniform horizontal rod AB of length l and mass m. The rod is pivoted at its centre O and can rotate freely in horizontal plane. The other ends of the two springs are fixed to rigid supports as shown in figure.

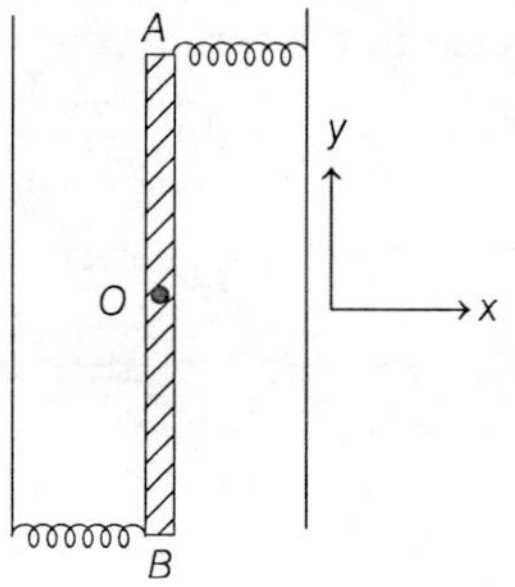

The rod is gently pushed through a small angle and released. The frequency of resulting oscillation is *[2019, 12 Jan Shift-I]*

(a) $\frac{1}{2\pi}\sqrt{\frac{2k}{m}}$ (b) $\frac{1}{2\pi}\sqrt{\frac{3k}{m}}$ (c) $\frac{1}{2\pi}\sqrt{\frac{6k}{m}}$ (d) $\frac{1}{2\pi}\sqrt{\frac{k}{m}}$

43 The mass and the diameter of a planet are three times the respective values for the earth. The period of oscillation of a simple pendulum on the earth is 2 s. The period of oscillation of the same pendulum on the planet would be *[2019, 11 Jan Shift-II]*

(a) $\frac{2}{\sqrt{3}}$ s (b) $\frac{3}{2}$ s (c) $2\sqrt{3}$ s (d) $\frac{\sqrt{3}}{2}$ s

44 A rod of mass M and length $2L$ is suspended at its middle by a wire. It exhibits torsional oscillations. If two masses each of m are attached at distance $L/2$ from its centre on both sides, it reduces the oscillation frequency by 20%. The value of ratio m/M is close to *[2019, 9Jan Shift-II]*

(a) 0.57 (b) 0.37 (c) 0.77 (d) 0.17

45 Two masses m and $m/2$ are connected at the two ends of a massless rigid rod of length l. The rod is suspended by a thin wire of torsional constant k at the centre of mass of the rod-mass system (see figure). Because of torsional constant k, the restoring torque is $\tau = k\theta$ for angular displacement θ. If the rod is rotated by θ_0 and released, the tension in it when it passes through its mean position will be *[2019, 9Jan Shift-I]*

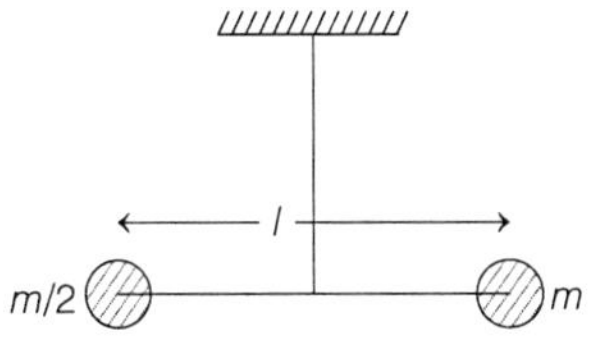

(a) $\frac{2k\theta_0^2}{l}$ (b) $\frac{k\theta_0^2}{l}$ (c) $\frac{3k\theta_0^2}{l}$ (d) $\frac{k\theta_0^2}{2l}$

46 An ideal gas enclosed in a vertical cylindrical container supports a freely moving piston of mass M. The piston and the cylinder have equal cross-sectional area A. When the piston is in equilibrium, the volume of the gas is V_0 and its pressure is p_0. The piston is slightly displaced from the equilibrium position and released. Assuming that the system is completely isolated from its surrounding, the piston executes a simple harmonic motion with frequency *[JEE Main 2013]*

(a) $\frac{1}{2\pi}\frac{A\gamma p_0}{V_0 M}$ (b) $\frac{1}{2\pi}\frac{V_0 M p_0}{A^2\gamma}$

(c) $\frac{1}{2\pi}\sqrt{\frac{A^2\gamma p_0}{MV_0}}$ (d) $\frac{1}{2\pi}\sqrt{\frac{MV_0}{A\gamma p_0}}$

47 The bob of a simple pendulum executes simple harmonic motion in water with a period t, while the period of oscillation of the bob is t_0 in air. Neglecting frictional force of water and given that the density of the bob is $(4/3) \times 1000 \text{ kgm}^{-3}$. What relationship between t and t_0 is true? *[AIEEE 2004]*

(a) $t = t_0$ (b) $t = \frac{t_0}{2}$ (c) $t = 2t_0$ (d) $t = 4t_0$

48 A particle at the end of a spring executes simple harmonic motion with a period t_1, while the corresponding period for another spring is t_2. If the period of oscillation with the two springs in series is T, then *[AIEEE 2004]*

(a) $T = t_1 + t_2$ (b) $T^2 = t_1^2 + t_2^2$

(c) $T^{-1} = t_1^{-1} + t_2^{-1}$ (d) $T^{-2} = t_1^{-2} + t_2^{-2}$

49 The length of a simple pendulum executing simple harmonic motion is increased by 21%. The percentage increase in the time period of the pendulum of increased length is *[AIEEE 2003]*

(a) 11% (b) 21% (c) 42% (d) 10%

Numerical Value Type Questions

50 The time period of simple harmonic motion of mass M in the given figure is $\pi\sqrt{\frac{\alpha M}{5k}}$, where the value of α is .... *[2024, 31 Jan Shift-II]*

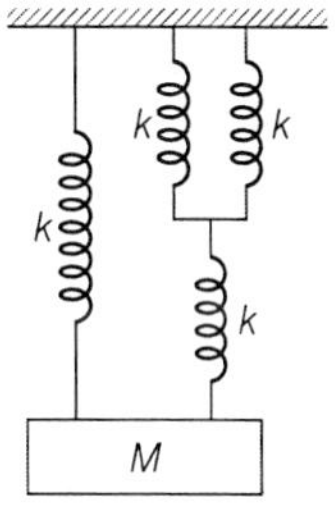

51 In the figure given below, a block of mass $M = 490$ g placed on a frictionless table is connected with two springs having same spring constant ($k = 2 \text{ N m}^{-1}$). If the block is horizontally displaced through X then the number of complete oscillations it will make in 14π s will be .... *[2023, 31 Jan Shift-I]*

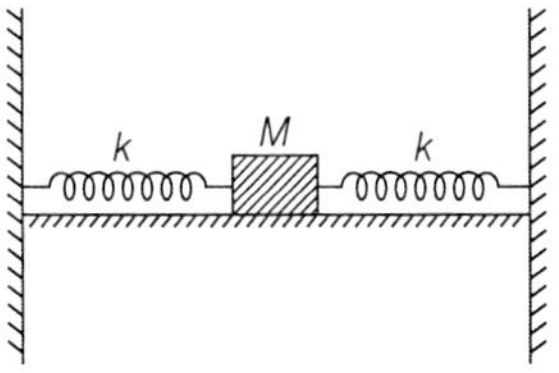

52 The general displacement of a simple harmonic oscillator is $x = A \sin \omega t$. Let T be its time period. The slope of its potential energy (U)-time (t) curve will be maximum when $t = T/\beta$. The value of β is .... *[2023, 30 Jan Shift-I]*

53

A block of a mass 2 kg is attached with two identical springs of spring constant 20 N/m each. The block is placed on a frictionless surface and the ends of the springs are attached to rigid supports (see figure). When the mass is displaced from its equilibrium position, it executes a simple harmonic motion. The time period of oscillation is $\pi/\sqrt{x}$ in SI unit. The value of x is .... *[2023, 24 Jan Shift-I]*

54 A rectangular block of mass 5 kg attached to a horizontal spiral spring executes simple harmonic motion of amplitude 1 m and time period 3.14 s. The maximum force exerted by spring on block is .... N. *[2023, 10 Jan Shift-II]*

55 The metallic bob of simple pendulum has the relative density 5. The time period of this pendulum is 10 s. If the metallic bob is immersed in water, then the new time period becomes $5\sqrt{x}$ s. The value of x will be .... *[2022, 29 July Shift-II]*

56 A mass 0.9 kg, attached to a horizontal spring, executes SHM with an amplitude A_1. When this mass passes through its mean position, then a smaller mass of 124 g is placed over it and both masses move together with amplitude A_2. If the ratio $\frac{A_1}{A_2}$ is $\frac{\alpha}{\alpha - 1}$, then the value of α will be .... *[2022, 27 July Shift-I]*

57 As per given figures, two springs of spring constants k and $2k$ are connected to mass m. If the period of oscillation in Fig. (*a*) is 3 s, then the period of oscillation in Fig. (*b*) will be $\sqrt{x}$ s. The value of x is .... *[2022, 26 July Shift-II]*

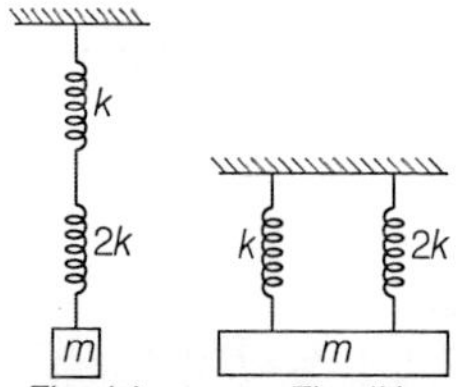

Fig. (a) Fig. (b)

58 A body is performing simple harmonic motion with an amplitude of 10 cm. The velocity of the body was tripled by air jet when it is at 5 cm from its mean position. The new amplitude of vibration is $\sqrt{x}$ cm. The value of x is .... *[2022, 29 June Shift-I]*

59 A particle of mass 1 kg is hanging from a spring of force constant 100 Nm^{-1}. The mass is pulled slightly downward and released, so that it executes free simple harmonic motion with time period T. The time when the kinetic energy and potential energy of the system will become equal, is T/x. The value of x is. *[2021, 31 Aug Shift-I]*

60 In the reported figure, two bodies A and B of masses 200 g and 800 g are attached with the system of springs. Springs are kept in a stretched position with some extension when the system is released. The horizontal surface is assumed to be frictionless. The angular frequency will be .... rad/s when $k = 20$ N/m. *[2021, 25 July Shift-I]*

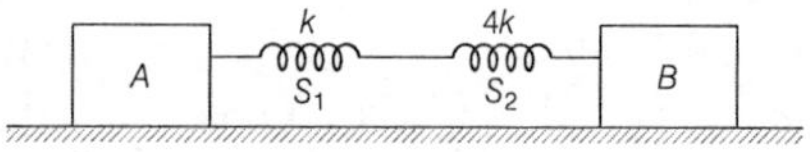

61 Consider two identical springs each of spring constant k and negligible mass compared to the mass M as shown. Fig.1 shows one of them and Fig.2 shows their series combination. The ratios of time period of oscillation of the two SHM is $T_b / T_a = \sqrt{x}$, where value of x is .... *[2021, 17 March Shift-I]*

(Round off to the nearest integer)

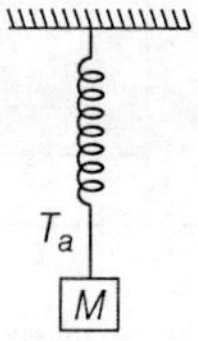

Fig. 1

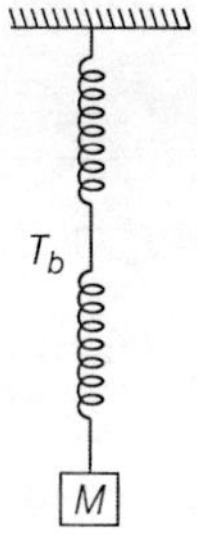

Fig. 2

62 Time period of a simple pendulum is T. The time taken to complete 5/8 oscillations starting from mean position is $\alpha/\beta\, T$. The value of α is .... *[2021, 26 Feb Shift-II]*

• Key Index

Topic 1

1 (d)	2 (a)	3 (a)	4 (b)	5 (c)	6 (b)	7 (a)	8 (a)	9 (d)	10 (c)
11 (b)	12 (b)	13 (b)	14 (c)	15 (b)	16 (a)	17 (a)	18 (b)	19 (c)	20 (c)
21 (c)	22 (d)	23 (b)	24 (c)	25 (b)	26 (d)	27 (d)	28 (d)	29 (b)	30 (d)
31 (c)	32 (c)	33 (d)	34 (d)	35 (a)	36 (d)	37 (b)	38 (c)	39 (a)	40 (d)
41 (b)	42 (a)	43 (a)	44 (a)	45 (22)	46 (17)	47 (6)	48 (12)	49 (10)	50 (3)
51 (7)	52 (2)	53 (9)	54 (12)	55 (4)	56 (99)	57 (2)	58 (88)	59 (40)	60 (1)
61 (1)	62 (2)	63 (2)	64 (6)	65 (3)					

Topic 2

1 (b)	2 (c)	3 (d)	4 (a)	5 (a)	6 (a)	7 (c)	8 (c)	9 (d)	10 (d)
11 (b)	12 (a)	13 (c)	14 (b)	15 (a)	16 (a)	17 (a)	18 (c)	19 (c)	20 (b)
21 (c)	22 (d)	23 (c)	24 (c)	25 (d)	26 (c)	27 (b)	28 (b)	29 (a)	30 (b)
31 (b)	32 (c)	33 (a)	34 (b)	35 (d)	36 (c)	37 (d)	38 (d)	39 (d)	40 (b)
41 (c)	42 (c)	43 (c)	44 (b)	45 (b)	46 (c)	47 (c)	48 (b)	49 (d)	50 (12)
51 (20)	52 (8)	53 (5)	54 (20)	55 (5)	56 (16)	57 (2)	58 (700)	59 (8)	60 (10)
61 (2)	62 (7)								

Detailed **Explanations**

Topic 1

01. *(d)* Particle perform SHM
Time period = 2s
Amplitude = 1 cm

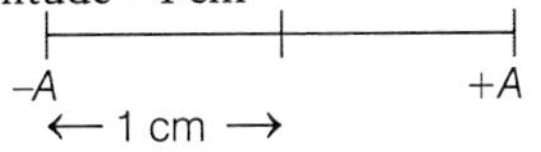

$$n = \frac{12.5}{2} = 6.25 \text{ cycle}$$

$\therefore \quad D = 4 \times 6 + 1 = 25$
$d = 1$
$$\frac{D}{d} = 25$$

02. *(a)* In simple harmonic motion, restoring force, $F = -kx$
where, k is force constant.
$\therefore \quad F \propto x$
Acceleration, $a = -\omega^2 x$
Here, negative sign indicates that acceleration acts in opposite direction to the displacement.
At extreme position, $x = A$
$\therefore \quad a_{max} = -\omega^2 A$
Hence, acceleration is maximum at extreme points.
velocity, $v = \omega\sqrt{A^2 - x^2}$
At mean position, $x = 0$.
$\therefore \quad v_{max} = \omega A$
Thus, from above discussion, we found that Statement (A), (B) and (C) are correct.

03. *(a)* Given, KE = PE

$$\frac{1}{2} m\omega^2 (A^2 - x^2) = \frac{1}{2} m\omega^2 x^2$$
$$A^2 - x^2 = x^2 \Rightarrow \frac{A^2}{2} = x^2$$
$$\Rightarrow \quad x = \frac{A}{\sqrt{2}}$$

04. *(b)* We know that in SHM,
TE = KE + PE, KE = TE – PE

But, $\text{KE} = \frac{1}{2} mv^2$ ...(i)

$$\text{KE} = \frac{1}{2} m\omega^2 (A^2 - x^2)$$
$$[\because v = \omega\sqrt{A^2 - x^2}]$$

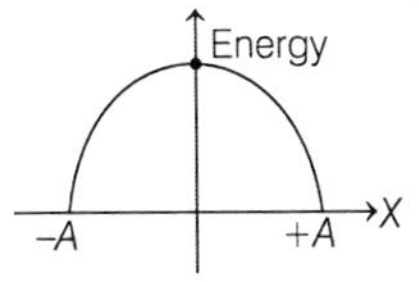

05. *(c)* Given, $\text{KE} = \frac{1}{2} k(A^2 - x^2)$
Graph between KE and x will be parabola.

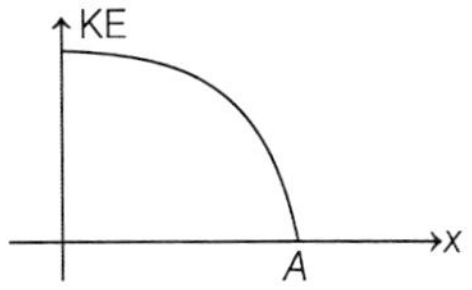

06. *(b)* Displacement of particle is given,
$$x = A\sin(\omega t + \delta) \quad \text{...(i)}$$
At $\quad t = 0, x = \frac{A}{2}$
So, from Eq. (i)
$$\frac{A}{2} = A\sin(\omega \times 0 + \delta)$$
$$\Rightarrow \quad \frac{1}{2} = \sin(\delta) \text{ or } \delta = \frac{\pi}{6} \text{ rad}$$

07. *(a)* In SHM, total Energy, $E = \frac{1}{2} kA^2$
which is equal to maximum potential energy.
Kinetic energy,
$$\text{KE} = \frac{1}{2} k(A^2 - x^2) \quad \text{... (i)}$$
Given that $E = 25 \text{ J} \Rightarrow x = \frac{A}{2}$
From Eq. (i),
$$\text{KE} = \frac{1}{2} k\left[A^2 - \left(\frac{A}{2}\right)^2\right]$$
$$= \frac{1}{2} k\frac{3A^2}{4} = \frac{3}{4} \times \left(\frac{1}{2} kA^2\right)$$
$$= \frac{3}{4} \times 25$$
$$\text{KE} = 18.75 \text{ J}$$

08. *(a)* In SHM,
$$\text{KE}_{av} = \text{PE}_{av} = \frac{1}{4} m\omega^2 a^2$$
$\therefore$ Total mechanical energy at any time = Sum of kinetic and potential energy
$= \frac{1}{2} m\omega^2 a^2 = \text{constant}.$
where, m = mass of body executing SHM,
ω = angular frequency
and $\quad a$ = amplitude of SHM.

> **• Concept Enhancer**
> In simple harmonic motion, kinetic energy and potential energy are periodic function of time and the time period of each being half the time period of oscillation.

09. *(d)* Given, time period, $T = \frac{\pi}{\omega}$

Time period, $T = \frac{2\pi}{\omega'}$
$$\frac{2\pi}{\omega'} = \frac{\pi}{\omega} \Rightarrow \omega' = 2\omega$$
Options (a) and (b) are incorrect.
Option (c),
$$\sin^2 2\omega t = \frac{1}{2}(2\sin^2 2\omega t)$$
$$= \frac{1}{2}(1 - \cos 4\omega t)$$
Hence, the angular frequency is 4ω.
Now, $3\cos\left(\frac{\pi}{4} - 2\omega t\right)$ has angular frequency of 2ω

10. *(c)* Potential energy of SHM
$$= \frac{1}{2} kx^2 = \frac{1}{2} m\omega^2 x^2$$
Kinetic energy of SHM
$$= \frac{1}{2} m(\omega\sqrt{A^2 - x^2})^2$$
Here, A is the amplitude of SHM.
According to question,
Potential energy of SHM = Kinetic energy of SHM
$$\frac{1}{2} m\omega^2 x^2 = \frac{1}{2} m\left(\omega\sqrt{A^2 - x^2}\right)^2$$
$$\Rightarrow \quad \omega^2 x^2 = \omega^2(\sqrt{A^2 - x^2})^2$$
$$\Rightarrow \quad \omega^2 x^2 = \omega^2(A^2 - x^2)$$
$$\Rightarrow \quad x = \pm\sqrt{\frac{A^2}{2}} \Rightarrow x = \pm\frac{A}{\sqrt{2}}$$

11. *(b)* Given, displacement-time equation,
$$Y = A\sin(\omega t + \phi_0)$$
At $t = 0, Y = A/2$
$\therefore \; Y = A/2 = A\sin(0 + \phi_0) = A\sin\phi_0$
So, $\sin\phi_0 = \frac{1}{2}$

$$\phi_0 = \frac{\pi}{6} \text{ or } \frac{5\pi}{6}$$

As particle is moving in negative direction

Hence, $\phi_0 > \pi$

So, $\phi_0 = \frac{5\pi}{6}$

12. *(b)* Since, the particle is executing SHM.

$$y = A\sin\omega t \Rightarrow \frac{y}{A} = \sin\omega t$$

and wave velocity equation will be

$$v_y = \frac{dy}{dt} = A\omega\cos\omega t$$

$$\Rightarrow \quad \frac{v_y}{A\omega} = \cos\omega t$$

Now, $\sin^2\omega t + \cos^2\omega t = 1$

$$\therefore \quad (y/A)^2 + (v_y/A\omega)^2 = 1$$

This equation is similar to the equation of ellipse.

13. *(b)* In simple harmonic motion,

$$x = A\sin\omega t$$

$$v = \omega\sqrt{A^2 - x^2}$$

or $\quad v = A\omega\cos\omega t$ and $a = -\omega^2 x$

or $\quad a = -\omega^2 A\sin\omega t$

Given, amplitude $A = 5$ cm and displacement $x = 4$ cm.

At this time (when $x = 4$ cm),

$$\Rightarrow \quad |v_{x=4}| = |a_{x=4}|$$

$$\text{or} \quad |\omega\sqrt{5^2 - 4^2}| = |-4\omega^2|$$

$$\Rightarrow \quad 3\omega = +4\omega^2$$

$$\Rightarrow \quad \omega = (3/4) \text{ rad/s}$$

So, time period, $T = \frac{2\pi}{\omega}$

$$\Rightarrow \quad T = \frac{2\pi}{3} \times 4 = \frac{8\pi}{3} \text{ s}$$

> • Concept **Enhancer**
>
> In simple harmonic motion, magnitude of velocity is maximum when magnitude of acceleration is minimum and *vice versa*.

14. *(c)* KE is maximum at mean position and minimum at extreme position $\left(\text{at } t = \frac{T}{4}\right)$.

15. *(b)* $\frac{aT}{x} = \frac{\omega^2 xT}{x} = \frac{4\pi^2}{T^2} \times T = \frac{4\pi^2}{T}$

$= \text{constant}$

16. *(a)* Average kinetic energy of particle

$$= \frac{1}{4} ma^2\omega^2 \qquad [\text{as } \omega = 2\pi\nu]$$

$$= \frac{1}{4} ma^2(2\pi\nu)^2 = \pi^2\nu^2 ma^2$$

17. *(a)* The maximum velocity of a particle performing SHM is given by $v = A\omega$, where A is the amplitude and ω is the angular frequency of oscillation.

$$\therefore \quad 4.4 = (7\times10^{-3}) \times 2\pi/T$$

$$\text{or} \quad T = \frac{7\times10^{-3}}{4.4} \times \frac{2\times22}{7} = 0.01 \text{ s}$$

18. *(b)* $\frac{d^2x}{dt^2} = -\alpha x \qquad \ldots\text{(i)}$

We know that for SHM

$$a = \frac{d^2x}{dt^2} = -\omega^2 x \qquad \ldots\text{(ii)}$$

From Eqs. (i) and (ii), we have

$$\omega^2 = \alpha \text{ or } \omega = \sqrt{\alpha}$$

$$\Rightarrow \quad \frac{2\pi}{T} = \sqrt{\alpha} \text{ or } T = \frac{2\pi}{\sqrt{\alpha}}$$

19. *(c)* In simple harmonic motion, when a particle is displaced to a position from its mean position, then its kinetic energy gets converted into potential energy. Hence, total energy of a particle remains constant or the total energy in simple harmonic motion does not depend on displacement x.

20. *(c)*

> **key idea** In simple harmonic motion, kinetic energy is maximum at mean position (due to its maximum velocity) and is minimum at extreme position (due to its zero velocity).

Kinetic energy,

$$\text{KE} = \frac{1}{2} m\omega^2(a^2 - x^2)$$

and potential energy $\text{PE} = \frac{1}{2} m\omega^2 x^2$

So, at mean position, $x = 0$

$$\therefore \quad \text{KE} = \frac{1}{2} m\omega^2 a^2 \text{ [maximum]}$$

$$\text{PE} = 0 \qquad \text{[minimum]}$$

21. *(c)* $x_1 = \sqrt{7}\sin 5t$

$$x_2 = 2\sqrt{7}\sin\left(5t + \frac{\pi}{3}\right)$$

from phase diagram

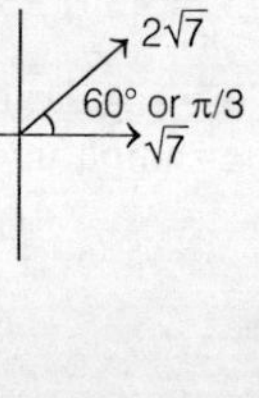

Amplitude of resultant SHM = 7

$$\phi = \tan^{-1}\left(\frac{2\sqrt{7}\times\frac{\sqrt{3}}{2}}{\sqrt{7} + 2\sqrt{7}\times\frac{1}{2}}\right)$$

$$= \tan^{-1}\frac{\sqrt{21}}{2\sqrt{7}} = \tan^{-1}\left(\frac{\sqrt{3}}{2}\right)$$

$$x_R = 7\sin(5t + \phi)$$

$$a_R = -7\times25\sin(5t + \phi)$$

$$a_{\max} = 175 \text{ cm/sec} = 175\times10^{-2} \text{ m/s}^2.$$

22. *(d)* $x = A\sin(\omega t + \phi)$

$$x_0 = A\sin\phi \qquad \ldots\text{(i)}$$

$$p = mA\,\omega\cos(\omega + \phi)$$

$$p_0 = mA\,\omega\cos\phi \qquad \ldots\text{(ii)}$$

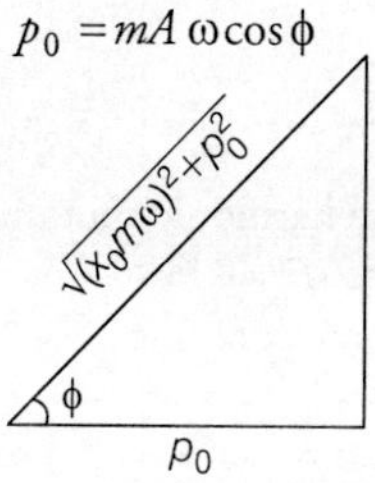

Dividing Eq. (ii) by Eq. (i), we get

$$\frac{mA\omega\cos\phi}{A\sin\phi} = \frac{p_0}{x_0}$$

$$m\omega\frac{\cos\phi}{\sin\phi} = \frac{p_0}{x_0}$$

$$\frac{\sin\phi}{m\omega\cos\phi} = \frac{x_0}{p_0}$$

$$\frac{\tan\phi}{m\omega} = \frac{x_0}{p_0} \Rightarrow \tan\phi = \left(\frac{x_0}{p_0}\right)m\omega$$

$$\sin\phi = \frac{x_0 m\omega}{\sqrt{(x_0 m\omega)^2 + p_0^2}}$$

$$A = \frac{x_0}{\sin\phi} = \frac{x_0}{\frac{x_0 m\omega}{\sqrt{(x_0 m\omega)^2 + p_0^2}}}$$

23. *(b)* Given, $x = A/2$

We know that, $\text{PE} = \frac{1}{2}kx^2$

$$= \frac{1}{2}k\left(\frac{A}{2}\right)^2 = \frac{1}{2}k\frac{A^2}{4} = \frac{kA^2}{8}$$

$$\text{KE} = \frac{1}{2}kA^2 - \frac{1}{2}kx^2$$

$$= \frac{1}{2}kA^2 - \frac{kA^2}{8} = \frac{3}{8}kA^2$$

According to question,

$$\therefore \quad \frac{\text{PE}}{\text{KE}} = \frac{\frac{kA^2}{8}}{\frac{3kA^2}{8}} = \frac{1}{3}$$

24. *(c)* Let equation of particle executing simple harmonic motion be

$$y = A\sin\omega t$$

At time t_1, $y = A/2$

$$\therefore \quad \frac{A}{2} = A\sin\omega t_1 \Rightarrow \frac{1}{2} = \sin\omega t_1$$

$$\Rightarrow \quad \frac{1}{2} = \sin(\omega\times 2) \qquad (\because t_1 = 2\text{ s})$$

$$\Rightarrow \quad \omega = \frac{\pi}{12} \qquad \ldots(\text{i})$$

Again, at time t_2, $y = A$

$$\therefore \quad A = A\sin\omega t_2 \Rightarrow 1 = \sin\omega t_2$$

$$\sin\frac{\pi}{2} = \sin\omega t_2 \Rightarrow t_2 = \frac{\pi}{2\omega}$$

$\therefore$ Time taken by particle in going from $A/2$ to A

$$t = t_2 - t_1 = \frac{\pi}{2\omega} - 2 = \frac{\pi}{2(\pi/12)} - 2$$

[From Eq. (i)]

$$= 4\text{ s}$$

25. *(b)* At $t = 1$ s, displacement of the particle,

$$x = \sin\pi\left(1 + \frac{1}{3}\right) = \sin\frac{4\pi}{3} = -\sin\left(\frac{\pi}{3}\right)$$

$$= -\frac{\sqrt{3}}{2}\text{ m}$$

So, $v = \omega\sqrt{A^2 - x^2} = \pi\sqrt{1 - \frac{3}{4}} = \frac{\pi}{2}$ m/s

$$= 1.57\text{ ms}^{-1} = 157\text{ cm s}^{-1}$$

26. *(d)* For a simple harmonic oscillator, its displacement is given by,

$$y = A\sin\omega t \qquad \ldots(\text{i})$$

At $t = 3$ s, it attains half of its amplitude. Thus,

$$A/2 = A\sin 3\omega$$

$$\Rightarrow \sin 3\omega = \frac{1}{2} \Rightarrow \sin 3\omega = \sin\frac{\pi}{6}$$

$$\Rightarrow \quad 3\omega = \frac{\pi}{6}$$

$$\Rightarrow \quad \omega = \frac{\pi}{18}$$

Since, $\frac{2\pi}{T} = \omega$, we get

$$\frac{2\pi}{T} = \frac{\pi}{18}$$

$$\therefore \quad T = 36\text{ s}$$

27. *(d)* We know that $\text{PE} = \frac{1}{2}kx^2$

$$\text{PE} = \frac{1}{2}kx_0^2\sin^2\omega t$$

Value of $\sin^2\omega t$ lies between 0 to 1. Thus, the value of potential energy will be greater than or equal to zero.

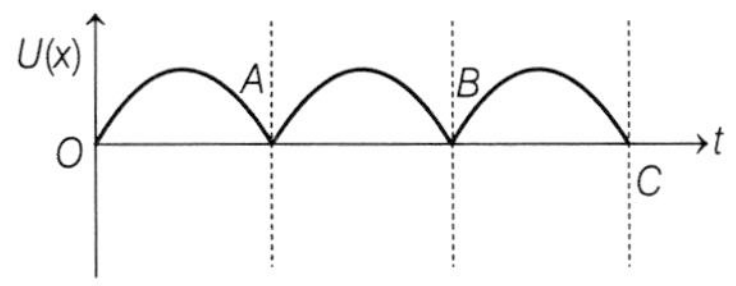

$$0 \le \text{PE}$$

This condition is only satisfied by graph (d).

28. *(d)* In first scenario, a particle is executing SHM of amplitude a and energy E.

$$\therefore \quad \text{Energy, } E = \frac{1}{2}ka^2 \qquad \ldots(\text{i})$$

In second scenario,

$$\frac{3E}{4} = \frac{1}{2}k(a^2 - y^2) \qquad \ldots(\text{ii})$$

where, y is the displacement.

On dividing Eqs. (i) and (ii), we get

$$\frac{E}{\frac{3E}{4}} = \frac{\frac{1}{2}ka^2}{\frac{1}{2}k(a^2 - y^2)} \Rightarrow \frac{4}{3} = \frac{a^2}{a^2 - y^2}$$

$$\Rightarrow 4a^2 - 4y^2 = 3a^2 \Rightarrow -4y^2 = -a^2$$

$$\Rightarrow \quad y^2 = \frac{a^2}{4} \Rightarrow y = \frac{a}{2}$$

29. *(b)* Given, wave displacement, $x = \frac{A}{2}$

As we know that,

$$\text{ME} = \text{KE} + \text{PE}$$

In SHM, $\text{KE} = \frac{1}{2}m\omega^2(A^2 - x^2)$

$$= \frac{1}{2}m\omega^2\left(A^2 - \frac{A^2}{4}\right)$$

$$= \frac{3}{8}m\omega^2A^2 \qquad \ldots(\text{i})$$

$$\therefore \quad \text{PE} = \frac{1}{2}m\omega^2\frac{A^2}{4} = \frac{1}{8}m\omega^2A^2$$

Now, ME = KE + PE

$$= \frac{3}{8}m\omega^2A^2 + \frac{1}{8}m\omega^2A^2$$

$$= \frac{4m\omega^2A^2}{8} = \frac{m\omega^2A^2}{2} \qquad \ldots(\text{ii})$$

On dividing Eq. (i) by Eq. (ii), we get

$$\frac{\text{KE}}{\text{ME}} = \frac{\frac{3}{8}m\omega^2A^2}{\frac{1}{2}m\omega^2A^2} = \frac{3}{4}$$

30. *(d)* Velocity of particle performing simple harmonic motion,

$$v = \pm\,\omega\sqrt{(A^2 - x^2)} \qquad \ldots(\text{i})$$

According to given situation,

$$v_1 = \omega\sqrt{A^2 - x_1^2} \qquad \ldots(\text{ii})$$

and

$$v_2 = \omega\sqrt{A^2 - x_2^2} \qquad \ldots(\text{iii})$$

On squaring Eq. (ii), we get

$$v_1^2 = \omega^2(A^2 - x_1^2) \qquad \ldots(\text{iv})$$

On squaring Eq. (iii), we get

$$v_2^2 = \omega^2(A^2 - x_2^2) \qquad \ldots(\text{v})$$

On subtracting Eq. (iv) from Eq. (v), we get

$$v_1^2 - v_2^2 = \omega^2(x_2^2 - x_1^2)$$

$$\Rightarrow \quad v_1^2 - v_2^2 = \left(\frac{2\pi}{T}\right)^2(x_2^2 - x_1^2)$$

$$\Rightarrow \quad T = 2\pi\sqrt{\frac{(x_2^2 - x_1^2)}{(v_1^2 - v_2^2)}}$$

31. *(c)*

> **key idea** If a particle executing SHM, graph obtained between its velocity and displacement is of elliptical shape.

Velocity of particle in SHM in terms of displacement x is given as

$$v = \omega\sqrt{A^2 - x^2}$$

$$\Rightarrow \quad v = A\omega\sqrt{1 - x^2/A^2}$$

$$\Rightarrow \quad \frac{v}{A\omega} = \sqrt{1 - x^2/A^2}$$

On squaring both side, we gets

$$\Rightarrow \quad \frac{v^2}{A^2\omega^2} = 1 - \frac{x^2}{A^2}$$

$$\Rightarrow \quad \frac{v^2}{A^2\omega^2} + \frac{x^2}{A^2} = 1 \qquad \ldots(\text{i})$$

As equation of ellipse is $\frac{x^2}{a^2} + \frac{y^2}{b^2} = 1$

Eq. (i) is similar to the equation of ellipse. So, the graph between velocity and displacement is an ellipse.

32. *(c)* The displacement-time graph is as shown below,

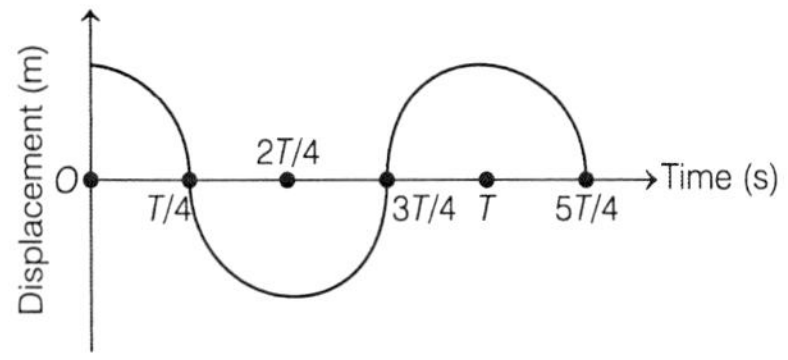

At $t = \frac{3T}{4}$, particle is at mean position.

At mean position of SHM, acceleration of particle is zero. i.e. force is zero.

Statement (A) is correct.

At $t = T$, particle is at extreme position. As direction is changing, so its acceleration is maximum. Statement (B) is correct.

At $t = T/4$, particle is at mean position, so whole of energy is kinetic. i.e. its speed is maximum. Statement (C) is correct.

Since, PE = KE at $t = \frac{T}{8}$, so Statement (D) is incorrect.

Hence, option (c) is correct.

33. *(d)*

> **key idea** The velocity of a particle executing SHM at any instant is defined as the time rate of change of its displacement at that instant.

Velocity, $v = \omega\sqrt{A^2 - x^2}$

Suppose that the new amplitude of the motion be A'.

Initial velocity of a particle performs SHM,

$$v^2 = \omega^2\left[A^2 - \left(\frac{2A}{3}\right)^2\right] \quad \text{... (i)}$$

Final velocity,

$$(3v)^2 = \omega^2\left[A'^2 - \left(\frac{2A}{3}\right)^2\right] \quad \text{...(ii)}$$

From Eqs. (i) and (ii), we get

$$\frac{1}{9} = \frac{A^2 - \frac{4A^2}{9}}{A'^2 - \frac{4A^2}{9}} \Rightarrow A' = \frac{7A}{3}$$

34. *(d)* In SHM, a particle starts from rest, we get

$$x = A\cos\omega t, \text{ at } t = 0, x = A$$

When, $t = t$, then $x = A - a$

When, $t = 2t$, then $x = A - 3a$

So, $A - a = A\cos\omega t$

$A - 3a = A\cos 2\omega t$

As, $\cos 2\omega\tau = 2\cos^2\omega t - 1$

$$\Rightarrow \frac{A-3a}{A} = 2\left(\frac{A-a}{A}\right)^2 - 1$$

$$\Rightarrow \frac{A-3a}{A} = \frac{2A^2 + 2a^2 - 4Aa - A^2}{A^2}$$

$$A^2 - 3aA = A^2 + 2a^2 - 4Aa$$

$$2a^2 = aA \Rightarrow A = 2a$$

Now, $A - a = A\cos\omega t \Rightarrow \cos\omega t = 1/2$

$$\frac{2\pi}{T} = \frac{\pi}{3}$$

$$\Rightarrow T = 6\pi$$

35. *(a)* Let $x_1 = A\sin(\omega t + \phi_1)$ and

$x_2 = A\sin(\omega t + \phi_2)$

$x_2 - x_1 = A[\sin(\omega t + \phi_2) - \sin(\omega t + \phi_1)]$

$$= 2A\cos\left(\frac{2\omega t + \phi_1 + \phi_2}{2}\right)\sin\left(\frac{\phi_2 - \phi_1}{2}\right)$$

The resultant motion can be treated as a simple harmonic motion with amplitude $2A\sin\left(\frac{\phi_2 - \phi_1}{2}\right)$.

Given, maximum distance between the particles

$$= X_0 + A$$

$\therefore$ Amplitude of resultant SHM

$$= X_0 + A - X_0 = A$$

$$\therefore \quad 2A\sin\left(\frac{\phi_2 - \phi_1}{2}\right) = A$$

$\Rightarrow$ Phase difference, $\phi_2 - \phi_1 = \pi/3$

36. *(d)* $x = x_0\cos\left(\omega t - \frac{\pi}{4}\right)$

Acceleration, $a = \frac{d^2x}{dt^2}$

$$= -\omega^2 x_0\cos\left(\omega t - \frac{\pi}{4}\right)$$

$$= \omega^2 x_0\cos\left(\omega t + \frac{3\pi}{4}\right)$$

So, $A = \omega^2 x_0$ and $\delta = \frac{3\pi}{4}$

37. *(b)* Here, $y = \sin^2\omega t$

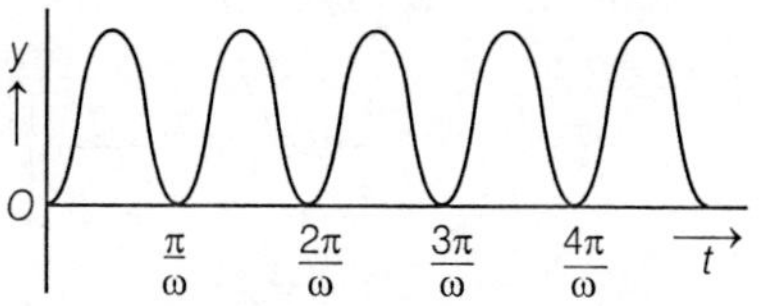

$$\frac{dy}{dt} = 2\omega\sin\omega t\cos\omega t = \omega\sin 2\omega t$$

$$\frac{d^2y}{dt^2} = 2\omega^2\cos 2\omega t$$

For SHM, $\frac{d^2y}{dt^2} \propto -y$

Hence, function is not SHM but periodic. From the *y-t* graph, time period is

$$T = \pi/\omega \cdot \quad \left[\because \frac{2\pi}{T} = 2\omega\right]$$

38. *(c)* $x = 4(\cos\pi t + \sin\pi t)$

$$= \frac{4}{\sqrt{2}} \times \sqrt{2}[\cos\pi t + \sin\pi t]$$

$$x = 4\sqrt{2}\left[\sin\pi t \cdot \frac{1}{\sqrt{2}} + \frac{1}{\sqrt{2}}\cos\pi t\right]$$

$$x = 4\sqrt{2}\left[\sin\pi t \cdot \cos\frac{\pi}{4} + \cos\pi t\sin\frac{\pi}{4}\right]$$

$$x = 4\sqrt{2}\sin\left[\pi t + \frac{\pi}{4}\right]$$

So, amplitude $= 4\sqrt{2}$

39. *(a)* Time period of small oscillation is

$$= 2\pi\sqrt{\frac{m_{total}}{K}}$$

Here, $m_{total} = M + m$

$$T = 2\pi\sqrt{\frac{M+m}{k}}$$

(A) is correct.

Force on the block is $= -kx$

$$F = -kx$$

$$\phi(M+m)a = -kx$$

$$a = \frac{-kx}{M+m}$$

{–ve sign is for direction only}

$$a = \frac{kx}{M+m}$$

(B) is correct.

Friction force on upper block $= ma$

$$f = m\left[\frac{kx}{M+m}\right]$$

(C) is incorrect.

Maximum amplitude before slipping

$$= \frac{\mu[M+m]}{k}g$$

Maximum friction $= f_{max} = \mu mg$

Friction required to keep box of mass m on box of mass $m = f_{req} = m\left[\frac{kA_{max}}{m+M}\right]$

$$f_{max} = f_{req}$$

$$\mu mg = \frac{mkA_{max}}{m+M}$$

$$\frac{\mu g}{k}[m+M] = A_{max}$$

(D) is correct.

Maximum static friction experienced by block of mass m is given by

$f = \mu mg$

(E) is incorrect.

Option (a) A, B, D correct only

40. *(d)* $x(t) = x_0\sin^2\left(\frac{t}{2}\right)$, where $x_0 = 1$ m

$$x = x_0\left(\frac{1-\cos t}{2}\right)$$

$$\frac{x}{x_0} = \frac{1-\cos t}{2}$$

$$\frac{x}{x_0} = \frac{1}{2} - \frac{1}{2}\cos t$$

$$\frac{x}{x_0}-\frac{1}{2}=-\frac{1}{2}\cos t$$

where $x_0 = 1$ m

$$\frac{x}{1}-\frac{1}{2}=-\frac{1}{2}\cos t$$

$$x-\frac{1}{2}=-\frac{1}{2}\cos t$$

Particle is oscillating between $x = 0$ to $x = 1$

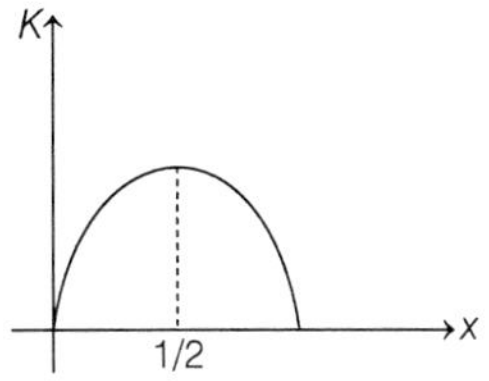

41. *(b)* Angular speed, $\omega=\frac{\pi/6}{0.1}=\frac{5\pi}{3}$ rad/s

Acceleration $\left(\frac{F}{m}\right)$ is maximum at point

$$M = A\omega^2 = 0.36\times\left(\frac{5\pi}{3}\right)^2=\pi^2$$

$= 9.87$ N/kg

42. *(a)* Time period of pendulum at temperature θ is given by

$$T_\theta = 2\pi\sqrt{\frac{l_\theta}{g}}=2\pi\sqrt{\frac{l_0(1+\alpha\theta)}{g}}$$

$$=2\pi\sqrt{\frac{l_0}{g}}(1+\alpha\theta)^{\frac{1}{2}}$$

$$T_\theta = T_0\left(1-\frac{1}{2}\alpha\theta\right)$$

[using binomial expansion]

Let the pendulum clock gives correct time at temperature θ_0.

$$\therefore \quad T_{\theta_0}=T_0\left(1+\frac{1}{2}\alpha\theta_0\right)$$

$$\frac{T_{40}-T_{\theta_0}}{T_0}=\frac{1}{2}\alpha(40-\theta_0)$$

or $\quad 12\,\text{s}=\alpha(40-\theta_0)24\,\text{h} \quad \ldots\text{(i)}$

$$\frac{T_{\theta_0}-T_{20}}{T_0}=\frac{1}{2}\alpha(\theta_0-20)$$

or $\quad 4\,\text{s}=\alpha(\theta_0-20)\,24\,\text{h} \quad \ldots\text{(ii)}$

From Eq. (i) and Eq. (ii) we get

$3(\theta_0-20)=(40-\theta_0)$

$3\theta_0+\theta_0=40+60$

$\theta_0 = 100/4 = 25°\text{C}$

From Eqs. (ii), we get

$4\,\text{s}=\alpha(25-20)\,(12\times 3600\,\text{s})$

$$\alpha=\frac{4}{5\times 12\times 3600}$$

$$=1.85\times 10^{-5}\,°\text{C}^{-1}$$

43. *(a)* KE of a body undergoing SHM is given by $\text{KE}=\frac{1}{2}m\omega^2A^2\cos^2\omega t$

and $\quad \text{KE}_{\text{max}}=\frac{m\omega^2A^2}{2}$

From given information,

$$\text{KE}=(\text{KE}_{\text{max}})\times\frac{75}{100}$$

$$\Rightarrow \quad \frac{m\omega^2A^2}{2}\cos^2\omega t=\frac{m\omega^2A^2}{2}\times\frac{3}{4}$$

or $\quad \cos\omega t=\pm\frac{\sqrt{3}}{2}$ or $\omega t=\frac{\pi}{6}$

or $\quad \frac{2\pi}{T}\times t=\frac{\pi}{6}$ or $t=\frac{T}{12}=\frac{1}{6}$ s

44. *(a)* Given, $y_1 = 0.1\sin\left(100\pi t+\frac{\pi}{3}\right)$

$\Rightarrow$ Velocity of particle 1 is given by,

$$v_1=\frac{dy_1}{dt}=0.1\times 100\pi\cos\left(100\pi t+\frac{\pi}{3}\right)$$

or $\quad v_1=10\pi\sin\left(100\pi t+\frac{\pi}{3}+\frac{\pi}{2}\right)$

or $\quad v_1=10\pi\sin\left(100\pi t+\frac{5\pi}{6}\right)$

and $\quad y_2 = 0.1\cos\pi t$

$\Rightarrow$ Velocity of particle 2,

$$\frac{dy_2}{dt}=v_2=-0.1\sin\pi t$$

or $\quad v_2=0.1\sin(\pi t+\pi)$

Hence, phase difference

$$\Delta\phi=\phi_1-\phi_2$$

$$=\left(100\pi t+\frac{5\pi}{6}\right)-(\pi t+\pi)$$

$$=\frac{5\pi}{6}-\pi=-\frac{\pi}{6} \quad [\text{at } t=0]$$

45. *(22)* Given, mass = 0.50 kg

Force, $F=(-50\ \text{Nm}^{-1})x$

Time period, $T=\frac{x}{35}$ s

We know that, $F=-kx$

$F=-50\ \text{Nm}^{-1}\ x\Rightarrow k=-50\ \text{Nm}^{-1}$

Time period, $\tau=2\pi\sqrt{\frac{m}{k}}=2\pi\sqrt{\frac{0.50}{50}}$

$$=\frac{2\pi}{10}=\frac{2\times 22}{10\times 7}=\frac{22}{35}\Rightarrow 22\times\frac{1}{35}\ \text{s}$$

Hence, $x = 22$

46. *(17)* We know that, $a=\omega^2x$

$16=\omega^2(4)$

$\omega^2=4\Rightarrow$ rad/s

$$v=\omega\sqrt{A^2-x^2}$$

$$A=\sqrt{\frac{v^2}{\omega^2}+x^2}$$

$$\Rightarrow \quad A=\sqrt{\frac{4}{4}+16}$$

$A=\sqrt{17}$ m $\quad\therefore\ x=17$

47. *(6)* Total energy of oscillating body

$$=\frac{1}{2}ma^2\omega^2=U+K$$

$$\Rightarrow \frac{1}{2}ma^2\omega^2=0.5+0.4=0.9\ \text{J} \quad \ldots\text{(i)}$$

Here, $m = 0.2$ kg, $\omega=2\pi f$

$$=2\pi\times\frac{25}{\pi}=50\ \frac{\text{rad}}{\text{s}}$$

So, from Eq. (i),

$$\frac{1}{2}\times 0.2\times a^2\times 50^2=0.9$$

$$\Rightarrow a^2=\frac{0.9}{50^2\times 0.1}=\left(\frac{3}{50}\right)^2$$

$$\Rightarrow \quad a=\frac{3}{50}\ \text{m}=\frac{3}{50}\times 100=6\ \text{cm}$$

48. *(12)* In SHM,

Maximum velocity is obtain at mean position,

$$v_{\text{max}}=\omega A=\frac{2\pi}{T}\times A=\frac{2\pi}{\pi}\times 0.06$$

$$=0.12\ \text{m/s}=12\ \text{cm/s}$$

49. *(10)* Given, displacement

$$x=10\sin\left(\omega t+\frac{\pi}{3}\right)$$

$\therefore$ Velocity of particle,

$$v=\frac{dx}{dt}=\frac{d}{dt}\left[10\sin\left(\omega t+\frac{\pi}{3}\right)\right]$$

$$=10\cos\left(\omega t+\frac{\pi}{3}\right)\cdot\omega$$

$$\Rightarrow \quad v=10\omega\cos\left(\frac{2\pi}{T}\cdot t+\frac{\pi}{3}\right)$$

At $t = 0$,

$$v=10\omega\cos\left(0+\frac{\pi}{3}\right)=10\times\frac{2\pi}{T}\left(\cos\frac{\pi}{3}\right)$$

$$=10\times\frac{2\times 3.14}{3.14}\times\frac{1}{2}=10\ \text{m/s}$$

50. *(3)* Frequency of oscillation of a string-mass system

$$f=\frac{1}{T}=\frac{1}{2\pi}\sqrt{\frac{k}{m}}$$

So, $\quad f_1=\frac{1}{2\pi}\sqrt{\frac{k}{m}}$

and $f_2 = \frac{1}{2\pi}\sqrt{\frac{k}{9m}} = \frac{1}{3}\cdot\frac{1}{2\pi}\sqrt{\frac{k}{m}}$

$= \frac{1}{3} f_1$

So, $f_1 / f_2 = 3$

51. *(7)*

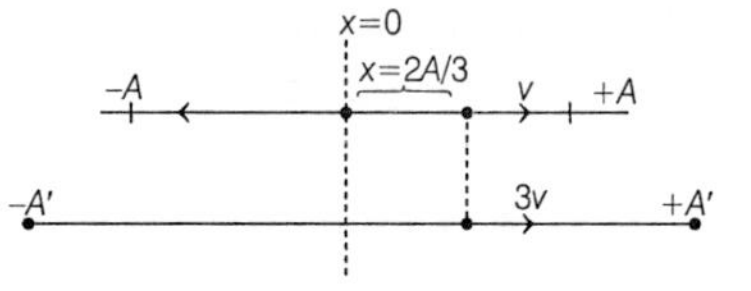

Velocity at $x = \frac{2A}{3}$ is,

$v = \omega\sqrt{A^2 - x^2} = \omega\sqrt{A^2 - \left(\frac{2A}{3}\right)^2}$

$= \omega A\left(\sqrt{\frac{5}{9}}\right)$

At $x = \frac{2A}{3}$, an impulse causes velocity to become three times.

$\therefore \quad v' = 3\omega A\frac{\sqrt{5}}{3} = \sqrt{5}\,\omega A$

As now, $\quad v' = \omega\sqrt{(A')^2 - x^2}$

$\Rightarrow \quad \sqrt{5}\,\omega A = \omega\sqrt{(A')^2 - \left(\frac{2A}{3}\right)^2}$

$\Rightarrow \quad 5A^2 = (A')^2 - \frac{4A^2}{9}$

$\Rightarrow \quad A'^2 = \left(5 + \frac{4}{9}\right)A^2 = \frac{49}{9}A^2$

or $\quad A' = \frac{7}{3}\text{A} = \frac{nA}{3}$ (given)

$\therefore \quad n = 7$

52. *(2)* In simple harmonic motion

$x = A\sin\omega t$

$x = A\sin\frac{2\pi}{T}t \quad \left(\therefore \omega = \frac{2\pi}{T}\right)$

when, $T = 6\pi$ s

then, $x = A\sin\left(\frac{2\pi}{6\pi}\cdot t\right)$

$x = A\sin\frac{t}{3} \quad \ldots\text{(i)}$

when, $x = A$, then $t = t_1$

Put value in Eq. (i),

$A = A\sin\frac{t_1}{3} \Rightarrow 1 = \sin\frac{t_1}{3}$

$\sin\frac{\pi}{2} = \sin\frac{t_1}{3} \Rightarrow t_1 = \frac{3\pi}{2}$

when, $x = \frac{\sqrt{3}}{2}A$, then $t = t_2$

Put value in Eq. (i),

$\frac{\sqrt{3}}{2}A = A\sin\frac{t_2}{3}$

$\sin\frac{\pi}{3} = \sin\frac{t_2}{3}$

$\Rightarrow \quad t_2 = \pi$

Time required, $\Delta t = t_1 - t_2$

$= \frac{3\pi}{2} - \pi = \frac{\pi}{2} = \frac{\pi}{x}$ s (given)

$\therefore \quad x = 2$

53. *(9)* Total kinetic energy, $E = \frac{1}{2}kA^2$

$U = \frac{1}{2}k\left(\frac{A}{3}\right)^2 = \frac{1}{2}\frac{kA^2}{9} = \frac{E}{9}$

$\text{KE} = E - \frac{E}{9} = \frac{8E}{9}$

$\frac{\text{Total KE}}{\text{KE}} = \frac{E}{8E/9} = \frac{9}{8} \Rightarrow x = 9$

54. *(12)* At mean position speed is maximum,

$v_{\max} = A\omega = 10$ cm / s

and using $v = \omega\sqrt{A^2 - x^2} = 5$ cm / s

$5 \text{ cm/s} = \omega\sqrt{A^2 - x^2} \quad \ldots\text{(i)}$

$\frac{10 \text{ cm / s}}{5 \text{ cm / s}} = \frac{A\omega}{\omega\sqrt{A^2 - x^2}} \quad \ldots\text{(ii)}$

$2 = \frac{A}{\sqrt{A^2 - x^2}}$

$\Rightarrow \quad 4 = \frac{A^2}{A^2 - x^2}$

$4A^2 - 4x^2 = A^2$

$3A^2 = 4x^2$

$\frac{3}{4}A^2 = x^2 \Rightarrow x = \frac{\sqrt{3}}{2}A$

$A = 4$ cm

$\Rightarrow \quad x = \frac{\sqrt{3}}{2} \times 4$

$x = 2\sqrt{3}$ cm

$x = \sqrt{12}$ cm

$\therefore \quad \alpha = 12$

> **• Concept Enhancer**
>
> Remember equation $[x(t)]$ of motion of a body performing simple harmonic motion will be periodic only when $x(t+T) = x(t)$, where T is time period of oscillation.

55. *(4)* Given, $y = A\cos 30°$

$= \frac{A\sqrt{3}}{2} = \frac{40\sqrt{3}}{2}$ cm $\quad [\because A = 40 \text{ cm}]$

$= 0.2\sqrt{3}$ m

Since, Kinetic energy,

$E_K = \frac{1}{2}m\omega^2(A^2 - y^2)$

$= \frac{1}{2}k(A^2 - y^2)$

$\Rightarrow \quad k = \frac{2E_K}{A^2 - y^2} = \frac{2 \times 200}{(0.4)^2 - (0.2\sqrt{3})^2}$

$= \frac{400}{0.16 - 0.12} = 1 \times 10^4$ N/m

$= 1 \times 10^x$ N/m (given)

$\therefore \quad x = 4$

56. *(99)* For pendulum

$T_{\max} = mg + \frac{mv^2}{L} \quad \ldots\text{(i)}$

Given $m = \frac{1}{4}$ kg, $L = 1$ m, $g = 9.8$ m/s^2

$A = \frac{1}{10}$ m

For SHM, $\text{KE}_{\max} = \frac{1}{2}mv^2 = \frac{1}{2}m\omega^2 A^2$

$\therefore \quad mv^2 = m\omega^2 A^2$

Using $\omega = \sqrt{\frac{g}{L}}$

$mv^2 = m\left(\sqrt{\frac{g}{l}}\right)^2 A^2 = \frac{mgA^2}{L} \quad \ldots\text{(ii)}$

Using Eqs. (ii) in (i)

$T_{\max} = mg + \frac{mgA^2}{L^2} = mg\left[1 + \frac{1}{10^2}\right]$

$= \frac{1}{4} \times 9.8 \times \frac{101}{100}$

$T_{\max} = \frac{98.98}{40} \Rightarrow x \cong 99$

57. *(2)* In SHM kinetic energy,

$\text{KE} = \frac{1}{2}k(A^2 - x^2)$

Potential energy, $U = \frac{1}{2}kx^2$

Given that, amplitude, $A = 3$ cm

$\text{KE} = U + 25\% \text{ of } U = U + 0.25\,U$

$\Rightarrow \quad \text{KE} = 1.25\,U$

$\Rightarrow \frac{1}{2}K(A^2 - x^2) = 1.25 \times \frac{1}{2}Kx^2$

$\Rightarrow A^2 - x^2 = 1.25x^2 \Rightarrow 2.25x^2 = A^2$

$x^2 = \frac{A^2}{2.25}$ or $x = \frac{A}{1.5}$

$\Rightarrow \quad x = \frac{3}{1.5} = 2$ cm

• Concept Enhancer

Kinetic energy and potential energy of a particle performing SHM is given as.

$KE = \frac{1}{2}k(A^2 - x^2)$ and $PE = \frac{1}{2}kx^2$

where, $k = m\omega^2$

where, A = amplitude

and x = displacement

58. *(88)* Given,

$$4v^2 = 50 - x^2 \Rightarrow v^2 = \frac{1}{4}(50 - x^2)$$

$$v^2 = \left(\frac{1}{2}\right)^2 [(\sqrt{50})^2 - x^2] \quad \ldots\text{(i)}$$

$$v^2 = \omega^2(A^2 - x^2) \quad \ldots\text{(ii)}$$

Thus, $\omega = \frac{1}{2} \Rightarrow \frac{2\pi}{T} = \frac{1}{2}$

$$T = 4\pi = 4 \times \frac{22}{7} = \frac{88}{7}\text{ s} = \frac{x}{7}\text{ s} \quad \text{[Given]}$$

$\therefore \quad x = 88$

59. *(40)* Given force on particle of 250 g executing SHM is,

$F = -25x$

As in SHM, $F = -kx$

So, $k = 25$ units.

Also, $\omega^2 = \frac{k}{m} = \frac{25}{250 \times 10^{-3}} = 100$

or $\omega = 10$ rad/s

Maximum speed of particle in SHM is

$v_{max} = A\omega$

Here, $v_{max} = 4$ m/s

So, $4 = A \times 10$

or $A = 0.4$ m $= 40$ cm

60. *(1)* Here, $A = 8$ cm, $T = 6$ s

we know that $x = A \sin \omega t$, when

$x = A/2$

$$\frac{A}{2} = A\cos\omega t \Rightarrow \cos\omega t = \frac{1}{2}$$

So, $\omega t = \frac{\pi}{3} \Rightarrow \frac{2\pi}{T} \times t = \frac{\pi}{3}$

$$t = \frac{T}{6} = \frac{6}{6} = 1\text{ s}$$

61. *(1)* Given, $y_1 = 10\sin\left(3\pi t + \frac{\pi}{3}\right)$

$A_1 = 10$

$y_2 = 5(\sin 3\pi t + \sqrt{3}\cos 3\pi t)$

$= 5\sin 3\pi t + 5\sqrt{3}\cos 3\pi t$

Amplitude of SHM y_2 is

$$A_2 = \sqrt{5^2 + (5\sqrt{3})^2}$$

$$= \sqrt{25 + 75} = \sqrt{100} = 10$$

$\therefore$ Ratio of amplitudes, $\frac{A_1}{A_2} = \frac{10}{10} = 1:1$

$= x:1$ [Given]

$\therefore \quad x = 1$

62. *(2)* Given, $x_1 = 5\sin\left(2\pi t + \frac{\pi}{4}\right)$.

$x_2 = 5\sqrt{2}(\sin 2\pi t + \cos 2\pi t)$.

Simplifying the second equation of simple harmonic motion as follows

$$x_2 = 5\sqrt{2}\left(\sin 2\pi t \times \frac{1}{\sqrt{2}} + \cos 2\pi t \times \frac{1}{\sqrt{2}}\right)\sqrt{2}$$

$$= 10\left(\sin 2\pi t \cos\frac{\pi}{4} + \cos 2\pi t \sin\frac{\pi}{4}\right)$$

$$= 10\sin\left(2\pi t + \frac{\pi}{4}\right)$$

Thus, $\frac{A_2}{A_1} = \frac{10}{5} = 2$

63. *(2)* In SHM

$$x(t) = A\sin(\omega t + \phi) \quad \ldots\text{(i)}$$

On differentiating equation with respect to t, we get

$$v(t) = A\omega\cos(\omega t + \phi) \quad \ldots\text{(ii)}$$

As per question, position of particle at $t = 0$ is 2 cm

Using Eq. (i), we get

$$2 = A\sin[\omega(0) + \phi]$$

$\Rightarrow \quad A\sin\phi = 2 \quad \ldots\text{(iii)}$

and velocity of particle at $t = 0$ is 2ω cms^{-1}

Using Eq. (ii), we get

$$2\omega = A\omega\cos[\omega(0) + \phi]$$

$\Rightarrow \quad A\cos\phi = 2 \quad \ldots\text{(iv)}$

On dividing Eq. (i) by Eq. (ii), we get

$$\frac{A\sin\phi}{A\cos\phi} = \frac{2}{2} \Rightarrow \tan\phi = 1$$

$\Rightarrow \quad \tan\phi = \tan 45°$

$\Rightarrow \quad \phi = 45° \quad [\because \tan 45° = 1]$

On putting the value of ϕ in Eq. (iii), we get

$$A\sin 45° = 2$$

$\Rightarrow \quad A\left(\frac{1}{\sqrt{2}}\right) = 2 \quad \left[\because \sin 45° = \frac{1}{\sqrt{2}}\right]$

$\Rightarrow \quad A = 2\sqrt{2} \quad \ldots\text{(v)}$

$= x\sqrt{2}$ cm [given]

$\therefore \quad x = 2$

64. *(6)* Given, the time period of the simple harmonic motion, $T = 2$ s

Displacement covered by the particle, $x = A/2$

Here, A is the amplitude of the particle. The general equation of the simple harmonic motion,

$$x = A\sin\omega t$$

$\Rightarrow \quad \frac{A}{2} = A\sin\left(\frac{2\pi}{T}t\right) \Rightarrow \frac{1}{2} = \sin\left(\frac{2\pi}{2}t\right)$

$\Rightarrow \quad \sin\frac{\pi}{6} = \sin(\pi t)$

$\Rightarrow \quad \frac{\pi}{6} = \pi t$

$\Rightarrow \quad t = \frac{1}{6} = \frac{1}{a}$ s [Given]

Hence, $a = 6$.

65. *(3)* Given, amplitude of particle in SHM $= a$

Maximum velocity, $v = A\omega$

Let displacement be y when speed,

$v_2 = v/2 = A\omega/2$

By using equation of wave velocity,

$$v_2^2 = \omega^2(A^2 - y^2)$$

$\because \quad \left(\frac{\omega A}{2}\right)^2 = \omega^2(A^2 - y^2)$

$\Rightarrow \quad \frac{\omega^2 A^2}{4} = \omega^2(A^2 - y^2)$

$\Rightarrow \quad \frac{A^2}{4} = A^2 - y^2$

$\Rightarrow \quad y = \frac{\sqrt{3}}{2}A = \frac{a\sqrt{x}}{2}$ [Given]

Hence, $x = 3$

Topic 2

01. *(b)* As we know

$$\omega = \sqrt{\frac{g}{l}}$$

$$\alpha = -\omega^2\theta$$

$$\alpha = -\frac{g}{l}\theta \;\; \alpha = \text{same} \quad \text{(given)}$$

$$\frac{g}{l_1}\theta_1 = \frac{g}{l_2}\theta_2$$

$$\theta_1 l_2 = \theta_2 l_1$$

02. *(c)* Time period $= 2\pi\sqrt{\frac{l}{g_{eff}}}$

As we go up, g_{eff} decreases, thus time period increases.

So, Assertion is true.

Reason is also true and explain the Assertion.

03. *(d)* Total mechanical energy of spring mass system,

$$E = \frac{1}{2} kA^2$$

Since, E does not depend on mass of oscillating particle, hence

New energy, $E' = E$

04. *(a)*

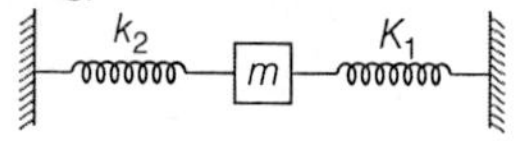

Both spring are in parallel

$k_{eq} = k_1 + k_2$

$$\because T = 2\pi\sqrt{\frac{m}{k_{eq}}} \Rightarrow T = 2\pi\sqrt{\frac{m}{k_1 + k_2}}$$

So, option (a) is correct.

05. *(a)* As we know, time period of simple pendulum is,

$$T = 2\pi\sqrt{\frac{L}{g}} \text{ or } T^2 = \frac{4\pi^2 L}{g}$$

or $\quad T^2 \propto L$

06. *(a)* We know that $\omega = \sqrt{k/m}$

For same spring, k remains constant.

$$\therefore \quad \frac{\omega_1}{\omega_2} = \sqrt{\frac{m_2}{m_1}}$$

Here, $m_1 = 1$ kg, $m_2 = 2$ kg

$$\therefore \quad \frac{\omega_1}{\omega_2} = \sqrt{\frac{2}{1}}$$

$$\frac{\omega_1}{\omega_2} = \sqrt{2} \Rightarrow \frac{\omega_2}{\omega_1} = \frac{1}{\sqrt{2}}$$

07. *(c)* Given, $m = 100 \text{ g} = 0.1 \text{ kg}$

$$R_e = 6400 \text{ km}$$
$$= 6.4 \times 10^6 \text{ m}$$

According to given situation, time period of oscillation of solid sphere,

$$T = 2\pi\sqrt{\frac{R_e}{g}} = 2 \times 3.14 = \sqrt{\frac{6.4 \times 10^6}{10}}$$

$$= 5028.6 \text{ s} = 83.8 \text{ min}$$
$$\simeq 84 \text{ min} \simeq 1 \text{ h } 24 \text{ min}$$

08. *(c)* For a pendulum, $T = 2\pi\sqrt{\frac{l}{g}}$, at mount Everest g reduces, so time period increases. Hence a pendulum clock will go slow (Assertion is false). As g decreases with height hence, its value at mount Everest is less than it's value on surface. (Reason is true).

09. *(d)* According to first situation.

$$T_0 = 2\pi\sqrt{\frac{l}{g}} \quad \ldots \text{(i)}$$

According to final situation, $l' = \frac{l}{16}$

$$T' = 2\pi\sqrt{\frac{l}{16 \times g}} = \frac{2\pi}{4}\sqrt{\frac{l}{g}} = \frac{T_0}{4}$$

[From Eq. (i)]

10. *(d)* We know that

$$v_{max} = A\omega$$

Given, $v_{max}(A) = v_{max}(B)$

$$\Rightarrow \quad A_1\omega_1 = A_2\omega_2$$

$$\frac{A_1}{A_2} = \frac{\omega_2}{\omega_1} \quad \ldots \text{(i)}$$

We know that $\omega = \sqrt{k/m}$

On substituting the value of ω in Eq. (i), we get

$$\frac{A_1}{A_2} = \sqrt{\frac{k_2}{k_1}}$$

11. *(b)* Statement I is false because time period of second's pendulum is always 2 s.

Therefore, time taken to move between two extreme positions will be $\frac{T}{2} = \frac{2}{2} = 1 \text{ s}$

12. *(a)* In series arrangement,

$$\frac{1}{k_{eq}} = \frac{1}{k_1} + \frac{1}{k_2}$$

$$\Rightarrow \quad \frac{1}{k_{eq}} = \frac{1}{k_1} + \frac{1}{k_1} = \frac{2}{k_1} \Rightarrow k_{eq} = \frac{k_1}{2}$$

As, $\quad T = 2\pi\sqrt{\frac{m}{k_1}}$

Where, m is mass of body connected with spring.

$$\Rightarrow \quad T \propto \sqrt{\frac{1}{k_1}} \text{ and } T' \propto \sqrt{\frac{2}{k_{eq}}}$$

$$\Rightarrow \quad T' = \sqrt{2}T$$

13. *(c)* Let spring constants of two springs be k_1 and k_2. Since, two springs are connected in parallel connection and parallel equivalent spring constant,

$k_{eq} = k_1 + k_2 \Rightarrow k_{eq} = 2k + 2k = 4k$

As, time period, $T = 2\pi\sqrt{\frac{m}{k_{eq}}}$

$$\Rightarrow \quad T = 2\pi\sqrt{\frac{m}{4k}} = \frac{2\pi}{2}\sqrt{\frac{m}{k}} = \pi\sqrt{\frac{m}{k}}$$

14. *(b)* As the child stands up, the effective length of pendulum decreases due to the reason that the centre of gravity rises up.

Hence, according to $T = 2\pi\sqrt{l/g}$, T will decrease.

15. *(a)* $g = \frac{GM}{R_e^2}$

Now, $M \to 4M$, $R \to 2R$

$$g' = \frac{G(4M)}{(2R_e)^2} = \frac{G(4M)}{4R_e^2}$$

$$g' = \frac{GM}{R_e^2} = g \Rightarrow T = 2\pi\sqrt{\frac{l}{g}}$$

A and R are correct.

but $T = 2\pi\sqrt{\frac{l}{g}}$ does not depend on mass ; R does not explain A.

16. *(a)* Time period of simple pendulum on the earth's surface,

$$T = 2\pi\sqrt{l/g} \quad \ldots \text{(i)}$$

where, l is length of pendulum.

At height $h = R$, gravitational acceleration g_h is given as

$$g_h = \frac{g}{\left(1 + \frac{h}{R}\right)^2} = \frac{g}{\left(1 + \frac{R}{R}\right)^2} = \frac{g}{4}$$

$\therefore$ New time period of simple pendulum at height h.

$$T' = 2\pi\sqrt{\frac{l}{g_h}} = 2\pi\sqrt{\frac{l}{g/4}} = 2 \times 2\pi\sqrt{\frac{l}{g}}$$

$= 2T$ [From Eq. (i)]

$= xT$ [Given]

$\therefore \quad x = 2$

17. *(a)* The situation given in the question can be depicted as shown below.

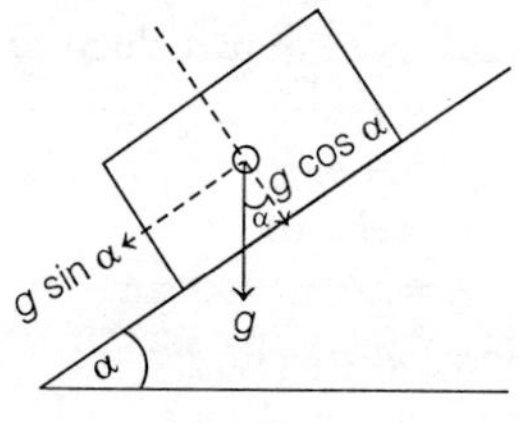

$\therefore g_{effective} = g \cos\alpha$

$$\therefore \quad T = 2\pi\sqrt{\frac{L}{g_{effective}}} = 2\pi\sqrt{\frac{L}{g\cos\alpha}}$$

18. *(c)* Given a massless spring, whose one end is fixed and other end is connected to a small object of mass (m).

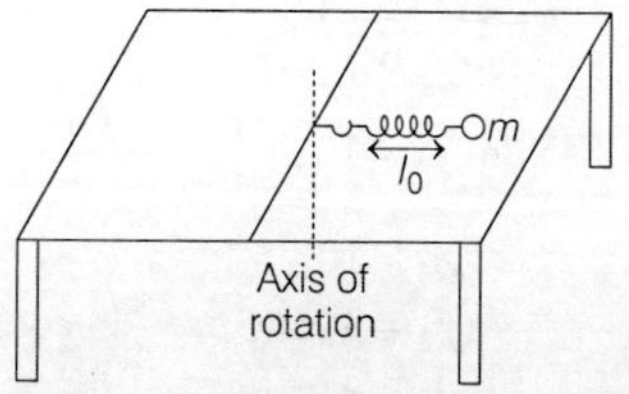

Spring remains horizontal on table and mass (m) revolves at constant angular

velocity (ω) about an axis passing through fixed end.
To find elongation of spring (say x)
Since, $m\omega^2 (l_0 + x) = kx$

$$\Rightarrow m\omega^2 l_0 + m\omega^2 x = kx$$

$$\Rightarrow m\omega^2 l_0 = kx - m\omega^2 x \Rightarrow x = \frac{m\omega^2 l_0}{(k - m\omega^2)}$$

19. *(c)* Time period of a pendulum is given by

$$T = 2\pi\sqrt{L/g} \quad \ldots\text{(i)}$$

Since, the lift is accelerated upwards with acceleration $\frac{g}{6}$, $a = g + \frac{g}{6} = \frac{7}{6}g$

New time period, $T' = 2\pi\sqrt{\frac{L}{(7/6)g}}$

$$= 2\pi\sqrt{\frac{6L}{7g}} = \sqrt{\frac{6}{7}} \times 2\pi\sqrt{\frac{L}{g}}$$

[From Eq. (i)]

$$T' = \sqrt{\frac{6}{7}}T$$

20. *(b)* According to given situation,

$$(v_{max})_1 = (v_{max})_2$$

$$\Rightarrow \omega_1 A_1 = \omega_2 A_2 \quad [\because v_{max} = \omega A]$$

$$\Rightarrow \frac{A_1}{A_2} = \frac{\omega_2}{\omega_1} = \frac{\sqrt{\frac{k_2}{m_2}}}{\sqrt{\frac{k_1}{m_1}}} \quad \left[\because \omega = \sqrt{\frac{k}{m}}\right]$$

$$= \sqrt{\frac{k_2}{k_1}}\sqrt{\frac{m_1}{m_2}} = \sqrt{\frac{9k}{2k}}\sqrt{\frac{0.05}{0.1}} = \frac{3}{2}$$

$$\therefore \quad A_1 : A_2 = 3 : 2$$

21. *(c)*

key idea You can use following formula to calculate time period of simple pendulum.

$$T = 2\pi\sqrt{l/g}$$

From the potential energy curve,

$$U_{max} = \frac{1}{2}kA^2 \Rightarrow 10 = \frac{1}{2}k(2)^2$$

$$\Rightarrow \quad k = 5\text{ N/m}$$

The length of the simple pendulum, $L = 4$ m
Time period of spring,

$$T = 2\pi\sqrt{\frac{m}{k}}$$

Time period of simple pendulum,

$$T = 2\pi\sqrt{\frac{l}{g}}$$

The time period of simple pendulum is same as the time period of the spring oscillation.

$$\Rightarrow \quad 2\pi\sqrt{\frac{l}{g}} = 2\pi\sqrt{\frac{m}{k}}$$

On substituting the values in the above equations, we get

$$2\pi\sqrt{\frac{4}{g}} = 2\pi\sqrt{\frac{5}{5}}$$

$$\Rightarrow \quad g = 4\text{ m/s}^2$$

22. *(d)* Time period of a simple pendulum,

$$T = 2\pi\sqrt{\frac{l}{g}} \quad \ldots\text{(i)}$$

When the lift moves upwards, then effective acceleration is

$$\Rightarrow \quad g_{eff} = g + a = g + \frac{g}{2} = \frac{3g}{2}$$

∴New time period,

$$T_1 = 2\pi\sqrt{\frac{l}{g_{eff}}} = 2\pi\sqrt{\frac{2l}{3g}}$$

$$\Rightarrow \quad T_1 = \sqrt{\frac{2}{3}}T \quad \text{[From Eq. (i)]}$$

23. *(c)* Let g_{eff} be the acceleration due to gravity.

∴Time period, $T = 2\pi\sqrt{l/g_{eff}}$

$$\Rightarrow \quad g_{eff} = 4\pi^2\frac{l}{T^2}$$

$$= 4\pi^2 \cdot \frac{2}{4} = 2\pi^2\text{ ms}^{-2}$$

24. *(c)* Time period of oscillation,

$$T = 2\pi\sqrt{\frac{M}{k_{eq}}}$$

$$\therefore \quad T = 2\pi\sqrt{\frac{M}{2k}} \quad [\because k_{eq} = k + k]$$

and frequency $(f) = \frac{1}{T} = \frac{1}{2\pi}\sqrt{\frac{2k}{M}}$

25. *(d)* Let extension in spring $= x$.
Then, total length of rotating spring $= l_0 + x$. Equating spring force and centripetal pull, we have

$$kx = m\omega^2(l_0 + x)$$

$$\Rightarrow \frac{x}{l_0} = \frac{m\omega^2}{k - m\omega^2} \Rightarrow \frac{x}{l_0} = \frac{m\omega^2}{k\left(1 - \frac{m\omega^2}{k}\right)}$$

As $k >> m\omega^2$, so $1 - \frac{m\omega^2}{k} \approx 1$

Hence, $\frac{x}{l_0} = \frac{m\omega^2}{k}$

26. *(c)* Initially, it is given that the unstretched length of the spring is l. When it is given an angular speed ω, then let x = stretched length of spring. Then, total length of the spring system while rotating will be $(l + x)$ as shown in the figure.

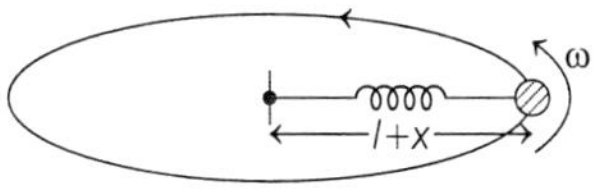

As we know, spring force will give the necessary centripetal force for rotation.

So, $\quad kx = m(l + x)\omega^2$

$$\Rightarrow (k - m\omega^2)x = ml\omega^2$$

$$\Rightarrow \quad x = \frac{ml\omega^2}{(k - m\omega^2)}$$

27. *(b)* Kinetic energy of a pendulum is maximum at its mean position.

$$K_{max} = \frac{1}{2}m\omega^2 a^2$$

$$\omega = \frac{2\pi}{T} = \frac{2\pi}{2\pi\sqrt{l/g}}$$

$$\text{or} \quad \omega = \sqrt{\frac{g}{l}} \quad \text{or} \quad \omega^2 = \frac{g}{l}$$

As amplitude is same in both cases so;

$$K_{max} \propto \omega^2 \quad \text{or} \quad K_{max} \propto \frac{1}{l}$$

[∵ g is constant]

According to given data,

$$K_1 \propto \frac{1}{l} \text{ and } K_2 \propto \frac{1}{2l}$$

$$\therefore \quad \frac{K_1}{K_2} = \left(\frac{1/l}{1/2l}\right) = 2 \quad \text{or} \quad K_1 = 2K_2$$

$$\Rightarrow \quad K_2 = K_1/2$$

28. *(b)* For a harmonic oscillator,

$$T = 2\pi\sqrt{\frac{m}{k}}$$

where, k = force constant and $T = \frac{1}{f}$

$$\therefore k = 4\pi^2 f^2 m$$

$$= 4 \times \left(\frac{22}{7}\right)^2 \times (10^{12})^2 \times \frac{108 \times 10^{-3}}{6.02 \times 10^{23}}$$

$$\Rightarrow \quad k = 7.1\text{ N/m}$$

29. *(a)* We know that time period,

$T = 2\pi\sqrt{L/g}$

When additional mass M is added to its bob

$$T_M = 2\pi\sqrt{\frac{L+\Delta L}{g}}$$

We know that

$$Y = \frac{Mg/A}{\Delta L/L} = \frac{MgL}{A\Delta L} \Rightarrow \Delta L = \frac{MgL}{AY}$$

$$\Rightarrow \therefore \quad T_M = 2\pi\sqrt{\frac{L + \frac{MgL}{AY}}{g}}$$

$$\Rightarrow \left(\frac{T_M}{T}\right)^2 = 1 + \frac{Mg}{AY}$$

or $$\frac{Mg}{AY} = \left(\frac{T_M}{T}\right)^2 - 1$$

or $$\frac{1}{Y} = \frac{A}{Mg}\left[\left(\frac{T_M}{T}\right)^2 - 1\right]$$

30. *(b)* Let at any instant, cube be at a depth x from the equilibrium position, then Net force acting on the cube = Upthrust on the portion of length x

$$\therefore \quad F = -\rho l^2 xg = -\rho l^2 g\, x \quad \text{...(i)}$$

$F \propto -x$ (SHM)

Negative sign shows that force is opposite to x. Hence, equation of SHM

$$F = -kx \quad \text{...(ii)}$$

On comparing Eqs. (i) and (ii), we get

$k = \rho l^2 g$

$$\therefore \quad T = 2\pi\sqrt{\frac{m}{k}} = 2\pi\sqrt{\frac{l^3 d}{\rho l^2 g}} = 2\pi\sqrt{\frac{ld}{\rho g}}$$

31. *(b)* Initial angular velocity of particle $= \omega_0$ and at any instant t, angular velocity $= \omega$

Therefore, for a displacement x, the resultant acceleration

$$f = (\omega_0^2 - \omega^2)x \quad \text{...(i)}$$

External force, $F = m(\omega_0^2 - \omega^2)x$ ...(ii)

Since, $F \propto \cos\omega t$ [Given]

From Eq. (ii), we get

$$m(\omega_0^2 - \omega^2)x \propto \cos\omega t \quad \text{...(iii)}$$

Now, from equation of simple harmonic motion,

$$x = A\sin(\omega t + \phi) \quad \text{...(iv)}$$

At $t = 0;\ x = A$

$$\therefore \quad A = A\sin(0 + \phi) \Rightarrow \phi = \frac{\pi}{2}$$

$$\therefore x = A\sin\left(\omega t + \frac{\pi}{2}\right) = A\cos\omega t \quad \text{...(v)}$$

Hence, from Eqs. (iii) and (v), we get

$$m(\omega_0^2 - \omega^2)A\cos\omega t \propto \cos\omega t$$

$$\Rightarrow \quad A \propto \frac{1}{m(\omega_0^2 - \omega^2)}$$

32. *(c)* As time period, $T = 2\pi\sqrt{\dfrac{M}{k}}$ ...(i)

$$T' = 2\pi\sqrt{\frac{M+m}{k}}$$

$$\Rightarrow \quad \frac{5T}{3} = 2\pi\sqrt{\frac{M+m}{k}} \quad \text{...(ii)}$$

On dividing Eq. (i) by Eq. (ii), we get

$$\frac{3}{5} = \sqrt{\frac{M}{M+m}} \quad \text{or} \quad \frac{9}{25} = \frac{M}{M+m}$$

or $9M + 9m = 25M$ or $16M = 9m$

or $$\frac{m}{M} = \frac{16}{9}$$

33. *(a)* $m = 10\text{ g} = 10 \times 10^{-3}$ kg

$A = h \times h = 10 \times 10\text{ cm}^2 = (0.1)^2\text{ m}^2$

When the body is floating

$mg = F_b$

$$mg = \rho_f gAx \quad \text{...(i)}$$

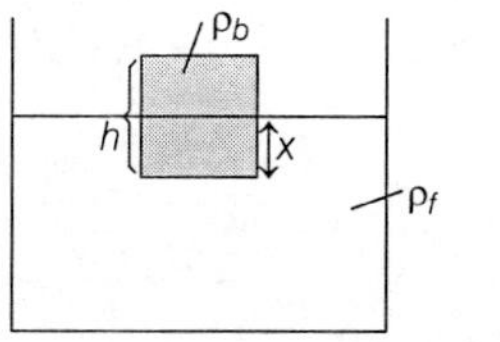

Let it is slightly displaced inside the filuid. Let at any instant, it is further y inside the fluid.

$$F = ma$$

$$F_b' - mg = ma$$

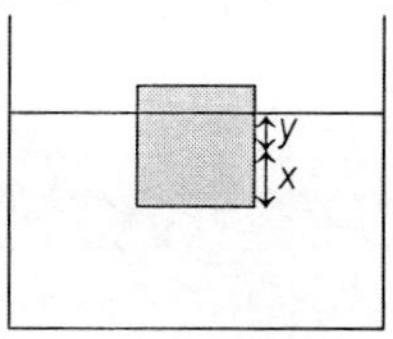

$$\rho_f A(x+y)g - mg = m\left(-\frac{d^2y}{dt^2}\right)$$

$$\rho_f Axg + \rho_f Ayg - mg = -\frac{md^2y}{dt^2}$$

Using Eq. (i)

$$\rho_f Ayg = -m\frac{d^2y}{dt^2}$$

$$\frac{d^2y}{dt^2} + \frac{\rho_f Agy}{m} = 0$$

$$\omega = \sqrt{\frac{\rho_f Ag}{m}}$$

For a cube/cuboidal block floating in a liquid, the time period is given by

$$T = 2\pi\sqrt{\frac{m}{\rho_f Ag}} = 2\pi\sqrt{\frac{10^{-2}}{10^3 \times (0.1)^2 \times 10}}$$

$$y\pi \times 10^{-2} = 2\pi\sqrt{10^{-4}}$$

$$y\pi \times 10^{-2} = 2\pi \times 10^{-2}$$

$$\Rightarrow \quad y = 2$$

34. *(b)*

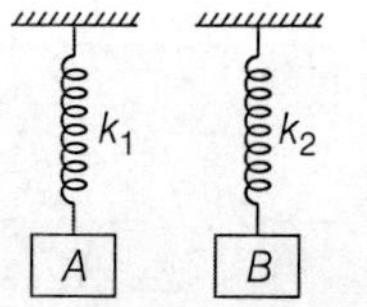

Angular velocity of SHM, $\omega = \sqrt{\dfrac{k}{m}}$

Angular velocity of 1st SHM,

$$\omega_1 = \sqrt{\frac{k_1}{m}}$$

Angular velocity of 2nd SHM,

$$\omega_2 = \sqrt{\frac{k_2}{m}}$$

Maximum velocity of SHM, $v = a\omega$

Maximum velocity of 1st SHM,

$v_1 = a\omega_1$

Maximum velocity of 2nd SHM,

$v_2 = a\omega_2$

According to question,

$$\frac{v_1}{v_2} = \frac{a\omega_1}{a\omega_2} = \sqrt{\frac{k_1}{m}} \times \sqrt{\frac{m}{k_2}}$$

$$\Rightarrow \quad \frac{v_1}{v_2} = \sqrt{\frac{k_1}{k_2}}$$

35. *(d)*

> **key idea** To solve this question, first use the relation for gravitation acceleration on going the above the earth's surface and use expression for time period of simple pendulum.

At height h, acceleration due to gravity is,

$$g_h = g \cdot \frac{R^2}{(R+h)^2}$$

At $h = 2R$,

$$g_h = g \cdot \frac{R^2}{(R+2R)^2} \Rightarrow g_h = \frac{g}{9}$$

For a second's pendulum time period is 2 s,

$$\therefore \quad T = 2\pi\sqrt{\frac{l}{g_h}} \Rightarrow \frac{T^2 \cdot g_h}{4\pi^2} = l$$

$$\text{or} \quad l = \frac{T^2 \cdot g}{4\pi^2 \times 9} \quad [\because g_h = g/9]$$

$$= \frac{4 \times \pi^2}{4\pi^2 \times 9} \Rightarrow l = \frac{1}{9}\text{ m}$$

36. *(c)* The time period of a simple pendulum is given by

$$T = 2\pi\sqrt{\frac{l}{g}}$$

Since, clocks are identical, so their lengths must be same.

$$\Rightarrow \quad \frac{T_1}{T_2} = \sqrt{\frac{g_2}{g_1}} \quad \ldots\text{(i)}$$

where, g_2 is the value of g at height h.

$$\text{As, } g_h = g\left(\frac{R_e}{R_e + h}\right)^2 \Rightarrow \frac{g_2}{g_1} = \left(\frac{R_e}{R_e + h}\right)^2$$

On substituting this value in Eq. (i), we get,

$$\frac{T_1}{T_2} = \sqrt{\left(\frac{R_e}{R_e + h}\right)^2} = \frac{R_e}{R_e + h}$$

$$\Rightarrow \quad \frac{4}{6} = \frac{6400}{6400 + h} \Rightarrow 6400 + h = 9600$$

$$\text{or} \quad h = 3200\text{ km}$$

37. *(d)* Time period of second pendulum,

$$T = 2\text{s}, \; g = \pi^2\text{ms}^{-2}$$

The value of gravitational acceleration at height, $h = 2R$, is given as

$$g' = \frac{g}{\left(1 + \frac{h}{R}\right)^2} = \frac{g}{\left(1 + \frac{2R}{R}\right)^2}$$

$$\Rightarrow \quad g' = \frac{g}{9} \quad \ldots\text{(i)}$$

We know that time period of second pendulum,

$$T = 2\pi\sqrt{\frac{l}{g'}}$$

$$\Rightarrow \quad 2 = 2\pi\sqrt{\frac{l}{g/9}} \quad \text{[Using Eq. (i)]}$$

$$\Rightarrow \quad 1 = \pi\sqrt{\frac{9l}{g}} \Rightarrow 1 = \pi^2 \times \frac{9l}{g}$$

$$\Rightarrow \quad l = \frac{g}{9\pi^2} = \frac{\pi^2}{9\pi^2} = \frac{1}{9}\text{ m}$$

38. *(d)* Given that a bob of mass m suspended by a thread of length l undergoes SHM with time period,

$$T_1 = T = 2\pi\sqrt{\frac{l}{g}} \quad \ldots\text{(i)}$$

In liquid, effective gravity,

$$g_{\text{eff}} = g\left(\frac{\sigma - \rho}{\sigma}\right)$$

where, ρ = density of liquid
and σ = density of body (bob).
Given, $\rho = \sigma/4 \Rightarrow \sigma = 4\rho$
On substituting this value, we get

$$g_{\text{eff}} = g\left(\frac{4\rho - \rho}{4\rho}\right) = \frac{3g}{4}$$

Now, the new time period of pendulum in liquid is

$$T' = 2\pi\sqrt{\frac{l'}{g_{\text{eff}}}} \quad \ldots\text{(ii)}$$

$$\text{Here,} \quad l' = l + \frac{l}{3} = \frac{4\ell}{3}$$

On substituting the values in Eq. (ii), we get

$$T' = 2\pi\sqrt{\frac{4l}{3} \times \frac{4}{3g}} = \left(2\pi\sqrt{\frac{l}{g}}\right)\frac{4}{3}$$

$$\Rightarrow \quad T' = \frac{4T}{3} \quad \text{[From Eq. (i)]}$$

39. *(d)* Given, radius of earth $= R$
Distance of chord from centre of earth $= R/2$

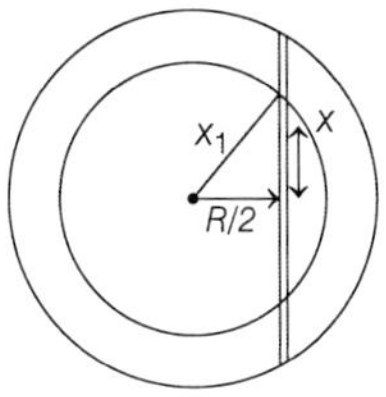

Let x_1 be the radius of inner circle and M be the mass of earth.
$\therefore$ m' (effective mass of earth)

$$= \frac{M}{\frac{4}{3}\pi R^3} \cdot \frac{4}{3}\pi x_1^3$$

$$\Rightarrow \quad m' = \frac{M}{R^3}x_1^3$$

Gravitational force

$$F = \frac{Gm'm}{x_1^2} = \frac{Gm}{x_1^2} \cdot \frac{M}{R^3}x_1^3$$

$$\Rightarrow m\omega^2 x_1 = \frac{GMmx_1}{R^3}$$

$$\Rightarrow \quad \omega = \sqrt{\frac{GM}{R^3}} \quad \ldots\text{(i)}$$

$$\text{Since,} \quad \omega = \frac{2\pi}{T} \text{ and } GM = gR^2$$

Substituting the above values in Eq. (i), we get

$$\Rightarrow \quad \frac{2\pi}{T} = \sqrt{\frac{gR^2}{R^3}} \Rightarrow T = 2\pi\sqrt{\frac{R}{g}}$$

40. *(b)* Given, initial amplitude $= A$
Velocity at mean position, $v = A\omega$
Applying conservation of momentum at mean position, we get

$$M_1v_1 = M_2v_2$$

$$\Rightarrow \quad MA\omega = (M + m)v'$$

$$\Rightarrow \quad v' = \frac{MA\omega}{M + m} = \frac{MA\sqrt{\frac{k}{M}}}{M + m}$$

$$\therefore \quad v' = A'\omega' = A'\sqrt{\frac{k}{M + m}}$$

$$\Rightarrow \quad A' = \frac{MA\sqrt{\frac{k}{M}}}{M + m} \times \sqrt{\frac{M + m}{k}}$$

$$A' = \sqrt{\frac{M}{M + m}}A$$

41. *(c)* Given that, displacement of particle at any instant

$$y(t) = y_0\sin^2\omega t,$$

$$y = y_0\left(\frac{1 - \cos 2\omega t}{2}\right)$$

$$y - \frac{y_0}{2} = -\frac{y_0}{2}\cos 2\omega t \quad \text{or}$$

$$Y = A\cos 2\omega t$$

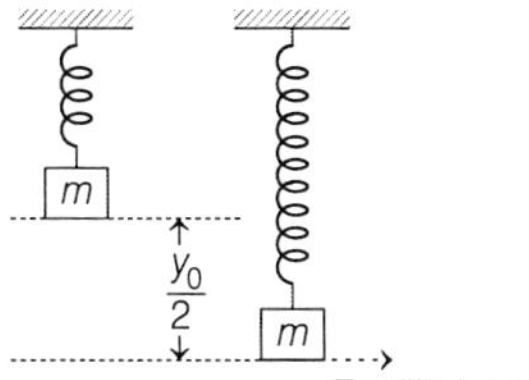

where, amplitude (displacement),

$$A = \frac{y_0}{2} \text{ and } 2\omega = \sqrt{k/m} \quad \ldots\text{(i)}$$

$$\text{Maximum displacement, } \frac{y_0}{2} = \frac{mg}{k}$$

$$\text{or} \quad k = \frac{2mg}{y_0} \quad \ldots\text{(ii)}$$

On putting the value of k from Eq. (ii) in Eq. (i), we get

$$2\omega = \sqrt{\frac{2mg}{y_0} \times \frac{1}{m}} \text{ or } \omega = \sqrt{\frac{g}{2y_0}}$$

42. *(c)* When a system oscillates, the magnitude of restoring torque of system is given by

$$\tau = C\theta \qquad \ldots(i)$$

Also, $\tau = I\alpha \qquad \ldots(ii)$

From Eqs. (i) and (ii),

$$\alpha = \frac{C}{I}\cdot\theta \qquad \ldots(iii)$$

and time period of oscillation of system will be $T = 2\pi\sqrt{\frac{I}{C}}$

In given case, magnitude of torque is

τ = Force × perpendicular distance

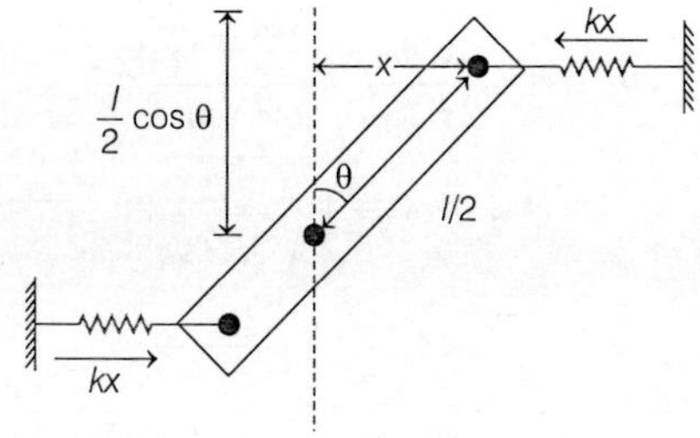

$$\tau = 2kx \times \frac{l}{2}\cos\theta$$

For small deflection,

$$\tau = \left(\frac{kl^2}{2}\right)\theta \qquad \ldots(iv)$$

∵ For small deflections,

$$\sin\theta = \frac{x}{\left(\frac{l}{2}\right)} \approx \theta \Rightarrow x = \frac{l\theta}{2}$$

Also, $\cos\theta \approx 1$

On comparing Eqs. (iv) and (i), we get

$$C = \frac{kl^2}{2}$$

$$\Rightarrow \quad \alpha = \frac{(kl^2/2)}{\left(\frac{1}{12}ml^2\right)}\cdot\theta \Rightarrow \alpha = \frac{6k}{m}\cdot\theta$$

Hence, time period of oscillation is

$$T = 2\pi\sqrt{\frac{m}{6k}}$$

Frequency of oscillation is given by

$$f = \frac{1}{T} = \frac{1}{2\pi}\sqrt{\frac{6k}{m}}$$

43. *(c)*

key idea You can calculate time period of a pendulum, with the help of following formula

$T = 2\pi\sqrt{l/g}$

where l is length of pendulum and g is acceleration due to gravity.

Period of motion of a pendulum is given by

$$T = 2\pi\sqrt{l/g} \qquad \ldots(i)$$

On the surface of earth,

$$\therefore \quad T_e = 2\pi\sqrt{\frac{l}{g_e}} \qquad \ldots(ii)$$

On the another planet,

$$\therefore \quad T_p = 2\pi\sqrt{l/g_p} \qquad \ldots(iii)$$

(∵ Pendulum is same, so l will be same)

From Eqs. (ii) and (iii), we get

$$\frac{T_e}{T_p} = \frac{2\pi\sqrt{l/g_e}}{2\pi\sqrt{l/g_p}} = \sqrt{\frac{g_p}{g_e}} \qquad \ldots(iv)$$

Now, $g_e = \frac{GM_e}{R_e^2}$ and $g_p = \frac{GM_p}{R_p^2}$

Given, $M_p = 3M_e$ and $R_p = 3R_e$

$$\therefore \quad g_p = \frac{G\times 3M_e}{9R_e^2} = \frac{1}{3}\cdot\frac{GM_e}{R_e^2} = \frac{1}{3}g_e$$

$$\Rightarrow \quad \frac{g_p}{g_e} = \frac{1}{3} \text{ or } \sqrt{\frac{g_p}{g_e}} = \frac{1}{\sqrt{3}} \qquad \ldots(v)$$

From Eqs. (iv) and (v), $T_p = \sqrt{3}\,T_e$

or $\quad T_p = 2\sqrt{3}$ s $\qquad$ (∵ $T_e = 2$ s)

44. *(b)* We know that in case of torsional oscillation frequency $\nu = \sqrt{k}/\sqrt{I}$

where, I is moment of inertia and k is torsional constant.

∴ According to question,

$$\nu_1 = \frac{\sqrt{k}}{\sqrt{\frac{M(2L)^2}{12}}}$$

(As, moment of inertia of a bar about COM is $I = \frac{ML^2}{12}$)

$$\text{or} \quad \nu_1 = \frac{\sqrt{k}}{\sqrt{ML^2/3}} \qquad \ldots(i)$$

When, two masses are attached at ends of rod. Then its moment of inertia is

$$\frac{M(2L)^2}{12} + 2m\left(\frac{L}{2}\right)^2$$

So, new frequency of oscillations is,

$$\nu_2 = \frac{\sqrt{k}}{\sqrt{\frac{M(2L)^2}{12} + 2m\left(\frac{L}{2}\right)^2}}$$

$$\nu_2 = \frac{\sqrt{k}}{\sqrt{\frac{ML^2}{3} + \frac{mL^2}{2}}} \qquad \ldots(ii)$$

As, $\nu_2 = 80\%$ of $\nu_1 = 0.8\nu_1$

$$\text{So,} \quad \frac{\sqrt{k}}{\sqrt{\frac{ML^2}{3} + \frac{mL^2}{2}}} = \frac{0.8\times\sqrt{k}}{\sqrt{\frac{ML^2}{3}}}$$

After solving it, we get,

$$m/M = 0.37$$

45. *(b)* Since in the given question, rotational torque, $\tau \propto$ angular displacement.

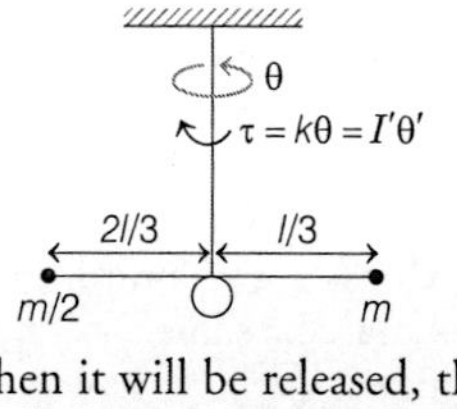

Thus, when it will be released, the system will execute SHM with a time period,

$$T = 2\pi\sqrt{I/k}$$

(where, I is moment of inertia and k is torsional constant) and the angular frequency is given as, $\omega = \sqrt{k/I}$.

If we know look at the top view of the above figure, we have

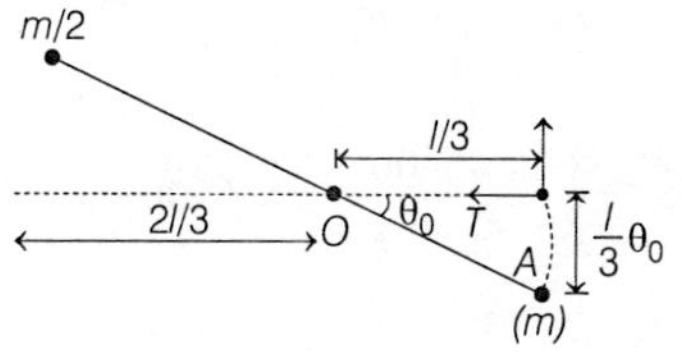

At some angular displacement θ_0, at point A the maximum velocity will be

$$F_c = \frac{mv_{\max}^2}{l/3}$$

$$v_{\max} = \frac{l}{3}\theta_0\omega = \frac{l}{3}\theta_0\sqrt{\frac{k}{I}} \qquad \ldots(i)$$

Then, tension in the rod when it passes through mean position will be

$$T = \frac{m\times v_{\max}^2}{\frac{l}{3}} = \frac{ml^2\theta_0^2 k\times 3}{9\times l\times I}$$

[Using Eq. (i)]

$$= \frac{ml\theta_0^2 k}{3I}$$

The moment of inertia I at point O,

$$= \frac{m}{2}\left(\frac{2l}{3}\right)^2 + m\left(\frac{l}{3}\right)^2$$

$$= \frac{2l^2m}{9} + \frac{ml^2}{9} = \frac{3ml^2}{9} = \frac{ml^2}{3}$$

$$\Rightarrow T = \frac{ml\theta_0^2 k\times 3}{3\times ml^2} = \frac{\theta_0^2 k}{l} = \frac{k\theta_0^2}{l}$$

46. *(c)* Pressure applied by piston,

$$\frac{Mg}{A} = p_0$$

$$Mg = p_0 A \quad \ldots\text{(i)}$$

As no exchange of heat takes place, the process is adiabatic.

$$p_0 V_0^\gamma = pV^\gamma,\ p_0 A x_0^\gamma = pA(x_0 - x)^\gamma$$

$$p = \frac{p_0 x_0^\gamma}{(x_0 - x)^\gamma}$$

Let piston be displaced by x.

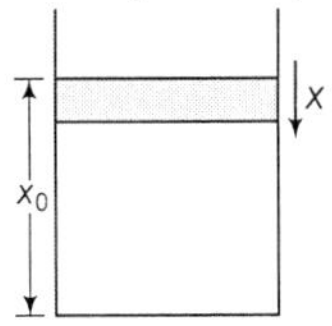

$$Mg - \left(\frac{p_0 x_0^\gamma}{(x_0 - x)^\gamma}\right) A = F_{\text{restoring}}$$

$$p_0 A\left(1 - \frac{x_0^\gamma}{(x_0 - x)^\gamma}\right) = F_{\text{restoring}}$$

$$[\because x_0 - x \approx x_0]$$

$$F = -\frac{\gamma p_0 Ax}{x_0}$$

$$\therefore \quad f = \frac{1}{2\pi}\sqrt{\frac{\gamma p_0 A}{x_0 M}} = \frac{1}{2\pi}\sqrt{\frac{\gamma p_0 A^2}{MV_0}}$$

47. *(c)* The time period of simple pendulum in air,

$$T = t_0 = 2\pi\sqrt{(l/g)} \quad \ldots\text{(i)}$$

In water, effective weight of bob,

w' = Weight of bob in air – Upthrust

$$\Rightarrow \rho V g_{\text{eff}} = mg - m'g$$

$$= \rho Vg - \rho' Vg = (\rho - \rho')Vg$$

where, ρ' = density of water,

ρ = density of bob.

$$\therefore \quad g_{\text{eff}} = \left(\frac{\rho - \rho'}{\rho}\right)g = \left(1 - \frac{\rho'}{\rho}\right)g$$

$$\therefore \quad t = 2\pi\sqrt{\left[\frac{l}{\left(1 - \frac{\rho'}{\rho}\right)g}\right]} \quad \ldots\text{(ii)}$$

$$\text{Thus,} \quad \frac{t}{t_0} = \sqrt{\left[\frac{1}{\left(1 - \frac{\rho'}{\rho}\right)}\right]}$$

$$= \sqrt{\left(\frac{1}{1 - \frac{1000}{(4/3)\times 1000}}\right)}$$

$$= \sqrt{\left(\frac{4}{4-3}\right)} = 2$$

or $\quad t = 2t_0$

48. *(b)* Time period of spring,

$$T = 2\pi\sqrt{\left(\frac{m}{k}\right)}$$

Here, k being the force constant of spring.

For first spring,

$$t_1 = 2\pi\sqrt{\left(\frac{m}{k_1}\right)} \quad \ldots\text{(i)}$$

For second spring,

$$t_2 = 2\pi\sqrt{\left(\frac{m}{k_2}\right)} \quad \ldots\text{(ii)}$$

In series, displacement is same,

$$\therefore \quad x = x_1 + x_2$$

$$\frac{-F}{k} = \frac{-F}{k_1} - \frac{F}{k_2} \Rightarrow k = \frac{k_1 k_2}{k_1 + k_2}$$

So, the effective force constant in their series combination is

$$k = \frac{k_1 k_2}{k_1 + k_2}$$

Time period of combination,

$$T = 2\pi\sqrt{\left[\frac{m(k_1 + k_2)}{k_1 k_2}\right]}$$

$$\Rightarrow \quad T^2 = \frac{4\pi^2 m(k_1 + k_2)}{k_1 k_2} \quad \ldots\text{(iii)}$$

From Eqs. (i) and (ii), we get

$$t_1^2 + t_2^2 = 4\pi^2\left(\frac{m}{k_1} + \frac{m}{k_2}\right)$$

$$\text{or} \quad t_1^2 + t_2^2 = 4\pi^2 m\left(\frac{1}{k_1} + \frac{1}{k_2}\right)$$

$$\text{or} \quad t_1^2 + t_2^2 = \frac{4\pi^2 m(k_1 + k_2)}{k_1 k_2}$$

$$\Rightarrow \quad t_1^2 + t_2^2 = T^2 \quad \text{[From Eq. (iii)]}$$

49. *(d)* $T = 2\pi\sqrt{\frac{l}{g}}$

$$T' = 2\pi\sqrt{\frac{l'}{g}} = 2\pi\sqrt{\frac{121l}{100g}}$$

$$\therefore l' = \frac{121}{100}l, \text{ So } T' = 1.1T$$

So, percentage increase in time period

$$= \frac{T' - T}{T} \times 100 = \frac{1.1T - T}{T} \times 100 = 10\%$$

50. *(12)*

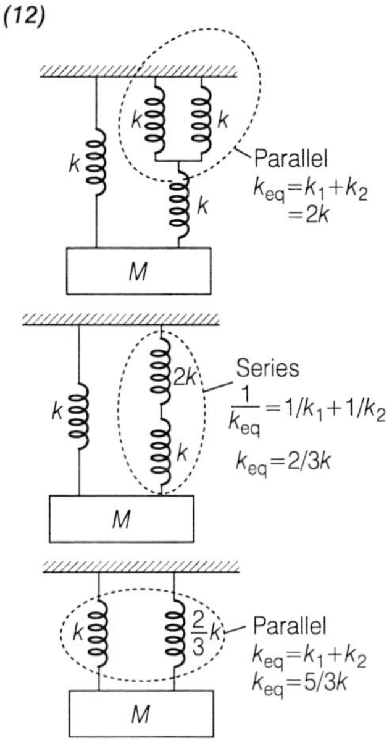

$$\text{Time period, } T = 2\pi\sqrt{\frac{m}{k_{\text{eq}}}} = 2\pi\sqrt{\frac{3m}{5k}}$$

$$= \pi\sqrt{\frac{12m}{5k}}$$

$$= \pi\sqrt{\frac{\alpha m}{5k}} \quad \text{(given)}$$

$$\therefore \quad \alpha = 12$$

51. *(20)* The time period of combination of springs and block is given as,

$$T = 2\pi\sqrt{m/k_{\text{eq}}}$$

In the given arrangement, for displacement of x of block both springs will have same deformation. So, both spring can be assumed in parallel.

In parallel combination,

$$k_{\text{eq}} = k_1 + k_2$$

$$= k + k = 2k = 2 \times 2$$

$$k_{\text{eq}} = 4 \text{ N/m}$$

Mass, $m = 490 \text{ g} = 0.49 \text{ kg}$

$$T = 2\pi\sqrt{\frac{m}{k_{\text{eq}}}} = 2\pi\sqrt{\frac{0.49}{4}} = 2\pi\sqrt{\frac{0.7 \times 0.7}{2 \times 2}}$$

$$= 2\pi \times \frac{0.7}{2} = 0.7\pi \text{ s}$$

No. of oscillations in 14π s,

$$N = \frac{14\pi}{T} = \frac{14\pi}{0.7\pi}$$

$$\Rightarrow \quad N = 20$$

52. *(8)* $U = \frac{1}{2} M\omega^2 A^2 \sin^2 \omega t$

So, slope $\frac{dU}{dt} = \frac{M\omega^3 A^2}{2} \sin 2\omega t$

It's value will be maximum, when

$\sin 2\omega t = 1$

$\Rightarrow \quad 2\omega t = \frac{\pi}{2}$

$\Rightarrow \quad 2 \times \frac{2\pi}{T} \times t = \frac{\pi}{2} \Rightarrow t = \frac{T}{8}$

So, $\beta = 8$

53. *(5)*

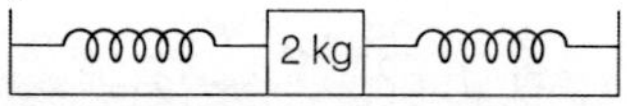

As, springs are in parallel arrangement, equivalent spring constant for system is;

$$k_{eq} = k_1 + k_2 = 40 \text{ Nm}^{-1}$$

So, time period of oscillation of mass m will be,

$$T = 2\pi\sqrt{\frac{m}{k_{eq}}}$$

$$= 2\pi\sqrt{\frac{2}{40}} = \pi\sqrt{\frac{4 \times 2}{40}} = \frac{\pi}{\sqrt{5}}$$

$$= \frac{\pi}{\sqrt{x}} \quad \text{[Given]}$$

$\therefore \quad x = 5$

54. *(20)* Given, time period,

$T = 3.14 \text{ s} = \pi \text{ s}$

$\Rightarrow \quad T = \frac{2\pi}{\omega} \Rightarrow \omega = 2$

Amplitude $= 1$ m

Mass $= 5$ kg

$\Rightarrow F_{max} = ma_{max} = m(A\omega^2)$

$= 5 \times 1 \times (2)^2 = 20$ N

55. *(5)* The given situation is shown below. Net force acting on the metallic bob in water,

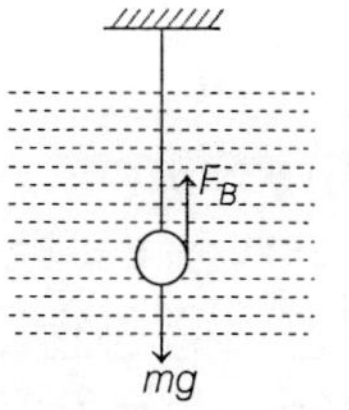

$mg' = mg - F_{upthrust}$

$\Rightarrow \quad mg' = mg - \rho_w Vg$

$\Rightarrow \quad g' = \frac{mg - \rho_w Vg}{m} = \frac{\rho_B Vg - \rho_w Vg}{\rho_B Vg}$

$= \left(1 - \frac{\rho_w}{\rho_B}\right) g = \left(1 - \frac{\rho_w}{5\rho_w}\right) g$

$\Rightarrow \quad g' = \frac{4}{5} g$

We know that, time period of simple pendulum,

$$T = 2\pi\sqrt{l / g}$$

$\Rightarrow \quad \frac{T'}{T} = \sqrt{\frac{g}{g'}} = \sqrt{\frac{g}{\frac{4}{5}g}} = \sqrt{\frac{5}{4}}$

$\Rightarrow \quad T = \sqrt{\frac{5}{4}} \times 10 = \frac{\sqrt{5}}{2} \times 10$ $\quad [\because T = 10 \text{ s}]$

$= \frac{\sqrt{x}}{2} \times 10$ $\quad$ [Given]

$\therefore \quad x = 5$

56. *(16)* Given, mass of the first body attached to spring, $M = 0.9$ kg Mass of small body attached to first body of mass M, $m = 124$ g $= 0.124$ kg

At mean position, $F_{net} = 0$

Therefore , by conservation of linear momentum, $p_1 = p_2$

As there is no loss of energy, total energy remains same

$\therefore \quad \frac{1}{2} kA^2 = \frac{p^2}{2m}$

$\left(\frac{A_1}{A_2}\right)^2 = \frac{p_1^2}{2m_1} \times \frac{2m_2}{p_2^2}$

$\left(\frac{A_1}{A_2}\right)^2 = \frac{m_2}{m_1} = \frac{1.024}{0.9}$

$\frac{A_1}{A_2} = \left(\frac{1024}{900}\right)^{1/2} = \frac{32}{30}$

$\frac{A_1}{A_2} = \frac{16}{15} = \frac{16}{16-1}$

$\therefore \quad \frac{A_1}{A_2} = \frac{\alpha}{\alpha - 1}$ $\quad$ (given)

then, $\alpha = 16$

57. *(2)* Case - I

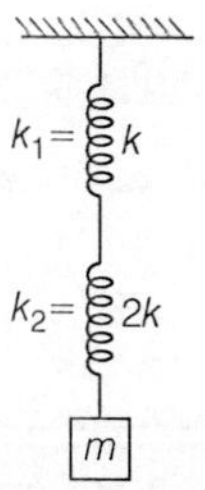

For series connection of spring equivalent force, constant k' is given as

$\frac{1}{k'} = \frac{1}{k_1} + \frac{1}{k_2} = \frac{1}{k} + \frac{1}{2k}$

$\Rightarrow \quad \frac{1}{k'} = \frac{3}{2k} \Rightarrow k' = \frac{2k}{3}$

Period of oscillation,

$T' = 2\pi\sqrt{m / k'}$

$\Rightarrow \quad 3 = 2\pi\sqrt{\frac{m}{\frac{2k}{3}}}$ $\quad [\because T' = 3 \text{ s}]$

$\Rightarrow \quad 3 = 2\pi\sqrt{\frac{3}{2}}\sqrt{\frac{m}{k}}$

$\Rightarrow \quad \sqrt{\frac{m}{k}} = \left(\sqrt{\frac{3}{2}}\right)\frac{1}{\pi}$ $\quad$ ...(i)

Case -2 For parallel combination,

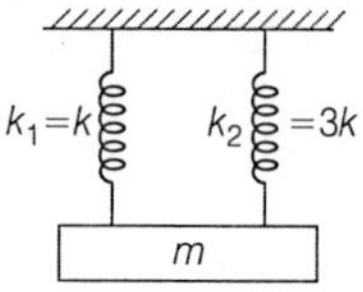

Equivalent force constant,

$k'' = k_1 + k_2$

$= k + 2k = 3k$

$\therefore$ Period of oscillation,

$T'' = 2\pi\sqrt{\frac{m}{k''}} = 2\pi\sqrt{\frac{m}{3k}} = \frac{2\pi}{\sqrt{3}}\sqrt{\frac{m}{k}}$

$\Rightarrow \quad T'' = \frac{2\pi}{\sqrt{3}}\left(\sqrt{\frac{3}{2}}\right)\frac{1}{\pi}$ $\quad$ [From Eq. (i)]

$= \sqrt{2} \text{ s} = \sqrt{x} \text{ s}$ $\quad$ [Given]

$x = 2$

58. *(700)* Initial amplitude of vibration, $a = 10$ cm $= 0.1$ m

Total energy of body = Kinetic energy at mean position

$\frac{1}{2} m\omega^2 a^2 = \frac{1}{2} m\omega^2 (0.1)^2$

Potential energy of body (at $x = 5$ cm),

PE ($x = 5$ cm), $U_1 = \frac{1}{2} m\omega^2 x^2$

$= \frac{1}{2} m\omega^2 \left(\frac{5}{100}\right)^2$

Also velocity at $x = 5$ cm is,

$v = \omega\sqrt{a^2 - x^2}$

$= \omega\sqrt{\left(\frac{10}{100}\right)^2 - \left(\frac{5}{100}\right)^2}$

$= \omega\sqrt{\frac{75}{(100)^2}} = \frac{\omega}{100} \times 5\sqrt{3}$

Air jet triples the velocity. So, velocity becomes, $v' = 3v$

or $\quad v' = \frac{\omega}{100} \times 5 \times 3\sqrt{3}$.

Now, total energy of body becomes,

$E_{total} = \text{PE} + \text{KE}$

$= \frac{1}{2} m\omega^2 \times \left(\frac{5}{100}\right)^2 + \frac{1}{2} m(v')^2$

$= \frac{1}{2} m\omega^2 \times \left(\frac{5}{100}\right)^2 + \frac{1}{2} m\omega^2$

$\times \left(\frac{5}{100}\right)^2 \times (3\sqrt{3})^2$

$$=\frac{1}{2}m\omega^2\times\frac{5^2}{100^2}(1+27)$$

$$=\frac{1}{2}m\omega^2\times\frac{25\times 28}{(100)^2}$$

So, amplitude a' is such that,

$$a'^2=\frac{25\times 28}{(100)^2}$$

or amplitude, $a'=\sqrt{\frac{25}{100}\times\frac{28}{100}}$

$$=\sqrt{\frac{7}{100}}\text{ m}=\sqrt{\frac{7}{100}}\times 100\text{ cm}$$

or $\quad a'=\sqrt{700}$ cm

Hence, $x=700$

59. *(8)* Given, mass of particle, $m=1$ kg

Spring constant, $k=100\text{ Nm}^{-1}$

Let time period is T.

Kinetic energy (KE) = Potential energy (PE) = E

KE in SHM = PE in SHM

$$\Rightarrow \frac{1}{2}m\omega^2(A^2-x^2)=\frac{1}{2}kx^2$$

$$\Rightarrow m\omega^2(A^2-x^2)=kx^2 \quad \ldots\text{(i)}$$

Since, force, $F=m\omega^2x=kx$

$$\Rightarrow k=m\omega^2$$

On substituting the value in Eq. (i), we get

$$\Rightarrow m\omega^2(A^2-x^2)=m\omega^2x^2$$

$$\Rightarrow A^2-x^2=x^2\Rightarrow A^2=2x^2$$

$$\Rightarrow x=A/\sqrt{2}$$

Since, wave displacement,

$$x=A\sin\omega t$$

$$\Rightarrow \frac{A}{\sqrt{2}}=A\sin\omega t\Rightarrow\sin\left(\frac{\pi}{4}\right)=\sin\omega t$$

$$\Rightarrow \omega t=\frac{\pi}{4}\Rightarrow\frac{2\pi}{T}\cdot t=\frac{\pi}{4}$$

$$\Rightarrow t=\frac{T}{8}=\frac{T}{x} \quad \text{[Given]}$$

$\therefore \quad x=8$

60. *(10)* Given, spring constants of springs k_1 and k_2 are k and $4k$.

where, $k=20$ N/m

$$m_A=200\text{ g}=200\times 10^{-3}\text{ kg}$$

$$m_B=800\text{ g}=800\times 10^{-3}\text{ kg}$$

As we know that

$$T=2\pi\sqrt{\frac{m_{eq}}{k_{eq}}} \quad \ldots\text{(i)}$$

where, equivalent or reduced mass,

$$m_{eq}=\frac{m_Am_B}{m_A+m_B}$$

$$\Rightarrow m_{eq}=\frac{(200\times 10^{-3})(800\times 10^{-3})}{200\times 10^{-3}+800\times 10^{-3}}$$

$$=16\times 10^{-2}\text{ kg}$$

and equivalent spring constant, in series

$$k_{eq}=\frac{k_1k_2}{k_1+k_2}=\frac{k\times 4k}{k+4k}=\frac{4k}{5}$$

$$=\frac{4}{5}\times 20=16\text{ N/m}$$

On putting these values in Eq. (i), we get

$$T=2\pi\sqrt{\frac{16\times 10^{-2}}{16}}$$

$$=2\pi\times 10^{-1}=\frac{2\pi}{10}$$

As, $\quad \omega=\frac{2\pi}{T}=\frac{2\pi}{2\pi}\times 10=10\text{ rads}^{-1}$

61. *(2)* Time period of figure 1 can be given as

$$T_a=2\pi\sqrt{\frac{M}{k}}$$

Time period can be given in figure 2 can be given as

$$T_b=2\pi\sqrt{\frac{M}{k_{eq}}}=2\pi\sqrt{\frac{M}{\frac{k}{2}}}\left(\because k_{eq}=\frac{k\times k}{k+k}\right)$$

Now, $\quad \frac{T_b}{T_a}=\frac{2\pi\sqrt{\frac{M}{k/2}}}{2\pi\sqrt{\frac{M}{k}}}$

$$\Rightarrow \frac{T_b}{T_a}=\sqrt{2}=\sqrt{x} \quad \text{[Given] ...(i)}$$

$\therefore \quad x=2$

62. *(7)* Given, angular displacement to complete $\frac{5}{8}\left(=\frac{1}{2}+\frac{1}{8}\right)$ rev

$$=\left(\pi+\frac{\pi}{6}\right)\text{rad}=\left(\frac{7\pi}{6}\right)\text{rad}$$

Since, $\quad \omega=\frac{2\pi}{T}=\frac{\theta}{t}$

$$\Rightarrow \theta=\frac{2\pi}{T}\cdot t\Rightarrow\frac{7\pi}{6}=\frac{2\pi}{T}t$$

$$\Rightarrow t=\frac{7T}{12}=\frac{\alpha}{\beta}T$$

Hence, $\alpha=7$

C-14

JEE Main Chapterwise Topicwise Physics

In JEE Main there is usually 0-1 question per paper are visible from this chapter. In past 3 years, 13 conceptual and 11 numerical questions were asked of moderate to difficult level from this chapter.

For detailed analysis scan QR code

WAVES

Topic 1 Basic of Waves and Progressive Waves

Level 1

01 The equation of a wave travelling on a string is , $y = \sin[20\pi x + 10\pi t]$, where x and t are distance and time in SI units. The minimum distance between two points having the same oscillating speed is *[2025, 7 April Shift-II]*

(a) 5.0 cm (b) 10 cm
(c) 20 cm (d) 2.5 cm

02 For a periodic motion represented by the equation $y = \sin\omega t + \cos\omega t$ the amplitude of the motion is *[2023, 10 April Shift-II]*

(a) 2 (b) $\sqrt{2}$ (c) 1 (d) 0.5

Imp. Speed of Transverse Waves

03 A steel wire with mass per unit length 7.0×10^{-3} kg m^{-1} is under tension of 70 N. The speed of transverse waves in the wire will be *[2023, 1 Feb Shift-I]*

(a) 10 m/s (b) 200π m/s
(c) 100 m/s (d) 50 m/s

04 A travelling wave is described by the equation $y(x, t) = [0.05 \sin(8x - 4t)]$ m.
The velocity of the wave is (all the quantities are in SI unit) *[2023, 24 Jan Shift-I]*

(a) 0.5 ms^{-1} (b) 8 ms^{-1}
(c) 2 ms^{-1} (d) 4 ms^{-1}

05 Which of the following equations correctly represents a travelling wave having wavelength $\lambda = 4.0$ cm, frequency $\nu = 100$ Hz and travelling in positive X-axis direction? *[2022, 30 June Shift-I]*

(a) $y = A\sin[(0.50\,\pi\,\text{cm}^{-1})x - (100\,\pi\text{s}^{-1})t]$

(b) $y = A\sin 2\pi[(0.25\,\text{cm}^{-1})x - (50\,\text{s}^{-1})t]$

(c) $y = A\sin\left[\left(\frac{2\pi}{4}\text{cm}^{-1}\right)x - \left(\frac{2\pi}{100}\text{s}^{-1}\right)t\right]$

(d) $y = A\sin\pi[(0.5\,\text{cm}^{-1})x - (200\,\text{s}^{-1})t]$

06 Which of the following equations represents a travelling wave? *[2021, 24 Feb Shift-II]*

(a) $y = A\sin(15x - 2t)$ (b) $y = Ae^{-x^2}(vt + \theta)$
(c) $y = Ae^{x}\cos(\omega t - \theta)$ (d) $y = A\sin x\cos\omega t$

07 A travelling harmonic wave is represented by the equation $y(x, t) = 10^{-3}\sin(50t + 2x)$, where x and y are in metre and t is in second. Which of the following is a correct statement about the wave? *[2019, 12 Jan Shift-I]*

(a) The wave is propagating along the negative X-axis with a speed of 25 ms^{-1}.
(b) The wave is propagating along the positive X-axis with a speed of 25 ms^{-1}.
(c) The wave is propagating along the positive X-axis with a speed of 100 ms^{-1}.
(d) The wave is propagating along the negative X-axis with a speed of 100 ms^{-1}

08 **Statement I** Two longitudinal waves given by equations– $y_1(x, t) = 2a\sin(\omega t - kx)$ and $y_2(x, t) = a\sin(2\omega t - 2kx)$ will have equal intensity.
Statement II Intensity of waves of given frequency in same medium is proportional to the square of amplitude only. *[AIEEE 2011]*

(a) Statement I is false but Statement II is true.
(b) Statement I is true but Statement II is false.
(c) Both Statement I and Statement II are true and Statement II is the correct explanation of Statement I.
(d) Both Statement I and Statement II are true but Statement II is not the correct explanation of Statement I.

09 The transverse displacement $y(x, t)$ of a wave on a string is given by $y(x, t) = e^{-(ax^2 + bt^2 + 2\sqrt{ab}\,xt)}$.
This represents a **[AIEEE 2011]**
(a) wave moving in $-x$ direction with speed $\sqrt{b/a}$
(b) standing wave of frequency $\sqrt{b}$
(c) standing wave of frequency $1/\sqrt{b}$
(d) wave moving in $+x$-direction with speed $\sqrt{\frac{a}{b}}$

10 The displacement y of a particle in a medium can be expressed as

$$y = 10^{-6} \sin\left(100t + 20x + \frac{\pi}{4}\right) \text{m}$$

where, t is in second and x in metre. The speed of the wave is **[AIEEE 2004]**
(a) 2000 ms^{-1} (b) 5 ms^{-1} (c) 20 ms^{-1} (d) $5\pi \text{ ms}^{-1}$

Level 2

11 Displacement of a wave is expressed as $x(t) = 5\cos\left(628t + \frac{\pi}{2}\right)$ m. The wavelength of the wave $\pi = 3.14$ when its velocity is 300 m/s is **[2025, 4 April Shift-II]**
(a) 0.33 m (b) 0.5 m (c) 3 m (d) 5 m

12 A sinusoidal wave of wavelength 7.5 cm travels a distance of 1.2 cm along the x-direction in 0.3 s. The crest P is at $x = 0$ at $t = 0$ s and maximum displacement of the wave is 2 cm. Which equation correctly represents this wave ? **[2025, 2 April Shift-II]**
(a) $y = 2\cos(0.83x - 3.35t)$ cm
(b) $y = 2\sin(0.83x - 3.5t)$ cm
(c) $y = 2\cos(3.35x - 0.83t)$ cm
(d) $y = 2\cos(0.13x - 0.5t)$ cm

13 The equation of a transverse wave travelling along a string is $y(x, t) = 4.0\sin[20 \times 10^{-3}x + 600t]$ mm, where x is in mm and t is in second. The velocity of the wave is **[2025, 23 Jan Shift-II]**
(a) -60 m/s (b) $+30$ m/s (c) $+60$ m/s (d) -30 m/s

Imp. Speed of Travelling Wave

14 In the wave equation, $y = 0.5\sin\frac{2\pi}{\lambda}(400t - x)$ m, the velocity of the wave will be **[2022, 28 July Shift-I]**
(a) 200 m/s (b) $200\sqrt{2}$ m/s
(c) 400 m/s (d) $400\sqrt{2}$ m/s

15 A transverse wave is represented by $y = 2\sin(\omega t - kx)$ cm. The value of wavelength (in cm) for which the wave velocity becomes equal to the maximum particle velocity, will be **[2022, 26 July Shift-II]**
(a) 4π (b) 2π (c) π (d) 2

16 A longitudinal wave is represented by $y = 10\sin 2\pi\left(nt - \frac{x}{\lambda}\right)$ cm. The maximum particle velocity will be four times the wave velocity, if the determined value of wavelength is equal to **[2022, 29 June Shift-I]**
(a) 2π (b) 5π (c) π (d) $5\pi/2$

17 A sound wave of frequency 245 Hz travels with the speed of 300 ms^{-1} along the positive X-axis. Each point of the wave moves to and fro through a total distance of 6 cm. What will be the mathematical expression of this travelling wave? **[2021, 17 March Shift-II]**
(a) $y(x, t) = 0.03[\sin 5.1x - (0.2 \times 10^3)t]$
(b) $y(x, t) = 0.06[\sin 5.1x - (1.5 \times 10^3)t]$
(c) $y(x, t) = 0.06[\sin 0.8x - (0.5 \times 10^3)t]$
(d) $y(x, t) = 0.03[\sin 5.1x - (1.5 \times 10^3)t]$

18 For a transverse wave travelling along a straight line, the distance between two peaks (crests) is 5 m, while the distance between one crest and one trough is 1.5 m. The possible wavelengths (in metre) of the waves are **[2020, 4 Sep Shift-I]**
(a) 1, 2, 3,...... (b) $\frac{1}{1}, \frac{1}{3}, \frac{1}{5}, \ldots$
(c) 1, 3, 5,..... (d) $\frac{1}{2}, \frac{1}{4}, \frac{1}{6}, \ldots$

19 Two identical strings X and Z made of same material have tensions T_X and T_Z in them. If their fundamental frequencies are 450 Hz and 300 Hz. respectively, then the ratio of T_X / T_Z is **[2020, 2 Sep Shift-I]**
(a) 1.25 (b) 2.25 (c) 0.44 (d) 1.5

20 A transverse wave travels on a taut steel wire with a velocity of v when tension in it is 2.06×10^4 N. When the tension is changed to T, the velocity changed to $v/2$. The value or T is close to **[2020, 8 Jan Shift-II]**
(a) 10.2×10^2 N (b) 5.15×10^3 N
(c) 2.50×10^4 N (d) 30.5×10^4 N

21 A progressive wave travelling along the positive x-direction is represented by $y(x, t) = A\sin(kx - \omega t + \phi)$. Its snapshot at $t = 0$ is given in the figure. **[2019, 12 April Shift-I]**

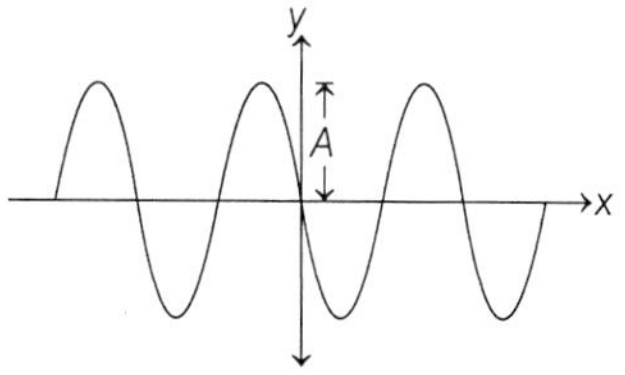

For this wave, the phase ϕ is
(a) $-\frac{\pi}{2}$ (b) π (c) 0 (d) $\frac{\pi}{2}$

Imp. Notes and Antinodes

22 A string of length 1 m and mass 5 g is fixed at both ends. The tension in the string is 8.0 N. The string is set into vibration using an external vibrator of frequency 100 Hz. The separation between successive nodes on the string is close to *[2019, 10 Jan Shift-I]*
(a) 16.6 cm (b) 33.3 cm
(c) 10.0 cm (d) 20.0 cm

23 A granite rod of 60 cm length is clamped at its middle point and is set into longitudinal vibrations. The density of granite is 2.7×10^3 kg/m^3 and its Young's modulus is 9.27×10^{10} Pa. What will be the fundamental frequency of the longitudinal vibrations? *[JEE Main 2018]*
(a) 5 kHz (b) 2.5 kHz (c) 10 kHz (d) 7.5 kHz

24 A uniform string of length 20 m is suspended from a rigid support. A short wave pulse is introduced at its lowest end. It starts moving up the string. The time taken to reach the support is (Take, $g = 10$ ms^{-2}) *[JEE Main 2016]*
(a) $2\pi\sqrt{2}$ s (b) 2 s (c) $2\sqrt{2}$ s (d) $\sqrt{2}$ s

25 A transverse wave is represented by
$$y = \frac{10}{\pi}\sin\left(\frac{2\pi}{T}t - \frac{2\pi}{\lambda}x\right) \text{cm}$$
For what value of the wavelength the wave velocity is twice the maximum particle velocity? *[JEE Main 2014]*
(a) 40 cm (b) 20 cm (c) 10 cm (d) 60 cm

26 In a transverse wave, the distance between a crest and neighbouring trough at the same instant is 4.0 cm and the distance between a crest and trough at the same place is 1.0 cm. The next crest appears at the same place after a time interval of 0.4 s. The maximum speed of the vibrating particles in the medium is *[JEE Main 2013]*
(a) $\frac{3\pi}{2}$ cm/s (b) $\frac{5\pi}{2}$ cm/s (c) $\frac{\pi}{2}$ cm/s (d) 2π cm/s

27 The disturbance $y(x, t)$ of a wave propagating in the positive x-direction is given by $y = \frac{1}{1+x^2}$ at time $t = 0$ and by $y = \frac{1}{[1+(x-1)^2]}$ at $t = 2$ s, where x and y are in meters. The shape of the wave disturbance does not change during the propagation. The velocity of wave in m/s is *[JEE Main 2012]*
(a) 2.0 (b) 4.0 (c) 0.5 (d) 1.0

28 The equation of a wave on a string of linear mass density 0.04 kg m^{-1} is given by
$$y = 0.02\ (\text{m}) \sin\left[2\pi\left(\frac{t}{0.04\ (s)} - \frac{x}{0.50\ (m)}\right)\right].$$
The tension in the string is *[AIEEE 2010]*
(a) 4.0 N (b) 12.5 N
(c) 0.5 N (d) 6.25 N

29 A wave travelling along the x-axis is described by the equation $y(x, t) = 0.005\cos(\alpha x - \beta t)$. If the wavelength and the time period of the wave are 0.08 m and 2.0 s respectively, then α and β in appropriate units are *[AIEEE 2008]*
(a) $\alpha = 25.00\,\pi$, $\beta = \pi$
(b) $\alpha = \frac{0.08}{\pi}$, $\beta = \frac{2.0}{\pi}$
(c) $\alpha = \frac{0.04}{\pi}$, $\beta = \frac{1.0}{\pi}$
(d) $\alpha = 12.50\,\pi$, $\beta = \frac{\pi}{2.0}$

Level 3

30 A heavy ball of mass M is suspended from the ceiling of a car by a light string of mass $m (m << M)$. When the car is at rest, the speed of transverse waves in the string is 60 ms^{-1}. When the car has acceleration a, the wave speed increases to 60.5 ms^{-1}. The value of a, in terms of gravitational acceleration g is closest to *[2019, 9 Jan Shift-I]*
(a) $g/20$ (b) $g/5$
(c) $g/30$ (d) $g/10$

Numerical Value Type Questions

31 The equation of wave is given by
$$Y = 10^{-2}\sin 2\pi(160t - 0.5x + \pi/4)$$
where x and Y are in meter and t in second. The speed of the wave is .... kmh^{-1}. *[2023, 11 April Shift-I]*

32 A transverse harmonic wave on a string is given by $y(x, t) = 5\sin(6t + 0.003x)$
where, x and y are in cm and t in sec. The wave velocity is .... ms^{-1}. *[2023, 10 April Shift-I]*

33 The mass per unit length of a uniform wire is 0.135 g/cm. A transverse wave of the form $y = -0.21\sin(x + 30t)$ is produced in it, where x is in metre and t is in second. Then, the expected value of tension in the wire is $x \times 10^{-2}$ N. Value of x is .... (Round-off to the nearest integer) *[2021, 26 Feb Shift-I]*

34 A wire of density 9×10^{-3} kg-cm^{-3} is stretched between two clamps 1 m apart. The resulting strain in the wire is 4.9×10^{-4}. The lowest frequency of the transverse vibrations in the wire is (Young's modulus of wire, $Y = 9 \times 10^{10}$ Nm^{-2}), (to the nearest integer) ....Hz . *[2020, 2 Sep Shift-II]*

Topic 2 Superposition and Reflection of Waves

Level 1

01 Consider the sound wave travelling in ideal gases of He, CH_4 and CO_2. All the gases have the same ratio p/ρ, where p is the pressure and ρ is the density. The ratio of the speed of sound through the gases $v_{He} : v_{CH_4} : v_{CO_2}$ is given by *[2025, 4 April Shift-I]*

(a) $\sqrt{\frac{5}{3}} : \sqrt{\frac{4}{3}} : \sqrt{\frac{7}{5}}$ (b) $\sqrt{\frac{7}{5}} : \sqrt{\frac{5}{3}} : \sqrt{\frac{4}{3}}$
(c) $\sqrt{\frac{5}{3}} : \sqrt{\frac{4}{3}} : \sqrt{\frac{4}{3}}$ (d) $\sqrt{\frac{4}{3}} : \sqrt{\frac{5}{3}} : \sqrt{\frac{7}{5}}$

Imp. Beat Frequency

02 The velocity of sound in a gas in which two wavelengths 4.08 m and 4.16 m produce 40 beats in 12 s, will be : *[2022, 28 June Shift-I]*

(a) 282.8 ms^{-1} (b) 175.5 ms^{-1}
(c) 353.6 ms^{-1} (d) 707.2 ms^{-1}

03 The equations of two waves are given by

$$y_1 = 5\sin 2\pi(x - vt)\,\text{cm}$$
$$y_2 = 3\sin 2\pi(x - vt + 1.5)\,\text{cm}$$

These waves are simultaneously passing through a string. The amplitude of the resulting wave is *[2022, 24 June Shift-I]*

(a) 2 cm (b) 4 cm (c) 5.8 cm (d) 8 cm

04 The correct figure that shows schematically, the wave pattern produced by superposition of two waves of frequencies 9 Hz and 11 Hz, is *[2019, 10 April Shift-II]*

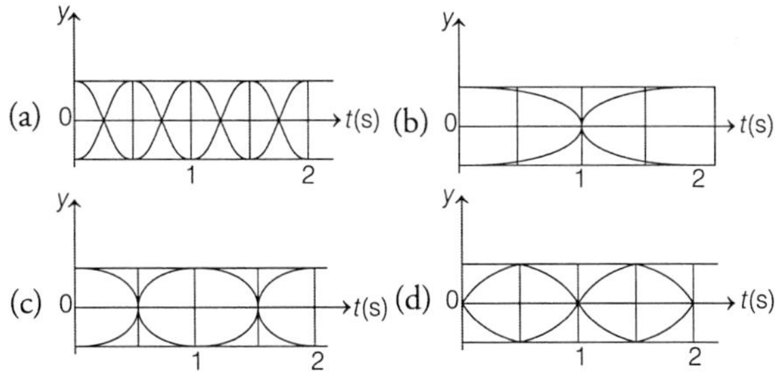

05 A pipe of length 85 cm is closed from one end. Find the number of possible natural oscillations of air column in the pipe whose frequencies lie below 1250 Hz. The velocity of sound in air is 340 m/s. *[JEE Main 2014]*

(a) 12 (b) 8 (c) 6 (d) 4

06 Three sound waves of equal amplitudes have frequencies $(\nu - 1)$, ν, $(\nu + 1)$. They superpose to give beat. The number of beats produced per second will be *[AIEEE 2009]*

(a) 4 (b) 3 (c) 2 (d) 1

Level 2

07 In an experiment with a closed organ pipe, it is filled with water by $\left(\frac{1}{5}\right)$th of its volume. The frequency of the fundamental note will change by *[2025, 4 April Shift-I]*

(a) – 20% (b) 20% (c) – 25% (d) 25%

08 A closed organ and an open organ tube are filled by two different gases having same bulk modulus but different densities ρ_1 and ρ_2, respectively. The frequency of 9th harmonic of closed tube is identical with 4th harmonic of open tube. If the length of the closed tube is 10 cm and the density ratio of the gases is $\rho_1 : \rho_2 = 1:16$, then the length of the open tube is *[2025, 22 Jan Shift-I]*

(a) $\frac{15}{7}$ cm (b) $\frac{20}{9}$ cm (c) $\frac{20}{7}$ cm (d) $\frac{15}{9}$ cm

09 The fundamental frequency of a closed organ pipe is equal to the first overtone frequency of an open organ pipe. If length of the open pipe is 60 cm, then the length of the closed pipe will be *[2024, 31 Jan Shift-I]*

(a) 60 cm (b) 30 cm (c) 45 cm (d) 15 cm

10 A tuning fork A of unknown frequency produces 5 beats/s with a fork of known frequency 340 Hz. When fork A is filled, the beat frequency decreases to 2 beats/s. What is the frequency of fork A? *[2021, 26 Feb Shift-II]*

(a) 342 Hz (b) 345 Hz (c) 335 Hz (d) 338 Hz

11 A student is performing the experiment of resonance column. The diameter of the column tube is 6 cm. The frequency of the tuning fork is 504 Hz. Speed of the sound at the given temperature is 336 m/s. The zero of the metre scale coincides with the top end of the resonance column tube. The reading of the water level in the column when the first resonance occurs is *[2021, 25 Feb, Shift-I]*

(a) 13 cm (b) 16.6 cm (c) 14.8 cm (d) 18.4 cm

12 A wire of length L and mass per unit length 6.0×10^{-3} kgm^{-1} is put under tension of 540 N. Two consecutive frequencies that it resonates at are 420 Hz and 490 Hz. Then, L in metres is *[2020, 9 Jan Shift-II]*

(a) 8.1 m (b) 2.1 m (c) 5.1 m (d) 1.1 m

Imp. Resonance Tube

13 A tuning fork of frequency 480 Hz is used in an experiment for measuring speed of sound (v) in air by resonance tube method. Resonance is observed to occur at two successive lengths of the air column $l_1 = 30$ cm and $l_2 = 70$ cm. Then, v is equal to *[2019, 12 April Shift-II]*

(a) 332 ms^{-1} (b) 384 ms^{-1}
(c) 338 ms^{-1} (d) 379 ms^{-1}

14 A string 2.0 m long and fixed at its ends is driven by a 240 Hz vibrator. The string vibrates in its third harmonic mode. The speed of the wave and its fundamental frequency is *[2019, 9 April Shift-II]*

(a) 180 m/s, 80 Hz (b) 320 m/s, 80 Hz
(c) 320 m/s, 120 Hz (d) 180 m/s, 120 Hz

15 A string is clamped at both the ends and it is vibrating in its 4th harmonic. The equation of the stationary wave is $Y = 0.3\sin(0.157x)\cos(200\pi t)$. The length of the string is (All quantities are in SI units) *[2019, 9 April Shift-I]*

(a) 60 m (b) 40 m (c) 80 m (d) 20 m

16 A pipe open at both ends has a fundamental frequency f in air. The pipe is dipped vertically in water, so that half of it is in water. The fundamental frequency of the air column is now *[JEE Main 2016, AIEEE 2012]*

(a) $\frac{f}{2}$ (b) $\frac{3f}{4}$ (c) $2f$ (d) f

17 A travelling wave represented by $y = A\sin(\omega t - kx)$ is superimposed on another wave represented by $y = A\sin(\omega t + kx)$. The resultant is *[AIEEE 2011]*

(a) a standing wave having nodes at

$$x = \left(n + \frac{1}{2}\right)\frac{\lambda}{2}, n = 0, 1, 2$$

(b) a wave travelling along $+x$ direction
(c) a wave travelling along $-x$ direction
(d) a standing wave having nodes at $x = \frac{n\lambda}{2}; n = 0, 1, 2$

18 While measuring the speed of sound by performing a resonance column experiment, a student gets the first resonance condition at a column length of 18 cm during winter. Repeating the same experiment during summer, she measures the column length to be x cm for the second resonance. Then, *[AIEEE 2008]*

(a) $18 > x$ (b) $x > 54$
(c) $54 > x > 36$ (d) $36 > x > 18$

19 A string is stretched between fixed points separated by 75.0 cm. It is observed to have resonant frequencies of 420 Hz and 315 Hz. There are no other resonant frequencies between these two. Then, the lowest resonant frequency for this string is *[AIEEE 2006]*

(a) 105 Hz (b) 1.05 Hz (c) 1050 Hz (d) 10.5 Hz

20 When two tuning forks (fork 1 and fork 2) are sounded simultaneously, 4 beats per second are heard. Now, some tape is attached on the prong of the fork 2. When the tuning forks are sounded again, 6 beats per second are heard. If the frequency of fork 1 is 200 Hz, then what will be the original frequency of fork 2? *[AIEEE 2005]*

(a) 200 Hz (b) 202 Hz
(c) 196 Hz (d) 204 Hz

21 A tuning fork of known frequency 256 Hz makes 5 beats per second with the vibrating string of a piano. The beat frequency decreases to 2 beats per second when the tension in the piano string is slightly increased. The frequency of the piano string before increasing the tension was *[AIEEE 2003]*

(a) $(256 + 2)$ Hz (b) $(256 - 2)$ Hz
(c) $(256 - 5)$ Hz (d) $(256 + 5)$ Hz

Level 3

22 The amplitude and phase of a wave that is formed by the superposition of two harmonic travelling waves,
$y_1(x, t) = 4\sin(kx - \omega t)$ and
$y_2(x, t) = 2\sin\left(kx - \omega t + \frac{2\pi}{3}\right)$, are :

(Take the angular frequency of initial waves same as ω) *[2025, 8 April Shift-II]*

(a) $\left[2\sqrt{3}, \frac{\pi}{6}\right]$ (b) $\left[6, \frac{\pi}{3}\right]$
(c) $\left[6, \frac{2\pi}{3}\right]$ (d) $\left[\sqrt{3}, \frac{\pi}{6}\right]$

23 Two harmonic waves moving in the same direction superimpose to form a wave $x = a\cos(1.5t)\cos(50.5t)$, where t is in seconds. Find the period with which they beat. (close to nearest integer) *[2025, 7 April Shift-I]*

(a) 6 s (b) 4 s (c) 2 s (d) 1 s

24 In a resonance tube experiment, when the tube is filled with water up to a height of 17.0 cm from bottom, it resonates with a given tuning fork. When the water level is raised, the next resonance with the same tuning fork occurs at a height of 24.5 cm. If the velocity of sound in air is 330 m/s, the tuning fork frequency is *[2020, 5 Sep Shift-I]*

(a) 1100 Hz (b) 3300 Hz
(c) 2200 Hz (d) 550 Hz

25 Three harmonic waves having equal frequency ν and same intensity I_0, have phase angles 0, $\pi/4$ and $-\frac{\pi}{4}$, respectively. When they are superimposed, the intensity of the resultant wave is closed to *[2020, 9 Jan Shift-I]*

(a) $0.2I_0$ (b) I_0
(c) $3I_0$ (d) $5.8I_0$

26 A wire of length $2L$, is made by joining two wires A and B of same length but different radii r and $2r$ and made of the same material. It is vibrating at a frequency such that the joint of the two wires forms a node. If the number of antinodes in wire A is p and that in B is q, then the ratio $p : q$ is *[2019, 8 April Shift-I]*

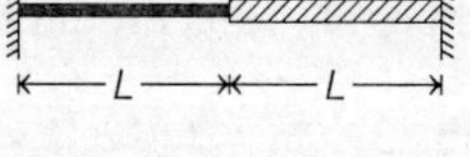

(a) 3 : 5 (b) 4 : 9 (c) 1 : 2 (d) 1 : 4

Imp. End Correction

27 A resonance tube is old and has jagged end. It is still used in the laboratory to determine velocity of sound in air. A tuning fork of frequency 512 Hz produces first resonance when the tube is filled with water to a mark 11 cm below a reference mark near the open end of the tube. The experiment is repeated with another fork of frequency 256 Hz which produces first resonance when water reaches a mark 27 cm below the reference mark. The velocity of sound in air, obtained in the experiment is close to *[2019, 12 Jan Shift-II]*

(a) 328 ms^{-1} (b) 341 ms^{-1}
(c) 322 ms^{-1} (d) 335 ms^{-1}

Numerical Value Type Questions

28 A closed and an open organ pipe have same lengths. If the ratio of frequencies of their seventh overtones is $\left(\frac{a-1}{a}\right)$, then the value of a is .... *[2024, 8 Jan Shift-I]*

29 Two open organ pipes of lengths 60 cm and 90 cm resonate at 6th and 5th harmonics respectively. The difference of frequencies for the given modes is .... Hz. (Velocity of sound in air $= 333$ m/s) *[2024, 06 April Shift-II]*

30 A sonometer wire of resonating length 90 cm has a fundamental frequency of 400 Hz when kept under some tension. The resonating-length of the wire with fundamental frequency of 600 Hz under same tension .... cm. *[2024, 5 April Shift-II]*

31 A tuning fork resonates with a sonometer wire of length 1 m stretched with a tension of 6 N. When the tension in the wire is changed to 54 N, the same tuning fork produces 12 beats per second with it. The frequency of the tuning fork is .... Hz. *[2024, 1 Feb Shift-I]*

32 In a closed organ pipe, the frequency of fundamental note is 30 Hz. A certain amount of water is now poured in the organ pipe, so that the fundamental frequency is increased to 110 Hz. If the organ pipe has a cross-sectional area of 2 cm^2, the amount of water poured to the organ tube is .... g. (Take speed of sound in air is 330 m/s) *[2024, 30 Jan Shift-I]*

33 A closed organ pipe 150 cm long gives 7 beats per second with an open organ pipe of length 350 cm, both vibrating in fundamental mode. The velocity of sound is .... m/s.

34 A guitar string of length 90 cm vibrates with a fundamental frequency of 120 Hz. The length of the string producing a fundamental frequency of 180 Hz will be .... cm. *[2023, 8 April, Shift-II]*

35 An organ pipe 40 cm long is open at both ends. The speed of sound in air is 360 ms^{-1}. The frequency of the second harmonic is .... Hz. *[2023, 8 April, Shift-I]*

36 The displacement equations of two interfering waves are given by $y_1 = 10 \sin\left(\omega t + \frac{\pi}{3}\right)$ cm,

$y_2 = 5[\sin \omega t + \sqrt{3} \cos \omega t]$ cm, respectively. The amplitude of the resultant wave is ...... cm. *[2023, 31 Jan, Shift-II]*

37 A wire of length of 30 cm, stretched between rigid supports, has it's nth and $(n+1)$th harmonics at 400 Hz and 450 Hz, respectively. If tension in the string is 2700 N, its linear mass density is .... kg/m. *[2022, 27 July Shift-II]*

38 A tuning fork of frequency 340 Hz resonates in the fundamental mode with an air column of length 125 cm in a cylindrical tube closed at one end. When water is slowly poured in it, the minimum height of water required for observing resonance once again is ... cm. (Velocity of sound in air is 340 ms^{-1}) *[2022, 28 June, Shift-II]*

39 A set of 20 tuning forks is arranged in a series of increasing frequencies. If each fork gives 4 beats with respect to the preceding fork and the frequency of the last fork is twice the frequency of the first, then the frequency of last fork is ...... Hz. *[2022, 26 June Shift-II]*

40 Two travelling waves of equal amplitudes and equal frequencies move in opposite directions along a string. They interfere to produce a stationary wave whose equation is given by

$$y = \left(10 \cos \pi x \sin \frac{2\pi t}{T}\right) \text{ cm.}$$

The amplitude of the particle at $x = 4/3$ cm will be ...... cm. *[2022, 24 June Shift-II]*

41 A wire having a linear mass density 9.0×10^{-4} kg/m is stretched between two rigid supports with a tension of 900 N. The wire resonates at a frequency of 500 Hz. The next higher frequency at which the same wire resonates is 550 Hz. The length of the wire is .... m. *[2021, 31 Aug, Shift-I]*

42 A tunning fork is vibrating at 250 Hz. The length of the shortest closed organ pipe that will resonate with the tuning fork will be .... cm. (Take, speed of sound in air as 340 ms^{-1}) *[2021, 27 Aug Shift-II]*

43 Two waves are simultaneously passing through a string and their equations are $y_1 = A_1 \sin k(x - vt), y_2 = A_2 \sin k(x - vt + x_0)$. Given amplitudes $A_1 = 12$ mm and $A_2 = 5$ mm, $x_0 = 3.5$ cm and wave number $k = 6.28 \text{ cm}^{-1}$. The amplitude of resulting wave will be ...... mm. *[2021, 26 Aug Shift-II]*

44 Two travelling waves produce a standing wave represented by equation,
$y = 1.0\ \text{mm}\cos(1.57\ \text{cm}^{-1})x\sin(78.5\ \text{s}^{-1})t$.

The node closest to the origin in the region $x > 0$ will be at $x =$ .... cm. *[2021, 26 Aug Shift-I]*

45 A closed organ pipe of length L and an open organ pipe contain gases of densities ρ_1 and ρ_2, respectively. The compressibility of gases are equal in both the pipes. Both the pipes are vibrating in their first overtone with same frequency. The length of the open pipe is $\frac{x}{3}L\sqrt{\frac{\rho_1}{\rho_2}}$ where x is ....

(Round off to the nearest integer) *[2021, 16 March Shift-II]*

46 A one metre long (both ends open) organ pipe is kept in a gas that has double the density of air at STP. Assuming the speed of sound in air at STP is 300 m/s, the frequency difference between the fundamental and second harmonic of this pipe is ...... Hz. *[2020, 8 Jan Shift-I]*

• Key Index

Topic 1

1 (a)	2 (b)	3 (c)	4 (a)	5 (d)	6 (a)	7 (a)	8 (d)	9 (a)	10 (b)
11 (c)	12 (a)	13 (d)	14 (c)	15 (a)	16 (b)	17 (d)	18 (b)	19 (b)	20 (b)
21 (b)	22 (d)	23 (a)	24 (c)	25 (a)	26 (b)	27 (c)	28 (d)	29 (a)	30 (b)
31 (1152)	32 (20)	33 (1215)	34 (35)						

Topic 2

1 (c)	2 (d)	3 (a)	4 (a)	5 (c)	6 (c)	7 (d)	8 (b)	9 (d)	10 (c)
11 (c)	12 (b)	13 (b)	14 (b)	15 (c)	16 (d)	17 (a)	18 (b)	19 (a)	20 (c)
21 (c)	22 (a)	23 (c)	24 (c)	25 (d)	26 (c)	27 (a)	28 (16)	29 (740)	30 (60)
31 (6)	32 (400)	33 (294)	34 (60)	35 (900	36 (20)	37 (3)	38 (50)	39 (152)	40 (5)
41 (10)	42 (34)	43 (7)	44 (1)	45 (4)	46 (106)				

Detailed **Explanations**

Topic 1

01. *(a)* $y = \sin(20\pi x + 10\pi t)$

$$k = 20\pi$$
$$\frac{2\pi}{\lambda} = 20\pi$$
$$\Rightarrow \quad \lambda = \frac{1}{10} = 0.1 \text{ m}$$

The minimum distance between two points having the same oscillating speed,

$$x = \frac{\lambda}{2} = \frac{0.1}{2} = 0.05 \text{ m} = 5 \text{ cm}$$

02. *(b)* Given, displacement,

$$y = \sin \omega t + \cos \omega t$$

i.e. $y = \sin \omega t + \sin\left(\omega t + \frac{\pi}{2}\right)$

Hence, phase difference, $\Delta\phi = \frac{\pi}{2}$

amplitude,

$$A = \sqrt{1^2 + 1^2 + 2 \times 1 \times 1 \times \cos\left(\frac{\pi}{2}\right)} = \sqrt{2}$$

03. *(c)* Given that, mass per unit length,

$$\mu = 7.0 \times 10^{-3} \text{ kg m}^{-1}$$

Tension in string, $T = 70$ N

Speed of transverse wave,

$$v = \sqrt{\frac{T}{\mu}} = \sqrt{\frac{70}{7.0 \times 10^{-3}}}$$
$$v = 100 \text{ m/s}$$

04. *(a)* Given, wave equation is ,

$$y = 0.05 \sin(8x - 4t),$$

Comparing it with general equation

$$\Rightarrow \quad y = A\sin(kx - \omega t)$$

We have $\quad k = 8$ and $\omega = 4$

So, wave speed is given by

$$v = \frac{\omega}{k} = \frac{4}{8} = 0.5 \text{ m/s}$$

05. *(d)* We know that,

$$k = \frac{2\pi}{\lambda} = \frac{2\pi}{4} = 0.5\,\pi \text{ cm}^{-1}$$
$$\omega = 2\pi\nu = 2\pi \times 100 = 200\,\pi\text{s}^{-1}$$

and equation of wave,

$$y = A\sin(kx - \omega t)$$

put the values of k and ω, then

$$y = A\sin \pi[(0.5)\text{ cm}^{-1}\, x - 200\text{ s}^{-1} t]$$

Hence, option (d) is correct answer.

06. *(a)* The equation of a travelling wave in standard form,

$$y = A\sin(\omega t \pm kx)$$

Only option (a), i.e.

$y = A\sin(15x - 2t)$ satisfies this equation.

07. *(a)* Wave equation is given as,

$$y = 10^{-3} \sin(50t + 2x)$$

Speed of wave is obtained by differentiating phase of wave.

Now, phase of wave from given equation is $\phi = 50t + 2x = \text{constant}$

Differentiating 'ϕ' w.r.t. t, we get

$$\frac{d}{dt}(50t + 2x) = \frac{d}{dt} \quad \text{(constant)}$$
$$\Rightarrow \quad 50 + 2\left(\frac{dx}{dt}\right) = 0$$
$$\Rightarrow \quad \frac{dx}{dt} = \frac{-50}{2} = -25 \text{ ms}^{-1}$$

So, wave is propagating in negative x–direction with a speed of 25 ms^{-1}.

> ***Alternate Solution***
>
> The general equation of a wave travelling in negative x-direction is given as
>
> $$y = a\sin(\omega t + kx) \quad \ldots\text{(i)}$$
>
> Given equation of wave is
>
> $$y = 10^{-3}\sin(50 + 2x) \quad \ldots\text{(ii)}$$
>
> Comparing Eqs. (i) and (ii), we get
>
> $$\omega = 50 \text{ and } k = 2$$
>
> Velocity of the wave, $v = \frac{\omega}{k} = \frac{50}{2} = 25$ m/s

08. *(d)* As Intensity,

$$I = \frac{1}{2}\rho\omega^2 A^2 v$$

Here, ρ = density of medium,
A = amplitude,
ω = angular frequency
and $\quad v$ = velocity of wave.

$\therefore$ Intensity depends on the amplitude, angular frequency as well as velocity of wave.

$$y_1 = 2a\sin(\omega t - kx)$$
$$I_1 = \frac{1}{2}\rho\omega^2(2a)^2\,\frac{\omega}{k}$$
$$= 2\frac{\rho\omega^3 a^2}{k}$$
$$y_2 = a\sin(2\omega t - 2kx)$$
$$I_2 = \frac{1}{2}\rho(2\omega)^2\,\frac{a^2 2\omega}{2k}$$
$$= \frac{2\rho\omega^3 a^2}{k}$$
$$I_1 = I_2$$

09. *(a)* $y(x, t) = e^{-(ax^2 + bt^2 + 2\sqrt{ab}\,xt)}$

$$= e^{-(\sqrt{a}\,x + \sqrt{b}\,t)^2}$$

It is a function of type $y = f(\omega t + kx)$

So, $y(x, t)$ represents wave travelling along $-x$-direction.

$\therefore$ Speed of wave $= \frac{\omega}{k} = \frac{\sqrt{b}}{\sqrt{a}} = \sqrt{\frac{b}{a}}$

10. *(b)* As given

$$y = 10^{-6}\sin\left(100t + 20x + \frac{\pi}{4}\right) \quad \ldots\text{(i)}$$

Comparing it with

$$y = a\sin(\omega t + kx + \phi), \quad \ldots\text{(ii)}$$

We obtain, $\omega = 100$ rad s^{-1}, $k = 20$ m^{-1}

$$\therefore \quad v = \frac{\omega}{k} = \frac{100}{20} = 5 \text{ ms}^{-1}$$

11. *(c)* Given displacement equation

$$x(t) = 5\cos\left(628t + \frac{\pi}{2}\right)\text{m}$$

and velocity $v = 300$ m/s

General equation of displacement,

$$x(t) = A\cos(\omega t + \phi)$$

On compairing,

$$\omega = 2\pi f = 628$$
$$\Rightarrow \quad f = \frac{628}{2\pi} \quad \ldots\text{(i)}$$

Also, $v = f\lambda$

$$\Rightarrow \quad \lambda = \frac{v}{f} \quad \ldots\text{(ii)}$$

Now, using Eqs. (i) and (ii)

$$\lambda = \frac{300 \times 2 \times 3.14}{628}$$
$$\lambda = 3 \text{ m}$$

12. *(a)* $\lambda = 7.5$ cm, difference, $x = 1.2$ cm

$$A = 2\text{ cm}$$

$\therefore$Velocity, $v = \frac{x}{t} = \frac{1.2}{0.3} = 4$ cm/s

Propagation constant,

$$k = \frac{2\pi}{\lambda} = \frac{2\pi}{7.5} = 0.83 \text{ cm}^{-1}$$

Also, $v = \frac{w}{k}$

$\Rightarrow w = vk = 4 \times \frac{2\pi}{7.5} = 3.35$ rad/s

$\therefore\ y = A\cos(kx - wt)$
$= 2\cos(0.83x - 3.35t)$ cm

13. *(d)* $y(x,t) = 4.0\sin(20\times10^{-3}x + 600t)$ mm

Compare $y(x,t) = A\sin(kx + \omega t)$
$A = 4.0,\ k = 20\times10^{-3},\ \omega = 600$

Velocity of wave $= \frac{\text{Coefficient of } t}{\text{Coefficient of } x}$
$= \frac{600}{20\times10^{-3}}$
$= 3\times10^4$ mm/s
$v = 30$ m/s

Also wave is travelling in negative x-direction.

14. *(c)* Wave equation is given as,

$$y = 0.5\sin\frac{2\pi}{\lambda}(400t - x)\text{ m}$$

$$\Rightarrow\quad y = 0.5\sin\left(\frac{800}{\lambda}\pi t - \frac{2\pi x}{\lambda}\right)\text{m}$$

Comparing above equation with wave equation,

$$y = A\sin(\omega t - kx)$$

We get, $\omega = \frac{800\pi}{\lambda}$ rad/s and $k = \frac{2\pi}{\lambda}$

$\therefore$ Wave velocity,

$$v = \frac{\omega}{k} = \frac{800\pi/\lambda}{2\pi/\lambda} = 400\text{ m/s}$$

15. *(a)* Here, $y = 2\sin(\omega t - kx)$

Maximum particle velocity, $v_m = A\omega$ and wave velocity,

$$v_p = \frac{\omega}{k}$$

$\because\quad v_p = v_m$

$$\frac{\omega}{k} = A\omega \Rightarrow k = \frac{1}{A} = \frac{2\pi}{\lambda}$$

$$\lambda = 2\pi A = 2\pi\times2 = 4\pi$$

16. *(b)* Wave equation is given as

$$y = 10\sin\left(2\pi nt - \frac{2\pi}{\lambda}\cdot x\right)$$

Wave speed $= \frac{\omega}{k} = \frac{2\pi n}{(2\pi/\lambda)} = n\lambda$

Maximum speed of oscillating particle
$= A\omega = 10(2\pi n)$

Given, maximum particle speed
$= 4\times$ wave speed

$\Rightarrow 10\times2\pi n = 4\times n\lambda$

$\Rightarrow \frac{10\times2\pi}{4} = \lambda$ or $\lambda = 5\pi$ units

17. *(d)* Given, the frequency of the sound wave, $\quad f = 245$ Hz

The speed of the travelling wave,
$v = 300$ m/s

As, total distance of to and fro motion is 6 cm.

Hence, the amplitude of the wave,
$A = 6/2 = 3$ cm $= 0.03$ m

As we know,
Angular frequency is given as
$\omega = 2\pi f$

Substituting the value of f, we get
$\omega = 2\pi(245)$
$\omega = 1.54\times10^3$ rad/s

Propagation constant is given as
$k = \frac{\omega}{v}$

Substituting the value of v and ω, we get

$$k = \frac{1.54\times10^3}{300}$$

$\Rightarrow\quad k = 5.1\text{ m}^{-1}$

General mathematical expression for a travelling wave along positive X-direction is given as

$$y = A\sin(kx - \omega t)$$

Substituting the values in the above equation, we get

$$y = 0.03\sin(5.1x - 1.5\times10^3 t)$$

18. *(b)* Distance between any two crests
$n\lambda = 5$ (n is an integer) ...(i)

Distance between any crest and any trough $(2m+1)\frac{\lambda}{2} = 1.5$
(m is an integer)

$\Rightarrow\quad (2m+1)\lambda = 3$ ...(ii)

Dividing Eq. (i) by Eq. (ii), we get

$\Rightarrow\quad \frac{n\lambda}{(2m+1)\lambda} = \frac{5}{3}$

$3n = 5(2m+1)$
$3n = 10m + 5$ ...(iii)

(I) $m = 1$ and $n = 5$
Now, on putting these values in Eqs. (i) and (ii), we get $\lambda = 1$

(II) $m = 4$ and $n = 15$
Now, on putting these values in Eqs. (i) and (ii), we get

$$\lambda = \frac{1}{3}$$

(III) $m = 7$ and $n = 25$
Now, on putting these values in Eqs. (i) and (ii), we get $\lambda = \frac{1}{5}$ and so on.

So, possible values of $\lambda = \frac{1}{1}, \frac{1}{3}, \frac{1}{5}, \ldots..$

Hence, correct option is (b).

19. *(b)* Fundamental frequency of oscillation of a taut string is

$$f = \frac{1}{2l}\sqrt{\frac{T}{\mu}}$$

Given, X and Z are identical strings.
$\therefore$ l and μ are same for both X and Z.

$$\Rightarrow\quad f \propto \sqrt{T} \Rightarrow \frac{f_X}{f_Z} = \sqrt{\frac{T_X}{T_Z}}$$

$$\Rightarrow\quad \frac{T_X}{T_Z} = \frac{f_X^2}{f_Z^2} = \left(\frac{450}{300}\right)^2 = \frac{9}{4} = 2.25$$

20. *(b)* Transverse wave speed over a string is given by

$$v = \sqrt{\frac{T}{\mu}} \quad \text{...(i)}$$

Here, when velocity is v, then tension, $T_1 = 2.06\times10^4$ N

Let when velocity is $v/2$, then tension is T, hence from Eq (i), we get

$$\frac{v}{\frac{v}{2}} = \sqrt{\frac{T_1}{T}} \Rightarrow T = \frac{T_1}{4} = \frac{2.06\times10^4}{4}$$

or $\quad T = 5.15\times10^3$ N

21. *(b)* From the given snapshot at $t = 0$,
$y = 0$ at $x = 0$
and $y = -$ve when x increases from zero.

Standard expression of any progressive wave is given by

$$y = A\sin(kx - \omega t + \phi)$$

Here, ϕ is the phase difference, we need to get at $t = 0$, $y = A\sin(kx + \phi)$

Clearly $\phi = \pi$, so that
$y = A\sin(kx + \pi)$
$y = -A\sin(kx)$

$\Rightarrow\quad y = 0$ at $x = 0$
and $\quad y = -$ve at $x > 0$

Which satisfies the given snapshot.

22. *(d)* Velocity v of the wave on the string $= \sqrt{\frac{T}{\mu}}$

Substituting the given values, we get

$$v = \sqrt{\frac{8}{5}} \times 1000 = 40 \text{ ms}^{-1}$$

Wavelength of the wave on the string,

$$\lambda = \frac{v}{f}$$

where, f = frequency of wave.

$$\Rightarrow \quad \lambda = \frac{40}{100} \text{ m} = 40 \text{ cm}$$

∴ Separation between two successive nodes is, $d = \frac{\lambda}{2} = \frac{40}{2} = 20.0$ cm

23. *(a)*

> **key idea** To find out the fundamental frequency of the longitudinal vibrations, you can use following formula,
>
> $$f = \frac{1}{2L}\sqrt{\frac{Y}{\rho}}$$

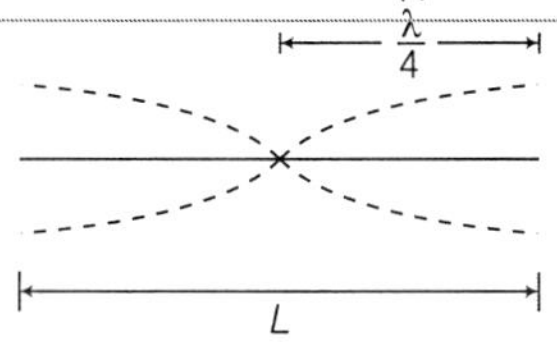

From vibration mode,

$$\frac{\lambda}{2} = L \Rightarrow \lambda = 2L$$

∴ Wave speed, $v = \sqrt{\frac{Y}{\rho}}$

So, frequency,

$$f = \frac{v}{\lambda} \Rightarrow f = \frac{1}{2L}\sqrt{\frac{Y}{\rho}}$$

$$= \frac{1}{2 \times 60 \times 10^{-2}}\sqrt{\frac{9.27 \times 10^{10}}{2.7 \times 10^{3}}}$$

$$\approx 5000 \text{ Hz}$$

$$f = 5 \text{ kHz}$$

24. *(c)* Tension T at a distance x from bottom is,

$$T = \frac{mgx}{l}$$

So, velocity at point P

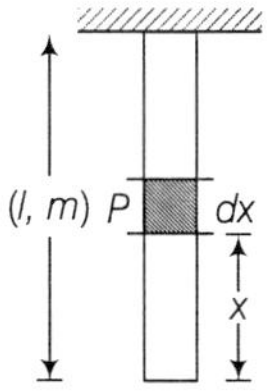

$$v = \sqrt{\frac{\frac{mgx}{l}}{m/l}}$$

i.e. $\quad v = \sqrt{gx}$

$$\frac{dx}{dt} = \sqrt{gx}$$

$$\Rightarrow \int_0^{20} \frac{dx}{\sqrt{x}} = \int_0^t \sqrt{g}\, dt$$

$$[2\sqrt{x}]_0^{20} = \sqrt{10}\, t \Rightarrow 2\sqrt{20} = \sqrt{10}\, t$$

$$t = 2\sqrt{2} \text{ s}$$

25. *(a)* Given, amplitude,

$$A = \frac{10}{\pi} \text{ cm}$$

Since, wave velocity = 2 × maximum particle velocity

i.e, $\quad \frac{\omega\lambda}{2\pi} = 2A\omega$

or, $\quad \lambda = 4\pi A$

$$= 4 \times \pi \times \frac{10}{\pi} = 40 \text{ cm}$$

26. *(b)* We have, $v = \frac{8}{0.4} = 20$ cm/s $\left[v = \frac{\lambda}{d}\right]$

and $\quad \lambda = 4 \times 2 = 8$ cm

∵ $\quad \omega = vk = 20 \times \frac{2\pi}{8} = 5\pi$ rad/s

So, $(v_{max})_{particle} = \omega A = 5\pi \times \frac{1}{2}$ $\quad [A = 0.5 \text{ cm}]$

$$= \frac{5\pi}{2} \text{ cm/s}$$

27. *(c)* The equation of wave at any time is obtained by putting $X = x - vt$

$Y(x, 0)$ at $t = 0$

$$= \frac{1}{1 + X^2} = \frac{1}{1 + (x - vt)^2} \quad \text{...(i)}$$

We know, at $t = 2$ s,

$$y = \frac{1}{1 + (x - 1)^2} \quad \text{...(ii)}$$

On comparing Eqs. (i) and Eq. (ii), we get

$$vt = 1 \Rightarrow v = 1/t$$

As $\quad t = 2$ sec

∴ $\quad v = \frac{1}{2} = 0.5$ m/s

28. *(d)* As $v = \sqrt{\frac{T}{\mu}}$

So, $T = \mu v^2 = \mu \frac{\omega^2}{k^2} = 0.04 \frac{(2\pi/0.04)^2}{(2\pi/0.50)^2}$

$$= 6.25 \text{ N}$$

29. *(a)* As general equation is

$$y(x, t) = a \cos(kx - \omega t)$$

where, $k = \frac{2\pi}{\lambda}$ and $\omega = \frac{2\pi}{T}$

∴ $\quad y(x, t) = 0.005 \cos(\alpha x - \beta t)$

$$\Rightarrow \quad \frac{2\pi}{\lambda} = \alpha \quad \text{and} \quad \frac{2\pi}{T} = \beta$$

So, $\quad \alpha = \frac{2\pi}{0.08} = 25\pi \text{ m}^{-1}$

and $\quad \beta = \frac{2\pi}{2} = \pi \text{ s}^{-1}$

30. *(b)* When the car is at rest, the situation can be shown in the figure below.

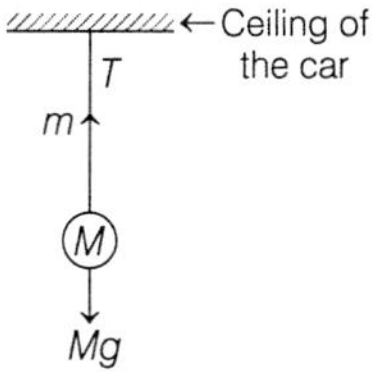

Since, $m << M$, then we can neglect the mass of the string. So, the initial velocity of the wave in the string can be given as,

$$v = \sqrt{\frac{T}{\mu}} = \sqrt{\frac{Mg}{\frac{M}{l}}} = 60 \text{ m/s} \quad \text{...(i)}$$

Now, when the car has acceleration a, then the situation can be shown in the figure given below.

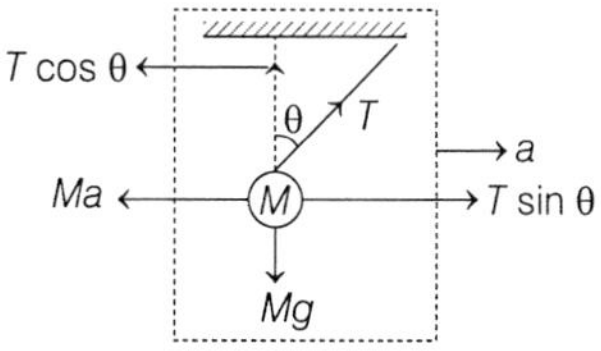

Resolving the components of 'T' along X-and Y-axis, we get

$$T \cos\theta = Mg \quad \text{...(ii)}$$

$$T \sin\theta = Ma \quad \text{...(iii)}$$

Squaring both sides of Eqs. (ii) and (iii) and adding them, we get

$$T^2(\sin^2\theta + \cos^2\theta) = M^2(g^2 + a^2)$$

$$\Rightarrow \quad T = M(g^2 + a^2)^{1/2} \quad \text{...(iv)}$$

Now, the velocity of the wave in the string would be given as

$$v' = \sqrt{\frac{T}{\mu}} = \sqrt{\frac{M(g^2 + a^2)^{\frac{1}{2}}}{\frac{M}{l}}}$$

$$= 60.5 \text{ m/s} \quad \text{...(v)}$$

Dividing Eq. (i) and Eq. (v), we get

$$\frac{v}{v'} = \frac{\sqrt{\frac{Mg}{\frac{M}{l}}}}{\sqrt{\frac{M(g^2 + a^2)^{\frac{1}{2}}}{\frac{M}{l}}}}$$

$$= \sqrt{\frac{g}{(g^2+a^2)^{\frac{1}{2}}}}$$

$$= \frac{60}{60.5} \text{ or } \frac{60.5}{60} = \sqrt{\frac{(g^2+a^2)^{\frac{1}{2}}}{g}}$$

Squaring both the sides, we get

$$\frac{(60.5)^2}{(60)^2} = \frac{(g^2+a^2)^{\frac{1}{2}}}{g}$$

or $$\left(1+\frac{0.5}{60}\right)^2 = \sqrt{1+\left(\frac{a}{g}\right)^2}$$

Using binomial expansion,

$$(1+x)^n = 1 + \frac{nx}{1!} + \ldots$$

On both sides, we get

$$1 + 2\times\frac{0.5}{60} = 1 + \frac{1}{2}\cdot\frac{a^2}{g^2}$$

$$\Rightarrow \quad 1+\frac{1}{60} = 1+\frac{1}{2}\cdot\frac{a^2}{g^2}$$

or $$\frac{a}{g} = \frac{1}{\sqrt{30}}$$

or $$a = \frac{g}{\sqrt{30}} = \frac{g}{5.4} \simeq \frac{g}{5}$$

> ***Alternate Solution***
>
> Here, $60 = \sqrt{\frac{mg}{\mu}}$ and $60.5 = \sqrt{\frac{m(g^2+a^2)^{1/2}}{\mu}}$
>
> so, $\frac{60.5}{60} = \left(\frac{g^2+a^2}{g^2}\right)^{1/4}$; $\left(1+\frac{0.5}{60}\right)^4 = \frac{g^2+a^2}{g^2}$
>
> $1+\frac{4\times0.5}{60} = \frac{g^2+a^2}{g^2} \Rightarrow g^2 + \frac{2g^2}{60} = g^2+a^2$
>
> $a = g\sqrt{\frac{2}{60}} = \frac{g}{\sqrt{30}} \simeq \frac{g}{5}$

31. *(1152)* Given,

$$Y = 10^{-2}\sin 2\pi(160t - 0.5x + \pi/4)$$

Speed of wave $= \frac{\omega}{k} = \frac{160}{0.5} = 320$ m/s

$$= 320 \times \frac{18}{5} = 1152 \text{ km/h}$$

32. *(20)* Transverse harmonic wave is,

$$y(x,t) = 5\sin(6t + 0.003x)$$

So here, $\omega = 6$ rad/s

and $k = 0.003\ \text{cm}^{-1}$

So, wave speed is given by

$$v = \frac{\omega}{k} = \frac{6}{0.003} \text{ cm/s}$$
$$= 2000 \text{ cm/s}$$
$$= 20 \text{ m/s}$$

33. *(1215)* Given, mass per unit length,

$$\mu = 0.135 \text{ g/cm}$$
$$= 1.35\times10^{-2} \text{ kg/m}$$

Transverse wave equation,

$$y = -0.21\sin(x + 30t)$$

From given equation,

$$\omega = 30 \text{ rad/s}, k = 1$$

Speed of wave,

$$v = \frac{\omega}{k} = \frac{30}{1} = 30 \text{ ms}^{-1}$$

Also, $v = \sqrt{\frac{T}{\mu}}$

$$\Rightarrow \quad T = v^2\mu$$
$$T = (30)^2 \times 1.35\times10^{-2}$$
$$= 900\times0.0135 = 12.15 \text{ N}$$
$$= 1215\times10^{-2} \text{ N}$$

Hence, $x = 1215$

34. *(35)* Density of wire,

$$d = 9\times10^{-3} \text{ kg/cm}^3$$
$$= 9\times10^3 \text{ kg/m}^3$$

Strain in the wire, $\varepsilon = 4.9\times10^{-4}$

Young's modulus of wire is

$$Y = 9\times10^{10} \frac{\text{N}}{\text{m}^2}$$

Lowest frequency of vibration in wire will be

$$f = \frac{1}{2L}\sqrt{\frac{T}{(M/L)}} \quad \ldots\text{(i)}$$

Now,

$$\frac{T}{M/L} = \frac{T}{V\rho/L} = \frac{T}{\left(\frac{LA\cdot\rho}{L}\right)} = \frac{T}{A\rho}$$

But $\frac{T}{A}$ = stress $= Y\times$ strain $= Y\times\varepsilon$

So, from Eq (i), frequency will be

$$f = \frac{1}{2L}\sqrt{\frac{Y\times\varepsilon}{\rho}}$$
$$= \frac{1}{2\times1}\times\sqrt{\frac{9\times10^{10}\times4.9\times10^{-4}}{9\times10^3}}$$
$$= 35 \text{ Hz}$$

Topic 2

01. *(c)* $v_{\text{sound}} = \sqrt{\frac{\gamma p}{\rho}}, \gamma = 1 + \frac{2}{f}$

$$\gamma_{He} = \frac{5}{3}, \gamma_{CH_4} = \gamma_{CO_2} \simeq 1.33 = \frac{4}{3}$$

So,

$$v_{He} : v_{CH_4} : v_{CO_2} :: \sqrt{\frac{5}{3}} : \sqrt{\frac{4}{3}} : \sqrt{\frac{4}{3}}$$

02. *(d)* As we know that, $f\times\lambda = v$

and beat frequency $= f_2 - f_1$

We have, $f_2 - f_1 = \frac{40}{12}$ [Given]

$$\Rightarrow \quad \frac{v}{4.08} - \frac{v}{4.16} = \frac{40}{12}$$

$$\Rightarrow \quad v = \frac{40\times4.08\times4.16}{12\times(4.16-4.08)}$$

$$\Rightarrow \quad v = 707.2 \text{ m/s}$$

03. *(a)* Equations of the two waves are given as

$$y_1 = 5\sin 2\pi(x - vt) \text{ cm}$$

$\therefore$ Amplitude, $A_1 = 5$ cm

and $y_2 = 3\sin 2\pi(x - vt + 1.5)$ cm

$\therefore$ Amplitude, $A_2 = 3$ cm

Phase difference between waves y_1 and y_2 is given as

$$\Delta\phi = 2\pi(1.5) = 3\pi$$

$\therefore$ Amplitude of resultant wave,

$$A = \sqrt{A_1^2 + A_2^2 + 2A_1A_2\cos(\Delta\phi)}$$
$$= \sqrt{A_1^2 + A_2^2 + 2A_1A_2(-1)}$$
$$[\because \cos 3\pi = \cos\pi = -1]$$
$$= \sqrt{(A_1 - A_2)^2}$$
$$= A_1 - A_2$$
$$= 5 - 3 = 2 \text{ cm}$$

04. *(a)* Beat frequency is given by,

$$f_{\text{beat}} = f_1 - f_2$$

Here, $f_1 = 11$ Hz and $f_2 = 9$ Hz

$\Rightarrow$ Beat frequency is,

$$f_{\text{beat}} = 11 - 9 = 2 \text{ Hz}$$

Hence, time period of beats or time interval between beats,

$$T = \frac{1}{f_{\text{beat}}}$$

$$\Rightarrow \quad T = \frac{1}{2} = 0.5 \text{ s}$$

So, resultant wave has a time period of 0.5 s which is correctly depicted in option (a) only.

05. *(c)* For closed organ pipe $= \frac{(2n+1)v}{4l}$

$(n = 0, 1, 2, \ldots)$

$$\frac{(2n+1)v}{4l} < 1250$$

$$(2n+1) < 1250 \times \frac{4 \times 0.85}{340}$$

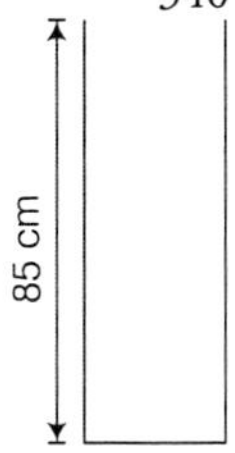

$$(2n+1) < 12.5$$

$\Rightarrow \quad 2n < 11.50$

$\Rightarrow \quad n < 5.25$

So, $\quad n = 0, 1, 2, 3, \ldots, 5$

So, we have 6 possibilities.

> **Alternate Solution**
>
> In closed organ pipe, fundamental node
>
> 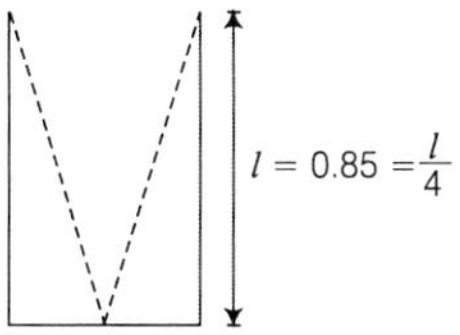
>
>
> *i.e.,* $\quad \lambda/4 = 0.85 \Rightarrow \lambda = 4 \times 0.85$
>
> As we know, $\nu = \frac{c}{\lambda} = \frac{340}{4 \times 0.85} = 100\text{ Hz}$
>
> $\therefore$ Possible frequencies $= 100$ Hz, 300 Hz, 500 Hz, 700 Hz, 900 Hz, 1100 Hz below 1250 Hz.

06. *(c)*

> **key idea** The number of beats produced per second can be find out using following formula
>
> f_b = difference between the frequency of interfering waves

Maximum number of beats

$= \nu + 1 - (\nu - 1) = 2$ beats/s

07. *(d)* $\lambda_1 = 4l$, $f = \frac{v}{\lambda_1}$

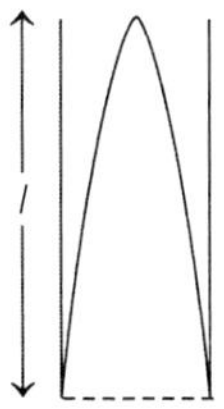

$$f_1 = \frac{v}{4l}$$

$$\lambda_2 = \frac{16l}{5}$$

after filled with water

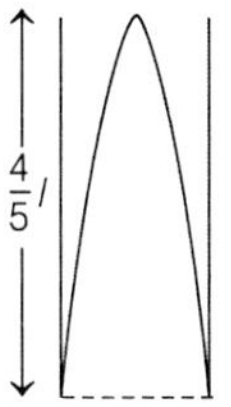

$$f_2 = \frac{v}{\lambda_2} = \frac{v}{\frac{16l}{5}} = \frac{5v}{16l}$$

$$\frac{\Delta f}{f} = \frac{f_2 - f_1}{f_1} = \frac{\frac{5v}{16l} - \frac{v}{4l}}{\frac{v}{4l}} = \frac{\frac{v}{16l}}{\frac{v}{4l}}$$

$$\frac{\Delta f}{f} = \frac{v}{16l} \times \frac{4l}{v} = \frac{1}{4}$$

$$\frac{\Delta f}{f} \times 100 = \frac{1}{4} \times 100 = 25\%$$

08. *(b)* Harmonic of closed pipe is given by $= \frac{nv_1}{4l_1}$

Frequency of 9th harmonic of closed pipe $= \frac{9v_1}{4l_1}$ $\quad \ldots$(i)

Frequency of 4th harmonic of open pipe $= \frac{2v_2}{l_2}$ $\quad \ldots$(ii)

As frequency of 9th and 4th harmonics are equal.

$$\frac{9v_1}{4l_1} = \frac{2v_2}{l_2}$$

$$\therefore \quad \frac{9}{4l_1}\sqrt{\frac{B}{\rho_1}} = \frac{2}{l_2}\sqrt{\frac{B}{\rho_2}} \quad \ldots\text{(iii)}$$

$$l_1 = 10 \text{ cm}, \rho_1 : \rho_2 = 1 : 16$$

Substituting all values in Eq.(iii)

$$l_2 = \frac{20}{9} \text{ cm}$$

09. *(d)* Fundamental frequency of closed organ pipe is,

$f_1 = \frac{v}{4L_1}$, v = Speed of sound

L_1 = Length of closed organ pipe

First overtone of open organ pipe has frequency,

$$f_2 = 2 \cdot \frac{v}{2L_2},$$

L_2 = Length of open organ pipe

Given, $f_1 = f_2$ and $L_2 = 60$ cm

$$\therefore \quad \frac{v}{4L_1} = 2 \cdot \frac{v}{2L_2}$$

$$\Rightarrow \quad L_1 = \frac{L_2}{4} = \frac{60}{4} = 15 \text{ cm}$$

10. *(c)* Given, $f = 340$ Hz

Let beat frequency of tuning fork A be f_A

and f_A' be the frequency of tuning fork A after filling.

In first case, beat frequency $= 5$

$\Rightarrow \quad f_A = 340 \pm 5$

$\Rightarrow \quad f_A = 345$ Hz

or $\quad$ 335 Hz

After filling tuning fork A,

$$f_A' = 340 \pm 2$$

$\Rightarrow \quad f_A' = 342$ or 338 Hz

As on filling, frequency of A increases. Hence, original frequency of tuning fork A is 335 Hz.

11. *(c)* $\lambda = \frac{v}{f} = \frac{336}{504}$ m $= \frac{2}{3}$ m

For end correction,

$$\therefore \quad \frac{\lambda}{4} = l + e = l + 0.3d$$

(d = diameter of column)

$$\Rightarrow \quad \frac{\frac{2}{3} \times 100}{4} = l + 0.3 \times 6 \Rightarrow 16.66 = l + 1.8$$

$\therefore \quad l = 14.8$ cm

12. *(b)* Resonant frequency for a stretched string is given by

$$f_n = nf_1 = \frac{n}{2l}\sqrt{\frac{T}{\mu}}$$

Given, two consecutive resonant frequencies are 420 Hz and 490 Hz.

So, $\quad \frac{n}{2l}\sqrt{\frac{T}{\mu}} = 420 \quad \ldots$(i)

and $\quad \frac{n+1}{2l}\sqrt{\frac{T}{\mu}} = 490 \quad \ldots$(ii)

Subtracting Eq. (ii) from Eq. (i), we get

$$\frac{1}{2l}\sqrt{\frac{T}{\mu}} = 490 - 420 = 70$$

Here, $\quad T = 540$ N,

$\mu = 6 \times 10^{-3}$ kgm^{-1}

So, $\quad l = \frac{1}{2 \times 70}\sqrt{\frac{T}{\mu}}$

$$= \frac{1}{2 \times 70} \times \sqrt{\frac{540}{6 \times 10^{-3}}}$$

$$= 2.14 \text{ m} \simeq 2.1 \text{ m}$$

13. *(b)* In a resonance tube apparatus, first and second resonance occur as shown below.

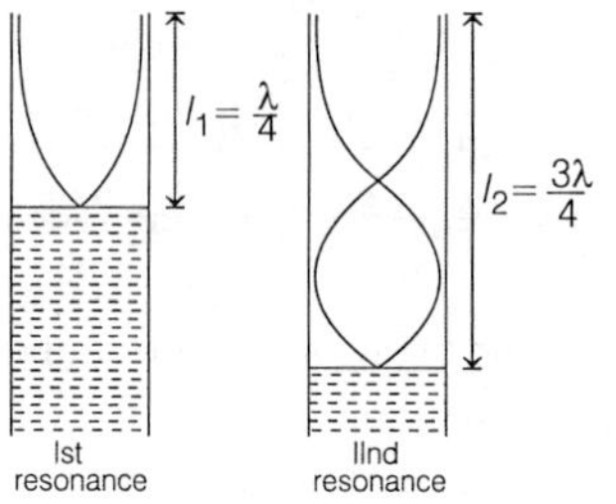

As in a stationary wave, distance between two successive nodes is $\lambda/2$ and distance of a node and an antinode is $\lambda/4$.

$$l_2 - l_1 = \frac{3\lambda}{4} - \frac{\lambda}{4} = \frac{\lambda}{2}$$

So, speed of sound, $v = f\lambda$

$$= f \times 2(l_2 - l_1)$$
$$= 480 \times 2 \times (70 - 30) \times 10^{-2}$$
$$= 384 \text{ ms}^{-1}$$

14. *(b)*

> **key idea** To find out speed of the wave, you can use following formula,
>
> Velocity of wave $= \frac{2f_n l}{n}$
>
> for third harmonic, $v = \frac{2f_3 l}{3}$
>
> For fundamental frequency, you can use
>
> $f = \frac{f_n}{n}$

Frequency of vibration of a string in nth harmonic is given by

$$f_n = n \cdot \frac{v}{2l} \quad \text{...(i)}$$

where, v = speed of sound and l = length of string.

Here, $f_3 = 240$ Hz, $l = 2$ m and $n = 3$

Substituting these values in Eq. (i), we get

$$\therefore \quad 240 = 3 \times \frac{v}{2 \times 2}$$

$$\Rightarrow \quad v = \frac{4 \times 240}{3} = 320 \text{ ms}^{-1}$$

Also, fundamental frequency is

$$f = \frac{f_n}{n} = \frac{f_3}{3} = \frac{240}{3} = 80 \text{ Hz}$$

15. *(c)* Given equation of stationary wave is

$$Y = 0.3 \sin (0.157x) \cos(200\pi t)$$

Comparing it with general equation of stationary wave, i.e. $Y = a \sin kx \cos \omega t$, we get

$$k = (2\pi/\lambda) = 0.157$$

$$\Rightarrow \quad \lambda = \frac{2\pi}{0.157} = 4\pi^2 \left(\because \frac{1}{2\pi} \approx 0.157\right) \quad \text{...(i)}$$

and $\omega = 200\,\pi = \frac{2\pi}{T} \Rightarrow T = \frac{1}{100}$ s

As the possible wavelength associated with nth harmonic of a vibrating string, i.e. fixed at both ends is given as

$$\lambda = \frac{2l}{n} \text{ or } l = n(\lambda/2)$$

Now, according to question, string is fixed from both ends and oscillates in 4th harmonic, so

$$4\left(\frac{\lambda}{2}\right) = l \Rightarrow 2\lambda = l$$

or $\quad l = 2 \times 4\pi^2 = 8\pi^2$ [using Eq. (i)]

Now, $\pi^2 \approx 10 \Rightarrow l \approx 80$ m

16. *(d)* For open ends, fundamental frequency f in air, we have

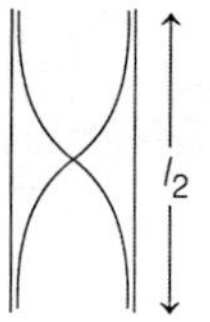

$$\frac{\lambda}{2} = l$$

$$\Rightarrow \quad \lambda = 2l \Rightarrow v = f\lambda$$

$$\Rightarrow \quad f = \frac{v}{\lambda} = \frac{v}{2l} \quad \text{...(i)}$$

When a pipe is dipped vertically in water, so that half of it is in water, we have

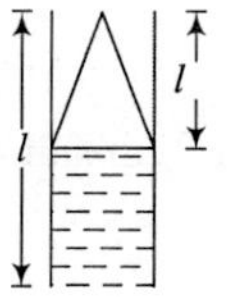

$$\frac{\lambda}{4} = \frac{l}{2}$$

$$\Rightarrow \quad \lambda = 2l \Rightarrow v = f'\lambda$$

$$\Rightarrow \quad f' = \frac{v}{\lambda} = \frac{v}{2l} = f \quad \text{...(ii)}$$

Thus, the fundamental frequency of the air column is now, $f = f'$

17. *(a)* By superposition principle,

$$y = y_1 + y_2 = A \sin(\omega t - kx) + A \sin(\omega t + kx)$$

$y = 2A \sin \omega t \cos kx$

Clearly, it is equation of standing wave for position of nodes

$$y = 0$$

i.e., $\quad x = (2n+1)\frac{\lambda}{4}$

$$\Rightarrow \quad \left(n + \frac{1}{2}\right)\frac{\lambda}{2}, n = 0, 1, 2, 3$$

18. *(b)* $l_1 = 18$ cm

$$f = \frac{v_1}{4l_1} \quad \text{[for first resonance]}$$

$$f = \frac{3v_2}{4l_2} \quad \text{[for second resonance]}$$

where, $l_2 = x$ according to given situation and also $v_1 < v_2$ as during summer temperature would be higher.

$$\Rightarrow \quad \frac{3v_2}{4l_2} = \frac{v_1}{4l_1} \text{ or } l_2 = 3l_1 \times \frac{v_2}{v_1}$$

$\Rightarrow x = 54 \times$ (A quantity greater than 1)

So, $x > 54$

19. *(a)* For string fixed at both the ends, resonant frequency are given by

$f = \frac{nv}{2L}$, where symbols have their usual meanings. It is given that 315 Hz and 420 Hz are two consecutive resonant frequencies, let these be nth and $(n+1)$th harmonics.

$$\text{Then, } 315 = \frac{nv}{2L} \quad \text{...(i)}$$

$$\text{and } 420 = \frac{(n+1)v}{2L} \quad \text{...(ii)}$$

Dividing Eq. (i) by Eq. (ii), we get

$$\frac{315}{420} = \frac{n}{n+1}$$

or $\quad n = 3$

From Eq. (i), lowest resonant frequency

$$f_0 = \frac{v}{2L} = \frac{315}{3} = 105 \text{ Hz}$$

20. *(c)* The frequency of fork 2

$$= 200 \pm 4$$
$$= 196 \text{ or } 204 \text{ Hz}$$

Since, on attaching the tape on the prong of fork 2, its frequency decreases but now the number of beats per second is 6 i.e., the frequency difference now increases.

It is possible only when before attaching the tape, the frequency of fork 2 is less than the frequency of tuning fork 1. Hence, the frequency of fork 2 is 196 Hz.

21. *(c)* $f_1 = 256$ Hz

For tuning fork, where $f_2 - f_1 = \pm 5$

f_2 = frequency of piano

$f_2 = (256 + 5)$ Hz or $(256 - 5)$ Hz

When tension is increased, the beat frequency decreases to 2 beats per second.

If we assume that the frequency of piano string is 261 Hz, then on increasing tension, frequency more than 261 Hz.

But it is given that beat frequency decreases to 2, therefore 261 Hz is not possible.

Hence, 251 Hz i.e., $(256-5)$ was the frequency of piano string before increasing tension.

22. *(a)*

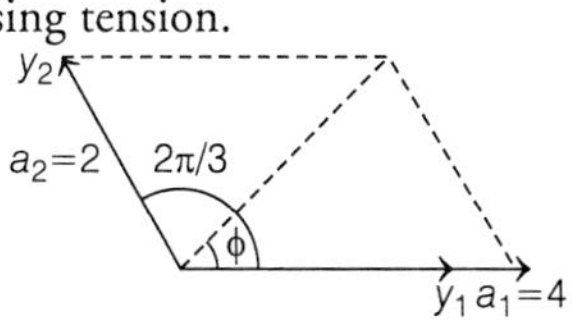

Phase difference between y_1 and y_2

$$\theta = \frac{2\pi}{3}$$

$\therefore$ Amplitude of resultant wave

$$A = \sqrt{a_1^2 + a_2^2 + 2a_1a_2\cos\theta}$$

$$= \sqrt{4^2 + 2^2 + 2\times4\times2\times\cos\frac{2\pi}{3}}$$

$$= \sqrt{16+4-8} = 2\sqrt{3}$$

Phase angle (ϕ) is given as,

$$\tan\phi = \frac{a_2\sin\theta}{a_1 + a_2\cos\theta} = \frac{2\sin\frac{2\pi}{3}}{4+2\cos\frac{2\pi}{3}}$$

$$= \frac{2\times\frac{\sqrt{3}}{2}}{4+2\left(-\frac{1}{2}\right)} = \frac{1}{\sqrt{3}} = \tan\frac{\pi}{6}$$

$$\therefore\ \phi = \frac{\pi}{6}$$

23. *(c)* $x = a\cos(1.5t)\cos(50.5t)$

$$x = \frac{2a}{2}\cos(1.5t)\cdot\cos(50.5t) \quad \ldots\text{(i)}$$

$$2\cos A\cos B = \cos(A+B) + \cos(A-B)$$

Eq. (i) can be written as

$$x = \frac{a}{2}\cos[(1.5+50.5)t + \frac{a}{2}\cos(50.5-1.5)]$$

$$x = \frac{a}{2}\cos(52t) + \frac{a}{2}\cos(49t)$$

$\therefore$Hence, $2\pi f_1 = 52 \quad 2\pi f_2 = 49$

$$f_1 = \frac{52}{2\pi} \qquad f_2 = \frac{49}{2\pi}$$

$$f_{\text{beat}} = f_1 - f_2$$

$$= \frac{52}{2\pi} - \frac{49}{2\pi} = \frac{3}{2\pi}\ \text{Hz}$$

$$T_{\text{beat}} = \frac{1}{f_{\text{beat}}} = \frac{2\pi}{3}\ \text{s}$$

$$= 2.09\ \text{s} \simeq 2\ \text{s}$$

24. *(c)* Let λ be the wavelength, L be the length of the tube.

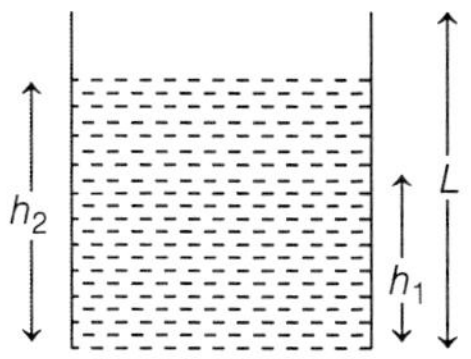

The resonance conditions for the given two cases are:

$$L - h_1 = \frac{n\lambda}{2} + \frac{\lambda}{4} \quad \ldots\text{(i)}$$

$$L - h_2 = (n-1)\frac{\lambda}{2} + \frac{\lambda}{4} \quad \ldots\text{(ii)}$$

Subtracting Eq. (ii) from Eq. (i), we get

$$(L-h_1) - (L-h_2)$$

$$= \frac{n\lambda}{2} + \frac{\lambda}{4} - (n-1)\frac{\lambda}{2} - \frac{\lambda}{4}$$

$$h_2 - h_1 = \lambda/2$$

Here, $h_1 = 17.0$ cm and $h_2 = 24.5$ cm

Substituting these values in above equation, we get

$$24.5 - 17.0 = \frac{\lambda}{2} \Rightarrow \frac{\lambda}{2} = 7.5\ \text{cm}$$

$$\Rightarrow \quad \lambda = 15.0\ \text{cm}$$

Now, the tuning fork frequency,

$$f = \frac{v}{\lambda} = \frac{330}{15\times10^{-2}}$$

$$= 2200\ \text{Hz}$$

Hence, correct option is (c).

25. *(d)* We can represent a wave vectorially with a phasor.

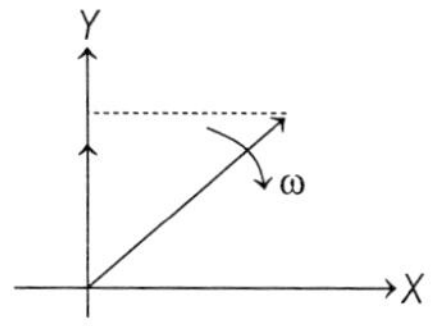

Here, we are given with following phasors.

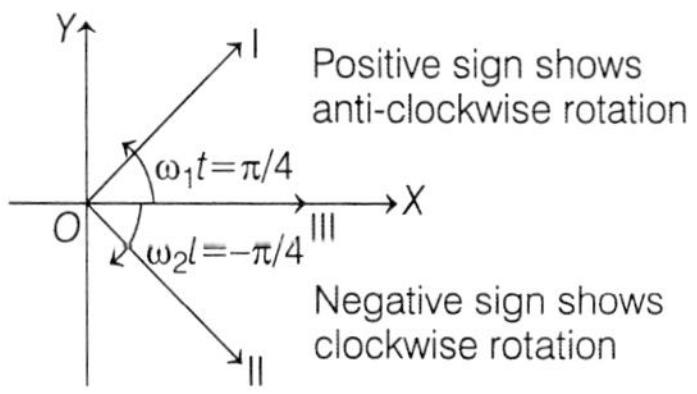

Each phasor is of same amplitude,

$$A^2 = I_0$$

From phasor diagram, amplitude of resultant of phasors I and II,

$$A' = A\cos\frac{\pi}{4} + A\cos\frac{\pi}{4}$$

$$= 2A\times\frac{1}{\sqrt{2}} = \sqrt{2}\,A$$

Now, this resultant phasor is along the direction of IIIrd phasor.

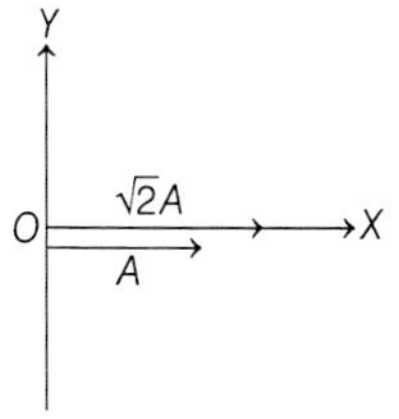

$\therefore$Resultant amplitude of these two will be

$$A_R = \sqrt{2}A + 1A = (\sqrt{2}+1)A$$

As intensity is proportional to square of amplitude, resultant intensity is given by

$$I_R = A_R^2 = (\sqrt{2}+1)^2 A^2$$

$$= (\sqrt{2}+1)^2 I_0$$

$$= (2+1+2\sqrt{2})I_0$$

$$= (3+2\sqrt{2})I_0$$

$$= (3+2\times1.4)I_0 = 5.8\,I_0$$

> **• Concept Enhancer**
>
> Length of phasor = Amplitude of wave and its projection on Y-axis gives variation in displacement.
>
> To find resultant amplitude obtained due to superposition of two or more waves, we can add phasors vectorially at any instant during their rotation.

26. *(c)*

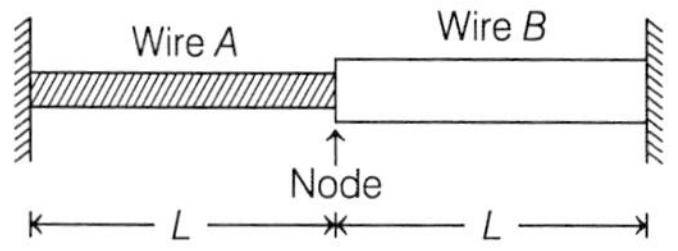

Let mass per unit length of wires are μ_A and μ_B, respectively.

$\because$ For same material, density is also same.

So, $\mu_A = \dfrac{\rho\pi r^2 L}{L} = \mu$

and $\mu_B = \dfrac{\rho 4\pi r^2 L}{L} = 4\mu$

Tension (T) in both connected wires are same.

So, speed of wave in wires are

$$v_A = \sqrt{\frac{T}{\mu_A}} = \sqrt{\frac{T}{\mu}}$$

and $v_B = \sqrt{\dfrac{T}{\mu_B}} = \sqrt{\dfrac{T}{4\mu}}$

$[\because \mu_A = \mu \text{ and } \mu_B = 4\mu]$

So, nth harmonic in such wires system is

$$f_{n\text{th}} = \frac{pv}{2L} \Rightarrow f_A = \frac{pv_A}{2L} = \frac{p}{2L}\sqrt{\frac{T}{\mu}}$$

(for p antinodes)

Similarly,

$$f_B = \frac{qv_B}{2L} = \frac{q}{2L}\sqrt{\frac{T}{4\mu}} = \frac{1}{2}\left(\frac{q}{2L}\sqrt{\frac{T}{\mu}}\right)$$

(for q antinodes)

As frequencies f_A and f_B are given equal.

So, $f_A = f_B$

$$\Rightarrow \frac{p}{2L}\sqrt{\frac{T}{\mu}} = \frac{q}{2}\left[\frac{1}{2L}\sqrt{\frac{T}{\mu}}\right]$$

$$\frac{p}{q} = \frac{1}{2} \Rightarrow p : q = 1 : 2$$

> • Concept **Enhancer**
>
> Exactly mid-way between two consecutive anti-node has a node and *vice-versa*.

27. *(a)* In first resonance,

length of air coloumn $= \frac{\lambda}{4}$.

So, $l_1 + e = \frac{\lambda}{4}$

or $11 \times 4 + 4e = \lambda$ $\quad [l_1 = 11 \text{ cm}]$

So, speed of sound is

$\Rightarrow \quad v = f_1\lambda = 512(44 + 4e) \quad \ldots(i)$

And in second case,

$$l_1' + e = \frac{\lambda'}{4}$$

or $27 \times 4 + 4e = \lambda'$

$\Rightarrow \quad v = f_2\lambda' = 256(108 + 4e) \ldots(ii)$

Dividing both Eqs. (i) and (ii), we get

$$1 = \frac{512(44 + 4e)}{256(108 + 4e)}$$

$\Rightarrow \quad e = 5$ cm

Substituting value of e in Eq. (i), we get

Speed of sound, $v = 512(44 + 4e)$

$= 512(44 + 4 \times 5)$

$= 512 \times 64 \text{ cm s}^{-1}$

$= 327.68 \text{ ms}^{-1} \approx 328 \text{ ms}^{-1}$

28. *(16)* Given, length of closed organ pipe (L_1) = Length of open organ pipe $(L_2) = L$ (say);

Now, For closed organ pipe, nth

Overtone, $f_n = (2n + 1)\frac{v}{4l}$

For open organ pipe, nth orvertone,

$$f_n = (n + 1)\frac{v}{2l}$$

$$\therefore \quad \frac{(f_n)_{\text{closed}}}{(f_n)_{\text{open}}} = \frac{\frac{15\,v}{ul}}{\frac{8\,v}{2l}}$$

$$= \frac{15}{4} \times \frac{2}{8} = \frac{15}{16}$$

29. *(740)* The difference in frequency in open organ pipe,

$$\Delta f = \frac{6v}{2 \times 0.6} - \frac{5v}{2 \times 0.9} \quad \left[\because f = \frac{nv}{2L}\right]$$

$$= \frac{6 \times 333}{2 \times 0.6} - \frac{5 \times 333}{2 \times 0.9}$$

$= 740$ Hz

30. *(60)* $(f_0)_1 l_1 = (f_0)_2 l_2$

$\Rightarrow 400 \times 90 = 600 \times l_2$

$$\Rightarrow l_2 = \frac{400 \times 90}{600} = 60 \text{ cm}$$

31. *(6)* Frequency of tuning fork

= Frequency of string

$$= f_1 = \frac{1}{2L}\sqrt{\frac{T}{\mu}}$$

$$= \frac{1}{2}\sqrt{\frac{6}{\mu}}$$

When tension is made 54 N, then 12 beats are produced.

Now, frequency of string is,

$$f_2 = \frac{1}{2}\sqrt{\frac{54}{\mu}} = \frac{3}{2}\sqrt{\frac{6}{\mu}} = 3f_1$$

Also, $3f_1 - f_1 = 12$

$\Rightarrow \quad 2f_1 = 12$

or $\quad f_1 = 6$ Hz

32. *(400)* Given,

For closed organ pipe fundamental frequency, $f = 30$ Hz

Let, length of closed organ pipe $= L$

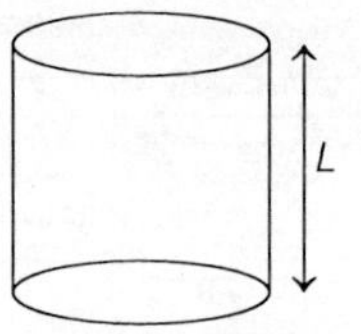

We know that,

Fundamental frequency $f = \frac{v}{4L}$

$$30 = \frac{330}{4L} \quad (v = 330 \text{ m/s given})$$

$$L = \frac{330}{30 \times 4} = \frac{11}{4} = 2.75 \text{ m}$$

Now, x g of water is added

Water filled upto height h.

New frequency $f' = 110$ Hz

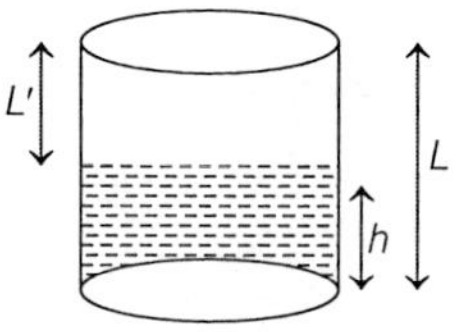

Again using formula Eq.(i)

$$f' = \frac{v}{4\,L'} \quad [L' = L - h]$$

$$110 = \frac{330}{4\,L'}$$

$$\Rightarrow \quad L' = \frac{330}{110 \times 4} = 0.75 \text{ cm}$$

$h = L - L'$

$\Rightarrow \quad h = 2.75 - 0.75 = 2$ m

Volume of liquid (water) filled

= area × height

$V = 2 \times 200 \text{ cm}^3$ ($a = 2 \text{ cm}^2$ given)

$= 400 \text{ cm}^3$

Amount of water (mass) = Volume × density of water

$= 400 \times 1 = 400$ g

33. *(294)*

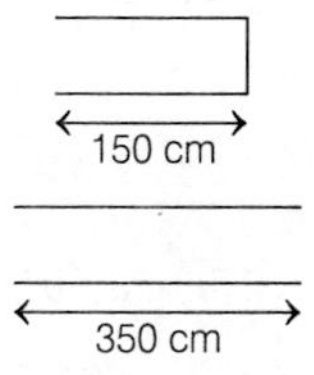

$$f_1 = \frac{v}{4l_1},\ f_2 = \frac{v}{2l_2}$$

$$f_1 = \frac{v \times 100}{4 \times 150},$$

$$f_2 = \frac{v \times 100}{2 \times 350}$$

$$f_1 = \frac{v}{6} \Rightarrow f_2 = \frac{v}{7}$$

$$\frac{v}{6} - \frac{v}{7} = 7 \Rightarrow \frac{v}{42} = 7$$

$v = 7 \times 42 \Rightarrow v = 294$ m/s

34. *(60)* Using $f = \frac{v}{2L}$, we get

$$120 = \frac{v}{2L} \quad \ldots(i)$$

and $$180 = \frac{v}{2L'} \quad \ldots(ii)$$

Dividing Eq. (ii) from Eq. (i), we get;

$$\frac{L'}{L} = \frac{120}{180} = \frac{2}{3}$$

$$\Rightarrow \quad L' = 90 \times \frac{2}{3} = 60 \text{ cm}$$

35. *(900)* For an open tube in second harmonic,

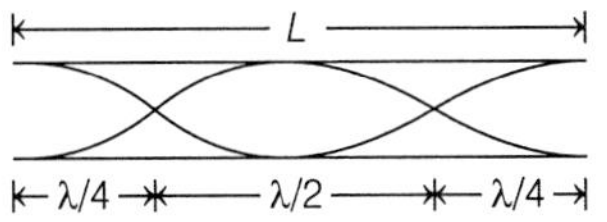

Wavelength of sound produced, in second harmonics,

$$\lambda = L$$

L = Length of tube

As, $\quad f \cdot \lambda = v$

(f = frequency, v = Speed of sound)

$$\Rightarrow \quad f = \frac{v}{\lambda} = \frac{v}{L} = \frac{360}{40 \times 10^{-2}}$$

$$= 900 \text{ Hz}$$

36. *(20)* Given, $y_1 = 10 \sin\left(\omega t + \frac{\pi}{3}\right)$ cm

$$y_2 = 5\,[\sin \omega t + \sqrt{3} \cos \omega t]$$

$$y_2 = 5\,[\sin \omega t + \sqrt{3} \cos \omega t] \times \frac{2}{2}$$

$$y_2 = 10\left[\sin \omega t \times \frac{1}{2} + \cos \omega t \times \frac{\sqrt{3}}{2}\right]$$

$$y_2 = 10\left[\sin \omega t \cos\frac{\pi}{3} + \cos \omega t \sin\frac{\pi}{3}\right]$$

$$y_2 = 10 \sin\left(\omega t + \frac{\pi}{3}\right) \text{ cm}$$

From given equation, $A_1 = 10$ cm, $A_2 = 10$ cm, $\phi = \frac{\pi}{3} - \frac{\pi}{3} = 0$

Amplitude of resultant wave,

$$R = \sqrt{A_1^2 + A_2^2 + 2A_1A_2 \cos\phi}$$

$$= \sqrt{(10)^2 + (10)^2 + 2(10)(10)\cos 0^\circ}$$

R = 20 cm

37. *(3)* We know, $\frac{nv}{2L} - f$

where, $\quad n$ = Harmonic number,
v = Speed of waves and
f = Frequency
$2L = 0.6$ m

$$\therefore \quad \frac{nv}{0.6} = 400 \text{ and } \frac{(n+1)v}{0.6} = 450$$

$$\Rightarrow \quad nv = 400 \times 0.6$$

and $nv + v = 450 \times 0.6 \Rightarrow nv = 240$

and $240 + v = 270$ m/s

$$\therefore \quad v = 30 \text{ m/s}$$

Now, we know that speed of wave in a string $= \sqrt{T/\mu}$

where, T = tension and $\mu = \frac{\text{mass}}{\text{length}}$

$$\therefore \quad v = \sqrt{\frac{T}{\mu}} \Rightarrow 30 = \sqrt{\frac{2700}{\mu}}$$

$$\Rightarrow \quad 900 = \frac{2700}{\mu}$$

$$\Rightarrow \quad \mu = 3 \text{ kg/m}$$

38. *(50)* In a closed organ tube, frequency of resonance of air column is

$$f = \frac{nv}{4L} \quad (\text{where } n = 1, 3, 5, 7 \ldots..)$$

Here, $f = 340$ Hz, $v = 340$ m/s

$L = 1.25$ m

So, $340 = \frac{n \times 340}{4 \times 1.25}$

$\Rightarrow n = 5$ (This is 3rd mode of vibration)

Hence, wavelength associated with this frequency,

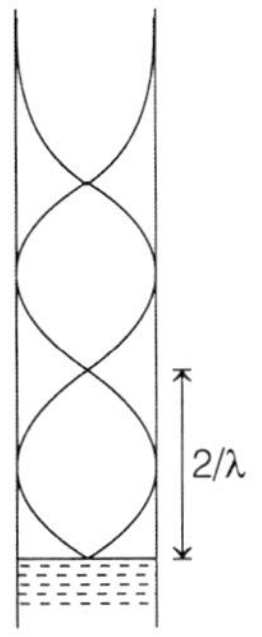

$$\lambda = 4\frac{L}{x}$$

$$\Rightarrow \quad \lambda = \frac{4L}{5} = \frac{4 \times 1.25}{5}$$

$$= 1 \text{ m}$$

or $\lambda = 100$ cm

Next mode of vibration occurs at a distance of half wavelength,

So, $h = \frac{100}{2} = 50$ cm

39. *(152)* Let f be frequency of first tuning fork.

According to given situation,

$$f_1 = f$$

Frequency of second tuning fork,

$$f_2 = f + 4$$

Frequency of third tuning fork,

$$f_3 = f + 2 \times 4$$

$$\vdots$$

$$f_{20} = f + (19 \times 4)$$

Since given that,

$$f_{20} = 2f_1 \Rightarrow f + (19 \times 4) = 2f$$

$$\Rightarrow \quad f = 76 \text{ Hz}$$

$\therefore$ Frequency of last tuning fork $= 2f$
$= 2 \times 76 = 152$ Hz

40. *(5)* Equation of stationary wave is given as

$$y = \left(10 \cos \pi x \sin \frac{2\pi t}{T}\right) \text{cm}$$

$\therefore$ Amplitude, $A = 10 \cos \pi x$

$\therefore$ Amplitude of particle at $x = \frac{4}{3}$ cm

$$A = 10 \cos \pi \times \frac{4}{3} = 10 \cos \frac{4\pi}{3}$$

$$= 10 \cos\left(\pi + \frac{\pi}{3}\right)$$

$$= 10\left(-\cos\frac{\pi}{3}\right) = 10\left(-\frac{1}{2}\right)$$

$$= -5 \text{ cm}$$

41. *(10)* Given that,

Linear mass density,

$$\mu = 9.0 \times 10^{-4} \frac{\text{kg}}{\text{m}}$$

Tension, $T = 900$ N

$\therefore$ Velocity of wire,

$$v = \sqrt{\frac{T}{\mu}} = \sqrt{\frac{900}{9 \times 10^{-4}}} = 10^3 \text{ m/s}$$

$$f_n = \frac{nv}{2l} = 500 \quad \ldots \text{(i)}$$

$$\text{And } f_{n+1} = \frac{(n+1)v}{2l} = 550 \quad \ldots \text{(ii)}$$

From Eqs. (i) and (ii),

$$f_{n+1} - f_n = \frac{v}{2l} = 50$$

$$\therefore \quad l = \frac{10^3}{2 \times 50} = 10 \text{ m}$$

42. *(34)* We know that, $f = \frac{nv}{4L}$

For L to be minimum, $n = 1$

$$\therefore \quad 250 = \frac{v}{4L} \Rightarrow 250 = \frac{340}{4L}$$

$$\Rightarrow \quad L = \frac{34}{4 \times 25} = 0.34 \text{ m}$$

$$\Rightarrow \quad L = 34 \text{ cm}$$

43. *(7)* Given, $y_1 = A_1 \sin k(x - vt)$

$$y_1 = 12 \sin 6.28\,(x - vt)$$

$$y_2 = 5 \sin 6.28\,(x - vt + 3.5)$$

We know,

Phase difference,

$$\Delta\phi = k(\Delta x) = 6.28 \times 3.5 = 7\pi$$

For resultant amplitude,

$$A^2 = A_1^2 + A_2^2 + 2A_1A_2 \cos\phi$$

$\Rightarrow$

$$A = \sqrt{(12)^2 + (5)^2 + 2(12)(5)\cos(7\pi)}$$

$$= \sqrt{144 + 25 - 120}$$

$$= \sqrt{49} = 7 \text{ mm}$$

44. *(1)* Given,

$$y = 1.0 \text{ mm} \cos(1.57 \text{ cm}^{-1})x \sin(78.5 \text{ s}^{-1})t$$

$\therefore \quad A = 1\text{ mm}; \quad k = 1.57\text{ cm}^{-1}$

$\omega = 78.5\text{ s}^{-1}$

At node, displacement, $y = 0$

$\cos(1.57\text{ cm}^{-1}x) = 0$

$\Rightarrow 1.57\text{ cm}^{-1}\, x = \frac{\pi}{2} \Rightarrow x = 1\text{ cm}$

45. *(4)* First overtone of open pipe $= \frac{v_2}{L_2}$

First overtone of closed pipe at one end

$$= \frac{3v}{4L_2}$$

As per question,

$$\frac{3v}{4L} = \frac{v_2}{L_2}$$

$$\Rightarrow \sqrt{\frac{B}{\rho_1}} \cdot \frac{3}{4L} = \sqrt{\frac{B}{\rho_2}} \cdot \frac{1}{L_2} \quad \left(\because V = \sqrt{\frac{B}{\rho}}\right)$$

$$\Rightarrow \quad L_2 = \frac{4L}{3}\sqrt{\frac{\rho_1}{\rho_2}} \quad \ldots\text{(i)}$$

According to question, the length of the open pipe is

$$L_2 = \frac{x}{3} L \sqrt{\frac{\rho_1}{\rho_2}} \quad \ldots\text{(ii)}$$

Comparing Eqs. (i) and (ii), we get

$$x = 4$$

• Concept Enhancer

In a closed pipe, only odd harmonics are present and all the even harmonics are absent here.

46. *(106)* Speed of sound in air by Laplace formula, $v = \sqrt{\frac{\gamma p}{\rho}}$

So, ratio of speed of sound in a gas to that of air,

$$\frac{v_{gas}}{v_{air}} = \sqrt{\frac{\rho_{air}}{\rho_{gas}}}$$

Here, $v_{air} = 300\text{ ms}^{-1}$ and $\frac{\rho_{air}}{\rho_{gas}} = \frac{1}{2}$

So, we have $\frac{v_{gas}}{300} = \frac{1}{\sqrt{2}}$

$$\Rightarrow \quad v_{gas} = \frac{300}{\sqrt{2}} = 150\sqrt{2}\text{ ms}^{-1}$$

Now, for a tube open at both ends.

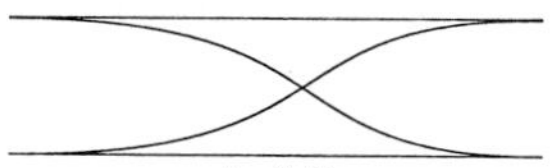

L=λ/2

$\therefore$ Fundamental frequency,

$$f_1 = \frac{v}{\lambda} = \frac{v}{2L}$$

Second harmonic frequency f_2 in tube,

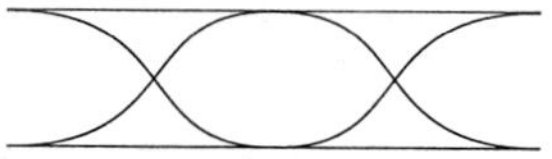

L=λ

$$f_2 = \frac{v}{\lambda}, \; f_2 = v/L$$

So, frequency difference,

$$\Delta f = f_2' - f_1 = \frac{v}{L} - \frac{v}{2L}$$

$$= \frac{v}{2L} = \frac{150\sqrt{2}}{2 \times 1} \quad (\because L = 1\text{ m})$$

$$= 75\sqrt{2} = 106.06\text{ Hz} \doteq 106\text{ Hz}$$

• Concept Enhancer

In an open pipe, all harmonics (even as well as odd) are present.

In JEE Main there is usually 1-2 question per paper are visible from this chapter. In past 3 years, 42 conceptual and 23 numerical questions were asked of moderate to difficult level from this chapter.
For detailed analysis scan QR code

ELECTRIC CHARGES AND FIELDS

Topic 1 Electric Charges and Coulomb's Law

Level 1

Imp. Surface Change Density

01 Electric charge is transferred to an irregular metallic disk as shown in figure. If $\sigma_1, \sigma_2, \sigma_3$ and σ_4 are charge densities at given points then, choose the correct answer from the options given below. *[2025, 8 April Shift-II]*

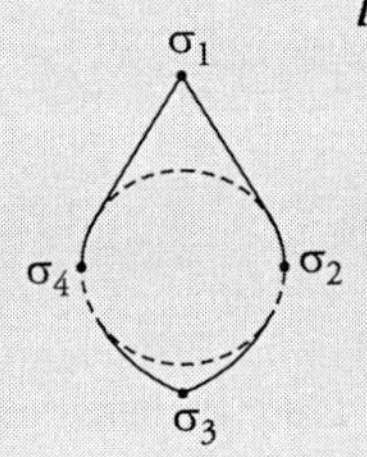

A. $\sigma_1 > \sigma_3, \sigma_2 = \sigma_4$ B. $\sigma_1 > \sigma_2, \sigma_3 > \sigma_4$
C. $\sigma_1 > \sigma_3 > \sigma_2 = \sigma_4$ D. $\sigma_1 < \sigma_3 < \sigma_2 = \sigma_4$
E. $\sigma_1 = \sigma_2 = \sigma_3 = \sigma_4$
(a) A and C Only (b) B and C Only
(c) D and E Only (d) A, B and C Only

02 A small uncharged conducting sphere is placed in contact with an identical sphere but having 4×10^{-8} C charge and then removed to a distance such that the force of repulsion between them is 9×10^{-3} N. The distance between them is

$\left(\text{Take } \frac{1}{4\pi\varepsilon_0} \text{ as } 9 \times 10^9 \text{ in SI units}\right)$ *[2025, 24 Jan Shift-II]*

(a) 2 cm (b) 3 cm (c) 1 cm (d) 4 cm

03 Two charges q and $3q$ are separated by a distance r in air. At a distance x from charge q, the resultant electric field is zero. The value of x is *[2024, 31 Jan Shift-I]*

(a) $\frac{r}{(1+\sqrt{3})}$ (b) $r(1+\sqrt{3})$ (c) $\frac{r}{3(1+\sqrt{3})}$ (d) $\frac{(1+\sqrt{3})}{r}$

04 A 10 μC charge is divided into two parts and placed at 1 cm distance so that the repulsive force between them is maximum. The charges of the two parts are *[2023, 13 April Shift-II]*

(a) 7 μC, 3 μC (b) 9 μC, 1 μC
(c) 5 μC, 5 μC (d) 8 μC, 2 μC

05 A charge of 4 μC is to be divided into two parts. The distance between the two divided charges is constant. The magnitude of the divided charges so that the force between them is maximum, will be. *[2022, 27 July Shift-II]*

(a) 1 μC and 3 μC (b) 2 μC and 2 μC
(c) 0 and 4 μC (d) 1.5 μC and 2.5 μC

06 Two particles A and B having charges 20 μC and $-5\ \mu$C, respectively are held fixed with a separation of 5 cm. At what position a third charged particle should be placed so that it does not experience a net electric force ? *[2021, 31 Aug Shift-I]*

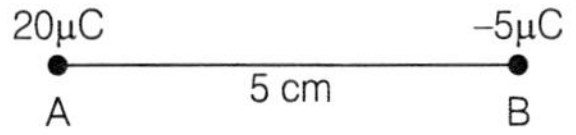

(a) At 5 cm from 20 μC on the left side of system
(b) At 5 cm from $-5\ \mu$C on the right side
(c) At 1.25 cm from $-5\ \mu$C between two charges
(d) At mid-point between two charges

07 A certain charge Q is divided into two parts q and $(Q - q)$. How should the charges Q and q be divided, so that q and $(Q - q)$ placed at a certain distance apart experience maximum electrostatic repulsion ? *[2021, 20 July Shift-I]*

(a) $Q = \frac{q}{2}$ (b) $Q = 2q$
(c) $Q = 4q$ (d) $Q = 3q$

Imp. Coulomb's Law

08 Three charges $+Q, q, +Q$ are placed respectively at distance 0, $d/2$ and d from the origin on the X-axis. If the net force experienced by $+Q$ placed at $x=0$ is zero, then value of q is **[2019, 9 Jan Shift-I]**
(a) $\frac{+Q}{2}$ (b) $\frac{+Q}{4}$ (c) $\frac{-Q}{2}$ (d) $\frac{-Q}{4}$

09 Shown in the figure are two point charges $+Q$ and $-Q$ inside the cavity of a spherical shell. The charges are kept near the surface of the cavity on opposite sides of the centre of the shell. If σ_1 is the surface charge on the inner surface and Q_1 net charge on it and σ_2 the surface charge on the outer surface and Q_2 net charge on it then

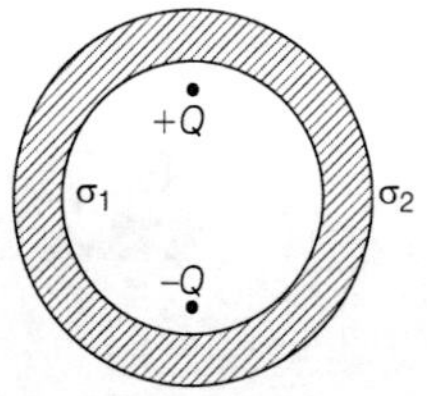

[JEE Main 2015]
(a) $\sigma_1 \neq 0, Q_1 \neq 0; \sigma_2 \neq 0, Q_2 \neq 0$
(b) $\sigma_1 \neq 0, Q_1 = 0; \sigma_2 \neq 0, Q_2 = 0$
(c) $\sigma_1 \neq 0, Q_1 = 0; \sigma_2 = 0, Q_2 = 0$
(d) $\sigma_1 = 0, Q_1 = 0; \sigma_2 = 0, Q_2 = 0$]

10 Two charges each equal to q, are kept at $x=-a$ and $x=a$ on the X-axis. A particle of mass m and charge $q_0 = q/2$ is placed at the origin. If charge q_0 is given, a small displacement $(y << a)$ along the Y-axis, the net force acting on the particle is proportional to **[JEE Main 2013]**
(a) y (b) $-y$ (c) $1/y$ (d) $-1/y$

11 If a charge q is placed at the centre of the line joining two equal charges Q such that the system is in equilibrium, then the value of q is **[AIEEE 2002]**
(a) $\frac{Q}{2}$ (b) $-\frac{Q}{2}$ (c) $\frac{Q}{4}$ (d) $-\frac{Q}{4}$

Level 2

12 Two identical conducting spheres P and S with charge Q on each, repel each other with a force 16 N. A third identical uncharged conducting sphere R is successively brought in contact with the two spheres. The new force of repulsion between P and S is **[2024, 6 April Shift-II]**
(a) 12 N (b) 4 N (c) 6 N (d) 1 N

Imp. Charge Density

13 Two isolated metallic solid spheres of radii R and $2R$ are charged such that both have same charge density σ. The spheres are then connected by a thin conducting wire. If the new charge density of the bigger sphere is σ'. The ratio σ'/σ is **[2023, 30 Jan Shift-I]**
(a) 5/3 (b) 9/4 (c) 4/3 (d) 5/6

14 If two charges q_1 and q_2 are separated with distance d and placed in a medium of dielectric constant k. What will be the equivalent distance between charges in air for the same electrostatic force? **[2023, 24 Jan Shift-I]**
(a) $d\sqrt{k}$ (b) $2d\sqrt{k}$ (c) $1.5d\sqrt{k}$ (d) $k\sqrt{d}$

15 Two identical metallic spheres A and B when placed at certain distance in air repel each other with a force of F. Another identical uncharged sphere C is first placed in contact with A and then in contact with B and finally placed at mid-point between spheres A and B. The force experienced by sphere C will be **[2022, 29 July Shift-II]**
(a) $3F/2$ (b) $3F/4$ (c) F (d) $2F$

16 Sixty four conducting drops each of radius 0.02 m and each carrying a charge of $5\mu C$ are combined to form a bigger drop. The ratio of surface density of bigger drop to the smaller drop will be **[2022, 26 Jan Shift-II]**
(a) 1 : 4 (b) 4 : 1 (c) 1 : 8 (d) 8 : 1

17 Two identical charged particles each having a mass 10 g and charge 2.0×10^{-7} C are placed on a horizontal table with a separation of L between them such that they stay in limited equilibrium. If the coefficient of friction between each particle and the table is 0.25, find the value of L. [Use, $g = 10\ ms^{-2}$] **[2022, 24 June Shift-II]**
(a) 12 cm (b) 10 cm (c) 8 cm (d) 5 cm

18 Find out the surface charge density at the intersection of point $X = 3$ m plane and X-axis, in the region of uniform line charge of 8 nC/m lying along the Z-axis in free space. **[2021, 16 March Shift-II]**
(a) $0.424\ nC\ m^{-2}$ (b) $47.88\ nC\ m^{-2}$
(c) $0.07\ nC\ m^{-2}$ (d) $4.0\ nC\ m^{-2}$

19 Two identical conducting spheres A and B carry equal charge. They are separated by a distance much larger than their diameters, and the force between them is F. A third identical conducting sphere, C is uncharged. Sphere C is first touched to A, then to B, and then removed. As a result the force between A and B would be equal to. **[JEE Main 2018]**
(a) $3F/8$ (b) $F/2$ (c) $3F/4$ (d) F

20 A charge Q is placed at each of the opposite corners of a square. A charge q is placed at each of the other two corners. If the net electrical force on Q is zero, then Q/q equals **[AIEEE 2009]**
(a) $-2\sqrt{2}$ (b) -1 (c) 1 (d) $-1/\sqrt{2}$

21 Four charges equal to $-Q$ are placed at the four corners of a square and a charge q is at its centre. If the system is in equilibrium, the value of q is **[AIEEE 2004]**
(a) $-\frac{Q}{4}(1+2\sqrt{2})$ (b) $\frac{Q}{4}(1+2\sqrt{2})$
(c) $-\frac{Q}{2}(1+2\sqrt{2})$ (d) $\frac{Q}{2}(1+2\sqrt{2})$

22 Three charges $-q_1, +q_2$ and $-q_3$ are placed as shown in the figure. The x-component of the force on $-q_1$ is proportional to *[AIEEE 2003]*

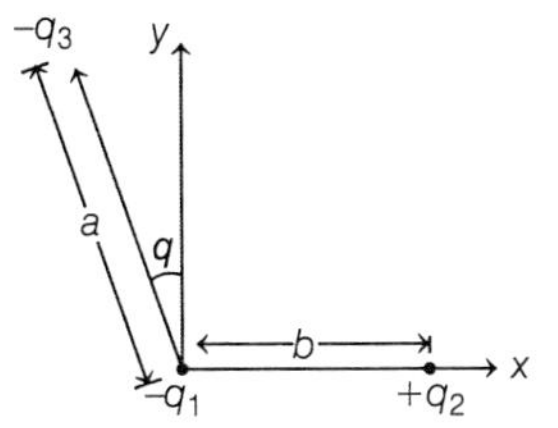

(a) $\frac{q_2}{b^2} - \frac{q_3}{a^2}\cos\theta$ (b) $\frac{q_2}{b^2} + \frac{q_3}{a^2}\sin\theta$

(c) $\frac{q_2}{b^2} + \frac{q_3}{a^2}\cos\theta$ (d) $\frac{q_2}{b^2} - \frac{q_3}{a^2}\sin\theta$

Level 3

23 In hydrogen like system the ratio of coulombian force and gravitational force between an electron and a proton is in the order of *[2024, 5 April Shift-I]*

(a) 10^{19} (b) 10^{39} (c) 10^{36} (d) 10^{29}

24 Two identical positive charge Q each are fixed at a distance of $2a$ apart from each other, Another point charge q_0 with mass m is placed at mid-point between two fixed charges. For a small displacement along the line joining the fixed charges, the charge q_0 executes SHM. The time period of ocscillation of charge q_0 will be *[2022, 27 July Shift-I]*

(a) $\sqrt{\frac{4\pi^3\varepsilon_o ma^3}{q_0 Q}}$ (b) $\sqrt{\frac{q_0 Q}{4\pi^3\varepsilon_o ma^3}}$

(c) $\sqrt{\frac{2\pi^3\varepsilon_o ma^3}{q_0 Q}}$ (d) $\sqrt{\frac{8\pi^3\varepsilon_o ma^3}{q_0 Q}}$

25 Two point charges Q each are placed at a distance d apart. A third point charge q is placed at a distance x from mid-point on the perpendicular bisector. The value of x at which charge q will experience the maximum coulomb's force is *[2022, 29 June Shift-II]*

(a) $x = d$ (b) $x = \frac{d}{2}$ (c) $x = \frac{d}{\sqrt{2}}$ (d) $x = \frac{d}{2\sqrt{2}}$

26 Three identical charged balls each of charge 2 C are suspended from a common point P by silk threads of 2 m each (as shown in figure). They form an equilateral triangle of side 1m.

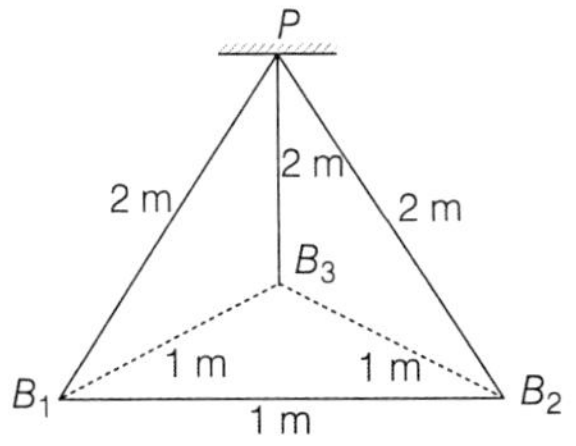

The ratio of net force on a charged ball to the force between any two charged balls will be *[2022, 27 June Shift-II]*

(a) 1 : 1 (b) 1 : 4 (c) $\sqrt{3} : 2$ (d) $\sqrt{3} : 1$

27 Two identical tennis balls each having mass m and charge q are suspended from a fixed point by threads of length l. What is the equilibrium separation when each thread makes a small angle θ with the vertical ? *[2021, 27 July Shift-I]*

(a) $d = \left(\frac{q^2 l}{2\pi\varepsilon_0 mg}\right)^{1/2}$ (b) $d = \left(\frac{q^2 l}{2\pi\varepsilon_0 mg}\right)^{1/3}$

(c) $d = \left(\frac{q^2 l^2}{2\pi\varepsilon_0 m^2 g}\right)^{1/3}$ (d) $d = \left(\frac{q^2 l^2}{2\pi\varepsilon_0 m^2 g^2}\right)^{1/3}$

28 Two electrons each are fixed at a distance $2d$. A third charge proton placed at the mid-point is displaced slightly by a distance $x(x << d)$ perpendicular to the line joining the two fixed charges. Proton will execute simple harmonic motion having angular frequency? (m = mass of charged particle) *[2021, 24 Feb Shift-II]*

(a) $\left(\frac{2q^2}{\pi\varepsilon_0 md^3}\right)^{1/2}$ (b) $\left(\frac{\pi\varepsilon_0 md^3}{2q^2}\right)^{1/2}$

(c) $\left(\frac{q^2}{2\pi\varepsilon_0 md^3}\right)^{1/2}$ (d) $\left(\frac{2\pi\varepsilon_0 md^3}{q^2}\right)^{1/2}$

29 Two identical charged spheres suspended from a common point by two massless strings of length l are initially a distance $d, (d << l)$ apart because of their mutual repulsion. The charge begins to leak from both the spheres at a constant rate. As a result, charges approach each other with a velocity v. Then, as a function of distance x between them, is *[AIEEE 2011]*

(a) $v \propto x^{-1}$ (b) $v \propto x^{1/2}$ (c) $v \propto x$ (d) $v \propto x^{-1/2}$

30 Two identical charged spheres are suspended by strings of equal lengths. The strings make an angle of 30° with each other. When suspended in a liquid of density 0.8 g cm^{-3}, the angle remains the same. If density of the material of the sphere is 1.6 g cm^{-3}, then dielectric constant of the liquid is *[AIEEE 2010]*

(a) 4 (b) 3 (c) 2 (d) 1

Numerical Value Type Questions

31 A positive ion A and a negative ion B has charges 6.67×10^{-19}C and 9.6×10^{-10}C and masses 19.2×10^{-27} kg and 9×10^{-27} kg respectively.

At an instant, the ions are separated by a certain distance r. At that instant the ratio of the magnitudes of electrostatic force to gravitational force is $P \times 10^{-13}$, where the value of P is ....

(Take $\frac{1}{4\pi\varepsilon_0} = 9 \times 10^9$ Nm^{2}C^{-2} and universal gravitational constant as 6.67×10^{-11} Nm^{2}kg^{-2})

[2025, 23 Jan Shift-I]

32 Three point charges q, $-2q$ and $2q$ are placed on X-axis at a distance $x = 0$, $x = 3/4R$ and $x = R$, respectively from origin as shown. If $q = 2 \times 10^{-6}$C and $R = 2$ cm, the magnitude of net force experienced by the charge $-2q$ is .... N. *[2023, 13 April Shift-II]*

33 As shown the figure, a configuration of two equal point charges ($q_0 = +2\mu$C) is placed on an inclined plane. Mass of each point charge is 20 g. Assume that there is no friction between charge and plane. For the system of two point charges to be in equilibrium (at rest) the height $h = x \times 10^{-3}$ m.
The value of x is .... *[2023, 11 April Shift-I]*

(Take, $\frac{1}{4\pi\varepsilon_0} = 9 \times 10^9$ Nm^{2}C^{-2}, $g = 10$ ms^{-2})

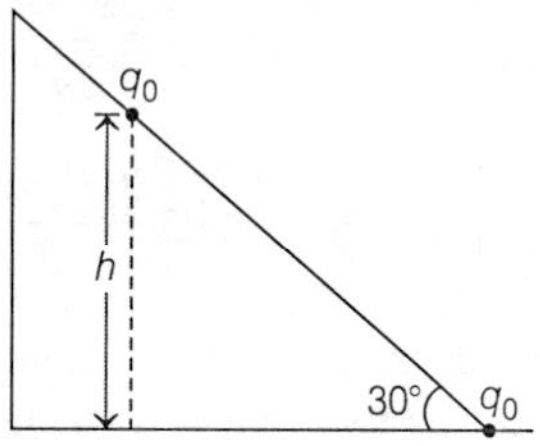

34 Two equal positive point charges are separated by a distance $2a$. The distance of a point from the centre of the line joining two charges on the equatorial line (perpendicular bisector) at which force experienced by a test charge q_0 becomes maximum is $a/\sqrt{x}$. The value of x is .... *[2023, 1 Feb Shift-I]*

35 A point charge $q_1 = 4q_0$ is placed at origin. Another point charge $q_2 = -q_0$ is placed at $x = 12$ cm. Charge of proton is q_0. The proton is placed on X-axis so that the electrostatic force on the proton is zero. In this situation, the position of the proton from the origin is .... cm. *[2023, 29 Jan Shift-I]*

36 Three point charges of magnitude 5 μC, 0.16 μC and 0.3 μC are located at the vertices A, B and C of a right angled triangle whose sides are $AB = 3$ cm, $BC = 3\sqrt{2}$ cm and $CA = 3$ cm and point A is the right angle corner. Charge at point A, experiences .... N of electrostatic force due to the other two charges.

[2022, 26 July Shift-II]

37 A particle of mass 1 mg and charge q is lying at the mid-point of two stationary particles kept at a distance 2 m when each is carrying same charge q. If the free charged particle is displaced from its equilibrium position through distance x ($x << 1$ m), the particle executes SHM. Its angular frequency of oscillation will be ...... $\times 10^5$ rad/s, if $q^2 = 10$ C^2.

[2021, 25 July Shift-I]

38 An infinite number of point charges, each carrying 1 μC charge, are placed along the Y-axis at $y = 1$ m, 2 m, 4 m, 8 m. The total force on a 1 C point charge, placed at the origin, is $x \times 10^3$ N. The value of x to the nearest integer, is ....

(Take, $\frac{1}{4\pi\varepsilon_0} = 9 \times 10^9$ N-m^2/C^2)

[2021, 18 March Shift-II]

Topic 2 Electric Field and Field Lines

Level 1

01 A point charge 2×10^{-2} C is moved from P to S in a uniform electric field of 30 NC^{-1} directed along positive X-axis. If coordinates of P and S are (1, 2, 0) m and (0, 0, 0) m respectively, the work done by electric field will be *[2023, 29 Jan Shift-II]*

(a) 1200 mJ (b) 600 mJ (c) −600 mJ (d) −1200 mJ

02 Two point charges A and B of magnitude $+8 \times 10^{-6}$C and -8×10^{-6} C, respectively are placed at a distance d apart. The electric field at the middle point O between the charges is 6.4×10^4 NC^{-1}. The distance d between the point charges A and B is

[2022, 28 June Shift-II]

(a) 2.0 m (b) 3.0 m (c) 1.0 m (d) 4.0 m

03 The three charges $q/2$, q and $q/2$ are placed at the corners A, B and C of a square of side a as shown in figure. The magnitude of electric field E at the corner D of the square, is *[2022, 28 June Shift-I]*

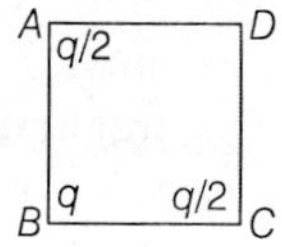

(a) $\frac{q}{4\pi\varepsilon_0 a^2}\left(\frac{1}{\sqrt{2}} + \frac{1}{2}\right)$ (b) $\frac{q}{4\pi\varepsilon_0 a^2}\left(1 + \frac{1}{\sqrt{2}}\right)$

(c) $\frac{q}{4\pi\varepsilon_0 a^2}\left(1 - \frac{1}{\sqrt{2}}\right)$ (d) $\frac{q}{4\pi\varepsilon_0 a^2}\left(\frac{1}{\sqrt{2}} - \frac{1}{2}\right)$

04 A vertical electric field of magnitude 4.9×10^5N/C just prevents a water droplet of a mass 0.1 g from falling. The value of charge on the droplet will be: (Given, $g = 9.8$ m/s^2) *[2022, 24 June Shift-I]*

(a) 1.6×10^{-9}C (b) 2.0×10^{-9}C
(c) 3.2×10^{-9}C (d) 0.5×10^{-9}C

05 A small point mass carrying some positive charge on it, is released from the edge of a table. There is an uniform electric field in this region in the horizontal direction. Which of the following options then correctly describe the trajectory of the mass? (Curves are drawn schematically and are not to scale) ***[2020, 2 Sep Shift-II]***

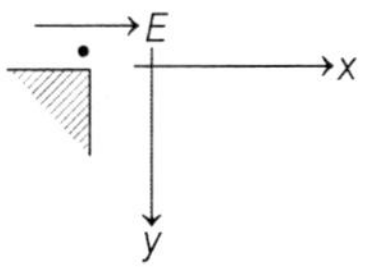

(a)
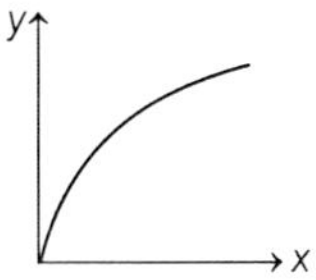

(b)
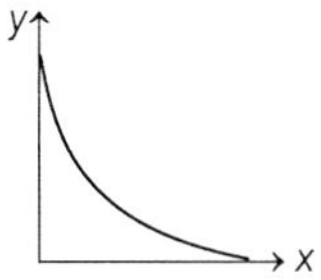

(c)
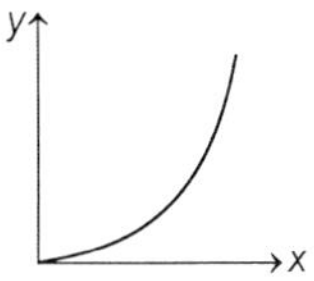

(d)
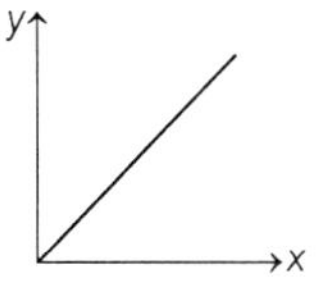

06 For a uniformly charged ring of radius R, the electric field on its axis has the largest magnitude at a distance h from its centre. Then, value of h is ***[2019, 9 Jan Shift-I]***

(a) $R/\sqrt{5}$ (b) $R/\sqrt{2}$ (c) R (d) $R\sqrt{2}$

07 A body of mass M and charge q is connected to a spring of spring constant k. It is oscillating along x-direction about its equilibrium position, taken to be at $x=0$, with an amplitude A. An electric field E is applied along the x-direction. Which of the following statements is correct ? ***[JEE Main 2018]***

(a) The total energy of the system is $\frac{1}{2}m\omega^2A^2+\frac{1}{2}\frac{q^2E^2}{k}$

(b) The new equilibrium position is at a distance $\frac{2qE}{k}$ from $x=0$

(c) The new equilibrium position is at a distance $\frac{qE}{2k}$ from $x=0$

(d) The total energy of the system is $\frac{1}{2}m\omega^2A^2-\frac{1}{2}\frac{q^2E^2}{k}$

08 A long cylindrical shell carries positive surface charge σ in the upper half and negative surface charge $-\sigma$ in the lower half. The electric field lines around the cylinder will look like figure given in (figures are schematic and not drawn to scale) ***[JEE Main 2015]***

(a)
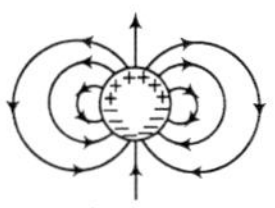

(c)
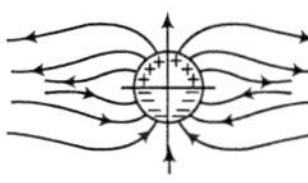

(b)
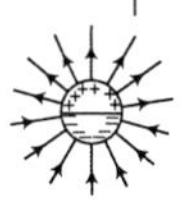

(d)

Level 2

09 A point charge $+q$ is placed at the origin. A second point charge $+9q$ is placed at $(d, 0, 0)$ in Cartesian coordinate system. The point in between them where the electric field vanishes is ***[2025, 2 April Shift-I]***

(a) $\left(\frac{3d}{4}, 0, 0\right)$ (b) $\left(\frac{d}{3}, 0, 0\right)$ (c) $\left(\frac{d}{4}, 0, 0\right)$ (d) $\left(\frac{4d}{3}, 0, 0\right)$

10 A particle of mass m and charge q is fastened to one end A of a massless string having equilibrium length l, whose other end is fixed at point O. The whole system is placed on a frictionless horizontal plane and is initially at rest. If uniform electric field is switched on along the direction as shown in figure, then the speed of the particle when it crosses the X-axis is ***[2025, 28 Jan Shift-I]***

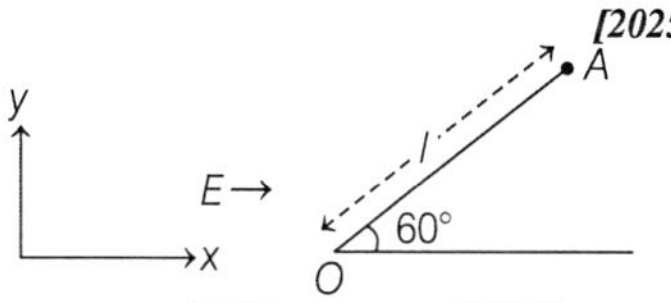

(a) $\sqrt{\frac{qEl}{m}}$ (b) $\sqrt{\frac{qEl}{2m}}$ (c) $\sqrt{\frac{2qEl}{m}}$ (d) $\sqrt{\frac{qEl}{4m}}$

11 An electron is made to enter symmetrically between two parallel and equally but oppositely charged metal plates, each of 10 cm length. The electron emerges out of the electric field region with a horizontal component of velocity 10^6 m/s. If the magnitude of the electric field between the plates is 9.1 V/cm, then the vertical component of velocity of electrons is (mass of electron $=9.1\times10^{-31}$ kg and charge of electron $=1.6\times10^{-19}$ C) ***[2025, 22 Jan Shift-I]***

(a) 16×10^4 m/s (b) 0

(c) 1×10^6 m/s (d) 16×10^6 m/s

Imp. Electric Field Intensity

12 A point charge of 10 μC is placed at the origin. At what location on the X-axis should a point charge of 40 μC be placed so that the net electric field is zero at $x=2$ cm on the X-axis ? ***[2023, 25 Jan Shift-II]***

(a) $x=8$ cm (b) $x=-4$ cm (c) $x=4$ cm (d) $x=6$ cm

13 Graphical variation of electric field due to a uniformly charged insulating solid sphere of radius R, with distance r from the centre O is represented by ***[2023, 8 April Shift-I]***

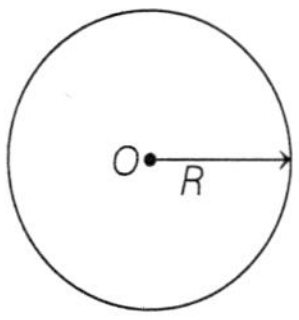

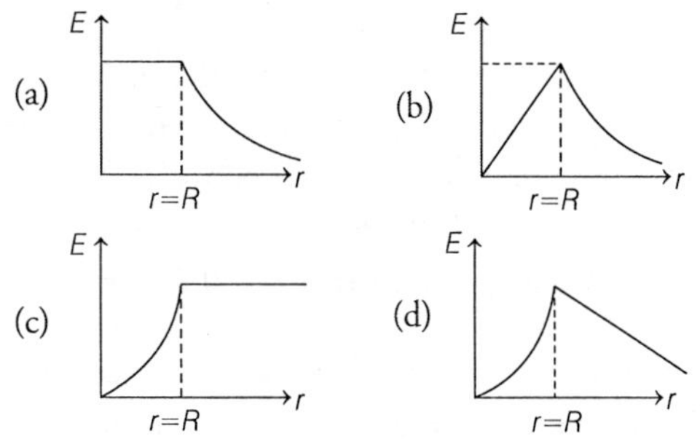

14 A positive charge particle of 100 mg is thrown in opposite direction to a uniform electric field of strength 1×10^5 NC^{-1}. If the charge on the particle is 40 μC and the initial velocity is 200 ms^{-1}, how much distance it will travel before coming to the rest momentarily **[2022, 29 June Shift-I]**

(a) 1 m (b) 5 m (c) 10 m (d) 0.5 m

15 What will be the magnitude of electric field at point O as shown in figure? Each side of the figure is l and perpendicular to each other. **[2021, 27 July Shift-II]**

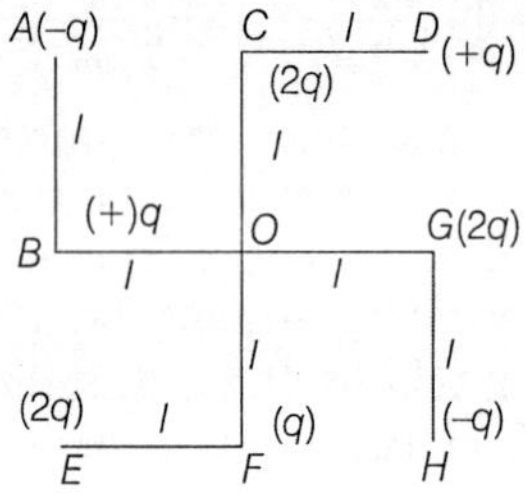

(a) $\frac{1}{4\pi\varepsilon_0}\frac{q}{l^2}$ (b) $\frac{1}{4\pi\varepsilon_0}\frac{q}{(2l^2)}(2\sqrt{2}-1)$

(c) $\frac{q}{4\pi\varepsilon_0 (2l)^2}$ (d) $\frac{1}{4\pi\varepsilon_0}\frac{2q}{2l^2}(\sqrt{2})$

16 A cube of side a has point charges $+Q$ located at each of its vertices except at the origin, where the charge is $-Q$. The electric field at the centre of cube is **[2021, 24 Feb Shift-I]**

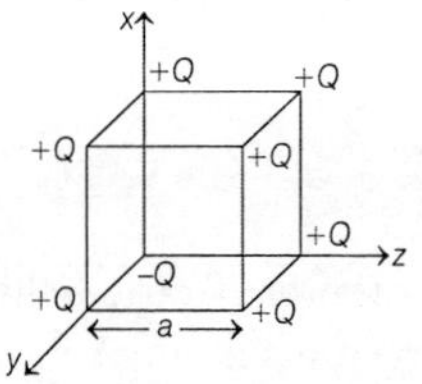

(a) $\frac{-Q}{3\sqrt{3}\pi\varepsilon_0 a^2}(\hat{x}+\hat{y}+\hat{z})$ (b) $\frac{Q}{3\sqrt{3}\pi\varepsilon_0 a^2}(\hat{x}+\hat{y}+\hat{z})$

(c) $\frac{-2Q}{3\sqrt{3}\pi\varepsilon_0 a^2}(\hat{x}+\hat{y}+\hat{z})$ (d) $\frac{2Q}{3\sqrt{3}\pi\varepsilon_0 a^2}(\hat{x}+\hat{y}+\hat{z})$

17 A uniformly charged disc of radius R having surface charge density σ is placed in the XY-plane with its centre at the origin. Find the electric field intensity along the Z-axis at a distance Z from origin **[2021, 27 Aug Shift-I]**

(a) $E = \frac{\sigma}{2\varepsilon_0}\left(1 - \frac{Z}{(Z^2+R^2)^{1/2}}\right)$

(b) $E = \frac{\sigma}{2\varepsilon_0}\left(1 + \frac{Z}{(Z^2+R^2)^{1/2}}\right)$

(c) $E = \frac{2\varepsilon_0}{\sigma}\left(\frac{1}{(Z^2+R)^{1/2}} + Z\right)$

(d) $E = \frac{\sigma}{2\varepsilon_0}\left(\frac{1}{(Z^2+R^2)^{1/2}} + \frac{1}{Z^2}\right)$

18 Figure shows a rod AB. which is bent in a 120° circular arc of radius R. A charge $(-Q)$ is uniformly distributed over rod AB. What is the electric field $\mathbf{E}$ at the centre of curvature O? **[2021, 27 Aug Shift-II]**

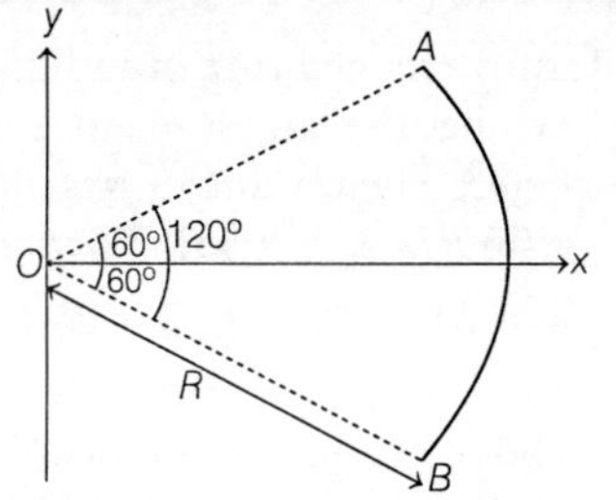

(a) $\frac{3\sqrt{3}Q}{8\pi\varepsilon_0 R^2}(\hat{i})$ (b) $\frac{3\sqrt{3}Q}{8\pi^2\varepsilon_0 R^2}(\hat{i})$

(c) $\frac{3\sqrt{3}Q}{16\pi^2\varepsilon_0 R^2}(\hat{i})$ (d) $\frac{3\sqrt{3}Q}{8\pi^2\varepsilon_0 R^2}(\hat{i})$

19 A particle of mass m and charge q is released from rest in a uniform electric field. If there is no other force on the particle, the dependence of its speed v on the distance x travelled by it is correctly given by (graphs are schematic and not drawn to scale) **[2020, 8 Jan Shift-II]**

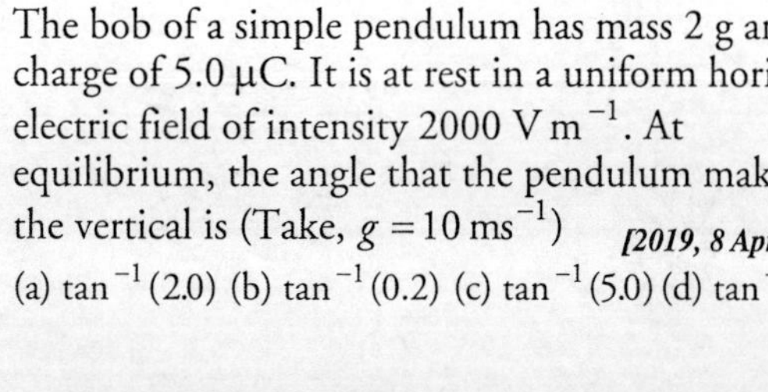

20 The bob of a simple pendulum has mass 2 g and a charge of 5.0 μC. It is at rest in a uniform horizontal electric field of intensity 2000 $V m^{-1}$. At equilibrium, the angle that the pendulum makes with the vertical is (Take, $g = 10$ ms^{-1}) **[2019, 8 April Shift-I]**

(a) $\tan^{-1}(2.0)$ (b) $\tan^{-1}(0.2)$ (c) $\tan^{-1}(5.0)$ (d) $\tan^{-1}(0.5)$

21 A wire of length L $(= 20$ cm$)$ is bent into a semicircular arc. If the two equal halves of the arc were each to be uniformly charged with charges $\pm Q$, $[|Q| = 10^3 \varepsilon_0$ Coulomb, where ε_0 is the permittivity (in SI units) of free space the net electric field at the centre O of the semicircular arc would be *[JEE Main 2015]*

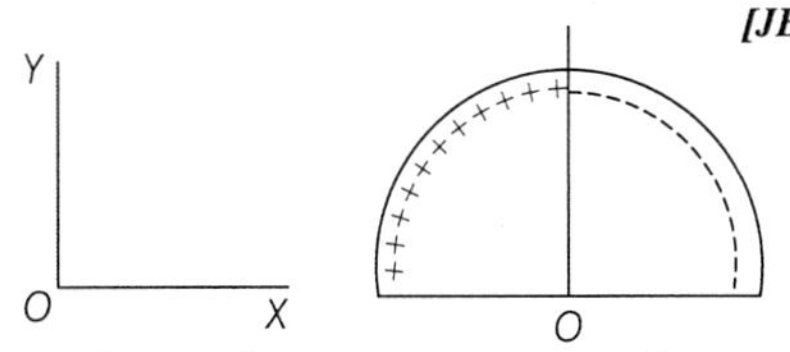

(a) $(50 \times 10^3 \text{ N/C})\hat{j}$ (b) $(50 \times 10^3 \text{ N/C})\hat{i}$
(c) $(25 \times 10^3 \text{ N/C})\hat{j}$ (d) $(25 \times 10^3 \text{ N/C})\hat{i}$

22 A thin semi-circular ring of radius r has a positive charge q distributed uniformly over it. The net field E at the centre O is *[AIEEE 2010]*

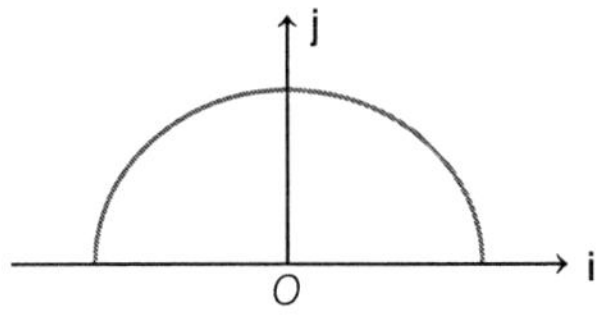

(a) $\dfrac{q}{4\pi^2 \varepsilon_0 r^2}\,\mathbf{j}$ (b) $-\dfrac{q}{4\pi^2 \varepsilon_0 r^2}\,\mathbf{j}$
(c) $-\dfrac{q}{2\pi^2 \varepsilon_0 r^2}\,\mathbf{j}$ (d) $\dfrac{q}{2\pi^2 \varepsilon_0 r^2}\,\mathbf{j}$

Level 3

23 Consider a circular loop that is uniformly charged and has a radius $a\sqrt{2}$. Find the position along the positive Z-axis of the cartesian coordinate system where the electric field is maximum if the ring was assumed to be placed in xy-plane at the origin *[2025, 2 April Shift-II]*

(a) $\dfrac{a}{\sqrt{2}}$ (b) a (c) $\dfrac{a}{2}$ (d) 0

24 An infinitely long positively charged straight thread has a linear charge density λ Cm^{-1}. An electron revolves along a circular path having axis along the length of the wire. The graph that correctly represents the variation of the kinetic energy of electron as a function of radius of circular path from the wire is *[2024, 4 April Shift-I]*

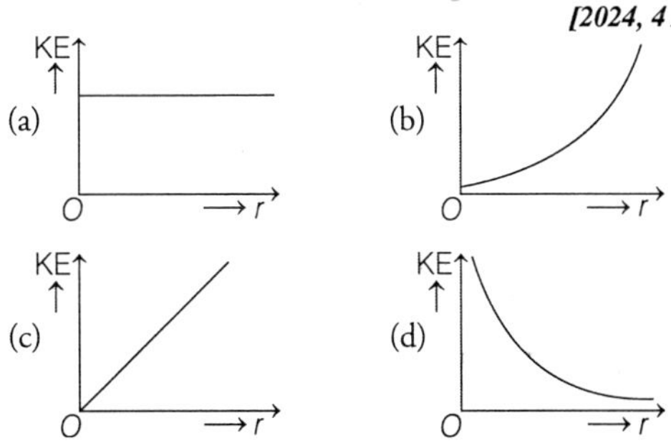

25 A particle of charge $-q$ and mass m moves in a circle of radius r around an infinitely long line charge of linear charge density $+\lambda$. Then, time period will be given as (Consider k as Coulomb's constant) *[2024, 30 Jan Shift-II]*

(a) $T = \dfrac{1}{2\pi r}\sqrt{\dfrac{m}{2k\lambda q}}$ (b) $T = \dfrac{1}{2\pi}\sqrt{\dfrac{2k\lambda q}{m}}$

(c) $T^2 = \dfrac{4\pi^2 m}{2k\lambda q} r^3$ (d) $T = 2\pi r\sqrt{\dfrac{m}{2k\lambda q}}$

26 A uniform electric field $E = (8\ m/e)$ V/m is created between two parallel plates of length 1 m as shown in figure, (where, m = mass of electron and e = charge of electron). An electron enters the field symmetrically between the plates with a speed of 2 m/s. The angle of the deviation θ of the path of the electron as it comes out of the field will be .... *[2022, 28 July Shift-II]*

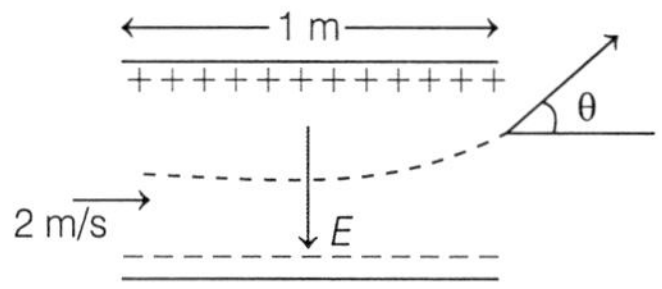

(a) $\tan^{-1}(4)$ (b) $\tan^{-1}(2)$
(c) $\tan^{-1}\left(\dfrac{1}{3}\right)$ (d) $\tan^{-1}(3)$

27 A long cylindrical volume contains a uniformly distributed charge of density ρ. The radius of cylindrical volume is R. A charge particle q revolves around the cylinder in a circular path. The kinetic energy of the particle is *[2022, 24 June Shift-II]*

(a) $\dfrac{\rho q R^2}{4\varepsilon_0}$ (b) $\dfrac{\rho q R^2}{2\varepsilon_0}$ (c) $\dfrac{q\rho}{4\varepsilon_0 R^2}$ (d) $\dfrac{4\varepsilon_0 R^2}{q\rho}$

28 An inclined plane making an angle of 30° with the horizontal is placed in a uniform horizontal electric field 200 N/C as shown in the figure. A body of mass 1 kg and charge 5 mC is allowed to slide down from rest at a height of 1m. If the coefficient of friction is 0.2, find the time taken by the body to reach the bottom.

[Take, $g = 9.8$ m/s^2, $\sin 30° = \dfrac{1}{2}$, $\cos 30° = \dfrac{\sqrt{3}}{2}$] *[2021, 26 Feb Shift-II]*

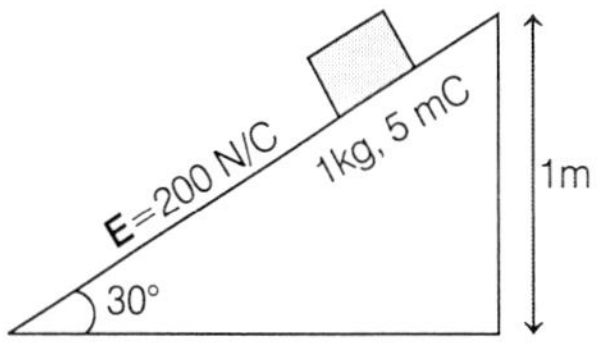

(a) 0.92 s (b) 0.46 s
(c) 2.3 s (d) 1.3 s

29 Find the electric field at point P (as shown in figure) on the perpendicular bisector of a uniformly charged thin wire of length L carrying a charge Q. The distance of the point P from the centre of the rod is $a = \frac{\sqrt{3}}{2}L$. **[2021, 26 Feb Shift I]**

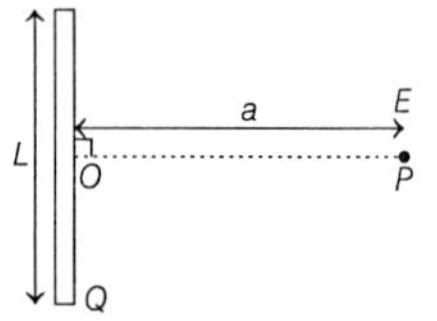

(a) $\frac{\sqrt{3}Q}{4\pi\varepsilon_0 L^2}$ (b) $\frac{Q}{3\pi\varepsilon_0 L^2}$ (c) $\frac{Q}{2\sqrt{3}\pi\varepsilon_0 L^2}$ (d) $\frac{Q}{4\pi\varepsilon_0 L^2}$

30 Consider a sphere of radius R which carries a uniform charge density ρ. if a sphere of radius $R/2$ is carved out of it, as shown, the ratio $\frac{|\mathbf{E}_A|}{|\mathbf{E}_B|}$ of magnitude of electric field $\mathbf{E}_A$ and $\mathbf{E}_B$, respectively, at points A and B due to the remaining portion is **[2020, 9 Jan Shift-I]**

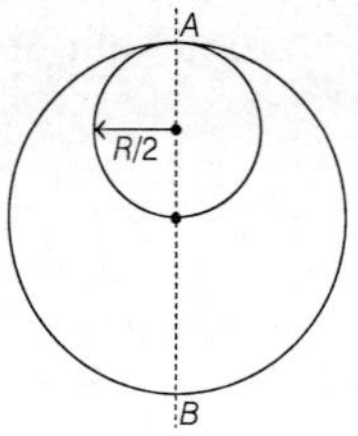

(a) $\frac{21}{34}$ (b) $\frac{18}{54}$ (c) $\frac{17}{54}$ (d) $\frac{18}{34}$

Imp. Electric Field Intensity Due to Electric Dipole

31 Four point charges $-q, +q, +q$ and $-q$ are placed on Y-axis at $y=-2d$, $y=-d$, $y=+d$ and $y=+2d$, respectively. The magnitude of the electric field E at a point on the X-axis at $x=D$, with $D>>d$, will behave as **[2019, 9 April Shift-II]**

(a) $E \propto \frac{1}{D}$ (b) $E \propto \frac{1}{D^3}$ (c) $E \propto \frac{1}{D^2}$ (d) $E \propto \frac{1}{D^4}$

Numerical Value Type Questions

32 An infinite plane sheet of charge having uniform surface charge density $+\sigma_s$ C/m^2 is placed on XY-plane. Another infinitely long line charge having uniform linear charge density $+\lambda_c$ C/m is placed at $z=4$ m plane and parallel to Y-axis. If the magnitude values $|\sigma_s|=2|\lambda_c|$ then at point (0, 0, 2), the ratio of magnitudes of electric field values due to sheet charge to that of line charge is $\pi\sqrt{n}:1$. The value of n is… . **[2024, 4 April Shift-I]**

33 An electron is moving under the influence of the electric field of a uniformly charged infinite plane sheet S having surface charge density $+\sigma$. The electron at $t=0$ is at a distance of 1 m from S and has a speed of 1 m/s. The maximum value of σ if the electron strikes S at $t=1$ s is $\alpha\left[\frac{m\varepsilon_0}{e}\right]\frac{\text{C}}{\text{m}^2}$, the value of α is ………. **[2024, 29 Jan Shift-I]**

34 A thin metallic wire having cross-sectional area of 10^{-4} m^2 is used to make a ring of radius 30 cm. A positive charge of 2π C is uniformly distributed over the ring, while another positive charge of 30 pC is kept at the centre of the ring. The tension in the ring is ……… N, provided that the ring does not get deformed (neglect the influence of gravity).

$\left(\text{given, } \frac{1}{4\pi\varepsilon_0} = 9 \times 10^9 \text{ SI units}\right)$ **[2024, 27 Jan Shift-I]**

35 A uniform electric field of 10 N/C is created between two parallel charged plates (as shown in figure). An electron enters the field symmetrically between the plates with a kinetic energy 0.5 eV. The length of each plate is 10 cm. The angle (θ) of deviation of the path of electron as it comes out of the field is …… . (in degree) **[2023, 25 Jan Shift-I]**

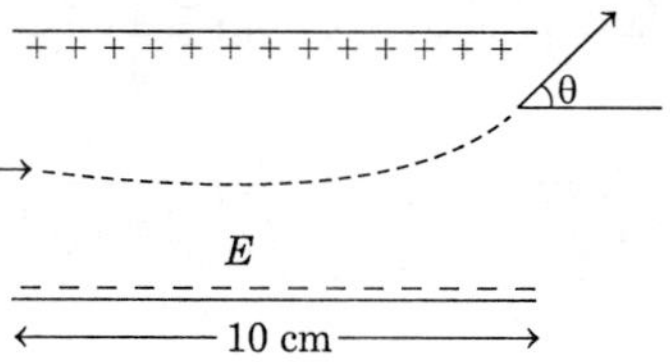

36 A stream of a positively charged particles having $\frac{q}{m} = 2 \times 10^{11}$ C/kg and velocity $\mathbf{v}_0 = 3 \times 10^7\,\hat{\mathbf{i}}$ m/s is deflected by an electric field $1.8\hat{\mathbf{j}}$ kV/m. The electric field exists in a region of 10 cm along x-direction. Due to the electric field, the deflection of the charge particles in the y-direction is ……… mm **[2023, 24 Jan Shift-I]**

Topic 3 Electric Dipole

Level 1

01 Two small spherical balls of mass 10 g each with charges -2μC and 2μC, are attached to two ends of very light rigid rod of length 20 cm. The arrangement is now placed near an infinite non-conducting charge sheet with uniform charge density of 100μC/m^2 such that length of rod makes an angle of 30° with electric field generated by charge sheet. Net torque acting on the rod is

(Take ε_0, 8.85×10^{-12} C^2/Nm^2) **[2025, 4 April Shift-I]**

(a) 2.24 Nm (b) 1.12 Nm
(c) 112 Nm (d) 11.2 Nm

02 Two point charges $-4\ \mu C$ and $4\ \mu C$ constituting an electric dipole, are placed at $(-9, 0, 0)$ cm and $(9, 0, 0)$ cm in a uniform electric field of strength $10^4\ NC^{-1}$. The work done on the dipole in rotating it from the equilibrium through 180° is *[2025, 23 Jan Shift-II]*
(a) 12.4 mJ (b) 16.4 mJ (c) 14.4 mJ (d) 18.4 mJ

Imp. Torque on Electric Dipole

03 Two charges each of magnitude 0.01 C and separated by a distance of 0.4 mm constitute an electric dipole. If the dipole is placed in an uniform electric field **E** of 10 dyne/C making 30° angle with **E**, the magnitude of torque acting on dipole is *[2023, 13 April Shift-I]*
(a) 4.0×10^{-10} Nm (b) 1.0×10^{-8} Nm
(c) 2.0×10^{-10} Nm (d) 1.5×10^{-9} Nm

04 The electric field due to a short electric dipole at a large distance r from centre of dipole on the equatorial plane varies with distance as *[2023, 15 April Shift-I]*
(a) $\frac{1}{r^2}$ (b) $\frac{1}{r^3}$ (c) $\frac{1}{r}$ (d) r

05 Two electric dipoles, A, B with respective dipole moments $\mathbf{d}_A = -4qa\,\hat{i}$ and $\mathbf{d}_B = -2qa\,\hat{i}$ are placed on the X-axis with a separation R, as shown in the figure

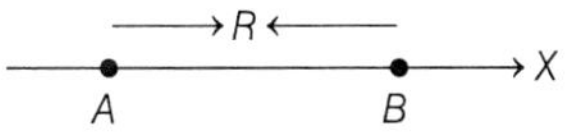

The distance from A at which both of them produce the same potential is *[2019, 10 Jan Shift-I]*
(a) $\frac{\sqrt{2}R}{\sqrt{2}+1}$ (b) $\frac{\sqrt{2}R}{\sqrt{2}-1}$ (c) $\frac{R}{\sqrt{2}+1}$ (d) $\frac{R}{\sqrt{2}-1}$

06 An electric dipole is placed at an angle of 30° to a non-uniform electric field. The dipole will experience *[AIEEE 2006]*
(a) a translational force only in the direction of the field
(b) a translational force only in a direction normal to the direction of the field
(c) a torque as well as a translational force
(d) a torque only

Level 2

07 An electric dipole of mass m, charge q and length l is placed in a uniform electric field $\mathbf{E} = E_0\hat{i}$. When the dipole is rotated slightly from its equilibrium position and released, the time period of its oscillations will be *[2025, 29 Jan Shift-I]*
(a) $\frac{1}{2\pi}\sqrt{\frac{2ml}{qE_0}}$ (b) $2\pi\sqrt{\frac{ml}{2qE_0}}$
(c) $2\pi\sqrt{\frac{ml}{qE_0}}$ (d) $\frac{1}{2\pi}\sqrt{\frac{ml}{2qE_0}}$

08 Given below are two statements.

Statement I A point charge is brought in an electric field. The value of electric field at a point near to the charge may increase, if the charge is positive.

Statement II An electric dipole is placed in a non-uniform electric field. The net electric force on the dipole will not be zero.

Choose the correct answer from the options given below. *[2022, 28 June Shift-I]*
(a) Both Statement I and Statement II are true.
(b) Both Statement I and Statement I are false.
(c) Statement I is true but Statement II is false.
(d) Statement I is false but Statement II is true.

09 Determine the electric dipole moment of the system of three charges, placed on the vertices of an equilateral triangle as shown in the figure. *[2019, 12 Jan Shift-I]*

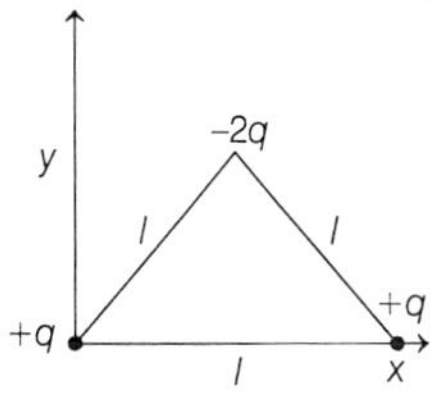

(a) $\sqrt{3}\,ql\,\frac{\hat{j}-\hat{i}}{\sqrt{2}}$ (b) $2ql\hat{j}$ (c) $-\sqrt{3}\,ql\,\hat{j}$ (d) $(ql)\frac{\hat{i}+\hat{j}}{\sqrt{2}}$

Level 3

10 A point particle of charge Q is located at P along the axis of an electric dipole 1 at a distance r as shown in the figure. The point P is also on the equatorial plane of a second electric dipole 2 at a distance r. The dipoles are made of opposite charge q separated by a distance $2a$. For the charge particle at P not to experience any net force, which of the following correctly describes the situation? *[2025, 23 Jan Shift-I]*

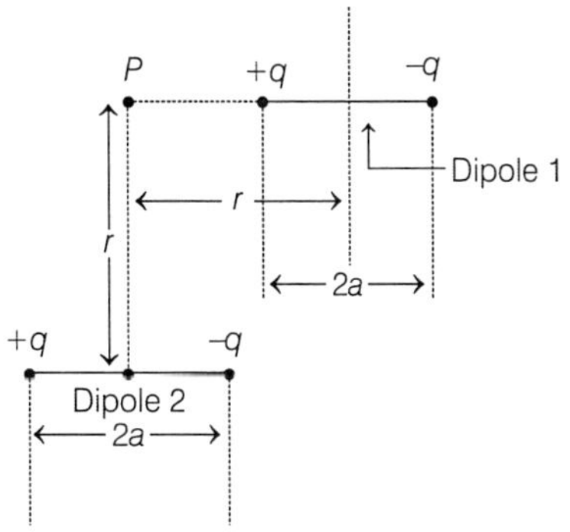

(a) $\frac{a}{r} \sim 20$ (b) $\frac{a}{r} \sim 10$ (c) $\frac{a}{r} \sim 0.5$ (d) $\frac{a}{r} \sim 3$

11 A dipole comprises of two charged particles of identical magnitude q and opposite in nature. The mass m of the positive charged particle is half of the mass of the negative charged particle. The two charges are separated by a distance l. If the dipole is placed in a uniform electric field E; in such a way that dipole axis makes a very small angle with the electric field, E. The angular frequency of the oscillations of the dipole when released is given by **[2023, 6 April Shift-II]**

(a) $\sqrt{\frac{8qE}{ml}}$ (b) $\sqrt{\frac{4qE}{3ml}}$ (c) $\sqrt{\frac{8qE}{3ml}}$ (d) $\sqrt{\frac{3qE}{2ml}}$

12 An electric dipole is formed by two equal and opposite charges q with separation d. The charges have same mass m. It is kept in a uniform electric field E. If it is slightly rotated from its equilibrium orientation, then its angular frequency ω is **[2019, 8 April Shift-II]**

(a) $\sqrt{\frac{2qE}{md}}$ (b) $2\sqrt{\frac{qE}{md}}$

(c) $\sqrt{\frac{qE}{md}}$ (d) $\sqrt{\frac{3qE}{2mL}}$

13 An electric dipole has a fixed dipole moment $\mathbf{p}$, which makes angle θ with respect to X-axis. When subjected to an electric field $\mathbf{E}_1 = E\hat{\mathbf{i}}$, it experiences a torque $\mathbf{T}_1 = \tau\hat{\mathbf{k}}$. When subjected to another electric field $\mathbf{E}_2 = \sqrt{3}E_1\hat{\mathbf{j}}$, it experiences a torque $\mathbf{T}_2 = -\mathbf{T}_1$. The angle θ is **[JEE Main 2017]**

(a) 45° (b) 60°

(c) 90° (d) 30°

Numerical Value Type Questions

14 An electric dipole of dipole moment 6×10^{-6} Cm is placed in uniform electric field of magnitude 10^6 V/m. Initially, the dipole moment is parallel to electric field. The work that needs to be done on the dipole to make its dipole moment opposite to the field, will be.......J. **[2025, 28 Jan Shift-II]**

15 The electric field at point P due to an electric dipole is E. The electric field at point R on equitorial line will be E/x. The value of x is ...... **[2024, 5 April Shift-II]**

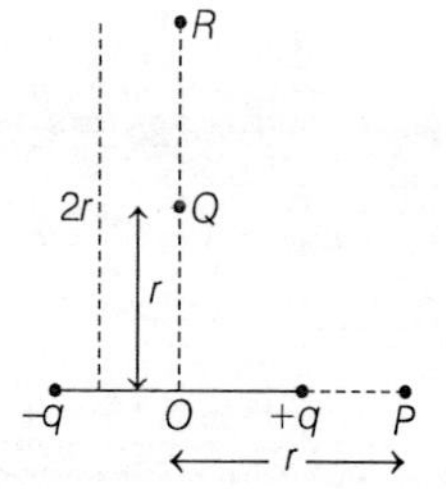

16 If the net electric field at point P along Y-axis is zero, then the ratio of $\left|\frac{q_2}{q_3}\right|$ is $\frac{8}{5\sqrt{x}}$, where x = ... **[2024, 8 April Shift-II]**

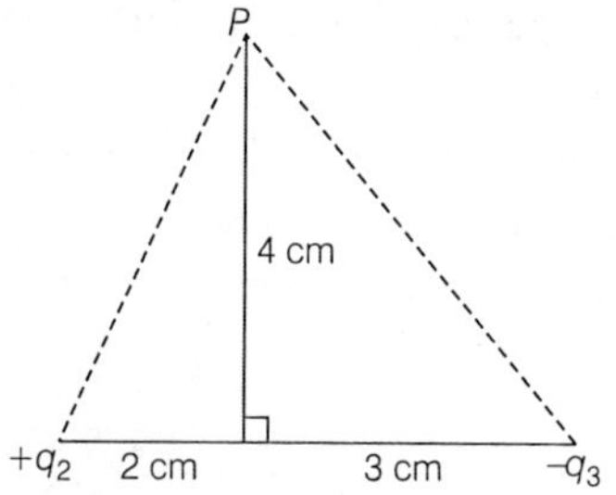

17 Two charges of $-4\,\mu$C and $+4\,\mu$C are placed at the points $A(1, 0, 4)$ m and $B(2, -1, 5)$ m located in an electric field $\mathbf{E} = 0.20\,\hat{\mathbf{i}}$ V/cm. The magnitude of the torque acting on the dipole is $8\sqrt{\alpha} \times 10^{-5}$ N-m, where α = ... **[2024, 27 Jan Shift-II]**

18 Two electric dipoles of dipole moments 1.2×10^{-30} C-m and 2.4×10^{-30} C-m are placed in two different uniform electric fields of strength 5×10^4 NC^{-1} and 15×10^4 NC^{-1}, respectively. The ratio of maximum torque experienced by the electric dipoles will be $1/x$ The value of x is .... **[2022, 28 July Shift-I]**

Topic 4 Electric Flux and Gauss's Law

Level 1

01 Consider two infinitely large plane parallel conducting plates as shown below. The plates are uniformly charged with a surface charge density $+\sigma$ and -2σ. The force experienced by a point charge $+q$ placed at the mid-point between two plates will be **[2025, 2 April Shift-I]**

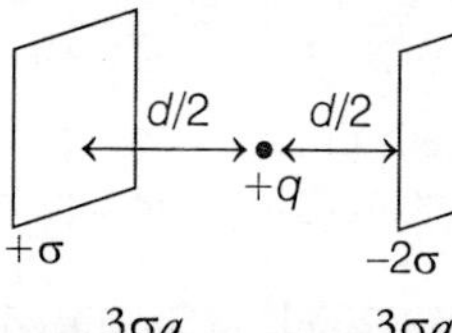

(a) $\frac{\sigma q}{2\varepsilon_0}$ (b) $\frac{3\sigma q}{2\varepsilon_0}$ (c) $\frac{3\sigma q}{4\varepsilon_0}$ (d) $\frac{\sigma q}{4\varepsilon_0}$

02 A point charge causes an electric flux of -2×10^4 Nm^{2}C^{-1} to pass through a spherical gaussian surface of 8.0 cm radius, centred on the charge. The value of the point charge is (Given, $\varepsilon_0 = 8.85 \times 10^{-12}$C^{2}N^{-1}m^{-2}) **[2025, 29 Jan Shift-II]**

(a) 15.7×10^{-8} C (b) 17.7×10^{-8} C

(c) -15.7×10^{-8} C (d) -17.7×10^{-8} C

03 A line charge of length $a/2$ is kept at the centre of an edge BC of a cube $ABCDEFGH$ having edge length a as shown in the figure. If the density of line charge is λC per unit length, then the total electric flux through all the faces of the cube will be .... (Take, ε_0 as the free space permittivity) **[2025, 22 Jan Shift-I]**

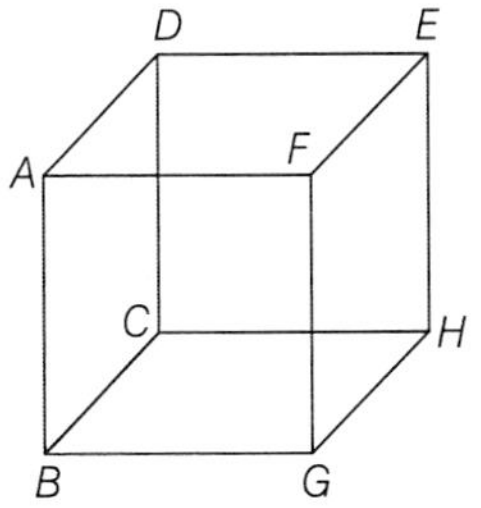

(a) $\frac{\lambda a}{4\varepsilon_0}$ (b) $\frac{\lambda a}{2\varepsilon_0}$ (c) $\frac{\lambda a}{16\varepsilon_0}$ (d) $\frac{\lambda a}{8\varepsilon_0}$

04 Five charges $+q, +5q, -2q, +3q$ and $-4q$ are situated as shown in the figure. The electric flux due to this configuration through the surface S is *[2024, 9 April Shift-II]*

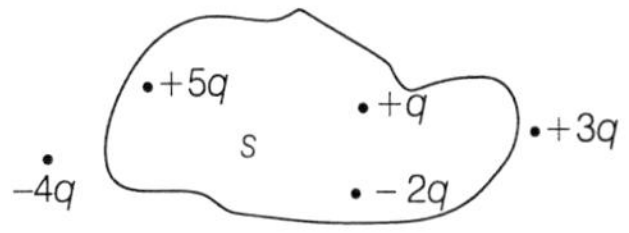

(a) $\frac{3q}{\varepsilon_0}$ (b) $\frac{5q}{\varepsilon_0}$ (c) $\frac{4q}{\varepsilon_0}$ (d) $\frac{q}{\varepsilon_0}$

Imp. Gauss's Law

05 σ is the uniform surface charge density of a thin spherical shell of radius R. The electric field at any point on the surface of the spherical shell is *[2024, 6 April Shift-I]*

(a) $\frac{\sigma}{4\varepsilon_0}$ (b) $\frac{\sigma}{\varepsilon_0}$ (c) $\frac{\sigma}{\varepsilon_0 R}$ (d) $\frac{\sigma}{2\varepsilon_0}$

06 A charge q is placed at the centre of one of the surface of a cube. The flux linked with the cube is *[2024, 4 April Shift-II]*

(a) $\frac{q}{8\varepsilon_0}$ (b) zero (c) $\frac{q}{2\varepsilon_0}$ (d) $\frac{q}{4\varepsilon_0}$

07 An electric field is given by $(6\hat{\mathbf{i}} + 5\hat{\mathbf{j}} + 3\hat{\mathbf{k}})$ N/C. The electric flux through a surface area $30\hat{\mathbf{i}}\ \text{m}^2$ lying in YZ-plane (in SI unit) is *[2024, 29 Jan Shift-II]*

(a) 90 (b) 150 (c) 60 (d) 180

08 Two charges of $5Q$ and $-2Q$ are situated at the points $(3a, 0)$ and $(-5a, 0)$, respectively. The electric flux through a sphere of radius $4a$ having centre at origin is *[2024, 29 Jan Shift-I]*

(a) $\frac{7Q}{\varepsilon_0}$ (b) $\frac{2Q}{\varepsilon_0}$ (c) $\frac{3Q}{\varepsilon_0}$ (d) $\frac{5Q}{\varepsilon_0}$

09 As shown in the figure, a point charge Q is placed at the centre of conducting spherical shell of inner radius a and outer radius b. The electric field due to charge Q in three different regions I, II and III is given by

(I : $r < a$, II : $a < r < b$, III : $r > b$ *[2023, 30 Jan Shift-II]*

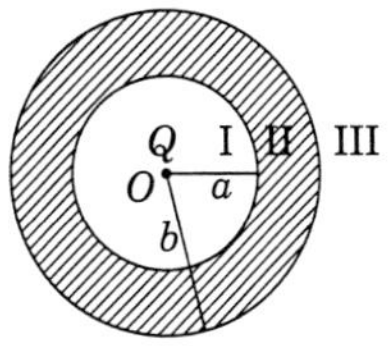

(a) $E_{\text{I}} \neq 0, E_{\text{II}} = 0, E_{\text{III}} = 0$
(b) $E_{\text{I}} = 0, E_{\text{II}} = 0, E_{\text{III}} \neq 0$
(c) $E_{\text{I}} \neq 0, E_{\text{II}} = 0, E_{\text{III}} \neq 0$
(d) $E_{\text{I}} = 0, E_{\text{II}} = 0, E_{\text{III}} = 0$

10 If a charge q is placed at the centre of a closed hemispherical non-conducting surface, the total flux passing through the flat surface would be *[2022, 27 June Shift-II]*

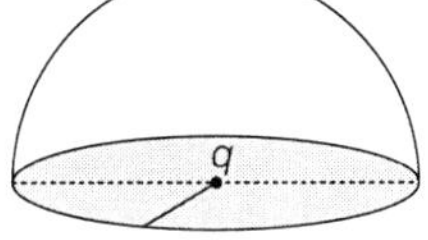

(a) $\frac{q}{\varepsilon_0}$ (b) $\frac{q}{2\varepsilon_0}$ (c) $\frac{q}{4\varepsilon_0}$ (d) $\frac{q}{2\pi\varepsilon_0}$

11 In the figure, a very large plane sheet of positive charge is shown. P_1 and P_2 are two points at distance l and $2l$ from the charge distribution. If σ is the surface charge density, then the magnitude of electric fields E_1 and E_2 at P_1 and P_2 respectively, are *[2022, 25 June Shift-I]*

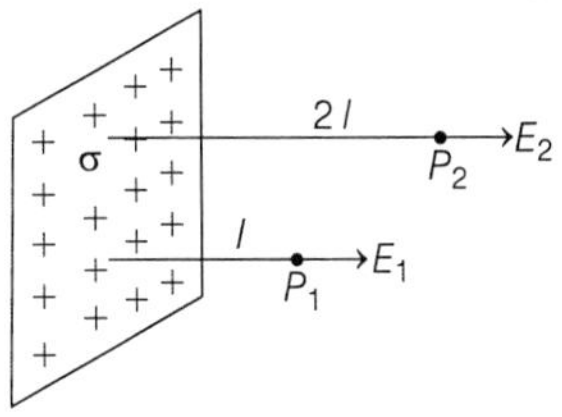

(a) $E_1 = \sigma/\varepsilon_0, E_2 = \sigma/2\varepsilon_0$ (b) $E_1 = 2\sigma/\varepsilon_0, E_2 = \sigma/\varepsilon_0$
(c) $E_1 = E_2 = \sigma/2\varepsilon_0$ (d) $E_1 = E_2 = \sigma/\varepsilon_0$

12 Given below are two statements, one is labelled as Assertion (A) and the other is labelled as Reason (R).

Assertion (A) If an electric dipole of dipole moment 30×10^{-5} C-m is enclosed by a closed surface, the net flux coming out of the surface will be zero.

Reason (R) Electric dipole consists of two equal and opposite charges.

In the light of the above statements, choose the most appropriate answer from the options given below. *[2021, 12 April Shift-I]*

(a) Both A and R are true and R is the correct explanation of A.
(b) A is false but R is true.
(c) Both A and R are true but R is not the correct explanation of A.
(d) A is true but R is false.

13 Given below are two statements.

Statement I An electric dipole is placed at the centre of a hollow sphere. The flux of electric field through the sphere is zero but the electric field is not zero anywhere in the sphere.

Statement II If R is the radius of a solid metallic sphere and Q be the total charge on it. The electric field at any point on the spherical surface of radius r ($< R$) is zero but the electric flux passing through this closed spherical surface of radius r is not zero.

In the light of the above statements choose the correct answer from the options given below. ***[2021, 26 Feb Shift-II]***

(a) Both Statement I and Statement II are true.
(b) Statement I is true but Statement II is false.
(c) Both Statement I and Statement II are false.
(d) Statement I is false but Statement II is true.

14 In finding the electric field using Gauss law the formula $|\mathbf{E}| = \dfrac{q_{enc}}{\varepsilon_0 |A|}$ is applicable. In the formula, ε_0 is permittivity of free space, A is the area of Gaussian surface and q_{enc} is charge enclosed by the Gaussian surface. This equation can be used in which of the following situation? ***[2020, 8 Jan Shift-I]***

(a) Only when the Gaussian surface is an equipotential surface and $|\mathbf{E}|$ is constant on the surface.
(b) Only when the Gaussian surface is an equipotential surface.
(c) For any choice of Gaussian surface.
(d) Only when $|\mathbf{E}| =$ constant on the surface.

15 This question has Statement I and Statement II. Of the four choices given after the statements, choose the one that best describes the two statements.

An insulating solid sphere of radius R has a uniform positive charge density ρ. As a result of this uniform charge distribution, there is a finite value of electric potential at the centre of the sphere, at the surface of the sphere and also at a point outside the sphere. The electric potential at infinite is zero.

Statement I When a charge q is taken from the centre to the surface of the sphere, its potential energy changes by $q\rho/3\varepsilon_0$.

Statement II The electric field at a distance r ($r < R$) from the centre of the sphere is $\rho r / 3\varepsilon_0$.

[AIEEE 2012]

(a) Statement I is false, Statement II is true.
(b) Statement I is true, Statement II is false.
(c) Statement I is true, Statement II is true, Statement II the correct explanation for Statement I.
(d) Statement I is true, Statement II is true, Statement II is not the correct explanation of Statement I.

16 A thin spherical shell of radius R has charge Q spread uniformly over its surface. Which of the following graphs most closely represents the electric field $E(r)$ produced by the shell in the range $0 \le r < x$, where r is the distance from the centre of the shell? ***[AIEEE 2008]***

(a)
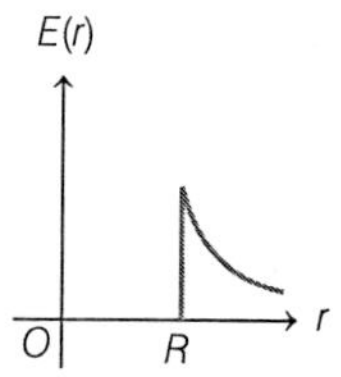

(b)
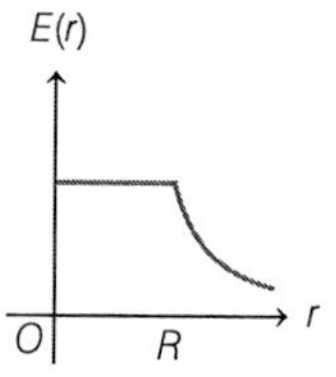

(c)
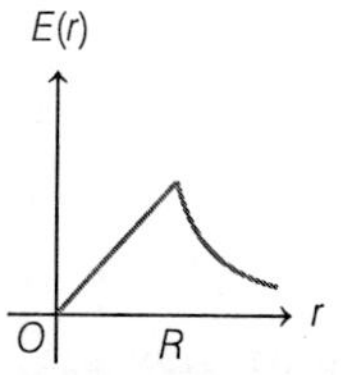

(d)
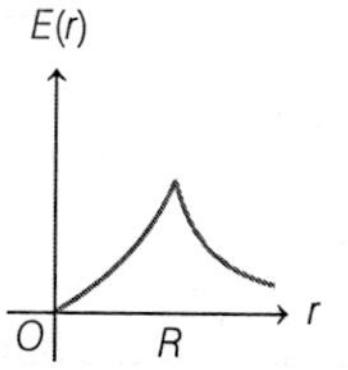

17 If the electric flux entering and leaving an enclosed surface respectively is ϕ_1 and ϕ_2, the electric charge inside the surface will be ***[AIEEE 2003]***

(a) $(\phi_2 - \phi_1)\varepsilon_0$ (b) $\dfrac{(\phi_1 + \phi_2)}{\varepsilon_0}$ (c) $\dfrac{(\phi_2 - \phi_1)}{\varepsilon_0}$ (d) $(\phi_1 + \phi_2)\varepsilon_0$

Level 2

18 A metallic ring is uniformly charged as shown in figure. AC and BD are two mutually perpendicular diameters. Electric field due to arc AB at O is E in magnitude. What would be the magnitude of electric field at O due to arc ABC? ***[2025, 4 April Shift-II]***

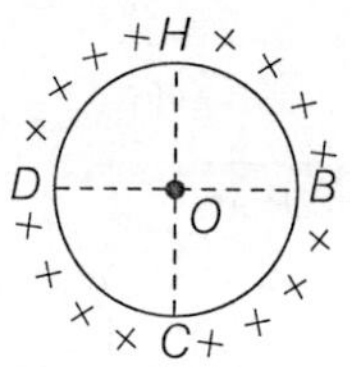

(a) $\sqrt{2}E$ (b) $2E$ (c) Zero (d) $E/2$

19 A small bob of mass 100 g and charge + 10 μC is connected to an insulating string of length 1 m. It is brought near to an infinitely long non-conducting sheet of charge density σ as shown in figure. If string subtends an angle of 45° with the sheet at equilibrium the charge density of sheet will be.

(Given, $\varepsilon_0 = 8.85 \times 10^{-12} \dfrac{\text{F}}{\text{m}}$ and acceleration due to gravity, $g = 10 \text{ m/s}^2$) ***[2025, 2 April Shift-I]***

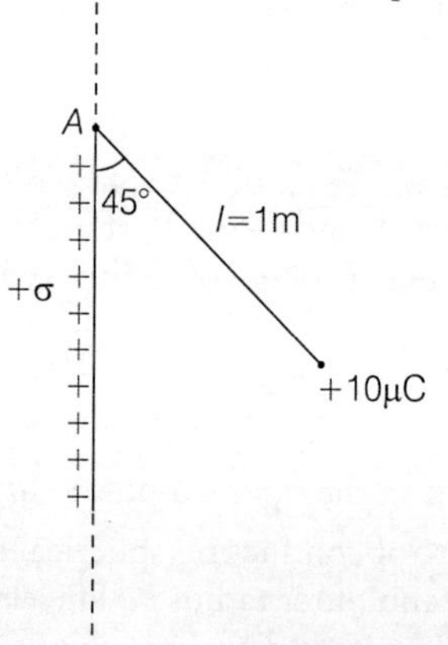

(a) 885 nC/m^2
(b) 0.885 nC/m^2
(c) 1.77 nC/m^2
(d) 17.7 nC/m^2

20 Two infinite identical charged sheets and a charged spherical body of charge density ρ are arranged as shown in figure. Then the correct relation between the electrical fields at A, B, C and D points is *[2025, 4 April Shift-I]*

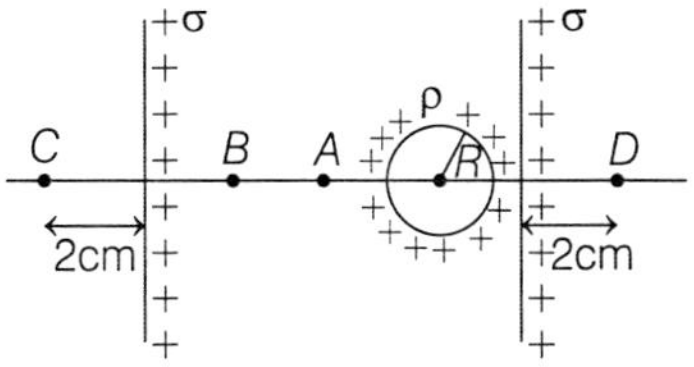

(a) $E_A > E_B$, $E_C = E_D$ (b) $E_A = E_B$, $E_C = E_D$
(c) $|E_A| = |E_B|$, $E_C > E_D$ (d) $E_C \neq E_D$, $E_A > E_B$

21 Given below are two statements. One is labelled as Assertion (A) and the other is labelled as Reason (R).

Assertion (A) The outer body of an air craft is made of metal which protects persons sitting inside from lightning-strikes.

Reason (R) The electric field inside the cavity enclosed by a conductor is zero. *[2025, 7 April Shift-II]*

In the light of the above statements, choose the most appropriate answer from the options given below:
(a) Both A and R are correct but R is not the correct explanation of A.
(b) A is not correct but R is correct.
(c) Both A and R are correct and R is the correct explanation of A.
(d) A is correct but R is not correct.

22 Match List-I with List-II. *[2025, 29 Jan Shift-I]*

List-I		List-II
(A) Electric field inside (distance $r > 0$ from centre) of a uniformly charged spherical shell with surface charge density σ and radius R.	(I)	σ / ε_0
(B) Electric field at distance $r > 0$ from a uniformly charged infinite plane sheet with surface charge density σ.	(II)	$\sigma / 2\varepsilon_0$
(C) Electric field outside (distance $r > 0$ from centre) of a uniformly charged spherical shell with surface charge density σ and radius R.	(III)	0
(D) Electric field between two oppositely charged infinite plane parallel sheets with uniform surface charge density σ.	(IV)	$\dfrac{\sigma}{\varepsilon_0 r^2}$

Choose the correct answer from the options given below.
(a) (A)-(II), (B)-(I), (C)-(IV), (D)-(III)
(b) (A)-(III), (B)-(II), (C)-(IV), (D)-(1)
(c) (A)-(IV), (B)-(II), (C)-(III), (D)-(I)
(d) (A)-(IV), (B)-(I), (C)-(III), (D)-(II)

23 C_1 and C_2 are two hollow concentric cubes enclosing charges $2Q$ and $3Q$ respectively as shown in figure. The ratio of electric flux passing through C_1 and C_2 is *[2024, 1 Feb Shift-II]*

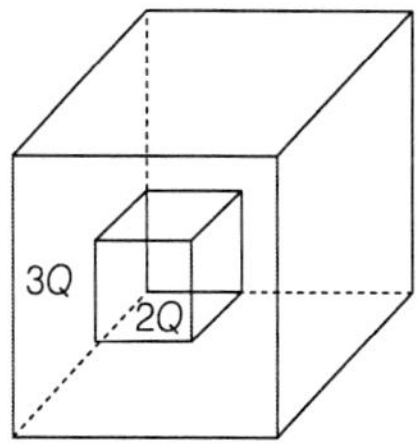

(a) 2 : 3 (b) 3 : 2 (c) 5 : 2 (d) 2 : 5

24 Let σ be the uniform surface charge density of two infinite thin plane sheets shown in figure. Then the electric fields in three different region E_{I}, E_{II} and E_{III} are *[2023, 1 Feb Shift-I]*

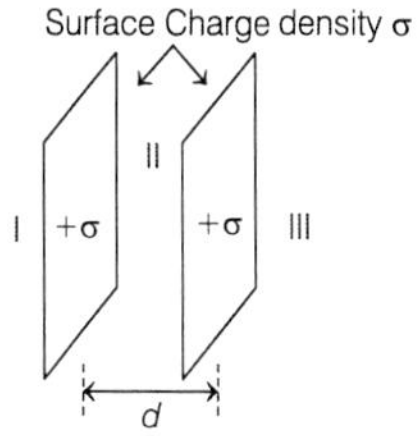

(a) $E_{\text{I}} = \dfrac{\sigma}{\varepsilon_0}\hat{n}$, $E_{\text{II}} = 0$, $E_{\text{III}} = \dfrac{\sigma}{\varepsilon_0}\hat{n}$

(b) $E_{\text{I}} = \dfrac{\sigma}{2\varepsilon_0}\hat{n}$, $E_{\text{II}} = 0$, $E_{\text{III}} = \dfrac{\sigma}{2\varepsilon_0}\hat{n}$

(c) $E_{\text{I}} = 0$, $E_{\text{II}} = \dfrac{\sigma}{\varepsilon_0}\hat{n}$, $E_{\text{III}} = 0$

(d) $E_{\text{I}} = \dfrac{2\sigma}{\varepsilon_0}\hat{n}$, $E_{\text{II}} = 0$, $E_{\text{III}} = \dfrac{2\sigma}{\varepsilon_0}\hat{n}$

Imp. Application of Gauss's Law

25 In a cuboid of dimension $2L \times 2L \times L$, a charge q is placed at the centre of the surface S having area of $4L^2$. The flux through the opposite surface to S is given by *[2023, 29 Jan Shift-I]*

(a) $\dfrac{q}{2\varepsilon_0}$ (b) $\dfrac{q}{12\varepsilon_0}$

(c) $\dfrac{q}{6\varepsilon_0}$ (d) $\dfrac{q}{3\varepsilon_0}$

26 A spherically symmetric charge distribution is considered with charge density varying as *[2022, 29 July Shift-I]*

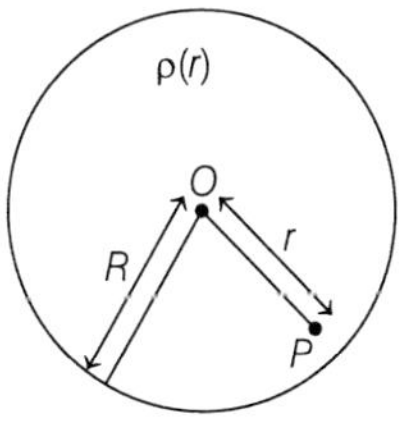

$$\rho(r) = \begin{cases} \rho_0\left(\dfrac{3}{4} - \dfrac{r}{R}\right) & \text{for } r \leq R, \\ & \text{For } z > R \end{cases}$$

where, $r(r < R)$ is the distance from the centre O (as shown in figure). The electric field at point P will be

(a) $\dfrac{\rho_0 r}{4\varepsilon_0}\left(\dfrac{3}{4} - \dfrac{r}{R}\right)$ (b) $\dfrac{\rho_0 r}{3\varepsilon_0}\left(\dfrac{3}{4} - \dfrac{r}{R}\right)$

(c) $\dfrac{\rho_0 r}{4\varepsilon_0}\left(1 - \dfrac{r}{R}\right)$ (d) $\dfrac{\rho_0 r}{5\varepsilon_0}\left(1 - \dfrac{r}{R}\right)$

27 A charge q is placed at one corner of a cube as shown in figure. The flux of electrostatic field E through the shaded area is *[2021, 25 Feb Shift-II]*

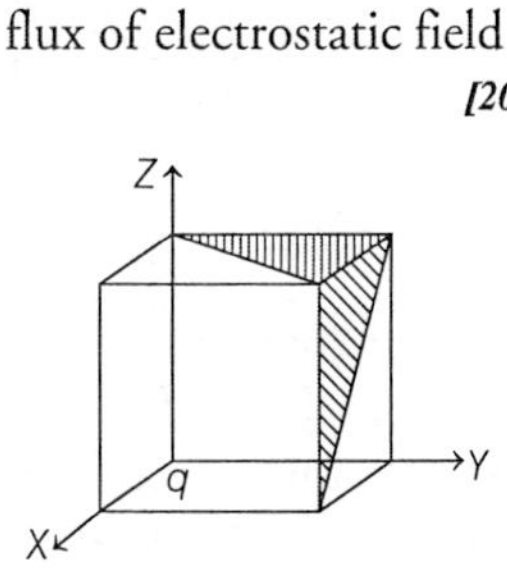

(a) $\frac{q}{48\varepsilon_0}$ (b) $\frac{q}{4\varepsilon_0}$ (c) $\frac{q}{8\varepsilon_0}$ (d) $\frac{q}{24\varepsilon_0}$

28 Shown in the figure is a shell made of a conductor. It has inner radius a and outer radius b and carries charge Q. At its centre is a dipole **p** as shown. In this case, *[2019, 12 April Shift-I]*

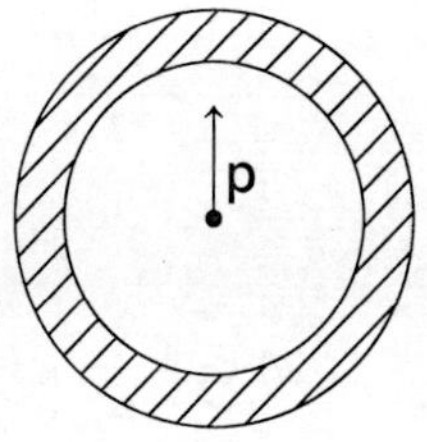

(a) surface charge density on the inner surface is uniform and equal to $\frac{(Q/2}{4\pi a^2}$

(b) electric field outside the shell is the same as that of a point charge at the centre of the shell

(c) surface charge density on the outer surface depends on $|\mathbf{p}|$

(d) surface charge density on the inner surface of the shell is zero everywhere

29 In a uniformly charged sphere of total charge Q and radius R, the electric field E is plotted as function of distance from the centre. The graph which would correspond to the above will be *[AIEEE 2012]*

(a)

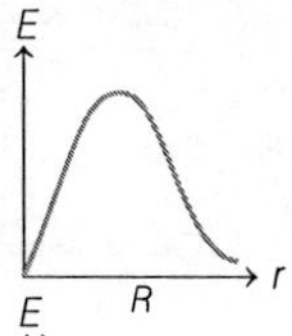

(c)

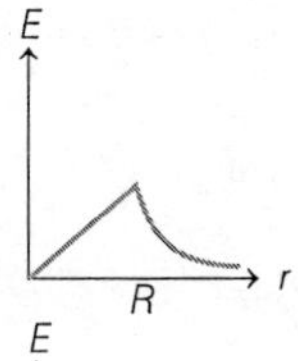

(b)

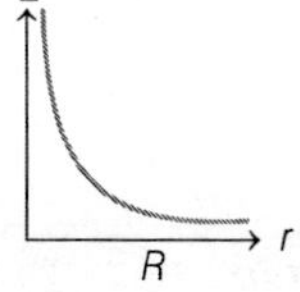

(d)

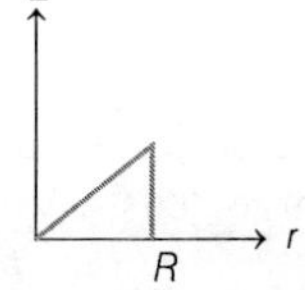

30 The electrostatic potential inside a charged spherical ball is given by $\phi = ar^2 + b$, where r is the distance from the centre and a, b are constants. Then, the charge density inside the ball is *[AIEEE 2011]*

(a) $-6a\,\varepsilon_0 r$ (b) $-24\,\pi a\,\varepsilon_0$
(c) $-6a\,\varepsilon_0$ (d) $-24\pi a\,\varepsilon_0 r$

31 Let there be a spherically symmetric charge distribution with charge density varying as

$\rho(r) = \rho_0\left(\frac{5}{4} - \frac{r}{R}\right)$ upto $r = R$ and $\rho(r) = 0$ for $r > R$,

where r is the distance from the origin. The electric field at a distance r, $(r < R)$ from the origin is given by *[AIEEE 2010]*

(a) $\frac{4\pi\rho_0 r}{3\varepsilon_0}\left(\frac{5}{3} - \frac{r}{R}\right)$ (b) $\frac{\rho_0 r}{4\varepsilon_0}\left(\frac{5}{3} - \frac{r}{R}\right)$

(c) $\frac{4\rho_0 r}{3\varepsilon_0}\left(\frac{5}{4} - \frac{r}{R}\right)$ (d) $\frac{\rho_0 r}{3\,\varepsilon_0}\left(\frac{5}{4} - \frac{r}{R}\right)$

32 Let $\rho(r) = \frac{Q}{\pi R^4} r$ be the charge density distribution for a solid sphere of radius R and total charge Q. For a point P inside the sphere at distance r_1 from the centre of the sphere, the magnitude of electric field is *[AIEEE 2009]*

(a) zero (b) $\frac{Q}{4\pi\varepsilon_0 r_1^2}$ (c) $\frac{Qr_1^2}{4\pi\varepsilon_0 R^4}$ (d) $\frac{Qr_1^2}{3\pi\varepsilon_0 R^4}$

Level 3

33 An infinitely long wire has uniform linear charge density $\lambda = 2\text{n C/m}$. The net flux through a Gaussian cube of side length $\sqrt{3}$ cm, if the wire passes through any two corners of the cube, that are maximally displaced from each other, would be x Nm^2C^{-1}, where x is

[Neglect any edge effects and use $\frac{1}{4\pi\varepsilon_0} = 9 \times 10^9$ SI units] *[2025, 8 April Shift-II]*

(a) $6.48\,\pi$ (b) $0.72\,\pi$ (c) $1.44\,\pi$ (d) $2.16\,\pi$

34 Let a total charge $2Q$ be distributed in a sphere of radius R with the charge density given by $\rho(r) = kr$, where r is the distance from the centre. Two charges A and B, of charge $-Q$ each, are placed on diametrically opposite points, at equal distance a, from the centre. If A and B do not experience any force, then *[2019, 12 April Shift-II]*

(a) $a = 8^{-1/4} R$ (b) $a = \frac{3R}{2^{1/4}}$

(c) $a = 2^{-1/4} R$ (d) $a = R/\sqrt{3}$

35 The region between two concentric spheres of radii a and b, respectively (see the figure), has volume charge density $\rho = A/r$, where A is a constant and r is the distance from the centre. At the centre of the spheres is a point charge Q. The value of A such that the electric field in the region between the spheres will be constant is *[JEE Main 2016]*

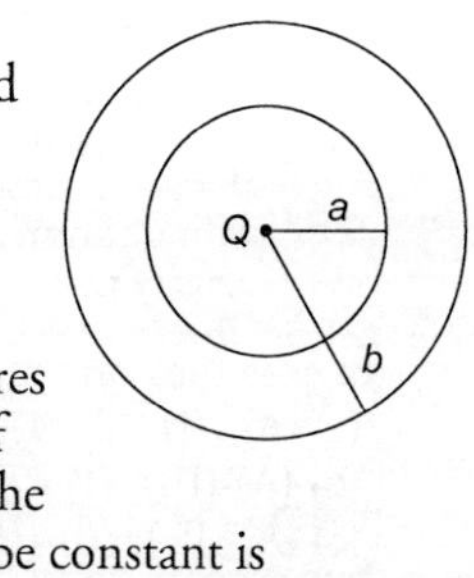

(a) $\frac{Q}{2\pi a^2}$ (b) $\frac{Q}{2\pi(b^2 - a^2)}$

(c) $\frac{2Q}{\pi(a^2 - b^2)}$ (d) $\frac{2Q}{\pi a^2}$

Numerical Value Type Questions

36 The electric field in a region is given by $\mathbf{E} = (2\hat{\mathbf{i}} + 4\hat{\mathbf{j}} + 6\hat{\mathbf{k}}) \times 10^3$ N/C. The flux of the field through a rectangular surface parallel to *xz*-plane is 6.0 $Nm^2 C^{-1}$. The area of the surface is .... cm^2 *[2025, 7 April Shift-II]*

37 A square loop of sides $a = 1$m is held normally in front of a point charge $q = 1$ C. The flux of the electric field through the shaded region is $\frac{5}{p} \times \frac{1}{\varepsilon_0} \frac{Nm^2}{C}$, where the value of p is .... *[2025, 24 Jan Shift-I]*

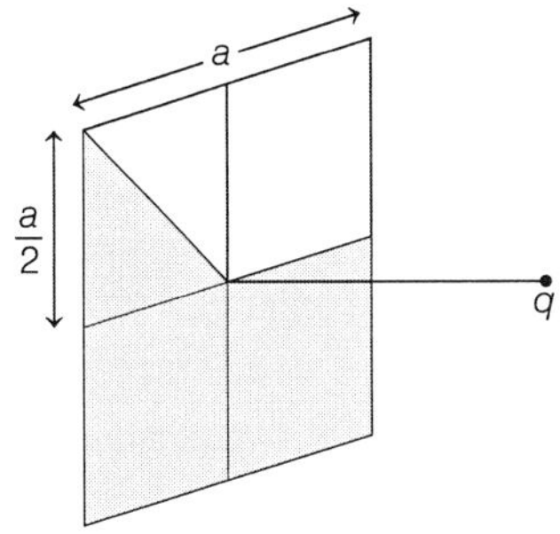

38 An electric field $\mathbf{E} = (2x\hat{\mathbf{i}})$ NC^{-1} exists in space. A cube of side 2m is placed in the space as per figure given below. The electric flux through the cube is .... Nm^2/C. *[2024, 9 April Shift-II]*

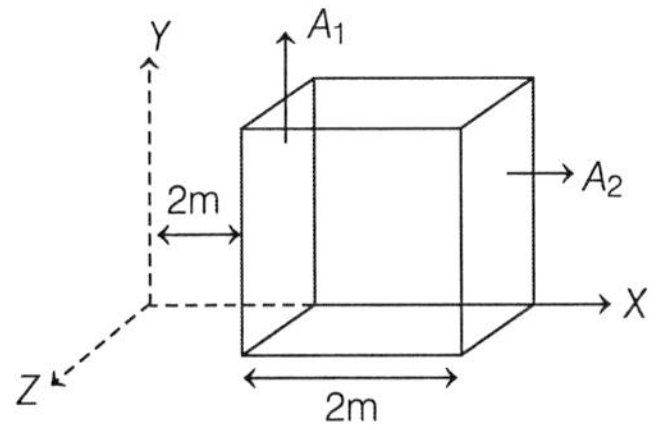

39 An electric field, $\mathbf{E} = \frac{2\hat{\mathbf{i}} + 6\hat{\mathbf{j}} + 8\hat{\mathbf{k}}}{\sqrt{6}}$ passes through the surface of $4\,m^2$ area having unit vector $\hat{\mathbf{n}} = \left(\frac{2\hat{\mathbf{i}} + \hat{\mathbf{j}} + \hat{\mathbf{k}}}{\sqrt{6}}\right)$. The electric flux for that surface is .... V m. *[2024, 8 April Shift-I]*

40 Three infinitely long charged thin sheets are placed as shown in figure. The magnitude of electric field at the point P is $x\sigma/\varepsilon_0$. The value of x is .... (all quantities are measured in SI units.) *[2024, 6 April Shift-I]*

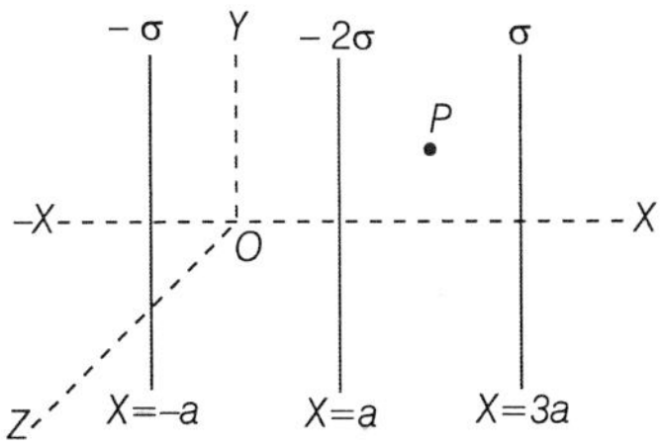

41 A cubical volume is bounded by the surfaces $x = 0$, $x = a$, $y = 0$, $y = a$, $z = 0$, $z = a$. The electric field in the region is given by $\mathbf{E} = E_0 x\hat{\mathbf{i}}$. Where, $E_0 = 4 \times 10^4\, NC^{-1}m^{-1}$. If $a = 2$ cm, the charge contained in the cubical volume is $Q \times 10^{-14}$ C. The value of Q is .... (Take, $\varepsilon_0 = 9 \times 10^{-12} C^2/N\text{-}m^2$) *[2023, Feb Shift-II]*

42 As shown in figure, a cuboid lies in a region with electric field $\mathbf{E} = 2x^2\,\hat{\mathbf{i}} - 4y\hat{\mathbf{j}} + 6\hat{\mathbf{k}}$ N/C. The magnitude of charge within the cuboid is $n\varepsilon_0 C$. The value of n is .... (If dimension of cuboid is $1 \times 2 \times 3\ m^3$) *[2023, 30 Jan Shift-II]*

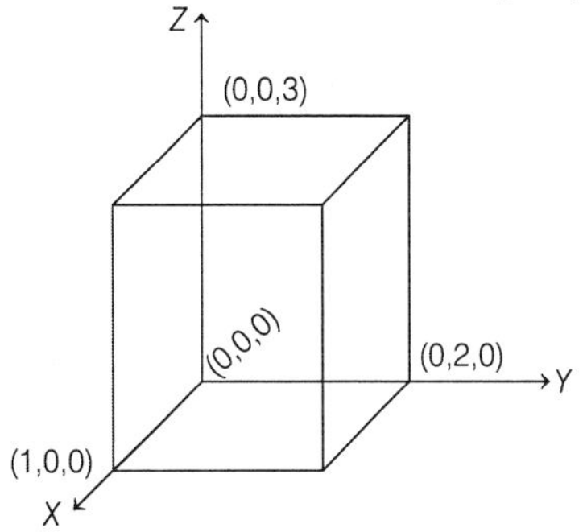

43 Expression for an electric field is given by $\mathbf{E} = 4000x^2\hat{\mathbf{i}}$ V/m. The electric flux through the cube of side 20 cm when placed in electric field (as shown in the figure) is ....V-cm. *[2023, 31 Jan Shift-I]*

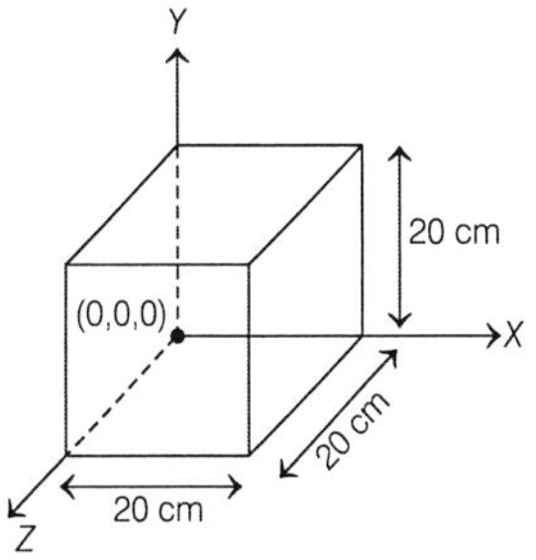

44 An electron revolves around an infinite cylindrical wire having uniform linear charge density 2×10^{-8} C-m^{-1} in circular path under the influence of attractive electrostatic field as shown in the figure. The velocity of electron with which it is revolving is ...... $\times 10^6 ms^{-1}$. Given, mass of electron $= 9 \times 10^{-31}$ kg *[2023, 10 April Shift-II]*

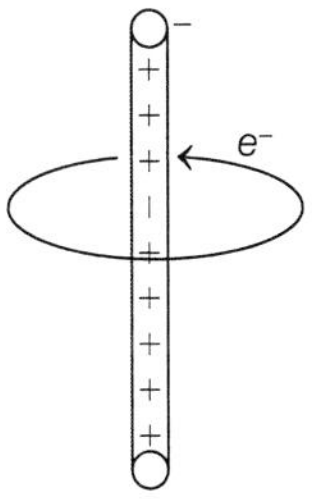

45 A long cylindrical volume contains a uniformly distributed charge of density ρ C-m^{-3}. The electric field inside the cylindrical volume at a distance $x = \frac{2\varepsilon_0}{\rho}$ m from its axis is .... Vm^{-1} [2022, 27 July Shift-I]

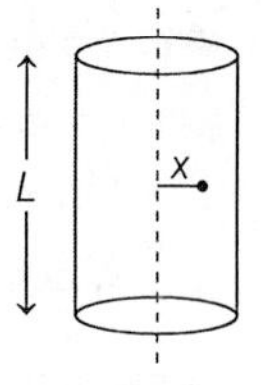

46 The volume charge density of a sphere of radius 6 m is 2 μC cm^{-3}. The number of lines of force per unit surface area coming out from the surface of the sphere is .... $\times 10^{10}$ NC^{-1}.
[Given, permittivity of vacuum, $\varepsilon_0 = 8.85 \times 10^{-12} C^2 N^{-1}$-m^{-2}]. [2022, 25 July shift-I]

47 The total charge enclosed in an incremental volume of $2 \times 10^{-9} m^3$ located at the origin is ...... nC, if electric flux density of its field is found as $D = e^{-x} \sin y\hat{i} - e^{-x} \cos y\hat{j} + 2z\hat{k}$ C/m^2. [2021, 22 July Shift-II]

48 A thin infinite sheet charge and an infinite line charge of respective charge densities $+\sigma$ and $+\lambda$ placed parallel at 5 m distance from each other. Point P and Q are at $3/\pi$ m and $4/\pi$ m perpendicular distance from line charge towards sheet charge, respectively. E_p and E_Q are the magnitudes of resultant electric field intensities at point P and Q, respectively. If $E_P/E_Q = 4/a$ for $2|\sigma| = |\lambda|$, then the value of a is .... [2021, 13 April Shift-I]

49 The electric field in a region is given by $E = \frac{2}{5}E_0\hat{i} + \frac{3}{5}E_0\hat{j}$ with $E_0 = 4.0 \times 10^3$ N/C. The flux of this field through a rectangular surface area 0.4 m^2 parallel to the YZ-plane is .... N-m^2 C^{-1}. [2021, 17 March Shift-II]

50 The electric field in a region is given $E = \left(\frac{3}{5}E_0\hat{i} + \frac{4}{5}E_0\hat{j}\right)\frac{N}{C}$. The ratio of flux of reported field through the rectangular surface of area 0.2 m^2 (parallel to YZ-plane) to that of the surface of area 0.3 m^2 (parallel to XZ-plane) is $a : b$, where a = ...
[Here, $\hat{i}$, $\hat{j}$ and $\hat{k}$ are unit vectors along X, Y and Z-axes, respectively] [2021, 25 Feb Shift-I]

Key **Index**

Topic 1

1 (d)	2 (a)	3 (a)	4 (c)	5 (b)	6 (b)	7 (b)	8 (d)	9 (c)	10 (a)
11 (d)	12 (c)	13 (d)	14 (a)	15 (b)	16 (b)	17 (a)	18 (a)	19 (a)	20 (a)
21 (b)	22 (b)	23 (b)	24 (a)	25 (d)	26 (d)	27 (b)	28 (c)	29 (d)	30 (c)
31 (*)	32 (5440)	33 (300)	34 (2)	35 (24)	36 (17)	37 (6000)	38 (12)		

Topic 2

1 (c)	2 (b)	3 (a)	4 (b)	5 (d)	6 (b)	7 (a)	8 (a)	9 (c)	10 (a)
11 (d)	12 (d)	13 (b)	14 (d)	15 (b)	16 (c)	17 (a)	18 (b)	19 (c)	20 (d)
21 (d)	22 (c)	23 (b)	24 (a)	25 (d)	26 (b)	27 (a)	28 (d)	29 (c)	30 (d)
31 (d)	32 (16)	33 (8)	34 (3)	35 (45)	36 (2)				

Topic 3

1 (b)	2 (c)	3 (c)	4 (b)	5 (a)	6 (c)	7 (b)	8 (a)	9 (c)	10 (*)
11 (d)	12 (a)	13 (b)	14 (12)	15 (16)	16 (5)	17 (2)	18 (6)		

Topic 4

1 (b)	2 (d)	3 (d)	4 (c)	5 (b)	6 (c)	7 (d)	8 (d)	9 (c)	10 (b)
11 (c)	12 (a)	13 (b)	14 (a)	15 (a)	16 (a)	17 (a)	18 (a)	19 (c)	20 (d)
21 (c)	22 (*)	23 (d)	24 (a)	25 (c)	26 (c)	27 (d)	28 (b)	29 (c)	30 (c)
31 (b)	32 (c)	33 (d)	34 (a)	35 (a)	36 (15)	37 (48)	38 (16)	39 (12)	40 (2)
41 (288)	42 (12)	43 (640)	44 (8)	45 (1)	46 (45)	47 (4)	48 (6)	49 (640)	50 (1)

Detailed **Explanations**

Topic 1

01. *(d)* Charge density (σ) is inversely proportional to the radius of curvature (ROC) i.e.

$$\sigma \propto \frac{1}{\text{ROC}}$$

$$\Rightarrow (\text{ROC})_1 < (\text{ROC})_3 < (\text{ROC})_2 = (\text{ROC})_4$$

$$\Rightarrow \sigma_1 > \sigma_3 > \sigma_2 = \sigma_4$$

Thus, statements A, B and C are correct.

02. *(a)*

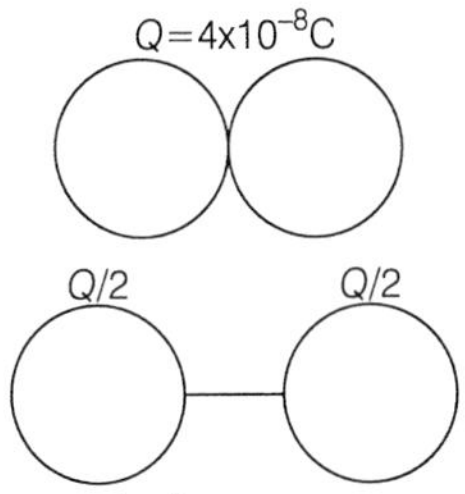

$$F = \frac{k\left(\frac{Q}{2}\right)(Q/2)}{r^2}$$

$$9\times10^{-3} = 9\times10^{9} \times \frac{4\times10^{-8}\times4\times10^{-8}}{4\times r^2}$$

$$r^2 = \frac{9\times10^{9}\times16\times10^{-16}}{4\times9\times10^{-3}} = 4\times10^{-4}$$

$$r = 2\times10^{-2}\text{ m}$$

$$\Rightarrow r = 2\text{ cm}$$

03. *(a)*

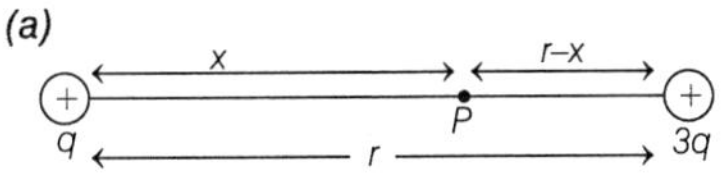

Net electric field at P is zero.

So, $$\frac{kq}{x^2} = \frac{k\cdot3q}{(r-x)^2}$$

$$\Rightarrow \frac{1}{x} = \frac{\sqrt{3}}{r-x}$$

$$\Rightarrow x = \left(\frac{1}{1+\sqrt{3}}\right)\cdot r$$

04. *(c)*

> **key idea** In this question, for maximum value of electrostatic repulsion, you put $\frac{dF}{dq} = 0$

Divide charge, $q = 10\mu\text{C}$ into two parts, $q_1 = x, q_2 = q - x$

$$F = \frac{k(x)(q-x)}{r^2}$$

For F to be maximum,

$$\frac{dF}{dx} = 0 \Rightarrow q - 2x = 0 \Rightarrow x = \frac{q}{2}$$

$$q_1 = x = \frac{q}{2} = \frac{10\mu\text{C}}{2} = 5\mu\text{C}$$

$$q_2 = q - x = 10\mu\text{C} - 5\mu\text{C} = 5\mu\text{C}$$

Option (c) is correct

05. *(b)* Let charge on one of the two charges is $q\,\mu C$, then using coulomb's law,

$$F = \frac{k(q)(4-q)}{r^2}$$

For maximum force, $\frac{dF}{dq} = 0$

$$\Rightarrow \frac{d}{dq}\left[\frac{k(q)(4-q)}{r^2} = 0\right]$$

$$\Rightarrow \frac{d}{dq}[4q - q^2] = 0 \Rightarrow 4 - 2q = 0$$

$$\Rightarrow \quad q = 2\mu\text{C}$$

06. *(b)* Let charge q be placed at P as shown below.

At point P forces due to $20\,\mu\text{C}$ and $-5\,\mu\text{C}$ should be in opposite direction.

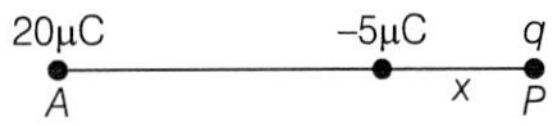

For net force **F = 0** and from Coulomb's law force

$$F = \frac{kq_1q_2}{r^2} \Rightarrow k\frac{20q}{(5+x)^2} = \frac{k5q}{x^2}$$

$$\Rightarrow x = 5\text{ cm}$$

07. *(b)* Let's say the charge q and $(Q-q)$ are at r distance from each other. This can be shown as

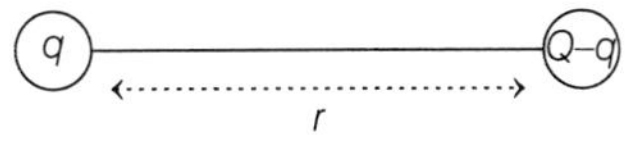

$$\therefore\ F = \frac{kq(Q-q)}{r^2} \Rightarrow F = \frac{k}{r^2}(qQ - q^2)$$

As we know that $\frac{dF}{dq} = 0$, for maximum force.

$$\Rightarrow \frac{dF}{dq} = \frac{d}{dq}\left[\frac{k}{r^2}(qQ - q^2)\right] = 0$$

$$\Rightarrow \frac{k}{r^2}(Q - 2q) = 0 \quad \Rightarrow \quad Q = 2q$$

08. *(d)* The given condition is shown in the figure given below,

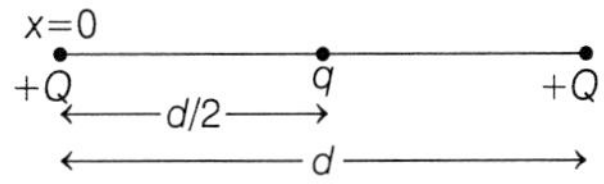

$\therefore$Net force on charge Q placed at origin i.e. at $x = 0$ in accordance with the principle of superposition can be given as

$$F_{\text{net}} = \frac{1}{4\pi\varepsilon_0}\cdot\frac{Q\times q}{\left(\frac{d}{2}\right)^2} + \frac{1}{4\pi\varepsilon_0}\cdot\frac{Q\times Q}{(d)^2}$$

Since, it has been given that

$$F_{\text{net}} = 0.$$

$$\Rightarrow \frac{1}{4\pi\varepsilon_0}\cdot\frac{Q\times q}{\left(\frac{d}{2}\right)^2} + \frac{1}{4\pi\varepsilon_0}\cdot\frac{Q\times Q}{(d)^2} = 0$$

or $$q = -\frac{Q}{4}$$

09. *(c)*

> **key idea** Charge distribution can not be zero when Q_{net} is zero. Because the distribution on the opposite surface is induced and relative closeness of the charge plays a significant role.

On outer surface there will be no charge.

There, fore $Q_2 = \sigma_2 = 0$

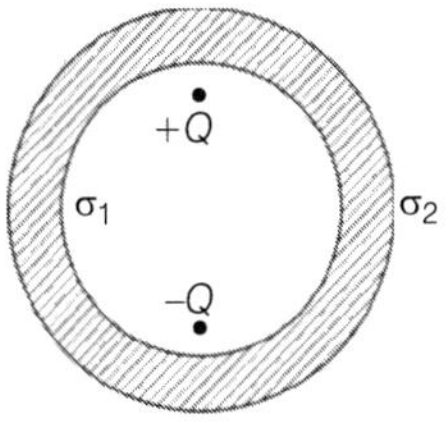

On inner surface total charge will be zero but charge distribution will be there so.

$$Q_1 = 0 \text{ and } \sigma_1 \neq 0$$

10. *(a)*

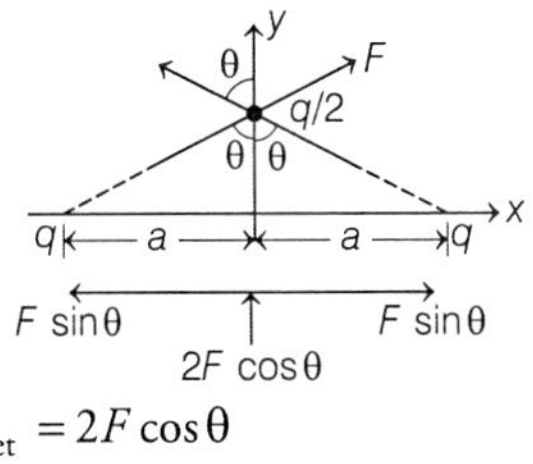

$$F_{\text{net}} = 2F\cos\theta$$

$$\therefore F_{net} = \frac{2kq\left(\frac{q}{2}\right)}{(\sqrt{y^2+a^2})^2} \cdot \frac{y}{\sqrt{y^2+a^2}}$$

$$F_{net} = \frac{2kq\left(\frac{q}{2}\right)y}{(y^2+a^2)^{3/2}}$$

$$\Rightarrow \frac{kq^2 y}{a^3} \propto y$$

11. *(d)* Let charge q be placed at mid-point of line AB as shown below.

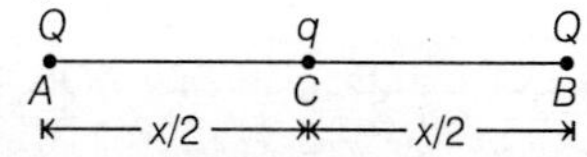

Also, $AB = x$[say]

$\therefore \quad AC = \frac{x}{2},$

$BC = \frac{x}{2}$

For the system to be in equilibrium,

$$F_{Qq} + F_{QQ} = 0$$

$$\Rightarrow \frac{1}{4\pi\varepsilon_0}\frac{Qq}{(x/2)^2} + \frac{1}{4\pi\varepsilon_0}\frac{QQ}{x^2} = 0$$

or $\quad q = -\frac{Q}{4}$

12. *(c)*

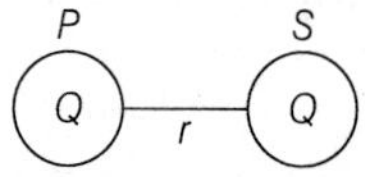

$$F_{PS} = \frac{1}{4\pi\varepsilon_0}\cdot\frac{Q^2}{r^2}$$

$$\Rightarrow \quad F_{PS} \propto Q^2 \quad \ldots(i)$$

$$\Rightarrow \quad F_{PS} = 16\text{ N}$$

When sphere P and R are brought in contact, then

Now, if sphere S and R are brought in contact, then

$\therefore$New force between sphere P and S,

$$F_{PS} \propto \frac{Q}{2} \times \frac{3Q}{4}$$

$$\Rightarrow \quad F_{PS} \propto \frac{3Q^2}{8}$$

$$F_{PS} = \frac{3}{8} \times 16 = 6\text{ N}$$

13. *(d)* Given, charge densities are equal before connection

$$\Rightarrow \sigma = \frac{Q_1}{4\pi R^2} = \frac{Q_2}{4\pi(2R)^2} \quad \left(\sigma = \frac{\text{charge}}{\text{area}}\right)$$

$$\Rightarrow 4Q_1 = Q_2$$

Charge on bigger solid sphere after connection,

$$\text{Also, } Q_2' = Q_T\left(\frac{R_2}{R_1+R_2}\right)$$

$$= (Q_1 + Q_2)\left(\frac{2R}{R+2R}\right)$$

$$= (Q_1 + Q_2)\left(\frac{2}{3}\right) = 5Q_1\left(\frac{2}{3}\right)$$

$$\sigma' = \frac{Q'_2}{A_2} = \frac{5Q_1(2/3)}{4\pi(2R)^2}$$

$$\text{But, } Q_1 = \frac{Q_2}{4} \Rightarrow \sigma' = \frac{10}{12}\cdot\frac{Q_2}{4\pi(2R)^2}$$

$$\therefore \quad \frac{\sigma'}{\sigma} = \frac{10}{12} = \frac{5}{6}$$

14. *(a)* Initially charges are placed in a medium of dielectric constant k.

$$F = \frac{q_1 q_2}{4\pi\varepsilon d^2} = \frac{q_1 q_2}{4\pi k\varepsilon_0 d^2} \quad \ldots \text{(i)}$$

Here, $k = \frac{\varepsilon}{\varepsilon_0}$

Without dielectric, when placed in air, force between charges,

$$F' = \frac{1}{4\pi\varepsilon_0}\frac{q_1 q_2}{r^2} \quad \ldots \text{(ii)}$$

On dividing Eq. (i) by Eq. (ii), we get

$$\frac{F}{F'} = \frac{1}{k} \times \frac{r^2}{d^2}$$

When, $F = F'$, we get

$$1 = \frac{r^2}{kd^2} \Rightarrow r^2 = kd^2$$

$$\Rightarrow \quad r = d\sqrt{k}$$

> **• Concept Enhancer**
>
> When two charges are placed in a dielectric medium except air, then electrostatic force between the charges always decreases.

15. *(b)* According to first condition,

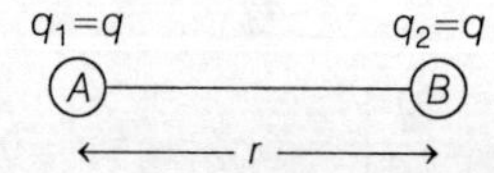

$$F = \frac{1}{4\pi\varepsilon_0}\cdot\frac{q_1 q_2}{r^2} = \frac{1}{4\pi\varepsilon_0}\cdot\frac{q\cdot q}{r^2}$$

$$F = \frac{1}{4\pi\varepsilon_0}\cdot\frac{q^2}{r^2} \quad \ldots\text{(i)}$$

When identical uncharged sphere C is placed in contact with sphere A, then

$$q_A = q_C = \frac{q}{2}$$

Now, C is placed in contact with B, then charge on sphere B and C

$$q_B' = q_C' = \frac{q + \frac{q}{2}}{2} = \frac{3q}{4}$$

Now, according to given condition, arrangement of charges are shown below.

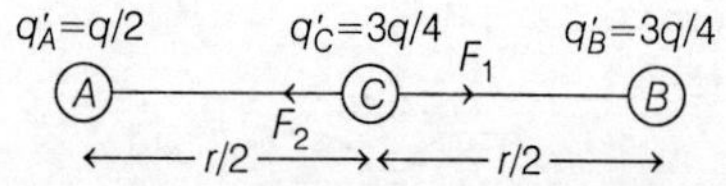

$\therefore$Net force experienced by sphere C,

$$F' = F_1 - F_2 = \frac{1}{4\pi\varepsilon_0}\cdot\frac{\frac{3q}{4}\cdot\frac{3q}{4}}{\left(\frac{r^2}{4}\right)}$$

$$-\frac{1}{4\pi\varepsilon_0}\cdot\frac{\frac{q}{2}\cdot\frac{3q}{4}}{\frac{r^2}{4}} = \frac{3}{4}\cdot\frac{1}{4\pi\varepsilon_0}\cdot\frac{q^2}{r^2}$$

$$\Rightarrow \quad F' = \frac{3F}{4} \quad \text{[From Eq. (i)]}$$

16. *(b)* If R be the radius of combined big drop and r be the radius of small drop. then

Volume of big drop

$= 64 \times$ Volume of small drop

$$\Rightarrow \frac{4}{3}\pi R^3 = 64 \times \frac{4}{3}\pi r^2$$

$$\Rightarrow \quad R = 4r = 4 \times 0.02 = 0.08\text{ m}$$

$[\because r = 0.02\text{ m}]$

Charge on combined drop,

$Q = 64 \times$ Charge on one small drop

$= 64q = 64 \times 5 \quad [\because q = 5\,\mu\text{C}]$

$= 320\,\mu\text{C}$

$$\therefore \frac{\sigma_{biggest}}{\sigma_{smaller}} = \frac{\frac{Q}{4\pi R^2}}{\frac{q}{4\pi r^2}} = \frac{Q}{q} \times \left(\frac{r}{R}\right)^2$$

$$= \frac{320 \times 10^{-6}}{5 \times 10^{-6}} \times \left(\frac{0.02}{0.08}\right)^2 = \frac{4}{1}$$

or 4 : 1

17. *(a)* Masses of each charged particle,

$$m_1 = m_2 = 10\text{g} = 0.01 \text{ kg}$$

and charges on them,

$$q_1 = q_2 = 2 \times 10^{-7} \text{C}$$
$$\mu = 0.25$$
$$g = 10 \text{ m/s}^2$$

The given situation is shown in the figure below,

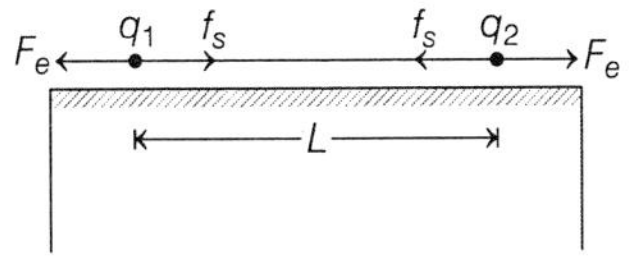

In the condition of limited equilibrium of either charge,

$$F_e = f_s$$
$$\frac{kq_1 q_2}{L^2} = \mu\, mg$$
$$\Rightarrow \frac{9 \times 10^9 \times 2 \times 10^{-7} \times 2 \times 10^{-7}}{L^2} = 0.25 \times 0.01 \times 10$$
$$\Rightarrow L = 0.12 \text{ m} = 12 \text{ cm}$$

18. *(a)* Given,

Linear charge density, $\lambda = 8$ nC/m $= 8 \times 10^{-9}$C/m

The relation between surface charge density and linear charge density can be given as

$$\frac{2k\lambda}{r} = \frac{\sigma}{\varepsilon_0} \quad \text{...(i)}$$
$$\Rightarrow \sigma = \frac{2k\lambda\varepsilon_0}{r}$$
$$= \frac{2 \times 9 \times 10^9 \times 8 \times 10^{-9} \times 8.85 \times 10^{-12}}{3}$$
$$= 0.424 \times 10^{-9} \text{ C m}^{-2}$$
$$= 0.424 \text{ nCm}^{-2}$$

19. *(a)* Initially force between spheres *A* and *B*,

$F = \frac{kq^2}{r^2}$. When *A* and *C* are touched, charge on both will be $\frac{q}{2}$. Again when *C* is touched with *B*, then charge on *B* is given by $q_B = \frac{\frac{q}{2} + q}{2} = \frac{3q}{4}$

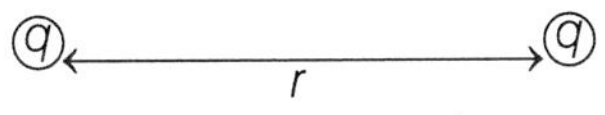

Required force between spheres *A* and *B* is given by

$$F' = \frac{kq_A q_B}{r^2} = \frac{k \times \frac{q}{2} \times \frac{3q}{4}}{r^2}$$
$$= \frac{3}{8}\frac{kq^2}{r^2} = \frac{3}{8}F$$

20. *(a)* Three forces F_{41}, F_{42} and F_{43} acting on *Q* as shown below.

Resultant of

$$F_{41} + F_{43} = \sqrt{2}F_{\text{each}} = \sqrt{2}\frac{1}{4\pi\varepsilon_0}\frac{Qq}{d^2}$$

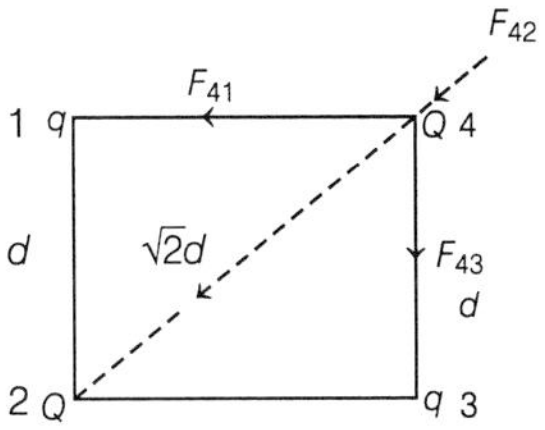

Resultant on *Q* becomes zero only when *q* charges are of negative nature.

$$\therefore \quad F = \frac{1}{4\pi\varepsilon_0}\frac{Q \times Q}{(\sqrt{2}\,d)^2}$$
$$\Rightarrow \sqrt{2}\frac{qQ}{d^2} = \frac{Q \times Q}{2d^2} \Rightarrow \sqrt{2} \times q = \frac{Q}{2}$$
$$\therefore \quad q = \frac{Q}{2\sqrt{2}} \text{ or } \frac{Q}{q} = -2\sqrt{2}$$

21. *(b)* The system is in equilibrium means the force experienced by each charge is zero.

It is clear that charge placed at centre would be in equilibrium for any value of *q*, so we are considering the equilibrium of charge placed at any corner.

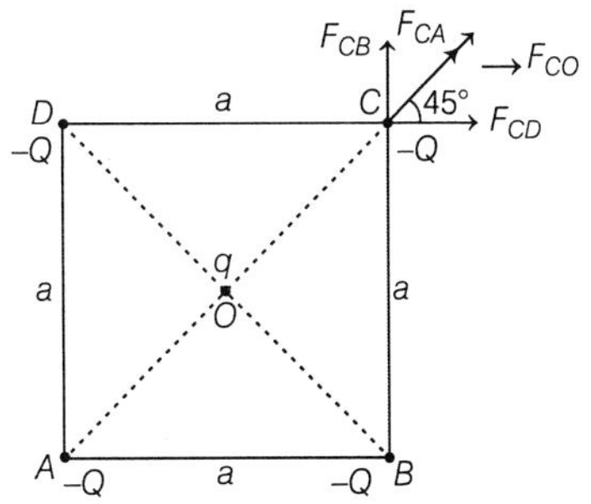

$$F_{CD} + F_{CA}\cos 45° + F_{CO}\cos 45° = 0$$
$$\Rightarrow \frac{1}{4\pi\varepsilon_0}\cdot\frac{(-Q)(-Q)}{a^2} + \frac{1}{4\pi\varepsilon_0}\frac{(-Q)(-Q)}{(\sqrt{2}\,a)^2} \times \frac{1}{\sqrt{2}} + \frac{1}{4\pi\varepsilon_0}\frac{(-Q)\,q}{(\sqrt{2}\,a/2)^2} \times \frac{1}{\sqrt{2}} = 0$$
$$\Rightarrow \frac{1}{4\pi\varepsilon_0}\frac{Q^2}{a^2} + \frac{1}{4\pi\varepsilon_0}\frac{Q^2}{2a^2}\cdot\frac{1}{\sqrt{2}} - \frac{1}{4\pi\varepsilon_0}\frac{2Qq}{a^2} \times \frac{1}{\sqrt{2}} = 0$$
$$\Rightarrow Q + \frac{Q}{2\sqrt{2}} - \sqrt{2}\,q = 0$$
$$\text{or} \quad q = (2\sqrt{2} + 1)\frac{Q}{4}$$

22. *(b)*

> **key idea** You can use parallelogram law along two perpendicular direction, to find the resultant force.

Force on $-q_1$,

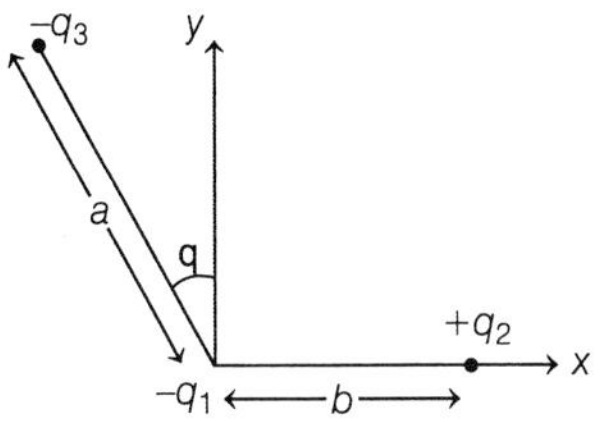

$$\text{F} = \frac{1}{4\pi\varepsilon_0}\frac{q_1 q_2}{b^2}\,\mathbf{i} + \frac{1}{4\pi\varepsilon_0}\frac{q_1 q_3}{a^2}[\sin\theta\,\mathbf{i} - \cos\theta\,\mathbf{j}]$$

From above, *x*-component of force is

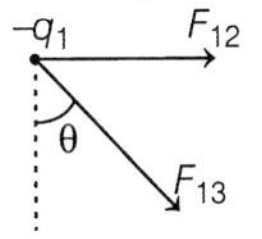

$$F_x = \frac{q_1}{4\pi\varepsilon_0}\left[\frac{q_2}{b^2} + \frac{q_3}{a^2}\sin\theta\right]$$
$$\Rightarrow F_x \propto \left[\frac{q_2}{b^2} + \frac{q_3}{a^2}\sin\theta\right]$$

23. *(b)* Coulombian force,

$$F_e = \frac{kQ_1Q_2}{r^2}$$
$$k = 9 \times 10^9 \text{ Nm}^2/\text{C}^2$$
$$F_e = \frac{9 \times 10^9 \times 1.6 \times 10^{-19} \times 1.6 \times 10^{-19}}{r^2} \quad \text{...(i)}$$

As $Q_1 = Q_2$ = charge on electron and proton (magnitude)

Similarly, for gravitational force,

$$F_g = \frac{Gm_1 m_2}{r^2}$$
$$= \frac{6.67 \times 10^{-11} \times 9.1 \times 10^{-31} \times 1.6 \times 10^{-27}}{r^2} \quad \text{...(ii)}$$

m_1 = mass of electron, m_2 = mass of proton

Taking ratio, $\frac{F_e}{F_g}$

$$= \frac{9\times10^9\times1.6\times10^{-19}\times1.6\times10^{-19}}{6.67\times10^{-11}\times9.1\times10^{-31}\times1.6\times10^{-27}}$$

$$= 0.23\times10^{40} \text{ or } 2.39\times10^{39}$$

So, the ratio is in order of 10^{39}.

24. *(a)* When point charge q_0 is displaced towards A, then force experienced by q_0 due to charge Q at A, is given by.

$$F_1 = \frac{kQq_0}{(a-x)^2}$$

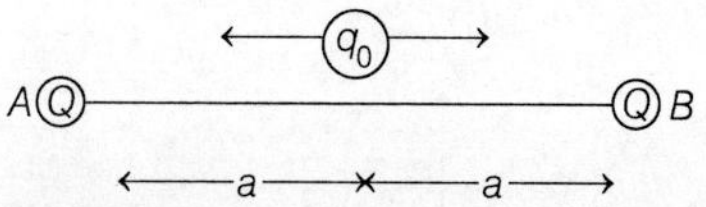

Similarly, Force experienced by point charge q_0 due to charge Q at point B is,

$$F_2 = \frac{kQq_0}{(a+x)^2},$$

Net force on q_0, $F = F_1 - F_2$

$$\therefore \quad F = \frac{kQq_0}{(a-x)^2} - \frac{kQq_0}{(a+x)^2}$$

$$= kQq_0\left[\frac{1}{(a-x)^2} - \frac{1}{(a+x)^2}\right]$$

$$= \frac{4kQqL}{a^2}x$$

Also, $F = kx = m\omega^2 x$

$$\therefore m\omega^2 x = \frac{4kQq_0}{a^3}x$$

or $\omega^2 = \frac{4kQq_0}{ma^3}$ or $\omega = \sqrt{\frac{4kQq_0}{ma^3}}$

So, time period,

$$T = 2\pi\sqrt{\frac{ma^3}{4kQq_0}} = 2\pi\sqrt{\frac{4\pi\varepsilon_0 ma^3}{4Qq_0}}$$

$$= 2\pi\sqrt{\frac{\pi\varepsilon_0 ma^3}{Qq_0}} = \sqrt{\frac{4\pi^3\varepsilon_0 ma^3}{Qq}}$$

25. *(d)* The given situation is shown below

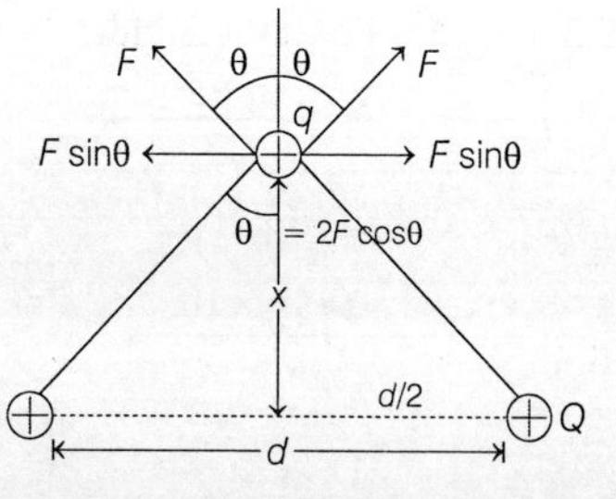

Net force on charge q = sum of components along perpendicular bisector

$$2F\cos\theta = \frac{2kQq}{\left(\frac{d^2}{4}+x^2\right)} \times \frac{x}{\left(\frac{d^2}{4}+x^2\right)^{1/2}}$$

$$\therefore \quad F_{\text{net}} = \frac{2kQq\cdot x}{\left(\frac{d^2}{4}+x^2\right)^{3/2}}$$

F_{net} is maximum, when $\frac{dF_{\text{net}}}{dx} = 0$

$$\Rightarrow \frac{d}{dx}\left(2kQq\cdot\frac{x}{\left(\frac{d^2}{4}+x^2\right)^{3/2}}\right) = 0$$

$$\Rightarrow \frac{\left(\frac{d^2}{4}+x^2\right)^{3/2}\cdot 1 - x\cdot\frac{3}{2}\left(\frac{d^2}{4}+x^2\right)^{1/2}\cdot 2x}{\left(\frac{d^2}{4}+x^2\right)^3} = 0$$

$$\Rightarrow x = \frac{d}{2\sqrt{2}}$$

• Concept Enhancer

Based on this concept, direct questions or indirect questions are asked, where force/electric field is maximum at equatorial line. You can use the above result to solve them.

26. *(d)* Given, 3 identical charged balls having charge 2 C each are suspended from a common point by silk threads of 2 m each (diagram shown), [each side of equilateral Δ is 1 m].

To find ratio of net force on a charged ball to the force between any two charged balls.

Let charges as 1, 2 and 3.

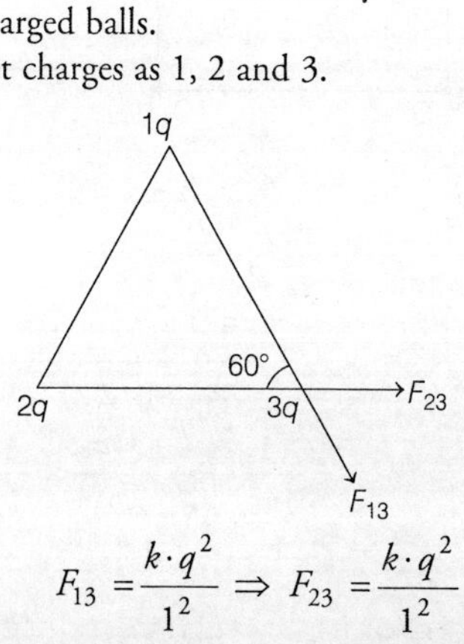

$$F_{13} = \frac{k\cdot q^2}{1^2} \Rightarrow F_{23} = \frac{k\cdot q^2}{1^2}$$

So, net force on charge 3

$$= \sqrt{F_{13}^2 + F_{23}^2 + (2\times F_{13}\times F_{23}\times\cos 60°)}$$

$$= \sqrt{\left(\frac{kq^2}{1^2}\right)^2 + \left(\frac{kq^2}{1^2}\right)^2 + \left(2\times\frac{kq^2}{1^2}\times\frac{kq^2}{1^2}\times\frac{1}{2}\right)}$$

$$\therefore \quad F_1 = F_{\text{net-3}} = \sqrt{3}\,\frac{kq^2}{1^2}. \qquad \text{...(i)}$$

Force between any two charges,

$$F_2 = \frac{kq^2}{1^2} \qquad \text{...(ii)}$$

On dividing Eq. (i) by Eq. (ii), we get

$$\frac{F_1}{F_2} = \sqrt{3} \text{ or } \sqrt{3}:1$$

27. *(b)* The given situation is shown below

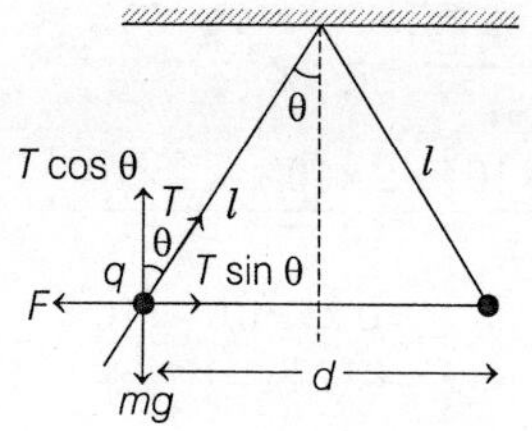

∵ According to Coulomb's law,

$$F = \frac{kq_1q_2}{r^2}$$

Here, $q_1 = q_2 = q$

∴ Force due to charge,

$$F = \frac{k\times q\times q}{d^2} \qquad \{\because r = d\}$$

$$\Rightarrow \quad F = \frac{kq^2}{d^2} \qquad \text{...(i)}$$

Using above diagram, we can write

$T\cos\theta = mg$ and $T\sin\theta = F$

$$\because \quad \tan\theta = \frac{\sin\theta}{\cos\theta} = \frac{F}{mg} = \frac{\frac{kq^2}{d^2}}{mg}$$

{Using Eq (i)}

$$\Rightarrow \quad \tan\theta = \frac{kq^2}{d^2mg} \qquad \text{...(ii)}$$

∵ We know that

$$\tan\theta \approx \sin\theta \approx \frac{d}{2l} \text{ for smaller value of } \theta \qquad \text{...(iii)}$$

∴ From Eqs. (ii) and (iii), we get

$$\frac{d}{2l} = \frac{kq^2}{d^2mg}$$

$$\Rightarrow \quad d^3mg = kq^2 2l \Rightarrow d^3 = \frac{kq^2 2l}{mg}$$

$$\because \quad k = \frac{1}{4\pi\varepsilon_0}$$

$$\Rightarrow \quad d^3 = \frac{q^2 2l}{4\pi\varepsilon_0 mg}$$

$$\Rightarrow \quad d = \left(\frac{q^2 l}{2\pi\varepsilon_0 mg}\right)^{1/3}$$

28 *(c)* The arrangement of charges is shown below

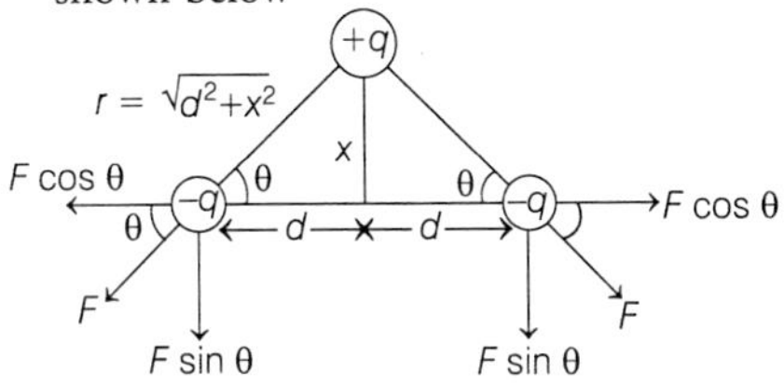

As we know that,

$$F = \frac{1}{4\pi\varepsilon_0}\frac{q_1 q_2}{r^2}$$

$$= \frac{1}{4\pi\varepsilon_0}\frac{q_1 q_2}{(d^2+x^2)} \quad \ldots (i)$$

Here, $q_1 = q_2 = q$

Force in SHM, $F = m\omega^2 x$ ...(ii)

Since, in order to have SHM, $+q$ should move downwards and force responsible for this will be only

$F' = F\sin\theta + F\sin\theta = 2F\sin\theta$ ...(iii)

Using Eqs. (ii) and (iii), we get

$$2F\sin\theta = m\omega^2 x$$

$$\Rightarrow \frac{2}{4\pi\varepsilon_0}\frac{q^2}{(d^2+x^2)}\sin\theta = m\omega^2 x$$

$$\Rightarrow \frac{2}{4\pi\varepsilon_0}\frac{q^2}{(d^2+x^2)}\cdot\frac{x}{(d^2+x^2)^{1/2}} = m\omega^2 x$$

$$\Rightarrow \omega = \left(\frac{1}{2\pi\varepsilon_0}\frac{q^2}{(d^2+x^2)^{3/2} m}\right)^{1/2}$$

As, $x << d$

$$\therefore \quad \omega = \left(\frac{1}{2\pi\varepsilon_0}\frac{q^2}{md^3}\right)^{1/2}$$

29. *(d)* At any instant,

$T\cos\theta = mg$ ...(i)

$T\sin\theta = F_e = \dfrac{kq^2}{x^2}$ ...(ii)

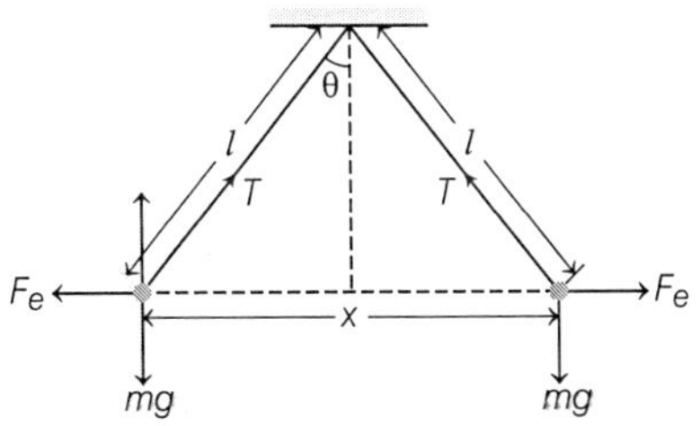

From Eqs. (i) and (ii), we get

$$kq^2/x^2 = mg\tan\theta$$

$$\Rightarrow \quad q^2 = \frac{mg}{k}\cdot\frac{x}{2l}\cdot x^2 \quad \left[\because \tan\theta \approx \frac{x}{2l}\right]$$

$$\Rightarrow \quad q^2 = \frac{mg}{2kl}x^3 \quad \ldots (iii)$$

$$\Rightarrow 2q\frac{dq}{dt} = \frac{3mg}{2kl}x^2\frac{dx}{dt}$$

$$\Rightarrow 2\left(\frac{mg}{2kl}x^3\right)^{1/2}\frac{dq}{dt} = \frac{3mg}{2kl}x^2 v$$

$$\left[\because q = \left(\frac{mg}{2kl}x^3\right)^{1/2}\right]$$

$$\Rightarrow vx^{1/2} = \text{constant} \Rightarrow v \propto x^{-1/2}$$

30. *(c)* From FBD of sphere, using Lami's theorem, we get

$F/mg = \tan\theta$ ...(i)

When suspended in liquid, as θ remains same.

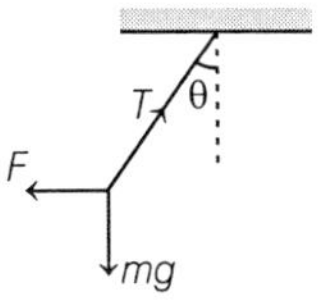

$$\therefore \quad \frac{F'}{mg\left(1-\frac{\rho}{d}\right)} = \tan\theta \quad \ldots (ii)$$

Using Eqs. (i) and (ii), we get

$$\frac{F}{mg} = \frac{F'}{mg\left(1-\frac{\rho}{d}\right)}, \text{ where } F' = \frac{F}{k}$$

$$\therefore \quad \frac{F}{mg} = \frac{F}{mgk\left(1-\frac{\rho}{d}\right)}$$

$$\text{or} \quad k = \frac{1}{1-\frac{\rho}{d}} = \frac{1}{\left(1-\frac{0.8}{1.6}\right)} = 2$$

31. *(*)* $P\times 10^{-13} - \dfrac{F_e}{F_G} = \dfrac{\dfrac{kq_1q_2}{r^2}}{\dfrac{Gm_1m_2}{r^2}}$

$$= \frac{9\times10^{9}\times 6.67\times10^{-19}\times 9.6\times10^{-10}}{6.67\times10^{-11}\times19.2\times10^{-27}\times 9\times10^{-27}}$$

$$= \frac{1}{2}\times10^{45}$$

32. *(5440)* Given,

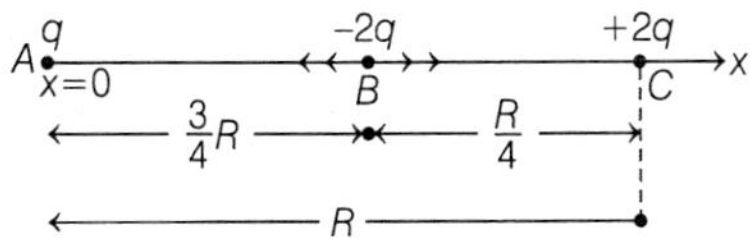

$$F_{BA} = \frac{kq\times 2q}{\left(\frac{3}{4}R\right)^2}$$

$$\Rightarrow F_{BA} = \frac{16k\times 2q^2}{9R^2}$$

$$F_{BA} = \frac{32kq^2}{9R^2}$$

$$F_{BC} = \frac{k4q^2}{(R/4)^2} = \frac{64kq^2}{R^2}$$

$$F_B = F_{BC} - F_{BA} = \frac{64kq^2}{R^2} - \frac{32kq^2}{9R^2}$$

$$= \frac{kq^2}{R^2}\left(64 - \frac{32}{9}\right)$$

$$= \frac{9\times10^9\times(2\times10^{-6})^2}{(2\times10^{-2})^2}\times\frac{544}{9}$$

$$= 5440 \text{ N}$$

33. *(300)* Point charge in equilibrium is at rest so,

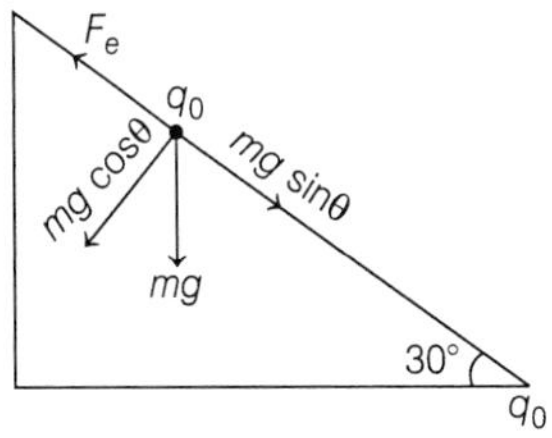

$$F_e = mg\sin\theta$$

$$\frac{kq_0q_0}{r^2} = mg\sin 30°$$

$$\frac{kq_0^2}{r^2} = \frac{mg}{2} \quad \left[\because r = \frac{h}{\sin 30°}\right]$$

$$\frac{kq_0^2}{\left(\frac{h}{\sin 30}\right)^2} = \frac{mg}{2}$$

$$\frac{9\times10^9\times(2\times10^{-6})^2}{4h^2} = \frac{20\times10^{-3}\times10}{2}$$

$$\frac{9\times4\times10^9\times10^{-12}}{4h^2} = 10^{-1}$$

$$h^2 = 9\times10^{-2}$$

$$h = 0.3 \text{ m} = 300\times10^{-3} \text{ m}$$

Hence, $x = 300$

34. *(2)* The situation given in question is as follows.

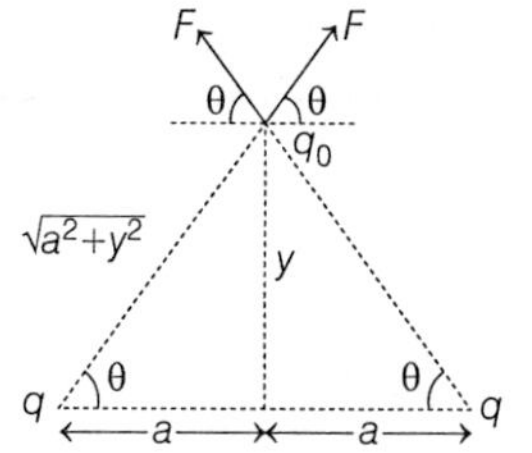

Resolving component of force acting on q_0 in x and y direction.

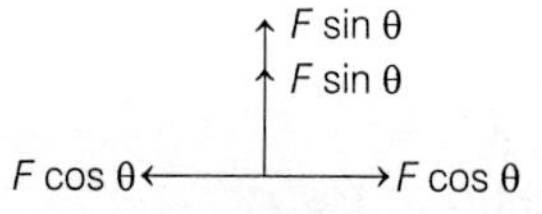

$$F_{\text{Net}} = 2F\sin\theta$$
$$= \frac{2kqq_0}{(\sqrt{a^2+y^2})^2}\frac{y}{\sqrt{a^2+y^2}}$$
$$= \frac{2kqq_0 y}{(a^2+y^2)^{3/2}}$$

For maximum value of F_{Net}, $\dfrac{dF_{\text{Net}}}{dy} = 0$

$$2kqq_0 \frac{d}{dy}\left(\frac{y}{(y^2+a^2)^{3/2}}\right) = 0$$
$$\Rightarrow \frac{d}{dy}\left(\frac{y}{(y^2+a^2)^{3/2}}\right) = 0$$
$$\Rightarrow (y^2+a^2)^{3/2}\frac{d}{dy}(y) - y\frac{d}{dy}(y^2+a^2)^{3/2} = 0$$
$$\Rightarrow (y^2+a^2)^{\frac{3}{2}}\cdot 1 - y\left\{\frac{3}{2}(y^2+a^2)^{\frac{1}{2}}\cdot 2y\right\} = 0$$
$$\Rightarrow (y^2+a^2)^{\frac{1}{2}}[(y^2+a^2) - 3y^2(1)] = 0$$
$$\Rightarrow (y^2+a^2)^{\frac{1}{2}}[a^2-2y^2] = 0$$
$$\Rightarrow a^2 - 2y^2 = 0 \text{ or } y = \frac{a}{\sqrt{2}} = \frac{a}{\sqrt{x}}$$

So, $x = 2$

35. *(24)*

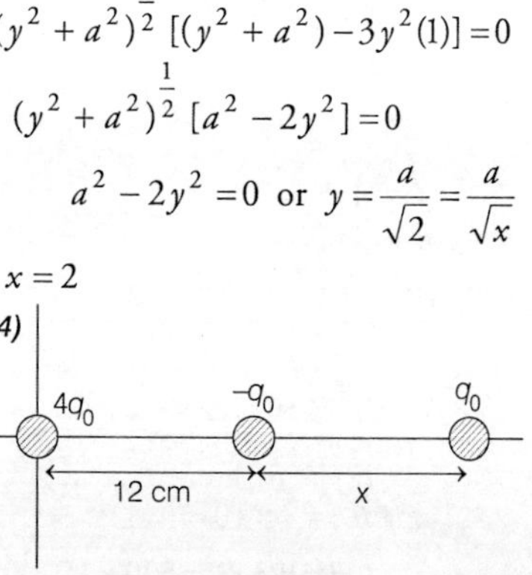

Force on proton of charge q_0 is zero in above shown situation

As, net force on q_0 is zero,

$$\frac{k4q_0(q_0)}{(12+x)^2\times 10^{-4}} = \frac{k(q_0)(q_0)}{x^2\times 10^{-4}}$$
$$\Rightarrow \quad 4x^2 = (12+x)^2 \Rightarrow x = 12 \text{ cm}$$

So, distance of proton from origin is $12+12 = 24$ cm

36. *(17)* The given situation is shown below

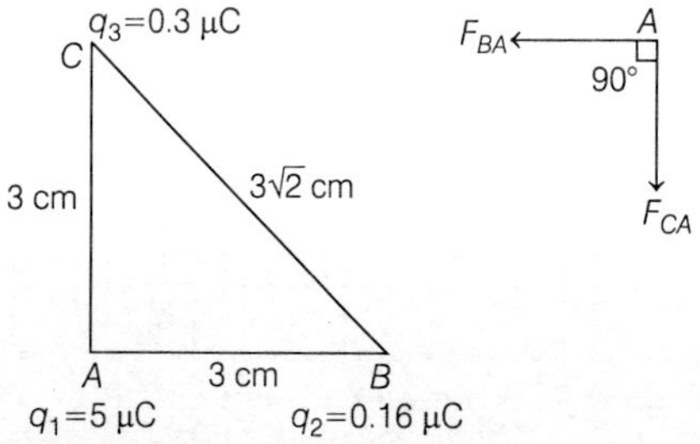

Here, $q_1 = 5\mu\text{C} = 5\times 10^{-6}\text{C}$

$$q_2 = 0.16\mu\text{C} = 1.6\times 10^{-7}\text{C}$$
$$q_3 = 0.3\mu\text{C} = 3\times 10^{-7}\text{C}$$

Force on charge q_1 due to charge q_2,

$$F_{BA} = 9\times 10^9\,\frac{q_1q_2}{(AB)^2}$$
$$= 9\times 10^9 \times \frac{5\times 10^{-6}\times 1.6\times 10^{-7}}{(3\times 10^{-2})^2} = 8\text{ N}$$

Force on charge q_1 due to charge q_3,

$$F_{CA} = 9\times 10^9\,\frac{q_1q_2}{(AC)^2}$$
$$= 9\times 10^9\times\frac{5\times 10^{-6}\times 3\times 10^{-7}}{(3\times 10^{-2})^2}$$
$$= 15\text{ N}$$

Net force experienced by charge on point A due to other two charges,

$$F = \sqrt{F_{CA}^2 + F_{BA}^2} = \sqrt{15^2+8^2}$$
$$= \sqrt{289} = 17\text{ N}$$

37. *(6000)* Given,

Mass, $m = 1$ mg $= 1\times 10^{-6}$kg,

Separation between two charges, d = 2m and $q^2 = 10\text{ C}^2$

When free charge is displaced from its position, then from given diagram,

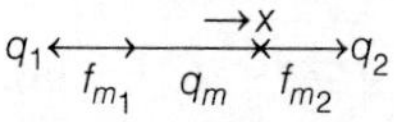

$$F_{\text{net}} = F_{m_1} - F_{m_2} = -m\omega^2 x$$

where, F_{net} = net electrostatic force,

F_{m_1}, F_{m_2} = force on m due to q_1 and q_2 respectively,

ω and x = angular frequency and equilibrium distance, respectively.

By using Coulomb's law,

$$F = \frac{kq_1q_2}{r^2}$$

where, k = Coulomb's constant

$$= 9\times 10^9\text{ N-m}^2\text{C}^{-2}$$

r = distance between two charges

$$= \frac{d}{2} = 1\text{ m}$$

$$\therefore\ F_{\text{net}} = \frac{kq_mq_1}{(r+x)^2} - \frac{kq_mq_2}{(r-x)^2} = -m\omega^2 x$$
$$\Rightarrow \frac{kq^2}{m}\left[\frac{(r-x)^2-(r+x)^2}{(r^2-x^2)^2}\right] = -\omega^2 x$$
$$(\because q_1 = q_2 = q_m = q)$$
$$\Rightarrow \frac{9\times 10^9\times 10}{10^{-6}}\left[\frac{-4rx}{r^4}\right] = -\omega^2 x$$
$$(\because x \ll 1\text{ m},\ \therefore r^2 - x^2 = r^2)$$
$$\Rightarrow \omega = \sqrt{36\times 10^{16}} = 6\times 10^8\text{ rads}^{-1}$$

From question $\omega = x\times 10^5$ rad/s

$$\therefore \qquad x = 6000$$

38. *(12)* The 1 C charge present in the origin and other 1μC charge are placed at 1 m, 2 m, 4 m, 8 m ...

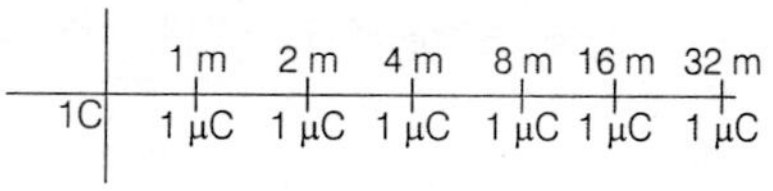

Using the Coulomb's law,

$$F = \frac{kq_1q_2}{r^2}$$
$$F = 9\times 10^9\times(1)\times 10^{-6}\left[1+\frac{1}{2^2}+\frac{1}{4^2}+\frac{1}{8^2}+\dots\right]$$
$$= 9\times 10^3\times\left[\frac{1}{1-\frac{1}{4}}\right]$$
$$= 12\times 10^3\text{ N}$$

• Concept Enhancer

Sum of infinite geometric series is given as,

$$S = \frac{a}{1-r}$$

where, $a \rightarrow$ first term, $t \rightarrow$ common ratio

for Ex. if series is given as

$$1+\frac{1}{x}+\frac{1}{x^2}+\frac{1}{x^2}\dots\infty$$

then, $a = 1$ and $r = \dfrac{1}{x}$

Topic 2

01. *(c)*

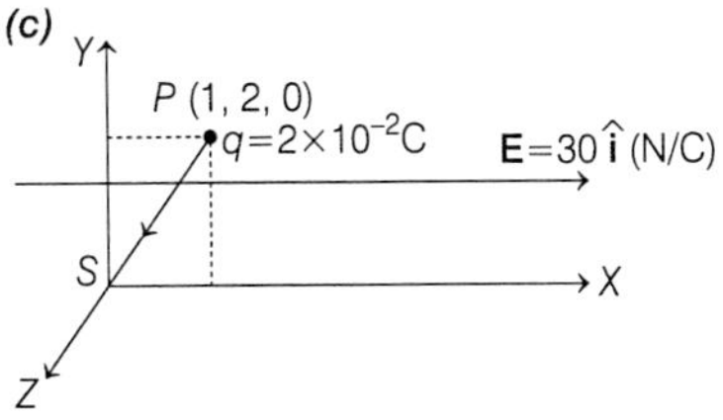

Work done $= \mathbf{F}\cdot\mathbf{s} = q\mathbf{E}\cdot\mathbf{s}$

$= (2\times10^{-2}\times30\hat{i})\cdot(-\hat{i}-2\hat{j})$

$= -60\times10^{-2}$ J

$= -600$ mJ

02. *(b)* The arrangement of charges is

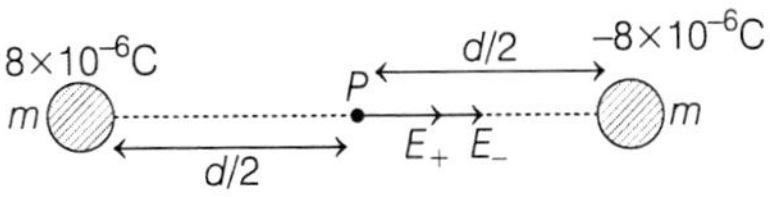

At mid-point net electric field is the sum of fields due to positive and negative charges.

$$\Rightarrow E_{net}(\text{at } P) = 2\times\frac{kq}{\left(\frac{d}{2}\right)^2}$$

$$= \frac{8\times9\times10^9\times8\times10^{-6}}{d^2}$$

Given, $E_{net} = 6.4\times10^4$ N/C

Hence, $d^2 = \dfrac{8\times9\times10^9\times8\times10^{-6}}{6.4\times10^4}$

$\Rightarrow \quad d^2 = 9$

$\Rightarrow \quad d = 3$ m

03. *(a)* Here, resultant of $\mathbf{E}_C$ and $\mathbf{E}_A$ will lie along $\mathbf{E}_B$.

So, $|\mathbf{E}_{net}| = |\mathbf{E}_B| + |\mathbf{E}_C + \mathbf{E}_A|$

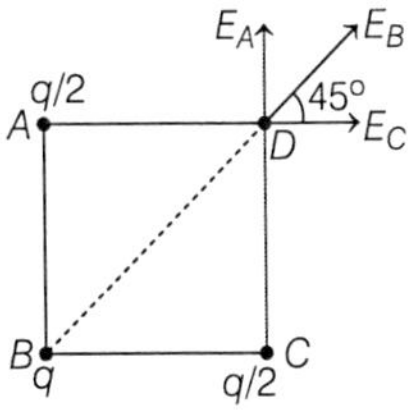

$$= \frac{kq}{(\sqrt{2}a)^2} + \left[\sqrt{2}\,\frac{kq}{2a^2}\right]$$

$$= \frac{kq}{a^2}\left[\frac{1}{2}+\frac{1}{\sqrt{2}}\right]$$

$$= \frac{q}{4\pi\varepsilon_0 a^2}\left[\frac{1}{\sqrt{2}}+\frac{1}{2}\right]$$

04. *(b)* As water droplet is in balanced condition.

So, $\quad \mathbf{F}_{net} = 0$

$\Rightarrow \quad mg = qE \Rightarrow q = mg/E$

$\Rightarrow \quad q = \dfrac{0.1\times10^{-3}\times9.8}{4.9\times10^5}$

$\Rightarrow \quad q = 2\times10^{-9}$ C

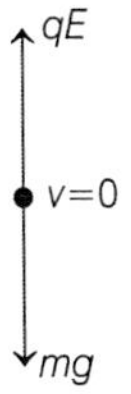

> **• Concept Enhancer**
>
> This is a famous Milikon's oil drop experiment. In this experiment a charged spherical oil drop is suspended in the presence of electri field.

05. *(d)* As gravitational force is considerably smaller than electric force. So, there is only one prominent acceleration in the direction of electric field.

Hence, path of charged particle of small mass is nearly a straight line.

06. *(b)* Electric field on the axis of a ring of radius R at a distance h from the centre is given as

$$E = \frac{kQh}{(h^2+R^2)^{3/2}}$$

Electric field will be maximum, if

$$\frac{dE}{dh} = 0$$

$$\Rightarrow \frac{d}{dh}\left[\frac{kQh}{(R^2+h^2)^{3/2}}\right] = 0$$

$$\Rightarrow \quad h = \frac{R}{\sqrt{2}}$$

07. *(a)* Equilibrium position will shift to point where resultant force = 0.

Hence,

$$kx_{eq} = qE \Rightarrow x_{eq} = \frac{qE}{k}$$

Total energy $= \dfrac{1}{2}m\omega^2A^2 + \dfrac{1}{2}kx_{eq}^2$

$$= \frac{1}{2}m\omega^2A^2 + \frac{1}{2}\frac{q^2E^2}{k}$$

08. *(a)* Field lines should originate from positive charge and terminate to negative charge.

Thus, (b) and (c) are not possible.

Electric field lines cannot form corners as shown in (d).

Thus, correct option is (a).

09. *(c)*

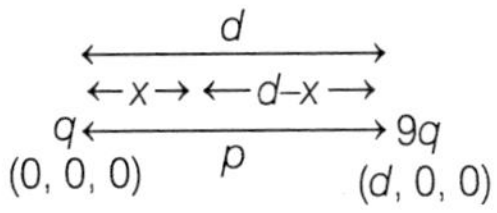

Let $E_p = E_q = E_{9q}$

$$\frac{k\cdot q}{x^2} = \frac{k\cdot 9q}{(d-x)^2}$$

$$\left(\frac{1}{x}\right)^2 = \left(\frac{3}{d-x}\right)^2$$

$$\frac{1}{x} = \frac{3}{d-x}$$

$$d - x = 3x$$

$$4x = d$$

$$x = \frac{d}{4}$$

Co-ordinate of $p\left(\dfrac{d}{4}, 0, 0\right)$

10. *(a)* Using the work-energy theorem, we have

$$W = \Delta KE$$

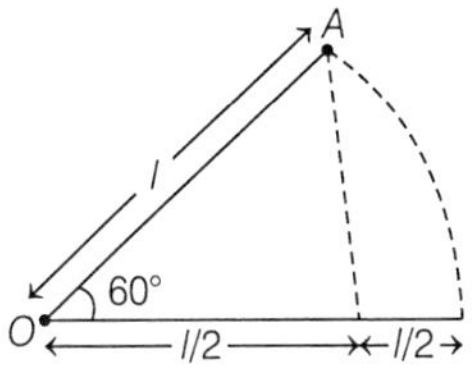

$$qE(l - l\cos60°) = \frac{1}{2}mv^2 - 0$$

$$qE\left(l - \frac{l}{2}\right) = \frac{1}{2}mv^2$$

$$qE\frac{l}{2} = \frac{1}{2}mv^2$$

$$\frac{qE\cdot l}{m} = v^2$$

$$\Rightarrow \quad v = \sqrt{\frac{qEl}{m}}$$

11. *(d)* The diagram for the situation is

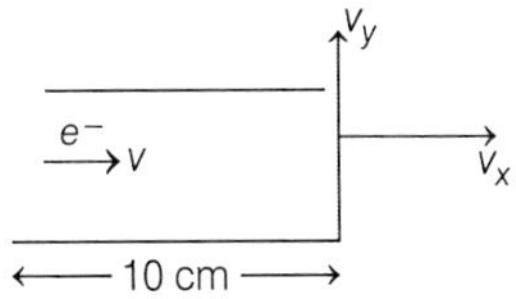

Time to travel 10 cm,

$$t = \frac{L}{v_x} = \frac{10\times10^{-2}}{10^6} = 10^{-7}\text{ s}$$

From 1st equation of motion

$$v_y = u_y + a_y t$$

$$v_y = 0 + \frac{eE}{m}\times10^{-7}$$

$$v_y = \frac{1.6\times10^{-19}}{9.1\times10^{-31}}\times9.1\times10^{+2}\times10^{-7}$$

$$v_y = 16\times10^6 \text{ m/s}$$

12. *(d)* Given,

$$q_0 = 10\,\mu C = 10^{-5} C$$

Let point charge of

$$q = 40\,\mu\, C = 4\times10^{-5}\, C$$

is placed at a distance x cm from origin O

According to given situation,

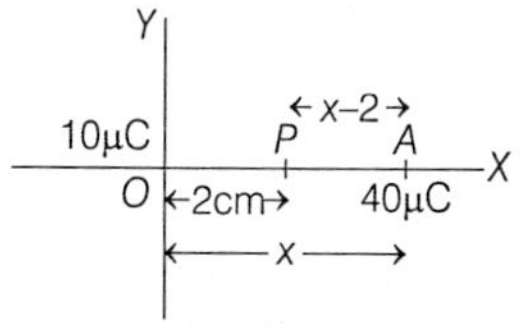

$$E_{OP} = E_{AP} \Rightarrow \frac{1}{4\pi\varepsilon_0}\cdot\frac{10^{-5}}{(2\times10^{-2})^2}$$

$$= \frac{1}{4\pi\varepsilon_0}\cdot\frac{4\times10^{-5}}{(x-2)^2\times(10^{-2})^2}$$

$$\Rightarrow \frac{1}{(2\times10^{-2})^2} = \frac{4}{(x-2)^2\times(10^{-2})^2}$$

$$\Rightarrow \frac{1}{2} = \frac{2}{x-2} \Rightarrow x - 2 = 4$$

$$\Rightarrow x = 6 \text{ cm}$$

13. *(b)* For an insulator charged sphere,

$$E = \frac{Qr}{4\pi\varepsilon_0 R^3} \quad \text{(Inside)}$$

$$E \propto r\ (r < R)$$

$$E = \frac{Q}{4\pi r^2} \quad \text{(Outside)}$$

and $E \propto \frac{1}{r^2}\ (r > R)$

So, graph between E and r will be same as given below

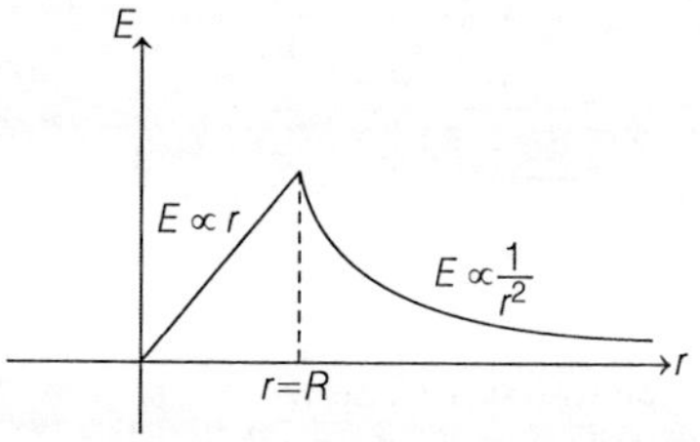

14. *(d)* The given situation is shown below

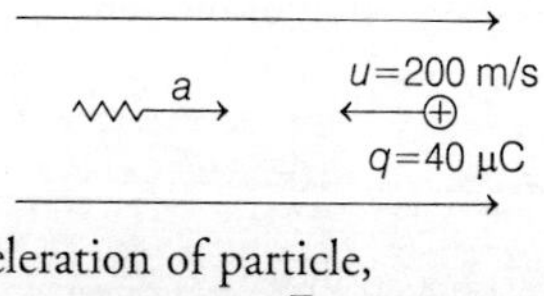

Acceleration of particle,

$$a = -\frac{F}{m} = -\frac{qE}{m}$$

$$= -\left(\frac{40\times10^{-6}\times1\times10^{5}}{100\times10^{-3}\times10^{-3}}\right)$$

$$= -4\times10^{4}\ \text{m/s}^2$$

Now, using $v^2 - u^2 = 2as$, we have

$$0 - (200)^2 = 2(-4\times10^4)\times s$$

or $s = 0.5$ m

15. *(b)* Here

$$E_A = \frac{kq}{(\sqrt{2}l)^2} = \frac{kq}{2l^2},$$

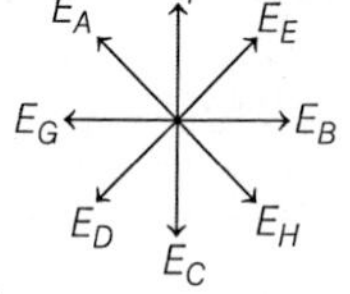

$$E_B = \frac{kq}{(l)^2}$$

E_A and E_H cancal out.

$$E_C = \frac{2kq}{l^2},\quad E_D = \frac{kq}{(\sqrt{2}l)^2}$$

$$E_E = \frac{2kq}{(2l)^2},\quad E_F = \frac{kq}{l^2},$$

$$E_G = \frac{2kq}{l^2},\quad E_H = \frac{kq}{(\sqrt{2}l)^2}$$

$$\therefore E_{\text{net}} = \sqrt{(E_C - E_F)^2 + (E_G - E_B)^2} - (E_E - E_D)$$

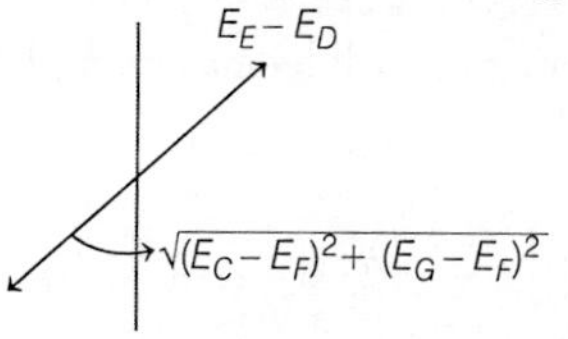

$$= \left\{\left(\frac{2kq}{l^2} - \frac{kq}{l^2}\right)^2 + \left(\frac{2kq}{l^2} - \frac{kq}{l^2}\right)^2\right\}^{1/2} - \left(\frac{kq^2}{l^2} - \frac{kq}{2l^2}\right)$$

$$= \frac{kq}{l^2}\sqrt{2} - \frac{kq}{2l^2} = \frac{kq}{2l^2}(2\sqrt{2}-1)$$

16. *(c)* We can replace $-Q$ charge at origin by $+Q$ and $-2Q$. Now, due to $+Q$ charge at every corner of cube, electric field at centre of cube is zero. So, net electric field at centre is only due to $-2Q$ charge at origin. Vector form of electric field strength,

$$\mathbf{E} = \frac{kq\mathbf{r}}{r^3}$$

Here, position vector,

$$r = \frac{a}{2}(\hat{\mathbf{x}} + \hat{\mathbf{y}} + \hat{\mathbf{z}})$$

$$\Rightarrow |\mathbf{r}| = \sqrt{\left(\frac{a}{2}\right)^2 + \left(\frac{a}{2}\right)^2 + \left(\frac{a}{2}\right)^2}$$

$$= \frac{\sqrt{3}\,a}{2}$$

$$\Rightarrow \mathbf{E} = \frac{1}{4\pi\varepsilon_0}\times\frac{(-2Q)\,a/2}{\left(\frac{a}{2}\sqrt{3}\right)^3}\cdot(\hat{\mathbf{x}} + \hat{\mathbf{y}} + \hat{\mathbf{z}})$$

$$\mathbf{E} = \frac{-2Q(\hat{\mathbf{x}} + \hat{\mathbf{y}} + \hat{\mathbf{z}})}{3\sqrt{3}\,\pi a^2\varepsilon_0}$$

17. *(a)* Consider a small ring of radius r and thickness dr on disc then charge, $dq = \sigma 2r\pi dr$, then

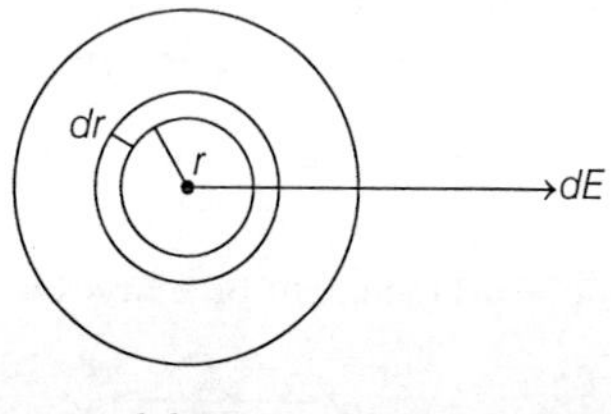

$$dE = \frac{kdqZ}{(r^2 + Z^2)^{3/2}} = \frac{k\sigma(2\pi r)drZ}{(r^2 + Z^2)^{3/2}}$$

On integrating both sides, we get

$$E = \int dE = \int_0^R \frac{k\sigma 2\pi r dr Z}{(Z^2 + r^2)}$$

$$= \frac{\sigma}{2\varepsilon_0}\left[1 - \frac{z}{\sqrt{R^2 + Z^2}}\right]$$

18. *(b)* Charge per unit length of the rod,

$$\lambda = \left(\frac{-Q}{R\theta}\right) = \left(\frac{-Q}{R\frac{2\pi}{3}}\right) = \frac{-3Q}{2\pi R}$$

$$\therefore \quad E = \int dE\cos\theta$$

$$\Rightarrow \quad E = \int_{-\pi/3}^{\pi/3}\frac{K\times Q}{\frac{2\pi}{3}R}\times\frac{Rd\theta}{R^2}\cos\theta$$

$$\Rightarrow \quad E = \frac{3}{2\pi}\frac{KQ}{R^2}[\sin\theta]_{-\pi/3}^{\pi/3}$$

$$= \frac{3}{2\pi}\frac{KQ}{R}\times\frac{2\sqrt{3}}{2}$$

$$\Rightarrow \quad E = \frac{3\sqrt{3}Q}{8\pi^2\varepsilon_0 R^2}(+\hat{\mathbf{i}})$$

19. *(c)*

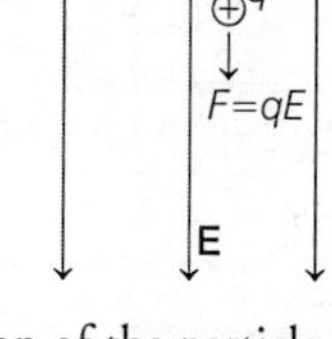

Acceleration of the particle,

$$a = \frac{F}{m} = \frac{q}{m}\cdot E$$

Velocity v and distance x can be related using

$$v^2 - u^2 = 2ax$$

$$\Rightarrow \quad v = \sqrt{2\left(\frac{q}{m}E\right)x} \quad (\because u = 0)$$

This equation resembles a parabola $y^2 = 4ax$. So, the graph between v and x will be as shown in option (c).

20. *(d)* The given situation is shown below.

At equilibrium,

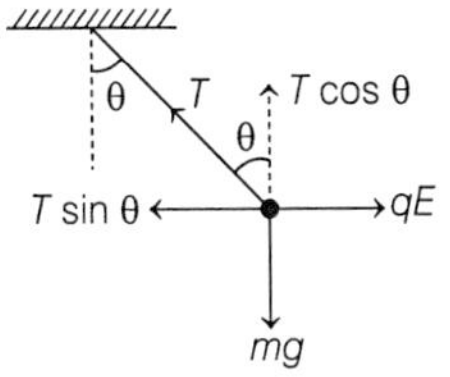

$T\cos\theta = mg$ ...(i)

and $T\sin\theta = qE$ ...(ii)

On dividing Eq. (ii) by Eq. (i), we get,

$$\tan\theta = \frac{qE}{mg}$$

$$\Rightarrow \quad \theta = \tan^{-1}\left(\frac{5\times10^{-6}\times2\times10^{3}}{2\times10^{-3}\times10}\right)$$

$$= \tan^{-1}(0.5)$$

21. *(d)* Given,

Length of wire, $L = 20$ cm

charge $\quad Q = 10^3\,\varepsilon_0$C

Electric field at the centre of the semicircular is

$$E = \frac{2k\lambda}{r} \Rightarrow E = \frac{2k\left(\frac{2Q}{\pi r}\right)}{r}\left[\text{As, } \lambda = \frac{2Q}{\pi r}\right]$$

$$= \frac{4kQ}{\pi r^2} = \frac{4kQ\pi^2}{\pi L^2} = \frac{4\pi kQ}{L^2}$$

$$\Rightarrow E = 4\pi \times \frac{1}{4\pi\varepsilon_0} \times \frac{10^3\,\varepsilon_0}{(20\times10^{-2})^2}$$

Electric field emerges from positive charge and terminates of negative charge. So, E by positive are at centre is away from arc and by –ve charge toward centre. Direction will be $+\hat{\mathbf{i}}$

$$= 25\times10^3\,\hat{\mathbf{i}}\text{ N/C}$$

22. *(c)* Linear charge density, $\lambda = \left(\frac{q}{\pi r}\right)$

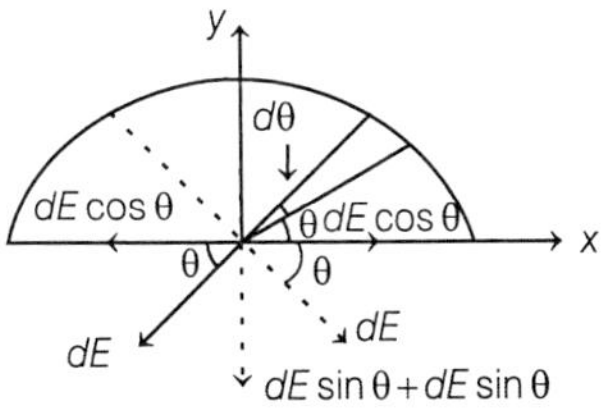

$$E = \int dE \sin\theta\,(-\mathbf{j})$$

$$= \int_0^{\pi}\int \frac{K\cdot dq}{r^2}\sin\theta\,(-\mathbf{j})$$

$$= \frac{K}{r^2}\int_0^{\pi}\frac{qr}{\pi r}\,d\theta\sin\theta\,(-\mathbf{j})$$

$$= \frac{K}{r^2}\frac{q}{\pi}\int_0^{\pi}\sin\theta\,d\theta\,(-\mathbf{j})$$

$$= \frac{1}{4\pi\varepsilon_0}\frac{q}{\pi r^2}[-\cos\theta]_0^{\pi}(-\mathbf{j})$$

$$= \frac{q}{2\pi^2\varepsilon_0 r^2}(-\mathbf{j})$$

23. *(b)* Electric field due to charged circular loop.

$$E = \frac{KQZ}{(Z^2+R^2)^{\frac{3}{2}}}$$

Electric field will be maximum if

$$\frac{dE}{dZ} = 0$$

$$\frac{d}{dZ}\left[\frac{KQZ}{(Z^2+R^2)^{\frac{3}{2}}}\right] = 0$$

$$KQ\frac{d}{dZ}\left[\frac{Z}{(Z^2+R^2)^{3/2}}\right] = 0$$

$\Rightarrow$

$$\frac{(Z^2+R^2)^{\frac{3}{2}}\cdot 1 - Z\left(\frac{3}{2}\right)(Z^2+R^2)^{\frac{1}{2}}(2Z)}{(Z^2+R^2)} = 0$$

$$\Rightarrow (Z^2+R^2)^{\frac{3}{2}} - 3Z^2(Z^2+R^2)^{\frac{1}{2}} = 0$$

$$\Rightarrow (Z^2+R^2)^{\frac{1}{2}}[Z^2+R^2-3Z^2] = 0$$

$$\Rightarrow R^2 - 2Z^2 = 0$$

$$\Rightarrow Z = \frac{R}{\sqrt{2}} = \frac{\sqrt{2}a}{\sqrt{2}} \quad [\because R = \sqrt{2}a]$$

$$= a$$

24. *(a)* Electric field at a distance r due to infinite long wire is

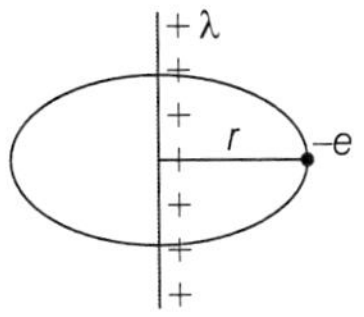

$$E = \frac{2k\lambda}{r} \quad \text{...(i)}$$

Force on electron

$$\Rightarrow \quad F = qE = eE \quad \text{...(ii)}$$

$$F = e\left(\frac{2k\lambda}{r}\right)$$

$$\Rightarrow \quad F = \frac{2k\lambda e}{r} \quad \text{...(iii)}$$

This force will provide the centripetal force,

$$\frac{mv^2}{r} = \frac{2k\lambda e}{r} \Rightarrow v = \sqrt{\frac{2k\lambda e}{m}}$$

Kinetic energy $\Rightarrow$ KE $= \frac{1}{2}mv^2$

$$\text{KE} = \frac{1}{2}m\left(\frac{2k\lambda e}{m}\right)$$

$= k\lambda e$ (where k is constant)

Kinetic energy will remain constant.

25. *(d)* Electric field of wire, $E = \dfrac{\lambda}{2\pi\varepsilon_0 r}$

$$= \frac{2k\lambda}{r}$$

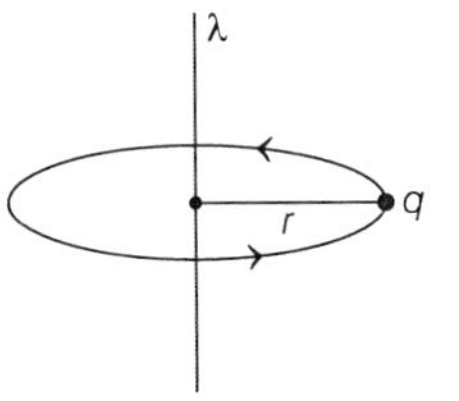

$$\therefore \quad F = qE = \frac{\lambda q}{2\pi\varepsilon_0 r} = \frac{2k\lambda q}{r}$$

Force on charged particle = Centripetal force

$$\Rightarrow \quad \frac{2k\lambda q}{r} = m\omega^2 r$$

$$\Rightarrow \quad \omega = \left(\frac{2k\lambda q}{mr^2}\right)^{\frac{1}{2}}$$

Hence, time period of rotation is,

$$T = \frac{2\pi}{\omega} = 2\pi r\sqrt{\frac{m}{2k\lambda q}}$$

26. *(b)* Given, electric field,

$$E = \frac{8m}{e}\text{ Vm}^{-1}$$

Length of plates, $l = 1$ m

Speed of electron, $v_x = 2$ ms^{-1}

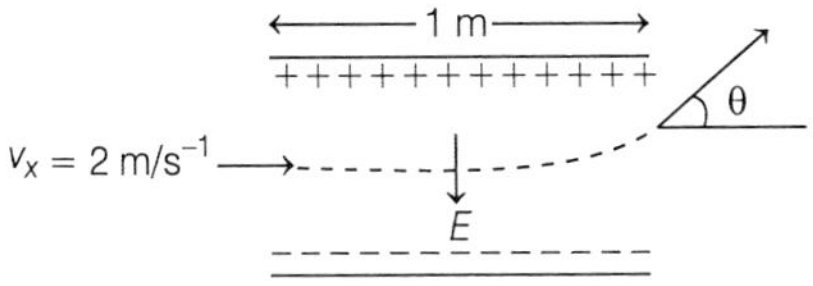

The time taken by the electron to come out of field undeflected is

$$t = \frac{l}{v_x} = \frac{1}{2} = 0.5\text{ s}$$

As, we know,

$$F = qE$$

Since, electron is accelerating opposite to the direction of **E**.

$$\Rightarrow m \times -a_y = e \times \frac{8m}{e}$$

$$\Rightarrow a_y = -8 \text{ ms}^{-2}$$

Using first equation of motion,

$$v_y = u_y + a_y t$$

$$\Rightarrow v_y = 0 + (-8)(0.5)$$

or $|v_y| = 4 \text{ ms}^{-1}$

As the electron is deflected by angle θ, so

$$\tan\theta = \frac{v_y}{v_x}$$

$$\Rightarrow \theta = \tan^{-1}\left(\frac{4}{2}\right)$$

$$= \tan^{-1}(2)$$

27. *(a)* Electric field due to cylinder of charge density ρ at a distance r from the axis of cylinder is given as

$$E = \frac{\rho R^2}{2\varepsilon_0 r}$$

∴Force on particle having charge q,

$$F = qE = \frac{q\rho R^2}{2\varepsilon_0 r}$$

The electric force on charge particle provides the required centripetal force.

i.e $\frac{mv^2}{r} = F \Rightarrow \frac{mv^2}{r} = \frac{q\rho R^2}{2\varepsilon_0 r}$

$$\Rightarrow mv^2 = \frac{q\rho R^2}{2\varepsilon_0} \quad \ldots\text{(i)}$$

∴Kinetic energy of charged particle,

$$K = \frac{1}{2}mv^2 = \frac{1}{2}\left(\frac{q\rho R^2}{2\varepsilon_0}\right)$$

[From Eq. (i)]

$$K = \frac{q\rho R^2}{4\varepsilon_0}$$

28. *(d)* Given, mass of block, $m = 1$ kg

Acceleration due to gravity,

$$g = 9.8 \text{ ms}^{-2}$$

Inclination, $\theta = 30°$

Electric field, $E = 200$ N/C

Coefficient of friction, $\mu = 0.2$

Charge, $q = 5 \text{ mC} = 5 \times 10^{-3}$ C

friction force, $f = \mu N$

Where, N be the normal reaction.

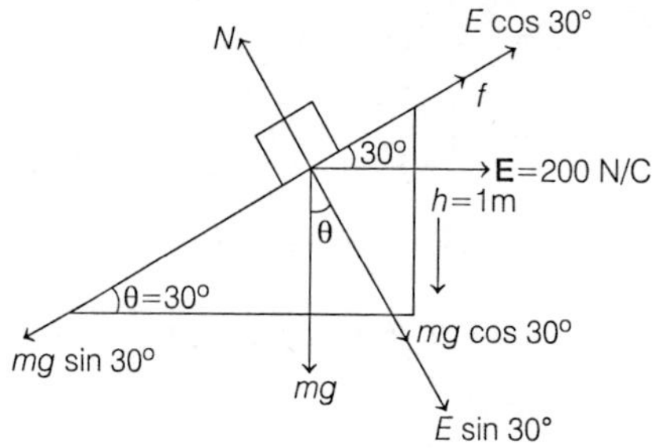

Force along the vertical of inclined will be zero.

$$\Rightarrow N = mg\cos 30° + qE\sin 30°$$

$$\Rightarrow N = 1 \times 9.8 \times \sqrt{3}/2 + 5 \times 10^{-3} \times 200 \times 1/2$$

$$= 8.99 \text{ N} \simeq 9 \text{ N}$$

$$\therefore f = \mu N = \frac{2}{10} \times 9 = 1.8 \text{ N}$$

Now, total force along the plane of incline, $mg\sin 30° - f - qE\cos 30° = ma$

$$\Rightarrow 1 \times 9.8 \times \frac{1}{2} - 1.8 - 5 \times 10^{-3} \times \frac{200\sqrt{3}}{2} = a$$

$$\Rightarrow a = 2.34 \text{ ms}^{-2}$$

Since, initial velocity of body,

$$u = 0 \text{ ms}^{-1}$$

and distance along incline, s

$$= h/\sin 30° = 1/\sin 30° = 2$$

By using second equation of motion,

$$s = ut + 1/2at^2$$

$$\Rightarrow 2 = 0 + 1/2 \times 2.34 \times t^2$$

$$\Rightarrow t^2 = \frac{4}{2.34}$$

$$\Rightarrow t = \frac{2}{\sqrt{2.34}} = 0.65 \times 2 = 1.3 \text{ s}$$

29. *(c)* Given, length of conductor $= L$

Charge on conductor $= Q$

According to figure,

$$OP = a = \frac{\sqrt{3}}{2}L, \; OQ = \frac{L}{2}$$

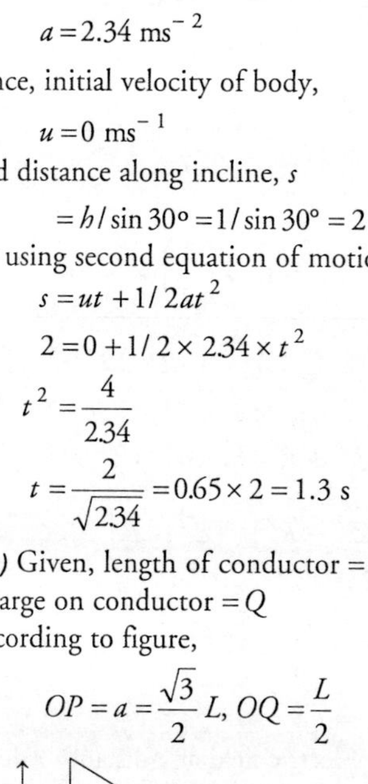

Let $PQ = r = \sqrt{OP^2 + OQ^2}$

$$\Rightarrow PQ = \sqrt{\left(\frac{\sqrt{3}}{2}L\right)^2 + \left(\frac{L}{2}\right)^2}$$

$$= \sqrt{\frac{3}{4}L^2 + \frac{L^2}{4}} = L$$

and E be the electric field at point P.

Since, E (due to finite wire)

$$= \frac{k\lambda}{a}(\sin\phi_1 + \sin\phi_2) \quad \ldots\text{(i)}$$

where, k = Coulomb's constant $= \frac{1}{4\pi\varepsilon_0}$

λ = linear charge density $= Q/L$

and $\sin\phi_1 = \sin\phi_2 = \frac{L/2}{L} = \frac{1}{2}$

On substituting the above value in Eq. (i), we get

$$E = \frac{k\lambda}{a} = \frac{1}{4\pi\varepsilon_0}\frac{Q}{L\cdot\frac{\sqrt{3}L}{2}}$$

$$= \frac{1}{2\sqrt{3}\pi\varepsilon_0}\frac{Q}{L^2}$$

30. *(d)* Electric field at point A,

$$E_A = \frac{k\rho \times \frac{4}{3}\pi\left(\frac{R}{2}\right)^3}{\left(\frac{R}{2}\right)^2}$$

$$= \frac{4\pi k\rho}{3}\cdot\frac{R}{2}$$

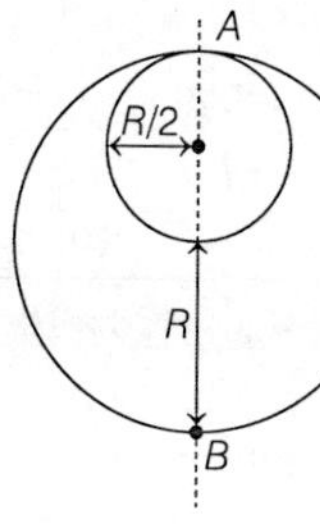

Electric field at point B,

$$E_B = \frac{k.\rho\frac{4}{3}\pi R^3}{R^2} - \frac{k\rho\frac{4}{3}\pi\left(\frac{R}{2}\right)^3}{\left(\frac{3R}{2}\right)^2}$$

$$E_B = k\rho\frac{4}{3}\pi\left(R - \frac{R}{18}\right)$$

$$= \frac{4\pi k\rho}{3} \times \frac{17R}{18}$$

$$\therefore \frac{E_A}{E_B} = \frac{R \times 18}{2 \times 17R} = \frac{18}{34}$$

31. *(d)*

key idea Remember $E \propto \frac{1}{r^2}$ for single charge

$E \propto \frac{1}{r^3}$ for dipole; $E \propto \frac{1}{r^4}$ for quadrupole

The given distribution of charges can be shown as the figure below

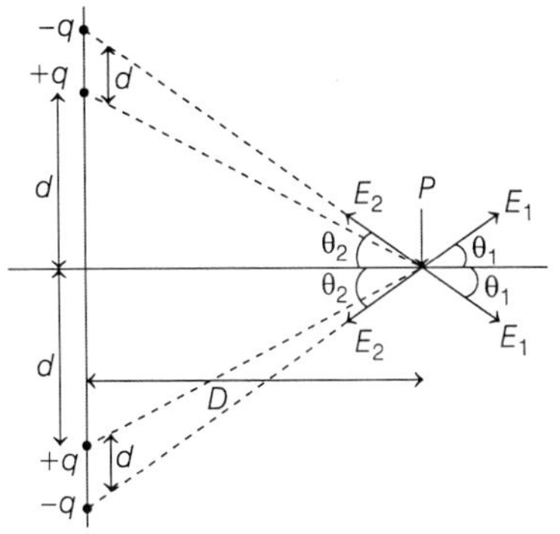

Electric field at point P,

$$E = E_1 \cos\theta_1 + E_1 \cos\theta_1 - E_2 \cos\theta_2 - E_2 \cos\theta_2$$

$$= 2E_1 \cos\theta_1 - 2E_2 \cos\theta_2$$

$$= \frac{2kq}{(d^2 + D^2)} \cos\theta_1 - \frac{2kq}{(2d)^2 + D^2} \cos\theta_2$$

As, $\cos\theta_1 = \frac{D}{(d^2 + D^2)^{1/2}}$

Similarly, $\cos\theta_2 = \frac{D}{[(2d)^2 + D^2]^{1/2}}$

$$= 2kqD[(d^2 + D^2]^{-3/2} - (4d^2 + D^2)^{-3/2}]$$

$$= \frac{2kqD}{D^3}\left[\left(1 + \frac{d^2}{D^2}\right)^{-3/2} - \left(1 + \frac{4d^2}{D^2}\right)^{-3/2}\right]$$

As $D >> d$, then by applying binomial approximation, we get

$$= \frac{2kq}{D^2}\left[1 - \frac{3}{2}\frac{d^2}{D^2} - \left(1 - \frac{3}{2}\left(\frac{4d^2}{D^2}\right)\right)\right]$$

$$= \frac{2kq}{D^2}\left[\frac{9d^2}{2D^2}\right]$$

$$= \frac{9kqd^2}{D^4}$$

$$\Rightarrow E \propto \frac{1}{D^4}$$

Alternate solution

Given charge distribution is as shown below

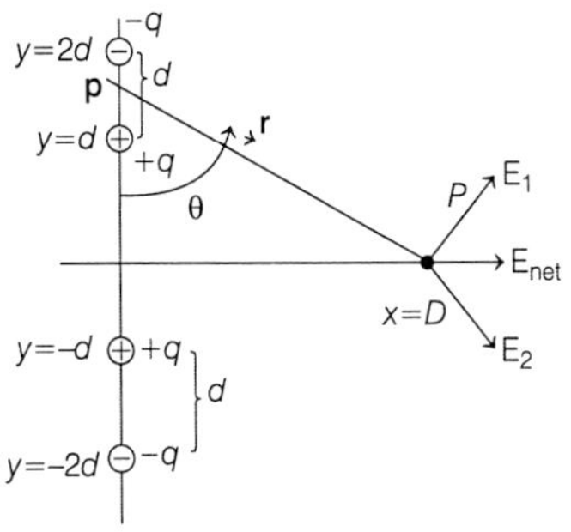

So, we can view above point charges as combination of pair of dipoles or a quadrupole.

By symmetry, the field components parallel to quadrupole cancels and the resultant perpendicular field is

$$E = \frac{2q}{4\pi\varepsilon_0}\left(\frac{1}{D^2} - \frac{D}{(D^2 + d^2)^{3/2}}\right)$$

$$= \frac{2q}{4\pi\varepsilon_0 D^2}\left(1 - \left(1 + \frac{d^2}{D^2}\right)^{-3/2}\right)$$

As, $\left(1 + \frac{d^2}{D^2}\right)^{-\frac{3}{2}} \approx \left(1 - \frac{3}{2}\frac{d^2}{D^2}\right)$

Using binomial expansion, we get,

$$E = \frac{3qd^2}{4\pi\varepsilon_0 D^4} \Rightarrow E \propto \frac{1}{D^4}$$

Note Dependence of field for a point charge is

$$E \propto \frac{1}{r^2}$$

For a dipole, it is, $E \propto \frac{1}{r^3}$

For a quadrupole, it is, $E \propto \frac{1}{r^4}$

32. *(16)* Given,

Surface charge density $= +\sigma_s$ C / m^2

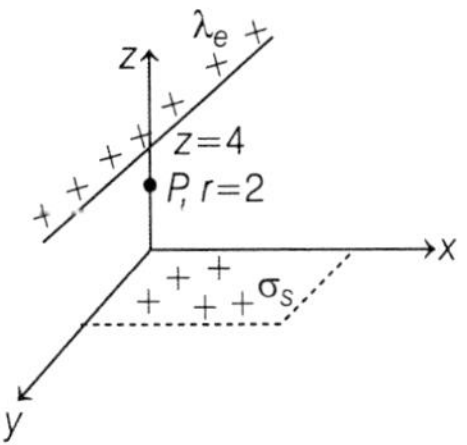

(electric field due to sheet)

Linear charge density $= +\lambda_e$ C / m

$$\sigma_s = 2\lambda_e$$

At point P, $E_s = \frac{\sigma_s}{2\varepsilon_0}$ ...(i)

$$E_l = \frac{\lambda_e}{2\pi r\varepsilon_0} \quad \text{...(ii)}$$

Now, From Eqs. (i) and (ii), We get

$$\frac{E_s}{E_l} = \frac{\sigma_s}{2\varepsilon_0} \times \frac{2\pi r\varepsilon_0}{\lambda_e} \; (r = 2)$$

$$\frac{E_s}{E_l} = \frac{4\pi}{1} \quad (\sigma_s = 2\lambda_e \text{ and } r = 2)$$

$$4\pi : 1 = \pi\sqrt{n} : 1$$

Hence, value of n is 16.

33. *(8)* Given, $u = 1$ m/s, $a = -\frac{\sigma e}{2\varepsilon_0 m}$

$$t = 1 \text{ s} \Rightarrow s = -1 \text{ m}$$

$$\Rightarrow \quad s = ut + \frac{1}{2}at^2$$

$$-1 = 1 \times 1 - \frac{1}{2}\frac{\sigma e}{2\varepsilon_0 m}(1)^2$$

$$-1 = 1 - \frac{1}{2}\frac{\sigma e}{2\varepsilon_0 m}$$

$$\frac{1}{2}\frac{\sigma e}{2\varepsilon_0 m} = 1 + 1 \Rightarrow \sigma = \frac{8\varepsilon_0 m}{e}$$

$$\alpha = 8$$

34. *(3)* Consider an arc of $d\theta$, force on arc is balanced by component of tension.

$$F_e = \frac{kQ}{R^2} \times \frac{q}{2\pi} d\theta$$

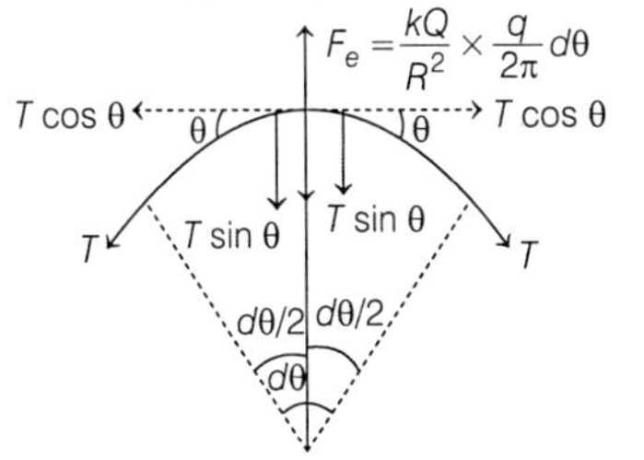

So, $2T \sin\frac{d\theta}{2} = F_e$

$$F_e = \frac{kQq}{R^2}\left(\frac{d\theta}{2\pi}\right)$$

$$2T \sin\left(\frac{d\theta}{2}\right) = \frac{kQq}{R^2}\left(\frac{d\theta}{2\pi}\right)$$

$$\Rightarrow \quad T = \frac{\frac{kQq}{R^2}\left(\frac{d\theta}{2\pi}\right)}{2\sin\left(\frac{d\theta}{2}\right)} = \left(\frac{\frac{kQq}{\pi R^2}}{2}\right)$$

As $\frac{\sin\left(\frac{d\theta}{2}\right)}{\left(\frac{d\theta}{2}\right)} = 1$

$$T = \frac{kQq}{2\pi R^2} = \frac{9\times10^9\times 2\pi\times 30\times10^{-12}}{2\pi\times(0.3)^2}$$

$$= \frac{9\times10^{-3}\times 30}{9\times10^{-2}} = 3\text{ N}$$

35. *(45)* Given, $E = 10$ N/C

Kinetic energy of electron,

$$K = 0.5\text{ eV} = 0.5\times1.6\times10^{-19}\text{ J}$$

$$= 8\times10^{-20}\text{ J}$$

$$l = 10\text{ cm} = 0.1\text{ m}$$

If θ be the angle of deviation of path as it comes out of the field, then

$$\tan\theta = \frac{eEl}{mv_0^2} = \frac{eEl}{2\times\frac{1}{2}mv_0^2}$$

$$= \frac{eEl}{2K} = \frac{1.6\times10^{-19}\times10\times0.1}{2\times8\times10^{-20}}$$

$$= \frac{16}{16} = 1$$

$$\theta = \tan^{-1}(1) = 45°$$

36. *(2)* Given situation is shown below,

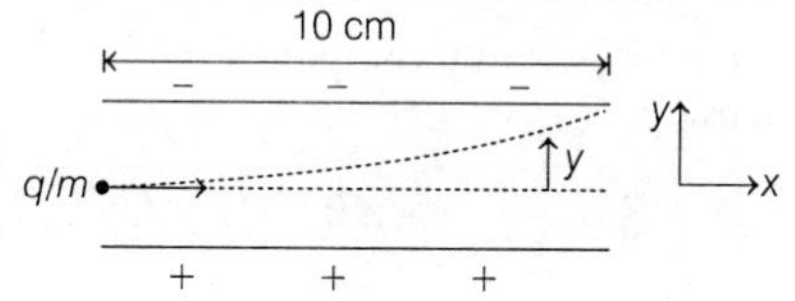

Velocity of particles is

$$v = 3\times10^7\text{ m/s}$$

Time taken by particle to cross a region of 10 cm in horizontal direction is

$$t = \frac{d}{v} = \frac{10\times10^{-2}}{3\times10^7} = \frac{10^{-8}}{3}\text{ s}$$

Acceleration of particle in y-direction

$$a = \frac{F}{m} = \frac{qE}{m} = \frac{q}{m}\cdot E$$

$$= 2\times10^{11}\times1.8\times10^3\text{ m/s}^2$$

$$= 3.6\times10^{14}\text{ m/s}^2$$

Deflection y of charged particle using

$$s_y = u_y t + \frac{1}{2}a_y t^2 \qquad [u_y = 0]$$

displacement in y-direction,

$$y = \frac{1}{2}\times3.6\times10^{14}\times\left(\frac{10^{-8}}{3}\right)^2$$

$$= \frac{3.6}{2\times9}\times10^{14-16}\text{ m}$$

$$= \frac{3.6}{18}\times10^{-2+3}\text{ mm} = 2\text{ mm}$$

Topic 3

01. *(b)* $E = \frac{\sigma}{2\varepsilon_0}$, $\tau = pE\sin\theta$

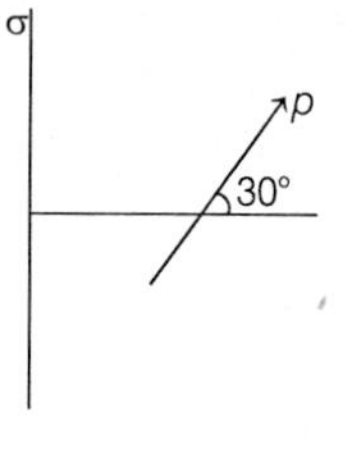

$$\tau = pE\sin\theta$$

$$= (2\times10^{-6})\left(\frac{2}{10}\right)\left(\frac{100\times10^{-6}}{2\times8.85\times10^{-12}}\right)\times\frac{1}{2}$$

$$= \frac{10}{8.85} = 1.12\text{ N-m}$$

02. *(c)* $U = -pE\cos\theta$

$$W_{ext} = \Delta U = U_f - U_i$$

$$= -pE\cos180° + pE\cos0°$$

$$= 2pE = 2qlE$$

$$= 2\times4\times10^{-6}\times18\times10^{-2}\times10^4$$

$$= 144\times10^{-4} = 14.4\text{ mJ}$$

03. *(c)* Electric dipole moment

$$p = q\times 2l$$

$$p = 0.01\times0.4\times10^{-3}$$

$$= 4\times10^{-6}\text{ C-m}$$

$$E = 10\times10^{-5}\text{ N/C}$$

$$|\tau| = |p\times E|$$

$$= 4\times10^{-6}\times10\times10^{-5}\sin30°$$

$$= 2\times10^{-10}\text{ N-m},$$

04. *(b)* Electric field due to electric dipole on equilotrial plane at a distance r from its centre is given as

$$E = \frac{1}{4\pi\varepsilon_0}\cdot\frac{p}{r^3} \Rightarrow E \propto \frac{1}{r^3}$$

05. *(a)*

> **key idea** As, dipole moments points in same direction
>
> 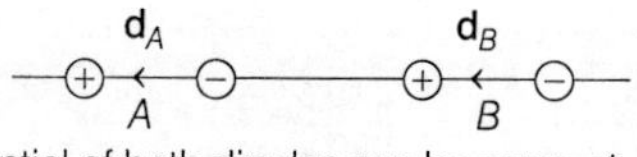
>
>
> So, potential of both dipoles can be same at some point between A and B.

Let potentials are same at P, distant x from B as shown below

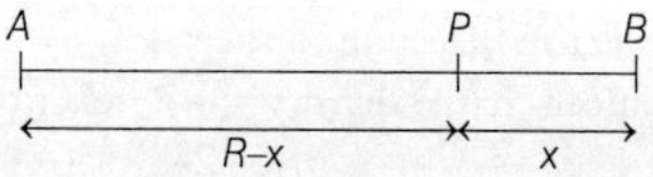

Then, $\frac{4qa}{(R-x)^2} = \frac{2qa}{(x)^2}$

$$2x^2 = (R-x)^2 \Rightarrow \sqrt{2}x = R - x$$

$$\Rightarrow \qquad x = \frac{R}{\sqrt{2}+1}$$

Distance from A is

$$R - x = R - \frac{R}{\sqrt{2}+1} = \frac{\sqrt{2}R}{\sqrt{2}+1}$$

06. *(c)* In a non-uniform electric field, the dipole may experience both non-zero torque as well as translational force. For example, as shown in figure

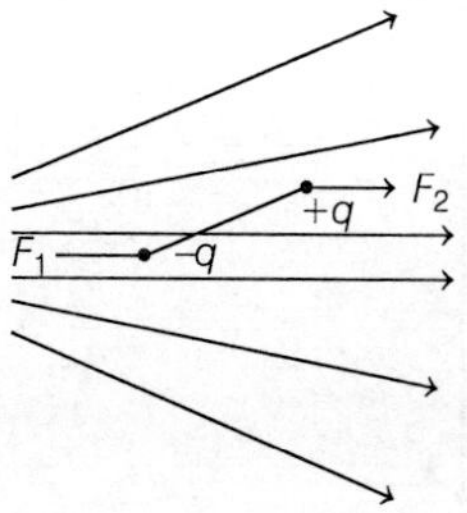

$F_1 \neq F_2$ as E is non-uniform. Therefore, torque would also be non-zero.

07. *(b)* Let the dipole is slightly rotated from its mean position

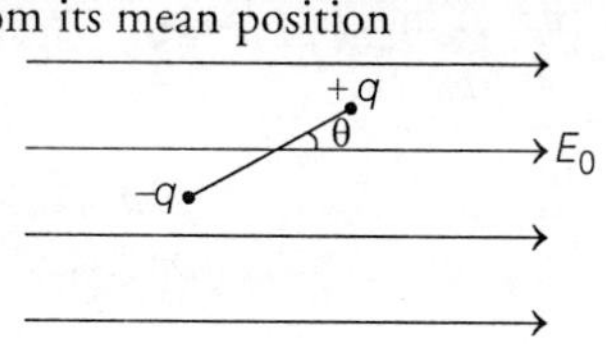

$$\tau = I\alpha$$

$$pE_0\sin\theta = 2\times m\left(\frac{l}{2}\right)^2\left(\frac{-d^2\theta}{dt^2}\right)$$

$$\frac{ml^2}{2}\frac{d^2\theta}{dt^2} + pE_0\sin\theta = 0$$

$$\frac{d^2\theta}{dt^2} + \frac{2pE_0\sin\theta}{ml^2} = 0$$

For θ to be small, $\sin\theta \simeq \theta$

$$\Rightarrow \qquad \frac{d^2\theta}{dt^2} + \frac{2pE_0\theta}{ml^2} = 0$$

$$\Rightarrow \omega = \sqrt{\frac{2pE_0}{ml^2}} = \sqrt{\frac{2qlE_0}{ml^2}} = \sqrt{\frac{2qE_0}{ml}}$$

$$T = 2\pi\sqrt{\frac{ml}{2qE_0}}$$

08. *(a)*

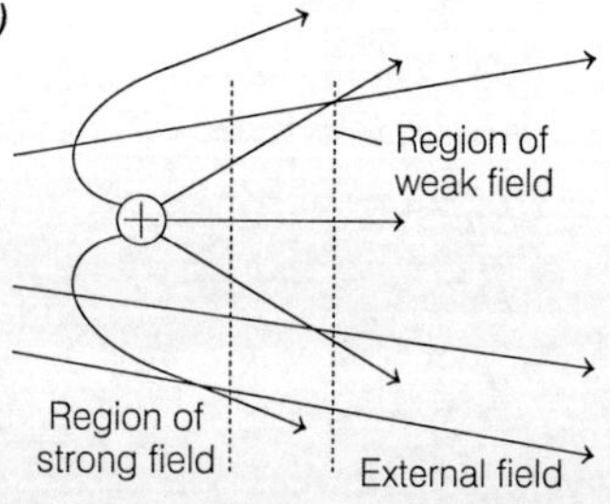

Statement I is correct as number density of field lines increases as we go near a positive charge placed in an electric field as shown above.

Statement II is also correct as net force on a dipole is non-zero in an non-uniform electric field as shown below.

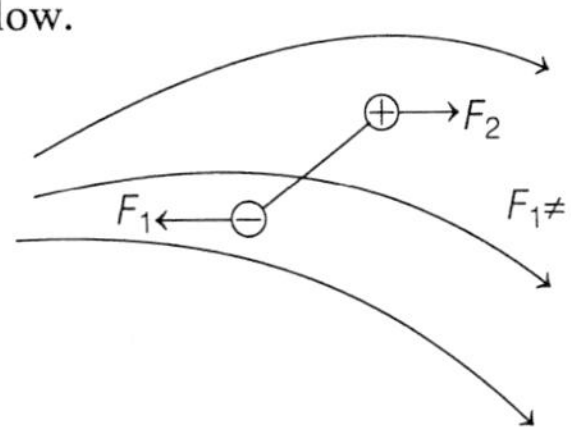

09. *(c)* Given system is equivalent to two dipoles inclined at 60° to each other as shown in the figure below

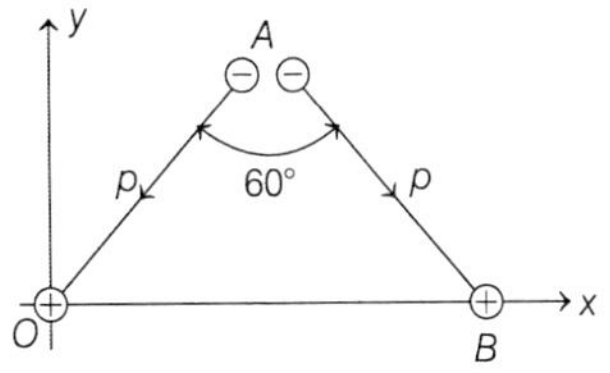

Now, magnitude of resultant of these dipole moments is

$$p_{\text{net}} = \sqrt{p^2 + p^2 + 2p \cdot p \cos 60^\circ}$$
$$= \sqrt{3}p = \sqrt{3}ql$$

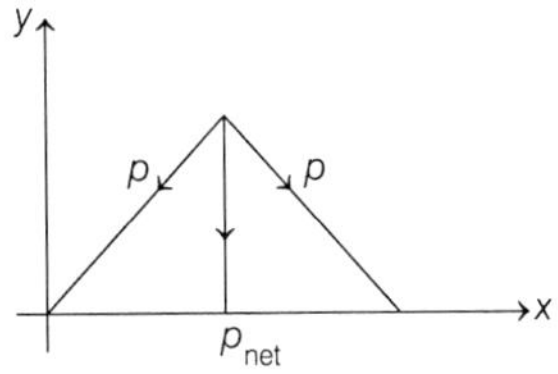

As, resultant is directed along negative y-direction

$$p_{\text{net}} = -\sqrt{3}p\hat{\mathbf{j}} = -\sqrt{3}ql\,\hat{\mathbf{j}}$$

10. (*)

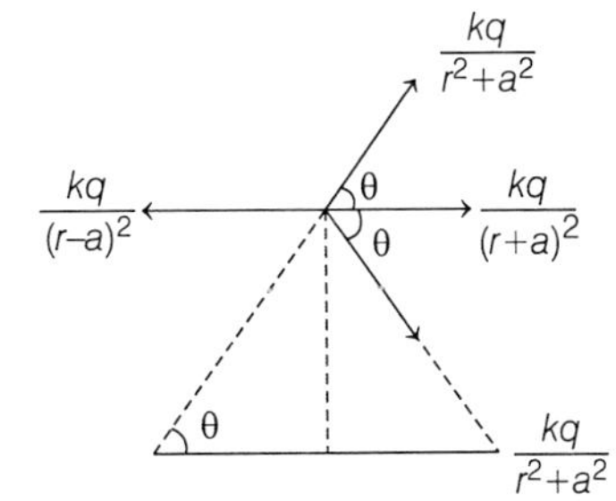

$$\frac{kq}{(r-a)^2} = \frac{kq}{(r+a)^2} + \frac{2kq}{(r^2+a^2)} \cos\theta$$

$$\frac{1}{(r-a)^2} = \frac{1}{(r+a)^2} + \frac{2a}{(r^2+a^2)^{\frac{3}{2}}}$$

$$\frac{1}{(r-a)^2} - \frac{1}{(r+a)^2} = \frac{2a}{(r^2+a^2)^{\frac{3}{2}}}$$

$$\frac{4ra}{(r^2-a^2)^2} = \frac{2a}{(r^2+a^2)^{\frac{3}{2}}}$$

$$\frac{2r}{(r^2-a^2)^2} = \frac{1}{(r^2+a^2)^{3/2}}$$

$$\frac{4r^2}{(r^2-a^2)^4} = \frac{1}{(r^2+a^2)^3}$$

$$4r^2(r^2+a^2)^3 = (r^2-a^2)^4$$

$$4r^8\left(1+\frac{a^2}{r^2}\right)^3 = r^8\left(1-\frac{a^2}{r^2}\right)^4$$

$$4\left(1+\frac{a^2}{r^2}\right)^3 = \left(1-\frac{a^2}{r^2}\right)^4$$

No option is matching.

> ***Alternate solution***
>
> Field on axis of a dipole ($2a << r$) is $E_{\text{axis}} = \frac{2kp}{r^3}$
>
> Field on equatorial axis of a dipole ($2a << r$) is $E_2 = E_{\text{equator}} = \frac{kp}{r^3}$
>
> Now, when charge Q at point P is placed, it is acted upon by two forces;
>
> 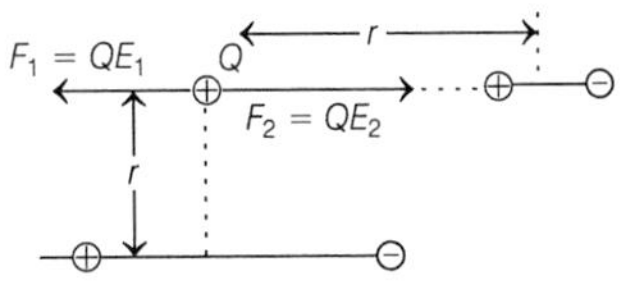
>
>
> Given net force on charge Q is zero.
>
> $\Rightarrow QE_1 = QE_2$
>
> But this is not possible for any value of a and r.

11. *(d)* In this case since masses of both end are not same, therefore we need to find centre of mass about which dipole will oscillate and then we will find moment of Inertia about this axis, to find torque and hence ω, COM will be $\frac{1}{3}$ from $2m$

and $\frac{2L}{3}$ from m

MI about this axis.

$$I = 2m \times \left(\frac{L}{3}\right)^2 + m\left(\frac{2L}{3}\right)^2$$

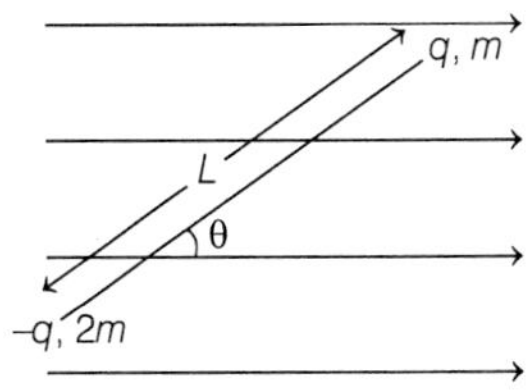

$$= \frac{2mL^2}{9} + \frac{4mL^2}{9} = \frac{6mL^2}{9} = \frac{2mL^2}{3},$$

Now, torque on the dipole,

$\tau = I\alpha = \sin\theta$

where, α is angular acceleration,

$$\therefore \quad \alpha = \frac{pE\sin\theta}{I}$$

$$\alpha = \left(\frac{pE}{I}\right)\theta$$

[For small angle $\sin\theta \approx \theta$]

$$\alpha = \left(\frac{qEL}{I}\right)\theta \qquad [\because p = qL]$$

$$\alpha = \left(\frac{3qEL}{2mL^2}\right)\theta \qquad \left[\because I = \frac{2mL^2}{3}\right]$$

$$\alpha = \left(\frac{3qE}{2mL}\right)\theta \qquad \ldots\text{(i)}$$

The above equation is similar to the equation for system excuting angular SHM.

Campare Eq. (i) with

$\alpha = \omega^2\theta$, then

$$\omega^2 = \frac{3qE}{2mL}$$

$$\Rightarrow \qquad \omega = \sqrt{\frac{3qE}{2mL}}$$

12. *(a)*

> **key idea** When an electric dipole is placed in an electric field **E** at some angle θ, then two forces equal in magnitude but opposite in direction acts on the +ve and –ve charges, respectively. These forces forms a couple which exert a torque, which is given as $\tau = \mathbf{p} \times \mathbf{E}$ where, **p** is dipole moment.

Torque on the dipole can also be given as

$$\tau = I\alpha = -pE\sin\theta$$

where, I is the moment of inertia and α is angular acceleration. For small angles, $\sin\theta \approx \theta$

$$\therefore \qquad \alpha = -\left(\frac{pE}{I}\right)\theta \qquad \ldots\text{(i)}$$

Moment of inertia of the given system is

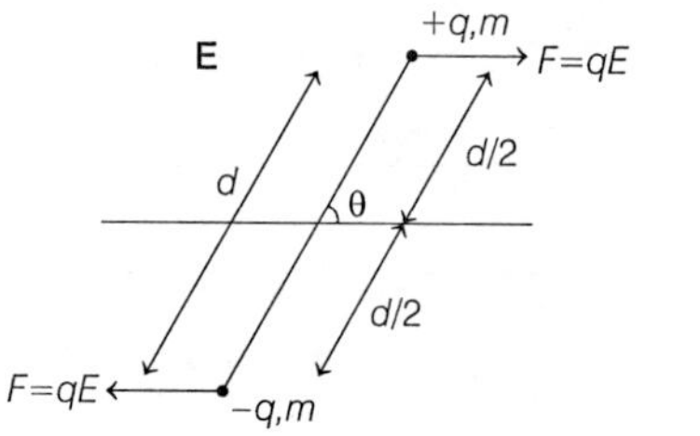

$$I = m\left(\frac{d}{2}\right)^2 + m\left(\frac{d}{2}\right)^2 = \frac{2md^2}{4} = \frac{md^2}{2}$$

Substituting the value of I in Eq. (i), we get

$$\Rightarrow \quad \alpha = -\left(\frac{2pE}{md^2}\right)\cdot\theta \quad \text{...(ii)}$$

The above equation is similar to the equation for a system executing angular SHM.

Comparing Eq. (ii), with the general equation of angular SHM, i.e.

$$\alpha = -\omega^2\theta$$

where, ω is the angular frequency, we get

$$\omega^2 = \frac{2pE}{md^2} \quad \text{or} \quad \omega = \sqrt{\frac{2pE}{md^2}}$$

As $\quad p = qd$

$$\therefore \quad \omega = \sqrt{\frac{2qdE}{md^2}} = \sqrt{\frac{2qE}{md}}$$

13. *(b)*

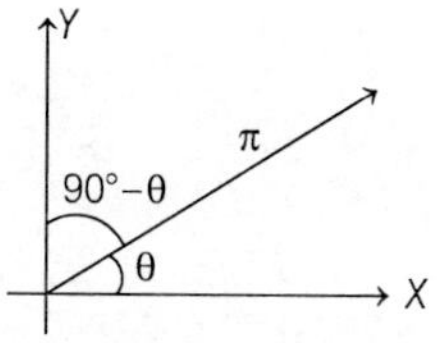

Torque applied on a dipole $\tau = pE\sin\theta$

For electric field, $E_1 = E\hat{i}$

Therefore, torque in this direction

$$\tau_1 = pE_1 \sin\theta$$

In electric field $E_2 = \sqrt{3}\,E\hat{j}$,

it means field is directed along positive Y-axis, so angle between dipole and field will be $90° - \theta$.

Torque in this direction

$$\tau_2 = pE\sin(90° - \theta)$$
$$= p\sqrt{3}\,E_1 \cos\theta$$

According to question

$$\tau_2 = -\tau_1 \Rightarrow |\tau_2| = |\tau_1|$$

$$\therefore pE_1 \sin\theta = p\sqrt{3}\,E_1 \cos\theta$$

$$\tan\theta = \sqrt{3} \Rightarrow \tan\theta = \tan 60°$$

$$\therefore \quad \theta = 60°$$

14. *(12)* $p = 6\times10^{-6}$ C-m

$$E = 10^6 \text{ V/m}$$
$$W = \Delta U = -pE(\cos\theta_f - \cos\theta_i)$$
$$W = -pE(\cos 180° - \cos 0°)$$
$$= -pE(-1-1) = 2pE = 12\text{J}$$

15. *(16)*

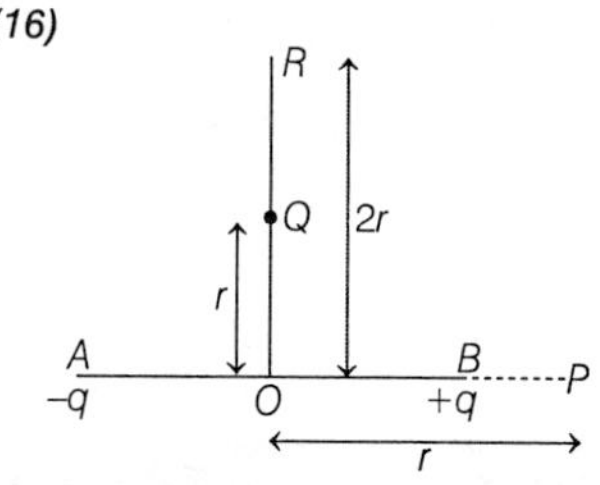

Electric field at point P due to electric dipole AB

$$E_p = \frac{1}{4\pi\varepsilon_0}\cdot\frac{2p}{r^3}$$

Similarly at point R,

$$E_R = \frac{1}{4\pi\varepsilon_0}\cdot\frac{p}{(2r)^3}$$

$$\therefore \quad \frac{E_R}{E_P} = \frac{\frac{1}{(2r)^3}}{\frac{2}{r^3}} = \frac{1}{16}$$

$$E_R = \frac{E_P}{16} = \frac{E}{16} = \frac{E}{x} \quad \text{(given)}$$

$$x = 16$$

16. *(5)*

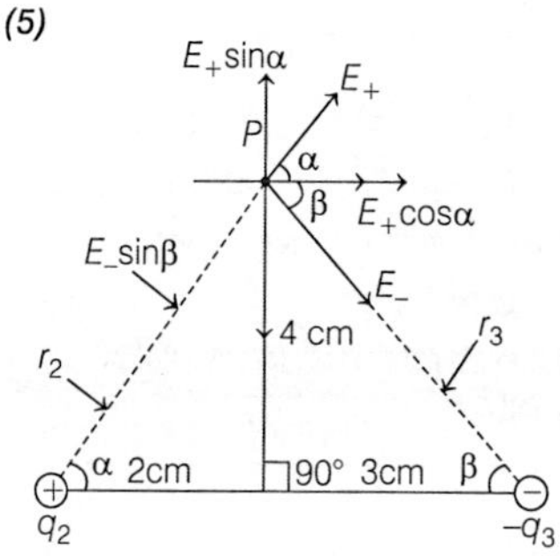

From geometry of figure ,

$$\sin\alpha = \frac{4}{\sqrt{4^2 + 2^2}} = \frac{4}{\sqrt{20}} = \frac{2}{\sqrt{5}}$$

$\sin\beta = 4/5$

Now net field along Y-axis at P will be zero when,

$$E_+ \sin\alpha = E_- \sin\beta$$

$$\frac{kq_2}{r_2^2}\cdot\sin\alpha = \frac{kq_3}{r_3^2}\cdot\sin\beta$$

$$\Rightarrow \frac{q_2}{q_3} = \left(\frac{r_2}{r_3}\right)^2 \cdot \frac{\sin\beta}{\sin\alpha}$$

$$= \frac{20}{25}\times\frac{4/5}{2/\sqrt{5}} = \frac{8}{5\sqrt{5}}$$

$$\therefore \quad x = 5$$

17. *(2)* Given,

$$q = 4\mu\text{C}$$
$$E = 0.20\hat{i}\text{ V/cm}$$
$$= 20\,\hat{i}\text{ V/m}$$

position vector of point A,

$$r_A = \hat{i} + 4\hat{k}$$

position vector of point B,

$$r_B = 2\hat{i} - \hat{j} + 5\hat{k}$$

Dipole moment,

$$\mathbf{p} = q[\mathbf{r}_B - \mathbf{r}_A]$$
$$= 4\times10^{-6}[\hat{i} - \hat{j} + \hat{k}]$$

Torque, $\tau = \mathbf{p}\times\mathbf{E}$

$$= 4\times10^{-6}[\hat{i} - \hat{j} + \hat{k}]\times 20\hat{i}$$
$$= 80\times10^{-6}[\hat{i}\times\hat{i} - \hat{j}\times\hat{i} + \hat{k}\times\hat{i}]$$
$$= 80\times10^{-6}[0 + \hat{k} + \hat{j}]$$
$$|\tau| = 80\times10^{-6}\sqrt{(1)^2 + (1)^2}$$
$$= 8\sqrt{2}\times10^{-5}\text{ N-m}$$
$$= 8\sqrt{\alpha}\times10^{-5}\text{ N-m}$$

$$\therefore \quad \alpha = 2$$

18. *(6)* Given, for two electric dipole,

$$p_1 = 1.2\times10^{-30}\text{ C-m}$$

and $\quad p_2 = 2.4\times10^{-30}$ C-m

$$E_1 = 5\times10^4\text{ NC}^{-1}$$
$$E_2 = 15\times10^4\text{ NC}^{-1}$$

We know that maximum torque on electric dipole in electric field,

$$\tau_{max} = pE$$

$$\Rightarrow \frac{(\tau_{max})_1}{(\tau_{max})_2} = \frac{p_1E_1}{p_2E_2}$$

$$= \frac{1.2\times10^{-30}\times5\times10^4}{2.4\times10^{-30}\times15\times10^4}$$

$$= \frac{1}{6} = \frac{1}{x} \quad \text{(given)}$$

$$\therefore \quad x = 6$$

Topic 4

01. *(b)*

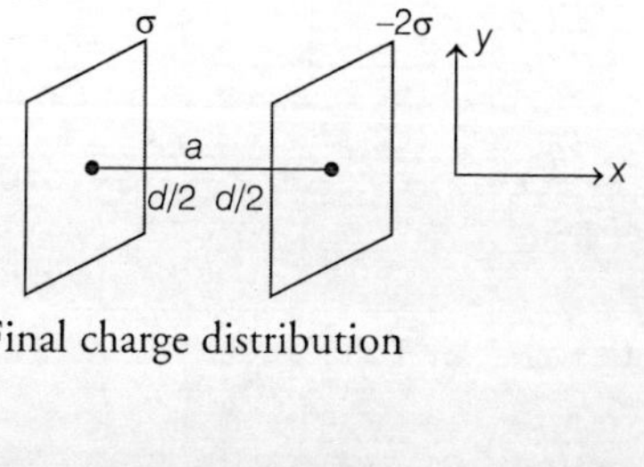

Final charge distribution

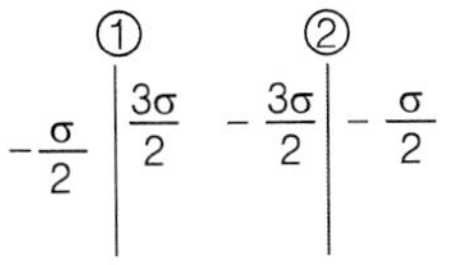

$$F = qE$$

$$F_{net} = \frac{3\sigma}{2\varepsilon_0} \cdot q$$

02. *(d)* By Gauss law

$$\phi = \frac{q_{in}}{\varepsilon_0}$$

$$\Rightarrow -2 \times 10^4 = \frac{q_{in}}{8.85 \times 10^{-12}}$$

$$\Rightarrow \quad q_{in} = -17.70 \times 10^{-8} \text{ C}$$

03. *(d)* As the charge is kept at edge's centre, then total electric flux is given by

$\phi = \frac{q}{\varepsilon_0}$ and $q = \frac{a\lambda}{8}$, where λ is linear change density.

$$\therefore \quad \phi = \frac{\lambda a}{8\varepsilon_0}$$

04. *(c)* Total flux $(\phi) = \frac{Q_{enclosed}}{\varepsilon_0}$

$$= \frac{+5q + q - 2q}{\varepsilon_0} = \frac{4q}{\varepsilon_0}$$

05. *(b)* According to Gauss's law,

$$\mathbf{E} \cdot d\mathbf{A} = \frac{q}{\varepsilon_0}$$

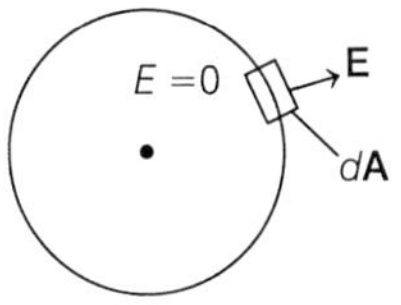

$$\Rightarrow \quad EdA \cos 0° = \frac{\sigma dA}{\varepsilon_0} \Rightarrow E = \frac{\sigma}{\varepsilon_0}$$

06. *(c)* According to Gauss's law,

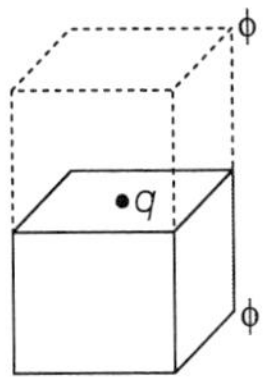

$$2\phi = \frac{q}{\varepsilon_0} \Rightarrow \phi = \frac{q}{2\varepsilon_0}$$

07. *(d)* Given,

$$\mathbf{E} = (6\hat{i} + 5\hat{j} + 3\hat{k}) \text{ N/C}$$

$$\mathbf{A} = 30\hat{i} \text{ m}^2$$

$\therefore$ Electric flux,

$$\phi = \mathbf{E} \cdot \mathbf{A} = (6\hat{i} + 5\hat{j} + 3\hat{k}) \cdot 30\hat{i}$$

$$= 180 \text{ N-m}^2/\text{C}$$

08. *(d)* Given,

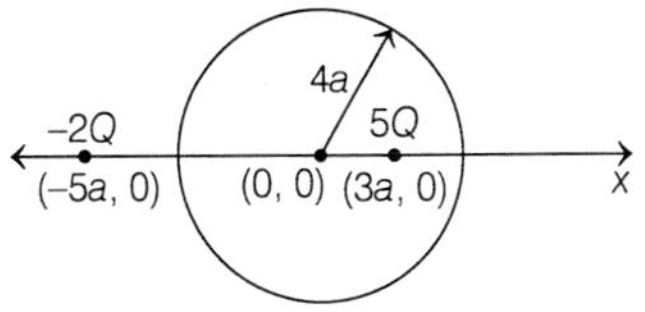

5Q charge is inside the spherical region

flux through sphere $= \frac{5Q}{\varepsilon_0}$

09. *(c)* Electric field in conducting materials is always zero. So, $E_{II} = 0$

If we make any Gaussian surface in region I and III, then it will enclose the charge Q.

Hence, by Gauss's law,

$$\oint E \cdot ds = \frac{Q \text{ enclosed}}{\varepsilon_0}$$

There will be electric field.

So, $E_I \neq 0$ and $E_{III} \neq 0$

10. *(b)* Given, a charge q is placed at centre of a closed hemispherical non-conducting surface. To find the total flux passing through flat surface. Flux passing through flat surface = Flux passing through curved surface

So, $\phi = \frac{q}{2\varepsilon_0}$

($\because$ For a complete sphere, flux $(\phi) = \frac{q}{\varepsilon_0}$)

11. *(c)* Electric field due to plane charge distribution does not depend on distances of different points near it i.e., it is same at all points near it, which is equal to $\frac{\sigma}{2\varepsilon_0}$

i.e. $\quad E_1 = E_2 = \frac{\sigma}{2\varepsilon_0}$

12. *(a)* Given, dipole moment,

$$\mathbf{p} = 30 \times 10^{-5} \text{ C-m}$$

Using Gauss's law, $\phi = \frac{Q_{in}}{\varepsilon_0}$

$$= 0 \quad (\because Q_{in} = 0 \text{ for dipole})$$

Thus, both A and R are true and R is the correct explanation of A.

13. *(b)* Net charge on electric dipole

$$= +q - q = 0$$

Hence, according to Gauss's law,

Electric flux, $\phi = \frac{q_{net}}{\varepsilon_0} = \frac{0}{\varepsilon_0} = 0$

Electric field due to electric dipole is non-zero and varies at point to point.

Hence, statement I is true.

Electric field due to charged solid sphere at a distance r from centre.

$$E = \frac{1}{4\pi\varepsilon_0} \cdot \frac{Qr}{R^3}$$

which is non-zero.

Hence, Statement II is false.

Hence, option (b) is the correct.

14. *(a)* Equation $|\mathbf{E}| = \frac{q_{en}}{\varepsilon_0 |A|}$

Given $\quad \int \mathbf{E} \cdot d\mathbf{A} = \frac{q_{en}}{\varepsilon_0}$

Now, in finding the electric field by above equation, the integral is easy to evaluate, if $|\mathbf{E}| =$ constant.

Also, if $|\mathbf{E}| =$ constant for the surface, then surface is equipotential.

15. *(a)* Statement I is dimensionally wrong while from Gauss' law,

$$E(4\pi r^2) = \frac{\rho \cdot \frac{4}{3}\pi r^3}{\varepsilon_0} \Rightarrow E = \frac{\rho r}{3\varepsilon_0}$$

Thus, Statement II is correct.

16. *(a)* For uniformly charged spherical shell, $\quad E = 0, \; r < R = \frac{Q}{4\pi\varepsilon_0 r^2}, \; r \geq R$

Thus, option (a) represents both electric field at $r < R$ and $r \geq R$.

17. *(a)* From Gauss' law,

Net flux $= \frac{\text{Charge enclosed}}{\varepsilon_0}$

$$\Rightarrow \quad \frac{q}{\varepsilon_0} = \phi_2 - \phi_1 \text{ or } q = (\phi_2 - \phi_1)\varepsilon_0$$

18. *(a)* Arc ABC,

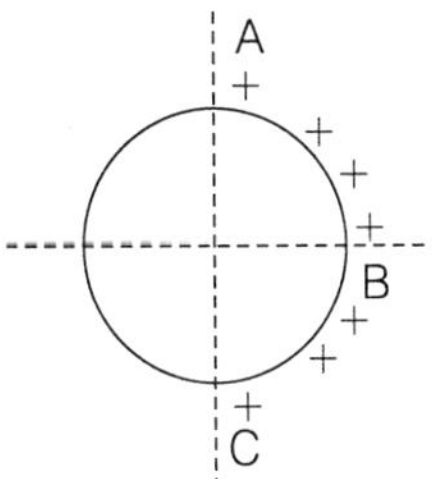

$$\mathrm{E} = \frac{2K\lambda}{R} \sin\left(\frac{\alpha}{2}\right) = \frac{2K\lambda}{R} \sin\left(\frac{180}{2}\right)$$

$$= \frac{2K\lambda}{R} \sin 90° = \frac{2K\lambda}{R} = \frac{\sqrt{2}}{\sqrt{2}} \frac{K\lambda}{R} \cdot 2$$

$$= \frac{2E \times \sqrt{2}}{\sqrt{2} \times \sqrt{2}} = \sqrt{2}E$$

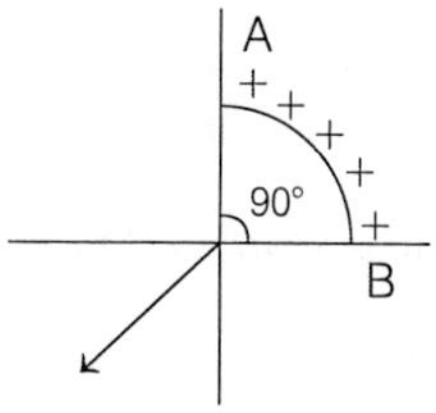

$$E = \frac{2K\lambda}{R}\sin\left(\frac{90}{2}\right) = \frac{2K\lambda}{R}\frac{1}{\sqrt{2}}$$

$$E = \sqrt{2}\frac{K\lambda}{R}$$

19. *(c)*

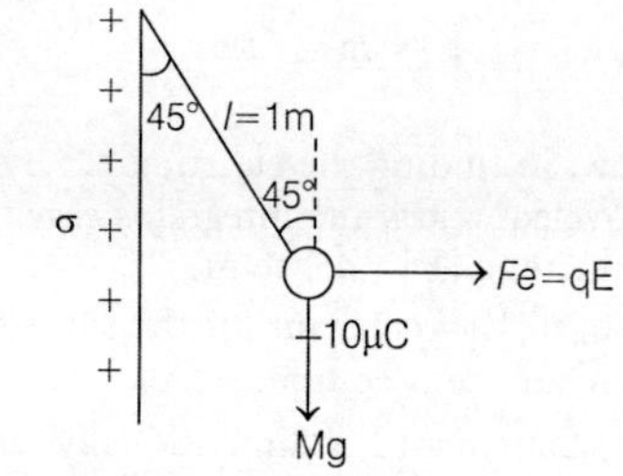

Given, $\varepsilon_0 = 8.85 \times 10^{-12}$

$m = 100 \text{ mg} = 100 \times 10^{-6}$ kg

$g = 10 \text{ m/s}^2$

$Q = 10\,\mu\text{C} = 10 \times 10^{-6}\text{C}$

$$E = \frac{\sigma}{2\varepsilon_0}$$

Initially

$\text{Fe} = mg$ $\quad qE = mg$

$qE = mg$ $\quad q\left(\frac{\sigma}{2\varepsilon_0}\right) = mg$

$$\sigma = \frac{2\varepsilon_0 mg}{q}$$

$$\sigma = \frac{2 \times 8.85 \times 10^{-12} \times 100 \times 10^{-6} \times 10}{10 \times 10^{-6}}$$

$\sigma = 17.7 \times 10^{-10}$ C/m^2

$\sigma = 17.7 \times 10^{-10}$ C/m^2 $= 1.77$ nC/m^2

20. *(d)*

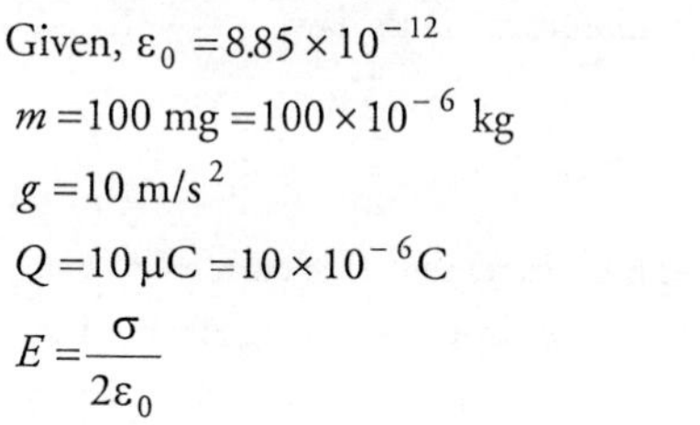

$$E = \frac{KQ}{R^2}, \rho = \frac{Q}{V}$$

$$\therefore \quad Q = \rho V = \rho\frac{4}{3}\pi R^3$$

$$E = \frac{K\rho\frac{4}{3}\pi R^3}{R^2}$$

$$= K\rho\frac{4}{3}\pi R^3$$

$$E = K\rho\frac{4}{3}\pi R$$

So $E_A > E_B > E_D > E_C$

$\mathbf{E_A} > \mathbf{E_B}$, $E_C \neq E_D$

21. *(c)* Electric field inside the cavity enclosed by a conductor is zero provided there are no charges placed inside the cavity.

This is because the charges on the conductor's surface will rearrange to cancel out any electric field inside the cavity, effectively creating a shield against external electric field.

22. *(*)* A. Electric field inside a spherical shell is always 0.

B. Electric field due to infinite plane sheet is $\sigma / 2\varepsilon_0$

C. Electric field outside a spherical shell is

$$\frac{q}{4\pi\varepsilon_0 r^2} \text{ or } \frac{\sigma}{\varepsilon_0}\frac{R^2}{r^2}$$

D. Electric field between two oppositely charged infinite sheet is σ / ε_0.

23. *(d)* Flux through any closed surface,

$$\phi_E = \frac{q_{\text{enclosed}}}{\varepsilon_0}$$

Flux through cube

$$C_1, \phi_1 = 2Q / \varepsilon_0$$

Flux through cube C_2,

$$\phi_2 = \frac{2Q + 3Q}{\varepsilon_0} = \frac{5Q}{\varepsilon_0}$$

Ratio, $\quad \dfrac{\phi_1}{\phi_2} = \dfrac{2Q/\varepsilon_0}{5Q/\varepsilon_0} = \dfrac{2}{5}$

• Concept Enhancer

Gauss' law is true for any closed surface, no matter what its shape or size be. It is applicable only for three dimensional closed surface.

24. *(a)* Electric field intensity due to an infinite thin plane sheets of uniform surface charge density σ is given,

as $\quad \mathbf{E} = \dfrac{\sigma}{2\varepsilon_0}\hat{\mathbf{n}}$

Situation given is shown below.

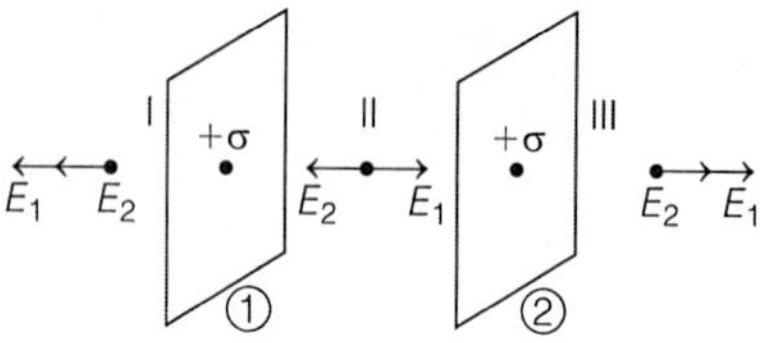

In region *I*, net field,

$$\mathbf{E}_\text{I} = \mathbf{E}_1 + \mathbf{E}_2$$

$$\mathbf{E}_\text{I} = \left(-\frac{\sigma}{2\varepsilon_0}\hat{\mathbf{n}}\right) + \left(-\frac{\sigma}{2\varepsilon_0}\hat{\mathbf{n}}\right) = -\frac{\sigma}{\varepsilon_0}\hat{\mathbf{n}}$$

Similarly, in region II and III,

$$\mathbf{E}_\text{II} = \left(-\frac{\sigma}{2\varepsilon_0}\hat{\mathbf{n}}\right) + \left(\frac{\sigma}{2\varepsilon_0}\hat{\mathbf{n}}\right) = 0$$

$$\mathbf{E}_\text{III} = \left(\frac{\sigma}{2\varepsilon_0}\hat{\mathbf{n}}\right) + \left(\frac{\sigma}{2\varepsilon_0}\hat{\mathbf{n}}\right) = \frac{\sigma}{\varepsilon_0}\hat{\mathbf{n}}$$

25. *(c)* Given, situation is as shown in figure.

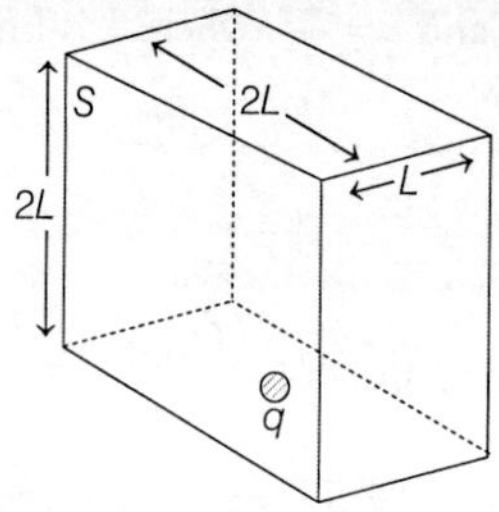

Now, we make cube complete by adding one more identical cuboid.

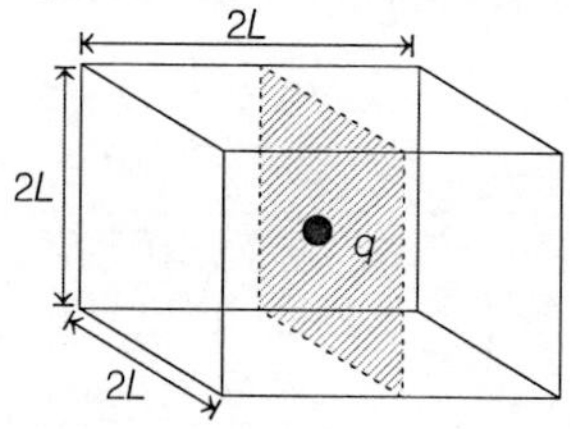

As, charge is placed symmetrically to above cube, flux through surface

(one of face $= \dfrac{1}{6}\left(\dfrac{q}{\varepsilon_0}\right)$

26. *(c)* Here, $E \cdot A = \dfrac{q}{\varepsilon_0}$

$$\Rightarrow E \cdot 4\pi r^2 = \frac{\int_0^r \rho_0\left(\frac{3}{4} - \frac{r}{R}\right)4\pi r^2\,dr}{\varepsilon_0}$$

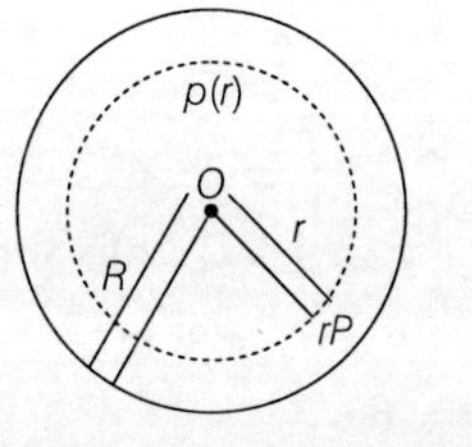

$$\Rightarrow E\cdot 4\pi r^2 = \frac{\rho_0 4\pi}{\varepsilon_0}\left(\frac{3}{4}\frac{r^3}{3} - \frac{r^4}{4R}\right)$$

$$\Rightarrow \quad E\cdot r^2 = \frac{\rho_0 r^3}{4\varepsilon_0}\left(1-\frac{r}{R}\right)$$

$$\Rightarrow \quad E = \frac{\rho_0 r}{4\varepsilon_0}\left(1-\frac{r}{R}\right)$$

27. *(d)*

> **key idea** You can use Gauss's law to calculate electric flux. According to Gauss's law, electric flux
>
> $\phi = \oint \mathbf{E}\cdot d\mathbf{S} = \frac{q}{\varepsilon_0}$,
>
> where q is total charge enclosed by that surface

Given, charge q is at one of the corner of the cube.

$\therefore$Contribution of q in cube will be

$q_{\text{enclosed}} = q/8$

As, only 3 faces of cube is allowing the flux lines to pass through it.

$$\therefore \text{Flux } (\phi) = \frac{q_{\text{enclosed}}}{\varepsilon_0}$$

$$= \frac{1}{3}\frac{q/8}{\varepsilon_0} = \frac{q}{24\varepsilon_0}$$

28. *(b)* Electric charge distribution at inner and outer surface of spherical shell due to the electric dipole can be shown as below

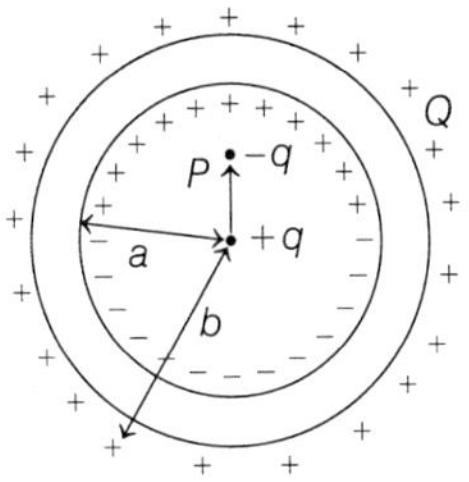

Here, we need to consider two different factors

(i) Charge on the spherical shell is $+Q$ which will be distributed on its outer surface as shown in figure.

(ii) Electric dipole will create non-uniform electric field inside the shell which will distribute the charges on inner surface as shown in figure.

But, its net contribution to the outer side of the shell will be zero as net charge of a dipole is zero.

$\therefore$Net charge on outer surface of shell will be $+Q$.

Hence, using (ii), option (a) is incorrect as field inside shell is not uniform.

Option (b) is correct, as net charge on outer surface is $+Q$ even in the presence of dipole.

Option (c) is incorrect, as surface charge density at outer surface is uniform $\left(=\frac{Q}{A}=\frac{Q}{4\pi b^2}\right)$.

Option (d) is incorrect, as surface charge density at inner surface is non-zero.

> ***Alternate Solution***
>
> Using Gauss' law at outer surface, let charge on dipole is q,
>
> $$\phi = \frac{\Sigma q}{\varepsilon_0} = E\cdot A \text{ or } E = \frac{1}{A\varepsilon_0}\Sigma q$$
>
> $$= \frac{(+Q+q-q)}{A\varepsilon_0} = \frac{Q}{A\varepsilon_0} = \frac{\sigma}{\varepsilon_0} = \text{constant}$$

29. *(c)* Electric field inside the uniformly charged sphere varies linearly,

$$E = \frac{kQ}{R^3}\cdot r, \ (r \le R),$$

while outside the sphere, it varies as inverse square of distance,

$$E = \frac{kQ}{r^2}; \ (r \ge R) \text{ which is correctly}$$

represented in option (c).

30. *(c)* Electric field,

$$E = -\frac{d\phi}{dr} = -2ar$$

By Gauss' theorem,

$$E\,(4\pi r^2) = \frac{q}{\varepsilon_0}$$

$$\Rightarrow \quad q = -8\pi\,\varepsilon_0\, ar^3$$

$$\rho = \frac{dq}{dV} = \frac{dq}{dr}\times\frac{dr}{dV}$$

$$= (-24\pi\varepsilon_0\, ar^2)\left(\frac{1}{4\pi r^2}\right)$$

$$= -6\,\varepsilon_0 a$$

31. *(b)* Apply Gauss's theorem, the total charge upto distance r can be calculated as followed

$$dq = 4\pi r^2\cdot dr\cdot\rho$$

$$= 4\pi r^2\cdot dr\cdot\rho_0\left[\frac{5}{4}-\frac{r}{R}\right]$$

$$= 4\pi\rho_0\left[\frac{5}{4}r^2\,dr - \frac{r^3}{R}\,dr\right]$$

$$\int dq = q = 4\pi\rho_0\int_0^r\left(\frac{5}{4}r^2\,dr - \frac{r^3}{R}\,dr\right)$$

$$q = 4\pi\rho_0\left[\frac{5}{4}\frac{r^3}{3} - \frac{1}{R}\frac{r^4}{4}\right]$$

As electric field intensity is given as,

$$E = \frac{kq}{r^2}$$

$$= \frac{1}{4\pi\varepsilon_0}\frac{1}{r^2}\cdot 4\pi\rho_0\left[\frac{5}{4}\left(\frac{r^3}{3}\right) - \frac{r^4}{4R}\right]$$

$$E = \frac{\rho_0 r}{4\varepsilon_0}\left[\frac{5}{3}-\frac{r}{R}\right]$$

32. *(c)*

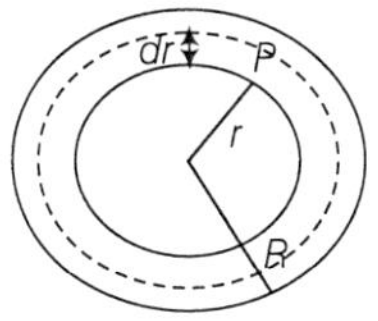

Uisng Gauss's law

$$E\cdot ds = \frac{q}{\varepsilon_0}$$

$$E4\pi r_1^2 = \frac{\int_0^{r_1}\frac{Q}{\pi R^4}r4\pi r^2\,dr}{\varepsilon_0}$$

$$\Rightarrow \quad E = \frac{Qr_1^2}{4\pi\varepsilon_0 R^4}$$

33. *(d)* Here, linear charge density

$\lambda = 2\text{n C/m} = 2\times10^{-9}$ C/m

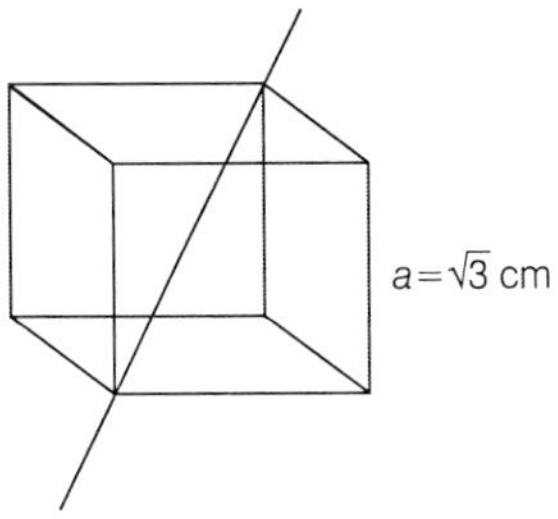

Length of wire inside the cube,

l = diagonal of cube $=\sqrt{3}a$

$=\sqrt{3}\times\sqrt{3}$ cm $=3\times10^{-2}$ m

$\therefore$Total charge enclosed by the cube

$$q = \lambda l = 2\times10^{-9}\times3\times10^{-2}$$

$$= 6\times10^{-11}\text{ C}$$

$$\therefore \text{Net flux, } \phi = \frac{q}{\varepsilon_0} = \frac{6\times10^{-11}}{\frac{1}{36\pi\times10^9}}$$

$$= 216\pi\times10^{-2}$$

$$= 2.16\pi\text{ Nm}^2\text{ C}^{-1}$$

$$\therefore \quad x = 2.16\,\pi$$

34. *(a)*

key idea Force on A is zero only when repulsion of A and B = attraction of positive charge distribution of radius a and charge A.

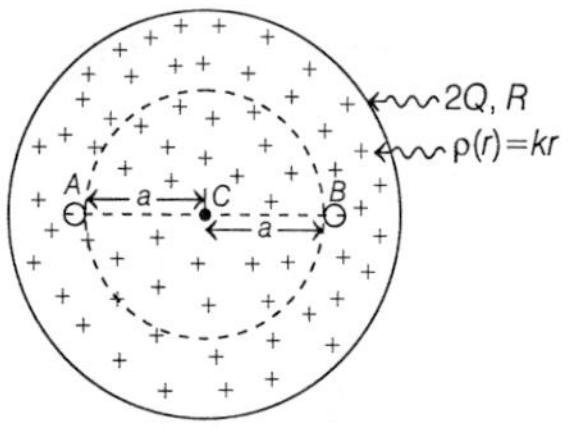

In given charge distribution, let r is radius of a shell of thickness dr.

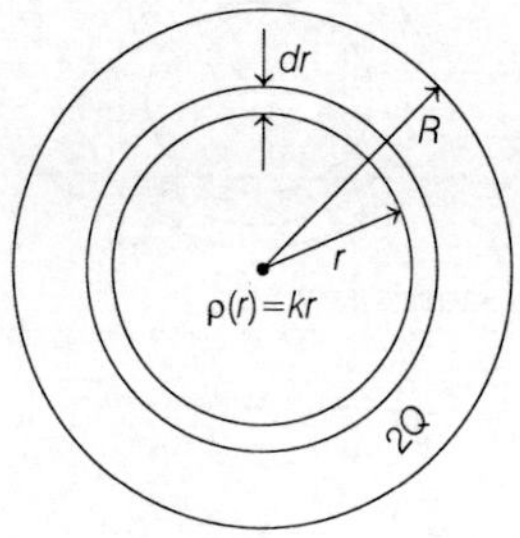

Charge dQ present in shell of thickness dr = Volume of shell × Volumetric charge density

$$\Rightarrow \quad dQ = (4\pi r^2 \times dr) \times (kr)$$
$$= 4\pi k r^3 \, dr$$

Total charge in sphere is

$$2Q = \int_0^R dQ = \int_0^R 4\pi k r^3 \, dr$$

$$\Rightarrow \quad 2Q = 4\pi k \left[\frac{r^4}{4}\right]_0^R$$

$$\Rightarrow \quad k = \frac{2Q}{\pi R^4}$$

Now, using Gauss' law, electric field on the surface of sphere of radius a is

$$E \int_0^a dA = \frac{1}{\varepsilon_0} \cdot \int_0^a (kr \, 4\pi r^2 \, dr)$$

$$\Rightarrow \quad E \cdot 4\pi a^2 = \frac{1}{\varepsilon_0} \cdot 4\pi k \left(\frac{a^4}{4}\right)$$

$$\Rightarrow \quad E = \frac{ka^2}{4\varepsilon_0} = \frac{2Q a^2}{4\pi\varepsilon_0 R^4}$$

Force of attraction on charge A (or B) due to this field is

$$F_1 = QE = \frac{2Q^2 a^2}{4\pi\varepsilon_0 R^4}$$

Force of repulsion on charge A due to B is

$$F_2 = \frac{1}{4\pi\varepsilon_0} \frac{Q^2}{(2a)^2} = \frac{1}{4\pi\varepsilon_0} \cdot \frac{Q^2}{4a^2}$$

If charge A(or B) does not feel any force, then,

$$F_1 = F_2$$

$$\Rightarrow \quad \frac{2Q^2 a^2}{4\pi\varepsilon_0 R^4} = \frac{1}{4\pi\varepsilon_0} \cdot \frac{Q^2}{4a^2}$$

$$\Rightarrow \quad 8a^4 = R^4 \Rightarrow a = 8^{-1/4} R$$

35. *(a)* As, Gaussian surface at distance r from centre,

$$\frac{Q + \int_a^r \frac{A}{r} 4\pi r^2 \, dr}{\varepsilon_0} = E 4\pi r^2$$

$$E \, 4\pi\varepsilon_0 r^2 = Q + A \frac{4\pi}{r^2}\left(\frac{r^2 - a^2}{2}\right)$$

$$E = \frac{1}{4\pi\varepsilon_0}\left[\frac{Q}{r^2} + A \, 2\pi\left(\frac{r^2 - a^2}{r^2}\right)\right]$$

$$E = \frac{1}{4\pi\varepsilon_0}\left(\frac{Q}{r^2} + A \, 2\pi - \frac{A \, 2\pi a^2}{r^2}\right)$$

$$E = \frac{1}{4\pi\varepsilon_0} \times A \times 2\pi$$

At the centre of the spheres is a point charge Q. The value of A such that the electric field in the region between the spheres will be constant is

As, $\quad Q = 2\pi A a^2$ i.e. $A = \dfrac{Q}{2\pi a^2}$

36. *(15)* $\mathbf{E} = (2\hat{\mathbf{i}} + 4\hat{\mathbf{j}} + 6\hat{\mathbf{k}}) \times 10^3$ N/C

The surface is parallel to XZ-plane, therefore the area vector is along $\hat{\mathbf{j}}$.

Flux = $\phi = \mathbf{E} \cdot \mathbf{A}$

$$6 = [2\hat{\mathbf{i}} + 4\hat{\mathbf{j}} + 6\hat{\mathbf{k}}] \times 10^3 A\hat{\mathbf{j}}$$

$$6 = 4A \times 10^3$$

$$\frac{6}{4 \times 10^3} = A$$

$$\Rightarrow \quad A = 1.5 \times 10^{-3} \text{ m}^2 = 15 \text{ cm}^2$$

37. *(48)* Considering charge at a distance of $\dfrac{a}{2}$ from the square.

Total flux through the square

$$= \frac{q}{\varepsilon_0}\left(\frac{1}{6}\right)$$

As flux is same in the plane, therefore flux is divided into equal parts

$\therefore$ Flux through shaded portion is $\dfrac{5}{8}$

$$= \frac{5}{8} \times \frac{q}{\varepsilon_0}\left(\frac{1}{6}\right)$$

$$= \frac{5}{48} \frac{1}{\varepsilon_0}$$

38. *(16)* Given, $\mathbf{E} = 2x\hat{\mathbf{i}}$ NC^{-1}

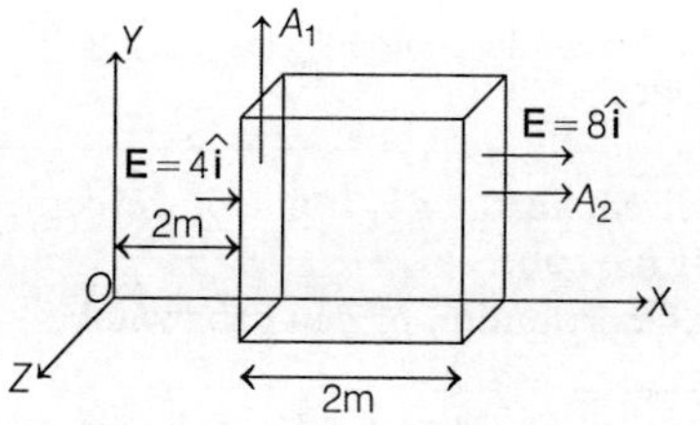

Total flux = Flux at A_2 − Flux at A_1

$$= E \cdot A_2 - E \cdot A_1$$
$$= 2 \times 4 \times 4 - 2 \times 2 \times 4$$
$$= 32 - 16 \text{ Nm}^2/\text{C}$$
$$= 16 \text{ Nm}^2/\text{C}$$

39. *(12)* Electric flux through a surface,

$$Q_E = \mathbf{E} \cdot \mathbf{A} = \mathbf{E} \cdot A(\hat{\mathbf{n}})$$

$$= \frac{2\hat{\mathbf{i}} + 6\hat{\mathbf{j}} + 8\hat{\mathbf{k}}}{\sqrt{6}} \cdot 4\left(\frac{2\hat{\mathbf{i}} + \hat{\mathbf{j}} + \hat{\mathbf{k}}}{\sqrt{6}}\right)$$

$$= \frac{4(4 + 6 + 8)}{6} = \frac{4 \times 18}{6} = 12 \text{ Vm}$$

40. *(2)* Electric field at point P

$$\mathbf{E}_p = \frac{\sigma}{2\varepsilon_0}(-\hat{\mathbf{i}}) + \frac{2\sigma}{2\varepsilon_0}(-\hat{\mathbf{i}}) + \frac{\sigma}{2\varepsilon_0}(-\hat{\mathbf{i}})$$

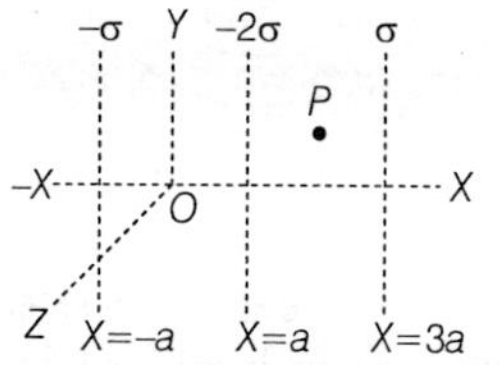

$$= \frac{2\sigma}{\varepsilon_0}(-\hat{\mathbf{i}})$$

$$\therefore \quad |\mathbf{E}_p| = \frac{2\sigma}{\varepsilon_0} = \frac{x\sigma}{\varepsilon_0} \quad \text{(given)}$$

$$\therefore \quad x = 2$$

41. *(288)* Drawing the cube given in question.

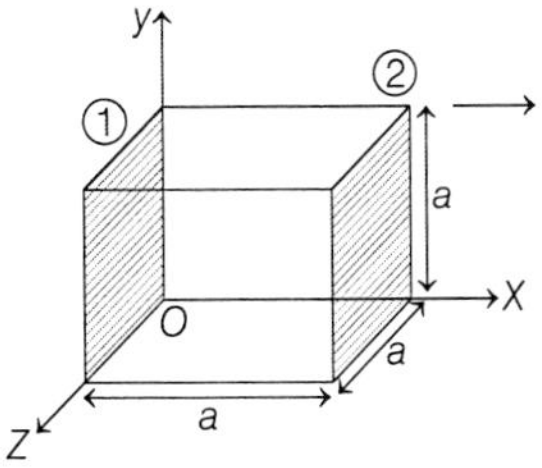

Since, electric field is along x-direction, so electric flux will be non zero only the shaded face.

Electric flux, $\phi = \mathbf{E} \cdot \mathbf{A} = EA\cos\theta \ (x = 0)$

At face 1, $\phi_1 = (0)a^2 = 0 \qquad (\because E = 0)$

At face 2, $\phi_2 = \mathbf{E} \cdot \mathbf{A}$

$$= (E_0 a\hat{\mathbf{i}}) \cdot (a^2 \hat{\mathbf{i}}) = E_0 a^3$$

$$= 4 \times 10^4 \times (2 \times 10^{-2})^3$$

$$\phi_2 = 32 \times 10^{-2} \text{ N/m}^2\text{-C}$$

Net flux, $\phi = \phi_2 - \phi_1$

$$= 32 \times 10^{-2} \text{ N/m}^2\text{-C}$$

From Gauss law, $\phi = \dfrac{q_{\text{enc}}}{\varepsilon_0}$

$$\Rightarrow q_{\text{enc}} = \phi\varepsilon_0 = 32 \times 10^{-2} \times 9 \times 10^{-12}$$

$$\Rightarrow q_{\text{enc}} = 288 \times 10^{-14} \text{ C}$$

42. *(12)* Given that

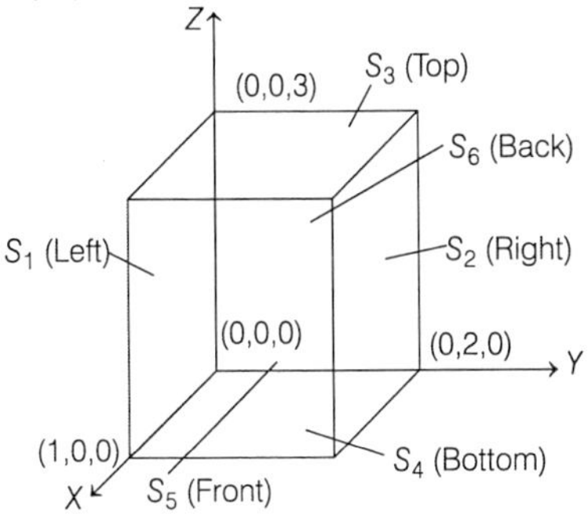

$$\mathbf{E} = (2x^2\hat{\mathbf{i}} - 4y\hat{\mathbf{j}} + 6\hat{\mathbf{k}}) \text{ N/C}$$

Flux passing through any surface like S_2 is calculated as

Area of surface $S_2 = 1 \times 3 = 3 \text{ m}^2$

Area vector of surface $S_2 = 3\hat{\mathbf{j}} \text{ m}^2$

$\therefore$ Flux passing through surface S_2,

$$\phi = \mathbf{E} \cdot \mathbf{A} = (2x^2\hat{\mathbf{i}} - 4y\hat{\mathbf{j}} + 6\hat{\mathbf{k}}) \cdot 3\hat{\mathbf{j}}$$

$$= -12y \text{ N-m}^2/\text{C} = -24 \text{ N-m}^2/\text{C}$$

$[\because y = 2]$

Similarly, for surfaces S_1, S_3, S_4 and S_6, $\phi = 0$ and $S_5 = 12 \text{ N-m}^2/\text{C}$.

Total flux passing through the cuboid,

$$\phi = \sum_{i=1}^{6} \phi_i \quad [\text{where} = \mathbf{E} . \mathbf{A}]$$

$$= -12 \text{ N-m}^2/\text{C}$$

By the Gauss's law,

$$\phi = \frac{q}{\varepsilon_0} \Rightarrow q = \phi\varepsilon_0 = -12\varepsilon_0 \text{ C}$$

$$|q| = n\varepsilon_0 \text{C} \qquad [\text{Given}]$$

$\therefore \quad n = 12$

43. *(640)*

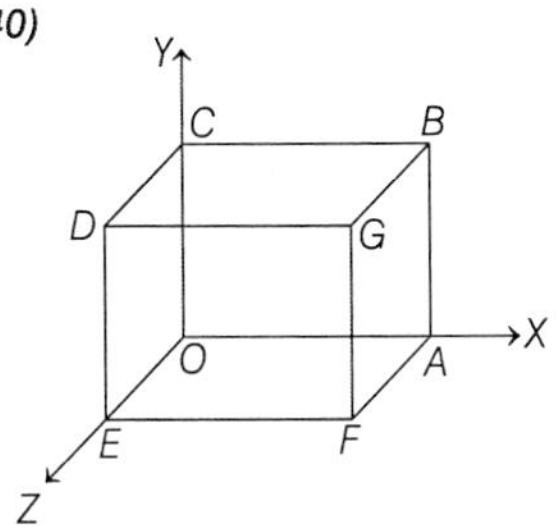

Since, the given electric field is along $\hat{\mathbf{i}}$ direction.

So, Electric flux through the faces which area vector is perpendicular to Electric field is zero.

$\therefore \ \phi_{OAEF} = \phi_{BCDG} = \phi_{BCDG} = \phi_{OAFE} = 0$

At face $ABGF$,

$$\mathbf{E} = 4000\, x^2\, \hat{\mathbf{i}}$$

$$\mathbf{E} = 4000(0.2)^2\, \hat{\mathbf{i}} = 160 \text{ V/m}$$

At face $OCDE$,

$$\mathbf{E} = 4000(0)^2 = 0 \text{ V/m}$$

Net flux, $\phi_{\text{cube}} = \phi_{ABGE} - \phi_{OCDE}$

$$= E_{ABGF} \times a^2 - E_{OCDE} \times a^2$$

$$= 160 \times (0.2)^2 - 0 = 6.4 \text{ V-m}$$

$$= 6.4 \times 100 \text{ V-cm}$$

$$\phi_{\text{cube}} = 640 \text{ V-cm}$$

44. *(8)* Given, charge density,

$$\lambda = \frac{q}{l} = 2 \times 10^{-8} \text{ Cm}^{-1}$$

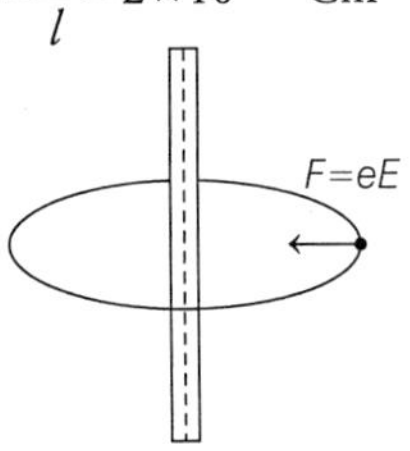

Forces of electron will be equilibrium when electrostatic force will be equal to centripetal force.

i.e., $\quad eE = \dfrac{mv^2}{r} \Rightarrow e \cdot \dfrac{2k\lambda}{r} = \dfrac{mv^2}{r}$

$$\Rightarrow \quad v = \sqrt{\frac{e \cdot 2k\lambda}{m}}$$

$$= \sqrt{\frac{1.6 \times 10^{-19} \times 2 \times 9 \times 10^9 \times 2 \times 10^{-8}}{9 \times 10^{-31}}}$$

$$= 8 \times 10^6 \text{ m/s}$$

45. *(1)* Electric field due to charged point at $x < R$ is given as

Charge on cylinder of radius, $x = \rho\pi x^2 h$

Using Gauss's law

$$E \cdot 2\pi x h = \pi\rho x^2 h\varepsilon_0$$

$$E = \frac{\rho x}{2\varepsilon_0} = \frac{\rho}{2\varepsilon_0} \times \frac{2\varepsilon_0}{\rho}$$

$$= 1 \qquad (\because x = 2\varepsilon_0/\rho)$$

46. *(45)* Number of field lines per unit area

$$= \frac{\text{Flux}}{\text{Area}} = \frac{Q}{\varepsilon_0 A} = \frac{\rho \times V}{\varepsilon_0 \times A} \qquad (\because Q = \rho \times V)$$

$$= \frac{\rho \times \frac{4}{3}\pi R^3}{\varepsilon_0 \times 4\pi R^2} = \frac{\rho \times R}{3\varepsilon_0}$$

$$= \frac{2 \times 10^{-6} \times 10^6 \times 6}{8.85 \times 10^{-12} \times 3}$$

$$= \frac{4}{8.85} \times 10^{12}$$

$$\approx 45 \times 10^{10} \text{ N/C}$$

47. *(4)* Given, incremental volume,

$dV = 2 \times 10^{-9} \text{m}^3$

Electric flux denisty,

$$D = e^{-x}\sin y\hat{\mathbf{i}} - e^{-x}\cos y\,\hat{\mathbf{j}} + 2z\hat{\mathbf{k}} \text{ C/m}^2$$

As we know that

$$\rho(x, y, z) = \nabla \cdot \mathbf{D} = \frac{d}{dx}(e^{-x}\sin y) - \frac{d}{dy}(e^{-x}\cos y) + \frac{d}{dz}2z$$

$$= -e^{-x}\sin y + e^{-x}\sin y + 2$$

$$\rho(0, 0, 0) = 2 = \frac{Q}{dV}$$

$$\Rightarrow \quad Q = 2\,dV = 2 \times 2 \times 10^{-9}$$

$$= 4 \times 10^{-9} = 4 \text{ nC}$$

48. *(6)* In the given question,

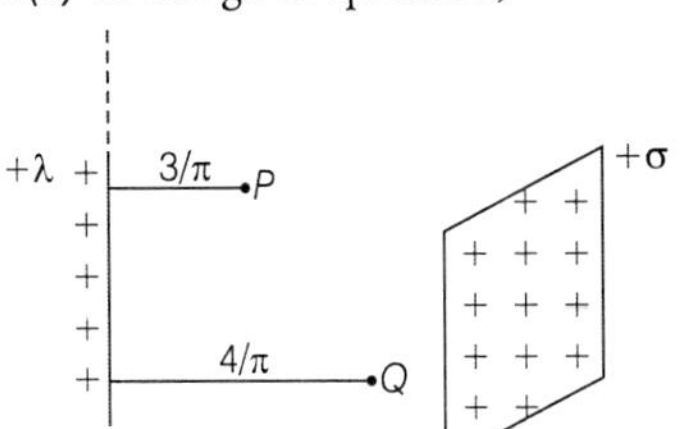

$$\frac{E_P}{E_Q} = \frac{\dfrac{\lambda}{2\pi\varepsilon_0\left(\dfrac{3}{\pi}\right)} - \dfrac{\sigma}{2\varepsilon_0}}{\dfrac{\lambda}{2\pi\varepsilon_0\left(\dfrac{4}{\pi}\right)} - \dfrac{\sigma}{2\varepsilon_0}}$$

$$= \frac{4}{3}\left(\frac{\lambda - 3\sigma}{\lambda - 4\sigma}\right)$$

$$= \frac{4}{3}\left(\frac{2\sigma - 3\sigma}{2\sigma - 4\sigma}\right)$$

$$= \frac{4}{3} \times \frac{1}{2}$$

$$= \frac{2}{3} = \frac{4}{6} = \frac{4}{a} \quad \text{[Given]}$$

$\therefore \quad a = 6$

49. *(640)* Given,

The electric field in the region,

$$\mathbf{E} = \frac{2}{5} E_0 \hat{\mathbf{i}} + \frac{3}{5} E_0 \hat{\mathbf{j}}$$

Here, $E_0 = 4 \times 10^3$ N/C

Area of the rectangular surface,

$$A = 0.4 \text{m}^2$$

The direction of electric field vector and area vector is same, so the angle between the electric field vector and area vector is 0.

As we know the expression of electric flux, $\phi = E \cdot A \cos\theta$ ...(i)

Here, E is the electric field vector and A is the surface area of the surfaces.

Consider the surface parallel to the Y-Z plane, so the area vector,

$$A = 0.4\mathbf{i} \text{ m}^2$$

Substituting the values in Eq. (i), we get

$$\phi = E \cdot A \cos 0° + EA \cos 90°$$

$$\Rightarrow \quad \phi = \frac{2}{5} E_0 (0.4) + 0$$

$$\Rightarrow \quad \phi = \frac{2}{5}(4 \times 10^3)(0.4)$$

$$= 640 \text{ Nm}^2\text{C}^{-1}$$

Hence, the electric flux of the surface parallel to the YZ-plane is $640 \text{ Nm}^2\text{C}^{-1}$.

50. *(1)* Given, $\mathbf{E} = \frac{3E_0}{5} \hat{\mathbf{i}} + \frac{4}{5} E_0 \hat{\mathbf{j}}$,

$$\mathbf{A}_1 = 0.2 \text{ m}^2 \hat{\mathbf{i}}$$

and $$\mathbf{A}_2 = 0.3 \text{ m}^2 \hat{\mathbf{j}}$$

Let ϕ_1 and ϕ_2 be the flux linked with area A_1 and A_2, respectively.

As we know that,

$$\phi = \oint \mathbf{E} \cdot d\mathbf{S} = \mathbf{E} \cdot \mathbf{A}$$

$$\Rightarrow \quad \phi_1 = (3/5 E_0 \hat{\mathbf{i}} + 4/5 E_0 \hat{\mathbf{j}}) \cdot 0.2 \hat{\mathbf{i}}$$

$$= 3/5\, E_0 \times 0.2$$

and similarly, $\phi_2 = 4/5\, E_0 \times 0.3$

Now, $$\frac{\phi_1}{\phi_2} = \frac{3/5 E_0 \times 0.2}{4/5\, E_0 \times 0.3}$$

$$= \frac{0.6}{1.2} = \frac{1}{2}$$

$\therefore \quad a = 1$

In JEE Main there is usually 1-2 question per paper is visible from this chapter. In past 3 years, 41 conceptual and 24 numerical questions were asked of moderate to difficult level from this chapter.
For detailed analysis scan QR code

ELECTROSTATICS POTENTIAL AND CAPACITANCE

Topic 1 Electrostatic Potential and Potential Energy

Level 1

01 The electrostatic potential on the surface of uniformly charged spherical shell of radius $R = 10$ cm is 120 V. The potential at the centre of shell, at a distance $r = 5$ cm from centre, and at a distance $r = 15$ cm from the centre of the shell respectively, are *[2025, 3 April Shift-I]*

(a) 40 V, 40 V, 80 V (b) 0 V, 120 V, 40 V
(c) 120 V, 120 V, 80 V (d) 0 V, 0 V, 80 V

02 Two metal spheres of radius R and $3R$ have same surface charge density σ. If they are brought in contact and then separated, the surface charge density on smaller and bigger sphere becomes σ_1 and σ_2, respectively. The ratio $\frac{\sigma_1}{\sigma_2}$ is *[2025, 8 April Shift-II]*

(a) 3 (b) $\frac{1}{3}$ (c) 9 (d) $\frac{1}{9}$

03 An electric dipole is placed at a distance of 2 cm from an infinite plane sheet having positive charge density σ_0. Choose the correct option from the following. *[2025, 29 Jan Shift-II]*

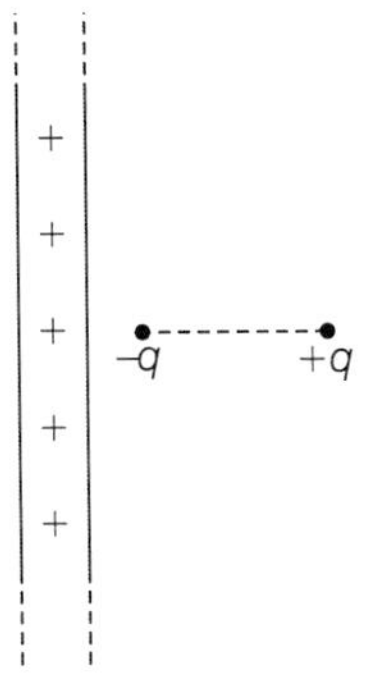

(a) Potential energy and torque both are maximum.
(b) Torque on dipole is zero and net force is directed away from the sheet.
(c) Torque on dipole is zero and net force acts towards the sheet.
(d) Potential energy of dipole is minimum and torque is zero.

04 Two charges 7 μC and −4 μC are placed at (−7 cm, 0, 0) and (7 cm, 0, 0) respectively. Given, $\varepsilon_0 = 8.85 \times 10^{-12} C^2N^{-1}m^{-2}$, the electrostatic potential energy of the charge configuration is *[2025, 23 Jan Shift-II]*

(a) −1.5 J (b) −1.2 J (c) −2.0 J (d) −1.8 J

Imp. Electric Potential Due to Electric Dipole

05 The electrostatic potential due to an electric dipole at a distance r varies as *[2024, 30 Jan Shift-I]*

(a) $1/r$ (b) $1/r^3$ (c) r (d) $1/r^2$

06 Given below are two statements : one is labelled as Assertion (A) and other is labelled as Reason (R).

Assertion (A) Work done by electric field on moving a positive charge on an equipotential surface is always zero. *[2024, 27 Jan Shift-II]*

Reason (R) Electric lines of forces are always perpendicular to equipotential surfaces.

In the light of the above statements, choose the **most appropriate** answer from the options given below

(a) (A) is not correct but (R) is correct.
(b) Both (A) and (R) correct but (R) is not the correct explanation of (A).
(c) Both (A) and (R) are correct and (R) is the correct explanation of (A).
(d) (A) is correct but (R) is not correct.

07 An electric charge 10^{-6} μC is placed at origin (0, 0) m of x-y coordinate system. Two points P and Q are situated at $(\sqrt{3}, \sqrt{3})$ m and $(\sqrt{6}, 0)$ m respectively. The potential difference between the points P and Q will be *[2024, 27 Jan Shift-I]*

(a) 3V (b) 0 V (c) $\sqrt{3}$ V (d) $\sqrt{6}$ V

08 Electric potential at a point P due to a point charge of 5×10^{-9} C is 50 V. The distance of P from the point charge is

(Assume, $\frac{1}{4\pi\varepsilon_0} = 9 \times 10^{+9}$ N-m^{2}C^{-2}) *[2023, 8 April Shift-II]*

(a) 90 cm (b) 3 cm (c) 0.9 cm (d) 9 cm

09 An α particle and a proton are accelerated from rest through the same potential difference. The ratio of linear momenta acquired by above two particle will be *[2022, 29 July Shift-II]*

(a) $\sqrt{2}:1$ (b) $2\sqrt{2}:1$ (c) $4\sqrt{2}:1$ (d) $8:1$

10 Given below are two statements.

Statement I Electric potential is constant within and at the surface of each conductor.

Statement II Electric field just outside a charged conductor is perpendicular to the surface of the conductor at every point.

In the light of the above statements, choose the most appropriate answer from the options given below. *[2022, 29 July Shift-I]*

(a) Both Statement I and Statement II are correct.
(b) Both Statement I and Statement II are incorrect.
(c) Statement I is correct but Statement II is incorrect.
(d) Statement I is incorrect but Statement II is correct.

11 Consider two charged metallic spheres S_1 and S_2 of radii R_1 and R_2, respectively. The electric fields E_1 (on S_1) and E_2 (on S_2) on their surfaces are such that $E_1 / E_2 = R_1 / R_2$. Then the ratio V_1(on S_1)/V_2 (on S_2) of the electrostatic potentials on each sphere is *[2020, 8 Jan Shift-II]*

(a) $\left(\frac{R_1}{R_2}\right)^3$ (b) R_2 / R_1 (c) R_1 / R_2 (d) $(R_1 / R_2)^2$

12 An electric field of 1000 V/m is applied to an electric dipole at angle of 45°. The value of electric dipole moment is 10^{-29} C-m. What is the potential energy of the electric dipole ? *[2019, 11 Jan Shift-II]*

(a) -9×10^{-20} J (b) -10×10^{-29} J
(c) -20×10^{-18} J (d) -7×10^{-27} J

13 Assume that an electric field $\mathbf{E} = 30x^2\,\hat{\mathbf{i}}$ exists in space. Then, the potential difference $V_A - V_O$, where V_O is the potential at the origin and V_A is the potential at $x = 2$ m, is *[JEE Main 2014]*

(a) 120 J (b) −120 J (c) −80 J (d) 80 J

14 Two points P and Q are maintained at the potentials of 10 V and −4 V, respectively. The work done in moving 100 electrons from P to Q is *[AIEEE 2009]*

(a) -19×10^{-17} J (b) 9.60×10^{-17} J
(c) -2.24×10^{-16} J (d) 2.24×10^{-16} J

15 The potential at a point x (measured in μm) due to some charges situated on the X-axis is given by $V(x) = 20/(x^2 - 4)$ V.

The electric field E at $x = 4$μm is given by *[AIEEE 2007]*

(a) $\frac{5}{3}$ V/μm and in the –ve x-direction
(b) $\frac{5}{3}$ V/μm and in the +ve x-direction
(c) $\frac{10}{9}$ V/μm and in the –ve x-direction
(d) $\frac{10}{9}$ V/μm and in the +ve x-direction

16 Charges are placed on the vertices of a square as shown. Let **E** be the electric field and V be the potential at the centre. If the charges on A and B are interchanged with those on D and C respectively, then *[AIEEE 2007]*

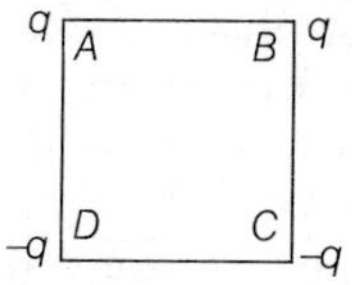

(a) **E** remains unchanged, V changes
(b) Both **E** and V change
(c) **E** and V remain unchanged
(d) **E** changes, V remains unchanged

17 Two spherical conductors A and B of radii 1 mm and 2 mm are separated by a distance of 5 cm and are uniformly charged. If the spheres are connected by a conducting wire, then in equilibrium condition, the ratio of the magnitude of the electric fields at the surfaces of spheres A and B is *[AIEEE 2006]*

(a) 4 : 1 (b) 1 : 2 (c) 2 : 1 (d) 1 : 4

18 A thin spherical conducting shell of radius R has a charge q. Another charge Q is placed at the centre of the shell. The electrostatic potential at a point P at a distance $R/2$ from the centre of the shell is *[AIEEE 2003]*

(a) $\frac{2Q}{4\pi\varepsilon_0 R}$ (b) $\frac{2Q}{4\pi\varepsilon_0 R} - \frac{2q}{4\pi\varepsilon_0 R}$
(c) $\frac{2Q}{4\pi\varepsilon_0 R} + \frac{q}{4\pi\varepsilon_0 R}$ (d) $\frac{(q+Q)}{4\pi\varepsilon_0}\frac{2}{R}$

19 On moving a charge of 20 C by 2 cm, 2 J of work is done, then the potential difference between the points is *[AIEEE 2002]*

(a) 0.1 V (b) 8 V
(c) 2 V (d) 0.5 V

Level 2

20 Given below are two statements : One is labelled as Assertion (A) and the other is labelled as Reason (R).

Assertion (A) Work done in moving a test charge between two points inside a uniformly charged spherical shell is zero, no matter which path is chosen.

Reason (R) Electrostatic potential inside a uniformly charged spherical shell is constant and is same as that on the surface of the shell.

In the light of the above statements, choose the correct answer from the options given below. *[2025, 8 April Shift-II]*

(a) A is true but R is false.
(b) A is false but R is true.
(c) Both A and R are true but R is not the correct explanation of A.
(d) Both A and R are true and R is the correct explanation of A.

21 Two large plane parallel conducting plates are kept 10 cm apart as shown in figure. The potential difference between them is V. The potential difference between the points A and B (shown in the figure) is *[2025, 2 April Shift-II]*

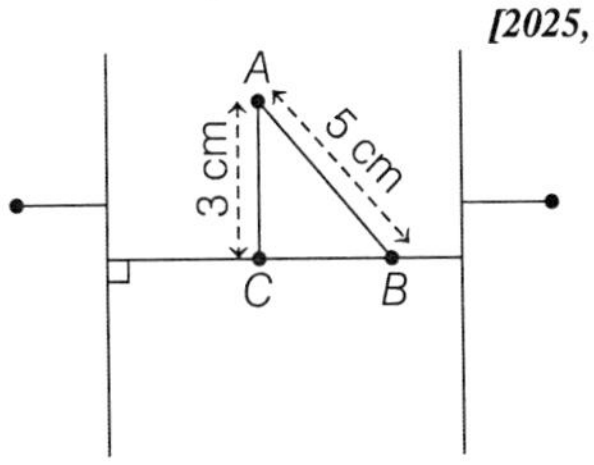

(a) 1 V (b) 3/4 V (c) 1/4 V (d) 2/5 V

22 Three infinitely long wires with linear charge density λ are placed along the X-axis, Y-axis and Z-axis, respectively. Which of the following denotes an equipotential surface ? *[2025, 28 Jan Shift-I]*

(a) xyz = constant
(b) $(x + y)(y + z)(z + x)$ = constant
(c) $xy + yz + zx$ = constant
(d) $(x^2 + y^2)(y^2 + z^2)(z^2 + x^2)$ = constant

23 For a short dipole at origin O, the dipole moment p is along X-axis, as shown in the figure. If the electric potential and electric field at A are V_0 and E_0, respectively, then the correct combination of the electric potential and electric field, respectively, at point B on the Y-axis is given by *[2025, 22 Jan Shift-II]*

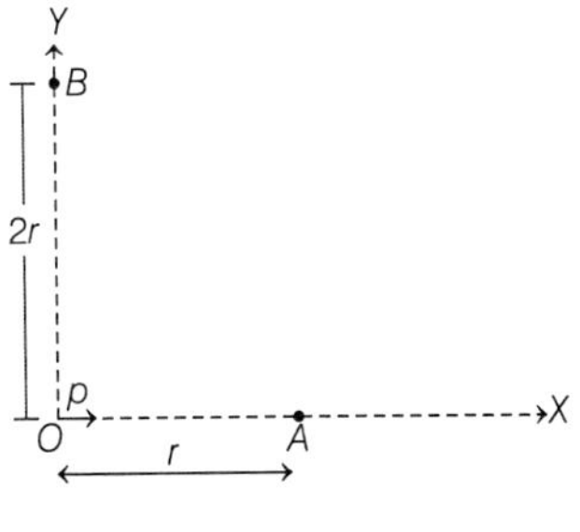

(a) $\frac{V_0}{2}$ and $\frac{E_0}{16}$ (b) zero and $\frac{E_0}{16}$
(c) zero and $\frac{E_0}{8}$ (d) V_0 and $\frac{E_0}{4}$

24 Two charged conducting spheres of radii a and b are connected to each other by a conducting wire. The ratio of charges of the two spheres respectively is *[2024, 8 April Shift-I]*

(a) $\frac{a}{b}$ (b) $\sqrt{ab}$
(c) $\frac{b}{a}$ (d) ab

25 For a uniformly charged thin spherical shell, the electric potential V radially away from the centre O of shell can be graphically represented as *[2023, 6 April Shift-I]*

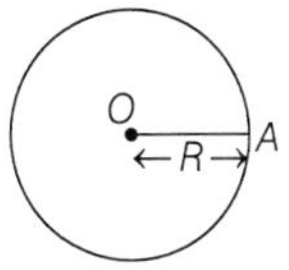

(a)
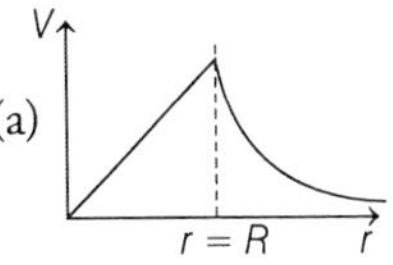

(b)
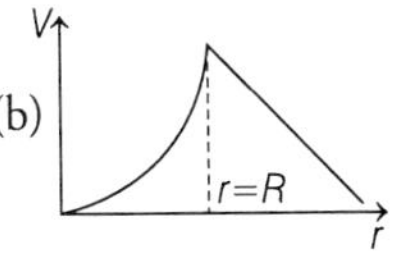

(c)
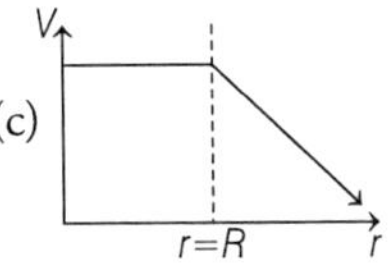

(d)
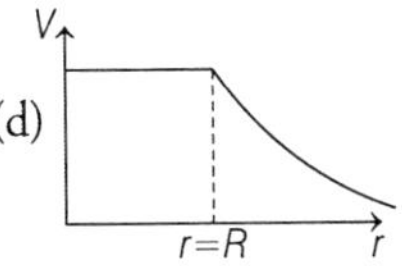

26 Which of the following correctly represents the variation of electric potential V of a charged spherical conductor of radius R with radial distance r from the centre? *[2023, 31 Jan Shift-I]*

(a)
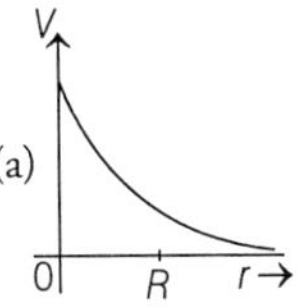

(b)
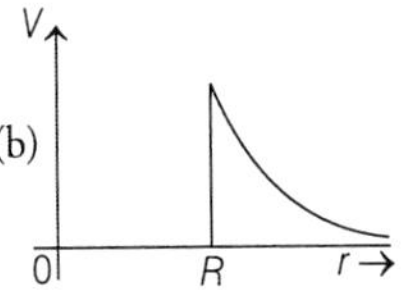

(c)
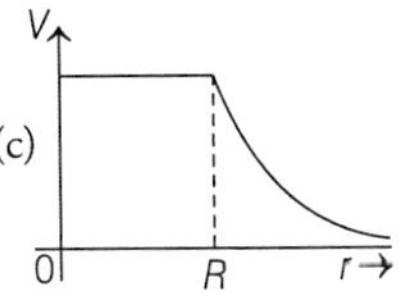

(d)
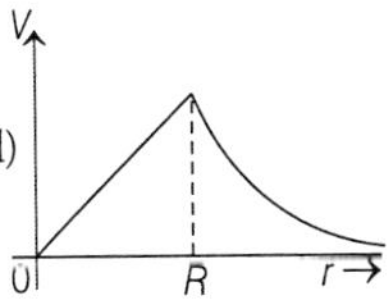

27 Two uniformly charged spherical conductors A and B of radii 5 mm and 10 mm are separated by a distance of 2 cm. If the spheres are connected by a conducting wire, then in equilibrium condition, the ratio of the magnitudes of the electric fields at the surface of the sphere A and B will be *[2022, 26 July Shift-II]*

(a) 1 : 2 (b) 2 : 1 (c) 1 : 1 (d) 1 : 4

28 If the electric potential at any point (x, y, z)m in space is given by $V = 3x^2$ V. The electric field at the point (1, 0, 3) m will be *[2022, 29 June Shift-II]*

(a) 3 Vm^{-1}, directed along positive X-axis.
(b) 3 Vm^{-1}, directed along negative X-axis.
(c) 6 Vm^{-1}, directed along positive X-axis.
(d) 6 Vm^{-1}, directed along negative X-axis.

29 The two thin coaxial rings, each of radius a and having charges $+Q$ and $-Q$, respectively are separated by a distance of s. The potential difference between the centres of the two rings is *[2021, 26 Aug Shift-II]*

(a) $\frac{Q}{2\pi\varepsilon_0}\left[\frac{1}{a}-\frac{1}{\sqrt{s^2+a^2}}\right]$ (b) $\frac{Q}{4\pi\varepsilon_0}\left[\frac{1}{a}-\frac{1}{\sqrt{s^2+a^2}}\right]$

(c) $\frac{Q}{4\pi\varepsilon_0}\left[\frac{1}{a}+\frac{1}{\sqrt{s^2+a^2}}\right]$ (d) $\frac{Q}{2\pi\varepsilon_0}\left[\frac{1}{a}+\frac{1}{\sqrt{s^2+a^2}}\right]$

30 A system of three charges are placed as shown in the figure

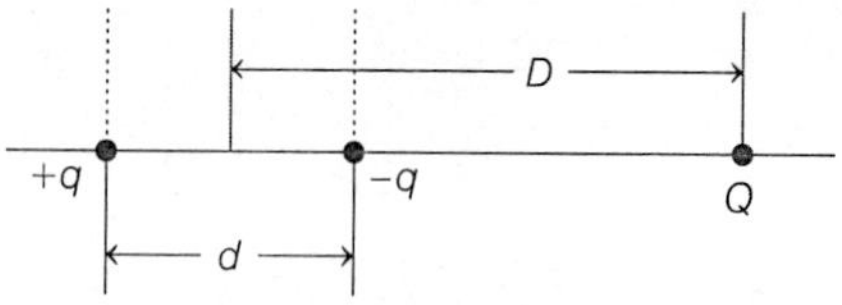

If $D >> d$, the potential energy of the system is best given by *[2019, 9 April Shift-I]*

(a) $\frac{1}{4\pi\varepsilon_0}\left[-\frac{q^2}{d}+\frac{2qQd}{D^2}\right]$ (b) $\frac{1}{4\pi\varepsilon_0}\left[+\frac{q^2}{d}+\frac{qQd}{D^2}\right]$

(c) $\frac{1}{4\pi\varepsilon_0}\left[-\frac{q^2}{d}-\frac{qQd}{2D^2}\right]$ (d) $\frac{1}{4\pi\varepsilon_0}\left[-\frac{q^2}{d}-\frac{qQd}{D^2}\right]$

31 There is a uniform spherically symmetric surface charge density at a distance R_0 from the origin. The charge distribution is initially at rest and starts expanding because of mutual repulsion. The figure that represent best the speed $V(R(t))$ of the distribution as a function of its instantaneous radius $R(t)$ is *[2019, 12 Jan Shift-I]*

(a)
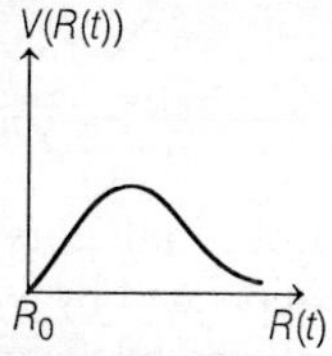

(b)
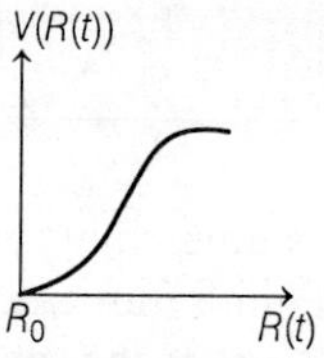

(c)
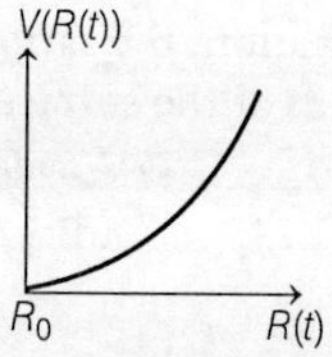

(d)
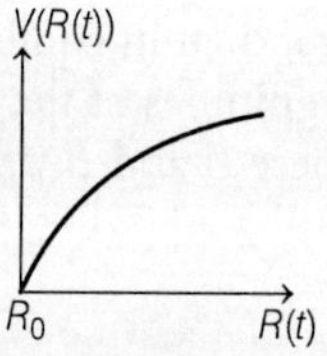

32 Three charges Q, $+q$ and $+q$ are placed at the vertices of a right angle isosceles triangle as shown below. The net electrostatic energy of the configuration is zero, if the value of Q is *[2019, 11 Jan Shift-I]*

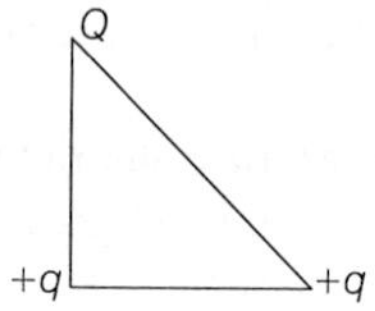

(a) $-2q$ (b) $\frac{-q}{1+\sqrt{2}}$ (c) $+q$ (d) $\frac{-\sqrt{2}q}{\sqrt{2}+1}$

33 Three concentric metal shells A, B and C of respective radii a, b and c $(a < b < c)$ have surface charge densities $+\sigma$, $-\sigma$ and $+\sigma$, respectively. The potential of shell B is *[JEE Main 2018]*

(a) $\frac{\sigma}{\varepsilon_0}\left[\frac{a^2-b^2}{a}+c\right]$ (b) $\frac{\sigma}{\varepsilon_0}\left[\frac{a^2-b^2}{b}+c\right]$

(c) $\frac{\sigma}{\varepsilon_0}\left[\frac{b^2-c^2}{b}+a\right]$ (d) $\frac{\sigma}{\varepsilon_0}\left[\frac{b^2-c^2}{c}+a\right]$

34 A charge Q is uniformly distributed over a long rod AB of length L as shown in the figure. The electric potential at the point O lying at distance L from the end A is *[JEE Main 2013]*

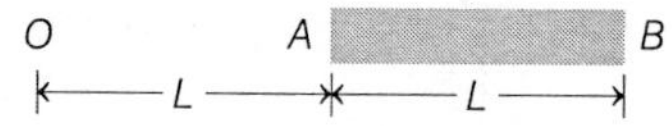

(a) $\frac{Q}{8\pi\varepsilon_0 L}$ (b) $\frac{3Q}{4\pi\varepsilon_0 L}$

(c) $\frac{Q}{4\pi\varepsilon_0 L\ln 2}$ (d) $\frac{Q\ln 2}{4\pi\varepsilon_0 L}$

35 Two positive charges of magnitude q are placed at the ends of a side 1 of a square of side $2a$. Two negative charges of the same magnitude are kept at the other corners. Starting from rest, if a charge Q moves from the middle of side 1 to the centre of square, its kinetic energy at the centre of square is *[AIEEE 2011]*

(a) $\frac{1}{4\pi\varepsilon_0}\frac{2qQ}{a}\left(1-\frac{1}{\sqrt{5}}\right)$ (b) zero

(c) $\frac{1}{4\pi\varepsilon_0}\frac{2qQ}{a}\left(1+\frac{1}{\sqrt{5}}\right)$ (d) $\frac{1}{4\pi\varepsilon_0}\frac{2qQ}{a}\left(1-\frac{2}{\sqrt{5}}\right)$

36 An electric charge 10^{-3} μC is placed at the origin (0, 0) of xy-coordinate system. Two points A and B are situated at $(\sqrt{2}, \sqrt{2})$ and (2, 0), respectively. The potential difference between the points A and B will be *[AIEEE 2007]*

(a) 9 V (b) zero
(c) 2 V (d) 4.5 V

37 Two insulating plates are both uniformly charged in such a way that the potential difference between them is $V_2 - V_1 = 20$ V (i.e., plate 2 is at a higher potential). The plates are separated by $d = 0.1$ m and can be treated as infinitely large. An electron is released from rest on the inner surface of plate 1. What is its speed when it hits plate 2?

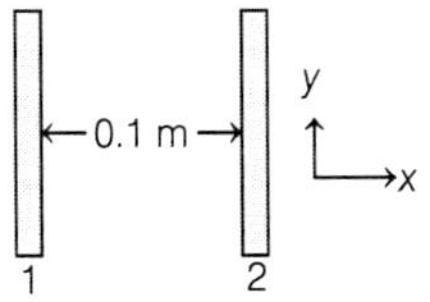

($e = 1.6 \times 10^{-19}$C, $m_0 = 9.11 \times 10^{-31}$ kg) *[AIEEE 2006]*

(a) 2.65×10^6 ms^{-1} (b) 7.02×10^{12} ms^{-1}
(c) 1.87×10^6 ms^{-1} (d) 32×10^{-19} ms^{-1}

Level 3

38 Two charges q_1 and q_2 are separated by a distance of 30 cm. A third charge q_3 initially at 'C' as shown in the figure, is moved along the circular path of radius 40 cm from C to D. If the difference in potential energy due to movement of q_3 from C to D is given by $\dfrac{q_3 K}{4\pi \in_0}$, the value of K is: *[2025, 7 April Shift-I]*

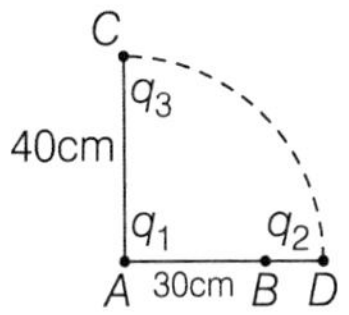

(a) $8q_1$ (b) $8q_2$ (c) $6q_2$ (d) $6q_1$

Imp. Electrostatic Potential Energy

39

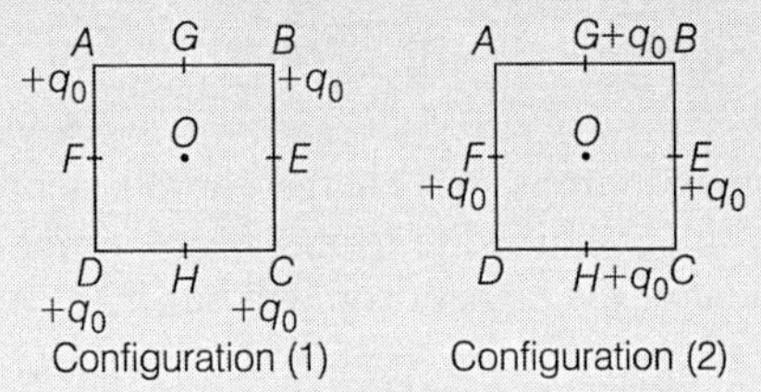

In the first configuration (1) as shown in the figure, four identical charges (q_0) are kept at the corners A, B, C and D of square of side length a. In the second configuration (2), the same charges are shifted to mid-points G, E, H and F, of the square. If $k = \dfrac{1}{4\pi\varepsilon_0}$, the difference between the potential energies of configuration (2) and (1) is given by *[2025, 24 Jan Shift-II]*

(a) $\dfrac{kq_0^2}{a}(3\sqrt{2} - 2)$ (b) $\dfrac{kq_0^2}{a}(4 - 2\sqrt{2})$
(c) $\dfrac{kq_0^2}{a}(4\sqrt{2} - 2)$ (d) $\dfrac{kq_0^2}{a}(3 - \sqrt{2})$

40 An electron rotates in a circle around a nucleus having positive charge Ze. Correct relation between total energy E of electron to its potential energy U is *[2024, 5 April Shift-I]*

(a) $E = 2U$ (b) $E = U$ (c) $2E = U$ (d) $2E = 3U$

41 The electric potential at the centre of two concentric half-rings of radii R_1 and R_2, having same linear charge density λ is *[2023, 24 Jan Shift-II]*

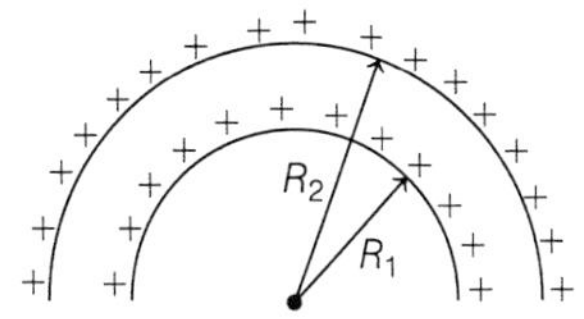

(a) $\dfrac{2\lambda}{\varepsilon_0}$ (b) $\dfrac{\lambda}{4\varepsilon_0}$ (c) $\dfrac{\lambda}{\varepsilon_0}$ (d) $\dfrac{\lambda}{2\varepsilon_0}$

42 Two point charges $4q$ and $-q$ are fixed on the X-axis at $x = \dfrac{-d}{2}$ and $x = \dfrac{d}{2}$, respectively. If a third point charge q is taken from the origin to $x = d$ along the semi-circle as shown in the figure, the energy of the charge will *[2020, 4 Sep Shift-I]*

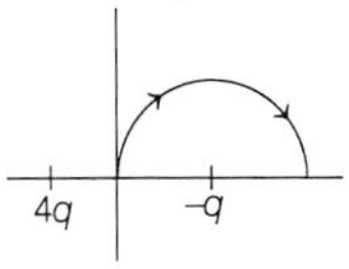

(a) increase by $\dfrac{2q^2}{3\pi\varepsilon_0 d}$

(b) decrease by $\dfrac{4q^2}{3\pi\varepsilon_0 d}$

(c) increase by $\dfrac{3q^2}{4\pi\varepsilon_0 d}$

(d) decrease by $\dfrac{q^2}{4\pi\varepsilon_0 d}$

43 A charge Q is distributed over two concentric conducting thin spherical shells radii r and R $(R > r)$. If the surface charge densities on the two shells are equal, the electric potential at the common centre is *[2020, 2 Sep Shift-II]*

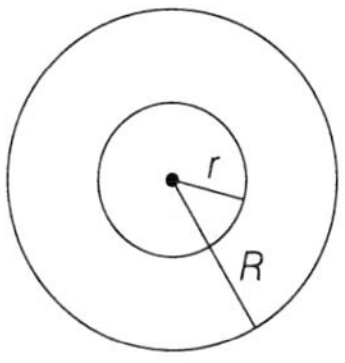

(a) $\dfrac{1}{4\pi\varepsilon_0} \dfrac{(R + r)}{2(R^2 + r^2)} Q$

(b) $\dfrac{1}{4\pi\varepsilon_0} \dfrac{(2R + r)}{(R^2 + r^2)} Q$

(c) $\dfrac{1}{4\pi\varepsilon_0} \dfrac{(R + 2r)Q}{2(R^2 + r^2)}$

(d) $\dfrac{1}{4\pi\varepsilon_0} \dfrac{(R + r)}{(R^2 + r^2)} Q$

Imp. Equipotential Surface

44 A charge Q is distributed over three concentric spherical shells of radii a, b, c $(a < b < c)$, such that their surface charge densities are equal to one another. The total potential at a point at distance r from their common centre, where $r < a$, would be **[2020, 10 Jan Shift-I]**

(a) $\frac{Q}{4\pi\varepsilon_0(a+b+c)}$ (b) $\frac{Q}{12\pi\varepsilon_0}\frac{ab+bc+ca}{abc}$

(c) $\frac{Q(a+b+c)}{4\pi\varepsilon_0(a^2+b^2+c^2)}$ (d) $\frac{Q(a^2+b^2+c^2)}{4\pi\varepsilon_0(a^3+b^3+c^3)}$

45 In free space, a particle A of charge $1\,\mu C$ is held fixed at a point P. Another particle B of the same charge and mass $4\,\mu g$ is kept at a distance of 1 mm from P. If B is released, then its velocity at a distance of 9 mm from P is $\left[\text{Take, } \frac{1}{4\pi\varepsilon_0} = 9\times 10^9\ \text{N - m}^2\text{C}^{-2}\right]$ **[2019, 10 April Shift-II]**

(a) 1.5×10^2 m/s (b) 3.0×10^4 m/s

(c) 1.0 m/s (d) 2.0×10^3 m/s

46 A uniformly charged ring of radius $3a$ and total charge q is placed in XY-plane centred at origin. A point charge q is moving towards the ring along the Z-axis and has speed v at $z = 4a$. The minimum value of v such that it crosses the origin is **[2019, 10 April Shift-I]**

(a) $\sqrt{\frac{2}{m}}\left(\frac{1}{5}\frac{q^2}{4\pi\varepsilon_0 a}\right)^{1/2}$ (b) $\sqrt{\frac{2}{m}}\left(\frac{4}{15}\frac{q^2}{4\pi\varepsilon_0 a}\right)^{1/2}$

(c) $\sqrt{\frac{2}{m}}\left(\frac{1}{15}\frac{q^2}{4\pi\varepsilon_0 a}\right)^{1/2}$ (d) $\sqrt{\frac{2}{m}}\left(\frac{2}{15}\frac{q^2}{4\pi\varepsilon_0 a}\right)^{1/2}$

47 A solid conducting sphere, having a charge Q, is surrounded by an uncharged conducting hollow spherical shell. Let the potential difference between the surface of the solid sphere and that of the outer surface of the hollow shell be V. If the shell is now given a charge of $-4Q$, the new potential difference between the same two surfaces is **[2019, 8 April Shift-I]**

(a) $-2V$ (b) $2V$ (c) $4V$ (d) V

48 Four equal point charges Q each are placed in the xy-plane at (0, 2), (4, 2), (4, –2) and (0, – 2). The work required to put a fifth charge Q at the origin of the coordinate system (in joule) will be **[2019, 10 Jan Shift-II]**

(a) $\frac{Q^2}{4\pi\varepsilon_0}$ (b) $\frac{Q^2}{4\pi\varepsilon_0}\left(1+\frac{1}{\sqrt{3}}\right)$

(c) $\frac{Q^2}{2\sqrt{2}\pi\varepsilon_0}$ (d) $\frac{Q^2}{4\pi\varepsilon_0}\left(1+\frac{1}{\sqrt{5}}\right)$

49 Two thin wire rings each having a radius R are placed at a distance d apart with their axes coinciding. The charges on the two rings are $+q$ and $-q$. The potential difference between the centres of the two rings is **[AIEEE 2005]**

(a) $\frac{qR}{4\pi\varepsilon_0 d^2}$ (b) $\frac{q}{2\pi\varepsilon_0}\left[\frac{1}{R}-\frac{1}{\sqrt{R^2+d^2}}\right]$

(c) zero (d) $\frac{q}{4\pi\varepsilon_0}\left[\frac{1}{R}-\frac{1}{\sqrt{R^2+d^2}}\right]$

Numerical Value Type Questions

50 At the centre of a half ring of radius $R = 10$ cm and linear charge density 4nCm^{-1}, the potential is $x\,\pi$V. The value of x is .... **[2024, 9 April Shift-I]**

51 The distance between charges $+q$ and $-q$ is $2l$ and between $+2q$ and $-2q$ is $4l$. The electrostatic potential at point P at a distance r from centre O is $-\alpha\left[\frac{ql}{r^2}\right]\times 10^9$ V, where the value of α is ....

(Use $\frac{1}{4\pi\varepsilon_0} = 9\times 10^9\,\text{Nm}^2\text{C}^{-2}$) **[2024, 31 Jan Shift-II]**

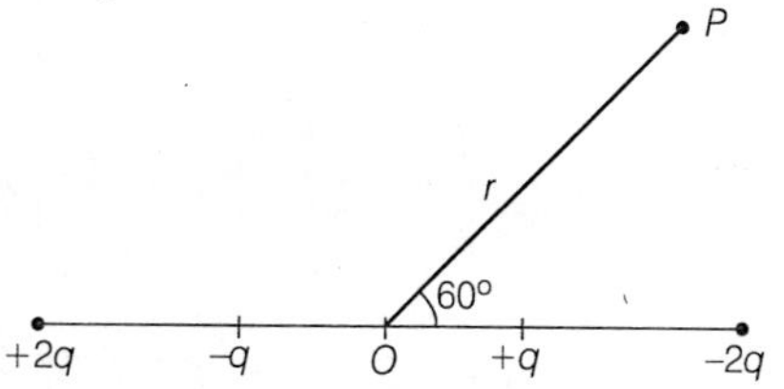

52 The electric potential at the surface of an atomic nucleus $(Z = 50)$ of radius 9×10^{-13} cm is .... $\times 10^6$ V. **[2024, 27 Jan Shift-II]**

53 64 identical drops each charged upto potential of 10mV are combined to form a bigger drop. The potential of the bigger drop will be .... mV. **[2023, 12 April Shift-I]**

54 Three concentric spherical metallic shells X, Y and Z of radius a, b and c, respectively [$a < b < c$] have surface charge densities $\sigma, -\sigma$ and σ, respectively. The shells X and Z are at same potential. If the radii of X and Y are 2 cm and 3 cm, respectively. The radius of shell Z is .... cm. **[2023, 10 April Shift-I]**

55 An electric dipole of dipole moment is 6.0×10^{-6}Cm placed in a uniform electric field of 1.5×10^3 NC^{-1} in such a way that dipole moment is along electric field. The work done in rotating dipole by 180° in this field will be .... mJ. **[2023, 8 April Shift-I]**

56 For a charged spherical ball, electrostatic potential inside the ball varies with r as $V = 2ar^2 + b$. Here, a and b are constant and r is the distance from the

centre. The volume charge density inside the ball is $-\lambda a\varepsilon$. The value of λ is ....
ε = permittivity of the medium [2023, 29 Jan Shift -II]

Imp. Potential at Drop

57 27 identical drops are charged at 22 V each. They combine to form a bigger drop. The potential of the bigger drop will be .... V. [2022, 25 June Shift-II]

58 27 similar drops of mercury are maintained at 10 V each. All these spherical drops combine into a single big drop. The potential energy of the bigger drop is ...... times that of a smaller drop. [2021, 26 Feb Shift-II]

59 512 identical drops of mercury are charged to a potential of 2 V each. The drops are joined to form a single drop. The potential of this drop is .... V. [2021, 25 Feb Shift-I]

Topic 2 Capacitors and Capacitance

Level 1

01 A parallel plate capacitor of capacitance 40 μF is connected to a 100 V power supply. Now, the intermediate space between the plates is filled with a dielectric material of dielectric constant $K = 2$.
Due to the introduction of dielectric material, the extra charge and the change in the electrostatic energy in the capacitor, respectively, are [2025, 22 Jan Shift-I]
(a) 4 mC and 0.2 J (b) 2 mC and 0.4 J
(c) 8mC and 2.0 J (d) 2 mC and 0.2 J

02 A capacitor of capacitance C is charged to a potential V. The flux of the electric field through a closed surface enclosing the positive plate of the capacitor is [2023, 11 April Shift-II]
(a) $\frac{CV}{2\varepsilon_0}$ (b) zero (c) $\frac{2CV}{\varepsilon_0}$ (d) $\frac{CV}{\varepsilon_0}$

03 Given below are two statements: One is labelled as Assertion (A) and the other is labelled as Reason (R).
Assertion (A) Two metallic spheres are charged to the same potential. One of them is hollow and another is solid and both have the same radii. Solid sphere will have lower charge than the hollow one.
Reason (R) Capacitance of metallic spheres depend on the radii of spheres.
In the light of the above statements, choose the correct answer from the options given below. [2023, 1 Feb Shift-II]
(a) A is true but R is false.
(b) A is false but R is true.
(c) Both A and R are true and R is the correct explanation of A.
(d) Both A and R are true but R is not the correct explanation of A.

Imp. Effect of Dielectric on Capacitance of Parallel Plate Capacitor

04 A slab of dielectric constant K has the same cross-sectional area as the plates of a parallel plate capacitor and thickness $\frac{3}{4}d$, where d is the separation of the plates. The capacitance of the capacitor when the slab is inserted between the plates will be
(Given, C_0 = capacitance of capacitor with air as medium between plates.) [2022, 28 July Shift-II]
(a) $\frac{4KC_0}{3+K}$ (b) $\frac{3KC_0}{3+K}$ (c) $\frac{3+K}{4KC_0}$ (d) $\frac{K}{4+K}$

05 A force of 10 N acts on a charged particle placed between two plates of a charged capacitor. If one plate of capacitor is removed, then the force acting on that particle will be [2022, 27 June Shift-I]
(a) 5 N (b) 10 N (c) 20 N (d) Zero

06 A parallel plate capacitor with plates of area 1 m^2 each, are at a separation of 0.1 m. If the electric field between the plates is 100 N/C, the magnitude of charge on each plate is
$\left(\text{Take, } \varepsilon_0 = 8.85 \times 10^{-12} \frac{C^2}{N-m^2}\right)$ [2019, 12 Jan Shift-II]
(a) 9.85×10^{-10} C (b) 8.85×10^{-10} C
(c) 7.85×10^{-10} C (d) 6.85×10^{-10} C

07 A parallel plate capacitor of capacitance 90 pF is connected to a battery of emf 20 V. If a dielectric material of dielectric constant $K = \frac{5}{3}$ is inserted between the plates, the magnitude of the induced charge will be [JEE Main 2018]
(a) 1.2 nC (b) 0.3 nC (c) 2.4 nC (d) 0.9 nC

08 A parallel plate capacitor is made of two circular plates separated by a distance of 5 mm with a dielectric of dielectric constant 2.2 between them. When the electric field in the dielectric is 3×10^4 V/m, the charge density of the positive plate will be close to [JEE Main 2014]
(a) 6×10^{-7} C/m^2 (b) 3×10^{-7} C/m^2
(c) 3×10^4 C/m^2 (d) 6×10^4 C/m^2

09 Two capacitors C_1 and C_2 are charged to 120 V and 200 V, respectively. It is found that by connecting them together, the potential on each one can be made zero. Then, [JEE Main 2013]
(a) $5C_1 = 3C_2$ (b) $3C_1 = 5C_2$
(c) $3C_1 + 5C_2 = 0$ (d) $9C_1 = 4C_2$

10 A sheet of aluminium foil of negligible thickness is introduced between the plates of a capacitor. The capacitance of the capacitor *[AIEEE 2003]*

(a) decreases (b) remains unchanged
(c) becomes infinite (d) increases

11 Capacitance (in F) of a spherical conductor having radius 1 m, is *[AIEEE 2002]*

(a) 1.1×10^{-10} (b) 10^{-6} (c) 9×10^{-9} (d) 10^{-3}

Level 2

12 A dipole with two electric charges of 2 μC magnitude each, with separation distance 0.5 μm, is placed between the plates of a capacitor such that its axis is parallel to an electric field established between the plates when a potential difference of 5 V is applied. Separation between the plates is 0.5 mm. If the dipole is rotated by 30° from the axis, it tends to realign in the direction due to a torque. The value of torque is *[2025, 7 April Shift-II]*

(a) 2.5×10^{-12} Nm (b) 5×10^{-9} Nm
(c) 5×10^{-3} Nm (d) 2.5×10^{-9} Nm

13 Using a battery, a 100 pF capacitor is charged to 60 V and then the battery is removed. After that, a second uncharged capacitor is connected to the first capacitor in parallel. If the final voltage across the second capacitor is 20 V, its capacitance is (in pF) *[2025, 3 April Shift-II]*

(a) 400 (b) 600 (c) 200 (d) 100

14 Given below are two statements : One is labelled as Assertion (A) and the other is labelled as Reason (R).

Assertion (A) Net dipole moment of a polar linear isotropic dielectric substance is not zero even in the absence of an external electric field.

Reason (R) In absence of an external electric field, the different permanent dipoles of a polar dielectric substance are oriented in random directions.

In the light of the above statements, choose the most appropriate answer from the options given below. *[2025, 3 April Shift-II]*

(a) A is not correct but R is correct.
(b) Both A and R are correct and R is the correct explanation of A.
(c) A is correct but R is not correct.
(d) Both A and R are correct but R is not the correct explanation of A.

15 A parallel plate capacitor was made with space two rectangular plates, each with a length of l = 3 cm and breath of b = 1 cm. The distance between the plates is 3 μm. Out of the following, which are the ways to increase the capacitance by a factor of 10 ? *[2025, 24 Jan Shift-I]*

A. $l = 30$ cm, $b = 1$ cm, $d = 1$ μm
B. $l = 3$ cm, $b = 1$ cm, $d = 30$ μm
C. $l = 6$ cm, $b = 5$ cm, $d = 3$ μm
D. $l = 1$ cm, $b = 1$ cm, $d = 10$ μm
E. $l = 5$ cm, $b = 2$ cm, $d = 1$ μm

Choose the correct answer from the options given below.

(a) C only (b) A only
(c) B and D only (d) C and E only

16 A capacitor has air as dielectric medium and two conducting plates of area 12 cm^2 and they are 0.6 cm apart. When a slab of dielectric having area 12 cm^2 and 0.6 cm thickness is inserted between the plates, one of the conducting plates has to be moved by 0.2 cm to keep the capacitance same as in previous case. The dielectric constant of the slab is (Given $\varepsilon_0 = 8.834\times 10^{-12}$ F/m) *[2024, 8 April Shift-II]*

(a) 1.33 (b) 0.66
(c) 1 (d) 1.50

17 Force between two point charges q_1 and q_2 placed in vacuum at r cm a part is F. Force between them when placed in a medium having dielectric constant $K = 5$ at $r/5$ cm apart will be *[2024, 31 Jan Shift-II]*

(a) $F/25$ (b) $F/5$
(c) $25F$ (d) $5F$

18 The distance between two plates of a capacitor is d and its capacitance is C_1, when air is the medium between the plates. If a metal sheet of thickness $2d/3$and of the same area as plate is introduced between the plates, the capacitance of the capacitor becomes C_2. The ratio $\frac{C_2}{C_1}$ is *[2023, 10 April Shift-II]*

(a) 3 : 1 (b) 1 : 1
(c) 4 : 1 (d) 2 : 1

19 A parallel plate capacitor has plate area 40 cm^2 and plates separation 2 mm. The space between the plates is filled with a dielectric medium of a thickness 1mm and dielectric constant 5. The capacitance of the system is *[2023, 25 Jan Shift-II]*

(a) $24\varepsilon_0$ F (b) $\frac{3}{10}\varepsilon_0$ F
(c) $\frac{10}{3}\varepsilon_0$ F (d) $10\varepsilon_0$ F

20 Two identical thin metal plates has charge q_1 and q_2 respectively, such that $q_1 > q_2$. The plates were brought close to each other to form a parallel plate capacitor of capacitance C. The potential difference between them is *[2022, 29 July Shift-II]*

(a) $\frac{(q_1 + q_2)}{C}$ (b) $\frac{(q_1 - q_2)}{C}$
(c) $\frac{(q_1 - q_2)}{2C}$ (d) $\frac{2(q_1 - q_2)}{C}$

21 A condenser of 2 μF capacitance is charged steadily from 0 to 5 C. Which of the following graph represents correctly the variation of potential difference V across its plates with respect to the charge Q on the condenser? *[2022, 25 July Shift-I]*

(a) 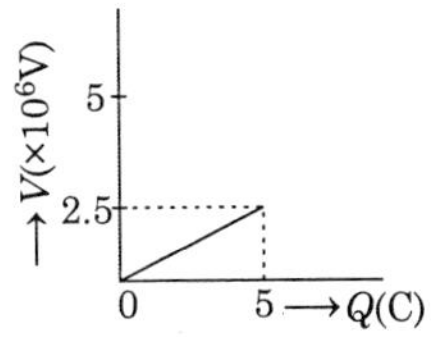

(b) 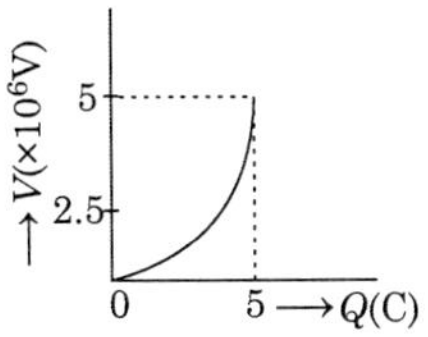

(c) 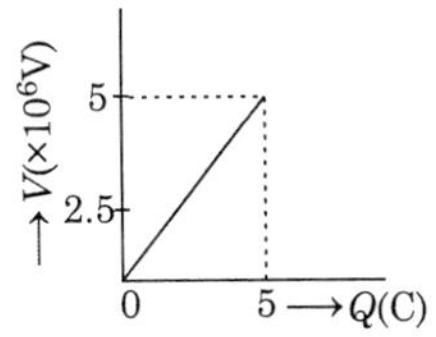

(d) 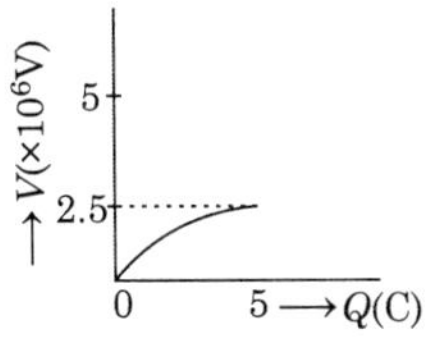

22 A parallel plate capacitor with plate area A and plate separation $d = 2$ m has a capacitance of 4 μF. The new capacitance of the system, if half of the space between them is filled with a dielectric material of dielectric constant, $K = 3$ (as shown in figure) will be *[2022, 26 June shift-II]*

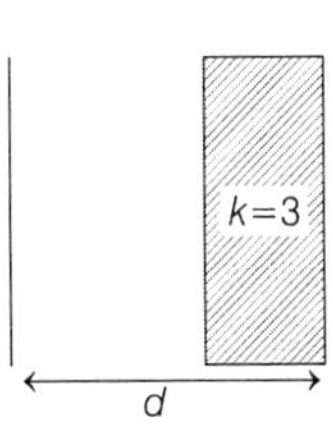

(a) 2 μF (b) 32 μF
(c) 6 μF (d) 8 μF

23 Two capacitors having capacitance C_1 and C_2, respectively are connected as shown in figure. Initially, capacitor C_1 is charged to a potential difference V volt by a battery. The battery is then removed and the charged capacitor C_1 is now connected to uncharged capacitor C_2 by closing the switch S. The amount of charge on the capacitor C_2, after equilibrium is *[2022, 26 June Shift-I]*

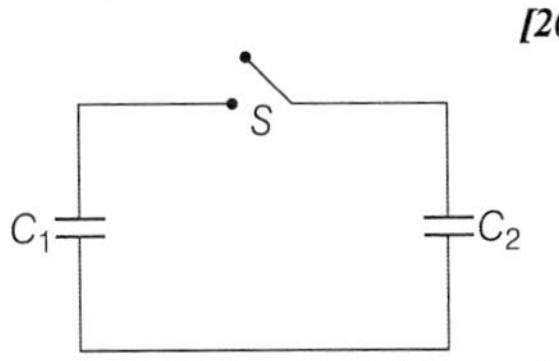

(a) $\dfrac{C_1 C_2}{(C_1 + C_2)}$ V (b) $\dfrac{(C_1 + C_2)}{C_1 C_2}$ V
(c) $(C_1 + C_2)$ V (d) $(C_1 - C_2)$ V

24 Two metallic plates form a parallel plate capacitor. The distance between the plates is d. A metal sheet of thickness $d/2$ and of area equal to area of each plate is introduced between the plates. What will be the ratio of the new capacitance to the original capacitance of the capacitor? *[2022, 25 June Shift-II]*

(a) 2 : 1 (b) 1 : 2 (c) 1 : 4 (d) 4 : 1

25 A parallel plate capacitor is formed by two plates, each of area 30π cm^2 separated by 1 mm. A material of dielectric strength 3.6×10^7 Vm^{-1} is filled between the plates. If the maximum charge that can be stored on the capacitor without causing any dielectric breakdown is 7×10^{-6} C, the value of dielectric constant of the material is

$$\left\{\text{Use, } \frac{1}{4\pi\varepsilon_0} = 9 \times 10^9 \text{ Nm}^2\text{C}^{-2}\right\}$$
[2022, 24 June Shift-I]

(a) 1.66 (b) 1.75 (c) 2.25 (d) 2.33

26 Two capacitors of capacities $2C$ and C are joined in parallel and charged upto potential V. The battery is removed and the capacitor of capacity C is filled completely with a medium of dielectric constant K. The potential difference across the capacitors will now be *[2021, 27 July Shift-I]*

(a $\dfrac{V}{K+2}$ (b) $\dfrac{V}{K}$ (c) $\dfrac{3V}{K+2}$ (d) $\dfrac{3V}{K}$

27 If q_f is the free charge on the capacitor plates and q_b is the bound charge on the dielectric slab of dielectric constant K placed between the capacitor plates, then bound charge q_b can be expressed as *[2021, 25 July Shift-II]*

(a) $q_b = q_f\left(1 - \dfrac{1}{\sqrt{K}}\right)$ (b) $q_b = q_f\left(1 - \dfrac{1}{K}\right)$

(c) $q_b = q_f\left(1 + \dfrac{1}{\sqrt{K}}\right)$ (d) $q_b = q_f\left(1 + \dfrac{1}{K}\right)$

28 A 10 μF capacitor is fully charged to a potential difference of 50 V. After removing the source voltage it is connected to an uncharged capacitor in parallel. Now, the potential difference across them becomes 20 V. The capacitance of the second capacitor is *[2020, 2 Sep Shift-II]*

(a) 15 μF (b) 30 μF
(c) 20 μF (d) 10 μF

29 A simple pendulum of length L is placed between the plates of a parallel plate capacitor having electric field E, as shown in figure. Its bob has mass m and charge q. The time period of the pendulum is given by *[2019, 10 April Shift-II]*

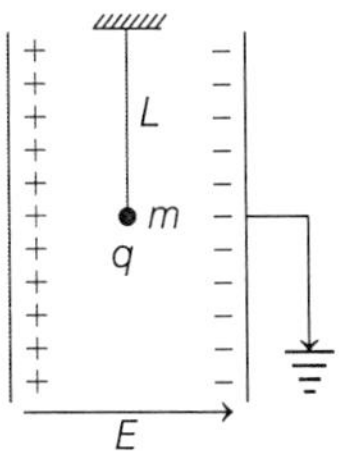

(a) $2\pi\sqrt{\dfrac{L}{\sqrt{g^2+\left(\dfrac{qE}{m}\right)^2}}}$ (b) $2\pi\sqrt{\dfrac{L}{\sqrt{g^2-\dfrac{q^2E^2}{m^2}}}}$

(c) $2\pi\sqrt{\dfrac{L}{\left(g+\dfrac{qE}{m}\right)}}$ (d) $2\pi\sqrt{\dfrac{L}{\left(g-\dfrac{qE}{m}\right)}}$

Level 3

30 A parallel plate capacitor is filled equally(half) with two dielectrics of dielectric constants ε_1 and ε_2, as shown in figures. The distance between the plates is d and area of each plate is A. If capacitance in first configuration and second configuration are C_1 and C_2 respectively, $\dfrac{C_1}{C_2}$ is *[2025, 3 April Shift-I]*

First Configuration

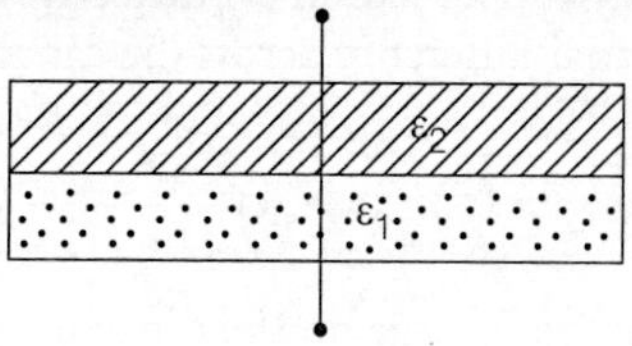

Second Configuration

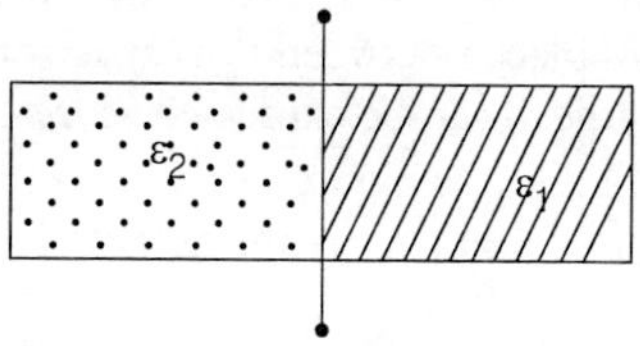

(a) $\dfrac{\varepsilon_1\varepsilon_2^2}{(\varepsilon_1+\varepsilon_2)^2}$ (b) $\dfrac{\varepsilon_1\varepsilon_2}{\varepsilon_1+\varepsilon_2}$

(c) $\dfrac{\varepsilon_0(\varepsilon_1+\varepsilon_2)}{2}$ (d) $\dfrac{4\varepsilon_1\varepsilon_2}{(\varepsilon_1+\varepsilon_2)^2}$

31 Two uniformly charged spherical conductors A and B of radii 5 mm and 10 mm are separated by a distance of 2 cm. If the spheres are connected by a conducting wire, then in equilibrium condition, the ratio of the magnitude of the electric fields at the surface of the sphere A and B will be *[2022, 26 July Shift-II]*

(a) 1 : 2 (b) 2 : 1 (c) 1 : 1 (d) 1 : 4

32 Capacitance of an isolated conducting sphere of radius R_1 becomes n times when it is enclosed by a concentric conducting sphere of radius R_2 connected to earth. *[2022, 25 July Shift-II]*

The ratio of their radii $\left(\dfrac{R_2}{R_1}\right)$ is

(a) $\dfrac{n}{n-1}$ (b) $\dfrac{2n}{2n+1}$ (c) $\dfrac{n+1}{n}$ (d) $\dfrac{2n+1}{n}$

33 A capacitor is made of two square plates each of side a making a very small angle α between them, as shown in figure. The capacitance will be close to *[2020, 8 Jan Shift-II]*

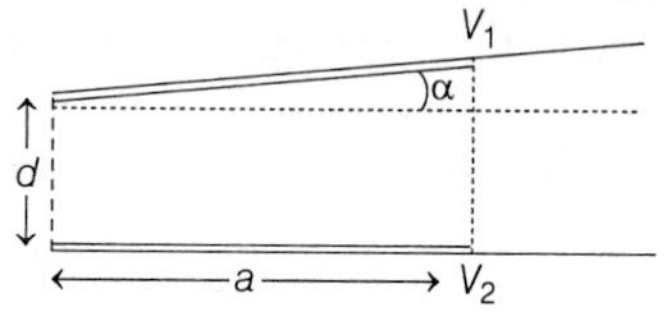

(a) $\dfrac{\varepsilon_0 a^2}{d}\left(1-\dfrac{\alpha a}{4d}\right)$ (b) $\dfrac{\varepsilon_0 a^2}{d}\left(1+\dfrac{\alpha a}{d}\right)$

(c) $\dfrac{\varepsilon_0 a^2}{d}\left(1-\dfrac{\alpha a}{2d}\right)$ (d) $\dfrac{\varepsilon_0 a^2}{d}\left(1-\dfrac{3\alpha a}{2d}\right)$

34 The space between the plates of a parallel plate capacitor is filled with a 'dielectric' whose 'dielectric constant' varies with distance as per the relation $K(x)=K_0+\lambda x$ ($\lambda=$ a constant).
The capacitance C, of the capacitor, would be related to its vacuum capacitance C_0 for the relation *[JEE Main 2014]*

(a) $C=\dfrac{\lambda d}{\ln(1+K_0\lambda d)}C_0$ (b) $C=\dfrac{\lambda}{d\ln(1+K_0\lambda d)}C_0$

(c) $C=\dfrac{\lambda d}{\ln(1+\lambda d/K_0)}C_0$ (d) $C=\dfrac{\lambda}{d\ln(1+K_0/\lambda d)}C_0$

Numerical Value Type Questions

35 A parallel plate capacitor has charge 5×10^{-6}C. A dielectric slab is inserted between the plates and almost fills the space between the plates. If the induced charge on one face of the slab is 4×10^{-6}C, then the dielectric constant of the slab is .... *[2025, 7 April Shift-I]*

36

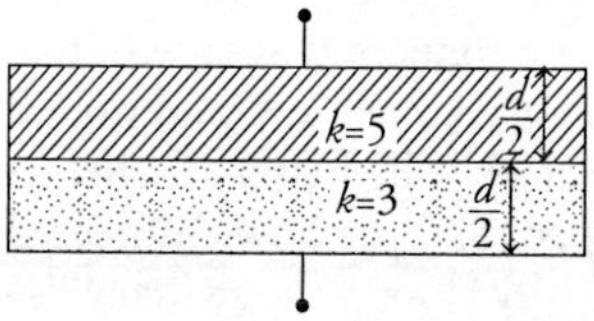

Space between the plates of a parallel plate capacitor of plate area 4 cm^2 and separation of (d) 1.77 mm, is filled with uniform dielectric materials with dielectric constants (3 and 5)as shown in figure. Another capacitor of capacitance 7.5 pF is connected in parallel with it. The effective capacitance of this combination is .... pF.
(Given $\epsilon_0=8.85\times10^{-12}$ F/m) *[2025, 8 April Shift-II]*

37 A parallel plate capacitor with plate separation 5 mm is charged up by a battery. It is found that on introducing a dielectric sheet of thickness 2mm, while keeping the battery connections intact, the capacitor draws 25% more charge from the battery than before. The dielectric constant of the sheet is .... *[2024, 31 Jan Shift-I]*

38 Two identical charged spheres are suspended by strings of equal lengths. The strings make an angle of 37° with each other. When suspended in a liquid of density 0.7 g/cm^3, the angle remains same. If density of material of the sphere is 1.4 g/cm^3, the dielectric constant of the liquid is .... $\left(\tan 37° = \frac{3}{4}\right)$ *[2024, 30 Jan Shift-II]*

39 A parallel plate capacitor with air between the plates has a capacitance of 15 pF. The separation between the plate becomes twice and the space between them is filled with a medium of dielectric constant 3.5. Then, the capacitance becomes $x/4$ pF. The value of x is .... *[2023, 24 Jan Shift-II]*

40 A capacitor has capacitance 5 μF when it's parallel plates are separated by air medium of thickness d. A slab of material of dielectric constant 1.5 having area equal to that of plates but thickness $d/2$ is inserted between the plates. Capacitance of the capacitor in the presence of slab will be ... μF. *[2023, 24 Jan Shift-II]*

41 A capacitor C_1 of capacitance 5 μF is charged to a potential of 30 V using a battery. The battery is then removed and the charged capacitor is connected to an uncharged capacitor C_2 of capacitance 10 μF as shown in figure.
When the switch is closed, charge flows between the capacitors. At equilibrium, the charge on the capacitor C_2 is .... μC. *[2022, 28 June Shift-II]*

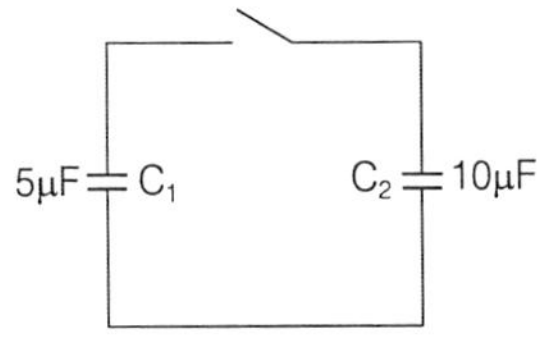

42 A parallel plate capacitor has plate area 100 m^2 and plate separation of 10 m. The space between the plates is filled upto a thickness 5 m with a material of dielectric constant of 10. The resultant capacitance of the system is x pF. The value of $\varepsilon_0 = 8.85 \times 10^{-12}$ C^2/N-m^2. The value of x to the nearest integer is .... *[2021, 18 March Shift-I]*

43 A 2 μF capacitor C_1 is first charged to a potential difference of 10 V using a battery. Then, the battery is removed and the capacitor is connected to an uncharged capacitor C_2 of 8 μF. The charge in C_2 on equilibrium condition is .... μC.
(Round off to the nearest integer) *[2021, 17 March Shift-II]*

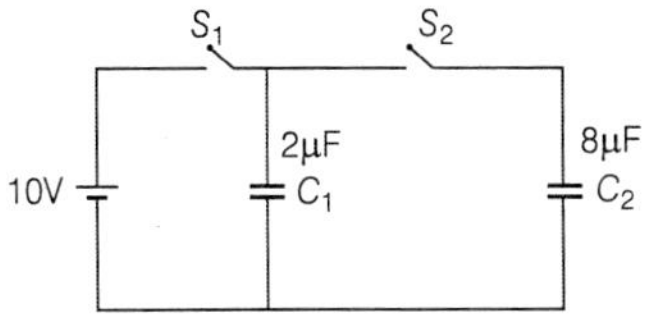

Topic 3 Combination of Capacitors and Energy Stored in a Capacitor

Level 1

01 Consider a parallel plate capacitor of area A (of each plate) and separation d between the plates. If E is the electric field and ε_0 is the permittivity of free space between the plates, then potential energy stored in the capacitor is *[2025, 24 Jan Shift-I]*

(a) $\frac{1}{2}\varepsilon_0 E^2 Ad$ (b) $\varepsilon_0 E^2 Ad$
(c) $\frac{1}{4}\varepsilon_0 E^2 Ad$ (d) $\frac{3}{4}\varepsilon_0 E^2 Ad$

Imp. Combination of Capacitors

02 The equivalent capacitance of the combination shown is *[2023, 10 April Shift-I]*

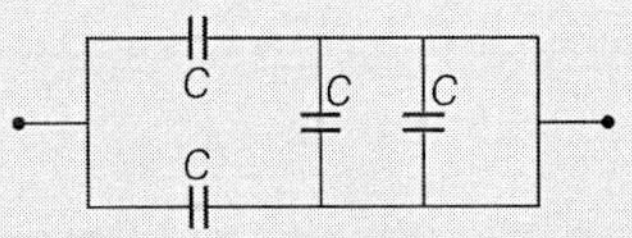

(a) $\frac{5}{3}C$ (b) $2C$ (c) $\frac{C}{2}$ (d) $4C$

03 The total charge on the system of capacitors $C_1 = 1\,\mu F, C_2 = 2\,\mu F, C_3 = 4\mu F$ and $C_4 = 3\mu F$ connected in parallel is
(Assume a battery of 20 V is connected to the combination) *[2022, 26 July Shift-I]*
(a) 200 μC (b) 200 C (c) 10 μC (d) 10 C

04 If the charge on a capacitor is increased by 2 C, the energy stored in it increases by 44%. The original charge on the capacitor is (in C) *[2022, 24 June Shift-II]*
(a) 10 (b) 20 (c) 30 (d) 40

05 A parallel plate capacitor is of area 6 cm^2 and a separation 3 mm. The gap is filled with three dielectric materials of equal thickness (see figure) with dielectric constants $K_1 = 10$, $K_2 = 12$ and $K_3 = 14$. The dielectric constant of a material which give same capacitance when fully inserted in above capacitor, would be *[2019, 10 Jan Shift-I]*

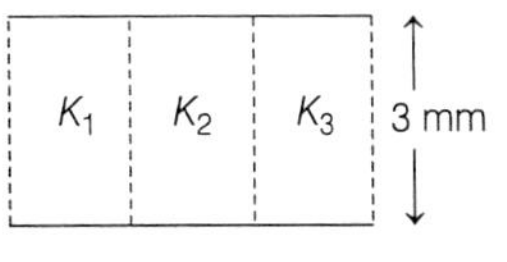

(a) 4 (b) 36 (c) 12 (d) 14

06 The equivalent capacitance between A and B in the circuit given below, is *[JEE Main 2018]*

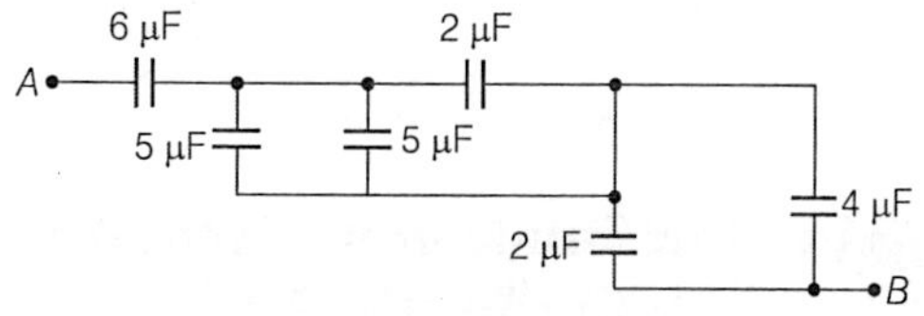

(a) $5.4\ \mu F$ (b) $4.9\ \mu F$
(c) $3.6\ \mu F$ (d) $2.4\ \mu F$

07 A battery is used to charge a parallel plate capacitor till the potential difference between the plates becomes equal to the electromotive force of the battery. The ratio of the energy stored in the capacitor and the work done by the battery will be *[AIEEE 2007]*

(a) 1 (b) 2
(c) 1/4 (d) 1/2

08 The work done in placing a charge of 8×10^{-18} C on a condenser of capacity $100\ \mu F$ is *[AIEEE 2003]*

(a) 16×10^{-32} J (b) 3.1×10^{-26} J
(c) 4×10^{-10} J (d) 32×10^{-32} J

09 If there are n capacitors in parallel connected to V volt source, then the energy stored is equal to *[AIEEE 2002]*

(a) CV (b) $1/2\, nCV^2$
(c) CV^2 (d) $\frac{1}{2n} CV^2$

Level 2

10 Three parallel plate capacitors C_1, C_2 and C_3 each of capacitance $5\ \mu F$ are connected as shown in figure. The effective capacitance between points A and B, when the space between the parallel plates of C_1 capacitor is filled with a dielectric medium having dielectric constant of 4, is *[2025, 4 April Shift-II]*

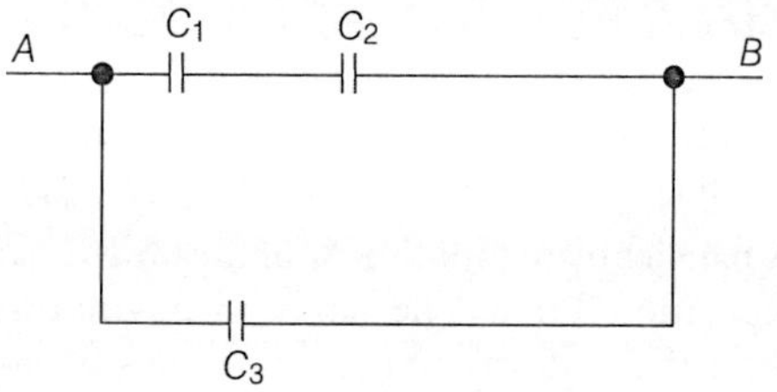

(a) $7.5\ \mu F$ (b) $9\ \mu F$
(c) $30\ \mu F$ (d) $22.5\ \mu F$

11 Two capacitors C_1 and C_2 are connected in parallel to a battery. Charge-time graph is shown below for the two capacitor. The energy stored with them are U_1 and U_2, respectively. Which of the given statements is true? *[2025, 28 Jan Shift-I]*

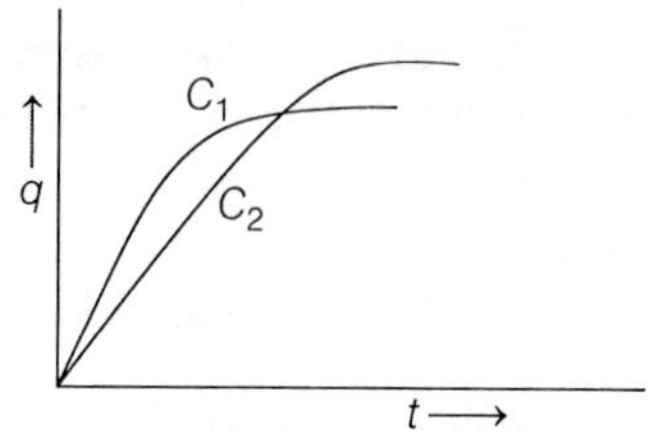

(a) $C_2 > C_1, U_2 < U_1$
(b) $C_1 > C_2, U_1 < U_2$
(c) $C_1 > C_2, U_1 > U_2$
(d) $C_2 > C_1, U_2 > U_1$

Imp. Energy Density

12 A parallel plate capacitor of capacitance $1\ \mu F$ is charged to a potential difference of 20 V. The distance between plates is $1\ \mu m$. The energy density between plates of capacitor is *[2025, 28 Jan Shift-II]*

(a) $1.8 \times 10^3\ J/m^3$ (b) $2 \times 10^2\ J/m^3$
(c) $1.8 \times 10^5\ J/m^3$ (d) $2 \times 10^{-4}\ J/m^3$

13 A capacitor of capacitance $100\ \mu F$ is charged to a potential of 12 V and connected to a 6.4 mH inductor to produce oscillations. The maximum current in the circuit would be *[2024, 29 Jan Shift-I]*

(a) 2.0 A (b) 1.5 A
(c) 1.2 A (d) 3.2 A

14 A parallel plate capacitor of capacitance 2F is charged to a potential V. The energy stored in the capacitor is E_1. The capacitor is now connected to another uncharged identical capacitor in parallel combination. The energy stored in the combination is E_2. The ratio E_2 / E_1 is *[2023, 11 April Shift-I]*

(a) 1 : 2 (b) 1 : 4
(c) 2 : 3 (d) 2 : 1

15 Two capacitors, each having capacitance $40\ \mu F$ are connected in series. The space between one of the capacitors is filled with dielectric material of dielectric constant K such that the equivalence capacitance of the system became $24\ \mu F$. The value of K will be *[2022, 28 July Shift-I]*

(a) 1.5 (b) 2.5
(c) 1.2 (d) 3

16 A parallel plate capacitor filled with a medium of dielectric constant 10, is connected across a battery and is charged. The dielectric slab is replaced by another slab of dielectric constant 15, then the energy of capacitor will *[2022, 29 June Shift-I]*

(a) increased by 50% (b) decreased by 15%
(c) increased by 25% (d) increased by 33%

17 The charge on capacitor of capacitance 15μF in the figure given below is **[2022, 26 June Shift-II]**

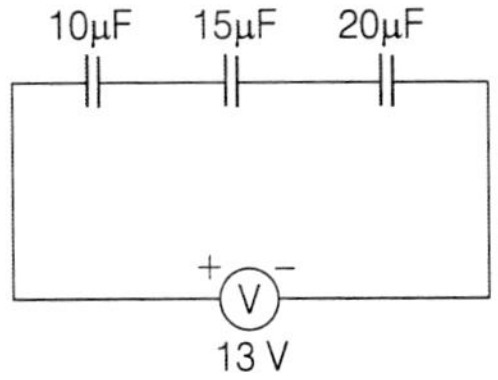

(a) 60 μC (b) 130 μC (c) 260 μC (d) 585 μC

18 Three capacitors $C_1 = 2\ \mu F$, $C_2 = 6\ \mu F$ and $C_3 = 12\ \mu F$ are connected as shown in figure. Find the ratio of the charges on capacitors C_1, C_2 and C_3, respectively. **[2021, 27 Aug Shift-II]**

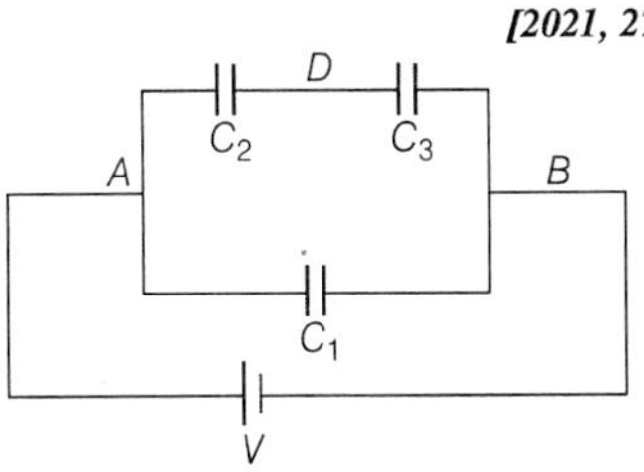

(a) 3 : 4 : 4 (b) 2 : 3 : 3 (c) 2 : 1 : 1 (d) 1 : 2 : 2

19 For changing the capacitance of a given parallel plate capacitor, a dielectric material of dielectric constant K is used, which has the same area as the plates of the capacitor. The thickness of the dielectric slab is $3/4d$, where d is the separation between the plates of parallel plate capacitor. The new capacitance (C') in terms of original capacitance (C_0) is given by the following relation **[2021, 16 March Shift-I]**

(a) $C' = \frac{3+K}{4K} C_0$ (b) $C' = \frac{4+K}{3} C_0$

(c) $C' = \frac{4K}{K+3} C_0$ (d) $C' = \frac{4}{3+K} C_0$

20 Two equal capacitors are first connected in series and then in parallel. The ratio of the equivalent capacities in the two cases will be **[2021, 24 Feb Shift-I]**

(a) 1 : 2 (b) 2 : 1 (c) 4 : 1 (d) 1 : 4

21 In the circuit shown in the figure, the total charge is 750 μC and the voltage across capacitor C_2 is 20 V. Then, the charge on capacitor C_2 is **[2020, 3 Sep Shift-I]**

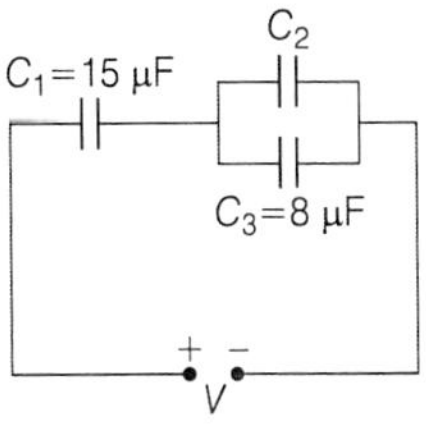

(a) 450 μC (b) 590 μC (c) 160 μC (d) 650 μC

22 Effective capacitance of parallel combination of two capacitors C_1 and C_2 is 10 μF. When these capacitors are individually connected to a voltage source of 1 V, the energy stored in the capacitor C_2 is 4 times that of C_1. If these capacitors are connected in series, then their effective capacitance will be **[2020, 8 Jan Shift-I]**

(a) 4.2 μF (b) 3.2 μF (c) 1.6 μF (d) 8.4 μF

23 In the given circuit, the charge on 4 μF capacitor will be **[2019, 12 April Shift-II]**

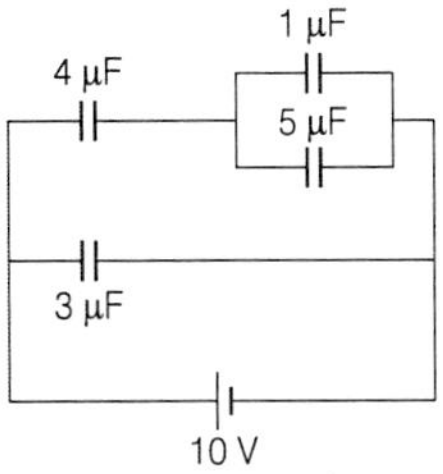

(a) 5.4 μC (b) 9.6 μC (c) 13.4 μC (d) 24 μC

24 The parallel combination of two air filled parallel plate capacitors of capacitance C and nC is connected to a battery of voltage, V. When the capacitors are fully charged, the battery is removed and after that a dielectric material of dielectric constant K is placed between the two plates of the first capacitor. The new potential difference of the combined system is **[2019, 9 April Shift-II]**

(a) $\frac{(n+1)V}{(K+n)}$ (b) $\frac{nV}{K+n}$ (c) V (d) $\frac{V}{K+n}$

25 In the circuit shown, find C if the effective capacitance of the whole circuit is to be 0.5 μF. All values in the circuit are in μF. **[2019, 12 Jan Shift-II]**

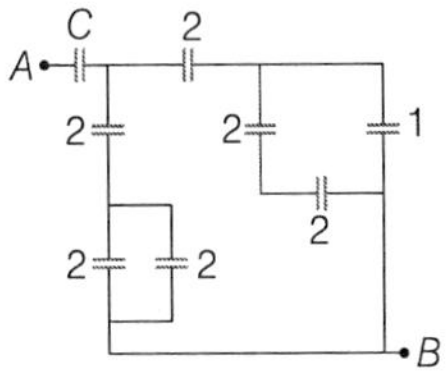

(a) $\frac{6}{5}$ μF (b) 4 μF (c) $\frac{7}{10}$ μF (d) $\frac{7}{11}$ μF

26 Seven capacitors, each of capacitance 2 μF are to be connected in a configuration to obtain an effective capacitance of $\left(\frac{6}{13}\right)$ μF. Which of the combinations shown in figures below will achieve the desired value? **[2019, 11 Jan Shift-II]**

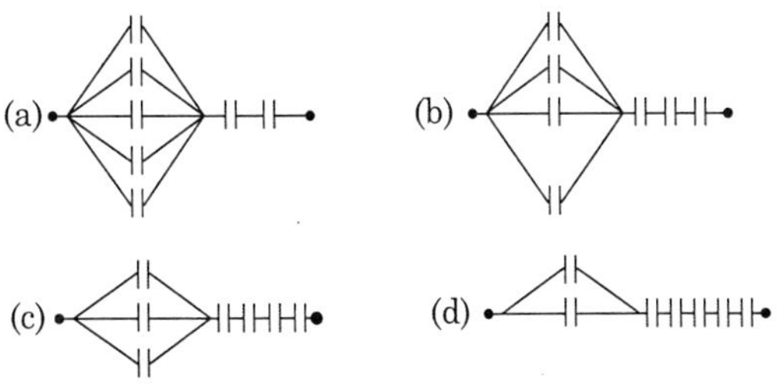

27 In the figure shown below, the charge on the left plate of the 10 μF capacitor is – 30 μC. The charge on the right plate of the 6 μF capacitor is *[2019, 11 Jan Shift-I]*

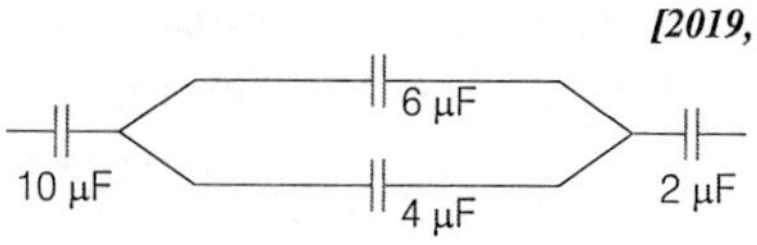

(a) + 12 μC (b) + 18 μC (c) – 12 μC (d) – 18 μC

28 A capacitance of 2 μF is required in an electrical circuit across a potential difference of 1kV. A large number of 1 μF capacitors are available which can withstand a potential difference of not more than 300 V. The minimum number of capacitors required to achieve this is *[JEE Main 2017]*

(a) 16 (b) 24 (c) 32 (d) 2

29 The energy stored in the electric field produced by a metal sphere is 4.5 J. If the sphere contains 4 μC charge, its radius will be

[Take, $\frac{1}{4\pi\varepsilon_0} = 9 \times 10^9$ N-m²/C²] *[JEE Main 2017]*

(a) 32 mm (b) 20 mm (c) 16 mm (d) 28 mm

30 A combination of capacitors is set-up as shown in the figure. The magnitude of the electric field, due to a point charge Q (having a charge equal to the sum of the charges on the 4 μF and 9 μF capacitors), at a point distance 30 m from it, would equal to *[JEE Main 2016]*

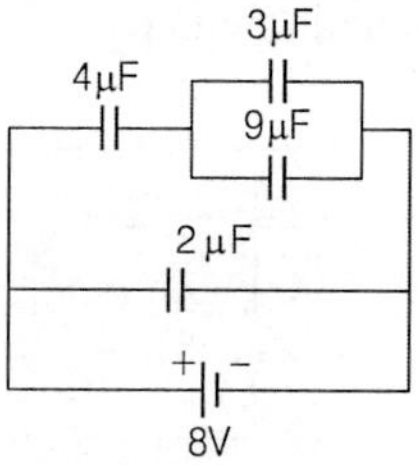

(a) 240 N/C (b) 360 N/C (c) 420 N/C (d) 480 N/C

31 In the given circuit, charge Q_2 on the 2 μF capacitor changes as C is varied from 1 μF to 3μF. Q_2 as a function of C is given properly by (figures are drawn schematically and are not to scale) *[JEE Main 2015]*

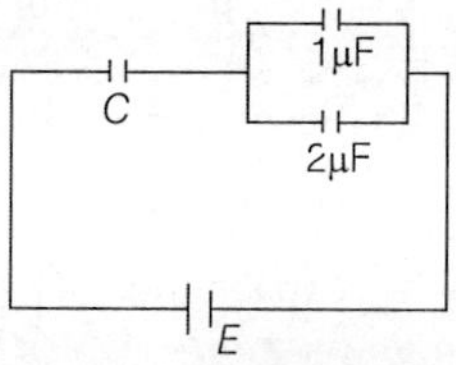

(a)

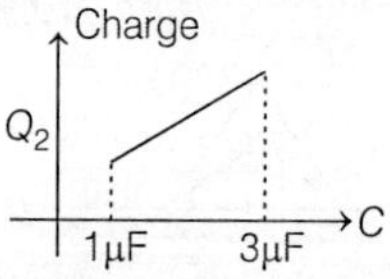

(b)

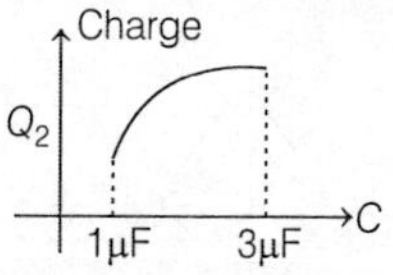

(c)

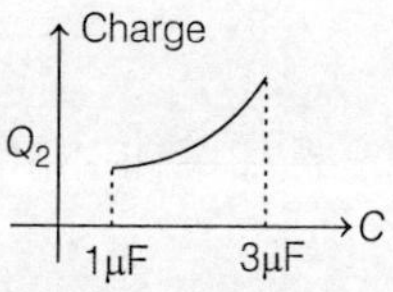

(d)

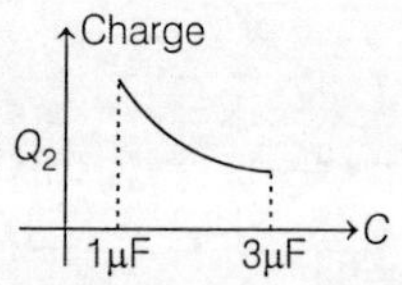

32 A parallel plate capacitor having a separation between the plates d, plate area A and material with dielectric constant K has capacitance C_0. Now one-third of the material is replaced by another material with dielectric constant $2K$, so that effectively there are two capacitors one with area 1/3 A, dielectric constant $2K$ and another with area $2/3\ A$ and dielectric constant K. If the capacitance of this new capacitor is C, then C/C_0 is *[JEE Main 2013]*

(a) 1 (b) 4/3 (c) 2/3 (d) 1/3

33 A parallel plate capacitor with air between the plates has a capacitance of 9 pF. The separation between its plates is d. The space between the plate is now filled with two dielectrics. One of the dielectrics has dielectric constant $K_1 = 3$ and thickness $d/3$ while the other one has dielectric constant $K_2 = 6$ and thickness $2d/3$. Capacitance of the capacitor is now *[AIEEE 2008]*

(a) 1.8 pF (b) 45 pF (c) 40.5 pF (d) 20.25 pF

34 A parallel plate capacitor is made by stacking n equally spaced plates connected alternatively. If the capacitance between any two adjacent plates is C, then the resultant capacitance is *[AIEEE 2005]*

(a) $(n-1)C$ (b) $(n+1)C$
(c) C (d) nC

35 A fully charged capacitor has a capacitance C. It is discharged through a small coil of resistance wire embedded in a thermally insulated block of specific heat capacity s and mass m. If the temperature of the block is raised by ΔT, the potential difference V across the capacitance is *[AIEEE 2005]*

(a) $\sqrt{\frac{2mC\Delta T}{s}}$ (b) $\frac{mC\Delta T}{s}$ (c) $\frac{ms\Delta T}{C}$ (d) $\sqrt{\frac{2ms\Delta T}{C}}$

Level 3

36 A capacitor $C_1 = 6\ \mu F$ is charged to a potential difference of $V_0 = 5$ V using a 5 V battery. The battery is removed and another capacitor $C_2 = 12\ \mu F$ is inserted in place of the battery. When the switch S is closed, the charge flows between the capacitors for some time until equilibrium condition is reached. What are the charges (q_1 and q_2) on the capacitors C_1 and C_2 when equilibrium condition is reached ? *[2025, 29 Jan Shift-II]*

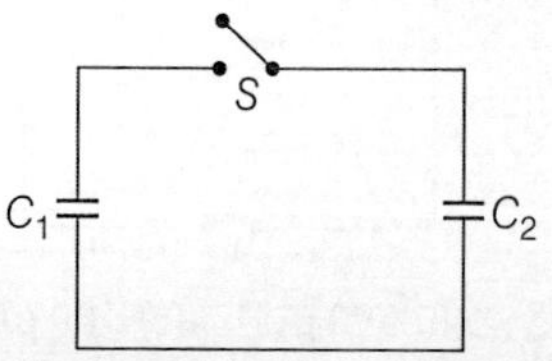

(a) $q_1 = 10\ \mu C$, $q_2 = 20\ \mu C$ (b) $q_1 = 30\ \mu C$, $q_2 = 15\ \mu C$
(c) $q_1 = 20\ \mu C$, $q_2 = 10\ \mu C$ (d) $q_1 = 15\ \mu C$, $q_2 = 30\ \mu C$

37 A capacitor is made of a flat plate of area A and a second plate having a stair-like structure as shown in figure. If the area of each stair is $\frac{A}{3}$ and the height is d, the capacitance of the arrangement is *[2024, 9 April Shift-I]*

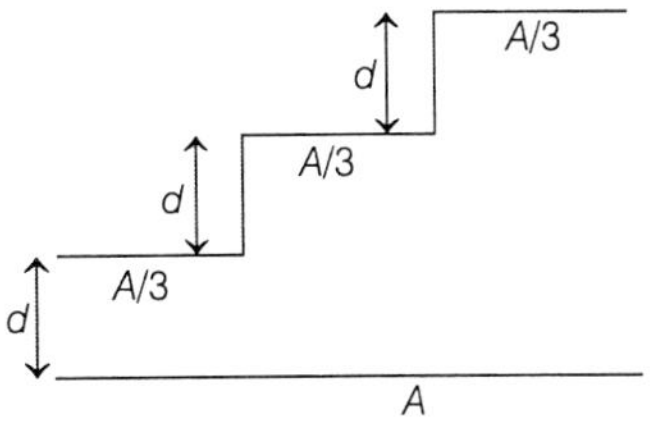

(a) $\frac{18\varepsilon_0 A}{11d}$ (b) $\frac{11\varepsilon_0 A}{18d}$ (c) $\frac{11\varepsilon_0 A}{20d}$ (d) $\frac{13\varepsilon_0 A}{17d}$

38 Two identical capacitors have same capacitance C. One of them is charged to the potential V and other to the potential $2V$. The negative ends of both are connected together. When the positive ends are also joined together, the decrease in energy of the combined system is *[2024, 1 Feb Shift-I]*

(a) $\frac{3}{4}CV^2$ (b) $\frac{1}{4}CV^2$ (c) $\frac{1}{2}CV^2$ (d) $2\,CV^2$

39 A source of potential difference V is connected to the combination of two identical capacitors as shown in the figure. When key K is closed, the total energy stored across the combination is E_1.

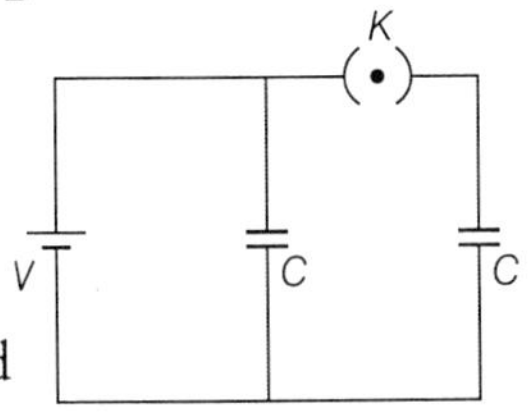

Now, key K is opened and dielectric of dielectric constant 5 is introduced between the plates of the capacitors. The total energy stored across the combination is now E_2. The ratio E_1/E_2 will be *[2022, 26 July Shift-II]*

(a) $\frac{1}{10}$ (b) $\frac{2}{5}$ (c) $\frac{5}{13}$ (d) $\frac{5}{26}$

40 A parallel-plate capacitor with plate area A has separation d between the plates. Two dielectric slabs of dielectric constant K_1 and K_2 of same area $A/2$ and thickness $d/2$ are inserted in the space between the plates. The capacitance of the capacitor will be given by *[2021, 26 Aug Shift-II]*

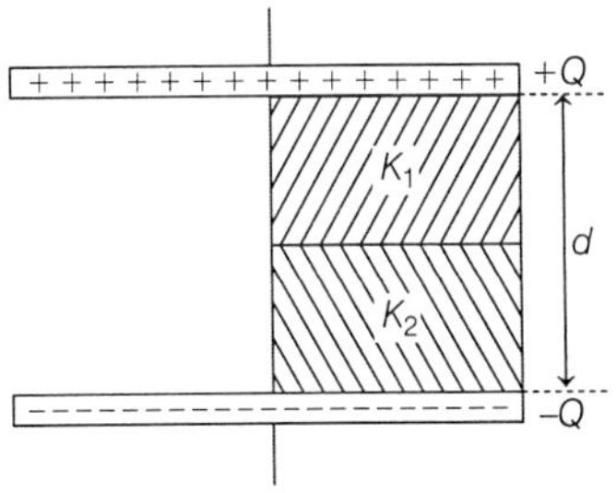

(a) $\frac{\varepsilon_0 A}{d}\left(\frac{1}{2}+\frac{K_1 K_2}{K_1+K_2}\right)$ (b) $\frac{\varepsilon_0 A}{d}\left(\frac{1}{2}+\frac{2(K_1+K_2)}{K_1 K_2}\right)$

(c) $\frac{\varepsilon_0 A}{d}\left(\frac{1}{2}+\frac{K_1+K_2}{K_1 K_2}\right)$ (d) $\frac{\varepsilon_0 A}{d}\left(\frac{1}{2}+\frac{K_1+K_2}{2(K_1+K_2)}\right)$

41 In the reported figure, a capacitor is formed by placing a compound dielectric between the plates of parallel plate capacitor. The expression for the capacity of the said capacitor will be (Take, area of plate $= A$) *[2021, 27 July Shift-II]*

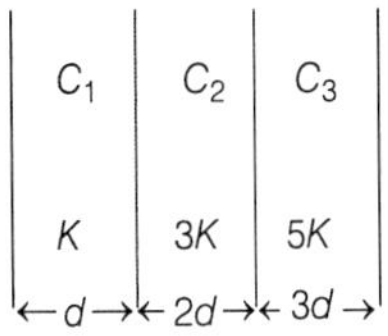

(a) $\frac{15}{34}\frac{K\varepsilon_0 A}{d}$ (b) $\frac{15}{6}\frac{K\varepsilon_0 A}{d}$ (c) $\frac{25}{6}\frac{K\varepsilon_0 A}{d}$ (d) $\frac{9}{6}\frac{K\varepsilon_0 A}{d}$

42 A parallel plate capacitor with plate area A and distance of separation d is filled with a dielectric. What is the capacity of the capacitor when permittivity of the dielectric varies as *[2021, 25 July Shift-I]*

$$\varepsilon(x)=\varepsilon_0+kx, \text{ for}\left(0<x\le\frac{d}{2}\right)$$

$$\varepsilon(x)=\varepsilon_0+k(d-x), \text{ for}\left(\frac{d}{2}\le x\le d\right)$$

(a) $\left(\varepsilon_0+\frac{kd}{2}\right)^{2/kA}$ (b) $\frac{kA}{2\ln\left(\frac{2\varepsilon_0+kd}{2\varepsilon_0}\right)}$

(c) zero (d) $\frac{kA}{2}\ln\left(\frac{2\varepsilon_0}{2\varepsilon_0-kd}\right)$

43 Consider the combination of two capacitors C_1 and C_2, with $C_2>C_1$, when connected in parallel, the equivalent capacitance is 15/4 times the equivalent capacitance of the same connected in series. Calculate the ratio of capacitors C_2/C_1. *[2021, 26 Feb Shift-I]*

(a) $\frac{15}{11}$ (b) $\frac{111}{80}$ (b) $\frac{29}{15}$ (d) $\frac{15}{4}$

44 A parallel plate capacitor has plate of length l, width w and separation of plates is d. It is connected to a battery of emf V. A dielectric slab of the same thickness d and of dielectric constant $K=4$ is being inserted between the plates of the capacitor. At what length of the slab inside plates, will the energy stored in the capacitor be two times the initial energy stored? *[2020, 5 Sep Shift-II]*

(a) $\frac{2l}{3}$ (b) $\frac{l}{3}$ (c) $\frac{l}{4}$ (d) $\frac{l}{2}$

45 Figure shows charge (q) *versus* voltage (V) graph for series and parallel combination of two given capacitors. The capacitances are **[2019, 10 April Shift-I]**

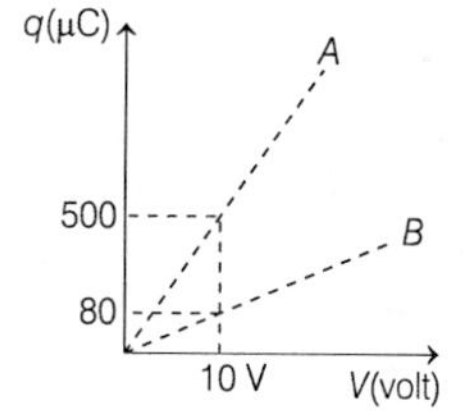

(a) 60 μF and 40 μF (b) 50 μF and 30 μF
(c) 20 μF and 30 μF (d) 40 μF and 10 μF

Imp. Energy Stored in Capacitor

46 A capacitor with capacitance 5 μF is charged to 5 μC. If the plates are pulled apart to reduce the capacitance to 2 μF, how much work is done? **[2019, 9 April Shift-I]**

(a) 6.25×10^{-6} J (b) 2.16×10^{-6} J
(c) 2.55×10^{-6} J (d) 3.75×10^{-6} J

47 A parallel plate capacitor with square plates is filled with four dielectrics of dielectric constants K_1, K_2, K_3, K_4 arranged as shown in the figure. The effective dielectric constant K will be

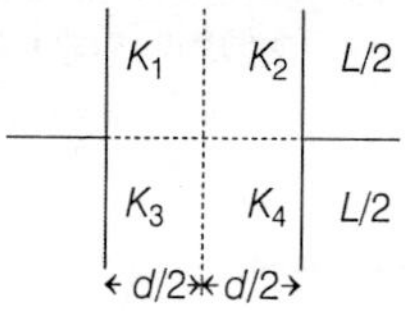

[2019, 9 Jan Shift-II]

(a) $K = \dfrac{(K_1 + K_2)(K_3 + K_4)}{2(K_1 + K_2 + K_3 + K_4)}$

(b) $K = \dfrac{(K_1 + K_2)(K_3 + K_4)}{K_1 + K_2 + K_3 + K_4}$

(c) $K = \dfrac{(K_1 + K_3)(K_2 + K_4)}{K_1 + K_2 + K_3 + K_4}$

(d) $K = \dfrac{(K_1 + K_4)(K_2 + K_3)}{2(K_1 + K_2 + K_3 + K_4)}$

Numerical Value Type Questions

48. Four capacitors each of capacitance 16 μF are connected as shown in the figure. The capacitance between points A and B is .... (in μF). **[2025, 4 April Shift-I]**

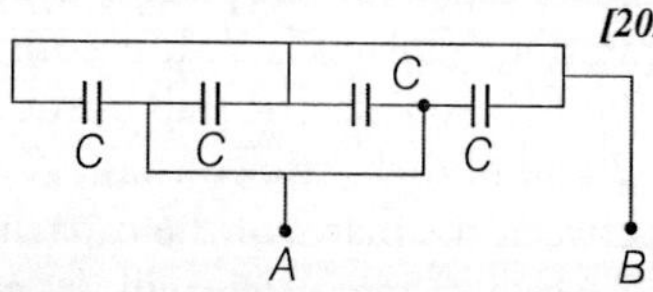

49 A capacitor of 10 μF capacitance whose plates are separated by 10 mm through air and each plate has area 4 cm^2 is now filled equally with two dielectric media of $K_1 = 2$, $K_2 = 3$ respectively as shown in figure. If new force between the plates is 8 N. The supply voltage is .... V. **[2024, 6 April Shift-II]**

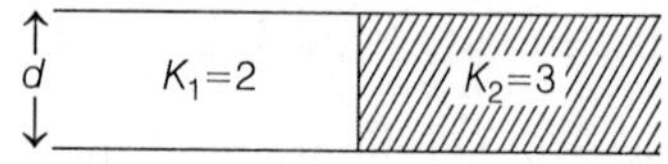

50 Three capacitors of capacitances 25 μF, 30 μF and 45 μF are connected in parallel to a supply of 100 V. Energy stored in the above combination is E. When these capacitors are connected in series to the same supply, the stored energy is $\dfrac{9}{x}E$. The value of x is .... **[2024, 5 April Shift-I]**

51 A capacitor of capacitance C and potential V has energy E. It is connected to another capacitor of capacitance $2C$ and potential $2V$. Then, the loss of energy is $\dfrac{x}{3}E$, where x is .... **[2024, 30 Jan Shift-I]**

52 In the given figure the total charge stored in the combination of capacitors is 100 μC. The value of x is .... **[2023, 15 April Shift-I]**

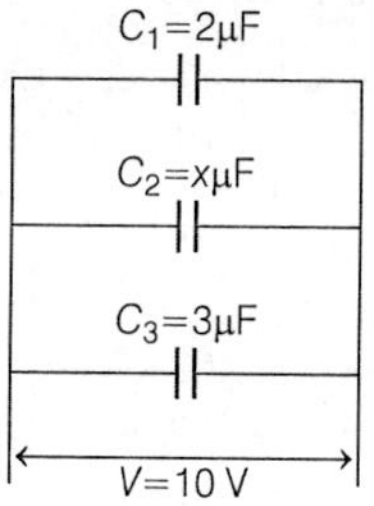

53 In the given circuit, $C_1 = 2$ μF, $C_2 = 0.2$ μF, $C_3 = 2$ μF, $C_4 = 4$ μF, $C_5 = 2$ μF, $C_6 = 2$ μF. The charge stored on capacitor C_4 is .... μC. **[2023, 11 April Shift-II]**

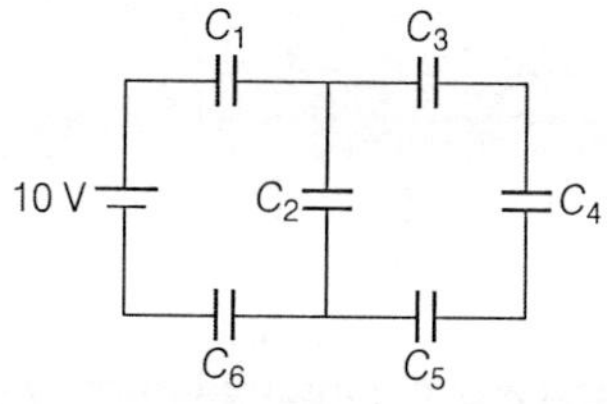

54 A 600 pF capacitor is charged by 200 V supply. It is then disconnected from the supply and is connected to another uncharged 600 pF capacitor. Electrostatic energy lost in the process is .... μJ. **[2023, 8 April Shift-II]**

55 As shown in the figure, two parallel plate capacitors having equal plate area of 200 cm² are joined in such a way that $a \neq b$. The equivalent capacitance of the combination is $x\,\varepsilon_0 F$. The value of x is **[2023, 6 April Shift-II]**

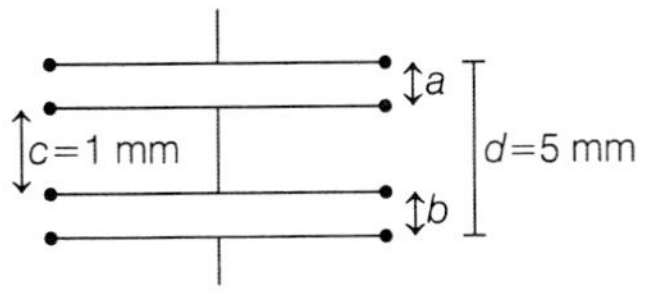

56 A parallel plate capacitor with plate area A and plate separation d is filled with a dielectric material of dielectric constant $K = 4$. The thickness of the dielectric material is x, where $x < d$.
Let C_1 and C_2 be the capacitance of the system for $x = \frac{1}{3}d$ and $x = \frac{2d}{3}$, respectively. If $C_1 = 2\,\mu F$ the value of C_2 is .... μF.
[2023, 6 April Shift-I]

57 Two parallel plate capacitors C_1 and C_2 each having capacitance of $10\,\mu F$ are individually charged by a 100 V DC source. Capacitor C_1 is kept connected to the source and a dielectric slab is inserted between its plates. Capacitor C_2 is disconnected from the source and then a dielectric slab is inserted in it. Afterwards the capacitor C_1 is also disconnected from the source and the two capacitors are finally connected in parallel combination. The common potential of the combination will be .... V.
(Assuming dielectric constant = 10) *[2023, 31 Jan Shift-II]*

58 A capacitor of capacitance $900\,\mu F$ is charged by a 100 V battery. The capacitor is disconnected from the battery and connected to another uncharged identical capacitor such that one plate of uncharged capacitor connected to positive plate and another plate of uncharged capacitor connected to negative plate of the charged capacitor. The loss of energy in this process is measured as $x \times 10^{-2}$ J. The value of x is ....
[2023, 30 Jan Shift-I]

59 A parallel plate capacitor with width 4 cm, length 8 cm and separation between the plates of 4 mm is connected to a battery of 20 V. A dielectric slab of dielectric constant 5 having length 1 cm, width 4 cm and thickness 4 mm is inserted between the plates of parallel plate capacitor.
The electrostatic energy of this system will be .... ε_0J.
(where, ε_0 is the permittivity of free space)
[2022, 27 July Shift-II]

60 A parallel plate capacitor is made up of stair like structure with a plate area A of each stair and that is connected with a wire of length b, as shown in the figure. The capacitance of the arrangement is $\frac{x}{15}\frac{\varepsilon_0 A}{b}$. The value of x is ....
[2022, 27 June Shift-II]

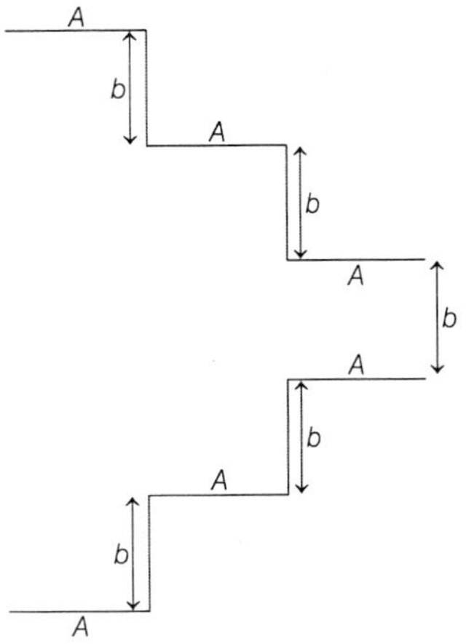

61 A parallel plate capacitor whose capacitance C is 14 pF is charged by a battery to a potential difference $V = 12$ V between its plates. The charging battery is now disconnected and a porcelain plate with $K = 7$ is inserted between the plates, then the plate would oscillate back and forth between the plates with a constant mechanical energy of .... pJ.
(Assume no friction) *[2021, 17 March Shift-I]*

62 Four identical rectangular plates with length, $l = 2$ cm and breadth, $b = 3/2$ cm are arranged as shown in figure. The equivalent capacitance between A and C is $x\varepsilon_0/d$. The value of x is ....
(Round off to the nearest integer)

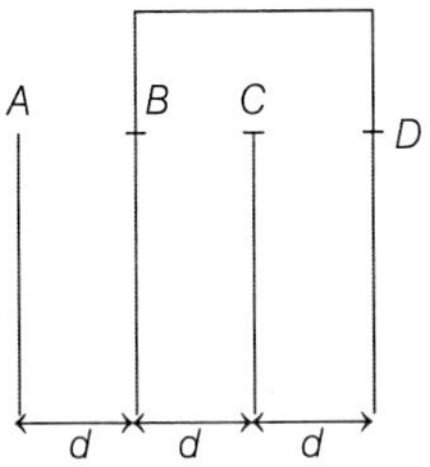

[2021, 17 March Shift-I]

63 A $5\,\mu F$ capacitor is charged fully by a 220V supply. It is then disconnected from the supply and is connected in series to another uncharged $2.5\,\mu F$ capacitor. If the energy change during the charge redistribution is $X/100$ J, then value of X to the nearest integer is .... *[2020, 2 Sep Shift-I]*

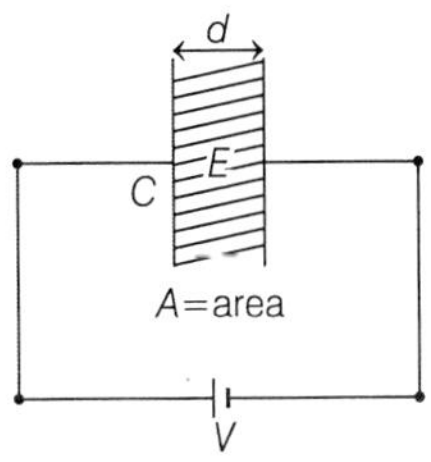

64 A 60 pF capacitor is fully charged by a 20 V supply. It is then disconnected from the supply and is connected to another uncharged 60 pF capacitor in parallel. The electrostatic energy that is lost in this process by the time, the charge is redistributed between them is (in nJ) *[2020, 7 Jan Shift-II]*

• Key **Index**

Topic 1

1 (c)	2 (a)	3 (d)	4 (d)	5 (d)	6 (b)	7 (b)	8 (a)	9 (b)	10 (a)
11 (d)	12 (d)	13 (c)	14 (d)	15 (d)	16 (d)	17 (c)	18 (c)	19 (a)	20 (d)
21 (d)	22 (d)	23 (b)	24 (a)	25 (d)	26 (c)	27 (b)	28 (d)	29 (a)	30 (d)
31 (d)	32 (d)	33 (b)	34 (d)	35 (a)	36 (b)	37 (a)	38 (b)	39 (a)	40 (c)
41 (d)	42 (b)	43 (d)	44 (c)	45 (d)	46 (d)	47 (d)	48 (d)	49 (b)	50 (36)
51 (27)	52 (8)	53 (160)	54 (5)	55 (18)	56 (12)	57 (198)	58 (243)	59 (128)	

Topic 2

1 (a)	2 (d)	3 (b)	4 (a)	5 (a)	6 (b)	7 (a)	8 (a)	9 (b)	10 (b)
11 (a)	12 (b)	13 (c)	14 (a)	15 (d)	16 (d)	17 (d)	18 (a)	19 (c)	20 (c)
21 (a)	22 (c)	23 (a)	24 (a)	25 (d)	26 (c)	27 (b)	28 (a)	29 (a)	30 (d)
31 (b)	32 (a)	33 (c)	34 (c)	35 (5)	36 (15)	37 (2)	38 (2)	39 (105)	40 (15)
41 (100)	42 (161)	43 (16)							

Topic 3

1 (a)	2 (b)	3 (a)	4 (a)	5 (c)	6 (d)	7 (d)	8 (d)	9 (b)	10 (b)
11 (d)	12 (a)	13 (b)	14 (b)	15 (a)	16 (a)	17 (a)	18 (d)	19 (c)	20 (d)
21 (b)	22 (c)	23 (d)	24 (a)	25 (d)	26 (c)	27 (b)	28 (c)	29 (c)	30 (c)
31 (b)	32 (b)	33 (c)	34 (a)	35 (d)	36 (a)	37 (b)	38 (b)	39 (c)	40 (a)
41 (a)	42 (b)	43 (*)	44 (b)	45 (d)	46 (d)	47 (*)	48 (64)	49 (80)	50 (86)
51 (2)	52 (5)	53 (5)	54 (6)	55 (5)	56 (3)	57 (55)	58 (2.25)	59 (240)	60 (23)
61 (864)	62 (2)	63 (4)	64 (6)						

Detailed **Explanations**

Topic 1

01. *(c)* We know that potential inside shell is equal to potential on the surface,

$$\therefore \quad V_{in} = V_{surface} = \frac{kq}{R} = 120 \text{ V}$$

$$\Rightarrow \quad kq = R \times 120 = 10 \times 10^{-2} \times 120$$

$$kq = 12$$

At, $r = 15$ cm $= 1.5 \times 10^{-1}$ m

$$V' = \frac{kg}{r} = \frac{12}{1.5 \times 10^{-1}} = 80 \text{ V}$$

02. *(a)* For metal sphere, potential is given as,

$$V = \frac{\sigma R}{\varepsilon_0}$$

$$\Rightarrow \quad V \propto \sigma R$$

After contact,

$$V_1 = V_2$$

$$\Rightarrow \quad \sigma_1 R_1 = \sigma_2 R_2$$

$$\Rightarrow \quad \frac{\sigma_1}{\sigma_2} = \frac{R_2}{R_1} = \frac{3R}{R} = 3$$

03. *(d)* Infinite plane sheet produces uniform electric field. Hence, both the charges of the dipole experience same force but in opposite direction. Further, this is the position of stable equilibrium. Hence, potential energy is minimum while the torque is zero.

04. *(d)* $U = \frac{kq_1 q_2}{r} = \frac{1}{4\pi\varepsilon_0} \frac{q_1 q_2}{r}$

$$= \frac{9 \times 10^9 \times (7 \times 10^{-6}) \times (-4 \times 10^{-6})}{(14) \times 10^{-2}}$$

$$= \frac{-9 \times 10^9 \times 28 \times 10^{-12}}{14 \times 10^{-2}}$$

$$= \frac{-252 \times 10^{-3}}{14 \times 10^{-2}} = \frac{-18 \times 10^{-3}}{10^{-2}} = -1.8 \text{ J}$$

05. *(d)* Electric potential due to electric dipole is given by $V = \frac{1}{4\pi\varepsilon_0} \frac{p\cos\theta}{r^2}$

where p is dipole moment.

$$\Rightarrow \quad V \propto \frac{1}{r^2}$$

06. *(b)* Electric lines of force are always perpendicular to equipotential surface

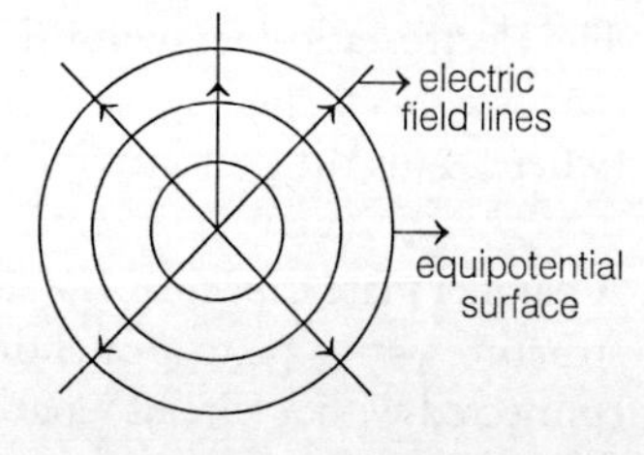

Work done by electric field on moving a positive charge on an equipotential surface is always zero.

07. *(b)* Potential due to a point charge $V = \frac{kq}{r}$.

Here, $OP = \sqrt{(\sqrt{3})^2 + (\sqrt{3})^2} = \sqrt{6}$

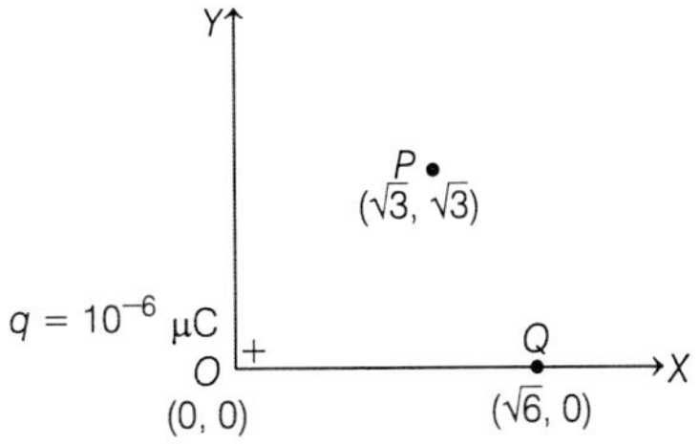

$\therefore \quad OP = OQ$

$\Rightarrow \quad V_P = V_Q$

Hence, $V_P - V_Q = 0$

08. *(a)* Potential due to a point charge at a distance r is,

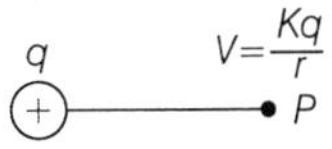

$$V = \frac{Kq}{r}$$

Here, $q = 5 \times 10^{-9}$ C

$V = 50$ V

$K = 9 \times 10^9$ N-m^2/C^2

So, $r = \frac{Kq}{V} = \frac{9 \times 10^9 \times 5 \times 10^{-9}}{50}$

$= \frac{45}{50} = 0.9$ m

$= 90$ cm

09. *(b)* We know that, linear momentum p and its kinetic energy E of moving charged particle accelerated with potential difference V is related as,

$$p = \sqrt{2mE} = \sqrt{2mqV} \quad [\because E = qV]$$

$\Rightarrow \quad p \propto \sqrt{mq}$

[For same potential difference]

$$\Rightarrow \quad \frac{p_\alpha}{p_p} = \sqrt{\frac{m_\alpha q_\alpha}{m_p q_p}} = \sqrt{\frac{4m_p \times 2q_p}{m_p \times q_p}}$$

$[\because m_\alpha = 4m_p$ and $q_\alpha = 2q_p]$

$$= \sqrt{4 \times 2} = \frac{2\sqrt{2}}{1}$$

$\therefore \ p_\alpha : p_p = 2\sqrt{2} : 1$

10. *(a)* We know that $E = \frac{-\partial V}{\partial r}$

$\because$ From inside till surface, $V =$ Constant

Since, electric field is zero along the surface of the conductor.

$\therefore$ E must be perpendicular to the surface of the conductors, so that its component along the surface of the conductors must be zero.

11. *(d)* Given, $\frac{E_1}{E_2} = \frac{R_1}{R_2}$ ...(i)

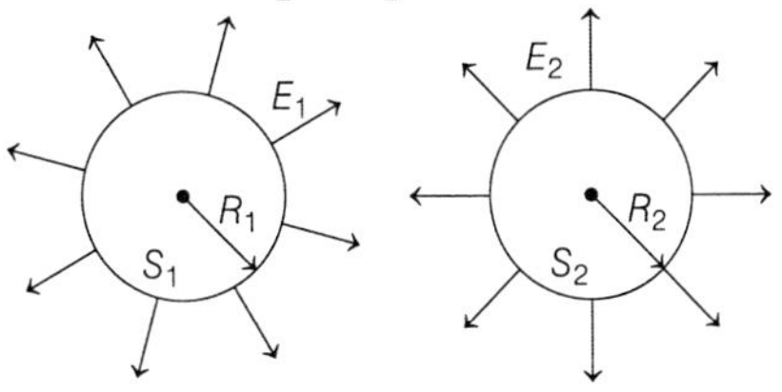

For a charged sphere, potential is given by

$$V = \frac{KQ}{R} = \frac{KQ}{R^2} \cdot R$$

or $\quad V = E \cdot R$

So, $\quad \frac{V_1}{V_2} = \frac{E_1 R_1}{E_2 R_2} = \frac{R_1}{R_2} \times \frac{R_1}{R_2}$ [from Eq. (i)]

$$\Rightarrow \quad \frac{V_1}{V_2} = \frac{R_1^2}{R_2^2} = \left(\frac{R_1}{R_2}\right)^2$$

> • Concept **Enhancer**
>
> The potential inside a metallic sphere is directly proportional to square of its radius, if $\frac{E_1}{E_2} = \frac{R_1}{R_2}$.

12. *(d)* Given, $\mathbf{E} = 1000$ V/m

$\theta = 45°$ and $\mathbf{p} = 10^{-29}$ C-m

We know that electric potential energy

$U = -\mathbf{p} \cdot \mathbf{E} = -pE \cos\theta$

$= -10^{-29} \times 1000 \times \cos 45°$

$\Rightarrow \ U \approx -7 \times 10^{-27}$ J

13. *(c)* As we know, potential difference $V_A - V_O$ is

$dV = -Edx$

$$\Rightarrow \int_{V_O}^{V_A} dV = -\int_0^2 30x^2 dx$$

$$V_A - V_O = -30 \times \left[\frac{x^3}{3}\right]_0^2$$

$= -10 \times [2^3 - (0)^3] = -10 \times 8 = -80$ J

14. *(d)* $W = QdV = Q(V_q - V_p)$

$= -100 \times (1.6 \times 10^{-19}) \times (-4 - 10)$

$= +100 \times 1.6 \times 10^{-19} \times 14$

$= +2.24 \times 10^{-16}$ J

15. *(d)* $E = -\frac{\partial V}{\partial x}\hat{\mathbf{i}} - \frac{\partial V}{\partial y}\hat{\mathbf{j}} - \frac{\partial V}{\partial z}\hat{\mathbf{k}}$

$$\Rightarrow \ E_x = -\frac{\partial V}{\partial x} = -\frac{d}{dx}\left[\frac{20}{x^2 - 4}\right]$$

$$= \frac{40x}{(x^2 - 4)^2}$$

$\Rightarrow E_x$ at $x = 4\,\mu$m $= \frac{10}{9}$ V/μm

and is along positive x-direction.

16. *(d)* Direction of E reverses while magnitude remains same and V remains unchanged because it is a scalar quantity.

17. *(c)* When the two conducting spheres are connected by a conducting wire, charge will flow from one sphere (having higher potential) to other (having lower potential) till both acquire the same potential. Therefore,

$$E = \frac{V}{r}$$

$$\Rightarrow \quad \frac{E_1}{E_2} = \frac{r_2}{r_1} = \frac{2}{1} = 2:1$$

18. *(c)* At P due to shell, potential

$$V_1 = \frac{q}{4\pi\varepsilon_0 R}$$

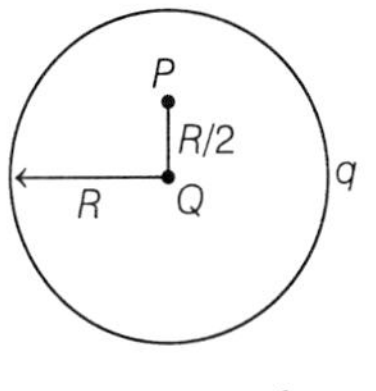

At P due to Q, potential

$$V_2 = \frac{Q}{4\pi\varepsilon_0 \frac{R}{2}} = \frac{2Q}{4\pi\varepsilon_0 R}$$

$\therefore$ Net potential at P,

$$V = V_1 + V_2 = \frac{q}{4\pi\varepsilon_0 R} + \frac{2Q}{4\pi\varepsilon_0 R}$$

19. *(a)* Potential difference between two points in an electric field is

$$V_A - V_B = \frac{W}{q_0}$$

where, W is work done by moving charge q_0 from point A to B.

Here, $\quad W = 2$ J, $q_0 = 20$ C

So, $V_A - V_B = \frac{2}{20} = 0.1$ V

20. *(d)* Electrostatic potential inside a uniformly charged spherical shell is constant and is same as that on the surface of the shell. Thus it works as a equipotential surface at which, work done to move a test charge between any two points, will be zero.

21. *(d)* Potential difference between points A and B,

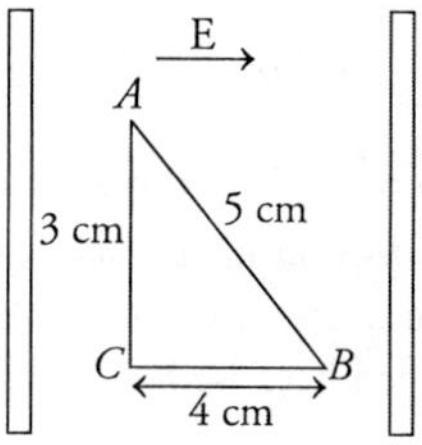

$$V_{AB} = E\cdot d = E(CB) = E\cdot 4$$
$$V_{AB} = 4E \quad ...(i)$$

Since, $V = E\cdot d$

$$V = E\times 10$$
$$\Rightarrow \quad E = \frac{V}{10} \quad ...(ii)$$

Now, from eqn (i) and (ii),

$$V_{AB} = 4\times\frac{V}{10} = \frac{2V}{5}$$

22. *(d)* $V = -\int \mathbf{E}\cdot d\mathbf{r} = -\int \frac{2k\lambda}{r} dr$

$$= -2k\lambda \ln r + c$$

Net potential due to all wires

$$V = -2k\lambda \ln\sqrt{x^2+y^2}$$
$$-2k\lambda \ln\sqrt{y^2+z^2} - 2k\lambda \ln\sqrt{z^2+x^2} + c$$

For V = constant

$$\sqrt{(x^2+y^2)(y^2+z^2)(z^2+x^2)} = c$$
$$\therefore \quad (x^2+y^2)(y^2+z^2)(z^2+x^2) = c$$

where c = constant

23. *(b)*

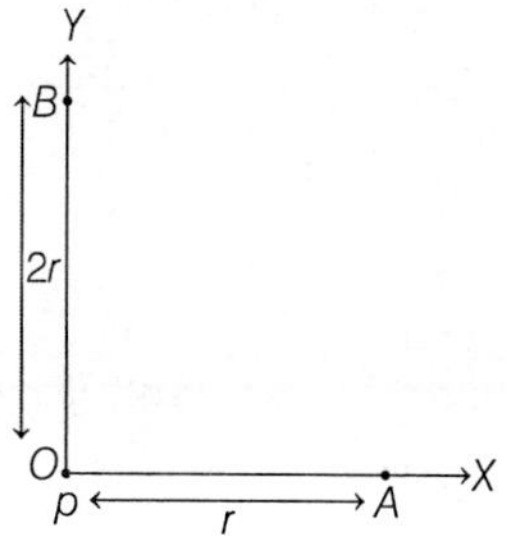

$$E_A = \frac{2kp}{r^3} = E_0 \Rightarrow V_A = \frac{kp}{r^2} = V_0$$
$$E_B = \frac{kp}{(2r)^3} = \frac{kp}{8r^3} = \frac{2kp}{16r^3}$$
$$E_B = \frac{E_0}{16} \Rightarrow V_B = 0$$

24. *(a)*

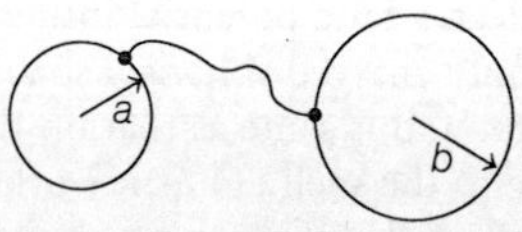

As spheres are connected they have same surface potential in equilibrium.

$$\Rightarrow \quad V_a = V_b$$
$$\Rightarrow \quad \frac{kq_A}{a} = \frac{kq_b}{b} \Rightarrow \frac{q_A}{q_B} = \frac{a}{b}$$

25. *(d)* For $r \le R$, $V = \frac{KQ}{R}$, constant everywhere inside

For $r > R$, $V = \frac{KQ}{r}$, Decreases with r

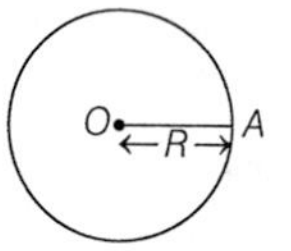

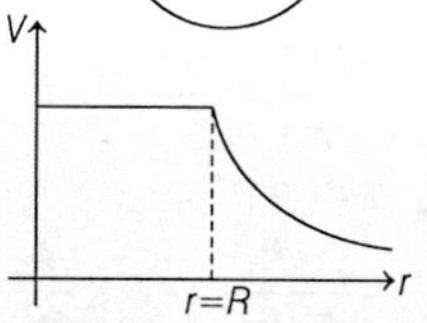

26. *(c)* Electric potential of a charged spherical conductor of radius (R) with radial distance is given as

$$V = \begin{cases} \frac{KQ}{R} & \text{If } r \le R \\ \frac{KQ}{r} & \text{If } r > R \end{cases}$$

Now, plotting graph between V and r

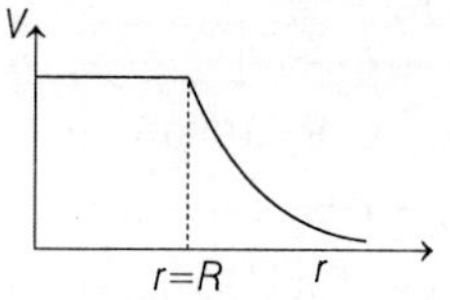

27. *(b)* When two conducting spheres are connected by a conducting wire, then charge will flow from the sphere of higher potential to the sphere of lower potential till both acquire the same potential.

$$\therefore \quad V_A = V_B$$
$$\text{or} \quad \frac{1}{4\pi\varepsilon_0}\frac{q_A}{r_A} = \frac{1}{4\pi\varepsilon_0}\frac{q_B}{r_B}$$
$$\text{or} \quad \frac{q_A}{q_B} = \frac{r_A}{r_B} = \frac{5}{10} \text{ or } \frac{q_A}{q_B} = \frac{1}{2}$$

Now, the ratio of electric fields,

$$\frac{E_1}{E_2} = \frac{K\frac{q_A}{r_A^2}}{K\frac{q_B}{r_B^2}}$$
$$\text{or} \quad \frac{E_1}{E_2} = \frac{q_A}{q_B}\times\frac{r_B^2}{r_A^2} = \frac{1}{2}\times\left(\frac{10}{5}\right)^2$$
$$\text{or} \quad \frac{E_1}{E_2} = \frac{2}{1}$$

28. *(d)*

key idea If the electric field is only function of x i.e. $E = f(x)$, then the direction of the electric field is along X-axis.

Electric field at some point (x, y, z),

$$E = -\frac{\partial V}{\partial x}\hat{\mathbf{i}} - \frac{\partial V}{\partial y}\hat{\mathbf{j}} - \frac{\partial V}{\partial z}\hat{\mathbf{k}}$$
$$= \frac{-\partial}{\partial x}(3x^2)\hat{\mathbf{i}} = -6x\hat{\mathbf{i}}$$

As, $\frac{\partial V}{\partial y} = \frac{\partial}{\partial y}3x^2 = 0$

Similarly,

$$\frac{\partial V}{\partial z} = 0$$

Hence, E (at 1, 0, 3) $= -6\hat{\mathbf{i}}$ V/m

So, direction of field will be along negative X-axis.

29. *(a)* The given situation is shown below.

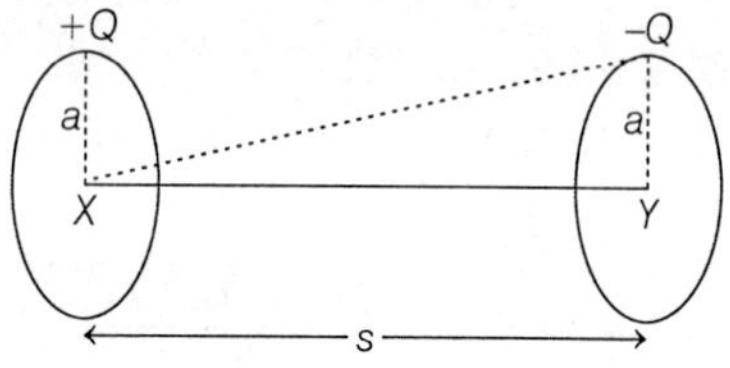

Potential at X,

$$V_X = \frac{KQ}{a} - \frac{KQ}{\sqrt{a^2+s^2}}$$

Potential at Y,

$$V_Y = -\frac{KQ}{a} + \frac{KQ}{\sqrt{a^2+s^2}}$$

$$\text{So, } V_X - V_Y = \frac{2KQ}{a} - \frac{2KQ}{\sqrt{a^2+s^2}}$$
$$= \frac{Q}{2\pi\varepsilon_0}\left[\frac{1}{a} - \frac{1}{\sqrt{s^2+a^2}}\right]$$

Concept Enhancer

If two charged rings are separated by a distance coaxially, then to find work done by moving a charge from centre of one ring to centre of other ring, use

$$W = q[V_1 - V_2]$$

where, V_1 = Potential at the centre of first ring due to second ring.

V_2 = Potential at the centre of second ring due to first ring.

30. *(d)*

key idea The potential energy of dipole and charge Q with dipole is

$(PT)_{total} = (PE)_{dipole} + (PE)_{dipole\ with\ Q}$

The system of two charges, i.e. $+q$ and $-q$ that are separated by distance d can be considered as a dipole. Thus, the charge Q would be at D distance from the centre of an electric dipole on its axial line.

So, the total potential energy of the system will be due to two components.

(1) Potential energy of dipole's own system

$$(PE)_1 = \frac{Kq_1q_2}{d} = -\frac{Kq^2}{d} \quad \ldots(i)$$

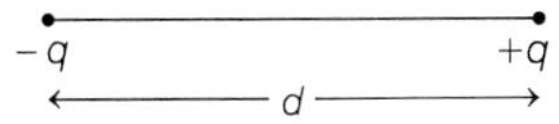

(2) Potential energy of charge Q and dipole system

$$(PE)_2 = -\frac{KQq}{D^2}\cdot d \quad \ldots(ii)$$

Hence, total potential energy of the system $(PE)_{total} = (PE)_1 + (PE)_2$

$$= -\frac{Kq^2}{d} - \frac{KQq}{D^2}\cdot d$$

$$\Rightarrow \quad (PE)_{total} = -\frac{1}{4\pi\varepsilon_0}\left[\frac{q^2}{d} + \frac{Qqd}{D^2}\right]$$

31. *(d)* At any instant t, total energy of charge distribution will be constant.

$$\text{i.e., } \frac{1}{2}mv^2 + \frac{KQ^2}{2R} = 0 + \frac{KQ^2}{2R_0}$$

$$\Rightarrow \frac{1}{2}mv^2 = \frac{KQ^2}{2R_0} - \frac{KQ^2}{2R}$$

$$\therefore \quad v = \sqrt{\frac{KQ^2}{m}\left(\frac{1}{R_0} - \frac{1}{R}\right)}$$

$$= c\sqrt{\frac{1}{R_0} - \frac{1}{R}}$$

where c is a constant $-\frac{KQ^2}{m}$,

Thus, the slope of speed versus R go on decreasing, it is correctly shown in option (d).

32. *(d)* Potential energy, $U = \frac{K\,q_1q_2}{r}$

Total energy of the charge system as shown in the figure below is

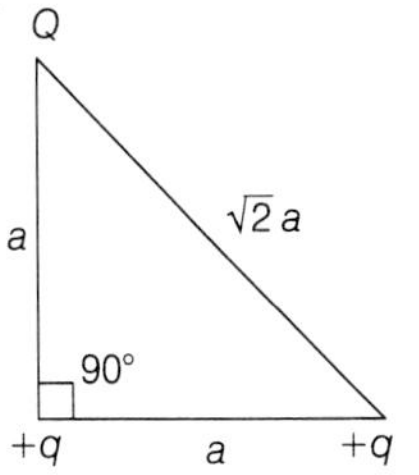

$$U = \frac{Kq^2}{a} + \frac{KQq}{a} + \frac{KQq}{\sqrt{2}a}$$

It is given that $\quad U = 0$

$$\therefore \quad \frac{Kq}{a}\left[q + Q + \frac{Q}{\sqrt{2}}\right] = 0$$

$$\Rightarrow \quad Q = \frac{-\sqrt{2}\times q}{(\sqrt{2}+1)}$$

33. *(b)*

key idea Potential of B = Potential due to charge on A + Potential due to charge on B + Potential due to charge on C.

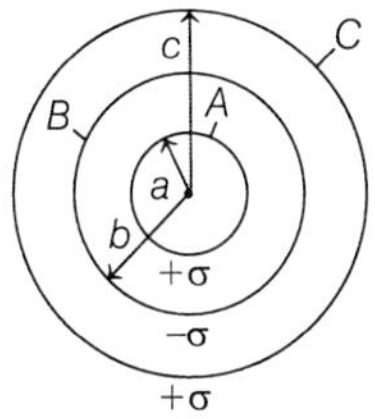

$$\therefore V_B = \frac{K(Q_A + Q_B)}{b} + \frac{KQ_C}{c}$$

$$= \frac{1}{4\pi\varepsilon_0}\left[\frac{\sigma 4\pi a^2}{b} - \frac{\sigma 4\pi b^2}{b} + \frac{\sigma 4\pi c^2}{c}\right]$$

$$= \frac{\sigma}{\varepsilon_0}\left(\frac{a^2 - b^2}{b} + \frac{c^2}{c}\right)$$

$$V_B = \frac{\sigma}{\varepsilon_0}\left(\frac{a^2 - b^2}{b} + c\right)$$

34. *(d)*

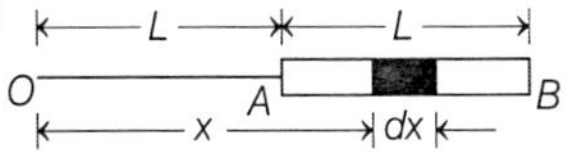

Let, an element of length dx of charge dq at distance x from point O.

$$dV = K\frac{dq}{x}$$

$$\text{where, } dq = \frac{Q}{L}dx$$

$$\therefore \quad V = \int_L^{2L}\frac{Kdq}{x} = \int_L^{2L}\frac{\left(\frac{Q}{L}\right)dx}{x}$$

$$= \frac{Q}{4\pi\varepsilon_0 L}\int_L^{2L}\left(\frac{1}{x}\right)dx$$

$$= \frac{Q}{4\pi\varepsilon_0 L}[\log_e x]_L^{2L}$$

$$= \frac{Q}{4\pi\varepsilon_0 L}[\log_e 2L - \log_e L]$$

$$= \frac{Q}{4\pi\varepsilon_0 L}\left[\log_e \frac{2L}{L}\right]$$

$$= \frac{Q}{4\pi\varepsilon_0 L}\ln(2)$$

35. *(a)* Initial potential energy of the system,

$$U_i = \frac{2KqQ}{a} + \frac{2K(-q)Q}{\sqrt{5}\,a}$$

$$= \frac{1}{4\pi\varepsilon_0}\cdot\frac{2qQ}{a}\left[1 - \frac{1}{\sqrt{5}}\right]$$

Final potential energy of the system,

$U_f = 0$

By conservation of energy,

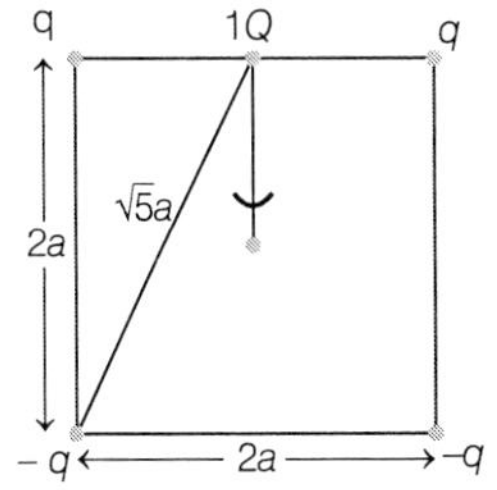

Gain in KE = Loss in PE

$$K = \frac{1}{4\pi\varepsilon_0}\cdot\frac{2qQ}{a}\left[1 - \frac{1}{\sqrt{5}}\right]$$

36. *(b)* Potential at A due to charge at O,

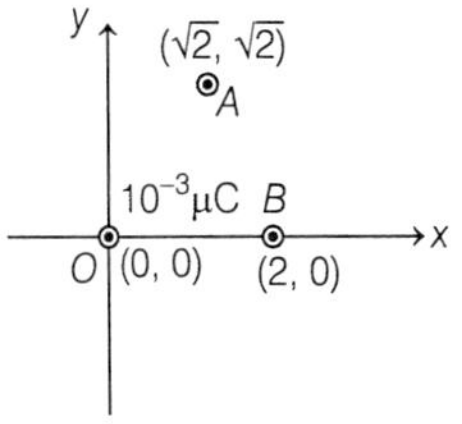

$$V_A = \frac{1}{4\pi\varepsilon_0}\frac{(10^{-3})}{OA}$$

$$= \frac{1}{4\pi\varepsilon_0}\cdot\frac{(10^{-3})}{\sqrt{(\sqrt{2})^2 + (\sqrt{2})^2}}$$

$$= \frac{1}{4\pi\varepsilon_0}\cdot\frac{(10^{-3})}{2}$$

Potential at B due to charge at O,

$$V_B = \frac{1}{4\pi\varepsilon_0}\cdot\frac{(10^{-3})}{OB} = \frac{1}{4\pi\varepsilon_0}\cdot\frac{(10^{-3})}{2}$$

So, $\quad V_A - V_B = 0$

37. *(a)* Since, $V_2 > V_1$, so electric field will point from plate 2 to plate 1.

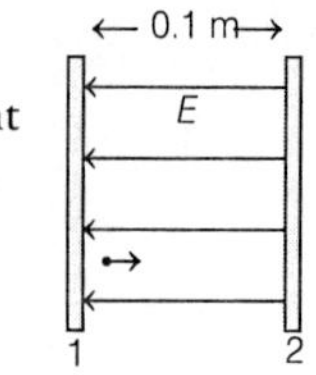

The electron will experience an electric force, opposite to the direction of electric field and hence move towards the plate 2. Use work-energy theorem to find speed of electron when it strikes the plate 2.

$$\therefore \quad \frac{m_e v^2}{2} - 0 = e(V_2 - V_1)$$

where, v is the required speed.

$$\Rightarrow \quad \frac{9.11 \times 10^{-31}}{2} v^2 = 1.6 \times 10^{-19} \times 20$$

or $$v = \sqrt{\frac{1.6 \times 10^{-19} \times 40}{9.11 \times 10^{-31}}}$$

$$= 2.65 \times 10^6 \text{ ms}^{-1}$$

38. *(b)* Potential at C

$$V_C = \frac{Kq_1}{0.4} + \frac{Kq_2}{0.5}$$

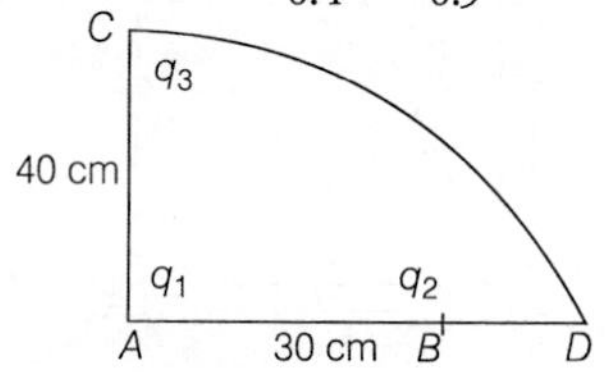

Potential at D

$$V_D = \frac{Kq_1}{0.4} + \frac{Kq_2}{0.1}$$

$$\Delta U = (V_D - V_C) q_3$$

$$= \left(\frac{Kq_2}{0.1} - \frac{Kq_2}{0.5}\right) \cdot (q_3)$$

$$\Delta U = 8\, Kq_2 q_3 = \frac{8 q_2 q_3}{4\pi\varepsilon_0}$$

$$\Delta U = \frac{8 q_2 q_3}{4\pi\varepsilon_0}$$

Value of K is $8q_2$.

39. *(a)*

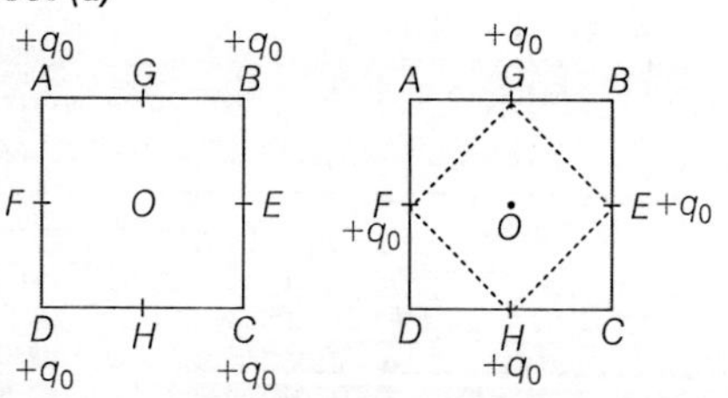

$$U_1 = \frac{4kq_0^2}{a} + \frac{2kq_0^2}{\sqrt{2}a} = \frac{kq_0^2}{a}(4 + \sqrt{2})$$

$$U_2 = \frac{kq_0^2}{\frac{a}{\sqrt{2}}}(4 + \sqrt{2}) = \frac{kq_0^2}{a}(4\sqrt{2} + 2)$$

$$\Delta U = U_2 - U_1 = \frac{kq_0^2}{a}(3\sqrt{2} - 2)$$

40. *(c)* The electrostatic force will provide necessary centripetal force for rotation.

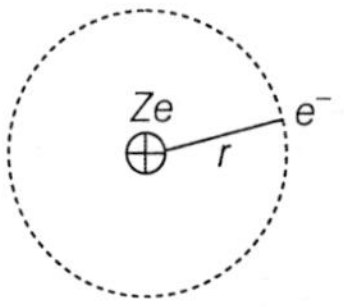

$$F_e = F_c$$

$$\frac{k(Ze)e}{r^2} = \frac{mv^2}{r} \quad \ldots\text{(i)}$$

Kinetic energy, $\text{KE} = \frac{1}{2}mv^2$

$$\Rightarrow \quad \text{KE} = \frac{1}{2}\frac{kZe(e)}{r} \quad \ldots\text{(ii)}$$

Potential energy, $\text{PE} = -\frac{k(Ze)e}{r} \quad \ldots\text{(iii)}$

$$\text{TE} = \text{PE} + \text{KE}$$

$$= \frac{k(Ze)(e)}{2r} - \frac{k(Ze)(e)}{r}$$

$$= -\frac{k(Ze)(e)}{2r} \quad \ldots\text{(iv)}$$

$$\text{TE} = \frac{\text{PE}}{2} \Rightarrow 2\text{TE} = \text{PE or } 2E = U$$

As total energy is denoted by E and potential energy is denoted by U.

> ***Alternative Solutions***
>
> When an electron is revolving in a circular path around the nucleus, then total energy of revolving electron in that orbit is always equal to half its potential energy in that orbit.
>
> i.e. $\quad E = \frac{U}{2} \Rightarrow U = 2E$

41. *(d)*

> **key idea** The electric potential of any continuous charge system is calculated by integration method.
>
> i.e., $V = \int dV = \int \frac{Kdq}{r}$

Potential at centre due to an elemental charge dq at centre of half-ring will be

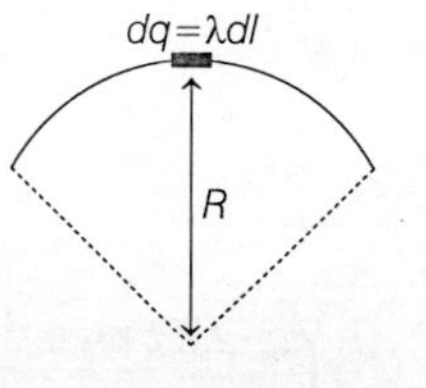

$$dV = \frac{Kdq}{R} = \frac{1}{4\pi\varepsilon_0} \cdot \frac{\lambda dl}{R}$$

So, potential due to both half-rings will be $V = \int_{\text{ring-1}} dV + \int_{\text{ring-2}} dV$

$$= \frac{\lambda}{4\pi\varepsilon_0 R_1}\int dl + \frac{\lambda}{4\pi\varepsilon_0 R_2}\int dl$$

$$= \frac{\lambda}{4\pi\varepsilon_0 R_1} \times \pi R_1 + \frac{\lambda}{4\pi\varepsilon_0 R_2} \times \pi R_2$$

$$= \frac{\lambda}{4\varepsilon_0} + \frac{\lambda}{4\varepsilon_0} = \frac{2\lambda}{4\varepsilon_0} = \frac{\lambda}{2\varepsilon_0}$$

42. *(b)*

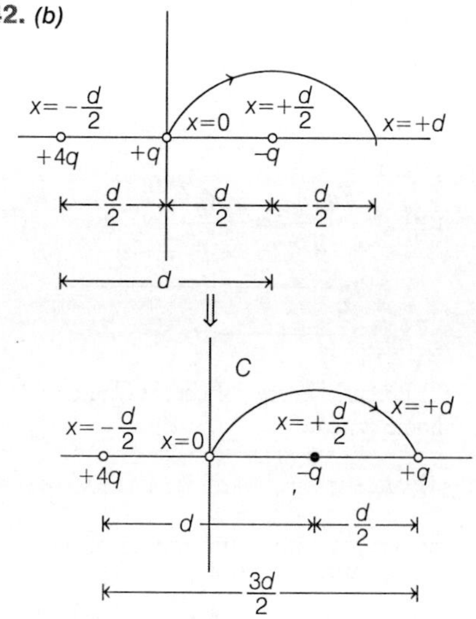

Change in potential energy,

$$\Delta U = U_f - U_i$$

$$\Delta U = \left[\frac{K(4q)(-q)}{d} + \frac{K(4q)(q)}{3d/2} + \frac{K(-q)(q)}{\frac{d}{2}}\right]$$

$$- \left[\frac{K(4q)(-q)}{d} + \frac{K(4q)(q)}{\frac{d}{2}} + \frac{K(-q)(q)}{\frac{d}{2}}\right]$$

$$= -\frac{4q^2}{3\pi\varepsilon_0 d}$$

So, ΔU will decrease by $\frac{4q^2}{3\pi\varepsilon_0 d}$.

43. *(d)* Given, total charge Q is uniformly distributed over concentric shells of radii r and R.

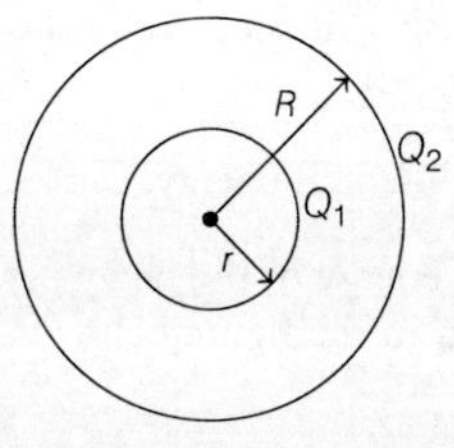

Let Q_1 and Q_2 be the charges over inner and outer shells, such that charge densities are equal on both shells.
Charge density of inner shell,

$$\sigma_1 = \frac{Q_1}{4\pi r^2}$$

Charge density of outer shell,

$$\sigma_2 = \frac{Q_2}{4\pi R^2}$$

but $\quad \sigma_1 = \sigma_2$

$$\Rightarrow \quad \frac{Q_1}{4\pi r^2} = \frac{Q_2}{4\pi R^2}$$

$$\Rightarrow \quad Q_2 = \frac{Q_1 R^2}{r^2}$$

Also, $Q_1 + Q_2 = Q$

$$\Rightarrow Q_1 + \frac{Q_1 R^2}{r^2} = Q \text{ or } Q_1 = \frac{Qr^2}{(r^2+R^2)}$$

and $\quad Q_2 = \dfrac{QR^2}{(r^2+R^2)}$

Potential due to Q_1 and Q_2 at common centre of shells is

$$V = V_{Q_1} + V_{Q_2} = \frac{kQ_1}{r} + \frac{kQ_2}{R}$$

$$= \frac{kQr}{r^2+R^2} + \frac{kQR}{r^2+R^2}$$

$$= \frac{1}{4\pi\varepsilon_0} \cdot \frac{Q(r+R)}{r^2+R^2}$$

• Concept Enhancer

Electric field inside a conducting spherical shell is zero and the potential is constant throughout inside the shell.

44. *(c)* Let q_1, q_2 and q_3 be the charge on the spherical shell with radii a, b and c, respectively.
Since, $\sigma_1 = \sigma_2 = \sigma_3$

$$\text{So, } \frac{q_1}{4\pi a^2} = \frac{q_2}{4\pi b^2} = \frac{q_3}{4\pi c^2}$$

$$\text{or} \quad q_1 = \frac{a^2}{b^2} q_2, \; q_3 = \frac{c^2}{b^2} q_2$$

Also, $\quad q_1 + q_2 + q_3 = Q$

$$\Rightarrow \frac{a^2}{b^2} q_2 + q_2 + \frac{c^2}{b^2} q_2 = Q$$

$$\Rightarrow \quad q_2 = \frac{b^2 Q}{b^2 + a^2 + c^2}$$

$$\text{Charge, } q_1 = \frac{a^2}{b^2}\left(\frac{b^2 Q}{b^2+a^2+c^2}\right)$$

$$\text{and} \quad q_3 = \frac{c^2}{b^2}\left(\frac{b^2 Q}{b^2+a^2+c^2}\right)$$

The potential at a point at distance r from common centre is

$$V = \frac{Kq_1}{a} + \frac{Kq_2}{b} + \frac{Kq_3}{c}$$

$$= Kq_2\left(\frac{a^2}{b^2}\frac{1}{a} + \frac{1}{b} + \frac{c^2}{b^2}\frac{1}{c}\right)$$

$$= \frac{1}{4\pi\varepsilon_0} \frac{b^2 Q}{a^2+c^2+b^2} \frac{a+b+c}{b^2}$$

$$= \frac{Q}{4\pi\varepsilon_0} \frac{a+b+c}{a^2+b^2+c^2}$$

45. *(d)* Given situation is shown in the figure below,

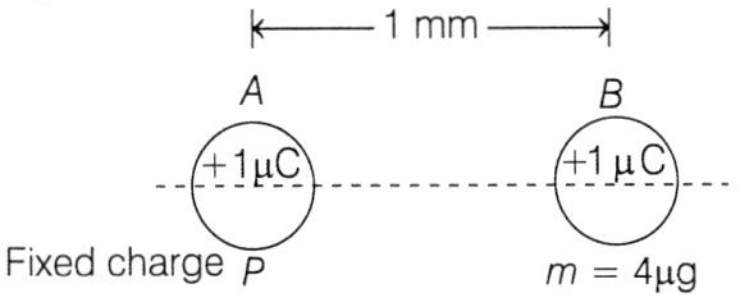

Now, potential energy of system of charges at separation of 1 mm is

$$U_1 = Kq_1 q_2 / r$$

Here, $q_1 = q_2 = 1 \times 10^{-6}$ C

$$r = 1 \text{ mm} = 1 \times 10^{-3} \text{ m}$$

$$\therefore \quad U_1 = \frac{9\times10^9 \times 1\times10^{-6} \times 1\times10^{-6}}{1\times10^{-3}}$$

$$= 9 \text{ J}$$

Potential energy of given system of charges at separation of 9 mm is

$$U_2 = \frac{Kq_1 q_2}{r}$$

$$= \frac{9\times10^9 \times (1\times10^{-6})^2}{9\times10^{-3}}$$

$$= 1 \text{ J}$$

By energy conservation,
Change in potential energy of system of A and B
= Kinetic energy of charged particle B

$$\Rightarrow U_1 - U_2 = \frac{1}{2} m_B v_B^2$$

where, m_B = mass of particle B

$$m_B = 4 \text{ μg}$$

$$- 4\times10^{-6} \times 10^{-3} \text{ kg}$$

$$= 4\times10^{-9} \text{ kg}$$

and v_B = velocity of particle B at separation of 9 mm

$$\Rightarrow \; 9 - 1 = \frac{1}{2} \times 4\times10^{-9} \times v_B^2$$

$$\Rightarrow \quad v_B^2 = 4\times10^9$$

$$\Rightarrow \quad v_B = 2\times10^3 \text{ m/s}$$

• Concept Enhancer

In the equations,

$$F = \frac{Kq_1q_2}{r^2} \text{ and } U = \frac{Kq_1q_2}{r}$$

In force (F) only magnitude of charges q_1, q_2 is taken, while in potential energy q_1 and q_2 are taken with sign.

46. *(d)* Potential at any point at distance x from the centre of the ring is given by

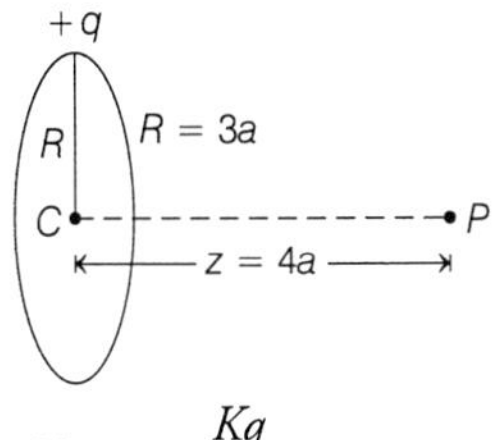

$$V_P = \frac{Kq}{\sqrt{R^2+z^2}}$$

Given, $R = 3a$ and $z = 4a$

$$\therefore \quad V_P = \frac{Kq}{\sqrt{9a^2+16a^2}} = \frac{Kq}{5a} \quad \ldots \text{(i)}$$

At centre,

$$x = 0$$

So, potential at centre is

$$V_C = \frac{Kq}{R} = \frac{Kq}{3a} \quad \ldots \text{(ii)}$$

Now, energy required to get this charge from $x = 4a$ to the centre is

$$\Delta U = q\,\Delta V = q\,[V_C - V_P]$$

$$= q\left[\frac{Kq}{3a} - \frac{Kq}{5a}\right]$$

$$\Delta U = \frac{2}{15}\frac{Kq^2}{a} \quad \ldots \text{(iii)}$$

This energy must be equal to (or less than) the kinetic energy of the charge,

$$\text{i.e. } \frac{1}{2}mv^2 \geq \frac{2}{15}\frac{Kq^2}{a}$$

So, minimum energy required is

$$\frac{1}{2}mv^2 = \frac{2}{15} \times \frac{1}{4\pi\varepsilon_0} \times \frac{q^2}{a}$$

(put $K = 1/4\pi\varepsilon_0$)

$\therefore$ Minimum velocity,

$$v^2 = \frac{2}{m} \times \frac{2}{4\pi\varepsilon_0} \times \frac{q^2}{15a}$$

$$\text{or} \quad v = \sqrt{\frac{2}{m}} \times \sqrt{\frac{2q^2}{4\pi\varepsilon_0 a \times 15}}$$

47. *(d)* Initially, when uncharged shell encloses charge Q, charge distribution due to induction will be as shown,

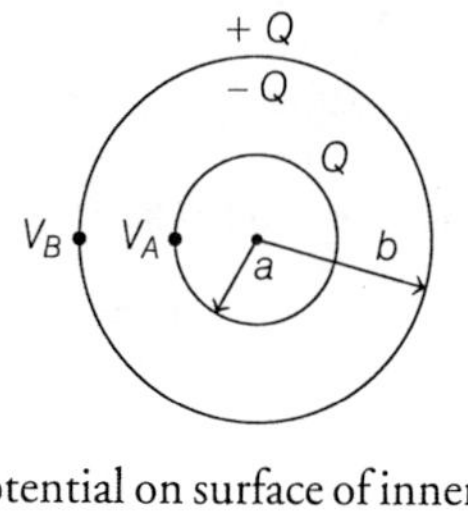

The potential on surface of inner shell is

$$V_A = \frac{KQ}{a} + \frac{K(-Q)}{b} + \frac{KQ}{b} \quad \ldots\text{(i)}$$

where, K = proportionality constant.
Potential on surface of outer shell is

$$V_B = \frac{KQ}{b} + \frac{K(-Q)}{b} + \frac{KQ}{b} \quad \ldots\text{(ii)}$$

Then, potential difference is

$$\Delta V_{AB} = V_A - V_B$$
$$= KQ\left(\frac{1}{a} - \frac{1}{b}\right)$$

Given, $\Delta V_{AB} = V$

So, $$KQ\left(\frac{1}{a} - \frac{1}{b}\right) = V \quad \ldots\text{(iii)}$$

Finally after giving charge $q = 4\,Q$ to outer shell, potential difference will be

$$\Delta V_{AB} = V_A - V_B$$
$$= \left(\frac{KQ}{a} + \frac{K(-4Q)}{b}\right) - \left(\frac{KQ}{b} + \frac{K(-4Q)}{b}\right)$$
$$= KQ\left(\frac{1}{a} - \frac{1}{b}\right) = V \quad \text{[From Eq. (iii)]}$$

48. *(d)* The four charges are shown in the figure below

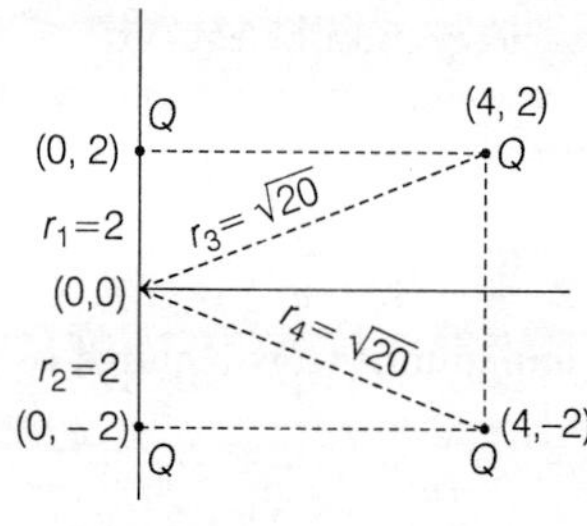

Electric potential at origin (0, 0) due to these charges can be found by scalar addition of electric potentials due to each charge.

$$\therefore \quad V = \frac{KQ}{r_1} + \frac{KQ}{r_2} + \frac{KQ}{r_3} + \frac{KQ}{r_4} \quad \ldots\text{(i)}$$

$$\Rightarrow \quad V = KQ\left[\frac{1}{2} + \frac{1}{2} + \frac{1}{\sqrt{20}} + \frac{1}{\sqrt{20}}\right]$$

$$= KQ\left[1 + \frac{1}{\sqrt{5}}\right]$$

$$\Rightarrow \quad V = KQ\frac{(\sqrt{5}+1)}{\sqrt{5}} \text{ volt} \quad \ldots\text{(ii)}$$

Now, if another charge Q is placed at origin, then work done to get the charge at origin

$$W = QV \quad \ldots\text{(iii)}$$

On putting the value of V from Eq. (ii) in Eq. (iii),

we get $$W = KQ^2\frac{(\sqrt{5}+1)}{\sqrt{5}} \text{ J}$$

or $$W = \frac{Q^2}{4\pi\varepsilon_0}\left(1 + \frac{1}{\sqrt{5}}\right) \text{J}$$

49. *(b)* V_A = Potential due to charge $+q$ on ring A
+ Potential due to charge $-q$ on ring B

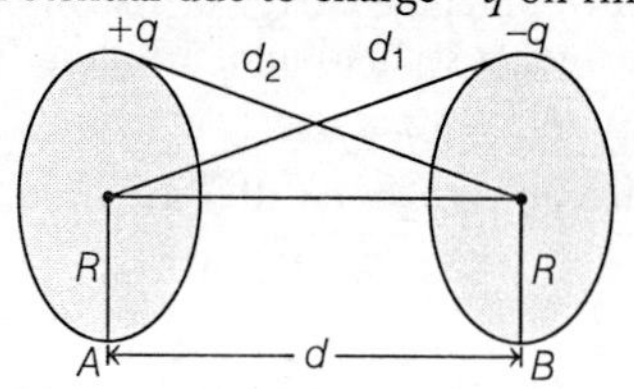

$$= \frac{1}{4\pi\varepsilon_0}\left(\frac{q}{R} - \frac{q}{d_1}\right)$$

$$= \frac{1}{4\pi\varepsilon_0}\left(\frac{q}{R} - \frac{q}{\sqrt{R^2+d^2}}\right)$$

$$[\because d_1 = \sqrt{R^2+d^2}] \quad \ldots\text{(i)}$$

Similarly,

$$V_B = \frac{1}{4\pi\varepsilon_0}\left(-\frac{q}{R} + \frac{q}{\sqrt{R^2+d^2}}\right)$$

Potential difference, $V_A - V_B$

$$= \frac{1}{4\pi\varepsilon_0}\left(\frac{q}{R} - \frac{q}{\sqrt{R^2+d^2}}\right) - \frac{1}{4\pi\varepsilon_0}\left(\frac{-q}{R} + \frac{q}{\sqrt{R^2+d^2}}\right)$$

$$= \frac{1}{4\pi\varepsilon_0}\frac{q}{R} + \frac{1}{4\pi\varepsilon_0}\frac{q}{R} - \frac{2}{4\pi\varepsilon_0}\frac{q}{\sqrt{R^2+d^2}}$$

$$= \frac{q}{2\pi\varepsilon_0}\left(\frac{1}{R} - \frac{1}{\sqrt{R^2+d^2}}\right)$$

50. *(36)* Potential at centre of half ring

$$V = \frac{kQ}{R} \Rightarrow V = \frac{k\lambda\pi R}{R}$$
$$V = k\lambda\pi$$
$$V = 9\times10^9 \times 4\times10^{-9}\,\pi$$
$$V = 36\pi \text{ V}$$
$$\therefore \quad x = 36$$

51. *(27)*

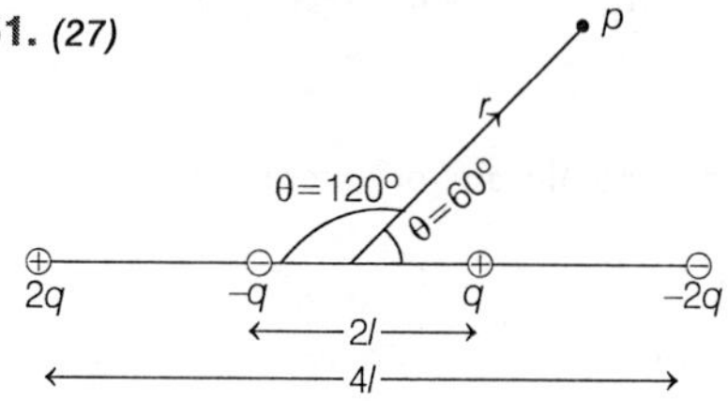

Electric potential due to pair $(-q, q)$ at P is,

$$V_1 = \frac{1}{4\pi\varepsilon_0}\frac{2lq}{r^2}\cos\theta$$

$$= \frac{1}{4\pi\varepsilon_0}\cdot\frac{2lq}{r^2}\cdot\cos 60° = \frac{1}{4\pi\varepsilon_0}\frac{lq}{r^2}$$

Electric potential due to pair $(-2q, 2q)$ at P is,

$$V_2 = \frac{1}{4\pi\varepsilon_0}\frac{(2\cdot 2l\cdot 2q)}{r^2}\cdot\cos 120°$$

$$= \frac{1}{4\pi\varepsilon_0}\cdot\frac{8ql}{r^2}\left(-\frac{1}{2}\right) = \frac{1}{4\pi\varepsilon_0}\left(\frac{-4lq}{r^2}\right)$$

So, net potential at $P = V_1 + V_2$

$$= \frac{1}{4\pi\varepsilon_0}\frac{ql}{r^2} + \frac{1}{4\pi\varepsilon_0}\left(\frac{-4ql}{r^2}\right)$$

$$= \frac{1}{4\pi\varepsilon_0}\times\frac{-3ql}{r^2} = 9\times10^9\times\frac{-3ql}{r^2}$$

$$= -27\frac{ql}{r^2}\times10^9 \text{ V}$$

$$= -\alpha\left[\frac{ql}{r^2}\right]\times10^9 \text{ V} \quad \text{(given)}$$

$$\therefore \quad \alpha = 27$$

52. *(8)* Given, $Z = 50$, $r = 9\times10^{-13}$ cm

$$V = \frac{KqZ}{r} = \frac{9\times10^9\times50\times1.6\times10^{-19}}{9\times10^{-13}\times10^{-2}}$$

$$V = 80\times10^{-10}\times10^{15} = 8\times10^6 \text{ V}$$

53. *(160)* Given, potential of small drop = 10 mV

Let q be the charge of each drop.

$$V = Kq/r$$

Now, for combination of 64 drop

$$64\times\frac{4}{3}\pi r^3 = \frac{4}{3}\pi R^3$$

$$\Rightarrow \quad R = 4r \text{ and } Q = 64q$$

Potential of bigger drop,

$$\frac{KQ}{R} = \frac{K64q}{4r} = \frac{16Kq}{r}$$

$$\Rightarrow \quad V = 160 \text{ mV}$$

54. *(5)*

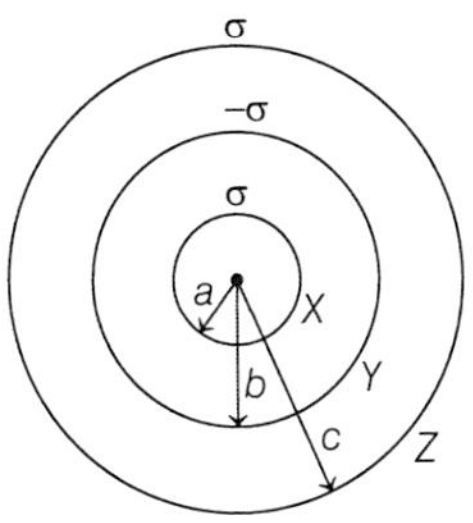

For a hollow spherical shell,

(i) Potential on the surface

$$=\frac{1}{4\pi\varepsilon_0}\cdot\frac{q}{R}$$

$$=\frac{1}{4\pi\varepsilon_0}\cdot\frac{4\pi R^2\cdot\sigma}{R},$$

$$R = \text{Radius of shell} = \frac{\sigma\cdot R}{\varepsilon_0}$$

(ii) Potential at an interior point = Potential at surface

(iii) Potential at an outer point

$$=\frac{Kq}{r}=\frac{1}{4\pi\varepsilon_0}\cdot\frac{\sigma\times 4\pi R^2}{\varepsilon}$$

$$=\frac{\sigma}{\varepsilon_0}\frac{R^2}{\varepsilon},\ \varepsilon = \text{distance from centre}$$

Now, potential of shell X,

$$V_X = V_{\text{interior of }Z} + V_{\text{interior of }Y} + V_{\text{surface of }X}$$

$$\frac{\sigma}{\varepsilon_0}\cdot c + \frac{(-\sigma)}{\varepsilon_0}\cdot b + \frac{\sigma}{\varepsilon_0}\cdot a$$

And potential of shell Z,

$$V_Z = V_{\text{exterior of }X} + V_{\text{exterior of }Y} + V_{\text{surface of }Z}$$

$$=\frac{\sigma}{\varepsilon_0}\cdot\frac{a^2}{c}+\left(\frac{-\sigma}{\varepsilon_0}\right)\cdot\frac{b^2}{c}+\frac{\sigma}{\varepsilon_0}\cdot c$$

Given, $V_X = V_Z$ and $a = 2$ cm, $b = 3$ cm

So we have;

$$\frac{\sigma}{\varepsilon_0}\cdot(c)-\frac{\sigma}{\varepsilon_0}\times(3\times10^{-2})+\frac{\sigma}{\varepsilon_0}\times 2\times10^{-2}$$

$$=\frac{\sigma}{\varepsilon_0}\times\frac{4\times10^{-4}}{c}-\frac{\sigma}{\varepsilon_0}\times\frac{9\times10^{-4}}{c}+\frac{\sigma}{\varepsilon_0}\cdot c$$

$$\Rightarrow c-3\times10^{-2}+2\times10^{-2}=\left(\frac{4\times10^{-4}-9\times10^{-4}}{c}\right)+c$$

$$\Rightarrow (-1\times10^{-2})c = -5\times10^{-4}$$

$$\Rightarrow \qquad c = 5\times10^{-2}$$

or radius of shell Z is 5 cm.

55. *(18)*

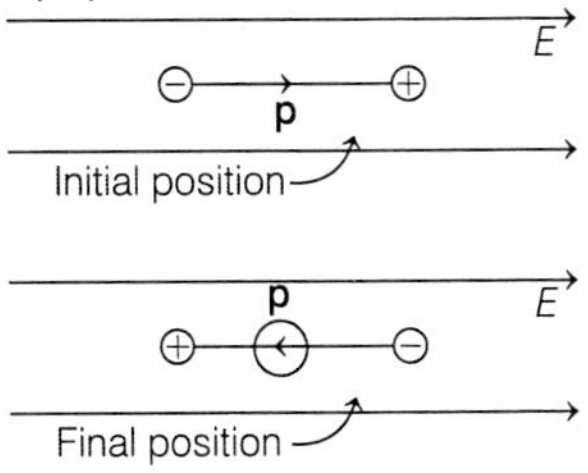

Work done in rotating a dipole is,

$$W_{\theta_1\to\theta_2} = -pE(\cos\theta_2-\cos\theta_1)$$

Here, $\theta_1 = 0°, \theta_2 = 180°$

$$P = 6\times10^{-6}\text{ Cm},$$

$$E = 1.5\times10^3\text{ NC}^{-1}$$

So, work done, W

$$=-6\times10^{-6}\times1.5\times10^3\times(\cos180°-\cos0°)$$

$$=18\times10^{-3}\text{ J} = 18\text{ mJ}$$

56. *(12)*

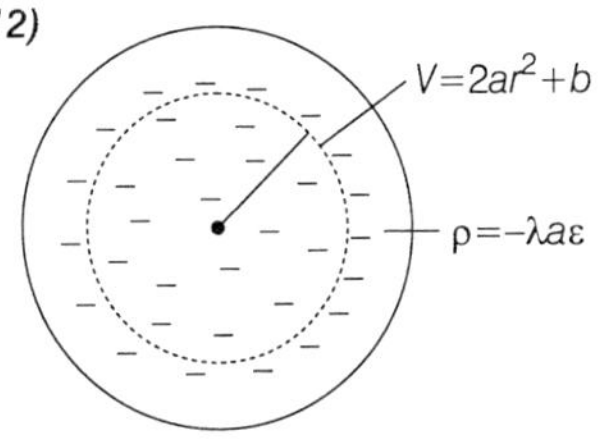

Given, volume charge density inside a charged spherical ball is $\rho = -\lambda a\varepsilon$

As, charge density is constant, charge enclosed in a Gauss' sphere of radius r is,

$$q=\rho\cdot V=-\lambda a\varepsilon\left(\frac{4}{3}\pi r^3\right)$$

by Gauss's law field on surface of Gauss's sphere is

$$E=\frac{1}{4\pi\varepsilon}\frac{q_{\text{enclosed}}}{r^2}$$

$$=\frac{1}{4\pi\varepsilon}\left(\frac{-\lambda a\varepsilon\frac{4}{3}\pi r^3}{r^2}\right)=-\frac{\lambda}{3}ar$$

Now, using $E=\frac{-dV}{dr}$, we get

$$V=\int_0^r dV=-\int_0^r(Edr)$$

or $$V=\int_0^r\frac{\lambda}{3}ar\,dr=\frac{\lambda a}{3}\cdot\left(\frac{r^2}{2}\right)$$

or $$V=\frac{\lambda}{6}ar^2 \qquad \ldots\text{(i)}$$

But potential is given,

$$V=2ar^2+b \qquad \ldots\text{(ii)}$$

On comparing Eq. (i) and, Eq. (ii) we get

$$\frac{\lambda}{6}=2 \quad\text{or}\quad \lambda=12$$

57. *(198)* Potential at each charged drop

$$V = 22\text{ V}$$

Let r and R be the radius of small and big drop, respectively.

∴ Volume of big drop = 27 × Volume of small drop

$$\Rightarrow \frac{4}{3}\pi R^3 = 27\times\frac{4}{3}\pi r^3$$

$$\Rightarrow \qquad R = 3r$$

∵ Potential at small drop

$$V = 22\text{ V}$$

or, $$\frac{1}{4\pi\varepsilon_0}\frac{q}{r}=22 \qquad \ldots\text{(i)}$$

Potential of bigger drop,

$$V'=\frac{1}{4\pi\varepsilon_0}\frac{27q}{R}$$

$$=\frac{1}{4\pi\varepsilon_0}\cdot\frac{27q}{3r}=9\cdot\frac{1}{4\pi\varepsilon_0}\cdot\frac{q}{r}$$

$$=9\times22 \qquad \text{[From Eq. (i)]}$$

$$=198\text{ V}$$

58. *(243)* Given,

$$n = 27$$

$$\Rightarrow \qquad V_1 = 10\text{ V}$$

Let q_1 be the charge of one drop, r_1 be its radius, r be the radius of bigger drop and q be its charge.

As, volume remains constant.

$$\therefore \quad 4/3\,\pi r^3 = 4/3\,\pi r_1^3\times27$$

$$\Rightarrow \qquad r = 3r_1$$

Electric potential $(V)=\frac{3Kq_1}{5r}$

where, K is Coulomb constant.

Therefore, potential energy of small drop,

$$U_S=\frac{3}{5}\frac{Kq_1^2}{r_1} \qquad \ldots\text{(i)}$$

and potential energy of bigger drop,

$$U_B=\frac{3}{5}\frac{K(q_1\times27)^2}{r}$$

$$U_B=\frac{3}{5}\frac{K(27)^2q_1^2}{3r_1} \qquad \ldots\text{(ii)}$$

On dividing Eq. (ii) by Eq. (i), we get

$$\frac{U_B}{U_s}=\frac{27\times27}{3}=243$$

$$\Rightarrow \quad U_B = 243\,U_s$$

59. *(128)* Given, number of mercury drops, $n = 512$

Voltage of each drop, $V = 2$ V

Let r, R be the radius of drop small and combined spherical drop, respectively.

Now, when all drops are joined into single drop, volume remains constant,

i.e., $512 \times \frac{4}{3}\pi r^3 = \frac{4}{3}\pi R^3$

$\Rightarrow \quad R = 8r$

$\because \quad V = \frac{Kq}{r} = 2 \quad \ldots(i)$

$\therefore \quad V_{net} = \frac{KQ}{R}$

where, Q be the charge of bigger sphere.

$\Rightarrow \quad V_{net} = \frac{K \times 512q}{8r} \quad \ldots(ii)$

On dividing Eq. (ii) by Eq. (i), we get

$$\frac{V_{net}}{V} = \frac{K \times 512\, q \times r}{8r \times Kq} = \frac{512}{8}$$

$\Rightarrow \quad V_{net} = 2 \times \frac{512}{8} = 128$ V

Topic 2

01. *(a)* The charge store in capacitor is

$$\Delta q = (KC - C)V$$
$$= 40 \times 10^{-6} \times 100$$
$$= 4 \times 10^{-3} = 4 \text{ mC}$$

Also, $\Delta U = \frac{1}{2}C'V^2 - \frac{1}{2}CV^2$

$$= \frac{1}{2}CV^2(2-1)$$
$$= \frac{1}{2} \times 40 \times 10^{-6} \times 10000$$
$$= 0.2 \text{ J}$$

02. *(d)*

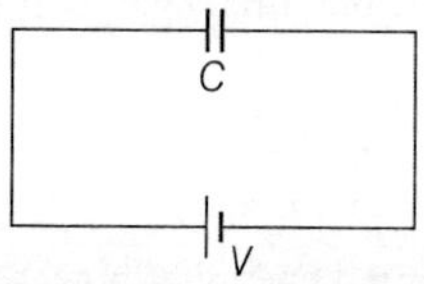

$$\phi = \frac{Q}{\varepsilon_0} \Rightarrow Q = CV$$
$$\phi = \frac{CV}{\varepsilon_0}$$

Option (d) is correct answer.

03. *(b)* The capacitance of spherical conductor of radius R is given as

$$C = 4\pi\varepsilon_0 R$$

So, capacitance does not depend on its charge, it depends only on the radius (R) of the conductor.

Charge on two sphere will be also same $Q = CV$, C is same and V is same for two spheres.

Therefore, Assertion is false and Reason is true.

04. *(a)* The capacitance of a parallel plate capacitor is given by

$$C_0 = \frac{\varepsilon_0 A}{d} \quad \ldots(i)$$

When the space between two parallel plates is partly filled with a dielectric slab of dielectric constant K, then capacitance becomes

$$C' = \frac{\varepsilon_0 A}{\left(d - t + \frac{t}{K}\right)}$$

where, t = thickness of dielectric slab $= \frac{3d}{4}$

$$\Rightarrow C' = \frac{\varepsilon_0 A}{\left(d - \frac{3d}{4} + \frac{3d}{4K}\right)} = \frac{\varepsilon_0 A}{\left(\frac{d}{4} + \frac{3d}{4K}\right)}$$
$$= \frac{4K\varepsilon_0 A}{(3+K)d}$$

$\therefore \quad C' = \frac{4KC_0}{3+K}$ [Using Eq. (i)]

05. *(a)*

Case 1 Force due to two charged plates in a capacitor

Case 2 Force due to single positively charged plate.

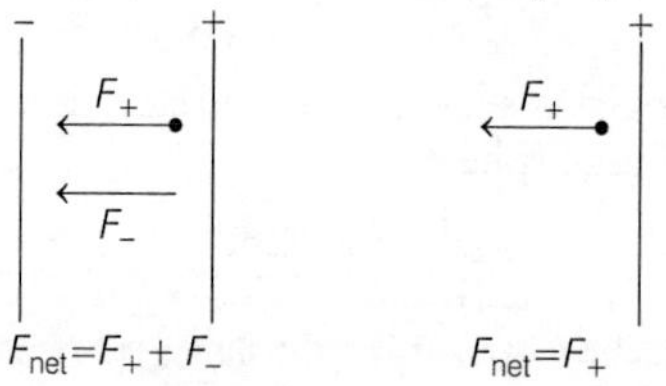

$$F_{new} = \frac{1}{2}F_{net} = \frac{1}{2} \times 10 = 5 \text{ N}$$

06. *(b)* If Q = charge on each plate, then

$$Q = CV = \frac{\varepsilon_0 A}{d} \cdot Ed = \varepsilon_0 AE$$

Here, $A = 1 \text{ m}^2, E = 100$ N/C

and $\varepsilon_0 = 8.85 \times 10^{-12} \frac{C^2}{N\text{-}m^2}$

So, by substituting given values, we get

$$Q = 8.85 \times 10^{-12} \times 1 \times 100$$
$$= 8.85 \times 10^{-10} \text{ C}$$

07. *(a)* Magnitude of induced charge is given by

$$Q' = (K-1)CV_0$$
$$= \left(\frac{5}{3} - 1\right)90 \times 10^{-12} \times 20$$
$$= 1.2 \times 10^{-9} \text{C} \Rightarrow Q' = 1.2 \text{ nC}$$

08. *(a)* Electric field inside dielectric,

$$\frac{\sigma}{K\varepsilon_0} = 3 \times 10^4$$
$$\Rightarrow \sigma = 2.2 \times 8.85 \times 10^{-12} \times 3 \times 10^4$$
$$= 6.6 \times 8.85 \times 10^{-8} = 5.841 \times 10^{-7}$$
$$= 6 \times 10^{-7} \text{ C/m}^2$$

09. *(b)*

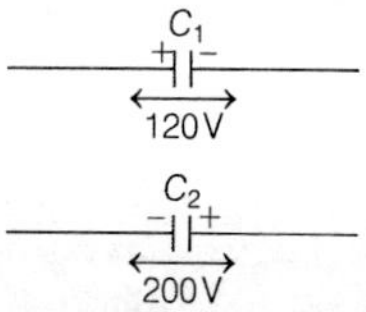

For potential to be made zero after connection, the charge of both capacitors are equal.

$\therefore \quad q_1 = q_2 \Rightarrow C_1V_1 = C_2V_2$

$\Rightarrow \quad 120\, C_1 = 200\, C_2 \Rightarrow 3C_1 = 5C_2$

10. *(b)*

Aluminium is a metal, so when we insert an aluminium foil, equal and opposite charges appear on its two surfaces. Since, it is of negligible thickness, it will not affect the capacitance.

> **Alternate Solution**
>
> From the formula, $C = \frac{\varepsilon_0 A}{d - t + \frac{t}{K}}$
>
> Here, $K = \infty$ and $t \to 0$
>
> So, $C = \frac{\varepsilon_0 A}{d+0} = \frac{\varepsilon_0 A}{d} = C_0$

11. *(a)* Capacitance of spherical conductor $= 4\pi\varepsilon_0 a$

where, a is radius of conductor.

Therefore, $C = \frac{1}{9 \times 10^9} \times 1$

$$= \frac{1}{9} \times 10^{-9} = 0.11 \times 10^{-9}$$
$$= 1.1 \times 10^{-10} \text{ F}$$

12. *(b)* Electric field

$$E = \frac{V}{d} = \frac{5}{5 \times 10^{-4}} = 10^4 \text{ V/m}$$

Electric dipole moment

$$p = q(2a) = 2 \times 10^{-6} \times 5 \times 10^{-7}$$

$= 10^{-12}$ C-m

$\therefore$ Torque, $\tau = pE\sin\theta$

$= 10^{-12} \times 10^4 \times \sin 30$

$= 0.5 \times 10^{-8}$

$= 5 \times 10^{-9}$ N-m

13. *(c)* Initial capacitance $= 100\,pF$

Initial voltage $= 60$ V

Initial charge

$Q_i = C_1V_1 = 100 \times 60 = 6000\,pC$

Now, new capacitor is added in parallel circuit, resulting in final voltage 20 V across both the capacitor.

$$Q_f = (C_1 + C_2)V_f$$

$$Q_f = (100 + C_2)20$$

According to Law of conservation of charge

$$Q_i = Q_f$$

$$6000 = 20\,[100 + C_2]$$

$$\frac{6000}{20} = 100 + C_2$$

$$200\,pF = C_2$$

14. *(a)* Net dipole moment of a polar linear isotropic dielectric substance is zero in the absence of an external electric field because polar dielectric are randomly oriented.

15. *(d)* The capacitance is given by the formula

$$C = \frac{A\varepsilon_0}{d}$$

Initial capacitance $= \dfrac{(l \times b)\varepsilon_0}{d}$

$$= \frac{3 \times 1}{3}\varepsilon_0 = \varepsilon_0$$

For option C, $l = 6$ cm, $b = 5$ cm, $d = 3$ cm.

$C = \dfrac{6 \times 5}{3} \times \varepsilon_0 = 10\varepsilon_0$ units

For option $E = l = 5$ cm $b = 2$ cm, $d = 1$ cm

$C = \dfrac{5 \times 2}{1}\varepsilon_0 = 10\varepsilon_0$ units

16. *(d)* Initially,

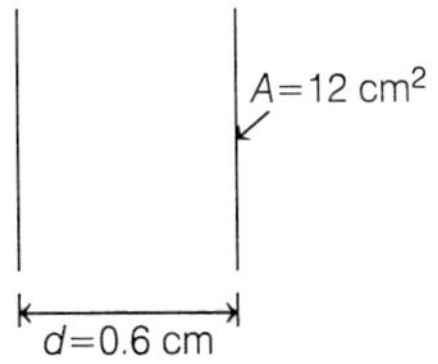

Finally,

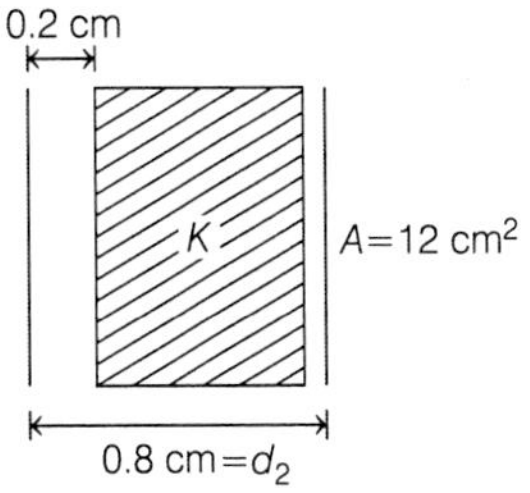

Given, capacitance is same in both cases.

$$\Rightarrow \quad \frac{\varepsilon_0 A}{d_1} = \frac{\varepsilon_0 A}{d_2 - t + \frac{t}{K}}$$

$$\Rightarrow \quad d_1 = d_2 - t + \frac{t}{K}$$

Here, $d_1 = 0.6$ cm

$d_2 = 0.8$ cm

$t = 0.6$ cm

So; $0.6 = 0.8 - 0.6 + \dfrac{0.6}{K}$

$$0.6 = 0.2 + \frac{0.6}{K} \Rightarrow 0.4 = \frac{0.6}{K}$$

$$K = \frac{6}{4} = \frac{3}{2} = 1.5$$

17. *(d)* Force between charges q_1 and q_2, distance r cm in vacuum is,

$$F = \frac{1}{4\pi\varepsilon_0}\frac{q_1q_2}{r^2 \times 10^{-4}} \quad \ldots\text{(i)}$$

Force between same charges in medium of dielectric constant $K = 5$, distance $\dfrac{r}{5}$ cm is, $F' = \dfrac{1}{4\pi\varepsilon_0 K}\dfrac{q_1q_2}{\left(\frac{r}{5}\right)^2 \times 10^{-4}}$

$$= 5 \cdot \frac{1}{4\pi\varepsilon_0}\frac{q_1q_2}{r^2 \times 10^{-4}}$$

$$= 5F$$

18. *(a)* Given, dielectric constant for metallic sheet, $K = \infty$

Thickness of (metallic plate), $t = \dfrac{2d}{3}$

Distance between two plates of capacitor $= d$

According to question, $C_1 = \dfrac{\varepsilon_0 A}{d}$

$$C_2 = \frac{\varepsilon_0 A}{d - t + \frac{t}{K}} = \frac{\varepsilon_0 A}{d - \frac{2d}{3} + 0} = 3C_1$$

$\Rightarrow C_2 / C_1 = 3$

19. *(c)* Given,

$$A = 40\text{ cm}^2 = 4 \times 10^{-3}\text{ m}^2$$

$$d = 2\text{ mm} = 2 \times 10^{-3}\text{ m}$$

Thickness $t = 1$ mm $= 1 \times 10^{-3}$ m

Dielectric constant, $K = 5$

$\therefore$ Capacitance, $C = \dfrac{\varepsilon_0 A}{(d - t) + \frac{t}{K}}$

$$= \frac{\varepsilon_0 \times 4 \times 10^{-3}}{(2 \times 10^{-3} - 1 \times 10^{-3}) + \frac{1 \times 10^{-3}}{5}}$$

$$= \frac{4\varepsilon_0 \times 10^{-3}}{10^{-3}\left(1 + \frac{1}{5}\right)} = 4\varepsilon_0 \times \frac{5}{6} = \frac{10\,\varepsilon_0}{3}\text{ F}$$

> **• Concept Enhancer**
>
> If a dielectric slab is inserted in parallel plate capacitor, then its capacitance will increase.

20. *(c)* The given situation is shown below.

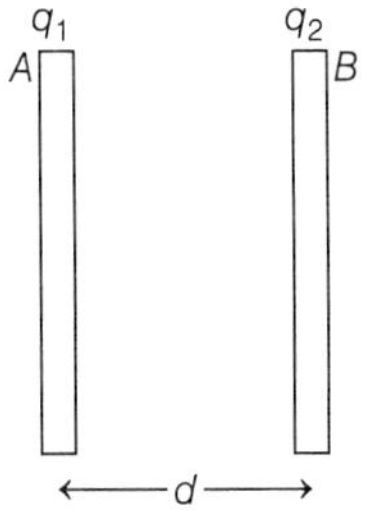

Net electric field inside the parallel plate capacitor.

$$E = \frac{\sigma_1}{2\varepsilon_0} - \frac{\sigma_2}{2\varepsilon_0} = \frac{1}{2\varepsilon_0}(\sigma_1 - \sigma_2)$$

$$= \frac{1}{2\varepsilon_0}\left[\frac{q_1}{A_1} - \frac{q_2}{A_2}\right]$$

$\therefore \quad A_1 = A_2 = A$

$$\Rightarrow \quad E = \frac{1}{2\varepsilon_0}\left[\frac{q_1}{A} - \frac{q_2}{A}\right] = \frac{q_1 - q_2}{2\varepsilon_0 A}$$

$\therefore$ Potential difference between the plates.

$$V = E \cdot d = \frac{(q_1 - q_2)d}{2\varepsilon_0 A}$$

$$= \frac{q_1 - q_2}{2\left(\frac{\varepsilon_0 A}{d}\right)} = \frac{q_1 - q_2}{2C}\left(\text{where, } \frac{\varepsilon_0 A}{d} = C\right)$$

21 *(a)* As, $C = \dfrac{Q}{V}$, $V = \dfrac{Q}{C}$

So, when capacitor is fully charged; given

$$Q = 5\text{ C and } C = 2 \times 10^{-6}\text{ F}$$

$$V = \frac{5}{2 \times 10^{-6}} = 2.5 \times 10^6\text{ V}$$

Also, $V = \left(\frac{1}{C}\right) \times Q$

This is equation of a straight line with slope $\frac{1}{C}$.

Hence, correct graph is

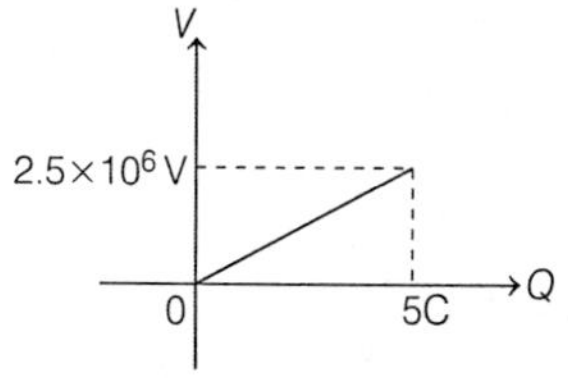

• Concept Enhancer

The slope of V-Q graph is reciprocal of the capacitance.

22. *(c)* Initial capacitance

$$C_1 = \frac{\varepsilon_0 A}{d} \quad \ldots(\text{i})$$

When dielectric material of dielectric constant ($K = 3$) is filled in half the space i.e. $t = d/2$, then new capacitance

$$C_2 = \frac{\varepsilon_0 A}{(d-t) + \left(\frac{t}{K}\right)} = \frac{\varepsilon_0 A}{\left(d - \frac{d}{2}\right) + \left(\frac{d/2}{K}\right)}$$

$$= \frac{\varepsilon_0 A}{\frac{d}{2} + \frac{d}{2K}} = \frac{\varepsilon_0 A}{d\left(\frac{1}{2} + \frac{1}{2K}\right)}$$

$$= \frac{C_1}{\frac{1}{2} + \frac{1}{2K}} \quad [\text{Using Eq. (i)}]$$

$$= \frac{4}{\frac{1}{2} + \frac{1}{2 \times 3}} = \frac{4}{4/6} = 6\,\mu\text{F}$$

23. *(a)* Initially when capacitor C_1 is charged, total charge Q flows through the circuit which charges the capacitor C_1 by V volt. Thus

$$Q = C_1 V$$

When battery is removed and C_2 is connected to C_1, the C_1 discharges until both capacitors acquire a common potential.

$$V_{\text{com}} = \left(\frac{C_1}{C_1 + C_2}\right) V$$

Therefore, the amount of charge on capacitor C_2 is given as

$$Q_2 = C_2 V_{\text{com}} = \frac{C_2 C_1}{C_1 + C_2} V$$

24. *(a)* Initial capacitance of parallel plate capacitor

$$C = \frac{\varepsilon_0 A}{d} \quad \ldots(\text{i})$$

Dielectric constant of metal sheet, $K = \infty$, when a metal sheet of thickness $t = \frac{d}{2}$ is introduced between the plates of capacitor, then new capacitance,

$$C' = \frac{\varepsilon_0 A}{(d-t) + \frac{t}{K}} = \frac{\varepsilon_0 A}{\left(d - \frac{d}{2}\right) + \frac{d/2}{\infty}}$$

$$= \frac{\varepsilon_0 A}{d/2 + 0} \Rightarrow C' = \frac{2\varepsilon_0 A}{d}$$

$$\Rightarrow C' = 2C \quad [\text{from Eq. (i)}]$$

$$\Rightarrow \frac{C'}{C} = \frac{2}{1} \Rightarrow C' : C = 2 : 1$$

25. *(d)* Area of each plate of capacitor,

$$A = 30\pi\,\text{cm}^2 = 30\pi \times 10^{-4}\ \text{m}^2$$

Separation between the plates,

$$d = 1\ \text{mm} = 10^{-3}\ \text{m}$$

Electric field, $E = 3.6 \times 10^7$ V/m

We know that,

$$E = \frac{\sigma}{K\varepsilon_0} = \frac{q/A}{K\varepsilon_0} \quad \left[\because \sigma = \frac{q}{A}\right]$$

$$\Rightarrow E = \frac{q}{AK\varepsilon_0} \Rightarrow K = \frac{q}{EA\varepsilon_0}$$

$$= \frac{7 \times 10^{-6}}{3.6 \times 10^7 \times 30\pi \times 10^{-4} \times \frac{1}{4\pi \times 9 \times 10^9}}$$

$$[\because q = 7 \times 10^{-6}\text{C}]$$

$$= 2.33$$

26. *(c)* Total charge on both capacitors,

$$Q = Q_1 + Q_2 = C_1 V + C_2 V$$
$$= 2CV + CV = 3CV$$
$$\Rightarrow Q = 3CV$$

When dielectric material is filled within the capacitor of capacitance C after removing the battery, then its charge will remain same but its capacitance becomes KC.

Now, potential difference across the capacitors will be common potential given as

$$V_C = \frac{Q}{KC + 2C} = \frac{3CV}{C(K+2)} = \frac{3V}{K+2}$$

27. *(b)* As we know that,

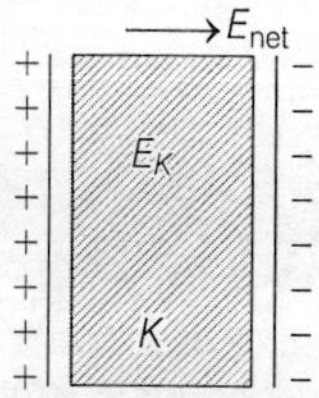

$$E_0 = \frac{\sigma_f}{\varepsilon_0}$$

and $E_K = \dfrac{\sigma_f}{K\varepsilon_0}$

where, E_0 is electric field without dielectric and E_K is electric field with dielectric.

σ = Surface charge density $= \frac{q}{A}$

ε_0 = Free space permittivity

$$\therefore \quad E_{\text{net}} = E_0 - E_K$$

$$\Rightarrow \quad \left(\frac{\sigma}{\varepsilon_0}\right)_b = \frac{\sigma_f}{\varepsilon_0} - \frac{\sigma_f}{K\varepsilon_0}$$

$$\Rightarrow \quad \frac{q_b}{A\varepsilon_0} = \frac{q_f}{A\varepsilon_0} - \frac{q_f}{K\varepsilon_0 A}$$

$$\Rightarrow \quad q_b = q_f\left(1 - \frac{1}{K}\right)$$

• Concept Enhancer

Dielectric constant of metals is very high comparable to dielectric constant of free space i.e., dielectric constant of metal is infinite and the net electric field inside the metal is zero.

28. *(a)* Initially we have a capacitor of capacitance, $C_1 = 10\,\mu$ F and $V_1 = 50$ V

Charge on capacitor is

$$Q = C_1 V_1 = 10\,\mu\text{F} \times 50\text{V} = 500\,\mu\text{C}$$

Now, another uncharged capacitor of capacity C_2 is connected in parallel with C_1 as shown in figure.

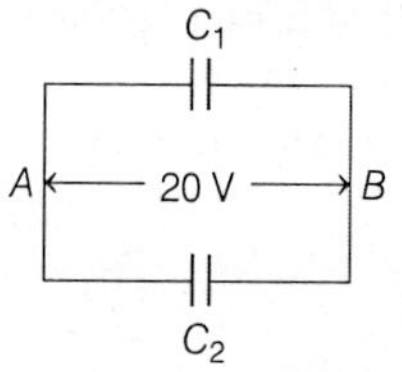

Given that after redistribution, voltage or common potential is 20 V.

Final charge on C_1 is

$$Q_1 = C_1 V = 10\,\mu\text{F} \times 20\text{V} = 200\,\mu\text{C}$$

So, another capacitor has a charge of

$$Q_2 = 500 - 200 = 300\,\mu\text{F}$$

Potential, $V = 20$ V

$$\therefore \text{Capacity, } C_2 = \frac{Q_2}{V}$$

$$= \frac{300\,\mu\text{F}}{20\,\text{V}} = 15\,\mu\text{F}$$

29. *(a)* When pendulum is oscillating between capacitor plates, it is subjected to two forces;

(i) Weight downwards $= mg$

(ii) Electrostatic force acting horizontally $= qE$

So, net acceleration of pendulum bob is resultant of accelerations produced by these two perpendicular forces.

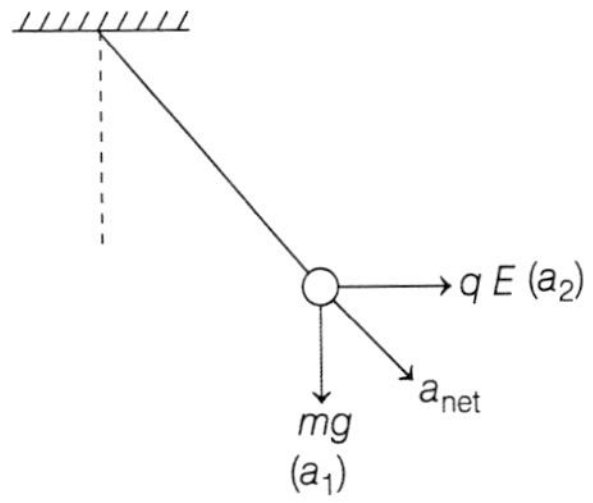

Net acceleration is,

$$a_{\text{net}} = \sqrt{a_1^2 + a_2^2} = \sqrt{g^2 + \left(\frac{qE}{m}\right)^2}$$

So, time period of oscillations of pendulum is

$$T = 2\pi\sqrt{\frac{l}{a_{\text{net}}}} = 2\pi\sqrt{\frac{L}{\sqrt{g^2 + \left(\frac{qE}{m}\right)^2}}}$$

30. *(d)* For the first configuration,

$$C' = \frac{\varepsilon_0 \varepsilon_2 A}{d/2} = \frac{2\varepsilon_0 \varepsilon_2 A}{d}$$

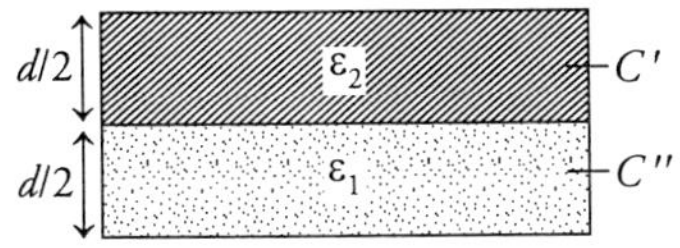

and $$C'' = \frac{\varepsilon_0 \varepsilon_1 A}{d/2} = \frac{2\varepsilon_0 \varepsilon_1 A}{d}$$

Since, C' and C'' are in series,

Thus, $$C_1 = \frac{2C_2' C''}{C' + C''} = \frac{2\varepsilon_1 \varepsilon_2 \varepsilon_0 A}{(\varepsilon_1 + \varepsilon_2) d} \quad \text{...(i)}$$

For the second configuration,

$$C' = \frac{\varepsilon_0 \varepsilon_1 \frac{A}{2}}{d} = \frac{\varepsilon_0 \varepsilon_1 A}{2d}$$

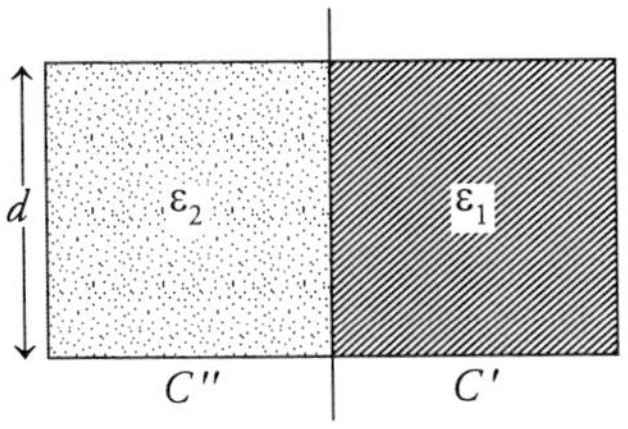

Similarly, $C'' = \dfrac{\varepsilon_0 \varepsilon_2 A}{2d}$

Since, C' and C'' are now parallel,

Thus, $C_2 = C' + C''$

$$C_2 = \frac{(\varepsilon_1 + \varepsilon_2)\varepsilon_0 A}{2d} \quad \text{...(ii)}$$

Using Eqs. (i) and (ii),

$$\frac{C_1}{C_2} = \frac{2\varepsilon_1 \varepsilon_2 \varepsilon_0 A}{(\varepsilon_1 + \varepsilon_2) d} \times \frac{2d}{(\varepsilon_1 + \varepsilon_2)\varepsilon_0 A}$$

$$= \frac{4\varepsilon_1 \varepsilon_2}{(\varepsilon_1 + \varepsilon_2)^2}$$

31. *(b)* Radii of two charged spherical conductors A and B are given as

$$R_A = 5 \text{ mm} = 5 \times 10^{-3} \text{ m}$$

and $R_B = 10 \text{ mm} = 10 \times 10^{-3}$ m

Separation between the spheres,

$$r = 2 \text{ cm} = 2 \times 10^{-2} \text{ m}$$

When both spheres are connected through a wire, then charge flows from one sphere of higher potential to sphere of lower potential till, while both spheres would reach at same common, potential (V).

After connection $$\frac{KQ}{R_1} = \frac{Kq}{k_2}$$

$$\frac{Q_1}{Q_2} = \frac{R_1}{R_2} \quad \text{...(i)}$$

But electric field on spheres A and B are given as,

$$E_A = \frac{\sigma}{\varepsilon_0} = \frac{q_A}{A\varepsilon_0} = \frac{q_A}{4\pi R_A^2 \varepsilon_0} = \frac{q_A}{4\pi\varepsilon_0 R_A^2}$$

Similarly, $$E_B = \frac{q_B}{4\pi\varepsilon_0 R_B^2}$$

$$\therefore \quad \frac{E_A}{E_B} = \frac{\dfrac{q_A}{4\pi\varepsilon_0 R_A^2}}{\dfrac{q_B}{4\pi\varepsilon_0 R_B^2}} = \frac{q_A}{q_B} \cdot \frac{R_B^2}{R_A^2}$$

$$= \frac{R_A}{R_B} \cdot \frac{R_B^2}{R_A^2} \text{ [From Eq. (i)]}$$

$$= \frac{R_B}{R_A} = \frac{10 \times 10^{-3}}{5 \times 10^{-3}} = \frac{2}{1}$$

$$\therefore E_A : E_B = 2:1$$

32. *(a)*

> **key idea** If capacitance of spherical shell is n times of the capacitance of isolated conducting sphere, then the ratio of their radii is $\left(\dfrac{n}{n-1}\right)$.

Capacitance of an isolated conducting sphere of radius R_1,

$$C_1 = 4\pi\varepsilon_0 R_1$$

Capacitance of spherical shell of radius R_1 and R_2, $C_2 = 4\pi\varepsilon_0 \dfrac{R_2 R_1}{R_2 - R_1}$

According to given situation,

$$C_2 = nC_1$$

$$\Rightarrow \frac{4\pi\varepsilon_0 R_1 R_2}{R_2 - R_1} = n \cdot 4\pi\varepsilon_0 R_1$$

$$\Rightarrow \frac{R_2}{R_2 - R_1} = n$$

$$\Rightarrow \frac{R_2 - R_1}{R_2} = \frac{1}{n} \Rightarrow 1 - \frac{R_1}{R_2} = \frac{1}{n}$$

$$\Rightarrow 1 - \frac{1}{n} = \frac{R_1}{R_2}$$

$$\Rightarrow \frac{n-1}{n} = \frac{R_1}{R_2} \Rightarrow \frac{R_2}{R_1} = \frac{n}{n-1}$$

33. *(c)* For given capacitor, consider a small width dx at a distance x from origin O.

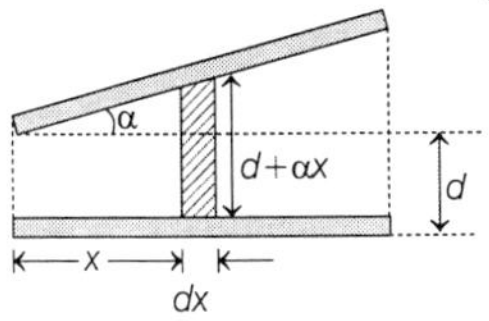

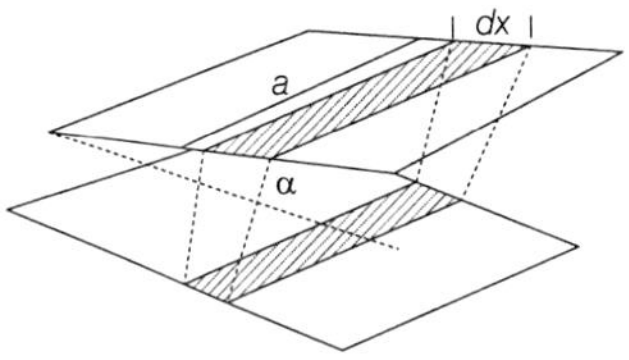

For this elemental capacitor, we have

Area of each plate $= a \times dx$

Distance between plates $= d + \alpha x$

Capacity of this elemental capacitor,

$$dC = \frac{\varepsilon_0 a dx}{d + \alpha x}$$

Total capacity of arrangement is sum of capacitances of all such elemental capacitors because they all are in parallel.

Hence, $$C_{\text{system}} = \int_0^a dC = \int_0^a \frac{\varepsilon_0 a}{d + \alpha x} \cdot dx$$

$$= \frac{\varepsilon_0 a}{\alpha} \ln\left(1 + \frac{\alpha a}{d}\right)$$

As, for small values of $a\alpha/d$,

$$\ln\left(1 + \frac{a\alpha}{d}\right) \approx \frac{a\alpha}{d} - \frac{1}{2}\left(\frac{a\alpha}{d}\right)^2$$

So, $$C_{\text{system}} = \frac{\varepsilon_0 a^2}{d}\left(1 - \frac{a\alpha}{2d}\right)$$

34. *(c)* The given situation is shown below.

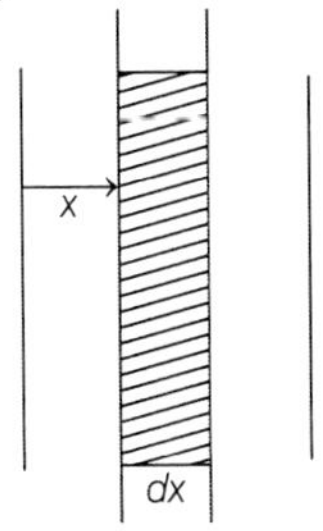

Let us take an elemental capacitor of width dx. Then,

$$dC = \frac{(K_0 + \lambda x)A\varepsilon_0 d}{dx\, d} = \frac{(K_0 + \lambda x)C_0 d}{dx} \quad \ldots(\text{i})$$

where, $C_0 = \frac{A\varepsilon_0}{d}$ (Vaccum)

Total capacitance $= C$

on integrating Eq. (i),

$$\Rightarrow \frac{1}{C} = \int_0^d \frac{dx}{(K_0 + \lambda x)C_0 d}$$

On integrating both sides, we get

$$\frac{1}{C} = \int_0^d \frac{dx}{(K_0 + \lambda x)C_0 d}$$

$$\Rightarrow \frac{1}{C} = \frac{1}{\lambda C_0 d}[\ln(K_0 + \lambda x)]_0^d$$

$$\Rightarrow \frac{1}{C} = \frac{1}{\lambda C_0 d}\left[\ln\left(\frac{K_0 + \lambda d}{K_0}\right)\right]$$

$$\Rightarrow C = \frac{\lambda C_0 d}{\ln\left(1 + \frac{\lambda d}{K_0}\right)}$$

35. *(5)* Charge on capacitor $= 5 \times 10^{-6}$ C

Induced charge on one face of dielectric

$= Q_i = 4 \times 10^{-6}$ C

$$Q_i = Q\left[1 - \frac{1}{K}\right]$$

$$K = \frac{1}{1 - \frac{Q_i}{Q}}$$

$$K = \frac{1}{1 - \frac{4 \times 10^{-6}}{5 \times 10^{-6}}} = 5$$

36. *(15)*

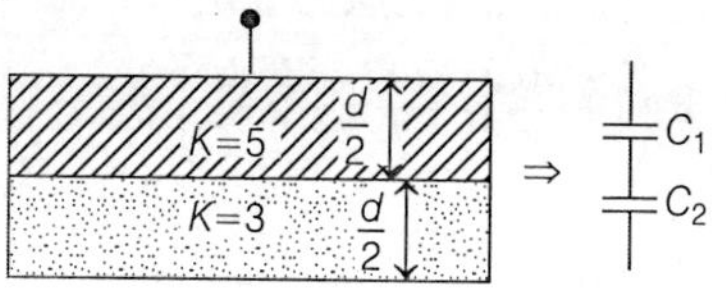

Now, $C_{eq} = \frac{C_1 C_2}{C_1 + C_2}$

where, $C_1 = \frac{K\varepsilon_0 A}{\frac{d}{2}}$

$$= \frac{5 \times 8.85 \times 10^{-12} \times 4 \times 10^{-4}}{\frac{1.77}{2} \times 10^{-3}}$$

$$= 20 \times 10^{-12}\ \text{F} = 20\ \text{pF}$$

and $C_2 = \frac{3 \times 8.85 \times 10^{-12} \times 4 \times 10^{-4}}{\frac{1.77}{2} \times 10^{-3}}$

$$= 12 \times 10^{-12}\ \text{F} = 12\ \text{pF}$$

$$\therefore C_{eq} = \frac{C_1 C_2}{C_1 + C_2} = \frac{20 \times 12}{20 + 12} = \frac{240}{32}$$

$$= 7.5\ \text{pF}$$

When another capacitance of 7.5 μF is connected parallel to it, then

$$C'_{eq} = 7.5 + 7.5 = 15\ \text{pF}$$

37. *(2)* As battery remains connected potential difference between plates remains same before and after insertion of dielectric. Capacitance before insertion is,

$$C_1 = \frac{\varepsilon_0 A}{d}$$

Capacity after insertion of dielectric is,

$$C_2 = \frac{\varepsilon_0 A}{\left(d - t + \frac{t}{K}\right)}$$

As charge stored is $Q = CV$

$\therefore$ Charge before insertion of dielectric

$$Q_1 = C_1 V = \frac{\varepsilon_0 A}{d} \cdot V$$

and charge after introduction of dielectric

$$Q_2 = C_2 V = \frac{\varepsilon_0 AV}{\left(d - t + \frac{t}{K}\right)}$$

Now given, $Q_2 = \frac{125}{100} Q_1$

$$\Rightarrow \frac{\varepsilon_0 AV}{\left(d - t + \frac{t}{K}\right)} = \frac{125}{100}\frac{\varepsilon_0 AV}{d}$$

$$\Rightarrow 125\left(d - t + \frac{t}{K}\right) = 100d$$

$$\Rightarrow 125\left(3 + \frac{2}{K}\right) = 100 \times 5$$

$$\Rightarrow K = 2$$

> **• Concept Enhancer**
>
> When a dielectric material is inserted between the space of parallel plate capacitor, then its capacitance increases. It is due to the fact that polarisation of the dielectric medium produces an opposing electric field, which decreases the potential between the plates.

38. *(2)*

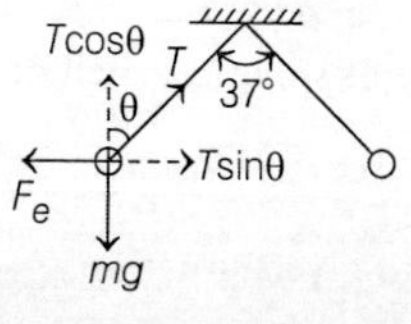

In air, $T\cos\theta = mg$ and $T\sin\theta = F_e$

$$\Rightarrow \tan\theta = \frac{F_e}{mg} = \frac{F_e}{V \cdot \rho \cdot g} \quad \ldots(\text{i})$$

In medium of dielectric constant K and density σ, weight of sphere

= Apparent weight

= True weight − Upthrust

$= mg - F_b$

$= V \cdot \rho \cdot g - V \cdot \sigma \cdot g$

$= Vg(\rho - \sigma)$

and electric force in dielectric is $\frac{F_e}{K}$

So, In medium we have,

$$\tan\theta = \frac{F_e / K}{V \cdot (\rho - \sigma) \cdot g} \quad \ldots(\text{ii})$$

From Eq. (i) and Eq. (ii), we have,

$$\frac{F_e}{V \cdot \rho \cdot g} = \frac{F_e}{V \cdot (\rho - \sigma) g \cdot K}$$

$$\Rightarrow K = \frac{\rho}{\rho - \sigma} = \frac{1.4}{1.4 - 0.7} = 2$$

39. *(105)* Initially, capacity of capacitor,

$$C = \frac{\varepsilon_0 A}{d} = 15\ \text{pF} \quad \text{[Given]}$$

When separation is doubled and space is filled with dielectric $K = 3.5$, capacity of capacitor becomes

$$C' = \frac{\varepsilon_0 KA}{2d} = \frac{K}{2}\left(\frac{\varepsilon_0 A}{d}\right)$$

$$= \frac{3.5}{2} \times 15\ \text{pF} = \frac{105}{4}\ \text{pF}$$

40. *(15)* Given, $C_0 = 5\ \mu\text{F} = 5 \times 10^{-6}$ F

Dielectric constant, $K = 1.5$

According to first condition,

$$C_0 = \frac{\varepsilon_0 A}{d} \quad \ldots(\text{i})$$

According to second condition,

Capacitance of the capacitor in the presence of slab,

$$C' = \frac{\varepsilon_0 KA}{d/2} = \frac{\varepsilon_0 A}{d} \times 2K = C_0 \times 2K$$

$$= 5 \times 10^{-6} \times 2 \times 1.5 = 15\ \mu\text{F}$$

41. *(100)* Given arrangement is

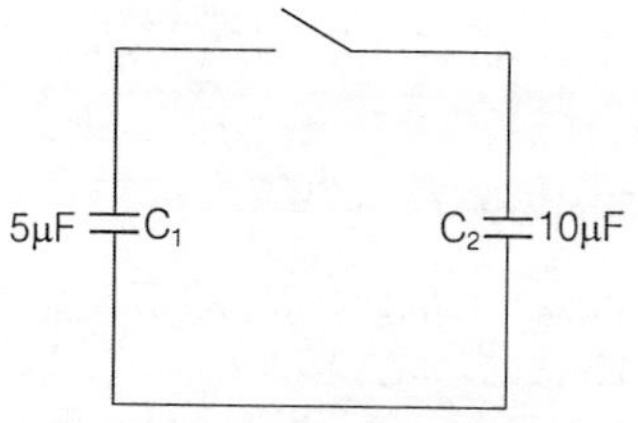

When, switch is closed, charge flows from capacitor C_1 to capacitor C_2 till both reaches some common potential.

Now, common potential

$$V_{com} = \frac{Q_1 + Q_2}{C_1 + C_2} = \frac{5 \times 30}{15} = 10\ \text{V}$$

Charge on capacitor C_2 after redistribution of charge is

$$Q_2' = C_2 \cdot V_{com}$$
$$= (10\times10^{-6})\times10 = 10^{-4}\ C$$
$$= 100\times10^{-6}\ C$$
$$= 100\ \mu C$$

42. *(161)* As we know, the expression of the capacitor with dielectric constant

$$C = \frac{A\varepsilon_0}{\left(d - t + \frac{t}{K}\right)}$$

$$\Rightarrow \quad C = \frac{100\times 8.85\times10^{-12}}{\left(10 - 5 + \frac{5}{10}\right)}$$

$$\Rightarrow \quad C = 161\times10^{-12}\ F = 161\ pF$$

43. *(16)* Let V is the voltage across each capacitor,

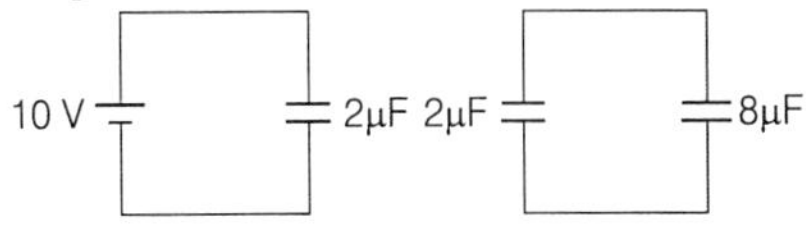

Now, after removing the battery the capacitor is connected.
So, using the law of conservation of charge,

$$2V + 8V = 2\times10 = 20$$
$$\Rightarrow \quad 10V = 20 \Rightarrow V = 2\ V$$

As we know, $Q = CV = 8\times 2 = 16\ \mu C$

Topic 3

01. *(a)* Potential energy per unit volume is given as $\frac{U}{V} = \frac{1}{2}\varepsilon_0 E^2$,
where E is electric field.

$$U = \frac{1}{2}\varepsilon_0 E_0^2 V = \left(\frac{1}{2}\varepsilon_0 E_0^2 A\cdot d\right)$$

Hence, option (a) is correct.

02. *(b)*

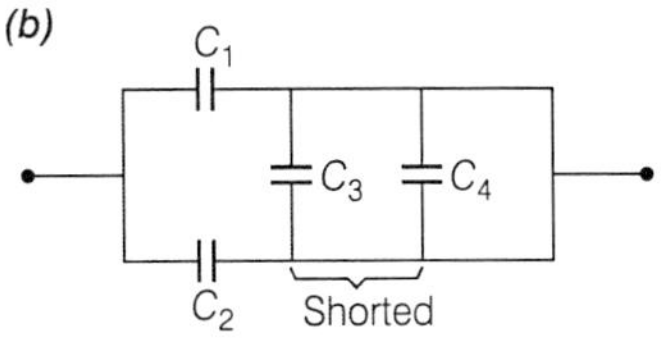

In given circuit, C_4 and C_3 are shorted so, capacitance of circuit is,

$$C_{CQ} = C_1 + C_2 = C + C = 2C$$

03. *(a)* Given,

$$C_1 = 1\ \mu F,\ C_2 = 2\ \mu F,$$
$$C_3 = 4\ \mu F$$
$$C_4 = 3\ \mu F,\ V = 20\ V$$

Since, all capacitances are connected in parallel, hence equivalent capacitance ,

$$C_{eq} = C_1 + C_2 + C_3 + C_4$$
$$= 1\mu F + 2\mu F + 4\mu F + 3\mu F$$
$$= 10\ \mu F$$

$\therefore$ Total charge, $Q = C_{eq}V$

$$= 10\times 20 = 200\ \mu C$$

04. *(a)* Let the original charge is Q.
So, initial energy stored,

$$U_i = \frac{Q^2}{2C}$$

Final energy stored, $\frac{(Q+2)^2}{2C} = U_f$

% change in energy

$$= \frac{U_f - U_i}{U_i}\times 100 = 44$$

$$\Rightarrow \quad \frac{(Q+2)^2 - Q^2}{Q^2} = 0.44$$

$$\Rightarrow \quad Q = 10\ C$$

05. *(c)* In the given arrangement, capacitor can be viewed as three-different capacitors connected in parallel as shown below,

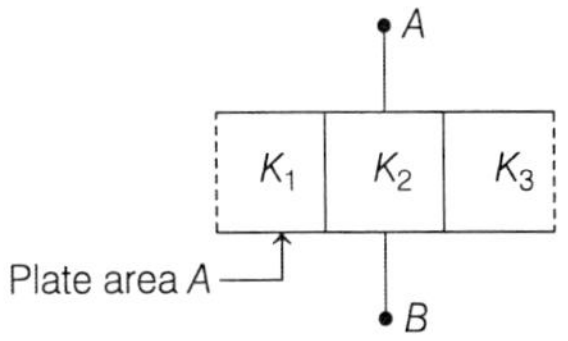

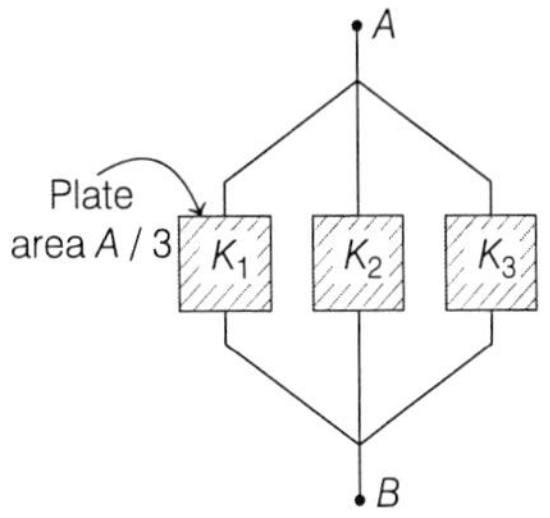

So, equivalent capacity of the system is $C_{eq} = C_1 + C_2 + C_3$

$$\Rightarrow \frac{K\varepsilon_0 A}{d} = \frac{K_1\varepsilon_0 A/3}{d} + \frac{K_2\varepsilon_0 A/3}{d} + \frac{K_3\varepsilon_0 A/3}{d}$$

$$\Rightarrow \quad K = \frac{K_1}{3} + \frac{K_2}{3} + \frac{K_3}{3}$$

Here $K_1 = 10$, $K_2 = 12$
and $K_3 = 14$

So, $K = \frac{10 + 12 + 14}{3}$

$$\Rightarrow \quad K = 12$$

06. *(d)* Given,

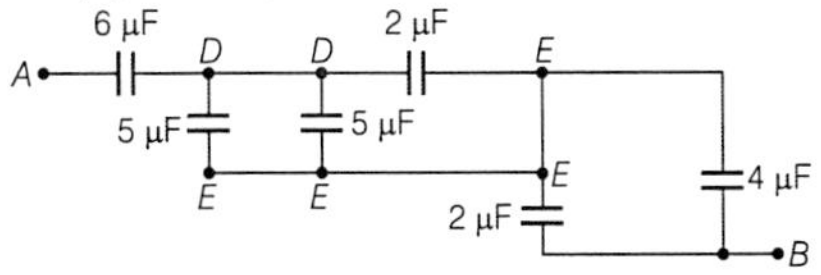

The equivalent capacitance is given by

$$\frac{1}{C_{eq}} = \frac{1}{6} + \frac{1}{12} + \frac{1}{6} = \frac{5}{12}$$

$$\Rightarrow C_{eq} = \frac{12}{5} = 2.4\ \mu F$$

07. *(d)* Ratio of energy stored in the capacitor and the work done by the battery

$$= \frac{\frac{1}{2}qV}{qV} = \frac{1}{2}$$

08. *(d)* $W = \frac{1}{2}\frac{q^2}{C}$

$$= \frac{1}{2}\times\frac{(8\times10^{-18})^2}{100\times10^{-6}}$$

$$= \frac{1}{2}\times\frac{64\times10^{-36}}{100\times10^{-6}} = 32\times10^{-32}\ J$$

09. *(b)* Energy stored by any system of capacitors $= \frac{1}{2}C_{net}V^2$

where, V is source voltage. Thus, n capacitors are connected in parallel.

So, $C_{net} = nC$

$$\therefore \quad E_{net} = \frac{1}{2}nCV^2$$

10. *(b)*

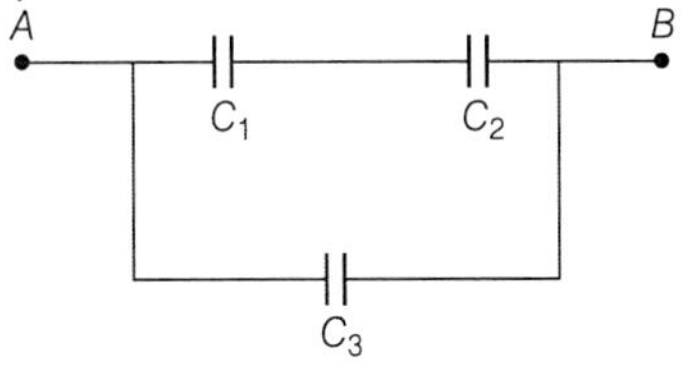

After dielectric

$C_1 = 4C$

$C_1 = 4\times 5 = 20\ \mu F$

$C_2 = C_3 = 5\ \mu F$

C_1 and C_2 arc in series
Which is parallel to C_3
so,

$$C_{eq} = \frac{C_1C_2}{C_1 + C_2} + C_3 = \frac{20\times5}{20+5} + 5$$

$$C_{eq} = 4 + 5 = 9\ \mu F$$

11. *(d)* Here, in this case both the capacitors C_1 and C_2 are connected in parallel to a battery.
In steady state,

$$\frac{C_1}{C_2} = \frac{Q_1}{Q_2} < 1 \Rightarrow C_2 > C_1$$

For energy stored in capacitor,

$$\frac{C_1}{C_2} = \frac{U_1}{U_2} < 1 \Rightarrow U_2 > U_1$$

12. *(a)* $C = 1\mu\text{F}, V = 20\text{V}, d = 1\,\mu\text{m}$

$$\text{Energy density} = \frac{1}{2}\varepsilon_0 E^2$$

$$V = Ed \Rightarrow E = \frac{V}{d}$$

$$\text{Energy density} = \frac{1}{2}\varepsilon_0 \left(\frac{V}{d}\right)^2$$

$$= \frac{1}{2}\varepsilon_0 \frac{V^2}{d^2} \qquad \left[\frac{V}{d} = 20 \times 10^{+6}\right]$$

$$= \frac{1}{2} \times 8.85 \times 10^{-12} \times (20 \times 10^{+6})^2$$

$$= 1.77 \times 10^3\ \text{J/m}^3 \cong 1.8 \times 10^3\ \text{J/m}^3$$

13. *(b)* By energy conservation,

$$\frac{1}{2}CV^2 = \frac{1}{2}LI_{\text{max}}^2$$

$$I_{\text{max}}^2 = \frac{CV^2}{L} \Rightarrow I_{\text{max}} = \sqrt{\frac{C}{L}}.V$$

$$C = 100\,\mu\text{F} = 100 \times 10^{-6}\ \text{F}$$

$$L = 6.4 \times 10^{-3}\ \text{H}, V = 12\ \text{V}$$

$$I_{\text{max}} = \sqrt{\frac{100 \times 10^{-6}}{6.4 \times 10^{-3}}}\ V$$

$$= \sqrt{\frac{100 \times 10^{-6}}{6.4 \times 10^{-3}}} \times 12$$

$$= \sqrt{\frac{100}{6400}} \times 12$$

$$= \frac{10}{80} \times 12 = \frac{120}{80} = \frac{12}{8} = \frac{3}{2}$$

$$I_{\text{max}} = 1.5\ \text{A}$$

14. *(b)* Given, $C = 2\text{F}$

$$\text{Energy stored, } E_1 = \frac{1}{2}CV^2 \qquad \ldots\text{(i)}$$

Now, common potential

$$V_C = \frac{C_1V_1 + C_2V_2}{C_1 + C_2} = \frac{CV + 0}{2C} = \frac{V}{2}$$

$$E_2 = \frac{1}{2}CV_C^2 = \frac{CV^2}{8} \qquad \ldots\text{(ii)}$$

$$\frac{\text{Eq. (ii)}}{\text{Eq. (i)}} = \frac{E_2}{E_1} = \frac{CV^2/8}{CV^2/2} = \frac{1}{4}$$

15. *(a)* According to first condition,

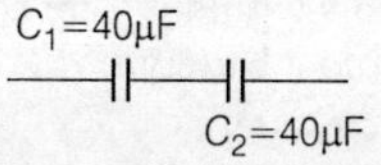

Equivalent capacitance,

$$C = \frac{C_1 \times C_2}{C_1 + C_2} = \frac{40 \times 40}{40 + 40} = 20\,\mu\text{F}$$

When the space between the one of the capacitor is filled dielectric material of dielectric constant K then.

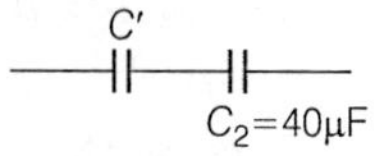

$$C' = C_1 K = K \times 40\mu\text{F} = 40\ K\mu\text{F}$$

$\therefore$ Equivalent capacitance,

$$C'' = \frac{C' \times C_2}{C' + C_2}$$

$$\Rightarrow \quad 24 = \frac{40K \times 40}{40K + 40} \Rightarrow 24 = \frac{40K}{K+1}$$

$$\Rightarrow 24K + 24 = 40K \Rightarrow 24 = 16K$$

$$\Rightarrow \quad K = \frac{24}{16} = 1.5$$

16. *(a)* Energy of a capacitor charged to potential V,

$$U = \frac{1}{2}CV^2 = \frac{1}{2}\frac{\varepsilon_0 A}{d} \cdot V^2 \cdot K$$

$$\left[\because C = \frac{\varepsilon_0 KA}{d}\right]$$

$$\Rightarrow \quad \frac{U_f}{U_i} = \frac{K_f}{K_i} = \frac{15}{10}$$ (as all other parameters remain constant)

Percentage change in energy,

$$\frac{U_f - U_i}{U_i} \times 100 = \frac{K_f - K_i}{K_i} \times 100$$

$$= \left[\frac{K_f}{K_i} - 1\right] \times 100$$

$$= \left[\frac{15}{10} - 1\right] \times 100$$

$$= \frac{15 - 10}{10} \times 100 = 50\%$$

17. *(a)* The given circuit diagram is shown below.

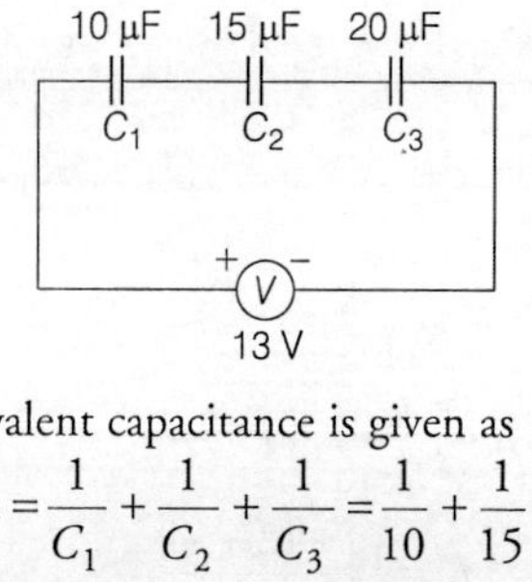

Equivalent capacitance is given as

$$\frac{1}{C_{\text{eq}}} = \frac{1}{C_1} + \frac{1}{C_2} + \frac{1}{C_3} = \frac{1}{10} + \frac{1}{15} + \frac{1}{20}$$

$$\Rightarrow C_{\text{eq}} = \frac{60}{13}\mu\text{F} = \frac{60}{13} \times 10^{-6}\text{F}$$

$\therefore$ Charge flow in the circuit,

$$Q = C_{\text{eq}}V = \frac{60}{13} \times 10^{-6} \times 13$$

$$= 60 \times 10^{-6}\text{C} = 60\ \mu\text{C}$$

Since, in series combination, charge on each capacitor is same, therefore charge on the capacitor of capacitance $15\ \mu\text{F}$ will be $60\ \mu\text{C}$.

18. *(d)* Equivalent capacitance of the series combination of C_2 and C_3

$$C' = \frac{C_2C_3}{C_2 + C_3} \Rightarrow C' = \frac{6 \times 12}{6 + 12} = 4\,\mu\text{F}$$

Charge on capacitors C_2 and C_3 is same.

$$\therefore \quad Q_2 = Q_3 = C'\text{V} = 4\,\mu\text{V}$$

Charge on capacitor

$$C_1, Q_1 = 2\mu\text{V}$$

Ratio of charge is given by

$$\therefore \quad Q_1 : Q_2 : Q_3 = 2 : 4 : 4$$

$$\text{or} \quad Q_1 : Q_2 : Q_3 = 1 : 2 : 2$$

19. *(c)* As per question, the figure can be shown for a parallel plate capacitors,

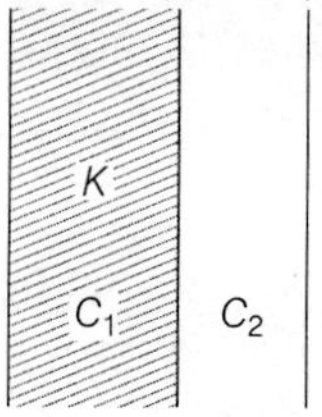

$$\text{Initially, capacitance, } \quad C_0 = \frac{\varepsilon_0 A}{d} \qquad \ldots\text{(i)}$$

In presence of dielectric medium

$$C = \frac{\varepsilon_0 KA}{d} \qquad \ldots\text{(ii)}$$

Also, from the figure capacitors C_1 and C_2 are in series.

$\therefore$ Equivalent capacitance is given by

$$\frac{1}{C_{\text{eq}}} = \frac{1}{C_1} + \frac{1}{C_2}$$

$$\Rightarrow \quad \frac{1}{C_{\text{eq}}} = \left(\frac{3d/4}{\varepsilon_0 KA}\right) + \left(\frac{d/4}{\varepsilon_0 A}\right)$$

[Using Eqs. (i) and (ii)]

$$\Rightarrow \quad \frac{1}{C_{\text{eq}}} = \frac{d}{4\varepsilon_0 A}\left(\frac{3+K}{K}\right)$$

$$\Rightarrow \quad C_{\text{eq}} = \frac{4KC_0}{(3+K)} \qquad \text{[Using Eq.(i)]}$$

20. *(d)* Given, $C_1 = C_2 = C$

When, both capacitors are connected in series,

$$\frac{1}{C_s} = \frac{1}{C} + \frac{1}{C} = \frac{2}{C} \Rightarrow C_s = \frac{C}{2}$$

When, both capacitors are connected in parallel,

$$C_p = C + C = 2C$$

$\therefore$The ratio of equivalent capacitance in series and parallel combination is

$$\frac{C_s}{C_p} = \frac{C/2}{2C} = \frac{1}{4} \Rightarrow C_s : C_p = 1:4$$

21. *(b)*

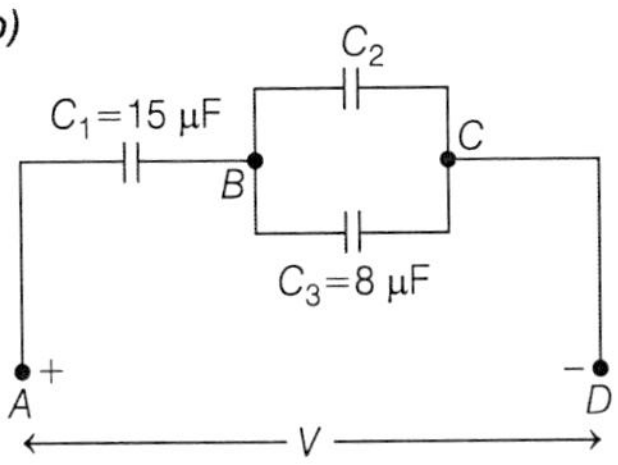

In given arrangement,

V_{BC} = potential drop across $C_2 = 20V$

As combination of capacitors C_2 and C_3 is in parallel, so potential drop across C_3 is also 20 V.

So, Q_3 = charge on $C_3 = C_3 \times V$

$$= 8 \times 10^{-6} \times 20 = 160\ \mu C$$

Also, total charge given by cell is

$$Q = 750\ \mu C$$

As combination of capacitors C_2 and C_3 is in series with capacitor C_1, so charge on C_1 = charge on combination of capacitors C_2 and C_3 = total charge given by cell.

Hence, $Q_2 + Q_3 = 750\ \mu C$

$\Rightarrow \quad Q_2 + 160 = 750$

$\Rightarrow \quad Q_2 = 750 - 160$

$\therefore \quad Q_2 = 590\ \mu C$

22. *(c)* Let capacitances of capacitors are C_1 and C_2.

Given, energy stored in capacitor 2 = 4 × energy stored in capacitor 1.

$\Rightarrow \quad \frac{1}{2} C_2 \times V^2 = 4 \times \frac{1}{2} C_1 \times V^2$

$\Rightarrow \quad C_2 = 4C_1$

Also, equivalent capacity in parallel combination is $10\ \mu F$.

$\Rightarrow \quad C_1 + C_2 = 10$

$\Rightarrow \quad 5C_1 = 10 \quad (\because C_2 = 4C_1)$

$\Rightarrow \quad C_1 = 2\ \mu F$

So, $\quad C_2 = 4 \times 2 = 8\ \mu F$

Equivalent capacity in series combination will be

$$C_{eq} = \frac{C_1 C_2}{C_1 + C_2} = \frac{8 \times 2}{8 + 2} = 1.6\ \mu F$$

23. *(d)* Given circuit is

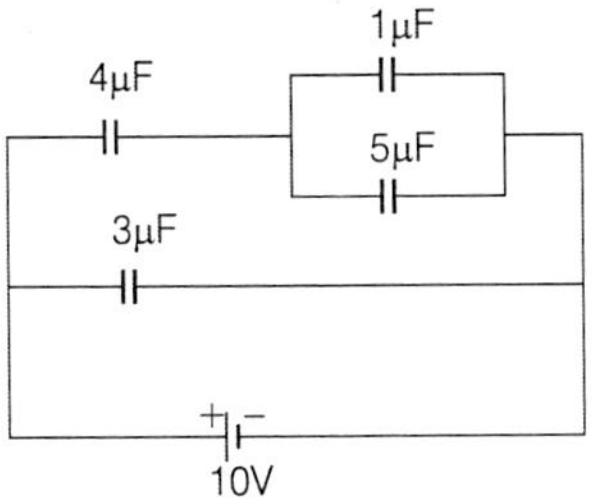

In parallel, $C_{eq} = 5 + 1 = 6\ \mu F$

and in series, $C'_{eq} = \frac{6 \times 4}{6 + 4} = 2.4\ \mu F$

This is equivalent to

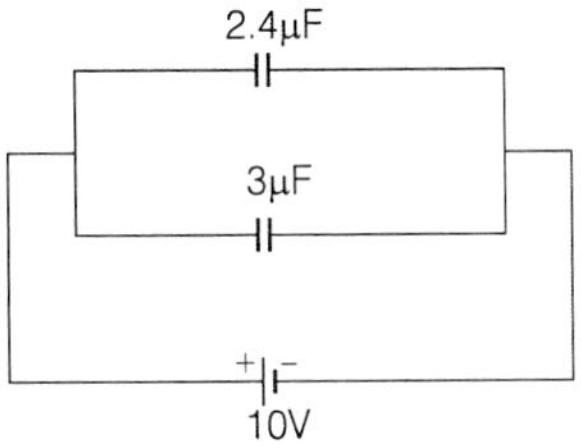

So, potential difference across upper branch = 10 V

Using $Q = C \times V$, charge delivered to upper branch is

$$Q = C'_{eq} \cdot V = 2.4\ \mu F \times 10\ V = 24\ \mu C$$

As we know, in series connection, same charge is shared by capacitors, so charge on 4 μF capacitor and 6 μF capacitor would be same, i.e. $Q'_{4\mu F} = 24\ \mu C$

24. *(a)*

> **key idea** When a dielectric slab is inserted to fill the space between the plates, then the charge on the capacitor plates remains same. However, the capacitance increase, i.e. $C = KC_0$, where C_0 is the capacitance of the capacitor without slab and K is the dielectric constant of slab.

When parallel combination is fully charged, charge on the combination is

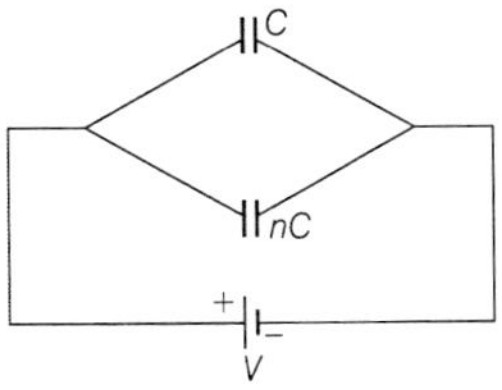

$$Q = C_{eq} V = C(1 + n) V$$

When battery is removed and a dielectric slab is placed between two plates of first capacitor, then charge on the system remains same. Now, equivalent capacitance after insertion of dielectric is

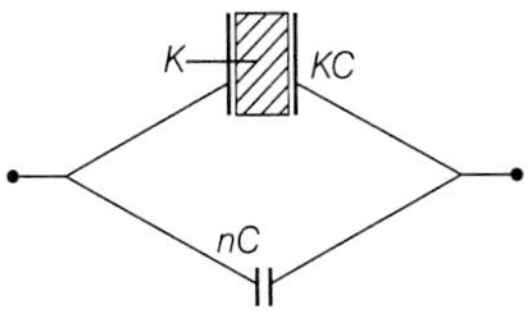

$$C_{eq} = KC + nC = (n + K) C$$

If potential value after insertion of dielectric is V', then charge on system is

$$Q' = C_{eq} V' = (n + K) CV'$$

As $\quad Q = Q'$, we have

$$C(1 + n)V = (n + K) CV'$$

$\therefore \quad V' = \frac{(1 + n) V}{(n + K)}$

25. *(d)*

> **key idea** Two capacitors C_1 and C_2, are connected in series, then their equivalent capacitance is
> $$C_{eq} = \frac{C_1 \times C_2}{C_1 + C_2}$$
> If they are connected in parallel, then their equivalent capacitance is $C_{eq} = C_1 + C_2$.

We simplify given circuit as

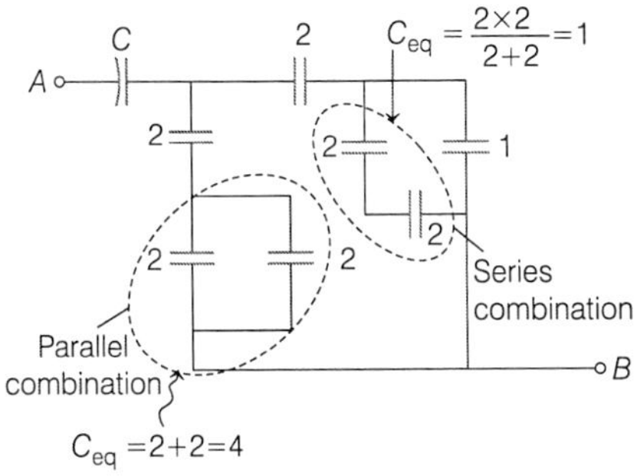

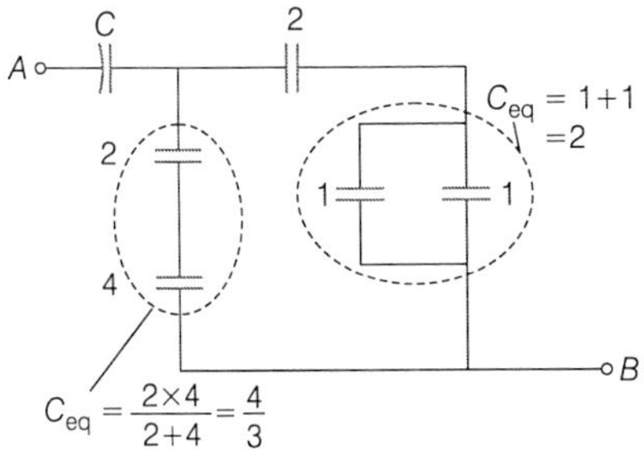

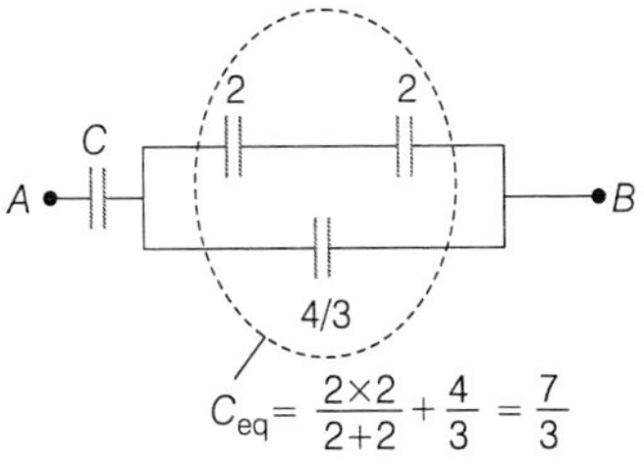

So, $C_{AB} = \dfrac{C \times 7/3}{C + 7/3} = \dfrac{1}{2}$ [given]

$\Rightarrow \dfrac{7C}{3} = \dfrac{C}{2} + \dfrac{7}{6}$

$\Rightarrow \left(\dfrac{14-3}{6}\right)C = \dfrac{7}{6} \Rightarrow C = \dfrac{7}{11}\,\mu F$

26. *(c)* Here, effective capacitance $= \left(\dfrac{6}{13}\right)\mu F$ In the given options some capacitors are joined in parallel and some are in series. For parallel combination, $C_P = C_1 + C_2 + C_3 + \ldots$

For series combination,

$$\frac{1}{C_S} = \frac{1}{C_1} + \frac{1}{C_2} + \frac{1}{C_3} \ldots$$

Here, capacity of each capacitor is $2\mu F$.

$\therefore$ Three capacitors must be in parallel to get $6\,\mu F$. Now, we consider the combinations with three capacitors in parallel (as shown in the figure below)

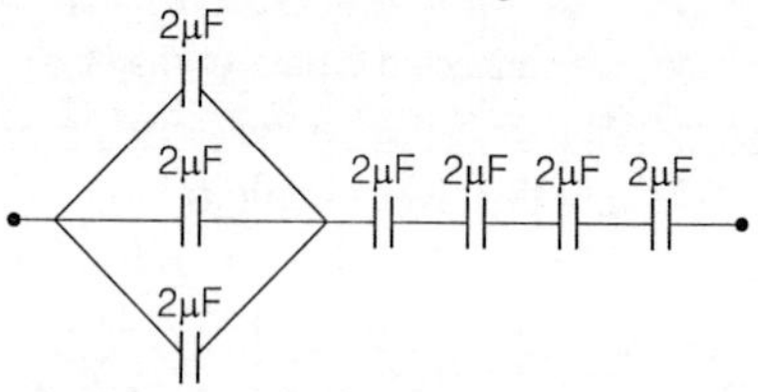

$$\therefore \frac{1}{C_{eq}} = \frac{1}{6} + \frac{1}{2} + \frac{1}{2} + \frac{1}{2} + \frac{1}{2}$$

$$= \frac{1}{6} + 2 = \frac{13}{6}$$

or $C_{eq} = \dfrac{6}{13}\,\mu F$

27. *(b)* Applying the concept of charge conservation on isolated plates of $10\,\mu F$, $6\,\mu F$ and $4\,\mu F$. Since, $6\,\mu F$ and $4\,\mu F$ are in parallel, so total charge on this combination will be $30\,\mu C$

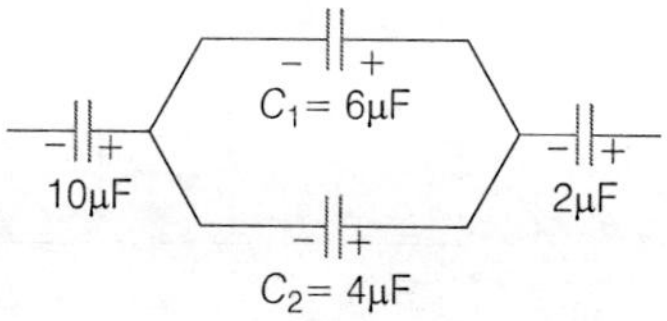

$\therefore$ Charge on $6\,\mu F$, capacitor

$$= \left(\frac{C_1}{C_1 + C_2}\right) q$$

$$= \frac{6}{6+4} \times 30$$

$$= 18\,\mu C$$

Since, the charge has been asked on the right plate of the capacitor. Thus, it would be $+18\,\mu C$.

> **Alternate Solution**
>
> Let charge on $6\mu F$ capacitor is $q\mu C$.
>
> Now, V at $6\mu F$ = V at $4\mu F$
>
> $\therefore \dfrac{q}{6\mu F} = \dfrac{30-q}{4\mu F}$ $\quad (\because V = q/C)$
>
> $\Rightarrow 4q = -6q + 180 \Rightarrow q = 18\mu C$

28. *(c)* As each capacitors cannot withstand more than 300 V, so there should be four capacitors in each row became in this condition 1 kV i.e. 1000 V will be divided by 4 (i.e. 250 not more than 300 V).

Now, equivalent capacitance of one row

$$= \frac{1}{4} \times 1\,\mu F = 0.25\,\mu F$$

[$\because$ in series combination, $C_{eq} = C/n$] we need equivalent of $2\,\mu F$, so let n such rows are needed

$\therefore n \times 0.25 = 2\,\mu F$

[$\because$ in parallel combination $C_{eq} = nC$]

$$n = \frac{2}{0.25} = 8$$

$\therefore$ Total number of capacitors = number of rows × number of capacitors in each row $= 8 \times 4 = 32$

29. *(c)* The energy stored in the electric field = 4.5 J

$$\Rightarrow \frac{Q^2}{2C} = 4.5 \text{ or } C = \frac{Q^2}{2 \times 4.5} \quad \ldots \text{(i)}$$

Capacitance of spherical conductor

$$= 4\pi\varepsilon_0 R$$

$$C = 4\pi\varepsilon_0 R = \frac{Q^2}{2 \times 4.5} \text{ [from Eq. (i)]}$$

$$R = \frac{1}{4\pi\varepsilon_0} \times \frac{(4 \times 10^{-6})^2}{2 \times 4.5}$$

$$= 9 \times 10^9 \times \frac{16}{9} \times 10^{-12}$$

$$= 16 \times 10^{-3}\ \text{m}$$

30. *(c)* Resultant circuit,

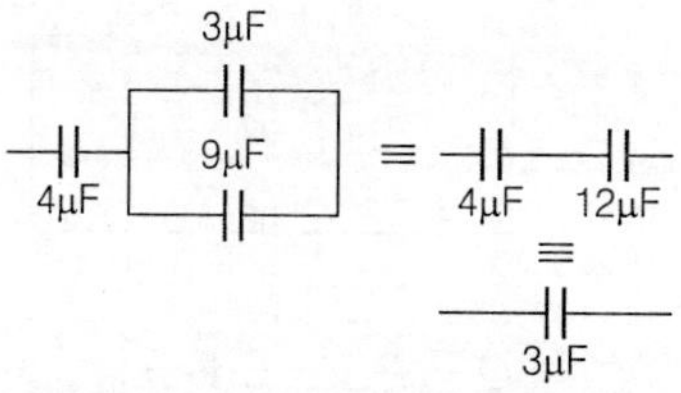

As, charge on $3\,\mu F = 3\,\mu F \times 8V = 24\mu C$

$\therefore$ Charge on $4\,\mu F$ = Charge on $12\,\mu F$ $= 24\,\mu C$

As charge on $4\,\mu F$ and $12\,\mu F$ will be same (in series)

$\therefore Q_1 = C_1 V_1 = C_2 V_2$

$\therefore V_1 = 6\,V$ for $4\,\mu F$

Charge on $3\mu F = 3\mu F \times 2V = 6\mu C$

Charge on $9\mu F = 9\mu F \times 2V = 18\,\mu C$

Charge on $4\,\mu F$ + Charge on $9\,\mu F$

$= (24 + 18)\mu C = 42\,\mu C$

$\therefore$ Electric field at a point distant 30 m

$$= \frac{9 \times 10^9 \times 42 \times 10^{-6}}{30 \times 30} = 420\,\text{N/C}$$

31. *(b)*

> **key idea** Assume negative terminal of the battery as grounded (0 V).

Suppose, potential of point x is V.

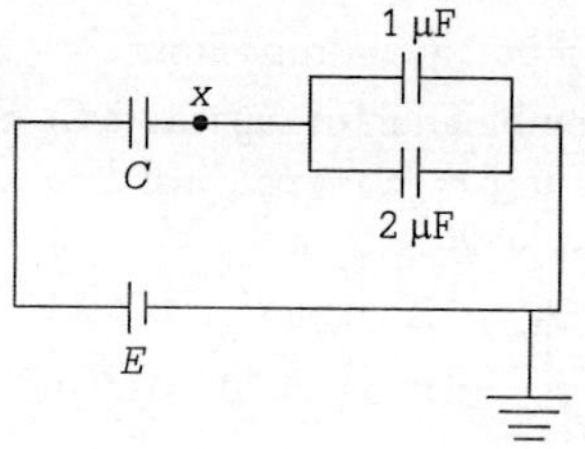

From the circuit diagram, we can write

$$Q_C = Q_1 + Q_2$$

or $C(E - V) = 1 \times V + 2 \times V$

or $V[C + 3] = CE$

or $V = \dfrac{CE}{3 + C}$

$$\therefore Q_2 = C_2(V) = \frac{2\,CE}{3 + C} = \frac{2E}{1 + 3/C}$$

As C_1 varied from $1\,\mu F$ to $3\,\mu F$, charge increases with decreasing slope.

Note As $C \to \infty$, $Q_2 \to 2E$ = constant

32. *(b)* For parallel plate capacitor with dielectric, $C = \dfrac{K\varepsilon_0 A}{d}$

For the given situation

$$C_{eq} = \frac{K\varepsilon_0 2A}{3d} + \frac{2K\varepsilon_0 A}{3d}$$

$$= \frac{4}{3}\frac{K\varepsilon_0 A}{d}$$

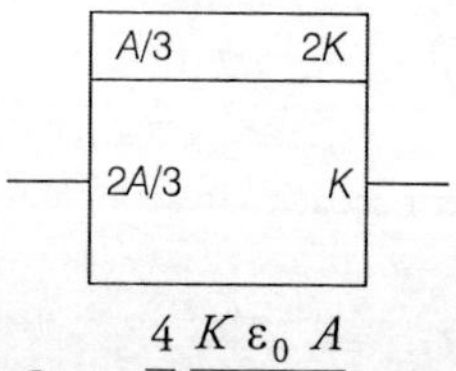

$$\therefore \frac{C_{eq}}{C} = \frac{\frac{4}{3}\frac{K\varepsilon_0 A}{d}}{\frac{K\varepsilon_0 A}{d}} = \frac{4}{3}$$

33. *(c)* As both capacitors are connected in series.

$$\therefore \quad C = \frac{C_1 C_2}{C_1 + C_2} \quad \ldots(\text{i})$$

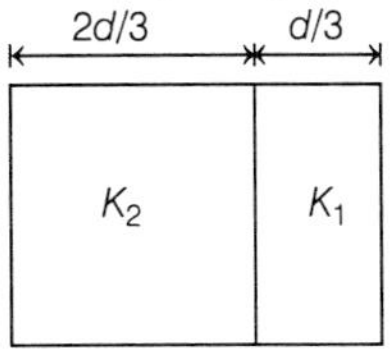

where, $$C_1 = \frac{K_1 \varepsilon_0 A}{d/3} \quad \ldots(\text{ii})$$

and $$C_2 = \frac{K_2 \varepsilon_0 A}{2d/3} \quad \ldots(\text{iii})$$

$$C = \frac{A\varepsilon_0}{d}\left[\frac{(3K_1)\left(\frac{3}{2}K_2\right)}{3K_1 + \frac{3}{2}K_2}\right]$$

$$= \frac{A\varepsilon_0}{d}\left[\frac{(9)(9)}{9+9}\right]$$

It is given that, $\varepsilon_0 A/d = 9$ pF

On substituting we get the result,

$$C = 40.5 \text{ pF}$$

34. *(a)* Each plate is taking part in the formation of two capacitors except the plates at the ends.

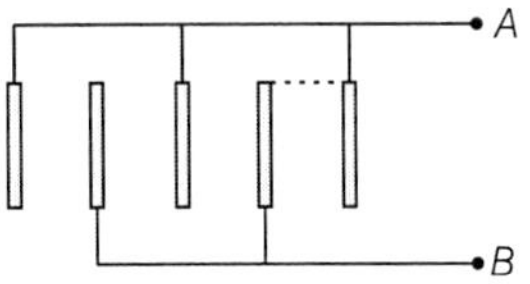

These capacitors are in parallel and n plates form $(n-1)$ capacitors.

Thus, equivalent capacitance between A and $B = (n-1)C$

35. *(d)* $$E = \left(\frac{1}{2}\right)CV^2 \quad \ldots(\text{i})$$

The energy stored in capacitor is lost in form of heat energy.

$$\therefore \quad H = ms\Delta T \quad \ldots(\text{ii})$$

From Eqs. (i) and (ii), we have

$$ms\,\Delta T = \left(\frac{1}{2}\right)CV^2 \text{ or } V = \sqrt{\frac{2ms\,\Delta T}{C}}$$

36. *(a)*

key idea When two capacitors at different potentials are joined, charge will flow from higher potential to lower potential. Charge flow occurs till both capacitors reaches to a common (same) potential.

$$V_{\text{Common}} = \frac{C_1V_1 + C_2V_2}{C_1 + C_2}$$

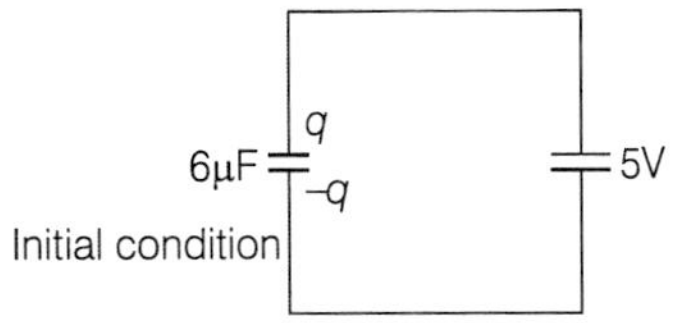

$$q = CV = 6 \times 5 = 30\ \mu\text{C}$$

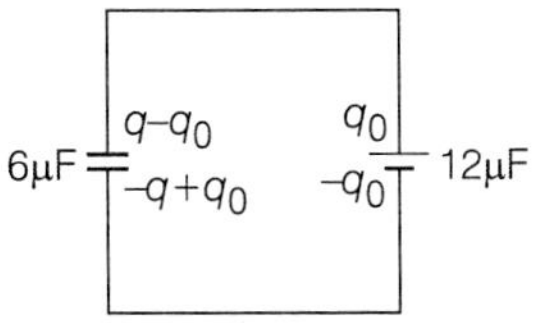

Let charge q_0 flows from C_1 to C_2.

Now apply KVL in the loop

$$\frac{q_0}{12} = \frac{q - q_0}{6}$$
$$6q_0 = 12q - 12q_0$$
$$18q_0 = 12q$$
$$q_0 = \frac{2}{3}q$$

According to the question,

$$q_1 = q - q_0$$
$$= q - \frac{2}{3}q = \frac{q}{3} = \frac{30}{3} = 10\ \mu\text{C}$$
$$q_2 = q_0 = \frac{2}{3}q = \frac{2}{3} \times 30 = 20\ \mu\text{C}$$

37. *(b)*

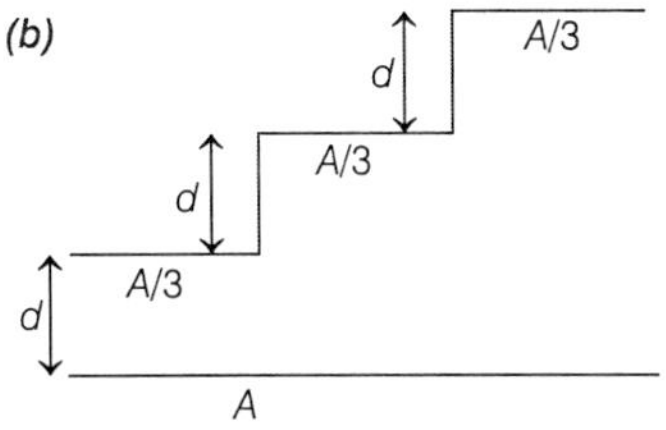

$$C_{\text{eq}} = C_1 + C_2 + C_3$$
$$C_{\text{eq}} = \frac{A\varepsilon_0}{3d} + \frac{A\varepsilon_0}{3(2d)} + \frac{A\varepsilon_0}{3(3d)}$$
$$C_{\text{eq}} = \frac{A\varepsilon_0}{3}\left(\frac{11}{6d}\right)$$
$$\Rightarrow C_{\text{eq}} = \frac{11A\varepsilon_0}{18d}$$

38. *(b)*

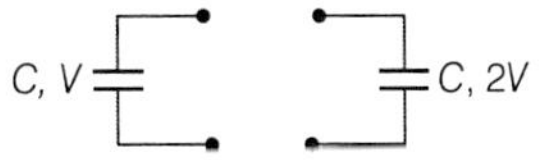

On joining both capacitors reaches a common potential given by,

$$V_C = \frac{Q_1 + Q_2}{C_1 + C_2}$$
$$= \frac{CV + 2CV}{C + C} = \frac{3V}{2}$$

Loss of energy = Initial energy − Final energy

$$= \frac{1}{2}CV^2 + \frac{1}{2}C(2V)^2 - \frac{1}{2}(C+C)\left(\frac{3V}{2}\right)^2$$
$$= \frac{1}{4}CV^2$$

39. *(c)* The given circuit diagram is shown below.

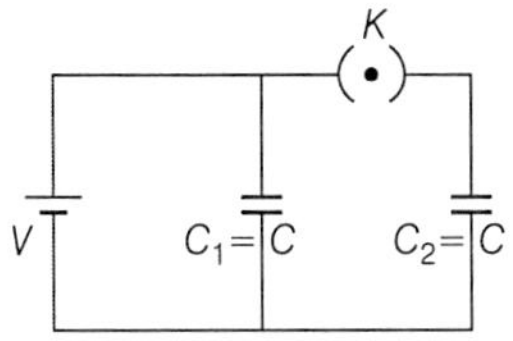

When key K is closed, then equivalent capacitance,

$$C' = C_1 + C_2 = C + C = 2C$$

$\therefore$ Total stored energy,

$$E_1 = \frac{1}{2}C'V^2 = \frac{1}{2}(2C)V^2 = CV^2$$

According to second condition, when key K is open, and material of dielectric constant 5 is introduced in each cpacitor, then

$C_1' = KC_1 = 5\,C$ and $C_2' = KC_2 = 5\,C$

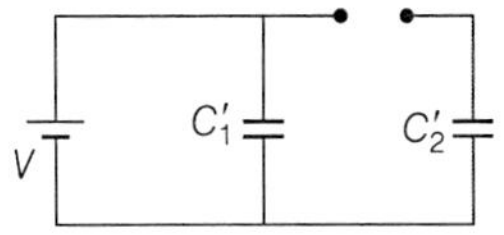

According to figure, energy stored in capacitor C_1'

$$e_1 = \frac{1}{2}C_1'V^2$$
$$= \frac{1}{2}(5C)V^2 = \frac{5}{2}CV^2$$

Energy stored on second capacitor,

$$e_2 = \frac{1}{2}\frac{q^2}{C_2'} = \frac{1}{2}\frac{(CV)^2}{5C} = \frac{CV^2}{10}$$

Total energy stored,

$$E_2 = e_1 + e_2 = \frac{5}{2}CV^2 + \frac{CV^2}{10}$$
$$= \frac{13}{5}CV^2$$

$$\because \quad \frac{E_1}{E_2} = \frac{CV^2}{\frac{13}{5}CV^2} = \frac{5}{13}$$

40. *(a)*

key idea If a parallel plate capacitor is inserted with different dielectric, then draw the equivalent circuit diagram to calculate equivalent capacitance.

Area of plates = A,

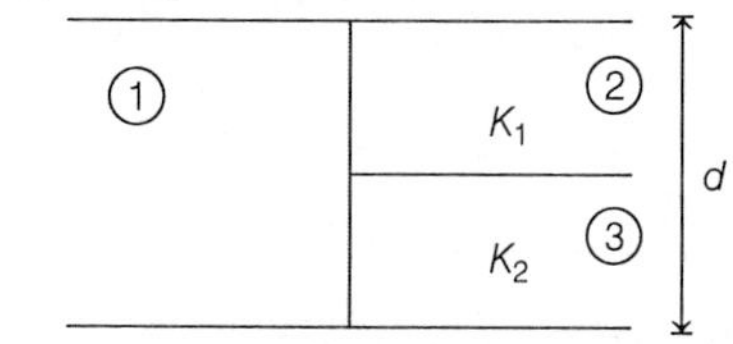

Now, there are three capacitance (C_1), (C_2) and (C_3) as shown below

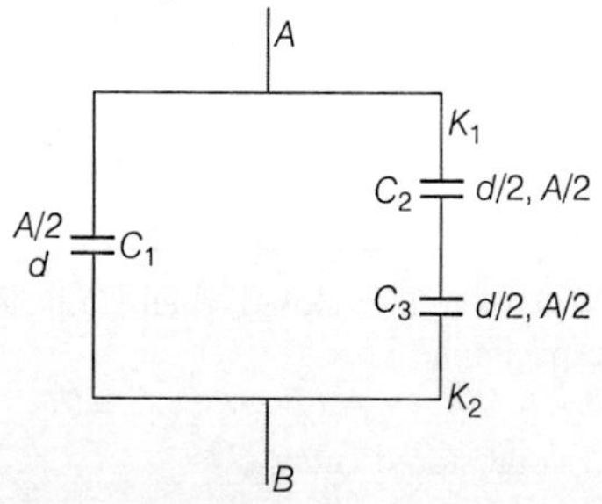

Here, $C_1 = \dfrac{\varepsilon_0 A}{2d}$

$$C_2 = \frac{K_1 \varepsilon_0 A/2}{d/2} = \frac{K_1 \varepsilon_0 A}{d}$$

$$C_3 = \frac{K_2 \varepsilon_0 A/2}{d/2} = \frac{K_2 \varepsilon_0 A}{d}$$

Now, C_2 and C_3 are in series.

∴ Equivalent capacitance is

$$\frac{1}{C_s} = \frac{1}{C_2} + \frac{1}{C_3} = \frac{d}{K_1 A \varepsilon_0} + \frac{d}{K_2 A \varepsilon_0}$$

$$= \frac{d}{A\varepsilon_0}\left[\frac{1}{K_1} + \frac{1}{K_2}\right] = \frac{d(K_1 + K_2)}{A\varepsilon_0 K_1 K_2}$$

$$\Rightarrow C_s = \frac{A\varepsilon_0 K_1 K_2}{d(K_1 + K_2)}$$

Now, C_s and C_1 are in parallel.

$$C_{eq} = \frac{A\varepsilon_0 K_1 K_2}{d(K_1 + K_2)} + \frac{\varepsilon_0 A}{2d}$$

$$\therefore C_{eq} = \frac{\varepsilon_0 A}{d}\left[\frac{1}{2} + \frac{K_1 K_2}{K_1 + K_2}\right]$$

41. *(a)* The value of C_1, C_2 and C_3 can be given as

$$C_1 = \frac{K\varepsilon_0 A}{d},\ C_2 = \frac{3K\varepsilon_0 A}{2d},\ C_3 = \frac{5K\varepsilon_0 A}{3d}$$

∵ In the above figure, we can see that C_1, C_2 and C_3 are in series combination.

$$\Rightarrow \frac{1}{C_{eq}} = \frac{1}{C_1} + \frac{1}{C_2} + \frac{1}{C_3}$$

$$\Rightarrow \frac{1}{C_{eq}} = \frac{1}{\frac{K\varepsilon_0 A}{d}} + \frac{1}{\frac{3K\varepsilon_0 A}{2d}} + \frac{1}{\frac{5K\varepsilon_0 A}{3d}}$$

$$\Rightarrow \frac{1}{C_{eq}} = \frac{d}{K\varepsilon_0 A} + \frac{2d}{3K\varepsilon_0 A} + \frac{3d}{5K\varepsilon_0 A}$$

$$\Rightarrow \frac{1}{C_{eq}} = \frac{d}{K\varepsilon_0 A}\left(1 + \frac{2}{3} + \frac{3}{5}\right)$$

$$\Rightarrow \frac{1}{C_{eq}} = \frac{d}{K\varepsilon_0 A}\left(\frac{15 + 10 + 9}{15}\right)$$

$$\Rightarrow \frac{1}{C_{eq}} = \frac{34d}{15K\varepsilon_0 A}$$

$$\Rightarrow C_{eq} = \frac{15}{34}\left(\frac{K\varepsilon_0 A}{d}\right)$$

42. *(b)* Permittivity of dielectric,

$$\varepsilon(x) = \varepsilon_0 + kx, \text{ for } \left(0 < x \le \frac{d}{2}\right)$$

$$\varepsilon(x) = \varepsilon_0 + k(d - x), \text{ for } \left(\frac{d}{2} \le x \le d\right)$$

As we know that

Capacitance with dielectric

$$(C_k) = kC_0 = \frac{k\varepsilon_0 A}{d}$$

Since, series capacitance adds up in reciprocal manner.

$$\frac{1}{C_{eq}} = \frac{dx}{A\varepsilon}$$

where, C_{eq} = equivalent capacitance

and dx = distance between plates.

$$\Rightarrow \frac{1}{C_{eq}} = \int_0^{\frac{d}{2}} \frac{dx}{A(\varepsilon_0 + kx)} + \int_{\frac{d}{2}}^{d} \frac{dx}{A[\varepsilon_0 + k(d - x)]}$$

$$= \frac{1}{Ak}[\ln(\varepsilon_0 + kx)]_0^{d/2} - \frac{1}{Ak}[\ln[\varepsilon_0 + k(d - x)]]_{d/2}^{d}$$

$$= \frac{1}{Ak}[\ln(\varepsilon_0 + kd/2) - \ln \varepsilon_0] - \frac{1}{Ak}\ln[\varepsilon_0 + k(d - d)] + \ln[\varepsilon_0 + k(d - d/2)]$$

$$= \frac{1}{Ak}\left[\ln\left(\varepsilon_0 + \frac{kd}{2}\right) - \ln \varepsilon_0 - \ln \varepsilon_0 + \ln\left(\varepsilon_0 + \frac{kd}{2}\right)\right]$$

$$= \frac{2}{Ak}\left[\ln\left(\varepsilon_0 + \frac{kd}{2}\right) - \ln \varepsilon_0\right]$$

$$= \frac{1}{Ak} 2\ln\left[\frac{\left(\varepsilon_0 + \frac{kd}{2}\right)}{\varepsilon_0}\right]$$

$$\Rightarrow C_{eq} = \frac{kA}{2\ln\left[\frac{\left(\varepsilon_0 + \frac{kd}{2}\right)}{\varepsilon_0}\right]}$$

$$= \frac{kA}{2\ln\left[\frac{2\varepsilon_0 + kd}{2\varepsilon_0}\right]}$$

43. *(*)* Let, equivalent capacitance of two capacitors C_1 and C_2 connected in parallel be C_a and equivalent capacitance of same, when connected in series be C_b.

According to given data,

$$C_a = \frac{15}{4} C_b \quad \ldots\text{(i)}$$

Since, equivalent capacitance in parallel combination,

$$C_{eq} = C_1 + C_2$$

$$\therefore \quad C_a = C_1 + C_2 \quad \ldots\text{(ii)}$$

And equivalent capacitance in series combination,

$$\frac{1}{C'_{eq}} = \frac{1}{C_1} + \frac{1}{C_2}$$

$$\therefore \quad \frac{1}{C_b} = \frac{1}{C_1} + \frac{1}{C_2} = \frac{C_2 + C_1}{C_1 C_2}$$

$$C_b = \frac{C_1 C_2}{C_1 + C_2} \quad \ldots\text{(iii)}$$

On substituting Eqs. (ii) and (iii) in Eq. (i), we get

$$C_1 + C_2 = \frac{15}{4}\left(\frac{C_1 C_2}{C_1 + C_2}\right)$$

$$\Rightarrow 4(C_1 + C_2)^2 = 15\, C_1 C_2$$

$$\Rightarrow 4C_1^2 + 4C_2^2 + 8\, C_1 C_2 = 15\, C_1 C_2$$

$$\Rightarrow 4C_1^2 + 4C_2^2 - 7C_1 C_2 = 0$$

On dividing both sides by C_1^2, we get

$$4 + 4\left(\frac{C_2}{C_1}\right)^2 - 7\frac{C_2}{C_1} = 0$$

$$\text{or} \quad 4\left(\frac{C_2}{C_1}\right)^2 - 7\left(\frac{C_2}{C_1}\right) + 4 = 0$$

If $\dfrac{C_2}{C_1} = x$, then $4x^2 - 7x + 4 = 0$

By using the concept of quadratic equation,

$$x = \frac{-b \pm \sqrt{b^2 - 4ac}}{2a} \Rightarrow x = \frac{7 \pm \sqrt{49 - 64}}{8}$$

$$\Rightarrow \quad x = \frac{C_2}{C_1} = \frac{7 \pm \sqrt{-15}}{8} = \frac{7 \pm \sqrt{15}\, i}{8}$$

44. *(b)* Let x be the required length of the slab.

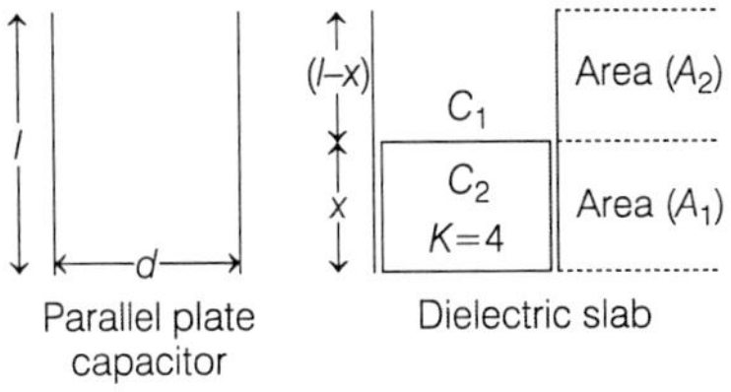

The capacitance of a parallel plate capacitor is given by

$$C = \frac{\varepsilon_0 A}{d}$$

Capacitance of parallel plate capacitor before inserting the dielectric slab,

$$C_{\text{initial}} = \frac{\varepsilon_0 A}{d} \Rightarrow C_{\text{initial}} = \frac{\varepsilon_0 lw}{d}$$

Now, capacitance of parallel plate capacitor after inserting the dielectric slab,

$$C_{\text{final}} = C_1 + C_2$$
$$= \frac{K\varepsilon_0 A_1}{d} + \frac{\varepsilon_0 A_2}{d}$$
$$= \frac{K\varepsilon_0 wx}{d} + \frac{\varepsilon_0 w(l-x)}{d}$$
$$= \frac{\varepsilon_0 w}{d}[Kx + (l-x)]$$
$$= \frac{\varepsilon_0 w}{d}[l + (K-1)x]$$

According to question,
Energy stored in capacitor = 2 (initial energy stored)

$$\Rightarrow \frac{1}{2}C_{\text{final}}V^2 = 2\left(\frac{1}{2}C_{\text{initial}}V^2\right)$$
$$\Rightarrow C_{\text{final}} = 2\,C_{\text{initial}}$$
$$\Rightarrow \frac{\varepsilon_0 w}{d}[l + (4-1)x] = \frac{2\varepsilon_0 wl}{d}$$
$$\Rightarrow \frac{\varepsilon_0 w}{d}[l + 3x] = \frac{2\varepsilon_0 wl}{d}$$
$$\Rightarrow l + 3x = 2l \Rightarrow 3x = l$$
$$\Rightarrow x = \frac{l}{3}$$

45. *(d)*

> **key idea** For a capacitor, $Q = CV$
> $\Rightarrow C = Q/V$
> Hence, slope of the Q-V curve should represent the capacitance.

In the given figure,
Slope of OA > Slope of OB
Since, we know that
net capacitance of parallel combination > net capacitance of series combination
∴ Parallel combination's capacitance,

$$C_P = C_1 + C_2 = \frac{500\,\mu\text{C}}{10\text{ V}} = 50\,\mu\text{F} \quad \ldots\text{(i)}$$

Series combination's capacitance,

$$C_S = \frac{C_1 C_2}{C_1 + C_2} = \frac{80\,\mu\text{C}}{10\text{ V}} = 8\,\mu\text{F} \quad \ldots\text{(ii)}$$

or $C_1 C_2 = 8 \times (C_1 + C_2)$
$= 8 \times 50\,\mu\text{F}$
$= 400\,\mu\text{F}$

[Using Eq. (i)]...(iii)

From Eqs. (i) and (iii), we get

$C_1 = 50 - C_2$ and $C_1 C_2 = 400$

$$\Rightarrow C_2(50 - C_2) = 400$$
$$\Rightarrow 50\,C_2 - C_2^2 = 400$$
$$\text{or } C_2^2 - 50\,C_2 + 400 = 0$$
$$\Rightarrow C_2 = \frac{+50 \pm \sqrt{2500 - 1600}}{2}$$
$$= \frac{+50 \pm 30}{2}$$
$$\Rightarrow C_2 = +40\,\mu\text{F or } +10\,\mu\text{F}$$

Also, $C_1 = 50 - C_2$

$$\Rightarrow C_1 = +10\,\mu\text{F or } +40\,\mu\text{F}$$

Hence, capacitance of two given capacitors is 10 μF and 40 μF.

46. *(d)* Potential energy stored in a capacitor is

$$U = \frac{1}{2}QV = \frac{1}{2}\frac{Q^2}{C}$$

So, initial energy of the capacitor,

$$U_i = \frac{1}{2}Q^2/C_1$$

Final energy of the capacitor,

$$U_f = \frac{1}{2}Q^2/C_2$$

As we know, work done,

$$W = \Delta U = U_f - U_i$$
$$= \frac{1}{2}Q^2\left[\frac{1}{C_2} - \frac{1}{C_1}\right]$$

Here, $Q = 5\,\mu\text{C} = 5 \times 10^{-6}$ C,
$C_1 = 5\,\mu\text{F} = 5 \times 10^{-6}$F,
$C_2 = 2\,\mu\text{F} = 2 \times 10^{-6}$ F

$$\Rightarrow \Delta U = \frac{1}{2} \times (5 \times 10^{-6})^2 \left[\frac{1}{2\times 10^{-6}} - \frac{1}{5\times 10^{-6}}\right]$$
$$\Rightarrow \Delta U = 3.75 \times 10^{-6}\text{ J}$$

47. *(*)* This capacitor system can be converted into two parts as shown in the figure.

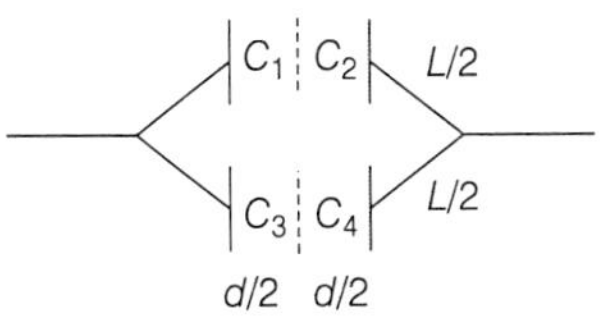

where, C_1, C_2, C_3 and C_4 are capacitance of the capacitor having dielectric constants K_1, K_2, K_3 and K_4, respectively.

Here, $$C_1 = \frac{K_1\varepsilon_0 A/2}{\frac{d}{2}} = \frac{K_1\varepsilon_0 A}{d}$$

Similarly, $$C_2 = \frac{K_2\varepsilon_0 A}{d},\ C_3 = \frac{K_3\varepsilon_0 A}{d}$$

and $$C_4 = \frac{K_4\varepsilon_0 A}{d}$$

Since, equivalent capacitance in series combination is

$$C_{\text{eq}} = \frac{C_1 \cdot C_2}{C_1 + C_2}$$

Here, C_1, C_2 and C_3, C_4 are in series combination.

$$\therefore (C_{\text{eq}})_{12} = \frac{C_1 \cdot C_2}{C_1 + C_2}$$
$$= \frac{\frac{K_1\varepsilon_0 A}{d}\cdot\frac{K_2\varepsilon_0 A}{d}}{\frac{K_1\varepsilon_0 A}{d} + \frac{K_2\varepsilon_0 A}{d}} = \frac{K_1 \cdot K_2}{K_1 + K_2}\cdot\frac{\varepsilon_0 A}{d}$$

Similarly, $$(C_{\text{eq}})_{34} = \frac{K_3 \cdot K_4}{K_3 + K_4}\cdot\frac{\varepsilon_0 A}{d}$$

Now, $(C_{\text{eq}})_{12}$ and $(C_{\text{eq}})_{34}$ are in parallel combination.

$$\therefore C_{\text{net}} = (C_{\text{eq}})_{12} + (C_{\text{eq}})_{34}$$
$$= \frac{K_1 \cdot K_2}{K_1 + K_2}\cdot\frac{\varepsilon_0 A}{d} + \frac{K_3 \cdot K_4}{K_3 + K_4}\cdot\frac{\varepsilon_0 A}{d}$$
$$\Rightarrow C_{\text{net}} = \left(\frac{K_1 \cdot K_2}{K_1 + K_2} + \frac{K_3 \cdot K_4}{K_3 + K_4}\right)\frac{\varepsilon_0 A}{d} \quad \ldots\text{(i)}$$

If K is effective dielectric constant, then

$$C_{\text{net}} = \frac{K\,\varepsilon_0 A}{d} \quad \ldots\text{(ii)}$$

From Eqs. (i) and (ii),

$$\frac{K\,\varepsilon_0 A}{d} = \left(\frac{K_1 \cdot K_2}{K_1 + K_2} + \frac{K_3 \cdot K_4}{K_3 + K_4}\right)\frac{\varepsilon_0 A}{d}$$

or $$K = \left(\frac{K_1 \cdot K_2}{K_1 + K_2} + \frac{K_3 \cdot K_4}{K_3 + K_4}\right)$$

48. *(64)*

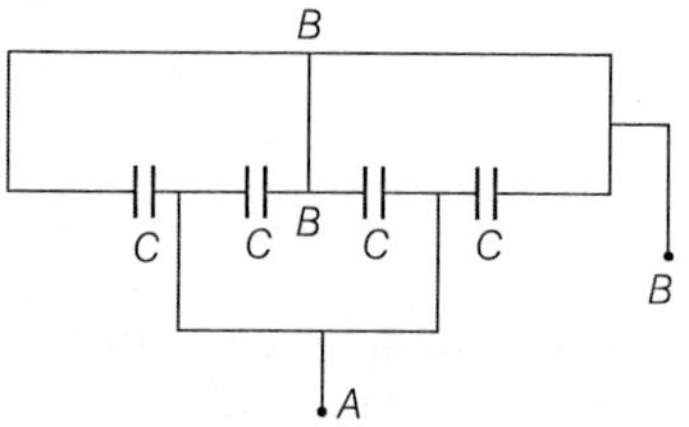

Redraw

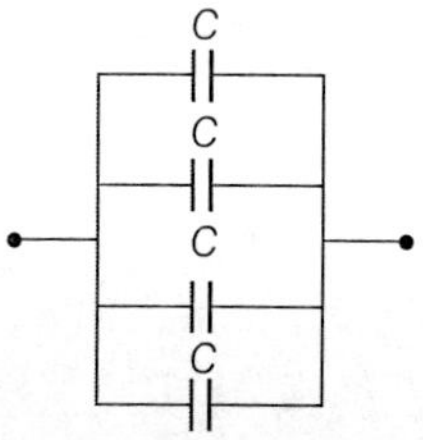

$C_{eq} = 4C = 4 \times 16\,\mu F = 64\,\mu F$

49. *(80)* The given situation is shown below.

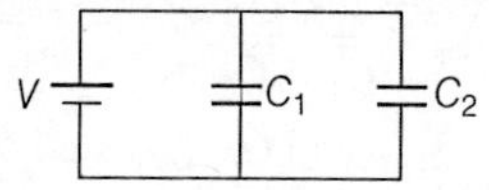

Given, $C = \dfrac{\varepsilon_0 A}{d} = 10\mu F$

$$C_{eq} = C_1 + C_2$$

$$= \frac{\varepsilon_0 (A/2) K_1 + \varepsilon_0 (A/2) K_2}{d}$$

$$= \frac{\varepsilon_0 A}{d}\left(\frac{K_1 + K_2}{2}\right)$$

$$= 10\left(\frac{2+3}{2}\right) = 25\,\mu F$$

Force between plates of capacitor,

$$F = \frac{CV^2}{2d} \quad \left(\because F = qE,\ E = \frac{V}{d},\ q = CV\right)$$

$$\Rightarrow \quad 8 = \frac{CV^2}{2 \times 10 \times 10^{-3}}$$

$$\Rightarrow \quad \frac{8 \times 2 \times 10^{-2}}{25 \times 10^{-6}} = V^2$$

$$\Rightarrow \quad V^2 = \frac{16}{25} \times 100 \times 100$$

$$V = 80\text{ V}$$

50. *(86)* In parallel combination, potential difference is same across all.

$$\text{Energy} = \frac{1}{2}(C_1 + C_2 + C_3)V^2$$

$$= \frac{1}{2}(25 + 30 + 45) \times (100)^2 \times 10^{-6}$$

$$E = 0.5\text{ J} \quad \ldots\text{(i)}$$

In series combination, charge is same for all.

$$\frac{1}{C_{eq}} = \frac{1}{C_1} + \frac{1}{C_2} + \frac{1}{C_3}$$

$$= \frac{1}{25} + \frac{1}{30} + \frac{1}{45}$$

$$C_{eq} = \frac{450}{43}\,\mu F \quad \ldots\text{(ii)}$$

$$\text{Energy} = \frac{Q^2}{2C_1} + \frac{Q^2}{2C_2} + \frac{Q^2}{2C_3}$$

$$= \frac{Q^2}{2}\left[\frac{1}{C_1} + \frac{1}{C_2} + \frac{1}{C_3}\right]$$

$$= \frac{V^2 C_{eq}}{2} \quad \ldots\text{(iii)}$$

$$= (100)^2 \times \frac{450}{43} \times 10^{-6} = \frac{4.5}{86}$$

Compare with given value,

$$\frac{4.5}{86} = \frac{9}{x}E = \frac{9}{x}(0.5)$$

$$\Rightarrow \quad x = 86$$

51. *(2)* Given,

For capacitor 1 capacitance $= C$

Potential $= V$, Energy $= E$

For capacitor 2

Capacitance $= 2C$

Potential $= 2V$

Energy of capacitor is given by

$$E = \frac{1}{2}CV^2 \quad \ldots\text{(i)}$$

Now, these two capacitor are connected

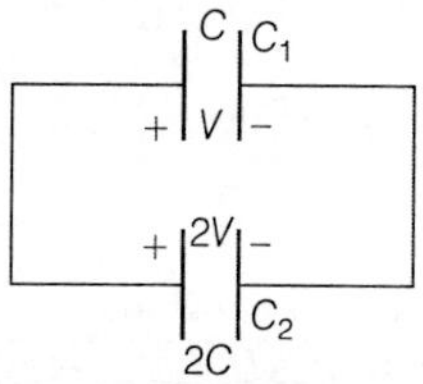

Energy loss when capacitor are connected is given by

$$U_{Loss} = \frac{1}{2}\frac{C_1 C_2}{C_1 + C_2}(V_1 - V_2)^2$$

$$= \frac{1}{2} \times \frac{2C^2}{3C}(2V - V)^2$$

$$= \frac{1}{3}CV^2 \quad \ldots\text{(ii)}$$

From Eq. (i), $U_{Loss} = \dfrac{2}{3}E$

Compare with term given in question

$$x = 2$$

52. *(5)* Total capacitance of the combination

$$C = C_1 + C_2 + C_3$$

$$= 2 + x + 3 = (5 + x)\,\mu F$$

Total charge, $Q = CV$

$$\Rightarrow \quad 100 = (5 + x) \times 10$$

$$\Rightarrow \quad x = 5\,\mu F$$

53. *(5)*

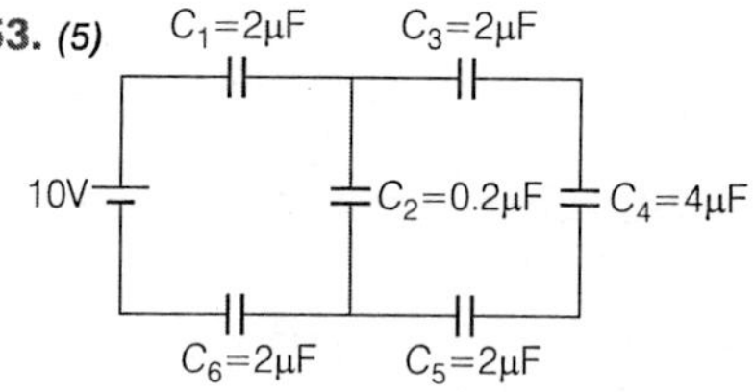

$$\frac{1}{C_{eq}} = \frac{1}{2} + \frac{1}{4} + \frac{1}{2}$$

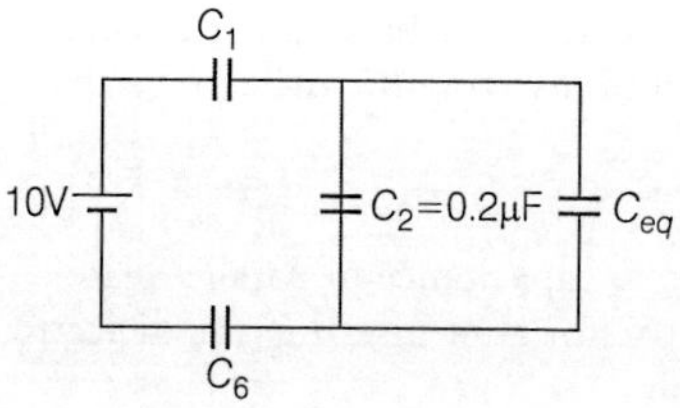

$$\frac{1}{C_{eq}} = \frac{5}{4} = \frac{4}{5} = 0.8$$

$$C'_{eq} = C_{eq} + C_2 = 0.8 + 0.2 = 1\mu F$$

$$\frac{1}{C_{eq}} = \frac{1}{2} + \frac{1}{2} + 1 \Rightarrow \frac{1}{C''_{eq}} = 2$$

$$C''_{eq} = \frac{1}{2} = 0.5\mu F$$

$$Q_4 = C''_{eq} \cdot V$$

$$= 0.5 \times 10 = 5\mu C$$

54. *(6)*

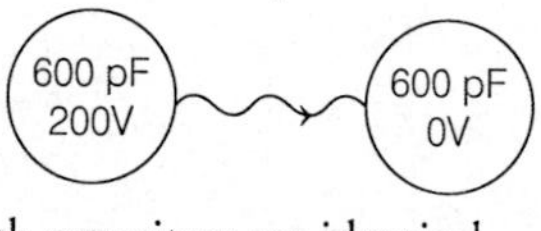

As both capacitors are identical, common potential after equilibrium is obtained is;

$$V' = \frac{Q_1 + Q_2}{C_1 + C_2} = \frac{C_1 V_1 + C_2 V_2}{C_1 + C_2}$$

$$= \frac{600 \times 200 + 600 \times 0}{600 + 600} = 100\text{ V}$$

Now, initial energy

$$= \frac{1}{2}C_1 V^2 = \frac{1}{2} \times 600 \times 10^{-12} \times 200 \times 200$$

$$= 12 \times 10^{-6}\text{ J}$$

$$\text{Final energy} = \frac{1}{2}(C_1 + C_2)V'^2$$

$$= \frac{1}{2} \times 1200 \times 10^{-12} \times (100)^2 = 6 \times 10^{-6}\text{ J}$$

$\therefore$ Energy cost

= Initial energy − Final energy

$= 6 \times 10^{-6}\text{ J} = 6\,\mu J$

55. *(5)* Distance between capacitor plates

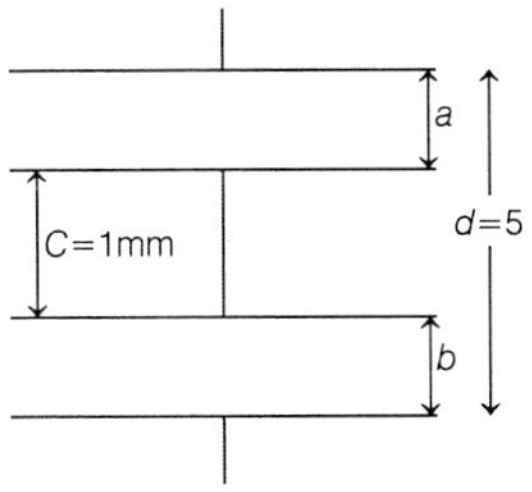

$$b = 5 - a - 1 = (4 - a)\text{ mm}$$

We know, $C = \varepsilon_0 A / d$

and in series arrangement

$$\frac{1}{C_{eq}} = \frac{1}{C_1} + \frac{1}{C_2}$$

$$\frac{1}{C_{eq}} = \frac{a}{\varepsilon_0 A} + \frac{4-a}{\varepsilon_0 A} = \frac{4(\text{mm})}{\varepsilon_0 A}$$

$$C_{eq} = \frac{\varepsilon_0 A}{4(\text{mm})}$$

given $A = 200\text{ cm}^2$

$$C_{eq} = \frac{\varepsilon_0 \times 200 \times 10^{-4}}{4 \times 10^{-3}}$$

$$C_{eq} = 50 \times 10^{-1} \varepsilon_0 = 5\varepsilon_0 \text{ F} \Rightarrow x = 5$$

56. *(3)*

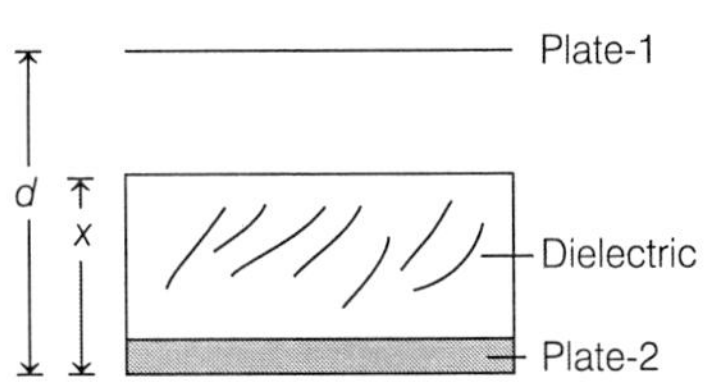

$$C_1 = \frac{\frac{\varepsilon_0 A}{\frac{2d}{3}} \times \frac{4\varepsilon_0 A}{\frac{d}{3}}}{\frac{\varepsilon_0 A}{\frac{2d}{3}} + \frac{4\varepsilon_0 A}{\frac{d}{3}}} = \frac{18}{\frac{3}{2} + 12} \cdot \frac{\varepsilon_0 A}{d}$$

$$= 18 \times \frac{2}{27} \frac{\varepsilon_0 A}{d} = \frac{4}{3} \frac{\varepsilon_0 A}{d}$$

According to question,

$$\frac{4}{3} \frac{\varepsilon_0 A}{d} = 2 \Rightarrow \frac{\varepsilon_0 A}{d} = \frac{3}{2} \qquad \text{...(i)}$$

$$\text{Now, } C_2 = \frac{\frac{\varepsilon_0 A}{\frac{d}{3}} \times \frac{4\varepsilon_0 A}{\frac{2d}{3}}}{\frac{\varepsilon_0 A}{\frac{d}{3}} + \frac{4\varepsilon_0 A}{\frac{2d}{3}}}$$

$$= \frac{18}{3+6} \frac{\varepsilon_0 A}{d} = 2 \times \frac{\varepsilon_0 A}{d} = 2 \times \frac{3}{2} = 3$$

57. *(55)* Given, situations in question are as follows

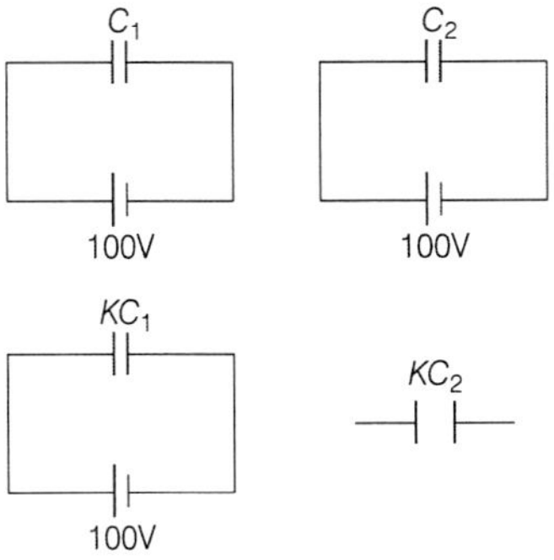

$$Q_1 = KC_1 \times 100$$
$$= 10 \times 10 \times 10^{-6} \times 100$$
$$Q_1 = 10^4 \ \mu\text{C}$$
$$Q_2 = 10^3 \times 10^{-6} = 10^3 \ \mu\text{C}$$

(As isolated capacitor)

Now, both capacitors are joined together as

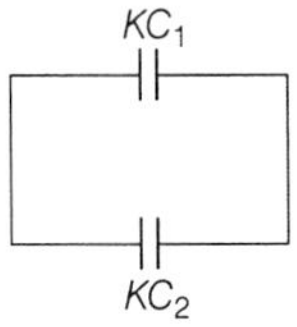

$$V_{\text{common}} = \frac{Q_1 + Q_2}{C_1' + C_2'}$$

$$= \frac{(10^4 + 10^3)\mu\text{C}}{(10 \times 10 + 10 \times 10)\mu\text{F}}$$

$$= \frac{10^3 \times 11}{200} = \frac{1100}{20} \text{ V}$$

$$V_{\text{common}} = 55 \text{ V}$$

58. *(2.25)*

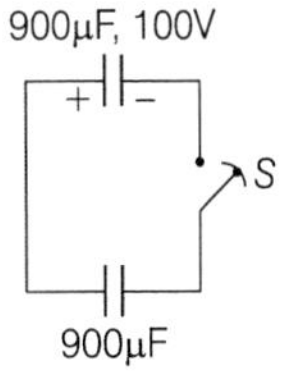

Common potential after redistribution of charges is

$$V' = \frac{C_1 V_1 + C_2 V_2}{C_1 + C_2}$$

$$= \frac{900 \times 10^{-6} \times 100 + 0}{1800 \times 10^{-6}}$$

$$= 50 \text{ V}$$

Initial energy stored

$$= \frac{1}{2} CV^2 = \frac{1}{2} \times 900 \times 10^{-6} \times (100)^2$$

$$= 4.5 \text{ J} = 450 \times 10^{-2} \text{ J}$$

Final energy stored

$$= \frac{1}{2}(C_1 + C_2)V'^2$$

$$= \frac{1}{2} \times 1800 \times 10^{-6} \times (50)^2$$

$$= 225 \times 10^{-2} \text{ J}$$

So, Loss of energy

$$= (450 - 225) \times 10^{-2} = 225 \times 10^{-2} \text{ J}$$

> ***Alternate Solution***
>
> Loss of energy is given as
>
> $$\Delta U = \frac{1}{2} \cdot \frac{C_1 C_2}{C_1 + C_2} (V_1 - V_2)^2$$
>
> $$= \frac{1}{2} \times \frac{900 \times 10^{-6} \times 900 \times 10^{-6} (100 - 0)^2}{(900 + + 900) \times 10^{-6}}$$
>
> $$= 2.25 \text{ J} = 225 \times 10^{-2} \text{ J}$$
>
> $\therefore \ x = 2.25$

59. *(240)* According to the question,

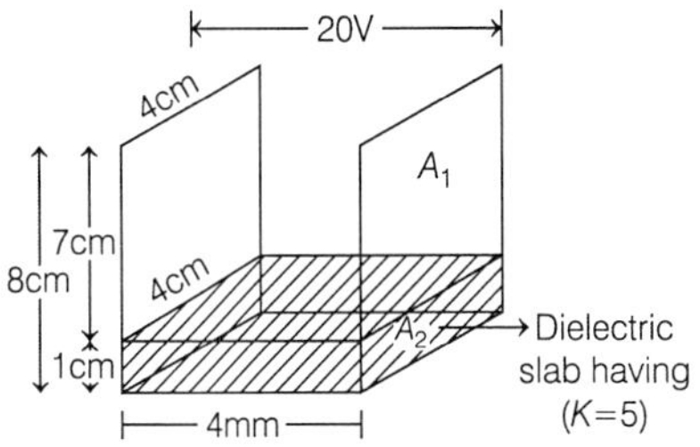

The above situation is a parallel combination of two capacitors,

$$\therefore \ C_{\text{equivalent}} = C_1 + C_2$$

$$C_{\text{equivalent}} = \frac{\varepsilon_0 A_1}{d} + \frac{K\varepsilon_0 A_2}{d} \qquad \text{...(i)}$$

(Since, on inserting dielectric the capacitance increases K times)

But, $A_1 = (7 \times 4)\text{ cm}^2 = 28 \times 10^{-4}\text{ m}^2$

and $A_2 = (1 \times 4)\text{ cm}^2 = 4 \times 10^{-4}\text{ m}^2$

∴On putting these values in Eq. (i), we get

$$C_{\text{equivalent}} = \frac{\varepsilon_0}{d}[A_1 + A_2 K]$$

$$= \frac{\varepsilon_0}{4 \times 10^{-3}}[(28 \times 10^{-4}) + (4 \times 10^{-4} \times 5)]$$

$$C_{\text{equivalent}} = \frac{\varepsilon_0}{4 \times 10^{-3}}[48 \times 10^{-4}]$$

$$= 12 \times 10^{-1} \varepsilon_0$$

∴The electrostatic energy of this system,

$$= \frac{1}{2} C_{\text{equivalent}} V^2$$

$$= \frac{1}{2} \times 12 \times 10^{-1} \times 20^2 \varepsilon_0$$

$$= \frac{1}{2} \times 12 \times 10^{-1} \times 400 \varepsilon_0$$

$$= 24 \times 10 \varepsilon_0 = 240 \varepsilon_0$$

60. *(23)* Given,

Plate Area of each stair $= A$

Capacitance of arrangement,

$$C = \frac{x}{15}\frac{\varepsilon_0 A}{b}$$

This circuit is equivalent to 3 capacitors in parallel.

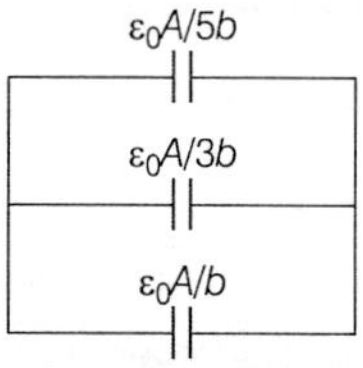

Capacitances are added to find equivalent capacity.

$$C_{eq} = \frac{\varepsilon_0 A}{b}\left(1 + \frac{1}{3} + \frac{1}{5}\right) = \frac{23}{15}\frac{\varepsilon_0 A}{b}$$

Comparing with given value, we get

$\Rightarrow \quad x = 23$

61. *(864)* Given, capacitance of capacitor, $C = 14$ pF

Potential difference, $V = 12$V

Energy of capacitor, $E_i = \frac{1}{2}CV^2$

$$= \frac{1}{2} \times 14 \times 12 \times 12 = 1008 \text{ pJ}$$

Now, the charging battery is disconnected and a porcelain plate with $K = 7$ is inserted between the plates

$$\therefore \quad E_f = \frac{E_i}{7} \Rightarrow E_f = \frac{1008}{7} \text{ pJ}$$

$$\Rightarrow \quad E_f = 144 \text{ pJ}$$

Mechanical energy with which the plate would oscillate back and forth between the plates will be

$$= (1008 - 144) \text{ pJ} = 864 \text{ pJ}$$

62. *(2)* The given figure can be shown with capacitors as

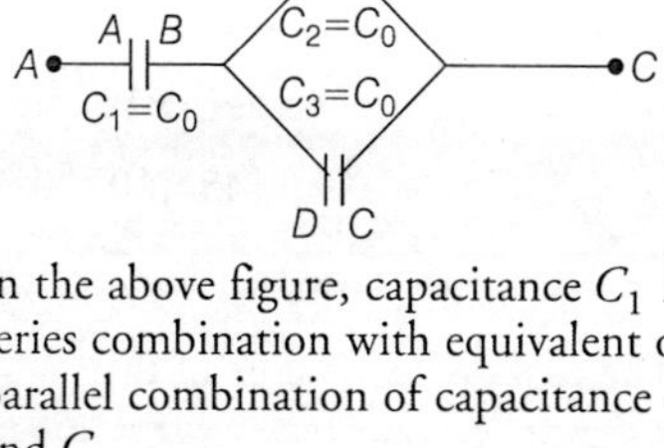

In the above figure, capacitance C_1 is in series combination with equivalent of parallel combination of capacitance C_2 and C_3.

$$C'_{eq} = C_2 + C_3$$

$$\Rightarrow \quad C'_{eq} = C_0 + C_0 = 2C_0$$

$$\Rightarrow \quad \frac{1}{C_{eq}} = \frac{1}{C_1} + \frac{1}{C'_{eq}}$$

$$= \frac{1}{C_0} + \frac{1}{2C_0} = \frac{3}{2C_0}$$

$$\Rightarrow \quad C_{eq} = \frac{2C_0}{3} \quad \ldots\text{(i)}$$

As, $C_0 = \frac{\varepsilon_0 A}{d} \quad \ldots$(ii)

From Eqs. (i) and (ii), we get

$$C_{eq} = \frac{2}{3}\left(\frac{\varepsilon_0 A}{d}\right)$$

$$\Rightarrow \quad C_{eq} = \frac{2\varepsilon_0}{3d}(l \times b) = \frac{2\varepsilon_0}{3d}\left(2 \times \frac{3}{2}\right)$$

$$\Rightarrow \quad C_{eq} = \frac{2\varepsilon_0}{d} \quad \ldots\text{(iii)}$$

According to question, the equivalent capacitance between A and C is $x\varepsilon_0 / d$. So, comparing it with Eq. (iii), we get $x = 2$

63. *(4)* Initial energy of charged capacitor,

$$U_1 = \frac{1}{2}C_1V_1^2$$

$$= \frac{1}{2} \times 5 \times 10^{-6} \times (220)^2$$

$$= 121 \times 10^{-3} \text{ J}$$

Common potential after redistribution of charges is

$$V = \frac{C_1V_1 + C_2V_2}{C_1 + C_2} \simeq \frac{C_1V_1}{C_1 + C_2}$$

$$\simeq \frac{5 \times 10^{-6} \times 220}{5 \times 10^{-6} + 2.5 \times 10^{-6}} = 220 \times \frac{2}{3} \text{ V}$$

Final stored energy is

$$U_2 = \frac{1}{2}(C_1 + C_2)V^2$$

$$= \frac{1}{2} \times 7.5 \times 10^{-6} \times \left(220 \times \frac{2}{3}\right)^2$$

$$\approx 80 \times 10^{-3} \text{ J}$$

Loss of energy

$$\Delta U = (121 - 80) \times 10^{-3} \text{ J}$$

$$= 41 \times 10^{-3} \text{ J} = \frac{41}{100} \text{ J}$$

$$\approx \frac{4}{100} \text{ J}$$

Given, $\Delta U = \frac{X}{100}$ J

$\therefore \quad X = 4$

64. *(6)* Initially a 60 pF capacitor is fully charged by a 20 V supply as shown in the figure.

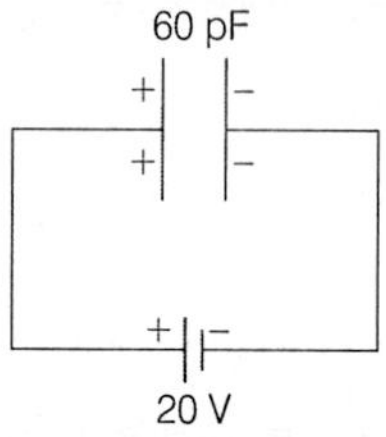

Its energy is,

$$U_1 = \frac{1}{2}CV^2$$

$$\Rightarrow \quad U_1 = \frac{1}{2} \times 60 \times 10^{-12} \times (20)^2$$

$$= 12 \times 10^{-9} \text{ J}$$

Then, it is connected to another uncharged 60 pF capacitor as shown in the figure.

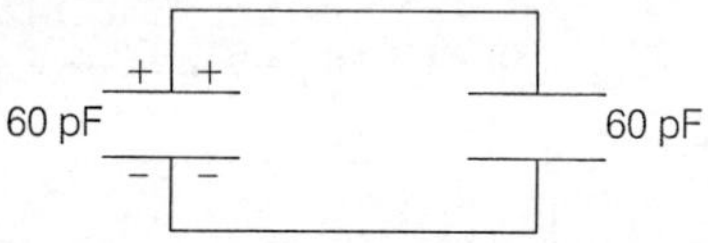

Transfer of charge from charged to uncharged capacitor occurs till both reaches a common potential.

If $\quad V =$ common potential, then

$$V = \frac{\text{total charge}}{\text{total capacitance}}$$

$$\Rightarrow \quad V = \frac{Q_1}{C_1 + C_2} = \frac{C_1V_1}{C_1 + C_2}$$

$$= \frac{60 \times 10^{-12} \times 20}{60 \times 10^{-12} + 60 \times 10^{-12}}$$

$$= 10 \text{ V}$$

Final energy of system,

$$U_2 = \frac{1}{2}(C_1 + C_2) \cdot V^2$$

$$= \frac{1}{2} \times 2 \times 60 \times 10^{-12} \times (10^2)$$

$$= 60 \times 10^{-10} = 6 \times 10^{-9} \text{ J}$$

Loss of energy, $\Delta U = U_1 - U_2$

$$= 12 \times 10^{-9} - 6 \times 10^{-9}$$

$$= 6 \times 10^{-9} \text{ J} = 6 \text{ nJ}$$

C-17

JEE Main Chapterwise Topicwise Physics

In JEE Main there is usually 1-2 question per paper are visible from this chapter. In past 3 years, 59 conceptual and 43 numerical questions were asked of moderate to difficult level from this chapter.

For detailed analysis scan QR code

CURRENT ELECTRICITY

Topic 1 Ohm's Law and Resistance

Level 1

01 A wire of resistance R is bent into a triangular pyramid as shown in figure with each segment having same length. The resistance between points A and B is $\frac{R}{n}$. The value of n is *[2025, 7 April Shift-I]*

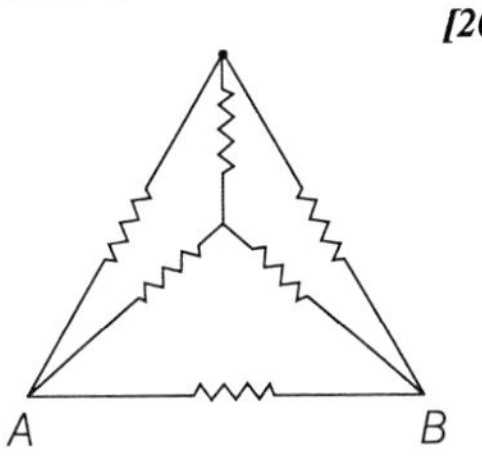

(a) 10 (b) 12
(c) 14 (d) 16

02 Current passing through a wire as function of time is given as $I(t) = 0.02t + 0.01$ A. The charge will flow through the wire from $t = 1$ s to $t = 2$ s is *[2025, 4 April Shift-I]*

(a) 0.04 C (b) 0.02 C
(c) 0.07 C (d) 0.06 C

Imp. Combination of Resistor

03 Find the equivalent resistance between two ends of the following circuit *[2025, 28 Jan Shift-I]*

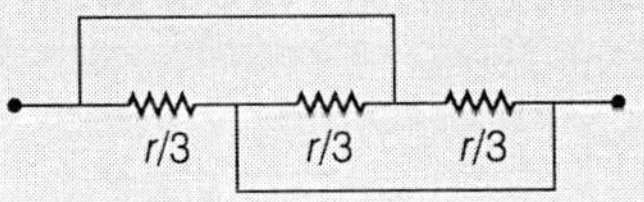

(a) $\frac{r}{9}$ (b) $\frac{r}{6}$ (c) r (d) $\frac{r}{3}$

04 Which of the following resistivity (ρ) *versus* temperature (T) curves is most suitable to be used in wire bound standard resistors? *[2025, 22 Jan Shift-I]*

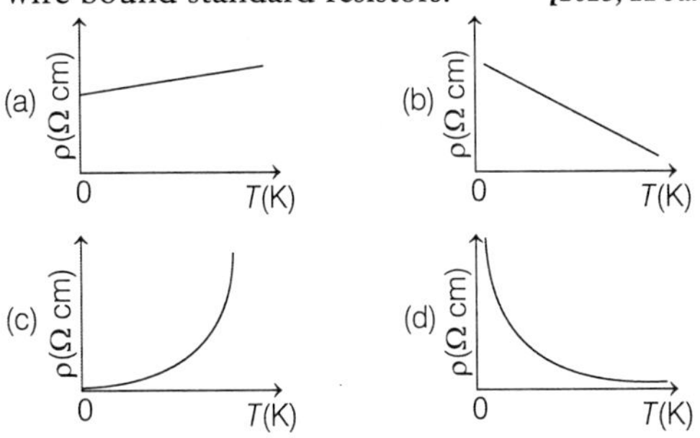

05 A wire of resistance R and length L is cut into 5 equal parts. If these parts are joined parallely, then resultant resistance will be *[2024, 27 Jan Shift-I]*

(a) $25R$ (b) $\frac{1}{25}R$ (c) $5R$ (d) $\frac{1}{5}R$

06 In a metallic conductor, under the effect of applied electric field, the free electrons of the conductor *[2023, 10 April Shift-II]*

(a) drift from higher potential to lower potential.
(b) move with the uniform velocity throughout from lower potential to higher potential
(c) move in the straight line paths in the same direction
(d) move in the curved paths from lower potential to higher potential

07 The drift velocity of electrons for a conductor connected in an electrical circuit is v_d. The conductor is now replaced by another conductor with same material and same length but double the area of cross-section. The applied voltage remains same. The new drift velocity of electrons will be *[2023, 31 Jan Shift-I]*

(a) v_d (b) $v_d / 2$
(c) $v_d / 4$ (d) $2v_d$

08 The resistance of a wire is 5 Ω. It's new resistance in ohm if stretched to 5 times of it's original length will be *[2023, 25 Jan Shift-II]*

(a) 5 (b) 25 (c) 125 (d) 625

09 Given below are two statements: one is labelled as Assertion (A) and the other is labelled as Reason (R).

Assertion (A) Alloys such as constantan and manganin are used in making standard resistance coils.

Reason (R) Constantan and manganin have very small value of temperature coefficient of resistance.

In the light of the above statements, choose the correct answer from the options given below. *[2022, 29 July Shift-II]*

(a) Both A and R are true and R is the correct explanation of A.
(b) Both A and R are true but R is not the correct explanation of A.
(c) A is true but R is false.
(d) A is false but R is true.

10 A wire of resistance R is drawn out so that its length is increased by twice of its original length. The ratio of new resistance to original resistance is *[2022, 28 July Shift-I]*

(a) 9 : 1 (b) 1 : 9
(c) 4 : 1 (d) 3 : 1

11 What will be the most suitable combination of three resistors $A = 2\Omega$, $B = 4\Omega$, $C = 6\Omega$, so that $\left(\frac{22}{3}\right)\Omega$ is equivalent resistance of combination? *[2022, June Shift-II]*

(a) Parallel combination of A and C connected in series with B.
(b) Parallel combination of A and B connected in series with C.
(c) Series combination of A and C connected in parallel with B.
(d) Series combination of B and C connected in series with A.

12 If you are provided a set of resistances $2\Omega, 4\Omega, 6\Omega$ and 8Ω. Connect these resistances, so as to obtain an equivalent resistance of $46/3\ \Omega$. *[2021, 26 Aug Shift-II]*

(a) 4 Ω and 6 Ω are in parallel with 2 Ω and 8 Ω in series.
(b) 6 Ω and 8 Ω are in parallel with 2 Ω and 4 Ω in series.
(c) 2 Ω and 6 Ω are in parallel with 4 Ω and 8 Ω in series.
(d) 2 Ω and 4 Ω are in parallel with 6 Ω and 8 Ω in series.

13 A current of 5 A is passing through a non-linear magnesium wire of cross-section 0.04 m^2. At every point, the direction of current density is at an angle of 60° with the unit vector of area of cross-section. The magnitude of electric field at every point of the conductor is

(Take, resistivity of magnesium, $\rho = 44 \times 10^{-8}\ \Omega$-m) *[2021, 20 July Shift-I]*

(a) 11×10^{-2} V/m (b) 11×10^{-7} V/m
(c) 11×10^{-5} V/m (d) 11×10^{-3} V/m

14 A current of 10 A exists in a wire of cross-sectional area of 5 mm^2 with a drift velocity of 2×10^{-3} ms^{-1}. The number of free electrons in each cubic metre of the wire is *[2021, 17 March Shift-I]*

(a) 2×10^6 (b) 625×10^{25}
(c) 2×10^{25} (d) 1×10^{23}

15 A conducting wire of length l, area of cross-section A and eectric resistivity ρ is connected between the terminals of a battery. A potential difference V is developed between its ends, causing an electric current. If the length of the wire of the same material is doubled and the area of cross-section is halved, the resultant current would be *[2021, 16 March Shift-I]*

(a) $\frac{1}{4}\frac{VA}{\rho l}$ (b) $\frac{3}{4}\frac{VA}{\rho l}$
(c) $\frac{1}{4}\frac{\rho l}{VA}$ (d) $4\frac{VA}{\rho l}$

16 Five equal resistances are connected in a network as shown in figure. The net resistance between the points A and B is *[2021, 26 Feb Shift-I]*

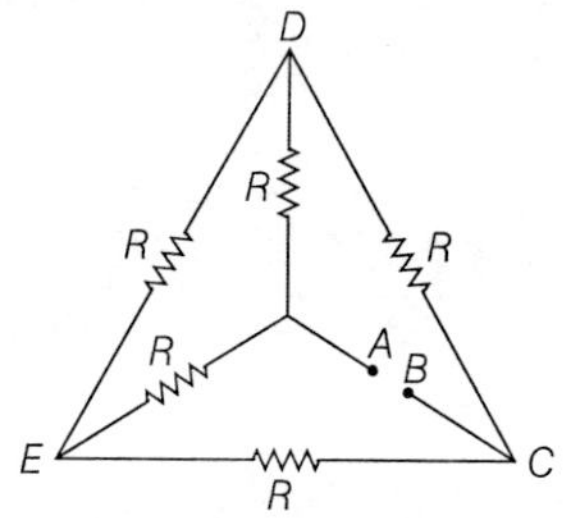

(a) R (b) $2R$
(c) $R/2$ (d) $3R/2$

17 Consider four conducting materials copper, tungsten, mercury and aluminium with resistivity ρ_C, ρ_T, ρ_M and ρ_A, respectively. Then, *[2020, 2 Sep Shift-I]*

(a) $\rho_A > \rho_T > \rho_C$ (b) $\rho_A > \rho_M > \rho_C$
(c) $\rho_C > \rho_A > \rho_T$ (d) $\rho_M > \rho_A > \rho_C$

18 A current of 5 A passes through a copper conductor (resistivity $= 1.7 \times 10^{-8}\ \Omega$-m) of radius of cross-section 5 mm. Find the mobility of the charges, if their drift velocity is 1.1×10^{-3} m/s. *[2019, 10 April Shift-I]*

(a) 1.5 m^2/V-s (b) 1.3 m^2/V-s
(c) 1.0 m^2/V-s (d) 1.8 m^2/V-s

19 Drift speed of electrons, when 1.5 A of current flows in a copper wire of cross-section 5 mm^2 is v. If the electron density in copper is 9×10^{28} /m^3, the value of v (in mm/s) is close to (Take, charge of electron $= 1.6 \times 10^{-19}$ C) *[2019, 9 Jan Shift-I]*

(a) 0.02 (b) 0.2
(c) 2 (d) 3

20 A copper rod of cross-sectional area A carries a uniform current I through it. At temperature T, if the volume charge density of the rod is ρ, how long will the charges take to travel a distance d ? *[JEE Main 2018]*

(a) $\frac{\rho dA}{I}$ (b) $\frac{\rho dA}{IT}$
(c) $\frac{2\rho dA}{I}$ (d) $\frac{2\rho dA}{IT}$

21 A uniform wire of length l and radius r has a resistance of $100\ \Omega$. It is recast into a wire of radius $r/2$. The resistance of new wire will be *[JEE Main 2017]*
(a) $400\ \Omega$ (b) $100\ \Omega$
(c) $200\ \Omega$ (d) $1600\ \Omega$

22 The temperature dependence of resistances of Cu and undoped Si in the temperature range 300-400 K, is best described by *[JEE Main 2016]*
(a) linear increase for Cu, linear increase for Si
(b) linear increase for Cu, exponential increase for Si
(c) linear increase for Cu, exponential decrease for Si
(d) linear decrease for Cu, linear decrease for Si

23 Suppose the drift velocity v_d in a material varied with the applied electric field E as $v_d \propto \sqrt{E}$. Then V-I graph for a wire made of such a material is best given by *[JEE Main 2015]*

(a)
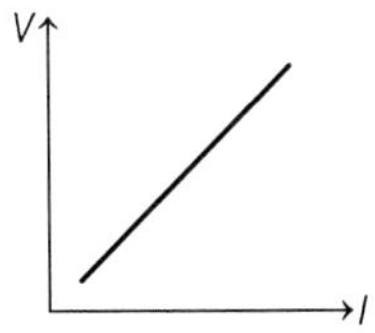

(b)
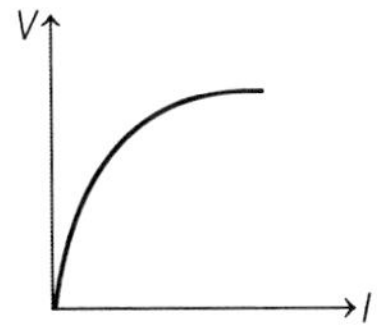

(c)
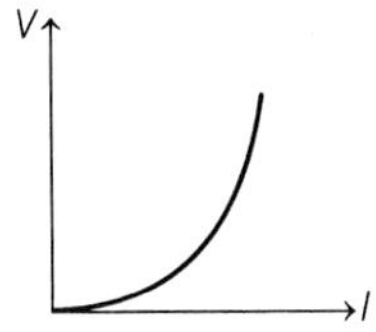

(d)
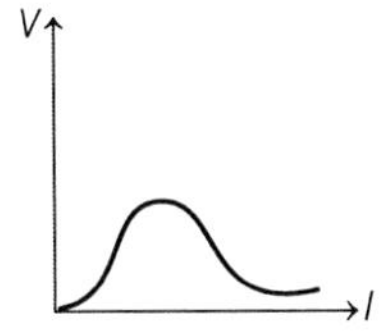

24 If $400\ \Omega$ of resistance is made by adding four $100\ \Omega$ resistance of tolerance 5%, then the tolerance of the combination is *[AIEEE 2011]*
(a) 20% (b) 5%
(c) 10% (d) 15%

25 If a wire is stretched to make it 0.1% longer, its resistance will *[AIEEE 2011]*
(a) increase by 0.2% (b) decrease by 0.2%
(c) decrease by 0.05% (d) increase by 0.05%

26 This question contains Statement I and Statement II. Of the four choices given after the statements, choose the one that best describes the two statements.

Statement I The temperature dependence of resistance is usually given as $R = R_0(1+\alpha\Delta T)$. The resistance of a wire changes from $100\ \Omega$ to $150\ \Omega$ when its temperature is increased from 27°C to 227°C. This implies that $\alpha = 2.5 \times 10^{-3}$ /°C.

Statement II $R = R_i(1+\alpha\Delta T)$ is valid only when the change in the temperature ΔT is small and $\Delta R = (R - R_0) << R_0$. *[AIEEE 2009]*
(a) Statement I is true, Statement II is false.
(b) Statement I is true, Statement II is true; Statement II is the correct explanation of Statement I.
(c) Statement I is true, Statement II is true; Statement II is not the correct explanation of Statement I.
(d) Statement I is false, Statement II is true.

27 The resistance of a wire is $5\ \Omega$ at 50°C and $6\ \Omega$ at 100°C. The resistance of the wire at 0°C will be *[AIEEE 2007]*
(a) $2\ \Omega$ (b) $1\ \Omega$
(c) $4\ \Omega$ (d) $3\ \Omega$

Imp. Temperature Coefficient of Resistance

28 The resistance of a bulb filament is $100\ \Omega$ at a temperature of 100°C. If its temperature coefficient of resistance is 0.005 per °C, its resistance will become $200\ \Omega$ at a temperature of *[AIEEE 2006]*
(a) 300°C (b) 400°C
(c) 500°C (d) 200°C

29 A 3 V battery with negligible internal resistance is connected in a circuit as shown in the figure. *[AIEEE 2003]*

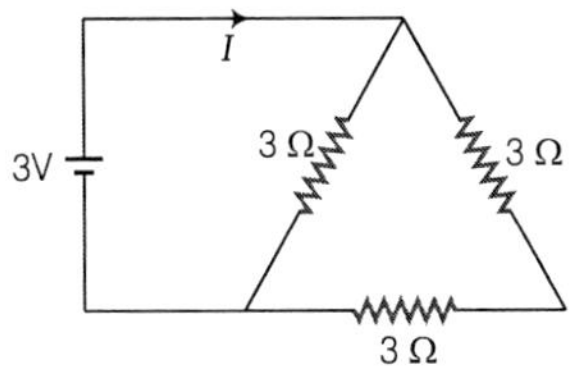

The current I in the circuit will be
(a) 1 A (b) 1.5 A (c) 2 A (d) $\frac{1}{3}$ A

Level 2

30 A wire of length 25 m and cross-sectional area 5 mm^2 having resistivity of 2×10^{-6} Ω-m is bent into a complete circle. The resistance between diametrically opposite points will be *[2025, 3 April Shift-I]*
(a) $12.5\ \Omega$ (b) $50\ \Omega$ (c) $25\ \Omega$ (d) $100\ \Omega$

Imp.

Combination of Resistors

31 From the combination of resistors with resistances values $R_1 = R_2 = R_3 = 5\,\Omega$ and $R_4 = 10\,\Omega$, which of the following combination is the best circuit to get on equivalent resistance of $6\,\Omega$? *[2025, 4 April Shift-II]*

(a)
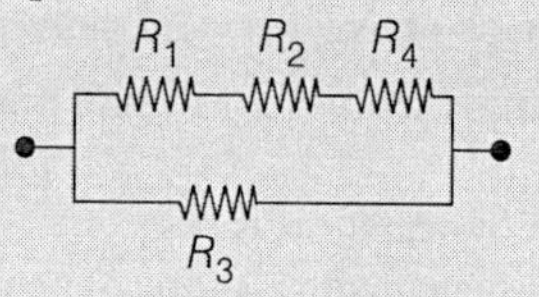

(b)
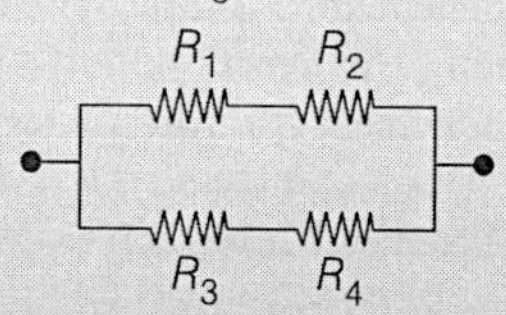

(c)
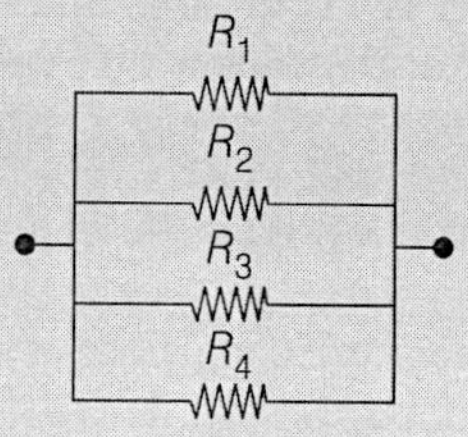

(d)
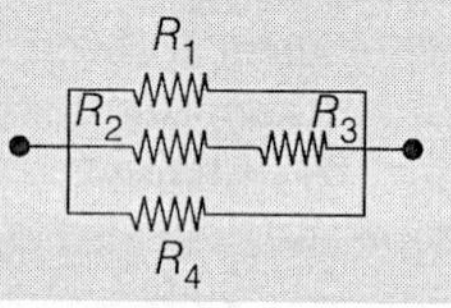

32 Identify the valid statements relevant to the given circuit at the instant when the key is closed. *[2025, 23 Jan Shift-I]*

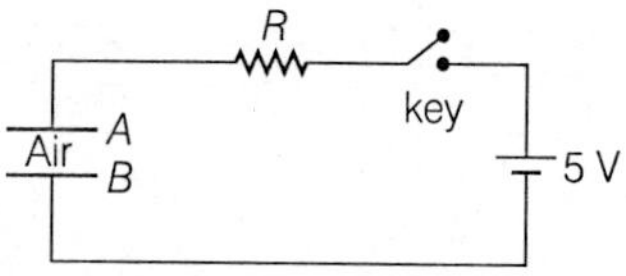

A. There will be no current through resistor R.
B. There will be maximum current in the connecting wires.
C. Potential difference between the capacitor plates A and B is minimum.
D. Charge on the capacitor plates is minimum.

Choose the correct answer from the options given below.

(a) C, D Only
(b) A, B, D Only
(c) A, C Only
(d) B, C, D Only

33 A wire of resistance R is bent into an equilateral triangle and an identical wire is bent into a square. The ratio of resistance between the two end points of an edge of the triangle to that of the square is *[2025, 28 Jan Shift-I]*

(a) $\frac{27}{32}$
(b) $\frac{9}{8}$
(c) $\frac{8}{9}$
(d) $\frac{32}{27}$

34 The equivalent resistance between A and B is *[2024, 9 April Shift-I]*

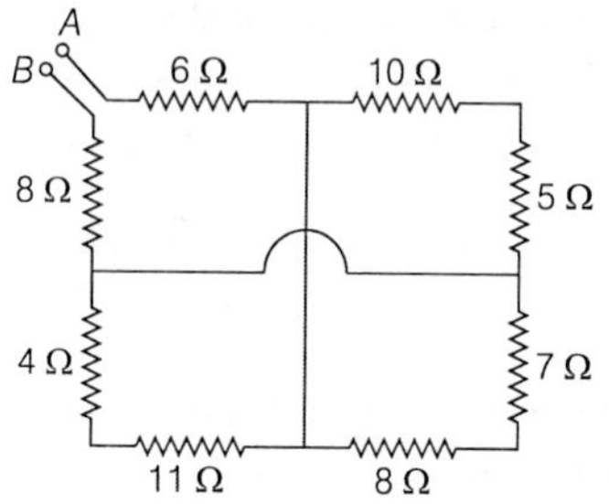

(a) $27\,\Omega$
(b) $18\,\Omega$
(c) $19\,\Omega$
(d) $25\,\Omega$

35 The electric current through a wire varies with time as $I = I_0 + \beta t$, where $I_0 = 20$ A and $\beta = 3$ A/s. The amount of electric charge crossed through a section of the wire in 20 s is *[2024, 29 Jan Shift-I]*

(a) 1600 C
(b) 800 C
(c) 80 C
(d) 1000 C

36 Given below are two statements.

Statement I The equivalent resistance of resistors in a series combination is smaller than least resistance used in the combination.

Statement II The resistivity of the material is independent of temperature. *[2023, 15 April Shift-I]*

In the light of above statements, choose the most appropriate answer from the options given below.

(a) Statement I is false but Statement II is true.
(b) Statement I is true but Statement II is false.
(c) Both Statement I and Statement II are true.
(d) Both Statement I and Statement II are false.

37 A wire of resistance $160\,\Omega$ is melted and drawn into a wire one-fourth of its length. The new resistance of the wire will be *[2023, 12 April Shift-I]*

(a) $16\,\Omega$
(b) $640\,\Omega$
(c) $10\,\Omega$
(d) $40\,\Omega$

38 The equivalent resistance between A and B as shown in figure is *[2023, 8 April Shift-II]*

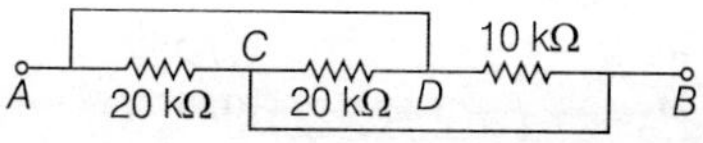

(a) $30\,k\Omega$
(b) $20\,k\Omega$
(c) $10\,k\Omega$
(d) $5\,k\Omega$

39 Equivalent resistance between the adjacent corners of a regular n-sided polygon of uniform wire of resistance R would be *[2023, 1 Feb Shift-II]*

(a) $\frac{(n-1)R}{n}$
(b) $\frac{(n-1)R}{(2n-1)}$
(c) $\frac{(n-1)R}{n^2}$
(d) $\frac{n^2R}{n-1}$

Imp. Equivalent Resistance

40 The equivalent resistance between A and B is .... *[2023, 30 Jan Shift-II]*

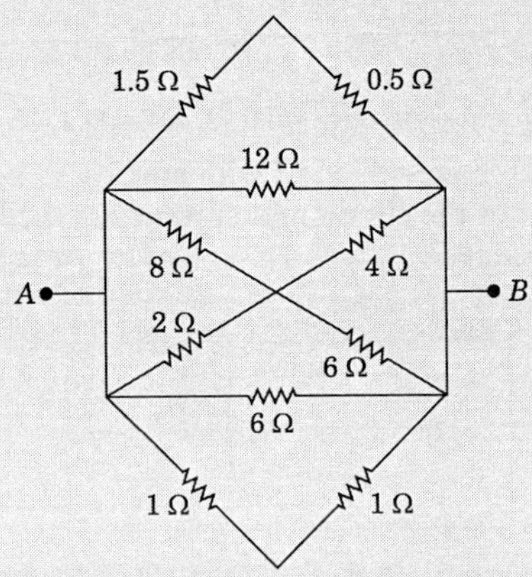

(a) $\frac{1}{2}\,\Omega$ (b) $\frac{2}{3}\,\Omega$ (c) $\frac{1}{3}\,\Omega$ (d) $\frac{3}{2}\,\Omega$

41 A uniform metallic wire carries a current 2 A, when 3.4 V battery is connected across it. The mass of uniform metallic wire is 8.92×10^{-3} kg, density is 8.92×10^{3} kg/m^3 and resistivity is 1.7×10^{-8} Ω-m. The length of wire is *[2023, 25 Jan Shift-I]*

(a) $l = 10$ m (b) $l = 100$ m (c) $l = 6.8$ m (d) $l = 5$ m

42 A 1 m long wire is broken into two unequal parts X and Y. The X part of the wire is stretched into another wire W. Length of W is twice the length of X and the resistance of W is twice that of Y. Find the ratio of length of X and Y. *2022, 29 July Shift-II]*

(a) 1 : 4 (b) 1 : 2 (c) 4 : 1 (d) 2 : 1

43 Two metallic wires of identical dimensions are connected in series. If σ_1 and σ_2 are the conductivities of these wires respectively, the effective conductivity of the combination is *[2022, 29 July Shift-I]*

(a) $\frac{\sigma_1\,\sigma_2}{\sigma_1+\sigma_2}$ (b) $\frac{2\sigma_1\,\sigma_2}{\sigma_1+\sigma_2}$ (c) $\frac{\sigma_1+\sigma_2}{2\sigma_1\,\sigma_2}$ (d) $\frac{\sigma_1+\sigma_2}{\sigma_1\,\sigma_2}$

44 Resistance of the wire is measured as 2 Ω and 3 Ω at 10° C and 30° C respectively. Temperature coefficient of resistance of the material of the wire is *[2022, 28 June Shift-II]*

(a) 0.033 °C^{-1} (b) −0.033 °C^{-1}
(c) 0.011 °C^{-1} (d) 0.055 °C^{-1}

45 (A) The drift velocity of electrons decreases with the increase in the temperature of conductor.
(B) The drift velocity is inversely proportional to the area of cross-section of given conductor.
(C) The drift velocity does not depend on the applied potential difference to the conductor.
(D) The drift velocity of electron is inversely proportional to the length of the conductor.
(E) The drift velocity increases with the increase in the temperature of conductor.

Choose the correct answer from the options given below. *[2022, 27 July Shift-II]*

(a) (A) and (B) only
(b) (A) and (D) only
(c) (B) and (E) only
(d) (B) and (C) only

46 The equivalent resistance between points A and B in the given network is *[2022, 26 June Shift-II]*

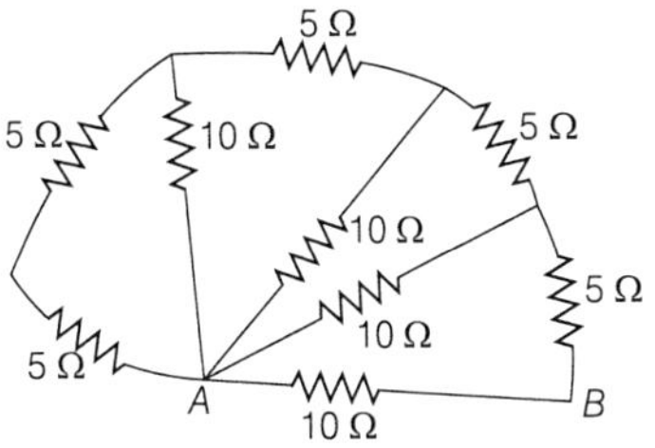

(a) 65 Ω (b) 20 Ω
(c) 5 Ω (d) 2 Ω

47 An aluminium wire is stretched to make its length, 0.4% larger. Then, percentage change in resistance is *[2022, 26 June Shift-I]*

(a) 0.4% (b) 0.2%
(c) 0.8% (d) 0.6%

48 The equivalent resistance of the given circuit between the terminals A and B is *[2021, 31 Aug Shift-II]*

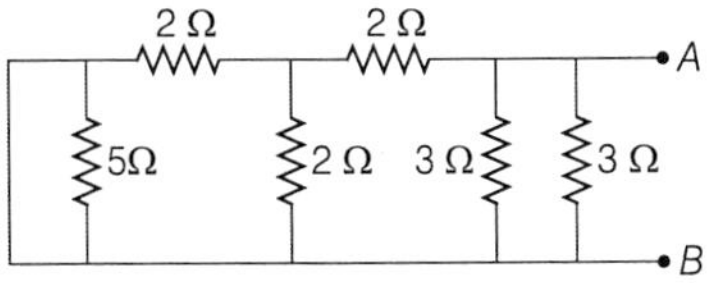

(a) 0 (b) 3 Ω
(c) 9 / 2 Ω (d) 1 Ω

49 What equal length of an iron wire and a copper-nickel alloy wire, each of 2 mm diameter connected parallel to give an equivalent resistance of 3 Ω ?
(Given, resistivities of iron and copper-nickel alloy wire are 12 μΩ cm and 51 μΩ cm res pectively) *[2021, 26 Aug Shift-I]*

(a) 82 m (b) 97 m
(c) 110 m (d) 90 m

50 The resistance of a conductor at 15°C is 16 Ω and at 100°C is 20 Ω. What will be the temperature coefficient of resistance of the conductor? *[2021, 27 July Shift-II]*

(a) 0.010°C^{-1} (b) 0.033°C^{-1}
(c) 0.003°C^{-1} (d) 0.042°C^{-1}

51 A copper Cu rod of length 25 cm and cross-sectional area 3 mm^2 is joined with a similar aluminium Al rod as shown in figure. Find the resistance of the combination between the ends A and B.
(Take, resistivity of copper $= 1.7 \times 10^{-8} \Omega$-m, resistivity of aluminium $= 2.6 \times 10^{-8} \Omega$-m) *[2021, 22 July Shift-II]*

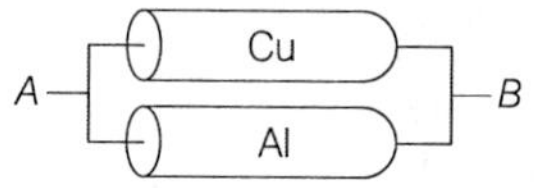

(a) 2.170 mΩ (b) 1.420 mΩ
(c) 0.0858 mΩ (d) 0.858 mΩ

52 A wire of 1 Ω has a length of 1 m. It is stretched till its length increases by 25 %. The percentage change in resistance to the nearest integer is *[2021, 26 Feb Shift-II]*
(a) 56% (b) 76%
(c) 12.5% (d) 25%

53 A current through a wire depends on time as $i = \alpha_0 t + \beta t^2$, where $\alpha_0 = 20$ A/s and $\beta = 8$ As^{-2}.
Find the charge crossed through a section of the wire in 15 s. *[2021, 24 Feb Shift-I]*
(a) 260 C (b) 2100 C
(c) 11250 C (d) 2250 C

54 Space between two concentric conducting spheres of radii a and b $(b > a)$ is filled with a medium of resistivity ρ. The resistance between the two spheres (in ohm) will be *[2019, 10 April Shift-II]*
(a) $\frac{\rho}{2\pi}\left(\frac{1}{a}+\frac{1}{b}\right)$ (b) $\frac{\rho}{4\pi}\left(\frac{1}{a}-\frac{1}{b}\right)$
(c) $\frac{\rho}{2\pi}\left(\frac{1}{a}-\frac{1}{b}\right)$ (d) $\frac{\rho}{4\pi}\left(\frac{1}{a}+\frac{1}{b}\right)$

55 A copper wire is stretched to make it 0.5% longer. The percentage change in its electrical resistance, if its volume remains unchanged is *[2019, 9 Jan Shift-I]*
(a) 2.0% (b) 1.0% (c) 0.5% (d) 2.5%

56 A heating element has a resistance of 100 Ω at room temperature. When it is connected to a supply of 220 V, a steady current of 2 A passes in it and temperature is 500°C more than room temperature. What is the temperature coefficient of resistance of the heating element? *[JEE Main 2018]*
(a) 1×10^{-4} °C^{-1} (b) 2×10^{-4} °C^{-1}
(c) 0.5×10^{-4} °C^{-1} (d) 5×10^{-4} °C^{-1}

57 The resistance of the series combination of two resistances is S. When they are joined in parallel, the total resistance is P. If $S = nP$, then the minimum possible value of n is *[AIEEE 2004]*
(a) 4 (b) 3
(c) 2 (d) 1

58 The length of a given cylindrical wire is increased by 100%. Due to the consequent decrease in diameter, the change in the resistance of the wire will be *[AIEEE 2003]*
(a) 200% (b) 100% (c) 50% (d) 300%

Level 3

59 In the given figure $R_1 = 10\,\Omega$, $R_2 = 8\Omega$, $R_3 = 4\Omega$ and $R_4 = 8\Omega$. Battery is ideal with emf 12 V. Equivalent resistant of the circuit and current supplied by battery are respectively. *[2024, 5 April Shift-I]*

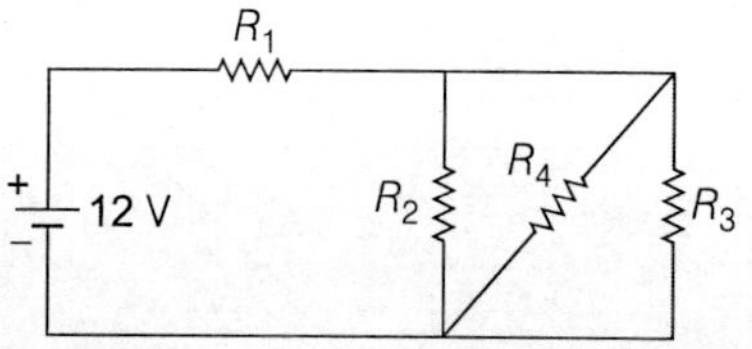

(a) 10.5 Ω and 1.14 A (b) 12 Ω and 1 A
(c) 12 Ω and 11.4 A (d) 10.5 Ω and 1 A

60 A potential divider circuit is shown in figure. The output voltage V_0 is *[2024, 30 Jan Shift-I]*

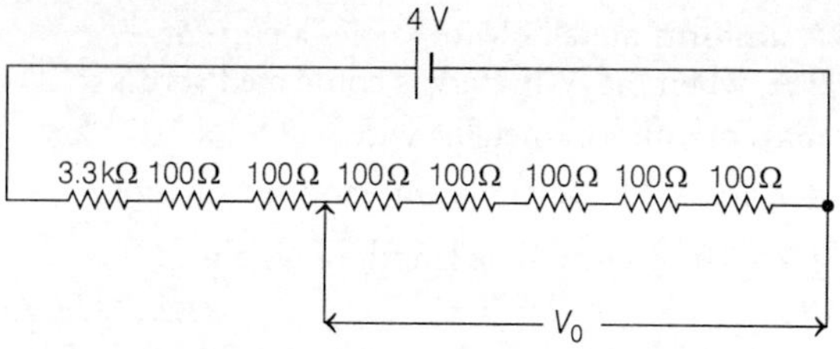

(a) 12 mV (b) 0.5 V
(c) 4 V (d) 2 mV

61 The resistance $R = \frac{V}{I}$ where $V = (200 \pm 5)$ V and $I = (20 \pm 0.2)$ A, the percentage error in the measurement of R is *[2024, 29 Jan Shift-I]*
(a) 3% (b) 7%
(c) 3.5% (d) 5.5%

62 In the given figure, a battery of emf E is connected across a conductor PQ of length l and different area of cross-sections having radii r_1 and $r_2 (r_2 < r_1)$. *[2021, 27 July Shift-I]*

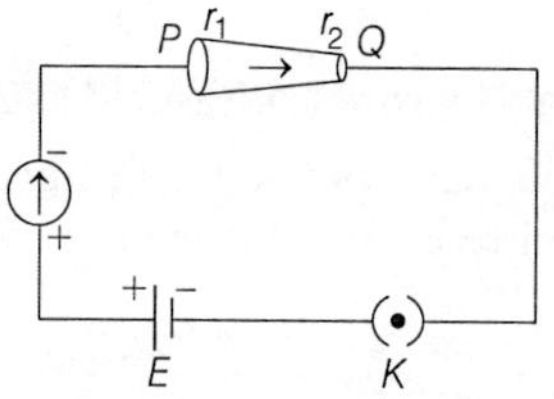

Choose the correct option as one moves from P to Q.
(a) Drift velocity of electron increases
(b) Electric field decreases
(c) Electron current decreases
(d) All of the above

63 Two conductors have the same resistance at 0°C but their temperature coefficients of resistance are α_1 and α_2. The respective temperature coefficients of their series and parallel combinations are nearly *[AIEEE 2010]*

(a) $\frac{\alpha_1 + \alpha_2}{2}, \alpha_1 + \alpha_2$ (b) $\alpha_1 + \alpha_2, \frac{\alpha_1 + \alpha_2}{2}$

(c) $\alpha_1 + \alpha_2, \frac{\alpha_1 \alpha_2}{\alpha_1 + \alpha_2}$ (d) $\frac{\alpha_1 + \alpha_2}{2}, \frac{\alpha_1 + \alpha_2}{2}$

Numerical Value Type Questions

64 A wire of resistance $9\,\Omega$ is bent to form an equilateral triangle. Then, the equivalent resistance across any two vertices will be....... ohm. *[2025, 24 Jan Shift-I]*

65 At steady state the charge on the capacitor, as shown in the circuit below, is .... μC. *[2025, 23 Jan Shift-II]*

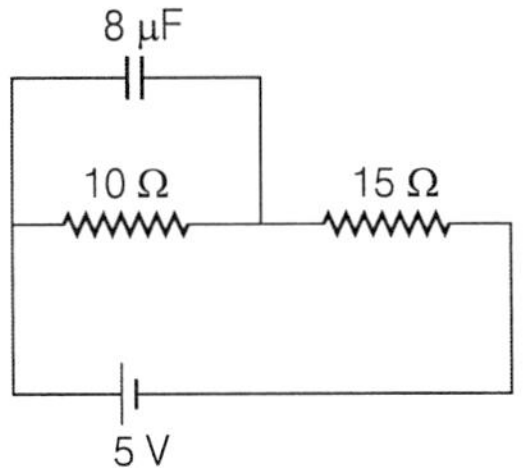

66 The net current flowing in the given circuit is ...... A. *[2025, 22 Jan Shift-II]*

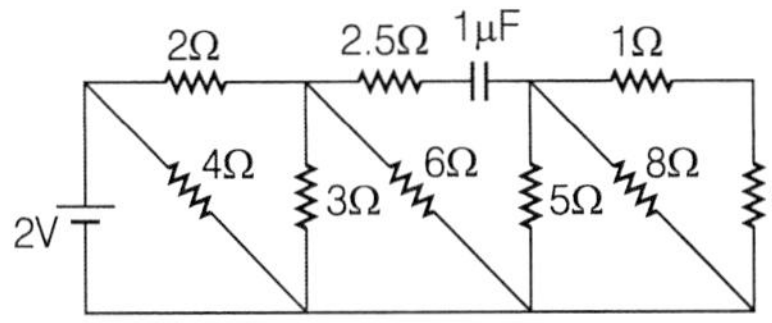

67 In the given figure an ammeter A consists of a $240\,\Omega$ coil connected in parallel to a $10\,\Omega$ shunt. The reading of the ammeter is .... mA. *[2024, 6 April Shift-II]*

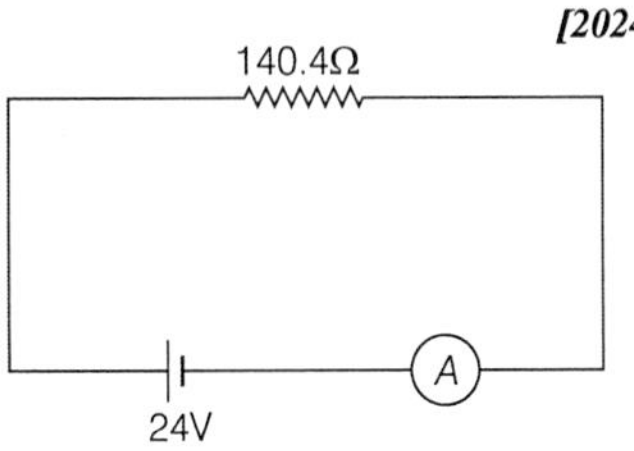

68 A wire of resistance R and radius r is stretched till its radius became $r/2$. If new resistance of the stretched wire is xR, then value of x is .... *[2024, 6 April Shift-I]*

69 A wire of resistance $20\,\Omega$ is divided into 10 equal parts, resulting pairs.A combination of two parts are connected in parallel and so on. Now, resulting pairs of parallel combination are connected in series. The equivalent resistance of final combination is .... Ω. *[2024, 5 April Shift-II]*

70 Two wires A and B are made up of the same material and have the same mass. Wire A has radius of 2.0 mm and wire B has radius of 4.0 mm. The resistance of wire B is 2Ω. The resistance of wire A is......... Ω. *[2024, 4 April Shift-II]*

71 The current in a conductor is expressed as $I = 3t^2 + 4t^3$, where I is in ampere and t is in second. The amount of electric charge that flows through a section of the conductor during $t = 1$ s to $t = 2$ s is .... C. *[2024, 1 Feb Shift-I]*

72 Two resistances of $100\,\Omega$ and $200\,\Omega$ are connected in series with a battery of 4 V and negligible internal resistance. A voltmeter is used to measure voltage across $100\,\Omega$ resistance, which gives reading as 1 V. The resistance of voltmeter must be .... Ω. *[2024, 30 Jan Shift-II]*

73 The current flowing through a conductor connected across a source is 2 A and 1.2 A at 0°C and 100°C respectively. The current flowing through the conductor at 50°C will be .... $\times 10^2$ mA. *[2023, 12 April Shift-I]*

74 A rectangular parallelopiped is measured as 1 cm $\times$ 1 cm $\times$ 100 cm. If its specific resistance is $3 \times 10^{-7}\,\Omega$-m, then the resistance between its two opposite rectangular faces will be .... $\times 10^{-7}\,\Omega$. *[2023, 10 April Shift-II]*

75 10 resistors each of resistance $10\,\Omega$ can be connected in such a way so as to get maximum and minimum equivalent resistance. The ratio of maximum and minimum equivalent resistance will be .... *[2023, 10 April Shift-I]*

76 The number density of free electrons in copper is nearly $8 \times 10^{28}\ m^{-3}$. A copper wire has its area of cross-section $= 2 \times 10^{-6}\ m^2$ and is carrying a current of 3.2 A. The drift speed of the electrons is .... $\times 10^{-6}\ ms^{-1}$. *[2023, 8 April Shift-II]*

77 A current of 2 A flows through a wire of cross-sectional area 25.0 mm^2. The number of free electrons in a cubic meter are 2.0×10^{28}. The drift velocity of the electrons is .... $\times 10^{-6}\ ms^{-1}$ (given, charge on electron $= 1.6 \times 10^{-19}$ C). *[2023, 8 April Shift-I]*

78 The length of a metallic wire is increased by 20% and its area of cross-section is reduced by 4%. The percentage change in resistance of the metallic wire is .... *[2023, 6 April Shift-I]*

79 In the given circuit, the equivalent resistance between the terminal A and B is .... Ω. *[2023, 25 Jan Shift-I]*

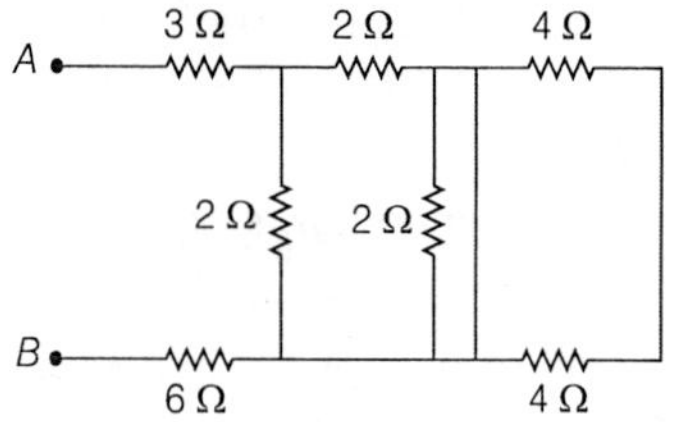

80 If a copper wire is stretched to increase its length by 20%, the percentage increase in resistance of the wire is .... %. *[2023, 24 Jan Shift-II]*

81 A hollow cylindrical conductor has length of 3.14 m, while its inner and outer diameters are 4 mm and 8 mm respectively. The resistance of the conductor is $n \times 10^{-3}$ Ω. If the resistivity of the material is 2.4×10^{-8} Ω-m, the value of n is .... *[2023, 24 Jan Shift-I]*

82 The current I flowing through the given circuit will be .... A. *[2022, 29 July Shift-I]*

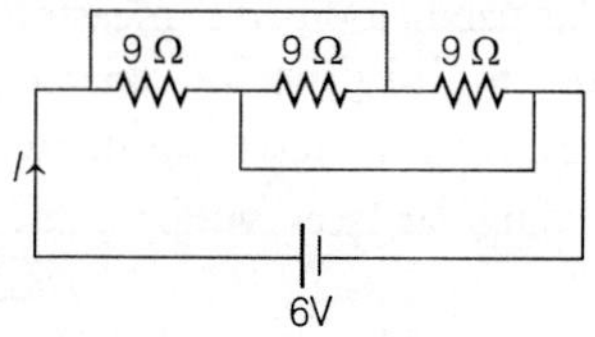

83 A 1 m long copper wire carries a current of 1 A. If the cross-section of the wire is 2.0 mm^2 and the resistivity of copper is 1.7×10^{-8} Ω-m, the force experienced by moving electron in the wire is ...... $\times 10^{-23}$ N. (Charge on electron $= 1.6 \times 10^{-19}$ C) *[2022, 27 July Shift-I]*

84 All resistance in figure are 1 Ω each. The value of current I is $a/5$ A. The value of a is .... *[2022, 28 June Shift-II]*

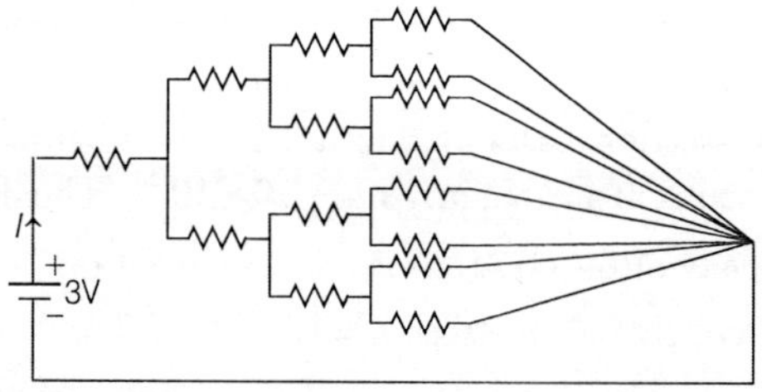

85 The current density in a cylindrical wire of radius 4 mm is 4×10^6 Am^{-2}. The current through the outer portion of the wire between radial distance $R/2$ and R is ...... π A. *[2022, 27 June Shift-I]*

86 The length of a given cylindrical wire is increased to double of its original length. The percentage increase in the resistance of the wire will be ......% *[2022, 25 June Shift-II]*

87 The total current supplied to the circuit as shown in figure by the 5 V battery is ... A. *[2022, 25 June Shift-I]*

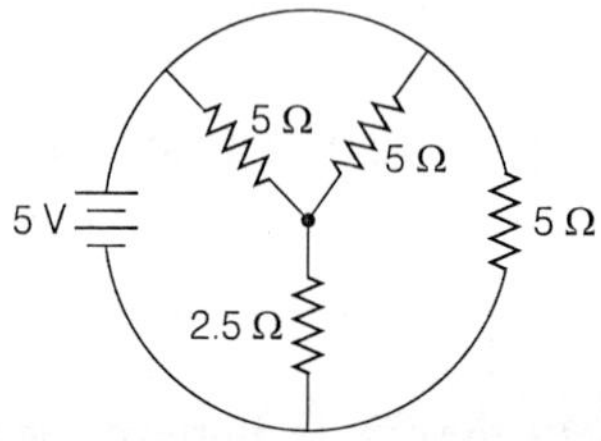

88 A square shaped wire with resistance of each side 3 Ω is bent to form a complete circle. The resistance between two diametrically opposite points of the circle in unit of Ω will be *[2021, 31 Aug Shift-I]*

89 A capacitor of 50 μF is connected in a circuit as shown in figure. The charge on the upper plate of the capacitor is .... μC. *[2021, 31 Aug Shift-I]*

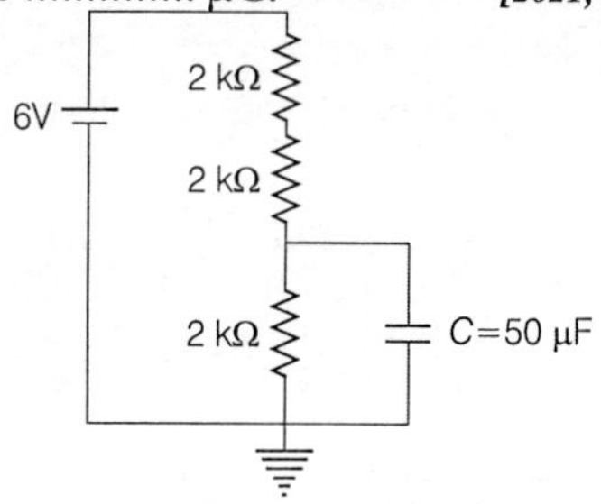

90 The ratio of the equivalent resistance of the network (shown in figure) between the points a and b when switch is open and switch is closed is $x : 8$. The value of x is *[2021, 27 Aug Shift-II]*

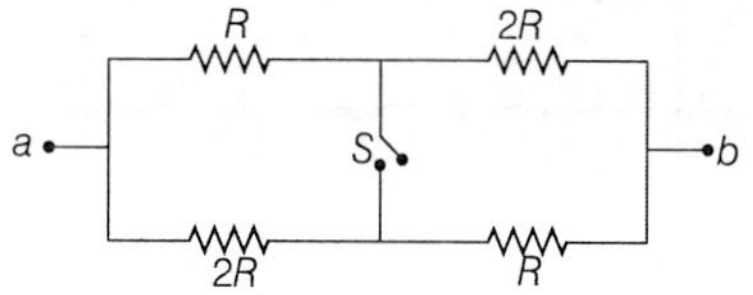

91 In Bohr's atomic model, the electron is assumed to revolve in a circular orbit of radius 0.5 Å. If the speed of electron is 2.2×10^6 m/s, then the current associated with the electron will be .... $\times 10^{-2}$ mA. [Take, π as 22/7] *[2021, 27 July Shift-I]*

92 A 16 Ω wire is bent to form a square loop. A 9 V supply having internal resistance of 1 Ω is connected across one of its sides. The potential drop across the diagonals of the square loop is ...... $\times 10^{-1}$ V. *[2021, 25 July Shift-II]*

93 In the given figure, switches S_1 and S_2 are in open condition. The resistance across ab when the switches S_1 and S_2 are closed is ...... Ω. *[2021, 20 July Shift-II]*

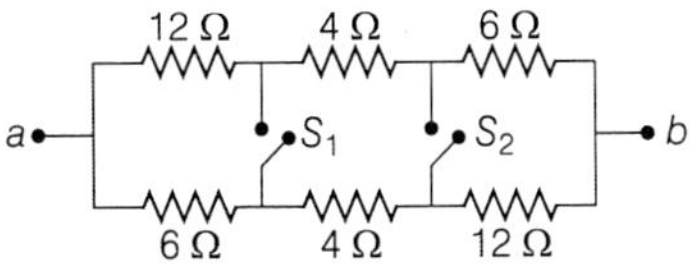

94 Two wires of same length and thickness having specific resistances $6\,\Omega$- cm and $3\,\Omega$-cm respectively are connected in parallel. The effective resistivity is $\rho\,\Omega$-cm. The value of ρ to the nearest integer, is .... *[2021, 18 March Shift-II]*

95 The voltage across the 10Ω resistor in the given circuit is x V.

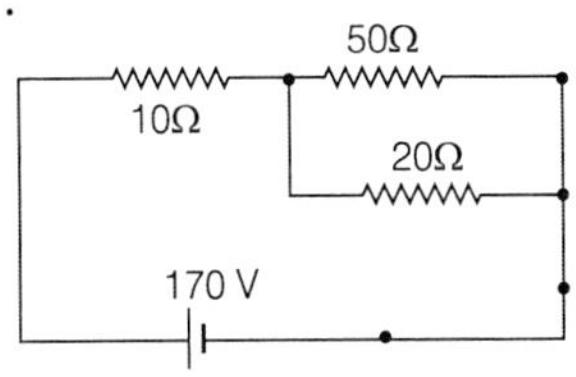

The value of x to the nearest integer is .... *[2021, 18 March Shift-I]*

96 The equivalent resistance of series combination of two resistors is s. When they are connected in parallel, the equivalent resistance is p. If $s = np$, then the minimum value for n is .... (Round off to the nearest integer) *[2021, 17 March Shift-I]*

97 In the figure given, the electric current flowing through the $5\,k\Omega$ resistor is x mA. *[2021, 16 March Shift-I]*

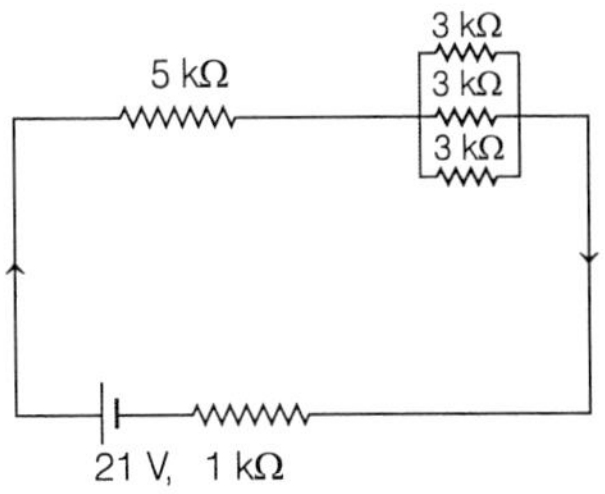

The value of x to the nearest Integer is ....

98 A current of 6 A enters one corner P of an equilateral triangle PQR having three wires of resistance $2\,\Omega$ each and leaves by the corner R. The current i_1 in ampere is .... *[2021, 25 Feb Shift-II]*

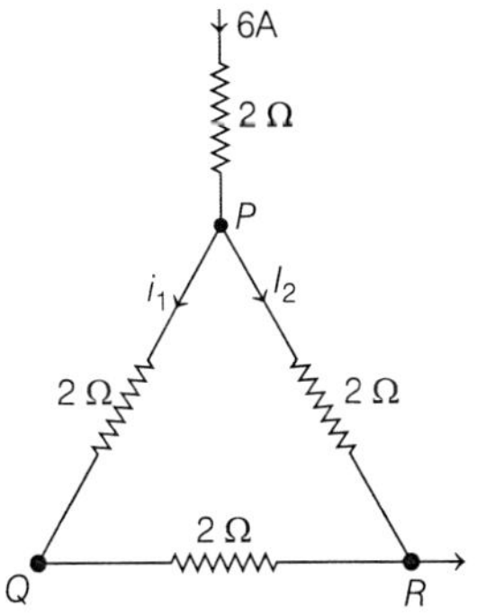

99 A cylindrical wire of radius 0.5 mm and conductivity 5×10^7 S/m is subjected to an electric field of 10 mV/m. The expected value of current in the wire will be $x^3\pi$ mA. The value of x is .... *[2021, 24 Feb Shift-II]*

Topic 2 Heating Effect of Current

Level 1

Imp. Electrical Power

01 There are n number of identical electric bulbs, each is designed to draw a power P independently from the mains supply. They are now joined in series across the mains supply. The total power drawn by the combination is *[2025, 4 April Shift-II]*
(a) P (b) P/n (c) P/n^2 (d) nP

02 The number of electrons flowing per second in the filament of a 110 W bulb operating at 220 V is (Given, $e = 1.6 \times 10^{-19}$ C) *[2024, 6 April Shift-II]*
(a) 1.25×10^{19} (b) 6.25×10^{17}
(c) 31.25×10^{17} (d) 6.25×10^{18}

03 An electric bulb rated 50 W-200 V is connected across a 100V supply. The power dissipation of the bulb is *[2024, 4 April Shift-II]*
(a) 50 W (b) 100 W (c) 25 W (d) 12.5 W

04 In a large building, there are 15 bulbs of 40 W, 5 bulbs of 100 W, 5 fans of 80 W and 1 heater of 1 kW. The voltage of the electric mains is 220 V. Then, the minimum current should be *[JEE Main 2014]*
(a) 8 A (b) 10 A (c) 12 A (d) 14 A

05 A heater coil is cut into two equal parts and only one part is now used in the heater. The heat generated will now be *[AIEEE 2005]*
(a) doubled (b) four times (c) one-fourth (d) halved

06 The resistance of hot tungsten filament is about 10 times the cold resistance. What will be the resistance of 100 W and 200 V lamp, when not in use? *[AIEEE 2005]*
(a) $40\,\Omega$ (b) $20\,\Omega$ (c) $400\,\Omega$ (d) $200\,\Omega$

07 The thermistors are usually made of *[AIEEE 2004]*
(a) metals with low temperature coefficient of resistivity
(b) metals with high temperature coefficient of resistivity
(c) metal oxides with high temperature coefficient of resistivity
(d) semiconducting materials having low temperature coefficient of resistivity

08 Time taken by a 836 W heater to heat 1 L of water from 10°C to 40°C is *[AIEEE 2004]*
(a) 50 s (b) 100 s
(c) 150 s (d) 200 s

09 A 220 V–1000 W bulb is connected across a 110 V mains supply. The power consumed will be *[AIEEE 2003]*

(a) 750 W (b) 500 W
(c) 250 W (d) 1000 W

10 If in the circuit, power dissipation is 150 W, then R is *[AIEEE 2002]*

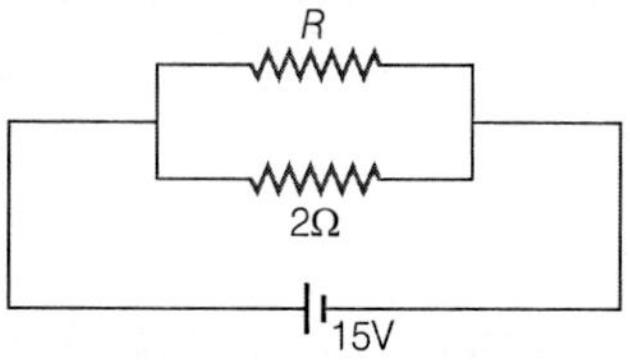

(a) 2 Ω (b) 6 Ω
(c) 5 Ω (d) 4 Ω

Level 2

11 A motor operating on 100 V draws a current of 1 A. If the efficiency of the motor is 91.6%, then the loss of power in units of cal/s is *[2025, 3 April Shift-II]*
(a) 2 (b) 4 (c) 6.2 (d) 8.4

12 Water boils in an electric kettle in 20 min after being switched on. Using the same main supply, the length of the heating element should be to times of its initial length if the water is to be boiled in 15 min. *[2024, 8 April Shift-II]*

(a) increased, 3/4 (b) decreased, 4/3
(c) decreased, 3/4 (d) increased, 4/3

13 The resistances of the platinum wire of a platinum resistance thermometer at the ice point and steam point are 8 Ω and 10 Ω respectively. After inserting in a hot bath of temperature 400°C, the resistance of platinum wire is *[2024, 4 April Shift-I]*
(a) 10 Ω (b) 2 Ω
(c) 8 Ω (d) 16 Ω

14 By what percentage will the illumination of the lamp decrease, if the current drops by 20%? *[2024, 31 Jan Shift-II]*

(a) 26% (b) 46%
(c) 36% (d) 56%

15 Two identical heater filaments are connected first in parallel and then in series. At the same applied voltage, the ratio of heat produced in same time for parallel to series will be *[2023, 11 April Shift-I]*
(a) 1 : 1 (b) 1 : 4
(c) 2 : 1 (d) 4 : 1

16 Ratio of thermal energy released in two resistors R and $3R$ connected in parallel in an electric circuit is *[2023, 29 Jan Shift-I]*

(a) 1 : 27 (b) 3 : 1 (c) 1 : 1 (d) 1 : 3

17 Due to cold weather, a 1 m water pipe of cross-sectional area 1 cm^2 is filled with ice at –10°C. Resistive heating is used to melt the ice. Current of 0.5 A is passed through a 4 kΩ resistance. Assuming that all the heat produced is used for melting, what is the minimum time required?
(Given, latent heat of fusion for water/ice $= 3.33 \times 10^5$ J kg^{-1}, specific heat of ice $= 2 \times 10^3$ J kg^{-1} and density of ice $= 10^3$ kg / m^3) *[2021, 1 Sep Shift-II]*

(a) 3.53 s (b) 0.353 s
(c) 35.3 s (d) 70.6 s

18 An electric bulb of 500 W at 100 V is used in a circuit having a 200 V supply. Calculate the resistance R to be connected in series with the bulb, so that the power delivered by the bulb is 500 W. *[2021, 26 Aug Shift-II]*

(a) 20 Ω (b) 30 Ω (c) 5 Ω (d) 10 Ω

19 In the given figure, the emf of the cell is 2.2 V and if internal resistance is 0.6 Ω. Calculate the power dissipated in the whole circuit *[2021, 26 Aug Shift-I]*

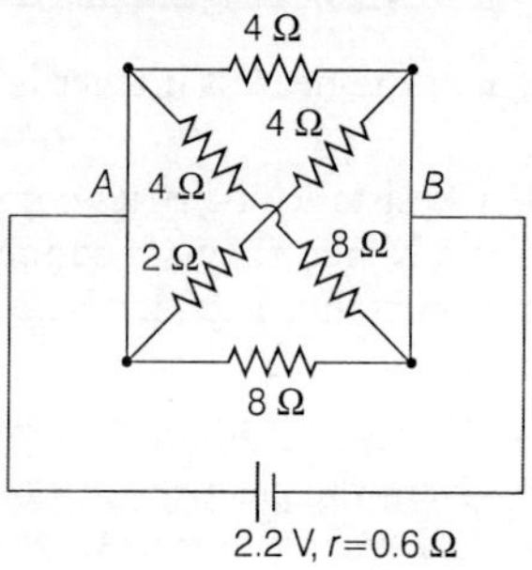

(a) 1.32 W (b) 0.65 W (c) 2.2 W (d) 4.4 W

20 A resistor develops 500 J of thermal energy in 20 s, when a current of 1.5 A is passed through it. If the current is increased from 1.5 A to 3 A, what will be the energy developed in 20 s ? *[2021, 16 March Shift-II]*
(a) 1500 J (b) 1000 J (c) 500 J (d) 2000 J

21 A battery of 3.0 V is connected to a resistor dissipating 0.5 W of power. If the terminal voltage of the battery is 2.5 V, the power dissipated within the internal resistance is *[2020, 4 Sep Shift-I]*
(a) 0.072 W (b) 0.125 W
(c) 0.50 W (d) 0.10 W

22 In a building, there are 15 bulbs of 45 W, 15 bulbs of 100 W, 15 small fans of 10 W and 2 heaters of 1 kW. The voltage of electric main is 220 V. The minimum fuse capacity (rated value) of the building will be *[2020, 7 Jan Shift-II]*

(a) 25 A (b) 10 A (c) 20 A (d) 15 A

23 Two equal resistances when connected in series to a battery consume electric power of 60 W. If these resistances are now connected in parallel combination to the same battery, the electric power consumed will be *[2019, 11 Jan Shift-I]*
(a) 60 W (b) 30 W
(c) 240 W (d) 120 W

24 Two electric bulbs marked 25 W-220 V and 100 W-220 V are connected in series to a 440 V supply. Which of the bulbs will fuse? *[AIEEE 2012]*

(a) Both (b) 100 W
(c) 25 W (d) None of these

25 An electric bulb is rated 220 V-100 W. The power consumed by it when operated on 110 V will be *[AIEEE 2006]*

(a) 75 W (b) 40 W (c) 25 W (d) 50 W

26 A wire when connected to 220 V mains supply has power dissipation P_1. Now, the wire is cut into two equal pieces which are connected in parallel to the same supply. Power dissipation in this case is P_2. Then, $P_2 : P_1$ is *[AIEEE 2002]*

(a) 1 (b) 4
(c) 2 (d) 3

Level 3

27 Two conductors have the same resistances at 0°C but their temperature coefficients of resistance are α_1 and α_2. The respective temperature coefficients for their series and parallel combinations are *[2024, 31 Jan Shift-I]*

(a) $\frac{\alpha_1+\alpha_2}{2}, \frac{\alpha_1+\alpha_2}{2}$ (b) $\frac{\alpha_1+\alpha_2}{2}, \alpha_1+\alpha_2$

(c) $\alpha_1+\alpha_2, \frac{\alpha_1+\alpha_2}{2}$ (d) $\alpha_1+\alpha_2, \frac{\alpha_1\alpha_2}{\alpha_1+\alpha_2}$

Imp. Heating Effect of Current

28 An electric toaster has resistance of 60 Ω at room temperature 27°C. The toaster is connected to a 220 V supply. If the current flowing through it reaches 2.75 A, the temperature attained by toaster is around (if, $\alpha = 2\times10^{-4}$/°C) *[2024, 30 Jan Shift-I]*

(a) 1667°C (b) 1694°C
(c) 694°C (d) 1235°C

29 Different combination of 3 resistors of equal resistance R are shown in the figures. The increasing order for power dissipation is *[2023, 13 April Shift-I]*

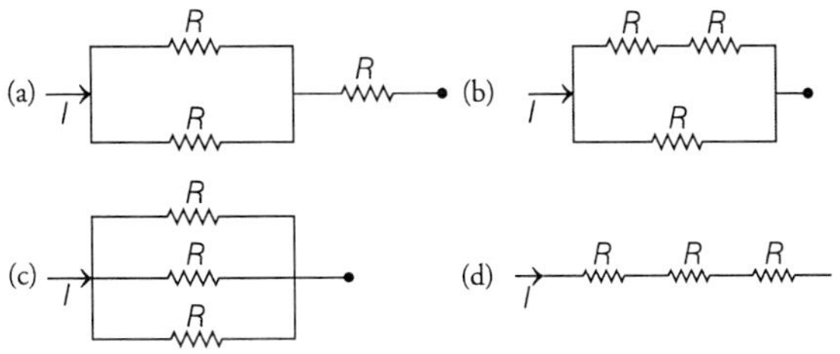

(a) $P_A < P_B < P_C < P_D$ (b) $P_C < P_D < P_A < P_B$
(c) $P_C < P_B < P_A < P_D$ (d) $P_B < P_C < P_D < P_A$

30 In the circuit shown, the resistance r is a variable resistance. If for $r = fR$, the heat generation in r is maximum, then the value of f is *[JEE Main 2016]*

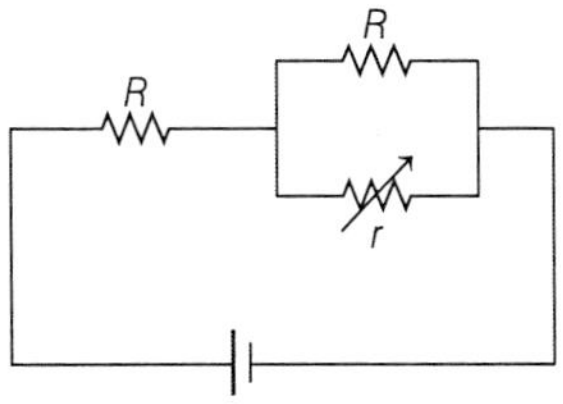

(a) 1/2 (b) 1
(c) 1/4 (d) 3/4

31 The supply voltage to room is 120 V. The resistance of the lead wires is 6 Ω. A 60 W bulb is already switched on. What is the decrease of voltage across the bulb, when a 240 W heater is switched on in parallel to the bulb ? *[JEE Main 2013]*

(a) Zero (b) 2.9 V (c) 13.3 V (d) 10. 4 V

Numerical Value Type Questions

32 At room temperature (27°C), the resistance of a heating element is 50 Ω. The temperature coefficient of the material is $2.4\times10^{-4}\,°C^{-1}$. The temperature of the element, when its resistance is 62 Ω is .... °C. *[2024, 9 April Shift-II]*

33 A heater is designed to operate with a power of 1000 W in a 100 V line. It is connected in combination with a resistance of 10 Ω and a resistance R, to a 100 V mains as shown in figure. For the heater to operate at 62.5 W, the value of R should be ...... Ω. *[2024, 8 April Shift-II]*

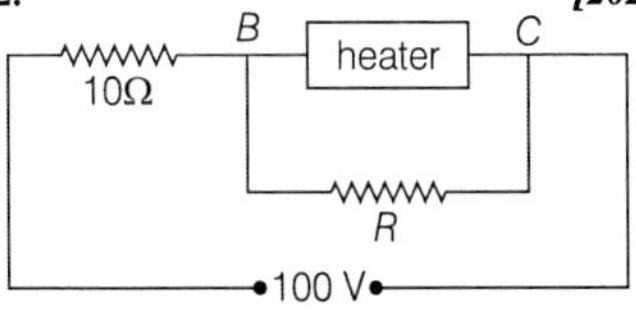

34 Resistance of a wire at 0°C, 100°C and t°C is found to be 10 Ω, 10.2 Ω and 10.95 Ω respectively. The temperature t in Kelvin scale is .... *[2024, 8 April Shift-I]*

Imp. Power Consumption in Electrical Circuit

35 In the following circuit, the battery has an emf of 2 V and an internal resistance of $\frac{2}{3}$ Ω. The power consumption in the entire circuit is .... W. *[2024, 31 Jan Shift-II]*

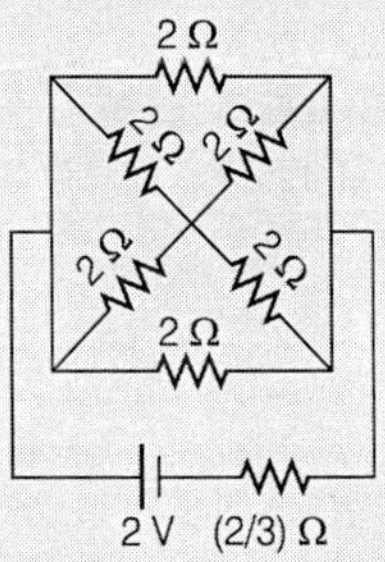

36 A potential V_0 is applied across a uniform wire of resistance R. The power dissipation is P_1. The wire is then cut into two equal halves and a potential of V_0 is applied across the length of each half. The total power dissipation ratio across two wires is P_2. The ratio $P_2 : P_1$ is $\sqrt{x} : 1$. The value of x is .... **[2023, 13 April Shift-I]**

37 An electrical bulb rated (220 V, 100 W) is connected in series with another bulb rated (220 V, 60 W). If the voltage across combination is 220 V, the power consumed by the 100 W bulb will be about .... W. **[2022, 28 July Shift-II]**

38 Two coils require 20 min and 60 min respectively to produce same amount of heat energy when connected separately to the same source. If they are connected in parallel arrangement to the same source; the time required to produce same amount of heat by the combination of coils will be .... min. **[2022, 29 June Shift-I]**

39 A resistor develops 300 J of thermal energy in 15 s when a current of 2 A is passed through it. If the current increases to 3 A, the energy developed in 10 s is .... J. **[2022, 25 June Shift-I]**

40 A uniform heating wire of resistance 36 Ω is connected across a potential difference of 240 V. The wire is then cut into half and potential difference of 240 V is applied across each half separately. The ratio of power dissipation in first case to the total power dissipation in the second case would be $1 : x$, where x is .... **[2021, 1 Sep Shift-II]**

41 A resistor dissipates 192 J of energy in 1 s when a current of 4 A is passed through it. Now, when the current is doubled, the amount of thermal energy dissipated in 5 s is .... J. **[2021, 31 Aug Shift-II]**

42 An electric bulb rated as 200 W at 100 V is used in a circuit having 200 V supply. The resistance R that must be put in series with the bulb so that the bulb delivers the same power is ....Ω **[2021, 25 July Shift-I]**

43 The energy dissipated by a resistor is 10 mJ in 1s when an electric current of 2 mA flows through it. The resistance is .... Ω. (Round off to the nearest integer) **[2021, 16 March Shift-II]**

Topic 3 Cells and Its Combination and Kirchhoff's Rules

Level 1

01 The battery of a mobile phone is rated as 4.2 V, 5800 mAh. How much energy is stored in it when fully charged? **[2025, 2 April Shift-I]**

(a) 24.4 kJ (b) 48.7 kJ
(c) 43.8 kJ (d) 87.7 kJ

02 Three voltmeters, all having different internal resistances are joined as shown in figure. When some potential difference is applied across A and B, their readings are V_1, V_2 and V_3. Choose the correct option. **[2024, 27 Jan Shift-II]**

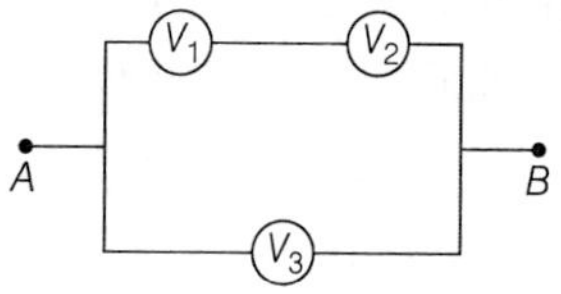

(a) $V_1 + V_2 > V_3$
(b) $V_1 \neq V_3 - V_2$
(c) $V_1 = V_2$
(d) $V_1 + V_2 = V_3$

03 A current of 15 mA flows in the circuit as shown in figure. The value of potential difference between the points A and B will be **[2022, 25 July Shift-II]**

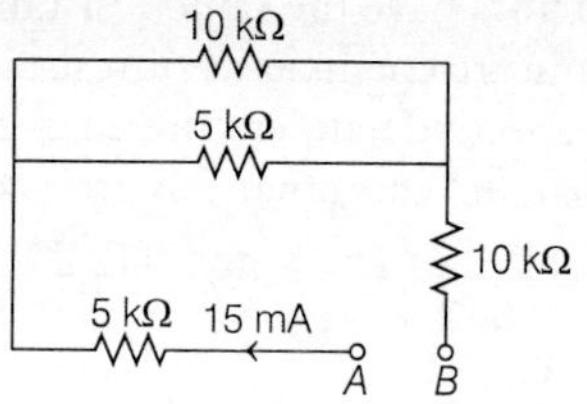

(a) 50 V (b) 75 V
(c) 150 V (d) 275 V

04 In the below circuit, the current in each resistance is **[JEE Main 2017]**

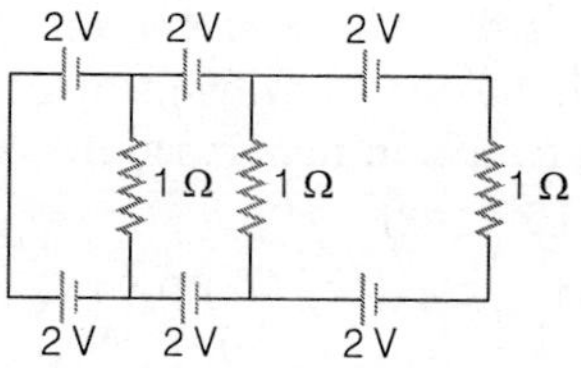

(a) 0.25 A (b) 0.5 A
(c) 0 A (d) 1 A

05 A 5 V battery with internal resistance 2 Ω and a 2 V battery with internal resistance 1Ω are connected to a 10 Ω resistor as shown in the figure.

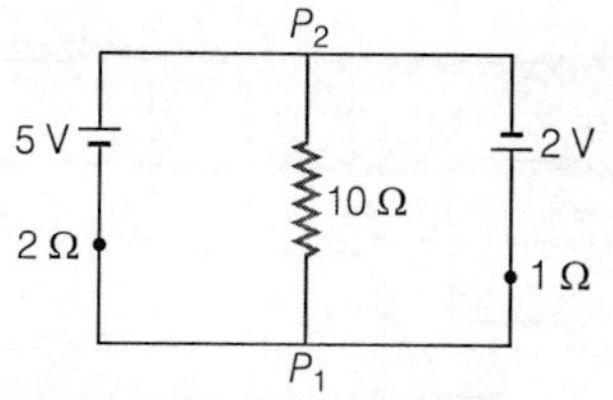

The current in the 10 Ω resistor is **[AIEEE 2008]**

(a) 0.27 A, P_2 to P_1
(b) 0.03 A, P_1 to P_2
(c) 0.03 A, P_2 to P_1
(d) 0.27 A, P_1 to P_2

06 The Kirchhoff's first law ($\Sigma i = 0$) and second law ($\Sigma iR = \Sigma E$) where, the symbols have their usual meanings, are respectively based on *[AIEEE 2006]*
(a) conservation of charge, conservation of momentum
(b) conservation of energy, conservation of charge
(c) conservation of momentum, conservation of charge
(d) conservation of charge, conservation of energy

07 An energy source will supply a constant current into the load, if its internal resistance is *[AIEEE 2005]*
(a) equal to the resistance of the load
(b) very large as compared to the load resistance
(c) zero
(d) non-zero but less than the resistance of the load

Level 2

08 Given below are two statements.
Statement I The equivalent emf of two non-ideal batteries connected in parallel is smaller than either of the two emfs.
Statement II The equivalent internal resistance of two non-ideal batteries connected in parallel is smaller than the internal resistance of either of the two batteries. *[2025, 22 Jan Shift-I]*
In the light of the above statements, choose the correct answer from the options given below.
(a) Both Statement I and Statement II are true.
(b) Both Statement I and Statement II are false.
(c) Statement I is true but Statement II is false.
(d) Statement I is false but Statement II is true.

09 In the given circuit, the terminal potential difference of the cell is *[2024, 8 April Shift-I]*

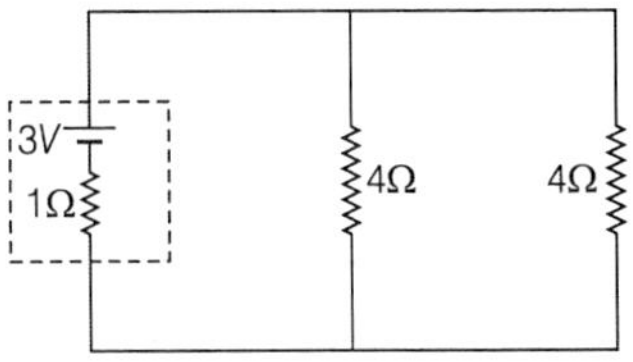

(a) 2 V (b) 1.5 V (c) 3 V (d) 4 V

Imp. Combination of Cells

10 The reading in the ideal voltmeter (V) shown in the given circuit diagram is *[2024, 1 Feb Shift-I]*

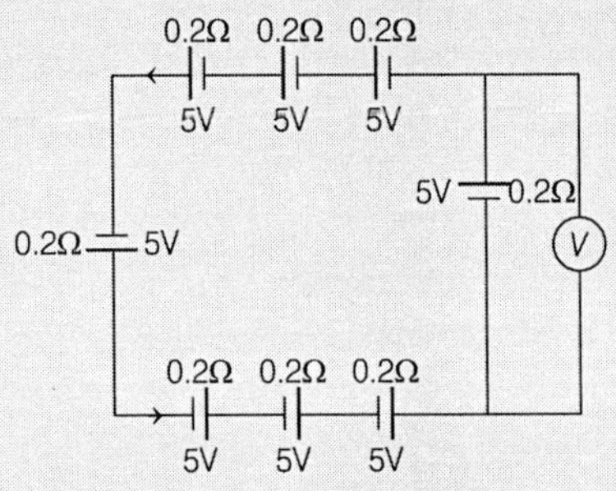

(a) 10 V (b) 0 V
(c) 5 V (d) 3 V

11 In the given circuit, the current in resistance R_3 is *[2024, 29 Jan Shift-II]*

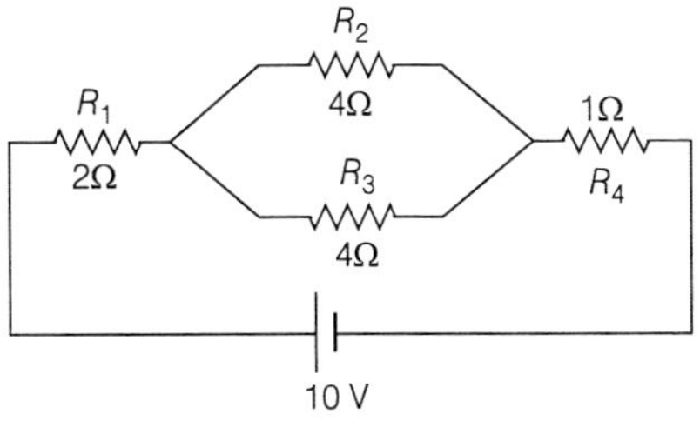

(a) 2.5 A (b) 1.5 A
(c) 2 A (d) 1 A

12 In the network shown below, the charge accumulated in the capacitor in steady state will be *[2023, 13 April Shift-II]*

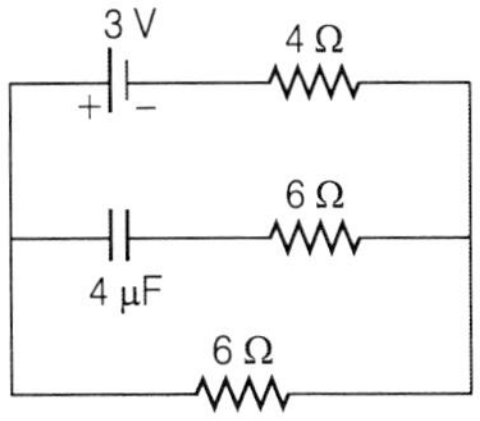

(a) 12 μC (b) 4.8 μC
(c) 7.2 μC (d) 10.3 μC

13

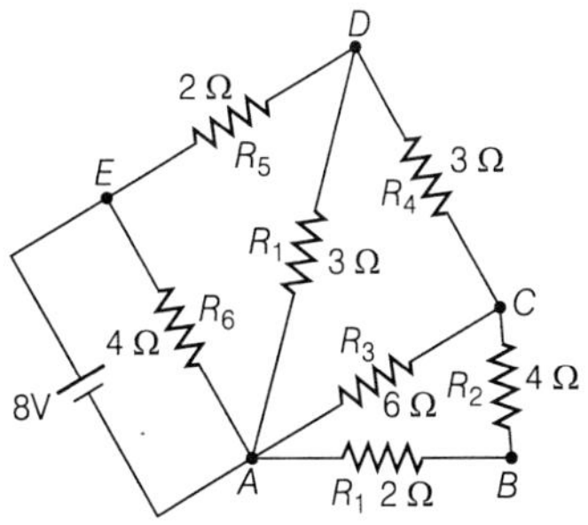

The current flowing through R_2 is *[2023, 11 April Shift-II]*
(a) $\frac{1}{3}$ A (b) $\frac{1}{4}$ A
(c) $\frac{2}{3}$ A (d) $\frac{1}{2}$ A

14 Figure shows a part of an electric circuit. The potentials at points a, b and c are 30 V, 12 V and 2 V respectively. The current through the 20 Ω resistor will be *[2023, 6 April Shift-II]*

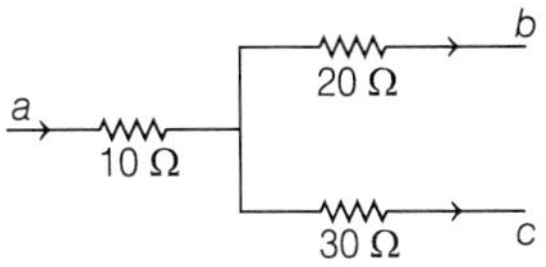

(a) 1.0 A (b) 0.4 A (c) 0.2 A (d) 0.6 A

15 As shown in the figure, a network of resistor is connected to a battery of 24V with an internal resistance of 3 Ω. The currents through the resistors

R_4 and R_5 are I_4 and I_5 respectively. The value of I_4 and I_5 are *[2023, 24 Jan Shift-I]*

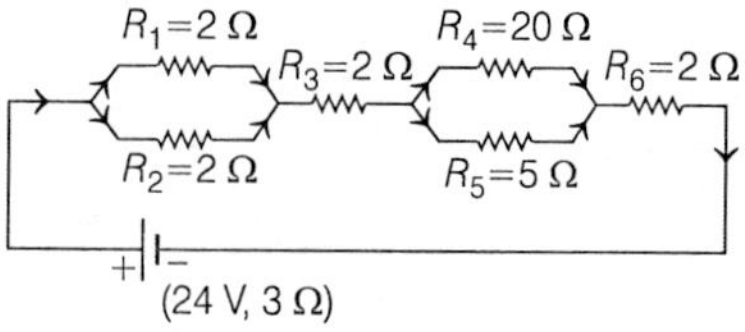

(a) $I_4 = \frac{8}{4}$ A and $I_5 = \frac{2}{5}$ A (b) $I_4 = \frac{2}{5}$ A and $I_5 = \frac{8}{5}$ A

(c) $I_4 = \frac{24}{5}$ A and $I_5 = \frac{6}{5}$ A (d) $I_4 = \frac{6}{5}$ A and $I_5 = \frac{24}{5}$ A

16 Two sources of equal emfs are connected in series. This combination is connected to an external resistance R. The internal resistances of the two sources are r_1 and $r_2 (r_1 > r_2)$. If the potential difference across the source of internal resistance r_1 is zero, then the value of R will be *[2022, 27 July Shift-I]*

(a) $r_1 - r_2$ (b) $\frac{r_1 r_2}{r_1 + r_2}$ (c) $\frac{r_1 + r_2}{2}$ (d) $r_2 - r_1$

17 The combination of two identical cells, whether connected in series or parallel combination provides the same current through an external resistance of 2 Ω. The value of internal resistance of each cell is *[2022, 29 June Shift-II]*

(a) 2 Ω (b) 4 Ω (c) 6 Ω (d) 8 Ω

18 Five identical cells each of internal resistance 1 Ω and emf 5V are connected in series and in parallel with an external resistance R. For what value of R, current in series and parallel combination will remain the same? *[2021, 27 Aug Shift-I]*

(a) 1 Ω (b) 25 Ω (c) 5 Ω (d) 10 Ω

19 The value of current in the 6 Ω resistance is *[2021, 20 July Shift-I]*

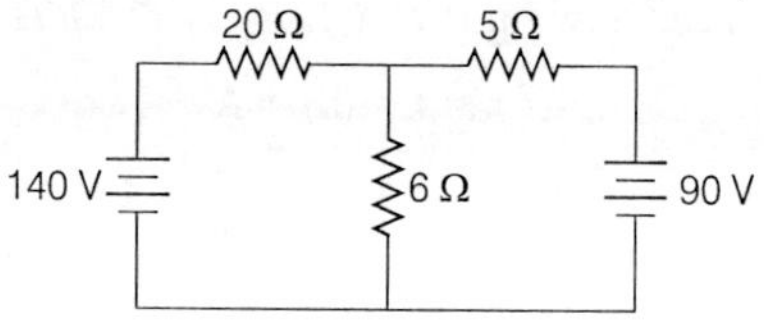

(a) 4 A (b) 8 A (c) 10 A (d) 6 A

20 Two cells of emf $2E$ and E with internal resistance r_1 and r_2 respectively are connected in series to an external resistor R (see figure). The value of R, at which the potential difference across the terminals of the first cell becomes zero is *[2021, 17 March Shift-II]*

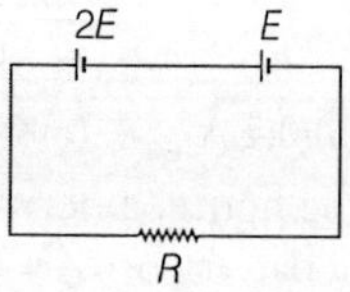

(a) $r_1 + r_2$ (b) $r_1/2 - r_2$ (c) $r_1/2 + r_2$ (d) $r_1 - r_2$

21 A cell E_1 of emf 6 V and internal resistance 2 Ω is connected with another cell E_2 of emf 4 V and internal resistance 8 Ω (as shown in the figure). The potential difference across points X and Y is *[2021, 24 Feb Shift-I]*

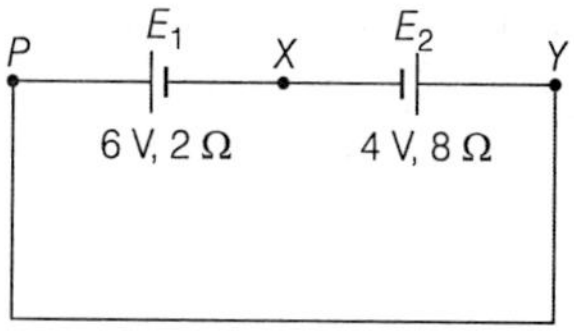

(a) 2.0 V (b) 3.6 V (c) 5.6 V (d) 10.0 V

Imp. Kirchoff's Law

23 In the given circuit, currents in different branches and value of one resistor are shown. Then, potential at point B with respect to the point A is *[2020, 5 Sep Shift-II]*

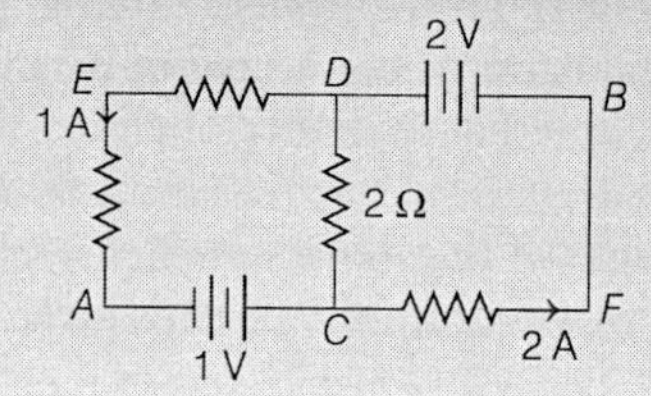

(a) +2 V (b) −2 V
(c) −1 V (d) +1 V

23 In the given circuit diagram, a wire is joining points B and D. The current in this wire is *[2020, 9 Jan Shift-I]*

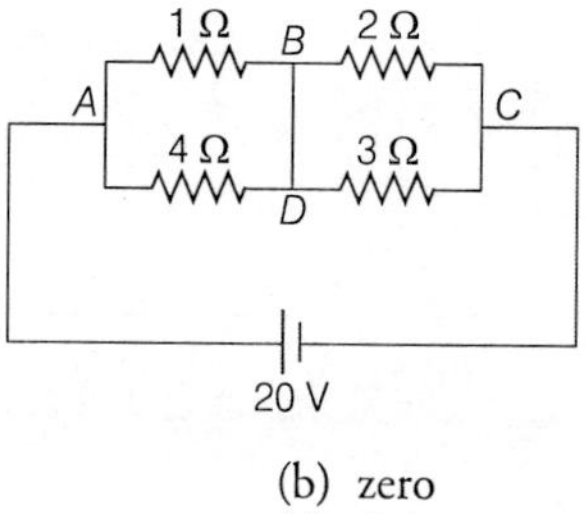

(a) 0.4 A (b) zero
(c) 2 A (d) 4 A

24 In the given circuit, an ideal voltmeter connected across the 10 Ω resistance reads 2 V. The internal resistance r, of each cell is *[2019, 10 April Shift-I]*

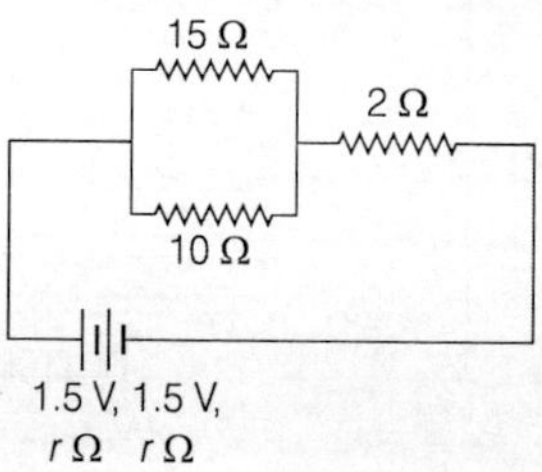

(a) 1.5 Ω (b) 0.5 Ω
(c) 1 Ω (d) 0 Ω

25 In the given circuit diagram, the currents $I_1 = -0.3$ A, $I_4 = 0.8$ A and $I_5 = 0.4$ A, are flowing as shown. The currents I_2, I_3 and I_6 respectively, are *[2019, 12 Jan Shift-II]*

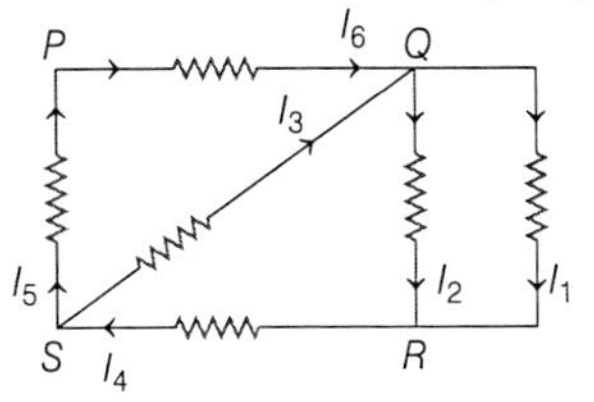

(a) 1.1 A, 0.4 A, 0.4 A (b) 1.1 A, −0.4 A, 0.4 A
(c) 0.4 A, 1.1 A, 0.4 A (d) −0.4 A, 0.4 A, 1.1 A

26 In the given circuit, the cells have zero internal resistance. The currents (in Ampere) passing through resistances R_1 and R_2 respectively are *[2019, 10 Jan Shift-I]*

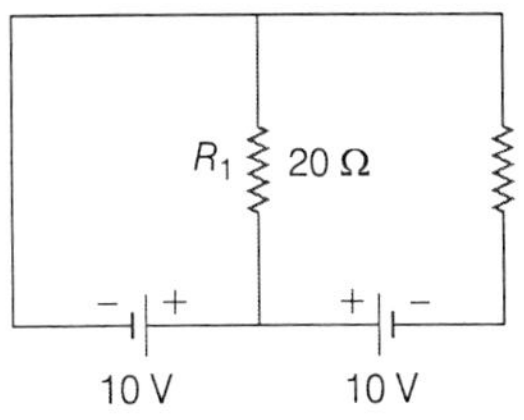

(a) 0.5, 0 (b) 1, 2 (c) 2, 2 (d) 0, 1

27 When the switch S in the circuit shown is closed, then the value of current i will be *[2019, 9 Jan Shift-I]*

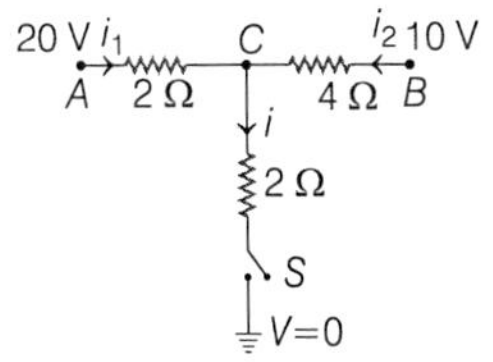

(a) 4 A (b) 3 A (c) 2 A (d) 5 A

28 Two batteries with emf 12 V and 13 V are connected in parallel across a load resistor of 10 Ω. The internal resistances of the two batteries are 1 Ω and 2 Ω, respectively. The voltage across the load lies between *[JEE Main 2018]*

(a) 11.6 V and 11.7 V (b) 11.5 V and 11.6 V
(c) 11.4 V and 11.5 V (d) 11.7 V and 11.8 V

29 In the given circuit diagram, when the current reaches steady state in the circuit, the charge on the capacitor of capacitance C will be *[JEE Main 2017]*

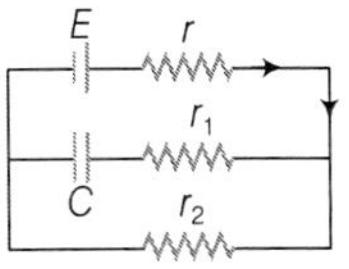

(a) $CE\dfrac{r_1}{(r_2+r)}$ (b) $CE\dfrac{r_2}{(r+r_2)}$

(c) $CE\dfrac{r_1}{(r_1+r)}$ (d) CE

30 In the circuit shown below, the current in the 1 Ω resistor is *[JEE Main 2015]*

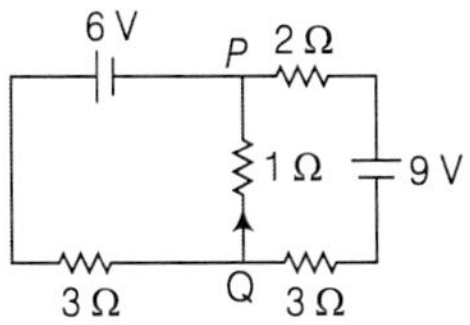

(a) 1.3 A, from P to Q (b) 0 A
(c) 0.13 A, from Q to P (d) 0.13 A, from P to Q

31 Two sources of equal emf are connected to an external resistance R. The internal resistances of the two sources are R_1 and R_2, $(R_2 > R_1)$. If the potential difference across the source having internal resistance R_2 is zero, then *[AIEEE 2005]*

(a) $R = \dfrac{R_2 \times (R_1 + R_2)}{(R_2 - R_1)}$ (b) $R = R_2 - R_1$

(c) $R = \dfrac{R_1 R_2}{(R_1 + R_2)}$ (d) $R = \dfrac{R_1 R_2}{(R_2 - R_1)}$

32 Two identical cells each of emf 1.5 V are connected in parallel across a parallel combination of two resistors each of resistance 20 Ω. A voltmeter connected in the circuit measures 1.2 V. The internal resistance of each cell is *[2002, June Shift I]*

(a) 2.5 Ω (b) 4 Ω (c) 5 Ω (d) 10 Ω

Level 3

33 The ratio of heat dissipated per second through the resistance 5 Ω and 10 Ω in the circuit given below is *[2024, 5 AprilShift -II]*

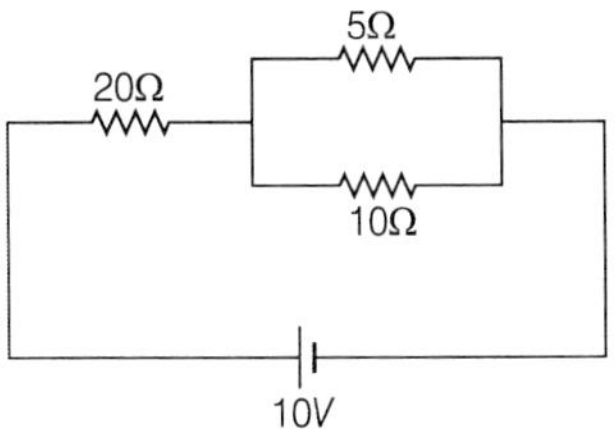

(a) 1 : 2 (b) 1 : 1 (c) 4 : 1 (d) 2 : 1

34 In this figure the resistance of the coil of galvanometer G is 2 Ω. The emf of the cell is 4 V. The ratio of potential difference across C_1 and C_2 is *[2023, 8 April Shift-I]*

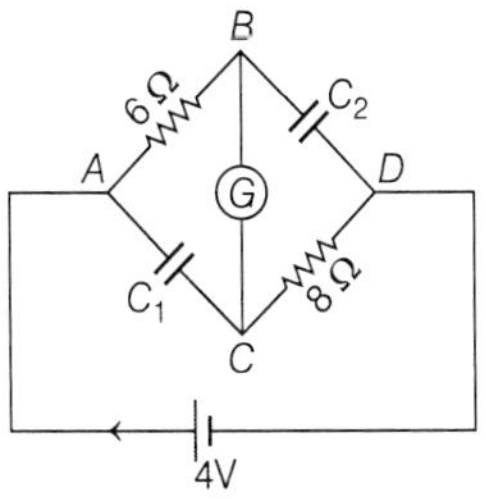

(a) $\dfrac{5}{4}$ (b) 1 (c) $\dfrac{4}{5}$ (d) $\dfrac{3}{4}$

35 A cell of emf 90 V is connected across series combination of two resistors each of 100 Ω resistance. A voltmeter of resistance 400 Ω is used to measure the potential difference across each resistor. The reading of the voltmeter will be *[2023, 24 Jan Shift-II]*

(a) 45 V` (b) 40 V (c) 80 V (d) 90 V

36

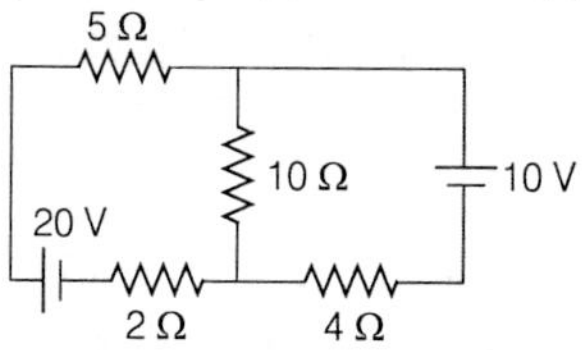

In above figure shown, the current in the 10 V battery is close to *[2020, 6 Sep Shift-II]*

(a) 0.71 A from positive to negative terminal
(b) 0.42 A from positive to negative terminal
(c) 0.21 A from positive to negative terminal
(d) 0.36 A from negative to positive terminal

37 A cell of internal resistance r drives current through an external resistance R. The power delivered by the cell to the external resistance will be maximum when *[2019, 8 April Shift-II]*

(a) $R = 2r$ (b) $R = r$ (c) $R = 0.001\,r$ (d) $R = 1000\,r$

38 In the given circuit, the internal resistance of the 18 V cell is negligible. If $R_1 = 400\ \Omega$, $R_3 = 100\ \Omega$ and $R_4 = 500\ \Omega$ and the reading of an ideal voltmeter across R_4 is 5 V, then the value of R_2 will be

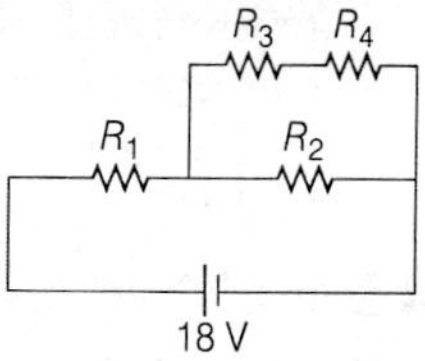

[2019, 9 Jan Shift-II]

(a) 550 Ω (b) 230 Ω (c) 300 Ω (d) 450 Ω

Numerical Value Type Questions

39 Two cells of emfs 1 V and 2 V and internal resistances 2 Ω and 1 Ω respectively, are connected in series with an external resistance of 6 Ω. The total current in the circuit is I_1. Now the same two cells in parallel configuration are connected to same external resistance. In this case, the total current drawn is I_2.

The value of $\left(\dfrac{I_1}{I_2}\right)$ is $\dfrac{x}{3}$. The value of x is .... *[2025, 3 April Shift-II]*

40 The current flowing through the 1Ω resistor is $\dfrac{n}{10}$ A.

The value of n is .... *[2024, 9 April Shift-I]*

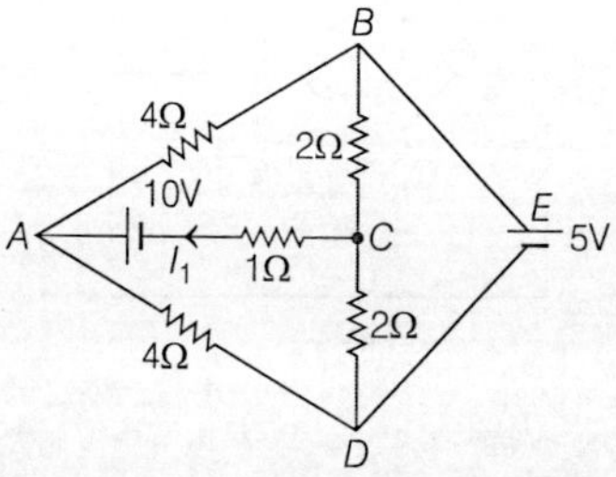

41 Twelve wires each having resistance 2Ω are joined to form a cube. A battery of 6 V emf is joined across point a and c. The voltage difference between e and f is...... V. *[2024, 4 April Shift-I]*

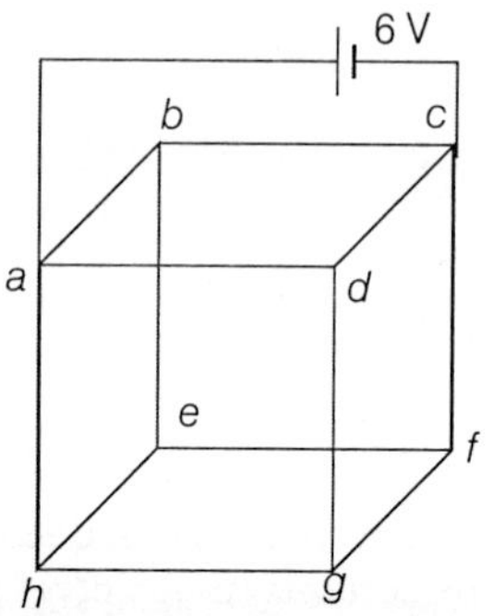

42 In the given circuit, the current flowing through the resistance 20 Ω is 0.3 A, while the ammeter reads 0.9 A. The value of R_1 is .... Ω. *[2024, 29 Jan Shift-II]*

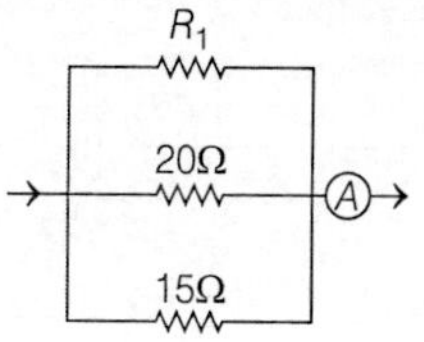

43 The charge accumulated on the capacitor connected in the following circuit is .... μC. (Given, $C = 150$ μF) *[2024, 27 Jan Shift-I]*

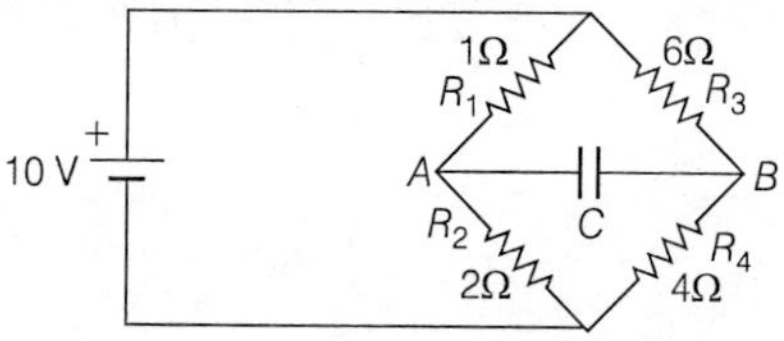

44 A network of four resistances is connected to 9 V battery, as shown in figure. The magnitude of voltage difference between the points A and B is .... V. *[2023, 15 April Shift-I]*

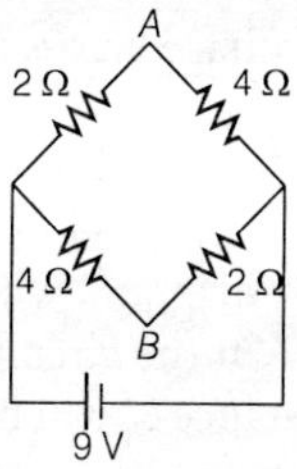

45 In the circuit shown, the energy stored in the capacitor is n μJ. The value of n is *[2023, 13 April Shift-II]*

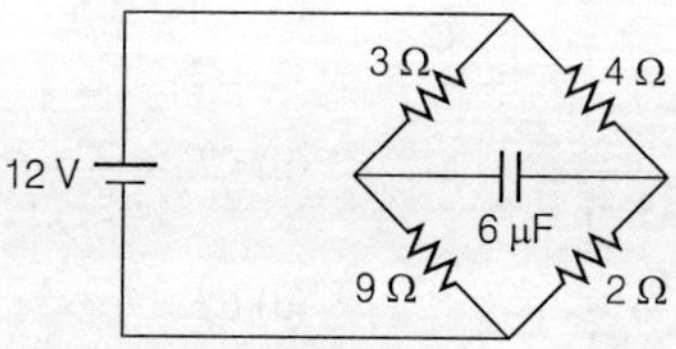

46 Two identical cells each of emf 1.5 V are connected in series across a $10\ \Omega$ resistance. An ideal voltmeter connected across $10\ \Omega$ resistance reads 1.5 V. The internal resistance of each cell is .... Ω.
[2023, 11 April Shift-II]

47 In the circuit diagram shown in figure given below, the current flowing through resistance 3Ω is $x/3$ A. The value of x is ....
[2023, 11 April Shift-I]

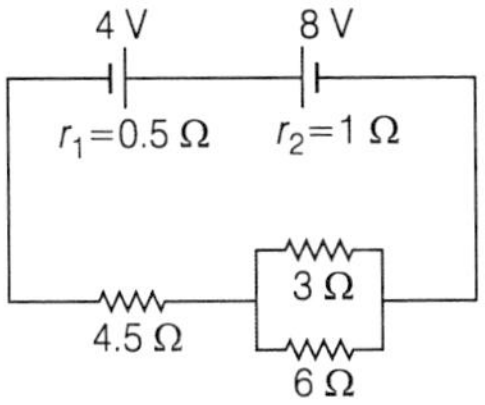

48 As shown in the figure, the voltmeter reads 2 V across $5\ \Omega$ resistor. The resistance of the voltmeter is ....Ω.
[2023, 6 April Shift-I]

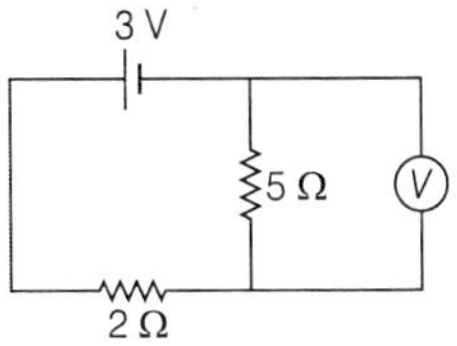

49 In the given circuit, the value of $\left|\dfrac{I_1 + I_3}{I_2}\right|$ is ....
[2023, 1 Feb Shift-II]

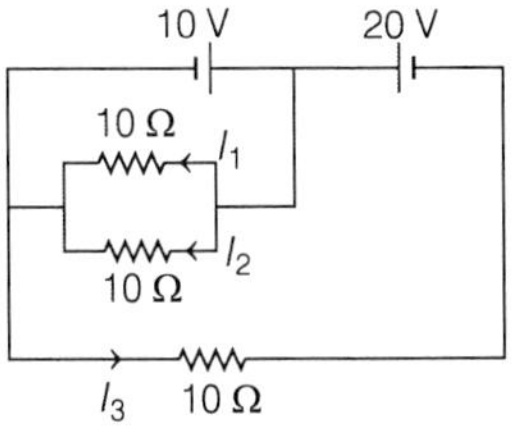

50 Two identical cells, when connected either in parallel or in series gives same current in an external resistance $5\ \Omega$. The internal resistance of each cell will be .... Ω.
[2023, 31 Jan Shift-I]

51 In the following circuit, the magnitude of current I_1, is .... A.
[2023, 30 Jan Shift-I]

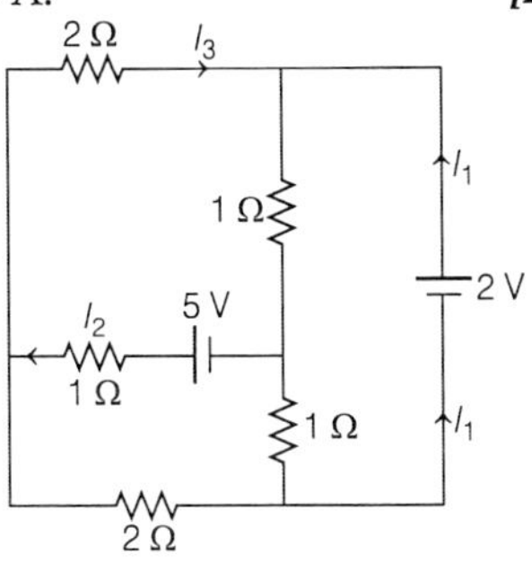

52 In the given figure, the value of V_0 will be .... V.
[2022, 25 July Shift-I]

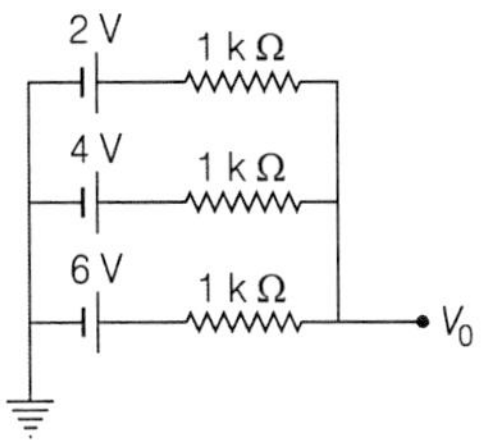

53 Two resistors are connected in series across a battery as shown in figure. If a voltmeter of resistance $2000\ \Omega$ is used to measure the potential difference across $500\ \Omega$ resistor, the reading of the voltmeter will be......V.
[2022, 29 June Shift-II]

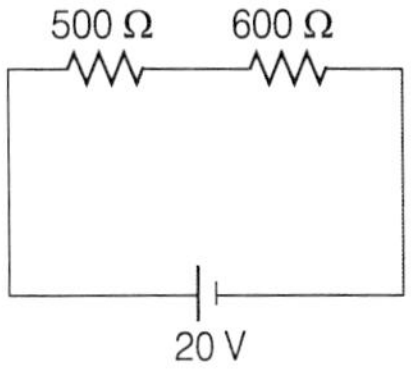

54 For the network shown below, the value of $V_B - V_A$ is V.
[2022, 29 June Shift-I]

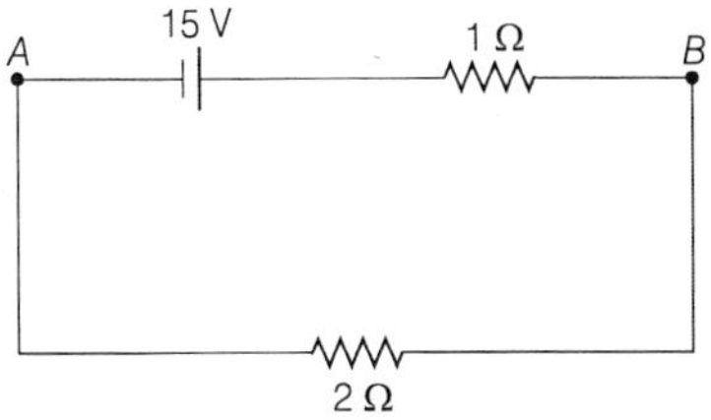

55 In an electric circuit, a cell of certain EMF provides a potential difference of 1.25 V across a load resistance of $5\ \Omega$. However, it provides a potential difference of 1 V across a load resistance of $2\ \Omega$. The emf of the cell is given by $\dfrac{x}{10}$ V. Then, the value of x is ....
[2021, 22 July Shift-II]

56 In an electrical circuit, a battery is connected to pass 20 C of charge through it in a certain given time. The potential difference between two plates of the battery is maintained at 15 V. The work done by the battery is .... J.
[2021, 26 Feb Shift-I]

Topic 4 Measuring Instruments

Level 1

01

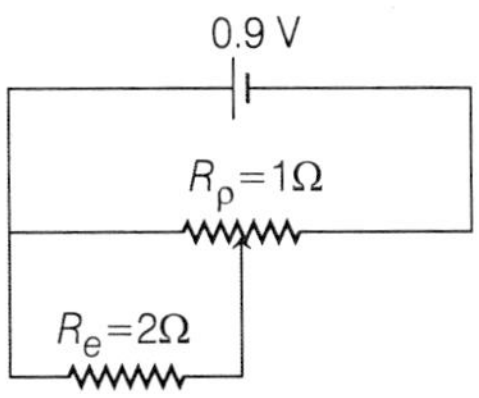

Sliding contact of a potentiometer is in the middle of the potentiometer wire having resistance $R_p = 1\Omega$ as shown in the figure. An external resistance of $R_e = 2\Omega$ is connected *via* the sliding contact. The electric current in the circuit is **[2025, 22 Jan Shift-I]**

(a) 1.35 A (b) 1.0 A
(c) 0.9 A (d) 0.3 A

02 The effective resistance between A and B, if resistance of each resistor is R, will be **[2024, 9 April Shift-II]**

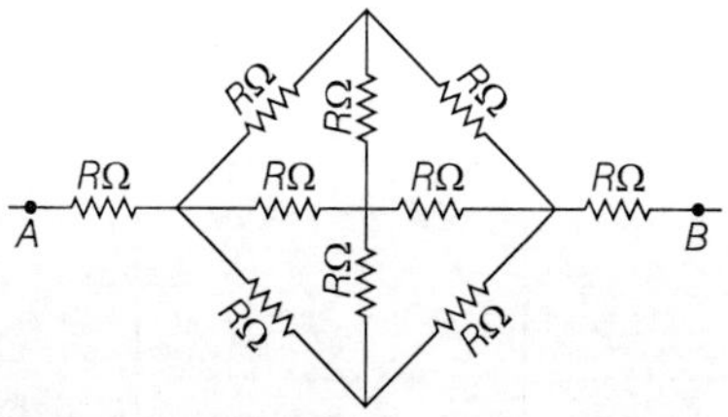

(a) $\frac{5R}{3}$ (b) $\frac{2}{3}R$
(c) $\frac{8R}{3}$ (d) $\frac{4R}{3}$

03 The resistance per centimeter of a meter bridge wire is r, with $X\ \Omega$ resistance in left gap. Balancing length from left end is at 40 cm with 25 Ω resistance in right gap. Now the wire is replaced by another wire of $2r$ resistance per centimeter. The new balancing length for same settings will be at **[2024, 31 Jan Shift-II]**

(a) 40 cm (b) 20 cm
(c) 80 cm (d) 10 cm

04 The equivalent resistance of the circuit shown below between points a and b is **[2023, 10 April Shift-I]**

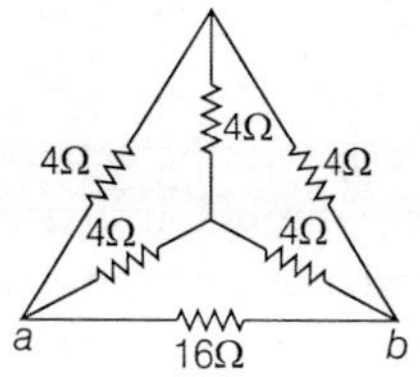

(a) 24 Ω
(b) 3.2 Ω
(c) 16 Ω
(d) 20 Ω

05 Which of the following statements is false? **[JEE Main 2017]**

(a) In a balanced Wheatstone bridge, if the cell and the galvanometer are exchanged, the null point is disturbed.
(b) A rheostat can be used as a potential divider.
(c) Kirchhoff's second law represents energy conservation.
(d) Wheatstone bridge is the most sensitive when all the four resistances are of the same order of magnitude.

Level 2

Imp. Wheatstone Bridge

06 The value of unknown resistance x for which the potential difference between B and D will be zero in the arrangement shown is **[2024, 6 April Shift-I]**

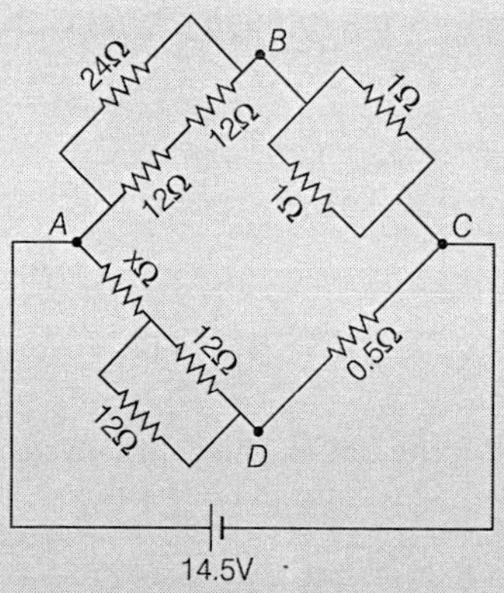

(a) 42 Ω (b) 3 Ω (c) 6 Ω (d) 9 Ω

07 To measure the internal resistance of a battery, potentiometer is used. For $R = 10\Omega$, the balance point is observed at $l = 500$ cm and for $R = 1\Omega$ the balance point is observed at $l = 400$ cm. The internal resistance of the battery is approximately. **[2024, 4 April Shift-I]**

(a) 0.1 Ω (b) 0.4 Ω (c) 0.2 Ω (d) 0.3 Ω

08 A wire of length 10 cm and radius $\sqrt{7} \times 10^{-4}$ m is connected across the right gap of a meter bridge. When a resistance of 4.5 Ω is connected on the left gap by using a resistance box, the balance length is found to be at 60 cm from the left end. If the resistivity of the wire is $R \times 10^{-7}\Omega$m, then value of R is **[2024, 27 Jan Shift-I]**

(a) 63 (b) 70 (c) 66 (d) 35

09 The equivalent resistance between A and B of the network shown in figure. **[2023, 1 Feb Shift-I]**

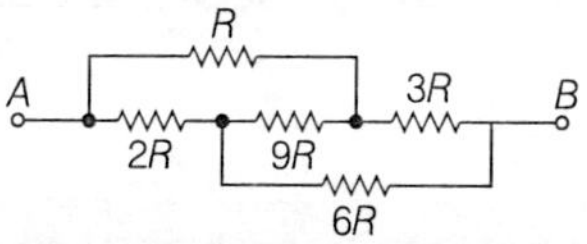

(a) $21\ R$ (b) $\frac{8}{3}R$ (c) $14\ R$ (d) $11\frac{2}{3}R$

10 The current I in the given circuit will be **[2022, 26 July Shift-I]**

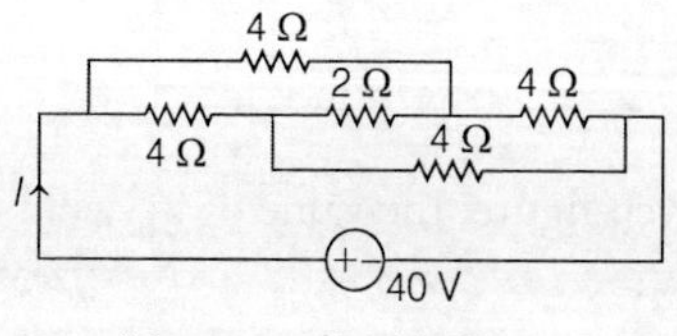

(a) 10 A (b) 20 A (c) 4 A (d) 40 A

11 Five equal resistances are connected in a network as shown in figure. The net resistance between the points A and B is *[2021, 26 Feb Shift-I]*

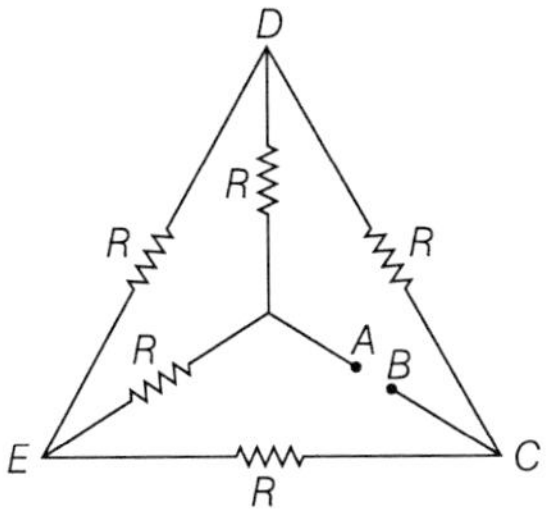

(a) $2R$ (b) $\frac{R}{2}$ (c) $\frac{3R}{2}$ (d) R

12 On interchanging the resistances, the balance point of a meter bridge shifts to the left by 10 cm. The resistance of their series combination is 1 kΩ. How much was the resistance on the left slot before interchanging the resistances? *[JEE Main 2018]*

(a) 990 Ω (b) 505 Ω
(c) 550 Ω (d) 910 Ω

13 Shown in the figure adjacent is a meter-bridge set up with null deflection in the galvanometer. The value of the unknown resistor R is *[AIEEE 2008]*

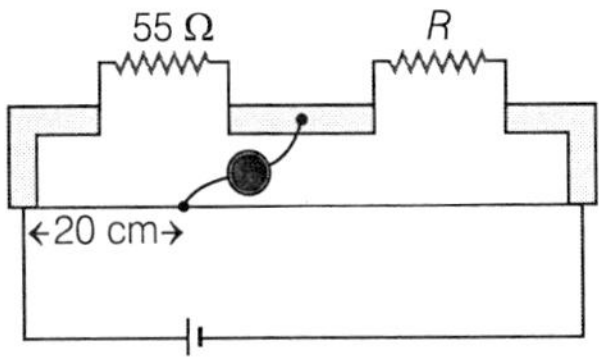

(a) 13.75 Ω (b) 220 Ω
(c) 110 Ω (d) 55 Ω

14 The current I drawn from the 5 V source will be *[AIEEE 2006]*

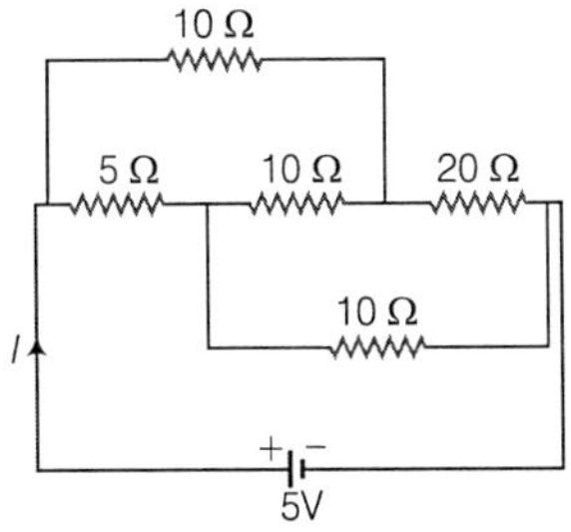

(a) 0.33 A (b) 0.5 A
(c) 0.67 A (d) 0.17 A

15 In a Wheatstone's bridge, three resistances P, Q and R are connected in the three arms and the fourth arm is formed by two resistances S_1 and S_2 connected in parallel. The condition for the bridge to be balanced will be *[AIEEE 2006]*

(a) $\frac{P}{Q} = \frac{2R}{S_1 + S_2}$ (b) $\frac{P}{Q} = \frac{R(S_1 + S_2)}{S_1 S_2}$

(c) $\frac{P}{Q} = \frac{R(S_1 + S_2)}{2 S_1 S_2}$ (d) $\frac{P}{Q} = \frac{R}{S_1 + S_2}$

Level 3

16 To measure the temperature coefficient of resistivity α of a semiconductor, an electrical arrangement shown in the figure is prepared. The arm BC is made up of the semiconductor. The experiment is being conducted at 25°C and resistance of the semiconductor arm is 3 mΩ. Arm BC is cooled at a constant rate of 2°C/s. If the galvanometer G shows no deflection after 10 s, then α is *[2024, 1 Feb Shift-II]*

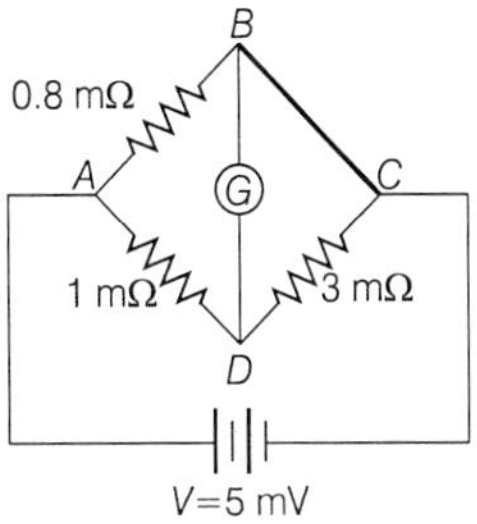

(a) -1.5×10^{-2} °C^{-1}
(b) -2.5×10^{-2} °C^{-1}
(c) -1×10^{-2} °C^{-1}
(d) -2×10^{-2} °C^{-1}

17 A galvanometer G of 2 Ω resistance is connected in the given circuit. The ratio of charge stored in C_1 and C_2 is *[2024, 1 Feb Shift-II]*

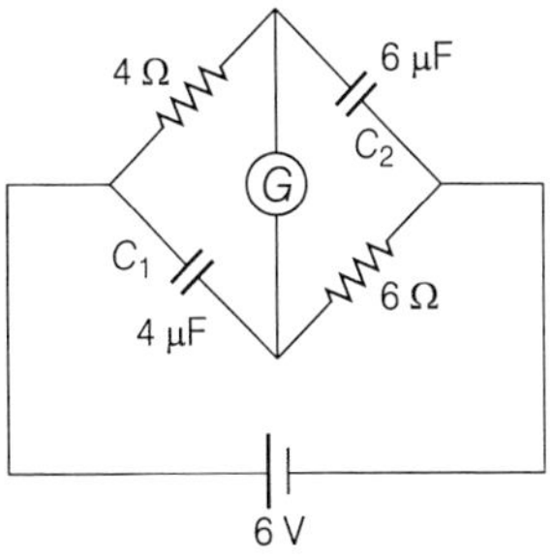

(a) 3/2 (b) 1/2
(c) 2/3 (d) 1

18 In a meter bridge when a resistance in the left gap is 2 Ω and unknown resistance in the right gap, the balance length is found to be 40 cm. On shunting the unknown resistance with 2 Ω, the balance length changes by *[2024, 1 Feb Shift-II]*

(a) 22.5 cm (b) 65 cm
(c) 62.5 cm (d) 20 cm

19 The four arms of a Wheatstone bridge have resistances as shown in the figure. A galvanometer of $15\,\Omega$ resistance is connected across BD. Calculate the current through the galvanometer when a potential difference of 10 V is maintained across AC. *[2021, 17 March Shift-II]*

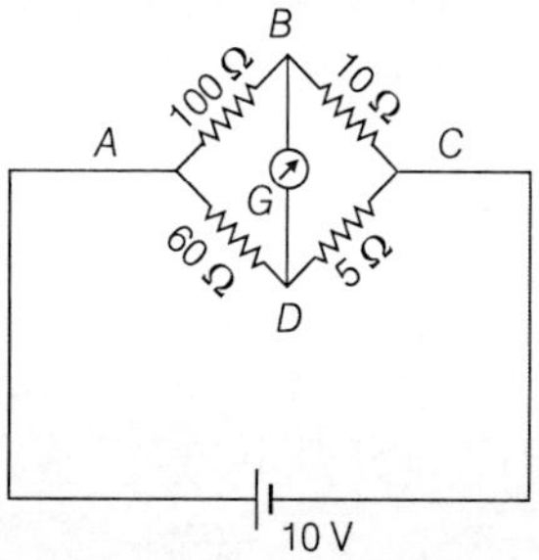

(a) 2.44 μA (b) 2.44 mA
(c) 4.87 mA (d) 4.87 μA

20 The resistance of the meter bridge AB in given figure is $4\,\Omega$. With a cell of emf $\varepsilon = 0.5$ V and rheostat resistance $R_h = 2\,\Omega$, the null point is obtained at some point J. When the cell is replaced by another one of emf $\varepsilon = \varepsilon_2$, the same null point J is found for $R_h = 6\,\Omega$. The emf ε_2 is *[2019, 11 Jan Shift-I]*

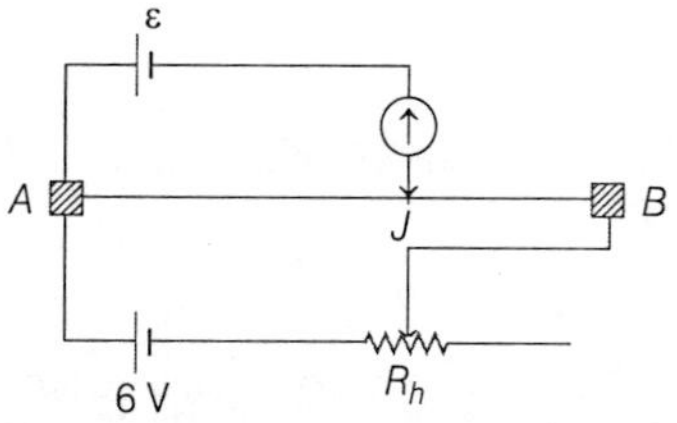

(a) 0.6 V (b) 0.3 V (c) 0.5 V (d) 0.4 V

21 In a meter bridge experiment, null point is obtained at 20 cm from one end of the wire when resistance X is balanced against another resistance Y. If $X < Y$, then where will be the new position of the null point from the same end, if one decides to balance a resistance of $4X$ against Y? *[AIEEE 2004]*

(a) 50 cm (b) 80 cm
(c) 40 cm (d) 70 cm

Numerical Value Type Questions

22 The value of current I in the electrical circuit as given below, when potential at A is equal to the potential at B, will be......... A. *[2025, 28 Jan Shift-II]*

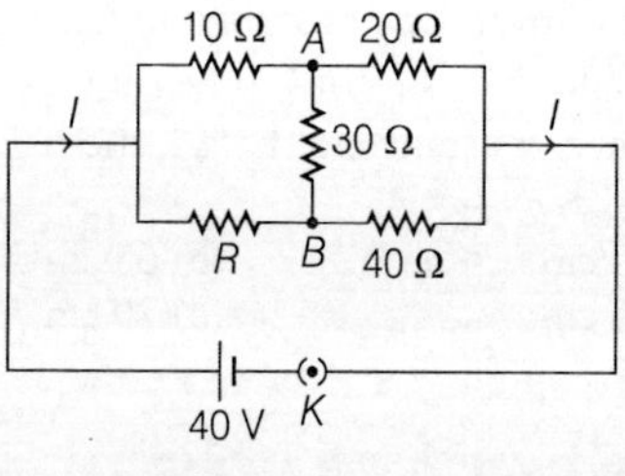

23 If the potential difference between B and D is zero, the value of x is $\frac{1}{n}\,\Omega$. The value of n is .... *[2023, 30 Jan Shift-II]*

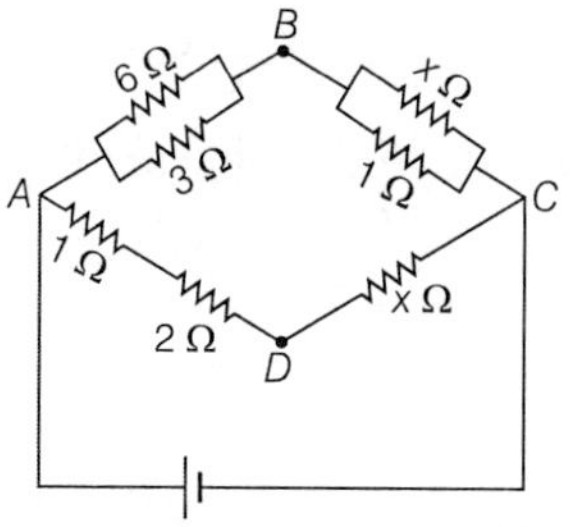

24 When two resistances R_1 and R_2 are connected in series and introduced into the left gap of a meter bridge and a resistance of $10\,\Omega$ is introduced into the right gap, a null point is found at 60 cm from left side. When R_1 and R_2 are connected in parallel and introduced into the left gap, a resistance of $3\,\Omega$ is introduced into the right-gap to get null point at 40 cm from left end. The product of $R_1 R_2$ is ...... Ω^2. *[2023, 29 Jan Shift-II]*

25 In a metre bridge experiment, the balance point is obtained if the gaps are closed by $2\,\Omega$ and $3\,\Omega$. A shunt of $X\Omega$ is added to $3\,\Omega$ resistor to shift the balancing point by 22.5 cm. The value of X is .... *[2023, 29 Jan Shift-I]*

26 In the given figure of meter bridge experiment, the balancing length AC corresponding to null deflection of the galvanometer is 40 cm. The balancing length, if the radius of the wire AB is doubled, will be .... cm. *[2022, 27 July Shift-II]*

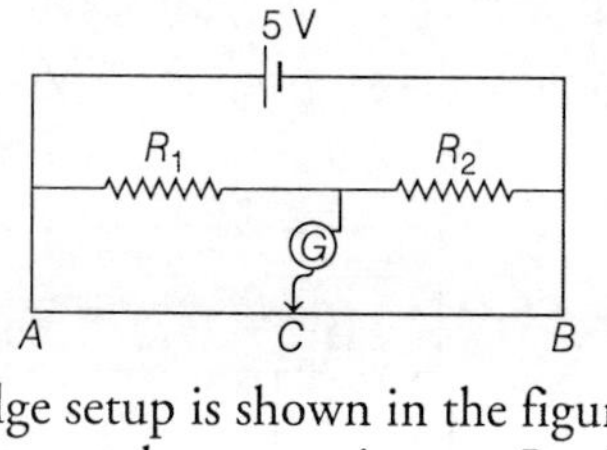

27 A meter bridge setup is shown in the figure. It is used to determine an unknown resistance R using a given resistor of $15\,\Omega$. The galvanometer G shows null deflection when tapping key is at 43 cm mark from end A. If the end correction for end A is 2 cm, then the determined value of R will be ...... Ω. *[2022, 28 June Shift-I]*

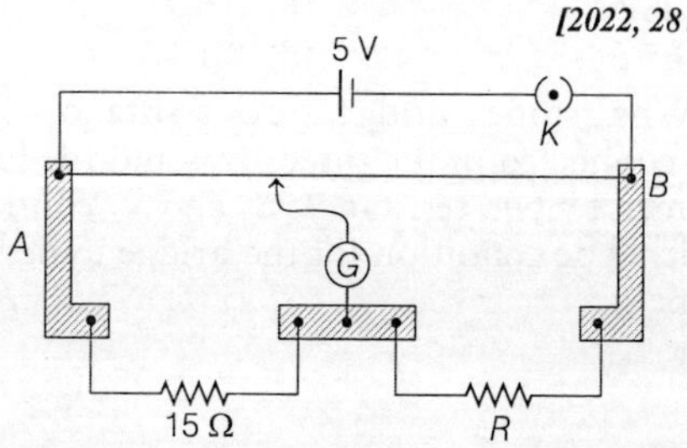

28 Current measured by the ammeter A in the reported circuit when no current flows through $10\,\Omega$ resistance, will be ...... A. *[2022, 28 June Shift-I]*

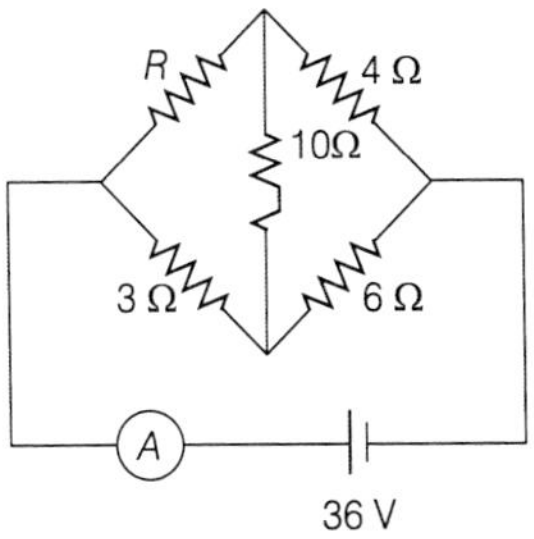

29 Consider a 72 cm long wire AB as shown in the figure. The galvanometer jockey is placed at P on AB at a distance x cm from A. The galvanometer shows zero deflection.

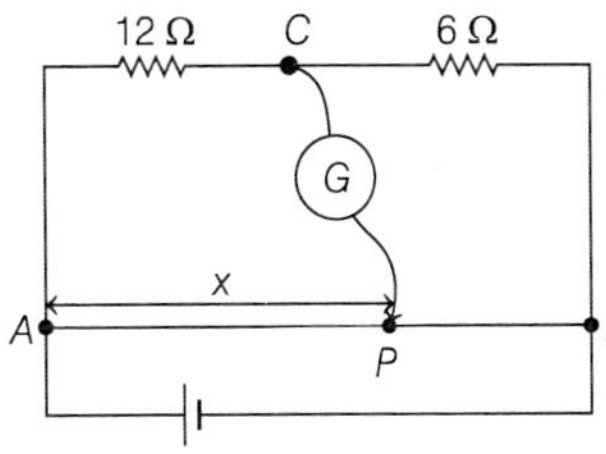

The value of x, to the nearest integer, is ....
[2021, 18 March Shift-II]

30 Four resistances of $15\,\Omega$, $12\,\Omega$, $4\,\Omega$ and $10\,\Omega$ respectively in cyclic order to form Wheatstone's network. The resistance that is to be connected in parallel with the resistance of $10\,\Omega$ to balance the network is .... Ω. *[2020, 8 Jan Shift-I]*

• Key **Index**

Topic 1

1	(b)	**2**	(a)	**3**	(a)	**4**	(a)	**5**	(b)	**6**	(d)	**7**	(a)	**8**	(c)	**9**	(a)	**10**	(a)
11	(b)	**12**	(d)	**13**	(c)	**14**	(b)	**15**	(a)	**16**	(a)	**17**	(d)	**18**	(c)	**19**	(a)	**20**	(a)
21	(d)	**22**	(c)	**23**	(c)	**24**	(b)	**25**	(a)	**26**	(a)	**27**	(c)	**28**	(b)	**29**	(b)	**30**	(*)
31	(b)	**32**	(d)	**33**	(d)	**34**	(c)	**35**	(d)	**36**	(d)	**37**	(c)	**38**	(d)	**39**	(c)	**40**	(b)
41	(a)	**42**	(b)	**43**	(b)	**44**	(a)	**45**	(b)	**46**	(c)	**47**	(c)	**48**	(d)	**49**	(b)	**50**	(c)
51	(d)	**52**	(a)	**53**	(c)	**54**	(b)	**55**	(b)	**56**	(b)	**57**	(a)	**58**	(d)	**59**	(b)	**60**	(b)
61	(c)	**62**	(a)	**63**	(d)	**64**	(2)	**65**	(16)	**66**	(1)	**67**	(160)	**68**	(16)	**69**	(5)	**70**	(32)
71	(22)	**72**	(200)	**73**	(15)	**74**	(3)	**75**	(100)	**76**	(125)	**77**	(25)	**78**	(25)	**79**	(10.33)	**80**	(44)
81	(2)	**82**	(2)	**83**	(136)	**84**	(8)	**85**	(48)	**86**	(300)	**87**	(2)	**88**	(3)	**89**	(100)	**90**	(9)
91	(0.112)	**92**	(45)	**93**	(10)	**94**	(2)	**95**	(70)	**96**	(4)	**97**	(3)	**98**	(2)	**99**	(5)		

Topic 2

1	(b)	**2**	(c)	**3**	(d)	**4**	(c)	**5**	(a)	**6**	(a)	**7**	(c)	**8**	(c)	**9**	(c)	**10**	(b)
11	(a)	**12**	(c)	**13**	(d)	**14**	(c)	**15**	(d)	**16**	(b)	**17**	(c)	**18**	(a)	**19**	(c)	**20**	(d)
21	(d)	**22**	(c)	**23**	(c)	**24**	(c)	**25**	(c)	**26**	(b)	**27**	(a)	**28**	(b)	**29**	(*)	**30**	(a)
31	(d)	**32**	(1027)	**33**	(5)	**34**	(748)	**35**	(3)	**36**	(16)	**37**	(14)	**38**	(15)	**39**	(450)	**40**	(4)
41	(3840)	**42**	(50)	**43**	(2500)														

Topic 3

1	(d)	**2**	(d)	**3**	(d)	**4**	(c)	**5**	(c)	**6**	(d)	**7**	(b)	**8**	(d)	**9**	(a)	**10**	(b)
11	(d)	**12**	(c)	**13**	(*)	**14**	(b)	**15**	(b)	**16**	(a)	**17**	(a)	**18**	(a)	**19**	(c)	**20**	(b)
21	(c)	**22**	(d)	**23**	(c)	**24**	(b)	**25**	(a)	**26**	(a)	**27**	(d)	**28**	(b)	**29**	(b)	**30**	(c)
31	(b)	**32**	(c)	**33**	(d)	**34**	(c)	**35**	(b)	**36**	(c)	**37**	(b)	**38**	(c)	**39**	(4)	**40**	(25)
41	(1)	**42**	(30)	**43**	(400)	**44**	(3)	**45**	(75)	**46**	(5)	**47**	(1)	**48**	(20)	**49**	(2)	**50**	(5)
51	(1.5)	**52**	(4)	**53**	(8)	**54**	(10)	**55**	(15)	**56**	(300)								

Topic 4

1	(b)	**2**	(c)	**3**	(a)	**4**	(b)	**5**	(a)	**6**	(c)	**7**	(d)	**8**	(c)	**9**	(b)	**10**	(a)
11	(d)	**12**	(c)	**13**	(b)	**14**	(b)	**15**	(b)	**16**	(c)	**17**	(b)	**18**	(a)	**19**	(c)	**20**	(b)
21	(a)	**22**	(2)	**23**	(2)	**24**	(30)	**25**	(2)	**26**	(40)	**27**	(19)	**28**	(10)	**29**	(48)	**30**	(10)

Detailed **Explanations**

Topic 1

01. *(b)*

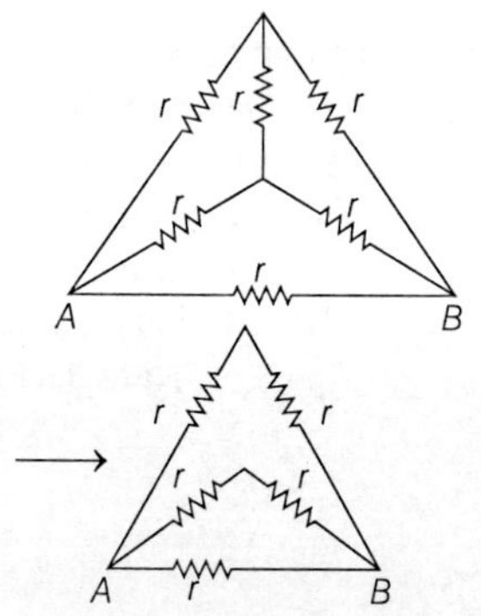

$$r = \frac{R}{6}$$

By balanced Wheatstone bridge

$$\frac{R_1}{R_2} = \frac{R_3}{R_4}$$

$$\frac{1}{R_{AB}} = \frac{1}{2r} + \frac{1}{2r} + \frac{1}{r}$$

$$\frac{1}{R_{AB}} = \frac{2}{r}$$

$$R_{AB} = \frac{r}{2} = \frac{R}{12}$$

02. *(a)* $I(t) = 0.02t + 0.01\text{ A}$

$$q = \int I dt$$

$$q = \int_1^2 (0.02t + 0.01)\, dt$$

$$q = \left[0.02\frac{t^2}{2} + 0.01t\right]_1^2$$

$$= 0.01(2)^2 + 0.01(2) - (0.01(1)^2 - 0.01(1))$$

$$= 0.04 + 0.02 - 0.01 - 0.01 = 0.04\text{ C}$$

03. *(a)* From the following circuit, all the resistances are parallel to each other, so it can be redrawn as

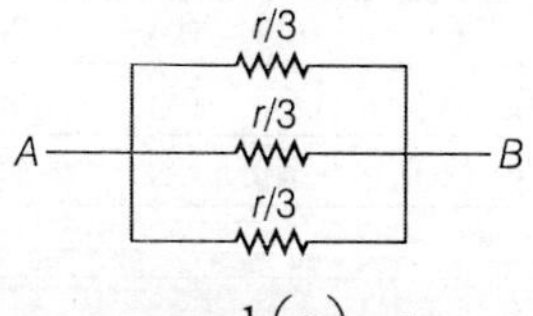

$$R_{\text{eq}} = \frac{1}{3}\left(\frac{r}{3}\right) = \frac{r}{9}$$

04. *(a)* For wire bound standard resistors, the value of resistivity changes very minimally when there is increase in temperature.

Hence, the correct graph is

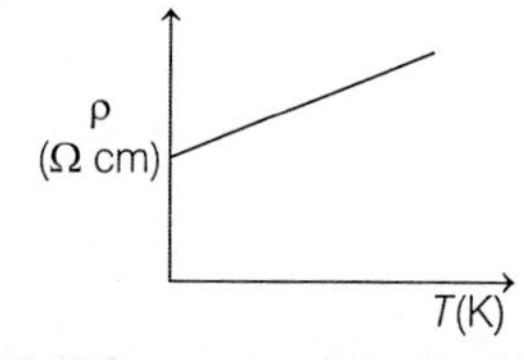

05. *(b)* As $R \propto l$

$\therefore$ Resistance of each part will be $\frac{R}{5}$.

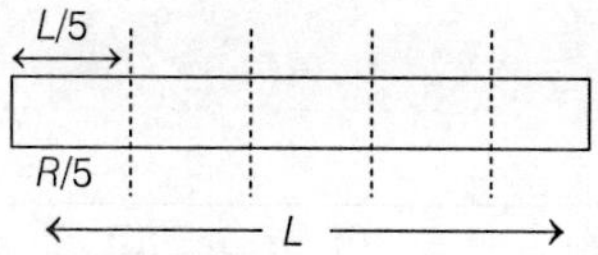

Parallel arrangement is,

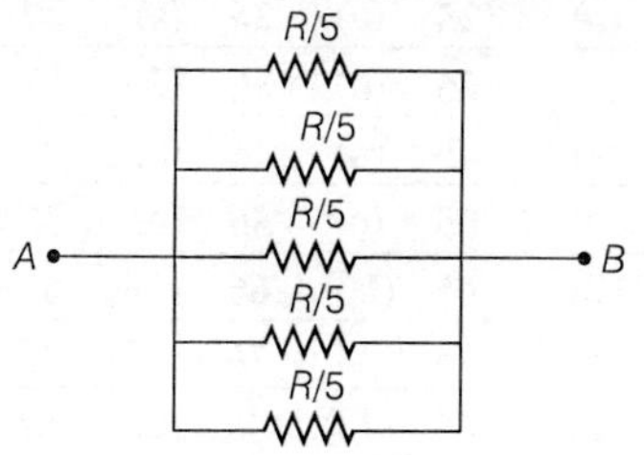

$$R_{\text{eq}} = \frac{R/5}{5} = \frac{R}{25}$$

06. *(d)* In the absence of electric field, electron masses move in straight lines between each successive collision. But under an external electric field, in a metallic conductor, electrons follow a curved path. Since electrons are negatively charged, therefore, they move from lower potential to higher potential.

07. *(a)* Drift velocity of electrons through conductor is given as,

$$v_d = \frac{I}{neA}$$

$$v_d = \frac{V}{neAR} = \frac{VA}{neA\rho l} = \frac{V}{ne\rho l}$$

$\because$ For same potential difference, same material, same length $v'_d = v_d$

Drift velocity is independent of area of cross-section so it will remain constant.

08. *(c)* Given, initial resistance of wire

$$R_1 = 5\,\Omega$$

If R_2 be the new resistance of the wire when wire is stretched to 5 times, then

$$R_2 = 5^2 R_1 \qquad (\because R_2 = n^2 R_1)$$

$$= 25 \times 5 = 125\,\Omega$$

> ***Alternate Solution***
>
> Given, $R = 5\Omega$
>
> We know that, $R = \frac{\rho L}{A}$
>
> Volume remain same after stretching
>
> $V = V' \Rightarrow AL = A'L'$
>
> $AL = A'(5L) \qquad (\because \text{Given, } L' = 5L)$
>
> $\frac{A}{5} = A'$
>
> So, new resistance $R' = \frac{\rho L'}{A'} = \frac{\rho 5L}{\frac{A}{5}} = 5 \times 5 \frac{\rho L}{A}$
>
> $= 25R = 25 \times 5 = 125\,\Omega$

09. *(a)* Constantan and manganin have very low value of temperature coefficient of resistance and high value of resistivity. Thus, their resistance changes very little with temperature. Hence, constantan and manganin are used in making standard resistance coils.

Thus, both A and R are true and R is the correct explanation of A.

10. *(a)* New length $= L + 2L = 3L$

Volume will remain same

i.e. $\qquad V = V'$

$\therefore \qquad A' = A/3$

Initial resistance, $R = \frac{\rho L}{A}$ ...(i)

New resistance, $R' = \frac{\rho L'}{A'}$ ...(ii)

Dividing Eq. (ii) by Eq. (i), we get

$$\frac{R'}{R} = \frac{L'}{A'} \times \frac{A}{L}$$

$$= \frac{3L}{A/3} \cdot \frac{A}{L} = 9$$

$\Rightarrow R' : R = 9 : 1$

> • Concept **Enhancer**
>
> We know that, when length of wire is increased to n times, then its resistance is increased by n^2 times.

11. *(b)* Equivalent resistances,

$$R_{eq} = (A \| B) + C$$
$$= \frac{A \times B}{A + B} + C = \frac{2 \times 4}{2+4} + 6 = \frac{22}{3}\Omega$$

12. *(d)* The required value of combination is $46/3\,\Omega$.

In order to achieve the above mentioned value of resistance from given resistances, we will connect $2\,\Omega$ and $4\,\Omega$ resistance in parallel, then join $6\,\Omega$ and $8\,\Omega$ resistance in series with the combination.

The circuit diagram for connection is shown below.

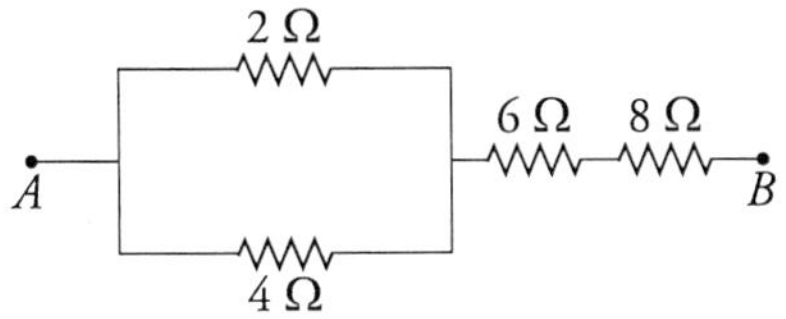

$$R_{eq} = (2\|4) + 6 + 8$$
$$= \frac{2\times 4}{2+4} + 14$$
$$= \frac{46}{3}\Omega$$

Thus, the resistances of $2\,\Omega$ and $4\,\Omega$ are in parallel and this combination is in series with 6Ω and 8Ω.

13. *(c)* Given, current, $I = 5$ A

Area of cross-section of wire,

$$A = 0.04\ \text{m}^2$$

We know that $J = \frac{I}{A}$

$\Rightarrow \quad I = JA$ or $\mathbf{I} = \mathbf{J}\cdot\mathbf{A}$

or $\quad I = JA\cos\theta$

where, J = current density

$$\Rightarrow \quad 5 = J\left(\frac{4}{100}\right)\times\cos(60°)$$

[$\because$ Given, $\theta = 60°$]

$$J = 500 \times \frac{1}{2} \qquad \left[\because \cos 60° = \frac{1}{2}\right]$$

$$\Rightarrow \quad J = 250\ \text{Am}^{-2}$$

The relation between electric field, current density and resistivity can be given as

$$\mathbf{E} = \rho\cdot\mathbf{J} = 44\times10^{-8}\times 250$$

[$\because$ Resistivity, $\rho = 44\times10^{-8}\ \Omega$-m]

$$= 11\times10^{-5}\ \text{V/m}$$

14. *(b)* Given, current, $I = 10$ A

Cross-sectional area,

$$A = 5\ \text{mm}^2 = 5\times10^{-6}\ \text{m}^2$$

Drift velocity, $v_d = 2\times10^{-3}\ \text{ms}^{-1}$

We know that, $I = neAv_d$

$$\Rightarrow \quad 10 = n\times1.6\times10^{-19}\times5\times10^{-6}\times2\times10^{-3}$$

$$\Rightarrow \quad n = \frac{10}{1.6\times10^{-19}\times5\times10^{-6}\times2\times10^{-3}}$$
$$= 0.625\times10^{28} = 625\times10^{25}$$

> **• Concept Enhancer**
>
> The value of current flowing through a conductor can be given by
>
> $I = neAv_d$
>
> where, n = number of free electrons
>
> and e = charge on an electron

15. *(a)* Initially, the resistance of wire is

$$R_1 = \rho L/A$$

In second case,

Length, $l' = 2l$

Area, $A' = A/2$

$$\therefore \quad R_2 = \frac{\rho l'}{A'} = \frac{\rho(2l)}{(A/2)} = \frac{4\rho l}{A} \quad \ldots(\text{i})$$

According to Ohm's law,

$$I = \frac{V}{R_2}$$

$$\Rightarrow I = \frac{V}{4\rho l/A} = \frac{1}{4}\frac{VA}{\rho l} \quad [\text{using Eq.(i)}]$$

16. *(a)* It is a balanced Wheatstone bridge, therefore resistance between E and D can be removed.

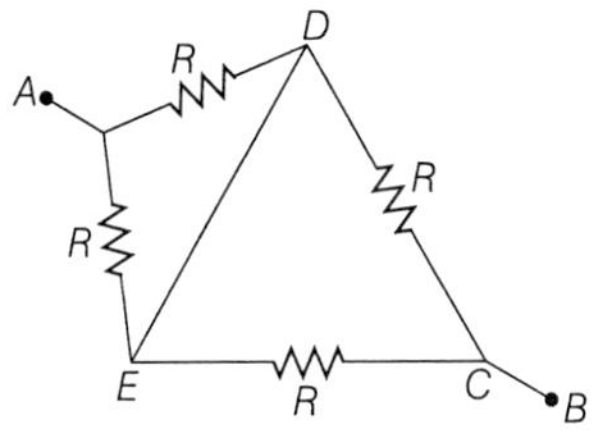

So, $\frac{1}{R_{eq}} = \frac{1}{2R} + \frac{1}{2R} = \frac{1}{R}$

$$\Rightarrow R_{AB} = R_{eq} = R$$

17. *(d)* Aluminium is more resistive than copper and mercury is most resistive of all. So, $\rho_M > \rho_A > \rho_C$

18. *(c)* Given, $I = 5$A,

$$\rho = 1.7\times10^{-8}\ \Omega\text{-m},$$
$$r = 5\ \text{mm} = 5\times10^{-3}\ \text{m},$$
$$v_d = 1.1\times10^{-3}\ \text{m/s}$$

Mobility of charges,

$$\mu = \frac{v_d}{E} \quad \ldots(\text{i})$$

$$\mu = \frac{v_d A}{\rho I} \qquad \left(\because E = \frac{\rho I}{A}\right)$$

$$= \frac{1.1\times10^{-3}\times\pi\times(5\times10^{-3})^2}{1.7\times10^{-8}\times5}$$
$$= \frac{86.35\times10^{-9}}{8.5\times10^{-8}} = 10.1\times10^{-1}$$

$$\Rightarrow \mu \approx 1\ \text{m}^2/\text{V-s}$$

19. *(a)* Relation between current I flowing through a conducting wire and drift velocity of electrons v_d is given as

$$I = neAv_d$$

where, n is the electron density and A is the area of cross-section of wire.

$$\Rightarrow \quad v_d = \frac{I}{neA}$$

On substituting the given values, we get

$$v = \frac{1.5}{9\times10^{28}\times1.6\times10^{-19}\times5\times10^{-6}}$$

$$v = \frac{1.5\times10^{-3}}{72}\ \text{m/s} = 0.2\times10^{-4}\ \text{m/s}$$

or $\quad v = 0.02$ mm/s

20. *(a)* Current flowing through copper rod is given by $I = neAv_d = \rho Av_d$

$$(\because \rho = ne)$$

$$v_d = \frac{I}{\rho A}$$

Time taken by charges to travel distanced d,

$$t = \frac{d}{v_d} = \frac{d}{(I/\rho A)} = \frac{\rho Ad}{I}$$

21. *(d)* Resistance of a wire of length l and radius r is given by

$$R = \frac{\rho l}{A} = \frac{\rho l}{A}\times\frac{A}{A} = \frac{\rho V}{A^2} = \frac{\rho V}{\pi^2 r^4}$$

$$(\because V = Al)$$

i.e., $R \propto \frac{1}{r^4}$

$$\therefore \quad \frac{R_1}{R_2} = \left(\frac{r_2}{r_1}\right)^4$$

Here, $R_1 = 100\,\Omega$, $r_1 = r$, $r_2 = \frac{r}{2}$, $R_2 = ?$

$$\therefore \quad R_2 = R_1\left(\frac{r_1}{r_2}\right)^4 = R_1\left(\frac{r}{r/2}\right)^4$$
$$= 16R_1 = 1600\,\Omega$$

22. *(c)* As, we know Cu is a conductor, so with increase in temperature, resistance will increase.

Then, Si is semiconductor, so with increase in temperature, resistance will decrease.

23. *(c)* Given,

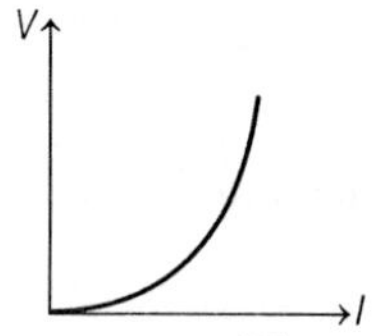

$$v_d \propto \sqrt{E}$$

We know, $I = neAv_d$ and $E = V/l$ or $E \propto V$

So $I \propto \sqrt{V},\ I^2 \propto V$

24. *(b)* Resistance of combination,

$$R_e = 4R$$

$$\Rightarrow \frac{\Delta R_e}{R_e} = \frac{\Delta R}{R} = \frac{5 \times 100}{100} = 5\%$$

25. *(a)* $R = \frac{\rho l}{A} = \frac{\rho l^2}{V}$ [$\because V =$ volume]

$$\therefore \frac{\Delta R}{R} = 2\frac{\Delta l}{l} = +\ 0.2\%$$

26. *(a)* $R = R_t\,(1 + \alpha\,\Delta T)$

Given, $R_t = 100\,\Omega$

$R = 150\,\Omega \Rightarrow \Delta T = 200°\,C$

$\therefore 150 = 100\,(1 + \alpha\ 200) \Rightarrow 0.5 = 200\,\alpha$

$$\Rightarrow \alpha = \frac{0.5}{200} = \frac{1}{400} = 25 \times 10^{-4}/°C$$

$$= 2.5 \times 10^{-3}/°C$$

$\therefore$Statement 1 is true.

27. *(c)* From $R_t = R_0(1 + \alpha t)$

$\therefore\ 5 = R_0(1 + 50\,\alpha)$ ...(i)

and $6 = R_0(1 + 100\,\alpha)$ ...(ii)

$$\therefore \frac{5}{6} = \frac{1 + 50\,\alpha}{1 + 100\,\alpha}$$

or $\alpha = \frac{1}{200}$

On putting the value of α in Eq. (i), we get

$$5 = R_0\left(1 + 50 \times \frac{1}{200}\right)$$

or $R_0 = 4\,\Omega$

28. *(b)* Let resistance of bulb filament be R_0 at 0°C, then from expression,

$R_\theta = R_0[1 + \alpha\Delta\theta]$ ...(i)

We have, $100 = R_0[1 + 0.005 \times 100]$ ...(ii)

and $200 = R_0[1 + 0.005 \times x]$

where, x is temperature in °C at which resistance become 200 Ω.

Dividing the above two equations, we get

$$\frac{200}{100} = \frac{1 + 0.005x}{1 + 0.005 \times 100}$$

$$\Rightarrow x = 400°C$$

29. *(b)* Resistance in the arms AC and BC are in series.

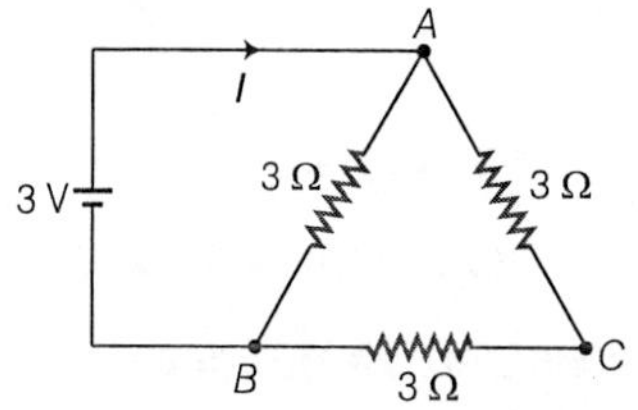

$\therefore\ R' = 3 + 3 = 6\,\Omega$

Now, R' and $3\,\Omega$ are in parallel.

$$\therefore R_{eq} = \frac{6 \times 3}{6 + 3} = 2\,\Omega$$

Now, $V = IR$

$$\Rightarrow I = \frac{3}{2} = 1.5\ A$$

30. (*) Here, $l = 25$ m, $A = 5\ mm^2$

$= 5 \times 10^{-6}\ m^2$

$\rho = 2 \times 10^{-6}\,\Omega m$

$$R = R_{wire} = \frac{\rho l}{A} = \frac{2 \times 10^{-6} \times 25}{5 \times 10^{-6}}$$

$R = 10\,\Omega$

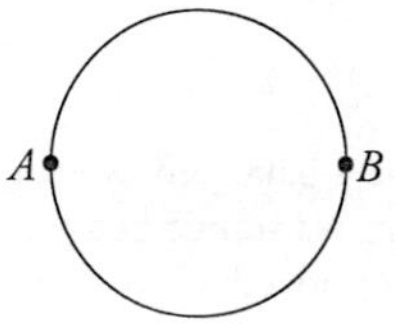

$\Rightarrow$

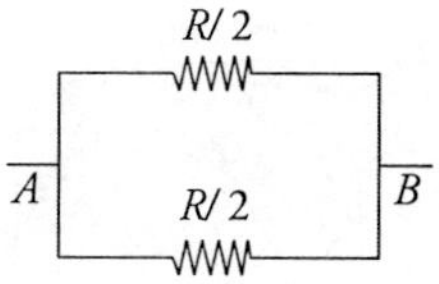

$$\therefore R_{eq} = \frac{\frac{R}{2} \times \frac{R}{2}}{\frac{R}{2} + \frac{R}{2}}$$

$$= \frac{R}{4} = \frac{10}{4} = 2.5\Omega$$

31. *(b)* Given, $R_1 = R_2 = R_3 = 5\Omega$

$R_4 = 10\,\Omega$

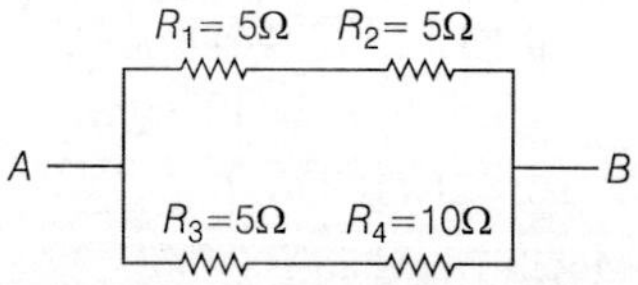

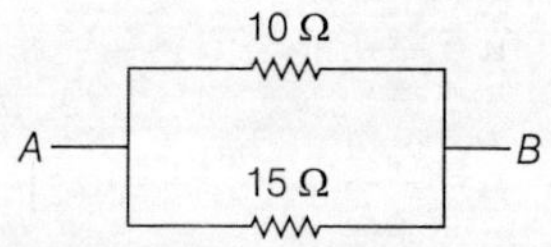

$$\frac{1}{R_{AB}} = \frac{1}{10} + \frac{1}{15} = \frac{15 + 10}{15 \times 10}$$

$$\frac{1}{R_{AB}} = \frac{25}{15 \times 10}$$

$$R_{AB} = \frac{15 \times 10}{25} = 6\Omega$$

32. *(d)* Initially capacitor behave as a short circuit.

So, current will be maximum. Charge on capacitor will be zero. Potential difference across capacitor will be zero. (B, C, D only)

33. *(d)* Let the resistance per unit length be λ, then

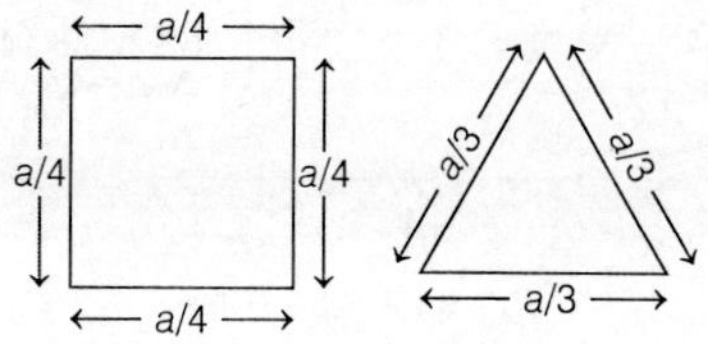

Square Triangle

$$R_{Triangle} = \frac{\left(\frac{2\lambda a}{3}\right)\left(\frac{\lambda a}{3}\right)}{\frac{2\lambda a}{3} + \frac{\lambda a}{3}} = \frac{2\lambda a}{9} \quad ...(i)$$

$$R_{Square} = \frac{\left(\frac{3\lambda a}{4}\right)\left(\frac{\lambda a}{4}\right)}{\frac{3\lambda a}{4} + \frac{\lambda a}{4}} = \frac{3\lambda a}{16} \quad ...(ii)$$

From Eqs. (i) and (ii), we get

$$\frac{R_{Triangle}}{R_{Square}} = \frac{2\lambda a}{9} \times \frac{16}{3\lambda a} = \frac{32}{27}$$

Alternate Solution

For a polygon of n sides, the equivalent resistance across two adjacent vertices is given by

$$R_{eq} = \frac{(n-1)R}{n^2}$$

where R is the total resistance.

So, according to question.

For triangle

$n = 3$ and $R = R$

$$(R_{eq})_t = \frac{(n-1)R}{n^2} = \frac{(3-1) \times R}{3^2} = \frac{2R}{9}$$

For square

$n = 4$ and $R = R$

$$(R_{eq})_{sq} = \frac{(n-1)R}{n^2} = \frac{(4-1)R}{4^2} = \frac{3R}{16}$$

$$\frac{(R_{eq})_t}{(R_{eq})_{sq}} = \frac{2R}{9 \times \frac{3R}{16}} = \frac{32}{27}$$

34. *(c)*

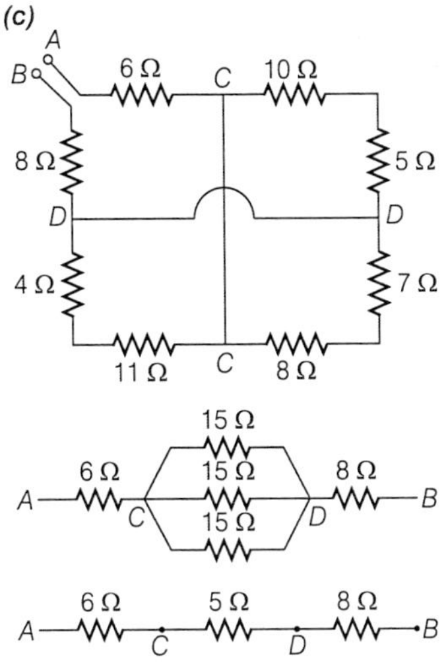

$R_{eq} = 6\Omega + 5\Omega + 8\Omega = 19\Omega$

35. *(d)* $I = I_0 + \beta t$, where $I_0 = 20$ A

and $\beta = 3$ A/s

So, $I = 20 + 3t$

$$\Rightarrow \frac{dq}{dt} = 20 + 3t$$

$$\Rightarrow \int dq = \int (20 + 3t)dt$$

$$q = \int_0^{20} 20\, dt + \int_0^{20} 3t\, dt$$

$$= \left[20t + \frac{3t^2}{2}\right]_0^{20}$$

$$q = 1000 \text{ C}$$

36. *(d)* In series combination, equivalent resistance,

$$R = R_1 + R_2 + R_3$$

Hence, the equivalent resistance of resistors in series combination is greater than largest resistance used in the combination.

Therefore, Statement I is incorrect.

Resistivity of the material depends on temperature.

For metals, it increases with increase in temperature and for semiconductor , it decreases with increase in temperature.

Thus, Statement II is also incorrect.

37. *(c)* Given, resistance of the wire,

$$R = 160\,\Omega$$

Let ρ be the resistivity of the wire.

Since, volume remains constant, thus

$$A_1 L_1 = A_2 L_2$$

$$\Rightarrow A_1 L = A_2 \frac{L}{4} \quad \left(\text{given, } L_2 = \frac{L_1}{4}\right)$$

$$\Rightarrow 4A_1 = A_2$$

Initially, $R_1 = \rho \frac{L_1}{A_1}$

Finally, $R_2 = \rho \frac{L_2}{A_2}$

$$\Rightarrow \frac{R_2}{R_1} = \frac{L_2 A_1}{L_1 A_2} = \frac{1}{4}\frac{A_1}{4A_1} = \frac{1}{16}$$

$$\Rightarrow R_2 = \frac{160}{16} = 10\,\Omega$$

38. *(d)* Given circuit is,

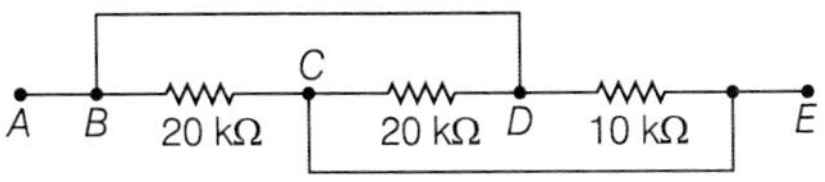

We can redraw above circuit as,

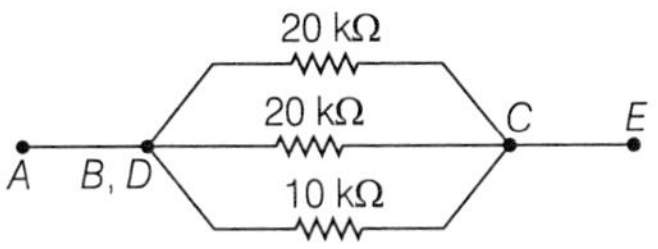

So, we have 3 resistances in parallel.

Equivalent resistance of circuit is;

$$\frac{1}{R_{eq}} = \frac{1}{R_1} + \frac{1}{R_2} + \frac{1}{R_3}$$

$$= \frac{1}{20} + \frac{1}{20} + \frac{1}{10} = \frac{1+1+2}{20}$$

So, $R_{eq} = \frac{20}{4}$ kΩ $= 5$ kΩ

39. *(c)* When a uniform wire of resistance R is shaped into a regular n-sided polygon, the resistance of each side will be

$$R_1 = R/n$$

Let, R_1 and R_2 be the resistance between adjacent corners of a regular polygon.

$\therefore$ The resistance of $(n-1)$ side,

$$R_2 = \left(\frac{n-1}{n}\right)R$$

Since, two parts are in parallel.

Therefore, $R_{eq} = \frac{R_1 R_2}{R_1 + R_2}$

$$= \frac{\left(\frac{R}{n}\right)\left(\frac{n-1}{n}\right)R}{\frac{R}{n} + \left(\frac{n-1}{n}\right)R}$$

$$\Rightarrow R_{eq} = \frac{(n-1)}{n^2}R$$

40. *(b)* The equivalent resistance between points A and B is given by

$$R_{AB} = \left(\frac{2\times12}{2+12}\right) \Big\| \left(\frac{8\times2}{8+2} + \frac{4\times6}{4+6}\right) \Big\| \left(\frac{6\times2}{6+2}\right)$$

$$= \left(\frac{12}{7}\right) \Big\| (4) \Big\| \left(\frac{3}{2}\right)$$

$$\Rightarrow \frac{1}{R_{AB}} = \frac{7}{12} + \frac{1}{4} + \frac{2}{3}$$

$$\therefore R_{AB} = \frac{2}{3}\,\Omega$$

41. *(a)* Given, $I = 2$ A, $V = 3.4$ V

$$m = 8.92 \times 10^{-3} \text{ kg},$$

$$d = 8.92 \times 10^3 \text{ kg/m}^3$$

Resistivity, $\rho = 1.7 \times 10^{-8}$ Ω-m

Resistance of wire,

$$R = \frac{V}{I} = \frac{3.4}{2} = 1.7\,\Omega$$

We know that, $R = \rho \cdot \frac{l}{A}$

$$\Rightarrow l = \frac{RA}{\rho} \quad \ldots\text{(i)}$$

$\therefore$Since, density,

$$d = \frac{m}{V} \Rightarrow d = \frac{m}{Al}$$

$$\Rightarrow A = \frac{m}{ld} \quad \ldots\text{(ii)}$$

From Eqs. (i) and (ii), we have

$$l = \frac{Rm}{\rho l d}$$

$$\Rightarrow l^2 = \frac{Rm}{\rho d} = \frac{1.7 \times 8.92 \times 10^{-3}}{1.7 \times 10^{-8} \times 8.92 \times 10^3}$$

$$= 10^2$$

$$\Rightarrow l^2 = 10^2 \Rightarrow l = 10 \text{ m}$$

42. *(b)* According to the question,

$$l_w = 2l_x \quad \ldots\text{(i)}$$

$$R_w = 2R_y$$

$$\Rightarrow \rho\frac{l_w}{A_w} = 2\rho\frac{l_y}{A_y} \quad \ldots\text{(ii)}$$

$$\Rightarrow \frac{l_w}{A_w} = \frac{2l_y}{A_y}, \frac{2l_x}{A_w} = \frac{2l_y}{A_y}$$

Also, $l_x + l_y + l_y = 1$ ... (iii)

and $A_x l_x = A_w l_w$ ...(iv)

From Eqs. (i) and (iv),

$$A_x l_x = A_w(2l_x), A_x = 2A_w$$

From (ii), $\frac{2l_x}{A_x/2} = \frac{2l_y}{A_y} \Rightarrow \frac{2l_x}{l_y} = \frac{A_x}{A_y}$ ...(v)

Since, $A_x = A_y$

$$\Rightarrow 2l_x = l_y, \frac{l_x}{l_y} = \frac{1}{2}$$

43. *(b)* According to the question

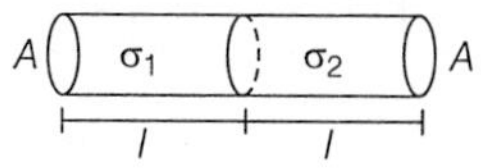

In series,

$$R_{eq} = R_1 + R_2 = \frac{1}{\sigma_1}\frac{l}{A} + \frac{1}{\sigma_2}\frac{l}{A}$$

$$\Rightarrow \quad \frac{1}{\sigma_{eq}}\frac{2l}{A} = \frac{l}{A}\left[\frac{\sigma_2 + \sigma_1}{\sigma_1\sigma_2}\right] \qquad \left[\because R = \frac{1}{\sigma}\frac{\ell}{A}\right]$$

$$\Rightarrow \quad \sigma_{eq} = \frac{2\sigma_1\sigma_2}{(\sigma_1 + \sigma_2)}$$

44. *(a)* Here, resistance of wire at 10 °C, $R_{10} = 2\,\Omega$

Resistance of wire at 30 °C, $R_{30} = 3\,\Omega$

Resistance at any temperature T is given by

$$R_T = R_0(1 + \alpha\Delta T)$$

$$\therefore \quad R_{10} = R_0(1 + 10\,\alpha)$$

or $\quad 2 = R_0(1 + 10\,\alpha) \quad$ ...(i)

Similarly, $R_{30} = R_0(1 + 30\,\alpha)$

or $\quad 3 = R_0(1 + 30\,\alpha) \quad$ ...(ii)

On dividing Eq. (i) by Eq. (ii), we get

$$\frac{2}{3} = \frac{(1 + 10\,\alpha)}{(1 + 30\,\alpha)}$$

$$\Rightarrow \quad 2 + 60\,\alpha = 3 + 30\,\alpha$$

$$\Rightarrow \quad \alpha = \frac{1}{30} = 0.033\,°\text{C}^{-1}$$

45. *(b)* We know, $v_d = \dfrac{eE\tau}{m}$

where, τ = relaxation time

$\because \quad v_d \propto \tau$

On increasing temperature, τ decreases. Thus, v_d decreases with increasing temperature.

$\therefore$ (A) is correct.

For a steady current the drift velocity is inversely proportional to the area of cross-ection but steady current is not mentioned.

We know, $E = \dfrac{V}{l}$

(where, V = potential difference)

$$\therefore \quad v_d = \frac{eV\tau}{ml}$$

$$v_d \propto V$$

$\therefore$ (C) is wrong.

$\therefore$ v_d is inversely proportional to the length of conductor.

Hence, (D) is correct.

46. *(c)* The given network is shown as

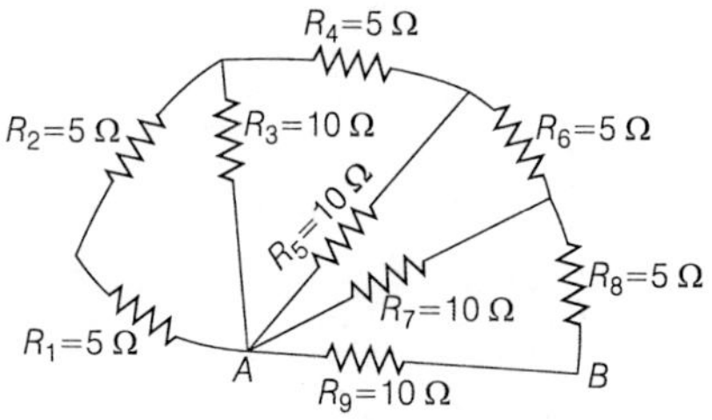

R_1 and R_2 are in series.

$$\therefore \quad R_{12} = R_1 + R_2 = 5 + 5 = 10\,\Omega$$

Now, R_{12} and R_3 are in parallel,

$$R_{123} = \frac{R_{12} \times R_3}{R_{12} + R_3} = \frac{10 \times 10}{10 + 10} = 5\,\Omega$$

Now, R_{123} and R_4 are in series,

$$R_{1234} = 5 + 5 = 10\,\Omega$$

Again, $R_{1234} \| R_5$

$$\therefore R_{12345} = \frac{R_{1234} \times R_5}{R_{1234} + R_5} = \frac{10 \times 10}{10 + 10} = 5\,\Omega$$

In similar way, we can calculate the equivalent resistance of next levels.

i.e. $R_{123456} = R_{12345} + R_6 = 5 + 5 = 10\,\Omega$

$$R_{1234567} = \frac{R_{123456} \times R_7}{R_{123456} + R_7} = \frac{10 \times 10}{10 + 10} = 5\,\Omega$$

$$\therefore R_{12345678} = R_{1234567} + R_8 = 5 + 5 = 10\,\Omega$$

$$\therefore R_{123456789} = \frac{R_{12345678} \times R_9}{R_{12345678} + R_9} = \frac{10 \times 10}{10 + 10} = 5\,\Omega$$

47. *(c)* Density,

$$d = \frac{\text{mass}}{\text{volume}} = \frac{\text{mass}}{\text{area} \times \text{length}}$$

$$d = \frac{m}{Al} \Rightarrow A = \frac{m}{ld}$$

$$\text{Resistance, } R = \frac{\rho l}{A} = \frac{\rho l}{m/ld} = \frac{\rho l^2 d}{m}$$

$$\therefore R = \left[\frac{\rho d}{m}\right] l^2 = \text{constant} \times l^2$$

$$\frac{\Delta R}{R} = 2 \times \frac{\Delta l}{l}$$

Percentage change in resistance is

$$\frac{\Delta R}{R} \times 100 = 2 \times \left[\frac{\Delta l}{l}\right] \times 100$$

Given, $\dfrac{\Delta l}{l} \times 100 = 0.4\%$

$$\therefore \quad \frac{\Delta R}{R} = 2 \times 0.4\% = 0.8\%$$

48. *(d)* In the given circuit, 5 Ω resistance is shorted. So, it can be discarded.

Now, we get a resolved circuit as shown below

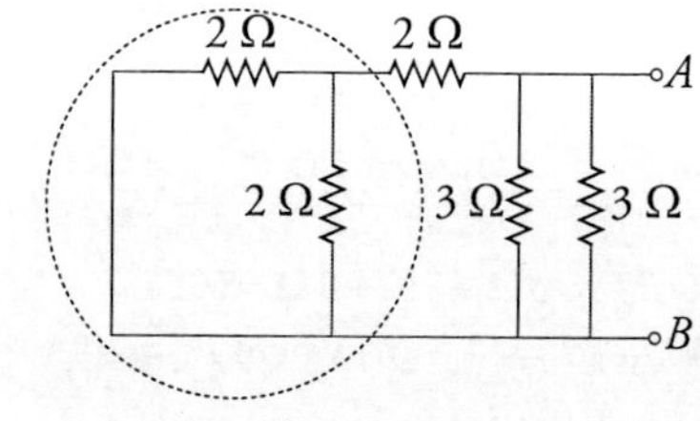

In parallel, $R_1 = \dfrac{2 \times 2}{2 + 2} = 1\,\Omega$

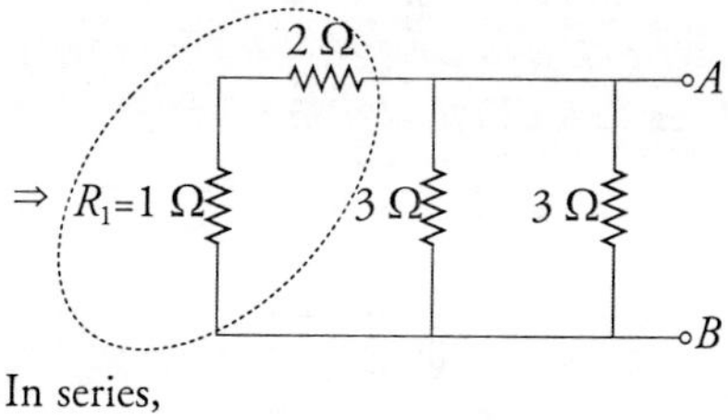

In series,

$$R_2 = R_1 + 2 = 1 + 2 = 3\,\Omega$$

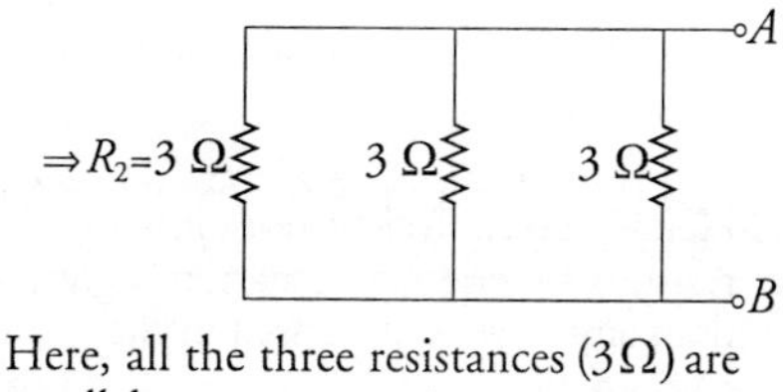

Here, all the three resistances (3Ω) are parallel.

$\therefore$ Equivalent resistance across A and B,

$$\frac{1}{R_{eq}} = \frac{1}{3} + \frac{1}{3} + \frac{1}{3} = 1$$

$$\Rightarrow \quad R_{eq} = 1\,\Omega$$

49. *(b)* Let the resistance of iron wire be R_1 and that of copper nickel alloy wire be R_2

$$r_1 = r_2 = 1\text{ mm} = 10^{-3}\text{ m}$$

$$\rho_1 = 12\,\mu\Omega\text{ cm} = 12 \times 10^{-8}\,\Omega\text{m}$$

$$\rho_2 = 51\,\mu\Omega\text{ cm} = 51 \times 10^{-8}\,\Omega\text{m}$$

For parallel combination,

$$R_{eq} = \frac{R_1 R_2}{R_1 + R_2} \text{ and } 3 = \frac{\dfrac{\rho_1 l}{\pi r_1^2}\dfrac{\rho_2 l}{\pi r_2^2}}{\dfrac{\rho_1 l}{\pi r_1^2} + \dfrac{\rho_2 l}{\pi r_2^2}}$$

$$\Rightarrow 3 = \frac{\frac{(12\times10^{-8})\,l}{\pi\times10^{-6}} \times \frac{51\times10^{-8}\,l}{\pi\times10^{-6}}}{\frac{12\times10^{-8}\,l}{\pi\times10^{-6}} + \frac{51\times10^{-8}\,l}{\pi\times10^{-6}}}$$

On solving, $l = 97$ m

50. *(c)* Given,

Resistance of conductor at 15°C $= 16\Omega$ and resistance of conductor at 100°C $= 20\Omega$

$$R = R_0(1+\alpha\Delta T) \quad \ldots\text{(i)}$$

$\therefore$ Using Eq. (i), we can write

$$16 = R_0[1+\alpha(15-T_0)] \quad \ldots\text{(ii)}$$

and $20 = R_0[1+\alpha(100-T_0)]$ ...(iii)

Assume that T_0 to be 0°C as per general convention and divide Eq. (ii) by Eq. (iii), we get

$$\frac{16}{20} = \frac{R_0(1+\alpha\times15)}{R_0(1+\alpha\times100)} \Rightarrow \frac{4}{5} = \frac{1+15\alpha}{1+100\alpha}$$

$$\Rightarrow 4 + 400\alpha = 5 + 75\alpha$$

$$\Rightarrow 400\alpha - 75\alpha = 5 - 4 \Rightarrow 325\alpha = 1$$

$$\Rightarrow \alpha = \frac{1}{325} = 0.003°\,C^{-1}$$

> **• Concept Enhancer**
>
> We know that,
>
> $R_t = R_0[1+\alpha(\Delta T)]$
>
> where, R_t = resistance of conductor at t°C,
>
> R_0 = resistance of conductor at 0°C,
>
> α = temperature coefficient of resistance
>
> and ΔT = temperature difference.

51. *(d)* We know that,

$$R = \rho l / A$$

$\therefore$ Resistance of copper (R_{Cu})

$$= \frac{1.7\times10^{-8}\times25\times10^{-2}}{3\times10^{-6}} = 14.167\times10^{-4}\,\Omega$$

Resistance of aluminium

$$(R_{Al}) = \frac{2.6\times10^{-8}\times25\times10^{-2}}{3\times10^{-6}} = 21.67\times10^{-4}\,\Omega$$

Hence, $\dfrac{1}{R_{eq}} = \dfrac{1}{R_{Cu}} + \dfrac{1}{R_{Al}}$

$$\Rightarrow R_{eq} = \frac{R_{Cu}\cdot R_{Al}}{R_{Cu}+R_{Al}}$$

$$\Rightarrow R_{eq} = \frac{14.167\times10^{-4}\times21.67\times10^{-4}}{10^{-4}(14.167+21.67)} = 0.857\text{ m}\Omega \simeq 0.858\text{ m}\Omega$$

52. *(a)* As we know, $R = \dfrac{\rho l}{A}$

Original length, $l_0 = 1$ m

Change in length, $l_1 = \dfrac{1\times25}{100} = 0.25$ m

New length = 1.25 m

As volume of wire remains constant, so

$$A_0 l_0 = A_1 l_1 \Rightarrow A_1 = \frac{l_0 A_0}{l_1}$$

If length becomes 1.25 m, area should decreases by 1.25.

$$R' = \frac{\rho(1.25)L}{A/1.25} = \frac{(1.25)^2\rho L}{A}$$

$$\therefore \quad R' = 1.5625\,R \quad \left[\because R = \frac{\rho L}{A}\right]$$

Therefore, increase in resistance

$$\frac{R'-R}{R}\times100 = \frac{1.56R - R}{R}\times100 = 56.25\% \simeq 56\%$$

53. *(c)* Given, $i = \alpha_0 t + \beta t^2$

where, $\alpha_0 = 20$ A/s, $\beta = 8$ A/s^2

We know that $\quad i = \dfrac{dq}{dt}$

$$\Rightarrow \frac{dq}{dt} = i = \alpha_0 t + \beta t^2 = 20t + 8t^2$$

$$\Rightarrow dq = (20t + 8t^2)dt$$

On integrating both sides, we get

$$\int_0^q dq = \int_0^{15}(20\,t + 8\,t^2)dt$$

$$q = \left[\frac{20t^2}{2} + \frac{8t^3}{3}\right]_0^{15} = 10\times(15)^2 + \frac{8}{3}\times(15)^3$$

$$\therefore \quad q = 11250\text{ C}$$

54. *(b)*

> **key idea** Resistance between surface of inner shell and a circumferential point of outer shell can be formed by finding resistance of a thin (differentially thin) shell in between these two shells. Then, this result can be integrated (summed up) to get resistance of the complete arrangement.

For a elemental shell of radius x and thickness dx,

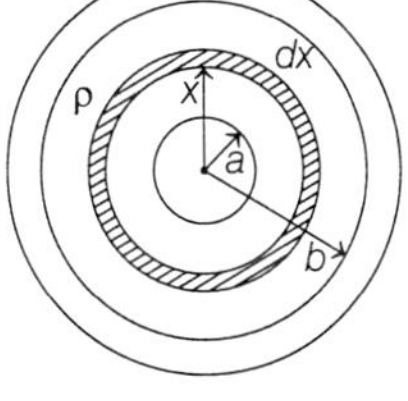

Resistance, $dR = \rho\dfrac{dl}{dA} \Rightarrow dR = \rho\dfrac{dx}{4\pi x^2}$

So, resistance of complete arrangement is

$$R = \int_a^b dR = \int_a^b \rho\frac{dx}{4\pi x^2} = \frac{\rho}{4\pi}\int_a^b x^{-2}dx$$

$$\Rightarrow R = \frac{\rho}{4\pi}\left(\frac{x^{-1}}{-1}\right)_a^b = \frac{\rho}{4\pi}\left(-\frac{1}{x}\right)_a^b$$

$$= \frac{\rho}{4\pi}\left(-\frac{1}{b} - \left(-\frac{1}{a}\right)\right) = \frac{\rho}{4\pi}\left(\frac{1}{a} - \frac{1}{b}\right)$$

55 *(b)* Electrical resistance of wire of length l, area of cross-section A and resistivity ρ is given as

$$R = \rho\frac{l}{A} \quad \ldots\text{(i)}$$

Since we know, volume of the wire is

$$V = A\times l \quad \ldots\text{(ii)}$$

$\therefore$ From Eqs. (i) and (ii), we get

$$R = \rho\frac{l^2}{V} \quad \ldots\text{(iii)}$$

As, the length has been increased to 0.5%.

$\therefore$ New length of the wire,

$l' = l + 0.5\%$ of $l = l + 0.005\,l = 1.005\,l$

But V and ρ remains unchanged.

So, new resistance,

$$R' = \frac{\rho[(1.005)\,l]^2}{V} \quad \ldots\text{(iv)}$$

On dividing Eq. (iv) and Eq. (iii), we get

$$\frac{R'}{R} = (1.005)^2$$

$\Rightarrow$ Percentage change in the resistance

$$= \left(\frac{R'}{R} - 1\right)\times100$$

$$= [(1.005)^2 - 1]\times100 = 1.0025\% \simeq 1\%$$

56. *(b)* Resistance after temperature increases by 500°C

$$R_T = \frac{\text{Voltage applied}}{\text{Current}} = \frac{220}{2} = 110\,\Omega$$

Also, $R_T = R_0(1+\alpha\Delta T)$

$$110 = 100(1+\alpha\times500)$$

$$\Rightarrow \alpha = \frac{10}{100\times500} = 2\times10^{-4}\,°C^{-1}$$

57. *(a)* Let resistances be R_1 and R_2.

Then, $S = R_1 + R_2$ and $P = \dfrac{R_1R_2}{R_1+R_2}$

$$\therefore (R_1+R_2) = \frac{n\times R_1R_2}{R_1+R_2} \quad [\because S = nP]$$

or $(R_1+R_2)^2 = nR_1R_2$

$$\Rightarrow n = \left[\frac{R_1^2 + R_2^2 + 2R_1R_2}{R_1R_2}\right]$$

$$= \left[\frac{R_1}{R_2} + \frac{R_2}{R_1} + 2\right] \quad ...(i)$$

We know,
Arithmetic Mean ≥ Geometric Mean

$$\frac{\frac{R_1}{R_2} + \frac{R_2}{R_1}}{2} \geq \sqrt{\frac{R_1}{R_2} \times \frac{R_2}{R_1}}$$

$$\Rightarrow \quad \frac{R_1}{R_2} + \frac{R_2}{R_1} \geq 2 \quad ...(ii)$$

From Eqs. (i) and (ii)
So, n (minimum value) = 2 + 2 = 4

58. *(d)* Given, $l' = l + 100\% l = 2l$

Initial volume = Final volume
i.e., $\pi r^2 l = \pi r'^2 l'$

or $r'^2 = \frac{r^2 l}{l'} = r^2 \times \frac{l}{2l}$ or $r'^2 = \frac{r^2}{2}$

$$\therefore \quad R' = \rho\frac{l'}{A'} = \rho\frac{2l}{\pi r'^2} \quad \left[\because R = \frac{\rho l}{A}\right]$$

$$= \frac{\rho \cdot 4l}{\pi r^2} = 4R$$

Thus, $\Delta R = R' - R = 4R - R = 3R$

$$\therefore \quad \% \Delta R = \frac{3R}{R} \times 100\% = 300\%$$

> ***Alternate Solution***
> By the shortcut formula, percentage change
> $$= x + y + \frac{xy}{100}$$
> Valid only for two variables in terms of percentage.
> As $\quad R \propto l^2 \%$
> $$\text{Increase in resistance} = 100 + 100 + \frac{100 \times 100}{100}$$
> $$= 300\%$$

59. *(b)* Given, $R_1 = 10\,\Omega$, $R_2 = 8\,\Omega$
$R_3 = 4\,\Omega$, $R_4 = 8\,\Omega$
$V = 12$ V

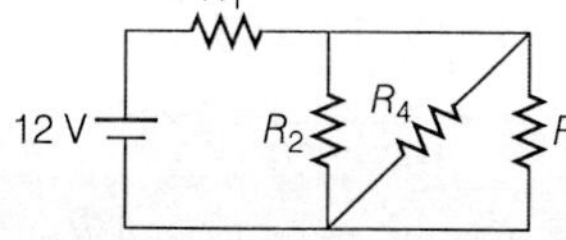

From circuit diagram, R_2, R_3 and R_4 are connected parallely.
∴ In parallel connection,

$$\frac{1}{R_{eq(234)}} = \frac{1}{R_2} + \frac{1}{R_3} + \frac{1}{R_4}$$

$$\frac{1}{R_{eq(234)}} = \frac{1}{8} + \frac{1}{4} + \frac{1}{8}$$

$$\Rightarrow \quad R_{eq(234)} = 2\,\Omega$$

Now, R_1 and $R_{eq(234)}$ are connected in series.

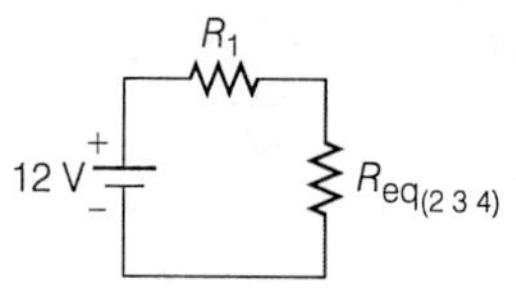

$$R_{eq} = R_1 + R_{eq(234)} = 2 + 10 = 12\,\Omega$$
$$R_{eq} = 12\,\Omega$$

$$\text{Current, } i = \frac{V}{R} = \frac{12}{12} = 1\text{ A}$$

60. *(b)*

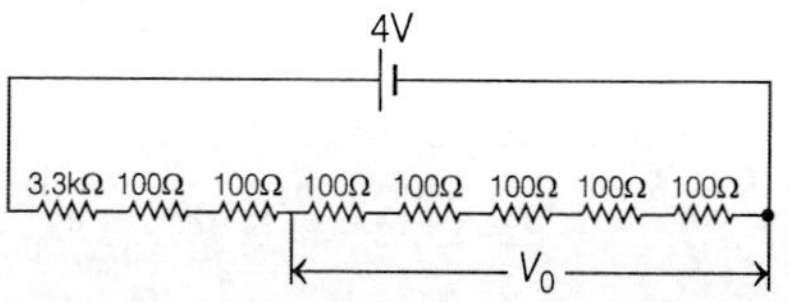

Total resistance in the circuit = Sum of all the resistances

$$R_{eq} = 100\,\Omega + 100\,\Omega + 100\,\Omega + 100\,\Omega + 100\,\Omega + 100\,\Omega + 100\,\Omega + 3.3\text{ k}\Omega$$

(All are connected in series)

$$R_{eq} = 4000\,\Omega$$

$$\text{Current } I = \frac{V}{R} \Rightarrow I = \frac{4}{4000}\text{ A}$$

Now, output voltage

$$V_0 = I \times R_{V_0}$$

[R_{V_0} is resistance across V_0]

$$V_0 = \frac{4}{4000} \times [100 + 100 + 100 + 100 + 100]$$

$$V_0 = \frac{4}{4000} \times 500 \Rightarrow V_0 = 0.5\text{ V}$$

> **• Concept Enhancer**
> Voltage division rule is given as
> $$V_1 = V \cdot \frac{R_1}{R_1 + R_2}, \text{ and } V_2 = V \cdot \frac{R_2}{R_1 + R_2}$$
> 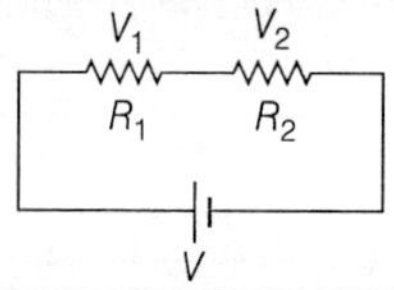
>

61. *(c)* $R = \frac{V}{I}$

$\Rightarrow V = (200 \pm 5)\text{ V} \Rightarrow I = (20 \pm 0.2)\text{ A}$

The percentage error in the measurement of R

$$R = V / I$$

Fractional error in R is

$$\frac{dR}{R} = \frac{dV}{V} + \frac{dI}{I}$$

$$\frac{dR}{R} = \frac{5}{200} + \frac{0.2}{20}$$

$$\frac{dR}{R} = \frac{5}{200} + \frac{0.2 \times 10}{20 \times 10}$$

$$= \frac{5}{200} + \frac{2}{200}$$

$$\frac{dR}{R} = \frac{7}{200}$$

Percentage error,

$$\frac{dR}{R} \times 100 = \frac{7}{200} \times 100 = 3.5\%$$

Percentage error = 3.5%

62. *(a)* If we consider only conductor PQ as follows

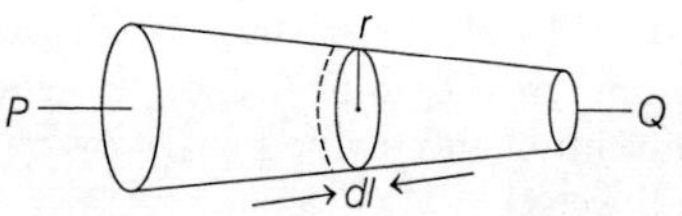

Let dl be the small element of conductor PQ.
∵ Current passing through PQ is constant.
So, resistance of this small element dl will be given as

$$dR = \rho dl / A \quad ...(i)$$

$$\because \quad \text{Current, } I = \frac{dV}{dR}$$

$$\Rightarrow \quad dV = IdR \quad ...(ii)$$

From Eqs. (i) and (ii), we get

$$dV = \frac{I\rho dl}{A} \Rightarrow dV = \frac{I\rho dl}{\pi r^2} \quad ...(iii)$$

$$\{\because A = \pi r^2\}$$

∵ We know that,
Electric field, $E = dV / dl \quad ...(iv)$
From Eqs. (iii) and (iv), we get

$$\Rightarrow \quad E = \frac{I\rho dl}{\pi r^2 dl}$$

$$\Rightarrow \quad E \propto \frac{1}{r^2} \quad ...(v)$$

Also, we know that drift velocity,

$$v_d = \frac{eE\tau}{m}$$

$$\Rightarrow \quad v_d \propto E \quad ...(vi)$$

∴ From Eqs. (v) and (vi), we can say
If r decreases, E will increase and due to this, v_d will increase.

63. *(d)* Let R_0 be the initial resistance of both conductors.
∴ At temperature θ, their resistances will be

$$R_1 = R_0(1 + \alpha_1 \theta)$$

and $\quad R_2 = R_0(1 + \alpha_2 \theta)$

For series combination,

$$R_s = R_1 + R_2$$

$$R_{s0}(1+\alpha_s\theta) = R_0(1+\alpha_1\theta) + R_0(1+\alpha_2\theta)$$

where, $R_{s0} = R_0 + R_0 = 2R_0$

$\therefore\ 2R_0(1+\alpha_s\theta) = 2R_0 + R_0\theta\,(\alpha_1+\alpha_2)$

or $\qquad \alpha_s = \dfrac{\alpha_1+\alpha_2}{2}$

For parallel combination,

$$R_p = \frac{R_1R_2}{R_1+R_2}$$

$$R_{p0}(1+\alpha_p\theta) = \frac{R_0(1+\alpha_1\theta)R_0(1+\alpha_2\theta)}{R_0(1+\alpha_1\theta)+R_0(1+\alpha_2\theta)}$$

where, $R_{p0} = \dfrac{R_0R_0}{R_0+R_0} = \dfrac{R_0}{2}$

$$\therefore\ \frac{R_0}{2}(1+\alpha_p\theta) = \frac{R_0^2(1+\alpha_1\theta+\alpha_2\theta+\alpha_1\alpha_2\theta^2)}{R_0(2+\alpha_1\theta+\alpha_2\theta)}$$

As α_1 and α_2 are small quantities.

$\therefore\ \alpha_1\alpha_2$ is negligible.

$$\text{or}\quad \alpha_p = \frac{\alpha_1+\alpha_2}{2+(\alpha_1+\alpha_2)\theta} = \frac{\alpha_1+\alpha_2}{2}\left[1-\left(\frac{\alpha_1+\alpha_2}{2}\right)\theta\right]$$

As $(\alpha_1+\alpha_2)^2$ is negligible.

$$\therefore\quad \alpha_p = \frac{\alpha_1+\alpha_2}{2}$$

64. *(2)* The resistor diagram is as follows,

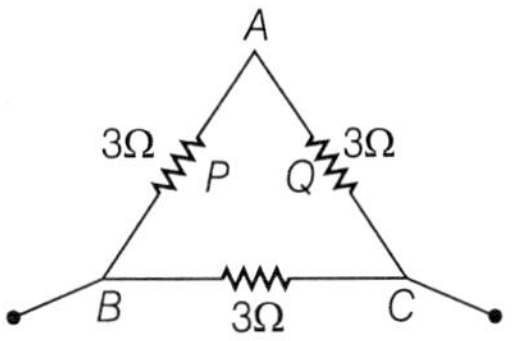

$\Rightarrow$ As 3Ω and 3Ω (P and Q) are in series.

$$R_{eq} = R_1 + R_2 = 3+3 = 6\,\Omega$$

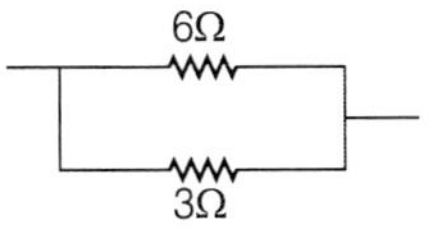

Now 6Ω and 3Ω are in parallel

$$R_{eq} = \frac{R_1\times R_2}{R_1+R_2} = \frac{6\times 3}{6+3} = 2\,\Omega$$

Time Saver Tech

For a polygon of n sides, the equivalent resistance across two adjacent vertices is given by

$$R_{eq} = \frac{(n-1)R}{n^2}$$

where, R is the total resestance.

So, according to question $n = 3$ and $R = 9\,\Omega$

$$R_{eq} = \frac{(n-1)R}{n^2} = \frac{(3-1)\times 9}{3^2} = 2\Omega$$

65. *(16)*

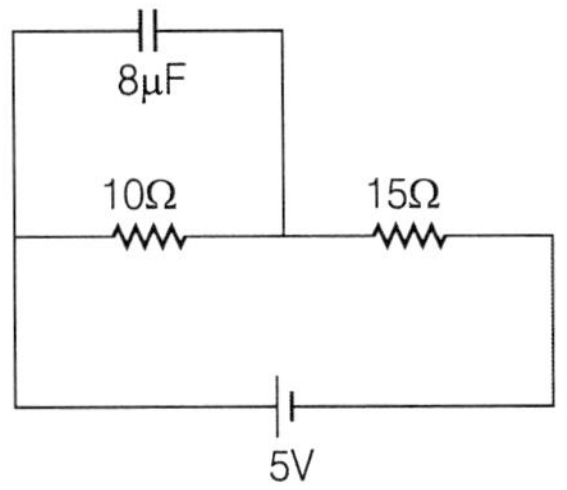

In steady state capacitor acts as open circuit

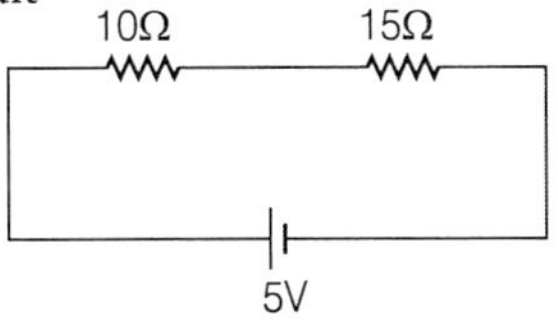

$$R_{eq} = 10 + 15 = 25\,\Omega$$

$$I = \frac{V}{R_{eq}} = \frac{5}{25} \text{ and } I = \frac{1}{5}\text{A}$$

Potential across 10Ω,

$$V_{10} = IR = \frac{1}{5}\times 10 = 2\text{V}$$

$$Q = CV = 8\times10^{-6}\times 2 = 16\times10^{-6} = 16\,\mu\text{C}$$

66. *(1)*

key idea A capacitor acts an open switch in a DC circuit.

$$R_{CB} = 2\Omega + 2\Omega = 4\Omega$$

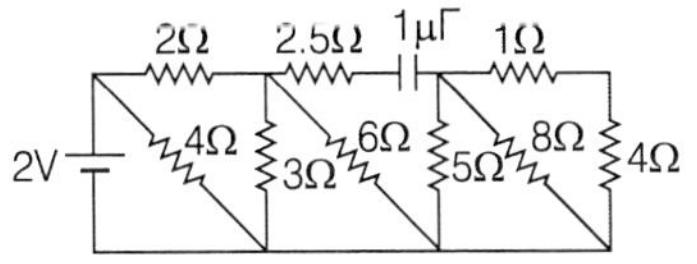

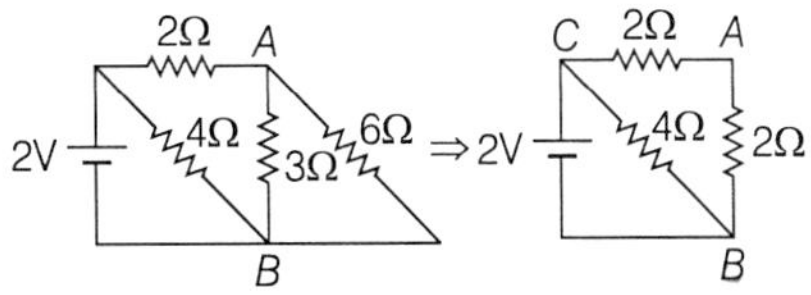

2Ω and 2Ω are in series

$$\frac{1}{R_{AB}} = \frac{1}{6}+\frac{1}{3} \Rightarrow \frac{1}{R_{AB}} = \frac{3}{6} \Rightarrow R_{AB} = 2$$

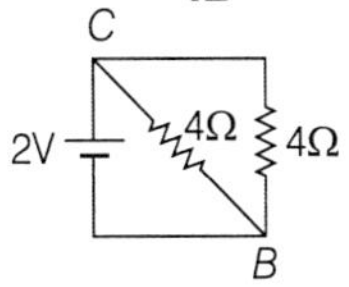

$$\frac{1}{R_{eq}} = \frac{1}{4}+\frac{1}{4} = \frac{2}{4} \Rightarrow R_{eq} = 2$$

$$I = V/R = 2/2 = 1\text{ A}$$

67. *(160)* The situation is shown below.

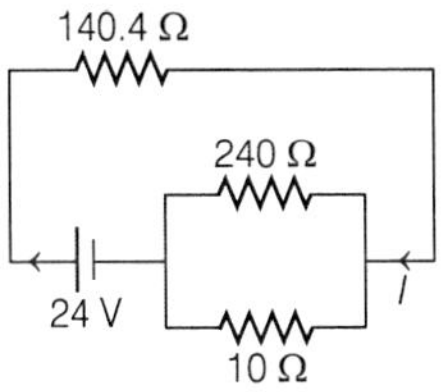

Now, equivalent resistance

$$R_{eq} = 140.4 + \frac{240\times10}{240+10} = 150\,\Omega$$

$\therefore$ Reading of ammeter,

$$I = \frac{V}{R_{eq}} = \frac{24}{150} = 0.16\text{ A} = 160\text{ mA}$$

68. *(16)* When wire is stretched, the volume remains same

$\therefore\quad V_1 = V_2$

$\pi r_1^2 l_1 = \pi r_2^2 l_2$

$$\Rightarrow\quad \frac{l_1}{l_2} = \left(\frac{r_2}{r_1}\right)^2 = \left(\frac{r/2}{r}\right)^2$$

$$\frac{l_1}{l_2} = \frac{1}{4} \qquad \ldots\text{(i)}$$

$$\therefore\quad \frac{R_2}{R_1} = \left(\frac{l_2}{l_1}\right)\left(\frac{r_1}{r_2}\right)^2 = (4)\left(\frac{r}{r/2}\right)^2 = 16$$

$\Rightarrow\quad R_2 = 16R_1 = 16R$

$= xR \qquad$ (given)

$\therefore\quad x = 16$

69. *(5)* Resistance of each part

$$R' = \frac{R}{10} = \frac{20}{10} = 2\,\Omega$$

Now the given situation is shown below

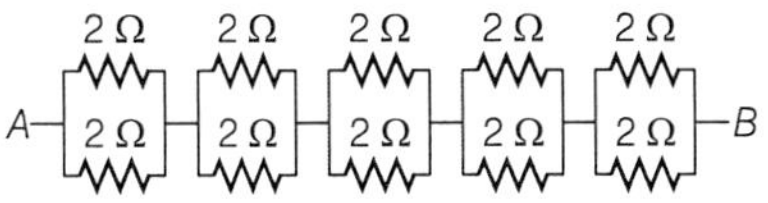

Equivalent resistance for parallel resistances is given by

$$R_{xy} = \frac{R_xR_y}{R_x+R_y}$$

Equivalent resistance,

$$R_{AB} = \frac{2\times 2}{2+2} + \frac{2\times 2}{2+2} + \frac{2\times 2}{2+2} + \frac{2\times 2}{2+2} + \frac{2\times 2}{2+2}$$

$$= 1+1+1+1+1 = 5\,\Omega$$

70. *(32)* Given, resistance of wire B,

$R_2 = 2\,\Omega$

mass of wire A = mass of wire B

$\Rightarrow \quad V_1 d = V_2 d \Rightarrow \pi r_1^2 l_1 = \pi r_2^2 l_2$

$$\Rightarrow \quad \frac{l_1}{l_2} = \left(\frac{r_2}{r_1}\right)^2 = \left(\frac{4}{2}\right)^2 \Rightarrow \frac{l_1}{l_2} = 4$$

$$\therefore \quad \frac{R_2}{R_1} = \frac{\rho\cdot\frac{l_2}{A_2}}{\rho\cdot\frac{l_1}{A_1}} = \frac{l_2}{l_1}\cdot\frac{A_1}{A_2}$$

$$= \frac{l_2}{l_1}\times\left(\frac{r_1}{r_2}\right)^2 = \frac{1}{4}\times\left(\frac{2}{4}\right)^2$$

$$= \frac{1}{4}\times\frac{1}{4}$$

$$\Rightarrow \quad \frac{R_2}{R_1} = \frac{1}{16}$$

$\Rightarrow \quad R_1 = 16R_2 = 16\times 2 = 32\,\Omega$

71. *(22)* Current in wire is,

$$I = \frac{dq}{dt} = 3t^2 + 4t^3$$

So, charge that flows from $t = 1\text{s}$ to $t = 2\text{s}$ is, $Q = \int dq = \int_{t=1}^{t=2} (3t^2 + 4t^3)\cdot dt$

$$= \left[\frac{3t^3}{3} + \frac{4t^4}{4}\right]_1^2$$

$$= (2^3 + 2^4) - (1^3 + 1^4)$$

$$= 22\,\text{C}$$

72. *(200)*

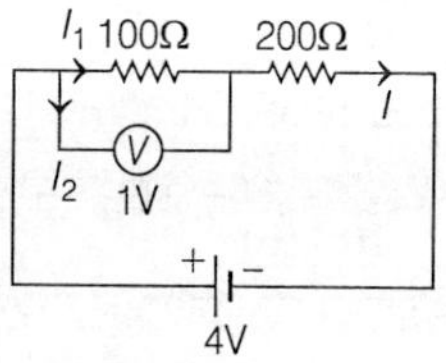

Given, potential drop across 100Ω resistance is 1 V.

So, potential drop across 200 Ω resistance = 3 V,

Current in circuit is,

$$I = \frac{V}{R} = \frac{3}{200}$$

$$= 1.5\times 10^{-3}\,\text{A} = 1.5\,\text{mA}$$

So, current in 100Ω resistance is,

$$I_1 = \frac{V}{R} = \frac{1}{100} = 10\,\text{mA}$$

∴Remaining 5mA current goes from voltmeter, using $V = IR_V$,

$\Rightarrow \quad 1 = 5\times 10^{-3}\times R_V$ or

$$R_V = \frac{1000}{5} = 200\,\Omega$$

73. *(15)* Given, current at 0°C = 2A

Current at 100°C = 1.2 A

For the same source we have

$i_0 R_0 = i_{100} R_{100}$

$\Rightarrow \quad 2R_0 = 1.2R_0(1+100\alpha) \quad \ldots$ (i)

$\Rightarrow 1 + 100\,\alpha = \frac{5}{3} \Rightarrow 100\,\alpha = \frac{2}{3}$

$\Rightarrow \quad 50\,\alpha = 1/3$

$\therefore \quad i_{50} R_{50} = i_0 R_0$ [Same source]

$$\Rightarrow \quad i_{50} = \frac{i_0 R_0}{R_{50}} = \frac{2\times R_0}{R_0(1+50\,\alpha)} = \frac{2}{1+\frac{1}{3}}$$

$$= 1.5\,\text{A} = 15\times 10^2\,\text{mA}$$

74. *(3)* Given, specific resistance,

$$\rho = 3\times 10^{-7}\,\Omega\text{-m}$$

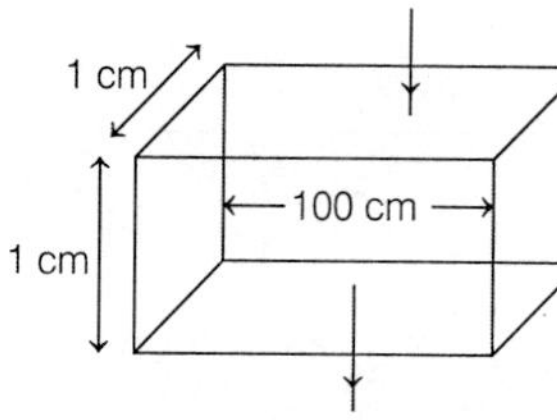

Resistance, $R = \rho\dfrac{l}{A}$.

Resistance through its two opposite faces will be

$$R = \frac{3\times 10^{-7}\times(1\times 10^{-2})}{100\times 1\times 10^{-4}}$$

$$= 3\times 10^{-7}\,\Omega$$

75. *(100)* Maximum resistance occurs when resistors are connected in series;

$$R_{\text{max}} = \Sigma R_i = 100\,\Omega$$

Minimum resistance occurs when resistors are connected in parallel.

$$\Rightarrow \quad \frac{1}{R_{\text{min}}} = \Sigma\frac{1}{R_i} = 1\,\Omega^{-1}$$

$$\Sigma\frac{1}{R_i} = 10\left(\frac{1}{10}\right) = 1\,\Omega$$

$\Rightarrow \quad R_{\text{min}} = 1\,\Omega$

Hence, $\dfrac{R_{\text{max}}}{R_{\text{min}}} = 100:1$

76. *(125)* Using $I = nv_d Ae$

We get, $\quad v_d = \dfrac{I}{neA}$

$$v_d = \frac{3.2}{8\times 10^{28}\times 1.6\times 10^{-19}\times 2\times 10^{-6}}$$

$$= \frac{2}{16\times 10^3} = 125\times 10^{-6}\,\text{m/s}$$

77. *(25)* Drift velocity of electrons in a conductor is,

$$I = nev_d A \quad \ldots\text{(i)}$$

Here, number of electrons per metre cube $n = 2\times 10^{28}\,\text{m}^{-3}$

Area of wire, $A = \pi r^2$

Charge of an electron, $e = 1.6\times 10^{-19}$ C

Current in wire, $I = 2$ A

So from Eq. (i) we have,

$$v_d = \frac{I}{neA}$$

$$= \frac{2}{2\times 10^{28}\times 1.6\times 10^{-19}\times 25\times 10^{-6}}$$

$$= 0.25\times 10^{-4}\,\text{ms}^{-1}$$

$$\Rightarrow v_d = 25\times 10^{-6}\,\text{ms}^{-1}$$

78. *(25)* $R = \dfrac{\rho l}{A}$

where $l' = 21.2\,l$, $A' = 0.96\,A$

$$R' = \frac{\rho\times 1.2\,l}{0.96A} = \frac{10}{8}R = 1.25R$$

It means 25% increase in resistance.

79. *(10.33)* Due to shunt, the 4Ω resistances will get short and so they will not contribute to the circuit.

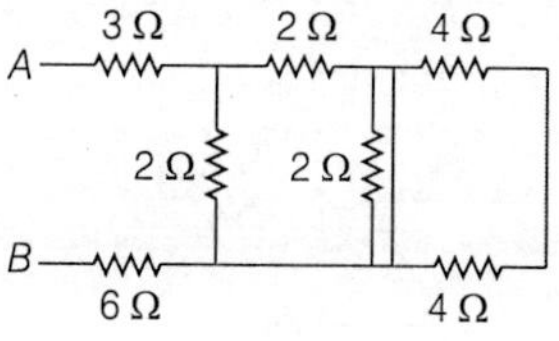

Thus,

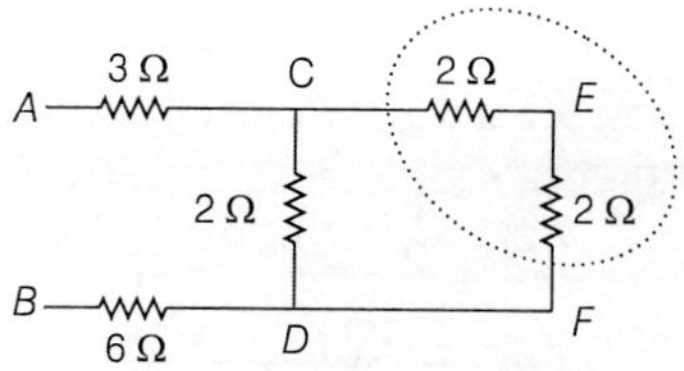

All resistances given in ohm (Ω)

$R_{CF} = 2+2 = 4\,\Omega$

$$\frac{1}{R_{CD}} = \frac{1}{R_{CF}} + \frac{1}{2} = \frac{1}{4} + \frac{1}{2} = \frac{3}{4}$$

$$R_{CD} = \frac{4}{3}\,\Omega \Rightarrow R_{AB} = R_{CD} + 3 + 6$$

$$= \frac{4+9+18}{3} = \frac{31}{3} = 10.33\,\Omega$$

80. *(44)* Resistance of stretched wire will be,

$$R' = \frac{\rho l'}{A'}$$

$$l' = 20\% + l = \frac{120}{100}l = \frac{6}{5}l$$

$$A = V/l$$

$$A' = V/l' = V/(6/5l) = \frac{5}{6}(V/l) = \frac{5}{6}A$$

$$R' = \rho\frac{l'}{A'} = \frac{\rho\cdot\frac{6}{5}l}{\frac{5}{6}A} = \frac{36}{25}\left(\rho\frac{l}{A}\right) = \frac{36}{25}R$$

So, Percentage increase in resistance will be $\frac{\Delta R}{R}\times 100 = \frac{R'}{R}\times 100$

$$= \left(\frac{36}{25} - 1\right)\times 100$$

$$= 11\times 4 = 44\%$$

> ***Alternate Solution***
>
> Let R be the initial value of resistance,
>
> New length of wire, $l' = l + \frac{20}{100}\times l = \frac{6l}{5} = nl$
>
> Here, $n = \frac{6}{5}$
>
> If length of wire increased by n times, then its resistance will be increased by n^2 times.
>
> i.e. $R' = n^2R = \left(\frac{6}{5}\right)^2 R = \frac{36}{25}R$
>
> ∴ Percentage increase in resistance
>
> $= \frac{R'-R}{R}\times 100 = \left(\frac{36}{25} - 1\right)\times 100 = 44\%$

81. *(2)*

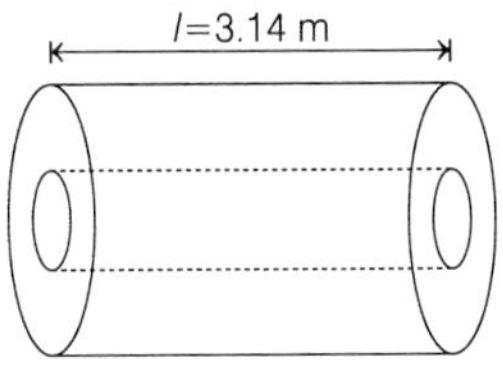

Inner diameter, $d_1 = 4$ mm $= 4\times 10^{-3}$ m

Outer diameter, $d_2 = 8$ mm $= 8\times 10^{-3}$ m

Resistance of conductor given is

$$R = \rho\frac{l}{A}$$

$$= \frac{2.4\times 10^{-8}\times 3.14}{\frac{\pi}{4}[(8\times 10^{-3})^2 - (4\times 10^{-3})^2]}$$

$$= 2\times 10^{-3}\,\Omega$$

Given, $R = n\times 10^{-3}\,\Omega$

So, $n = 2$

82. *(2)* Equivalent resistance of given circuit is given as

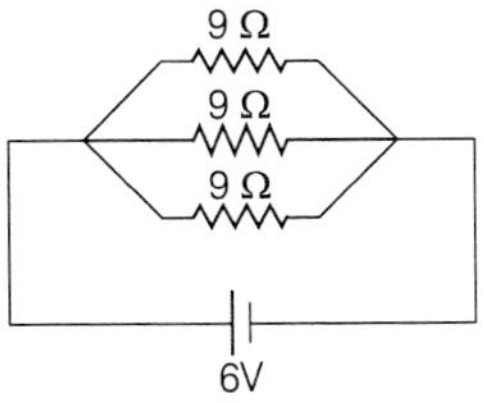

$$R_{eq} = \frac{9}{3}\Omega = 3\Omega,$$

Now, $V = IR_{eq}$, $I = \frac{V}{R_{eq}} = \frac{6}{3} = 2$ A

83. *(136)* Given, $l = 1$ m, $I = 1$ A

$$A = 2\text{ mm}^2 = 2\times 10^{-6}\text{ m}^2$$

$$\rho = 1.7\times 10^{-8}\,\Omega\text{-m}$$

We know that

Current density,

$$\Rightarrow \quad J = \frac{E}{\rho} \qquad \left(\because \sigma = \frac{1}{\rho}\right)$$

$$\Rightarrow \quad \frac{I}{A} = \frac{E}{\rho} \qquad \left(\because J = \frac{I}{A}\right)$$

$$\Rightarrow \quad E = I\rho/A \qquad \ldots\text{(i)}$$

∴ Force on electron,

$$F = eE \qquad \text{[From Eq. (i)]}$$

$$= eI\rho/A$$

$$= \frac{1.6\times 10^{-19}\times 1\times 1.7\times 10^{-8}}{2\times 10^{-6}}$$

$$= 136\times 10^{-23}\text{ N}$$

84. *(8)*

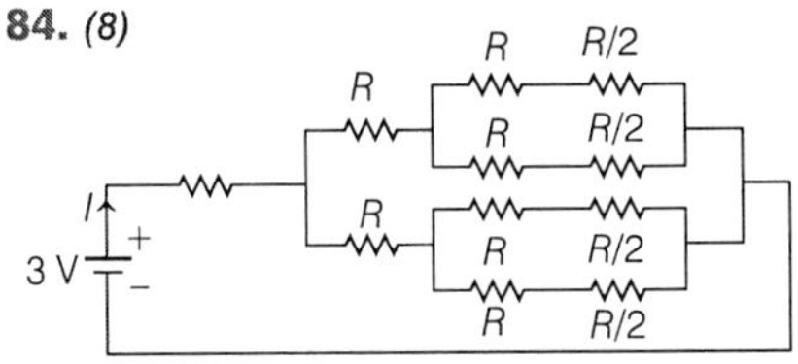

Equivalent resistance of the given circuit will be, $R_{eq} = \frac{15R}{8}$

Resistance of each resistor, $R = 1\Omega$

$$\therefore \quad R_{eq} = \frac{15}{8}\Omega$$

Now, current in the circuit,

$$I = \frac{V}{R_{eq}} = \frac{3\times 8}{15} = \frac{8}{5}\text{ A}$$

So, $a = 8$

85. *(48)* Given, current density,

$$\sigma_i = 4\times 10^6\text{Am}^{-2}$$

$$\sigma_i = \frac{I}{A} = 4\times 10^6$$

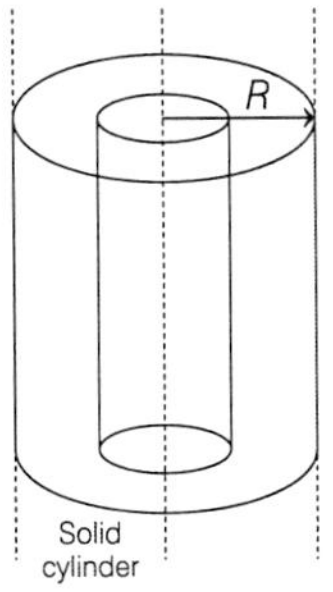

As we know, $(dV)_{\text{Cylinder}} = 2\pi\cdot rdr\cdot l$

Current for a small portion of cylinder is given as,

$$dI = \sigma_i\left(\frac{dV}{l}\right)$$

$$\int_0^I dI = \int_{R/2}^{R} 4\times 10^6\cdot 2\pi rdr$$

$$I = 8\pi\times 10^6\int_{R/2}^{R} r\cdot dr$$

$$I = 8\pi\times 10^6\times\left[\frac{R^2}{2}\right]_{R/2}^{R}$$

$$= \frac{8\pi\times 10^6}{8}[4R^2 - R^2]$$

$$I = 3\pi\times 10^6\times(4\times 10^{-3})^2$$

$$= 48\pi\text{A}$$

86. *(300)* Let initial length $= l$, final length $= 2l$

As the length is doubled, so new resistance,

$$R' = n^2R = (2)^2R = 4R$$

Percentage resistance,

$$= \frac{R'-R}{R}\times 100 = \frac{4R-R}{R}\times 100 = 300\,\%$$

87. *(2)* Circuit diagram is shown below

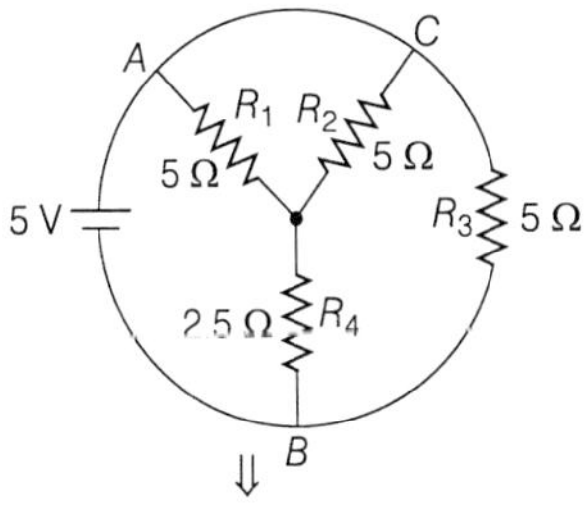

⇓

$$R_{12} = R_1 \,||\, R_2 = \frac{R_1 R_2}{R_1 + R_2}$$

$$= \frac{5\times 5}{5+5} = 2.5\,\Omega$$

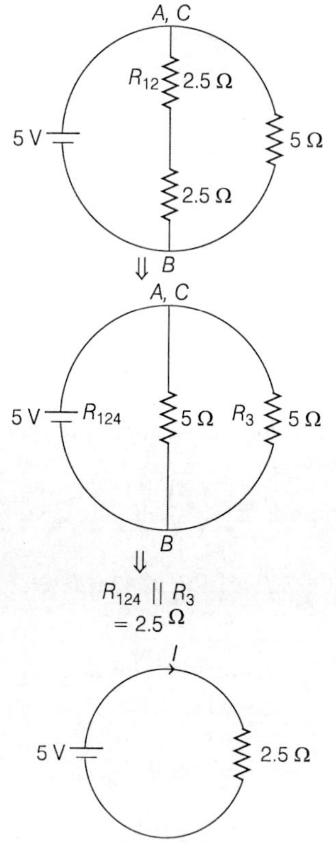

$$\therefore I = \frac{5}{2.5} = 2\,A$$

88. *(3)* The perimeter of square $= 4a$

(a = side of square)

If the radius of shape of circle be r, then

$\therefore \quad 2\pi r = 4a$

$\Rightarrow \quad r = \frac{4a}{2\pi} \Rightarrow r = \frac{2a}{\pi}$

Since, resistance of each side of square $= 3\,\Omega$

$\therefore$Total resistance of square $= 4 \times 3 = 12\,\Omega$

i.e. resistance of length $2\pi r = 12\,\Omega$

$\Rightarrow$ Resistance of $\pi r = 6\,\Omega$

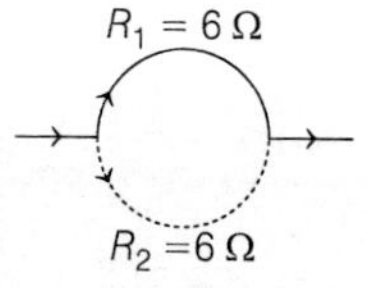

Now, equivalent resistance of circle diametrically opposite,

$$R_{eq} = \frac{R_1 R_2}{R_1 + R_2} = \frac{6 \times 6}{6 + 6} = 3\,\Omega$$

89. *(100)*

key idea In steady state, capacitor will act as open circuit.

According to given circuit diagram,

$C = 50\,\mu F = 50 \times 10^{-6} F$, $V = 6\,V$

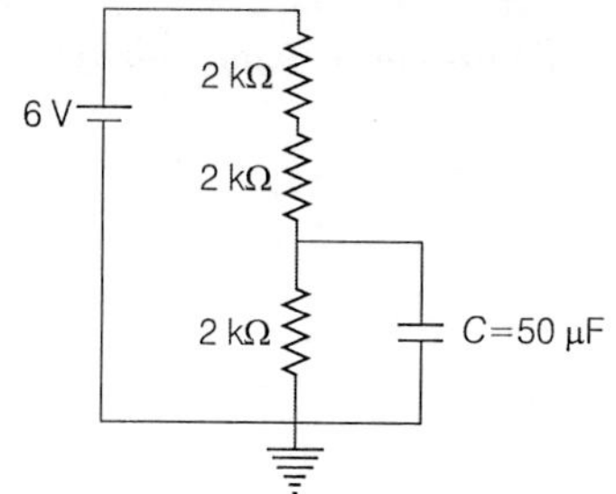

$\therefore$Equivalent resistance

$$R_{eq} = (2 + 2 + 2)\,k\Omega = 6\,k\Omega$$

Circuit current, $I = \frac{V}{R_{eq}}$

$$= \frac{6}{6 \times 1000} = 10^{-3}\,A$$

$\therefore$Voltage across $2\,k\Omega = I \times 2 \times 10^3$

$$= 10^{-3} \times 2 \times 10^3 = 2\,V$$

Now, charge on capacitor,

$$q = CV = 50 \times 10^{-6} \times 2$$

$$= 100 \times 10^{-6}\,C = 100\,\mu C$$

90. *(9)* According to given circuit diagram When switch is open, combination R_1 and R_2 will be in series and also combination R_3 and R_4 will be in series and these branches will be in parallel.

$\therefore$Equivalent resistance,

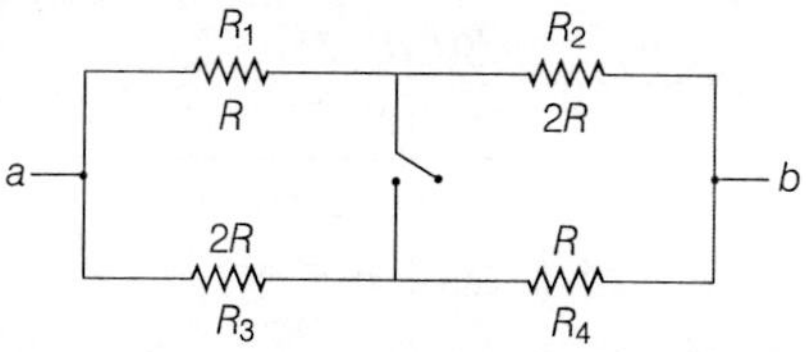

$$(R_{eq}) = \frac{3R \cdot 3R}{3R + 3R} = \frac{9R^2}{6R}$$

$$= \frac{3}{2}R \quad \ldots(i)$$

When switch is closed, combination R_1 and R_3 will be in parallel and also, combination R_2 and R_4 will be in parallel. After that, both will be in series.

$\therefore$Equivalent resistance,

$$R'_{eq} = \frac{R \cdot 2R}{R + 2R} + \frac{R \cdot 2R}{R + 2R}$$

$$= \frac{2R^2}{3R} + \frac{2R^2}{3R}$$

$$= \frac{2R}{3} + \frac{2R}{3} = \frac{4R}{3} \quad \ldots(ii)$$

Now, dividing Eq. (i) by Eq. (ii), we get

$$\frac{R_{eq}}{R_{eq}'} = \frac{3\frac{R}{2}}{4\frac{R}{3}} = \frac{9}{8} \Rightarrow x = 9$$

91. *(0.112)* Radius of circular orbit,

$$r = 0.5\,Å = 0.5 \times 10^{-10}\,m$$

Velocity of electron,

$$v = 2.2 \times 10^6\,ms^{-1}$$

$\because$ We know that,

Current, $I = \frac{q}{t}$

where, q = charge and t = time taken.

$$\Rightarrow \quad I = \frac{e}{t} \quad \ldots(i) \{\because q = e\}$$

As electron is moving in circular orbit, so it means electron is tracing a distance of $2\pi r$.

$$\because \quad \text{Time} = \frac{\text{Distance}}{\text{Speed}} \Rightarrow t = \frac{2\pi r}{v} \quad \ldots(ii)$$

From Eqs. (i) and (ii), we get

$$I = \frac{e}{\frac{2\pi r}{v}} \Rightarrow I = \frac{ev}{2\pi r}$$

$$\Rightarrow \quad I = \frac{1.6 \times 10^{-19} \times 2.2 \times 10^6}{2 \times \frac{22}{7} \times 0.5 \times 10^{-10}}$$

$$= 0.112 \times 10^{-2}\,mA$$

92. *(45)* Given, resistance of wire $= 16\,\Omega$

Let length of wire $= 4\,L$

Hence, side of square, $a = L$

Resistance of one side $= \frac{16}{4} = 4\,\Omega$

Supply voltage $V = 9\,V$ and internal resistance $= 1\,\Omega$

Current flowing through circuit be I and voltage across diagonal be V'.

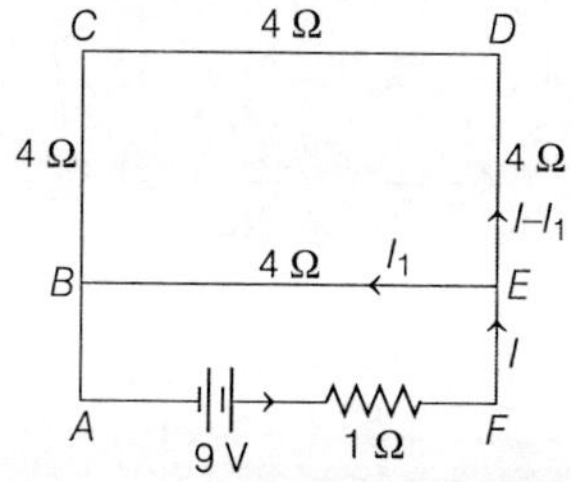

Three 4Ω resistance are in series with 4Ω and $1\,\Omega$ resistance in parallel.

Equivalent resistance R_{eq}

$$= \frac{12 \times 4}{12 + 4} + 1 = \frac{48}{16} + 1 = 4\,\Omega$$

As we know that

By Ohm's law, $V = IR$

$$\Rightarrow \quad I = \frac{V}{R} \Rightarrow I = \frac{9}{4}\,A$$

Now current through BE be I_1 and $EDCB$ is $(I - I_1)$.

∵ In parallel connection,
Voltage across BE (V_1) = Voltage across $EDCB$ (V_2)

$$\therefore I_1 \times 4 = \left(\frac{9}{4} - I_1\right)12 \Rightarrow I_1 = \left(\frac{9}{4} - I_1\right)3$$

$$\Rightarrow \quad I_1 = \frac{27}{16}\text{ A}$$

Equivalent resistance across diagonal,

$$\frac{1}{R_d} = \frac{1}{4+4} + \frac{1}{4+4} = \frac{1}{8} + \frac{1}{8}$$

$$R_d = \frac{8}{2} = 4\ \Omega$$

$$\text{and} \quad I_2 = I - I_1 = \frac{9}{4} - \frac{27}{16} = \frac{9}{16}\text{ A}$$

$$\therefore \quad V' = (4+4)\, I_2$$

$$= 8 \times \frac{9}{16} = 4.5\text{ V} = 45 \times 10^{-1}\text{ V}$$

93. *(10)* When switches S_1 and S_2 are closed, then the given figure will be represented as follows

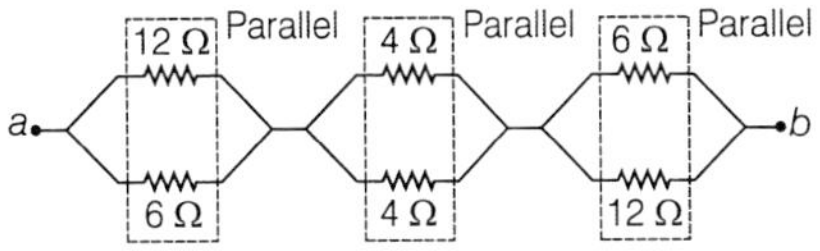

∴ The resistance across ab will be given as

$$R_{ab} = \frac{12 \times 6}{12+6} + \frac{4 \times 4}{4+4} + \frac{6 \times 12}{6+12}$$

$$\Rightarrow R_{ab} = \frac{72}{18} + 2 + \frac{72}{18}$$

$$\Rightarrow R_{ab} = 4 + 2 + 4 = 10\,\Omega$$

94. *(2)* Given, specific resistance for wire 1,

$$\rho_1 = 6\,\Omega\text{-cm}$$

Specific resistance for wire 2,

$$\rho_2 = 3\,\Omega\text{-cm}$$

Resistance, $R = \rho l / A$

For parallel connections, $R_{net} = \dfrac{R_1 R_2}{R_1 + R_2}$

$$\Rightarrow \frac{\rho l}{A} = \frac{\dfrac{\rho_1 l}{A} \times \dfrac{\rho_2 l}{A}}{\dfrac{\rho_1 l}{A} + \dfrac{\rho_2 l}{A}} \Rightarrow \rho = \frac{\rho_1 \rho_2}{\rho_1 + \rho_2}$$

$$= \frac{6 \times 3}{6+3} = 2\,\Omega\text{-cm}$$

95. *(70)* According to the question,

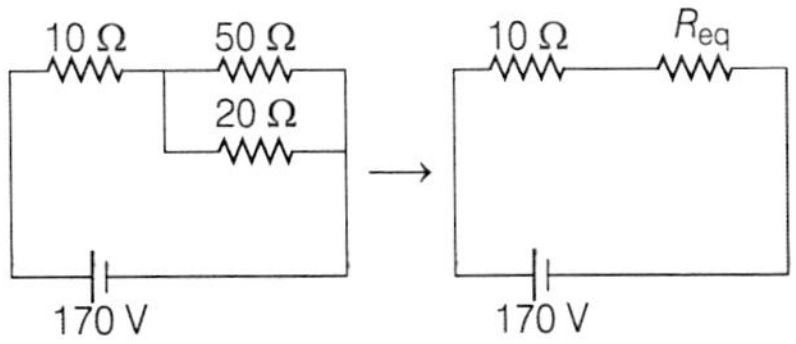

Equivalent resistance of the circuit,

$$R_{eq} = \frac{50 \times 20}{50 + 20} \Rightarrow R_{eq} = \frac{100}{7}\,\Omega$$

According to voltage division rule, voltage across the $10\,\Omega$ resistance of the circuit,

$$V_{10\,\Omega} = 170 \times \left(\frac{10}{10 + \dfrac{100}{7}}\right)$$

$$V_{10\,\Omega} = 70\text{ V}$$

The value of x to the nearest integer is 70.

96. *(4)* Let two resistors have resistances R_1 and R_2, respectively.
As per question, equivalent resistance of series combination is s

$$\Rightarrow \quad s = R_1 + R_2 \quad \ldots\text{(i)}$$

and equivalent resistance of parallel combination is p

$$\Rightarrow \quad p = \frac{R_1 R_2}{R_1 + R_2} \quad \ldots\text{(ii)}$$

According to the question,

$$s = np \quad \ldots\text{(iii)}$$

From Eqs. (i),(ii) and (iii), we get

$$R_1 + R_2 = n\frac{R_1 R_2}{(R_1 + R_2)}$$

$$\Rightarrow n(R_1 R_2) = (R_1 + R_2)^2$$

$$\Rightarrow \quad n = \frac{(R_1 + R_2)^2}{R_1 R_2}$$

For n to be minimum, $R_1 = R_2 = R$

$$\Rightarrow \quad n = \frac{(R+R)^2}{R \cdot R} \Rightarrow n = 4$$

97. *(3)* According to the figure given in question, all $3\,\Omega$ resistances are in parallel combination. So, their equivalent resistance is

$$\frac{1}{R_{\text{parallel}}} = \frac{1}{3} + \frac{1}{3} + \frac{1}{3}$$

$$\frac{1}{R_{\text{parallel}}} = \frac{3}{3}$$

$$\Rightarrow \quad R_{\text{parallel}} = 1\text{ k}\Omega$$

Equivalent circuit

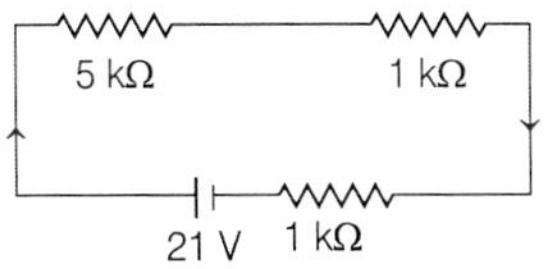

$5\text{ k}\Omega$ and $1\text{ k}\Omega$ are in series to the equivalent of all $3\,\Omega$ resistances.

$$\therefore \quad R_{net} = 5 + 1 + R_{\text{parallel}} \quad \ldots\text{(i)}$$

$$\Rightarrow \quad R_{net} = (5 + 1 + 1)\text{ k}\Omega$$

$$= 7\text{ k}\Omega = 7 \times 10^3\,\Omega$$

∴ The value of electric current flowing through $5\text{ k}\Omega$ resistor will be

$$I = \frac{V}{R_{net}} = \frac{21}{7 \times 10^3} = 3 \times 10^{-3}\text{ A}$$

$$= 3\text{ mA}$$

98. *(2)* Let resistances be R_1, R_2, R_3 and R_4 and I_1 current is passing through R_4 as shown in figure
∴ $I_2 = (6 - I_1)$ is passing through R_2
As, same current is flowing through R_4 and R_3,
∴ R_4 and R_3 are in series.

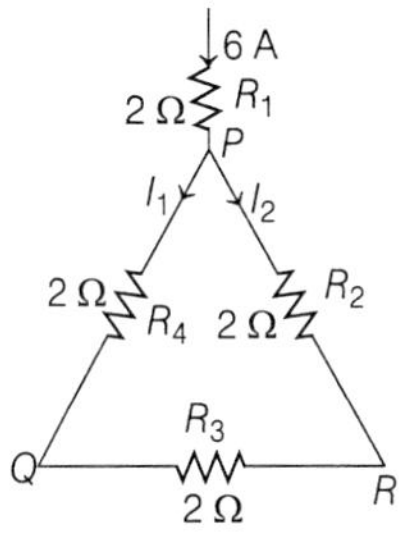

and series equivalent resistance,

$$R_{eq} = R_4 + R_3$$

$$\therefore \quad R_{eq} = 2 + 2 = 4\,\Omega$$

Voltage through R_{eq} and R_2 will be same.

$$\Rightarrow \quad I_1 R_{eq} = I_2 R_2$$

$$\Rightarrow \quad I_1 4 = (6 - I_1)\, 2$$

$$\Rightarrow \quad 2I_1 = 6 - I_1$$

$$\Rightarrow \quad I_1 = 2\text{ A}$$

99. *(5)* Given, radius of cylindrical wire, $r = 0.5\text{ mm} = 0.5 \times 10^{-3}$ m

Conductivity, $\sigma = 5 \times 10^7$ S/m

Electric field, $E = 10$ mV/m $= 10 \times 10^{-3}$ V/m

We know that current density,

$$\therefore \quad J = \sigma E = 5 \times 10^7 \times 10 \times 10^{-3}$$

$$= 5 \times 10^5\text{ A/m}^2$$

Also, $J = I / A \Rightarrow I = JA$

$$\Rightarrow \quad I = 5 \times 10^5 \times \pi \times (0.5 \times 10^{-3})^2$$

$$= 5 \times 10^5 \times \pi \times 25 \times 10^{-8}$$

$$= 125\,\pi \times 10^{-3}$$

$$\Rightarrow \quad x^3\ \pi\text{mA} = 125\,\pi\text{ mA}$$

$$\Rightarrow \quad x = 5$$

Topic 2

01. *(b)* $P = \dfrac{V^2}{R}$

$R_S = R_1 + R_2 + R_3 + \ldots + R_n$

$\dfrac{V^2}{P_S} = \dfrac{V^2}{P} + \dfrac{V^2}{P} + \ldots + \dfrac{V^2}{P_n}$

$P_S = \dfrac{P}{n}$

02. *(c)*

> **key idea** To solve such type of problem,
> (i) Calculate the value of current using $I = P/V$ and
> (ii) Using this value of I, you can calcualte number of electrons using
> $n = It/e$

We know that,
Power $P = VI$

$\Rightarrow \quad I = \dfrac{P}{V} = \dfrac{110}{220} = 0.5$ A

Since, $I = \dfrac{ne}{t}$

$\Rightarrow \quad \dfrac{n}{t} = \dfrac{I}{e} = \dfrac{0.5}{1.6 \times 10^{-19}}$

$= 31.25 \times 10^{17}$ electrons/second

03. *(d)* For bulb, $P = 50$ W, $V = 200$ V

$\therefore$ Resistance of the bulb,

$R = \dfrac{V^2}{P} \quad \left[\because P = \dfrac{V^2}{R}\right]$

$= \dfrac{200 \times 200}{50} = 800\,\Omega$

$\therefore$ Power dissipation when bulb is connected with $V' = 100$ V supply,

$P' = \dfrac{(V')^2}{R} = \dfrac{100 \times 100}{800} = 12.5$ W

04. *(c)* Total power P consumed
$= (15 \times 40) + (5 \times 100) + (5 \times 80) + (1 \times 1000)$
$= 2500$ W

As, we know that
$P = VI$

$\Rightarrow \quad I = \dfrac{2500}{220} = \dfrac{125}{11} = 11.3$ A

Hence, minimum capacity should be 12 A.

05. *(a)* Heat generated,

$H_1 = \dfrac{V^2}{R} t$ and $H_2 = \dfrac{V^2}{(R/2)} t$

$\therefore \quad \dfrac{H_2}{H_1} = 2$

or $\quad H_2 = 2H_1$

06. *(a)* $P = V^2/R$

$\therefore \quad R_{hot} = \dfrac{V^2}{P} = \dfrac{200 \times 200}{100} = 400\,\Omega$

$R_{cold} = \dfrac{400}{10} = 40\,\Omega$

07. *(c)* This implies that they have negative and high temperature coefficient of resistivity. They are usually made of metal oxides with high temperature coefficient of resistivity.

> **• Concept Enhancer**
> Thermistors are the resistors made up of semiconductors whose resistance decreases with the increase in temperature.

08. *(c)* Let time taken in boiling the water by the heater be t seconds. Then,

$Q = ms\,\Delta T \Rightarrow \dfrac{Pt}{J} = ms\,\Delta T$

$\therefore \quad \dfrac{836}{4.2} t = 1 \times 1000(40 - 10)$

or $\quad \dfrac{836}{4.2} t = 1000 \times 30$

or $\quad t = \dfrac{1000 \times 30 \times 4.2}{836} = 150$ s

09. *(c)* $R = \dfrac{V^2}{P} = \dfrac{(220)^2}{1000}$

where, V and P are denoting rated voltage and power, respectively.

$\therefore P_{consumed} = \dfrac{V^2_{applied}}{R}$

$= \dfrac{110 \times 110}{220 \times 220} \times 1000 = 250$ W

10. *(b)* $P = \dfrac{V^2}{R_{net}}$

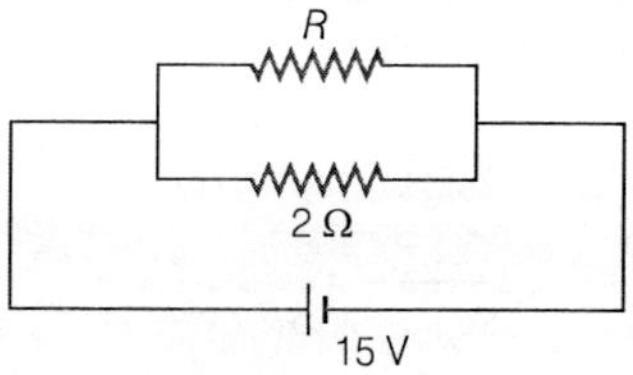

$\therefore \quad 150 = \dfrac{(15)^2}{\dfrac{R \cdot 2}{2 + R}} \Rightarrow R = 6\,\Omega$

11. *(a)* $I = 1$ A
We know that

efficiency, $\eta = \dfrac{P_{out}}{P_{in}}$

$\Rightarrow \quad 0.916 = \dfrac{P_{out}}{VI}$

$0.916 = \dfrac{P_{out}}{100 \times 1}$

$\Rightarrow \quad P_{out} = 91.6$ Watt

Loss of power $= 100 - 91.6$
$= 8.4$ W $= 8.4$ J/s
$= \dfrac{8.4}{4.2}$ cal/s $= 2$ cal/s

12. *(c)* Let initial resistance is R and length is l.

We know power $P = \dfrac{V^2}{R}$ and $R = \dfrac{\rho l}{A}$

$\therefore \quad P \propto \dfrac{1}{l}$

For case I $t = t_1 = 20$ min
and
For case II $t = t_2 = 15$ min

$P_1 \propto \dfrac{1}{t_1}$ and $P_2 \propto \dfrac{1}{t_2}$

$\Rightarrow \quad \dfrac{P_1}{P_2} = \dfrac{t_2}{t_1} = \dfrac{15}{20} = \dfrac{l_2}{l_1} \Rightarrow l_2 = \dfrac{3}{4} l_1$

Hence, length is decreased by 3/4 times.

13. *(d)* Given,
At $t = 0°$C, $R_0 = 8\,\Omega$
$t = 100°$C, $R = 10\,\Omega$
Using the relation,

$R = R_0(1 + \alpha t)$

Here, $t = \Delta t = (100°\text{C})$

$10 = 8(1 + \alpha \times 100)$

$\alpha = 1/400$

Now with $t = 400°$C

$R = R_0(1 + \alpha \times 400)$

$R = 8\left[1 + \dfrac{1}{400} \times 400\right]$

$R = 16\Omega$

14. *(c)* Illumination $\propto$ Power

Now, $P_i = I^2 R \Rightarrow P_f = \left(\dfrac{80}{100} I\right)^2 R$

$= 0.64 P_i$

Percentage drop in power (illumination)

$= \dfrac{P_i - P_f}{P_i} \times 100$

$= \dfrac{P_i(1 - 0.64)}{P_i} \times 100$

$= 0.36 \times 100 = 36\%$

15. *(d)* Given, $H = \frac{V^2}{R_{eq}} \cdot t$

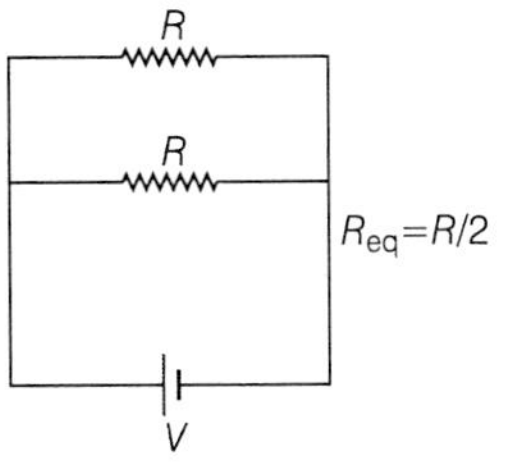

$$\frac{1}{R_{eq}} = \frac{1}{R} + \frac{1}{R}, \; R_{eq} = \frac{R}{2}$$

$$H_1 = \frac{2V^2}{R} \cdot t \qquad \dots \text{(i)}$$

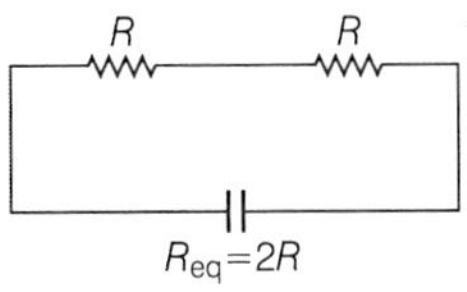

$$H_2 = \frac{V^2}{2R} \cdot t \qquad \dots \text{(ii)}$$

$$\frac{H_1}{H_2} = \frac{\frac{2V^2}{R} \cdot t}{\frac{V^2}{2R} \cdot t} \Rightarrow \frac{H_1}{H_2} = \frac{4}{1}$$

16. *(b)*

key idea If two resistors R_1 and R_2 are connected in parallel as shown in figure,

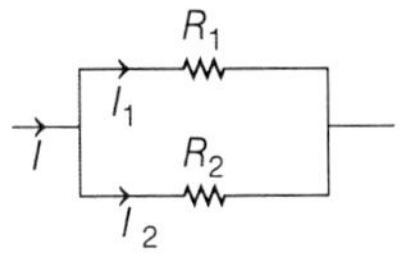

then you can find the value of I_1 and I_2 using current division rule as given below.

$$I_1 = \left(\frac{R_2}{R_1 + R_2}\right) I \text{ and } I_2 = \left(\frac{R_1}{R_1 + R_2}\right) I$$

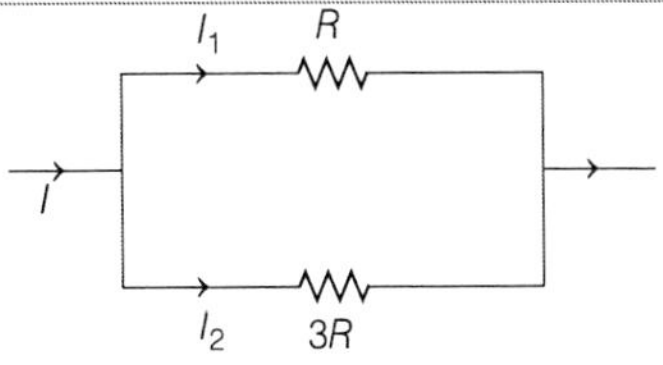

According to current division rule,

$$I_1 = \frac{3R}{(R + 3R)} I = \frac{3I}{4}$$

and $$I_2 = \frac{RI}{(R + 3R)} = \frac{I}{4}$$

Hence, ratio of thermal energies released in resistors is

$$\frac{H_1}{H_2} = \frac{I_1^2 R}{I_2^2 \times 3R} = \frac{\frac{9}{16}}{\frac{1}{16} \times 3} = 3:1$$

17. *(c)* Mass of ice $= \rho A l$

$$= 10^3 \times 10^{-4} \times 1 = 0.1 \text{ kg}$$

Energy required to melt the ice

$$= ms\Delta T + mL$$
$$= 0.1 \times 2 \times 10^3 (10) + 0.1 \times 3.33 \times 10^5$$
$$= 2 \times 10^3 + 3.33 \times 10^4 = 3.53 \times 10^4 \text{ J} \qquad \dots \text{(i)}$$

Also, heat produced due to electric current

$$H = i^2 Rt = (0.5)^2 \times 4 \times 10^3 \times t$$
$$= 10^3 \times t \qquad \dots \text{(ii)}$$

From Eqs. (i) and (ii), we get

$$3.53 \times 10^4 = 10^3 t$$
$$\therefore \quad t = 35.3 \text{ s}$$

18. *(a)* Given, power rating of bulb,

$$P_B = 500 \text{ W}$$

Voltage across bulb,

$$V_B = 100 \text{ V}$$

Supply voltage,

$$V_S = 200 \text{ V}$$

If a resistance R is attached in series with the bulb, then the voltage across resistance will be 100 V.

Now, current flowing in circuit when bulb delivers power of 500 W is given as

$$P_B = V_B I$$
$$\Rightarrow \quad 500 = 100 \times I \Rightarrow I = 5 \text{ A}$$

Same amount of current will flow from the resistance as it is connected in series.

Using Ohm's law,

$$V = IR$$
$$\Rightarrow \quad 100 = 5 \times R$$
$$\Rightarrow \quad R = 20\,\Omega$$

Thus, the resistance connected in series is 20 Ω.

19. *(c)* The given circuit diagram can be drawn as

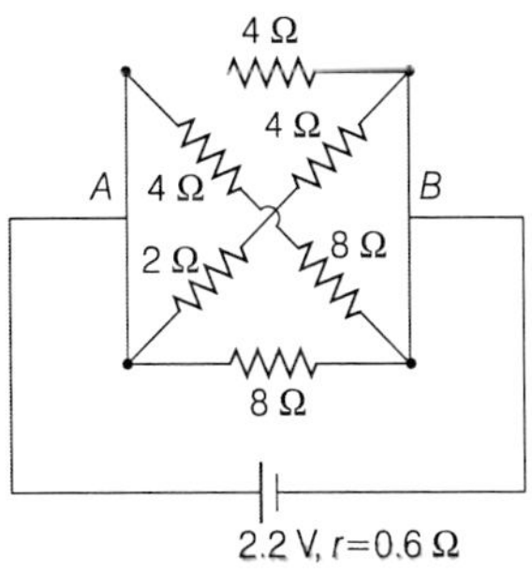

Redrawing the above circuit diagram as

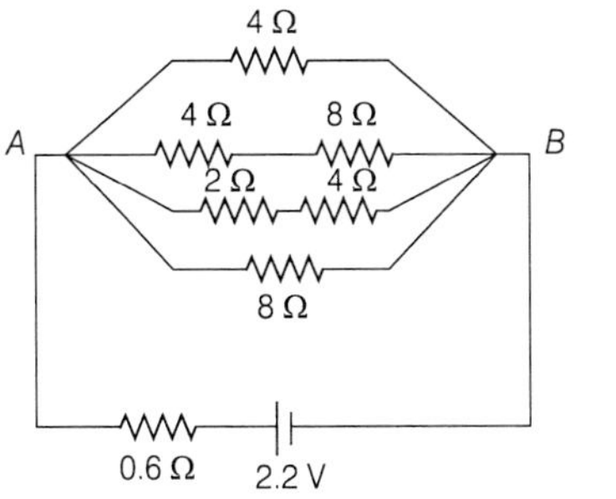

⇓

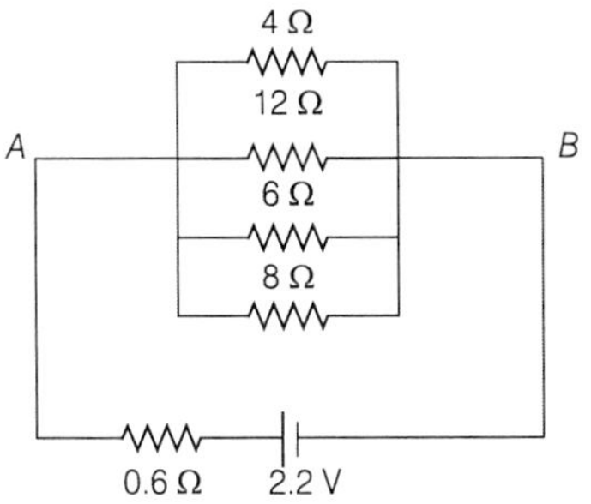

Equivalent resistance of the circuit between point A and B is given as

$$\frac{1}{R_{eq}} = \frac{1}{R_1} + \frac{1}{R_2} + \frac{1}{R_3} + \frac{1}{R_4}$$
$$= \frac{1}{4} + \frac{1}{12} + \frac{1}{6} + \frac{1}{8}$$
$$\Rightarrow \quad R_{eq} = 1.6\,\Omega$$

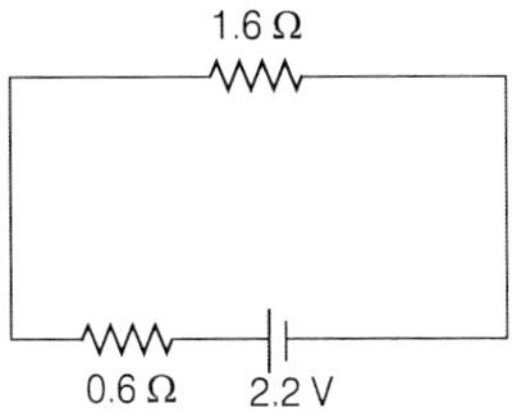

$$R_{total} = 1.6 + 0.6 = 2.2\,\Omega$$

$$P = \frac{V^2}{R_{total}} = \frac{2.2 \times 2.2}{2.2} = 2.2 \text{ W}$$

20. *(d)* Given, Heat energy,

$$H_1 = 500 \text{ J}$$

Initial current, $I_1 = 1.5$ A, final current,

$$I_2 = 3 \text{ A}$$

and time, $t = 20$ s

According to Joule's law of heating,

$$H = I^2 Rt$$
$$\Rightarrow \quad H_1 = I_1^2 Rt \qquad \dots \text{(i)}$$
and $$H_2 = I_2^2 Rt \qquad \dots \text{(ii)}$$

[∵ Resistance and time being the same in both cases]

Dividing Eq. (i) by Eq. (ii), we get

$$\Rightarrow \frac{H_1}{H_2} = \frac{I_1^2 Rt}{I_2^2 Rt} \Rightarrow \frac{H_1}{H_2} = \left(\frac{I_1}{I_2}\right)^2$$

$$\Rightarrow \frac{H_1}{H_2} = \left(\frac{1.5}{3}\right)^2 = \left(\frac{15}{30}\right)^2$$

$$\Rightarrow \frac{H_1}{H_2} = \left(\frac{1}{2}\right)^2$$

$$\Rightarrow \frac{H_1}{H_2} = \frac{1}{4} \Rightarrow \frac{500}{H_2} = \frac{1}{4}$$

$$\Rightarrow H_2 = 500 \times 4$$

$$\Rightarrow H_2 = 2000 \text{ J}$$

21. *(d)* Using Kirchhoff's loop law,

$$V = E - ir$$
$$2.5 = 3 - ir$$
$$ir = 0.5 \quad \ldots(\text{i})$$

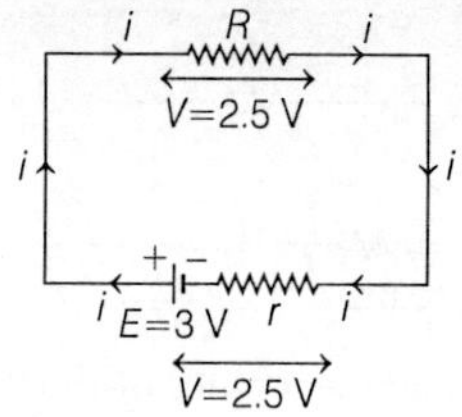

Now, from Ohm's law,

$$V = iR, \; 2.5 = iR, \quad \ldots(\text{ii})$$

and given that

$$P_R = 0.5, \; i^2 R = 0.5 \quad \ldots(\text{iii})$$

On dividing Eq. (i) by Eq (ii), we get

$$\frac{ir}{iR} = \frac{0.5}{2.5} \Rightarrow \frac{r}{R} = \frac{1}{5} \quad \ldots(\text{iv})$$

Now, $\frac{P_r}{P_R} = \frac{i^2 r}{i^2 R}$

$$\Rightarrow \frac{P_r}{0.5} = \frac{r}{R} \Rightarrow \frac{P_r}{0.5} = \frac{1}{5}$$

$$\Rightarrow P_r = 0.1 \text{ W} = 0.10 \text{ W}$$

22 *(c)* Power of 15 bulbs (each of 45 W),

$$P_1 = 15 \times 45 = 675 \text{ W}$$

Power of 15 bulbs (each of 100 W),

$$P_2 = 15 \times 100 = 1500 \text{ W}$$

Power of 15 fans (each of 10 W),

$$P_3 = 15 \times 10 = 150 \text{ W}$$

Power of 2 heaters (each of 1 kW),

$$P_4 = 2 \times 1000 = 2000 \text{ W}$$

Total power usage of building,

$$P = P_1 + P_2 + P_3 + P_4 = 4325 \text{ W}$$

Using $P = VI$, current drawn from mains supply of 220 V,

$$I = \frac{P}{V} = \frac{4325}{220} = 19.66 \text{ A}$$

$\therefore$ Minimum fuse capacity required is 20 A.

23 *(c)* Let P_1 and P_2 be the individual electric powers of the two resistances, respectively.

In series combination, power is

$$P_0 = \frac{P_1 P_2}{P_1 + P_2} = 60 \text{ W}$$

Since, the resistances are equal and in series,

$$P_1 = P_2 = 120 \text{ W}$$

In parallel combination, power is

$$P = P_1 + P_2 = 120 + 120 = 240 \text{ W}$$

> ***Alternate Solution***
>
> Let R be the resistance.
>
> $\therefore$ Net resistance in series $= R + R = 2R$
>
> $P = \frac{V^2}{2R} = 60 \text{ W} \Rightarrow \frac{V^2}{R} = 120 \text{ W}$
>
> New resistance in parallel $= \frac{R \times R}{R + R} = R/2$
>
> $P' = \frac{V^2}{R/2} = 2\left(\frac{V^2}{R}\right) = 240 \text{ W}$

24. *(c)* As the rated power of 25 W is less than 100 W, it implies that 25 W bulb has higher resistance. As in series connection, current through both the bulbs is same but heating in 25 W bulb is more than that of 100 W bulb. So, 25 W bulb will get fused.

25. *(c)* Resistance of electric bulb,

$R = \frac{V^2}{P_{\text{rated}}}$, where subscripts denote for rated parameters.

$$\Rightarrow R = \frac{(220)^2}{100}$$

Power consumed at 110 V,

$$P_{\text{consumed}} = V^2 / R$$

$$\therefore P_{\text{consumed}} = \frac{(110)^2}{(220)^2/100} = 25 \text{ W}$$

26. *(b)* Case I

Using the formula, $P = \frac{V^2}{R}$ $\quad \ldots(\text{i})$

where, R is resistance of wire, V is voltage across wire and P is power dissipation in wire

and $\quad R = \frac{\rho l}{A}$ $\quad \ldots(\text{ii})$

From Eqs. (i) and (ii), we have

$$P_1 = \frac{V^2}{\rho l / A}$$

or $\quad P_1 = \frac{V^2}{\rho l} \cdot A \quad \ldots(\text{iii})$

Case II

Let R_2 be net resistance of two wires in parallel, then

$$R_2 = \frac{R \times R}{R + R} = \frac{R}{2}$$

where, R is the resistance of half wire.

$$\Rightarrow R_2 = \frac{\rho \cdot \left(\frac{l}{2}\right)}{A \cdot 2} = \frac{\rho l}{4A}$$

or $\quad P_2 = \frac{V^2}{\rho l} \cdot 4A \quad \ldots(\text{iv})$

Hence, from Eqs. (iii) and (iv), we get

$$\frac{P_1}{P_2} = \frac{1}{4}$$

$$\Rightarrow \frac{P_2}{P_1} = \frac{4}{1}$$

27. *(a)* In series,

$$R_{\text{eq}} = R_1 + R_2$$

$$\Rightarrow 2R(1 + \alpha_{\text{eq}} \Delta T) = R(1 + \alpha_1 \Delta T) + R(1 + \alpha_2 \Delta T)$$

$$\Rightarrow 2\alpha_{\text{eq}} = \alpha_1 + \alpha_2$$

or $\quad \alpha_{\text{eq}} = \frac{\alpha_1 + \alpha_2}{2}$

In parallel,

$$\frac{1}{R_{\text{eq}}} = \frac{1}{R_1} + \frac{1}{R_2}$$

or $\quad \frac{dR_{\text{eq}}}{R_{\text{eq}}^2} = \frac{dR_1}{R_1^2} + \frac{dR_2}{R_2^2}$

$$\frac{\alpha_{\text{eq}} R_{\text{eq}} \Delta T}{(R_{\text{eq}})^2} = \frac{\alpha_1 R_1 \Delta T}{R_1^2} + \frac{\alpha_2 R_2 \Delta T}{R_2^2}$$

$$(\because R = R_0(1 + \alpha T))$$

or $\quad \frac{\alpha_{\text{eq}}}{R_{\text{eq}}} = \frac{\alpha_1}{R_1} + \frac{\alpha_2}{R_2}$

$$\Rightarrow \frac{2\alpha_{\text{eq}}}{R} = \frac{\alpha_1 + \alpha_2}{R}$$

$\therefore \quad R_{\text{eq}} = \frac{R}{2}$ and $R_1 = R_2 = R$

So, $\quad \alpha_{\text{eq}} = \frac{\alpha_1 + \alpha_2}{2}$

> **Time Saver Tech**
>
> Always remember, in problems having two conductors with coefficient of thermal resistances α_1 and α_2, respectively and also having same resistance at 0°C, then equivalent value of coefficient of thermal resistance in series and parallel combination is same which is equal to the average value of α_1 and α_2.

28. *(b)* Given,

Resistance of toaster $= 60\,\Omega = R_0$

Supply voltage = 220 V

Current reaching at toaster $I_1 = 2.75$ A

Then, final resistance of toaster

$$R_F = \frac{V\text{ (supply)}}{\text{Current}}$$

$$= \frac{220}{2.75} = 80\,\Omega$$

Now, resistance dependency on temperature is given by

$$R_t = R_0(1+\alpha t) \qquad [\because R_t = R_F]$$

$$80 = 60\,[1+\alpha\,(T-27)]$$

(Room temperature = 27° given)

$$\alpha = 2\times10^{-4}\,/^\circ\text{C (given)}$$

$$80 = 60\,[1+2\times10^{-4}\,(T-27)]$$

$$\Rightarrow \quad \frac{4}{3} - 1 = 2\times10^{-4}\,(T-27)$$

$$\Rightarrow \quad \frac{10000}{6} + 27 = T$$

$$\Rightarrow \quad T = 1694^\circ\text{C}$$

29. *(*)* Suppose battery of emf ε is applied across each circuit.

$$R_{eq} = \frac{3R}{2}$$

(A) 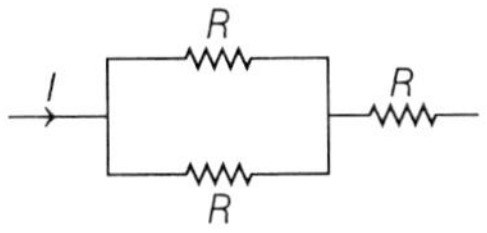

$$P_A = \frac{\varepsilon^2}{R_{eq}} = \frac{\varepsilon^2}{\frac{3R}{2}} = \frac{2\varepsilon^2}{3R} = 0.67\frac{\varepsilon^2}{R}$$

(B) 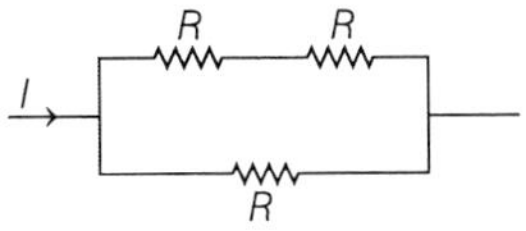

$$P_B = \frac{\varepsilon^2}{\frac{2R}{3}} = \frac{3}{2R}\varepsilon^2 = 1.5\frac{\varepsilon^2}{R}$$

(C) 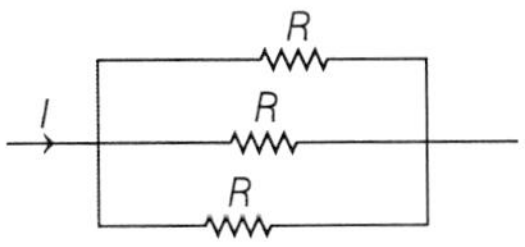

$$P_C = \frac{\varepsilon^2}{R_{eq}}$$

$$R_{eq} = \frac{R}{3},\; P_C = \frac{3\varepsilon^2}{R}$$

(D) 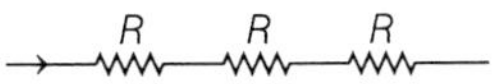

$$P_D = \frac{\varepsilon^2}{3R} \Rightarrow P_D = 0.33\frac{\varepsilon^2}{R}$$

From above we find,

$$P_D < P_A < P_B < P_C.$$

30. *(a)* Let the source voltage be V.

Equivalent resistance of the circuit when $r = fR$

$$R_{eq} = R + \frac{r\times R}{r+R} = R + \frac{fR}{f+1}$$

$$= \frac{(2f+1)R}{(f+1)}$$

∴Current in the circuit,

$$I = \frac{V}{R_{eq}} = \frac{V(f+1)}{R(2f+1)}$$

Current in the resistance $r(= fR)$

$$I_2 = \frac{I}{f+1} = \frac{V}{R(2f+1)}$$

Now, heat generated per unit time in r

$$H = I_2^2 r = \frac{V^2 f}{R(2f+1)^2}$$

For maximum H, $\dfrac{dH}{df} = 0$

$$\Rightarrow \quad \frac{V^2}{R}\left[\frac{1}{(2f+1)^2} - \frac{4f}{(2f+1)^3}\right] = 0$$

or $2f + 1 - 4f = 0$

$$\Rightarrow \quad f = \frac{1}{2}$$

31. *(d)* $\quad P = \dfrac{V^2}{R}$

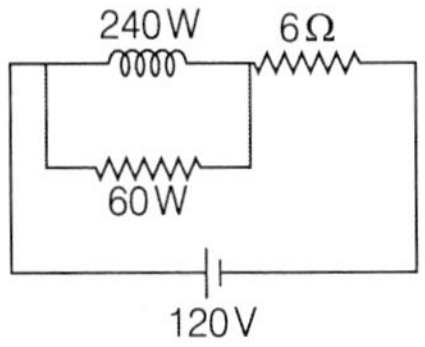

Resistance of the bulb,

$$R = \frac{120\times120}{60} = 240\,\Omega$$

$$R_{eq} = (240+6) = 246\,\Omega$$

$$\Rightarrow \quad I_1 = \frac{V}{R_{eq}} = \frac{120}{246}$$

$$V_1 = I_1R_1 = \frac{120}{246}\times240$$

$$= 117.073\text{ V}$$

Resistance of the heater

$$= \frac{V^2}{P} = \frac{120\times120}{240} = 60\,\Omega$$

As bulb and heater are connected in parallel.

$$\text{Net resistance} = \frac{240\times60}{300} = 48\,\Omega$$

Total resistance, $R_2 = 48 + 6 = 54\,\Omega$

Total current, $I_2 = V/R_2 = 120/54$

Potential across heater = Potential across bulb

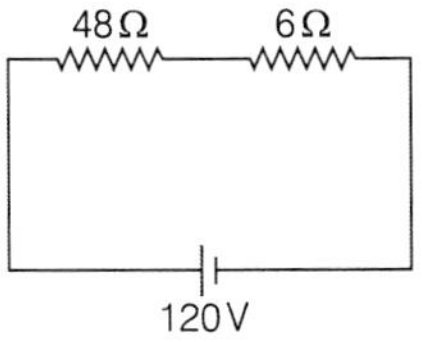

$$V_2 = \frac{120}{54}\times48 = 106.66\text{ V}$$

$$V_1 - V_2 = 117.073 - 106.66$$

$$= 10.4\text{ V}$$

32. *(1027)* Given, room temperature,

$$T_0 = 27^\circ\text{C}$$

Resistance at room temperature,

$$R_0 = 50\,\Omega$$

Resistance at given temperature T,

$$R_T = 62\,\Omega$$

Temperature coefficient

$$\alpha = 2.4\times10^{-4}\;^\circ\text{C}^{-1}.$$

The relation between resistance and temperature is given by

$$R = R_0\,[1+\alpha(T-T_0)]$$

$$\Rightarrow \quad 62 = 50[1+2.4\times10^{-4}(T-27)]$$

$$\Rightarrow \quad \frac{62}{50} = [1 + 2.4\times10^{-4}\,T - 27 \times 2.4\times10^{-4}]$$

$$\Rightarrow \quad 1.24 = 1 + 2.4\times10^{-4}\,T - 27\times2.4\times10^{-4}$$

$$\Rightarrow \quad 0.24 + 64.8\times10^{-4} = 2.4\times10^{-4}\,T$$

$$\Rightarrow \quad 2400\times10^{-4} + 64.8\times10^{-4} = 2.4\times10^{-4}\,T$$

$$\Rightarrow \quad \frac{2464.8}{2.4} = T$$

$$\Rightarrow \quad T = 1027\,^\circ\text{C}$$

33. *(5)*

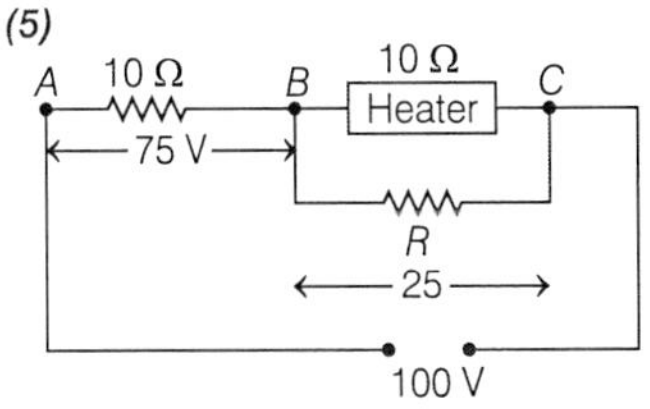

Heater has a rated power of 1000 W in 100 V line.

$$P = V^2/R \Leftrightarrow R = V^2/P$$

$$\Rightarrow \quad R = \frac{100\times100}{1000} = 10\,\Omega$$

So, heater has a resistance of $10\,\Omega$.

For heater to operate at 62.5 W in given circuit.

$$P = V^2/R \text{ or } V^2 = P \times R$$

$\Rightarrow V^2 = 62.5 \times 10 = 625$

or $V = \sqrt{625} = 25$ V

Hence, $V_{BC} = 25$ V

So, $V_{AB} = 75$ V

As $V_{AB} = I_{AB} \cdot R_{AB}$

$75 = I_{AB} \cdot 10$

or $I_{AB} = \frac{75}{10} = \frac{15}{2} = 7.5$ A

Now, $V_{BC} = I_{BC} \cdot R_{BC}$

$\Rightarrow \quad 25 = 7.5\left(\frac{R_1 R_2}{R_1 + R_2}\right)$

$\Rightarrow \quad 25 = 7.5\left(\frac{R \times 10}{10 + R}\right)$

$10 + R = 3R \Rightarrow 2R = 10$ or $R = 5\,\Omega$

34. *(748)* Resistance at 100°C is

$R_{100} = R_0(1 + \alpha\Delta T)$

$\Rightarrow 10.2 = 10[1 + \alpha(100 - 0)]$

$\Rightarrow 10.2 = 10 + \alpha \times 1000 \Rightarrow \frac{0.2}{1000} = \alpha$

or temperature coefficient is,

$\alpha = 2 \times 10^{-4}\,°C^{-1}$

Now, at temperature t, resistance is $10.95\,\Omega$ then,

$R_t = R_0[1 + \alpha(t - 0)]$

$\Rightarrow 10.95 = 10 + 10 \times \alpha \times t$

$\Rightarrow \frac{0.95}{10 \times 2 \times 10^{-4}} = t \Rightarrow \frac{95 \times 10^2}{10 \times 2} = t$

or $t = \frac{950}{2} = 475°C$

So; $t = 475 + 273 = 748$ K

35. *(3)* The given circuit is

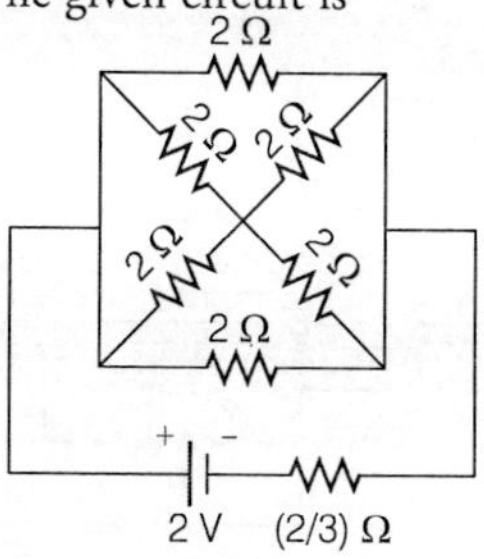

Equivalent circuit is

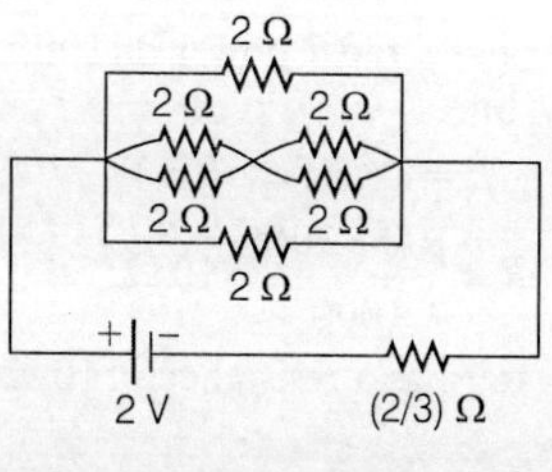

$\Rightarrow$

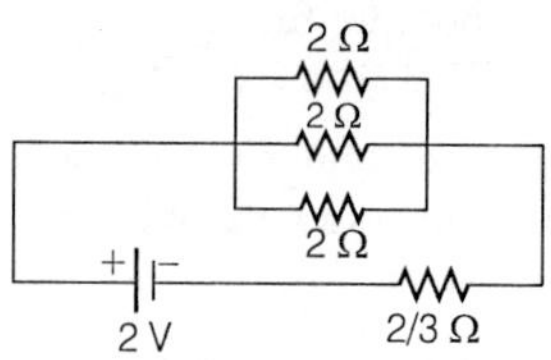

So, $R_{eq} = \frac{2}{3} + \frac{2}{3} = \frac{4}{3}\,\Omega$

Power, $P = \frac{V^2}{R} = \frac{4}{4/3} = 3$ W

36. *(16)* According to first condition,

$P_1 = \frac{V_0^2}{R}$

According to second condition, when wire is cut into two halves, then

$P_2 = \frac{V_0^2}{\frac{R}{2}} + \frac{V_0^2}{\frac{R}{2}} \Rightarrow P_2 = \frac{4V_0^2}{R} = 4P_1$

$\Rightarrow \quad \frac{P_2}{P_1} = \frac{4}{1} = \frac{\sqrt{x}}{1}$ (given)

$\therefore \quad \sqrt{x} = 4 \Rightarrow x = 16$

37. *(14)* Given, voltage,

$V = 220$ V

Power of 1st bulb, $P_1 = 100$ W

Power of 2nd bulb, $P_2 = 60$ W

Resistance of 1st bulb, $R_1 = \frac{V^2}{P_1}$ $\left[\because P = \frac{V^2}{R}\right]$

$= \frac{(220)^2}{100} = 484\,\Omega$

Resistance of 2nd bulb,

$R_2 = \frac{V^2}{P_2} = \frac{(220)^2}{60} = 806.67\,\Omega$

Since, two bulbs are connected in series, so current will be same in both and given by

$I = \frac{V}{R_{net}} = \frac{V}{R_1 + R_2}$

$= \frac{220}{(484 + 806.67)} = 0.17$ A

So, power consumed by 100 W bulb is

$P_1 = I^2 R_1$

$= (0.17)^2 \times 484 \simeq 14$ W

38. *(15)* When the coils are connected in parallel, the time taken to produce same amount of heat when connected separately to the same source, is

$\frac{1}{t_p} = \frac{1}{t_1} + \frac{1}{t_2}$ or

$t_p = \frac{t_1 t_2}{t_1 + t_2} = \frac{20 \times 60}{20 + 60}$

$\therefore t_p = 15$ min

39. *(450)* Here, heat, $H = 300$ J, time, $t = 15$ s, current, $i = 2$ A

New current, $i' = 3$ A, New time,

$t' = 10$ s

$H = i^2 Rt$ ...(i)

$H' = i' Rt'$ ...(ii)

On dividing Eq. (i) by (ii), we get

$\frac{300}{H'} = \left(\frac{2}{3}\right)^2 \times \frac{15}{10};$

$\frac{300}{H'} = \frac{4}{9} \times \frac{15}{10};$

$H' = \frac{300 \times 9 \times 10}{4 \times 15}$

Thus, $H' = 450$ J

40. *(4)* For case I,

The potential difference of the uniform wire, $V = 240$ V

The resistance of the uniform wire,

$R_1 = 36\,\Omega$

The power dissipation in the first case,

$P_1 = \frac{V^2}{R_1} = \frac{(240)^2}{36}$

For case II,

The resistance of each half,

$R_2 = \frac{R_1}{2} = \frac{36}{2} = 18\,\Omega$

$P_2 = \frac{V^2}{R_2} + \frac{V^2}{R_2} = \frac{(240)^2}{18} + \frac{(240)^2}{18}$

$= \frac{(240)^2}{9}$

Thus, the ratio of the total power dissipation in the first case to the second case,

$\frac{P_1}{P_2} = \frac{(240)^2/36}{(240)^2/9}$

$\Rightarrow \quad \frac{P_1}{P_2} = \frac{1}{4}$

On comparing with, $\frac{P_1}{P_2} = \frac{1}{x}$

The value of the $x = 4$.

41. *(3840)* Given that, initial current,

$I_1 = 4$ A

Final current, $I_2 = 2I_1 = 8$ A

Initial heat dissipated,

$H_1 = 192$ J

Initial time, $t_1 = 1\,\text{s}$
Final time, $t_2 = 5\,\text{s}$
Let final heat dissipated $= H_2$
By Joule's law of heating,

$$H \propto I^2 RT$$

Since, resistance remains same at initial and final condition,

$$\therefore \quad \frac{H_2}{H_1} = \frac{I_2^2 Rt_2}{I_1^2 Rt_1} = \frac{I_2^2 t_2}{I_1^2 t_1}$$

On substituting the given values, we get

$$\frac{H_2}{192} = \left(\frac{8}{4}\right)^2 \times \frac{5}{1}$$

$$\Rightarrow \quad H_2 = 3840\,\text{J}$$

42. *(50)* Rating of bulb = 200 W at 100 V
Supply voltage, $V = 200\,\text{V}$
Resistance of bulb,

$$R' = \frac{100 \times 100}{200} = 50\Omega$$

Let current be I, when resistance R is connected in series with the bulb.

Using, $P = VI$

$$200 = 100 \times I$$
$$\Rightarrow \quad I = 2\,\text{A}$$

Now, $V = I(R + R')$

$$200 = 2(R + 50)$$
$$\Rightarrow \quad R = 50\,\Omega$$

43. *(2500)* Given, energy dissipated by a resistor,

$$H = 10\text{ mJ} = 10 \times 10^{-3}\text{ J}$$

Time, $t = 1\text{ s}$
Electric current,

$$I = 2\text{ mA} = 2 \times 10^{-3}\text{ A}$$

Resistance, $R = ?$
According to Joule's law of heating,

$$H = I^2 Rt$$

$$\Rightarrow \quad R = \frac{H}{I^2 T} \quad \ldots\text{(i)}$$

On substituting the given values in Eq. (i), we get

$$R = \frac{10 \times 10^{-3}}{(2 \times 10^{-3})^2 \times 1}$$

$$\Rightarrow \quad R = \frac{10^{-2}}{4 \times 10^{-6}}$$

$$\Rightarrow \quad R = 0.25 \times 10^4$$

$$\Rightarrow \quad R = 2500\,\Omega$$

Topic 3

01. *(d)* Given, $V = 4.2\text{ V}$

$\therefore$ Energy supplied by battery $= Vq$

$$= Vq \Rightarrow 4.2 \times 5800 \times 3600 \times 10^{-3}$$
$$= 87696\text{ J}$$

$\therefore$ Energy stored in the battery when fully charged

$$= 87.696\text{ kJ} \simeq 87.7\text{ KJ}$$

02. *(d)* By series and parallel combination,

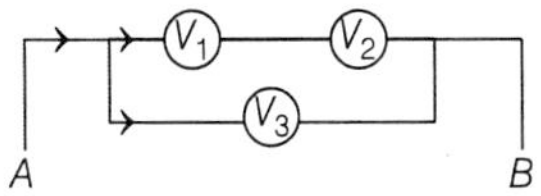

$$V_1 + V_2 = V_3$$

Hence, option $V_3 = V_1 + V_2$ is correct option.

03. *(d)* Let the current in 5 kΩ is i and in 10 kΩ is $(15 - i)$

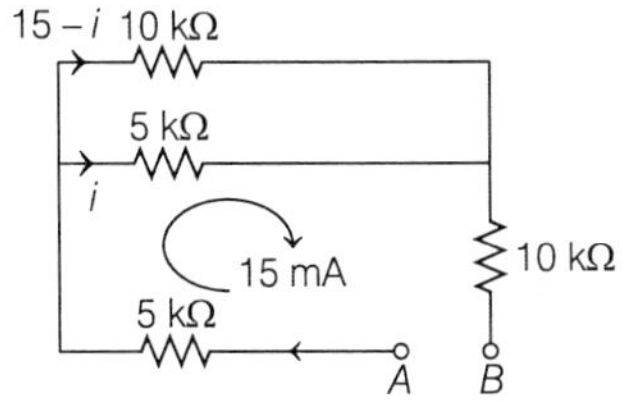

So, $i \times 5 = (15 - i) \times 10$

$$i = 10\text{ mA}$$

Using KVL in lower loop,

$$V_A - 15 \times 5 - 10 \times 5 - 10 \times 15 - V_B = 0$$
$$V_A - V_B = 275\text{ V}$$

04. *(c)* Each resistance is connected with two cells combined in opposite direction, so potential drop across each resistor is zero. Hence, the current through each of resistor is zero.

05. *(c)*

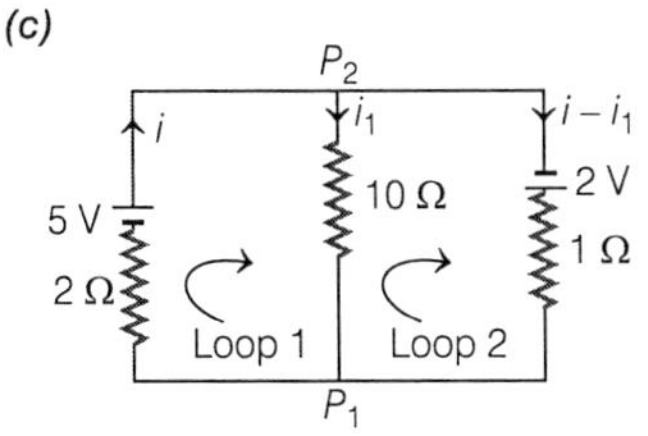

Applying Kirchhoff's law in loop 1

$$-10i_1 \quad 2i + 5 = 0 \quad \ldots\text{(i)}$$

Applying Kirchhoff's law in loop 2

$$10i_1 + 2 - 1 \times (i - i_1) = 0$$
$$10i_1 + 2 - i + i_1 = 0$$
$$11i_1 - i = -2$$
$$22i_1 - 2i = -4 \quad \ldots\text{(ii)}$$

Solving Eqs. (i) and (ii)

$$i_1 = \frac{1}{32}\text{ A} = 0.03\text{A from } P_2 \text{ to } P_1$$

06. *(d)* Kirchhoff's Ist law or KCL states that the algebraic sum of current meeting at any junction is equal to zero.
It is based on the conservation of charge.
Kirchhoff's IInd law or KVL states that algebraic sum of changes in potential around any closed resistor loop must be zero. Thus, we can say that KVL is based on conservation of energy.

07. *(b)* $I = \dfrac{E}{R + r}$

As r (internal resistance) and E are constant values and R is variable, so for I to be constant, we have to remove R from equation,
If $r >> R$, then $R + r \approx r$,
So, $I = \dfrac{E}{r}$, which is constant.

08. *(d)* Statement I is false but statement II is correct. The correct form of statement I is
When two non-ideal batteries are connected in parallel, then equivalent emf is lies between E_1 and E_2

$$E_{\text{eq}} = \frac{\frac{E_1}{r_1} + \frac{E_2}{r_2}}{\frac{1}{r_1} + \frac{1}{r_2}} \Rightarrow \frac{E_1 r_2 + E_2 r_1}{r_1 + r_2}$$

09. *(a)* Given circuit is

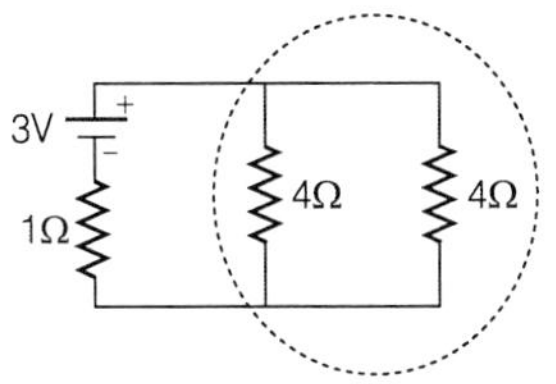

Parallel, $R_{\text{eq}} = \dfrac{R_1 R_2}{R_1 + R_2}$

$$= \frac{4 \times 4}{4 + 4} = 2\,\Omega$$

Total resistance of circuit $= 3\,\Omega$

Current through cell, $I = \dfrac{V}{R} = \dfrac{3}{3} = 1\text{ A}$

Terminal potential difference

$V = E - Ir \Rightarrow V = 3 - 1 \times 1 = 2\text{ V}$

10. *(b)*

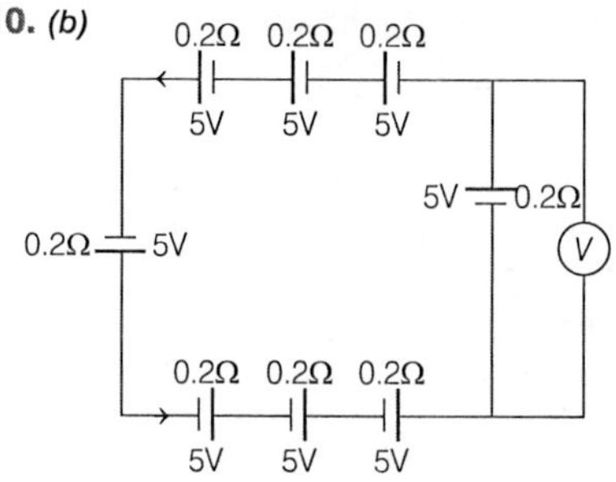

All cells are identical $r = 0.2\,\Omega$, $V = 5$ V
Above circuit can be reduced to,

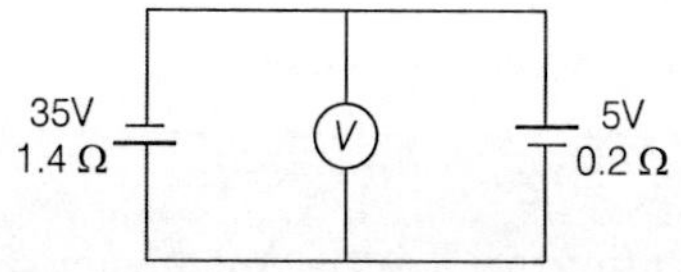

So, potential drop across parallel combination of above cells is,

$$V = \frac{E_2 r_1 - E_1 r_2}{r_1 + r_2} = \frac{5 \times 1.4 - 35 \times 0.2}{1.4 + 0.2}$$

$$= 0 \text{ V}$$

11. *(d)* $R_{eq} = R_1 + R_4 + \dfrac{R_2 R_3}{R_2 + R_3}$

$$= 2 + 1 + \frac{4 \times 4}{4 + 4} = 5\,\Omega$$

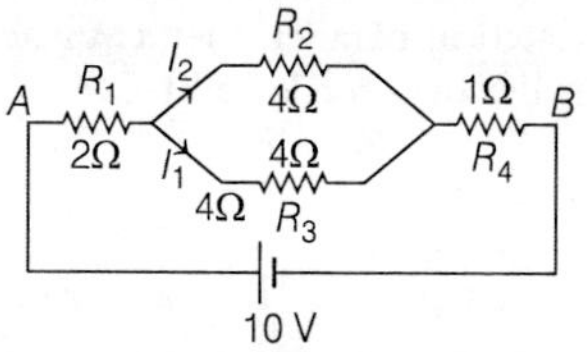

$\therefore \quad I = \dfrac{V}{R} = \dfrac{10}{5} = 2\text{A}$

Since, $R_2 = R_3 = 4\,\Omega$

$\therefore \quad I_1 = I_2 = \dfrac{I}{2} = \dfrac{2}{2} = 1\text{ A}$

> **• Concept Enhancer**
> In series combination, current through each resistor is same whereas in parallel combination, voltage across each resistor is same.

12. *(c)*

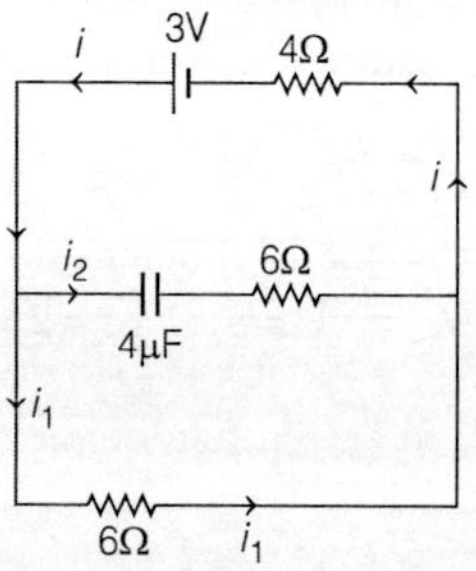

In steady state, no current will pass through capacitor, hence capacitor will act as open switch so $i_2 = 0$

$$i_1 = \frac{3}{6+4} = \frac{3}{10}\text{A}$$

Potential difference on 6Ω resistor

$$= 6 \times \frac{3}{10} = 1.8 \text{ V}$$

capacitor will have same potential so charge

$$Q = CV$$
$$= 4 \times 1.8 = 7.2\,\mu\text{C}$$

13. *(*)*

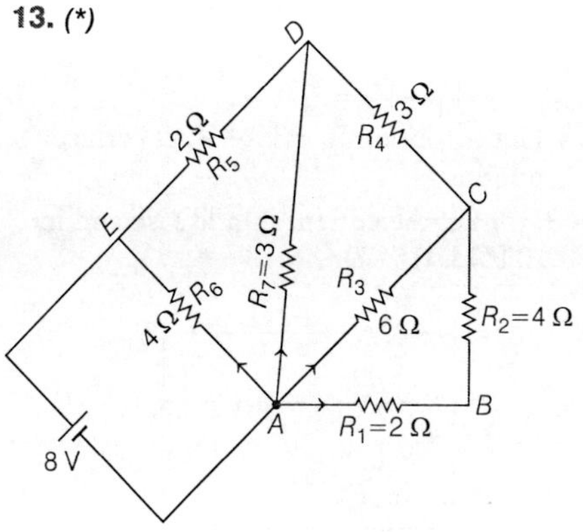

$$\frac{1}{R_{eq}} = \frac{1}{(4+2)} + \frac{1}{(3+3)} + \frac{1}{6} + \frac{1}{(2+4)}$$

$$= \frac{4}{6} \text{ or } R_{eq} = \frac{3}{2}$$

$$i = \frac{V}{R_{eq}} = \frac{8}{3} \times 2 = \frac{16}{3}\text{ A}$$

Since, resistance in all arms are same, current through all arms is the same.

$$\therefore \quad i_c = \frac{i}{4} = \frac{16}{3} \times \frac{1}{4} = \frac{4}{3}\text{ A}$$

Current through $R_2 = \dfrac{4}{3}$ A

14. *(b)* Given,

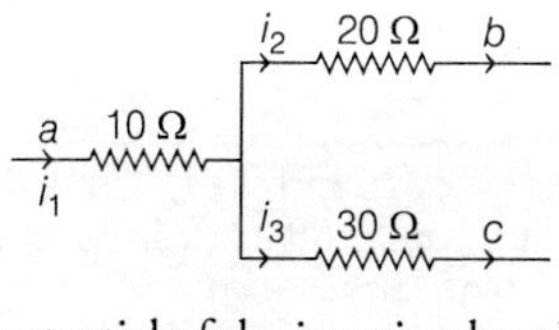

Let potential of the junction be x V.
using junction law $i_1 + i_2 + i_3 = 0$

$$\text{or } \frac{x-30}{10} + \frac{x-12}{20} + \frac{x-2}{30} = 0$$

$$\Rightarrow \frac{1}{60}[6x - 180 + 3x - 36 + 2x - 4] = 0$$

$$\frac{1}{60}[11x - 220] = 0 \Rightarrow x = \frac{220}{11} = 20 \text{ V}$$

Current through $20\,\Omega$ is $= \dfrac{x-12}{20}$

$$i_2 = \frac{20-12}{20} = 0.4 \text{ A}$$

15. *(b)* Given circuit is

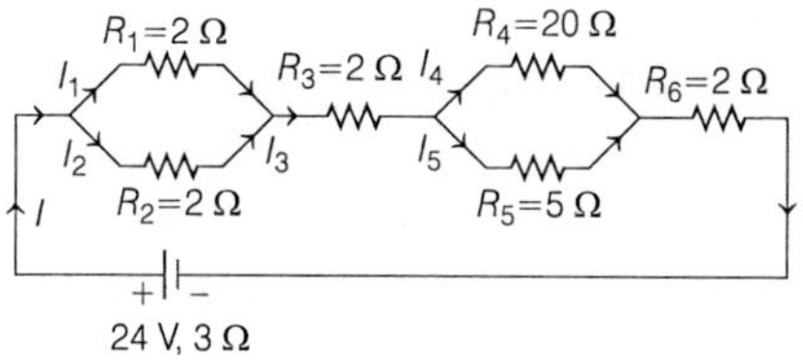

Total resistance of circuit

$$= (2 \| 2) + 2 + (20 \| 5) + 2 + 3$$

$$= \left(\frac{2 \times 2}{2+2}\right) + 2 + \left(\frac{20 \times 5}{20+5}\right) + 2 + 3$$

$$= 3 + 4 + 5 = 12\,\Omega$$

$$I = \frac{24}{12} = 2\text{ A}$$

So, current through resistor R_4,

$$I_4 = \left(\frac{5}{25}\right) \times 2 = \frac{2}{5}\text{ A}$$

And current through resistor R_5,

$$I_5 = \frac{20}{25} \times 2 = \frac{8}{5}\text{ A}$$

16. *(a)* The circuit diagram for the given condition is shown below.

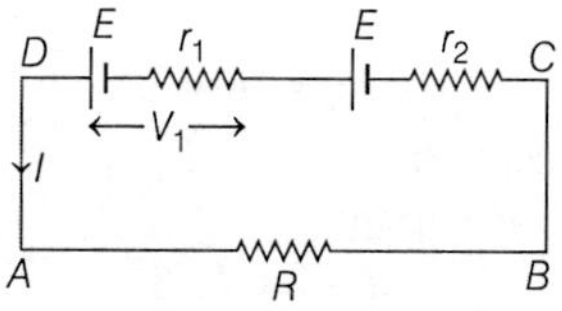

Given, $V_1 = 0$
i.e $E - Ir_1 = 0$

$$E = Ir_1 \quad \ldots\text{(i)}$$

On applying KVL in loop $ABCDA$.

$$RI + Ir_2 - E + Ir_1 - E = 0$$
$$I(R + r_1 + r_2) = 2E$$
$$\Rightarrow \quad I(R + r_1 + r_2) = 2 \times Ir_1 \quad \text{[From Eq. (i)]}$$
$$\Rightarrow \quad R + r_1 + r_2 = 2r_1$$
$$\Rightarrow \quad R = 2r_1 - r_1 - r_2$$
$$\Rightarrow \quad R = r_1 - r_2$$

17. *(a)* Let the internal resistance of cell is r and emf is ε. External resistance, $R = 2\,\Omega$

When, they are in series

$$i = \frac{2\varepsilon}{R + 2r} \quad \ldots\text{(i)}$$

When, they are in parallel

$$i' = \frac{\varepsilon}{R + r/2} \quad \ldots \text{(ii)}$$

From (i) and (ii), $i = i'$

$$\Rightarrow \frac{2\varepsilon}{R + 2r} = \frac{\varepsilon}{R + r/2}$$

$$\Rightarrow 2R + r = R + 2r;\ R = r \Rightarrow r = 2\,\Omega$$

18. *(a)* When all cells are connected in parallel as shown below.

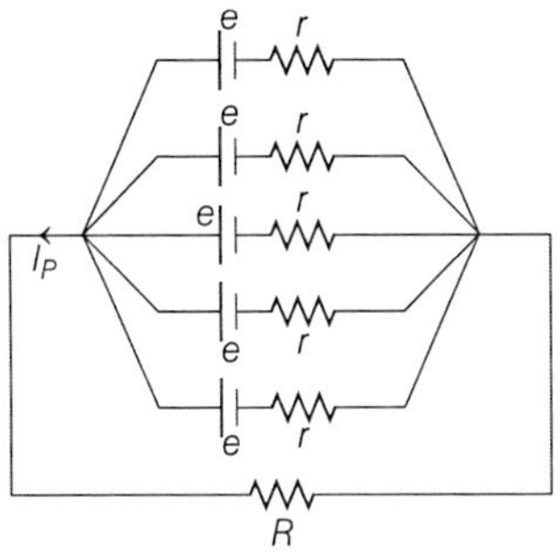

Potential will remain same as, $V_P = 5\,\text{V}$

Net resistance in parallel combination will be given as

$$R_p = R + \left[\frac{1}{\frac{1}{r} + \frac{1}{r} + \frac{1}{r} + \frac{1}{r} + \frac{1}{r}}\right]$$

$$= R + \frac{r}{5}$$

$$R_p = R + \frac{1}{5} \quad (\because r = 1\,\Omega) \ldots \text{(i)}$$

When all cells are connected in series as shown below

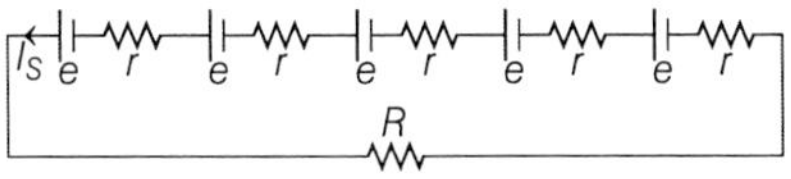

The net potential will increase as cells are connected in series,

$V_s = 5 + 5 + 5 + 5 + 5 = 25\,\text{V}$

Net resistance

$$R_s = r + r + r + r + r + R = 5r + R$$

$$R_s = 5 + R \quad (\because r = 1\,\Omega) \ldots \text{(ii)}$$

Since, $I = V/R$

As current in both series and parallel combination is same,

$$\frac{V_p}{R_p} = \frac{V_s}{R_s}$$

$$\Rightarrow \frac{5}{R + \frac{1}{5}} = \frac{25}{5 + R}$$

[From Eqs. (i) and (ii)]

$$\Rightarrow 25 + 5R = 25R + 5$$

$$\Rightarrow R = 1\,\Omega$$

19. *(c)* The given figure can be represented as

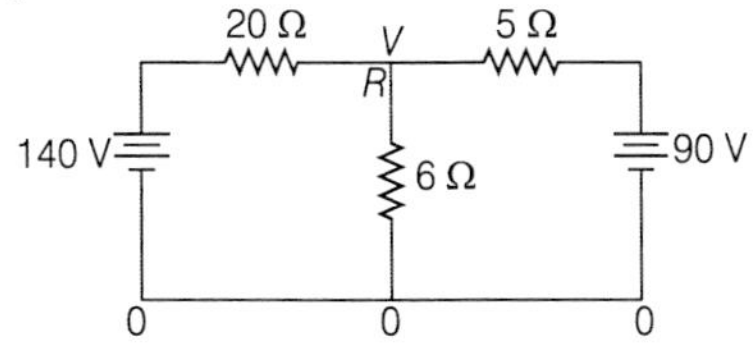

From the above figure, it can be clearly seen that the voltage at point R is assumed as V.

Therefore, applying Kirchhoff's current law at point R, we can write

$$\frac{V - 0}{6} + \frac{V - 90}{5} + \frac{V - 140}{20} = 0$$

$$\Rightarrow \frac{V}{6} + \frac{V - 90}{5} + \frac{V - 140}{20} = 0$$

$$\Rightarrow \frac{10V + 12V - 1080 + 3V - 420}{60} = 0$$

$$\Rightarrow 25V = 1500 \Rightarrow V = 60\,\text{V}$$

Therefore, current through $6\,\Omega$ resistor is

$$I = V/R = 60/6 = 10\,\text{A}$$

20. *(b)* The given circuit can be drawn as

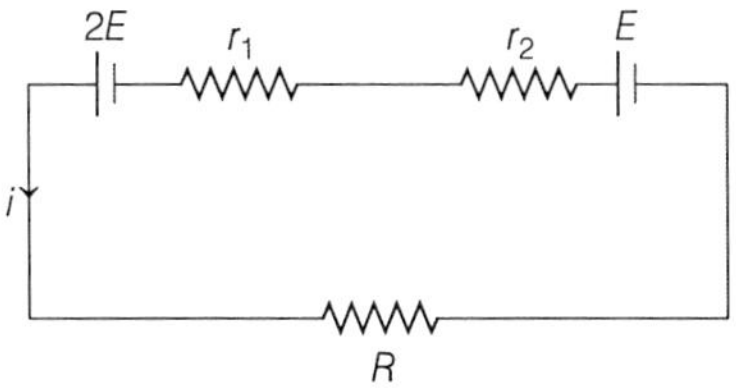

Since, in series combination, the current through each resistance remains same. So, equivalent resistance of the circuit is given as

$$R_{\text{equivalent}} = R + r_1 + r_2$$

and equivalent emf,

$$E_{\text{equivalent}} = 2E + E = 3E$$

From Ohm's law, $I = \dfrac{E_{\text{equivalent}}}{R_{\text{equivalent}}}$

$$\Rightarrow I = \frac{3E}{R + r_1 + r_2}$$

When potential difference is zero across the first cell.

Then, $2E = Ir_1$

On substituting the values in the above equation, we get

$$2E = \frac{3E}{R + r_1 + r_2} r_1$$

$$2R + 2r_1 + 2r_2 = 3r_1$$

$$\Rightarrow R = \frac{r_1 - 2r_2}{2}$$

$$\Rightarrow R = \frac{r_1}{2} - r_2$$

21. *(c)* The circuit can be shown as below

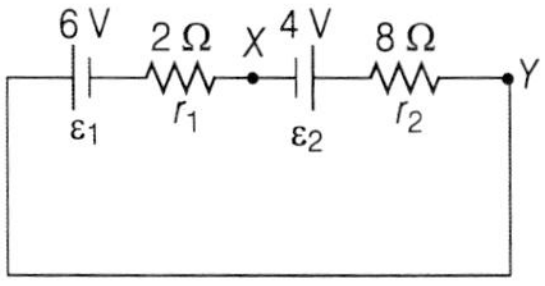

The current through the circuit,

$$I = \frac{\varepsilon_1 - \varepsilon_2}{r_1 + r_2} = \frac{6 - 4}{10} = \frac{1}{5}\,\text{A}$$

$\therefore$ Potential difference across points X and Y is

$$V_{XY} = \varepsilon_2 + Ir_2$$

$$= 4 + \frac{1}{5} \times 8 = 5.6 = 5.6\,\text{V}$$

22. *(d)* The given circuit can be drawn as,

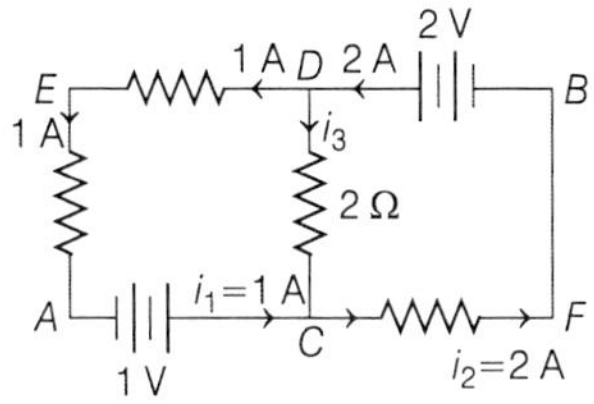

Current in branch DC (using KCl at point C),

$$i_1 + i_3 = i_2$$

$$2 - 1 - i_3 = 0$$

$$\Rightarrow i_3 = 1\,\text{A}$$

Now, while moving from A to B *via* C and D, the potential along $ACDB$,

$$V_A + 1 + 2 \times i_3 - 2 = V_B$$

$$\Rightarrow V_B - V_A = 1\,\text{V}$$

23. *(c)* In given circuit, current distribution following Kirchhoff's law will be as shown in the figure.

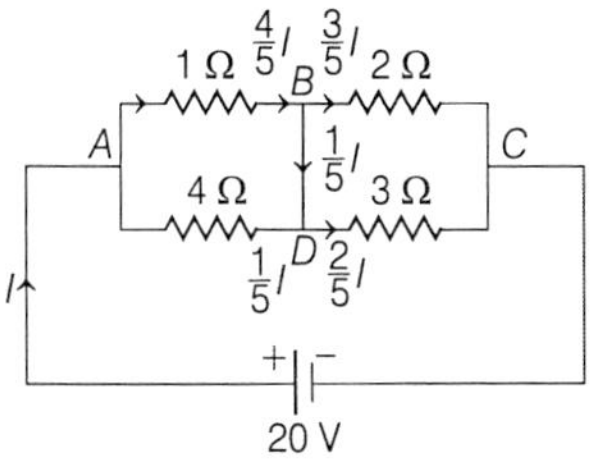

Since, the current flows in inverse ratio of the resistance of branch.

Now, total circuit resistance,

$$R_{\text{eq}} = (1\,\Omega \,||\, 4\,\Omega) + (2\,\Omega \,||\, 3\,\Omega)$$

$$= \left(\frac{1 \times 4}{1 + 4}\right) + \left(\frac{2 \times 3}{2 + 3}\right) = \frac{4}{5} + \frac{6}{5}$$

$$= \frac{10}{5}\,\Omega = 2\,\Omega$$

So, current drawn from cell,

$$I = \frac{V}{R_{\text{eq}}} = \frac{20\,\text{V}}{2\,\Omega} = 10\,\text{A}$$

Hence, current through BD arm is (refer to circuit diagram),

$$I_{BD} = \frac{I}{5} = \frac{10}{5} = 2\,\text{A}$$

24. *(b)* For the given circuit

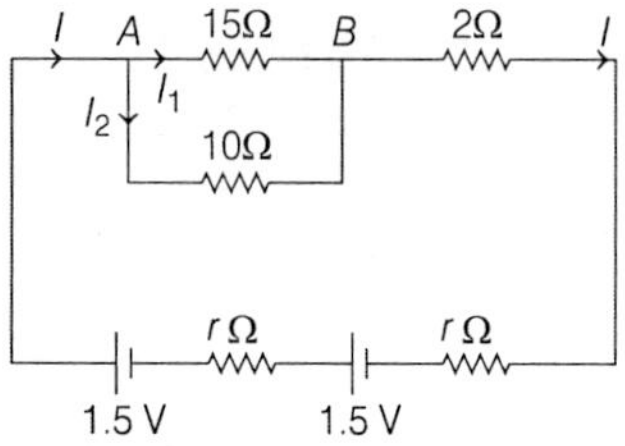

Given, $V_{AB} = 2\,\text{V}$

$\therefore$Current in circuit, $I = I_1 + I_2$

$$= \frac{2}{15} + \frac{2}{10} \quad [\because V = IR \text{ or } I = V/R]$$

$$= \frac{4+6}{30} = \frac{1}{3}\,\text{A} \quad \dots\text{(i)}$$

Also, voltage drop across $(r + r)$ resistors is

= voltage of the cell − voltage drop across AB

$= 3 - 2 = 1\,\text{V}$

Using $V = IR$ over the entire circuit

$$\Rightarrow \quad 1 = I(2 + 2r)$$

$$= \frac{1}{3}(2 + 2r) \quad \text{[Using Eq. (i)]}$$

$$\Rightarrow \quad 3 = 2 + 2r \quad \text{or} \quad 2r = 1\,\Omega$$

or $\quad r = \frac{1}{2}\Omega = 0.5\,\Omega$

> ***Alternate Solution***
>
> Equivalent resistance between A and B is
>
> $$\frac{1}{R_{AB}} = \frac{1}{10} + \frac{1}{15} = \frac{1}{6} \Rightarrow R_{AB} = 6\,\Omega$$
>
> 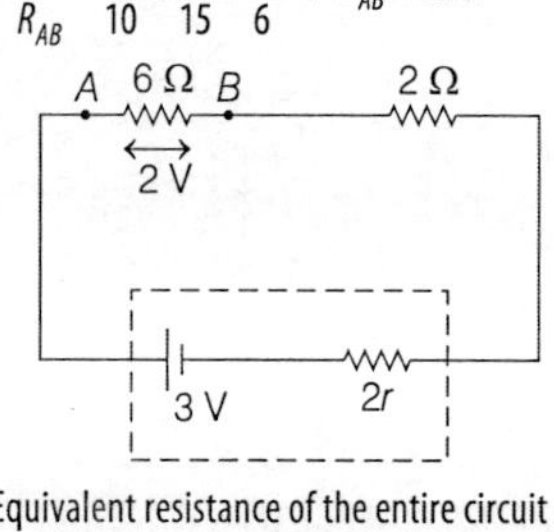
>
>
> $\therefore$Equivalent resistance of the entire circuit is,
>
> $R_{eq} = 6\,\Omega + 2\,\Omega + 2r = 8 + 2r$
>
> Now, current passing through the circuit is given as, $I = \frac{E_{net}}{R + r_{eq}} = \frac{E_{net}}{R_{eq}}$
>
> where, R is external resistance, r_{eq} is net internal resistance and E_{net} is the emf of the cells.
>
> Here, $\quad E_{net} = 1.5 + 1.5 = 3\,\text{V}$
>
> $r_{eq} = r + r = 2r \Rightarrow I = 3/(8 + 2r)$
>
> Also, reading of the voltmeter, $V = 2\text{V} = I \cdot R_{AB}$
>
> $$2 = \left(\frac{3}{8 + 2r}\right) \times 6$$
>
> $\Rightarrow \quad 8 + 2r = 9$ or $r = 1/2 = 0.5\,\Omega$

25. *(a)* Given circuit with currents as shown in the figure below, [In the question $I_1 = 0.3$ A is given, due to it we change the direction of I_1, in this figure]

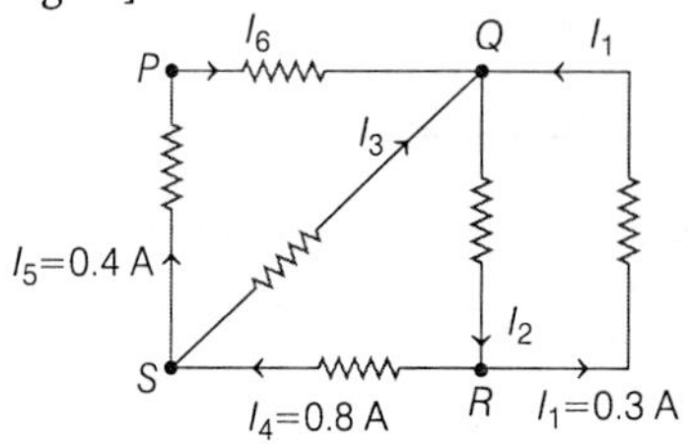

From Kirchhoff's junction rule, $\Sigma I = 0$

At junction S,

$$I_4 = I_5 + I_3$$

$$\Rightarrow \quad 0.8 = 0.4 + I_3$$

$$\Rightarrow \quad I_3 = 0.4\,\text{A}$$

At junction P, $I_5 = I_6$

$$\Rightarrow \quad I_6 = 0.4\,\text{A}$$

At junction Q,

$$I_2 = I_1 + I_3 + I_6$$

$$= 0.3 + 0.4 + 0.4 = 1.1\,\text{A}$$

26. *(a)* By Kirchhoff's loop rule in the given loop $ABEFA$, we get

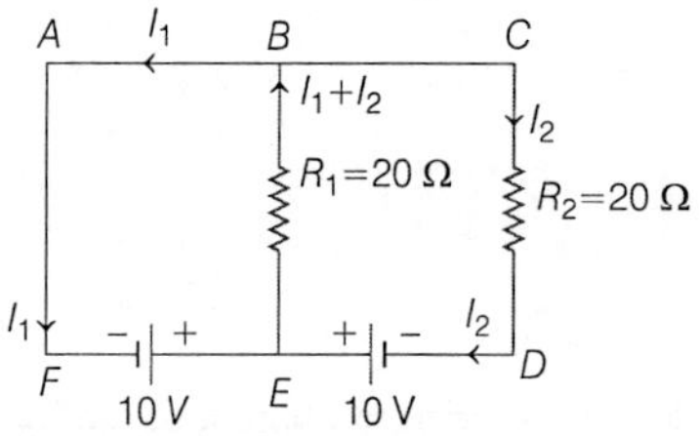

$$10 - (I_1 + I_2)R_1 = 0$$

$$\Rightarrow \quad 10 - (I_1 + I_2)\,20 = 0$$

or $\quad I_1 + I_2 = \frac{1}{2} \quad \dots\text{(i)}$

And from loop $BCDEB$, we get

$$10 - (I_1 + I_2)R_1 - I_2R_2 = 0$$

$$\Rightarrow \quad 10 - (I_1 + I_2 + I_2)\,20 = 0$$

$$\Rightarrow \quad I_1 + 2I_2 = \frac{1}{2} \quad \dots\text{(ii)}$$

On solving Eqs. (i) and (ii), we get

$$I_2 = 0 \quad \text{and} \quad I_1 = 0.5\,\text{A}$$

27. *(d)* When the switch S is closed, the circuit, hence formed is given in the figure below,

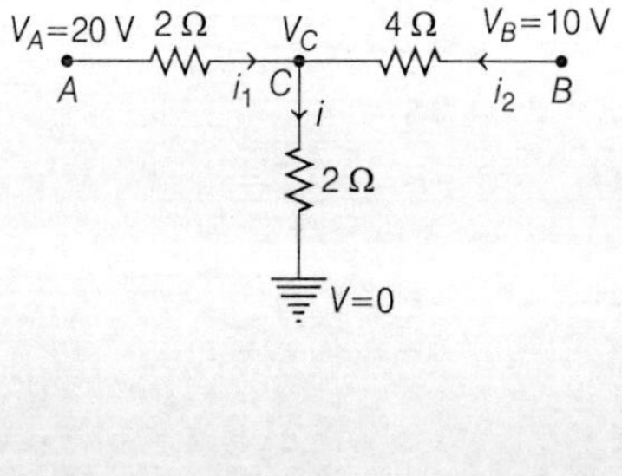

Since, in the above circuit, that point is C

$$\therefore \quad i_1 + i_2 = i$$

$$\Rightarrow \quad \frac{V_A - V_C}{2} + \frac{V_B - V_C}{4} = \frac{V_C - V}{2}$$

($\because$ Using Ohm's law, $V = iR$)

or $\quad \frac{20 - V_C}{2} + \frac{10 - V_C}{4} = \frac{V_C - 0}{2}$

$$\Rightarrow 20 - V_C + \frac{10 - V_C}{2} = V_C$$

$$\Rightarrow 40 - 2V_C + 10 - V_C = 2V_C$$

$$V_C = 10\,\text{V}$$

$\therefore$ The current,

$$i = \frac{V_C}{2}$$

$$= \frac{10}{2} = 5\,\text{A}$$

> **• Concept Enhancer**
>
> Kirchhoff's current law, states that the sum of all the currents directed towards a point in a circuit is equal to the sum of all the currents directed away from that point.

28. *(b)* For parallel combination of cells,

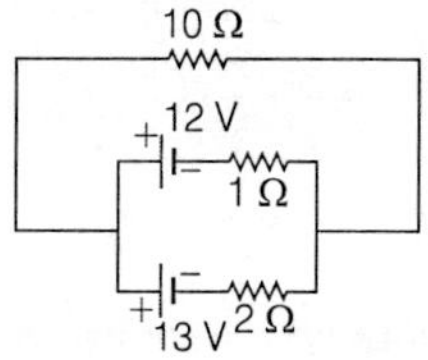

$$E_{eq} = \frac{\frac{E_1}{r_1} + \frac{E_2}{r_2}}{\frac{1}{r_1} + \frac{1}{r_2}}$$

$$\therefore E_{eq} = \frac{\frac{12}{1} + \frac{13}{2}}{\frac{1}{1} + \frac{1}{2}} = \frac{37}{3}\,\text{V}$$

Potential drop across $10\,\Omega$ resistance,

$$V = \left(\frac{E}{R_{total}}\right) \times 10$$

$$= \frac{37/3}{\left(10 + \frac{2}{3}\right)} \times 10$$

$$= 11.56\,\text{V}$$

$\therefore \quad V = 11.56\,\text{V}$

Hence, load lies between 11.5 V and 11.6 V

Alternate Solution

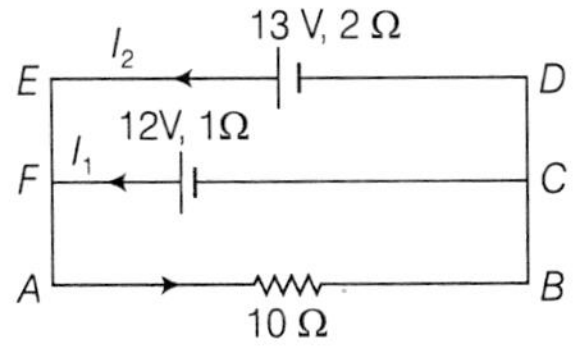

On applying KVL, in loop *ABCFA*, we get

$$-12 + 10(I_1 + I_2) + 1 \times I_1 = 0$$

$$\Rightarrow 12 = 11I_1 + 10I_2 \quad \ldots\text{(i)}$$

Similarly,

In loop *ABDEA*,

$$-13 + 10(I_1 + I_2) + 2 \times I_2 = 0$$

$$\Rightarrow \quad 13 = 10I_1 + 12I_2 \quad \ldots\text{(ii)}$$

On solving Eqs. (i) and (ii), we get

$$I_1 = \frac{7}{16}\text{A},\ I_2 = \frac{23}{32}\text{A}$$

∴Voltage drop across 10 Ω resistance is

$$V = 10\left(\frac{7}{16} + \frac{23}{32}\right) = 11.56\text{ V}$$

29. *(b)* In steady state, no current flows through the capacitor. So, resistance r_p becomes ineffective.

So, the current in circuit,

$$I = \frac{E}{r + r_2 \text{ (Total Resistance)}}$$

∵ Potential drop across capacitor = Potential drop across $r_2 = Ir_2 = \dfrac{Er_2}{r + r_2}$

∴ Stored charge of capacitor,

$$Q = CV = CE\frac{r_2}{r + r_2}$$

30. *(c)*

key idea Connect point Q to ground and apply KCL.

Consider the grounded circuit as shown below.

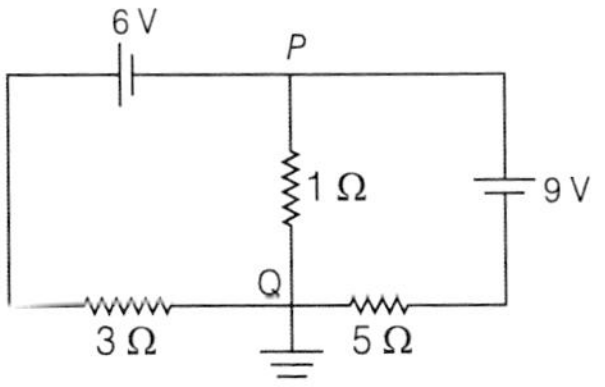

Applying KCL at point *Q*, we can write

Incoming current at Q = outgoing current from Q

$$\Rightarrow \quad \frac{V+6}{3} + \frac{V}{1} = \frac{9-V}{5}$$

$$\text{or } V\left[\frac{1}{3} + \frac{1}{5} + 1\right] = \frac{9}{5} - 2 \text{ or } V = \frac{-3}{23}$$

$$= -0.13\text{ V}$$

Thus, current in the 1 Ω resistance is 0.13 A, from *Q* to *P*.

31. *(b)* As R_1, R_2 and R in series.

$$R_{eq} = R_1 + R_2 + R$$

$$\therefore \text{ Net current, } I = \frac{2E}{R_1 + R_2 + R}$$

According to the question,

$$-(V_A - V_B) = E - IR_2$$

$$\therefore \quad 0 = E - IR_2$$

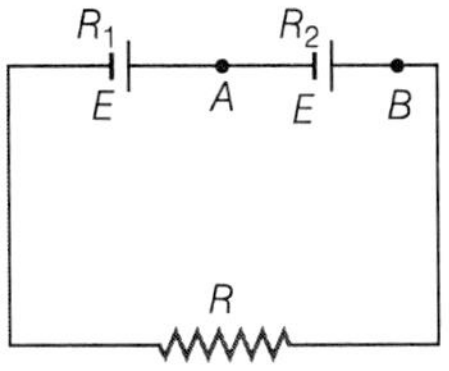

$$\text{or} \quad E = IR_2$$

$$\text{or} \quad E = \frac{2E}{R_1 + R_2 + R}R_2$$

$$\text{or} \quad R_1 + R_2 + R = 2R_2$$

$$\text{or} \quad R = R_2 - R_1$$

32. *(c)* The circuit diagram for the given situation is shown below

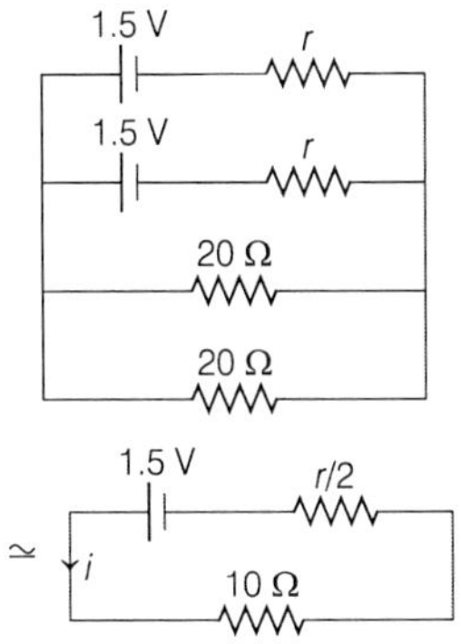

We know that,

$$V = E - i\frac{r}{2}$$

$$\Rightarrow \quad 1.2 = 1.5 - i\left(\frac{r}{2}\right)$$

$$\Rightarrow \quad ir/2 = 1.5 - 1.2 = 0.3$$

$$\Rightarrow \quad ir = 0.6 \quad \ldots\text{(i)}$$

Again, current flowing through the circuit,

$$i = \frac{\text{total emf}}{\text{total resistance}}$$

$$\Rightarrow \quad \frac{0.6}{r} = \frac{1.5}{10 + \dfrac{r}{2}} \quad \text{[From Eq. (i)]}$$

fter solving, we get $r = 5\Omega$

33. *(d)*

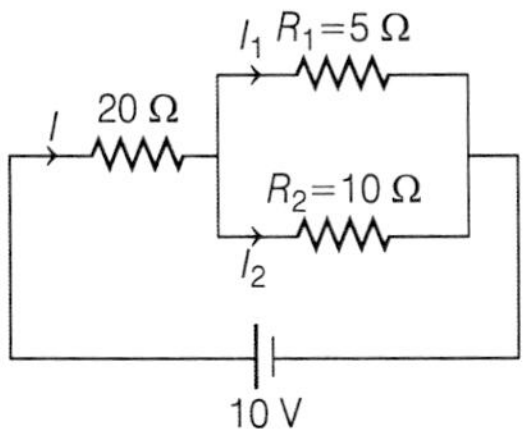

Equivalent resistance of the circuit,

$$R = 20 + \frac{5 \times 10}{5 + 10} = 20 + \frac{10}{3} = \frac{70}{3}\Omega$$

$$\therefore \quad I = \frac{V}{R} = \frac{10}{70/3} = \frac{3}{7}\text{A}$$

∴According to current division rule, current through 10Ω arm,

$$I_1 = I \cdot \frac{10}{10+5} = \frac{3}{7} \times \frac{2}{3} = \frac{2}{7}\text{A}$$

and current through 5Ω, arm,

$$I_2 = I \cdot \frac{5}{10+5} = \frac{3}{7} \times \frac{1}{3} = \frac{1}{7}\text{A}$$

∴Ratio of heat dissipation per second through 5 Ω to 10 Ω resistance

$$= \frac{I_1^2 R_1}{I_2^2 R_2} = \frac{(2/7)^2 \times 5}{(1/7)^2 \times 10} = \frac{2}{1} \text{ or } 2 : 1$$

34. *(c)* In steady state capacitors acts like open switch so, we have following circuit

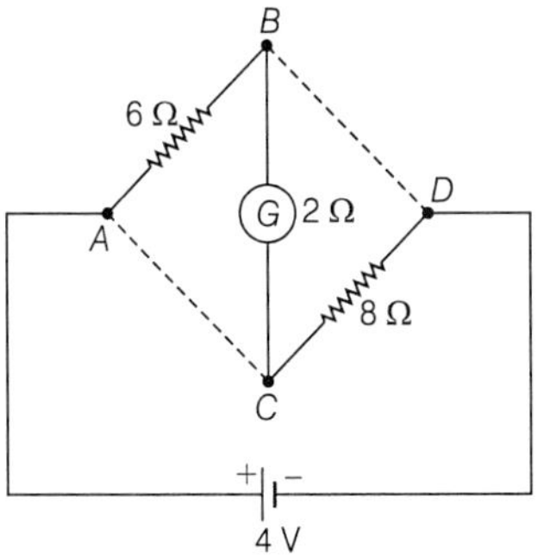

Hence current in circuit is

$$I = \frac{V}{R_1 + R_2 + R_3} = \frac{4}{6+2+8} = \frac{1}{4}\text{A}$$

Taking *A* at 4V; potential of point *C* is

$$V_C = 4 - \frac{1}{4}(6+2) = 2\text{ V}$$

And potential of point *B* is,

$$V_B = 4 - \frac{1}{4} \times 6 = \frac{5}{2}\text{ V}$$

$$\text{Hence, } \Delta V_{C_1} = 4 - 2 = 2\text{ V}$$

$$\text{and } \Delta V_{C_2} = \frac{5}{2} - 0 = \frac{5}{2}\text{ V}$$

$$\text{Hence, Ratio } \frac{\Delta V_{C_1}}{\Delta V_{C_2}} = \frac{2}{5/2} = \frac{4}{5}$$

35. *(b)*

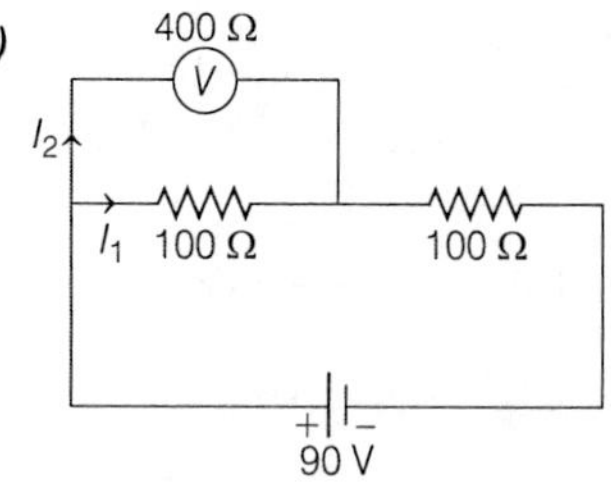

Total resistance of circuit with voltmeter connected across any of resistor will be

$$R_{eq} = (400 || 100) + 100$$

$$= \frac{400 \times 100}{400 + 100} + 100 = 180\,\Omega$$

Current through cell,

$$I = \frac{V}{R_{eq}} = \frac{90}{180} = \frac{1}{2}\,\text{A}$$

Potential drop across any of resistor will be

$$V = IR = \frac{1}{2} \times 100 = 50\text{ V}$$

If I_1 and I_2 are currents through resistor and voltmeter then,

$$400 I_2 = 100 I_1 \text{ or } 4I_2 = I_1$$

Also, $I_1 + I_2 = 0.5$

$\Rightarrow \quad 4I_2 + I_2 = 0.5$

$\Rightarrow \quad I_2 = 0.1\text{ A}$

$\therefore \quad I_1 = 0.5 - 0.1 = 0.4\text{ A}$

$\therefore$Reading of voltmeter

$$V = -I_2 R_{400\,\Omega}$$

$$= 0.1 \times 400 = 40\text{ V}$$

36. *(c)* Assume the current in branches as shown in figure,

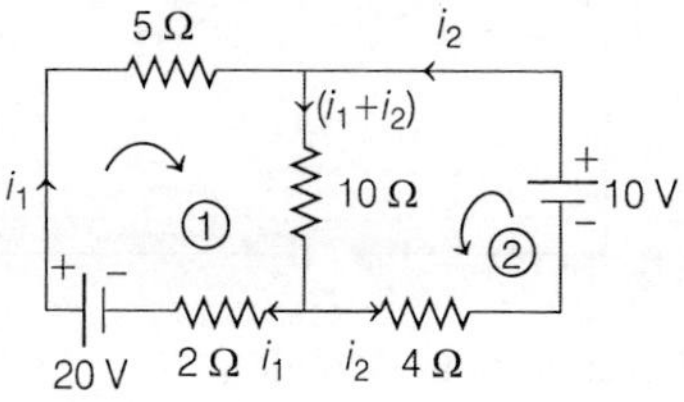

On applying KVL in loop 1, we get

$$+20 - 5i_1 - 10(i_1 + i_2) - 2i_1 = 0$$

$$17i_1 + 10i_2 = 20 \quad \ldots\text{(i)}$$

On applying KVL in loop 2, we get

$$+10 - 10(i_1 + i_2) - 4i_2 = 0$$

$$5i_1 + 7i_2 = 5 \quad \ldots\text{(ii)}$$

On solving Eqs. (i) and (ii),

we get

$$i_2 = -0.214\text{ A}$$

or $\quad i_2 \simeq -0.21\text{ A}$

It means the direction of current is from positive to negative terminal of battery inside it.

Alternate Solution

The circuit can be reduced to

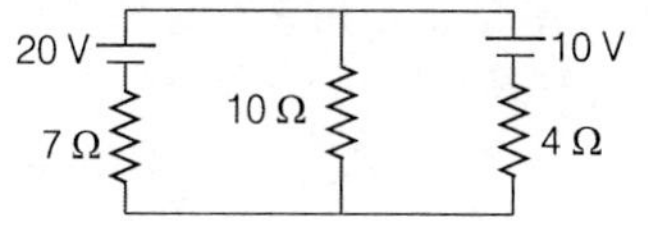

Using voltage division rule, voltage across 10 Ω resistor,

$$V' = \frac{20 \times 10}{10 + 7} = \frac{200}{17}\text{ V}$$

and resistance, $R' = \frac{7 \times 10}{7 + 10} = \frac{70}{17}\,\Omega$

Now, circuit becomes

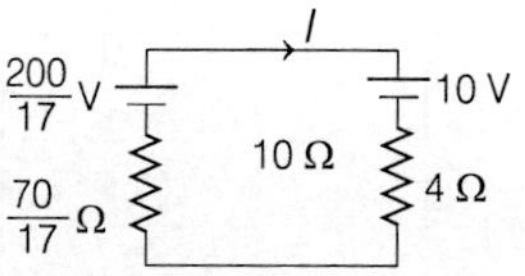

$$\therefore I = \frac{V_{eq}}{R_{eq}} = \frac{\frac{200}{17} - 10}{\frac{70}{17} + 4} = \frac{30}{138}\text{ A} = 0.21\text{ A}$$

37. *(b)* Given circuit is shown in the figure below.

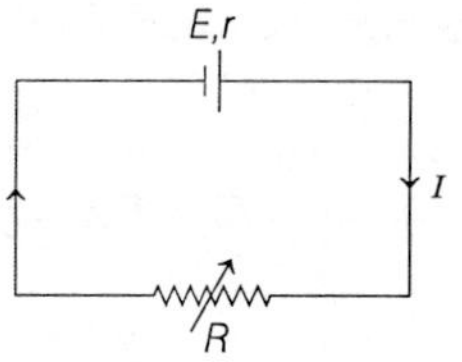

Net current, $\quad I = \frac{E}{R + r} \quad \ldots\text{(i)}$

Power across R is given as

$$P = I^2 R = \left(\frac{E}{R + r}\right)^2 \cdot R$$

[Using Eq. (i)]

For the maximum power,

$$\frac{dP}{dR} = 0$$

$$\Rightarrow \frac{dP}{dR} = \frac{d}{dR}\left(\left(\frac{E}{R + r}\right)^2 \cdot R\right)$$

$$= E^2 \frac{d}{dR}\left(\frac{R}{(R + r)^2}\right)$$

$$= E^2 \left[\frac{(R + r)^2 \times 1 - 2R \times (R + r)}{(R + r)^4}\right] = 0$$

$\Rightarrow \quad (R + r)^2 = 2R(R + r)$

or $\quad R + r = 2R$

$\Rightarrow \quad r = R$

$\therefore$The power delivered by the cell to the external resistance is maximum when $R = r$.

38. *(c)* According to question, the voltage across R_4 is 5 V, then the current across it

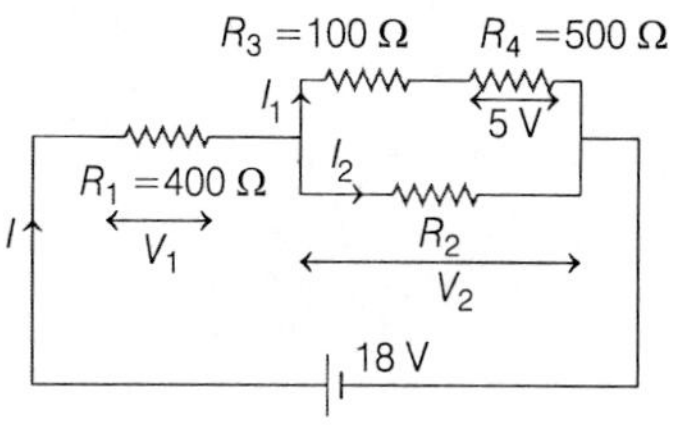

According to Ohm's law,

$\Rightarrow \quad V = IR$

$\Rightarrow \quad 5 = I_1 \times R_4$

$\Rightarrow \quad 5 = I_1 \times 500$

$$I_1 = \frac{5}{500} = \frac{1}{100}\,\text{A}$$

The potential difference across series combination of R_3 and R_4,

$\Rightarrow \quad V_2 = (R_3 + R_4)I$

$$= 600 \times \frac{1}{100} = 6\text{ V}$$

So, potential difference (across R_1)

$$V_1 = 18 - 6 = 12\text{ V}$$

Current through R_1 is,

$$I = \frac{V_1}{R_1} = \frac{12}{400} = \frac{3}{100}\,\text{A}$$

So, current through R_2 is,

$$I_2 = I - I_1 = \frac{3}{100} - \frac{1}{100}\,\text{A} = \frac{2}{100}\,\text{A}$$

Now, from $V = IR$, we have,

$$R_2 = \frac{V_2}{I_2} = \frac{6}{(2/100)} = 300\,\Omega$$

39. *(4)* Cell A : EMF $\Sigma_1 = 1\text{ V}$

Internal resistance $r_1 = 2\Omega$

Cell B : EMF $= 2$ V

Internal resistance $r_2 = 1\Omega$

External resistance : $R = 6\Omega$

$\Rightarrow$ Series combination

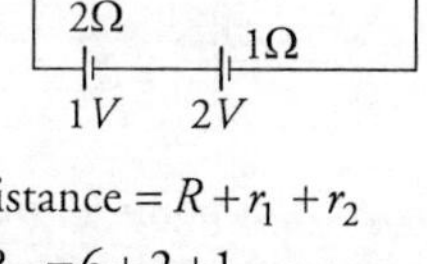

Net Resistance $= R + r_1 + r_2$

$$R_S = 6 + 2 + 1$$

$$R_S = 9\Omega$$

Net EMF $= E_S = 1 + 2 = 3V$

$$\therefore \text{Current } I_1 = \frac{E_S}{R_S} = \frac{3}{9}A = \frac{1}{3}A$$

Parallel combination

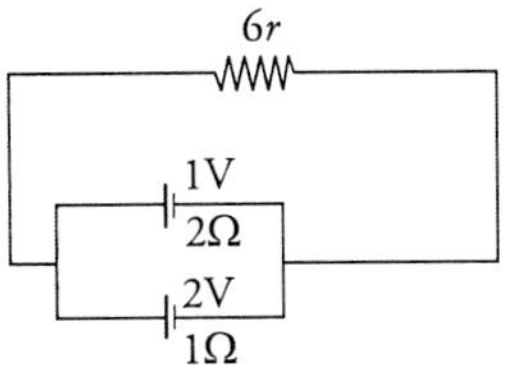

Internal resistance in parallel combination of cell $= \frac{1}{r_1} + \frac{1}{r_2}$

$$\frac{1}{r} = \frac{1}{2} + \frac{1}{1} = \frac{3}{2} \Rightarrow r = \frac{2}{3}\Omega$$

Net resistance of parallel combination of cell in circuit

$$R_{net} = \frac{2}{3} + 6 = \frac{20}{3}\Omega$$

Net EMF of cells in parallel combination is given by

$$E_{net} = \frac{E_1 r_2 + E_2 r_1}{r_1 + r_2} = \frac{1.1 + 2.2}{2 + 1}$$

$$\Rightarrow \quad \frac{1+4}{3} \Rightarrow \frac{5}{3}\text{ V}$$

$\therefore$ Current in circuit $I_2 = \frac{E_{net}}{R_{net}}$

$$\Rightarrow \frac{5/3}{20/3} = I_2 = \frac{1}{4}A$$

Now $\frac{I_1}{I_2} = \frac{1/3}{1/4} = \frac{4}{3} = \frac{x}{3}$

$\therefore \quad x = 4$

40. *(25)*

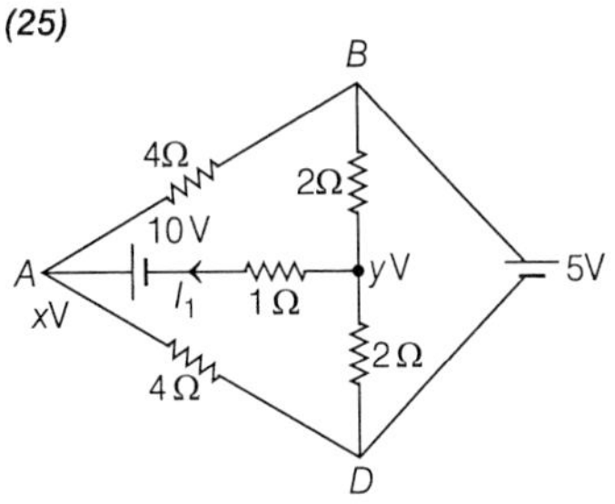

$$\frac{y-5}{2} + \frac{y-0}{2} + \frac{y-x+10}{1} = 0$$

$$y - 5 + y + 2y - 2x + 20 = 0$$

$$4y - 2x + 15 = 0 \quad \ldots\text{(i)}$$

$$\frac{x-5}{4} + \frac{x-0}{4} + \frac{x-10-y}{1} = 0$$

$$x - 5 + x + 4x - 40 - 4y = 0$$

$$6x - 4y - 45 = 0 \quad \ldots\text{(ii)}$$

From Eqs. (i) and (ii)

$$x = \frac{15}{2}, y = 0$$

$$i = \frac{y - x + 10}{1} = \frac{0 - 7.5 + 10}{1} = 2.5\text{ A}$$

$\therefore \quad n = 25$

41. *(1)* From symmetry,

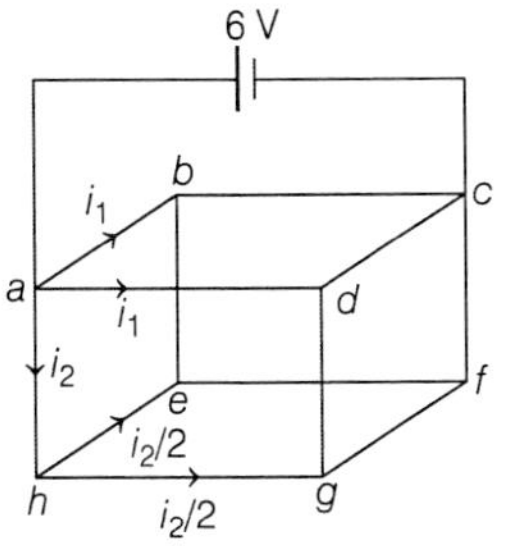

Current through $e - b$ and $g - d = 0$

$$R_{eq} = \frac{3}{4} \times R = \frac{3}{2}\Omega$$

Current through battery $= \frac{6 \times 2}{3} = 4$ A

$$i_2 = \frac{2}{8} \times 4 = 1\text{A}$$

$$\Delta V \text{ across } e - f = \frac{i_2}{2} \times R = \frac{1}{2} \times 2 = 1\text{ V}$$

42. *(30)* Given, $i_{20} = 0.3$A

All resistors are connected in parallel, so potential difference

$$V = 0.3 \times 20 = 6\text{ V}$$

Current through 15Ω resistor

$$i_{15} = \frac{6}{15} = 0.4\text{ A}$$

Ammeter reading, $i_A = 0.9$ A

$\because$ From KCL, current through R_1

$$i_1 = i_A - i_{15} - i_{20} = 0.9 - 0.4 - 0.3 = 0.2$$

$$\therefore \quad R_1 = \frac{V}{i_1} = \frac{6}{0.2} = 30\,\Omega$$

43. *(400)*

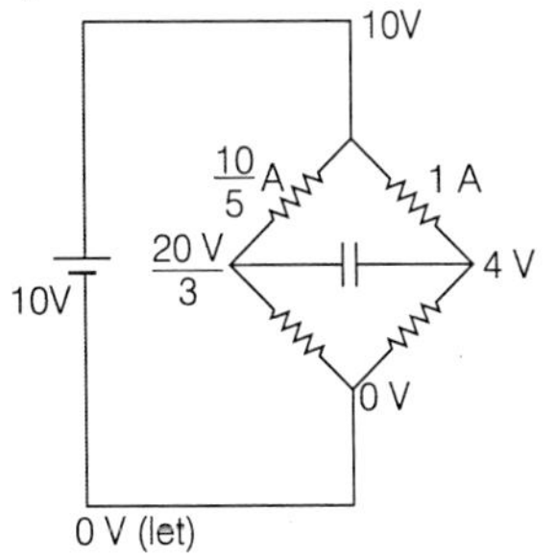

$$\Delta V|_{capacitor} = \left| 4 - \frac{20}{3} \right| = \frac{8}{3}\text{ V}$$

$$\therefore \quad q = \frac{8}{3} \times 150 = 400\,\mu\text{C}$$

44 *(3)* Applying KVL in loop *ADBCA*,

$$2I_1 + 4I_1 - 2I_2 - 4I_2 = 0$$

$$\Rightarrow \quad I_1 = I_2 \quad \ldots\text{(i)}$$

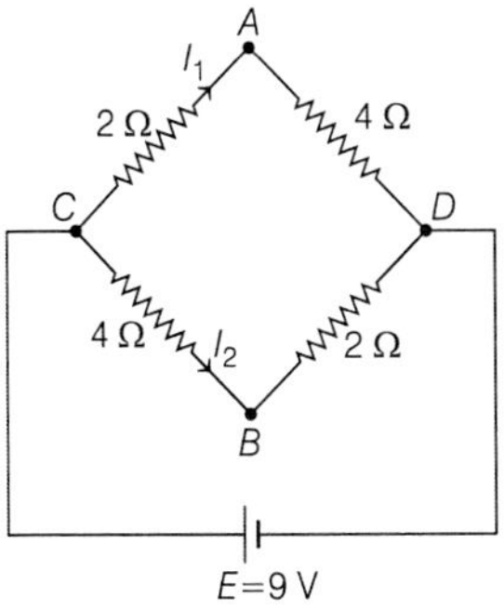

Applying KVL in loop *CBDEC*, we have

$$4I_2 + 2I_2 - 9 = 0$$

$$\Rightarrow \quad I_2 = \frac{9}{6}$$

$$I_2 = 1.5\text{ A}$$

$\therefore$ From Eq. (i), $I_1 = 1.5$ A

$$\therefore \quad V_A = 9 - I_1 \times 2 = 9 - 1.5 \times 2 = 6\text{ V}$$

$$V_B = 9 - I_2 \times 4$$

$$9 - 1.5 \times 4 = 3\text{ V}$$

$$\therefore \quad V_A - V_B = 6 - 3 = 3\text{ V}$$

45 *(75)*

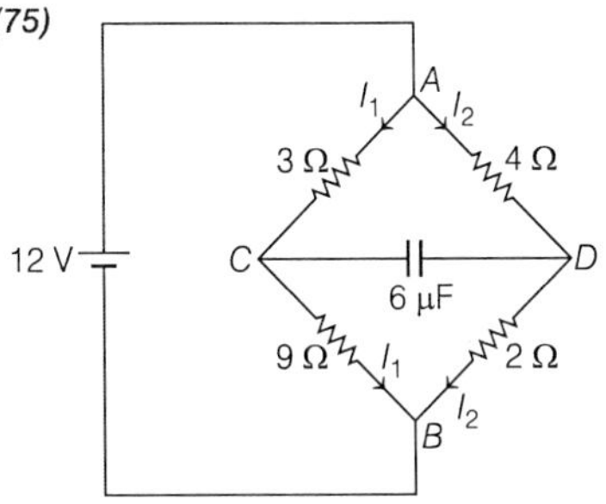

$$I_1 = \frac{12}{3+9} = 1\text{ A}$$

$$\Rightarrow I_2 = \frac{12}{4+2} = 2\text{ A}$$

$$V_A - V_C = 3I_1 = 3 \times 1 = 3\text{ V}$$

$$V_A - V_D = 4 \times I_2 = 4 \times 2 = 8\text{ V}$$

$$V_C - V_D = 5\text{ V}$$

$$U = \frac{1}{2}CV^2 = \frac{1}{2} \times 6 \times 5^2$$

$$\{C = 6\,\mu\text{F}, V_{CD} = 5\text{ V}\}$$

$$U = 75\,\mu\text{J}$$

46. *(5)*

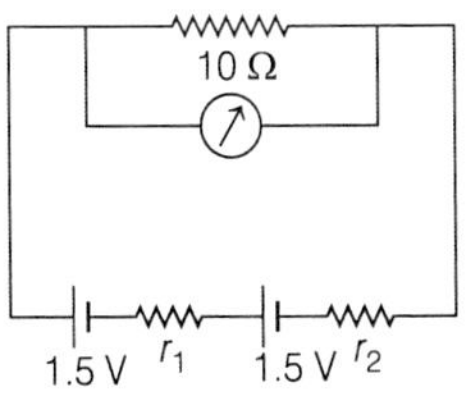

$$r = \left(\frac{E}{V} - 1\right)R$$

$$r = \left(\frac{3.0}{1.5} - 1\right) \times 10 = \left(\frac{30}{15} - 1\right) \times 10$$

$$= (2-1) \times 10$$

$$V_{eq} = 10$$

$$r_1 = 5\Omega \text{ or } r_2 = 5\Omega$$

47 *(1)*

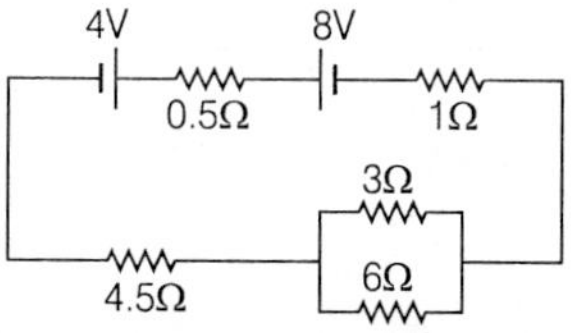

$$R_{eq} = 0{\cdot}5 + 1 + 4{\cdot}5 + 2$$

$$R_{eq} = 6 + 2 = 8\,\Omega$$

$$I = \frac{8-4}{8} = \frac{1}{2}\text{ A} = 0.5\text{ A}$$

Current through resistance 3Ω is I_1 and that through resistance 6Ω is I_2

$$I_1 : I_2 = \frac{1}{3} : \frac{1}{6}$$

$$I_1 : I_2 = 2:1$$

and $\quad I_1 + I_2 = 0{\cdot}5\text{ A}$

$$I_1 = \frac{2}{3} \times 0{\cdot}5 = \frac{1}{3}\text{ A}$$

So, $\quad \frac{1}{3} = \frac{x}{3} \Rightarrow x = 1$

48 *(20)*

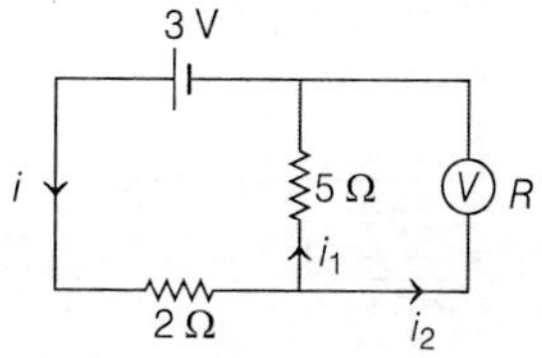

Given, the potential across 5Ω and voltmeter is 2 V. To find resistance R of voltmeter.

Let current in 5Ω be i_1 and R be V/i_2, potential across 2Ω will be 1 V and current $i = \frac{1}{2}$ A

Using junction law, $i = i_1 + i_2$

$$\frac{1}{2} = \frac{2}{5} + \frac{2}{R} \Rightarrow \frac{2}{R} = \frac{1}{2} - \frac{2}{5} = \frac{1}{10}$$

$$R = 20\,\Omega$$

49. *(2)* The given circuit may be reduced in following form.

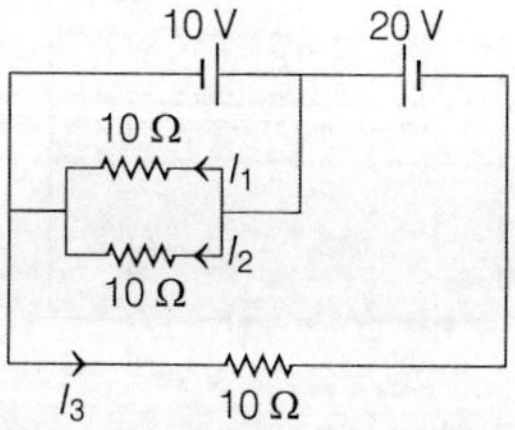

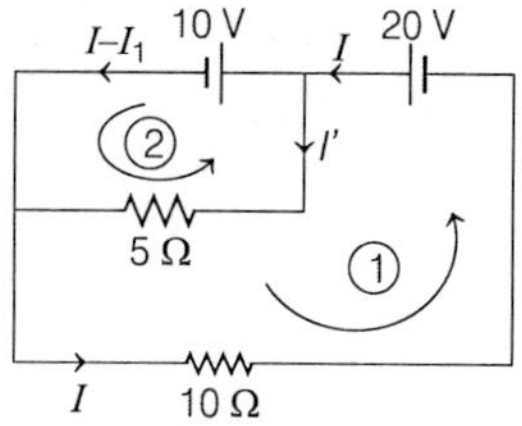

On applying KVL in loop-1 along the direction shown, we get

$$20 - 10 - 10I = 0$$

$$\Rightarrow \qquad I = 1\text{ A}$$

On applying KVL in loop-2 along the direction shown, we get

$$-10 + 5I' = 0$$

$$\Rightarrow \qquad I' = 2\text{ A}$$

On comparing, we get $I_3 = 1\text{ A}$ $[I_3 = I]$

$$= I_1 = \frac{I'}{2} = 1\text{ A}$$

So, the value of $\left|\frac{I_1 + I_3}{I_2}\right| = \left|\frac{1+1}{1}\right| = 2$

50. *(5)* The given condition is as follows

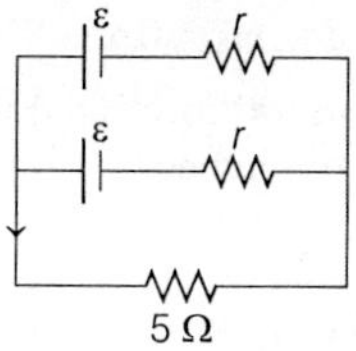

In parallel combination of cell,

$$r_{eq} = \frac{r_1 r_2}{r_1 + r_2} = \frac{r \times r}{r + r} = \frac{r}{2}$$

$$\varepsilon_{eq} = \frac{\varepsilon_1 r_2 + \varepsilon_2 r_1}{r_1 + r_2}$$

$$= \frac{\varepsilon \times r + \varepsilon \times r}{r + r} = \varepsilon$$

$$I_1 = \frac{\varepsilon_{eq}}{R + \frac{r}{2}} = \frac{\varepsilon}{5 + \frac{r}{2}}$$

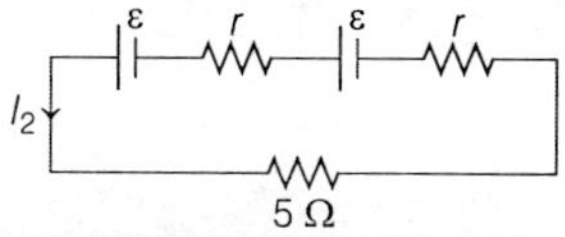

In series, $r_{eq} = r_1 + r_2 = r + r = 2r$

$$\varepsilon_{eq} = \varepsilon_1 + \varepsilon_2 = \varepsilon + \varepsilon = 2\varepsilon$$

$$I_2 = \frac{\varepsilon_{eq}}{R + r_{eq}} = \frac{2\varepsilon}{5 + 2r}$$

Current through load,

According to question, $I_1 = I_2$

$$\Rightarrow \quad \frac{\varepsilon}{5 + \frac{r}{2}} = \frac{2\varepsilon}{5 + 2r}$$

$$\text{or} \quad 5 + \frac{r}{2} = \frac{5 + 2r}{2}$$

$$\Rightarrow \quad 10 + r = 5 + 2r \Rightarrow r = 5\,\Omega$$

51 *(1.5)*

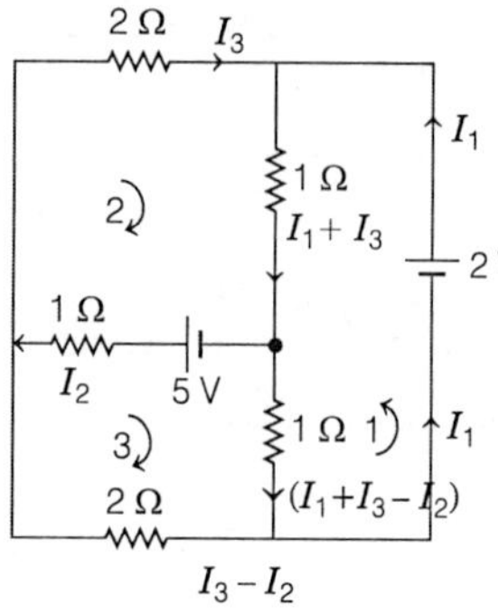

For first loop,

$$(I_1 + I_3) + (I_1 + I_3 - I_2) = 2$$

$$\Rightarrow \quad 2I_1 - I_2 + 2I_3 = 2 \qquad \ldots \text{(i)}$$

For second loop,

$$2I_3 + I_1 + I_3 + I_2 = 5$$

$$I_1 + I_2 + 3I_3 = 5 \qquad \ldots \text{(ii)}$$

For third loop,

$$I_2 - 5 = 2(I_3 - I_2) + (I_1 + I_3 - I_2)$$

$$\Rightarrow \quad I_1 - 4I_2 + 3I_3 = -5 \qquad \ldots \text{(iii)}$$

On subtracting Eq. (iii) from Eq. (ii), we get

$$5I_2 = 10 \Rightarrow I_2 = 2\text{ A}$$

On putting in Eq. (i) and Eq. (ii), we get

$$\Rightarrow \quad 2I_1 + 2I_3 = 4 \Rightarrow [I_1 + I_3 = 2]$$

$$\Rightarrow \quad I_1 = 2 - I_3 \qquad \ldots \text{(iv)}$$

and $\quad I_1 + 3I_3 = 3$

$$\Rightarrow \quad 2 - I_3 + 3I_3 = 3 \quad \text{[From Eq. (iv)]}$$

$$\Rightarrow \quad 2 + 2I_3 = 3 \Rightarrow I_3 = \frac{1}{2}\text{ A}$$

$\therefore$ From Eq. (iv), we get

$$\Rightarrow \quad I_1 = \frac{3}{2}\text{ A} = 1.5\text{ A}$$

52. *(4)* The given circuit diagram is shown below

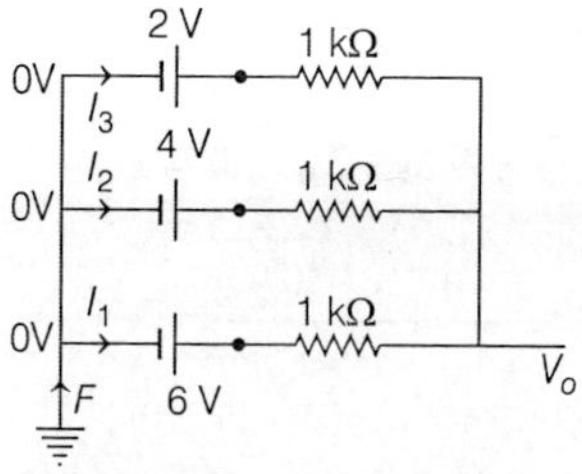

Using Kirchhoff's junction rule,

$$I_1 + I_2 + I_3 = 0$$

$$\frac{6 - V_0}{1 \times 10^3} + \frac{4 - V_0}{1 \times 10^3} + \frac{2 - V_0}{1 \times 10^3} = 0$$

$$\Rightarrow \quad 12 - 3V_0 = 0$$

$$\Rightarrow \quad V_0 = 4\text{ V}$$

53. *(8)* Given arrangement is as shown

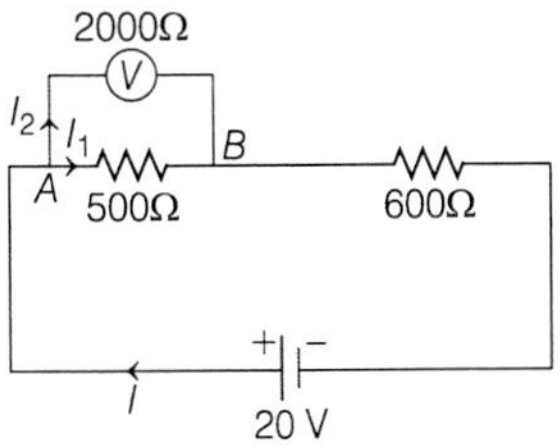

Total circuit resistance after connecting voltmeter is

$$R = (2000\,\Omega || 500\,\Omega) + 600\Omega$$
$$= \frac{2000 \times 500}{2000 + 500} + 600$$
$$= 400 + 600 = 1000\Omega$$

Current through cell, $I = \frac{V}{R}$

$$= \frac{20}{1000} = 20 \text{ mA} \quad \ldots \text{(i)}$$

Now in part AB,

$$I_2 \times 2000 = I_1 \times 500$$
$$\Rightarrow \quad 4I_2 = I_1 \quad \ldots \text{(ii)}$$

Also, $\quad I_1 + I_2 = 20 \text{ mA} \quad \ldots \text{(iii)}$

From Eqs. (ii) and (iii), we get

$$5I_2 = 20 \text{ mA} \Rightarrow I_2 = 4 \text{ mA}$$

and $\quad I_1 = 4I_2 = 16 \text{ mA}$

Voltmeter is measuring potential difference across 500 Ω resistor

So, $\quad V_{AB} = I_1 \times 500$
$$= 16 \times 10^{-3} \times 500 = 8 \text{ V}$$

54. *(10)* The circuit diagram is shown below

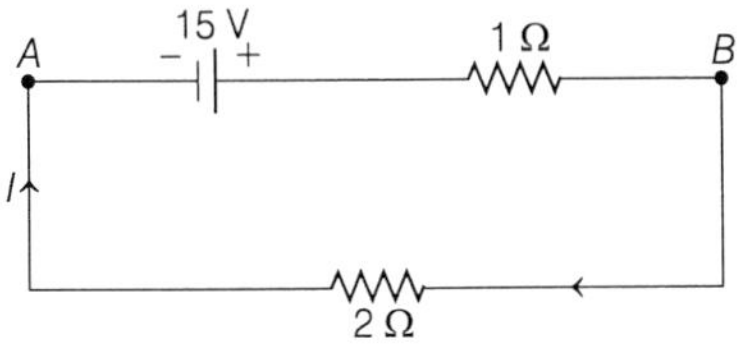

Current in circuit, $I = \frac{V}{R_{\text{total}}} = \frac{15}{3} = 5$ A

Now potential drop in 2 Ω resistance $= V_2 = IR = 5 \times 2 = 10$ V

Hence, $V_{BA} = V_B - V_A = V_2 = 10$ V

55. *(15)* Let emf of cell = ε

Potential difference across the terminal of cell, $V_1 = 1.25$ V

Load resistance, $R_{L_1} = 5\Omega$

when load resistance, $R_{L_2} = 2\Omega$,

then $V_2 = 1$ V

I_1, I_2 be the current through load in the two above mentioned cases and r be internal resistance of cell.

As we know that,

By using Kirchhoff's voltage law,

$$\varepsilon - V = I_1 R_{L_1}, \ \varepsilon = I_1 R_{L_1} + V$$
$$\varepsilon = I_1 (R_{L_1} + r) \quad [\because V = I_1 r]$$
$$\Rightarrow \quad I_1 = \frac{\varepsilon}{R_{L_1} + r} = \frac{\varepsilon}{5 + r}$$
$$\because \quad V_1 = I_1 R_{L_1}$$
$$\therefore \quad 1.25 = \frac{\varepsilon}{5 + r} \times 5$$
$$\Rightarrow \quad \frac{\varepsilon}{5 + r} = 0.25 = \frac{25}{100} = \frac{1}{4}$$
$$\Rightarrow \quad 4\varepsilon = 5 + r \quad \ldots \text{(i)}$$

and $\quad V_2 = I_2 R_{L_2} = \frac{\varepsilon}{2 + r} 2 = 1$

$$\Rightarrow \quad 2\varepsilon = 2 + r \quad \ldots \text{(ii)}$$

On subtracting Eq. (i) and Eq. (ii), we get

$$2\varepsilon = 3$$
$$\varepsilon = \frac{3}{2} = 1.5 \text{ V} = \frac{15}{10} \text{ V}$$

Hence, $\quad x = 15$

56. *(300)* Given, charge passing through circuit,

$$q = 20 \text{ C}$$

Potential difference between two plates,

$$V = 15 \text{ V}$$

Let W be the amount of work done by battery.

$$\therefore \quad W = qV = 20 \times 15 = 300 \text{ J}$$

Topic 4

01. *(b)* R_{eq} of the given circuit is,

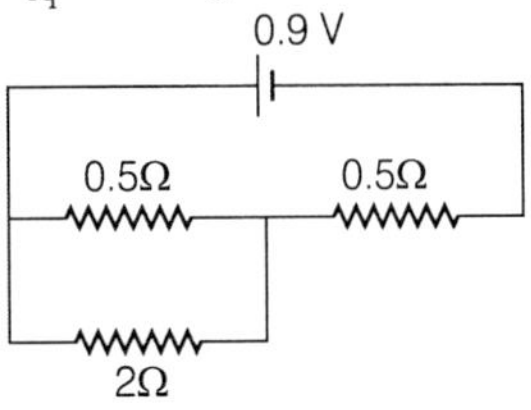

$$R_{\text{eq}} = 0.5 + \frac{0.5 \times 2}{2 + 0.5}$$

As for series $R_{\text{eq}} = R_1 + R_2$ and for parallel connection $R_{\text{eq}} = \frac{R_1 \times R_2}{R_1 + R_2}$

$$\therefore \quad R_{\text{eq}} = \frac{45}{50} = \frac{9}{10} = 0.9\Omega$$

Current is given by Ohm's law

$$i = \frac{V}{R} = \frac{0.9}{0.9} = 1\text{A}$$

02. *(c)* The given configuration has a balanced Wheatstone bridge. Thus, equivalent circuit is

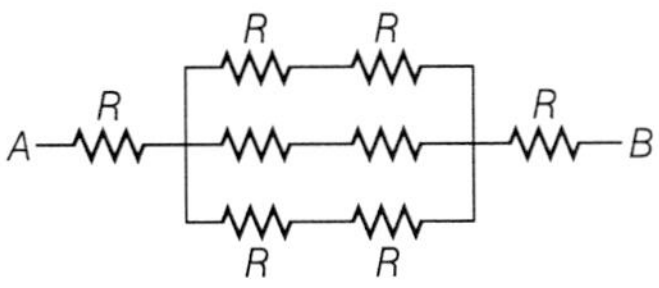

The equivalent resistance of the box,

$$\frac{1}{r_1} = \frac{1}{2R} + \frac{1}{2R} + \frac{1}{2R}$$
$$\frac{1}{r_1} = \frac{3}{2R} \Rightarrow r_1 = \frac{2R}{3}$$

Net resistance,

$$r = R + \frac{2R}{3} + R = \frac{8R}{3}\Omega$$

03. *(a)*

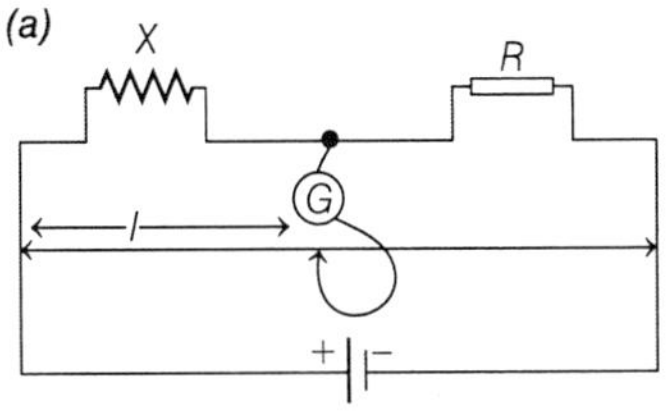

Balance length is independent of the resistance per metre length of meter bridge. So, new balance length remains same.

04. *(b)* Given circuit can be redrawn as;

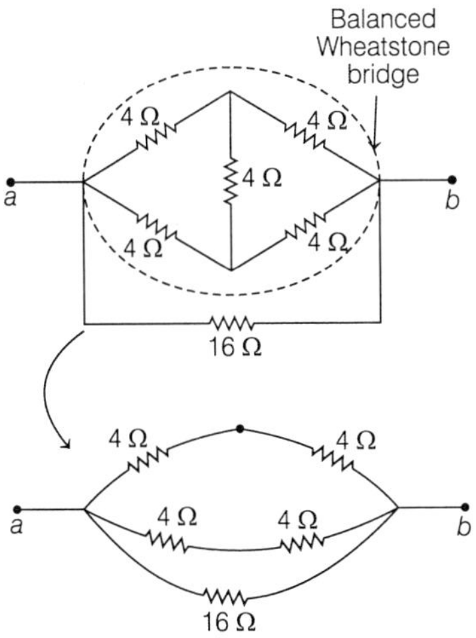

So, we have two 8 Ω and one 16 Ω resistor in parallel.

If R_{eq} is equivalent resistance then,

$$\frac{1}{R_{\text{eq}}} = \frac{1}{8} + \frac{1}{8} + \frac{1}{16} = \frac{5}{16}$$
$$\Rightarrow \quad R_{\text{eq}} = \frac{16}{5} = 3.2\Omega$$

05. *(a)* In a balanced Wheatstone bridge, there is no effect on the position of null point, if we exchange the battery and galvanometer. So, option (a) is incorrect.

06. *(c)*

> **key idea** Use condition of balanced Wheatstone bridge to solve this problem.

Since, $V_B - V_D = 0$

$\therefore \quad V_B = V_D$

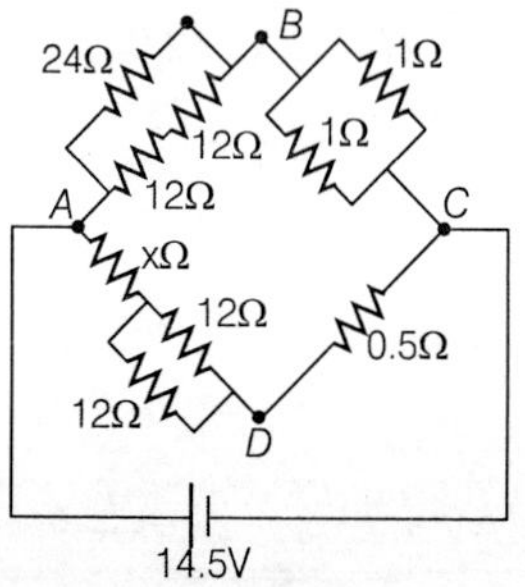

i.e., This is the condition of balanced Wheatstone bridge

$\therefore \quad \dfrac{R_{AB}}{R_{BC}} = \dfrac{R_{AD}}{R_{DC}}$

$\Rightarrow \quad \dfrac{\dfrac{24\times(12+12)}{24+(12+12)}}{\dfrac{1\times1}{1+1}} = \dfrac{x+\dfrac{12\times12}{12+12}}{0.5}$

$\Rightarrow \quad \dfrac{12}{0.5} = \dfrac{x+6}{0.5}$

$\Rightarrow \quad x = 6\ \Omega$

07. *(d)* Using principle of potentiometer,
Balance point $\propto$ Potential

$l_1 \propto \dfrac{ER_1}{R_1 + r}$ and $l_2 \propto \dfrac{ER_2}{R_2 + r}$

where r is internal resistance of battery.

$\Rightarrow \quad \dfrac{l_1}{l_2} = \dfrac{ER_1}{R_1 + r} \times \dfrac{R_2 + r}{ER_2}$

$\dfrac{l_1}{l_2} = \dfrac{R_1}{R_2} \times \dfrac{(R_2 + r)}{(R_1 + r)}$

Given, $l_1 = 500$ cm,

$l_2 = 400$ cm,

$R_1 = 10\ \Omega$, $R_2 = 1\ \Omega$

$\dfrac{500}{400} = \dfrac{10(1+r)}{1(10+r)}$

$2 = 7r$

$r = 0.3\ \Omega$

08. *(c)*

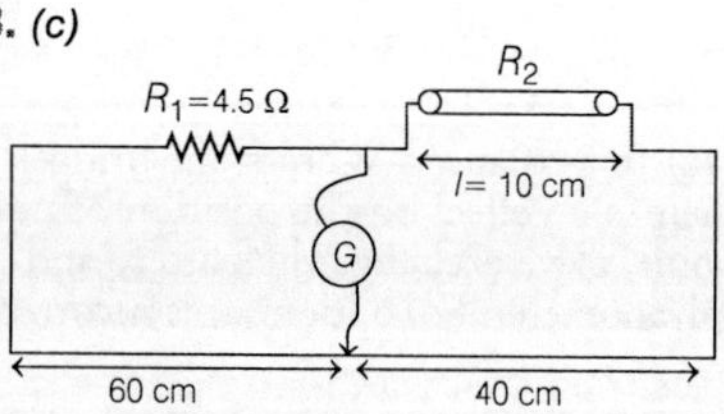

$r = \sqrt{7} \times 10^{-4}$ m

$l = 10$ cm $= 10 \times 10^{-2}$ m $= \dfrac{1}{10}$ m

Under balance condition $\dfrac{R_1}{l_1} = \dfrac{R}{l_2}$

Here, $R_1 = 4.5\Omega$, $l_1 = 60$ cm

$l_2 = 100 - l_1 = 40$ cm

$R_2 = R$

$\dfrac{l_1}{l_2} = \dfrac{R_1}{R_2} \Rightarrow \dfrac{60}{40} = \dfrac{4.5\ \Omega}{R}$

$R = 3\ \Omega$

$R = \rho \dfrac{l}{A} \Rightarrow 3 = \rho \times \dfrac{1}{\pi r^2}$

$3 = \rho \times \dfrac{1}{10 \times \pi \times 7 \times 10^{-8}}$

$\Rightarrow \quad \rho = 21\pi \times 10^{-7} = 21\pi \times 10^{-7}$

$\rho = 21 \times \dfrac{22}{7} \times 10^{-7} = 66 \times 10^{-7}$

$\rho = 66 \times 10^{-7}$

$\Rightarrow \quad R = 66$

09. *(b)* The given network of resistances can be rearranged as shown in figure.

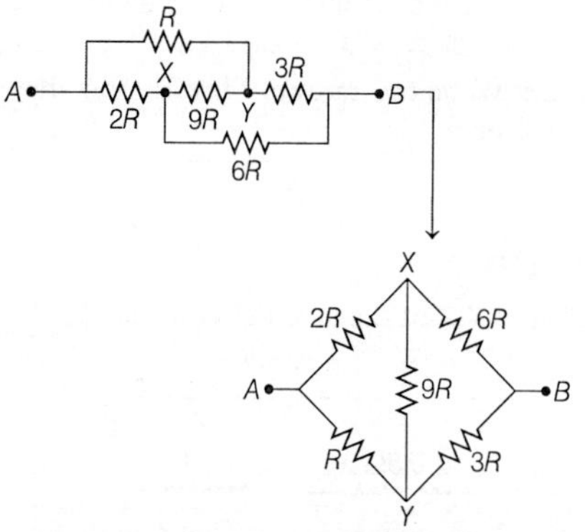

Above arrangement of wheatstone bridge is under balanced condition. So, no current will flow through resistor $9R$ connected between X and Y and can be removed.

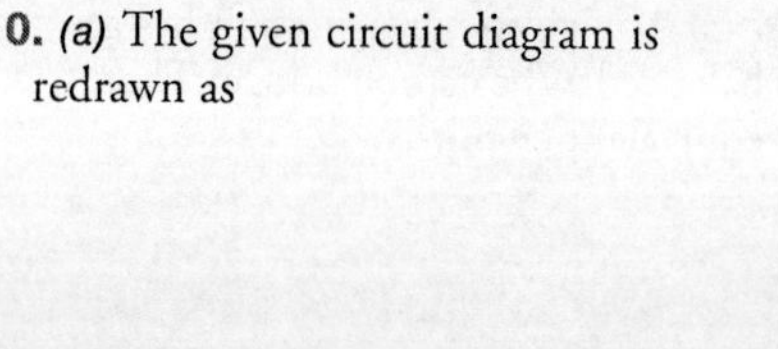

$R_{eq} = \dfrac{4R \times 8R}{4R + 8R} = \dfrac{8}{3}R$

10. *(a)* The given circuit diagram is redrawn as

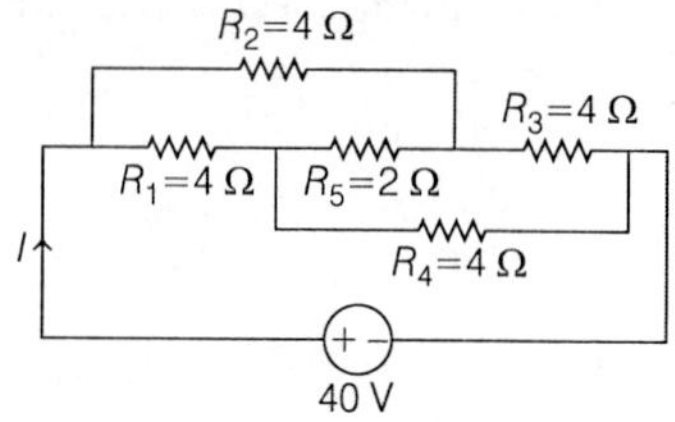

The above circuit diagram is redrawn as,

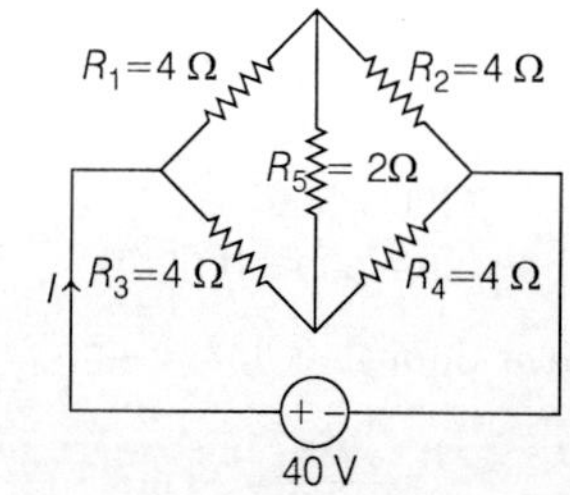

Since, $\dfrac{R_1}{R_2} = \dfrac{R_3}{R_4} = \dfrac{4}{4} = \dfrac{1}{1}$

$\therefore$ Given circuit diagram is a balanced Wheatstone bridge circuit. Hence, R_5 is ineffective.

i.e. Now, redrawing the given circuit as

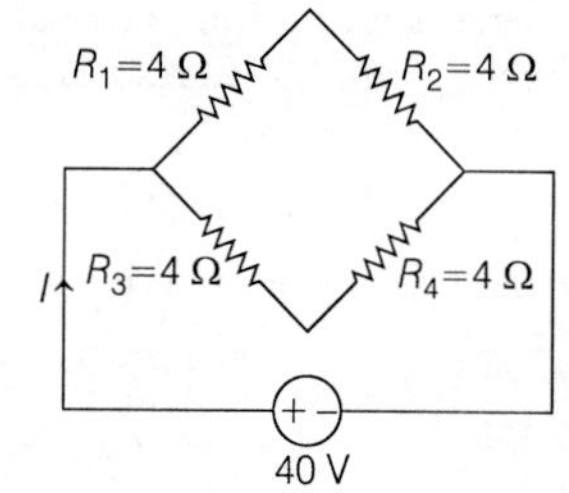

$\therefore$ Equivalent resistance of the circuit,

$R = (R_1 + R_2) || (R_3 + R_4)$

$= (4+4) || (4+4) = 8 || 8 = \dfrac{8\times8}{8+8} = 4\Omega$

$\therefore \quad$ Current, $I = \dfrac{V}{R} = \dfrac{40}{4} = 10$ A

11. *(d)* Given, all resistances have same resistance R.

Now, we can redraw the circuit as below.

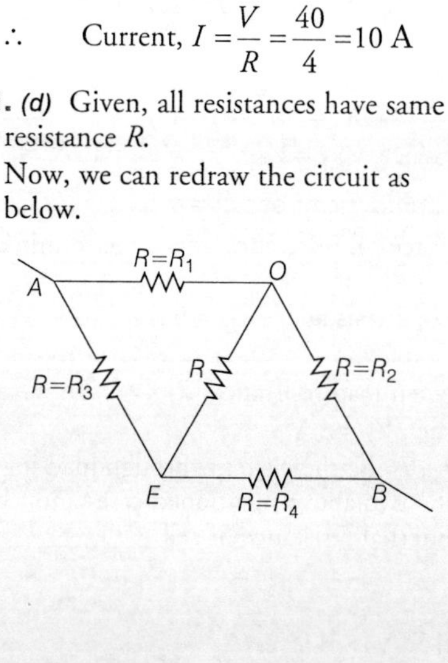

Let resistances be R_1, R_2, R_3 and R_4.

$\because \quad \dfrac{R_1}{R_3} = \dfrac{R_2}{R_4}$

So, circuit will behave as a Wheatstone bridge and no current will flow through middle resistor.

$\therefore \quad R_{eq} = \dfrac{(R_1 + R_2)(R_3 + R_4)}{(R_1 + R_2) + (R_3 + R_4)}$

$= \dfrac{(R + R)(R + R)}{(R + R) + (R + R)} = R$

12. *(c)* We have, $X + Y = 1000\,\Omega$

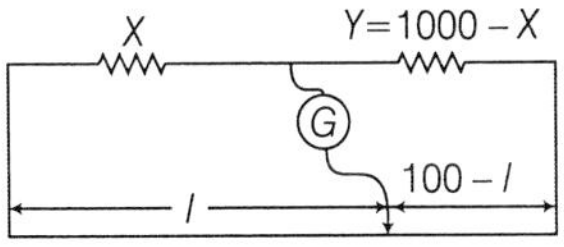

Initially, $\dfrac{X}{l} = \dfrac{1000 - X}{100 - l}$ ...(i)

When, X and Y are interchanged, then

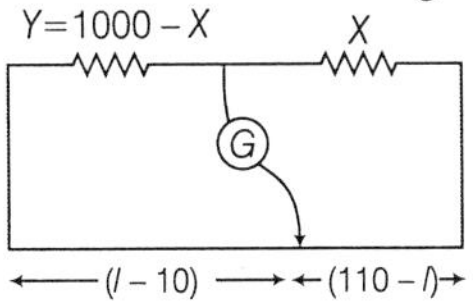

$\dfrac{1000 - X}{l - 10} = \dfrac{X}{100 - (l - 10)}$

or $\dfrac{1000 - X}{l - 10} = \dfrac{X}{110 - l}$ ...(ii)

From Eqs. (i) and (ii), we get

$\dfrac{100 - l}{l} = \dfrac{l - 10}{110 - l}$

$\therefore \quad l = 55$ cm

On substituting the value of l in Eq. (i), we get

$\dfrac{X}{55} = \dfrac{1000 - X}{100 - 55}$

$\Rightarrow \quad 20X = 11000$

$\therefore \quad X = 550\,\Omega$

13. *(b)* When bridge is balanced, ratio of resistance in opposite arms is equal.

$P/Q = R/S$

Given, $P = 55\,\Omega$, $Q = R$

As $R \propto l$

$R \propto l_1$, where $l_1 = 20$ cm

$S \propto l_2$, where $l_2 = 100 - 20 = 80$ cm

$\dfrac{R}{S} = \dfrac{l_1}{l_2} = \dfrac{20}{80}$

$\therefore \quad 55/R = 20/80$

or $R = 220\,\Omega$

14. *(b)* The given circuit can be redrawn as

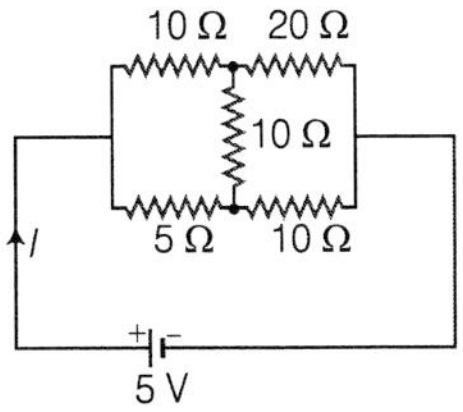

Which is a balanced Wheatstone's bridge and hence no current flows in the middle resistor, so equivalent circuit would be as shown below.

$\Rightarrow$

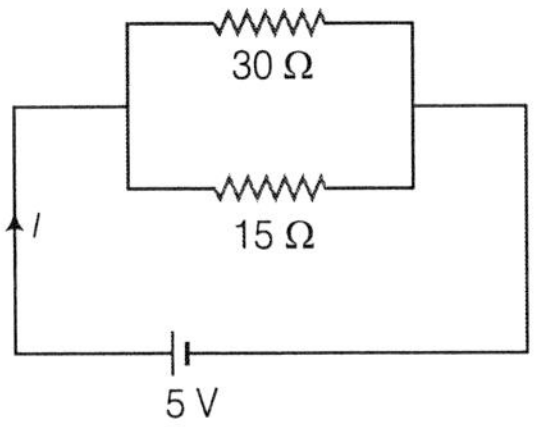

30 Ω || 15 Ω = 10 Ω

$\Rightarrow$

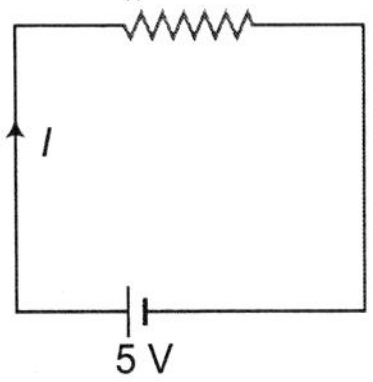

$I = \dfrac{V}{R} = \dfrac{5}{10} = 0.5$ A

15. *(b)* For balanced Wheatstone's bridge,

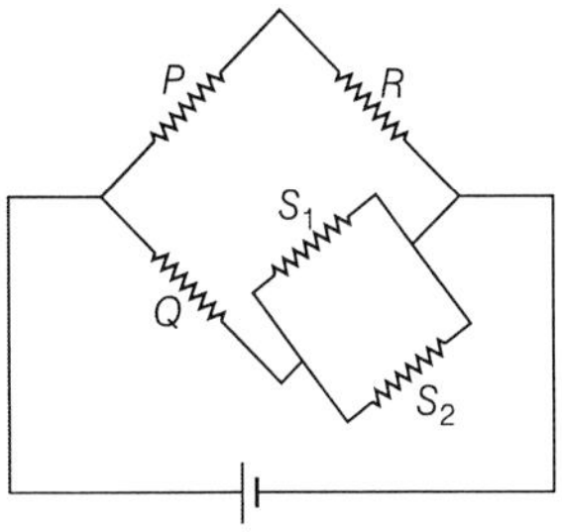

$P/Q = R/S$

Here, $S = S_1 \,||\, S_2 = \dfrac{S_1 S_2}{S_1 + S_2}$

$\Rightarrow \quad \dfrac{P}{Q} = \dfrac{R(S_1 + S_2)}{S_1 S_2}$

16. *(c)*

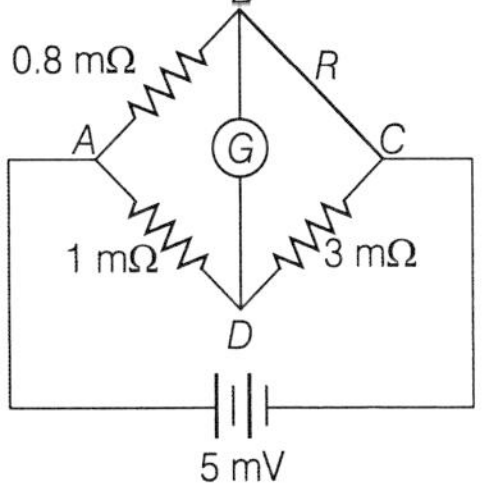

Arm BC is cooled at 2° C/ S from initial temperature of 25°C for 10s.

Temperature of BC at balance point is,

$T_2 = T_1 - 2 \times 10 = 25 - 20 = 5$°C

Initial resistance of arm $BC = R_{T_1}$

$= 3$ m Ω at 25°C

Final resistance of BC at balance point is,

$R_{T_2} = R_{T_1}(1 + \alpha\Delta T)$

or $R_{T_2} = 3(1 + \alpha(20))$

At balance condition, $\dfrac{0.8}{1} = \dfrac{R_{T2}}{3}$

$\Rightarrow \quad R_{T_2} = \dfrac{3 \times 0.8}{1}$ m Ω $= 2.4$ m Ω

$\Rightarrow 3(1 + \alpha(20)) = 2.4 \Rightarrow 1 + \alpha(20) = 0.8$

$\Rightarrow \alpha = \dfrac{-0.2}{20} = -0.01 = -1 \times 10^{-2}$ °C^{-1}

17. *(b)* Given, circuit is

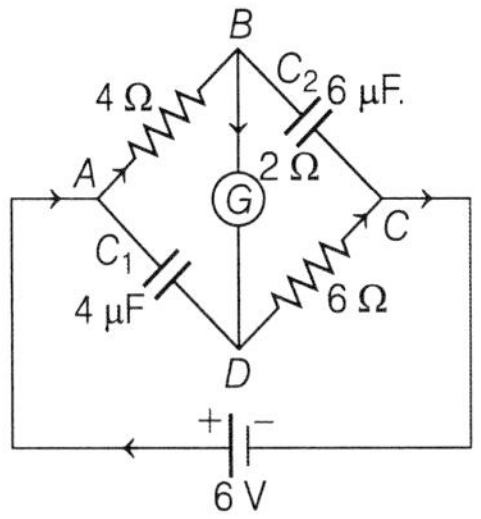

No, current flows through capacitors.

Current through path, $ABDC$ is $I = \dfrac{V}{R_{eq}}$

$\Rightarrow \quad I = \dfrac{6}{4 + 2 + 6} = \dfrac{1}{2}$ A

Now, taking $V_A = 6$ and $V_C = 0$,

Potential drops across,

$AB = I_{AB} \times R_{AB} = V_{AB}$

$\Rightarrow \quad V_{AB} = \dfrac{1}{2} \times 4 = 2$ V

$V_{BD} = I_{BD} \times R_{BD} = \dfrac{1}{2} \times 2 = 1$ V

$V_{DC} = I_{DC} \times R_{DC} = \dfrac{1}{2} \times 6 = 3$ V

Now, $V_A - V_B = V_{AB} = 2$ V

$\Rightarrow \quad 6 - V_B = 2$ or $V_B = 4$ V

and $V_{BD} = 1 \Rightarrow V_B - V_D = 1$

$\Rightarrow \quad 4 - V_D = 1$

$V_D = 3$ V

So, potential drops across capacitors C_1 and C_2 are $V_{AD} = V_A - V_D$

$= 6 - 3 = 3$ V

and $V_{BC} = V_B - V_C$

$V_{BC} = 4 - 0 = 4$ V

Charge stored in a capacitor, $Q = CV$

$\therefore$Ratio of charge stored $\dfrac{Q_1}{Q_2} = \dfrac{C_1 V_1}{C_2 V_2}$

$= \dfrac{4 \times 3}{6 \times 4} = \dfrac{1}{2}$

18. *(a)*

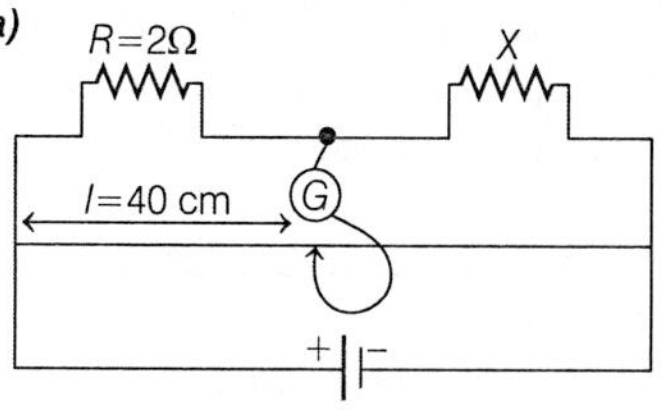

Initially, $\dfrac{R}{l} = \dfrac{X}{100 - l} \Rightarrow \dfrac{2}{40} = \dfrac{X}{60}$

$\Rightarrow \quad X = 3\Omega$

When X is shunted with 2Ω.

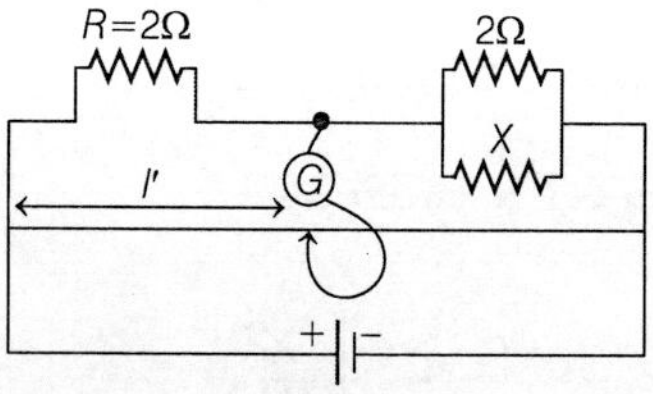

$$\frac{R}{l'} = \frac{\left(\dfrac{2X}{2+X}\right)}{100 - l'}$$

Substituting R and X,

$$\Rightarrow \quad \frac{2}{l'} = \frac{6}{5(100 - l')} \Rightarrow 3l' = 500 - 5l'$$

$$\Rightarrow \quad 8l' = 500 \Rightarrow l' = 62.5 \text{ cm}$$

Balance length change
$= 62.5 - 40 = 22.5$ cm

> **• Concept Enhancer**
>
> Coefficient of thermal resistance of meter bridge wire is very low and its resistivity is very high so that dimensions of wire remains unchanged due to less heating effect.

19. *(c)* As, A is directly connected to the positive terminal of the battery,
$V_A = 10$ V and $V_C = 0$

By nodal analysis at B,

$$\frac{V_B - 10}{100} + \frac{V_B - V_D}{15} + \frac{V_B - 0}{10} = 0$$

$$53V_B - 20V_D = 30 \quad \ldots \text{(i)}$$

By nodal analysis at D,

$$\frac{V_D - 10}{60} + \frac{V_D - V_B}{15} + \frac{V_D - 0}{5} = 0$$

$$-4V_B + 17V_D = 10 \quad \ldots \text{(ii)}$$

On solving Eqs. (i) and (ii) by substitution method, we get

$$V_D = 0.792 \text{ V} \Rightarrow V_B = 0.865 \text{ V}$$

The current through the galvanometer,

$$I = \frac{V_B - V_D}{R}$$

On substituting the values in the above equation, we get

$$I = \frac{0.865 - 0.792}{15} \Rightarrow I = 4.87 \text{ mA}$$

20. *(b)* Let length of null point J be x and length of the potentiometer wire be L.
In first case, current in the circuit

$$I_1 = \frac{6}{4+2} = 1 \text{ A}$$

$\therefore$ Potential gradient $= I \times R = \dfrac{1 \times 4}{L}$

$\Rightarrow$ Potential difference in part AJ

$$= \frac{1 \times 4}{L} \times x = \varepsilon_1$$

Given, $\varepsilon_1 = 0.5 = \dfrac{4x}{L}$ or $\dfrac{x}{L} = \dfrac{1}{8}$ $\quad \ldots$(i)

In second case, current in the circuit

$$I_2 = \frac{6}{4+6} = 0.6 \text{ A}$$

$\therefore$ Potential gradient $= \dfrac{0.6 \times 4}{L}$

$\Rightarrow$ Potential difference in part AJ

$$= \frac{0.6 \times 4}{L} \times x = \varepsilon_2$$

$\Rightarrow \quad \varepsilon_2 = \dfrac{0.6 \times 4}{L} \times \dfrac{L}{8}$ $\quad$ [Using Eq. (i)]

$\Rightarrow \quad \varepsilon_2 = 0.3$ V

21. *(a)* Meter bridge is an arrangement which works on Wheatstone's principle, so the balancing condition is

$$\frac{R}{S} = \frac{l_1}{l_2}$$

Where, $l_2 = 100 - l_1$

Case I $\quad R = X, S = Y, l_1 = 20$ cm,

$$l_2 = 100 - 20 = 80 \text{ cm}$$

$\therefore \quad \dfrac{X}{Y} = \dfrac{20}{80}$ $\quad \ldots$(i)

Case II Let the position of null point be obtained at a distance l from same end.

$\therefore \quad R = 4X, \quad S = Y,$

$$l_1 = l, l_2 = 100 - l$$

So, from Eq. (i), we get

$$\frac{4X}{Y} = \frac{l}{100 - l}$$

$$\frac{X}{Y} = \frac{l}{4(100 - l)} \quad \ldots \text{(ii)}$$

Therefore, from Eqs. (i) and (ii), we get

$$\frac{l}{4(100 - l)} = \frac{20}{80} \Rightarrow \frac{l}{4(100 - l)} = \frac{1}{4}$$

or $\quad l = 100 - l$ or $2l = 100$

Hence, $\quad l = 50$ cm

22 *(2)*

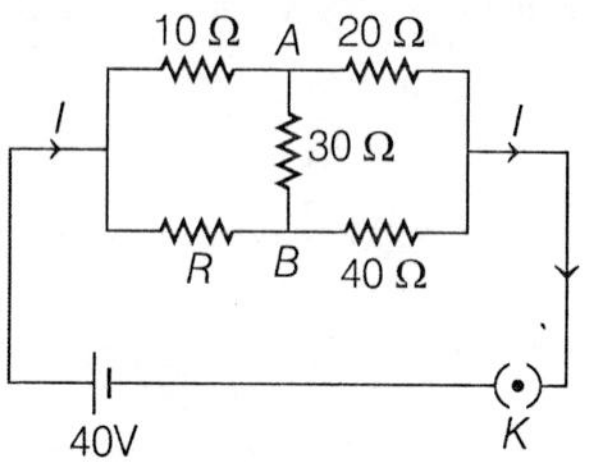

$V_A = V_B$, the bridge is in balanced condition

$$\frac{10}{R} = \frac{20}{40} \Rightarrow R = 20\Omega$$

$$I = \frac{V}{R} = \frac{40}{20} = 2\text{A}$$

23. *(2)* The potential difference between point B and D is zero, hence it is the balanced Wheatstone bridge circuit.

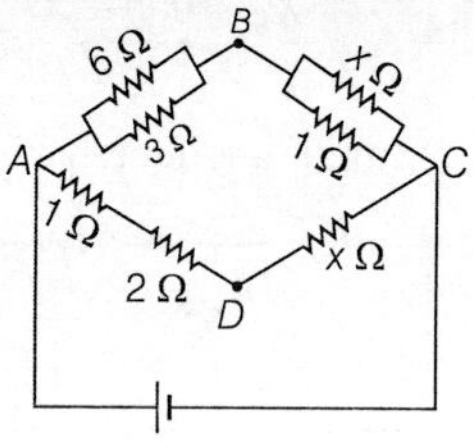

$$\therefore \quad \frac{R_{AB}}{R_{BC}} = \frac{R_{AD}}{R_{DC}}$$

$$\Rightarrow \quad \frac{\dfrac{6 \times 3}{6+3}}{\dfrac{x \times 1}{x+1}} = \frac{1+2}{x} \Rightarrow \frac{2(x+1)}{x} = \frac{3}{x}$$

$$\Rightarrow \quad 2x^2 + 2x - 3x = 0 \Rightarrow 2x^2 - x = 0$$

$$\Rightarrow \quad x(2x - 1) = 0 \Rightarrow x = 1/2 = 1/n$$

[Given]

$\therefore \quad n = 2$

> **• Concept Enhancer**
>
> The Wheatstone bridge is used for the precise measurement of low resistance. when four resistances P, Q, R and S are connected as shown in the circuit diagram, then under balanced condition, no current flows through galvanometer G. In this situation.

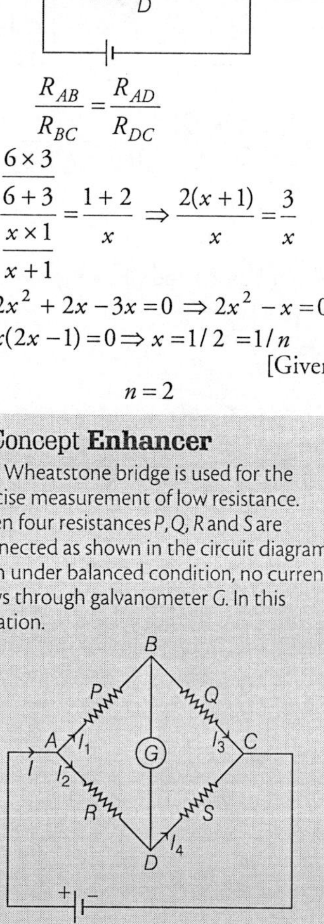

$$\frac{P}{Q} = \frac{R}{S}$$

24. *(30)* Given situation I is

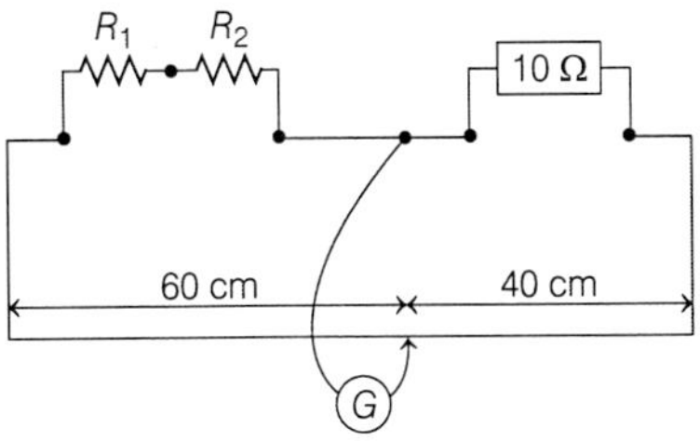

And situation II is

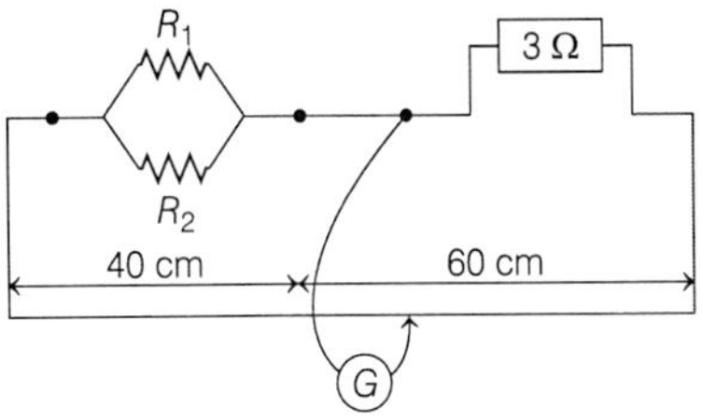

From situation I in balance

$$\frac{R_1 + R_2}{60} = \frac{10}{40} \quad \ldots\text{(i)}$$

And from situation II in balance

$$\frac{\left(\frac{R_1 R_2}{R_1 + R_2}\right)}{40} = \frac{3}{60} \quad \ldots\text{(ii)}$$

From Eq. (i), we get

$$R_1 + R_2 = 15\,\Omega$$

On substituting this in Eq. (ii), we get

$$\frac{R_1 R_2}{(15)} = 2 \text{ or } R_1 R_2 = 30\,\Omega$$

25. *(2)*

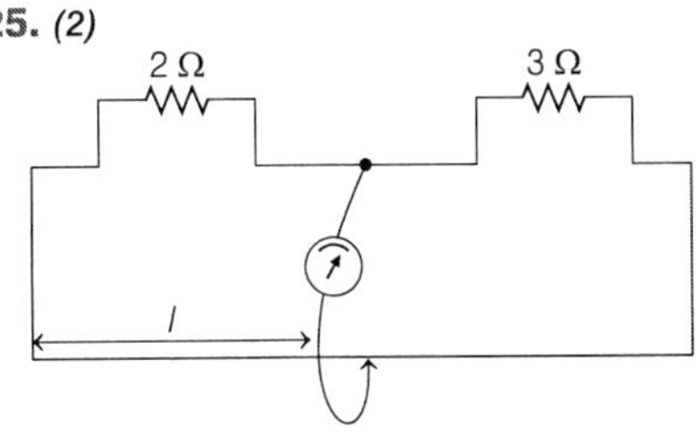

From Ist condition given,

$$\frac{2}{l} = \frac{3}{100 - l} \Rightarrow l = 40 \text{ cm}$$

Now, 3Ω resistor is added with a shunt of $X\,\Omega$.

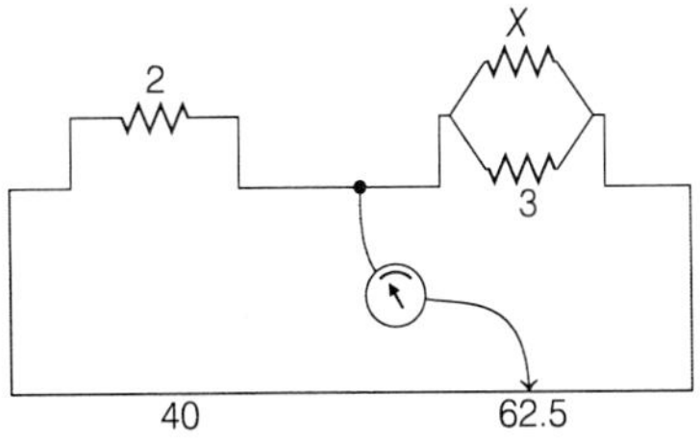

Balance length in above condition is $40 + 22.5 = 62.5$ cm

$$\text{So, } \frac{2}{62.5} = \frac{R}{(100 - 62.5)} \quad \ldots\text{(i)}$$

Here, R = Equivalent resistance of 3Ω and $X\,\Omega$ in parallel

$$= \frac{3X}{3 + X}$$

So, $R = 1.2\,\Omega$ [From Eq. (i)]

$$\Rightarrow \frac{3X}{3 + X} = 1\cdot 2$$

$$\Rightarrow 1\cdot 8X = 3\cdot 6$$

or $X = 2\,\Omega$

26. *(40)* We know, in a balanced meter bridge,

$$\frac{R_1}{R_2} = \frac{AC}{CB} \quad \ldots\text{(i)}$$

$$\Rightarrow \frac{R_1}{R_2} = \frac{40}{CB}$$

As we can clearly see from Eq. (i) that the condition for a balanced Wheatstone bridge (balanced meter bridge) is independent of the radius of wire AB, therefore the balancing length will not change and will remain 40 cm only.

27 *(19)* Balance length is given 43 cm. So, other point's length will be

$$l_2 = (100 - l_1)$$
$$= 100 - 43 = 57\text{cm}$$

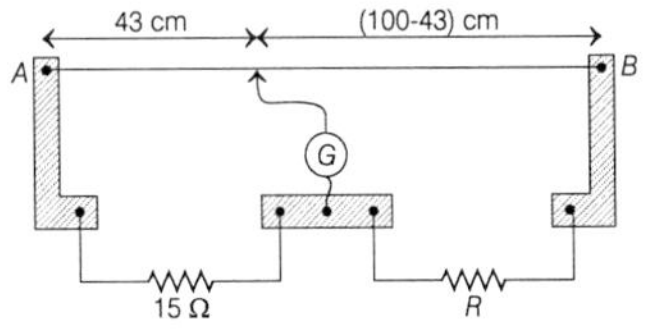

Now under balanced condition,

$$\frac{15}{l_1} = \frac{R}{l_2} \quad \text{(when there is no end error)}$$

$$\frac{15}{l_1 + e_1} = \frac{R}{l_2 + e_2} \quad \text{(when there are errors}$$

e_1 at end A and e_2 at end B

Here, $l_1 = 43$ cm, $l_2 = 100 - 43 = 57$ cm

$e_1 = 2$ cm, $e_2 = 0$ (not given)

So, we have

$$\frac{15}{43 + 2} = \frac{R}{57}$$

$$\Rightarrow R = \frac{15 \times 57}{(43 + 2)} = 19\,\Omega$$

28. *(10)*

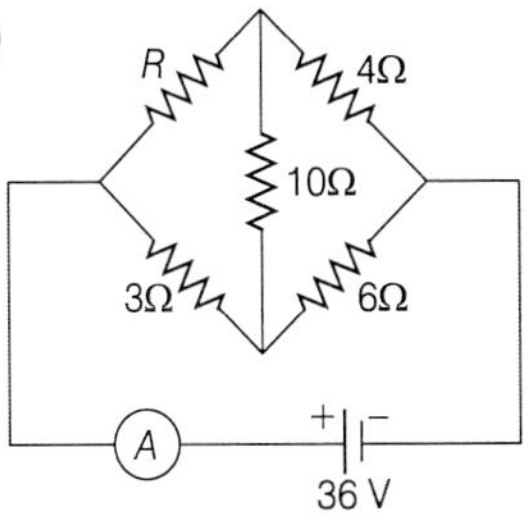

When no current flows through 10Ω resistor, then given Wheatstone bridge will be balanced and so ratio of resistances are equal.

$$\Rightarrow \frac{R}{3} = \frac{4}{6} \Rightarrow R = 2\,\Omega$$

So, total resistance of circuit will be $(2 + 4) || (3 + 6)$

$$\Rightarrow R_{\text{total}} = \frac{6 \times 9}{6 + 9} = \frac{18}{5}\,\Omega$$

Current through ammeter,

$$I = \frac{V}{R_{\text{total}}} = \frac{36}{\frac{18}{5}} = 10 \text{ A}$$

29. *(48)* At the balanced condition of the Wheatstone bridge,

$$\frac{R}{S} = \frac{L_1}{L - L_1}$$

$$\Rightarrow \frac{12}{6} = \frac{x}{72 - x}$$

$$\Rightarrow x = 48 \text{ cm}$$

∴The galvanometer jockey is placed at P on AB at a distance of 48 cm from the A.

30 *(10)* Cyclic order and resistance $X\,\Omega$ which is connected to obtain balance condition is as shown in the figure

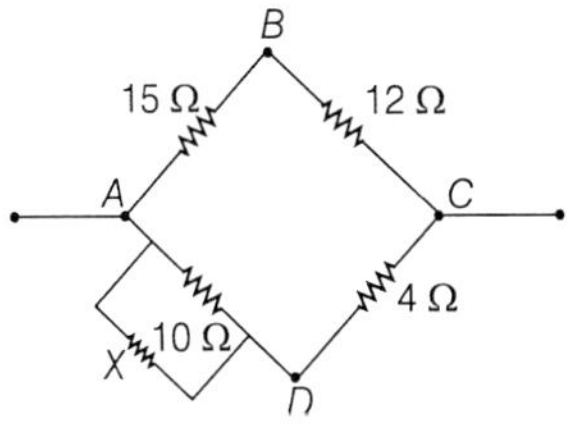

For balancing condition,

$$\frac{15}{12} = \frac{\left(\frac{10X}{10 + X}\right)}{4}$$

$$\Rightarrow \frac{10X}{10 + X} = 5$$

$$\Rightarrow X = 10\,\Omega$$

C-18

JEE Main Chapterwise
Topicwise Physics

In JEE Main there is usually 1-2 question per paper is visible from this chapter. In past 3 years, 53 conceptual and 28 numerical questions were asked of moderate to difficult level from this chapter.

For detailed analysis scan QR code

MOVING CHARGES AND MAGNETISM

Topic 1 Biot Savart's Law and Ampere's Circuital Law

Level 1

Imp. Magnetic Field on the Axis of Circular Coil

01 Let B_1 the magnitude of magnetic field at center of a circular coil of radius R carrying current I. Let B_2 be the magnitude of magnetic field at an axial distance x from the center. For $x:R=3:4$, $\frac{B_2}{B_1}$ is

[2025, 2 April Shift-I]

(a) 64 : 125 (b) 25 : 16 (c) 4 : 5 (d) 16 : 25

02 Given below are two statements.

Statement I When currents vary with time, Newton's third law is valid only if momentum carried by the electromagnetic field is taken into account.

Statement II Ampere's circuital law does not depend on Biot-Savart's law.

In the light of the above statements, choose the correct answer from the options given below.

[2024, 9 April Shift-I]

(a) Both Statement I and Statement II are true.
(b) Statement I is false but Statement II is true.
(c) Statement I is true but Statement II is false.
(d) Both Statement I and Statement II are false.

03 In a coaxial straight cable, the central conductor and the outer conductor carry equal currents in opposite directions. The magnetic field is zero

[2024, 5 April Shift-I]

(a) outside the cable
(b) in between the two conductors
(c) inside the inner conductor
(d) inside the outer conductor

04 An electron is projected with uniform velocity along the axis inside a current carrying long solenoid. Then, *[2024, 4 April Shift-I]*

(a) the electron path will be circular about the axis.
(b) the electron will experience a force at 45° to the axis and execute a helical path.
(c) the electron will be accelerated along the axis.
(d) the electron will continue to move with uniform velocity along the axis of the solenoid.

05 An electron is moving along the positive X-axis. If the uniform magnetic field is applied parallel to the negative Z-axis, then *[2023, 13 April Shift-II]*

A. the electron will experience magnetic force along positive Y-axis.
B. the electron will experience magnetic force along negative Y-axis.
C. the electron will not experience any force in magnetic field.
D. the electron will continue to move along the positive X-axis.
E. the electron will move along circular path in magnetic field.

Choose the correct answer from the options given below.

(a) C and D only (b) B and D only
(c) B and E only (d) A and E only

06 A charge particle moving in magnetic field B, has the components of velocity along B as well as perpendicular to B. The path of the charge particle will be *[2023, 8 April Shift-I]*

(a) circular path
(b) helical path with the axis along magnetic field B
(c) helical path with the axis perpendicular to the direction of magnetic field B
(d) straight along the direction of magnetic field B

Imp. Ampere's Circuital Law

07 A long straight wire of circular cross-section (radius a) is carrying steady current I. The current I is uniformly distributed across this cross-section. The magnetic field is *[2023, 6 April Shift-I]*

(a) directly proportional to r in the region $r < a$ and inversely proportional to r in the region $r > a$
(b) inversely proportional to r in the region $r < a$ and uniform throughout in the region $r > a$
(c) uniform in the region $r < a$ and inversely proportional to distance r from the axis, in the region $r > a$
(d) zero in the region $r < a$ and inversely proportional to r in the region $r > a$

08 As shown in the figure, a long straight conductor with semi-circular arc of radius $\pi / 10$ m is carrying current $I = 3$ A. The magnitude of the magnetic field at the centre O of the arc is (The permeability of the vacuum $= 4\pi \times 10^{-7}$ NA^{-2}) *[2023, 1 Feb Shift-II]*

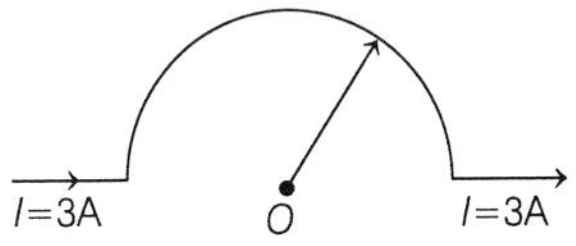

(a) $3\,\mu$T (b) $1\,\mu$T
(c) $4\,\mu$T (d) $6\,\mu$T

09 A long solenoid is formed by winding 70 turns cm^{-1}. If 2.0 A current flows, then the magnetic field produced inside the solenoid is ....
$(\mu_0 = 4\pi \times 10^{-7}$ T-m A$^{-1})$ *[2023, 29 Jan Shift-II]*

(a) 352×10^{-4} T (b) 176×10^{-4} T
(c) 1232×10^{-4} T (d) 88×10^{-4} T

10 A single current carrying loop of wire carrying current I flowing in anti-clockwise direction seen from + ve z-direction and lying in XY-plane is shown in figure. The plot of $\hat{j}$ component of magnetic field (B_y) at a distance a (less than radius of the coil) and on YZ-plane *versus* z-coordinate looks like *[2023, 29 Jan Shift-I]*

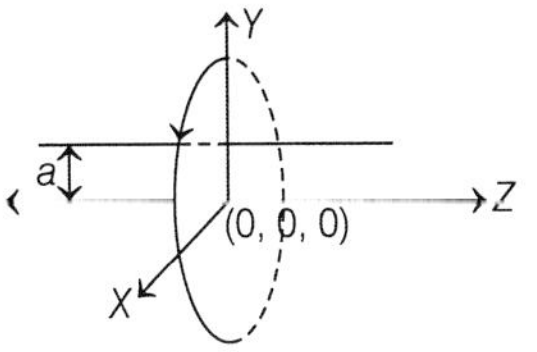

(a)
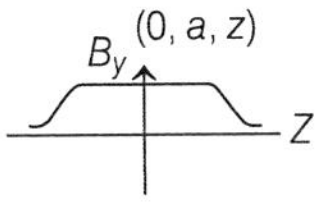

(b)
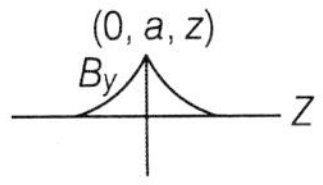

(c)
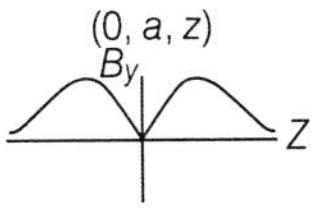

(d)
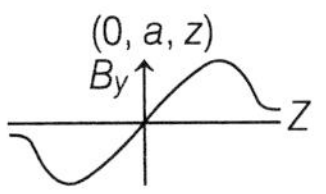

11 A solenoid of 1200 turns is wound uniformly in a single layer on a glass tube 2m long and 0.2 m in diameter. The magnetic intensity at the centre of the solenoid when a current of 2A flows through it is *[2023, 25 Jan Shift-I]*

(a) 1A m^{-1}
(b) 1.2×10^3 A m^{-1}
(c) 2.4×10^3 A m^{-1}
(d) 2.4×10^{-3} A m^{-1}

12 An infinitely long hollow conducting cylinder with radius R carries a uniform current along its surface. Choose the correct representation of magnetic field B as a function of radial distance r from the axis of cylinder. *[2022, 28 June Shift-I]*

(a)
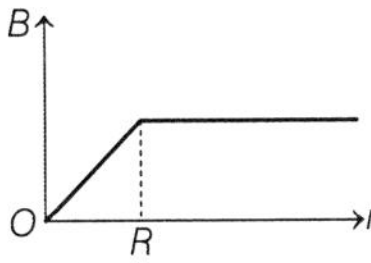

(b)
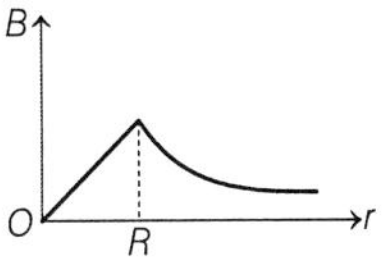

(c)
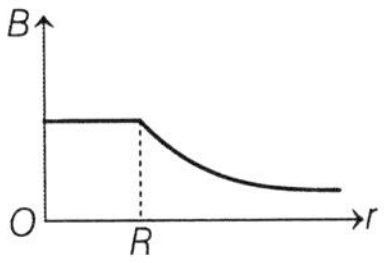

(d)
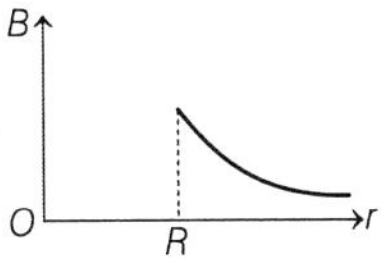

13 A long solenoid carrying a current produces a magnetic field B along its axis. If the current is doubled and the number of turns per cm is halved, the new value of magnetic field will be equal to *[2022, 25 June Shift-II]*

(a) B (b) $2\,B$
(c) $4\,B$ (d) $B/2$

14 Figures A and B shown two long straight wires of circular cross- section a and b (with $a < b$), carrying current I which is uniformly distributed across the cross-section. The magnitude of magnetic field B varies with radius r and can be represented as *[2021, 27 July Shift-II]*

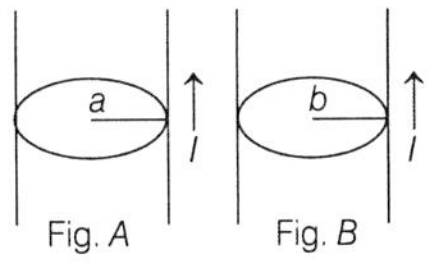

Fig. A Fig. B

(a)
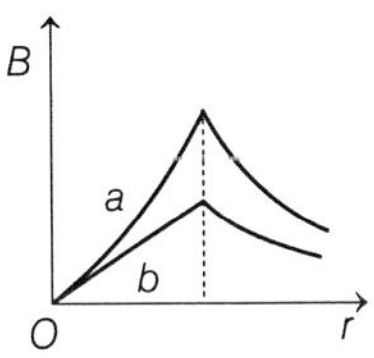

(b)
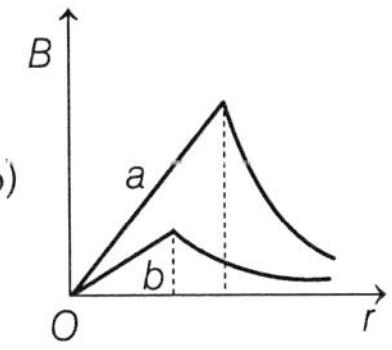

(c)
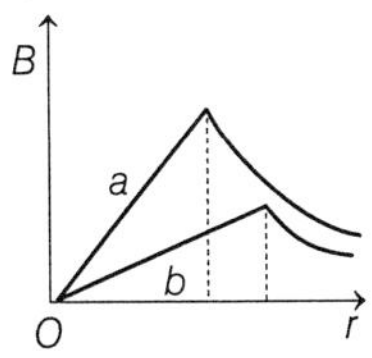

(d)
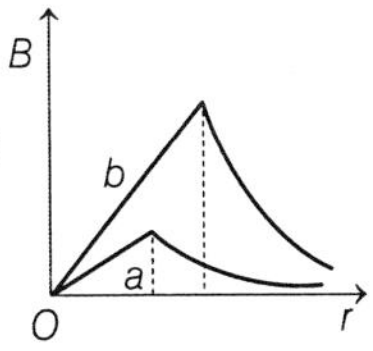

15 Four identical long solenoids A, B, C and D are connected to each other as shown in the figure. If the magnetic field at the centre of A is 3 T, the field at the centre of C would be (Assume that, the magnetic field is confined with in the volume of respective solenoid) *[2021, 18 March Shift-I]*

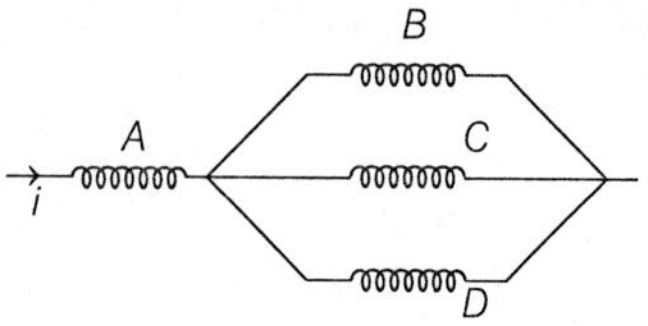

(a) 12T (b) 6T (c) 9T (d) 1T

16 A current I flows along the length of an infinitely long, straight, thin walled pipe. Then, *[AIEEE 2007]*
(a) the magnetic field is zero only on the axis of the pipe.
(b) the magnetic field is different at different points inside the pipe.
(c) the magnetic field at any point inside the pipe is zero.
(d) the magnetic field at all points inside the pipe is the same but not zero.

Imp. Magnetic Field Due to Solenoid

17 A long solenoid has 200 turns per cm and carries a current I. The magnetic field at its centre is 6.28×10^{-2} Wb/m^2. Another long solenoid has 100 turns per cm and it carries a current $I/3$. The value of the magnetic field at its centre is *[AIEEE 2006]*
(a) 1.05×10^{-2} Wb/m^2 (b) 1.05×10^{-5} Wb/m^2
(c) 1.05×10^{-3} Wb/m^2 (d) 1.05×10^{-4} Wb/m^2

18 If in a circular coil A of radius R, current I is flowing and in another coil B of radius $2R$ a current $2I$ is flowing, then the ratio of the magnetic fields, B_A and B_B produced by them will be *[AIEEE 2002]*
(a) 1 (b) 2 (c) 1/2 (d) 4

Level 2

19

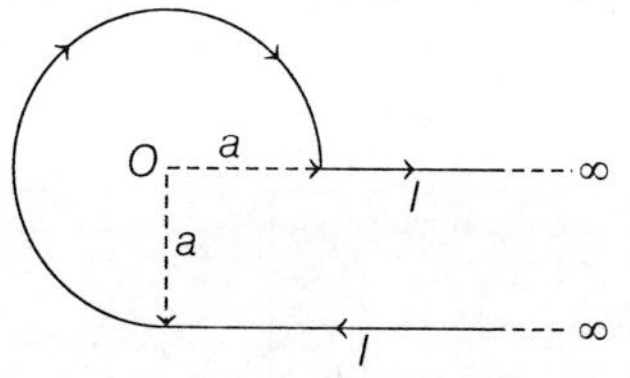

An infinite wire has a circular bend of radius a and carrying a current I as shown in figure. The magnitude of magnetic field at the origin O of the arc is given by *[2025, 28 Jan Shift-II]*

(a) $\frac{\mu_0}{4\pi}\frac{I}{a}\left[\frac{3\pi}{2}+1\right]$ (b) $\frac{\mu_0}{4\pi}\frac{I}{a}\left[\frac{\pi}{2}+1\right]$

(c) $\frac{\mu_0}{4\pi}\frac{I}{a}\left[\frac{3\pi}{2}+2\right]$ (d) $\frac{\mu_0}{2\pi}\frac{I}{a}\left[\frac{\pi}{2}+2\right]$

20 Two insulated circular loop A and B of radius a carrying a current of I in the anti-clockwise direction as shown in the figure. The magnitude of the magnetic induction at the centre will be *[2024, 30 Jan Shift-I]*

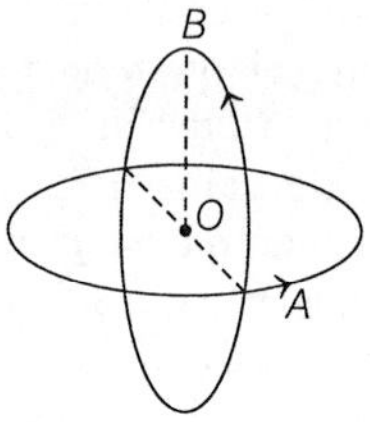

(a) $\frac{\sqrt{2}\mu_0 I}{a}$ (b) $\frac{2\mu_0 I}{a}$ (c) $\frac{\mu_0 I}{\sqrt{2}a}$ (d) $\frac{\mu_0 I}{2a}$

21 An element $\Delta l = \Delta x\hat{i}$ is placed at the origin and carries a large current $I = 10$ A. The magnetic field on the Y-axis at a distance of 0.5 m from the elements Δx of 1 cm length is *[2023, 6 April Shift-I]*

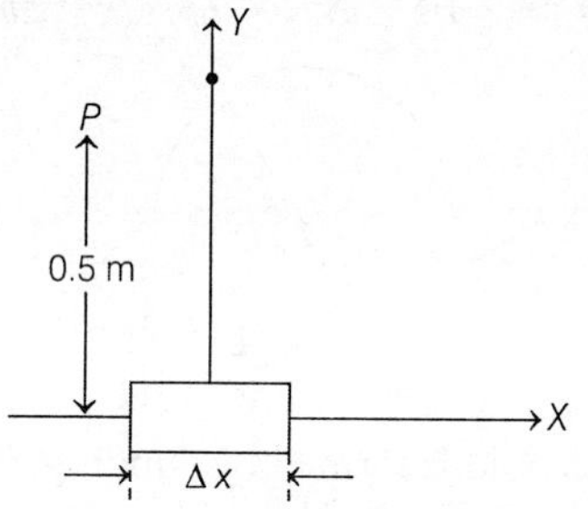

(a) 12×10^{-8} T (b) 8×10^{-8} T
(c) 4×10^{-8} T (d) 10×10^{-8} T

22 The source of time varying magnetic field may be
A. a permanent magnet
B. an electric field changing linearly with time
C. direct current
D. a decelerating charge particle
E. an antenna fed with a digital signal
Choose the correct answer from the options given below *[2023, 13 April Shift-I]*
(a) D only (b) B and D only
(c) C and E only (d) A only

23 Given below are two statements: One is labelled as Assertion (A) and the other is labelled as Reason (R)
Assertion (A) An electric fan continues to rotate for some time after the current is switched off.
Reason (R) Fan continues to rotate due to inertia of motion. *[2023, 10 April Shift-II]*
In the light of above statements. choose the most appropriate answer from the options given below
(a) A is correct but R is not correct.
(b) A is not correct but R is correct.
(c) Both A and R are correct and R is the correct explanation of A.
(d) Both A and R are correct but R is not the correct explanation of A.

Imp. Magnetic Field Due to Semicircular arc

24 Find the magnetic field at the point P in figure. The curved portion is a semicircle connected to two long straight wires. **[2023, 1 Feb Shift-II]**

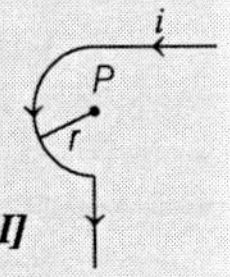

(a) $\frac{\mu_0 i}{2r}\left(\frac{1}{2}+\frac{1}{2\pi}\right)$ (b) $\frac{\mu_0 i}{2r}\left(1+\frac{1}{\pi}\right)$

(c) $\frac{\mu_0 i}{2r}\left(\frac{1}{2}+\frac{1}{\pi}\right)$ (d) $\frac{\mu_0 i}{2r}\left(1+\frac{2}{\pi}\right)$

25 A long conducting wire having a current I flowing through it, is bent into a circular coil of N turns. Then it is bent into a circular coil of n turns. The magnetic field is calculated at the centre of coils in both the cases. The ratio of the magnetic field in first case to that of second case is **[2023, 31 Jan Shift-II]**

(a) $N^2 : n^2$ (b) $N : n$ (c) $n^2 : N^2$ (d) $n : N$

26 A circular loop of radius r is carrying current I A. The ratio of magnetic field at the centre of circular loop and at a distance r from the centre of the loop on its axis is **[2023, 24 Jan Shift-I]**

(a) $1 : \sqrt{2}$ (b) $1 : 3\sqrt{2}$ (c) $3\sqrt{2} : 2$ (d) $2\sqrt{2} : 1$

27 The magnetic field at the center of current carrying circular loop is B_1. The magnetic field at a distance of $\sqrt{3}$ times radius of the given circular loop from the centre on its axis is B_2. The value of $\frac{B_1}{B_2}$ will be **[2022, 28 July Shift-II]**

(a) 9 : 4 (b) $12 : \sqrt{5}$ (c) 8 : 1 (d) $5 : \sqrt{3}$

28 The electric current in a circular coil of 2 turns produces a magnetic induction B_1 at its centre. The coil is unwound and is rewound into a circular coil of 5 turns and the same current produces a magnetic induction B_2 at its centre. The ratio of B_2 / B_1 is **[2022, 25 July Shift-II]**

(a) $\frac{5}{2}$ (b) $\frac{25}{4}$ (c) $\frac{5}{4}$ (d) $\frac{25}{2}$

29 A coil of n number of turns wound tightly in the form of a spiral with inner and outer radii r_1 and r_2, respectively. When a current of strength I is passed through the coil, the magnetic field at its centre will be **[2022, 30 June Shift I]**

(a) $\frac{\mu_0 nI}{2(r_2 - r_1)}$ (b) $\frac{\mu_0 nI}{r_2}$

(c) $\frac{\mu_0 nI}{r_2 - r_1}\log_e \frac{r_1}{r_2}$ (d) $\frac{\mu_0 nI}{2(r_2 - r_1)}\log_e \frac{r_2}{r_1}$

30 A long straight wire with a circular cross-section having radius R, is carrying a steady current I. The current I is uniformly distributed across this cross-section. Then, the variation of magnetic field due to current I with distance r ($r < R$) from its centre will be **[2022, 25 June Shift-I]**

(a) $B \propto r^2$ (b) $B \propto r$ (c) $B \propto \frac{1}{r^2}$ (d) $B \propto \frac{1}{r}$

31 A hairpin like shape as shown in figure is made by bending a long current carrying wire. What is the magnitude of a magnetic field at point P which lies on the centre of the semicircle? **[2021, 17 March Shift-II]**

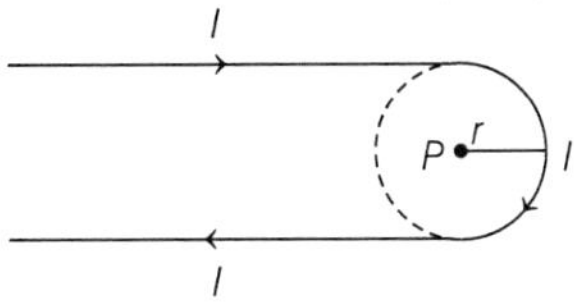

(a) $\frac{\mu_0 I}{4\pi r}(2-\pi)$ (b) $\frac{\mu_0 I}{4\pi r}(2+\pi)$

(c) $\frac{\mu_0 I}{2\pi r}(2+\pi)$ (d) $\frac{\mu_0 I}{2\pi r}(2-\pi)$

32 As shown in the figure, two infinitely long, identical wires are bent by 90° and placed in such a way that the segments LP and QM are along the X-axis, while segments PS and QN are parallel to the Y-axis. If $OP = OQ = 4$ cm and the magnitude of the magnetic field at O is 10^{-4} T and the two wires carry equal currents (see figure), the magnitude of the current in each wire and the direction of the magnetic field at O will be (Take, $\mu_0 = 4\pi \times 10^{-7}$ NA^{-2}) **[2019, 12 Jan Shift-I]**

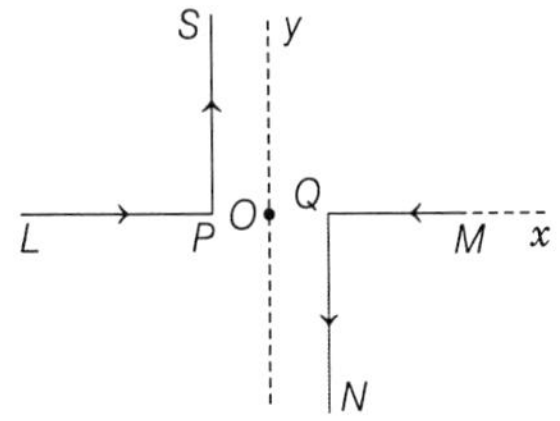

(a) 40 A, perpendicular out of the page
(b) 20 A, perpendicular into the page
(c) 20 A, perpendicular out of the page
(d) 40 A, perpendicular into the page

33 A current I flows in an infinitely long wire with cross-section in the form of a semi circular ring of radius R. The magnitude of the magnetic induction along its axis is **[AIEEE 2011]**

(a) $\frac{\mu_0 I}{2\pi^2 R}$ (b) $\frac{\mu_0 I}{2\pi R}$ (c) $\frac{\mu_0 I}{4\pi R}$ (d) $\frac{\mu_0 I}{\pi^2 R}$

34 Two long parallel wires are at a distance $2d$ apart. They carry steady equal current flowing out of the plane of the paper as shown. The variation of the magnetic field along the line xx' is given by **[AIEEE 2010]**

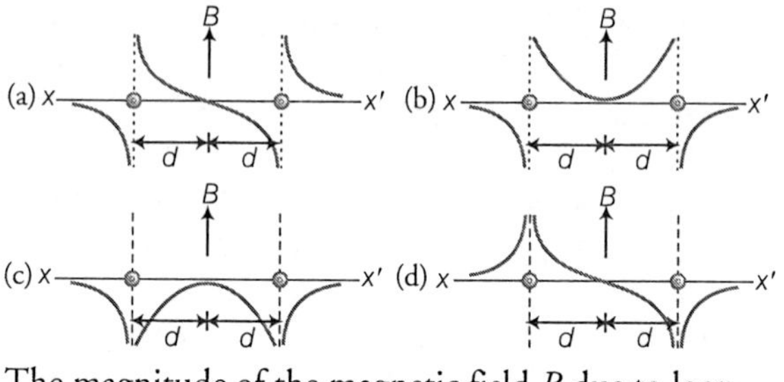

35 The magnitude of the magnetic field B due to loop $ABCD$ at the origin O is **[AIEEE 2009]**

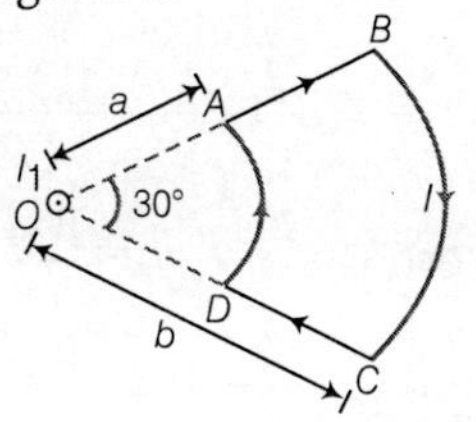

(a) zero (b) $\frac{\mu_0 I(b-a)}{24ab}$

(c) $\frac{\mu_0 I}{4\pi}\left[\frac{b-a}{ab}\right]$ (d) $\frac{\mu_0 I}{4\pi}\left[2(b-a)+\frac{\pi}{3}(a+b)\right]$

36 Two concentric coils each of radius equal to 2π cm are placed at right angles to each other. 3 A and 4A are the currents flowing in each coil, respectively. The magnetic induction in Wb/m^2 at the centre of the coils will be ($\mu_0 = 4\pi \times 10^{-7}$ Wb/Am) **[AIEEE 2005]**

(a) 12×10^{-5} (b) 10^{-5} (c) 5×10^{-5} (d) 7×10^{-5}

37 A long wire carries a steady current. It is bent into a circle of one turn and the magnetic field at the centre of the coil is B. It is then bent into a circular loop of n turns. The magnetic field at the centre of the coil will be **[AIEEE 2004]**

(a) nB (b) $n^2 B$
(c) $2nB$ (d) $2n^2 B$

Level 3

38 Figure shows a current carrying square loop $ABCD$ of edge length is a lying in a plane. If the resistance of the ABC part is r and that of ADC part is $2r$, then the magnitude of the resultant magnetic field at centre of the square loop is **[2025, 8 April Shift-II]**

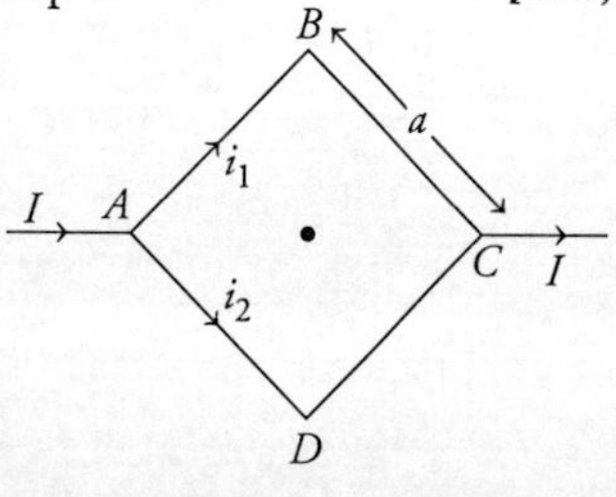

(a) $\frac{\mu_0 I}{2\pi a}$ (b) $\frac{\sqrt{2}\mu_0 I}{3\pi a}$

(c) $\frac{3\pi\mu_0 I}{\sqrt{2}a}$ (d) $\frac{2\mu_0 I}{3\pi a}$

39 Consider a long straight wire of a circular cross-section (radius a) carrying a steady current I. The current is uniformly distributed across this cross-section. The distances from the centre of the wire's cross-section at which the magnetic field (inside the wire, outside the wire) is half of the maximum possible magnetic field, anywhere due to the wire, will be **[2025, 29 Jan Shift-I]**

(a) $[a/4, 3a/2]$ (b) $[a/2, 3a]$
(c) $[a/4, 2a]$ (d) $[a/2, 2a]$

40 A long straight wire of a circular cross-section with radius a carries a steady current I. The current I is uniformly distributed across this cross-section. The plot of magnitude of magnetic field B with distance r from the centre of the wire is given by **[2025, 24 Jan Shift-II]**

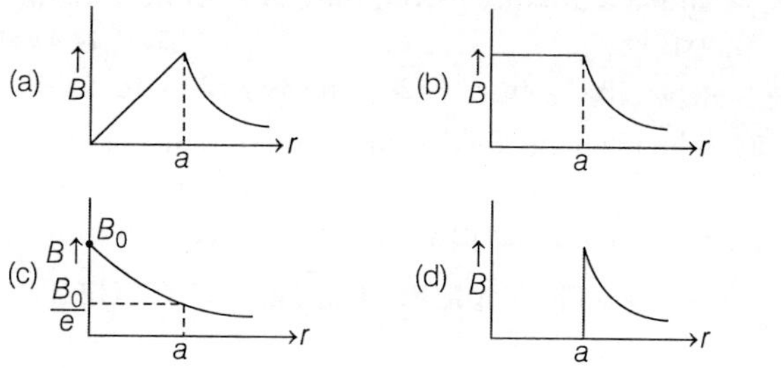

41 A long straight wire of radius a carries a steady current I. The current is uniformly distributed across its cross-section. The ratio of the magnetic field at $a/2$ and $2a$ from axis of the wire is **[2024, 8 April Shift-II]**

(a) 4 : 1 (b) 3 : 4
(c) 1 : 4 (d) 1 : 1

Imp. Magnetic Field Due to Straight Wire

42 As shown in the figure, a current of 2 A flowing in an equilateral triangle of side $4\sqrt{3}$ cm. The magnetic field at the centroid O of the triangle is **[2023, 30 Jan Shift-I]**

(Neglect the effect of earth's magnetic field)

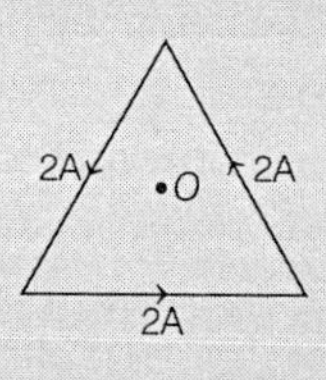

(a) $\sqrt{3} \times 10^{-4}$ T
(b) $4\sqrt{3} \times 10^{-4}$ T
(c) $3\sqrt{3} \times 10^{-5}$ T
(d) $4\sqrt{3} \times 10^{-5}$ T

43 Match List I with List II.

List I (Current configuration)	List II (Magnitude of Magnetic Field at point O)
A. 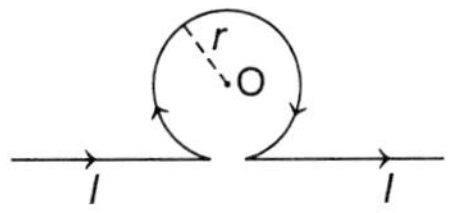	I. $B_0 = \frac{\mu_0 I}{4\pi r}(\pi + 2)$
B. 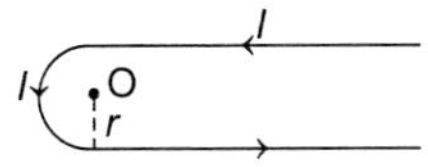	II. $B_0 = \frac{\mu_0}{4}\frac{I}{r}$
C. 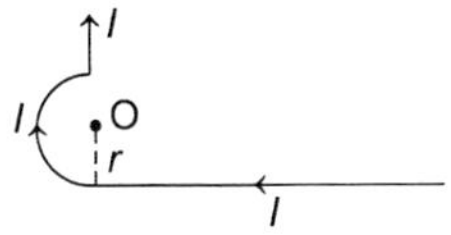	III. $B_0 = \frac{\mu_0 I}{2\pi r}(\pi - 1)$
D. 	IV. $B_0 = \frac{\mu_0 I}{4\pi r}(\pi + 1)$

[2023, 25 Jan Shift-I]

Choose the correct answer from the options given below.

(a) A-III, B-I, C-IV, D-II (b) A-III, B-IV, C-I, D-II
(c) A-I, B-III, C-IV, D-II (d) A-II, B-I, C-IV, D-III

44 The magnetic field at the centre of a circular coil of radius r, due to current I flowing through it, is B. The magnetic field at a point along the axis at a distance $r/2$ from the centre is *[2022, 24 June Shift-I]*

(a) $B/2$ (b) $2B$
(c) $\left(\frac{2}{\sqrt{5}}\right)^3 B$ (d) $\left(\frac{2}{\sqrt{3}}\right)^3 B$

45 A coil having N turns is wound tightly in the form of a spiral with inner and outer radii a and b, respectively. Find the magnetic field at centre, when a current I passes through coil *[2021, 31 Aug Shift-I]*

(a) $\frac{\mu_0 IN}{2(b-a)}\ln\left(\frac{b}{a}\right)$ (b) $\frac{\mu_0 I}{8}\ln\left[\frac{a+b}{a-b}\right]$
(c) $\frac{\mu_0 I}{4(a-b)}\ln\left[\frac{1}{a}-\frac{1}{b}\right]$ (d) $\frac{\mu_0 I}{8}\ln\left(\frac{a-b}{a+b}\right)$

46 A coaxial cable consists of an inner wire of radius a surrounded by an outer shell of inner and outer radii b and c, respectively. The inner wire carries an electric current i_0, which is distributed uniformly across cross-sectional area. The outer shell carries an equal current in opposite direction and distributed uniformly. What will be the ratio of the magnetic field at a distance x from the axis when (i) $x < a$ and (ii) $a < x < b$? *[2021, 27 Aug Shift-II]*

(a) $\frac{x^2}{a^2}$ (b) $\frac{a^2}{x^2}$ (c) $\frac{x^2}{b^2 - a^2}$ (d) $\frac{b^2 - a^2}{x^2}$

47 Magnetic fields at two points on the axis of a circular coil at a distance of 0.05 m and 0.2 m from the centre are in the ratio 8 : 1. The radius of coil is *[2021, 25 Feb Shift-I]*

(a) 0.2 m (b) 0.1 m
(c) 0.15 m (d) 1.0 m

48 A thin ring of 10 cm radius carries a uniformly distributed charge. The ring rotates at a constant angular speed of 40π rad s^{-1} about its axis, perpendicular to its plane. If the magnetic field at its centre is 3.8×10^{-9} T, then the charge carried by the ring is close to ($\mu_0 = 4\pi \times 10^{-7}$ N/A^2). *[2019, 12 April Shift-I]*

(a) 2×10^{-6} C
(b) 3×10^{-5} C
(c) 4×10^{-5} C
(d) 7×10^{-6} C

Imp. Magnetic Field Due to Circular Arc

49 A current loop, having two circular arcs joined by two radial lines as shown in the figure. It carries a current of 10 A. The magnetic field at point O will be close to *[2019, 9 Jan Shift-I]*

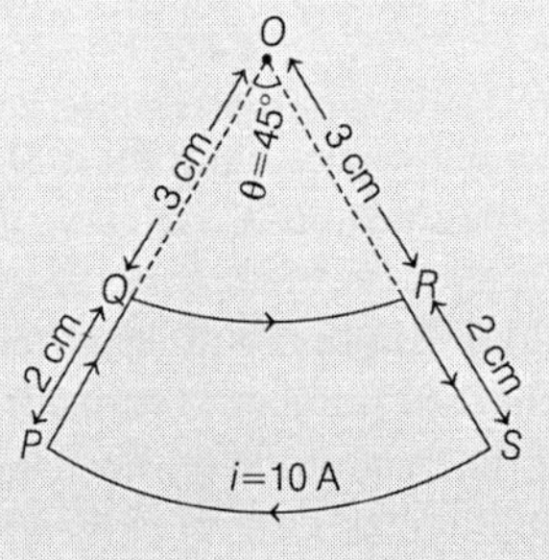

(a) 1.0×10^{-5} T (b) 1.0×10^{-7} T
(c) 1.5×10^{-7} T (d) 1.5×10^{-5} T

50 Two identical wires A and B, each of length l, carry the same current I. Wire A is bent into a circle of radius R and wire B is bent to form a square of side a. If B_A and B_B are the values of magnetic field at the centres of the circle and square respectively, then the ratio $\frac{B_A}{B_B}$ is *[JEE Main 2016]*

(a) $\frac{\pi^2}{8}$ (b) $\frac{\pi^2}{16\sqrt{2}}$ (c) $\frac{\pi^2}{16}$ (d) $\frac{\pi^2}{8\sqrt{2}}$

51 A long straight wire of radius a carries a steady current I. The current is uniformly distributed across its cross-section. The ratio of the magnetic field at $a/2$ and $2a$ is *[AIEEE 2007]*

(a) 1/4 (b) 4
(c) 1 (d) 1/2

52 The magnetic field due to a current carrying circular loop of radius 3 cm at a point on the axis at a distance of 4 cm from the centre is 54 μT. What will be its value at the centre of the loop ? *[AIEEE 2004]*

(a) 250 μT (b) 150 μT
(c) 125 μT (d) 75 μT

Numerical Value Type Questions

Imp. Magnetic Field Due to Semicircular Current Loop

53 A loop *ABCDA*, carrying current $I = 12$ A is placed in a plane, consists of two semi-circular segments of radius $R_1 = 6\pi$ m and $R_2 = 4\pi$ m. The magnitude of the resultant magnetic field at center *O* is $k \times 10^{-7}$ T. The value of *k* is ....
(Given $\mu_0 = 4\pi \times 10^{-7}$ Tm A^{-1}) *[2025, 3 April Shift-I]*

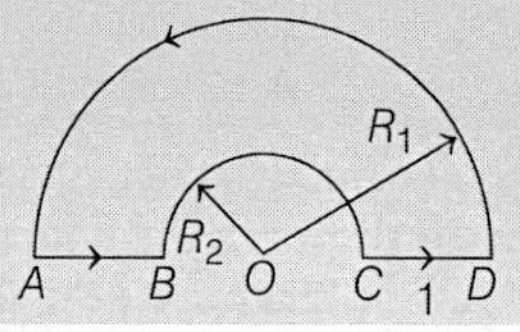

54 The magnetic field inside a 200 turns solenoid of radius 10 cm is 2.9×10^{-4} T.
If the solenoid carries a current of 0.29 A, then the length of the solenoid is.............π cm *[2025, 29 Jan Shift-II]*

55 A solenoid of length 0.5 m has a radius of 1 cm and is made up of *m* number of turns. It carries a current of 5 A. If the magnitude of the magnetic field inside the solenoid is 6.28×10^{-3} T, then the value of *m* is .... *[2024, 5 April Shift-II]*

56 Two parallel long current carrying wire separated by a distance 2*r* are shown in the figure. The ratio of magnetic field at *A* to the magnetic field produced at *C* is *x*/7. *[2024, 4 April Shift-II]*
The value of *x* is...... .

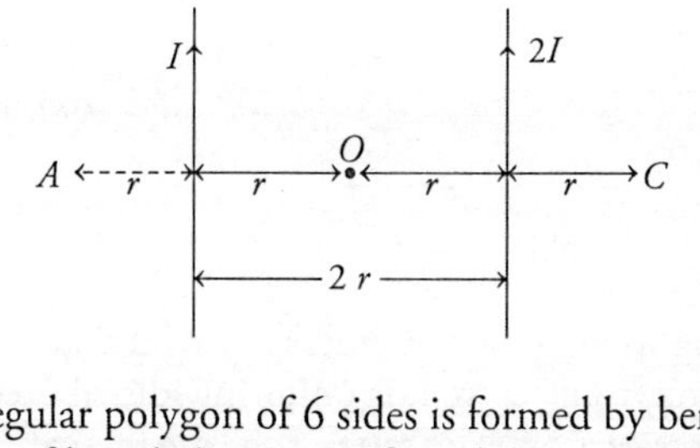

57 A regular polygon of 6 sides is formed by bending a wire of length 4π metre. If an electric current of $4\pi\sqrt{3}$ A is flowing through the sides of the polygon, the magnetic field at the centre of the polygon would be $x \times 10^{-7}$ T. The value of *x* is .... *[2024, 1 Feb Shift-I]*

58 Two circular coils *P* and *Q* of 100 turns each have same radius of π cm. The current in *P* and *R* are 1 A and 2 A respectively. *P* and *Q* are placed with their planes mutually perpendicular with their centres coincide. The resultant magnetic field induction at the centre of the coils is $\sqrt{x}$ mT, where $x =$ ....
[Use, $\mu_0 = 4\pi \times 10^{-7}$ T mA^{-1}] *[2024, 31 Jan Shift-II]*

59 The current of 5 A flows in a square loop of sides 1 m is placed in air. The magnetic field at the centre of the loop is $x\sqrt{2} \times 10^{-7}$ T. The value of *x* is .... *[2024, 30 Jan Shift-II]*

60 Two long, straight wires carry equal currents in opposite directions as shown in figure. The separation between the wires is 5.0 cm. The magnitude of the magnetic field at a point *P* mid-way between the wires is .... μT. *[2024, 27 Jan Shift-I]*
(Given, $\mu_0 = 4\pi \times 10^{-7}$ TmA^{-1})

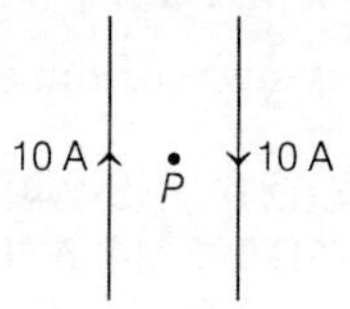

61 An electron in a hydrogen atom revolves around its nucleus with a speed of $6.76 \times 10^6 ms^{-1}$ in an orbit of radius 0.52 Å. The magnetic field produced at the nucleus of the hydrogen atom is .... T. *[2023, 15 April Shift-I]*

62 The ratio of magnetic field at the centre of a current carrying coil of radius *r* from the magnetic field at distance *r* from the centre of coil on its axis is $\sqrt{x} : 1$. The value of *x* is .... *[2023, 8 April Shift-II]*

63 Two identical circular wires of radius 20 cm and carrying current $\sqrt{2}$ A are placed in perpendicular planes as shown in figure. The net magnetic field at the centre of the circular wires is .....$\times 10^{-8}$ T.
(Take, $\pi = 3.14$) *[2023, 6 April Shift-I]*

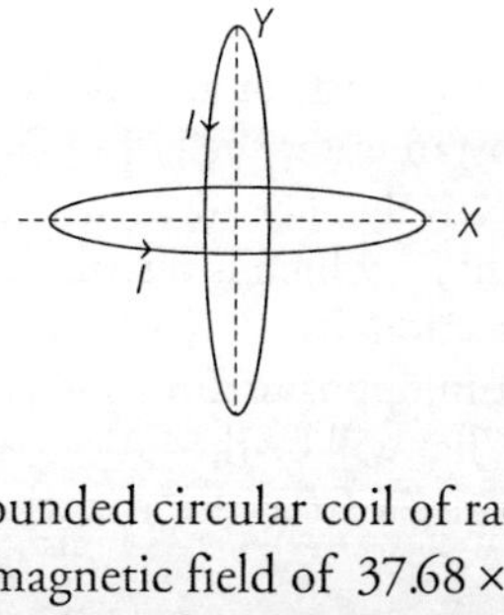

64 A closely wounded circular coil of radius 5 cm produces a magnetic field of 37.68×10^{-4} T at its centre. The current through the coil is .... A.
(Given, number of turns in the coil is 100 and $\pi = 3.14$) *[2022, 29 July Shift-I]*

Topic 2 Magnetic Force and Motion of Charged Particle in Magnetic Field

Level 1

01 Given below are two statements. One is labelled as Assertion (A) and the other is labelled as Reason (R)

Assertion (A) A electron in a certain region of uniform magnetic field is moving with constant velocity in a straight line path.

Reason (R) The magnetic field in that region is along the direction of velocity of the electron.

In the light of the above statements, choose the correct answer from the options given below. *[2025, 24 Jan Shift-II]*

(a) A is true but R is false.
(b) Both A and R are true and R is the correct explanation of A.
(c) Both A and R are true but R is not the correct explanation of A.
(d) A is false but R is true.

02 The electrostatic force $\mathbf{F}_1$ and magnetic force $\mathbf{F}_2$ acting on a charge q moving with velocity v can be written. *[2024, 5 April Shift-II]*

(a) $\mathbf{F}_1 = q\mathbf{E}, \mathbf{F}_2 = q(\mathbf{B}\times\mathbf{v})$
(b) $\mathbf{F}_1 = q\mathbf{v}, \mathbf{E}\cdot\mathbf{F}_2 = q(\mathbf{B}\cdot\mathbf{v})$
(c) $\mathbf{F}_1 = q\mathbf{E}, \mathbf{F}_2 = q(\mathbf{v}\times\mathbf{B})$
(d) $\mathbf{F}_1 = q\mathbf{B}, \mathbf{F}_2 = q(\mathbf{B}\times\mathbf{v})$

03 An electron is allowed to move with constant velocity along the axis of current carrying straight solenoid. *[2023, 11 April Shift-II]*

A. The electron will experience magnetic force along the axis of the solenoid.
B. The electron will not experience magnetic force.
C. The electron will continue to move along the axis of the solenoid.
D. The electron will be accelerated along the axis of the solenoid.
E. The electron will follow parabolic path-inside the solenoid.

Choose the correct answer from the options given below.

(a) B and C only (b) A and D only
(c) B and E only (d) B, C and D only

Imp. Force on Charged Particle in Magnetic Field

04 A charge particle is moving in a uniform magnetic field $(2\hat{i}+3\hat{j})$ T. If it has an acceleration of $(\alpha\hat{i}-4\hat{j})$ m/s^2, then the value of α will be *[2022, 26 July Shift-I]*

(a) 3 (b) 6
(c) 12 (d) 2

05 An electron with energy 0.1 keV moves at right angle to the earth's magnetic field of 1×10^{-4} Wbm^{-2}. The frequency of revolution of the electron will be (Take, mass of electron $=9.0\times10^{-31}$ kg) *[2022, 25 July Shift-II]*

(a) 1.6×10^5 Hz (b) 5.6×10^5 Hz
(c) 2.8×10^6 Hz (d) 1.8×10^6 Hz

06 A proton and an alpha particle of the same velocity enter in a uniform magnetic field which is acting perpendicular to their direction of motion. The ratio of the radii of the circular paths described by the alpha particle and proton is *[2022, 26 June Shift-I]*

(a) 1 : 4 (b) 4 : 1
(c) 2 : 1 (d) 1 : 2

07 Two ions having same mass have charges in the ratio 1 : 2. They are projected normally in a uniform magnetic field with their speeds in the ratio 2 : 3. The ratio of the radii of their circular trajectories is *[2021, 25 July Shift-II]*

(a) 1 : 4 (b) 4 : 3 (c) 3 : 1 (d) 2 : 3

08 A negative test charge is moving near a long straight wire carrying a current. The force acting on the test charge is parallel to the direction of the current. The motion of the charge is: *[JEE Main 2017]*

(a) away from the wire
(b) towards the wire
(c) parallel to the wire along the current
(d) parallel to the wire opposite to the current

09 A proton (mass m) accelerated by a potential difference V flies through a uniform transverse magnetic field B. The field occupies a region of space by width d. If α be the angle of deviation of proton from initial direction of motion (see figure), the value of $\sin\alpha$ will be *[JEE Main 2015]*

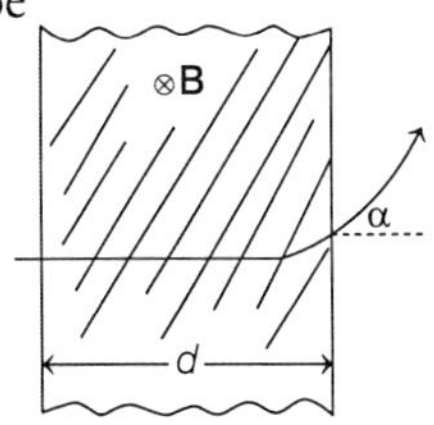

(a) $qV\sqrt{\dfrac{Bd}{2m}}$ (b) $\dfrac{B}{2}\sqrt{\dfrac{qd}{mV}}$

(c) $\dfrac{B}{d}\sqrt{\dfrac{q}{2mV}}$ (d) $Bd\sqrt{\dfrac{q}{2mV}}$

10 An electric charge $+q$ moves with velocity $\mathbf{v}=3\,\mathbf{i}+4\,\mathbf{j}+\mathbf{k}$, in an electromagnetic field given by $\mathbf{E}=3\,\mathbf{i}+\mathbf{j}+2\,\mathbf{k}$, $\mathbf{B}=\mathbf{i}+\mathbf{j}-3\,\mathbf{k}$. The y component of the force experienced by $+q$ is *[AIEEE 2011]*

(a) $2\,q$ (b) $11\,q$ (c) $5\,q$ (d) $3\,q$

11 A charged particle with charge q enters a region of constant, uniform and mutually orthogonal fields E and B with a velocity v perpendicular to both E and B and comes out without any change in magnitude or direction of v. Then, *[AIEEE 2007]*

(a) $\mathbf{v} = \mathbf{E} \times \frac{\mathbf{B}}{B^2}$
(b) $\mathbf{v} = \mathbf{B} \times \frac{\mathbf{E}}{B^2}$
(c) $\mathbf{v} = \mathbf{E} \times \frac{\mathbf{B}}{E^2}$
(d) $\mathbf{v} = \mathbf{B} \times \frac{\mathbf{E}}{E^2}$

12 A charged particle moves through a magnetic field perpendicular to its direction. Then, *[AIEEE 2007]*

(a) the momentum changes but the kinetic energy is constant
(b) both momentum and kinetic energy of the particle are not constant
(c) both momentum and kinetic energy of the particle are constant
(d) kinetic energy changes but the momentum is constant

13 In a region, steady and uniform electric and magnetic fields are present. These two fields are parallel to each other. A charged particle is released from rest in this region. The path of the particle will be a *[AIEEE 2006]*

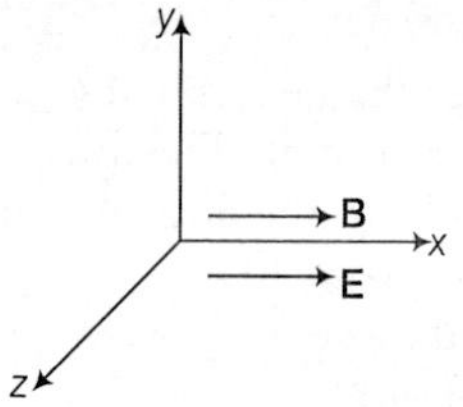

(a) helix
(b) straight line
(c) ellipse
(d) circle

14 A charged particle of mass m and charge q travels on a circular path of radius r that is perpendicular to a magnetic field B. The time taken by the particle to complete one revolution is *[AIEEE 2005]*

(a) $\frac{2\pi mq}{B}$
(b) $\frac{2\pi q^2 B}{m}$
(c) $\frac{2\pi qB}{m}$
(d) $\frac{2\pi m}{qB}$

15 A uniform electric field and a uniform magnetic field are acting along the same direction in a certain region. If an electron is projected along the direction of the fields with a certain velocity, then *[AIEEE 2005]*

(a) its velocity will decrease
(b) its velocity will increase
(c) it will turn towards right of direction of motion
(d) it will turn towards left of direction of motion

16 If an electron and a proton having same momenta enter perpendicularly to a magnetic field, then *[AIEEE 2002]*

(a) curved path of electron and proton will be same (ignoring the sense of revolution)
(b) they will move undeflected
(c) curved path of electron is more curved than that of proton
(d) path of proton is more curved

17 The time period of a charged particle undergoing a circular motion in a uniform magnetic field is independent of its *[AIEEE 2002]*

(a) speed
(b) mass
(c) charge
(d) magnetic induction

Level 2

18 Given below are two statements. one is labelled as Assertion (A) and the other is labelled as Reason (R).

Assertion (A) If Oxygen ion (O^{-2}) and Hydrogen ion (H^+) enter normal to the magnetic field with equal momentum, then the path of O^{-2} ion has a smaller curvature than that of H^+.

Reason (R) A proton with same linear momentum as an electron will form a path of smaller radius of curvature on entering a uniform magneuc field perpendicularily. *[2025, 3 April Shift-II]*

In the light of the above statements, choose the correct answer from the options given below.

(a) Both A and R are true and R is the correct explanation of A.
(b) A is false but R is true.
(c) Both A and R are true but R is not the correct explanation of A.
(d) A is true but R is false.

19 A particle of charge q, mass m and kinetic energy E enters in magnetic field perpendicular to its velocity and undergoes a circular arc of radius (r). Which of the following curves represents the variation of r with E? *[2025, 7 April Shift-I]*

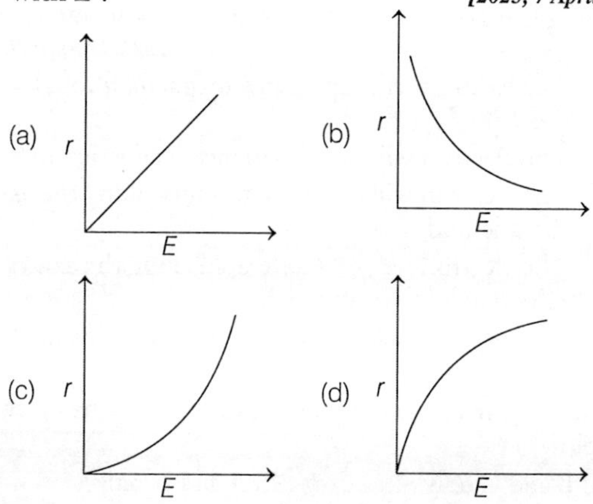

Imp. Motion of Charge Particle in Uniform Magnetic Field

20 A proton and a deutron ($q = +e$, $m = 2.0$ u) having same kinetic energies enter a region of uniform magnetic field **B**, moving perpendicular to **B**. The ratio of the radius r_d of deutron path to the radius r_p of the proton path is *[2024, 9 April Shift-II]*

(a) 1 : 2
(b) 1 : 1
(c) $1 : \sqrt{2}$
(d) $\sqrt{2} : 1$

21 Two particles X and Y having equal charges are being accelerated through the same potential difference. Thereafter they enter normally in a region of uniform magnetic field and describes circular paths of radii R_1 and R_2 respectively. The mass ratio of X and Y is *[2024, 29 Jan Shift-II]*

(a) $\left(\frac{R_2}{R_1}\right)$ (b) $\left(\frac{R_1}{R_2}\right)^2$ (c) $\left(\frac{R_2}{R_1}\right)^2$ (d) $\left(\frac{R_1}{R_2}\right)$

22 A proton moving with a constant velocity passes through a region of space without any change in its velocity. If **E** and **B** represent the electric and magnetic fields respectively, then the region of space may have

A. $E = 0, B = 0$ B. $E = 0, B \neq 0$
C. $E \neq 0, B = 0$ D. $E \neq 0, B \neq 0$

Choose the most appropriate answer from the options given below. *[2024, 27 Jan Shift-I]*

(a) A, B and D only (b) A, C and D only
(c) A, B and C only (d) B, C and D only

23 A beam of light travelling along X-axis is described by the electric field $E_y = 900 \sin \omega(t - x/c)$. The ratio of electric force to magnetic force on a charge q moving along Y-axis with a speed of 3×10^7 ms^{-1} will be [Given, speed of light $= 3 \times 10^8$ ms^{-1}) *[2022, 27 July Shift-I]*

(a) 1 : 1 (b) 1 : 10
(c) 10 : 1 (d) 1 : 2

24 Given below are two statements.

Statement I The electric force changes the speed of the charged particle and hence changes its kinetic energy, whereas the magnetic force does not change the kinetic energy of the charged particle.

Statement II The electric force accelerates the positively charged particle perpendicular to the direction of electric field. The magnetic force accelerates the moving charged particle along the direction of magnetic field.

In the light of the above statements, choose the most appropriate answer from the options given below. *[2022, 29 June Shift-II]*

(a) Both Statement I and Statement II are correct.
(b) Both Statement I and Statement II are incorrect.
(c) Statement I is correct but statement II is incorrect.
(d) Statement I is incorrect but statement II is correct.

25 A charged particle moves along circular path in a uniform magnetic field in a cyclotron. The kinetic energy of the charged particle increases to 4 times its initial value. What will be the ratio of new radius to the original radius of circular path of the charged particle *[2022, 29 June Shift-I]*

(a) 1 : 1 (b) 1 : 2
(c) 2 : 1 (d) 1 : 4

26 Given below are two statements, one is labelled as Assertion (A) and the other is labelled as Reason (R).

Assertion (A) In a uniform magnetic field, speed and energy remains the same for a moving charged particle.

Reason (R) Moving charged particle experiences magnetic force perpendicular to its direction of motion. *[2022, 24 June Shift-I]*

(a) Both A and R are true and R is the correct explanation of A.
(b) Both A and R are true but R is not the correct explanation of A.
(c) A is true but R is false.
(d) A is false but R is true.

27 A deuteron and an α-particle having equal kinetic energy enter perpendicular into a magnetic field. Let r_d and r_α be their respective radii of circular path. The value of $\frac{r_d}{r_\alpha}$ is equal to *[2021, 20 July Shift-I]*

(a) $1/\sqrt{2}$ (b) $\sqrt{2}$ (c) 1 (d) 2

28 A proton and an α-particle, having kinetic energies K_p and K_α, respectively, enter into a magnetic field at right angles. The ratio of the radii of trajectory of proton to that of α-particle is 2 : 1. The ratio of $K_p : K_\alpha$ is *[2021, 18 March Shift-II]*

(a) 1 : 8 (b) 8 : 1
(c) 1 : 4 (d) 4 : 1

29 A particle of charge q and mass m is moving with a velocity $-v\hat{\mathbf{i}}(v \neq 0)$ towards a large screen placed in the YZ-plane at a distance d. If there is a magnetic field $\mathbf{B} = B_0\hat{\mathbf{k}}$, the minimum value of v for which the particle will not hit the screen is *[2020, 6 Sep Shift-I]*

(a) $\frac{qdB_0}{3m}$ (b) $\frac{2qdB_0}{m}$ (c) $\frac{qdB_0}{m}$ (d) $\frac{qdB_0}{2m}$

30 A particle of mass m and charge q has an initial velocity $\mathbf{v} = v_0\hat{\mathbf{j}}$. If an electric field $\mathbf{E} = E_0\hat{\mathbf{i}}$ and magnetic field $\mathbf{B} = B_0\hat{\mathbf{i}}$ act on the particle, its speed will double after a time *[2020, 7 Jan Shift-II]*

(a) $\frac{3mv_0}{qE_0}$ (b) $\frac{\sqrt{3}mv_0}{qE_0}$ (c) $\frac{2mv_0}{qE_0}$ (d) $\frac{\sqrt{2}mv_0}{qE_0}$

31 A proton and an α-particle (with their masses in the ratio of 1 : 4 and charges in the ratio of 1 : 2) are accelerated from rest through a potential difference V. If a uniform magnetic field B is set up perpendicular to their velocities, the ratio of the radii $r_p : r_\alpha$ of the circular paths described by them will be *[2019, 12 Jan Shift-I]*

(a) $1 : \sqrt{2}$ (b) $1 : \sqrt{3}$
(c) 1 : 3 (d) 1 : 2

32 A particle of mass m and charge q is in an electric and magnetic field is given by $\mathbf{E}=2\hat{\mathbf{i}}+3\hat{\mathbf{j}}, \mathbf{B}=4\hat{\mathbf{j}}+6\hat{\mathbf{k}}$. The charged particle is shifted from the origin to the point $P(x=1, y=1)$ along a straight path. The magnitude of the total work done is *[2019, 11 Jan Shift-II]*

(a) $(0.35)\,q$ (b) $(0.15)\,q$
(c) $(2.5)\,q$ (d) $5\,q$

Imp. Radius of Circular Path of Charged Particles in Magnetic Field

33 An electron, a proton and an alpha particle having the same kinetic energy are moving in circular orbits of radii r_e, r_p, r_α respectively, in a uniform magnetic field B. The relation between r_e, r_p, r_α is *[JEE Main 2018]*

(a) $r_e > r_p = r_\alpha$ (b) $r_e < r_p = r_\alpha$
(c) $r_e < r_p < r_\alpha$ (d) $r_e < r_\alpha < r_p$

34 Proton, deuteron and alpha particles of same kinetic energy are moving in circular trajectories in a constant magnetic field. The radii of proton, deuteron and alpha particle are respectively, r_p, r_d and r_α. Which one of the following relation is correct? *[AIEEE 2012]*

(a) $r_\alpha = r_p = r_d$ (b) $r_\alpha = r_p < r_d$
(c) $r_\alpha > r_d > r_p$ (d) $r_\alpha = r_d > r_p$

35 A particle of charge -16×10^{-18} C moving with velocity 10 ms^{-1} along the X-axis enters a region where a magnetic field of induction B is along the Y-axis and an electric field of magnitude 10^4 Vm^{-1} is along the negative Z-axis. If the charged particle continues moving along the X-axis, the magnitude of B is *[AIEEE 2003]*

(a) 10^3 Wb/m^2 (b) 10^5 Wb/m^2
(c) 10^{16} Wb/m^2 (d) 10^{-3} Wb/m^2

Level 3

36 Uniform magnetic fields of different strengths (B_1 and B_2) both normal to the plane of the paper exist as shown in the figure. A charged particle of mass m and charge q, at the interface at an instant, moves into the region 2 with velocity v and returns to the interface. It continues to move into region 1 and finally reaches the interface. What is the displacement of the particle during this movement along the interface?

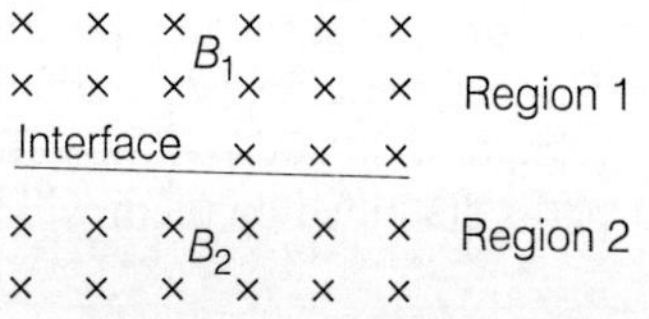

(Consider the velocity of the particle to be normal to the magnetic field and $B_2 > B_1$) *[2025, 7 April Shift-I]*

(a) $\dfrac{mv}{qB_1}\left(1-\dfrac{B_1}{B_2}\right)$ (b) $\dfrac{mv}{qB_1}\left(1-\dfrac{B_2}{B_1}\right)\times 2$
(c) $\dfrac{mv}{qB_1}\left(1-\dfrac{B_1}{B_2}\right)\times 2$ (d) $\dfrac{mv}{qB_1}\left(1-\dfrac{B_2}{B_1}\right)$

37 Consider a long thin conducting wire carrying a uniform current I. A particle having mass M and charge q is released as a distance a from the wire with a speed v_0 along the direction of current in the wire. The particle gets attracted to the wire due to magnetic force. The particle turns round when it is at distance x from the wire. The value of x is [μ_0 is vacuum permeability] *[2025, 28 Jan Shift-I]*

(a) $a\left[1-\dfrac{mv_0}{q\mu_0 I}\right]$ (b) $\dfrac{a}{2}$
(c) $ae^{-\frac{4\pi m v_0}{q\mu_0 I}}$ (d) $a\left[1-\dfrac{mv_0}{2q\mu_0 I}\right]$

38 A velocity selector consists of electric field $\mathbf{E}=E\,\hat{\mathbf{k}}$ and magnetic field $\mathbf{B}=B\,\hat{\mathbf{j}}$ with $B=12$ mT. The value of E required for an electron of energy 728 eV moving along the positive X-axis to pass undeflected is (Given, mass of electron $=9.1\times10^{-31}$ kg) *[2022, 26 July Shift-II]*

(a) 192 kVm^{-1} (b) 192 mVm^{-1}
(c) 9600 kVm^{-1} (d) 16 kVm^{-1}

39 Two charged particles, having same kinetic energy, are allowed to pass through a uniform magnetic field perpendicular to the direction of motion. If the ratio of radii of their circular path is 6 : 5 and their respective masses ratio is 9 : 4. Then, the ratio of their charges will be *[2022, 25 July Shift-I]*

(a) 8 : 5 (b) 5 : 4
(c) 5 : 3 (d) 8 : 7

40 A proton, a deuteron and an α-particle with same kinetic energy enter into a uniform magnetic field at right angle to magnetic field. The ratio of the radii of their respective circular paths is *[2022, 24 June Shift-II]*

(a) $1:\sqrt{2}:\sqrt{2}$ (b) $1:1:\sqrt{2}$
(c) $\sqrt{2}:1:1$ (d) $1:\sqrt{2}:1$

41 Two ions of masses 4 amu and 16 amu have charges $+2e$ and $+3e$, respectively. These ions pass through the region of constant perpendicular magnetic field. The kinetic energy of both ions is same. Then, *[2021, 27 Aug Shift-I]*

(a) lighter ion will be deflected less than heavier ion
(b) lighter ion will be deflected more than heavier ion
(c) both ions will be deflected equally
(d) no ion will be deflected

42 A proton, a deuteron and an α-particle are moving with same momentum in a uniform magnetic field. The ratio of magnetic forces acting on them is ...... and their speed is ...... in the ratio. *[2021, 25 Feb Shift-I]*
(a) 1 : 2 : 4 and 2 : 1 : 1
(b) 2 : 1 : 1 and 4 : 2 : 1
(c) 4 : 2 : 1 and 2 : 1 : 1
(d) 1 : 2 : 4 and 1 : 1 : 2

43 A proton, an electron and a helium nucleus, have the same energy. They are in circular orbits in a plane due to magnetic field perpendicular to the plane. Let r_p, r_e and r_{He} be their respective radii, then *[2019, 10 April Shift-I]*
(a) $r_e < r_p = r_{He}$ (b) $r_e > r_p = r_{He}$
(c) $r_e < r_p < r_{He}$ (d) $r_e > r_p > r_{He}$

44 The region between $y = 0$ and $y = d$ contains a magnetic field $\mathbf{B} = B\hat{\mathbf{k}}$. A particle of mass m and charge q enters the region with a velocity $\mathbf{v} = v\hat{\mathbf{i}}$. If $d = \frac{mv}{2qB}$ then the acceleration of the charged particle at the point of its emergence at the other side is *[2019, 11 Jan Shift-II]*
(a) $\frac{qvB}{m}\left(\frac{\sqrt{3}}{2}\hat{\mathbf{i}} + \frac{1}{2}\hat{\mathbf{j}}\right)$ (b) $\frac{qvB}{m}\left(\frac{1}{2}\hat{\mathbf{i}} - \frac{\sqrt{3}}{2}\hat{\mathbf{j}}\right)$
(c) $\frac{qvB}{m}\left(\frac{-\hat{\mathbf{j}} + \hat{\mathbf{i}}}{\sqrt{2}}\right)$ (d) $\frac{qvB}{m}\left(\frac{\hat{\mathbf{i}} + \hat{\mathbf{j}}}{\sqrt{2}}\right)$

45 A particle having the same charge as of electron moves in a circular path of radius 0.5 cm under the influence of a magnetic field of 0.5 T. If an electric field of 100 V/m makes it to move in a straight path, then the mass of the particle is (Take, charge of electron = 1.6×10^{-19} C) *[2019, 9 Jan Shift-II]*
(a) 1.6×10^{-19} kg
(b) 1.6×10^{-27} kg
(c) 9.1×10^{-31} kg
(d) 2.0×10^{-24} kg

Numerical Value Type Questions

46 A particle of charge 1.6 μC and mass 16 μg is present in a strong magnetic field of 6.28 T. The particle is then fired perpendicular to magnetic field. The time required for the particle to return to original location for the first time is .... s. ($\pi = 3.14$) *[2025, 4 April Shift-II]*

47 A lightly wound long solenoid carries a current of 1.5 A. An electron is executing uniform circular motion inside the solenoid with a time period of 75 ns. The number of turns per metre in the solenoid is ....
[Take mass of electron $m_e = 9 \times 10^{-31}$ kg, charge of electron $|q_e| = 1.6 \times 10^{-19}$ C,
$\mu_0 = 4\pi \times 10^{-7} \frac{N}{A^2}$, $1\,ns = 10^{-9}$ s] *[2025, 24 Jan Shift-II]*

48 A proton is moving undeflected in a region of crossed electric and magnetic fields at a constant speed of 2×10^5 ms^{-1}. When the electric field is switched off, the proton moves along a circular path of radius 2 cm. The magnitude of electric field is $x \times 10^4$ N/C. The value of x is ....
Take the mass of the proton = 1.6×10^{-27} kg. *[2025, 22 Jan Shift-II]*

49 An electron moves through a uniform magnetic field $\mathbf{B} = B_0\hat{\mathbf{i}} + 2B_0\hat{\mathbf{j}}$ T. At a particular instant of time, the velocity of electron is $\mathbf{u} = 3\hat{\mathbf{i}} + 5\hat{\mathbf{j}}$ m/s. If the magnetic force acting on electron is $F = 5e\hat{\mathbf{k}}$ N, where e is the charge of electron, then the value of B_0 is .... T. *[2024, 31 Jan Shift-I]*

50 A charge of 4.0 μC is moving with a velocity of 4.0×10^{-6} ms^{-1} along the positive Y-axis under a magnetic field $\mathbf{B}$ of strength $(2\hat{\mathbf{k}})$ T. The force acting on the charge is $x\hat{\mathbf{i}}$ N. The value of x is .... *[2024, 29 Jan Shift-II]*

51 A proton with a kinetic energy of 2.0 eV moves into a region of uniform magnetic field of magnitude $\frac{\pi}{2} \times 10^{-3}$ T. The angle between the direction of magnetic field and velocity of proton is 60°. The pitch of the helical path taken by the proton is ...... cm. (Take, mass of proton = 1.6×10^{-27} kg and charge on proton = 1.6×10^{-19} C). *[2023, 6 April Shift-II]*

52 A charge particle of 2 μC accelerated by a potential difference of 100 V enters a region of uniform magnetic field of magnitude 4 mT at right angle to the direction of field. The charge particle completes semicircle of radius 3 cm inside magnetic field. The mass of the charge particle is .... $\times 10^{-18}$ kg. *[2023, 1 Feb Shift-I]*

53 A singly ionised magnesium atom ($A = 24$) ion is accelerated to kinetic energy 5 keV and is projected perpendicularly into a magnetic field B of the magnitude 0.5 T. The radius of path formed will be .... cm. *[2022, 28 June Shift-I]*

54 A deuteron and a proton moving with equal kinetic energy enter into to a uniform magnetic field at right angle to the field. If r_d and r_p are the radii of their circular paths respectively, then the ratio r_d / r_p will be $\sqrt{x} : 1$ where x is .... *[2022, 27 June Shift-II]*

Topic 3 Force and Torque on Current Carrying Conductor

Level 1

01 A uniform magnetic field of 2×10^{-3} T acts along positive y-direction. A rectangular loop of sides 20 cm and 10 cm with current of 5A is in YZ-plane. The current is in anti-clockwise sense with reference to negative X-axis. Magnitude and direction of the torque is *[2024. 31 Jan Shift-II]*

(a) 2×10^{-4} N-m along positive z-direction
(b) 2×10^{-4} N-m along negative z-direction
(c) 2×10^{-4} N-m along positive x-direction
(d) 2×10^{-4} N-m along positive y-direction

02 A wire X of length 50 cm carrying a current of 2 A is placed parallel to a long wire Y of length 5 m. The wire Y carries a current of 3 A. The distance between two wires is 5 cm and currents flow in the same direction. The force acting on the wire Y is *[2022, 29 July Shift-II]*

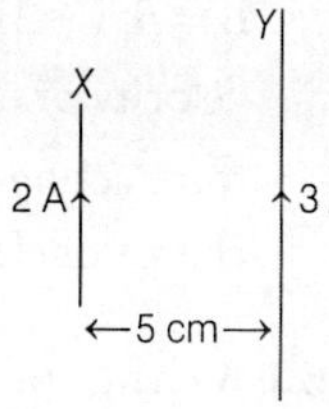

(a) 1.2×10^{-5} N directed towards wire X
(b) 1.2×10^{-4} N directed away from wire X
(c) 1.2×10^{-4} N directed towards wire X
(d) 2.4×10^{-5} N directed towards wire X

03 As shown in the figure, a metallic rod of linear density 0.45 kg m^{-1} is lying horizontally on a smooth incline plane which makes an angle of 45° with the horizontal. The minimum current flowing in the rod required to keep it stationary, when 0.15 T magnetic field is acting on it in the vertical upward direction, will be (Use, $g = 10$ m/s^2) *[2022, 28 July Shift-I]*

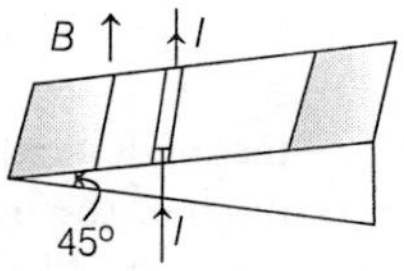

(a) 30 A (b) 15 A (c) 10 A (d) 3 A

Imp. Force Between Two Parallel Current Carrying Conductor

04 Two long current carrying conductors are placed parallel to each other at a distance of 8 cm between them. The magnitude of magnetic field produced at mid-point between the two conductors due to current flowing in them is 300 μT. The equal current flowing in the two conductors is *[2022, 29 June Shift-II]*

(a) 30A in the same direction.
(b) 30A in the opposite direction.
(c) 60A in the opposite direction.
(d) 300A in the opposite direction.

05 Two parallel, long wires are kept 0.20 m apart in vacuum, each carrying current of x A in the same direction. If the force of attraction per metre of each wire is 2×10^{-6} N, then the value of x is approximately *[2022, 28 June Shift-II]*

(a) 1 (b) 2.4 (c) 1.4 (d) 2

06 A loop of flexible wire of irregular shape carrying current is placed in an external magnetic field. Identify the effect of the field on the wire. *[2021, 18 March Shift-I]*

(a) Loop assumes circular shape with its plane normal to the field.
(b) Loop assumes circular shape with its plane parallel to the field.
(c) Wire gets stretched to become straight.
(d) Shape of the loop remains unchanged.

07 A circular coil having N turns and radius r carries a current I. It is held in the XZ-plane in a magnetic field $B\hat{\mathbf{i}}$. The torque on the coil due to the magnetic field (in N-m) is *[2019, 8 April Shift-I]*

(a) $\dfrac{Br^2 I}{\pi N}$ (b) $B\pi r^2 IN$ (c) $\dfrac{B\pi r^2 I}{N}$ (d) zero

08 Two coaxial solenoids of different radii carry current I in the same direction. Let F_1 be the magnetic force on the inner solenoid due to the outer one and F_2 be the magnetic force on the outer solenoid due to the inner one. Then, *[JEE Main 2015]*

(a) $F_1 = F_2 = 0$
(b) F_1 is radially inwards and F_2 is radially outwards
(c) F_1 is radially inwards and $F_2 = 0$
(d) F_1 is radially outwards and $F_2 = 0$

09 A rectangular loop of sides 10 cm and 5 cm carrying a current I of 12 A is placed in different orientations as shown in the figures below.

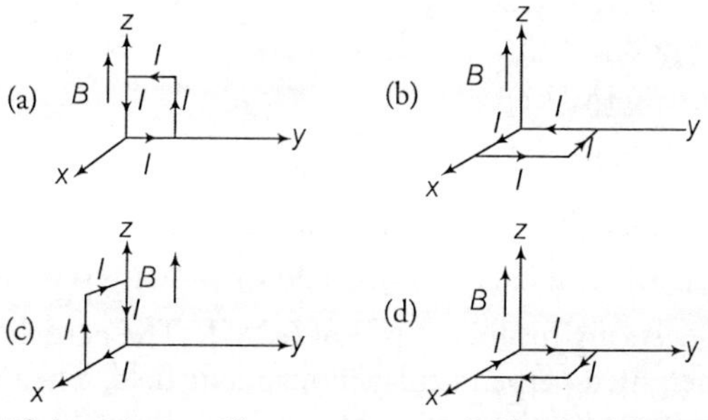

If there is a uniform magnetic field of 0.3 T in the positive z-direction in which orientations the loop would be in (i) stable equilibrium and (ii) unstable equilibrium ? *[JEE Main 2015]*

(a) (a) and (b) respectively
(b) (a) and (c) respectively
(c) (b) and (d) respectively
(d) (b) and (c) respectively

10 Due to the presence of the current I_1 at the origin in the figure, *[AIEEE 2009]*

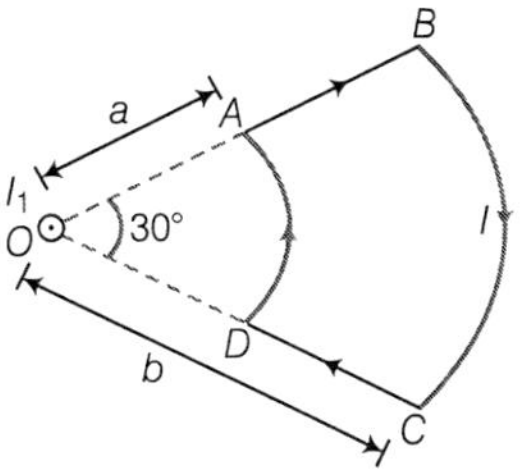

(a) the forces on AB and DC are zero
(b) the forces on AD and BC are zero
(c) the magnitude of the net force on the loop is given by
$$\frac{\mu_0 I I_1}{4\pi}\left[2(b-a)+\frac{\pi}{3}(a+b)\right]$$
(d) the magnitude of the net force on the loop is given by
$$\frac{\mu_0 I I_1}{24ab}(b-a)$$

11 Two thin, long, parallel wires, separated by a distance d carry a current of I ampere in the same direction. They will *[AIEEE 2005]*

(a) attract each other with a force of $\frac{\mu_0 I^2}{(2\pi d)}$

(b) repel each other with a force of $\frac{\mu_0 I^2}{(2\pi d)}$

(c) attract each other with a force of $\frac{\mu_0 I^2}{(2\pi d^2)}$

(d) repel each other with a force of $\frac{\mu_0 I^2}{(2\pi d^2)}$

12 If a current is passed through a spring, then the spring will *[AIEEE 2002]*

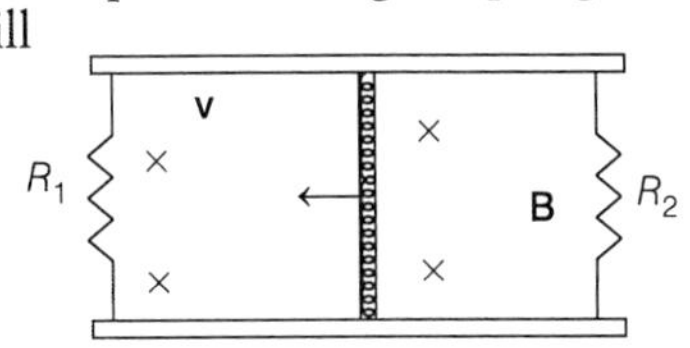

(a) expand (b) compress
(c) remain same (d) None of these

13 Wires 1 and 2 carrying currents I_1 and I_2, respectively are inclined at an angle θ to each other. What is the force on a small element dl of wire 2 at a distance r from wire 1 (as shown in figure) due to the magnetic field of wire 1? *[AIEEE 2002]*

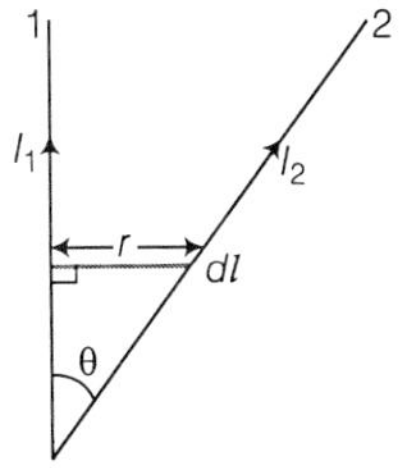

(a) $\frac{\mu_0}{2\pi r} I_1 I_2 \, dl \tan\theta$ (b) $\frac{\mu_0}{2\pi r} I_1 I_2 \, dl \sin\theta$

(c) $\frac{\mu_0}{2\pi r} I_1 I_2 \, dl \cos\theta$ (d) $\frac{\mu_0}{4\pi r} I_1 I_2 \, dl \sin\theta$

Level 2

14 A rigid wire consists of a semicircular portion of radius R and two straight sections. The wire is partially immersed in a perpendicular magnetic field $B = B_0\hat{j}$ as shown in figure. The magnetic force on the wire if it has a current i is *[2024, 31 Jan Shift-I]*

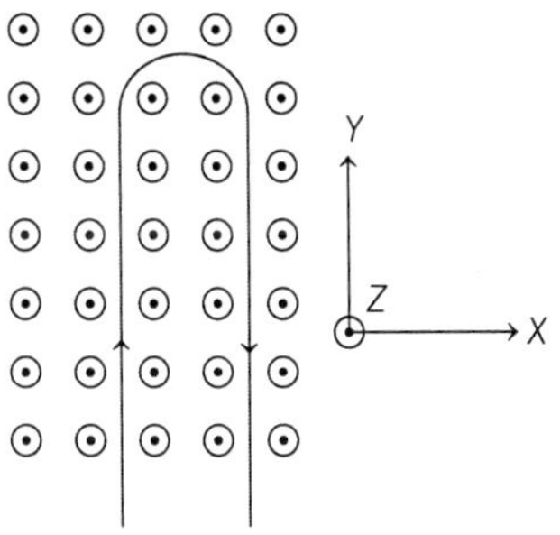

(a) $iBR\hat{j}$ (b) $-iBR\hat{j}$ (c) $2iBR\hat{j}$ (d) $-2iBR\hat{j}$

15 Two long straight wires P and Q carrying equal current 10 A each were kept parallel to each other at 4 cm distance. Magnitude of magnetic force experienced by 10 cm length of wire P is F_1. If distance between wires is halved and currents on them are doubled, force F_2 on 10 cm length of wire P will be *[2023, 1 Feb Shift-II]*

(a) $8F_1$ (b) $10F_1$ (c) $F_1/10$ (d) $F_1/8$

16 A triangular shaped wire carrying 10 A current is placed in a uniform magnetic field of 0.5 T, as shown in figure. The magnetic force on segment CD is (Given, $BC = CD = BD = 5$ cm). *[2022, 28 July Shift-II]*

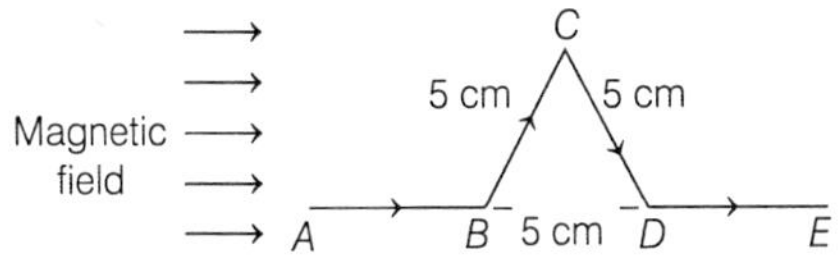

(a) 0.126 N (b) 0.312 N (c) 0.216 N (d) 0.245 N

17 A square loop of side $2a$ and carrying current I is kept in xy-plane with its centre at origin. A long wire carrying the same current I is placed parallel to the Z-axis and passing through the point $(0, b, 0)$, $(b >> a)$. The magnitude of the torque on the loop about Z-axis is given by *[2020, 5 Sep Shift-I]*

(a) $\frac{2\mu_0 I^2 a^3}{\pi b^2}$ (b) $\frac{2\mu_0 I^2 a^2}{\pi b}$

(c) $\frac{\mu_0 I^2 a^3}{2\pi b^2}$ (d) $\frac{\mu_0 I^2 a^2}{2\pi b}$

18 A wire carrying current I is bent in the shape $ABCDEFA$ as shown, where rectangle $ABCDA$ and $ADEFA$ are perpendicular to each other. If the sides of the rectangles are of lengths a and b, then the magnitude and direction of magnetic moment of the loop $ABCDEFA$ is *[2020, 2 Sep Shift-II]*

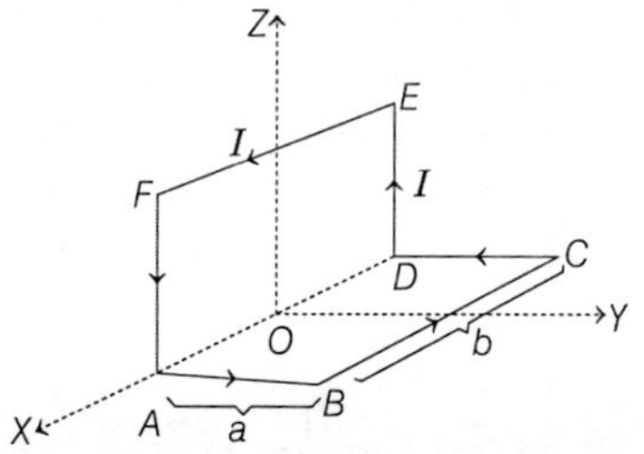

(a) abI, along $\left(\frac{\hat{j}}{\sqrt{2}}+\frac{\hat{k}}{\sqrt{2}}\right)$ (b) $\sqrt{2}\,abI$, along $\left(\frac{\hat{j}}{\sqrt{2}}+\frac{\hat{k}}{\sqrt{2}}\right)$

(c) $\sqrt{2}\,abI$, along $\left(\frac{\hat{j}}{\sqrt{5}}+\frac{2\hat{k}}{\sqrt{5}}\right)$ (d) abI, along $\left(\frac{\hat{j}}{\sqrt{5}}+\frac{2\hat{k}}{\sqrt{5}}\right)$

19 A square loop is carrying a steady current I and the magnitude of its magnetic dipole moment is m. If this square loop is changed to a circular loop and it carries the same current, the magnitude of the magnetic dipole moment of circular loop (in A-m) will be *[2019, 10 April Shift-II]*

(a) $\frac{4m}{\pi}$ (b) $\frac{3m}{\pi}$ (c) $\frac{2m}{\pi}$ (d) $\frac{m}{\pi}$

Imp. Torque on current Loop in Magnetic Field

20 A rectangular coil (dimension 5 cm × 2.5 cm) with 100 turns, carrying a current of 3A in the clockwise direction, is kept centred at the origin and in the X-Z plane. A magnetic field of 1 T is applied along X-axis. If the coil is tilted through 45° about Z-axis, then the torque on the coil is *[2019, 9 April Shift-I]*

(a) 0.27 N-m (b) 0.38 N-m
(c) 0.42 N-m (d) 0.55 N-m

21 The dipole moment of a circular loop carrying a current I, is m and the magnetic field at the centre of the loop is B_1. When the dipole moment is doubled by keeping the current constant, the magnetic field at the centre of the loop is B_2. The ratio B_1/B_2 is *[JEE Main 2018]*

(a) 2 (b) $\sqrt{3}$
(c) $\sqrt{2}$ (d) $1/\sqrt{2}$

22 Two long current carrying thin wires, both with current I, are held by insulating threads of length L and are in equilibrium as shown in the figure, with threads making an angle θ with the vertical. If wires have mass λ per unit length then, the value of I is (g = gravitational acceleration) *[JEE Main 2015]*

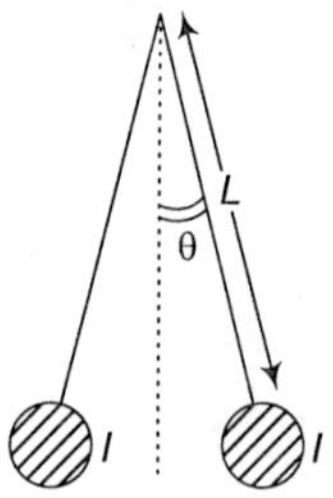

(a) $\sin\theta\sqrt{\frac{\pi\lambda gL}{\mu_0\cos\theta}}$ (b) $2\sin\theta\sqrt{\frac{\pi\lambda gL}{\mu_0\cos\theta}}$

(c) $2\sqrt{\frac{\pi gL}{\mu_0}\tan\theta}$ (d) $\sqrt{\frac{\pi\lambda gL}{\mu_0}\tan\theta}$

23 A conductor lies along the Z-axis at $-1.5 \le z < 1.5$ m and carries a fixed current of 10.0 A in $-a_z$ direction (see figure). For a field $\mathbf{B} = 3.0\times10^{-4}\,e^{-0.2x}\,a_y$ T, find the power required to move the conductor at constant speed to $x = 2.0$ m, $y = -0$ in 5×10^{-3} s. Assume parallel motion along the X-axis. *[JEE Main 2014]*

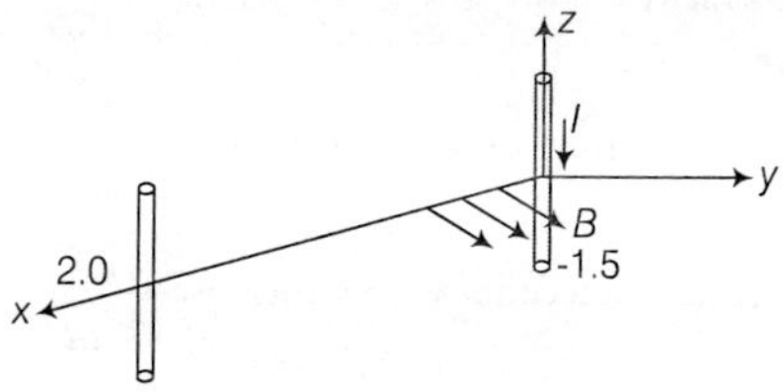

(a) 1.57 W (b) 2.97 W (c) 14.85W (d) 29.7 W

24 A thin circular disk of radius R is uniformly charged with density $\sigma > 0$ per unit area. The disk rotates about its axis with a uniform angular speed ω. The magnetic moment of the disk is *[AIEEE 2011]*

(a) $2\pi R^4\sigma\omega$ (b) $\pi R^4\sigma\omega$

(c) $\frac{\pi R^4}{2}\sigma\omega$ (d) $\frac{\pi R^4}{4}\sigma\omega$

25 Two long conductors, separated by a distance d carry currents I_1 and I_2 in the same direction. They exert a force F on each other. Now, the current in one of them is increased to two times and its direction is reversed. The distance is also increased to $3d$. The new value of the force between them is *[AIEEE 2004]*

(a) $-2F$ (b) $\frac{F}{3}$ (c) $-\frac{2F}{3}$ (d) $-\frac{F}{3}$

Level 3

26 Two long parallel conductors S_1 and S_2 are separated by a distance 10 cm and carrying currents of 4 A and 2 A, respectively. The conductors are placed along X-axis in XY-plane. There is a point P located between the conductors (as shown in figure).
A charge particle of 3π coulomb is passing through

the point P with velocity. $\mathbf{v} = (2\hat{\mathbf{i}} + 3\hat{\mathbf{j}})$ m/s, where $\hat{\mathbf{i}}$ and $\hat{\mathbf{j}}$ represent unit vector along x and y axis, respectively. The force acting on the charge particle is $4\pi \times 10^{-5}(-x\hat{\mathbf{i}} + 2\hat{\mathbf{j}})$ N. The value of x is

[2022, 27 June Shift-II]

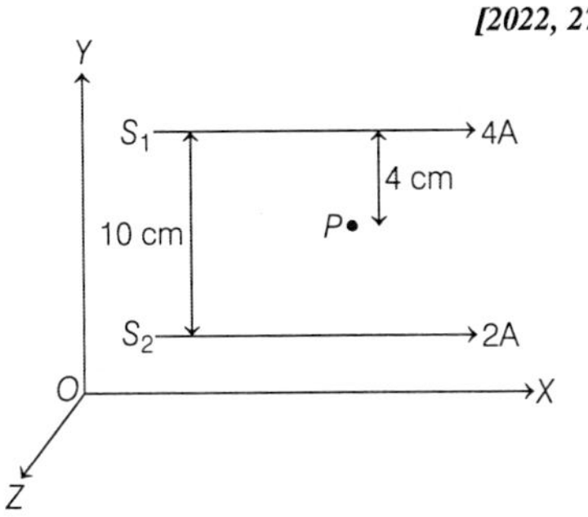

(a) 2 (b) 1
(c) 3 (d) −3

27 A circular coil has moment of inertia 0.8 kg m^2 around any diameter and is carrying current to produce a magnetic moment of 20 Am^2. The coil is kept initially in a vertical position and it can rotate freely around a horizontal diameter. When a uniform magnetic field of 4 T is applied along the vertical, it starts rotating around its horizontal diameter. The angular speed of the coil, acquired after rotating by 60°, will be *[2020, 4 Sep Shift-II]*

(a) 10 rad s^{-1} (b) 20π rad s^{-1}
(c) 10π rad s^{-1} (d) 20 rad s^{-1}

28 Two wires A and B are carrying currents I_1 and I_2 as shown in the figure. The separation between them is d. A third wire C carrying a current I is to be kept parallel to them at a distance x from A such that the net force acting on it is zero. The possible values of x are *[2019, 10 April Shift-I]*

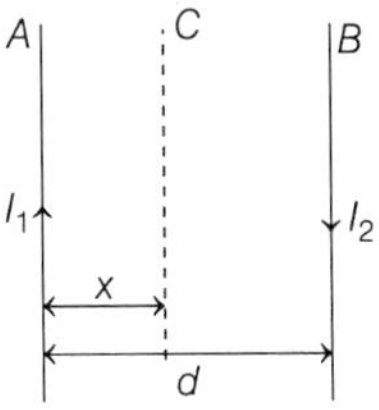

(a) $x = \left(\dfrac{I_2}{I_1 + I_2}\right) d$ and $x = \left(\dfrac{I_2}{I_1 - I_2}\right) d$

(b) $x = \left(\dfrac{I_1}{I_1 - I_2}\right) d$ and $x = \left(\dfrac{I_2}{I_1 + I_2}\right) d$

(c) $x = \left(\dfrac{I_1}{I_1 + I_2}\right) d$ and $x = \left(\dfrac{I_2}{I_1 - I_2}\right) d$

(d) $x = \pm \dfrac{I_1 d}{(I_1 - I_2)}$

29 A rigid square loop of side a and carrying current I_2 is lying on a horizontal surface near a long current I_1 carrying wire in the same plane as shown in figure. The net force on the loop due to the wire will be

[2019, 9 April Shift-I]

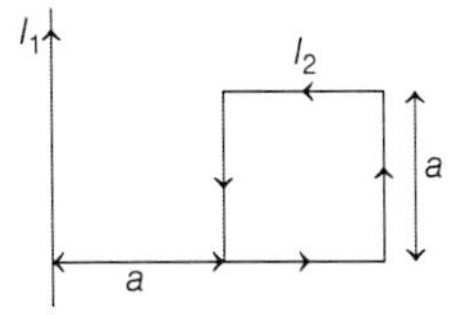

(a) repulsive and equal to $\dfrac{\mu_0 I_1 I_2}{2\pi}$

(b) attractive and equal to $\dfrac{\mu_0 I_1 I_2}{3\pi}$

(c) zero

(d) repulsive and equal to $\dfrac{\mu_0 I_1 I_2}{4\pi}$

30 A thin strip 10 cm long is on an U-shaped wire of negligible resistance and it is connected to a spring of spring constant 0.5Nm^{-1} (see figure). The assembly is kept in a uniform magnetic field of 0.1 T. If the strip is pulled from its equilibrium position and released, the number of oscillations performs before its amplitude decreases by a factor of e is N. If the mass of the strip is 50 grams, its resistance 10 Ω and air drag negligible, N will be close to

[2019, 8 April Shift-I]

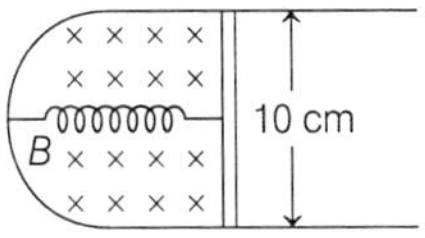

(a) 1000 (b) 50000 (c) 5000 (d) 10000

31 An insulating thin rod of length l has a linear charge density $\rho(x) = \rho_0 \dfrac{x}{l}$ on it. The rod is rotated about an axis passing through the origin ($x = 0$) and perpendicular to the rod. If the rod makes n rotations per second, then the time averaged magnetic moment of the rod is *[2019, 10 Jan Shift-I]*

(a) $n\rho l^3$ (b) $\pi n \rho l^3$ (c) $\dfrac{\pi}{3} n \rho l^3$ (d) $\dfrac{\pi}{4} n \rho l^3$

32 An infinitely long current-carrying wire and a small current- carrying loop are in the plane of the paper as shown. The radius of the loop is a and distance of its centre from the wire is $d (d >> a)$. If the loop applies a force F on the wire, then *[2019, 9 Jan Shift-I]*

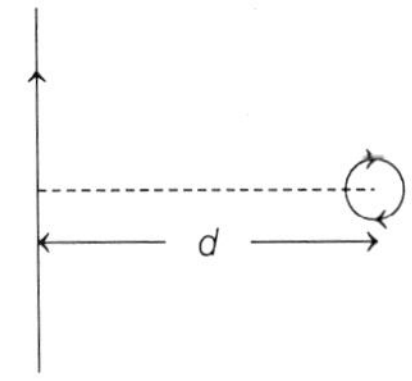

(a) $F \propto \left(\dfrac{a^2}{d^3}\right)$ (b) $F = 0$

(c) $F \propto \left(\dfrac{a}{d}\right)$ (d) $F \propto \left(\dfrac{a}{d}\right)^2$

Numerical Value Type Questions

33 A 4.0 cm long straight wire carrying a current of 8A is placed perpendicular to a uniform magnetic field of strength 0.15 T. The magnetic force on the wire is .... mN. *[2025, 3 April Shift-I]*

34 Two long parallel wires X and Y, separated by a distance of 6 cm, carry currents of 5 A and 4 A, respectively, in opposite directions as shown in the figure. Magnitude of the resultant magnetic field at point P at a distance of 4 cm from wire Y is $x \times 10^{-5}$ T.

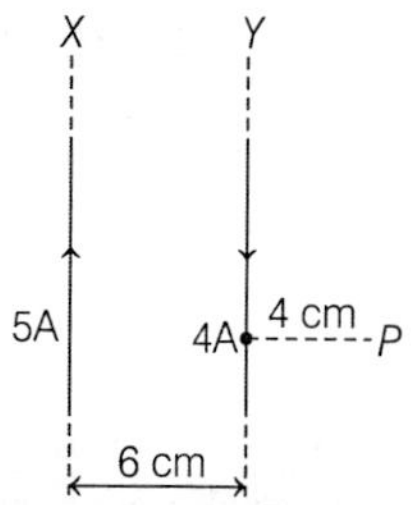

The value of x is .... Take permeability of free space as $\mu_0 = 4\pi \times 10^{-7}$ SI units. *[2025, 22 Jan Shift-II]*

35 A current of 5 A exists in a square loop of side $\frac{1}{\sqrt{2}}$ m. Then, the magnitude of the magnetic field B at the centre of the square loop will be $p \times 10^{-6}$T, where value of p is .... [Take, $\mu_0 = 4\pi \times 10^{-7}$ mA^{-1}] *[2025, 24 Jan Shift-I]*

36 A square loop of edge length 2 m carrying current of 2A is placed with its edges parallel to the XY-axis. A magnetic field is passing through the XY-plane and expressed as $\mathbf{B} = B_0(1+4x)\hat{\mathbf{k}}$, where $B_0 = 5$ T. The net magnetic force experienced by the loop is .... N. *[2024, 9 April Shift-I]*

37 A 2A current carrying straight metal wire of resistance 1 Ω, resistivity 2×10^{-6}Ωm, area of cross-section 10 mm^2 and mass 500 g is suspended horizontally in mid air by applying a uniform magnetic field B. The magnitude of B is .... $\times 10^{-1}$ T (given, $g = 10$ m/s^2) *[2024, 5 April Shift-I]*

38 The magnetic field existing in a region is given by $\mathbf{B} = 0.2(1+2x)\hat{\mathbf{k}}$T. A square loop of edge 50 cm carrying 0.5 A current is placed in XY-plane with its edges parallel to the XY-axes, as shown in figure. The magnitude of the net magnetic force experienced by the loop is...... mN. *[2024, 4 April Shift-I]*

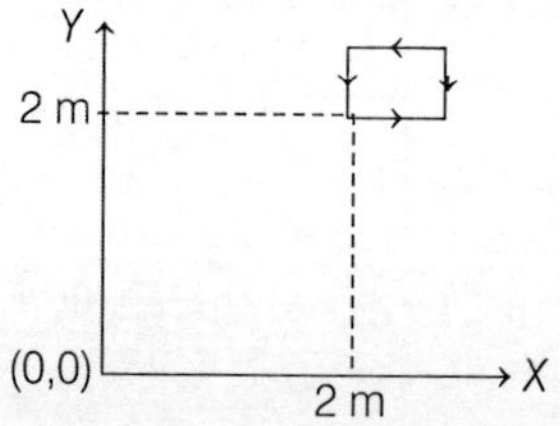

39 The horizontal component of earth's magnetic field at a place is 3.5×10^{-5} T. A very long straight conductor carrying current of $\sqrt{2}$ A in the direction from South East to North West is placed. The force per unit length experienced by the conductor is .... $\times 10^{-6}$ N/m. *[2024, 30 Jan Shift-I]*

40 A straight wire AB of mass 40 g and length 50 cm is suspended by a pair of flexible leads in uniform magnetic field of magnitude 0.40 T as shown in the figure. The magnitude of the current required in the wire to remove the tension in the supporting leads is .... A. (Take, $g = 10$ ms^{-2}) *[2023, 13 April Shift-II]*

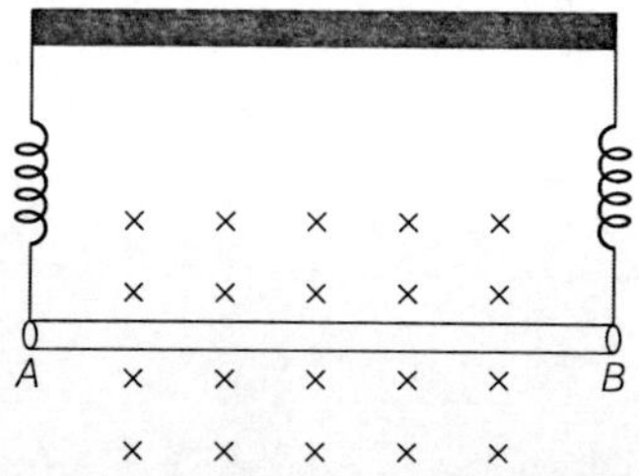

41 Two 10 cm long, straight wires, each carrying a current of 5 A are kept parallel to each other. If each wire experienced a force of 10^{-5} N, then separation between the wires is .... cm. *[2022, 26 June Shift-II]*

42 A uniform conducting wire of length is 24 a, and resistance R is wound up as a current carrying coil in the shape of an equilateral triangle of side a and then in the form of a square of side a. The coil is connected to a voltage source V_0. The ratio of magnetic moment of the coils in case of equilateral triangle to that for square is $1:\sqrt{y}$. The value of y is .... *[2021, 27 Aug Shift-I]*

43 A coil in the shape of an equilateral triangle of side 10 cm lies in a vertical plane between the pole pieces of permanent magnet producing a horizontal magnetic field 20 mT. The torque acting on the coil when a current of 0.2 A is passed through it and its plane becomes parallel to the magnetic field will be $\sqrt{x} \times 10^{-5}$ Nm. The value of x is.......... . *[2021, 26 Aug Shift-II]*

Topic 4 Moving Coil Galvanometer

Level 1

01 A galvanometer having a coil of resistance 30 Ω need 20 mA of current for full-scale deflection. If a maximum current of 3 A is to be measured using this galvanometer, the resistance of the shunt to be added to the galvanometer should be $\frac{30}{X}$Ω, where X is *[2025, 23 Jan Shift-I]*

(a) 149 (b) 298
(c) 596 (d) 447

02 A current of 200 μA deflects the coil of a moving coil galvanometer through 60°. The current to cause deflection through $\pi / 10$ rad is *[2024, 27 Jan Shift-II]*
(a) 180 μA (b) 120 μA (c) 30 μA (d) 60 μA

03 Given below are two statements.
Statement I If the number of turns in the coil of a moving coil galvanometer is doubled then the current sensitivity becomes double.
Statement II Increasing current sensitivity of a moving coil galvanometer by only increasing the number of turns in the coil will also increase its voltage sensitivity in the same ratio.
In the light of the above statements, choose the correct answer from the options given below. *[2023, 10 April Shift-I]*
(a) Statement I is false but Statement II is true.
(b) Statement I is true but Statement II is false.
(c) Both Statement I and Statement II are false.
(d) Both Statement I and Statement II are true.

04 Certain galvanometers have a fixed core made of non magnetic metallic material. The function of this metallic material is *[2023, 8 April Shift-I]*
(a) to produce large deflecting torque on the coil
(b) to bring the coil to rest quickly
(c) to make the magnetic field radial
(d) to oscillate the coil in magnetic field for longer period of time

05 For a moving coil galvanometer, the deflection in the coil is 0.05 rad when a current of 10 mA is passed through it. If the torsional constant of suspension wire is 4.0×10^{-5} N-m rad^{-1} the magnetic field is 0.01 T and the number of turns in the coil is 200, the area of each turn (in cm^2) is *[2023, 25 Jan Shift-II]*
(a) 2.0 (b) 1.5 (c) 0.5 (d) 1.0

06 The current sensitivity of a galvanometer can be increased by
A. decreasing the number of turns
B. increasing the magnetic field
C. decreasing the area of the coil
D. decreasing the torsional constant of the spring
Choose the most appropriate answer from the options given below. *[2022, 28 July Shift-I]*
(a) B and C only (b) C and D only
(c) A and C only (d) B and D only

07 Consider a galvanometer shunted with 5 Ω resistance and 2% of current passes through it. What is the resistance of the given galvanometer? *[2021, 31 Aug Shift-I]*
(a) 300 Ω (b) 344 Ω (c) 245 Ω (d) 226 Ω

08 For full scale deflection of total 50 divisions, 50 mV voltage is required in galvanometer. The resistance of galvanometer if its current sensitivity is 2 div/mA will be *[2021, 27 Aug Shift-II]*
(a) 1 Ω (b) 5 Ω (c) 4 Ω (d) 2 Ω

09 A galvanometer having a coil resistance 100 Ω gives a full scale deflection when a current of 1 mA is passed through it. What is the value of the resistance which can convert this galvanometer into a voltmeter giving full scale deflection for a potential difference of 10 V? *[2020, 8 Jan Shift-II]*
(a) 8.9 kΩ (b) 10 kΩ
(c) 9.9 kΩ (d) 7.9 kΩ

10 A moving coil galvanometer allows a full scale current of 10^{-4} A. A series resistance of 2 MΩ is required to convert the above galvanometer into a voltmeter of range 0-5 V. Therefore, the value of shunt resistance required to convert the above galvanometer into an ammeter of range 0.10 mA is *[2019, 10 April Shift-I]*
(a) 100 Ω (b) 500 Ω
(c) 200 Ω (d) 10 Ω

Imp. Moving Coil Galvanometer as Ammeter

11 The resistance of a galvanometer is 50 Ω and the maximum current which can be passed through it is 0.002 A. What resistance must be connected to it in order to convert it into an ammeter of range 0-0.5 A? *[2019, 9 April Shift-II]*
(a) 0.2 Ω (b) 0.5 Ω (c) 0.002 Ω (d) 0.02 Ω

12 A moving coil galvanometer has resistance 50 Ω and it indicates full deflection at 4 mA current. A voltmeter is made using this galvanometer and a 5 kΩ resistance. The maximum voltage, that can be measured using this voltmeter, will be close to *[2019, 9 April Shift-I]*
(a) 40 V (b) 10 V
(c) 15 V (d) 20 V

13 When a current of 5 mA is passed through a galvanometer having a coil of resistance 15 Ω, it shows full scale deflection. The value of the resistance to be put in series with the galvanometer to convert it into a voltmeter of range 0-10 V is *[JEE Main 2017]*
(a) 2.045×10^3 Ω (b) 2.535×10^3 Ω
(c) 4.005×10^3 Ω (d) 1.985×10^3 Ω

14 A galvanometer having a coil resistance of 100 Ω gives a full scale deflection when a current of 1 mA is passed through it. The value of the resistance which can convert this galvanometer into ammeter giving a full scale deflection for a current of 10 A, is *[JEE Main 2016]*
(a) 0.01 Ω (b) 2 Ω
(c) 0.1 Ω (d) 3 Ω

15 An ammeter reads upto 1 A. Its internal resistance is 0.81 Ω. To increase the range to 10 A, the value of the required shunt is *[AIEEE 2003]*
(a) 0.03 Ω (b) 0.3 Ω
(c) 0.9 Ω (d) 0.09 Ω

16 If an ammeter is to be used in place of a voltmeter, then we must connect with the ammeter a *[AIEEE 2002]*
(a) low resistance in parallel (b) high resistance in parallel
(c) high resistance in series (d) low resistance in series

Level 2

17 In a moving coil galvanometer, two moving coils M_1 and M_2 have the following particulars :
$R_1 = 5\,\Omega, N_1 = 15, A_1 = 3.6 \times 10^{-3}\,m^2, B_1 = 0.25\,T$
$R_2 = 7\,\Omega, N_2 = 21, A_2 = 1.8 \times 10^{-3}\,m^2, B_2 = 0.50\,T$
Assuming that torsional constant of the springs are same for both coils, what will be the ratio of voltage sensitivity of M_1 and M_2? *[2025, 2 April Shift-II]*
(a) 1 : 3 (b) 1 : 1
(c) 1 : 4 (d) 1 : 2

18 Consider a moving coil galvanometer (MCG).
A. The torsional constant in moving coil galvanometer has dimentions $[ML^2T^{-2}]$.
B. Increasing the current sensitivity may not necessarily increase the voltage sensitivity.
C. If we increase number of turns (N) to its double ($2N$), then the voltage sensitivity doubles.
D. MCG can be converted into an ammeter by introducing a shunt resistance of large value in parallel with galvanometer.
E. Current sensitivity of MCG depends inversely on number of turns of coil.
Choose the correct answer from the options given below. *[2025, 23 Jan Shift-I]*
(a) B, D, E Only (b) A, D, Only
(c) A, B Only (d) A, B, E Only

19 A galvanometer has a coil of resistance 200 Ω with a full scale deflection at 20 μA. The value of resistance to be added to use it as an ammeter of range 0-20 mA is *[2024, 9 April Shift-I]*
(a) 0.40 Ω (b) 0.20 Ω
(c) 0.50 Ω (d) 0.10 Ω

20 A galvanometer of resistance 100 Ω when connected in series with 400 Ω measures a voltage of upto 10 V. The value of resistance required to convert the galvanometer into ammeter to read upto 10 A is $x \times 10^{-2}\,\Omega$. The value of x is *[2024, 5 April Shift-II]*
(a) 800 (b) 20 (c) 2 (d) 200

21 In an ammeter, 5% of the main current passes through the galvanometer. If resistance of the galvanometer is G, the resistance of ammeter will be *[2024, 1 Feb Shift-II]*
(a) $\frac{G}{200}$ (b) 199 G (c) 200 G (d) $\frac{G}{199}$

22 A galvanometer has a resistance of 50 Ω and it allows maximum current of 5 mA. It can be converted into voltmeter to measure upto 100 V by connecting in series a resistor of resistance *[2024, 1 Feb Shift-I]*
(a) 19950 Ω (b) 5975 Ω
(c) 19500 Ω (d) 20050 Ω

23 A galvanometer having coil resistance 10 Ω shows a full scale deflection for a current of 3 mA. For it to measure a current of 8 A, the value of the shunt should be *[2024, 29 Jan Shift-I]*
(a) $4.85 \times 10^{-3}\,\Omega$ (b) $3 \times 10^{-3}\,\Omega$
(c) $3.75 \times 10^{-3}\,\Omega$ (d) $2.75 \times 10^{-3}\,\Omega$

24 The deflection in moving coil galvanometer falls from 25 divisions to 5 division when a shunt of 24 Ω is applied. The resistance of galvanometer coil will be *[2024, 29 Jan Shift-I]*
(a) 48 Ω (b) 96 Ω
(c) 12 Ω (d) 100 Ω

25 For designing a voltmeter of range 50 V and an ammeter of range 10 mA using a galvanometer which has a coil of resistance 54 Ω showing a full scale deflection for 1 mA as in figure. *[2023, 15 April Shift-I]*

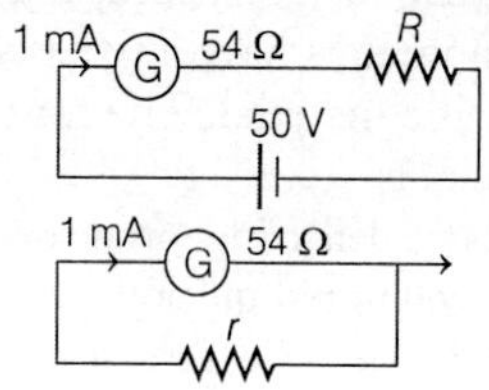

A. for voltmeter $R \approx 50\,k\Omega$
B. for ammeter $r \approx 0.2\,\Omega$
C. for ammeter $r = 6\,\Omega$
D. for voltmeter $R \approx 5\,k\Omega$
E. for voltmeter $R \approx 500\,\Omega$
Choose the correct answer from the options given below.
(a) C and D (b) C and E (c) A and B (d) A and C

26 The current sensitivity of moving coil galvanometer is increased by 25%. This increase is achieved only by changing the number of turns of coil and area of cross-section of the wire while keeping the resistance of galvanometer coil constant. The percentage change in the voltage sensitivity will be *[2023, 11 April Shift-I]*
(a) + 25% (b) – 25% (c) – 50% (d) zero

27 A 72 Ω galvanometer is shunted by a resistance of 8 Ω. The percentage of the total current which passes through the galvanometer is *[2022, 27 June Shift-I]*
(a) 0.1% (b) 10% (c) 25% (d) 0.25%

28 If n represents the actual number of deflections in a converted galvanometer of resistance G and shunt resistance S. Then, the total current I when its figure of merit is K will be *[2022, 25 June Shift-II]*
(a) $\frac{KS}{(S+G)}$ (b) $\frac{(G+S)}{nKS}$
(c) $\frac{nKS}{(G+S)}$ (d) $\frac{nK(G+S)}{S}$

Imp. Moving Coil Galvanometer as Voltmeter

29 A galvanometer of resistance G is converted into a voltmeter of range 0-1V by connecting a resistance R_1 in series with it. The additional resistance that should be connected in series with R_1 to increase the range of the voltmeter to 0-2V will be *[2020, 5 Sep Shift-I]*

(a) $R_1 + G$ (b) R_1 (c) G (d) $R_1 - G$

30 A galvanometer having a resistance of 20 Ω and 30 divisions on both sides has figure of merit 0.005 ampere/division. The resistance that should be connected in series such that it can be used as a voltmeter upto 15 V is *[2019, 11 Jan Shift-II]*

(a) 100 Ω (b) 125 Ω
(c) 120 Ω (d) 80 Ω

31 A moving coil galvanometer has a coil with 175 turns and area 1 cm^2. It uses a torsion band of torsion constant 10^{-6} N-m/rad. The coil is placed in a magnetic field B parallel to its plane. The coil deflects by 1° for a current of 1 mA. The value of B (in tesla) is approximately *[2019, 9 April Shift-II]*

(a) 10^{-3} (b) 10^{-4} (c) 10^{-1} (d) 10^{-2}

32 A moving coil galvanometer has 150 equal divisions. Its current sensitivity is 10 divisions per milliampere and voltage sensitivity is 2 divisions per millivolt. In order that each division reads 1 V, the resistance in Ohms needed to be connected in series with the coil will be *[AIEEE 2005]*

(a) 10^3 (b) 10^5
(c) 99995 (d) 9995

Level 3

33 A moving coil galvanometer, having a resistance G, produces full scale deflection when a current I_g flows through it. This galvanometer can be converted into (i) an ammeter of range 0 to $I_0 (I_0 > I_g)$ by connecting a shunt resistance R_A to it and (ii) into a voltmeter of range 0 to V $(V = GI_0)$ by connecting a series resistance R_V to it. Then, *[2019, 12 April Shift-II]*

(a) $R_A R_V = G^2 \left(\frac{I_0 - I_g}{I_g}\right)$ and $\frac{R_A}{R_V} = \left(\frac{I_g}{(I_0 - I_g)}\right)^2$

(b) $R_A R_V = G^2$ and $\frac{R_A}{R_V} = \left(\frac{I_g}{I_0 - I_g}\right)^2$

(c) $R_A R_V = G^2 \left(\frac{I_g}{I_0 - I_g}\right)$ and $\frac{R_A}{R_V} = \left(\frac{I_0 - I_g}{I_g}\right)^2$

(d) $R_A R_V = G^2$ and $\frac{R_A}{R_V} = \frac{I_g}{(I_0 - I_g)}$

34 A galvanometer of resistance 100 Ω has 50 divisions on its scale and has sensitivity of 20 μA/division. It is to be converted to a voltmeter with three ranges of 0-2 V, 0-10 V and 0-20 V. The appropriate circuit to do so is *[2019, 12 April Shift-I]*

(a)
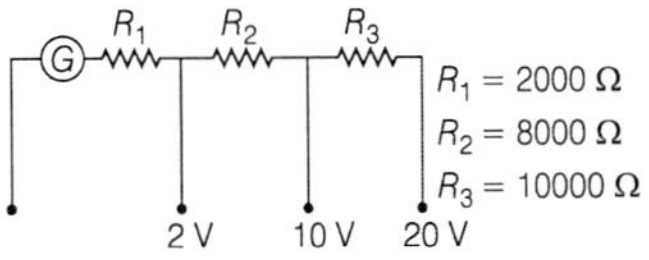

(b)
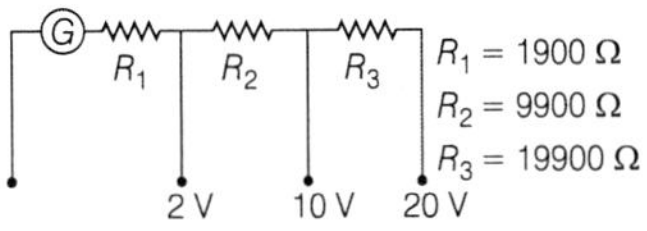

(c)
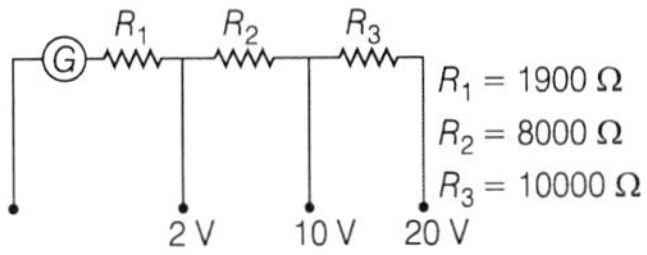

(d)
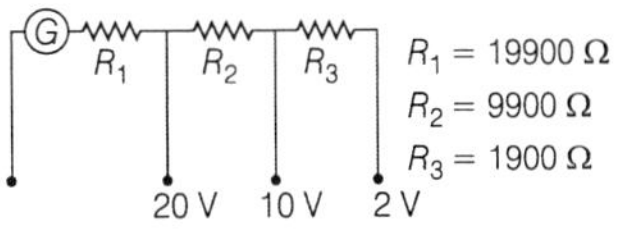

Numerical Value Type Questions

35. In the figure shown below, a resistance of 150.4 Ω is connected in series to an ammeter A of resistance 240 Ω. A shunt resistance of 10 Ω is connected in parallel with the ammeter. The reading of the ammeter is .... mA. *[2025, 3 April Shift-II]*

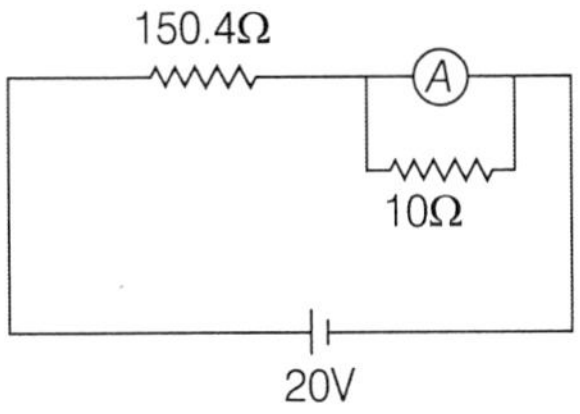

36 A moving coil galvanometer has 100 turns and each turn has an area of 2.0 cm^2. The magnetic field produced by the magnet is 0.01 T and the deflection in the coil is 0.05 radian when a current of 10 mA is passed through it. The torsional constant of the suspension wire is $x \times 10^{-5}$ N-m/rad. The value of x is .... *[2024, 1 Feb Shift-II]*

37 A galvanometer coil has 500 turns and each turn has an average area of 3×10^{-4} m^2. If a torque of 1.5 N-m is required to keep this coil parallel to a magnetic field when a current of 0.5 A is flowing through it, the strength of the field (in T) is .... *[2020, 3 Sep Shift-II]*

Key Index

Topic 1

1 (a)	**2** (c)	**3** (a)	**4** (d)	**5** (c)	**6** (b)	**7** (a)	**8** (a)	**9** (b)	**10** (d)
11 (b)	**12** (d)	**13** (a)	**14** (c)	**15** (d)	**16** (c)	**17** (a)	**18** (a)	**19** (a)	**20** (c)
21 (c)	**22** (b)	**23** (c)	**24** (a)	**25** (a)	**26** (d)	**27** (c)	**28** (b)	**29** (d)	**30** (b)
31 (b)	**32** (b)	**33** (d)	**34** (a)	**35** (b)	**36** (c)	**37** (b)	**38** (b)	**39** (d)	**40** (a)
41 (d)	**42** (c)	**43** (a)	**44** (c)	**45** (a)	**46** (a)	**47** (b)	**48** (b)	**49** (a)	**50** (d)
51 (c)	**52** (a)	**53** (1)	**54** (8)	**55** (500)	**56** (5)	**57** (72)	**58** (20)	**59** (40)	**60** (160)
61 (40)	**62** (8)	**63** (628)	**64** (3)						

Topic 2

1 (b)	**2** (c)	**3** (a)	**4** (b)	**5** (c)	**6** (c)	**7** (b)	**8** (b)	**9** (d)	**10** (b)
11 (a)	**12** (a)	**13** (b)	**14** (d)	**15** (a)	**16** (a)	**17** (a)	**18** (d)	**19** (d)	**20** (d)
21 (b)	**22** (a)	**23** (c)	**24** (c)	**25** (c)	**26** (a)	**27** (b)	**28** (d)	**29** (c)	**30** (b)
31 (a)	**32** (d)	**33** (b)	**34** (b)	**35** (a)	**36** (c)	**37** (c)	**38** (a)	**39** (b)	**40** (d)
41 (b)	**42** (b)	**43** (c)	**44** (*)	**45** (d)	**46** (0.01)	**47** (250)	**48** (2)	**49** (5)	**50** (32)
51 (40)	**52** (144)	**53** (10)	**54** (2)						

Topic 3

1 (a)	**2** (a)	**3** (a)	**4** (b)	**5** (c)	**6** (a)	**7** (b)	**8** (a)	**9** (c)	**10** (b)
11 (a)	**12** (b)	**13** (c)	**14** (d)	**15** (a)	**16** (c)	**17** (b)	**18** (b)	**19** (a)	**20** (a)
21 (c)	**22** (b)	**23** (b)	**24** (d)	**25** (c)	**26** (c)	**27** (*)	**28** (d)	**29** (d)	**30** (c)
31 (d)	**32** (d)	**33** (48)	**34** (1)	**35** (8)	**36** (160)	**37** (5)	**38** (50)	**39** (35)	**40** (2)
41 (5)	**42** (3)	**43** (3)							

Topic 4

1 (a)	**2** (d)	**3** (b)	**4** (b)	**5** (d)	**6** (d)	**7** (c)	**8** (d)	**9** (c)	**10** (*)
11 (a)	**12** (d)	**13** (d)	**14** (a)	**15** (d)	**16** (c)	**17** (b)	**18** (c)	**19** (b)	**20** (b)
21 (*)	**22** (a)	**23** (c)	**24** (b)	**25** (a)	**26** (a)	**27** (b)	**28** (d)	**29** (a)	**30** (d)
31 (a)	**32** (d)	**33** (b)	**34** (c)	**35** (5)	**36** (4)	**37** (20)			

Detailed **Explanations**

Topic 1

01. *(a)*

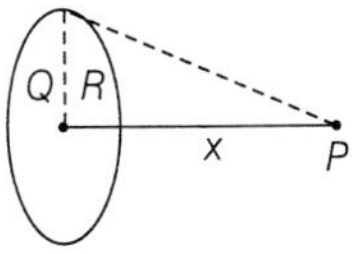

$x : R = 3 : 4$

$\sin\theta = \frac{4}{5}$

$$B_1 = \frac{\mu_0 I}{2R},\ B_2 = B_1 \sin^3\theta$$

$$\frac{B_2}{B_1} = \sin^3\theta = \left(\frac{4}{5}\right)^3 = \frac{64}{125}$$

02. *(c)* Statement I is true but Statement II is false.

Ampere's circuital law is not independent of Biot-Savart's law.

03. *(a)*

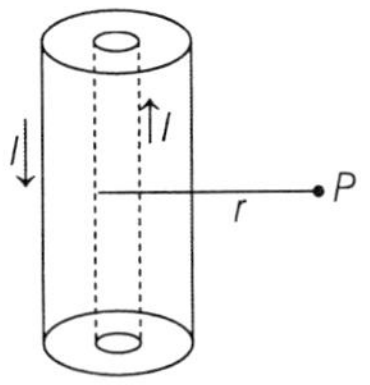

According to Ampere's circuit law,

$$\oint \mathbf{B}\cdot d\mathbf{l} = \mu_0 (I_{\text{enclosed}})$$

As $I_{\text{enclosed}} = 0$ (net current)

$\Rightarrow \quad B = 0$

Hence, magnetic field outside is zero.

04. *(d)* Since, electron is moving uniformly along magnetic field inside a current carrying solenoid

i.e. $\quad \theta = 0°$

Then using Lorentz force formula,

$$F_{\text{net}} = qvB\sin\theta = 0$$

Hence, electron will move in direction of its initial velocity without changing its direction motion and speed.

05. *(c)* An electron is moving along the positive X-axis. If the uniform magnetic field is applied parallel to the negative Z-axis then magnetic force

$$\mathbf{F} = -e(\mathbf{v}\times\mathbf{B})$$

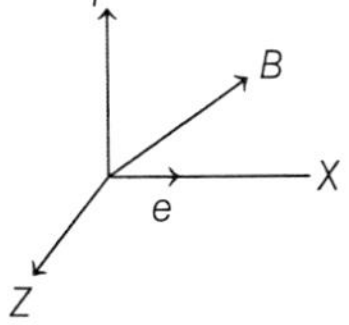

So, the electron will experience magnetic force along –ve Y-axis $(-\hat{j})$ and the e^- will move along circular path in magnetic field.

So, option (c) is correct answer.

06. *(b)*

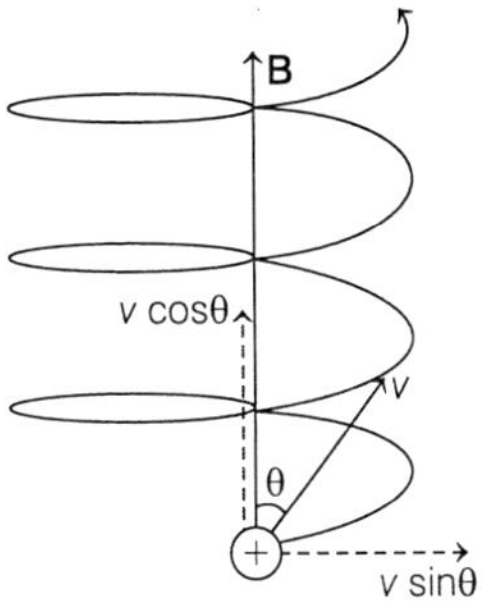

When velocity **v** is inclined at angle θ with **B**, components of **v** along **B** is $v\cos\theta$ and perpendicular to **B** is $v\sin\theta$.

Due to $v\sin\theta$, the charged particle rotates in a circular path.

Also due to $v\cos\theta$, it travels along **B**. So path is helical.

07. *(a)* In this case of solid infinite current carrying wire.

Using ampere circuital law,

Case I If $r \le R$, $B = \frac{\mu_0 i}{2\pi R^2}\cdot r \Rightarrow B \propto r$

Case II $r > R$, $\quad B = \frac{\mu_0 i}{2\pi r} \Rightarrow B \propto \frac{1}{r}$

Option (a) is correct.

08. *(a)*

> **key idea** You can use the following two facts for the calculation of magnetic field.
>
> (i) Magnetic field due to semicircular current loop, $B_1 = \frac{\mu_0 I}{4r}$
>
> (ii) Magnetic field due to current carrying wire at a point on its axis is zero.

Given that, $R = \frac{\pi}{10}$ m, $I = 3$A,

$\mu_0 = 4\pi\times10^{-7}$ N/A^2.

The point O lies on the line of wire-1 and wire-2. So, magnetic field is zero due to wire-1 but due to semi-circular arc, it is non zero.

Magnetic field due to semi-circular arc,

$$B = \frac{\mu_0 I}{4R} = \frac{4\pi\times10^{-7}\times3}{4\times\frac{\pi}{10}}$$

$$= 3\times10^{-6}\ \text{T}$$

$\Rightarrow \quad B = 3\ \mu\text{T}$

09. *(b)* Magnetic field inside a solenoid,

$$B = \mu_0 n I$$

Here, $n = 70$ turns/cm $= 7000$ turns/m

and $\quad I = 2$A

So, $\quad B = 4\pi\times10^{-7}\times7000\times2$

$= 175.84\times10^{-4}$ T

$\approx 176\times10^{-4}$ T

10. *(d)* The y-component of magnetic field as we move from centre to circumference of coil, first increases, then decreases.

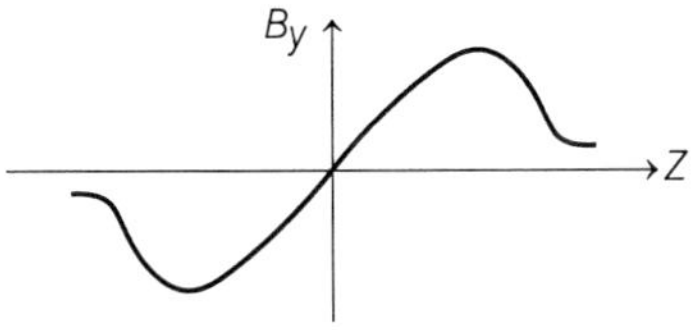

11. *(b)* Given, number of turns in solenoid,

$$N = 1200$$

Length of glass tube,

$$l = 2\ \text{m}$$

$\therefore$ Number of turns per unit length,

$$n = \frac{N}{l} = \frac{1200}{2} = 600\ \text{turns/m}$$

$$I = 2\ \text{A}$$

$\therefore$ Magnetic field intensity at the centre of the solenoid

$$H = \frac{B}{\mu_0} = \frac{\mu_0 ni}{\mu_0} = ni$$

$$H = ni = 600\times2 = 1200$$

$$H = 1.2\times10^3\ \text{Am}^{-1}$$

> • Concept **Enhancer**
>
> Magnetic field intensity due to current carrying solenoid is given as,
>
> $B = \mu_0 ni$
>
> where, n is number of turns per unit length and i is current flowing through the solenoid.

12. *(d)*

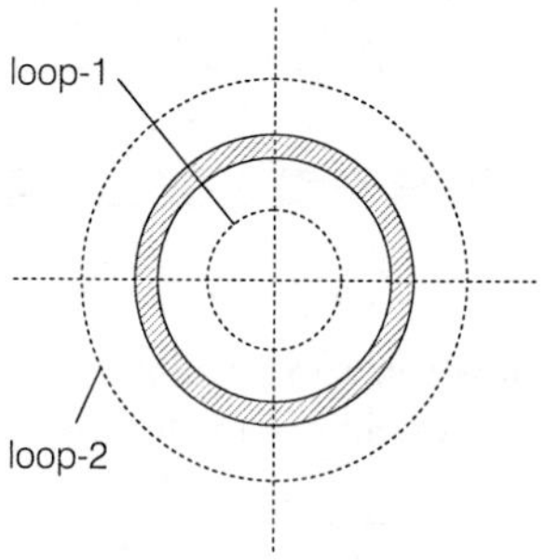

Consider two loops, loop-1 and loop-2. For loop-1, there is no enclosed current. So, following Ampere's law,

$$\oint \mathbf{B}\cdot d\mathbf{l} = \mu_0 I_{\text{net}}$$

$$\Rightarrow \quad B_{\text{in}} = 0$$

Also, applying ampere circuital law for loop-2.

$$\oint \mathbf{B}\cdot d\mathbf{l} = B\times 2\pi r = \mu_0 I_{\text{net}}$$

$$\Rightarrow \quad B_{\text{out}} = \frac{\mu_0 I}{2\pi r} \text{ or } B_{\text{out}} \propto \frac{1}{r}$$

So, B varies with distance from centre as

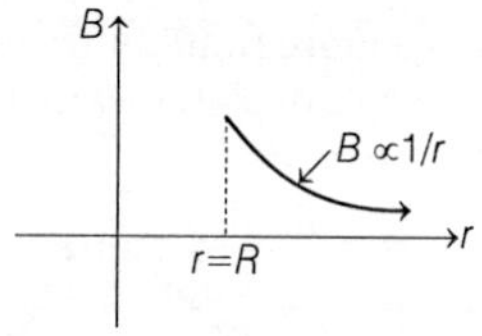

13. *(a)* Magnetic field due to a long current carrying solenoid

$$B = \mu_0 nI \qquad \ldots\text{(i)}$$

When $I' = 2I$ and $n' = \frac{n}{2}$, then new value of magnetic field.

$$B' = \mu_0 n' I' = \mu_0\left(\frac{n}{2}\right)\cdot 2I$$

$$= \mu_0 nI = B \quad \text{[From Eq. (i)]}$$

$$B' = B$$

14. *(c)* We know that magnetic field inside circular cross-section carrying current I (which is uniformly distributed across the cross-section) is given as

$$B = \frac{\mu_0 Ix}{2\pi r^2}$$

$$\therefore \quad B \propto x$$

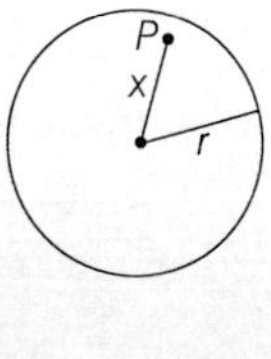

Magnetic field outside the circular cross-section is given as

$$B = \frac{\mu_0 I}{2\pi r} \Rightarrow B \propto \frac{1}{r}$$

Hence, graph between B and r showing in option (c) is correct.

15. *(d)* Given, magnetic field at the centre at A is 3T. Since, the solenoid B, C and D are connected in parallel arrangement.

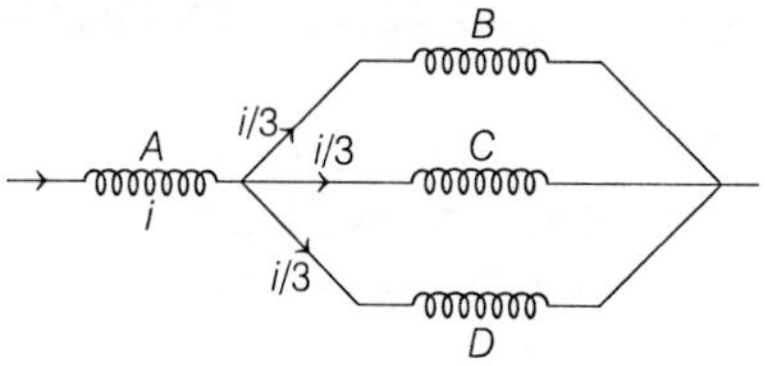

We know that the magnetic field in the circuit is directly proportional to the current in the circuit.

$$B \propto i$$

$$\Rightarrow \quad \frac{B_A}{B_C} = \frac{I_A}{I_C} \Rightarrow \frac{3T}{B_C} = \frac{i}{i/3}$$

$$\Rightarrow \quad B_C = 1\,\text{T}$$

16. *(c)* By using Ampere's circuital law, the magnetic field at any point inside the pipe is zero.

17. *(a)* Magnetic field due to a long solenoid is given by,

$$B = \mu_0 nI$$

From given data,

$$6.28\times 10^{-2} = \mu_0 \times 200\times 10^2 \times I \ \ldots\text{(i)}$$

$$\text{and } B = \mu_0 \times 100\times 10^2 \times \left(\frac{I}{3}\right) \quad \ldots\text{(ii)}$$

On solving Eqs. (i) and (ii), we get

$$B \approx 1.05\times 10^{-2}\ \text{Wb}/\text{m}^2$$

18. *(a)* Magnetic field in circular coil A is

$$B_A = \frac{\mu_0 NI}{2R}$$

where, R is radius and I is current flowing in coil. Similarly,

$$B_B = \frac{\mu_0 N(2I)}{2\cdot(2R)} = \frac{\mu_0 NI}{2R}$$

$$\therefore \quad \frac{B_A}{B_B} = 1$$

19. *(a)*

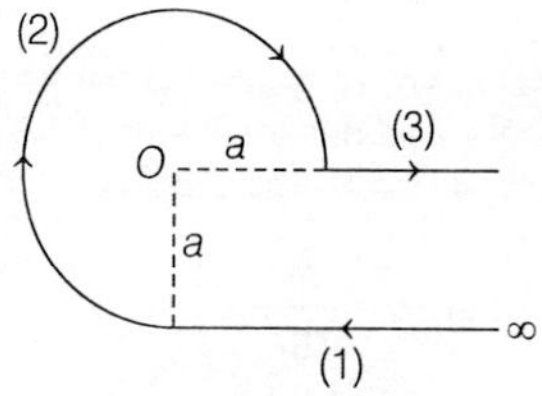

$$\mathbf{B}_1 = \frac{\mu_0 I}{4\pi a} \otimes$$

$$\mathbf{B}_2 = \frac{\mu_0}{4\pi}\frac{I}{a}\left(\frac{3\pi}{2}\right) \otimes$$

$$\mathbf{B}_3 = 0$$

$$\mathbf{B}_{\text{net}} = \mathbf{B}_1 + \mathbf{B}_2 + \mathbf{B}_3$$

$$= \frac{\mu_0 I}{4\pi a} + \frac{\mu_0}{4\pi}\frac{I}{a}\left(\frac{3\pi}{2}\right) + 0$$

$$= \frac{\mu_0 I}{4\pi a}\left[1 + \frac{3\pi}{2}\right]$$

20. *(c)* Given, $I = 10\text{A}$

$$r = 0.5\text{ m}$$

$$\therefore \quad d\mathbf{B} = \frac{\mu_0}{4\pi}\frac{I(d\mathbf{l}\times\mathbf{r})}{r^3}$$

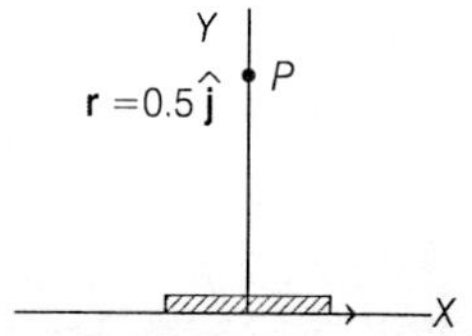

$$\Delta l = \Delta x\hat{\mathbf{i}} \Rightarrow dl = \frac{1}{100}\hat{\mathbf{i}}$$

$$= 10^{-7} \times \frac{10\left(\frac{1}{100}\hat{\mathbf{i}} \times 0.5\hat{\mathbf{j}}\right)}{(0.5)^3}$$

$$d\mathbf{B} = 4\times 10^{-8}\,\text{T}\,\hat{\mathbf{k}}$$

$$\therefore \quad |d\mathbf{B}| = 4\times 10^{-8}\ \text{T}$$

21. *(c)* Magnetic field at centre O due to current carrying loop is given by

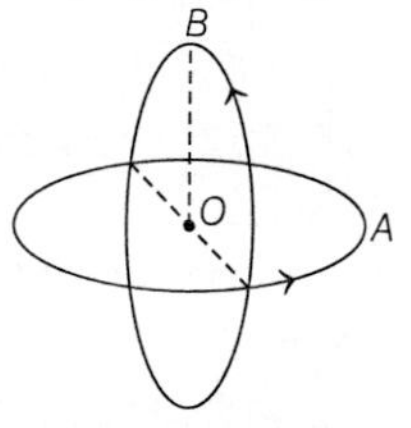

$$B = \frac{\mu_0 I}{2a} \qquad \ldots\text{(i)}$$

(I is current in loop, a is radius)

In this case, we have 2 perpendicular identical loops,

So, net magnetic field will be given by

$$B_{\text{net}} = \sqrt{B_1^2 + B_2^2}$$

$$B_{\text{net}} = \sqrt{2}\,B \ [\because B_1 = B_2 = B]$$

$$B_{\text{net}} = \frac{\sqrt{2}\,\mu_0 I}{2a} \Rightarrow B_{\text{net}} = \frac{\mu_0 I}{\sqrt{2}a}$$

22. *(b)* **For B** An electric field changing linearly with time can be a source of time varying magnetic field just as a time varying magnetic field produces a changing electric field.

For D A decelerating charged particle is nothing more than a negatively accelerating charged particle. An accelerating charged particle produces a magnetic field. Hence, option (b) is correct.

23. *(c)* Both A and R are correct and R is the correct explanation of A.

An electric fan rotates due to magnetic field produced inside the coil of fan. Magnetic field is produced due to current.

When the fan is turned off, it still continues to rotate. It is due to inertia of motion however because of the presence of air resistance it eventually slows down and eventually stops after few seconds.

24. *(a)* Magnetic field at point P,

$$\mathbf{B}_P = \mathbf{B}_1 + \mathbf{B}_2 + \mathbf{B}_3$$

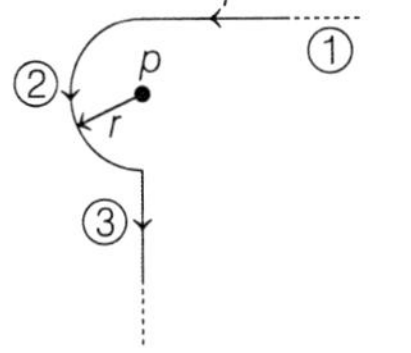

$$\mathbf{B}_P = \frac{\mu_0 i}{4\pi R} \odot + \frac{\mu_0 i}{4R} \odot + 0$$

$$\mathbf{B}_P = \frac{\mu_0 i}{4R}\left(\frac{1}{\pi} + 1\right) \odot$$

or $$\mathbf{B}_P = \frac{\mu_0 i}{2R}\left(\frac{1}{2\pi} + \frac{1}{2}\right)$$

25. *(a)* Magnetic field at the centre of current carrying coil of R having N number of turns is given as,

$$B = \frac{\mu_0 NI}{2R}$$

For two given situations

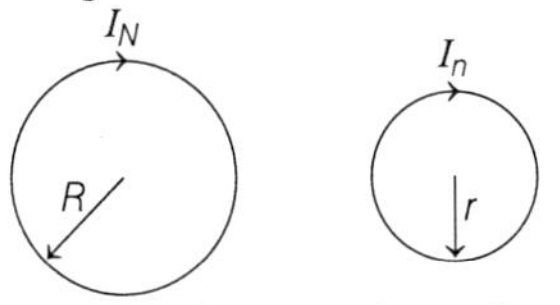

$l_1 = N \times 2\pi R$ $\quad$ $l_2 = n \times 2\pi r$

Length remains same, $l_1 = l_2$

$$N \times 2\pi R = n \times 2\pi r \text{ or } \frac{r}{R} = \frac{N}{n}$$

Since, $B_1 = \dfrac{\mu_0 NI}{2R}$ and $B_2 = \dfrac{\mu_0 nI}{2r}$

$$\therefore \frac{B_1}{B_2} = \frac{\frac{\mu_0 NI}{2R}}{\frac{\mu_0 nI}{2r}} = \frac{N}{n}\left(\frac{r}{R}\right) = \left(\frac{N}{n}\right)\left(\frac{N}{n}\right)$$

$$\Rightarrow \frac{B_1}{B_2} = \left(\frac{N}{n}\right)^2$$

26. *(d)* Magnetic field at centre,

$$B_0 = \frac{\mu_0 I}{2r}$$

Magnetic field at axial point A,

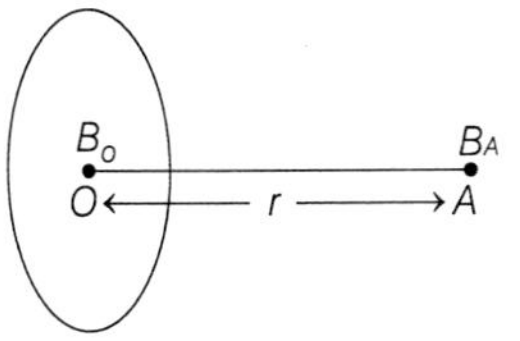

$$B_A = \frac{\mu_0 I r^2}{2(x^2 + r^2)^{3/2}}$$

Here, $x = r$ [Given]

$$\Rightarrow B_A = \frac{\mu_0 I}{4\sqrt{2}r}$$

So, $$\frac{B_0}{B_A} = \frac{4\sqrt{2}}{2} = \frac{2\sqrt{2}}{1} \text{ or } 2\sqrt{2} : 1$$

• Concept Enhancer

Magnetic field due to current I carrying circular loop of radius r on its axis at a distance x from the centre is given as,

$$B = \frac{\mu_0 I r^2}{2(x^2 + r^2)^{3/2}}$$

27. *(c)* We have, $B_1 = \dfrac{\mu_0 I}{2R}$ ... (i)

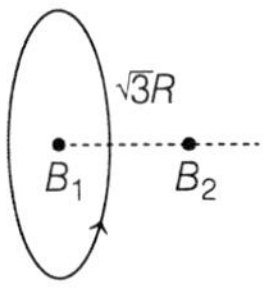

$$B_2 = \frac{\mu_0 I R^2}{2(R^2 + 3R^2)^{3/2}}$$

$$= \frac{1}{8}\left(\frac{\mu_0 I}{2R}\right) \quad \text{...(ii)}$$

On dividing Eq. (i) by Eq. (ii), we get

$$\Rightarrow \frac{B_1}{B_2} = \frac{8}{1}$$

28. *(b)* **Case - I** Magnetic field at the centre of circular current carrying coil having $2(N_1 = 2)$ turns is given as

$$B_1 = \frac{\mu_0 N_1 I}{2r_1}, \quad B_1 = \frac{\mu_0 \times 2 \times I}{2r_1}$$

$$\Rightarrow B_1 = \frac{2\mu_0 I}{2r_1} \quad \text{...(i)}$$

Case -II Magnetic field at the centre of circular current carrying coil having 5 turns $(N_2 = 5)$ is given as

$$B_2 = \frac{\mu_0 N_2 I}{2r_2}$$

$$\Rightarrow B_2 = \frac{5\mu_0 I}{2r_2} \quad \text{...(ii)}$$

On dividing Eq. (ii) by Eq. (i), we get

$$\frac{B_2}{B_1} = \frac{5r_1}{2r_2} \quad \text{...(iii)}$$

According to given situation

$$l_1 = l_2$$

$$\Rightarrow 2 \times 2\pi r_1 = 5 \times 2\pi r_2$$

$$\Rightarrow 2r_1 = 5r_2 \Rightarrow \frac{r_1}{r_2} = \frac{5}{2}$$

From Eq. (iii), $\dfrac{B_2}{B_1} = \dfrac{5}{2} \times \dfrac{5}{2} = \dfrac{25}{4}$

29. *(d)* Magnetic field at centre due to this strip dr is given as

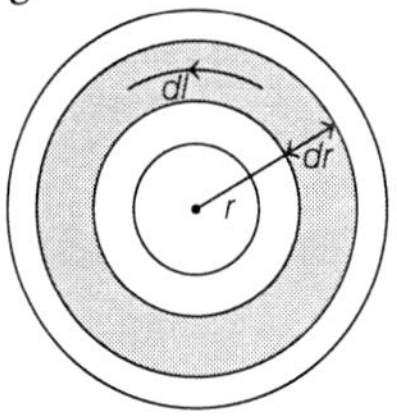

$$dB = \frac{\mu_0}{2r} dI = \frac{\mu_0}{2r} \cdot \frac{nI}{r_2 - r_1} \cdot dr$$

$$\text{So, } B_{\text{net}} = \int dB = \frac{\mu_0 nI}{2(r_2 - r_1)} \int_{r_1}^{r_2} \frac{dr}{r}$$

$$= \frac{\mu_0 nI}{2(r_2 - r_1)} \log_e \left(\frac{r_2}{r_1}\right)$$

30. *(b)*

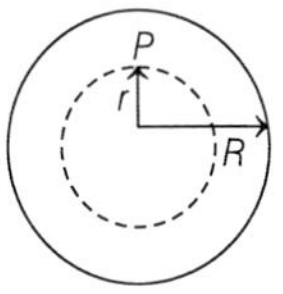

Current density, $J = \dfrac{I}{\pi R^2}$

$\therefore$ Current enclosed (i) in area πr^2

$$= J\pi r^2 = \frac{I}{\pi R^2} \times \pi r^2 = \frac{Ir^2}{R^2}$$

Now, by using Ampere's circuital law, we get $\oint \mathbf{B} \cdot d\mathbf{l} = \mu_0 i$

$$\Rightarrow \oint B \cdot dl \cos 0° = \mu_0 I \frac{r^2}{R^2}$$

$$\Rightarrow B \oint dl = \mu_0 I \frac{r^2}{R^2}$$

[$\because |\mathbf{B}|$ = constant on amperian loop]

$$\Rightarrow B \cdot 2\pi r = \mu_0 I \frac{r^2}{R^2}$$

$$\Rightarrow B = \frac{\mu_0 I}{2\pi} \frac{r^2}{rR^2} = \left(\frac{\mu_0 I}{2\pi R^2}\right) r$$

$$\therefore B \propto r$$

31. *(b)* For the straight wire,

$$B_1 = B_3 = \frac{\mu_0 I}{4\pi r}$$

For the semicircular wire, the expression of the magnetic field,

$$B_2 = \frac{\mu_0 I}{4r}$$

Hence, the total magnetic field at the point P, $B = B_1 + B_2 + B_3$

$$B = \frac{\mu_0 I}{4\pi r} + \frac{\mu_0 I}{4r} + \frac{\mu_0 I}{4\pi r}$$

$$\Rightarrow \quad B = \frac{\mu_0 I}{4\pi r}(1 + 1 + \pi)$$

$$\Rightarrow \quad B = \frac{\mu_0 I}{4\pi r}(2 + \pi)$$

32. *(b)* There is no magnetic field along axis of a current-carrying wire.

Also, magnetic field near one of end of an infinitely long wire is $\frac{\mu_0 I}{4\pi r}$ tesla.

Hence, magnetic field due to segments LP and MQ at O is zero.

Using right hand rule, we can check that magnetic field due to segments PS and QN at O is in same direction perpendicularly into the plane of paper.

Hence, $B_O = B_{PS} + B_{QN}$

$$= \frac{\mu_0 i}{4\pi r} + \frac{\mu_0 i}{4\pi r} = \frac{\mu_0 i}{2\pi r}$$

So, $$i = \frac{2\pi r B_0}{\mu_0}$$

Here, $r = OP = OQ = 4$ cm

and $B_0 = 10^{-4}$ T.

On substituting values, we get

$$\Rightarrow \quad i = \frac{2\pi \times 4 \times 10^{-2} \times 10^{-4}}{4\pi \times 10^{-7}} = 20\text{ A}$$

Also, magnetic field points perpendicular into the plane of paper.

33. *(d)* Consider that the wire is made up of large number of thin wires of infinite length.

Let these wires each be of thickness dl subtending an angle $d\theta$ at centre.

Current through this wire,

$$dI = \frac{d\theta}{\pi} I$$

$\therefore$ Magnetic field at centre due to this portion,

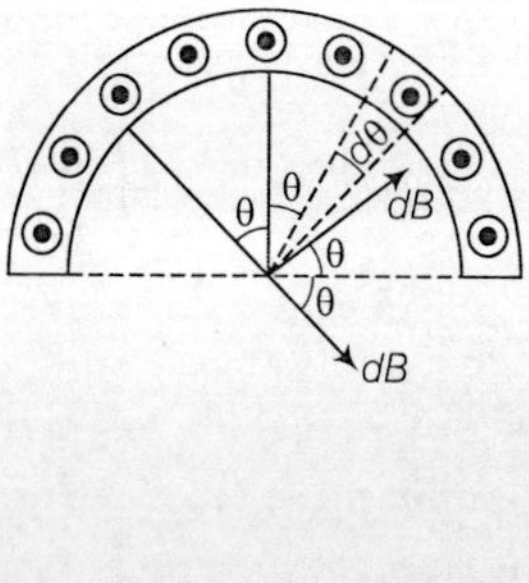

$$dB = \frac{\mu_0}{4\pi} \cdot \frac{2dI}{R} = \frac{\mu_0 I}{2\pi^2 R} d\theta$$

Net magnetic field at the centre,

$$B = \int_{-\pi/2}^{\pi/2} dB \cos\theta = \frac{\mu_0 I}{2\pi^2 R}$$

$$\int_{-\pi/2}^{\pi/2} \cos\theta \, d\theta$$

$$= \frac{\mu_0 I}{\pi^2 R}$$

34. *(a)* The magnetic field in between because of each wire will be in opposite direction.

$$\therefore B_{\text{in between}} = \frac{\mu_0 i}{2\pi x}\mathbf{j} - \frac{\mu_0 i}{2\pi(2d - x)}(-\mathbf{j})$$

$$= \frac{\mu_0 i}{2\pi}\left[\frac{1}{x} - \frac{1}{2d - x}\right](\mathbf{j})$$

At $x = d$, $B_{\text{in between}} = 0$

For $x < d$, $B_{\text{in between}} = (\mathbf{j})$

For $x > d$, $B_{\text{in between}} = (-\mathbf{j})$

Towards x, net magnetic field will add up and direction will be $(-\mathbf{j})$.

Towards x', net magnetic field will add up and direction will be $(\mathbf{j})$.

35. *(b)* Net magnetic field due to loop $ABCD$ at O is

$$B = B_{AB} + B_{BC} + B_{CD} + B_{DA}$$

$$= 0 + \frac{\mu_0 I}{4\pi a} \times \frac{\pi}{6} + 0 - \frac{\mu_0 I}{4\pi b} \times \frac{\pi}{6}$$

$$= \frac{\mu_0 I}{24a} - \frac{\mu_0 I}{24b} = \frac{\mu_0 I}{24ab}(b - a)$$

36. *(c)* $$B_P = \frac{\mu_0 I_2}{2R} = \frac{4\pi \times 10^{-7} \times 4}{2 \times 0.02\pi}$$

$$= 4 \times 10^{-5}\text{ Wb/m}^2$$

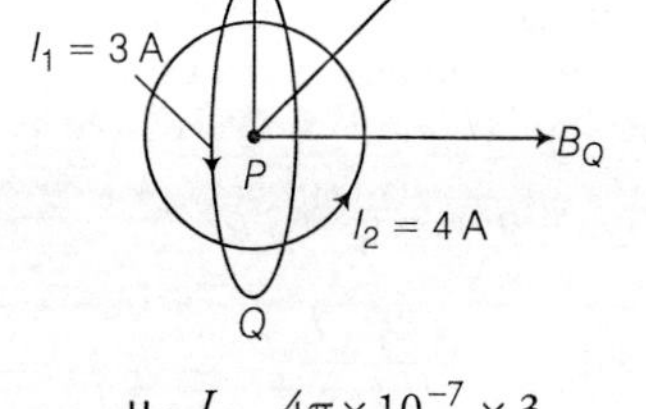

and $$B_Q = \frac{\mu_0 I_1}{2R} = \frac{4\pi \times 10^{-7} \times 3}{2 \times 0.02\pi}$$

$$= 3 \times 10^{-5}\text{ Wb/m}^2$$

$$\therefore \quad B = \sqrt{B_P^2 + B_Q^2}$$

$$= \sqrt{(4 \times 10^{-5})^2 + (3 \times 10^{-5})^2}$$

$$= 5 \times 10^{-5}\text{ Wb/m}^2$$

37. *(b)* The magnetic field at the centre of circular coil is,

$$B = \frac{\mu_0 I}{2r}$$

where r = radius of circle $= l / 2\pi$

$[\because l = 2\pi r]$

$$\therefore \quad B = \frac{\mu_0 I}{2} \times \frac{2\pi}{l} = \frac{\mu_0 I\pi}{l} \quad \ldots\text{(i)}$$

When wire of length l is bent into a circular loops of n turns, then

$$l = n \times 2\pi r'$$

$$\Rightarrow \quad r' = \frac{l}{n \times 2\pi}$$

Thus, new magnetic field,

$$B' = \frac{\mu_0 nI}{2r'} = \frac{\mu_0 nI}{2} \times \frac{n \times 2\pi}{l}$$

$$= \frac{\mu_0 I\pi}{l} \times n^2 = n^2 B$$

[From Eq. (i)]

38. *(b)* The given situation is shown below.

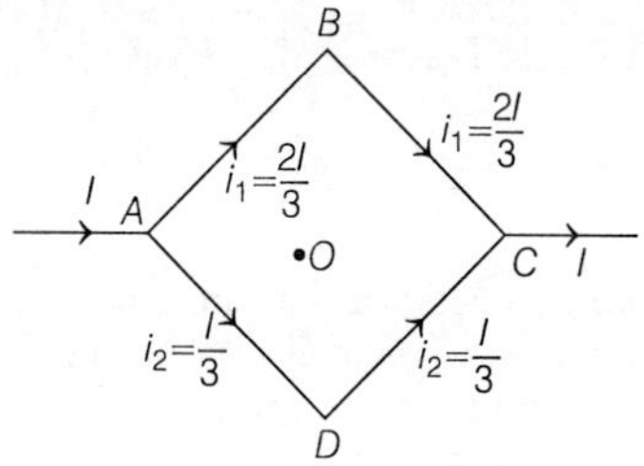

Magnetic field at O due to wire AB

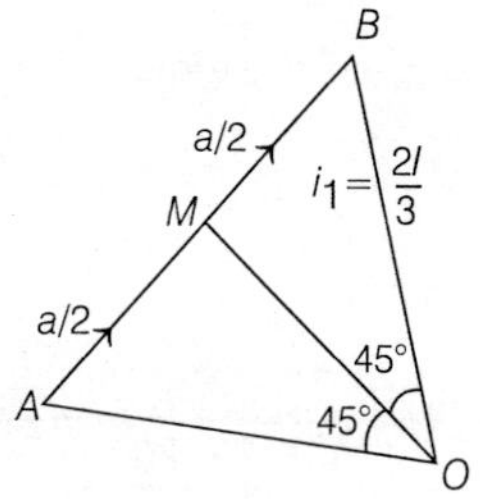

$$\frac{OM}{MB} = \cot 45°$$

$$\Rightarrow OM = MB = \frac{a}{2}$$

$$\mathbf{B}_{AB} = \frac{\mu_0}{4\pi}\frac{i_1}{\frac{a}{2}}(\sin 45° + \sin 45°)$$

$$= \frac{\mu_0}{4\pi} \cdot \frac{2I \cdot 2}{3a}\left(\frac{1}{\sqrt{2}} + \frac{1}{\sqrt{2}}\right)$$

$$= \frac{\mu_0}{4\pi} \cdot \frac{4\sqrt{2}I}{3a}(-\hat{\mathbf{k}})$$

Similarly, $\mathbf{B}_{BC} = \frac{\mu_0}{4\pi} \cdot \frac{4\sqrt{2}I}{3a}(-\hat{\mathbf{k}})$

$$\mathbf{B}_{AD} = \frac{\mu_0}{4\pi} \frac{2\sqrt{2}I}{3a} (\hat{\mathbf{k}})$$

and $\mathbf{B}_{DC} = \frac{\mu_0}{4\pi} \cdot \frac{2\sqrt{2}I}{3a} (\hat{\mathbf{k}})$

$\therefore$ Net magnetic field at centre of loop,

$$\mathbf{B} = \mathbf{B}_{AB} + \mathbf{B}_{BC} + \mathbf{B}_{AD} + \mathbf{B}_{DC}$$

$$= \frac{-\sqrt{2}\mu_0 I}{3\pi a} \hat{\mathbf{k}}$$

$$\therefore B = |\mathbf{B}| = \frac{\sqrt{2}\,\mu_0 I}{3\pi a}$$

39. *(d)*

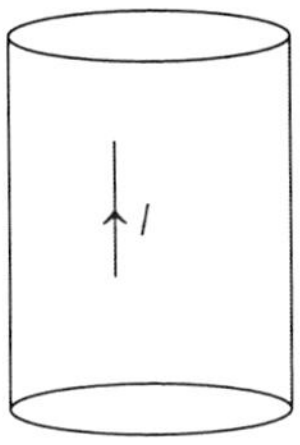

Due to a thick wire, maximum magnetic

Field is at its surface, $B_{max} = \frac{\mu_0 I}{2\pi a}$

Internal magnetic field is, $B_i = \frac{\mu_0 Ir}{2\pi a^2}$

For $\quad B_i = \frac{1}{2} B_{max}$

$$\frac{\mu_0 Ir}{2\pi a^2} = \frac{1}{2} \times \frac{\mu_0 I}{2\pi a} \Rightarrow r = \frac{a}{2}$$

External magnetic field is

$$B_0 = \frac{\mu_0 I}{2\pi r}$$

For $\quad B_0 = \frac{1}{2} \times B_{max}$

$$\frac{\mu_0 I}{2\pi r} = \frac{1}{2} \times \frac{\mu_0 I}{2\pi a}$$

$$r = 2a$$

40. *(a)*

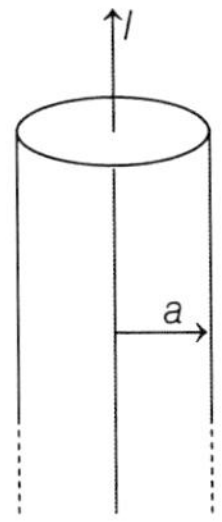

$$B_{in} = \frac{\mu_0 Ir}{2\pi a^2}$$

$B_{in} \propto r \rightarrow$ Straight line

$$B_{out} = \frac{\mu_0 I}{2\pi r}$$

$B_{out} \propto \frac{1}{r} \rightarrow$ Rectangular hyperbola

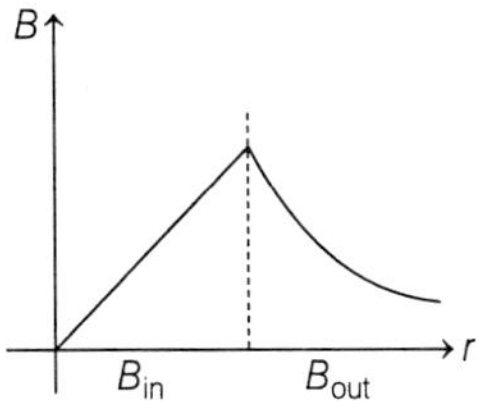

41. *(d)* Magnetic field outside a wire is,

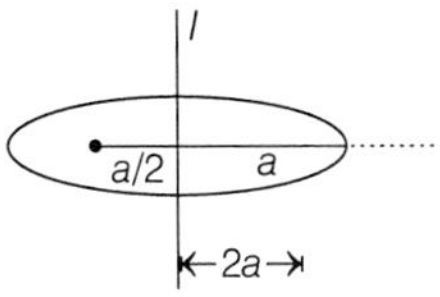

$$B_1 = \frac{\mu_0 i}{2\pi r_1} \quad \ldots(i)$$

Magnetic field inside of a wire is,

$$B_2 = \frac{\mu_0}{2\pi} \cdot \frac{ir_2}{R^2}$$

Here, $r_1 = 2a$, $r_2 = \frac{a}{2}$ and $R = a$

$$\therefore \quad B_1 = \frac{\mu_0 i}{2\pi(2a)} = \frac{\mu_0 i}{4\pi a} \quad \ldots(ii)$$

and $B_2 = \frac{\mu_0 i\left(\frac{a}{2}\right)}{2\pi a^2} = \frac{\mu_0 i}{4\pi a} \quad \ldots(iii)$

From Eq. (ii) and Eq. (iii)

So, $\frac{B_1}{B_2} = 1:1$

42. *(c)* Given, current, $I = 2$ A

Side of equilateral triangle,

$$a = 4\sqrt{3} \text{ cm}$$

$$\therefore \quad QM = \frac{QR}{2} = \frac{a}{2} = 2\sqrt{3} \text{ cm}$$

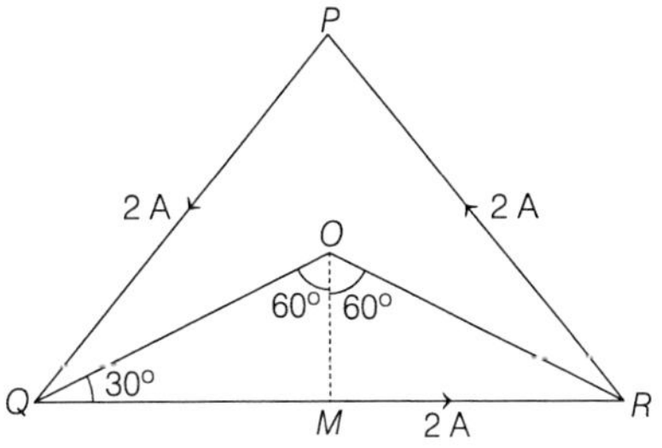

In ΔQOM

$$\tan 30° = \frac{OM}{QM} \Rightarrow \frac{1}{\sqrt{3}} = \frac{OM}{2\sqrt{3}}$$

$\Rightarrow \quad OM = 2$ cm

Magnetic field at O due to current carrying wire QR,

$$B_{QR} = \frac{\mu_0}{4\pi} \cdot \frac{I}{OM} (\sin 60° + \sin 60°)$$

$$= \frac{\mu_0}{4\pi} \cdot \frac{2}{2 \times 10^{-2}} \left[\frac{\sqrt{3}}{2} + \frac{\sqrt{3}}{2}\right]$$

$$= \sqrt{3} \times 10^{-5} \text{ T}$$

Similarly, $B_{RP} = B_{PQ} = B_{QR}$

$$= \sqrt{3} \times 10^{-5} \text{ T}$$

Net magnetic field at 'O' due to all three wires, $B = B_{QR} + B_{PQ} + B_{RP}$

$$= 3B_{QR} = 3\sqrt{3} \times 10^{-5} \text{ T}$$

43. *(a)*

key idea Here, you can use the following two formulae for the calculation of magnetic field.

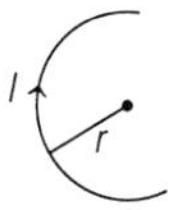

(i) Magnetic field due to semicircular current loop, $\boldsymbol{B}_1 = \frac{\mu_0 \boldsymbol{I}}{4\boldsymbol{r}}$

(ii) Magnetic field due to long straight wire at its one end,

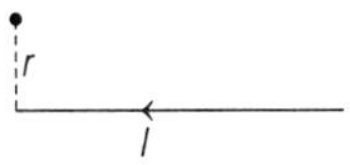

$$B_2 = \frac{\mu_0 I}{4\pi r}$$

A $\rightarrow$ III

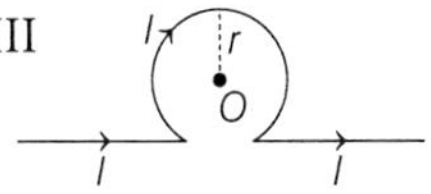

Magnetic field at centre O,

B_0 = Magnetic field due to circular wire $- 2 \times$ magnetic field due to straight wire

$$= \frac{\mu_0 I}{2r} - 2 \times \frac{\mu_0 I}{4\pi r} = \frac{\mu_0 I}{2\pi r}(\pi - 1)$$

B $\rightarrow$ I

Magnetic field at point O

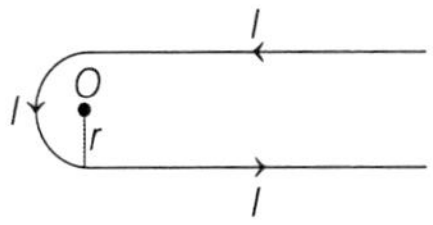

$B_0 = 2 \times$ Magnetic field due to straight wire + Magnetic field due to semi-circular wire

$$= 2 \times \frac{\mu_0 I}{4\pi r} + \frac{1}{2} \cdot \frac{\mu_0 I}{2r} = \frac{\mu_0 I}{4\pi r}(2 + \pi)$$

C $\rightarrow$ IV

Magnetic field at O

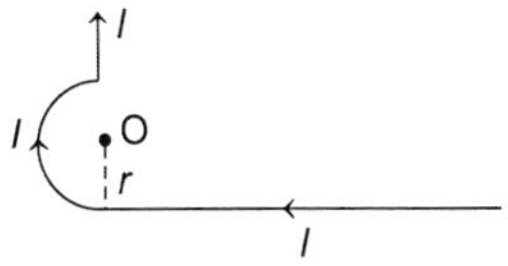

B_0 = Magnetic field due to straight wire + Magnetic field due to semi-circular wire, $= \frac{\mu_0 I}{4\pi r} + \frac{1}{2} \cdot \frac{\mu_0 I}{2r} = \frac{\mu_0 I}{4\pi r}(1 + \pi)$

D → II,

Magnetic field due to straight wire will be zero.

$\therefore B_0 =$ Magnetic field due to semi-circular wire.

$$= \frac{1}{2} \times \frac{\mu_0 I}{2r} = \frac{\mu_0 I}{4r}$$

44. *(c)*

> **key idea** To find the value of magnetic field at the centre, you can use following formula,
>
> $$B_c = \frac{\mu_0 I}{2r}$$
>
> and for magnetic field on the axis at a distance d from centre, you can use following formula.
>
> $$B = \frac{\mu_0 I r^2}{2(d^2 + r^2)^{3/2}}$$

Magnetic field due to current carrying circular loop at the centre,

$$B_c = \frac{\mu_0 I}{2r} \quad \ldots \text{(i)}$$

Magnetic field on the axis at a distance x from the centre

$$B = \frac{\mu_0 I r^2}{2(x^2 + r^2)^{3/2}}$$

Here, $x = \frac{r}{2}$,

Hence, $$B' = \frac{\mu_0 I r^2}{2\left[\left(\frac{r}{2}\right)^2 + r^2\right]^{3/2}}$$

$$= \frac{\mu_0 I r^2}{2\left(\frac{5r^2}{4}\right)^{3/2}} = \frac{\mu_0 I}{2r} \cdot \left(\frac{4}{5}\right)^{3/2}$$

$$= \frac{\mu_0 I}{2r} \cdot \left(\frac{2}{\sqrt{5}}\right)^3$$

$$= B_c \left(\frac{2}{\sqrt{5}}\right)^3 \quad \text{[From Eq. (i)]}$$

$$= \left(\frac{2}{\sqrt{5}}\right)^3 B \quad [\because B_c = B]$$

45. *(a)* Consider an element of thickness dr at a distance r from the centre.

Now, number of turns in dr element,

$$dN = \frac{N}{(b-a)} dr$$

Magnetic field due to this element at the centre,

$$dB = \frac{\mu_0}{2} \frac{dN \cdot I}{r}$$

$$\therefore \quad dB = \frac{1}{2} \frac{\mu_0 I}{r} \frac{N}{(b-a)} dr$$

$\therefore$ Total magnetic field at the centre,

$$B = \int_a^b dB = \int_a^b \frac{\mu_0 IN}{2(b-a)} \frac{dr}{r}$$

$$= \frac{N}{2} \frac{\mu_0 I}{(b-a)} \ln r \Big|_a^b$$

$$= \frac{N}{2} \frac{\mu_0 I}{(b-a)} \ln\left(\frac{b}{a}\right)$$

46. *(a)* Given, radius of inner wire of co-axial cable $= a$

Inner radius of outer shell $= b$

Outer radius of outer shell $= c$

Let, magnetic field at distance $x < a = B_1$

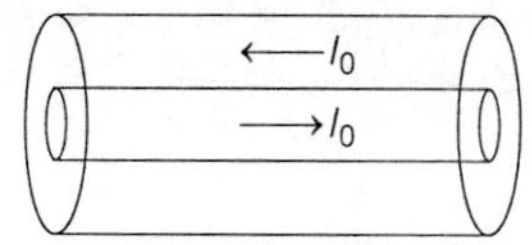

and magnetic field at distance

$$a < x < b = B_2$$

By using Ampere circuital law,

$$\oint B \cdot dl = n\mu_0 I_{enc}$$

where, $dl =$ perimeter of small circular element,

$n =$ number of turns,

$\mu_0 =$ free space permeability

and I = current.

$$\therefore \quad B_1 \cdot 2\pi x = n\mu_0 I_{enc}$$

$$\Rightarrow \quad B_1 = \frac{\mu_0}{2\pi x} \frac{I_0}{\pi a^2} \cdot \pi x^2$$

$$= \frac{\mu_0}{2\pi} \frac{I_0 x}{a^2} \quad \ldots \text{(i)}$$

Now, for $B_2 (a < x < b)$

$$B_2 \cdot 2\pi x = \mu_0 I_0$$

$$B_2 = \frac{1}{2\pi} \frac{\mu_0 I_0}{x} \quad \ldots \text{(ii)}$$

$\therefore$ From Eqs. (i) and (ii), we get

$$\frac{B_1}{B_2} = \frac{\frac{\mu_0 I_0 x}{2\pi a^2}}{\frac{\mu_0 I_0}{2\pi x}} = \frac{x^2}{a^2}$$

47. *(b)* Given, $d_1 = 0.05$ m, $d_2 = 0.2$ m

and $B_1 : B_2 = 8 : 1$

Let radius of coil be r.

As we know that magnetic field at a point on axis of a coil,

$$B = \frac{\mu_0}{2} \cdot \frac{Ir^2}{(d_1^2 + r^2)^{3/2}}$$

$$\therefore \quad \frac{B_1}{B_2} = \frac{r^2 \cdot (d_2^2 + r^2)^{3/2}}{(d_1^2 + r^2)^{3/2} \cdot r^2}$$

$$= \frac{(d_2^2 + r^2)^{3/2}}{(d_1^2 + r^2)^{3/2}}$$

$$\Rightarrow \quad \frac{8}{1} = \left(\frac{d_2^2 + r^2}{d_1^2 + r^2}\right)^{3/2}$$

$$\Rightarrow \quad \frac{2^3}{1} = \left(\frac{d_2^2 + r^2}{d_1^2 + r^2}\right)^{3/2}$$

$$\Rightarrow \quad \frac{4}{1} = \frac{d_2^2 + r^2}{d_1^2 + r^2}$$

$$\Rightarrow \quad 4d_1^2 + 4r^2 = d_2^2 + r^2$$

$$\Rightarrow 4(0.05)^2 + 4r^2 = (0.2)^2 + r^2$$

$$\Rightarrow \quad 3r^2 = 0.03 \Rightarrow r = 0.1 \text{ m}$$

48. *(b)* Given, $\mu_0 = 4\pi \times 10^{-7}$ N/A^2,

$$\omega = 40\pi \text{ rad/s}$$

$$B_{\text{at centre}} = 3.8 \times 10^{-9} \text{ T}$$

and $R = 10$ cm $= 0.1$ m

Now, we know that

$$B = \frac{\mu_0 I}{2r} \quad \ldots \text{(i)}$$

$$\therefore \quad I = \frac{Q}{T} \quad \ldots \text{(ii)}$$

Here, $$T = \frac{2\pi}{\omega} \Rightarrow I = \frac{Q\omega}{2\pi} \quad \ldots \text{(iii)}$$

By putting the value of I from Eq. (iii) to Eq. (i), we get

$$B = \frac{\mu_0 Q\omega}{2r \times 2\pi} \text{ or } Q = \frac{2Br \times 2\pi}{\mu_0 \omega}$$

$$Q = \frac{2 \times 3.8 \times 10^{-9} \times 0.1 \times 2\pi}{4\pi \times 10^{-7} \times 40\pi}$$

or $Q = 3 \times 10^{-5}$ C

49. *(a)*

> **key idea** When a point P lies on the axial position of current-carrying conductor, then magnetic field at P is always zero.

From the given figure as shown below.

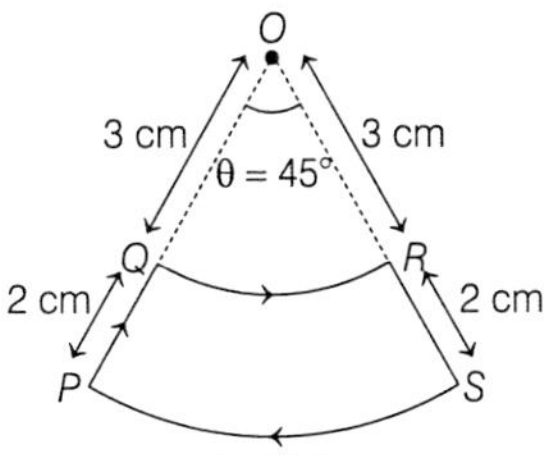

The magnetic field at point O due to wires PQ and RS will be zero.

Magnetic field due to arc QR at point O will be

$$B_1 = \frac{\theta}{2\pi}\left(\frac{\mu_0 i}{2a}\right)$$

$$\Rightarrow \quad B_1 = \frac{\pi}{2\pi \times 4}\left(\frac{\mu_0 \times 10}{2 \times 3 \times 10^{-2}}\right)$$

$$= \frac{5 \times \mu_0 \times 10^2}{24}$$

Direction of field B_1 will be coming out of the plane of figure.

Similarly, field at point O due to arc SP will be

$$B_2 = \frac{\pi}{4}\left(\frac{1}{2\pi}\right)\left[\frac{\mu_0 \times 10}{2 \times (2+3) \times 10^{-2}}\right]$$

$$= \frac{\mu_0 \times 5}{2 \times 20 \times 10^{-2}} = \frac{\mu_0 \times 10^2}{8}$$

Direction of B_2 is going into the plane of the figure.

∴The resultant field at O is

$$B = B_1 - B_2$$

$$= \frac{1}{2}\left(\frac{5 \times \mu_0}{12 \times 10^{-2}} - \frac{\mu_0}{4 \times 10^{-2}}\right)$$

$$= \frac{1}{2}\left(\frac{2\mu_0}{12 \times 10^{-2}}\right)$$

$$= \frac{4\pi \times 10^{-7}}{12 \times 10^{-2}}$$

$$\cong 1 \times 10^{-5}\ \text{T}$$

50. *(d)* Magnetic field in case of circle of radius R, we have

$$B_A = \frac{\mu_0 I}{2R}$$

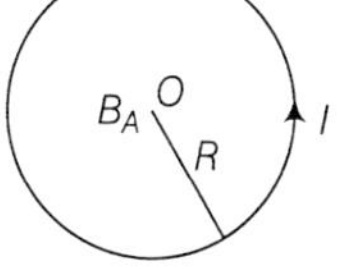

As, $2\pi R = l$, (where l is length of a wire)

$$R = \frac{l}{2\pi}$$

$$\Rightarrow B_A = \frac{\mu_0 I}{2 \times \frac{l}{2\pi}} = \frac{\mu_0 I\pi}{l} \qquad \text{...(i)}$$

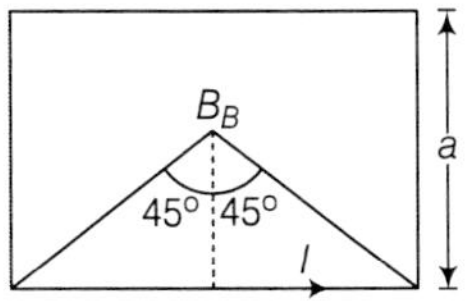

Magnetic field in case of square of side a, we get

$$B_B = 4 \times \frac{\mu_0}{4\pi} \times \frac{I}{\left(\frac{a}{2}\right)}\left(\frac{1}{\sqrt{2}} + \frac{1}{\sqrt{2}}\right)$$

$$\Rightarrow \quad B_B = \frac{4I\mu_0}{\pi a\sqrt{2}}$$

$$= \frac{\mu_0 2\sqrt{2} I}{a\pi}$$

As, $\quad 4a = l,\ a = l/4$

$$\Rightarrow \quad B_B = \frac{8\sqrt{2}\,\mu_0 I}{\pi l} \qquad \text{...(ii)}$$

On dividing Eq. (i) by Eq. (ii), we get

$$\frac{B_A}{B_B} = \frac{\pi^2}{8\sqrt{2}}$$

51. *(c)* Current density,

$$J = \frac{I}{\pi a^2}$$

From Ampere's circuital law,

$$\oint B \cdot dl = \mu_0 \cdot I_{\text{enclosed}}$$

For $r < a$, $B \times 2\pi r = \mu_0 \times J \times \pi r^2$

$$\Rightarrow \quad B = \frac{\mu_0 I}{\pi a^2} \times \frac{r}{2}$$

At $r = \frac{a}{2}$, $B_1 = \frac{\mu_0 I}{4\pi a}$

For $r > a$, $B \times 2\pi r = \mu_0 I$

$$\Rightarrow \quad B = \frac{\mu_0 I}{2\pi r}$$

At $r = 2a$, $B_2 = \frac{\mu_0 I}{4\pi a}$

$$\therefore \quad \frac{B_1}{B_2} = 1$$

52. *(a)* The magnetic field at a point on the axis of a circular loop at a distance x from the centre is

$$B = \frac{\mu_0 I R^2}{2(R^2 + x^2)^{3/2}} \qquad \text{...(i)}$$

Given,

$B = 54\ \mu\text{T}$, $x = 4$ cm,

$R = 3$ cm

On putting the given values in Eq. (i), we get

$$54 = \frac{\mu_0 I \times (3)^2}{2(3^2 + 4^2)^{3/2}}$$

$$\Rightarrow \quad 54 = \frac{9\mu_0 I}{2(25)^{3/2}} = \frac{9\mu_0 I}{2 \times (5)^3}$$

$$\therefore \quad \mu_0 I = \frac{54 \times 2 \times 125}{9}$$

$$\Rightarrow \quad \mu_0 I = 1500\ \mu\text{T-cm} \qquad \text{...(ii)}$$

Now, putting $x = 0$ in Eq. (i), magnetic field at the centre of loop is

$$B = \frac{\mu_0 I R^2}{2R^3}$$

$$= \frac{\mu_0 I}{2R} = \frac{1500}{2 \times 3} \quad \text{[From Eq. (ii)]}$$

$$= 250\ \mu\text{T}$$

53. *(1)* Magnetic field due to straight wire AB and CD at centre O is zero.

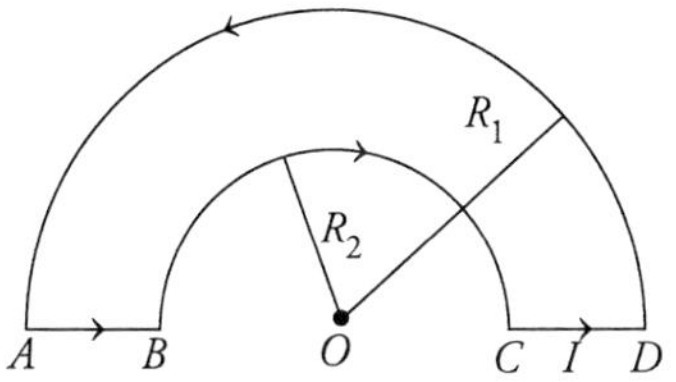

∴Net magnetic field, at O

$$B_0 = |B_{R_1} - B_{R_2}|$$

$$= \frac{\mu_0 I}{4R_2} - \frac{\mu_0 \cdot I}{4R_1}$$

$$= \frac{\mu_0 I}{4}\left(\frac{1}{R_2} - \frac{1}{R_1}\right)$$

$$= \frac{4\pi \times 10^{-7} \times 12}{4}\left(\frac{1}{4\pi} - \frac{1}{6\pi}\right)$$

$$= 12\pi \times 10^{-7}\left(\frac{1}{12\pi}\right)$$

$$= 1 \times 10^{-7}\ \text{T}$$

$$= k \times 10^{-7}\ \text{T} \qquad \text{(given)}$$

$\therefore k = 1$

54. *(8)* $N = 200$, $B = 2.9 \times 10^{-4}$ T, $I = 0.29$ A

For a solenoid

$$B = \mu_0 \frac{N}{l} I$$

$$\Rightarrow 2.9 \times 10^{-4} = \frac{4\pi \times 10^{-7} \times 200 \times 0.29}{l}$$

$$\Rightarrow \quad l = \frac{4\pi \times 10^{-7} \times 2 \times 29}{2.9 \times 10^{-4}}$$

$$l = 80\pi \times 10^{-3} \text{ m} = 8\pi \text{ cm}$$

55. *(500)* Number of turns per unit length, $n = \frac{m}{l} = \frac{m}{0.5} = 2m$

Since, $B = \mu_0 nI$

$\Rightarrow \quad 6.28 \times 10^{-3} = 4\pi \times 10^{-7} \times 2m \times 5$

$\therefore \quad 10^{-3} = 2 \times 10^{-6} m$

$\Rightarrow \quad m = \frac{1}{2 \times 10^{-3}} = 500$

56. *(5)*

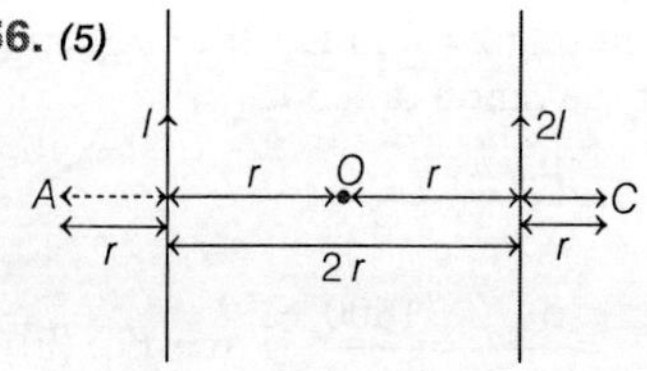

Net magnetic field at point A,

$$B_1 = \frac{\mu_0}{2\pi} \cdot \frac{I}{r} + \frac{\mu_0}{2\pi} \cdot \frac{2I}{(2r + r)}$$

$$= \frac{\mu_0}{2\pi} \cdot \frac{I}{r}\left(1 + \frac{2}{3}\right)$$

$$= \frac{\mu_0 I}{2\pi r}\left(\frac{5}{3}\right)$$

Net magnetic field at point C,

$$B_2 = \frac{\mu_0}{2\pi} \cdot \frac{I}{(r + 2r)} + \frac{\mu_0}{2\pi} \cdot \frac{2I}{r}$$

$$= \frac{\mu_0 I}{2\pi r}\left(\frac{1}{3} + 2\right)$$

$$= \frac{\mu_0 I}{2\pi r} \cdot \left(\frac{7}{3}\right)$$

$$\therefore \quad \frac{B_1}{B_2} = \frac{\frac{\mu_0 I}{2\pi r}\left(\frac{5}{3}\right)}{\frac{\mu_0 I}{2\pi r}\left(\frac{7}{3}\right)} = \frac{5}{7} = \frac{x}{7} \quad \text{(given)}$$

$\therefore \quad x = 5$

57. *(72)*

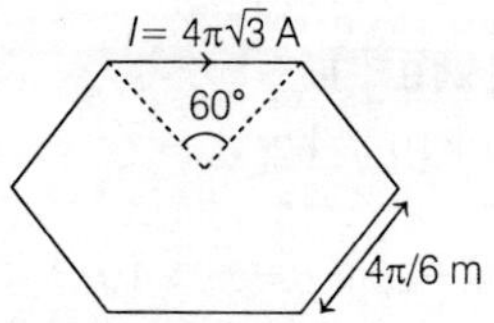

Using Biot-Savart's formula, magnetic field due to a finite wire is,

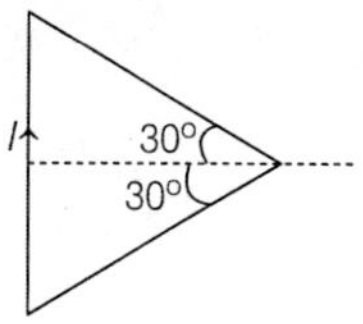

$$B = \frac{\mu_0}{4\pi} \cdot \frac{I}{r} \, 2 \sin \phi$$

Here, $I = 4\pi\sqrt{3}$ A, $\phi = 30°$

Distance r = Altitude of equilateral triangle of side $4\pi/6$

$$\therefore \quad r = \frac{\sqrt{3} \times \text{side}}{2} = \frac{4\pi\sqrt{3}}{12} \text{ m}$$

So, magnetic field due to all 6 wires along sides of hexagon is,

$$B = 6 \times \frac{\mu_0}{4\pi} \times \frac{I}{r} \times 2 \sin \phi$$

$$= 6 \times 10^{-7} \times \frac{4\pi\sqrt{3}}{4\pi \frac{\sqrt{3}}{12}} \times 2 \times \sin 30°$$

$$= 72 \times 10^{-7} \text{ T}$$

58. *(20)*

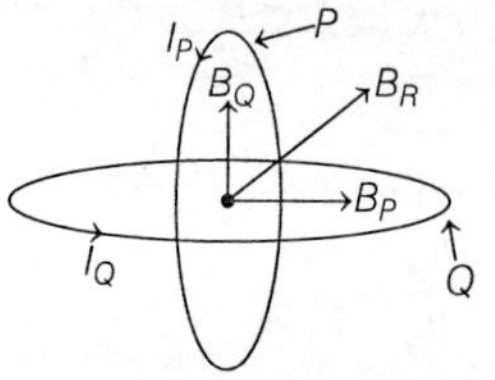

Resultant magnetic field's magnitude is,

$$B_R = \sqrt{B_P^2 + B_Q^2}$$

Now, $B_P = \frac{\mu_0 N i_P}{2 r_P} = \frac{\mu_0 \times 100 \times 1}{2\pi \times 10^{-2}}$

$$= \frac{\mu_0}{4\pi} \times 2 \times 10^4$$

$$= 2 \times 10^4 \times 10^{-7} = 2 \times 10^{-3} \text{ T}$$

and $B_Q = \frac{\mu_0 N I_Q}{2 r_Q} = \frac{\mu_0 \times 100 \times 2}{2 \times \pi \times 10^{-2}}$

$$= \frac{\mu_0}{4\pi} \times 4 \times 10^4 = 4 \times 10^{-3} \text{ T}$$

So, resultant magnetic field is,

$$B_R = \sqrt{2^2 + 4^2} \times 10^{-3} = \sqrt{20} \text{ mT}$$

$$= \sqrt{x} \text{ mT} \quad \text{(given)}$$

$\therefore \quad x = 20$

59. *(40)* Magnetic field at centre is sum of magnetic fields produced by each side of the cube (obtained by Biot-Savart's law)

So, $B_{\text{centre}} = 4 \times \frac{\mu_0 I}{4\pi d} (\sin 45° + \sin 45°)$

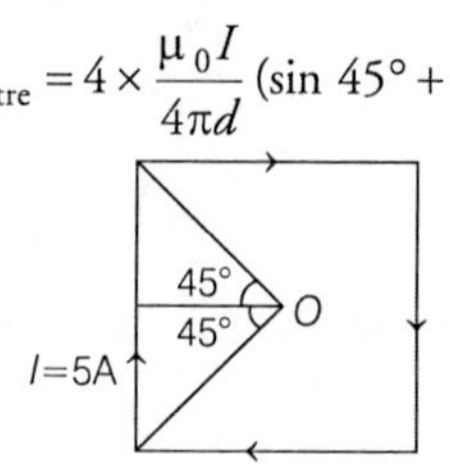

$$= \frac{4 \times 10^{-7} \times 5}{0.5} \times \frac{2}{\sqrt{2}}$$

$$= 4\sqrt{2} \times 10^{-6} \text{ T}$$

$$= 40\sqrt{2} \times 10^{-7} \text{ T}$$

$$= x\sqrt{2} \times 10^{-7} \text{ T (given)}$$

$\therefore \quad x = 40$

60. *(160)* As magnetic fields of both wires are in same direction,

$$B = \frac{\mu_0 I}{2\pi D} \times 2 = \frac{4\pi \times 10^{-7} \times 10}{2\pi \times \frac{5}{2} \times 10^{-2}} \times 2$$

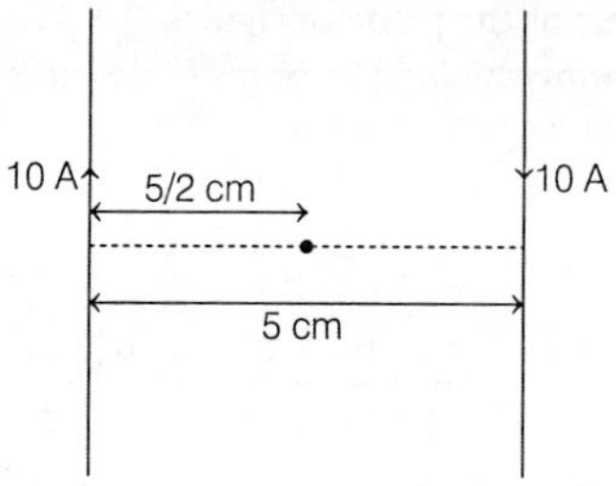

$$= 2 \times 2 \times 2 \times 2 \times 10^{-7} \times 10^2$$

$$= 16 \times 10^{-5} = 160 \times 10^{-6}$$

$$B = 160 \, \mu\text{T}$$

61. *(40)* Given,

$$v = 6.76 \times 10^6 \text{ ms}^{-1}$$

$$r = 0.52 \text{ Å} = 0.52 \times 10^{-10} \text{ m}$$

Current associated with revolving electron,

$$I = \frac{e}{T} = \frac{e}{\left(\frac{2\pi r}{v}\right)} = \frac{ev}{2\pi r}$$

$\therefore$ Magnetic field,

$$B = \frac{\mu_0 I}{2r} = \frac{\mu_0 ev}{4\pi r^2} = \left(\frac{\mu_0}{4\pi}\right) \cdot \frac{ev}{r^2}$$

$$= \frac{10^{-7} \times 1.6 \times 10^{-19} \times 6.76 \times 10^6}{(0.52 \times 10^{-10})^2}$$

$$= 40 \text{ T}$$

62. *(8)* Using $B_{axis} = \dfrac{\mu_0 I r^2}{2(r^2 + x^2)^{3/2}}$

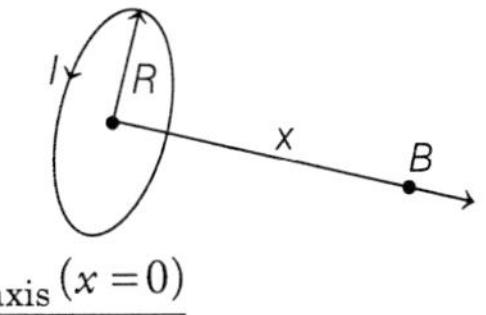

So, $\dfrac{B_{axis}(x=0)}{B_{axis}(x=r)}$

$$= \frac{\mu_0 I r^2}{2(r^2+0)^{3/2}} \times \frac{2(r^2+r^2)^{3/2}}{\mu_0 I r^2}$$

$$= \frac{2^{3/2} r^3}{r^3} = \frac{\sqrt{8}}{1}$$

So, $x = 8$

63. *(628)*

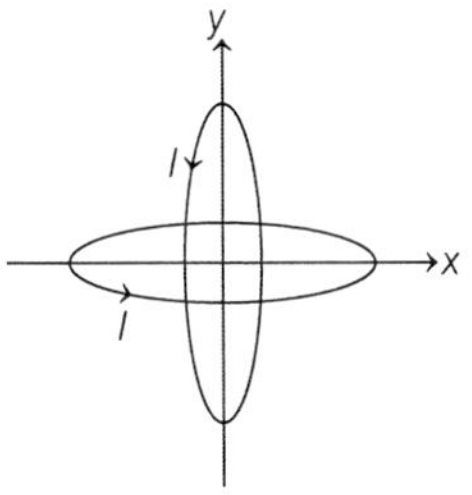

$$\mathbf{B}_{net} = \frac{\mu_0 i}{2r}\hat{\mathbf{i}} + \frac{\mu_0 i}{2r}\hat{\mathbf{j}} = \frac{\mu_0 i}{2r}\sqrt{2}$$

$$= 4\pi \times 10^{-7} \times \sqrt{2} \times \sqrt{2} \times \frac{1}{2 \times 0.2}$$

$$= 2 \times 3.14 \times 10^{-6} = 628 \times 10^{-8}\ \text{T}$$

64. *(3)* We know that formula for B_{centre} of a coil is

$$B_{centre} = \frac{\mu_0 NI}{2R} \quad \ldots \text{(i)}$$

Given, $B_{centre} = 37.68 \times 10^{-4}$ T

$R = 5$ cm $= 0.05$ m

$\pi = 3.14$

$N = 100$ turns

$\therefore$ On putting all values in Eq. (i) we get,

$$37.68 \times 10^{-4} = \frac{100 \times 4\pi \times 10^{-7} \times I}{2 \times 0.05}$$

$$(\because \mu_0 = 4\pi \times 10^{-7}\,\text{TmA}^{-1})$$

$\Rightarrow \quad I = 3\text{A}$

Topic 2

01. *(b)* $\mathbf{F} = q(\mathbf{v} \times \mathbf{B})$

$\mathbf{F} = 0$

$\mathbf{v} \,||\, \mathbf{B}$

$\theta = 0°$ or $180°$

Both A and R are true.

02. *(c)* Electrostatic force, $\mathbf{F}_1 = q\mathbf{E}$ and magnetic force $\mathbf{F}_2 = q(\mathbf{v} \times \mathbf{B})$

03. *(a)* An electron is allowed to move with constant velocity along the axis of current carrying straight solenoid.
The electron will not experience magnetic force.
The electron will continue to move along the axis of the solenoid.

• Concept Enhancer

When a charged particle moves parallel to the magnetic field, the magnetic force acting on the particle is zero.

04. *(b)* We know that

$$F = q(\mathbf{v} \times \mathbf{B})$$

$$\Rightarrow \quad m\mathbf{a} = q(\mathbf{v} \times \mathbf{B})$$

$\therefore$ $\mathbf{a} \perp \mathbf{B}$, so, $\mathbf{a} \cdot \mathbf{B} = 0$

$$(\alpha\hat{\mathbf{i}} - 4\hat{\mathbf{j}})(2\hat{\mathbf{i}} + 3\hat{\mathbf{j}}) = 0$$

$$\Rightarrow \quad 2\alpha - 12 = 0$$

$$\Rightarrow \quad \alpha = 6$$

05. *(c)* We know that, frequency of revolution of the electron,

$$f = \frac{1}{T} = \frac{eB}{2\pi m} \quad \left(\because T = \frac{2\pi m}{qB}\right)$$

$$= \frac{1.6 \times 10^{-19} \times 10^{-4}}{2\pi \times 9 \times 10^{-31}}$$

$$= 2.8 \times 10^{6}\ \text{Hz}$$

06. *(c)* Radius of a charged particle in uniform magnetic field is given by,

$$R = mv/qB$$

For α particle, $R_\alpha = \dfrac{m_\alpha v}{q_\alpha B}$

For proton, $R_p = \dfrac{m_p v}{q_p \cdot B}$

$$\Rightarrow \quad \frac{R_\alpha}{R_p} = \frac{m_\alpha / q_\alpha}{m_p / q_p} = \frac{2}{1}$$

07. *(b)* Let masses of ions be m_1 and m_2 respectively, charges be q_1 and q_2 respectively and speed of charges be v_1 and v_2.

Given, $m_1 = m_2 = m$

$q_1 : q_2 = 1 : 2$

$v_1 : v_2 = 2 : 3$

As we know that $r = \dfrac{mv}{qB}$

$$\therefore \quad \frac{r_1}{r_2} = \frac{m_1 v_1}{q_1 B} \times \frac{q_2 B}{m_2 v_2}$$

$$= \frac{m}{m} \times \frac{2}{1} \times \frac{2}{3} = \frac{4}{3}$$

08. *(b)* The force is parallel to the direction of current in magnetic field,

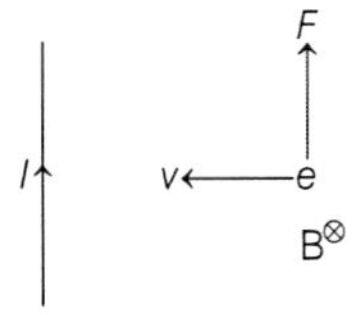

Since, $F = q(\mathbf{v} \times \mathbf{B})$, hence according to Fleming's left hand rule, the direction of motion of charge is towards the wire.

09. *(d)* From figure, $\sin\alpha = d/R$

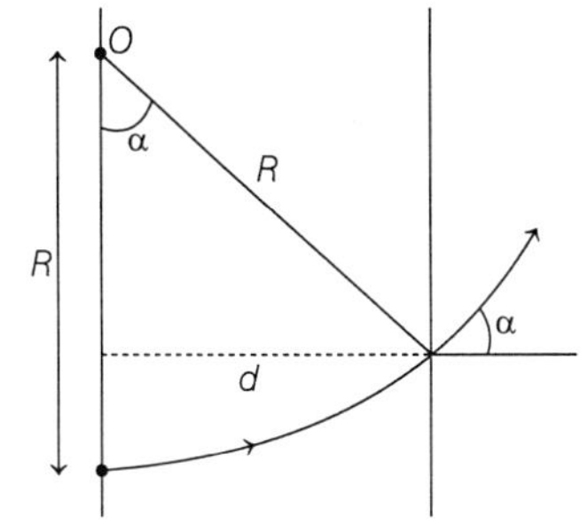

And we know, $\dfrac{mv^2}{R} = qvB \Rightarrow R = \dfrac{mv}{qB}$

$$\therefore \quad \sin\alpha = \frac{dqB}{mv}$$

$$\Rightarrow \sin\alpha = Bd\sqrt{\frac{q}{2mV}} \quad \left(\because qV = \frac{1}{2}mv^2\right)$$

10. *(b)* By Lorentz force, net force

$$\mathbf{F} = q(\mathbf{E} + \mathbf{v} \times \mathbf{B})$$

$$= q[(3\hat{\mathbf{i}} + \hat{\mathbf{j}} + 2\hat{\mathbf{k}}) + (3\hat{\mathbf{i}} + 4\hat{\mathbf{j}} + \hat{\mathbf{k}}) \times (\hat{\mathbf{i}} + \hat{\mathbf{j}} - 3\hat{\mathbf{k}})]$$

$$= q[-10\hat{\mathbf{i}} + 11\hat{\mathbf{j}} + \hat{\mathbf{k}}]$$

$\therefore \; F_y = 11q$

11. *(a)* As **v** of charged particle is remaining constant, it means force acting on charged particle is zero.

$$F = q\mathbf{E} + q(\mathbf{v} \times \mathbf{B}) = 0$$

$$q\mathbf{E} - q(\mathbf{v} \times \mathbf{B}) = 0$$

So, $\quad q(\mathbf{v} \times \mathbf{B}) = q\mathbf{E} \Rightarrow \mathbf{v} \times \mathbf{B} = \mathbf{E}$

$\Rightarrow \quad (\mathbf{v} \times \mathbf{B}) \times \mathbf{B} = \mathbf{E} \times \mathbf{B}$

$(\mathbf{B} \cdot \mathbf{B})\mathbf{v} - (\mathbf{v} \cdot \mathbf{B})\mathbf{B} = \mathbf{E} \times \mathbf{B}$

$\Rightarrow \quad \mathbf{v} = \dfrac{\mathbf{E} \times \mathbf{B}}{B^2}$

12. *(a)* In case of motion of a charged particle perpendicular to the magnetic field, i.e., displacement, the work done $W = \int \mathbf{F} \cdot d\mathbf{s} = \int F ds \cos\theta = 0$ [as $\theta = 90°$]
and by work-energy theorem,
$W = \Delta$ KE, the kinetic energy and hence speed v remains constant. But **v** changes, so momentum changes.

13. *(b)* Let E and B be along x-axis. When a charged particle is released from rest, it will experience an electric force along the direction of electric field or opposite to the direction of electric field depending on the nature of charge.

Due to this force, it acquires some velocity along x-axis. Due to this motion of charge, magnetic force cannot have non-zero value because angle between **v** and **B** would be either 0° or 180°.

So, only electric force is acting on particle and hence it will move along a straight line.

14. *(d)* We know that or $r = mv / qB$

The time taken by the particle to complete one revolution,

$$T = \frac{2\pi r}{v} = \frac{2\pi mv}{vqB} = \frac{2\pi m}{qB}$$

15. *(a)* When **E**, **v** and **B** are all along same direction, then magnetic force experienced by electron is zero while electric force is acting opposite to velocity of electron, so velocity of electron will decrease.

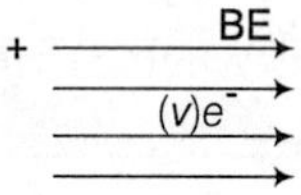

16. *(a)* Since, momenta are same and masses of electron and proton are different, so they will attain different velocities and hence experience different forces. But radius of circular path depends on momenta, so both will be moving on same trajectory (curved path).

17. *(a)* Time period T is

$$T = \frac{2\pi r}{v} = \frac{2\pi}{v}\left(\frac{mv}{qB}\right) = \frac{2\pi m}{qB}$$

So, T is independent of its speed.

18. *(d)* We know that,

$$r = \frac{mv}{Bq} = \frac{p}{Bq}$$

⇒ For equal momentum,

$$r \propto \frac{1}{q}$$

i.e., $r_{O^{-2}} < r_{H^+}$

19. *(d)*

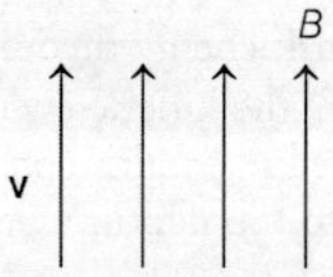

$$\frac{mv^2}{r} = qvB$$

$$mv = qBr$$

$$E = \frac{1}{2}mv^2 = \frac{1}{2}m\left(\frac{qBr}{m}\right)^2$$

$$= \frac{1}{2}m\frac{q^2B^2r^2}{m^2} = \frac{1}{2} \times \frac{q^2B^2r^2}{m}$$

$$E = \left(\frac{q^2B^2}{2m}\right)r^2 \Rightarrow E \propto r^2$$

or $r^2 \propto E$

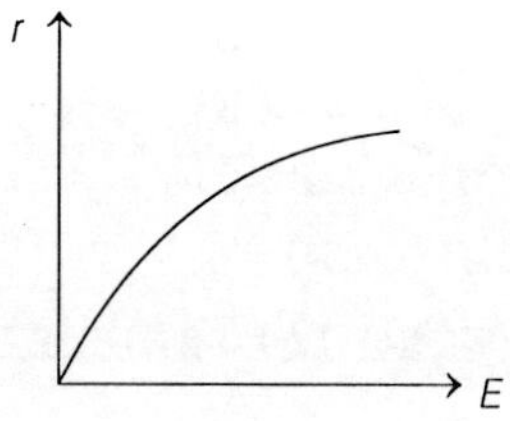

20. *(d)* We know that, in uniform magnetic field.

$$R = \frac{\sqrt{2mK}}{qB}$$

Since, same K (kinetic energy)

$$R \propto \frac{\sqrt{m}}{q}$$

$$\therefore \quad \frac{R_d}{R_p} = \sqrt{\frac{m_d}{m_p}} \times \frac{q_p}{q_d}$$

$$\therefore \quad \frac{R_d}{R_p} = \sqrt{\frac{2m_p}{m_p}} \times \frac{e}{e}$$

$$\therefore \quad \frac{R_d}{R_p} = \frac{\sqrt{2}}{1} \text{ or } \sqrt{2}:1$$

21. *(b)* When a charge particle is accelerated with potential difference V, then

$$\frac{1}{2}mv^2 = qV \Rightarrow v = \sqrt{\frac{2qV}{m}}$$

∴ When charged particle enters into magnetic field B, then

$$R = \frac{mv}{Bq} = \frac{m}{Bq} \times \sqrt{\frac{2qV}{m}}$$

$$= \frac{1}{B}\sqrt{\frac{2mV}{q}}$$

$$\therefore \quad R \propto \sqrt{m}$$

Thus, $\frac{R_1}{R_2} = \sqrt{\frac{m_x}{m_y}} \Rightarrow \left(\frac{R_1}{R_2}\right)^2 = \frac{m_x}{m_y}$

$$\Rightarrow \quad \frac{m_x}{m_y} = \left(\frac{R_1}{R_2}\right)^2$$

22. *(a)* As, $F_e = qE$ and $F_B = Bqv$

When $E = 0$ and $B = 0$, there is no force on charged particle.

When $E = 0$ and **v** || **B**, there is no force on charged particle.

When $F_e = F_B$ or $v = \frac{E}{B}$ the net force on charged particle is zero.

(A) $E = 0, B = 0$ (B) $E = 0, B \neq 0$

(D) $E \neq 0, B \neq 0$

Option (a) A, B, D correct.

23. *(c)* Given, velocity, $v = 3 \times 10^7$ ms^{-1}

We have, $\frac{F_e}{F_B} = \frac{qE_0}{qvB_0} = \frac{E_0}{vB_0} = \frac{c}{v}$ $\left(\because \frac{E_0}{B_0} = c\right)$

$$= \frac{3 \times 10^8}{3 \times 10^7} = \frac{10}{1}$$

24. *(c)* Statement I is correct.

As electric force and acceleration,

$$F = qE \text{ and } a = \frac{q}{m} \cdot E$$

Also magnetic force in a perpendicular magnetic field,

$$F = q(\mathbf{v} \times \mathbf{B}) \Rightarrow \mathbf{v} \perp \mathbf{F}$$

So, work done by force,

$$W = (\mathbf{F} \cdot \mathbf{v}) = \text{zero}$$

As work done is zero, so no change in kinetic energy occurs.

Therefore, Statement II is incorrect as electric force on a positive particle is along the direction of electric field.

25. *(c)* Kinetic energy of a particle rotating in a magnetic field is

$$\text{KE} = \frac{1}{2}mv^2 = \frac{B^2q^2r^2}{2m}$$

$$\left[\because r = \frac{mv}{Bq} \Rightarrow v = \frac{rBq}{m}\right]$$

$$\Rightarrow \text{KE} \propto r^2 \Rightarrow \frac{K_f}{K_i} = \left(\frac{r_f}{r_i}\right)^2$$

Given, $K_f = 4K_i$

$$\Rightarrow \quad \left(\frac{r_f}{r_i}\right)^2 = 4 \Rightarrow r_f = 2r_i$$

or $\quad \frac{r_f}{r_i} = \frac{2}{1}$

26. *(a)* In a uniform magnetic field, a moving charged particle experiences a magnetic force perpendicular to its direction of motion, which is called **Lorentz force** and is given as

$$\mathbf{F}=q(\mathbf{v}\times\mathbf{B}) \text{ i.e. } \mathbf{F}\perp\mathbf{v}$$

i.e. force on charged particle always remains perpendicular to its velocity i.e. its displacement. Hence, work done

$$=\mathbf{F}\cdot\mathbf{s}=Fs\cos 90°=0$$

Hence, according to work-energy theorem, $\Delta E_K = W = 0$

i.e. kinetic energy = constant.

Hence, in uniform magnetic field, speed and energy remains the same for moving charged particle.

Therefore, both A and R are correct and R is the correct explanation of A.

27. *(b)* Since, kinetic energy, $K=\frac{1}{2}mv^2$

$$\Rightarrow \quad mv^2 = 2K \Rightarrow v^2 = \frac{2K}{m} \qquad \text{...(i)}$$

We know that

$$r=\frac{mv}{Bq} \qquad \text{...(ii)}$$

From Eqs. (i) and (ii), we get

$$r=\frac{m\sqrt{\frac{2K}{m}}}{Bq} \Rightarrow r=\frac{\sqrt{2Km}}{Bq} \qquad \text{...(iii)}$$

Since, K and B are same for both deuteron and α-particle.

Therefore, from Eq. (iii), we get

$$r \propto \frac{\sqrt{m}}{q}$$

$$\therefore \quad \frac{r_d}{r_\alpha}=\sqrt{\frac{m_d}{m_\alpha}}\cdot\frac{q_\alpha}{q_d}=\sqrt{\frac{2}{4}}\left(\frac{2}{1}\right)$$

$$\left[\because m_d = 2m_p \text{ and } m_\alpha = 4m_p \atop q_\alpha = 2e \text{ and } q_d = e.\right]$$

$$\frac{r_d}{r_\alpha}=\sqrt{2}$$

28. *(d)* We know that

$$r = mv/qB$$

$$\Rightarrow \quad r=\frac{p}{qB} \qquad (\because p = mv)$$

$$\because \quad \frac{r_p}{r_\alpha}=\frac{p_p}{q_p}\times\frac{q_\alpha}{p_\alpha}$$

$$\Rightarrow \quad \frac{2}{1}=\frac{+2e}{+e}\times\frac{p_p}{p_\alpha} \qquad \left[\because \frac{r_p}{r_\alpha}=\frac{2}{1}\right]$$

$$\Rightarrow \quad \frac{p_p}{p_\alpha}=1$$

Since, kinetic energy and momentum are related as $K=\frac{p^2}{2m}$

$$\therefore \quad \frac{K_p}{K_\alpha}=\frac{p_p^2 m_\alpha}{m_p p_\alpha^2}$$

$$\Rightarrow \quad \frac{K_p}{K_\alpha}=\frac{m_\alpha}{m_p}=\frac{4m_p}{m_p}$$

$$\Rightarrow \quad \frac{K_p}{K_\alpha}=\frac{4}{1}$$

29. *(c)* As we know that, if a charge particle moves in a uniform magnetic field, then its path is always circular. Considering the positive charge the direction of magnetic force acting on it, has been shown by dotted line.

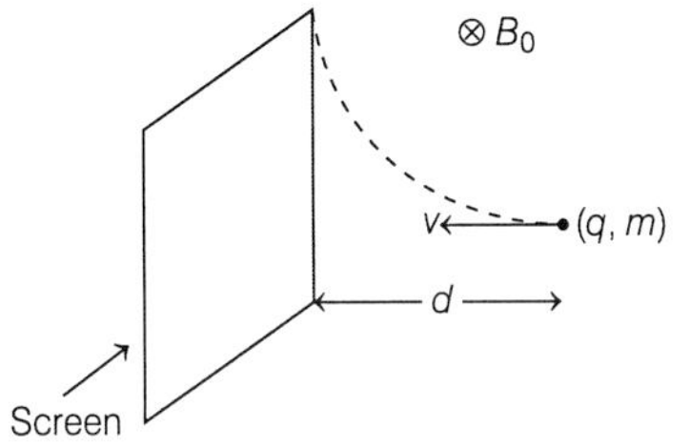

As, charge particle should not hit the screen this means radius of circular path must be less than distance d.

i.e. $R \le d \Rightarrow \frac{mv}{qB_0} \le d \qquad \left[\because R=\frac{mv}{qB_0}\right]$

or $\quad v \le \frac{qB_0 d}{m}$

Maximum value of $v=\frac{qB_0 d}{m}$

Note In this problem, minimum value has been asked which is clearly an error, so read it maximum value.

30. *(b)* Acceleration produced by electric field,

$$\mathbf{a}=\frac{qE_0}{m}\hat{\mathbf{i}}$$

After time t, velocity of the particle,

$$\mathbf{v}=\mathbf{u}+\mathbf{a}t$$

$$\Rightarrow \quad \mathbf{v}=v_0\hat{\mathbf{j}}+\left(\frac{qE_0 t}{m}\right)\hat{\mathbf{i}} \qquad [\because u = v_0\hat{\mathbf{j}}]$$

So, speed of the particle,

$$|\mathbf{v}|=\sqrt{v_0^2+\left(\frac{qE_0 t}{m}\right)^2}$$

Now, given $|\mathbf{v}| = 2v_0$ after time t, so

$$2v_0=\sqrt{v_0^2+\left(\frac{qE_0 t}{m}\right)^2}$$

$$\Rightarrow 4v_0^2 = v_0^2+\left(\frac{qE_0 t}{m}\right)^2$$

$$\Rightarrow 3v_0^2=\left(\frac{qE_0 t}{m}\right)^2$$

So, time interval, $t=\frac{\sqrt{3}\,mv_0}{qE_0}$.

31. *(a)* Radius of path of charged particle q in a uniform magnetic field B of mass 'm' moving with velocity v is

$$r=\frac{mv}{Bq}=\frac{m\sqrt{(2qV/m)}}{Bq}$$

$$\Rightarrow \quad r \propto \frac{\sqrt{m}}{\sqrt{q}} \text{ so, required ratio is}$$

$$\Rightarrow \quad \frac{r_p}{r_\alpha}=\sqrt{\frac{m_p}{m_\alpha}}\times\sqrt{\frac{q_\alpha}{q_p}}=\sqrt{\frac{1}{4}}\times\sqrt{\frac{2}{1}}$$

$$=\frac{\sqrt{2}}{2}=\frac{1}{\sqrt{2}}$$

32. *(d)* Here, $\mathbf{E}=2\hat{\mathbf{i}}+3\hat{\mathbf{j}}$, $\mathbf{B}=4\hat{\mathbf{j}}+6\hat{\mathbf{k}}$,

q = charge on a particle.

Initial position, $r_1 = (0, 0)$

Final position, $r_2 = (1, 1)$

Net force experienced by charge particle inside electromagnetic field is

$$\mathbf{F}_{\text{net}} = q\mathbf{E}+q(\mathbf{v}\times\mathbf{B}) = q(2\hat{\mathbf{i}}+3\hat{\mathbf{j}})$$

[Here, $\mathbf{v}\times\mathbf{B}=0$]

$$=(2q\hat{\mathbf{i}}+3q\hat{\mathbf{j}})$$

$$\therefore \quad dW = \mathbf{F}_{\text{net}}\cdot d\mathbf{r}$$

$$\Rightarrow \int dW = \int_{r_1}^{r_2}(2q\hat{\mathbf{i}}+3q\hat{\mathbf{j}})\cdot(dx\,\hat{\mathbf{i}}+dy\,\hat{\mathbf{j}})$$

[here, $d\mathbf{r}=dx\,\hat{\mathbf{i}}+dy\,\hat{\mathbf{j}}$]

$$\Rightarrow \quad W = 2q\int_0^1 dx + 3q\int_0^1 dy$$

or $\quad W = 2q + 3q$ or $W = 5q$

33. *(b)* From $Bqv=\frac{mv^2}{r}$, we have

$$r=\frac{mv}{Bq}=\frac{\sqrt{2mK}}{Bq}$$

where, K is the kinetic energy.

As, kinetic energies of particles are same

$$r \propto \frac{\sqrt{m}}{q}$$

$$\Rightarrow r_e : r_p : r_\alpha = \frac{\sqrt{m_e}}{e}:\frac{\sqrt{m_p}}{e}:\frac{\sqrt{4m_p}}{2e}$$

Clearly, $r_p = r_\alpha$ and r_e is least

$[\because m_e < m_p]$

So, $\quad r_p = r_\alpha > r_e$

34. *(b)* For charged particle in magnetic field, radius

$$r=\frac{mv}{qB}=\frac{\sqrt{2Km}}{qB}$$

[as $mv = p \Rightarrow K = \frac{1}{2}mv^2 = \frac{m^2v^2}{2m}$

$= p^2/2m \; p = \sqrt{2\,Km}]$

$\Rightarrow \quad r \propto \frac{\sqrt{m}}{q}$ or $m_d = 2m_p$

and $q_d = q_p$; $m_\alpha = 4m_p$ and $q_\alpha = 2q_p$

$\Rightarrow r_p : r_d : r_\alpha = \frac{\sqrt{m_p}}{q_p} : \frac{\sqrt{2m_p}}{q_p} : \frac{\sqrt{4m_p}}{2q_p}$

$= 1 : \sqrt{2} : 1 \Rightarrow r_\alpha = r_p < r_d$

35. *(a)* The force on a particle is

So, $\mathbf{F} = q(\mathbf{E} + \mathbf{v} \times \mathbf{B})$ or $\mathbf{F} = \mathbf{F}_e + \mathbf{F}_m$

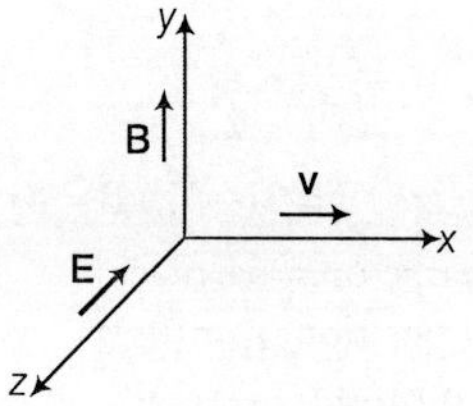

$\therefore \quad \mathbf{F}_e = q\mathbf{E} = -16 \times 10^{-18} \times 10^4 (-\hat{\mathbf{j}})$

$= 16 \times 10^{-14} \; \hat{\mathbf{k}}$

and $\mathbf{F}_m = -16 \times 10^{-18} (10\,\hat{\mathbf{i}} \times B\,\hat{\mathbf{j}})$

$= -16 \times 10^{-17} \times B(+\hat{\mathbf{k}})$

$= -16 \times 10^{-17} \, B\,\hat{\mathbf{k}}$

Since, particle will continue to move along + *X*-axis, so resultant force is equal to zero.

$\mathbf{F}_e + \mathbf{F}_m = 0$

$\Rightarrow 16 \times 10^{-14} = 16 \times 10^{-17} B$

$\Rightarrow \quad B = \frac{16 \times 10^{-14}}{16 \times 10^{-17}} = 10^3$

or $\quad B = 10^3 \text{ Wb/m}^2$

36. *(c)* As **v** is perpendicular is **B**, so charge particle will move in circular path,

Radius $R = \frac{mv}{qB}$

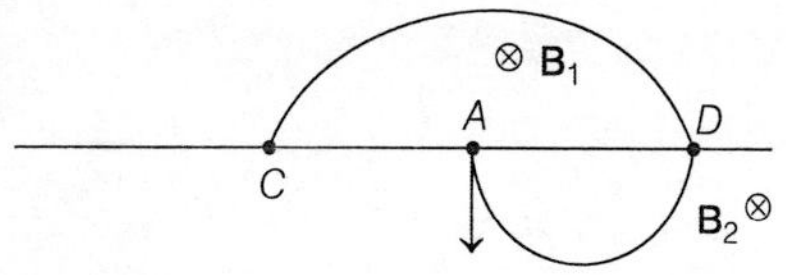

Starting point $= A$

Ending point $= C$

Net displacement $= AC$

$AC = CD - AD$

$AC = \frac{2mv}{qB_1} - \frac{2mv}{qB_2}$

$AC = \frac{2mv}{qB_1}\left(1 - \frac{B_1}{B_2}\right)$

$= \frac{mv}{qB_1}\left(1 - \frac{B_1}{B_2}\right) \times 2$

37. *(c)*

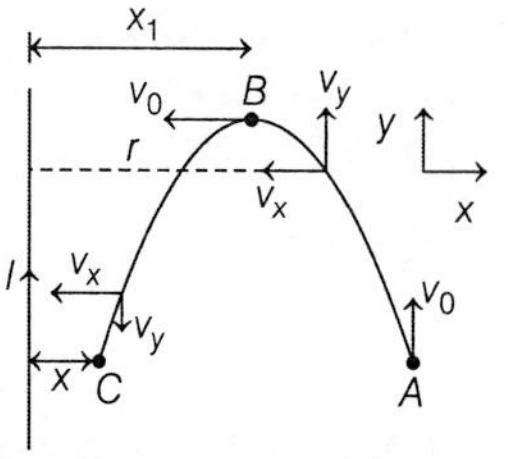

$A \to B$

$\mathbf{v} = -v_x\hat{\mathbf{i}} + v_y\hat{\mathbf{j}}$

$\mathbf{B} = \frac{\mu_0 I}{2\pi r}(-\hat{\mathbf{k}})$

$\mathbf{F} = q(\mathbf{v} \times \mathbf{B}) = \frac{\mu_0 Iq}{2\pi r}(-v_x\hat{\mathbf{j}} - v_y\hat{\mathbf{i}})$

$a_x = -\frac{\mu_0 Iq}{2\pi m} \cdot \frac{v_y}{r} \Rightarrow a_y = -\frac{\mu_0 Iq}{2\pi m} \cdot \frac{v_x}{r}$

$\frac{v_x dv_x}{dr} = -\frac{\mu_0 Iq}{2\pi m} \frac{v_y}{r}$

$\frac{v_x dv_x}{v_y} = -\frac{\mu_0 Iq}{2\pi m} \frac{dr}{r}$

$\int_0^{v_0} \frac{v_x dv_x}{\sqrt{v_0^2 - v_x^2}} = -\frac{\mu_0 I_q}{2\pi m} \int_a^{x_1} \frac{dr}{r}$

Let $z^2 = v_0^2 - v_x^2$

$2z\,dz = -2v_x dv_x$

$z\,dz = -v_x dv_x$

$\frac{v_x dv_x}{\sqrt{v_0^2 - v_x^2}} = -\frac{z\,dz}{z} = -dz$

Then integral becomes

$-\int_{v_0}^{0} dz = -\frac{\mu_0 Iq}{2\pi m} \ln \frac{x_1}{a}$

$v_0 = -\frac{\mu_0 Iq}{2\pi m} \ln \frac{x_1}{a}$

$x_1 = ae^{\frac{-2\pi m v_0}{\mu_0 Iq}}$ ...(i)

For $B \to C$, $\mathbf{v} = -v_x\hat{\mathbf{i}} - v_y\hat{\mathbf{j}}$

$\mathbf{B} = \frac{\mu_0 I}{2\pi r}(-\hat{\mathbf{k}})$

$\mathbf{F} = q(\mathbf{v} \times \mathbf{B}) = \frac{\mu_0 Iq}{2\pi r}(-v_x\hat{\mathbf{j}} + v_y\hat{\mathbf{i}})$

$a_x = +\frac{\mu_0 Iq}{2\mu m} \frac{v_y}{r}, \; a_y = -\frac{\mu_0 Iq}{2\pi m} \cdot \frac{v_x}{r}$

$\frac{v_x dv_x}{dr} = \frac{\mu_0 Iq}{2\pi m} \frac{v_y}{r}$

$\Rightarrow \int_{v_0}^{0} \frac{v_x dv_x}{\sqrt{v_0^2 - v_x^2}} = \frac{\mu_0 Iq}{2\pi m} \int_{x_1}^{x} \frac{dr}{r}$

$\frac{\mu_0 Iq}{2\pi m} \ln \frac{x}{x_1} = -\int_0^{v_0} dt = -v_0$

$x = x_1 e^{\frac{-2\pi m v_0}{\mu_0 Iq}}$ ...(ii)

From Eqs. (i) and (ii), $x = ae^{\frac{-4\pi m v_0}{\mu_0 Iq}}$

38. *(a)* Given, $\mathbf{E} = E\hat{\mathbf{k}}$

and magnetic field, $\mathbf{B} = B\hat{\mathbf{j}}$

$= 12 \times 10^{-3}\,\hat{\mathbf{j}}\text{T}$ $(\because B = 12 \text{ mT})$

Kinetic energy of electron,

$K = 728 \text{ eV} = 728 \times 1.6 \times 10^{-19}$ J

$= 1.16 \times 10^{-16}$ J

We know that electron will move along positive *X*-axis without deflection, thus,

$v = E/B$

Since, $K = \frac{1}{2}mv^2$

$\Rightarrow \quad K = \frac{1}{2}m\left(\frac{E}{B}\right)^2 \Rightarrow K = \frac{1}{2}m\frac{E^2}{B^2}$

$\Rightarrow \quad E^2 = \frac{2KB^2}{m} \quad \Rightarrow \quad E = \sqrt{\frac{2KB^2}{m}}$

$= \sqrt{\frac{2 \times 1.16 \times 10^{-16} \times (12 \times 10^{-3})^2}{9.1 \times 10^{-31}}}$

$= 192 \times 10^3 \text{ Vm}^{-1}$

$= 192 \text{ kVm}^{-1}$

39. *(b)* Given, $K_A = K_B$

$\Rightarrow \quad \frac{1}{2}m_A v_A^2 = \frac{1}{2}m_B v_B^2$

and $\quad \frac{m_A}{m_B} = \frac{9}{4}$, also $\frac{r_A}{r_B} = \frac{6}{5}$

Now, $\frac{1}{2}m_A v_A^2 = \frac{1}{2}m_B v_B^2$

$\Rightarrow \quad \left(\frac{v_A}{v_B}\right)^2 = \frac{m_B}{m_A} = \frac{4}{9}$ or $\frac{v_A}{v_B} = \frac{2}{3}$

Since, $\quad r = \frac{mv}{qB}$

So, $\quad \frac{r_A}{r_B} = \frac{m_A v_A}{q_A \cdot B} \times \frac{q_B \cdot B}{m_B v_B}$

$\Rightarrow \quad \frac{r_A}{r_B} = \frac{m_A}{m_B} \cdot \frac{v_A}{v_B} \cdot \frac{q_B}{q_A}$

On substituting values $\frac{r_A}{r_B}$, $\frac{m_A}{m_B}$ and $\frac{v_A}{v_B}$, we get

$$\frac{6}{5}=\frac{9}{4}\times\frac{2}{3}\times\frac{q_B}{q_A}$$

$$\Rightarrow \quad \frac{q_A}{q_B}=\frac{9}{4}\times\frac{2}{3}\times\frac{5}{6}\Rightarrow\frac{q_A}{q_B}=\frac{5}{4}$$

40. *(d)* When a charged particle enters into magnetic field perpendicular to it, then it moves on circular path, and radius r of circular path is given as,

$$r=\frac{mv}{Bq}=\frac{p}{Bq}$$

[where, momentum $p=mv$]

$$=\frac{\sqrt{2mK}}{Bq}$$

$\left[\because p=\sqrt{2mK} \text{ where, } K=\text{kinetic energy}\right]$

$$\Rightarrow \quad r\propto\frac{\sqrt{m}}{q}$$

$$\therefore r_p:r_d:r_\alpha=\frac{\sqrt{m_p}}{q_p}:\frac{\sqrt{m_d}}{q_d}:\frac{\sqrt{m_\alpha}}{q_\alpha}$$

$\left[\text{Since, } m_d=2m_p, m_\alpha=4m_p, q_p=q_d=e \text{ and } q_\alpha=2e\right]$

$$\therefore r_p:r_d:r_\alpha=\frac{\sqrt{m_p}}{e}:\frac{\sqrt{2m_p}}{e}:\frac{\sqrt{4m_p}}{2e}$$

$$=1:\sqrt{2}:1$$

41. *(b)* Given, mass of 1st ion,

$$m_1=4 \text{ amu}$$

Mass of 2nd ion, $m_2=16$ amu

Charge on 1st ion, $q_1=+2e$

Charge on 2nd ion, $q_2=+3e$

Kinetic energy of both ion is same,

$$KE_1=KE_2$$

The radius of path traced by a charged particle in magnetic field is given as

$$r=\frac{\sqrt{2m\text{KE}}}{qB}$$

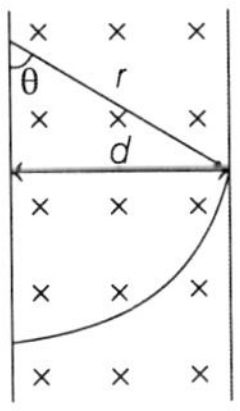

$$\Rightarrow \quad r\propto\frac{\sqrt{m}}{q}$$

$$\Rightarrow \quad \frac{r_1}{r_2}=\sqrt{\frac{m_1}{m_2}}\times\left(\frac{q_2}{q_1}\right)$$

$$\Rightarrow \frac{r_1}{r_2}=\sqrt{\frac{4}{16}}\times\left(\frac{+3e}{2e}\right)=\frac{1}{2}\times\frac{3}{2}=\frac{3}{4}$$

$$\Rightarrow r_1=\frac{3}{4}r_2\Rightarrow r_1<r_2$$

Now, the deflection of ion in region with constant magnetic field as shown in figure is denoted by θ.

From diagram, $\sin\theta=\frac{d}{r}$

$$\theta\propto\frac{1}{r} \quad [\because d=\text{constant}]$$

As, $r_1<r_2$ from above expression $\theta_1>\theta_2$.

Thus, deflection of lighter ion will be more than the deflection of heavier ion.

42. *(b)* Let F_p, F_d and F_α be the forces and v_p, v_d and v_α be the velocities of proton, deuteron and α-particle, respectively.

Since, $\quad F=Bqv$

On dividing and multiplying F by m, we get

$$F=Bqv\frac{m}{m}$$

$$\Rightarrow \quad F=B\frac{q}{m}p \quad (\because p=mv)$$

$$\Rightarrow \quad F\propto q/m \; (\because p \text{ and } B \text{ are same})$$

$$\therefore F_p:F_d:F_\alpha=\frac{+q}{m}:\frac{+q}{2m}:\frac{+2q}{4m}$$

$$=1:1/2:1/2=2:1:1$$

and $\quad F=Bqv\Rightarrow v\propto F/q$

$$\therefore v_p:v_d:v_\alpha=\frac{F_p}{q_p}:\frac{F_d}{q_d}:\frac{F_\alpha}{q_\alpha}$$

$$=\frac{2x}{q}:\frac{x}{q}:\frac{x}{2q}$$

$$=2:1:1/2=4:2:1$$

43. *(c)* When a moving charged particle is placed in a magnetic field **B**.

Then, the net magnetic force acting on it is

$$\mathbf{F}_m=q(\mathbf{v}\times\mathbf{B})$$

$$\mathbf{F}_m=qvB\sin\theta$$

Here, $\quad \theta=90°$

$$\mathbf{F}_m=qvB$$

Also, due to this net force, the particle transverses in a circular path, whose necessary centripetal force is being provided by F_m, i.e.

$$\frac{mv^2}{r}=qvB$$

$$\Rightarrow \quad r=\frac{mv^2}{qvB}=\frac{mv}{qB}\Rightarrow r\propto m$$

So, for electron,

$$r_e=\frac{m_ev}{eB} \text{ or } r_e\propto m_e$$

For proton, $\quad r_p=\frac{m_pv}{eB}$

or $\quad r_p\propto m_p$

For He-particle,

$$r_{\text{He}}=\frac{4m_pv}{2eB}=\frac{2m_pv}{eB}$$

Clearly, $\quad r_{\text{He}}>r_p \quad (\because r_{\text{He}}=2r_p)$

and we know that $m_p>m_e$

$\left[\because m_p\approx10^{-27}\text{ kg}, m_e\approx10^{-31}\text{ kg}\right]$

$$\therefore \quad r_p>r_e$$

$$\Rightarrow \quad r_{\text{He}}>r_p>r_e$$

44. *(*)* Situation given in question is shown below;

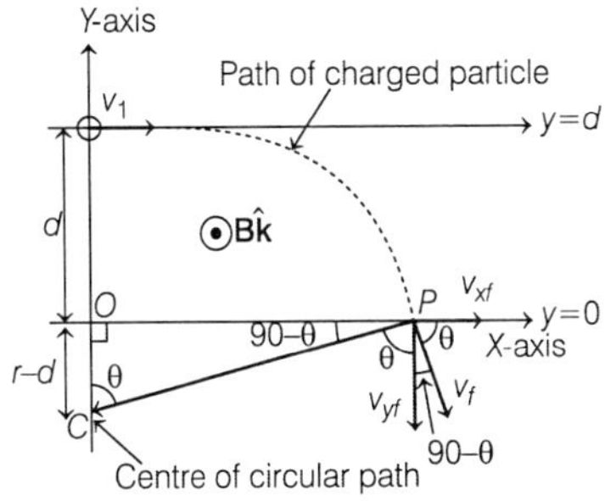

Path taken by particle of charge q and mass m is a circle of radius r where,

$$r=\frac{mv}{Bq}$$

Here, final velocity

$$\mathbf{v}_f=\mathbf{v}_x\hat{\mathbf{i}}+\mathbf{v}_{yf}(-\hat{\mathbf{j}})$$

$$=v\cos60°\hat{\mathbf{i}}-v\sin60°\hat{\mathbf{j}}$$

$$=v\left(\frac{1}{2}\hat{\mathbf{i}}-\frac{\sqrt{3}}{2}\hat{\mathbf{j}}\right)$$

So, change of velocity of charged particle is

$$\Delta\mathbf{v}=\mathbf{v}_f-\mathbf{v}_i=v\left(\frac{1}{2}\hat{\mathbf{i}}-\frac{\sqrt{3}}{2}\hat{\mathbf{j}}\right)-v\hat{\mathbf{i}}$$

$$=-v\left(\frac{1}{2}\hat{\mathbf{i}}+\frac{\sqrt{3}}{2}\hat{\mathbf{j}}\right)$$

If $t=$ time taken by charged particle to cross region of magnetic field then,

$$t=\frac{\text{distance } OP}{\text{speed in direction } OP}$$

$$=\frac{r\times\frac{\sqrt{3}}{2}}{v}=\frac{\frac{mv}{Bq}\times\frac{\sqrt{3}}{2}}{v}$$

$$=\frac{\sqrt{3}m}{2Bq}$$

So, acceleration of charged particle at the point its emergence is;

$$\text{Acceleration } \mathbf{a} = \frac{\Delta \mathbf{v}}{\Delta t} = \frac{-v\left(\frac{1}{2}\hat{\mathbf{i}} + \frac{\sqrt{3}}{2}\hat{\mathbf{j}}\right)}{\frac{\sqrt{3}}{2}\frac{m}{Bq}}$$

$$= \frac{-Bqv}{m}\left(\frac{\hat{\mathbf{i}}}{\sqrt{3}} + \hat{\mathbf{j}}\right) \text{ ms}^{-2}$$

None of the option given matches with answer.

45. *(d)* According to given condition, when a particle having charge same as electron move in a magnetic field on circular path, then the force always acts towards the centre and perpendicular to the velocity.

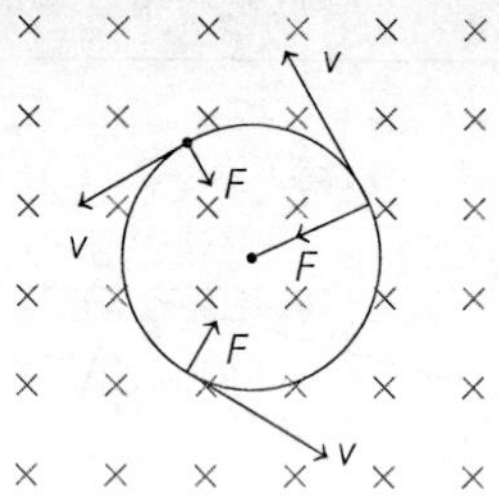

Here, $B = 0.5$ T

$R =$ radius of circular path $= 5$ cm

Now, the magnetic force is

$$F_m = q(v \times B)$$
$$= q\,vB \sin 90°$$
$$F_m = qvB \quad \ldots\text{(i)}$$

When the electric field is applied and the particle moves in a straight path, then this is the case of velocity selector.

$$\therefore \quad F_e = qE \quad \ldots\text{(ii)}$$

In velocity selector,

$$F_m = F_e$$
$$qvB = qE \quad \ldots\text{(iii)}$$

Initially particle moves under the magnetic field, so the radius of circular path taken by the particle is

$$R = \frac{mv}{qB} \quad \ldots\text{(iv)}$$

From Eqs. (iii) and (iv), we get

$$m = \frac{qB^2R}{E}$$

$$m = \frac{1.6 \times 10^{-19} \times 0.25 \times 0.5 \times 10^{-2}}{10^2}$$

$$m = 2 \times 10^{-24} \text{ kg}$$

46. *(0.01)*

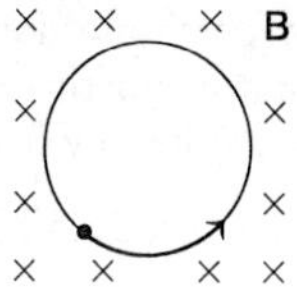

$$B = 6.28 \text{ T}$$
$$m = 16\,\mu\text{g}$$
$$m = 16 \times 10^{-6} \times 10^{-3} \text{ kg}$$
$$q = 1.6 \times 10^{-6} \text{ C}$$
$$T = \frac{2\pi m}{qB}$$
$$= \frac{2\pi \times 16 \times 10^{-6} \times 10^{-3}}{1.6 \times 10^{-6} \times 6.28} = 10^{-2}$$
$$T = 0.01 \text{ sec}$$

47. *(250)* Since time period of revolving charge is $\frac{2\pi m}{qB}$.

where $B =$ magnetic field

Due to solenoid $B = \mu_0 nI$

$$T = \frac{2\pi m}{q(\mu_0 nI)}$$

$$75 \times 10^{-9} = \frac{2\pi \times 9 \times 10^{-31}}{1.6 \times 10^{-19} \times 4\pi \times 10^{-7} \times n \times 1.5}$$

$$\Rightarrow \quad n = 250$$

48. *(2)* For uniform speed, $v = E/B$

Radius of path when electric field is switched off ;

$$R = \frac{mv}{eB} \Rightarrow R = \frac{mv^2}{eE}$$

$$E = \frac{mv^2}{eR}$$

$$= \frac{1.6 \times 10^{-27} \times 4 \times 10^{10}}{1.6 \times 10^{-19} \times 2 \times 10^{-2}}$$

$$= 2 \times 10^4 \text{ N/C}$$

$$x = 2$$

49. *(5)* Force on electron is given by

$$\mathbf{F} = q(\mathbf{v} \times \mathbf{B})$$

or $$\mathbf{F} = e \cdot \begin{vmatrix} \hat{\mathbf{i}} & \hat{\mathbf{j}} & \hat{\mathbf{k}} \\ 3 & 5 & 0 \\ B_0 & 2B_0 & 0 \end{vmatrix}$$

$$= e \cdot (6B_0 - 5B_0)\,\hat{\mathbf{k}}$$

$$= eB_0\hat{\mathbf{k}} \quad \ldots\text{(i)}$$

Given, $\mathbf{F} = 5e\hat{\mathbf{k}}$ ...(ii)

So, from Eqs. (i) and (ii),

$$B_0 = 5 \text{ T}$$

50. *(32)* $F = q(\mathbf{v} \times \mathbf{B})$

$$= 4 \times 10^{-6}(4 \times 10^6\,\hat{\mathbf{j}} \times 2\hat{\mathbf{k}})$$
$$= 4 \times 10^{-6} \cdot 8 \times 10^6\,\hat{\mathbf{i}} = 32\hat{\mathbf{i}} \text{ N}$$
$$= x\,\hat{\mathbf{i}} \text{ N} \quad \text{(given)}$$
$$\therefore \quad x = 32$$

51. *(40)* $B = \frac{\pi}{2} \times 10^{-3}$ (given)

$$\text{KE} = \frac{1}{2}mv^2 \text{ and } v = \sqrt{\frac{2\text{KE}}{m}}$$

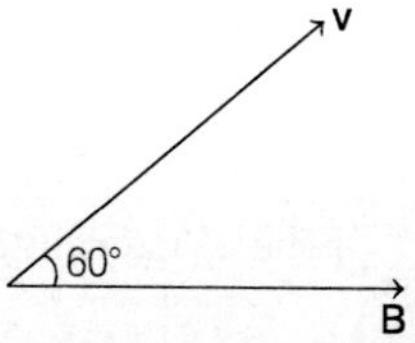

Pitch $= v\cos 60° \times$ time period of one rotation

$$= v\cos 60° \times \frac{2\pi m}{eB}$$

$$= \sqrt{\frac{2 \times 2 \times 1.6 \times 10^{-19}}{1.6 \times 10^{-27}}} \times \cos 60° \times \frac{2\pi \times 1.6 \times 10^{-27}}{1.6 \times 10^{-19} \times \frac{\pi}{2} \times 10^{-3}}$$

$$= 2 \times 10^4 \times \frac{1}{2} \times 4 \times 10^{-5}$$

$$= 4 \times 10^{-1} \text{ m} = 40 \text{ cm}$$

52. *(144)* The given situation in question is as follows.

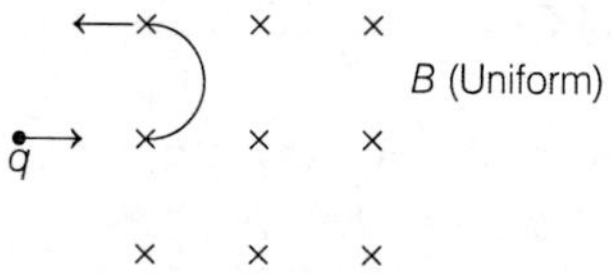

Kinetic energy of charged particle, $\text{KE} = qV$

$$\Rightarrow \quad \text{KE} = 2 \times 10^{-6} \times 100$$
$$= 2 \times 10^{-4} \text{ J}$$

Radius of circular path, $R = mv/qB$

$$R = \frac{p}{qB} = \frac{\sqrt{2(\text{KE})m}}{qB} \text{ or } m = \frac{q^2B^2R^2}{2(\text{KE})}$$

$$= \frac{(2 \times 10^{-6})^2 \times (4 \times 10^{-3})^2 \times (3 \times 10^{-2})^2}{2 \times 2 \times 10^{-4}}$$

$$m = 144 \times 10^{-18} \text{ kg}$$

53. *(10)* We know that

$$r = \frac{mV}{qB} = \frac{\sqrt{2m(\text{KE})}}{qB}$$

$$= \sqrt{\frac{2 \times 24 \times 1.67 \times 10^{-27} \times 5000 \times 1.6 \times 10^{-19}}{1.6 \times 10^{-19} \times 0.5}}$$

$= 0.1\text{ m} = 10\text{ cm}$

54. *(2)* Given, a deutron and a proton have equal KE enter into a uniform magnetic field.

$\theta = 90°$ $[\because \mathbf{v} \perp \mathbf{B}]$

$$\frac{r_d}{r_p} = \frac{\sqrt{x}}{1}$$

Since, radius of circular path,

$$r = \frac{\sqrt{2mK}}{qB}$$

($\because$ KE and B are same)

$$\Rightarrow \quad \frac{r_d}{r_p} = \frac{\sqrt{m_d}}{\sqrt{m_p}} \times \frac{q_d}{q_p} = \sqrt{2}$$

Since, $m_d = 2m_p$ and $q_p = q_d$

Comparing with given value, we get

$\therefore \quad x = 2$

Topic 3

01. *(a)*

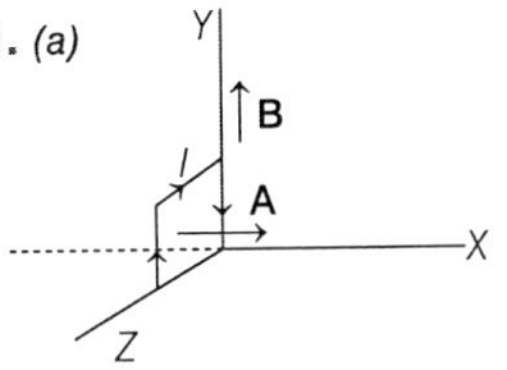

Torque on loop,

$= \mathbf{m} \times \mathbf{B} = I\mathbf{A} \times \mathbf{B}$

$= (5 \times 0.02\,\hat{\mathbf{i}}) \times (2 \times 10^{-3}\,\hat{\mathbf{j}})$

$= 2 \times 10^{-4}\,\hat{\mathbf{k}}$ N-m

02. *(a)* The given situation is shown below.

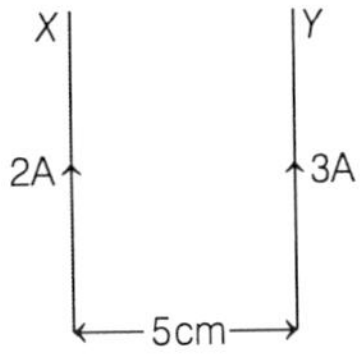

$L_X = 50\text{ cm} = 0.5\text{ m} \Rightarrow I_X = 2\text{ A}$

$L_Y = 5\text{ m} \Rightarrow \quad I_Y = 3\text{ A}$

$r = 5\text{ cm} = 5 \times 10^{-2}$ m

Magnitude of force acting on both wires X and Y will be same and given as

$$F = \frac{\mu_0 I_X I_Y}{2\pi r} \times L_X$$

$$= 2 \times 10^{-7} \times \frac{2 \times 3 \times 0.5}{5 \times 10^{-2}} = 1.2 \times 10^{-5}\text{ N}$$

$\therefore$ Force acting on the wire Y will be 1.2×10^{-5} N directed towards wire X because both wires having current in the same direction, hence they attract to each other.

03. *(a)* The given situation is shown below

Force on metalic rod, $F = ILB$

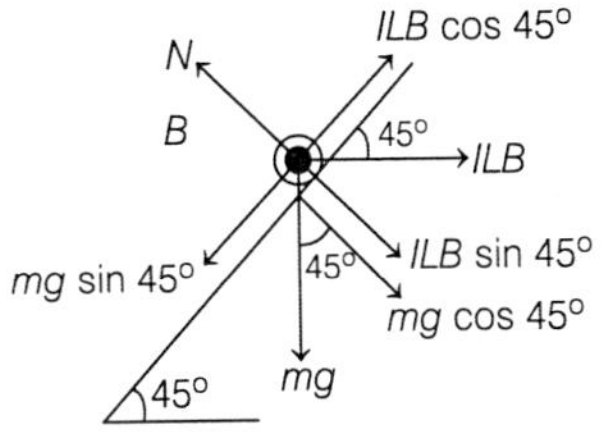

Here, $mg \sin 45° = ILB \cos 45°$

$$\therefore I = \left(\frac{m}{L}\right)\frac{g}{B} = \frac{(0.45)(10)}{0.15} = 30\text{ A}$$

04. *(b)* Here, current should be in opposite direction

$$B_p = B_{1p} + B_{2p},\ B_p = 2\frac{\mu_0 I}{2\pi r}$$

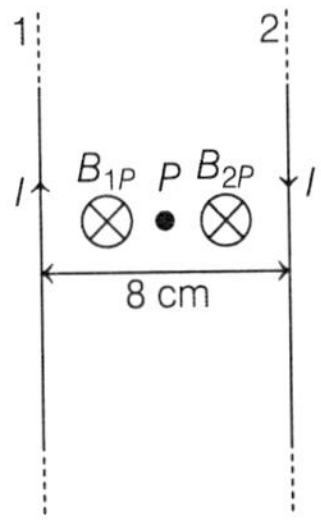

$$300 \times 10^{-6} = 2 \times 2 \times 10^{-7} \times \frac{I}{4 \times 10^{-2}}$$

$3 \times 10^{-4} = 10^{-5} \times I \Rightarrow I = 30\text{ A}$

05. *(c)* Force per unit length between two current carrying wire is given by

$$\frac{F}{l} = \frac{\mu_0 I_1 I_2}{2\pi d} \quad \ldots\text{(i)}$$

Here, $\dfrac{F}{l} = 2 \times 10^{-6}$ N/m

$I_1 = I_2 = x\text{A},\ d = 0.20$ m

So, from Eq. (i),

$$2 \times 10^{-6} = \frac{4\pi \times 10^{-7} x^2}{2\pi \times 0.2}$$

$\Rightarrow \quad x^2 = 2 \Rightarrow x = \sqrt{2} = 1.4$

06. *(a)* A loop of wire will act likc magnetic dipole. In equilibrium, direction of dipole will be parallel to direction of magnetic field and its plane remains normal to the field.

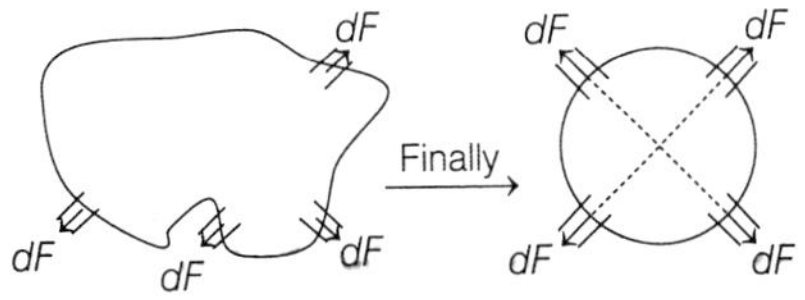

Now the force on each element of wire due to magnetic field will be perpendicular to length of element and outward in direction. So, loop will become circular.

Hence, the correct option is (a).

07. *(b)* According to the question, the situation can be drawn as follows

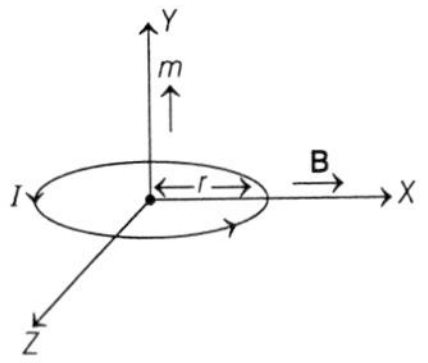

Let the current I be flowing in anti-clockwise direction, then the magnetic moment of the coil is $m = NIA$

Its direction is perpendicular to the area of the coil and is along Y-axis.

Then, torque on the current coil is

$\tau = \mathbf{m} \times \mathbf{B} = mB \sin 90° = NIAB$

$= NI\pi r^2 B$(N-m)

08. *(a)* Consider the two coaxial solenoids. Due to one of the solenoids magnetic field at the centre of the other can be assumed to be constant.

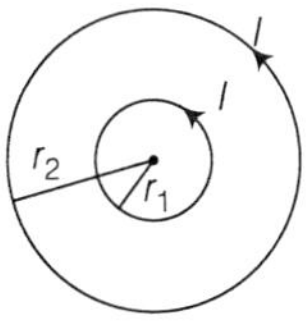

Due to symmetry, forces on upper and lower part of a solenoid will be equal and opposite and hence resultant is zero.

Therefore, $F_1 = F_2 = 0$

09. *(c)* Since, **B** is uniform only torque acts on a current carrying loop.

As, $\tau = \mathbf{M} \times \mathbf{B} \Rightarrow |\tau| = |\mathbf{M}||\mathbf{B}| \sin\theta$

For orientation shown in (b)

$\theta = 0^0$, $\tau = 0$ (stable equilibrium) and for

(d) $\theta = \pi$, $\tau = 0$ (unstable equilibrium)

10. *(b)* The forces on AD and BC are zero because magnetic field due to a straight wire on AD and BC is parallel to elementary length of the loop.

11. *(a)* The force per unit length between the two wires is

$$\frac{F}{l} = \frac{\mu_0}{4\pi} \cdot \frac{2I^2}{d} = \frac{\mu_0 I^2}{2\pi d}$$

The force will be attractive as current directions in both are same.

12. *(b)*

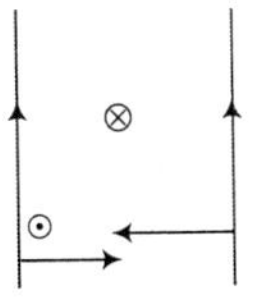

Due to flow of current in same direction in two adjacent sides, an attractive magnetic force will be produced due to which spring will get compressed.

13. *(c)* The component $dl \cos\theta$ of element dl is parallel to the length of the wire 1. Hence, force on this elemental component

$$F = \frac{\mu_0}{4\pi} \cdot \frac{2I_1 I_2}{r}(dl \cos\theta)$$
$$= \frac{\mu_0 I_1 I_2 \, dl \cos\theta}{2\pi r}$$

14. *(d)* Force on wire $= i(\mathbf{l} \times \mathbf{B})$

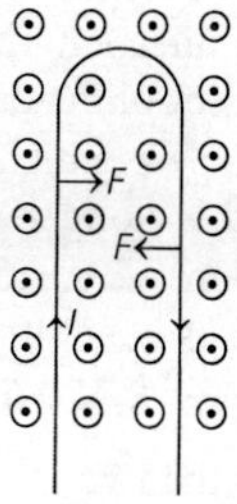

Force due to long straight wire cancle each other.
Here $\mathbf{l}$ is effective length vector,

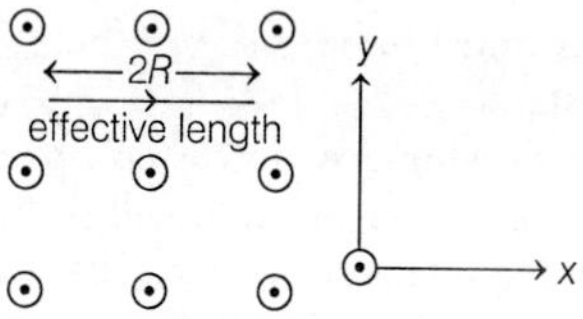

$$\mathbf{l} = (2R)\hat{\mathbf{i}}$$

So, force on wire is,

$$\mathbf{F} = i(2R\hat{\mathbf{i}} \times B_0 \hat{\mathbf{k}})$$
$$= 2iRB_0(\hat{\mathbf{i}} \times \hat{\mathbf{k}}) = (2iRB_0)(-\hat{\mathbf{j}})$$

Note In this question magnetic field in diagram is shown towards Z-axis but written $B = B_0\hat{\mathbf{j}}$.

15. *(a)* Force between two wires carrying current I_1 and I_2 is

$$F = \frac{\mu_0 I_1 I_2}{2\pi d} . l$$

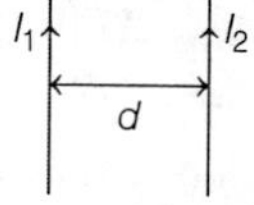

Here, $I_1 = I_2 = 10$ A, $d = 4 \times 10^{-2}$ m,
$l = 10 \times 10^{-2}$ m

So, $F_1 = \frac{\mu_0 I_1 I_2 l}{2\pi d}$

If distance between wires is halved and currents are doubled.

Then, $F_2 = \frac{\mu_2 I_1' I_2' l}{2\pi d'} = \frac{\mu_0 4 I_1 I_2 l}{2\pi \frac{d}{2}}$

$$= 8\left(\frac{\mu_0 I_1 I_2 l}{2\pi d}\right) = 8F_1$$

• Concept Enhancer

Force between two parallel current carrying wire per unit length is calculated by using following formula.

$$\frac{F}{l} = \frac{\mu_0 I_1 I_2}{2\pi d}$$

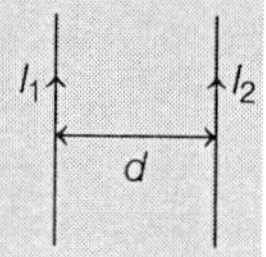

16. *(c)* Given, $i = 10$ A, $B = 0.5$ T

Since, $BC = CD = BD = l$ (say)$=5$ cm, so ΔBCD is an equilateral triangle and angle between CD and BD is 60°. Also, as magnetic field is along BD, so angle between **B** and **l** is 60°.

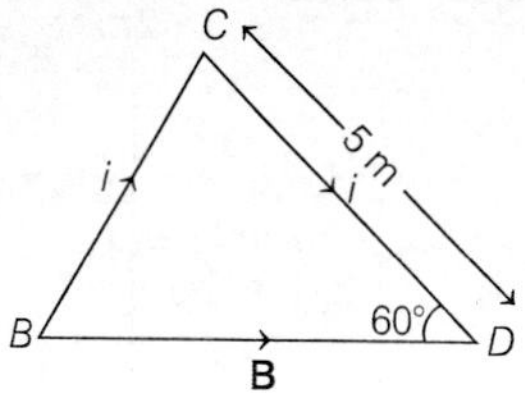

$\therefore$ Force of segment CD, $F = i\mathbf{l} \times \mathbf{B}$

$$= ilB \sin\theta$$
$$= 10 \times 5 \times 10^{-2} \times 0.5 \times \sin 60°$$
$$= 0.25 \times \sqrt{3}/2 = 0.216 \text{ N}$$

17. *(b)* The situation given in question is shown in the following figure

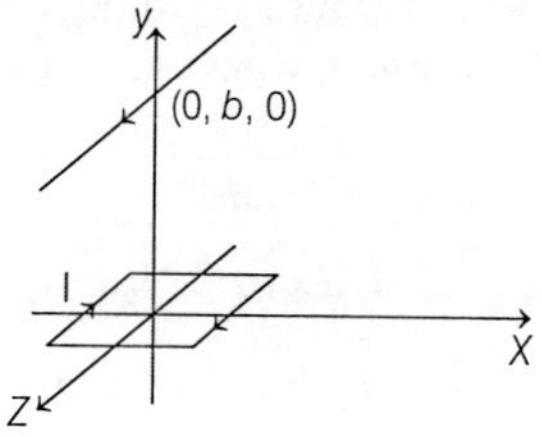

Force acting towards wire,

$$F_1 = \frac{\mu_0 I^2}{2\pi r} 2a$$

Force acting away from wire,

$$F_2 = \frac{\mu_0 I^2}{2\pi r} 2a$$

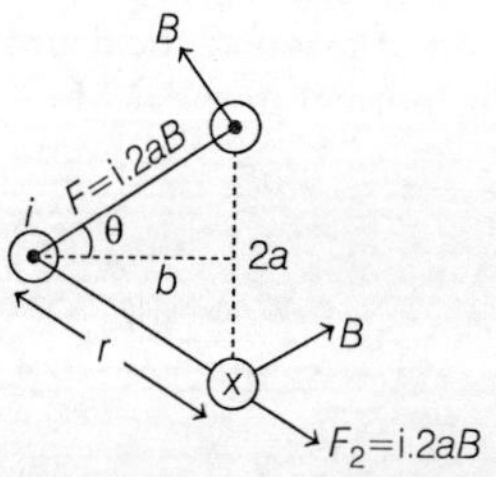

Torque acting on the loop about Z-axis, $\tau = F_1 a \cos\theta + F_2 a \cos\theta$

$$= 2\left(\frac{\mu_0 I^2 2a}{2\pi r}\right) a \cdot \frac{b}{r} \quad \left(\because \cos\theta = \frac{b}{r}\right)$$
$$= \frac{2\mu_0 I^2 a^2 b}{\pi(a^2 + b^2)} \quad (\because r = \sqrt{a^2 + b^2})$$
$$= \frac{2\mu_0 I^2 a^2 b}{\pi b^2} \quad (\because b >>> a)$$
$$= \frac{2\mu_0 I^2 a^2}{\pi b}$$

Hence, correct option is (b).

18. *(b)* Magnetic moment is a vector quantity, $\mathbf{m} = NI\mathbf{A}$, with its direction parallel to area vector.
Given rectangular loop can be viewed as a combination of two loops $ABCDA$ and $ADEFA$. So, **m** (loop)
$= \mathbf{m}\,(ABCDA) + \mathbf{m}\,(ADEFA)$
The area vector of $ABCDA$ is along Z-axis and that of $ADEFA$ is along Y-axis as shown in figure,

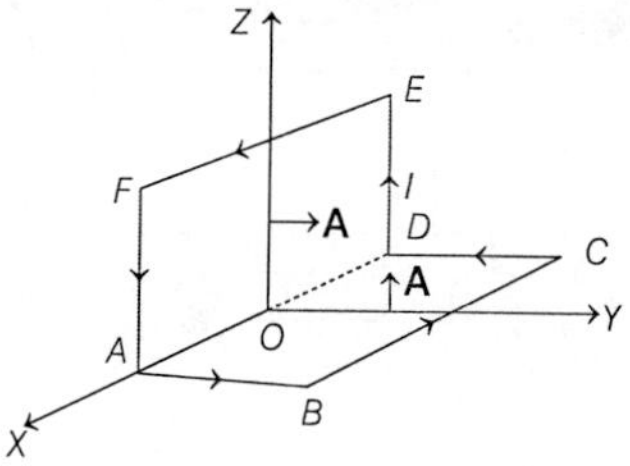

Now, $\mathbf{m}\,(ABCDA) = IA\hat{\mathbf{k}} = I(ab)\hat{\mathbf{k}}$

and $\mathbf{m}\,(ADEFA) = IA\hat{\mathbf{j}} = I(ab)\hat{\mathbf{j}}$

$\therefore$ $\mathbf{m}$ (loop) $= Iab\hat{\mathbf{k}} + Iab\hat{\mathbf{j}}$
$= Iab(\hat{\mathbf{k}} + \hat{\mathbf{j}})$

$$= (Iab\sqrt{2})\left(\frac{\hat{\mathbf{k}}}{\sqrt{2}} + \frac{\hat{\mathbf{j}}}{\sqrt{2}}\right)$$

So, magnitude of $\mathbf{m} = \sqrt{2} Iab$
and direction of $\mathbf{m}$ is along $\frac{1}{\sqrt{2}}(\hat{\mathbf{j}} + \hat{\mathbf{k}})$ z.

19. *(a)*

key idea Magnetic dipole moment of a current carrying loop is $m = IA$ (A - m²)
where, I = current in loop and A = area of loop.

Let the given square loop has side a, then $m = Ia^2$

When square is converted into a circular loop of radius r,

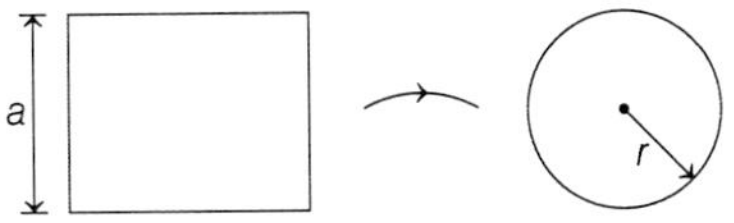

Then, wire length will be same in both area,

$\Rightarrow \quad 4a = 2\pi r$

$\Rightarrow \quad r = \dfrac{4a}{2\pi} = \dfrac{2a}{\pi}$

Hence, area of circular loop formed is,

$A' = \pi r^2$

$= \pi\left(\dfrac{2a}{\pi}\right)^2$

$= \dfrac{4a^2}{\pi}$

Magnitude of magnetic dipole moment of circular loop will be

$m' = IA' = I\dfrac{4a^2}{\pi}$

Ratio of magnetic dipole moments of both shapes is,

$\dfrac{m'}{m} = \dfrac{I\cdot\dfrac{4a^2}{\pi}}{Ia^2} = \dfrac{4}{\pi} \Rightarrow m' = \dfrac{4m}{\pi}$ (A-m)

20. *(a)* Given,

Area of the rectangular coil,

$A = 5\text{ cm} \times 2.5\text{ cm}$

$\Rightarrow \quad A = 12.5\text{ cm}^2 = 12.5 \times 10^{-4}\text{ m}^2$

Number of turns, $N = 100$ turns

Current through the coil, $I = 3$ A

Magnetic field applied, $B = 1$ T

Angle between the magnetic field and area vector of the coil, $\theta = 45°$

As we know that when a coil is tilted by an angle θ in the presence of some external magnetic field, then the net torque experienced by the coil is,

$\tau = \mathbf{M} \times \mathbf{B} = NI(\mathbf{A} \times \mathbf{B})$

$= NIAB\sin\theta$

On substituting the given values, we get

$\tau = 100 \times 3 \times 12.5 \times 10^{-4} \times 1 \times \sin 45°$

$\tau = 0.707 \times 100 \times 3 \times 12.5 \times 10^{-4}$ N-m

$= 2.651 \times 10^{-1}$ N-m

≈ 0.27 N-m

21. *(c)* Initially, $m = I\pi R^2$

and $\quad B_1 = \dfrac{\mu_0 I}{2R_1}$

Finally, $\quad m' = 2m = I\pi R_2^2$

$\Rightarrow \quad 2I\pi R_1^2 = I\pi R_2^2$

or $\quad R_2 = \sqrt{2}R_1$

So, $\quad B_2 = \dfrac{\mu_0 I}{2(R_2)} = \dfrac{\mu_0 I}{2\sqrt{2}R_1}$

Hence, ratio $\dfrac{B_1}{B_2} = \dfrac{\left(\dfrac{\mu_0 I}{2R_1}\right)}{\left(\dfrac{\mu_0 I}{2\sqrt{2}R_1}\right)} = \sqrt{2}$

$\therefore$ Ratio $\dfrac{B_1}{B_2} = \sqrt{2}$

22. *(b)* Consider free body diagram of the wire. As the wires are in equilibrium, they must carry current in opposite direction.

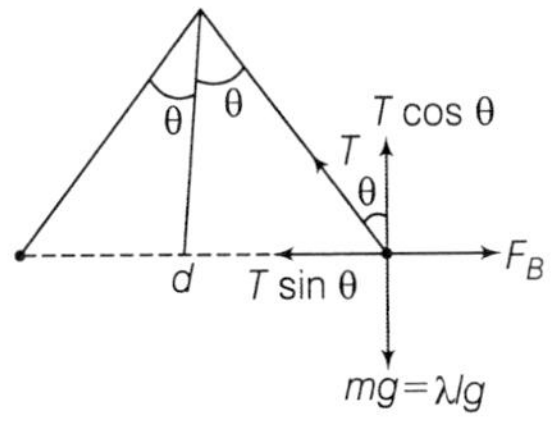

Here, $F_B = \dfrac{\mu_0 I^2 l}{2\pi d}$, where l is length of each wire and d is separation between wires.

From figure, $d = 2L\sin\theta$

$T\cos\theta = mg = \lambda lg$ (in vertical direction)...(i)

$T\sin\theta = F_B = \dfrac{\mu_0 I^2 l}{4\pi L\sin\theta}$ (in horizontal direction) ...(ii)

From Eqs. (i) and (ii), we get

$\dfrac{T\sin\theta}{T\cos\theta} = \dfrac{\mu_0 I^2 l}{4\pi L\sin\theta \times \lambda lg}$

$\therefore \quad I = \sqrt{\dfrac{4\pi\lambda Lg\sin^2\theta}{\mu_0\cos\theta}}$

$= 2\sin\theta\sqrt{\dfrac{\pi\lambda Lg}{\mu_0\cos\theta}}$

23. *(b)* Force exerted on a current carrying conductor,

$F_{ext} = B(x)\,IL$

Average power $= \dfrac{\text{Work done}}{\text{Time taken}}$

$P = \dfrac{1}{t}\int_0^2 F_{ext}\cdot dx = \dfrac{1}{t}\int_0^2 B(x)\,IL\,dx$

$= \dfrac{1}{5\times10^{-3}}\int_0^2 3\times10^{-4}e^{-0.2x} \times 10 \times 3\,dx$

$= 9\,[1 - e^{-0.4}] = 9\left[1 - \dfrac{1}{e^{0.4}}\right]$

$= 2.967 \approx 2.97$ W

24. *(d)* Let us consider the disc to be made up of large number of concentric elementary rings.

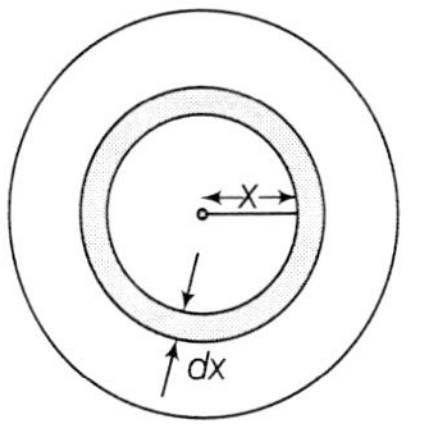

Consider one such ring of radius x and thickness dx. Charge on this elementary ring,

$dq = \sigma \times 2\pi x dx = 2\pi\sigma x dx$

Current associated with this elementary ring,

$dI = \dfrac{dq}{dt} = dq \times f = \sigma\omega x dx$

$[\because f$ is frequency and $\omega = 2\pi f$]

Magnetic moment of this elementary ring,

$dM = dI\,\pi x^2 = \pi\sigma\omega x^3\,dx$

$\therefore$ Magnetic moment of the entire disc,

$M = \int_0^R dM = \pi\sigma\omega\int_0^R x^3\,dx$

$= \dfrac{1}{4}\pi R^4\,\sigma\omega$

25. *(c)* Force acting between two current carrying conductors

$F = \dfrac{\mu_0}{2\pi}\dfrac{I_1 I_2}{d}\,l$ ...(i)

where, d = distance between the conductors

l = length of small conductor

Again, $F' = \dfrac{\mu_0}{2\pi}\dfrac{(-2I_1)(I_2)}{(3d)}\cdot l$

$= -\dfrac{\mu_0}{2\pi}\dfrac{2I_1I_2}{3d}\cdot l$ ...(ii)

Thus, from Eqs. (i) and (ii), we get

$\dfrac{F'}{F} = -\dfrac{2}{3} \Rightarrow F' = -\dfrac{2}{3}F$

26. *(c)*

key idea To solve this question, you can use following steps.

(1) Find out the value of magnetic field at point P using following formula :

$\mathbf{B}_p = \dfrac{\mu_0}{2\pi}\left[\dfrac{I_1}{r_1} - \dfrac{I_2}{r_2}\right](-\hat{\mathbf{k}})$

(2) Now, determine the value of force acting on the charge particle using following formula :

$\mathbf{F} = q(\mathbf{v} \times \mathbf{B}_p)$

According to given situation, magnetic field at point P.

$$\mathbf{B}_p = \frac{\mu_0}{2\pi}\left[\frac{I_1}{r_1} - \frac{I_2}{r_2}\right](-\hat{\mathbf{k}})$$

$$= \frac{2\times10^{-7}}{10^{-2}}\left[\frac{4}{4} - \frac{2}{6}\right](-\hat{\mathbf{k}})$$

$$= 2\times10^{-5}\left[1-\frac{1}{3}\right](-\hat{\mathbf{k}}) = \frac{4}{3}\times10^{-5}(-\hat{\mathbf{k}})$$

So, $\mathbf{F} = q(\mathbf{v}\times\mathbf{B}_p)$

$$= 3\pi\left[(2\hat{\mathbf{i}}+3\hat{\mathbf{j}})\times\left(-\frac{4}{3}\times10^{-5}\hat{\mathbf{k}}\right)\right]$$

$$= 3\pi\left[\frac{8}{3}\hat{\mathbf{j}} - 4\hat{\mathbf{i}}\right]\times10^{-5}$$

$$= 4\pi\times10^{-5}[-3\hat{\mathbf{i}}+2\hat{\mathbf{j}}]$$

$$= 4\pi\times10^{-5}(-x\hat{\mathbf{i}}+2\hat{\mathbf{j}})$$

Hence, $x = 3$

27. *(*)*

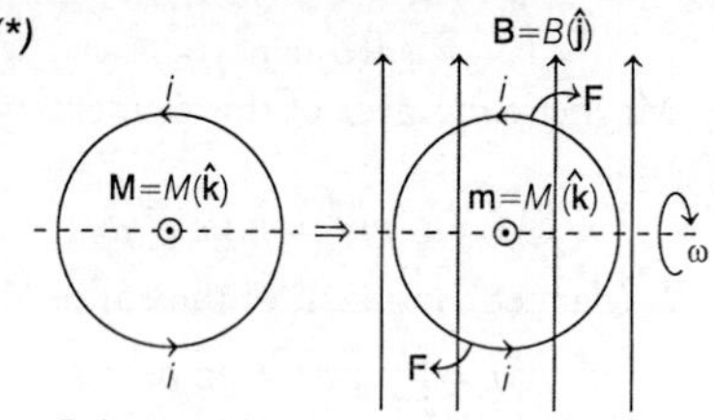

Before applying the magnetic field   After applying the magnetic field

Given that $I_{dia} = 0.8$ kg-m^2,

$M = 20$ A-m^2, $B = 4$ T

$\theta =$ angle between M and B after rotation of 60° by coil $= 90° - 60° = 30°$ and $\omega = ?$

Applying the law of conservation of energy,

$$(TE)_{initial} = (TE)_{final} \Rightarrow U_i + K_i = U_f + K_f$$

$$0 + 0 = -MB\cos\theta + \frac{1}{2}I_{dia}\omega^2$$

$$0 = -20\times4\cos30° + \frac{1}{2}\times0.8\times\omega^2$$

$$0 = -20\times4\times\frac{\sqrt{3}}{2} + 0.4\,\omega^2$$

$$0 = -40\sqrt{3} + 0.4\,\omega^2$$

$$\Rightarrow 40\sqrt{3} = 0.4\,\omega^2 \Rightarrow \omega^2 = \frac{40\sqrt{3}}{0.4}$$

$$= 100\sqrt{3} = 100\times1.732 = 173.2$$

$$\omega = \sqrt{173.2} = 13.16 \text{ rad/s}$$

Answer is unavailable in options. So, bonus marks should be provided.

28. *(d)* Net force on the third wire, carrying current I in the following first case is

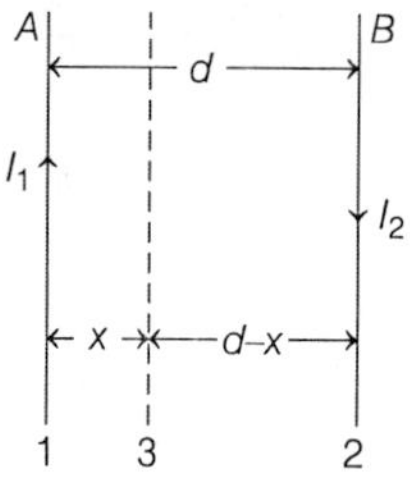

$$F_{13} + F_{23} = 0$$

Using thumb rule, direction of **B** at inside region of wires A and B will be same.

$$\therefore \frac{\mu_0 I_1 I}{2\pi x} + \frac{\mu_0 I_2 I}{2\pi(d-x)} = 0$$

$$\Rightarrow \frac{I_1}{x} + \frac{I_2}{d-x} = 0 \Rightarrow \frac{I_1}{x} = \frac{I_2}{x-d}$$

or $\quad (x-d)I_1 = xI_2$

$$\Rightarrow x(I_1 - I_2) = dI_1$$

$$\Rightarrow \quad x = \frac{I_1}{(I_1 - I_2)}\cdot d \quad \dots \text{(i)}$$

Second case of balanced force can be as shown

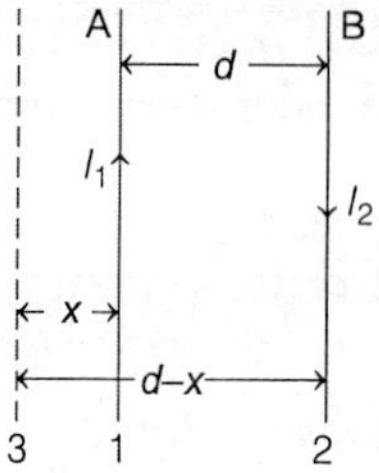

Using thumb rule, directions of **B** at any point on wires A and B will be opposite, so net force,

$$\frac{\mu_0 I_1 I}{2\pi x} - \frac{\mu_0 I_2 I}{2\pi(d+x)} = 0$$

or $\quad \frac{I_1}{x} - \frac{I_2}{(d+x)} = 0 \Rightarrow \frac{I_1}{x} = \frac{I_2}{d+x}$

$$\Rightarrow (d+x)I_1 = xI_2$$

$$\Rightarrow (I_2 - I_1)x = dI_1$$

$$\Rightarrow \quad x = -\frac{I_1}{(I_1 - I_2)}\cdot d \quad \dots \text{(ii)}$$

From Eqs. (i) and (ii), it is clear that

$$x = \pm\frac{I_1}{(I_1 - I_2)}d$$

29. *(d)*

> **key idea** Net force experienced by two wires separated by same distance is attractive, if current flow in them in same direction. However, this force is repulsive in nature, if current in them flows in opposite direction.

Force on a wire 1 in which current I_1 is flowing due to another wire 2 which are separated by a distance r is given as

$$\mathbf{F} = I_1(\mathbf{l}\times\mathbf{B}_2)$$

or $F = \frac{\mu_0 I_1 I_2}{2\pi r}\cdot l\sin\theta \quad \left[\because \mathbf{B}_2 = \frac{\mu_0 I_2}{2\pi r}\right]$

Thus, the given square loop can be drawn as shown below

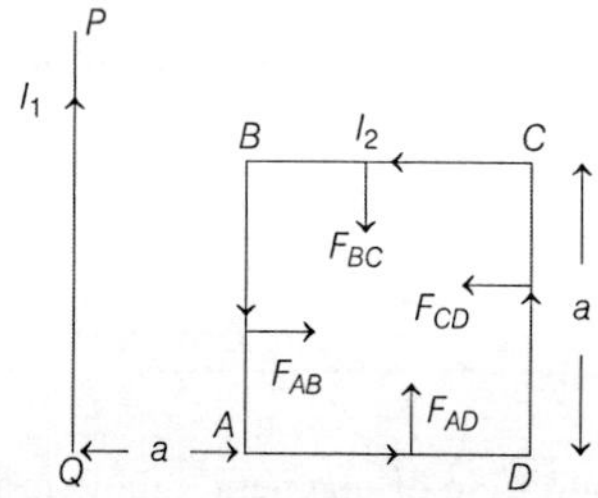

$$F_{AB} = \frac{\mu_0 I_1}{2\pi a}\cdot I_2 a$$

(away from wire PQ)

$F_{BC} = F_{AD} = 0 \qquad [\because \theta = 0°]$

$$F_{CD} = \frac{\mu_0 I_1}{2\pi(2a)}\cdot I_2 a = \frac{\mu_0 I_1}{4\pi a}\cdot I_2 a$$

(towards the wire PQ)

$$\therefore \quad F_{net} = F_{AB} - F_{CD}$$

$$= \frac{\mu_0 I_1 I_2}{2\pi} - \frac{\mu_0 I_1 I_2}{4\pi}$$

(away from wire)

$$= \frac{\mu_0 I_1 I_2}{4\pi}$$

(repulsive in nature)

30. *(c)* There are two forces on slider.

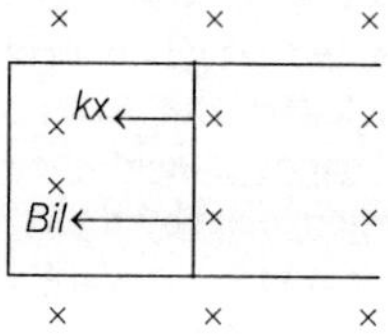

Spring force $= kx$

where, $k =$ spring constant.

As the slider is kept in a uniform magnetic field $B = 0.1$ T, hence it will experience a force, i.e.

Magnetic force $= Bil$

where, $l =$ length of the strip.

Now, using $F_{net} = ma$

We have,

$$(-kx) + (-Bil) = ma$$

$$\Rightarrow -kx - Bil - ma = 0$$

$$\Rightarrow -kx - \frac{B^2 l^2}{R}\cdot v - ma = 0 \quad \left[\because i = \frac{Blv}{R}\right]$$

and acceleration, $a = \dfrac{d^2x}{dt^2}$

Hence, the modified equation becomes

$$\Rightarrow \quad \frac{md^2x}{dt^2} + \frac{B^2l^2}{R}\left(\frac{dx}{dt}\right) + kx = 0$$

This is the equation of damped simple harmonic motion.

So, amplitude of oscillation varies with time as

$$A = A_0 e^{-\frac{B^2l^2}{2Rm}\cdot t}$$

Now, when amplitude is $\dfrac{A_0}{e}$, then

$$\frac{A_0}{e} = \frac{A_0}{e^{\frac{B^2l^2}{2Rm}\cdot t}} \quad \text{[As given]}$$

$$\Rightarrow \quad \left(\frac{B^2l^2}{2Rm}\right)t = 1$$

or $$t = \frac{2Rm}{B^2l^2}$$

According to the question, magnetic field $B = 0.1$ T, mass of strip $m = 50 \times 10^{-3}$ kg, resistance $R = 10\,\Omega$,

$$l = 10 \text{ cm} = 10 \times 10^{-2} \text{ m}$$

$$\therefore \quad t = \frac{2Rm}{B^2l^2} = \frac{2 \times 10 \times 50 \times 10^{-3}}{(0.1)^2 \times (10 \times 10^{-2})^2}$$

$$= \frac{1}{10^{-4}} = 10000 \text{ s}$$

Given, spring constant, $k = 0.5 \text{ Nm}^{-1}$

Also, time period of oscillation is

$$T = 2\pi\sqrt{\frac{m}{k}} = 2\pi\sqrt{\frac{50 \times 10^{-3}}{0.5}}$$

$$= \frac{2\pi}{\sqrt{10}} \approx 2\text{ s}$$

So, number of oscillations is

$$N = \frac{t}{T} = \frac{10000}{2} = 5000$$

31. *(d)*

> **key idea** A rotating charge constitutes a current. Hence, a rotating charged rod behaves like a current carrying coil. If charge q rotates with a frequency n, then equivalent current is $I = qn$ and magnetic moment associated with this current is $M = IA$ where, A = area of coil or area swept by rotating rod.

Let dq be the charge on dx length of rod at a distance x from origin as shown in the figure below.

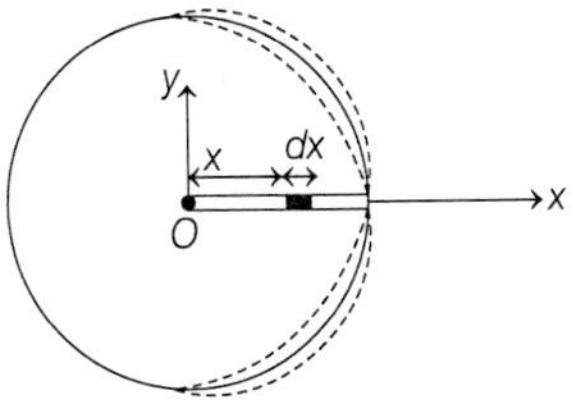

The magnetic moment dm of this portion dx is given as

$$dm = (dI)\,A$$

$$dm = ndqA$$

$$[\because I = qn, \ \therefore dI = n \cdot dq]$$

$$= n\,\rho\,dx\,A$$

Where, ρ = charge density of rod $= \rho_0 \dfrac{x}{l}$.

$$\text{So, } dm = \frac{n\rho_0 x\,dx\,\pi x^2}{l} = \frac{\pi n \rho_0}{l} \cdot x^3 \cdot dx$$

Total magnetic moment associated with rotating rod is sum of all the magnetic moments of all such differentiable elements of the rod. So, magnetic moment associated with complete rod is

$$M = \int_{x=0}^{x=l} dm = \int_0^l \frac{\pi n \rho_0}{l} \cdot x^3\,dx$$

$$= \frac{\pi n \rho_0}{l} \cdot \int_0^l x^3\,dx = \frac{\pi n \rho_0}{l}\left[\frac{x^4}{4}\right]_0^l$$

$$= \frac{\pi n \rho_0 l^3}{4} \text{ at } x = l \Rightarrow \rho = \rho_0$$

$$\therefore M = \frac{\pi}{4} n\rho l^3$$

32. *(d)* Suppose the effective length of this dipole be 'l'. Thus, the top view of the condition can be shown in the figure given below.

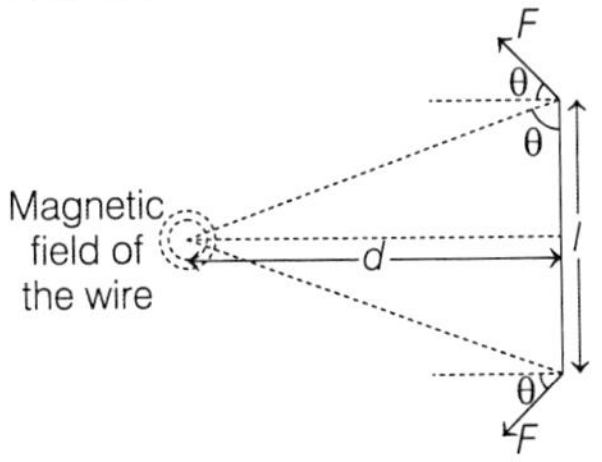

Now, the net force on the loop

$$F_{net} = 2F\cos\theta = 2mB\cos\theta$$

From the figure, get

$$\cos\theta = \frac{\frac{l}{2}}{\sqrt{d^2 + \frac{l^2}{4}}}$$

$$\Rightarrow \quad F_{net} = \frac{2mB\,l}{2\sqrt{d^2 + \frac{l^2}{4}}} \quad \text{...(i)}$$

Since, $M = IA = I\pi r^2 = ml$ ...(ii)

and magnetic field due to a straight infinitely long current-carrying conductor at a distance

x is $$B = \frac{\mu_0 I'}{2\pi x} \quad \text{...(iii)}$$

$\therefore$ Using Eqs. (ii) and (iii), rewriting Eq. (i), we get

$$F_{net} = \frac{I\pi a^2 \mu_0 I'}{2\pi\left(\sqrt{d^2 + \frac{l^2}{4}}\right)^2}$$

$$\Rightarrow \quad F_{net} \propto \frac{a^2}{\left(d^2 + \frac{l^2}{4}\right)}$$

l can be neglected as the loop is kept at very large distance.

or $$F_{net} \propto \left(\frac{a}{d}\right)^2$$

33. *(48)* Magnetic force on the wire,

$$F = IBl = 8 \times 0.15 \times 4 \times 10^{-2}$$

$$= 4.8 \times 10^{-2} \text{ N} = 48 \times 10^{-3} \text{ N} = 48 \text{ mN}$$

34. *(1)*

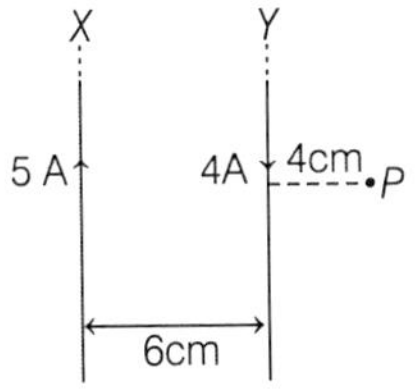

$$B_1 = \frac{\mu_0 I_1}{2\pi r_1}$$

$$r_1 = 6\text{cm} + 4\text{cm} = 10\text{cm} = 0.1 \text{ m}$$

$$B_2 = \frac{\mu_0 I_2}{2\pi r_2} \Rightarrow r_2 = 4 \text{ cm} = 0.04 \text{ cm} \quad \odot$$

$$B_1 = \frac{\mu_0(5)}{2\pi(0.1)} \otimes \Rightarrow B_2 = \frac{\mu_0(4)}{2\pi \times 0.04}$$

$$\text{B}_{net} = \text{B}_1 - \text{B}_2$$

$$= \frac{\mu_0(5)}{2\pi(0.1)} - \frac{\mu_0(4)}{2\pi(0.04)} = \frac{-100\mu_0}{4\pi}$$

$$= -100 \times 10^{-7} = -1 \times 10^{-5} \text{ T}$$

$$x = 1$$

35. *(8)* Magnetic field due to single side is

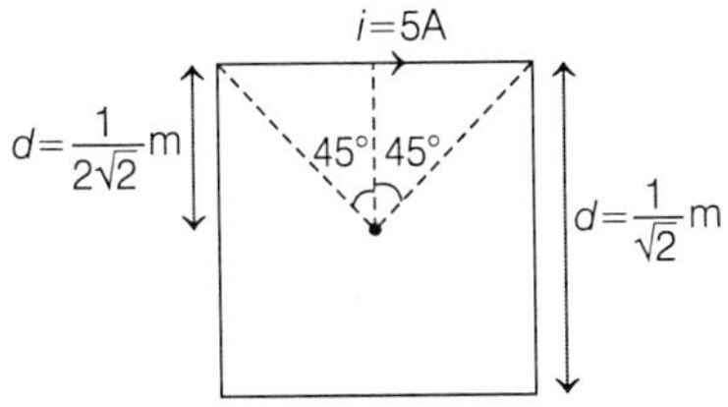

$$B = \frac{\mu_0 i}{4\pi d}[\sin\theta_1 + \sin\theta_2]$$

$$= \frac{10^{-7} \times 5 \times 2}{\frac{1}{2\sqrt{2}}} \times \frac{1}{\sqrt{2}} = 2 \times 10^{-6}\ \text{T}$$

$\therefore B_{net}$ at centre $O = 4B = 8 \times 10^{-6}$ T

$p = 8$

36. *(160)*

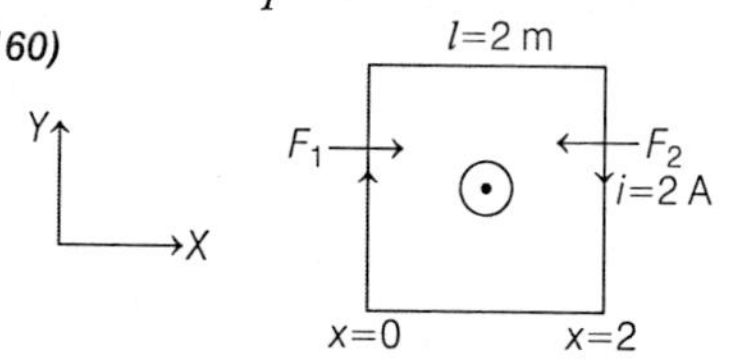

$B(x=0) = B_0$

$B(x=2) = 9B$

$F = ilB$

$F_1 = ilB_0$ and $F_2 = 9ilB_2$

$F = F_2 - F_1 = 8ilB_0$

$= 8 \times 2 \times 2 \times 5 = 160$ N

37. *(5)* Resistance of wire, $R = \frac{\rho L}{A}$ ...(i)

ρ = resistivity $= 2 \times 10^{-6}\ \Omega$m,

$A = 10\ \text{mm}^2 = 10^{-5}\ \text{m}^2$

Put the values in Eq. (i),

$$R = \frac{2 \times 10^{-6} \times L}{10^{-5}}$$

$L = 5$ m

As it is suspended, gravitational force is balanced by magnetic force.

$mg = Bil$

$$B = \frac{mg}{il}$$

$$= \frac{500 \times 10^{-3} \times 10}{2 \times 5}$$

$B = 5 \times 10^{-1}$ T

Hence, magnetic field is 5×10^{-1} T.

38. *(50)* Force on segment parallel to X-axis will cancel each other.

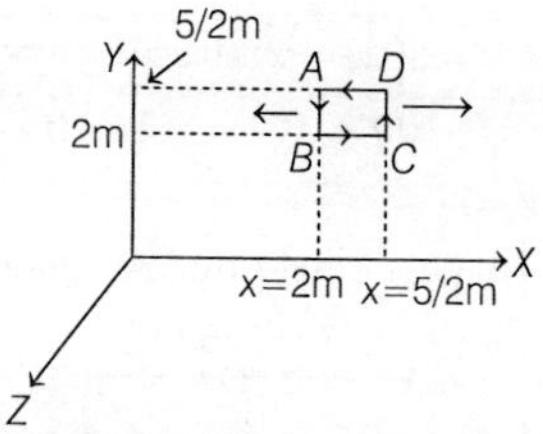

$F_{BC} + F_{DA} = 0$

Hence, F_{net} will be due to portion parallel to Y-axis.

$F_{AB} = ilB = 0.5 \times 0.5 \times 5$

$= 1.25\ \text{N} \times 0.2 = 0.25$ N

$F_{CD} = 0.5 \times 0.5 \times 6$

$= 1.5 \times 0.2 = 0.3$ N

$F_{net} = 0.3\ \text{N} - 0.25$ N

$= 50 \times 10^{-3}\ \text{N} = 50$ mN

39. *(35)* Given, $B_H = 3.5 \times 10^{-5}$ T

Current in wire, $i = \sqrt{2}$ A

$\theta = 45°$

Now, Force per unit length is given by

$$F = i\,lB\sin\theta,\ \frac{F}{l} = iB\sin\theta$$

$$= \sqrt{2} \times 3.5 \times 10^{-5} \times \frac{1}{\sqrt{2}} \quad (\because \theta = 45°)$$

$= 3.5 \times 10^{-5}$ N/m

$= 35 \times 10^{-6}$ N/m

Hence, missing value is 35.

40. *(2)*

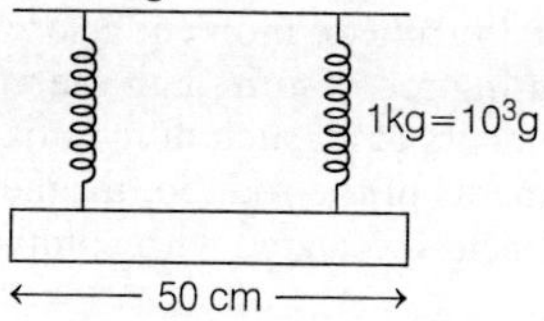

$1\ \text{kg} = 10^3$ g

$m = 40\text{g},\ l = 50$ cm

$$m = \frac{40}{1000}\ \text{kg} = 4 \times 10^{-2}\ \text{kg}$$

$l = 50$ cm

$1\ \text{cm} = 10^{-2}$ m

$l = 50\ \text{cm} = 50 \times 10^{-2}\ \text{m} = 0.5$ m

$\because\ iBl = mg$

$$i = \frac{mg}{Bl}$$

$$i = \frac{4 \times 10^{-2} \times 10}{0.40 \times 0.5} = \frac{10^{-1}}{0.5 \times 0.10}$$

$$= \frac{1}{5 \times 0.10} = \frac{1}{0.50} = \frac{100}{50} = 2\ \text{A}$$

41. *(5)* Given, $i_1 = i_2 = 5$ A

$l = 10\ \text{cm} = 0.1$ m

$F = 10^{-5}$ N

We have $\frac{F}{l} = \frac{\mu_0}{2\pi}\frac{i_1 i_2}{r}$

$$\Rightarrow \quad r = \frac{\mu_0}{2\pi} i_1 i_2 \times \frac{l}{F}$$

$$\Rightarrow \quad r = 2 \times 10^{-7} \times 25 \times \frac{0.1}{10^{-5}}$$

$= 50 \times 10^{-7} \times 10^4$

$= 50 \times 10^{-3}\ \text{m} = 5$ cm

42. *(3)* Length of conducting wire, $l = 24a$

Resistance of wire is R.

Ratio of magnetic moment of coil for equilateral triangle to square is $1 : \sqrt{y}$.

Side of equilateral triangle is a and also side of square is a. Number of turns in equilateral triangle will be

$$n_T = \frac{24a}{3a} = 8$$

[$\because$ Length of one turn is $3a$]

Number of turns in square will be

$$n_S = \frac{24a}{4a} = 6$$

[$\because$ Length of one turn is $4a$]

Magnetic moment of coil is given as

$M = nIA$

The ratio of magnetic moment for triangular coil to square coil is given as

$$\frac{M_T}{M_s} = \frac{n_T I A_T}{n_S I A_S}$$

[$\because I$ is same in both coils]

On substituting the values in above expression, we get

$$\frac{1}{\sqrt{y}} = \frac{\left(8 \times \frac{\sqrt{3}}{4} a^2\right)}{6 \times a^2}$$

$$= \frac{8\sqrt{3}}{24} = \frac{1}{\sqrt{3}}$$

$\Rightarrow \quad \sqrt{y} = \sqrt{3} \Rightarrow y = 3$

Hence, the value of y is 3.

43. *(3)* Side of equilateral triangular coil,

$a = 10\ \text{cm} = 0.1$ m

Horizontal magnetic field,

$B = 20\ \text{mT} = 20 \times 10^{-3}$ T

Current in coil,

$i = 0.2$ A

Torque acting on coil,

$\tau = \sqrt{x} \times 10^{-5}$ N-m

Torque acting on coil can be calculated as

$\tau = MB\sin\theta$

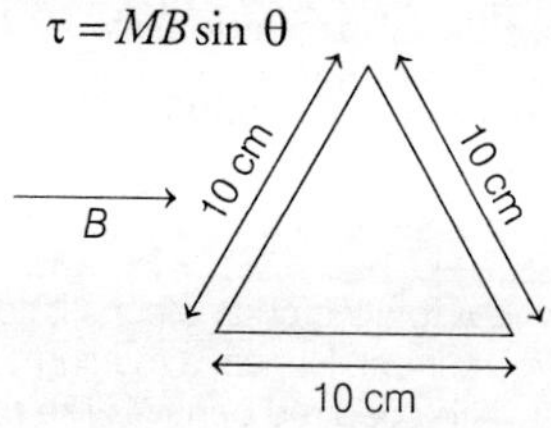

Here, M is pole strength and B is magnetic field.

Magnetic pole strength, $M = iA$

$$\tau = i\left(\frac{\sqrt{3}}{4}a^2\right)B\sin\theta$$

$$\left[\because A = \frac{\sqrt{3}}{4}a^2 \quad \text{(for equilateral triangle)}\right]$$

$= 0.2 \times \frac{\sqrt{3}}{4} \times (0.1)^2 \times 20 \times 10^{-3}$

$\times \sin 90° \quad [\because \mathbf{A} \perp \mathbf{B}]$

$= \sqrt{3} \times 10^{-5}$ N-m

After comparing obtained value with the given value, $x = 3$.

Topic 4

01. *(a)* $I_g G = (I - I_g) S$ or $I_g G \approx IS$

$$S = \frac{I_g G}{I} = \frac{(20 \times 10^{-3})(30\Omega)}{3\text{A}}$$

$$= \frac{600 \times 10^{-3}}{3} = \frac{6 \times 10^{-1}}{3}$$

$$= \frac{6}{30} = \frac{6 \times 5}{30 \times 5} = \frac{30}{150}$$

$X \approx 150$, $X = 149$

02. *(d)* Given, $\theta_1 = 60° = \frac{\pi}{3}$ rad,

$$\theta_2 = \frac{\pi}{10} \text{ rad}$$

In a galvanometer,

$\therefore \quad \theta \propto I$

$$\therefore \quad \frac{I_2}{I_1} = \frac{\theta_2}{\theta_1} = \frac{\frac{\pi}{10}}{\frac{\pi}{3}} = \frac{3}{10}$$

$$\Rightarrow \quad I_2 = \frac{3}{10} I_1 = \frac{3}{10} \times 200 = 60 \,\mu\text{A}$$

03. *(b)* Current sensitivity of a galvanometer,

$$S_i = \frac{\theta}{I} = \frac{BNA}{k}$$

B = magnetic field of pole pieces;

N = number of turns of coil

A = area of coil, k = spring constant.

So, when number of turns is doubled, S_i is also doubled.

Now voltage sensitivity of galvanometer is,

$$S_V = \frac{\theta}{IR} = \frac{BNA}{kR}$$

As N/R remains constant, voltage sensitivity does not increases.

04. *(b)* In a galvanometer to damp out the oscillations of the coil, core is either of a non-magnetic material or a magnetic core is covered with a sheet of a non-magnetic material.

Eddy currents induced in the non-magnetic material converts oscillation energy into heat.

In this way oscillation of coil dies out quickly.

05. *(d)* Given, deflection in the coil,

$\theta = 0.05$ rad

Current, $I = 10 \text{ mA} = 10^{-2}$ A

Magnetic field,

$B = 0.01$ T, $n = 200$

Torsional constant,

$C = 4 \times 10^{-5}$ N-m rad^{-1}

We know that,

$$\frac{\theta}{I} = \frac{nBA}{C} \Rightarrow A = \frac{C\theta}{nBI}$$

$$= \frac{4 \times 10^{-5} \times 0.05}{200 \times 0.01 \times 10^{-2}}$$

$= 1 \times 10^{-4}$ m^2

$= 1 \times 10^{-4} \times 10^4$ cm^2 $= 1$ cm^2

06. *(d)* We know that, current sensitivity of a galvanometer is given as

$$G = \frac{\theta}{i} = \frac{NBA}{K}$$

where, N = number of turns in the coil, B = magnetic field, A = area of loop

K = Galvanometer constant or current reduction factor or torsional constant of the spring.

Hence, G will be increased when either B increases or K decreases.

07. *(c)* Given,

Shunt resistance, $R_{sn} = 5\,\Omega$

Let, galvanometer resistance = G

Total, current = I

Voltage across galvanometer (V_G)

= Voltage across shunt resistance R_{sn}

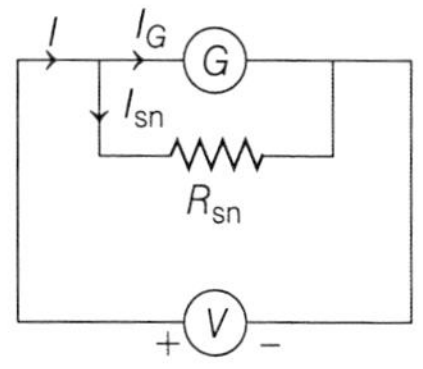

$I_G \times G = I_{sn} \times R_{sn}$

$\Rightarrow 0.02I \times G = 0.98I \times 5$

$$\Rightarrow \quad G = \frac{98 \times 5}{2} = 49 \times 5$$

$= 245\,\Omega$

08. *(d)* Given, number of divisions,

$n = 50$

Voltage, $V = 50 \text{ mV} = 50 \times 10^{-3}$ V

$$\text{Current sensitivity, } S_i = \frac{2 \text{ div}}{\text{mA}} = \frac{2 \text{ div}}{10^{-3} \text{ A}}$$

$$\therefore \text{Current, } I = \frac{n}{S_i} = \frac{50 \times 10^{-3}}{2}$$

$= 25 \times 10^{-3}$ A

By using Ohm's law,

$$\text{Resistance } (R) = \frac{\text{Potential } (V)}{\text{Current } (I)}$$

$$= \frac{50 \times 10^{-3}}{25 \times 10^{-3}} = 2\Omega$$

09. *(c)* Here,

I_g = full scale deflection current

$= 1\text{mA} = 1 \times 10^{-3}$ A

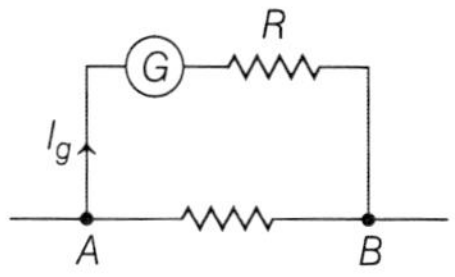

V_{AB} = Full scale deflection potential

$= 10$ V

G = Galvanometer resistance $= 100\,\Omega$

From the circuit diagram,

$I_g (G + R) = V$

$\Rightarrow 10^{-3} (100 + R) = 10$

$\Rightarrow \quad R = 10000 - 100$

$= 9900\,\Omega = 9.9$ kΩ

10. *(*)* Given data,

$I = 10^{-4}$ A, $R_S = 2\,\text{M}\Omega = 2 \times 10^6\,\Omega$

$V_{max} = 5$ V

Let internal resistance of galvanometer is R_G.

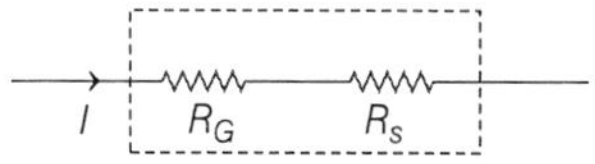

Voltmeter

Then, $I \times R_S + I \times R_G = V_{max}$

$\Rightarrow \quad 2 \times 10^6 \times 10^{-4} + 10^{-4} \times R_G = 5$

$\Rightarrow \quad 10^{-4} R_G = 5 - 200 = -195$

or $\quad R_G = -195 \times 10^4\,\Omega$

Resistance cannot be negative.

$\therefore$ No option is correct.

11. *(a)*

> **key idea** An ammeter is a type of galvanometer with a shunt connected in parallel to the galvanometer.

Ammeter circuit is shown in the figure below

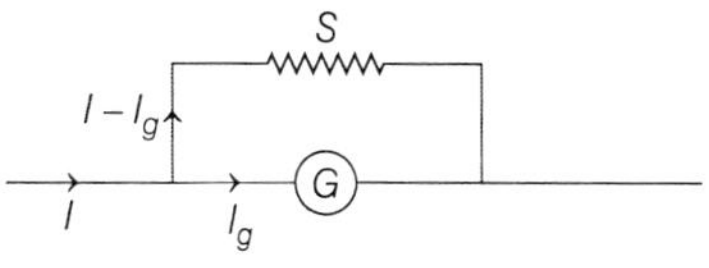

So, $I_g G = (I - I_g) S$

Here, $I_g = 0.002$ A,

$I = 0.5$ A, $G = 50\,\Omega$

So, shunt resistance required is

$$S = \frac{I_g G}{I - I_g} = \frac{0.002 \times 50}{(0.5 - 0.002)} \approx 0.2\,\Omega$$

12. *(d)* Given, resistance of galvanometer,

$$R_g = 50\,\Omega$$

Current, $I_g = 4\text{ mA} = 4 \times 10^{-3}$ A

Resistance used in converting a galvanometer in voltmeter, R

$$= 5\,\text{k}\Omega = 5 \times 10^3\,\Omega$$

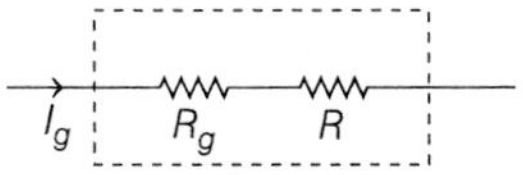

Voltmeter

∴ Maximum current in galvanometer is

$$I_g = \frac{E}{R + R_g}$$

$$\therefore E = I_g (R + R_g)$$
$$= 4 \times 10^{-3} \times (5 \times 10^3 + 50)$$
$$= 5050 \times 4 \times 10^{-3}$$
$$= 20.2\text{ V} \simeq 20\text{ V}$$

13. *(d)* Suppose, a resistance R_s is connected in series with galvanometer to convert it into voltmeter.

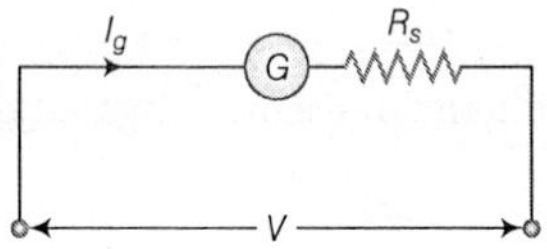

$$I_g (G + R_s) = V \Rightarrow R = \frac{V}{I_g} - G$$

$$\Rightarrow \quad R_s = 1985 = 1.985\text{ k}\Omega$$

or $\quad R_s = 1.985 \times 10^3\,\Omega$

14. *(a)* Maximum voltage that can be applied across the galvanometer coil

$$= 100\,\Omega \times 10^{-3}\text{ A} = 0.1\text{ V}.$$

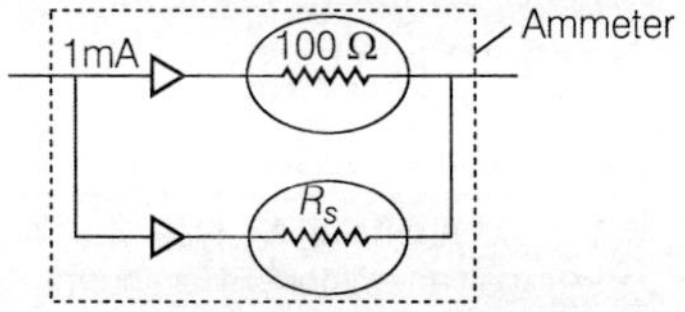

If R_s is the shunt resistance, then

$$R_s \times 10\text{ A} = 0.1\text{ V} \Rightarrow R_s = 0.01\,\Omega$$

15. *(d)* To increase the range of ammeter, we have to connect a small resistance in parallel (shunt), let its value be R.

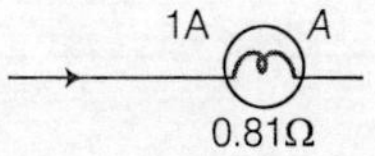

Apply KCL at junction to divide the current.

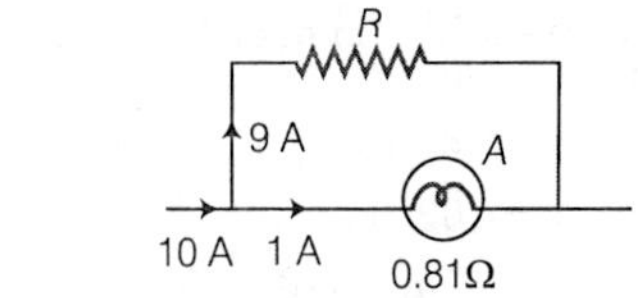

Voltage across R = Voltage across ammeter

i.e., $\quad 9R = 0.81 \times 1$

or $\quad R = \frac{0.81}{9} = 0.09\,\Omega$

16. *(c)* A voltmeter is a high resistance device and is always connected in parallel with the circuit. While an ammeter is a low resistance device and is always connected in series with the circuit.

So, to use ammeter in place of voltmeter, a high resistance must be connected in series with the ammeter to make its resistance high.

17. *(b)* We know that,

Voltage sensitivity, $S_V = \frac{\theta}{V} = \frac{NAB}{KR}$

$$\Rightarrow S_V \propto \frac{NAB}{R}$$

$$\Rightarrow \frac{(S_V)_1}{(S_V)_2} = \frac{N_1 A_1 B_1 R_2}{N_2 A_2 B_2 R_1}$$

$$= \frac{15 \times 3.6 \times 10^{-3} \times 0.25 \times 7}{21 \times 1.8 \times 10^{-3} \times 0.50 \times 5}$$

$$\therefore (S_V)_1 : (S_V)_2 = 1:1$$

18. *(c)* (A) $\tau = C\theta$

$$\tau = [\text{ML}^2\text{T}^{-2}]$$
$$[\text{ML}^2\text{T}^{-2}] = C[1]$$
$$C = [\text{ML}^2\text{T}^{-2}]$$

(B) Current senstivity $= \frac{\theta}{I}$

$$= \frac{BNA}{C}$$

Voltage sensitivity $= \frac{BNA}{RC}$

R also depends on N.

(C) Voltage sensitivity does not change on increasing N or $\frac{N}{R}$ = constant.

(D) MCG cannot be converted into an ammeter by introducing a shunt resistance of large value in parallel with galvanometer

(E) $CS = \frac{NAB}{C}$

$$CS \propto N$$

Ans. A, B only correct.

19. *(b)* For a galvanometer; $i_g R_g = i_s R_s$

where g represent galvanometer and s represent shunt,

$$\because i_g R_g = (i - i_g) R_s$$

Here, $i = 20$ mA, $i_g = 20\,\mu$A

$$\Rightarrow R_s = \left(\frac{20 \times 10^{-6}}{20 \times 10^{-3} - 20 \times 10^{-6}} \right) \times 200$$

$$= \frac{200}{100 - 1} \simeq 0.2\,\Omega$$

20. *(b)* $I_g = \frac{V}{R_g + R_S} = \frac{10}{100 + 400}$

$$I_g = 20 \times 10^{-3}\text{ A}$$

When galvanometer is used as ammeter, then

$$I_g R_g = (I - I_g) S$$

where S is shunt resistance.

$$20 \times 10^{-3} \times 100 = 10S$$

$$\Rightarrow \quad S = 20 \times 10^{-2}\,\Omega$$

Hence, value of $x = 20$

21. *(*)*

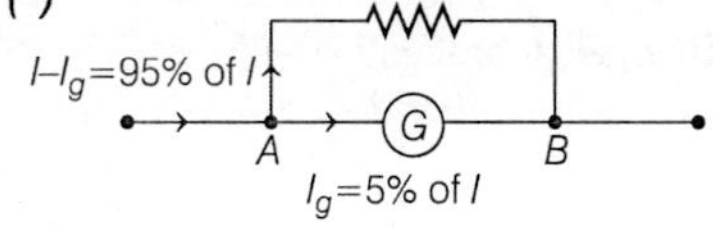

Ammeter given is shown above

Let $\quad I$ = input current.

Then, potential drop across galvanometer (or shunt S) is

$$V_{AB} = I_g G = (I - I_g) S$$

$$\Rightarrow \frac{5}{100} I \cdot G = \frac{95}{100} I \cdot S$$

$$\Rightarrow \quad S = \frac{5}{95} G = \frac{1}{19} G$$

Resistance of ammeter = resistance of parallel combination of G and S

$$= \frac{GS}{G + S} = \frac{G \times \frac{1}{19} G}{G + \frac{1}{19} G} = \frac{G}{20}$$

22. *(a)*

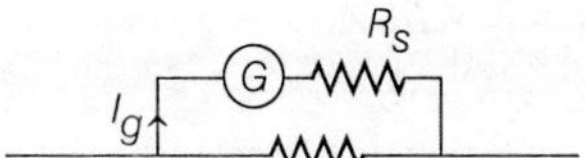

For voltmeter,

$$V_{AB} = I_g (G + R_S)$$

Given, $V_{AB} = 100$ V,

$I_g = 5$ mA

$\Rightarrow \quad G = 50\,\Omega$

So, $\quad 100 = 5 \times 10^{-3} (50 + R_S)$

$$\Rightarrow \frac{100}{5 \times 10^{-3}} = 50 + R_S$$

or $R_S = 20000 - 50 = 19950\,\Omega$

23. *(c)* Given, $G = 10\Omega$, $I_g = 3$ mA

$I = 8$ A

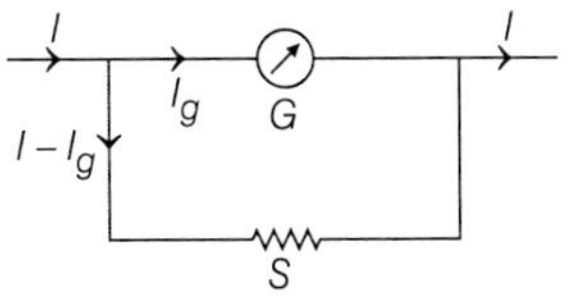

$$I_g G = (I - I_g)S$$

$$S = \frac{I_g G}{I - I_g} = \frac{3\times10^{-3}\times10}{8 - 3\times10^{-3}}$$

$$S = \frac{3\times10^{-3}\times10}{8 - 0.003}$$

$$= 3.75\times10^{-3}\,\Omega$$

24. *(b)* Let $x =$ current/division

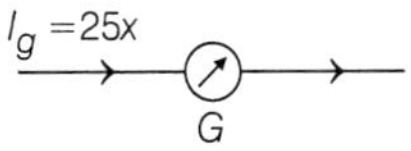

After applying shunt,

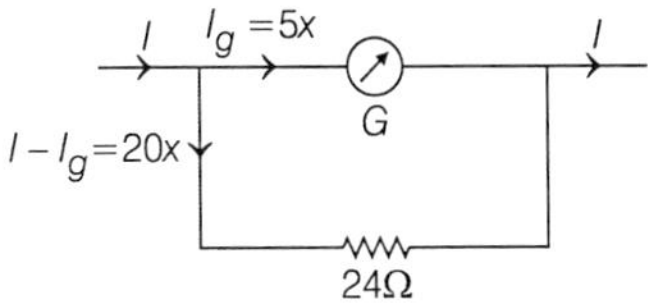

$$5x\times G = 20x\times 24$$

$$G = 4\times 24 = 96\Omega$$

25. *(a)* For ammeter, shunt resistance,

$$r = \frac{GI_g}{I - I_g} = \frac{54\times1\times10^{-3}}{(10-1)\times10^{-3}} = 6\,\Omega$$

For voltmeter, series resistance,

$$R = \frac{V}{I_g} - G = \frac{50}{1\times10^{-3}} - 54$$

$$= 5000 - 54$$

$$= 4946\,\Omega \simeq 5000\,\Omega \simeq 5\text{ k}\Omega$$

26. *(a)* Given, $\tau = mB$

$A =$ Area of coil

$B =$ magnetic field

$$K\theta = IANB$$

$$\frac{\theta}{I} = \frac{ANB}{K}$$

$$\frac{\theta}{I} = \text{Current sensitivity.}$$

$$1.25\left(\frac{\theta}{I}\right)_1 = \left(\frac{\theta}{I}\right)_2 \qquad \ldots\text{(i)}$$

$$1.25\left[\frac{AN_1B}{K}\right] = \left[\frac{AN_2B}{K}\right]$$

$$1.25 = \frac{N_2}{N_1}$$

$$\frac{N_2}{N_1} = \frac{125}{100} = \frac{5}{4} \qquad \ldots\text{(ii)}$$

$$R = \rho\frac{l}{a} = \text{constants}$$

Voltage sensitivity

$$= \frac{\theta}{V} = \frac{\theta}{IR} = \frac{\text{Current sensitivity}}{R}$$

$$R = \text{Constant}$$

Voltage sensitivity $\propto$ current sensitivity since, the current sensitivity is increased by 25%, the voltage sensitivity will also be increased by same percentage i.e., 25%.

27. *(b)*

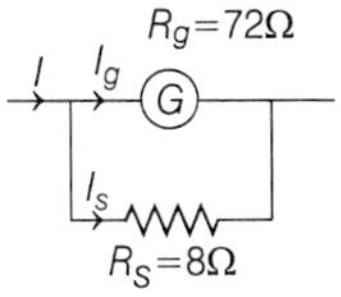

From KVL, $I = I_g + I_S$

where, I_g is the current in galvanometer and I_S is the current in shunt

$$1 = \frac{I_g}{I} + \frac{I_S}{I} \Rightarrow 1 = \frac{I_g}{I} + \left(\frac{R_g}{R_S}\times\frac{I_g}{I}\right)$$

$$\left(\because \frac{I_S}{I_g} = \frac{R_g}{R_S}\text{ So, } I_S = \frac{R_g}{R_S}\times I_g\right)$$

$$1 = \frac{I_g}{I}\left(1 + \frac{R_g}{R_S}\right) = \frac{I_g}{I}\left(1 + \frac{72}{8}\right)$$

$$1 = \frac{I_g}{I}(1+9) \Rightarrow \frac{I_g}{I} = \frac{1}{10}\text{ or } 10\%$$

28. *(d)* According to given situation, circuit diagram of galvanometer is shown below

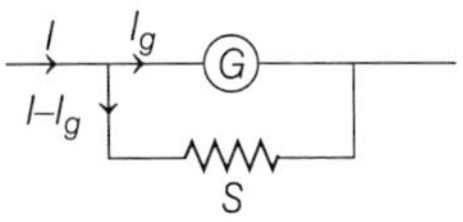

Figure of merit,

$$K = \frac{I_g}{n} \Rightarrow I_g = Kn \qquad \ldots\text{(i)}$$

From the above circuit diagram,

$$I_g G = (I - I_g)S$$

$$\Rightarrow I_g G = IS - I_g S$$

$$\Rightarrow IS = I_g G + I_g S$$

$$\Rightarrow IS = I_g(G+S) \Rightarrow I = \frac{I_g}{S}(G+S)$$

$$\Rightarrow \quad = \frac{nK}{S}(G+S) \qquad \text{[From Eq. (i)]}$$

29. *(a)* According to question,

From Ohm's law, $V = IR$

$$\therefore \quad i_g(R_1 + G) = 1\text{V} \qquad \ldots\text{(i)}$$

Let x be the required resistance.

Now, $i_g(R_1 + G + x) = 2$V

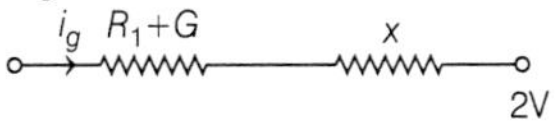

Using Eq. (i) in above equation, we get

$$\frac{1}{(R_1+G)}(R_1 + G + x) = 2$$

$$R_1 + x + G = 2(R_1 + G)$$

$$\Rightarrow \quad x = R_1 + G$$

Hence, correct option is (a).

30. *(d)* Resistance of galvanometer, $(G) = 20\Omega$

Number of divisions on both side $= 30$

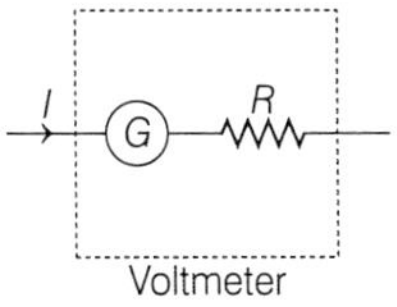

Figure of merit $= 0.005$ ampere/division

$\therefore$ Full scale deflection current $(I_g) =$ Number of divisions × figure of merit

$$\Rightarrow \quad I_g = 30\times0.005 = 0.15\text{ A}$$

Now, for measuring 15V by this galvanometer, let we use a resistance of R ohm in series with the galvanometer.

$\therefore$ Effective resistance of voltmeter,

$$R_{\text{eff}} = R + 20$$

As, maximum potential measured by voltmeter.

$$V = I_g\cdot R_{\text{eff}}\text{ or } 15 = 0.15(R + 20)$$

$$\text{or} \quad R + 20 = 100\text{ or } R = 80\Omega$$

31. *(a)* In a moving coil galvanometer in equilibrium, torque on coil due to current is balanced by torque of torsion band.

As, torque on coil,

$$\tau = \mathbf{M}\times\mathbf{B} = NIAB\sin\alpha$$

$$\therefore \quad \alpha = 90° \Rightarrow \tau = NIBA$$

Torque of torsion band, $T = k\theta$

where, $k =$ torsion constant of torsion band

and $\theta =$ deflection of coil in radians or angle of twist of restoring torque.

$$\therefore \quad BINA = k\theta\text{ or } B = \frac{k\theta}{INA} \qquad \ldots\text{(i)}$$

Here, $k = 10^{-6}$ N - m / rad,

$I = 1\times10^{-3}$ A, $N = 175$

$A = 1\text{ cm}^2 = 1\times10^{-4}\text{ m}^2$

$$\theta = 1° = \frac{\pi}{180}\text{ rad}$$

On substituting values in Eq. (i), we get

$$B = \frac{10^{-6}\times22}{1\times10^{-3}\times175\times7\times180\times10^{-4}}$$

$$= 0.998\times10^{-3} \simeq 10^{-3}\text{ T}$$

32. *(d)* Full scale deflection current

$$=\frac{150}{10}\text{ mA}=15\text{ mA}$$

Full scale deflection voltage

$$=\frac{150}{2}\text{ mV}=75\text{ mV}$$

Galvanometer resistance,

$$G=\frac{75\text{ mV}}{15\text{ mA}}=5\,\Omega$$

Required full scale deflection voltage,

$$V=1\times150=150\text{ V}$$

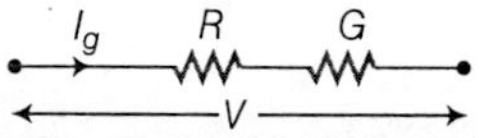

Let resistance to be connected in series be R.

$\Rightarrow \quad V=I_g(R+G)$

$\therefore \quad 150=15\times10^{-3}\,(R+5)$

or $\quad 10^4=R+5$

or $\quad R=10000-5=9995$

33. *(b)* To use galvanometer as an ammeter, a low resistance in parallel is used.

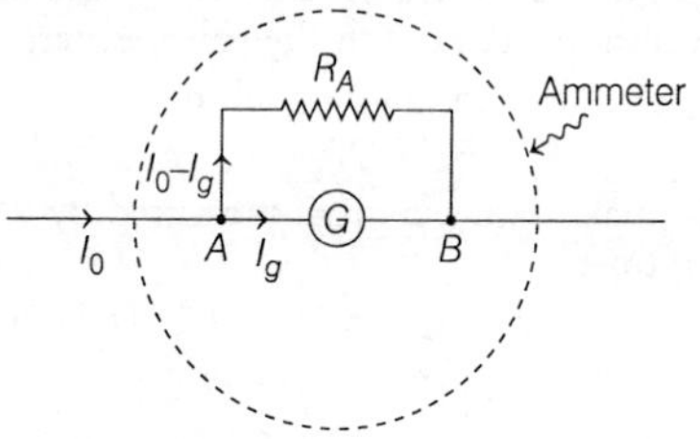

In ammeter, if I_g = full scale deflection current, then equating potential drops across points marked AB, we have

$$V_{AB}=I_gG=(I_0-I_g)R_A$$

$$\Rightarrow \quad R_A=\frac{I_gG}{I_0-I_g} \quad \ldots(\text{i})$$

Here, G = resistance of galvanometer coil. When a galvanometer is used as a voltmeter, a high resistance (R_V) in series is used.

Equating potential across point AB,

$$V_{AB}=(G+R_V)\,I_g$$

But $\quad V_{AB}=I_0G \quad$ [Given]

So, $\quad I_0G=(G+R_V)\,I_g$

$$\Rightarrow \quad R_V=\frac{(I_0-I_g)G}{I_g} \quad \ldots(\text{ii})$$

From Eqs. (i) and (ii), we have

$$\frac{R_A}{R_V}=\frac{\left(\dfrac{I_gG}{I_0-I_g}\right)}{\dfrac{(I_0-I_g)G}{I_g}}$$

$$=\frac{I_g^2}{(I_0-I_g)^2}$$

and

$$R_A\times R_V=\frac{I_gG}{(I_0-I_g)}\times\frac{(I_0-I_g)\,G}{I_g}$$

$$=G^2$$

34. *(c)* Given, divisions in scale of galvanometer, $n=50$

Sensitivity of galvanometer,

$$\frac{I_g}{n}=20\ \mu\text{A/division}$$

$\therefore$ Current in galvanometer,

$$I_g=\frac{I_g}{n}\times n=20\,\mu\text{A}\times50$$

$\Rightarrow \quad I_g=1000\ \mu\text{A}=1\text{ mA}$

Now, for R, it should be converted into 2V voltmeter.

$\therefore V_1=I_g(R_1+G)\Rightarrow 2=10^{-3}(R_1+100)$

$\Rightarrow 2000=R_1+100$

$\therefore \quad R_1=1900\Omega \quad \ldots(\text{i})$

For R_2, it should be converted into 10V voltmeter.

$\therefore \quad V_2=I_g\,[(R_1+R_2)+G]$

$\Rightarrow \quad 10=10^{-3}\,[(R_1+R_2)+100]$

$\Rightarrow 10000=R_1+R_2+100=2000+R_2$

$\therefore \quad R_2=8000\Omega \quad \ldots(\text{ii})$

For R_3, it should be converted into 20V voltmeter.

$\therefore \quad V_3=I_g[(R_1+R_2+R_3+G]$

$\Rightarrow \quad 20=10^{-3}\,[1900+8000+R_3+100]$

$\Rightarrow 20000=R_3+10000$

$\therefore \quad R_3=10000\,\Omega \quad \ldots(\text{iii})$

From Eqs. (i), (ii) and (iii), it is clear that option (c) is correct.

35. *(5)*

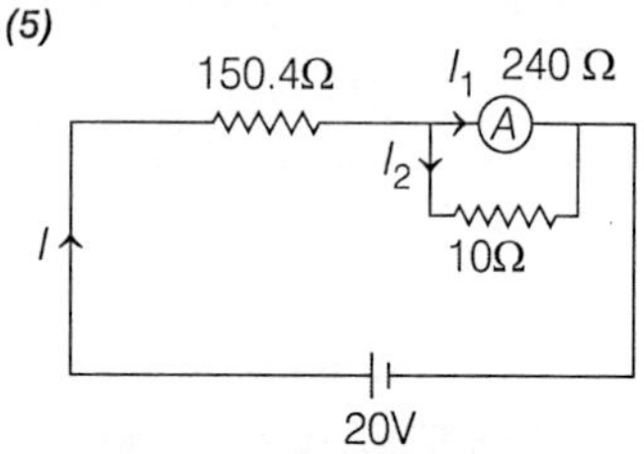

$$R_{eq}=150.4+\frac{240\times10}{240+10}$$

$R_{eq}=150.4+9.6$

$R_{eq}=160\Omega$

$$I=\frac{120}{160}=0.125\text{ A}$$

Voltage across ammeter,

$V_{\text{ammeter}}=I\times9.6$

$=0.125\times9.6=1.25\text{ V}$

Current across ammeter,

$$I_{\text{ammeter}}=\frac{V_{\text{ammeter}}}{R_{\text{ammeter}}}=\frac{1.2}{240}$$

$=5\times10^{-3}\text{ A}=5\text{ mA}$

36. *(4)* In a galvanometer

$$Q\propto I \text{ or } kQ=BINA$$

Substituting given values,

$$k(0.05)=0.01\times(10\times10^{-3})\times(100)\times(2\times10^{-4})$$

$$\Rightarrow \quad k=4\times10^{-5}\,\frac{\text{N-m}}{\text{rad}}$$

37. *(20)* Torque,

$$\tau=BINA\sin\theta$$

Here, $\quad \theta=90°$

$\Rightarrow \quad \sin\theta=1,\ \tau=1.5\text{ N-m}$

$A=3\times10^{-4}\text{ m}^2,$

$N=500 \quad\text{and } I=0.5\text{ A}$

So, $\quad B=\dfrac{\tau}{INA}$

$$=\frac{1.5}{0.5\times500\times3\times10^{-4}}=20\text{ T}$$

In JEE Main there is usually 0-1 question per paper is visible from this chapter. In past 3 years, 11 conceptual and 4 numerical questions were asked of moderate to difficult level from this chapter.
For detailed analysis scan QR code

MAGNETISM AND MATTER

Topic 1 Bar Magnet and Magnetic Dipole Moment

Level 1

01 The magnetic moments associated with two closely wound circular coils A and B of radius $r_A = 10$ cm and $r_B = 20$ cm respectively, are equal if (where N_A, I_A and N_B, I_B are number of turn and current of A and B respectively) *[2023, 30 Jan Shift-I]*

(a) $2N_AI_A = N_BI_B$
(b) $4N_AI_A = N_BI_B$
(c) $N_A = 2N_B$
(d) $N_AI_A = 4N_BI_B$

Imp. Oscillation of Bar Magnet in Uniform Magnetic Field

02 A small circular loop of conducting wire has radius a and carries current i. It is placed in a uniform magnetic field B perpendicular to its plane such that when rotated slightly about its diameter and released, it starts performing simple harmonic motion of time period T. If the mass of the loop is m, then *[2020, 9 Jan Shift-II]*

(a) $T = \sqrt{\frac{\pi m}{iB}}$ (b) $T = \sqrt{\frac{2m}{iB}}$
(c) $T = \sqrt{\frac{2\pi m}{iB}}$ (d) $T = \sqrt{\frac{\pi m}{2iB}}$

03 A magnetic needle of magnetic moment 6.7×10^{-2} Am^2 and moment of inertia 7.5×10^{-6} kg m^2 is performing simple harmonic oscillations in a magnetic field of 0.01 T. Time taken for 10 complete oscillations is *[JEE Main 2017]*

(a) 8.89 s (b) 6.98 s
(c) 8.76 s (d) 6.65 s

04 A magnetic dipole in a constant magnetic field has *[JEE Main 2017]*

(a) maximum potential energy when the torque is maximum
(b) zero potential energy when the torque is minimum.
(c) zero potential energy when the torque is maximum
(d) minimum potential energy when the torque is maximum.

05 The magnetic lines of force inside a bar magnet *[JEE Main 2013]*

(a) are from North-pole to South-pole of the magnet
(b) do not exist
(c) depend upon the area of cross-section of the bar magnet
(d) are from South-pole to North-pole of the magnet

06 A bar magnet of length 6 cm magnetic moment of $4\,\text{J T}^{-1}$. Find the strength of magnetic field at a distance of 200 cm from the centre of the magnet along its equatorial line. *[AIEEE 2012]*

(a) 4×10^{-8} T (b) 3.5×10^{-8} T
(c) 5×10^{-8} T (d) 3×10^{-8} T

07 A thin circular disc of radius R is uniformly charged with density $\sigma > 0$ per unit area. The disc rotates about its axis with a uniform angular speed ω. The magnetic moment of the disc is *[AIEEE 2011]*

(a) $\pi R^4 \sigma \omega$ (b) $\frac{\pi R^4}{2} \sigma \omega$
(c) $\frac{\pi R^4}{4} \sigma\omega$ (d) $2\pi R^4 \sigma\omega$

08 A magnetic needle is kept in a non-uniform magnetic field. It experiences *[AIEEE 2005]*

(a) a torque but not a force
(b) Neither a force nor a torque
(c) a force and a torque
(d) a force but not a torque

Level 2

09 Given below are two statements. One is labelled as Assertion (A) and the other is labelled as Reason (R).

Assertion (A) Magnetic monopoles do not exist.

Reason (R) Magnetic field lines are continuous and form closed loops.

In the light of the above statements, choose the most appropriate answer from the options given below: *[2025, 7 April, Shift-II]*

(a) A is not correct but R is correct.
(b) A is correct but R is not correct.
(c) Both A and R are correct and R is the correct explanation of A.
(d) Both A and R are correct but R is not the correct explanation of A.

Imp. Potential Energy of Magnetic Dipole

10 A magnetic dipole experiences a torque of $80\sqrt{3}$ N-m when placed in uniform magnetic field in such a way that dipole moment makes angle of 60° with magnetic field. The potential energy of the dipole is *[2025, 3 April, Shift-II]*

(a) 80 J (b) –80 J (c) –60 J (d) $-40\sqrt{3}$ J

11

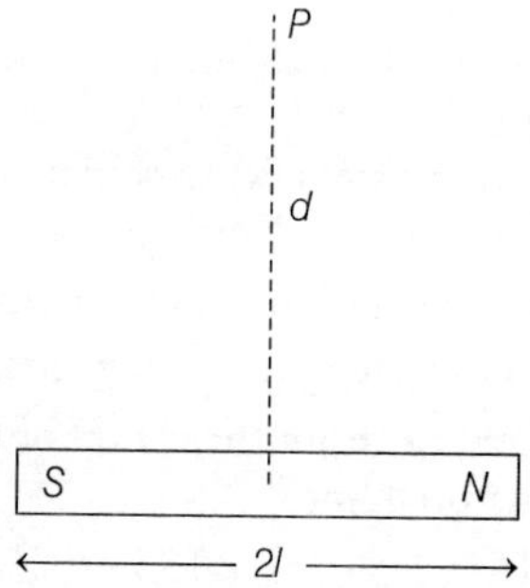

A bar magnet has total length $2l = 20$ units and the field point P is at a distance $d = 10$ units from the centre of the magnet. If the relative uncertainly of length measurement is 1%, then uncertainty of the magnetic field at point P is *[2025, 28 Jan Shift-II]*

(a) 5% (b) 3% (c) 4% (d) 10%

Imp. Work Done in Rotating the Bar Magnet

12 The magnetic moment of a bar magnet is 0.5 Am^2. It is suspended in a uniform magnetic field of 8×10^{-2} T. The work done in rotating it from its most stable to most unstable position is *[2024, 4 April Shift-II]*

(a) 8×10^{-2} J (b) 4×10^{-2} J
(c) 16×10^{-2} J (d) zero

13 A small bar magnet placed with its axis at 30° with an external field of 0.06 T experiences a torque of 0.018 Nm. The minimum work required to rotate it from its stable to unstable equilibrium position is *[2020, 4 Sep Shift-I]*

(a) 9.2×10^{-3} J (b) 7.2×10^{-2} J
(c) 6.4×10^{-2} J (d) 11.7×10^{-3} J

14 Two short bar magnets of length 1 cm each have magnetic moments 1.20 A m^2 and 1.00 A m^2, respectively. They are placed on a horizontal table parallel to each other with their North poles pointing towards the South. They have a common magnetic equator and are separated by a distance of 20.0 cm. The value of the resultant horizontal magnetic induction at the mid-point O of the line joining their centres is close to
(Horizontal component of Earth's magnetic induction is 3.6×10^{-5} Wb/m^2) *(JEE Main 2013)*

(a) 5.80×10^{-4} Wb/m^2 (b) 3.6×10^{-5} Wb/m^2
(c) 2.56×10^{-4} Wb/m^2 (d) 3.50×10^{-4} Wb/m^2

15 The length of a magnet is large compared to its width and breadth. The time period of its oscillation in a vibration magnetometer is 2s. The magnet is cut along its length into three equal parts and these parts are then placed on each other with their like poles together. The time period of this combination will be *[AIEEE 2004, 2003]*

(a) $2\sqrt{3}$ s (b) $\frac{2}{3}$ s (c) 2 s (d) $\frac{2}{\sqrt{3}}$ s

16 A magnetic needle lying parallel to a magnetic field requires W units of work to turn it through 60°. The torque needed to maintain the needle in this position will be *[AIEEE 2003]*

(a) $\sqrt{3}\,W$ (b) $\sqrt{2}\,W$ (c) $\frac{\sqrt{3}}{2}W$ (d) $2W$

17 A thin rectangular magnet suspended freely has a period of oscillation equal to T. Now, it is broken into two equal halves (each having half of the original length) and one piece is made to oscillate freely in the same field. If its period of oscillation is T', the ratio T'/T is *[AIEEE 2003]*

(a) $\frac{1}{2\sqrt{2}}$ (b) $\frac{1}{2}$ (c) 2 (d) $\frac{1}{4}$

Level 3

18 A circular coil has moment of inertia 0.8 kg m^2 around any diameter and is carrying current to produce a magnetic moment of 20 Am^2. The coil is kept initially in a vertical position and it can rotate freely around a horizontal diameter. When a uniform magnetic field of 4 T is applied along the vertical, it starts rotating around its horizontal diameter. The angular speed the coil acquires after rotating by 60° will be *[2020, 4 Sep Shift-II]*

(a) 10 rad s^{-1} (b) 10 π rad s^{-1}
(c) 20 π rad s^{-1} (d) 20 rad s^{-1}

19 A magnetic dipole is acted upon by two magnetic fields which are inclined to each other at an angle of 75°. One of the fields has a magnitude of 15 mT. The dipole attains stable equilibrium at an angle of 30° with this field. The magnitude of the other field (in mT) is close to *[JEE Main 2016]*

(a) 1 (b) 11
(c) 36 (d) 1060

Numerical Value Type Questions

Imp. Magnetic Dipole Moment

20 A straight magnetic strip has a magnetic moment of 44 Am^2. If the strip is bent in a semicircular shape, its magnetic moment will be .... Am^2. (Given, $\pi = 22/7$) *[2024, 9 April Shift-II]*

21 A coil having 100 turns, area of 5×10^{-3} m^2, carrying current of 1 mA is placed in uniform magnetic field of 0.20 T such a way that plane of coil is perpendicular to the magnetic field. The work done in turning the coil through 90° is .... μJ. *[2024, 6 April Shift-II]*

22 A circular coil having 200 tuns, 2.5×10^{-4} m^2 area and carrying 100 μA current is placed in a uniform magnetic field of 1T. Initially the magnetic dipole moment **M** was directed along **B**. Amount of work, required to rotate the coil through 90° from its initial orientation such that **M** becomes perpendicular to **B** is .... μ J. *[2024, 6 April Shift-I]*

23 Two short magnetic dipoles m_1 and m_2 each having magnetic moment of 1 Am^2 are placed at point O and P, respectively. The distance between OP is 1 metre. The torque experienced by the magnetic dipole m_2 due to the presence of m_1 is .... $\times 10^{-7}$ N m. *[2021, 26 Aug Shift-I]*

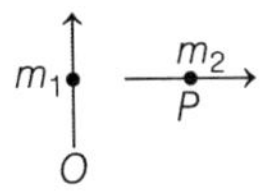

24 In a uniform magnetic field, the magnetic needle has a magnetic moment 9.85×10^{-2} A-m^2 and moment of inertia 5×10^{-6} kg-m^2. If it performs 10 complete oscillations in 5 s, then the magnitude of the magnetic field is .... mT. [Take, π^2 as 9.85] *[2021, 27 July Shift-I]*

Topic 2 Magnetic Materials

Level 1

01 The relationship between the magnetic susceptibility (χ) and the magnetic permeability (μ) is given by (μ_0 is the permeability of free space and μ_r is relative permeability) *[2025, 2 April Shift-I]*

(a) $\chi = \frac{\mu}{\mu_0} - 1$ (b) $\chi = \frac{\mu_r}{\mu_0} + 1$
(c) $\chi = \mu_r + 1$ (d) $\chi = 1 - \frac{\mu}{\mu_0}$

02 Paramagnetic substances
A. align themselves along the directions of external magnetic field.
B. attract strongly towards external magnetic field.
C. has susceptibility little more than zero.
D. move from a region of strong magnetic field to weak magnetic field.
Choose the most appropriate answer from the options given below. *[2024, 8 April Shift-I]*

(a) B, D Only (b) A, C Only
(c) A, B, C, D (d) A, B, C Only

03 Given below are two statements.
Statement I The diamagnetic property depends on temperature.
Statement II The induced magnetic dipole moment in a diamagnetic sample is always opposite to the magnetisic field.
In the light of above statements, choose the most appropriate answer from the options given below. *[2023, 12 April Shift-II]*

(a) Both Statement I and Statement II are false.
(b) Statement I is false but Statement II is true.
(c) Both Statement I and Statement II are true.
(d) Statement I is true but Statement II is false.

Imp. Properties of Magnetic Materials

04 In a ferromagnetic material, below the Curie temperature, a domain is defined as *[2021, 25 Feb Shift-II]*
(a) a macroscopic region with zero magnetisation
(b) a macroscopic region with saturation magnetisation
(c) a macroscopic region with randomly oriented magnetic dipoles
(d) a macroscopic region with consecutive magnetic dipoles oriented in opposite direction

05 A soft ferromagnetic material is placed in an external magnetic field. The magnetic domains *[2021, 24 Feb Shift-II]*
(a) decrease in size and changes orientation
(b) increase in size but no change orientation
(c) may increase or decrease in size and change its orientation
(d) have no relation with external magnetic field

06 A perfectly diamagnetic sphere has a small spherical cavity at its centre,which is filled with a paramagnetic substance. The whole system is placed in a uniform magnetic field **B** Then, the field inside the paramagnetic substance is *[2020, 3 Sep Shift-II]*

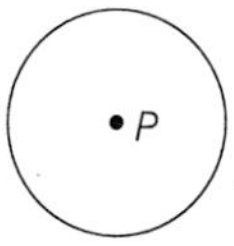

(a) much larger than $|\mathbf{B}|$ and parallel to $\mathbf{B}$
(b) $\mathbf{B}$
(c) zero
(d) much larger than $|\mathbf{B}|$ but opposite to $\mathbf{B}$

07 A bar magnet is demagnetised by inserting it inside a solenoid of length 0.2 m, 100 turns and carrying a current of 5.2 A. The coercivity of the bar magnet is *[2019, 9 Jan Shift-I]*

(a) 1200 A/m (b) 285 A/m
(c) 2600 A/m (d) 520 A/m

08 Curie temperature is the temperature above which *[AIEEE 2003]*

(a) a ferromagnetic material becomes paramagnetic
(b) a paramagnetic material becomes diamagnetic
(c) a ferromagnetic material becomes diamagnetic
(d) a paramagnetic material becomes ferromagnetic

Level 2

09 The percentage increase in magnetic field (B) when space within a current carrying solenoid is filled with magnesium (magnetic susceptibility $\chi_{Mg} = 1.2 \times 10^{-5}$) is *[2025, 7 April, Shift-I]*

(a) $\frac{6}{5} \times 10^{-3}\%$ (b) $\frac{5}{6} \times 10^{-5}\%$
(c) $\frac{5}{3} \times 10^{-5}\%$ (d) $\frac{5}{6} \times 10^{-4}\%$

10 A solenoid having area A and length l is filled with a material having relative permeability 2. The magnetic energy stored in the solenoid is *[2025, 2 April, Shift-II]*

(a) $\frac{B^2 Al}{\mu_0}$ (b) $B^2 Al$
(c) $\frac{B^2 Al}{2\mu_0}$ (d) $\frac{B^2 Al}{4\mu_0}$

Imp. Curie Temperature

11 **Statement I** The ferromagnetic property depends on temperature. At high temperature, a ferromagnet becomes a paramagnet.
Statement II At high temperature, the domain wall area of a ferromagnetic substance increases.
In the light of the above statements, choose the most appropriate answer from the options given below. *[2021, 22 July Shift-II]*

(a) Statement I is true but Statement II is false.
(b) Both Statement I and Statement II are true.
(c) Both Statement I and Statement II are false.
(d) Statement I is false but Statement II is true.

12 An iron rod of volume 10^{-3} m^3 and relative permeability 1000 is placed as core in a solenoid with 10 turns/cm. If a current of 0.5 A is passed through the solenoid, then the magnetic moment of the rod will be *[2020, 5 Sep Shift-II]*

(a) 50×10^2 A-m^2
(b) 5×10^2 A-m^2
(c) 500×10^2 A-m^2
(d) 0.5×10^2 A-m^2

13 Needles N_1, N_2 and N_3 are made of a ferromagnetic, paramagnetic and diamagnetic substance respectively. A magnet when brought close to them will *[AIEEE 2006]*

(a) attract N_1 and N_2 strongly but repel N_3
(b) attract N_1 strongly, N_2 weakly and repel N_3 weakly
(c) attract N_1 strongly, but repel N_2 and N_3 weakly
(d) attract all three of them

Level 3

14 Match List I with List II.

	List I (Y vs X)		List II (Shape of Graph)
A.	Y = magnetic susceptibility X = magnetising field	I.	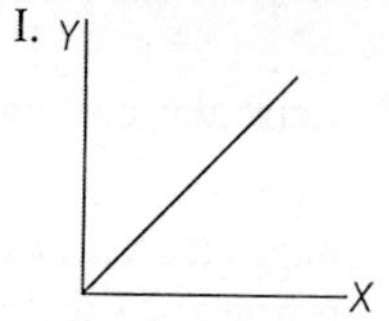
B.	Y = magnetic field X = distance from centre of a current carrying wire for $x < a$ (where a = radius of wire)	II.	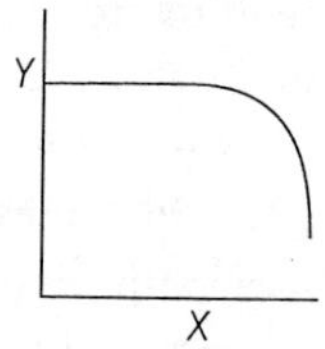
C.	Y = magnetic field X = distance from centre of a current carrying wire for $x > a$ (where a = radius of wire)	III.	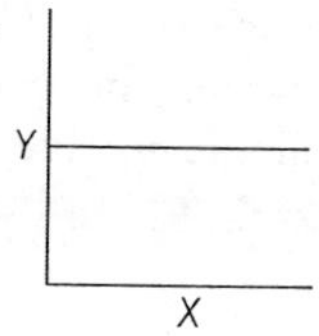
D	Y = magnetic field inside solenoid X = distance from centre	IV.	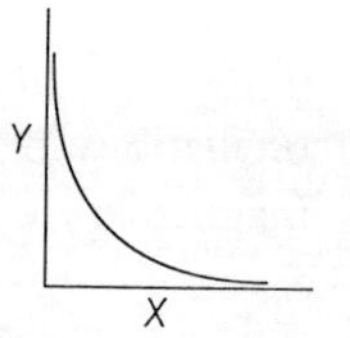

Choose the correct answer from the options given below. *[2024, 6 April Shift-II]*

	A	B	C	D
(a)	IV	I	III	II
(b)	III	I	IV	II
(c)	III	IV	I	II
(d)	I	III	II	IV

15 Which of the following statements are correct?

A. Electric monopoles do not exist, whereas magnetic monopoles exist.

B. Magnetic field lines due to a solenoid at its ends and outside cannot be completely straight and confined.

C. Magnetic field lines are completely confined within a toroid.

D. Magnetic field lines inside a bar magnet are not parallel.

E. $\chi = -1$ is the condition for a perfect diamagnetic material, where χ is its magnetic susceptibility.

Choose the correct answer from the options given below. *[2021, 18 March Shift-II]*

(a) C and E (b) B and D
(c) A and B (d) B and C

Numerical Value Type Questions

16 The coercivity of a magnet is 5×10^3 A/m.
The amount of current required to be passed in a solenoid of length 30 cm and the number of turns 150, so that the magnet gets demagnetised when inside the solenoid is .... A. *[2024, 8 April Shift-II]*

• Key **Index**

Topic 1

1 (d)	**2** (c)	**3** (d)	**4** (c)	**5** (d)	**6** (c)	**7** (c)	**8** (c)	**9** (c)	**10** (b)
11 (b)	**12** (a)	**13** (b)	**14** (c)	**15** (b)	**16** (a)	**17** (b)	**18** (a)	**19** (b)	**20** (28)
21 (100)	**22** (5)	**23** (1)	**24** (8)						

Topic 2

1 (a)	**2** (b)	**3** (b)	**4** (b)	**5** (c)	**6** (c)	**7** (c)	**8** (a)	**9** (a)	**10** (d)
11 (a)	**12** (b)	**13** (b)	**14** (b)	**15** (a)	**16** (10)				

Detailed **Explanations**

Topic 1

01. *(d)* Magnetic moment of a circular coil is, $M = NIA = NI(\pi r^2)$

When 2 coils A and B have same magnetic moments then,

$$M_A = M_B$$

$$\Rightarrow N_A I_A(\pi r_A^2) = N_B I_B(\pi r_B^2)$$

$$\Rightarrow N_A I_A (10)^2 = N_B I_B (20)^2$$

(Given, $r_A = 10$ cm and $r_B = 20$ cm)

$$\Rightarrow \quad N_A I_A = 4 N_B I_B$$

02. *(c)* Time period,

$$T = 2\pi\sqrt{\frac{I}{\mu B}}$$

where, μ = magnetic dipole

Moment of dipole = Current × Area

$$= \pi a^2 i$$

I = Moment of inertia of a coil

$$= \frac{ma^2}{2}$$

So, $T = 2\pi\sqrt{\frac{ma^2/2}{\pi a^2 iB}} = 2\pi\sqrt{\frac{m}{2\pi iB}}$

$$= \sqrt{\frac{2\pi m}{iB}}$$

03. *(d)* Time period of oscillation is

$$T = 2\pi\sqrt{\frac{I}{MB}}$$

$$\Rightarrow \quad T = 2\pi\sqrt{\frac{7.5\times10^{-6}}{6.7\times10^{-2}\times0.01}}$$

$$= 0.665 \text{ s}$$

Time taken for 10 oscillations

$$= 10 \times T = 6.65 \text{ s}$$

04. *(c)* Potential energy,

$$U = -MB\cos\theta$$

Torque, $\tau = MB\sin\theta$

Torque will be maximum (τ_{max}) when

$$\theta = 90°$$

Thus, potential energy,

$$U = -MB\cos 90° = 0$$

05. *(d)* Inside bar magnet, magnetic lines of force are from South to North.

06. *(c)* Strength of magnetic field on equitorial line

$$B = \frac{\mu_0}{4\pi}\frac{M}{(r^2 + l^2)^{3/2}}$$

where $M = 4$ J/T, $r = 200$ cm,

$$l = \frac{6 \text{ cm}}{2} = 3\times10^{-2} \text{ m}$$

$$B = \frac{10^{-7}\times4}{[2^2 + (3\times10^{-2})^2]^{\frac{3}{2}}}$$

$$= 5\times10^{-8} \text{ T}$$

07. *(c)* Consider the disc is made up of large number of rings.

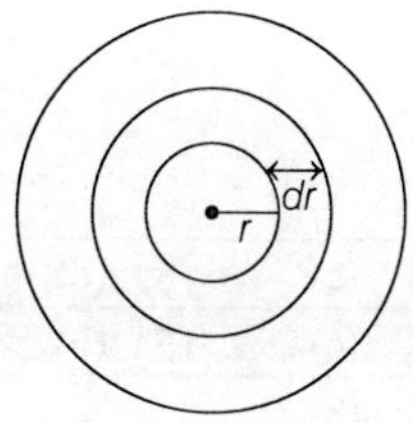

Let one such ring of radius r and thickness dr

Charge on this elementary ring

$dq = \sigma \times 2\pi r dr$ (σ = surface charge density)

Current associated with this ring

$$dI = \frac{dq}{dt} = dqf \; [f \text{ is frequency, } f = \frac{1}{T}]$$

$$dI = \sigma 2\pi f \, rdr$$

$$= \sigma\omega rdr \qquad (\omega = 2\pi f)$$

Magnetic moment of this ring,

$$dM = dI \cdot dA$$

$$= \sigma\omega rdr \times \pi r^2$$

$$dM = \pi\sigma\omega r^3 \, dr$$

$$\int_0^M dM = \pi\sigma\omega\int_0^R r^3 \, dr$$

$$M = \frac{1}{4}\pi R^4 \sigma\omega$$

08. *(c)* Magnetic needle is placed in non-uniform magnetic field. It experiences force and torque both due to unequal forces acting on poles.

> **• Concept Enhancer**
>
> When a magnetic needle is placed in a uniform magnetic field, then it will experience only torque.

09. *(c)* Magnetic field lines are always continuous, forming closed loops, indicating that they always originate from north pole and enter into south pole. This contrasts with the behaviour of magnetic field lines in the presence of an electric monopole, where they would diverge from a single source.

The concept of magnetic monopoles is also related to the fact that any magnet always has both a north and south pole.

10. *(b)* Torque on magnetic dipole

$$\tau = MB\sin\theta$$

$$\Rightarrow \quad 80\sqrt{3} = MB\sin 60°$$

$$\Rightarrow \quad 80\sqrt{3} = MB\frac{\sqrt{3}}{2}$$

$$\Rightarrow \quad MB = 160$$

Potential energy,

$$U = -MB\cos\theta$$

$$= -160\cos 60°$$

$$= -160 \times \frac{1}{2}$$

$$= -80 \text{ J}$$

11. *(b)* $B = \frac{\mu_0}{4\pi}\frac{m}{r^3}$

$$B \propto \frac{1}{r^3}$$

$$\frac{\Delta r}{r}\times100 = 1\%$$

$$100\times\frac{\Delta B}{B} = 3\times\left(\frac{\Delta r}{r}\right)\times100$$

% uncertainty in $B = 3\%$

12. *(a)* Given, magnetic moment,

$$M = 0.5 \text{ Am}^2$$

$$B = 8\times10^{-2} \text{ T}$$

∴ For most stable state, $\theta_1 = 0°$ and for most unstable stable, $\theta_2 = 180°$

∴ Work done,

$$W = MB(\cos 0° - \cos 180°)$$

$$= 0.5\times8\times10^{-2}(1+1)$$

$$= 8\times10^{-2} \text{ J}$$

13. *(b)* Torque on a bar magnet,

$$\tau = MB\sin\theta$$

$$0.018 = M \times 0.06 \times \sin 30°$$

$$0.018 = M\times0.06\times\frac{1}{2}$$

$$M = 0.6 \text{ A-m}^2$$

Now, work done to rotate bar magnet,

$$W = MB[\cos\theta_1 - \cos\theta_2]$$

Here, $\theta_1 = 0°$ and $\theta_2 = 180°$

So, $W = 0.6\times0.06[\cos(0°) - \cos(180°)]$

$$= 0.036[(1) - (-1)]$$

$$= 0.036(2) = 0.072 \text{ J}$$

$$= 7.2\times10^{-2} \text{J}$$

14. *(c)* As the point O lies on equatorial line with respect to both the magnets. Therefore, the net magnetic field at point O is

$$\mathbf{B}_{net} = B_1 + B_2 + B_H$$

$$\mathbf{B}_{net} = \frac{\mu_0}{4\pi}\frac{M_1}{r^3} + \frac{\mu_0}{4\pi}\frac{M_2}{r^3} + B_H$$

$$= \frac{\mu_0}{4\pi r^3}(M_1 + M_2) + B_H$$

$$\Rightarrow \mathbf{B}_{net} = \frac{10^{-7}}{(10\times10^{-2})^3}[1.2+1] + 3.6\times10^{-5}$$

$$= 2.56\times10^{-4}\ \text{Wb/m}^2$$

15. *(b)* In initial condition,

$$T = 2\pi\sqrt{\frac{I}{MB}} = 2\text{ s, where } I = \frac{1}{12}ml^2$$

When the magnet is cut into three pieces the pole strength will remain the same as m and moment of inertia of each part will be.

$$I' = \frac{1}{12}\left(\frac{m}{3}\right)\left(\frac{l}{3}\right)^2 \times 3 = \frac{I}{9}$$

$\therefore$ $M = \text{Pole strength } (m) \times l$

$\therefore$ New magnetic moment,

$$M' = m\times\left(\frac{l}{3}\right)\times 3 = ml = M$$

New time period,

$$T = 2\pi\sqrt{\frac{I'}{M'B}} = 2\pi\sqrt{\frac{I}{9MB}}$$

$$\Rightarrow T' = \frac{T}{\sqrt{9}} = \frac{2}{3}\text{ s}$$

16. *(a)* Workdone to turn a magnetic needle in magnetic field from angle θ_1 to θ_2 is given by

$$W = MB(\cos\theta_1 - \cos\theta_2)$$

Here, $\theta = 0°, \theta_2 = 60°$

$$\therefore\ W = MB(\cos 0° - \cos 60°)$$

$$= MB\left(1 - \frac{1}{2}\right) = \frac{MB}{2}$$

$\therefore$ Torque, $\tau = MB\sin\theta = MB\sin 60°$

$$= \sqrt{3}\frac{MB}{2} = \sqrt{3}W$$

17. *(b)* When magnet is divided into two equal parts, the magnetic dipole moment,

$$M' = \text{Pole strength} \times \frac{l}{2} = \frac{M}{2}$$

[$\because$ pole strength remains same]

Also, the mass of magnet becomes half i.e., $m' = m/2$

Moment of inertia of magnet, $I = \dfrac{ml^2}{12}$

New moment of inertia,

$$I' = \frac{1}{12}\left(\frac{m}{2}\right)\left(\frac{l}{2}\right)^2 = \frac{ml^2}{12\times 8}$$

$$\therefore\ I' = I/8$$

Now, $T = 2\pi\sqrt{\left(\dfrac{I}{MB}\right)}$

$$T' = 2\pi\sqrt{\left(\frac{I'}{M'B}\right)} = 2\pi\sqrt{\left(\frac{I/8}{MB/2}\right)}$$

$$\therefore\ T' = \frac{T}{2} \Rightarrow \frac{T'}{T} = \frac{1}{2}$$

18. *(a)* Given, $I = 0.8\ \text{kg m}^2$

$M = 20\ \text{Am}^2$

Rotational kinetic energy of circular coil, $\text{KE} = \frac{1}{2}I\omega^2$

Potential energy of bar magnet $= -MB\cos\phi$

According to law of conservation of energy,

$$\frac{1}{2}I\omega^2 = U_{in} - U_f = 0 - (-MB\cos 60°)$$

$$[\because U_i = -MB\cos 90° = 0]$$

$$\Rightarrow \frac{MB}{2} = \frac{1}{2}I\omega^2$$

$$\Rightarrow \frac{20\times 4}{2} = \frac{1}{2}(0.8)\omega^2$$

$$\Rightarrow 100 = \omega^2 \Rightarrow \omega = 10\text{ rad/s}$$

19. *(b)* The magnetic dipole reaches in stable equilibrium under the influence of these two fields making an angle $\theta_1 = 30°$ with B_1 and $\theta_2 = 75° - 30° = 45°$ with B_2.

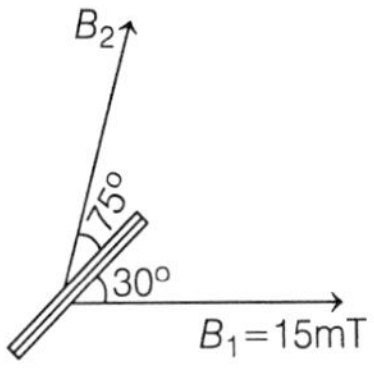

For stable equilibrium, net torque acting on dipole must be zero, i.e,

$\tau_1 + \tau_2 = 0$ or $\tau_1 = \tau_2$

or $mB_1\sin\theta_1 = mB_2\sin\theta_2$

$$\Rightarrow B_2 = B_1\times\frac{\sin\theta_1}{\sin\theta_2} = 15\times\frac{\sin 30°}{\sin 45°}$$

$$= 15\times\frac{1}{2}\times\sqrt{2}$$

$$= 10.6\text{ mT} \approx 11\text{ mT}$$

20. *(28)* Magnetic moment of straight strip, $M = 44\ \text{Am}^2$

When it is bent in semicircular shape.

$$l = \pi r \text{ or } r = l/\pi$$

Now, the distance between two poles,

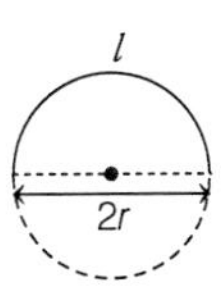

$$2r = 2l/\pi$$

Its new magnetic moment is,

$$M' = m\times 2r = m\times\frac{2l}{\pi}$$

$$= \frac{2}{\pi}(m\times l) \quad (\because m\times l = M)$$

$$= \frac{2}{\pi}\times 44 = \frac{2}{22}\times 44\times 7$$

$$= 7\times 4 = 28\ \text{Am}^2$$

21. *(100)* Work done in turning coil through 90°

$= MB(\cos 0° - \cos 90°)$

$= MB(1-0) = MB$ $\quad [M = NIA]$

$= NIAB$

$= 100\times1\times10^{-3}\times5\times10^{-3}\times0.2$

$= 10^{-4}\ \text{J} = 100\times10^{-6}\ \text{J} = 100\ \mu\text{J}$

22. *(5)* Amount of work done,

$W = MB\cos 0° - MB\cos 90°$

$= MB - 0 = MB$

$= NIAB$ $\quad [\because M = NIA]$

$= 200\times100\times10^{-6}\times2.5\times10^{-4}\times1$

$= 5\times10^{-6}\ \text{J} = 5\ \mu\text{J}$

23. *(1)* Here, $M_1 = 1\ \text{A-m}^2 = M_2$

$OP = 1\text{ m}$

Torque, $\tau = \mathbf{M}_2\times\mathbf{B}_1$

B_1 is the magnetic field at point P due to short magnet M_1.

$$\tau = M_2B_1\sin 90°$$

$$= 1\times\frac{\mu_0}{4\pi}\cdot\frac{M_1}{(1)^3}\times 1,$$

$$\Rightarrow \tau = 10^{-7}\text{N m}$$

24. *(8)*

> **key idea** You can find the time peirod of oscillation of a freely suspended bar magnet in magnetic field B using following formula.
>
> $$T = 2\pi\sqrt{\frac{I}{MB}}$$
>
> where, $I \rightarrow$ moment of inertia, $M \rightarrow$ magnetic dipole moment

Given,

Magnetic moment,

$$M = 9.85\times10^{-2}\ \text{A-m}^2$$

Moment of inertia,

$$I = 5\times10^{-6}\ \text{kg m}^2$$

$\because$ We know that

$$T = 2\pi\sqrt{I/MB}$$

$$\Rightarrow \quad \frac{5}{10} = 2\pi\sqrt{\frac{5\times 10^{-6}}{9.85\times 10^{-2}\times B}}$$

$$\Rightarrow \quad B = 8\times 10^{-3}\ \text{T}$$

Topic 2

01. *(a)* We have, $\mu_r = (1+\chi)$

$$\mu_r = 1+\chi$$
$$\mu_r - 1 = \chi$$
$$\mu = \mu_0\mu_r$$
$$\Rightarrow \quad \mu_r = \frac{\mu}{\mu_0}$$
$$\chi = \mu_r - 1$$
$$\chi = \frac{\mu}{\mu_0} - 1$$

02. *(b)* Paramagnetic substances have a weak net dipole moment of their own. So, they tend to align with external magnetic field and their susceptibility is a small positive member.

03. *(b)* **Statement 1** False, diamagnetism is an intrinsic property of material that arises due to presence of completely filled electron shells, not due to temperature.
Statement 2 True, diamagnetic material have negative magnetic susceptibility, which means that they, oppose the applied magnetic field.
Therefore, Statement I is false and Statement II is true.

04. *(b)* Since, in ferromagnetic material, with increase in temperature susceptibility decreases,
∴Ferromagnetic material below Curie temperature will show saturation magnetisation.

05. *(c)* Soft ferromagnetic materials can magnetised and demagnetised easily. These materials, when placed in an external magnetic field, experiences net dipole moment, hence their orientation changes. When they are aligned in direction of field, the orientation grow and when aligned in opposite direction, size reduces.

06. *(c)* As, the sphere is perfectly diamagnetic, it completely expels field as shown below:

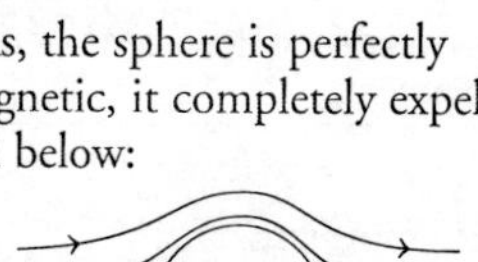

Therefore, there will be no magnetic field inside the cavity of sphere.

So, the paramagnetic sample at the spherical cavity at the centre of sphere remains uneffected by external field and field inside paramagnetic sample is zero.

07. *(c)* For a solenoid magnetic induction,

$$B = \mu_0 nI \qquad \text{...(i)}$$

Also, $\quad B = \mu_0 H \qquad \text{...(ii)}$

∴ From Eqs. (i) and (ii), we get

$$\mu_0 nI = \mu_0 H$$

or $\quad H = nI = \frac{N}{l}I$

$$H = \frac{100}{0.2}\times 5.2$$
$$= 2600\ \text{A/m}$$

08. *(a)* At Curie temperature, a ferromagnetic substance transits into paramagnetic substance.

09. *(a)* % change in

$$B = \frac{B_{\text{new}} - B_{\text{old}}}{B_{\text{old}}}\times 100\%$$
$$= \frac{\mu ni - \mu_0 ni}{\mu_0 ni}\times 100\%$$
$$= \frac{\mu - \mu_0}{\mu_0}\times 100\%$$
$$= \frac{\mu_0\mu_r - \mu_0}{\mu_0}\times 100\%$$
$$= (\mu_r - 1)\times 100\%$$
$$= \chi_n \times 100$$
$$= 1.2\times 10^{-3}\%$$
$$= \frac{6}{5}\times 10^{-3}\%$$

10. *(d)* We know that magnetic energy stored per unit volume.

$$\frac{U}{V} = \frac{B^2}{4\mu_0}$$
$$\Rightarrow \quad U = \frac{B^2}{4\mu_0}\cdot V$$
$$= \frac{B^2}{4\mu_0}\cdot Al = \frac{B^2 Al}{4\mu_0}$$

11. *(a)* Given, As we know that ferromagnetism of any object decreases with increase in temperature and above Curie temperature, ferromagnetic material become paramagnetic material. Hence, Statement I is true.
But, in case of ferromagnetic material below Curie temperature, atoms are alligned and parallel.
But above Curie temperature, atoms losses their ordered magnetic moments. Hence, Statement II is false.

12. *(b)* Magnetic moment of an iron core solenoid,

$$M = \mu_r NiA$$
$$M = \mu_r Ni\frac{V}{l} \qquad \left\{\because A = \frac{V}{l}\right\}$$

Substituting all the given values in above equation, we get

$$M = 1000\times \frac{10}{10^{-2}}\times 0.5\times 10^{-3}$$
$$(\because V = 10^{-3}\ \text{m}^3)$$
$$= 500\ \text{Am}^2 = 5\times 10^2\ \text{A-m}^2$$

13. *(b)* Ferromagnetic substances have strong tendency to get magnetised (induced magnetic moment) in the same direction as that of applied magnetic field, so magnet attracts N_1 strongly. Paramagnetic substances get weakly magnetised (magnetic moment induced is small) in the same direction as that of applied magnetic field, so magnet attracts N_2 weakly.
Diamagnetic substances also get weakly magnetised when placed in an external magnetic field but in opposite direction and hence N_3 is weakly repelled by magnet.

14. *(b)* Magnetic susceptibility (χ) of a magnetic material does not depend on magnetising field (H)

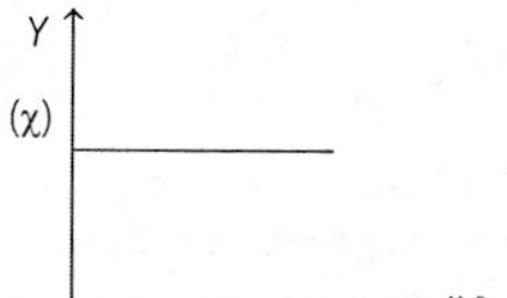

∴ A → III

Magnetic field due to current carrying wire for $x < a$

$$B = \frac{\mu_0 Ix}{2\pi a^2}$$
$$\Rightarrow \quad B \propto x$$

∴ B → I

Magnetic field due to a current carrying wire for $x > a$

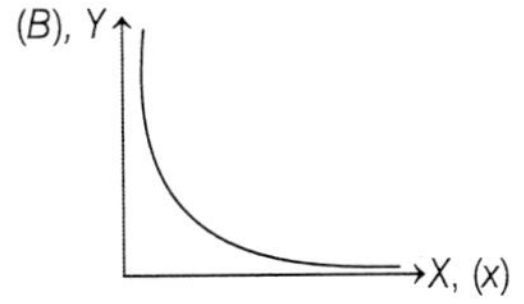

$\therefore \quad B = \frac{\mu_0 I}{2\pi x} \Rightarrow B \propto \frac{1}{x}$

$\therefore$ C → IV

Magnetic field inside a solenoid is constant and becomes zero outside the solenoid.

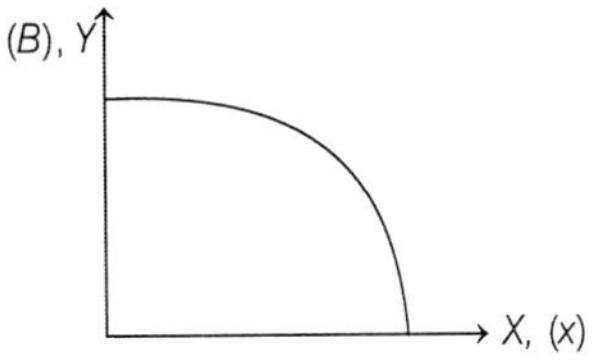

$\therefore$ D → II

15. *(a)* **Statement** (A) Magnetic monopoles do not exist, they always exist in pairs as South and North poles but electric monopoles exist.
So, Statement A is incorrect.

Statement (B) Solenoids have a both ends, so magnetic field lines cannot be confined. Inside the solenoids, the magnetic field lines are parallel to the length of the conductor. So, Statement B is incorrect.

Statement (C) Toroid has no ends. It is closed loop conductor, so the magnetic field lines are completely confined. So, the Statement C is correct.

Statement (D) Inside the bar magnet, the magnetic field lines are parallel. So, the Statement D is incorrect.

Statement (E) Magnetic susceptibility of the diamagnetic material is –1.
So, the Statement E is correct.
Hence, the correct Statements are C and E.

• Concept Enhancer

Among the three types of magnetic materials (Diamagnetic, paramagnetic and ferromagnetic), only diamagnetic substance has negative value of susceptibility.

16. *(10)* Coercivity produced by solenoid is,

$$H = nI = 5 \times 10^3 \ \frac{\text{A}}{\text{m}}$$

$$\Rightarrow \quad \frac{N}{l} \times I = 5 \times 10^3$$

Here, $\quad N = 150, l = 30 \times 10^{-2}$ m

$$\Rightarrow \frac{150}{30 \times 10^{-2}} \times I = 5 \times 10^3$$

$$\Rightarrow \quad I = \frac{5 \times 10^3 \times 30 \times 10^{-2}}{150}$$

$$= 10 \text{ A}$$

C-20

JEE Main Chapterwise
Topicwise Physics

In JEE Main there is usually 0-1 question per paper is visible from this chapter. In past 3 years, 20 conceptual and 23 numerical questions were asked of moderate to difficult level from this chapter.

For detailed analysis scan QR code

ELECTROMAGNETIC INDUCTION

Topic 1 Magnetic Flux, Faraday's and Lenz's Laws

Level 1

01 A coil of area A and N turns is rotating with angular velocity ω in a uniform magnetic field $\mathbf{B}$ about an axis perpendicular to $\mathbf{B}$. Magnetic flux ϕ and instant emf ε across it, at an instant when $\mathbf{B}$ is parallel to the plane of coil, are *[2025, 29 Jan Shift-I]*

(a) $\phi = AB,\ \varepsilon = NAB\omega$ (b) $\phi = 0,\ \varepsilon = NAB\omega$
(c) $\phi = AB,\ \varepsilon = 0$ (d) $\phi = 0,\ \varepsilon = 0$

02 A rectangular loop of length 2.5 m and width 2 m is placed at 60° to a magnetic field of 4 T. The loop is removed from the field in 10 s. The average emf induced in the loop during this time is *[2024, 27 Jan Shift-I]*

(a) -2 V (b) -1 V (c) $+2$ V (d) $+1$ V

03 The induced emf can be produced in a coil by

A. moving the coil with uniform speed inside uniform magnetic field
B. moving the coil with non uniform speed inside uniform magnetic field
C. rotating the coil inside the uniform magnetic field
D. changing the area of the coil inside the uniform magnetic field

Choose the correct answer from the options given below. *[2023, 6 April Shift-I]*

(a) B and C only (b) A and C only
(c) B and D only (d) C and D only

04 A coil is placed in a time varying magnetic field. If the number of turns in the coil were to be halved and the radius of wire is doubled, the electrical power dissipated due to the current induced in the coil would be (Assume the coil to be shorted circuited) *[2022, 28 June, Shift-II]*

(a) halved (b) quadrupled
(c) the same (d) doubled

Imp. Induced EMF

05 The magnetic flux through a coil perpendicular to its plane is varying according to the relation, $\phi = (5t^3 + 4t^2 + 2t - 5)$ Wb. If the resistance of the coil is 5 Ω, then the induced current through the coil at $t = 2$ s will be *[2022, 26 June Shift-I]*

(a) 15.6 A (b) 16.6 A
(c) 17.6 A (d) 18.6 A

06 A coil is placed in a magnetic field $\mathbf{B}$ as shown below.

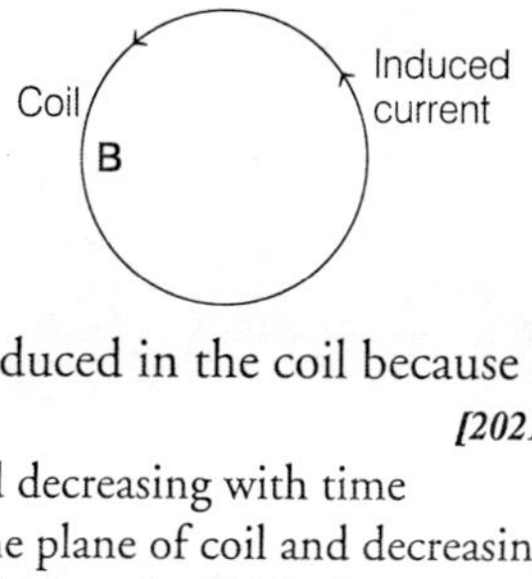

A current is induced in the coil because $\mathbf{B}$ is *[2021, 31 Aug Shift-II]*

(a) outward and decreasing with time
(b) parallel to the plane of coil and decreasing with time
(c) outward and increasing with time
(d) parallel to the plane of coil and increasing with time

07 A conducting bar of length L is free to slide on two parallel conducting rails as shown in the figure.

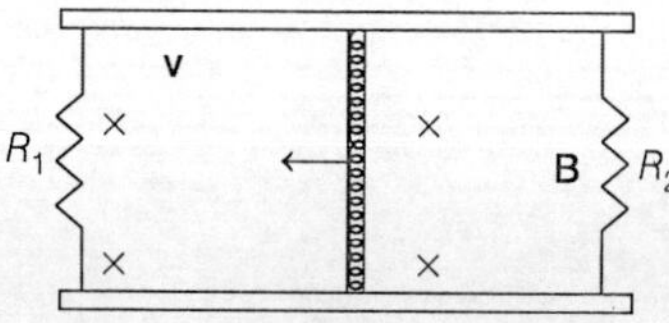

Two resistors R_1 and R_2 are connected across the ends of the rails. There is a uniform magnetic field $\mathbf{B}$

pointing into the page. An external agent pulls the bar to the left at a constant speed v.
The correct statement about the directions of induced currents I_1 and I_2 flowing through R_1 and R_2 respectively is **[2021, 16 March Shift-I]**
(a) both I_1 and I_2 are in anti-clockwise direction.
(b) both I_1 and I_2 are in clockwise direction.
(c) I_1 is in clockwise direction and I_2 is in anti-clockwise direction.
(d) I_1 is in anticlockwise direction and I_2 is in clockwise direction.

08 Consider a circular coil of wire carrying constant current I, forming a magnetic dipole. The magnetic flux through an infinite plane that contains the circular coil and excluding the circular coil area is given by ϕ_i. The magnetic flux through the area of the circular coil area is given by ϕ_o. Which of the following option is correct? **[2020, 7 Jan Shift-I]**
(a) $\phi_i > \phi_o$ (b) $\phi_i < \phi_o$ (c) $\phi_i = \phi_o$ (d) $\phi_i = -\phi_o$

09 At the centre of a fixed large circular coil of radius R, a much smaller circular coil of radius r is placed. The two coils are concentric and are in the same plane. The larger coil carries a current I. The smaller coil is set to rotate with a constant angular velocity ω about an axis along their common diameter. Calculate the emf induced in the smaller coil after a time t of its start of rotation. **[JEE Main 2018]**
(a) $\frac{\mu_0 I}{4R}\omega\pi r^2 \sin\omega t$ (b) $\frac{\mu_0 I}{4R}\omega r^2 \sin\omega t$
(c) $\frac{\mu_0 I}{2R}\omega\pi r^2 \sin\omega t$ (d) $\frac{\mu_0 I}{4R}\omega\pi r^2 \sin\omega t$

10 The flux linked with a coil at any instant t is given by $\phi = 10t^2 - 50t + 250$. The induced emf at $t = 3$ s is **[AIEEE 2006]**
(a) −190 V (b) −10 V (c) 10 V (d) 190 V

Level 2

11

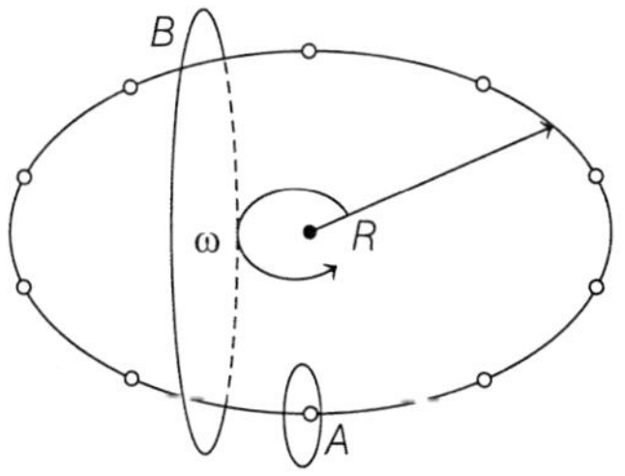

N equally spaced charges each of value q, are placed on a circle of radius R. The circle rotates about its axis with an angular velocity ω as shown in the figure. A bigger amperian loop B encloses the whole circle where as a smaller amperian loop A encloses a small segment. The difference between enclosed currents, $I_A - I_B$, for the given amperian loops is **[2025, 24 Jan Shift-II]**
(a) $\frac{2\pi}{N}q\omega$ (b) $\frac{N}{2\pi}q\omega$
(c) $\frac{N}{\pi}q\omega$ (d) $\frac{N^2}{2\pi}q\omega$

12 A coil is places perpendicular to a magnetic field of 5000 T. When the field is changed to 3000 T in 2s, an induced emf of 22 V is produced in the coil. If the diameter of the coil is 0.02 m, then the number of turns in the coil is **[2024, 31 Jan, Shift-I]**
(a) 140 (b) 35 (c) 7 (d) 70

13 A square loop of area 25 cm^2 has a resistance of 10 Ω. The loop is placed in uniform magnetic field of magnitude 40.0 T. The plane of loop is perpendicular to the magnetic field. The work done in pulling the loop out of the magnetic field slowly and uniformly in 1.0 sec. will be **[2023, 29 Jan, Shift-II]**
(a) 2.5×10^{-3} J (b) 1.0×10^{-3} J
(c) 5×10^{-3} J (d) 1.0×10^{-4} J

14 A conducting circular loop of radius $10/\sqrt{\pi}$ cm is placed perpendicular to a uniform magnetic field of 0.5 T. The magnetic field is decreased to zero in 0.5 s at a steady rate. The induced emf in the circular loop at 0.25 s is **[2023, 22 Jan, Shift-I]**
(a) emf = 5 mV (b) emf = 100 mV
(c) emf = 1 mV (d) emf = 10 mV

15 A uniform magnetic field **B** exists in a direction perpendicular to the plane of a square loop made of a metal wire. The wire has a diameter of 4 mm and a total length of 30 cm. The magnetic field changes with time at a steady rate $dB/dt = 0.032$ Ts^{-1}. The induced current in the loop is close to
(Take, resistivity of the metal wire $= 1.23 \times 10^{-8}$ Ωm) **[2020, 3 Sep Shift-II]**
(a) 0.61 A (b) 0.43 A (c) 0.53 A (d) 0.34 A

16 An elliptical loop having resistance R, of semi-major axis a and semi-minor axis b is placed in a magnetic field as shown in the figure.

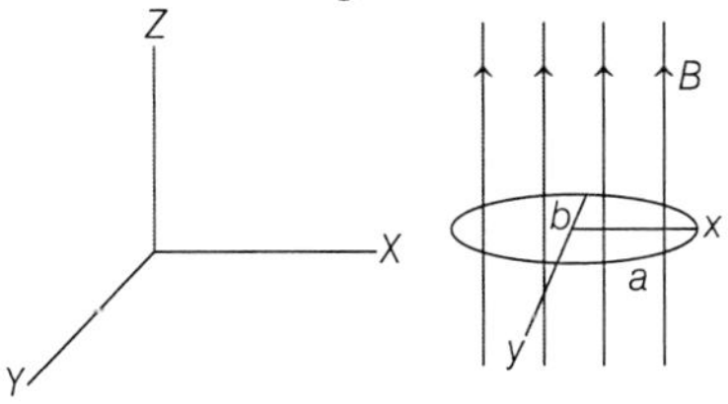

If the loop is rotated about the X-axis with angular frequency ω, the average power loss in the loop due to Joule heating is **[2020, 3 Sep, Shift-I]**
(a) $\frac{\pi^2 a^2 b^2 B^2 \omega^2}{2R}$ (b) zero
(c) $\frac{\pi abB\omega}{R}$ (d) $\frac{\pi^2 a^2 b^2 B^2 \omega^2}{R}$

17 A planar loop of wire rotates in a uniform magnetic field. Initially at $t = 0$, the plane of the loop is perpendicular to the magnetic field. If it rotates with a period of 10 s about an axis in its plane, then the magnitude of induced emf will be maximum and minimum respectively at *[2020, 7 Jan Shift-II]*
(a) 2.5 s and 7.5 s (b) 2.5 s and 5.0 s
(c) 5.0 s and 10.0 s (d) 5.0 s and 7.5 s

18 A long solenoid of radius R carries a time (t) dependent current $I(t) = I_0(1-t)$. A ring of radius $2R$ is placed coaxially near its middle. During the time interval $0 \le t \le 1$, the induced current I_R and the induced EMF V_R in the ring change as *[2020, 7 Jan, Shift-I]*
(a) direction of I_R remains unchanged and V_R is zero at $t = 0.25$
(b) direction of I_R remains unchanged and V_R is maximum at $t = 0.5$
(c) at $t = 0.25$, direction of I_R reverses and V_R is maximum
(d) at $t = 0.5$, direction of I_R reverses and V_R is zero.

19 In a coil of resistance 100 Ω, a current is induced by changing the magnetic flux through it as shown in the figure. The magnitude of change in flux through the coil is *[JEE Main 2017]*

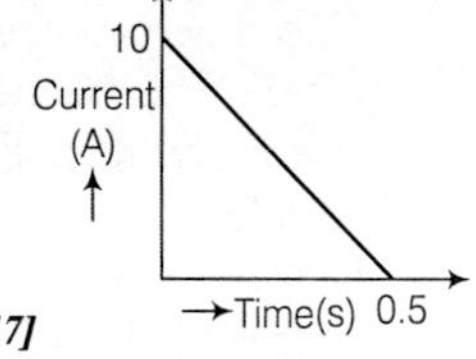

(a) 225 Wb (b) 250 Wb (c) 275 Wb (d) 200 Wb

20 A circular loop of radius 0.3 cm lies parallel to a much bigger circular loop of radius 20 cm. The centre of the small loop is on the axis of the bigger loop. The distance between their centres is 15 cm. If a current of 2.0 A flows through the smaller loop, then the flux linked with bigger loop is *[JEE Main 2013]*
(a) 9.2×10^{-11} Wb (b) 6×10^{-11} Wb
(c) 3.3×10^{-11} Wb (d) 6.6×10^{-9} Wb

21 A coil having n turns and resistance $R\,\Omega$ is connected with a galvanometer of resistance $4R\,\Omega$. This combination is moved for time t seconds from a magnetic field W_1 weber to W_2 weber. The induced current in the circuit is *[AIEEE 2004]*
(a) $\dfrac{W_2 - W_1}{5Rnt}$ (b) $-\dfrac{n(W_2 - W_1)}{5Rt}$
(c) $-\dfrac{(W_2 - W_1)}{Rnt}$ (d) $-\dfrac{n(W_2 - W_1)}{Rt}$

Level 3

22 At time $t = 2$ magnetic field of 1000 gauss is passing perpendicularly through the area defined by the closed loop as shown in the figure. If the magnetic field reduces linearly to 500 gauss, in the next 5 s, then induced emf in the loop is *[2020, 8 Jan Shift-I]*

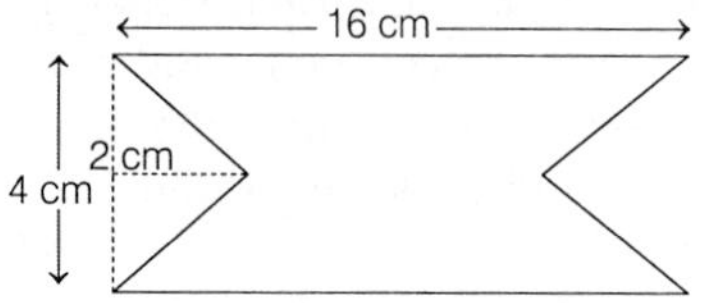

(a) 48 μV (b) 28 μV (c) 56 μV (d) 36 μV

23 A conducting metal circular-wire-loop of radius r is placed perpendicular to a magnetic field which varies with time as $B = B_0 e^{-t/\tau}$, where B_0 and τ are constants, at time $t = 0$. If the resistance of the loop is R, then the heat generated in the loop after a long time ($t \to \infty$) is *[JEE Main 2016]*
(a) $\dfrac{\pi^2 r^4 B_0^4}{2\tau R}$ (b) $\dfrac{\pi^2 r^4 B_0^2}{2\tau R}$
(c) $\dfrac{\pi^2 r^4 B_0^2 R}{\tau}$ (d) $\dfrac{\pi^2 r^4 B_0^2}{\tau R}$

24 In a uniform magnetic field of induction B, a wire in the form a semi-circle of radius r rotates about the diameter of the circle with angular frequency ω. The axis of rotation is perpendicular to the field. If the total resistance of the circuit is R, the mean power generated per period of rotation is *[AIEEE 2004]*
(a) $\dfrac{B\pi r^2 \omega}{2R}$ (b) $\dfrac{(B\pi r^2 \omega)^2}{8R}$
(c) $\dfrac{(B\pi r^2 \omega)^2}{2R}$ (d) $\dfrac{(B\pi r \omega^2)^2}{8R}$

Numerical Value Type Questions

25 An insulated copper wire of 100 turns is wrapped around a wooden cylindrical core of the cross-sectional area 24 cm^2. The two ends of the wire are connected to a resistor. The total resistance in the circuit is 12 Ω. If an externally applied uniform magnetic field in the core along its axis changes from 1.5 T in one direction to 1.5 T in the opposite direction, the charge flowing through a point in the circuit during the change of magnetic field will be ………… mC. *[2023, 13 April shift-II]*

26 The magnetic field B crossing normally a square metallic plate of area 4m^2 is changing with time as shown in figure. The magnitude of induced emf in the plate during $t = 2$ s to $t = 4$ s, is ………… mV. *[2023, 11 April shift-I]*

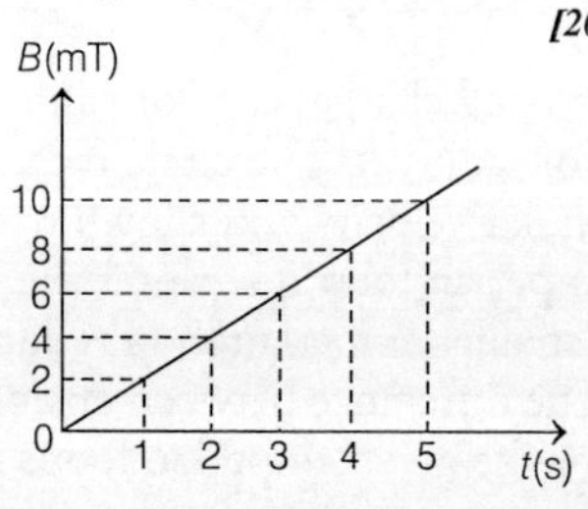

27 A square loop of side 2.0 cm is placed inside a long solenoid that has 50 turns per centimetre and carries a sinusoidally varying current of amplitude 2.5 A and angular frequency 700 rad s^{-1}. The central axes of the loop and solenoid coincide. The amplitude of the emf induced in the loop is $x \times 10^{-4}$ V. The value of x is .... (Take, $\pi = 22/7$) *[2023, 10 April Shift-II]*

28 A conducting circular loop is placed in XY-plane in presence of magnetic field $\mathbf{B} = (3t^3 \hat{j} + 3t^2 \hat{k})$ in SI unit. If the radius of the loop is 1 m, the induced emf in the loop, at time $t = 2$ s is $n\pi$V. The value of n is .... *[2022, 27 July shift-II]*

29 In a coil of resistance 8 Ω, the magnetic flux due to an external magnetic field varies with time as $\phi = \frac{2}{3}(9 - t^2)$. The value of total heat produced in the coil, till the flux becomes zero, will be .... J. *[2022, 26 July Shift-II]*

Imp. Induced Currect

30 In the given figure, the magnetic flux through the loop increases according to the relation $\phi_B(t) = 10t^2 + 20t$, where ϕ_B is in milliwebers and t is in seconds. The magnitude of current through $R = 2\Omega$ resistor at $t = 5$ s is ..... mA. *[2021, 20 July Shift-II]*

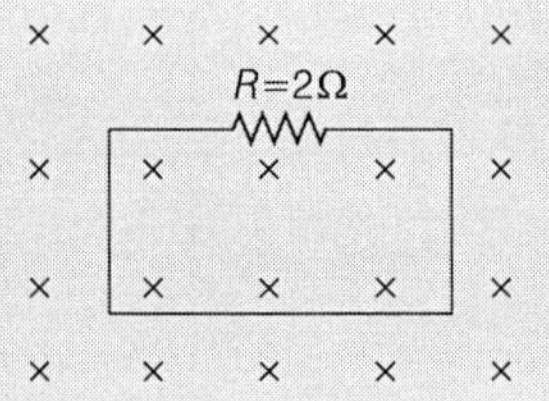

Topic 2 Motional EMF and Eddy Current

Level 1

01 A uniform magnetic field of 0.4 T acts perpendicular to a circular copper disc 20 cm in radius. The disc is having a uniform angular velocity of 10 π rad s^{-1} about an axis through its centre and perpendicular to the the disc. What is the potential difference developed between the axis of the disc and the rim ? ($\pi = 3.14$) *[2025, 28 Jan Shift-II]*

(a) 0.2512 V (b) 0.0628 V (c) 0.1256 V (d) 0.5024 V

02 A square loop of side 15 cm being moved towards right at a constant speed of 2 cm/s as shown in figure. The front edge enters the 50 cm wide magnetic field at $t = 0$. The value of induced emf in the loop at $t = 10$ s will be *[2024, 9 April Shift-II]*

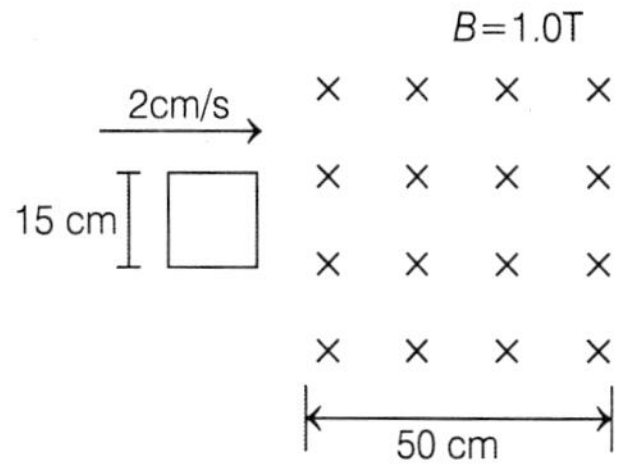

(a) zero (b) 3 mV (c) 0.3 mV (d) 4.5 mV

03 Given below are two statements, One is labelled as Assertion (A) and the other is labelled as Reason (R).

Assertion (A) A bar magnet dropped through a metallic cylindrical pipe takes more time to come down compared to a non-magnetic bar with same geometry and mass.

Reason (R) For the magnetic bar, Eddy currents are produced in the metallic pipe which oppose the motion of the magnetic bar.

In the light of the above statements, choose the most appropriate answer from the options given below. *[2023, 11 April Shift-II]*

(a) Both A and R are true but R is not the correct explanation of A.
(b) A is false but R is true.
(c) A is true but R is false.
(d) Both A and R are true and R is the correct explanation of A.

04 An emf of 0.08 V is induced in a metal rod of length 10 cm held normal to a uniform magnetic field of 0.4 T, when moves with a velocity of *[2023, 8 April Shift-II]*

(a) 0.5 ms^{-1} (b) 2 ms^{-1} (c) 20 ms^{-1} (d) 3.2 ms^{-1}

05 A wire of length 1 m moving with velocity 8 m/s at right angles to a magnetic field of 2 T. The magnitude of induced emf between the ends of wire will be *[2023, 25 Jan, Shift-II]*

(a) 8 V (b) 12 V (c) 16 V (d) 20 V

Imp. Motional EMF

06 A metallic rod of length L is rotated with an angular speed of ω normal to a uniform magnetic field B about an axis passing through one end of rod as shown in figure. The induced emf will be *[2023, 24 Jan, Shift-II]*

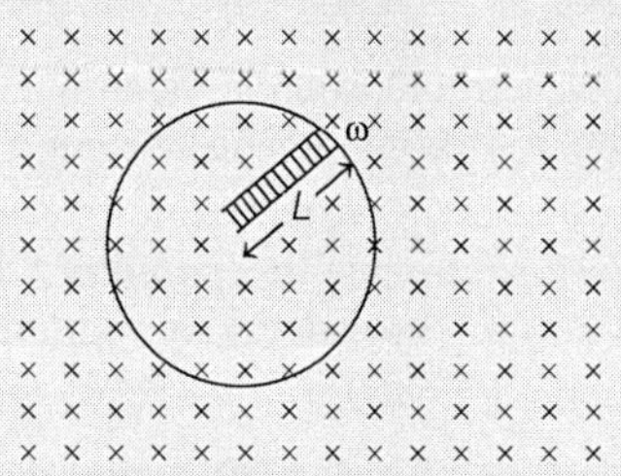

(a) $\frac{1}{2}BL^2\omega$ (b) $\frac{1}{4}BL^2\omega$ (c) $\frac{1}{2}B^2L^2\omega$ (d) $\frac{1}{4}B^2L\omega$

07 A metallic conductor of length 1m rotates in a vertical plane parallel to East-West direction about one of its end with angular velocity 5 rad/s. If the horizontal component of earth's magnetic field is 0.2×10^{-4} T, then emf induced between the two ends of the conductor is *[2022, 26 June Shift-II]*
(a) 5 μV (b) 50 μV (c) 5 mV (d) 50 mV

08 A constant magnetic field of 1 T is applied in the $x > 0$ region. A metallic circular ring of radius 1 m is moving with a constant velocity of 1 m/s along the X-axis. At $t = 0$ s, the centre of O of the ring is at $x = -1$ m. What will be the value of the induced emf in the ring at $t = 1$ s?
(Assume the velocity of the ring does not change.) *[2021, 27 Aug Shift-II]*

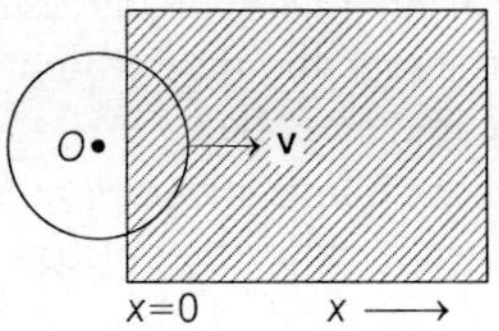

(a) 1 V (b) 2π V (c) 2 V (d) 0

Imp. Motional EMF

09 An aeroplane with its wings spread 10 m, is flying at a speed of 180 km/h in a horizontal direction. The total intensity of Earth's field at that part is 2.5×10^{-4} Wb/m^2 and the angle of dip is 60°. The emf induced between the tips of the plane wings will be *[2021, 26 Feb Shift-II]*
(a) 108.25 mV (b) 54.125 mV
(c) 88.37 mV (d) 62.50 mV

10 A coil is suspended in a uniform magnetic field with the plane of the coil parallel to the magnetic lines of force. When a current is passed through the coil, it starts oscillating; it is very difficult to stop. But, if an aluminium plate is placed near to the coil, it stops. This is due to *[AIEEE 2012]*
(a) development of air current when the plate is placed
(b) induction of electrical charge on the plate
(c) shielding of magnetic lines of force as aluminium is a paramagnetic material
(d) electromagnetic induction in the aluminium plate giving rise to electromagnetic damping

11 A horizontal straight wire 20 m long extending from East to West is falling with a speed of 5.0 m/s, at right angles to the horizontal component of the earth's magnetic field 0.30×10^{-4} Wb/m^2. The instantaneous value of the emf induced in the wire will be *[AIEEE 2011]*
(a) 6.0 mV (b) 3 mV
(c) 4.5 mV (d) 1.5 mV

12 A boat is moving due East in a region where the earth's magnetic field is 5.0×10^{-5} NA^{-1} m^{-1} due North and horizontal. The boat carries a vertical aerial 2m long. If the speed of the boat is 1.50 ms^{-1}, the magnitude of the induced emf in the wire of aerial is *[AIEEE 2011]*
(a) 0.75 mV (b) 0.50 mV (c) 0.15 mV (d) 1 mV

Imp. Motional EMF

13 A metal conductor of length 1 m rotates vertically about one of its ends at angular velocity 5 rad/s. If the horizontal component of earth's magnetic field is 0.2×10^{-4} T, then the e.m.f. developed between the two ends of the conductor is *[AIEEE 2004]*
(a) 5 μV (b) 50 μV (c) 5 mV (d) 50 mV

Level 2

14 A rectangular metallic loop is moving out of a uniform magnetic field region to a field free region with a constant speed. When the loop is partially inside the magnetic field, the plot of magnitude of induced emf (ε) with time (t) is given by *[2025, 22 Shift -II]*

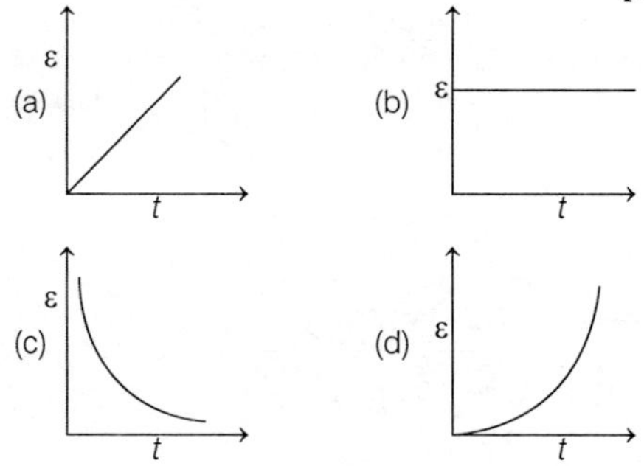

15 A square loop of side 20 cm and resistance 1Ω is moved towards right with a constant speed v_0. The right arm of the loop is in a uniform magnetic field of 5 T. The field is perpendicular to the plane of the loop and is going into it. The loop is connected to a network of resistors each of value 4 Ω. What should be the value of v_0, so that a steady current of 2 mA flows in the loop ? *[2021, 1 Sep Shift-II]*

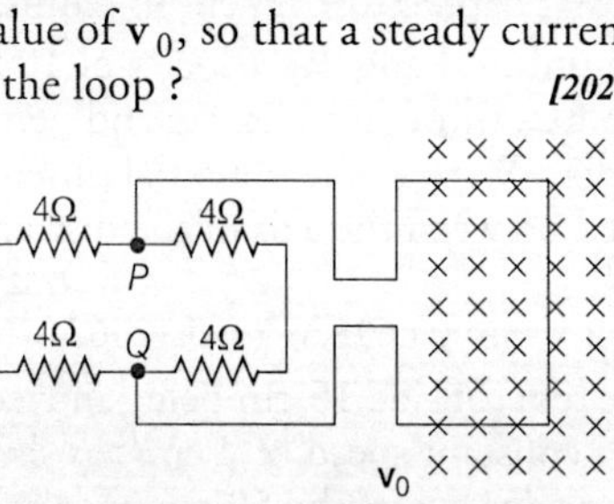

(a) 1 m/s (b) 1 cm/s
(c) 10^2 m/s (d) 10^{-2} cm/s

16 The magnetic field in a region is given by $\mathbf{B} = B_0\left(\frac{x}{a}\right)\hat{\mathbf{k}}$. A square loop of side d is placed with its edges along the X and Y-axes. The loop is moved with a constant velocity $\mathbf{v} = v_0\hat{\mathbf{i}}$. The emf induced in the loop is **[2021, 16 March Shift-II]**

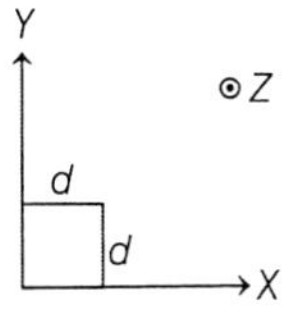

(a) $\frac{B_0 v_0^2 d}{2a}$ (b) $\frac{B_0 v_0 d}{2a}$

(c) $\frac{B_0 v_0 d^2}{a}$ (d) $\frac{B_0 v_0 d^2}{2a}$

17 Consider a thin metallic sheet perpendicular to the plane of the paper moving with speed v in a uniform magnetic field B going into the plane of the paper (see figure). If charge densities σ_1 and σ_2 are induced on the left and right surfaces, respectively of the sheet, then (ignore fringe effects) **[JEE Main 2016]**

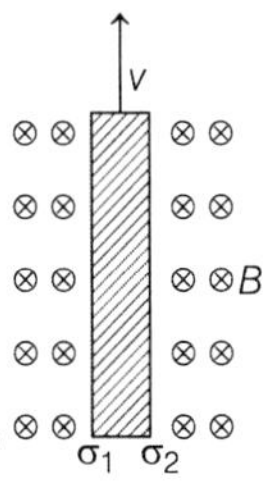

(a) $\sigma_1 = \frac{-\varepsilon_0 vB}{2}, \sigma_2 = \frac{\varepsilon_0 vB}{2}$

(b) $\sigma_1 = \varepsilon_0 vB, \sigma_2 = -\varepsilon_0 vB$

(c) $\sigma_1 = \frac{\varepsilon_0 vB}{2}, \sigma_2 = \frac{-\varepsilon_0 vB}{2}$

(d) $\sigma_1 = \sigma_2 = \varepsilon_0 vB$

18 A metallic rod of length l is tied to a string of length $2l$ and made to rotate with angular speed ω on a horizontal table with one end of the string fixed. If there is a vertical magnetic field B in the region, the emf induced across the ends of the rod is **[JEE Main 2013]**

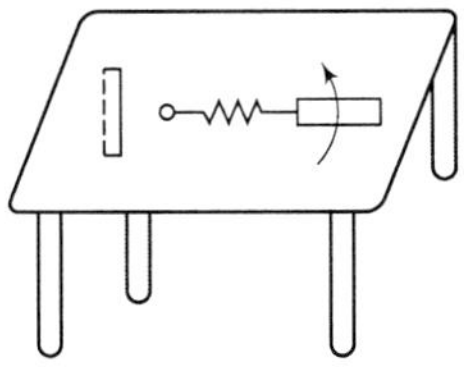

(a) $\frac{2B\omega l^3}{2}$ (b) $\frac{3B\omega l^3}{2}$

(c) $\frac{4B\omega l^2}{2}$ (d) $\frac{5B\omega l^2}{2}$

19 One conducting U-tube can slide inside another as shown in figure, maintaining electrical contacts between the tubes. The magnetic field B is perpendicular to the plane of the figure. If each tube moves towards the other at a constant speed v, then the emf induced in the circuit in terms of B, l and v where l is the width of each tube, will be **[AIEEE 2005]**

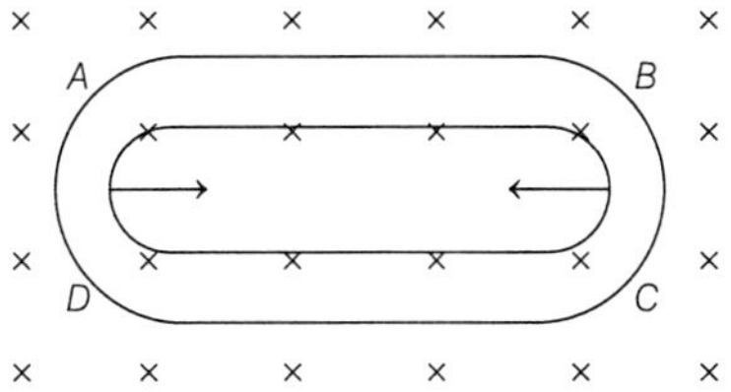

(a) zero (b) $2Blv$

(c) Blv (d) $-Blv$

20 A conducting square loop of side L and resistance R moves in its plane with a uniform velocity v perpendicular to one of its sides. A magnetic induction B constant in time and space, pointing perpendicular and into the plane at the loop exists everywhere with half the loop outside the field, as shown in figure. The induced emf is **[AIEEE 2002]**

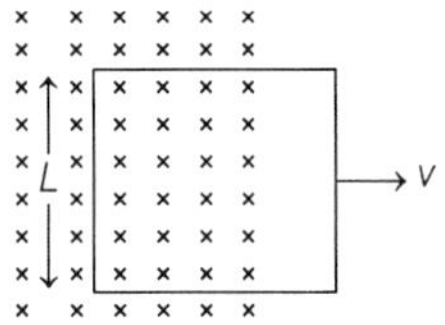

(a) zero (b) RvB

(c) vBL/R (d) vBL

Level 3

21 A bar magnet is passing through a conducting loop of radius R with velocity v. The radius of the bar magnet is such that it just passes through the loop. The induced emf in the loop can be represented by the approximate curve **[2021, 26 Aug. Shift-I]**

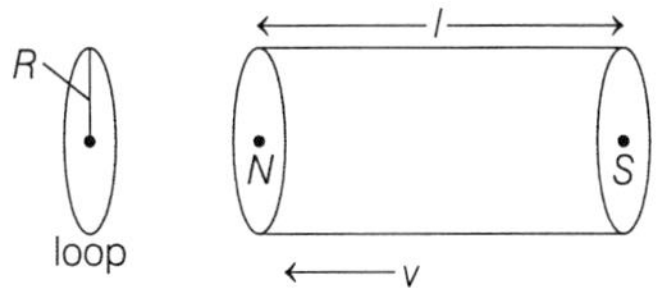

(a)
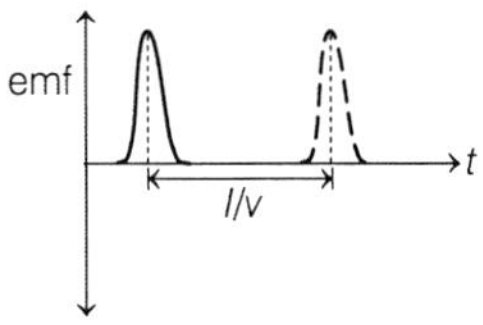

(b)
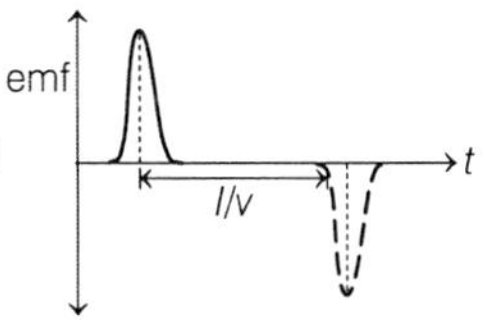

(c)
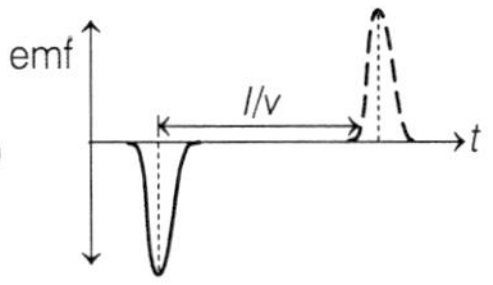

(d)
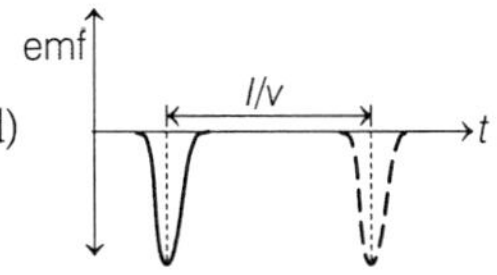

22 A rectangular loop has a sliding connector PQ of length l and resistance $R\,\Omega$ and it is moving with a speed v as shown. The set-up is placed in a uniform magnetic field going into the plane of the paper. The three currents I_1, I_2 and I are **[AIEEE 2010]**

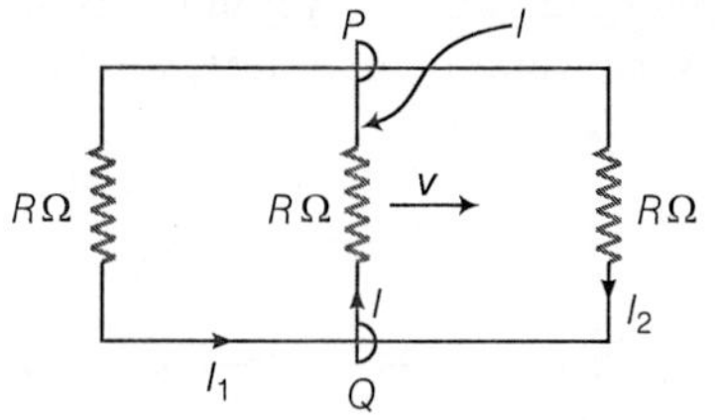

(a) $I_1 = -I_2 = \frac{Blv}{R}, I = \frac{2Blv}{R}$ (b) $I_1 = I_2 = \frac{Blv}{3R}, I = \frac{2Blv}{3R}$

(c) $I_1 = I_2 = I = \frac{Blv}{R}$ (d) $I_1 = I_2 = \frac{Blv}{6R}, I = \frac{Blv}{3R}$

Numerical Value Type Questions

23 Conductor wire $ABCDE$ with each arm 10 cm in length is placed in magnetic field of $\frac{1}{\sqrt{2}}$ T, perpendicular to its plane. When conductor is pulled towards right with constant velocity of 10 cm/s, induced emf between points A and E is .... mV.

[2025, 4 April Shift-I]

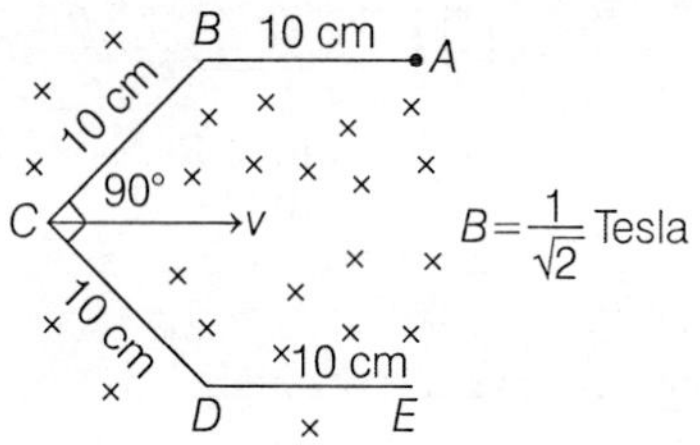

24

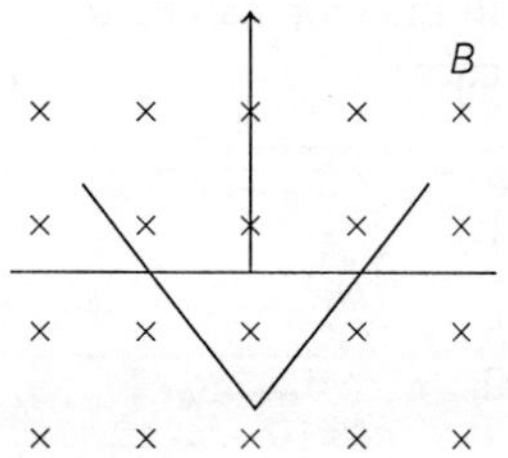

A conducting bar moves on two conducting rails as shown in the figure. A constant magnetic field B exists into the page. The bar starts to move from the vertex at time $t = 0$ with a constant velocity. If the induced EMF is $E \propto t^n$, then value of n is..........

[2025, 28 Jan Shift-II]

25 A square loop $PQRS$ having 10 turns, area 3.6×10^{-3} m^2 and resistance $100\,\Omega$ is slowly and uniformly being pulled out of a uniform magnetic field of magnitude $B = 0.5$ T as shown. Work down in pulling the loop out of the field in 1.0 s is ..... $\times 10^{-6}$ J.

[2024, 8 April Shift-I]

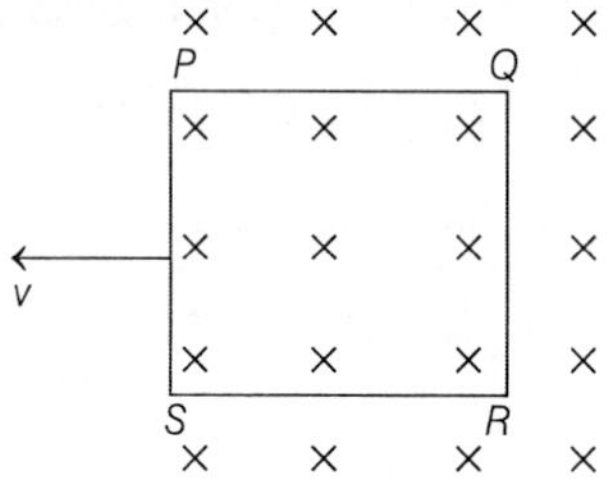

26 A rod of length 60 cm rotates with a uniform angular velocity 20 rad s^{-1} about its perpendicular bisector. in a uniform magnetic field 0.5 T. The direction of magnetic field is parallel to the axis of rotation. The potential difference between the two ends of the rod is......V. *[2024, 4 April Shift-II]*

27 A rectangular loop of sides 12 cm and 5 cm, with its sides parallel to the X-axis and Y-axis respectively, moves with a velocity of 5 cm/s in the positive X-axis direction, in a space containing a variable magnetic field in the positive z-direction. The field has a gradient of 10^{-3} T/cm along the negative x-direction and it is decreasing with time at the rate of 10^{-3} T/s. If the resistance of the loop is 6 mΩ, the power dissipated by the loop as heat is .... $\times 10^{-9}$ W.

[2024, 1 Feb Shift-I]

28 A ceiling fan having 3 blades of length 80 cm each is rotating with an angular velocity of 1200 rpm. The magnetic field of earth in that region is 0.5 G and angle of dip is 30°. The emf induced across the blades is $N\pi \times 10^{-5}$ V. The value of N is ....

[2024, 30 Jan Shift-I]

29 A 20 cm long metallic rod is rotated with 210 rpm about an axis normal to the rod passing through its one end. The other end of the rod is in contact with a circular metallic ring. A constant and uniform magnetic field 0.2 T parallel to the axis exists everywhere. The emf developed between the centre and the ring is .... mV.

$\left(\text{take, } \pi = \frac{22}{7}\right)$ *[2023, 15 April Shift-I]*

30 A conducting circular loop is placed in a uniform magnetic field of 0.4 T with its plane perpendicular to the field. Somehow, the radius of the loop starts expanding at a constant rate of 1 mm/s. The magnitude of induced emf in the loop at an instant when the radius of the loop is 2 cm will be .... μV. *[2023, 12 April Shift-I]*

31 A metallic cube of side 15 cm moving along Y-axis at a uniform velocity of 2 ms^{-1}. In a region of uniform magnetic field of magnitude 0.5T directed along Z-axis.

In equilibrium the potential difference between the face of higher and lower potential developed because of the motion through the field will be .... mV. *[2023, 11 April Shift-II]*

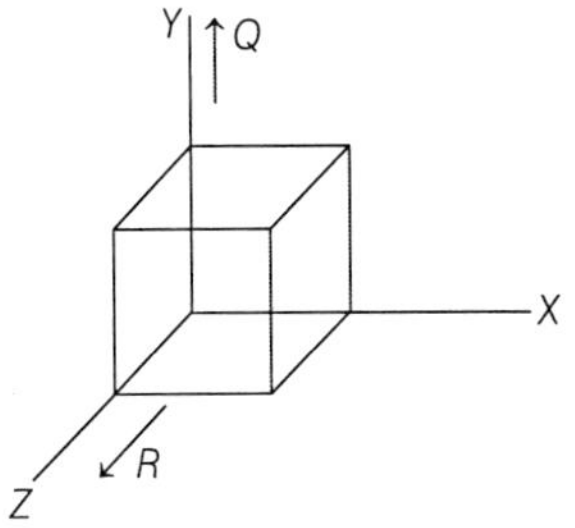

32 A 1 m long metal rod XY completes the circuit as shown in figure. The plane of the circuit is perpendicular to the magnetic field of flux density 0.15 T. If the resistance of the circuit is 5 Ω, the force needed to move the rod in direction, as indicated, with a constant speed of 4 m/s will be .... 10^{-3} N. *[2023, 10 April Shift-I]*

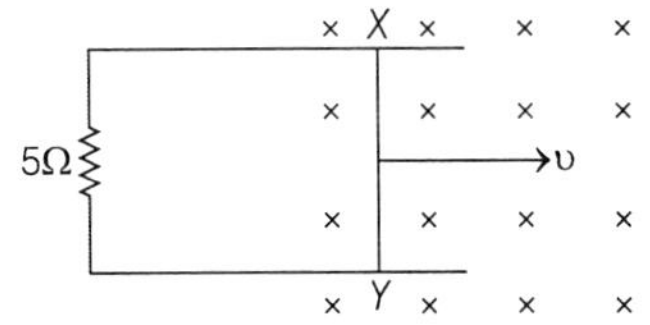

33 A certain elastic conducting material is stretched into a circular loop. It is placed with its plane perpendicular to a uniform magnetic field $B = 0.8$ T. When released, the radius of the loop starts shrinking at a constant rate of 2 cms^{-1}. The induced emf in the loop at an instant when the radius of the loop is 10 cm will be ...... mV. *[2023, 29 Jan, Shift-I]*

34 A metallic rod of length 20 cm is placed in North-South direction and is moved at a constant speed of 20 m/s towards East. The horizontal component of the Earth's magnetic field at that place is 4×10^{-3} T and the angle of dip is 45°. The emf induced in the rod is ...... mV. *[2022, 27 June Shift-II]*

35 A circular coil of 1000 turns each with area 1 m^2 is rotated about its vertical diameter at the rate of one revolution per second in a uniform horizontal magnetic field of 0.07T. The maximum voltage generation will be ...... V. *[2022, 24 June Shift-II]*

36 A circular coil of radius 8.0 cm and 20 turns is rotated about its vertical diameter with an angular speed of 50 rad s^{-1} in a uniform horizontal magnetic field of 3.0×10^{-2}T. The maximum emf induced in the coil will be ...... $\times 10^{-2}$ V.

(Rounded off to the nearest integer) *[2021, 26 Aug Shift-I]*

Imp. Energy Dissipation

37 A circular conducting coil of radius 1 m is being heated by the change of magnetic field **B** passing perpendicular to the plane in which the coil is laid. The resistance of the coil is 2 μΩ. The magnetic field is slowly switched off such that its magnitude changes in time as

$$B = \frac{4}{\pi} \times 10^{-3} \left(1 - \frac{t}{100}\right) \text{T}$$

The energy dissipated by the coil before the magnetic field is switched off completely is $E =$ ...... mJ. *[2021, 25 July Shift-I]*

Topic 3 Inductor, Inductance (Self and Mutual) and *L*-*R* Circuit

Level 1

01 Consider I_1 and I_2 are the currents flowing simultaneously in two nearby coils 1 and 2, respectively. If L_1 = self-inductance of coil 1, M_{12} = mutual inductance of coil 1 with respect to coil 2, then the value of induced emf in coil 1 will be *[2025, 29 Jan Shift-I]*

(a) $\varepsilon_1 = -L_1 \dfrac{dI_1}{dt} - M_{12} \dfrac{dI_1}{dt}$

(b) $\varepsilon_1 = -L_1 \dfrac{dI_1}{dt} - M_{12} \dfrac{dI_2}{dt}$

(c) $\varepsilon_1 = -L_1 \dfrac{dI_1}{dt} + M_{12} \dfrac{dI_2}{dt}$

(d) $\varepsilon_1 = -L_1 \dfrac{dI_2}{dt} - M_{12} \dfrac{dI_1}{dt}$

02 A 12 V battery connected to a coil of resistance 6 Ω through a switch, drives a constant current in the circuit. The switch is opened in 1 ms. The emf induced across the coil is 20 V. The inductance of the coil is *[2023, 15 April Shift-I]*

(a) 5 mH (b) 10 mH (c) 12 mH (d) 8 mH

03 Two coils of self-inductance L_1 and L_2 are connected in series combination having mutual inductance of the coils as M. The equivalent self-inductance of the combination will be *[2022, 26 June, Shift-II]*

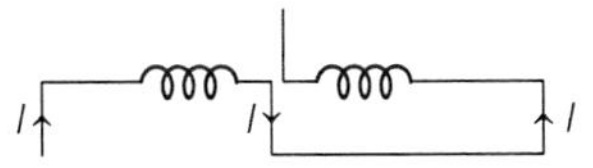

(a) $\dfrac{1}{L_1} + \dfrac{1}{L_2} + \dfrac{1}{M}$ (b) $L_1 + L_2 + M$

(c) $L_1 + L_2 + 2M$ (d) $L_1 + L_2 - 2M$

04 Figure shows a circuit that contains four identical resistors with resistance $R = 2.0\,\Omega$, two identical inductors with inductance $L = 2.0$ mH and an ideal battery with emf $E = 9$ V. The current i just after the switch S is closed will be *[2021, 24 Feb, Shift-II]*

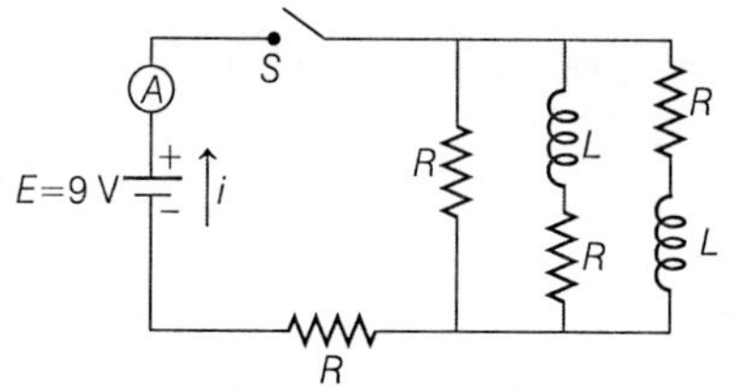

(a) 3.37 A (b) 3.0 A (c) 2.25 A (d) 9 A

05 Consider the L-R circuit shown in the figure. If the switch S is closed at $t = 0$, then the amount of charge that passes through the battery between $t = 0$ and $t = L/R$ is *[2019, 12 April, Shift-II]*

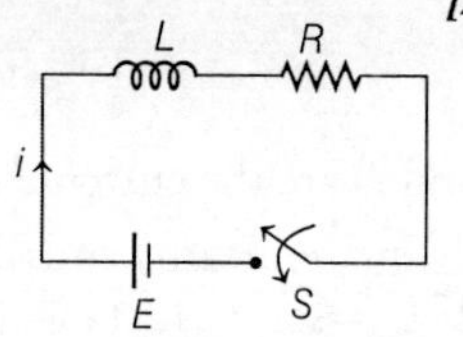

(a) $\frac{7.3\,EL}{R^2}$ (b) $\frac{2.7\,EL}{R^2}$ (c) $\frac{EL}{7.3\,R^2}$ (d) $\frac{EL}{2.7\,R^2}$

06 In the figure shown, a circuit contains two identical resistors with resistance $R = 5\,\Omega$ and an inductance with $L = 2$ mH. An ideal battery of 15 V is connected in the circuit. *[2019, 12 Jan Shift-I]*

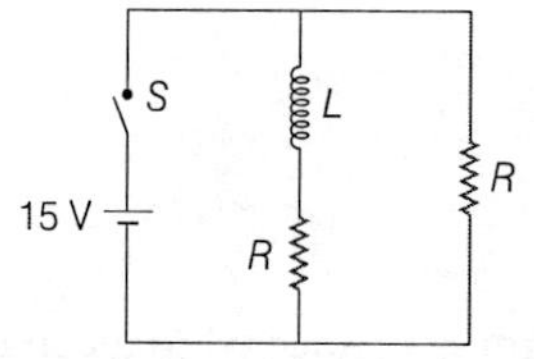

What will be the current through the battery long after the switch is closed?

(a) 6 A (b) 3 A
(c) 5.5 A (d) 7.5 A

07 There are two long coaxial solenoids of same length l. The inner and outer coils have radii r_1 and r_2 and number of turns per unit length n_1 and n_2, respectively. The ratio of mutual inductance to the self-inductance of the inner coil is *[2019, 11 Jan Shift-I]*

(a) $\frac{n_2}{n_1} \cdot \frac{r_1}{r_2}$ (b) $\frac{n_2}{n_1} \cdot \frac{r_2^2}{r_1^2}$ (c) $\frac{n_2}{n_1}$ (d) $\frac{n_1}{n_2}$

08 Two coaxial solenoids are made by winding thin insulated wire over a pipe of cross-sectional area $A = 10\text{ cm}^2$ and length $= 20$ cm. If one of the solenoids has 300 turns and the other 400 turns, their mutual inductance is ($\mu_0 = 4\pi \times 10^{-7}$ T m A^{-1}) *[AIEEE 2008]*

(a) $2.4\,\pi \times 10^{-4}$ H (b) $2.4\,\pi \times 10^{-5}$ H
(c) $4.8\,\pi \times 10^{-4}$ H (d) $4.8\,\pi \times 10^{-5}$ H

09 Two coils are placed close to each other. The mutual inductance of the pair of coils depends on *[AIEEE 2003]*

(a) the rates at which currents are changing in the two coils
(b) relative position and orientation of the two coils
(c) the materials of the wires of the coils
(d) the currents in the two coils

Level 2

Imp. Self Inductance

10 Regarding self-inductance. *[2025, 23 Jan Shift-I]*

A. The self-inductance of the coil depends on its geometry.
B. Self-inductance does not depend on the permeability of the medium.
C. Self-induced emf opposes any change in the current in a circuit.
D. Self-inductance is electromagnetic analogue of mass in mechanics.
E. Work needs to be done against self-induced emf in establishing the current.

Choose the correct answer from the options given below.

(a) A, B, C, E only (b) A, B, C, D only
(c) A, C, D, E only (d) B, C, D, E only

11 In a coil, the current changes from −2A to +2A in 0.2 s and induces an emf of 0.1 V. The self inductance of the cell is *[2024, 6 April, Shift-II]*

(a) 4 mH (b) 5 mH
(c) 1 mH (d) 2.5 mH

12 Two conducting circular loops A and B are placed in the same plane with their centres coinciding as shown in figure. The mutual inductance between them is *[2024, 5 April, Shift-I]*

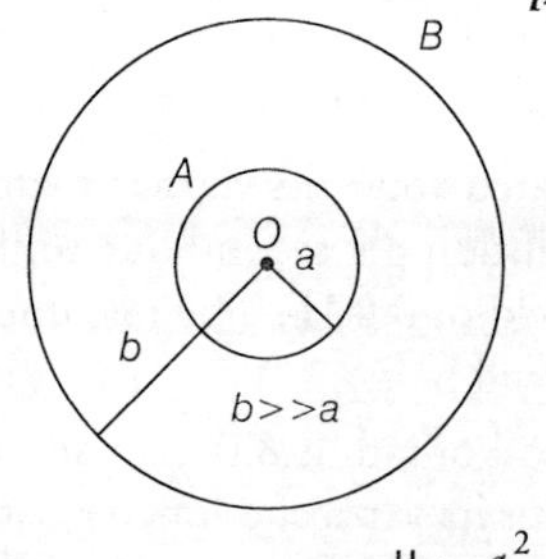

(a) $\frac{\mu_0 \pi b^2}{2a}$ (b) $\frac{\mu_0}{2\pi} \cdot \frac{a^2}{b}$

(c) $\frac{\mu_0 \pi a^2}{2b}$ (d) $\frac{\mu_0}{2\pi} \cdot \frac{b^2}{a}$

Imp. Mutual Induction

13 Find the mutual inductance in the arrangement, when a small circular loop of wire of radius R is placed inside a large square loop of wire of side L ($L >> R$). The loops are coplanar and their centres coincide **[2023, 29 Jan, Shift-I]**

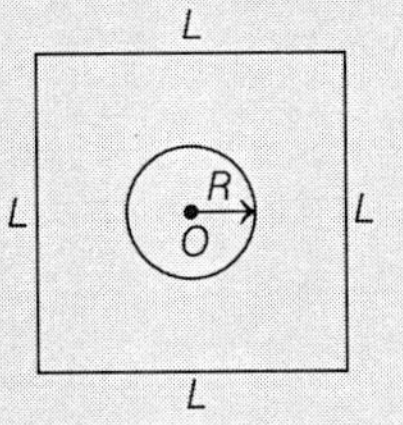

(a) $M = \dfrac{2\sqrt{2}\mu_0 R^2}{L}$ (b) $M = \dfrac{2\sqrt{2}\mu_0 R}{L^2}$

(c) $M = \dfrac{\sqrt{2}\mu_0 R}{L^2}$ (d) $M = \dfrac{\sqrt{2}\mu_0 R^2}{L}$

14 A coil of inductance 1 H and resistance 100 Ω is connected to a battery of 6 V. Determine approximately

(A) The time elapsed before the current acquires half of its steady - state value.

(B) The energy stored in the magnetic field associated with the coil at an instant 15 ms after the circuit is switched on. (Given, $\ln 2 = 0.693$, $e^{-3/2} = 0.25$) **[2022, 29 July, Shift-I]**

(a) $t = 10$ ms, $U = 2$ mJ
(b) $t = 10$ ms, $U = 1$ mJ
(c) $t = 7$ ms, $U = 1$ mJ
(d) $t = 7$ ms, $U = 2$ mJ

15 A small square loop of wire of side l is placed inside a large square loop of wire $L (L >> l)$. Both loops are coplanar and their centres coincide at point O as shown in figure. The mutual inductance of the system is **[2022, 25 July shift-I]**

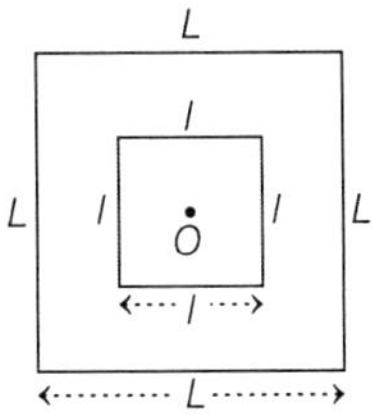

(a) $\dfrac{2\sqrt{2}\mu L^2}{\pi l}$ (b) $\dfrac{\mu_0 l^2}{2\sqrt{2}\pi L}$ (c) $\dfrac{2\sqrt{2}\mu_0 l^2}{\pi L}$ (d) $\dfrac{\mu_0 L^2}{2\sqrt{2}\pi l}$

16 A small square loop of side a and one turn is placed inside a larger square loop of side b and one turn ($b >> a$). The two loops are coplanar with their centres coinciding. If a current i is passed in the square loop of side b, then the coefficient of mutual inductance between the two loops is **[2021, 31 Aug. Shift-I]**

(a) $\dfrac{\mu_0}{4\pi} 8\sqrt{2} \dfrac{a^2}{b}$ (b) $\dfrac{\mu_0}{4\pi} \dfrac{8\sqrt{2}}{a}$

(c) $\dfrac{\mu_0}{4\pi} 8\sqrt{2} \dfrac{b^2}{a}$ (d) $\dfrac{\mu_0}{4\pi} \dfrac{8\sqrt{2}}{b}$

17 An inductor coil stores 64 J of magnetic field energy and dissipates energy at the rate of 640 W when a current of 8 A is passed through it. If this coil is joined across an ideal battery, find the time constant of the circuit in seconds. **[2021, 26 Aug. Shift-I]**

(a) 0.4 (b) 0.8 (c) 0.125 (d) 0.2

18 The time taken for the magnetic energy to reach 25% of its maximum value, when a solenoid of resistance R, inductance L is connected to a battery, is **[2021, 18 March Shift-II]**

(a) $\dfrac{L}{R}\ln 5$ (b) infinite (c) $\dfrac{L}{R}\ln 2$ (d) $\dfrac{L}{R}\ln 10$

19 A series L-R circuit is connected to a battery of emf V. If the circuit is switched on at $t = 0$, then the time at which the energy stored in the inductor reaches $(1/n)$ times of its maximum value, is **[2020, 4 Sep, Shift-II]**

(a) $\dfrac{L}{R}\ln\left(\dfrac{\sqrt{n}+1}{\sqrt{n}-1}\right)$ (b) $\dfrac{L}{R}\ln\left(\dfrac{\sqrt{n}-1}{\sqrt{n}}\right)$

(c) $\dfrac{L}{R}\ln\left(\dfrac{\sqrt{n}}{\sqrt{n}-1}\right)$ (d) $\dfrac{L}{R}\ln\left(\dfrac{\sqrt{n}}{\sqrt{n}+1}\right)$

20 As shown in the figure, a battery of emf ε is connected to an inductor L and resistance R in series. The switch is closed at $t = 0$. The total charge that flows from the battery, between $t = 0$ and $t = t_c$ (t_c is the time constant of the circuit) is **[2020, 8th Jan, Shift-II]**

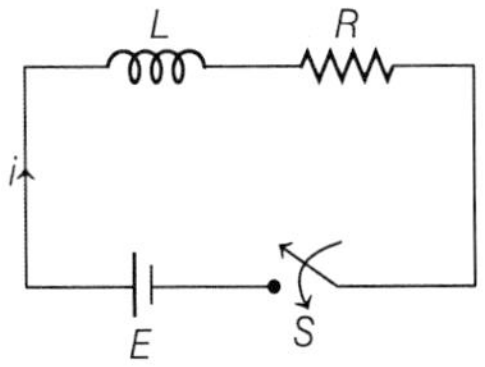

(a) $\dfrac{\varepsilon L}{R^2}\left(1 - \dfrac{1}{e}\right)$ (b) $\dfrac{\varepsilon L}{R^2}$

(c) $\dfrac{\varepsilon R}{eL^2}$ (d) $\dfrac{\varepsilon L}{eR^2}$

21 A coil of self inductance 10 mH and resistance 0.1 Ω is connected through a switch to a battery of internal resistance 0.9 Ω. After the switch is closed, the time taken for the current to attain 80% of the saturation value is [take $\ln 5 = 1.6$] **[2019, 10 April, Shift-II]**

(a) 0.016 s (b) 0.324 s
(c) 0.002 s (d) 0.103 s

22 Two coils P and Q are separated by some distance. When a current of 3 A flows through coil P, a magnetic flux of 10^{-3} Wb passes through Q. No current is passed through Q. When no current passes through P and a current of 2 A passes through Q, the flux through P is **[2019, 9 April Shift-II]**

(a) 6.67×10^{-3} Wb (b) 6.67×10^{-4} Wb
(c) 3.67×10^{-3} Wb (d) 3.67×10^{-4} Wb

23 A 20 H inductor coil is connected to a 10 Ω resistance in series as shown in figure. The time at which rate of dissipation of energy (Joule's heat) across resistance is equal to the rate at which magnetic energy is stored in the inductor, is *[2019, 8 April, Shift-I]*

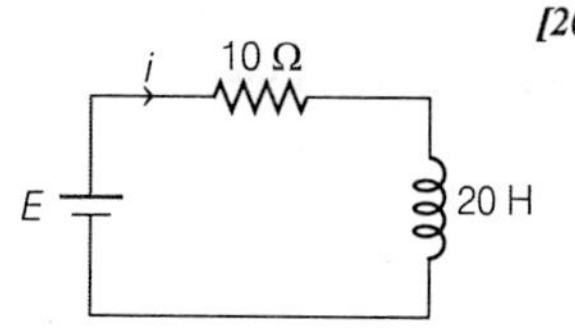

(a) $\frac{1}{2}\ln 2$ (b) $\ln 2$ (c) $\frac{2}{\ln 2}$ (d) $2\ln 2$

24 A small circular loop of wire of radius a is located at the centre of a much larger circular wire loop of radius b. The two loops are in the same plane. The outer loop of radius b carries an alternating current $I = I_0 \cos(\omega t)$. The emf induced in the smaller inner loop is nearly *[JEE Main 2017]*

(a) $\frac{\pi\mu_0 I_0}{2}\cdot\frac{a^2}{b}\omega\cos(\omega t)$ (b) $\frac{\pi\mu_0 I_0 b^2}{a}\omega\cos(\omega t)$

(c) $\frac{\pi\mu_0 I_0}{2}\cdot\frac{a^2}{b}\omega\sin(\omega t)$ (d) $\pi\mu_0 I_0\frac{a^2}{b}\omega\sin(\omega t)$

25 An inductor ($L = 0.03$ H) and a resistor ($R = 0.15$ kΩ) are connected in series to a battery of 15 V emf in a circuit shown below. The key K_1 has been kept closed for a long time. Then at $t = 0$, K_1 is opened and key K_2 is closed simultaneously. At $t = 1$ ms, the current in the circuit will be ($e^5 \cong 150$) *[JEE Main 2015]*

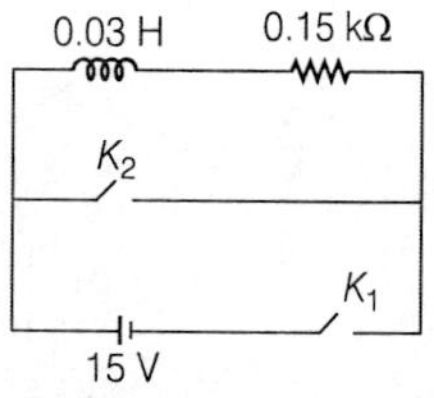

(a) 6.7 mA (b) 0.67 mA (c) 100 mA (d) 67 mA

26 In the circuit shown here, the point C is kept connected to point A till the current flowing through the circuit becomes constant. Afterward, suddenly, point C is disconnected from point A and connected to point B at time $t = 0$. Ratio of the voltage across resistance and the inductor at $t = L/R$ will be equal to *[JEE Main 2014]*

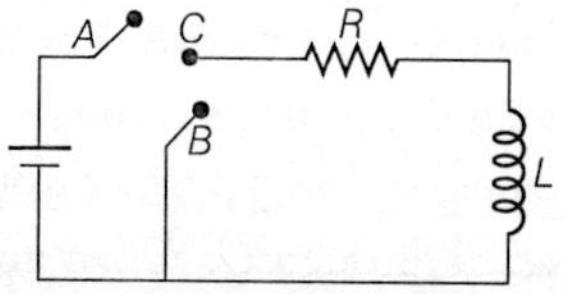

(a) $\frac{1-e}{e}$ (b) $\frac{e}{1-e}$

(c) 1 (d) −1

Imp. Growth of Current in *LR* Circuit

27 An ideal coil of 10 H is connected in series with a resistance of 5 Ω and a battery of 5 V, 2 s after the connection is made, the current flowing in ampere in the circuit is *[AIEEE 2007]*

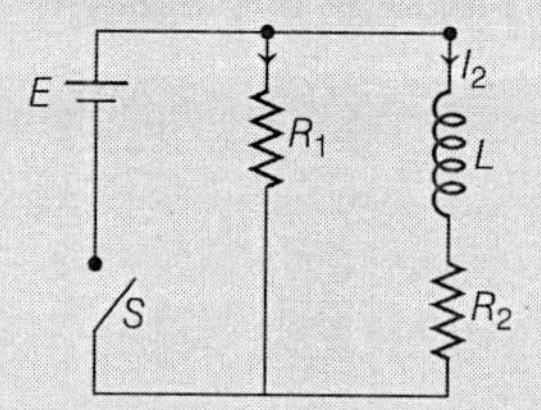

(a) $(1 - e^{-1})$ (b) $(1 - e)$

(c) e (d) e^{-1}

28 An inductor ($L = 100$ mH), a resistor ($R = 100$ Ω) and a battery ($E = 100$ V) are initially connected in series as shown in the figure. After a long time the battery is disconnected after short circuiting the points A and B. The current in the circuit 1 ms after the short circuit is *[AIEEE 2006]*

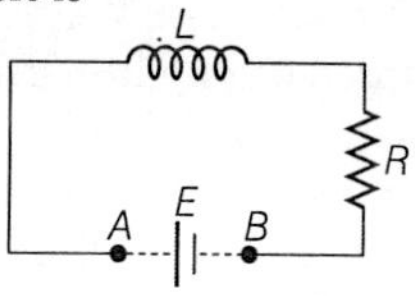

(a) 1 A (b) $(1/e)$ A

(c) e A (d) 0.1 A

29 A coil of inductance 300 mH and resistance 2 Ω is connected to a source of voltage 2 V. The current reaches half of its steady state value in *[JEE Main 2005]*

(a) 0.15 s (b) 0.3 s

(c) 0.05 s (d) 0.1 s

30 The inductance between A and D is

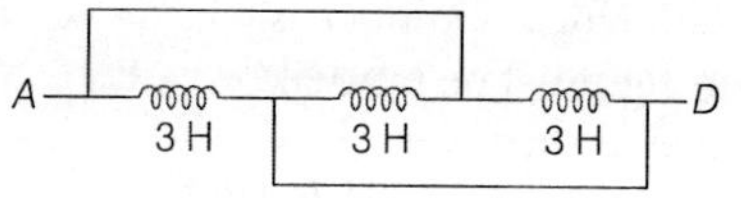

[AIEEE 2002]

(a) 3.66 H (b) 9 H (c) 0.66 H (d) 1 H

Level 3

31 A very long solenoid of radius R is carrying current $I(t) = kte^{-\alpha t}$ ($k > 0$), as a function of time ($t \geq 0$). Counter clockwise current is taken to be positive. A circular conducting coil of radius $2R$ is placed in the equatorial plane of the solenoid and concentric with the solenoid. The current induced in the outer coil is correctly depicted, as a function of time, by *[2019, 9 April Shift-II]*

(a) 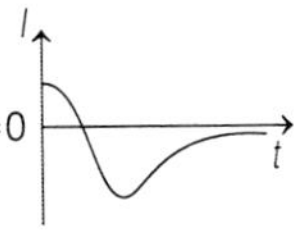

(b) 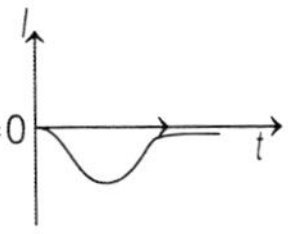

(c) 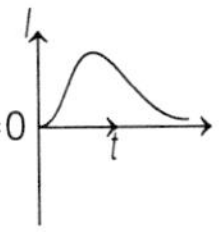

(d) 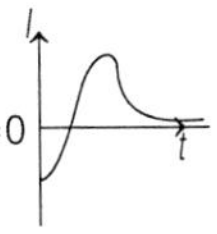

32 The self-induced emf of a coil is 25 V. When the current in it is changed at uniform rate from 10 A to 25 A in 1s, the change in the energy of the inductance is *[2019, 10 Jan Shift-II]*

(a) 437.5 J (b) 740 J (c) 637.5 J (d) 540 J

Numerical Value Type Questions

33

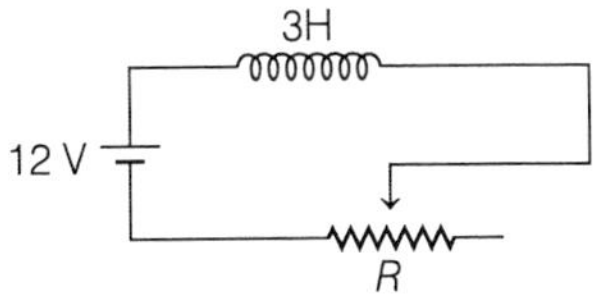

In the given circuit the sliding contact is pulled outwards such that electric current in the circuit changes at the rate of 8 A/s. At an instant when R is $12\,\Omega$, the value of the current in the circuit will be .... A. *[2025, 23 Jan Shift-I]*

Imp. Induced EMF Across Inductor

34 The current in an inductor is given by $I = (3t + 8)$ where t is in second. The magnitude of induced emf produced in the inductor is 12 mV. The self-inductance of the inductor .... mH. *[2024, 5 April Shift-II]*

35 The magnetic flux ϕ (in weber) linked with a closed circuit of resistance $8\,\Omega$ varies with time (in seconds) as $\phi = 5t^2 - 36t + 1$. The induced current in the circuit at $t = 2$ s is .... A. *[2024, 31 Jan Shift-II]*

36 A small square loop of wire of side l is placed inside a large square loop of wire of side $L (L = l^2)$. The loops are coplanar and their centers coincide. The value of the mutual inductance of the system is $\sqrt{x} \times 10^{-7}$ H, where $x = \ldots$. *[2024, 31 Jan Shift-I]*

37 A square loop of side 10 cm and resistance $0.7\,\Omega$ is placed vertically in east-west plane. A uniform magnetic field of 0.20 T is set up across the plane in north-east direction. The magnetic field is decreased to zero in 1 s at a steady rate. Then, magnitude of induced emf is $\sqrt{x} \times 10^{-3}$ V. The value of x is .... *[2024, 29 Jan Shift-I]*

38 Two coils have mutual inductance 0.002 H. The current changes in the first coil according to the relation $i = i_0 \sin \omega t$, where $i_0 = 5$A and $\omega = 50\pi$ rad/s. The maximum value of emf in the second coil is $\frac{\pi}{\alpha}$ V. The value of α is .... *[2024, 27 Jan Shift-I]*

39 Two concentric circular coils with radii 1 cm and 1000 cm, and number of turns 10 and 200 respectively are placed coaxially with centers coinciding. The mutual inductance of this arrangement will be .... $\times 10^{-8}$ H.

(Take, $\pi^2 = 10$) *[2023, 6 April Shift-II]*

40

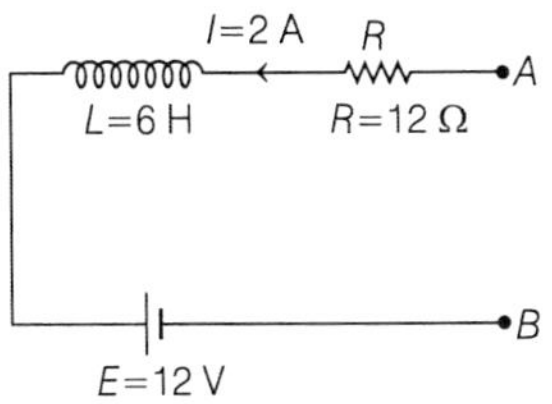

[2023, 30 Jan Shift-II]

As per the given figure, if $dI/dt = -1$ A/s, then the value of V_{AB} at this instant will be .... V.

41 Three identical resistors with resistance $R = 12\,\Omega$ and two identical inductors with self inductance $L = 5$ mH are connected to an ideal battery with emf of 12 V as shown in figure. The current through the battery long after the switch has been closed will be .... A. *[2023, 24 Jan, Shift-II]*

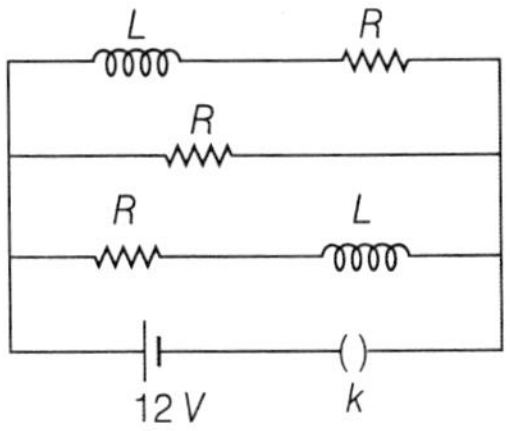

42 For the given circuit, the current through battery of 6 V just after closing the switch S will be .... A. *[2022, 28 July Shift-II]*

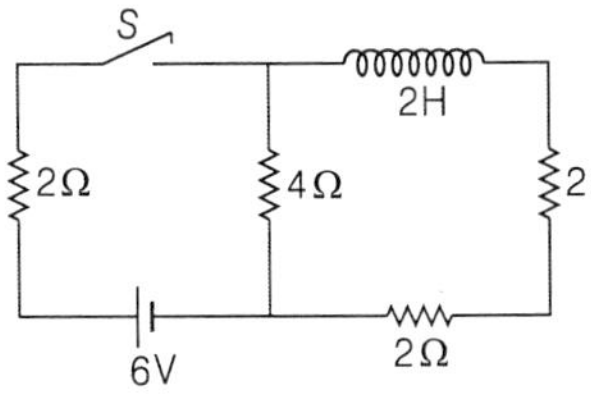

43 A $10\,\Omega$, 20 mH coil carrying constant current is connected to a battery of 20 V through a switch. Now, after switch is opened current becomes zero in 100 μs. The average emf, induced in the coil is .... V. *[2022, 26 June Shift-I]*

44 The current in a coil of self inductance 2.0 H is increasing according to $I = 2 \sin t^2$ A. The amount of energy spent during the period when current changes from 0 to 2 A is ... J. *[2022, 25 June Shift-I]*

45 A coil of inductance 2 H having negligible resistance is connected to a source of supply whose voltage is given by V = 3t V (where, t is in second). If the voltage is applied when $t = 0$, then the energy stored in the coil after 4 s is .... J. *[2021, 25 Feb Shift-I]*

Key Index

Topic 1

1 (b)	2 (d)	3 (d)	4 (d)	5 (a)	6 (a)	7 (c)	8 (d)	9 (c)	10 (b)
11 (b)	12 (d)	13 (b)	14 (d)	15 (a)	16 (a)	17 (b)	18 (d)	19 (b)	20 (a)
21 (b)	22 (c)	23 (b)	24 (b)	25 (60)	26 (8)	27 (44)	28 (12)	29 (2)	30 (60)

Topic 2

1 (a)	2 (a)	3 (d)	4 (b)	5 (c)	6 (a)	7 (b)	8 (c)	9 (a)	10 (d)
11 (b)	12 (c)	13 (b)	14 (b)	15 (b)	16 (c)	17 (b)	18 (d)	19 (b)	20 (d)
21 (c)	22 (b)	23 (10)	24 (1)	25 (3)	26 (0)	27 (216)	28 (32)	29 (88)	30 (50)
31 (150)	32 (18)	33 (10)	34 (16)	35 (440)	36 (60)	37 (80)			

Topic 3

1 (b)	2 (b)	3 (d)	4 (c)	5 (d)	6 (a)	7 (c)	8 (a)	9 (b)	10 (c)
11 (b)	12 (c)	13 (a)	14 (c)	15 (c)	16 (a)	17 (d)	18 (c)	19 (c)	20 (d)
21 (a)	22 (b)	23 (d)	24 (c)	25 (b)	26 (d)	27 (a)	28 (b)	29 (d)	30 (d)
31 (d)	32 (a)	33 (3)	34 (4)	35 (2)	36 (128)	37 (2)	38 (2)	39 (4)	40 (30)
41 (3)	42 (1)	43 (400)	44 (4)	45 (144)					

Detailed **Explanations**

Topic 1

01. *(b)* Flux, $\phi = NBA\cos\theta$

For $\theta = 90°$

$\phi = 0$

Induced emf,

$$e = \frac{-d\phi}{dt} = -\frac{d}{dt}(NBA\cos\theta)$$

Here, $\theta = \omega t$

$$\Rightarrow \quad e = \frac{-d}{dt}(NBA\cos\omega t)$$

$$= +NBA\sin\omega t \times \omega$$

$$e = NBA\omega\sin\omega t$$

For $\theta = 90° = \omega t \Rightarrow e = NBA\omega$

02. *(d)* $E_{avg} = \frac{\Delta\phi}{\Delta t} = \frac{BA\cos\theta}{10}$

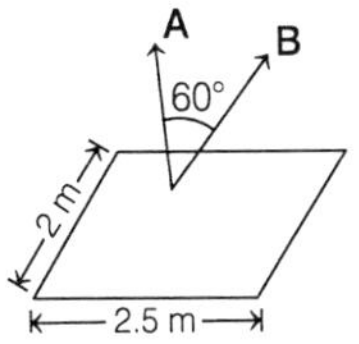

$$= 4 \times 2 \times \frac{5}{2} \times \frac{\cos 60°}{10}$$

$$= 4 \times 5 \times \frac{1}{10 \times 2} = 1\text{ V}$$

= 1 V

03. *(d)* Induced emf can be induced in coil by changing magnetic flux as

$$\phi = \mathbf{B}\cdot\mathbf{A} = BA\cos\theta$$

By rotating coil, angle θ changes and hence flux changes. By changing area, magnetic flux also changes.

04. *(d)* As the number of turns are halved, so length of wire is halved and radius is doubled, then area will be 4 times the previous one i.e. $A_f = 4A_i$

Previous resistance = R;

New resistance = $\frac{R}{8}$

Previous emf = ε, New emf = $\frac{\varepsilon}{2}$

$$P_i = \frac{\varepsilon^2}{R} \text{ and } P_f = 2P_i$$

05. *(a)*

> **key idea** If induced emf is a fuction of time i.e. $e = f(t)$, then magnetic flux is calculated by
>
> $$\phi = \int e\,dt$$

Given, flux,

$$\phi = (5t^3 + 4t^2 + 2t - 5)\text{ Wb}$$

$$t = 2\text{ s}, R = 5\Omega$$

Induced emf,

$$e = \frac{d\phi}{dt} = 15t^2 + 8t + 2$$

At, $t = 2$ s,

$$e = (15 \times 2^2) + (8 \times 2) + 2$$

$$= 78\text{ V}$$

We have, $e = IR$

$\therefore$ Induced current,

$$I = \frac{e}{R} = \frac{78}{5} = 15.6\text{ A}$$

06. *(a)* As the induced current is anti-clockwise which is also making N-pole on the outward face.

According to the Lenz's law, the induced current always opposes the nature by which it is produced.

Hence, induced current is in the direction, so the strength of N-pole is maintained as it is decreasing.

07. *(c)* According to Lenz's law, "An induced current flows in a direction in such a way that it always opposes the cause that induced it".

When the bar slides toward the loop (First) area will be decreased and due to this magnetic flux in it decreases

As, $\phi \propto A$

Due to this emf will be induced

By Right hand thumb rule, the direction of current I_1 comes out to be clockwise.

As area of loop 2 increases due to which magnetic flux.

Again apply Right hand thumb rules the direction of current I_2 comes out to be anti-clockwise.

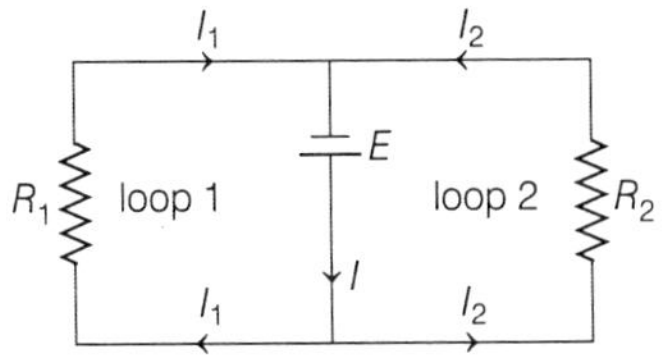

08. *(d)* We are given with following situation,

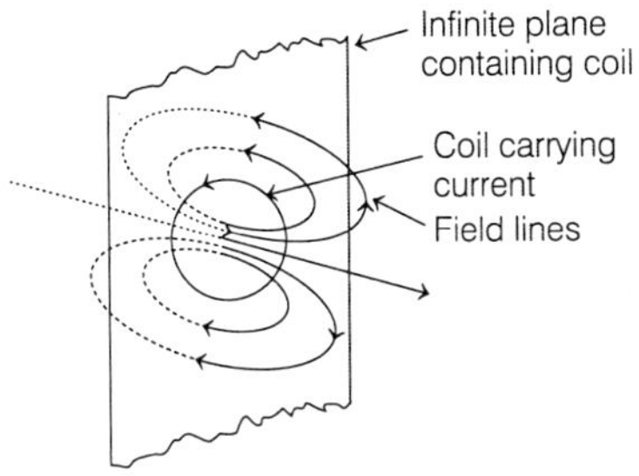

From the diagram of field lines, we can observe that no matter howmany field lines emitted from coil, all of them goes back into the infinite plane only.

So, magnetic flux emanating from coil is equal and opposite to the flux linked with infinite plane.

So, $\phi_i = -\phi_o$

09. *(c)* Emf induced in smaller coil is given by

$$\varepsilon = -\frac{d\phi}{dt} = -B \times A\frac{d}{dt}(\cos\theta)$$

$$[\because \phi = BA\cos\theta]$$

$$= \frac{\mu_0 I}{2R}(\pi r^2)(\sin\theta) \times \frac{d\theta}{dt}$$

$$= \frac{\mu_0 I}{2R}\omega\pi r^2\sin\omega t$$

($\because B$ is assumed to be constant in small region.)

10. *(b)* $\phi = 10t^2 - 50t + 250$

From Faraday's law of electromagnetic induction,

$$e = -\frac{d\phi}{dt}$$

$$\therefore \quad e = -[10 \times 2t - 50]$$

$$\therefore \; e|_{t=3\text{ s}} = -[10 \times 6 - 50] = -10\text{ V}$$

11. *(b)*

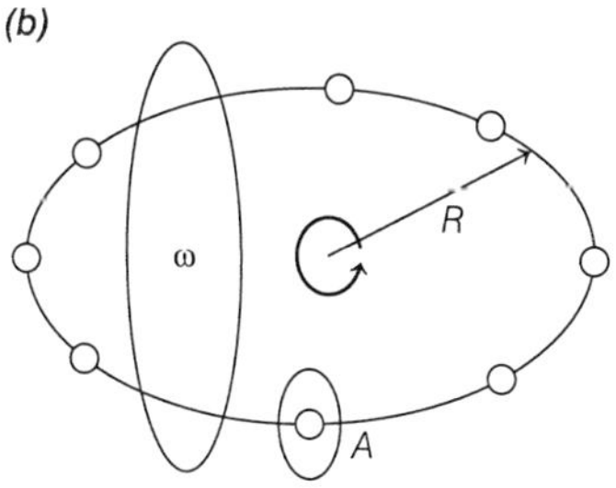

$$I_A = \frac{Nq}{\frac{2\pi}{\omega}}$$

$$I_A = \frac{Nq\omega}{2\pi}$$
$$I_B = 0$$
$$I_A - I_B = \frac{Nq\omega}{2\pi}$$

12. *(d)* EMF produced in the coil is given by,

$$\varepsilon = \frac{d}{dt}\phi_B = NA\cdot\frac{dB}{dt} = NA\cdot\frac{\Delta B}{\Delta t}$$
$$= N\times\frac{\pi D^2}{4}\left(\frac{\Delta B}{\Delta t}\right)$$

Here, change in magnetic field

$$\Delta B = B_2 - B_1$$
$$= 5000 - 3000 = 2000\text{ T}$$

Time interval, $\Delta t = 2$ s

Diameter, $D = 0.02$ m

Emf induced $= 22$ V

So, we have,

$$22 = \frac{N\,\pi\,(0.02)^2\times 2000}{4\times 2}$$

$\Rightarrow$ $N = 70$ turns

13. *(b)*

key idea The work done in pulling the loop out of region of magnetic field is calculated by

$W = \frac{V^2}{R}t$ where,

$= \left(\frac{d\phi}{dt}\right)^2\frac{t}{R}$ $\quad\left(\because V = \frac{d\phi}{dt}\right)$

$= B^2\left(\frac{dA}{dt}\right)^2\frac{t}{R}$ $\quad[\because \phi = (B.A)]$

Work done in pulling the loop out of region of magnetic field,

$$W = V\cdot q = VIt$$
$$\Rightarrow\quad W = \frac{V^2}{R}t$$
$$\Rightarrow\quad = \left(\frac{d}{dt}\phi\right)^2\cdot\frac{t}{R} = \left(\frac{d}{dt}BA\right)^2\cdot\frac{t}{R}$$
$$\Rightarrow\quad W = B^2\cdot\frac{A^2}{t^2}\cdot\frac{t}{R} = \frac{B^2A^2}{tR}$$

Here, $B = 40$ T, $A = 25\times10^{-4}$ m^2, $t = 1$ s, $R = 10\,\Omega$

$\therefore$ Work done

$$= W = \frac{(40)^2\times(25\times10^{-4})^2}{1\times10}$$
$$= 1.0\times10^{-3}\text{ J}$$

14. *(d)* It is given,

In 0.5 s, magnetic field of 0.5 T becomes zero.

So, rate of change of magnetic field,

$$\frac{\Delta B}{\Delta t} = \frac{0-0.5}{0.5} = -1\text{ T/s}$$

(Negative sign shows field is reducing)

Hence, magnetic field at 0.25 s will be

$$\frac{\Delta B}{0.25} = -1\text{ T/s} \Rightarrow \frac{B_2 - B_1}{0.25} = -1$$

or $\quad B_2 = 0.5 - 0.25 = 0.25$ T

So, magnitude of induced emf in coil at $t = 0.25$ s will be

$$\varepsilon = \frac{d}{dt}\phi = \left(\frac{dB}{dt}\right)\cdot A$$
$$= \frac{\Delta B}{\Delta t}\cdot\pi r^2 = \frac{0.25}{0.25}\times\pi\times\left(\frac{10}{\sqrt{\pi}}\times10^{-2}\right)^2$$
$$= 10^{-2}\text{ V} = 10\times10^{-3}\text{ V} = 10\text{ mV}$$

15. *(a)* Induced emf in square loop,

$$E = A\left(\frac{dB}{dt}\right)$$
$$= l^2\left(\frac{dB}{dt}\right) = \left(\frac{L}{4}\right)^2\left(\frac{dB}{dt}\right)$$

Here, $L = 30$ cm

$$= 30\times10^{-2}\text{ m}$$
$$\frac{dB}{dt} = 0.032\text{ Ts}^{-1}$$

$\therefore$ $\quad E = 1.8\times10^{-4}$ V

Induced current

$$= \frac{E}{R} = \frac{E}{\left(\rho\frac{L}{A}\right)} = \frac{EA}{\rho L}$$

Here, $\rho = 1.23\times10^{-8}\,\Omega$-m

$$L = 30\times10^{-2}\text{ m}$$
$$A = \pi r^2 = \pi(2\times10^{-3})^2$$
$$= 12.56\times10^{-6}\text{ m}^2$$

$$\text{So, } I = \frac{1.8\times10^{-4}\times1.256\times10^{-5}}{1.23\times10^{-8}\times30\times10^{-2}}$$
$$= 0.61\text{ A}$$

16. *(a)* Let resistance $= R$

The induced emf is given by

$$e = NBA\omega\cos\omega t$$

or $\quad e = AB\omega\cos\omega t \quad [\because N = 1]$

Average power loss in the loop,

$$P_{av} = \left(\frac{e^2}{R}\right) = \left(\frac{A^2B^2\omega^2\cos^2\omega t}{R}\right)$$

The average value of $\cos^2\omega t = \frac{1}{2}$

$$\therefore\quad P_{av} = \frac{A^2B^2\omega^2}{2R}$$

Substituting, $A = \pi ab$

$$\therefore\quad P_{av} = \frac{\pi^2a^2b^2B^2\omega^2}{2R}$$

17. *(b)* Induced emf in the wire loop,

$$e = BA\omega\sin\omega t$$

Induced emf is maximum when $\sin\omega t = \pm1$

$$\Rightarrow\quad \omega t = \frac{\pi}{2}$$
$$\Rightarrow\quad \frac{2\pi}{T}\cdot t = \frac{\pi}{2}$$

Here, $\quad T = 10$ s

So, $\quad t = \frac{10}{4} = 2.5$ s

and induced emf is minimum when $\sin\omega t = 0$

$$\Rightarrow\quad \omega t = \pi \text{ or } \frac{2\pi}{T}\cdot t = \pi$$
$$\Rightarrow\quad t = \frac{10}{2} = 5\text{ s}$$

18. *(d)* Given, current, $I = I_0t(1-t)$

$I = I_0t - I_0t^2$, Also $\phi = BA$, $\phi = \mu_0 nIA$

Induced emf,

$$V_R = -\frac{d\phi}{dt} = -\mu_0 nAI_0(1-2t)$$

At $t = 0.5$, $V_R = 0$, For $t < 0.5$, $V_R < 0$

Thus, in time interval, $0\le t\le1$, current will reverse its direction.

19. *(b)* Induced constant, $I = \frac{e}{R}$

Here, e = induced emf $= \frac{d\phi}{dt}$

$$I = \frac{e}{R} = \left(\frac{d\phi}{dt}\right)\cdot\frac{1}{R}$$
$$d\phi = IRdt$$
$$\phi = \int IRdt$$

$\therefore$ Here, R is constant

$$\therefore\quad \phi = R\int Idt$$

$$\int I\cdot dt = \text{Area under } I\text{-}t \text{ graph}$$
$$= \frac{1}{2}\times10\times0.5 = 2.5$$

$$\therefore\quad \phi = R\times2.5 = 100\times2.5$$
$$= 250\text{ Wb}$$

20. *(a)* The magnetic field due to the bigger circular loop is

$$B = \frac{\mu_0 IR^2}{2\,(x^2+R^2)^{3/2}}$$

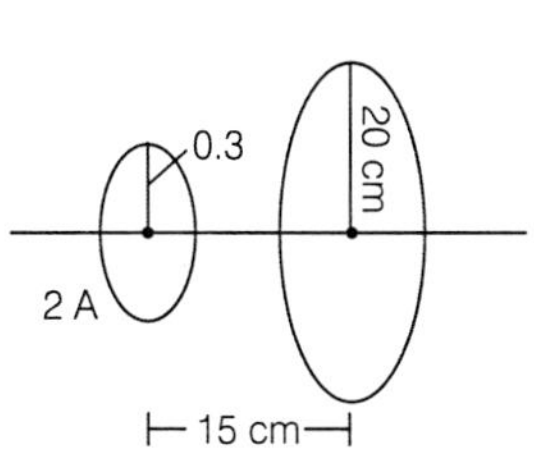

Given, $I = 2$ A, $R = 20$ cm

$= 20 \times 10^{-2}$ m,

$x = 15$ cm

$= 15 \times 10^{-2}$ m

$$B = \frac{\mu_0 \times 2 \times (20 \times 10^{-2})^2}{[(0.2)^2 + (0.15)^2]^{3/2}}$$

Flux linked, $\phi = BA$

$A = \pi r^2 = \pi \times (0.3 \times 10^{-2})^2$

$$\therefore \phi = \frac{\mu_0 \times 2 \times (20 \times 10^{-2})^2}{2[(0.2)^2 + (0.15)^2]^{3/2}} \times \pi \times (0.3 \times 10^{-2})^2$$

$\phi = 9.2 \times 10^{-11}$ Wb

• Concept Enhancer

A current carrying circular loop is equivalent to a bar magnet whose magnetic moment is given by $M = NIA$

21. *(b)* The rate of change of flux or emf induced in the coil is $e = -n\frac{d\phi}{dt}$.

$\therefore$ Induced current,

$$I = \frac{e}{R'} = -\frac{n}{R'}\frac{d\phi}{dt} \quad \ldots(i)$$

Given, $R' = R + 4R = 5R$,

$d\phi = W_2 - W_1$, $dt = t$

[here, W_1 and W_2 are flux associated with one turn]

Putting the given values in Eq. (i), we get

$$I = -\frac{n}{5R}\frac{(W_2 - W_1)}{t}$$

22. *(c)*

key idea If B_1 and B_2 are the initial and final magnetic field, Δt and A are time interval and area respectively given in question, then induced emf is calculated by, $E = \left|\frac{B_2 - B_1}{\Delta t}\right| \cdot A$

Induced emf in the loop,

$$E = -\frac{\Delta\phi}{\Delta t} = -\left(\frac{\Delta B}{\Delta t}\right)(A) \quad [\because \phi = BA]$$

$$= -\left(\frac{B_2 - B_1}{\Delta t}\right)A \quad \ldots(i)$$

Here, $B_2 = 500\text{ G} = 500 \times 10^{-4}$ T,

$B_1 = 1000\text{ G} = 1000 \times 10^{-4}$ T,

$\Delta t = 5$ s

$\therefore$ A = area of loop

= Area of rectangle − Area of two triangles

$$= \left(16 \times 4 - 2 \times \frac{1}{2} \times 4 \times 2\right)\text{cm}^2$$

$= 56 \times 10^{-4}\text{ m}^2$

Using Eq. (i), we get

$$E = \frac{(1000 - 500) \times 10^{-4} \times 56 \times 10^{-4}}{5}$$

$= 56 \times 10^{-6}\text{ V} = 56\ \mu\text{V}$

23. *(b)*

key idea If $B = f(t)$, then the heat generated in the loop is calculated by $H = \int \frac{E^2}{R} dt$, where $E = -A\frac{dB}{dt}$

Here, magnetic field, $B = B_0 e^{-\frac{t}{\tau}}$

Area of the circular loop, $A = \pi r^2$

Flux linked with the loop at any time t,

$$\phi = BA = \pi r^2 B_0 e^{-\frac{t}{\tau}}$$

Emf induced in the loop,

$$\varepsilon = -\frac{d\phi}{dt} = \pi r^2 B_0 \frac{1}{\tau} e^{-\frac{t}{\tau}}$$

Net heat generated in the loop,

$$H = \int_0^\infty \frac{\varepsilon^2}{R} dt = \frac{\pi^2 r^4 B_0^2}{\tau^2 R}\int_0^\infty e^{-\frac{2t}{\tau}} dt$$

$$= \frac{\pi^2 r^4 B_0^2}{\tau^2 R} \times \frac{1}{\left(-\frac{2}{\tau}\right)} \times \left[e^{-\frac{2t}{\tau}}\right]_0^\infty$$

$$= \frac{-\pi^2 r^4 B_0^2}{2\tau^2 R} \times \tau(0 - 1) = \frac{\pi^2 r^4 B_0^2}{2\tau R}$$

24. *(b)* Magnetic flux linked,

$$\phi = BA\cos\omega t = \frac{B\pi r^2 \cos\omega t}{2}$$

[Semi circle]

$\therefore$ Induced emf,

$$e = \frac{-d\phi}{dt} = \frac{-1}{2}B\pi r^2 \omega \sin\omega t$$

$$\therefore \text{Power, } P = \frac{e^2}{R} = \frac{B^2\pi^2 r^4 \omega^2 \sin^2\omega t}{4R}$$

$$= \frac{(B\pi r^2 \omega)^2}{4R}\sin^2\omega t$$

$\because \langle \sin^2\omega t \rangle = 1/2$

$\therefore$ Mean power generated

$$= \frac{(B\pi r^2\omega)^2}{4R} \times \frac{1}{2} = \frac{(B\pi r^2\omega)^2}{8R}$$

25. *(60)* $\Delta Q = \frac{\Delta\phi}{R} = \frac{2NBA}{R}$

$$= \frac{2 \times 100 \times 1.5 \times 24 \times 10^{-4}}{12}$$

$= 6 \times 10^{-2}\text{ C} = 60$ mC

26. *(8)* We know that magnitude of induced emf,

$$|e| = \frac{d\phi}{dt}$$

$$e = \frac{dBA}{dt} = \frac{AdB}{dt}$$

$e = A \times$ slope of B-t curve

$$= A\left[\frac{8 - 4}{4 - 2}\right] = 4 \times \frac{4}{2} = 8\text{ V}$$

27 *(44)* Given, side of square loop,

$a = 2.0\text{ cm} = 2 \times 10^{-2}$ m

Area of loop $= (2 \times 10^{-2})^2$

$= 4 \times 10^{-4}\text{ m}^2$

Amplitude of current = 2.5 A

Angular frequency = 700 rad/s

No. of turns in solenoid

= 50 turns per cm

$= \frac{50}{10^{-2}}$ turns per m

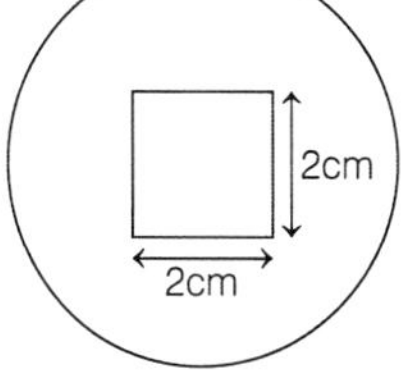

B due to selenoid $= \mu_0 nI$

$\phi_{\text{through square}} = \mu_0 nI \times A$

Emf induced, $\varepsilon = \mu_0 nA \times \frac{dI}{dt}$

$= \mu_0 nA I_0 \omega\cos\omega t$

$[\because I = I_0 \sin\omega t]$

$|\varepsilon| = \mu_0 nAI_0\omega$

$$= 4\pi \times 10^{-7} \times \frac{50}{10^{-2}} \times 4 \times 10^{-4} \times 2.5 \times 700$$

$= 44 \times 10^{-4}$ V

28. *(12)*

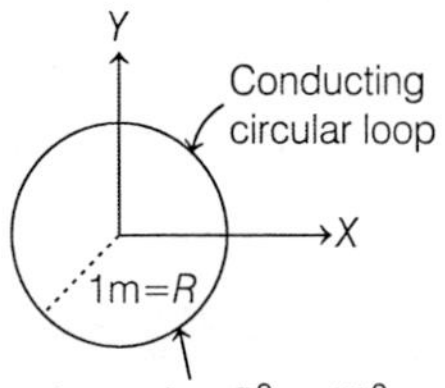

Area$=A=\pi R^2=\pi(1)^2$
or $\mathbf{A}=\pi\hat{\mathbf{k}}$

Given, $\mathbf{B}=(3t^3\hat{\mathbf{j}}+3t^2\hat{\mathbf{k}})\text{T}$

$\therefore$Flux through loop,

$$\phi_B=\mathbf{B}\cdot\mathbf{A}=(3t^3\hat{\mathbf{j}}+3t^2\hat{\mathbf{k}})\cdot(\pi\hat{\mathbf{k}})$$
$$=3\pi t^2\ \text{Wb}$$

We know,

Induced emf, $\varepsilon=-\dfrac{Nd\phi_B}{dt}$

(where, $N=$ number of turns)

$\Rightarrow\quad \varepsilon=-1d\dfrac{3\pi t^2}{dt}=-3\pi\dfrac{dt^2}{dt}$ $\quad(\because N=1)$

$\Rightarrow\quad \varepsilon=-3\pi\times 2t=-6\pi t\ \text{V}$

Now, at $t=2\text{s}$

$\varepsilon=-6\times\pi\times 2=-12\pi\,\text{V}$

$\therefore\quad n\pi=-12\pi$

$\Rightarrow\quad n=-12\Rightarrow|n|=12$

29. *(2)* Given, resistance of coil,

$R=8\Omega$

Magnetic flux, $\phi=\dfrac{2}{3}(9-t^2)$

According to Faraday's law of electromagnetic induction, induced emf in the coil,

i.e $=-\dfrac{d\phi}{dt}=\dfrac{-d}{dt}\dfrac{2}{3}(9-t^2)=-\dfrac{2}{3}\times(-2t)$

$=\dfrac{4}{3}t$

Heat produced during, $t=0$ to $t=3$ s

$$H=\int_0^3\frac{e^2}{R}dt=\int_0^3\frac{1}{8}\left(\frac{4}{3}t\right)^2dt$$
$$=\int_0^3\frac{2}{9}t^2dt=\frac{2}{9}\left[\frac{t^3}{3}\right]_0^3$$
$$=\frac{2}{9}\times\frac{27}{3}=2\,\text{J}$$

30. *(60)* As per question, flux through the loop increases according to

$$\phi_B(t)=10t^2+20t \qquad \ldots\text{(i)}$$

where, ϕ_B is in milliwebers and t in seconds.

Differentiate Eq. (i) with respect to t, we get,

Induced emf$|\text{E}|=\dfrac{d\phi}{dt}=(20t+20)\,\text{mV}$

Current, $i=\dfrac{|E|}{R}$

$\Rightarrow\quad i=\dfrac{20t+20}{2}$

$\Rightarrow\quad i=(10t+10)\text{mA}$

$\therefore$Current, $=(10\times5+10)$ mA $[\because t=5\text{s}]$

$=(50+10)$ mA,

$i=60$ mA

Topic 2

01. *(a)* $B=0.4$ T, $R=20$cm and $\omega=10\pi$ rad/s

$$E=\frac{1}{2}B\omega R^2$$
$$=\frac{1}{2}\times0.4\times10\times3.14\times(20\times10^{-2})^2$$
$$=0.2512\ \text{V}$$

02. *(a)* At $t=0$, the distance of the square loop inside the magnetic field will be 20 cm.

Side of the loop $=15$ cm

Now the whole loop is inside at $t=10$ s

Hence, $d\phi=0$

i.e. $e=-\dfrac{d\phi}{dt}=0$

03. *(d)* Moving a magnet in conducting pipe causes change in flux and hence an emf is induced. This emf causes eddy current in conducting pipe in such a way that it tries to oppose the change in flux, therefore magnet slows down. So, both A and R are true and R is the correct explanation of A.

04. *(b)* When a rod is moving perpendicular to an uniform magnetic field, magnitude of induced emf is

$E=Blv$

Here $\quad E=0.08v$

$l=10\text{ cm}=10\times10^{-2}$ m

$v=?,\ B=0.4$ T

So, velocity v is

$$v=\frac{E}{Bl}=\frac{0.08}{0.4\times10\times10^{-2}}$$
$$=2\ \text{m/s}$$

05. *(c)* Given, length of wire, $l=1$ m,

Velocity, $v=8$ m/s

$\theta=90°$

Magnetic field, $B=2$ T

$\therefore$ Induced emf, $e=Bvl\sin\theta$

$=2\times8\times1\times\sin90°=16$ V

06. *(a)* Induced emf in a rotating rod

$$=\frac{d\phi}{dt}=B\left(\frac{dA}{dt}\right)=\frac{B\cdot A}{T}$$
$$=B\cdot\frac{\pi L^2}{2\pi/\omega}=\frac{1}{2}BL^2\omega$$

• Concept Enhancer

When a metallic rod of length L is rotated with an angular speed ω normal to uniform magnetic field B about an axis passing through the one end of the rod, then induced emf across its ends is given as,

$$e=\frac{1}{2}BL^2\omega$$

07. *(b)* Given,

$l=1$ m, $\omega=5$ rad/s

$B_H=0.2\times10^{-4}$ T

Induced emf between the two ends of conductor due to its rotation in B_H,

$$e=\frac{B_H\omega l^2}{2}=\frac{0.2\times10^{-4}\times5\times1^2}{2}$$
$$=5\times10^{-5}\ \text{V}=50\times10^{-6}\ \text{V}=50\,\mu\text{V}$$

08. *(c)* Given, magnetic field, $B=1$ T

Radius of ring, $R=1$ m

Velocity of ring, $v=1$ m/s

Time taken, $t=1$ s

Let emf be ε.

Since, $\varepsilon=Blv\sin\theta$

$\therefore\quad =1\times2R\times1\times\sin90°=2R=2$ V

09. *(a)* Given, length of aeroplane wing,

$l=10$ m

Speed of aeroplane,

$v=180$ km/h $=50$ ms^{-1}

Magnetic flux density,

$B=2.5\times10^{-4}$ Wb/m^2

Angle of dip, $\theta=60°$

The emf induced between the tips of plane wings,

$$\varepsilon=Blv\sin\theta$$
$$=2.5\times10^{-4}\times10\times50\sin60°$$
$$=2.5\times5\times10^{-2}\times\sqrt{3}/2$$
$$=6.25\times1.732\times10^{-2}$$
$$=108.25\times10^{-3}\ \text{V}=108.25\ \text{mV}$$

10. *(d)* According to Lenz's law, electromagnetic induction takes place in the aluminium plate due to which eddy current is developed which oppose the motion or vibrations of coil.

This causes loss in energy which results in damping of oscillatory motion of the coil.

11. *(b)* Induced emf, $e = B_H\, lv$

$$= 0.30 \times 10^{-4} \times 20 \times 5.0 \quad [\theta = 90°]$$

$$= 3\text{ mV}$$

12. *(c)* Induced emf,

$$e = B_H\, lv = 5.0 \times 10^{-5} \times 2 \times 1.50$$

$$= 0.15 \times 10^{-3}\text{ V} = 0.15\text{ mV}$$

13. *(b)* Induced emf,

$$e = \frac{1}{2} B\omega l^2 = \frac{1}{2} \times (0.2 \times 10^{-4})(5)(1)^2$$

$\therefore$ Induced emf,

$$e = \frac{10^{-4}}{2} = \frac{100 \times 10^{-6}}{21} = 50\ \mu\text{V}$$

14. *(b)*

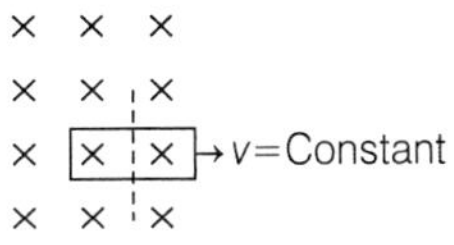

$B =$ Constant

Motional emf $\Rightarrow \varepsilon = Blv =$ Constant

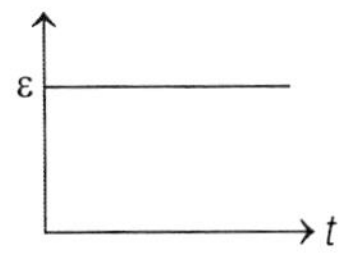

15. *(b)* According to given circuit diagram, equivalent resistance between point P and Q.

$$R_{PQ} = (4+4) \,||\, (4+4) = \frac{8 \times 8}{8+8} = 4\,\Omega$$

The equivalent circuit can be drawn as,

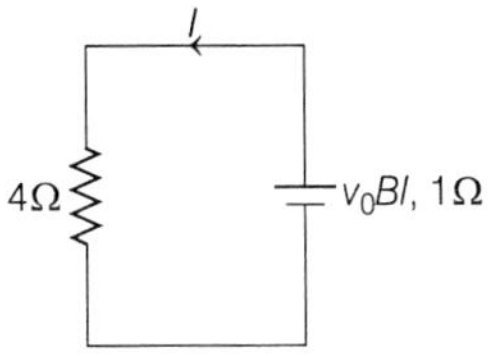

Equivalent resistance,

$R_{eq} = 4 + 1 = 5\,\Omega$

Magnetic field, $B = 5$ T

The side of the square loop,

$l = 20\text{ cm} = 0.20\text{ m}$

The steady value of the current,

$I = 2\text{ mA} = 2 \times 10^{-3}$ A

Induced emf, $e = Bv_0 l \quad [\theta = 90°]$

Induced current, $I = e / R_{eq}$

Substituting the values in the above equation, we get

$$2 \times 10^{-3} = \frac{5 \times v_0 \times 0.2}{5}$$

$$\Rightarrow \quad v_0 = 10^{-2}\text{ m/s} = 1\text{ cm/s}$$

$\therefore$ The value of $v_0 = 1$ cm/s, so that a steady current of 2 mA flows in the loop.

16. *(c)* The given situation can be shown as

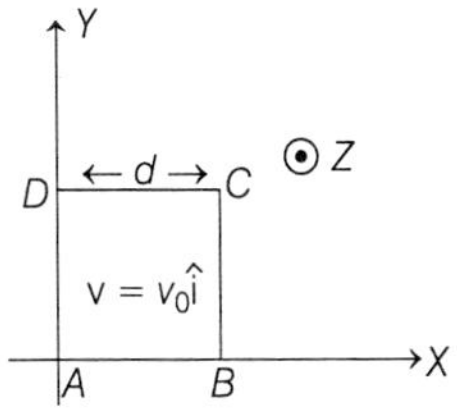

From the above figure, it can be seen that emf induced inside DC and AB will be zero. It is because **v** is parallel to length of conductor along X-axis.

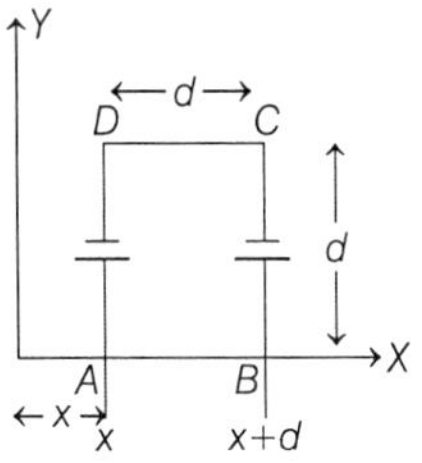

$\therefore$ Net emf induced $= E_{BC} - E_{AD}$

$$[E = \mathbf{v} \times \mathbf{B} \cdot \mathbf{d}]$$

$$= \frac{B_0 (x+d) v_0 d}{a} - \frac{B_0 x v_0 d}{a}$$

$$= \frac{B_0 v_0 d^2}{a}$$

17. *(b)* Magnetic force on electron in the metal sheet is given by

$\mathbf{F}_m = -e(\mathbf{v} \times \mathbf{B})$

At equilibrium,

$F_m = F_e$ (induced)

and $\sigma_2 = -\sigma_1$

$$evB = e\frac{\sigma}{\varepsilon_0}$$

$$\Rightarrow \quad \sigma = \varepsilon_0 vB = \sigma_1$$

$$\sigma_2 = -\varepsilon_0 vB$$

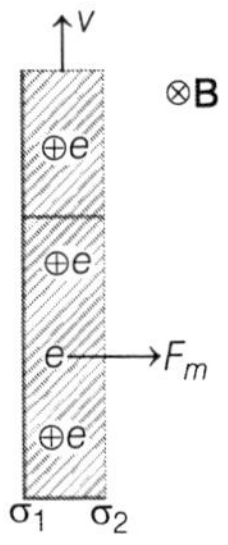

18. *(d)* $\because$ Induced emf is rate of change of magnetic flux.

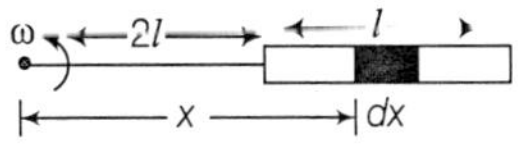

Consider a element of length dx at a distance x from the fixed end of the string. Emf induced in the element

$$d_E = B(\omega x)dx$$

Thus, emf induced across the ends of the rod

$$E = \int_{2l}^{3l} B\omega x\, dx$$

$$= B\omega \left[\frac{x^2}{2}\right]_{2l}^{3l} = \frac{5B\omega l^2}{2}$$

19. *(b)* Relative velocity of one tube w.r. to another $= v_{rel}$, then

$$\varepsilon = B \times \text{width of tube} \times v_{rel}$$

$$= B \times l \times [v - (-v)] = 2Blv$$

20. *(d)*

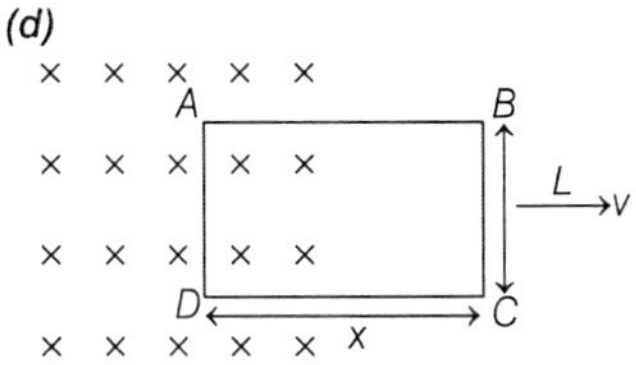

As BC is outside the magnetic field,

$$\therefore \quad e = \frac{-d\phi}{dt} = \frac{dB \cdot A}{dt} = 0 \quad [B = 0]$$

Emf due to AD,

$$e = -B\frac{dA}{dt} = -\frac{Bd(L \times x)}{dt} = -BLv$$

$$|e| = BLv$$

emf due to AB and CD will be zero as they are not carrying any flux.

21. *(c)* Given, radius of loop is R and velocity of the bar magnet is v.

According to Faraday's law of electromagnetic induction, the emf induced in loop is given by

$$e = -\frac{Nd\phi}{dt} \quad \text{...(i)}$$

As initially the bar magnet is moving towards the coil, so the rate of change of magnetic flux linked with the coil will increase. Thus, initially the emf induced in coil will decrease as per Eq. (i) (due to –ve sign).

When magnet reaches the middle point in the coil, at that moment the emf induced will be equal to zero because in this situation, change in magnetic flux associated with bar magnet becomes zero. Now, after that when magnet moves forward, the magnetic flux again starts changing but the polarity of magnetic flux change will be reversed. So, the emf will now increase in opposite direction.

The graph for change in emf of coil is shown below.

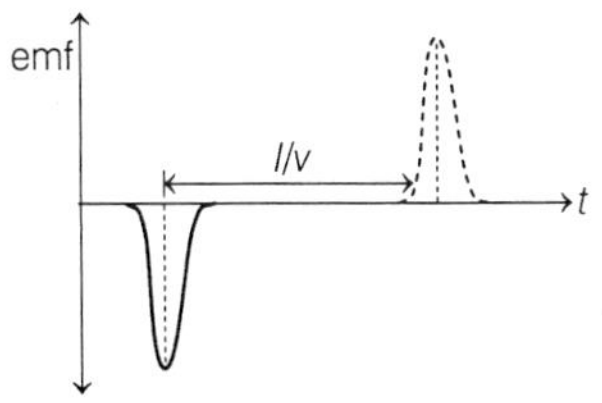

22. *(b)*

> **key idea** If a circuit contains a conductor of length l and moving with velocity v in a uniform magnetic field B, then solve the problem apply KVL as usual by taking potential difference across the conductor as Bvl.

A moving conductor is equivalent to a battery of emf $= vBl$ [angle in 90° between velocity and B] [motion emf]

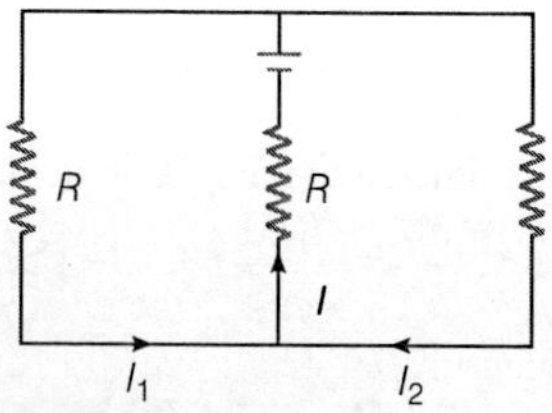

Equivalent circuit,

$$I = I_1 + I_2 \quad \text{...(i)}$$

Applying Kirchhoff's law,

$$I_1 R + IR - vBl = 0 \quad \text{...(ii)}$$

$$I_2 R + IR - vBl = 0 \quad \text{...(iii)}$$

Adding Eqs. (ii) and (iii), we get

$$I_1 R + IR - vBl + I_2 R + IR - vBl = 0$$

$$\Rightarrow \quad 2IR + IR = 2vBl \quad \text{[From Eq. (i)]}$$

$$\Rightarrow \quad I = \frac{2vBl}{3R}$$

Subtracting Eq.(iii) from Eq.(ii), we get

$$I_1 R + IR - vBl = 0$$

$$\underline{{}_{-}I_2 R +_{-} IR -_{+} vBl =_{-} 0}$$

$$I_1 R - I_2 R = 0$$

$$\Rightarrow \quad R(I_1 - I_2) = 0$$

$$\Rightarrow \quad I_1 - I_2 = 0 \quad [\because R \neq 0]$$

$$\Rightarrow \quad I_1 = I_2 \quad \text{...(iv)}$$

From Eq. (i), we get

$$I = 2I_1 \quad [\because I_1 = I_2]$$

$$\Rightarrow \quad 2I_1 = I$$

$$\Rightarrow \quad I_1 = \frac{I}{2} = \frac{2vBl}{2 \times 3R} = \frac{vBl}{3R}$$

$$\therefore \quad I_1 = I_2 = \frac{vBl}{3R}$$

23. *(10)*

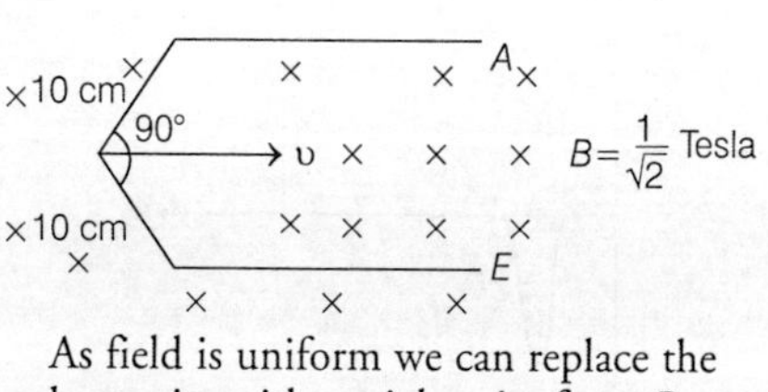

As field is uniform we can replace the bent wire with straight wire from B to D So, emf

$$\varepsilon = Bvl_{BD} \sin$$

$$= \frac{1}{\sqrt{2}} \times 0.1 \times 10\sqrt{2} \times 10^{-2} \times \sin 90°$$

$$= 10 \times 10^{-3} \times 1 = 10 \text{ mV}$$

24. *(1)*

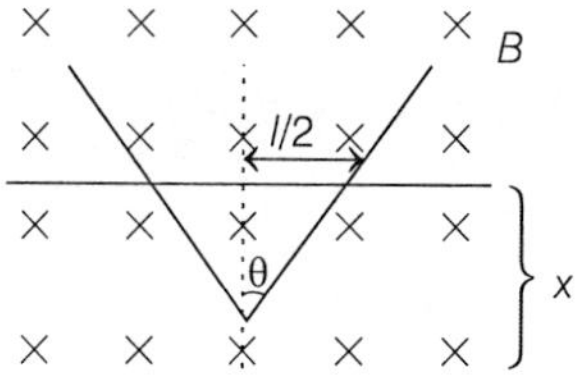

From the figure,

$$\tan\theta = \frac{l}{2x}$$

$$l = 2x \tan\theta$$

$$E = Blv = 2x \tan\theta vB \quad (x = vt)$$

$$E = 2\tan\theta v^2 tB$$

$$E \propto t$$

25. *(3)*

> **key idea** To calculate work done W is pulling the loop out of the field, use two steps as given below.
>
> **Step 1** Calculate induced emf using $e = \Delta\phi/\Delta t$
>
> **Step 2** Find work done, using $W = P \times t$ where, $P = e^2/R$

Induced emf in loop

$$e = \frac{\Delta\phi}{\Delta t} = \frac{NBA}{t}$$

Induced power, $P = \frac{e^2}{R} = \frac{N^2 B^2 A^2}{t^2 \times R}$

Work done $= P \times t$

$$= \frac{N^2 B^2 A^2}{t^2 \times R} \times t = \frac{N^2 B^2 A^2}{t \times R}$$

$$= \frac{10^2 \times \left(\frac{1}{2}\right)^2 \times (3.6 \times 10^{-3})^2}{1 \times 100}$$

$$= \frac{3.6 \times 3.6 \times 10^{-6}}{4} = 3.24 \times 10^{-6} \text{ J}$$

$$\approx 3 \times 10^{-6} \text{ J}$$

26. *(0)* We know that emf induced across the rotating rod, $\varepsilon = \frac{B\omega l^2}{2}$

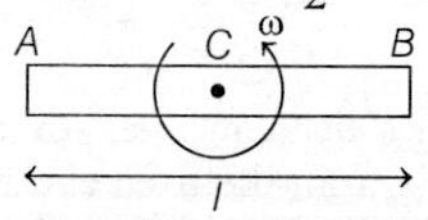

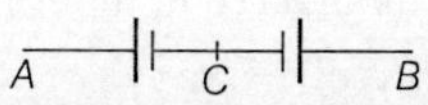

$\therefore$ Induced emf in rod AC = induced emf in rod BC

$$= \frac{B\omega (l/2)^2}{2} = \frac{B\omega l^2}{8}$$

Also, by symmetry, $V_A - V_C = V_B - V_C$

$$\Rightarrow \quad V_A - V_B = 0$$

27. *(216)*

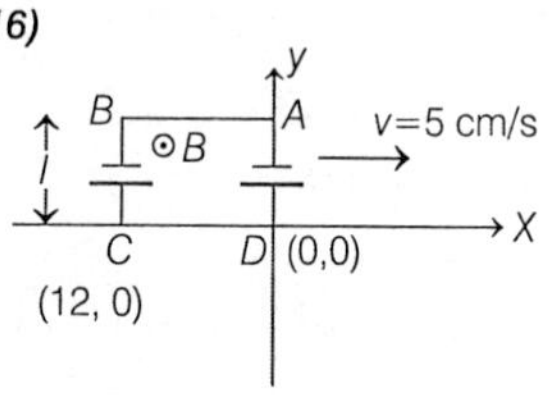

Rate of change of flux due to time variation $= A \cdot \frac{dB}{dt}$

$$= \frac{12}{100} \times \frac{5}{100} \times 10^{-3} \cdot \left(\text{T} \cdot \frac{\text{m}^2}{\text{s}}\right)$$

$$= 60 \times 10^{-7} \frac{\text{Wb}}{\text{s}}$$

Rate of change of flux due to motion of the loop in a non-uniform magnetic field.

$$= A \cdot v \cdot \frac{dB}{dx}$$

$$= \frac{12}{100} \times \frac{5}{100} \times \frac{5}{100} \times \frac{10^{-3}}{\frac{1}{100}}$$

$$= 60 \times 5 \times 10^{-7} = 300 \times 10^{-7} \frac{\text{Wb}}{\text{s}}$$

So, induced emf

$$E = 60 \times 10^{-7} + 300 \times 10^{-7}$$

$$= 360 \times 10^{-7} \text{ V}$$

Power dissipated as heat,

$$= \frac{E^2}{R} = \frac{(360 \times 10^{-7})^2}{6 \times 10^{-3}}$$

$$= 216 \times 10^{-9} \text{ W}$$

28. *(32)* Given,

Length of blade, $l = 80$ cm

Rotational speed = 1200 rpm

Angular speed, $\omega = 2\pi \times f$

$$= \frac{2\pi}{60} \times 1200 \text{ rad/s}$$

$$\omega = 40\pi \text{ rad/s}$$

Induced emf, $\varepsilon = \frac{1}{2} B_V \omega l^2 \quad \text{...(i)}$

$$B_V = B_H \sin\theta$$

$$B_H = 0.5 \text{G} = 0.5 \times 10^{-4} \text{ T}$$

$$\theta = 30°, \sin 30° = \frac{1}{2}$$

$$B_V = \frac{1}{4} \times 10^{-4}$$

$$\varepsilon = \frac{1}{2} \times \frac{1}{4} \times 10^{-4} \times 40\pi \times (80)^2 \times 10^{-4}$$

$$\varepsilon = 3.2\pi \times 10^{-4} \text{ V} = 32\pi \times 10^{-5} \text{ V}$$

29. *(88)* Given, Angular speed of metallic rod.

$$\omega = 2\pi \times \frac{210}{60} \text{ rad/s}$$
$$= 7\pi \text{ rad/s}$$

Length of rod, $l = 20$ cm $= 0.2$ m

Magnetic field, $B = 0.2$ T

The emf induced between the centre and the ring,

$$e = \frac{1}{2} B\omega l^2$$
$$= \frac{1}{2} \times 0.2 \times 7\pi \times (0.2)^2$$
$$= 0.088 \text{ V} = 88 \text{ mV}$$

30. *(50)* Magnetic field, $B = 0.4$ T

Expansion in the radius of loop

$$\left(\frac{dr}{dt}\right) = 1 \text{ mm/s} = 10^{-3} \text{ m/s}$$

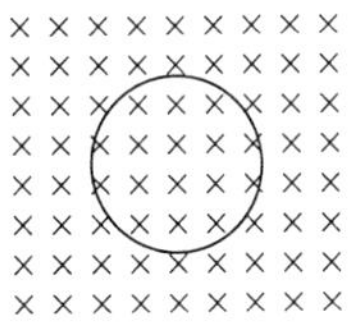

Thus, $\frac{dA}{dt} = 2\pi r \frac{dr}{dt}$

$$\text{emf induced, } \varepsilon = \left(-\frac{d\phi}{dt}\right) = \frac{BdA}{dt}$$
$$= 0.4\, 2\pi r \frac{dr}{dt}$$
$$|\varepsilon|_{r = 2\text{cm}} = 0.4 \times 2 \times \pi \times 2 \times 10^{-2} \times 10^{-3} \text{ V}$$
$$= 16\pi \mu \text{ V} = 50.24 \mu\text{V} \simeq 50 \mu \text{ V}$$

31. *(150)* Given,

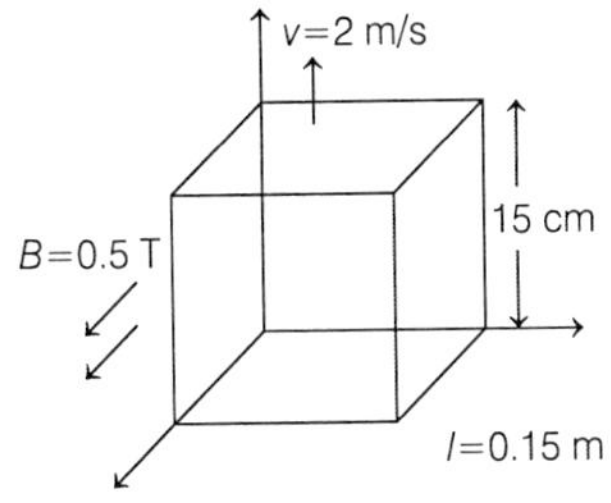

$$l = 0.15 \text{ m}$$
$$\text{Emf} = Bvl = 0.5 \times 2 \times 0.15$$
$$= 0.15 \text{V} = 150 \text{ mV}$$

32. *(18)* Force on rod is given by

$$F = \frac{B^2 l^2 v}{R}$$

Substituting value we get

$$F = \frac{(0.15)^2 (1^2) \times 4}{5}$$
$$= 18 \times 10^{-3} \text{ N}$$

33. *(10)*

key idea If a loop is shrinking with velocity v in a uniform magnetic field perpendicularly, then the induced emf in the coil is calculated by

$$e = 2\pi r B v$$

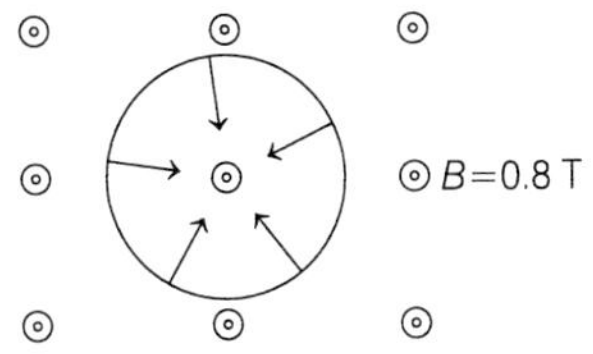

For loop given that, $\frac{dr}{dt} = 2$ cm/ s

Magnitude of induced emf in loop,

$$E = \frac{d\phi}{dt} = \frac{d}{dt} BA = B \cdot \frac{d}{dt} \pi r^2 = 2\pi B r \cdot \frac{dr}{dt}$$

Here, $B = 0.8$ T, $r = 10$ cm, $\frac{dr}{dt} = 2$ cm/s

So,

$$E = 2 \times \pi \times 0.8 \times 10 \times 10^{-2} \times 2 \times 10^{-2}$$
$$= 100 \times 10^{-4} \text{ V}$$
$$= 10 \text{ mV}$$

34. *(16)*

key idea If the speed of a rod is inclined to magnetic field with an angle θ, then the motional emf is

$$e = Bvl \sin\theta$$

Given, a metallic rod, $l = 20$ cm

It is placed in N-S direction

$$v = 20 \text{ m/s (towards East)}$$
$$B_H = 4 \times 10^{-3} \text{ T}$$
$$\theta = 45°$$

To find emf induced in rod,

$$E = Blv \sin \theta$$
$$= 4 \times 10^{-3} \times \frac{20}{100} \times 20$$
$$(\because \theta = 90° \Rightarrow \sin 90° = 1)$$
$$= 16 \times 10^{-3} \text{ V} = 16 \text{ mV}$$

35. *(440)* Given, number of turns in coil

$$N = 1000$$

Area, $A = 1 \text{ m}^2$

Angular velocity, $\omega = 2\pi \times$ number of revolutions per second

$$= 2\pi \times 1$$
$$\Rightarrow \quad \omega = 2\pi = 2 \times 3.14 = 6.28 \text{ rad/s}$$

$\therefore$ Maximum voltage produced due to revolution of coil,

$$V_{\max} = NBA\omega$$
$$= 1000 \times 0.07 \times 1 \times 6.28$$
$$= 439.6 \simeq 440 \text{ V}$$

36. *(60)* Given, radius of circular coil,

$$r = 8.0 \text{ cm} = 0.08 \text{ m}$$

Number of turns, $N = 20$

Angular speed of coil, $\omega = 50 \text{ rad s}^{-1}$

Magnetic field, $B = 3.0 \times 10^{-2}$ T

The maximum emf induced in a coil can be expressed as $\varepsilon = \omega NBA$

Here, A is the area of coil.

$$\varepsilon = NB\,\omega(\pi r^2)$$

Substituting the given values, we get

$$\varepsilon = 20 \times 3.0 \times 10^{-2} \times 50 \times \pi \times (0.08)^2$$
$$= 60.3 \times 10^{-2} \text{ V} \simeq 60 \times 10^{-2} \text{ V}$$

37. *(80)* Given,

Radius of coil, $r = 1$ m

Resistance of coil,

$$R = 2\mu\,\Omega = 2 \times 10^{-6} \Omega$$

Magnetic field,

$$B = \frac{4}{\pi} \times 10^{-3} \left(1 - \frac{t}{100}\right) \text{T}$$

As we know that,

$$\text{Induced emf}(\varepsilon) = -\frac{d\phi}{dt} = -A\frac{dB}{dt}$$
$$(\because \phi = \text{B} \cdot \text{A})$$

where, $\frac{d\phi}{dt}$ = rate of charge of flux

A = area of coil

$\frac{dB}{dt}$ = rate of change of magnetic field strength

$$\Rightarrow \quad \varepsilon = -(\pi r^2) \frac{d}{dt}\left[\frac{4}{\pi} \times 10^{-3}\left(1 - \frac{t}{100}\right)\right]$$
$$= 4 \times 10^{-5} \text{ V}$$

If $B = 0$,

Then,

$$\frac{4}{\pi} \times 10^{-3}\left(1 - \frac{t}{100}\right) = 0 \Rightarrow t = 100 \text{ s}$$

Since, Energy (E) = Power (P) × Time (t)

Also, $P = \frac{\varepsilon^2}{R}$

$$\Rightarrow \quad E = \frac{\varepsilon^2}{R} \times t$$
$$= \frac{(4 \times 10^{-5})^2}{2 \times 10^{-6}} \times 100 = 80 \text{ mJ}$$

Topic 3

01. *(b)* $\phi_1 = L_1 I_1 + M_{12} I_2$

$$e_1 = -\frac{d\phi_1}{dt} = -\frac{d}{dt}[L_1 I_1 + M_{12} I_2]$$
$$e_1 = -L_1 \frac{dI_1}{dt} - M_{12}\frac{dI_2}{dt}$$

02. *(b)* The steady state current flowing in the circuit until opening the switch is,

$$I = \frac{E}{R} = \frac{12}{6} = 2\text{ A}$$

The final current becomes zero, when the switch is opened.

We know that induced emf,

$$e = -L\frac{dI}{dt}$$

$$20 = -L\left[\frac{0-(2)}{1\times 10^{-3}}\right]$$

$$\Rightarrow \quad 20 = L \times 2 \times 10^{3}$$

$$\Rightarrow \quad L = 10^{-2}\text{ H} = 10\text{ mH}$$

03. *(d)* Here, the two coils of inductance L_1 and L_2 and the mutual inductance between them is M. So, the current in both the coil is opposite to each other. The equivalent self inductance is given by $L_{eq} = L_1 + L_2 - 2M$

So, $L_{eq} = L_1 + L_2 - 2M$

04. *(c)* Given, $R = 2.0\ \Omega$,

$$L = 2.0\text{ mH} = 2\times 10^{-3}\text{ H},\ E = 9\text{ V}$$

When the switch is closed at $t = 0$ in *L-R* circuit. The inductors will offer infinite resistance (behaves as an open circuit). Therefore, the equivalent circuit can be redrawn as follows

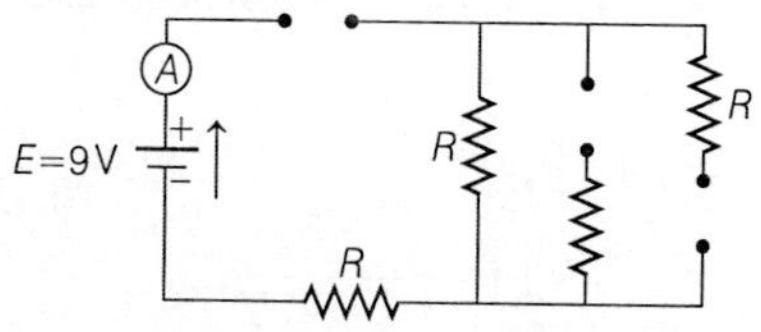

Current through the circuit is

$$i = \frac{E}{R_{total}} = \frac{E}{R+R} = \frac{9}{2R}$$

$$= \frac{9}{2\times 2} = \frac{9}{4} = 2.25\text{ A}$$

05. *(d)* As, $i = dq/dt$

$$\therefore \quad q = \int i\,dt = \int_0^{L/R} \frac{E}{R}[1 - e^{-Rt/L}]\,dt$$

$$q = \frac{EL}{R^2}\frac{1}{e} \Rightarrow q = \frac{EL}{2.7R^2}$$

06. *(a)*

key idea At steady state, i.e., after a long time, the maximum current flows through the inductor. Hence, it behaves as a short circuit, i.e. a piece of wire.

After a sufficiently long time, in steady state, resistance offered by inductor is zero. So, circuit is reduced to

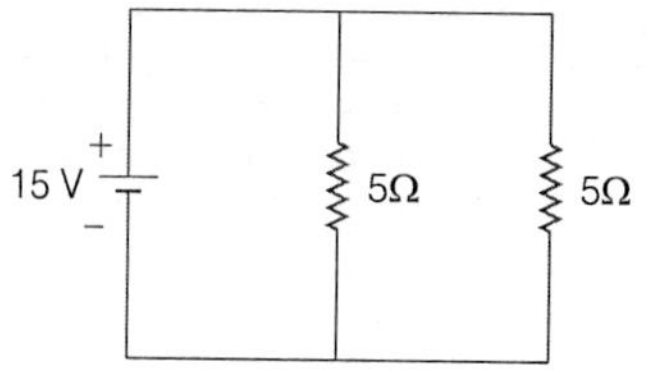

$\therefore$ Current in circuit is

$$I = \frac{E}{R_{eq}} = \frac{15}{\left(\frac{5\times 5}{5+5}\right)} = \frac{15\times 2}{5} = 6\text{ A}$$

• Concept Enhancer

At steady state i.e. after a very long time, the maximum charge flows through a capacitor in *R-C* circuit. The capacitor becomes saturated and it behaves as an open circuit.

07. *(c)* Mutual inductance for a coaxial solenoid of radius r_1 and r_2 and number of turns n_1 and n_2, respectively is given as,

$M = \mu_0 n_1 n_2\ \pi r_1^2\ l$ (for internal coil of radius r_1)

Self inductance for the internal coil,

$$L = \mu_0 n_1^2 \pi r_1^2 l$$

$$\therefore \quad \frac{M}{L} = \frac{n_1 n_2}{n_1^2} = \frac{n_2}{n_1}$$

• Concept Enhancer

The self inductance of a coil depends on shape, size, material and the medium in which coil is placed.

08. *(a)* Mutual inductance,

$$M = \mu_0 n_1 n_2 \pi r_1^2 l$$

From, $\phi_2 = \pi r_1^2(\mu_0 n i) n_2 l$

$$A = \pi r_1^2 = 10\text{ cm}^2, l = 20\text{ cm},$$

$$N_1 = 300, N_2 = 400$$

$$M = \frac{\mu_0 N_1 N_2 A}{l}$$

$$= \frac{4\pi\times 10^{-7}\times 300\times 400\times 10\times 10^{-4}}{0.20}$$

$$= 2.4\pi\times 10^{-4}\text{ H}$$

09. *(b)* Mutual inductance M between two coils is given two coils is given by

$$M = \mu_0 n_1 n_2 \pi r_1^2 L$$

where, n_1, n_2 are number of turns, r_1 is the radius of coil and L is the length. From the above formula, it is clear that mutual inductance depends on the distance between the coils and geometry (πr^2 = area) of two coils.

10. *(c)* $L = \dfrac{\mu_0 N^2 A}{l}$

Self-inductance of the coil depends on its geometry.

In electromotive analysis, "self-inductance is considered the electrical equivalent of mass" in mechanics.

Work needs to be done against the self induced emf because this opposing force tried to resist the change in current flow.

A, C, D, E only correct.

11. *(b)* We know that, emf, $\varepsilon = -\dfrac{LdI}{dt}$

$$\Rightarrow \quad |\varepsilon| = \frac{LdI}{dt} \Rightarrow |\varepsilon| = L\frac{\Delta I}{\Delta t}$$

$$\Rightarrow \quad 0.1 = L\frac{[2-(-2)]}{0.2} \Rightarrow L = \frac{0.1\times 0.2}{4}$$

$$= 5\times 10^{-3}\text{ H} = 5\text{ mH}$$

12. *(c)* Flux, ϕ = mutual inductance × current

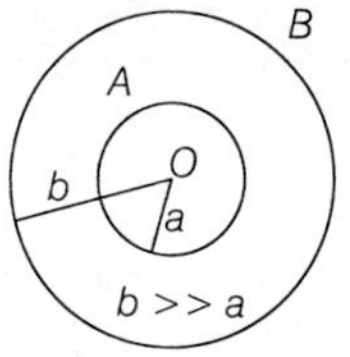

$$\phi = Mi = B\cdot A$$

$$Mi = \frac{\mu_0 i}{2b}\cdot \pi a^2 \left[B = \frac{\mu_0 i}{2r} \text{ and } A = \pi a^2\right]$$

$$M = \frac{\mu_0 \pi a^2}{2b}$$

13. *(a)*

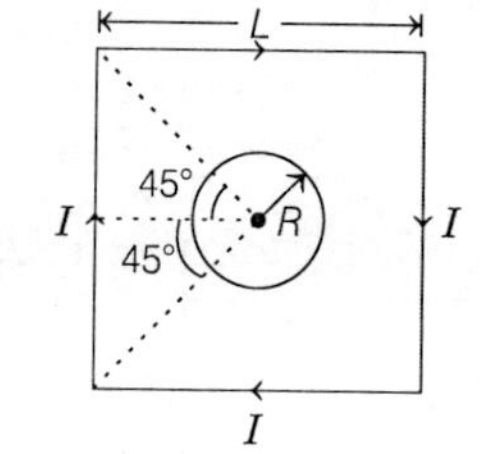

Mutual inductance,

$$M = \frac{\text{Number of Flux Linkages}}{\text{Current}}$$

$$= \frac{B_1 A_2}{I_1}$$

B_1 = Field of square due to current I in it

$$= \left(\frac{\mu_0}{4\pi}\times\frac{2I\sin 45°}{L/2}\right)\times 4 = \frac{\mu_0 2\sqrt{2}I}{\pi L}$$

$$\text{So, } M = \frac{\frac{\mu_0 2\sqrt{2}I}{\pi L}\times \pi R^2}{I} = \frac{\mu_0 2\sqrt{2}R^2}{L}$$

14. *(c)*

> **key idea** The potential energy stored in a inductor during its charging in *L-R* circuit is calculated by $U = \frac{1}{2}Li^2$
> where, $i = i_0\left(i - e^{-\frac{Rt}{L}}\right)$

(A) Given,

$$L = 1\text{ H}, R = 100\,\Omega, V = 6\text{ V}$$

In *L-R* circuit,

$$\because \quad i = i_0\left(1 - e^{-\frac{Rt}{L}}\right)$$

When current acquires half of its steady-state value.

i.e., $i = \frac{i_0}{2}$

$$\frac{i_0}{2} = i_0\left(1 - e^{-\frac{Rt}{L}}\right) \quad \left(\because i = \frac{i_0}{2}\right)$$

$$e^{\frac{Rt}{L}} = 2; \frac{Rt}{L} = \ln 2; t = \frac{L}{R}\ln 2;$$

$$t = \frac{1}{100} \times 0.693 = 7\text{ ms}$$

(B) Now, current in circuit at an instant 15 ms

$$i = i_0\left(1 - e^{-\frac{Rt}{L}}\right); i = \frac{V}{R}\left(1 - e^{-\frac{Rt}{L}}\right)$$

$$i = \frac{6}{100}\left(1 - e^{\frac{-100 \times 15 \times 10^{-3}}{1}}\right);$$

$$i = \frac{6}{100}\left(1 - e^{\frac{-3}{2}}\right)$$

$$i = \frac{6}{100}(1 - 0.25) \Rightarrow i = 0.045$$

Now energy stored,

$$U = \frac{1}{2}Li^2 = \frac{1}{2} \times 1 \times (0.045)^2, \; U = 1\text{ mJ}$$

15. *(c)*

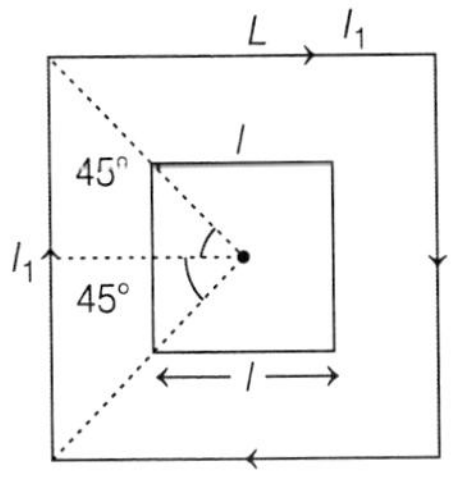

Magnetic field due to one of side at near centre is

$$B_1 = \frac{\mu_0}{4\pi} \cdot \frac{2I_1 \sin 45°}{L/2}$$

Hence, net magnetic field at centre is

$$B = 4B_1 = \frac{\mu_0 I_1 2\sqrt{2}}{4\pi L} \times 4$$

Flux linked with inner coil,

$$\phi_B = B \cdot A = \frac{\mu_0 I_1 \times 2\sqrt{2} \times 4}{4\pi L} \times l^2$$

Mutual inductance of coils is

$$M = \frac{\phi_B}{I_1} = \frac{\mu_0 \times 2\sqrt{2} \times 4 \times l^2}{4\pi L}$$

$$M = \frac{\mu_0 \times 2\sqrt{2} \times l^2}{\pi L} = \frac{2\sqrt{2}\,\mu_0 l^2}{\pi L}$$

16. *(a)* Given, side of one square loop $= a$

Side of 2nd square loop $= b$

$$b >> a$$

Current $= i$

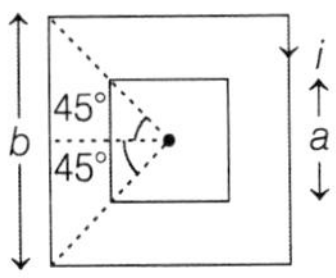

Since, $\phi = Mi$

$$M = \frac{\phi}{i} = \frac{B \cdot A}{i}$$

$B =$ magnetic field and $A =$ area.

Mutual inductance by 4 straight conductor,

$$M = \frac{4B \cdot A}{i}$$

$$M = \frac{4\left[\frac{\mu_0}{4\pi}\frac{i}{b/2}(\sin \pi/4 + \sin \pi/4)\right]a^2}{i}$$

$$= \left(\frac{\mu_0}{4\pi} 8\sqrt{2}\frac{a^2}{b}\right)$$

17. *(d)* Magnetic energy is stored in an inductor, whereas energy is dissipated through resistor.

Given, $U_m = 64$ J

$$i = 8\text{ A}$$

and $\quad P = 640$ W

As, magnetic energy,

$$U_m = \frac{1}{2}Li^2 \Rightarrow 64 = \frac{1}{2} \times L \times 64$$

$$\Rightarrow \quad L = 2\text{ H}$$

Energy dissipated, $P = i^2 R$

$$\Rightarrow \quad 640 = 64\,R \Rightarrow R = 10\,\Omega$$

Time constant of *R-L* circuit is given by

$$\tau = \frac{L}{R}$$

$$\Rightarrow \tau = \frac{2}{10} = \frac{1}{5} = 0.2\text{ s}$$

18. *(c)*

> **key idea** The time for which the current in *L-R* circuit increases to *I* is given by
> $t = \frac{L}{R}\ln\left(\frac{I_0}{I_0 - I}\right)$

The expression of the magnetic energy stored in the solenoid,

$$U = (LI^2)/2$$

The maximum value of the magnetic energy stored in the solenoid,

$$U_0 = LI_0^2/2$$

Given, $\quad U = 25\% \times U_0$

$$\Rightarrow \quad \frac{LI^2}{2} = \frac{1}{4} \times \frac{LI_0^2}{2} \Rightarrow I = \frac{I_0}{2}$$

Therefore, using the formula for the decay current in *L-R* circuit,

$$I = I_0(1 - e^{-t/\tau}) \Rightarrow \frac{I_0}{2} = I_0(1 - e^{-t/\tau})$$

$$e^{-t/\tau} = 1/2 \Rightarrow t = \tau \ln 2$$

$$\Rightarrow \quad t = \frac{L}{R}\ln 2$$

19. *(c)* The energy stored in the inductor is given by

$$U = \frac{1}{2}LI^2 \Rightarrow U \propto I^2 = \frac{U_0}{n}$$

So, $$\frac{U}{U_0} = \left(\frac{I}{I_0}\right)^2, \frac{1}{n} = \left(\frac{I}{I_0}\right)^2 \quad \ldots\text{(i)}$$

$$\frac{I}{I_0} = 1 - e^{-t/z} = 1 - e^{-Rt/L}$$

From Eq. (i),

$$1 - e^{-\frac{Rt}{L}} = \frac{1}{\sqrt{n}}, 1 - \frac{1}{\sqrt{n}} = e^{-\frac{Rt}{L}}$$

$$\frac{\sqrt{n} - 1}{\sqrt{n}} = e^{\frac{-Rt}{L}}, \; t = \frac{L}{R}\ln\left(\frac{\sqrt{n}}{\sqrt{n} - 1}\right)$$

20. *(d)*

> **key idea** The charge flowing through an *LR* circuit is calculated by
> $q = \int_{t_1}^{t_2} i\,dt$, where,
> $i = i_0(1 - e^{-t/\tau})$ and $\tau = \frac{L}{R}$

In case of charging of *L-R* circuit, $i = i_0(1 - e^{-t/\tau})$, where $\tau = L/R = t$

So, charge, $q = \int_0^{t_c} i\,dt = \int_0^{t_c} i_0(1 - e^{t/t_c})dt$

$$= \frac{\varepsilon}{R}\int_0^{t_c}(1 - e^{-t/t_c})dt = \frac{\varepsilon}{R}[t + t_c e^{-t/t_c}]_0^{t_c}$$

$= \frac{\varepsilon}{R}[(t_c + t_c e^{-t_c/t_c}) - (0 + t_c e^{-0})]$

$= \frac{\varepsilon}{R}[(t_c + t_c e^{-1}) - t_c] = \frac{\varepsilon}{R} t_c e^{-1}$

$= \frac{\varepsilon}{R} \times \frac{L}{Re} \Rightarrow q = \frac{\varepsilon L}{eR^2}$

21. *(a)* In an L-R-circuit the current at any time t is given as

$$I = I_0(1 - e^{-t/\tau}) \quad \text{...(i)}$$

Given, $I = \frac{80}{100} I_0$

$\therefore \frac{80}{100} I_0 = I_0(1 - e^{-t/\tau})$ [Using Eq. (i)]

$\Rightarrow e^{-tR/L} = 5$ or $t = \frac{L}{R}\ln 5$

$\Rightarrow \quad t = \frac{10 \times 10^{-3}}{0.1 + 0.9}(1.6) = 0.016$ s

22. *(b)*

> **key idea** The mutual inductance of a coil 1 due to coil 2 is given by
> $M_{12} = \frac{N_1\phi_2}{I_2}$, where I_2 = current flowing in coil 2.

As, coefficient of mutual induction is same for both coils

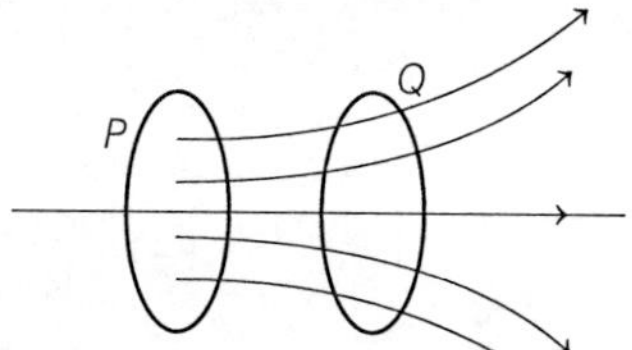

$\therefore \quad M_{PQ} = M_{QP}$

$\Rightarrow \frac{N_P \phi_{PQ}}{I_Q} = \frac{N_Q \phi_{QP}}{I_P}$ ... (i)

Here, $N_P = N_Q = 1$,

$\phi_{PQ} = ?$, $\phi_{QP} = 10^{-3}$ Wb

$I_Q = 2$ A, $I_P = 3$ A

Substituting values in Eq (i), we get

$\phi_{PQ} = \frac{I_Q \cdot \phi_{QP}}{I_P} = \frac{2}{3} \times 10^{-3}$

$= 0.667 \times 10^{-3} = 6.67 \times 10^{-4}$ Wb

23. *(d)* For L-R circuit

$i = \frac{E}{R}(I - e^{-t/\tau})$

where $\tau = \frac{L}{R} = \frac{20}{10} = 2$

Rate of magnetic energy stored in inductor $= i \times L \times \frac{di}{dt}$ ...(i)

Rate of dissipation of energy across resistance $= i^2 R$ ... (ii)

Equating both (given is question)

$i^2 R = iL\frac{di}{dt}$

$R \times \frac{E}{R}\left(I - e^{-\frac{Rt}{L}}\right) = L \times \frac{E}{R} \times \frac{R}{L} e^{-\frac{Rt}{L}}$

$\Rightarrow \quad 2e^{-\frac{Rt}{L}} = 1 \Rightarrow \frac{R}{L}t = \ln 2$

$t = \frac{L}{R}\ln 2 \Rightarrow t = 2\ln 2$

24. *(c)* The induced emf, $\varepsilon = -MI$ ...(i)

where mutual inductance, $M = \frac{\mu_0 \pi a^2}{2b}$

The current is given by, $I = I_0 \cos(\omega t)$

Putting these values in Eq. (i),

$\varepsilon = \frac{-\mu_0 \pi a^2}{2b} \frac{d}{dt}(I_0 \cos(\omega t))$

$= \frac{\mu_0 \pi a^2}{2b} I_0 \omega \sin(\omega t)$

$= \frac{\pi \mu_0 I_0}{2} \frac{a^2}{b} \omega \sin(\omega t)$

25. *(b)* When key K_1 is kept closed, a steady current $I_0 \left(= \frac{\varepsilon}{R}\right)$ flows through the circuit.

When K_1 is opened and K_2 is closed, current at time t in the circuit given by,

$I = I_0 e^{-t/\tau} = \frac{\varepsilon}{R} e^{-\frac{tR}{L}} \quad \left(\because \tau = \frac{L}{R}\right)$

Here, $\varepsilon = 15$ V, $R = 0.15$ kΩ $= 150\,\Omega$

$L = 0.03$ H, $t = 1$ ms $= 10^{-3}$ s

$\therefore \quad I = \frac{15}{150} e^{-\left(\frac{10^{-3} \times 150}{0.03}\right)} = \frac{e^{-5}}{10} = \frac{1}{10e^5}$

$= \frac{1}{10 \times 150} = 6.67 \times 10^{-4}$ A $= 0.67$ mA

26. *(d)* Initially, current in the circuit $= I_0$

After time t current is $I = I_0 e^{(-t/\tau)}$

$\therefore$ Voltage drop across the resistance,

$V_R = IR = V_0 e^{-t/\tau}$ ... (i)

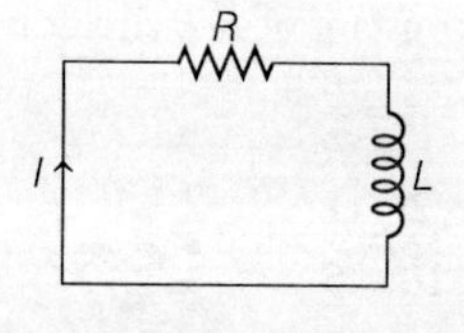

Voltage across the inductor,

$V_L = -L\frac{dI}{dt} = L\left[-\frac{I_0}{\tau} e^{(-t/\tau)}\right]$

$\Rightarrow V_L = I_0 R e^{-t/\tau} = -V_0 e^{-t/\tau}$ ... (ii)

From Eqs. (i) and (ii), $\frac{V_R}{V_L} = -1$

27. *(a)* During the growth of current in L-R circuit, current at any time t is given by

$I = I_0\left(1 - e^{-\frac{R}{L}t}\right)$

or $I = \frac{E}{R}\left(1 - e^{-\frac{R}{L}t}\right) = \frac{5}{5}\left(1 - e^{-\frac{5}{10} \times 2}\right)$

$I = (1 - e^{-1})$

28. *(b)* Maximum current in L-R circuit,

$I_0 = \frac{E}{R} = \frac{100}{100} = 1$ A

The current decays for 1 millisecond $= 1 \times 10^{-3}$ s

$I = (1)e^{\left(\frac{(-1 \times 10^{-3}) \times 100}{100 \times 10^{-3}}\right)}$ or $I = e^{-1} = \frac{1}{e}$ A

29. *(d)* During growth of charge in an inductance, the current at any time t is

$I = I_0(1 - e^{-Rt/L})$

or $\quad \frac{I_0}{2} = I_0\,(1 - e^{-Rt/L})$

or $e^{-Rt/L} = \frac{1}{2} = 2^{-1}$ or $\frac{Rt}{L} = \ln 2$

$\Rightarrow \quad t = \frac{L}{R}\ln 2 = \frac{300 \times 10^{-3}}{2} \times (0.693)$

or $t = 0.1$ s

30. *(d)*

> **key idea** For a short circuit problem, the potential difference across the wire is zero, meaning the potential throughant the entire short circuit is the same.

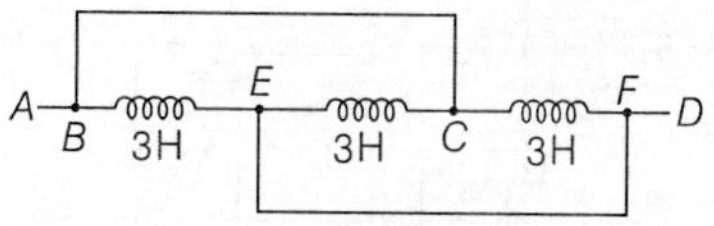

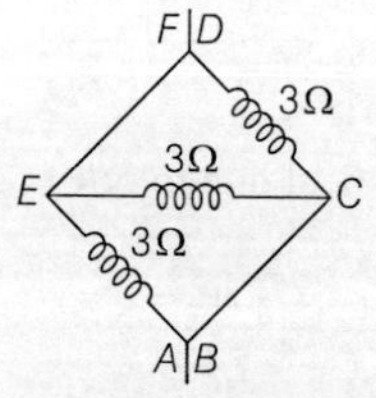

Here, inductors are in parallel.

$\therefore \quad \frac{1}{L} = \frac{1}{3} + \frac{1}{3} + \frac{1}{3}$ or $L = 1$ H

31. *(d)* Magnetic flux associated with the outer coil is

$\phi_{outer} = \mu_0 \pi NR \cdot I$

$= \mu_0 N\pi R(kte^{-\alpha t}) = Cte^{-\alpha t}$

where, $C = \mu_0 N\pi Rk =$ constant

Induced emf, $e = \frac{-d\phi_{outer}}{dt}$

$= -Ce^{-\alpha t} + (-\alpha C t e^{-\alpha t})$

$\therefore$ Induced current, $I = \frac{e}{\text{Resistance}}$

$\Rightarrow$ At $t = 0$, $I = -$ve

$\therefore$ The correct graph representing this condition is given in option (d).

Alternate Solution

Given solenoid is shown below as,

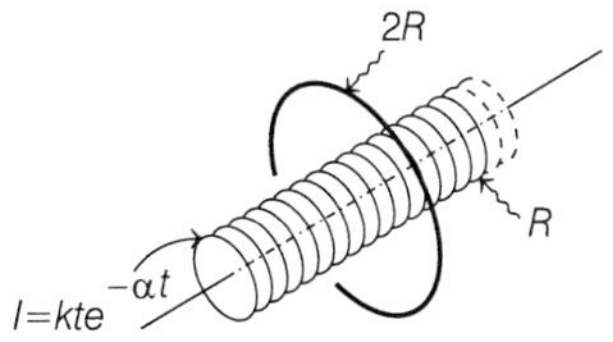

At $t = 0$, current in solenoid

$= I(t=0) = k(0)e^{-\alpha \cdot 0} = 0$

Graph of $e^{\alpha t}$ and kt *versus* time can be shown as,

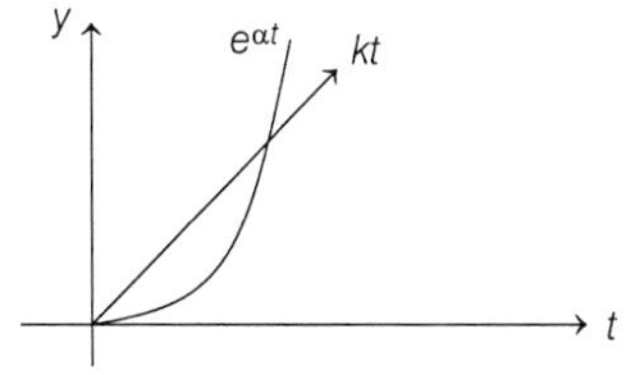

As, $I = \frac{kt}{e^{\alpha t}}$; Initially, $kt > e^{\alpha t}$

So, current is increasing in magnitude.

Finally, after a short time $kt < e^{\alpha t}$. So, current is decreasing in magnitude.

But in both cases, it remains positive or counter clockwise. So, current induced is at first anti-clockwise (following Lenz's law) and then it becomes clockwise and finally reduces to zero as $t \longrightarrow \infty$.

So, correct graph of induced current is

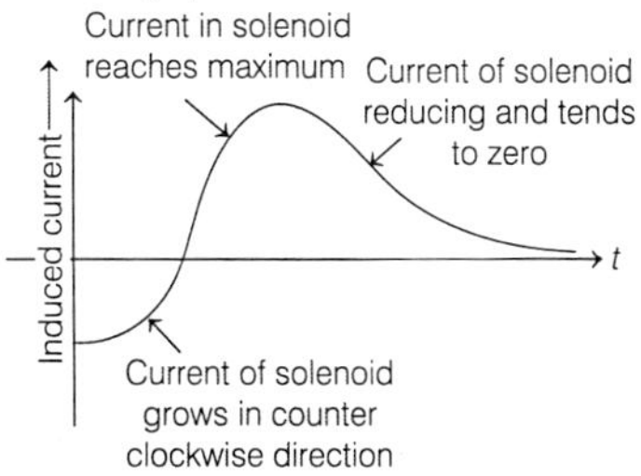

32. *(a)* Energy stored in an inductor of inductance L and current I is given by

$$E = \frac{1}{2} LI^2$$

When current is being changed from I_1 to I_2, change in energy will be

$$\Delta E = E_2 - E_1$$

$$= \frac{1}{2} LI_2^2 - \frac{1}{2} LI_1^2 \quad \ldots \text{(i)}$$

As I_1 and I_2 are given, we need to find value of L.

Now, induced emf in a coil is

$$\varepsilon = L \frac{dI}{dt}$$

Here, $\varepsilon = 25$ V,

$dI = I_2 - I_1 = (25 - 10) = 15$ A

and $dt = 1$ s

$\Rightarrow \quad 25 = L \times \frac{15}{1}$ or $L = \frac{25}{15} = \frac{5}{3}$ H

Putting values of L, I_1 and I_2 in Eq. (i), we get

$$\Delta E = \frac{1}{2} \times \frac{5}{3} \times [25^2 - 10^2] = \frac{1}{2} \times \frac{5}{3} \times 525$$

$$\Delta E = 437.5 \text{ J}$$

33. *(3)*

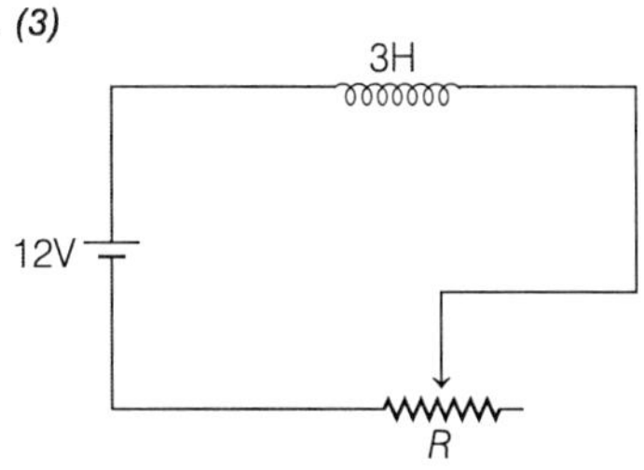

$$\varepsilon - L\frac{dI}{dt} - IR = 0$$

$$12 - 3(-8) - I \times 12 = 0$$

$$I = 3$$

34. *(4)* Given, $I = 3t + 8$

Induced emf, $e = 12$ mV $= 12 \times 10^{-3}$ V

Since, $|e| = L\frac{dI}{dt}$

$\Rightarrow 12 \times 10^{-3} = L \times 3$

$\Rightarrow \quad L = 4 \times 10^{-3}$ H $= 4$ mH

35. *(2)* Induced current,

$$I = -\frac{e}{R} = -\left(\frac{\frac{d}{dt}\phi_B}{R}\right) = -\left(\frac{10t - 36}{8}\right)$$

Induced current at $t = 2$s is,

$$I = -\left(\frac{10 \times 2 - 36}{8}\right) = 2 \text{ A}$$

36. *(128)* For two concentric coplanar square loops,

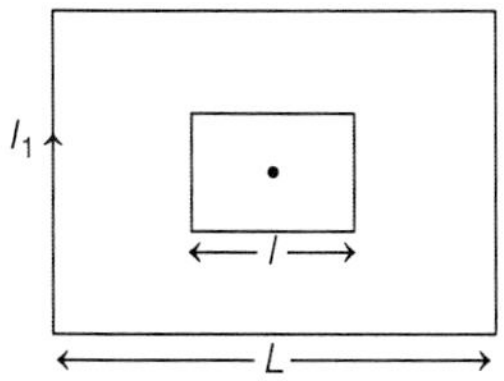

Magnetic field at centre due to outer loop is,

$$B_1 = \frac{\mu_0}{4\pi} \cdot \frac{8\sqrt{2}I_1}{L}$$

So, coefficient of mutual inductance is,

$$M = \frac{B_1 A_2}{I_1} = \frac{\mu_0}{4\pi} \frac{8\sqrt{2}}{L} \times l^2$$

Here, $\frac{\mu_0}{4\pi} = 10^{-7}$ and $l^2 = L$

So, $M = 8\sqrt{2} \times 10^{-7}$ H

Now given,

$M = \sqrt{x} \times 10^{-7} = 8\sqrt{2} \times 10^{-7}$

$\therefore \quad \sqrt{x} = 8\sqrt{2}$

or $\quad x = 64 \times 2 = 128$

37. *(2)*

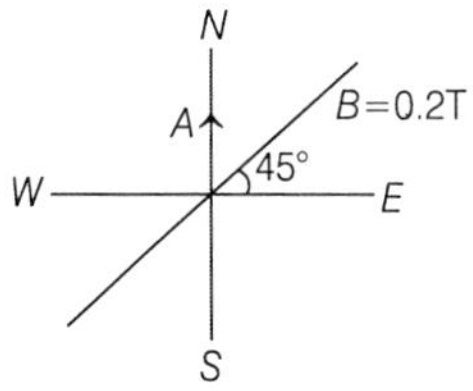

$$\mathbf{A} = (0.1)^2 \hat{\mathbf{j}}, \quad \mathbf{B} = \frac{0.2}{\sqrt{2}}\hat{\mathbf{i}} + \frac{0.2}{\sqrt{2}}\hat{\mathbf{j}}$$

Magnitude of induced emf,

$$e = \frac{\Delta\phi}{\Delta t} = \frac{\mathbf{BA} - 0}{1} = \sqrt{2} \times 10^{-3} \text{ V}$$

$$x = 2$$

38. *(2)* $L = 0.002$ H, $i = i_0 \sin \omega t$, $i_0 = 5$A, $\omega = 50\pi$ rad/s

Induced emf is, $E = -L\frac{di}{dt}$

Also, $\frac{di}{dt} = \omega L \cos \omega t$

$$E = -i_0 \omega L \cos \omega t$$

When, $\cos \omega t = -1$, E will be maximum

$E_{max} = -5 \times 50\pi \times 0.002 \times (-1)$

$= 5 \times 50\pi \times 0.002$

$= \frac{5 \times 50\pi \times 2}{1000} = \frac{\pi}{2}$ V

$\therefore \quad \alpha = 2$

39. *(4)* Given, $a = 1000$ cm, $b = 1$ cm

$b << a$

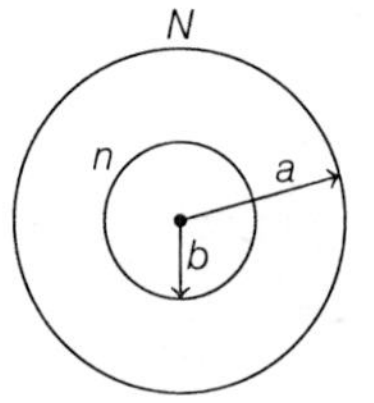

We will take larger coil as primary

$$B = \frac{\mu_0 i_p N}{2a}$$

Flux passing through secondary coil,

$$\phi_s = BA = \frac{\mu_0 i_p N}{2a} \times \pi b^2 \times n$$

Mutual Inductancel,

$$M = \frac{\phi_s}{i_p} \Rightarrow M = \frac{\mu_0 N n \pi b^2}{2a}$$

$$M = \frac{4\pi \times 10^{-7} \times 200 \times 10 \times \pi \times 1 \times 10^{-4}}{2 \times 1000 \times 10^{-2}}$$

$$= 4\pi^2 \times 10^{-9}$$

or $M = 4 \times 10^{-8}$ H (using $\pi^2 = 10$)

40. *(30)* $V_A - IR - \frac{LdI}{dt} - 12 = V_B$

$\Rightarrow \quad V_A - V_B = 2 \times 12 + 6(-1) + 12$

$\Rightarrow \quad V_A - V_B = 24 - 6 + 12$

$\Rightarrow \quad V_A - V_B = 30$ V

$\Rightarrow \quad \Delta V = 30$ V

41. *(3)* After a long time, induced emf in the coil will be zero.

So, we have equivalent circuit,

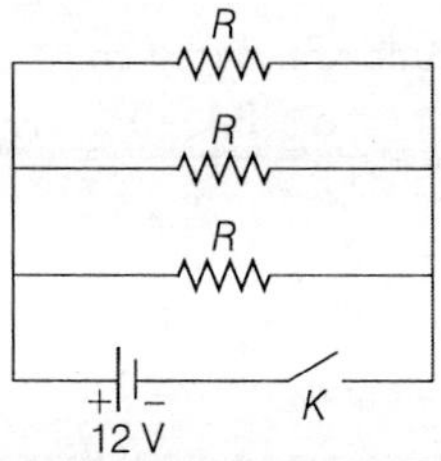

Equivalent resistance R_{eq} of circuit is given by,

$$\frac{1}{R_{eq}} = \frac{1}{R_1} + \frac{1}{R_2} + \frac{1}{R_3} = \frac{3}{12} = \frac{1}{4}$$

or $\quad R_{eq} = 4\Omega$

Circuit current, $I = \frac{V}{R_{eq}} = \frac{12}{4} = 3$ A

42. *(1)* Just after closing the switch S, the current in the inductor branch is zero for this short duration of time at $t = 0$. So, the circuit reduced to as shown below.

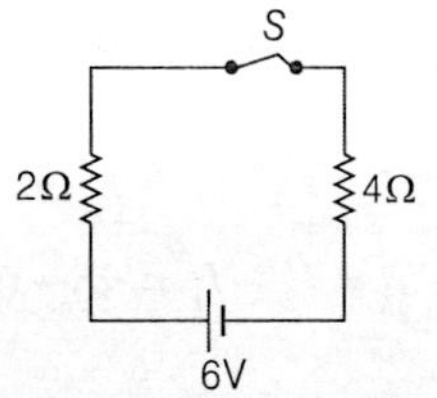

The equivalent resistance of circuit,

$$R_{eq} = 4 + 2 = 6\Omega$$

Current through battery,

$$I = \frac{V}{R_{eq}} = \frac{6}{6} = 1 \text{ A}$$

43. *(400)* Given, resistance,

$R = 10\,\Omega$,

Inductance, $\quad L = 20$ mH

Voltage, $\quad V = 20$ V, $\Delta t = 100$ μs

Induced emf, $e = \frac{\Delta\phi}{\Delta t} = L\frac{dI}{dt} \quad [\because \phi = LI]$

$$e = L\frac{\Delta V}{R \cdot \Delta t}$$

$$= \frac{20 \times 10^{-3} \times 20 \times 10^6}{10 \times 100}$$

$\Rightarrow \quad e = 400$ V

44. *(4)* Self inductance of coil,

$L = 2.0$ H

Current, $\quad I = 2\sin t^2$

At $\quad t = 0, I = 0$A

When, $\quad I = 2$A

Then, $\quad 2 = 2\sin t^2$

$\Rightarrow \quad \sin t^2 = 1 = \sin \pi/2$

$\Rightarrow \quad t^2 = \pi/2 \Rightarrow t = \sqrt{\pi/2}$

Since, $\quad I = 2\sin t^2$

$dI = 4t \cos t^2 dt$

$\therefore$ Amount of energy spent,

$$E = \int LI\, dI$$

$$= \int_0^{\sqrt{\pi/2}} 2 \times 2\sin t^2 \times 4t \cos t^2 dt$$

$$= \int_0^{\sqrt{\pi/2}} 8 \times 2t \sin t^2 \cos t^2 dt$$

$$= 8\int_0^{\sqrt{\pi/2}} t \sin(2t^2)\, dt$$

Put, $2t^2 = x$	when $t = 0$, $x = 0$
$4tdt = dx$	when $t = \sqrt{\pi/2}$
$tdt = \frac{dx}{4}$	$x = 2 \times \frac{\pi}{2} = \pi$

$$= 8\int_0^{\pi} \sin x \frac{dx}{4} = \frac{8}{4}\int_0^{\pi} \sin x\, dx$$

$$= 2\int_0^{\pi} \sin x\, dx$$

$$= 2 \cdot 2\int_0^{\pi/2} \sin x\, dx$$

$$= 4[-\cos x]_0^{\pi/2}$$

$$= 4[-\cos \pi/2 + \cos 0] = 4 \text{ J}$$

45. *(144)* Given, inductance of coil,

$L = 2$H

Supply voltage, $V = 3t$ V

Let E be the energy stored in the coil.

Since, emf $V = L \cdot \frac{dI}{dt} \Rightarrow 3t = L\frac{dI}{dt}$

$\Rightarrow \quad 3\,tdt = LdI$

On integrating both sides, we get

$$3\frac{t^2}{2} = LI$$

At $t = 4$ s,

$$\frac{3}{2} \times 4^2 = LI \Rightarrow \frac{3}{2} \times \frac{16}{L} = I$$

$\Rightarrow \quad 24/L = I \quad \ldots$(i)

As, $\quad E = 1/2\, LI^2$

$\therefore \quad E = 1/2\, L\frac{24^2}{L^2}$ [From Eq. (i)]

$$= \frac{1}{2}\frac{24^2}{L} = \frac{24^2}{2 \times 2} = 144 \text{ J}$$

C-21

JEE Main Chapterwise Topicwise Physics

In JEE Main there is usually 0-1 question per paper is visible from this chapter. In past 3 years, 31 conceptual and 24 numerical questions were asked of easy to moderate level from this chapter.

For detailed analysis scan QR code

ALTERNATING CURRENT

Topic 1 AC Circuits and Power in AC Circuits

Level 1

01 An AC current is represented as

$$i = 5\sqrt{2} + 10\cos\left(650\pi t + \frac{\pi}{6}\right)\text{A}$$

The r.m.s value of the current is *[2025, 7 April, Shift-I]*

(a) 10 A (b) 50 A

(c) 100 A (d) $5\sqrt{2}$ A

02 An alternating current is represented by the equation, $i = 100\sqrt{2}\sin(100\pi t)$ A. The rms value of current and the frequency of the given alternating current are *[2025, 4 April, Shift-I]*

(a) $\frac{100}{\sqrt{2}}$ A, 100 Hz (b) $100\sqrt{2}$ A, 100 Hz

(c) $50\sqrt{2}$ A, 50 Hz (d) 100 A, 50 Hz

03 An electric bulb rated as 100 W-200 V is connected to an AC source of rms voltage 220 V. The peak value of current through the bulb is *[2025, 3 April, Shift-II]*

(a) 2.2 A (b) 0.32 A

(c) 0.64 A (d) 0.45 A

04 A series L-C-R circuit is connected to an alternating source of emf E. The current amplitude at resonant frequency is I_0. If the value of resistance R becomes twice of its initial value, then amplitude of current at resonance will be *[2025, 22 Jan Shift-II]*

(a) $2I_0$ (b) I_0 (c) $\frac{I_0}{\sqrt{2}}$ (d) $\frac{I_0}{2}$

05 In an AC circuit, the instantaneous current is zero, when the instantaneous voltage is maximum. In this case, the source may be connected to

A. pure inductor

B. pure capacitor

C. pure resistor

D. combination of an inductor and capacitor.

Choose the correct answer from the options given below. *[2024, 4 April Shift-I]*

(a) A and B only (b) A, B and D only

(c) A, B and C only (d) B, C and D only

06 Given below are two statements.

Statement I An AC circuit undergoes electrical resonance if it contains either a capacitor or an inductor.

Statement II An AC circuit containing a pure capacitor or a pure inductor consumes higher power due to its non-zero power factor.

In the light of the above statements, choose the correct answer from the options given below. *[2023, 13 April Shift-II]*

(a) Statement I is true but Statement II is false.

(b) Both Statement I and Statement II are true.

(c) Statement I is false but Statement II is true.

(d) Both Statement I and Statement II are false.

Imp. Power Factor L-C-R Circuit at Resonance

07 Given below are two statements.

Statement I When the frequency of an AC source in a series L-C-R circuit increases, the current in the circuit first increases, attains a maximum value and then decreases.

Statement II In a series L-C-R circuit, the value of power factor at resonance is one.

In the light of the above statements, choose the most appropriate answer from the options given below. *[2023, 12 April Shift-I]*

(a) Both Statement I and Statement II are true.

(b) Statement I is incorrect but Statement II is true.

(c) Statement I is correct but Statement II is false.

(d) Both Statement I and Statement II are false.

08 Given below are two statements.

Statement I Maximum power is dissipated in a circuit containing an inductor, a capacitor and a resistor connected in series with an AC source, when resonance occurs.

Statement II Maximum power is dissipated in a circuit containing pure resistor due to zero phase difference between current and voltage.
In the light of the above statements, choose the correct answer from the options given below :
[2023, 10 April Shift-I]

(a) Both Statement I and Statement II are false.
(b) Statement I is false but Statement II is true.
(c) Statement I is true but Statement II is false.
(d) Both Statement I and Statement II are true.

09 A capacitor of capacitance 150.0 μF is connected to an alternating source of emf given by $E = 36\sin(120\pi t)$ V. The maximum value of current in the circuit is approximately equal to *[2023, 6 April Shift-II]*

(a) 2 A (b) $\sqrt{2}$ A
(c) $2\sqrt{2}$ A (d) $1/\sqrt{2}$ A

10 In the given circuit, rms value of current I_{rms} through the resistor R is *[2023, 30 Jan Shift-II]*

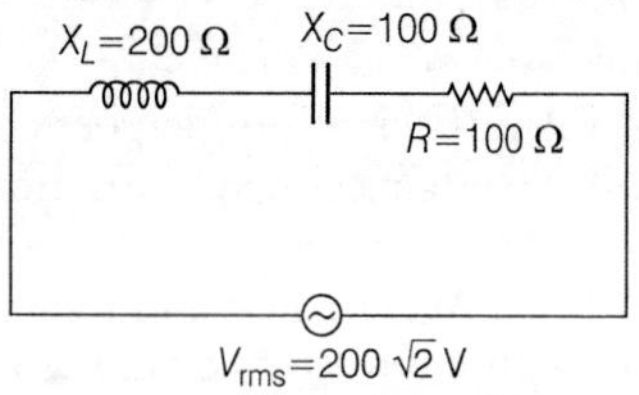

(a) 20 A (b) $2\sqrt{2}$ A
(c) 2 A (d) $\frac{1}{2}$ A

11 For the given figures, choose the correct options. *[2023, 29 Jan Shift-II]*

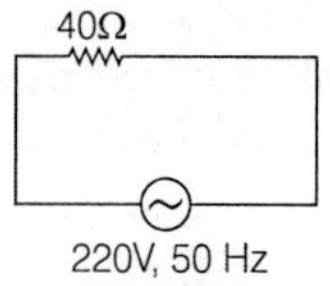

Fig. (a)

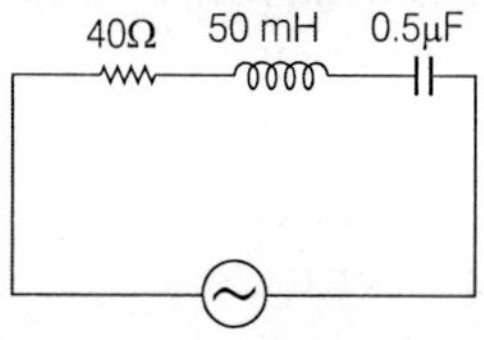

Fig. (b)

(a) The rms current in circuit (b) can be larger than that in (a).
(b) The rms current in Fig. (a) is always equal to that in Fig. (b).
(c) At resonance, current in (b) is less than that in (a).
(d) The rms current in circuit (b) can never be larger than that in (a).

Imp. Resonance Frequency

12 In an L-C oscillator, if values of inductance and capacitance become twice and eight times respectively, then the resonant frequency of oscillator becomes x times its initial resonant frequency ω_0. The value of x is *[2023, 25 Jan Shift-I]*

(a) 1/16 (b) 16
(c) 4 (d) 1/4

13 The equation of current in a purely inductive circuit is $5\sin(49\pi t - 30°)$. If the inductance is 30 mH then the equation for the voltage across the inductor, will be $\left(\text{Let, } \pi = \frac{22}{7}\right)$ *[2022, 28 July, Shift-I]*

(a) $1.47\sin(49\pi t - 30°)$ (b) $1.47\sin(49\pi t + 60°)$
(c) $23.1\sin(49\pi t - 30°)$ (d) $23.1\sin(49\pi t + 60°)$

14 When you walk through a metal detector carrying a metal object in your pocket, it raises an alarm. This phenomenon works on *[2022, 25 July Shift-II]*

(a) Electromagnetic induction
(b) Resonance in AC circuits
(c) Mutual induction in AC circuits
(d) Interference of electromagnetic waves

15 To increase the resonant frequency in series L-C-R circuit, *[2022, 25 July Shift-I]*

(a) source frequency should be increased
(b) another resistance should be added in series with the first resistance.
(c) another capacitor should be added in series with the first capacitor.
(d) the source frequency should be decreased.

Imp. Wattless Current

16 If wattless current flows in the AC circuit, then the circuit is *[2022, 25 June Shift-I]*

(a) purely resistive circuit (b) purely inductive circuit
(c) L-C-R series circuit (d) R-C series circuit only

17 Given below are two statements.
Statement I The reactance of an AC circuit is zero. It is possible that the circuit contains a capacitor and an inductor.
Statement II In AC circuit, the average power delivered by the source never becomes zero.
In the light of the above statements, choose the correct option. *[2022, 24 June Shift-II]*

(a) Both Statement I and Statement II are true.
(b) Both Statement I and Statement II are false.
(c) Statement I is true but Statement II is false.
(d) Statement I is false but Statement II is true.

18 A 100 Ω resistance, a 0.1 μF capacitor and an inductor are connected in series across a 250 V supply at variable frequency. Calculate the value of inductance of inductor at which resonance will occur. Given that the resonant frequency is 60 Hz. *[2021, 27 July Shift-II]*

(a) 0.70 H (b) 70.3 mH
(c) 7.03×10^{-5} H (d) 70.3 H

19 In a circuit consisting of a capacitance and a generator with alternating emf $E_g = E_{g0} \sin \omega t$, V_C and I_C are the voltage and current. Correct phasor diagram for such circuit is *[2021, 22 July Shift-II]*

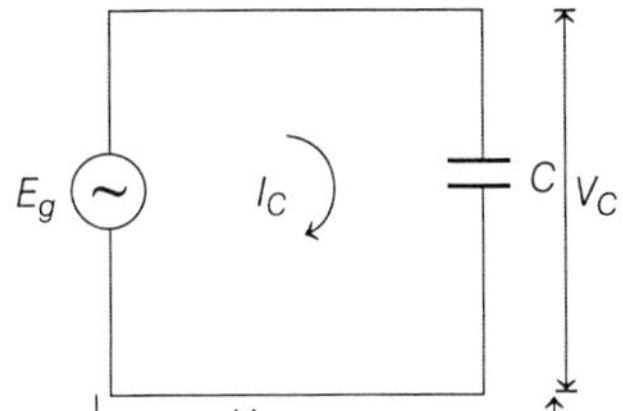

(a)
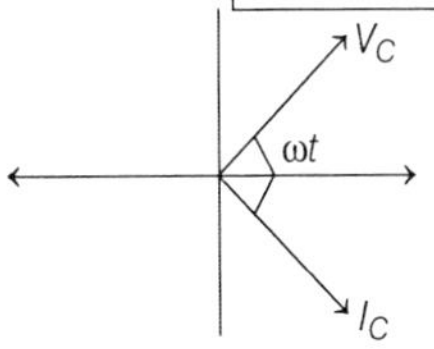

(b)
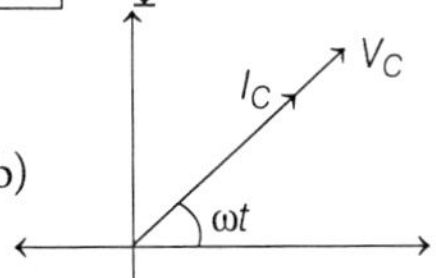

(c)
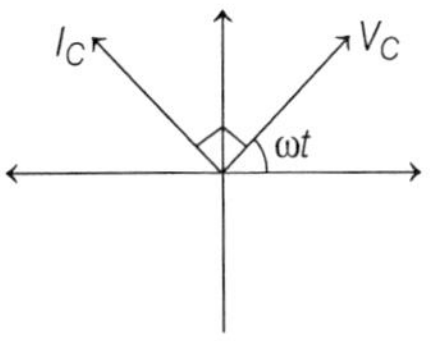

(d)
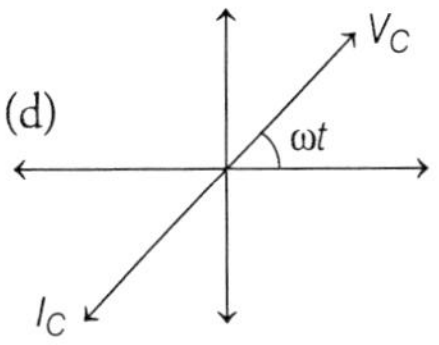

20 Match List I with List II.

List I	List II
A. $\omega L > \frac{1}{\omega C}$	(i) Current is in phase with EMF
B. $\omega L = \frac{1}{\omega C}$	(ii) Current lags behind the applied EMF
C. $\omega L < \frac{1}{\omega C}$	(iii) Maximum current occurs
D. Resonant frequency	(iv) Current leads the EMF

Choose the correct answer from the options given below. *[2021, 22 July Shift-II]*

Codes

	A	B	C	D		A	B	C	D
(a)	(ii)	(i)	(iv)	(iii)	(b)	(ii)	(i)	(iii)	(iv)
(c)	(iii)	(i)	(iv)	(ii)	(d)	(iv)	(iii)	(ii)	(i)

21 In a series *L-C-R* circuit, the inductive reactance X_L is 10 Ω and the capacitive reactance X_C is 4 Ω. The resistance *R* in the circuit is 6 Ω. The power factor of the circuit is *[2021, 18 March Shift-II]*

(a) $\frac{1}{2}$ (b) $\frac{1}{2\sqrt{2}}$ (c) $\frac{1}{\sqrt{2}}$ (d) $\frac{\sqrt{3}}{2}$

22 Match List I with List II.

List-I	List-II
A. Phase difference between current and voltage in a purely resistive AC circuit	1. π/ 2, current leads voltage
B. Phase difference between current and voltage in a pure inductive AC circuit	2. zero
C. Phase difference between current and voltage in a pure capacitive AC circuit	3. π/ 2, current lags voltage
D. Phase difference between current and voltage in an *L-C-R* series circuit	4. $\tan^{-1}\left(\frac{X_C - X_L}{R}\right)$

Choose the most appropriate answer from the options given below. *[2021, 17 March Shift-II]*

	A	B	C	D		A	B	C	D
(a)	1	3	4	2	(b)	2	4	3	1
(c)	2	3	4	1	(d)	2	3	1	4

23 An alternating current is given by the equation $i = i_1 \sin \omega t + i_2 \cos \omega t$. The rms current will be *[2021, 26 Feb Shift-I]*

(a) $\frac{1}{\sqrt{2}}(i_1^2 + i_2^2)^{1/2}$ (b) $\frac{1}{\sqrt{2}}(i_1 + i_2)^2$
(c) $\frac{1}{2}(i_1^2 + i_2^2)^{1/2}$ (d) $\frac{1}{\sqrt{2}}(i_1 + i_2)$

24 In the figure shown, a circuit contains two identical resistors with resistance $R = 5\,\Omega$ and an inductance with $L = 2$ mH. An ideal battery of 15 V is connected in the circuit. What will be the current through the battery long after the switch is closed ? *[2019, 12 Jan Shift-I]*

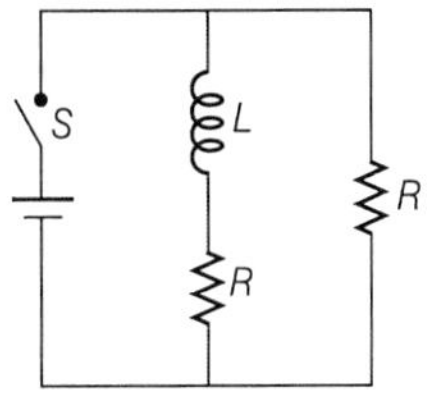

(a) 5.5 A (b) 7.5 A (c) 3 A (d) 6 A

25 A sinusoidal voltage $V(t) = 100 \sin(500t)$ is applied across a pure inductance of $L = 0.02$ H. The current through the coil is *[JEE Main 2014]*

(a) $10 \cos(500t)$ (b) $-10 \cos(500t)$
(c) $10 \sin(500t)$ (d) $-10 \sin(500t)$

26 In an AC circuit, the voltage applied is $E = E_0 \sin \omega t$. The resulting current in the circuit is $I = I_0 \sin\left(\omega t - \frac{\pi}{2}\right)$. The power consumption in the circuit is given by *[AIEEE 2007]*

(a) $P = \frac{E_0 I_0}{\sqrt{2}}$ (b) $P =$ zero
(c) $P = \frac{E_0 I_0}{2}$ (d) $P = \sqrt{2} E_0 I_0$

27 In a series resonant L-C-R circuit, the voltage across R is 100 V and $R = 1\,k\Omega$ with $C = 2\,\mu F$. The resonant frequency ω is 200 rad/s. At resonance, the voltage across L is *[AIEEE 2006]*

(a) 2.5×10^{-2} V (b) 40 V
(c) 250 V (d) 4×10^{-3} V

28 A circuit has a resistance of $12\,\Omega$ and an impedance of $15\,\Omega$. The power factor of the circuit will be *[AIEEE 2005]*

(a) 0.8 (b) 0.4 (c) 1.25 (d) 0.125

29 The phase difference between the alternating current and emf is $\pi/2$. Which of the following cannot be the constituent of the circuit ? *[AIEEE 2005]*

(a) C alone (b) R, L
(c) L, C (d) L alone

30 Alternating current cannot be measured by DC ammeter because *[AIEEE 2004]*

(a) AC cannot pass through DC ammeter
(b) AC changes direction
(c) average value of current for complete cycle is zero
(d) DC ammeter will get damaged

31 In an L-C-R circuit, capacitance is changed from C to $2C$. For the resonant frequency to remains unchanged, the inductance should be changed from L to *[AIEEE 2004]*

(a) $4L$ (b) $2L$ (c) $\frac{L}{2}$ (d) $\frac{L}{4}$

32 The power factor of an AC circuit having resistance R and inductance L (connected in series) and an angular velocity ω is *[AIEEE 2002]*

(a) $\frac{R}{\omega L}$ (b) $\frac{R}{(R^2 + \omega^2 L^2)^{1/2}}$
(c) $\frac{\omega L}{R}$ (d) $\frac{R}{(R^2 - \omega^2 L^2)^{1/2}}$

Level 2

33 An alternating current is given by $I = I_A \sin \omega t + I_B \cos \omega t$. The rms current will be *[2025, 24 Jan Shift-I]*

(a) $\frac{\sqrt{I_A^2 + I_B^2}}{2}$ (b) $\frac{|I_A + I_B|}{\sqrt{2}}$
(c) $\sqrt{I_A^2 + I_B^2}$ (d) $\sqrt{\frac{I_A^2 + I_B^2}{2}}$

34 Given below are two statements. One is labelled as Assertion (A) and the other is labelled as Reason (R).
Assertion (A) Choke coil is simply a coil having a large inductance but a small resistance. Choke coils are used with fluorescent mercury-tube fittings. If household electric power is directly connected to a mercury tube, the tube will be damaged.
Reason (R) By using the choke coil, the voltage across the tube is reduced by a factor $(R / \sqrt{R^2 + \omega^2 L^2})$, where ω is frequency of the supply across resistor R and inductor L. If the choke coil were not used, the voltage across the resistor would be the same as the applied voltage.
In the light of the above statements, choose the most appropriate answer from the options given below. *[2025, 29 Jan Shift-I]*

(a) A is false but R is true.
(b) Both A and R are true and R is the correct explanation of A.
(c) Both A and R are true but R is not the correct explanation of A.
(d) A is true but R is false.

35 A bulb and a capacitor are connected in series across an AC supply. A dielectric is then placed between the plates of the capacitor. The glow of the bulb *[2024, 9 Jan Shift-I]*

(a) remains same (b) increases
(c) becomes zero (d) decreases

36 A L-C-R circuit is at resonance for a capacitor C, inductance L and resistance R. Now, the value of resistance is halved keeping all other parameters same. The current amplitude at resonance will be now *[2024, 8 April Shift-I]*

(a) zero (b) double
(c) same (d) halved

37 Given below are two statements.
Statement I In an L-C-R series circuit, current is maximum at resonance.
Statement II Current in a purely resistive circuit can never be less than that in a series L-C-R circuit when connected to same voltage source.
In the light of the above statements, choose the correct options given below. *[2024, 6 April Shift-I]*

(a) Statement I is false but Statement II is true.
(b) Both Statement I and Statement II are true.
(c) Both Statement I and Statement II are false.
(d) Statement I is true but Statement II is false.

38 A series L-C-R circuit is subjected to an AC signal of 200V, 50 Hz. If the voltage across the inductor ($L = 10$ mH) is 31.4 V, then the current in this circuit is .... *[2024, 5 April Shift-II]*

(a) 68 A (b) 10 A
(c) 63 A (d) 10 mA

39 An alternating voltage of amplitude 40 V and frequency 4 kHz is applied directly across the capacitor of $12\,\mu F$. The maximum displacement current between the plates of the capacitor is nearly. *[2024, 5 April Shift-I]*

(a) 8 A (b) 10 A
(c) 12 A (d) 13 A

40 Match List I with List II.

	List I		List II
A.	Purely capacitive circuit	I.	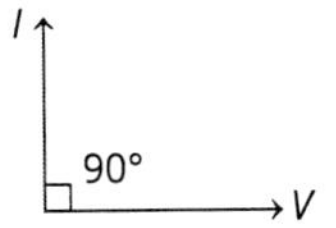
B.	Purely inductive circuit	II.	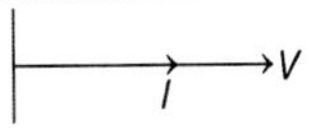
C	*L-C-R* series at resonance	III.	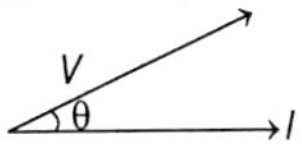
D.	*L-C-R* series circuit	IV.	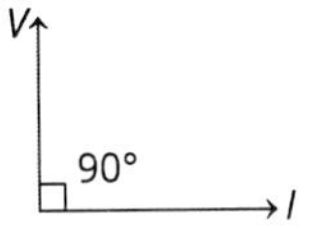

Choose the correct answer from the options given below. *[2024, 4 April Shift-II]*

(a) A- IV, B- I, C- II, D- III (b) A- IV, B- I, C- III, D- II
(c) A- I, B- IV, C- III, D- II (d) A- I, B- IV, C- II, D- III

Imp. L-C-R Series Resonance Circuit

41 In series *L-C-R* circuit, the capacitance is changed from C to $4C$. To keep the resonance frequency unchanged, the new inductance should be *[2024, 1 Feb Shift-I]*

(a) increased by $2L$ (b) increased to $4L$
(c) reduced by $\frac{1}{4}L$ (d) reduced by $\frac{3}{4}L$

42 An AC voltage $V = 20 \sin 200\pi t$ is applied to a series *L-C-R* circuit which drives a current $I = 10 \sin\left(200\pi t + \frac{\pi}{3}\right)$. The average power dissipated is *[2024, 31 Jan Shift-II]*

(a) 173.2 W (b) 200 W
(c) 21.6 W (d) 50 W

Imp. Average Power Dissipation

43 In an AC circuit, voltage and current are given by $V = 100 \sin (100t)$ V and $I = 100 \sin\left(100t + \frac{\pi}{3}\right)$ mA respectively. The average power dissipated in one cycle is *[2024, 29 Jan Shift-II]*

(a) 25 W (b) 2.5 W (c) 5 W (d) 10 W

44 In a series *L-R* circuit with $X_L = R$, power factor is P_1. If a capacitor of capacitance C with $X_C = X_L$ is added to the circuit, the power factor becomes P_2. The ratio of P_1 to P_2 will be *[2023, 30 Jan Shift-I]*

(a) $1:1$ (b) $1:2$
(c) $1:3$ (d) $1:\sqrt{2}$

45 An alternating emf $E = 440 \sin 100\pi t$ is applied to a circuit containing an inductance of $\sqrt{2}/\pi$ H. If an AC ammeter is connected in the circuit, its reading will be *[2022, 29 July Shift-I]*

(a) 4.4 A (b) 1.55 A (c) 2.2 A (d) 3.11 A

46 A direct current of 4 A and an alternating current of peak value 4 A flow through resistance of 3 Ω and 2 Ω, respectively. The ratio of heat produced in the two resistances in same interval of time will be *[2022, 27 July Shift-I]*

(a) 3 : 2 (b) 3 : 1 (c) 3 : 4 (d) 4 : 3

47 An AC source rated 220 V, 50 Hz is connected to a resistor. The time taken by the current to change from its maximum to the rms value is *[2021, 18 March Shift-I]*

(a) 2.5 ms (b) 25 ms (c) 2.5 s (d) 0.25 ms

48 What happens to the inductive reactance and the current in a purely inductive circuit, if the frequency is halved? *[2021, 17 March Shift-II]*

(a) Both inductive reactance and current will be halved.
(b) Inductive reactance will be halved and current will be doubled.
(c) Inductive reactance will be doubled and current will be halved.
(d) Both inductive reactance and current will be doubled.

49 Figure shows a circuit that contains four identical resistors with resistance $R = 2.0\ \Omega$, two identical inductors with inductance $L = 2.0$ mH and an ideal battery with electromotive force $E = 9$ V. The current i just after the switch S is closed will be *[2021, 24 Feb Shift-II]*

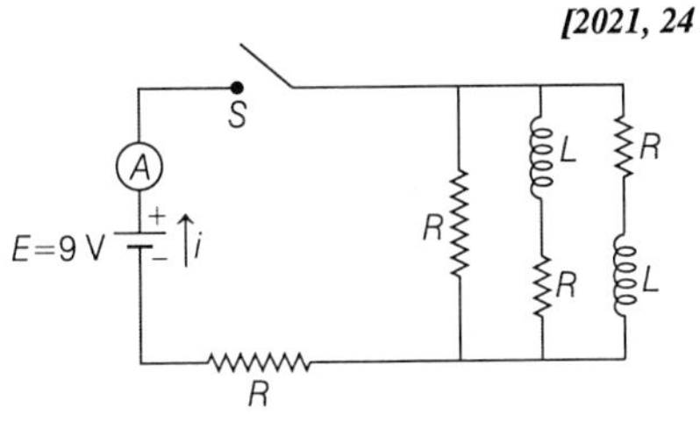

(a) 2.25 A (b) 3.0 A
(c) 3.37 A (d) 9 A

50 In *L-C* circuit, the inductance $L = 40$ mH and capacitance $C = 100\ \mu$F. If a voltage $V(t) = 10 \sin(314t)$ is applied to the circuit, the current in the circuit is given as *[2020, 9 Jan Shift-II]*

(a) $0.52 \cos 314t$ (b) $0.52 \sin 314t$
(c) $10 \cos 314t$ (d) $5.2 \cos 314t$

51 An *L-C-R* circuit behaves like a damped harmonic oscillator. Comparing it with a physical spring-mass damped oscillator having damping constant b, the correct equivalence would be *[2020, 7 Jan Shift-I]*

(a) $L \leftrightarrow \frac{1}{b}, C \leftrightarrow \frac{1}{m}, R \leftrightarrow \frac{1}{k}$ (b) $L \leftrightarrow m, C \leftrightarrow \frac{1}{k}, R \leftrightarrow b$
(c) $L \leftrightarrow m, C \leftrightarrow k, R \leftrightarrow b$ (d) $L \leftrightarrow k, C \leftrightarrow b, R \leftrightarrow m$

52 In an AC circuit, the instantaneous emf and current are given by $e = 100 \sin 30t$, $i = 20 \sin\left(30t - \frac{\pi}{4}\right)$.

In one cycle of AC, the average power consumed by the circuit and the wattless current are, respectively *[JEE Main 2018]*

(a) 50 , 10 (b) $\frac{1000}{\sqrt{2}}$, 10

(c) $\frac{50}{\sqrt{2}}$, 0 (d) 50 , 0

53 A metal wire of linear mass density of 9.8 g/m is stretched with a tension of 10 kg-wt between two rigid supports 1 m apart. The wire passes at its middle point between the poles of a permanent magnet and it vibrates in resonance when carrying an alternating current of frequency n. The frequency n of the alternating source is *[AIEEE 2003]*

(a) 50 Hz (b) 100 Hz (c) 200 Hz (d) 25 Hz

Level 3

54. A coil of negligible resistance is connected in series with 90 Ω resistor across 120 V, 60 Hz supply. A voltmeter reads 36 V across resistance. Inductance of the coil is *[2024, 8 April Shift-II]*

(a) 0.76 H (b) 2.86 H
(c) 0.91 H (d) 0.286 H

55 An alternating voltage $V(t) = 220 \sin 100\pi t$ volt is applied to a purely resistive load of 50 Ω. The time taken for the current to rise from half of the peak value to the peak value is *[2024, 30 Jan Shift-II]*

(a) 2.2 ms (b) 5 ms (c) 7.2 ms (d) 3.3 ms

56 A series L-R circuit connected with an AC source $E = (25 \sin 1000\,t)$ V has a power factor of $1/\sqrt{2}$. If the source of emf is changed to $E = (20 \sin 2000\,t)$ V, the new power factor of the circuit will be *[2024, 30 Jan Shift-I]*

(a) $\frac{1}{\sqrt{3}}$ (b) $\frac{1}{\sqrt{7}}$ (c) $\frac{1}{\sqrt{2}}$ (d) $\frac{1}{\sqrt{5}}$

57 A circuit element X when connected to an AC supply of peak voltage 100 V gives a peak current of 5 A which is in phase with the voltage. A second element Y when connected to the same AC supply also gives the same value of peak current which lags behind the voltage by $\frac{\pi}{2}$. If X and Y are connected in series to the same supply, what will be the rms value of the current in ampere ? *[2022, 29 July Shift-II]*

(a) $\frac{10}{\sqrt{2}}$ (b) $\frac{5}{\sqrt{2}}$

(c) $5\sqrt{2}$ (d) $\frac{5}{2}$

58 A resistance of 40 Ω is connected to a source of alternating current rated 220 V, 50 Hz. Find the time taken by the current to change from its maximum value to the rms value *[2022, 24 June Shift-I]*

(a) 2.5 ms (b) 1.25 ms
(c) 2.5 s (d) 0.25 s

59 In the given circuit the AC source has $\omega = 100$ rad s^{-1}. Considering the inductor and capacitor to be ideal, what will be the current I flowing through the circuit? *[2021, 26 Aug Shift-II]*

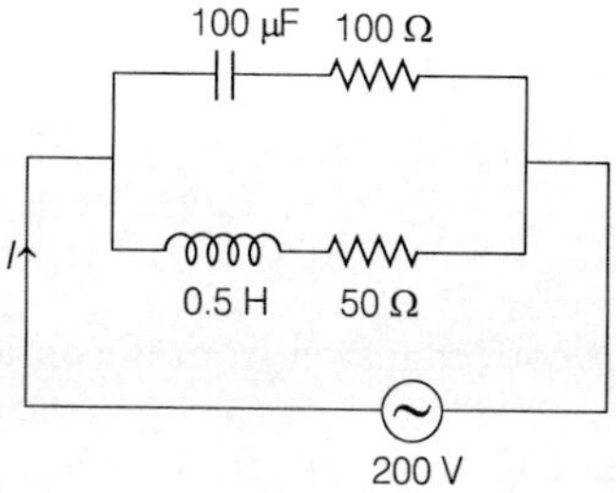

(a) 5.9 A (b) 4.24 A (c) 0.94 A (d) 6 A

Imp. Average Power

60 A series L-C-R circuit driven by 300 V at a frequency of 50 Hz contains a resistance $R = 3$ kΩ, an inductor of inductive reactance $X_L = 250\pi\,\Omega$ and an unknown capacitor. The value of capacitance to maximise the average power should be (Take, $\pi^2 = 10$) *[2021, 26 Aug Shift-I]*

(a) 4 μF (b) 25 μF (c) 400 μF (d) 40 μF

61 A 0.07 H inductor and a 12 Ω resistor are connected in series to a 220 V, 50 Hz AC source. The approximate current in the circuit and the phase angle between current and source voltage are, respectively. [Take, π as $\frac{22}{7}$] *[2021, 27 July Shift-I]*

(a) 8.8 A and $\tan^{-1}\left(\frac{11}{6}\right)$ (b) 88 A and $\tan^{-1}\left(\frac{11}{6}\right)$

(c) 0.88 A and $\tan^{-1}\left(\frac{11}{6}\right)$ (d) 8.8 A and $\tan^{-1}\left(\frac{6}{11}\right)$

Imp. Characteristic of L-C-R Series Circuit

62 For a series L-C-R circuit with $R = 100\,\Omega$, $L = 0.5$ mH and $C = 0.1$ pF connected across 220 V-50 Hz AC supply, the phase angle between current and supplied voltage and the nature of the circuit is *[2021, 20 July Shift-II]*

(a) 0°, resistive circuit
(b) ≈ 90°, predominantly inductive circuit
(c) 0° resonance circuit
(d) ≈ 90°, predominantly capacitive circuit

63 AC voltage $V(t) = 20 \sin \omega t$ V of frequency 50 Hz is applied to a parallel plate capacitor. The separation between the plates is 2 mm and the area is 1 m^2. The amplitude of the oscillating displacement current for the applied AC voltage is …… .
(Take, $\varepsilon_0 = 8.85 \times 10^{-12} \ C^2/N\text{-}m^2$) *[2021, 20 July Shift-I]*

(a) 21.14 μA (b) 83.37 μA
(c) 27.79 μA (d) 55.58 μA

64 An AC current is given by $I = I_1 \sin \omega t + I_2 \cos \omega t$. A hot wire ammeter will give a reading *[2021, 17 March Shift-I]*

(a) $\sqrt{\dfrac{I_1^2 - I_2^2}{2}}$ (b) $\sqrt{\dfrac{I_1^2 + I_2^2}{2}}$ (c) $\dfrac{I_1 + I_2}{\sqrt{2}}$ (d) $\dfrac{I_1 + I_2}{2\sqrt{2}}$

65 Find the peak current and resonant frequency of the following circuit (as shown in figure). *[2021, 26 Feb Shift-II]*

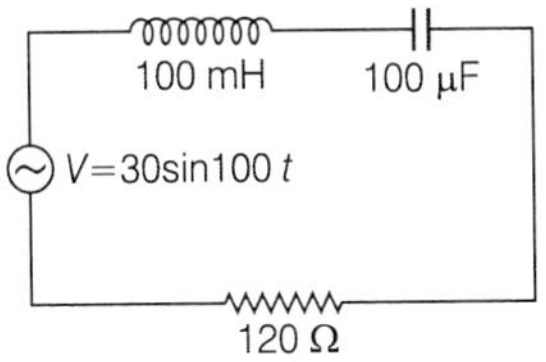

(a) 0.2 A and 50 Hz (b) 0.2 A and 100 Hz
(c) 2 A and 100 Hz (d) 2 A and 50 Hz

66 The angular frequency of alternating current in an L-C-R circuit is 100 rad/s. The components connected are shown in the figure. Find the value of inductance of the coil and capacity of condenser. *[2021, 25 Feb Shift-I]*

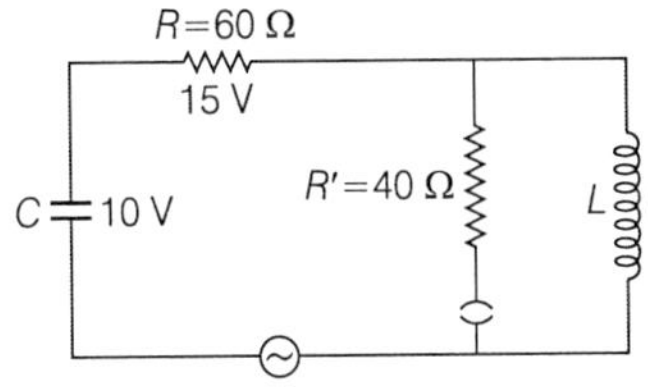

(a) 0.8 H and 150 μF (b) 0.8 H and 250 μF
(c) 1.33 H and 250 μF (d) 1.33 H and 150 μF

67 The current i at time $t = 0$ and $t = \infty$, respectively for the given circuit is *[2021, 25 Feb Shift-I]*

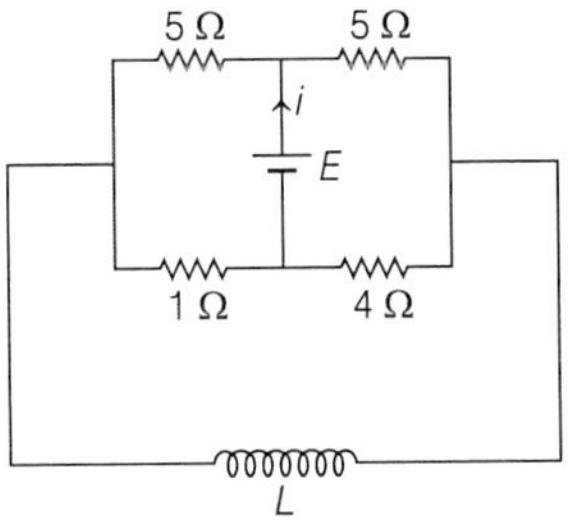

(a) $\dfrac{18E}{55}, \dfrac{5E}{18}$ (b) $\dfrac{10E}{33}, \dfrac{5E}{18}$ (c) $\dfrac{5E}{18}, \dfrac{18E}{55}$ (d) $\dfrac{5E}{18}, \dfrac{10E}{33}$

68 A circuit connected to an AC source of emf $e = e_0 \sin(100t)$ with t in seconds, gives a phase difference of $\pi/4$ between the emf e and current i. Which of the following circuits will exhibit this? *[2019, 8 April Shift-II]*

(a) R-C circuit with $R = 1 \text{ k}\Omega$ and $C = 1 \ \mu\text{F}$
(b) R-L circuit with $R = 1 \text{ k}\Omega$ and $L = 1$ mH
(c) R-C circuit with $R = 1 \text{ k}\Omega$ and $C = 10 \ \mu\text{F}$
(d) R-L circuit with $R = 1 \text{ k}\Omega$ and $L = 10$ mH

Numerical Value Type Questions

69 An inductor of reactance 100 Ω, a capacitor of reactance 50 Ω, and a resistor of resistance 50 Ω' are connected in series with an AC source of 10 V, 50 Hz. Average power dissipated by the circuit is ………… W. *[2025, 7 April Shift-II]*

70 For ac circuit shown in figure, $R = 100 \text{ k}\Omega$ and $C = 100$ pF and the phase difference between V_{in} and $(V_B - V_A)$ is 90°. The input signal frequency is 10^x rad/sec. where 'x' is

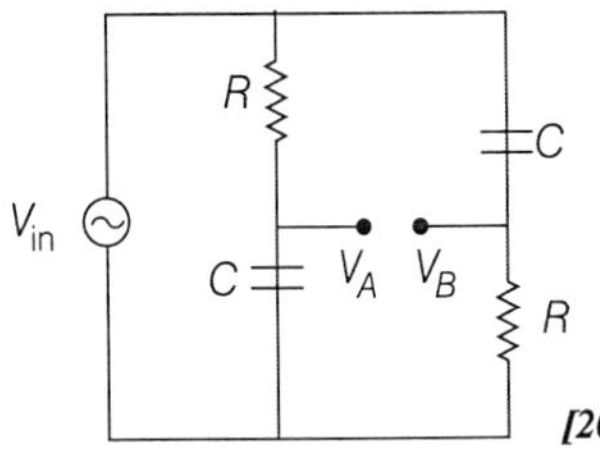

[2025, 7 April Shift-I]

71 An inductor of self-inductance 1 H is connected in series with a resistor of 100π Ω and an AC supply of 100π V, 50 Hz. Maximum current flowing in the circuit is …… A. *[2025, 4 April Shift-II]*

72 In a series L-C-R circuit, a resistor of 300 Ω, a capacitor of 25 nF and an inductor of 100 mH are used. For maximum current in the circuit, the angular frequency of the AC source is ….$\times 10^4$ rad s^{-1}. *[2025, 23 Jan Shift-II]*

73 A capacitor of reactance $4\sqrt{3}$ Ω and a resistor of resistance 4 Ω are connected in series with an AC source of peak value $8\sqrt{2}$ V. The power dissipation in the circuit is ………… W. *[2024, 9 April Shift-II]*

74 When a coil is connected across a 20 V DC supply, it draws a current of 5 A. When it is connected across 20 V, 50 Hz AC supply, it draws a current of 4 A. The self inductance of the coil is ………. mH. (Take, $\pi = 3$) *[2024, 9 April Shift-I]*

75 An alternating emf $E = 110\sqrt{2} \sin 100t$ V is applied to a capacitor of 2μF, the rms value of current in the circuit is ………. mA. *[2024, 8 April Shift-II]*

Imp. Resonance Frequency

76 For a given series *L-C-R* circuit it is found that maximum current is drawn when value of variable capacitance is 2.5 nF. If resistance of 200 Ω and 100 mH inductor is being used in the given circuit. The frequency of AC source is .... $\times 10^3$ Hz (Given, $\pi^2 = 10$) **[2024, 6 April Shift-II]**

77 When a DC voltage of 100 V is applied to an inductor, a DC current of 5A flows through it. When an AC voltage of 200V peak value is connected to inductor, its inductive reactance is found to be $20\sqrt{3}\Omega$. The power dissipated in the circuit is .... W. **[2024, 6 April Shift-I]**

78 An AC source is connected in given series *L–C–R* circuit. The rms potential difference across the capacitor of 20 μF is .... V. **[2024, 5 April Shift-I]**

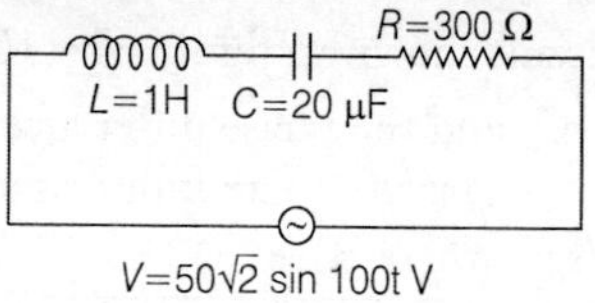

79 A alternating current at any instant is given by $i = \left[6 + \sqrt{56}\sin\left(100\pi t + \frac{\pi}{3}\right)\right]$ A. The rms value of the current is...... A. **[2024, 4 April Shift-I]**

Imp. Power Factor

80 A series *L-C-R* circuit with $L = \frac{100}{\pi}$ mH, $C = \frac{10^{-3}}{\pi}$ F and $R = 10\,\Omega$, is connected across an AC source of 220 V, 50 Hz supply. The power factor of the circuit would be .... **[2024, 27 Jan Shift-II]**

81 A series *L-C-R* circuit is connected to an AC source of 220 V, 50 Hz. The circuit contain a resistance $R = 100\,\Omega$ and an inductor of inductive reactance $X_L = 79.6\,\Omega$. The capacitance of the capacitor needed to maximise the average rate at which energy is supplied will be .... μF. **[2023, 1 Feb Shift-I]**

82 A series *L-C-R* circuit consists of $R = 80\,\Omega$, $X_L = 100\,\Omega$, and $X_C = 40\,\Omega$. The input voltage is $2500\cos(100\,\pi t)$ V. The amplitude of current, in the circuit is .... A. **[2023, 31 Jan Shift-II]**

83 An inductor of 0.5 mH, a capacitor of 20 μF and resistance of 20 Ω are connected in series with a 220 V AC source. If the current is in phase with the emf, the amplitude of current of the circuit is $\sqrt{x}$ A. The value of x is .... **[2023, 31 Jan Shift-I]**

84 An inductor of inductance 2 μH is connected in series with a resistance, a variable capacitor and an AC source of frequency 7 kHz. The value of capacitance for which maximum current is drawn into the circuit is $\frac{1}{x}$ F, where the value of x is .... $\left(\text{Take}, \pi = \frac{22}{7}\right)$ **[2023, 29 Jan Shift-II]**

85 A series *L-C-R* circuit is connected to an AC source of 220 V, 50 Hz. The circuit contains a resistance $R = 80\,\Omega$, an inductor of inductive reactance $X_L = 70\Omega$, and a capacitor of capacitive reactance $X_C = 130\Omega$. The power factor of circuit is $x / 10$. The value of x is .... **[2023, 25 Jan Shift-II]**

86 An *L-C-R* series circuit of capacitance 62.5 nF and resistance of 50 Ω, is connected to an AC source of frequency 2.0 kHz. For maximum value of amplitude of current in circuit, the value of inductance is .... mH. (Take, $\pi^2 = 10$) **[2023, 25 Jan Shift-I]**

87 To light a 50 W, 100 V lamp is connected in series with a capacitor of capacitance $\frac{50}{\pi\sqrt{x}}$ μF with 200 V, 50Hz AC source. The value of x will be .... **[2022, 27 July Shift-I]**

88 An inductor of 0.5 mH, a capacitor of 200 μF and a resistor of 2Ω are connected in series with a 220 V AC source. If the current is in phase with the emf, the frequency of AC source will be ...... $\times 10^2$ Hz. **[2022, 29 June Shift-II]**

89 In the given circuit, the magnitude of V_L and V_C are twice that of V_R. Given that $f = 50$ Hz, the inductance of the coil is $1/K\pi$ mH. The value of K is ....

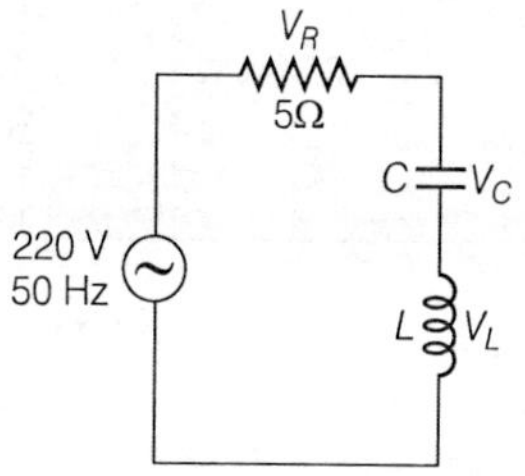

[2022, 28 June Shift-II]

90 A telegraph line of length 100 km has a capacity of 0.01 μF/km and it carries an alternating current at 0.5 kilo cycle per second. If minimum impedance is required, then the value of the inductance that needs to be introduced in series is ...... mH. (If, $\pi = \sqrt{10}$) **[2022, 28 June Shift-I]**

91 A 110 V, 50 Hz, AC source is connected in the circuit (as shown in figure). The current through the resistance 55Ω at resonance in the circuit, will be .... A. **[2022, 26 June Shift-I]**

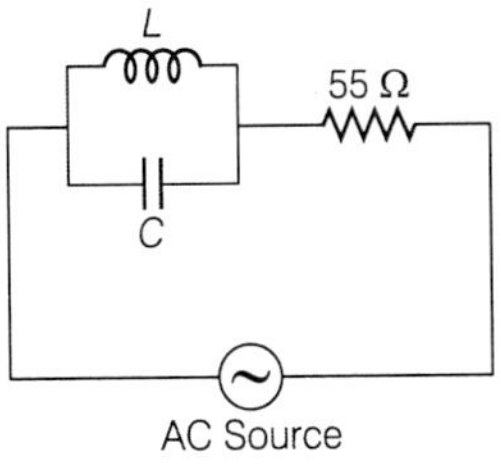

92 In a series L-C-R circuit, the inductance, capacitance and resistance are $L = 100$ mH, $C = 100$ μF and $R = 10\Omega$, respectively. They are connected to an AC source of voltage 220V and frequency of 50 Hz. The approximate value of current in the circuit will be .... A. *[2022, 25 June Shift-II]*

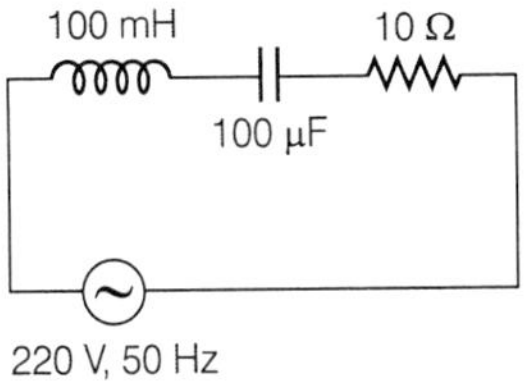

93 As shown in the figure an inductor of inductance 200 mH is connected to an AC source of emf 220 V and frequency 50 Hz. The instantaneous voltage of the source is 0 V when the peak value of current is $\sqrt{a}/\pi$A. The value of a is .... *[2022, 24 June Shift-I]*

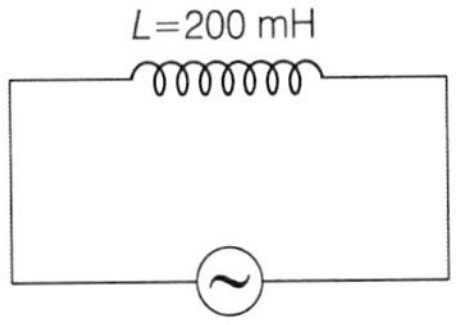

94 At very high frequencies, the effective impedance of the given circuit will be ....Ω. *[2021, 31 Aug Shift-II]*

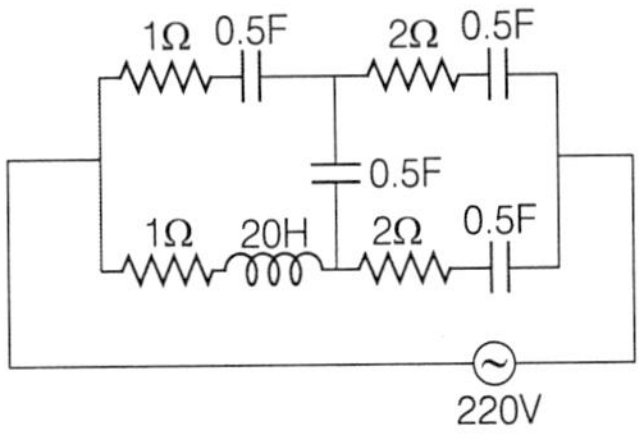

95 An AC circuit has an inductor and a resistor of resistance R in series, such that $X_L = 3R$. Now, a capacitor is added in series such that $X_C = 2R$. The ratio of new power factor with the old power factor of the circuit is $\sqrt{5}:x$ The value of x is .... *[2021, 27 Aug Shift-II]*

96 The alternating current is given by

$$i = \left\{\sqrt{42}\sin\left(\frac{2\pi}{T}t\right) + 10\right\} \text{A}$$

The rms value of this current is ...... A. *[2021, 27 Aug Shift-I]*

97 Two circuits are shown in the figures (a) and (b). At a frequency of .... rad/s, the average power dissipated in one cycle will be same in both the circuits. *[2021, 25 July Shift-II]*

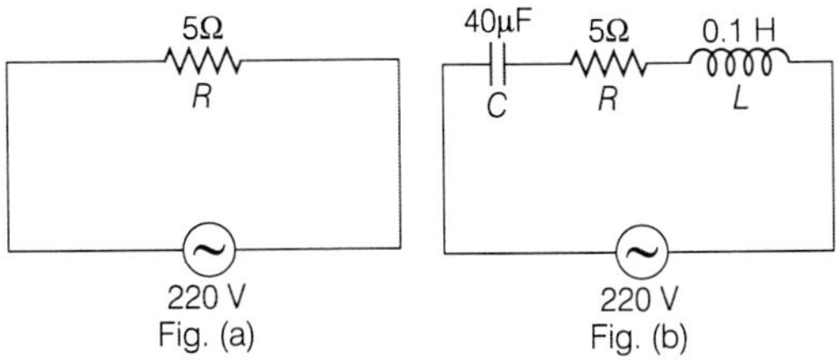

98 A series L-C-R circuit of $R = 5\,\Omega$, $L = 20$ mH and $C = 0.5$ μF is connected across an AC supply of 250 V, having variable frequency. The power dissipated at resonance condition is ...... $\times 10^2$ W. *[2021, 20 July Shift-II]*

99 In an L-C-R series circuit, an inductor 30 mH and a resistor 1Ω are connected to an AC source of angular frequency 300 rad/s. The value of capacitance for which, the current leads the voltage by 45° is $\frac{1}{x} \times 10^{-3}$ F. Then, the value of x is .... *[2021, 20 July Shift-I]*

100 A sinusoidal voltage of peak value 250 V is applied to a series L-C-R circuit, in which $R = 8\,\Omega$, $L = 24$ mH and $C = 60$ μF. The value of power dissipated at resonant condition is x kW. The value of x to the nearest integer is .... *[2021, 16 March Shift-I]*

101 A transmitting station releases waves of wavelength 960 m. A capacitor of 2.56 μF is used in the resonant circuit. The selfinductance of coil necessary for resonance is .... $\times 10^{-8}$ H. *[2021, 25 Feb Shift-I]*

Topic 2 Growth and Decay of Current, Energy Store in AC Circuit

Level 1

01

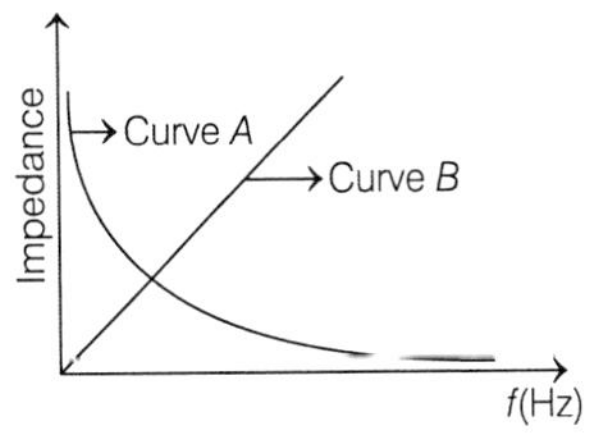

As per the given graph, choose the correct representation for curve A and curve B.
{Where, X_C = reactance of pure capacitive circuit connected with AC source
X_L = Reactance of pure inductive circuit connected with AC source
R = Impedance of pure resistive circuit connected

with AC source.
Z = Impedance of the *L-C-R* series circuit}
[2023, 11 April Shift-I]

(a) $A = X_L, B = R$ (b) $A = X_L, B = Z$
(c) $A = X_C, B = X_L$ (d) $A = X_C, B = R$

02 The current flowing through an ac circuit is given by $I = 5\sin(120\pi t)$A
How long will the current take to reach the peak value starting from zero? **[2022, 27 June Shift-I]**

(a) $\frac{1}{60}$ s (b) 60 s (c) $\frac{1}{120}$ s (d) $\frac{1}{240}$ s

03 A capacitor of capacitance $C = 1\,\mu F$ is suddenly connected to a battery of 100 V through a resistance $R = 100\,\Omega$. The time taken for the capacitor to be charged to get 50 V is **[2021, 27 July Shift-I]**

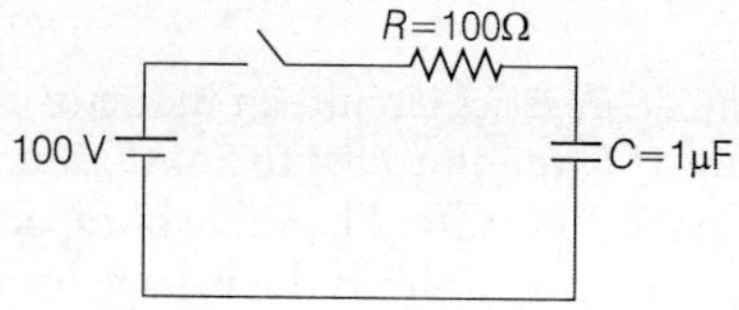

[Take, In 2 = 0.69]

(a) 1.44×10^{-4} s (b) 3.33×10^{-4} s
(c) 0.69×10^{-4} s (d) 0.30×10^{-4} s

04 An *R-C* circuit as shown in the figure is driven by an AC source generating a square wave. The output wave pattern monitored by CRO would look close to **[2021, 16 March Shift-I]**

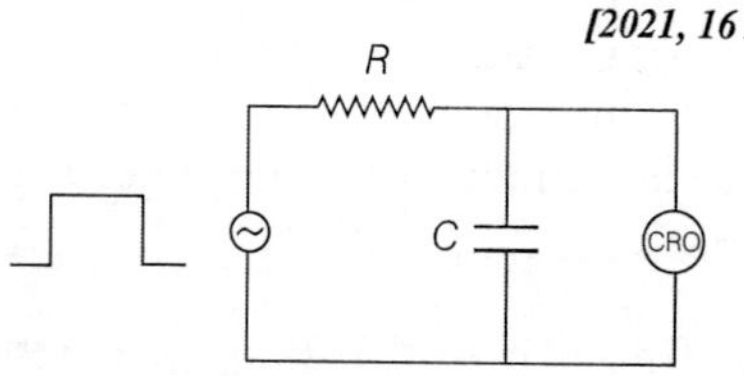

(a) 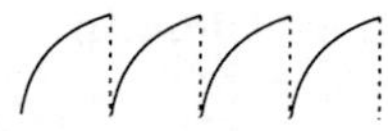 (b) 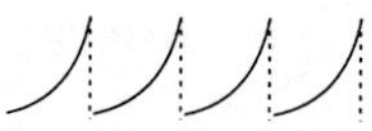

(c)  (d) 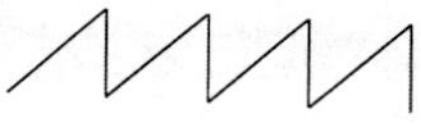

05 In an *L-C-R* circuit as shown below, both switches are open initially. Now, switch S_1 is closed and S_2 kept open (q is charge on the capacitor and $\tau = RC$ is capacitance time constant). Which of the following statement is correct? **[JEE Main 2013]**

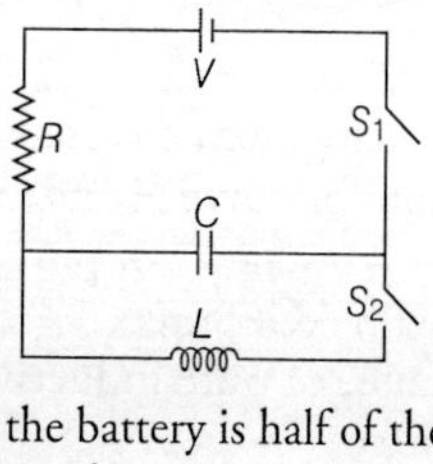

(a) Work done by the battery is half of the energy dissipated in the resistor
(b) At $t = \tau$, $q = CV/2$
(c) At $t = 2\tau$, $q = CV(1 - e^{-2})$
(d) At $t = \tau/2$, $q = CV(1 - e^{-1})$

Level 2

06 In *L-C-R* circcuit as shown below, both switches are open initally. Now switch S_1 is closed, S_2 kept open. (q is charge on the capacitor and $\tau = RC$ is capacitive time constant)

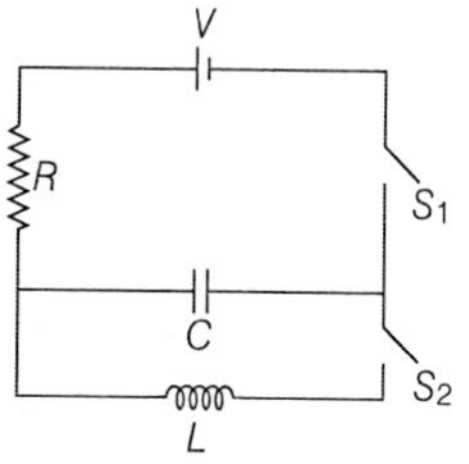

Which of the following statement is correct ? **[JEE Main 2013]**

(a) At $t = \tau/2q - CV(1 - e^{-1})$
(b) Work done by the battery is half of the energy dissipated in the resistor
(c) At $t = \tau, q - CV/2$
(d) At $t = 2\tau, q = CV(t - e^{-2})$

07 A resistor R and 2 μF capacitor in series is connected through a switch to 200 V direct supply. Across the capacitor is a neon bulb that lights up at 120 V. Calculate the value of R to make the bulb light up 5 s after the switch has been closed (Take, $\log_{10} 2.5 = 0.4$) **[AIEEE 2011]**

(a) $1.7 \times 10^5\,\Omega$ (b) $2.7 \times 10^6\,\Omega$
(c) $3.3 \times 10^7\,\Omega$ (d) $1.3 \times 10^4\,\Omega$

Level 3

08 An *L-C-R* circuit is equivalent to a damped pendulum. In an *L-C-R* circuit, the capacitor is charged to Q_0 and then connected to the L and R as shown below.

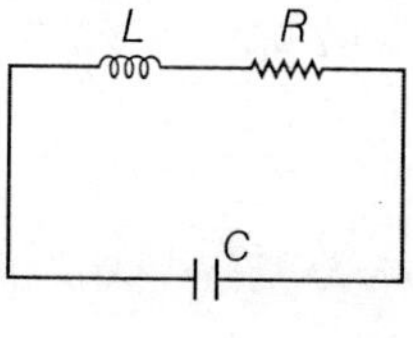

If a student plots graphs of the square of maximum charge Q_{max}^2 on the capacitor with time t for two different values L_1 and L_2 ($L_1 > L_2$) of L, then which of the following represents this graph correctly? (plots are schematic and not drawn to scale) **[JEE Main 2015]**

(a) 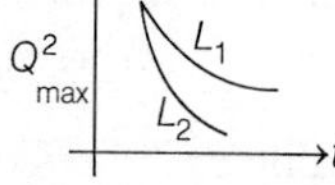

(b) 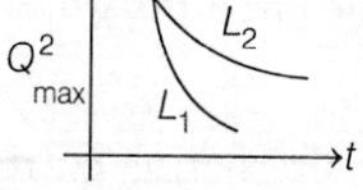

(c) 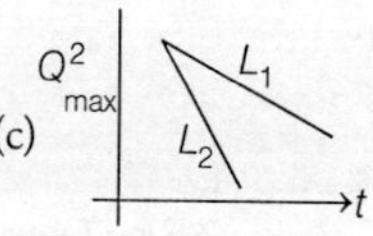

(d) 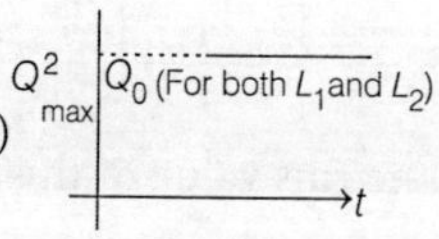

09 Let C be the capacitance of a capacitor discharging through a resistor R. Suppose t_1 is the time taken for the energy stored in the capacitor to reduce to half its initial value and t_2 is the time taken for the charge to reduce to one-fourth its initial value. Then, the ratio t_1 / t_2 will be *[AIEEE 2010]*

(a) 1 (b) $\frac{1}{2}$ (c) $\frac{1}{4}$ (d) 2

Numerical Value Type Questions

10 The electric field between the two parallel plates of a capacitor of 1.5 μF capacitance drops to one-third of its initial value in 6.6 μs when the plates are connected by a thin wire. The resistance of this wire is .... Ω. (Given, $\log 3 = 1.1$) *[2024, 5 April Shift-I]*

11 An oscillating L-C circuit consists of a 75 mH inductor and a 1.2 μF capacitor. If the maximum charge to the capacitor is 2.7 μC. The maximum current in the circuit will be .... mA. *[2023, 8 April Shift-I]*

Imp. Energy Stored in AC Circuit

12 A capacitor of capacitance 500μF is charged completely usinga DC supply of 100 V. It is now connected to an inductor of inductance 50 mH to form an L-C circuit. The maximum current in L-C circuit will be .... A. *[2022, 29 July, Shift-II]*

13 An AC source is connected to an inductance of 100 mH, a capacitance of 100μ F and a resistance of 120Ω as shown in figure. The time in which the resistance having a thermal capacity 2N/°C will get heated by 16°C is .... s. *[2022, 28 June Shift-I]*

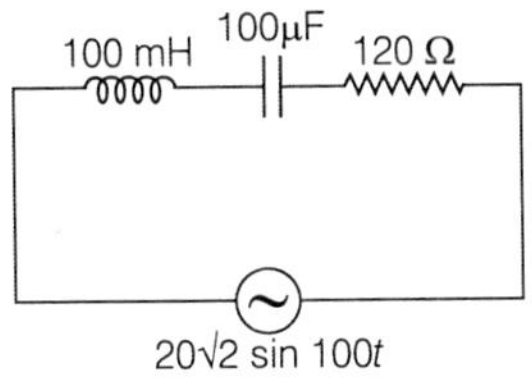

Topic 3 AC Generator and Transformer

Level 1

Imp. Transformer

01 Primary side of a transformer is connected to 230 V, 50 Hz supply. Turns ratio of primary to secondary winding is 10 : 1. Load resistance connected to secondary side is 46 Ω. The power consumed in it is *[2024, 27, Jan Shift-II]*

(a) 11.5 W (b) 10.0 W (c) 12.5 W (d) 12.0 W

02 Match List I with List II.

	List I		List II
A	AC generator	(i)	Detects the presence of current in the circuit
B	Galvanometer	(ii)	Converts mechanical energy into electrical energy
C	Transformer	(iii)	Works on th principle of resonance in AC circuit
D	Metal detector	(iv)	Changes an alternating voltage for smaller or greater value

Choose the correct answer fromthe options given below *[2022, 25, June Shift-I]*

(a) (A)-(ii), B-(i), C-(iv),D-(iii)
(b) (A)-(ii), B-(i), C-(iii),D-(iv)
(c) (A)-(iii), B-(iv), C-(ii),D-(i)
(d) (A)-(iii), B-(i), C-(ii),D-(iv)

03 For the given circuit, comment on the type of transformer used. *[2021, 16 March Shift-II]*

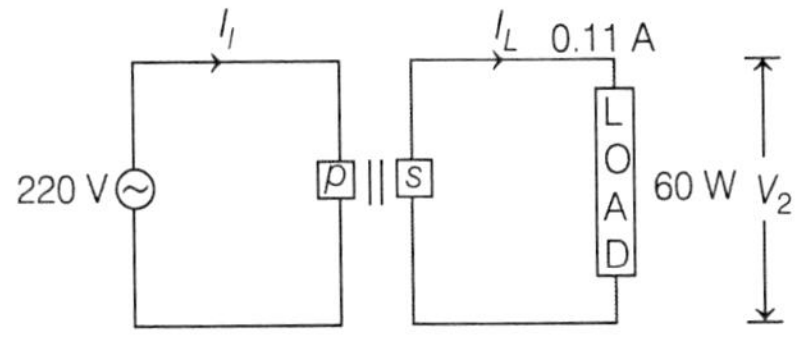

(a) Auxilliary transformer
(b) Auto transformer
(c) Step up transformer
(d) Step down transformer

04 In an AC generator, a coil with N turns, all of the same area A and total resistance R, rotates with frequency ω in a magnetic field B. The maximum value of emf generated in the coil is *[AIEEE 2006]*

(a) $NABR\,\omega$ (b) NAB
(c) $NABR$ (d) $NAB\,\omega$

05 The core of any transformer is laminated so as to

(a) reduce the energy loss due to eddy currents *[AIEEE 2003]*
(b) make it light weight
(c) make it robust and strong
(d) increase the secondary voltage

06 In a transformer, number of turns in the primary are 140 and that in the secondary are 280. If current in primary is 4 A, then that in the secondary is *[AIEEE 2002]*

(a) 4 A (b) 2 A (c) 6 A (d) 10 A

Level 2

07 A transformer has an efficiency of 80% and works at 10 V and 4 kW. If the secondary voltage is 240 V, then the current in the secondary coil is *[2024, 1 Feb Shift-II]*

(a) 15.1 A (b) 13.33 A
(c) 1.59 A (d) 1.33 A

08 Primary coil of a transformer is connected to 220 V AC. Primary and secondary turns of the transforms are 100 and 10 respectively. Secondary coil of transformer is connected to two series resistances shown in figure. The output voltage (V_0) is

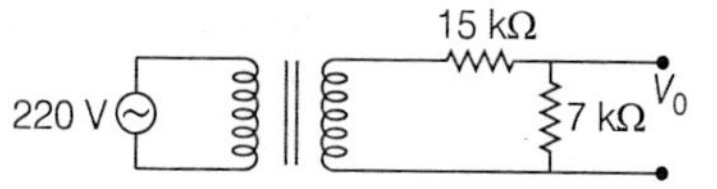

[2024, 30 Jan Shift-I]

(a) 15 V (b) 44 V
(c) 7 V (d) 22 V

09 A transformer operating at primary voltage 8 kV and secondary voltage 160 V serves a load of 80 kW. Assuming the transformer to be ideal with purely resistive load and working on unit power factor, the loads in the primary and secondary circuits would be *[2022, 28 July Shift-II]*

(a) 800 Ω and 1.06 Ω (b) 10 Ω and 500 Ω
(c) 800 Ω and 0.32 Ω (d) 1.06 Ω and 500 Ω

Imp. Efficiency of Transformer

10 An electrical power line, having a total resistance of 2Ω, delivers 1 kW at 220 V. The efficiency of the transmission line is approximately *[2020, 5 Sep Shift-I]*

(a) 91% (b) 85% (c) 96% (d) 72%

Level 3

11 A transformer consisting of 300 turns in the primary and 150 turns in the secondary gives output power of 2.2 kW. If the current in the secondary coil is 10 A, then the input voltage and current in the primary coil are *[2019, 10 April Shift-I]*

(a) 440 V and 5 A (b) 220 V and 20 A
(c) 220 V and 10 A (d) 440 V and 20 A

12 A power transmission line feeds input power at 2300 V to a stepdown transformer with its primary windings having 4000 turns. The output power is delivered at 230 V by the transformer. If the current in the primary of the transformer is 5A and its efficiency is 90%, the output current would be *[2019, 9 Jan Shift-II]*

(a) 45 A (b) 50 A
(c) 25 A (d) 35 A

Numerical Value Type Questions

13 A coil of 200 turns and area 0.20 m^2 is rotated at half a revolution per second and is placed in uniform magnetic field of 0.01 T perpendicular to axis of rotation of the coil. The maximum voltage generated in the coil is $\frac{2\pi}{\beta}$ V. The value of β is .... *[2024, 1 Feb Shift-II]*

14 An ideal transformer with purely resistive load operates at 12 kV on the primary side. It supplies electrical energy to a number of nearby houses at 120 V. The average rate of energy consumption in the houses served by the transformer is 60 kW. The value of resistive load R_S required in the secondary circuit will be .... mΩ. *[2023, 6 April Shift-I]*

15 A square shaped coil of area 70 cm^2 having 600 turns rotates in a magnetic field of 0.4 Wbm^{-2}, about an axis which is parallel to one of the side of the coil and perpendicular to the direction of field. If the coil completes 500 revolution in a minute. The instantaneous emf when the plane of the coil is inclined at 60° with the field, will be .... V. $\left(\text{Take}, \pi = \frac{22}{17}\right)$ *[2023, 1 Feb Shift-II]*

16 In an AC generator, a rectangular coil of 100 turns each having area 14×10^{-2} m^2 is rotated at 360 rev/min about an axis perpendicular to a uniform magnetic field of magnitude 3.0 T. The maximum value of the emf produced will be .... V. $\left(\text{Take}, \pi = \frac{22}{7}\right)$ *[2023, 30 Jan Shift-I]*

17 A common transistor radio set requires 12 V (DC) for its operation. The DC source is constructed by using a transformer and a rectifier circuit, which are operated at 220 V(AC) on standard domestic AC supply. The number of turns of secondary coil are 24, then the number of turns of primary are ... . *[2021, 24 Feb Shift-I]*

Key Index

Topic 1

1 (a)	**2** (d)	**3** (c)	**4** (d)	**5** (b)	**6** (d)	**7** (a)	**8** (d)	**9** (a)	**10** (c)
11 (d)	**12** (d)	**13** (d)	**14** (b)	**15** (c)	**16** (b)	**17** (c)	**18** (d)	**19** (c)	**20** (a)
21 (c)	**22** (d)	**23** (a)	**24** (d)	**25** (b)	**26** (b)	**27** (c)	**28** (a)	**29** (c)	**30** (c)
31 (c)	**32** (b)	**33** (d)	**34** (b)	**35** (b)	**36** (b)	**37** (b)	**38** (b)	**39** (c)	**40** (d)
41 (d)	**42** (d)	**43** (b)	**44** (d)	**45** (c)	**46** (b)	**47** (a)	**48** (b)	**49** (a)	**50** (a)
51 (b)	**52** (b)	**53** (a)	**54** (a)	**55** (d)	**56** (d)	**57** (d)	**58** (a)	**59** (*)	**60** (a)
61 (a)	**62** (d)	**63** (c)	**64** (b)	**65** (a)	**66** (b)	**67** (d)	**68** (c)	**69** (1)	**70** (5)
71 (1)	**72** (2)	**73** (4)	**74** (10)	**75** (22)	**76** (10)	**77** (250)	**78** (50)	**79** (8)	**80** (1)
81 (40)	**82** (25)	**83** (242)	**84** (3872)	**85** (8)	**86** (100)	**87** (3)	**88** (5)	**89** (0)	**90** (100)
91 (0)	**92** (22)	**93** (242)	**94** (2)	**95** (1)	**96** (11)	**97** (500)	**98** (125)	**99** (3)	**100** (4)
101 (10)									

Topic 2

1 (c)	**2** (d)	**3** (c)	**4** (c)	**5** (c)	**6** (d)	**7** (b)	**8** (a)	**9** (c)	**10** (4)
11 (9)	**12** (10)	**13** (15)							

Topic 3

1 (a)	**2** (a)	**3** (c)	**4** (d)	**5** (a)	**6** (b)	**7** (b)	**8** (c)	**9** (c)	**10** (c)
11 (a)	**12** (a)	**13** (5)	**14** (240)	**15** (44)	**16** (1584)	**17** (440)			

Detailed **Explanations**

Topic 1

01. *(a)* $i = 5\sqrt{2} + 10\cos\left(650\pi t + \frac{\pi}{6}\right)$

$$i^2 = 50 + 100\cos^2\left(650\pi t + \frac{\pi}{6}\right)$$
$$+ (2)(5\sqrt{2})(10)\cos\left(650\pi t + \frac{\pi}{6}\right)$$

As we know, average value of $\sin^2\theta$ or $\cos^2\theta = \frac{1}{2}$ amd average value of $\sin\theta$ or $\cos\theta = 0$

$$<i^2> = 50 + \frac{100}{2} + 0$$
$$<i^2> = 100$$
$$<i> = 10\text{ A}$$

02. *(d)* $i = 100\sqrt{2}\sin(100\pi t)$, $i_0 = 100\sqrt{2}$,

$$\omega = 100\pi$$
$$i_{\text{rms}} = \frac{i_0}{\sqrt{2}}$$
$$i_{\text{rms}} = \frac{100\sqrt{2}}{\sqrt{2}} = 100\text{ A}$$
$$f = \frac{\omega}{2\pi} = \frac{100\pi}{2\pi} = 50\text{ Hz}$$

03. *(c)* We know that,

$$P = V_{\text{rms}} \times I_{\text{rms}}$$
$$100 = 220 \times I_{\text{rms}}$$
$$\Rightarrow \quad I_{\text{rms}} = \frac{100}{220} = \frac{5}{11}$$
$$\therefore \quad I_0 = \sqrt{2} \times I_{\text{rms}}$$
$$= \sqrt{2} \times \frac{5}{11}$$
$$= 0.64\text{ A}$$

04. *(d)* Initially $I_0 = \varepsilon_m / R$

R become twice of initial value

$R \to 2R$

Finally $I_0' = \frac{\varepsilon_m}{2R} = \frac{I_0}{2}$

05. *(b)* Statement given in A, B and D are correct, while C is incorrect.

When instantaneous current is zero and instantaneous voltage is maximum, this means that the phase difference between current and voltage is $\frac{\pi}{2}$.

This can be achieved in pure inductor, pure capacitor and combination of pure inductor and capacitor.

In resistor circuit phase difference is zero.

06. *(d)* **Statement I – False**

Reason - Electrical resonance occurs in AC circuit when capacitive and inductive reactance are equal ($X_L = X_C$), causing the impedance of circuit minimum. This happen in series or parallel *L-R-C* circuit.

Statement II – False

An AC circuit containing a pure capacitor or inductor will have a power factor 0, not a non zero power factor. Power factor of a capacitor is – 1 and + 1 for inductor, but only considering the reactive components power factor is zero.

07. *(a)* As we know that the impedance of an *L-C-R* circuit is,

$$Z = \sqrt{R^2 + (X_L - X_C)^2}$$
$$= \sqrt{R^2 + \left(2\pi fL - \frac{1}{2\pi fC}\right)^2}$$

As, f increases, Z decreases and hence current is increases.

At $f = f_r$ (resonance), Z is minimum and current is maximum. After $f > f_r$, Z increases and current decreases.

Power factor

$$\cos\phi = \frac{\text{Resistance } (R)}{\text{Total impedance } (Z)}$$

At resonance, $\cos\phi = \frac{R}{R} = 1$

Thus, both Statements are true.

08. *(d)* Power dissipated in an *L-C-R* circuit is $P = V_{\text{rms}} I_{\text{rms}} \cos\phi$

Now, at resonance $\phi = 0°$

So, $\cos\phi = 1$ and power dissipation is maximum at resonance.

09. *(a)* Given, alternating AC source

$$E = 36\sin(120\pi t)\text{ V}$$
$$C = 150\mu\text{F}$$

using $Q = CV \Rightarrow Q = CE_0 \sin\omega t$

Current, $i = \frac{dQ}{dt} = CE_0\omega\cos\omega t$

Maximum value of current $i_0 = CE_0\omega$

or $\quad i_0 = 150 \times 10^{-6} \times 36 \times 120\pi$

$$= 2.03\text{ A} \approx 2\text{A}$$

10. *(c)* Given, $X_L = 200\,\Omega$,

$$X_C = 100\,\Omega,\ R = 100\,\Omega$$
$$V_{\text{rms}} = 200\sqrt{2}\text{ V}$$

Impedance,

$$Z = \sqrt{R^2 + (X_L - X_C)^2}$$
$$= \sqrt{(100)^2 + (200 - 100)^2}$$
$$= \sqrt{100^2 + 100^2} = 100\sqrt{2}\,\Omega$$
$$\therefore \quad I_{\text{rms}} = \frac{V_{\text{rms}}}{Z} = \frac{200\sqrt{2}}{100\sqrt{2}} = 2\text{ A}$$

11. *(d)* In circuit (b),

Inductive reactance

$$X_L = 2\pi fL$$
$$= 2\pi \times 50 \times 50 \times 10^{-3}\,\Omega$$
$$= 15.7\,\Omega$$

And capacitive reactance,

$$X_C = \frac{1}{2\pi fC} = \frac{1}{2\pi \times 50 \times 0.5 \times 10^{-6}}$$
$$= 6366\,\Omega$$

As $X_L \neq X_C$, there is no resonance in circuit (b). So, impedance of circuit (b) is always greater than (a).

$\Rightarrow I_{\text{rms}}$ is always smaller in circuit (b).

• Concept Enhancer

The reactance of inductor (X_L) is directly proportional to frequency whereas, the reactance of a capacitor (X_C) is inversely proportional to the frequency.

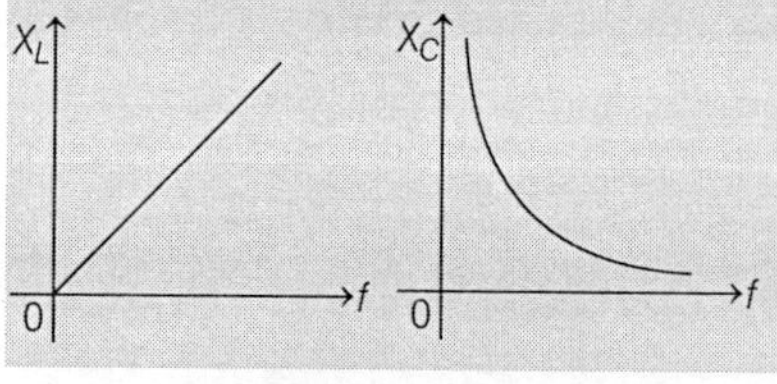

12. *(d)* In *L-C* oscillator, resonance frequency is given as,

$$\omega_0 = \frac{1}{\sqrt{LC}} \quad \ldots\text{(i)}$$

When $L' = 2L$ and $C' = 8C$,

then $\omega_0' = \frac{1}{\sqrt{L'C'}}$

$$= \frac{1}{\sqrt{2L \times 8C}}$$

$$= \frac{1}{4\sqrt{LC}}$$

$$= \frac{\omega_0}{4} \quad \text{[From Eq. (i)]}$$

$$= \left(\frac{1}{4}\right)\omega_0 = x\omega_0 \quad \text{(Given)}$$

$$\Rightarrow \quad x = \frac{1}{4}$$

13. *(d)* Voltage amplitude across the inductor is given by

$$V_0 = i_0 X_L$$
$$= i_0(\omega L)$$
$$= (5)(49\pi)(30 \times 10^{-3})$$
$$= 23.1 \text{V}$$

Voltage will lead current by 90°. [in inductive circuit]

Therefore, the equation for the voltage across the inductor will be

$$V = 23.1 \sin(49\pi t + 60°)$$

14. *(b)* When a person walks through a metal detector carrying a metal object in his pocket, then impedance of AC circuit changes. It results in appereciable change in the AC current due to its resonance condition, which causes production of sound in the form of a beep.

Thus, metal detector works on the principle of resonance in AC circuits.

15. *(c)* Resonant frequency of an *L-C-R* circuit is given by

$$\omega_r = \frac{1}{\sqrt{LC}}$$

So, resonance frequency does not depend on resistance of circuit and frequency of supply.

To increase resonance frequency, we have to reduce inductance or capacitance of circuit.

As capacitance reduces when capacitors are connected in series (i.e. $\frac{1}{C_{series}} = \frac{1}{C_1} + \frac{1}{C_2}$), we can add another capacitor in series to increase resonant frequency.

> **• Concept Enhancer**
>
> At resonance in *L-C-R* series circuit, the amplitude of current will be maximum.
>
> i.e. $I_0 = \frac{E_0}{\sqrt{R^2 + \left(\omega L - \frac{1}{\omega C}\right)^2}}$
>
> For I_0 to be maximum,
>
> $\omega L = \frac{1}{\omega C}$ or $X_L = X_C$
>
> Hence, at resonance, $X_L = X_C$

16. *(b)* For wattless current,

$$P_{avg} = 0$$
$$\Rightarrow \quad \cos\phi = 0$$
$$\phi = \pi/2$$

We know that, for purely inductive circuit, phase difference between alternating current and alternating voltage is $\pi/2$ because alternating voltage in purely inductive circuit leads $\frac{\pi}{2}$ angle from alternating current.

> **• Concept Enhancer**
>
> In an AC circuit, $P_{ac} = V_{rms} \cdot I_{rms} \cos\phi$
>
> or $P_{ac} = \frac{1}{2} I_0 V_0 \cos\phi$

17. *(c)* An *L-C* resonance circuit have zero reactance since, $X_L = X_C$.

Therefore, it is possible that the circuit contains a capacitor and an inductor.

In purely inductive or capacitive circuit, voltage lags or leads the current by 90°.

So, power in both circuit over one full periodic cycle will be equal to zero.

Thus, Statement I is true but Statement II is false.

18. *(d)* Given,

Resistance, $R = 100\,\Omega$

Capacitance, $C = 0.1\,\mu\text{F}$

Voltage, $V = 250$ V

Resonant frequency, $f_0 = 60$ Hz

We know that,

Resonant frequency, $f_0 = \frac{1}{2\pi\sqrt{LC}}$

$$L = \frac{1}{4\pi^2 f_0^2 C} = \frac{1}{4\pi^2 \times (60)^2 \times 0.1 \times 10^{-6}}$$

$$= \frac{10^5}{4\pi^2 \times 36} = 70.3 \text{ H}$$

19. *(c)* Given, input alternating emf

$$(E_g) = E_{g0} \sin \omega t$$

As we know that, capacitive current leads voltage by $\frac{\pi}{2}$ rad.

$\therefore$ Phasor diagram will be

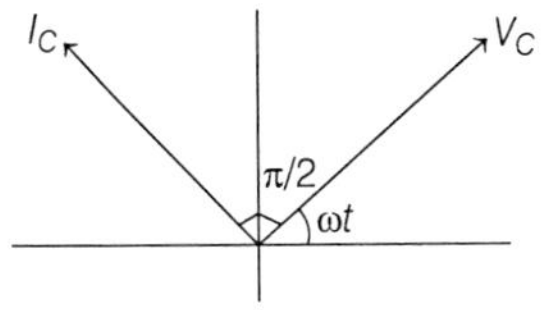

where, V_C and I_C are voltage and current in capacitive circuit.

20. *(a)* (A) $\omega L > \frac{1}{\omega C}$

Since, inductive resistance > capacitive resistance, hence inductive current lag behind applied voltage.

Hence, A → (ii)

(B) $\omega L = \frac{1}{\omega C}$

Since, inductive resistance = capacitive resistance

Hence, this circuit behaves like a *L-C-R* series resonant circuit. Therefore, $Z = R$ and phase difference = 0.

$\therefore$ Circuit will be purely resistive and resistive current remains in phase with voltage.

Hence, B → (i)

(C) $\omega L < \frac{1}{\omega C}$

Since, inductive resistance < capacitive resistance, hence capacitive current lead with respect to applied voltage.

Hence, C → (iv)

(D) As we know that, at resonance impedance $(Z) = R$ (resistance)

$\therefore$ Circuit current will be maximum.

Hence D → (iii)

Hence, option (a) is the correct.

21. *(c)* Given, inductive reactance, $X_L = 10\,\Omega$

Capacitive reactance, $X_C = 4\,\Omega$

Resistance, $R = 6\,\Omega$

Therefore, power factor = $\cos\phi$

$$= \frac{R}{Z} = \frac{R}{\sqrt{R^2 + (X_L - X_C)^2}}$$

$$= \frac{6}{\sqrt{6^2 + (10-4)^2}} = \frac{1}{\sqrt{2}}$$

22. *(d)* A. In a purely resistive AC circuit, the phase difference between the current and voltage is zero.

B. In a purely inductive AC circuit, the current lags the voltage, so the phase difference between the current and voltage is $\pi/2$.

C. In a purely capacitive AC circuit the current leads the voltage, so the phase difference between the current and voltage is $\pi/2$.

D. The phase difference between current and voltage in an L-C-R series circuit is $\phi = \tan^{-1}\frac{(X_C - X_L)}{R}$

$\therefore$ The correct match is

A-(2), B-(3), C-(1), D-(4).

23. *(a)* Given, $i = i_1 \sin \omega t + i_2 \cos \omega t$

Let I_{rms} be the rms current.

$$\therefore I_{\text{rms}} = \left(\frac{i_1^2 + i_2^2}{2}\right)^{1/2}$$

$$\Rightarrow I_{\text{rms}} = \frac{1}{\sqrt{2}}(i_1^2 + i_2^2)^{1/2}$$

24. *(d)* Long time after switch is closed, the inductor will be idle so, the equivalent diagram will be as below

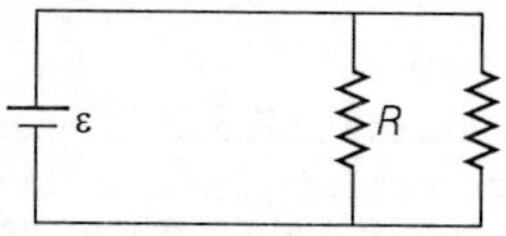

Current is given by

$$I = \frac{\varepsilon}{\left(\frac{R \times R}{R+R}\right)} = \frac{2\varepsilon}{R} = \frac{2 \times 15}{5} = 6\text{A}$$

25. *(b)* In a pure inductive circuit current always lags behind the emf by phase angle $\pi/2$.

If $V(t) = V_0 \sin \omega t$

then $I = I_0 \sin\left(\omega t - \frac{\pi}{2}\right)$

Now, given $V(t) = 100 \sin (500t)$

and $I_0 = \frac{E_0}{\omega L} = \frac{100}{500 \times 0.02}$ $[\because L = 0.02\text{ H}]$

$$I_0 = 10 \sin\left(500t - \frac{\pi}{2}\right)$$

$$\Rightarrow I_0 = -10\cos(500t)$$

26. *(b)* For given circuit, current is lagging the voltage by $\frac{\pi}{2}$, so circuit is purely inductive and there is no power consumption in the circuit. The work done by battery is stored as magnetic energy in the inductor.

27. *(c)* At resonance, $\omega L = \frac{1}{\omega C}$

Current flowing through the circuit,

$$I = \frac{V_R}{R} = \frac{100}{1000} = 0.1\text{ A}$$

So, voltage across L is given by

$$V_L = I X_L = I\omega L$$

But $\omega L = 1/\omega C$

$$\therefore V_L = \frac{I}{\omega C} = V_C$$

$$= \frac{0.1}{200 \times 2 \times 10^{-6}} = 250\text{ V}$$

28. *(a)* Power factor,

$$P = \cos\phi = \frac{R}{Z} = \frac{12}{15} = \frac{4}{5} = 0.8$$

29. *(c)*

(a) In a circuit having C alone, the voltage lags the current by $\pi/2$.

(b) In circuit containing R and L, the voltage leads the current by $\pi/2$.

(c) In L-C circuit, the phase difference between current and voltage can have any value between 0 to $\pi/2$ depending on the values of L and C.

(d) In a circuit containing L alone, the voltage leads the current by $\pi/2$.

Hence, option (c) is correct.

30. *(c)* The full cycle of alternating current consists of two half cycles. For one half, current is positive and for second half, current is negative. Therefore, for an AC cycle, the net value of current average out to zero. While the DC ammeter, read the average value. Hence, the alternating current cannot be measured by DC ammeter.

31. *(c)*

> **key idea** For resonance condition, apply
> LC = constant
> or $L_1C_1 = L_2C_2$

In the condition of resonance,

$$X_L = X_C \text{ or } \omega L = \frac{1}{\omega C} \quad \ldots\text{(i)}$$

Since, resonance frequency remains unchanged, so

$\sqrt{LC}$ = constant or LC = constant

$\therefore$ $L_1C_1 = L_2C_2$

or $L \times C = L_2 \times 2C$

or $L_2 = L/2$

32. *(b)* From the relation,

$$\tan\phi = \omega L/R$$

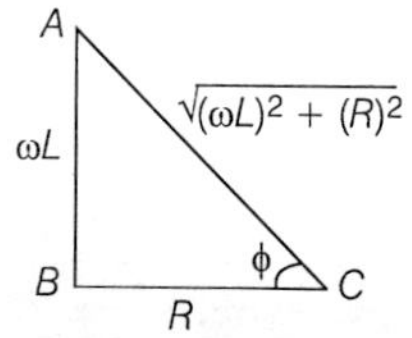

Power factor,

$$\cos\phi = \frac{1}{\sqrt{1+\tan^2\phi}}$$

$$= \frac{1}{\sqrt{1+(\omega L/R)^2}}$$

$$= \frac{R}{\sqrt{R^2 + \omega^2 L^2}}$$

> **• Concept Enhancer**
>
> From impedance Z we can draw impedance triangle.
>
> 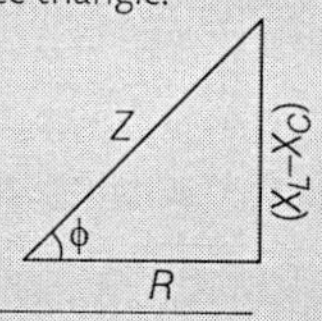
>
>
> i.e. $Z = \sqrt{R^2 + (X_L - X_C)^2}$
>
> $\tan\phi = \frac{X_L - X_C}{R}$
>
> $\therefore Z = R \cdot \sqrt{1+\tan^2\phi}$
>
> or Power factor, $\cos\phi = \frac{R}{Z} = \frac{1}{\sqrt{1+\tan^2\phi}}$

33. *(d)* Given current value is,

$$I = I_A \sin \omega t + I_B \cos \omega t$$

I_{rms} is given by

$$I_{\text{rms}} = \sqrt{\frac{\int_0^T I^2 dt}{\int_0^T dt}}$$

$$= \sqrt{\frac{\int I_A^2 \sin^2 \omega t dt + \int I_B^2 \cos^2 \omega t dt + 2\int I_A I_B \sin \omega t \cos \omega t dt}{\int dt}}$$

$$= \sqrt{\frac{I_A^2 \times \frac{T}{2} + I_B^2 \times \frac{T}{2} + I_A I_B \times 0}{T}}$$

$$= \sqrt{\frac{I_A^2 + I_B^2}{2}}$$

34. *(b)* Inductor in an AC circuit simply provides potential difference without consuming any energy. So, an inductor in the form of choke coil is connected in series with fluoroscent - mercury tube fitting to reduce the voltage across it and protect it. So, Assertion is true.
Voltage across tube is same as source voltage in absence of inductor.
Let it be V. On adding inductor,
Voltage across the inductor,

$$V_L = I\,X_L = \frac{V}{Z} X_L$$

New voltage across tube is

$$V_R' = \sqrt{V^2 - V_L^2}$$

$$= \sqrt{V^2 - \frac{V^2 X_L^2}{Z^2}} = V\sqrt{\frac{Z^2 - X_L^2}{Z^2}}$$

$$V_R' = \frac{VR}{Z} \Rightarrow \frac{V_R'}{V} = \frac{R}{Z} = \frac{R}{\sqrt{R^2 + X_L^2}}$$

Hence, Reason is true and explains the Assertion.

35. *(b)*

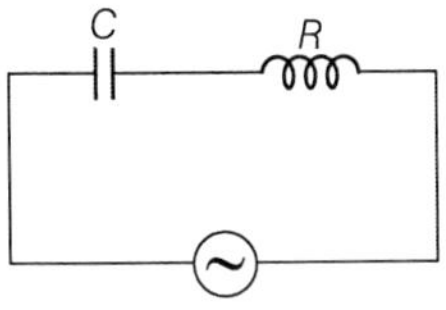

$$Z = \sqrt{R^2 + X_C^2}$$

$$\Rightarrow X_C = \frac{1}{\omega C}$$

Due to dielectric

$C\uparrow \Rightarrow X_C\downarrow \Rightarrow Z\downarrow$

So, current is increases is and thus bulb will glow more brighter.

36. *(b)* As *L-C-R* is given in resonance

$\therefore X_L = X_C$ and circuit impedance

$$Z = \sqrt{(X_L - X_C)^2 + R^2} = R$$

So, current amplitude

$$I_1 = \frac{V_0}{Z'} = \frac{V_0}{R} \text{ (Initially)}$$

When resistance is halved, current amplitude,

$$I_2 = \frac{V_0}{Z'} = \frac{V_0}{R/2} = 2\frac{V_0}{R} = 2I_1$$

So, current is doubled.

Time Saver Tech

In *L-C-R* series resonance circuit,

Current amplitude, $I_0 \propto \frac{1}{R}$

$$\therefore \quad \frac{I_{0_2}}{I_{0_1}} = \frac{R_1}{R_2} = \frac{R}{R/2} = 2 \Rightarrow I_{0_2} = 2I_{0_1}$$

37. *(b)* Given, *L-C-R* series circuit is at resonance,

$$\therefore \text{Impedance, } Z = \sqrt{R^2 + (X_L - X_C)^2}$$

$$Z_{min} = R$$

Thus, Current is maximum at resonance in *L-C-R* circuit.
For same supply voltage V.
Current in purely resistive circuit.

$$I' = V/R$$

and current in *L-C-R* series circuit

$$I' = \frac{V}{Z} = \frac{V}{\sqrt{R^2 + (X_L - X_C)^2}}$$

Since, $Z \geq R$

$\therefore \quad I'' \leq I'$

Thus, both Statements I and II are true.

• Concept Enhancer

At the condition of resonance, only resistive components are connected in the circuit. Therefore, no voltage drops across capacitive and inductive elements. So, a series *L-C-R* resonant circuit is also known as voltage amplification circuit.

38. *(b)* In a series *L-C-R* circuit,

$V_{rms} = 200$ V, $f = 50$ Hz, $V_L = 31.4$ V

$\Rightarrow IX_L = 31.4$

$$\Rightarrow \quad I = \frac{31.4}{X_L} = \frac{31.4}{\omega L} = \frac{31.4}{2\pi f L}$$

$$= \frac{31.4}{2 \times 3.14 \times 50 \times 10 \times 10^{-3}}$$

$$= \frac{1}{10^{-1}} = 10 \text{ A}$$

39. *(c)* Given, voltage amplitude,

$$V_0 = 40 \text{ V}$$

Frequency, $f = 4$ kHz
Capacitance, $C = 12\,\mu$F
Displacement current is same as conduction current in capacitor.
Capacitive reactance,

$$X_C = \frac{1}{\omega C} = \frac{1}{2\pi f C}$$

$$= \frac{1}{2\pi \times 4 \times 10^3 \times 12 \times 10^{-6}} = 3.317\,\Omega$$

$$\text{Current, } I = \frac{V}{X_C} = \frac{40}{3.317} = 12 \text{ A}$$

40. *(d)* In purely capacitive circuit, alternating current leads 90° from alternating voltage.

$\therefore \quad A \rightarrow I$

In purely inductive circuit, alternating voltage leads 90° from alternating current.

Thus, $\quad B \rightarrow IV$

In *L-C-R* series circuit at resonance,

$$X_L = X_C$$

$$\therefore \cos\phi = \frac{R}{Z} = \frac{R}{\sqrt{R^2 + (X_L - X_C)^2}}$$

$$= R/R = 1 = \cos 0°$$

$\therefore \quad \phi = 0$

i.e. at resonance, alternating voltage and current are same phase.

Thus, $\quad C \rightarrow II$

In *L-C-R* series circuit, when $V_L > V_C$ alternating voltage leads ϕ angle with alternating current.

Thus, $\quad D \rightarrow III$

41. *(d)* Resonant frequency is given by

$$\omega_r = 1/\sqrt{LC}$$

If capacitance is changed to $4C$, then

$$\omega_r = 1/\sqrt{L'4C}$$

For same resonant frequency

$$L'4C = LC$$

or $$L' = \frac{1}{4} L$$

So, inductance must be reduced by $\frac{3}{4}L$.

42. *(d)* In *L-C-R* circuit,

$$V = 20 \sin 200\pi t$$

$$I = 10 \sin\left(200\pi t + \frac{\pi}{3}\right)$$

Average power dissipated per cycle

$$= V_{rms}\, I_{rms} \cos\phi$$

$$= \frac{V_{max}}{\sqrt{2}} \cdot \frac{I_{max}}{\sqrt{2}} \cdot \cos\phi$$

$$= \frac{20 \times 10}{2} \times \cos\frac{\pi}{3} = 100 \times \frac{1}{2} = 50 \text{ W}$$

43. *(b)* In an AC circuit,

$$V = 100 \sin(100t) \text{ V}$$

$$I = 100 \sin(100t + \pi/3) \text{ mA}$$

$\therefore$Phase difference between V and I

$$\phi = \pi/3$$

$\therefore$Average power dissipation

$$P_{av} = V_{rms} \times I_{rms} \times \cos\phi$$

$$= \frac{100}{\sqrt{2}} \times \frac{100}{\sqrt{2}} \times 10^{-3} \times \cos\frac{\pi}{3}$$

$$= 2.5 \text{ W}$$

44. *(d)*

key idea Power factor of *L-R* circuit can be calculated with the help of impedance triangle given below.

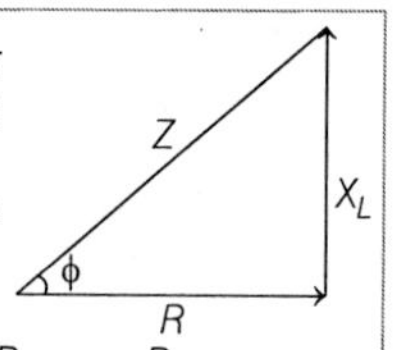

Power factor = $\cos\phi = \frac{R}{Z} = \frac{R}{(R^2 + X_L^2)}$

Power factor, $\cos\phi = R/Z$,

So, power factor,

$$P_1 = \frac{R}{\sqrt{X_L^2 + R^2}} = \frac{R}{\sqrt{R^2 + R^2}} \quad (\because X_L = R)$$

So, $P_1 = \frac{1}{\sqrt{2}}$

Now, When a capacitor is added in circuit such that, $X_L = X_C$

Then, power factor will be

$$\Rightarrow \quad P_2 = \frac{R}{\sqrt{R^2 + (X_L - X_C)^2}}$$

$$= \frac{R}{\sqrt{R^2}} = \frac{R}{R} = 1$$

Hence ratio is, $\frac{P_1}{P_2} = \frac{1}{\sqrt{2}}$

45. *(c)* Given, $E = 440 \sin 100\,\pi t$ V

$$E_0 = 440 \text{ V and } \omega = 100\pi$$

$$L = \frac{\sqrt{2}}{\pi} \text{ H}$$

We know, the reading of an AC ammeter gives I_{rms}.

So, let's calculate I_{rms}.

$$X_L = \omega L = 100\pi \frac{\sqrt{2}}{\pi}$$

$$= 100\sqrt{2}\,\Omega$$

∴Peak value of current,

$$I_0 = \frac{E_0}{X_L} = \frac{440}{100\sqrt{2}}$$

$$\Rightarrow \quad I_0 = \frac{4.4}{\sqrt{2}} = \frac{4.4\sqrt{2}}{2} = 2.2\sqrt{2}\text{A}$$

$$I_{rms} = \frac{I_0}{\sqrt{2}} = \frac{2.2\sqrt{2}}{\sqrt{2}} = 2.2 \text{ A}$$

46. *(b)* Given, for direct current,

$$I_1 = 4 \text{ A}, R_1 = 3\Omega$$

For alternating peak current,

$$I_0 = 4 \text{ A}$$

$$I_{rms} = \frac{I_0}{\sqrt{2}} = \frac{4}{\sqrt{2}} = 2\sqrt{2} \text{ A}$$

Resistance, $R_2 = 2\Omega$

$$\frac{\text{Heat produced across } R_1 (H_1)}{\text{Heat produced across } R_2 (H_2)} = \frac{I_1^2 R_1 t}{I_2^2 R_2 t}$$

$$\Rightarrow \quad \frac{H_1}{H_2} = \frac{I_1^2 R_1}{(I_{rms})^2 R_2} = \frac{4^2 \times 3}{(2\sqrt{2})^2 \times 2}$$

$$= \frac{4 \times 4 \times 3}{8 \times 2} = \frac{3}{1}$$

$$\therefore \; H_1 : H_2 = 3:1$$

47. *(a)* Given, the frequency of the AC source, $f = 50$ Hz

Angular frequency of the circuit,

$$\omega = 2\pi f$$

$$\Rightarrow \quad \omega = 2\pi(50) = 100\pi$$

The general expression of the current in AC circuit,

$$I_{rms} = I_0 \sin \omega t$$

$$\Rightarrow \quad \frac{I_0}{\sqrt{2}} = I_0 \sin(100\,\pi t)$$

$$\Rightarrow \quad \sin(100\,\pi t) = 1/\sqrt{2}$$

$$\Rightarrow \quad 100\,\pi t = \sin^{-1}(1/\sqrt{2})$$

$$\Rightarrow \quad 100\,\pi t = \frac{\pi}{4} \Rightarrow t = \frac{1}{400}\text{ s}$$

$$\Rightarrow \quad t = 2.5 \times 10^{-3}\text{ s} = 2.5 \text{ ms}$$

48. *(b)* As we know, the inductive reactance is directly proportional to the frequency of the AC circuit

i.e., $\quad X_L = \omega L$

$\Rightarrow \quad X_L = 2\,\pi f L \quad (\because \omega = 2\pi f)$

Here, f is the frequency of the AC circuit,

L is the inductive resistance

and X_L is the inductive reactance.

When the frequency of an AC circuit is halved, then the inductive reactance of the circuit is also halved.

i.e. $\quad X_L' = \frac{X_L}{2}$

Using Ohm's law, $I = \frac{V}{X_L}$

When the frequency is halved, then the current

$$I' = \frac{V}{X_L'}$$

$$\Rightarrow I' = \frac{V}{X_L/2} \quad \Rightarrow I' = 2I$$

The current becomes doubled.

49. *(a)* Given, resistance, $R = 2\Omega$

Inductance, $L = 2$ mH, emf, $E = 9$ V and let i be the current.

∵ At $t = 0$ when switch is closed, inductors behave as open circuit.

∴Effective circuit will be

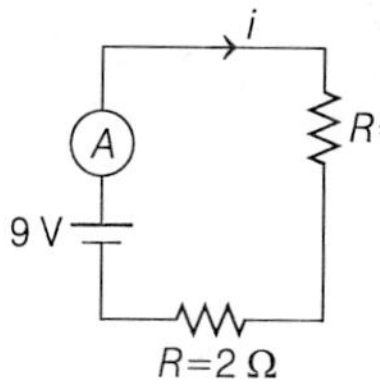

By using Ohm's law, $V = i\,R_{eq}$

$$\Rightarrow \quad i = V/R_{eq}$$

where, R_{eq} is equivalent resistance of series resistors,

i.e., $\quad R_{eq} = R + R = 2R$

$$= 2 \times 2 = 4\Omega$$

$$\therefore \quad i = \frac{9}{4} = 2.25 \text{ A}$$

50. *(a)* Impedance of given ciruit,

$$Z = \sqrt{(X_C - X_L)^2} = X_C - X_L$$

$$= \frac{1}{\omega C} - \omega L$$

$$= \frac{1}{314 \times 100 \times 10^{-6}} - 314 \times 40 \times 10^{-3}$$

$$= 19.28\Omega$$

As $X_C > X_L$, circuit is capacitive, hence current in circuit leads emf by $\pi/2$ radians.

Current in circuit is given by

$$I = I_{max} \sin\left(\omega t + \frac{\pi}{2}\right)$$

$$= \frac{V_{max}}{Z} \cos \omega t = \frac{10}{19.28} \times \cos 314t$$

or $\quad I = 0.52 \cos 314t$

51. *(b)*

key idea *L-C* oscillation of *L-C-R* circuit is equivalent to the oscillation of spring mass system in mechanical system.

where, *L* is equivalent to mass *m*, *C* is equivalent to $1/k$

R is equivalent to damping constant *b*.

For an *L-C-R* circuit,

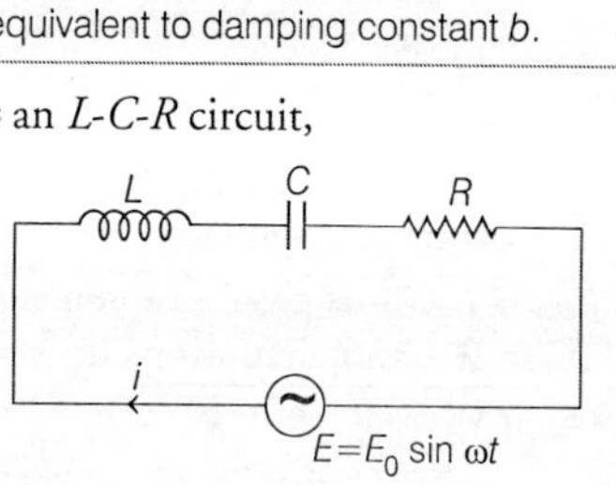

Oscillation of charge q is given by

$$L\frac{d^2q}{dt^2} + R \cdot \frac{dq}{dt} + \frac{1}{C}q = E_0 \sin \omega t \quad \ldots\text{(i)}$$

Now, for a damped harmonic oscillator, we have

$$m\frac{d^2x}{dt^2} + b \cdot \frac{dx}{dt} + kx = F_0 \cos \omega t \quad \ldots\text{(ii)}$$

On comparing Eqs. (i) and (ii), we get the following analogy,

$$L \equiv m,\ R \equiv b \text{ and } \frac{1}{C} \equiv k$$

or $L \leftrightarrow m,\ C \leftrightarrow \frac{1}{k}$ and $R \leftrightarrow b$

52. *(b)* Given, $e = 100 \sin 30t$

and $i = 20 \sin\left(30t - \frac{\pi}{4}\right)$

$\therefore$ Average power,

$$P_{av} = V_{rms} I_{rms} \cos\phi$$

$$= \frac{100}{\sqrt{2}} \times \frac{20}{\sqrt{2}} \times \cos\frac{\pi}{4} = \frac{1000}{\sqrt{2}}\ \text{W}$$

Wattless currents, $I = I_{rms} \sin\phi$

$$= \frac{20}{\sqrt{2}} \times \sin\frac{\pi}{4} = \frac{20}{2} = 10\ \text{A}$$

$$\therefore\quad P_{av} = \frac{1000}{\sqrt{2}}\ \text{W}$$

and $I_{wattless} = 10$ A

53. *(a)* The wire will vibrate with the same frequency as that of the source. this can be considered as an example of forced vibration.

$$T = 10 \times 9.8\ \text{N} = 98\ \text{N}$$

$$m = 9.8 \times 10^{-3}\ \text{kg m}^{-1}$$

Frequency of wire,

$$f = \frac{1}{2L}\sqrt{\left(\frac{T}{m}\right)} = \frac{1}{2 \times 1}\sqrt{\left(\frac{98}{9.8 \times 10^{-3}}\right)}$$

$$= 50\ \text{Hz}$$

At resonance condition, frequency of alternating source, n = frequency of wire = 50 Hz

54. *(a)*

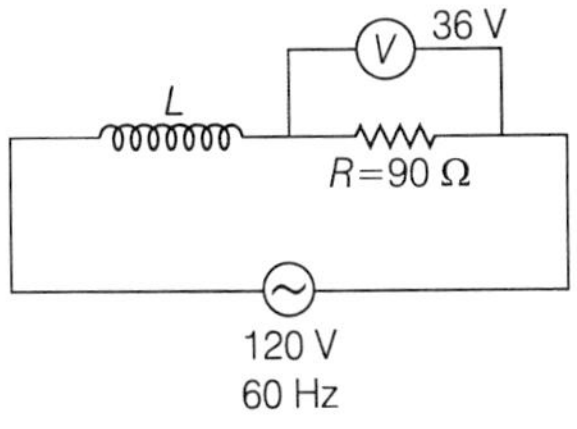

According to the question,

$\because\quad \omega = 2\pi f$

$\Rightarrow\quad \omega = 2 \times 3.14 \times 60 = 376.8$ rad/s

$$V = I_{rms} R$$

$$\Rightarrow\quad 36 = I_{rms} R$$

$$I_{rms} = \frac{V}{\sqrt{X_L^2 + R^2}} \text{ for } L\text{-}R \text{ circuit}$$

$$36 = \frac{120}{\sqrt{X_L^2 + R^2}} \times R \qquad \ldots\text{(i)}$$

$$R = 90\,\Omega \Rightarrow 36 = \frac{120 \times 90}{\sqrt{X_L + 90^2}}$$

$$\sqrt{X_L^2 + 90^2} = 300$$

$$X_L = 286.18$$

$$\Rightarrow\quad \omega L = 286.18$$

$$L = \frac{286.18}{\omega} \Rightarrow L = \frac{286.18}{376.8}$$

$$L = 0.76\ \text{H}$$

55. *(d)* For a resistive load, supply is

$$V = 220 \sin(100\pi t)$$

So, current is

$$I = \frac{V}{R},\ R = 50\,\Omega$$

$$\Rightarrow\quad I = \frac{220}{50}\sin(100\pi t)$$

$$\Rightarrow\quad I = 4.4 \sin(100\pi t)$$

Now, current is peak (maximum) when

$$\sin\theta = 1 \text{ or } 100\pi t_1 = \frac{\pi}{2} \Rightarrow t_1 = \frac{1}{200}\ \text{s}$$

Current is half of peak value when $\sin\theta = 1/2$

$$\text{or}\quad 100\pi t_2 = \frac{\pi}{6} \Rightarrow t_2 = \frac{1}{600}\ \text{s}$$

So, time interval, $t_1 - t_2 = \frac{1}{200} - \frac{1}{600}$

$$= \frac{1}{300}\ \text{s} = \frac{1000}{3000}\ \text{ms} = 3.3\ \text{ms}$$

56. *(d)* Given,

$$E_1 = (25 \sin(1000\,t))\ \text{V}$$

Power factor, $P_1 = \frac{1}{\sqrt{2}}$

Now, $\quad E_2 = (20 \sin(2000\,t))$ V

Power factor of L-R circuit is given by

$$P_1 = \frac{R}{\sqrt{R^2 + X_L^2}} \Rightarrow \frac{1}{\sqrt{2}} = \frac{R}{\sqrt{R^2 + X_L^2}}$$

$$\Rightarrow\quad 2R^2 = R^2 + X_L^2$$

$$R = X_L \qquad \ldots\text{(i)}$$

Now, $\omega = 2\omega$ in E_2 and $X_L \propto \omega$

$$P_2 = \frac{R}{\sqrt{R^2 + 4X_L^2}}$$

From Eq. (i), we get

$$P_2 = \frac{R}{\sqrt{R^2 + 4R^2}} \Rightarrow P_2 = \frac{1}{\sqrt{5}}$$

57. *(d)* Given, for the first element X,

Peak voltage, $V_0 = 100$ V.

Peak current, $I_0 = 5$A

$$\therefore\quad R = \frac{V_0}{I_0} = \frac{100}{5} = 20\,\Omega$$

(Since, current is in phase with voltage, hence element X is resistor).

For second element Y,

Current lags behind the voltage, hence Y is inductor.

$$\therefore\quad X_L = \frac{V_0}{I_0} = \frac{100}{5} = 20\,\Omega$$

When X (resistor) and Y (inductor) are connected in series with supply voltage $V_0 = 100$ V, then

Impedance of the circuit,

$$Z = \sqrt{R^2 + X_L^2}$$

$$= \sqrt{20^2 + 20^2} = 20\sqrt{2}\,\Omega$$

$\therefore$ rms value of alternating current,

$$I_{rms} = \frac{V_{rms}}{Z} = \frac{\frac{V_0}{\sqrt{2}}}{20\sqrt{2}} = \frac{\frac{100}{\sqrt{2}}}{20\sqrt{2}} = \frac{5}{2}\ \text{A}$$

> **• Concept Enhancer**
>
> The impedance of R-L circuit is the vector sum of resistance R and impedance X_L of an inductor.
>
> i.e. $\quad Z = \sqrt{R^2 + X_L^2}$

58. *(a)* Given, $R = 40\ \Omega$,

$$V_{rms} = 220\ \text{V and } f = 50\ \text{Hz}$$

Instantaneous value of alternating current,

$$I = I_0 \sin\omega t$$

For maximum value of alternating current, $\quad I = I_0$

$$\therefore\quad I_0 = I_0 \sin\omega t$$

$$\Rightarrow\quad \sin\omega t = 1$$

$$\Rightarrow\quad \omega t = \frac{\pi}{2} \Rightarrow t = \frac{\pi}{2\omega}$$

If t' be the time at rms value, then

$$I_{rms} = I_0 \sin\omega t'$$

$$\Rightarrow\quad \frac{I_0}{\sqrt{2}} = I_0 \sin\omega t' \Rightarrow \frac{1}{\sqrt{2}} = \sin\omega t'$$

$$\Rightarrow\quad \frac{\pi}{4} = \omega t' \Rightarrow\quad t' = \frac{\pi}{4\omega}$$

$\therefore$ Time taken to change the value of current from maximum value to rms value

$$= t - t' = \frac{\pi}{2\omega} - \frac{\pi}{4\omega} = \frac{\pi}{4\omega} = \frac{\pi}{4 \times 2\pi f}$$

$$= \frac{1}{8f} = \frac{1}{8 \times 50} = \frac{1}{400}\ \text{s}$$

$$= 2.5 \times 10^{-3}\ \text{s} = 2.5\ \text{ms}$$

59. *(*)* Given, angular frequency,

$$\omega = 100\ \text{rad/s}$$

Capacitance of capacitor,

$$C = 100\ \mu\text{F} = 100 \times 10^{-6}\ \text{F}$$

Inductance of inductor coil, $L = 0.5$ H

Resistance in upper branch, $R_1 = 100\,\Omega$

Resistance in lower branch, $R_2 = 50\,\Omega$

In the given circuit consider current in upper branch be i_1 and current flowing in lower branch be i_2. The net current flowing in circuit will be I.

Impedance of upper branch can be calculated as

$$Z_1 = \sqrt{X_C^2 + R_1^2} = \sqrt{\left(\frac{1}{\omega C}\right)^2 + R_1^2}$$

$$= \sqrt{\left(\frac{1}{100 \times 100 \times 10^{-6}}\right)^2 + 100^2}$$

$$= \sqrt{100^2 + 100^2} = 100\sqrt{2}\,\Omega$$

Impedance of lower branch can be calculated as

$$= \sqrt{(\omega L)^2 + R_2^2}$$

$$= \sqrt{(100 \times 0.5)^2 + 50^2}$$

$$= 50\sqrt{2}\,\Omega$$

Current flowing in upper branch,

$$i_1 = \frac{V}{Z_1} = \frac{200}{100\sqrt{2}} = \sqrt{2}\text{A}$$

Phase of current in upper branch,

$$\cos\phi_1 = \frac{R_1}{Z_1} = \frac{100}{100\sqrt{2}} = \frac{1}{\sqrt{2}}$$

$$\Rightarrow \phi_1 = 45°$$

Thus, in upper branch, current leads voltage by 45° as capacitor is present.

Current flowing in lower branch i_2 is

$$i_2 = \frac{V}{Z_2} = \frac{200}{50\sqrt{2}} = 2\sqrt{2}\text{A}$$

Phase of current in lower branch is

$$\cos\phi_2 = \frac{R_2}{Z_2} = \frac{50}{50\sqrt{2}}$$

$$\Rightarrow \quad \phi_2 = 45°$$

Thus, in lower branch current lags voltage by 45° as inductor is present.

Thus, the net current,

$$I = \sqrt{i_1^2 + i_2^2}$$

$$I = \sqrt{(\sqrt{2})^2 + (2\sqrt{2})^2}$$

$$I = \sqrt{10} = 3.16\text{A}$$

Thus, no option in the given question is correct.

60. *(a)*

key idea To solve an AC circuit containing L, R and C, first individually calculate the impedance of L and C, then calculate total impedance by

$$Z = \sqrt{R^2 + (X_L - X_C)^2}$$

The circuit diagram can be drawn as,

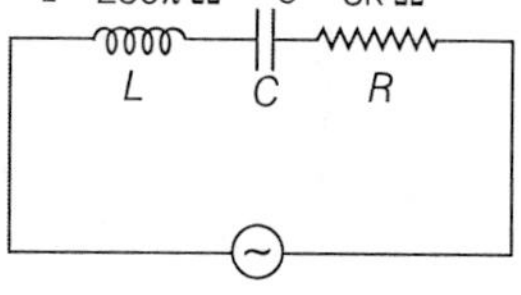

Average power of an L-C-R circuit is given by

$$P_{av} = V_{rms}\, i_{rms} \cos\phi$$

For P_{av} to be maximum,

$$\cos\phi = 1 \qquad \ldots(i)$$

Now, $Z = \sqrt{(X_L - X_C)^2 + R^2}$

As, $\cos\phi = 1 = \frac{R}{Z}$ [using Eq, (i)]

$$\Rightarrow \quad R = Z$$

$$\Rightarrow \quad R = \sqrt{(X_L - X_C)^2 + R^2}$$

On squaring both sides, we get

$$R^2 = (X_L - X_C)^2 + R^2$$

$$\Rightarrow \quad X_L = X_C$$

Since, $X_C = \frac{1}{\omega C} = \frac{1}{2\pi f C}$

and $X_L = 250\pi\Omega$ [Given]

$$\Rightarrow \quad 250\pi = \frac{1}{2\pi \times 50 \times C}$$

$$\Rightarrow \quad C = \frac{1}{2\pi^2 \times 50 \times 250}$$

$$\Rightarrow \quad C = 4 \times 10^{-6}\text{ F} = 4\,\mu\text{F}$$

61. *(a)* Given,

Inductance, $L = 0.07$ H

Resistance, $R = 12\,\Omega$

Voltage, $V = 220$ V

Frequency, $f = 50$ Hz

∵ We know that,

Angular frequency, $\omega = 2\pi f$

$$= 2\pi \times 50 = 100\pi\text{ Hz}$$

∵ Inductive reactance, $X_L = \omega L$

$$= 100\pi \times 0.07$$

$$= 100 \times \frac{22}{7} \times \frac{7}{100} = 22\,\Omega$$

∵ A resistor and an inductor are connected in series,

∴ Impedance,

$$Z = \sqrt{R^2 + X_L^2} = \sqrt{12^2 + 22^2}$$

$$= \sqrt{144 + 484} = \sqrt{628} \approx 25\,\Omega$$

∴ Current,

$$I = \frac{V}{Z} = \frac{220}{25} = 8.8\text{ A}$$

$$\Rightarrow \quad I = 8.8\text{A}$$

Also, phase angle between current and source voltage can be given as

$$\tan\phi = \left(\frac{X_L}{R}\right)$$

$$\Rightarrow \quad \tan\phi = \frac{22}{12}$$

$$\Rightarrow \quad \phi = \tan^{-1}\left(\frac{11}{6}\right)$$

62. *(d)* According to question, there is a series L-C-R circuit.

Resistance, $R = 100\,\Omega$

Capacitance, $C = 0.1\text{ pF} = 0.1 \times 10^{-12}$ F

Inductance, $L = 0.5\text{ mH} = 0.5 \times 10^{-3}$ H

Frequency, $f = 50$ Hz

Voltage, $V = 220$ V

∵ Inductive reactance, $X_L = \omega L$

$$= 2\pi f L$$

$$= 2\pi \times 50 \times 0.5 \times 10^{-3}$$

$$\Rightarrow \quad X_L = 50\pi \times 10^{-3}\,\Omega$$

∵ Capacitive reactance,

$$X_C = \frac{1}{\omega C} = \frac{1}{2\pi f C}$$

$$\Rightarrow \quad X_C = \frac{1}{2\pi \times 50 \times 10^{-12} \times 0.1}$$

$$\Rightarrow \quad X_C = \frac{1}{100\pi \times \frac{1}{10} \times 10^{-12}}$$

$$\Rightarrow \quad X_C = \frac{10^{11}}{\pi}\Omega$$

On comparing X_L with X_C, we can write

$$X_C >> X_L$$

and $|X_C - X_L| >> R$

which means that the circuit is predominantly capacitive circuit and the phase angle between current and emf is 90°.

63. *(c)* Given,

AC voltage, $V(t) = 20 \sin \omega t$ V.

Frequency, $f = 50$ Hz

Separation between the plates,

$$d = 2\text{ mm} = 2 \times 10^{-3}\text{ m}$$

Area, $A = 1\text{ m}^2$

As, $C = \frac{\varepsilon_0 A}{d}$

where, ε_0 = electrical permittivity of free space

$$= 8.854 \times 10^{-12}\text{ N}^{-1}\text{ C}^2\text{m}^{-2}$$

$$\Rightarrow \quad C = \frac{\varepsilon_0 \times 1}{2 \times 10^{-3}} \qquad \ldots(i)$$

Capacitive reactance,

$$X_C = 1/\omega C \qquad \ldots(ii)$$

From Eqs. (i) and (ii), we get

$$X_C = \frac{2\times10^{-3}}{2\times 50\pi\times \varepsilon_0} \qquad (\because \omega = 2\pi f)$$

$$\Rightarrow \quad X_C = \frac{2\times10^{-3}}{25}\times 9\times10^{9}$$

$$\Rightarrow \quad X_C = \frac{18}{25}\times10^{6}\ \Omega$$

By using Ohm's law,

As, $I_0 = \frac{V_0}{X_C} = \frac{20\times 25}{18}\times10^{-6}$

$= 27.78\times 10^{-6}$

$\Rightarrow \quad I_0 = 27.78\ \mu\text{A}$

$\therefore$ The amplitude of the oscillating displacement current for applied AC voltage will be approximately 27.79 μA.

64. *(b)* Given,

$$I = I_1 \sin \omega t + I_2 \cos\omega t \qquad \ldots(\text{i})$$

We know that rms value of current is given by

$$I_{\text{rms}} = \sqrt{\frac{\int I^2 dt}{T}}$$

$$\Rightarrow \quad I_{\text{rms}} = \sqrt{\frac{\int_0^T (I_1 \sin \omega t + I_2 \cos\omega t)^2\, dt}{T}}$$

[Using Eq. (i)]

Squaring on both sides of the above equation, we get

$$(I_{\text{rms}})^2 = \int_0^T \frac{(I_1 \sin \omega t + I_2 \cos\omega t)^2}{T}\, dt$$

$$\Rightarrow (I_{\text{rms}})^2 = \int_0^T \frac{(I_1^2 \sin^2 \omega t + I_2 \cos^2 \omega t + 2I_1 I_2 \sin\omega t \cos\omega t)\, dt}{T}$$

$$\Rightarrow \quad I_{\text{rms}} = \sqrt{\frac{I_1^2}{2} + \frac{I_2^2}{2} + 0}$$

$$\Rightarrow \quad I_{\text{rms}} = \sqrt{\frac{I_1^2 + I_2^2}{2}}$$

> **• Concept Enhancer**
>
> Heat consumed by a constant current (I_{rms}) is equal to heat consumed by instantaneous current in the same time interval.
>
> i.e. $I_{\text{rms}}^2 Rt = \int_0^t I^2 R dt$

65. *(a)* Given, inductance,

$L = 100$ mH

$= 100\times10^{-3}$ H

Capacitance, $C = 100\ \mu$F

$= 100\times10^{-6}$ F

Resistance, $R = 120\ \Omega$

Maximum voltage, $V_0 = 30$ V

Angular frequency,

$\omega = 100\ \text{rads}^{-1}$

Impedance, $Z = \sqrt{(X_L - X_C)^2 + R^2}$

Here, $X_L = \omega L = 100\times100\times10^{-3}$

$= 10\ \Omega$

$\therefore \quad X_C = \frac{1}{\omega C} = \frac{1}{100\times100\times10^{-6}}$

$= 100\ \Omega$

$\Rightarrow \quad Z = \sqrt{(X_C - X_L)^2 + R^2}$

$= \sqrt{(100-10)^2 + (120)^2}$

$= \sqrt{(90)^2 + 120^2} = 150\ \Omega$

As we know that, peak current,

$$I_0 = \frac{V_0}{Z} = \frac{30}{150} = 0.2\text{ A}$$

Resonance frequency,

$$\therefore \quad f = \frac{1}{2\pi\sqrt{LC}}$$

$$= \frac{1}{2\pi\sqrt{100\times10^{-3}\times100\times10^{-6}}}$$

$= 50$ Hz

66. *(b)* Given, angular frequency,

$\omega = 100\ \text{rad s}^{-1}$

$R = 60\Omega,\ V_R = 15$ V,

$R' = 40\Omega,\ V_{R'} = V_L = 20$ V

and $V_C = 10$ V

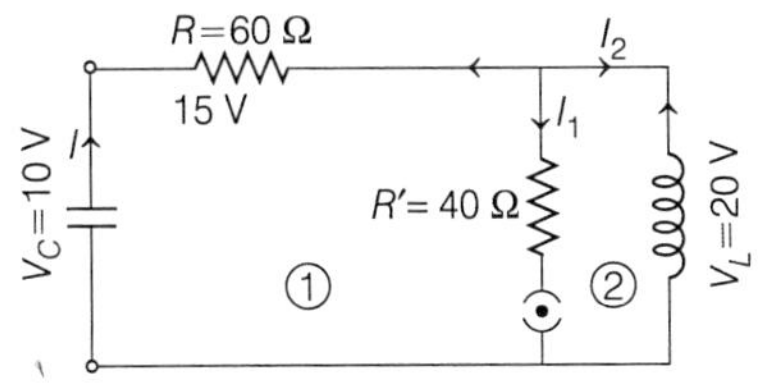

By using Ohm's law,

$V = IR \ \Rightarrow\ I = V/R$

$\Rightarrow \quad I = 15/60 = 1/4$ A $\quad \ldots(\text{i})$

and $I_1 = \frac{V_{R'}}{R'} = 20/40 = 1/2$ A $\quad \ldots(\text{ii})$

As, $X_C = \frac{V_C}{I} = \frac{10}{1/4} = 40\ \Omega$

and $X_C = \frac{1}{\omega C}$

$\Rightarrow \quad C = \frac{1}{X_C\omega} = \frac{1}{40\times100}$

$= 0.25\times10^{-3}$ F $= 0.25$ mF

$= 250\ \mu$F

By using KCL in loop 2,

$I_2 = I - I_1 = 1/4 - 1/2 = -1/4$ A

$\therefore \quad X_L = \frac{V_L}{|I_2|} = \frac{20}{1/4} = 80\Omega \Rightarrow \omega L = 80$

$\Rightarrow \quad L = \frac{80}{\omega} = \frac{80}{100} = 0.8$ H

67. *(d)* As we know that at time $t = 0$, inductor acts as open circuit.

Then, the circuit becomes

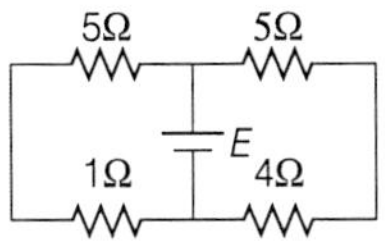

Therefore,

$$R_{\text{eq}} = \frac{(5+1)(5+4)}{(5+1)+(5+4)}$$

$$= \frac{6\times9}{6+9} = \frac{54}{15} = \frac{18}{5}$$

By using Ohm's law,

$V = IR_{\text{eq}} \qquad [\because V = E]$

$\Rightarrow \quad I = \frac{E\times5}{18} = \frac{5E}{18}$

At $t = \infty$, inductor will act as short circuit. It is shown below

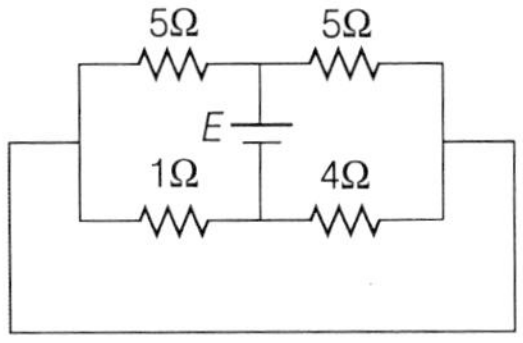

Therefore,

$$R_{\text{eq}} = \frac{5\times5}{5+5} + \frac{1\times4}{1+4}$$

$$= \frac{25}{10} + \frac{4}{5} = \frac{5}{2} + \frac{4}{5} = \frac{33}{10}\ \Omega$$

and $\quad I = \frac{E}{\frac{33}{10}} = \frac{10E}{33}$

68. *(c)* Given, phase difference, $\phi = \pi/4$

As we know, for *R-L* or *R-C* circuit,

$$\tan\phi = \frac{X_C \text{ or } X_L}{\text{Resistance}\,(R)}$$

$$\tan\frac{\pi}{4} = \frac{X_C \text{ or } X_L}{R}$$

$$1 = \frac{X_C \text{ or } X_L}{R} \Rightarrow R = X_C \text{ or } X_L$$

Also, given $e = e_0 \sin(100\,t)$

Comparing the above equation with general equation of emf, i.e.

$e = e_0 \sin\omega t$, we get

$\omega = 100$ rad/s $= 10^2$ rad/s

Now, checking option wise,

For *R-C* circuit, with

$R = 1\,k\Omega = 10^3\ \Omega$

and $C = 1\mu F = 10^{-6} F$

So, $X_C = \frac{1}{\omega C} = \frac{1}{10^2 \times 10^{-6}} = 10^4\ \Omega$

$\Rightarrow \quad R \neq X_C$

For R - L circuit, with

$R = 1\,k\Omega = 10^3\,\Omega$

and $L = 1\ mH = 10^{-3} H$

So, $X_L = \omega L = 10^2 \times 10^{-3} = 10^{-1}\Omega$

$\Rightarrow \quad R \neq X_L$

For R - C circuit, with

$R = 1\,k\Omega = 10^3\,\Omega$ and $C = 10\mu F$

$= 10 \times 10^{-6} F = 10^{-5} F$

So, $X_C = \frac{1}{10^2 \times 10^{-5}} = 10^3\,\Omega \Rightarrow R = C$

For R - L circuit, with

$R = 1\,k\Omega = 10^3\,\Omega$ and $L = 10\ mH$

$= 10 \times 10^{-3}\ H = 10^{-2} H$

$X_L = 10^2 \times 10^{-2} = 1\Omega \Rightarrow R \neq X_L$

Alternate Solution

Since, $\tan\frac{\pi}{4} = 1 = \frac{X_C \text{ or } X_L}{R}$

$\therefore$ For R-C circuit, we have

$1 = \frac{1}{C\omega R}$ or $\omega = \frac{1}{CR}$ ...(i)

Similarly, for R-L circuit, we have

$1 = \frac{\omega L}{R} \Rightarrow \omega = \frac{R}{L}$ ...(ii)

It is given in the question that, $\omega = 100$ rad/s

Thus, again by substituting the given values of R, C or L option wise in the respective Eqs. (i) and (ii), we get that only for option (c),

$\omega = \frac{1}{CR} = \frac{1}{10 \times 10^{-6} \times 10^3}$ or $\omega = 100$ rad/s

69. *(1)* Average power dissipated

$P_{avg} = V_{rms} \times I_{rms} \times \cos\phi$

$= V_{rms} \times \frac{V_{rms}}{Z} \times \frac{R}{Z}$

$= \frac{V_{rms}^2 R}{Z^2}$

$= \frac{V_{rms}^2 R}{R^2 + (X_L - X_C)^2}$

$= \frac{10^2 \times 50}{50^2 + (100 - 50)^2}$

$= \frac{5 \times 10^3}{2500 + 2500}$

$= \frac{5 \times 10^3}{5000} = 1\ W$

70. *(5)*

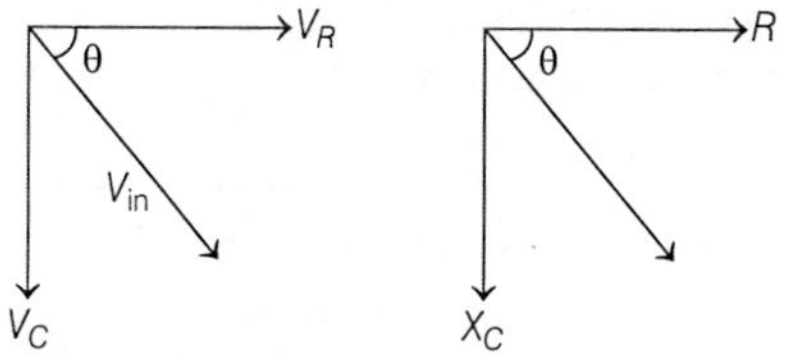

$V_A - V_B$

Given $R = 100\ k\Omega = 100 \times 10^3\ \Omega$

$C = 100\ pF = 100 \times 10^{-12}\ F$

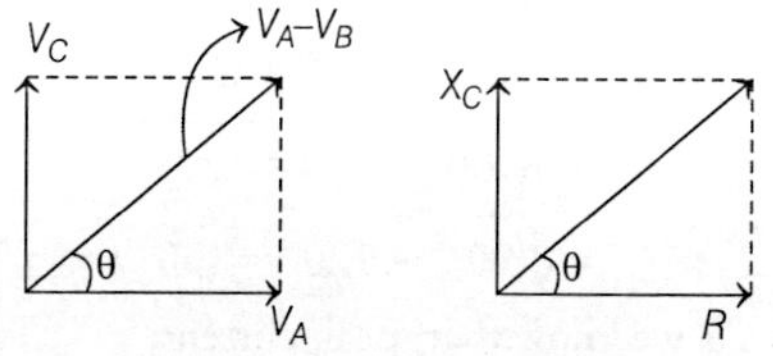

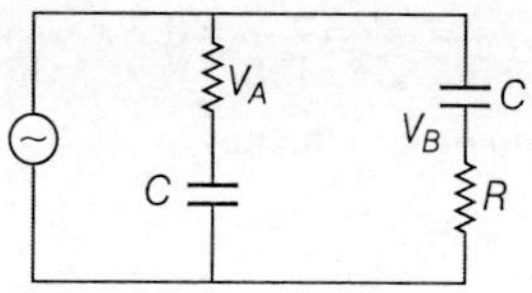

$\theta = 45°$

$\theta + \theta = 90°$

$\tan\theta = \frac{X_C}{R}$

$\Rightarrow \tan 45° = \frac{X_C}{R}$

$\Rightarrow \quad X_C = R$

$X_C = R = \frac{1}{\omega C}$

$\omega = \frac{1}{RC} = \frac{1}{100 \times 10^3 \times 100 \times 10^{-12}}$

$= \frac{10^{12}}{10^7} = 10^5$

$\omega = 10^5$

$x = 5$

71. *(1)* $R = 100\pi$

$X_L = \omega L$

$Z = \sqrt{R^2 + (X_L)^2}$

$= \sqrt{R^2 + (\omega L)^2}$

$= \sqrt{(100\pi)^2 + (2\pi + 50 \times 1)^2}$

$= \sqrt{(100\pi)^2 + (100\pi)^2}$

$= \sqrt{2} \times 100\pi$

$I_{rms} = \frac{V}{Z} = \frac{100\pi}{\sqrt{2} \times 100\pi} = \frac{1}{\sqrt{2}}$

$I_{max} = \sqrt{2}\, I_{rms} = \sqrt{2} \times \frac{1}{\sqrt{2}}$

$= 1\ A$

72. *(2)* The current is maximum at resonance,

$\omega = \frac{1}{\sqrt{LC}} = \frac{1}{\sqrt{(100 \times 10^{-3})(25 \times 10^{-9})}}$

$= \sqrt{\frac{1}{10^{-10} \times 25}} = \sqrt{\frac{10^{10}}{25}} = \frac{10^5}{5}$

$= 2 \times 10^4$ rad/s

73. *(4)* Given, capacitor reactance,

$X_C = 4\sqrt{3}\Omega$

Resistance, $R = 4\Omega$

Peak voltage, $V_0 = 8\sqrt{2}$ V

Impedance, $Z = \sqrt{X_C^2 + R^2}$

$= \sqrt{(4\sqrt{3})^2 + (4)^2}$

$= \sqrt{48 + 16} = \sqrt{64} = 8\Omega$

Peak current, $I_0 = \frac{V_0}{Z} = \frac{8\sqrt{2}}{8} = \sqrt{2}$ A

Power dissipated, $P = V_{rms} I_{rms} \cos\phi$

$= \frac{8\sqrt{2}}{\sqrt{2}} \times \frac{\sqrt{2}}{\sqrt{2}} \times \frac{R}{Z}$

$= 8 \times \frac{4}{8} = 4$ W

74. *(10)* Case I

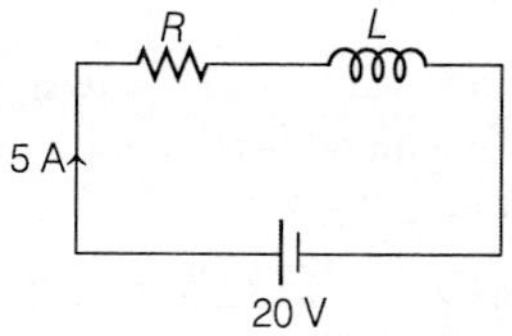

$i = \frac{20}{R} \Rightarrow 5 = \frac{20}{R} \Rightarrow R = 4\Omega$

Case 2

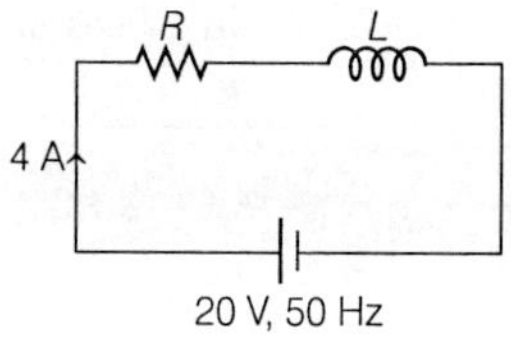

$i = \frac{20}{Z} \Rightarrow 4 = \frac{20}{Z} = \frac{20}{\sqrt{R^2 + X_L^2}}$

$\sqrt{R^2 + X_L^2} = 5 \Rightarrow R^2 \times X_L^2 = 25$

$(4)^2 + X_L^2 = 25$

$X_L^2 = 25 - 16 = 9$

$X_L = 3\Omega$

$L = \frac{X_L}{2\pi f} = \frac{3}{2\pi \times 50}$

$L \simeq \frac{1000}{100}$ mH

$\simeq 10$ mH

75. *(22)*

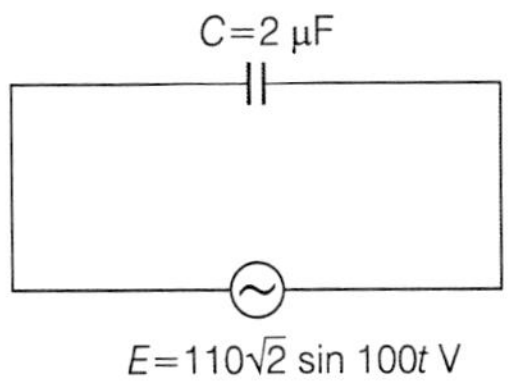

By comparing with standard equation of emf

$$E = E_0 \sin(\omega t)$$

we get, $\omega = 100 \dfrac{\text{rad}}{\text{s}}$

Capative reactance,

$$X_C = \frac{1}{C\omega} = \frac{1}{2 \times 10^{-6} \times 100}$$

$$= \frac{10000}{2} = 5000\,\Omega$$

We know that,

$$I_0 = \frac{E_0}{X_C} \Rightarrow I_0 = \frac{110\sqrt{2}}{5000}$$

$$\Rightarrow \quad I_{\text{rms}} = \frac{I_0}{\sqrt{2}} = \frac{110}{5000}$$

$$= 2.2 \times 10^{-2}\,\text{A} = 22 \times 10^{-3}\,\text{A}$$

$$= 22\,\text{mA}$$

76. *(10)* For the case of maximum current, *L-C-R* circuit should be in resonance.

$\therefore$ Frequency, $f = \dfrac{1}{2\pi\sqrt{LC}}$

$$= \frac{1}{2 \times \pi\sqrt{100 \times 10^{-3} \times 2.5 \times 10^{-9}}}$$

$$= \frac{1}{2\pi \times 5\sqrt{10^{-11}}} = \frac{\sqrt{10^{11}}}{10\pi}$$

$$= \frac{10^5\sqrt{10}}{10\pi} \qquad [\because \sqrt{10} = \pi]$$

$$= 10^4\ \text{Hz}$$

$$= 10 \times 10^3\ \text{Hz}$$

77. *(250)* For DC supply of 100 V, resistance of inductor,

$$R = \frac{V}{I} = \frac{100}{5} = 20\,\Omega$$

$\therefore$ For AC voltage

$$X_L = 20\sqrt{3}\,\Omega$$

$$\therefore \quad Z = \sqrt{R^2 + X_L^2}$$

$$= \sqrt{20^2 + (20\sqrt{3})^2}$$

$$= \sqrt{1600} = 40\,\Omega$$

$\therefore$ Power dissipated in the circuit,

$$P = i_{\text{rms}}^2 R = \left(\frac{V_{\text{rms}}}{Z}\right)^2 \times R$$

$$= \frac{\left(\dfrac{200}{\sqrt{2}}\right)^2}{40^2} \times 20 = 250\ \text{W}$$

78. *(50)* $V = 50\sqrt{2} \sin(100t)$ V $\quad$ ...(i)

From Eq. (i), $\omega = 100$

Inductive reactance, $X_L = \omega L$

$\therefore X_L = 100 \times 1 = 100\,\Omega$ $\quad$ ...(ii)

$[L = 1\,\text{H given}]$

Capacitive inductance, $X_C = \dfrac{1}{\omega C}$

$$= \frac{1}{100 \times 20 \times 10^{-6}} = 500\,\Omega \quad \text{...(iii)}$$

Total impedance of *L-C-R* circuit,

$$Z = \sqrt{(X_L - X_C)^2 + R^2}$$

$$= \sqrt{(100 - 500)^2 + 300^2} \Rightarrow Z = 500\,\Omega$$

$$i_{\text{rms}} = \frac{V_{\text{rms}}}{Z} = \frac{50}{500} = 0.1\,\text{A} \quad \text{..(iv)}$$

rms voltage across capacitor,

$$V_{\text{rms}} = X_C i_{\text{rms}} = 500 \times 0.1 = 50\,\text{V}$$

$$V_{\text{rms}} = 50\,\text{V}$$

79. *(8)* Given,

$$i = \left[6 + \sqrt{56} \sin\left(100\pi t + \frac{\pi}{3}\right)\right]$$

$$i_1 = 6\,\text{A}, i_2 = \sqrt{56}\,\text{A}$$

$$i_{\text{rms}} = \sqrt{i_1^2 + \frac{i_2^2}{2}} = \sqrt{36 + \frac{56}{2}} = 8\,\text{A}$$

80. *(1)*

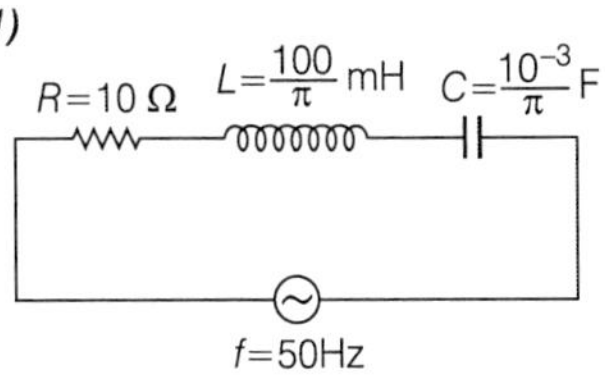

$$X_L = \omega L = 2\pi f \cdot L$$

$$X_L = \frac{100}{\pi} \times 2\pi \times 50 \times 10^{-3}$$

$$X_L = 100 \times 100 \times 10^{-3} = 10\,\Omega$$

$$X_C = \frac{1}{\omega C} = \frac{1}{2\pi f \cdot C} = \frac{1}{2\pi \times 50 \times \dfrac{10^{-3}}{\pi}}$$

$$= \frac{1}{100 \times 10^{-3}}$$

$$X_C = \frac{1}{10^{-1}} = 10\,\Omega$$

Power factor

$$\Rightarrow \quad \cos\phi = \frac{R}{Z} = \frac{R}{\sqrt{R^2 + (X_L - X_C)^2}}$$

$$= \frac{10}{\sqrt{100 + (10 - 10)^2}} = \frac{10}{10} = 1$$

> **• Concept Enhancer**
> Series *L-C-R* circuit draw maximum current when $X_L = X_C$. In this case power dissipation in the circuit is maximum because power factor is unity.

81. *(40)* Given that,

$$V = 220\,\text{V}, f = 50\,\text{Hz}, R = 100\,\Omega$$

$$X_L = 79.6\,\Omega$$

or $\quad C = \dfrac{1}{100\pi \times 79.6} - X_L = \dfrac{1}{\omega C}$

$$\Rightarrow \quad C = 40 \times 10^{-6}\,\text{F}$$

82. *(25)* Given that,

$$R = 80\,\Omega, X_L = 100\,\Omega,$$

$$X_C = 40\,\Omega, V_0 = 2500\,\text{V}$$

Amplitude of current,

$$I_0 = \frac{V_0}{Z} = \frac{V_0}{\sqrt{R^2 + (X_L - X_C)^2}}$$

$$= \frac{2500}{\sqrt{(80)^2 + (100 - 40)^2}}$$

$$= \frac{2500}{\sqrt{(80)^2 + (60)^2}} = \frac{2500}{100}$$

$$I_0 = 25\,\text{A}$$

83. ***(242)***

> **key idea** When the current is in the phase with the emf in *L-C-R* series circuit, then circuit is in resonant condition. In this situation you can find amplitudes of current directly with the help of following formula.
>
> i.e., $\quad I_0 = \sqrt{2} \times \dfrac{V_{\text{rms}}}{R}$

Given that, current in phase with the emf, then electrical circuit is in resonance.

$$\text{Current, } I_{\text{rms}} = \frac{V_{\text{rms}}}{R} = \frac{220}{20} = 11\text{A}$$

Amplitude of current,

$$I_0 = \sqrt{2} \times I_{\text{rms}}$$

$$= \sqrt{2} \times 11 = \sqrt{2} \times \sqrt{121}$$

$$I_0 = \sqrt{242}\,\text{A}$$

84. *(3872)* Current drawn in *L-C-R* series circuit is maximum at resonance.

And at resonance,

$$\omega = \frac{1}{\sqrt{LC}}$$

Substituting given values,

$$2\pi(7000) = \frac{1}{\sqrt{2 \times 10^{-6} \times C}}$$

$$\Rightarrow \quad C = \frac{1}{3872}\,\text{F}$$

85. *(8)* In series L-C-R circuit.

$V = 220\text{ V}, f = 50\text{ Hz}, R = 80\,\Omega,$
$X_L = 70\,\Omega, X_C = 130\,\Omega$

$\therefore$ Power factor,

$$\cos\phi = \frac{R}{Z} = \frac{R}{\sqrt{R^2 + (X_C - X_L)^2}}$$

$$= \frac{80}{\sqrt{80^2 + (130-70)^2}}$$

$$= \frac{80}{\sqrt{80^2 + 60^2}} = \frac{80}{100} = \frac{8}{10} = \frac{x}{10} \quad \text{(given)}$$

$\therefore \quad x = 8$

86. *(100)* In L-C-R series circuit,

$C = 62.5\text{ nF} = 6.25 \times 10^{-8}\text{ F}$
$R = 50\,\Omega$
$f = 2\text{ kHz} = 2 \times 10^3\text{ Hz}$

At resonance condition, maximum value of current flows.

Hence, resonance frequency,

$$f = \frac{1}{2\pi\sqrt{LC}} \Rightarrow L = \frac{1}{4\pi^2 f^2 C}$$

$$= \frac{1}{4 \times 10 \times (2 \times 10^3)^2 \times (6.25) \times 10^{-8}}$$

$$= \frac{1}{10}\text{ H} = 100\text{ mH}$$

87. *(3)* Power, $P = \dfrac{V^2}{R} \Rightarrow R = \dfrac{V^2}{P}$

$\Rightarrow$ Resistance, $R = \dfrac{100 \times 100}{50}$

$\Rightarrow \quad R = 200\Omega$

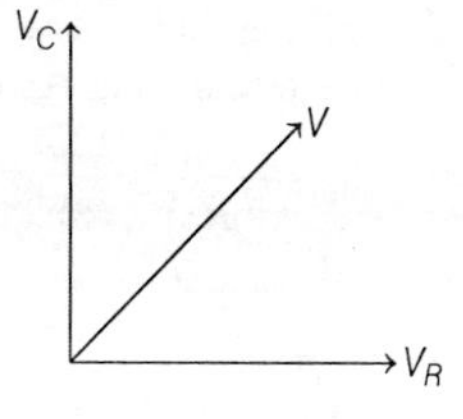

By phasor Diagram

$V_R^2 + V_C^2 = V^2$

$\Rightarrow (100)^2 + V_C^2 = (200)^2$

$\Rightarrow \quad V_C = 100\sqrt{3}$

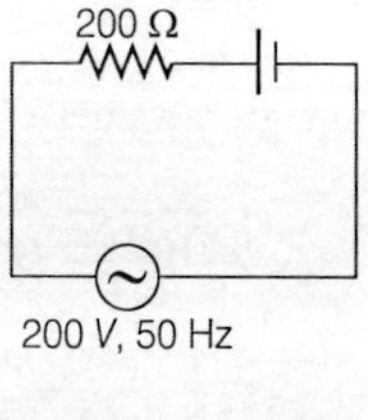

Now, current, $I = \dfrac{100}{200} = \dfrac{1}{2}\text{ A}$

As, $\quad V_C = I \times X_C,$

$V_C = 100\sqrt{3}$ and $I = \dfrac{1}{2}\text{ A}$

So, capacitive reactance, $X_C = 200\sqrt{3}$

$\Rightarrow \quad 200\sqrt{3} = \dfrac{1}{\omega C}$

$\Rightarrow \quad C = \dfrac{1}{200\sqrt{3} \times 2\pi \times 50} = \dfrac{50}{\pi\sqrt{3}}\,\mu\text{F}$

So, $\quad x = 3$

88. *(5)* As current is in phase with emf, so the circuit is in resonance, and frequency of source = frequency of resonance

$$= \frac{1}{2\pi\sqrt{LC}}$$

$$= \frac{1}{2 \times \pi \times \sqrt{0.5 \times 10^{-3} \times 200 \times 10^{-6}}}$$

$$= 0.5 \times 10^3 = 5 \times 10^2\text{ Hz}$$

89. *(0)* We have,

Inductive reactance, $X_L = \omega L$

$\Rightarrow \quad X_L = 2\pi f L$

Now, $\quad V_L = 2V_R$

$\Rightarrow \quad iX_L = 2iR \Rightarrow X_L = 2R$

So, $\quad 2R = 2\pi f L \Rightarrow R = \pi f L$

$\Rightarrow \quad 5 = \pi \times 50 \times L$

$\Rightarrow \quad L = \dfrac{1}{10\pi}\text{H} = \dfrac{1000}{10\pi}\text{ mH} = \dfrac{1}{0.1\,\pi}\text{ mH}$

So, $K = 0.01 \approx 0$

90. *(100)* Total capacity of line = Capacity per km × Length in km

$= 0.01 \times 10^{-6} \times 100 = 1\,\mu\text{F}$

Angular frequency of current,

$\omega = 2\pi f$
$= 2 \times \pi \times 0.5 \times 10^3 = 2 \times \pi \times 500$
$= 1000\pi\text{ rad/s}$

For impedance to be minimum circuit should be operating under resonance condition,

$\Rightarrow \quad \omega L = \dfrac{1}{\omega C}$

Or inductance of circuit,

$$L = \frac{1}{\omega^2 C}$$

$$\Rightarrow L = \frac{1}{10^6 \pi^2 \times 10^{-6}} = \frac{1}{10}\text{ H}$$

$= 100\text{ mH}$

91. *(0)* This is a case of parallel L-C-R circuit

The impedance relation is

$$\frac{1}{Z} = \sqrt{\left(\frac{1}{X_L} - \frac{1}{X_C}\right)^2}$$

At resonance,

$X_L = X_C$. So $Z \to \infty$

$I = \dfrac{V}{Z} \to 0$

92. *(22)* Given,

$L = 100\text{ mH},$
$C = 100\mu\text{F}, R = 10\Omega,$
$V = 220\text{ V}, f = 50\text{ Hz}$

Impedance,

$$Z = \sqrt{R^2 + (X_L - X_C)^2}$$

$$= \sqrt{10^2 + \left(2\pi \times 50 \times 0.1 - \frac{1}{2\pi \times 50 \times 10^{-4}}\right)^2}$$

$= 10\Omega$

So, current, $I = \dfrac{E}{Z} = \dfrac{220}{10} = 22\text{A}$

93. *(242)* Given, $L = 200\text{ mH} = 0.2\text{ H}$

$V_{\text{rms}} = 220\text{ V}$
$f = 50\text{ Hz}$

$\therefore$ Inductive reactance,

$X_L = 2\pi f L$
$= 2\pi \times 50 \times 0.2$
$= 20\pi\Omega$

Peak current,

$$I_0 = \frac{V_0}{X_L} = \frac{\sqrt{2}\,V_{\text{rms}}}{20\,\pi} = \frac{\sqrt{2} \times 220}{20\,\pi}$$

$$I_0 = \frac{11\sqrt{2}}{\pi}$$

$$\Rightarrow I_0 = \frac{\sqrt{242}}{\pi}\text{ A} \quad \ldots\text{(i)}$$

Since, in inductive circuit, there is a phase difference of $\dfrac{\pi}{2}$ between alternating current and voltage. Hence, when instantaneous voltage is zero, then current reaches at its peak value, which is given by Eq. (i),

$$I_0 = \frac{\sqrt{242}}{\pi} = \frac{\sqrt{a}}{\pi}\text{ A} \quad \text{(Given)}$$

$\therefore \quad a = 242$

94. *(2)* We know that, at very high frequency capacitive reactance becomes negligible i.e. short circuit (SC) and inductive reactance becomes very large i.e. open circuit (OC).

i.e. $\quad X_C \to 0$

and $\quad X_L \to \infty.$

Now, the circuit can be rearranged as shown in figure.

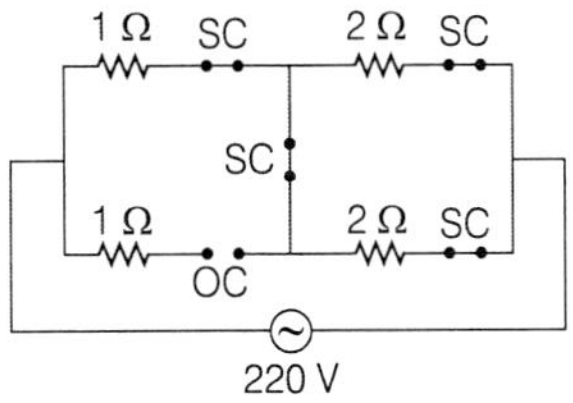

Final circuit is

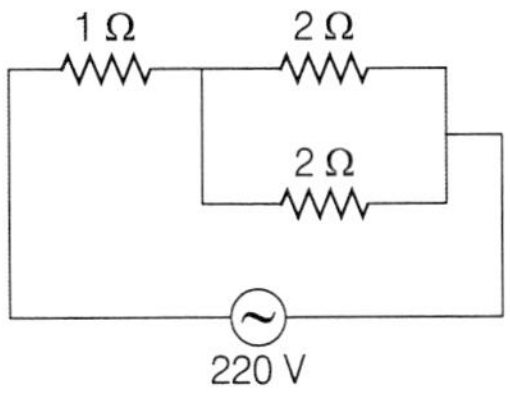

Hence, equivalent resistance,

$$R_{eq} = 1 + \frac{2 \times 2}{2+2} = 2\Omega$$

Thus, correct answer is 2.

95. *(1)* Given, resistance of resistor $= R$

Inductance, $X_L = 3R$

Capacitance $X_C = 2R$

As we know that,

Power factor, $\cos\phi = \frac{R}{Z}$

where, Z is impedance,

$$Z = \sqrt{R^2 + (X_L - X_C)^2}$$

For **case I**, an inductance connected in series with resistance

$$\cos\phi_1 = \frac{R}{Z_1} = \frac{R}{\sqrt{R^2 + (X_L - 0)^2}}$$

$$= \frac{R}{\sqrt{R^2 + (3R)^2}} = \frac{R}{R\sqrt{10}} = \frac{1}{\sqrt{10}} \quad \ldots\text{(i)}$$

For **case II**, A capacitor is also connected in series with resistance

$$\therefore \quad \cos\phi_2 = \frac{R}{Z_2}$$

$$\Rightarrow \cos\phi_2 = \frac{R}{\sqrt{R^2 + (3R - 2R)^2}}$$

$$= \frac{R}{\sqrt{R^2 + R^2}}$$

$$= \frac{R}{R\sqrt{2}} = \frac{1}{\sqrt{2}} \quad \ldots\text{(ii)}$$

Now, ratio of Eqs. (ii) and (i), we get

$$\frac{\cos\phi_2}{\cos\phi_1} = \frac{\frac{1}{\sqrt{2}}}{\frac{1}{\sqrt{10}}} = \sqrt{\frac{10}{2}} = \sqrt{5} : 1$$

$$\therefore \quad x = 1$$

96. *(11)* Given, equation of alternating current,

$$i = \left[\sqrt{42}\sin\left(\frac{2\pi}{T}t\right) + 10\right] \text{A}$$

From given equation, we get

$$i = i_1 + i_2$$

where, $i_1 = \sqrt{42}\sin\left(\frac{2\pi}{T}t\right)$ A

and $\quad i_2 = 10$ A

Now, i_1 is oscillating current, whereas i_2 is direct current and its value does not change with time.

$$(i_1)_{\text{rms}} = \frac{\sqrt{42}}{\sqrt{2}} = \sqrt{21}\text{ A} \Rightarrow (i_2)_{\text{rms}} = 10\text{ A}$$

We know that,

$$i_{\text{rms}}^2 = (i_1)_{\text{rms}}^2 + (i_2)_{\text{rms}}^2$$

Substituting the values, we get

$$i_{\text{rms}} = \sqrt{(\sqrt{21})^2 + 10^2} = \sqrt{121}$$

$$\Rightarrow \quad i_{\text{rms}} = 11\text{ A}$$

Thus, rms value of given equation of current is 11 A.

97. *(500)* For circuit (*a*), $P_a = V_{\text{rms}}^2 / R$

For circuit (*b*), $P_b = \frac{V_{\text{rms}}^2}{Z^2} \times R$

Given that $P_a = P_b$

$$\Rightarrow \frac{V_{\text{rms}}^2}{R} = \frac{V_{\text{rms}}^2}{Z^2} \times R \Rightarrow R^2 = Z^2$$

$$\Rightarrow \quad 25 = \left(\sqrt{(X_C - X_L)^2 + (5)^2}\right)^2$$

$$\Rightarrow \quad 25 = (X_C - X_L)^2 + 25$$

$$\Rightarrow X_C = X_L \Rightarrow \frac{1}{\omega C} = \omega L$$

$$\Rightarrow \omega^2 = \frac{1}{LC} = \frac{10^6}{0.1 \times 40}$$

$$\Rightarrow \quad \omega = 500 \text{ rad/s}$$

98. *(125)* Given, resistance, $R = 5\Omega$

Inductance, $L = 20$ mH

Capacitance, $C = 0.5\ \mu$F

Voltage, $V = 250$ V

$$I_{\text{rms}} = \frac{V}{Z} = \frac{V}{R}$$

[At resonance, $Z = R$]

Power dissipated, $P = I_{\text{rms}}^2 R$

$$\text{or} \quad P = \frac{V^2}{R} = \frac{250 \times 250}{5}$$

$$= 12500 \text{ J/s} = 125 \times 10^2 \text{ W}$$

99. *(3)* Given, Inductance, $L = 30$ mH

Resistance, $R = 1\Omega$

Angular frequency, $\omega = 300$ rad/s

We know that in *L-C-R* circuit,

$$\tan\phi = \frac{X_C - X_L}{R}$$

where, $\phi =$ phase angle $= 45°$

$X_C =$ capacitive reactance $= \frac{1}{\omega C}$

$X_L =$ inductive reactance $= \omega L$

$$\Rightarrow \quad \tan 45° = \frac{X_C - X_L}{R}$$

$$\Rightarrow \quad X_C - X_L = R \quad [\because \tan 45° = 1]$$

$$\Rightarrow \quad \frac{1}{\omega C} - \omega L = R$$

$$\Rightarrow \frac{1}{\omega C} - 300 \times 30 \times 10^{-3} = 1$$

$$\Rightarrow \quad \frac{1}{\omega C} = 10 \Rightarrow \omega C = \frac{1}{10}$$

$$\Rightarrow \quad C = \frac{1}{10\omega}$$

$$\Rightarrow \quad C = \frac{1}{10 \times 300}$$

$$\Rightarrow \quad C = \frac{1}{3} \times 10^{-3}\text{ F} = \frac{1}{x} \times 10^{-3}$$

Hence, $x = 3$

100. *(4)* Given, $V_0 = 250$ V,

$R = 8\Omega$, $L = 24$ mH and $C = 60\ \mu$F

We know that at resonance power ,

$$P = \frac{V_{\text{rms}}^2}{R}$$

$$\Rightarrow \quad P = \frac{(250/\sqrt{2})^2}{8}$$

$$\left[\because V_{\text{rms}} = \frac{V_0}{\sqrt{2}} = \frac{250}{\sqrt{2}}\text{ V}\right]$$

$$= (250)^2/16$$

$$= \frac{62500}{16} = 3906.25 \text{ W}$$

$$\approx 4 \text{ kW}, x = 4$$

101. *(10)* Given, wavelength of transmission signal,

$$\lambda = 960 \text{ m}$$

Capacitance, $C = 2.56\ \mu$F

$$= 2.56 \times 10^{-6}\text{ F}$$

As we know resonance frequency,

$$f = \frac{1}{2\pi\sqrt{LC}}$$

Also, frequency f

$$\therefore \quad \frac{v}{\lambda} = \frac{1}{2\pi\sqrt{LC}}$$

$$\Rightarrow \sqrt{LC} = \frac{\lambda}{v \times 2\pi}$$

On squaring both sides, we get

$$\Rightarrow \quad LC = \frac{\lambda^2}{v^2 \times 4\pi^2}$$

$$\Rightarrow \quad L = \frac{\lambda^2}{v^2 \times 4\pi^2 \times C}$$

$$\Rightarrow \quad L = \frac{(960)^2}{(3\times10^8)^2 \times 4\pi^2 \times 2.56\times10^{-6}}$$

$$\therefore \quad L = 10\times10^{-8}\ \text{H}$$

Topic 2

01. *(c)*

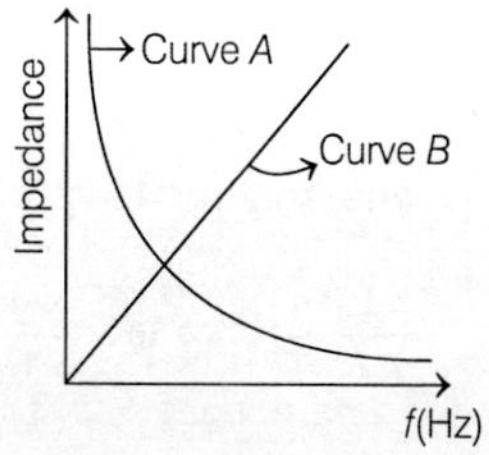

$X_L = \omega L = 2\pi fL \qquad [X_L \propto f]$

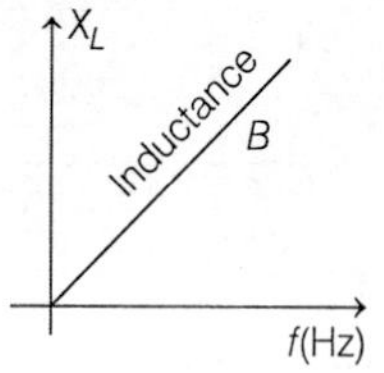

$X_C = \dfrac{1}{\omega C} = \dfrac{1}{2\pi fC} \qquad \left[X_C \propto \dfrac{1}{f}\right]$

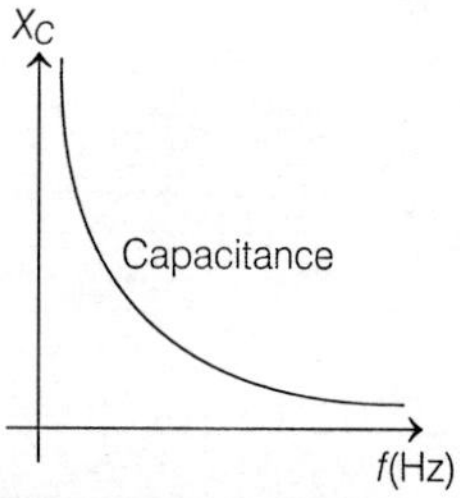

$R =$ constant

$A \to X_C \qquad X_C = A$

$B \to X_L \qquad X_L = B$

Option (c) is correct.

02. *(d)* Given, $\omega = 120\pi$

and $T = \dfrac{2\pi}{\omega} = \dfrac{2\pi}{120\pi}$

$= \dfrac{\pi}{60\pi} = \dfrac{1}{60}$

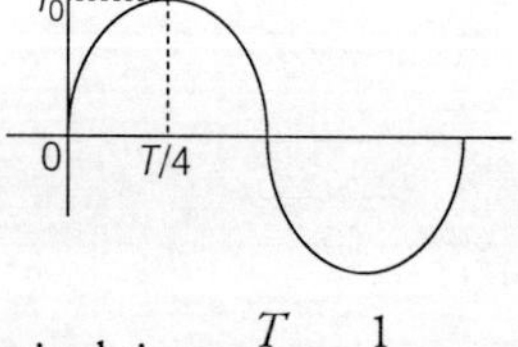

So, required time $= \dfrac{T}{4} = \dfrac{1}{240}$ s

03. *(c)* We know that in R-C circuit

$$V = V_0\left(1 - e^{-\left(\frac{t}{RC}\right)}\right)$$

where, V = voltage across the capacitor,

V_0 = supply voltage,

t = elapsed time since the application of voltage

and RC = time constant of the RC-charging circuit

$$\Rightarrow 50 = 100\left(1 - e^{-\frac{t}{RC}}\right) \Rightarrow \frac{1}{2} = 1 - e^{-\frac{t}{RC}}$$

By solving we get

$$\Rightarrow \qquad 2 = e^{\frac{t}{RC}} \qquad \text{...(i)}$$

Taking log on both sides of Eq. (i), we get

$$\ln 2 = \ln\left(e^{\frac{t}{RC}}\right)$$

$$\Rightarrow \qquad t = RC \ln 2$$

$$= 100\times10^{-6}\times(0.69)$$

$$\Rightarrow \qquad t = 0.69\times10^{-4}\ \text{s}$$

04. *(c)* When square wave is applied at the input, then

For charging, the capacitor

$$Q_1 = Q(1 - e^{\frac{-t}{RC}})$$

Similarly, for discharging the capacitor,

$$Q_2 = Q_{\max}(e^{\frac{-t}{RC}})$$

In this manner, charging and discharging exponentially with time will keep on happening alternatively. Therefore, the output wave pattern monitored by CRO would look close to

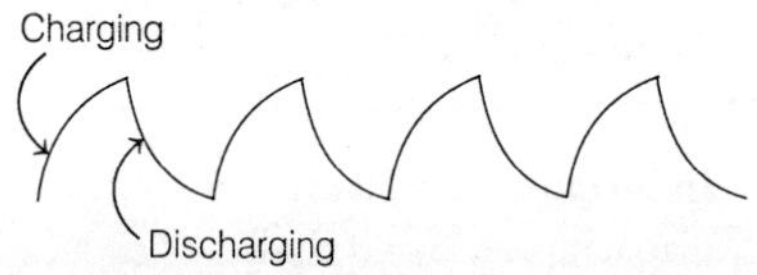

05. *(c)* For charging capacitor, q is given as

$$q = q_0(1 - e^{-t/\tau}) = CV(1 - e^{-t/\tau})$$

At $\quad t = 2\tau,$

$$q = CV\left(1 - e^{\frac{-2\tau}{\tau}}\right)$$

$$\Rightarrow \qquad q = CV(1 - e^{-2})$$

06. (d) As switch S_1 is closed and switch S_2 is kept open. Now, capacitor is charging through a resistor R. Charge on a capacitor at any time t is

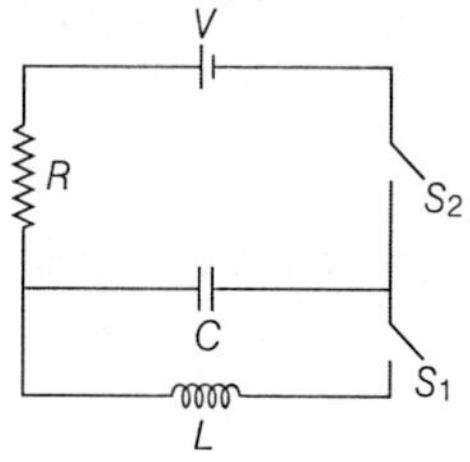

$q = q_0(1 - e^{-t/\tau}),\ q = CV(1 - e^{-t/\tau})$

[As $q_0 = CV$]

At, $t = \dfrac{\tau}{2}$

$q = CV(1 - e^{-\tau/2\tau}) = CV(1 - e^{-1/2})$

At $t = \tau$

$q = CV(1 - e^{-\tau/\tau}) = CV(1 - e^{-1})$

At $t = 2\tau$

$q = CV(1 - e^{-2\tau/\tau}) = CV(1 - e^{-2})$

07. *(b)* Neon bulb is filled with gas, so its resistance is infinite, hence no current flows through it.

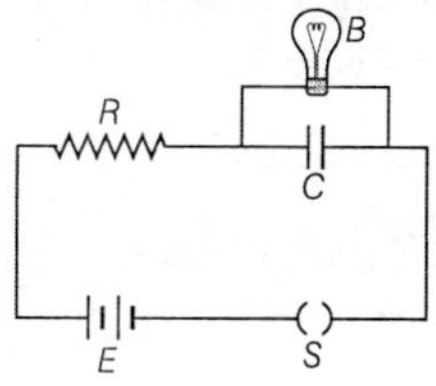

Now, $V_c = E(1 - e^{-t/RC})$

$\Rightarrow \quad 120 = 200(1 - e^{-t/RC})$

$\Rightarrow e^{-t/RC} = \dfrac{2}{5} \Rightarrow t = RC\ \ln 2.5$

$\Rightarrow \quad R = \dfrac{t}{C\ln 2.5} = \dfrac{t}{2.303\,C\log 2.5}$

$= 2.7\times10^6\,\Omega$

08. *(a)* Considert the L-C-R circuit at any time t

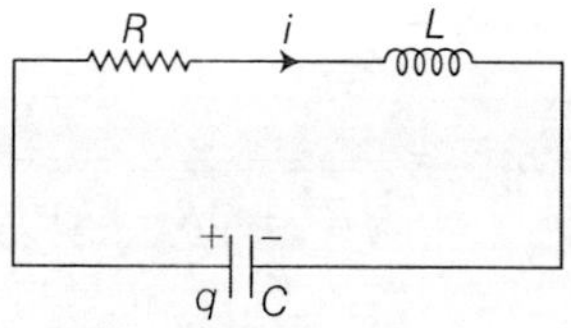

Now, applying KVL

We have $\dfrac{q}{C} - iR - \dfrac{Ldi}{dt} = 0$

As current is decreasing with time, we can write $i = -\dfrac{dq}{dt}$

$$\Rightarrow \qquad \frac{q}{C} + \frac{dq}{dt}R + \frac{Ld^2q}{dt^2} = 0$$

or $\frac{d^2q}{dt^2}+\frac{R}{L}\frac{dq}{dt}+\frac{q}{LC}=0$

This equation is equivalent to that of a damped oscillator

Thus, we can write the solution as

$$Q_{max}(t)=Q_0\cdot e^{-Rt/2L}$$

or $Q_{max}^2=Q_0^2\,e^{-\frac{Rt}{L}}$

As $L_1>L_2$.

Therefore, damping is faster for L_2.

Hence, option (a) is correct.

> ***Alternate Solution***
>
> Inductance is inertia of circuit. It means inductance opposes the flow of charge, more inductance means decay of charge is slow.
>
> In option (a), in a given time to, $Q_1^2>Q_2^2$.
>
> 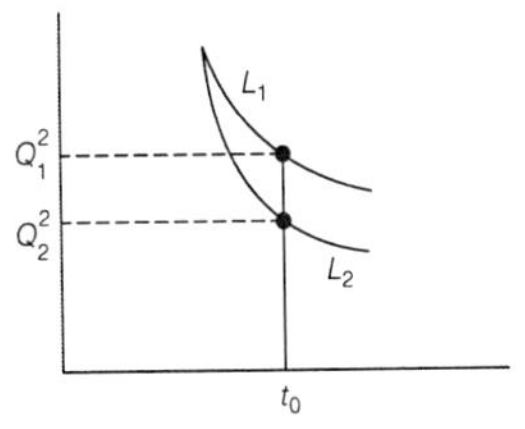
>
>
> So, $L_1>L_2$. Hence, option (a) is correct.

09. *(c)* Energy stored in capacitor,

$$U=\frac{1}{2}\frac{q^2}{C}=\frac{1}{2C}(q_0e^{-t/\tau})^2$$

$$=\frac{q_0^2}{2C}e^{-2t/\tau}\quad[\text{where, }\tau=CR]$$

$$U=U_ie^{-2t/\tau},$$

$$\frac{1}{2}U_i=U_ie^{-2t_1/\tau}\quad[\text{Given}]$$

$$\frac{1}{2}=e^{-2t_1/\tau}$$

$$\Rightarrow\quad t_1=\frac{\tau}{2}\ln 2$$

Now, $q=q_0e^{-t/\tau}$,

$$\frac{1}{4}q_0=q_0e^{-t_2/\tau}\quad\left[q=\frac{1}{4}q_0\right]$$

$$t_2=\tau\ln 4=2\tau\ln 2$$

$$\therefore\quad\frac{t_1}{t_2}=\frac{1}{4}$$

10. *(4)* According to given data in question,

E_0 changes to $\frac{E_0}{3}$

[E_0 is initial value of electric field]

$\therefore q_0$ changes to $\frac{q_0}{3}$

Also, $\frac{q_0}{3}=q_0e^{-t/RC}\quad[\because q=q_0e^{-t/RC}]$

$$\ln\left(\frac{1}{3}\right)=-\frac{t}{RC}\Rightarrow R=\frac{t}{C\ln 3}\quad\ldots\text{(i)}$$

Put $t=6.6\,\mu s$, $C=1.5\,\mu F$, $R=4\,\Omega$

Hence, resistance of the wire is $4\,\Omega$.

11. *(9)* In an oscillatory L-C circuit initial energy stored is,

$$U_{E\max}=\frac{1}{2}\frac{Q_{max}^2}{C}$$

and maximum energy of inductor when capacitor is fully empty and current in circuit is maximum is,

$$U_{B_{max}}=\frac{1}{2}LI_{max}^2$$

Assuming N_0 loss of energy,

$$U_{E_{max}}=U_{B_{max}}$$

$$\frac{1}{2}\frac{Q_{max}^2}{C}=\frac{1}{2}LI_{max}^2\Rightarrow I_{max}^2=\frac{Q_{max}^2}{LC}$$

$$\Rightarrow\quad I_{max}=Q_{max}/\sqrt{LC}$$

$$=\frac{2.7\times10^{-6}}{\sqrt{75\times10^{-3}\times1.2\times10^{-6}}}$$

$$=\frac{2.7\times10^{-6}}{\sqrt{9\times10^{-8}}}=\frac{2.7\times10^{-6}}{3\times10^{-4}}$$

$$=0.9\times10^{-2}=9\text{ mA}$$

12. *(10)* Given, inductance, $L=50$ mH

$$=50\times10^{-3}\text{ H}$$

Capacitance, $C=500\times10^{-6}$ F

Energy stored in capacitor given by

$$U=\frac{1}{2}CV^2$$

$$=\frac{1}{2}\times500\times10^{-6}\times10^4=\frac{5}{2}\text{J}$$

Current will be maximum when capacitor is equal to inductor.

i.e., $\frac{1}{2}LI^2=\frac{5}{2}$

$$\Rightarrow\quad I=\sqrt{\frac{5}{L}}=\sqrt{\frac{5}{50\times10^{-3}}}$$

$$=10\text{ A}$$

13. *(15)* We know, impedance

$$Z=\sqrt{R^2+(X_L-X_C)^2}$$

$$=\sqrt{120^2+(10-100)^2}$$

$$=150\Omega$$

$$\left[\begin{array}{l}\because X_L=\omega L=100\times100\times10^{-3}=10\Omega\\ X_C=\frac{1}{\omega_c}=\frac{1}{100\times100\times10^{-6}}=100\Omega\end{array}\right]$$

So, $i_{rms}=\frac{V_{rms}}{Z}=\frac{20}{150}=\frac{2}{15}$ A

Now,

Heat lost $=ms\Delta T=i_{rms}^2R\Delta t$

$$\Rightarrow\quad\Delta t=\frac{(ms)\Delta T}{i_{rms}^2R}=\frac{2\times16}{\frac{4}{225}\times120}=15\text{ s}$$

Topic 3

01. *(a)* $\frac{N_1}{N_2}=\frac{V_1}{V_2}\Rightarrow\frac{10}{1}=\frac{230}{V_2}$

$$\Rightarrow\quad V_2=23\text{ V}$$

$$P_2=\frac{V_2^2}{R_2}=\frac{23\times23}{46}=\frac{23}{2}=11.5\text{ W}$$

02. *(a)* AC Generator: It is a device which converts mechanical energy into electrical energy.

Galvanometer : It shows deflection when current passes through it, so it is used to show presence of current in the wire.

Transformer : It is used to step up or stepdown the voltage by principle of mutual inductance.

Metal detector : It contain inductor coil and use principle of induction and resonance in AC circuit.

> • Concept **Enhancer**
>
> AC generator works on the principle of electromagnetic induction by producing alternating current
>
> $i=i_0\sin\omega t$
>
> and emf, $e=NAB\omega\sin\omega t$

03. *(c)* Voltage across secondary coil,

$$=\frac{\text{Power across load}}{\text{Current passing through load}}$$

$$\Rightarrow\quad V_2=\frac{P}{I_L}=\frac{60}{0.11}\Rightarrow V_2=545.45\text{ V}$$

Voltage across primary coil,

$V_1=220\text{ V}\Rightarrow V_2>V_1$

It means that stepup transformer is used.

04. *(d)* The emf generated would be maximum when flux (cutting) would be maximum i.e., angle between area vector of coil and magnetic field is 0°. The emf generated is given by (as a function of time)

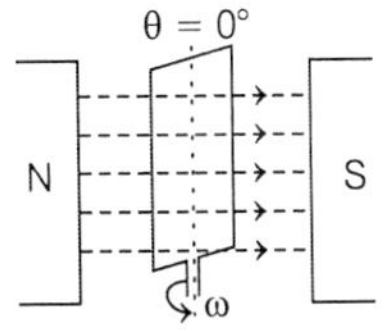

Flux, $\phi=NBA\cos\theta\quad[\theta=\omega t]$

$$\phi = NBA\cos\omega t$$

$$e = -\frac{d\phi}{dt} = NBA\omega \sin\omega t$$

$$e_{max} = NBA\omega \qquad [\omega t = 90°]$$

05. *(a)* The core of transformer is laminated to reduce energy loss due to eddy currents because induction is reduced by laminating.

06. *(b)* Given, $I_p = 4$ A, $N_p = 140$

and $N_s = 280$

From the formula,

$$\frac{I_p}{I_s} = \frac{N_s}{N_p} \text{ or } \frac{4}{I_s} = \frac{280}{140}$$

So, $I_s = 2$ A

07. *(b)* Given, input to transformer is 10 V, 4 kW and secondary voltage is 240 V, efficiency = 80%

Now, efficiency $= \frac{P_{out}}{P_{in}}$

$$\Rightarrow \frac{P_{out}}{P_{in}} = \frac{80}{100} \Rightarrow \frac{V_{out} \cdot I_{out}}{P_{in}} = \frac{80}{100}$$

$$\Rightarrow \frac{240 \times I_{out}}{4 \times 1000} = \frac{80}{100}$$

$$\Rightarrow I_{out} = \frac{80 \times 4 \times 1000}{240 \times 100} = 13.33 \text{ A}$$

08. *(c)* Given,

$V_{in} = 220$ V, $N_P = 100$, $N_S = 10$

Using transformer equation

$$\frac{V_{in}}{V_S} = \frac{N_P}{N_S}$$

$$\Rightarrow V_S = \frac{V_{in} \times N_S}{N_P} = \frac{220 \times 10}{100} = 22 \text{ V}$$

Now, two resistance are connected in series.

$$R_{eq} = R_1 + R_2$$

$$\Rightarrow R_{eq} = 15 + 7 = 22 \text{ k}\Omega$$

$$I_0 = \frac{V_S}{R_{eq}} = \frac{22}{22} = 1 \text{ mA}$$

$$\Rightarrow V_0 = I_0 \times R_2$$

$$V_0 = 1 \text{ mA} \times 7 \text{ k}\Omega = 7 \text{ V}$$

09. *(c)* Given, $V_1 = 8$ kV $= 8000$ V

$V_2 = 160$ V

$P_2 = 80$ kW $= 80 \times 1000$ W

Since, transformer is ideal with purely resistive load and have unity power factor, so

$$P_1 = P_2 = 80000 \text{ W}$$

$$\Rightarrow V_1 I_1 = V_2 I_2 = 80000 \text{ W}$$

$$\Rightarrow I_1 = \frac{80000}{V_1} = \frac{80000}{8000} = 10 \text{ A}$$

and $I_2 = \frac{80000}{V_2} = \frac{80000}{160} = 500$ A

Primary load,

$$R_1 = \frac{V_1}{I_1} = \frac{8000}{10} = 800\Omega$$

and secondary load,

$$R_2 = \frac{V_2}{I_2} = \frac{160}{500} = 0.32\Omega$$

10. *(c)* Given,

$P = 1$ kW $= 1000$ W

$V = 220$ V, $R = 2\Omega$

Current, $I = \frac{P}{V} = \frac{1000}{220}$ A

Power loss, $P_{loss} = I^2 R$

$$= \left(\frac{1000}{220}\right)^2 \times 2$$

$$= 41.32 \text{ W}$$

Now, the efficiency of transmission line,

$$\eta = \left(\frac{P}{P + P_{loss}}\right) \times 100$$

$$= \left(\frac{1000}{1000 + 41.32}\right) \times 100$$

$$\eta = 96.03\% \approx 96\%$$

Hence, correct option is (c).

11. *(a)* Given,

Number of turns in primary,

$N_1 = 300$

Number of turns in secondary,

$N_2 = 150$

Output power,

$P_2 = 2.2$ kW $= 2.2 \times 10^3$ W

Current in secondary coil,

$I_2 = 10$A

Output power,

$P_2 = I_2 V_2$

$$\Rightarrow V_2 = \frac{P_2}{I_2} = \frac{2.2 \times 10^3}{10} = 220 \text{ V} \quad \text{... (i)}$$

We know that

$$\frac{N_1}{N_2} = \frac{\text{Input voltage}}{\text{Output voltage}}$$

$$\frac{N_1}{N_2} = \frac{V_1}{V_2} \Rightarrow V_1 = \left(\frac{N_1}{N_2}\right) V_2$$

$$\Rightarrow V_1 = \left(\frac{300}{150}\right) \times (220 \text{ V})$$

[using Eq. (i)]

$V_1 = 440$ V ....(ii)

Again, $\frac{V_1}{V_2} = \frac{I_2}{I_1} = \frac{220}{440} \times 10$

[using Eqs. (i) and (ii)]

$\Rightarrow I_1 = 5$ A

12. *(a)* For a transformer, there are two circuits which have N_p and N_s (number of coil turns), I_p and I_S (currents) respectively as shown below.

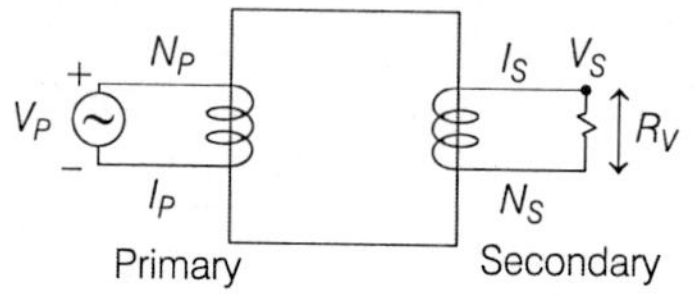

Here, input voltage,

$V_p = 2300$ V

Number of turns in primary coil,

$N_P = 4000$

Output voltage, $V_S = 230$ V

Output power, $P_S = V_S \cdot I_S$

Input power, $P_P = V_P I_P$

$\therefore$ The efficiency of the transformer is

$$\eta = \frac{\text{Output (secondary) power}}{\text{Input (primary) power}}$$

$$\Rightarrow \eta = \frac{V_S \cdot I_S}{V_P \cdot I_P} \times 100$$

$$\Rightarrow \eta = \frac{(230)(I_S)}{(2300)(5)} \times 100$$

$$90 = \frac{230\, I_S}{(2300) \times 5} \times 100$$

$\Rightarrow I_S = 45$ A

13. *(5)* Given, Angular speed of coil, $\omega = \frac{1}{2}$ rotation per second $\omega = \pi(s^{-1})$

$B = 0.01$ T, $N = 200$, $A = 0.2 \text{ m}^2$

Maximum voltage generated in a coil

$V = NBA\,\omega$

$$\Rightarrow \frac{2\pi}{\beta} = 200 \times 0.01 \times 0.2 \times \pi$$

$$\Rightarrow \frac{2\pi}{\beta} = \frac{2\pi}{5} \Rightarrow \beta = 5$$

14. *(240)* Given, $\frac{V_S}{V_P} = \frac{N_S}{N_P}$

$$\frac{120}{12000} = \frac{N_S}{N_P}$$

$$\frac{N_S}{N_P} = \frac{1}{100} \quad \text{...(i)}$$

For an ideal transformer, input power = Output power

and power is given by

$$P = VI$$

$$I_P V_P = I_S V_S = 60000 \text{ W}$$

$$12000 \times I_P = 60000$$

$$I_P = 5 \text{ A}$$

Now, $R_P = \frac{V_P}{I_P} = \frac{12000}{5} = 2400\,\Omega$

For secondary coil,

$$P_S = V_S \cdot I_S$$

$$I_S = \frac{P_S}{V_S} = \frac{60000}{120} = 500\text{A}$$

$$\frac{V_P}{V_S} = \frac{I_P R_P}{I_S R_S}$$

$$\Rightarrow R_S = \frac{I_P}{I_S} \times \frac{V_S}{V_P} \times R_P$$

$$= \frac{I_P}{I_S} \times \frac{N_S}{N_P} \times R_P$$

$$= \frac{5}{500} \times \frac{1}{100} \times 2400$$

$$R_S = 240 \times 10^{-3}\,\Omega = 240\text{ m}\Omega$$

15. *(44)* Given that, area,

$$A = 70 \text{ cm}^2 = 70 \times 10^{-4} \text{ m}^2$$

Number of turns, $N = 600$

Magnetic field, $B = 0.4 \text{ Wbm}^{-2}$

Angular velocity,

$$\omega = 500 \text{ rpm} = 500 \times \frac{\pi}{30} \text{ rad/s}$$

$$\Rightarrow \quad \omega = \frac{50\pi}{3} \text{ rad/s}$$

The angle between the normal to coil and the direction of magnetic field,

$$\theta = 90° - 60° = 30°$$

Induced emf in rotating coil is given as

$$e = N\omega BA \sin\theta$$

$$= 600 \times \frac{50\pi}{3} \times 0.4 \times 70 \times 10^{-4} \times \sin 30°$$

$$\Rightarrow \quad e = 44 \text{ V}$$

16. *(1584)* Number of turns in rectangular coil,

$$N = 100$$

Area, $A = 14 \times 10^{-2} \text{ m}^2$

Rotation frequency,

$$f = 360 \text{ rev/min} = \frac{360}{60} = 6 \text{ rev/s}$$

Magnetic field, $B = 3.0 \text{ T}$

∴ Maximum value of emf produced,

$$e = NBA\omega$$

$$= NBA(2\pi f)$$

$$= 100 \times 3 \times 14 \times 10^{-2} \times 2 \times \frac{22}{7} \times 6$$

$$= 1584 \text{ V}$$

17. *(440)*

> **key idea** For a transformer
>
> $P_{input} = P_{output}$ (in ideal case)
>
> $\Rightarrow V_P I_P = V_S I_S$

As we know,

In a transformer,

$$\frac{N_P}{N_S} = \frac{V_P}{V_S}$$

Since, $N_S = 24$,

$V_P = 220 \text{ V}$

and $V_S = 12 \text{ V}$

$$\frac{N_P}{24} = \frac{220}{12}$$

$$\Rightarrow \quad N_P = \frac{220 \times 24}{12} = 440$$

In JEE Main there is usually 0-1 question per paper is visible from this chapter. In past 3 years, 44 conceptual and 7 numerical questions were asked of moderate to difficult level from this chapter.
For detailed analysis scan QR code

ELECTROMAGNETIC WAVES

Topic 1 Displacement Current and Properties of EM Waves

Level 1

Imp. Displacement Current

01 If ε_0 denotes the permittivity of free space and ϕ_E is the flux of the electric field through the area bounded by the closed surface, then dimensions of $\left(\varepsilon_0 \frac{d\phi_E}{dt}\right)$ are that of *[2025, 7 April Shift-I]*
(a) electric field (b) electric charge
(c) electric potential (d) electric current

02 A plane electromagnetic wave of frequency 20 MHz travels in free space along the $+x$-direction. At a particular point in space and time, the electric field vector of the wave is $E_y = 9.3\ \text{Vm}^{-1}$. Then, the magnetic field vector of the wave at that point is *[2025, 23 Jan Shift-II]*
(a) $B_z = 1.55 \times 10^{-8}$ T
(b) $B_z = 3.1 \times 10^{-8}$ T
(c) $B_z = 9.3 \times 10^{-8}$ T
(d) $B_z = 6.2 \times 10^{-8}$ T

03 A plane electromagnetic wave propagates along the $+x$-direction in free space. The components of the electric filed **E** and magnetic field, **B** vectors associated with the wave in cartesian frame are *[2025, 29 Jan Shift-II]*
(a) E_x, B_y (b) E_y, B_z
(c) E_z, B_y (d) E_y, B_x

04 In the given electromagnetic wave $E_y = 600 \sin(\omega t - kx)\ \text{Vm}^{-1}$, intensity of the associate light beam is (in W/m^2) (Given, $\varepsilon_0 = 9 \times 10^{-12}\ \text{C}^2\text{N}^{-1}\text{m}^{-2}$) *[2024, 6 April Shift-II]*
(a) 486 (b) 972
(c) 243 (d) 729

05 If frequency of electromagnetic wave is 60 MHz and it travels in air along z-direction, then the corresponding electric and magnetic field vectors will be mutually perpendicular to each other and the wavelength of the wave (in m) is *[2024, 1 Feb Shift-II]*
(a) 2 (b) 2.5 (c) 10 (d) 5

06 Given below are two statements.
Statement I Electromagnetic waves carry energy as they travel through space and this energy is equally shared by the electric and magnetic fields.
Statement II When electromagnetic waves strike a surface, a pressure is exerted on the surface.
In the light of the above statements, choose the most appropriate answer from the options given below. *[2024, 31 Jan Shift-II]*
(a) Both Statement I and Statement II are correct.
(b) Both Statement I and Statement II are incorrect.
(c) Statement I is correct but Statement II is incorrect.
(d) Statement I is incorrect but Statement II is correct.

07 An object is placed in a medium of refractive index 3. An electromagnetic wave of intensity $6 \times 18^8\ \text{W/m}^2$ falls normally on the object and it is absorbed completely. The radiation pressure on the object would be (speed of light in free space $= 3 \times 10^8$ m/s) *[2024, 27 Jan Shift-II]*
(a) $2\ \text{Nm}^{-2}$ (b) $18\ \text{Nm}^{-2}$
(c) $36\ \text{Nm}^{-2}$ (d) $6\ \text{Nm}^{-2}$

Imp. Intensity of EM Wave

08 A plane electromagnetic wave propagating in x-direction is described by
$E_y = (200\ \text{Vm}^{-1}) \sin[1.5 \times 10^7 t - 0.05x]$.
The intensity of the wave is
(Use, $\varepsilon_0 = 8.85 \times 10^{-12}\ \text{C}^2\text{N}^{-1}\ \text{m}^{-2}$) *[2024, 27 Jan Shift-I]*
(a) $53.1\ \text{Wm}^{-2}$ (b) $106.2\ \text{Wm}^{-2}$
(c) $26.6\ \text{Wm}^{-2}$ (d) $35.4\ \text{Wm}^{-2}$

21 In free space, an electromagnetic wave of 3 GHz frequency strikes over the edge of an object of size $\lambda/100$, where λ is the wavelength of the wave in free space. The phenomenon, which happens there will be *[2022, 26 June Shift-I]*
(a) reflection (b) refraction (c) diffraction (d) scattering

22 An expression for oscillating electric field in a plane electromagnetic wave is given as
$E_z = 300 \sin(5\pi \times 10^3 x - 3\pi \times 10^{11} t)\ \text{Vm}^{-1}$.
Then, the value of magnetic field amplitude will be (Given, speed of light in vacuum $c = 3 \times 10^8\ \text{ms}^{-1}$) *[2022, 30 June Shift-I]*
(a) 1×10^{-6} T (b) 5×10^{-6} T
(c) 18×10^9 T (d) 21×10^9 T

23 The relative permittivity of distilled water is 81. The velocity of light in it will be
(Take, $\mu_r = 1$) *[2021, 27 July Shift-I]*
(a) 4.33×10^7 m/s (b) 2.33×10^7 m/s
(c) 3.33×10^7 m/s (d) 5.33×10^7 m/s

24 In an electromagnetic wave, the electric field vector and magnetic field vector are given as $\mathbf{E} = E_0 \hat{\mathbf{i}}$ and $\mathbf{B} = B_0 \hat{\mathbf{k}}$, respectively. The direction of propagation of electromagnetic wave is along *[2021, 20 July Shift-II]*
(a) $\hat{\mathbf{k}}$ (b) $\hat{\mathbf{j}}$
(c) $(-\hat{\mathbf{k}})$ (d) $(-\hat{\mathbf{j}})$

25 A plane electromagnetic wave propagating along y-direction can have the following pair of electric field **E** and magnetic field **B** components. *[2021, 18 March Shift-II]*
(a) E_y, B_y or E_z, B_z (b) E_y, B_x or E_x, B_y
(c) E_x, B_z or E_z, B_x (d) E_x, B_y or E_y, B_x

26 A plane electromagnetic wave of frequency 25 GHz is propagating in vacuum along the z-direction. At a particular point in space and time, the magnetic field is given by $\mathbf{B} = 5 \times 10^{-8}\,\hat{\mathbf{j}}$T. The corresponding electric field **E** is (speed of light, $c = 3 \times 10^8\ \text{ms}^{-1}$) *[2020, 8 Jan Shift-II]*
(a) $-1.66 \times 10^{-16}\,\hat{\mathbf{i}}$ V/m (b) $1.66 \times 10^{-16}\,\hat{\mathbf{i}}$ V/m
(c) $-15\,\hat{\mathbf{i}}$ V/m (d) $15\,\hat{\mathbf{i}}$ V/m

27 If the magnetic field of a plane electromagnetic wave is given by
$B = 100 \times 10^{-6} \sin\left[2\pi \times 2 \times 10^{15}\left(t - \frac{x}{c}\right)\right]$ then the maximum electric field associated with it is (Take, the speed of light $= 3 \times 10^8$ m/s) *[2019, 10 Jan Shift-I]*
(a) 6×10^4 N/C (b) 4×10^4 N/C
(c) 3×10^4 N/C (d) 4.5×10^4 N/C

28 Magnetic field in a plane electromagnetic wave is given by $\mathbf{B} = B_0 \sin(kx + \omega t)\,\hat{\mathbf{j}}$T
Expression for corresponding electric field will be (where, c is speed of light.) *[JEE Main 2017]*
(a) $\mathbf{E} = -B_0 c \sin(kx + \omega t)\,\hat{\mathbf{k}}$ V/m
(b) $\mathbf{E} = B_0 c \sin(kx - \omega t)\,\hat{\mathbf{k}}$ V/m
(c) $\mathbf{E} = \frac{B_0}{c} \sin(kx + \omega t)\,\hat{\mathbf{k}}$ V/m
(d) $\mathbf{E} = B_0 c \sin(kx + \omega t)\,\hat{\mathbf{k}}$ V/m

29 The magnetic field in a travelling electromagnetic wave has a peak value of 20 nT. The peak value of electric field strength is *[JEE Main 2013]*
(a) 12 V/m (b) 3 V/m
(c) 6 V/m (d) 9 V/m

Imp. Energy Density

30 The rms value of the electric field of the light coming from the sun is 720 NC^{-1}. The average total energy density of the electromagnetic wave is *[AIEEE 2006]*
(a) $4.58 \times 10^{-6}\ \text{Jm}^{-3}$ (b) $6.37 \times 10^{-9}\ \text{Jm}^{-3}$
(c) $81.35 \times 10^{-12}\ \text{Jm}^{-3}$ (d) $3.3 \times 10^{-3}\ \text{Jm}^{-3}$

31 Electromagnetic waves are transverse in nature is evident by *[AIEEE 2002]*
(a) polarisation (b) interference
(c) reflection (d) diffraction

Level 2

32 The radiation pressure exerted by a 450 W light source on a perfectly reflecting surface placed at 2m away from it, is
(a) 6×10^{-8} Pa (b) 1.5×10^{-8} Pa
(c) 0 (d) 3×10^{-8} Pa

33 Due to presence of an EM wave whose electric component is given by $E = 100 \sin(\omega t - kx)\,\text{NC}^{-1}$ a cylinder of length 200 cm holds certain amount of em energy inside it. If another cylinder of same length but half diameter than previous one holds same amount of em energy, the magnitude of the electric field of the corresponding EM wave should be modified as *[2025, 28 Jan Shift-I]*
(a) $25 \sin(\omega t - kx)\,\text{NC}^{-1}$ (b) $50 \sin(\omega t - kx)\,\text{NC}^{-1}$
(c) $200 \sin(\omega t - kx)\,\text{NC}^{-1}$ (d) $400 \sin(\omega t - kx)\,\text{NC}^{-1}$

34 The electric field of an electromagnetic wave in free space is
$E = 57 \cos[7.5 \times 10^6 t - 5 \times 10^{-3}(3x + 4y)]\,(4\hat{\mathbf{i}} - 3\hat{\mathbf{j}})$ N/C
The associated magnetic field in Tesla is *[2025, 23 Jan Shift-I]*

21 In free space, an electromagnetic wave of 3 GHz frequency strikes over the edge of an object of size $\lambda/100$, where λ is the wavelength of the wave in free space. The phenomenon, which happens there will be *[2022, 26 June Shift-I]*

(a) reflection (b) refraction (c) diffraction (d) scattering

22 An expression for oscillating electric field in a plane electromagnetic wave is given as

$E_z = 300 \sin(5\pi \times 10^3 x - 3\pi \times 10^{11} t)\ \text{Vm}^{-1}$.

Then, the value of magnetic field amplitude will be (Given, speed of light in vacuum $c = 3 \times 10^8\ \text{ms}^{-1}$) *[2022, 30 June Shift-I]*

(a) 1×10^{-6} T (b) 5×10^{-6} T
(c) 18×10^9 T (d) 21×10^9 T

23 The relative permittivity of distilled water is 81. The velocity of light in it will be
(Take, $\mu_r = 1$) *[2021, 27 July Shift-I]*

(a) 4.33×10^7 m/s (b) 2.33×10^7 m/s
(c) 3.33×10^7 m/s (d) 5.33×10^7 m/s

24 In an electromagnetic wave, the electric field vector and magnetic field vector are given as $\mathbf{E} = E_0 \hat{\mathbf{i}}$ and $\mathbf{B} = B_0 \hat{\mathbf{k}}$, respectively. The direction of propagation of electromagnetic wave is along *[2021, 20 July Shift-II]*

(a) $\hat{\mathbf{k}}$ (b) $\hat{\mathbf{j}}$
(c) $(-\hat{\mathbf{k}})$ (d) $(-\hat{\mathbf{j}})$

25 A plane electromagnetic wave propagating along y-direction can have the following pair of electric field **E** and magnetic field **B** components. *[2021, 18 March Shift-II]*

(a) E_y, B_y or E_z, B_z (b) E_y, B_x or E_x, B_y
(c) E_x, B_z or E_z, B_x (d) E_x, B_y or E_y, B_x

26 A plane electromagnetic wave of frequency 25 GHz is propagating in vacuum along the z-direction. At a particular point in space and time, the magnetic field is given by $\mathbf{B} = 5 \times 10^{-8} \hat{\mathbf{j}}$T. The corresponding electric field **E** is (speed of light, $c = 3 \times 10^8\ \text{ms}^{-1}$) *[2020, 8 Jan Shift-II]*

(a) $-1.66 \times 10^{-16} \hat{\mathbf{i}}$ V/m (b) $1.66 \times 10^{-16} \hat{\mathbf{i}}$ V/m
(c) $-15 \hat{\mathbf{i}}$ V/m (d) $15 \hat{\mathbf{i}}$ V/m

27 If the magnetic field of a plane electromagnetic wave is given by

$B = 100 \times 10^{-6} \sin\left[2\pi \times 2 \times 10^{15}\left(t - \frac{x}{c}\right)\right]$ then the maximum electric field associated with it is (Take, the speed of light $= 3 \times 10^8$ m/s) *[2019, 10 Jan Shift-I]*

(a) 6×10^4 N/C (b) 4×10^4 N/C
(c) 3×10^4 N/C (d) 4.5×10^4 N/C

28 Magnetic field in a plane electromagnetic wave is given by $\mathbf{B} = B_0 \sin(kx + \omega t)\ \hat{\mathbf{j}}$T

Expression for corresponding electric field will be (where, c is speed of light.) *[JEE Main 2017]*

(a) $\mathbf{E} = -B_0 c \sin(kx + \omega t)\ \hat{\mathbf{k}}$ V/m
(b) $\mathbf{E} = B_0 c \sin(kx - \omega t)\ \hat{\mathbf{k}}$ V/m
(c) $\mathbf{E} = \frac{B_0}{c} \sin(kx + \omega t)\ \hat{\mathbf{k}}$ V/m
(d) $\mathbf{E} = B_0 c \sin(kx + \omega t)\ \hat{\mathbf{k}}$ V/m

29 The magnetic field in a travelling electromagnetic wave has a peak value of 20 nT. The peak value of electric field strength is *[JEE Main 2013]*

(a) 12 V/m (b) 3 V/m
(c) 6 V/m (d) 9 V/m

Imp. Energy Density

30 The rms value of the electric field of the light coming from the sun is 720 NC^{-1}. The average total energy density of the electromagnetic wave is *[AIEEE 2006]*

(a) $4.58 \times 10^{-6}\ \text{Jm}^{-3}$ (b) $6.37 \times 10^{-9}\ \text{Jm}^{-3}$
(c) $81.35 \times 10^{-12}\ \text{Jm}^{-3}$ (d) $3.3 \times 10^{-3}\ \text{Jm}^{-3}$

31 Electromagnetic waves are transverse in nature is evident by *[AIEEE 2002]*

(a) polarisation (b) interference
(c) reflection (d) diffraction

Level 2

32 The radiation pressure exerted by a 450 W light source on a perfectly reflecting surface placed at 2m away from it, is *[2025, 3 April Shift-I]*

(a) 6×10^{-8} Pa (b) 1.5×10^{-8} Pa
(c) 0 (d) 3×10^{-8} Pa

33 Due to presence of an EM wave whose electric component is given by $E = 100 \sin(\omega t - kx)\text{NC}^{-1}$ a cylinder of length 200 cm holds certain amount of em energy inside it. If another cylinder of same length but half diameter than previous one holds same amount of em energy, the magnitude of the electric field of the corresponding EM wave should be modified as *[2025, 28 Jan Shift-I]*

(a) $25 \sin(\omega t - kx)\ \text{NC}^{-1}$ (b) $50 \sin(\omega t - kx)\ \text{NC}^{-1}$
(c) $200 \sin(\omega t - kx)\ \text{NC}^{-1}$ (d) $400 \sin(\omega t - kx)\ \text{NC}^{-1}$

34 The electric field of an electromagnetic wave in free space is

$E = 57 \cos[7.5 \times 10^6 t - 5 \times 10^{-3}(3x + 4y)]$ $(4\hat{\mathbf{i}} - 3\hat{\mathbf{j}})$ N/C

The associated magnetic field in Tesla is *[2025, 23 Jan Shift-I]*

(a) $\mathbf{B} = \frac{57}{3\times10^8}\cos[7.5\times10^6 t - 5\times10^{-3}(3x+4y)](5\hat{\mathbf{k}})$

(b) $\mathbf{B} = -\frac{57}{3\times10^8}\cos[7.5\times10^6 t - 5\times10^{-3}(3x+4y)](5\hat{\mathbf{k}})$

(c) $\mathbf{B} = -\frac{57}{3\times10^8}\cos[7.5\times10^6 t - 5\times10^{-3}(3x+4y)](\hat{\mathbf{k}})$

(d) $\mathbf{B} = \frac{57}{3\times10^8}\cos[7.5\times10^6 t - 5\times10^{-3}(3x+4y)](\hat{\mathbf{k}})$

35 The magnetic field in a plane electromagnetic wave is $B_y = 3.5\times10^{-7}\sin(1.5\times10^3 x + 0.5\times10^{11} t)$ T. The corresponding electric field will be *[2024, 9 April Shift-II]*

(a) $E_z = 1.17\sin(1.5\times10^3 x + 0.5\times10^{11} t)\ \text{Vm}^{-1}$

(b) $E_y = 1.17\sin(1.5\times10^3 x + 0.5\times10^{11} t)\ \text{Vm}^{-1}$

(c) $E_z = 105\sin(1.5\times10^3 x + 0.5\times10^{11} t)\ \text{Vm}^{-1}$

(d) $E_y = 10.5\sin(1.5\times10^3 x + 0.5\times10^{11} t)\ \text{Vm}^{-1}$

36 A plane EM wave is propagating along x-direction. It has a wavelength of 4 mm. If electric field is in y-direction with the maximum magnitude of 60 Vm^{-1}, the equation for magnetic field is *[2024, 9 April Shift-I]*

(a) $B_z = 2\times10^{-7}\sin\left[\frac{\pi}{2}\times10^3(x - 3\times10^8 t)\right]\hat{\mathbf{k}}$ T

(b) $B_x = 60\sin\left[\frac{\pi}{2}(x - 3\times10^8 t)\right]\hat{\mathbf{i}}$ T

(c) $B_z = 2\times10^{-7}\sin\left[\frac{\pi}{2}(x - 3\times10^8 t)\right]\hat{\mathbf{k}}$ T

(d) $B_z = 60\sin\left[\frac{\pi}{2}(x - 3\times10^8 t)\right]\hat{\mathbf{k}}$ T

37 A plane progressive wave is given by $y = 2\cos 2\pi(330t - x)$ m. The frequency of the wave is *[2024, 8 April Shift-II]*

(a) 330 Hz (b) 340 Hz
(c) 165 Hz (d) 660 Hz

38 Electromagnetic waves travel in a medium with speed of 1.5×10^8 m/s. The relative permeability of the medium is 2.0. The relative pemittivity will be *[2024, 6 April Shift-I]*

(a) 1 (b) 4 (c) 2 (d) 5

39 The electric field in an electromagnetic wave is given by $\mathbf{E} = \hat{\mathbf{i}}\,40\cos\omega\left(t - \frac{z}{c}\right)\text{NC}^{-1}$. The magnetic field induction of this wave is (in SI unit) *[2024, 4 April Shift-I]*

(a) $\mathbf{B} = \hat{\mathbf{i}}\frac{40}{c}\cos\omega\left(t - \frac{z}{c}\right)$ (b) $\mathbf{B} = \hat{\mathbf{k}}\frac{40}{c}\cos\omega\left(t - \frac{z}{c}\right)$

(c) $\mathbf{B} = \hat{\mathbf{j}}\frac{40}{c}\cos\omega\left(t - \frac{z}{c}\right)$ (d) $\mathbf{B} = \hat{\mathbf{j}}\,40\cos\omega\left(t - \frac{z}{c}\right)$

Imp. Energy Density of EM Waves

40 In a plane EM wave, the electric field oscillates sinusoidally at a frequency of 5×10^{10} Hz and an amplitude of 50 Vm^{-1}. The total average energy density of the electromagnetic field of the wave is [Use, $\varepsilon_0 = 8.85\times10^{-12}\ \text{C}^2/\text{N-m}^2$] *[2024, 31 Jan Shift-I]*

(a) $2.212\times10^{-10}\ \text{Jm}^{-3}$ (b) $1.106\times10^{-8}\ \text{Jm}^{-3}$
(c) $4.425\times10^{-8}\ \text{Jm}^{-3}$ (d) $2.212\times10^{-8}\ \text{Jm}^{-3}$

41 Match List I with List II.

List I	List II
A. Gauss's law of magnetostatics	I. $\oint \mathbf{E}\cdot d\mathbf{A} = \frac{1}{\varepsilon_0}\int \rho dV$
B. Faraday's law of electromagnetic induction	II. $\oint \mathbf{B}\cdot d\mathbf{A} = 0$
C. Ampere's law	III. $\int \mathbf{E}\cdot d\mathbf{l} = \frac{-d}{dt}\int \mathbf{B}\cdot d\mathbf{A}$
D. Gauss's law of electrostatics	IV. $\oint \mathbf{B}\cdot d\mathbf{l} = \mu_0 I$

Choose the correct answer from the options given below. *[2024, 30 Jan Shift-II]*

(a) A-III, B-IV, C-I, D-II (b) A-I, B-III, C-IV, D-II
(c) A-II, B-III, C-IV, D-I (d) A-IV, B-II, C-III, D-I

42 The electric field of an electromagnetic wave in free space is represented as $\mathbf{E} = E_0\cos(\omega t - kz)\,\hat{\mathbf{i}}$. The corresponding magnetic induction vector will be *[2024, 30 Jan Shift-I]*

(a) $\mathbf{B} = E_0 c\cos(\omega t - kz)\,\hat{\mathbf{j}}$ (b) $\mathbf{B} = \frac{E_0}{c}\cos(\omega t + kz)\,\hat{\mathbf{j}}$

(c) $\mathbf{B} = \frac{E_0}{c}\cos(\omega t - kz)\,\hat{\mathbf{j}}$ (d) $\mathbf{B} = E_0 c\cos(\omega t + kz)\,\hat{\mathbf{j}}$

43 A plane electromagnetic wave of frequency 35 MHz travels in free space along the x-direction. At a particular point (in space and time) $\mathbf{E} = 9.6\,\hat{\mathbf{j}}$ V/m. The value of magnetic field at this point is *[2024, 29 Jan Shift-II]*

(a) $3.2\times10^{-8}\ \hat{\mathbf{k}}$ T (b) $9.6\times10^{-8}\ \hat{\mathbf{k}}$ T
(c) $3.2\times10^{-8}\ \hat{\mathbf{i}}$ T (d) $9.6\ \hat{\mathbf{j}}$ T

44 Match List I with List II

List I	List II
A. $\oint \mathbf{B}\cdot d\mathbf{l} = \mu_0 i_0 + \mu_0\varepsilon_0\frac{d\phi_E}{dt}$	I. Gauss's law for electricity
B. $\oint \mathbf{E}\cdot d\mathbf{l} = \frac{d\phi_B}{dt}$	II. Gauss's law for magnetism
C. $\oint \mathbf{E}\cdot d\mathbf{A} = \frac{Q}{\varepsilon_0}$	III. Faraday's law
D. $\oint \mathbf{B}\cdot d\mathbf{A} = 0$	IV. Ampere-Maxwell's law

Choose the correct answer from the options given below. *[2024, 29 Jan Shift-I]*

(a) A-I, B-II, C-III, D-IV (b) A-IV, B-III, C-I, D-II
(c) A-II, B-III, C-I, D-IV (d) A-IV, B-I, C-III, D-II

45 The electric field in an electromagnetic wave is given as $E = 20\sin\omega\left(t - \frac{x}{c}\right)\hat{j}\ NC^{-1}$,

where ω and c are angular frequency and velocity of electromagnetic wave, respectively. The energy contained in a volume of $5 \times 10^{-4}\ m^3$ will be

(Given, $\varepsilon_0 = 8.85 \times 10^{-12} C^2/Nm^2$) *[2023, 11 April Shift-I]*

(a) 28.5×10^{-13} J (b) 17.7×10^{-13} J
(c) 8.85×10^{-13} J (d) 88.5×10^{-13} J

46 The ratio of average electric energy density and total average energy density of electromagnetic wave is *[2023, 1 Feb Shift-II]*

(a) 1/2 (b) 2 (c) 1 (d) 3

47 Given below are two statements.

Statement I Electromagnetic waves are not deflected by electric and magnetic field.

Statement II The amplitude of electric field and the magnetic field in electromagnetic waves are related to each other as $E_0 = \sqrt{\frac{\mu_0}{\varepsilon_0}} B_0$.

In the light of the above statements, choose the correct answer from the options given below. *[2023, 29 Jan Shift-II]*

(a) Statement I is false but Statement II is true.
(b) Statement I is true but Statement II is false.
(c) Both Statement I and Statement II are true.
(d) Both Statement I and Statement II are false.

48 Which of the following are true ?

A. Speed of light in vacuum is dependent on the direction of propagation.
B. Speed of light in a medium is independent the wavelength of light.
C. The speed of light is independent the motion of the source.
D. The speed of light in a medium is independent of intensity.

Choose the correct answer from the options given below. *[2023, 29 Jan Shift-I]*

(a) B and C (b) B and D (c) C and D (d) A and C

49 Match List I with List II.

List I	List II
A. Gauss's law in Electrostatics	I. $\oint E \cdot dl = -\frac{d\phi_B}{dt}$
B. Faraday's law	II. $\oint B \cdot dA = 0$
C. Gauss's law in magnetism	III. $\oint B \cdot dl = \mu_0 i_0 + \mu_0\varepsilon_0 \frac{d\phi_E}{dt}$
D. Ampere-Maxwell law	IV. $\oint E \cdot dS = \frac{q}{\varepsilon_0}$

Choose the correct answer from the options given below. *[2023, 25 Jan Shift -II]*

(a) A-III, B-IV, C-I, D-II (b) A-IV, B-I, C-II, D-III
(c) A-I, B-II, C-III, D-IV (d) A-II, B-III, C-IV, D-I

50 Sunlight falls normally on a surface of area 36 cm^2 and exerts an average force of 7.2×10^{-9} N within a time period of 20 min. Considering a case of complete absorption, the energy flux of incident light is *[2022, 28 July Shift-II]*

(a) 25.92×10^2 W/cm^2 (b) 8.64×10^{-6} W/cm^2
(c) 6.0 W/cm^2 (d) 0.06 W/cm^2

51 As shown in the figure, after passing through the medium 1. The speed of light v_2 in medium 2 will be (Given, $c = 3 \times 10^8 ms^{-1}$) *[2022, 28 July Shift-I]*

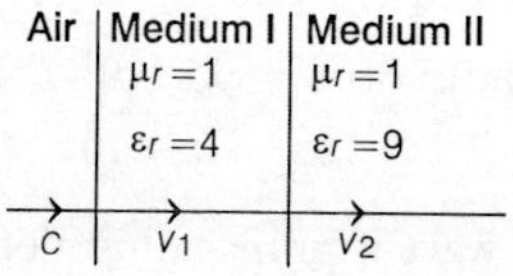

(a) 1.0×10^8 m s^{-1} (b) 0.5×10^8 m s^{-1}
(c) 1.5×10^8 m s^{-1} (d) 3.0×10^8 m s^{-1}

52 A beam of light travelling along X-axis is described by the electric field $E_y = 900\sin\omega(t - x/c)$. The ratio of electric force to magnetic force on a charge q moving along Y-axis with a speed of 3×10^7 ms^{-1} will be (Given, speed of light $= 3 \times 10^8$ ms^{-1}) *[2022, 27 July Shift-I]*

(a) 1 : 1 (b) 1 : 10
(c) 10 : 1 (d) 1 : 2

53 The oscillating magnetic field in a plane electromagnetic wave is given by $B_y = 5 \times 10^{-6} \sin 1000\pi(5x - 4 \times 10^8 t)$ T. The amplitude of electric field will be *[2022, 26 July Shift-II]*

(a) 15×10^2 Vm^{-1} (b) 5×10^{-6} Vm^{-1}
(c) 16×10^{12} Vm^{-1} (d) 4×10^2 Vm^{-1}

54 The magnetic field of a plane electromagnetic wave is given by $\mathbf{B} = 2 \times 10^{-8}\sin(0.5 \times 10^3 x + 1.5 \times 10^{11} t)\hat{j}$T The amplitude of the electric field would be *[2022, 26 July Shift-I]*

(a) 6 Vm^{-1} along X-axis
(b) 3 Vm^{-1} along Z-axis
(c) 6 Vm^{-1} along Z-axis
(d) 2×10^{-8} Vm^{-1} along Z-axis

55 A RADAR sends an electromagnetic signal of electric field $E_0 = 2.25$ V/m and magnetic field $B_0 = 1.5 \times 10^{-8}$ T which strikes a target on line of sight at a distance of 3 km in a medium. After that a pail of signal (echo) reflects back towards the RADAR with same velocity and by same path. If the signal was transmitted at time $t = 0$ from RADAR, then after how much time echo will reach to the RADAR? [2022, 28 June Shift-I]

(a) 2.0×10^{-5} s (b) 4.0×10^{-5} s
(c) 1.0×10^{-5} s (d) 8.0×10^{-5} s

56 The electromagnetic waves travel in a medium at a speed of 2.0×10^8 m/s. The relative permeability of the medium is 1.0. The relative permittivity of the medium will be [2022, 25 June Shift-II]

(a) 2.25 (b) 4.25 (c) 6.25 (d) 8.25

57 The electric field in an electromagnetic wave is given by $E = 56.5 \sin \omega(t - x/c)$ NC^{-1}. Find the intensity of the wave, if it is propagating along X-axis in the free space. (Given, $\varepsilon_0 = 8.85 \times 10^{-12}$ $C^2N^{-1}m^{-2}$) [2022, 25 June Shift-I]

(a) 5.65 Wm^{-2} (b) 4.24 Wm^{-2}
(c) 1.9×10^{-7} Wm^{-2} (d) 56.5 Wm^{-2}

58 A plane electromagnetic wave travels in a medium of relative permeability 1.61 and relative permittivity 6.44. If magnitude of magnetic intensity is 4.5×10^{-2} Am^{-1} at a point, what will be the approximate magnitude of electric field intensity at that point? (Given, permeability of free space $\mu_0 = 4\pi \times 10^{-7}$ NA^{-2}, speed of light in vacuum, $c = 3 \times 10^8$ ms^{-1}) [2022, 24 June Shift-I]

(a) 16.96 Vm^{-1} (b) 2.25×10^{-2} Vm^{-1}
(c) 8.48 Vm^{-1} (d) 6.75×10^6 Vm^{-1}

Imp. Relation μ_0, ε_0 and c

59 Electric field of plane electromagnetic wave propagating through a non–magnetic medium is given by $E = 20\cos(2 \times 10^{10} t - 200x)$ V/m. The dielectric constant of the medium is equal to (Take, $\mu_r = 1$) [2021, 1 Sep Shift-II]

(a) 9 (b) 2 (c) 1/3 (d) 3

60 Electric field in a plane electromagnetic wave is given by $E = 50\sin(500x - 10 \times 10^{10} t)$ V/m.

The velocity of electromagnetic wave in this medium is (Given, c = speed of light in vacuum) [2021, 27 Aug Shift-I]

(a) $\frac{3}{2}c$ (b) c (c) $\frac{2}{3}c$ (d) $\frac{c}{2}$

61 A light beam is described by $E = 800 \sin(\omega t - x/c)$. An electron is allowed to move normal to the propagation of light beam with a speed 3×10^7 ms^{-1}. What is the maximum magnetic force exerted on the electron? [2021, 26 Aug Shift-II]

(a) 1.28×10^{-17} N (b) 1.28×10^{-18} N
(c) 12.8×10^{-17} N (d) 12.8×10^{-18} N

62 A linearly polarised electromagnetic wave in vacuum is $E = 3.1 \cos[(1.8)z - (5.4 \times 10^6)t]\hat{i}$ N/C is incident normally on a perfectly reflecting wall at $z = a$. Choose the correct option. [2021, 25 July Shift-I]

(a) The wavelength is 5.4 m.
(b) The frequency of electromagnetic wave is 54×10^4 Hz.
(c) The transmitted wave will be $3.1 \cos[(1.8)z - (5.4 \times 10^6)t]\hat{}$ N/C
(d) The reflected wave will be $3.1 \cos[(1.8)z + (5.4 \times 10^6)t]\hat{}$ N/C

63 Intensity of sunlight is observed as 0.092 Wm^{-2} at a point in free space. What will be the peak value of magnetic field at that point? ($\varepsilon_0 = 8.85 \times 10^{-12} C^{-2}N^{-1}m^{-2}$) [2021, 22 July Shift-II]

(a) 2.77×10^{-8} T (b) 1.96×10^{-8} T
(c) 8.31 T (d) 5.88 T

64 A plane electromagnetic wave of frequency 100 MHz is travelling in vacuum along the x-direction. At a particular point in space and time, $\mathbf{B} = 2.0 \times 10^{-8}\ \hat{\mathbf{k}}$T (where, $\hat{\mathbf{k}}$ is unit vector along z-direction). What is E at this point? [2021, 18 March Shift-I]

(a) 0.6 $\hat{\mathbf{j}}$ V/m (b) 6.0 $\hat{\mathbf{k}}$ V/m
(c) 6.0 $\hat{\mathbf{j}}$ V/m (d) 0.6 $\hat{\mathbf{k}}$ V/m

65 A plane electromagnetic wave of frequency 500 MHz is travelling in vacuum along y-direction. At a particular point in space and time, $\mathbf{B} = 8.0 \times 10^{-8}\ \hat{\mathbf{z}}$T. The value of electric field at this point is (speed of light = 3×10^8 ms^{-1}, $\hat{\mathbf{x}}, \hat{\mathbf{y}}, \hat{\mathbf{z}}$ are unit vectors along x, y and z-direction). [2021, 16 March Shift-I]

(a) $-24\hat{\mathbf{x}}$ V/m (b) $2.6\hat{\mathbf{x}}$ V/m
(c) $24\hat{\mathbf{x}}$ V/m (d) $-2.6\hat{\mathbf{x}}$ V/m

66 For an electromagnetic wave travelling in free space, the relation between average energy densities due to electric U_e and magnetic U_m fields is [2021, 16 March Shift-I]

(a) $U_e = U_m$ (b) $U_e > U_m$ (c) $U_e < U_m$ (d) $U_e \neq U_m$

67 For a plane electromagnetic wave, the magnetic field at a point x and time t is $\mathbf{B}(x, t) = [1.2 \times 10^{-7} \sin(0.5 \times 10^3 x + 1.5 \times 10^{11} t)\ \hat{\mathbf{k}}]$ T

The instantaneous electric field E corresponding to B is (Given, speed of light, $c = 3 \times 10^8$ m/s) [2020, 6 Sep Shift-II]

(a) $\mathbf{E}(x, t) = [-36\sin(0.5 \times 10^3 x + 1.5 \times 10^{11} t)\hat{\mathbf{j}}]\frac{V}{m}$
(b) $\mathbf{E}(x, t) = [36\sin(1 \times 10^3 x + 0.5 \times 10^{11} t)\hat{\mathbf{j}}]\frac{V}{m}$
(c) $\mathbf{E}(x, t) = [36\sin(0.5 \times 10^3 x + 1.5 \times 10^{11} t)\hat{\mathbf{k}}]\frac{V}{m}$
(d) $\mathbf{E}(x, t) = [36\sin(1 \times 10^3 x + 1.5 \times 10^{11} t)\hat{\mathbf{j}}]\frac{V}{m}$

68 If the magnetic field in a plane electromagnetic wave is given by
$\mathbf{B} = 3\times10^{-8}\sin(1.6\times10^{3}x + 48\times10^{10}t)\hat{j}\,\text{T}$, then what will be expression for electric field? *[2020, 7 Jan Shift-I]*

(a) $\mathbf{E} = 60\sin(1.6\times10^{3}x + 48\times10^{10}t)\hat{k}\ \text{V/m}$
(b) $\mathbf{E} = 3\times10^{-8}\sin(1.6\times10^{3}x + 48\times10^{10}t)\hat{j}\ \text{V/m}$
(c) $\mathbf{E} = 3\times10^{-8}\sin(1.6\times10^{3}x + 48\times10^{10}t)\hat{j}\ \text{V/m}$
(d) $\mathbf{E} = 9\sin(1.6\times10^{3}x + 48\times10^{10}t)\hat{k}\ \text{V/m}$

69 An electromagnetic wave is represented by the electric field $\mathbf{E} = E_0\,\hat{n}\sin[\omega t + (6y - 8z)]$. Taking unit vectors in x, y and z- directions to be $\hat{i}, \hat{j}, \hat{k}$, the direction of propagation $\hat{s}$, is *[2019, 12 April Shift-I]*

(a) $\hat{s} = \dfrac{3\hat{i} - 4\hat{j}}{5}$ (b) $\hat{s} = \dfrac{-4\hat{k} + 3\hat{j}}{5}$
(c) $\hat{s} = \left(\dfrac{-3\hat{j} + 4\hat{k}}{5}\right)$ (d) $\hat{s} = \dfrac{4\hat{j} - 3\hat{k}}{5}$

70 Light is incident normally on a completely absorbing surface with an energy flux of $25\ \text{W cm}^{-2}$. If the surface has an area of $25\ \text{cm}^2$, the momentum transferred to the surface in 40 min time duration will be *[2019, 10 April Shift-II]*

(a) 3.5×10^{-6} N-s (b) 6.3×10^{-4} N-s
(c) 1.4×10^{-6} N-s (d) 5.0×10^{-3} N-s

71 A 27 mW laser beam has a cross-sectional area of $10\ \text{mm}^2$. The magnitude of the maximum electric field in this electromagnetic wave is given by [Take, permittivity of space, $\varepsilon_0 = 9\times10^{-12}$ SI units and speed of light, $c = 3\times10^{8}$ m/s] *[2019, 11 Jan Shift-II]*

(a) 1 kV/m (b) 0.7 kV/m (c) 2 kV/m (d) 1.4 kV/m

72 A plane electromagnetic wave of frequency 50 MHz travels in free space along the positive x-direction. At a particular point in space and time, $\mathbf{E} = 6.3\,\hat{j}$ V/m. The corresponding magnetic field $\mathbf{B}$, at that point will be *[2019, 9 Jan Shift-I]*

(a) $18.9\times10^{8}\,\hat{k}$ T (b) $6.3\times10^{-8}\,\hat{k}$ T
(c) $18.9\times10^{-8}\,\hat{k}$ T (d) $2.1\times10^{-8}\,\hat{k}$ T

73 An EM wave from air enters a medium. The electric fields are $\mathbf{E}_1 = E_{01}\hat{x}\cos\left[2\pi\nu\left(\dfrac{z}{c} - t\right)\right]$ in air and $\mathbf{E}_2 = E_{02}\hat{x}\cos[k(2z - ct)]$ in medium, where the wave number k and frequency ν refer to their values in air. The medium is non-magnetic.
If ε_{r_1} and ε_{r_2} refer to relative permittivities of air and medium respectively, which of the following options is correct? *[JEE Main 2018]*

(a) $\dfrac{\varepsilon_{r_1}}{\varepsilon_{r_2}} = 4$ (b) $\dfrac{\varepsilon_{r_1}}{\varepsilon_{r_2}} = 2$
(c) $\dfrac{\varepsilon_{r_1}}{\varepsilon_{r_2}} = \dfrac{1}{4}$ (d) $\dfrac{\varepsilon_{r_1}}{\varepsilon_{r_2}} = \dfrac{1}{2}$

74 A red LED emits light at 0.1W uniformly around it. The amplitude of the electric field of the light at a distance of 1 m from the diode is *[JEE Main 2015]*

(a) 5.48 V/m (b) 7.75 V/m
(c) 1.73 V/m (d) 2.45 V/m

75 An electromagnetic wave travelling in the x-direction has frequency of 2×10^{14} Hz and electric field amplitude of $27\ \text{V m}^{-1}$. From the options given below, which one describes the magnetic field for this wave? *[JEE Main 2015]*

(a) $\mathbf{B}(x, t) = (3\times10^{-8}\,\text{T})\hat{j}\sin[2\pi(1.5\times10^{-8}x - 2\times10^{14}t)]$
(b) $\mathbf{B}(x, t) = (9\times10^{-8}\,\text{T})\hat{k}\sin[2\pi(1.5\times10^{-6}x - 2\times10^{14}t)]$
(c) $\mathbf{B}(x, t) = (9\times10^{-8}\,\text{T})\hat{i}\sin[2\pi(1.5\times10^{-8}x - 2\times10^{14}t)]$
(d) $\mathbf{B}(x, t) = (9\times10^{-8}\,\text{T})\hat{j}\sin[1.5\times10^{-6}x - 2\times10^{14}t]$

76 For plane electromagnetic waves propagating in the z-direction, which one of the following combination gives the correct possible direction for electric (E) and magnetic (B) field, respectively? *[JEE Main 2015]*

(a) $(\hat{i} + 2\hat{j})$ and $(2\hat{i} - \hat{j})$ (b) $(-2\hat{i} - 3\hat{j})$ and $(3\hat{i} - 2\hat{j})$
(c) $(2\hat{i} + 3\hat{j})$ and $(\hat{i} + 2\hat{j})$ (d) $(3\hat{i} + 4\hat{j})$ and $(4\hat{i} - 3\hat{j})$

77 An electromagnetic wave in vacuum has the electric and magnetic fields $\mathbf{E}$ and $\mathbf{B}$, which are always perpendicular to each other. The direction of polarisation is given by $\mathbf{X}$ and that of wave propagation by $\mathbf{k}$. Then, *[AIEEE 2012]*

(a) $\mathbf{X} \parallel \mathbf{B}$ and $\mathbf{k} \parallel \mathbf{B}\times\mathbf{E}$ (b) $\mathbf{X} \parallel \mathbf{E}$ and $\mathbf{k} \parallel \mathbf{E}\times\mathbf{B}$
(c) $\mathbf{X} \parallel \mathbf{B}$ and $\mathbf{k} \parallel \mathbf{E}\times\mathbf{B}$ (d) $\mathbf{X} \parallel \mathbf{E}$ and $\mathbf{k} \parallel \mathbf{B}\times\mathbf{E}$

78 An electromagnetic wave of frequency $\nu = 3.0$ MHz passes from vacuum into a dielectric medium with permittivity $\varepsilon = 4.0$. Then, *[AIEEE 2004]*

(a) wavelength is doubled and the frequency remains unchanged
(b) wavelength is doubled and frequency becomes half
(c) wavelength is halved and frequency remains unchanged
(d) wavelength and frequency both remain unchanged

79 A radiation of energy E falls normally on a perfectly reflecting surface. The momentum transferred to the surface is *[AIEEE 2004]*

(a) $\dfrac{E}{c}$ (b) $\dfrac{2E}{c}$
(c) Ec (d) $\dfrac{E}{c^2}$

Level 3

Imp. Characteristic of EM Wave

80 The magnetic field of an EM wave is given by $\mathbf{B} = \left(\frac{\sqrt{3}}{2}\hat{i} + \frac{1}{2}\hat{j}\right) 30 \sin\left[\omega\left(t - \frac{z}{c}\right)\right]$ (SI Units).

The corresponding electric field in SI unit is *[2025, 28 Jan Shift-II]*

(a) $\mathbf{E} = \left(\frac{\sqrt{3}}{2}\hat{i} - \frac{1}{2}\hat{j}\right) 30c \sin\left[\omega\left(t + \frac{z}{c}\right)\right]$

(b) $\mathbf{E} = \left(\frac{1}{2}\hat{i} - \frac{\sqrt{3}}{2}\hat{j}\right) 30c \sin\left[\omega\left(t - \frac{z}{c}\right)\right]$

(c) $\mathbf{E} = \left(\frac{1}{2}\hat{i} + \frac{\sqrt{3}}{2}\hat{j}\right) 30c \sin\left[\omega\left(t + \frac{z}{c}\right)\right]$

(d) $\mathbf{E} = \left(\frac{3}{4}\hat{i} + \frac{1}{4}\hat{j}\right) 30c \cos\left[\omega\left(t - \frac{z}{c}\right)\right]$

81 Average force exerted on a non-reflecting surface at normal incidence is 2.4×10^{-4} N. If $360\ \text{W/cm}^2$ is the light energy flux during span of 1 h 30 min. Then, the area of the surface is *[2024, 8 April Shift-II]*

(a) $0.2\ \text{m}^2$ (b) $0.1\ \text{m}^2$
(c) $0.02\ \text{m}^2$ (d) $20\ \text{m}^2$

82 A point source of 100 W emits light with 5% efficiency. At a distance of 5 m from the source, the intensity produced by the electric field component is *[2023, 30 Jan Shift-II]*

(a) $\frac{1}{10\pi}\frac{\text{W}}{\text{m}^2}$ (b) $\frac{1}{20\pi}\frac{\text{W}}{\text{m}^2}$
(c) $\frac{1}{2\pi}\frac{\text{W}}{\text{m}^2}$ (d) $\frac{1}{40\pi}\frac{\text{W}}{\text{m}^2}$

83 An EM wave propagating in the x-direction has a wavelength of 8 mm. The electric field vibrating in y-direction has maximum magnitude of $60\ \text{Vm}^{-1}$. Choose the correct equations for electric and magnetic fields, if the EM wave is propagating in vacuum : *[2022, 28 June Shift-II]*

(a) $E_y = 60 \sin\left[\frac{\pi}{4} \times 10^3 (x - 3 \times 10^8 t)\right] \hat{j}\ \text{Vm}^{-1}$
$E_z = 2 \sin\left[\frac{\pi}{4} \times 10^3 (x - 3 \times 10^8 t)\right] \hat{k}\text{T}$

(b) $E_y = 60 \sin\left[\frac{\pi}{4} \times 10^3 (x - 3 \times 10^8 t)\right] \hat{j}\text{Vm}^{-1}$
$B_z = 2 \times 10^{-7} \sin\left[\frac{\pi}{4} \times 10^3 (x - 3 \times 10^8 t)\right] \hat{k}\text{T}$

(c) $E_y = 2 \times 10^{-7} \sin\left[\frac{\pi}{4} \times 10^3 (x - 3 \times 10^8 t)\right] \hat{j}\ \text{Vm}^{-1}$
$B_z = 60 \sin\left[\frac{\pi}{4} \times 10^3 (x - 3 \times 10^8 t)\right] \hat{k}\text{T}$

(d) $E_y = 2 \times 10^{-7} \sin\left[\frac{\pi}{4} \times 10^4 (x - 4 \times 10^8 t)\right] \hat{j}\ \text{Vm}^{-1}$
$B_z = 60 \sin\left[\frac{\pi}{4} \times 10^4 (x - 4 \times 10^8 t)\right] \hat{k}\text{T}$

84 An electric bulb is rated as 200 W. What will be the peak magnetic field at 4 m distance produced by the radiations coming from this bulb? Consider this bulb as a point source with 3.5% efficiency. *[2022, 24 June Shift-II]*

(a) 1.19×10^{-8} T (b) 1.71×10^{-8} T
(c) 0.84×10^{-8} T (d) 3.36×10^{-8} T

85 The magnetic field vector of an electromagnetic wave is given by $\mathbf{B} = B_0 \frac{\hat{i} + \hat{j}}{\sqrt{2}} \cos(kz - \omega t)$ T, where $\hat{i}, \hat{j}$ represents unit vector along X and Y-axis, respectively. At $t = 0$, two electric charges q_1 of 4π C and q_2 of 2π C located at $\left(0, 0, \frac{\pi}{k}\right)$ and $\left(0, 0, \frac{3\pi}{k}\right)$ respectively, have the same velocity of $0.5\, c\, \hat{i}$. (where, c is the velocity of light). The ratio of the force acting on charge q_1 to q_2 is *[2021, 31 Aug Shift-I]*

(a) $2\sqrt{2} : 1$ (b) $1 : \sqrt{2}$ (c) $2 : 1$ (d) $\sqrt{2} : 1$

86 The electric field of a plane electromagnetic wave is given by $\hat{E} = E_0(\hat{x} + \hat{y}) \sin(kz - \omega t)$.

Its magnetic field will be given by *[2020, 4 Sep Shift-II]*

(a) $\frac{E_0}{c}(\hat{x} + \hat{y}) \sin(kz - \omega t)$ (b) $\frac{E_0}{c}(-\hat{x} + \hat{y}) \sin(kz - \omega t)$
(c) $\frac{E_0}{c}(\hat{x} - \hat{y}) \sin(kz - \omega t)$ (d) $\frac{E_0}{c}(\hat{x} - \hat{y}) \cos(kz - \omega t)$

87 The electric field of a plane electromagnetic wave is given by $\mathbf{E} = E_0 \hat{i} \cos(kz) \cos(\omega t)$

The corresponding magnetic field $\mathbf{B}$ is then given by *[2019, 10 April Shift-I]*

(a) $\mathbf{B} = \frac{E_0}{c} \hat{j} \sin(kz) \sin(\omega t)$

(b) $\mathbf{B} = \frac{E_0}{c} \hat{j} \sin(kz) \cos(\omega t)$

(c) $\mathbf{B} = \frac{E_0}{c} \hat{k} \sin(kz) \cos(\omega t)$

(d) $\mathbf{B} = \frac{E_0}{c} \hat{j} \cos(kz) \sin(\omega t)$

88 The magnetic field of an electromagnetic wave is given by $\mathbf{B} = 1.6 \times 10^{-6} \cos(2 \times 10^7 z + 6 \times 10^{15} t)(2\hat{i} + \hat{j})\ \text{Wbm}^{-2}$.

The associated electric field will be [2019, 8 April Shift-II]

(a) $E = 4.8 \times 10^2 \cos(2 \times 10^7 z - 6 \times 10^{15} t)$ Vm^{-1}

(b) $E = 4.8 \times 10^2 \cos(2 \times 10^7 z - 6 \times 10^{15} t)(2\hat{j} + \hat{i})$ Vm^{-1}

(c) $E = 4.8 \times 10^2 \cos(2 \times 10^7 z + 6 \times 10^{15} t)(\hat{i} - 2\hat{j})$ Vm^{-1}

(d) $E = 4.8 \times 10^2 \cos(2 \times 10^7 z + 6 \times 10^{15} t)(-\hat{i} + 2\hat{j})$ Vm^{-1}

89 An electromagnetic wave of intensity 50 Wm^{-2} enters in a medium of refractive index n without any loss. The ratio of the magnitudes of electric fields and the ratio of the magnitudes of magnetic fields of the wave before and after entering into the medium are respectively, given by [2019, 11 Jan Shift-I]

(a) $\left(\frac{1}{\sqrt{n}}, \sqrt{n}\right)$ (b) $(\sqrt{n}, \sqrt{n})$

(c) $\left(\frac{1}{\sqrt{n}}, \frac{1}{\sqrt{n}}\right)$ (d) $\left(\sqrt{n}, \frac{1}{\sqrt{n}}\right)$

90 A light wave is incident normally on a glass slab of refractive index 1.5. If 4% of light gets reflected and the amplitude of the electric field of the incident light is 30 V/m, then the amplitude of the electric field for the wave propogating in the glass medium will be [2019, 12 Jan Shift-I]

(a) 30 V/m (b) 6 V/m (c) 10 V/m (d) 24 V/m

Numerical Value Type Questions

91 If an optical medium possesses a relative permeability of $\frac{10}{\pi}$ and relative permittivity of $\frac{1}{0.0885}$, then the velocity of light is greater in vacuum than that in this medium by .... times. [2025, 4 April, Shift-II]

$\mu_0 = 4\pi \times 10^{-7} \mu / m, \varepsilon_0 = 8.85 \times 10^{-12} F / m$

$C = 3 \times 10^8$ m / s

Imp. Displacement Current

92 A parallel plate capacitor consisting of two circular plate of radius 10 cm is being charged by a constant current of 0.15 A. If the rate of change of potential difference between the plates is 7×10^8 V/s, then the integer value of the distance between the parallel plates is $\left(\text{Take, } \varepsilon_0 = 9 \times 10^{12} \frac{F}{m}, \pi = \frac{22}{7}\right)$........ μm. [2025, 29 Jan Shift-II]

93 A time varying potential difference is applied between the plates of a parallel plate capacitor of capacitance 25 μF. The dielectric constant of the medium between the capacitor plates is 1. It produces an instantaneous displacement current of 0.25 mA in the intervening space between the capacitor plates, the magnitude of the rate of change of the potential difference will be...Vs^{-1}. [2025, 23 Jan Shift-II]

94 A parallel plate capacitor of area $A = 16$ cm^2 and separation between the plates 10 cm, is charged by a DC current. Consider a hypothetical plane surface of area $A_0 = 3.2$ cm^2 inside the capacitor and parallel to the plates. At an instant, the current through the circuit is 6A. At the same instant the displacement current through A_0 is .... mA. [2025, 22 Jan Shift-II]

95 An electron with kinetic energy 5 eV enters a region of uniform magnetic field of 3 μT perpendicular to its direction. An electric field E is applied perpendicular to the direction of velocity and magnetic field. The value of E. So, that electron moves along the same path is .... NC^{-1}.

(Given, mass of electron $= 9 \times 10^{-31}$ kg, electric charge $= 1.6 \times 10^{-19}$ C) [2024, 8 April Shift-I]

96 In a medium, the speed of light wave decreases to 0.2 times to its speed in free space. The ratio of relative permittivity to the refractive index of the medium is $x : 1$. The value of x is ....

(Given, speed of light in free space $= 3 \times 10^8$ m s^{-1} and for the given medium $\mu_r = 1$) [2023, 31 Jan Shift-I]

Imp. Radiation Pressure Over a Reflecting Surface

97 A point source of light is placed at the centre of curvature of a hemispherical surface. The source emits a power of 24 W. The radius of curvature of hemisphere is 10 cm and the inner surface is completely reflecting. The force on the hemisphere due to the light falling on it is .... $\times 10^{-8}$ N. [2023, 30 Jan Shift-I]

98 Nearly 10% of the power of a 110 W light bulb is converted to visible radiation. The change in average intensities of visible radiation, at a distance of 1 m from the bulb to a distance of 5 m is $a \times 10^{-2}$ W/m^2. The value of a will be [2022, 29 July Shift-II]

99 The displacement current of 4.425 μA is developed in the space between the plates of parallel plate capacitor when voltage is changing at a rate of 10^6 Vs^{-1}. The area of each plate of the capacitor is 40 cm^2. The distance between each plate of the capacitor is $x \times 10^{-3}$ m. The value of x is ....

(Permittivity of free space, $\varepsilon_0 = 8.85 \times 10^{-12}$ C^2N^{-1}-m^{-2}) [2022, 29 June Shift-II]

100 The intensity of the light from a bulb incident on a surface is 0.22 W/m^2. The amplitude of the magnetic field in this light wave is ...... $\times 10^{-9}$ T.

(Given, permittivity of vacuum, $\varepsilon_0 = 8.85 \times 10^{-12}\ C^2\ N^{-1}\text{-}m^{-2}$, speed of light in vacuum, $c = 3 \times 10^8\ ms^{-1}$) *[2022, 29 June Shift-I]*

101 The electric field in an electromagnetic wave is given by $E = (50\ NC^{-1}) \sin \omega\left(t - \frac{x}{c}\right)$.
The energy contained in a cylinder of volume V is 5.5×10^{-12} J. The value of V is .... cm^3.
(Given $\varepsilon_0 = 8.85 \times 10^{-12}\ C^2N^{-1}\ m^{-2}$). *[2021, 31 Aug Shift-I]*

102 A plane electromagnetic wave with frequency of 30 MHz travels in free space. At particular point in space and time, electric field is 6 V/m. The magnetic field at this point will be $x \times 10^{-8}$ T. The value of x is .... *[2021, 27 Aug Shift-II]*

103 The electric field intensity produced by the radiation coming from a 100 W bulb at a distance of 3 m is E. The electric field intensity produced by the radiation coming from 60 W at the same distance is $\sqrt{\frac{x}{5}}\, E$, where the value of x is .... *[2021, 17 March Shift-II]*

104 Seawater at a frequency $f = 9 \times 10^2$ Hz, has permittivity $\varepsilon = 80\varepsilon_0$ and resistivity $r = 0.25\ \Omega$-m. Imagine a parallel plate capacitor is immersed in seawater and is driven by an alternating voltage source $V(t) = V_0 \sin(2\pi ft)$. Then, the conduction current density becomes 10^x times the displacement current density after time $t = \frac{1}{800}$ s. The value of x is .... $\left(\text{Take, } \frac{1}{4\pi\varepsilon_0} = 9 \times 10^9 N\text{-}m^2C^{-2}\right)$ *[2021, 17 March Shift-II]*

105 If 2.5×10^{-6} N average force is exerted by a light wave on a non-reflecting surface of 30 cm^2 area during 40 min of time span, the energy flux of light just before it falls on the surface is .... W/cm^2. (Round off to the nearest integer) (Assume complete adsorption and normal incidence conditions are there) *[2021, 17 March Shift-I]*

106 A radiation is emitted by 1000 W bulb and it generates an electric field and magnetic field at P, placed at a distance of 2 m. The efficiency of the bulb is 1.25%. The value of peak electric field at P is $x \times 10^{-1}$ V/m. Value of x is ....
(Rounded-off to the nearest integer)
[Take, $\varepsilon_0 = 8.85 \times 10^{-12}\ C^2N^{-1}m^{-2}$, $c = 3 \times 10^8\ ms^{-1}$] *[2021, 26 Feb Shift-I]*

107 An electromagnetic wave of frequency 3 GHz enters a dielectric medium of relative electric permittivity 2.25 from vacuum. The wavelength of this wave in that medium will be .... $\times 10^{-2}$ cm. *[2021, 24 Feb Shift-II]*

108 An electromagnetic wave of frequency 5 GHz, is travelling in a medium whose relative electric permittivity and relative magnetic permeability both are 2. Its velocity in this medium is ...... $\times 10^7$ m/s. *[2021, 24 Feb Shift-I]*

109 Suppose that intensity of a laser is $\left(\frac{315}{\pi}\right) W/m^2$.
The rms electric field (in V/m) associated with this source is close to the nearest integer is ..... .
(Take, $\varepsilon_0 = 8.86 \times 10^{-12}\ C^2N^{-1}m^{-2}$ and $c = 3 \times 10^8\ ms^{-1}$) *[2020, 6 Sep Shift-I]*

Topic 2 EM Spectrum

Level 1

01 Arrange the following in the ascending order of wavelength (λ)
(A) Microwaves (λ_1) (B) Ultraviolet rays (λ_2)
(C) Infrared rays (λ_3) (D) X-rays (λ_4)
Choose the most appropriate answer from the options given below. *[2025, 24 Jan Shift-II]*
(a) $\lambda_3 < \lambda_4 < \lambda_2 < \lambda_1$ (b) $\lambda_4 < \lambda_2 < \lambda_3 < \lambda_1$
(c) $\lambda_4 < \lambda_3 < \lambda_2 < \lambda_1$ (d) $\lambda_4 < \lambda_3 < \lambda_1 < \lambda_2$

02 Match List I with List II.

List I (EM-Wave)	List II (Wavelength Range)
A. Infrared	I. $< 10^{-3}$ nm
B. Ultraviolet	II. 400 nm to 1 nm
C. X-rays	III. 1 mm to 700 nm
D. Gamma rays	IV. 1 nm to 10^{-3} nm

Choose the correct answer from the options given below. *[2024, 5 April Shift-II]*
(a) A-IV, B-III, C-II, D-I
(b) A-II, B-I, C-IV, D-III
(c) A-III, B-II, C-IV, D-I
(d) A I, B III, C II, D IV

03 Arrange the following in the ascending order of wavelength.
A. Gamma rays (λ_1) B. X-rays (λ_2)
C. Infrared waves (λ_3) D. Microwaves (λ_4)
Choose the most appropriate answer from the options given below. *[2024, 4 April Shift-II]*
(a) $\lambda_2 < \lambda_1 < \lambda_4 < \lambda_3$ (b) $\lambda_1 < \lambda_2 < \lambda_3 < \lambda_4$
(c) $\lambda_4 < \lambda_3 < \lambda_1 < \lambda_2$ (d) $\lambda_4 < \lambda_3 < \lambda_2 < \lambda_1$

04 Given below are two statements, one is labelled as Assertion (A) and the other is labelled as Reason (R).

Assertion (A) EM waves used for optical communication have longer wavelengths than that of microwave, employed in Radar technology.

Reason (R) Infrared EM waves are more energetic than microwaves, (used in Radar).

In the light of the above statements, choose the most appropriate answer from the options given below. *[2023, 12 April Shift-I]*

(a) Both A and R are true and R is the correct explanation of A.
(b) Both A and R are true but R is not the correct explanation of A.
(c) A is true but R is false.
(d) A is false but R is true.

05 Which is the correct ascending order of wavelength? *[2022, 26 June Shift-II]*

(a) $\lambda_{\text{visible}} < \lambda_{\text{X-ray}} < \lambda_{\text{gamma-ray}} < \lambda_{\text{microwave}}$
(b) $\lambda_{\text{gamma-ray}} < \lambda_{\text{X-ray}} < \lambda_{\text{visible}} < \lambda_{\text{microwave}}$
(c) $\lambda_{\text{X-ray}} < \lambda_{\text{gamma-ray}} < \lambda_{\text{visible}} < \lambda_{\text{microwave}}$
(d) $\lambda_{\text{microwave}} < \lambda_{\text{visible}} < \lambda_{\text{gamma-ray}} < \lambda_{\text{X-ray}}$

Imp. Electromagnetic Spectrum

06 Choose the correct option relating wavelengths of different parts of electromagnetic wave spectrum *[2020, 4 Sep Shift-I]*

(a) $\lambda_{\text{X-rays}} < \lambda_{\text{microwaves}} < \lambda_{\text{radio waves}} < \lambda_{\text{visible}}$
(b) $\lambda_{\text{visible}} > \lambda_{\text{X-rays}} > \lambda_{\text{radio waves}} > \lambda_{\text{microwaves}}$
(c) $\lambda_{\text{radio waves}} > \lambda_{\text{microwaves}} > \lambda_{\text{visible}} > \lambda_{\text{X-rays}}$
(d) $\lambda_{\text{visible}} < \lambda_{\text{microwaves}} < \lambda_{\text{radio waves}} < \lambda_{\text{X-rays}}$

07 Arrange the following electromagnetic radiations per quantum in the order of increasing energy.

A. Blue light B. Yellow light
C. X-ray D. Radio wave *[JEE Main 2016]*

(a) D, B, A, C (b) A, B, D, C
(c) C, A, B, D (d) B, A, D, C

08 Which of the following are not electromagnetic waves? *[AIEEE 2002]*

(a) cosmic rays (b) γ-rays
(c) β-rays (d) X-rays

Level 2

09 Match List I with List II.

List I	List II
A. Microwave	I. 400 nm to 1 nm
B. Ultraviolet	II. 1 nm to 10^{-3} nm
C. X-Ray	III. 1 mm to 700 nm
D. Infrared	IV. 0.1 m to 1 mm

Choose the correct answer from the options given below. *[2023, 15 April Shift-I]*

(a) A-IV, B-I, C-II, D-III
(b) A-I, B-IV, C-II, D-III
(c) A-IV, B-II, C-I, D-III
(d) A-IV, B-I, C-III, D-II

10 Match List I with List II.

List I	List II
A. Microwave	I. Radioactive decay of the nucleus
B. Gamma rays	II. Rapid acceleration and deceleration of electron in aerials
C. Radio waves	III. Inner shell electrons
D. X-rays	IV. Klystron valve

Choose the correct answer from the options given below *[2023, 1 Feb Shift-I]*

(a) A-IV, B-I, C-II, D-III
(b) A-I, B-II, C-III, D-IV
(c) A-IV, B-III, C-II, D-I
(d) A-I, B-III, C-IV, D-II

11 Match List I with List II

List I	List II
(A) UV-rays	(i) Diagnostic tool in medicine
(B) X-rays	(ii) Water purification
(C) Microwave	(iii) Communication radar
(D) Infrared wave	(iv) Improving visibility in foggy days

Choose the correct answer from the given below *[2022, 29 July Shift-I]*

(a) (A)-(iii), (B)-(ii), (C)-(i), (D)-(iv)
(b) (A)-(ii), (B)-(i), (C)-(iii), (D)-(iv)
(c) (A)-(ii), (B)-(iv), (C)-(iii), (D)-(i)
(d) (A)-(iii), (B)-(i), (C)-(ii), (D)-(iv)

12 Match List I with List II.

List I	List II
A. Source of microwave frequency	1. Radioactive decay of nucleus
B. Source of infrared frequency	2. Magnetron
C. Source of gamma rays	3. Inner shell electrons
D. Source of X-rays	4. Vibration of atoms and molecules

Choose the correct answer from the options given below. *[2021, 24 Feb Shift-II]*

	A	B	C	D
(a)	1	2	3	4
(b)	4	3	2	1
(c)	2	4	3	1
(d)	2	4	1	3

13 The correct match between the entries in table are *[2020, 5 Sep Shift-II]*

Column I (Radiation)	Column II (Wavelength)
(A) Microwaves	(I) 100 m
(B) γ-rays	(II) 10^{-15} m
(C) AM radiowaves	(III) 10^{-10} m
(D) X-rays	(IV) 10^{-3} m

	A	B	C	D
(a)	II	I	IV	III
(b)	I	III	IV	II
(c)	III	II	I	IV
(d)	IV	II	I	III

Imp. Application of EM Waves

14 Match List I (Electromagnetic wave type) with List II (Its association/application) and select the correct option from the choices given below the lists. *[JEE Main 2014]*

List I	List II
A. Infrared waves	1. To treat muscular strain
B. Radio waves	2. For broadcasting
C. X-rays	3. To detect fracture of bones
D. Ultraviolet	4. Absorbed by the ozone layer of the atmosphere

Codes

	A	B	C	D		A	B	C	D
(a)	4	3	2	1	(b)	1	2	4	3
(c)	3	2	1	4	(d)	1	2	3	4

Key Index

Topic 1

1 (d)	**2** (b)	**3** (b)	**4** (a)	**5** (d)	**6** (a)	**7** (d)	**8** (a)	**9** (c)	**10** (d)
11 (c)	**12** (b)	**13** (b)	**14** (c)	**15** (a)	**16** (a)	**17** (c)	**18** (b)	**19** (a)	**20** (c)
21 (d)	**22** (b)	**23** (c)	**24** (d)	**25** (c)	**26** (d)	**27** (c)	**28** (d)	**29** (c)	**30** (a)
31 (a)	**32** (a)	**33** (c)	**34** (b)	**35** (c)	**36** (a)	**37** (a)	**38** (c)	**39** (c)	**40** (b)
41 (c)	**42** (c)	**43** (a)	**44** (b)	**45** (c)	**46** (a)	**47** (b)	**48** (c)	**49** (b)	**50** (d)
51 (a)	**52** (c)	**53** (d)	**54** (c)	**55** (b)	**56** (a)	**57** (b)	**58** (c)	**59** (a)	**60** (c)
61 (a)	**62** (d)	**63** (a)	**64** (c)	**65** (a)	**66** (a)	**67** (a)	**68** (d)	**69** (c)	**70** (d)
71 (d)	**72** (d)	**73** (c)	**74** (d)	**75** (*)	**76** (b)	**77** (b)	**78** (c)	**79** (b)	**80** (b)
81 (c)	**82** (d)	**83** (b)	**84** (b)	**85** (c)	**86** (b)	**87** (a)	**88** (c)	**89** (d)	**90** (d)
91 (6)	**92** (1320)	**93** (10)	**94** (1200)	**95** (4)	**96** (5)	**97** (4)	**98** (84)	**99** (8)	**100** (43)
101 (500)	**102** (2)	**103** (3)	**104** (6)	**105** (25)	**106** (137)	**107** (667)	**108** (15)	**109** (194)	

Topic 2

1 (b)	**2** (c)	**3** (b)	**4** (d)	**5** (b)	**6** (c)	**7** (a)	**8** (c)	**9** (a)	**10** (a)
11 (b)	**12** (d)	**13** (d)	**14.** (d)						

Detailed Explanations

Topic 1

01. *(d)* We know that formula for displacement current is

$$I_d = \varepsilon_0 \frac{d\phi_E}{dt}$$

02. *(b)* $E_y = cB_z$

$$B_z = \frac{E_y}{c} = \frac{9.3}{3\times 10^8}$$

$$B_z = 3.1\times 10^{-8}\ \text{T}$$

03. *(b)* $\hat{E}\times\hat{B}$ gives the direction of the wave propagation.
Since, wave is propagating in positive x-direction.
$\Rightarrow$ Electric field is in y-direction (E_y).
and magnetic field is in z-direction (B_z).

04. *(a)* Given,

$$E_y = 600\sin(\omega t - kx)\ \text{Vm}^{-1}$$

$$\therefore\quad E_0 = 600\ \text{Vm}^{-1}$$

Intensity, $I = \frac{1}{2}\varepsilon_0 E_0^2 c$

$$= \frac{1}{2}\times 9\times 10^{-12}\times 600^2\times 3\times 10^8$$

$$= 4.86\times 10^2$$

$$= 486\ \text{W/m}^2$$

05. *(d)* Frequency, $f = 60\times 10^6$ Hz

Velocity, $c = 3\times 10^8$ m/s

$$\text{Wavelength} = \frac{c}{f} = \frac{3\times 10^8}{60\times 10^6}$$

$$= 0.5\times 10 = 5\ \text{m}$$

06. *(a)* Both Statements are correct as energy is equally shared by the electric and magnetic fields and radiation pressure is caused due to momentum transferred by EM waves.

07. *(d)* Given, $I = 6\times 10^8\ \text{W/m}^2$

$$\mu = 3$$

Radiation pressure

$$= \frac{I}{v} = \frac{I\mu}{c} = \frac{6\times 10^8\times 3}{3\times 10^8}$$

$$= 6\ \text{N/m}^2$$

08. *(a)* EM wave

$$E = 200\sin[1.5\times 10^7 t - 0.05x]\ \frac{\text{N}}{\text{C}}$$

$$I = \frac{1}{2}\varepsilon_0 E_0^2\cdot c = \frac{1}{2}\varepsilon_0 E_0^2\cdot\left(\frac{\omega}{k}\right)$$

$$= \frac{1}{2}\times 8.85\times 10^{-12}\times[200]^2\times\frac{1.5\times 10^7}{0.05}$$

$$I = 53.1\ \text{W/m}^2$$

09. *(c)* $\mathbf{E}\times\mathbf{B} = \hat{c}$, option (c) is correct

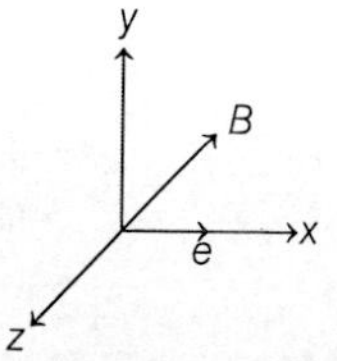

The direction of propagation of electromagnetic wave is negative Y-axis $(-\hat{j})$

10. *(d)* $\oint \mathbf{E}\cdot d\mathbf{l} = -\frac{\delta\phi_B}{\delta t}$

11. *(c)* Given, $\mathbf{E} = 6.6\hat{j}$ V/m

$$|\mathbf{E}| = c|\mathbf{B}|$$

$$|\mathbf{B}| = \frac{|\mathbf{E}|}{c} = \frac{6.6}{3\times 10^8}$$

$$= 2.2\times 10^{-8}\hat{k}\text{T}$$

$$\mathbf{E}\times\mathbf{B} = \hat{c}$$

12. *(b)* Given, magnitude of magnetic field

$$|B| = 6\times 10^{-7}\ \text{T}$$

We know that $\left|\frac{E}{B}\right| = c$

$$\Rightarrow\quad E = Bc = 6\times 10^{-7}\times 3\times 10^8$$

$$= 180\ \text{Vm}^{-1}$$

13. *(b)* Energy of electromagnetic wave remains constant with time. But frequency of change of magnetic energy into electric and electric into magnetic occurs at frequency double the frequency of wave.
Energy of an electromagnetic waves

$$U = \frac{\varepsilon_0 E^2}{2} + \frac{B^2}{2\mu_0}$$

$$= \frac{\varepsilon_0 E^2}{2} + \frac{\varepsilon_0 E^2}{2} = \varepsilon_0 E^2\ \left(\because B = \frac{E}{c}\right)$$

$$E = E_0\cos(\omega t - kx)$$

$$\therefore\quad U = U_0\cos(2\omega t - k'x) + C$$

14. *(c)* Given,

$$E = E_0\sin(\omega t - kx)$$

$$B = B_0\sin(\omega t - kx)$$

In EM waves, average electric energy density is equal to average magnetic energy density.

$$\frac{1}{4}\varepsilon_0 E_0^2 = \frac{1}{4\mu_0}B_0^2$$

15. *(a)* Electromagnetic wave is transporting energy in the negative z-direction. This means that EM wave is propagating along negative z-direction $(-\hat{k})$. Direction of electric field is given along $+\ y$-direction $(\hat{j})$.
Since, electric field vector and magnetic field vector are perpendicular to each other and also perpendicular to the direction of propagation of EM wave.
If $\hat{x}$ be the unit vector along the direction of magnetic field, then

$$\hat{j}\times\hat{x} = -\hat{k}$$

Clearly, $\hat{x} = \hat{i}$

16. *(a)* As, $\frac{E_0}{B_0} = c = \frac{\omega}{k}$

$$\Rightarrow\quad kE_0 = \omega B_0$$

17. *(c)* Magnetic field vector is perpendicular to both direction of propagation of electromagnetic wave and direction of electric field vector.
Also, $E/B = c$ and $\omega/k = c$
So, magnetic field vector can be given by

$$B = \frac{1}{\omega}(\mathbf{k}\times\mathbf{E})\ \left(\because B = \frac{E}{c} = \frac{E}{\omega/k} = \frac{kE}{\omega}\right)$$

18. *(b)* We know that energy of an electromagnetic wave is equally divided

i.e. $U_B = U_E \Rightarrow \frac{1}{4}\frac{B_0^2}{\mu_0} = \frac{1}{4}\varepsilon_0 E_0^2$

Also, electric field, magnetic field and directions of propagation of wave are mutually perpendicular to each other.

19. *(a)* $B_0 = \frac{E_0}{c} = \frac{540}{3\times 10^8}$

$$= 180\times 10^{-8}\ \text{T} = 18\times 10^{-7}\ \text{T}$$

20. *(c)* In air, the speed of light is given by

$c = \frac{1}{\sqrt{\mu_0\varepsilon_0}}$ and in material medium, the speed of light is given by

$$v = \frac{1}{\sqrt{\mu_0\varepsilon_0\mu_r\varepsilon_r}}$$

So, Statement II is false.

21. *(d)* Given frequency,

$$f = 3\,\text{GHz} = 3\times 10^9\ \text{Hz}$$

Wavelength, $\lambda = \dfrac{c}{f}$

$$= \frac{3\times 10^8}{3\times 10^9} = \frac{1}{10}\ \text{m}$$

The size of the object is given by,

$$\frac{\lambda}{100} \approx \left(\frac{1}{10}\right)^3$$

This satisfies the condition of Rayleigh's scattering.

22. *(b)* We know that,

$$B_0 = \frac{E_0}{v} = \frac{E_0 k}{\omega} = \frac{300\times 5\pi\times 10^3}{3\pi\times 10^{11}}$$

$$= 5\times 10^{-6}\ \text{T}$$

23. *(c)* We know that $v = \dfrac{c}{\sqrt{\varepsilon_r \mu_r}}$

$$\Rightarrow \quad v = \frac{3\times 10^8}{\sqrt{1\times 81}} \qquad [\because \mu_r = 1]$$

$$= \frac{3\times 10^8}{9} = \frac{10^8}{3}$$

$$= 3.33\times 10^7\ \text{ms}^{-1}$$

24. *(d)* In an electromagnetic wave, the electric field vector and magnetic field vector are given as

$$\mathbf{E} = E_0\hat{\mathbf{i}} \qquad \text{...(i)}$$

and $$\mathbf{B} = B_0\hat{\mathbf{k}} \qquad \text{...(ii)}$$

$\because$ Direction of electromagnetic wave is given by direction of $\mathbf{E}\times\mathbf{B}$ because it is perpendicular to both $\mathbf{E}$ and $\mathbf{B}$ vectors.

So, unit vector in direction of $\mathbf{E}\times\mathbf{B}$ will be

$$= \frac{\mathbf{E}\times\mathbf{B}}{|\mathbf{E}\times\mathbf{B}|} \qquad \text{...(iii)}$$

From Eqs. (i), (ii) and (iii), we get

$$\frac{E_0\hat{\mathbf{i}}\times B_0\hat{\mathbf{k}}}{E_0 B_0 \sin 90^\circ} = \hat{\mathbf{i}}\times\hat{\mathbf{k}} = -\hat{\mathbf{j}}$$

25. *(c)* Electric field, magnetic field and the direction of wave propagation are mutually perpendicular to each other.

For electromagnetic waves,

$$\mathbf{E}\times\mathbf{B} = \mathbf{c}$$

The direction of wave propagation is along $+y$-direction.

Therefore, the possible direction of electric field and magnetic field are (E_x, B_z) and (E_z, B_x).

26. *(d)* Using $\mathbf{E} = \mathbf{B}\times\mathbf{v}$, we have

$$\mathbf{E} = (5\times 10^{-8}\hat{\mathbf{j}})\times(3\times 10^8\hat{\mathbf{k}})$$

$$= 15\hat{\mathbf{i}}\ \text{Vm}^{-1}$$

27. *(c)* For an electromagnetic wave,

$$E_{\text{max}} = B_{\text{max}}\times c$$

$$\Rightarrow \quad E_{\text{max}} = 100\times 10^{-6}\times 3\times 10^8$$

$$= 3\times 10^4\ \frac{\text{N}}{\text{C}}$$

28. *(d)* Given, $\mathbf{B} = B_0 \sin(kx + \omega t)\,\hat{\mathbf{j}}\text{T}$

The relation between electric and magnetic field is,

$$E = Bc$$

The electric field component is perpendicular to the direction of propagation and the direction of magnetic field. Therefore, the electric field component along Z-axis is obtained as, $E = cB_0 \sin(kx + \omega t)\,\hat{\mathbf{k}}$.

29. *(c)* $E_0 = B_0 c = (20\times 10^{-9}\ \text{T})$

$$\times (3\times 10^8\ \text{m s}^{-1})$$

$$= 6\ \text{V/m}$$

30. *(a)* Total average energy $= \varepsilon_0 E_{\text{rms}}^2$

$$= 8.85\times 10^{-12}\times (720)^2$$

$$= 4.58\times 10^{-6}\ \text{Jm}^{-3}$$

31. *(a)* Polarisation using polariser and analyser, we get change in intensity of light at different rotations, hence we conclude that light waves are transverse in nature.

32. *(a)* Radiation pressure

$$p = \frac{2I}{c}$$

where, $$I = \frac{\text{Power}}{\text{Area}}$$

$$= \frac{450}{4\pi r^2}$$

$$= \frac{450}{4\pi\times 4}$$

$$= \frac{450}{16\pi}$$

$$\therefore \quad p = \frac{2\times 450}{16\pi\times 3\times 10^8}$$

$$= \frac{150}{8\pi\times 10^8}$$

$$\simeq 6\times 10^{-8}\ \text{Pa}$$

33. *(c)* For EM wave,

$$E = 100 \sin(\omega t - kx)$$

Energy density, $P_{\text{av}} = \dfrac{1}{2}\varepsilon_0 (E_0)^2$

Initial volume, $V_0 = \pi R^2 l$

Volume in next situation

$$= V = \frac{\pi R^2 l}{4} = \frac{V_0}{4}$$

$$U_0 = \frac{1}{2}\varepsilon_0 (E_0)^2 \times V_0$$

$$= \frac{1}{2}\varepsilon_0 (E_1{}^2)\cdot\frac{V_0}{4}$$

$$E_0{}^2 = \frac{E_1{}^2}{4} \Rightarrow E_1 = 2E_0 = 200\ \text{V/m}$$

So, the electric field of the corresponding EM-wave is modified as $200 \sin(\omega t - kx)$ N/C.

34. *(b)* $E = 57\cos[7.5\times 10^6 t - 5\times 10^{-3}$ $(3x + 4y)]\,(4\hat{\mathbf{i}} - 3\hat{\mathbf{j}})$ N/C

$$\mathbf{k} = 3\hat{\mathbf{i}} + 4\hat{\mathbf{j}}$$

$$\hat{\mathbf{k}} = \frac{3\hat{\mathbf{i}} + 4\hat{\mathbf{j}}}{\sqrt{(3)^2 + (4)^2}} = \frac{3\hat{\mathbf{i}} + 4\hat{\mathbf{j}}}{5}$$

$$\mathbf{E} = 4\hat{\mathbf{i}} - 3\hat{\mathbf{j}}$$

$$\hat{\mathbf{E}} = \frac{4\hat{\mathbf{i}} - 3\hat{\mathbf{j}}}{\sqrt{(4)^2 + (3)^2}} = \frac{4\hat{\mathbf{i}} - 3\hat{\mathbf{j}}}{5}$$

$$\hat{\mathbf{B}} = \hat{\mathbf{k}}\times\hat{\mathbf{E}}$$

$$\hat{\mathbf{B}} = -\hat{\mathbf{z}}$$

$$B_0 = \frac{E_0}{c} = \frac{57}{3\times 10^8}$$

$$\mathbf{B} = \frac{-57}{3\times 10^8}[\cos(7.5\times 10^6 t - 5\times 10^{-3}(3x + 4y))\,5\hat{\mathbf{k}}]$$

35. *(c)* Given,

$$B_y = (3.5\times 10^{-7})\sin(1.5\times 10^3 x + 0.5\times 10^{11} t)\ \text{T}$$

We know that, $\mathbf{E}\times\mathbf{B}$ = direction of propagation of wave i.e, x-direction. Hence, E must be in z-direction.

Now we know that,

$$\left|\frac{E_0}{B_0}\right| = c$$

$$E_0 = B_0 c = 3.5\times 10^{-7}\times 3\times 10^8$$

$$= 105\ \text{Vm}^{-1}$$

$$\therefore E_z = E_0 \sin(1.5\times 10^3 x + 0.5\times 10^{11} t)$$

$$= 105 \sin(1.5\times 10^3 x + 0.5\times 10^{11} t)\ \text{V/m}$$

36. *(a)* We know that, $\left|\frac{E_0}{B_0}\right| = c$

$\Rightarrow \quad 60 = B_0 \times 3 \times 10^8$

$B_0 = 2 \times 10^{-7} \quad \Rightarrow c = f\lambda$

$3 \times 10^8 = f \times 4 \times 10^{-3} \Rightarrow f = \frac{3}{4} \times 10^{11}$

$\omega = 2\pi f = 2\pi \times \frac{3}{4} \times 10^{11}$

$\omega = \frac{\pi}{2} \times 10^3 c$

$k = \frac{2\pi}{\lambda} = \frac{2\pi}{4 \times 10^{-3}} = \frac{\pi}{2} \times 10^3$

$\therefore B_z = 2 \times 10^{-7} \sin \left[\frac{\pi}{2} \times 10^3 (x - 3 \times 10^8 t)\right] \hat{k}\,\text{T}$

37. *(a)* Wave equation is,

$y = 2\cos 2\pi(330t - x)$

$y = 2\cos(2\pi \times 330t - 2\pi x)$

Compare with standard wave equation,

$y = A\cos(\omega t - kx)$

So, angular frequency

$\omega = 2\pi \times 330 \frac{\text{rad}}{\text{s}}$

$\therefore$ Linear frequency is,

$f = \frac{\omega}{2\pi} = \frac{2\pi \times 330}{2\pi}$

$= 330$ Hz

38. *(c)* Speed of EM wave in medium is,

$v = \frac{1}{\sqrt{\mu\varepsilon}} = \frac{1}{\sqrt{\mu_0 \mu_r\, \varepsilon_0 \varepsilon_r}}$

$= \frac{1}{\sqrt{\mu_0 \varepsilon_0}} \cdot \frac{1}{\sqrt{\mu_r \varepsilon_r}}$

$v = \frac{c}{\sqrt{\mu_r \varepsilon_r}}$

$\Rightarrow \quad \mu_r \varepsilon_r = \frac{c^2}{v^2}$

$\Rightarrow \quad 2 \times \varepsilon_r = \frac{(3 \times 10^8)^2}{(1.5 \times 10^8)^2}$

$\Rightarrow \quad 2 \times \varepsilon_r = 4$

$\Rightarrow \quad \varepsilon_r = 2$

39. *(c)* Given, $\mathbf{E} = \hat{\mathbf{i}}\, 40 \cos \omega\left(t - \frac{z}{c}\right) \frac{\text{N}}{\text{C}}$

Compare it with standard equation,

$\mathbf{E} = \mathbf{E}_0 \cos \omega t$

$\Rightarrow \quad E_0 = 40\hat{\mathbf{i}}$

Using the relation for magnitude,

$B_0 = \frac{E_0}{c}$

and as $c = \mathbf{E} \times \mathbf{B}$, B should be along Y-axis.

$\therefore \quad \mathbf{B} = \hat{\mathbf{j}} \frac{40}{c} \cos \omega\left(t - \frac{z}{c}\right)$

Hence, correct option is *(c)*.

40. *(b)* Average energy density of an electromagnetic wave is,

$u_{\text{avg}} = u_e + u_B$

$= 2u_e = \varepsilon_0 E^2$

$E =$ RMS value of electric field $= \frac{E_0}{\sqrt{2}}$

So, $\quad u_{\text{avg}} = \varepsilon_0 \left(\frac{E_0}{\sqrt{2}}\right)^2 = \frac{1}{2}\varepsilon_0 \cdot E_0^2$

$= \frac{1}{2} \times 8.85 \times 10^{-12} \times (50)^2$

$= 1.106 \times 10^{-8}\ \text{Jm}^{-3}$

41. *(c)* Gauss's law of magnetism,

$\oint \mathbf{B} \cdot d\mathbf{A} = 0,$

Faraday's law of EMI,

$\oint \mathbf{E} \cdot d\mathbf{l} = -\frac{d}{dt}\phi_B$

$= -\frac{d}{dt}\oint \mathbf{B} \cdot d\mathbf{A}$

Ampere's law, $\oint \mathbf{B} \cdot d\mathbf{l} = \mu_0 I$

Gauss's law of electrostatics,

$\oint \mathbf{E} \cdot d\mathbf{A} = \frac{q_{\text{net}}}{\varepsilon_0} = \frac{1}{\varepsilon_0}\oint \rho \cdot dV$

(For volumetric charge distribution)

Correct option (c) A-II, B-III, C-IV, D-I

42. *(c)* Electric field of EM wave is given by $\mathbf{E} = E_0 \cos(\omega t - kz)\,\hat{\mathbf{i}}$, the direction of

$\hat{\mathbf{E}} = \hat{\mathbf{i}},\ \hat{\mathbf{c}} = \hat{\mathbf{k}}$

We know that

$\hat{\mathbf{c}} = \hat{\mathbf{E}} \times \hat{\mathbf{B}}$ and $B_0 = \frac{E_0}{c}$

then direction of magnetic induction

$\hat{\mathbf{B}} = \hat{\mathbf{c}} \times \hat{\mathbf{E}} = \hat{\mathbf{k}} \times \hat{\mathbf{i}} = \hat{\mathbf{j}}$

therefore magnetic induction vector is,

$\mathbf{B} = \frac{E_0}{c} \cos(\omega t - kz)\,\hat{\mathbf{j}}$

• Concept Enhancer

During propagation of EM wave, electric field vector and magnetic field vector are perpendicular to each other and also perpendicular to direction of propagation of EM wave.

43. *(a)* Given,

$f = 35$ MHz $= 3.5 \times 10^7$ Hz

$\mathbf{E} = 9.6\,\hat{\mathbf{j}}$ V/m

$\therefore \quad E = 9.6$ V/m

$B = \frac{E}{c} = \frac{9.6}{3 \times 10^8}$

$= 3.2 \times 10^{-8}$ T

Since, EM wave is travelling along X-axis i.e., along $\hat{\mathbf{i}}$.

Thus, magnetic field must be directed along $\hat{\mathbf{k}}$ because $\mathbf{E}$ and $\mathbf{B}$ are perpendicular to each other and also perpendicular to the direction of propagation of EM wave.

$\therefore \quad \mathbf{B} = 3.2 \times 10^{-8}\,\hat{\mathbf{k}}$ T

44. *(b)* (1) Ampere-Maxwell's law,

$\oint \mathbf{B} \cdot d\mathbf{l} = \mu_0 i_c + \mu_0 \varepsilon_0 d\frac{\phi_E}{dt}$

(2) Faraday's law, $\oint \mathbf{E} \cdot d\mathbf{l} = d\frac{\phi_B}{dt}$

(3) Gauss's law for electricity,

$\Rightarrow \quad \oint \mathbf{E} \cdot d\mathbf{A} = \frac{Q}{\varepsilon_0}$

(4) Gauss's law for magnetism,

$\oint \mathbf{B} \cdot d\mathbf{A} = 0$

A-IV, B-III, C-I, D-II

45. *(c)* Given,

$\mathbf{E} = 20 \sin \omega\left(t - \frac{x}{c}\right)\hat{\mathbf{j}}$ ...(i) [Given]

$\mathbf{E} = E_0 \sin \omega\left(t - \frac{x}{c}\right)$ ...(ii)

On comparing Eqs. (i) and (ii), we get

$E_0 = 20$

Energy density $= \frac{1}{2}\varepsilon_0 E_0^2$

$= \frac{1}{2} \times 8.85 \times 10^{-12} \times 400\ \text{J/m}^3$

Energy contained in the given volumes

$= 200 \times 8.85 \times 10^{-12} \times 5 \times 10^{-4}$

$= 8.85 \times 10^{-12} \times 10^{-4} \times 1000$

$= 8.85 \times 10^{-13}$ J

46. *(a)* For electromagnetic wave,

Average energy density,

$U_E = \frac{1}{2}\varepsilon_0 E_{\text{rms}}^2 = \frac{1}{2}\varepsilon_0 \left(\frac{E_0}{\sqrt{2}}\right)^2$

$U_E = \frac{1}{4}\varepsilon_0 E_0^2$

where, E_0 is peak value of Electric field.
Total average energy density,

$$U_T = \frac{1}{2}\varepsilon_0 E_0^2$$

Ratio, $\frac{U_E}{U_T} = \frac{1}{2}$

Alternate Solution

When an electromagnetic wave propagates through a medium, then average electric energy density (U_E) is equal to average magnetic energy density (U_B).

$\therefore$ Total average energy density, $U_T = U_E + U_B$

Thus, $\frac{U_E}{U_T} = \frac{U_E}{U_E + U_B} = \frac{U_E}{2U_E} = \frac{1}{2}$

47. *(b)* Electromagnetic waves are photon stream and they are not deflected by electric or magnetic fields.

Also, $\frac{E_0}{B_0} = c = \frac{1}{\sqrt{\varepsilon_0 \mu_0}}$

$\therefore$ Statement I is true but Statement II is false.

48. *(c)* Speed of light does not depend on the direction of propagation in vacuum. Speed of light depends on wavelength of light in a medium. Speed of light is independent of motion of source or intensity of source.

49. *(b)* A $\rightarrow$ IV, B $\rightarrow$ I, C $\rightarrow$ II, D $\rightarrow$ III

According to Gauss's law in electrostatics,

$$\phi = \oint \mathbf{E} \cdot d\mathbf{S} = \frac{q}{\varepsilon_0}$$

According to Faraday's law,

$$\oint \mathbf{E} \cdot d\mathbf{l} = -\frac{d\phi_B}{dt}$$

According to Gauss's law in magnetism,

$$\oint \mathbf{B} \cdot d\mathbf{A} = 0$$

According to Ampere-Maxwell law,

$$\oint \mathbf{B} \cdot d\mathbf{l} = \mu_0 (I_c + I_d) = \mu_0 I_c + \mu_0 \varepsilon_0 \frac{d\phi_E}{dt}$$

50. *(d)*

key idea In the case of complete absorption, you can find the energy flux of incident light using the following expression.

$$p = \frac{F}{A} = \frac{I}{c}$$

where, $p \rightarrow$ pressure, $F \rightarrow$ force, $A \rightarrow$ area, $I \rightarrow$ energy flux of incident light.

In case of complete absorption, the pressure exerted by incident light is

$$p = I/c,$$

where, I = Energy flux of incident light.

$$\Rightarrow \frac{F}{A} = \frac{I}{c}$$

$$\Rightarrow I = \frac{Fc}{A} = \frac{7.2\times10^{-9}\times3\times10^8}{36\times10^{-4}}$$

$$= 6\times10^2 \text{ W/m}^2 = 0.06 \text{ W/cm}^2$$

51. *(a)* We know that

$$\mu_2 = \frac{c_{air}}{v_2}, \frac{\mu_2}{\mu_{air}} = \frac{c}{v_2},$$

μ_2 is the refractive index of medium 2

$$\frac{\sqrt{\mu_{r_2}\varepsilon_{r_2}}}{1} = \frac{c}{v_2} \Rightarrow \frac{\sqrt{1\times9}}{1} = \frac{c}{v_2}$$

$$\Rightarrow v_2 = \frac{C}{3} = \frac{3\times10^8}{3} = 1\times10^8 \text{ m/s}$$

52. *(c)* Electric field component

$$E_y = 900 \sin\omega\left(t - \frac{x}{c}\right)$$

Speed of charge moving along Y-axis,

$$v = 3\times10^7 \text{ m/s}$$

$$\therefore \frac{\text{Electric force }(F_e)}{\text{Magnetic force }(F_m)} = \frac{qE}{Bqv}$$

$$\Rightarrow \frac{F_e}{F_m} = \frac{E}{Bv} = \frac{E/B}{v} = \frac{c}{v} \quad \left(\because \frac{E}{B} = c\right)$$

$$= \frac{3\times10^8}{3\times10^7} = \frac{10}{1}$$

$$\Rightarrow F_e : F_m = 10 : 1$$

53. *(d)* Amplitude of magnetic field,

$B_0 = 5\times10^{-6}$ T

Here, speed of EM wave, $v = \omega/k$

$$= \frac{1000\pi\times4\times10^8}{1000\pi\times5}$$

$$= 0.8\times10^8 \text{ m/s}$$

$$= 8\times10^7 \text{ m/s}$$

We know that $E_0 = B_0 v$

$$= 5\times10^{-6}\times8\times10^7 = 400 \text{ V/m}$$

$$= 4\times10^2 \text{ V/m}$$

54. *(c)* The amplitude of magnetic field vector, $B_0 = 2\times10^{-8}$ T

$$v = \frac{\omega}{R} = \frac{1.5\times10^{11}}{0.5\times10^3}$$

We know that

$$E_0 = B_0 c$$

(where, c is speed of light)

$$= 2\times10^{-8}\times3\times10^8 = 6 \text{ V/m}$$

From the given expression of magnetic field vector **B** of EM wave, it is clear that EM wave is travelling along X-axis and magnetic field vector vibrates along y-direction, hence electric field vector will vibrate along z-direction because in EM wave, **E** and **B** are perpendicular to each other and also perpendicular to direction of motion of EM wave.

55. *(b)*

key idea In such type of question, you can first calculate the speed of EM wave using, $v = \frac{E_0}{B_0}$ and after that you can find the time t taken by signal using the following formula $t = \frac{2d}{v}$, where $d \rightarrow$ distance.

Speed of Electromagnetic signal in medium,

$$v = \frac{E_0}{B_0} = \frac{2.25}{1.5\times10^{-8}}$$

$$= 1.5\times10^8 \text{ m/s}$$

Time taken by signal for to and fro from RADAR,

$$t = \frac{2d}{v} = \frac{2\times3\times10^3 \text{ m}}{1.5\times10^8 \text{ m/s}}$$

$$= 4\times10^{-5} \text{ s}$$

56. *(a)* We know that the refractive index of medium, $\mu = \sqrt{\mu_r \varepsilon_r}$

$$\Rightarrow \frac{c}{v} = \sqrt{\mu_r \varepsilon_r} \quad \left[\because \mu = \frac{c}{v}\right]$$

$$\Rightarrow \frac{3\times10^8}{2\times10^8} = \sqrt{1\times\varepsilon_r}$$

$$\Rightarrow 1.5 = \sqrt{\varepsilon_r} \Rightarrow 2.25 = \varepsilon_r$$

$$\Rightarrow \varepsilon_r = 2.25$$

57. *(b)* Intensity of EM wave is given by

$$I = \frac{1}{2}\varepsilon_0 E^2 c$$

$$= \frac{1}{2}\times8.85\times10^{-12}\times(56.5)^2 \times3\times10^8 = 4.24 \text{ Wm}^{-2}$$

58. *(c)* We know that

$$c = \frac{1}{\sqrt{\mu_0\varepsilon_0}} \text{ and } v = \frac{1}{\sqrt{\mu\varepsilon}}$$

$$\therefore \frac{c}{v} = \sqrt{\frac{\mu\varepsilon}{\mu_0\varepsilon_0}} = \sqrt{\left(\frac{\mu}{\mu_0}\right)\left(\frac{\varepsilon}{\varepsilon_0}\right)}$$

$$= \sqrt{\mu_r \varepsilon_r}$$

$$\Rightarrow \frac{c}{v} = \sqrt{1.61\times6.44}$$

$$\Rightarrow v = \frac{c}{\sqrt{1.61 \times 6.44}} \Rightarrow v = \frac{3 \times 10^8}{\sqrt{1.61 \times 6.44}}$$

But speed of electromagnetic wave,

$$v = \frac{E}{B} \Rightarrow E = B.v$$

$$B = \mu H = \mu_0 \mu_r H$$

$$B = 4\pi \times 10^{-7} \times 1.61 \times 4.5 \times 10^{-2}$$

$$E = 4.5 \times 10^{-2} \times \frac{3 \times 10^8 \times 4\pi \times 10^{-7} \times 1.61}{\sqrt{1.61 \times 6.44}}$$

$$= 8.48 \text{ Vm}^{-1}$$

59. *(a)* Comparing given equation with the standard equation,

$$E = E_0 \cos(\omega t - kx) \text{ V/m},$$

we get

$$k = 200$$

$$\omega = 2 \times 10^{10} \text{ rad/s}$$

$\therefore$ Speed of the wave,

$$v = \frac{\omega}{k} = \frac{2 \times 10^{10}}{200} = 10^8 \text{ m/s}$$

Refractive index,

$$\mu = \frac{c}{v} = \frac{3 \times 10^8}{10^8} = 3$$

As we know,

$$\mu = \sqrt{\varepsilon_r \mu_r}$$

$$3 = \sqrt{\varepsilon_r (1)} \Rightarrow \varepsilon_r = 9$$

60. *(c)* From the given expression of electric field,

$$E_0 = 50 \text{ V/m}, \omega = 10 \times 10^{10} \text{ rad s}^{-1},$$

$$k = 500 \text{ rad m}^{-1}$$

Speed of wave in medium will be

$$v = \frac{\omega}{k} = \frac{10 \times 10^{10}}{500} = 2 \times 10^8 \text{ ms}^{-1}$$

$$= \frac{2}{3} \times 3 \times 10^8 \text{ m/s} = \frac{2}{3} c$$

61. *(a)* Here, $E_0 = 800 \text{ Vm}^{-1}$

As we know, the magnitude of magnetic field is given as

$$E_0 = B_0 c \Rightarrow 800 = B_0 \times 3 \times 10^8$$

$$\Rightarrow B_0 = \frac{800}{3 \times 10^8} \text{ T}$$

$$\therefore F_{max} = evB_0$$

$$= (1.6 \times 10^{-19}) \times (3 \times 10^7) \times \frac{800}{3 \times 10^8}$$

(Given, $v = 3 \times 10^7 \text{ ms}^{-1}$)

$$F_{max} = 12.8 \times 10^{-18} \text{N}$$

or $= 1.28 \times 10^{-17}$

62. *(d)* Given,

Linearly polarised electromagnetic wave,

$$E = 3.1 \cos [(1.8) z - (5.4 \times 10^6) t] \hat{i} \text{ N/C}$$

The given wave is travelling along +ve X-axis.

Angular frequency, $\omega = 5.4 \times 10^6 \text{ rad s}^{-1}$

Wave constant, $k = 1.8 \text{ m}^{-1}$

$$k = \frac{2\pi}{\lambda} \Rightarrow \lambda = \frac{2\pi}{R} = \frac{2\pi}{1.8} = 3.4 \text{ m}$$

$$f = \frac{\omega}{2\pi} = \frac{5.4 \times 10^6}{2\pi} = 85 \times 10^4 \text{ Hz}$$

As, we know that,

If incident wave is $x = A \sin(\omega t - kx)$

Then, for normal incidence, reflected wave $x_r = A \sin(\omega t + kx)$

Hence, reflected wave will be

$$E_r = 3.1 \cos [(1.8) z + (5.4 \times 10^6) t] \hat{i} \text{ N/C}.$$

63. *(a)* Given, intensity of sunlight,

$$I = 0.092 \text{ W/m}^2$$

As we know that

$$I = \frac{\varepsilon_0 c}{2} E_0^2$$

$$\Rightarrow E_0 = \sqrt{2 \frac{I}{\varepsilon_0 c}}$$

$$\text{Since, } B_0 = \frac{E_0}{c} = \frac{1}{c} \sqrt{\frac{2I}{\varepsilon_0 c}}$$

$$= \frac{1}{3 \times 10^8} \sqrt{\frac{2 \times 0.092}{8.85 \times 10^{-12} \times 3 \times 10^8}}$$

$$= 2.77 \times 10^{-8} \text{ T}$$

64. *(c)* Given, $f = 100$ MHz

$$B = (2.0 \times 10^{-8}) \hat{k} \text{T}$$

The speed of the light in free space,

$$c = (3 \times 10^8) \hat{i}$$

(The direction of wave propagates in x-direction)

Magnitude of electric field,

$$E = Bc = 2 \times 10^{-8} \times 3 \times 10^8$$

$$\Rightarrow E = 6 \text{ V/m}$$

If $\hat{n}$ be the direction of E, then $\hat{i} = \hat{n} \times \hat{k}$

$$\Rightarrow \hat{n} = \hat{j}$$

Hence, the electric field at this point is $6\hat{j}$ V/m.

65. *(a)* we know that, $E_0 = cB_0$

$$= 3 \times 10^8 \times 8 \times 10^{-8} = 24 \text{ V/m}$$

$\therefore$ Direction of propagation is given by $\hat{E} \times \hat{B}$.

As, the wave is travelling in y-direction and the magnetic field is in z-direction.

$$\Rightarrow \hat{E} \times \hat{z} = \hat{y} \Rightarrow \hat{E} = -\hat{x}$$

$\therefore$ The value of electric field will be $-24\hat{x}$ V/m.

66. *(a)* We know that

$$U_e = \frac{1}{2} \varepsilon_0 E^2 \quad \text{...(i)}$$

$$\text{and } U_m = \frac{B^2}{2\mu_0} \quad \text{...(ii)}$$

Also, $E = cB$ $\left(\because c = \frac{1}{\sqrt{\mu_0 \varepsilon_0}} \right)$

$$\Rightarrow E = \frac{B}{\sqrt{\mu_0 \varepsilon_0}}$$

$$\Rightarrow U_e = \frac{1}{2} \varepsilon_0 \times \frac{1}{\mu_0 \varepsilon_0} B^2$$

$$\Rightarrow U_e = \frac{B^2}{2\mu_0}$$

$$\Rightarrow U_e = U_m \quad \text{[Using Eq. (ii)]}$$

67. *(a)* Given that $\mathbf{B}(x, t)$

$$= [1.2 \times 10^{-7} \sin(0.5 \times 10^3 x + 1.5 \times 10^{11} t) \hat{k}] \text{ T}$$

E_0 = (maximum value of B_0) × (speed of light)

$$= B_0 \times c = 1.2 \times 10^{-7} \times 3 \times 10^8$$

$$= 36 \text{ V/m}$$

Since, wave is propagating in negative x-direction and magnetic field is in z-direction. So, electric field vector must be in negative y-direction as direction of wave propagation is in the direction of $\mathbf{E} \times \mathbf{B}$.

So, electric field at a point x and time t is given by

$$\mathbf{E}(x, t) = -36 \sin(0.5 \times 10^3 x + 1.5 \times 10^{11} t) \hat{j} \text{V/m}$$

Here, phase of **E** will be same that of **B**.

68. *(d)* It is given that

$$\mathbf{B} = 3 \times 10^{-8} \sin(1.6 \times 10^3 x + 48 \times 10^{10} t) \hat{j} \text{ T}$$

$$= B_0 \sin(kx + \omega t) \hat{j} \text{ T}$$

Here, $k = 1.6 \times 10^3$ units

and $\omega = 48 \times 10^{10}$ units

Then, it can be implied that electromagnetic wave is travelling along X-axis (this is obtained from phase

$kx + \omega t$) and its magnetic vector **B** is directed along $+Y$-axis (this is indicated by unit vector $\hat{j}$).

Hence, electric field vector is along $+Z$-axis and its magnitude is given by

$E_0 = c \cdot B_0 = 3 \times 10^8 \times 3 \times 10^{-8} = 9$ V/m

$\therefore$ Expression for **E**

$$= 9 \sin(kx + \omega t)\hat{k}$$

$$= 9 \sin(1.6 \times 10^3 x + 48 \times 10^{10} t)\hat{k} \text{ V/m}$$

69. *(c)*

> **key idea** You can use standard expression of electromagnetic wave, which is given by
>
> $\mathbf{E} = E_0\hat{n}[\sin(\omega t - \mathbf{k}\cdot\hat{r})]$ ...(i)
>
> Here, **k** is the propagation vector. Direction of propagation in this case is $\hat{\mathbf{k}}$.

Given expression of electromagnetic wave,

$$\mathbf{E} = E_0\hat{n}\sin[\omega t + (6y - 8z)]$$

$$\Rightarrow \mathbf{E} = E_0\hat{n}\sin[\omega t - (8z - 6y)] \quad \text{...(ii)}$$

On comparing Eq. (ii) with Eq. (i), we get

$$\mathbf{k}\cdot\hat{r} = 8z - 6y \quad \text{...(iii)}$$

Here, $\hat{r} = x\hat{i} + y\hat{j} + z\hat{k}$

and $\mathbf{k} = k_x\hat{i} + k_y\hat{j} + k_z\hat{k}$

$$\therefore \mathbf{k}\cdot\hat{r} = xk_x + yk_y + zk_z \quad \text{...(iv)}$$

From Eqs. (iii) and (iv), we get

$xk_x = \text{zero} \Rightarrow k_x = 0$

$yk_y = -6y \Rightarrow k_y = -6$

$zk_z = 8z \Rightarrow k_z = 8$

Hence, $\mathbf{k} = -6\hat{j} + 8\hat{k}$

So, direction of propagation,

$$\hat{s} = \hat{k} = \frac{\mathbf{k}}{|\mathbf{k}|} = \frac{-6\hat{j} + 8\hat{k}}{\sqrt{6^2 + 8^2}}$$

$$= \frac{-6\hat{j} + 8\hat{k}}{10} = \frac{-3\hat{j} + 4\hat{k}}{5}$$

70. *(d)* Momentum transferred to the surface is

$$\Delta p = F \times \Delta t = \frac{IA}{c} \times \Delta t$$

$$\Rightarrow \Delta p = \frac{25 \times 25 \times 2400}{3 \times 10^8}$$

$$= 5 \times 10^{-3} \text{ N-s}$$

71. *(d)* Intensity of electromagnetic wave is given by the relation,

$$I = \frac{1}{2} nc\,\varepsilon_0 E^2$$

where, n is refractive index, for air, $n = 1$.

$$\therefore I = \frac{1}{2} c \cdot \varepsilon_0 E^2 \quad \text{...(i)}$$

Also, $I = \dfrac{P}{A}$ ...(ii)

From Eqs. (i) and (ii), we get

$$\frac{1}{2} c\,\varepsilon_0 E^2 = \frac{P}{A}$$

or $E^2 = \dfrac{2P}{Ac\,\varepsilon_0}$

or $E = \sqrt{\dfrac{2 \times 27 \times 10^{-3}}{10 \times 10^{-6} \times 3 \times 10^8 \times 9 \times 10^{-12}}}$

$$\simeq 1.4 \times 10^3 \text{ V/m}$$

$$= 1.4 \text{ k V/m}$$

72. *(d)* $\because |\mathbf{B}| = \dfrac{|\mathbf{E}|}{c} = \dfrac{63}{3 \times 10^8}$

$$= 2.1 \times 10^{-8}$$

As here, direction of **E** is $+y$-direction i.e. $\hat{j}$ and direction of propagation of wave is in $+x$-direction i.e., $\hat{i}$

$\therefore$ If **B** has to be perpendicular to **E** and direction of propagation, then it must in $+z$-direction i.e., $\hat{k}$.

Thus, magnetic field,

$$\mathbf{B} = 2.1 \times 10^{-8}\,\hat{k} \text{ T}.$$

73. *(c)*

> **key idea** You can calculate speed of progressive wave using, $v = \dfrac{\omega}{k}$

As electric field in air is,

$$E_1 = E_{01}\hat{x}\cos\left(\frac{2\pi\nu z}{c} - 2\pi\nu t\right)$$

$$\therefore \text{Speed in air} = \frac{2\pi\nu}{\left(\frac{2\pi\nu}{c}\right)} = c$$

Also, $c = \dfrac{1}{\sqrt{\mu_0 \varepsilon_{r_1} \varepsilon_0}}$ ...(i)

In medium,

$$E_2 = E_{02}x\cos(2kz - kct)$$

$$\therefore \text{Speed in medium} = \frac{kc}{2k} = \frac{c}{2}$$

Also, $\dfrac{c}{2} = \dfrac{1}{\sqrt{\mu_0 \varepsilon_{r_2} \varepsilon_0}}$ ...(ii)

As medium is non-magnetic medium,

$$\mu_{\text{medium}} = \mu_{\text{air}}$$

On dividing Eq. (i) by Eq. (ii), we get

$$2 = \sqrt{\frac{\varepsilon_{r_2}}{\varepsilon_{r_1}}} \Rightarrow \frac{\varepsilon_{r_1}}{\varepsilon_{r_2}} = \frac{1}{4}$$

74. *(d)* Intensity of light, $I = u_{av}c$

Also, $I = \dfrac{P}{4\pi r^2}$ and $u_{av} = \dfrac{1}{2}\varepsilon_0 E_0^2$

Therefore,

$$\therefore \frac{P}{4\pi r^2} = \frac{1}{2}\varepsilon_0 E_0^2 c \text{ or } E_0 = \sqrt{\frac{2P}{4\pi\varepsilon_0 r^2 c}}$$

Here, $P = 0.1$ W, $r = 1$ m,

$c = 3 \times 10^8$ m s^{-1}

$$\therefore E_0 = \sqrt{\frac{2 \times 0.1 \times 9 \times 10^9}{1^2 \times 3 \times 10^8}}$$

$$= \sqrt{6} = 2.45 \text{ V m}^{-1}$$

75. *(*)* $\nu = 2 \times 10^{14}$ Hz, $E_0 = 27$ V m^{-1}

We know, $B_0 = \dfrac{E_0}{c}$,

$$B_0 = \frac{27}{3 \times 10^8} = 9 \times 10^{-8} \text{ T}$$

Wavelength,

$$\lambda = \frac{c}{\nu} = \frac{3 \times 10^8}{2 \times 10^{14}} = 1.5 \times 10^{-6} \text{ m},$$

$$B = B_0 \sin 2\pi\left(\frac{x}{\lambda} - \nu t\right)$$

$$B = (9 \times 10^{-8} \text{ T})$$

$$\sin 2\pi\left(\frac{x}{1.5 \times 10^{-6}} - 2 \times 10^{14} t\right)$$

Oscillation of B can be along either $\hat{j}$ or $\hat{k}$ direction.

*None of the given options is correct.

76. *(b)* For electromagnetic wave, direction of propagation, **E** and **B** are mutually perpendicular to each other. According to question, $\mathbf{E} \times \mathbf{B}$ = direction of propagation = $+z$-direction only option (b) satisfies both conditions (i) $\mathbf{E}\cdot\mathbf{B} = 0$ (ii) $(\mathbf{E} \times \mathbf{B})$ directed along the z-axis.

77. *(b)* In electromagnetic wave, the direction of propagation of wave, electric field and magnetic field are mutually perpendicular, i.e., wave propagates perpendicular to **E** and **B** or along $\mathbf{E} \times \mathbf{B}$.

While polarisation of wave takes place parallel to electric field vector.

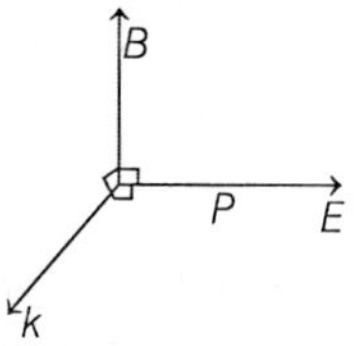

78. *(c)* In vacuum, $\varepsilon_0 = 1$

In medium, $\varepsilon = 4$

$\therefore$ Refractive index, $\mu = \sqrt{\dfrac{\varepsilon}{\varepsilon_0}} = \sqrt{\dfrac{4}{1}} = 2$

Wavelength, $\lambda' = \dfrac{\lambda}{\mu} = \dfrac{\lambda}{2}$

and wave velocity, $v = \dfrac{c}{\mu} = \dfrac{c}{2}\left[\because \mu = \dfrac{c}{v}\right]$

Hence, it is clear that wavelength and velocity will become half but frequency remains unchanged when the wave is passing through any medium.

79. *(b)* Initial momentum of surface,

$p_i = E/c$

Since, the surface is perfectly reflecting, so the same momentum will be reflected completely.

Final momentum, $p_f = E/c$ [negative value]

$\therefore$ Change in momentum,

$$\Delta p = p_f - p_i = -\frac{E}{c} - \frac{E}{c} = -\frac{2E}{c}$$

Thus, momentum transferred to the surface is

$$\Delta p' = |\Delta p| = \frac{2E}{c}$$

80. *(b)*

$$\mathbf{B} = \left(\frac{\sqrt{3}}{2}\hat{\mathbf{i}} + \frac{1}{2}\hat{\mathbf{j}}\right)30\sin\left[\omega\left(t - \frac{z}{c}\right)\right]$$

$$B_0 = 30\sqrt{\frac{3}{4} + \frac{1}{4}} = 30$$

$$E_0 = c\,|B_0| = 30c$$

Now **B** lies in *XY*-plane and wave travels in +*z*-direction by the righ hand rule, *E* must lie in *XY*-plane and perpendicular to **B**.

$$\mathbf{E} = \mathbf{B} \times \mathbf{c}$$

$$= \left(\frac{\sqrt{3}}{2}\hat{\mathbf{i}} + \frac{1}{2}\hat{\mathbf{j}}\right) \times (\hat{\mathbf{k}})$$

$$= \left(-\frac{\sqrt{3}}{2}\hat{\mathbf{j}} + \frac{1}{2}\hat{\mathbf{i}}\right)$$

$$= \left(\frac{1}{2}\hat{\mathbf{i}} - \frac{\sqrt{3}}{2}\hat{\mathbf{j}}\right)$$

$$\therefore \mathbf{E} = \left(\frac{1}{2}\hat{\mathbf{i}} - \frac{\sqrt{3}}{2}\hat{\mathbf{j}}\right)30c\sin\left[\omega\left(t - \frac{z}{c}\right)\right]$$

81. *(c)* For a non-reflecting (100% absorbing) surface, radiation pressure is given by,

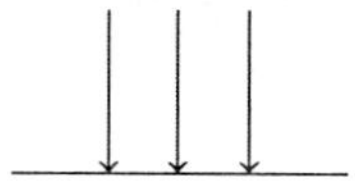

$$p = \frac{I}{c}$$

Here, p = pressure $= \dfrac{F}{A}$

F = Force, A = Area

and, Intensity $I = \dfrac{\text{Energy}}{\text{Area} \times \text{Time}}$ = Energy flux

and c = speed of light

So we have,

$$p = \frac{I}{c} \Rightarrow \frac{F}{A} = \frac{I}{c}$$

$\Rightarrow$ Area of surface, $A = \dfrac{F \times c}{I}$

Substituting given values,

$$F = 2.4 \times 10^{-4}\ \text{N}$$

$$I = 360\frac{\text{W}}{\text{cm}^2} = 360 \times 10^4\ \frac{\text{W}}{\text{m}^2}$$

$$c = 3 \times 10^8\ \frac{\text{m}}{\text{s}}$$

$$\therefore \text{Area, } A = \frac{2.4 \times 10^{-4} \times 3 \times 10^8}{360 \times 10^4}$$

$$= \frac{2.4 \times 3}{360} = \frac{24 \times 3}{36 \times 100}$$

$$= \frac{2}{100} = 0.02\ \text{m}^2$$

82. *(d)* A point source of 100 W emits light with 5% efficiency.

$\therefore$ Efficient power produced by point source,

$$P' = 5\%\text{ of }100\text{ W} = \frac{5}{100} \times 100 = 5\text{ W}$$

At a distance, $r = 5$ m, intensity produced by point source.

$$I = \frac{P'}{4\pi r^2} = \frac{5}{4\pi \times 5^2} = \frac{1}{20\pi}$$

Since, $U_E = U_B$

Hence, intensity produced by the electric field component.

$$I_E = \frac{I}{2} = \frac{1}{2} \times \frac{1}{20\pi}$$

$$= \frac{1}{40\pi}\ (\text{W/m}^2)$$

83. *(b)* We know that $c = \dfrac{E_0}{B_0}$

$$\Rightarrow \quad B_0 = \frac{E_0}{c} = \frac{60}{3 \times 10^8} = 2 \times 10^{-7}\ \text{T}$$

$\hat{\mathbf{E}} \times \hat{\mathbf{B}}$ must be in direction of propagation.

Hence, direction of **B** is along $\hat{\mathbf{k}}$

$$k = \frac{2\pi}{\lambda} = \frac{2\pi}{8} \times 10^3 = \frac{\pi}{4} \times 10^3\ \text{m}^{-1}$$

So, $E_y = 60\sin\left[\dfrac{\pi}{4} \times 10^3 (x - ct)\right]\hat{\mathbf{j}}$ V/m

$B_z = 2 \times 10^{-7}\sin\left[\dfrac{\pi}{4} \times 10^3 (x - ct)\right]\hat{\mathbf{k}}$T

[where, $c = 3 \times 10^8$ m/s]

84. *(b)* Power of electric bulb, $P = 200$ W

Efficiency,

$$\eta = 3.5\% = 0.035$$

$$r = 4\text{ m}$$

$\therefore$ Intensity of radiation at $r = 4$ m

$$I = \frac{\text{Power}}{\text{Area}} \times \eta = \frac{200}{4\pi r^2} \times 0.035$$

$$= \frac{200}{4 \times 3.14 \times 4^2} \times 0.035$$

$$= 3.48 \times 10^{-2}\ \text{W/m}^2$$

But, $I = \dfrac{cB_0^2}{2\mu_0}$

$$\Rightarrow B_0^2 = \frac{2\mu_0 I}{c} \Rightarrow B = \sqrt{\frac{2\mu_0 I}{c}}$$

$$= \sqrt{\frac{2 \times 4\pi \times 10^{-7} \times 3.48 \times 10^{-2}}{3 \times 10^8}}$$

$$= 1.71 \times 10^{-8}\ \text{T}$$

85. *(c)* The equation of magnetic field vector of an electromagnetic field is

$$\mathbf{B} = \frac{B_0}{\sqrt{2}}(\hat{\mathbf{i}} + \hat{\mathbf{j}})\cos(kz - \omega t)\text{T}$$

Electric charges

$$q_1 \text{ at } P\left(0, 0, \frac{\pi}{k}\right) = 4\pi C$$

$$q_2 \text{ at } Q\left(0, 0, \frac{3\pi}{k}\right) = 2\pi C$$

At $t = 0$ and $P\left(0, 0, \dfrac{\pi}{k}\right)$, where $z = \dfrac{\pi}{k}$,

Magnetic field vector,

$$\mathbf{B} = \frac{B_0}{\sqrt{2}}(\hat{\mathbf{i}} + \hat{\mathbf{j}})\cos\left(k\frac{\pi}{k} - 0\right)\text{T}$$

$$= \frac{B_0}{\sqrt{2}}(\hat{\mathbf{i}} + \hat{\mathbf{j}})\cos\pi\text{T}$$

$$= -\frac{B_0}{\sqrt{2}}(\hat{i}+\hat{j})\text{T}$$

and at $t=0$ and $Q\left(0,0,\frac{3\pi}{k}\right)$, where $z=\frac{3\pi}{k}$.

Magnetic field vector,

$$\mathbf{B} = \frac{B_0}{\sqrt{2}}(\hat{i}+\hat{j})\cos\left(k\frac{3\pi}{k}-0\right)$$

$$= -\frac{B_0}{\sqrt{2}}(\hat{i}+\hat{j})\text{T}$$

[$\because$ Using $\cos\pi = \cos 3\pi = -1$]

Velocity of charges q_1 and q_2, $\mathbf{v} = 0.5c\hat{i}$

$\therefore$ $\mathbf{F} = q(\mathbf{v}\times\mathbf{B})$

$\therefore$ For q_1,

$$\mathbf{F}_1 = 4\pi\left[0.5c\hat{i}\times\left\{-\frac{B_0}{\sqrt{2}}(\hat{i}+\hat{j})\right\}\right]\text{N}$$

$$\mathbf{F}_1 = -\frac{2\pi c B_0}{\sqrt{2}}\hat{k}\text{ N}$$

Similarly, for q_2

$$\Rightarrow \mathbf{F}_2 = 2\pi\left[0.5c\hat{i}\times\left\{-\frac{B_0}{\sqrt{2}}(\hat{i}+\hat{j})\right\}\right]\text{N}$$

$$\Rightarrow \mathbf{F}_2 = -\frac{\pi c B_0}{\sqrt{2}}\hat{k}\text{ N}$$

Hence, ratio of magnitudes of $\mathbf{F}_1$ to $\mathbf{F}_2$,

$$\frac{F_1}{F_2} = \frac{\frac{2\pi c B_0}{\sqrt{2}}}{\frac{\pi c B_0}{\sqrt{2}}} = \frac{2}{1} \Rightarrow F_1 : F_2 = 2:1$$

86. *(b)* Electric field vector of the given EM wave is

$$\mathbf{E} = E_0(\hat{\mathbf{x}}+\hat{\mathbf{y}})\sin(kz-\omega t)$$

Unit vector of electric field,

$$\hat{\mathbf{E}} = \frac{\hat{\mathbf{x}}+\hat{\mathbf{y}}}{\sqrt{1^2+1^2}} = \frac{\hat{\mathbf{x}}+\hat{\mathbf{y}}}{\sqrt{2}}$$

Wave travels along positive z-direction. So, unit vector into the direction of propagation of EM wave $\hat{\mathbf{c}} = \hat{\mathbf{z}}$.

Amplitude of magnetic field,

$$B_0 = E_0/c$$

and unit vector of magnetic field $= \hat{\mathbf{B}}$

So, the expected equation of magnetic field vector will be

$$\mathbf{B} = B_0(\hat{\mathbf{B}})\sin(kz-\omega t)$$

$$= \frac{E_0}{c}(\hat{\mathbf{B}})\sin(kz-\omega t) \quad \ldots\text{(i)}$$

Also, $\hat{\mathbf{E}}\times\hat{\mathbf{B}} = \hat{\mathbf{c}} = \hat{\mathbf{z}}$ ...(ii)

Now, let's check the given alternatives

(a) $\mathbf{B} = \frac{E_0}{c}(\hat{\mathbf{x}}+\hat{\mathbf{y}})\sin(kz-\omega t)$

Here, $\hat{\mathbf{B}} = \frac{\hat{\mathbf{x}}+\hat{\mathbf{y}}}{\sqrt{1^2+1^2}} = \frac{\hat{\mathbf{x}}+\hat{\mathbf{y}}}{\sqrt{2}}$

So, $\hat{\mathbf{E}}\times\hat{\mathbf{B}} = \left(\frac{\hat{\mathbf{x}}+\hat{\mathbf{y}}}{\sqrt{2}}\right)\times\left(\frac{\hat{\mathbf{x}}+\hat{\mathbf{y}}}{\sqrt{2}}\right)$

$$= \frac{1}{2}[(\hat{\mathbf{x}}\times\hat{\mathbf{x}})+(\hat{\mathbf{x}}\times\hat{\mathbf{y}})+(\hat{\mathbf{y}}\times\hat{\mathbf{x}})+(\hat{\mathbf{y}}\times\hat{\mathbf{y}})]$$

$$= \frac{1}{2}[(0)+(\hat{\mathbf{z}})+(-\hat{\mathbf{z}})+(0)]$$

$$= \frac{1}{2}[\hat{\mathbf{z}}-\hat{\mathbf{z}}] = 0$$

which is impossible for an EM wave. So, this option is wrong.

(b) $\mathbf{B} = \frac{E_0}{c}(-\hat{\mathbf{x}}+\hat{\mathbf{y}})\sin(kz-\omega t)$

Here, $\hat{\mathbf{B}} = \frac{-\hat{\mathbf{x}}+\hat{\mathbf{y}}}{\sqrt{1^2+1^2}} = \frac{-\hat{\mathbf{x}}+\hat{\mathbf{y}}}{\sqrt{2}}$

So, $\hat{\mathbf{E}}\times\hat{\mathbf{B}} = \left(\frac{\hat{\mathbf{x}}+\hat{\mathbf{y}}}{\sqrt{2}}\right)\times\left(-\frac{\hat{\mathbf{x}}+\hat{\mathbf{y}}}{\sqrt{2}}\right)$

$$\frac{1}{2}[-(\hat{\mathbf{x}}\times\hat{\mathbf{x}})+(\hat{\mathbf{x}}\times\hat{\mathbf{y}})-(\hat{\mathbf{y}}\times\hat{\mathbf{x}})+(\hat{\mathbf{y}}\times\hat{\mathbf{y}})]$$

$$= \frac{1}{2}[-(0)+(\hat{\mathbf{z}})-(-\hat{\mathbf{z}})+(0)]$$

$$= \frac{1}{2}[\hat{\mathbf{z}}+\hat{\mathbf{z}}] = \frac{1}{2}(2\hat{\mathbf{z}}) = \hat{\mathbf{z}}$$

which satisfies for Eq. (ii). So, this option is correct.

(c) $\mathbf{B} = \frac{E_0}{c}(\hat{\mathbf{x}}-\hat{\mathbf{y}})\sin(kz-\omega t)$

Here, $\hat{\mathbf{B}} = \frac{\hat{\mathbf{x}}-\hat{\mathbf{y}}}{\sqrt{1^2+1^2}} = \frac{\hat{\mathbf{x}}-\hat{\mathbf{y}}}{\sqrt{2}}$

So, $\hat{\mathbf{E}}\times\hat{\mathbf{B}} = \left(\frac{\hat{\mathbf{x}}+\hat{\mathbf{y}}}{\sqrt{2}}\right)\times\left(\frac{\hat{\mathbf{x}}-\hat{\mathbf{y}}}{\sqrt{2}}\right)$

$$= \frac{1}{2}[(\hat{\mathbf{x}}\times\hat{\mathbf{x}})-(\hat{\mathbf{x}}\times\hat{\mathbf{y}})+(\hat{\mathbf{y}}\times\hat{\mathbf{x}})-(\hat{\mathbf{y}}\times\hat{\mathbf{y}})]$$

$$= \frac{1}{2}[(0)-(\hat{\mathbf{z}})+(-\hat{\mathbf{z}})-(0)]$$

$$= \frac{1}{2}[-\hat{\mathbf{z}}-\hat{\mathbf{z}}] = \frac{1}{2}(-2\hat{\mathbf{z}}) = -\hat{\mathbf{z}}$$

which doesn't satisfy for Eq. (ii). So, this option is wrong.

(d) $\mathbf{B} = \frac{E_0}{c}(\hat{\mathbf{x}}-\hat{\mathbf{y}})\cos(kz-\omega t)$

This option is wrong, as the function in eq. (i) is a sine function, not a cosine function.

Hence, option (b) is correct.

87. *(a)* We know that

$$(\mathbf{E}\times\mathbf{B}) \,||\, v$$

According to given situation, direction of **B** must be Y-axis.

From Maxwell's equation,

$$\nabla\times\mathbf{E} = -\frac{\partial\mathbf{B}}{\partial t}$$

Here, $\frac{\partial\mathbf{E}}{\partial Z} = -\frac{\partial B}{\partial t}$ ...(i)

and $B_0 = \frac{E_0}{c}$ ...(ii)

Given, $\mathbf{E} = E_0\,\hat{i}\cos kz\cos\omega t$

$$\Rightarrow \frac{-\partial\mathbf{E}}{\partial Z} = kE_0\sin kz\cos\omega t$$

$\therefore$ Using Eq. (i), we get

$$\frac{\partial\mathbf{B}}{\partial t} = kE_0\sin kz\cos\omega t$$

Integrating both sides of the above equation w.r.t. t, we get

$$\Rightarrow \mathbf{B} = \frac{k}{\omega}E_0\sin kz\sin\omega t$$

$$= \frac{E_0}{c}\sin kz\sin\omega t$$

$$\Rightarrow \mathbf{B} = \frac{E_0}{c}\sin(kz)\sin(\omega t)\,\hat{j}$$

• Concept Enhancer

For an electromagnetic wave, its electric field vector **E** and magnetic field vector **B** is mutually perpendicular to each other and also to its direction of propagation.

88 *(c)* Given,

$$\mathbf{B} = 1.6\times10^{-6}\cos(2\times10^{7}z + 6\times10^{15}t)(2\hat{i}+\hat{j})\text{ Wbm}^{-2}$$

From the given equation, it can be said that the electromagnetic wave is propagating negative z-direction, i.e. $-\hat{k}$.

Equation of associated electric field will be $\mathbf{E} = (|\mathbf{B}|c)\cos(kz+\omega t)\cdot\hat{\mathbf{n}}$

where, $\hat{\mathbf{n}}$ = a vector perpendicular to **B**.

So, $|\mathbf{E}| = |\mathbf{B}|c$

$$= 1.6\times10^{-6}\times3\times10^{8}$$

$$= 4.8\times10^{2}\text{ V/m}$$

Since, we know that for an electromagnetic wave, E and B are mutually perpendicular to each other.

So, $\mathbf{E}\cdot\mathbf{B}=0$

From the given options, when

$$\hat{\mathbf{n}}=\hat{\mathbf{i}}-2\hat{\mathbf{j}}$$

$$\mathbf{E}\cdot\mathbf{B}=(2\hat{\mathbf{i}}+\hat{\mathbf{j}})\cdot(\hat{\mathbf{i}}-2\hat{\mathbf{j}})=0$$

Also, when $\hat{\mathbf{n}}=-\hat{\mathbf{i}}+2\hat{\mathbf{j}}$

$$\mathbf{E}\cdot\mathbf{B}=(2\hat{\mathbf{i}}+\hat{\mathbf{j}})\cdot(-\hat{\mathbf{i}}+2\hat{\mathbf{j}})=0$$

But, we also know that the direction of propagation of electromagnetic wave is perpendicular to both E and B, i.e. it is in the direction of $\mathbf{E}\times\mathbf{B}$.

Again, when $\hat{\mathbf{n}}=\hat{\mathbf{i}}-2\hat{\mathbf{j}}$

$$\mathbf{E}\times\mathbf{B}=(2\hat{\mathbf{i}}+\hat{\mathbf{j}})\times(\hat{\mathbf{i}}-2\hat{\mathbf{j}})=-\hat{\mathbf{k}}$$

and when $\hat{\mathbf{n}}=-\hat{\mathbf{i}}+2\hat{\mathbf{j}}$

$$\mathbf{E}\times\mathbf{B}=(2\hat{\mathbf{i}}+\hat{\mathbf{j}})\times(-\hat{\mathbf{i}}+2\hat{\mathbf{j}})=\hat{\mathbf{k}}$$

But, it is been given in the question that the direction of propagation of wave is in $-\hat{\mathbf{k}}$.

Thus, associated electric field will be

$\mathbf{E}=4.8\times10^{2}$

$\cos(2\times10^{7}z+6\times10^{15}t)(\hat{\mathbf{i}}-2\hat{\mathbf{j}})\ \text{Vm}^{-1}$

89. *(d)* $c=\dfrac{1}{\sqrt{\mu_0\varepsilon_0}}=\dfrac{E_0}{B_0}$ ...(i)

Now, when it enters in a medium of refractive index n, its speed is given as,

$$v=\frac{1}{\sqrt{\mu\varepsilon}}=\frac{1}{\sqrt{K\varepsilon_0\mu}}=\frac{E}{B} \quad \text{...(ii)}$$

where, K is dielectric strength of the medium.

($\because$ For a transparent medium, $\mu_0\approx\mu$)

Using Eqs. (i) and (ii), we get

$$\frac{v}{c}=\frac{1}{\sqrt{K}} \quad \text{...(iii)}$$

Also, refractive index of medium is n and is given as

$$\frac{c}{v}=n$$

or $$\frac{v}{c}=\frac{1}{n} \quad \text{...(iv)}$$

$\therefore$ From Eqs. (iii) and (iv), we get

$$n=\sqrt{K}$$

or $$K=n^2 \quad \text{...(v)}$$

The intensity of a EM wave is given as,

$$I=\frac{1}{2}\varepsilon_0E_0^2c$$

and in the medium, it is given as

$$I'=\frac{1}{2}K\varepsilon_0E^2v$$

It is given that, $I=I'$

$$\Rightarrow \frac{1}{2}\varepsilon_0E_0^2c=\frac{1}{2}K\varepsilon_0E^2v$$

or $$\left(\frac{E_0}{E}\right)^2=\frac{Kv}{c} \quad \text{...(vi)}$$

From Eqs. (iv), (v) and (vi), we get

$$\left(\frac{E_0}{E}\right)^2=\frac{n^2}{n}=n$$

or $$\frac{E_0}{E}=\sqrt{n}$$

Similarly, $$\frac{1}{2}\frac{B_0^2}{\mu_0}\cdot c=\frac{1}{2}\cdot\frac{B^2}{\mu_0}v$$

$$\Rightarrow \frac{B_0}{B}=\frac{1}{\sqrt{n}}$$

> ***Alternate Solution***
>
> We know that,
>
> $\left(\dfrac{E_0}{B_0}\right)_{\text{air/vacuum}}=c$ and $\left(\dfrac{E}{B}\right)_{\text{medium}}=v$
>
> Also, $n=\dfrac{c}{v}$
>
> $\dfrac{E_0/B_0}{E/B}=\dfrac{c}{v}=n\Rightarrow\dfrac{E_0/E}{B_0/B}=n$
>
> This is possible only if $\dfrac{E_0}{E}=\sqrt{n}$ and $\dfrac{B_0}{B}=\dfrac{1}{\sqrt{n}}$.

90. *(d)* Energy of a light wave $\propto$ Intensity of the light wave

Since, intensity $=\varepsilon vA^2$

where, ε is the permittivity of the medium in which light is travelling with velocity v and A is its amplitude.

Since, only 4% of the energy of the light gets reflected.

$\therefore$ 96% of the energy of the light is transmitted.

$E_{\text{transmitted}}(E_t)=96\%$ of $E_{\text{incident}}(E_i)$

$$\varepsilon_r vA_t^2=\frac{96}{100}\times\varepsilon_0\times c\times A_i^2$$

$$A_t^2=\frac{96}{100}\cdot\frac{\varepsilon_0}{\varepsilon_r}\cdot\frac{c}{v}\cdot A_i^2$$

$$A_t^2=\frac{96}{100}\cdot\frac{v^2}{c^2}\cdot\frac{c}{v}\cdot A_i^2\quad\left[\because\sqrt{\frac{\varepsilon_0}{\varepsilon_r}}=\frac{v}{c}\right]$$

$$A_t^2=\frac{96}{100}\cdot\frac{v}{c}\cdot A_i^2$$

$$=\frac{96}{100}\times\frac{1}{3/2}\times30^2$$

$$\left[\because\mu=\frac{c}{v}=1.5\right]$$

$$A_t=\sqrt{\frac{64}{100}\times30^2}$$

$$\Rightarrow A_t=24\ \text{V/m}$$

91. *(6)* Velocity of light in of μ and ε is

$$\left[c=\frac{1}{\sqrt{\mu_0\varepsilon_0}}\right]$$

$$v_m=\frac{1}{\sqrt{\mu\varepsilon}}=\frac{1}{\sqrt{\mu_0\mu_r}}\times\frac{1}{\sqrt{\varepsilon_0\varepsilon_r}}$$

$$=\frac{1}{\sqrt{\mu_r\varepsilon_r}}\times\frac{1}{\sqrt{\mu_0\varepsilon_0}}$$

$$=\frac{c}{\sqrt{\mu_r\varepsilon_r}}$$

$$=\frac{c}{\sqrt{\frac{10}{\pi}\times\frac{1}{0.0885}}}=\frac{c}{\sqrt{36}}=\frac{c}{6}$$

92. *(1320)* $\dfrac{dV}{dt}=7\times10^8$ V/s, $i=0.15$ A,

$r=10$ cm $=0.1$ m

From displacement current,

$$i_d=C\frac{dV}{dt}$$

$$\Rightarrow 0.15=\frac{A\varepsilon_0}{d}\times7\times10^8$$

$$\Rightarrow d=\frac{A\varepsilon_0\times7\times10^8}{0.15}$$

$$=\frac{\pi\times(0.1)^2\times9\times10^{-12}\times7\times10^8}{0.15}$$

$$=\frac{22}{7}\times\frac{10^{-2}\times9\times10^{-4}\times7}{15\times10^{-2}}$$

$$=\frac{22\times9}{15}\times10^{-4}=13.2\times10^{-4}\ \text{m}$$

$$=1320\ \mu\text{m}$$

93. *(10)* $V=\dfrac{Q}{C}$

$$\frac{dV}{dt}=\frac{1}{C}\cdot\frac{dQ}{dt}$$

$$\frac{dV}{dt}=\frac{1}{C}(I)=\frac{0.25\times10^{-3}}{25\times10^{-6}}$$

$$=\frac{1000}{100}=10\ \text{Vs}^{-1}$$

94. *(1200)* Given, $A_0 = 3.2\text{ cm}^2$

$I = 6\text{A}$

Displacement current,

$$J_d = I / A$$

$$\text{Current density} = \frac{\text{Current}}{\text{Area}} = \frac{6}{16}$$

Current through small area

$$= J_d \times A_0 = \frac{6}{16} \times 3.2$$

$$= 1.2\text{ A} = 1200\text{ mA}$$

95. *(4)* For no deflection,

$$v = \frac{E}{B} \quad \ldots\text{(i)}$$

and $v = \sqrt{\frac{2K}{m}}$

Using given data,
We have,

$$v = \sqrt{\frac{2 \times 5 \times 1.6 \times 10^{-19}}{9 \times 10^{-31}}}$$

So, from Eq. (i),

$$v^2 = \frac{E^2}{B^2}$$

or $E^2 = v^2 \times B^2$

$$= \frac{2 \times 5 \times 1.6 \times 10^{-19}}{9 \times 10^{-31}} \times (3 \times 10^{-6})^2$$

$$= \frac{2 \times 5 \times 1.6 \times 9}{9} \times 10^{-19-12+31}$$

$\Rightarrow E^2 = 16$

or $E = 4\,\frac{\text{N}}{\text{C}}$

96. *(5)* Given that speed in medium,

$$v = 0.2\,c$$

Refractive index, $n = \frac{c}{v}$

$$n = \frac{c}{0.2c} = 5$$

$\therefore \quad n = \sqrt{\mu_r \varepsilon_r}$

or $\quad n^2 = \mu_r \varepsilon_r$ or $\frac{\varepsilon_r}{n} = \frac{n}{\mu_r} = \frac{5}{1}$

On comparing with $\frac{\varepsilon_r}{n} = \frac{x}{1}$, we get

$$x = 5$$

97. *(4)* Radiation pressure over a perfectly reflecting surface is given by $p = \frac{2I}{c}$

Now, Intensity $I = \frac{\text{Energy}}{\text{Area} \times \text{time}}$

$$= \frac{\text{Power}}{\text{Area}}$$

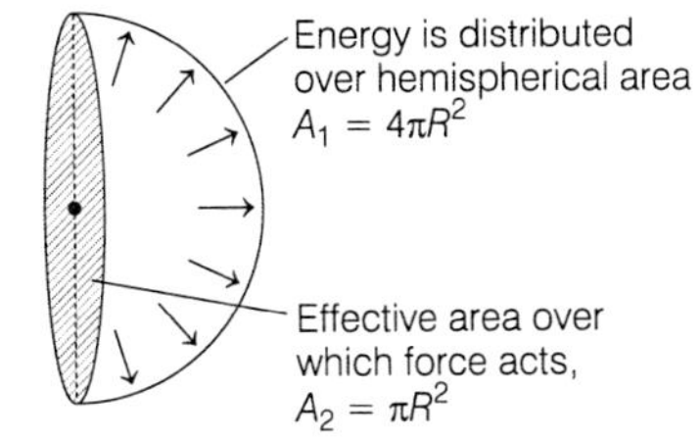

So, force on hemispherical surface,

$$F = p \times A_2$$

$$= \frac{2I}{c} \times A_2 = \frac{2(p / A_1) \times A_2}{c}$$

$$= \frac{2 \times 24 \times \pi R^2}{4\pi R^2 \times 3 \times 10^8} = 4 \times 10^{-8}\text{ N}$$

98. *(84)* Effective power of light bulb converted into visible radiation,

$$P' = 10\%\text{ of } 110\text{ W}$$

$$= \frac{10}{100} \times 110\text{ W} = 11\text{ W}$$

Change in average intensities of visible radiation,

$$I_1 - I_2 = \frac{P'}{4\pi r_1^2} - \frac{P'}{4\pi r_2^2}$$

Here, $r_1 = 1$ m and $r_2 = 5$ m

$$= \frac{P'}{4\pi}\left(\frac{1}{r_1^2} - \frac{1}{r_2^2}\right)$$

$$= \frac{11}{4\pi}\left(\frac{1}{1^2} - \frac{1}{5^2}\right)$$

$$= \frac{11}{4 \times \frac{22}{7}}\left(\frac{24}{25}\right) = \frac{7}{8} \times \frac{24}{25} = \frac{21}{25} = 0.84$$

$$= 84 \times 10^{-2}\text{ W/m}^2$$

$$= a \times 10^{-2}\text{ W/m}^2 \quad \text{[Given]}$$

$\therefore \quad a = 84$

99. *(8)* Displacement current is given by

$$I_d = \varepsilon_0 \frac{d\phi_E}{dt} = \varepsilon_0 \frac{d}{dt}(EA)$$

$$= \varepsilon_0 A \cdot \frac{d}{dt}\left(\frac{V}{d}\right)$$

or $I_d = \frac{\varepsilon_0 A}{d}\left(\frac{d}{dt}V\right) \quad \ldots\text{(i)}$

Here, given
$dV / dt = 10^6\text{ Vs}^{-1}$

$$I_d = 4.425 \times 10^{-6}\text{ A}$$

$$A = 40 \times 10^{-4}\text{ m}^2$$

$$\varepsilon_0 = 8.85 \times 10^{-12}\text{ C}^2\text{N}^{-1}\text{m}^{-2}$$

On substituting values in Eq. (i), we get

$$d = \frac{8.85 \times 10^{-12} \times 40 \times 10^{-4} \times 10^6}{4.425 \times 10^{-6}}$$

$$= 8 \times 10^{-3}\text{ m}$$

so, $\quad x = 8$

100. *(43)* Intensity I and magnitude of electric field, E are related as

$$I = \frac{1}{2}\varepsilon_0 E^2 \cdot c$$

$$\Rightarrow \quad E^2 = \frac{2I}{\varepsilon_0 \times c}$$

Also, $E / B = c \Rightarrow E = B \cdot c$

So, $\quad B^2 = \frac{2I}{\varepsilon_0 \times c^3}$

On substituting values, we get

$$\Rightarrow B^2 = \frac{2 \times 0.22}{8.85 \times 10^{-12} \times 27 \times 10^{24}}$$

$$\Rightarrow \quad B = \sqrt{\left(\frac{2 \times 0.22}{8.85 \times 27 \times 10^{12}}\right)}$$

$$= 0.429 \times 10^{-7}$$

$$= 42.9 \times 10^{-9}\text{ T}$$

or $\quad B \approx 43 \times 10^{-9}\text{ T}$

101. *(500)* As we know that

$$\frac{\text{Energy }(U)}{\text{Volume }(V)} = \frac{1}{2}\varepsilon_0 E_0^2$$

$$\therefore \quad V = \frac{2U}{\varepsilon_0 E_0^2}$$

$$= \frac{2 \times 5.5 \times 10^{-12}}{8.854 \times 10^{-12} \times (50)^2}$$

$$= \frac{11}{8.854 \times 50^2} = 4.97 \times 10^{-4}$$

$$= 497 \times 10^{-6}\text{m}^3 = 500\text{ cm}^3$$

102. *(2)* We know that,

$$B = \frac{E}{c} \Rightarrow B = \frac{6}{3 \times 10^8}$$

$$\Rightarrow \quad B = 2 \times 10^{-8}\text{ T} \quad \therefore x = 2$$

103. *(3)* Intensity

$$I = \frac{\text{Energy}}{\text{Area} \times \text{Time}} = \frac{\text{Power}}{\text{Area}}$$

$$= U_{\text{avg}}\, c = \frac{1}{2} c\varepsilon_0 E_0^2$$

Here,

U_{avg} = Average energy density,

c = Speed of radiation in air (or vacuum),

ε_0 = Permittivity of the free space

and E_0 = Peak value of the electric field.

$$\Rightarrow \quad P \propto E_0^2$$

$\Rightarrow \quad \dfrac{P_1}{P_2} = \dfrac{E_1^2}{E_2^2}$

$\Rightarrow \quad \dfrac{E_1}{E_2} = \sqrt{\dfrac{P_1}{P_2}}$

$\Rightarrow \quad \dfrac{E_1}{E_2} = \sqrt{\dfrac{60}{100}}$

$\Rightarrow \quad E_1 = \sqrt{\dfrac{3}{5}} E_2 = \sqrt{\dfrac{3}{5}} E \quad (\because E_2 = E)$

On comparing with $\sqrt{\dfrac{x}{5}} E$, we get, $x = 3$

104. *(6)* As we know, conduction current,

$$I_c = \frac{[v_0 \sin(2\pi f t)]}{\rho \dfrac{d}{A}}$$

$$\Rightarrow \quad I_c = \frac{A[v_0 \sin(2\pi f t)]}{\rho d} \quad \ldots(i)$$

As we know the displacement current equation,

$$I_d = \frac{\varepsilon_0 \varepsilon_r A}{d} \frac{d}{dt}[v_0 \sin(2\pi f t)]$$

$$I_d = \frac{\varepsilon_0 \varepsilon_r A}{d} v_0 (2\pi f)[\cos(2\pi f t)] \quad \ldots(ii)$$

On dividing Eq. (ii) by Eq. (i), we get

$$\frac{I_d}{I_c} = \varepsilon_0 \varepsilon_r \times \rho 2\pi f \cot(2\pi f t)$$

On substituting the values in the above equation, we get

$$\frac{I_d}{I_c} = \frac{1}{4\pi \times 9 \times 10^9} \times (80) \times 2\pi \times 900(0.25) \cot\left(2\pi(900)\frac{1}{800}\right)$$

$$\frac{I_d}{I_c} = \frac{1}{10^6} \Rightarrow I_c = 10^6 I_d$$

On comparing with 10^x, we get $x = 6$.

105. *(25)* Given,

$F = 2.5 \times 10^{-6}$N,

$A = 30\ \text{cm}^2, I = ?$

We know that energy flux, $P = \dfrac{I}{c}$

or $\quad \dfrac{F}{A} = \dfrac{I}{c} \quad \left[\because P = \dfrac{F}{A}\right]$

$\Rightarrow \quad F = \dfrac{I}{c} A$

$\Rightarrow \quad I = \dfrac{F \times c}{A}$

$\Rightarrow \quad I = \dfrac{2.5 \times 10^{-6} \times 3 \times 10^8}{30}$

$= 25\ \text{W cm}^{-2}$

106. *(137)* Given, Power of bulb,

$P = 1000$ W

Distance, $d = 2$ m

Efficiency of bulb = 1.25%,

$\therefore$ Intensity,

$$I = \frac{\text{Power}(P)}{\text{Area}(A)} = \frac{1}{2}\varepsilon_0 E^2 c$$

$$\Rightarrow \frac{1.25}{100} \times \frac{1000}{4\pi(2)^2} = \frac{1}{2} \times 8.854 \times 10^{-12} \times E^2 \times 3 \times 10^8$$

$$\Rightarrow E = \sqrt{\frac{12.5}{16\pi} \times \frac{2}{8.854 \times 10^{-4} \times 3}}$$

$= 13.7\text{NC}^{-1}$

$= 137 \times 10^{-1}\ \text{NC}^{-1}$

or Vm^{-1}

$\therefore x = 137$

107. *(667)* Since, $f = c/\lambda$

$$\lambda = \frac{c}{f} = \frac{3 \times 10^8}{3 \times 10^9} = 0.1\ \text{m}$$

$\because \lambda_m$ (wavelength of wave in a medium) $= \lambda / \mu$

and as we know that $\mu = \sqrt{\mu_r \varepsilon_r}$

As, dielectric is non-magnetic,

$\mu_r = 1 \Rightarrow \mu = \sqrt{2.25} = 1.5$

$\Rightarrow \quad \lambda_m = \dfrac{0.1}{1.5} = \dfrac{1}{15} = 0.0667$ m

$= 6.67$ cm

$= 667 \times 10^{-2}$ cm

108. *(15)* Given, $\mu_r = \varepsilon_r = 2$

Speed of electromagnetic wave v is given by

$$v = \frac{c}{n}$$

where, n = refractive index

$= \sqrt{\mu_r \varepsilon_r} = \sqrt{4} = 2$

$\Rightarrow \quad v = \dfrac{3 \times 10^8}{2} = 15 \times 10^7$ m/s

$\because x \times 10^7 = 15 \times 10^7$

$\Rightarrow \quad x = 15$

109. *(194)* Given, intensity,

$$I = \left(\frac{315}{\pi}\right) \text{W/m}^2$$

As, intensity, $I = \dfrac{1}{2}\varepsilon_0 E_0^2 c$

or $\quad E_0 = \sqrt{\dfrac{2I}{\varepsilon_0 c}}$

$$= \sqrt{\frac{2 \times \left(\dfrac{315}{\pi}\right)}{8.86 \times 10^{-12} \times 3 \times 10^8}}$$

$E_0 \approx 275$ V/m

$\therefore \quad E_{\text{rms}} = \dfrac{E_0}{\sqrt{2}}$

$= \dfrac{275}{\sqrt{2}} \simeq 194$ V/m

Topic 2

01. *(b)* γ-rays, X-rays, UV-rays, visible rays, IR, micro wave, radio wave $\lambda\uparrow$

$\lambda_4 < \lambda_2 < \lambda_3 < \lambda_1$

02. *(c)* Wavelength range for infrared wave is 1 mm to 700 nm.

Hence, $\quad$ A → III

Wavelength range for ultraviolet wave is 400 nm to 1 nm.

$\therefore \quad$ B → II

Wavelength range for X-rays is 1 nm to 10^{-3} nm.

$\therefore \quad$ C → IV

Wavelength range for Gamma rays is less than 10^{-3} nm.

$\therefore \quad$ D → I

03. *(b)*

$\lambda_{\gamma\text{-rays}} < \lambda_{\text{X-rays}} < \lambda_{\text{infrared}} < \lambda_{\text{microwave}}$

$\Rightarrow \lambda_1 < \lambda_2 < \lambda_3 < \lambda_4$

04. *(d)* Optical communication is performed in the frequency range of 1 THz to 1000 THz.

From microwave to UV range.

So, EM wave used for optical communication have shorter wavelength than that of microwaves used in RADAR.

Also, $\nu_{\text{infrared}} > \nu_{\text{microwave}}$

$\therefore$ Infrared EM waves are more energetic than microwave.

Hence, A is false and R is true.

05. *(b)* The increasing order of wavelength of EM wave is γ-rays < X-ray < UV < Visible < IR < Micro < Radio

So, $\lambda_{\text{gamma}} < \lambda_{\text{X-ray}} < \lambda_{\text{visible}} < \lambda_{\text{microwave}}$

06. *(c)* $\lambda_{\text{X-rays}} = 0.01$ Å to 100 Å

$\lambda_{\text{visible}} = 3800$ Å to 7000 Å

$\lambda_{\text{microwaves}} = 10^7$ Å to 3×10^9 Å

$\lambda_{\text{radio waves}} = 10^7$ Å to 10^{15} Å

So, it is clear from the values that

$\lambda_{\text{radio waves}} > \lambda_{\text{microwaves}} > \lambda_{\text{visible}} > \lambda_{\text{X-rays}}$

07. *(a)* As, we know energy liberated,

$$E = \frac{hc}{\lambda} \text{ i.e. } E \propto \frac{1}{\lambda}$$

So, lesser the wavelength, greater will be energy liberated by electromagnetic radiations per quantum.

As, order of wavelength is given by

X-ray, VIBGYOR, Radio waves

(C) (A) (B) (D)

∴Order of electromagnetic radiations per quantum.

$\Rightarrow$ D < B < A < C

08. *(c)* Cosmic rays, γ-rays and X-rays are part of electromagnetic spectrum, while β-rays are emitted by radioactive elements. Hence, β-rays is not electromagnetic waves. It is of electrons.

09. *(a)*

EM Wave	Wavelength range
Microwave	0.1 m to 1 mm
Ultraviolet	400 nm to 1 nm
X-rays	1 nm to 10^{-3} nm
Infrared	1 mm to 700 nm

Hence,

A → IV, B → I, C → II, D → III

10. *(a)*

S.No.	Radiation	Production
A.	Microwave	IV. Special electronic devices such as klystron tube.
B.	Gamma rays	I. Due to decay of radioactive nuclei
C.	Radiowaves	II. Oscillating circuits
D.	X-rays	III. Due to transition of inner shell electrons

Thus, correct sequence from List I to List II is A-IV, B-I, C-II, D-III.

11. *(b)* (A)-(ii) UV rays are used in water purification.

(B)-(i) X-rays are used in diagnostic tool in medicine.

(C)-(iii) Microwaves are used in communication and Radar system.

(D)-(iv) Infrared waves are used in improving visibility in foggy days.

12. *(d)* As we know that;

A. Source of microwave frequency is magnetron.

B. Source of infrared frequency is vibration of atoms and molecules.

C. Source of gamma ray is radioactive decay of nucleus.

D. Source of X-ray is transition of electron in inner shells.

∴ The correct match is

A → (2), B → (4), C → (1), D → (3).

13. *(d)* Energy is inversely proportional to wavelength,

i.e., $E = \frac{hc}{\lambda} \Rightarrow E \propto \frac{1}{\lambda}$

Energies of given radiations have the following order:

$E_{\gamma\text{-rays}} > E_{\text{X-rays}} > E_{\text{microwaves}} > E_{\text{AM radiowaves}}$

$\therefore \lambda_{\gamma\text{-rays}} < \lambda_{\text{X-ray}} < \lambda_{\text{microwaves}} < \lambda_{\text{AM radiowaves}}$

Thus, microwaves $\rightarrow 10^{-3}$ m

γ-rays $\rightarrow 10^{-15}$ m

AM radiowaves $\rightarrow 100$ m

X-rays $\rightarrow 10^{-10}$ m

∴(A) → (IV), (B) → (II), (C) → (I), (D) → (III).

Hence, correct option is (d).

14. *(d)* (A) Infrared waves are used to treat muscular strain.

(B) Radiowaves are used for broadcasting purposes.

(C) X-rays are used to detect fracture of bones.

(D) Ultraviolet rays are absorbed by ozone.

C-23

JEE Main Chapterwise Topicwise Physics

In JEE Main there is usually 1-2 question per paper is visible from this chapter. In past 3 years, 61 conceptual and 24 numerical questions were asked of easy to moderate level from this chapter.

For detailed analysis scan QR code

RAY OPTICS

Topic 1 Reflection of Light

Level 1

01 A finite size object is placed normal to the principal axis at a distance of 30 cm from a convex mirror of focal length 30 cm. A plane mirror is now placed in such a way that the image produced by both the mirrors coincide with each other. The distance between the two mirrors is *[2025, 4 April Shift-II]*

(a) 45 cm (b) 15 cm
(c) 7.5 cm (d) 22.5 cm

02 A concave mirror of focal length f in air is dipped in a liquid of refractive index μ. Its focal length in the liquid will be *[2025, 23 Jan Shift-II]*

(a) $f/(\mu-1)$ (b) f/μ
(c) f (d) μf

03 When one light ray is reflected from a plane mirror with 30° angle of reflection, the angle of deviation of the ray after reflection is *[2023, 11 April Shift-II]*

(a) 140° (b) 110°
(c) 120° (d) 130°

04 An object is placed at a distance of 12 cm infront of a plane mirror. A virtual and erect image is formed by the mirror. Now, the mirror is moved by 4 cm towards the stationary object. The distance by which the position of image would be shifted, will be *[2023, 10 April Shift-I]*

(a) 4 cm towards mirror (b) 8 cm away from mirror
(c) 2 cm towards mirror (d) 8 cm towards mirror

05 The light rays from an object have been reflected towards an observer from a standard flat mirror, the image observed by the observer are

A. real B. erect
C. smaller in size than object D. laterally inverted

Choose the most appropriate answer from the options given below. *[2023, 25 Jan Shift-II]*

(a) B and C only (b) B and D only
(c) A, C and D only (d) A and D only

06 A short straight object of height 100 cm lies before the central axis of a spherical mirror, whose focal length has absolute value $f = 40$ cm. The image of object produced by the mirror is of height 25 cm and has the same orientation of the object. One may conclude from the information. *[2021, 26 Feb Shift-I]*

(a) Image is real, same side of concave mirror
(b) Image is virtual, opposite side of concave mirror
(c) Image is real, same side of convex mirror
(d) Image is virtual, opposite side of convex mirror

Imp. Image formation by two plane mirrors

07 To get three images of a single object, one should have two plane mirrors at an angle of *[AIEEE 2003]*

(a) 60° (b) 90°
(c) 120° (d) 30°

08 If two mirrors are kept at 60° to each other, then the number of images formed by them is *[AIEEE 2002]*

(a) 5 (b) 6 (c) 7 (d) 8

Level 2

09 When an object is placed 40 cm away from a spherical mirror an image of magnification $\frac{1}{2}$ is produced. To obtain an image with magnification of $\frac{1}{3}$ the object is to be moved *[2025, 4 April Shift-I]*

(a) 20 cm towards the mirror.
(b) 20 cm away from the mirror.
(c) 40 cm away from the mirror.
(d) 80 cm away from the mirror.

10 Two identical objects are placed in front of convex mirror and concave mirror having same radii of curvature of 12 cm, at same distance of 18 cm from the respective mirrors. The ratio of sizes of the images formed by convex mirror and by concave mirror is *[2025, 2 April Shift-II]*

(a) 3 (b) $\frac{1}{3}$ (c) $\frac{1}{2}$ (d) 2

11 A mirror is used to produce an image with magnification of $\frac{1}{4}$. If the distance between object and its image is 40 cm, then the focal length of the mirror is .... *[2025, 7 April Shift-II]*

(a) 15 cm (b) 10 cm
(c) 10.7 cm (d) 12.7 cm

12 A concave mirror produces an image of an object such that the distance between the object and image is 20 cm. If the magnification of the image is −3, then the magnitude of the radius of the curvature of the mirror is *[2025, 28 Jan Shift-II]*

(a) 7.5 cm (b) 30 cm
(c) 3.75 cm (d) 15 cm

13 If the distance between object and its two times magnified virtual image produced by a curved mirror is 15 cm, the focal length of the mirror must be *[2024, 29 Jan Shift-II]*

(a) −12 cm (b) (10/3) cm
(c) −10 cm (d) 15 cm

14 A convex mirror of radius of curvature 30 cm forms an image that is half the size of the object. The object distance is *[2024, 29 Jan Shift-I]*

(a) − 45 cm (b) 15 cm
(c) 45 cm (d) −15 cm

15 Two objects A and B are placed at 15 cm and 25 cm from the pole in front of a concave mirror having radius of curvature 40 cm. The distance between images formed by the mirror is *[2023, 1 Feb Shift-II]*

(a) 60 cm (b) 40 cm
(c) 160 cm (d) 100 cm

16 Two plane mirrors M_1 and M_2 are at right angle to each other shown. A point source P is placed at a and $2a$ meter away from M_1 and M_2, respectively. The shortest distance between the images thus formed is (Take, $\sqrt{5} = 2.3$)

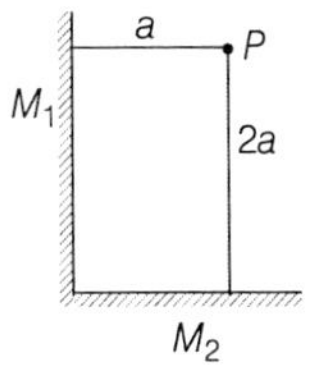

[2021, 31 Aug Shift-I]

(a) 3 a (b) 4.6 a
(c) 2.3 a (d) $2\sqrt{10}\ a$

17 The incident ray, reflected ray and the outward drawn normal are denoted by the unit vectors **a**, **b** and **c**, respectively. Then, choose the correct relation for these vectors. *[2021, 26 Feb Shift-II]*

(a) $\mathbf{b} = \mathbf{a} + 2\mathbf{c}$ (b) $\mathbf{b} = 2\mathbf{a} + \mathbf{c}$
(c) $\mathbf{b} = \mathbf{a} - 2(\mathbf{a} \cdot \mathbf{c})\mathbf{c}$ (d) $\mathbf{b} = \mathbf{a} - \mathbf{c}$

Imp. Magnification Produced by Spherical Mirror

18

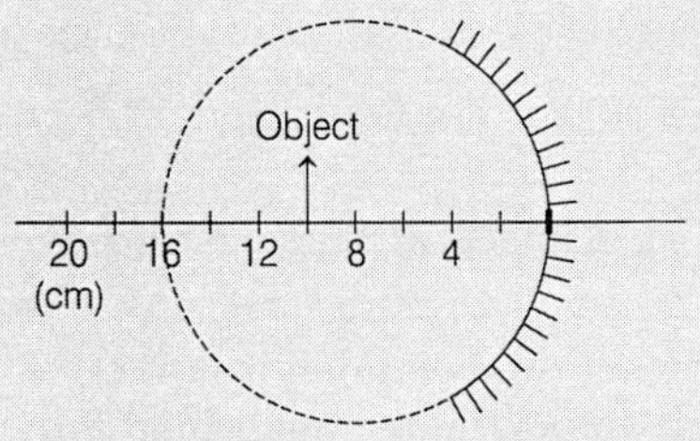

A spherical mirror is obtained as shown in the above figure from a hollow glass sphere. If an object is in front of the mirror, what will be the nature and magnification of the image of the object? (Figure drawn as schematic and not to scale) *[2020, 2 Sep Shift-I]*

(a) Erect, virtual and unmagnified
(b) Inverted, real and magnified
(c) Erect, virtual and magnified
(d) Inverted, real and unmagnified

19 A concave mirror for face viewing has focal length of 0.4 m. The distance at which you hold the mirror from your face in order to see your image upright with a magnification of 5 is *[2019, 9 April Shift-I]*

(a) 0.16 m (b) 1.60 m (c) 0.32 m (d) 0.24 m

20 Two plane mirrors are inclined to each other such that a ray of light incident on the first mirror M_1 and parallel to the second mirror M_2 is finally reflected from the second mirror M_2 parallel to the first mirror (M_1). The angle between the two mirrors will be *[2019, 9 Jan Shift-II]*

(a) 75° (b) 90° (c) 60° (d) 45°

Level 3

21 An object is placed beyond the centre of curvature C of the given concave mirror. If the distance of the object is d_1 from C and the distance of the image formed is d_2 from C, the radius of curvature of this mirror is *[2021, 27 Aug Shift-I]*

(a) $\frac{2d_1 d_2}{d_1 - d_2}$ (b) $\frac{2d_1 d_2}{d_1 + d_2}$ (c) $\frac{d_1 d_2}{d_1 + d_2}$ (d) $\frac{d_1 d_2}{d_1 - d_2}$

22 Car B overtakes another car A at a relative speed of 40 ms^{-1}. How fast will the image of car B appear to move in the mirror of focal length 10 cm fitted in car A, when the car B is 1.9 m away from the car A? *[2021, 26 Aug Shift I]*

(a) 4 ms^{-1} (b) 0.2 ms^{-1} (c) 40 ms^{-1} (d) 0.1 ms^{-1}

23 A car is fitted with a convex side-view mirror of focal length 20 cm. A second car 2.8 m behind the first car is overtaking the first car at a relative speed of 15 m/s. The speed of the image of the second car as seen in the mirror of the first one is *[AIEEE 2011]*

(a) $\frac{1}{15}$ m/s (b) 10 m/s (c) 15 m/s (d) $\frac{1}{10}$ m/s

Numerical Value Type Questions

24 Distance between object and its image (magnified by $-\frac{1}{3}$) is 30 cm. The focal length of the mirror used is $\left(\frac{x}{4}\right)$ cm, where magnitude of value of x is .... *[2025, 4 April Shift-I]*

25 The driver sitting inside a parked car is watching vehicles approaching from behind with the help of his side view mirror, which is a convex mirror with radius of curvature $R = 2$ m. Another car approaches him from behind with a uniform speed of 90 km/h. When the car is at a distance of 24 m from him, the magnitude of the acceleration of the image of the car in the side view mirror is a. The value of $100a$ is ....m/s^2. *[2025, 22 Jan Shift-I]*

26 As shown in the figure, a plane mirror is fixed at a height of 50 cm from the bottom of tank containing water $\left(\mu = \frac{4}{3}\right)$. The height of water in the tank is 8 cm. A small bulb is placed at the bottom of the water tank. The distance of image of the bulb formed by mirror at the bottom of the tank is .... cm. *[2023, 11 April Shift-II]*

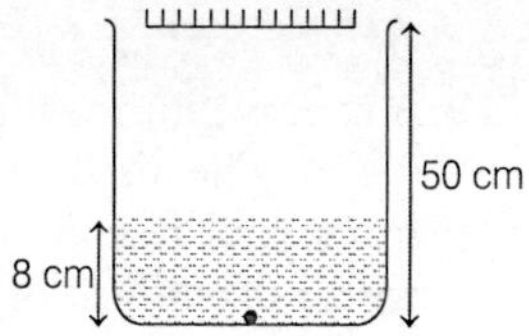

27 Two vertical parallel mirrors A and B are separated by 10 cm. A point object O is placed at a distance of 2 cm from mirror A. The distance of the second nearest image behind mirror A from the mirror A is .... cm. *[2023, 8 April Shift-I]*

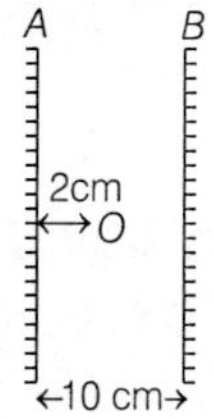

28 A thin cylindrical rod of length 10 cm is placed horizontally on the principle axis of a concave mirror of focal length 20 cm. The rod is placed in such a way that mid-point of the rod is at 40 cm from the pole of mirror. The length of the image formed by the mirror will be $x/3$ cm. The value of x is .... *[2023, 1 Feb Shift-I]*

Imp. Mirror Formula

29 An object O is placed at a distance of 100 cm infront of a concave mirror of radius of curvature 200 cm as shown in the figure. The object starts moving towards the mirror at a speed 2 cm/s. The position of the image from the mirror after 10 s will be at ...... cm. *[2022, 28 July Shift-II]*

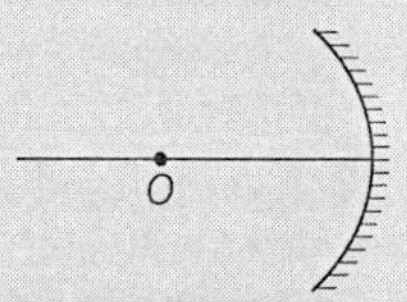

30 When an object is kept at a distance of 30 cm from a concave mirror, the image is formed at a distance of 10 cm from the mirror. If the object is moved with a speed of 9 cms^{-1}, the speed (in cms^{-1}) with which image moves at that instant is .... *[2020, 3 Sep Shift-II]*

Topic 2 Refraction, TIR and Prism

Level 1

01 A monochromatic light of frequency 5×10^{14} Hz travelling through air, is incident on a medium of refractive index '2'. Wavelength of the refracted light will be *[2025, 3 April Shift-II]*
(a) 600 nm (b) 500 nm (c) 400 nm (d) 300 nm

02 Consider following statements for refraction of light through prism, when angle of deviation is minimum.
A. The refracted ray inside prism becomes parallel to the base.
B. Larger angle prisms provide smaller angle of minimum deviation.
C. Angle of incidence and angle of emergence becomes equal.
D. There are always two sets of angle of incidence for which deviation will be same except at minimuin deviation setting.
E. Angle of refraction becomes double of prism angle.
Choose the correct answer from the options given below. *[2025, 3 April Shift-I]*
(a) A, B and E Only (b) B, D and E Only
(c) A, C and D Only (d) B, C and D Only

03

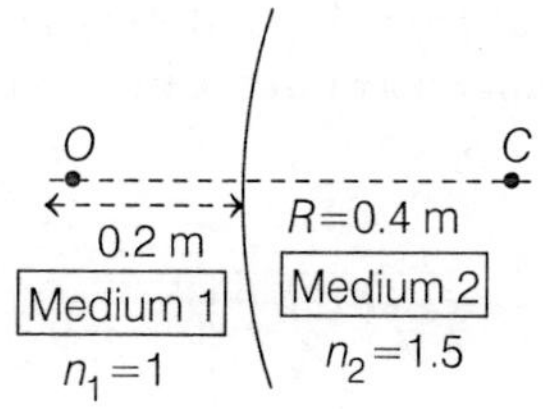

A spherical surface separates two media of refractive indices 1 and 1.5 as shown in figure. Distance of the image of an object O, is (C is the center of curvature of the spherical surface and R is the radius of curvature) *[2025, 2 April Shift-I]*
(a) 0.4 m right to the spherical surface
(b) 0.4 m left to the spherical surface
(c) 0.24 m left to the spherical surface
(d) 0.24 m right to the spherical surface

04 What is the lateral shift of a ray refracted through a parallel sided glass slab of thickness h in terms of the angle of incidence i and angle of refraction r, if the glass slab is placed in air medium? *[2025, 23 Jan Shift-I]*

(a) $\dfrac{h\tan(i-r)}{\tan r}$ (b) $\dfrac{h\cos(i-r)}{\sin r}$

(c) $\dfrac{h\sin(i-r)}{\cos r}$ (d) h

Imp. Angle of Minimum Deviation

05 The refractive index of the material of a glass prism is $\sqrt{3}$. The angle of minimum deviation is equal to the angle of the prism. What is the angle of the prism? *[2025, 23 Jan Shift-II]*

(a) 58° (b) 48° (c) 60° (d) 50°

06 A thin prism P_1 with angle 4° made of glass having refractive index 1.54, is combined with another thin prism P_2 made of glass having refractive index 1.72 to get dispersion without deviation. The angle of the prism P_2 in degrees is *[2025, 28 Jan Shift-I]*

(a) 3 (b) 16/3 (c) 4 (d) 1.5

07 Given below are two statements.

Statement I When the white light passed through a prism. the red light bends lesser than yellow and violet.

Statement II The refractive indices are different for different wavelengths in dispersive medium.

In the light of the above statements, choose the correct answer from the options given below. *[2024, 5 April Shift-II]*

(a) Statement I is true but Statement II is false.
(b) Both Statement I and Statement II are true.
(c) Statement I is false but Statement II is true.
(d) Both Statement I and Statement II are false.

08 A monochromatic light wave with wavelength λ_1 and frequency v_1 in air enters another medium. If the angle of incidence and angle of refraction at the interface are 45° and 30° respectively, then the wavelength λ_2 and frequency v_2 of the refracted wave are *[2023, 6 April Shift-I]*

(a) $\lambda_2=\lambda_1, v_2=\sqrt{2}v_1$ (b) $\lambda_2=\sqrt{2}\lambda_1, v_2=v_1$

(c) $\lambda_2=\dfrac{1}{\sqrt{2}}\lambda_1, v_2=v_1$ (d) $\lambda_2=\lambda_1, v_2=\dfrac{1}{\sqrt{2}}v_1$

09 Light enters from air into a given medium at an angle of 45° with interface of the air-medium surface. After refraction, the light ray is deviated through an angle of 15° from its original direction. The refractive index of the medium is *[2022, 29 July Shift-II]*

(a) 1.732 (b) 1.333 (c) 1.414 (d) 2.732

10 Consider a light travelling in air is incident into a medium of refractive index $\sqrt{2n}$. The incident angle is twice that of refracting angle. Then, the angle of incidence will be *[2022, 27 June Shift-I]*

(a) $\sin^{-1}(\sqrt{n})$ (b) $\cos^{-1}\left(\sqrt{\dfrac{n}{2}}\right)$

(c) $\sin^{-1}(\sqrt{2n})$ (d) $2\cos^{-1}\left(\sqrt{\dfrac{n}{2}}\right)$

11 The expected graphical representation of the variation of angle of deviation δ with angle of incidence i in a prism is *[2021, 27 July Shift-II]*

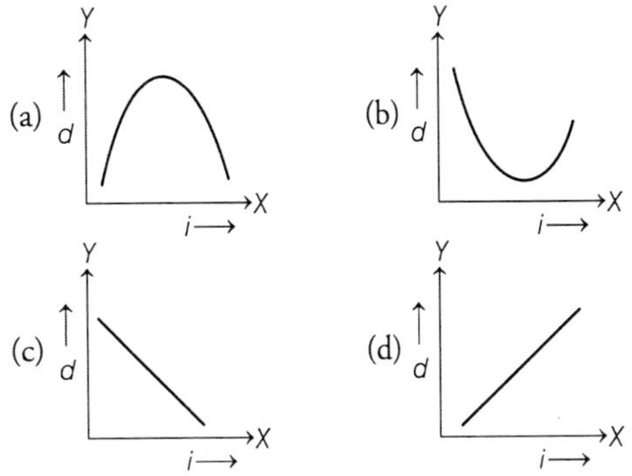

12 A ray of laser of a wavelength 630 nm is incident at an angle of 30° at the diamond-air interface. It is going from diamond to air. The refractive index of diamond is 2.42 and that of air is 1. Choose the correct option. *[2021, 25 July Shift-I]*

(a) Angle of refraction is 24.41°
(b) Angle of refraction is 30°
(c) Refraction is not possible
(d) Angle of refraction is 53.4°

13 Three rays of light, namely red (R), green (G) and blue (B) are incident on the face PQ of a right angled prism PQR as shown in figure.

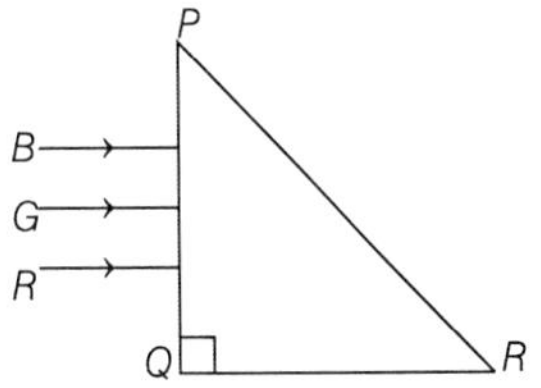

The refractive indices of the material of the prism for red, green and blue wavelength are 1.27, 1.42 and 1.49, respectively. The colour of the ray(s) emerging out of the face PR is *[2021, 18 March Shift-II]*

(a) green (b) red
(c) blue and green (d) blue

14 Red light differs from blue light as they have *[2021, 16 March Shift-II]*

(a) different frequencies and different wavelengths
(b) different frequencies and same wavelengths
(c) same frequencies and same wavelengths
(d) same frequencies and different wavelengths

15 The angle of deviation through a prism is minimum when

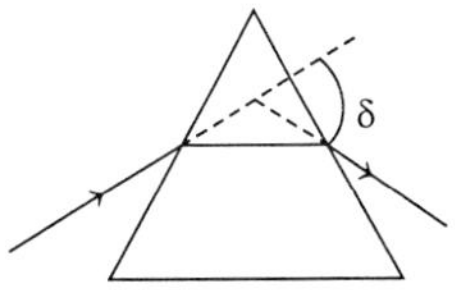

A. incident ray and emergent ray are symmetric to the prism

B. the refracted ray inside the prism becomes parallel to its base

C. angle of incidence is equal to that of the angle of emergence

D. angle of emergence is double the angle of incidence

Choose the correct answer from the options given below. *[2021, 16 March Shift-I]*

(a) Statements (A), (B) and (C) are true.
(b) Only statement (D) is true.
(c) Only statements (A) and (B) are true.
(d) Statements (B) and (C) are true.

16 As the beam enters the medium, it will *[AIEEE 2010]*

(a) diverge
(b) converge
(c) diverge near the axis and converge near the periphery
(d) travel as a cylindrical beam

17 A transparent solid cylinder rod has a refractive index of $2/\sqrt{3}$. It is surrounded by air. A light ray is incident at the mid-point of one end of the rod as shown in the figure. The incident angle θ for which the light ray grazes along the wall of the rod is

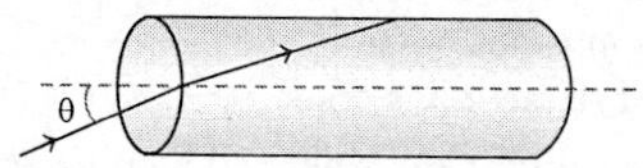

[AIEEE 2009]

(a) $\sin^{-1}\left(\frac{1}{2}\right)$ (b) $\sin^{-1}\left(\frac{\sqrt{3}}{2}\right)$

(c) $\sin^{-1}\left(\frac{2}{\sqrt{3}}\right)$ (d) $\sin^{-1}\left(\frac{1}{\sqrt{3}}\right)$

18 The refractive index of glass is 1.520 for red light and 1.525 for blue light. Let D_1 and D_2 be angles of minimum deviations for red and blue light respectively in a prism of this glass. Then, *[AIEEE 2006]*

(a) $D_1 < D_2$
(b) $D_1 = D_2$
(c) can be less than or greater than D_2 depending upon the angle of prism
(d) $D_1 > D_2$

19 Which of the following is used in optical fibres ? *[AIEEE 2002]*

(a) Total internal reflection (b) Scattering
(c) Diffraction (d) Refraction

Level 2

20 Given below are two statements. One is labelled as Assertion (A) and the other is labelled as Reason (R).

Assertion (A) Refractive index of glass is higher than that of air.

Reason (R) Optical density of a medium is directly proportionate to its mass density which results in a proportionate refractive index. *[2025, 7 April Shift-II]*

In the light of the above statements, choose the most appropriate answer from the options given below.

(a) Both A and R are correct but R is not the correct explanation of A.
(b) Both A and R are correct and R is the correct explanation of A.
(c) A is correct but R is not correct.
(d) A is not correct but R is correct.

21 A transparent film of refractive index, 2.0 is coated on a glass slab of refractive index, 1.45. What is the minimum thickness of transparent film to be coated for the maximum transmission of green light of wavelength 550 nm? [Assume that the light is incident nearly perpendicular to the glass surface.] *[2025, 22 Jan Shift-II]*

(a) 94.8 nm (b) 275 nm
(c) 68.7 nm (d) 137.5 nm

22 At the interface between two materials having refractive indices n_1 and n_2, the critical angle for reflection of an em wave is θ_{1C}. The n_2 material is replaced by another material having refractive index n_3 such that the critical angle at the interface between n_1 and n_3 materials is θ_{2C}. If $n_3 > n_2 > n_1$, $\frac{n_2}{n_3} = \frac{2}{5}$ and $\sin\theta_{2C} - \sin\theta_{1C} = 1/2$, then θ_{1C} is *[2025, 29 Jan Shift-I]*

(a) $\sin^{-1}\left(\frac{5}{6n_1}\right)$ (b) $\sin^{-1}\left(\frac{2}{3n_1}\right)$

(c) $\sin^{-1}\left(\frac{1}{3n_1}\right)$ (d) $\sin^{-1}\left(\frac{1}{6n_1}\right)$

23 A spherical surface of radius of curvature R, separates air from glass (refractive index = 1.5). The centre of curvature is in the glass medium. A point object O placed in air on the optic axis of the surface. So, that its real image is formed at I inside glass. The line OI intersects the spherical surface at P and $PO = PI$. The distance PO equals to *[2025, 23 Jan Shift-I]*

(a) $5R$ (b) $1.5\ R$ (c) $3\ R$ (d) $2\ R$

24 Critical angle of incidence for a pair of optical media is 45°. The refractive indices of first and second media are in the ratio *[2024, 8 April Shift-I]*

(a) 1 : 2 (b) $1:\sqrt{2}$ (c) 2 : 1 (d) $\sqrt{2}:1$

Imp. Refraction Through Prism

25 The refractive index of a prism with apex angle A is $\cot\frac{A}{2}$. The angle of minimum deviation is *[2024, 31 Jan Shift-I]*

(a) $\delta_m = 180° - 3A$ (b) $\delta_m = 180° - A$
(c) $\delta_m = 180° - 2A$ (d) $\delta_m = 180° - 4A$

26 A vessel of depth d is half filled with oil of refractive index n_1 and the other half is filled with water of refractive index n_2. The apparent depth of this vessel when viewed from above will be *[2023, 13 April Shift-I]*

(a) $\frac{2d(n_1 + n_2)}{n_1 n_2}$ (b) $\frac{d n_1 n_2}{2(n_1 + n_2)}$ (c) $\frac{d n_1 n_2}{(n_1 + n_2)}$ (d) $\frac{d(n_1 + n_2)}{2 n_1 n_2}$

27 The critical angle for a denser-rarer interface is 45°. The speed of light in rarer medium is 3×10^8 m/s. The speed of light in the denser medium is *[2023, 11 April Shift-I]*

(a) 5×10^7 m/s (b) 2.12×10^8 m/s
(c) $\sqrt{2} \times 10^8$ m/s (d) 3.12×10^7 m/s

Imp. Apparent Depth

28 A microscope is focussed on an object at the bottom of a bucket. If liquid with refractive index 5 / 3 is poured inside the bucket, then microscope have to be raised by 30 cm to focus the object again. The height of the liquid in the bucket is *[2023, 31 Jan Shift-II]*

(a) 75 cm (b) 50 cm (c) 12 cm (d) 18 cm

29 A thin prism P_1 with an angle 6° and made of glass of refractive index 1.54 is combined with another prism P_2 made from glass of refractive index 1.72 to produce dispersion without average deviation. The angle of prism P_2 is *[2023, 30 Jan Shift-II]*

(a) 4.5° (b) 1.3° (c) 7.8° (d) 6°

30 Light travels in two media M_1 and M_2 with speeds 1.5×10^8 ms^{-1} and 2.0×10^8 ms^{-1}, respectively. The critical angle between them is *[2022, 26 July Shift-II]*

(a) $\tan^{-1}\left(\frac{3}{\sqrt{7}}\right)$ (b) $\tan^{-1}\left(\frac{2}{3}\right)$
(c) $\cos^{-1}\left(\frac{3}{4}\right)$ (d) $\sin^{-1}\left(\frac{2}{3}\right)$

31 The speed of light in media A and B are 2.0×10^{10} cm/s and 1.5×10^{10} cm/s, respectively. A ray of light enters from the medium B to A at an incident angle θ. If the ray suffers total internal reflection, then *[2022, 29 June Shift-II]*

(a) $\theta = \sin^{-1}\left(\frac{3}{4}\right)$ (b) $\theta > \sin^{-1}\left(\frac{2}{3}\right)$
(c) $\theta < \sin^{-1}\left(\frac{3}{4}\right)$ (d) $\theta > \sin^{-1}\left(\frac{3}{4}\right)$

32 The refracting angle of a prism is A and refractive index of the material of the prism is $\cot A/2$. Then the angle of minimum deviation will be: *[2022, 28 June Shift-I]*

(a) $180 - 2A$ (b) $90 - A$
(c) $180 + 2A$ (d) $180 - 3A$

33 A light wave travelling linearly in a medium of dielectric constant 4, incident on the horizontal interface separating medium with air. The angle of incidence for which the total intensity of incident wave will be reflected back into the same medium, will be (Given, relative permeability of medium, $\mu_r = 1$) *[2022, 25 June Shift-I]*

(a) 10° (b) 20° (c) 30° (d) 60°

34 The difference of speed of light in the two media A and B, $(v_A - v_B)$ is 2.6×10^7 m/s. If the refractive index of medium B is 1.47, then the ratio of refractive index of medium B to medium A is (Given, speed of light in vacuum $c = 3 \times 10^8$ ms^{-1}) *[2022, 25 June Shift-I]*

(a) 1.303 (b) 1.318
(c) 1.13 (d) 0.12

35 A glass tumbler having inner depth of 17.5 cm is kept on a table. A student starts pouring water ($\mu = 4/3$) into it while looking at the surface of water from above. When he feels that the tumbler is half filled, he stops pouring water. Up to what height, the tumbler is actually filled? *[2021, 1 Sep Shift-II]*

(a) 11.7 cm (b) 10 cm
(c) 7.5 cm (d) 8.75 cm

Imp. Total Internal Reflection

36 A ray of light entering from air into a denser medium of refractive index 4 / 3, as shown in figure. The light ray suffers total internal reflection at the adjacent surface as shown. The maximum value of angle θ should be equal to *[2021, 25 July Shift-II]*

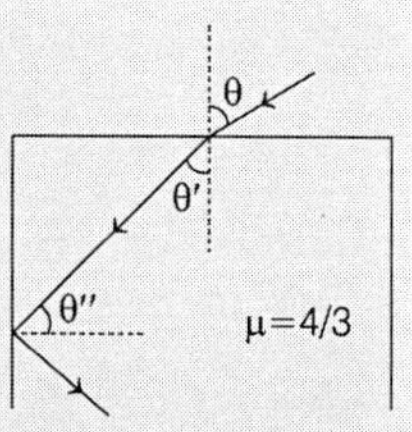

(a) $\sin^{-1}\frac{\sqrt{7}}{3}$ (b) $\sin^{-1}\frac{\sqrt{5}}{4}$
(c) $\sin^{-1}\frac{\sqrt{7}}{4}$ (d) $\sin^{-1}\frac{\sqrt{5}}{3}$

37 A ray of light passes from a denser medium to a rarer medium at an angle of incidence i. The reflected and refracted rays make an angle of 90° with each other. The angle of reflection and refraction are respectively, r and r'. The critical angle is given by *[2021, 22 July Shift-II]*

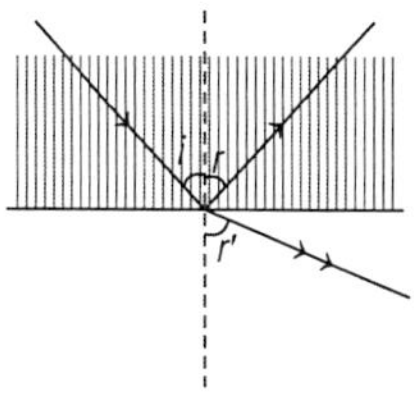

(a) $\sin^{-1}(\tan r')$ (b) $\tan^{-1}(\sin i')$
(c) $\sin^{-1}(\tan r)$ (d) $\sin^{-1}(\cot r)$

38 The critical angle of a medium for a specific wavelength, if the medium has relative permittivity 3 and relative permeability 4 / 3 for this wavelength, will be *[2020, 8 Jan Shift-I]*

(a) 45° (b) 60°
(c) 15° (d) 30°

39 A monochromatic light is incident at a certain angle on an equilateral triangular prism and suffers minimum deviation. If the refractive index of the material of the prism is $\sqrt{3}$, then the angle of incidence is *[2019, 11 Jan Shift-II]*

(a) 45° (b) 90°
(c) 60° (d) 30°

40 A fish looking up through the water sees the outside world, contained in a circular horizon. If the refractive index of water is 4/3 and the fish is 12 cm below the water surface, the radius of this circle (in cm) is *[AIEEE 2005]*

(a) $36\sqrt{7}$ (b) $36/\sqrt{7}$
(c) $36\sqrt{5}$ (d) $4\sqrt{5}$

41 A light ray is incident perpendicular to one face of a 90° prism and is totally internally reflected at the glass-air interface. If the angle of reflection is 45°, we conclude that for the refractive index n as *[AIEEE 2004]*

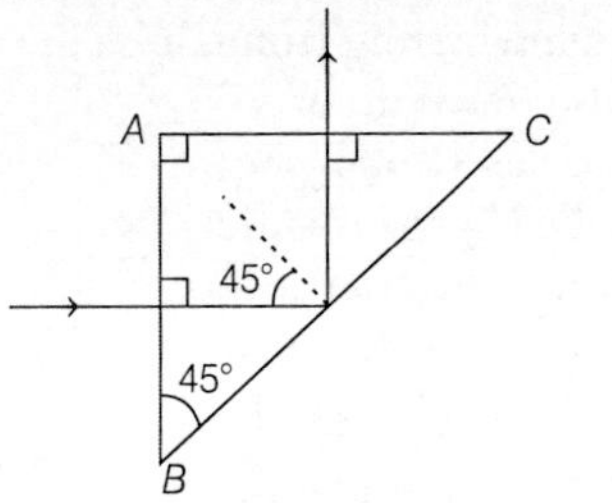

(a) $n<\frac{1}{\sqrt{2}}$ (b) $n>\sqrt{2}$
(c) $n>\frac{1}{\sqrt{2}}$ (d) $n<\sqrt{2}$

Level 3

42 A transparent block A having refractive index $\mu = 1.25$ is surrounded by another medium of refractive index $\mu = 1.0$ as shown in figure. A light ray is incident on the flat face of the block with incident angle θ as shown in figure. What is the maximum value of θ for which light suffers total internal reflection at the top surface of the block? *[2025, 7 April Shift-II]*

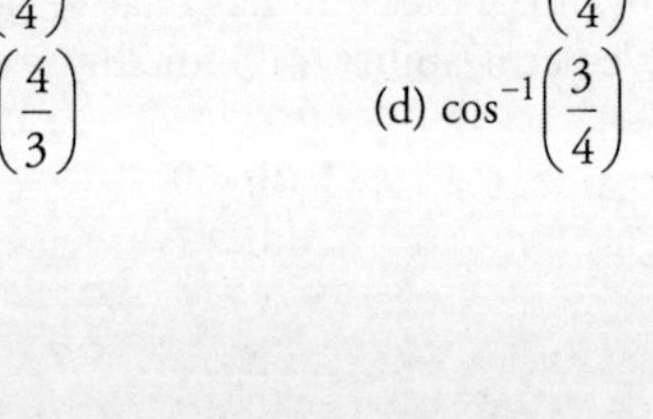

(a) $\sin^{-1}\left(\frac{3}{4}\right)$ (b) $\tan^{-1}\left(\frac{3}{4}\right)$
(c) $\tan^{-1}\left(\frac{4}{3}\right)$ (d) $\cos^{-1}\left(\frac{3}{4}\right)$

43 In a long glass tube, mixture of two liquids A and B with refractive indices 1.3 and 1.4 respectively, forms a convex refractive meniscus towards A. If an object placed at 13 cm from the vertex of the meniscus in A forms an image with a magnification of -2, then the radius of curvature of meniscus is *[2025, 28 Jan Shift-II]*

(a) 2/3 cm (b) 4/3 cm (c) 1/3 cm (d) 1 cm

44 A hemispherical vessel is completely filled with a liquid of refractive index μ. A small coin is kept at the lowest point O of the vessel as shown in figure. The minimum value of the refractive index of the liquid, so that a person can see the coin from point E (at the level of the vessel) is .... *[2025, 28 Jan Shift-I]*

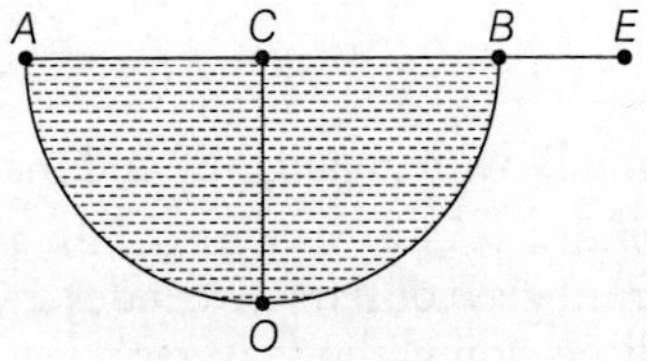

(a) $\frac{\sqrt{3}}{2}$ (b) $\frac{3}{2}$ (c) $\sqrt{3}$ (d) $\sqrt{2}$

45

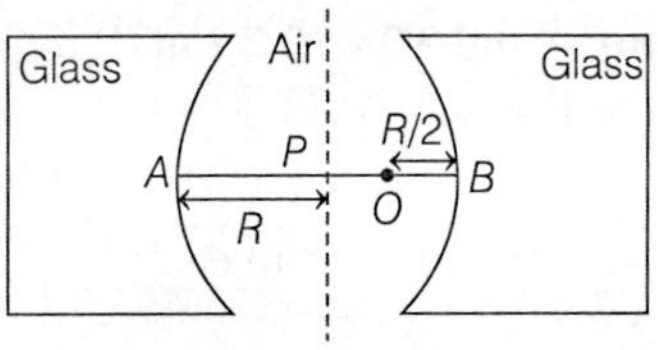

Two concave refracting surfaces of equal radii of curvature and refractive index 1.5 face each other in air as shown in figure. A point object O is placed mid-way, between P and B. The separation between the images of O, formed by each refracting surface is *[2025, 29 Jan Shift-II]*

(a) $0.214R$ (b) $0.411R$ (c) $0.124R$ (d) $0.114R$

46 If the refractive index of the material of a prism is $\cot(A/2)$, where A is the angle of prism, then the angle of minimum deviation will be *[2024, 27 Jan Shift-I]*

(a) $\pi - 2A$ (b) $\frac{\pi}{2} - A$ (c) $\pi - A$ (d) $\frac{\pi}{2} - 2A$

47 In figure, the optical fibre is $l = 2$ m long and has a diameter of $d = 20\ \mu$m. If a ray of light is incident on one end of the fibre at angle $\theta_1 = 40°$, the number of reflections it makes before emerging from the other end is close to, (refractive index of fibre is 1.31 and $\sin 40° = 0.64$) *[2019, 8 April Shift-I]*

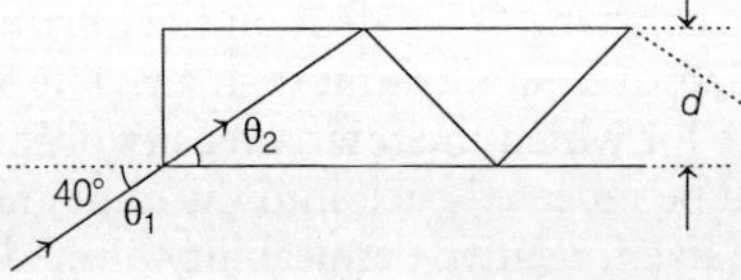

(a) 55000 (b) 66000 (c) 45000 (d) 57000

48 In an experiment for determination of refractive index of glass of a prism by i-δ, plot, it was found that a ray incident at an angle 35° suffers a deviation of 40° and that it emerges at an angle 79°. In that case, which of the following is closest to the maximum possible value of the refractive index? *[JEE Main 2016]*

(a) 1.5 (b) 1.6
(c) 1.7 (d) 1.8

49 Monochromatic light is incident on a glass prism of angle A. If the refractive index of the material of the prism is μ, a ray incident at an angle θ, on the face AB would get transmitted through the face AC of the prism provided *[JEE Main 2015]*

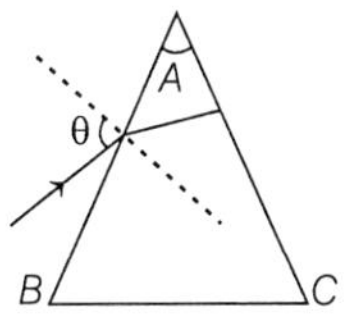

(a) $\theta > \sin^{-1}\left[\mu \sin\left(A - \sin^{-1}\left(\frac{1}{\mu}\right)\right)\right]$

(b) $\theta < \sin^{-1}\left[\mu \sin\left(A - \sin^{-1}\left(\frac{1}{\mu}\right)\right)\right]$

(c) $\theta > \cos^{-1}\left[\mu \sin\left(A + \sin^{-1}\left(\frac{1}{\mu}\right)\right)\right]$

(d) $\theta < \cos^{-1}\left[\mu \sin\left(A + \sin^{-1}\left(\frac{1}{\mu}\right)\right)\right]$

Numerical Value Type Questions

50 A ray of light suffers minimum deviation when incident on a prism having angle of the prism equal to 60°. The refractive index of the prism material is $\sqrt{2}$. The angle of incidence (in degrees) is .... *[2025, 2 April, Shift-II]*

51 Light from a point source in air falls on a spherical glass surface (refractive index, $\mu = 1.5$ and radius of curvature $= 50$ cm). The image is formed at a distance of 200 cm from the glass surface inside the glass. The magnitude of distance of the light source from the glass surface is .... m. *[2025, 3 April, Shift-II]*

52 A container contains a liquid with refractive index of 1.2 up to a height of 60 cm and another liquid having refractive index 1.6 is added to height H above first liquid. If viewed from above, the apparent shift in the position of bottom of container is 40 cm. The value of H is .... cm. (Consider liquids are immisible) cm. *[2025, 7 April, Shift-I]*

Imp. Refraction through Glass Slab

53 Two light beams fall on a transparent material block at point 1 and 2 with angle θ_1 and θ_2, respectively, as shown in figure. After refraction, the beams intersect at point 3 which is exactly on the interface at other end of the block. Given the distance between 1 and 2, $d = 4\sqrt{3}$ cm and $\theta_1 = \theta_2 = \cos^{-1}(n_2/2n_1)$, where refractive index of the block $n_2 >$ refractive index of the outside medium n_1, then the thickness of the block is .... cm. *[2025, 29 Jan Shift-I]*

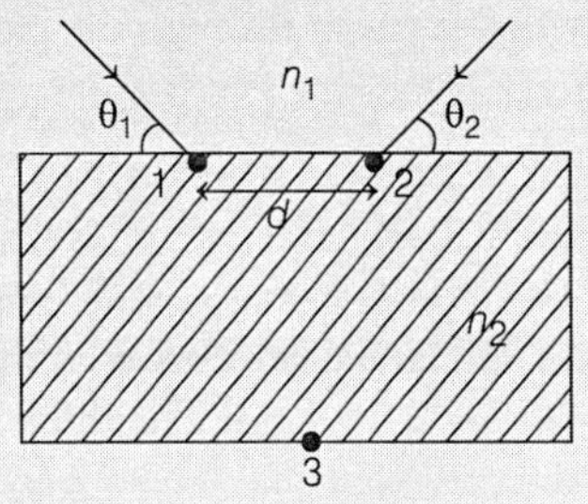

54 A light ray is incident on a glass slab of thickness $4\sqrt{3}$ cm and refractive index $\sqrt{2}$. The angle of incidence is equal to the critical angle for the glass slab with air. The lateral displacement of ray after passing through glass slab is......cm.
(given, $\sin 15° = 0.25$) *[2024, 4 April Shift-II]*

55 Two immiscible liquids of refractive indices $\frac{8}{5}$ and $\frac{3}{2}$, respectively are put in a beaker as shown in the figure. The height of each column is 6 cm. A coin is placed at the bottom of the beaker. For near normal vision, the apparent depth of the coin is $\frac{\alpha}{4}$ cm. The value of α is .... *[2024, 27 Jan Shift-I]*

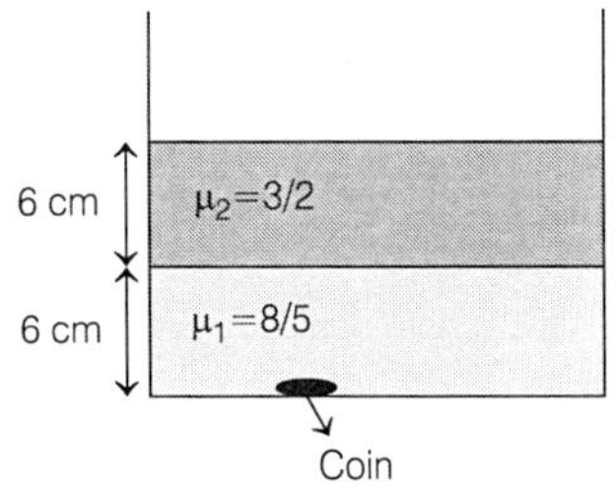

56 The refractive index of a transparent liquid filled in an equilateral hollow prism is $\sqrt{2}$. The angle of minimum deviation for the liquid will be .... *[2023, 15 April Shift-I]*

57 A fish rising vertically upward with a uniform velocity of 8 ms^{-1}, observes that a bird is diving vertically downward towards the fish with the velocity of 12 ms^{-1}. If the refractive index of water is 4/3, then the actual velocity of the diving bird to pick the fish, will be .... ms^{-1}. *[2023, 13 April Shift-I]*

58 A pole is vertically submerged in a swimming pool, such that it gives a length of shadow 2.15 m within water when sunlight is incident at an angle of 30° with the surface of water. If swimming pool is filled to a height of 1.5 m, then the height of the pole above the water surface in centimeters is ($n_W = 4/3$) .... ***[2023, 6 April Shift-I]***

59 A ray of light is incident from air on a glass plate having thickness $\sqrt{3}$ cm and refractive index $\sqrt{2}$. The angle of incidence of a ray is equal to the critical angle for glass-air interface. The lateral displacement of the ray when it passes through the plate is .... $\times 10^{-2}$ cm, (Given, $\sin 15° = 0.26$) ***[2022, 25 Jan Shift-I]***

60 A ray of light is incident at an angle of incidence of 60° on a glass slab of refractive index $\sqrt{3}$. After refraction, the light ray emerges out from other parallel faces and lateral shift between incident ray and emergent ray is $4\sqrt{3}$ cm. The thickness of the glass slab is ...... cm. ***[2022, 24 June Shift-II]***

61 Cross-section view of a prism is the equilateral triangle ABC is shown in the figure. The minimum deviation is observed using this prism when the angle of incidence is equal to the prism angle. The time taken by light to travel from P (mid-point of BC) to A is ..... $\times 10^{-10}$ s. (Given, speed of light in vacuum $= 3 \times 10^8$ m/s and $\cos 30° = \sqrt{3}/2$) ***[2021, 31 Aug Shift-II]***

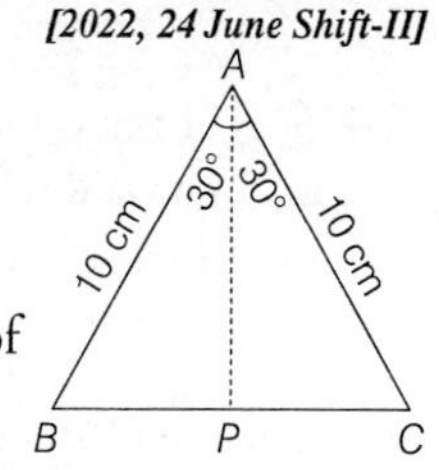

62 A prism of refractive index n_1 and another prism of refractive index n_2 are stuck together (as shown in the figure). n_1 and n_2 depend on λ, the wavelength of light, according to the relation

$$n_1 = 1.2 + \frac{10.8 \times 10^{-14}}{\lambda^2} \text{ and } n_2 = 1.45 + \frac{1.8 \times 10^{-14}}{\lambda^2}$$

The wavelength for which rays incident at any angle on the interface BC pass through without bending at that interface will be .... nm. ***[2021, 27 July Shift-I]***

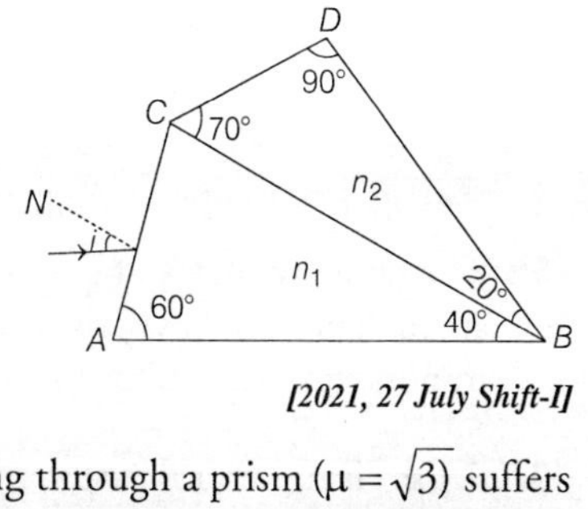

63 A ray of light passing through a prism ($\mu = \sqrt{3}$) suffers minimum deviation. It is found that the angle of incidence is double the angle of refraction within the prism. Then, the angle of prism is .... (in degrees). ***[2021, 22 July Shift-II]***

64 A deviation of 2° is produced in the yellow ray when prism of crown and flint glass are achromatically combined. Taking dispersive powers of crown and flint glass are 0.02 and 0.03, respectively and refractive index for yellow light for these glasses are 1.5 and 1.6, respectively. The refracting angles for crown glass prism will be ......° (in degree).
(Round off to the nearest integer) ***[2021, 16 March Shift-II]***

65 A prism of angle $A = 1°$ has a refractive index $\mu = 1.5$. A good estimate for the minimum angle of deviation (in degree) is close to $N/10$. Value of N is .... ***[2020, 5 Sep Shift-II]***

Topic 3 Lenses

Level 1

01 The radii of curvature for a thin convex lens are 10 cm and 15 cm respectively. The focal length of the lens is 12 cm. The refractive index of the lens material is ***[2025, 3 April, Shift-I]***
(a) 1.2 (b) 1.8 (c) 1.4 (d) 1.5

Imp. Lens Maker's Formula

02 A plano-convex lens having radius of curvature of first surface 2 cm exhibits focal length of f_1 in air. Another plano-convex lens with first surface radius of curvature 3 cm has focal length of f_2 when it is immersed in a liquid of refractive index 1.2. If both the lenses are made of same glass of refractive index 1.5, the ratio of f_1 and f_2 will be ***[2025, 24 Jan Shift-I]***
(a) 1 : 3 (b) 2 : 3 (c) 1 : 2 (d) 3 : 5

03 What is the relative decrease in focal length of a lens for an increase in optical power by 0.1 D from 2.5 D? ['D' stands for dioptre] ***[2025, 24 Jan Shift-I]***
(a) 0.1 (b) 0.01
(c) 0.04 (d) 0.40

Imp. Focal Length After Cutting Part of a Lens

04 Two identical symmetric double convex lenses of focal length f are cut into two equal parts L_1, L_2 by AB plane and L_3, L_4 by XY-plane as shown in figure, respectively. The ratio of focal lengths of lenses L_1 and L_3 is ***[2025, 29 Jan Shift-II]***

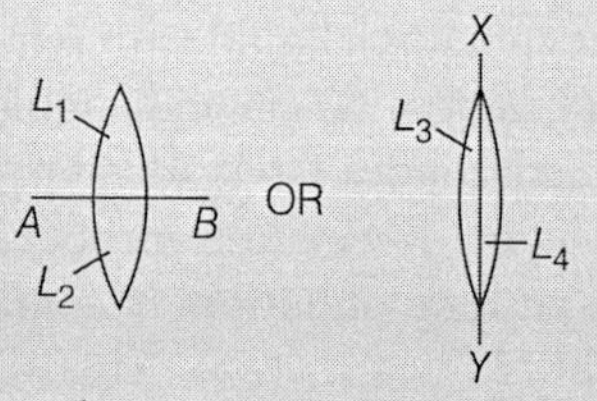

(a) 1 : 1 (b) 1 : 2 (c) 1 : 4 (d) 2 : 1

05 The following figure represents two biconvex lenses L_1 and L_2 having focal length 10 cm and 15 cm respectively. The distance between L_1 and L_2 is **[2024, 9 April Shift-II]**

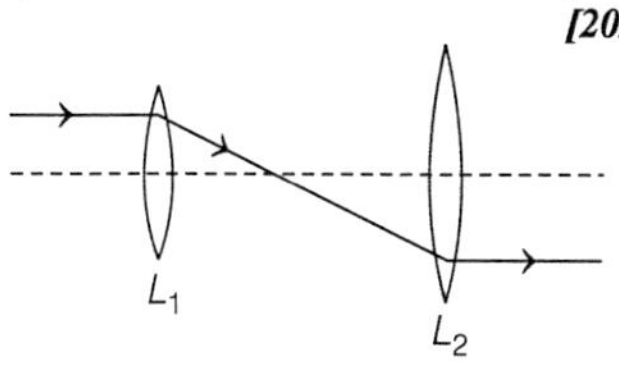

(a) 15 cm (b) 10 cm
(c) 35 cm (d) 25 cm

06 Given below are two statements
Statement I When an object is placed at the centre of curvature of a concave lens, image is formed at the centre of curvature of the lens on the other side.
Statement II Concave lens always forms a virtual and erect image.
In the light of the above statements, choose the correct answer from the options given below. **[2024, 9 April Shift-I]**
(a) Statement I is true but Statement II is false.
(b) Both Statement I and Statement II are true.
(c) Both Statement I and Statement II are false.
(d) Statement I is false but Statement II is true.

07 Light emerges out of a convex lens when a source of light kept at its focus. The shape of wavefront of the light is **[2024, 5 April Shift-I]**
(a) cylindrical (b) plane
(c) spherical (d) both spherical and cylindrical

Imp. Cutting of Lens

08 A convex lens has power P. It is cut into two halves along its principal axis. Further one piece (out of the two halves) is cut into two halves perpendicular to the principal axis (as shown in figure). Choose the incorrect option for the reported pieces. **[2022, 27 June Shift-II]**

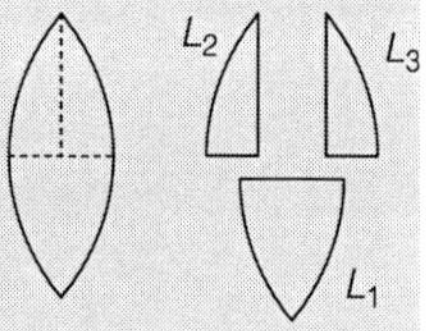

(a) Power of $L_1 = \frac{P}{2}$ (b) Power of $L_2 = \frac{P}{2}$
(c) Power of $L_3 = \frac{P}{2}$ (d) Power of $L_1 = P$

09 An object is placed at the focus of concave lens having focal length f. What is the magnification and distance of the image from the optical centre of the lens? **[2021, 31 Aug Shift-I]**
(a) $1, \infty$ (b) Very high, ∞
(c) $\frac{1}{2}, \frac{f}{2}$ (d) $\frac{1}{4}, \frac{f}{4}$

10 When monochromatic red light is used instead of blue light in a convex lens, its focal length will **[AIEEE 2011]**
(a) not depend on colour of light (b) increase
(c) decrease (d) remain same

11 In an optics experiment, with the position of the object fixed, a student varies the position of a convex lens and for each position, the screen is adjusted to get a clear image of the object. A graph between the object distance u and the image distance v, from the lens, is plotted using the same scale for the two axis. A straight line passing through the origin and making an angle of 45° with the x-axis meets the experimental curve at P. The coordinates of P will be **[AIEEE 2009]**
(a) $(2f, 2f)$ (b) $\left(\frac{f}{2}, \frac{f}{2}\right)$ (c) (f, f) (d) $(4f, 4f)$

12 A student measures the focal length of a convex lens by putting an object pin at a distance u from the lens and measuring the distance v of the image pin. The graph between u and v plotted by the student should look like **[AIEEE 2008]**

(a)
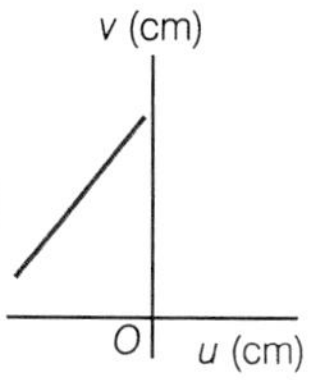

(b)
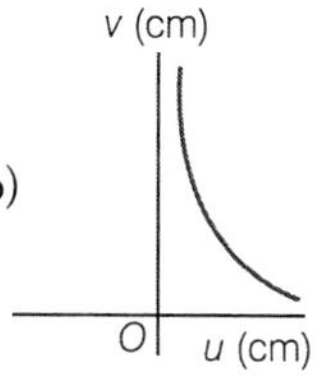

(c)
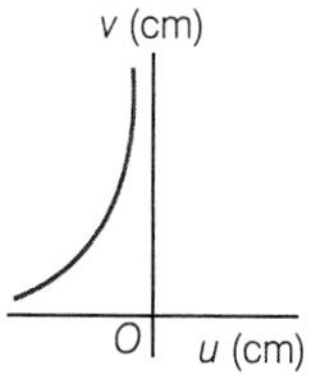

(d)
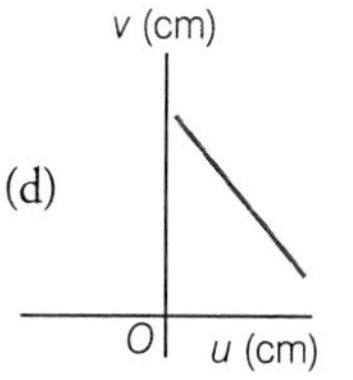

13 Two lenses of power –15D and +5D are in contact with each other. The focal length of the combination is **[AIEEE 2007]**
(a) –20 cm (b) –10 cm (c) +20 cm (d) +10 cm

Level 2

14 Two thin convex lenses of focal lengths 30 cm and 10 cm are placed coaxially 10 cm apart. The power of this combination is **[2025, 7 April Shift-I]**
(a) 1 D (b) 10 D (c) 5 D (d) 20 D

15 A lens having refractive index 1.6 has focal length of 12 cm, when it is in air. Find the focal length of the lens when it is placed in water.
(Take refractive index of water as 1.28) **[2025, 7 April Shift-I]**
(a) 555 mm (b) 288 mm
(c) 655 mm (d) 355 mm

16 A concave-convex lens of refractive index 1.5 and the radii of curvature of its surfaces are 30 cm and 20 cm, respectively. The concave surface is upwards and is filled with a liquid of refractive index 1.3. The focal length of the liquid-glass combination will be **[2025, 8 April Shift-II]**
(a) $\frac{800}{11}$ cm (b) $\frac{700}{11}$ cm c) $\frac{600}{11}$ cm (d) $\frac{500}{11}$ cm

17 A convex lens made of glass (refractive index = 1.5) has focal length 24 cm in air. When it it totally immersed in water (refractive index = 1.33), its focal length changes to *[2025, 29 Jan Shift-II]*

(a) 24 cm (b) 96 cm (c) 48 cm (d) 72 cm

18 Let u and v be the distances of the object and the image from a lens of focal length f. The correct graphical representation of u and v for a convex lens when $|u| > f$, is *[2025, 29 Jan Shift-I]*

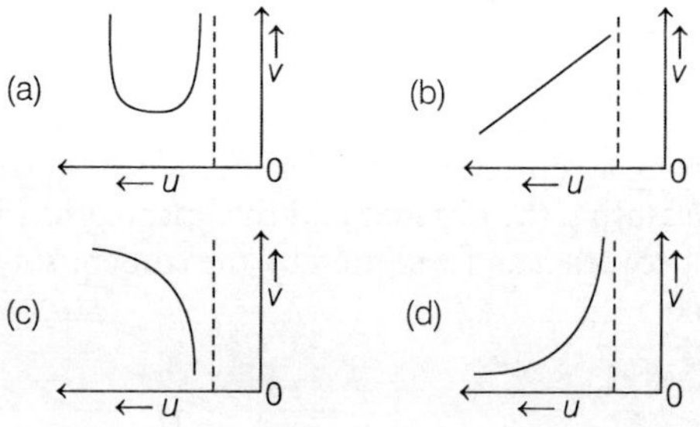

19 A thin plano-convex lens made of glass of refractive index 1.5 is immersed in a liquid of refractive index 1.2. When the plane side of the lens is silver coated for complete reflection, the lens immersed in the liquid behaves like a concave mirror of focal length 0.2 m. The radius of curvature of the curved surface of the lens is *[2025, 24 Jan Shift-I]*

(a) 0.20 m (b) 0.10 m
(c) 0.25 m (d) 0.15 m

20 A symmetric thin biconvex lens is cut into four equal parts by two planes AB and CD as shown in figure. If the power of original lens is 4D, then the power of a part of the divided lens is *[2025, 22 Jan Shift-II]*

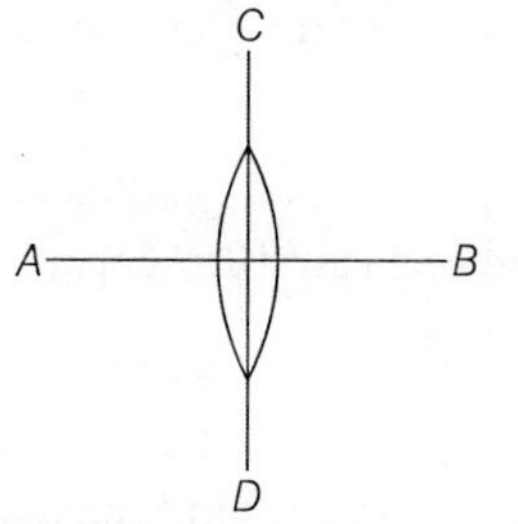

(a) D (b) 2D (c) 4D (d) 8D

21 For the thin convex lens, the radii of curvature are at 15 cm and 30 cm respectively. The focal length the lens is 20 cm. The refractive index of the material is *[2024, 6 April Shift-II]*

(a) 1.2 (b) 1.5 (c) 1.8 (d) 1.4

22 An effective power of a combination of 5 identical convex lenses which are kept in contact along the principal axis is 25 D. Focal length of each of the convex lens is *[2024, 4 April Shift-I]*

(a) 25 cm (b) 500 cm
(c) 50 cm (d) 20 cm

Imp. Lens Maker's Formula

23 A biconvex lens of refractive index 1.5 has a focal length of 20 cm in air. Its focal length when immersed in a liquid of refractive index 1.6 will be *[2024, 29 Jan Shift-I]*

(a) −16 cm (b) +160 cm (c) −160 cm (d) +16 cm

24 A person has been using spectacles of power −1.0 dioptre for distant vision and a separate reading glass of power 2.0 dioptres. What is the least distance of distinct vision for this person *[2023, 30 Jan Shift-I]*

(a) 50 cm (b) 30 cm
(c) 10 cm (d) 40 cm

25 The power of a lens (biconvex) is $1.25\ m^{-1}$ in particular medium. Refractive index of the lens is 1.5 and radii of curvature are 20 cm and 40 cm, respectively. The refractive index of surrounding medium is *[2022, 28 July Shift-II]*

(a) 1.0 (b) 9/7
(c) 3/2 (d) 4/3

26 The refractive index of a converging lens is 1.4. What will be the focal length of this lens if it is placed in a medium of same refractive index ? (Assume the radii of curvature of the faces of lens are R_1 and R_2 respectively) *[2021, 16 March Shift-II]*

(a) 1 (b) Infinite
(c) $\dfrac{R_1 R_2}{R_1 - R_2}$ (d) Zero

27 For a concave lens of focal length f, the relation between object and image distances u and v, respectively, from its pole can best be represented by ($u = v$ is the reference line) *[2020, 5 Sep Shift-I]*

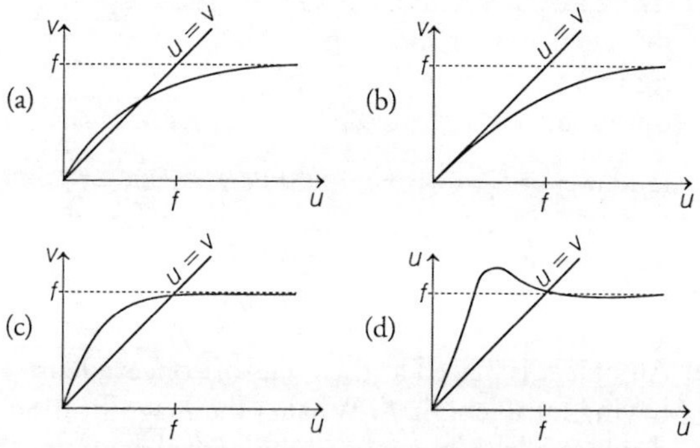

28 A plano-convex lens of refractive index μ_1 and focal length f_1 is kept in contact with plano-concave lens of refractive index μ_2 and focal length f_2. If the radius of curvature of their spherical faces is R each and $f_1 = 2f_2$, then μ_1 and μ_2 are related as *[2019, 10 Jan Shift-I]*

(a) $3\mu_2 - 2\mu_1 = 1$ (b) $2\mu_2 - \mu_1 = 1$
(c) $2\mu_1 - \mu_2 = 1$ (d) $\mu_1 + \mu_2 = 3$

29 A convex lens of focal length 20 cm produces images of the same magnification 2 when an object is kept at two distances x_1 and x_2 ($x_1 > x_2$) from the lens. The ratio of x_1 and x_2 is *[2019, 9 April Shift-II]*
(a) 5 : 3 (b) 2 : 1 (c) 4 : 3 (d) 3 : 1

30 A convex lens (of focal length 20 cm) and a concave mirror, having their principal axis along the same lines, are kept 80 cm apart from each other. The concave mirror is to the right of the convex lens.
When an object is kept at a distance of 30 cm to the left of the convex lens, its image remains at the same position even if the concave mirror is removed. The maximum distance of the object for which this concave mirror, by itself would produce a virtual image would be *[2019, 8 April Shift-II]*
(a) 25 cm (b) 20 cm (c) 10 cm (d) 30 cm

31 A plano-convex lens (focal length f_2, refractive index μ_2, radius of curvature R) fits exactly into a plano-concave lens (focal length f_1, refractive index μ_1, radius of curvature R). Their plane surfaces are parallel to each other. Then, the focal length of the combination will be *[2019, 12 Jan Shift-II]*
(a) $f_1 - f_2$ (b) $\dfrac{R}{\mu_2 - \mu_1}$
(c) $f_1 + f_2$ (d) $\dfrac{2f_1 f_2}{f_1 + f_2}$

Imp. Imersion of Lens in Liquid

32 Formation of real image using a biconvex lens is shown below. If the whole set up is immersed in water without disturbing the object and the screen positions, what will one observe on the screen? *[2019, 12 Jan Shift-II]*

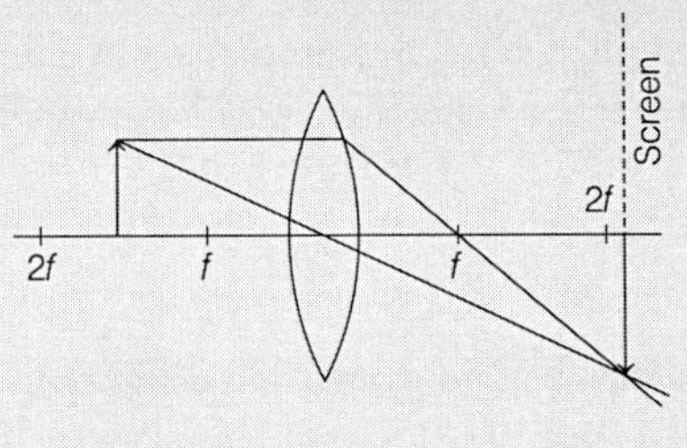

(a) No change (b) Magnified image
(c) Image disappears (d) Erect real image

33 What is the position and nature of image formed by lens combination shown in figure? (where, f_1 and f_2 are focal lengths) *[2019, 12 Jan Shift-I]*

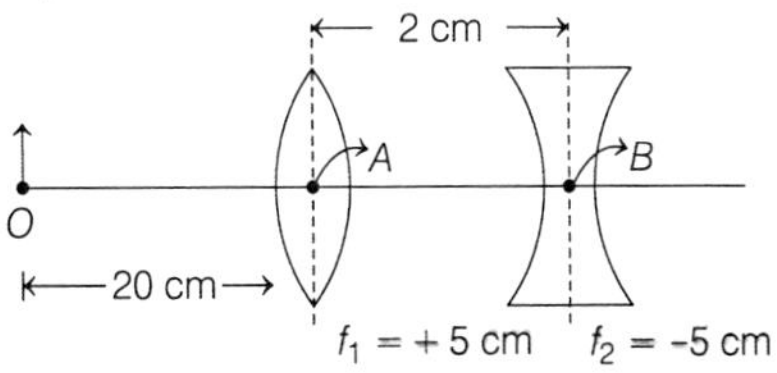

(a) 20/3 cm from point B at right, real
(b) 70 cm from point B at right, real
(c) 40 cm from point B at right, real
(d) 70 cm from point B at left, virtual

34 An object is at a distance of 20 m from a convex lens of focal length 0.3 m. The lens forms an image of the object. If the object moves away from the lens at a speed of 5 m/s, the speed and direction of the image will be *[2019, 11 Jan Shift-I]*
(a) 3.22×10^{-3} m/s towards the lens
(b) 0.92×10^{-3} m/s away from the lens
(c) 2.26×10^{-3} m/s away from the lens
(d) 1.16×10^{-3} m/s towards the lens

35 A diverging lens with magnitude of focal length 25 cm is placed at a distance of 15 cm from a converging lens of magnitude of focal length 20 cm. A beam of parallel light falls on the diverging lens. The final image formed is *[JEE Main 2017]*
(a) virtual and at a distance of 40 cm from convergent lens
(b) real and at a distance of 40 cm from the divergent lens
(c) real and at a distance of 6 cm from the convergent lens
(d) real and at a distance of 40 cm from convergent lens

36 Diameter of a plano-convex lens is 6 cm and thickness at the centre is 3 mm. If speed of light in material of lens is 2×10^8 m/s, the focal length of the lens is *[JEE Main 2013]*
(a) 15 cm (b) 20 cm
(c) 30 cm (d) 10 cm

Level 3

37 A convex lens of focal length 30 cm is placed in contact with a concave lens of focal length 20 cm. An object is placed at 20 cm to the left of this lens system. The distance of the image from the lens in cm is .... *[2025, 8 April, Shift-II]*
(a) 30 (b) 45
(c) 15 (d) $\dfrac{60}{7}$

38 A slanted object AB is placed on one side of convex lens as shown in the diagram. The image is formed on the opposite side. Angle made by the image with principal axis is *[2025, 2 April, Shift-I]*

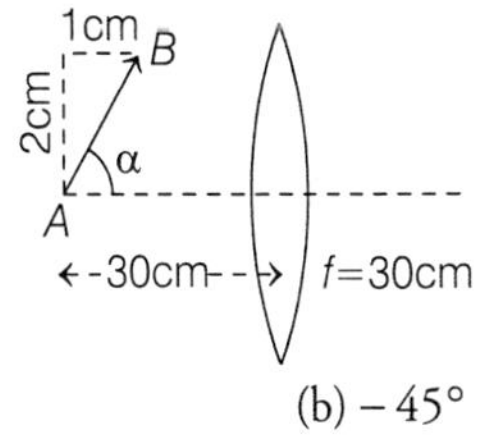

(a) + 45° (b) – 45°
(c) $-\dfrac{\alpha}{2}$ (d) $-\alpha$

39 In the diagram given below, there are three lenses formed. Considering negligible thickness of each of them as compared to $|R_1|$ and $|R_2|$, i.e., the radii of curvature for upper and lower surface of the glass lens, the power of the combinations is *[2025, 22 Jan Shift-I]*

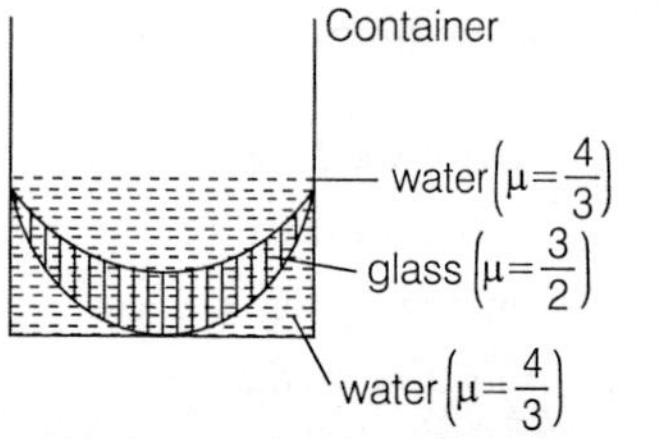

(a) $-\frac{1}{6}\left(\frac{1}{|R_1|}-\frac{1}{|R_2|}\right)$ (b) $\frac{1}{6}\left(\frac{1}{|R_1|}-\frac{1}{|R_2|}\right)$

(c) $\frac{1}{6}\left(\frac{1}{|R_1|}+\frac{1}{|R_2|}\right)$ (d) $-\frac{1}{6}\left(\frac{1}{|R_1|}+\frac{1}{|R_2|}\right)$

40 A photograph of a landscape is captured by a drone camera at a height of 18 km. The size of the camera film is 2 cm × 2 cm and the area of the landscape photographed is 400 km^2. The focal length of the lens in the drone camera is *[2025, 24 Jan Shift-II]*

(a) 1.8 cm (b) 0.9 cm

(c) 2.8 cm (d) 2.5 cm

41 Given is a thin convex lens of glass (refractive index μ) and each side having radius of curvature R. One side is polished for complete reflection. At what distance from the lens an object be placed on the optic axis, so that the image gets formed on the object itself? *[2025, 22 Jan Shift-I]*

(a) μR (b) $\frac{R}{\mu}$

(c) $\frac{R}{(2\mu-1)}$ (d) $\frac{R}{(2\mu-3)}$

42 Given a thin convex lens (refractive index μ_2), kept in a liquid (refractive index $\mu_1, \mu_1 < \mu_2$) having radii of curvatures $|R_1|$ and $|R_2|$. Its second surface is silver polished. Where should an object be placed on the optic axis, so that a real and inverted image is formed at the same place? *[2025, 23 Jan Shift-I]*

(a) $\frac{\mu_1 |R_1|\cdot|R_2|}{\mu_2 (|R_1|+|R_2|)-\mu_1 |R_2|}$

(b) $\frac{\mu_1 |R_1|\cdot|R_2|}{\mu_1 (|R_1|+|R_2|)-\mu_2 |R_2|}$

(c) $\frac{\mu_1 |R_1|\cdot|R_2|}{\mu_1 (2|R_1|+|R_2|)-\mu_2\sqrt{|R_1|\cdot|R_2|}}$

(d) $\frac{(\mu_2+\mu_1)|R_1|}{(\mu_2-\mu_1)}$

43 The position of the image formed by the combination of lenses is *[2024, 8 April Shift-II]*

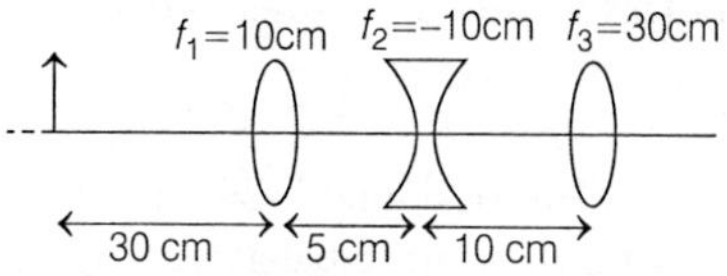

(a) 30 cm (left of third lens)
(b) 30 cm (right of third lens)
(c) 15 cm (right of second lens)
(d) 15 cm (left of second lens)

44 Find the distance of the image from object O, formed by the combination of lenses in the figure. *[2021, 8 April Shift-II]*

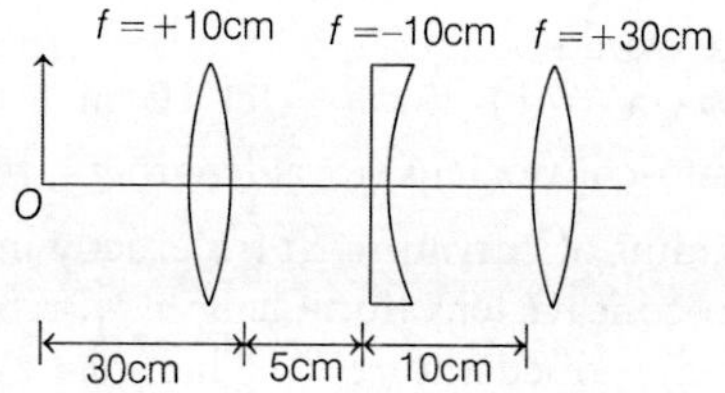

(a) 75 cm (b) 10 cm

(c) 20 cm (d) infinity

45 Curved surfaces of a plano-convex lens of refractive index μ_1 and a plano-concave lens of refractive index μ_2 have equal radius of curvature as shown in figure. Find the ratio of radius of curvature to the focal length of the combined lenses. *[2021, 27 Aug Shift-I]*

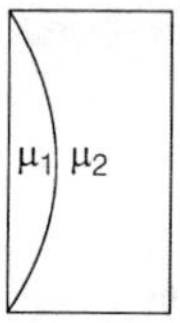

(a) $\frac{1}{\mu_2-\mu_1}$ (b) $\mu_1-\mu_2$

(c) $\frac{1}{\mu_1-\mu_2}$ (d) $\mu_2-\mu_1$

46 A thin lens made of glass (refractive index = 1.5) of focal length f = 16 cm is immersed in a liquid of refractive index 1.42. If its focal length in liquid is f_1, then the ratio f_1 / f is closest to the integer *[2020, 7 Jan Shift-II]*

(a) 1 (b) 5

(c) 9 (d) 17

47 An upright object is placed at a distance of 40 cm in front of a convergent lens of focal length 20 cm. A convergent mirror of focal length 10 cm is placed at a distance of 60 cm on the other side of the lens. The position and size of the final image will be *[2019, 8 April Shift-I]*

(a) 20 cm from the convergent mirror, same size as the object
(b) 40 cm from the convergent mirror, same size as the object
(c) 40 cm from the convergent lens, twice the size of the object
(d) 20 cm from the convergent mirror, twice size of the object

48 A thin convex lens made from crown glass $\left(\mu = \frac{3}{2}\right)$ has focal length f. When it is measured in two different liquids having refractive indices 4/3 and 5/3. It has the focal lengths f_1 and f_2, respectively. The correct relation between the focal lengths is *[JEE Main 2014]*
(a) $f_1 = f_2 < f$
(b) $f_1 > f$ and f_2 becomes negative
(c) $f_2 > f$ and f_1 becomes negative
(d) f_1 and f_2 both become negative

49 A plano-convex lens of refractive index 1.5 and radius of curvature 30 cm is silvered at the curved surface. Now, this lens has been used to form the image of an object. At what distance from this lens may an object be placed in order to have a real image of the size of the object. *[AIEEE 2004]*
(a) 20 cm (b) 30 cm
(c) 60 cm (d) 80 cm

Numerical Value Type Questions

50 The distance between object and its 3 times magnified virtual image as produced by a convex lens is 20 cm. The focal length of the lens used is .... cm. *[2024, 1 Feb Shift-I]*

51 The distance between object and its two times magnified real image as produced by a convex lens is 45 cm. The focal length of the lens used is ...... cm. *[2024, 30 Jan Shift-I]*

52 A biconvex lens of focal length 10 cm is cut in two identical parts along a plane perpendicular to the principal axis. The power of each lens after cut is .... D. *[2023, 13 April Shift-II]*

53 Two convex lenses of focal length 20 cm each are placed coaxially with a separation of 60 cm between them. The image of distant object formed by the combination is at .... cm from the first lens. *[2023, 12 April Shift-I]*

54 The radius of curvature of each surface of a convex lens having refractive index 1.8 is 20 cm. The lens is now immersed in a liquid of refractive index 1.5. The ratio of power of lens in air to its power in the liquid will be $x : 1$. The value of x is .... *[2023, 11 April Shift-I]*

55 A point object, 'O' is placed infront of two thin symmetrical coaxial convex lenses L_1 and L_2 with focal length 24 cm and 9 cm, respectively. The distance between two lenses is 10 cm and the object is placed 6 cm away from lens L_1 as shown in the figure. The distance between the object and the image formed by the system of two lenses is .... cm. *[2023, 10 April Shift-II]*

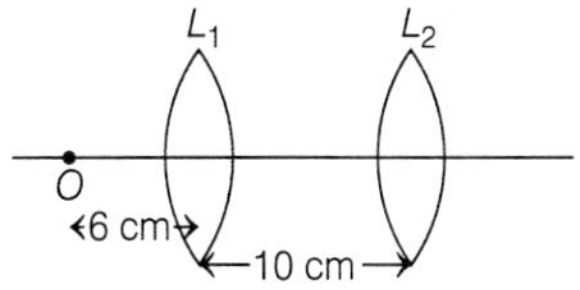

Imp. Refraction Through Convex Spherical Surface

56 Two transparent media having refractive indices 1.0 and 1.5 are separated by a spherical refracting surface of radius of curvature 30 cm. The centre of curvature of surface is towards denser medium and a point object is placed on the principle axis in rarer medium at a distance of 15 cm from the pole of the surface. The distance of image from the pole of the surface is .... cm. *[2023, 8 April Shift-II]*

57 An object is placed on the principal axis of convex lens of focal length 10 cm as shown. A plane mirror is placed on the other side of lens at a distance of 20 cm. The image produced by the plane mirror is 5 cm inside the mirror. The distance of the object from the lens is .... cm. *[2023, 25 Jan Shift-II]*

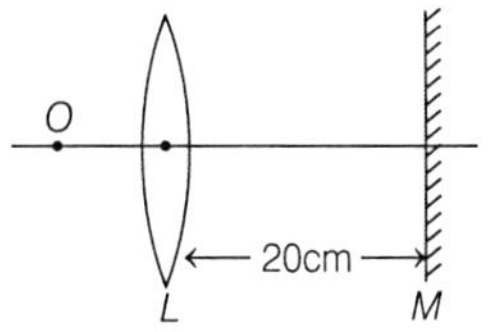

58 A convex lens of refractive index 1.5 and focal length 18 cm in air is immersed in water. The change in focal length of the lens will be .... cm.
(Given, refractive index index of water $= 4/3$) *[2023, 24 Jan Shift-II]*

59 As shown in the figure, a combination of a thin plano concave lens and a thin plano convex lens is used to image an object placed at infinity. The radius of curvature of both the lenses is 30 cm and refractive index of the material for both the lenses is 1.75. Both the lenses are placed at distance of 40 cm from each other. Due to the combination, the image of the object is formed at distance
$x =$ .... cm, from concave lens. *[2023, 24 Jan Shift-I]*

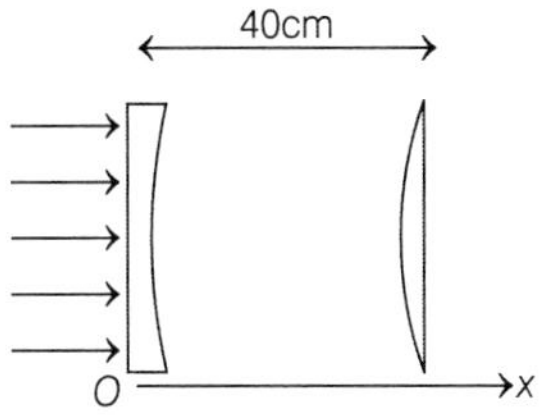

60 An object is placed at a distance of 12 cm from a convex lens. A convex mirror of focal length 15 cm is placed on other side of lens at 8 cm as shown in the figure. Image of object coincides with the object.

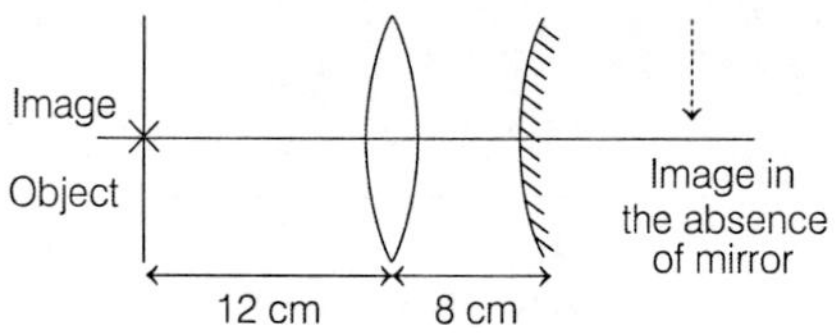

When the convex mirror is removed, a real and inverted image is formed at a position. The distance of the image from the object will be .... cm. *[2021, 26 Aug Shift-II]*

61 The same size images are formed by a convex lens when the object is placed at 20 cm or at 10 cm from the lens. The focal length of convex lens is .... cm. *[2021, 25 Feb Shift-I]*

Topic 4 Optical Instruments

Level 1

01 In a reflecting telescope, a secondary mirror is used to *[2023, 8 April Shift-I]*
(a) remove spherical aberration
(b) reduce the problem of mechanical support
(c) make chromatic aberration zero
(d) move the eyepiece outside the telescopic tube

02 The image formed by an objective of a compound microscope is *[AIEEE 2003]*
(a) virtual and diminished
(b) real and diminished
(c) real and enlarged
(d) virtual and enlarged

Level 2

03 Given below are two statements: one is labelled as Assertion (A) and the other is labelled as Reason (R).
Assertion (A) For a simple microscope, the angular size of the object equals the angular size of the image.
Reason (R) Magnification is achieved as the small object can be kept much closer to the eye than 25 cm and hence, it subtends a large angle.
In the light of the above statements, choose the most appropriate answer from the options given below. *[2021, 26 Feb Shift-II]*
(a) A is true but R is false.
(b) Both A and R are true but R is not the correct explanation of A.
(c) Both A and R are true and R is the correct explanation of A.
(d) A is false but R is true.

Imp. Magnifying Power of Telescope

04 The magnifying power of a telescope with tube length 60 cm is 5. What is the focal length of its eyepiece? *[2020, 8 Jan Shift-I]*
(a) 10 cm (b) 20 cm (c) 30 cm (d) 40 cm

Numerical Value Type Questions

05 An object viewed from a near point distance of 25 cm, using a microscopic lens with magnification 6, gives an unresolved image. A resolved image is observed at infinite distance with a total magnification double the earlier using an eyepiece along with the given lens and a tube of length 0.6 m, if the focal length of the eyepiece is equal to .... cm. *[2021, 20 July Shift-I]*

06 In a compound microscope, the magnified virtual image is formed at a distance of 25 cm from the eye-piece. The focal length of its objective lens is 1 cm. If the magnification is 100 and the tube length of the microscope is 20 cm, then the focal length of the eye-piece lens (in cm) is .... *[2020, 4 Sep Shift-I]*

07 A compound microscope consists of an objective lens of focal length 1 cm and an eye piece of focal length 5 cm with a separation of 10 cm. The distance between an object and the objective lens, at which the strain on the eye is minimum is $n/40$ cm. The value of n is .... *[2020, 5 Sep Shift-I]*

• Key **Index**

Topic 1

1 (b)	**2** (c)	**3** (c)	**4** (d)	**5** (b)	**6** (d)	**7** (b)	**8** (a)	**9** (d)	**10** (c)
11 (c)	**12** (d)	**13** (c)	**14** (d)	**15** (c)	**16** (b)	**17** (c)	**18** (d)	**19** (c)	**20** (c)
21 (a)	**22** (d)	**23** (a)	**24** (45)	**25** (8)	**26** (48)	**27** (18)	**28** (32)	**29** (400)	**30** (1)

Topic 2

1 (d)	**2** (c)	**3** (b)	**4** (c)	**5** (c)	**6** (a)	**7** (b)	**8** (c)	**9** (c)	**10** (d)
11 (b)	**12** (c)	**13** (b)	**14** (a)	**15** (a)	**16** (b)	**17** (d)	**18** (a)	**19** (a)	**20** (c)
21 (d)	**22** (*)	**23** (a)	**24** (d)	**25** (c)	**26** (d)	**27** (b)	**28** (a)	**29** (a)	**30** (a)
31 (d)	**32** (a)	**33** (d)	**34** (c)	**35** (b)	**36** (a)	**37** (c)	**38** (d)	**39** (c)	**40** (b)
41 (b)	**42** (a)	**43** (a)	**44** (d)	**45** (d)	**46** (a)	**47** (d)	**48** (a)	**49** (a)	**50** (45)
51 (4)	**52** (80)	**53** (6)	**54** (2)	**55** (31)	**56** (30)	**57** (3)	**58** (50)	**59** (52)	**60** (12)
61 (5)	**62** (600)	**63** (60)	**64** (12)	**65** (5)					

Topic 3

1 (d)	**2** (a)	**3** (c)	**4** (b)	**5** (d)	**6** (d)	**7** (b)	**8** (a)	**9** (c)	**10** (b)
11 (a)	**12** (c)	**13** (b)	**14** (b)	**15** (b)	**16** (c)	**17** (b)	**18** (d)	**19** (b)	**20** (b)
21 (b)	**22** (d)	**23** (c)	**24** (b)	**25** (b)	**26** (b)	**27** (b)	**28** (c)	**29** (d)	**30** (c)
31 (b)	**32** (c)	**33** (b)	**34** (d)	**35** (d)	**36** (c)	**37** (c)	**38** (b)	**39** (a)	**40** (a)
41 (c)	**42** (a)	**43** (b)	**44** (a)	**45** (b)	**46** (c)	**47** (a)	**48** (b)	**49** (a)	**50** (15)
51 (10)	**52** (5)	**53** (100)	**54** (4)	**55** (34)	**56** (30)	**57** (30)	**58** (54)	**59** (120)	**60** (50)
61 (15)									

Topic 4

1 (d)	**2** (c)	**3** (c)	**4** (a)	**5** (25)	**6** (6)	**7** (50)

Detailed **Explanations**

Topic 1

01. *(b)*

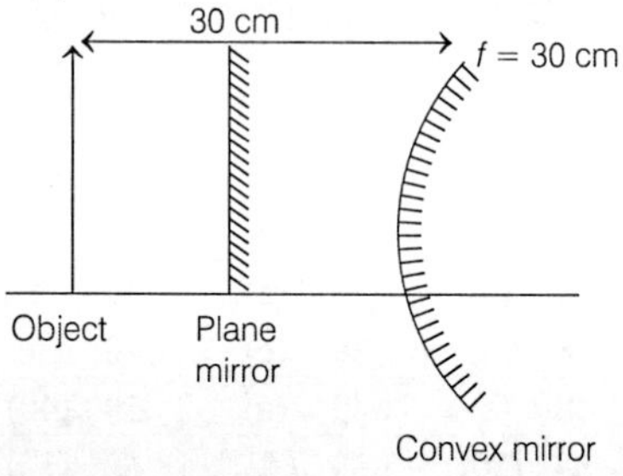

For convex mirror,

$$\frac{1}{v}+\frac{1}{u}=\frac{1}{f}$$

$$\frac{1}{v}-\frac{1}{30}=\frac{1}{30}$$

$$\frac{1}{v}=\frac{1}{30}+\frac{1}{30}=\frac{2}{30}$$

$$\frac{1}{v}=\frac{1}{15}$$

$$v=15 \text{ cm}$$

02. *(c)* Focal length of spherical mirror does not depends on medium in which it placed.

03. *(c)* Given, $i=30°$

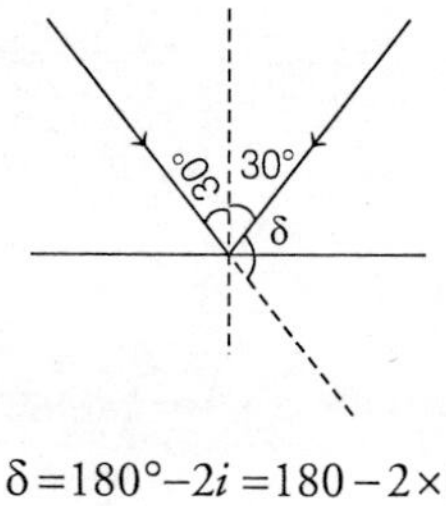

$$\delta=180°-2i=180-2\times30°$$
$$=180-60°=120°$$

04. *(d)* In a plane mirror, image is always at same distance from object.

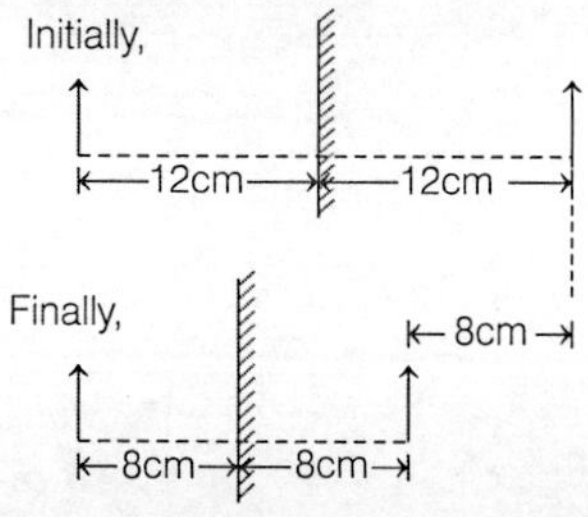

From above diagrams, distance by which image moved is 8 cm towards mirror (from it's previous position).

05. *(b)* When the light rays from an object have been reflected towards an observer from a standard flat mirror, the image observed by the observer is virtual, **erect**, same size as the object and **laterally inverted.**

06. *(d)* Given, $h_o=100$ cm; $f=40$ cm

$$h_i=25 \text{ cm}$$

Nature of image is erect means virtual. As, h_i is less than h_o, so mirror used is convex mirror.

Hence, (d) is correct option, i.e. image is virtual, opposite and mirror is convex.

07. *(b)* Number of images, $n=\frac{360°}{\theta}-1$

where, θ is angle between mirrors.

$$\therefore \quad 3=\frac{360°}{\theta}-1 \quad \text{or} \quad \theta=90°$$

08. *(a)* Number of images, $n=\frac{360°}{\theta}-1$

where, θ is angle between mirrors.

Thus,

$$\theta=60° \qquad \text{[Given]}$$

So, number of images, $n=\frac{360°}{60°}-1=5$

> **• Concept Enhancer**
>
> The number of image formed by the two plane mirrors inclined at angle θ to each other is given as:
>
> **Case I** if $\frac{360°}{\theta}$ = even, then no. of images $=\frac{360°}{\theta}-1$
>
> **Case 2** if $\frac{360°}{\theta}$ = odd, then, no. of images $=\frac{360°}{\theta}$

09. *(d)* $m=\frac{1}{2}$ given, $u=-40$ cm

$$m=\frac{f}{f-u}$$

$$\frac{1}{2}=\frac{f}{f-(-40)}$$

$$\frac{1}{2}=\frac{f}{f-4}$$

$$\frac{1}{2}=\frac{f}{f+40}$$

$$f+40=2f$$
$$f=40 \text{ cm}$$

Now $m=\frac{1}{3}$, $u=?$

$$m=\frac{f}{f-u}$$

$$\frac{1}{3}=\frac{40}{40-u}$$

$$40-u=120 \Rightarrow u=-80 \text{ cm}$$

10. *(c)* For concave mirror,

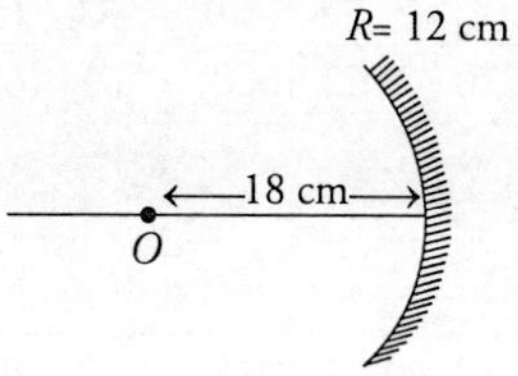

$$m_1=\frac{f}{u-f}$$

$$=\frac{6}{18-6}=\frac{1}{2}$$

For convex mirror,

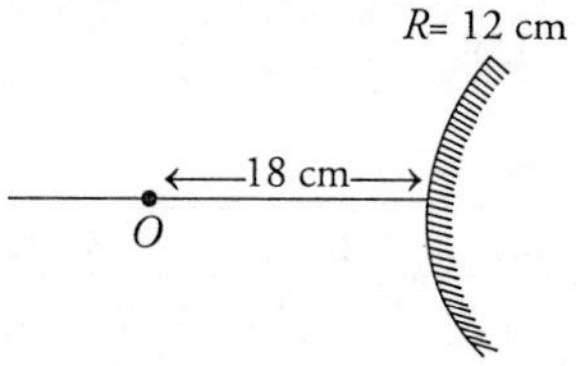

$$m_2=\frac{f}{u+f}$$

$$=\frac{6}{18+6}=\frac{1}{4}$$

$$\therefore \quad \frac{m_2}{m_1}=\frac{\frac{1}{4}}{\frac{1}{2}}=\frac{1}{2}$$

$$\Rightarrow \quad \frac{m_2}{m_1}=\frac{1}{2}$$

11. *(c)* Magnification produced by the mirror

$$m=-\frac{v}{u}=-\left(\frac{v}{-u}\right)=\frac{v}{u}$$

$$\Rightarrow \quad m=\frac{v}{u}$$

$$\Rightarrow \quad \frac{1}{4}=\frac{v}{u}$$

$$\Rightarrow \quad u=4v$$

Here, $u+v=40$

$\Rightarrow \quad 4v+v=40$

$\Rightarrow \quad 5v=40 \Rightarrow v=8$ cm

$\therefore \quad u=4v=4\times 8=32$ cm

Now, using mirror formula,

$$\frac{1}{v}+\frac{1}{u}=\frac{1}{f}$$

$$\Rightarrow \frac{1}{8}-\frac{1}{32}=\frac{1}{f}$$

$$\Rightarrow \frac{3}{32}=\frac{1}{f}$$

$$f=\frac{32}{3}\text{ cm}=10.7\text{ cm}$$

12. *(d)*

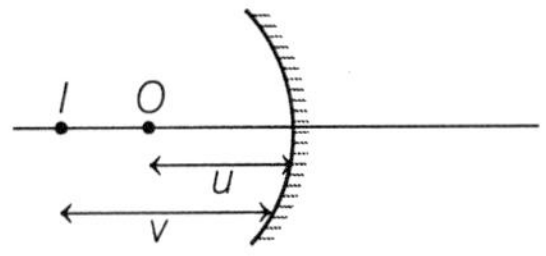

$$=-\frac{v}{u}=-3$$

and $v-u=20$ cm

$$f=\frac{vu}{v+u}=\frac{(-30)(-10)}{-30-10}=\frac{300}{-40}$$

$$\therefore f=-\frac{30}{4}=-\frac{15}{2}$$

$R=-15$ cm $\quad [2f=R]$

$\therefore$ Magnitude of R is 15 cm.

13. *(c)* Given,

Image distance, $v=15$ cm

Here, $m=2$

i.e., $\frac{-v}{u}=2 \Rightarrow v=-2u$

Distance between object and image

$$-u+v=15\text{ cm}$$

$$-u+(-2u)=15\text{ cm}$$

$\Rightarrow \quad u=-5$ cm

$\therefore \quad v=-2\times -5=10$ cm

Using mirror formula,

$$\Rightarrow \frac{1}{f}=\frac{1}{v}+\frac{1}{u}$$

$$\therefore \frac{1}{f}=\frac{1}{10}+\frac{1}{-5}=-\frac{1}{10}$$

$$f=-10\text{ cm}$$

14. *(d)*

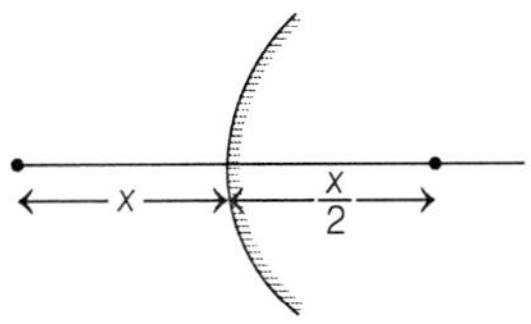

$R=30$ cm

$$f=\frac{R}{2}=\frac{30}{2}=+15\text{ cm}$$

Magnification $(m)=\pm\frac{1}{2}$

For convex mirror, virtual image is formed for real object.

Therefore, m is + ve

$$m=\frac{f}{f-u} \Rightarrow +\frac{1}{2}=\frac{f}{f-u}$$

$$\Rightarrow +\frac{1}{2}=\frac{15}{15-u}$$

$$15-u=30$$

$$15-30=u \Rightarrow u=-15\text{ cm}$$

15. *(c)* The given situation in question is as follows.

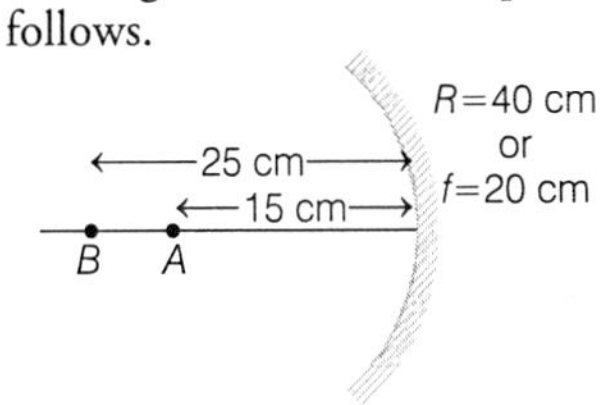

Using mirror formula, $\frac{1}{v}+\frac{1}{u}=\frac{1}{f}$

For point A, $u=-15$ cm, $f=-20$ cm

$$\frac{1}{v_A}+\frac{1}{-15}=\frac{1}{-20}$$

$$\frac{1}{v_A}=\frac{1}{15}-\frac{1}{20}=\frac{4-3}{60}=\frac{1}{60}$$

$$v_A=+60\text{ cm}$$

For point B, $u=-25$ cm, $f=-20$ cm

$$\frac{1}{v_B}+\frac{1}{-25}=\frac{1}{-20}$$

$$\Rightarrow \frac{1}{v_B}=\frac{1}{25}-\frac{1}{20}=\frac{4-5}{100}=-\frac{1}{100}$$

$\Rightarrow \quad v_B=-100$ cm

Since, the sign of image formed by concave mirror is opposite i.e., both images lie opposite to pole.

Hence, the distance between images formed by the mirror is

$$d=|v_A-v_B|$$

$$=|60-(-100)|$$

$\Rightarrow \quad d=160$ cm

16. *(b)*

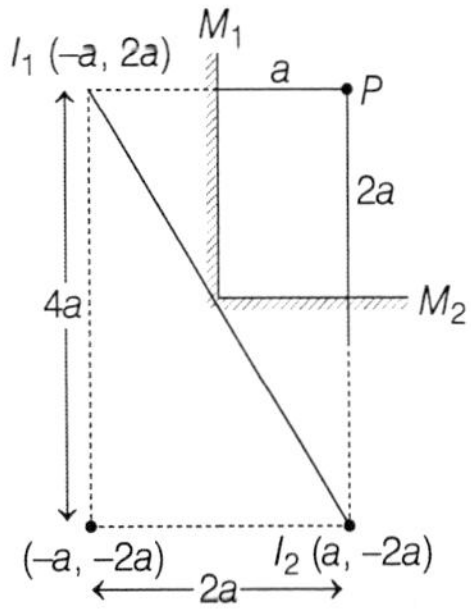

According to the given figure, we have two plane mirror, placed at 90° to each other.

Let I_1 is the image of P due to mirror M_1 and I_2 is the image of P due to mirror M_2

Therefore, the shortest distance between images i.e., between I_1 and I_2

$$I_1I_2=\sqrt{(4a)^2+(2a)^2}$$

$$=\sqrt{16a^2+4a^2}=a\sqrt{20}=a2\sqrt{5}$$

$$=a\times 2\times 2.3=4.6a$$

> **• Concept Enhancer**
>
> Image formed by plane mirror are virtual, erect, same size and at the same distance behind the miror.

17. *(c)* Taking component of **a**, **b** and **c** along $\hat{i}$, $\hat{j}$ and $\hat{k}$.

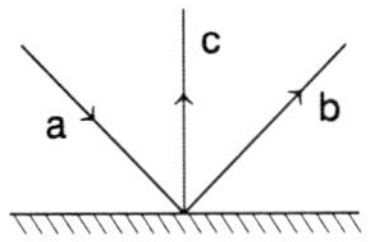

$\therefore \mathbf{a}=\sin\theta\hat{i}-\cos\theta\hat{j}$

$\mathbf{b}=\sin\theta\hat{i}+\cos\theta\hat{j}$

$\mathbf{c}=\hat{j}$

Now, on solving option (c), we get

$$\mathbf{a}-2(\mathbf{a}\cdot\mathbf{c})\mathbf{c}=(\sin\theta\hat{i}-\cos\theta\hat{j})-2[(\sin\theta\hat{i}-\cos\theta\hat{j})\cdot\hat{j}]\hat{j}$$

$$=\sin\theta\hat{i}+\cos\theta\hat{j}=\mathbf{b}$$

18. *(d)*

> **key idea** To solve these types of questions, you can use the following two formulae (i) $\frac{1}{v}+\frac{1}{u}=\frac{1}{f}$; $m=\frac{-v}{u}$
>
> Where, symbols have their usual meanings.

Radius of curvature of given concave mirror $R=-8$ cm

So, focal length, $f=\frac{R}{2}=-\frac{8}{2}=-4$ cm

Object distance, $u=-10$ cm

Using mirror formula,

$$\frac{1}{v}=\frac{1}{f}-\frac{1}{u}=-\frac{1}{4}-\frac{1}{-10}=-\frac{3}{20}$$

$\Rightarrow \quad v=-6.67$ cm (real)

Also, magnification,

$$m=-\frac{v}{u}=-\frac{(-6.67)}{(-10)}=-0.667$$

As m is negative and less than one, so image is inverted and diminished or unmagnified.

19. *(c)* Given, $f=-0.4$ m

Magnification $=5$

We know that $m = -\dfrac{\text{image distance}}{\text{object distance}}$

$\Rightarrow \quad \dfrac{v}{u} = -5$ or $v = -5u$

Using mirror formula, $\dfrac{1}{v}+\dfrac{1}{u}=\dfrac{1}{f}$

$$\frac{1}{-5u}+\frac{1}{u} = -\frac{1}{0.4} \Rightarrow \frac{4}{5u} = -\frac{1}{0.4}$$

$$\Rightarrow \quad u = -\frac{1.6}{5} = -0.32 \text{ m}$$

> ***Alternate Solution***
> Magnification produced by a mirror can also be given as $m = \dfrac{f}{f-u}$
> Substituting the given values, we get
> $5 = \dfrac{-0.4}{-0.4-u}$ or $u = -0.32$ m

20. *(c)* The given situation is shown below,

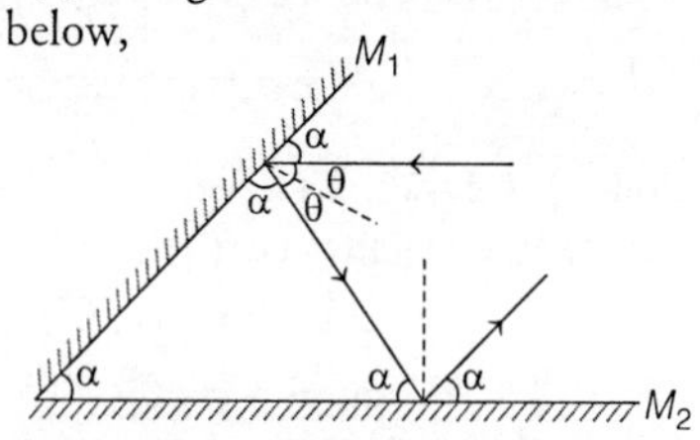

$$\alpha+\alpha+\alpha = 180°$$
$$3\alpha = 180° \Rightarrow \alpha = 60°$$

21. *(a)* The given situation is shown in the following ray diagram.

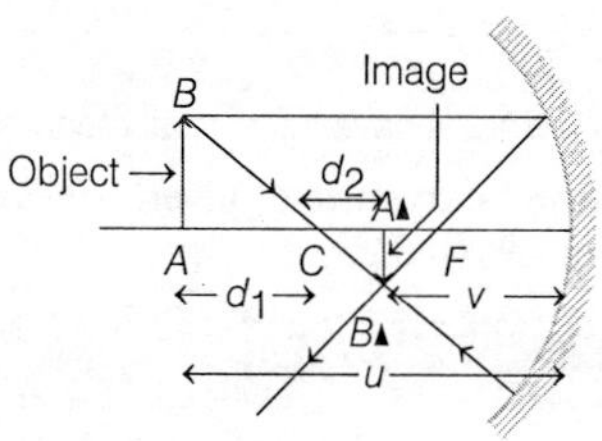

Consider radius of curvature of mirror is R.

Object distance, $u = -(R+d_1)$

Image distance, $v = -(R-d_2)$

Focal length of mirror, $f = \dfrac{-R}{2}$

Now, from mirror formula, we have

$$\frac{1}{f} = \frac{1}{v}+\frac{1}{u} = \frac{u+v}{uv} \Rightarrow uv = f(u+v)$$

On substituting the values in above expression, we get

$$[-(R+d_1)]\,[-(R-d_2)] = \frac{R}{2}(R+d_1+R-d_2)$$

$$\Rightarrow 2(R^2 - d_2R + d_1R - d_1d_2)$$

$$= 2R^2 + d_1R - d_2R$$

$$R(d_1-d_2) = 2d_1d_2$$

$$\Rightarrow \quad R = \frac{2d_1d_2}{d_1-d_2}$$

> **• Concept Enhancer**
> When the object is placed beyond the centre of curvature of the concave mirror, then a real and diminished image is formed between centre of curvature and focus.

22. *(d)* According to the question,
Velocity of car B w.r.t. mirror of car A,

$$v_{BM} = 40 \text{ m/s}$$

Distance of car B from mirror of car A,

$$u = 1.9 \text{ m} = 190 \text{ cm}$$

Focal length of mirror, $f = 10$ cm

We have to find the velocity of image of car B in the mirror, i.e. v_{IM}

We know that

$$v_{IM} = -m^2 v_{BM} \qquad \ldots\text{(i)}$$

Where, m is the magnification produced by mirror of car A,

$$\Rightarrow m = \frac{f}{f-u} = \frac{10}{10-(-190)} = \frac{10}{200} = \frac{1}{20}$$

On substituting the values in Eq. (i), we get

$$v_{IM} = -\left(\frac{1}{20}\right)^2 \times 40$$

$$= -\frac{1}{400}\times 40 = -0.1 \text{ m/s}$$

Here, negative sign shows that, the speed of image w.r.t. mirror of car A is in opposite direction.

23. *(a)* For the mirror, $\dfrac{1}{u}+\dfrac{1}{v}=\dfrac{1}{f}$

Differentiate this equation w.r.t. t, we get

$$-\frac{1}{u^2}\frac{du}{dt} - \frac{1}{v^2}\frac{dv}{dt} = 0 \Rightarrow \frac{dv}{dt} = -\frac{v^2}{u^2}\left(\frac{du}{dt}\right)$$

But $\dfrac{v}{u} = \dfrac{f}{u-f}$

$$\therefore \quad \frac{dv}{dt} = -\left(\frac{f}{u-f}\right)^2\left(\frac{du}{dt}\right)$$

$$= -\left(\frac{0.2}{-2.8-0.2}\right)^2 \times 15$$

$$= \frac{1}{15} \text{ ms}^{-1}$$

24. *(45)* $m = -\dfrac{1}{3}$

$$m = -\frac{(-v)}{(-u)}$$

$$-\frac{1}{3} = -\frac{v}{u}$$

$$v = \frac{u}{3}$$

Distance between object and image

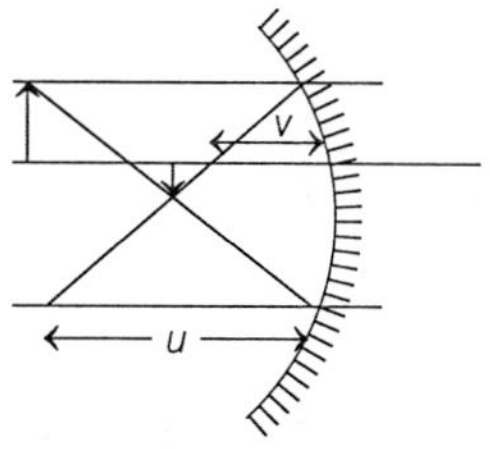

$$u - v = 30 \qquad \ldots(1)$$

$$u - \frac{u}{3} = 30$$

$$\frac{2u}{3} = 30$$

$u = 45$, put value in Eq. (1)

$v = 15$

$$\frac{1}{f} = \frac{1}{v}+\frac{1}{u} = \frac{-1}{15} - \frac{1}{45}$$

$$\frac{1}{f} = \frac{-60}{15\times 45} = \frac{-4}{45} \text{ or } |f| = \frac{45}{4}$$

25. *(8)* Using mirror formula

$$v = \frac{uf}{u-f} = \frac{-24\times 1}{-24-1} = \frac{24}{25}$$

$$m = \frac{-v}{u} = \frac{24}{25(-24)} = \frac{1}{25}$$

Using the formula

$$\frac{1}{v}-\frac{1}{u}=\frac{1}{f} \qquad \ldots\text{(i)}$$

$$v_I = -m^2 v_o = \frac{-1}{(25)^2}\times 25 = \frac{-1}{25}$$

Differentiating Eq. (i)

$$\frac{-1}{v^2}\left(\frac{dv}{dt}\right) + \frac{1}{u^2}\left(\frac{du}{dt}\right) = 0 \qquad \ldots\text{(ii)}$$

$$\left[\frac{dv}{dt} = v_I, \frac{du}{dt} = v_o\right]$$

Again Differentiating Eq. (ii)

$$\frac{+2}{v^3}(v_I)^2 - \frac{1}{v^2}\cdot a_I - \frac{2}{u^3}(v_o)^2 + \frac{1}{u^2}a_o = 0$$

$$\frac{a_I}{v^2} = \frac{2}{v^3}\cdot v_I^2 - \frac{2}{u^3}\cdot v_o^2$$

Substituting the value

$$a_I = \frac{2}{24.25} - \frac{2}{24}$$

$$100a_I = \frac{2}{25} \times 100 = 8$$

26. *(48)* Given, $\Delta x = \left(1 - \frac{1}{\mu}\right)d$

Image distance formed in plane mirror

$$= h - \Delta x$$

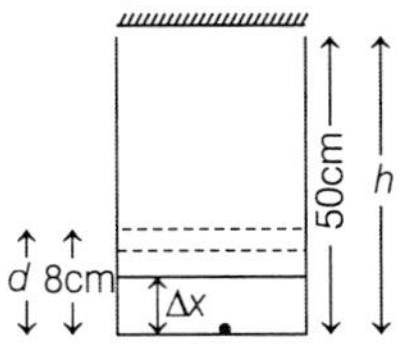

$$= h - \left(1 - \frac{1}{\mu}\right)d$$

$$= h - d + \frac{d}{\mu}$$

$$= 50 - 8 + \frac{8}{4/3}$$

$$= 50 - 8 + \frac{8}{4} \times 3 \text{ cm}$$

$$= 56 - 8 = 48 \text{ cm}$$

27. *(18)*

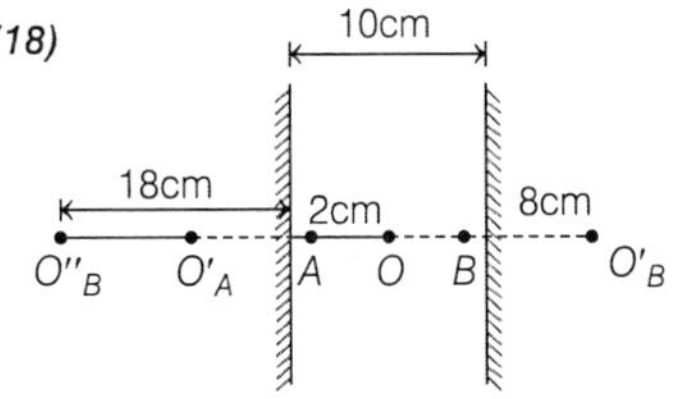

Reflection of O in mirror B is image $O'B$ at a distance of 18 cm from A. Image of $O'B$ is at $O''B$ which is also at a distance of 18 cm from A.

28. *(32)* Given that $l = 10$ cm

$f = 20$ cm, $d = 40$ cm

According to question,

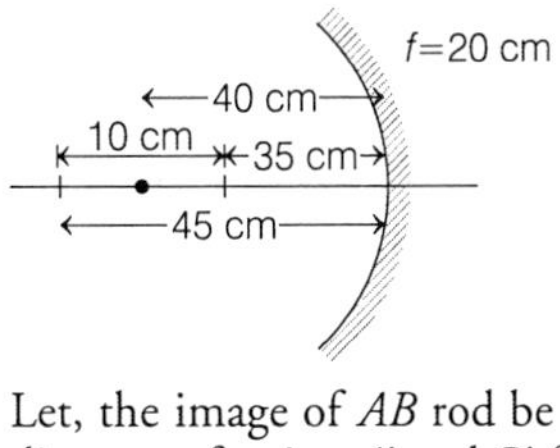

Let, the image of AB rod be $A'B'$ and distance of point A' and B' from mirror be v_A' and v_B' respectively.

From mirror formula,

For point A, $\frac{1}{v_A'} + \frac{1}{(-45)} = \frac{1}{-20}$

$$\frac{1}{v_A'} - \frac{1}{45} = \frac{1}{-20} \Rightarrow v_A' = -36 \text{ cm}$$

For point B, $\frac{1}{v_B'} + \frac{1}{(-35)} = \frac{1}{-20}$

$$\Rightarrow \frac{1}{v_B'} - \frac{1}{35} = \frac{1}{-20} \Rightarrow v_B' = \frac{-140}{3} \text{ cm}$$

Length of rod $= |v_A' - v_B'|$

$$= \left| -36 + \frac{140}{3} \right| = \frac{32}{3} \text{ cm}$$

So, $x = 32$

29. *(400)* Given, $u = 100$ cm

$$R = -200 \text{ cm}$$

$$\therefore \quad f = \frac{R}{2} = -100 \text{ cm}$$

Speed of object $= 2$ cm / s

After 10 s distance moved by object towards mirror,

$$s = v \times t = 2 \times 10 = 20 \text{ cm}$$

Distance of object,

$$u' = 100 - 20 = 80 \text{ cm}$$

$$\Rightarrow \frac{1}{v} + \frac{1}{u} = \frac{1}{f}$$

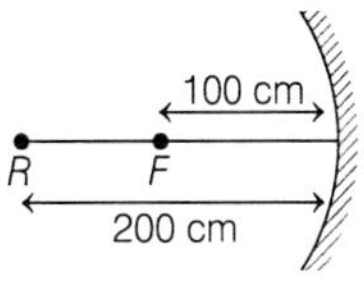

$$\Rightarrow -\frac{1}{100} = \frac{1}{v} + \frac{1}{-80}$$

$$\Rightarrow \frac{-1}{100} + \frac{1}{80} = \frac{1}{v}$$

$$\Rightarrow v = 400 \text{ cm}$$

30. *(1)* Given, object distance,

$$u = -30 \text{ cm},$$

Image distance, $v = -10$ cm

and speed of object, $v_o = 9 \text{ cms}^{-1}$

Let v_i be the speed of image.

Magnification of concave mirror,

$$m = -\frac{v}{u} = \frac{-10}{30} = -\frac{1}{3}$$

$$\text{Now,} v_i = v_o(m^2) = 9 \times \left(\frac{-1}{3}\right)^2 = 1 \text{ cms}^{-1}$$

Topic 2

01. *(d)* $\lambda_{\text{air}} = \frac{c}{v} = \frac{3 \times 10^8}{5 \times 10^{14}}$

$$= 0.6 \times 10^{-6} \text{ m}$$

Wavelength of refracted light in medium,

$$\lambda_{\text{med.}} = \frac{\lambda_{\text{air}}}{\mu} = \frac{0.6 \times 10^{-6}}{2}$$

$$= 0.3 \times 10^{-6} = 300 \times 10^{-9} \text{m}$$

$$= 300 \text{ nm}$$

02. *(c)* For the minimum deviation through the prism, refracted rays inside the prism becomes parallel to the base

For the case minimum deviation,

$$i = e$$

For angles of deviation other than the minimum, there are usually two angles of incidence giving the same deviation.

03. *(b)*

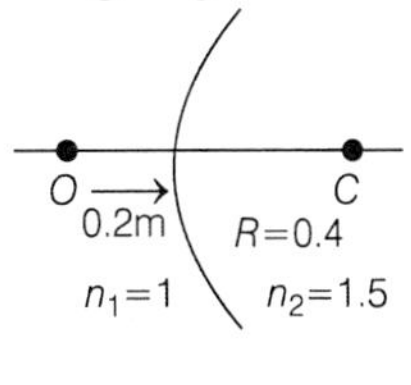

$$\frac{\mu_2}{v} - \frac{\mu_1}{u} = \frac{\mu_2 - \mu_1}{R}$$

$$\frac{1.5}{v} - \frac{1}{-(0.2)} = \frac{1.5 - 1}{0.4}$$

$$\frac{1.5}{v} = \frac{0.5}{0.4} - \frac{1}{0.2}$$

$$\frac{1.5}{v} = -\frac{1.5}{0.4}$$

$$v = -0.4 \text{ m}$$

04. *(c)*

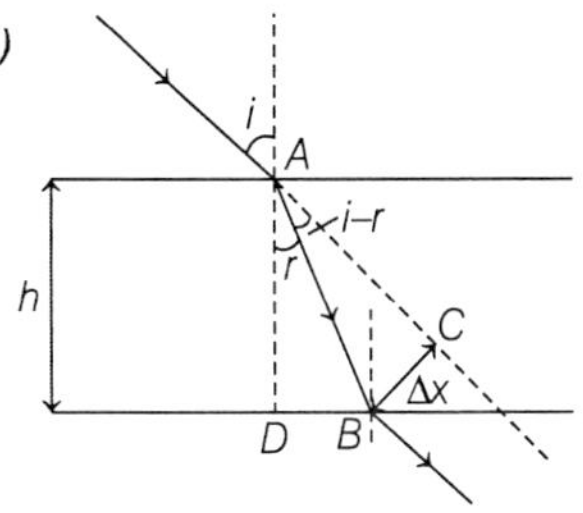

In triangle ABC

$$\sin(i - r) = \frac{\Delta x}{AB}$$

$$\Delta x = AB \sin(i - r) \quad \ldots\text{(i)}$$

In ΔABD

$$\cos r = \frac{AD}{AB}$$

$$AB = \frac{AD}{\cos r} \quad \ldots\text{(ii)}$$

Put the value in Eq. (i)

$$\Delta x = \frac{AD}{\cos r} \cdot \sin(i - r) \; [AD = h]$$

$$\Delta x = \frac{h \sin(i - r)}{\cos r}$$

05. *(c)* $\mu = \dfrac{\sin\left(\dfrac{A + \delta_{\min}}{2}\right)}{\sin\left(\dfrac{A}{2}\right)}$

$$\delta_{\min} = A$$

$$\sqrt{3} = \frac{\sin\left(\frac{A + A}{2}\right)}{\sin\left(\frac{A}{2}\right)} = \frac{\sin\left(\frac{2A}{2}\right)}{\sin\left(\frac{A}{2}\right)}$$

$$= \frac{2 \sin\left(\frac{A}{2}\right) \cdot \cos\left(\frac{A}{2}\right)}{\sin\left(\frac{A}{2}\right)}$$

$$\sqrt{3} = 2 \cos\left(\frac{A}{2}\right)$$

$$\frac{\sqrt{3}}{2} = \cos\left(\frac{A}{2}\right)$$

$$\Rightarrow \frac{A}{2} = 30° \Rightarrow A = 60°$$

06. *(a)* Angle of deviation for thin prism

$$\delta = A(n-1)$$

For no deviation, $\delta_1 = \delta_2$

$$4°(1.54-1) = (1.72-1)A$$

$$\frac{4 \times 0.54}{0.72} = A \Rightarrow A = 3°$$

07. *(b)* When the white light passed through a prism, then red light bends least compared to all other colour of light. As, $\lambda_{red} > \lambda_{yellow} > \lambda_{violet}$

Light ray with longer wavelength bends less. Hence, Statement I is true.

The refractive indexes are different for different wavelengths of dispersive medium.

08. *(c)* $1 \sin 45° = \mu \sin 30°$

$$\frac{1}{\sqrt{2}} = \mu \times \frac{1}{2}$$

$$\mu = \sqrt{2} \quad \text{...(i)}$$

Now, $\frac{\mu_1}{\mu_2} = \frac{v_2}{v_1} = \frac{\lambda_2}{\lambda_1}$ ...(ii)

Using Eqs. (i) and (ii)

$$\lambda_2 = \frac{1}{\sqrt{2}}\lambda_1$$

Now, for relation between frequencies

Frequency, $(\nu) = \frac{v}{\lambda}$

$$\frac{\nu_1}{\nu_2} = \frac{v_1}{v_2} \times \frac{\lambda_2}{\lambda_1} = 1$$

$$\nu_1 = \nu_2$$

09. *(c)* Angle of deviation, $\delta = 15°$

$\therefore$ Angle of refraction, $r = i - \delta$

$$= 45° - 15° = 30°$$

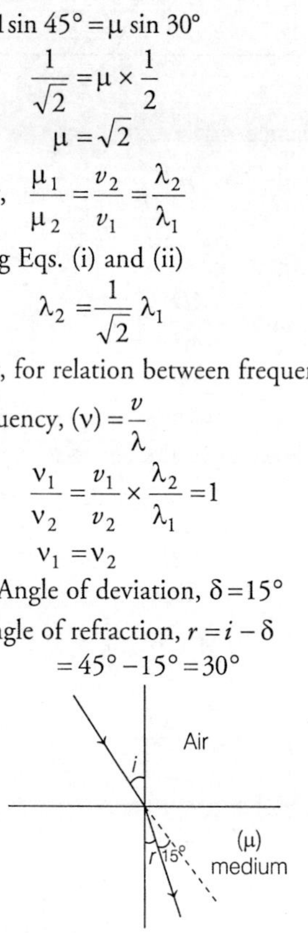

According to Snell's law,

$$\mu = \frac{\sin i}{\sin r} = \frac{\sin 45°}{\sin 30°} = \frac{1/\sqrt{2}}{1/2} = \frac{2}{\sqrt{2}}$$

$$= \sqrt{2} = 1.414$$

10. *(d)* Angle of incidence, $i = 2r$

According to snell's law,

As, $\mu = \frac{\sin i}{\sin r} \Rightarrow \sqrt{2n} = \frac{\sin 2r}{\sin r}$

$$= \frac{2 \sin r \cos r}{\sin r} = 2 \cos r$$

or $\cos r = \frac{\sqrt{2n}}{2} = \sqrt{\frac{n}{2}}$ or $r = \cos^{-1}\left(\sqrt{\frac{n}{2}}\right)$

Since, angle of incidence $i = 2r$.

Hence, $i = 2\cos^{-1}\left(\sqrt{\frac{n}{2}}\right)$

11. *(b)* If the angle of incidence is increased gradually, then the angle of deviation first decreases, attains a minimum value (δ_m), and then again starts increasing.

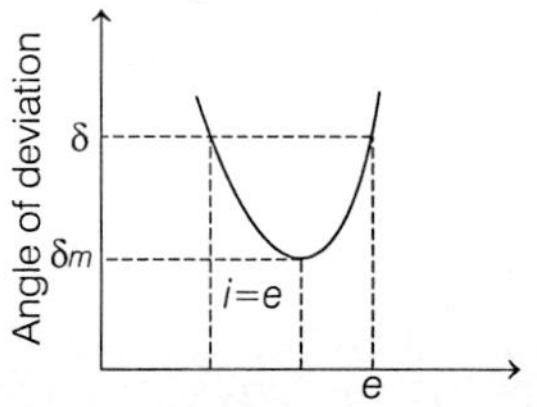

12. *(c)*

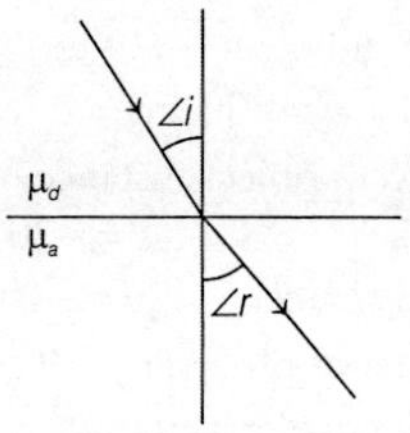

By using Snell's law, $\frac{\mu_a}{\mu_d} = \frac{\sin i}{\sin r}$

$$\Rightarrow \quad \sin r = \frac{\mu_d}{\mu_a} \sin i = 2.42 \times \sin 30°$$

$$= 2.42 \times \frac{1}{2} = 1.21 > 1$$

But the maximum value of $\sin\theta$ is 1.

$\therefore$ In above case, refraction is not possible.

13. *(b)* From the given figure,

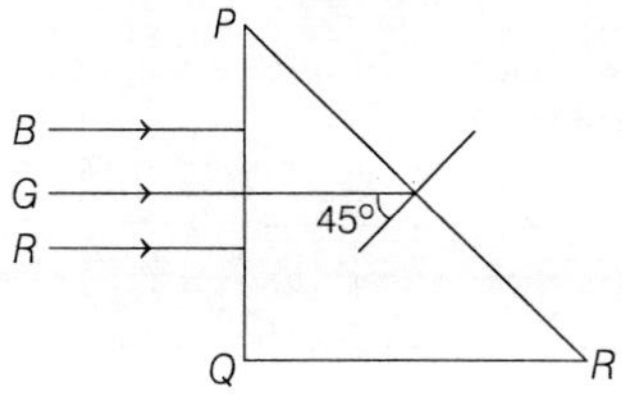

We know that,

$$\mu = \frac{1}{\sin i_C} \Rightarrow \mu = \frac{1}{\sin 45°}$$

$$\Rightarrow \quad \mu = 1.414$$

As $\mu_R < \mu$ while μ_G and $\mu_B > \mu$, so only red colour will be transmitted through face *PR* while green and blue rays will suffer total internal reflection.

> **Concept Enhancer**
>
> The rays will emerge out from the prism only when the angle of incidence is less than the critical angle of glass air interface.

14. *(a)* Since, $\lambda\nu = c =$ constant

Red light and blue light have different wavelengths and different frequencies but same speed.

15. *(a)* The propagation of light ray from a prism is shown below

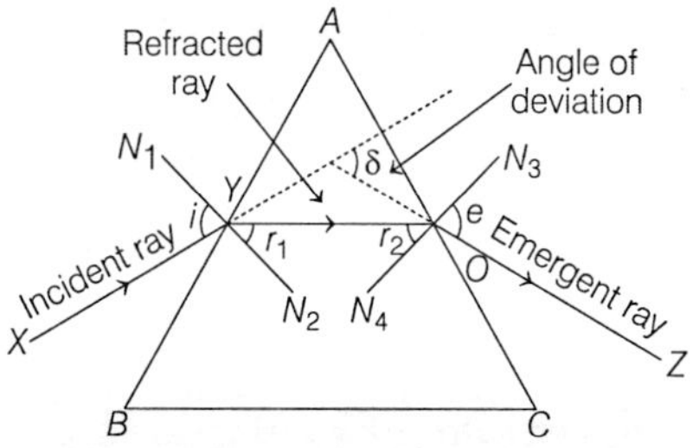

From the above figure, we can say that minimum value of angle of deviation can only be achieved when

A. incident ray and emergent ray are symmetric to the prism.

B. the refracted ray inside the prism becomes parallel to its base.

C. angle of incidence is equal to angle of emergence.

$\therefore$ Statement (A), (B) and (C) are true.

> **Concept Enhancer**
>
> Refracted ray inside the prism is parallel to the base only for equilateral and isosceles prism.

16. *(b)* As intensity is maximum at axis. Therefore, μ will be maximum and speed will be minimum on the axis of the beam. Hence, beam will converge.

17. *(d)* $\sin C = \frac{\sqrt{3}}{2}$ ...(i)

$$\sin r = \sin(90° - C) = \cos C = \frac{1}{2}$$

$$\frac{\sin\theta}{\sin r} = \frac{\mu_2}{\mu_1}$$

$$\Rightarrow \quad \sin\theta = \frac{2}{\sqrt{3}} \times \frac{1}{2}$$

$$\Rightarrow \quad \theta = \sin^{-1}\left(\frac{1}{\sqrt{3}}\right)$$

18. *(a)* $D = (\mu - 1)A$

For blue light, μ is greater than that for red light, so $D_2 > D_1$.

> **Alternate Solution**
>
> $D_1 = (1.520 - 1)A$, $D_2 = (1.525 - 1)A$
>
> $\Rightarrow \frac{D_1}{D_2} < 1 \Rightarrow D_2 > D_1$

19. *(a)* Optical fibres work on the principle of total internal reflection.

> **• Concept Enhancer**
> The core should have greater refractive index than cladding for TIR to take place because TIR will happen only light travels from denser medium to rarer medium.

20. *(c)* Optical density is a measure of how much a medium slows down the speed of light and is directly related to the refractive index of medium. Mass density is the ratio of mass to volume, and while a denser material might have a higher value of refractive index, it's not a simple direct proportionality. Thus, optical density of a medium is not directly proportionate to its mass density.
Hence, Assertion is correct but Reason is not correct.

21. *(d)*

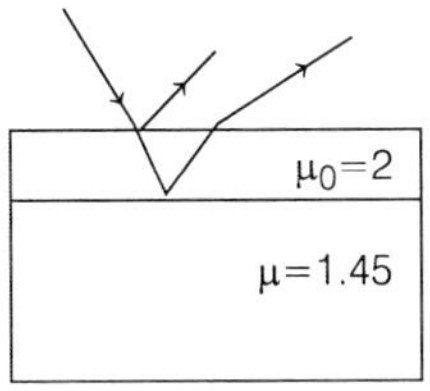

$$\Delta p = 2\mu_0 t = n\lambda$$

$$t = \frac{n\lambda}{2\mu_0} \quad \therefore t_{\min} = \frac{\lambda}{2\mu_0}$$

$$t_{\min} = \frac{550}{2 \times 2} = 137.5 \text{ nm}$$

22. *(*)*

> **key idea** Whenever the difference in the refractive index is large, the value of critical angle is small. In the question, difference between n_3 and n_1 is more as compared to n_2 and n_1 therefore $\theta_{2c} < \theta_{1c}$.

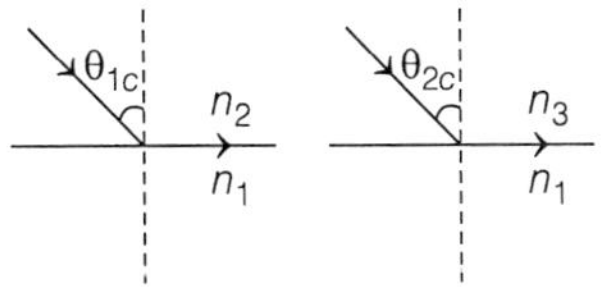

$$\sin\theta_{1C} = \frac{n_1}{n_2} \quad \ldots(i)$$

$$\sin\theta_{2C} = \frac{n_1}{n_3} \quad \ldots(ii)$$

Divide Eq. (ii) by Eq. (i)

$$\frac{\sin\theta_{2C}}{\sin\theta_{1C}} = \frac{n_1}{n_3} \times \frac{n_2}{n_1}$$

$$= \frac{n_2}{n_3} = \frac{2}{5}$$

$$\Rightarrow \quad \sin\theta_{2C} = \frac{2}{5}\sin\theta_{1C}$$

As given in the question

$$\sin\theta_{2C} - \sin\theta_{1C} = \frac{1}{2}$$

$$\Rightarrow \quad \frac{2}{5}\sin\theta_{1C} - \sin\theta_{1C} = \frac{1}{2}$$

$$\Rightarrow \quad \frac{-3}{5}\sin\theta_{1C} = \frac{1}{2}$$

$$\Rightarrow \quad \sin\theta_{1C} = \frac{-5}{6}$$

$$\Rightarrow \quad \theta_{1C} = \sin^{-1}\left(\frac{-5}{6}\right)$$

23. *(a)*

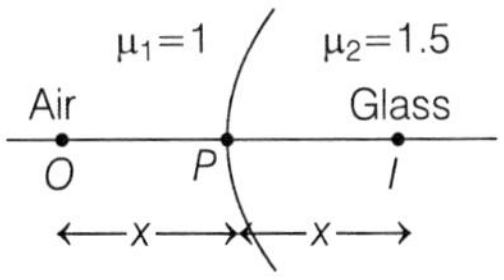

$$PO = u = -x$$
$$PI = v = x$$
$$PO = PI$$

$$\frac{\mu_2}{v} - \frac{\mu_1}{u} = \frac{\mu_2 - \mu_1}{R}$$

$$\frac{1.5}{x} + \frac{1}{x} = \frac{1}{2R}$$

$$\frac{5}{2x} = \frac{1}{2R}$$

$$x = 5R$$

24. *(d)* We are given with following situation,

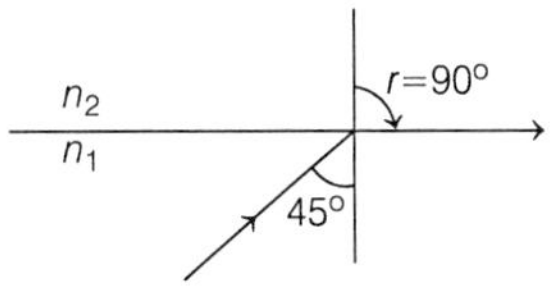

Now, refractive index of medium 2 with respect to medium 1,

$$n = \frac{n_2}{n_1} = \frac{\sin i}{\sin r} = \frac{\sin 45^\circ}{\sin 90^\circ} = \frac{1}{\sqrt{2}}$$

$$\therefore \quad \frac{n_1}{n_2} = \frac{\sqrt{2}}{1}$$

25. *(c)* Given, $\mu = \cot\frac{A}{2}$

So, $$\cot\frac{A}{2} = \frac{\sin\left(\frac{A+\delta_m}{2}\right)}{\sin\left(\frac{A}{2}\right)}$$

$$\Rightarrow \quad \cos\frac{A}{2} = \sin\left(\frac{A+\delta_m}{2}\right)$$

$$\Rightarrow \sin\left(90^\circ - \frac{A}{2}\right) = \sin\left(\frac{A+\delta_m}{2}\right)$$

or $$90^\circ - \frac{A}{2} = \frac{A+\delta_m}{2}$$

$$\Rightarrow \quad 180^\circ = 2A + \delta_m$$

$$\Rightarrow \quad \delta_m = 180^\circ - 2A$$

26. *(d)* The given situation is

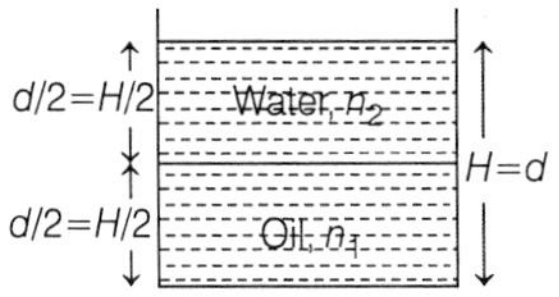

$$d_{\text{app}} = \frac{\frac{d}{2}}{n_2} + \frac{\frac{d}{2}}{n_1} = \frac{d}{2}\left[\frac{1}{n_2} + \frac{1}{n_1}\right]$$

$$= \frac{d}{2}\left[\frac{n_1 + n_2}{n_1 n_2}\right]$$

27. *(b)* $\sin i_c = \frac{\mu_r}{\mu_d}$

$$i_c = 45 \qquad \text{[given]}$$

$$\sin 45^\circ = \frac{\mu_r}{\mu_d} \Rightarrow \frac{\mu_r}{\mu_d} = \frac{1}{\sqrt{2}}$$

We know,

$$V \propto \frac{1}{\mu} \Rightarrow \frac{V_d}{V_r} = \frac{\mu_r}{\mu_d}$$

$$\frac{V_d}{3\times 10^8} = \frac{1}{\sqrt{2}} \Rightarrow V_d = \frac{3\times 10^8}{\sqrt{2}}$$

$$V_d = 3 \times 0.7 \times 10^8$$

$$V_d = 2{\cdot}12 \times 10^8 \text{ m/s}$$

28. *(a)* Let height of liquid filled in bucket be h.
To see image, a microscope should be focussed at that point where image is formed.
Apparent depth,

$$d_{\text{app}} = \frac{d}{\mu} = \frac{h}{5/3} = \frac{3h}{5}$$

Shift in position of image = Original position − Apparent depth

$$\text{Shift} = h - \frac{3h}{5} = \frac{2h}{5}$$

Given that Shift = 30 cm

$$\therefore \quad 30 = \frac{2}{5}h$$

$$\Rightarrow \quad h = \frac{30 \times 5}{2} = 75 \text{ cm}$$

29. *(a)* We know that for small angle prism,

$$\delta = (\mu - 1)\, A,$$

According to given situation, net deviation

$$A_1(\mu_1 - 1) - A_2(\mu_2 - 1) = 0$$

$$\Rightarrow A_2 = \frac{A_1(\mu_1 - 1)}{\mu_2 - 1} = \frac{6(1.54 - 1)}{1.72 - 1}$$

$$= 6 \times \frac{0.54}{0.72} = 4.5°$$

30. *(a)*

> **key idea** When a ray of light incident from denser madium to rarer medium at angle of incidence $i \geq i_c$, then refractive index (μ) of denser medium and i_c are related as, $\sin i_c = \frac{1}{\mu}$.

Speed of light in medium M_1 and M_2 are given as

$v_1 = 1.5 \times 10^8$ m/s, $v_2 = 2 \times 10^8$ m/s If i_c be the critical angle for incident light in medium M_1, then

$$\sin i_c = \frac{1}{{}^2\mu_1} \qquad \dots(i)$$

$$\text{But } {}^2\mu_1 = \frac{v_2}{v_1} = \frac{2 \times 10^8}{1.5 \times 10^8} = \frac{4}{3}$$

From Eq. (i), we get

$$\sin i_c = \frac{1}{4/3} \Rightarrow \sin i_c = \frac{3}{4}$$

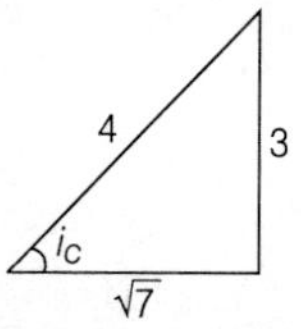

$$\therefore \quad \tan i_c = \frac{3}{\sqrt{7}}$$

$$\Rightarrow \quad i_c = \tan^{-1}(3/\sqrt{7})$$

31. *(d)* The given situation is shown below

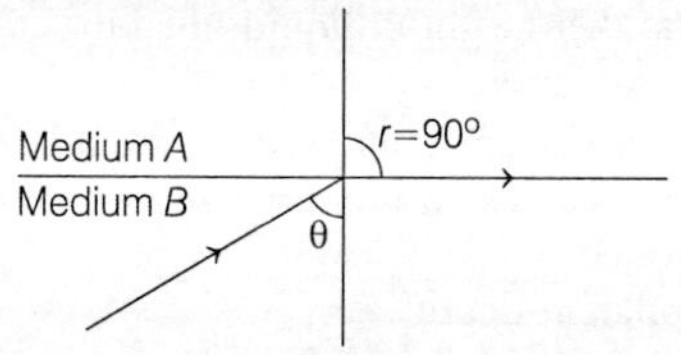

As total internal reflection occurs, so θ_{min} = critical incidence angle

$$\Rightarrow \frac{\sin \theta_{min}}{\sin r} = n_{AB} = \frac{v_B}{v_A}$$

$$\Rightarrow \sin \theta_{min} = \sin r \times \frac{v_B}{v_A}$$

Here, $r = 90°$, $v_B = 1.5 \times 10^{10}$ cm/s and $v_A = 2 \times 10^{10}$ cm/s

$$\text{So, } \sin \theta_{min} = \frac{\sin 90° \times 1.5 \times 10^{10}}{2 \times 10^{10}}$$

$$\Rightarrow \sin \theta_{min} = \frac{3}{4} \text{ or } \theta_{min} = \sin^{-1}\left(\frac{3}{4}\right)$$

$$\Rightarrow \quad \theta \geq \sin^{-1}\left(\frac{3}{4}\right) \text{ for TIR}$$

32. *(a)* By refraction formula through prism, we get

$$n = \frac{\sin\left(\frac{A + \delta_m}{2}\right)}{\sin\left(\frac{A}{2}\right)} = \cot\left(\frac{A}{2}\right) \quad \text{[Given]}$$

$$\Rightarrow \sin\left(\frac{A + \delta_m}{2}\right) = \frac{\cos\left(\frac{A}{2}\right)}{\sin\left(\frac{A}{2}\right)} \times \sin\left(\frac{A}{2}\right)$$

$$\text{or} \quad \sin\left(\frac{A + \delta_m}{2}\right) = \cos\left(\frac{A}{2}\right)$$

$$\Rightarrow \quad \sin\left(\frac{A + \delta_m}{2}\right) = \sin\left(\frac{\pi}{2} - \frac{A}{2}\right)$$

$$\Rightarrow \quad \frac{A + \delta_m}{2} = \frac{\pi}{2} - \frac{A}{2}$$

$$\text{or} \quad \delta_m = 180° - 2A$$

33. *(d)* We know that for total internal reflection, $i > i_c$

$$\therefore \quad \sin i > \sin i_c \qquad \dots(i)$$

$$\text{But} \quad \sin i_c = \frac{\mu_{rarer}}{\mu_{denser}} \qquad \dots(ii)$$

From Eqs. (i) and (ii), we get

$$\sin i > \frac{\mu_{rarer}}{\mu_{denser}} \qquad \dots(iii)$$

$$\text{Since,} \quad \mu = \sqrt{\mu_r \cdot \varepsilon_r}$$

$$= \sqrt{1 \cdot \varepsilon_r} = \sqrt{\varepsilon_r}$$

$$\therefore \frac{\mu_{rarer}}{\mu_{denser}} = \sqrt{\frac{(\varepsilon_r)_{rarer}}{(\varepsilon_r)_{denser}}} = \sqrt{\frac{1}{4}} = \frac{1}{2}$$

$$\therefore \text{From Eq. (iii), } \sin i > \frac{1}{2}$$

$$\Rightarrow \quad \sin i > \sin 30°$$

$$\Rightarrow \quad i > 30°$$

$\therefore$ $i = 60°$ [From the given options]

34. *(c)* Here, $v_A - v_B = 2.6 \times 10^7$ m/s, $\mu_B = 1.47$, $c = 3 \times 10^8$ m/s

We know that

$$v_A = \frac{c}{\mu_A}, \; v_B = \frac{c}{\mu_B};$$

$$\because \quad v_A - v_B = c\left(\frac{1}{\mu_A} - \frac{1}{\mu_B}\right)$$

$$\Rightarrow 2.6 \times 10^7 = 3 \times 10^8 \left(\frac{1}{\mu_A} - \frac{1}{\mu_B}\right);$$

$$\Rightarrow \quad \frac{0.26}{3} = \frac{1}{\mu_B} - \frac{1}{1.47} \Rightarrow \mu_B = 1.30$$

$$\text{So,} \quad \frac{\mu_B}{\mu_A} = \frac{1.97}{1.30} = 1.13$$

35. *(b)* Let us draw the diagram of glass tumbler.

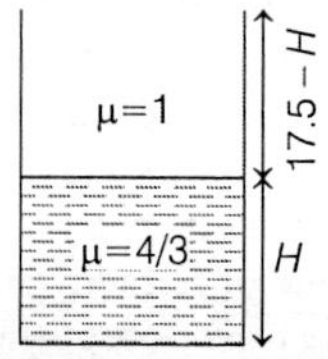

As we know that

$$\mu_{water} = \frac{H_{real}}{H_{appearent}}$$

$$\Rightarrow \quad \frac{4}{3} = \frac{H}{H_{appearent}}$$

$$\Rightarrow H_{appearent} = 3H/4$$

The height of air observed by observer $= 17.5 - H$

According to the above figure,

$$\frac{3H}{4} = 17.5 - H$$

$$\Rightarrow \quad H = 10.11 \text{ cm} \simeq 10 \text{ cm}$$

36. *(a)* Given, refractive index, $\mu = \frac{4}{3}$

By using Snell's law,

$$\mu = \frac{\sin i}{\sin r} \Rightarrow \frac{4}{3} = \frac{\sin \theta}{\sin \theta'} \qquad \dots(i)$$

From diagram, $\theta'' = 90 - \theta'$

For total internal reflection, $\frac{1}{\mu} = \sin \theta''$

$$\Rightarrow \sin(90 - \theta') = \frac{3}{4} \Rightarrow \cos \theta' = \frac{3}{4}$$

$$\text{Since,} \quad \sin \theta' = \sqrt{1 - \cos^2 \theta'}$$

$$\Rightarrow \quad \sin \theta' = \sqrt{1 - \left(\frac{3}{4}\right)^2}$$

$$= \sqrt{\frac{7}{16}} = \frac{\sqrt{7}}{4}$$

On putting this value in Eq. (i), we get

$$\frac{4}{3} = \frac{\sin \theta \times 4}{\sqrt{7}}$$

$$\Rightarrow \quad \sin \theta = \frac{\sqrt{7}}{3}$$

$$\Rightarrow \; \theta = \sin^{-1}\left(\frac{\sqrt{7}}{3}\right)$$

37. *(c)* According to given situation

$$r + r' = 90°$$

$$\Rightarrow \quad r' = 90° - r = 90° - i \quad \ldots(i)$$

By using Snell's law

$n_1 \sin i = n_2 \sin r'$,

From Eq. (i), we have

$$n_1 \sin i = n_2 \sin(90° - i)$$

$$n_1 \sin i = n_2 \sin(90° - i)$$

$$n_1 \sin i = n_2 \cos i$$

$$\Rightarrow \quad \tan i = n_2 / n_1 \quad \ldots(ii)$$

Also $\sin C = n_2 / n_1$ where, C is critical angle.

From Eq. (ii), we get

$$\sin C = \tan i$$

$$C = \sin^{-1}(\tan i) = \sin^{-1}(\tan r)$$

38. *(d)* Speed of light in a medium is given by

$$v = \frac{1}{\sqrt{\varepsilon\mu}} = \frac{1}{\sqrt{\varepsilon_0 \mu_0}} \times \frac{1}{\sqrt{\varepsilon_r \mu_r}}$$

$$= \frac{c}{\sqrt{\varepsilon_r \mu_r}}$$

Here, $\varepsilon_r = 3$ and $\mu_r = \frac{4}{3}$

$$\therefore \quad \frac{v}{c} = \frac{1}{\sqrt{\varepsilon_r \mu_r}}$$

$$= \frac{1}{\sqrt{3 \times \frac{4}{3}}} = \frac{1}{2}$$

So, refractive index of medium w.r.t. vacuum or air, $n = c / v = 2$

Hence, critical angle,

$$i_c = \sin^{-1}\left(\frac{1}{n}\right) = \sin^{-1}\frac{1}{2} = 30°$$

39. *(c)* Given, refractive index of material of prism $n = \sqrt{3}$, prism angle $A = 60°$

Using prism formula,

$$n = \frac{\sin\left(\frac{A+\delta}{2}\right)}{\sin\left(\frac{A}{2}\right)}$$

$$\Rightarrow \sqrt{3} = \frac{\sin\left(\frac{60+\delta}{2}\right)}{\sin 30°}$$

$$\Rightarrow \sin\left(\frac{60+\delta}{2}\right) = \frac{\sqrt{3}}{2}$$

$$\Rightarrow \sin\left(\frac{60+\delta}{2}\right) = \sin 60°$$

$$\text{or} \quad \frac{60+\delta}{2} = 60$$

$$\Rightarrow \quad \delta = 60°$$

Incident angle, $i = \frac{60+\delta}{2} = 60°$

> ***Alternate Solution***
>
> For minimum deviation, ray should pass symmetrically (i.e. parallel to the base of the equilateral prism)
>
> From geometry of given figure, we have, $r = 30°$
>
> 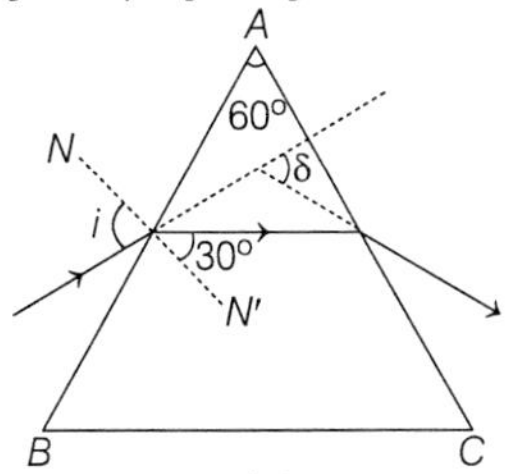
>
>
> Using Snell's law, $n = \frac{\sin i}{\sin r}$
>
> $\sin i = n \sin r = \sqrt{3} \sin 30°$
>
> $\Rightarrow \quad \sin i = \frac{\sqrt{3}}{2}$ or $i = 60°$

40. *(b)* The situation is shown in figure.

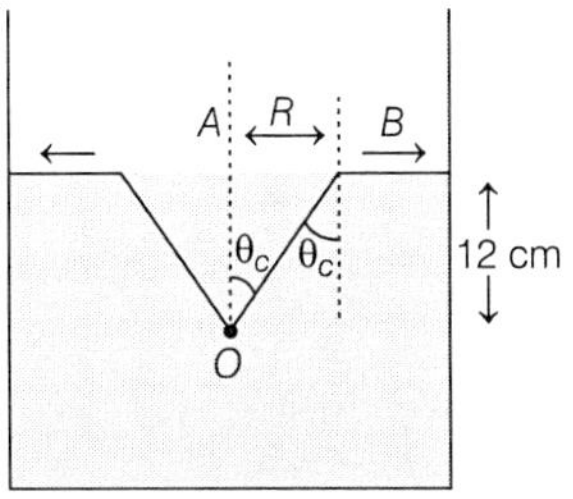

$$\sin\theta_C = \frac{1}{\mu} \Rightarrow \tan\theta_C = \frac{AB}{OA}$$

$$\therefore \quad AB = OA \tan\theta_C$$

$$= OA \frac{\sin\theta_c}{\cos\theta_c} = \frac{\sin\theta_c}{\sqrt{1 - \sin^2\theta_c}} \times OA$$

$$= \frac{1/\mu}{\sqrt{1 - 1/\mu^2}} \times OA = \frac{OA}{\sqrt{\mu^2 - 1}}$$

$$\text{or} \quad AB = \frac{OA}{\sqrt{\mu^2 - 1}} = \frac{12}{\sqrt{\left(\frac{4}{3}\right)^2 - 1}} = \frac{36}{\sqrt{7}}$$

41. *(b)* For total internal reflection

i.e., $C < i$

or $C < 45°$ $\quad [\because \angle i = 45°]$

But $n = \frac{1}{\sin C}$

$$\Rightarrow \quad C = \sin^{-1}\left(\frac{1}{n}\right)$$

$$\text{or} \quad \sin^{-1}\left(\frac{1}{n}\right) < 45° \quad \text{or} \quad \frac{1}{n} < \sin 45°$$

$$\text{or } n > \frac{1}{\sin 45°} \quad \text{or} \quad n > \frac{1}{(1/\sqrt{2})} \quad \text{or } n > \sqrt{2}$$

42. *(a)*

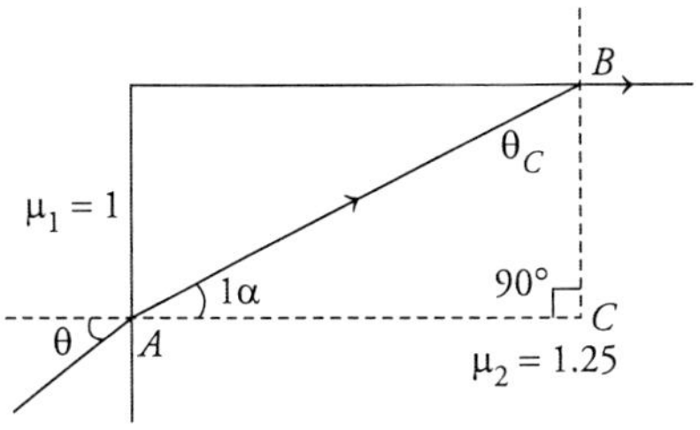

In ΔABC, $r + \theta_c = 90°$

$$r = 90° - \theta_c \quad \ldots(i)$$

Now, applying Snell's law at point A

$$\mu_1 \sin\theta = \mu_2 \sin r$$

$$\mu_1 \sin\theta = \mu_2 \sin[90° - \theta_c]$$

$$\mu_1 \sin\theta = \mu_2 \cos\theta_c$$

$$\sin\theta = \frac{\mu_2}{\mu_1} \cos\theta_c \quad \ldots(ii)$$

Applying Snell's law at point B

$$\mu_2 \sin\theta_c = \mu_1 \sin 90°$$

$$\sin\theta_c = \frac{\mu_1}{\mu_2}$$

$$\cos\theta_c = \sqrt{1 - \left(\frac{\mu_1}{\mu_2}\right)^2} \quad \ldots(iii)$$

Using Eqs. (ii) and (iii),

$$\sin\theta = \frac{\mu_2}{\mu_1}\sqrt{1 - \left[\frac{\mu_1}{\mu_2}\right]^2}$$

$$\sin\theta = \frac{1.25}{1}\sqrt{0.36}$$

$$\sin\theta = 1.25 \times 0.6 = 0.75 = \frac{3}{4}$$

$$\theta = \sin^{-1}\left(\frac{3}{4}\right)$$

43. *(a)* $u = -13$ cm

$n_1 = 1.3$, $n_2 = 1.4$

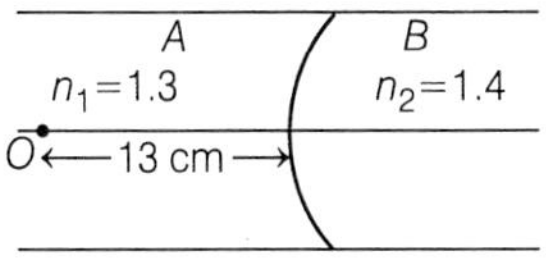

$$m = \frac{\frac{v}{n_2}}{\frac{u}{n_1}}$$

$$-2 = \frac{v}{1.4} \times \frac{1.3}{(-13)} \Rightarrow \frac{v}{1.4} = +\frac{2 \times 13}{1.3}$$

$$\frac{1.4}{v} = \frac{1.3}{2 \times 13} \quad \ldots(i)$$

$$\frac{n_2}{v} - \frac{n_1}{u} = \frac{n_2 - n_1}{R}$$

$$\frac{1.4}{v} - \frac{1.3}{(-1.3)} = \frac{1.4 - 1.3}{R}$$

$$\frac{1.4}{v} + \frac{1.3}{1.30} = \frac{0.1}{R}$$

$$\frac{1.4}{v} = \frac{1}{10R} - \frac{1}{10}$$

$$\frac{1.4}{v} = \frac{1 - R}{10R} \quad \ldots(ii)$$

Put Eq. (i) in Eq. (ii)

$$\frac{1.3}{2 \times 1.3} = \frac{1 - R}{10R}$$

$$+13R = 26 - 26R$$

$$39R = 26$$

$$R = \frac{26}{39} = \frac{2}{3} = \frac{2}{3} \text{ cm}$$

44. *(d)*

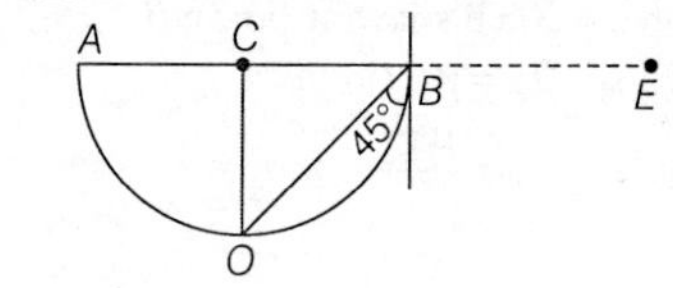

In this case,

Critical angle, $\theta_c = 45°$

Using the condition, we have

$$\sin \theta_c = \frac{1}{\mu}$$

$$\mu = \frac{1}{\sin 45°} = \sqrt{2}$$

45. *(d)*

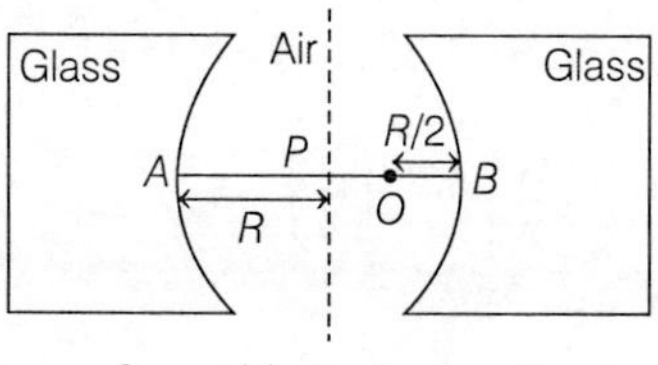

Image formed by a single refracting surface is given by

$$\frac{n_2}{v} - \frac{n_1}{u} = \frac{n_2 - n_1}{R} \quad \ldots(i)$$

For surface A,

$$n_1 = 1, n_2 = 1.5$$

$$u = \frac{-3R}{2}, R = -R$$

Applying Eq. (i)

$$\frac{1.5}{v_A} - \frac{1}{\frac{-3R}{2}} = \frac{1.5 - 1}{-R}$$

$$\Rightarrow \frac{1.5}{v_A} = \frac{-0.5}{R} - \frac{2}{3R}$$

$$\Rightarrow \frac{1.5}{v_A} = \frac{-1.5 - 2}{3R}$$

$$\Rightarrow \frac{1.5}{v_A} = \frac{-3.5}{3R}$$

$$\Rightarrow v_A = \frac{-4.5R}{3.5} = \frac{-9R}{7}$$

Similarly for surface B,

$$n_1 = 1, n_2 = 1.5$$

$$u = \frac{-R}{2}, R = -R$$

Applying Eq. (i)

$$\frac{1.5}{v_B} - \frac{1}{\frac{-R}{2}} = \frac{1.5 - 1}{-R}$$

$$\Rightarrow \quad \frac{1.5}{v_B} = \frac{-0.5}{R} - \frac{2}{R} \Rightarrow \frac{1.5}{v_B} = -\frac{2.5}{R}$$

$$\Rightarrow \quad v_B = \frac{-1.5R}{2.5} = \frac{-3R}{5}$$

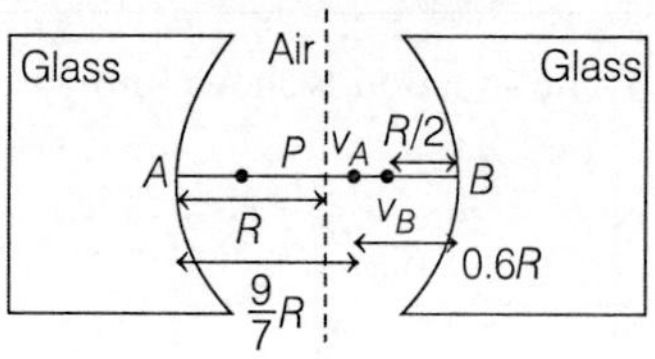

Distance between v_A and v_B is given by

$$d = 2R - (|v_A| + |v_B|)$$

$$= 2R - \left(\frac{9R}{7} + \frac{3R}{5}\right) = 2R - \frac{9R}{7} - \frac{3R}{5}$$

$$= \left(\frac{70 - 45 - 21}{35}\right)R = \frac{4R}{35}$$

$$= 0.114R$$

46. *(a)* By Snell's law

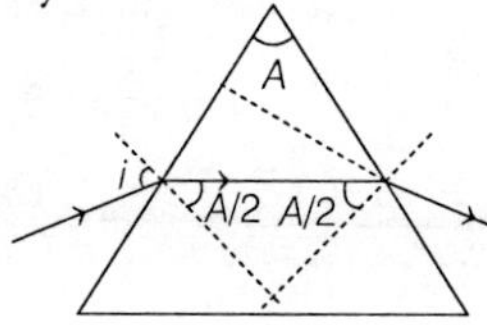

$$1 \sin i = \mu \sin \frac{A}{2}$$

$$\sin i = \mu \sin\left(\frac{A}{2}\right) \text{ and } \mu = \cot \frac{A}{2}$$

$$\sin i = \cot\left(\frac{A}{2}\right) \cdot \sin\left(\frac{A}{2}\right)$$

$$\sin i = \frac{\cos\left(\frac{A}{2}\right)}{\sin\left(\frac{A}{2}\right)} \cdot \sin\left(\frac{A}{2}\right)$$

$$\sin i = \cos\left(\frac{A}{2}\right)$$

$$\sin i = \sin\left(\frac{\pi}{2} - \frac{A}{2}\right) \Rightarrow i = \frac{\pi}{2} - \frac{A}{2}$$

$$2i = \pi - A$$

$$\delta_{min} = 2i - A = \pi - A - A = \pi - 2A$$

Alternate Solution

$$\mu = \frac{\sin\left(\frac{A + \delta_{min}}{2}\right)}{\sin\left(\frac{A}{2}\right)} \Rightarrow \cot \frac{A}{2} = \frac{\sin\left(\frac{A + \delta_{min}}{2}\right)}{\sin\left(\frac{A}{2}\right)}$$

$$\frac{\cos \frac{A}{2}}{\sin\left(\frac{A}{2}\right)} = \frac{\sin\left(\frac{A + \delta_{min}}{2}\right)}{\sin\left(\frac{A}{2}\right)}$$

$$\cos\left(\frac{A}{2}\right) = \sin\left(\frac{\pi}{2} - \frac{A}{2}\right) = \sin\left(\frac{A + \delta_{min}}{2}\right)$$

$$\frac{\pi}{2} - \frac{A}{2} = \frac{A + \delta_{min}}{2} \Rightarrow \frac{\pi}{2} - \frac{A}{2} = \frac{A}{2} + \frac{\delta_{min}}{2}$$

$$\Rightarrow \quad \delta_{min} = (\pi - 2A)$$

47. *(d)* Total internal reflection occurs through given glass rod as shown in figure.

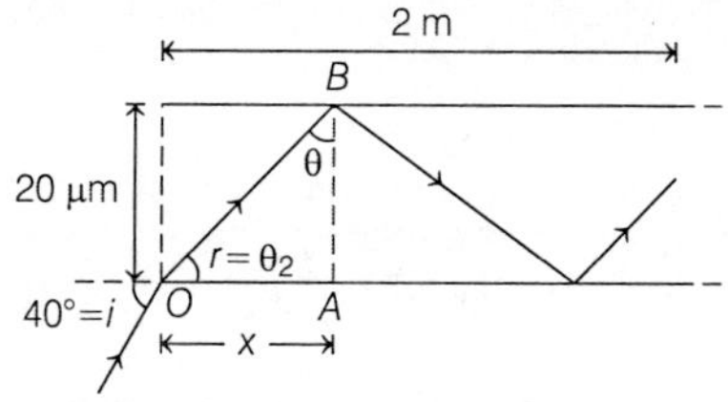

From Snell's law, $n_1 \sin i = n_2 \sin r$

where, $n_1 = 1$, $n_2 = 1.31$ and $i = 40°$

So, we get,

$$1 \sin 40° = 1.31 \sin r$$

$$\Rightarrow \quad \sin r = \frac{0.64}{1.31} = 0.49 \approx 0.5$$

So, $\quad r = 30°$

From ΔOAB, $\theta = 90 - r = 60°$

Now, $\tan \theta = \dfrac{x}{20 \ \mu m}$

$$\Rightarrow \quad x = 20\sqrt{3} \ \mu m$$

$$[\because \tan 60° = \sqrt{3}]$$

One reflection occurs in $20\sqrt{3}$ μm.

∴ Total number of reflections occurring in 2 m

$$n = \frac{2m}{20\sqrt{3} \ \mu m}$$

$$= \frac{2}{20\sqrt{3} \times 10^{-6}}$$

$$= 57735 \text{ reflections}$$

$$\approx 57000 \text{ reflections}$$

48. *(a)* Given, $i = 35°, \delta = 40°, e = 79°$.

So, angle of deviation by a glass prism,

$$\delta = i + e - A$$

$\Rightarrow \quad 40° = 35° + 79° - A$

i.e., Angle of prism $\Rightarrow A = 74°$.

Such that $r_1 + r_2 = A = 74°$.

We know that,

$$n = \frac{\sin\left(\frac{A + \delta m}{2}\right)}{\sin\left(\frac{A}{2}\right)} \quad \ldots(i)$$

Let us put $\mu = 1.5$ in eq. (i), we get

$$1.5 = \frac{\sin\left(\frac{A + \delta_{min}}{2}\right)}{\sin A/2}$$

$$\Rightarrow \quad 1.5 = \frac{\sin\left(\frac{74° + \delta_{min}}{2}\right)}{\sin 37°}$$

$$\Rightarrow \quad 0.9 = \sin\left(37° + \frac{\delta_{min}}{2}\right)$$

($\because \sin 37° \approx 0.6$)

$$\sin 64° = \sin\left(37° + \frac{\delta_{min}}{2}\right)$$

($\because \sin 64° = 0.9$)

$$37° + \frac{\delta_{min}}{2} = 64° \Rightarrow \delta_{min} \approx 54°$$

This angle is greater than the 40° deviation angle already given. For greater μ, deviation will be even higher. Hence, μ of the given prism should be less than 1.5. Hence, the closest option will be 1.5.

49. *(a)*

> **key idea** The ray will get transmitted through face AC, if $i_{AC} < i_C$

Consider the ray diagram as shown below.

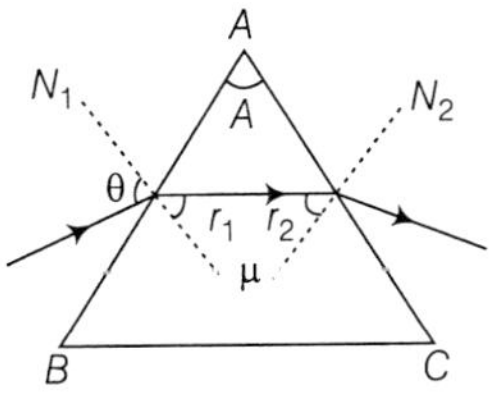

For transmission of light through face AC

$$i_{AC} < i_C \quad \text{or} \quad A - r_1 < i_C$$

or $\quad \sin(A - r_1) < \sin i_C \quad$ or

$$\sin(A - r_1) < \frac{1}{\mu}$$

$$A - r_1 < \sin^{-1}\left(\frac{1}{\mu}\right)$$

or $\quad \sin r_1 > \sin\left[A - \sin^{-1}\left(\frac{1}{\mu}\right)\right]$

Now, applying Snell's law at the face AB

$$1 \times \sin\theta = \mu \sin r_1$$

or $\quad \sin r_1 = \dfrac{\sin\theta}{\mu}$

$$\Rightarrow \quad \frac{\sin\theta}{\mu} > \sin\left[A - \sin^{-1}\left(\frac{1}{\mu}\right)\right]$$

or $\theta > \sin^{-1}\left[\mu \sin\left\{A - \sin^{-1}\left(\frac{1}{\mu}\right)\right\}\right]$

50. *(45)* We know that,

$$\mu = \frac{\sin\left(\frac{A + \delta_m}{2}\right)}{\sin\frac{A}{2}}$$

$$\Rightarrow \quad \sqrt{2} = \frac{\sin\left(\frac{60° + \delta_m}{2}\right)}{\sin\frac{60°}{2}}$$

$$\Rightarrow \sin\left(\frac{60° + \delta_m}{2}\right) = \sqrt{2}\sin 30°$$

$$= \sqrt{2} \times \frac{1}{2}$$

$$= \frac{1}{\sqrt{2}}$$

$$\Rightarrow \sin\left(\frac{60° + \delta_m}{2}\right) = \sin 45°$$

$$\Rightarrow \quad \frac{60° + \delta_m}{2} = 45°$$

$$\Rightarrow \quad 60° + \delta_m = 90°$$

$$\Rightarrow \quad \delta_m = 30°$$

$\therefore$ Angle of incidence,

$$i = \frac{A + \delta_m}{2}$$

$$= \frac{60° + 30°}{2} = 45°$$

51. *(4)*

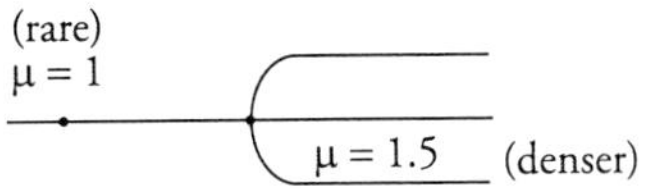

$$\mu_1 = 1$$
$$\mu_2 = 1.5$$
$$R = 50$$
$$v = 200$$
$$u = ?$$

$$\frac{\mu_2}{v} - \frac{\mu_1}{v} = \frac{\mu_2 - \mu_1}{R}$$

$$\frac{1.5}{200} - \frac{1}{u} = \frac{1.5 - 1}{50}$$

$$\frac{1.5}{200} - \frac{1}{u} = \frac{0.5}{50}$$

$$\frac{1.5}{200} - \frac{0.5}{50} = \frac{1}{u}$$

$$\frac{1.5 - 2.0}{200} = \frac{1}{u}$$

$$\frac{-0.5}{200} = \frac{1}{u}$$

$$u = -400 \text{ cm} = -4 \text{ m}$$

$$n = 4$$

52. *(80)*

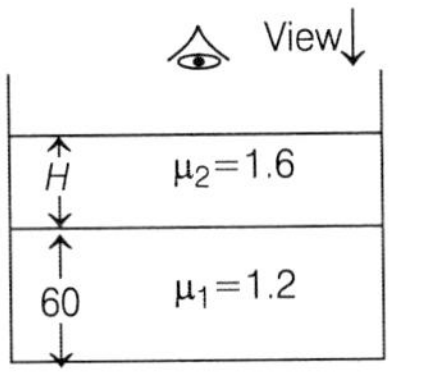

y = apparent depth of bottom

$$y = \frac{H}{\mu_2} + \frac{60}{\mu_1}$$

$$\frac{y}{1} = \frac{H}{1.6} + \frac{60}{1.2}$$

Shift $= 40$

$$H + 60 - y = 40$$

$$H + 60 = 40 + y$$

$$H + 60 = 40 + \frac{H}{1.6} + \frac{60}{1.2}$$

[put value of y]

$$H + 60 = 40 + \frac{H}{1.6} + 50$$

$$H - \frac{H}{1.6} = 90 - 60$$

$$\frac{0.6H}{1.6} = 30$$

$$\frac{6}{16}H = 30$$

$$H = 80 \text{ cm}$$

53. *(6)* Given

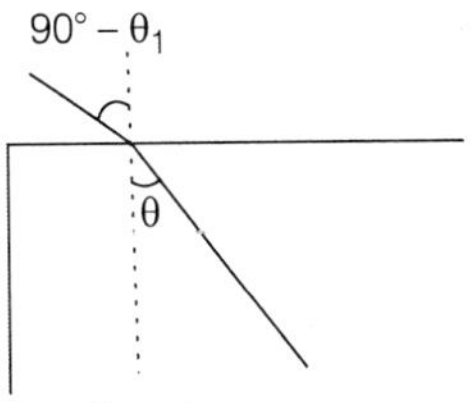

$$\theta_1 = \cos^{-1}\left(\frac{n_2}{2n_1}\right) \Rightarrow \cos\theta_1 = \frac{n_2}{2n_1}$$

By Snell's law

$$\frac{\sin(90° - \theta_1)}{\sin\theta} = \frac{n_2}{n_1}$$

$$\frac{\cos\theta_1}{\sin\theta} = \frac{n_2}{n_1}$$

$$\frac{n_2}{2n_1 \times \sin\theta} = \frac{n_2}{n_1} \Rightarrow \sin\theta = \frac{1}{2}$$

or $\theta = 30^\circ$

Now, $\tan\theta = \dfrac{d}{2\times t}$

$$t = \frac{d}{2\tan\theta} = \frac{4\sqrt{3}}{2\times\tan 30^\circ} = 6\text{ cm}$$

54. *(2)* From the given situation,

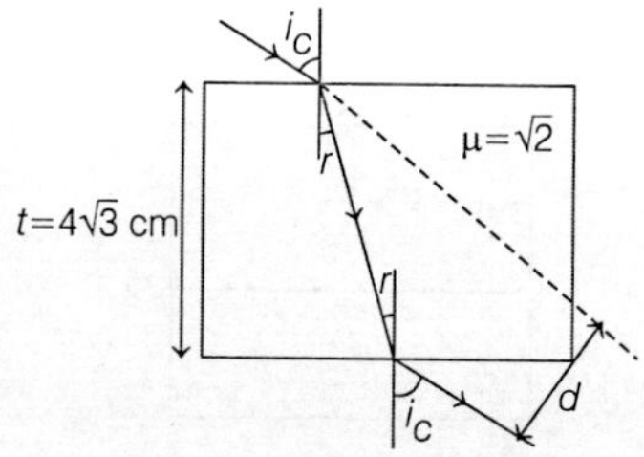

$$\sin i_C = \frac{1}{\mu} = \frac{1}{\sqrt{2}} = \sin 45^\circ$$

∴ $i_C = 45^\circ$

But $\sin i_C = \mu \sin r$

⇒ $\sin 45^\circ = \sqrt{2}\sin r$

⇒ $\dfrac{1}{\sqrt{2}} = \sqrt{2}\sin r$

⇒ $\sin r = \dfrac{1}{2} = \sin 30^\circ$

∴ $r = 30^\circ$

∴ Lateral displacement,

$$d = t\sin(i_C - r)\sec r$$
$$= 4\sqrt{3}\sin(45^\circ - 30^\circ)\sec 30^\circ$$
$$= 4\sqrt{3}\sin 15^\circ \times \frac{2}{\sqrt{3}}$$
$$= 8\sin 15^\circ = 8\times 0.25 = 2\text{ cm}$$

55. *(31)*

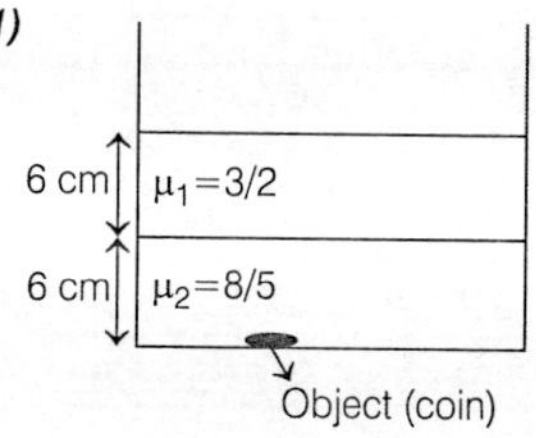

$$\text{Apparent depth} = \frac{h_1}{\mu_{\text{rel }1}} + \frac{h_2}{\mu_{\text{rel }2}}$$

$$= \frac{6}{3/2} + \frac{6}{8/5} = 4 + \frac{15}{4}$$

$$= \frac{31}{4}\text{ cm} = \frac{\alpha}{4}\text{ cm (given)}$$

$\alpha = 31$

56. *(30)* Given, for equilateral prism, $A = 60^\circ, \mu = \sqrt{2}$

We know that, $\mu = \dfrac{\sin\left(\dfrac{A+\delta_m}{2}\right)}{\sin\dfrac{A}{2}}$

$$\Rightarrow \sqrt{2} = \frac{\sin\left(\dfrac{A+\delta_m}{2}\right)}{\sin 30^\circ}$$

$$\Rightarrow \sin\left(\frac{A+\delta_m}{2}\right) = \frac{1}{\sqrt{2}} = \sin 45^\circ$$

$$\Rightarrow \frac{A+\delta_m}{2} = 45^\circ$$

$$\Rightarrow \delta_m = 90^\circ - A$$
$$= 90 - 60^\circ = 30^\circ$$

57. *(3)* Effective distance of bird as observed by the fish $d_{\text{eff}} = d_1 + \mu d$

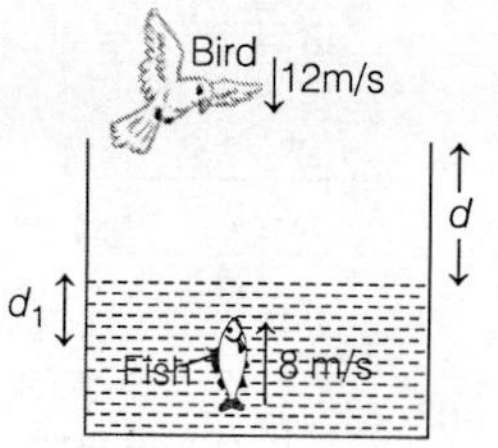

Suppose v be the actual velocity of bird.

⇒ $v_{\text{eff}} = v_1 + \mu v$

⇒ $12 = 8 + \dfrac{4}{3}v$

⇒ $v = 3$ m/s

58. *(50)* According to Snell's law,

$$\sin 60^\circ = \frac{4}{3}\sin r$$

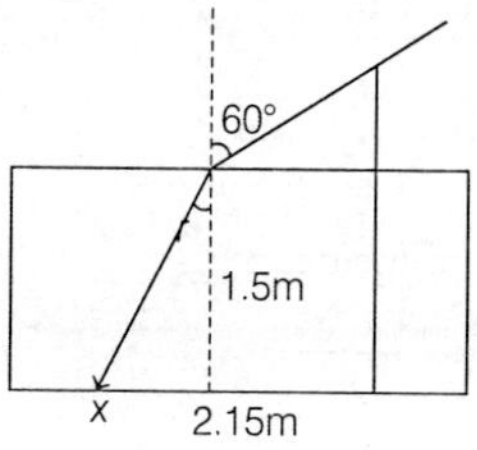

$$\sin r = \frac{3}{4}\times\frac{\sqrt{3}}{2} = \frac{3\sqrt{3}}{8} \quad \text{...(i)}$$

$$\cos r = \sqrt{1 - \frac{27}{64}} = \frac{\sqrt{37}}{8} = 0.75$$

$$\tan r = \sqrt{\frac{27}{37}} \Rightarrow \frac{x}{1.5} = 0.85$$

$$x = 0.85\times 1.5 = 1.275\text{ m}$$

$$\tan 30^\circ = \frac{y}{2.15 - 1.275} = \frac{y}{0.875}$$

$$y = \frac{0.875}{1.732} = 0.50$$

So length of pole above water surface $= 0.50\text{ m} = 50\text{ cm}$

59. *(52)* The given situation is shown below.

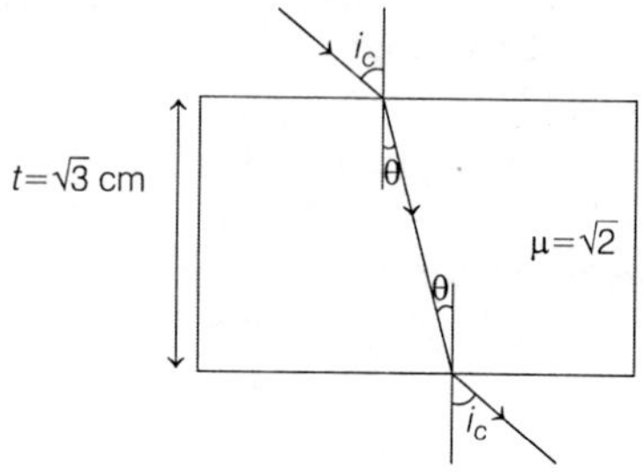

We know that $\sin i_c = \dfrac{1}{\mu} = \dfrac{1}{\sqrt{2}} = \sin 45^\circ$

∴ $i_c = 45^\circ$

Again, $\dfrac{\sin i_c}{\sin\theta} = \mu$ [From Snell's law]

⇒ $\sin i_c = \mu\sin\theta \Rightarrow \sin 45^\circ = \sqrt{2}\sin\theta$

⇒ $\sin\theta = \dfrac{1}{2} = \sin 30^\circ \Rightarrow \theta = 30^\circ$

∴ Lateral displacement,

$$x = t\sin(i - r)\sec r$$
$$= \sqrt{3}\sin(45^\circ - 30^\circ)\sec 30^\circ$$
$$= \sqrt{3}\sin 15^\circ\left(\frac{2}{\sqrt{3}}\right)$$
$$= 52\times 10^{-2}\text{ cm}$$

60. *(12)* If t is the thickness of the glass slab, then

$$l = t\sin i\left[1 - \frac{\cos i}{\sqrt{\mu^2 - \sin^2 i}}\right]$$

$$\Rightarrow 4\sqrt{3} = t\sin 60^\circ \left[1 - \frac{\cos 60^\circ}{\sqrt{(\sqrt{3})^2 - \sin^2 60^\circ}}\right]$$

$$= \frac{\sqrt{3}}{2}t\left[1 - \frac{\frac{1}{2}}{\sqrt{3 - \frac{3}{4}}}\right]$$

$$= \frac{\sqrt{3}}{2}t\left[1 - \frac{\frac{1}{2}}{3/2}\right] = \frac{\sqrt{3}}{2}t\left[\frac{2}{3}\right]$$

$$\Rightarrow 4\sqrt{3} = \frac{t}{\sqrt{3}} \Rightarrow t = 12\text{ cm}$$

61. *(5)* At minimum deviation,

Angle of incident = Angle of prism

i.e. $i = A = 60^\circ$

Angle of refraction,

$$r = \frac{A}{2} = \frac{60^\circ}{2} = 30^\circ$$

Then, by Snell's law of refraction,

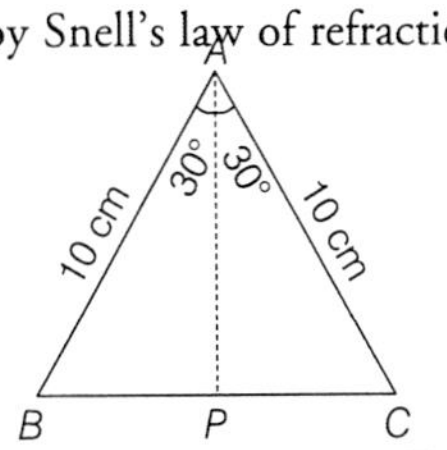

$$\mu = \frac{\sin i}{\sin r} = \frac{\sin 60°}{\sin 30°} = \frac{\frac{\sqrt{3}}{2}}{\frac{1}{2}} = \sqrt{3}$$

Now, in ΔABP, $AP = AB \cos 30°$

$$= \frac{10\sqrt{3}}{2} = 5\sqrt{3} \text{ cm}$$

$$= 5\sqrt{3} \times 10^{-2} \text{ m}$$

i.e. optical distance travelled by light along AP,

$$d = \mu \times AP = \sqrt{3} \times 5\sqrt{3} \times 10^{-2}$$

$$= 15 \times 10^{-2} \text{ m}$$

Now, the time taken by light ($c = 3 \times 10^8$ m/s) to travel from P to A,

$$t = \frac{d}{c} = \frac{15 \times 10^{-2}}{3 \times 10^8} \text{ s} = 5 \times 10^{-10} \text{ s}$$

62. *(600)* $\because$ For no deviation, $n_1 = n_2$

$$\Rightarrow 1.2 + \frac{10.8 \times 10^{-14}}{\lambda^2} = 1.45 + \frac{1.8 \times 10^{-14}}{\lambda^2}$$

$$\Rightarrow 0.25 = \frac{9 \times 10^{-14}}{\lambda^2}$$

$$\Rightarrow \lambda^2 = \frac{9 \times 10^{-14}}{0.25} \Rightarrow \lambda = \frac{3}{5} \times 10^{-6}$$

$$\Rightarrow \lambda = 6 \times 10^{-7} \text{ m}$$

$$\Rightarrow \lambda = 600 \text{ nm}$$

63. *(60)*

> **key idea** If μ is refractive index of material of prism, then from Snell's law
>
> $$\mu = \frac{\sin i}{\sin r} = \frac{\sin(A + \delta_m)/2}{\sin A/2} \quad \ldots \text{(i)}$$
>
> where, A is angle of prism and δ_m is minimum deviation through prism.

Given, let refractive index of prism,

$$\mu = \sqrt{3}$$

In the case of minimum deviation,

$$i = 2r = A \quad \ldots \text{(i)}$$

As we know that

$$i + e = A + \delta_m \quad [\because i = e]$$

$$2i = A + \delta_m$$

$$2A = A + \delta_m \Rightarrow A = \delta_m$$

By using prism formula,

$$\mu = \frac{\sin\left(\frac{A + \delta_m}{2}\right)}{\sin\left(\frac{A}{2}\right)}$$

$$\Rightarrow \sqrt{3} = \frac{\sin\left(\frac{A + A}{2}\right)}{\sin\frac{A}{2}} = \frac{\sin A}{\sin\frac{A}{2}}$$

$$\Rightarrow \sqrt{3} \sin\frac{A}{2} = \sin A = \sin 2\frac{A}{2}$$

$$= 2 \sin\frac{A}{2} \cos\frac{A}{2}$$

$$\Rightarrow \frac{\sqrt{3}}{2} = \cos\frac{A}{2} \Rightarrow \cos 30° = \cos\frac{A}{2}$$

$$\therefore A/2 = 30° \Rightarrow A = 60°$$

64. *(12)* This is a case of achromatic combination.

$$\therefore \quad \theta_{\text{net}} = 0$$

$$\Rightarrow \quad \theta_1 - \theta_2 = 0$$

$$\Rightarrow \quad \theta_1 = \theta_2$$

$$\Rightarrow \quad \omega_1 \delta_1 = \omega_2 \delta_2 \quad [\because \theta = \omega\delta] \ldots \text{(i)}$$

and $\quad \delta_{\text{net}} = \delta_1 - \delta_2 = 2° \quad \ldots \text{(ii)}$

From Eqs. (i) and (ii), we get

$$\delta_1 - \frac{\omega_1 \delta_1}{\omega_2} = 2°$$

$$\Rightarrow \delta_1\left(1 - \frac{\omega_1}{\omega_2}\right) = 2°$$

$$\Rightarrow \delta_1\left(1 - \frac{2}{3}\right) = 2°$$

$$\Rightarrow \quad \delta_1 = 6°$$

Also, $\quad \delta_1 = (\mu_1 - 1)A_1$

$$\Rightarrow \quad 6° = (1.5 - 1)\, A_1$$

$$\Rightarrow \quad A_1 = 12°$$

65. *(5)* Deviation for small-angled prism is given by $\delta = (\mu - 1)A$

Given, $A = 1°$, $\mu = 1.5$

On substituting these values in above equation, we get,

$$\delta = (1.5 - 1)1 \quad \delta = 0.5$$

According to question, $\delta = N/10$

$$\Rightarrow \quad 0.5 = \frac{N}{10}$$

$$\Rightarrow \quad N = 5$$

Topic 3

01. *(d)* Using lens Maker's formula,

$$\frac{1}{f} = (\mu - 1)\left(\frac{1}{R_1} - \frac{1}{R_2}\right)$$

$$\Rightarrow \quad \frac{1}{12} = (\mu - 1)\left(\frac{1}{10} - \frac{1}{-15}\right)$$

$$\Rightarrow \quad \frac{1}{12} = (\mu - 1)\frac{1}{6}$$

$$\Rightarrow \quad (\mu - 1) = \frac{1}{2}$$

$$\Rightarrow \quad \mu = \frac{3}{2} = 1.5$$

02. *(a)* Using lens Maker formula

$$\frac{1}{f} = ({}_1\mu_2 - 1)\left[\frac{1}{R_1} - \frac{1}{R_2}\right]$$

$$\frac{1}{f_1} = (1.5 - 1)\left[\frac{1}{2} - 0\right]$$

$$f_1 = 4 \text{ cm}$$

For f_2, $\dfrac{1}{f_2} = \left(\dfrac{1.5}{1.2} - 1\right)\left(\dfrac{1}{3} - 0\right)$

$$f_2 = 12 \text{ cm}$$

$$f_1 : f_2 \Rightarrow 4 : 12 = 1 : 3$$

03. *(c)* Focal length is given by,

$$f = \frac{1}{P}$$

For small change in P

$$\frac{\Delta f}{f} = -\frac{\Delta P}{P}$$

As given, $\Delta P = 0.1$ D and $P = 2.5$ D

$$\Rightarrow \frac{\Delta f}{f} = -\frac{0.1}{2.5} = -0.04$$

So, relative decrease in focal length is 0.04.

04. *(b)* Since, lens is cut horizontally, hence, focal length remains same. So, focal length of L_1 is f

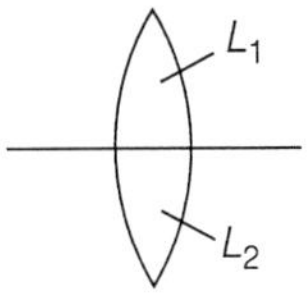

In second lens, it is cut vertically, hence, focal length becomes $2f$.

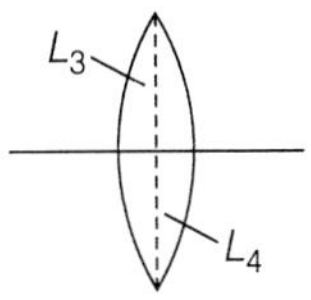

$$\Rightarrow \quad \frac{f_{L_1}}{f_{L_3}} = \frac{f}{2f} = \frac{1}{2}$$

05. *(d)* Given, $f_1 = 10$ cm

$$f_2 = 15 \text{ cm}$$

From the diagram it is clear that rays from lens L_1 converges at a fixed point f_1 and from where the lens L_2 diverges the rays parallel to principal axis.

i.e. $\quad f_1 + f_2 = 10 + 15 = 25$ cm

06. *(d)* Statement I is false but Statement II is true.

If an object is at the centre of curvature of convex lens, image is formed at the centre of curvature of the lens on the other side. But it is not true for concave lens. Concave lens always form a virtual and erect image for real objects.

07. *(b)*

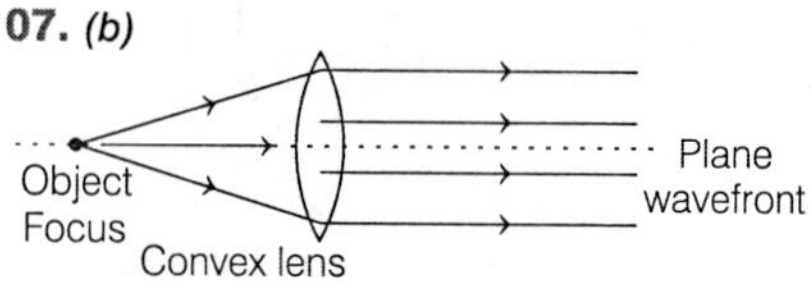

When light emerges out of convex lens (object is kept at focus), the shape of wavefront of light is plane.

> **• Concept Enhancer**
>
> Based on the path followed by the rays emitting from a source, there are three types of wavefront : spherical, cylindrical and plane wavefront. For a point source, we obtain a spherical wavefront. For a linear source, we obain a cylindrical wavefront. Plane wavefront are generated from a very distant source.

08. *(a)* When the lens is cut horizontally, the focal length of the lens does not change. Hence, no change in power but when the lens is cut vertically then focal length of the lens gets doubled and hence power of the lens gets halved.

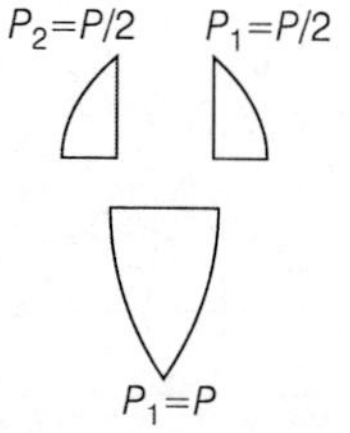

09. *(c)* Given, focal length of concave lens, $f' = -f$

By using lens formula,

$$\frac{1}{f'} = \frac{1}{v} - \frac{1}{u}$$

$$\Rightarrow \quad \frac{1}{v} = \frac{1}{f'} + \frac{1}{u}$$

$$= -\frac{1}{f} - \frac{1}{f} = -2/f$$

$$\Rightarrow \quad v = -f/2$$

$$\therefore \quad |v| = \frac{f}{2}$$

$$\therefore \quad m = \frac{v}{u} = \frac{-f/2}{-f} = \frac{1}{2}$$

10. *(b)* $\frac{1}{f} = (\mu - 1)\left(\frac{1}{R_1} - \frac{1}{R_2}\right)$

Also, by Cauchy's formula,

$$\mu = A + \frac{B}{\lambda^2} + \frac{C}{\lambda^4} + \ldots \Rightarrow \quad \mu \propto \frac{1}{\lambda}$$

As $\quad \lambda_{\text{blue}} < \lambda_{\text{red}} \Rightarrow \mu_{\text{blue}} > \mu_{\text{red}}$

Hence, $\quad f_{\text{red}} > f_{\text{blue}}$

11. *(a)* It is possible when object kept at centre of curvature.

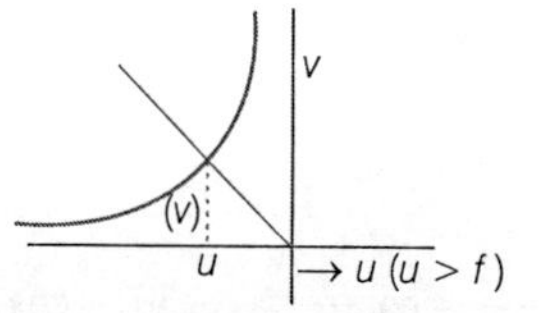

For $\quad u = v, u = 2f, v = 2f$

12. *(c)* Experimental observation at $v = \infty, u = -f$

At $\; u = \infty, v = f$

$$\frac{1}{v} - \frac{1}{u} = \frac{1}{f} = \text{constant}$$

So, (c) is the correct graph.

13. *(b)* Power of a lens is reciprocal of its focal length. Power of combined lens is

$$P = P_1 + P_2 = -15 + 5 = -10 \text{ D}$$

$$\therefore \quad f = \frac{1}{P} = \frac{100}{-10} \text{ cm or } f = -10 \text{ cm}$$

14. *(b)* $f_1 = 30$ cm, $f_2 = 10$ cm

$$\frac{1}{f_{\text{eq}}} = \frac{1}{f_1} + \frac{1}{f_2} - \frac{d}{f_1 f_2}$$

d = distance between lens

$$= \frac{1}{0.3} + \frac{1}{0.1} - \frac{0.1}{(0.3)(0.1)}$$

$$\frac{1}{f_{\text{eq}}} = \frac{1}{0.1}, f_{\text{eq}} = 0.1$$

$$\text{Power} = \frac{1}{f_{\text{eq}}} = \frac{1}{0.1} = \frac{10}{1} = 10 \text{D}$$

15. *(b)* As we know,

$$\frac{1}{f} = \left(\frac{\mu_L}{\mu_m} - 1\right)\left(\frac{1}{R_1} - \frac{1}{R_2}\right)$$

For air $\mu_m = 1$

$$\frac{1}{12} = [1.6 - 1]\left[\frac{1}{R_1} - \frac{1}{R_2}\right]$$

$$\frac{1}{12} = \frac{6}{10}\left[\frac{1}{R_1} - \frac{1}{R_2}\right]$$

$$\frac{1}{R_1} - \frac{1}{R_2} = \frac{10}{72}$$

For water

$$\frac{1}{f} = \left(\frac{1.6}{1.28} - 1\right)\left(\frac{1}{R_1} - \frac{1}{R_2}\right)$$

$$\frac{1}{f} = \frac{32}{128} \times \frac{10}{72}$$

$$\frac{1}{f} = \frac{1}{4} \times \frac{10}{72}$$

$$f = 28.8 \text{ cm}$$

$$f = 288 \text{ mm}$$

16. *(c)* $\mu_g = 1.5$, $R_1 = 30$ cm

$R_2 = 20$ cm

$\mu_l = 1.3$

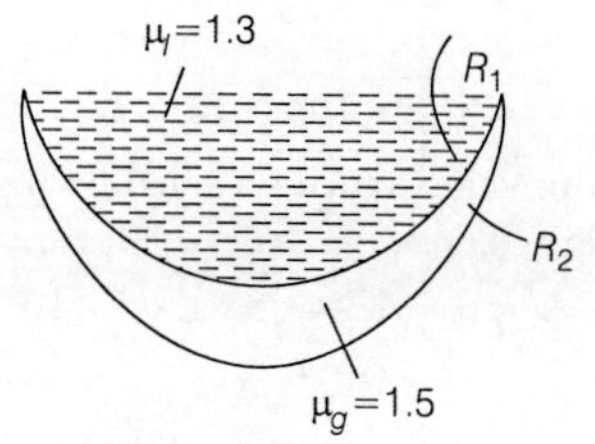

Focal length of liquid glass combination is given as

$$\frac{1}{f} = \frac{1}{f_1} + \frac{1}{f_2}$$

$$= ({}^a\mu_l - 1)\left(\frac{1}{\infty} - \frac{1}{-30}\right)$$

$$+ ({}^a\mu_g - 1)\left(\frac{1}{-30} - \frac{1}{-20}\right)$$

$$= (1.3 - 1)\left(\frac{1}{30}\right) + (1.5 - 1)\left(\frac{1}{60}\right)$$

$$= \frac{0.3}{30} + \frac{0.5}{60}$$

$$= \frac{1}{100} + \frac{1}{120} = \frac{6+5}{600}$$

$$\Rightarrow \frac{1}{f} = \frac{11}{600}$$

$$\Rightarrow f = \frac{600}{11} \text{ cm}$$

17. *(b)* Focal length of a lens

$$\frac{1}{f} = \left(\frac{n_2}{n_1} - 1\right)\left(\frac{1}{R_1} - \frac{1}{R_2}\right)$$

For lens in air

$$n_1 = 1, n_2 = 1.5, f = f_a$$

$$\frac{1}{f_a} = \left(\frac{1.5}{1} - 1\right)\left(\frac{1}{R_1} - \frac{1}{R_2}\right)$$

$$= \frac{1}{2}\left(\frac{1}{R_1} - \frac{1}{R_2}\right) \quad \ldots \text{(i)}$$

For lens in water

$$n_1 = 1.33 = \frac{4}{3}, n_2 = 1.5 = \frac{3}{2}, f = f_w$$

$$\frac{1}{f_w} = \left(\frac{3}{2 \times \frac{4}{3}} - 1\right)\left(\frac{1}{R_1} - \frac{1}{R_2}\right)$$

$$= \frac{1}{8}\left(\frac{1}{R_1} - \frac{1}{R_2}\right) \qquad \text{...(ii)}$$

Divide Eq. (i) by Eq. (ii)

$$\frac{\frac{1}{f_a}}{\frac{1}{f_w}} = \frac{\frac{1}{2}\left(\frac{1}{R_1} - \frac{1}{R_2}\right)}{\frac{1}{8}\left(\frac{1}{R_1} - \frac{1}{R_2}\right)}$$

$$\Rightarrow \quad f_w = 4f_a$$

$$= 4 \times 24 = 96 \text{ cm}$$

18. *(d)* As u increases, v decreases, while u is –ve and v is +ve.

Further, relationship between u and v is not a straight line as seen from the following equation.

$$\frac{1}{v} - \frac{1}{u} = \frac{1}{f}$$

Thus

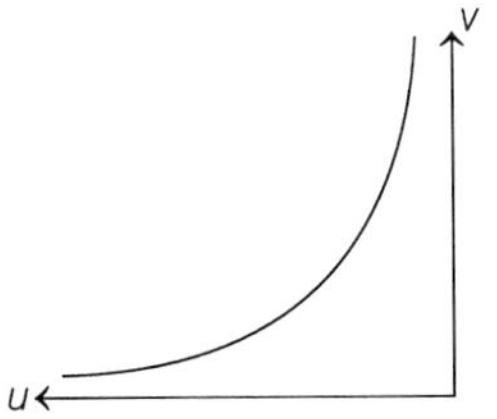

19. *(b)*

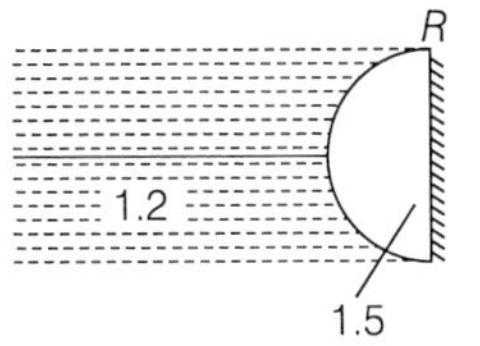

$\mu_2 = 1.5, \mu_1 = 1.2, R_1 = R, R_2 \to \infty$

Using thin lens formula

$$\frac{1}{f_{\text{lens}}} = \left(\frac{\mu_2}{\mu_1} - 1\right)\left(\frac{1}{R_2} - \frac{1}{R_2}\right)$$

$$\Rightarrow \quad \frac{1}{f_{\text{lens}}} = \left(\frac{1.5}{1.2} - 1\right)\left(\frac{1}{R} - \frac{1}{\infty}\right)$$

$$\Rightarrow \quad \frac{1}{f_{\text{lens}}} = \frac{1}{4R} \qquad \text{...(i)}$$

Now, lens is silvered, thus its focal length is

$$\frac{1}{f_{\text{lens}}} = \frac{2}{R} = \frac{2}{\infty} = 0$$

By combination of less

$$\frac{1}{f_{\text{eq}}} = \frac{1}{f_{\text{lens}}} - \frac{1}{f'} + \frac{1}{f_{\text{lens}}}$$

$$\frac{1}{f_{\text{eq}}} = \frac{2}{f_{\text{lens}}} = \frac{2}{4R}$$

$$f_{\text{eq}} = 2R \Rightarrow R = 0.1$$

20. *(b)*

$$\frac{1}{f_1} = (\mu - 1)\frac{2}{R} \Rightarrow P = 4\text{D}$$

$$\Rightarrow$$

$$\frac{1}{f_2} = (\mu - 1)\frac{1}{R} = \frac{P}{2}$$

$$= \frac{4\text{D}}{2} = 2\text{D}$$

$$P' = 2\text{D}$$

21. *(b)* By lens Maker's formula,

$$\frac{1}{f} = (\mu - 1)\left(\frac{1}{R_1} - \frac{1}{R_2}\right)$$

$$\frac{1}{20} = (\mu - 1)\left[\frac{1}{15} - \frac{1}{(-30)}\right]$$

$$\frac{1}{20} = (\mu - 1)\left(\frac{3}{30}\right)$$

$$\Rightarrow \quad \mu - 1 = \frac{1}{2} \Rightarrow \mu = \frac{3}{2} = 1.5$$

22. *(d)* We know that,

$P_{\text{eq}} = \Sigma P_i$ (equivalent power of combination of lens)

All lenses are identical.

$\therefore 5P = 25\text{ D} \Rightarrow P = 5\text{ D}$

Power of 1 lens = 5 D

Using relation, $P = \frac{1}{f}$

$$f = \frac{1}{P} = \frac{1}{5}\text{ m} = 20 \text{ cm}$$

Hence, focal length is 20 cm.

23. *(c)* Given, $\mu_a = 1.5$

$$\mu_m = 1.6$$

$$f_a = 20 \text{ cm}$$

Ratio of focal lengths in air and in medium using lens Maker's formula is,

As $\quad \dfrac{f_m}{f_a} = \dfrac{(\mu_a - 1)\mu_m}{(\mu_a - \mu_m)}$

$$\frac{f_m}{20} = \frac{(1.5 - 1)1.6}{(1.5 - 1.6)}$$

$$\frac{f_m}{20} = -\frac{0.5 \times 1.6}{0.1}$$

$$f_m = -\frac{20 \times 0.5 \times 1.6}{0.1}$$

$$f_m = -\frac{32 \times 5}{1} = -160 \text{ cm}$$

24. *(b)* Given, power of lens used for distant ($u = \infty$) vision is,

$P_1 = -1.0\text{ D} \Rightarrow$ Focal length of lens used $f_1 = -100$ cm

$$\left(\because f\text{ (cm)} = \frac{100}{P\text{(in D)}}\right)$$

So, image of distant object must be formed with in the far point (point upto which person can see).

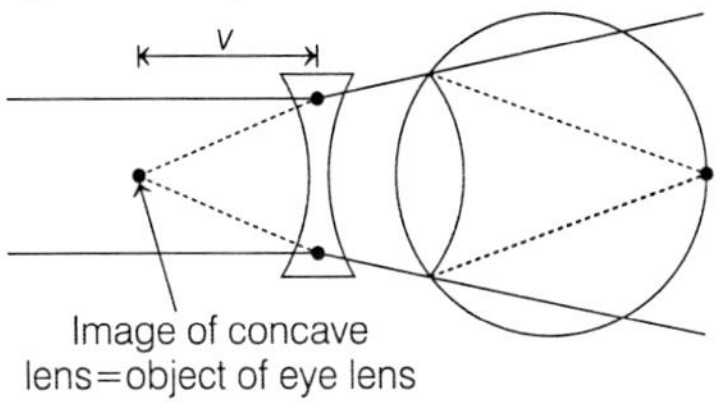

For concave lens,

$$\frac{1}{v} - \frac{1}{u} = \frac{1}{f_1}$$

$$u = \infty, v = ?, f_1 = -100 \text{ cm}$$

$$\Rightarrow \quad \frac{1}{v} = -\frac{1}{100}$$

or $\quad v = -100$ cm.

Now, person is using a lens of power + 2D for reading.

Focal length of convex lens used

$$= f_2 = \frac{100}{P\text{(in D)}} = +50 \text{ cm}$$

Now, if distance of distinct vision is x, then image of x must be formed at 100 cm from eye.

$$u = -x, v = -100 \text{ cm}$$

$$f = 50 \text{ cm}$$

$$\frac{1}{-100} - \frac{1}{-x} = \frac{1}{50}$$

$$\Rightarrow \quad \frac{1}{x} = \frac{1}{50} + \frac{1}{100} = \frac{3}{100}$$

$$x = \frac{100}{3} = 33.33 \text{ cm.}$$

Nearest answer is 30 cm.

25. *(b)* Given, power $P = 1.25\text{ m}^{-1}$

Refractive index of lens, $\mu_l = 1.5$

Radii of curvature of lens are

$R_1 = 20$ cm, $R_2 = -40$ cm

Using Lens Makers formula,

$$\frac{1}{f} = \left(\frac{\mu_l}{\mu_s} - 1\right)\left(\frac{1}{R_1} - \frac{1}{R_2}\right)$$

where, μ_s = Refractive index of surrounding medium

$$\Rightarrow \frac{1.25}{100} = \left(\frac{1.5}{\mu_s} - 1\right)\left(\frac{1}{20} + \frac{1}{40}\right)\left(\because f = \frac{1}{P}\right)$$

$$\Rightarrow \frac{1}{80} = \left(\frac{1.5}{\mu_s} - 1\right)\frac{6}{80}$$

$$\Rightarrow \frac{1.5}{\mu_s} - 1 = \frac{1}{6} \text{ or } \mu_s = \frac{6}{7} \times 1.5 = \frac{9}{7}$$

26. *(b)* Consider a convex lens of refractive index μ_2 and μ_1 is the refractive index of the medium in which it is placed.

i.e., $\mu_2 = \mu_1$ ... (i)

According to the Lens Maker's formula,

$$\frac{1}{f} = \left[\frac{\mu_1}{\mu_2} - 1\right]\left[\frac{1}{R_1} - \frac{1}{R_2}\right] \quad \text{...(ii)}$$

$$\frac{1}{f} = [1-1]\left[\frac{1}{R_1} - \frac{1}{R_2}\right] = 0$$

$$\Rightarrow \frac{1}{f} = 0$$

$$\Rightarrow f = \text{Infinite}$$

27. *(b)* Using lens formula,

$$\frac{1}{f} = \frac{1}{v} - \frac{1}{u} \Rightarrow \frac{1}{v} = \frac{1}{f} + \frac{1}{u}$$

$$\Rightarrow v = \frac{uf}{u+f}$$

Now, considering two cases:

Case I If $v = u$,

$$\Rightarrow u = \frac{uf}{u+f} \Rightarrow u + f = f$$

$$\Rightarrow u = 0$$

So, the curve will pass through origin.

Case II If $u = \infty$

$$\Rightarrow v = f$$

So, as u tends towards infinity, v tends towards focus and all other distances will be less than focal length f and it will never intersect $u = v$ line.

28. *(c)* Given, $f_1 = 2f_2$

$$\Rightarrow \frac{1}{|f_1|} = \frac{1}{|2f_2|} \quad \text{...(i)}$$

Using lens Maker's formula, we get

$$\frac{1}{f_1} = (\mu_1 - 1)\left(\frac{1}{R_1} - \frac{1}{R_2}\right)$$

$$\frac{1}{f_2} = (\mu_2 - 1)\left(\frac{1}{R_1'} - \frac{1}{R_2'}\right)$$

$$\Rightarrow \left|(\mu_1 - 1)\left(\frac{1}{\infty} - \frac{1}{-R}\right)\right|$$

$$= \left|\frac{1}{2}(\mu_2 - 1)\left(\frac{1}{-R} - \frac{1}{\infty}\right)\right|$$

[$\because$ Using Eq. (i)]

$$\Rightarrow \frac{\mu_1 - 1}{R} = \frac{\mu_2 - 1}{2R}$$

$$\Rightarrow 2\mu_1 - \mu_2 = 1$$

29. *(d)* Since, image formed by a convex lens can be real or virtual in nature.

Thus, $m = +2$ or -2.

First let us take image to be real in nature, then $m = -2 = \frac{v}{u}$, where v is image distance and u is object distance.

$$\Rightarrow v = -2x_1 \quad [\text{Taking } u = x_1]$$

Now, by using lens equation,

$$\frac{1}{v} - \frac{1}{u} = \frac{1}{f}$$

$$\Rightarrow \frac{1}{-2x_1} - \frac{1}{x_1} = \frac{1}{20} \Rightarrow \frac{-3}{2x_1} = \frac{1}{20}$$

$$\Rightarrow x_1 = -30 \text{ cm}$$

Now, let us take image to be virtual in nature, then

$$m = 2 = v/u$$

$$\Rightarrow v = 2x_2 \quad [\text{Taking, } u = x_2]$$

Again, by using lens equation,

$$\frac{1}{v} - \frac{1}{u} = \frac{1}{f}$$

$$\Rightarrow \frac{1}{2x_2} - \frac{1}{x_2} = \frac{1}{20}$$

$$\Rightarrow \frac{-1}{2x_2} = \frac{1}{20}$$

$$\Rightarrow x_2 = -10 \text{ cm}$$

So, the ratio of x_1 and x_2 is

$$\frac{x_1}{x_2} = \frac{-30}{-10} = 3:1$$

Alternate Solution

Magnification for a lens can also be written as

$$m = \left(\frac{f}{f+u}\right)$$

When $m = -2$ (for real image)

$$-2 = \frac{f}{f + x_1}$$

$$\Rightarrow -2f + (-2)x_1 = f \text{ or } x_1 = -\frac{3f}{2}$$

Similarly, when $m = +2$ (for virtual image)

$$+2 = \frac{f}{f + x_2}$$

$$\Rightarrow 2f + 2x_2 = f \text{ or } x_2 = \frac{-f}{2}$$

Now, the ratio of x_1 and x_2 is

$$\frac{x_1}{x_2} = \frac{\frac{-3f}{2}}{-f/2} = \frac{3}{1}$$

30. *(c)* The given situation can be drawn as shown below

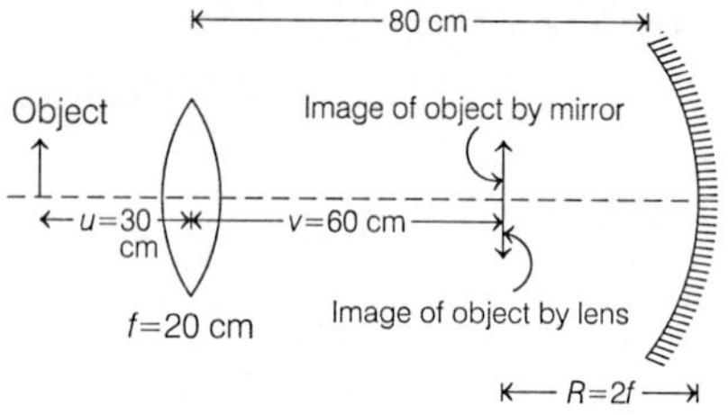

For lens formula, $\frac{1}{v} - \frac{1}{u} = \frac{1}{f}$

On substituting given values, we get

$$\frac{1}{v} - \frac{1}{-30} = \frac{1}{20}$$

$$\Rightarrow v = 60 \text{ cm}$$

So, this image is at a distance of $80 - 60 = 20$ cm from the mirror.

Image position on removing mirror does not change means, image formed by the mirror coincides with image formed by the lens.

This condition is only possible if any object that has been placed in front of concave mirror is at centre of curvature i.e., at $2f$. So, radius of curvature of mirror is $R = 20$ cm

$\therefore$ Focal length of mirror,

$$f = \frac{R}{2} = 10 \text{ cm}$$

As, for virtual image, the object is to be kept between pole and focus of the mirror.

$\therefore$ The maximum distance of the object for which this concave mirror by itself produces a virtual image would be 10 cm.

31. *(b)* Given combination is as shown below

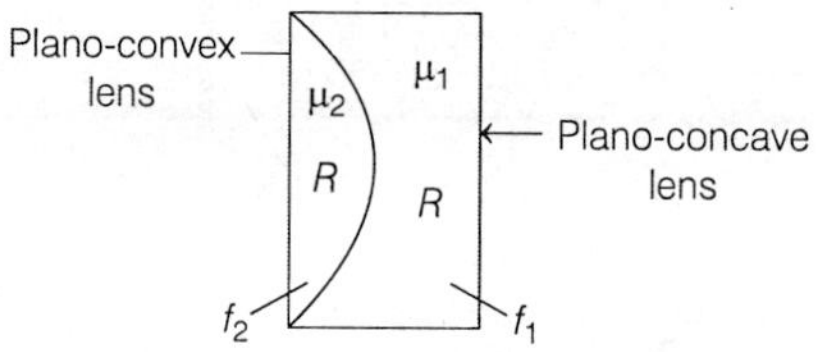

As lenses are in contact, equivalent focal length of combination is

$$\frac{1}{f_{eq}} = \frac{1}{f_1} + \frac{1}{f_2}$$

Using lens Maker's formula,

$$\frac{1}{f} = (\mu - 1)\left(\frac{1}{R_1} - \frac{1}{R_2}\right)$$

Here,

$$\frac{1}{f_1} = (\mu_1 - 1)\left(\frac{1}{-R} - \frac{1}{\infty}\right) = \frac{(1-\mu_1)}{R}$$

and $\dfrac{1}{f_2} = (\mu_2 - 1)\left(\dfrac{1}{\infty} - \dfrac{1}{(-R)}\right)$

$= \dfrac{(\mu_2 - 1)}{R}$

$\therefore \quad \dfrac{1}{f_{eq}} = \left(\dfrac{\mu_2 - 1}{R}\right) + \left(\dfrac{1 - \mu_1}{R}\right)$

$= \dfrac{\mu_2 - 1 + 1 - \mu_1}{R} = \dfrac{\mu_2 - \mu_1}{R}$

So, $f_{eq} = \dfrac{R}{\mu_2 - \mu_1}$.

32. *(c)* When given set up is immersed in water, focal length of lens increases,

Note: $\dfrac{f_{liquid}}{f_{air}} = \dfrac{n_{ga} - 1}{n_{gl} - 1} = \dfrac{n_{ga} - 1}{\left(\dfrac{n_{ga}}{n_{la}} - 1\right)}$

Now, $n_{ga} = \dfrac{3}{2}$ and $n_{la} = \dfrac{4}{3}$

$\therefore f_{liquid} = f_{air}\left(\dfrac{\frac{1}{2}}{\frac{1}{8}}\right) = 4 f_{air}$

As focal length increases there is no focussing of image on screen

$\therefore$ image will disappear.

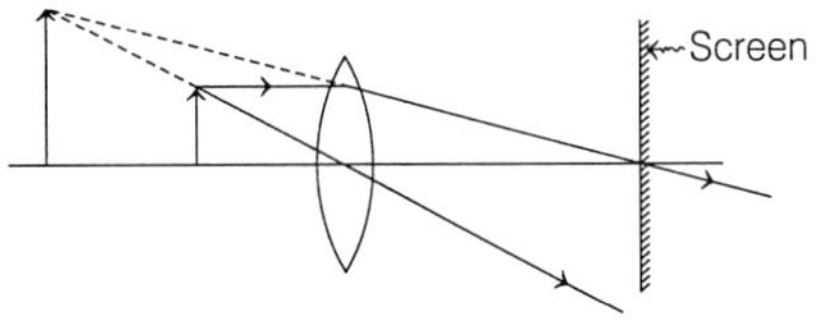

Actually a virtual image is formed on same side of object.

33. *(b)* For lens A, object distance,

$u = -20$ cm

Focal length, $f = +5$ cm

From lens formula, $\dfrac{1}{v} - \dfrac{1}{u} = \dfrac{1}{f}$

We have, $\dfrac{1}{v} = \dfrac{1}{5} - \dfrac{1}{20}$

$\dfrac{1}{v} = \dfrac{20 - 5}{20 \times 5} = \dfrac{15}{100}$,

$v = \dfrac{20}{3}$ cm

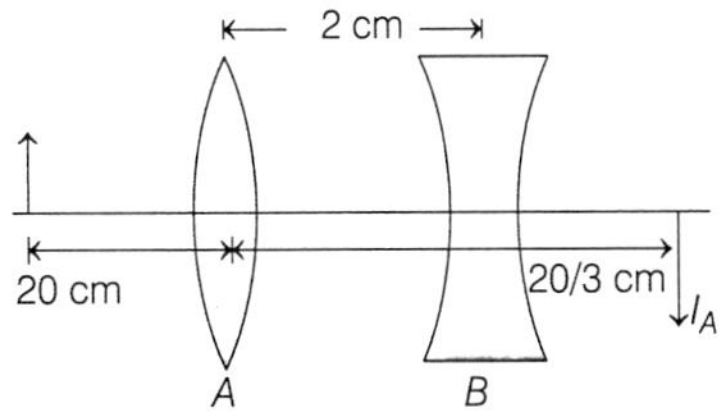

For lens B, image of A is object for B.

$\therefore \quad u = \dfrac{20}{3} - 2 = +\dfrac{14}{3}$ cm

$f = -5$ cm

Now, from lens formula, we have

$\dfrac{1}{v} - \dfrac{1}{u} = \dfrac{1}{f}$

$\Rightarrow \quad \dfrac{1}{v} = \dfrac{1}{-5} + \dfrac{3}{14}$

$\dfrac{1}{v} = \dfrac{15 - 14}{5 \times 14}$

$\Rightarrow \quad v = 70$ cm

Hence, image is on right of lens B and is real in nature.

34. *(d)* Lens formula is given as

$\dfrac{1}{f} = \dfrac{1}{v} - \dfrac{1}{u}$ ... (i)

$\Rightarrow \quad \dfrac{1}{v} = \dfrac{1}{f} + \dfrac{1}{u} \Rightarrow \dfrac{uf}{u + f} = v$

$\Rightarrow \quad \dfrac{v}{u} = \dfrac{f}{u + f}$ ...(ii)

Now, by differentiating Eq. (i), we get

$0 = -\dfrac{1}{v^2} \cdot \dfrac{dv}{dt} + \dfrac{1}{u^2} \cdot \dfrac{du}{dt}$

[$\because f$ (focal length of a lens is constant)]

or $\dfrac{dv}{dt} = \dfrac{v^2}{u^2} du/dt$

$\Rightarrow \quad \dfrac{dv}{dt} = \left(\dfrac{f}{u + f}\right)^2 \dfrac{du}{dt}$ [Using Eq. (ii)]

Given, $f = 0.3$ m, $u = -20$ m,

$du/dt = 5$ m/s

$\therefore \quad \dfrac{dv}{dt} = \left(\dfrac{0.3}{0.3 - 20}\right)^2 \times 5 = \left(\dfrac{3}{197}\right)^2 \times 5$

$= 1.16 \times 10^{-3}$ m/s

35. *(d)* Focal length of diverging lens is 25 cm.

As the rays are coming parallel, so the image (I_1) will be formed at the focus of diverging lens i.e. at 25 cm towards left of diverging lens.

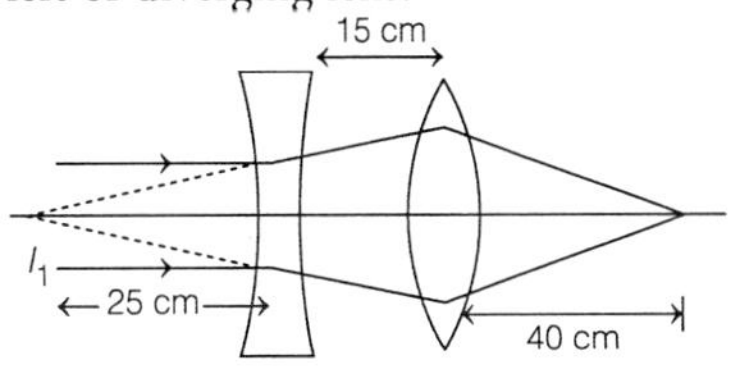

The image (I_1) will work as object for converging lens.

For converging lens, distance of object u (i.e. distance of I_1) $= -(25 + 15)$

$= -40$ cm

$f = 20$ cm

$\therefore$ From len's formula $\dfrac{1}{f} = \dfrac{1}{v} - \dfrac{1}{u}$

$\dfrac{1}{20} = \dfrac{1}{v} - \dfrac{1}{-40} \Rightarrow \dfrac{1}{v} = \dfrac{1}{20} - \dfrac{1}{40}$

$\dfrac{1}{v} = \dfrac{1}{40} \Rightarrow v = 40$ cm

v is positive so image will be real and will form at right side of converging lens at 40 cm.

36. *(c)*

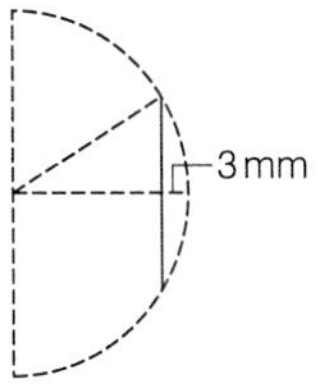

Refractive endex, $\eta = \dfrac{3 \times 10^8}{2 \times 10^8}$

$\therefore \quad n = 3/2 \quad [\because n = c/v]$

$3^2 + (R - 3 \text{ mm})^2 = R^2$

$\Rightarrow 3^2 + R^2 - 2R(3 \text{ mm}) + (3 \text{ mm})^2 = R^2$

$\Rightarrow \quad R \approx 15$ cm

$\therefore \quad \dfrac{1}{f} = \left(\dfrac{3}{2} - 1\right)\left(\dfrac{1}{15}\right)$

$\left[\because \dfrac{1}{f} = (n - 1)\left(\dfrac{1}{R_1} - \dfrac{1}{R_2}\right) \text{ and } R_1 = \infty, R_2 = -15\right]$

$\Rightarrow \quad f = 30$ cm

37. *(c)* For convex lens, $f_1 = 30$ cm

For concave lens, $f_2 = -20$ cm

Combined focal length

$\dfrac{1}{f} = \dfrac{1}{f_1} + \dfrac{1}{f_2}$

$= \dfrac{1}{30} - \dfrac{1}{20}$

$= \dfrac{2 - 3}{60}$

$= -\dfrac{1}{60}$

$\therefore \quad f = -60$ cm

When object is placed at a distance of 20 cm, i.e. $u = -20$ cm.

Using lens formulae,

$\dfrac{1}{v} - \dfrac{1}{u} = \dfrac{1}{f}$

$\Rightarrow \dfrac{1}{v} - \dfrac{1}{-20} = \dfrac{1}{-60}$

$$\Rightarrow \quad \frac{1}{v} = -\frac{1}{20} - \frac{1}{60}$$

$$\Rightarrow \quad v = -\frac{60}{4}$$

$$v = -15 \text{ cm}$$

$\therefore$ Distance of image from lens is 15 cm.

38. *(b)*

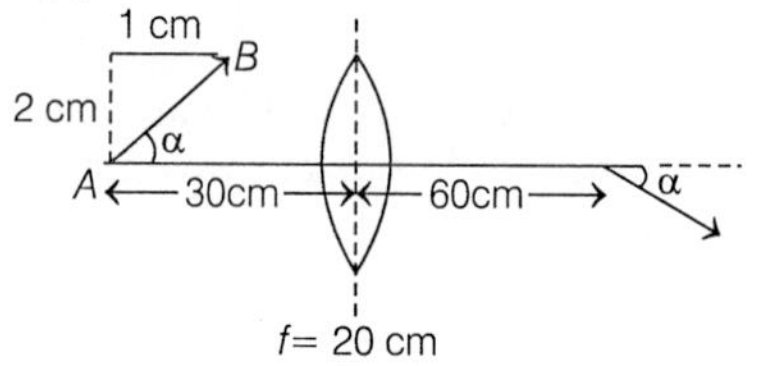

Location of image of A

$$\frac{1}{v} - \frac{1}{u} = \frac{1}{f}$$

$$\frac{1}{v} - \frac{1}{-30} = \frac{1}{20}$$

$$\frac{1}{v} = \frac{1}{20} - \frac{1}{30} = \frac{30-20}{30 \times 20}$$

$$\frac{1}{v} = \frac{10}{600} = \frac{1}{60}$$

$$v = 60 \text{ cm}$$

$$\therefore \quad m = 2$$

Since size of object is small w.r.t to the location hence,

$$dv = m^2 du$$

$$dv = 4 \times 1 = 4 \text{ cm}$$

$$h_I = mh_O$$

$$= h_i(d_y)$$

$$= 2 \times 2 = 4 \text{ cm}$$

$\therefore$ Angle made with principle axis $= -45°$

39. *(a)* There are three lens in the given diagram,

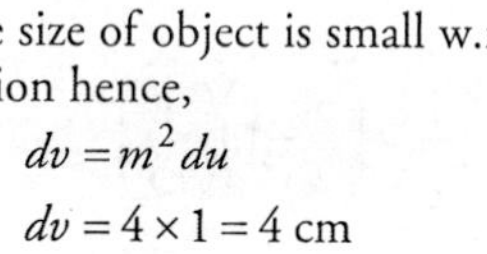

Where R is radius of curvature and P is power of lens

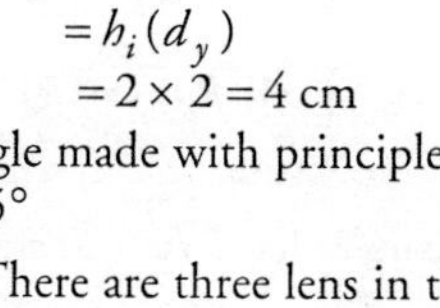

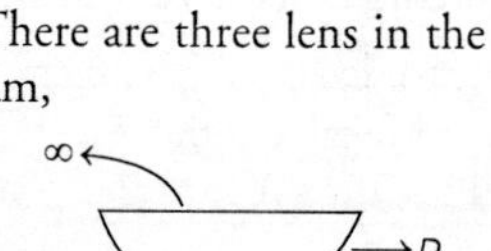

Total power is given by

$$P_{total} = P_1 + P_2 + P_3$$

$$\Rightarrow P_1 = \left[\frac{4}{3} - 1\right]\left[\frac{1}{\infty} - \frac{1}{|R_1|}\right]$$

$$\Rightarrow P_1 = \left[\frac{1}{3|R_1|}\right]$$

$$P_2 = \frac{1}{2}\left[\frac{1}{-|R_1|} - \frac{1}{-|R_2|}\right]$$

$$P_3 = \frac{1}{3}\left[\frac{1}{-|R_2|} - \frac{1}{\infty}\right] \Rightarrow \frac{-1}{3|R_2|}$$

$$P_{total} = P_1 + P_2 + P_3 = \frac{-1}{6}\left[\frac{1}{|R_1|} - \frac{1}{|R_2|}\right]$$

40. *(a)*

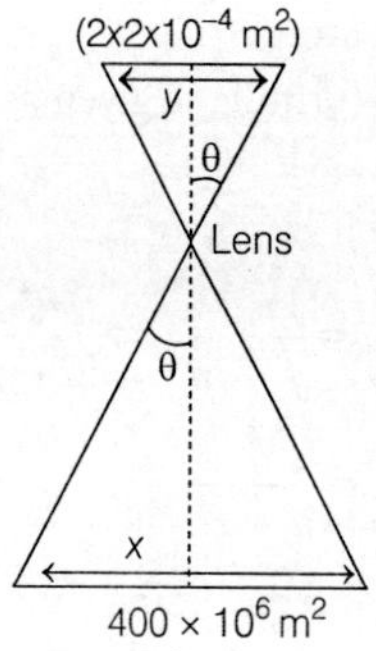

$$H = 18 \text{ km}$$

Size of camera film $= 2 \text{ cm} \times 2 \text{ cm}$

$$A_{image} = 400 \text{ km}^2$$

$$x = 20 \times 10^3 \text{ m}$$

$$x = 2 \times 10^4 \text{ m}$$

$$\frac{2 \times 10^4}{2 \times 10^{-2}} = \frac{x}{y} = 10^6 = \frac{18 \text{ km}}{f}$$

$$f = \frac{18 \text{ km}}{10^6}$$

$$f = \frac{18 \times 10^3 \text{ m}}{10^6} = 18 \times 10^{-3}$$

$$f = 18 \text{ mm} = 1.8 \text{ cm}$$

41. *(c)*

> **key idea** On silvering surface of a lens it behaves like a mixture. As light ray passes twice through lens in process of image formation from a silvered lens, we have
>
> $$\frac{1}{F} = \frac{2}{f_l} + \frac{1}{f_m}$$
>
> 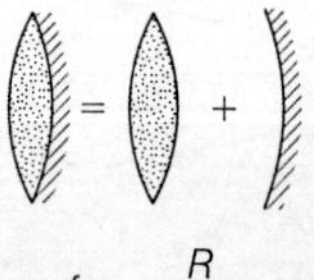
>
>
> Here, $f_l = \dfrac{R}{2(\mu - 1)}$ and $f_m = \dfrac{R}{2}$

When object's image is lies on object itself that means, the combination of two lens acts as mirror.

$$P_{eq} = 2P_l + P_m$$

$$\frac{-1}{f_{eq}} = \frac{2}{f_l} - \frac{1}{f_m}$$

$$= \frac{4(\mu - 1)}{R} - \frac{2}{-R}$$

$$= \frac{1}{R}[4\mu - 4 + 2]$$

$$\frac{1}{f_{eq}} = \frac{-1}{R}(4\mu - 2) \quad \ldots \text{(i)}$$

$$\Rightarrow \quad f_{eq} = \frac{R}{2}$$

$$R = 2f_{eq} = \frac{-2R}{4\mu - 2}$$

$$\text{or} \quad \frac{-R}{(2\mu - 1)}$$

42. *(a)*

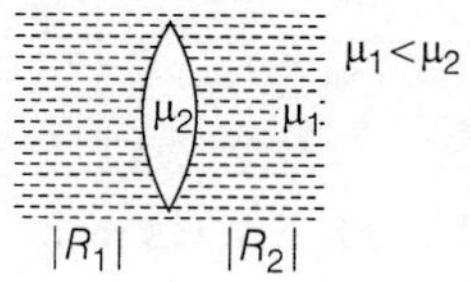

$$\frac{1}{f_{eq}} = \frac{2}{f_L} - \frac{1}{f_m}$$

$$f_m = -\frac{|R|}{2}$$

$$\frac{1}{f_L} = \left(\frac{\mu_2}{\mu_1} - 1\right)\left(\frac{1}{R_1} + \frac{1}{R_2}\right)$$

$$\frac{1}{f_{eq}} = 2\left(\frac{\mu_2 - \mu_1}{\mu_1}\right)\left(\frac{R_1 + R_2}{R_1 R_2}\right) + \frac{2}{R_2}$$

$$= \frac{2}{R_2}\left[\frac{(\mu_2 - \mu_1)(R_1 + R_2) + \mu_1 R_1}{\mu_1 R_1}\right]$$

$$= \frac{2}{R_2}\left[\frac{\mu_2 R_1 + \mu_2 R_2 - \mu_1 R_1 - \mu_1 R_2 + \mu_1 R_1}{\mu_1 R_1}\right]$$

$$\frac{1}{f_{eq}} = \frac{2[\mu_2 R_1 + \mu_2 R_2 - \mu_1 R_2]}{\mu_1 R_1 R_2}$$

For same size of image, $u = 2f$

$$u = \frac{\mu_1 R_1 R_2}{\mu_2 R_1 + \mu_2 R_2 - \mu_1 R_2}$$

$$= \frac{\mu_1 |R_1||R_2|}{\mu_2 (|R_1| + |R_2|) - \mu_1 |R_2|}$$

Option (a) is correct.

43. *(b)*

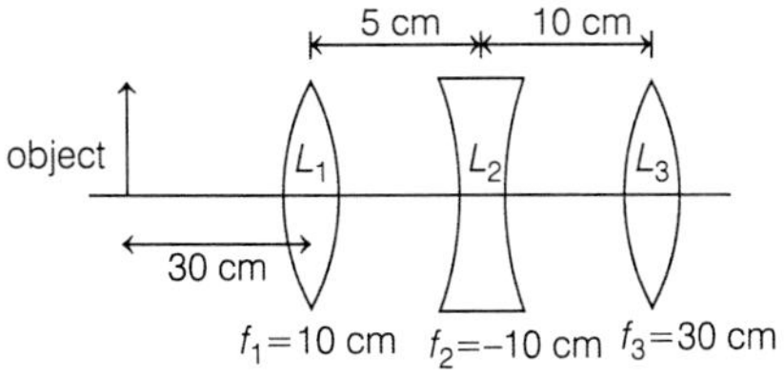

For lens L_1,

$$u = -30 \text{ cm}, f = 10 \text{ cm}$$

Using, $\frac{1}{v} - \frac{1}{u} = \frac{1}{f}$, we have,

$$\frac{1}{v} - \frac{1}{(-30)} = \frac{1}{10}$$

$$\Rightarrow \quad \frac{1}{v} = \frac{1}{10} - \frac{1}{30} = \frac{2}{30} = \frac{1}{15}$$

$$\Rightarrow \quad v = 15 \text{ cm}$$

This image acts like object for L_2.

$\Rightarrow$ For L_2, $u = +10$ cm,

$$f = -10 \text{ cm}$$

$$\frac{1}{v} - \frac{1}{u} = \frac{1}{f} \quad \Leftrightarrow \quad \frac{1}{v} - \frac{1}{10} = \frac{1}{-10}$$

$$\Rightarrow \quad \frac{1}{v} = 0 \text{ or } v = \infty$$

For L_2, rays from L_1, are directed at its second focal point

∴ rays becomes parallel after refraction.

Hence, for L_3,

$u = \infty$ (parallel rays falls on L_3)

$\Rightarrow$ Images is formed at focus of L_3, at 30 cm on right of L_3.

44. *(a)* Given, focal length of first lens,

$f_1 = +10$ cm

Focal length of second lens,

$f_2 = -10$ cm

Focal length of third lens, $f_3 = +30$ cm

Object distance from first lens,

$u_1 = -30$ cm

Using lens formula for first lens, we get

$$\frac{1}{f_1} = \frac{1}{v_1} - \frac{1}{u_1}$$

$$\Rightarrow \quad \frac{1}{-10} = \frac{1}{v_1} - \frac{1}{-30}$$

$$\Rightarrow \quad \frac{1}{v_1} = \frac{1}{10} - \frac{1}{30} = \frac{2}{30}$$

$$\Rightarrow \quad v_1 = 15 \text{ cm}$$

Thus, the image formed by first lens is 15 cm to the right of first lens.

The object distance for second lens will be $u_2 = 15 \text{ cm} - 5 \text{ cm} = 10$ cm

Using lens formula for first lens, we get

$$\frac{1}{f_2} = \frac{1}{v_2} - \frac{1}{u_2} \Rightarrow \frac{1}{-10} = \frac{1}{v_2} - \frac{1}{10}$$

$$\Rightarrow \quad v_2 = \infty$$

Thus, the object distance for third lens, $u_3 = \infty$.

Using lens formula for third lens, we get

$$\frac{1}{f_3} = \frac{1}{v_3} - \frac{1}{u_3}$$

$$\Rightarrow \quad \frac{1}{30} = \frac{1}{v_3} - \frac{1}{\infty} \Rightarrow \frac{1}{v_3} = \frac{1}{30}$$

$$v_3 = 30 \text{ cm}$$

The final image will form 30 cm from the third lens.

The distance of final image from object will be $30 + 10 + 5 + 30 = 75$ cm.

Thus, the distance of final image formed from object O is 75 cm.

45. *(b)* According to given diagram, Refractive indices are μ_1 and μ_2. Radius of two lenses are

$R_1 = R_2 = R$

By using lens Maker's formula,

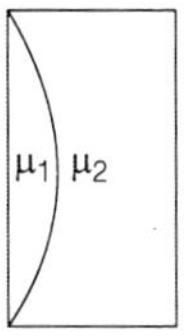

$$\frac{1}{f} = (\mu - 1)\left(\frac{1}{R_1} - \frac{1}{R_2}\right)$$

Where, f = focal length.

Now, for plano-convex lens

$R_1 = \infty$

and $R_2 = -R$

$$\frac{1}{f_1} = (\mu_1 - 1)\left(\frac{1}{\infty} - \frac{1}{-R}\right)$$

$$\Rightarrow \quad \frac{1}{f_1} = (\mu_1 - 1)\left(\frac{1}{R}\right) \quad \ldots\text{(i)}$$

For plano-concave lens, $R_1 = -R$ and $R_2 = \infty$

$$\frac{1}{f_2} = (\mu_2 - 1)\left(\frac{1}{-R} - \frac{1}{\infty}\right)$$

$$\frac{1}{f_2} = (\mu_2 - 1)\left(-\frac{1}{R}\right) \quad \ldots\text{(ii)}$$

Therefore, combined focal length

$$\frac{1}{f} = \frac{1}{f_1} + \frac{1}{f_2}$$

$$= (\mu_1 - 1)\left(\frac{1}{R}\right) + (\mu_2 - 1)\left(-\frac{1}{R}\right)$$

$$= \frac{\mu_1}{R} - \frac{1}{R} - \frac{\mu_2}{R} + \frac{1}{R}$$

$$\frac{1}{f} = \frac{\mu_1 - \mu_2}{R}$$

$$\therefore \quad \frac{R}{f} = \mu_1 - \mu_2$$

> **• Concept Enhancer**
>
> If two lenses of focal lengths f_1 and f_2 are kept in contact, then their combined focal length is given as, $\frac{1}{f} = \frac{1}{f_1} + \frac{1}{f_2}$

46. *(c)* For a glass-lens in air, focal length,

$$\frac{1}{f_{\text{air}}} = \frac{1}{f} = \left(\frac{n_g}{n_a} - 1\right)\left(\frac{1}{R_1} - \frac{1}{R_2}\right) \quad \ldots\text{(i)}$$

When, lens is dipped in a liquid, its focal length,

$$\frac{1}{f_{\text{liquid}}} = \frac{1}{f_1} = \left(\frac{n_g}{n_l} - 1\right)\left(\frac{1}{R_1} - \frac{1}{R_2}\right) \quad \ldots\text{(ii)}$$

From Eqs. (i) and (ii), we get

$$\frac{f_1}{f} = \frac{\left(\frac{n_g}{n_a} - 1\right)}{\left(\frac{n_g}{n_l} - 1\right)}$$

As $f = 16$ cm, $n_g = 1.5$,

$n_l = 1.42$, we have

$$\frac{f_1}{f} = \frac{\left(\frac{1.5}{1} - 1\right)}{\left(\frac{1.5}{1.42} - 1\right)}$$

$$\Rightarrow \quad \frac{f_1}{f} = \frac{\left(\frac{3}{2} - 1\right)}{\left(\frac{150}{142} - 1\right)} = \frac{\left(\frac{1}{2}\right)}{\left(\frac{4}{71}\right)} = \frac{71}{8} \simeq 9$$

47. *(a)* In given system of lens and mirror, position of object O in front of lens is at a distance $2f$.

i.e. $u = 2f = 40$ cm

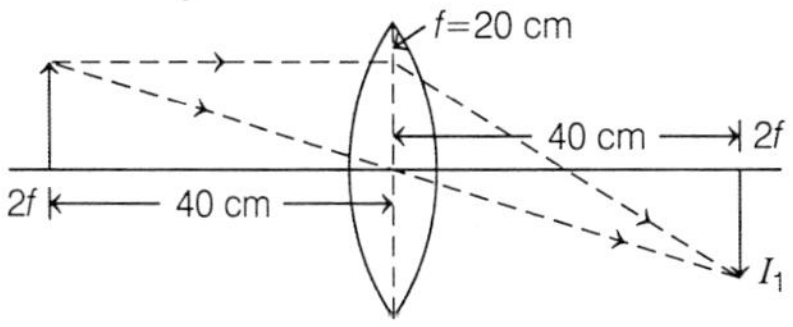

So, image (I_1) formed is real, inverted and at a distance,

$v = 2f = 2 \times 20 = 40$ cm (behind lens)

magnification, $m_1 = \frac{v}{u} = \frac{40}{40} = 1$

Thus, size of image is same as that of object.

This image (I_1) acts like a real object for mirror.

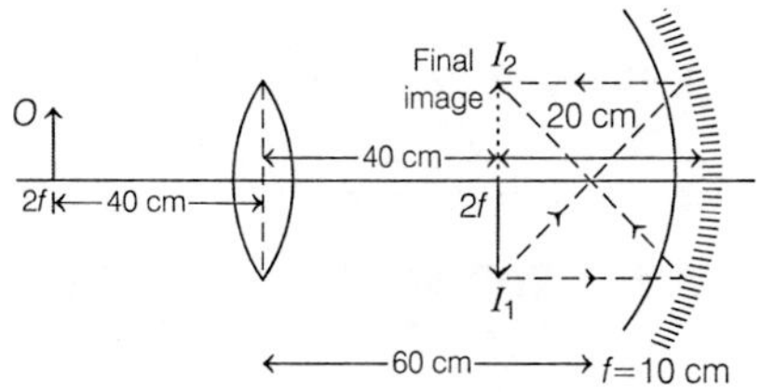

As object distance for mirror is

$$u = C = 2f = -20 \text{ cm}$$

Where, C = centre of curvature.

So, image (I_2) formed by mirror is at $2f$.

$\therefore$ For mirror $v = 2f = 2(-10)$
$= -20$ cm

Magnification,

$$m_2 = -\frac{v}{u} = -\frac{(-20)}{(-20)} = -1$$

Thus, image size is same as that of object.

Image obtained after refraction from lenes and reflection from mirror matches with option (a).

48. *(b)* According to lens maker's formula, when the lens in the air

$$\frac{1}{f} = \left(\frac{3}{2} - 1\right)\left(\frac{1}{R_1} - \frac{1}{R_2}\right)$$

$$\Rightarrow \quad \frac{1}{f} = \frac{1}{2x} \Rightarrow f = 2x$$

$$\left(\text{Here, } \frac{1}{x} = \frac{1}{R_1} - \frac{1}{R_2}\right)$$

In case of liquid,where refractive indices are 4/3 and 5/3, we get

Focal lenght in first liquid,

$$\frac{1}{f_1} = \left(\frac{\mu_s}{\mu_{l_1}} - 1\right)\left(\frac{1}{R_1} - \frac{1}{R_2}\right)$$

$$\frac{1}{f_1} = \left(\frac{3/2}{4/3} - 1\right)\frac{1}{x} \Rightarrow f_1 \text{ is positive.}$$

Nature of lens does not change.

$$\frac{1}{f_1} = \frac{1}{8x} = \frac{1}{4(2x)} = \frac{1}{4f} \Rightarrow f_1 = 4f$$

Focal length in second liquid,

$$\frac{1}{f_2} = \left(\frac{\mu_s}{\mu_{l_2}} - 1\right)\left(\frac{1}{R_1} - \frac{1}{R_2}\right)$$

$$\frac{1}{f_2} = \left(\frac{3/2}{5/3} - 1\right)\left(\frac{1}{x}\right)$$

$\Rightarrow f_2$ is negative.

Nature of lens will change *i.e.*, convex behave as concave.

• Concept Enhancer

When a lens of refractive index μ_g is kept in a liquid of refractive index μ_l, then if

(i) $\mu_g > \mu_l$, then there is no change in nature of lens.

(ii) $\mu_g < \mu_l$, then nature of lens will be changed.

(iii) $\mu_g = \mu_l$, then lens will behave like a plane glass.

49. *(a)* Let the focal length of silvered lens be F.

$$\frac{1}{F} = \frac{1}{f} + \frac{1}{f} + \frac{1}{f_m} = \frac{2}{f} + \frac{1}{f_m}$$

where, f = focal length of lens before silvering,

f_m = focal length of spherical mirror.

$$\frac{1}{F} = \frac{2}{f} + \frac{2}{R} \qquad [\because R = 2f_m] \ldots\text{(i)}$$

$$\text{Now, } \frac{1}{f} = (\mu - 1)\left(\frac{1}{R_1} - \frac{1}{R_2}\right) \qquad \ldots\text{(ii)}$$

Here, $R_1 = \infty$, $R_2 = 30$ cm

$$\therefore \quad \frac{1}{f} = (1.5 - 1)\left(\frac{1}{\infty} - \frac{1}{30}\right)$$

$$\text{or} \quad \frac{1}{f} = -\frac{0.5}{30} = -\frac{1}{60} \text{ or } f = -60 \text{ cm}$$

Hence, from Eq. (i), we get

$$\frac{1}{F} = \frac{2}{60} + \frac{2}{30} = \frac{6}{60}$$

$$F = 10 \text{ cm}$$

Again given that

Size of object = Size of image

i.e., $\quad O = I$

$$\therefore \quad m = -\frac{v}{u} = \frac{I}{O} \Rightarrow \frac{v}{u} = -1$$

or $\quad v = -u$

Thus, from lens formula,

$$\frac{1}{F} = \frac{1}{v} - \frac{1}{u} \Rightarrow \frac{1}{10} = \frac{1}{-u} - \frac{1}{u}$$

$$\frac{1}{10} = -\frac{2}{u}$$

$$\therefore \quad u = -20 \text{ cm}$$

• Concept Enhancer

A plano-convex lens behaves as a concave mirror, if its one surface (curved) is silvered. The rays refracted from plane surface are reflected from curved surface and again refract from plane surface. Therefore, in this lens two refractions and one reflection occur.

50. *(15)* Given, magnification $m = 3$

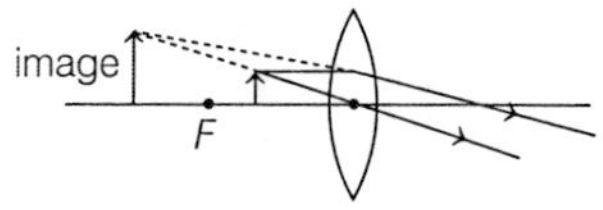

Magnification, $m = v/u$

$$v = 3u$$

$$v - u = 20 \text{ cm}$$

$$2u = 20 \text{ cm}$$

$$u = 10 \text{ cm}$$

$$\text{Now using, } \frac{1}{v} - \frac{1}{u} = \frac{1}{f}$$

$$\frac{1}{(-30)} - \frac{1}{(-10)} = \frac{1}{f}$$

$$\frac{1}{-30} + \frac{1}{10} = \frac{1}{f} \quad = \frac{1}{f} = \frac{2}{30} = \frac{1}{5}$$

$$f = 15 \text{ cm}$$

51. *(10)* Given,

Distance between object and image = 45 cm

Magnification = 2 (real image)

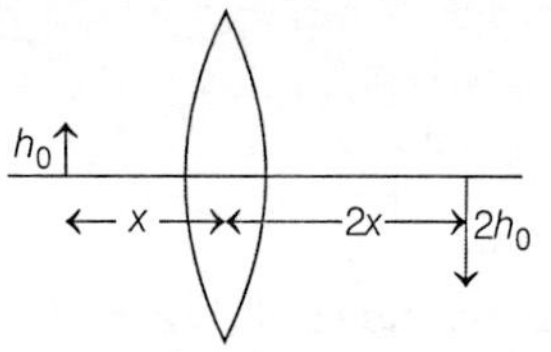

Let object height = h_0

$$\text{then, magnification} = \frac{\text{Image height}}{\text{Object height}}$$

Image height = $2h_0$

$$\text{Also, magnification} = \frac{\text{Image distance}}{\text{Object distance}}$$

Let object distance = x

Image distance = $2x$

$$x + 2x = 45 \text{ (given)}$$

$x = 15$ cm, (object distance) = u

Image distance, $2x = 30$ cm $= v$

Using lens formula,

$$\frac{1}{v} - \frac{1}{u} = \frac{1}{f}$$

$$\frac{1}{30} - \frac{1}{-15} = \frac{1}{f} \Rightarrow \frac{1+2}{30} = \frac{1}{f}$$

$$f = 10 \text{ cm}$$

52. *(5)* Given $f = 10$ cm $= 0.1$ m

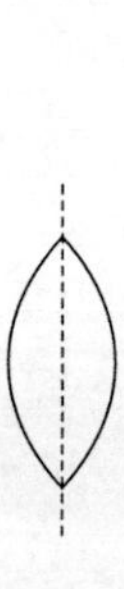

$$P_1 + P_2 = P = \frac{1}{f}$$

$$2P_1 = \frac{1}{0.1}$$

$$P_1 = \frac{1}{0.2} \Rightarrow P_1 = \frac{10}{2}$$

$$\Rightarrow \quad P_1 = 5\text{D}$$

53. *(100)* According to question, the situation described is given below

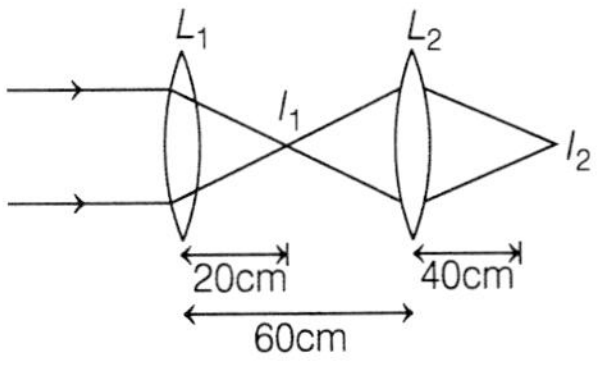

For L_1; $\frac{1}{v}-\frac{1}{u}=\frac{1}{f} \Rightarrow \frac{1}{v}-\frac{1}{\infty}=\frac{1}{f}$

$\therefore \quad v=f$

For L_2; $I_1 \to$ object and $I_2 \to$ image

$u=-(60-20)=-40$ cm

$\Rightarrow \quad u=-40$ cm and $f=20$ cm

$$\frac{1}{v}-\frac{1}{u}=\frac{1}{f} \Rightarrow \frac{1}{v}-\frac{1}{(-40)}=\frac{1}{20}$$

$$\Rightarrow \quad \frac{1}{v}=\frac{1}{20}-\frac{1}{40}$$

$$\Rightarrow \quad \frac{1}{v}=\frac{1}{40}$$

$\therefore \quad v=40$ cm

Hence, the distance from

$L_1=60+40=100$ cm

54. *(4)* $\frac{1}{f}=\left(\frac{\mu_l}{\mu_m}-1\right)\left(\frac{1}{R}-\frac{1}{-R}\right)$

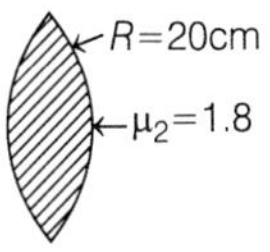

$$P_I=\frac{2}{R}\left[\frac{1\cdot8}{1}-1\right]$$

$$P_I=\frac{2}{R}\times0\cdot8$$

$$P_I=\frac{1\cdot6}{R} \quad \ldots(i)$$

Now, $P_2=\frac{2}{R}\left[\frac{1\cdot8}{1\cdot5}-1\right]$

$$P_2=\frac{2}{R}\left[\frac{0\cdot3}{1\cdot5}\right]$$

$$=\frac{2}{R}\times\frac{3}{15}=\frac{2}{5R}$$

$$\frac{P_{air}}{P_{liquid}}=\frac{P_1}{P_2}=\frac{\left(\frac{1\cdot6}{R}\right)}{\left(\frac{0\cdot4}{R}\right)}=\frac{4}{1}=\frac{x}{1}$$

55. *(34)* Given focal length of $L_1=24$ cm

Focal length of $L_2=9$ cm

Distance b/w two lenses $=10$ cm

According to question, for L_1

$$\frac{1}{v}+\frac{1}{6}=\frac{1}{24} \Rightarrow v=-8 \text{ cm}$$

For L_2

$$\frac{1}{v}+\frac{1}{18}=\frac{1}{9} \Rightarrow \frac{1}{v}=\frac{1}{18}$$

$\Rightarrow \quad v=18$ cm

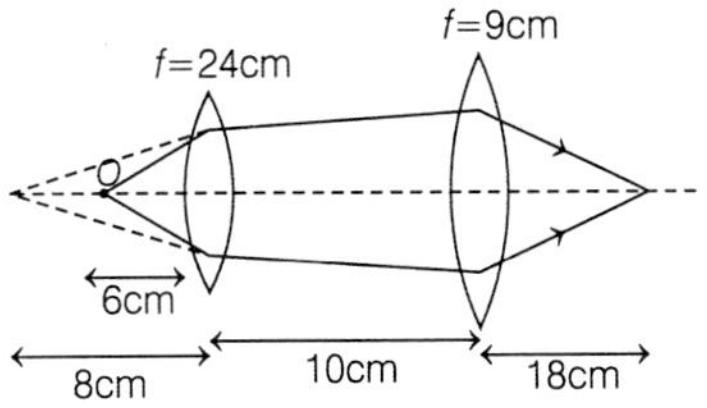

Distance between the object and image is $=18+10+6=34$ cm

56. *(30)* Given, $n_1=1$

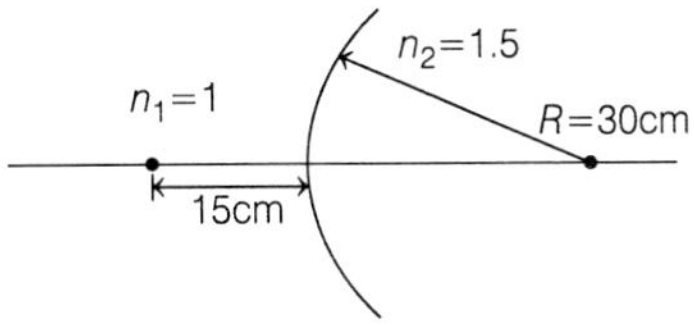

Using refraction formula,

$$\frac{n_2}{v}-\frac{n_1}{u}=\frac{n_2-n_1}{R}$$

We get, $\frac{1.5}{v}-\frac{1}{(-15)}=\frac{1.5-1}{30}$

$\Rightarrow \quad v=-30$ cm

(Negative sign shows image is formed on same side as that of object)

57. *(30)* Focal length of convex lens,

$f=10$ cm

Since, image produced by the plane mirror is 5 cm inside the mirror. Hence, distance image formed by the convex mirror

$v=20-5=15$ cm

$\therefore$ By lens formula,

$$\frac{1}{v}-\frac{1}{u}=\frac{1}{f}$$

$$\Rightarrow \quad \frac{1}{u}=\frac{1}{v}-\frac{1}{f}=\frac{1}{15}-\frac{1}{10}$$

$$=-\frac{1}{30}$$

$\Rightarrow \quad u=-30$ cm

Thus, distance of the object from the lens is 30 cm.

58. *(54)* Focal length of a lens in a liquid f_l can be determined by

$$\frac{f_l}{f_a}=\frac{{}_a\mu_g-1}{{}_l\mu_g-1}=\frac{\frac{\mu_g}{\mu_a}-1}{\frac{\mu_g}{\mu_l}-1}$$

Here, $f_a=$ focal length of lens in air $=18$ cm

$\mu_g=1.5; \mu_l=4/3$

So, $f_l=\frac{f_a\left(\frac{\mu_g}{\mu_a}-1\right)}{\left(\frac{\mu_g}{\mu_l}-1\right)}=\frac{18\times\left(\frac{3}{2}-1\right)}{\left(\frac{3/2}{4/3}-1\right)}$

$=72$ cm

$\therefore$ Change in focal length

$=72-18=54$ cm

59. *(120)* Using lens maker's formula,

$$\frac{1}{f}=(\mu-1)\left(\frac{1}{R_1}-\frac{1}{R_2}\right)$$

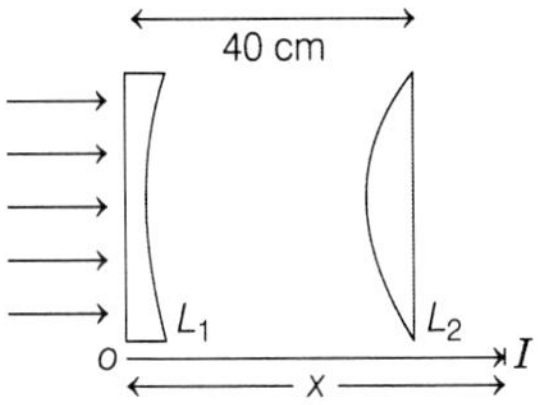

For lens, $L_1 \Rightarrow R_1=\infty$,

$R_2=30$ cm

$\mu=1.75$

So, $\frac{1}{f_1}=0.75\left(-\frac{1}{30}\right)=-\frac{1}{40}$

or $f_1=-40$ cm

For lens L_2, $R_1=+30$ cm, $R_2=\infty$ and $\mu=1.75$

So, $\frac{1}{f_2}=0.75\left(\frac{1}{30}\right)=40$ cm

Now, in given situation, for lens L_1,

$u=\infty, f_1=-40$ cm, $v=?$

Using, $\frac{1}{v}-\frac{1}{u}=\frac{1}{f} \Rightarrow \frac{1}{v}=\frac{1}{-40}$

or $v=-40$ cm.

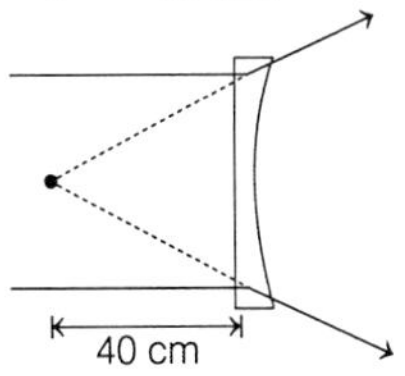

Now for lens L_2,

$u=-80$ cm, $v=?$, $f=40$ cm

So, $\frac{1}{v}-\frac{1}{-80}=\frac{1}{40}$

or $\frac{1}{v}=\frac{1}{40}-\frac{1}{80}=\frac{1}{80}$

or $v=80$ cm

So, distance of image from plano concave (L_1) lens is $80+40=120$ cm.

60. *(50)* When the ray on mirror falls directly perpendicular on the surface of it.

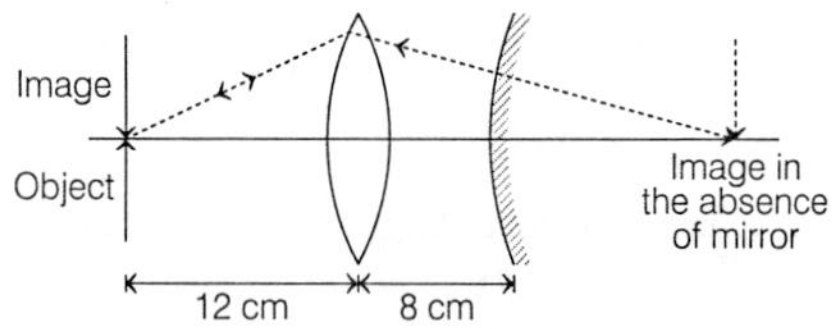

The light will converge at centre of curvature of mirror and due to that after removing the mirror the light will converge at same position.

The focal length of mirror,

$f = 15$ cm So, radius of curvature of mirror,

$$2f = 30 \text{ cm}$$

The image distance from the object after convex mirror is removed will be calculated as,

12 cm + 8 cm + 30 cm = 50 cm.

Thus, the image will formed at a distance of 50 cm from object.

61. *(15)* Let v be the position of image, h_i be the height of image and h_o be the height of object.

Given, $h_i = h_o$

Since, magnification,

$$m = h_i / h_o = h_i / h \quad \ldots(\text{i})$$

By using lens formula,

$$1/f = 1/v - 1/u$$

$$\Rightarrow \quad 1/v = 1/f + 1/u$$

$$\Rightarrow \quad v = \frac{fu}{u+f}$$

$$\therefore \quad m = \frac{v}{u} = \frac{f}{f+u}$$

Now, from Eq. (i), m can be ± 1.

$$\text{For,} \quad m = +1 = \frac{f}{-10+f} \quad \ldots(\text{ii})$$

For $m = -1$

$$= \frac{f}{-20+f} \quad \ldots(\text{iii})$$

On dividing Eq. (iii) by Eq. (ii), we get

$$-1 = \frac{-10+f}{-20+f}$$

$$\Rightarrow \quad 20 - f = -10 + f$$

$$\Rightarrow \quad 30 = 2f$$

$$\Rightarrow \quad f = 15 \text{ cm}$$

Topic 4

01. *(d)*

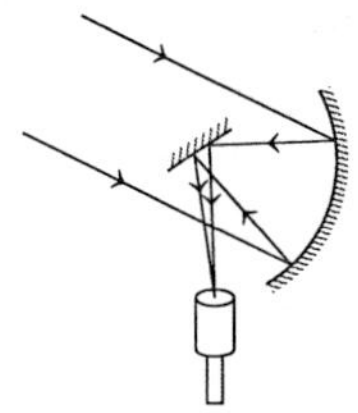

In a reflecting telescope secondary mirror turns light towards eyepiece which is outside the telescope tube.

02. *(c)* Objective of compound microscope is a convex lens. Convex lenses form real and enlarged image when an object is placed near its' focus.

03. *(c)* The formation of image with simple microscope is shown below.

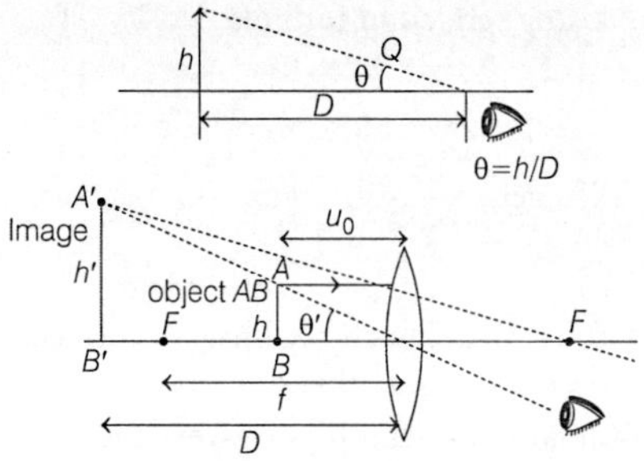

Here, $\theta' = \frac{h}{u_0} = \frac{h'}{D} = \frac{h'}{25}$

Where, $D = 25$ cm (least distance of distinct vision)

Here, θ' is same for both object and image, hence Assertion is true.

Magnification, $m = \frac{\theta'}{\theta} = \frac{D}{u_0}$

Hence, if $u_0 < D$ (25 cm), hence the value of θ' will obtain large.

So, option (c) is the correct.

04. *(a)* For a telescope in normal setting,

$$f_o + f_e = L$$

(length of the tube of telescope)

$$\text{and} \quad \frac{f_o}{f_e} = m \quad \text{(magnification)}$$

According to the given values in the question, we have

$$f_o + f_e = 60 \text{ cm and } \frac{f_o}{f_e} = 5$$

$$\Rightarrow \quad f_o = 5 f_e$$

On solving the above equations, we have

$$f_e = 10 \text{ cm}$$

05. *(25)* Given, magnification, $M = 6$

Since, we know that

$$M = 1 + \frac{D}{f_0}$$

where, D = least distance of distinct vision = 25 cm

and f_0 = focal length of objective lens.

$$\Rightarrow \quad 6 = 1 + \frac{D}{f_0} \Rightarrow 6 = 1 + \frac{25}{f_0}$$

$$\Rightarrow \quad 5 = \frac{25}{f_0} \Rightarrow f_0 = 5 \text{ cm}$$

For compound microscope, magnifying power is given by

$$M = \frac{l \cdot D}{f_0 f_e} = 2M_{\text{simple microscope}}$$

and l = length of the given tube = 0.6 m

$$\Rightarrow \quad 12 = \frac{60 \times 25}{5 \cdot f_e}$$

[$\because$ Magnification is doubled]

$$\Rightarrow \quad f_e = 25 \text{ cm}$$

06. *(6)* When the final image is formed at least distance of distinct vision (i.e., at 25 cm) from eye-piece of a compound microscope, then the magnification is given by

$$m = -\frac{v_0}{u_0}\left(1 + \frac{D}{f_e}\right)$$

$$\Rightarrow \quad m = \frac{-L}{f_0}\left(1 + \frac{D}{f_e}\right)$$

$$\Rightarrow \quad -100 = -\frac{20}{1}\left(1 + \frac{25}{f_e}\right)$$

$$\Rightarrow \quad 5 = 1 + \frac{25}{f_e} \Rightarrow 4 = \frac{25}{f_e}$$

$$\Rightarrow \quad f_e = \frac{25}{4} = 6.25 \simeq 6 \text{ cm}$$

07. *(50)* Strain on the eye is minimum so the intermediate image is formed at focal length of eye-piece.

For objective,

$$v = 5 \text{ cm}, f = 1 \text{ cm}$$

$\therefore$ Using lens formula,

$$\frac{1}{v} - \frac{1}{u} = \frac{1}{f} \Rightarrow \frac{1}{5} - \frac{1}{u} = \frac{1}{1}$$

$$\Rightarrow \quad u = \frac{-5}{4} \text{ cm}$$

$$\Rightarrow \quad |u| = \frac{5}{4} \text{ cm} \Rightarrow \frac{n}{40} = \frac{5}{4}$$

$$\Rightarrow \quad n = \frac{5}{4} \times 40 \Rightarrow n = 50$$

C-24

JEE Main Chapterwise Topicwise Physics

In JEE Main there is usually 0-1 question per paper is visible from this chapter. In past 3 years, 29 conceptual and 19 numerical questions were asked of moderate to difficult level from this chapter.

For detailed analysis scan QR code

WAVE OPTICS

Topic 1 Huygens Principle and Interference of Light

Level 1

01 Two plane polarised light waves combine at a certain point whose electric field components are

$$E_1 = E_0 \sin \omega t$$

$$E_2 = E_0 \sin\left(\omega t + \frac{\pi}{3}\right)$$

Find the amplitude of the resultant wave. *[2025, 7 April Shift-I]*

(a) E_0 (b) $0.9\ E_0$ (c) $1.7\ E_0$ (d) $3.4\ E_0$

02 In a Young's double slit experiment, the source is white light. One of the slits is covered by red filter and another by a green filter. In this case *[2025, 8 April Shift-II]*

(a) there shall be no interference fringes.
(b) there shall be an interference pattern for red distinct from that for green.
(c) there shall be an interference pattern, where each fringe's pattern centre is green and outer edges is red.
(d) there shall be alternate interference fringes of red and green.

03 Two monochromatic light beams have intensities in the ratio 1:9. An interference pattern is obtained by these beams. The ratio of the intensities of maximum to minimum is *[2025, 3 April Shift-II]*

(a) 9 : 1 (b) 8 : 1 (c) 3 : 1 (d) 4 : 1

04 In a Young's double slit experiment, the slits are separated by 0.2 mm. If the slits separation is increased to 0.4 mm, the percentage change of the fringe width is *[2025, 4 April Shift-I]*

(a) 0% (b) 100%
(c) 50% (d) 25%

05 Given below are two statements. One is labelled as Assertion (A) and the other is labelled as Reason (R).

Assertion (A) In Young's double slit experiment, the fringes produced by red light are closer as compared to those produced by blue light.

Reason (R) The fringe width is directly proportional to the wavelength of light.

In the light of the above statements, choose the correct answer from the options given below.

(a) A is false but R is true. *[2025, 22 Jan Shift-II]*
(b) A is true but R is false.
(c) Both A and R are true and R is the correct explanation of A.
(d) Both A and R are true but R is not the correct explanation of A.

06 The Young's double slit interference experiment is performed using light consisting of 480 nm and 600 mm wavelengths to form interference patterns. The least number of the bright fringes of 480 nm light that are required for the first coincidence with the bright fringes formed by 600 nm light is *[2025, 24 Jan Shift-I]*

(a) 5 (b) 4 (c) 6 (d) 8

07 The ratio of intensities at two points P and Q on the screen in a Young's double slit experiment where phase difference between two waves of same amplitude are $\frac{\pi}{3}$ and $\frac{\pi}{2}$, respectively are *[2023, 10 April Shift-II]*

(a) 2 : 3 (b) 3 : 1
(c) 3 : 2 (d) 1 : 3

08 The width of fringe is 2 mm on the screen in a double slits experiment for the light of wavelength of 400 nm. The width of the fringe for the light of wavelength 600 nm will be *[2023, 8 April Shift-II]*

(a) 1.33 mm (b) 2 mm
(c) 4 mm (d) 3 mm

Imp. Position of Fringes in YDSE

09 In Young's double slits experiment, the position of 5th bright fringe from the central maximum is 5 cm. The distance between slits and screen is 1 m and wavelength of used monochromatic light is 600 nm. The separation between the slits is *[2023, 25 Jan Shift-I]*

(a) 36 μm (b) 48 μm
(c) 12 μm (d) 60 μm

10 In Young's double slit experiment, the fringe width is 12 mm. If the entire arrangement is placed in water of refractive index $4/3$, then the fringe width becomes (in mm) *[2022, 26 July Shift-I]*
(a) 16 (b) 9
(c) 48 (d) 12

11 Using Young's double slit experiment, a monochromatic light of wavelength 5000 Å produces fringes of fringe width 0.5 mm. If another monochromatic light of wavelength 6000 Å is used and the separation between the slits is doubled, then the new fringe width will be *[2022, 29 June Shift-I]*
(a) 0.5 mm (b) 1.0 mm (c) 0.6 mm (d) 0.3 mm

12 The interference pattern is obtained with two coherent light sources of intensity ratio 4 : 1 and the ratio $\frac{I_{max} + I_{min}}{I_{max} - I_{min}}$ is $\frac{5}{x}$. Then, the value of x will be equal to *[2022, 25 June Shift-II]*
(a) 3 (b) 4 (c) 2 (d) 1

13 The two light beams having intensities I and $9I$ interfere to produce a fringe pattern on a screen. The phase difference between the beams is $\pi/2$ at point P and π at point Q. Then the difference between the resultant intensities at P and Q will be *[2022, 25 June Shift-I]*
(a) $2I$ (b) $6I$ (c) $5I$ (d) $7I$

14 The light waves from two coherent sources have same intensity $I_1 = I_2 = I_0$. In interference pattern the intensity of light at minima is zero. What will be the intensity of light at maxima ? *[2021, 27 Aug Shift-II]*
(a) I_0 (b) $2I_0$ (c) $5I_0$ (d) $4I_0$

15 In Young's double slit experiment, if the source of light changes from orange to blue, then *[2021, 27 July Shift-I]*
(a) the central bright fringe will become a dark fringe.
(b) the distance between consecutive fringes will decrease.
(c) the distance between consecutive fringes will increase.
(d) the intensity of the minima will increase.

16 In Young's double slit experiment the two slits are separated by 2 mm and the screen is placed one meter away. When a light of wavelength 500 nm is used, the fringe separation will be *[2021, 26 Feb Shift-I]*
(a) 0.75 mm (b) 0.25 mm
(c) 1 mm (d) 0.50 mm

Imp. Maximum and Minimum Intensities

17 Two coherent light sources having intensity in the ratio $2x$ produce an interference pattern. The ratio $\frac{I_{max} - I_{min}}{I_{max} + I_{min}}$ will be *[2021, 25 Feb Shift-I]*
(a) $\frac{2\sqrt{2x}}{x+1}$ (b) $\frac{\sqrt{2x}}{2x+1}$ (c) $\frac{\sqrt{2x}}{x+1}$ (d) $\frac{2\sqrt{2x}}{2x+1}$

18 If the source of light used in a Young's double slit experiment is changed from red to violet, then *[2021, 24 Feb Shift-II]*
(a) the consecutive fringe lines will come closer
(b) the central bright fringe will become a dark fringe
(c) the fringes will become brighter
(d) the intensity of minima will increase

19 Two light waves having the same wavelength λ in vacuum are in phase initially. Then, the first wave travels a path L_1 through a medium of refractive index n_1 while the second wave travels a path of length L_2 through a medium of refractive index n_2. After this, the phase difference between the two waves is *[2020, 3 Sep Shift-II]*
(a) $\frac{2\pi}{\lambda}(n_1 L_1 - n_2 L_2)$ (b) $\frac{2\pi}{\lambda}\left(\frac{L_1}{n_1} - \frac{L_1}{n_2}\right)$
(c) $\frac{2\pi}{\lambda}\left(\frac{L_2}{n_1} - \frac{L_1}{n_2}\right)$ (d) $\frac{2\pi}{\lambda}(n_2 L_1 - n_1 L_2)$

20 Interference fringes are observed on a screen by illuminating two thin slits 1 mm apart with a light source ($\lambda = 632.8$ nm). The distance between the screen and the slits is 100 cm. If a bright fringe is observed on a screen at a distance of 1.27 mm from the central bright fringe, then the path difference between the waves, which are reaching this point from the slits is close to *[2020, 2 Sep Shift-I]*
(a) 2 nm (b) 1.27 μm (c) 2.87 nm (d) 2.05 μm

Imp. Shifting of Fringes

21 In double slit experiment, when a thin film of thickness t having refractive index μ is introduced in front of one of the slits, the maximum at the centre of the fringe pattern shifts by one fringe width. The value of t is (λ is the wavelength of the light used) *[2019, 12 April Shift-I]*
(a) $\frac{2\lambda}{(\mu-1)}$ (b) $\frac{\lambda}{2(\mu-1)}$ (c) $\frac{\lambda}{(\mu-1)}$ (d) $\frac{\lambda}{(2\mu-1)}$

22 Light of wavelength 550 nm falls normally on a slit of width 22.0×10^{-5} cm. The angular position of the second minima from the central maximum will be (in radians) *[JEE Main 2018]*
(a) $\frac{\pi}{8}$ (b) $\frac{\pi}{12}$ (c) $\frac{\pi}{6}$ (d) $\frac{\pi}{4}$

23 On a hot summer night, the refractive index of air is smallest near the ground and increases with height from the ground. When a light beam is directed horizontally, the Huygens principle leads us to conclude that as it travels, the light beam *[JEE Main 2015]*
(a) becomes narrower
(b) goes horizontally without any deflection
(c) bends downwards
(d) bends upwards

24 Two coherent point sources S_1 and S_2 are separated by a small distance d as shown. The fringes obtained on the screen will be *[JEE Main 2013]*

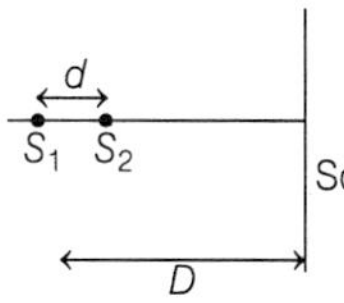

(a) concentric circles (b) points
(c) straight lines (d) semi-circles

25 n identical waves each of intensity I_0 interfere with each other. The ratio of maximum intensities if the interference is (i) coherent and (ii) incoherent is *[JEE Main 2013]*

(a) n^2 (b) $\frac{1}{n}$ (c) $\frac{1}{n^2}$ (d) n

26 Two coherent plane light waves of equal amplitude makes a small angle α (<< 1) with each other. They fall almost normally on a screen. If λ is the wavelength of light waves, the fringe width Δx of interference patterns of the two sets of waves on the screen is *[JEE Main 2012]*

(a) $\frac{2\lambda}{\alpha}$ (b) $\frac{\lambda}{\alpha}$ (c) $\frac{\lambda}{(2\alpha)}$ (d) $\frac{\lambda}{\sqrt{\alpha}}$

27 In a Young's double slit experiment, the two slits act as coherent sources of waves of equal amplitude A and wavelength λ. In another experiment with the same arrangement, the two slits are made to act as incoherent sources of waves of same amplitude and wavelength. If the intensity at the middle point of the screen in the first case is I_1 and in the second case is I_2, then the ratio $\frac{I_1}{I_2}$ is *[AIEEE 2011]*

(a) 4 (b) 2 (c) 1 (d) 0.5

28 The initial shape of the wavefront of the beam is *[AIEEE 2010]*

(a) convex
(b) concave
(c) convex near the axis and concave near the periphery
(d) planar

29 A mixture of light, consisting of wavelength 590 nm and an unknown wavelength, illuminates Young's double slit and gives rise to two overlapping interference patterns on the screen. The central maximum of both lights coincide. Further, it is observed that the third bright fringe of known light coincides with the 4th bright fringe of the unknown light. From this data, the wavelength of the unknown light is *[AIEEE 2009]*

(a) 393.4 nm (b) 885.0 nm
(c) 442.5 nm (d) 776.8 nm

30 A Young's double slit experiment uses a monochromatic source. The shape of the interference fringes formed on a screen is *[AIEEE 2005]*

(a) hyperbola (b) circle
(c) straight line (d) parabola

31 The maximum number of possible interference maxima for slit-separation equal to twice the wavelength in Young's double-slit experiment, is *[AIEEE 2004]*

(a) infinite (b) five (c) three (d) zero

Level 2

32 Width of one of the two slits in a Young's double slit interference experiment is half of the other slit. The ratio of the maximum to the minimum intensity in the interference pattern is *[2025, 3 April Shift-II]*

(a) $9:1$
(b) $(2\sqrt{2}+1):(2\sqrt{2}-1)$
(c) $(3+2\sqrt{2}):(3-2\sqrt{2})$
(d) $3:1$

33 A light wave is propagating with plane wave fronts of the type $x+y+z=$ constant. The angle made by the direction of wave propagation with the X-axis is *[2025, 2 April Shift-I]*

(a) $\cos^{-1}\left(\sqrt{\frac{2}{3}}\right)$ (b) $\cos^{-1}\left(\frac{1}{\sqrt{3}}\right)$
(c) $\cos^{-1}\left(\frac{1}{3}\right)$ (d) $\cos^{-1}\left(\frac{2}{3}\right)$

34 Young's double slit inteference apparatus is immersed in a liquid of refractive index 1.44. It has slit separation of 1.5 mm. The slits are illuminated by a parallel beam of light whose wavelength in air is 690 nm. The fringe width on a screen placed behind the plane of slits at a distance of 0.72 m, will be *[2025, 24 Jan Shift-II]*

(a) 0.23 mm (b) 0.33 mm
(c) 0.46 mm (d) 0.63 mm

35 The width of one of the two slits in Young's double slit experiment is d while that of the other slit is xd. If the ratio of the maximum of the minimum intensity in the interference pattern on the screen is $9:4$, then what is the value of x?
(Assume that the field strength varies according to the slit width) *[2025, 23 Jan Shift-II]*

(a) 2 (b) 5
(c) 4 (d) 3

36 The width of one of the two slits in a Young's double slit experiment is 4 times that of the other slit. The ratio of the maximum and the minimum intensity in the interference pattern is *[2024, 4 April Shift-I]*

(a) $1:1$ (b) $9:1$
(c) $4:1$ (d) $16:1$

37 In Young's double slit experiment, two slits are illuminated with a light of wavelength 800 nm. The line joining A_1P is perpendicular to A_1A_2 as shown in the figure. If the first minimum is detected at P, the value of slits separation a will be

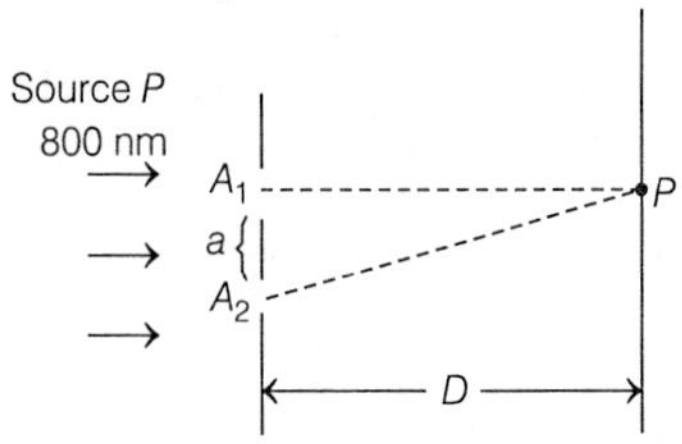

The distance of screen from slits, $D = 5$ cm [2023, 29 Jan Shift-I]

(a) 0.5 mm (b) 0.1 mm
(c) 0.4 mm (d) 0.2 mm

38 In Young's double slit experiment performed using a monochromatic light of wavelength λ, when a glass plate ($\mu = 1.5$) of thickness $x\lambda$ is introduced in the path of one of the interfering beams, the intensity at the position where the central maxima occurred previously remains unchanged. The value of x will be [2022, 28 June Shift-II]

(a) 3 (b) 2 (c) 1.5 (d) 0.5

39 Two light beams of intensities in the ratio of 9 : 4 are allowed to interfere. The ratio of the intensity of maxima and minima will be [2022, 24 June Shift-II]

(a) 2 : 3 (b) 16 : 81
(c) 25 : 169 (d) 25 : 1

40 In Young's double slit experiment, the distance between the slits varies in time as $d(t) = d_0 + a_0 \sin \omega t$, where d_0, ω and a_0 are constants. The difference between the largest fringe width and the smallest fringe width obtained over time is given as [2021, 25 July Shift-I]

(a) $\frac{2\lambda D(d_0)}{(d_0^2 - a_0^2)}$ (b) $\frac{2\lambda D a_0}{(d_0^2 - a_0^2)}$ (c) $\frac{\lambda D}{d_0^2} a_0$ (d) $\frac{\lambda D}{d_0 + a_0}$

41 In Young's double slit arrangement, slits are separated by a gap of 0.5 mm, and the screen is placed at a distance of 0.5 m from them. The distance between the first and the third bright fringe formed when the slits are illuminated by a monochromatic light of 5890 Å is [2021, 18 March Shift-I]

(a) 1178×10^{-9} m (b) 1178×10^{-6} m
(c) 1178×10^{-12} m (d) 5890×10^{-7} m

42 In Young's double slit experiment, the width of one of the slit is three times the other slit. The amplitude of the light coming from a slit is proportional to the slit-width. Find the ratio of the maximum to the minimum intensity in the interference pattern. [2021, 24 Feb Shift-I]

(a) 4 : 1 (b) 2 : 1 (c) 1 : 4 (d) 3 : 1

43 In Young's double slit experiment, light of 500 nm is used to produce an interference pattern. When the distance between the slits is 0.05 mm, the angular width (in degree) of the fringes formed on the distance screen is close to [2020, 3 Sep Shift-I]

(a) 0.17° (b) 0.57° (c) 1.7° (d) 0.07°

44 In Young's double slit experiment, 16 fringes are observed in a certain segment of the screen when light of wavelength 700 nm is used. If the wavelength of light is changed to 400 nm, the number of fringes observed in the same segment of the screen would be [2020, 2 Sep Shift-II]

(a) 24 (b) 30 (c) 18 (d) 28

45 In Young's double slit experiment, the separation between the slits is 0.15 mm. In the experiment, a source of light of wavelength 589 nm is used and the interference pattern is observed on a screen kept 1.5 m away. The separation between the successive bright fringes on the screen is [2020, 7 Jan Shift-II]

(a) 4.9 mm (b) 3.9 mm (c) 6.9 mm (d) 5.9 mm

Imp. Fringe Width

46 The figure shows a Young's double slit experimental setup. It is observed that when a thin transparent sheet of thickness t and refractive index μ is put in front of one of the slits, the central maximum gets shifted by a distance equal to n fringe widths. If the wavelength of light used is λ, t will be [2019, 9 April Shift-I]

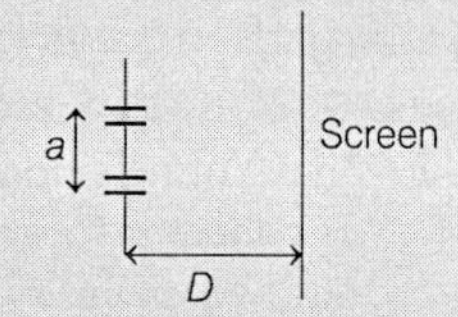

(a) $\frac{2nD\lambda}{a(\mu - 1)}$ (b) $\frac{2D\lambda}{a(\mu - 1)}$ (c) $\frac{D\lambda}{a(\mu - 1)}$ (d) $\frac{nD\lambda}{a(\mu - 1)}$

47 In Young's double slit experiment, the path difference at a certain point on the screen between two interfering waves is 1/8th of wavelength. The ratio of the intensity at this point to that at the centre of a bright fringe is close to [2019, 11 Jan Shift-I]

(a) 0.80 (b) 0.74 (c) 0.94 (d) 0.85

48 In Young's double slit experiment, slits are separated by 0.5 mm and the screen is placed 150 cm away. A beam of light consisting of two wavelengths, 650 nm and 520 nm, is used to obtain interference fringes on the screen. The least distance from the common central maximum to the point where the bright fringes due to both the wavelengths coincide, is [JEE Main 2017]

(a) 7.8 mm (b) 9.75 mm
(c) 15.6 mm (d) 1.56 mm

49 In Young's double slit experiment, the distance between slits and the screen is 1.0 m and monochromatic light of 600 nm is being used. A person standing near the slits is looking at the fringe

pattern. When the separation between the slits is varied, the interference pattern disappears for a particular distance d_0 between the slits. If the angular resolution of the eye is $(1/60)°$, the value of d_0 is close to **[JEE Main 2016]**
(a) 1 mm (b) 3 mm (c) 2 mm (d) 4 mm

50 Two monochromatic light beams of intensity 16 and 9 units are interfering. The ratio of intensities of bright and dark parts of the resultant pattern is **[JEE Main 2014]**
(a) 16/9 (b) 4/3 (c) 7/1 (d) 49/1

51 In Young's double slit experiment, one of the slit is wider than other, so that amplitude of the light from one slit is double of that from other slit. If I_m is the maximum intensity, the resultant intensity I when they interfere at phase difference ϕ, is given by **[AIEEE 2012]**
(a) $\frac{I_m}{9}(4+5\cos\phi)$ (b) $\frac{I_m}{3}\left(1+2\cos^2\frac{\phi}{2}\right)$
(c) $\frac{I_m}{5}\left(1+4\cos^2\frac{\phi}{2}\right)$ (d) $\frac{I_m}{9}\left(1+8\cos^2\frac{\phi}{2}\right)$

52 In Young's double slit experiment, the intensity at a point where the path difference is $\frac{\lambda}{6}$ (λ being the wavelength of the light used) is I. If I_0 denotes the maximum intensity, then I/I_0 is equal to **[AIEEE 2007]**
(a) $1/\sqrt{2}$ (b) $\sqrt{3}/2$
(c) 1/2 (d) 3/4

Level 3

Imp. Young's Double Slit Experiment

53 Given below are two Statements. One is labelled as **Assertion** (A) and the other is labelled as **Reason** (R).
Assertion (A) If Young's double slit experiment is performed in an optically denser medium than air, then the consecutive fringes come closer.
Reason (R) The speed of light reduces in an optically denser medium than air while its frequency does not change. **[2025, 22 Jan Shift-I]**
In the light of the above statements, choose the most appropriate answer from the options given below.
(a) (A) is true but (R) is false.
(b) Both (A) and (R) are true but (R) is not the correct explanation of (A).
(c) Both (A) and (R) are true and (R) is the correct explanation of (A).
(d) (A) is false but (R) is true.

54 In Young's double slit experiment, light from two identical sources are superimposing on a screen. The path difference between the two lights reaching at a point on the screen is $7\lambda/4$. The ratio of intensity of fringe at this point with respect to the maximum intensity of the fringe is **[2024, 29 Jan Shift-II]**
(a) $\frac{3}{4}$ (b) $\frac{1}{3}$
(c) $\frac{1}{2}$ (d) $\frac{1}{4}$

Imp. Maximum and Minimum Intensities

55 Two coherent sources of light interfere. The intensity ratio of two sources is 1 : 4. For this interference pattern if the value of $\frac{I_{max}+I_{min}}{I_{max}-I_{min}}$ is equal to $\frac{2\alpha+1}{\beta+3}$, then α/β will be **[2022, 27 July Shift-II]**
(a) 1.5 (b) 2 (c) 0.5 (d) 1

56 In Young's double slit experiment, the ratio of the slit's width is 4 : 1. The ratio of the intensity of maxima to minima, close to the central fringe on the screen, will be **[2019, 10 April Shift-II]**
(a) 4 : 1 (b) 25 : 9
(c) 9 : 1 (d) $(\sqrt{3}+1)^4$:16

57 In Young's double slit experiment with slit separation 0.1 mm, one observes a bright fringe at angle $1/40$ rad by using light of wavelength λ_1. When the light of the wavelength λ_2 is used a bright fringe is seen at the same angle in the same set up. Given that λ_1 and λ_2 are in visible range (380 nm to 740 nm), their values are **[2019, 10 Jan Shift-I]**
(a) 380 nm, 525 nm (b) 400 nm, 500 nm
(c) 380 nm, 500 nm (d) 625 nm, 500 nm

58 Two coherent sources produce waves of different intensities which interfere. After interference, the ratio of the maximum intensity to the minimum intensity is 16. The intensity of the waves are in the ratio **[2019, 9 Jan Shift-I]**
(a) 16 : 9 (b) 5 : 3
(c) 25 : 9 (d) 4 : 1

Numerical Value Type Questions

59 Two coherent monochromatic light beams of intensities $4I$ and $9I$ are superimposed. The difference between the maximum and minimum intensities in the resulting interference pattern is xI. The value of x is .... **[2025, 3 April Shift-I]**

60 A double slit interference experiment performed with a light of wavelength 600 nm forms an interference fringe pattern on a screen with 10th bright fringe

having its centre at a distance of 10 mm from the central maximum. Distance of the centre of the same 10th bright fringe from the central maximum when the source of light is replaced by another source of wavelength 660 nm would be .... mm.
[2025, 28 Jan Shift-I]

61 A thin transparent film with refractive index 1.4, is held on circular ring of radius 1.8 cm. The fluid in the film evaporates such that transmission through the film at wavelength 560 nm goes to a minimum every 12 s. Assuming that the film is flat on its two sides, the rate of evaporation is .... $\pi \times 10^{-13} m^3/s$.
[2025, 28 Jan Shift-II]

62 Monochromatic light of wavelength 500 nm is used in Young's double slit experiment. An interference pattern is obtained on a screen. When one of the slits is covered with a very thin glass plate (refractive index = 1.5), the central maximum is shifted to a position previously occupied by the 4th bright fringe. The thickness of the glass-plate is .... μm.
[2024, 9 April Shift-II]

63 Two slits are 1 mm apart and the screen is located 1 m away from the slits. A light of wavelength 500 nm is used. The width of each slit to obtain 10 maxima of the double slit pattern within the central maximum of the single slit pattern is .... $\times 10^{-4}$ m.
[2024, 8 April Shift-II]

64 Two coherent monochromatic light beams of intensities I and $4I$ are superimposed. The difference between maximum and minimum possible intensities in the resulting beam is xI. The value of x is ....
[2024, 6 April Shift-II]

65 In Young's double slit experiment, carried out with light of wavelength 5000 Å, the distance between the slits is 0.3 mm and the screen is at 200 cm from the slits. The central maximum is at $x = 0$ cm. The value of x for third maxima is .... mm.
[2024, 5 April Shift-I]

66 Two wavelengths λ_1 and λ_2 are used in Young's double slit experiment. $\lambda_1 = 450$ nm and $\lambda_2 = 650$ nm. The minimum order of fringe produced by λ_2 which overlaps with the fringe produced by λ_1 is n. The value of n is...... .
[2024, 4 April Shift-I]

67 Two waves of intensity ratio 1:9 cross each other at a point. The resultant intensities at that point, when (a) waves are incoherent is I_1 (b) waves are coherent is I_2 and differ in phase by 60°. If $\frac{I_1}{I_2} = \frac{10}{x}$, then $x =$....... .
[2024, 31 Jan Shift-I]

68 A beam of light consisting of two wavelengths 7000 Å and 5500 Å is used to obtain interference pattern in Young's double slit experiment. The distance between the slits is 2.5 mm and the distance between the plane of slits and the screen is 150 cm. The least distance from the central fringe, where the bright fringes due to both the wavelengths coincide, is $n \times 10^{-5}$ m. The value of n is ...... *[2023, 6 April Shift-II]*

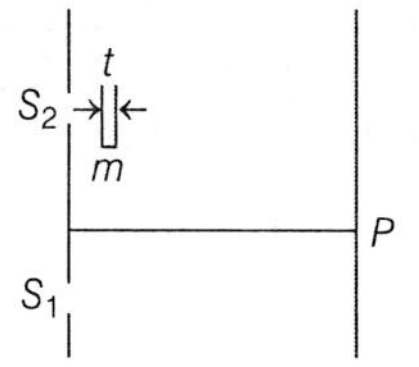

69 As shown in the figure, in Young's double slit experiment, a thin plate of thickness $t = 10$ μm and refractive index $\mu = 1.2$ is inserted infront of slit S_1. The experiment is conducted in air ($\mu = 1$) and uses a monochromatic light of wavelength $\lambda = 500$ nm, due to the insertion of the plate, central maxima is shifted by a distance of $x\beta_0$. β_0 is the fringe-width before the insertion of the plate. The value of the x is ....
[2023, 1 Feb Shift-II]

70 Two light waves of wavelengths 800 and 600 nm are used in Young's double slit experiment to obtain interference fringes on a screen placed 7 m away from plane of slits. If the two slits are separated by 0.35 mm, then shortest distance from the central bright maximum to the point where the bright fringes of the two wavelength coincide will be .... mm.
[2023, 31 Jan Shift-II]

71 Two light beams of intensities $4I$ and $9I$ interfere on a screen. The phase difference between these beams on the screen at point A is zero and at point B is π. The difference of resultant intensities, at the point A and B, will be .... I.
[2022, 29 July Shift-I]

72 In Young's double slit experiment, a laser light of 560 nm produces an interference pattern with consecutive bright fringes separation of 7.2 mm. Now, another light is used to produce an interference pattern with consecutive bright fringes separation of 81 mm. The wavelength of second light is .... nm.
[2022, 28 July Shift-I]

73 Two beams of light having intensities I and $4I$ interference to produce a fringe pattern on a screen. The phase difference between the two beams are $\pi/2$ and $\pi/3$ at points A and B, respectively. The difference between the resultant intensities at the two points is xI. The value of x will be ....
[2022, 27 July Shift-I]

74 In a double slit experiment with monochromatic light, fringes are obtained on a screen placed at some distance from the plane of slits. If the screen is moved by 5×10^{-2} m towards the slits, the change in fringe width is 3×10^{-3} cm. If the distance between the slits is 1 mm, then the wavelength of the light will be .... nm *[2022, 29 June Shift-II]*

75 In Young's double slit experiment the two slits are 0.6 mm distance apart. Interference pattern is observed on a screen at a distance 80 cm from the slits. The first dark fringe is observed on the screen directly opposite to one of the slits. The wavelength of light will be ...... nm. *[2022, 27 June Shift-I]*

76 The width of one of the two slits in a Young's double slit experiment is three times the other slit. If the amplitude of the light coming from a slit is proportional to the slit-width, the ratio of minimum to maximum intensity in the interference pattern is $x:4$ where x is .... *[2021, 1 Sep Shift-II]*

77 White light is passed through a double slit and interference is observed on a screen 1.5 m away. The separation between the slits is 0.3 mm. The first violet and red fringes are formed 2.0 mm and 3.5 mm away from the central white fringes. The difference in wavelengths of red and violet light is .... nm. *[2021, 26 Aug Shift-I]*

78 A fringe width of 6 mm was produced for two slits separated by 1 mm. The screen is placed 10 m away. The wavelength of light used is x nm. The value of x to the nearest integer is .... *[2021, 16 March Shift-I]*

79 In a Young's double slit experiment, 15 fringes are observed on a small portion of the screen when light of wavelength 500 nm is used. Ten fringes are observed on the same section of the screen when another light source of wavelength λ is used. Then, the value of λ is (in nm) .... *[2020, 9 Jan Shift-II]*

Topic 2 Diffraction and Doppler Effect of Light

Level 1

01 A single slit of width a is illuminated by a monochromatic light of wavelength 600 nm. The value of a for which first minimum appears at $\theta = 30°$ on the screen will be *[2023, 15 April Shift-I]*

(a) 1.2 μm (b) 3 μm
(c) 0.6 μm (d) 1.8 μm

02 For a specific wavelength 670 nm of light coming from a galaxy moving with velocity v, the observed wavelength is 670.7 nm. The value of v is *[2022, 26 June Shift-II]*

(a) 3×10^{8} ms^{-1}
(b) 3×10^{10} ms^{-1}
(c) 3.13×10^{5} ms^{-1}
(d) 4.48×10^{5} ms^{-1}

03 Consider the diffraction pattern obtained from the sunlight incident on a pinhole of diameter 0.1μm. If the diameter of the pinhole is slightly increased, it will affect the diffraction pattern such that *[2021, 25 Feb Shift-II]*

(a) its size increases and intensity increases
(b) its size increases, but intensity decreases
(c) its size decreases, but intensity increases
(d) its size decreases and intensity decreases

04 In a double-slit experiment, green light 5303 Å falls on a double slit having a separation of 19.44 μm and a width of 4.05 μm. The number of bright fringes between the first and the second diffraction minima is *[2019, 11 Jan Shift-II]*

(a) 10 (b) 04
(c) 05 (d) 09

Level 2

05 A microwave of wavelength 2.0 cm falls normally on a slit of width 4.0 cm. The angular spread of the central maxima of the diffraction pattern obtained on a screen 1.5 m away from the slit, will be *[2024, 1 Feb Shift-II]*

(a) 15° (b) 30°
(c) 60° (d) 45°

06 A monochromatic light of wavelength 6000 Å is incident on the single slit of width 0.01 mm. If the diffraction pattern is formed at the focus of the convex lens of focal length 20 cm, the linear width of the central maximum is *[2024, 1 Feb Shift-I]*

(a) 120 mm (b) 12 mm
(c) 60 mm (d) 24 mm

Imp. Single Slit Diffraction Pattern

07 The diffraction pattern of a light of wavelength 400 nm diffracting from a slit of width 0.2 mm is focused on the focal plane of a convex lens of focal length 100 cm. The width of the 1st secondary maxima will be *[2024, 30 Jan Shift-I]*

(a) 2 cm (b) 0.2 mm (c) 2 mm (d) 0.02 mm

08 A single slit of width b is illuminated by a coherent monochromatic light of wavelength λ. If the second and fourth minima in the diffraction pattern at a distance 1 m from the slit are at 3 cm and 6 cm, respectively from the central maximum, what is the width of the central maximum? (i.e. distance between first minimum on either side of the central maximum) *[JEE Main 2017]*

(a) 6.0 cm (b) 1.5 cm (c) 4.5 cm (d) 3.0 cm

09 The box of a pin hole camera, of length L, has a hole of radius a. It is assumed that when the hole is illuminated by a parallel beam of light of wavelength λ the spread of the spot (obtained on the opposite wall of the camera) is the sum of its geometrical spread and the spread due to diffraction. The spot would then have its minimum size (say b_{min}) when *[JEE Main 2016]*

(a) $a = \frac{\lambda^2}{L}$ and $b_{min} = \left(\frac{2\lambda^2}{L}\right)$

(b) $a = \sqrt{\lambda L}$ and $b_{min} = \left(\frac{2\lambda^2}{L}\right)$

(c) $a = \sqrt{\lambda L}$ and $b_{min} = \sqrt{4\lambda L}$

(d) $a = \frac{\lambda^2}{L}$ and $b_{min} = \sqrt{4\lambda L}$

Level 3

10 The angular width of the central maximum in a single slit diffraction pattern is 60°. The width of the slit is 1 μm. The slit is illuminated by monochromatic plane waves. If another slit of same width is made near it, Young's fringes can be observed on a screen placed at a distance 50 cm from the slits. If the observed fringe width is 1 cm, what is slit separation distance? (i.e. distance between the centres of each slit.) *[JEE Main 2018]*

(a) 25 μm (b) 50 μm
(c) 75 μm (d) 100 μm

Numerical Value Type Questions

11 In a Young's double slit experiment, two slits are located 1.5 mm apart. The distance of screen from slits is 2 m and the wavelength of the source is 400 nm. If the 20 maxima of the double slit pattern are contained within the central maximum of the single slit diffraction pattern, then the width of each slit is $x \times 10^{-3}$ cm, where x -value is .... *[2025, 4 April Shift-II]*

12 If the measured angular separation between the second minimum to the left of the central maximum and the third minimum to the right of the central maximum is 30° in a single slit diffraction pattern recorded using 628 nm light, then the width of the slit is .... μm. *[2025, 2 AprilShift-I]*

Imp. Angular Position of Maxima and Minima

13 A parallel beam of monochromatic light of wavelength 600 nm passes through single slit of 0.4 mm width. Angular divergence corresponding to second order minima would be .... $\times 10^{-3}$ rad. *[2024, 8 April Shift-I]*

14 In a single slit diffraction pattern, a light of wavelength 6000 Å is used. The distance between the first and third minima in the diffraction pattern is found to be 3 mm, when the screen in placed 50 cm away from slits. The width of the slit is ...... $\times 10^{-4}$ m. *[2024, 29 Jan Shift-II]*

15 A parallel beam of monochromatic light of wavelength 5000 Å is incident normally on a narrow slit of width 0.001 mm. The light is focused by convex lens on screen, placed on its focal plane. The first minima will be formed for the angle of diffraction of .... (degree). *[2024, 27 Jan Shift-II]*

16 Sodium light of wavelengths 650 nm and 655 nm is used to study diffraction at a single slit of aperture 0.5 mm. The distance between the slit and the screen is 2.0 m. The separation between the positions of the first maxima of diffraction pattern obtained in the two cases is ...... $\times 10^{-5}$ m. *[2022, 24 June Shift-I]*

17 Orange light of wavelength 6000×10^{-10} m illuminates a single slit of width 0.6×10^{-4} m. The maximum possible number of diffraction minima produced on both sides of the central maximum is .... *[2020, 4 Sep Shift-II]*

Topic 3 Polarisation

Level 1

01 When unpolarized light is incident at an angle of 60° on a transparent medium from air, the reflected ray is completely polarised. The angle of refraction in the medium is *[2024, 31 Jan Shift-II]*

(a) 90° (b) 45° (c) 60° (d) 30°

02 When a polaroid sheet is rotated between two crossed polaroids then the transmitted light intensity will be maximum for a rotation of *[2024, 27 Jan Shift-II]*

(a) 45° (b) 60° (c) 90° (d) 30°

03 Two polaroids A and B are placed in such a way that the pass-axis of polaroids are perpendicular to each other. Now, another polaroid C is placed between A and B bisecting angle between them. If intensity of unpolarised light is I_0 then intensity of transmitted light after passing through polaroid B will be *[2023, 31 Jan Shift-I]*

(a) $I_0/8$ (b) $I_0/2$ (c) zero (d) $I_0/4$

04 Unpolarised light of intensity I is incident on a system of two polarisers, A followed by B. The intensity of emergent light is $I/2$. If a third polariser C is placed between A and B, the intensity of emergent light is reduced to $I/3$. The angle between the polarisers A and C also between C and B is θ. Then *[JEE Main 2018]*

(a) $\cos\theta = \left(\frac{1}{3}\right)^{1/2}$ (b) $\cos\theta = \left(\frac{2}{3}\right)^{1/4}$

(c) $\cos\theta = \left(\frac{2}{3}\right)^{1/2}$ (d) $\cos\theta = \left(\frac{1}{3}\right)^{1/4}$

05 A beam of unpolarised light of intensity I_0 is passed through a polaroid A and then through another polaroid B which is oriented so that its principal plane makes an angle of 45° relative to that of A. The intensity of the emergent light is *[JEE Main 2013]*

(a) I_0 (b) $\frac{I_0}{2}$ (c) $\frac{I_0}{4}$ (d) $\frac{I_0}{8}$

06 **Statement I** On viewing the clear, blue portion of the sky through a Calcite crystal, the intensity of transmitted light varies as the crystal is rotated.

Statement II The light coming from the sky is polarised due to scattering of sun light by particles in the atmosphere. The scattering is largest for blue light. *[AIEEE 2011]*

(a) Statement I is false, Statement II is true.
(b) Statement I is true, Statement II is false.
(c) Statement I is true, Statement II is true; Statement II is the correct explanation of Statement I.
(d) Statement I is true, Statement II is true; Statement II is not the correct explanation of Statement I.

07 When an unpolarised light of intensity I_0 is incident on a polarising sheet, the intensity of the light which does not get transmitted is *[AIEEE 2005]*

(a) $\frac{1}{2}I_0$ (b) $\frac{1}{4}I_0$ (c) zero (d) I_0

08 The angle of incidence at which reflected light is totally polarised for reflection from air to glass (refractive index n), is *[AIEEE 2004]*

(a) $\sin^{-1}(n)$ (b) $\sin^{-1}\left(\frac{1}{n}\right)$
(c) $\tan^{-1}\left(\frac{1}{n}\right)$ (d) $\tan^{-1}(n)$

Level 2

09 Two polarisers P_1 and P_2 are placed in such a way that the intensity of the transmitted light will be zero. A third polariser P_3 is inserted in between P_1 and P_2 at particular angle between P_2 and P_3. The transmitted intensity of the light passing the through all three polarisers maximum. The angle between the polarisers P_2 and P_3 is *[2025, 4 April shift-II]*

(a) $\frac{\pi}{4}$ (b) $\frac{\pi}{6}$
(c) $\frac{\pi}{8}$ (d) $\frac{\pi}{3}$

10 A beam of unpolarised light of intensity I_0 is passed through a polaroid A and then through another polaroid B which is oriented, so that its principal plane makes an angle of 45°, relative to that of A. The intensity of emergent light is *[2024, 30 Jan shift-II]*

(a) $I_0/8$ (b) $I_0/4$
(c) $I_0/2$ (d) I_0

11 n polarising sheets are arranged such that each makes an angle 45° with the preceeding sheet. An unpolarised light of intensity I is incident into this arrangement. The output intensity is found to be $I/64$. The value of n will be *[2023, 1 Feb Shift-I]*

(a) 3 (b) 6
(c) 4 (d) 5

Imp. Brewster's Law

12 Given below are two statements.

Statement I If the Brewster's angle for the light propagating from air to glass is θ_B, then the Brewster's angle for the light propagating from glass to air is $\frac{\pi}{2}-\theta_B$.

Statement II The Brewster's angle for the light propagating from glass to air is $\tan^{-1}(\mu_g)$, where μ_g is the refractive index of glass. In the light of the above statements, choose the correct answer from the options given below. *[2023, 24 Jan shift-I]*

(a) Statement I is false but Statement II is true.
(b) Statement I is true but Statement II is false.
(c) Both Statement I and Statement II are false.
(d) Both Statement I and Statement II is true.

13 Unpolarised light of intensity I passes through an ideal polariser A. Another identical polariser B is placed behind A. The intensity of light beyond B is found to be $I/2$. Now, another identical polariser C is placed between A and B. The intensity beyond B is now found to be I/8. The angle between polariser A and C is *[JEE Main 2018]*

(a) 0° (b) 30°
(c) 45° (d) 60°

14 Two beams, A and B, of plane polarised light with mutually perpendicular planes of polarisation are seen through a polaroid. From the position, when the beam A has maximum intensity (and beam B has zero intensity), a rotation of polaroid through 30° makes the two beams appear equally bright. If the initial intensities of the two beams are I_A and I_B respectively, then I_A/I_B equals *[JEE Main 2014]*

(a) 3 (b) 3/2
(c) 1 (d) 1/3

Level 3

15 In a Young's double slit experiment, three polarisers are kept as shown in the figure. The transmission axes of P_1 and P_2 are orthogonal to each other. The polariser P_3 covers both the slits with its transmission axis at 45° to those of P_1 and P_2. An unpolarised light of wavelength λ and intensity I_0 is incident on P_1 and P_2. The intensity at a point after P_3, where the path difference between the light waves from S_1 and S_2 is $\lambda/3$, is *[2025, 24 Jan Shift-II]*

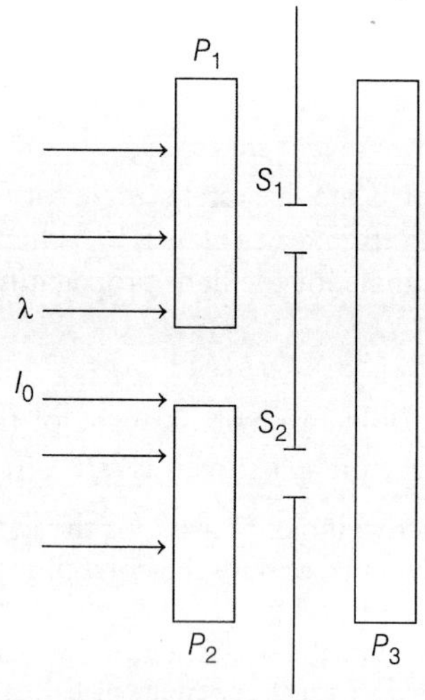

(a) $I_0/3$ (b) I_0 (c) $I_0/2$ (d) $I_0/4$

16 A beam of plane polarised light of large cross-sectional area and uniform intensity of 3.3 Wm^{-2} falls normally on a polariser, which rotates about its axis with an angular speed of 31.4 rad/s. The energy of light passing through the polariser per revolution is close to (Take, cross-sectional area $= 3 \times 10^{-4}\ m^2$) *[2020, 4 Sep Shift-I]*

(a) 1.5×10^{-4} J (b) 5.0×10^{-4} J
(c) 1.0×10^{-4} J (d) 1.0×10^{-5} J

17 A polariser-analyser set is adjusted such that the intensity of light coming out of the analyser is just 10% of the original intensity. Assuming that the polariser-analyser set does not absorb any light, the angle by which the analyser need to be rotated further to reduce the output intensity to be zero, is *[2020, 7 Jan Shift-I]*

(a) 71.6° (b) 90° (c) 45° (d) 18.4°

18 A system of three polarisers P_1, P_2, P_3 is set up such that the pass axis of P_3 is crossed with respect to that of P_1. The pass axis of P_2 is inclined at 60° to the pass axis of P_3. When a beam of unpolarised light of intensity I_0 is incident on P_1, the intensity of light transmitted by the three polarisers is I. The ratio (I_0/I) equals (nearly) *[2019, 12 April Shift-II]*

(a) 5.33 (b) 16.00 (c) 10.67 (d) 1.80

19 Consider a tank made of glass (refractive index 1.5) with a thick bottom. It is filled with a liquid of refractive index μ. A student finds that, irrespective of what the incident angle i (see figure) is for a beam of light entering the liquid, the light reflected from the liquid glass interface is never completely polarized. For this to happen, the minimum value of μ is *[2019, 9 Jan Shift-I]*

(a) $\sqrt{\frac{5}{3}}$ (b) $\frac{4}{3}$ (c) $\frac{3}{\sqrt{5}}$ (d) $\frac{5}{\sqrt{3}}$

Numerical Value Type Questions

20 Unpolarised light of intensity 32 Wm^{-2} passes through the combination of three polaroids such that the pass axis of the last polaroid is perpendicular to that of the pass axis of first polaroid. If intensity of emerging light is 3 Wm^{-2}, then the angle between pass axis of first two polaroids is .... °. *[2023, 10 April Shift-I]*

21 Unpolarised light is incident on the boundary between two dielectric media, whose dielectric constants are 2.8 (medium-1) and 6.8 (medium-2), respectively. To satisfy the condition, so that the reflected and refracted rays are perpendicular to each other, the angle of incidence should be $\tan^{-1}\left(1+\frac{10}{\theta}\right)^{1/2}$. The value of θ is ....

(Given for dielectric media, $\mu_r = 1$) *[2023, 29 Jan Shift-II]*

22 A source of light is placed in front of a screen. Intensity of light on the screen is I. Two polaroids P_1 and P_2 are so placed in between the source of light and screen that the intensity of light on screen is $I/2$. P_2 should be rotated by an angle of .... (degrees), so that the intensity of light on the screen becomes $3I/8$. *[2021, 26 Aug Shift-II]*

23 An unpolarised light beam is incident on the polariser of a polarisation experiment and the intensity of light beam emerging from the analyser is measured as 100 lumens. Now, if the analyser is rotated around the horizontal axis (direction of light) by 30° in clockwise direction, the intensity of emerging light will be ...... lumens. *[2021, 24 Feb Shift-I]*

• Key **Index**

Topic 1

1 (c)	**2** (a)	**3** (d)	**4** (c)	**5** (a)	**6** (a)	**7** (c)	**8** (d)	**9** (d)	**10** (b)
11 (d)	**12** (b)	**13** (b)	**14** (d)	**15** (b)	**16** (b)	**17** (d)	**18** (a)	**19** (a)	**20** (b)
21 (c)	**22** (c)	**23** (d)	**24** (a)	**25** (d)	**26** (b)	**27** (b)	**28** (d)	**29** (c)	**30** (a)
31 (b)	**32** (c)	**33** (b)	**34** (a)	**35** (b)	**36** (b)	**37** (d)	**38** (b)	**39** (d)	**40** (b)
41 (b)	**42** (a)	**43** (b)	**44** (d)	**45** (d)	**46** (*)	**47** (d)	**48** (a)	**49** (c)	**50** (d)
51 (d)	**52** (d)	**53** (c)	**54** (c)	**55** (b)	**56** (c)	**57** (d)	**58** (c)	**59** (24)	**60** (11)
61 (54)	**62** (4)	**63** (2)	**64** (8)	**65** (10)	**66** (9)	**67** (13)	**68** (462)	**69** (4)	**70** (48)
71 (24)	**72** (630)	**73** (2)	**74** (600)	**75** (450)	**76** (1)	**77** (300)	**78** (600)	**79** (750)	

Topic 2

1 (a)	**2** (c)	**3** (c)	**4** (b)	**5** (c)	**6** (d)	**7** (c)	**8** (d)	**9** (c)	**10** (a)
11 (6)	**12** (15)	**13** (3)	**14** (2)	**15** (30)	**16** (3)	**17** (98)			

Topic 3

1 (d)	**2** (a)	**3** (a)	**4** (b)	**5** (c)	**6** (c)	**7** (a)	**8** (d)	**9** (a)	**10** (b)
11 (b)	**12** (b)	**13** (c)	**14** (d)	**15** (d)	**16** (c)	**17** (d)	**18** (c)	**19** (c)	**20** (30)
21 (7)	**22** (30)	**23** (75)							

Detailed **Explanations**

Topic 1

01. *(c)* $E_1 = E_0 \sin(\omega t)$ $\qquad \phi = \frac{\pi}{3}$

$$E_2 = E_0 \sin\left(\omega t + \frac{\pi}{3}\right) \qquad A_1 = E_0,$$

$A_2 = E_0$

$$E_R = \sqrt{(E_0)^2 + (E_0)^2 + 2(E_0)(E_0)\cos\frac{\pi}{3}}$$

$$= \sqrt{E_0^2 + E_0^2 + E_0^2} = E_0\sqrt{3}$$

$$= 1.73 E_0$$

02. *(a)* For interference to occur, the two waves must be coherent-they must have the same frequency and a constant phase difference. Sine red and green light are of different frequencies, they are not coherent. Hence, there shall be no interference fringe.

03. *(d)* $\frac{I_{max}}{I_{min}} = \frac{(\sqrt{I_1} + \sqrt{I_2})^2}{(\sqrt{I_1} - \sqrt{I_2})^2}$

$$= \frac{(\sqrt{I_2})^2\left[\sqrt{\frac{I_1}{I_2}} + 1\right]^2}{(\sqrt{I_2})^2\left(\sqrt{\frac{I_1}{I_2}} - 1\right)^2}$$

$$= \frac{\left(\sqrt{\frac{I_1}{I_2}} + 1\right)^2}{\left(\sqrt{\frac{I_1}{I_2}} - 1\right)^2}$$

$$= \frac{\left(\sqrt{\frac{1}{9}} + 1\right)^2}{\left(\sqrt{\frac{1}{9}} - 1\right)^2}$$

$$= \frac{\left(\frac{4}{3}\right)^2}{\left(-\frac{2}{3}\right)^2} = 4$$

$\therefore I_{max} : I_{min} = 4 : 1$

04. *(c)* $\beta = \frac{D\lambda}{d} \propto \frac{1}{d}$

If d is doubled then β is half so 50% decrement will occur in fringe width.

05. *(a)* In YDSE fringe width

$$\beta = \frac{\lambda D}{d} \text{ and } \lambda_R > \lambda_B$$

$$\beta_R > \beta_B$$

> **• Concept Enhancer**
>
> On immersing the apparatus in liquid (denser medium), then wavelength decreases and hence fringe width also decreases.
>
> i.e. $\quad \beta_{med} = \frac{\beta_{air}}{\mu}$

06. *(a)* The fringe width is given by

$$\frac{n_1\lambda_1 D}{d} = \frac{n_2\lambda_2 D}{d}$$

$$\Rightarrow \qquad n_1 \times 480 = n_2 \times 600$$

$$(n_1)_{min} = 5$$

07. *(c)* For same amplitude,

$$I_1 = I_2 = ka^2 = I_0$$

When, $\phi_1 = \pi/3$

$$I_{0_1} = I_1 + I_2 + 2\sqrt{I_1 I_2}\cos\phi$$

$$I_{0_1} = I_0 + I_0 + 2I_0 \cos\frac{\pi}{3} = 3I_0$$

When, $\phi_2 = \pi/2$

$$I_{0_2} = I_1 + I_2 + 2\sqrt{I_1 I_2}\cos\phi$$
$$= 2I_0 + 2I_0 \cos\frac{\pi}{2} = 2I_0$$

Hence ratio, $\frac{I_{0_1}}{I_{0_2}} = \frac{3}{2}$

08. *(d)* Fringe width in *YDSE*, $\beta = \frac{\lambda D}{d}$

So, (In same experimental set up)

$$\frac{\beta_2}{\beta_1} = \lambda_2/\lambda_1 \Rightarrow \beta_2 = \beta_1 \cdot \frac{\lambda_2}{\lambda_1}$$

Here $\beta_1 = 2$ mm

$\lambda_2 = 600$ nm, $\lambda_1 = 400$ nm.

$$\therefore \quad \beta_2 = 2\text{mm} \times \frac{600\text{ nm}}{400\text{ nm}} = \frac{2\times 6}{4}\text{ mm}$$
$$= 3\text{ mm}$$

09. *(d)* Given, $D = 1$ m

$\lambda = 600\text{ nm} = 6\times 10^{-7}$ m

$x_5 = 5\text{ cm} = 5\times 10^{-2}$ m

We know that position of *n*th bright fringe, $x_n = nD\lambda/d$

Here, $n = 5$,

$$\therefore \quad x_5 = \frac{5D\lambda}{d} \Rightarrow d = \frac{5D\lambda}{x_5}$$
$$= \frac{5\times 1\times 6\times 10^{-7}}{5\times 10^{-2}} = 60\ \mu\text{m}$$

> • Concept **Enhancer**
>
> In Young's double slit experiment, the position of *n*th bright fringe from the central maximum is given by
>
> $$x_n = \frac{nD\lambda}{d}$$
>
> where, the symbols have their usual meanings.

10. *(b)* In Young's double slit experiment,

Fringe width, $\beta = 12\text{ mm} = 1.2\times 10^{-2}$ m

When entire arrangement of Young's double slit experiment is placed in water of refractive index, $\mu = \frac{4}{3}$, then

$$\text{Fringe width, } \beta' = \frac{\beta}{\mu} = \frac{1.2\times 10^{-2}}{4/3}$$
$$= 0.9\times 10^{-2}\text{ m} = 9\times 10^{-3}\text{ m} = 9\text{ mm}$$

11. *(d)* Fringe width, $\beta = \lambda D/d$

$$\Rightarrow \quad \frac{\beta_2}{\beta_1} = \frac{\lambda_2}{\lambda_1}\times\frac{d_1}{d_2} \quad \ldots\text{(i)}$$

Here, $d_2 = 2d_1$ and $\lambda_1 = 5000$ Å

$\lambda_2 = 6000$ Å, and $\beta_1 = 0.5$ mm

Hence, from Eq. (i), we get

$$\beta_2 = 0.5\times\frac{6000}{5000}\times\frac{1}{2}$$
$$\Rightarrow \beta_2 = 0.3\text{ mm}$$

12. *(b)* According to given situation,

$$\frac{I_1}{I_2} = \frac{4}{1}, \text{ So, } \frac{I_1}{I_2} = \left(\frac{A_1}{A_2}\right)^2$$
$$\left(\frac{A_1}{A_2}\right)^2 = \frac{4}{1} \Rightarrow \frac{A_1}{A_2} = \frac{2}{1},$$
$$\frac{A_{max}}{A_{min}} = \frac{2+1}{2-1} = \frac{3}{1} \Rightarrow \frac{I_{max}}{I_{min}} = \left(\frac{A_{max}}{A_{min}}\right)^2 = \frac{9}{1}$$
$$\Rightarrow \frac{I_{max}+I_{min}}{I_{max}-I_{min}} = \frac{5}{4} = \frac{5}{x}$$

So, $\quad x = 4$

13. *(b)* Given,

$$I_1 = I, I_2 = 9I, \phi_P = \frac{\pi}{2}, \phi_Q = \pi$$
$$I_P = I + 9I + 2\sqrt{I\times 9I}\cos\frac{\pi}{2} = 10I$$
$$I_Q = I_1 + I_2 + 2\sqrt{I_1 I_2}\cos\pi;$$
$$I_Q = I + 9I - 2\sqrt{I\times 9I} = 4I$$

Then, the difference between the resultant intensities at *P* and *Q* will be $10I - 4I = 6I$.

14. *(d)* Given, intensity of two light waves

$$I_1 = I_2 = I_0 \Rightarrow I_{min} = 0$$

Since, $\quad I_{min} = (\sqrt{I_1} - \sqrt{I_2})^2$

$$0 = (\sqrt{I_1} - \sqrt{I_2})^2$$

$\therefore \quad I_1 = I_2 = I_0$

and $\quad I_{max} = (\sqrt{I_1} + \sqrt{I_2})^2$

$$= (\sqrt{I_0} + \sqrt{I_0})^2$$
$$= (2\sqrt{I_0})^2 = 4I_0$$

15. *(b)* We know that in Young's double slit experiment, fringe width, $\beta = \frac{D\lambda}{d}$

$$\beta \propto \lambda$$

Since, $\quad \lambda_{blue} < \lambda_{orange}$

Thus, $\quad \beta_{blue} < \beta_{orange}$

$\therefore$ In Young's double slit experiment, if the source of light changes from orange to blue, then distance between the consecutive fringes will decrease.

16. *(b)* Fringe width,

$$\beta = \frac{D\lambda}{d} = \frac{1\times 500\times 10^{-9}}{2\times 10^{-3}} = 2.5\times 10^{-4}\text{ m}$$
$$\Rightarrow \quad \beta = 0.25\text{ mm}$$

17. *(d)* Given, $\frac{I_1}{I_2} = 2x \Rightarrow I_1 = 2I_2 x$

As we know,

$$I_{max} = (\sqrt{I_1} + \sqrt{I_2})^2$$

and $\quad I_{min} = (\sqrt{I_1} - \sqrt{I_2})^2$

$$\therefore \quad \frac{I_{max} - I_{min}}{I_{max} + I_{min}} = \frac{4\sqrt{I_1 I_2}}{2(I_1 + I_2)}$$

(using dividendo rule)

$$= \frac{2\sqrt{2I_2 x\cdot I_2}}{2I_2 x + I_2} = \frac{2\sqrt{2x}}{2x+1}$$

18. *(a)* According to Young's double slit experiment,

The distance of *n*th bright fringe from the centre, $y_n = \frac{n\lambda D}{d}$

Since, $\quad \lambda_{violet} < \lambda_{red}$

$\therefore \quad y_{violet} < y_{red}$

$\therefore$ Consecutive fringe lines will come closer.

19. *(a)* Optical path difference for first ray which travels a path L_1 through a refractive index $n_1 = n_1 L_1$

L_2 in refractive index, $n_2 = n_2 L_2$

Path difference $= n_1 L_1 - n_2 L_2$

$$\text{Phase difference} = \frac{2\pi}{\lambda}\times\text{ path difference}$$
$$= 2\frac{\pi}{\lambda}[n_1 L_1 - n_2 L_2]$$

20. *(b)* Given, setup of Young's double slit experiment is as shown

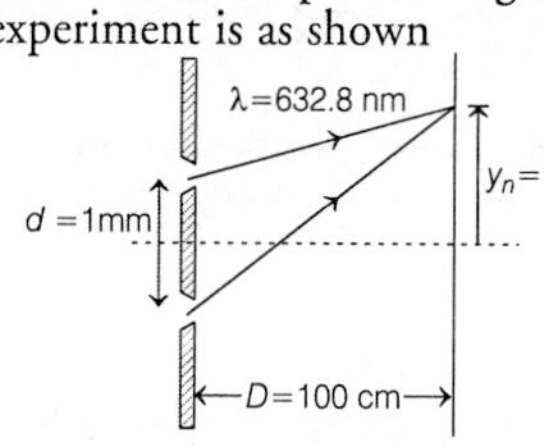

Distance of a bright fringe from central bright fringe is

$$y_n = \frac{n\lambda D}{d}$$

where, $n\lambda$ = path difference between rays from lower and upper slit $= \frac{y_n d}{D}$

$$= \frac{1.27\times 10^{-3}\times 10^{-3}}{100\times 10^{-2}}$$
$$= 1.27\times 10^{-6}\text{ m} = 1.27\ \mu\text{m}$$

21. *(c)* As we know,

Path difference introduced by thin film,

$$\Delta = (\mu - 1)t \quad \ldots\text{(i)}$$

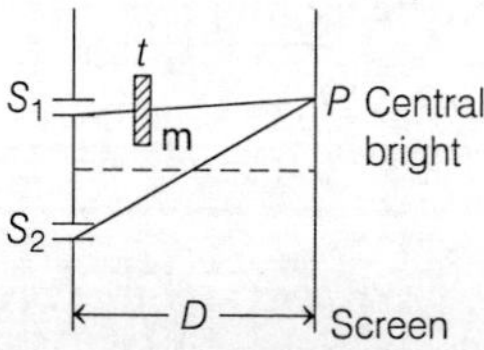

and if fringe pattern shifts by one fringe width, then path difference,

$$\Delta = 1\times\lambda = \lambda \quad \ldots\text{(ii)}$$

So, from Eqs. (i) and (ii), we get

$$(\mu-1)t=\lambda \Rightarrow t=\frac{\lambda}{\mu-1}$$

Alternate Solution

Path difference introduced by the thin film of thickness t and refractive index μ is given by

$$\Delta=(\mu-1)t$$

$\therefore$ Position of the fringe is

$$x=\frac{\Delta D}{d}=\frac{(\mu-1)tD}{d} \quad \ldots(i)$$

Fringe width of one fringe is given by

$$\beta=\frac{\lambda D}{d} \quad \ldots(ii)$$

Given that $x=\beta$, so from Eqs. (i) and (ii), we get

$$\Rightarrow \frac{(\mu-1)tD}{d}=\frac{\lambda D}{d} \Rightarrow (\mu-1)t=\lambda \text{ or } t=\frac{\lambda}{(\mu-1)}$$

22. *(c)* We know that if the angular position of second minima from central maxima is θ then,

$$\sin\theta=\frac{2\lambda}{a}=\frac{2\times550\times10^{-9}}{22\times10^{-7}}=\frac{1}{2}$$

$$\theta=\pi/6 \text{ rad}$$

23. *(d)*

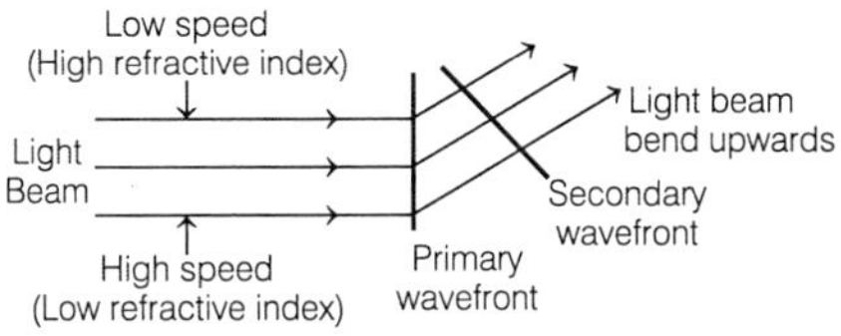

Due to speed variation a tilted wavefront occurs and light beam bend upwards.

24. *(a)* When the screen is placed perpendicular to the line joining the sources the fringes will be concentric circles.

25. *(d)* We know that,

Maximum intensity for two waves in case of coherent interference is

$$I_{max}=(\sqrt{I_1}+\sqrt{I_2})^2$$

$$(\because \theta=2n\pi, \cos\theta=1)$$

Similarly, for n identical waves

$$I_{max}=(\sqrt{I_0}+\sqrt{I_0}+\ldots)^2=n^2I_0 \ \ldots(i)$$

Maximum intensity for two waves in case of incoherent interference is

$$I'_{max}=I_1+I_2 \quad (\because \langle\cos\theta\rangle=0)$$

$\therefore$ For n identical waves

$$I'_{max}=nI_0 \quad \ldots(ii)$$

Divide Eq. (i) and (ii)

$$\frac{I_{max}}{I'_{max}}=\frac{n^2I_0}{nI_0}=n$$

26. *(b)* We know that $\beta=\frac{\lambda D}{d}$

Now, for small angle α

$$\sin\alpha=\frac{d}{D} \Rightarrow \frac{1}{\sin\alpha}=\frac{D}{d}$$

$$\therefore \beta=\frac{\lambda}{\sin\alpha}\approx\frac{\lambda}{\alpha}$$

27. *(b)* For coherent sources,

$$I_1=4I_0\cos^2\frac{\phi}{2}=4I_0$$

For incoherent sources,

$$I_2=I_0+I_0=2I_0 \quad \therefore \frac{I_1}{I_2}=2$$

28. *(d)* For a parallel beam, wavefront will be planar.

29. *(c)* $3\lambda_1=4\lambda_2$ $\quad \left[\because \frac{3D\lambda_1}{d}=4\frac{D\lambda_2}{d}\right]$

$$\Rightarrow \lambda_2=\frac{3}{4}\lambda_1=\frac{3}{4}\times590=\frac{1770}{4}$$

$$=442.5 \text{ nm}$$

30. *(a)* The shape of the interference fringes formed on a screen is hyperbola line.

31. *(b)* For possible interference maxima on the screen, the condition is

$$d\sin\theta=n\lambda \quad \ldots(i)$$

Given, d = slit-width = 2λ

$$2\lambda\sin\theta=n\lambda \quad \text{or} \quad 2\sin\theta=n$$

The maximum value of $\sin\theta$ is 1, hence

$$n=2\times1=2$$

Thus, Eq. (i) must be satisfied by 5 integer values *i.e.*, –2, –1, 0, 1, 2. Hence, the maximum number of possible interference maxima is 5.

32. *(c)* Since, $I \propto$ width

Thus, $I_1=I_0$ and $I_2=2I_0$

$$\Rightarrow \frac{I_2}{I_1}=2$$

$$\therefore \frac{I_{max}}{I_{min}}=\frac{(\sqrt{I_1}+\sqrt{I_2})^2}{(\sqrt{I_1}-\sqrt{I_2})^2}$$

$$=\frac{(\sqrt{I_1})^2\left[1+\sqrt{\frac{I_2}{I_1}}\right]^2}{(\sqrt{I_1})^2\left[1-\sqrt{\frac{I_2}{I_1}}\right]^2}=\frac{[1+\sqrt{2}]^2}{(1-\sqrt{2})^2}$$

$$=\frac{1+2+2\sqrt{2}}{1+2-2\sqrt{2}}=\frac{3+2\sqrt{2}}{3-2\sqrt{2}}$$

33. *(b)* The direction of propagation of light is perpendicular to the wave front and is symmetric about X, Y and Z-axis.

$\therefore$ Angle made by the light with X, Y and Z-axis is same.

$\therefore \cos\alpha=\cos\beta=\cos\gamma,$

$X+Y+Z=\cos\pi$

also $\cos^2\alpha+\cos^2\beta+\cos^2\gamma=1$

$$\alpha=\cos^{-1}\left(\frac{1}{\sqrt{3}}\right)$$

34. *(a)*

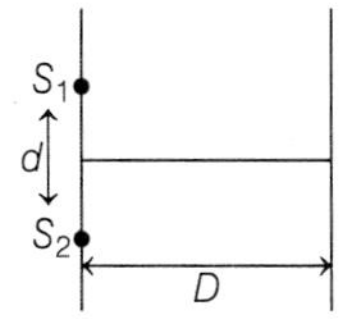

$$d=1.5 \text{ mm}=1.5\times10^{-3} \text{ m}$$

$$\lambda=690 \text{ nm}=690\times10^{-9} \text{ m}$$

$$D=0.72 \text{ m}$$

Fringe width, $\beta=?$

$$=\frac{690\times10^{-9}\times0.72}{1.44\times1.5\times10^{-3}} \quad \left[\beta_m=\frac{\beta_{air}}{\mu}\right]$$

$$\left[\beta_{air}=\frac{\lambda D}{d}\right]$$

$$=0.23\times10^{-3}=0.23 \text{ mm}$$

35. *(b)* $\frac{I_{max}}{I_{min}}=\frac{9}{4}$ (given)

$$I_{max}=(\sqrt{I_1}+\sqrt{I_2})^2$$

$$I_{min}=(\sqrt{I_1}-\sqrt{I_2})^2$$

Put the value

$$\frac{(\sqrt{I_1}+\sqrt{I_2})^2}{(\sqrt{I_1}-\sqrt{I_2})^2}=\frac{9}{4}$$

$$\frac{\sqrt{I_1}+\sqrt{I_2}}{\sqrt{I_1}-\sqrt{I_2}}=\frac{3}{2} \quad \ldots(i)$$

Given intensity is directly proportional to the square of the slit width

$$I \propto d^2$$

Thus, according to question $\sqrt{I_1}=xd$, $\sqrt{I_2}=d$ put the value in Eq. (i)

$$\frac{xd+d}{xd-d}=\frac{3}{2}$$

$$\frac{d(x+1)}{d(x-1)}=\frac{3}{2}$$

$$2(x+1)=3(x-1)$$

$$2x+2=3x-3$$

$$5=3x-2x$$

$$x=5$$

36. *(b)* Given,

$$\frac{W_1}{W_2}=4$$

We know that,

$$\frac{W_1}{W_2}=\frac{I_1}{I_2}=\left(\frac{a_1}{a_2}\right)^2$$

$$\Rightarrow 4=\left(\frac{a_1}{a_2}\right)^2$$

$$\Rightarrow \frac{a_1}{a_2}=2$$

$\therefore \quad \dfrac{I_{max}}{I_{min}} = \dfrac{(a_1 + a_2)^2}{(a_1 - a_2)^2}$

$= \dfrac{a_2^2\left(\dfrac{a_1}{a_2}+1\right)^2}{a_2^2\left(\dfrac{a_1}{a_2}-1\right)^2} = \dfrac{(2+1)^2}{(2-1)^2} = \dfrac{9}{1}$

$\therefore \; I_{max} : I_{min} = 9:1$

37. *(d)* In given YDSE, first minimum occurs at P. When wavelength of light used is

$\lambda = 800$ nm

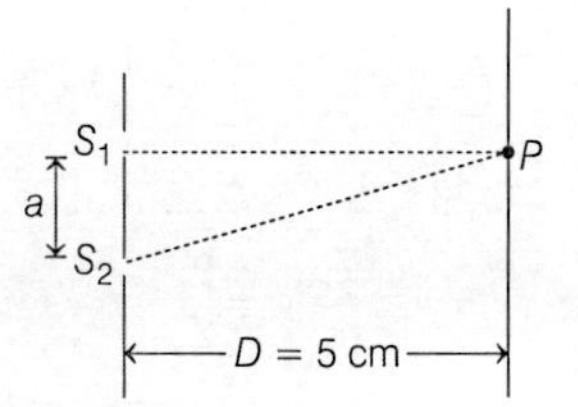

For 1st minima,
Path difference,

$S_2P - S_1P = \sqrt{(5+a)^2} - 5 = \dfrac{\lambda}{2}$

Taking cm unit for length

$\Rightarrow \sqrt{5^2 + a^2} - 5 = \dfrac{\lambda}{2}$

$5\left(1 + \dfrac{a^2}{5^2}\right)^{1/2} - 5 = \dfrac{\pi}{2}$

$5\left(1 + \dfrac{1}{2} \times \dfrac{a^2}{5^2}\right) - 5 = \dfrac{\lambda}{2}$

$\dfrac{a^2}{2 \times 5} = \dfrac{\lambda}{2}$

$\Rightarrow \quad a = 0.2$ mm

38. *(b)* In YDSE, maximum occurs at same position. This means extra path difference due to glass plate is a multiple of wavelength λ.

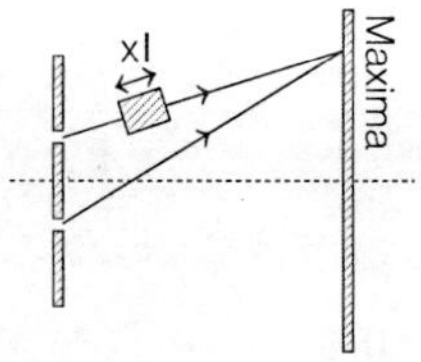

Now, extra path difference due to glass plate is

$\Delta L = (\mu - 1)t \quad (\because t = \text{thickness} = x\lambda)$

$\Rightarrow \Delta L = (1.5 - 1)x\lambda$

$\Rightarrow \Delta L = x\lambda / 2$

Now, for given case,

$\Delta L = \dfrac{x\lambda}{2} = n\lambda$

(As maxima stays at same place)

$\Rightarrow \quad x = 2n$

As, $n = 0, 1, 2, 3$ ......
So, value of x can be
$x = 0, 2, 4, 6$ ...... etc.

• Concept Enhancer

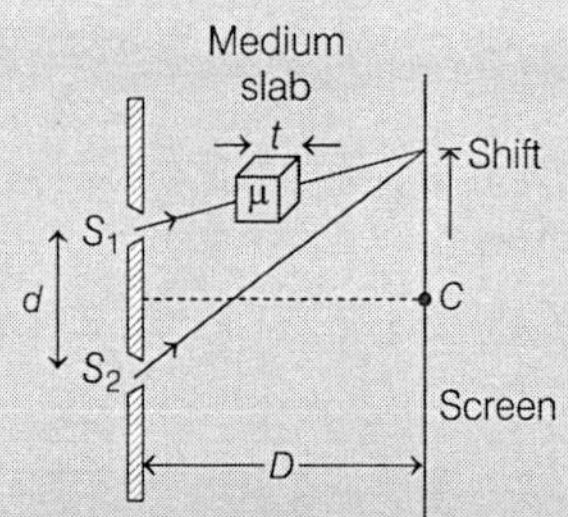

A transparent medium slab is kept in path of light beam from one of the slits in YDSE. Fringe pattern will gets shifted to that side.

Remember three formulae

- Additional path difference $= (\mu - 1)t$
- Fringe shift $= \dfrac{D}{d}(\mu - 1)t = \dfrac{\beta}{\lambda}(\mu - 1)t$
- Shift in number of n fringes, $n = \dfrac{(\mu - 1)t}{\lambda}$ or $t = \dfrac{n\lambda}{(\mu - 1)}$

39. *(d)* Given, $I_1 : I_2 = 9:4$

$\dfrac{I_{max}}{I_{min}} = \dfrac{(\sqrt{I_1} + \sqrt{I_2})^2}{(\sqrt{I_1} - \sqrt{I_2})^2} = \left(\dfrac{3+2}{3-2}\right)^2$

$= 25:1$

40. *(b)* Given, separation between slits,

$d(t) = d_0 + a_0 \sin \omega t$

As we know that,
Fringe width $(\beta) = \lambda D / d$

$\Rightarrow \quad \beta = \dfrac{\lambda D}{d_0 + a_0 \sin \omega t}$

Now, maximum fringe width

$\beta_{max} = \dfrac{\lambda D}{d_0 - a_0} \quad (\text{for } \sin \omega t = -1)$

and minimum fringe width

$\beta_{min} = \dfrac{\lambda D}{d_0 + a_0} \quad (\text{for } \sin \omega t = +1)$

$\therefore \beta_{max} - \beta_{min} = \lambda D \left[\dfrac{1}{d_0 - a_0} - \dfrac{1}{d_0 + a_0}\right]$

$= \lambda D \left[\dfrac{2a_0}{d_0^2 - a_0^2}\right]$

41. *(b)* Given, The distance of sources from the screen, $D = 0.5$ m
The distance between the slits,
$d = 0.5$ mm $= 0.0005$ m
The wavelength of the monochromatic light,

$\lambda = 5890$ Å $= 5890 \times 10^{-10}$ m

The distance between the first and the third bright fringe formed, $y = 2\beta$

As we know, $\beta = \dfrac{\lambda D}{d}$

On substituting the values in the above equation, we get

$y = 2\left(\dfrac{\lambda D}{d}\right) = 2\left(\dfrac{5890 \times 10^{-10} \times 0.5}{0.0005}\right)$

$= 1178 \times 10^{-6}$ m

42. *(a)* Given, amplitude $\propto$ width of slit

$\Rightarrow \quad A_2 = 3A_1$

We know that, $\dfrac{I_{max}}{I_{min}} = \dfrac{(\sqrt{I_1} + \sqrt{I_2})^2}{(\sqrt{I_1} - \sqrt{I_2})^2}$

$\because$ Intensity,
$I \propto A^2$

$\therefore \quad \dfrac{I_{max}}{I_{min}} = \dfrac{(A_1 + A_2)^2}{(|A_1 - A_2|)^2}$

$= \left(\dfrac{A_1 + 3A_1}{|A_1 - 3A_1|}\right)^2 = \left(\dfrac{4A_1}{2A_1}\right)^2 = \dfrac{4}{1}$

$\therefore \; I_{max} : I_{min} = 4:1$

43. *(b)*

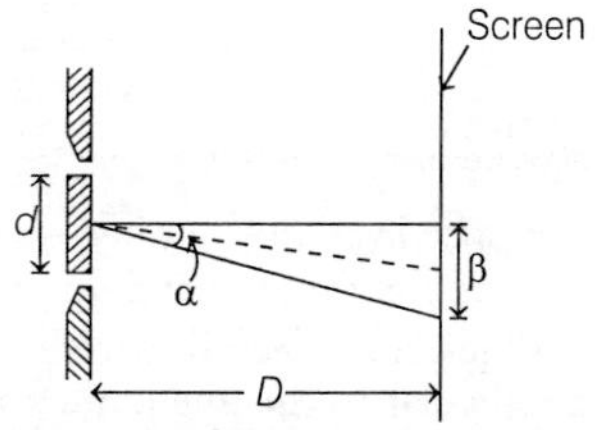

Angular width of central maxima in YDSE is given by,

$\alpha = \dfrac{\beta}{D} = \dfrac{\lambda}{d}$ (in radian)

$= \dfrac{\lambda}{d} \times \dfrac{180}{\pi}$ (in degree)

Here, $\lambda = 500$ nm $= 500 \times 10^{-9}$ m
and $d = 0.05$ mm $= 0.05 \times 10^{-3}$ m.
Hence, angular width is

$\alpha = \dfrac{500 \times 10^{-9}}{0.05 \times 10^{-3}} \times \dfrac{180}{\pi} = 0.57°$

44. *(d)* In YDSE, let segment length is $2y$ as shown in figure.

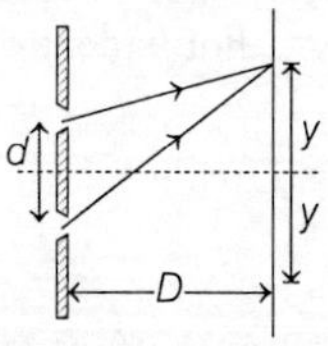

Then,

$2y = 2 \times N \times \dfrac{\lambda D}{d}$

where, N = number of fringes in segment length $2y$.

$\Rightarrow \quad N\lambda = yd / D$

So, for same segment length $N\lambda$ is constant.

$\Rightarrow \quad N_1\lambda_1 = N_2\lambda_2$

Given, $N_1 = 16$, $\lambda_1 = 700$ nm

$N_2 = ?$, $\lambda_2 = 400$ nm

$$\Rightarrow \quad N_2 = \frac{N_1 \lambda_1}{\lambda_2} = 16 \times \frac{700}{400} = 28$$

$\therefore$ 28 fringes are observed in same.

45. *(d)* In YDSE, separation between successive bright fringes is width of one dark fringe.

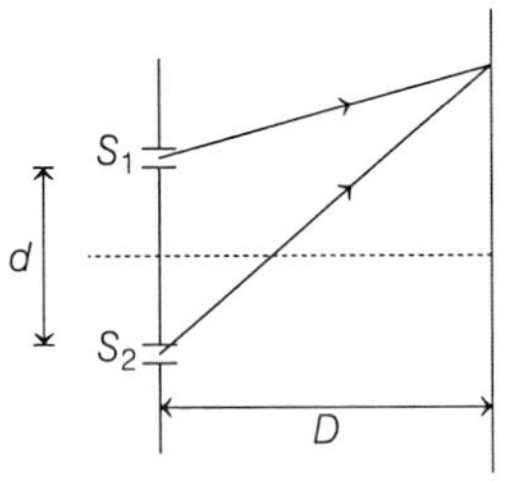

As, fringe width in YDSE,

$$\beta = \frac{\lambda D}{d} \quad \ldots \text{(i)}$$

Here, $\lambda = 589 \text{ nm} = 589 \times 10^{-9}$ m,

$D = 1.5$ m,

$d = 0.15 \text{ mm} = 0.15 \times 10^{-3}$ m

On substituting the given values in Eq. (i), we get

$$\beta = \frac{589 \times 10^{-9} \times 1.5}{0.15 \times 10^{-3}} = 589 \times 10^{-5} \text{ m}$$

$$= 5.89 \text{ mm or } 5.9 \text{ mm}$$

46. *(*)* Path difference introduced by a slab of thickness t and refractive index μ is given by, $\Delta = (\mu - 1)t$

Position of the fringe is

$$x = \frac{\Delta D}{d} = \frac{(\mu - 1)t\,D}{d}$$

Also, fringe width is given by

$$\beta = \lambda D / d$$

According to the question,

$$n\beta = x$$

$$\Rightarrow \quad \frac{n\lambda D}{d} = (\mu - 1)t\frac{D}{d}$$

$$\Rightarrow \quad n\lambda = (\mu - 1)t$$

$$\text{or} \quad t = \frac{n\lambda}{(\mu - 1)}$$

So, no option is correct.

> **Alternate Solution**
>
> Path difference, $\Delta = \frac{xd}{D} = (\mu - 1)t$ or $t = \frac{xd}{(\mu - 1)D}$
>
> Also, $\beta = \frac{\lambda D}{d} \Rightarrow \frac{d}{D} = \frac{\lambda}{\beta}$
>
> $\therefore \quad t = \frac{x\lambda}{(\mu - 1)\beta} = \frac{n\lambda D \times \lambda}{d(\mu - 1)\beta} \quad [\because x = n\beta]$
>
> $\Rightarrow \quad t = \frac{n\lambda}{(\mu - 1)}$
>
> So, no option is correct.

47. *(d)* Let intensity of each wave is I_0.

Then, intensity at the centre of bright fringe will be $4I_0$.

Given, path difference, $\Delta x = \lambda / 8$

$\therefore$ Phase difference, $\phi = \Delta x \times \frac{2\pi}{\lambda}$

$$\Rightarrow \quad \phi = \frac{\lambda}{8} \times \frac{2\pi}{\lambda} \text{ or } \phi = \pi / 4$$

Intensity of light at this point,

$$I' = I_0 + I_0 + 2I_0 \cos\left(\frac{\pi}{4}\right)$$

$$= 2I_0 + \sqrt{2}\,I_0 = 3.41\,I_0$$

Now, $\frac{I'}{4I_0} = \frac{3.41}{4} = 0.85$

48. *(a)* Let n_1th fringe formed due to first wavelength and n_2th fringe formed due to second wavelength coincide i.e. their distance from common central maxima will be same

i.e. $\quad y_{n_1} = y_{n_2} \Rightarrow \frac{n_1 \lambda_1 D}{d} = \frac{n_2 \lambda_2 D}{d}$

$$\Rightarrow \quad \frac{n_1}{n_2} = \frac{\lambda_2}{\lambda_1} = \frac{520}{650} = \frac{4}{5}$$

$\because \quad n_1 = 4$ and $n_2 = 5$

Hence, distance of the point of coincidence from the central maxima is

$$y = \frac{n_1 \lambda_1 D}{d} = \frac{n_2 \lambda_2 D}{d}$$

$$= \frac{4 \times 650 \times 10^{-9} \times 1.5}{0.5 \times 10^{-3}} = 7.8 \text{ mm}$$

49. *(c)* For a particular distance d_0 between the slits, the eye is not able to resolve two consecutive bright fringes.

Now, $\theta = \frac{\beta}{D}$ but $\beta = \frac{\lambda D}{d_0} \quad \left(\because \theta = \frac{\lambda}{d_0}\right)$

$$\text{or} \quad d_0 = \frac{\lambda}{\theta} = \frac{600 \times 10^{-9} \text{ m}}{\frac{1}{60} \times \frac{\pi}{180} \text{ rad}}$$

$$= 2.06 \times 10^{-3} \text{ m} \approx 2 \text{ mm}$$

50. *(d)* We know that

Intensity $\propto$ (amplitude)^2

$$\frac{I_1}{I_2} = \frac{16}{9} = \frac{a_1^2}{a_2^2} \Rightarrow a_1 = 4;\ a_2 = 3$$

Therefore, the ratio of intensities of bright and dark parts

$$\frac{I_{\text{Bright}}}{I_{\text{Dark}}} = \frac{(a_1 + a_2)^2}{(a_1 - a_2)^2} = \frac{(4+3)^2}{(4-3)^2} = \frac{49}{1}$$

51. *(d)* Given, $a_1 = 2a_2$

$$\Rightarrow \quad I_1 = 4I_2 = 4I_0 \quad [\because I \propto a^2]$$

$$\therefore \quad I_{\max} = (\sqrt{I_1} + \sqrt{I_2})^2$$

$$= (3\sqrt{I_2})^2 = 9I_2 = 9I_0$$

Now, $I = I_1 + I_2 + 2\sqrt{I_1 I_2} \cos\phi$

$$= 4I_0 + I_0 + 2\sqrt{4I_0 \cdot I_0} \cos\phi$$

$$= 5I_0 + 4I_0 \cos\phi$$

$$= \frac{I_m}{9}(5 + 4\cos\phi) \quad \left[\because I_0 = \frac{I_m}{9}\right]$$

$$= \frac{I_m}{9}[1 + 4(1 + \cos\phi)]$$

$$= \frac{I_m}{9}(1 + 8\cos^2\phi/2)$$

$$\left[\because \cos\phi = 2\cos^2\frac{\phi}{2} - 1\right]$$

52. *(d)* Phase difference $= \frac{2\pi}{\lambda} \times$ Path difference

i.e., $\quad \phi = \frac{2\pi}{\lambda} \times \frac{\lambda}{6} = \frac{\pi}{3}$

As, $\quad I = I_{\max} \cos^2\left(\frac{\phi}{2}\right)$

or $\quad \frac{I}{I_{\max}} = \cos^2\left(\frac{\phi}{2}\right)$

or $\quad \frac{I}{I_0} = \cos^2\left(\frac{\pi}{6}\right) = \frac{3}{4} \quad [\because I_{\max} = I_0]$

53. *(c)* Both A and R are true and R is the correct explanation of A.

$\beta \cdot$(fringe width) $= \frac{\lambda D}{d}$.

In denser medium, λ decrease therefore β will also decreases.

$\therefore$ Fringe will come closer.

Also $\mu = \frac{c}{v}$ or $v = \frac{c}{\mu}$

Frequency remains same.

$$\Rightarrow \quad \mu = \frac{\lambda_{\text{vaccum}} \times f}{\lambda_{\text{medium}} \times f}$$

$$\Rightarrow \quad \lambda_{\text{medium}} = \frac{\lambda_{\text{vaccum}}}{\mu}$$

54. *(c)* Given,

Path difference,

$$\Delta x = \frac{7\lambda}{4}$$

If I be the intensity of two identical source, then

$$I_{\max} = I_1 + I_2 + 2\sqrt{I_1 I_2} \cos\phi$$

$$= I + I + 2I\cos 0° = 4I$$

When, $\Delta x = \frac{7\lambda}{4}$

Then, $\Delta\phi' = \frac{2\pi}{\lambda} \times \Delta x = \frac{2\pi}{\lambda} \times \frac{7\lambda}{4} = \frac{7}{2}\pi$

∴ Intensity

$$I' = I_1 + I_2 + 2\sqrt{I_1 I_2}\cos\Delta\phi'$$
$$= I + I + 2\sqrt{I\cdot I}\cos\frac{7\pi}{2}$$
$$= 2I + 2I\cos\left(2\pi + \frac{3\pi}{2}\right)$$
$$= 2I + 2I\cos\left(\frac{3\pi}{2}\right)$$
$$= 2I + 2I\cos(\pi + \pi/2)$$
$$= 2I + 2I(-\cos\pi/2) = 2I$$
$$\therefore \quad \frac{I'}{I_{max}} = \frac{2I}{4I} = \frac{1}{2}$$

55. *(b)*

> **key idea** In YDSE, $I_{max} = (\sqrt{I_1} + \sqrt{I_2})^2$
> and $I_{min} = (\sqrt{I_1} - \sqrt{I_2})^2$
> So, $\dfrac{I_{max} - I_{min}}{I_{max} + I_{min}} = \dfrac{2\sqrt{I_1 I_2}}{(I_1 + I_2)}$

Given, $I_1 : I_2 = 1:4$

$\Rightarrow \quad I_1 = k$ and $I_2 = 4k$

$$\therefore \frac{I_{max} + I_{min}}{I_{max} - I_{min}} = \frac{I_1 + I_2}{2\sqrt{I_1}\sqrt{I_2}}$$
$$= \frac{K + 4K}{2\sqrt{k}\sqrt{4k}} = \frac{5}{4}$$

Now, since given that

$$\frac{I_{max} + I_{min}}{I_{max} - I_{min}} = \frac{5}{4} = \frac{2\alpha + 1}{\beta + 3}$$

$\Rightarrow \quad 2\alpha + 1 = 5 ; \beta + 3 = 4$

$\Rightarrow \quad \alpha = \frac{4}{2} = 2$; and $\beta = 1$

$\alpha/\beta = 2$

56. *(c)* In YDSE, ratio of intensities of maxima and minima is given by

$$\frac{I_{max}}{I_{min}} = \frac{(\sqrt{I_1} + \sqrt{I_2})^2}{(\sqrt{I_1} - \sqrt{I_2})^2}$$
$$\Rightarrow \frac{I_{max}}{I_{min}} = \frac{(a_1 + a_2)^2}{(a_1 - a_2)^2}$$
$$\Rightarrow \frac{I_{max}}{I_{min}} = \frac{(\sqrt{W_1} + \sqrt{W_2})^2}{(\sqrt{W_1} - \sqrt{W_2})^2}$$

where, W_1 and W_2 are the widths of slits, respectively.

Here, $\left(\dfrac{W_1}{W_2}\right) = \left(\dfrac{a_1}{a_2}\right)^2 = \dfrac{4}{1} \Rightarrow \sqrt{\dfrac{W_1}{W_2}} = \dfrac{2}{1}$

So, we have

$$\frac{I_{max}}{I_{min}} = \left(\frac{\sqrt{W_1} + \sqrt{W_2}}{\sqrt{W_1} - \sqrt{W_2}}\right)^2 = \left(\frac{\sqrt{\frac{W_1}{W_2}} + 1}{\sqrt{\frac{W_1}{W_2}} - 1}\right)^2$$
$$= \left(\frac{2+1}{2-1}\right)^2 = 9:1$$

57. *(d)* In a YDSE, path difference between two rays, reaching at some common point P located at angular position θ as shown in the figure below is

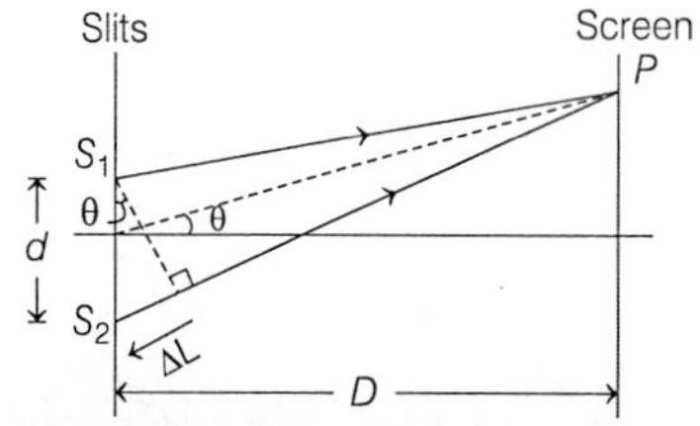

$\Delta L = d\sin\theta$

For small value of θ,

$$\sin\theta \approx \theta$$

So, path difference $= \Delta L = d\theta$

For a bright fringe at same angular position θ, both of the rays from slits S_1 and S_2 are in phase.

Hence, path difference is an integral multiple of wavelength of light used.

i.e., $\Delta L = n\lambda$ or $d\theta = n\lambda \Rightarrow \lambda = \dfrac{d\theta}{n}$

Here, $\theta = \dfrac{1}{40}$ rad, $d = 0.1$ mm

Hence, $\lambda = \dfrac{0.1}{40n}$ mm $= \dfrac{0.1\times10^{-3}}{40n}$ m

$$= \frac{0.1\times10^{-3}\times10^{9}}{40n}\text{ nm}$$
$$\Rightarrow \quad \lambda = \frac{2500}{n}\text{ nm}$$

So, with light of wavelength λ_1, we get

$$\lambda_1 = \frac{2500}{n_1}\text{(nm)}$$

and with light of wavelength λ_2, we get

$$\lambda_2 = \frac{2500}{n_2}\text{(nm)}$$

Now, choosing different integral values for n_1 and n_2, (i.e., $n_1, n_2 = 1, 2, 3 \ldots$ etc) we find that for

$$n_1 = 4,$$
$$\lambda_1 = \frac{2500}{4} = 625\text{ nm}$$

and for $n_2 = 5$, $\lambda_2 = \dfrac{2500}{5} = 500$ nm

These values lie in given interval 500 nm to 625 nm.

58. *(c)* Let the intensity of two coherent sources be I_1 and I_2, respectively.

It is given that $= \dfrac{I_{max}}{I_{min}} = \dfrac{16}{1}$

Since, we know,

$$I_{max} = (\sqrt{I_1} + \sqrt{I_2})^2$$

and $\quad I_{min} = (\sqrt{I_1} - \sqrt{I_2})^2$

∴ We can write,

$$\frac{I_{max}}{I_{min}} = \frac{(\sqrt{I_1} + \sqrt{I_2})^2}{(\sqrt{I_1} - \sqrt{I_2})^2} = 16$$
$$\Rightarrow \frac{\sqrt{I_1} + \sqrt{I_2}}{\sqrt{I_1} - \sqrt{I_2}} = \frac{\sqrt{16}}{1} = \frac{4}{1}$$
$$\sqrt{I_1} + \sqrt{I_2} = 4\sqrt{I_1} - 4\sqrt{I_2}$$
$$\Rightarrow \quad 5\sqrt{I_2} = 3\sqrt{I_1}, \ \frac{\sqrt{I_1}}{\sqrt{I_2}} = \frac{5}{3}$$

On squaring both the sides, we get

$$\frac{I_1}{I_2} = \frac{25}{9}$$

59. *(24)*

$$I_{max} = (\sqrt{I_1} + \sqrt{I_2})^2$$
$$= (\sqrt{4I} + \sqrt{9I})^2$$
$$= (2\sqrt{I} + 3\sqrt{I})^2$$
$$= 4I + 9I + 12I = 25I$$
$$I_{min} = (\sqrt{I_1} - \sqrt{I_2})^2$$
$$= (\sqrt{4I} - \sqrt{9I})^2$$
$$= 4I + 9I - 2\sqrt{4I\times9I} = I$$
$$\therefore I_{max} - I_{min} = 25I - I$$
$$= 24I = xI \text{ (given)}$$

Thus, $x = 24$

60. *(11)* Distance of 10th bright fringe from central maxima is

$$y = 10\frac{\lambda D}{d}$$
$$\frac{y_2}{y_1} = \frac{\lambda_2}{\lambda_1}$$

(For different wavelengths)

$$y_2 = \frac{660\text{nm}}{600\text{nm}}\times 10\text{ mm}$$
$$y_1 = 11\text{ mm}$$

61. *(54)* Maxima condition,

$$2\mu t = n\lambda$$
$$t = \frac{n\lambda}{2\mu},$$
$$t = \frac{\lambda}{2\mu}, \frac{2\lambda}{2\mu}$$

$(n = 1, 2, \ldots\ldots)$

Minima condition

$$2\mu t = (2n-1)\frac{\lambda}{2}, \ldots\ldots$$
$$t = \frac{(2n-1)\lambda}{2\mu\times2} = \frac{(2n-1)\lambda}{4\mu}$$

$$t = \frac{\lambda}{4\mu}, \frac{3\lambda}{4\mu} [n = 1, 2, \ldots\ldots]$$

$$\Delta t = \frac{3\lambda}{4\mu} - \frac{\lambda}{4\mu}$$

$$= \frac{2\lambda}{4\mu} = \frac{\lambda}{2\mu}$$

Rate of evaporation $= \frac{A \cdot \Delta t}{\text{time}}$

$$= \frac{\pi r^2 \times \Delta t}{t}$$

$$= \frac{\pi \times (1.8 \times 10^{-2})^2 \times 2 \times 560 \times 10^{-9}}{4 \times 1.4 \times 2}$$

$$= 54\pi \times 10^{-13} \text{ m}^3/\text{s}$$

62. *(4)* Given, wavelength, $\lambda = 500$ nm

$$= 500 \times 10^{-9} \text{ m}$$

Refractive index, $\mu = 1.5$
Fringe shift $= 4\beta$
where β is the fringe width.
When a slit is covered by a transparent medium, the fringe shift is given by
$n\lambda = (\mu - 1)t$
Here, t = thickness of the slab
$4 \times 500 \times 10^{-9} = (1.5 - 1) \times t$

$$\Rightarrow t = \frac{2 \times 10^{-6}}{0.5} = 4 \times 10^{-6}$$

$$= 4 \,\mu\text{m}$$

63. *(2)* Width of central maxima

$$\alpha = \frac{2\lambda D}{a}, \; a = \text{Slit width}$$

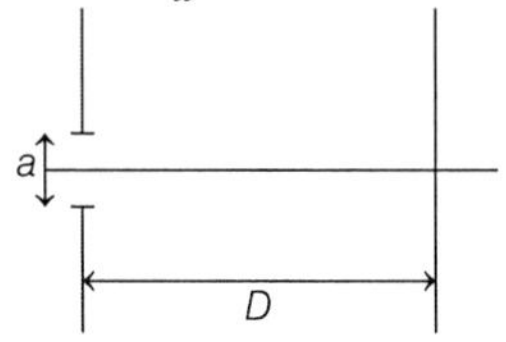

Width of maxima or bright fringe in YDSE is,

$$\beta = \frac{\lambda D}{d}$$

When 10 maxima lies in central maxima then

$$\therefore \quad \alpha = 10\beta \Rightarrow \frac{2\lambda D}{a} = \frac{10\lambda D}{d}$$

$$d = 5a \Rightarrow a = d/5$$

$$a = \frac{10 \times 10^{-4} \text{ m}}{5} \Rightarrow a = 2 \times 10^{-4} \text{ m}$$

64. *(8)* We know that,

$$I_{\text{max}} = (\sqrt{I} + \sqrt{4I})^2$$

$$= I + 4I + 2 \times \sqrt{I} \times \sqrt{4I} = 9I$$

$$I_{\text{min}} = (\sqrt{4I} - \sqrt{I})^2$$

$$= 4I + I - 2\sqrt{4I} \cdot \sqrt{I}$$

$$= 5I - 4I = I$$

$$\therefore I_{\text{max}} - I_{\text{min}} = 9I - I = 8I = xI \text{ (given)}$$

$$\therefore \quad x = 8$$

65. *(10)* Given, wavelength, $\lambda = 5000$ Å,
$d = 0.3$ mm, $D = 200$ cm

Fringe width, $\beta = \frac{\lambda D}{d}$

$$\beta = \frac{5000 \times 10^{-10} \times 200 \times 10^{-2}}{0.3 \times 10^{-3}}$$

$$\beta = \frac{10 \times 10^{-3}}{3} \text{ m} \quad \ldots\text{(i)}$$

Now, for 3rd maxima, $x_3 = 3\beta$

$$\Rightarrow \quad x_3 = \frac{3 \times 10 \times 10^{-3}}{3} = 10 \times 10^{-3} \text{ m}$$

$$x_3 = 10 \text{ mm}$$

66. *(9)* Given, $\lambda_1 = 450$ nm, $\lambda_2 = 650$ nm

For overlap,

$$n_1\lambda_1 = n_2\lambda_2$$

$$\Rightarrow \quad \frac{\lambda_2}{\lambda_1} = \frac{n_1}{n_2} \Rightarrow \frac{n_1}{n_2} = \frac{650}{450} = \frac{13}{9}$$

$$n_2 = 9$$

The minimum order of fringe produced is 9.

67. *(13)* Given intensity ratio of waves A and B is, $\frac{I_A}{I_B} = \frac{1}{9}$

So, if $I_A = x'$, $I_B = 9x'$
Now, I_1 = resultant intensity due to incoherent addition

$$= x' + 9x' = 10x'.$$

and I_2 resultant intensity due to coherent addition ($\Delta\phi = 60°$)

$$= I_A + I_B + 2\sqrt{I_A}\sqrt{I_B} \cos\phi$$

$$= x' + 9x' + 2\sqrt{x'}\sqrt{9x'} \cos 60°$$

$$= x' + 9x' + 3x' = 13x'$$

Ratio, $\frac{I_1}{I_2} = \frac{10x'}{13x'} = \frac{10}{x}$ or $x = 13$

68. *(462)* Let n_1 maxima of 7000 Å coincide. with n_2 maxima of 5500 Å, therefore $y_{n_1} = y_{n_2}$

$$\Rightarrow \quad n_1 \frac{D\lambda_1}{d} = n_2 \frac{D\lambda_2}{d}$$

$$\Rightarrow \quad \frac{n_1}{n_2} = \frac{\lambda_2}{\lambda_1} = \frac{5500}{7000} = \frac{11}{14}$$

Therefore, 11th maxima of 7000 Å will coincide with 14th maxima of 5500 A°. Least distance from central bright fringe

$$y = n_1\beta_1$$

$$y = \frac{n_1\lambda_1 D}{d}$$

$$= \frac{11 \times 7000 \times 10^{-10}}{2.5 \times 10^{-3}} \times 150 \times 10^{-2}$$

$$= \frac{11 \times 7 \times 5}{2.5} \times 10^{-5} \text{ m}$$

$$y = 462 \times 10^{-5} \text{ m}$$

Therefore, $n = 462$

69. *(4)* Given that thickness,

$$t = 10 \times 10^{-6} \text{ m}$$

$$\mu = 1.2$$

$$\lambda = 500 \times 10^{-9} \text{ m}$$

When, the glass slab is inserted infront of one slit then the shift of central fringe is obtained by

$$\Delta W = \frac{(\mu - 1)tD}{d}$$

Given that $\Delta W = x\beta_0 = x\frac{\lambda D}{d}$

$$\therefore \quad \frac{(\mu - 1)tD}{d} = \frac{x\lambda D}{d} \Rightarrow x = \frac{(\mu - 1)t}{\lambda}$$

$$= \frac{(1.2 - 1) \times 10^{-5}}{500 \times 10^{-9}} = 4$$

70. *(48)* Given that $d = 0.35$ mm,

$$D = 7 \text{ m}, \lambda_1 = 800 \text{ nm},$$

$$\lambda_2 = 600 \text{ nm}$$

If bright fringes of the two wavelength coincide, then their position on screen are same.

i.e., $\quad y_1 = y_2$

$$n_1 \frac{\lambda_1 D}{d} = \frac{n_2\lambda_2 D}{d} \Rightarrow n_1\lambda_1 = n_2\lambda_2$$

or $\quad \frac{n_1}{n_2} = \frac{\lambda_2}{\lambda_1} = \frac{600}{800} = \frac{3}{4}$

3rd maxima of λ_1 and 4th maxima of λ_2 will coincide.
Shortest distance from central point,

$$y = \frac{n_1\lambda_1 D}{d} = \frac{3 \times 800 \times 10^{-9} \times 7}{0.35 \times 10^{-3}}$$

$$y = 48 \times 10^{-3} \Rightarrow y = 48 \text{ mm}$$

71. *(24)* Given,

$$I_1 = 4I \Rightarrow I_2 = 9I$$

At point A, $\Delta\phi_A = 0$
At point B, $\Delta\phi_B = \pi$
To find, $(I_A)_{\text{resultant}} - (I_B)_{\text{resultant}}$
But we know that, in interference when $\Delta\phi = 0$, then resultant intensity in maximum i.e.

$$(I_A)_{\text{resultant}} = I_{\text{max}}$$

and when $\Delta\phi = \pi$ then resultant intensity is minimum i.e.

$$(I_B)_{\text{resultant}} = I_{\text{min}}$$

$\therefore$We have to find, $(I_{\text{max}} - I_{\text{min}})$

Now,

$$I_{\text{max}} = (\sqrt{I_1} + \sqrt{I_2})^2 = (\sqrt{4I} + \sqrt{9I})^2$$

$$= 4I + 9I + 12I$$

$$\Rightarrow I_{\text{max}} = 25I$$

Also,

$$I_{min} = (\sqrt{I_1} - \sqrt{I_2})^2 = (\sqrt{4I} - \sqrt{9I})^2$$
$$= 4I + 9I - 12I = I$$
$$\Rightarrow \quad I_{min} = I$$
$$\therefore I_{max} - I_{min} = 25I - I = 24I$$

72. *(630)* In Young's double slit experiment, wavelength of laser light,
$\lambda_1 = 560 \text{ nm} = 5.6 \times 10^{-7}$ m

Seperation of consecutive bright fringes is equal to fringe width β.

$$\therefore \beta = D\lambda / d \Rightarrow \frac{\beta_1}{\beta_2} = \frac{\lambda_1}{\lambda_2} \quad \ldots (i)$$

$\Rightarrow$ Here, $\beta_1 = 7.2 \text{ nm} = 7.2 \times 10^{-3}$ m

and $\beta_2 = 8.1 \text{ mm} = 8.1 \times 10^{-3}$ m

On putting these values in Eq. (i), we get

$$\frac{7.2 \times 10^{-3}}{8.1 \times 10^{-3}} = \frac{5.6 \times 10^{-7}}{\lambda_2}$$
$$\Rightarrow \frac{8}{9} = \frac{5.6 \times 10^{-7}}{\lambda_2}$$
$$\Rightarrow \lambda_2 = \frac{9}{8} \times 5.6 \times 10^{-7}$$
$$= 6.3 \times 10^{-7} \text{ m}$$
$$= 630 \times 10^{-9} \text{ m} = 630 \text{ nm}$$

73. *(2)* Given, at point A,
Phase difference

$$\phi_1 = \frac{\pi}{2}, I_1 = I \text{ and } I_2 = 4I$$

$\therefore$Resultant intensity at point A,

$$I_A = I_1 + I_2 + 2\sqrt{I_1 I_2} \cos \phi_1$$
$$= I + 4I + 2\sqrt{I \times 4I} \cos \frac{\pi}{2}$$
$$= 5I + 0 \Rightarrow I_A = 5I$$

Similarly, for point B

$$I_1 = I, I_2 = 4I \text{ and } \phi_2 = \frac{\pi}{3}$$

Resultant intensity at this condition,

$$I_B = I_1 + I_2 + 2\sqrt{I_1 I_2} \cos \phi_2$$
$$= I + 4I + 2\sqrt{I \times 4I} \cos \frac{\pi}{3}$$
$$= 5I + 2 \times 2I \times \frac{1}{2} = 7I$$
$$\therefore I_B - I_A = 7I - 5I = 2I = xI \quad \text{[Given]}$$
$$x = 2$$

74. *(600)* We know that,

$$\beta = \frac{\lambda D}{d} \Rightarrow \Delta\beta = \lambda \frac{\Delta D}{d}$$
$$\lambda = \frac{\Delta\beta d}{\Delta D} = \frac{3 \times 10^{-5} \times 10^{-3}}{5 \times 10^{-2}}$$
$$= 0.6 \times 10^{-6} = 600 \times 10^{-9} \text{m}$$
$$= 600 \text{ nm}$$

75. *(450)* Given, fringe width, $\beta = 0.6$ mm
$= 0.6 \times 10^{-3}$ m

According to question,

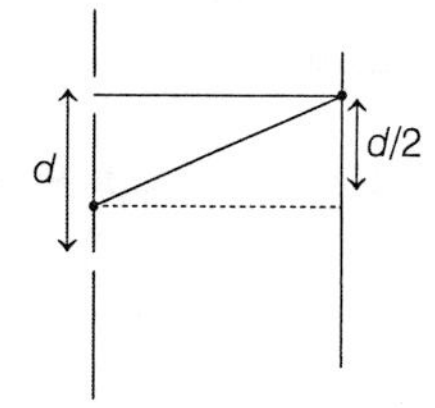

$$y_{dark} = \frac{d}{2} \Rightarrow \frac{\lambda D}{2d} = \frac{d}{2}$$
$$\lambda = \frac{d^2}{D} = \frac{(6 \times 10^{-4})^2}{0.8} = \frac{36 \times 10^{-8}}{8 \times 10^{-1}}$$
$$= 4.5 \times 10^{-7} \text{ m}$$
$$\therefore \quad \lambda = 450 \text{ nm}$$

76. *(1)*

key idea To solve this question, you should use following relation

Intensity in YDSE $\propto$ slit width $\propto$ (Amplitude)^2

Given, $b_1 = 3b_2$

Here, b_1 = width of the one of the two slit and b_2 = width of the other slit.

As we know that,

Intensity, $I \propto$ (Amplitude)^2

$$\Rightarrow \frac{I_1}{I_2} = \left(\frac{b_1}{b_2}\right)^2 \Rightarrow \frac{I_1}{I_2} = \left(\frac{3b_2}{b_2}\right)^2$$
$$I_1 = 9I_2$$

As we know, the ratio of the minimum intensity to the maximum intensity in the interference pattern,

$$\frac{I_{min}}{I_{max}} = \left(\frac{\sqrt{I_1} - \sqrt{I_2}}{\sqrt{I_1} + \sqrt{I_2}}\right)^2$$

On substituting the values in the above equations, we get

$$\frac{I_{min}}{I_{max}} = \frac{(\sqrt{9I_2} - \sqrt{I_2})^2}{(\sqrt{9I_2} + \sqrt{I_2})^2} = \left(\frac{3-1}{3+1}\right)^2$$
$$\frac{I_{min}}{I_{max}} = \frac{1}{4}$$

Comparing with, $\dfrac{I_{min}}{I_{max}} = \dfrac{x}{4}$

The value of $x = 1$.

77. *(300)* In a Young's double slit experiment for interference pattern, the position of bright fringe is given by

$$y_n = \frac{n\lambda D}{d}$$

Here, $D = 1.5$ m, $d = 0.3 \times 10^{-3}$ m

and $n = 1$

$$\lambda = \frac{y_n d}{D}$$

For first violet,

$$y_n = 2 \text{ mm} = 2 \times 10^{-3} \text{ m}$$
$$\therefore \quad \lambda_{violet} = 2 \times 10^{-3} \left(\frac{d}{D}\right)$$

For first red $y_n = 3.5 \text{ mm} = 3.5 \times 10^{-3}$ m

$$\lambda_{red} = 3.5 \times 10^{-3} \left(\frac{d}{D}\right)$$

The difference in wavelengths of red and violet light is

$$\Delta\lambda = \lambda_{red} - \lambda_{violet}$$
$$= 3.5 \times 10^{-3} \left(\frac{d}{D}\right) - 2 \times 10^{-3} \left(\frac{d}{D}\right)$$
$$= \frac{d}{D}(1.5) \times 10^{-3}$$
$$= \frac{0.3 \times 10^{-3}}{1.5} \times 1.5 \times 10^{-3}$$
$$= 0.3 \times 10^{-6}$$
$$= 0.3 \times 10^{-6} \times 10^3 \times 10^{-3}$$
$$= 300 \times 10^{-9} = 300 \text{ nm}$$

78. *(600)* Given,

$$d = 1 \text{ mm} = 10^{-3} \text{ m}$$
$$\beta = 6 \text{ mm} = 6 \times 10^{-3} \text{ m}$$

Distance between the screen and the slit,

$$D = 10 \text{ m}$$

As, $\beta = \dfrac{D\lambda}{d}$

$$\Rightarrow \lambda = \frac{\beta d}{D} = \frac{6 \times 10^{-3} \times 10^{-3}}{10}$$
$$= 6 \times 10^{-7} \text{ m} = 600 \text{ nm}$$

On comparing with the given value in the question i.e., x nm, the value of $x = 600$.

79. *(750)* Given, in a YDSE,

$$\Rightarrow \frac{15 \times 500 \times D}{d} = \frac{10 \times \lambda \times D}{d}$$
$$\Rightarrow \lambda = \frac{15 \times 500}{10} = 750 \text{ nm}$$

Topic 2

01. *(a)* Given, $\lambda = 600 \text{ nm} = 6 \times 10^{-7}$ m

According to given situation,
For nth minima of diffraction pattern
$n\lambda = a \sin \theta$
Here, $n = 1$ so,

$$a = \frac{\lambda}{\sin \theta} = \frac{6 \times 10^{-7}}{\sin 30°}$$
$$= 12 \times 10^{-7} \text{ m} = 1.2 \times 10^{-6} \text{ m}$$
$$= 1.2 \,\mu\text{m}$$

02. *(c)* Given, $\lambda_{actual} = 670$ nm

and $\lambda_{observed} = 670.7$ nm

Change in wavelength,

$$\Delta\lambda = \lambda_{observed} - \lambda_{actual} = 670.7 - 670 = 0.7 \text{ nm} = 7 \times 10^{-10} \text{ m}$$

According to Doppler's shift, we have

$$\frac{\Delta\lambda}{\lambda_{actual}} = \frac{v}{c}$$

$$\Rightarrow \frac{7\times10^{-10}}{670\times10^{-9}} = \frac{v}{3\times10^{8}}$$

$$\Rightarrow \frac{7}{6700} = \frac{v}{3\times10^{8}}$$

$$\Rightarrow v = \frac{21\times10^{8}}{6700} = 3.13\times10^{5} \text{ m/s}$$

03. *(c)* Given, diameter of pinhole,

$$a = 0.1\ \mu\text{m} = 0.1\times10^{-6} \text{ m}$$

∵ Path difference,

$$\Delta x = a\sin\phi = n\lambda \quad \ldots\text{(i)}$$

where, ϕ is the phase difference and λ be the wavelength.

As, $I = 4I_0 \cos^2\phi$

and $\sin\phi = \frac{n\lambda}{a}$ [From Eq. (i)]

If a increases $\leftrightarrow$ $\sin\phi$ or ϕ decreases

As ϕ decreases $\leftrightarrow$ $\cos\phi$ increases

Hence, on decreasing diameter of pinhloe, the size of diffraction maxima decreases and intensity increases.

04. *(b)* For diffraction location of first minima, $y_1 = D\lambda / a$

Location of second minima, $y_2 = \frac{2D\lambda}{a}$

So, $\Delta y = y_2 - y_1 = \frac{D\lambda}{a}$

Let n bright fringes occurs in this region then

$$\Delta y = n\beta \Rightarrow \frac{\lambda D}{a} = \frac{n\lambda D}{d}$$

$$n = \frac{d}{a} = \frac{19.44}{4.05} = 4.8$$

So only 4 bright fringes are observed in this region.

05. *(c)* Central maxima lies between two first minimas on either side of centre line.

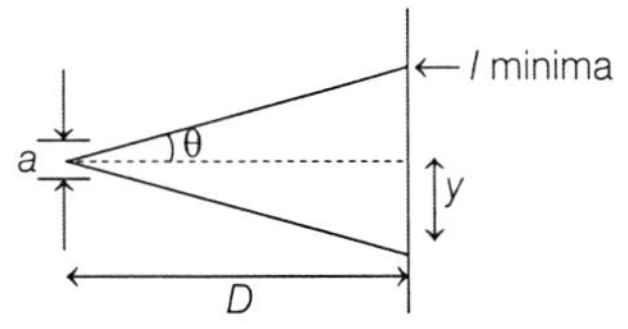

Angular width of central maxima is 2θ

where, $\sin\theta = \frac{\lambda}{a} \Rightarrow \theta = \sin^{-1}\frac{\lambda}{a}$

So, angular width $= 2\theta = 2\sin^{-1}\frac{\lambda}{a}$

$$= 2\sin^{-1}\left(\frac{2}{4}\right) = 2\sin^{-1}\left(\frac{1}{2}\right)$$

$$= 2\times 30° = 60°$$

06. *(d)* Width of central maxima in case of diffraction is,

$$2y = \frac{2\lambda D}{a}$$

As a lens is used,

Here, $D = f = 20$ cm $= 20\times10^{-2}$ m

Wavelength, $\lambda = 6000$ Å $= 6000\times10^{-10}$ m

Width of slit $= a = 0.01$ mm $= 0.01\times10^{-3}$ m

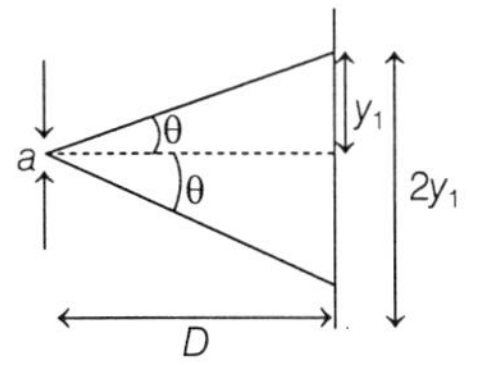

So, width of central maxima

$$= \frac{2D\lambda}{a} = \frac{2\times6000\times10^{-10}\times20\times10^{-2}}{0.01\times10^{-3}}$$

$$= 24\times10^{-3} \text{ m} = 24 \text{ mm}$$

07. *(c)* Given,

Slit width $(a) = 0.2$ mm $= 0.2\times10^{-3}$ m

Wavelength of light,

$\lambda = 400$ nm $= 400\times10^{-9}$ m

Screen distance $(D) = 100\times10^{-2}$ m

In diffraction pattern for width of 1st secondary maxima,

$$\text{Width} = \frac{\lambda D}{a} = \frac{400\times10^{-9}\times100\times10^{-2}}{0.2\times10^{-3}}$$

$$= 2\times10^{-3} \text{ m} = 2 \text{ mm}$$

08. *(d)* For single slit diffraction,

$$\sin\theta = \frac{n\lambda}{b}$$

Position of nth minima from central maxima $= \frac{n\lambda D}{b}$

When, $n = 2$, then $x_2 = \frac{2\lambda D}{b} = 0.03$ ... (i)

When, $n = 4$, then $x_4 = \frac{4\lambda D}{b} = 0.06$ ... (ii)

On subtracting Eq. (i) from Eq (ii), we get

$$x_4 - x_2 = \frac{4\lambda D}{b} - \frac{2\lambda D}{b} = 0.03$$

or $\frac{\lambda D}{b} = \frac{0.03}{2}$,

then width of central maximum

$$= \frac{2\lambda D}{b} = 2\times\frac{0.03}{2} = 0.03 \text{ m} = 3 \text{ cm}$$

09. *(c)* In diffraction, first minima, we have , $\sin\theta = \frac{\lambda}{a}$

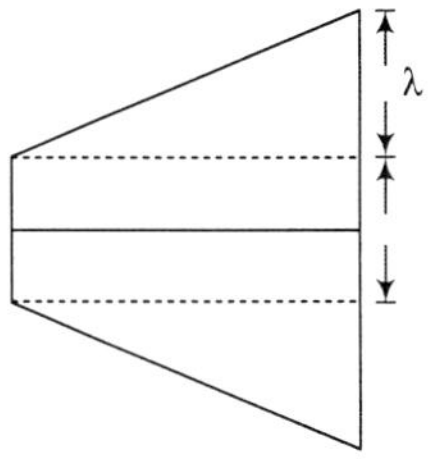

So, size of a spot $b = 2a + \frac{2L\lambda}{a}$ ...(i)

Then, minimum size of a spot, we get

$$\frac{\partial b}{\partial a} = 0$$

$$\Rightarrow 1 - \frac{L\lambda}{a^2} = 0$$

$$\Rightarrow a = \sqrt{\lambda L} \quad \ldots\text{(ii)}$$

So, $b_{min} = 2\sqrt{\lambda L} + 2\sqrt{\lambda L}$

[By substituting the value of a from Eq. (ii) in Eq. (i)]

$$= 4\sqrt{\lambda L}$$

So, the radius of the spot,

$$b_{min} = \frac{4}{2}\sqrt{\lambda L} = \sqrt{4\lambda L}$$

10. *(a)* Angular width of diffraction pattern $= 60°$

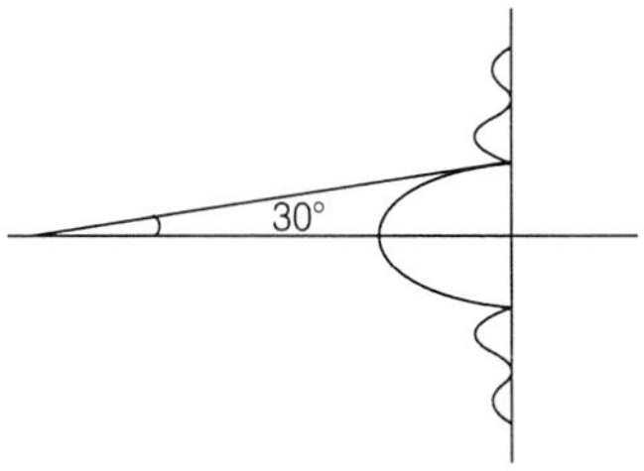

For first minima, $\frac{a}{2}\sin\theta = \frac{\lambda}{2}$,

[Here, $a = 10^{-6}$ m, $\theta = 30°$]

$$\Rightarrow \lambda = 10^{-6}\times\sin 30°$$

$$\Rightarrow \lambda = \frac{10^{-6}}{2} \text{ m}$$

Now, in case of interference caused by bringing second slit,

$\therefore$ Fringe width, $\beta = \frac{\lambda D}{d}$

[Here, $\lambda = \frac{10^{-6}}{2}$ m, $\beta = 1$ cm $= \frac{1}{100}$ m,

$d = ?$ and $D = 50$ cm $= \frac{50}{100}$ m]

So, $d = \frac{\lambda D}{\beta} = \frac{10^{-6} \times 50}{2 \times \frac{1}{100} \times 100}$

$= 25 \times 10^{-6}$ m or $d = 25\ \mu$m

11. *(15)* Width of 20 maxima of double slit = Width of central maxima of single slit

$\frac{20\lambda D}{d} = \frac{2\lambda D}{a}$

$\frac{10}{d} = \frac{1}{a}$

$a = \frac{d}{10} = \frac{1.5 \times 10^{-1}}{10} = 15 \times 10^{-3}$ cm

On value of x is 15.

12. *(6)*

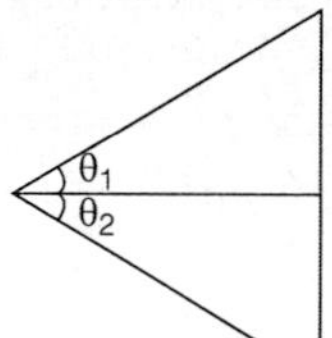

$\theta_1 = \sin^{-1}\left(\frac{2\lambda}{9}\right)$

$\theta_2 = \sin^{-1}\left(\frac{3\lambda}{9}\right)$

$\theta_1 + \theta_2 = 30° \rightarrow$ given

$\theta_1 + \theta_2 = \sin^{-1}\left(\frac{2\lambda}{a}\right) + \sin^{-1}\left(\frac{3\lambda}{a}\right)$

If we taken approximate,

$\frac{\pi}{6} = \frac{2\lambda}{a} + \frac{3\lambda}{a}$

$\frac{\pi}{6} = \frac{5\lambda}{a}$

$\frac{\pi}{6} = \frac{5}{a}$ (628 nm)

$a = 6\ \mu$m

13. *(3)* For second minima, $a \sin\theta = 2\lambda$

$\Rightarrow 0.4 \times 10^{-3} \times \sin\theta = 2 \times 600 \times 10^{-9}$

$\Rightarrow \sin\theta = \frac{2 \times 600 \times 10^{-9}}{0.4 \times 10^{-3}}$

Using $\sin\theta = \theta$, we get

or $\theta = \frac{2 \times 6}{4} \times 10^{-3} = 3 \times 10^{-3}$ rad

Angular divergence, $2\theta = 6 \times 10^{-3}$ rad

14. *(2)* Given,

$D = 50$ cm $= 0.5$ m

$\lambda = 6000$ Å $= 6 \times 10^{-7}$ m

$\Delta x = 3$ mm $= 3 \times 10^{-3}$ m

Position of nth minima

$x_n = \frac{nD\lambda}{d}$

Distance between 3rd order minima and Ist order minimum,

$\Delta x = x_3 - x_1 = \frac{(3-1)\lambda D}{d}$

$\Delta x = \frac{2\lambda D}{d}$

$\Rightarrow d = \frac{2\lambda D}{\Delta x} = \frac{2 \times 6 \times 10^{-7} \times 0.5}{3 \times 10^{-3}}$

$= 2 \times 10^{-4}$ m

15. *(30)* $a \sin\theta = n\lambda$

For 1st minima,

$\sin\theta = \frac{1 \times 5000 \times 10^{-10}}{0.001 \times 10^{-3}}$

$\sin\theta = 5 \times 10^{-1} \Rightarrow \sin\theta = 0.5$

$\sin\theta = \frac{1}{2} \Rightarrow \theta = 30°$

16. *(3)* Given, $\lambda_1 = 650$ nm $= 650 \times 10^{-9}$ m

$\lambda_2 = 655$ nm $= 655 \times 10^{-9}$ m

Width of aperture,

$a = 0.5$ mm $= 5 \times 10^{-4}$ m

Distance between screen and source, $D = 2$ m

We know that position of first maxima, $a \sin\theta = 3\lambda/2$

Since, $\sin\theta \simeq \theta = \frac{y}{D}$

$\therefore a \cdot \frac{y}{D} = \frac{3\lambda}{2} \Rightarrow y = \frac{3\lambda D}{2a}$

$\therefore y_1 = \frac{3\lambda_1 D}{2a}$ and $y_2 = \frac{3\lambda_2 D}{2a}$

$\therefore \Delta y = y_2 - y_1 = \frac{3\lambda_2 D}{2a} - \frac{3\lambda_1 D}{2a}$

$= \frac{3D}{2a}(\lambda_2 - \lambda_1)$

$= \frac{3 \times 2}{2 \times 5 \times 10^{-4}}(655 \times 10^{-9} - 650 \times 10^{-9})$

$= 3 \times 10^{-5}$ m

17. *(198)* In single slit diffraction, if width of slit is a, wavelength of light is λ and angular position on screen is θ, then for nth minima, the condition is

$a\sin(\theta) = n\lambda \Rightarrow \sin(\theta) = \frac{n\lambda}{a}$ ...(i)

$\because \sin(\theta) < 1 \Rightarrow \frac{n\lambda}{a} < 1$

$\Rightarrow n < \frac{a}{\lambda}$

$\Rightarrow n < \frac{0.6 \times 10^{-4}}{6000 \times 10^{-10}}$

$\Rightarrow n < \frac{10^{-4}}{10^4 \times 10^{-10}}$

$\Rightarrow n < 10^2 \Rightarrow n < 100$

So, maximum possible number of minima on one side of central maximum = 99.

Therefore, maximum possible number of minima on both sides of central maximum = 2 × 99 = 198

Topic 3

01. *(d)* According to Brewster's rule, when polarisation of reflected ray occurs, then reflected and refracted rays are perpendicular.

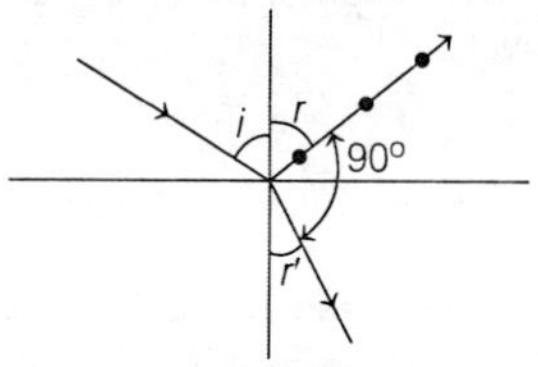

So, $r + r' = 90°$

$\Rightarrow r' = 90° - r = 90° - i \quad (\because i = r)$

$= 90° - 60° = 30°$

02. *(a)* When a polaroid sheet is rotated through an angle ϕ between two crossed polaroids, then intensity of transmitted light.

$I = I_0 \cos^2\phi \cdot \sin^2\phi$

$\therefore I_{max}$ at $\phi = 45°$

$\phi = 45°$

03. *(a)* As, we know that if an unpolarised light is converted into plane polarised light then intensity becomes half of incident intensity

i.e., $I = \frac{I_0}{2}$,

and from law of Malus, $I = I_0 \cos^2\theta$

After polaroid A,

$I = \frac{I_0}{2}$

After polaroid C,

$I = \left(\frac{I_0}{2}\right)\cos^2 45° = \frac{I_0}{4}$

After polaroid B,

$I = \frac{I_0}{4}\cos^2 45° = \frac{I_0}{8}$

> **• Concept Enhancer**
> Intensity of plane polarised light that passes through the analyser is directly proportional to the square of the cosine of the angle between the plane of polarizer and the transmission axis of the analyser.

04. *(b)* Using Malus' law,

$$I_C = \left(\frac{I}{2}\cos^2\theta\right)$$

and $I_B = I_C \cos^2\theta = \frac{I}{2}\cos^4\theta$

$$\frac{I}{3} = \frac{I}{2}\cos^4\theta \qquad \left(\because I_B = \frac{I}{3}\right)$$

$$\cos^4\theta = \frac{2}{3};\ \cos\theta = \left(\frac{2}{3}\right)^{1/4}$$

05. *(c)* Relation between intensities is

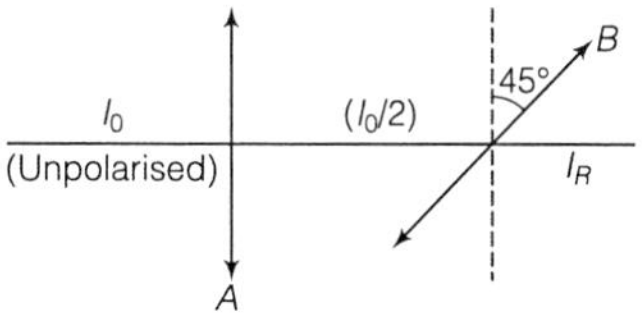

$$I_R = \left(\frac{I_0}{2}\right)\cos^2(45°) = \frac{I_0}{2}\times\frac{1}{2} = \frac{I_0}{4}$$

06. *(c)* Scattering for blue light is largest and it is polarised by scattering.
Also, for polarised light,

$$I = I_0\cos^2\theta$$

07. *(a)* Intensity of polarised light $= I_0/2$

$\therefore$ Intensity of untransmitted light

$$= I_0 - \frac{I_0}{2} = \frac{I_0}{2}$$

08. *(d)* The particular angle of incidence for which reflected light is totally polarised for reflection from air to glass, is called the angle of polarisation (i_p) or **Brewster's law.**

Accordingly, $n = \tan i_p \Rightarrow i_p = \tan^{-1}(n)$

where, n is refractive index of glass.

09. *(a)*

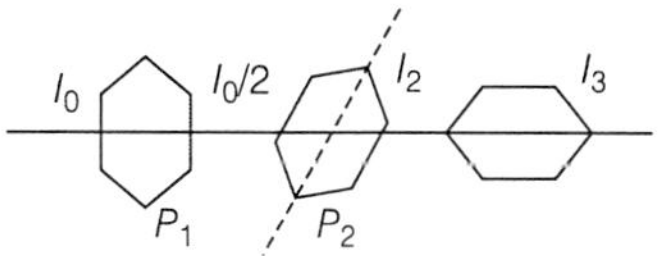

$$I_1 = \frac{I_0}{2}$$

$$I_2 = I_1\cos^2\theta = \frac{I_0}{2}\cos^2\theta$$

$$I_3 = \frac{I_2}{2}\cos^2(90-\theta)$$

$$= \left(\frac{I_0}{2}\cos^2\theta\right)\times\frac{1}{2}\sin^2\theta$$

$$= \frac{I_0}{4}\sin^2\theta\cdot\cos^2\theta\times\frac{4}{4}$$

$$= \frac{I_0}{4}\times\frac{4}{4}[\sin 2\theta]^2$$

for max I_3,

$$\theta = 45° = \frac{\pi}{4}$$

10. *(b)*

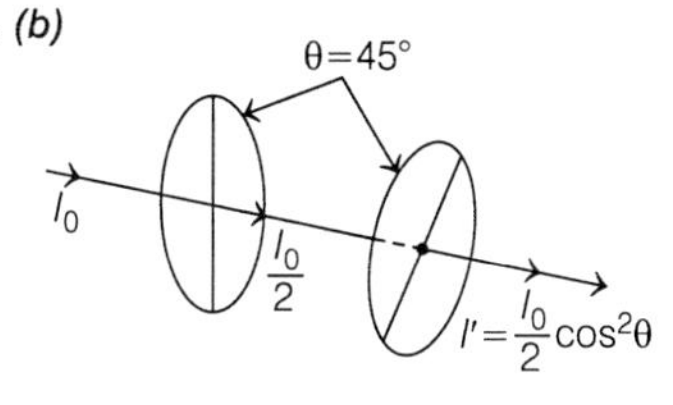

Intensity of light after 1st polaroid $= \frac{I_0}{2}$

By Malus' law,

Intensity of light after second polaroid is,

$$I' = \frac{I_0}{2}\times\cos^2\theta$$

$$= \frac{I_0}{2}\times\cos^2 45°$$

$$= \frac{I_0}{2}\times\frac{1}{2} = \frac{I_0}{4}$$

11. *(b)* If an unpolarised light is converted into plane polarised light, then its intensity becomes half. $I = \frac{I_0}{2}$, where I_0 is intensity of unpolarised light.

According to Malus,

$$I = I_0\cos^2\theta,$$

$$I = \frac{I_0}{2}$$

$$[\cos^2 45\times\cos^2 45\times\ldots\times(n-1)\text{ times}]$$

$$\Rightarrow \quad \frac{I_0}{64} = \frac{I_0}{2}\times\left(\frac{1}{2}\right)^{n-1}$$

$$\Rightarrow \quad \frac{1}{32} = \frac{1}{2^{n-1}}$$

$$\Rightarrow \quad \frac{1}{2^5} = \frac{1}{2^{n-1}}$$

$$\therefore \quad n-1 = 5$$

or $\quad n = 6$

12. *(b)*

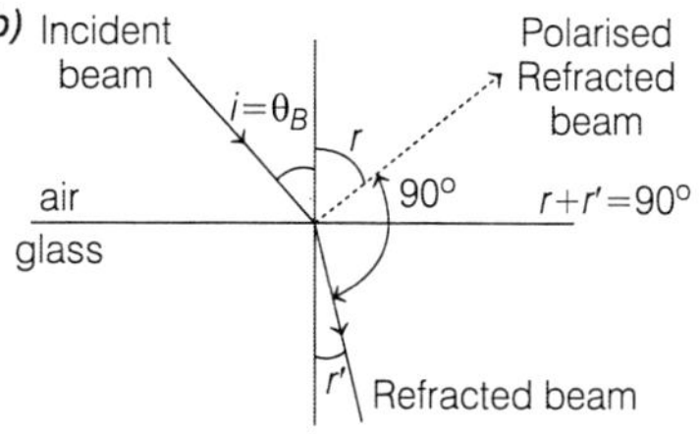

As, angle of incidence
= angle of reflection

So, $\theta_B + r' = 90°$

or $\quad r' = 90 - \theta_B = \frac{\pi}{2} - \theta_B$

So, Statement I is correct.

Also, for light propagating from air to glass,

$$\mu_g = \frac{\sin i}{\sin r} = \frac{\sin\theta_B}{\sin\left(\frac{\pi}{2}-\theta_B\right)}$$

$$= \frac{\sin\theta_B}{\cos\theta_B} = \tan\theta_B$$

or θ_B(air to glass) $= \tan^{-1}(\mu_g)$

So, θ_B(glass to air) $= \tan^{-1}\left(\frac{1}{\mu_g}\right)$

Hence, Statement II is incorrect.

13. *(c)*

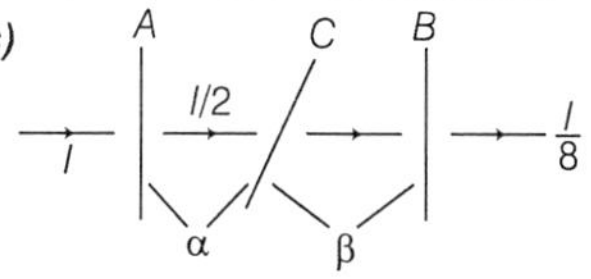

Using Malus's law, intensity available after $C = \frac{I}{2}\times\cos^2\alpha$

and intensity available after

$$= \frac{I}{2}\cos^2\alpha\times\cos^2\beta = \frac{I}{8}\text{ (given)}$$

So, $\frac{I}{2}\times\cos^2\alpha\cdot\cos^2\beta = \frac{I}{8}$

$$\Rightarrow \quad \cos^2\alpha\cdot\cos^2\beta = 1/4$$

This is satisfied with $\alpha = 45°$ and $\beta = 45°$

So, angle between A and C is 45°.

14. *(d)* By law of Malus i.e., $I = I_0\cos^2\theta$

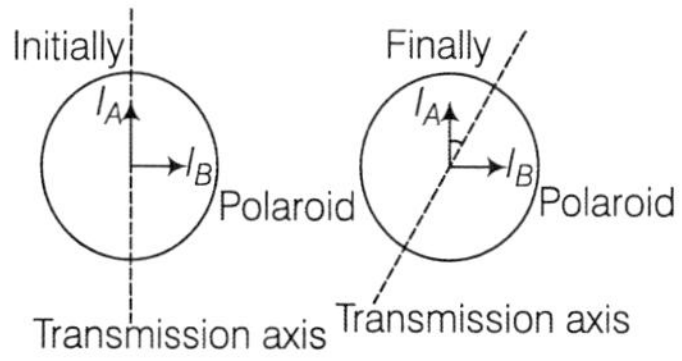

Now, $\quad I_{A'} = I_A\cos^2 30°,$

$$I_{B'} = I_B\cos^2 60°$$

As, $\quad I'_A = I'_B$

$$I_A\cos^2 30° = I_B\cos^2 60°$$

$$\Rightarrow \quad I_A\frac{3}{4} = I_B\frac{1}{4}$$

$$\Rightarrow \quad \frac{I_A}{I_B} = \frac{1}{3}$$

15. *(d)*

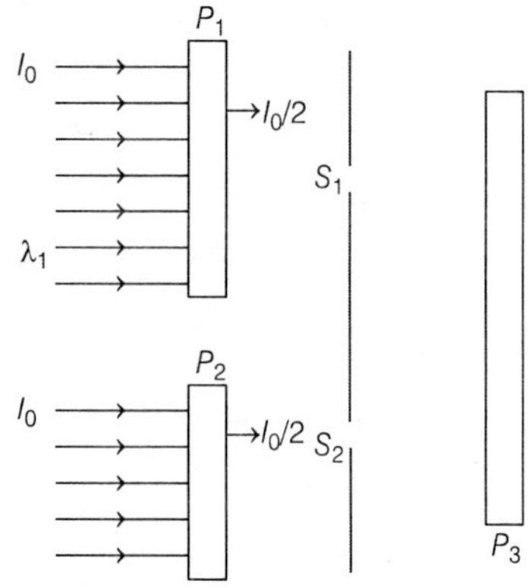

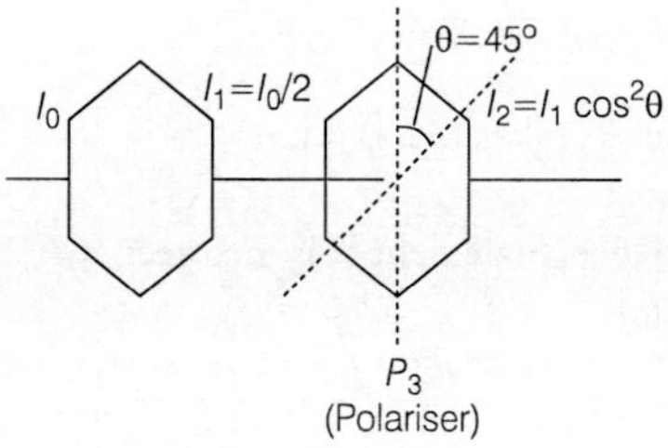

$$I_2=I_1\cos^2\theta=\frac{I_0}{2}\cos^2 45°$$

$$=\frac{I_0}{2}\left(\frac{1}{\sqrt{2}}\right)^2=\frac{I_0}{4}$$

Now at a point where path difference is $\frac{\lambda}{3}$ phase difference

$$\Delta\phi=\frac{2\pi}{\lambda}\Delta x$$

$$\Delta\phi=\frac{2\pi}{\lambda}\cdot\frac{\lambda}{3}=\frac{2\pi}{3}$$

$$I_R=\frac{I_0}{4}+\frac{I_0}{4}+2\sqrt{\frac{I_0}{4}\times\frac{I_0}{4}}\cos 120°$$

$$=\frac{I_0}{4}$$

16. *(c)* Intensity falling on polariser,

$$I_0=3.3\ \text{W/m}^2$$

Area of polariser, $A=3\times10^{-4}\ \text{m}^2$

Angular speed of rotation of polariser,

$$\omega=31.4\ \text{rad/s}$$

Now, average energy is given by

$$\langle E\rangle=\langle IAt\rangle$$

$$\langle E\rangle=\langle I_0\cos^2\theta\times A\times t\rangle$$

$$\langle E\rangle=I_0 At\langle\cos^2\theta\rangle$$

$$\langle E\rangle=I_0 At\left(\frac{1}{2}\right)\quad\left[\because\langle\cos^2\theta\rangle=\frac{1}{2}\right]$$

Now, the calculation is about one revolution, then

Time taken = Time period

$$t=T=\frac{2\pi}{\omega}=\frac{2\pi}{31.4}$$

$$=\frac{2\pi}{10\pi}=\frac{1}{5}\ \text{s}$$

Now, $\langle E\rangle=I_0 At\left(\frac{1}{2}\right)$

Substituting all the values in it, we get

$$\langle E\rangle=3.3\times3\times10^{-4}\times\frac{1}{5}\times\left(\frac{1}{2}\right)$$

$$=0.99\times10^{-4}\ \text{J}\approx1\times10^{-4}\ \text{J}$$

Hence, correct option is (c).

17 *(d)* By Malus' law,

Output intensity obtained from analyser,

$$I'=I_0\cos^2\theta$$

where θ = angle between polariser and analyser.

Now, given I' = 10% of initial intensity

$$\Rightarrow\quad I'=\frac{1}{10}I_0=I_0\times\cos^2\theta$$

$$\Rightarrow\quad \cos^2\theta=\frac{1}{10}\Rightarrow\cos\theta=\frac{1}{\sqrt{10}}$$

$$\Rightarrow\quad \theta=\cos^{-1}\frac{1}{\sqrt{10}}\approx71.6°$$

Now, output intensity is zero when angle between polariser and analyser is 90°.

So, analyser must be rotated by $90-71.6=18.40$ to reduce output intensity to zero.

• Concept Enhancer

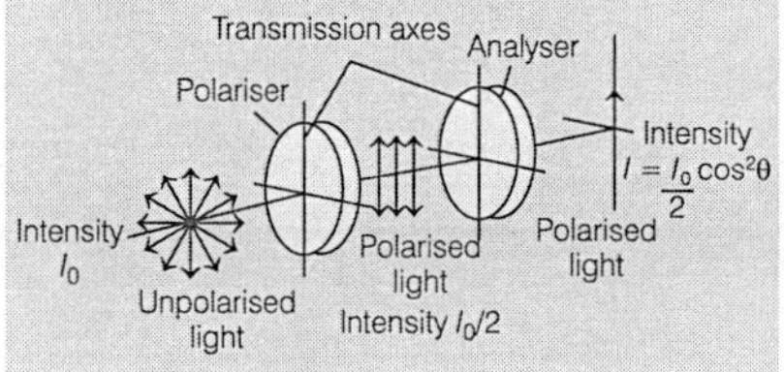

If an unpolarised light is converted into plane polarised light (say by passing it through a polaroid or a Nicol-prism), its intensity becomes half, $I=\frac{I_0}{2}$

The intensity of the polarised light transmitted through the analyser varies as the square of the cosine of the angle between the plane of transmission of the analyser and the plane of the polariser.

$$I=\frac{I_0}{2}\cos^2\theta$$

18. *(c)*

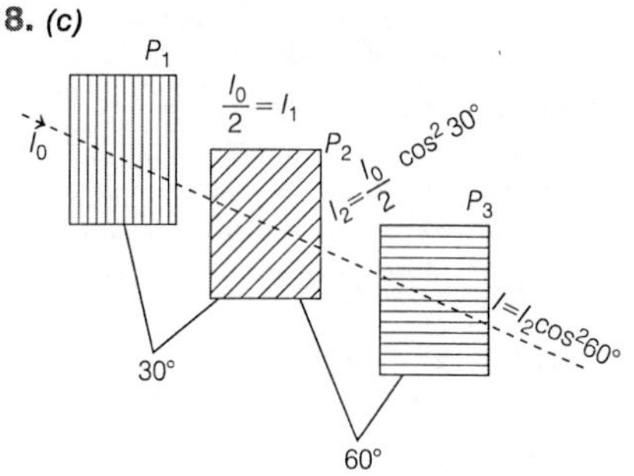

When unpolarised light passes through polaroid P_1, intensity obtained is

$$I_1=\frac{I_0}{2}$$

where, I_0 = intensity of incident light.

Now, this transmitted light is polarised and it passes through polariser P_2.

So, intensity I_2 transmitted is obtained by Malus law,

$$\Rightarrow\quad I_2=I_1\cos^2\theta$$

As angle of pass axis of P_1 and P_3 is 90° and angle of pass axis of P_2 and P_3 is 60°, so angle between pass axis of P_1 and P_2 is $(90°-60°)=30°$.

$$\text{So, } I_2=\frac{I_0}{2}\cos^2 30°=\frac{I_0}{2}\times\left(\frac{\sqrt{3}}{2}\right)^2=\frac{3}{8}I_0$$

When this light passes through the third polariser P_3, intensity I transmitted is again obtained by Malus law,

$$I=I_2\cos^2 60°$$

$$=\left(\frac{3}{8}I_0\right)\cos^2 60°$$

$$=\frac{3}{8}I_0\times\left(\frac{1}{2}\right)^2=\frac{3}{32}I_0$$

$$\text{So, ratio }\frac{I_0}{I}=\frac{32}{3}=10.67$$

19. *(c)* We known that $\frac{\sin i}{\sin r}=\mu$

and $\tan r=1.5/\mu$

$$\frac{\sin i}{\sin r}=\frac{1.5}{\tan r}=\frac{1.5}{\sin r}\times\cos r$$

$$\Rightarrow\quad \sin i=1.5\cos r$$

$$\Rightarrow\quad \frac{\sin i}{\cos r}=1.5$$

$$\frac{\sin i}{\sqrt{1-\sin^2 r}}=1.5\Rightarrow\frac{\sin i}{\sqrt{1-\frac{\sin^2 i}{\mu^2}}}=1.5$$

$$\Rightarrow\quad \frac{\mu\sin i}{\sqrt{\mu^2-\sin^2 i}}=1.5$$

$$\Rightarrow \frac{\mu^2 \sin^2 i}{\mu^2 - \sin^2 i} = 2.25$$

$$\mu^2 \sin^2 i = 2.25\mu^2 - 2.25 \sin^2 i$$

$$\Rightarrow (\mu^2 + 2.25)\sin^2 i = 2.25\mu^2$$

$$\sin^2 i = \frac{2.25\mu^2}{\mu^2 + 2.25}$$

For i to be imaginary,

$$2.25\mu^2 > \mu^2 + 2.25$$

$$\Rightarrow \quad 1.25\mu^2 > 2.25$$

$$\mu > \sqrt{\frac{2.25}{1.25}}$$

$$\Rightarrow \mu > \sqrt{\frac{9}{5}}$$

$$\mu > \frac{3}{\sqrt{5}}, \text{ minimum value of } \mu = \frac{3}{\sqrt{5}}$$

20. *(30)* The given situation is shown below

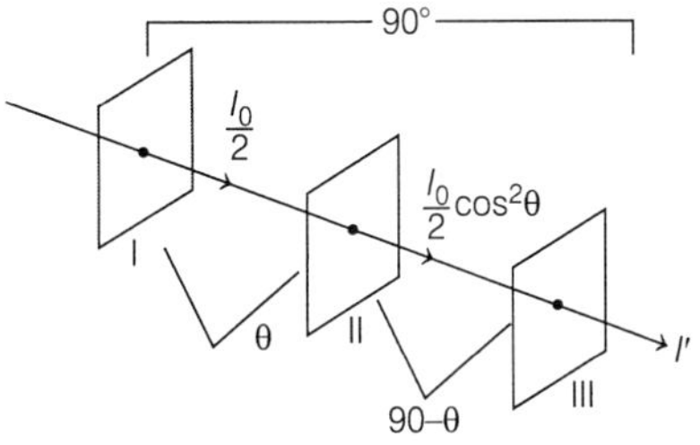

Output intensity from IIIrd polaroid using Malus Law, is

$$I' = \frac{I_0}{2} \cos^2 \theta \cdot \cos^2 (90 - \theta)$$

$$= \frac{I_0}{2} \cos^2 \theta \cdot \sin^2 \theta$$

$$\text{or,} \quad 3 = \frac{32}{2} \cdot \cos^2 \theta \sin^2 \theta$$

$$\Rightarrow \quad 3 = 4\,(4 \cos^2 \theta \sin^2 \theta)$$

$$\Rightarrow \quad \frac{3}{4} = (\sin 2\theta)^2$$

$$\text{or } \sin 2\theta = \frac{\sqrt{3}}{2}$$

$$\Rightarrow \quad 2\theta = 60°$$

$$\text{or} \quad \theta = 30°$$

21. *(7)* $\mu_1 = \sqrt{2.8}$

$$\mu_2 = \sqrt{6.8}$$

$$\mu_1 \sin i = \mu_2 \cos i$$

$$\tan i = \frac{\mu_2}{\mu_1} = \sqrt{\frac{6.8}{2.8}}$$

$$\Rightarrow \tan i = \left(\frac{2.8 + 4}{2.8}\right)^{\frac{1}{2}}$$

$$i = \tan^{-1}\left(1 + \frac{10}{7}\right)^{\frac{1}{2}} \Rightarrow \theta = 7$$

22. *(30)* Given, intensity of light falling on screen is I.
Intensity of light when P_1 and P_2 placed between screen and source is $I/2$.

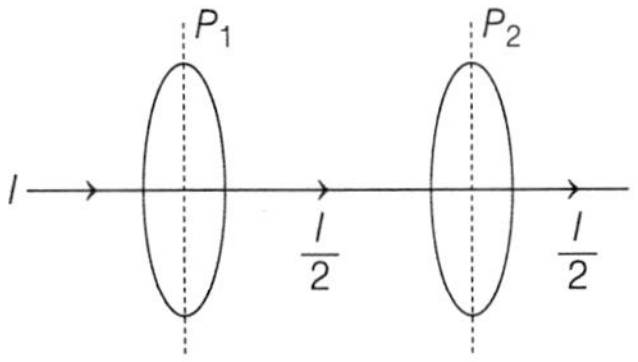

Initially let us consider the angle between analyser and polariser is 0°. Then, the intensity of transmitted light is $I/2$. Thus, intensity of light become $I/2$ after transmitting through 1st polariser as shown in the figure.
If we rotate the polarise P_2 by angle ϕ, then the resulting intensity of light on screen will be $3I/8$.

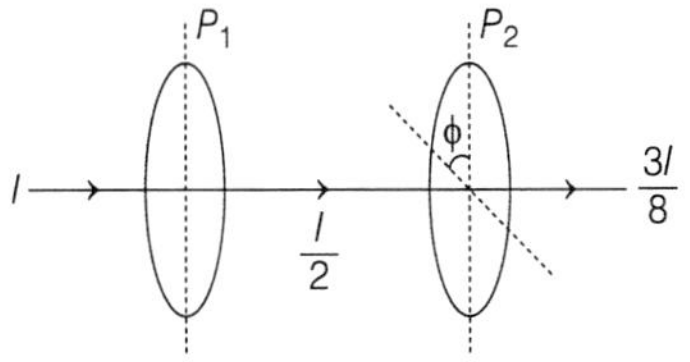

Now, for second condition, apply Malus's law,

$$\frac{3I}{8} = \frac{I}{2} \cos^2 \phi$$

$$\cos^2 \phi = \frac{3}{4}$$

$$\Rightarrow \quad \cos \phi = \frac{\sqrt{3}}{2}$$

$$\phi = 30°.$$

23. *(75)* Given, $I_0 = 100$ lumens

When the analyser is rotated through an angle θ, the intensity of light will becomes

$$I = I_0 \cos^2 \theta = 100 \times \cos^2 30°$$

$$= 100 \times \left(\frac{\sqrt{3}}{2}\right)^2$$

$$= 75 \text{ lumens}$$

In JEE Main there is usually 1-2 question per paper is visible from this chapter. In past 3 years, 60 conceptual and 4 numerical questions were asked of moderate to difficult level from this chapter.

For detailed analysis scan QR code

DUAL NATURE OF RADIATION AND MATTER

Topic 1 Photoelectric Effect

Level 1

Imp. Work Function

01 The work function of a metal is 3 eV. The colour of the visible light that is required to cause emission of photoelectrons is *[2025, 3 April Shift-I]*
(a) Red (b) Blue
(c) Green (d) Yellow

02 Given below are two statements. One is labelled as Assertion (A) and the other is labelled as Reason (R).
Assertion (A) Emission of electrons in photoelectric effect can be suppressed by applying a sufficiently negative electron potential to the photoemissive substance.
Reason (R) A negative electric potential, which stops the emission of electrons from the surface of a photoemissive substance, varies linearly with frequency of incident radiation.
In the light of the above statements, choose the most appropriate answer from the options given below. *[2025, 29 Jan Shift-I]*
(a) Both A and R are true and R is the correct explanation of A.
(b) Both A and R are true but R is not the correct explanation of A.
(c) (A) is false but R is true.
(d) (A) is true but R is false.

03 In an experiment with photoelectric effect, the stopping potential, *[2025, 29 Jan Shift-II]*
(a) increases with increase in the intensity of the incident light
(b) decreases with increase in the intensity of the incident light
(c) increases with increase in the wavelength of the incident light
(d) is $(1/e)$ times the maximum kinetic energy of the emitted photoelectrons

04 The work functions of Cesium (Cs) and Lithium (Li) metals are 1.9 eV and 2.5 eV respectively. If we incident a light of wavelength 550 nm on these two metal surfaces, then photoelectric effect is possible for the case of *[2025, 22 Jan Shift-I]*
(a) Both Cs and Li (b) Neither Cs nor Li
(c) Cs only (d) Li only

05 UV light of 4.13 eV is incident on a photosensitive metal surface having work function 3.13 eV. The maximum kinetic energy of ejected photoelectrons will be *[2024, 9 April Shift-II]*
(a) 3.13 eV (b) 7.26 eV
(c) 4.13 eV (d) 1 eV

06 When UV light of wavelength 300 nm is incident on the metal surface having work function 2.13 eV, electron emission takes place. The stopping potential is (Given, $hc = 1240$ eV-nm) *[2024, 6 April Shift-II]*
(a) 4 V (b) 4.1 V (c) 1.5 V (d) 2 V

07 In photoelectric experiment energy of 2.48 eV irradiates a photo sensitive material. The stopping potential was measured to be 0.5 V. Work function of the photo sensitive material is *[2024, 6 April Shift-I]*
(a) 0.5 eV (b) 2.48 eV (c) 1.68 eV (d) 1.98 eV

08 Which of the following phenomena does not explain by wave nature of light? *[2024, 6 April Shift-I]*
A. Reflection B. Diffraction
C. Photoelectric effect D. Interference
E. Polarisation
Choose the most appropriate answer from the options given below
(a) C only (b) B, D only
(c) A, C only (d) E only

09 Which of the following statement is not true about stopping potential V_0 ? *[2024, 5 April Shift-II]*

(a) It depends upon frequency of the incident light.
(b) It is $1/e$ times the maximum kinetic energy of electrons emitted.
(c) It increases with increase in intensity of the incident light.
(d) It depends on the nature of emitter material.

10 Which figure shows the correct variation of applied potential difference V with photoelectric current I at two different intensities of light ($I_1 < I_2$) of same wavelengths? *[2024, 4 April Shift-I]*

(a)
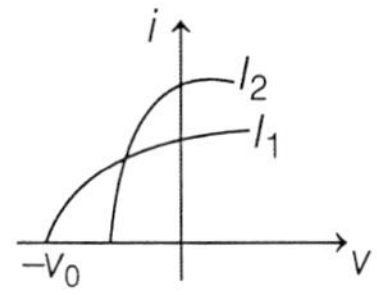

(b)
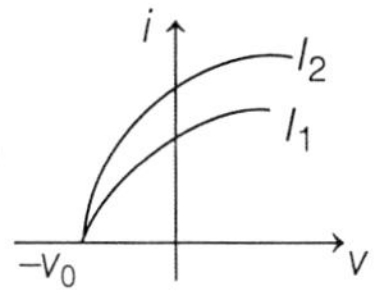

(c)
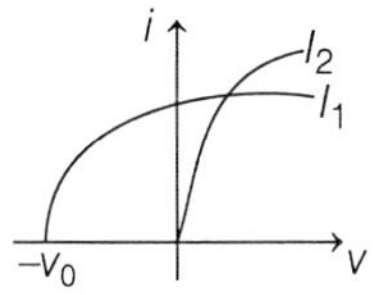

(d)
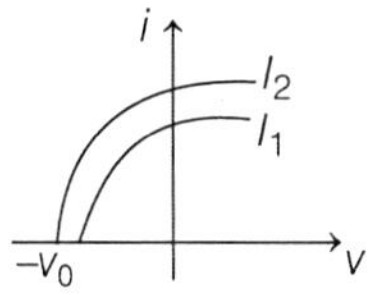

11 In a photoelectric effect experiment, a light of frequency 1.5 times the threshold frequency is made to fall on the surface of photosensitive material. Now, if the frequency is halved and intensity is doubled, the number of photo electrons emitted will be *[2024, 31 Jan Shift-II]*

(a) doubled (b) halved (c) quadrupled (d) Zero

12 For the photoelectric effect, the maximum kinetic energy (E_K) of the photoelectrons is plotted against the frequency (ν) of the incident photons as shown in figure. The slope of the graph gives *[2024, 30 Jan Shift-II]*

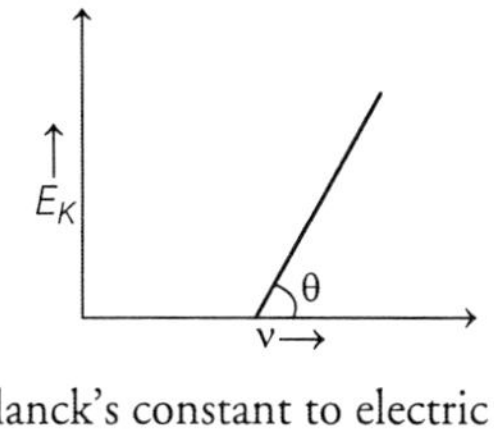

(a) ratio of Planck's constant to electric charge
(b) charge of electron
(c) Planck's constant
(d) work function of the metal

13 The work function of a substance is 3.0 eV. The longest wavelength of light that can cause the emission of photoelectrons from this substance is approximately *[2024, 30 Jan Shift-I]*

(a) 200 nm (b) 215 nm
(c) 414 nm (d) 400 nm

14 The threshold frequency of a metal with work function 6.63 eV is *[2024, 27 Jan Shift-II]*

(a) 16×10^{12} Hz (b) 1.6×10^{15} Hz
(c) 16×10^{15} Hz (d) 1.6×10^{12} Hz

15 A convex lens of focal length 40 cm forms an image of an extended source of light on a photoelectric cell. A current I is produced. The lens is replaced by another convex lens having the same diameter but focal length 20 cm. The photoelectric current now is *[2024, 27 Jan Shift-I]*

(a) I (b) $4I$ (c) $\frac{I}{2}$ (d) $2I$

16 Given below are two statements.

Statement I Out of microwaves, infrared rays and ultraviolet rays, ultraviolet rays are the most effective for the emission of electrons from a metallic surface.

Statement II Above the threshold frequency, the maximum kinetic energy of photoelectrons is inversely proportional to the frequency of the incident light. *[2023, 13 April Shift-II]*

In the light of the above statements, choose the correct answer from the options given below.

(a) Both Statement I and Statement II are false.
(b) Both Statement I and Statement II are true.
(c) Statement I is false but Statement II is true.
(d) Statement I is true but Statement II is false.

17 The difference between threshold wavelengths for two metal surfaces A and B having work function $\phi_A = 9$ eV and $\phi_B = 4 \cdot 5$ eV in nm is (Given, $hc = 1242$ eV-nm) *[2023, 13 April Shift-I]*

(a) 264 (b) 276
(c) 540 (d) 138

Imp. Calculation of Work Function using (ν-V) Graph

18 The variation of stopping potential V_0 as a function of the frequency ν of the incident light for a metal is shown in figure. The work function of the surface is *[2023, 10 April Shift-II]*

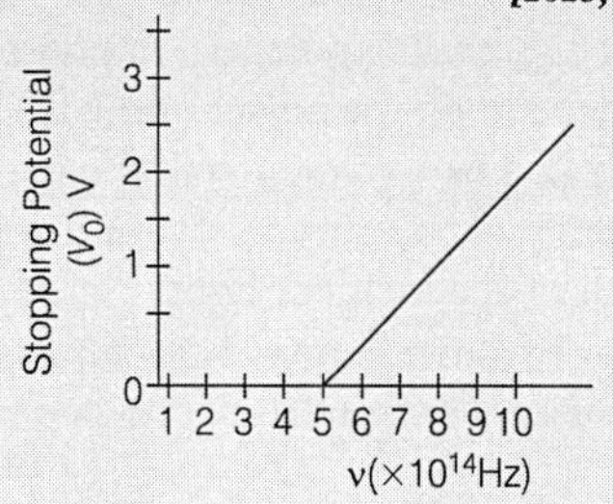

(a) 1.36 eV (b) 2.98 eV
(c) 18.6 eV (d) 2.07 eV

19 In photoelectric effect *[2023, 8 April Shift-II]*

A. The photocurrent is proportional to the intensity of the incident radiation
B. Maximum Kinetic energy with which photoelectrons are emitted depends on the intensity of incident light.
C. Maximum KE with which photoelectrons are emitted depends on the frequency of incident light.

D. The emission of photoelectrons require a minimum threshold intensity of incident radiation.
E. Maximum KE of the photoelectrons is independent of the frequency of the incident light.

Choose the correct answer from the options given below
(a) A and B only
(b) B and C only
(c) A and C only
(d) A and E only

20 The work functions of aluminium and gold are 4.1 eV and and 5.1 eV respectively. The ratio of the slope of the stopping potential *versus* frequency plot for gold to that of aluminium is *[2023, 6 April Shift-II]*
(a) 1.24 (b) 1.5
(c) 1 (d) 2

21 If the two metals A and B are exposed to radiation of wavelength 350 nm. The work functions of metals A and B are 4.8 eV and 2.2 eV. Then, choose the correct option. *[2023, 31 Jan Shift-II]*
(a) Both metals A and B will emit photoelectrons
(b) Metal B will not emit photoelectrons
(c) Both metals A and B will not emit photo electrons
(d) Metal A will not emit photoelectrons

22 The threshold wavelength for photoelectric emission from a material is 5500 Å. Photoelectrons will be emitted, when this material is illuminated with monochromatic radiation from a
A. 75 W infrared lamp B. 10 W infrared lamp
C. 75 W ultraviolet lamp D. 10W ultraviolet lamp

Choose the correct answer from the options given below. *[2023, 29 Jan Shift-I]*
(a) B and C only (b) C and D only
(c) A and D only (d) C only

23 Given below are two statements.

Statement I Stopping potential in photoelectric effect does not depend on the power of the light source.

Statement II For a given metal, the maximum kinetic energy of the photoelectron depends on the wavelength of the incident light.

In the light of above statements, choose the most appropriate answer from the options given below. *[2023, 25 Jan Shift-II]*
(a) Both Statement I and Statement II are incorrect.
(b) Both Statement I and Statement II are correct.
(c) Statement I is correct but Statement II is incorrect.
(d) Statement I is incorrect but Statement II is correct.

24 From the photoelectric effect experiment, following observations are made. Identify which of these are correct?
A. The stopping potential depends only on the work function of the metal.
B. The saturation current increases as the intensity of incident light increases.
C. The maximum kinetic energy of a photoelectron depends on the intensity of the incident light.
D. Photoelectric effect can be explained using the wave theory of light.

Choose the correct answer from the options given below. *[2023, 24 Jan Shift-I]*
(a) B only (b) B and C only
(c) A, B and D (d) A, C and D

25 Two streams of photons, possessing energies equal to five and ten times the work function of metal are incident on the metal surface successively. The ratio of maximum velocities of the photoelectron emitted, in the two cases respectively, will be *[2022, 28 July Shift-II]*
(a) 1 : 2 (b) 1 : 3 (c) 2 : 3 (d) 3 : 2

26 With reference to the observations in photoelectric effect, identify the correct statements from below
(A) The square of maximum velocity of photoelectrons varies linearly with frequency of incident light.
(B) The value of saturation current increases on moving the source of light away from the metal surface.
(C) The maximum kinetic energy of photoelectrons decreases on decreasing the power of LED (light emitting diode) source of light.
(D) The immediate emission of photoelectrons out of metal surface cannot be explained by particle nature of light/electromagnetic waves.
(E) Existence of threshold wavelength cannot be explained by wave nature of light/electromagnetic waves.

Choose the correct answer from the options given below. *[2022, 27 July Shift-II]*
(a) (A) and (B) only (b) (A) and (E) only
(c) (C) and (E) only (d) (D) and (E) only

Imp. Photoelectric Effect

27 Given below are two statements. One is labelled as Assertion (A) and the other is labelled as Reason (R).

Assertion (A) The photoelectric effect does not take place, if the energy of the incident radiation is less than the work function of a metal.

Reason (R) Kinetic energy of the photoelectrons is zero, if the energy of the incident radiation is equal to the work function of a metal.

In the light of the above statements choose the most appropriate answer from the options given below. *[2022, 29 June Shift-I]*
(a) Both A and R are correct and R is the correct explanation of A.
(b) Both A and R are correct but R is not the correct explanation of A.
(c) A is correct but R is not correct.
(d) A is not correct but R is correct.

28 In a photoelectric experiment, increasing the intensity of incident light *[2021, 27 Aug Shift-I]*

(a) increases the number of photons incident and also increases the KE of the ejected electrons.
(b) increases the frequency of photons incident and increases the KE of the ejected electrons.
(c) increases the frequency of photons incident and the KE of the ejected electrons remains unchanged.
(d) increases the number of photons incident and the KE of the ejected electrons remains unchanged.

29 Two identical photocathodes receive the light of frequencies f_1 and f_2, respectively. If the velocities of the photoelectrons coming out are v_1 and v_2 respectively, then *[2021, 17 March Shift-II]*

(a) $v_1^2 - v_2^2 = \frac{2h}{m}[f_1 - f_2]$ (b) $v_1^2 + v_2^2 = \frac{2h}{m}[f_1 + f_2]$

(c) $v_1 - v_2 = \left[\frac{2h}{m}(f_1 + f_2)\right]^{\frac{1}{2}}$ (d) $v_1 - v_2 = \left[\frac{2h}{m}(f_1 - f_2)\right]^{\frac{1}{2}}$

30 The stopping potential in the context of photoelectric effect depends on the following property of incident electromagnetic radiation *[2021, 16 March Shift-I]*

(a) phase (b) intensity
(c) amplitude (d) frequency

31 In a photoelectric effect experiment the threshold wavelength of light is 380 nm. If the wavelength of incident light is 260 nm, the maximum kinetic energy of emitted electrons will be

$\left(\text{Given, } E \text{ (in eV)} = \frac{1237}{\lambda \text{ (in nm)}}\right)$ *[2019, 10 April Shift-I]*

(a) 15.1 eV (b) 4.5 eV
(c) 1.5 eV (d) 3.0 eV

32 The electric field of light wave is given as

$E = 10^{-3}\cos\left(\frac{2\pi x}{5\times10^{-7}} - 2\pi\times6\times10^{14}t\right)\hat{x}\,\frac{N}{C}$. This

light falls on a metal plate of work function 2 eV. The stopping potential of the photoelectrons is

$\left(\text{Given, } E \text{ (in eV)} = \frac{12375}{\lambda \text{ (in Å)}}\right)$ *[2019, 9 April Shift-I]*

(a) 0.48 V (b) 0.72 V
(c) 2.48 V (d) 2.0 V

33 In a photoelectric experiment, the wavelength of the light incident on a metal is changed from 300 nm to 400 nm. The decrease in the stopping potential is close to ($hc = 1240$ nm eV) *[2019, 11 Jan, Shift-II]*

(a) 2.0 V (b) 0.5 V
(c) 1.0 V (d) 1.5 V

34 The anode voltage of a photocell is kept fixed. The wavelength λ of the light falling on the cathode is gradually changed. The plate current I of photocell varies as follows *[JEE Main 2013]*

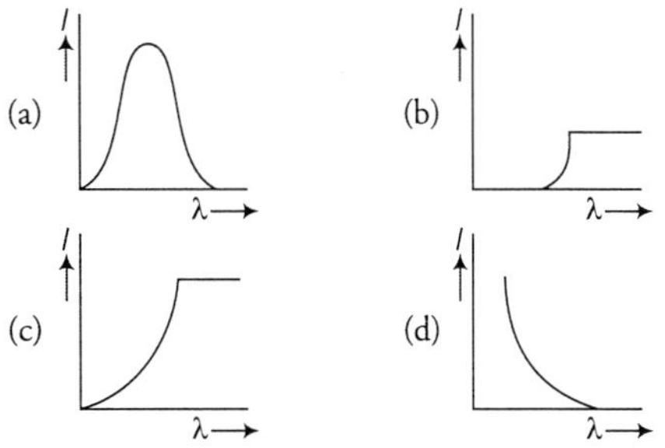

35 The anode voltage of a photocell is kept fixed. The wavelength λ of the light falling on the cathode is gradually changed. The plate current I of the photocell varies as follows *[AIEEE 2006]*

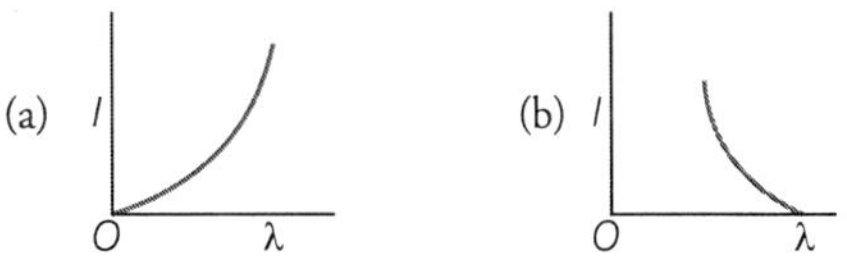

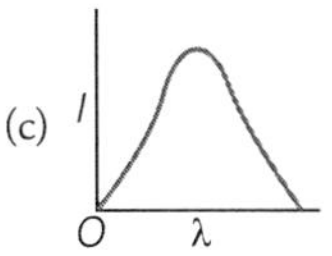

(d) None of these

36 The time taken by a photoelectron to come out after the photon strikes is approximately *[AIEEE 2006]*

(a) 10^{-4} s (b) 10^{-10} s
(c) 10^{-16} s (d) 10^{-1} s

37 According to Einstein's photoelectric equation, the plot of the kinetic energy of the emitted photoelectrons from a metal *versus* the frequency of the incident radiation gives a straight line whose slope *[AIEEE 2004]*

(a) depends on the nature of the metal used
(b) depends on the intensity of the radiation
(c) depend both on the intensity of the radiation and the metal used
(d) is the same for all metals and independent of the intensity of the radiation

38 Sodium and copper have work functions 2.3 eV and 4.5 eV, respectively. Then, the ratio of the wavelengths is nearest to *[AIEEE 2002]*

(a) 1 : 2 (b) 4 : 1
(c) 2 : 1 (d) 1 : 4

Level 2

39 Given below are two statements. One is labelled as Assertion (A) and the other is labelled as Reason (R).

Assertion (A) In photoelectric effect, on increasing the intensity of incident light the stopping potential increases.

Reason (R) Increase in intensity of light increases the rate of photoelectrons emitted, provided the frequency of incident light is greater than threshold frequency. *[2025, 4 April Shift-I]*

In the light of the above statements, choose the correct answer from the options given below.
(a) Both A and R are true and R is the correct explanation of A.
(b) Both A and R are true but R is not the correct explanation of A.
(c) (A) is false but R is true.
(d) (A) is true but R is false.

40 A monochromatic light is incident on a metallic plate having work function ϕ. An electron, emitted normally to the plate from a point A with maximum kinetic energy, enters a constant magnetic field, perpendicular to the initial velocity of electron. The electron passes through a curve and hits back the plate at a point B. The distance between A and B is:
(Given, The magnitude of charge of an electron is e and mass is m, h is Planck's constant and c is velocity of light. Take the magnetic field exists throughout the path of electron) **[2025, 2 April, Shift-I]**

(a) $2\sqrt{m\left(\frac{hc}{\lambda}-\phi\right)}/eB$ (b) $\sqrt{2m\left(\frac{hc}{\lambda}-\phi\right)}/eB$

(c) $\sqrt{8m\left(\frac{hc}{\lambda}-\phi\right)}/eB$ (c) $\sqrt{m\left(\frac{hc}{\lambda}-\phi\right)}/eB$

41 In photoelectric effect an EM wave is incident on a metal surface and electrons are ejected from the surface. If the work function of the metal is 2.14 eV and stopping potential is 2 V, what is the wavelength of the EM wave?
(Given, $hc = 1242$ eV-nm where h is the Planck's constant and c is the speed of light in vacuum) **[2025, 23 Jan, Shift-II]**
(a) 400 nm (b) 300 nm
(c) 600 nm (d) 200 nm

42 In photoelectric effect, the stopping potential (V_0) *versus* frequency (ν) curve is plotted.
(h is the Planck's constant and ϕ_0 is work function of metal)
(A) V_0 *versus* ν is linear.
(B) The slope of V_0 *versus* ν curve $= \frac{\phi_0}{h}$.
(C) h constant is related to the slope of V_0 *versus* ν line.
(D) The value of electric charge of electron is not required to determine h using the V_0 *versus* ν curve.
(E) The work function can be estimated without knowing the value of h.
Choose the correct answer from the options given below. **[2025, 24 Jan, Shift-II]**
(a) A, C and E only (b) C and D only
(c) A, B and C only (d) D and E only

43 Given below are two statements, one is labelled as Assertion (A) and the other is labelled as Reason (R).
Assertion (A) Number of photons increases with increase in frequency of light.
Reason (R) Maximum kinetic energy of emitted electrons increases with the frequency of incident radiation. **[2024, 4 April Shift-II]**
In the light of the above statements, choose the most appropriate answer from the options given below.
(a) Both A and R are correct but R is not the correct explanation of A.
(b) Both A and R are correct and R is the correct explanation of A.
(c) A is not correct but R is correct.
(d) A is correct but R is not correct.

44 Given below are two statements.

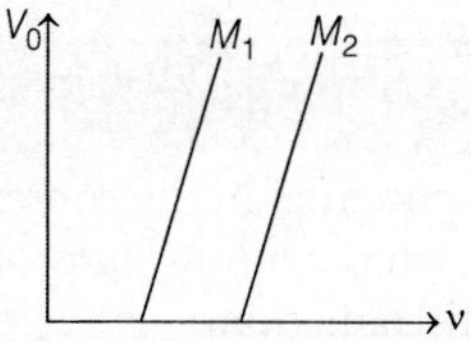

Statement I Figure shows the variation of stopping potential with frequency ν for the two photosensitive material M_1 and M_2. The slope gives value of h/e, where h is Planck's constant, e is the charge of electron.
Statement II M_2 will emit photoelectrons of greater kinetic energy for the incident radiation having same frequency. **[2024, 5 April Shift-I]**
In the light of the above statements, choose the most appropriate answer from the options given below.
(a) Statement I is incorrect but Statement II is correct.
(b) Both Statement I and Statement II are incorrect.
(c) Both Statement I and Statement II are correct.
(d) Statement I is correct and Statement II is incorrect.

45 When a metal surface is illuminated by light of wavelength λ the stopping potential is 8V. When the same surface is illuminated by light of wavelength 3λ, stopping potential is 2V. The threshold wavelength for this surface is **[2024, 31 Jan Shift-I]**
(a) 9λ (b) 5λ (c) 4.5λ (d) 3λ

46 A metallic surface is illuminated with radiation of wavelength λ, the stopping potential is V_0. If the same surface is illuminated with radiation of wavelength 2λ, the stopping potential becomes $V_0/4$. The threshold wavelength for this metallic surface will be **[2023, 11 April Shift-I]**
(a) 3λ (b) $\frac{3}{2}\lambda$
(c) $\frac{\lambda}{4}$ (d) 4λ

47 The threshold frequency of a metal is f_0. When the light of frequency $2f_0$ is incident on the metal plate, the maximum velocity of photoelectrons is v_1. When the frequency of incident radiation is increased to $5f_0$. The maximum velocity of photoelectrons emitted is v_2. The ratio of v_1 to v_2 is *[2023, 1 Feb, Shift-II]*

(a) $\frac{v_1}{v_2}=\frac{1}{4}$ (b) $\frac{v_1}{v_2}=\frac{1}{2}$ (c) $\frac{v_1}{v_2}=\frac{1}{8}$ (d) $\frac{v_1}{v_2}=\frac{1}{16}$

48 The kinetic energy of emitted electron is E when the light incident on the metal has wavelength λ. To double the kinetic energy, the incident light must have wavelength *[2022, 29 July Shift-I]*

(a) $\frac{hc}{E\lambda - hc}$ (b) $\frac{hc\lambda}{E\lambda + hc}$ (c) $\frac{h\lambda}{E\lambda + hc}$ (d) $\frac{hc\lambda}{E\lambda - hc}$

49 A metal exposed to light of wavelength 800 nm and emits photoelectrons with a certain kinetic energy. The maximum kinetic energy of photoelectron doubles when light of wavelength 500 nm is used. The work function of the metal is
(Take, $hc = 1230$ eV-nm) *[2022, 25 July Shift-I]*

(a) 1.537 eV (b) 2.46 eV (c) 0.615 eV (d) 1.23 eV

50 The electric field at a point associated with a light wave is given by

$$E = 200\,[\sin(6\times10^{15})\,t + \sin(9\times10^{15})\,t]\,\text{Vm}^{-1}$$

(Given, $h = 4.14 \times 10^{-15}$ eV-s)
If this light falls on a metal surface having a work function of 2.50 eV, the maximum kinetic energy of the photoelectrons will be *[2022, 29 June Shift-II]*

(a) 1.90 eV (b) 3.27 eV
(c) 3.60 eV (d) 3.42 eV

51 Let K_1 and K_2 be the maximum kinetic energies of photoelectrons emitted when two monochromatic beams of wavelength λ_1 and λ_2, respectively are incident on a metallic surface. If $\lambda_1 = 3\lambda_2$, then *[2022, 28 June shift-II]*

(a) $K_1 > \frac{K_2}{3}$ (b) $K_1 < \frac{K_2}{3}$ (c) $K_1 = \frac{K_2}{3}$ (d) $K_2 = \frac{K_1}{3}$

52 The light of two different frequencies whose photons have energies 3.8 eV and 1.4 eV, respectively, illuminate a metallic surface whose work function is 0.6 eV successively. The ratio of maximum speeds of emitted electrons for the two frequencies, respectively will be *[2022, 24 June Shift-II]*

(a) 1 : 1 (b) 2 : 1 (c) 4 : 1 (d) 1 : 4

53 In a photoelectric experiment ultraviolet light of wavelength 280 nm is used with lithium cathode having work-function $\phi = 2.5$ eV. If the wavelength of incident light is switched to 400 nm, find out the change in the stopping potential.
($h = 6.63\times10^{-34}$ Js, and $c = 3\times10^8$ ms^{-1}) *[2021, 26 Aug Shift-I]*

(a) 1.3 V (b) 1.1 V (c) 1.9 V (d) 0.6 V

Imp. Stopping Potential

54 The stopping potential for electrons emitted from a photosensitive surface illuminated by light of wavelength 491 nm is 0.710 V. When the incident wavelength is changed to a new value, the stopping potential is 1.43 V. The new wavelength is *[2021, 25 Feb Shift-II]*

(a) 309 nm (b) 329 nm
(c) 382 nm (d) 400 nm

55 In a photoelectric effect experiment, the graph of stopping potential V *versus* reciprocal of wavelength $1/\lambda$ obtained is shown in the figure. As the intensity of incident radiation is increased *[2020, 4 Sep Shift-II]*

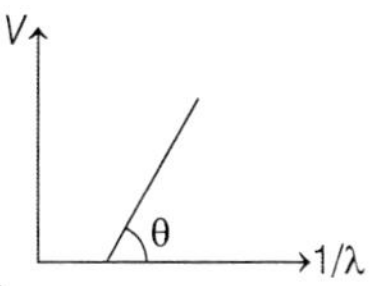

(a) graph does not change
(b) straight line shifts to left
(c) slope of the straight line get more steep
(d) straight line shifts to right

56 The following figure shows few data points in a photoelectric effect experiment for a certain metal. The minimum energy for ejection of electron from its surface is (Take, Planck's constant, $h = 6.62\times10^{-34}$ Js)

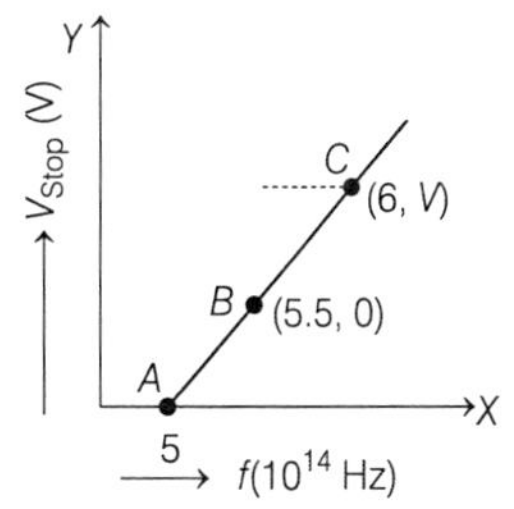

[2020, 4 Sep Shift-I]

(a) 1.93 eV (b) 2.59 eV
(c) 2.27 eV (d) 2.10 eV

57 The stopping potential V_0 (in volt) as a function of frequency (ν) for a sodium emitter, is shown in the figure. The work function of sodium, from the data plotted in the figure, will be
[Take, Planck's constant $(h) = 6.63\times10^{-34}$ J-s, electron charge, $e = 1.6\times10^{-19}$ C] *[2019, 12 April Shift-I]*

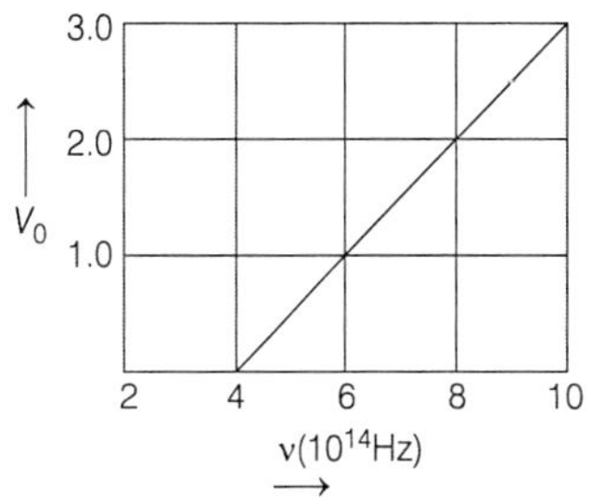

(a) 1.82 eV (b) 1.66 eV
(c) 1.95 eV (d) 2.12 eV

58 The magnetic field associated with a light wave is given at the origin, by
$B = B_0 [\sin(3.14\times10^7)ct + \sin(6.28\times10^7)ct]$.
If this light falls on a silver plate having a work function of 4.7 eV, what will be the maximum kinetic energy of the photoelectrons?
(Take, $c = 3\times10^8\ \text{ms}^{-1}$ and $h = 6.6\times10^{-34}$ J-s)
[2019, 9 Jan Shift-II]
(a) 7.72 eV (b) 6.82 eV (c) 8.52 eV (d) 12.5 eV

59 Radiation of wavelength λ is incident on a photocell. The fastest emitted electron has speed v. If the wavelength is changed to $3\lambda/4$, the speed of the fastest emitted electron will be *[JEE Main 2016]*
(a) $> v\left(\frac{4}{3}\right)^{1/2}$ (b) $< v\left(\frac{4}{3}\right)^{1/2}$ (c) $= v\left(\frac{4}{3}\right)^{1/2}$ (d) $= v\left(\frac{3}{4}\right)^{1/2}$

60 **Statement I** When ultraviolet light is incident on a photocell, its stopping potential is V_0 and the maximum kinetic energy of the photoelectrons is K_{max}. When the ultraviolet light is replaced by X-rays, both V_0 and K_{max} increase.
Statement II Photoelectrons are emitted with speeds ranging from zero to a maximum value, because of the range of frequencies present in the incident light.
[AIEEE 2010]
(a) Statement I is true, Statement II is true; Statement II is the correct explanation of Statement I.
(b) Statement I is true, Statement II is true; Statement II is not the correct explanation of Statement I.
(c) Statement I is true, Statement II is false.
(d) Statement I is false, Statement II is true.

61 The surface of a metal is illuminated with the light of 400 nm. The kinetic energy of the ejected photoelectrons was found to be 1.68 eV. The work function of the metal is ($hc = 1240$ eV-nm)
[AIEEE 2009]
(a) 3.09 eV (b) 1.42 eV (c) 151 eV (d) 1.68 eV

Level 3

62 A metal surface is illuminated by a radiation of wavelength 4500 Å. The ejected photo-electron enters a constant magnetic field of 2 mT making an angle of 90° with the magnetic field. If it starts revolving in a circular path of radius 2 mm, the work function of the metal is approximately.
[2022, 26 June Shift-II]
(a) 1.36 eV (b) 1.69 eV (c) 2.78 eV (d) 2.23 eV

63 A monochromatic neon lamp with wavelength of 670.5 nm illuminates a photo-sensitive material which has a stopping voltage of 0.48 V. What will be the stopping voltage if the source light is changed with another source of wavelength of 474.6 nm ?
[2021, 27 Aug Shift-II]
(a) 0.96 V (b) 1.25 V (c) 0.24 V (d) 1.5 V

64 When radiation of wavelength λ is incident on a metallic surface, the stopping potential of ejected photoelectrons is 4.8 V. If the same surface is illuminated by radiation of double the previous wavelength, then the stopping potential becomes 1.6 V. The threshold wavelength of the metal is
[2021, 25 July Shift-II]
(a) 2λ (b) 4λ (c) 8λ (d) 6λ

65 Radiation with wavelength 6561 Å falls on a metal surface to produce photoelectrons. The electrons are made to enter a uniform magnetic field of 3×10^{-4} T. If the radius of the largest circular path followed by the electrons is 10 mm, the work function of the metal is close to *[2020, 9 Jan Shift-I]*
(a) 0.8 eV (b) 1.1 eV
(c) 1.8 eV (d) 1.6 eV

Numerical Value Type Questions

Imp. Einstein Equation of Photoelectric Effect

66 A light source of wavelength λ illuminates a metal surface and electrons are ejected with maximum kinetic energy of 2 eV. If the same surface is illuminated by a light source of wavelength $\lambda/2$, then the maximum kinetic energy of ejected electrons will be (the work function of metal is 1 eV)
[2025, 22 Jan Shift-II]
(a) 3 eV (b) 6 eV (c) 2 eV (d) 5 eV

67 An atom absorbs a photon of wavelength 500 nm and emits another photon of wavelength 600 nm. The net energy absorbed by the atom in this process is $n\times10^{-4}$ eV. The value of n is ....
[Assume the atom to be stationary during the absorption and emission process]
(Take, $h = 6.6\times10^{-34}$ J-s and $c = 3\times10^8$ m/s).
[2023, 13 April Shift-II]

68 When light of frequency twice the threshold frequency is incident on the metal plate, the maximum velocity of emitted electron is v_1. When the frequency of incident radiation is increased to five times the threshold value, the maximum velocity of emitted electron becomes v_2. If $v_2 = x\, v_1$, the value of x will be .... *[2022, 24 June Shift-I]*

69 A light beam of wavelength 500 nm is incident on a metal having work function of 1.25 eV, placed in a magnetic field of intensity B. The electrons emitted perpendicular to the magnetic field B, with maximum kinetic energy are bent into circular arc of radius 30 cm. The value of B is ...... $\times10^{-7}$ T.
(Take, $hc = 20\times10^{-26}$ J-m, mass of electron $= 9\times10^{-31}$ kg) *[2021, 25 July Shift-II]*

70 A certain metallic surface is illuminated by monochromatic radiation of wavelength λ. The stopping potential for photoelectric current for this radiation is $3V_0$. If the same surface is illuminated with a radiation of wavelength 2λ, the stopping potential is V_0. The threshold wavelength of this surface for photoelectric effect is ...... λ. *[2021, 20 July Shift-II]*

71 Two stream of photons, possessing energies equal to twice and ten times the work function of metal are incident on the metal surface successively. The value of ratio of maximum velocities of the photoelectrons emitted in the two respective cases is $x : y$. The value of x is .... *[2021, 26 Feb Shift-II]*

72 The surface of a metal is illuminated alternately with photons of energies $E_1 = 4$ eV and $E_2 = 2.5$ eV respectively. The ratio of maximum speeds of the photoelectrons emitted in the two cases is 2. The work function of the metal (in eV) is .... *[2020, 5 Sep Shift-II]*

73 When radiation of wavelength λ is used to illuminate a metallic surface, the stopping potential is V. When the same surface is illuminated with radiation of wavelength 3λ, the stopping potential is $V/4$. If the threshold wavelength for the metallic surface is $n\lambda$, then value of n will be .... *[2020, 2 Sep Shift-I]*

Topic 2 Particle Nature of Light-The Photon

Level 1

01 Which of the following phenomena can not be explained by wave theory of light ? *[2025, 28 Jan Shift-II]*

(a) Refraction of light (b) Diffraction of light
(c) Reflection of light (d) Compton effect

02 Given below are two statements. One is labelled as Assertion (A) and the other is labelled as Reason (R).

Assertion (A) Electromagnetic waves carry energy but not momentum.

Reason (R) Mass of a photon is zero.

In the light of the above statements, choose the most appropriate answer from the options given below. *[2025, 29 Jan Shift-I]*

(a) A is false but R is true.
(b) A is true but R is false.
(c) Both A and R are true but R is not the correct explanation of A.
(d) Both A and R are true and R is the correct explanation of A.

03 Monochromatic light of frequency 6×10^{14} Hz is produced by a laser. The power emitted is 2×10^{-3} W. How many photons per second on an average, are emitted by the source? (Given, $h = 6.63 \times 10^{-34}$ Js) *[2024, 1 Feb Shift-II]*

(a) 6×10^{15} (b) 7×10^{16}
(c) 5×10^{15} (d) 9×10^{18}

04 Two sources of light emit with a power of 200 W. The ratio of number of photons of visible light emitted by each source having wavelengths 300 nm and 500 nm, respectively, will be *[2024, 29 Jan Shift-II]*

(a) 1 : 5 (b) 5 : 3
(c) 3 : 5 (d) 1 : 3

05 Given below are two statements

Statement I Two photons having equal linear momentum have equal wavelengths.

Statement II If the wavelength of photon is decreased, then the momentum and energy of a photon will also decrease.

In the light of the above statements, choose the correct answer from the options given below. *[2021, 24 Feb Shift-I]*

(a) Both Statement I and Statement II are true.
(b) Both Statement I and Statement II are false.
(c) Statement I is true but Statement II is false.
(d) Statement I is false but Statement II is true.

06 If a source of power 4 kW produces 10^{20} photons/second, the radiation belong to a part of the spectrum called *[AIEEE 2010]*

(a) X-rays (b) ultraviolet rays
(c) microwaves (d) γ-rays

Level 2

07 If a source of electromagnetic radiation having power 15 kW produces 10^{16} photons per second, the radiation belongs to a part of spectrum is (Take, Planck's constant $h = 6 \times 10^{-34}$ Js) *[2023, 31 Jan Shift-I]*

(a) radiowaves (b) gamma rays
(c) ultraviolet rays (d) microwaves

08 A small object at rest, absorbs a light pulse of power 20 mW and duration 300 ns. Assuming speed of light as 3×10^8 m/s, the momentum of the object becomes equal to *[2023, 30 Jan Shift-I]*

(a) 1×10^{-17} kg-m/s (b) 2×10^{-17} kg-m/s
(c) 0.5×10^{-17} kg-m/s (d) 3×10^{-17} kg-m/s

09 Two sources of light emit X-rays of wavelength 1 nm and visible light of wavelength 500 nm, respectively. Both the sources emit light of the same power 200 W. The ratio of the number density of photons of X-rays to the number density of photons of the visible light of the given wavelengths is *[2020, 3 Sep Shift-II]*

(a) 1/250 (b) 500 (c) 250 (d) 1/500

10 A 2 mW laser operates at a wavelength of 500 nm. The number of photons that will be emitted per second is [Given, Planck's constant $h = 6.6 \times 10^{-34}$ Js, speed of light $c = 3.0 \times 10^8$ m/s]

[2019, 10 April Shift-2]

(a) 1×10^{16} (b) 5×10^{15}
(c) 1.5×10^{16} (d) 2×10^{16}

Level 3

11 A parallel beam of light of wavelength 900 nm and intensity 100 Wm^{-2} is incident on a surface perpendicular to the beam. The number of photons crossing through area of 1 cm^2 perpendicular to the beam in one second is *[2022, 26 July Shift-I]*

(a) 3×10^{16} (b) 4.5×10^{16}
(c) 4.5×10^{17} (d) 4.5×10^{20}

Numerical Value Type Questions

12 The ratio of the power of a light source S_1 to that the light source S_2 is 2. S_1 is emitting 2×10^{15} photons per second at 600 nm. If the wavelength of the source S_2 is 300 nm, then the number of photons per second emitted by S_2 is .... $\times 10^{14}$.

[2025, 24 July Shift-II]

Imp. Intensity of Radiation

13 A beam of electromagnetic radiation of intensity 6.4×10^{-5} W/cm^2 is comprised of wavelength $\lambda = 310$ nm. It falls normally on a metal (work function $\phi = 2$ eV) of surface area of 1 cm^2. If one in 10^3 photons ejects an electron, total number of electrons ejected in 1 s is 10^x. ($hc = 1240$ eV-nm, 1 eV = 1.6×10^{-19} J), then x is ....

[2020, 7 Jan Shift-I]

Topic 3 Matter Waves

Level 1

Imp. de-Broglie Wavelength

01 If λ and K are de-Broglie wavelength and kinetic energy, respectively, of a particle with constant mass. The correct graphical representation for the particle will be *[2025, 29 Jan Shift-I]*

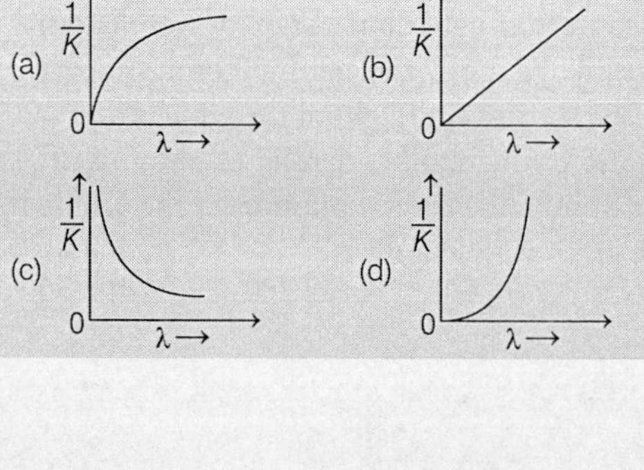

02 A proton of mass m_p has same energy as that of a photon of wavelength λ. If the proton is moving at non-relativistic speed, then ratio of its de-Broglie wavelength to the wavelength of photon is *[2025, 28 Jan Shift-I]*

(a) $\frac{1}{c}\sqrt{\frac{E}{2m_p}}$ (b) $\frac{1}{2c}\sqrt{\frac{E}{m_p}}$ (c) $\frac{1}{c}\sqrt{\frac{E}{m_p}}$ (d) $\frac{1}{c}\sqrt{\frac{2E}{m_p}}$

03 If the total energy transferred to a surface in time t is 6.48×10^5 J, then the magnitude of the total momentum delivered to this surface for complete absorption will be *[2024, 30 Jan Shift-II]*

(a) 4.32×10^{-3} kg m/s (b) 2.16×10^{-3} kg m/s
(c) 1.58×10^{-3} kg m/s (d) 2.46×10^{-3} kg m/s

04 A proton, an electron and an alpha particle have the same energies. Their de-Broglie wavelengths will be compared as *[2024, 9 April Shift-I]*

(a) $\lambda_e > \lambda_\alpha > \lambda_p$ (b) $\lambda_p < \lambda_e < \lambda_\alpha$
(c) $\lambda_\alpha < \lambda_p < \lambda_e$ (d) $\lambda_p > \lambda_e > \lambda_\alpha$

05 The de-Broglie wavelength of an electron having kinetic energy E is λ. If the kinetic energy of electron becomes $E/4$, then its de-Broglie wavelength will be *[2023, 15 April Shift-I]*

(a) $\lambda/2$ (b) $\lambda/\sqrt{2}$ (c) 2λ (d) $\sqrt{2}\lambda$

06 A proton and an α-particle are accelerated from rest by 2V and 4V potentials, respectively. The ratio of their de-Broglie wavelength is *[2023, 12 April Shift-I]*

(a) 16 : 1 (b) 2 : 1 (c) 8 : 1 (d) 4 : 1

07 The ratio of the de-Broglie wavelength of proton and electron having same kinetic energy (Assume, $m_p = m_e \times 1849$) *[2023, 11 April Shift-II]*

(a) 1 : 43 (b) 1 : 62 (c) 1 : 30 (d) 2 : 43

08 Proton p and electron e will have same de-Broglie wavelength when the ratio of their momentum is (assume, $m_p = 1849\, m_e$) *[2023, 8 April Shift-I]*

(a) 1 : 1849 (b) 43 : 1 (c) 1 : 1 (d) 1 : 43

09 A proton moving with one-tenth of velocity of light has a certain de-Broglie wavelength of λ. An alpha particle having certain kinetic energy has the same de-Broglie wavelength λ. The ratio of kinetic energy of proton and that of alpha particle is *[2023, 1 Feb, Shift-I]*

(a) 2 : 1 (b) 1 : 2 (c) 4 : 1 (d) 1 : 4

10 The ratio of de-Broglie wavelength of an α particle and a proton accelerated from rest by the same potential is $1/\sqrt{m}$, the value of m is *[2023, 29 Jan, Shift-II]*

(a) 16 (b) 8 (c) 4 (d) 2

11 Electron beam used in an electron microscope, when accelerated by a voltage of 20 kV, has a de-Broglie wavelength of λ_0. If the voltage is increased to 40 kV, then the de-Broglie wavelength associated with the electron beam would be *[2023, 25 Jan, Shift-I]*

(a) $9\lambda_0$ (b) $\frac{\lambda_0}{\sqrt{2}}$ (c) $\frac{\lambda_0}{2}$ (d) $3\lambda_0$

12 An α-particle and a proton are accelerated from rest through the same potential difference. The ratio of linear momentum acquired by above two particles will be *[2022, 29 July, Shift-II]*
(a) $\sqrt{2}:1$ (b) $2\sqrt{2}:1$ (c) $4\sqrt{2}:1$ (d) $8:1$

13 The equation $\lambda = \frac{1.227}{x}$ nm can be used to find the de-Broglie wavelength of an electron. In this equation, x stands for
where, m = mass of electron
p = momentum of electron,
K = kinetic energy of electron and
V = accelerating potential in volts for electron. *[2022, 28 July Shift-I]*
(a) $\sqrt{mK}$ (b) $\sqrt{p}$ (c) $\sqrt{K}$ (d) $\sqrt{V}$

14 The ratio of wavelength of proton and deuteron accelerated by potential V_p and V_d is $1:\sqrt{2}$. Then, the ratio of V_p to V_d will be *[2022, 25 July, Shift-II]*
(a) 1 : 1 (b) $\sqrt{2}:1$ (c) 2 : 1 (d) 4 : 1

15 The de-Broglie wavelengths for an electron and a photon are λ_e and λ_p, respectively. For the same kinetic energy of electron and photon, which of the following represents the correct relation between the de-Broglie wavelengths of two? *[2022, 28 June Shift-I]*
(a) $\lambda_p \propto \lambda_e^2$ (b) $\lambda_p \propto \lambda_e$
(c) $\lambda_p \propto \sqrt{\lambda_e}$ (d) $\lambda_p \propto \sqrt{\frac{1}{\lambda_e}}$

16 An α-particle and a carbon 12 atom has same kinetic energy K. The ratio of their de-Broglie wavelengths $(\lambda_\alpha : \lambda_{c12})$ is *[2022, 27 June, Shift-I]*
(a) $1:\sqrt{3}$ (b) $\sqrt{3}:1$ (c) $3:1$ (d) $2:\sqrt{3}$

Imp. Comparision of de-Broglie Wavelength

17 A proton, a neutron, an electron and an α-particle have same energy. If $\lambda_p, \lambda_n, \lambda_e$ and λ_α are the de-Broglie's wavelengths of proton, neutron, electron and α particle respectively, then choose the correct relation from the following. *[2022, 25 June Shift-II]*
(a) $\lambda_p = \lambda_n > \lambda_e > \lambda_\alpha$ (b) $\lambda_\alpha < \lambda_n < \lambda_p < \lambda_e$
(c) $\lambda_e < \lambda_p = \lambda_n > \lambda_\alpha$ (d) $\lambda_e = \lambda_p = \lambda_n = \lambda_\alpha$

18 An electron of mass m_e and a proton of mass m_p are accelerated through the same potential difference. The ratio of the de-Broglie wavelength associated with the electron to that with the proton is *[2021, 22 July Shift-II]*
(a) $\frac{m_p}{m_e}$ (b) 1 (c) $\sqrt{\frac{m_p}{m_e}}$ (d) $\frac{m_e}{m_p}$

19 A particle is travelling 4 times as fast as an electron. Assuming the ratio of de-Broglie wavelength of a particle to that of electron is 2 : 1, the mass of the particle is *[2021, 18 March Shift-I]*
(a) 1/16 times the mass of electron
(b) 8 times the mass of electron
(c) 16 times the mass of electron
(d) 1/8 times the mass of electron

20 The de-Broglie wavelength associated with an electron and a proton were calculated by accelerating them through same potential of 100 V. What should nearly be the ratio of their wavelengths? (m_p = 1.00727 u, m_e = 0.00055 u) *[2021, 16 March Shift-II]*
(a) 1860 : 1 (b) $(1860)^2 : 1$
(c) 41.4 : 1 (d) 43 : 1

21 An electron of mass m_e and a proton of mass $m_p = 1836\, m_e$ are moving with the same speed. The ratio of their de-Broglie wavelength $\frac{\lambda_{\text{electron}}}{\lambda_{\text{proton}}}$ will be *[2021, 25 Feb Shift-II]*
(a) 1 (b) 1836
(c) 1/1836 (d) 918

22 An α-particle and a proton are accelerated from rest by a potential difference of 200 V. After this, their de-Broglie wavelengths are λ_α and λ_p, respectively. The ratio λ_p/λ_α is *[2021, 25 Feb Shift-I]*
(a) 3.8 (b) 8
(c) 7.8 (d) 2.8

23 The de-Broglie wavelength of a proton and α-particle are equal. The ratio of their velocities is *[2021, 24 Feb Shift-II]*
(a) 4 : 3 (b) 4 : 1
(c) 4 : 2 (d) 1: 4

24 Assuming the nitrogen molecule is moving with rms velocity at 400 K, the de-Broglie wavelength of nitrogen molecule is close to
(Given, weight of N_2 molecule $= 4.64 \times 10^{-26}$ kg, Boltzmann constant $= 1.38 \times 10^{-23}$ J/K and Planck's constant $= 6.63 \times 10^{-34}$ J-s) *[2020, 6 Sep Shift-II]*
(a) 0.24Å (b) 0.20Å
(c) 0.34Å (d) 0.44Å

25 An electron (of mass m) and a photon have the same energy E in the range of a few electron volt. The ratio of the de Broglie wavelength associated with the electron and the wavelength of the photon is (c = speed of light in vacuum) *[2020, 7 Jan Shift-II]*
(a) $\left(\frac{E}{2m}\right)^{1/2}$ (b) $\frac{1}{c}\left(\frac{2E}{m}\right)^{1/2}$
(c) $c\,(2mE)^{1/2}$ (d) $\frac{1}{c}\left(\frac{E}{2m}\right)^{1/2}$

26 After absorbing a slowly moving neutron of mass m_N (momentum ~ 0), a nucleus of mass M breaks into two nuclei of masses m_1 and $5m_1$, $(6m_1 = M + m_N)$, respectively. If the de-Broglie wavelength of the nucleus with mass m_1 is λ, then de-Broglie wavelength of the other nucleus will be *[AIEEE 2011]*

(a) 25λ (b) 5λ (c) $\lambda/5$ (d) λ

27 Photon of frequency ν has a momentum associated with it. If c is the velocity of light, the momentum is *[AIEEE 2007]*

(a) ν/c (b) $h\nu c$ (c) $h\nu/c^2$ (d) $h\nu/c$

28 Formation of covalent bonds in compounds exhibits *[AIEEE 2002]*

(a) wave nature of electron
(b) particle nature of electron
(c) both wave and particle nature of electron
(d) None of the above

Level 2

29 An electron with mass m with an initial velocity $(t = 0)$ $\mathbf{v} = v_0\hat{\mathbf{i}} (v_0 > 0)$ enters a magnetic field $\mathbf{B} = B_0\hat{\mathbf{j}}$. If the initial de-Broglie wavelength at $t = 0$ is λ_0, then its value after time 't' would be *[2025, 2 April Shift-II]*

(a) λ_0 (b) $\dfrac{\lambda_0}{\sqrt{1+\dfrac{e^2B_0^2t^2}{m^2}}}$

(c) $\dfrac{\lambda_0}{\sqrt{1-\dfrac{e^2B_0^2t^2}{m^2}}}$ (d) $\lambda_0\sqrt{1+\dfrac{e^2B_0^2t^2}{m^2}}$

30 A photoemissive substance is illuminated with a radiation of wavelength λ_i so that it releases electrons with de-Broglie wavelength λ_e. The longest wavelength of radiation that can emit photoelectron is λ_0. Expression for de-Broglie wavelength is given by (m = mass of the electron, h = Planck's constant and c = speed of light) *[2025, 7 April Shift-II]*

(a) $\lambda_e = \sqrt{\dfrac{h\lambda_0}{2mc}}$ (b) $\lambda_e = \sqrt{\dfrac{h\lambda_i}{2mc}}$

(c) $\lambda_e = \dfrac{h}{\sqrt{2mc\left(\dfrac{1}{\lambda_i}-\dfrac{1}{\lambda_0}\right)}}$ (d) $\lambda_e = \sqrt{\dfrac{h}{2mc\left(\dfrac{1}{\lambda_i}-\dfrac{1}{\lambda_0}\right)}}$

31 An electron in the ground state of the hydrogen atom has the orbital radius of 5.3×10^{-11} m while that for the electron in third excited state is 8.48×10^{-10} m. The ratio of the de-Broglie wavelengths of electron in the ground state to that in the excited state is *[2025, 22 Jan Shift-I]*

(a) 3 (b) 4 (c) 16 (d) 9

32 An electron of mass m with an initial velocity $\mathbf{v} = v_0\hat{\mathbf{i}} (v_0 > 0)$ enters an electric field $\mathbf{E} = -E_0\hat{\mathbf{k}}$. If the initial de-Broglie wavelength is λ_0, the value after time t would be *[2025, 24 Jan Shift-I]*

(a) λ_0 (b) $\dfrac{\lambda_0}{\sqrt{1-\dfrac{e^2E_0^2t^2}{m^2v_0^2}}}$

(c) $\dfrac{\lambda_0}{\sqrt{1+\dfrac{e^2E_0^2t^2}{m^2v_0^2}}}$ (d) $\lambda_0\sqrt{1+\dfrac{e^2E_0^2t^2}{m^2v_0^2}}$

33 A proton and an electron have the same de-Broglie wavelength. If K_p and K_e be the kinetic energies of proton and electron respectively, then choose the correct relation *[2024, 8 April Shift-II]*

(a) $K_p = K_e$ (b) $K_p < K_e$ (c) $K_p = K_e^2$ (d) $K_p > K_e$

34 The de-Broglie wavelengths of a proton and an α-particle are λ and 2λ, respectively. The ratio of the velocities of proton and α-particle will be *[2024, 1 Feb Shift-I]*

(a) 1 : 2 (b) 1 : 8 (c) 8 : 1 (d) 4 : 1

35 The de-Broglie wavelength of an electron is the same as that of a photon. If velocity of electron is 25% of the velocity of light, then the ratio of KE of electron and KE of photon will be *[2024, 29 Jan Shift-I]*

(a) 8/1 (b) 1/8 (c) 1/4 (d) 1/1

36 The de-Broglie wavelength of a molecule in a gas at room temperature 300 K is λ_1. If the temperature of the gas is increased to 600 K, then the de-Broglie wavelength of the same gas molecule becomes *[2023, 10 April Shift-I]*

(a) $\dfrac{1}{\sqrt{2}}\lambda_1$ (b) $2\lambda_1$ (c) $\sqrt{2}\lambda_1$ (d) $\dfrac{1}{2}\lambda_1$

37 The kinetic energy of an electron, α-particle and a proton are given as $4K$, $2K$ and K respectively. The de-Broglie wavelength associated with electron λ_e, α-particle (λ_α) and the proton λ_p are as follows *[2023, 6 April Shift-I]*

(a) $\lambda_\alpha = \lambda_p > \lambda_e$ (b) $\lambda_\alpha < \lambda_p < \lambda_e$
(c) $\lambda_\alpha > \lambda_p > \lambda_e$ (d) $\lambda_\alpha = \lambda_p < \lambda_e$

38 An electron accelerated through a potential difference V_1 has a de-Broglie wavelength of λ. When the potential is changed to V_2, its de-Broglie wavelength increases by 50%. The value of $\left(\dfrac{V_1}{V_2}\right)$ is equal to *[2023, 30 Jan, Shift-II]*

(a) 4 (b) $\dfrac{9}{4}$ (c) 3 (d) $\dfrac{3}{2}$

39 The temperature of an ideal gas in 3-dimensions is 300 K. The corresponding de-Broglie wavelength of the electron approximately at 300 K, is
[m_e = mass of electron $= 9 \times 10^{-31}$ kg
h = Planck's constant $= 6.6 \times 10^{-34}$ Js
k_B = Boltzmann constant $= 1.38 \times 10^{-23}\ \text{JK}^{-1}$]
[2021, 1 Sep Shift-II]
(a) 6.26 nm (b) 8.46 nm
(c) 2.26 nm (d) 3.25 nm

40 Consider two separate ideal gases of electrons and protons having same number of particles. The temperature of both the gases are same. The ratio of the uncertainty in determining the position of an electron to that of a proton is proportional to *[2021, 31 Aug Shift-II]*
(a) $\left(\frac{m_p}{m_e}\right)^{3/2}$ (b) $\sqrt{\frac{m_e}{m_p}}$
(c) $\sqrt{\frac{m_p}{m_e}}$ (d) $\frac{m_p}{m_e}$

41 A moving proton and electron have the same de-Broglie wavelength. If K and p denote the KE and momentum, respectively. Then, choose the correct option. *[2021, 31 Aug Shift-I]*
(a) $K_p < K_e$ and $p_p = p_e$ (b) $K_p = K_e$ and $p_p = p_e$
(c) $K_p < K_e$ and $p_p < p_e$ (d) $K_p > K_e$ and $p_p = p_e$

42 An electron moving with speed v and a photon moving with speed c, have same de-Broglie wavelength. The ratio of kinetic energy of electron to that of photon is *[2021, 25 July Shift-II]*
(a) $\frac{3c}{v}$ (b) $\frac{v}{3c}$ (c) $\frac{v}{2c}$ (d) $\frac{2c}{v}$

43 An electron having de-Broglie wavelength λ is incident on a target in a X-ray tube. Cut-off wavelength of emitted X-ray is *[2021, 20 July Shift-II]*
(a) zero (b) $\frac{2m^2c^2\lambda^2}{h^2}$ (c) $\frac{2mc\lambda^2}{h}$ (d) $\frac{hc}{mc}$

44 An electron of mass m and magnitude of charge $|e|$ initially at rest gets accelerated by a constant electric field **E**. The rate of change of de-Broglie wavelength of this electron at time t ignoring relativistic effects is *[2020, 9 Jan Shift-II]*
(a) $\frac{h}{|e|E\sqrt{t}}$ (b) $-\frac{h}{|e|Et}$ (c) $\frac{-h}{|e|Et^2}$ (d) $\frac{|e|Et}{h}$

45 A particle moving with kinetic energy E has de-Broglie wavelength λ. If energy ΔE is added to its energy, the wavelength become $\lambda/2$. Value of ΔE is *[2020, 9 Jan Shift-I]*
(a) $2E$ (b) $4E$
(c) $3E$ (d) E

46 If the kinetic energy of a free electron doubles, its de-Broglie wavelength changes by the factor *[AIEEE 2005]*
(a) $\frac{1}{2}$ (b) 2 (c) $\frac{1}{\sqrt{2}}$ (d) $\sqrt{2}$

Level 3

47 A proton and an electron are associated with same de-Broglie wavelength. The ratio of their kinetic energies is
(Assume $h = 6.63 \times 10^{-34}$ J-s, $m_e = 9.0 \times 10^{-31}$ kg and $m_p = 1836$ times m_e) *[2024, 8 April Shift-I]*
(a) $1:\frac{1}{1836}$ (b) $1:\sqrt{1836}$ (c) $1:1836$ (d) $1:\frac{1}{\sqrt{1836}}$

48 An electron (mass m) with an initial velocity $\mathbf{v} = v_0\hat{\mathbf{i}} (v_0 > 0)$ is moving in an electric field $\mathbf{E} = E_0\hat{\mathbf{i}} (E_0 > 0)$ where E_0 is constant. If at $t = 0$ de-Broglie wavelength is $\lambda_0 = \frac{h}{mv_0}$, then its de-Broglie wavelength after time t is given by *[2022, 27 July Shift-I]*
(a) λ_0 (b) $\lambda_0\left(1 + \frac{eE_0t}{mv_0}\right)$
(c) $\lambda_0 t$ (d) $\frac{\lambda_0}{\left(1 + \frac{eE_0t}{mv_0}\right)}$

49 The de-Broglie wavelength of a particle having kinetic energy E is λ. How much extra energy must be given to this particle, so that the de-Broglie wavelength reduces to 75% of the initial value ? *[2021, 26 Aug Shift-II]*
(a) $1/9\,E$ (b) $7/9\,E$ (c) E (d) $16/9\,E$

50 Two particles move at right angle to each other. Their de-Broglie wavelengths are λ_1 and λ_2, respectively. The particles suffer perfectly inelastic collision. The de-Broglie wavelength λ of the final particle, is given by *[2019, 8 April Shift-I]*
(a) $\frac{1}{\lambda^2} = \frac{1}{\lambda_1^2} + \frac{1}{\lambda_2^2}$ (b) $\lambda = \sqrt{\lambda_1\lambda_2}$
(c) $\lambda = \frac{\lambda_1 + \lambda_2}{2}$ (d) $\frac{2}{\lambda} = \frac{1}{\lambda_1} + \frac{1}{\lambda_2}$

51 A particle A of mass m and initial velocity v collides with a particle B of mass $m/2$ which is at rest. The collision is head on, and elastic. The ratio of the de-Broglie wavelengths λ_A to λ_B after the collision is *[JEE Main 2017]*
(a) $\frac{\lambda_A}{\lambda_B} = 2$ (b) $\frac{\lambda_A}{\lambda_B} = \frac{2}{3}$
(c) $\frac{\lambda_A}{\lambda_B} = \frac{1}{2}$ (d) $\frac{\lambda_A}{\lambda_B} = \frac{1}{3}$

Numerical Value Type Questions

52 An electron is released from rest near an infinite non-conducting sheet of uniform charge density $-\sigma$. The rate of change of de-Broglie wavelength associated with the electron varies inversely as nth power of time. The numerical value of n is ....
[2025, 8 April Shift-II]

53 A particle of mass 9.1×10^{-31} kg travels in a medium with a speed of 10^6 m/s and a photon of a radiation of linear momentum 10^{-27} kg-m/s travels in vacuum. The wavelength of photon is .... times the wavelength of the particle.
[2021, 27 July Shift-I]

Key Index

Topic 1

1 (b)	2 (b)	3 (d)	4 (c)	5 (d)	6 (d)	7 (d)	8 (a)	9 (c)	10 (b)
11 (d)	12 (c)	13 (c)	14 (b)	15 (a)	16 (d)	17 (d)	18 (d)	19 (c)	20 (c)
21 (d)	22 (b)	23 (d)	24 (a)	25 (c)	26 (b)	27 (b)	28 (d)	29 (a)	30 (d)
31 (c)	32 (a)	33 (c)	34 (d)	35 (b)	36 (b)	37 (d)	38 (c)	39 (c)	40 (c)
41 (b)	42 (a)	43 (c)	44 (d)	45 (a)	46 (a)	47 (b)	48 (b)	49 (c)	50 (d)
51 (b)	52 (b)	53 (a)	54 (c)	55 (a)	56 (c)	57 (b)	58 (a)	59 (a)	60 (c)
61 (b)	62 (a)	63 (b)	64 (b)	65 (b)	66 (d)	67 (4125)	68 (2)	69 (125)	70 (4)
71 (1)	72 (2)	73 (9)							

Topic 2

1 (d)	2 (*)	3 (c)	4 (c)	5 (c)	6 (a)	7 (b)	8 (b)	9 (d)	10 (b)
11 (b)	12 (5)	13 (11)							

Topic 3

1 (d)	2 (a)	3 (b)	4 (c)	5 (c)	6 (d)	7 (a)	8 (c)	9 (c)	10 (b)
11 (b)	12 (b)	13 (d)	14 (d)	15 (a)	16 (b)	17 (b)	18 (c)	19 (d)	20 (d)
21 (b)	22 (d)	23 (b)	24 (a)	25 (d)	26 (d)	27 (d)	28 (a)	29 (a)	30 (d)
31 (*)	32 (c)	33 (b)	34 (c)	35 (b)	36 (a)	37 (b)	38 (b)	39 (a)	40 (c)
41 (a)	42 (c)	43 (c)	44 (c)	45 (c)	46 (c)	47 (c)	48 (d)	49 (b)	50 (a)
51 (a)	52 (2)	53 (910)							

Detailed **Explanations**

Topic 1

01. *(b)* We know that,

$$(E_k)_{max} = \frac{hc}{\lambda} - \phi_0$$

For the photoelectric emission,

$$\frac{hc}{\lambda} > \phi_0$$

$$\Rightarrow \quad \lambda < \frac{hc}{\phi_0}$$

$$\Rightarrow \quad \lambda < \frac{1242}{3} \text{ nm}$$

$$\Rightarrow \quad \lambda < 414 \text{ nm}$$

Since, $\lambda_{blue} < 414$ nm

Thus, blue colour of visible light is required to cause photoelectric emission.

02. *(b)* A sufficiently negative potential which stops the photoelectric effect is called as stopping potential. So, assertion is correct.

Stopping potential varies linearly with frequency as given by the equation.

$$V_0 = \frac{h\nu}{e} - \frac{\phi_0}{e}$$

Hence, reason is also correct.

03. *(d)* In photoelectric effect, stopping potential is independent of the intensity of light while it decreases with the increase of wavelength. Hence, options (a), (b) and (c) are wrong.

Now,

$$V_0 = \frac{KE_{max}}{e}$$

Hence, option (d) is correct.

04. *(c)* For photoelectric effect, incident energy must be equal to or greater than the work function.

For wavelength, $\lambda = 550$ nm

incident energy

$$E = \frac{1240}{\lambda} = \frac{1240}{550} \simeq 2.25 \text{ eV}$$

Hence, only cesium will show photoelectric effect.

05. *(d)* Energy of UV radiation,

$$E = 4.13 \text{ eV}$$

Work function of metal, $\phi = 3.13$ eV

According of Einstein's photoelectric equation, $K_{max} = E - h\nu_0$

$$\therefore \quad h\nu_0 = \phi = 4.13 - 3.13 = 1.00 \text{ eV}$$

06. *(d)* Given, $\lambda = 300$ nm

$$\phi_0 = 2.13 \text{ eV}$$

According to Einstein's equation,

$$\frac{hc}{\lambda} - \phi_0 = K_{max} \Rightarrow \frac{hc}{\lambda} - \phi_0 = eV_0$$

$$\Rightarrow \left(\frac{1240}{300} - 2.13\right) \text{eV} = eV_0 \Rightarrow V_0 = 2 \text{ V}$$

07. *(d)* According to Einstein's photo electric equation.

$$E_{max} = h\nu - \phi_0$$

$$\Rightarrow \quad \phi_0 = h\nu - E_{max} = h\nu - eV_0$$

$$= (2.48 - 0.5) \text{ eV} = 1.98 \text{ eV}$$

08. *(a)* Photoelectric effect explains particle nature of light while the phenomenon of diffraction, reflection, interference and polarisation explain wave nature of light.

09. *(c)* Stopping potential does not depend on the intensity of incident light. It depends on the frequency and nature of emitter material.

Since, $eV_0 = K_{max}$

$$\therefore \; V_0 = \frac{1}{e}(K_{max})$$

10. *(b)* Given lights are of same wavelength, so the stopping potential will remain same.

Since $I_2 > I_1$ (Intensity), this signifies that saturation current corresponding to I_2 will be greater than that corresponding to I_1.

11. *(d)* Initially, frequency of incident light, $f_1 = 1.5 f_0$. If frequency is halved, then new incident frequency

$$f' = \frac{1}{2} \times 1.5 \times f_0 = 0.75 \, f_0$$

As this frequency is lower than threshold frequency f_0, no photoemission will occur.

12. *(c)* By Einstein's photoelectric equation, $K_{max} = h\nu - \phi_0$

Comparing with $y = mx + c$

Here, K_{max} is on Y-axis and frequency ν on X-axis.

So, slope is h (Planck's constant)

13. *(c)* Given,

Work function, $\phi = 3.0$ eV

Work function is given by formula

$$\phi = \frac{hc}{\lambda} \text{ or } \lambda = \frac{hc}{\phi}$$

$$\lambda = \frac{6.64 \times 10^{-34} \times 3 \times 10^{8}}{3 \times 1.6 \times 10^{-19}}$$

$$\lambda = 414 \times 10^{-9} \text{ m}$$

$$\Rightarrow \quad \lambda = 414 \text{ nm}$$

14. *(b)* Given, work function

$\phi = 6.63 \text{ eV} = 6.63 \times 1.6 \times 10^{-19}$ V

Work function

$$\phi = h\nu_0$$

$$\nu_0 = \frac{\phi}{h} = \frac{6.63 \times 1.6 \times 10^{-19}}{6.62 \times 10^{-34}}$$

$$\nu_0 = 1.6 \times 10^{15} \text{Hz}$$

15. *(a)*

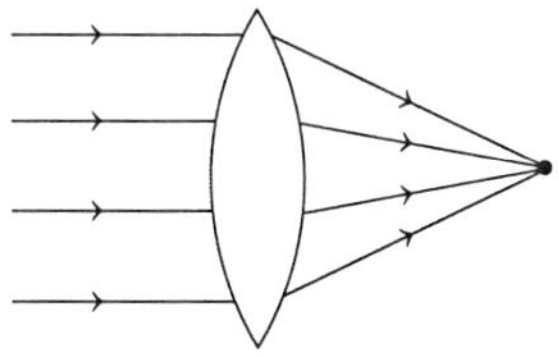

Intensity of image formed by lens does not depends on focal length of lens.

It only depends on diameter of lens.

As intensity remains same in both cases, so there is no change in photo current.

16. *(d)* Statement I is true but Statement II is false.

Out of microwaves, infrared rays and ultraviolet rays, ultraviolet rays are the most effective for the emission of electron from metallic surface.

Option (d) is correct.

Explanation UV rays have maximum frequency hence are most effective for emission of electrons from the metallic surface.

$$KE_{max} = hf - hf_0$$

Thus, $KE_{max} \propto f$

17. *(d)* Given, $\phi_A = 9$ eV $\Rightarrow \phi_B = 4.5$ eV

Threshold wavelength of metal surface, A

$$\lambda_A = \frac{1242}{\phi_A} \text{ nm} = \frac{1242}{9} \text{ nm}$$

Similarly, $\quad \lambda_B = 1242/4.5$

$$\lambda_B - \lambda_A = \frac{1242}{4.5} - \frac{1242}{9}$$

$$= 276 - 138 = 138 \text{ nm}$$

Alternate Solution

Given, $\phi_A = 9$ eV, $\phi_B = 4.5$ eV

We know, $E = \frac{hc}{\lambda} - \phi_0$ or for $E = 0$

$$\therefore \quad \phi_0 = \frac{hc}{\lambda} \Rightarrow \lambda = \frac{hc}{\phi}$$

$$\lambda_A = \frac{hc}{\phi}, \lambda_B = \frac{hc}{\phi_B}$$

$$\lambda_B - \lambda_A = \frac{hc}{\phi_B} - \frac{hc}{\phi_A} = hc\left\{\frac{1}{4.5} - \frac{1}{9}\right\}$$

$$= 1242\left\{\frac{1}{9}\right\} = 138 \text{ nm}$$

$$\Delta\lambda = 138 \text{ nm}$$

18. *(d)* From Einstein's photoelectric equation,

we have, $eV_0 = h\nu - \phi \Rightarrow 0 = h\nu - \phi$

$$\Rightarrow \phi = h\nu = 6.6 \times 10^{-34} \times 5 \times 10^{14}$$

$$= 33 \times 10^{-20} \text{ J}$$

$$= \frac{33 \times 10^{-20}}{1.6 \times 10^{-19}} \text{ eV}$$

$$= 2.07 \text{ eV}$$

19. *(c)* In photoemission,

Intensity $\propto$ photocurrent

Maximum KE of emitted electrons depends on frequency of incident light.

Photoemission is possible only when frequency is greater than threshold frequency. So, Statements A and C are correct.

20. *(c)* Using $KE_{max} = eV_s = hf - \phi_0$

When, ϕ_0 = work function

V_s = Stopping potential

f = frequency

$$V_s = \frac{h}{e} f - \frac{\phi_0}{e}$$

Therefore, the slope m will be same for all graphs and will be independent of ϕ_0.

21. *(d)* Energy corresponding to incident radiation,

$$E = \frac{hc}{\lambda}$$

$$= \frac{1240}{\lambda(\text{in nm})} = \frac{1240}{350}$$

$$E = 3.54 \text{ eV}$$

Work function for metal-A,

$$\phi_A = 4.8 \text{ eV}$$

and for metal-B, $\phi_B = 2.2$ eV

If energy of incident radiation is greater than work function of metal then photoelectron will emit. For metal-B, $E > \phi_B$ and for metal-A, $E < \phi_A$.

So, metal-A will not emit photoelectron but metal-B will emit.

22. *(b)* Photoemission occurs when $\lambda < \lambda_0$. Infrared wavelength is more than 7500 Å. So, no photoemission occurs from an infrared bulb.

Wavelength for ultraviolet is less than 4500 Å, so photoemission occurs from an ultraviolet lamp.

23. *(d)* If power of light source is increased, then frequency of light source increases. Thus, stopping potential is increased also. Hence, statement I is incorrect.

For a given metal, the maximum kinetic energy of the photoelectron depends on the wavelength (or frequency) of the incident radiation.

24. *(a)* Photoelectric effect cannot be explained by wave theory of light.

Maximum kinetic energy of photoelectrons depends on frequency of incident radiation.

Stopping potential depends on incident frequency and work function.

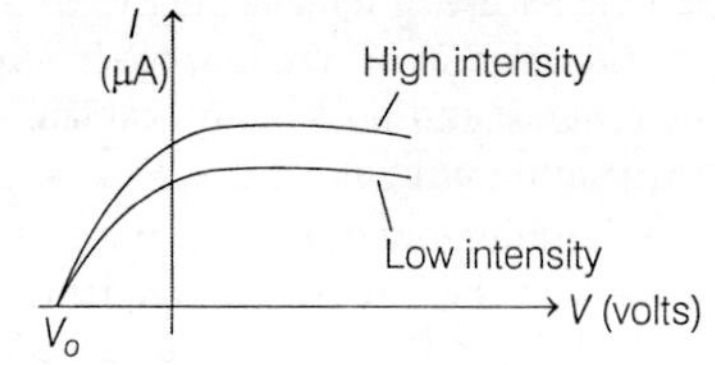

As intensity increases, saturation value of photocurrent increases.

Stopping potential,

$$V_0 = \frac{h\nu}{e} - \frac{\phi_0}{e} = \frac{1}{e}(h\nu - \phi_0)$$

Thus, stopping potential depends both on work function and frequency of incident radiation.

25. *(c)* According to Einstein's equation,

$$KE_{max} = h\nu - \phi, \; h\nu = KE_{max} + \phi$$

For 1st metal, $h\nu = 5\phi$

Therefore, $5\phi = KE_1 + \phi$

$$\Rightarrow \quad KE_1 = 4\phi \quad \ldots (i)$$

For IInd metal, $10\phi = KE_2 + \phi$

$$\Rightarrow \quad KE_2 = 9\phi \quad \ldots (ii)$$

From Eqs. (i) and (ii), we get

$$\Rightarrow \quad \frac{KE_1}{KE_2} = \frac{4\phi}{9\phi} \Rightarrow \frac{\frac{1}{2}mv_1^2}{\frac{1}{2}mv_2^2} = \frac{4}{9}$$

$$\Rightarrow \quad \frac{v_1^2}{v_2^2} = \frac{4}{9}; \; v_1 : v_2 = 2 : 3$$

26. *(b)* According to Einstein's equation,

$$K_{max} = E - \phi, \Rightarrow \frac{1}{2}mv_{max}^2 = h\nu - \phi \quad \ldots (i)$$

Clearly, from above equation, v_{max}^2 varies linearly with v. Threshold wavelength depends upon the work function and hence, nature of material.

27. *(b)* The minimum energy required to eject electrons from a metal surface, must be equal to the work function of that metal. Thus, according to Einstein's photoelectric equation, kinetic energy of the emitted electrons is $KE = h\nu - \phi_0$

If $h\nu = \phi_0$, then $KE = 0$

28. *(d)* As per Einstein's theory of photoelectric effect when a photon strikes the metal surface it ejects an electron. If we increase the intensity of light, the number of photons incident on surface will increase and thus, the number of electrons ejected will increase proportionally and the photocurrent will also increase.

Kinetic energy of ejected electrons is given as, $\frac{1}{2}m_e v_e^2 = eV_0$

Here, V_0 is stopping potential.

The stopping potential of surface is constant and it will not increase with increase in intensity of light, so the kinetic energy of ejected electrons will not increase.

Thus, increasing the intensity of light will increase the number of photons incident and the kinetic energy will remain unchanged.

> **• Concept Enhancer**
>
> The value of saturation current depends on the intensity of incident radiation but it is independent of the frequency of the incident radiation. The photoelectric emission starts instantaneously without any apparent time log (~ 10^{-8}s), even when the incident radiation has low intensity.

29. *(a)* We know that, the photoelectric effect equation, $KE = hf - \phi$

For two identical photocathodes,

$$\frac{1}{2}mv_1^2 = hf_1 - \phi \quad \ldots (i)$$

$$\frac{1}{2}mv_2^2 = hf_2 - \phi \quad \ldots (ii)$$

Since, the material of photocathodes is same, so the value of the work-function ϕ is same.

Subtracting Eq. (ii) from Eq. (i) we get

$$\frac{1}{2}mv_1^2 - \frac{1}{2}mv_2^2 = hf_1 - hf_2$$

$$v_1^2 - v_2^2 = \frac{2h}{m}(f_1 - f_2)$$

30. *(d)* The stopping potential V_0 is related to the frequency of incident radiations by following relation

$$eV_0 = h\nu - \phi \qquad (\phi = \text{work-function})$$

The stopping potential in the context of photoelectric effect depends only on the frequency ν of incident electromagnetic radiation.

31. *(c)* According to photoelectric equation,

$$K_{max} = h(\nu_i - \nu_0)$$

$$= h\left(\frac{c}{\lambda_i} - \frac{c}{\lambda_0}\right) = hc\left(\frac{\lambda_0 - \lambda_i}{\lambda_0 \lambda_i}\right)$$

$$= 1237\left(\frac{380 - 260}{380 \times 260}\right) \text{eV} = 1.5 \text{ eV}$$

32. *(a)* From given wave equation we have

$$k = \frac{2\pi}{\lambda} = \frac{2\pi}{5 \times 10^{-7}}$$

or $\lambda = 5 \times 10^{-7}$ m

$\lambda = 5000$ Å

Now stopping potential in eV,

$$= \left(\frac{12375}{5000}\right) - 2$$

$$= 2.48 - 2$$

$$= 0.48 \text{ eV}$$

33. *(c)* Given,

$$E_1 = \frac{1240 \text{ eV - nm}}{300 \text{ nm}} = 4.13 \text{ eV}$$

and $E_2 = \dfrac{1240 \text{ eV - nm}}{400 \text{ nm}} = 3.10$ eV

According to Einstein's photoelectric equation,

$$K_{max_1} = E_1 - \phi_0, \; K_{max_2} = E_2 - \phi_0$$

and $K_{max_1} - K_{max_2} = E_1 - E_2 = 1.03$ eV

$V_1 - V_2 = 1.03$ V $\qquad (\because K = eV)$

34. *(d)* As λ is increased, there will be a value of λ above which photoelectron will cease to come out, so photocurrent will become zero.

• Concept Enhancer

Maximum wavelength of incident radiation above which no emission of photoelectrons take place called threshold wavelength.

35. *(b)* According to the photoelectric effect in a photocell, if a light of wavelength λ is incident on a cathode, then electrons are emitted, which constitute the photoelectric current.

As λ is increased then it cross the threshold wavelength, above it photocurrent will become zero.

36. *(b)* The photoelectric effect is an instantaneous phenomenon (experimentally proved). It takes approximate time of the order of 10^{-10} s.

37. *(d)* Einstein's photoelectric equation is

$$\text{KE}_{max} = h\nu - \phi \qquad \text{...(i)}$$

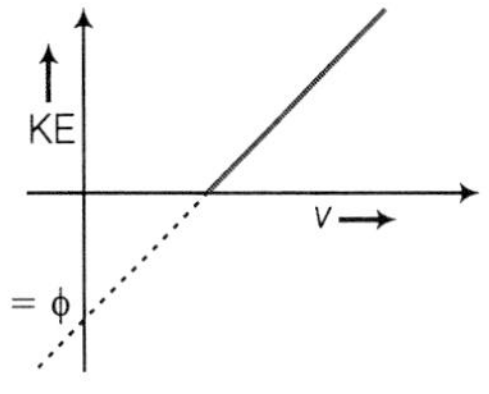

The equation of line is

$$y = mx + c \qquad \text{...(ii)}$$

Comparing above two equations, we get

$$m = h, \quad c = -\phi$$

Hence, slope of graph is equal to Planck's constant (non-variable) and does not depend on intensity of radiation. But interception c depends on work function ϕ.

38. *(c)* Work function, $W = hc/\lambda$

Therefore, $\dfrac{\lambda_{Na}}{\lambda_{Cu}} = \dfrac{W_{Cu}}{W_{Na}}$

$$= \frac{4.5}{2.3} = 2 \text{ (nearly)}$$

39. *(c)* $V_s = \dfrac{h\nu - \phi}{e}$

So stopping potential does not depends on intensity.

$$I = \frac{nh\nu}{A}$$

On increasing intensity no. of photons per see, n increases so the no. of electros increases

A → False

R → True

40. *(c)* $\text{KE}_{max} = \dfrac{hc}{\lambda} - \phi$, $R = \dfrac{p}{qB}$

$$p = \sqrt{2m \, \text{KE}_{max}}$$

$$p = \sqrt{2m\left(\frac{hc}{\lambda} - \phi\right)}$$

$$d_{A \to B} = 2R$$

$$d_{A \to B} = \frac{2p}{qB}$$

$$= \frac{2\sqrt{2m\left(\frac{hc}{\lambda} - \phi\right)}}{eB}$$

$$= \frac{\sqrt{8m\left(\frac{hc}{\lambda} - \phi\right)}}{eB}$$

41. *(b)* $E = 2 + 2.14 = 4.14$ eV

$$E = \frac{hc}{\lambda}$$

$$\lambda = \frac{hc}{E} = \frac{1242}{4.14} = 300 \text{ nm}$$

42. *(a)* $h\nu = \phi + \text{KE}_{max}$

$$\text{KE}_{max} = eV_0$$

$$h\nu = \phi + eV_0$$

$$V_0 = \frac{h\nu}{e} - \frac{\phi}{e}$$

It is in the form of $y = mx + c$

(A) V_0 versus ν is linear.

(B) Slope $= \dfrac{h}{e}$

(C) h = constant

(D) Incorrect

(E) $\phi = eV_0$

(A), (C) and (E) only are correct.

43. *(c)* Number of photons does not depend on the frequency of light. Number of photons increases with increase of intensity of light.

Maximum kinetic energy of emitted photo electrons, $K_{max} = h(\nu - \nu_0)$

Thus, K_{max} increases with increase of frequency of incident radiation.

Thus, A is not correct but R is correct.

44. *(d)* Statement I is correct but Statement II is incorrect. The correct form is,

We know from work function equation, $eV_0 = h\nu - \phi$

$$V_0 = \frac{h}{e}\nu - \frac{\phi}{e}$$

Now, material M_2 has higher work function, so it will have lower kinetic energy.

Hence, option (d) is correct.

45. *(a)* Relation of stopping potential and threshold wavelength is,

$$eV_0 = \frac{hc}{\lambda} - \frac{hc}{\lambda_0}$$

Given, when wavelength is λ, $V_0 = 8$ V

$$e \cdot 8 = \frac{hc}{\lambda} - \frac{hc}{\lambda_0} \qquad \text{...(i)}$$

and when wavelength is 3λ, stopping potential is 2V,

$$e\cdot 2=\frac{hc}{3\lambda}-\frac{hc}{\lambda_0} \quad \ldots(ii)$$

Dividing Eq. (i) by Eq. (ii), we get

$$4=\frac{\frac{hc}{\lambda}-\frac{hc}{\lambda_0}}{\frac{hc}{3\lambda}-\frac{hc}{\lambda_0}} \Rightarrow \frac{4}{3}\frac{hc}{\lambda}-\frac{4hc}{\lambda_0}=\frac{hc}{\lambda}-\frac{hc}{\lambda_0}$$

So, $\lambda_0=9\lambda$

46. *(a)* Given, $E=\text{KE}+\phi_0$

Now, $$\frac{hc}{\lambda}=eV_0+\phi_0 \quad \ldots(i)$$

and $$\frac{hc}{2\lambda}=\frac{eV_0}{4}+\phi_0 \quad \ldots(ii)$$

Eq. (ii) × 4

$$\frac{hc}{2\lambda}\times 4=\left(\frac{eV_0}{4}+\phi_0\right)\times 4$$

$$\frac{2hc}{\lambda}=eV_0+4\phi_0$$

Putting value of $eV_0=\frac{hc}{\lambda}-\phi_0$ [From Eq. (i)]

$$\frac{2hc}{\lambda}=\frac{hc}{\lambda}-\phi+4\phi_0$$

$$\Rightarrow \frac{2hc}{\lambda}-\frac{hc}{\lambda}=3\phi_0$$

$$\frac{hc}{\lambda}=3\phi_0 \Rightarrow \frac{hc}{\lambda}=\frac{3hc}{\lambda_0}\left[\phi=\frac{hc}{\lambda_0}\right]$$

$$\frac{1}{\lambda}=\frac{3}{\lambda_0} \Rightarrow \lambda_0=3\lambda$$

Option (a) is correct.

47. *(b)*

key idea When a light beam of frequency (f) more than threshold frequency (f_0) is incident on the metal surface, then electrons are emitted from the metal surface and maximum kinetic energy of these emitted electrons are calculated by the following equation.

$$\frac{1}{2}mv_{max}^2=hf-hf_0$$

Using the photoelectric equation

$$hf=hf_0+eV_0 \text{ or } hf=hf_0+\frac{1}{2}mv_{max}^2$$

In case (i) $f=2f_0, v_{max}=v_1$

$$h(2f_0)=hf_0+\frac{1}{2}mv_1^2$$

$$\frac{1}{2}mv_1^2=hf_0 \quad \ldots(i)$$

In case (ii) $f=5f_0$

$$h(5f_0)=hf_0+\frac{1}{2}mv_2^2$$

$$\Rightarrow \frac{1}{2}mv_2^2=4hf_0 \quad \ldots(ii)$$

On dividing Eq. (i) with Eq. (ii), we get

$$\left(\frac{v_1}{v_2}\right)^2=\frac{1}{4} \text{ or } \frac{v_1}{v_2}=\frac{1}{2}$$

48. *(b)* By the photoelectric equation, Incident photon energy = Work function + Kinetic energy

$$\Rightarrow \frac{hc}{\lambda}=\phi+E \quad \ldots(i)$$

Now, when E is doubled,

$$\therefore \frac{hc}{\lambda'}=\phi+2E \quad \ldots(ii)$$

Now, subtracting Eq. (i) from Eq. (ii),

Then, $E=\frac{hc}{\lambda'}-\frac{hc}{\lambda}$

$$\Rightarrow E=hc\left[\frac{1}{\lambda'}-\frac{1}{\lambda}\right] \Rightarrow \frac{E}{hc}=\frac{1}{\lambda'}-\frac{1}{\lambda}$$

$$\Rightarrow \frac{1}{\lambda'}=\frac{E}{hc}+\frac{1}{\lambda} \Rightarrow \frac{1}{\lambda'}=\frac{E\lambda+hc}{hc\lambda}$$

$$\Rightarrow \lambda'=\frac{hc\lambda}{E\lambda+hc}$$

49. *(c)* Using Einstein's photoelectric equation,

$$\frac{hc}{\lambda}=K_{max}+\phi_0, \text{ we have,}$$

$$\frac{hc}{800}=K+\phi_0 \quad \ldots(i)$$

and $$\frac{hc}{500}=2K+\phi_0 \quad \ldots(ii)$$

Multiplying Eq. (i) by 2 and subtracting from Eq. (ii), we get

$$\frac{hc}{500}-\frac{2\times hc}{800}=(2K+\phi_0)-2(K+\phi_0)$$

$$\Rightarrow \frac{hc}{100}\left(\frac{1}{5}-\frac{1}{4}\right)=-\phi_0$$

or $$-\phi_0=\frac{1230 \text{ eV-nm}}{100 \text{ nm}}\left(\frac{4-5}{5\times 4}\right)$$

$$\Rightarrow \phi_0=0.615 \text{ eV}$$

50. *(d)*

key idea To solve this question, first calculate the frequency of photo having largest energy using $f=\frac{\omega_{max}}{2\pi}$ and after that use Einstein's photoelectric equation to calculate maximum kinetic energy of the photoelectrons.

Given, $h=4.14\times 10^{-15}$ eV

$$=4.14\times 10^{-15}\times 1.6\times 10^{-19} \text{ J-s}$$
$$=6.624\times 10^{-34} \text{ J-s}$$

incident electric field, $E=200$ $[\sin(6\times 10^{15})t+\sin(9\times 10^{15})t]$ Vm^{-1}

So, $\omega_{max}=9\times 10^{15}$ rad/s

Frequency of largest energy photon,

$$f=\frac{\omega_{max}}{2\pi}=\frac{9\times 10^{15}}{2\pi} \text{ Hz}$$

Using Einstein's photoelectric equation, we have

$$K_{max}=hf-\phi_0$$
$$=\frac{6.624\times 10^{-34}\times 9\times 10^{15}}{2\pi\times 1.6\times 10^{-19}}-2.5$$
$$=5.93\times 10^{-34+15+19}-2.5$$
$$=5.93-2.5=3.43 \text{ eV} \simeq 3.42 \text{ eV}$$

51. *(b)*

key idea For the calculation of maximum kinetic energy of emitted photoelectron, you can use Einstein photoelectric equation given as,

$$K_{max}=h\nu-\phi_0=\frac{hc}{\lambda}-\phi_0$$

where, symbols are there usual meaning.

By Einstein's photoelectric equation, we have

$$K_{max}=\frac{hc}{\lambda}-\phi_0$$

So, in given situation,

$$K_1=\frac{hc}{\lambda_1}-\phi_0 \quad \ldots(i)$$

and $$K_2=\frac{hc}{\lambda_2}-\phi_0, \quad \ldots(ii)$$

Given $\lambda_1=3\lambda_2$, so from Eq. (i), We have

$$K_1=\frac{hc}{3\lambda_2}-\phi_0 \quad \ldots(iii)$$

Multiply Eq. (iii) by 3 and subtracting from Eq. (ii), we get

$$K_2-3K_1=\left(\frac{hc}{\lambda_2}-\phi_0\right)-3\left(\frac{hc}{3\lambda_2}-\phi_0\right)$$

$$\Rightarrow K_2-3K_1=2\phi_0 \Rightarrow 3K_1=K_2-2\phi_0$$

$$\Rightarrow 3K_1<K_2 \text{ or } K_1<K_2/3$$

52. *(b)* According to Einsteins's photoelectric equation,

$$\frac{1}{2}mv^2=E_{photon}-\phi,$$

For the first photon having energy $E_1=3.8$ eV,

$$\frac{1}{2}mv_1^2=E_1-\phi \Rightarrow \frac{1}{2}mv_1^2=3.8-0.6$$

$$\frac{1}{2}mv_1^2=3.2 \text{ eV} \quad \ldots(i)$$

Similarly, for the second photon,
$E_2 = 1.4$ eV

$$\therefore \quad \frac{1}{2}mv_2^2 = E_2 - \phi = 1.4 - 0.6$$

$$\Rightarrow \quad \frac{1}{2}mv_2^2 = 0.8 \text{ eV} \quad \ldots \text{(ii)}$$

Dividing Eq. (i) by Eq. (ii), we get

$$\frac{\frac{1}{2}mv_1^2}{\frac{1}{2}mv_2^2} = \frac{3.2}{0.8} \Rightarrow \left(\frac{v_1}{v_2}\right)^2 = (2)^2$$

$$\Rightarrow \quad \frac{v_1}{v_2} = \frac{2}{1} \text{ or } v_1 : v_2 = 2:1$$

53. *(a)* By Einstein's photoelectric equation,

$$hf = \phi + (\text{KE})_{\text{max}}$$

$$\Rightarrow \quad \frac{hc}{\lambda} = \phi + eV_0 \Rightarrow eV_0 = \frac{hc}{\lambda} - \phi$$

Here, $\lambda = 280$ nm and $\phi = 2.5$ eV

$$\Rightarrow e(V_0)_1 = \frac{1240}{280} - 2.5$$

$$\Rightarrow e(V_0)_1 = 1.93 \text{ eV} \Rightarrow (V_0)_1 = 1.93 \text{ V}$$

Similarly, $e(V_0)_2 = \frac{1240}{400} - 2.5$

$$\Rightarrow \quad (V_0)_2 = 0.6 \text{ V}$$

$\therefore$ Change in stopping potential,

$$\Delta V = (V_0)_1 - (V_0)_2$$
$$= 1.93 - 0.6 = 1.33 \simeq 1.3 \text{ V}$$

54. *(c)* Let λ_2 will be the new incident wavelength of electrons.

Stopping potential $(V_1) = 1.43$V

As, energy $(E) = \frac{1240}{\lambda_1} = \phi_0 + eV$

where, ϕ_0 is work-function and V is stopping potential

$$\therefore \quad E_1 = \frac{1240}{491} = \phi_0 + 0.71 \quad \ldots \text{(i)}$$

$$\text{and} \quad E_2 = \frac{1240}{\lambda_2} = \phi_0 + 1.43 \quad \ldots \text{(ii)}$$

Now, subtracting Eqs. (i) from (ii), we get

$$E_2 - E_1 = 1240\left(\frac{1}{\lambda_2} - \frac{1}{491}\right) = 0.72$$

$$\Rightarrow \quad \frac{1}{\lambda_2} = 0.00058 + 0.00204 = 0.00262$$

$$\therefore \quad \lambda_2 = 381.7 \text{ nm} \simeq 382 \text{ nm}$$

55. *(a)* From photoelectric equation,

$$\phi = \phi_0 + eV \Rightarrow \frac{hc}{\lambda} = \phi_0 + eV$$

$$eV = \frac{hc}{\lambda} - \phi_0 \Rightarrow V = \frac{hc}{\lambda e} - \frac{\phi_0}{e}$$

$$V = \left(\frac{hc}{e}\right)\left(\frac{1}{\lambda}\right) - \frac{\phi_0}{e}$$

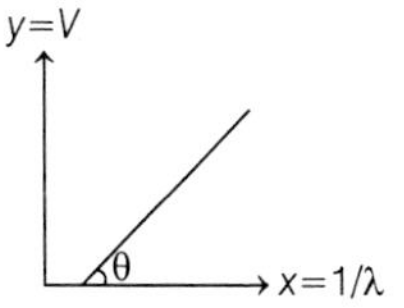

On comparing the above equation with equation of straight line, i.e.,

$$y = mx + c, \text{ we get}$$

Slope, $\quad m = hc/e$

y-intercept $c = -\frac{\phi_0}{e}$

Now, if we change intensity of incident radiation, then h, c, e and ϕ_0 will not change. Therefore, neither slope of the graph, nor the y-intercept cut by graph will change on changing the intensity of incident radiation.

So, graph will not change.

> • Concept **Enhancer**
>
> If we change intensity of incident radiation, then neither slope of the graph nor the y-intercept cut by graph will change.

56. *(c)* From Einstein's photoelectric equation,

$$\phi = \phi_0 + eV_0$$

$$\Rightarrow \quad hf = \phi_0 + eV_0 \quad \ldots \text{(i)}$$

From graph, take point B

$$(5.5 \times 10^{14} \text{ Hz, 0 V})$$

On satisfying eq. (i) with point B, we get

$$6.62 \times 10^{-34} \times 5.5 \times 10^{14}$$
$$= \phi_0 + 1.6 \times 10^{-19} \times 0$$

$$36.41 \times 10^{-20} = \phi_0 + 0$$

$$\phi_0 = 36.41 \times 10^{-20} \text{ J}$$

$$= \frac{36.41 \times 10^{-20}}{1.6 \times 10^{-19}} \text{ eV}$$

$$= 2.275 \text{ eV} \simeq 2.27 \text{ eV}$$

> • Concept **Enhancer**
>
> The stopping potential varies linearly with the frequency of incident radiation for a given photosensitive material.

57. *(b)* Using Einstein's relation of photoelectric effect,

$$(\text{KE})_{\text{max}} = eV_0 = h\nu - h\nu_0 = h\nu - W$$
$$[\because W = h\nu_0]$$

$$\text{or} \quad V_0 = \frac{h}{e}\nu - \frac{W}{e}$$

From graph at

$$V_0 = 0 \text{ and } \nu = 4 \times 10^{14} \text{ Hz}$$

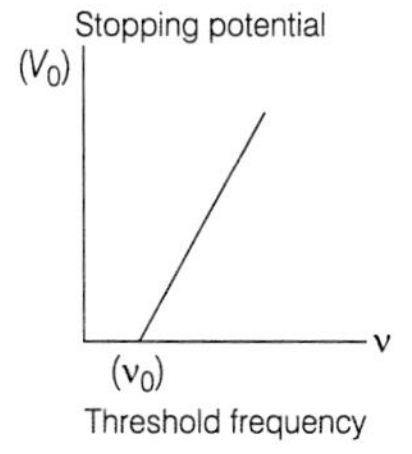

$$\therefore \quad 0 = \frac{6.63 \times 10^{-34}}{e} \times 4 \times 10^{14} - \frac{W}{e}$$

$$\Rightarrow \quad \frac{W}{e} = \frac{6.63 \times 10^{-34} \times 4 \times 10^{14}}{e} \text{ J}$$

$$\text{or} \quad W = 6.63 \times 4 \times 10^{-20} \text{ J}$$

$$\text{or} \quad W = \frac{6.63 \times 4 \times 10^{-20}}{1.6 \times 10^{-19}}$$
$$= 1.657 \text{ eV}$$

$$\therefore \quad W = 1.66 \text{ eV}$$

> ***Alternate Solution***
>
> From graph, threshold frequency,
>
> $\nu_0 = 4 \times 10^{14}$ Hz (where, $V_0 = 0$)
>
> $\therefore$Work function, $W = h\nu_0$
>
> $\Rightarrow \quad W = 6.63 \times 10^{-34} \times 4 \times 10^{14}$ J
>
> $\Rightarrow \quad W = \frac{6.63 \times 4 \times 10^{-20}}{1.6 \times 10^{-19}}$ eV
>
> $= 1.657 \text{ eV} \approx 1.66 \text{ eV}$

58. *(a)* Given, $B = B_0[(\sin(3.14 \times 10^7 ct) + \sin(6.28 \times 10^7 ct)]$

Here, the photoelectric effect produced by the angular frequency,

$$\omega = 6.28 \times 10^7 \times c$$

$$\Rightarrow \quad \omega = 6.28 \times 10^7 \times 3 \times 10^8$$

$$\omega = 2\pi \times 10^7 \times 3 \times 10^8 \text{ rad/s} \quad \ldots \text{(i)}$$

Using Eqs. (i)

$$h\nu = \frac{h\omega}{2\pi}$$

$$= \frac{h \times 2\pi \times 10^7 \times 3 \times 10^8}{2\pi}$$

$$= 12.4 \text{ eV}$$

Therefore, according to Einstein equation for photoelectric effect

$$E = h\nu = \phi + \text{KE}_{\text{max}}$$

$$\Rightarrow \text{KE}_{\text{max}} = E - \phi$$

(where, ϕ = work-function = 4.7 eV)

$$\text{KE}_{\text{max}} = 12.4 - 4.7 = 7.7 \text{ eV}$$

$$\text{or} \quad \text{KE}_{\text{max}} = 7.7 \text{ eV}$$

59. *(a)*

> **key idea** According to the law of conservation of energy, i.e. Energy of a photon ($h\nu$) = Work function (ϕ) + Kinetic energy of the photoelectron $\left(\frac{1}{2}mv^2{}_{max}\right)$

According to Einstein's photoelectric emission of light, $E = (\text{KE})_{max} + \phi$

As, $\frac{hc}{\lambda} = (\text{KE})_{max} + \phi$

If the wavelength of radiation is changed to $\frac{3\lambda}{4}$, then

$$\Rightarrow \quad \frac{4}{3}\frac{hc}{\lambda} = \left[\frac{4}{3}(\text{KE})_{max} + \frac{\phi}{3}\right] + \phi$$

For fastest emitted electron,

$$(\text{KE})_{max} = \frac{1}{2}mv'^2 + \phi$$

$$\Rightarrow \quad \frac{1}{2}mv'^2 = \frac{4}{3}\left(\frac{1}{2}mv^2\right) + \frac{\phi}{3}$$

i.e. $\quad v' > v\left(\frac{4}{3}\right)^{1/2}$

60. *(c)* Since, the frequency of ultraviolet light is less than the frequency of X-rays, the energy of each incident photon will be more for X-rays

$$\text{KE}_{photoelectron} = h\nu - \phi$$

Stopping potential is to stop the fastest photoelectron

$$V_0 = \frac{h\nu}{e} - \frac{\phi}{e}$$

So, KE_{max} and V_0 both increases.But KE ranges from zero to KE_{max}, because of loss of energy due to subsequent collisions before getting ejected but not due to range of frequencies in the incident light.

> **Concept Enhancer**
>
> One photon can emit only one electron from a photosensitive material, no matter how much its energy is.

61. *(b)* $\text{KE}_{max} = eV_0$

$$\Rightarrow \quad \frac{1}{2}mv^2 = eV_0 = 1.68 \text{ eV}$$

$$\Rightarrow \quad h\nu = \frac{hc}{\lambda} = \frac{1240 \text{ eV-nm}}{400 \text{ nm}} = 3.1 \text{ eV}$$

$$\Rightarrow \quad 3.1 \text{ eV} = W_0 + 1.68 \text{ eV}$$

[from Einstein equation, $E = W_0 + K_{max}$]

$$W_0 = 1.42 \text{ eV}$$

62. *(a)* Radius of circular path,

$$R = \frac{mv}{Bq} \Rightarrow v = \frac{RBq}{m}$$

Kinetic energy of electron,

$$K = \frac{1}{2}mv^2 = \frac{1}{2}m\left(\frac{RBq}{m}\right)^2 = \frac{1}{2}m\frac{R^2B^2q^2}{m^2} = \frac{R^2B^2q^2}{2m}$$

$$= \frac{(2\times10^{-3})^2\times(2\times10^{-3})^2 \times (1.6\times10^{-19})^2}{2\times9.1\times10^{-31}}$$

$$= 2.25\times10^{-19} \text{ J}$$

$$= \frac{2.25\times10^{-19}}{1.6\times10^{-19}} \text{ eV} = 1.4 \text{ eV}$$

Energy of radiation,

$$E = \frac{hc}{\lambda} = \frac{6.62\times10^{-34}\times3\times10^8}{4.5\times10^{-7}} \text{ J}$$

$$= \frac{6.62\times10^{-34}\times3\times10^8}{4.5\times10^{-7}\times1.6\times10^{-19}} \text{ eV} = 2.76 \text{ eV}$$

$\therefore$ Work function,

$$\phi = E - K = 2.76 - 1.4 = 1.36 \text{ eV}$$

63. *(b)* $\lambda_1 = 670.5$ nm,

$V_1 = 0.48$ V

$\lambda_2 = 474.6$ nm

By using Einstein photoelectric equation, $eV_1 = \frac{hc}{\lambda_1} - \phi_0 \quad \ldots\text{(i)}$

ϕ_0 is work function

and similarly, $eV_2 = \frac{hc}{\lambda_2} - \phi_0$

$$\Rightarrow \quad \phi_0 = \frac{hc}{\lambda_2} - eV_2 \quad \ldots\text{(ii)}$$

From Eqs. (i) and (ii), we get

$$\therefore \quad \frac{hc}{\lambda_1} - eV_1 = \frac{hc}{\lambda_2} - eV_2$$

$$\Rightarrow eV_1 - eV_2 = hc\left(\frac{1}{\lambda_1} - \frac{1}{\lambda_2}\right)$$

$$\Rightarrow \quad V_1 - V_2 = \frac{hc}{e}\left(\frac{1}{\lambda_1} - \frac{1}{\lambda_2}\right)$$

$$\Rightarrow \quad V_2 = V_1 + \frac{hc}{e}\left(\frac{1}{\lambda_2} - \frac{1}{\lambda_1}\right)$$

$$= 0.48 + \frac{6.63\times10^{-34}\times3\times10^8}{1.6\times10^{-19}} \times \frac{1}{10^{-9}} \times \left(\frac{1}{474.6} - \frac{1}{670.5}\right) = 1.25 \text{ V}$$

64. *(b)* Given,

In Case 1

Wavelength of light $= \lambda$

Stopping potential of ejected electron,

$$V_{01} = 4.8 \text{ V}$$

In Case 2

Let wavelength of light be $\lambda_2 = 2\lambda$

Stopping potential,

$$V_{02} = 1.6\text{V}$$

By using equation for photoelectric effect, $E = h\nu - h\nu_0$

$$E = \frac{hc}{\lambda} - \frac{hc}{\lambda_0} = hc\left[\frac{1}{\lambda} - \frac{1}{\lambda_0}\right]$$

$$\Rightarrow \quad V = \frac{hc}{e}\left(\frac{1}{\lambda} - \frac{1}{\lambda_0}\right) \quad [\because E = Ve]$$

For case I

$$4.8 = \frac{hc}{e}\left(\frac{1}{\lambda} - \frac{1}{\lambda_0}\right) \quad \ldots\text{(i)}$$

For case II

and $\quad 1.6 = \frac{hc}{e}\left(\frac{1}{2\lambda} - \frac{1}{\lambda_0}\right) \quad \ldots\text{(ii)}$

On dividing Eq. (i) by Eq. (ii), we get

$$\frac{4.8}{1.6} = \frac{\frac{hc}{e}\left(\frac{1}{\lambda} - \frac{1}{\lambda_0}\right)}{\frac{hc}{e}\left(\frac{1}{2\lambda} - \frac{1}{\lambda_0}\right)}$$

$$\Rightarrow \quad 3 = \frac{\left(\frac{1}{\lambda} - \frac{1}{\lambda_0}\right)}{\left(\frac{1}{2\lambda} - \frac{1}{\lambda_0}\right)} \Rightarrow \lambda_0 = 4\lambda$$

65. *(b)* Radius of largest circular path,

$$r = \frac{mv_{max}}{Be}$$

$\because$ Velocity of most energetic electron

$v_{max} = \frac{Ber}{m}$, Kinetic energy,

$$K_{max} = \frac{1}{2}mv_{max}^2 = \frac{1}{2}m\cdot\frac{B^2e^2r^2}{m^2}$$

$$\Rightarrow \quad K_{max} = \frac{1}{2}\left(\frac{B^2e^2r^2}{m}\right) \text{J}$$

Kinetic energy of emitted electron (in electron volts),

$$K_{max}\text{ (eV)} = \frac{K_{max}\text{ (in joule)}}{e}$$

$\therefore \quad K_{max}\ (eV) = \frac{1}{2}\left(\frac{B^2er^2}{m}\right)$

Here given,

$B = 3\times10^{-4}$ T, $e = 1.6\times10^{-19}$ C,

$r = 10$ mm $= 10^{-2}$ m

and $\quad m = 9.1\times10^{-31}$ kg

So,

$$K_{max} = \frac{(3\times10^{-4})^2\times(1.6\times10^{-19})\times(10^{-2})^2}{2\times(9.1\times10^{-31})}$$

$$= \frac{9\times1.6}{2\times9.1}\times10^{31-31} \approx 0.8 \text{ eV}$$

$\therefore$ Energy of incident photon

$$= hf = \frac{hc}{\lambda} = \frac{12400 \text{ eV - Å}}{6561 \text{ Å}} \approx 1.9 \text{ eV}$$

(**Note** $hc = 1240$ eV- nm = 12400 eV-Å)

So, work function of metal,

ϕ_0 = incident energy − kinetic energy of emitted electron

$= 1.9 - 0.8 = 1.1$ eV

> **• Concept Enhancer**
>
> When emitted photoeletrons enters into magnetic field (*B*) perpendicular to it, then these electrons move on circular path in magnetic field. Maximum radius (*r*) of circular path is calculated by, $r = \frac{mv_{max}}{Be}$
>
> where, m = mass of electron and
> v_{max} = maximum velocity of emitted electrons

66. *(d)* $\frac{hc}{\lambda} = \phi + KE_{max}$

Given $KE_{max} = 2$eV

Work function (ϕ) = 1eV

$\frac{hc}{\lambda} = \phi + KE_{max} = 1 + 2 = (1+2)$ eV

$\frac{hc}{\lambda} = 3$ eV $\quad \ldots$(i)

According to 2nd condition

$\lambda \to \frac{\lambda}{2}$

Then, $\frac{hc}{\frac{\lambda}{2}} = \phi + KE_{max}$

$2\frac{hc}{\lambda} = 1 \text{ eV} + KE_{max}$

As, $\frac{hc}{\lambda} = 3$eV

$2(3)\text{eV} = 1 \text{ eV} + KE_{max}$

$6 - 1 = KE_{max}$

$KE_{max} = 5$eV

67. *(4125)* $E = E_1 - E_2 = hc\left(\frac{1}{\lambda_1} - \frac{1}{\lambda_2}\right)$

$$E = 6.6\times10^{-34}\times3\times10^8\times10^9\left(\frac{1}{500} - \frac{1}{600}\right)$$

$= 6.6\times10^{-20}\text{J} = 4.125\times10^{-1}$ eV

$= 4125\times10^{-4}$ eV

68. *(2)* Let ν_0 be the value of threshold frequency.

According to Einstein's photoelectric equation, $\frac{1}{2}mv^2 = h\nu - h\nu_0$

For the first situation, $\nu_1 = 2\nu_0$

i.e. $\quad \frac{1}{2}mv_1^2 = h\nu_1 - h\nu_0$

$= h\cdot2\nu_0 - h\nu_0 = h\nu_0$

$\Rightarrow \quad \frac{1}{2}mv_1^2 = h\nu_0 \quad \ldots$ (i)

For the second situation,

$\nu_2 = 5\nu_0$,

Then $\frac{1}{2}mv_2^2 = h\nu_2 - h\nu_0 = h5\nu_0 - h\nu_0$

$\Rightarrow \quad \frac{1}{2}mv_2^2 = 4h\nu_0 \quad \ldots$ (ii)

Dividing Eq. (ii) by Eq. (i), we get

$$\frac{\frac{1}{2}mv_1^2}{\frac{1}{2}mv_2^2} = \frac{h\nu_0}{4h\nu_0} = \frac{1}{4}$$

$$\left(\frac{v_1}{v_2}\right)^2 = \left(\frac{1}{2}\right)^2 \Rightarrow \frac{v_1}{v_2} = \frac{1}{2}$$

$\Rightarrow \quad v_2 = 2v_1 = xv_1 \quad$ (given)

$\therefore \quad x = 2$

69. *(125)* According to Einstein's photoelectric equation,

$$KE = E_i - \phi_0 = \frac{hc}{\lambda} - \phi_0$$

$$= \frac{20\times10^{-26}}{500\times10^{-9}\times1.6\times10^{-19}} - 1.25$$

$= 0.025\times10^2 - 1.25$

$= 2.5 - 1.25$

$= 1.25$ eV

$\because \quad KE = \frac{p^2}{2m}$

where, p is momentum.

$\Rightarrow \ p = \sqrt{2mKE}$

$= \sqrt{2\times9\times10^{-31}\times1.25\times1.6\times10^{-19}}$

$= 6\times10^{-25}$ kg-ms^{-1}

Now, $r = mv/qB$

Here r is radius, mv is momentum and B is magnetic field.

$\Rightarrow \quad B = \frac{p}{qr} = \frac{6\times10^{-25}}{1.6\times10^{-19}\times30\times10^{-2}}$

$= 0.125\times10^{-4} = 125\times10^{-7}$ T

70. *(4)* We know that, according to Einstein's photoelectric equation,

$KE = \frac{hc}{\lambda} - \phi$ or $eV_0 = \frac{hc}{\lambda} - \phi \quad \ldots$(i)

As per question,

$\Rightarrow \quad e(3V_0) = \frac{hc}{\lambda} - \phi \quad \ldots$(ii)

and when the same surface is illuminated with a radiation of wavelength 2λ, stopping potential becomes V_0.

$\Rightarrow \quad eV_0 = \frac{hc}{2\lambda} - \phi \quad \ldots$(iii)

On dividing Eq. (ii) by Eq. (iii), we get

$$\frac{e(3V_0)}{eV_0} = \frac{\frac{hc}{\lambda} - \phi}{\frac{hc}{2\lambda} - \phi}$$

On solving, we will get

$\phi = \frac{hc}{4\lambda} \quad \ldots$(iv)

$\because$ We know that,

$\phi = hc/\lambda_t \quad \ldots$(v)

where, λ_t is the threshold wavelength.

$\therefore$From Eqs. (iv) and (v), we get

$\lambda_t = 4\lambda$

71. *(1)* Given, first incident energy, $E_{i_1} = 2\phi_0$ (work function ϕ_0), Second incident energy, $E_{i_2} = 10\ \phi_0$

If v_1 and v_2 be the respective maximum velocities in two cases, then by using the concept of photoelectric effect,

$KE = E_i - \phi_0$

$\therefore \quad 1/2\ mv_1^2 = 2\phi_0 - \phi_0 = \phi_0 \quad \ldots$(i)

and $1/2mv_2^2 = 10\ \phi_0 - \phi_0 = 9\ \phi_0 \quad \ldots$(ii)

On dividing Eq. (i) by Eq. (ii), we get

$$\left(\frac{v_1}{v_2}\right)^2 = \frac{1}{9}$$

$\Rightarrow \quad \frac{v_1}{v_2} = \frac{1}{3} = \frac{x}{y}$

$\therefore \quad x = 1$

72. *(2)* Let the work function of the metal be ϕ.

$$KE = \frac{hc}{\lambda} - \phi$$

For case I, $\frac{1}{2}mv_1^2 = 4 - \phi$ ...(i)

For case II, $\frac{1}{2}mv_2^2 = 2.5 - \phi$ ...(ii)

Given, $v_1 / v_2 = 2$

From eqs. (i) and (ii), we get

$$\frac{v_1^2}{v_2^2} = \frac{4-\phi}{2.5-\phi}$$

$$\Rightarrow (2)^2 = \frac{4-\phi}{2.5-\phi}$$

$$\Rightarrow 4(2.5-\phi) = 4-\phi$$

$$\Rightarrow 10 - 4\phi = 4 - \phi$$

$$\Rightarrow 3\phi = 6 \Rightarrow \phi = \frac{6}{3} = 2 \text{ eV}$$

73. *(9)* With radiation wavelength λ and work function ϕ_0, stopping potential is V.

$$\Rightarrow \frac{hc}{\lambda} - \phi_0 = eV \quad ...(i)$$

When wavelength is 3λ, stopping potential is $V/4$.

$$\Rightarrow \frac{hc}{3\lambda} - \phi_0 = \frac{eV}{4} \quad ...(ii)$$

Now, multiply Eq. (ii) by 3 and subtracting Eq (i) from it, we get

$$\Rightarrow \left(\frac{hc}{\lambda} - 3\phi_0\right) - \left(\frac{hc}{\lambda} - \phi_0\right) = \frac{3 \text{ eV}}{4} - \text{eV}$$

$$\Rightarrow 2\phi_0 = \frac{1}{4}eV \Rightarrow eV = 8\phi_0$$

Putting this value in Eq. (i), we get

$$\frac{hc}{\lambda} - \phi_0 = 8\phi_0$$

$$\Rightarrow \frac{hc}{\lambda} = 9\phi_0$$

$$\Rightarrow \frac{hc}{\lambda} = 9\frac{hc}{\lambda_0}$$

{As $\phi_0 = \frac{hc}{\lambda_0}$, where λ_0 = threshold wavelength}

$$\Rightarrow \lambda_0 = 9\lambda \quad \therefore n = 9$$

Topic 2

01. *(d)* Wave theory of light cannot explain Compton effect, or particle nature of light.

Wave theory of light can explain

→ Reflection of light

→ Diffraction of light

→ Refraction of light

• Concept Enhancer

When a high frequency photon scatters due to an interaction with a charged particle, there is a decrease in the energy of the photon and thus, an increase in its wavelength. This trade off between wavelength and energy in response to collision is the Compton effect.

02. *(*)* Electromagnetic waves carry both energy and momentum.

So, Assertion is false.

Rest mass of the photon is zero but mass of the photon is not zero.

So, Reason is also false.

03. *(c)* Power radiated $= \frac{\text{Energy emitted}}{\text{Time}}$

$$\Rightarrow P = \frac{N \times E_1}{t} = \frac{N}{t} \times hf \Rightarrow P = n \times hf$$

where $n = N/t$

= number of photons per second

$$\therefore \quad n = \frac{P}{hf} = \frac{2\times10^{-3}}{(6.63\times10^{-34}\times6\times10^{14})}$$

$$= 5\times10^{15} \text{ s}^{-1}$$

04. *(c)* $P = 200$ W

Number of photons emitted per second

$$n = \frac{P}{E} = \frac{P}{(hc/\lambda)}$$

$$n = P\lambda / hc \Rightarrow n \propto \lambda$$

$$\Rightarrow \frac{n_1}{n_2} = \frac{\lambda_1}{\lambda_2} = \frac{300}{500} = \frac{3}{5}$$

$$\therefore \quad n_1 : n_2 = 3:5$$

05. *(c)* As we know, $\lambda = \frac{h}{p} = \frac{h}{\sqrt{2mK}}$

If linear momentum of two photons are equal, then their wavelengths are also equal. Also, if the wavelength is decreased, then the momentum and energy of photon will increase.

06. *(a)* As power = Energy/time

or $4\times10^3 = Nhf / 10^{20}$

$$\left[\because \frac{\text{energy}}{\text{time}} = \text{power}\right]$$

$$f = \frac{4\times10^3}{10^{20}\times6.023\times10^{-34}}$$

$$f = 6.64\times10^{16} \text{ Hz}$$

The obtained frequency lies in the band of X-rays.

07. *(b)* Given that, power, $P = 15$ kW

$$= 15\times10^3 \text{ W}$$

Photon per second, $n = 10^{16}$ photon/s

Planck's constant, $h = 6\times10^{-34}$ J-s.

As we know that power,

$$P = \frac{E}{t} = n\frac{hc}{\lambda}$$

$$\Rightarrow \lambda = \frac{nhc}{P} = \frac{10^{16}\times6\times10^{-34}\times3\times10^8}{15\times10^3}$$

$$= 1.2\times10^{-13} = 0.0012 \text{ Å}$$

This wavelength corresponds to gamma rays.

08. *(b)* According to mass - energy equation is

$$E = mc^2 = pc$$

$$\Rightarrow p = \frac{E}{c} = \frac{P\cdot t}{c} \quad [\because E = P\times t]$$

Substituting values, we get

$$p = \left(\frac{20\times10^{-3}\times300\times10^{-9}}{3\times10^8}\right)$$

$$= 2\times10^{-17} \text{ kg-ms}^{-1}$$

09. *(d)* Power of source $= \frac{\text{Energy emitted}}{\text{Time}}$

$$= \frac{\text{Number of photons} \times \text{Energy of 1 photon}}{\text{Time}}$$

= Number density of photons × Energy of 1 photon

$$= n\cdot\frac{hc}{\lambda}$$

Now, $P_{X\text{-rays}} = P_{\text{visible light}}$

$$\Rightarrow n_X\left(\frac{hc}{\lambda_X}\right) = n_V\left(\frac{hc}{\lambda_V}\right)$$

$$\Rightarrow \frac{n_V}{n_X} = \frac{\lambda_V}{\lambda_X} = \frac{500}{1} = 500$$

So, $\frac{\lambda_X}{\lambda_V} = \frac{1}{500}$

10. *(b)* Power of laser is given as

$$P = \frac{\text{Energy}}{\text{Time}}$$

$$= \frac{\text{Number of photons emitted} \times \text{Energy of one photon}}{\text{Time}}$$

$$\Rightarrow P = \frac{NE}{t} = \left(\frac{N}{t}\right)\cdot E$$

So, number of photons emitted per second

$$= \frac{N}{t} = \frac{P}{E}$$

$$= \frac{P}{hc/\lambda} = \frac{P\lambda}{hc} \quad \left[\because E = h\nu = \frac{hc}{\lambda}\right]$$

Here, $h = 6.6 \times 10^{-34}$ J-s,

$\lambda = 500 \text{ nm} = 500 \times 10^{-9}$ m

$c = 3 \times 10^8 \text{ ms}^{-1}$

$P = 2\text{mW} = 2 \times 10^{-3}$ W

$$\therefore \quad \frac{N}{t} = \frac{2 \times 10^{-3} \times 500 \times 10^{-9}}{6.6 \times 10^{-34} \times 3 \times 10^8}$$

$\approx 5 \times 10^{15}$ photons per second

11. *(b)* Given, wavelength of parallel beam of light,

$\lambda = 900 \text{ nm} = 9 \times 10^{-7}$ m

Intensity of incident radiation,

$I = 100 \text{ Wm}^{-2}$

Area of surface, $A = 1 \text{ cm}^2 = 10^{-4} \text{ m}^2$

$\therefore$ The energy of each photon,

$$E_1 = \frac{hc}{\lambda} = \frac{6.63 \times 10^{-34} \times 3 \times 10^8}{9 \times 10^{-7}}$$

$= 2.21 \times 10^{-19}$ J

We know that, intensity of light wave,

$I = \frac{P}{A}$

$\Rightarrow \quad P = IA = 100 \times 10^{-4} = 10^{-2}$ W

$= 10^{-2}$ J/s

$\therefore$ Total energy of light beam in one second, $E = P \times t = 10^{-2} \times 1 = 10^{-2}$ J

Number of photons

$$= \frac{\text{Total energy } (E)}{\text{Energy of one photon } (E_1)}$$

$$= \frac{10^{-2}}{2.21 \times 10^{-19}} = 4.5 \times 10^{16}$$

12. *(5)* Since power emitted by source is given as

$$= \frac{\text{Total energy emitted}}{\text{Time}}$$

$$= \frac{(E_1 \text{ photon}) \times \text{Number of photons } (N)}{t}$$

$P_1 = (E_1)n,$

$$n = \frac{N}{t} = \frac{\text{Number of photon}}{\text{second}}$$

$$\frac{P_1}{P_2} = \frac{(E_1)(n_1)}{(E_2)(n_2)} = \frac{\left(\frac{hc}{\lambda_1}\right)(n_1)}{\left(\frac{hc}{\lambda_2}\right)(n_2)}$$

$$\frac{P_1}{P_2} = \left(\frac{\lambda_2}{\lambda_1}\right)\left(\frac{n_1}{n_2}\right)$$

Substituting the given values, we get

$$2 = \frac{300}{600} \times \frac{2 \times 10^{15}}{n_2}$$

$n_2 = \frac{1}{2} \times 10^{15} = 5 \times 10^{14}$ photon/second

13. *(11)* Given, intensity of falling radiation, $I = 6.4 \times 10^{-5} \text{ Wcm}^{-2}$

Wavelength of radiation, $\lambda = 310$ nm

Work function of metal, $\phi = 2$ eV

Area of metal surface, $A = 1 \text{ cm}^2$

Energy of one photon of radiation,

$$E_1 = \frac{hc}{\lambda} = \frac{1240 \text{ eV-nm}}{\lambda \text{ (nm)}}$$

or $E_1 = \frac{1240}{310} \text{eV} = \frac{1240}{310} \times 1.6 \times 10^{-19}$ J

$= 6.4 \times 10^{-19}$ J

So, photon flux of radiation falling over surface = number of photons falling over surface per second

$$= \frac{\text{Total incident energy}}{\text{Energy of one photon}} = \frac{I \times A}{E_1}$$

$$= \frac{6.4 \times 10^{-5} \times 1}{6.4 \times 10^{-19}} = 10^{14} \text{ photons/second}$$

As only one in 10^3 photons are able to emit an electron, number of ejected electrons in 1 s

$$= \frac{10^{14}}{10^3} = 10^{11} \Rightarrow x = 11$$

Topic 3

01. *(d)* For a particle, de-Broglie wavelength is given as

$$\lambda = \frac{h}{\sqrt{2mK}}$$

$$2mK = \frac{h^2}{\lambda^2} \Rightarrow \frac{1}{2mK} = \frac{\lambda^2}{h^2}$$

The graph between $\frac{1}{K}$ and λ is upward parabola.

02. *(a)* Given,

Energy of photon = Kinetic energy of proton of mass m_p

$$\frac{1}{2} m_p v^2 = E_0$$

$$\Rightarrow \quad p = \sqrt{2m_p E_0}$$

$$\text{So,} \quad \lambda_p = \frac{h}{\sqrt{2m_p E_0}} \quad \ldots\text{(i)}$$

For photon, $\frac{hc}{\lambda} = E_0$

$$\Rightarrow \quad \lambda_{\text{Ph}} = \frac{hc}{E_0} \quad \ldots\text{(ii)}$$

Using Eqs. (i) and (ii), we get

$$\frac{\lambda_p}{\lambda_{\text{ph}}} = \frac{h}{\sqrt{2m_p E}} \times \frac{E}{hc} = \frac{1}{c}\sqrt{\frac{E}{2m_p}}$$

03. *(b)* Momentum of photon is,

$$\text{Momentum} = \frac{\text{Total energy}}{\text{Speed}} \quad (\because E = pc)$$

(In case of 100% absorption of energy)

$$\therefore \text{ Momentum} = \frac{6.48 \times 10^5}{3 \times 10^8}$$

$= 2.16 \times 10^{-3}$ kg m/s

04. *(c)* de-Broglie's wavelength is given by

$$\lambda_{DB} = \frac{h}{p} = \frac{h}{\sqrt{2mK}}$$

$$\lambda_{DB} \propto \frac{1}{\sqrt{m}} \Rightarrow \lambda_\alpha < \lambda_p < \lambda_e$$

05. *(c)* de-Broglie wavelength of electron in terms of kinetic energy K,

$$\lambda = \frac{h}{\sqrt{2mK}} \Rightarrow \lambda \propto \frac{1}{\sqrt{K}}$$

$$\therefore \quad \frac{\lambda_1}{\lambda_2} = \sqrt{\frac{K_2}{K_1}}$$

If $K_2 = \frac{E}{4}$ and $K_1 = E$

(Given in question)

$$\therefore \quad \frac{\lambda_1}{\lambda_2} = \sqrt{\frac{E/4}{E}} = \left(\frac{1}{4}\right)^{1/2} = \frac{1}{2}$$

or $\quad \lambda_2 = 2\lambda_1$

06. *(d)* de-Broglie wavelength is given by

$$\lambda = \frac{h}{mv} = \frac{h}{\sqrt{2mq\Delta v}}$$

According to question,

$$\frac{\lambda_\alpha}{\lambda_p} = \sqrt{\frac{m_p v_p q_p}{m_\alpha v_\alpha q_\alpha}}$$

$$\Rightarrow \quad \frac{\lambda_\alpha}{\lambda_p} = \sqrt{\frac{1 \times 2 \times 1}{4 \times 4 \times 2}} = \frac{1}{4}$$

$$\Rightarrow \quad \lambda_p : \lambda_\alpha = 4:1$$

07. *(a)* $p = \sqrt{2\text{KE}m}$

$$\lambda = \frac{h}{p}$$

$$\Rightarrow \quad \frac{\lambda_p}{\lambda_e} = \frac{p_e}{p_p} = \sqrt{\frac{2\text{KE}m_e}{2\text{KE}m_p}}$$

$$= \sqrt{\frac{m_e}{m_p}} = \sqrt{\frac{1}{1849}} = \frac{1}{43}$$

Option (a) is correct answer.

08. *(c)* Given de-Broglie wavelength of proton and electron have same value;

$$\lambda_p = \lambda_e$$

$$\frac{h}{p_p} = \frac{h}{p_e} \quad \{p = \text{momentum}, \; h = \text{Planck's constant}\}$$

$$\Rightarrow \quad p_p = p_e$$

$$\therefore \quad \text{Ratio } \frac{p_p}{p_e} = 1:1$$

09. *(c)* de-Broglie wavelength of a moving particle with kinetic energy KE is given as

$$\lambda = \frac{h}{\sqrt{2m\text{KE}}}$$

$$\text{or} \quad \text{KE} = \frac{h^2}{2m\lambda^2}$$

For same wavelength, KE $\propto \frac{1}{m}$

$\because$ Mass of α-particle = 4 (mass of proton)

$$\frac{\text{KE}_p}{\text{KE}_\alpha} = \frac{m_\alpha}{m_p} = \frac{4m_p}{m_p} = \frac{4}{1}$$

• Concept Enhancer

When a charged particle of charge q is moving with kinetic energy K, then de-Broglie wavelength associated with charged particle is given as,

$$\lambda = \frac{h}{\sqrt{2mK}}$$

10. *(b)* de-Broglie wavelength of a charged particle accelerated by a potential V,

$$\lambda = \frac{h}{p} = \frac{h}{\sqrt{2mqV}}$$

$$\frac{\lambda_\alpha}{\lambda_\beta} = \sqrt{\frac{m_p}{m_\alpha} \cdot \frac{q_p}{q_\alpha}} = \sqrt{\frac{m_p}{4m_p} \cdot \frac{q_p}{2q_p}}$$

$$= 1/\sqrt{8} = 1/\sqrt{m} \quad \text{(given)}$$

$$\therefore \quad m = 8$$

11. *(b)* de-Broglie wavelength in terms of accelerated potential difference

$$\lambda = \frac{12.27}{\sqrt{V}}$$

$$\Rightarrow \quad \lambda \propto \frac{1}{\sqrt{V}}$$

$$\Rightarrow \quad \frac{\lambda_2}{\lambda_1} = \sqrt{\frac{V_1}{V_2}} = \sqrt{\frac{20 \text{ kV}}{40 \text{ kV}}}$$

Here, $V_1 = 20$ kV and $V_2 = 40$ kV

$$= \sqrt{\frac{1}{2}} = \frac{1}{\sqrt{2}}$$

$$\Rightarrow \quad \lambda_2 = \frac{\lambda_1}{\sqrt{2}} = \frac{\lambda_0}{\sqrt{2}} \quad [\because \lambda_1 = \lambda_0]$$

12. *(b)* For α-particle,

$$K_\alpha = \frac{p_\alpha^2}{2m_\alpha} = (2e)V \quad \ldots \text{(i)}$$

$$\text{For a proton, } K_p = \frac{p_p^2}{2m_p} = eV \quad \ldots \text{(ii)}$$

From Eqs. (i) and (ii),

$$\left(\frac{p_\alpha}{p_p}\right)^2 = \frac{2m_\alpha}{m_p} \quad \left[\because \frac{m_\alpha}{m_p} = 4\right]$$

$$\therefore \left(\frac{p_\alpha}{p_p}\right)^2 = 2(4)$$

$$\Rightarrow \quad \frac{p_\alpha}{p_p} = \sqrt{8} = 2\sqrt{2}$$

13. *(d)* We know de-Broglie wavelength of electron accelerating under the effect of potential difference V is given as

$$\lambda = \frac{1.227}{\sqrt{V}} = \frac{1.227}{x} \quad \text{(given)}$$

$$\therefore \quad x = \sqrt{V}$$

14. *(d)* The de-Broglie wavelength is given by, $\lambda = \frac{h}{\sqrt{2mqV}}$

$$\Rightarrow \frac{\lambda_p}{\lambda_d} = \sqrt{\frac{m_d q_d V_d}{m_p q_p V_p}}$$

$$\Rightarrow \frac{1}{\sqrt{2}} = \sqrt{\frac{2m_p \times q.V_d}{m_p \times q \times V_p}} \quad [\therefore m_d = 2m_p]$$

$$\Rightarrow \frac{1}{2} = 2.\frac{V_d}{V_p} \Rightarrow \frac{V_p}{V_d} = \frac{4}{1}$$

15. *(a)* de-Broglie wavelengths of a photon and an electron of same kinetic energy E are

$$\text{for photon, } \lambda_p = \frac{h}{p} = \frac{hc}{E_p}$$

$$\Rightarrow \quad E_p = hc/\lambda_p \quad \ldots \text{(i)}$$

$$\text{for electron, } \lambda_e = \frac{h}{p} = \frac{h}{\sqrt{2mE_e}}$$

$$\Rightarrow \quad E_e = \frac{h^2}{2m\lambda_e^2} \quad \ldots \text{(ii)}$$

As $E_p = E_e$. So, from Eqs. (i) and (ii), we get

$$\frac{hc}{\lambda_p} = \frac{h^2}{2m\lambda_e^2}$$

$$\text{or} \quad \lambda_p \propto \lambda_e^2$$

16. *(b)* de-Broglie wavelength, $\lambda = \frac{h}{\sqrt{2mK}}$

For the same kinetic energy, $\lambda \propto \frac{1}{\sqrt{m}}$

$$\therefore \quad \frac{\lambda_\alpha}{\lambda_c} = \sqrt{\frac{m_c}{m_\alpha}} = \sqrt{\frac{12}{4}} = \frac{\sqrt{3}}{1} \text{ or } \sqrt{3}:1$$

17. *(b)*

key idea You can compare de-Broglie wavelengths of different moving particles with same energy (E) with the help of following formula,

$$\lambda = \frac{h}{\sqrt{2mE}}, \text{ where, } m \text{ is the mass of moving particle.}$$

de-Broglie wavelength of moving particle having energy E is given as

$$\lambda = \frac{h}{\sqrt{2mE}}$$

Since, E is same for proton, neutron, electron and α-particle.

$$\therefore \quad \lambda \propto \frac{1}{\sqrt{m}}$$

Since, $m_\alpha > m_n > m_p > m_e$

$$\therefore \quad \lambda_e > \lambda_p > \lambda_n > \lambda_\alpha$$

18. *(c)*

key idea You can calculate de-Broglie wavelength of moving charged particle accelerated through potential difference V with the help of following formula.

$$\lambda = \frac{h}{\sqrt{2mqV}}$$

where, symbols have their usual meanings.

de-Broglie wavelength,

$$\lambda = \frac{h}{p} = \frac{h}{\sqrt{2mE}} = \frac{h}{\sqrt{2mqV}}$$

$$\therefore \quad \lambda \propto \frac{1}{\sqrt{mq}}$$

$$\text{Hence, } \frac{\lambda_e}{\lambda_p} = \frac{\sqrt{m_p q_p}}{\sqrt{m_e q_e}} = \frac{\sqrt{m_p}}{\sqrt{m_e}} \quad [\because q_p = q_e]$$

19. *(d)* Given, $\frac{\lambda_p}{\lambda_e} = \frac{2}{1}, \frac{v_e}{v_p} = \frac{1}{4}$

We know that, de-Broglie wavelength,

$$\lambda = h/mv$$

$$\Rightarrow \frac{\lambda_p}{\lambda_e} = \frac{m_e v_e}{m_p v_p} \Rightarrow \frac{2}{1} = \frac{m_e}{m_p}\left(\frac{1}{4}\right)$$

$$\Rightarrow m_p = \frac{1}{8} m_e$$

20. *(d)* Given, mass of proton,
$m_p = 1.00727$ u
Mass of electron, $m_e = 0.00055$ u
Potential, $V = 100$ V
As, de-Broglie wavelength can be given as

$$\lambda = \frac{h}{\sqrt{2mqV}}$$

$$\Rightarrow \quad \lambda \propto \frac{1}{\sqrt{m}}$$

$$\Rightarrow \quad \frac{\lambda_1}{\lambda_2} = \sqrt{\frac{m_2}{m_1}}$$

$$\Rightarrow \quad \frac{\lambda_e}{\lambda_p} = \sqrt{\frac{m_p}{m_e}} = \sqrt{\frac{1.00727 \text{ u}}{0.00055 \text{ u}}}$$

$$\frac{\lambda_e}{\lambda_p} = \sqrt{1831.4} = 42.79$$

or $\frac{\lambda_e}{\lambda_p} \approx \frac{43}{1}$ or $\lambda_e : \lambda_p = 43:1$

21. *(b)* As, wavelength $\lambda = \frac{h}{p} = \frac{h}{mv}$

where, h is Planck's constant and p is momentum.

$$\therefore \quad \lambda \propto \frac{1}{m}$$

$$\Rightarrow \quad \frac{\lambda_e}{\lambda_p} = \frac{m_p}{m_e} = \frac{1836\, m_e}{m_e} = 1836$$

22. *(d)* de-Broglie wavelength,

$$\lambda = \frac{h}{\sqrt{2mqV}}$$

$$\Rightarrow \quad \frac{\lambda_p}{\lambda_\alpha} = \sqrt{\frac{m_\alpha q_\alpha}{m_p q_p}} = \sqrt{\frac{4m_p \cdot 2e}{m_p \cdot e}} = \frac{2\sqrt{2}}{1} = 2 \times 1.414 = 2.8$$

23. *(b)* Given, $\lambda_p = \lambda_\alpha$

As we know that, $\lambda = h/p$

$$\therefore \quad \frac{h}{p_p} = \frac{h}{p_\alpha} \Rightarrow p_p = p_\alpha$$

$$\Rightarrow \quad m_p v_p = m_\alpha v_\alpha$$

$$\Rightarrow \quad m_p v_p = 4m_p v_\alpha \qquad (\because m_\alpha = 4m_p)$$

$$\Rightarrow v_p / v_\alpha = 4/1 \text{ or } 4:1$$

24. *(a)* The de-Broglie wavelength,

$$\lambda = \frac{h}{\sqrt{3k_B T m}}$$

$$= \frac{6.63 \times 10^{-34}}{\sqrt{3 \times 1.38 \times 10^{-23} \times 400 \times 4.64 \times 10^{-26}}}$$

$$\lambda \simeq 0.24 \text{ Å}$$

25. *(d)* Here, energy of electron $= E$

$$\Rightarrow 1/2mv^2 = E$$

$$\Rightarrow \quad m^2v^2 = 2mE$$

$$\Rightarrow \quad p^2 = 2mE$$

or momentum of electron,

$$p = \sqrt{2mE}$$

So, de-Broglie wavelength

$$\lambda_e = \frac{h}{p} = \frac{h}{\sqrt{2mE}}$$

Now, photon energy $= E$

$$\Rightarrow \quad hf = E \Rightarrow \frac{hc}{\lambda_p} = E$$

or wavelength of photon, $\lambda_p = hc/E$
Hence, ratio

$$\frac{\lambda_e}{\lambda_p} = \frac{\left(\frac{h}{\sqrt{2mE}}\right)}{\left(\frac{hc}{E}\right)} = \frac{1}{c}\left(\frac{E}{2m}\right)^{1/2}$$

26. *(d)* de-Broglie wavelength, $\lambda = \frac{h}{mv} = \frac{h}{p}$

where, $p =$ momentum
By conservation of momentum,

$$p_1 + p_2 = 0 \text{ or } p_1 = p_2 \Rightarrow \frac{h}{p_1} = \frac{h}{p_2}$$

$$\therefore \quad \lambda_1 = \lambda_2 = \lambda \qquad (\because \lambda = h/p)$$

27. *(d)* The momentum of the photon,

$$p = h/\lambda = h\nu/c$$

28. *(a)* Formation of covalent bonds due to the wave nature of particles is done in compounds.

29. *(a)* Here, velocity of electron,

$$\mathbf{v} = V_0\hat{\mathbf{i}}$$

and magnetic field, $B = \mathbf{B}_0\hat{\mathbf{j}}$

Since, $\mathbf{v} \perp \mathbf{B}$, these electron will move on circular path in magnetic field and work done by magnetic field on electron will be zero. i.e. no change in kinetic energy of electron (According to work-energy theorem).
So, speed of electron remains same with time.
Therefore, there will be no change in de-Broglie wavelength with time.

30. *(d)* According to Einstein's photoelectric equation,

$$E_k = E - \phi_0 \qquad \ldots(i)$$

de-Broglie wavelength

$$\lambda_e = \frac{h}{\sqrt{2mE_k}}$$

$$\Rightarrow \quad E_k = \frac{h^2}{2m\lambda_e^2}$$

$\therefore$ From Eq. (i), we get

$$\frac{h^2}{2m\lambda_e^2} = \frac{hc}{\lambda_i} - \frac{hc}{\lambda_0}$$

$$\Rightarrow \quad \lambda_e = \sqrt{\frac{h}{2mc\left(\frac{1}{\lambda_i} - \frac{1}{\lambda_0}\right)}}$$

31. (*) According to de - Broglie's equation,

$$\lambda = \frac{h}{p} \Rightarrow \frac{h}{mv} \qquad \ldots(i)$$

From Bohr's 2nd postulate,

$$mvr = \frac{nh}{2\pi}$$

$$mv = \frac{nh}{2\pi r} \qquad \ldots(ii)$$

Substitute the value of mv from Eq (i)

$$\lambda = \frac{2\pi rh}{nh} \Rightarrow \lambda \propto \frac{r}{n}$$

$$\frac{\lambda_1}{\lambda_4} = \frac{r_1 n_4}{r_4 n_1} \qquad \ldots(ii)$$

Substituting all the values

$$\therefore \quad \frac{\lambda_1}{\lambda_4} = \frac{5.3 \times 10^{-11} \times 4}{1 \times 8.48 \times 10^{-10}} = \frac{1}{4}$$

32. *(c)* Initial velocity of electron $= v_0\hat{\mathbf{i}}$

Electric field $\mathbf{E} = -E_0\hat{\mathbf{k}}$

Velocity when it enters the field

$$\mathbf{v} = \mathbf{v}_0\hat{\mathbf{i}} + \frac{E_0 e}{m} t\,\hat{\mathbf{k}}$$

$$|\mathbf{v}| = \sqrt{v_0^2 + \frac{E_0^2 e^2 t^2}{m^2}}$$

Now, according to de-Broglie

$$\lambda_0 = \frac{h}{mv_0}$$

and

$$\lambda' = \frac{h}{mv_0\sqrt{1 + \frac{E_0^2 e^2 t^2}{v_0^2 m^2}}}$$

$$\lambda' = \frac{\lambda_0}{\sqrt{1 + \frac{E_0^2 e^2 t^2}{v_0^2 m^2}}}$$

33. *(b)* de-Broglie wavelength is given by

$$\lambda = \frac{h}{p} = \frac{h}{\sqrt{2mK}}$$

Given, $\lambda_{\text{proton}} = \lambda_{\text{electron}}$

$$\Rightarrow \frac{h}{\sqrt{2m_pK_p}} = \frac{h}{\sqrt{2m_eK_e}}$$

$$\Rightarrow m_pK_p = m_eK_e$$

$$\Rightarrow \frac{K_p}{K_e} = \frac{m_e}{m_p} = \frac{1}{1836}$$

$$\therefore m_p \approx 1836 m_e$$

$$\Rightarrow \frac{K_p}{K_e} < 1 \text{ or } K_p < K_e$$

34. *(c)* de-Broglie wavelength,

$$\lambda = \frac{h}{p} = \frac{h}{mv}$$

For proton, de-Broglie wavelength,

$$\lambda_1 = \frac{h}{m_p v_p} = \lambda \text{ (given)}$$

For α-particle, de-Broglie wavelength

$$\lambda_2 = \frac{h}{m_\alpha v_\alpha} = 2\lambda \text{ (given)}$$

Dividing, we get

$$\frac{\lambda_1}{\lambda_2} = \frac{m_\alpha v_\alpha}{m_p v_p} = \frac{1}{2}$$

or $$\frac{4 \times v_\alpha}{1 \times v_p} = \frac{1}{2} \quad \left(\text{As, } \frac{m_\alpha}{m_p} = 4\right)$$

$$\Rightarrow \frac{v_\alpha}{v_p} = \frac{1}{8} \Rightarrow \frac{v_p}{v_\alpha} = \frac{8}{1}$$

• Concept Enhancer

The de-Broglie waves cannot be electromagnetic in nature because electromagnetic waves are produced by motion of charge particle.

35. *(b)* For photon,

$$E_p = \frac{hc}{\lambda_p} \Rightarrow \lambda_p = \frac{hc}{E_p}$$

For electron,

$$\lambda_e = \frac{h}{m_e v_e} = \frac{hv_e}{2K_e}$$

$$\Rightarrow v_e = 0.25c$$

$$\lambda_e = \frac{h \times 0.25c}{2K_e} = \frac{hc}{8K_e}$$

$$\lambda_p = \lambda_e$$

$$\frac{hc}{E_p} = \frac{hc}{8K_e} \Rightarrow \frac{K_e}{E_p} = \frac{1}{8}$$

36. *(a)* For a gas molecule at absolute temperature T, $v_{\text{rms}} \propto \sqrt{T}$

Hence, momentum of gas molecule,

$$p \propto \sqrt{T}$$

$$\Rightarrow \frac{p_2}{p_1} = \sqrt{\frac{T_2}{T_1}} = \sqrt{\frac{600}{300}}$$

$$\Rightarrow p_2 = \sqrt{2}p_1 \text{ or } \frac{p_1}{p_2} = \frac{1}{\sqrt{2}}$$

So ratio of de-Broglie wavelength is

$$\frac{\lambda_2}{\lambda_1} = \frac{h/p_2}{h/p_1} = \frac{p_1}{p_2} = \frac{1}{\sqrt{2}} \Rightarrow \lambda_2 = \frac{\lambda_1}{\sqrt{2}}$$

37. *(b)* According to de-Broglie, momentum $p = h/\lambda$,
where h is Planck's constant
and λ = wavelength

$$p = \sqrt{2m\text{KE}}$$

Now, $$\lambda = \frac{h}{p} = \frac{h}{\sqrt{2m\text{KE}}}$$

$$\lambda_e = \frac{h}{\sqrt{2m_e\text{KE}_e}} = \frac{h}{\sqrt{2m_e \times 4K}} = \frac{h}{\sqrt{8m_eK}}$$

$$\lambda_p = \frac{h}{\sqrt{2m_p\text{KE}_p}} = \frac{h}{\sqrt{2m_pK}}$$

$$\lambda_\alpha = \frac{h}{\sqrt{2m_\alpha\text{KE}_\alpha}} = \frac{h}{\sqrt{2m_\alpha \cdot 2K}}$$

$$= \frac{h}{\sqrt{2 \times 4m_p \times 2K}} = \frac{h}{\sqrt{16m_pK}}$$

Form above $\lambda_\alpha < \lambda_p < \lambda_e$
Option (b) is correct.

38. *(b)* de-Broglie wavelength λ is related with potential difference as

$$\lambda = \frac{h}{\sqrt{2mqV}}$$

$$\Rightarrow \frac{\lambda_1}{\lambda_2} = \sqrt{\frac{V_2}{V_1}} \quad \text{...(i)}$$

Here, $\lambda_1 = \lambda$, $\lambda_2 = \lambda_1 + 50\%$ of λ_1

$$= \lambda_1 + \frac{\lambda_1}{2} = \frac{3\lambda}{2}$$

∴From Eq. (i), we have

$$\frac{V_2}{V_1} = \left(\frac{\lambda_1}{\lambda_2}\right)^2 = \left(\frac{\lambda}{3\lambda/2}\right)^2 = \left(\frac{2}{3}\right)^2 = \frac{4}{9}$$

$$\Rightarrow \frac{V_1}{V_2} = \frac{9}{4}$$

39. *(a)* As we know that, de-Broglie wavelength,

$$\lambda = \frac{h}{mv} = \frac{h}{\sqrt{2mE}} \quad \text{...(i)}$$

Here, E is the kinetic energy.

$$E = \frac{3k_BT}{2}$$

Substituting value of E in Eq. (i), we get,

$$\lambda = \frac{h}{\sqrt{3mk_BT}}$$

Substituting the given values in the above equation, we get

$$\lambda = \frac{6.6 \times 10^{-34}}{\sqrt{3 \times 9 \times 10^{-31} \times 1.38 \times 10^{-23} \times 300}}$$

$$= 6.26 \text{ nm}$$

40. *(c)* At same temperature kinetic energy of proton and electron are same

$$\text{KE}_p = \text{KE}_e \Rightarrow \frac{p_p^2}{2m_p} = \frac{p_e^2}{2m_e}$$

$$\frac{p_p^2}{p_e^2} = \frac{m_p}{m_e} \quad \text{...(i)}$$

Using uncertainty in determining the position, $\Delta x \approx \frac{h}{\Delta p}$

$$\therefore \text{Ratio, } \frac{\Delta x_e}{\Delta x_p} = \frac{\Delta p_p}{\Delta p_e} = \sqrt{\frac{m_p}{m_e}}$$

[from Eq. (i)]

41. *(a)* According to question,

$$\lambda_p = \lambda_e = \lambda$$

By using de-Broglie wavelength,

$$\lambda = \frac{h}{p} = \frac{h}{\sqrt{2mK}}$$

where, m is mass and h is Planck's constant.

$$\therefore p = h/\lambda \quad \text{... (i)}$$

From Eq. (i)

$$p_e = p_p \text{ as } \lambda_e = \lambda_p$$

Again, $$K = h^2/2m\lambda^2 \quad \text{... (ii)}$$

From Eq. (ii)

$$\therefore \frac{K_e}{K_p} = \frac{m_p}{m_e} \times \frac{\lambda_p}{\lambda_e}$$

$$\Rightarrow \frac{K_e}{K_p} = \frac{m_p}{m_e} \quad [\because \lambda_e = \lambda_p]$$

$$\because m_p > m_e$$

$$\Rightarrow K_e > K_p$$

42. *(c)* Given, $\lambda_e = \lambda_p$ ...(i)

As we know that,
de-Broglie wavelength,

$$\lambda = \frac{h}{p} = \frac{h}{m_p} \quad \text{...(ii)}$$

Energy of photon, $E_p = \frac{hc}{\lambda_p}$

Using Eq. (ii)

$$E_p = \frac{hcm_pc}{h} = m_pc^2 \quad \text{...(iii)}$$

Similarly, $$E_e = \frac{1}{2}m_ev^2 \quad \text{...(iv)}$$

Dividing Eqs. (iv) by (iii), we get

$$\frac{E_e}{E_p}=\frac{\frac{1}{2}m_e v^2}{m_p c^2} \quad \dots(v)$$

From Eqs. (i) and (ii), we get

$$p_e=p_p$$

$$\Rightarrow \quad m_e v=m_p c \Rightarrow \frac{m_e}{m_p}=\frac{c}{v}$$

Putting this value in Eq. (v), we get

$$\frac{E_e}{E_p}=\frac{1}{2}\frac{c}{v}\cdot\frac{v^2}{c^2}=\frac{v}{2c}$$

43. *(c)* Since, we know that de-Broglie wavelength of an electron can be given as

$$\lambda=h/mv \quad \dots(i)$$

Also, we know that,

$$\text{Kinetic energy, } K=\frac{p^2}{2m} \quad \dots(ii)$$

where, p is the linear momentum.

∵ Linear momentum,

$$p=\frac{h}{\lambda} \quad \dots(iii)$$

From Eq. (ii) and Eq. (iii), we get

$$K=\frac{h^2}{2\lambda^2 m}$$

$$\Rightarrow \quad h\nu=\frac{h^2}{2\lambda^2 m} \quad \left\{\because K=h\nu=\frac{hc}{\lambda}\right\}$$

$$\Rightarrow \quad \frac{hc}{\lambda_C}=\frac{h^2}{2\lambda^2 m}$$

where, λ_C = cut-off wavelength.

$$\Rightarrow \quad \lambda_C=\frac{2m\lambda^2 c}{h}$$

44. *(c)* Initial acceleration of electron,

$$a=\frac{F}{m}=\frac{|e|\mathbf{E}}{m}$$

Velocity of electron after time t,

$$v=u+at=\frac{|e|\mathbf{E}}{m}\cdot t$$

Momentum of electron, $mv=|e|\mathbf{E}\cdot t$

de-Broglie wavelength λ associated with electron,

$$\lambda=\frac{h}{p}=\frac{h}{|e|\mathbf{E}\cdot t}$$

Rate of change of de Broglie wavelength with time,

$$\frac{d\lambda}{dt}=\frac{-h}{|e|\mathbf{E}t^2}$$

Here, negative sign shows wavelength decreases with time.

45. *(c)* de-Broglie wavelength of a moving particle is given by

$$\lambda=\frac{h}{p}=\frac{h}{\sqrt{2mK}}$$

where, K = kinetic energy of the particle.

Now, in given condition,

$$\lambda=\frac{h}{\sqrt{2mE}} \quad \dots(i)$$

$$\text{and} \quad \frac{\lambda}{2}=\frac{h}{\sqrt{2m(E+\Delta E)}} \quad \dots(ii)$$

On dividing Eq. (i) by Eq. (ii), we get

$$2=\frac{\sqrt{E+\Delta E}}{\sqrt{E}} \Rightarrow \Delta E=3E$$

46. *(c)* We know that, $\lambda=\dfrac{h}{mv}$

$$\text{and} \quad K=\frac{1}{2}mv^2=\frac{(mv)^2}{2m}$$

$$\Rightarrow \quad mv=\sqrt{2mK}$$

$$\text{Thus,} \quad \lambda=\frac{h}{\sqrt{2mK}}$$

$$\Rightarrow \quad \lambda \propto \frac{1}{\sqrt{K}}$$

$$\Rightarrow \quad \frac{\lambda_2}{\lambda_1}=\frac{\sqrt{K_1}}{\sqrt{K_2}}=\frac{\sqrt{K_1}}{\sqrt{2K_1}}$$

$$[\because K_2=2K_1]$$

$$\text{or} \quad \frac{\lambda_2}{\lambda_1}=\frac{1}{\sqrt{2}} \text{ or } \lambda_2=\frac{\lambda_1}{\sqrt{2}}$$

47. *(c)* de-Broglie wavelength is, $\lambda=\dfrac{h}{p}$,

$$\left[p^2=m^2v^2=2m\left(\frac{1}{2}mv^2\right)=2mK\right]$$

$$=\frac{h}{\sqrt{2mK}}$$

Given, $\lambda_{\text{electron}}=\lambda_{\text{proton}}$

$$\Rightarrow \quad \frac{h}{\sqrt{2m_eK_e}}=\frac{h}{\sqrt{2m_pK_p}}$$

$$\Rightarrow \quad m_eK_e=m_pK_p$$

$$\Rightarrow \quad \frac{K_e}{K_p}=\frac{m_p}{m_e}=\frac{1836\,m_e}{m_e}$$

$$\Rightarrow \quad \frac{K_e}{K_p}=1836 \text{ or } \frac{K_p}{K_e}=\frac{1}{1836}$$

48. *(d)* Initial velocity of electron,

$$\mathbf{v}=v_0\hat{\mathbf{i}} \quad (v_0>0)$$

$$\text{Electric field, } \mathbf{E}=E_0\hat{\mathbf{i}} \quad (E_0>0)$$

At time $t=0$, de-Broglie wavelength,

$$\lambda_0=\frac{h}{mv_0} \quad \dots(i)$$

∴ Acceleration on electron,

$$\mathbf{a}=\frac{eE_0}{m}(-\hat{\mathbf{i}}) \Rightarrow \mathbf{a}=\frac{-eE_0}{m}\hat{\mathbf{i}}$$

∴ Velocity of electron after time t,

$$v(t)=v_0+|a|t=v_0+\frac{eE_0}{m}t$$

∴ de-Broglie wavelength,

$$\lambda'=\frac{h}{m[v(t)]}=\frac{h}{m\left(v_0+\frac{eE_0}{m}t\right)}$$

$$=\frac{h}{mv_0+eE_0t}$$

$$\therefore \quad \frac{\lambda_0}{\lambda'}=\frac{\frac{h}{mv_0}}{\frac{h}{(mv_0+eE_0t)}}$$

$$=\frac{mv_0+eE_0t}{mv_0}=1+\frac{eE_0t}{mv_0}$$

$$\therefore \quad \lambda'=\frac{\lambda_0}{\left(1+\frac{eE_0t}{mv_0}\right)}$$

49. *(b)* Given, initial kinetic energy,

$$E_1=E$$

Initial de-Broglie wavelength,

$$\lambda_1=\lambda$$

Consider the wavelength of particle changes to 75% of λ after providing energy ΔE to the particle.

Hence final wavelength of particle,

$$\lambda_2=0.75\lambda$$

Final energy, $E_2=E+\Delta E$

Relationship between de-Broglie wavelength and energy of particle is given as

$$\lambda=\frac{h}{\sqrt{2mE}}$$

Therefore, we get $\lambda \propto \dfrac{1}{\sqrt{E}}$

Thus, we can write the relationship as

$$\frac{\lambda_1}{\lambda_2}=\frac{\sqrt{E_2}}{\sqrt{E_1}}$$

$$\Rightarrow \frac{\lambda}{0.75\lambda}=\frac{\sqrt{E+\Delta E}}{\sqrt{E}} \Rightarrow \frac{4}{3}=\frac{\sqrt{E+\Delta E}}{\sqrt{E}}$$

Squaring both sides of above equation, we get

$$\frac{16}{9}=\frac{E+\Delta E}{E}$$

$$\Rightarrow 16E=9E+9\Delta E$$

$$\Rightarrow \Delta E=\frac{7}{9}E$$

50. *(a)* Given, de-Broglie wavelengths for particles are λ_1 and λ_2.

So, $\lambda_1 = \frac{h}{p_1}$ and $\lambda_2 = \frac{h}{p_2}$

and momentum of particles are

$$p_1 = \frac{h}{\lambda_1} \text{ and } p_2 = \frac{h}{\lambda_2}$$

Given that, particles are moving perpendicular to each other and collide inelastically.

So, they move as a single particle.

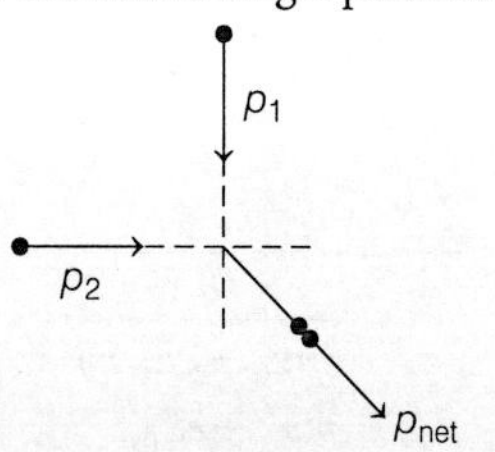

So, by conservation of momentum and vector addition law, net momentum after collision,

$$p_{net} = \sqrt{p_1^2 + p_2^2 + 2p_1p_2 \cos 90^\circ}$$

$$= \sqrt{p_1^2 + p_2^2}$$

Since, $p_1 = \frac{h}{\lambda_1}$ and $p_2 = \frac{h}{\lambda_2}$

So, $$p_{net} = \sqrt{\frac{h^2}{\lambda_1^2} + \frac{h^2}{\lambda_2^2}} \quad \ldots(i)$$

Let the de-Broglie wavelength after the collision is λ_{net}, then

$$p_{net} = \frac{h}{\lambda_{net}} \quad \ldots(ii)$$

From Eqs. (i) and (ii), we get

$$\frac{h}{\lambda_{net}} = \sqrt{\frac{h^2}{\lambda_1^2} + \frac{h^2}{\lambda_2^2}}$$

$$\Rightarrow \frac{1}{\lambda_{net}^2} = \frac{1}{\lambda_1^2} + \frac{1}{\lambda_2^2}$$

$$\Rightarrow \frac{1}{\lambda^2} = \frac{1}{\lambda_1^2} + \frac{1}{\lambda_2^2} \quad [\because \lambda_{net} = \lambda]$$

51. *(a)* For elastic collision,

$$p_{\text{before collision}} = p_{\text{after collision}}.$$

$$mv = mv_A + \frac{m}{2} v_B$$

$$2v = 2v_A + v_B \quad \ldots(i)$$

Now, coefficient of restitution,

$$e = \frac{v_B - v_A}{u_A - u_B}$$

Here, $u_B = 0$ (Particle at rest) and for elastic collision $e = 1$, $u_A = v$, then

$$\therefore \quad 1 = \frac{v_B - v_A}{v}$$

$$\Rightarrow \quad v = v_B - v_A \quad \ldots(ii)$$

From Eqs. (i) and (ii), we get

$$v_A = \frac{v}{3} \text{ and } v_B = \frac{4v}{3}$$

Hence, $$\frac{\lambda_A}{\lambda_B} = \frac{\left(\frac{h}{mv_A}\right)}{\frac{h}{\frac{m}{2}.v_B}}$$

$$= \frac{v_B}{2v_A} = \frac{4/3}{2/3} = 2$$

52. *(2)* Let de-Broglie wavelength is λ and momentum of electron at any time t is p.

$$\therefore \quad p = \frac{h}{\lambda}$$

$$\Rightarrow \quad \frac{dp}{dt} = -\frac{h}{\lambda^2}\frac{d\lambda}{dt}$$

$$\Rightarrow \quad F = -\frac{h}{\lambda^2}\frac{d\lambda}{dt}$$

$$\Rightarrow \quad ma = -\frac{h}{\lambda^2}\frac{d\lambda}{dt}$$

$$\Rightarrow \quad ma = -\frac{h}{\left(\frac{h}{p}\right)^2}\frac{d\lambda}{dt}$$

$$\Rightarrow \quad a = -\frac{p^2}{mh}\frac{d\lambda}{dt}$$

$$\Rightarrow \quad a = -\frac{mv^2}{h}\frac{d\lambda}{dt}$$

$$\Rightarrow \quad \frac{d\lambda}{dt} = \frac{-ah}{mv^2} \quad \ldots(i)$$

Since, $a = \frac{qE}{m} = \frac{e}{m} \cdot \frac{\sigma}{2\varepsilon_0} = \frac{\sigma e}{2m\varepsilon_0}$

Again, $v = u + at \Rightarrow at = \frac{\sigma et}{2m\varepsilon_0} \quad \ldots(ii)$

Now, from Eqs. (i) and (ii), we get

$$\frac{d\lambda}{dt} = \frac{-\frac{\sigma eh}{2m\varepsilon_0}}{m \cdot \left(\frac{\sigma et}{2m\varepsilon_0}\right)^2} = \frac{-2h\varepsilon_0}{\sigma et^2}$$

$$\Rightarrow \frac{d\lambda}{dt} \propto \frac{1}{t^2}, \therefore n = 2$$

53. *(910)* We know that, de-Broglie wavelength $\lambda = h/mv$

$\therefore$For particle,

$$\lambda_{pa} = \frac{h}{mv} = \frac{6.6 \times 10^{-34}}{9.1 \times 10^{-31} \times 10^6} \quad \ldots(i)$$

and for photon,

$$\lambda_{ph} = \frac{h}{p} = \frac{6.6 \times 10^{-34}}{10^{-27}} \quad \ldots(ii)$$

From Eqs. (i) and (ii), we get

$$\Rightarrow \frac{\lambda_{ph}}{\lambda_{pa}} = \frac{\frac{6.6 \times 10^{-34}}{10^{-27}}}{\frac{6.6 \times 10^{-34}}{9.1 \times 10^{-31} \times 10^6}}$$

$$\Rightarrow \frac{\lambda_{ph}}{\lambda_{pa}} = 910 \Rightarrow \lambda_{ph} = 910\,(\lambda_{pa})$$

C-26

JEE Main Chapterwise Topicwise Physics

In JEE Main there is usually 0-1 question per paper is visible from this chapter. In past 3 years, 28 conceptual and 19 numerical questions were asked of moderate to difficult level from this chapter.

For detailed analysis scan QR code

ATOMS

Topic 1 Alpha-Particle Scattering and Rutherford Model of Atom

Level 1

01 Given below are two statements:

Statement I Most of the mass of the atom and all its positive charge are concentrated in a tiny nucleus and the electrons revolve around it, is Rutherford's model.

Statement II An atom is a spherical cloud of positive charges with electrons embedded in it, is a special case of Rutherford's model. *[2024, 29 Jan Shift-II]*

In the light of the above statements, choose the **most appropriate** from the options given below

(a) Statement I is true but Statement II is false
(b) Both Statement I and Statement II are false
(c) Both Statement I and Statement II are true
(d) Statement I is false but Statement II is true

02 Choose the correct option from the following options given below *[2022, 24 June Shift-I]*

(a) In the ground state of Rutherford's model electrons are in stable equilibrium. While in Thomson's model electrons always experience a net-force.
(b) An atom has a nearly continuous mass distribution in a Rutherford's model but has a highly non-uniform mass distribution in Thomson's model.
(c) A classical atom based on Rutherford's model is doomed to collapse.
(d) The positively charged part of the atom possesses most of the mass in Rutherford's model but not in Thomson's model.

03 The graph which depicts the results of Rutherford gold foil experiment with α-particles is

θ : Scattering angle

Y : Number of scattered α-particles detected

(plots are schematic and not to scale) *[2020, 8 Jan Shift-I]*

(a)
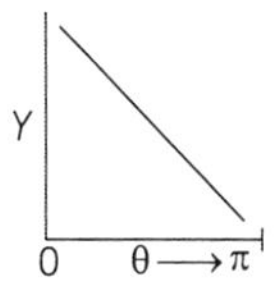

(b)
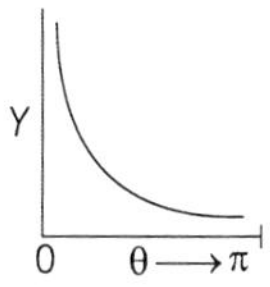

(c)
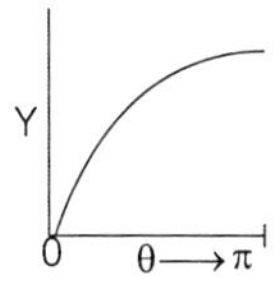

(d)
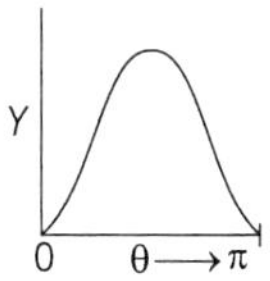

04 An alpha nucleus of energy $(1/2)\ mv^2$ bombards a heavy nuclear target of charge Ze. Then, the distance of closest approach for the alpha nucleus will be proportional to *[AIEEE 2006]*

(a) v^2 (b) $1/m$
(c) $1/v^4$ (d) $1/Ze$

Level 2

05 An α-particle of energy 5 MeV is scattered through $180°$ by a fixed uranium nucleus. The distance of the closest approach is of the order of *[AIEEE 2004]*

(a) 1 Å (b) 10^{-10} cm
(c) 10^{-12} cm (d) 10^{-15} cm

Numerical Value Type Questions

Imp. Impact Parameter

06 $\sqrt{d_1}$ and $\sqrt{d_2}$ are the impact parameters corresponding to scattering angles 60° and 90° respectively, when an α-particle is approaching a gold nucleus. For $d_1 = x\, d_2$, the value of x will be .... *[2022, 29 June Shift-I]*

Topic 2 Bohr's Model and Hydrogen Spectrum

Level 1

01 In a hydrogen like ion, the energy difference between the 2nd excitation energy state and ground is 108.8 eV. The atomic number of the ion is *[2025, 7 April, Shift-I]*
(a) 3 (b) 1 (c) 2 (d) 4

02 Given below are two statements. One is labelled as Assertion (A) and the other is labelled as Reason (R).
Assertion (A) The Bohr model is applicable to hydrogen and hydrogen-like atoms only.
Reason (R) The formulation of Bohr model does not include repulsive force between electrons. *[2025, 3 April, Shift-II]*
In the light of the above statements, choose the correct answer from the options given below.
(a) Both A and R are true and R is not the correct explanation of A.
(b) Both A and R are true but R is the correct explanation of A.
(c) A is false but R is true.
(d) A is true but R is false.

03 The frequency of revolution of the electron in Bohr's orbit varies with n, the principal quantum number is *[2025, 28 Jan Shift-II]*
(a) $\frac{1}{n^3}$ (b) $\frac{1}{n^2}$ (c) $\frac{1}{n}$ (d) $\frac{1}{n^4}$

04 The number of spectral lines emitted by atomic hydrogen that is in the 4th energy level, is *[2025, 29 Jan Shift-II]*
(a) 3 (b) 1 (c) 6 (d) 0

05 A hydrogen atom in ground state is given an energy of 10.2 eV. How many spectral lines will be emitted due to transition of electrons ? *[2024, 9 April Shift-II]*
(a) 1 (b) 6 (c) 10 (d) 3

06 The angular momentum of an electron in a hydrogen atom is proportional to (where r is the radius of orbit of electron) *[2024, 5 April Shift-II]*
(a) r (b) $\frac{1}{r}$ (c) $\frac{1}{\sqrt{r}}$ (d) $\sqrt{r}$

07 According to Bohr's theory, the moment of momentum of an electron revolving in 4th orbit of hydrogen atom is *[2024, 4 April Shift-II]*
(a) $\frac{h}{\pi}$ (b) $2\frac{h}{\pi}$ (c) $8\frac{h}{\pi}$ (d) $\frac{h}{2\pi}$

08 The ratio of the magnitude of the kinetic energy to the potential energy of an electron in the 5th excited state of a hydrogen atom is *[2024, 30 Jan Shift-II]*
(a) 1 (b) 1/4 (c) 1/2 (d) 4

09 The energy of He^+ ion in its first excited state is, (The ground state energy for the hydrogen atom is -13.6 eV) *[2023, 11 April Shift-II]*
(a) -27.2 eV (b) -3.4 eV (c) -54.4 eV (d) -13.6 eV

10 The angular momentum for the electron in Bohr's orbit is L. If the electron is assumed to revolve in second orbit of hydrogen atom, then the change in angular momentum will be *[2023, 10 April Shift-I]*
(a) L (b) $\frac{L}{2}$ (c) Zero (d) $2L$

11 The energy levels of an hydrogen atom are shown below. The transition corresponding to emission of shortest wavelength is *[2023, 6 April Shift-I]*

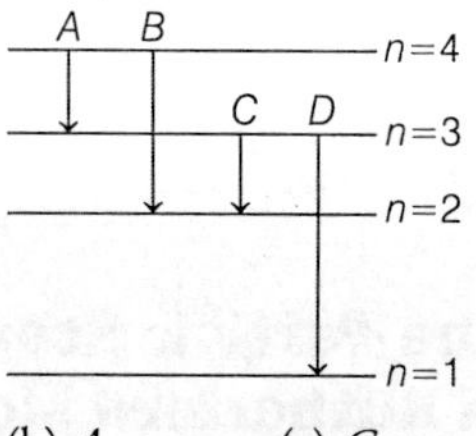

(a) D (b) A (c) C (d) B

12 Speed of an electron in Bohr's 7th orbit for H-atom is 3.6×10^6 m/s. The corresponding speed of the electron in 3rd orbit, (in m/s) is *[2023, 30 Jan Shift-I]*
(a) 3.6×10^6 (b) 1.8×10^6
(c) 8.4×10^6 (d) 7.5×10^6

13 The momentum of an electron revolving in nth orbit is given by (symbols have their usual meanings) *[2022, 25 July Shift-I]*
(a) $\frac{nh}{2\pi r}$ (b) $\frac{nh}{2r}$
(c) $\frac{nh}{2\pi}$ (d) $\frac{2\pi r}{nh}$

14 The magnetic moment of an electron e revolving in an orbit around nucleus with an orbital angular momentum is given by *[2022, 25 July Shift-I]*
(a) $\mu_L = \frac{eL}{2m}$ (b) $\mu_L = -\frac{eL}{2m}$
(c) $\mu_l = -\frac{eL}{m}$ (d) $\mu_l = -\frac{2eL}{m}$

15 Given below are two statements
Statement I In hydrogen atom, the frequency of radiation emitted when an electron jumps from lower energy orbit E_1 to higher energy orbit E_2, is given as $hf = E_1 - E_2$.
Statement II The jumping of electron from higher energy orbit E_2 to lower energy orbit E_1 is associated with frequency of radiation given as $f = (E_2 - E_1)/h$.
This condition is Bohr's frequency condition.
In the light of the above statements, choose the correct options. *[2022, 27 June Shift-II]*
(a) Both Statement I and Statement II are true.
(b) Both Statement I and Statement II are false.
(c) Statement I is true but Statement II is false.
(d) Statement I is false but Statement II is true.

16 A hydrogen atom is in ground state absorbs 10.2 eV of energy. The angular momentum of electron of the hydrogen atom will increase by the value of (given, Planck's constant $= 6.6 \times 10^{-34}$ J-s) *[2022, 27 June Shift-I]*

(a) 2.10×10^{-34} J-s (b) 1.05×10^{-34} J-s
(c) 3.15×10^{-34} J-s (d) 4.2×10^{-34} J-s

17 The ratio for the speed of the electron in the 3rd orbit of He^+ to the speed of the electron in the 3rd orbit of hydrogen atom will be *[2022, 25 June Shift-I]*

(a) 1 : 1 (b) 1 : 2 (c) 4 : 1 (d) 2 : 1

18 The energy required to remove the electron from a singly ionised helium atom is 2.2 times the energy required to remove and electron from helium atom. The total energy required to ionise the helium atom completely is *[JEE Main 2018]*

(a) 34 eV (b) 79 eV (c) 10 eV (d) 109 eV

19 Some energy levels of a molecule are shown in the figure. The ratio of the wavelengths $r = \lambda_1 / \lambda_2$ is given by *[JEE Main 2017]*

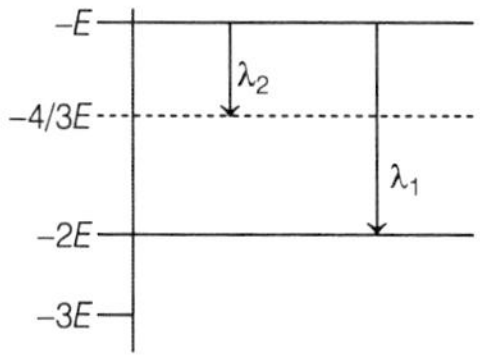

(a) $r = \frac{2}{3}$ (b) $r = \frac{3}{4}$ (c) $r = \frac{1}{3}$ (d) $r = \frac{4}{3}$

20 The acceleration of an electron in the first orbit of hydrogen atom ($n = 1$) is *[JEE Main 2017]*

(a) $\frac{h^2}{\pi^2 m^2 r^3}$ (b) $\frac{h^2}{4\pi^2 m^2 r^3}$ (c) $\frac{h^2}{4\pi m^2 r^3}$ (d) $\frac{h^2}{8\pi^2 m^2 r^3}$

Imp. Hydrogen Spectrum

21 Hydrogen ($_1H^1$), deuterium ($_1H^2$), singly ionised helium $(_2He^4)^+$ and doubly ionised lithium $(_3Li^8)^{++}$ all have one electron around the nucleus. Consider an electron transition form $n = 2$ to $n = 1$. If the wavelengths of emitted radiation are $\lambda_1, \lambda_2, \lambda_3$ and λ_4, respectively for four elements, then approximately which one of the following is correct? *[JEE Main 2014]*

(a) $4\lambda_1 = 2\lambda_2 = 2\lambda_3 - \lambda_4$ (b) $\lambda_1 = 2\lambda_2 = 2\lambda_3 - \lambda_4$
(c) $\lambda_1 = \lambda_2 = 4\lambda_3 = 9\lambda_4$ (d) $\lambda_1 = 2\lambda_2 = 3\lambda_3 = 4\lambda_4$

22 In a hydrogen like atom electron make transition from an energy level with quantum number n to another with quantum number $(n - 1)$. If $n >> 1$, the frequency of radiation emitted is proportional to *[JEE Main 2013]*

(a) $\frac{1}{n}$ (b) $\frac{1}{n^2}$ (c) $\frac{1}{n^3/2}$ (d) $\frac{1}{n^3}$

23 Hydrogen atom is excited from ground state to another state with principal quantum number equal to 4. Then, the number of spectral lines in the emission spectra will be *[JEE Main 2013]*

(a) 2 (b) 3 (c) 5 (d) 6

24 The transition from the state $n = 4$ to $n = 3$ in a hydrogen like atom results in ultraviolet radiation. Infrared radition will be obtained in the transition from *[AIEEE 2009]*

(a) $2 \to 1$ (b) $3 \to 2$ (c) $4 \to 2$ (d) $5 \to 4$

25 Suppose an electron is attracted towards the origin by a force k/r, where k is a constant and r is the distance of the electron from the origin. By applying Bohr model to this system, the radius of the nth orbital of the electron is found to be r_n and the kinetic energy of the electron to be T_n. Then, which of the following is true? *[AIEEE 2008]*

(a) $T_n \propto \frac{1}{n^2}, r_n \propto n^2$ (b) T_n is independent of n, $r_n \propto n$
(c) $T_n \propto \frac{1}{n}, r_n \propto n$ (d) $T_n \propto \frac{1}{n}, r_n \propto n^2$

26 Which of the following transitions in hydrogen atoms emit photons of highest frequency? *[AIEEE 2007]*

(a) $n = 2$ to $n = 6$ (b) $n = 6$ to $n = 2$
(c) $n = 2$ to $n = 1$ (d) $n = 1$ to $n = 2$

Imp. Ionisation Energy of Excited State

27 If the binding energy of the electron in a hydrogen atom is 13.6 eV, the energy required to remove the electron from the first excited state of Li^{2+} is *[AIEEE 2003]*

(a) 30.6 eV (b) 13.6 eV
(c) 3.4 eV (d) 122.4 eV

28 If 13.6 eV energy is required to ionise the hydrogen atom, then the energy required to remove an electron from $n = 2$ is *[AIEEE 2002]*

(a) 10.2 eV (b) zero (c) 3.4 eV (d) 6.8 eV

Level 2

29 For a hydrogen atom, the ratio of the largest wavelength of Lyman series to that of the Balmer series is *[2025, 7 April Shift-I]*

(a) 3 : 4 (b) 5 : 36 (c) 5 : 27 (d) 27 : 5

30 Considering Bohr's atomic model for hydrogen atom

(a) the energy of H atom in ground state is same as energy of He^+ ion in its first excited state.
(b) the energy of H atom in ground state is same as that for Li^{++} ion in its second excited state.
(c) the energy of H atom in its ground state is same as that of He^+ ion for its ground state.

(d) the energy of He^+ ion in its first excited state is same as that for Li^{++} ion in its ground state.

Choose the correct answer from the options given below *[2025, 2 April, Shift-I]*

(a) (A), (D) only (b) (B), (D) only
(c) (A), (C) only (d) (A), (B) only

31 Given below are two statements.

Statement I The dimensions of Planck's constant and angular momentum are same.

Statement II In Bohr's model electron revolve around the nucleus only in those orbits for which angular momentum is integral multiple of Planck's constant. *[2025, 4 April, Shift-II]*

In the light of the above statements, choose the most appropriate answer from the options given below.

(a) Both Statement I and Statement II are correct.
(b) Statement I is incorrect but Statement II is correct.
(c) Both Statement I and Statement II are incorrect.
(d) Statement I is correct but Statement II is incorrect.

32 Considering the Bohr model of hydrogen like atoms, the ratio of the radius of 5th orbit of the electron in Li^{2+} and He^+ is *[2025, 4 April, Shift-I]*

(a) $\frac{9}{4}$ (b) $\frac{4}{9}$ (c) $\frac{2}{3}$ (d) $\frac{3}{2}$

33 During the transition of electron from state A to state C of a Bohr atom, the wavelength of emitted radiation is 2000 Å and it becomes 6000 Å when the electron jumps from state B to state C. Then, the wavelength of the radiation emitted during the transition of electrons from state A to state B is *[2025, 24 Jan Shift-I]*

(a) 3000 Å (b) 6000 Å
(c) 2000 Å (d) 4000 Å

34 The longest wavelength associated with Paschen series is (Given, $R_H = 1.097 \times 10^7$ Si unit) *[2024, 6 April Shift-II]*

(a) 1.094×10^{-6} m (b) 3.646×10^{-6} m
(c) 1.876×10^{-6} m (d) 2.973×10^{-6} m

35 The minimum energy required by a hydrogen atom in ground state to emit radiation in Balmer series is nearly: *[2024, 1 Feb Shift-I]*

(a) 13.6 eV (b) 12.1 eV (c) 1.5 eV (d) 1.9 eV

36 If the wavelength of the first member of Lyman series of hydrogen is λ. The wavelength of the second member will be *[2024, 31 Jan Shift-I]*

(a) $\frac{27}{5}\lambda$ (b) $\frac{5}{27}\lambda$ (c) $\frac{27}{32}\lambda$ (d) $\frac{32}{27}\lambda$

37 Given below are two statements:

Statement I Planck's constant and angular momentum have same dimensions.

Statement II Linear momentum and moment of force have same dimensions. *[2024, 27 Jan Shift-I]*

In the light of the above statements, choose the correct answer from the options given below

(a) Both Statement I and Statement II are false.
(b) Both Statement I and Statement II are true.
(c) Statement I is true but Statement II is false.
(d) Statement I is false but Statement II is true.

38 A 12.5 eV electron beam is used to bombard gaseous hydrogen at room temperature. The number of spectral lines emitted will be *[2023, 12 April Shift-I]*

(a) 1 (b) 3 (c) 2 (d) 4

39 A small particle of mass m moves in such a way that its potential energy $U = \frac{1}{2} m\omega^2 r^2$, where ω is constant and r is the distance of the particle from origin. Assuming Bohr's quantisation of momentum and circular orbit, the radius of nth orbit will be proportional to *[2023, 6 April Shift-II]*

(a) $\frac{1}{n}$ (b) n (c) $\sqrt{n}$ (d) n^2

40 The energy levels of an atom is shown in figure.

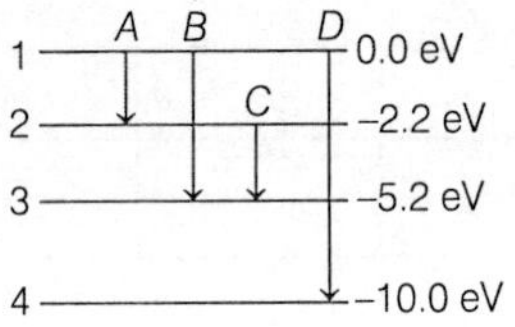

Which one of these transitions will result in the emission of a photon of wavelength 124.1 nm ?
(Given, $h = 6.62 \times 10^{-34}$ J-s) *[2023, 25 Jan Shift-II]*

(a) A (b) C (c) D (d) B

41 Hydrogen atom from excited state comes to the ground state by emitting a photon of wavelength λ. The value of principal quantum number n of the excited state will be (R = Rydberg constant) *[2022, 25 July Shift-II]*

(a) $\sqrt{\frac{\lambda R}{\lambda - 1}}$ (b) $\sqrt{\frac{\lambda R}{\lambda R - 1}}$ (c) $\sqrt{\frac{R}{\lambda R - 1}}$ (d) $\sqrt{\frac{\lambda R^2}{\lambda R - 1}}$

42 In Bohr's atomic model of hydrogen, let K, P and E are the kinetic energy, potential energy and total energy of the electron, respectively. Choose the correct option when the electron undergoes transitions to a higher level. *[2022, 24 June Shift-II]*

(a) All K, P and E increase.
(b) K decreases, P and E increase.
(c) P decreases, K and E increase.
(d) K increases, P and E decrease.

43 A particular hydrogen like ion emits radiation of frequency 2.92×10^{15} Hz when it makes transition from $n = 3$ to $n = 1$. The frequency in Hz of radiation emitted in transition from $n = 2$ to $n = 1$ will be *[2021, 26 Aug Shift-I]*

(a) 0.44×10^{15} (b) 6.57×10^{15}
(c) 4.38×10^{15} (d) 2.46×10^{15}

44 The wavelength of the photon emitted by a hydrogen atom when an electron makes a transition from $n=2$ to $n=1$ state is *[2021, 25 Feb Shift-II]*
(a) 121.8 nm (b) 194.8 nm
(c) 490.7 nm (d) 913.3 nm

45 In the given figure, the energy levels of hydrogen atom have been shown alongwith some transitions marked *A, B, C, D* and *E*. The transitions *A, B* and *C* respectively represent. *[2021, 24 Feb Shift-I]*

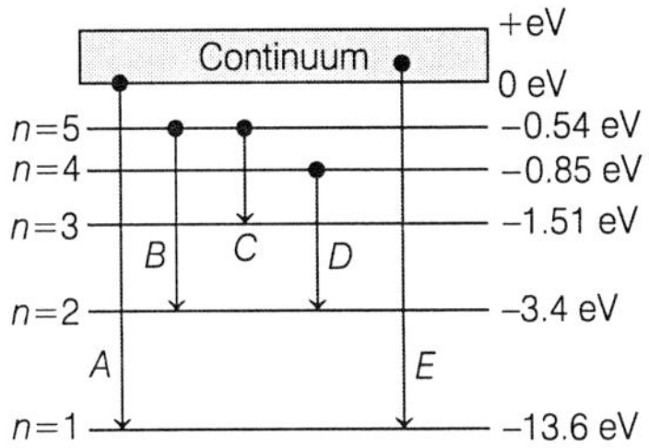

(a) The first member of the Lyman series, third member of Balmer series and second member of Paschen series.
(b) The ionisation potential of hydrogen, second member of Balmer series and third member of Paschen series.
(c) The series limit of Lyman series, second member of Balmer series and second member of Paschen series.
(d) The series limit of Lyman series, third member of Balmer series and second member of Paschen series.

46 The energy required to ionise a hydrogen like ion in its ground state is 9 Rydbergs. What is the wavelength of the radiation emitted when the electron in this ion jumps from the second excited state to the ground state ? *[2020, 9 Jan Shift-II]*
(a) 8.6 nm (b) 24.2 nm
(c) 11.4 nm (d) 35.8 nm

47 The time period of revolution of electron in its ground state orbit in a hydrogen atom is 1.6×10^{-16} s. The frequency of revolution of the electron in its first excited state (in Hz) is *[2020, 7 Jan Shift-I]*
(a) 1.6×10^{14} (b) 5.6×10^{12}
(c) 6.2×10^{15} (d) 7.8×10^{14}

48 The electron in a hydrogen atom first jumps from the third excited state to the second excited state and subsequently to the first excited state. The ratio of the respective wavelengths λ_1/λ_2 of the photons emitted in this process is *[2019, 12 April Shift-II]*
(a) 20/7 (b) 27/5
(c) 7/5 (d) 9/7

49 Consider an electron in a hydrogen atom, revolving in its second excited state (having radius 4.65 Å). The de-Broglie wavelength of this electron is *[2019, 12 April Shift-II]*
(a) 3.5 Å (b) 6.6 Å
(c) 12.9 Å (d) 9.7 Å

50 In a hydrogen like atom, when an electron jumps from the *M*-shell to the *L*-shell, the wavelength of emitted radiation is λ. If an electron jumps from *N*-shell to the *L*-shell, the wavelength of emitted radiation will be *[2019, 11 Jan Shift-II]*
(a) $\frac{27}{20}\lambda$ (b) $\frac{25}{16}\lambda$
(c) $\frac{20}{27}\lambda$ (d) $\frac{16}{25}\lambda$

51 A hydrogen atom, initially in the ground state is excited by absorbing a photon of wavelength 980 Å. The radius of the atom in the excited state in terms of Bohr radius a_0 will be
(Take, $hc=12500$ eV-Å) *[2019, 11 Jan Shift-I]*
(a) $4a_0$ (b) $9a_0$
(c) $16a_0$ (d) $25a_0$

52 If the series limit frequency of the Lyman series is ν_L, then the series limit frequency of the Pfund series is *[JEE Main 2018]*
(a) $25\nu_L$ (b) $16\nu_L$ (c) $\frac{\nu_L}{16}$ (d) $\frac{\nu_L}{25}$

53 The radiation corresponding to $3\to2$ transition of hydrogen atom falls on a metal surface to produce photoelectrons. These electrons are made to enter a magnetic field of 3×10^{-4} T. If the radius of the largest circular path followed by these electrons is 10.0 mm, the work function of the metal is close to *[JEE Main 2014]*
(a) 1.6 eV (b) 1.8 eV
(c) 1.1 eV (d) 0.8 eV

54 A diatomic molecule is made of two masses m_1 and m_2 which are separated by a distance *r*. If we calculate its rotational energy by applying Bohr's rule of angular momentum quantisation, its energy will be given by (*n* is an integer) *[AIEEE 2012]*
(a) $\frac{(m_1+m_2)^2n^2\hbar^2}{2m_1^2m_2^2r^2}$ (b) $\frac{n^2\hbar^2}{2(m_1+m_2)r^2}$
(c) $\frac{2n^2\hbar^2}{(m_1+m_2)r^2}$ (d) $\frac{(m_1+m_2)n^2\hbar^2}{2m_1m_2r^2}$

Level 3

55 An electron projected perpendicular to a uniform magnetic field *B* moves in a circle. If Bohr's quantisation is applicable, then the radius of the electronic orbit in the first excited state is *[2025, 22 Jan Shift-I]*
(a) $\sqrt{\frac{h}{2\pi eB}}$ (b) $\sqrt{\frac{2h}{\pi eB}}$
(c) $\sqrt{\frac{h}{\pi eB}}$ (d) $\sqrt{\frac{4h}{\pi eB}}$

56 The ratio of the shortest wavelength of Balmer series to the shortest wavelength of Lyman series for hydrogen atom is *[2024, 6 April Shift-I]*
(a) 4 : 1 (b) 1 : 4 (c) 2 : 1 (d) 1 : 2

Imp. Energy Released During Transition from Higher Energy State to Lower Energy State

57 An electron of a hydrogen like atom, having $Z = 4$, jumps from 4th energy state to 2nd energy state. The energy released in this process, will be (given, $Rch = 13.6$ eV)
where, R = Rydberg constant, c = Speed of light in vacuum and h = Planck's constant *[2023, 1 Feb Shift-II]*
(a) 13.6 eV (b) 10.5 eV (c) 40.8 eV (d) 3.4 eV

58 Find the ratio of energies of photons produced due to transition of an electron of hydrogen atom from its (i) second permitted energy level to the first level, and (ii) the highest permitted energy level to the first permitted level. *[2022, 29 July Shift-I]*
(a) 3 : 4 (b) 4 : 3 (c) 1 : 4 (d) 4 : 1

59 The radiation corresponding to $3 \to 2$ transition of a hydrogen atom falls on a gold surface to generate photoelectrons. These electrons are passed through a magnetic field of 5×10^{-4} T. Assume that the radius of the largest circular path followed by these electrons is 7 mm, the work-function of the metal is (Take, mass of electron $= 9.1 \times 10^{-31}$ kg) *[2021, 20 July Shift-I]*
(a) 1.36 eV (b) 1.88 eV (c) 0.16 eV (d) 0.82 eV

60 In a hydrogen atom, electron makes a transition from $(n+1)$th level to the nth level. If $n >> 1$, the frequency of radiation emitted is proportional to *[2020, 2 Sep Shift-II]*
(a) $1/n$ (b) $1/n^3$ (c) $1/n^2$ (d) $1/n^4$

61 In Li^{++}, electron in first Bohr orbit is excited to a level by a radiation of wavelength λ. When the ion gets de-excited to the ground state in all possible ways (including intermediate emissions), a total of six spectral lines are observed. What is the value of λ ? [Take, $h = 6.63 \times 10^{-34}$ J-s, $c = 3 \times 10^8$ ms^{-1}] *[2019, 10 April Shift-II]*
(a) 9.4 nm (b) 12.3 nm (c) 10.8 nm (d) 11.4 nm

62 A particle of mass m moves in a circular orbit in a central potential field $U(r) = \frac{1}{2}kr^2$. If Bohr's quantisation conditions are applied, radii of possible orbitals and energy levels vary with quantum number n as *[2019, 12 Jan Shift-I]*
(a) $r_n \propto n$, $E_n \propto n$ (b) $r_n \propto n^2$, $E_n \propto \frac{1}{n^2}$
(c) $r_n \propto \sqrt{n}$, $E_n \propto n$ (d) $r_n \propto \sqrt{n}$, $E_n \propto \frac{1}{n}$

63 An electron from various excited states of hydrogen atom emit radiation to come to the ground state. Let λ_n, λ_g be the de-Broglie wavelength of the electron in the nth state and the ground state, respectively. Let Λ_n be the wavelength of the emitted photon in the transition from the nth state to the ground state. For large n, (A, B are constants) *[JEE Main 2018]*
(a) $\Lambda_n \approx A + \frac{B}{\lambda_n^2}$ (b) $\Lambda_n \approx A + B\lambda_n^2$
(c) $\Lambda_n^2 \approx A + B\lambda_n^2$ (d) $\Lambda_n^2 \approx \lambda$

Numerical Value Type Questions

64 Assuming the validity of Bohr's atomic model for hydrogen like ions the radius of Li^{++} ion in its ground state is given by $\frac{1}{x}a_0$, where x = ...
(where a_0 is the first Bohr's radius) *[2025, 2 April Shift-II]*

65 An electron in the hydrogen atom initially in the fourth excited state makes a transition to nth energy state by emitting a photon of energy 2.86 eV. The integer value of n will be .... *[2025, 3 April Shift-II]*

66 In an alpha particle scattering experiment distance of closest approach for the α particle is 4.5×10^{-14} m. If target nucleus has atomic number 80. Then, maximum velocity of α-particle is .... $\times 10^5$ m/s approximately.
$\left(\frac{1}{4\pi\varepsilon_0} = 9 \times 10^9 \text{ SI unit, mass of } \alpha \text{ particle} = 6.72 \times 10^{-27} \text{ kg}\right)$
[2024, 8 April Shift-I]

67 Radius of a certain orbit of hydrogen atom is 8.48 Å. If energy of electron in this orbit is E/x, then x = ... (Given $a_0 = 0.529$ Å, E = energy of electron in ground state). *[2024, 6 April Shift-I]*

68 A particular hydrogen-like ion emits the radiation of frequency 3×10^{15} Hz when it makes transition from $n = 2$ to $n = 1$. The frequency of radiation emitted in transition from $n = 3$ to $n = 1$ is $\frac{x}{9} \times 10^{15}$ Hz, when x = ... *[2024, 1 Feb Shift-II]*

69 A electron of hydrogen atom on an excited state is having energy $E_n = -0.85$ eV. The maximum number of allowed transitions to lower energy level is .... *[2024, 30 Jan Shift-I]*

70 Hydrogen atom is bombarded with electrons accelerated through a potential difference of V, which causes excitation of hydrogen atoms. If the experiment is being performed at $T = 0$ K, the minimum potential difference needed to observe any

Balmer series lines in the emission spectra will be $\frac{\alpha}{10}$ V, where α = ... *[2024, 29 Jan Shift-II]*

71 When a hydrogen atom going from $n=2$ to $n=1$ emits a photon, its recoil speed is $\frac{x}{5}$ m/s. Where x = ... (Use, mass of hydrogen atom $= 1.6 \times 10^{-27}$ kg) *[2024, 29 Jan Shift-I]*

72 If Rydberg's constant is R, the longest wavelength of radiation in Paschen series will be $\alpha / 7R$, where α = ... *[2024, 27 Jan Shift-II]*

73 As per given figure A, B and C are the first, second and third energy levels of hydrogen atom, respectively. If the ratio of the two wavelengths $\left(\text{i.e.} \frac{\lambda_1}{\lambda_2}\right)$ is $\frac{7}{4n}$, then the value of n will be .... *[2023, 15 April Shift-I]*

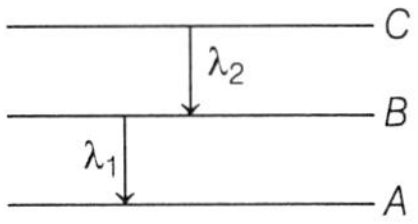

74 The radius of 2nd orbit of He^+ of bohr's model is r_1 and that of fourth orbit of Be^{3+} is represented as r_2. Now, the ratio r_2 / r_1 is $x : 1$. The value of x is .... *[2023, 13 April Shift-I]*

75 If 917 Å be the lowest wavelength of Lyman series then the lowest wavelength of Balmer series will be .... Å. *[2023, 10 April Shift-II]*

76 The ratio of wavelength of spectral lines H_α and H_β in the Balmer series is $\frac{x}{20}$. The value of x is .... *[2023, 10 April Shift-II]*

77 Experimentally it is found that 12.8 eV energy is required to separate a hydrogen atom into a proton and an electron. So, the orbital radius of the electron in a hydrogen atom $\frac{9}{x} \times 10^{-19}$ m. The value of the x is ..... .

$(1 \text{ eV} = 1.6 \times 10^{-19} \text{ J}, \frac{1}{4\pi\varepsilon_0} = 9 \times 10^9 \text{ Nm}^2/\text{C}^2$ and electronic charge $= 1.6 \times 10^{-19}$ C) *[2023, 6 April Shift-II]*

78 The radius of fifth orbit of the Li^{++} is ...... $\times 10^{12}$ m. Take, radius of hydrogen atom = 0.51 Å *[2023, 6 April Shift-I]*

79 A light of energy 12.75 eV is incident on a hydrogen atom in its ground state. The atom absorbs the radiation and reaches to one of its excited states. The angular momentum of the atom in the excited state is $\frac{x}{\pi} \times 10^{-17}$ eV-s. The value of x is ....
(Use $h = 4.14 \times 10^{-15}$ eV-s, $c = 3 \times 10^8 \text{ ms}^{-1}$). *[2023, 1 Feb Shift-I]*

80 If the binding energy of a ground state electron in a hydrogen atom is 13.6 eV , then the energy required to remove the electron from the second excited state of Li^{2+} will be $x \times 10^{-1}$ eV. The value of x is .... *[2023, 31 Jan Shift -II]*

81 For a hydrogen atom, λ_1 and λ_2 are the wavelengths corresponding to the transitions 1 and 2, respectively as shown in figure. The ratio of λ_1 and λ_2 is $x / 32$. The value of x is .... *[2023, 31 Jan Shift-I]*

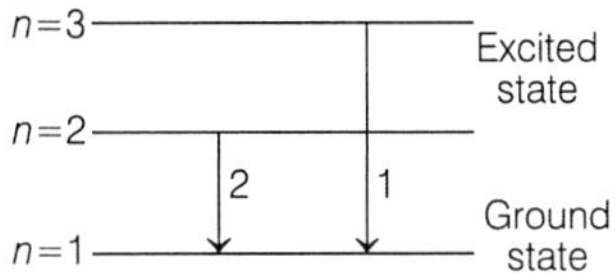

82 The wavelength of the radiation emitted is λ_0 when an electron jumps from the second excited state to the first excited state of hydrogen atom. If the electron jumps from the third excited state to the second orbit of the hydrogen atom the wavelength of the radiation emitted will be $\frac{20}{x}\lambda_0$. The value of x is ...... *[2023, 25 Jan Shift-I]*

83 x different wavelengths may be observed in the spectrum from a hydrogen sample if the atoms are excited to states with principal quantum number $n = 6$? The value of x is .... *[2021, 27 Aug Shift-II]*

84 The K_α X-ray of molybdenum has wavelength 0.071 nm. If the energy of a molybdenum atom with a K electron knocked out is 27.5 keV, the energy of this atom when an L electron is knocked out will be .... keV. (Round off to the nearest integer) [Take, $h = 4.14 \times 10^{-15}$ eV-s, $c = 3 \times 10^8 \text{ ms}^{-1}$] *[2021, 27 July Shift-II]*

85 In Bohr's atomic model, the electron is assumed to revolve in a circular orbit of radius 0.5 Å, If the speed of electron is 2.2×10^6 m / s, then the current associated with the electron will be ......$\times 10^{-2}$ mA $\left[\text{Take, } \pi - \frac{22}{7}\right]$ *[2021, 27 July Shift-I]*

86 The first three spectral lines of H-atom in the Balmer series are given $\lambda_1, \lambda_2, \lambda_3$ considering the Bohr atomic model, the wavelengths of first and third spectral lines $\left(\frac{\lambda_1}{\lambda_3}\right)$ are related by a factor of approximately $x \times 10^{-1}$. The value of x to the nearest integer, is .... *[2021, 16 March Shift-I]*

87 A particle of mass 200 MeV/c^2 collides with a hydrogen atom at rest. Soon after the collision, the particle comes to rest and the atom recoils and goes to its first excited state. The initial kinetic energy of the particle (in eV) is $N/4$. The value of N is .... (Given, the mass of the hydrogen atom to be 1 GeV/c^2) *[2020, 5 Sep Shift-I]*

88 In the line spectra of hydrogen atom, difference between the largest and the shortest wavelengths of the Lyman series is 304 Å. The corresponding difference for the Paschan series (in Å) is .... *[2020, 4 Sep Shift-I]*

89 The first member of the Balmer series of hydrogen atom has a wavelength of 6561 Å. The wavelength of the second member of the Balmer series (in nm) is .... *2020, 8 Jan Shift-II]*

• Key **Index**

Topic 1

1	(a)	2	(c)	3	(b)	4	(b)	5	(c)	6	(3)

Topic 2

1	(a)	2	(b)	3	(a)	4	(c)	5	(a)	6	(d)	7	(b)	8	(c)	9	(d)	10	(a)
11	(a)	12	(c)	13	(a)	14	(b)	15	(d)	16	(b)	17	(d)	18	(b)	19	(c)	20	(b)
21	(c)	22	(d)	23	(d)	24	(d)	25	(b)	26	(c)	27	(a)	28	(c)	29	(c)	30	(d)
31	(d)	32	(c)	33	(a)	34	(c)	35	(b)	36	(c)	37	(c)	38	(b)	39	(c)	40	(c)
41	(b)	42	(b)	43	(d)	44	(a)	45	(d)	46	(c)	47	(d)	48	(a)	49	(d)	50	(c)
51	(c)	52	(d)	53	(c)	54	(d)	55	(c)	56	(a)	57	(c)	58	(a)	59	(d)	60	(b)
61	(c)	62	(c)	63	(a)	64	(3)	65	(2)	66	(156)	67	(16)	68	(32)	69	(6)	70	(102)
71	(17)	72	(144)	73	(9.8)	74	(2)	75	(3668)	76	(27)	77	(16)	78	(425)	79	(828)	80	(136)
81	(27)	82	(35)	83	(15)	84	(10)	85	(112)	86	(15)	87	(51)	88	(10553)	89	(486)		

Detailed **Explanations**

Topic 1

01. *(a)* According to Rutherford model, an atom consists of positively particles concentrated at the centre known as nucleus.

The size of nucleus is very small as compared to the size of atom.

Thus, Statement I is true but Statement II is false.

02. *(c)* According to Rutherford model, electrons revolve around the nucleus in circular orbit i.e. electrons accelerate around the nucleus.

The acceleration would cause the electron to radiate energy in the form of electromagnetic waves, which would imply that the electron would continually lose energy.

Thus, their velocity would decrease, causing it to spiral and eventually fall into the nucleus.

This would imply that atoms are naturally unstable which is not possible because atoms are naturally stable.

Therefore, classical atom based on Rutherford's model is doomed to collapse.

03. *(b)* In Rutherford's experiment, number of particles scattered at large angles is very less and most of the particles are scattered at small angles. Hence, graph of Y = number of α-particles and θ = scattering angle is as shown in the figure below.

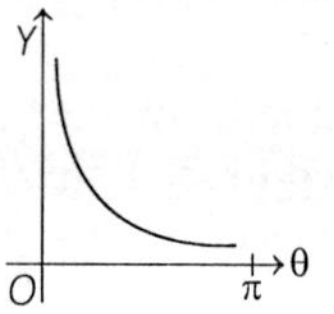

04. *(b)* Let, at distance of closest approach distance be r, then from work-energy theorem,

$$0-\frac{mv^2}{2}=-\frac{1}{4\pi\varepsilon_0}\frac{Ze\times 2e}{r}$$

$$\text{or } r=\frac{1}{4\pi\varepsilon_0}\frac{4Ze^2}{mv^2}$$

So, $$r\propto\frac{1}{m}$$

Hence, option (b) is correct.

05. *(c)* For closest approach,

$\therefore \quad KE = PE,$

$$5 \text{ MeV} = \frac{1}{4\pi\varepsilon_0} \frac{q_1 q_2}{r}$$

$$\text{or } 5 \times 10^6 \times e = \frac{(9 \times 10^9) \times (92e)(2e)}{r}$$

$$\text{or} \quad r = \frac{9 \times 10^9 \times 92 \times 2 \times e}{5 \times 10^6}$$

$$= \frac{9 \times 10^9 \times 92 \times 2 \times (1.6 \times 10^{-19})}{5 \times 10^6}$$

$$r = 5.3 \times 10^{-14} \text{ m} = 5.3 \times 10^{-12} \text{ cm}$$

06. *(3)*

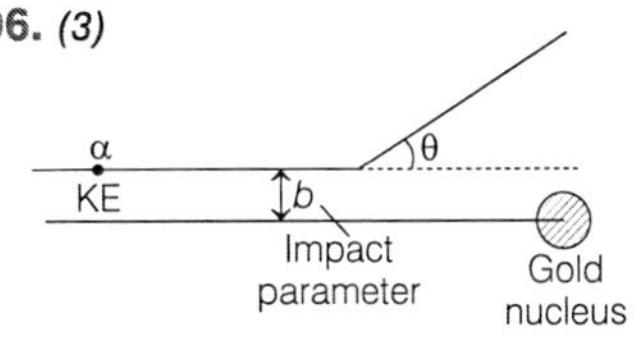

Impact parameter b is given by

$$b = \frac{Ze^2 \cdot \cot(\theta/2)}{4\pi\varepsilon_0 \cdot (KE_\alpha)} \text{ or } b \propto \cot\frac{\theta}{2}$$

With given values, we get

$$\therefore \frac{\sqrt{d_1}}{\sqrt{d_2}} = \frac{\cot(60°/2)}{\cot(90°/2)} = \frac{\cot 30°}{\cot 45°}$$

$$\text{or} \quad \frac{\sqrt{d_1}}{\sqrt{d_2}} = \frac{\sqrt{3}}{1} \Rightarrow d_1 = 3d_2$$

Hence, value of x is 3.

Topic 2

01. *(a)* $\Delta E = 108.8$ eV given

$$\Delta E = 13.6 Z^2 \left(\frac{1}{n_1^2} - \frac{1}{n_2^2} \right)$$

$$= 13.6 Z^2 \left(\frac{1}{1} - \frac{1}{9} \right)$$

$$108.8 = 13.6 Z^2 \left(1 - \frac{1}{9} \right)$$

$$108.8 = \frac{13.6 Z^2}{9} \times 8$$

$$\Rightarrow \quad 13.6 = \frac{13.6 Z^2}{9}$$

$$Z = 3$$

02. *(b)* The formulation of Bohr model does not include repulsive force between electrons. The Bohr model is applicable to hydrogen and hydrogen-like atoms only.

03. *(a)* n = Principal quantum number

$$f_n = \frac{4\pi^2 k^2 e^4 m}{h^3} \times \frac{Z^2}{n^3}$$

$$f_n \propto \frac{Z^2}{n^3}$$

04. *(c)* Total number of spectral line

$$= \frac{(n-1)n}{2}$$

4th energy level means $n = 4$

$$\Rightarrow \text{Total number of lines} = \frac{(4-1)4}{2}$$

$$= 6$$

05. *(a)* Energy given, $E = 10.2$ eV

Energy of the electron

$$= -13.6 + 10.2 \text{ eV} = -3.4 \text{ eV}$$

This corresponds to the energy of the first excited state $n = 2$.

Therefore, only one spectral line i.e., $n_2 \to n_1$ will be emitted.

06. *(d)* Angular momentum,

$$L = \frac{nh}{2\pi} \Rightarrow L \propto n \quad \ldots \text{(i)}$$

Since, $r \propto n^2 \Rightarrow r = Cn^2$

$$\Rightarrow \quad n = \sqrt{\frac{r}{C}} \Rightarrow n \propto \sqrt{r} \quad \ldots \text{(ii)}$$

From Eqs. (i) and (ii), we get

$$L \propto \sqrt{r}$$

07. *(b)* Moment of momentum (or angular momentum) of electron revolving in nth orbit of hydrogen atom,

$$L_n = nh/2\pi$$

For 4th orbit, $n = 4$

$$\therefore \quad L_4 = \frac{4h}{2\pi} = \frac{2h}{\pi}$$

08. *(c)* Relation of kinetic, potential and total energy is given by

$$KE = |TE| = \left| \frac{PE}{2} \right|$$

This relation is in accordance with Bohr model.

Therefore, ratio of kinetic and potential energy is

$$\left| \frac{KE}{PE} \right| = \frac{1}{2}$$

Hence, ratio of magnitude's of KE and PE of an electron is 1/2.

09. *(d)* For He^+

$$E = -13.6 \frac{Z^2}{n^2}$$

$$= -13.6 \times \frac{4}{4} = -13.6 \text{ eV}$$

10. *(a)* Bohr's orbit is orbit of ground state of atom.

So, in first orbit angular momentum

$$L = h/2\pi$$

In second orbit angular momentum

$$L' = 2h/2\pi$$

Hence, change in angular momentum is

$$\Delta L = L' - L = \frac{2h}{2\pi} - \frac{h}{2\pi} = \frac{h}{2\pi} \text{ or } L$$

11. *(a)* $E = hc/\lambda$

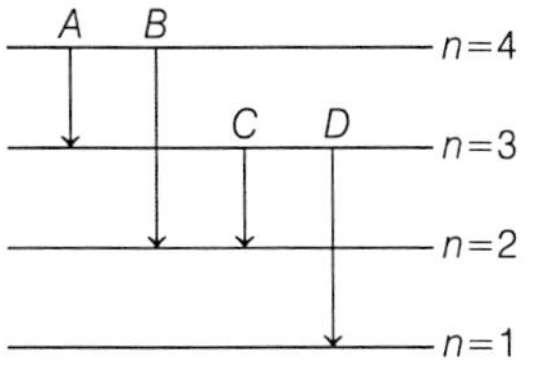

$$\lambda = hc/E$$

For λ_{min}, E must be maximum and E is maximum for (D).

Hence, option (a) is correct.

12. *(c)* Speed of an electron in nth Bohr's orbit is given as, $v_n \propto \frac{1}{n}$

$$\Rightarrow \quad \frac{v_7}{v_3} = \frac{3}{7} \Rightarrow \frac{3.6 \times 10^6}{v_3} = \frac{3}{7}$$

$$\Rightarrow \quad v_3 = \frac{7}{3} \times 3.6 \times 10^6$$

$$\Rightarrow \quad v_3 = 8.4 \times 10^6 \text{ ms}^{-1}$$

13. *(a)* We know that angular momentum

$$L = nh/2\pi$$

As, $L = mvr \Rightarrow mvr = nh/2\pi$

$\Rightarrow mv = nh/2\pi r$

or $\quad p = mv = nh/2\pi r$

14. *(b)* Orbital angular momentum of revolving electron.

$$\mu_L = I \cdot A = \frac{qA}{t}$$

$$\mu_L = \frac{-e.v}{2\pi r} . \pi r^2 = \frac{-evr}{2},$$

$$\mu_L = \frac{-e}{2} . \frac{L}{m} = \frac{-eL}{2m} \quad (\because L = mvr)$$

15. *(d)* When an electron jumps from lower orbit E_1 to higher orbit E_2, then it absorbs energy. If f be the frequency of absorbed photon, then

$$hf = E_2 - E_1 \Rightarrow f = (E_2 - E_1)/h$$

So, only Statement II is correct.

16. *(b)* Change in energy,

$$\Delta E = 10.2 \text{ eV}$$

If transition occurs from ground state i.e, $\quad n = 1$ to $n = 2$, then change in momentum,

$$\Delta L = \frac{2h}{2\pi} - \frac{h}{2\pi} = \frac{h}{2\pi} = \frac{6.6 \times 10^{-34}}{2 \times \pi}$$

$$= 1.05 \times 10^{-34} \text{ J-s}$$

17. *(d)* Speed of electron in nth orbit is given as

$$v = \frac{e}{2\varepsilon_0 h}\left(\frac{Z}{n}\right)$$

[where, Z = atomic number]

$$\Rightarrow \quad v \propto \frac{Z}{n}$$

$$\Rightarrow \frac{v_{He^+}}{v_H} = \frac{Z_{He^+}}{Z_H} \times \frac{n_H}{n_{He^+}}$$

$[\because Z_{He^+} = 2; Z_H = 1]$

$$= \frac{2}{1} \times \frac{3}{3} = \frac{2}{1}$$

$\therefore \; v_{He^+} : v_H = 2 : 1$

18. *(b)* It is known that energy required to remove an electron from singly ionised helium atom = 54.4 eV

Energy required to remove the electron from helium atom = x eV

According to given situation,

54.4 eV = $2.2x \Rightarrow x = 24.73$ eV

Total energy required to ionise helium atom = 54.4 + 24.73 = 79.13 eV

$\simeq$ 79 eV

19. *(c)* We have, $\lambda = hc / \Delta E$

$\therefore$ So, ratio of wavelengths

$$\frac{\lambda_1}{\lambda_2} = \frac{hc / \Delta E_1}{hc / \Delta E_2} = \frac{\Delta E_2}{\Delta E_1} = \frac{\left(\frac{4}{3}E - E\right)}{2E - E} = \frac{1}{3}$$

20. *(b)* For revolving electron in the 1st orbit.

$$\frac{mv^2}{r} = \frac{1}{4\pi\varepsilon_0}\frac{e^2}{r^2} \quad \dots \text{(i)}$$

and $\quad mvr = h / 2\pi \quad \dots$(ii)

Form Eqs. (i) and (ii), we get

$$m^2v^2r^2 = h^2 / 4\pi^2$$

On dividing both sides by r^3, we get

$$\frac{m^2v^2}{r} = \frac{h^2}{4\pi^2 r^3} \Rightarrow \frac{v^2}{r} = \frac{h^2}{4\pi^2 r^3 m^2}$$

This is required acceleration of the electron.

21. *(c)* For hydrogen atom, we get

$$\frac{1}{\lambda} = R Z^2\left(\frac{1}{1^2} - \frac{1}{2^2}\right)$$

$$\Rightarrow \quad \frac{1}{\lambda_1} = R(1)^2\left(\frac{3}{4}\right) \quad \dots \text{(i)}$$

$$\Rightarrow \quad \frac{1}{\lambda_2} = R(1)^2\left(\frac{3}{4}\right) \quad \dots \text{(ii)}$$

$$\Rightarrow \quad \frac{1}{\lambda_3} = R(2)^2\left(\frac{3}{4}\right) \quad \dots \text{(iii)}$$

$$\Rightarrow \quad \frac{1}{\lambda_4} = R(3)^2\left(\frac{3}{4}\right) \quad \dots \text{(iv)}$$

Using Eqs. (i), (ii), (iii) and (iv), we get

$$\lambda_1 = \lambda_2 = 4\lambda_3 = 9\lambda_4$$

22. *(d)* $\therefore \Delta E = h\nu$

$$\Rightarrow \quad \nu = \frac{\Delta E}{h} = k\left[\frac{1}{(n-1)^2} - \frac{1}{n^2}\right]$$

$$= \frac{k(2n-1)}{n^2(n-1)^2} \approx \frac{2k}{n^3} \propto \frac{1}{n^3}$$

23. *(d)* In emission spectrum, number of bright lines is given by

$$\frac{n(n-1)}{2} = \frac{4(4-1)}{2} = 6$$

24. *(d)* IR corresponds to least value of $\left(\frac{1}{n_1^2} - \frac{1}{n_2^2}\right)$

i.e., from Paschen, Brackett and Pfund series. Thus, the transition corresponds to $5 \rightarrow 4$.

25. *(b)* $\frac{mv^2}{r_n} = \frac{k}{r_n}$ [Given]

So, $\frac{1}{2}mv_n^2 = T_n = \frac{k}{2}$

and $\quad mvr_n = nh / 2\pi$

[From Bohr's theory]

On solving, $r_n \propto n$ and T_n is independent of n.

26. *(c)* Emission spectrum would rises when electron makes a jump from higher energy level to lower energy level.

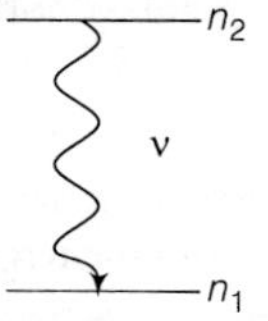

Frequency of emitted photon is proportional to change in energy of two energy levels, i.e.,

$$\nu = RcZ^2\left(\frac{1}{n_1^2} - \frac{1}{n_2^2}\right)$$

Here, energy difference is more in $n = 2$ to $n = 1$. So, option (c) is correct.

27. *(a)* $E = -Z^2 \frac{13.6}{n^2}$ eV

For first excited state, $E_2 = -3^2 \times \frac{13.6}{4}$

[$\because n = 2$, for Ist excited state]

$= -30.6$ eV

Ionisation energy for first excited state of Li^{2+} is 30.6 eV.

28. *(c)* Energy required to remove an electron from nth orbit is

$$E_n = -\frac{13.6}{n^2}$$

Here, $n = 2$

Therefore, $E_2 = -\frac{13.6}{2^2} = -3.4$ V

Energy required $= -E_2 = 3.4$ eV

29. *(c)* For hydrogen atom

Lyman series

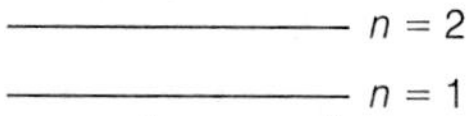

$$\frac{1}{\lambda_1} = R\left[\frac{1}{n_1^2} - \frac{1}{n_2^2}\right]$$

$$\frac{1}{\lambda_1} = R\left[\frac{1}{1} - \frac{1}{(2)^2}\right]$$

$$\frac{1}{\lambda_1} = R\left[1 - \frac{1}{4}\right]$$

$$\frac{1}{\lambda_1} = \frac{3R}{4}$$

$$\lambda_1 = \frac{4}{3R}$$

and Balmer series

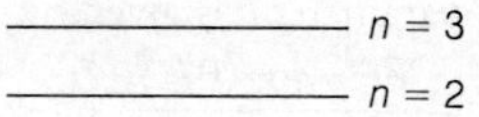

$$\frac{1}{\lambda_2} = R\left[\frac{1}{(2)^2} - \frac{1}{(3)^2}\right] = \frac{5R}{36},$$

$$\lambda_2 = \frac{36}{5R} \Rightarrow \frac{\lambda_1}{\lambda_2} = \frac{5}{27}$$

30. *(d)* According to Bohr's atomic model

$$E \propto \frac{Z^2}{n^2}, \; Z_H = 1$$

$$Z_{He^+} = 2, Z_{Li^{+2}} = 3$$

1st excited state $n = 2$

2nd excited state $n = 3$

(A) and (B) only

31. *(d)* $E = h\nu \quad E \rightarrow$ Energy

$[ML^2T^{-2}] = h[T^{-1}]$

$h \rightarrow$ Planck's constant

$\nu \rightarrow$ Frequency

$$[h] = \frac{[ML^2T^{-2}]}{[T^{-1}]}$$

$$= [ML^2T^{-1}]$$

$$L = MvR$$

$$[L] = [MLT^{-1}L]$$

$$= [ML^2T^{-1}]$$

$$L = \frac{nh}{2\pi}$$

L is integral multiple of $\frac{h}{2\pi}$.

Statement-I is correct.

Statement-II is incorrect.

32. *(c)* $R_n = \frac{n^2}{Z}R_0$

For $Li^{2+} \rightarrow Z = 3$

$$R_5 = R_0 \frac{n^2}{Z} = R_0 \frac{(5)^2}{3} = R_0 \frac{25}{3}$$

For $He^+ \rightarrow Z = 2$

$$R_5 = R_0 \frac{n^2}{Z} = R_0 \frac{25}{2}$$

$$\therefore \quad \frac{R_{Li^{2+}}}{R_{H^+}} = \frac{R_0 \times \frac{25}{3}}{R_0 \times \frac{25}{2}}$$

$$= \frac{25}{3} \times \frac{2}{25} = \frac{2}{3}$$

33. *(a)* The energy transition from A to C is given by

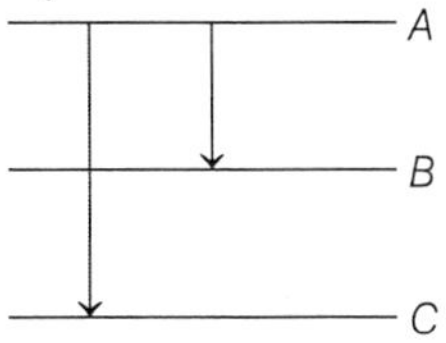

$$E_A - E_C = \frac{hc}{\lambda}$$

where, $\lambda = 2000$ Å ... (i)

Similarly, when transition take place from A to B

$$E_A - E_B = [E_A - E_C) - (E_B - E_C)$$

$$\frac{hc}{\lambda_{AB}} = \frac{hc}{2000 \text{Å}} - \frac{hc}{6000 \text{ Å}}$$

$$\therefore \lambda_{AB} = 3000 \text{ Å}$$

Alternate Solution

The relationship between the wavelengths for step-by-step jump of electron and direct jump of electron is given by

$$\frac{1}{\lambda} = \frac{1}{\lambda_1} + \frac{1}{\lambda_2} + \frac{1}{\lambda_3} + \ldots\ldots$$

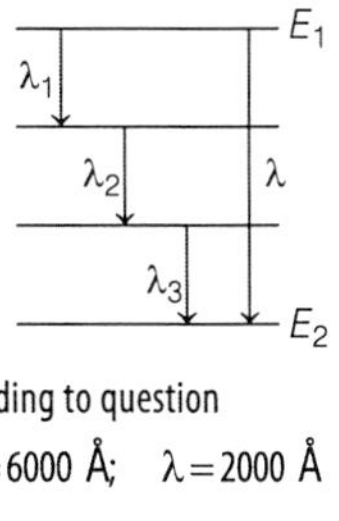

So, according to question

$\lambda_1 = 6000$ Å; $\lambda = 2000$ Å

$\lambda_2 = ?$

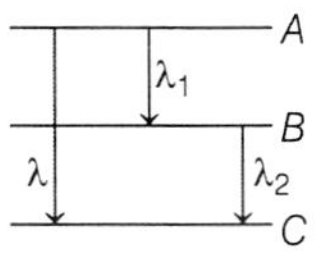

Using the formula

$$\frac{1}{\lambda} = \frac{1}{\lambda_1} + \frac{1}{\lambda_2}$$

$$\frac{1}{2000} = \frac{1}{6000} + \frac{1}{\lambda_2} \Rightarrow \lambda_2 = 3000 \text{Å}$$

34. *(c)* Wavelength of spectral lines in hydrogen atom is given by,

$$\frac{1}{\lambda} = R\left(\frac{1}{n_1^2} - \frac{1}{n_2^2}\right)$$

For longest wavelength of Paschen series

$n_1 = 3, n_2 = 4$

$$\therefore \quad \frac{1}{\lambda} = R\left(\frac{1}{3^2} - \frac{1}{4^2}\right)$$

$$\lambda = \frac{144}{7R} = \frac{144}{7 \times 1.097 \times 10^7}$$

$$= 1.876 \times 10^{-6} \text{ m}$$

35. *(b)* First transition of Balmer series is from $n = 3$ to $n = 2$ state.

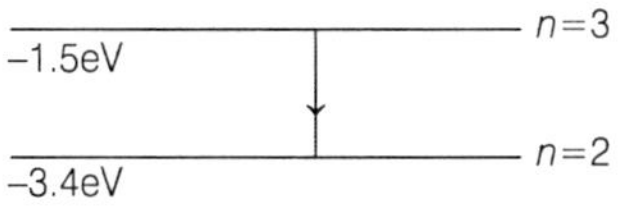

If H atom electron is in ground state, it can reach $n = 3$ state when it absorbs an energy,

$$\Delta E = E_3 - E_1 = -1.51 - (-13.6)$$
$$= 13.6 - 1.51$$
$$= 12.09 \text{ eV} \approx 12.1 \text{ eV}$$

36. *(c)* For Lyman series,

$$\frac{1}{\lambda} = R\left(\frac{1}{1^2} - \frac{1}{n^2}\right)$$

For first member, transition is from $n = 2$ to $n = 1$.

$$\therefore \quad \frac{1}{\lambda_1} = R\left(1 - \frac{1}{2^2}\right) = \frac{3R}{4}$$

For second member, transition is from $n = 3$ to $n = 1$.

$$\therefore \quad \frac{1}{\lambda_2} = R\left(1 - \frac{1}{9}\right) = \frac{8R}{9}$$

$$\text{So, } \frac{\frac{1}{\lambda_1}}{\frac{1}{\lambda_2}} = \frac{\lambda_2}{\lambda} = \frac{3R/4}{8R/9} \Rightarrow \lambda_2 = \frac{27}{32}\lambda$$

37. *(c)* Angular momentum of an electron in Bohr's atom is,

$$L = \frac{nh}{2\pi}$$

As n = a whole number, 2π = constant

$\therefore$ $[L] = [h]$

Statement I is true.

Linear momentum

$$p = mv \Rightarrow [p] = [\text{MLT}^{-1}]$$

Moment of force,

$$\tau = F \cdot d \Rightarrow [\tau] = [\text{ML}^2\text{T}^{-2}]$$

$\therefore$ $[p] \neq [\tau]$

Statement II is false.

38. *(b)* Given, energy of the electron beam $= 12.5$ eV

According to Bohr's postulate, an electron makes jump to higher energy orbital if it absorbs a photon of energy equal to different between the energies of an excited state and the ground state. Assuming that collided electron takes energy equal to 10.2 eV or 12.09 eV from incoming electron beam. The maximum excited state is $n = 3$. So number of spectral lines is $\frac{3(3-1)}{2} = 3$.

39. *(c)* $F = -\frac{dU}{dr} = -m\omega^2 r$

$$|F| = \frac{mv^2}{r} = m\omega^2 r$$

$$v^2 = \omega^2 r^2 \Rightarrow v = \omega r$$

Bohr's postulate,

$$mvr = \frac{nh}{2\pi} \Rightarrow m\omega r^2 = \frac{nh}{2\pi}$$

$$r \propto \sqrt{n}$$

40. *(c)*

> **key idea** If an atom jumps from higher energy state to lower energy state and emits photon of wavelength λ. Then, you can calculate energy of photon during transition using the following formula.
>
> i.e. $E = \frac{hc}{\lambda}$

Given, wavelength of photon,

$\lambda = 124.1$ nm $= 1.24 \times 10^{-7}$ m

$h = 6.62 \times 10^{-34}$ J-s

$\therefore$ Energy of photon,

$$E = \frac{hc}{\lambda} = \frac{6.62 \times 10^{-34} \times 3 \times 10^8}{1.24 \times 10^{-7}}$$

$$= 1.6 \times 10^{-18} \text{ J}$$

$$= \frac{1.6 \times 10^{-18}}{1.6 \times 10^{-19}} \text{ eV} = 10 \text{ eV}$$

Energy of photon in transition D,

$$E' = 0 - (-10) = 10 \text{ eV}$$

41. *(b)* Wavelength emitted by photon,

$$\frac{1}{\lambda} = R\left(\frac{1}{n_1^2} - \frac{1}{n_2^2}\right), n_1 = 1, n_2 = n$$

$$\frac{1}{\lambda} = R\left[\frac{1}{1^2} - \frac{1}{n^2}\right]; \frac{1}{\lambda R} = 1 - \frac{1}{n^2} \ldots \text{(i)}$$

$$\Rightarrow \frac{1}{n^2} = 1 - \frac{1}{\lambda R} = \frac{\lambda R - 1}{\lambda R};$$

$$n^2 = \frac{\lambda R}{\lambda R - 1} \Rightarrow n = \sqrt{\frac{\lambda R}{\lambda R - 1}}$$

42. *(b)* In Bohr's hydrogen atom

$$K = \frac{13.6}{n^2}, P = \frac{-27.2}{n^2}$$

$$E = \frac{-13.6}{n^2}$$

When electron undergoes transitions to a higher energy level, then n increases, thus we see that K decreases, P and E increase.

• Concept Enhancer

Total energy of revolving electron in a stationary orbit of hydrogen atom is negative which means the electron is bound to the nucleus and not free to leave it.

43. *(d)* The frequency of the emitted radiation

$$f = RZ^2c\left(\frac{1}{n_2} - \frac{1}{n_1}\right)$$

For H-atom, $Z = 1$

So, for transition from $n = 3$ to $n = 1$,

$$f_1 = Rc\left(1 - \frac{1}{3^2}\right) \quad \ldots(i)$$

Similarly for transition from $n = 2$ to $n = 1$,

$$f_2 = Rc\left(1 - \frac{1}{2^2}\right) \quad \ldots(ii)$$

On dividing Eq. (i) by Eq. (ii), we get

$$\frac{f_1}{f_2} = \frac{Rc\left(1 - \frac{1}{9}\right)}{Rc\left(1 - \frac{1}{4}\right)} = \frac{\frac{8}{9}}{\frac{3}{4}} = \frac{32}{27}$$

Here, $f_1 = 2.92 \times 10^{15}$ Hz

$$\Rightarrow \frac{2.92 \times 10^{15}}{f_2} = \frac{32}{27}$$

$$f_2 = \frac{27 \times 2.92 \times 10^{15}}{32} = 2.46 \times 10^{15} \text{ Hz}$$

44. *(a)* Given, electron makes a transition from $n = 2$ to $n = 1$. From Bohr's hydrogen spectrum,

$$\frac{1}{\lambda} = R\left[\frac{1}{n_f^2} - \frac{1}{n_i^2}\right]$$

$$\Rightarrow \frac{1}{\lambda} = 1.097 \times 10^7\left[\frac{1}{1^2} - \frac{1}{2^2}\right]$$

$$= 1.097 \times 10^7\left[\frac{1}{1} - \frac{1}{4}\right] = \frac{3}{4} \times 1.097 \times 10^7$$

$$\lambda = \frac{4}{3 \times 1.097 \times 10^7} = 1.215 \times 10^{-7} \text{m}$$

$$= 121.5 \times 10^{-9} \text{ m} \simeq 121.8 \text{ nm}$$

45. *(d)* In transition A, hydrogen atom comes from higher energy state $n = \infty$ to lower energy state $n = 1$. Hence, transition A represents series limit of Lyman series. In transition B, hydrogen atom comes from higher energy state $n = 5$ to lower energy state $n = 2$. Hence, transition B represents 3rd line of Balmer series.

In transition C, hydrogen atom comes from higher energy state $n = 5$ to lower energy state $n = 3$. Hence, transition C represents 2nd line of Paschen series. Hence, option (d) is correct.

46. *(c)* As, ionisation energy of ground state = – (ground state energy)

$$\Rightarrow \quad 9 \times 13.6 = -E_1$$

or $\quad E_1 = -9 \times 13.6$ eV

and energy of nth state for the given atom will be

$$E_n = \frac{E_1}{n^2} = \frac{-9 \times 13.6}{n^2} \text{ eV}$$

In transition from second excited state $(n = 3)$ to ground state $(n = 1)$, energy released will be

$$\Delta E = E(n = 3) - E(n = 1)$$

$$= \left(\frac{-9 \times 13.6}{3^2}\right) - \left(\frac{-9 \times 13.6}{1^2}\right)$$

or $\quad \Delta E = 9 \times 13.6\left(1 - \frac{1}{9}\right)$

$$= 8 \times 13.6 \text{ eV}$$

Thus,

$$\lambda = \frac{hc}{\Delta E} = \frac{1240 \text{ (eV - nm)}}{\Delta E \text{ (eV)}}$$

$$\Rightarrow \quad \lambda = \frac{1240}{8 \times 13.6} = 11.39 \text{ nm} \approx 11.4 \text{ nm}$$

47. *(d)* Time period of revolution of an electron in its nth state (or nth orbit) of H-atom,

$$T_n = \frac{2\pi r_n}{v_n} = \frac{2\pi r_1 \cdot n^2}{(v_1 / n)}$$

where, r_1 and v_1 are radius and velocity of electron in ground state.

So, $\quad T_n = \frac{2\pi r_1}{v_1} \cdot n^3$ or $T_n = T_1 \cdot n^3$

where, T_1 = time period of revolution in ground state.

Now, in first excited state $n = 2$, so time period,

$$T_2 = T_1 \times 2^3 = 8 \times 1.6 \times 10^{-16} \text{ s}$$

$$\therefore \; f_2 = \frac{1}{T_2} = \frac{1}{8 \times 1.6 \times 10^{-16}}$$

$$= 7.8 \times 10^{14} \text{ Hz}$$

48. *(a)*

key idea When an electron transits from an initial energy state n_i to final energy state n_f, then you can calculate wavelength of emitted photon with the help of the following formula.

$$\frac{1}{\lambda} = R\left(\frac{1}{n_i^2} - \frac{1}{n_f^2}\right)$$

Wavelength λ of emitted photon

$$\frac{1}{\lambda} = R\left(\frac{1}{n_f^2} - \frac{1}{n_i^2}\right)$$

In transition from $n = 4$ to $n = 3$,

$$\therefore \frac{1}{\lambda_1} = R\left(\frac{1}{3^2} - \frac{1}{4^2}\right) = R\left(\frac{7}{9 \times 16}\right) \ldots(i)$$

In transition from $n = 3$ to $n = 2$, we have

$$\frac{1}{\lambda_2} = R\left(\frac{1}{2^2} - \frac{1}{3^2}\right) = R\left(\frac{5}{9 \times 4}\right) \ldots(ii)$$

So, from Eqs. (i) and (ii), the ratio of $\frac{\lambda_1}{\lambda_2}$ is

$$\frac{\lambda_1}{\lambda_2} = \frac{\left(\frac{9 \times 16}{7R}\right)}{\left(\frac{9 \times 4}{5R}\right)} = \frac{20}{7}$$

49. *(d)* By Bohr's IInd postulate, for revolving electron,

Angular momentum $= \frac{nh}{2\pi} \Rightarrow mvr_n = \frac{nh}{2\pi}$

$\Rightarrow$ Momentum of electron,

$$p = mv = \frac{nh}{2\pi r_n}$$

de-Broglie wavelength associated with electron is

$$\lambda_n = h / p = 2\pi r_n / n$$

Given, $\quad n = 3, r_n = 4.65$ Å

$$\therefore \quad \lambda_n = \frac{(2 \times \pi \times 4.65)}{3} \approx 9.7 \text{ Å}$$

50. *(c)* For hydrogen or hydrogen like atoms, we know that

$$\frac{1}{\lambda} = RZ^2\left[\frac{1}{n_1^2} - \frac{1}{n_2^2}\right] \quad \ldots(i)$$

When electron jumps from M-shell to the L - shell, then

$n_1 = 2$ (for L - shell)

$n_2 = 3$ (for M - shell)

$\therefore$ Eq (i) becomes

$$\frac{1}{\lambda} = RZ^2\left[\frac{1}{2^2} - \frac{1}{3^2}\right] = \frac{5}{36} RZ^2] \quad \ldots(ii)$$

Now, electron jumps from N-shell to the L - shell, for this

$n_1 = 2$ (for L - shell)

$n_2 = 4$ (for N - shell)

$\therefore$ Eq (i) becomes

$$\frac{1}{\lambda'} = RZ^2\left[\frac{1}{2^2} - \frac{1}{4^2}\right]$$

$$= \frac{3}{16} RZ^2 \quad \text{...(iii)}$$

Now, we divide Eq (ii) by Eq (iii)

$$\frac{\lambda'}{\lambda} = \left(\frac{5}{36} RZ^2\right) \div \left(\frac{3}{16} RZ^2\right) = \frac{20}{27}$$

or $\lambda' = \frac{20}{27}\lambda$

51. *(c)*

> **key idea** Energy associated with an electron in n th Bohr's orbit is directly calculated by
> $$E_n = \frac{-13.6}{n^2} \text{ eV}$$

We know that $\Delta E = hc/\lambda$.

Here, $hc = 12500$ eV Å and $\lambda = 980$ Å

$\Rightarrow E_n - E_1 = 12.76$ eV

Since, the energy associated with an electron in nth Bohr's orbit is given as,

$$E_n = \frac{-13.6}{n^2} \text{ eV} \quad \text{...(i)}$$

$$\Rightarrow \quad E_n = E_1 + 12.76 \text{ eV} \quad \text{...(ii)}$$

$$= \frac{-13.6}{(1)^2} + 12.76 = -0.84$$

On putting this value in Eq. (i), we get

$$\Rightarrow \quad n^2 = \frac{-13.6}{-0.84} = 16$$

$$\Rightarrow \quad n = 4$$

and radius of nth orbit, $r_n = n^2 a_0$

$$\Rightarrow \quad r_n = 16\, a_0$$

52. *(d)* For Lyman series,

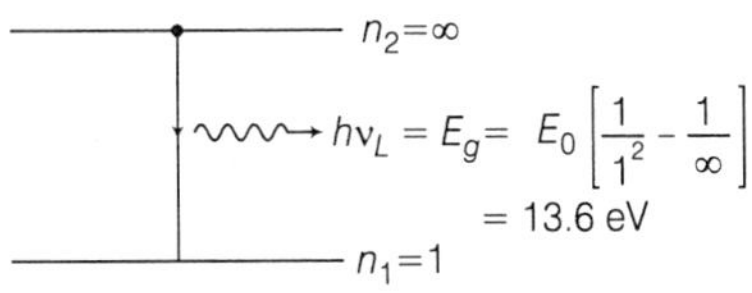

$$h\nu_L = 13.6 \quad \text{...(i)}$$

In Pfund series

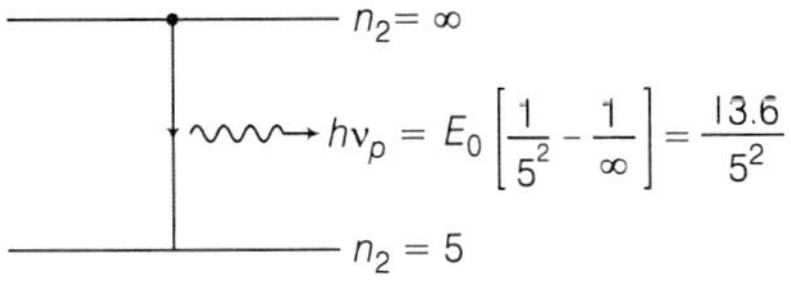

$$h\nu_p = 13.6/5^2 \quad \text{...(ii)}$$

From Eqs. (i) and (ii), we get

$$25h\nu_p = h\nu_L$$

$$\Rightarrow \quad \nu_p = \frac{\nu_L}{25}$$

53. *(c)* We know that radius of a charged particle moving in a uniform magnetic field, $R = \frac{mv}{qB}$

$$\text{or } R^2 = \frac{m^2v^2}{q^2B^2} = \frac{2m\left(\frac{1}{2}mv^2\right)}{q^2B^2} = \frac{2m(\text{KE})}{q^2B^2}$$

$$\Rightarrow \text{KE} = \frac{q^2B^2R^2}{2m}$$

$$\text{KE}_{\text{max}} = \frac{q^2B^2R^2_{\text{max}}}{2m} = 0.80 \text{ eV}$$

Energy of photon corresponding transition from orbit $3 \to 2$ in hydrogen atom.

$$E = 13.6\left(\frac{1}{2^2} - \frac{1}{3^2}\right) = 1.89 \text{ eV}$$

Using Einstein photoelectric equation,

$$E = \text{KE}_{\text{max}} + \phi$$

$$\Rightarrow \quad 1.89 = 0.8 + \phi \Rightarrow \phi = 1.09 \approx 1.1 \text{ eV}$$

54. *(d)* Rotational kinetic energy of the two body system rotating about their centre of mass is

$$\text{RKE} = \frac{1}{2}\mu\omega^2 r^2$$

where, $\mu = \frac{m_1 m_2}{m_1 + m_2}$ = reduced mass

and angular momentum,

$$L = \mu\omega r^2 = nh/2\pi$$

$$\therefore \text{RKE} = \frac{1}{2}\mu\omega^2 r^2 = \frac{1}{2}\mu \cdot \left(\frac{nh}{2\pi\mu r^2}\right)^2 r^2$$

$$= \frac{n^2h^2}{8\pi^2\mu r^2} = \frac{n^2\hbar^2}{2\mu r^2} = \frac{(m_1 + m_2)n^2\hbar^2}{2m_1 m_2 r^2}$$

55. *(c)* Radius, $R = \frac{mv}{qB}$ ...(i)

and $\quad mvR = \frac{nh}{2\pi}$ ...(ii)

Put the value of mv from Eq.(i) in Eq. (ii)

$$mv = ReB$$

$$\Rightarrow ReB(R) = \frac{nh}{2\pi}$$

$$R^2 eB = \frac{nh}{2\pi}$$

$$\Rightarrow \quad R^2 = \frac{nh}{2\pi eB}$$

$$R = \sqrt{\frac{nh}{2\pi eB}}$$

For first excited state $n = 2$

$$\therefore \quad R = \sqrt{\frac{2h}{2\pi eB}} = \sqrt{\frac{h}{\pi eB}}$$

56. *(a)* We know that wavelength of spectral lines in H-atom is given by,

$$\frac{1}{\lambda} = R\left(\frac{1}{n_1^2} - \frac{1}{n_2^2}\right)$$

For shortest wavelength of Balmer series,

$$n_1 = 2 \text{ and } n_2 = \infty$$

$$\frac{1}{\lambda_B} = R\left(\frac{1}{2^2} - \frac{1}{\infty}\right) = \frac{R}{4}$$

$$\Rightarrow \quad \lambda_B = \frac{4}{R} \quad \text{...(i)}$$

For shortest wavelength of Lyman series, $n_1 = 1$ and $n_2 = \infty$

$$\therefore \quad \frac{1}{\lambda_L} = R\left(\frac{1}{1^2} - \frac{1}{\infty}\right) = R$$

$$\therefore \quad \lambda_L = \frac{1}{R} \quad \text{...(ii)}$$

From Eqs. (i) and (ii), we get

$$\frac{\lambda_B}{\lambda_L} = \frac{4/R}{1/R} = 4$$

$$\therefore \quad \lambda_B : \lambda_L = 4:1$$

57. *(c)* The energy of an electron in nth orbit is given as $E_n = -Rhc\frac{Z^2}{n^2}$

Electron is making an transition from 4th energy state to 2nd energy state, then energy released

$$\Delta E = E_4 - E_2$$

$$= -Rhc\frac{Z^2}{4^2} - \left(-Rhc\frac{Z^2}{2^2}\right)$$

$$\Rightarrow \quad \Delta E = Rhc\left(\frac{1}{2^2} - \frac{1}{4^2}\right)Z^2$$

For hydrogen like atom, $Z = 4$

$$\Delta E = 13.6\left(\frac{1}{4} - \frac{1}{16}\right) \times (4)^2$$

$$= 13.6 \times \left(\frac{3}{16}\right) \times (4)^2$$

$$\Rightarrow \quad \Delta E = 40.8 \text{ eV}$$

> **Alternate Solution**
>
> When an electron of a hydrogen like atom, jumps from the energy state n_2 to energy state n_1, then released energy,
>
> $$\Delta E = E_2 - E_1 = \frac{-13.6Z^2}{n_2^2} - \left[\frac{-13.6Z^2}{n_1^2}\right]$$
>
> $$= 13.6Z^2\left[\frac{1}{n_1^2} - \frac{1}{n_2^2}\right] = 13.6 \times 4^2\left[\frac{1}{2^2} - \frac{1}{4^2}\right]$$
>
> [Here, $Z = 4, n_1 = 2, n_2 = 4$]
>
> $= 40.8$ eV

58. *(a)* We know that the energy of an electron in the nth orbit

$$E_n = \frac{-13.6}{n^2}\ \text{eV}$$

(i) Energy of photon produced in transition from $n=2$ to $n=1$ energy level,

$$E_2 - E_1 = \left[\frac{-13.6}{4} - \left(\frac{-13.6}{1}\right)\right]\text{eV}$$

$$= \frac{13.6}{4} \times 3\ \text{eV}$$

$$\Rightarrow E_2 - E_1 = \frac{3}{4} \times 13.6\ \text{eV} \quad \ldots \text{(i)}$$

(ii) Energy of photon produced in transition from

$$n = \infty \text{ to } n = 1$$

($\because$ Highest energy level is $n = \infty$)

$$\Rightarrow E_\infty - E_1 = \left[\frac{-13.6}{\infty} - \left(\frac{-13.6}{1}\right)\right]\text{eV}$$

$$= 13.6\ \text{eV} \quad \ldots \text{(ii)}$$

$\therefore$ From Eqs. (i) and (ii),

$$\frac{E_2 - E_1}{E_\infty - E_1} = \frac{\frac{3}{4} \times 13.6\ \text{eV}}{13.6\ \text{eV}} = \frac{3}{4} \text{ or } 3:4$$

59. *(d)* Energy of photon can be given as

$$E_p = 13.6\left[\frac{1}{n_1^2} - \frac{1}{n_2^2}\right]\text{eV}$$

where, n_1 = lower energy level and n_2 = higher energy level.

As per question, $n_1 = 2$, $n_2 = 3$

$$\therefore\ E_p = 13.6\left[\frac{1}{(2)^2} - \frac{1}{(3)^2}\right]$$

$$= 13.6\left[\frac{1}{4} - \frac{1}{9}\right] = 13.6\left[\frac{9-4}{36}\right]$$

$$= 1.89\ \text{eV}$$

For gold plate,

$$\phi = E_P - \text{KE}_{\text{max}} \quad \ldots \text{(i)}$$

[Given, $B = 5 \times 10^{-4}$ T, $r = 7$ mm $= 7 \times 10^{-3}$ m, $q = 1.6 \times 10^{-19}$ C and $m = 9.1 \times 10^{-31}$ kg]

Therefore, velocity of photoelectrons will be

$$v = \frac{Bqr}{m}$$

$$= \frac{5 \times 10^{-4} \times 1.6 \times 10^{-19} \times 7 \times 10^{-3}}{9.1 \times 10^{-31}}$$

$$= 6.15 \times 10^5\ \text{ms}^{-1}$$

Kinetic energy will be

$$\therefore\ \text{KE}_{\text{max}} = \frac{1}{2} mv^2$$

$$= \frac{1 \times 9.1 \times 10^{-31} \times (6.15 \times 10^5)^2}{2 \times 1.6 \times 10^{-19}}\ \text{eV}$$

$$= 1.075\ \text{eV}$$

From Eq. (i),

$$\phi = (1.89 - 1.075)\ \text{eV} = 0.82\ \text{eV}$$

60. *(b)* Frequency of emitted radiation during transition from $(n+1)$th level to the nth level in hydrogen atom is

$$f = \Delta E / h \qquad \{\because \Delta E = hf\}$$

$$= \frac{1}{h}[E_{n+1} - E_n]$$

$$= \frac{1}{h}\left[\frac{-13.6}{(n+1)^2} - \left(\frac{-13.6}{n^2}\right)\right]$$

$$= \frac{1}{h} \times 13.6 \times \left(\frac{1}{n^2} - \frac{1}{(n+1)^2}\right)$$

$$= \frac{13.6}{h}\left\{\frac{2n+1}{n^2(n+1)^2}\right\}$$

$$\Rightarrow f = \frac{13.6}{h}\left\{\frac{n\left(2 + \frac{1}{n}\right)}{n^4\left(1 + \frac{1}{n}\right)^2}\right\} \quad \ldots \text{(i)}$$

If $n >> 1$, $\frac{1}{n}$ and $\frac{1}{n^2} \approx 0$

So, neglecting $\frac{1}{n}$ and $\frac{1}{n^2}$ in Eq (i), we get

$$f = \frac{13.6}{h} \times \left(\frac{2n}{n^4}\right) = \frac{2 \times 13.6}{h} \times \frac{1}{n^3}$$

$$\Rightarrow f \propto \frac{1}{n^3}$$

61. *(c)*

> **key idea** You can directly calculate the wavelength of incident photon by using the following formula.
>
> $$\lambda = \frac{hc}{\Delta E}$$
>
> where, ΔE is energy absorbed by electron from the incident photon.

Number of spectral lines produced as an excited electron falls to ground state $(n=1)$ is,

$$N = \frac{n(n-1)}{2}$$

$$\therefore\ 6 = \frac{n(n-1)}{2} \qquad [\because N = 6]$$

$$\Rightarrow n = 4$$

So, Li^{++} electron is in it's 3rd excited state.

Now, using the expression of energy of an electron in nth energy level,

$$E_n = -\frac{13.6 Z^2}{n^2}\ \text{eV}$$

where, Z is the atomic number.

$\therefore$ Energy levels of Li^{++} electron are as shown

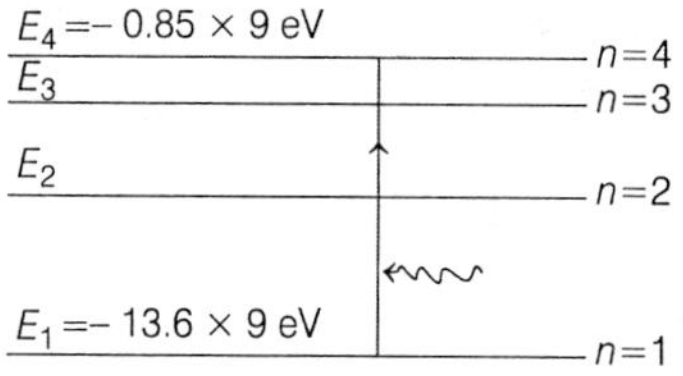

So, energy absorbed by electron from incident photon of wavelength λ is

$$\Delta E = \frac{hc}{\lambda}$$

$$\Rightarrow (13.6 \times 9 - 0.85 \times 9) = \frac{hc}{\lambda}$$

$$\Rightarrow \quad \lambda = \frac{hc}{9(13.6 - 0.85)}$$

$$\Rightarrow \quad \lambda = \frac{1240\ \text{eV -nm}}{9 \times 12.75\ \text{eV}} = 10.8\ \text{nm}$$

62. *(c)* As, for conservative fields

$$\text{F} = -\left(\frac{d\text{U}}{d\text{r}}\right)$$

$\therefore$ Magnitude of force on particle is

$$\Rightarrow \quad F = \frac{dU}{dr} = \frac{d}{dr}\left(\frac{1}{2}kr^2\right) \Rightarrow F = kr$$

This force is acting like centripetal force.

$$\therefore \quad \frac{mv^2}{r} = kr \quad \ldots \text{(i)}$$

So, for nth orbit,

$$\Rightarrow \quad m^2 v_n^2 = mkr_n^2$$

$$\Rightarrow \quad \frac{n^2 h^2}{4\pi^2 r^2} = mkr_n^2 \quad \left[\because v_n = \frac{nh}{2\pi mr}\right]$$

Therefore, $\quad r_n^4 \propto n^2 \Rightarrow r_n^2 \propto n$

So, $\quad r_n \propto \sqrt{n}$

Energy of particle is

$$E_n = \text{PE} + \text{KE} = \frac{1}{2}kr_n^2 + \frac{1}{2}mv_n^2$$

$$= \frac{1}{2}kr_n^2 + \frac{1}{2}kr_n^2 \quad \text{[Using Eq. (i)]}$$

$$= kr_n^2$$

So, energy, $E_n \propto r_n^2 \Rightarrow E_n \propto n$

63. *(a)* If wavelength of emitted photon in de-excitation is Λ_n;

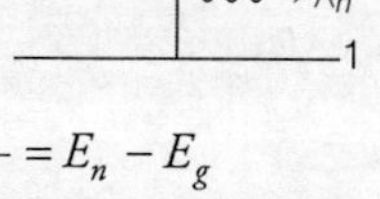

Then, $\frac{hc}{\Lambda_n} = E_n - E_g$

$$\frac{hc}{\Lambda_n} = \frac{p_n^2}{2m} - \frac{p_g^2}{2m}, \qquad \left[\because E = \frac{p^2}{2m}\right]$$

As energies are negative, we get

$$\frac{hc}{\Lambda_n} = \frac{p_g^2}{2m} - \frac{p_n^2}{2m}$$

$$= \frac{p_g^2}{2m}\left(1 - \left(\frac{p_n}{p_g}\right)^2\right)$$

$$= \frac{h^2}{2m\lambda_g^2}\left(1 - \frac{\lambda_g^2}{\lambda_n^2}\right)$$

$$\left[\because p \propto \lambda^{-1}, p = \frac{h}{\lambda}\right]$$

$$\Rightarrow \Lambda_n = \frac{2m\lambda_g^2 c}{h}\left(1 - \frac{\lambda_g^2}{\lambda_n^2}\right)^{-1}$$

$$\Rightarrow \Lambda_n = \frac{2m\lambda_g^2 c}{h}\left(1 + \frac{\lambda_g^2}{\lambda_n^2}\right)$$

$$[\because (1-x)^{-n} = 1 + nx]$$

$$\Rightarrow \Lambda_n \simeq A + \frac{B}{\lambda_n^2}$$

where, $A = \left[\frac{2mc\lambda_g^2}{h}\right]$

and $B = \left[\frac{2mc\lambda_g^4}{h}\right]$ are constants.

64. *(3)* We know that,

$$r = r_0 \frac{n^2}{Z}$$

For Li^{++}, $Z = 3$ and $x = 1$

$$\therefore \quad r = r_0 \cdot \frac{1^2}{3} = \frac{r_0}{3}$$

Here, $r_0 = a_0$

$$\therefore \quad r = \frac{a_0}{3}$$

$$= \frac{a_0}{x} \qquad \text{[given]}$$

$$\therefore \quad x = 3$$

65. *(2)* We know that

$$E = 13.6\left(\frac{1}{n_2^2} - \frac{1}{n_1^2}\right)$$

$$\Rightarrow \quad 2.86 = 13.6\left(\frac{1}{n^2} - \frac{1}{5^2}\right)$$

$\Rightarrow$ Solving, we get

$$n^2 = 4$$

$$\Rightarrow \quad n = 2$$

66. *(156)*

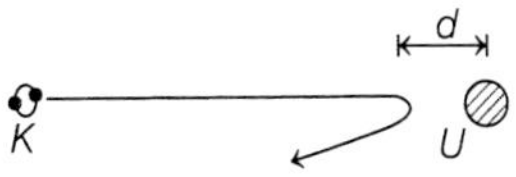

At distance of closest approach, all of kinetic energy becomes potential energy

$$\Rightarrow \frac{1}{2}mv^2 = \frac{kq_1q_2Z}{r}$$

$$\Rightarrow \left(\frac{1}{2} \times 6.72 \times 10^{-27}\right)v^2$$

$$= \frac{(9 \times 10^9 \times 2 \times 1.6 \times 10^{-19} \times 80 \times 1.6 \times 10^{-19})}{4.5 \times 10^{-14}}$$

$$v^2 = \frac{9 \times 10^9 \times 2 \times 80 \times (1.6 \times 10^{-19})^2 \times 2}{6.72 \times 10^{-27} \times 4.5 \times 10^{-14}}$$

$$= \frac{9 \times 4 \times 80 \times (1.6)^2}{6.72 \times 4.5} \times 10^{-38+9+27+14}$$

$$\Rightarrow v^2 = 243.089 \times 10^{12}$$

or $v = 15.59 \times 10^6$ m/s $\simeq 156 \times 10^5$ m/s

67. *(16)*

> **key idea** To solve this problem, use the following two steps.
> (i) Find the position of orbit using $r_n = 0.529\, n^2$ and
> (ii) For energy of electron, are $E = \frac{-13.6 Z^2}{n^2}$

Radius of orbit,

$$r_n = 8.48 \text{ Å} \Rightarrow a_0 = 0.59 \text{ Å}$$

We know that, for H-atom

$$r = \frac{a_0 n^2}{Z} \quad [Z = 1 \text{for H-atom}]$$

$$r_n = 0.529\, n^2$$

$$\Rightarrow 8.48 = 0.529\, n^2$$

$$\Rightarrow n^2 = 16 \quad \Rightarrow n = 4$$

Also, $E \propto 1/n^2$

$$\Rightarrow \quad \frac{E_2}{E_1} = \left(\frac{n_1}{n_2}\right)^2 = \left(\frac{1}{4}\right)^2 = \frac{1}{16}$$

$$\therefore \quad E_2 = \frac{E_1}{16} = \frac{E}{16} = \frac{E}{x} \qquad \text{(given)}$$

$$\therefore \quad x = 16$$

68. *(32)* We know that

$$E = 13.6\left[\frac{1}{n_1^2} - \frac{1}{n_2^2}\right]$$

Energy released in transition,

$$E_1 = 13.6\left[\frac{1}{1} - \frac{1}{2^2}\right] \text{eV}$$

$$h \cdot f_1 = \left(13.6 \times \frac{3}{4}\right) \text{eV} \qquad \ldots\text{(i)}$$

Similarly, $E_2 = 13.6\left[\frac{1}{1} - \frac{1}{3^2}\right]$ eV

$$h \cdot f_2 = 13.6 \times \frac{8}{9} \text{ eV} \qquad \ldots\text{(ii)}$$

From Eqs. (i) and (ii), we get

$$\frac{f_2}{f_1} = \frac{32}{27}$$

$$\therefore \quad f_2 = \frac{32}{27} \times f_1 = \frac{32}{27} \times 3 \times 10^{15} \text{ Hz}$$

$$= \frac{32}{9} \times 10^{15} \text{ Hz}$$

$$= \frac{x}{9} \times 10^{15} \text{ Hz} \qquad \text{(given)}$$

$$\therefore \quad x = 32$$

69. *(6)* Given,

For H-atom, $E_n = -0.85$ eV

We know nth energy level of atom is given by

$$E = \frac{-13.6 \text{ eV } Z^2}{n^2} \quad (Z = 1 \text{ for H-atom})$$

$$-0.85 \text{ eV} = \frac{-13.6 \text{ eV}}{n^2}$$

$$\Rightarrow \quad n = 4$$

Number of allowed transitions

$$= \frac{n(n-1)}{2} = \frac{4(3)}{2} = 6$$

Hence, maximum number of allowed transition is 6.

70. *(102)* For Balmer series $n = 2$

$$\Delta E = E_2 - E_1 = 13.6\left(1 - \frac{1}{4}\right)$$

$$= \frac{3}{4} \times 13.6 = 10.2 \text{ eV}$$

$$\therefore V = 10.2 \text{ V}$$

$$= \frac{102}{10} \text{ V} = \frac{\alpha}{10} \text{ V} \qquad \text{[given]}$$

$$\therefore \quad \alpha = 102$$

> **• Concept Enhancer**
> In hydrogen spectrum, Lyman series occurs in ultraviolet region, Balmer series occurs in visible region while Pfund, Brackett and Paschen series lies in infrared region.

71. *(17)*

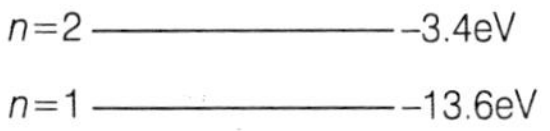

$$\Delta E = 10.2 \text{ eV}$$

Recoil speed, $v = \frac{\Delta E}{mc}$

$$= \frac{10.2\text{ eV}}{1.6\times10^{-27}\times3\times10^{8}}$$

$$= \frac{10.2\times1.6\times10^{-19}}{1.6\times10^{-27}\times3\times10^{8}}$$

$$v = 3.4\text{ m/s} = \frac{17}{5}\text{ m/s}$$

$$\Rightarrow \quad x = 17$$

72. *(144)* $\frac{1}{\lambda} = R\left(\frac{1}{n_1^2} - \frac{1}{n_2^2}\right)$

For longest wavelength of Paschen series,

$$n_1 = 3 \text{ and } n_2 = 4$$

$$\therefore \quad \frac{1}{\lambda_2} = R\left(\frac{1}{3^2} - \frac{1}{4^2}\right)$$

$$= R\left(\frac{7}{144}\right)$$

$$\therefore \quad \lambda_L = \frac{144}{7R} = \frac{\alpha}{7R} \quad \text{(given)}$$

$$\therefore \quad \alpha = 144$$

73. *(9.8)* According to given situation,

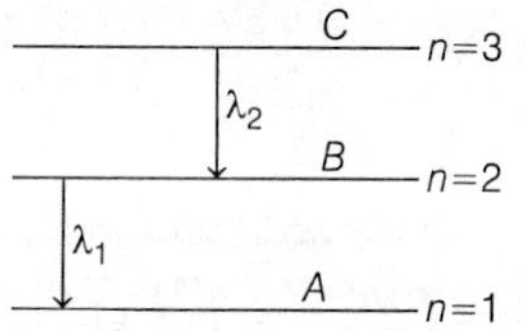

$$E_B - E_A = \frac{hc}{\lambda_1}$$

$$\Rightarrow \frac{-13.6}{2^2} - \frac{(-13.6)}{1^2} = \frac{hc}{\lambda_1}$$

$$\Rightarrow 10.2 = \frac{hc}{\lambda_1}$$

$$\Rightarrow \quad hc/\lambda_1 = 10.2 \quad \ldots\text{(i)}$$

Similarly, $E_C - E_B = \frac{hc}{\lambda_2}$

$$\Rightarrow \frac{-13.6}{3^2} - \left(-\frac{13.6}{2^2}\right) = \frac{hc}{\lambda_2}$$

$$\Rightarrow \quad 13.6\left(\frac{1}{4} - \frac{1}{9}\right) = \frac{hc}{\lambda_2}$$

$$\Rightarrow \quad 13.6\times\frac{5}{36} = \frac{hc}{\lambda_2} \Rightarrow 1.89 = \frac{hc}{\lambda_2}$$

$$\Rightarrow \quad \frac{hc}{\lambda_2} = 1.89 \quad \ldots\text{(ii)}$$

On dividing Eqs. (ii) by Eq. (i), we get

$$\frac{hc/\lambda_1}{hc/\lambda_2} = \frac{10.2}{1.89} \Rightarrow \frac{\lambda_2}{\lambda_1} = 5.6$$

$$\Rightarrow \quad \frac{\lambda_1}{\lambda_2} = \frac{1}{5.6} = \frac{7}{4\times9.8} = \frac{7}{4n} \quad \text{[Given]}$$

$$\therefore \quad n = 9.8$$

74. *(2)* We know that radius of nth orbit of H-atom is given as,

$$r_n = r_0 \times \frac{n^2}{Z}$$

For 2nd orbit of He^+,

$$Z = 2, n = 2$$

$$\therefore \quad r_1 = r_0 \cdot \frac{2^2}{2} = 2r_0$$

$$\Rightarrow \quad r_1 = 2r_0 \quad \ldots\text{(i)}$$

Similarly, for fourth orbit Be^{3-}

$$Z = 4, x = 4$$

$$\therefore \quad r_2 = r_0 \times \frac{4^2}{4} = 4r_0$$

$$\therefore \quad \frac{r_1}{r_2} = \frac{2r_0}{4r_0} = \frac{1}{2}$$

$$\Rightarrow \quad \frac{r_2}{r_1} = \frac{2}{1} = \frac{x}{1} \quad \text{[Given]}$$

$$\therefore \quad x = 2$$

75. *(3668)* Given, lowest wavelength of Lyman series = 917 Å

For lowest wavelength of lyman series

$$\frac{1}{\lambda} = RZ^2\left(\frac{1}{1^2} - \frac{1}{\infty^2}\right) = RZ^2$$

For lowest wavelength of Balmer series,

$$\frac{1}{\lambda'} = RZ^2\left(\frac{1}{2^2} - \frac{1}{\infty^2}\right) = \frac{RZ^2}{4}$$

$$\Rightarrow \quad \lambda' = \frac{4}{RZ^2} = 4\times917 = 3668\text{ Å}$$

76. *(27)* Wavelength of emitted radiation is

$$\frac{1}{\lambda} = R\left(\frac{1}{n_1^2} - \frac{1}{n_2^2}\right)$$

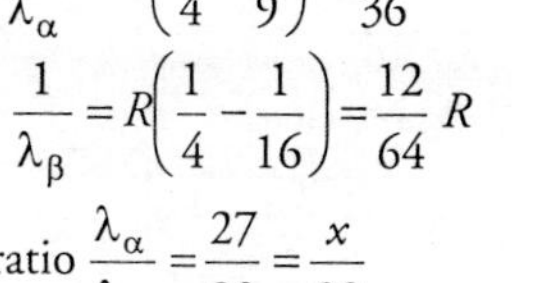

For H_α line, $n_2 = 3, n_1 = 2$

For H_β line, $n_2 = 4, n_1 = 2$

So, $\frac{1}{\lambda_\alpha} = R\left(\frac{1}{4} - \frac{1}{9}\right) = \frac{5R}{36}$

and $\frac{1}{\lambda_\beta} = R\left(\frac{1}{4} - \frac{1}{16}\right) = \frac{12}{64}R$

So, ratio $\frac{\lambda_\alpha}{\lambda_\beta} = \frac{27}{20} = \frac{x}{20}$

Therefore, $x = 27$

77. *(16)* Using, $E = \frac{ke^2}{2r}$, $r = \frac{ke^2}{2E}$

Given, $E = 12.8\text{ eV} = 12.8\times e$

$$r = \frac{9\times10^9\times e^2}{2\times12.8e}$$

$$= \frac{9\times10^9\times1.6\times16^{-19}}{2\times12.8}$$

$$= \frac{9\times10^{-10}}{2\times\frac{12.8}{1.6}} = \frac{9\times10^{-10}}{\frac{25.6}{1.6}}$$

$$= \frac{9\times10^{-10}}{16} = \frac{9}{x}\times10^{-10}\text{ m}$$

$$x = 16$$

78. *(425)* $r_n = \frac{0.51n^2}{Z}$ Å

For Li^{++}, $\quad Z = 3$

$$r_5 = 0.51\times\frac{25}{3}\times10^{-10}\text{ m}$$

$$= 17\times25\times10^{-12}\text{ m}$$

$$= 425\times10^{-12}\text{ m}$$

79. *(828)* Energy of photon, $E = 12.75$ eV

Let, the electron makes transition from ground states ($n = 1$) to ($n - 1$)th excited state (or nth shell), then we can write.

$$E = 13.6\left(\frac{1}{1^2} - \frac{1}{n^2}\right)\times Z^2 \text{ (in eV)}$$

For hydrogen atom, $Z = 1$,

$$E = 13.6\left(\frac{1}{1} - \frac{1}{n^2}\right)$$

$$12.75 = 13.6\left(1 - \frac{1}{n^2}\right)$$

$$\Rightarrow \quad n = 4$$

Angular momentum,

$$L = \frac{nh}{2\pi} = 4\times\frac{4.14\times10^{-15}}{2\pi}\text{ eV-s}$$

$$\Rightarrow \quad L = 828\times10^{-17}\text{ eV-s}$$

80. *(136)* Binding energy of an electron is negative of total energy in that state.

BE = − Total energy

$$= -\left(-13.6\frac{Z^2}{n^2}\right) = 13.6\frac{Z^2}{n^2}\text{ eV}$$

For Li^{++}, $Z = 3$, $n = 2 + 1 = 3$

$$\text{BE} = 13.6\frac{(3)^2}{(3)^2} = 13.6 = 136\times10^{-1}\text{eV}$$

On comparing, $x = 136$

81. *(27)*

key idea When an electron in hydrogen atom jumps from higher energy state n_2 to lower energy state n_1, then during this transition, you can calculate the wavelength corresponding to above transition using the following formula

$$\frac{1}{\lambda} = R\left(\frac{1}{n_1^2} - \frac{1}{n_2^2}\right)$$

For hydrogen atom, when electron makes an transition from nth level to ground state, then wavelength of emitted photon is given as

$$\frac{1}{\lambda} = R\left(\frac{1}{1^2} - \frac{1}{n^2}\right)$$

For transition-1,

$$\frac{1}{\lambda_1} = R\left(\frac{1}{1^2} - \frac{1}{3^2}\right)$$

$$\Rightarrow \quad \frac{1}{\lambda_1} = R\left(1 - \frac{1}{9}\right) = \frac{8R}{9}$$

$$\text{or} \quad \lambda_1 = \frac{9}{8R} \quad \ldots\text{(i)}$$

For transition-2,

$$\frac{1}{\lambda_2} = R\left(\frac{1}{1^2} - \frac{1}{2^2}\right)$$

$$\Rightarrow \quad \frac{1}{\lambda_2} = R\left(1 - \frac{1}{4}\right) = \frac{3}{4}R$$

$$\Rightarrow \quad \lambda_2 = \frac{4}{3R} \quad \ldots\text{(ii)}$$

On dividing Eq. (i) and Eq. (ii), we get

$$\frac{\lambda_1}{\lambda_2} = \frac{27}{32} = \frac{x}{32} \quad \text{[Given]}$$

$$\therefore \quad x = 27$$

82. *(35)* According to first condition,

$$\frac{hc}{\lambda_0} = \frac{-13.6}{3^2} - \left(\frac{-13.6}{2^2}\right)$$

$$\Rightarrow \quad \frac{hc}{\lambda_0} = \frac{13.6 \times 5}{36} \quad \ldots\text{(i)}$$

According to second condition,

$$\frac{hc}{\lambda'} = \frac{-13.6}{4^2} - \left(\frac{-13.6}{3^2}\right)$$

$$= \frac{13.6}{9} - \frac{13.6}{16}$$

$$\frac{hc}{\lambda'} = \frac{13.6 \times 7}{16 \times 9} \quad \ldots\text{(ii)}$$

On dividing Eq. (i) by Eq. (ii), we get

$$\frac{hc/\lambda_0}{hc/\lambda'} = \frac{13.6 \times 5}{36} \times \frac{16 \times 9}{13.6 \times 7}$$

$$\Rightarrow \quad \frac{\lambda'}{\lambda_0} = \frac{4}{7} = \frac{20}{35}$$

$$\Rightarrow \quad \lambda' = \frac{20}{35}\lambda_0 = \frac{20}{x}\lambda_0 \quad \text{[Given]}$$

$$\therefore \quad x = 35$$

83. *(15)*

> **key idea** If an atom is excited to nth energy level, then number of possible wavelength observed are calculated with the help of following formula.
>
> Number of possible wavelength, $N = \frac{n(n-1)}{2}$

Given, principal quantum number,

$$n = 6$$

Number of wavelength,

$$N = \frac{n(n-1)}{2} = \frac{6(6-1)}{2} = 3(5) = 15$$

$$\therefore \quad x = 15$$

84. *(10)* As per question, K_α X-ray of molybdenum, has wavelength 0.071 nm.

Since, we know that

$$K_\alpha = \frac{hc}{\lambda} = \frac{4.14 \times 10^{-15} \times 3 \times 10^8}{0.071 \times 10^{-9}}$$

$$= 17.49 \text{ keV}$$

$\therefore$ Energy difference between L and K will be,

$$E_K - E_L = K_\alpha \Rightarrow E_K - E_L = 17.49$$

$$\Rightarrow \quad E_L = E_K - 17.49$$

$$\Rightarrow \quad E_L = 27.5 - 17.49$$

$$\Rightarrow \quad E_L \approx 10 \text{ keV}$$

$\therefore$ The energy of molybdenum atom when an L-electron is knoked out will be 10 keV.

85. *(112)* Radius, $r = 0.5 \text{Å} = 0.5 \times 10^{-10}$ m

Speed, $v = 2.2 \times 10^6$ m/s

Current associated due to moving electron,

$$I = \frac{e}{T} = \frac{e\omega}{2\pi} \Rightarrow I = \frac{e}{2\pi} \cdot \frac{v}{r}$$

$$= \frac{1.6 \times 10^{-19} \times 2.2 \times 10^6 \times 7}{2 \times 22 \times 0.5 \times 10^{-10}}$$

$$\therefore \quad I = 112 \times 10^{-2} \text{ mA}$$

86. *(15)* Since, we know that

$$\frac{1}{\lambda} = RZ^2\left(\frac{1}{n_1^2} - \frac{1}{n_2^2}\right) \quad \ldots\text{(i)}$$

Therefore, for 1st spectral line of Balmer series,

$$n_1 = 2 \text{ and } n_2 = 3$$

$$\frac{1}{\lambda_1} = RZ^2\left(\frac{1}{2^2} - \frac{1}{3^2}\right)$$

$$\Rightarrow \quad \frac{1}{\lambda_1} = RZ^2\left(\frac{5}{36}\right) \quad \ldots\text{(ii)}$$

Similarly, for 3rd spectral line,

$$n_1 = 2 \text{ and } n_2 = 5$$

$$\frac{1}{\lambda_3} = RZ^2\left(\frac{1}{2^2} - \frac{1}{5^2}\right)$$

$$\Rightarrow \quad \frac{1}{\lambda_3} = RZ^2\left(\frac{21}{100}\right) \quad \ldots\text{(iii)}$$

Now, dividing Eq. (iii) by Eq. (ii), we get

$$\frac{\frac{1}{\lambda_3}}{\frac{1}{\lambda_1}} = \frac{RZ^2\left(\frac{21}{100}\right)}{RZ^2\left(\frac{5}{36}\right)} \Rightarrow \frac{\lambda_1}{\lambda_3} = \frac{21}{100} \times \frac{36}{5}$$

$$\Rightarrow \quad \lambda_1 / \lambda_3 = 1.512 \simeq 15 \times 10^{-1}$$

$$\simeq x \times 10^{-1} \quad \text{[Given]}$$

87. *(51)* Mass of particle

$$= 200 \text{ MeV/c}^2 = m$$

Mass of hydrogen $= 1 \text{ GeV/c}^2$

$m_{\text{Hydrogen}} = 5$ m

Using law of conservation of linear momentum,

$$mv + 0 = 0 + 5mv' \Rightarrow v' = v/5$$

When the atom goes to its first excited state,

Energy dissipated

$$= \left[\frac{-13.6}{(2)^2}\right] \text{eV} - \left[\frac{-13.6}{(1)^2}\right] \text{eV}$$

$$= -3.4 + 13.6 = 10.2 \text{ eV} \quad \ldots\text{(i)}$$

Loss of kinetic energy

$$= \frac{1}{2}mv^2 - \frac{1}{2}(5m)\left(\frac{v}{5}\right)^2 = \frac{1}{2}mv^2\left(1 - \frac{1}{5}\right)$$

$$= \frac{4}{5}\left(\frac{mv^2}{2}\right) = \frac{4}{5}K \quad \ldots\text{(ii)}$$

From Eqs. (i) and (ii), we get

$$\frac{4}{5}K = 10.2 \text{ eV}$$

$$\Rightarrow \quad K = \frac{10.2 \times 5}{4} \text{ eV} = \frac{51.0}{4} \text{ eV}$$

$$\Rightarrow \quad \frac{N}{4} = \frac{51}{4} \text{ eV}$$

$$\Rightarrow \quad N = 51$$

88. *(10553)* As, $\frac{1}{\lambda} = R\left(\frac{1}{n_i^2} - \frac{1}{n_f^2}\right)$

$$\Rightarrow \quad \lambda = \frac{1}{R\left(\frac{1}{n_f^2} - \frac{1}{n_i^2}\right)} \quad \ldots\text{(i)}$$

For Lyman series

Largest wavelength

$$(n_i = 2 \text{ and } n_f = 1)$$

$$\lambda_1 = \frac{1}{R\left(\frac{1}{1^2} - \frac{1}{2^2}\right)} = \frac{1}{R\left(\frac{3}{4}\right)} = \frac{4}{3R}$$

Shortest wavelength

($n_i = \infty$ and $n_f = 1$)

$$\lambda_2 = \frac{1}{R\left(\frac{1}{1^2} - \frac{1}{\infty^2}\right)} = \frac{1}{R(1)} = \frac{1}{R}$$

$$\Delta\lambda = \lambda_1 - \lambda_2$$

$$\Rightarrow \quad 304 = \frac{4}{3R} - \frac{1}{R}$$

$$\Rightarrow \quad 304 = \frac{1}{3R} \Rightarrow R = \frac{1}{3 \times 304}$$

$$\Rightarrow \quad R = \frac{1}{912} \quad \text{...(ii)}$$

For Paschan series

Largest wavelength ($n_i = 4$ and $n_f = 3$)

$$\lambda_3 = \frac{1}{R\left(\frac{1}{3^2} - \frac{1}{4^2}\right)} = \frac{1}{R\left(\frac{7}{144}\right)} = \frac{144}{7R}$$

Shortest wavelength ($n_i = \infty$ and $n_f = 3$)

$$\lambda_4 = \frac{1}{R\left(\frac{1}{3^2} - \frac{1}{\infty^2}\right)} = \frac{1}{R\left(\frac{1}{9}\right)} = \frac{9}{R}$$

Now, the difference between the largest and the shortest wavelengths of the Paschan series,

$$\Delta\lambda = \lambda_3 - \lambda_4 = \frac{144}{7R} - \frac{9}{R}$$

$$= \frac{144 - 63}{7R} = \frac{81}{7R}$$

$$= \frac{81}{7\,(1/912)} \quad \text{[From Eq. (i)]}$$

$$= \frac{81 \times 912}{7}$$

$$= 10553.14 \text{ Å}$$

$$\simeq 10553 \text{ Å}$$

89. *(486)* Balmer series of H-atom spectrum occurs when an excited electron jumps to second level or $n = 2$ state.

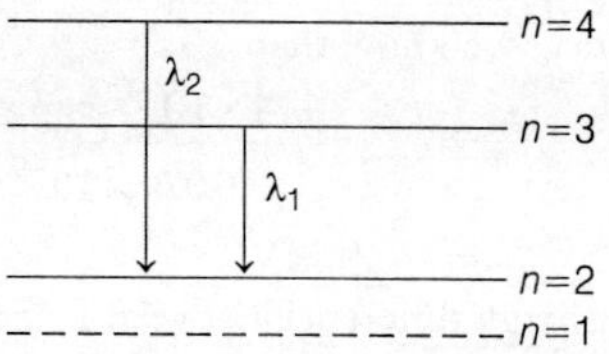

Emitted wavelengths are given by

$$\frac{1}{\lambda} = R\left(\frac{1}{2^2} - \frac{1}{n^2}\right)$$

where, $n = 3, 4, 5 \ldots \infty$.

First member is obtained in transition $n = 3$ to $n = 2$.

$$\Rightarrow \quad \frac{1}{\lambda_1} = R\left(\frac{1}{2^2} - \frac{1}{3^2}\right) \quad \text{...(i)}$$

Second member is obtained in transition $n = 4$ to $n = 2$.

$$\Rightarrow \quad \frac{1}{\lambda_2} = R\left(\frac{1}{2^2} - \frac{1}{4^2}\right) \quad \text{...(ii)}$$

From Eqs. (i) and (ii), we have

$$\frac{\frac{1}{\lambda_1}}{\frac{1}{\lambda_2}} = \frac{\left(\frac{1}{2^2} - \frac{1}{3^2}\right)}{\left(\frac{1}{2^2} - \frac{1}{4^2}\right)}$$

$$\Rightarrow \quad \frac{\lambda_2}{\lambda_1} = \frac{\left(\frac{1}{4} - \frac{1}{9}\right)}{\left(\frac{1}{4} - \frac{1}{16}\right)}$$

$$= \frac{5}{36} \times \frac{64}{12}$$

$$\Rightarrow \quad \frac{\lambda_2}{\lambda_1} = \frac{5}{9} \times \frac{4}{3} = \frac{20}{27}$$

$$\Rightarrow \quad \lambda_2 = \lambda_1 \times \frac{20}{27} = 6561 \times \frac{20}{27}$$

$$= 4860 \text{ Å} = 486 \text{ nm}$$

C-27

JEE Main Chapterwise Topicwise Physics

In JEE Main there is usually 0-1 question per paper is visible from this chapter. In past 3 years, 22 conceptual and 14 numerical questions were asked of moderate to difficult level from this chapter.

For detailed analysis scan QR code

NUCLEI

Topic 1 Nucleus

Level 1

01 For a nucleus of mass number A and radius R, the mass density of nucleus can be represented as *[2025, 8 April Shift-II]*

(a) $A^{\frac{1}{3}}$ (b) independent of A

(c) A^3 (d) $A^{\frac{2}{3}}$

02 From the statements given below

A. The angular momentum of an electron in nth orbit is an integral multiple of $\hbar$.

B. Nuclear forces do not obey inverse square law.

C. Nuclear forces are spin dependent.

D. Nuclear forces are central and charge independent.

E. Stability of nucleus is inversely proportional to the value of packing fraction.

Choose the correct answer from the options given below. *[2024, 1 Feb Shift-II]*

(a) A, B, C, E only (b) B, C, D, E only

(c) A, C, D, E only (d) A, B, C, D only

03 Given below are two statements, one is labelled as Assertion (A) and the other is labelled as Reason (R).

Assertion (A) The binding energy per nucleon is practically independent of the atomic number for nuclei of mass number in the range 30 to 170.

Reason (R) Nuclear force is short ranged.

In the light of the above statements, choose the most appropriate answer from the options given below. *[2023, 13 April Shift-II]*

(a) Both A and R are true and R is the correct explanation of A.

(b) Both A and R are true but R is not the correct explanation of A.

(c) A is true but R is false.

(d) A is false but R is true.

04 Given below are two statements, one is labelled as Assertion (A) and the other is labelled as Reason (R).

Assertion (A) The nuclear density of nuclides ${}_{5}^{10}B$, ${}_{3}^{6}Li$, ${}_{26}^{56}Fe$, ${}_{10}^{20}Ne$ and ${}_{83}^{209}Bi$ can be arranged as $\rho_{Bi}^{N} > \rho_{Fe}^{N} > \rho_{Ne}^{N} > \rho_{B}^{N} > \rho_{Li}^{N}$

Reason (R) The radius R of nucleus is related to its mass number A as $R = R_0 A^{1/3}$, where R_0 is a constant.

In the light of the above statements, choose the correct answer from the options given below. *[2023, 30 Jan Shift-II]*

(a) A is true but R is false.

(b) Both A and R are true but R is not the correct explanation of A.

(c) A is false but R is true.

(d) Both A and R are true and R is the correct explanation of A.

05 Mass numbers of two nuclei are in the ratio of 4 : 3. Their nuclear densities will be in the ratio of *[2022, 26 July Shift-II]*

(a) 4 : 3 (b) $\left(\frac{3}{4}\right)^{1/3}$ (c) 1 : 1 (d) $\left(\frac{4}{3}\right)^{1/3}$

06 The decay of a proton to neutron is *[2021, 18 March Shift-II]*

(a) not possible as proton mass is less than the neutron mass

(b) possible only inside the nucleus

(c) not possible but neutron to proton conversion is possible

(d) always possible as it is associated only with β^+ decay

Level 2

07 Given below are two statements. One is labelled as Assertion (A) and the other is labelled as Reason (R).

Assertion (A) The density of the copper (${}_{29}^{64}Cu$) nucleus is greater than that of the carbon (${}_{6}^{12}C$) nucleus.

Reason (R) The nucleus of mass number A has a radius proportional to $A^{1/3}$. *[2025, 7 April Shift-II]*

In the light of the above statements, choose the most appropriate answer from the options given below:

(a) Both A and R are correct but R is not the correct explanation of A.
(b) Both A and R are correct and R is the correct explanation of A.
(c) A is not correct but R is correct.
(d) A is correct but R is not correct.

08 Choose the correct nuclear process from the below options [p: proton, n : neutron, e^- : electron, e^+ : positron, ν: neutrino, $\bar{\nu}$: anti-neutrino] *[2025, 18 March Shift-II]*

(a) $n \rightarrow p + e^- + \nu$ (b) $n \rightarrow p + e^+ + \bar{\nu}$
(c) $n \rightarrow p + e^+ + \nu$ (d) $n \rightarrow p + e^- + \bar{\nu}$

Imp. Radius of Nucleus

09 The mass number of nucleus having radius equal to half of the radius of nucleus with mass number 192 is *[2024, 31 Jan Shift-II]*

(a) 20 (b) 32 (c) 24 (d) 40

10 Read the following statements.

A. Volume of the nucleus is directly proportional to the mass number.
B. Volume of the nucleus is independent of mass number.
C. Density of the nucleus is directly proportional to the mass number.
D. Density of the nucleus is directly proportional to the cube root of the mass number
E. Density of the nucleus is independent of the mass number.

Choose the correct option from the following options. *[2022, 29 July Shift-II]*

(a) A and D only (b) A and E only
(c) B and E only (d) A and C only

11 Which of the following figures represents the variation of $\left(\frac{R}{R_0}\right)$ with ln A (If R = radius of a nucleus and A = its mass number)? *[2022, 25 June Shift-II]*

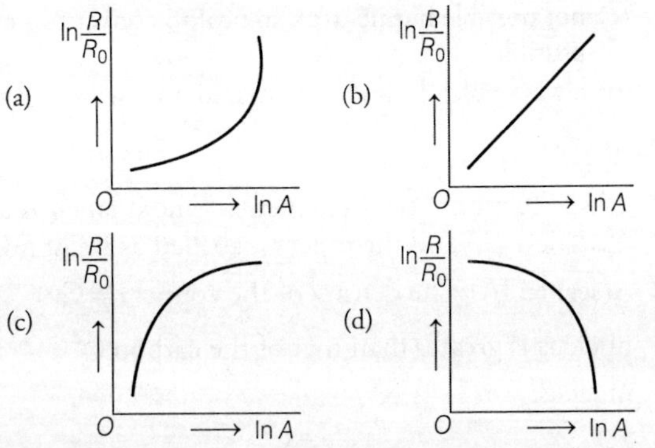

Numerical Value Type Questions

12 The radius of a nucleus of mass number 64 is 4.8 fermi. Then the mass number of another nucleus having radius of 4 fermi is $\frac{1000}{x}$, where x is .... *[2024, 1 Feb Shift-I]*

13 A nucleus has mass number A_1 and volume V_1. Another nucleus has mass number A_2 and volume V_2. If relation between mass number is $A_2 = 4A_1$, then $\frac{V_2}{V_1}$ = ... *[2024, 31 Jan Shift-II]*

Imp. Nuclear Density

14 Assume that protons and neutrons have equal masses. Mass of a nucleon is 1.6×10^{-27} kg and radius of nucleus is $1.5 \times 10^{-15} A^{1/3}$ m. The approximate ratio of the nuclear density and water density is $n \times 10^{13}$. The value of n is .... *[2023, 24 Jan Shift-I]*

Topic 2 Nuclear Fission and Fusion and Binding Energy

Level 1

01 Energy released when two deuterons ($_1H^2$) fuse to form a helium nucleus ($_2He^4$) is :
(Given : Binding energy per nucleon of $_1H^2$ = 1.1 MeV and binding energy per nucleon of $_2He^4$ = 7.0 MeV) *[2025, 2 April Shift-II]*

(a) 8.1 MeV (b) 23.6 MeV (c) 5.9 MeV (d) 26.8 MeV

02 Given below are two statements. One is labelled as Assertion (A) and the other is labelled as Reason (R).

Assertion (A) The binding energy per nucleon is found to be practically independent of the atomic number A, for nuclei with mass numbers between 30 and 170.

Reason (R) Nuclear force is long range.

In the light of the above statements, choose the correct answer from the options given below. *[2025, 23 Jan Shift-II]*

(a) Both A and R are true but R is not the correct explanation of A.
(b) A is false but R is true.
(c) A is true but R is false.
(d) Both A and R are true and R is the correct explanation of A.

03 A nucleus at rest disintegrates into two smaller nuclei with their masses in the ratio of 2 : 1. After disintegration they will move *[2024, 9 April Shift-II]*

(a) in opposite directions with speed in the ratio of 1 : 2 respectively.
(b) in opposite directions with the same speed.

(c) in the same direction with same speed.
(d) in opposite directions with speed in the ratio of 2 : 1 respectively.

04 The energy equivalent of 1 g of substance is **[2024, 9 April Shift-I]**
(a) 5.6×10^{26} MeV (b) 5.6 eV
(c) 11.2×10^{24} MeV (d) 5.6×10^{12} MeV

Imp. Binding Energy

05 If M_0 is the mass of isotope ${}^{12}_{5}\text{B}$, M_p and M_n are the masses of proton and neutron, then nuclear binding energy of isotope is **[2024, 8 April Shift-II]**
(a) $(M_0-12M_n)c^2$ (b) $(5M_p+7M_n-M_0)c^2$
(c) $(M_0-5M_p-7M_n)c^2$ (d) $(M_0-5M_p)c^2$

06 In a hypothetical fission reaction
${}_{92}X^{236}\rightarrow{}_{56}Y^{141}+{}_{36}Z^{92}+3R$ **[2024, 8 April Shift-II]**
The identity of emitted particles (R) is
(a) proton (b) γ-radiations
(c) neutron (d) electron

07 Which of the following nuclear fragments corresponding to nuclear fission between neutron (${}^{1}_{0}n$) and uranium isotope (${}^{235}_{92}\text{U}$) is correct? **[2024, 4 April Shift-I]**
(a) ${}^{144}_{56}\text{Ba}+{}^{89}_{36}\text{Kr}+4{}^{1}_{0}n$
(b) ${}^{140}_{56}\text{Xe}+{}^{94}_{38}\text{Sr}+3{}^{1}_{0}n$
(c) ${}^{153}_{51}\text{Sb}+{}^{99}_{41}\text{Nb}+3{}^{1}_{0}n$
(d) ${}^{144}_{56}\text{Ba}+{}^{89}_{36}\text{Kr}+3{}^{1}_{0}n$

08 The atomic mass of ${}_{6}\text{C}^{12}$ is 12.000000 u and that of ${}_{6}\text{C}^{13}$ is 13.003354 u. The required energy to remove a neutron from ${}_{6}\text{C}^{13}$, if mass of neutron is 1.008665 u, will be **[2024, 27 Jan Shift-II]**
(a) 4.95 MeV (b) 62.5 MeV
(c) 49.5 MeV (d) 6.25 MeV

09 The mass of proton, neutron and helium nucleus are respectively 1.0073 u, 1.0087 u and 4.0015 u. The binding energy of helium nucleus is **[2023, 1 Feb Shift-I]**
(a) 14.2 MeV (b) 28.4 MeV
(c)7.1 MeV (d) 56.8 MeV

Direction (Q. Nos. 9 and 10) are based on the following paragraph.

A nucleus of mass $M+\Delta m$ is at rest and decays into two daughter nuclei of equal mass $M/2$ each. Speed of light is c.

10 The binding energy per nucleon for the parent nucleus is E_1 and that for the daughter nuclei is E_2. Then, **[AIEEE 2010]**
(a) $E_2=2E_1$ (b) $E_1>E_2$
(c) $E_2>E_1$ (d) $E_1=2E_2$

11 The speed of daughter nuclei is **[AIEEE 2010]**
(a) $c\dfrac{\Delta m}{M+\Delta m}$ (b) $c\sqrt{\dfrac{2\Delta m}{M}}$
(c) $c\sqrt{\dfrac{\Delta m}{M}}$ (d) $c\sqrt{\dfrac{\Delta m}{M+\Delta m}}$

12 The below is a plot of binding energy per nucleon E_b, against the nuclear mass M, A, B, C, D, E, F correspond to different nuclei. Consider four reactions.
(i) $A+B\rightarrow C+\varepsilon$ (ii) $C\rightarrow A+B+\varepsilon$
(iii) $D+E\rightarrow F+\varepsilon$ and (iv) $F\rightarrow D+E+\varepsilon$

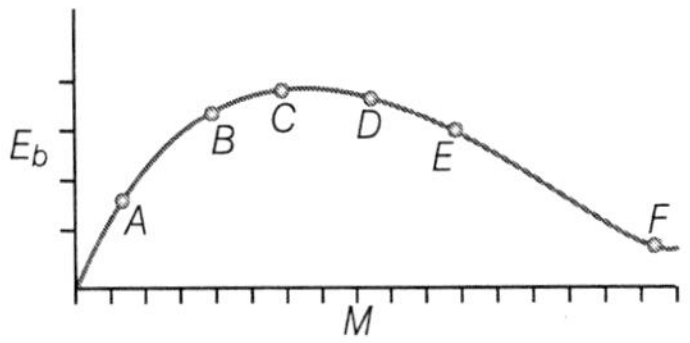

where, ε is the energy released. In which reactions ε is positive? **[AIEEE 2009]**
(a) (i) and (iv) (b) (i) and (iii)
(c) (ii) and (iv) (d) (ii) and (iii)

13 **Statement I** Energy is released when heavy nuclei undergo fission or light nuclei undergo fusion.
Statement II For heavy nuclei, binding energy per nucleon increases with increasing Z while for light nuclei, it decreases with increasing Z. **[AIEEE 2008]**
(a) Statement I is true, Statement II is true; Statement II is not correct explanation of Statement I.
(b) Statement I is true, Statement II is false.
(c) Statement I is false, Statement II is true.
(d) Statement I is true, Statement II is true; Statement II is correct explanation of Statement I.

14 If M_o is the mass of an oxygen isotope ${}_{8}\text{O}^{17}$, M_p and M_n are the masses of a proton and a neutron, respectively, the nuclear binding energy of the isotope is **[AIEEE 2007]**
(a) $(M_o-8M_p)c^2$ (b) $(M_o-8M_p-9M_n)c^2$
(c) M_oc^2 (d) $(M_o-17M_n)c^2$

15 If the binding energy per nucleon in ${}^{7}_{3}\text{Li}$ and ${}^{4}_{2}\text{He}$ nuclei are 5.60 MeV and 7.06 MeV respectively, then in the reaction $p+{}^{7}_{3}\text{Li}\longrightarrow 2{}^{4}_{2}\text{He}$
energy of proton must be **[AIEEE 2006]**
(a) 28.24 MeV (b) 17.28 MeV
(c) 1.46 MeV (d) 39.2 MeV

16 A nuclear transformation is denoted by $X(n,\alpha)\rightarrow{}^{7}_{3}\text{Li}$. Which of the following is the nucleus of element X? **[AIEEE 2005]**
(a) ${}^{12}_{6}\text{C}$ (b) ${}^{10}_{5}\text{B}$
(c) ${}^{9}_{5}\text{B}$ (d) ${}^{11}_{4}\text{Be}$

17 In the nuclear fusion reaction, ${}_1^2H + {}_1^3H \longrightarrow {}_2^4He + n$ given that the repulsive potential energy between the two nuclei is 7.7×10^{-14} J, the temperature at which the gases must be heated to initiate the reaction is nearly [Boltzmann's constant, $k_B = 1.38 \times 10^{-23}$ J/K] *[AIEEE 2003]*

(a) 10^7 K (b) 10^5 K (c) 10^3 K (d) 10^9 K

Level 2

18 A radioactive material P first decays into Q and then Q decays to non-radioactive material R. Which of the following figure represents time dependent mass of P, Q and R? *[2025, 4 April Shift-II]*

(a)
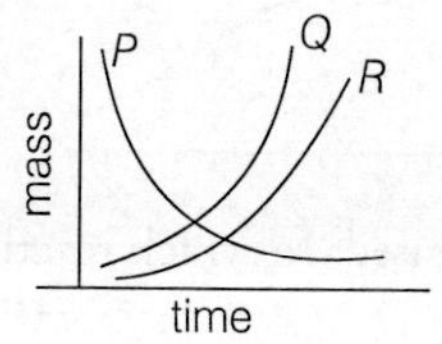

(b)
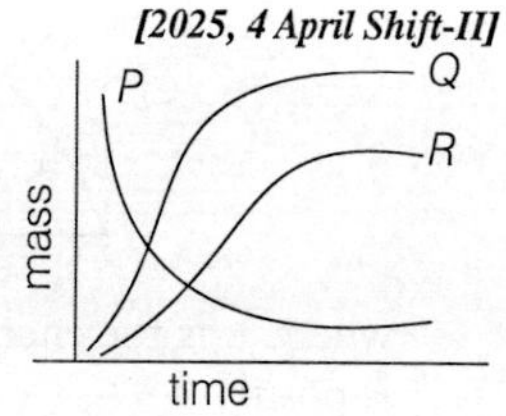

(c)
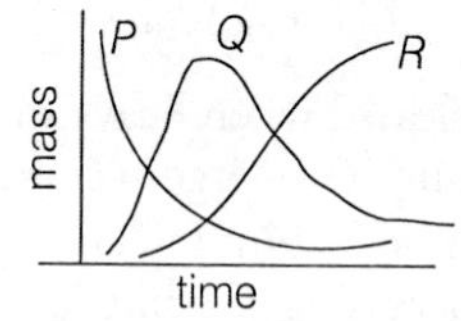

(d)
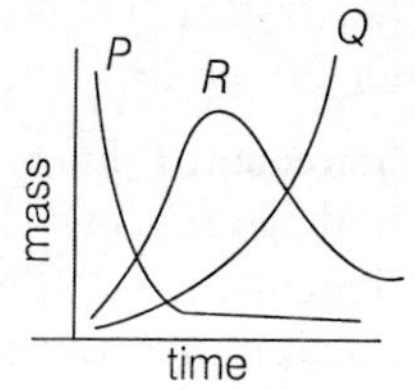

19 Binding energy of a certain nucleus is 18×10^8 J. How much is the difference between total mass of all the nucleons and nuclear mass of the given nucleus? *[2024, 8 April Shift-I]*

(a) 20 µg (b) 2 µg (c) 0.2 µg (d) 10 µg

20 In a nuclear fission reaction of an isotope of mass M, three similar daughter nuclei of same mass are formed. The speed of a daughter nuclei in terms of mass defect ΔM will be *[2024, 30 Jan Shift-II]*

(a) $c\sqrt{\frac{2\Delta M}{M}}$ (b) $\sqrt{\frac{2c\Delta M}{M}}$ (c) $c\sqrt{\frac{3\Delta M}{M}}$ (d) $\frac{\Delta Mc^2}{3}$

Imp. Energy Released in Chain Reaction

21 The explosive in a Hydrogen bomb is a mixture of ${}_1H^2$, ${}_1H^3$ and ${}_3Li^6$ in some condensed form. The chain reaction is given by

$$ {}_3Li^6 + {}_0n^1 \longrightarrow {}_2He^4 + {}_1H^3 $$
$$ {}_1H^2 + {}_1H^3 \longrightarrow {}_2He^4 + {}_0n^1 $$

During the explosion the energy released is approximately [Given : M (Li) = 6.01690 amu, M (${}_1H^2$) = 2.01471 amu, M (${}_2He^4$) = 4.00388 amu, and 1 amu = 931.5 MeV] *[2024, 29 Jan Shift-I]*

(a) 12.64 MeV (b) 28.12 MeV
(c) 22.22 MeV (d) 16.48 MeV

22 Nucleus A is having mass number 220 and its binding energy per nucleon is 5.6 MeV. It splits in two fragments B and C of mass numbers 105 and 115. The binding energy of nucleons in B and C is 6.4 MeV per nucleon. The energy Q released per fission will be *[2022, 24 June Shift-I]*

(a) 0.8 MeV (b) 275 MeV
(c) 220 MeV (d) 176 MeV

23 Find the binding energy per nucleon for ${}_{50}^{120}Sn$. Mass of proton $m_p = 1.00783$ u, mass of neutron $m_n = 1.00867$ u and mass of tin nucleus $m_{Sn} = 119.902199$ u. (Take, 1 u = 931 MeV) *[2020, 4 Sep Shift-II]*

(a) 9.0 MeV (b) 7.5 MeV
(c) 8.0 MeV (d) 8.5 MeV

Imp. Nuclear Fission

24 Consider the nuclear fission $Ne^{20} \longrightarrow 2He^4 + C^{12}$ Given that the binding energy/nucleon of Ne^{20}, He^4 and C^{12} are respectively, 8.03 MeV, 7.07 MeV and 7.86 MeV. Identify the correct statement. *[2019, 10 Jan Shift-I]*

(a) Energy of 3.6 MeV will be released.
(b) Energy of 12.4 MeV will be supplied.
(c) 8.3 MeV energy will be released.
(d) Energy of 9.72 MeV has to be supplied.

25 Assume that a neutron breaks into a proton and an electron. The energy released during this process is (mass of neutron = 1.6750×10^{-27} kg, mass of proton = 1.6725×10^{-27} kg, mass of electron = 9×10^{-31} kg) *[AIEEE 2012]*

(a) 0.73 MeV (b) 7.10 MeV
(c) 6.30 MeV (d) 5.4 MeV

26 When ${}_3^7Li$ nuclei are bombarded by protons, and the resultant nuclei are ${}_4^8Be$, the emitted particles will be *[AIEEE 2006]*

(a) alpha particles (b) beta particles
(c) gamma photons (d) neutrons

27 The binding energy per nucleon of deuteron ${}_1^2H$ and helium nucleus ${}_2^4He$ is 1.1 MeV and 7 MeV, respectively. If two deuteron nuclei react to form a single helium nucleus, then the energy released is *[AIEEE 2004]*

(a) 13.9 MeV
(b) 26.9 MeV
(c) 23.6 MeV
(d) 19.2 MeV

Level 3

28 Match the List-I with List-II. *[2025, 3 April Shift-I]*

	List-I		List-II
A.	${}_0^1n + {}_{92}^{235}U \rightarrow {}_{54}^{140}Xe + {}_{38}^{94}Sr + 2{}_0^1n$	1.	Chemical reaction
B.	$2H_2 + O_2 \rightarrow 2H_2O$	2.	Fusion with + ve Q value
C.	${}_1^2H + {}_1^2H \rightarrow {}_2^3He + {}_0^1n$	3.	Fission
D.	${}_1^1H + {}_1^3H \rightarrow {}_1^2H + {}_1^2H$	4.	Fusion, with – ve Q value

Choose the correct answer from the options given below.
(a) A-II, B-I, C-III, D-IV
(b) A-III, B-I, C-IV, D-II
(c) A-III, B-I, C-II, D-IV
(d) A-II, B-I, C-IV, D-III

29 The energy released in the fusion of 2 kg of hydrogen deep in the sum is E_H and the energy released in the fission of 2 kg of ${}^{235}U$ is E_U. The ratio $\frac{E_H}{E_U}$ is approximately , (Consider the fusion reaction as $4{}_1^1H + 2e^- \longrightarrow {}_2^4He + 2v + 6\gamma + 26.7$ MeV, energy released in the fission reaction of ${}^{235}U$ is 200 MeV per fission nucleus and $N_A = 6.023 \times 10^{23}$) *[2024, 9 April Shift-II]*
(a) 7.62 (b) 9.13
(c) 15.04 (d) 25.6

30 A nucleus with mass number 184 initially at rest emits an α-particle. If the Q-value of the reaction is 5.5 MeV, calculate the kinetic energy of the α-particle. *[2021, 22 July Shift-II]*
(a) 5.0 MeV (b) 5.5 MeV
(c) 0.12 MeV (d) 5.38 MeV

31 A nucleus of mass M emits γ- ray photon of frequency ν. The loss of internal energy by the nucleus is [Take, c is the speed of electromagnetic wave.] *[2021, 20 July Shift-I]*
(a) $h\nu$ (b) zero
(c) $h\nu\left[1 - \frac{h\nu}{2Mc^2}\right]$ (d) $h\nu\left[1 + \frac{h\nu}{2Mc^2}\right]$

32 You are given that mass of ${}_3^7Li = 7.0160$ u, mass of ${}_2^4He = 4.0026$ u and mass of ${}_1^1H = 1.0079$ u. When 20 g of ${}_3^7Li$ is converted into ${}_2^4He$ by proton capture, the energy liberated (in kWh), is [mass of nucleon = 1 GeV/c^2] *[2020, 6 Sep Shift-I]*
(a) 4.5×10^5 (b) 8×10^6
(c) 6.82×10^5 (d) 1.33×10^6

Numerical Value Type Questions

33 A star has 100% helium composition. It starts to convert three ${}^4He$ into one ${}^{12}C$ via triple alpha process as ${}^4He + {}^4He + {}^4He \longrightarrow {}^{12}C + Q$. The mass of the star is 2.0×10^{32} kg and it generates energy at the rate of 5.808×10^{30} W. The rate of converting these ${}^4He$ to ${}^{12}C$ is $n \times 10^{42}$ s^{-1}, where n is ………… *[2024, 9 April Shift-I]*
[Take, mass of ${}^4He = 4.0026$ u, mass of ${}^{12}C = 12$ u]

Imp. Nuclear Fusion

34 If three helium nuclei combine to form a carbon nucleus then the energy released in this reaction is ……….. $\times 10^{-2}$ MeV. (Given 1 u = 931 MeV/c^2, atomic mass of helium = 4.002603u) *[2024, 5 April Shift-I]*

35 The disintegration energy Q for the nuclear fission of ${}^{235}U \rightarrow {}^{140}Ce + {}^{94}Zr + n$ is……… MeV.
Given atomic masses of
${}^{235}U$: 235.0439 u : ${}^{140}Ce$: 139.91054 u.
${}^{94}Zr$: 93.9063u : n : 1.0086 u.
Value of $c^2 = 931$ MeV/u *[2024, 4 April Shift-II]*

36 The mass defect in a particular reaction is 0.4 g. The amount of energy liberated is $n \times 10^7$ kWh, where n = ……. (Speed of light $= 3 \times 10^8$ m/s) *[2024, 31 Jan Shift-I]*

37 In a nuclear fission process, a high mass nuclide ($A \approx 236$) with binding energy 7.6 MeV/Nucleon dissociated into middle mass nuclides ($A \approx 118$), having binding energy of 8.6 MeV/Nucleon. The energy released in the process would be ……….. MeV. *[2024, 27 Jan Shift-I]*

38 ${}_{92}^{238}A \longrightarrow {}_{90}^{234}B + {}_2^4D + Q$
In the given nuclear reaction, the approximate amount of energy released will be
[Given, mass of ${}_{92}^{238}A = 238.05079 \times 931.5$ MeV$/c^2$,
mass of ${}_{92}^{234}B = 234.04363 \times 931.5$ MeV$/c^2$,
mass of ${}_2^4D = 4 \cdot 00260 \times 931 \cdot 5$ MeV$/c^2$] *[2023, 13 April Shift-I]*

39 A common example of a alpha decay is
$${}_{92}^{238}U \longrightarrow {}_{90}^{234}Th + {}_2He^4 + Q$$
Given, ${}_{92}^{238}U = 238.05060$ u, ${}_{90}^{234}Th = 234.04360$ u,
$${}_2^4He = 4.00260 \text{ u and } 1u = 931.5 \frac{MeV}{c^2}$$
The energy released Q during the alpha decay of ${}_{92}^{238}U$ is ………… MeV *[2023, 12 April Shift-I]*

40 A nucleus with mass number 242 and binding energy per nucleon as 7.6 MeV breaks into two fragment each with mass number 121. If each fragment nucleus has binding energy per nucleon as 8.1 MeV, the total gain in binding energy is .... MeV.

[2023, 8 April Shift-I]

41 Nucleus A having $Z = 17$ and equal number of protons and neutrons has 1.2 MeV binding energy per nucleon. Another nucleus B of $Z = 12$ has total 26 nucleons and 1.8 MeV binding energy per nucleons. The difference of binding energy of B and A will be .... MeV.

[2023, 1 Feb Shift-II]

42 A nucleus disintegrates into two smaller parts, which have their velocities in the ratio 3 : 2. The ratio of their nuclear sizes will be $(x/3)^{1/3}$. The value of x is ....

[2023, 25 Jan Shift-II]

43 The energy released per fission of nucleus of ^{240}X is 200 MeV. The energy released, if all the atoms in 120 g of pure ^{240}X undergo fission is .... $\times 10^{25}$ MeV. (Given, $N_A = 6 \times 10^{23}$)

[2023, 24 Jan Shift-II]

44 From the given data, the amount of energy required to break the nucleus of aluminium $^{27}_{13}$Al is .... $x \times 10^{-3}$ J.

Mass of neutron = 1.00866 u

Mass of proton = 1.00726 u

Mass of aluminium nucleus = 27.18846 u

(Assume 1 u corresponds to x joule of energy)

(Round off to the nearest integer) *[2021, 25 July Shift-II]*

Key Index

Topic 1

1 (b)	2 (a)	3 (a)	4 (c)	5 (c)	6 (b)	7 (c)	8 (d)	9 (c)	10 (b)
11 (b)	12 (27)	13 (4)	14 (11)						

Topic 2

1 (b)	2 (c)	3 (a)	4 (a)	5 (b)	6 (c)	7 (d)	8 (a)	9 (b)	10 (c)
11 (b)	12 (a)	13 (b)	14 (b)	15 (b)	16 (b)	17 (d)	18 (c)	19 (a)	20 (a)
21 (c)	22 (d)	23 (d)	24 (d)	25 (a)	26 (c)	27 (c)	28 (c)	29 (a)	30 (d)
31 (d)	32 (d)	33 (5)	34 (727)	35 (208)	36 (1)	37 (236)	38 (4.25)	39 (4)	40 (121)
41 (6)	42 (2)	43 (6)	44 (27)						

Detailed **Explanations**

Topic 1

01. *(b)* Density of nucleus is independent of mass number A.

02. *(a)* Nuclear force does not depends on the distance between nucleons but on their relative orientation. So, it is not a central force.

03. *(a)* Both A and R are true.
$\Rightarrow$ For nuclei with atomic number 30 to 170 there is a saturation in nuclear force per nucleon as nucleons being added at large distance do not make much impact on the nuclear forces at a specific region.

04. *(c)* Nuclear density is independent of mass number, hence A is incorrect. The radius R of nucleus is related to its mass number A as $R = R_0 A^{1/3}$

where, R_0 is constant.

05. *(c)*

key idea Nuclear density is independent of mass number.

Nuclear density is independent of mass number of nuclei, hence option (c) is correct.

06. *(b)* The decay of a proton to neutron is possible only inside the nucleus because the mass of proton is less than the mass of neutron. This mass difference is compensated with the binding energy of nucleus. This decay is not possible outside the nucleus because particle cannot be decayed into the greater mass.

7. *(c)* Nuclear density $\rho = \dfrac{m}{V}$

$$= \frac{mA}{\frac{4}{3}\pi R^3}$$

$$= \frac{mA}{\frac{4}{3}\pi (R_0 A^{1/3})^3} \quad [\because R = R_0 A^{1/3}]$$

$$= \frac{mA}{\frac{4}{3}\pi R_0^3 A}$$

$$= \frac{3m}{4\pi R_0^3}$$

$\therefore$ ρ is independent mass number (A), i.e. ρ is almost constant.

8. *(d)*

key idea A neutron outside the nucleus is unstable having mean life of 1000 s.
So, if breaks into proton and electron given by the relation

$$n \rightarrow p + e^- + \bar{\nu}$$

A proton outside the nucleus is stable but inside the nucleus it can break into neutron and position for the stablity of nucleus.

$$p \rightarrow n + e^+ + \nu$$

In the following options, (b) and (c) are not following charge conservation and in neutron decay, anti-neutrino ($\bar{\nu}$) is released

$$n \rightarrow p + e^- + \bar{\nu}$$

09. *(c)* Radius of nucleus is, $R = R_0 A^{1/3}$

For nucleus of mass number 192, radius is

$$R = R_0\ (192)^{1/3} \quad \ldots(i)$$

For nucleus of radius $R/2$ let mass number is A', then

$$R/2 = R_0 A'^{1/3} \quad \ldots(ii)$$

Dividing Eq. (i) by Eq. (ii), we get

$$2A'^{1/3} = 192^{1/3}$$

$$\Rightarrow \quad 8A' = 192$$

$$\Rightarrow \quad A' = 192/8 = 24$$

10. *(b)* We know that radius of nucleus,

$$R = R_0 A^{1/3} \quad \ldots(i)$$

$$\therefore \quad V = \frac{4}{3}\pi R^3 = \frac{4}{3}\pi (R_0 A^{1/3})^3$$

[From Eq. (i)]

$$\Rightarrow \quad V = \frac{4}{3}\pi R_0^3 A$$

$$\Rightarrow \quad V \propto A$$

Hence, Statement A is correct and Statement B is incorrect.

If m is the average mass per nucleon, then mass of nucleus, $M = mA$

$\therefore$ Density of nucleus,

$$\rho = \frac{M}{V} = \frac{mA}{\frac{4}{3}\pi R_0^3 A},\ \rho = \frac{3m}{4\pi R_0^3}$$

From above equation of density of nucleus, we see that density of nucleus is independent of mass number (A). Hence, Statement E is correct and Statements C and D are incorrect.

11. *(b)* Radius of nucleus is given by

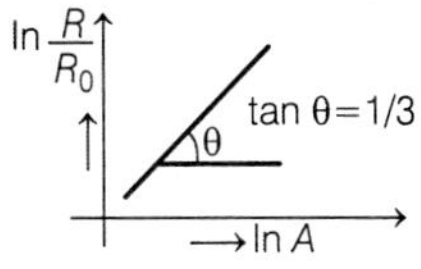

$$R = R_0 A^{1/3}$$

$$\Rightarrow \quad R/R_0 = A^{1/3}$$

$$\Rightarrow \quad \ln\frac{R}{R_0} = \frac{1}{3}\ln A$$

On comparing it with the equation of straight line $y = mx + c$, we get $m = 1/3$ and $c = 0$

12. *(27)* Using $R = R_0 A^{1/3}$, we have

$$4.8 = R_0 (64)^{1/3} \quad \ldots(i)$$

Also, $$4 = R_0 A^{1/3} \quad \ldots(ii)$$

Dividing Eq. (ii) by Eq. (i) we get,

$$\frac{4}{4.8} = \left(\frac{A}{64}\right)^{1/3}$$

$$\frac{40}{48} = \left(\frac{A}{64}\right)^{1/3}$$

$$\Rightarrow \quad \frac{125}{216} = \frac{A}{64}$$

$$\Rightarrow \quad A = \frac{125 \times 64}{216} = \frac{8000}{216} = \frac{1000}{27}$$

$$= \frac{1000}{x}\ \text{(given)} \quad \therefore\ x = 27$$

13. *(4)* Radius of a nucleus, $R = R_0 A^{1/3}$

So, volume is $V = \dfrac{4}{3}\pi R_0^3 A$

Now, ratio $\dfrac{V_2}{V_1}$ is, $\dfrac{V_2}{V_1} = \dfrac{A_2}{A_1} = \dfrac{4A_1}{A_1} = 4$

14. *(11)* Mass of one nucleon

$$= 1.6 \times 10^{-27}\ \text{kg}$$

Volume of nucleus $= \dfrac{4}{3}\pi R^3$

$$= \frac{4}{3}\pi (1.5 \times 10^{-15})^3 A$$

where, A = number of nucleons in nucleus. So, Nuclear density

$$= \frac{\text{Mass}}{\text{Volume}} = \frac{A \times 1.6 \times 10^{-27}}{\frac{4}{3}\pi \times (1.5 \times 10^{-15})^3 \times A}$$

$$= \frac{1.6 \times 10^{-27}}{\frac{4}{3}\pi \times (1.5)^3 \times 10^{-45}} \text{ kg/ m}^3$$

$$= 1.1 \times 10^{17} \text{ kg / m}^3$$

Hence, ratio of nuclear density and density of water $= 1000 \text{ kg/ m}^3$ will be

$$\frac{\rho_{\text{nucleus}}}{\rho_{\text{water}}} = \frac{1.1 \times 10^{17}}{1000} = 11 \times 10^{13}$$

$\therefore \quad n \times 10^{13} \Rightarrow n = 11$

Topic 2

01. *(b)* ${}_1H^2 + {}_1H^2 \longrightarrow {}_1He^4$

Energy released

$= (\text{BE})_{\text{reactant}} - (\text{BE})_{\text{product}}$

$= 1.1 \times 2 + 1.1 \times 2 - 7 \times 4$

$= -23.6 \text{ MeV}$

Negative sign indicates that energy is released.

02. *(c)* Binding energy per nucleon is found to be practically independent of the atomic number A, for nuclei with mass number between 30 and 170. Nuclear force is short range force. A is true but R is false.

03. *(a)* By conservation of momentum,

$p_{\text{initial}} = p_{\text{final}}$

$0 = m_1 v_1 + m_2 v_2$

As, $\frac{m_1}{m_2} = \frac{2}{1}$ given $\Rightarrow \frac{v_1}{v_2} = -\left[\frac{1}{2}\right]$

Thus, the velocity ratio will be $1:2$ in opposite directions.

04. *(a)* From mass energy equivalence relation, we have, $E = Mc^2$

$E = 1 \times 10^{-3} \times (3 \times 10^8)^2 \text{ J}$

$= 1 \times 10^{-3} \times 9 \times 10^6 \times 6.24 \times 10^{18} \text{ eV}$

$E = 56.169 \times 10^{31} \text{ eV}$

$E = 5.6 \times 10^{26} \text{ MeV}$

05. *(b)* Nuclear BE $=$ Mass defect $\times c^2$

$=$ (Mass of nucleons $-$ mass of nucleus) c^2

$= (ZM_p + NM_n - M_0)c^2$

Here, $Z = 5$, $N = 12 - 5 = 7$

$\therefore$ Binding energy is,

$\text{BE} = (5M_p + 7M_n - M_0)c^2$

06. *(c)* Given reaction is,

${}^{236}_{92}X \longrightarrow {}^{141}_{56}Y + {}^{92}_{36}Z + 3R$

Both A and Z should be conserved,

We have, $3A = 236 - 141 - 92$

$A = 1$

and $\quad 3Z = 92 - 56 - 36 \Rightarrow Z = 0$

So, particle is, ${}^1_0R$ which is a neutron.

07. *(d)* The complete nuclear fission reaction of U^{235} is,

${}^1_0n + {}^{235}_{92}U \rightarrow {}^{144}_{56}Ba + {}^{89}_{36}Kr + 3\,{}^1_0n$

Here, atomic and mass number are balanced. Hence option (d) is correct.

08. *(a)* Given, ${}_6C^{12} = 12.000000$ u

${}_6C^{13} = 13.003354$ u, ${}_0n^1 = 1.008665$ u

Mass defect,

$\Delta M = [12.000000 + 1.008665] - 13.003354$

$= 0.005311$ u

The required energy,

$\Delta E = 0.005311 \times 931.5$

$= 4.9471 \approx 4.95$ MeV

09. *(b)* Binding energy $= \Delta mc^2$

$= [Zm_p + (A - Z)m_n - m_{He}]c^2$

$= [Zm_p + (A - Z)m_n - m_{He}] \times 931.5$ MeV

$= (2 \times 1.0073 + (4 - 2) \times 1.0087 - 4.00154) \times 931.5$

Binding energy $= 28.4$ MeV

10. *(c)* After decay, the daughter nuclei will be more stable, hence binding energy per nucleon will be more than that of their parent nucleus.

11. *(b)* By conservation of momentum,

$0 = \frac{M}{2}v_1 - \frac{M}{2}v_2$

$v_1 = v_2 \quad$ ...(i)

$\Delta mc^2 = \frac{1}{2} \cdot \frac{M}{2} \cdot v_1^2 + \frac{1}{2} \cdot \frac{M}{2} \cdot v_2^2 \quad$ ...(ii)

$\Delta mc^2 = \frac{M}{2}v_1^2 \Rightarrow \frac{2\Delta mc^2}{M} = v_1^2$

$\Rightarrow v_1 = c\sqrt{2\Delta m / M}$

12. *(a)* 1st reaction is fusion and 4th reaction is fission.

13. *(b)* Statement I is true and Statement II is false which can be directly concluded from binding energy/ nucleon curve.

14. *(b)* Binding energy,

$\text{BE} = (M_{\text{nucleus}} - M_{\text{nucleons}})c^2$

$= (M_o - 8M_p - 9M_n)c^2$

15. *(b)* From the equation,

Energy of proton $+ (7 \times 5.60)$

$= 2 \times [4 \times 7.06]$

$\therefore$ Energy of proton $= 17.28$ MeV

16. *(b)* The notation $X(n, \alpha) \rightarrow {}^7_3Li$ means X atom bombarded with neutrons and breaks into Lithium and α-particle.

${}^A_ZX + {}^1_0n \longrightarrow {}^7_3Li + {}^4_2He$

It implies that, $A + 1 = 7 + 4$

$\Rightarrow \quad A = 10$

and $\quad Z + 0 = 3 + 2$

$\Rightarrow \quad Z = 5$

17. *(d)* $\frac{3}{2}k_BT = 7.7 \times 10^{-14}$ J

or $\quad T = \frac{2 \times 7.7 \times 10^{-14}}{3 \times 1.38 \times 10^{-23}}$

$= 3.7 \times 10^9$ K

18. (3) $P \rightarrow Q \rightarrow R$

graph between radioactive material mass and time

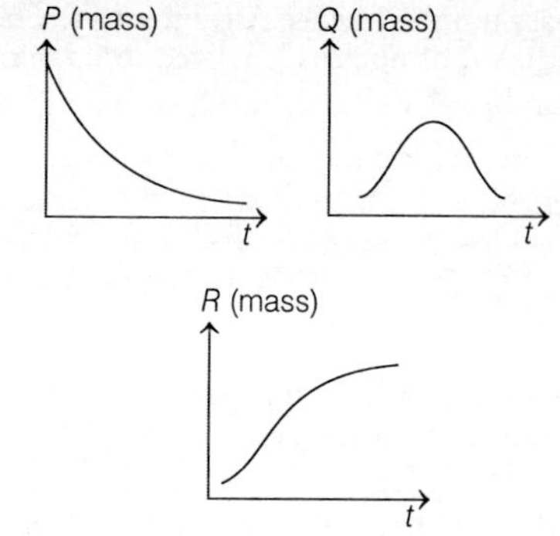

19. *(a)* Binding energy, $E = \Delta M \cdot c^2 \quad$ ...(i)

$\Delta M =$ mass defect, $c = 3 \times 10^8$ m/s

$E = 18 \times 10^8$ J

Substituting values in Eq. (i) we get,

$\Delta M = \frac{E}{c^2} = \frac{18 \times 10^8}{(3 \times 10^8)^2}$

$= 2 \times 10^{-8} \text{ kg} = 2 \times 10^{-8} \times 10^3 \text{ g}$

$= 2 \times 10^{-5} \text{ g} = 20 \times 10^{-6} \text{ g} = 20 \mu\text{g}$

20. *(a)*

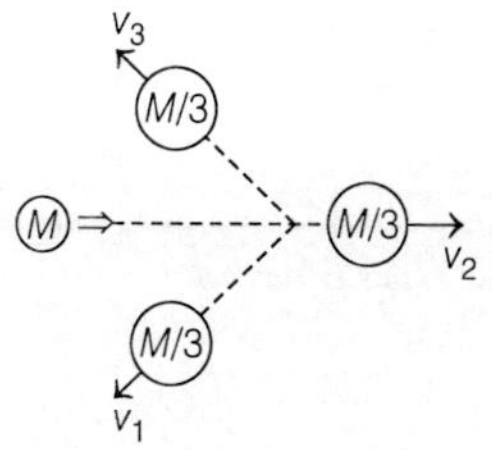

Energy released in process of fission is,

$E = \Delta Mc^2$

$\Delta M =$ mass defect

This energy appears as KE of 3 daughter nuclei, each of equal mass, so they must have equal speeds.

$$\therefore \Delta M \cdot c^2 = 3 \times \frac{1}{2}\left(\frac{M}{3}\right) \cdot v^2$$

$$\Rightarrow \quad v = c \cdot \sqrt{2\Delta M / M}$$

21. *(c)* In given question,

$${}_3\text{Li}^6 + {}_0n^1 \rightarrow {}_2\text{He}^4 + {}_1\text{H}^3$$

$${}_1\text{H}^2 + {}_1\text{H}^3 \rightarrow {}_2\text{He}^4 + {}_0n^1$$

So net reaction is,

$${}_3\text{Li}^6 + {}_1\text{H}^2 \rightarrow 2({}_2\text{He}^4)$$

Energy released in process, $Q = \Delta mc^2$

$$Q = [m(\text{Li}) + m({}_1\text{H}^2) - 2\,m({}_2\text{He}^4)] \times 931.5 \text{ MeV}$$

$$Q = [6.01690 + 2.01471 - 2 \times 4.00388] \times 931.5 \text{ MeV}$$

$Q = 22.216$ MeV, $Q = 22.22$ MeV

22. *(d)* Binding energy per nucleon of nucleus A, $E_A = 5.6 \text{ MeV} \Rightarrow M_A = 220$

Binding energy per nucleon of nucleus B and C,

$$E_B = E_C = E = 6.4 \text{ MeV}$$
$$M_B = 105, M_C = 115$$

$\therefore$ Released energy per fission

$Q =$ Total binding energy of products Nuclei – Total binding energy of reactants nuclei

$= (M_B + M_C)E - M_A \times E_A$

$= (105 + 115)(6.4) - 220 \times 5.6$

$= 176$ MeV

23. *(d)* $m_{\text{theoretical}} = Zm_p + (A - Z)m_n$

$= 50(1.00783\text{ u}) + (120 - 50)\ (1.00867\text{ u})$

$= 50.391500\text{ u} + 70.606900\text{ u}$

$= 120.998400$ u

$m_{\text{experimental}} = 119.902199$ u

So, mass defect,

$\Delta m = m_{\text{theoretical}} - m_{\text{experimental}}$

$= 120.998400\text{ u} - 119.902199\text{ u}$

$= 1.096201$ u

Now, total binding energy,

$\text{BE} = \Delta m \times 931$ MeV

$= 1.096201 \times 931$ MeV

$= 1020.563131$ MeV

Therefore, binding energy per nucleon,

$$(\text{BE})_{\text{per nucleon}} = \frac{\text{BE}}{\text{Total number of nucleons}}$$

$$= \frac{1020.563131}{120}$$

$= 8.50469275833$ MeV

≈ 8.5 MeV

24. *(d)*

> **key idea** You can calculate absorbed or released energy in nuclear reaction with the help of following formula.
>
> $\Delta Q =$ Binding energy of product – Binding energy of nuclei

Energy absorbed or released in a nuclear reaction is given by;

$$\Delta Q = \text{Binding energy of products} - \text{Binding energy of reactants.}$$

If energy is absorbed, ΔQ is negative and if it is positive then energy is released. Also, Binding energy = Binding energy per nucleon × Number of nucleons.

Here, binding energy of products

$= 2 \times (\text{BE of He}^4) + (\text{BE of C}^{12})$

$= 2\ (4 \times 7.07) + (12 \times 7.86)$

$= 150.88$ MeV

and binding energy of reactants

$= 20 \times 8.03 = 160.6$ MeV

So, $\Delta Q = (\text{BE})_{\text{Products}} - (\text{BE})_{\text{reactants}}$

$= 150.88 - 160.6 = -9.72$ MeV.

As ΔQ is negative

$\therefore$ energy of 9.72 Mev is absorbed in the reaction.

$\therefore$ Option (d) is correct.

25. *(a)* Energy released = Mass defect $\times c^2$

$= (m_n - m_p - m_e) \times c^2$

[$\because$ energy released mass of reactant – mass of product $\times c^2$]

$$= \frac{(1.6750 \times 10^{-27} - 1.6725 \times 10^{-27} - 9 \times 10^{-31})}{1.66 \times 10^{-27}} \times 931.5 \text{ MeV} \quad [\because 1\text{amu} = 931.5 \text{ MeV}]$$

$$= \frac{16 \times 10^{-31}}{1.66 \times 10^{-27}} \times 931.5 \text{ MeV}$$

$= 0.9$ MeV

which is nearest to 0.73 MeV

26. *(c)* The nuclear reaction can be represented as

$${}^7_3\text{Li} + {}^1_1\text{H} \longrightarrow {}^8_4\text{Be} + {}^A_Z X$$

Applying conservation of atomic number (charge), $3 + 1 = 4 + Z \Rightarrow Z = 0$

Applying conservation of atomic mass,

$$7 + 1 = 8 + A \Rightarrow A = 0$$

Thus, the emitted radiation is γ-photons $({}^0_0 X)$.

27. *(c)* As given

$${}^2_1\text{H} + {}^2_1\text{H} \longrightarrow {}^4_2\text{He} + \text{energy}$$

The binding energy per nucleon of a deuteron $({}^2_1\text{H}) =$ 1.1 MeV

$\therefore$ Total binding energy of one deuteron nucleus $= 2 \times 1.1 = 2.2$ MeV

The binding energy per nucleon of helium $({}^4_2\text{He}) = 7$ MeV

$\therefore$ Total binding energy

$= 4 \times 7 = 28$ MeV

Hence, energy released in the above process

$= 28 - 2 \times 2.2$

$= 28 - 4.4$

$= 23.6$ MeV

28. *(c)*

(A) ${}^1_0n + {}^{235}_{92}\text{U} \rightarrow {}^{140}_{54}\text{Xe} + {}^{94}_{38}\text{Sr} + 2\,{}^1_0n$ is a fission process of Uranium $[{}^{235}_{92}\text{U}]$

B. $2\text{H}_2 + \text{O}_2 \rightarrow 2\text{H}_2\text{O}$ is a chemical reaction

C. ${}^1_1\text{H} + {}^2_1\text{H} \rightarrow {}^3_2\text{He} + {}^1_0n$

Here fusion of Hydrogen and Deuterium is taking place with the Q value.

D. ${}^1_1\text{H} + {}^3_1\text{H} \rightarrow {}^2_1\text{H} + {}^2_1\text{H}$

Here resulting products are less stable and have lower binding energy per nucleon. Thus, energy must be supplied to make this reaction happen.

Therefore, in this case fusion is taking place with Q value

A $\rightarrow$ III

B $\rightarrow$ I

C $\rightarrow$ II

D $\rightarrow$ IV

29. *(a)* 1 mol of ${}^{235}\text{U} = 235$ g

$\Rightarrow \quad 1N_A / 235 = 2\text{g}$

$$\Rightarrow \quad 2\text{ kg} = \frac{2 \times 10^3 \times 6.023 \times 10^{23}}{235}$$

$= 51.25 \times 10^{23}$ atoms

Energy released in 2 kg ${}^{235}\text{U} = 51.25 \times 10^{23} \times 200$ MeV

1 mole of ${}^1_1\text{H} = 1$ g

$\Rightarrow$ 2 kg of Hydrogen $= 2 \times 10^3 N_A$

$= 2 \times 10^3 \times 6.023 \times 10^{23}$ atoms

Now, 4 Hydrogen produces 26.7 MeV.

Thus, 2 kg hydrogen produce energy

$$= \frac{2 \times 10^3}{4} \times 6.023 \times 10^{23} \times 26.7 \text{ MeV}$$

$$\frac{E_\text{H}}{E_\text{U}} = \frac{3.0115 \times 10^3 \times 10^{23} \times 26.7}{51.25 \times 10^{23} \times 200} = 7.84$$

Which is closest of 7.62 option (a).

30. *(d)* Given, initial mass number be A is 184 and Q value is 5.5 MeV.

Equation of α-emission is

$$X_Z^A \longrightarrow Y_{Z-2}^{A-4} + \alpha + Q$$

$$X_Z^{184} \longrightarrow Y_{Z-2}^{180} + \alpha + Q$$

$p_y, p_\alpha, m_y, m_\alpha, \text{KE}_Y$ and KE_α be the momentum, mass, kinetic energy of Y and α, respectively.

By using conservation of momentum,

Momentum of α = Momentum of Y

$$\Rightarrow \quad p_\alpha = p_Y$$

$$\therefore \quad \sqrt{2m_\alpha E_\alpha} = \sqrt{2m_y E_y}$$

$$[\because \text{Momentum, } p = \sqrt{2mE}]$$

$$\Rightarrow E_\alpha = \frac{180}{4} E_Y = 45 E_Y$$

$$\Rightarrow E_Y = \frac{E_\alpha}{45}$$

By using conservation of energy,

$$Q = \text{KE}_Y + \text{KE}_\alpha$$

$$= \frac{\text{KE}_\alpha}{45} + \text{KE}_\alpha$$

$$\Rightarrow 5.5 = \text{KE}_\alpha (46/45)$$

$$\Rightarrow \text{KE}_\alpha = \frac{5.5 \times 45}{46} = 5.38 \text{ MeV}$$

31. *(d)* Energy of γ-ray, $E_\gamma = h\nu$

and momentum of γ-ray, $p_\gamma = h/\lambda$ ...(i)

As, $\quad p_\gamma = E_\gamma / c = h\nu / c$ ...(ii)

From Eqs. (i) and (ii), we get

$$p_\gamma = \frac{h\nu}{c} = \frac{h}{\lambda} \quad \left[\because \lambda = \frac{c}{\nu}\right]$$

Since, during the emission of γ-ray photon, momentum is conserved.

$$\therefore p_\gamma + p_{\text{decayed nuclei}} = 0$$

$$\Rightarrow p_\gamma = p_{\text{decayed nuclei}}$$

$$\Rightarrow h\nu / c = p_{\text{decayed nuclei}} \quad \text{...(iii)}$$

Kinetic energy of decayed nuclei,

$$\text{KE} = \frac{1}{2}\left(\frac{p^2_{\text{decayed nuclei}}}{M}\right)$$

$$\text{K.E} = \frac{(p^2_{\text{decayed nuclei}})}{2M} \quad \text{...(iv)}$$

From Eqs. (iii) and (iv), we get

$$\text{KE} = \frac{1}{2M}\left[\frac{h\nu}{c}\right]^2$$

$\therefore$ Loss in internal energy

$$= E_\gamma + \text{KE}_{\text{decayed nuclei}}$$

$$= h\nu + \frac{1}{2M}\left[\frac{h\nu}{c}\right]^2$$

$$= h\nu\left[1 + \frac{h\nu}{2Mc^2}\right]$$

32. *(d)*

> **key idea** In such type of question, calculate mass defect (Δm) in terms of atomic mass unit (amu). After that, you can calculate released energy with the help of following formula.
>
> $\Delta Q = \Delta m \times 931$ MeV

Given that $m(^7_3\text{Li}) = 7.0160$ u

$m(^4_2\text{He}) = 4.0026 \text{ u} \Rightarrow m(^1_1\text{H}) = 1.0079$ u

Nuclear reaction :

$$^7_3\text{Li} + ^1_1\text{H} \longrightarrow ^4_2\text{He} + ^4_2\text{He}$$

Energy released in one nuclear reaction,

$$Q = \Delta mc^2 = \Delta m \times 931 \text{ MeV}$$

$$= (7.0160 + 1.0079 - 2 \times 4.0026) \times 931$$

$$= 17.41 \text{ MeV}$$

Number of atoms in 20 g of Li

$$= (20/7) \times 6.023 \times 10^{23} = 1.72 \times 10^{24}$$

$\therefore$ Total energy liberated

$$= Q \times \text{Total number of Li atoms}$$

$$= (17.41 \text{ MeV}) \times (1.72 \times 10^{24}) \times 1.6 \times 10^{-19}$$

$$= 4.79 \times 10^{12} \text{ J}$$

Energy (in kWh)

$$= \frac{4.79 \times 10^{12}}{3.6 \times 10^6} \quad (\because 1 \text{ kWh} = 3.6 \times 10^6 \text{ J})$$

$$= 1.33 \times 10^6 \text{ kWh}$$

33. *(5)* $^4\text{He} + ^4\text{He} + ^4\text{He} \longrightarrow ^{12}\text{C} + Q$

Power generated $= \dfrac{N}{t} Q$

where $N \rightarrow$ Number of reactions

$$Q = (3m_{\text{He}} - m_{\text{C}})c^2$$

$$= (3 \times 4.0026 - 12)(3 \times 10^8)^2$$

$$Q = 7.266 \text{ MeV}$$

$$\frac{N}{t} = \frac{\text{Power}}{Q} = \frac{5.808 \times 10^{30}}{7.266 \times 10^6 \times 1.6 \times 10^{-19}}$$

$$N/t = 5 \times 10^{42} \text{s}^{-1} \quad \therefore \quad n = 5$$

34. *(727)* The complete nuclear fusion reaction is, $3\,^4_2\text{He} \rightarrow ^{12}_6\text{C} + \gamma$-rays

Mass defect $= \Delta m = (3m_{\text{He}} - m_{\text{C}})$

$\Delta m = (3 \times 4.002603 - 12) = 0.007809$ u

Energy released,

$$\Delta E = \Delta m \times 931 = 7.27 \text{ MeV}$$

$$\Delta E = 727 \times 10^{-2} \text{ MeV}$$

35. *(208)* Mass of $\text{U}^{235} = 235.0439$ u

Mass of $^{140}\text{Ce} = 139.91054$ u

Mass of $^{94}\text{Zr} = 93.9063$ u

$m(n) = 1.0086$ u

$\therefore$ Mass defect $= m(^{235}\text{U}) - [m(^{140}\text{Ce}) + m(^{94}\text{Zr}) + m(n)]$

$(\Delta m) = 235.0439$ u $- (139.91054 \text{ u} + 93.9063 \text{ u} + 1.0086 \text{ u})$

$\Rightarrow \Delta m = 0.2236$ u

$\therefore$ Disintegration energy,

$$Q = \Delta m c^2 = 0.2236 \times 931$$

$$= 208.2 \text{ MeV} \simeq 208 \text{ MeV}$$

36. *(1)* Energy released $= mc^2$

$$= 0.4 \times 10^{-3} \times (3 \times 10^8)^2$$

$$= 3.6 \times 10^{13} \text{ J} = \frac{3.6 \times 10^{13}}{3.6 \times 10^6} \text{ kWh}$$

$$(\because 1 \text{ kWh} = 3.6 \times 10^6 \text{ J})$$

$$= 1 \times 10^7 \text{ kWh} \therefore n = 1$$

37. *(236)* Given in the question,

$$\text{X}^{236} \rightarrow 2\text{X}^{118}$$

before after

Initially, $\quad (\text{BE})_i = 7.6$ MeV/nucleon $= 7.6 \times 236$ MeV

Finally, $\quad (\text{BE})_f = 8.6$ MeV/nucleon $= 2 \times 8.6 \times 118$ MeV

Heat released $(Q) = (\text{BE})_f - (\text{BE})_i$

$$= 8.6 \times 2 \times 118 - 7.6 \times 236$$

$$= 2029.6 - 1793.6 = 236 \text{ MeV}$$

38. *(4.25)* In the given nuclear reaction, mass defect,

$$\Delta m = [m(^{238}_{92}A) - m(^{234}_{90}B) - m(^4_2D)]$$

$$= 238.05079 - 234.04363 - 4.00260$$

$$= 0.004564 \text{ amu}$$

$\therefore$ Amount of energy released.

$$\Delta Q = \Delta m \times 931.5 \text{ MeV}$$

$$= 0.004564 \times 931.5 = 4.25 \text{ MeV}$$

39. *(4)* Nuclear reaction is as follows

$$^{238}_{92}\text{U} \longrightarrow ^{234}_{90}\text{Th} + ^4_2\text{He} + Q$$

Energy released $(Q) = \Delta m \times 931.5$ MeV

$$= (m_{\text{U}} - m_{\text{Th}} - m_{\text{He}})_{\text{amu}} \times 931.5 \text{ MeV}$$

$$= (238.05060 - 234.04360 - 4.00260) \times 931.5 \text{ MeV}$$

$$= 0.0044 \times 931.5 \text{ MeV} = 4.0986 \text{ MeV}$$

40. *(121)*

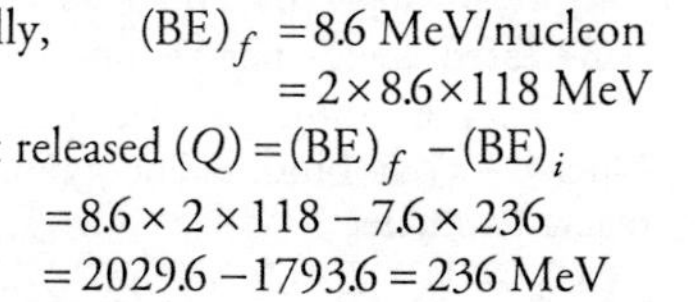

Total initial BE $= 242 \times 7.6$

$$= 1839.2 \text{ MeV}$$

Total final BE $=121\times 2\times 8.1=1960.2$
Gain in BE $=1960.2-1839.2$
$=121$ MeV

41. *(6)* For nucleus A,
Number of proton, $Z=17$ and number of neutrons, $N=17$ (Given, $Z=N$)
$A=Z+N=17+17=34$
Binding energy per nucleon
$=BE/A=1.2$ MeV
$(\text{BE})_A=1.2\times A=1.2\times 34=40.8$ MeV
For nucleus B,
Number of protons, $Z=12$,
Number of nucleons, $A=26$
$\text{BE}/A=1.8$ MeV
$(\text{BE})_B=1.8\times A=1.8\times 26=46.8$ MeV
Therefore, difference in binding energy of B and A is $\Delta\text{BE}=\text{BE}_B-\text{BE}_A$
$\Rightarrow \Delta\text{BE}=46.8-40.8$
$\Rightarrow \Delta\text{BE}=6$ MeV

42. *(2)* When nucleus disintegrates into two smaller parts of masses m_1 and m_2, then according to conservation of linear momentum, $m_1v_1-m_2v_2=0$
$\Rightarrow \quad m_1/m_2=v_2/v_1=2/3$
$[\because v_1:v_2=3:2\Rightarrow v_2/v_1=2/3]$
$\Rightarrow \quad m_1/m_2=2/3\Rightarrow \rho_1V_1/\rho_2V_2=2/3$
$\Rightarrow \quad V_1/V_2=2/3 \qquad (\because \rho_1=\rho_2)$

$$\Rightarrow \frac{\frac{4}{3}\pi r_1^3}{\frac{4}{3}\pi r_2^3}=\frac{2}{3}\Rightarrow\left(\frac{r_1}{r_2}\right)^3=\frac{2}{3}$$

$$\Rightarrow \left(\frac{r_1}{r_2}\right)=\left(\frac{2}{3}\right)^{1/3}=\left(\frac{x}{3}\right)^{1/3} \quad \text{[given]}$$

$\therefore \quad x=2$

> **Alternate Solution**
> When a nucleus disintegrates into two nuclei having velocity ratio $v_1:v_2$, then the ratio of their nuclear sizes (or radii) is given as
> $r_1/r_2=(v_2/v_1)^{1/3}=(2/3)^{1/3} \quad [\because v_1:v_2=3:2]$
> $\therefore \quad x=2$

43. *(6)* Molar mass of given isotope,
$^{240}X=240$ g
So, in 240 g number of nucleus present
$=N_A=6\times 10^{23}$ (as given in question)
$\therefore$ In 120 g, number of nucleus present
$=N_A/2=3\times 10^{23}$
1 nucleus gives 200 MeV of energy in fission. So, total energy obtained by fission of 120 g of ^{240}X will be
$3\times 10^{23}\times 200=6\times 10^{25}\approx 6\times 10^{25}$ MeV

44. *(27)* As we know that,
Binding energy, $\text{BE}=\Delta mc^2$
Mass defect, $(\Delta m)=Zm_p+(A-Z)m_n-m$
where, Z is atomic number or number of proton and $A-Z$ is number of neutron.
$\therefore \Delta m=13\times 1.00726+14\times 1.00866$
-27.18846
$=13.09438+14.12124-27.18846$
$=0.02716$ u
Given that $1\text{ u}=x$ J,
$E=27.16\times 10^{-3}$ J

In JEE Main there are usually 0-2 questions per paper are visible from this chapter. In past 3 years, 57 conceptual and 1 numerical questions were asked of moderate to difficult level from this chapter.

For detailed analysis scan QR code

SEMICONDUCTOR ELECTRONICS : MATERIALS, DEVICES & SIMPLE CIRCUITS

Topic 1 Semiconductor and *p-n* Junction Diode

Level 1

01 The output voltage in the following circuit is (Consider ideal diode case) *[2025, 8 April Shift-II]*

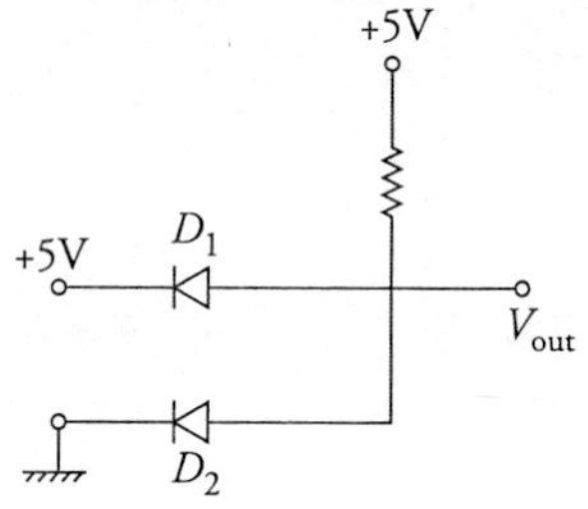

(a) −5 V (b) 10 V (c) 0 V (d) + 5 V

Imp. Biasing of *p-n* Junction Diode

02 Which of the following circuits represents a forward biased diode? *[2025, 22 Jan Shift-I]*

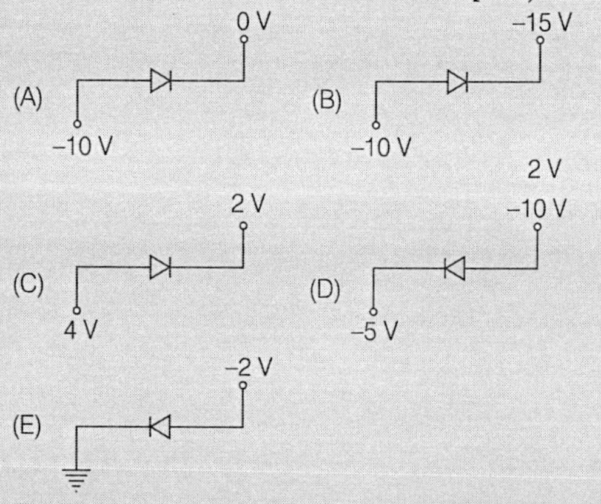

Choose the correct answer from the options given below:

Options

(a) (B), (C) and (E) only (b) (*C*) and (*E*) only
(c) (B), (D) and (E) only (d) (A) and (D) only

03 What is the current through the battery in the circuit shown below? *[2025, 23 Jan Shift-II]*

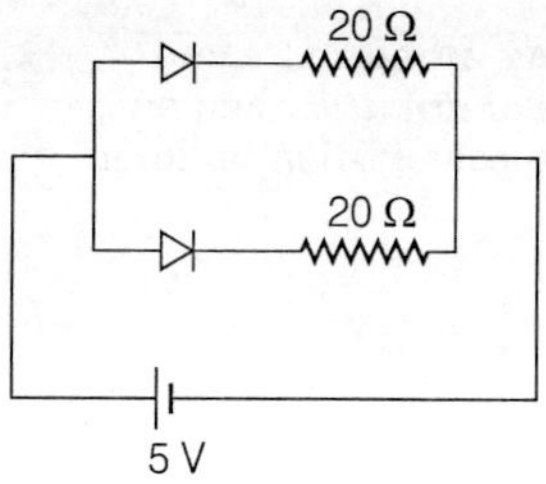

(a) 1.5 A (b) 1.0 A
(c) 0.5 A (d) 0.25 A

04 The *I-V* characteristics of an electronic device shown in the figure. The device is *[2024, 9 April Shift-II]*

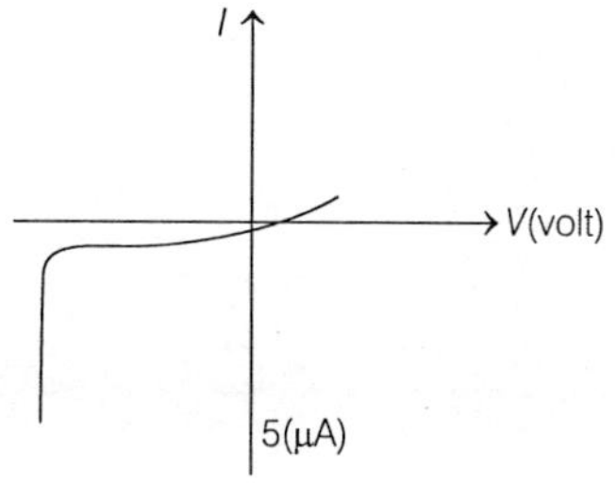

(a) a diode which can be used as a rectifier
(b) a transistor which can be used as an amplifier
(c) a zener diode which can be used as a voltage regulator
(d) a solar cell

05 A light emitting diode (LED) is fabricated using GaAs semiconducting, material whose band gap is 1.42 eV. The wavelength of light emitted from the LED is *[2024, 9 April Shift-I]*

(a) 875 nm (b) 1243 nm
(c) 650 nm (d) 1400 nm

06 Which of the diode circuit shows correct biasing used for the measurement of dynamic resistance of *p-n* junction diode? *[2024, 4 April Shift-II]*

(a) 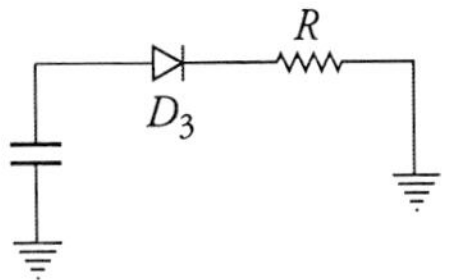

(b) 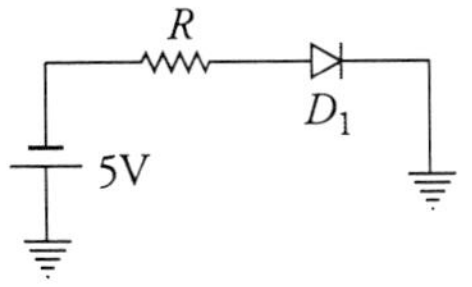

(c) 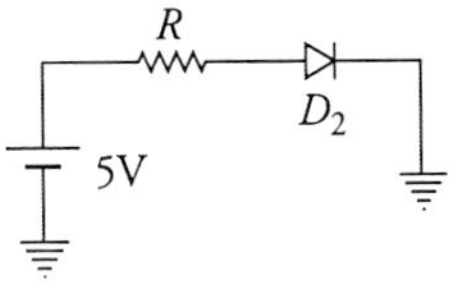

(d) 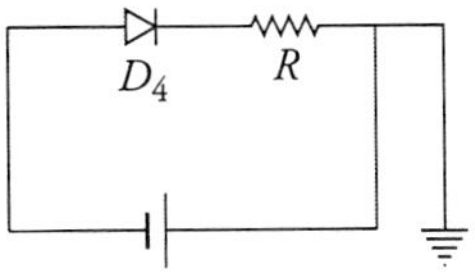

Imp. Biasing of Diodes

07 Which of the following circuits is reverse-biased? *[2024, 27 Jan Shift-I]*

(a) 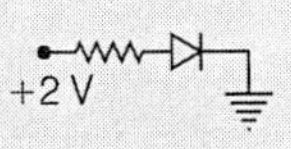

(b) 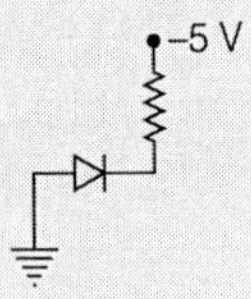

(c) 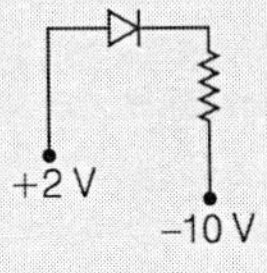

(d) 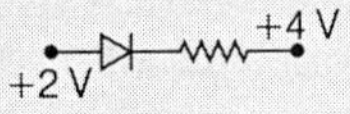

08 Given below are two statements. One is labelled as Assertion (A) and the other is labelled as Reason (R).

Assertion (A) Diffusion current in a *p-n* junction is greater than the drift current in magnitude if the junction is forward biased.

Reason (R) Diffusion current in a *p-n* junction is from the *n*-side to the *p*-side, if the junction is forward biased.

In the light of the above statements, choose the most appropriate answer from the options given below. *[2023, 6 April Shift-II]*

(a) Both A and R are correct and R is the correct explanation of A.
(b) A is not correct but R is correct.
(c) A is correct but R is not correct.
(d) Both A and R are correct but R is not the correct explanation of A.

09 The resistivity ρ of semiconductor varies with temperature. Which of the following curve represents the correct behaviour? *[2023, 6 April Shift-I]*

(a) 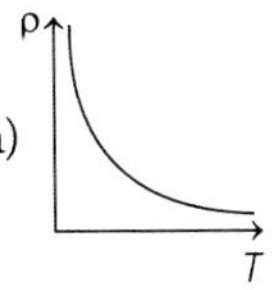

(b) 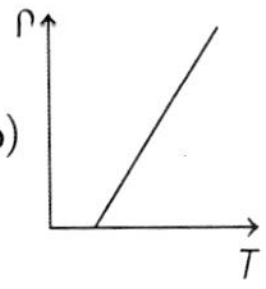

(c) 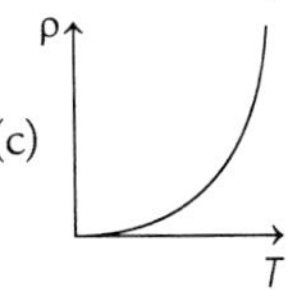

(d) 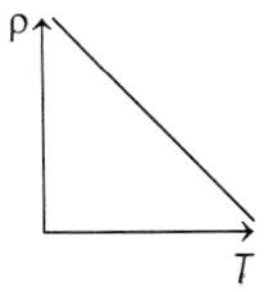

10 Choose the correct statement about zener diode. *[2023, 1 Feb Shift-II]*

(a) It works as a voltage regulator in reverse bias and behaves like simple *p-n* junction diode in forward bias.
(b) It works as a voltage regulator only in forward bias.
(c) It works as a voltage regulator in forward bias and behaves like simple *p-n* junction diode in reverse bias.
(d) It works as a voltage regulator in both forward and reverse bias.

11 Match List I with List II.

	List I		List II
A.	Intrinsic semiconductor	I.	Fermi-level near the valence band
B.	*n*-type semiconductor	II.	Fermi-level in the middle of valence and conduction band
C.	*p*-type semiconductor	III.	Fermi-level near the conduction band
D.	Metals	IV.	Fermi-level inside the conduction band

Choose the correct answer from the options given below. *[2023, 1 Feb, Shift-I]*

(a) A-I, B-II, C-III, D-IV
(b) A-III, B-I, C-II, D-IV
(c) A-II, B-III, C-I, D-IV
(d) A-II, B-I, C-III, D-I

12 The effect of increase in temperature on the number of electrons in conduction band n_e and resistance of a semiconductor will be as *[2023, 31 Jan Shift-I]*

(a) n_e increases, resistance decreases
(b) Both n_e and resistance decrease
(c) n_e decreases. resistance increases
(d) Both n_e and resistance increase

13 Which one of the following statement is not correct in the case of light emitting diodes ?

A. It is a heavily doped *p-n* junction.
B. It emits light only when it is forward biased.
C. It emits light only when it is reverse biased.
D. The energy of the light emitted is equal to or slightly less than the energy gap of the semiconductor used.

Choose the correct answer from the options given below *[2023, 29 Jan Shift-I]*

(a) B (b) C
(c) A (d) C and D

14 Given below are two statements.

One is labelled as Assertion A and the other is labelled as Reason R.

Assertion (A) Photodiodes are preferably operated in reverse bias condition for light intensity measurement.

Reason (R) The current in the forward bias is more than the current in the reverse bias for a *p-n* junction diode.

In the light of the above statements, choose the correct answer from the options given below. *[2023, 24 Jan Shift-I]*

(a) Both A and R are true and R is the correct explanation of A.
(b) Both A and R are true but R is not the correct explanation of A.
(c) A is true but R is false.
(d) A is false but R is true.

Imp. *V-I* Characteristics of Solar Cell

15 Identify the solar cell characteristics from the following options *[2022, 28 July Shift-I]*

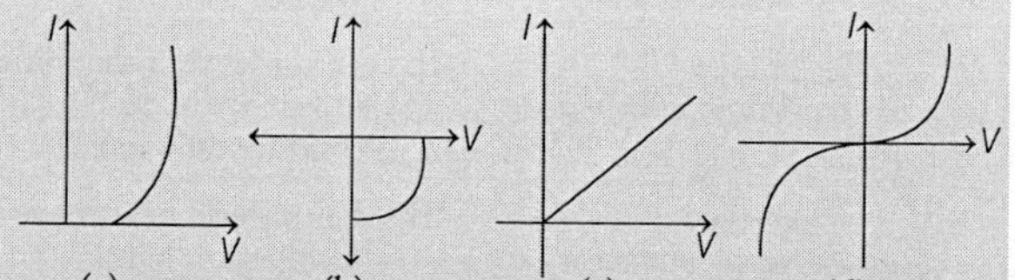

(a) (b) (c) (d)

16 For using a multimeter to identify diode from electrical components, choose the correct statement out of the following about the diode. *[2022, 28 June Shift-I]*

(a) It is two terminal device which conducts current in both directions.
(b) It is two terminal device which conducts current in one direction only.
(c) It does not conduct current gives initial deflection which decays to zero.
(d) It is three terminal device which conducts current in one direction only between central terminal and either of the remaining two terminals.

17 The photodiode is used to detect the optical signals. These diodes are preferably operated in reverse biased mode because *[2022, 25 June, Shift-I]*

(a) fractional change in majority carries produce higher forward bias current
(b) fractional change in majority carriers produce higher reverse bias current
(c) fractional change in minority carriers produce higher forward bias current
(d) fractional change in minority carriers produce higher reverse bias current.

18 With increasing biasing voltage of a photodiode, the photocurrent magnitude *[2020, 5 Sep Shift-I]*

(a) increases initially and after attaining certain value, it decreases
(b) increases initially and saturates finally
(c) remains constant
(d) increases linearly

19 Mobility of electrons in a semiconductor is defined as the ratio of their drift velocity to the applied electric field. If for an *n*-type semiconductor, the density of electrons is 10^{19} m^{-3} and their mobility is 1.6 m^2/(V-s), then the resistivity of the semiconductor (since, it is an *n*-type semiconductor contribution of holes is ignored) is close to *[2019, 9 Jan Shift-I]*

(a) 2 Ω-m (b) 0.2 Ω-m (c) 0.4 Ω-m (d) 4 Ω-m

20 The reading of the ammeter for a silicon diode in the given circuit is *[JEE Main 2018]*

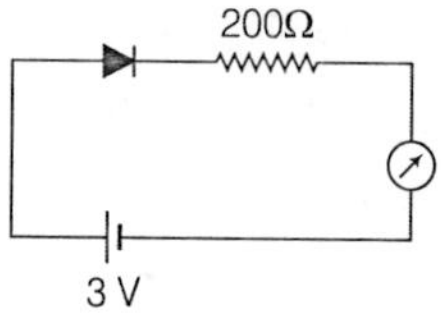

(a) 0 (b) 15 mA
(c) 11.5 mA (d) 13.5 mA

21 What is the conductivity of a semiconductor sample having electron concentration of 5×10^{18} m^{-3}, hole concentration of 5×10^{19} m^{-3}, electron mobility of 2.0 m^{2}V^{-1} s^{-1} and hole mobility of 0.01 m^2 V^{-1}s^{-1}? *[JEE Main 2017]*

(Take, charge of electron as 1.6×10^{-19} C)

(a) 1.68 (Ω-m)$^{-1}$ (b) 1.83 (Ω-m)$^{-1}$
(c) 0.59 (Ω-m)$^{-1}$ (d) 1.20 (Ω-m)$^{-1}$

22 Identify the semiconductor devices whose characteristics are as given below, in the order (A),(B),(C),(D). *[JEE Main 2016]*

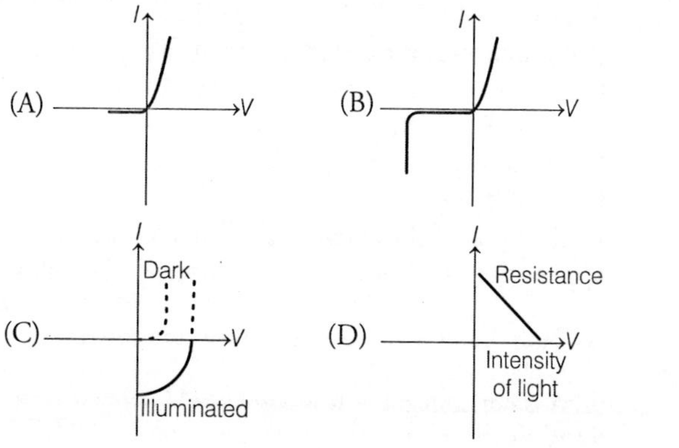

(a) Simple diode, zener diode, Solar cell, Light dependent resistance
(b) zener diode, Simple diode, Light dependent resistance, Solar cell
(c) Solar cell, Light dependent resistance, zener diode, Simple diode
(d) zener diode, Solar cell, Simple diode, Light dependent resistance

23 The temperature dependence of resistance of Cu and undoped Si in the temperature range 300-400 K, is best described by *[JEE Main 2016]*

(a) linear increase for Cu, exponential decrease of Si
(b) linear decrease for Cu, linear decrease for Si
(c) linear increase for Cu, linear increase for Si
(d) linear increase for Cu, exponential increase for Si

24 In an unbiased p-n junction, electrons diffuse from n-region to p-region because *[JEE Main 2015]*

(a) holes in p-region attract them
(b) electrons travel across the junction due to potential difference
(c) only electrons move from n to p region and not the vice-versa
(d) electron concentration in n-region is more compared to that in p-region

25 The forward biased diode connection is *[JEE Main 2014]*

(a) 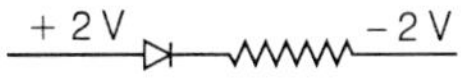

(c) 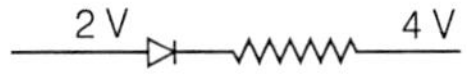

(b) 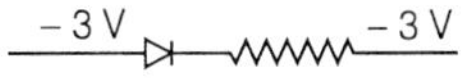

(d) 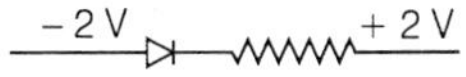

26 A p-n junction D shown in the figure can act as a rectifier. An alternating current source V is connected in the circuit.

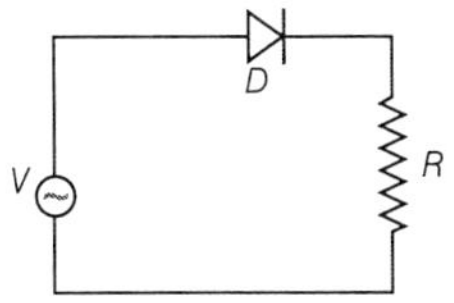

[AIEEE 2009]

(a) 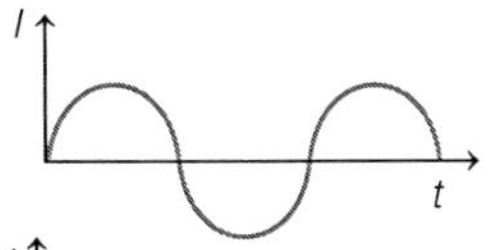

(b) 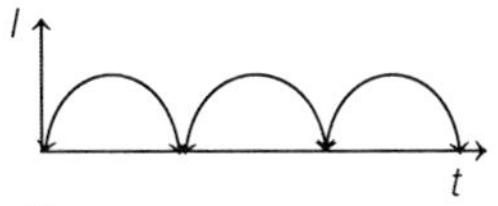

(c) 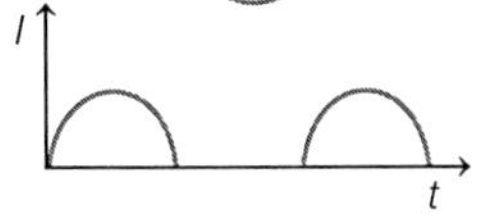

(d) 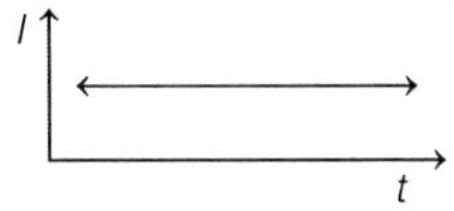

27 Carbon, silicon and germanium have four valence electrons each. At room temperature, which one of the following statements is most appropriate? *[AIEEE 2007]*

(a) The number of free conduction electrons is significant in C but small in Si and Ge
(b) The number of free conduction electrons is negligibly small in all the three
(c) The number of free electrons for conduction is significant in all the three
(d) The number of free electrons for conduction is significant only in Si and Ge but small is C

28 In the following, which one of the diodes is reverse biased? *[AIEEE 2006]*

(a) 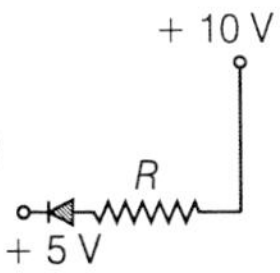

(b) 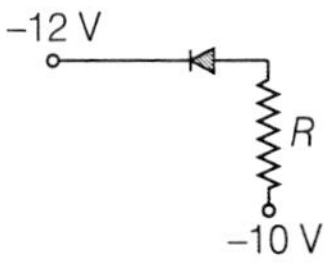

(c) 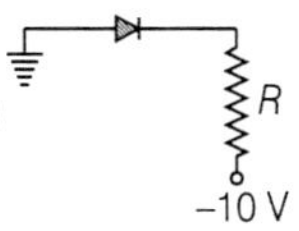

(d) 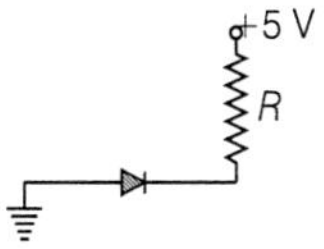

29 The electrical conductivity of a semiconductor increases when electromagnetic radiation of wavelength shorter than 2480 nm, is incident on it. The band gap in eV for the semiconductor is *[AIEEE 2005]*

(a) 1.1 eV (b) 2.5 eV (c) 0.5 eV (d) 0.7 eV

30 In a full wave rectifier circuit operating from 50 Hz mains frequency, the fundamental frequency in the ripple would be *[AIEEE 2005]*

(a) 50 Hz (b) 25 Hz (c) 100 Hz (d) 70.7 Hz

31 A piece of copper and another of germanium are cooled from room temperature to 77 K, the resistance of *[AIEEE 2004, 03]*

(a) each of them increases
(b) each of them decreases
(c) copper decreases and germanium increases
(d) copper increases and germanium decreases

32 When p-n junction diode is forward biased, then *[AIEEE 2004]*

(a) the depletion region is reduced and barrier height is increased
(b) the depletion region is widened and barrier height is reduced
(c) both the depletion region and barrier height are reduced
(d) both the depletion region and barrier height are increased

33 In the middle of the depletion layer of reverse biased p-n junction, the *[AIEEE 2003]*

(a) electric field is zero
(b) potential is maximum
(c) electric field is maximum
(d) potential is zero

34 The energy band gap is maximum in *[AIEEE 2002]*

(a) metals (b) superconductors
(c) insulators (d) semiconductors

35 By increasing the temperature, the specific resistance of a conductor and a semiconductor *[AIEEE 2002]*

(a) increases for both
(b) decreases for both
(c) increases, decreases respectively
(d) decreases, increases respectively

36 At absolute zero, Si acts as *[AIEEE 2002]*

(a) non-metal (b) metal
(c) insulator (d) None of these

Level 2

37 A Zener diode with 5 V zener voltage is used to regulate an unregulated dc voltage input of 25 V. For a 400 Ω resistor connected in series, the Zener current is found to be 4 times load current. The load current (I_L) and load resistance (R_L) are *[2025, 2 April Shift-I]*

(a) $I_L = 10$ mA; $R_L = 500\,\Omega$
(b) $I_L = 0.02$ mA; $R_L = 250\,\Omega$
(c) $I_L = 10$ A; $R_L = 0.5\,\Omega$
(d) $I_L = 20$ mA; $R_L = 250\,\Omega$

38 In the following circuit, the reading of the ammeter will be (Take, Zener breakdown voltage = 4 V) *[2025, 7 April Shift-I]*

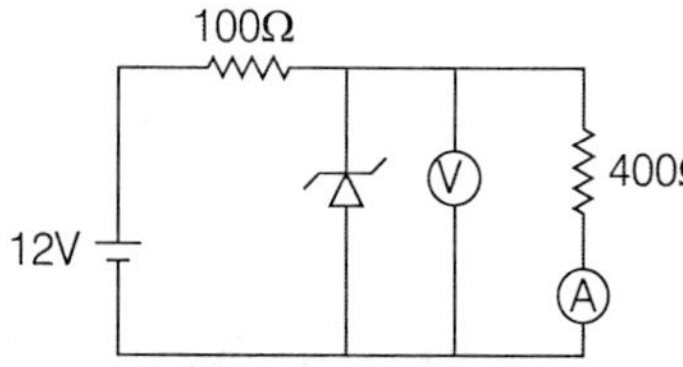

(a) 80 mA (b) 60 mA
(c) 10 mA (d) 24 mA

39 Consider a n-type semiconductor in which n_e and n_h are number of electrons and holes, respectively.
(A) Holes are minority carriers
(B) The dopant is a pentavalent atom
(C) $n_e n_h \neq n_i^2$
(where n_i is number of electrons or holes in semiconductor when it is intrinsic form)
(D) $n_e n_h \geq n_i^2$
(E) The holes are not generated due to the donors
Choose the correct answer from the options given below. *[2025, 4 April Shift-II]*
(a) (A), (C), (E) only (b) (A), (B), (E) only
(c) (A), (C), (D) only (d) (A), (B), (C) only

40

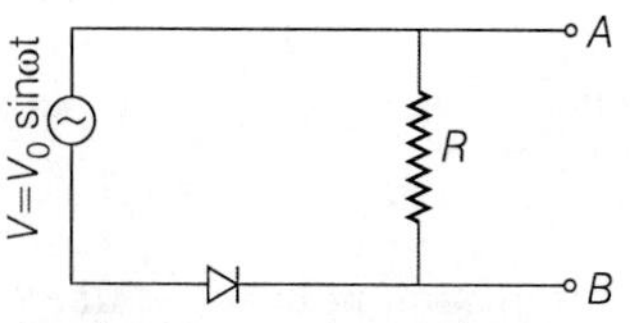

In the circuit shown here, assuming threshold voltage of diode is negligibly small, then voltage V_{AB} is correctly represented by *[2025, 28 Jan Shift-II]*

(a) V_{AB} would be zero at all time

(b)
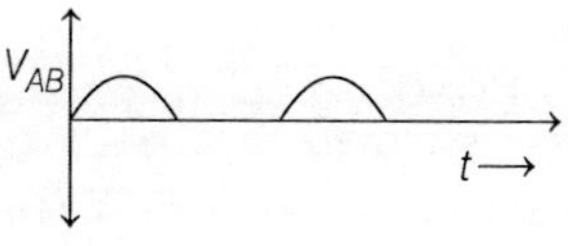

(c)
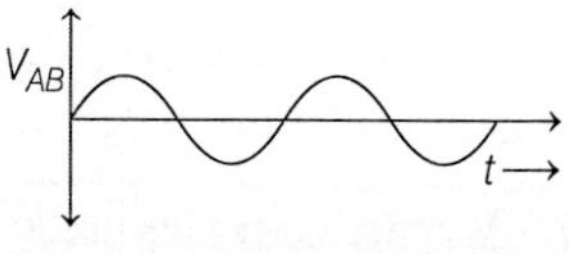

(d)
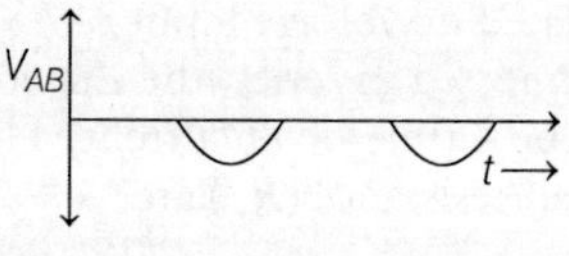

Imp. Light Emitting Diodes

41 Consider the following statements. *[2025, 24 Jan Shift-I]*
A. The junction area of solar cell is made very narrow compared to a photo diode.
B. Solar cells are not connected with any external bias.
C. LED is made of lightly doped p-n junction.
D. Increase of forward current results in continuous increase of LED light intensity.
E. LEDs have to be connected in forward bias for emission of light.
Choose the correct answer from the options given below.
(a) B, D, E Only (b) A, C Only
(c) B, E Only (d) A, C, E Only

42 The acceptor level of a p-type semiconductor is 6 eV. The maximum wavelength of light which can create a hole would be (Given, $hc = 1242$ eV-nm) *[2024, 6 April Shift-II]*
(a) 414 nm (b) 207 nm
(c) 407 nm (d) 103.5 nm

43 The value of net resistance of the network as shown in the given figure is *[2024, 4 April Shift-I]*

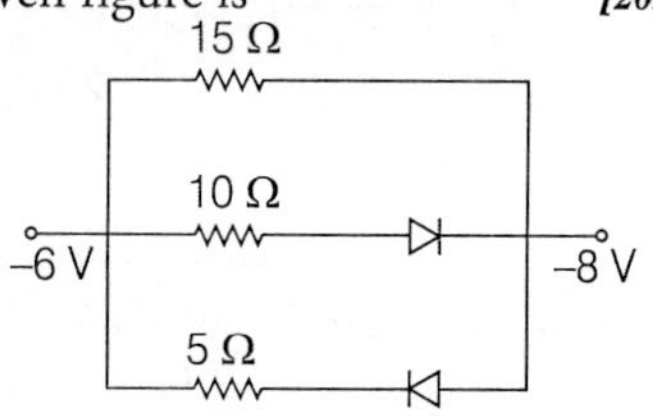

(a) $6\,\Omega$ (b) $\left(\frac{15}{4}\right)\Omega$ (c) $\left(\frac{5}{2}\right)\Omega$ (d) $\left(\frac{30}{11}\right)\Omega$

44 In the given circuit, the voltage across load resistance (R_L) is *[2024, 30 Jan Shift-II]*

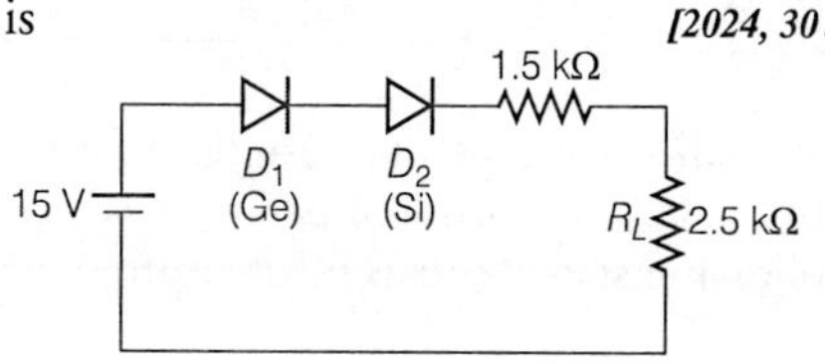

(a) 8.75 V (b) 14.00 V (c) 8.50 V (d) 9.00 V

Imp. Zener Diode

45 A Zener diode of breakdown voltage 10 V is used as a voltage regulator as shown in the figure. The current through the Zener diode is *[2024, 30 Jan Shift-I]*

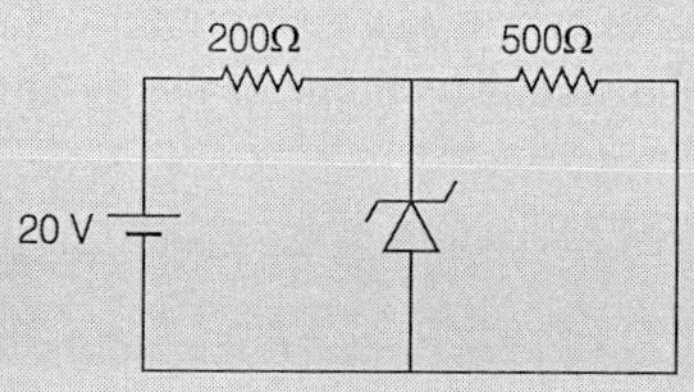

(a) 30 mA (b) 20 mA (c) 0 (d) 50 mA

46 In the given circuit, the breakdown voltage of the Zener diode is 3.0 V. What is the value of I_z? *[2024, 29 Jan Shift-I]*

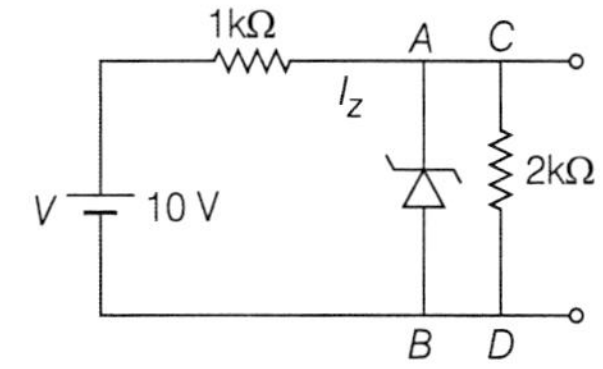

(a) 5.5 mA (b) 3.3 mA
(c) 10 mA (d) 7 mA

47 If each diode has a forward bias resistance of 25 Ω in the below circuit. *[2023, 10 April Shift-II]*

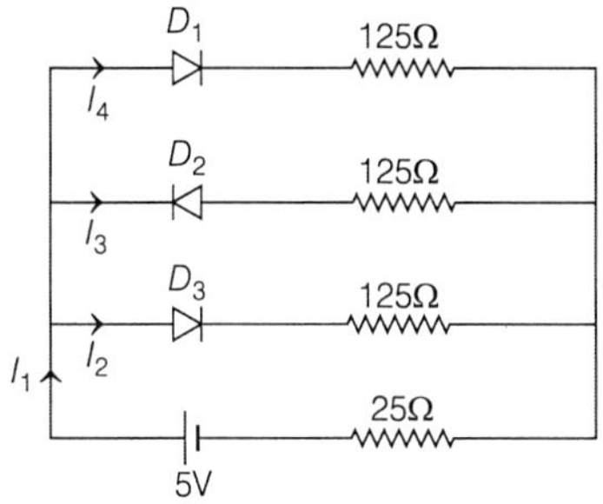

Which of the following options is correct?

(a) $\frac{I_1}{I_2}=1$ (b) $\frac{I_2}{I_3}=1$

(c) $\frac{I_3}{I_4}=1$ (d) $\frac{I_1}{I_2}=2$

Imp.

Zener Diode as Regulator

48 A zener diode of power rating 1.6 W is to be used as voltage regulator. If the zener diode has a breakdown of 8 V and it has to regulate voltage fluctuating between 3 V and 10 V. The value of resistance R for safe operation of diode will be *[2023, 10 April Shift-I]*

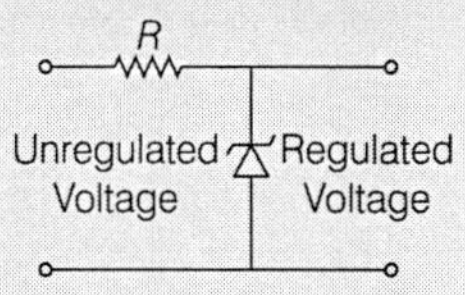

(a) 13.3 Ω (b) 10 Ω (c) 13 Ω (d) 12 Ω

49 **Statement I** When a Si sample is doped with Boron, it becomes p-type and when doped by Arsenic it becomes n-type semiconductor such that p-type has excess holes and n-type has excess electrons.

Statement II When such p-type and n-type semiconductors are fused to make a junction, a current will automatically flow which can be detected with an externally connected ammeter.

In the light of above statements, choose the most appropriate answer from the options given below. *[2023, 25 Jan Shift-II]*

(a) Statement I is correct but Statement II is incorrect.
(b) Statement I is incorrect but Statement II is correct.
(c) Both Statement I and Statement II are incorrect.
(d) Both Statement I and Statement II are correct.

50 Given below are two statements. One is labelled as Assertion A and the other is labelled as Reason R

Assertion A Photodiodes are used in forward bias usually for measuring the light intensity.

Reason R For a p-n junction diode at applied voltage V, the current in the forward bias is more than the current in the reverse bias for $|V_z| > \pm V \geq |V_0|$, where V_0 is the threshold voltage and V_z is the breakdown voltage.

In the light of the above statements, choose the correct answer from the options given below. *[2023, 25 Jan Shift-I]*

(a) Both A and R are true but R is not the correct explanation of A.
(b) A is true but R is false.
(c) Both A and R are true and R is correct explanation of A.
(d) A is false but R is true.

51 In the given figure, each diode has a forward bias resistance of 30 Ω and infinite resistance in reverse bias. The current I_1 will be *[2021, 1 Sep Shift-II]*

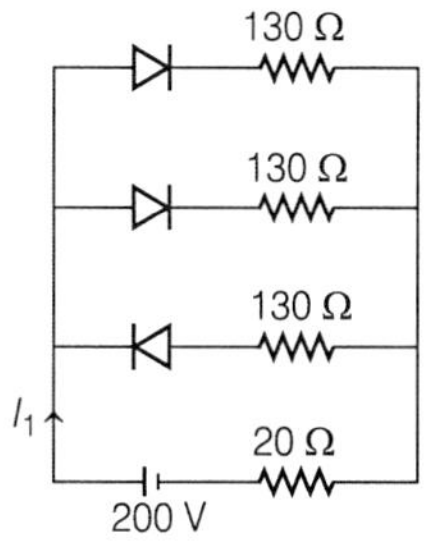

(a) 3.75 A (b) 2.35 A (c) 2 A (d) 2.73 A

52 **Statement I** By doping silicon semiconductor with pentavalent material, the electrons density increases.

Statement II The n-type semiconductor has net negative charge.

In the light of the above statements, choose the most appropriate answer from the options given below. *[2021, 26 Aug Shift-I]*

(a) Statement I is true but Statement II is false.
(b) Statement I is false but Statement II is true.
(c) Both Statement I and Statement II are true.
(d) Both Statement I and Statement II are false.

53 Consider a situation in which reverse biased current of a particular p-n junction increases when it is exposed to a light of wavelength ≤ 621 nm. During this process, enhancement in carrier concentration takes place due to generation of hole-electron pairs. The value of band gap is nearly *[2021, 22 July Shift-II]*

(a) 2 eV (b) 4 eV
(c) 1 eV (d) 0.5 eV

54 For the circuit shown below, calculate the value of I_z. *[2021, 20 July Shift-I]*

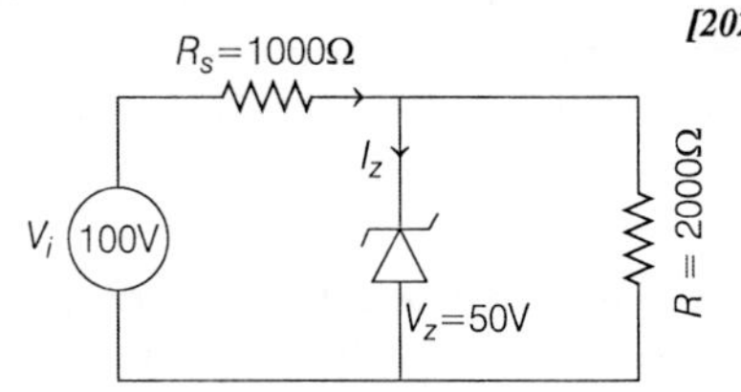

(a) 25 mA (b) 0.15 A (c) 0.1 A (d) 0.05 A

55 For extrinsic semiconductors when doping level is increased, *[2021, 25 Feb Shift-II]*

(a) Fermi level of p-type semiconductor will go upward and Fermi level of n-type semiconductors will go downward
(b) Fermi level of p-type semiconductors will go downward and Fermi level of n-type semiconductor will go upward
(c) Fermi level of p and n-type semiconductors will not be affected
(d) Fermi level of both p-type and n-type semiconductors will go upward for $T > T_F$ K and downward for $T < T_F$ K, where T_F is Fermi temperature

56 LED is constructed from GaAsP semiconducting material. The energy gap of this LED is 1.9 eV. Calculate the wavelength of light emitted and its colour. [$h = 6.63 \times 10^{-34}$ Js and $c = 3 \times 10^{8}$ ms^{-1}] *[2021, 26 Feb Shift-I]*

(a) 1046 nm and red colour (b) 654 nm and orange colour
(c) 1046 nm and blue colour (d) 654 nm and red colour

57 A 5 V battery is connected across the points X and Y. Assume D_1 and D_2 to be normal silicon diodes. Find the current supplied by the battery, if the positive terminal of the battery is connected to point X. *[2021, 25 Feb Shift-I]*

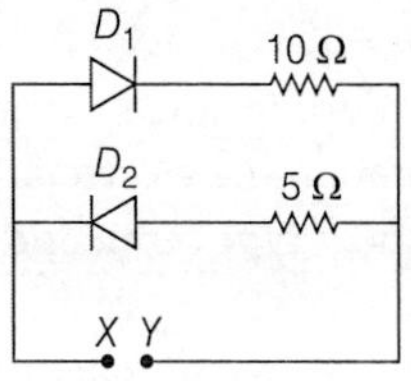

(a) ~ 0.5 A (b) ~ 1.5 A
(c) ~ 0.86 A (d) ~ 0.43 A

58 Zener breakdown occurs in a p-n junction having p and n both *[2021, 24 Feb Shift-II]*

(a) lightly doped and have wide depletion layer
(b) heavily doped and have narrow depletion layer
(c) lightly doped and have narrow depletion layer
(d) heavily doped and have wide depletion layer

59 When a diode is forward biased, it has a voltage drop of 0.5 V. The safe limit of current through the diode is 10 mA. If a battery of emf 1.5 V is used in the circuit, the value of minimum resistance to be connected in series with the diode, so that the current does not exceed the safe limit is *[2020, 3 Sep Shift-I]*

(a) 300 Ω (b) 50 Ω
(c) 100 Ω (d) 200 Ω

60 The current i in the network is *[2020, 9 Jan Shift-II]*

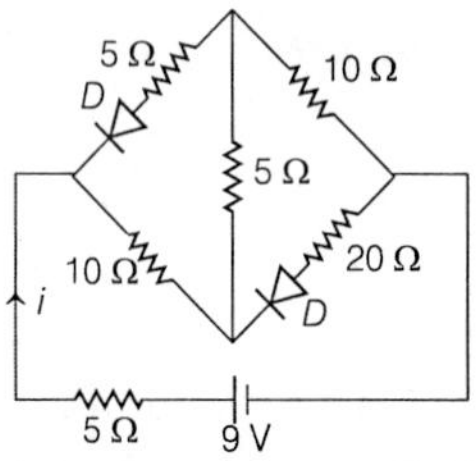

(a) 0.2A (b) 0 A (c) 0.6 A (d) 0.3 A

61 The reverse breakdown voltage of a zener diode is 5.6 V in the given circuit. *[2019, 8 April Shift-I]*

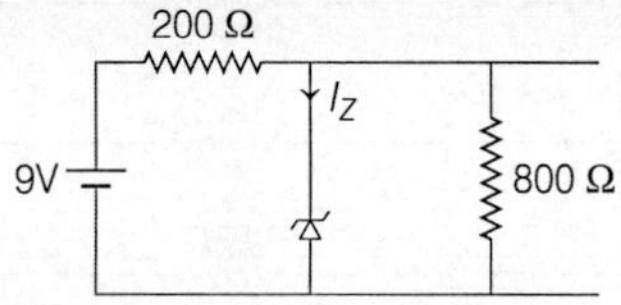

The current I_Z through the zener is
(a) 10 mA (b) 17 mA
(c) 15 mA (d) 7 mA

62 The circuit shown below contains two ideal diodes, each with a forward resistance of 50 Ω. If the battery voltage is 6 V, the current through the 100 Ω resistance (in ampere) is *[2019, 11 Jan Shift-II]*

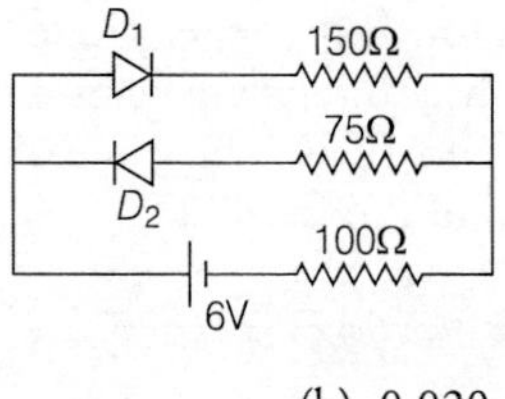

(a) 0.027 (b) 0.020
(c) 0.030 (d) 0.036

Imp. Zener Current

63 In the given circuit, the current through zener diode is close to *[2019, 11 Jan Shift-I]*

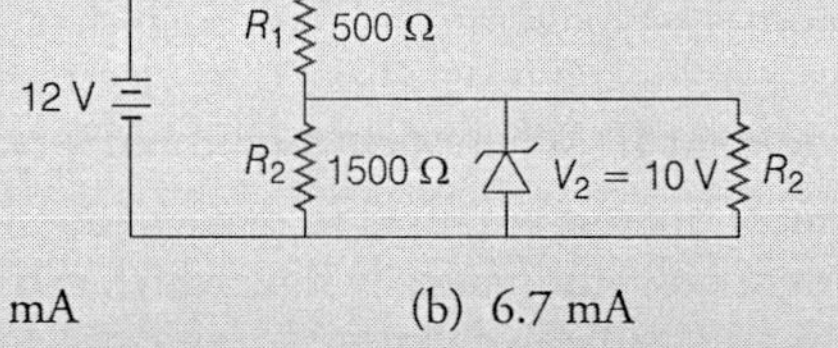

(a) 6.0 mA (b) 6.7 mA
(c) 0 (d) 4.0 mA

64 At 0.3 V and 0.7 V, the diodes Ge and Si become conductor respectively. In given figure, if ends of diode Ge overturned, the change in potential V_0 will be *[2019, 9 Jan Shift-II]*

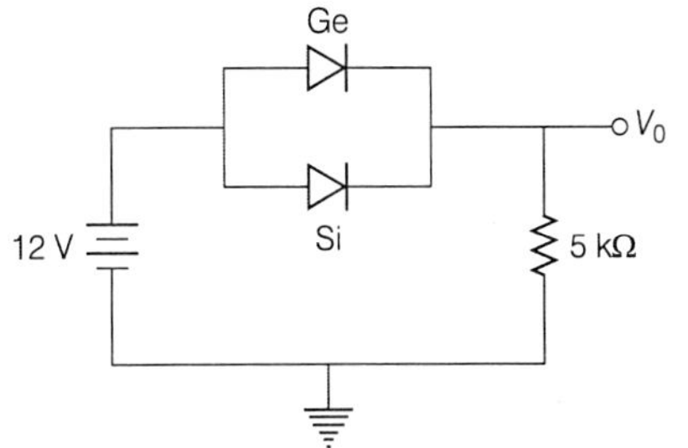

(a) 0.2 V (b) 0.6V
(c) 0.4 V (d) 0.8V

65 For LED's to emit light in visible region of electromagnetic light, it should have energy band gap in the range of *[JEE Main 2014]*
(a) 0.1 eV to 0.4 eV (b) 0.5 eV to 0.8 eV
(c) 0.9 eV to 1.6 eV (d) 1.7 eV to 3.0 eV

66 If in a *p-n* junction diode, a square input signal of 10 V is applied as shown. *[AIEEE 2007]*

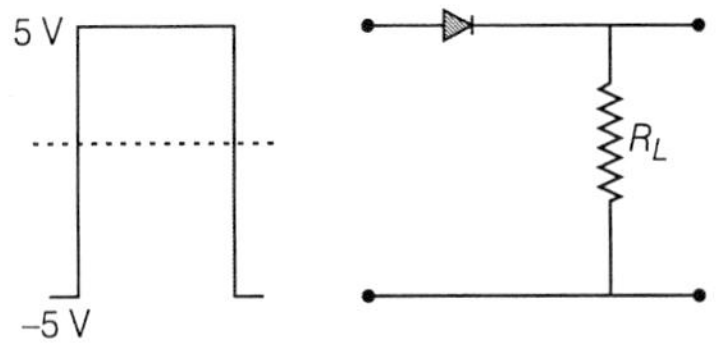

Then, the output signal across R_L will be

(a)
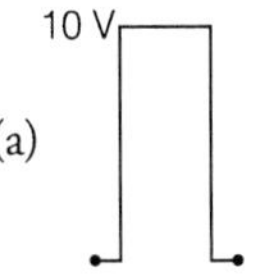

(b)
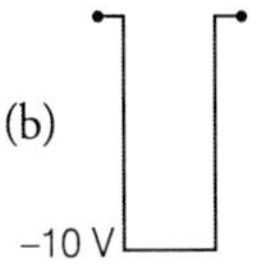

(c)
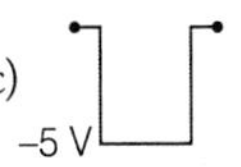

(d)
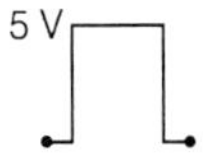

67 If the lattice constant of this semiconductor is decreased, then which of the following is correct? *[AIEEE 2006]*

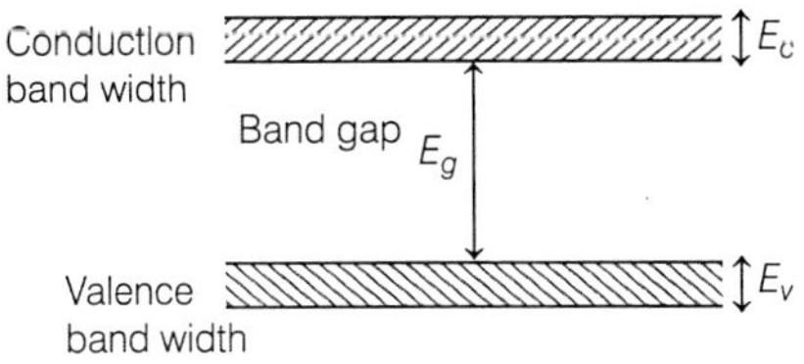

(a) All E_c, E_g, E_v increase
(b) E_c and E_v increase but E_g decreases
(c) E_c and E_v decrease but E_g increases
(d) All E_c, E_g, E_v decrease

68 If the ratio of the concentration of electrons to that of holes in a semiconductor is 7/5 and the ratio of currents is 7/4, then what is the ratio of their drift velocities? *[AIEEE 2006]*
(a) 5/8 (b) 4/5 (c) 5/4 (d) 4/7

69 The circuit has two oppositely connected ideal diodes in parallel. What is the current flowing in the circuit? *[AIEEE 2006]*

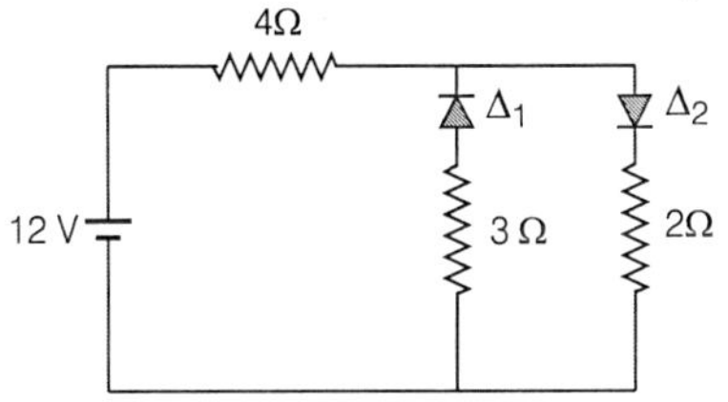

(a) 1.71 A (b) 2.00 A (c) 2.31 A (d) 1.33 A

70 The difference in the variation of resistance with temperature in a metal and a semiconductor arises essentially due to the difference in the *[AIEEE 2003]*
(a) crystal structure
(b) variation of the number of charge carriers with temperature
(c) type of bonding
(d) variation of scattering mechanism with temperature

Level 3

71 Refer to the circuit diagram given in the figure, which of the following observations are correct? *[2025, 23 Jan Shift-I]*
A. Total resistance of circuit is 6 Ω.
B. Current in ammeter is 1 A.
C. Potential across *AB* is 4 V.
D. Potential across *CD* is 4 V.
E. Total resistance of the circuit is 8 Ω.
Choose the correct answer from the options given below.

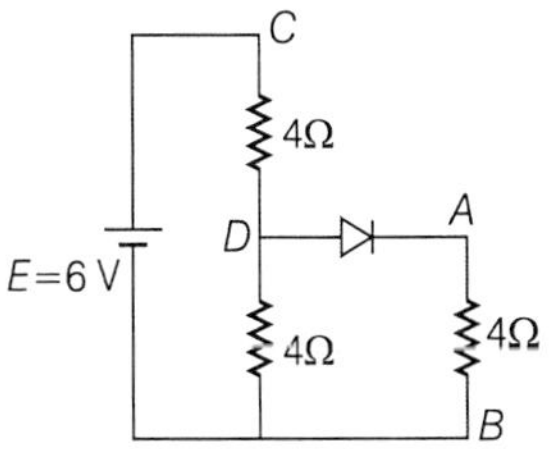

(a) B, C and E Only
(b) A, B and D Only
(c) A, C and D Only
(d) A, B and C Only

72 In the given circuit if the power rating of Zener diode is 10 mW, the value of series resistance R_S to regulate the input unregulated supply is *[2024, 1 Feb Shift-I]*

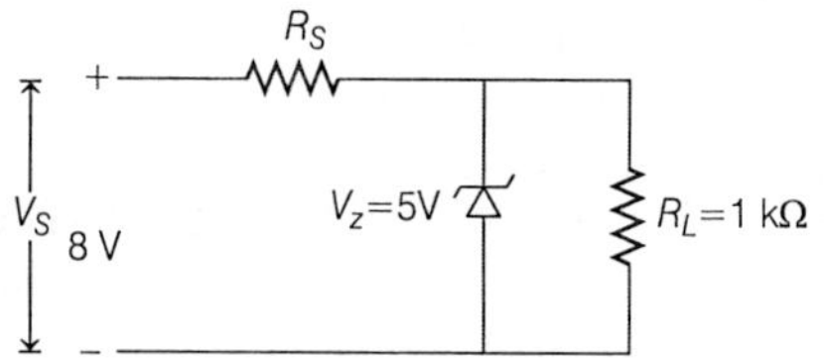

(a) 5 kΩ (b) 1 kΩ (c) 10 kΩ (d) 10 Ω

73 **Statement I** To get a steady DC output from the pulsating voltage received from a full wave rectifier we can connect a capacitor across the output parallel to the load R_L.

Statement II To get a steady DC output from the pulsating voltage received from a full wave rectifier we can connect an inductor in series with R_L.

In the light of the above statements, choose the most appropriate answer from the options given below. *[2021, 31 Aug Shift-II]*

(a) Statement I is true but Statement II is false.
(b) Statement I is false but Statement II is true.
(c) Both Statement I and Statement II are false.
(d) Both Statement I and Statement II are true.

Imp.

Half Wave Rectifier

74 Choose the correct wave form that can represent the voltage across R of the following circuit, assuming the diode is ideal one. *[2021, 31 Aug Shift-I]*

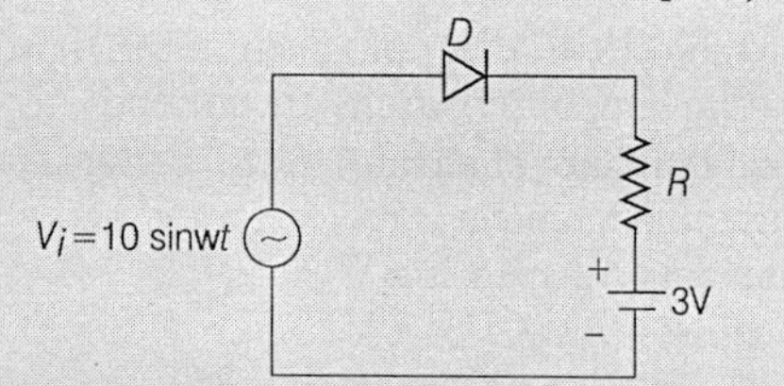

(a)
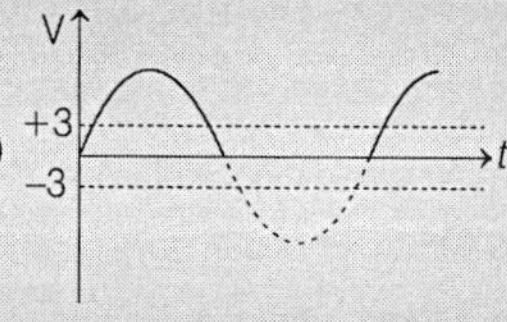

(b)
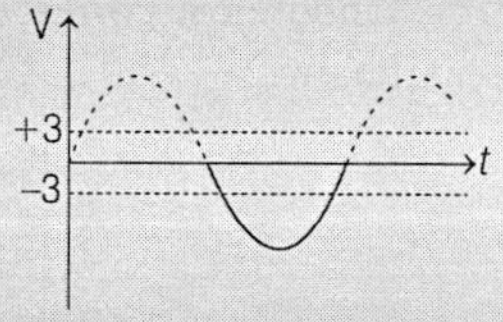

(c)
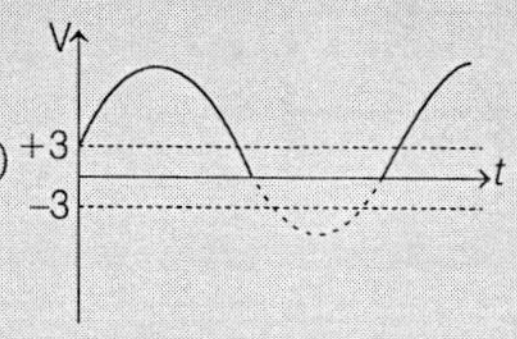

(d)
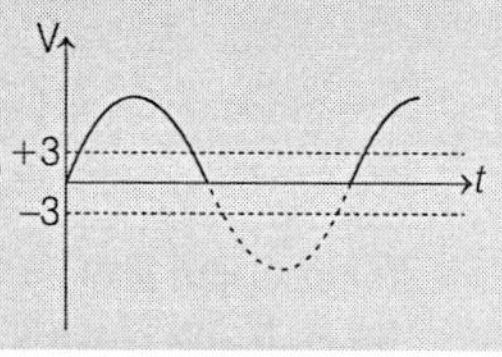

75 In the figure, potential difference between A and B is *[2020, 7 Jan Shift-II]*

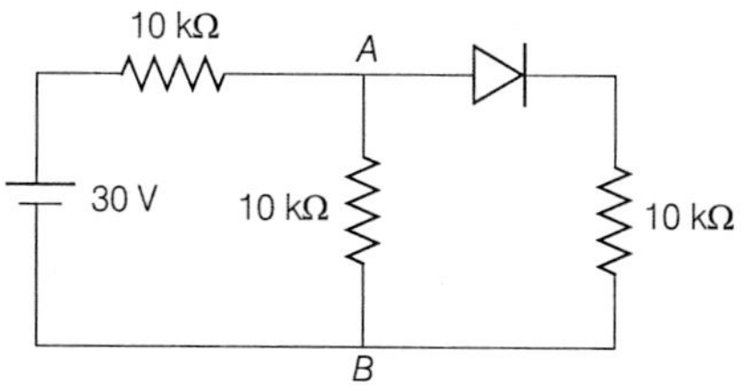

(a) zero (b) 10 V
(c) 5 V (d) 15 V

76 Figure shows a DC voltage regulator circuit, with a zener diode of breakdown voltage = 6 V.
If the unregulated input voltage varies between 10 V to 16 V, then what is the maximum zener current ? *[2019, 12 April Shift-II]*

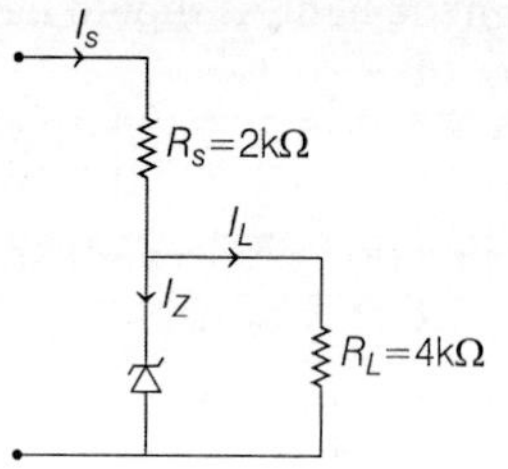

(a) 2.5 mA (b) 1.5 mA
(c) 7.5 mA (d) 3.5 mA

77 The figure represents a voltage regulator circuit using a zener diode. The breakdown voltage of the zener diode is 6 V and the load resistance is $R_L = 4$ kΩ. The series resistance of the circuit is $R_i = 1$ kΩ. If the battery voltage V_B varies from 8V to 16V, what are the minimum and maximum values of the current through zener diode ? *[2019, 10 April Shift-II]*

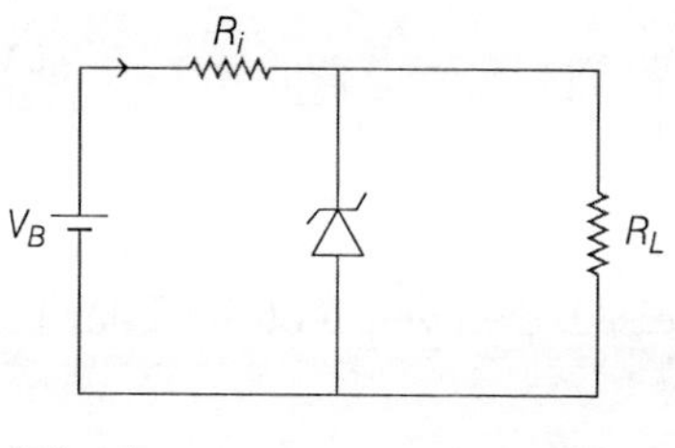

(a) 1.5 mA, 8.5 mA
(b) 1 mA, 8.5 mA
(c) 0.5 mA, 8.5 mA
(d) 0.5 mA, 6 mA

Numerical Value Type Questions

78 A potential divider circuit is connected with a DC source of 20 V, a light emitting diode of glow in voltage 1.8 V and a zener diode of breakdown voltage of 3.2 V. The length (PR) of the resistive wire is 20 cm. The minimum length of PQ to just glow the LED is .... cm. *[2024, 8 April Shift-II]*

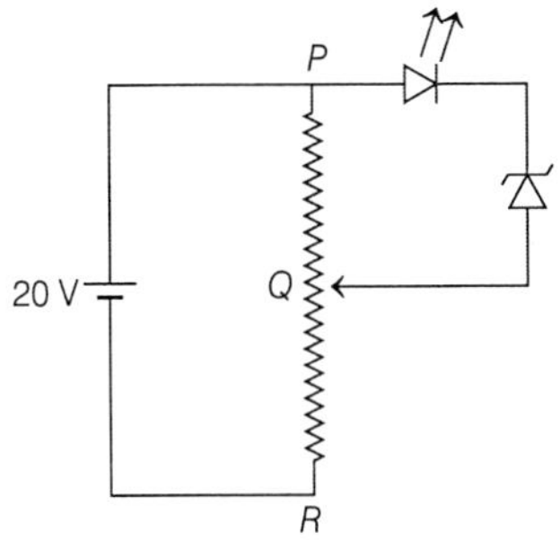

79 A 8 V zener diode along with a series resistance R is connected across a 20 V supply (as shown in the figure). If the maximum zener current is 25 mA, then the minimum value of R will be .... Ω. *[2022, 29 July Shift-II]*

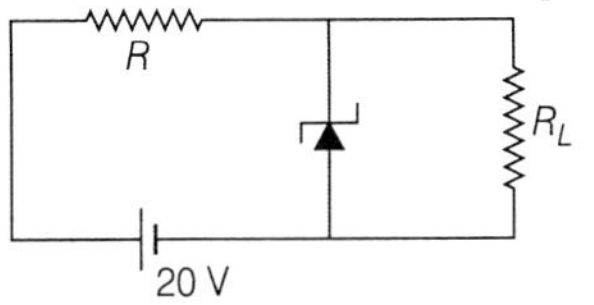

80 In the circuit shown below, maximum zener diode current will be ......mA.

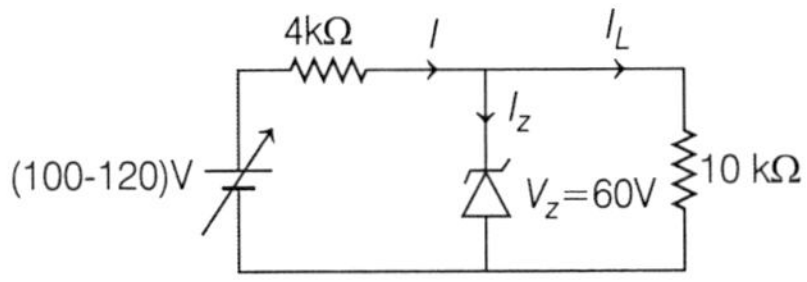

[2022, 26 July Shift-I]

81 Two ideal diodes are connected in the network as shown in figure. The equivalent resistance between A and B is .... Ω. *[2022, 25 July Shift-II]*

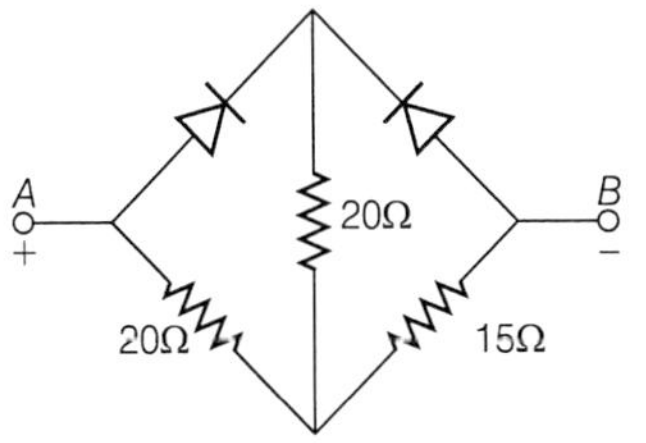

82 If the potential barrier across a p-n junction is 0.6 V. Then, the electric field intensity in the depletion region having the width of 6×10^{-6} m, will be .... $\times 10^5$ N/C. *[2022, 29 June Shift-I]*

Imp. Band Gap Energy

83 The energy band gap of semiconducting material to produce violet (wavelength = 4000 Å) LED is .... eV. (Round off to the nearest integer) *[2022, 25 July, Shift-I]*

84 The circuit diagram used to study the characteristics curve of a zener diode is connected to variable power supply (0-15 V) as shown in figure. A zener diode with maximum potential $V_z = 10$ V and maximum power dissipation of 0.4 W is connected across a potential divider arrangement. The value of resistance R_P connected in series with the zener diode to protect it from the damage is .... Ω. *[2022, 30 June Shift-I]*

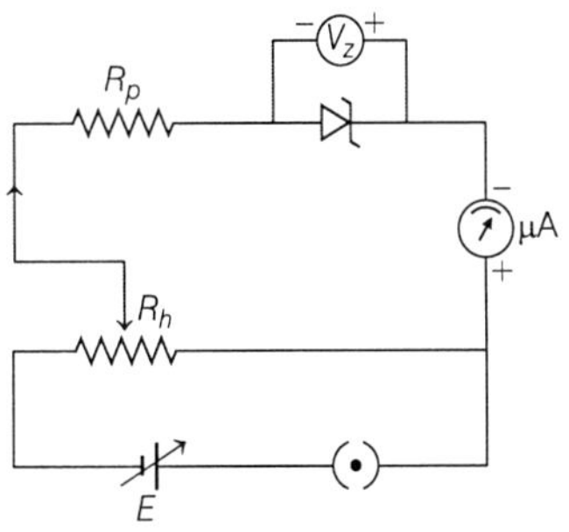

85 The cut-off voltage of the diodes (shown in figure) in forward bias is 0.6 V. The current through the resistor of 40 Ω is .... mA. *[2022, 27 June Shift-II]*

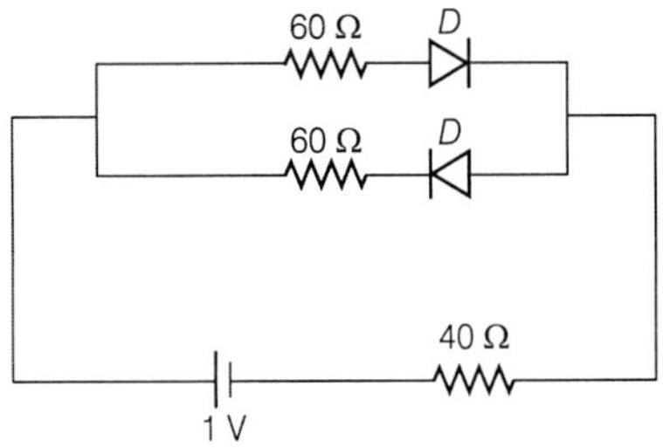

86 As per the given circuit, the value of current through the battery will be .... A. *[2022, 26 June Shift-I]*

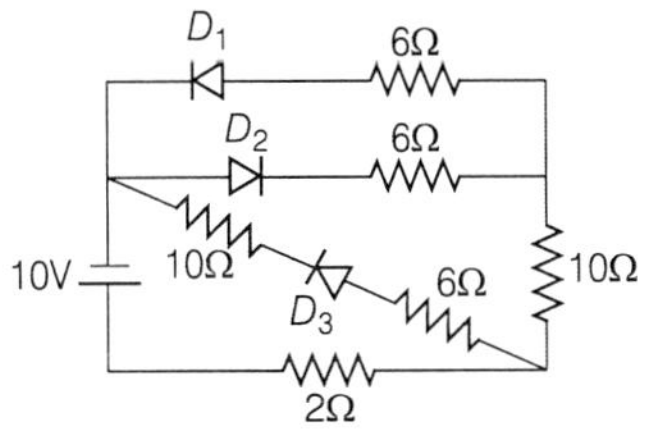

87 In the given circuit, the value of current I_L will be ...... mA. (When, $R_L = 1\,k\Omega$) *[2022, 24 June Shift-II]*

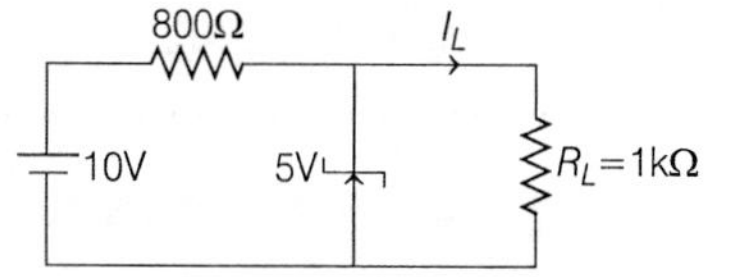

88 A zener diode of power rating 2W is to be used as a voltage regulator. If the Zener diode has a breakdown of 10 V and it has to regulate voltage fluctuated between 6 V and 14 V, the value of R_s for safe operation should be .... Ω *[2021, 27 Aug Shift-II]*

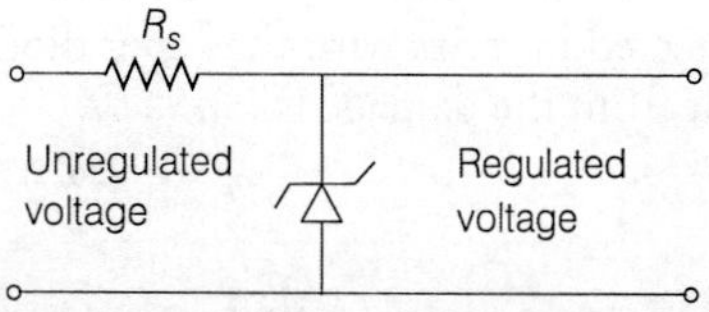

89 For the given circuit, the power across Zener diode is .... mW. *[2021, 26 Aug Shift-II]*

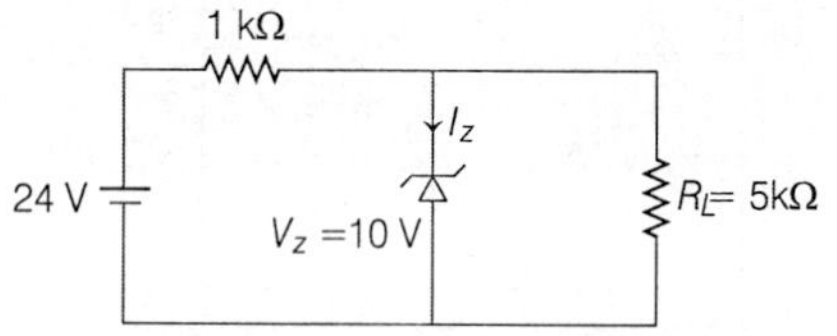

90 In a semiconductor, the number density of intrinsic charge carriers at 27°C is $1.5 \times 10^{16}/\text{m}^3$. If the semiconductor is doped with impurity atom, the hole density increases to $4.5 \times 10^{22}/\text{m}^3$. The electron density in the doped semiconductor is ...... $\times 10^9/\text{m}^3$. *[2021, 25 July Shift-II]*

91 In a given circuit diagram, a 5 V Zener diode along with a series resistance is connected across a 50 V power supply. The minimum value of the resistance required, if the maximum Zener current is 90 mA will be .... Ω. *[2021, 22 July Shift-II]*

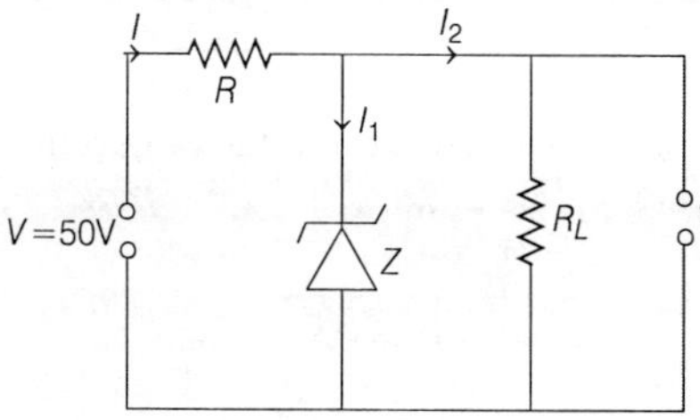

92 A Zener diode having Zener voltage 8 V and power dissipation rating of 0.5 W is connected across a potential divider arranged with maximum potential drop across Zener diode is as shown in the diagram. The value of protective resistance R_p is ....Ω. *[2021, 20 July Shift-II]*

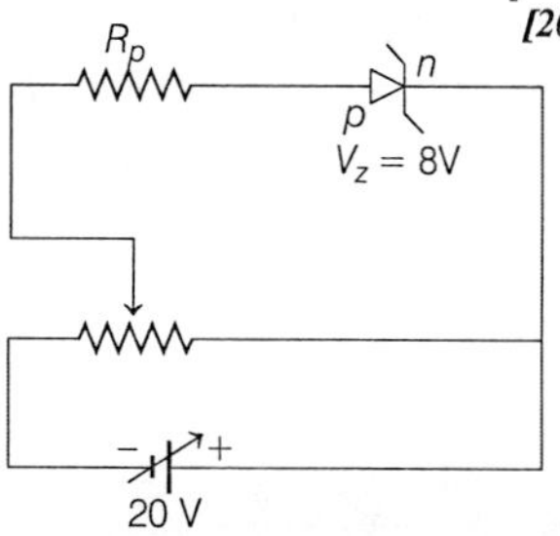

93 The value of power dissipated across the Zener diode ($V_z = 15$ V) connected in the circuit as shown in the figure is $x \times 10^{-1}$ W.

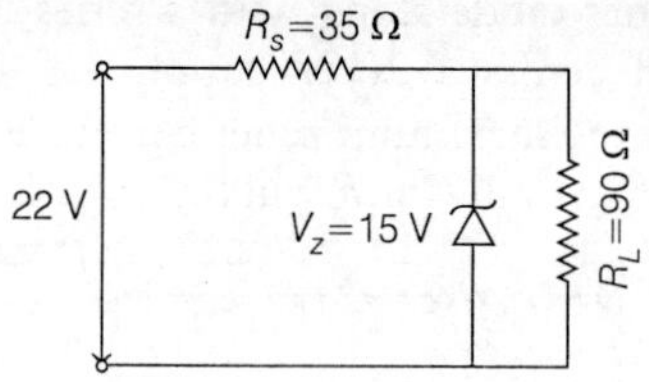

The value of x, to the nearest integer, is .... *[2021, 16 March Shift-I]*

94 Both the diodes used in the circuit shown are assumed to be ideal and have negligible resistance when these are forward biased. Built in potential in each diode is 0.7 V. For the input voltages shown in the figure, the voltage (in volts) at point A is .... *[2020, 9 Jan Shift-I]*

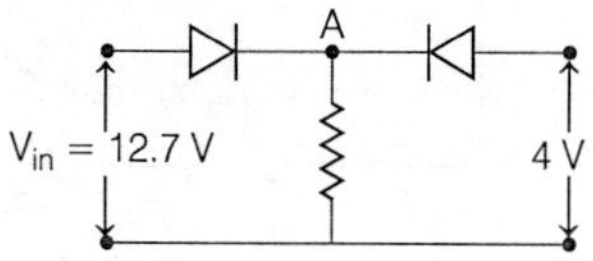

Topic 2 Digital Circuits

01 The truth table corresponding to the circuit given below is *[2025, 3 April Shift-II]*

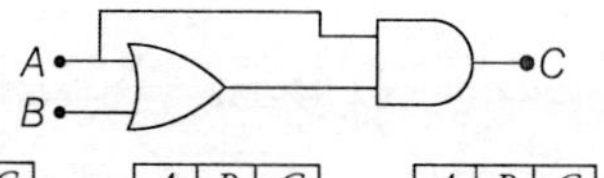

(a)

A	B	C
0	0	1
0	1	0
1	0	0
1	1	0

(b)

A	B	C
0	0	1
1	0	0
0	1	0
1	1	0

(c)

A	B	C
0	0	0
0	1	0
1	0	1
1	1	1

(d)

A	B	C
0	0	0
1	0	0
0	1	0
1	1	1

02 Choose the correct logic circuit for the given truth table having inputs A and B. *[2025, 3 April Shift-I]*

Inputs		Output
A	B	Y
0	0	0
0	1	0
1	0	1
1	1	1

(a) 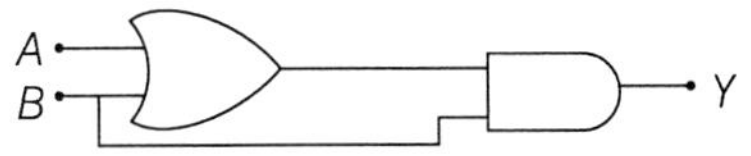

(b) 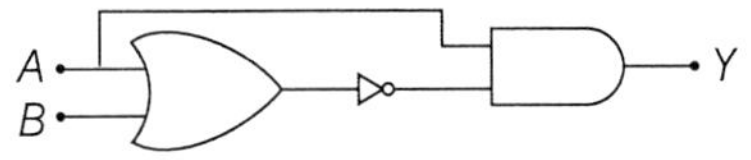

(c) 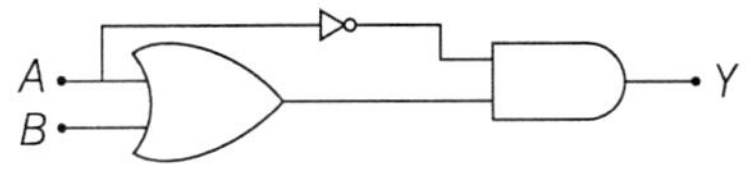

(d) 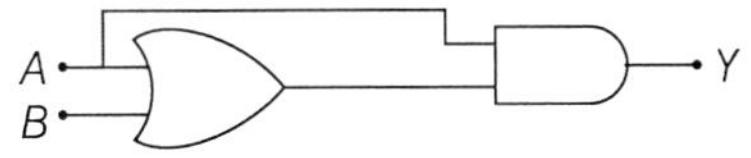

03 The output of the circuit is low (zero) for

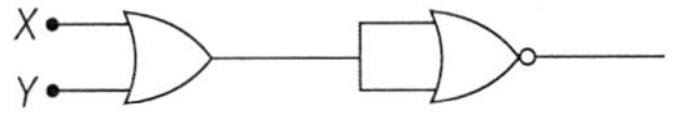

(a) $X=0, Y=0$
(b) $X=0, Y=1$
(c) $X=1, Y=0$
(d) $X=1, Y=1$

Choose the correct answer from the options given below. *[2025, 24 Jan Shift-II]*

(a) B, C and D only
(b) A, B and D only
(c) A, C and D only
(d) A, B and C only

04 The output Y of logic circuit given below is 0 only when *[2024, 5 April Shift-II]*

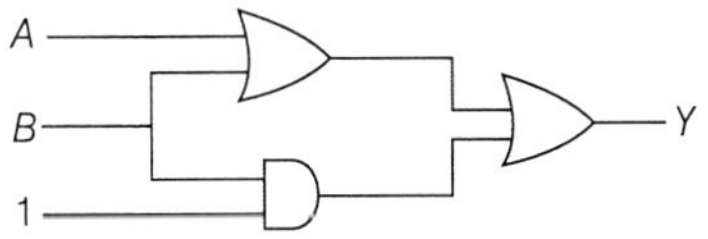

(a) $A=1, B=0$ (b) $A=0, B=0$
(c) $A=0, B=1$ (d) $A=1, B=1$

Imp. Logic Gates

05 The output from a NAND gate havign inputs A and B given below will be, *[2023, 13 April Shift-II]*

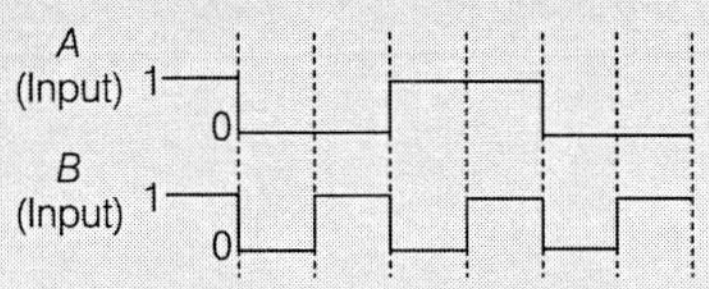

(a) 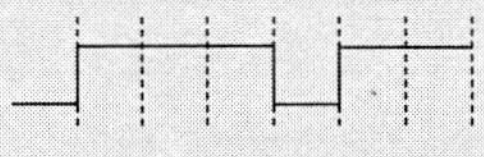 (b) 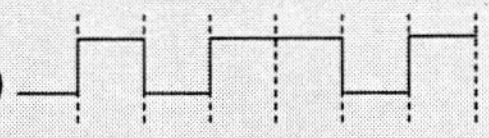

(c) 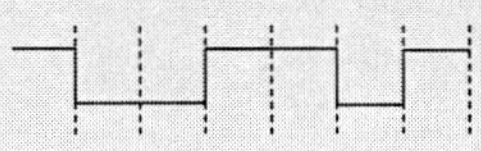 (d) 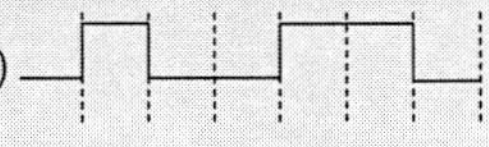

06 Name the logic gate equivalent to the diagram attached. *[2023, 6 April Shift-I]*

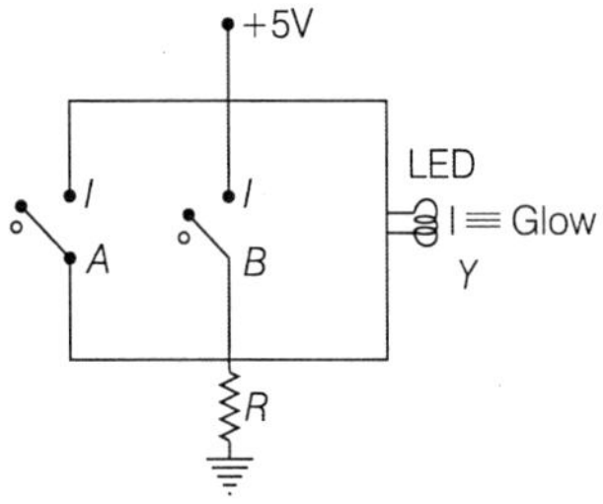

(a) OR (b) NAND (c) NOR (d) AND

07 The logic gate equivalent of the given circuit diagram is *[2023, 24 Jan Shift-II]*

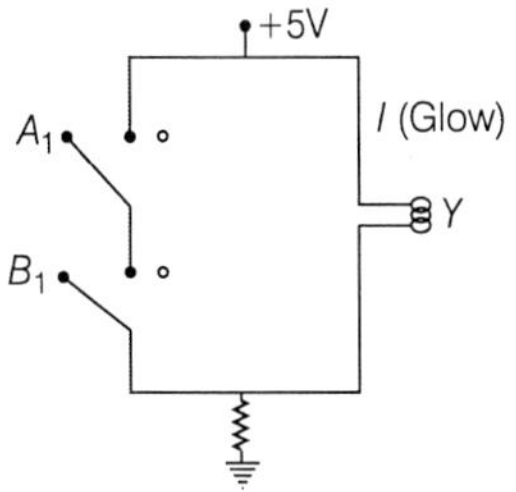

(a) NAND (b) AND (c) NOR (d) OR

08 A logic gate circuit has two inputs A and B and output Y. The voltage waveforms of A, B and Y are shown below. *[2022, 27 July Shift-I]*

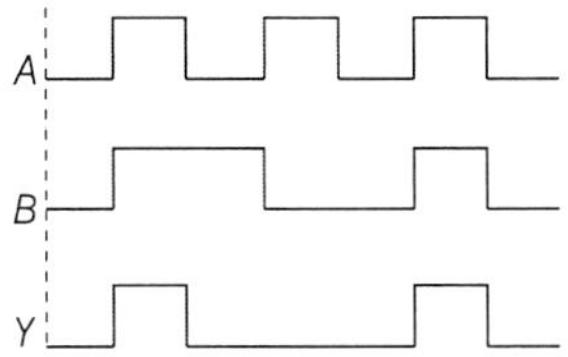

The logic gate circuit is

(a) AND gate (b) OR gate
(c) NOR gate (d) NAND gate

09 Identify the correct logic gate for the following output Y of two inputs A and B. *[2022, 27 June Shift-I]*

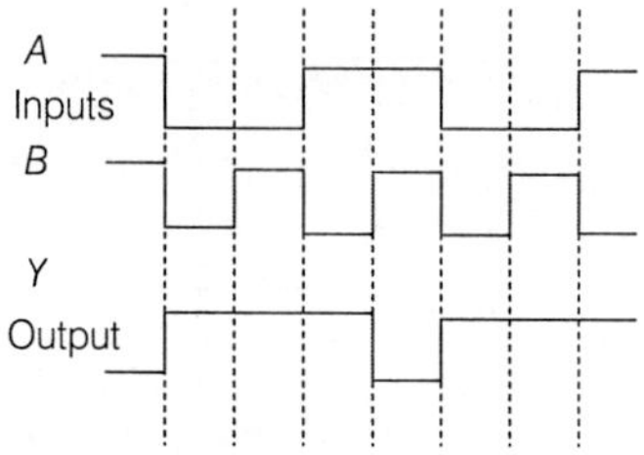

(a) 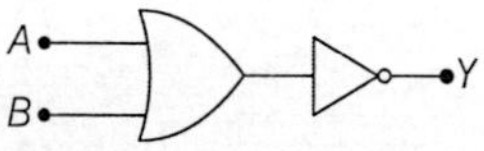

(b) 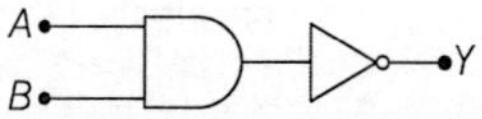

(c) 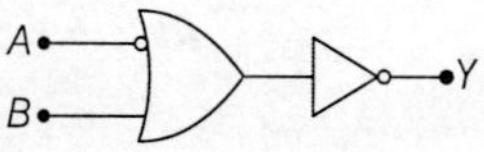

(d) 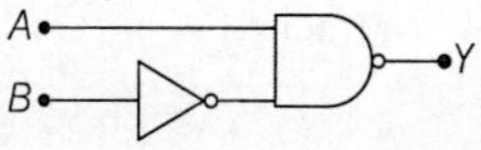

10 Identify the logic operation carried out. *[2021, 25 July Shift-I]*

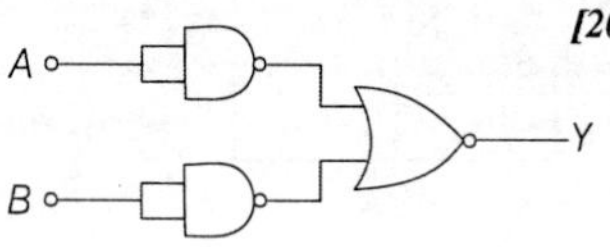

(a) OR (b) AND
(c) NOR (d) NAND

11

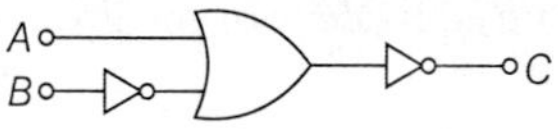

The logic circuit shown above is equivalent to *[2021, 24 Feb Shift-II]*

(a) 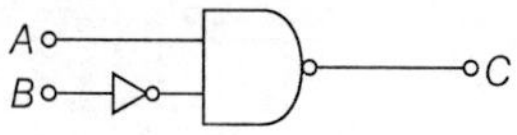

(b) 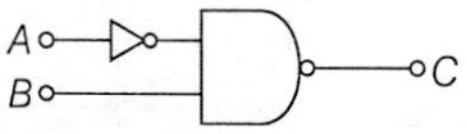

(c) 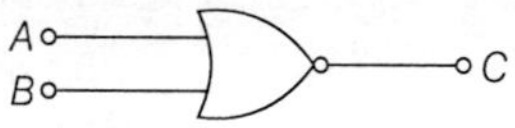

(d) 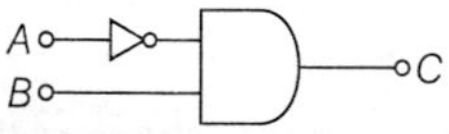

12 Which of the following gives a reversible operation? *[2020, 7 Jan Shift-I]*

(a) 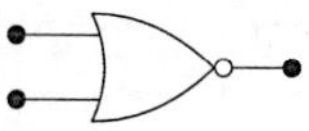
(b) 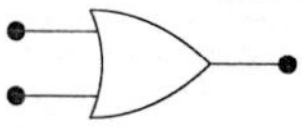
(c) 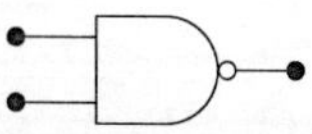
(d) 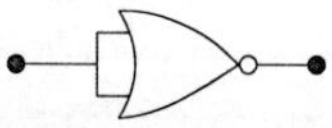

13 The logic gate equivalent to the given logic circuit is *[2019, 9 April Shift-II]*

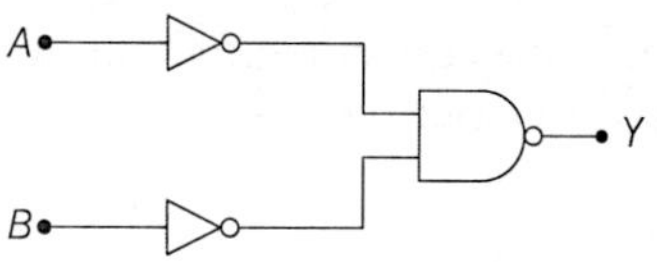

(a) NOR (b) NAND
(c) OR (d) AND

14 The truth table given in figure represents

A	B	Y
0	0	0
0	1	1
1	0	1
1	1	1

[JEE Main 2016]

(a) OR gate (b) NAND gate
(c) AND gate (d) NOR gate

15 If a, b, c and d are inputs to a gate and x is its output, then, as per the following time graph, the gate is *[JEE Main 2016]*

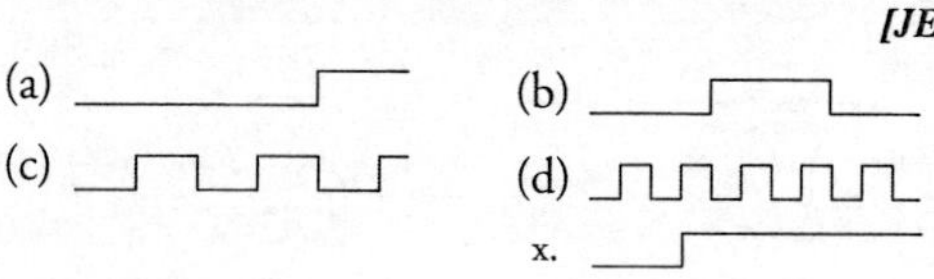

(a) NOT (b) AND
(c) OR (d) NAND

16 Which of the following circuits correctly represents the following truth table? *[JEE Main 2013]*

A	B	C
0	0	0
0	1	0
1	0	1
1	1	0

(a) 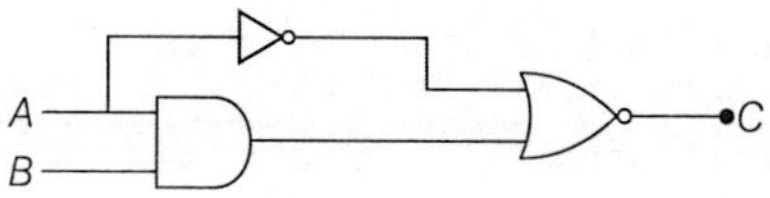

(b) 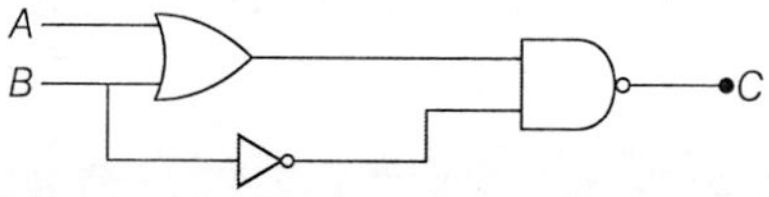

(c) 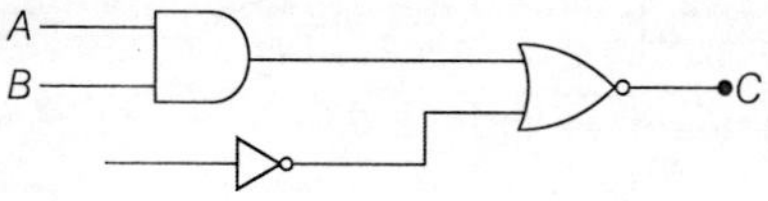

(d) 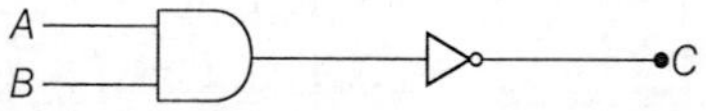

17 Which one of the following is the boolean expression for NOR gate? *[AIEEE 2012]*

(a) $Y = \overline{A + B}$ (b) $Y = \overline{A . B}$
(c) $Y = A . B$ (d) $Y = \overline{A}$

18 The figure shows a combination of two NOT gates and a NOR gate. *[AIEEE 2012]*

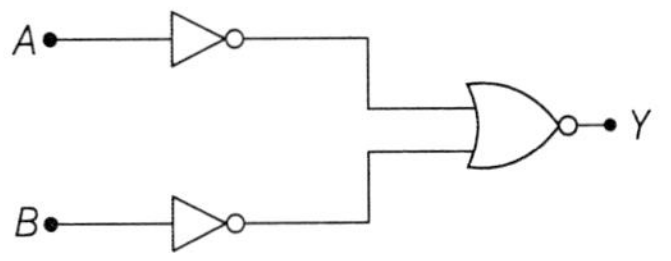

The combination is equivalent to a

(a) NAND gate (b) NOR gate
(c) AND gate (d) OR gate

19 The output of an OR gate is connected to both the inputs of a NAND gate. The combination will serve as a *[AIEEE 2011]*

(a) OR gate (b) NOT gate
(c) NOR gate (d) AND gate

20 The combination of gates shown below yields *[AIEEE 2010]*

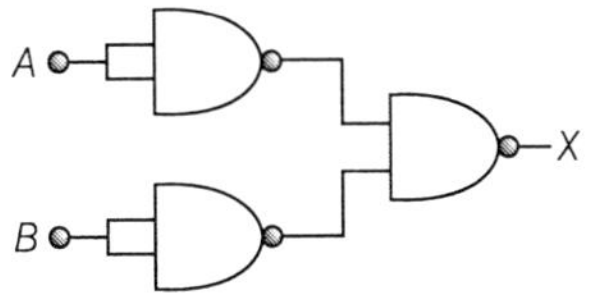

(a) OR gate (b) NOT gate
(c) XOR gate (d) NAND gate

Level 2

21 Consider the following logic circuit. *[2025, 7 April Shift-II]*

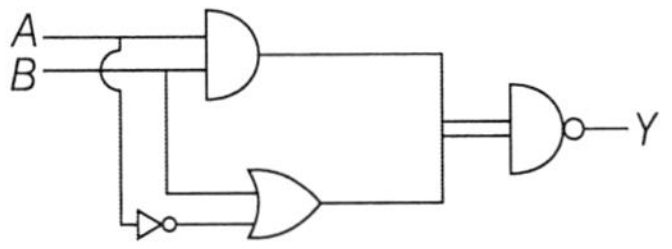

The output is $Y = 0$ when

(a) $A = 1$ and $B = 0$ (b) $A = 0$ and $B = 1$
(c) $A = 0$ and $B = 0$ (d) $A = 1$ and $B = 1$

22

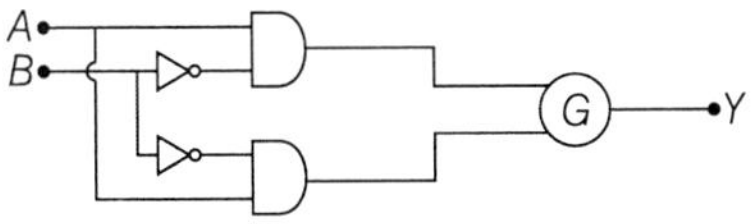

A	B	Y
0	0	1
0	1	0
1	0	0
1	1	1

To obtain the given truth table, following logic gate should be placed at *G*. *[2025, 22 Jan Shift-II]*

(a) AND gate
(b) OR gate
(c) NAND gate
(d) NOR gate

23 Which of the following circuits has the same output as that of the given circuit? *[2025, 28 Jan Shift-I]*

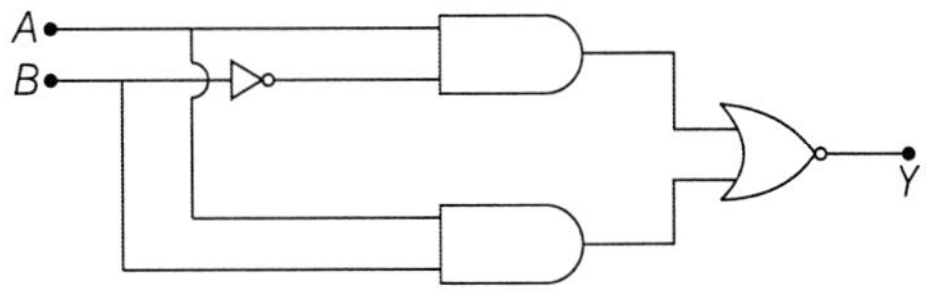

(a) 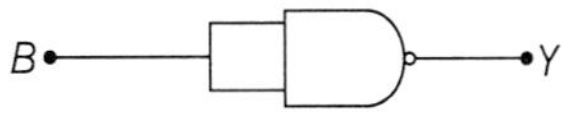

(b) 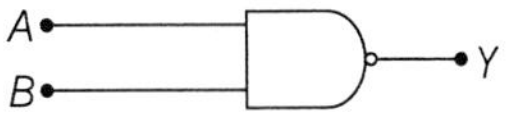

(c) 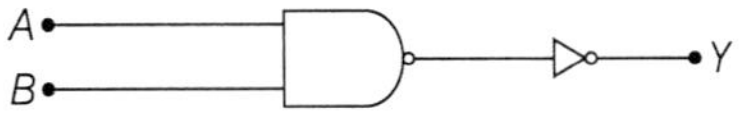

(d) 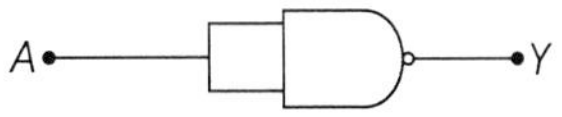

Imp. Truth Table of Logic Circuit

24 The correct truth table for the following logic circuit is *[2024, 6 April Shift-I]*

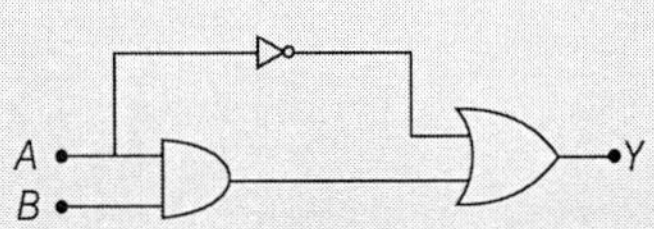

(a)

A	B	Y
0	0	0
0	1	1
1	0	0
1	1	1

(b)

A	B	Y
0	0	1
0	1	1
1	0	0
1	1	0

(c)

A	B	Y
0	0	1
0	1	1
1	0	0
1	1	1

(d)

A	B	Y
0	0	0
0	1	0
1	0	0
1	1	1

25 Following gates section is connected in a complete suitable circuit. *[2024, 5 April Shift-I]*

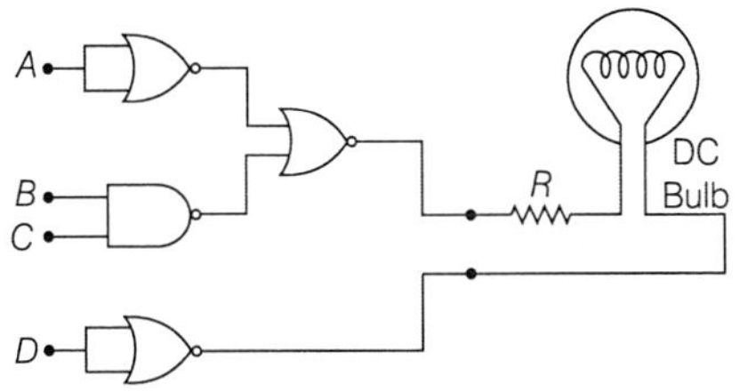

For which of the following combination, bulb will glow (ON)

(a) A = 1, B = 0, C = 0, D = 0
(b) A = 0, B = 1, C = 1, D = 1
(c) A = 0, B = 0, C = 0, D = 1
(d) A = 1, B = 1, C = 1, D = 0

26 Identify the logic operation performed by the given circuit. *[2024, 31 Jan Shift-I]*

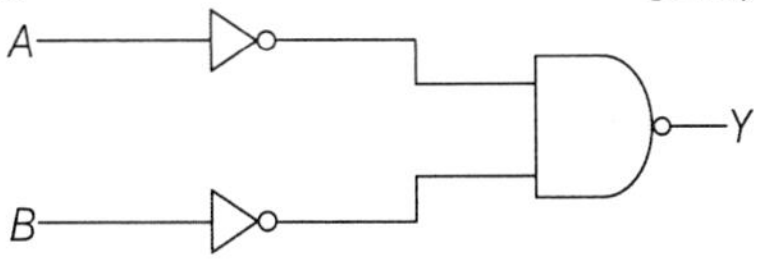

(a) NAND (b) AND (c) NOR (d) OR

27 The truth table for this given circuit is *[2024, 29 Jan Shift-II]*

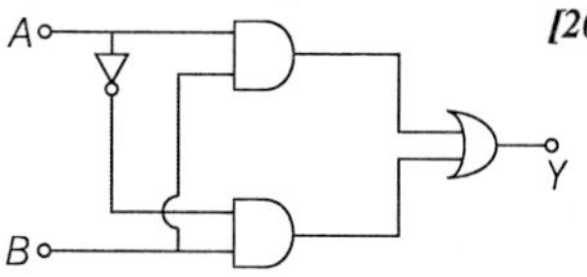

(a)

A	B	Y
0	0	0
0	1	1
1	0	0
1	1	1

(b)

A	B	Y
0	0	1
0	1	0
1	0	1
1	1	0

(c)

A	B	Y
0	0	1
0	1	1
1	0	1
1	1	0

(d)

A	B	Y
0	0	0
0	1	0
1	0	0
1	1	1

28 The truth table of the given circuit diagram is *[2024, 27 Jan Shift-II]*

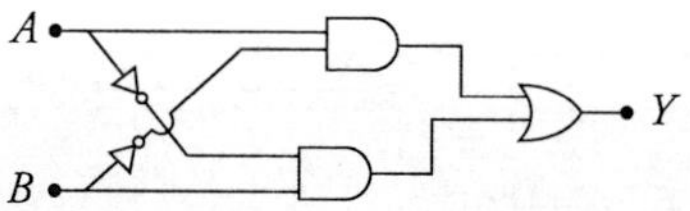

(a)

A	B	Y
0	0	0
0	1	0
1	0	0
1	1	1

(b)

A	B	Y
0	0	1
0	1	1
1	0	1
1	1	0

(c)

A	B	Y
0	0	1
0	1	0
1	0	0
1	1	1

(d)

A	B	Y
0	0	0
0	1	1
1	0	1
1	1	0

Imp. Output Waveforms

29 For the following circuit and given inputs A and B, choose the correct option for output Y. *[2023,13 April Shift-I]*

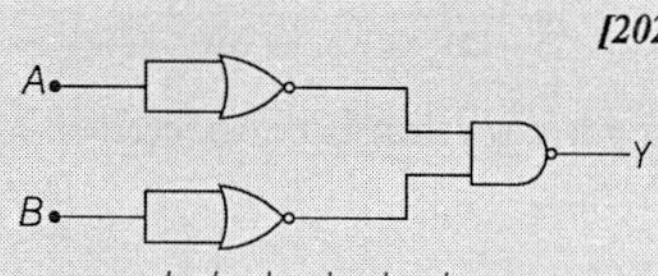

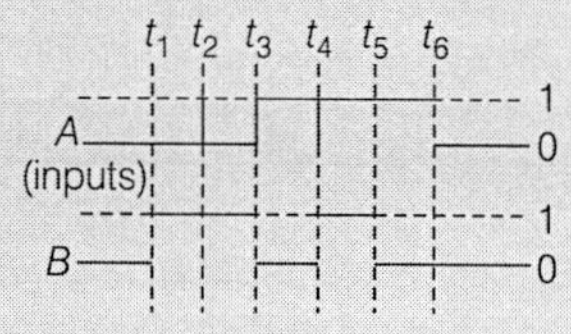

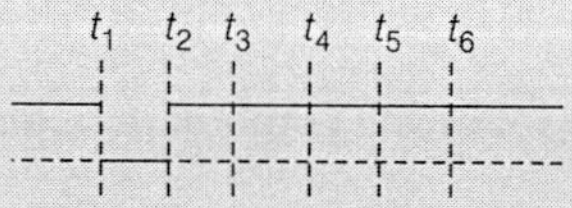

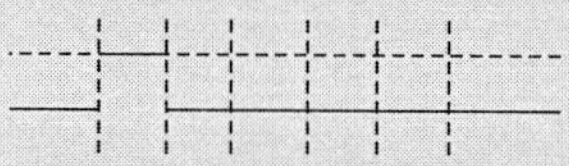

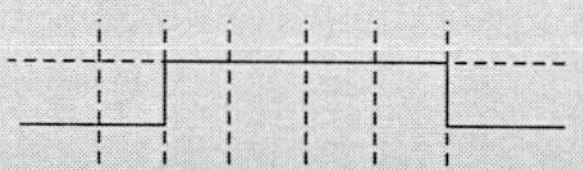

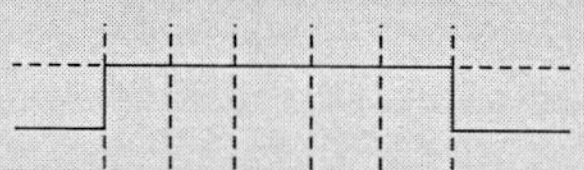

30 The logic operations performed by the given digital circuit is equivalent to *[2023, 11 April Shift-I]*

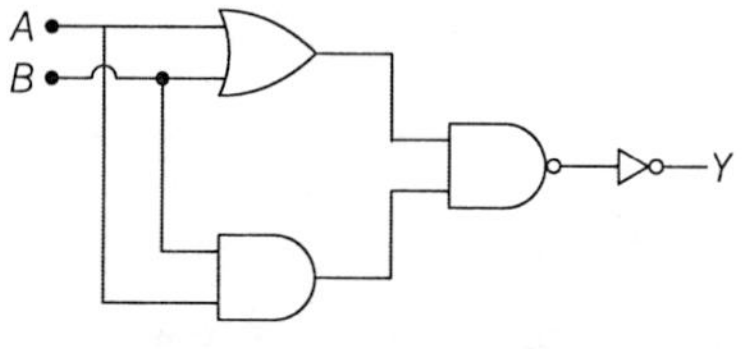

(a) NAND (b) OR (c) NOR (d) AND

31 In the following logic circuit, the sequence of the inputs A, B are (0, 0), (0,1), (1, 0) and (1, 1). The output Y for this sequence will be *[2021, 31 Aug Shift-I]*

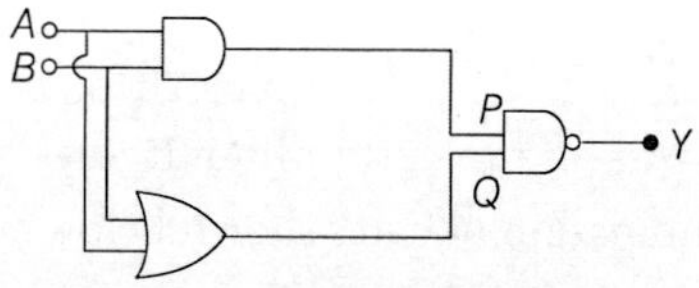

(a) 1, 0, 1, 0 (b) 0, 1, 0, 1
(c) 1, 1, 1, 0 (d) 0, 0, 1, 1

32 Identify the logic operation carried out by the given circuit. *[2021, 26 Aug Shift-I]*

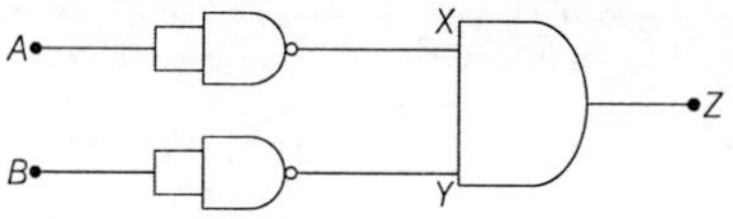

(a) OR (b) AND (c) NOR (d) NAND

33 Find the truth table for the function Y of A and B represented in the following figure. *[2021, 27 July Shift-II]*

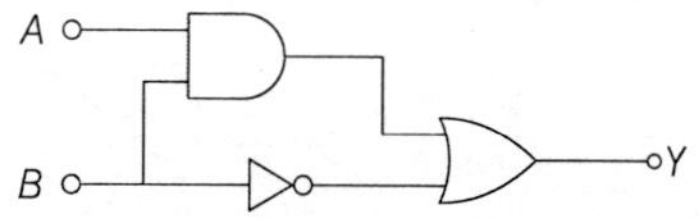

(a)

A	B	Y
0	0	1
0	1	1
1	0	0
1	1	0

(b)

A	B	Y
0	0	1
0	1	0
1	0	1
1	1	1

(c)

A	B	Y
0	0	0
0	1	0
1	0	0
1	1	1

(d)

A	B	Y
0	0	0
0	1	1
1	0	1
1	1	1

34 The output of the given combination gates represents *[2021, 17 March Shift-I]*

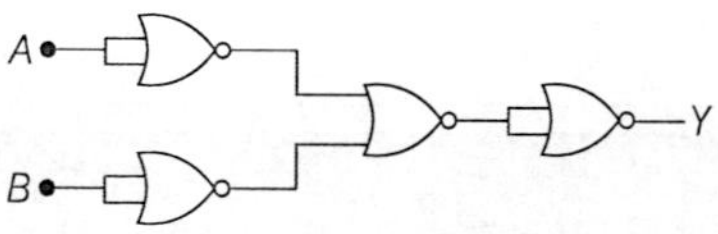

(a) XOR gate (b) NAND gate
(c) AND gate (d) NOR gate

35 The output of the given logic circuit is *[2019, 12 Jan Shift-I]*

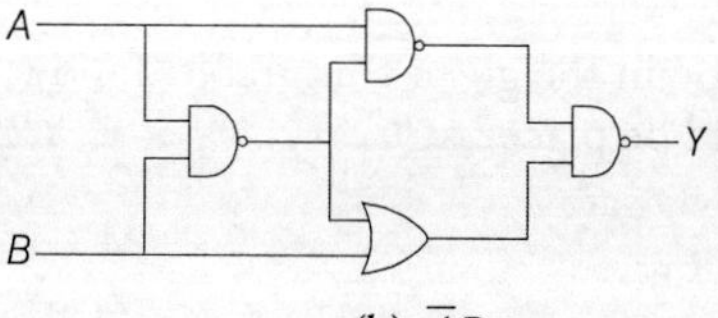

(a) $A\overline{B}$ (b) $\overline{A}B$
(c) $AB+\overline{AB}$ (d) $A\overline{B}+\overline{A}B$

Imp. NAND gate as Universal gate

36 Truth table for system of four NAND gates as shown in figure is *[AIEEE 2012]*

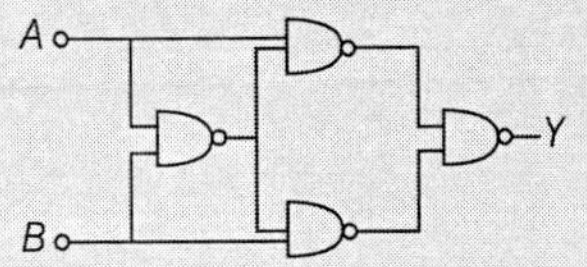

(a)

A	B	Y
0	0	0
0	1	1
1	0	1
1	1	0

(b)

A	B	Y
0	0	1
0	1	1
1	0	0
1	1	0

(c)

A	B	Y
0	0	0
0	1	0
1	0	1
1	1	1

(d)

A	B	Y
0	0	1
0	1	0
1	0	0
1	1	1

Level 3

37 The Boolean expression $Y = A\overline{B}C + \overline{A}\overline{C}$ can be realised with which of the following gate configurations. *[2025, 4 April Shift-I]*

A. One 3-input AND gate, 3 NOT gates and one 2-input OR gate, One 2-input AND gate.

B. One 3-input AND gate, 1 NOT gate, One 2-input NOR gate and one 2-input OR gate.

C. 3-input OR gate, 3 NOT gates and one 2-input AND gate.

Choose the correct answer from the options given below.

(a) B, C Only
(b) A, B Only
(c) A, C Only
(d) A, B, C Only

38 In the digital circuit shown in the figure, for the given inputs the P and Q values are *[2025, 2 April Shift-II]*

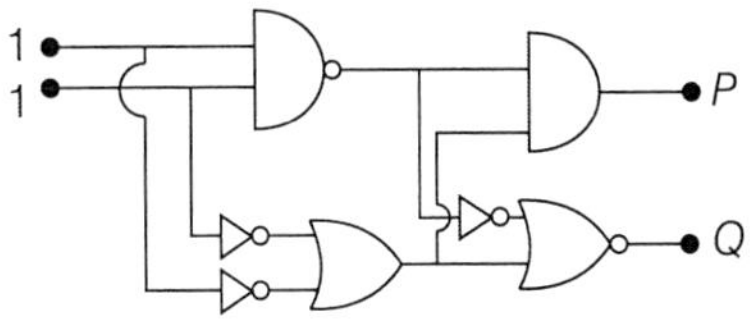

(a) $P = 1, Q = 1$ (b) $P = 0, Q = 0$
(c) $P = 1, Q = 0$ (d) $P = 0, Q = 1$

39

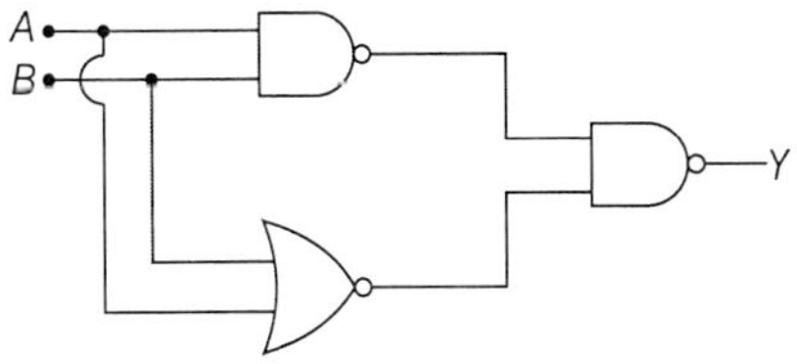

For the circuit shown above, equivalent gate is *[2025, 29 Jan Shift-I]*

(a) AND gate (b) NOT gate
(c) NAND gate (d) OR gate

40 The truth table for the circuit given below is *[2025, 29 Jan Shift-II]*

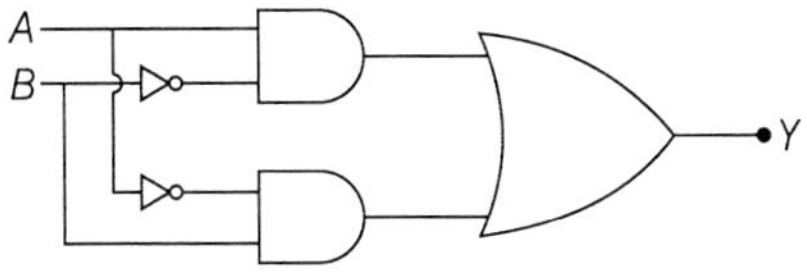

(a)

A	B	Y
0	0	0
1	0	0
1	1	0
0	1	1

(b)

A	B	Y
0	0	0
1	1	1
1	0	1
0	1	1

(c)

A	B	Y
0	0	0
1	0	1
0	1	0
1	1	0

(d)

A	B	Y
0	0	0
0	1	1
1	0	1
1	1	0

41

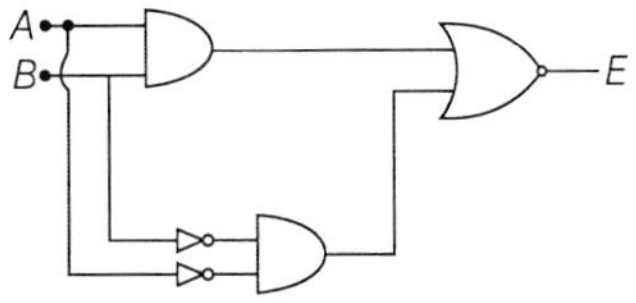

A	B	E
0	0	0
0	1	X
1	0	Y
1	1	0

In the truth table of the above circuit the value of X and Y are *[2024, 9 April Shift-II]*

(a) 0, 0 (b) 1, 1
(c) 0, 1 (d) 1, 0

42 The output Y of following circuit for given inputs is *[2024, 8 April Shift-I]*

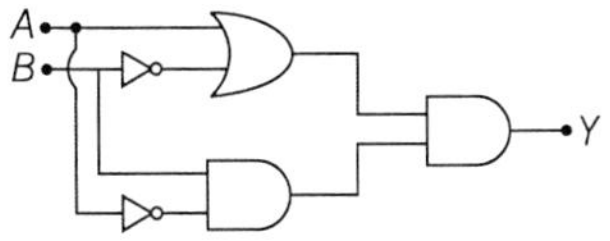

(a) 0 (b) $A \cdot B(A + B)$
(c) $\overline{A} \cdot B$ (d) $A \cdot B$

43

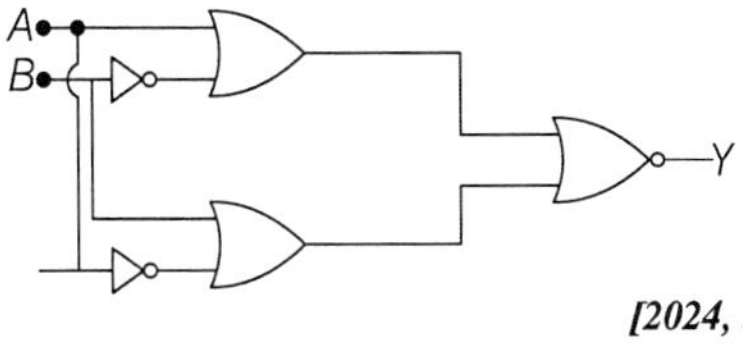

[2024, 31 Jan Shift-II]

The output of the given circuit diagram is

(a)

A	B	Y
0	0	0
1	0	0
0	1	0
1	1	1

(b)

A	B	Y
0	0	0
1	0	0
1	1	0
0	1	0

(c)

A	B	Y
0	0	0
1	0	0
0	1	1
1	1	0

(d)

A	B	Y
0	0	0
1	0	1
0	1	1
1	1	0

44 The output Y for the inputs A and B of circuit is given by *[2023, 30 Jan Shift-II]*

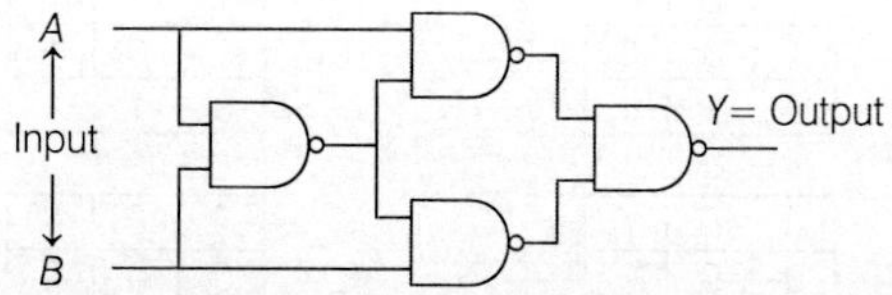

Truth table of the shown circuit is

(a)

A	B	Y
0	0	0
1	0	0
0	1	0
1	1	1

(b)

A	B	Y
0	0	0
1	0	0
0	1	0
1	1	0

(c)

A	B	Y
0	0	0
1	0	0
0	1	1
1	1	0

(d)

A	B	Y
0	0	0
1	0	1
0	1	1
1	1	0

45 The output waveform of the given logical circuit for the following inputs A and B as shown below, is *[2023, 30 Jan Shift-I]*

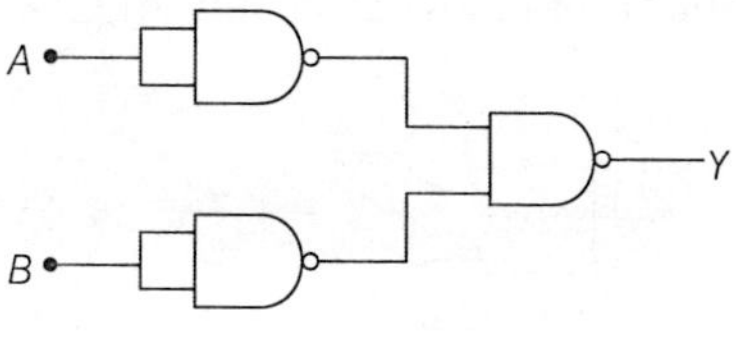

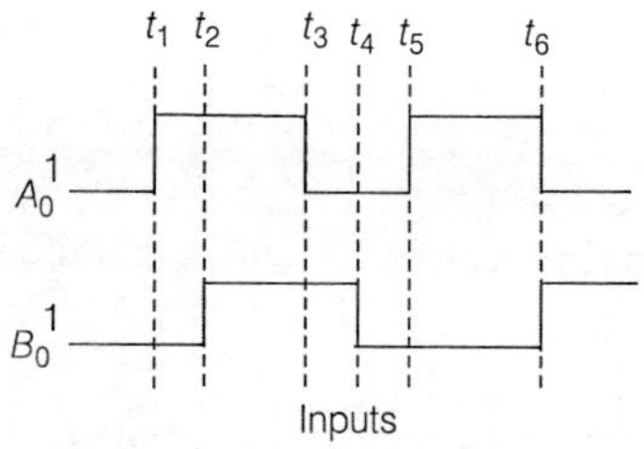

(a)
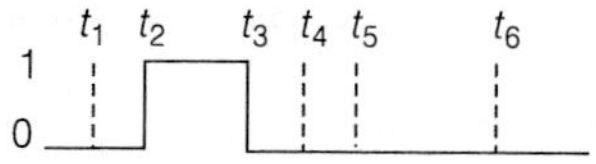

(b)
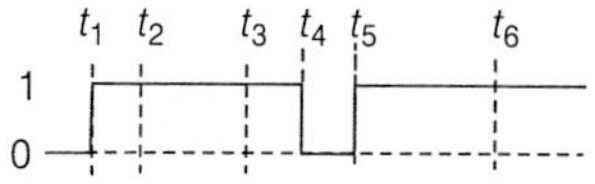

(c)
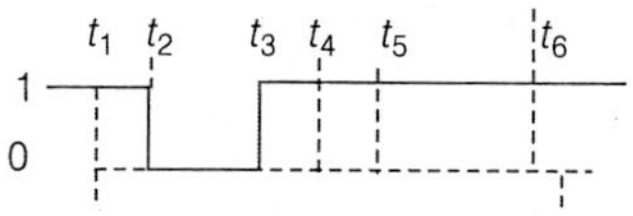

(d)
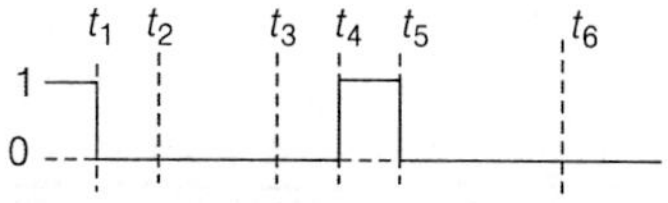

Imp. Ex-OR Gate

46 For the given logic gates combination, the correct truth table will be *[2023, 29 Jan Shift-II]*

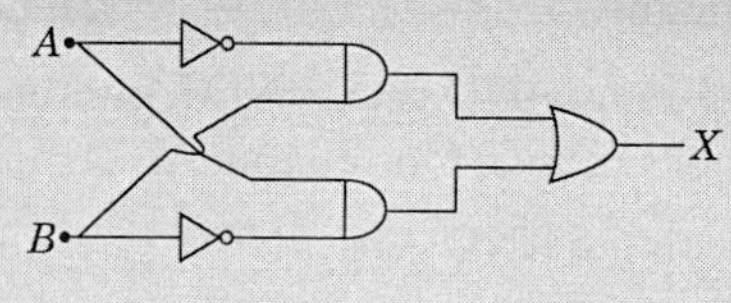

(a)

A	B	X
0	0	1
0	1	0
1	0	1
1	1	0

(b)

A	B	X
0	0	1
0	1	0
1	0	0
1	1	0

(c)

A	B	X
0	0	0
0	1	1
1	0	1
1	1	1

(d)

A	B	X
0	0	0
0	1	1
1	0	1
1	1	0

47 If V_A and V_B are the input voltages (either 5V or 0V) and V_0 is the output voltage then the two gates represented in the following circuit A and B are *[2021, 31 Aug Shift-II]*

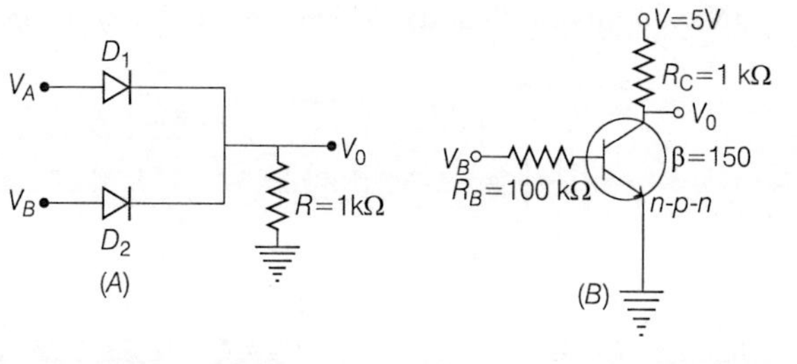

(a) AND and OR gate
(b) OR and NOT gate
(c) NAND and NOR gate
(d) AND and NOT gate

48 Which one of the following will be the output of the given circuit ? *[2021, 17 March Shift-II]*

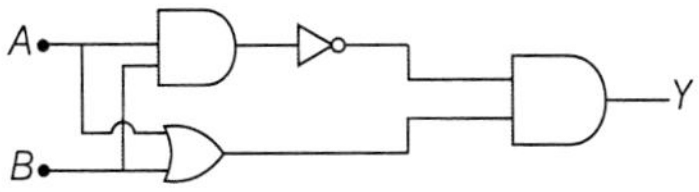

(a) NOR gate (b) NAND gate
(c) AND gate (d) XOR gate

49 Draw the output signal Y in the given combination of gates *[2021, 26 Feb Shift-II]*

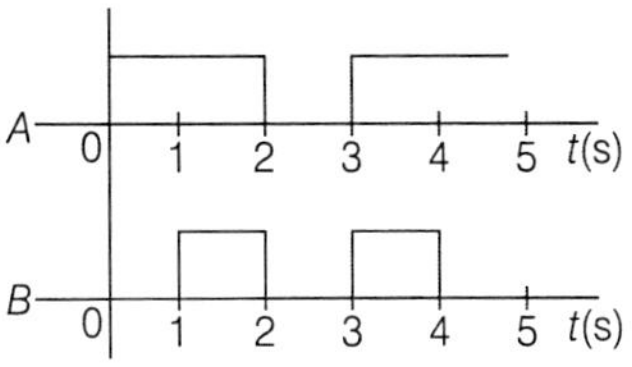

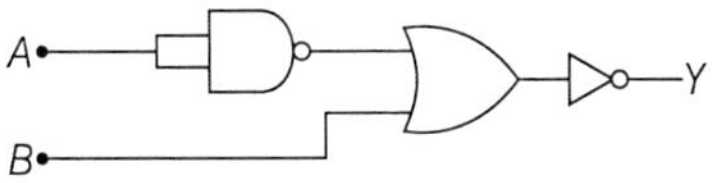

(a)
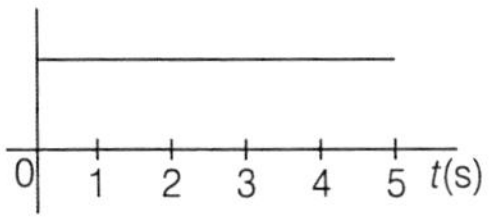

(b)
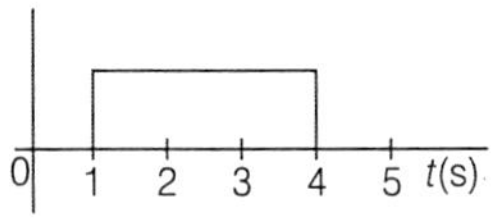

(c)
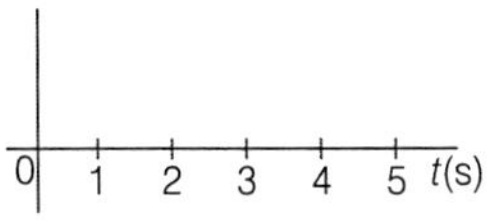

(d)
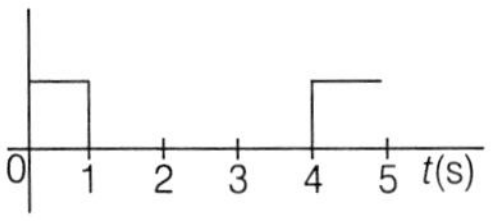

Imp. Ex-NOR Gate

50 The truth table for the followng logic circuit is *[2021, 25 Feb Shift-II]*

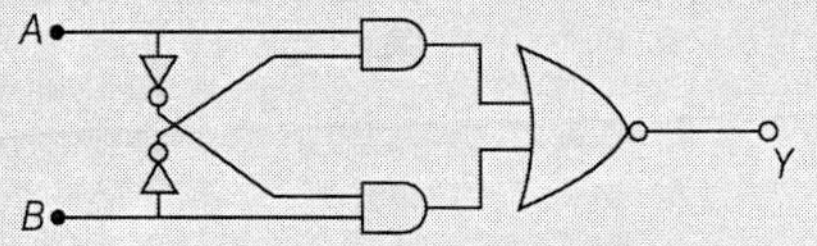

(a)

A	B	Y
0	0	0
0	1	1
1	0	1
1	1	0

(b)

A	B	Y
0	0	1
0	1	0
1	0	0
1	1	1

(c)

A	B	Y
0	0	0
0	1	0
1	0	1
1	1	0

(d)

A	B	Y
0	0	0
0	1	1
1	0	0
1	1	1

51 To get output 1 at R, for the given logic gate circuit, the input values must be *[2019, 10 Jan Shift-I]*

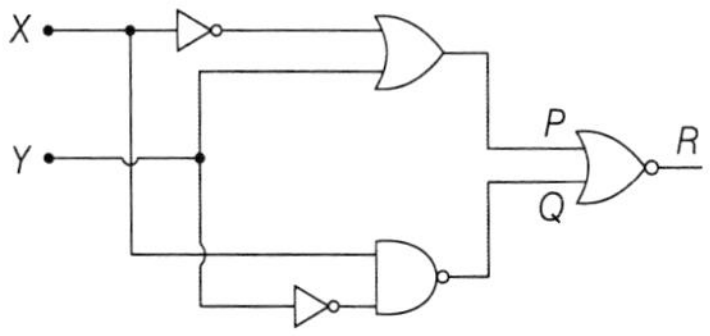

(a) $X=0, Y=0$ (b) $X=1, Y=0$
(c) $X=1, Y=1$ (d) $X=0, Y=1$

Numerical Value Type Questions

52 In the logic circuit shown in the figure, if input A and B are 0 to 1 respectively, the output at Y would be x. The value of x is .... *[2021, 16 March Shift-I]*

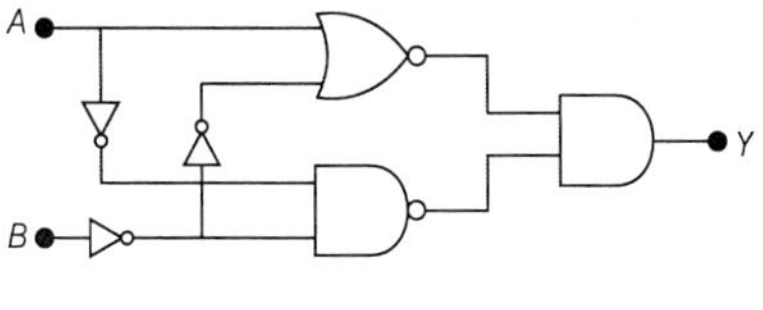

Key Index

Topic 1

1 (c)	**2** (a)	**3** (c)	**4** (c)	**5** (a)	**6** (c)	**7** (d)	**8** (c)	**9** (a)	**10** (a)
11 (c)	**12** (a)	**13** (b)	**14** (b)	**15** (b)	**16** (b)	**17** (d)	**18** (b)	**19** (c)	**20** (c)
21 (a)	**22** (a)	**23** (a)	**24** (d)	**25** (a)	**26** (c)	**27** (d)	**28** (d)	**29** (c)	**30** (c)
31 (c)	**32** (c)	**33** (a)	**34** (c)	**35** (c)	**36** (c)	**37** (a)	**38** (c)	**39** (b)	**40** (d)
41 (c)	**42** (b)	**43** (a)	**44** (a)	**45** (a)	**46** (a)	**47** (d)	**48** (b)	**49** (a)	**50** (d)
51 (c)	**52** (a)	**53** (a)	**54** (a)	**55** (b)	**56** (d)	**57** (a)	**58** (b)	**59** (c)	**60** (d)
61 (a)	**62** (b)	**63** (c)	**64** (c)	**65** (d)	**66** (d)	**67** (c)	**68** (c)	**69** (b)	**70** (b)
71 (b)	**72** (*)	**73** (d)	**74** (c)	**75** (b)	**76** (d)	**77** (c)	**78** (5)	**79** (480)	**80** (9)
81 (25)	**82** (1)	**83** (3)	**84** (125)	**85** (4)	**86** (1)	**87** (5)	**88** (20)	**89** (120)	**90** (5)
91 (500)	**92** (192)	**93** (5)	**94** (12)						

Topic 2

1 (c)	**2** (d)	**3** (a)	**4** (b)	**5** (a)	**6** (c)	**7** (a)	**8** (a)	**9** (b)	**10** (b)
11 (d)	**12** (d)	**13** (c)	**14** (a)	**15** (c)	**16** (a)	**17** (a)	**18** (c)	**19** (c)	**20** (a)
21 (d)	**22** (*)	**23** (d)	**24** (c)	**25** (a)	**26** (d)	**27** (a)	**28** (d)	**29** (d)	**30** (d)
31 (c)	**32** (c)	**33** (b)	**34** (b)	**35** (a)	**36** (a)	**37** (b)	**38** (b)	**39** (d)	**40** (d)
41 (b)	**42** (a)	**43** (b)	**44** (a)	**45** (b)	**46** (d)	**47** (b)	**48** (d)	**49** (d)	**50** (b)
51 (b)	**52** (0)								

Detailed **Explanations**

Topic 1

01. *(c)* From the given circuit diagram, diode D_1 is reverse biased and diode D_2 is in forward biased. There, current flows through diode D_2. Hence voltage drop on resistor will be 5V. Thus, $V_{out} = 0$.

02. *(a)*

> **key idea** To check weather a diode is in forward bias or in reverse bias, check sign of potentials of *p* and *n* side,
>
> If $V_p - V_n = \begin{cases} +, \text{Forward bias} \\ -, \text{Reverse bias} \end{cases}$

For forward bias $V_p - V_n$ is positive.
Among the given circuits, only B, C and E are forward bias.
In forward biasing, positive or high concentration part is connect to high potential and *N* part of semiconductor is connect to low potential.

03. *(c)*

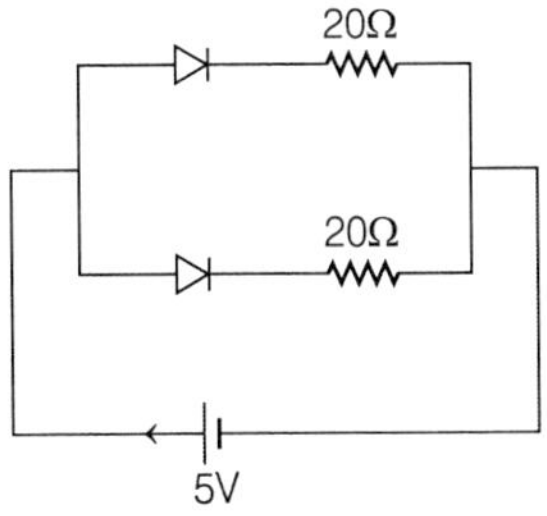

Both diode are forward bias,
In forward bias diode is act as conducting path
So, resistor are in parallel

$$\frac{1}{R_{eq}} = \frac{1}{20} + \frac{1}{20} \Rightarrow R_{eq} = 10\,\Omega$$

$$I = \frac{V}{R_{eq}} = \frac{5}{10} = 0.5\text{ A}$$

04. *(c)* The device shows a rapid current increase in reverse biasing after a certain point. This is a zener diode operating in reverse biased connection. It can also be used as a voltage regulator as a small voltage difference creates a large change in current after breakdown voltage.

05. *(a)* We know that,

$$E = \frac{hc}{\lambda}$$

$$\therefore \lambda = \frac{hc}{E} = \frac{1242\text{ eV}}{1.42\text{ eV}}\text{nm} \quad (\because hc = 1242\text{ eV})$$

$$= 875\text{ nm (approx.)}$$

06. *(c)* For the measurement of dynamic resistance of *p-n* junction diode, diode must be connected in forward bias. A diode is said to be forward biased if *p*-side is connected with higher potential than *n*-side. Thus, diode shown in option (c) is connected in forward biased, hence it is used for the measurement of dynamic resistance.

07. *(d)* A diode is said to be in reverse biased when *p*-side is connected to less voltage than *n*-side. i.e.
$V_p - V_n = 2 - 4 = -2(-\text{ve})$
So, diode is in reverse bias.
Hence, circuit is in reverse bias.

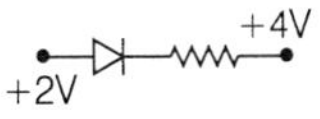

08. *(c)* A is correct but R is incorrect as per as theory of *p-n* junction diodes. For a *p-n* junction diode, the current in the forward bias is greater than the current in the reverse bias when the applied voltage is more than the threshold voltage and less than the breakdown voltage.

09. *(a)* A semiconductor starts conducting more as the temperature increases. It means resistance decreases with increase in temperature. So, if temperature increases its resistivity decreases.

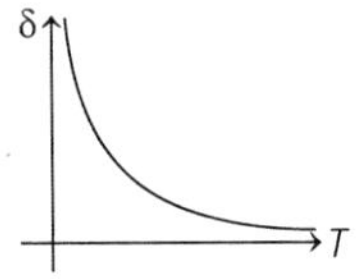

Also, $\rho = \frac{m}{ne^2\tau}$

As temperature increases, τ decreases but n increases and n is dominant over τ, thus reducing overall resistivity of the semiconductor.

10. *(a)* Zener diode is a reverse biased heavily doped *p-n* junction diode. It is operated in breakdown region.
Zener diode is designed to operate in the reverse breakdown voltage continuously without being damaged.

11. *(c)* The highest energy level that an electron can occupy at the absolute zero temperature is known as the Fermi level. The fermi level lies between the valence band and conduction band because at absolute zero temperature, the electrons are all in the lowest energy state.
In intrinsic semiconductor, Fermi level is between the two bands.
In *p*-type, Fermi level is closer to valence band.
In *n*-type, Fermi level is closer to conduction band.
In metal, Fermi level is inside conduction band.
Thus, correct sequence is
A - II, B- III, C - I, D- IV.

12. *(a)* As temperature increases, more electrons are excited to conduction band and hence conductivity increases, therefore resistance decreases.

13. *(b)* An LED is a heavily doped *p-n* junction. When an electron transits from conduction band to valence band in forward bias, it emits energy which is nearly equal to the energy gap in form of visible radiation.
So, an LED cannot be used in reverse bias.

14. *(b)* A photodiode is operated in reverse bias because photocurrent is more significant in reverse bias as compared to the forward bias. For a *p-n* junction diode, the current in the forward bias is greater than the current in the reverse bias when the applied voltage is more than the threshold voltage and less than the breakdown voltage.

15. *(b)* The *V-I* curve for solar cell is in 4th quadrant, because current is negative here signifies that current is supplied by cell not drawn.
I-V characteristics of solar cell is as shown below.

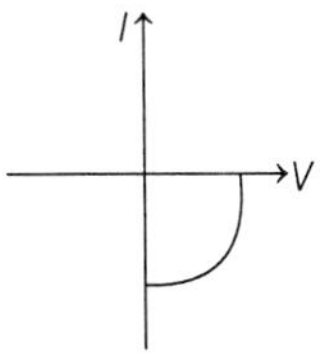

• **Concept Enhancer**

Solar cell is based on photovoltaic effect. It converts radiation into electrical energy.

16. *(b)* Diode is two terminal device which conducts current in forward bias only. It will not conduct in reverse bias.

17. *(d)* Photodiode is used in reverse bias because very small change in minority charge carriers produces high value of reverse bias current.

18. *(b)* The *I-V* characteristic curve for a photodiode is shown below.

With increasing biasing voltage of a photodiode, the photocurrent first increases and then becomes saturated, i.e., after certain potential, electron-hole pair production will be nearly constant.

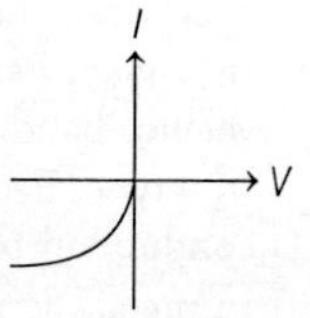

19. *(c)* We know that conductivity

$$\sigma = n_e e \mu_e$$
$$\sigma = 10^{19} \times 1.6 \times 10^{-19} \times 1.6$$
$$= 2.56$$

As, resistivity, $\rho = \frac{1}{\sigma} = \frac{1}{2.56}$

or $\rho = 0.39 \simeq 0.4\ \Omega\text{-m}$

20. *(c)*

key idea Potential drop in a silicon diode in forward bias is around 0.7 V.

In given circuit, potential drop across 200 Ω resistor is

$$I = \Delta V_{net} / R = 3 - 0.7/200$$

$\Rightarrow$ $I = 0.0115\ \text{A} \Rightarrow I = 11.5\ \text{mA}$

21. *(a)* Here, $n_e = 5 \times 10^{18}\ \text{m}^{-3}$

$n_h = 5 \times 10^{19}\ \text{m}^{-3}$

$\mu_e = 2.0\ \text{m}^2\ \text{V}^{-1}\ \text{s}^{-1}$

$\mu_h = 0.01\ \text{m}^2\ \text{V}^{-1}\text{s}^{-1}$

The conductivity of semiconductor,

$\sigma = e(n_e \mu_e + n_h \mu_h) = (1.6 \times 10^{-19})$

$(5 \times 10^{18} \times 2 + 5 \times 10^{19} \times 0.01)$

$$= 1.6 \times 1.05$$
$$= 1.68\ (\Omega\text{m})^{-1}$$

• **Concept Enhancer**

The electrical conductivity of semiconductors increases as temperature increases.

22. *(a)* So, Simple diode works in breakdown region → *(a)* forward bias

Zener diode → *(b)*

Works in breakdown region

Solar cell → *(c)* reverse bias

Light dependent resistance → *(d)* forward bias

23. *(a)* The relation between temperature and resistance for Cu and undoped Si are given below, this is because Cu is metal and resistance increase with increase in temperature as $R = R_0(1 + \alpha T)$ while Si is semiconductor and R decreases as T increases.

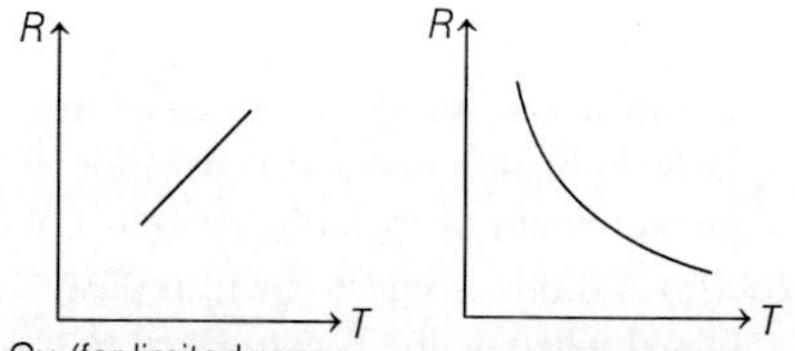

Cu (for limited range of temperature) Undoped Si

24. *(d)* Because electron concentration in n-region is more as compared to that in p-region.

• **Concept Enhancer**

In *p-n* junction diode, holes/electrons try to diffuse from their side to other side. Only these electrons/holes crosses the junction, which have high kinetic energy.

25. *(a)* For forward bias, p-side must be at higher potential than n-side.

So,

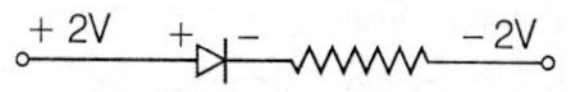

is forward biased.

26. *(c)* Given figure is half-wave rectifier. [Since, diode allows the signal to pass during positive half only i.e., forward bias]

27. *(d)* As Si and Ge are semiconductor at room temperature (300 K) the energy band gap is low, as at result of which electron in covalent bond break and move to conduction band.

A hole is also created in VB. Therefore Si and Ge has higher number of free electron. While in carbon the energy gap is very high compare to Si and Ge, so very few electron moves to VB.

28. *(d)* For forward bias, p is + ve, n is – ve.

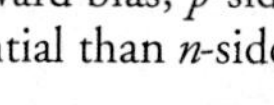

For reverse biasing of an ideal diode, the potential of n-side should be higher than potential of p-side. Only option (d) is satisfying the criterion for reverse biasing.

29. *(c)* $E_g = h\nu = \frac{hc}{\lambda}$

$$= \left(\frac{6.63 \times 10^{-34} \times 3 \times 10^8}{2480 \times 10^{-9} \times 1.6 \times 10^{-19}}\right) \text{eV}$$

$[\because 1\ \text{eV} = 1.6 \times 10^{-19}\ \text{J}]$

$$= 0.5\ \text{eV}$$

30. *(c)* Given, $f = 50$ Hz

$\Rightarrow$ $T = \frac{1}{50}$

For full wave rectifier,

$$T_1 = \frac{T}{2} = \frac{1}{100}$$

$\therefore$ $f_1 = 100$ Hz $[\because f = 1/T]$

31. *(c)* We know that resistance of conductor is directly proportional to temperature (i.e., $R \propto \Delta t$), while resistance of semiconductor is inversely proportional to temperature.

$\left(\text{i.e., } R \propto \frac{1}{\Delta t}\right)$.

Since, copper is pure conductor and germanium is a semiconductor due to decrease in temperature, resistance of conductor decreases while that of semiconductor increases.

32. *(c)* In forward bias, the majority charge carriers are repelled by applied field (both n and p-side). This causes the narrowing of depletion region and reduction in barrier height.

33. *(a)* Due to the reverse biasing, the width of depletion region increases and current flowing through the diode is almost zero. In this case, electric field is almost zero at the middle of the depletion region.

34. *(c)* Metals have minimum energy gap and insulators have maximum energy gap and semi-conductors have energy gap lying between insulators and metals.

35. *(c)* If we increase the temperature, the specific resistance (resistivity) of conductor increases while that of semiconductor decreases.

36. *(c)* At very low temperature electrons does not have enough energy to cross band gap (or 1.1 eV) in Si. Therefore, there are no free charge carriers at absolute zero.

37. *(a)*

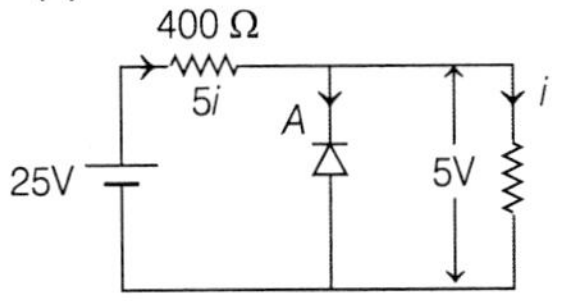

From circuit diagram

$$5I_L = \frac{20}{400} = \frac{1}{20}$$

$$I_L = \frac{1}{20 \times 5} = \frac{1}{100} \text{ A}$$

$$I_L = 10 \text{ mA}$$

$= \text{Load current}$

Also, $V_L = 5\text{V}$

$$\therefore \quad R_L = \frac{V_L}{I_L} = \frac{5}{10 \times 10^{-3}} = 500\Omega$$

38. *(c)*

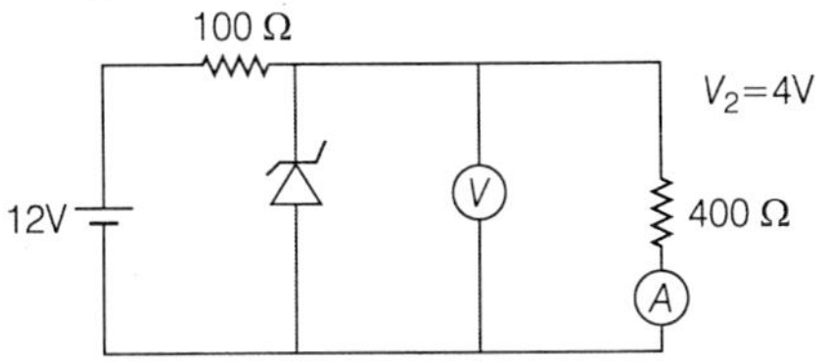

$$V_1 = \frac{400\Omega}{100 + 400} \times 12 \text{ V}$$

$$= \frac{4}{5} \times 12 = \frac{48}{5} \text{ V}$$

Here, $V_1 > V_Z$ (V_Z = Zener diode voltage)

So Zener breakdown will be take place.

So, voltage across 400Ω will be 4 V.

$$I = \frac{4}{400} \text{ A} = \frac{1}{100} \text{ A} = 10 \text{ mA}$$

39. *(b)* (A) *n*-type semiconductor holes are minority charge carriers and e^- are majority charge carriers.

(B) Dopant are pentavalent atoms.

(C) $n_e \cdot n_n = n_i^2$ for intrinsic semiconductor

(E) In *n*-type semiconductor primary source of holes generated due to thermal excitation.

(A), (B), (E) are correct.

40. *(d)*

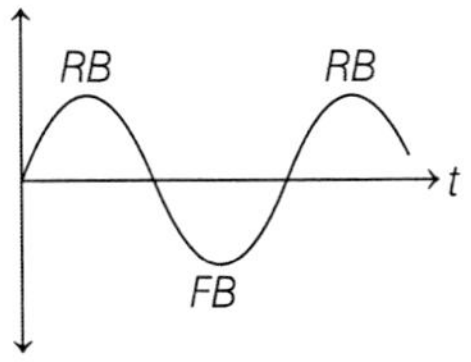

$V = V_0 \sin \omega t$

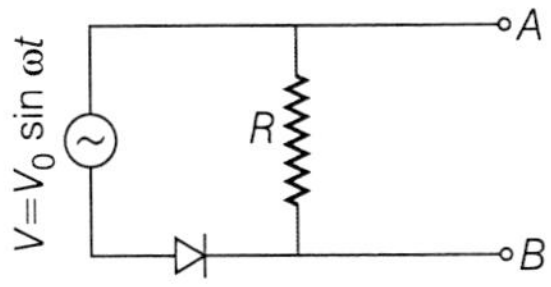

For first half cycle → *RB*

For second half cycle → *FB*

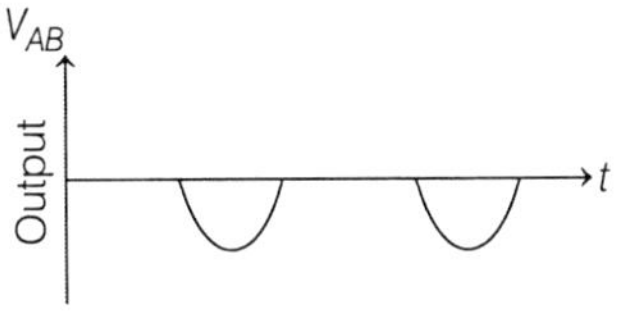

41. *(c)* Statements given in B and E are correct, while A, C and D are incorrect.

Junction area of solar cell is made broad compared to photodiode.

LED is made of heavily doped *p - n* junction.

Increase of forward current does not always result in increase in light intensity.

42. *(b)* Given, acceptor energy level,

$$E_g = 6 \text{ eV}$$

Let λ be the maximum wavelength of light which can create a hole then,

$$E = \frac{1242}{\lambda(\text{nm})}$$

$$\Rightarrow \quad \lambda = \frac{1242}{E} \text{ nm} = \frac{1242}{6} = 207 \text{ nm}$$

43. *(a)* Diode 2 is connected in reverse bias, so no current will flow.

Diode 1 is forward bias. Now drawing the simplified circuit

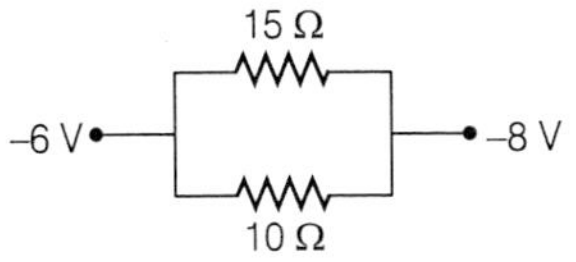

15 Ω and 10 Ω are in parallel connection.

$$R_{eq} = \frac{R_1 R_2}{R_1 + R_2}$$

$$= \frac{15 \times 10}{15 + 10} = 6\,\Omega$$

Hence, option (a) is correct.

44. *(a)*

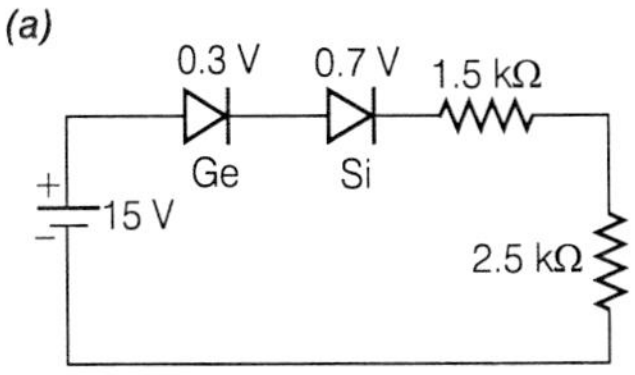

As both diodes are in forward bias, current will flow.

Potential drops of diodes are 0.7 V and 0.3 V in forward bias.

So, current in circuit is

$$I = \frac{V}{R} = \frac{15 - 0.3 - 0.7}{(2.5 + 1.5) \times 10^3}$$

$$= \frac{7}{2} \times 10^{-3} \text{ A} = \frac{7}{2} \text{ mA}$$

So, potential drop across 2.5 kΩ resistor is,

$$V = IR = \frac{7}{2} \times 10^{-3} \times 2.5 \times 10^3 = 8.75 \text{ V}$$

• **Concept Enhancer**

While deciding, that the diode is in forward bias, a confusion arises specially when both terminals of diode are connected with only positive voltage or only negative voltage. So, always remember a diode is said to be in forward bias if its *p*-side is connected with more positive voltage than its *n*-side.

45. *(a)* Given,

Zener breakdown voltage = 10 V

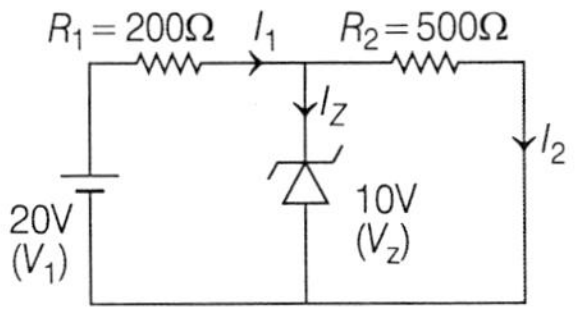

Using junction rule,

$$I_Z = I_1 - I_2$$

$$I_1 = \frac{V_1}{R_1} = \frac{20 - 10}{200} \text{ A} = \frac{10}{200} \text{ A}$$

$$I_2 = \frac{V_1}{R_2} = \frac{10}{500} \text{ A}$$

$$\therefore \quad I_Z = I_1 - I_2 = \frac{10}{200} - \frac{10}{500}$$

$$I_Z = \frac{3}{100} \text{ A} = 30 \text{ mA}$$

46. *(a)*

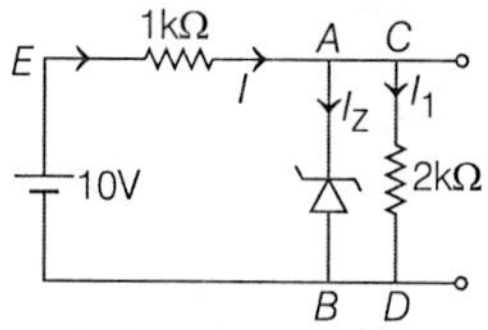

Given in question, $V_Z = 3.0$ V

Let potential at $B = 0$ V

Let potential at E, $V_E = 10$ V

$$V_C = V_A = 3\text{ V}$$

$$I_Z + I_1 = I,$$

$$I_Z = I - I_1$$

$$I = \frac{10-3}{1000} = \frac{7}{1000}\text{ A}$$

$$I_1 = \frac{V_C - V_D}{2\text{k}\Omega} = \frac{3-0}{2000}\text{ A}$$

$$I_Z = \frac{7-1.5}{1000} = 5.5\text{ mA}$$

• Concept Enhancer

Zener diode is used as a shunt voltage regulator. It exhibits a controlled breakdown. Both Avalanche and zener breakdown occur in zener diode.

47. *(d)* Given, forward bias resistance

$$r = 25\,\Omega$$

Diode D_2 is in reverse biased, thus $I_3 = 0$.

Since, resistance in each branch is same

Thus, $I_2 = I_4$

From KCL,

$$I_2 + I_4 + I_3 = I_1$$

$$\Rightarrow \quad 2I_2 = I_1$$

$$\Rightarrow \quad \frac{I_1}{I_2} = 2$$

48. *(b)*

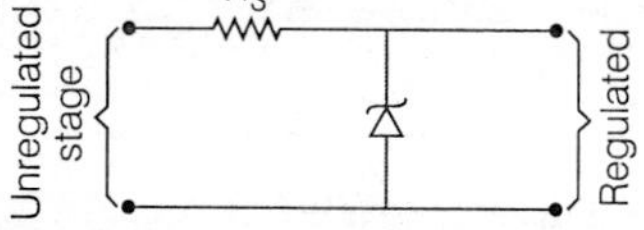

Power rating of zener $= 1.6$ W

Max voltage across zener $= 8$ V

Using $P = VI$; for zener diode,

$$I = \frac{P}{V} = \frac{1.6}{8} = 0.2\text{ A}$$

Hence, for safe operation 2V (excess voltage of 10 V) potential drop must occur in series resistor R_S.

For R_S, $V = 2$ V, $I = 0.2$ A

Using, $V = IR$,

$$R = \frac{2}{0.2} = 10\,\Omega$$

49. *(a)* When Si is doped with trivalent impurities like Boron, then a p-type semiconductor is formed. In p-type semiconductor, excess number of holes are found because holes are majority carrier in p-type semiconductor.

When Si is doped with pentavalent impurities (like Arsenic), then an n-type semiconductor is formed.

In n-type semiconductor, excess no. of electrons are found because these are majority carrier. Thus, Statement I is correct.

When p-type semiconductor and n-type semiconductor are fused to make a junction, a potential barrier is produced across the junction which stops the flow of current.

Thus, no current is detected by externally connected ammeter.

Hence, Statement II is incorrect.

50. *(d)* Photodiodes are used in reverse bias for measuring the light intensity.

For a p-n junction diode, the current in the forward bias is greater than the current in the reverse bias when the applied voltage is more than the threshold voltage and less than the breakdown voltage.

51. *(c)*

key idea To solve this question, you can use following steps

(1) Find out the equivalent resistance of this circuit.

(2) Using Ohm's law you get total current of this circuit. $I = \frac{V}{R_{eq}}$

In the given circuit,

Diode D_1 and D_2 are in forward bias whereas diode D_3 is in reverse bias.

$$R_{eq} = \frac{(130+30)\times(130+30)}{(130+30)+(130+30)} + 20$$

$$= 80 + 20 = 100\,\Omega$$

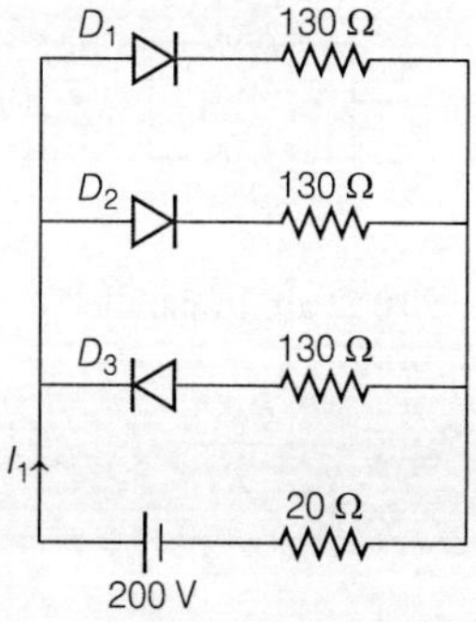

Current, $I = \frac{V}{R_{eq}} = \frac{200}{100} = 2$ A

52. *(a)* We know that pentavalent impurities have excess free electrons. So, when a silicon semiconductor is doped with pentavalent impurities, the electron density increases. But the whole semiconductor is electrically neutral, therefore net charge on n-type semiconductor is zero.

Hence, Statement I true but Statement II is false.

53. *(a)* Given, $\lambda = 621 \times 10^{-9}$ m

As we know that the band gap energy is given by

$$E = \frac{hc}{\lambda}\text{ J} \Rightarrow E = \frac{hc}{\lambda_e}\text{ eV} \quad \ldots\text{(i)}$$

$$E = \frac{6.63\times10^{-34}\times3\times10^{8}}{621\times10^{-9}\times1.6\times10^{-19}}$$

$$= 0.02001 \times 10^{2}$$

$$= 2.001\text{ eV} \simeq 2\text{eV}$$

54. *(a)* Consider the given figure and draw the direction of current as follows

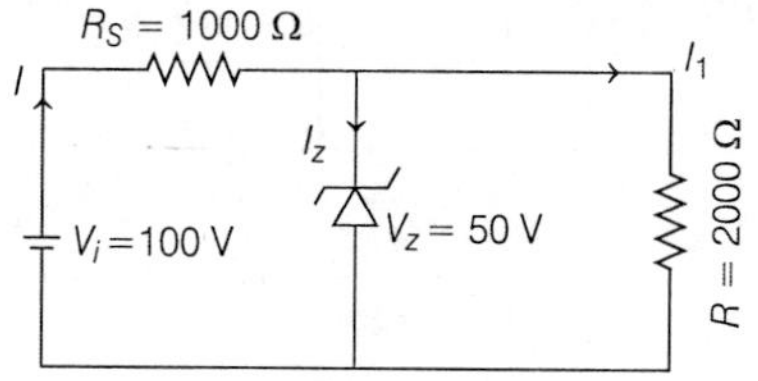

$\therefore$ Current, $I = \frac{V_i - V_Z}{R_S}$

$$\Rightarrow \quad I = \frac{100-50}{1000} = 50\text{ mA} \quad \ldots\text{(i)}$$

[$\because$ Resistance and Zener diode are in parallel, therefore voltages across them are same]

and current, $I_1 = \frac{50}{2000} = 25$ mA ...(ii)

By using Kirchhoff's current law,

$$\because \quad I = I_1 + I_Z$$

$$\Rightarrow \quad I_Z = I - I_1 \quad \ldots\text{(iii)}$$

From Eq. (i) and (ii) and, we get

$$\Rightarrow \quad I_Z = (50 - 25)\text{ mA}$$

$$I_Z = 25\text{ mA}$$

55. *(b)* In case of n-type semiconductor, the energy level diagram will be

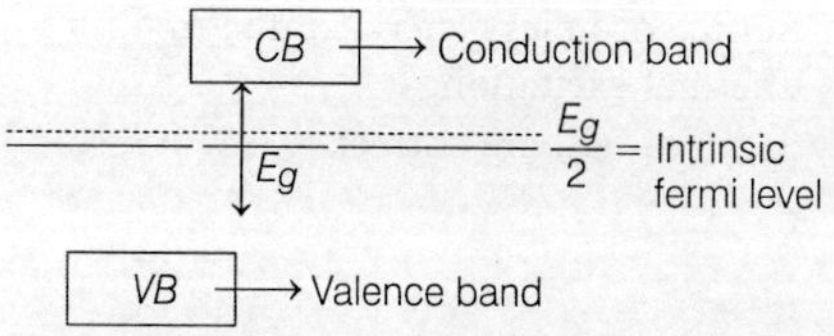

In case of n-type semiconductor $n > p$, so the fermi level will go upward. Similarly, in case of p-type semiconductor $p > n$, so the Fermi level will go downward.

56. *(d)* Given, energy gap of LED,

$$E = 1.9 \text{ eV}$$

Speed of light in free space,

$$c = 3 \times 10^8 \text{ ms}^{-1}$$

Planck's constant, $h = 6.63 \times 10^{-34}$ J-s

As we know that $E = \dfrac{hc}{\lambda}$

$$\Rightarrow \quad \lambda = \frac{hc}{E}$$

$$\Rightarrow \quad \lambda = \frac{6.63 \times 10^{-34} \times 3 \times 10^8}{1.9 \times 1.6 \times 10^{-19}}$$

$$= 654 \times 10^{-9} \text{ m}$$

$$= 654 \text{ nm}$$

As, wavelength of red light is 600 nm.

$\therefore$ Required wavelength will be of red colour.

57. *(a)* Given, supply voltage, $V = 5$ V

The circuit diagram, when positive terminal of the battery is connected to X is as shown below

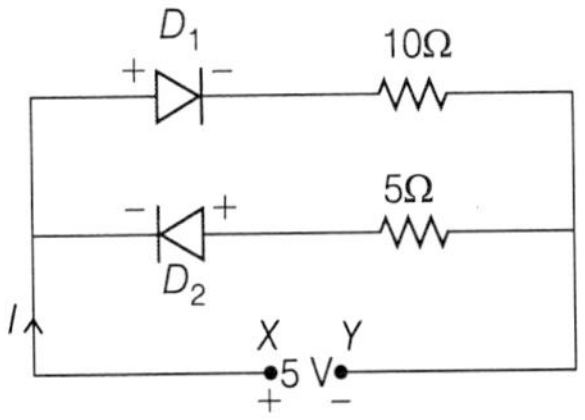

Let I current is coming from battery.

$\therefore D_1$ will act as closed circuit as forward biased and D_2 will act as open circuit as reverse biased.

D_2 is in reverse bias, so it will not pass current. Since, potential drop across the diode is not given

$$V = iR \Rightarrow i = \frac{V}{R} = \frac{5}{10}$$

$$i = 0.5 \text{ A}$$

58. *(b)* As we know that Zener breakdown takes place, when we supply reverse bias voltage to Zener diode. Due to heavily dopping, the electrons in the valence band of p-type region can jump easily to the conduction band of n-type region, hence due to high electric field, zener breakdown occur.

Hence, Zener breakdown is easily observed in Zener diode which is heavily doped and having narrow depletion region.

59. *(c)* Given circuit is

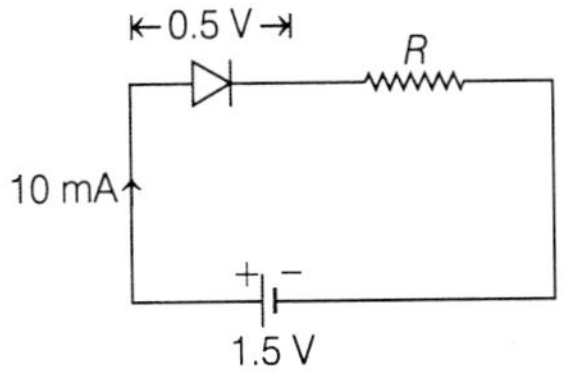

Now, by Kirchhoff 's loop rule,

$$1.5 - 0.5 - R \times 10 \times 10^{-3} = 0$$

$$\Rightarrow \quad R = 100\,\Omega$$

60. *(d)* As both diodes are in reverse bias, so given circuit is equivalent to the circuit as shown in the figure.

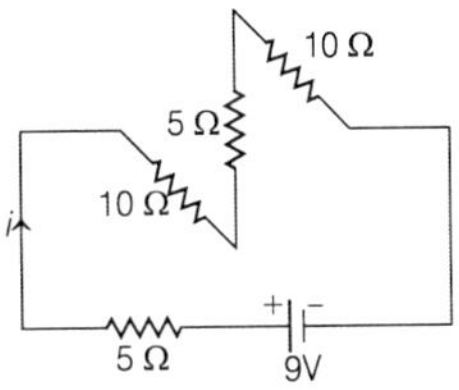

Equivalent resistance of circuit,

$$R = 5 + 10 + 5 + 10 = 30\Omega$$

$\therefore$ Circuit current,

$$I = \frac{V}{R} = \frac{9}{30} = 0.3 \text{ A}$$

61. *(a)* Given circuit is Zener diode circuit

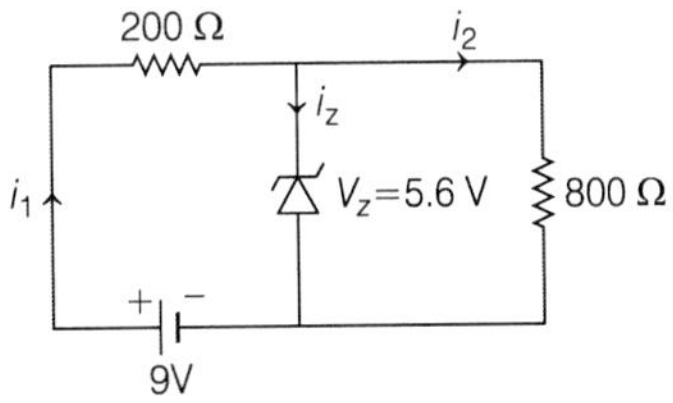

Where, potential drop across 800 Ω resistance = potential drop across Zener diode = 5.6 V

So, current, $i_2 = \dfrac{V}{R} = \dfrac{5.6}{800} = 7$mA

Now, potential drop across 200 Ω resistance $= 9 - 5.6 = 3.4$ V

Current, $i_1 = \dfrac{V}{R} = \dfrac{3.4}{200} = 17$ mA

So, $i_z = i_1 - i_2 = 17 - 7 = 10$ mA

62. *(b)* In this circuit, D_1 is forward biased and D_2 is reversed biased.

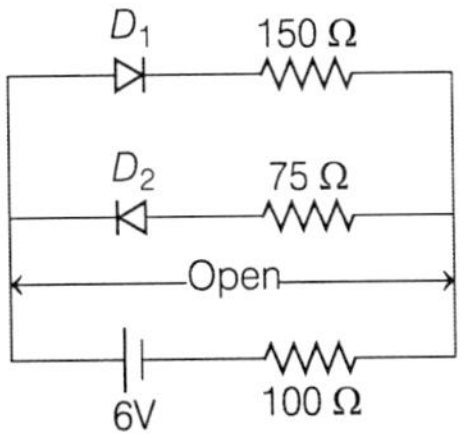

Resistance of D_1 is 50Ω.

$\therefore$ Net resistance of the circuit,

$$R_{net} = 50 + 150 + 100 = 300\,\Omega$$

$\therefore$ Current through the $100\,\Omega$ resistance

$$= \frac{V}{R_{net}} = \frac{6}{300} = 0.020 \text{ A}$$

63. *(c)*

> **key idea** When the applied reverse voltage (V) reaches the breakdown voltage of the Zener diode, then only a large amount of current is flown through it, otherwise it is approximately zero.

In the given situation, if we consider that Zener diode is at breakdown. Then, potential drop across 1500 Ω resistances will be 10 V. So, potential drop at 500 Ω resistor will be 2 V.

$\therefore$ Current in $R_1 = \dfrac{2}{500} = 4\text{mA} = I_1$ (say)

Current in each

$$R_2 = \frac{10}{750} = \frac{2}{150} = 13.33 \text{ mA} = I_2 \text{ (say)}$$

$\Rightarrow$ which is not possible.

So, Zener diode will never reach to its breakdown.

$\therefore$ Current flowing through a reverse biased Zener diode $= 0$.

64. *(c)* Initially Ge and Si are both forward biased. So, current will effectively pass through Ge diode with a voltage drop of 0.3 V.

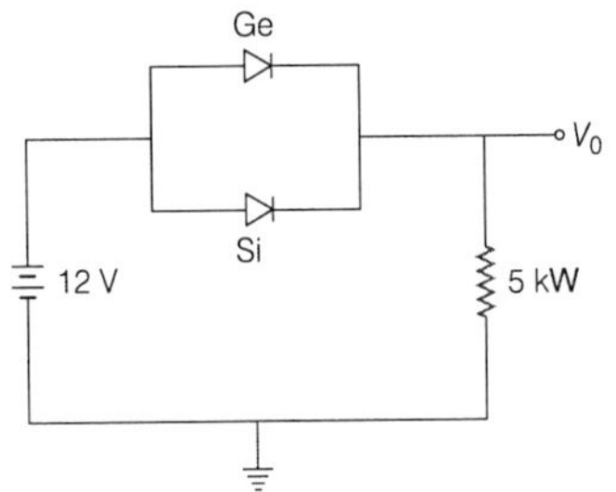

$\therefore$ Initial output voltage,

$$V_0 = 12 - 0.3$$

$$= 11.7 \text{ V}$$

If Ge is reversed biased, then only Si diode will work. In this condition, output voltage

$$V_0 = 12 - 0.7 = 11.3 \text{ V}$$

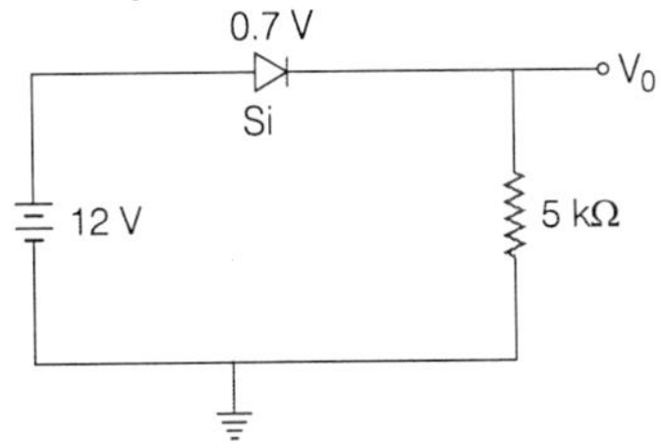

$\therefore$ Change in output voltage

$= 11.7 - 11.3$

$= 0.4$ V

65. *(d)* For visible region

$\lambda = (4 \times 10^{-7} - 7 \times 10^{-7})$ m

$\because \quad E_g = \dfrac{hc}{\lambda}$

$E_{gL} = \dfrac{6.6 \times 10^{-34} \times 3 \times 10^{8}}{4 \times 10^{-7} \times 1.6 \times 10^{-19}}$

$= 3.09 \text{ eV} \simeq 3 \text{ eV}$

$E_{gL} = \dfrac{6.6 \times 10^{-34} \times 3 \times 10^{8}}{7 \times 10^{-7}}$

$= \dfrac{19.8 \times 10^{-26}}{7 \times 10^{-7}}$ J

$\simeq \dfrac{2.8 \times 10^{-19}}{1.6 \times 10^{-19}}$

$\Rightarrow \quad E_{gL} = 1.75$ eV

$\therefore$ Energy band gap given in option (d) is correct.

66. *(d)* During – ve cycle, diode will not allow the signal to pass.

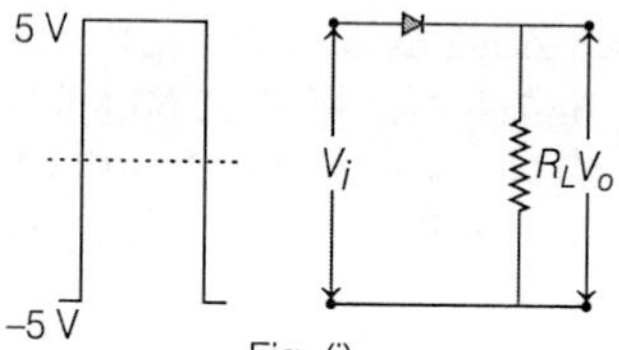

Fig. (i)

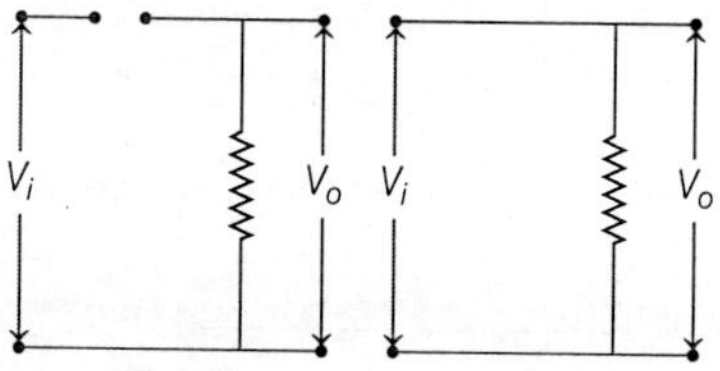

Fig. (ii) Fig. (iii)

For $V_i < 0$, the diode is reverse biased and hence offer infinite resistance, so circuit would be like as shown in Fig. (ii) and $V_o = 0$.

For $V_i > 0$, the diode is forward biased and circuit would be as shown in Fig. (iii) and $V_o = V_i$.

Hence, option (*d*) is correct.

67. *(c)* Lattice constant is the side length of the cube. Lattice represents the distance between molecules and/or atom which form the crystal.

So, decrease in lattice constant means the electron are more tightly bound to atom and hence require more energy to remove leading to increase in band gap energy while due to decrease in lattice constant, valence and conduction band also decreases.

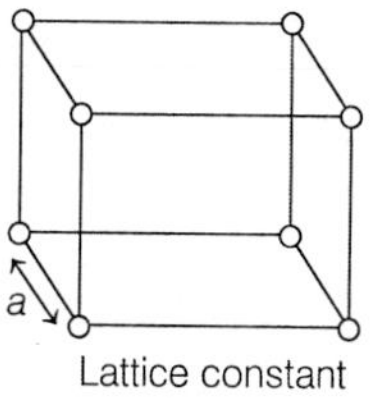

Lattice constant

68. *(c)* $I = n_e A v_d$

$\Rightarrow \quad \dfrac{I_e}{I_h} = \dfrac{n_e \times (v_d)_e}{n_h \times (v_d)_h}$

Here, $\dfrac{n_e}{n_h} = \dfrac{7}{5}, \quad \dfrac{I_e}{I_h} = \dfrac{7}{4}$

$\therefore \quad \dfrac{7}{4} = \dfrac{7}{5} \times \dfrac{(v_d)_e}{(v_d)_h}$

or $\quad \dfrac{(v_d)_e}{(v_d)_h} = \dfrac{5}{7} \times \dfrac{7}{4} = \dfrac{5}{4}$

69. *(b)* In the given, circuit diode D_1 is reverse biased while D_2 is forward biased, so the circuit can be redrawn as

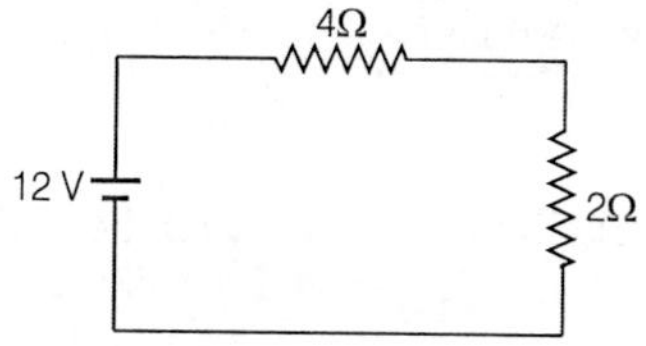

[$\because$ For ideal diodes, reverse biased means open and forward biased means short]

Apply KVL to get current following through the circuit

$-12 + 4I + 2I = 0$

or $\quad I = \dfrac{12}{6} = 2$ A

70. *(b)* As we know for metal the resistance increases with increase in temperature

$R = R_0(1 + \alpha\Delta T)$

In metal with increase in temperature, the vibration of metal ions increases. As atom start to vibrate with more amplitude, the collision between electrons also increase. These collision reduce the energy of electron which means resistivity increases.

While in semiconductor the band gap is very less. When temperature increases electron easing more to CB by overcoming the band gap. So, with increase in temperature the density of charge carrier also increases.

71. *(b)*

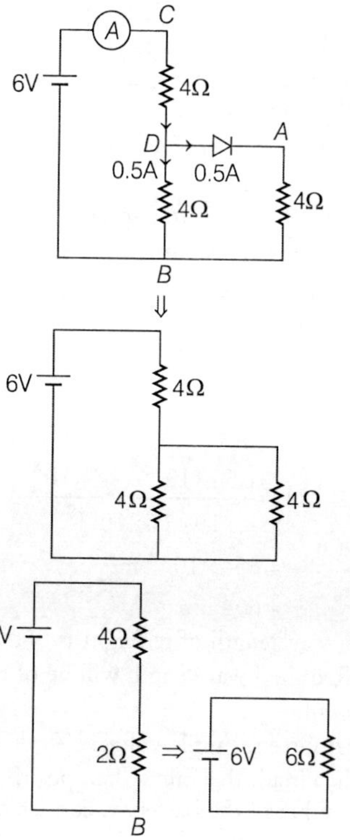

Current through ammeter = 1 A

$R_{net} = 6\Omega$

$V_{AB} = 0.5 \times 4 = 2$ V

$V_{CD} = 1 \times 4 = 4$ V

A, B and D are correct.

72. *(*)* Given voltage regulator circuit is,

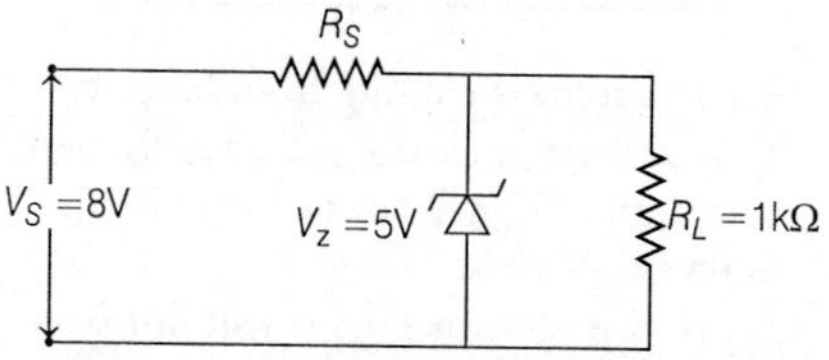

Given, power (rated) for Zener,

$V_z I_z = 10$ mW

Current through Zener diode

$I_z = \dfrac{P}{V_z} = \dfrac{10}{5} \times 10^{-3}$

$= 2 \times 10^{-3}$ A

Current through load resistor

$I_L = \dfrac{V}{R_L} = \dfrac{5}{10^3}$

$= 5 \times 10^{-3}$ A

Current through R_S is,

$I_S = I_Z + I_L$

$= 7 \times 10^{-3}$ A

Potential drop across, R_S

$= 8 - 5 = 3$ V

So, $R_S = \dfrac{V}{I} = \dfrac{3}{7 \times 10^{-3}}$

$= \dfrac{3000}{7} \simeq 428\,\Omega$

(No option is matching)

73. *(d)* According to statement I and II, full wave rectifier with capacitor and inductor are as shown in figure.

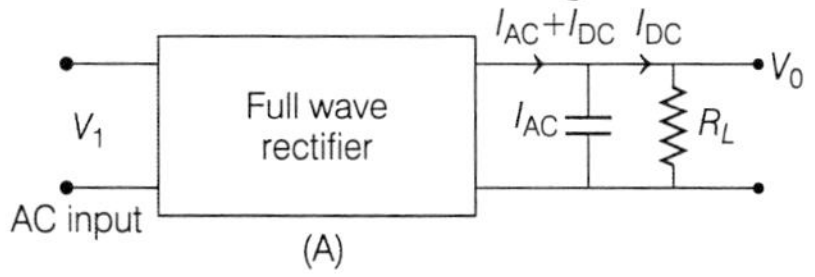

(A)

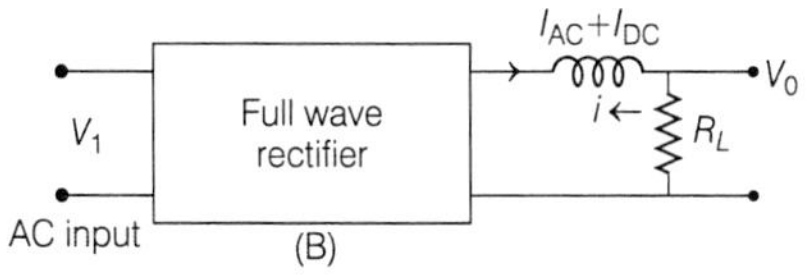

(B)

In figure (A), the capacitor is connected in parallel to load resistance (R_L), as it can block DC current while it conducts AC current.

$\therefore$DC current can pass only through R_L which gives DC output.

In figure (B), the AC current is time varying current which can produce an opposition in current due to inductor that cancels the effect of AC in the load resistance (R_L). Hence, pure DC output is obtained across R_L.

Hence, statement I and Statement II are true.

74. *(a)* According to given circuit diagram,

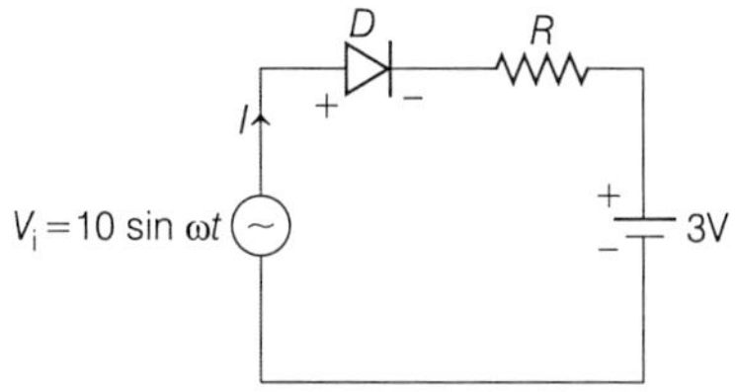

Let I current is flowing through the circuit.

$\therefore$Kirchhoff's voltage loop equation will be

$10 \sin \omega t - IR + 3 = 0$

$\Rightarrow \qquad IR = V = 10 \sin \omega t + 3$

$\therefore$At time, $t = 0$, $V = 0 + 3 = +3\text{V}$ ... (i)

$t = \dfrac{T}{4}$,

$V = 10 \sin\left(\dfrac{2\pi}{T}\right)\cdot\dfrac{T}{4} + 3$

$= 10 + 3 = 13\text{ V}$ ...(ii)

$t = T/2$ to T, diode is in reverse bias.

$\therefore \qquad V = 0$ ...(iii)

We see that at $t = 0$, $V = +3$

and $\quad T/2 \le t \le T, V = 0$

75. *(b)* In ideal condition, a forward biased *p-n* junction acts like a conductive path of zero resistance.

So, given circuit is equivalent to circuit shown below

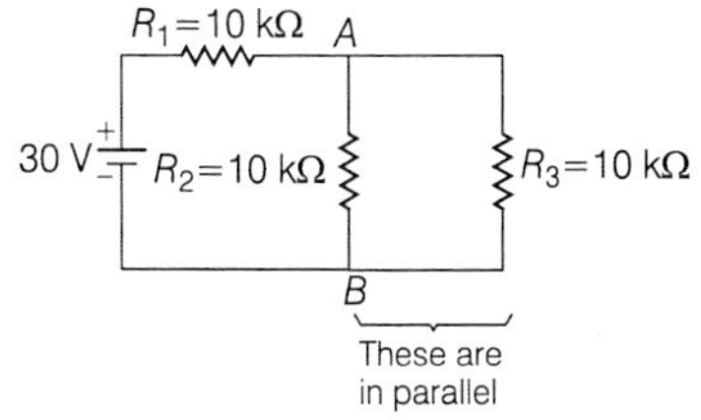

$R_{eq} = \dfrac{10 \times 10}{10 + 10} + 10$

$= 15\text{ k}\Omega$

So, total current in the circuit,

$I = \dfrac{V}{R_{eq}} = \dfrac{30}{15\text{ k}\Omega}$

$= 2 \times 10^{-3}$ A

Now, potential drop across R_1, (10 kΩ),

$V_{R_1} = IR_1 = 2 \times 10^{-3} \times 10 \times 10^3$

$= 20$ V

$V_{AB} = 30 - 20 = 10$ V

This can be seen in the figure below.

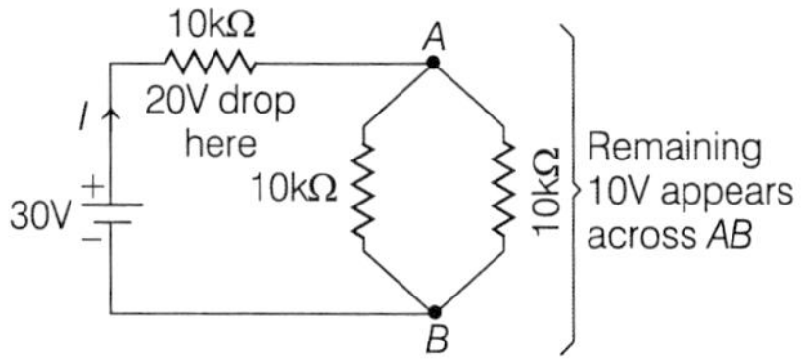

76. *(d)* In given voltage regulator circuit breakdown of Zener occurs at 6 V.

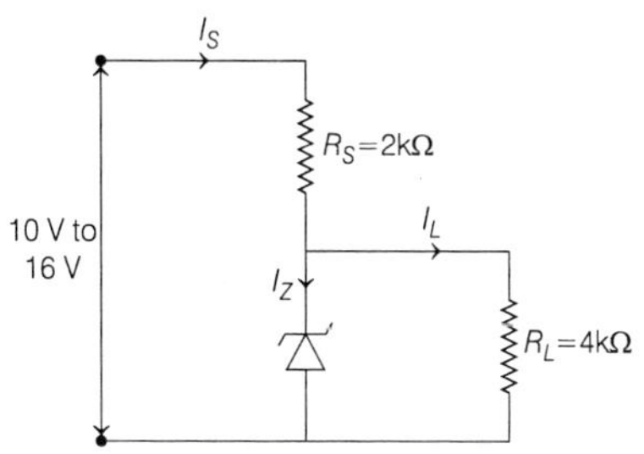

After breakdown, voltage across load resistance ($R_L = 4\text{k}\Omega$) is,

$V = V_Z = 6$ V

$\therefore \quad I_L = \dfrac{V_Z}{R_L} = \dfrac{6}{4000}$

$= 1.5 \times 10^{-3}$ A

$V_S = V - V_Z$

$= 16 - 6 = 10$ V

So, current across series resistance is

$\therefore \; I_S = \dfrac{V_S}{R_S} = \dfrac{10}{2 \times 10^3} = 5 \times 10^{-3}$ A

So, current across Zener diode is

$I_Z = I_S - I_L = 5 \times 10^{-3} - 1.5 \times 10^{-3}$

$= 3.5 \times 10^{-3}$ A $= 3.5$ mA

77. *(c)* In given voltage regulator circuit,

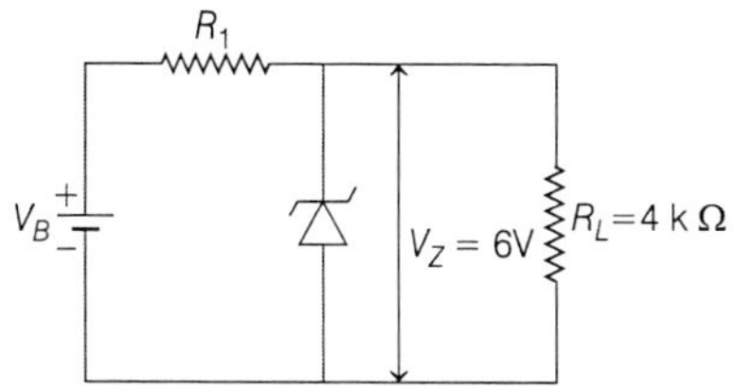

$i_L = \dfrac{V_Z}{R_L} = \dfrac{6}{4 \times 10^3}$

$= 1.5 \times 10^{-3}$ A

Now, when $V_B = 8$ V

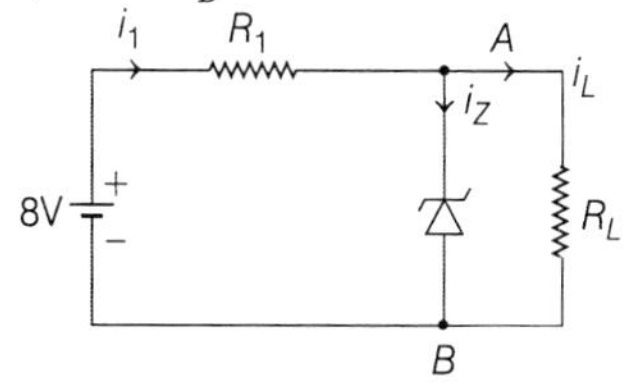

Potential drop across $R_1 = 8 - 6 = 2$ V

So, current through R_1 is

$i_1 = \dfrac{V}{R_1} = \dfrac{2}{1 \times 10^3}$

$= 2 \times 10^{-3}$ A

So, current through Zener diode is

$i_Z = i_1 - i_L$

$= 2 \times 10^{-3} - 1.5 \times 10^{-3}$

$= 0.5 \times 10^{-3}$ A

$= 0.5$ mA

Similarly, when $V_B = 16$ V

$V_{R_1} = 16 - 6 = 10$ V

$\therefore \qquad i_1 = \dfrac{10}{1 \times 10^3}$

$= 10 \times 10^{-3}$ A

Hence, $i_Z = i_1 - i_L$

$= 10 \times 10^{-3} - 1.5 \times 10^{-3}$

$= 8.5 \times 10^{-3}$ A

$= 8.5$ mA

78. *(5)* Potential drop across PQ length of wire $= 1.8 + 3.2 = 5.0$ V

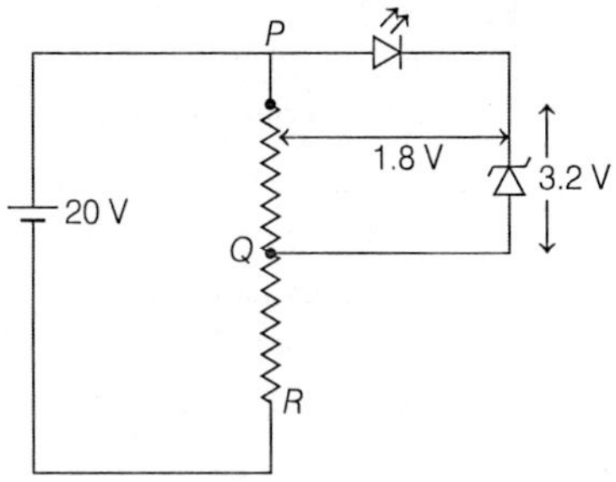

Now, $PR = 20$ cm and $V_{PR} = 20$ V

$\therefore$ Potential drop per unit length

$$(\text{per cm}) = \frac{20\text{ V}}{20\text{ cm}} = 1\text{ V/cm}$$

So, potential drop of 5 V occurs at length PQ of 5 cm.

> • Concept **Enhancer**
>
> Zener diode is utilized as a shunt voltage regulator. It is connected in parallel with the load, operating in reverse bias. Once the Zener diode exceeds its breakdown voltage, it maintains a constant voltage across the load.

79. *(480)* The circuit diagram for the given situation is shown below.

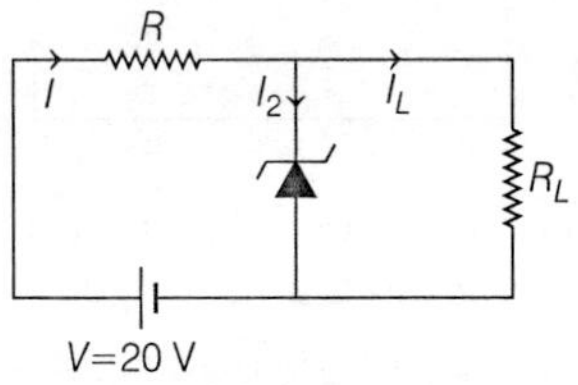

$$V_Z = 8\text{V}$$

$$I = \frac{V - V_Z}{R} = \frac{20-8}{R} \Rightarrow I = \frac{12}{R}\text{A}$$

$$\therefore (I_Z)_{\max} = I \Rightarrow 25 \times 10^{-3} = \frac{12}{R}$$

$$\Rightarrow \quad R = \frac{12}{25 \times 10^{-3}} = 480\,\Omega$$

80. *(9)* The given circuit diagram is represented as

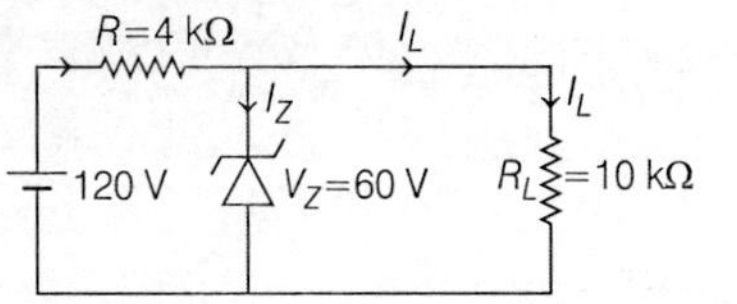

For maximum zener diode current

$$\therefore \quad I = \frac{120 - V_Z}{R} = \frac{120-60}{4 \times 10^3}$$

$$= 15 \times 10^{-3}\text{ A} = 15\text{ mA}$$

$$I_L = \frac{V_L}{R_L} = \frac{V_Z}{R_L} \quad (\because V_L = V_Z)$$

$$= \frac{60}{10 \times 10^3} = 6 \times 10^{-3}\text{ A} = 6\text{ mA}$$

$\therefore$ Zener diode current,

$$I_Z = I - I_L = 15 - 6 = 9\text{ mA}$$

81. *(25)* Here, left diode is forward biased and right diode is reversed biased. So,

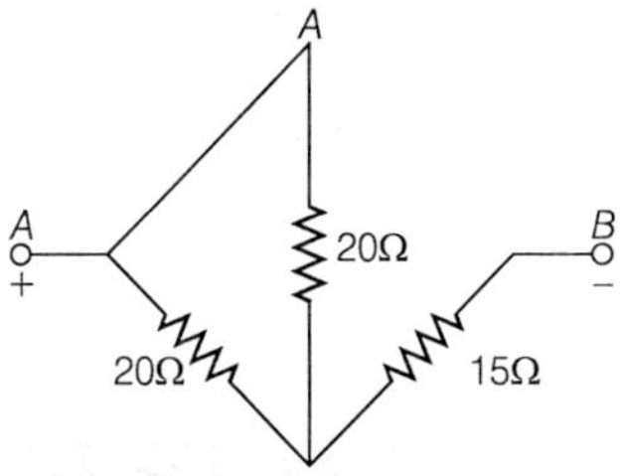

Forward biased diode will act as wire while the reverse biased will act as open circuit

$$\therefore \quad \frac{1}{R'} = \frac{1}{20} + \frac{1}{20}$$

$$R' = 10\,\Omega$$

$\therefore$ Equivalent resistance

$$R = 10 + 15 = 25\,\Omega$$

82. *(1)* We know that

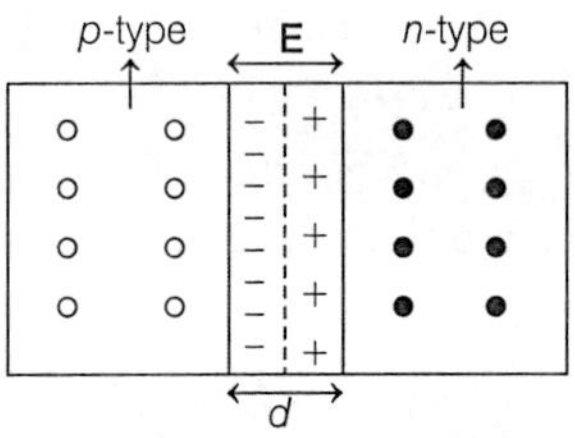

$$|\text{E}| = \frac{V}{d}$$

$$\Rightarrow \quad |\text{E}| = \frac{0.6}{6 \times 10^{-6}}\text{ N/C}$$

$$\Rightarrow \quad |\text{E}| = 1 \times 10^5\text{ N/C}$$

83. *(3)* $\because$ Given, $\lambda = 4000$ Å

$$E(\text{eV}) = \frac{12400}{\lambda(\text{Å})}$$

$$= \frac{12400}{4000} \simeq 3\text{ eV}$$

84. *(125)* $\because$ Power, $P_{zm} = I_{zm} V_z$

$$I_{zm} = \frac{P_{zm}}{V_z} = \frac{0.4}{10} = 0.04\text{ A}$$

$$\text{So,} \quad R_P = \frac{E_{\max} - V_z}{I_{zm}} = \frac{15-10}{0.04}$$

$$= \frac{5}{0.04} = 125\,\Omega$$

85. *(4)* Here, diode in upper branch is in forward biasing and diode in lower branch is in reverse biasing.

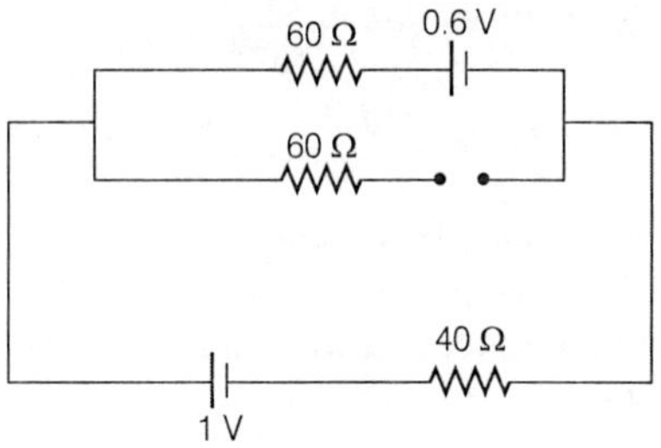

$$R_{\text{eq}} = 60 + 40 = 100\,\Omega$$

$$V_{\text{net}} = 1 - 0.6 = 0.4\text{ V}$$

$$\text{So,} \quad i = \frac{V_{\text{net}}}{R_{\text{eq}}} = \frac{0.4}{100} A = 4\text{ mA}$$

86. *(1)* In the given circuit, the diode D_1 and D_3 are forward biased, while D_2 is reverse biased. Hence, no current flows from D_2. The resultant circuit can be represented as,

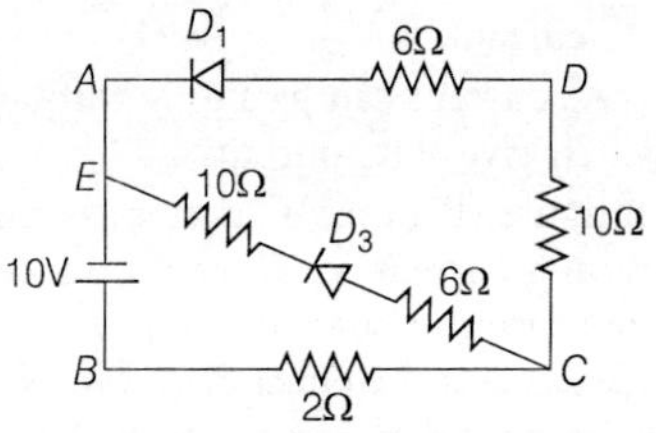

The resultant resistance over the loop, *AECDA*

$$\frac{1}{R'} = \frac{1}{16} + \frac{1}{16}$$

$$\Rightarrow \quad R' = \frac{16}{2} = 8\,\Omega$$

The effective resistance is

$$R = 8 + 2 = 10\,\Omega$$

$\therefore$ Current through circuit,

$$I = \frac{V}{R} = \frac{10}{10} = 1\text{A}$$

87. *(5)* The given circuit diagram is shown below.

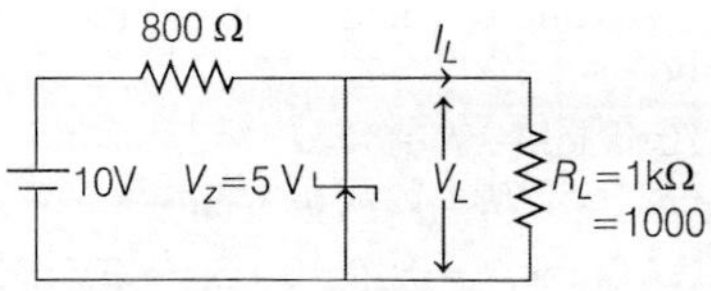

From the above diagram, voltage across R_L is equal V_z

i.e. $\quad V_L = V_Z = 5\text{V}$

$\Rightarrow \quad V_L = 5\text{ V}$

$\Rightarrow I_L R_L = 5$

$$I_L = \frac{5}{R_L} = \frac{5}{1000}$$

$= 5 \times 10^{-3}$ A = 5 mA

88. *(20)* Given, power, $P = 2$ W

Zener breakdown voltage, $V_z = 10$ V

Voltage range, (V_1 to V_2) = 6 V and 14 V

Let resistance be R.

Since, $P = VI_z$

Zener current, $I_z = \frac{P}{V_z}$

$= \frac{2}{10} = 0.2$ A

For safe operation of Zener diode,

$I_z = I_s = 0.2$A

$\therefore V_2 - V_z = I_s R_s$

$\Rightarrow 14 - 10 = I_s R_s$

$\Rightarrow R_s = \frac{4}{0.2} = \frac{40}{2} = 20\,\Omega$

89. *(120)* The circuit diagram is shown below

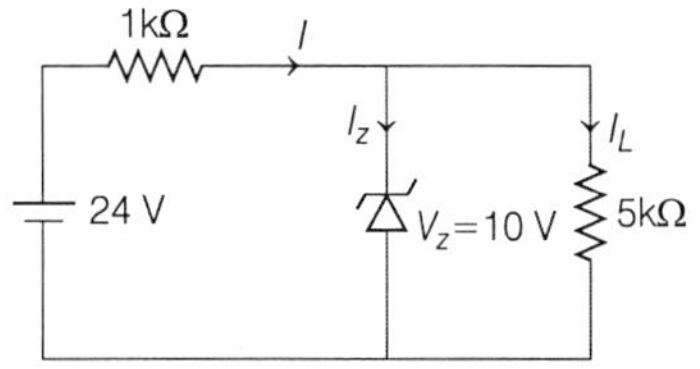

The supply voltage, $V_s = 24$ V

$V_z = 10$ V

Current I_L flowing through the load resistance can be calculated as

$I_L = \frac{10}{5 \times 10^3} = 2 \times 10^{-3}$ A = 2 mA

Since, the voltage drop across 5 kΩ is 10 V, then the voltage drop across 1 kΩ resistance, $V' = 24 - 10 = 14$ V

Current flowing through 1 kΩ resistance can be calculated as

$I = \frac{V'}{1 \times 10^3} = \frac{14}{1 \times 10^3}$

$= 14 \times 10^{-3}$ A = 14 mA

From circuit diagram, $I = I_z + I_L$

$I_z = I - I_L = 14 - 2 = 12$ mA

Power across Zener diode is

$P = V_z \times I_z = 10 \times 12 = 120$ mW

Thus, the power across the Zener diode is 120 mW.

90. *(5)* Given, number density of intrinsic charge carrier, $n_i = 1.5 \times 10^{16} / \text{m}^3$

Number density of hole,

$n_h = 4.5 \times 10^{22}\ \text{m}^{-3}$

Let number density of electron be n_e.

As we know that,

$n_i^2 = n_e n_h$

$\Rightarrow n_e = \frac{n_i^2}{n_h} = \frac{(1.5 \times 10^{16})^2}{4.5 \times 10^{22}}$

$= \frac{1.5 \times 1.5 \times 10^{32}}{4.5 \times 10^{22}}$

$= 0.5 \times 10^{10} = 5 \times 10^9\ \text{m}^{-3}$

91. *(500)* Given, Zener voltage, $V_z = 5$ V

Supply voltage, $V = 50$ V

Zener current, $I_1 = 90$ mA

$= 90 \times 10^{-3}$ A

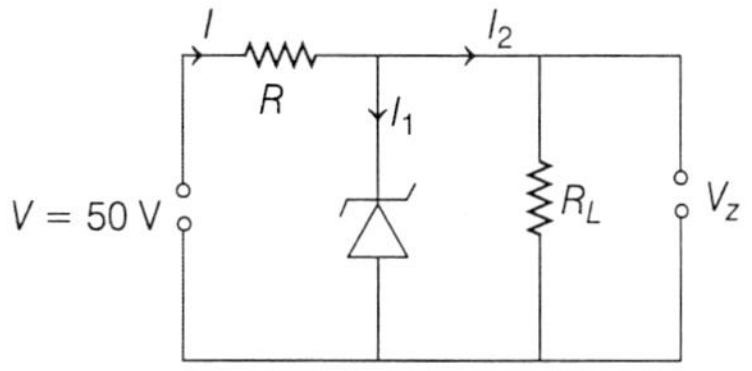

$\because R = \frac{V - V_z}{I}$

$\because R = \frac{50 - 5}{90 \times 10^{-3}} = \frac{45}{9} \times 10^2 = 500\,\Omega$

As maximum current is I_1, so resistance of Zener diode will be minimum.

By using Kirchhoff's current law,

$I = I_1 + I_2$

$\because I_1$ is maximum,

$\therefore I_2$ will be minimum

Hence, minimum value of load resistance R_L will be 500 Ω.

92. *(192)* Given, $V_z = 8$V, $P = 0.5$ W

$\because$ Power, $P = VI \Rightarrow 0.5 = 8 \times I_z$

$\Rightarrow I_z = \frac{0.5}{8} = \frac{5}{80} = \frac{1}{16}$ A

$\because E = V + IR_p$

Where, E is emf and V is terminal voltage.

$\Rightarrow 20 = 8 + \frac{1}{16} \times R_p \Rightarrow 20 - 8 = \frac{R_p}{16}$

$\Rightarrow 12 = \frac{R}{16} \Rightarrow R_p = 12 \times 16$

$\Rightarrow R_p = 192\,\Omega$

93. *(5)* Consider the figure with current in the directions shown below.

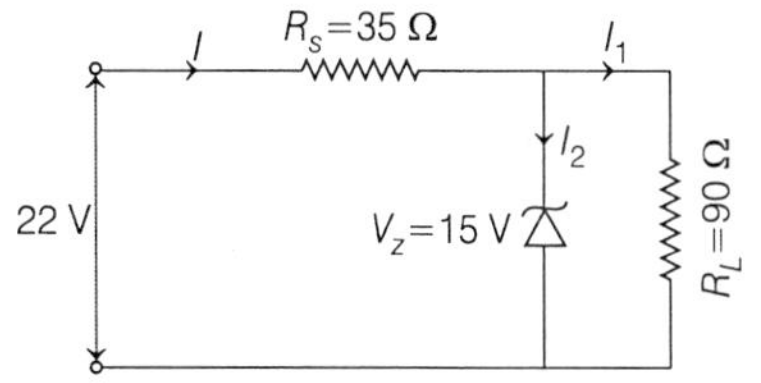

Voltage across R_s, $V_s = 22 - 15 = 7$ V

Current through R_s, $I = \frac{V_s}{R_S}$

$\Rightarrow I = \frac{7}{35} = \frac{1}{5}$ A

Current through 90 Ω resistance,

$I_1 = \frac{V_z}{R_L} = \frac{15}{90} = \frac{1}{6}$ A

$\therefore$ Current through Zener diode,

$I_2 = \frac{1}{5} - \frac{1}{6} = \frac{1}{30}$ A

$\therefore$ Power through Zener diode,

$P = V_z I_2 = 15 \times \frac{1}{30}$

$= \frac{1}{2} = 0.5$ W

$= 5 \times 10^{-1}$ W $= x \times 10^{-1}$ W [Given]

So, the value of x is 5.

94. *(12)* We have given circuit,

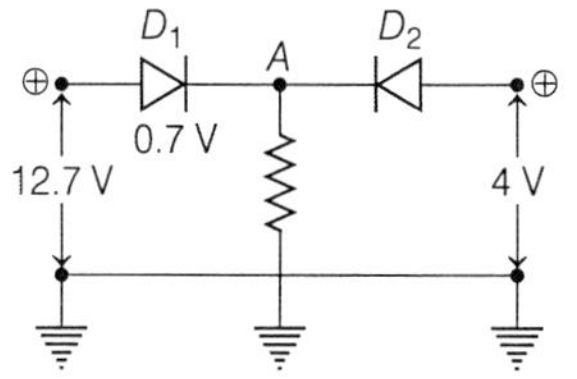

Diode D_1 is in forward bias and D_2 is in reverse bias, so D_1 conducts and D_2 resists.

$\therefore$ Potential at $A = 12.7 - 0.7 = 12$ V

Topic 2

01. *(c)*

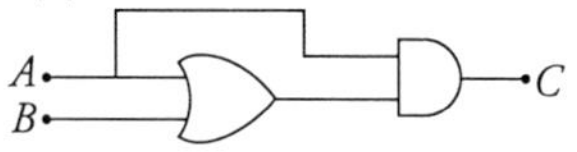

Output $C = A(A + B)$

$= A^2 + AB$

$= A + AB$

$= A(1 + B) = A$

Truth table

A	B	C = A
0	0	0
0	1	0
1	0	1
1	1	1

02. *(d)* From the logic circuit shown in option (d)

$Y = (A + B)A$

This output satisfies the given truth table.

03. *(a)*

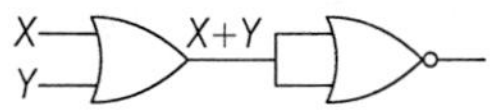

Combination of OR gate and NOT gate = NOR gate

X	Y	Z
0	0	1
0	1	0
1	0	0
1	1	0

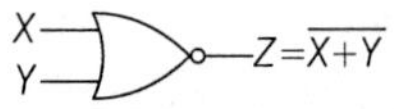

B, C and D are correct.

• Concept Enhancer

Using NAND and NOR gates, we can construct any type of logic gate, hence, NAND and NOR gate are also known as universal gate.

04. *(b)*

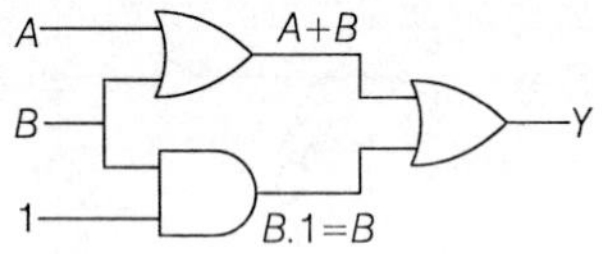

$$Y = A + B + B = A + B$$

Hence, output Y will be zero only when $A = 0$, $B = 0$

05. *(a)* We know that output of NAND gate is high only when one or both input are low (zero) and low when both inputs are high (1). Thus, option (a) is correct.

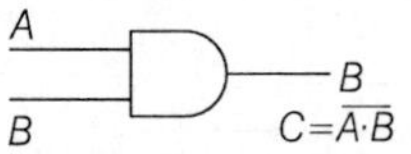

06. *(c)* Truth table for the given logic gate is

A	B	Y
0	0	1
0	1	0
1	0	0
1	1	0

Output Y is output of NOR gate.

07. *(a)* Circuit will not work when both A and B are closed.

A	B	Output
0	0	1
1	0	0
0	1	0
1	1	0

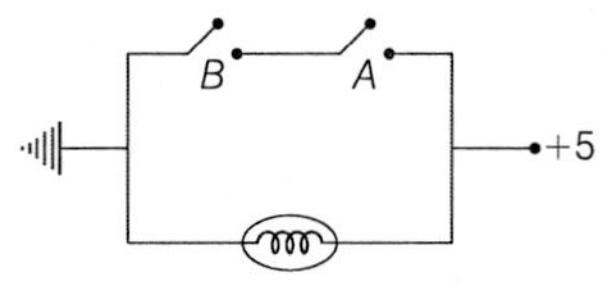

So above circuit is logic equivalent of NOT + AND = NAND

08. *(a)* From the given input and output waveforms, we observe that output Y is high only when both the inputs A and B are individually high. This is possible only in AND gate.

09. *(b)* We have truth table as shown

A	B	Y
0	0	1
0	1	1
1	0	1
1	1	0

Which resembles to NAND gate, so correct option is (b).

10. *(b)* By using de-Morgan's theorem,

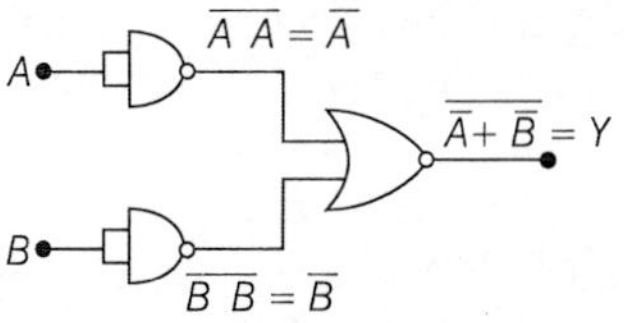

$$Y = \overline{\overline{A} + \overline{B}}$$
$$= \overline{\overline{A}} \cdot \overline{\overline{B}} = A \cdot B$$

So, the given circuit perform AND operation.

11. *(d)* The logic circuit is given as

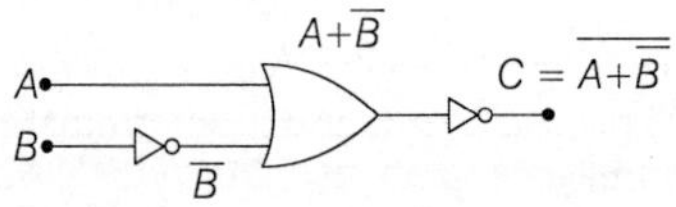

By using de-Morgan's theorem,

$$C = \overline{A + \overline{B}}$$
$$= \overline{A} \cdot \overline{\overline{B}} = \overline{A} \cdot B$$

This relation can be shown by the circuit drawn below

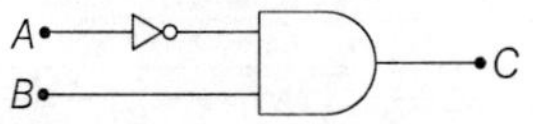

Hence, correct option is (d).

12. *(d)* Given gates and their truth tables are as below.

1. 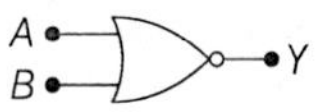 NOR gate

A	B	Y
0	0	1
1	0	0
0	1	0
1	1	0

2. 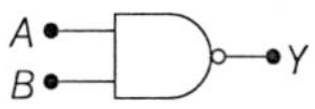 NAND gate

A	B	Y
0	0	1
1	0	1
0	1	1
1	1	0

3. 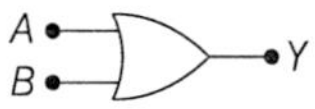 OR gate

A	B	Y
0	0	0
1	0	1
0	1	1
1	1	1

4. 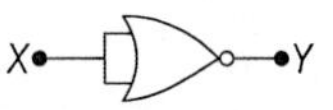 One input NOR gate

X	Y
0	1
1	0

Clearly, 4th gate converts a low voltage (0) into a high voltage (1) and *vice-versa*. So, 4th gate or one input NOR gate gives a reversible operation.

13. *(c)* Truth table for given combination of logic gates is

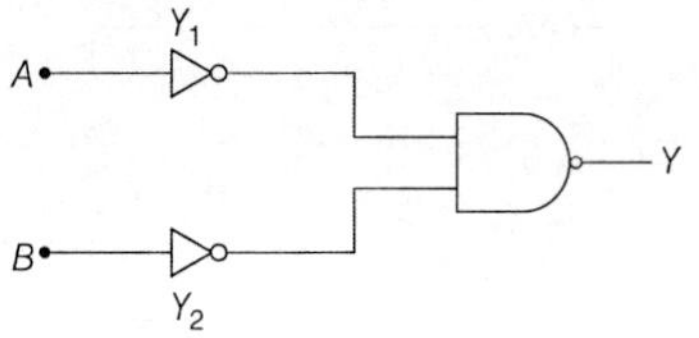

A	B	$Y_1 = \overline{A}$	$Y_2 = \overline{B}$	$Y = \overline{Y_1 \cdot Y_2}$
0	0	1	1	0
1	0	0	1	1
0	1	1	0	1
1	1	0	0	1

Output Y resembles output of an OR gate. So, given combination acts like an OR gate.

Alternate Solution

The given logic gate circuit can be drawn as shown below

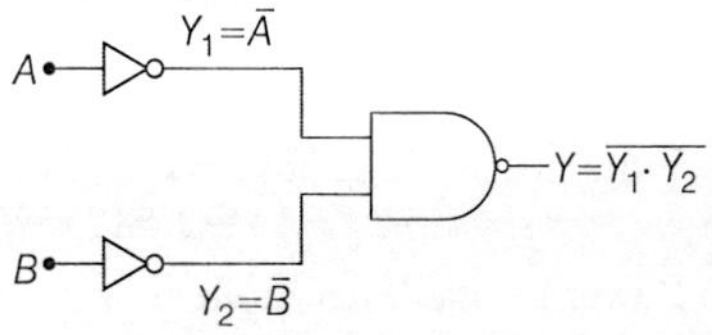

Here, $Y = \overline{Y_1 \cdot Y_2} = \overline{\overline{A} \cdot \overline{B}}$

Using de-Morgan's theorem, i.e. $\overline{x \cdot y} = \overline{x} + \overline{y}$

$\therefore \quad Y = \overline{\overline{A}} + \overline{\overline{B}} = A + B \qquad [\because \overline{\overline{x}} = x]$

This represents the boolean expression for OR gate.

14. *(a)* Here, from truth table we observe that

$$Y = A + B$$

Therefore, it represent OR gate.

15. *(c)* Output of OR gate is 0 when all inputs are 0 and output is 1 when atleast one of the inputs is 1.
Observing output x, it is 0 when all inputs are 0 and it is 1 when atleast one of the inputs is 1.
$\therefore$ The gate is OR.

Alternative Solution

OR gate

a	b	c	d	x
0	0	0	0	0
0	0	0	1	1
0	0	1	0	1
0	0	1	1	1
0	1	0	0	1
0	1	0	1	1
0	1	1	0	1
0	1	1	1	1
1	0	0	0	1

16. *(a)* Truth table for circuit shown is option (a) is

$\overline{Y + \overline{A}} = C$ $\quad [Y = A \cdot B]$

A	B	Y	$Y + \overline{A}$	C
0	0	0	1	0
0	1	0	1	0
1	0	0	0	1
1	1	1	0	0

17. *(a)* The Boolean expression for NOR gate is written as, $= \overline{\;+\;}$

18. *(c)* From the diagram,

$$Y = \overline{\overline{A} + \overline{B}} = \overline{\overline{A}} \cdot \overline{\overline{B}} = AB$$

= Output of AND gate

19. *(c)*

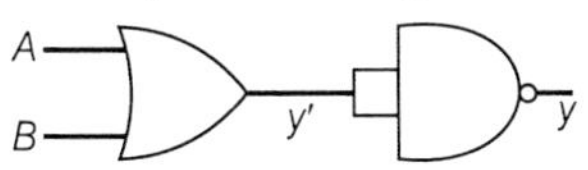

$Y' = A + B$ and $Y = \overline{Y'} = \overline{A + B}$

i.e., output of a NOR gate.

20. *(a)* Truth table for given combination is

A	B	X
0	0	0
1	0	1
0	1	1
1	1	1

This comes out to be truth table of OR gate.

Alternate method

$$Y = \overline{\overline{A} \cdot \overline{B}} = \overline{\overline{A + B}} = A + B$$

OR gate

21. *(d)*

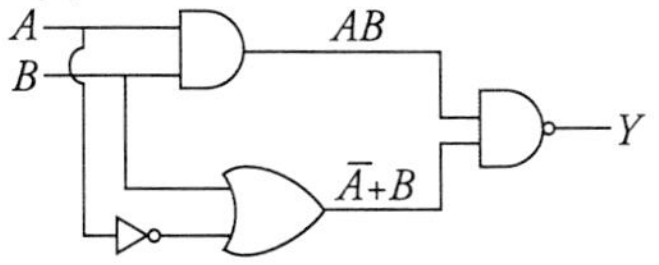

Output,

$$Y = \overline{AB(\overline{A} + B)}$$
$$= \overline{AB} + \overline{(\overline{A} + B)}$$
$$= \overline{AB} + \overline{\overline{A}} \cdot \overline{B}$$
$$= \overline{AB} + A\overline{B}$$
$$= \overline{A} + \overline{B} + A\overline{B}$$
$$= \overline{A} + \overline{B}\,(1 + A)$$
$$= \overline{A} + \overline{B} \quad [1 + A = 1]$$

When $A = 1$ and $B = 1$,

Then, $Y = \overline{1} + \overline{1}$
$= 0 + 0 = 0$

22. *(*)*

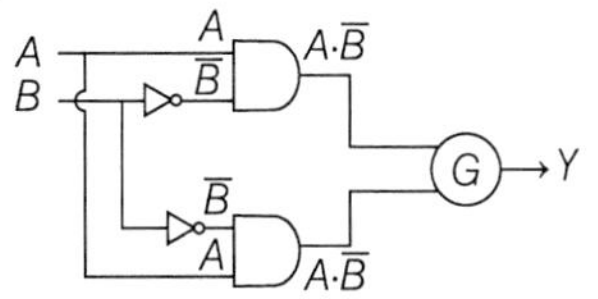

A	B	Y
0	0	1
0	1	0
1	0	0
1	1	1

$$Y = \overline{\overline{A\overline{B}} + \overline{\overline{A}B}} = \overline{\overline{A\overline{B}}} \cdot \overline{\overline{\overline{A}B}}$$
$$= (\overline{A} + B)(A + \overline{B})$$

Thus, output G is NOT gate.
Hence, no option is correct.
(Bonus marks)

23. *(d)* From the given circuit, the expression of Boolean algebra is

$$Y = \overline{A\overline{B} + AB}$$
$$= \overline{A\,(B + \overline{B})}$$
$$\Rightarrow Y = \overline{A}$$

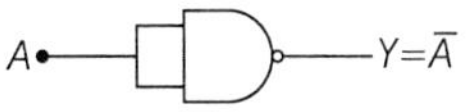

24. *(c)*

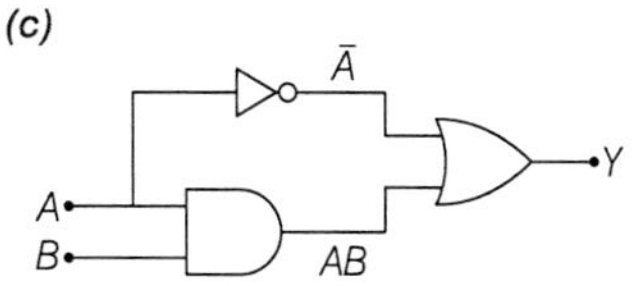

$Y = \overline{A} + AB$

$\therefore$Truth table is

A	B	$\overline{A}$	AB	$Y = \overline{A} + AB$
0	0	1	0	1
0	1	1	0	1
1	0	0	0	0
1	1	0	1	1

25. *(a)*

key idea While solving such question, keep in mind that bulb will glow if it has potential drop on it. For this, one end of the bulb must be at high (1) and other must be at low (0).

The bulb will glow only if there is potential drop across it. One terminal must be high, other should be low.
If $A = 1$, then Ist NOR gate will give 0 output. If B and $C = 0$, then Ist NAND gate will give 1 output.

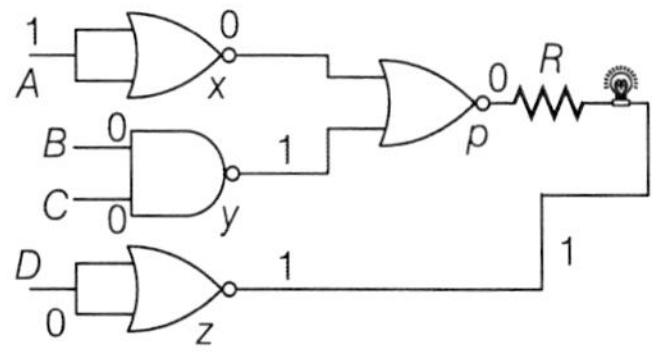

$D = 0$, only then the bulb will glow.
X is first NOR gate, and
Y is first NAND gate.
Hence, $A = 1$, $B = 0$, $C = 0$, $D = 0$.

• Concept Enhancer

Using NAND and NOR gates, we can construct any type of logic gate, hence, NAND and NOR gate are also known as universal gate.

26. *(d)*

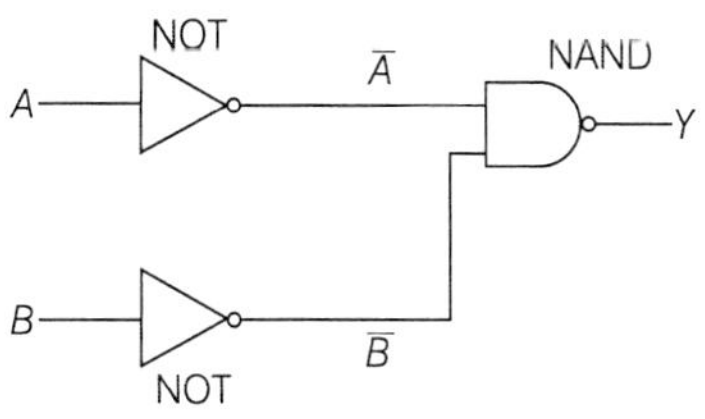

$$\therefore Y = \overline{\overline{A} \cdot \overline{B}} = \overline{\overline{A}} + \overline{\overline{B}} = A + B$$

= output of OR gate

Alternate Solution

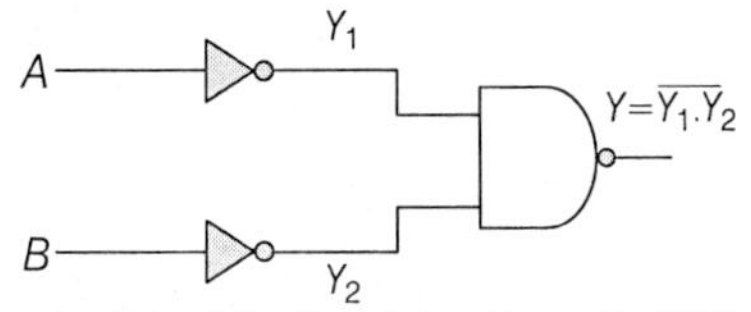

A	B	$Y_1=\overline{A}$	$Y_2=\overline{B}$	Y_1Y_2	$Y=\overline{Y_1Y_2}$
0	0	1	1	1	0
0	1	1	0	0	1
1	0	0	1	0	1
1	1	0	0	0	1

This truth table is simillar to OR gate.

27. *(a)* The given circuit is

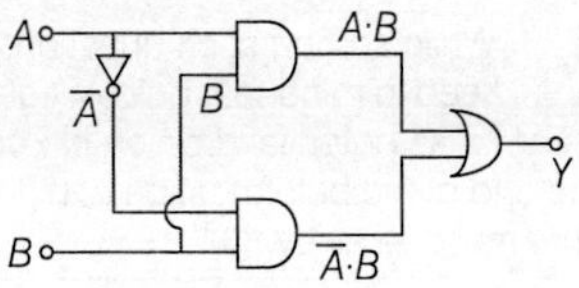

$\therefore \quad Y = AB + \overline{A}B$

$= B\cdot(A+\overline{A}) = B$

$\therefore$Truth table will be

A	B	$Y=B$
0	0	0
0	1	1
1	0	0
1	1	1

28. *(d)*

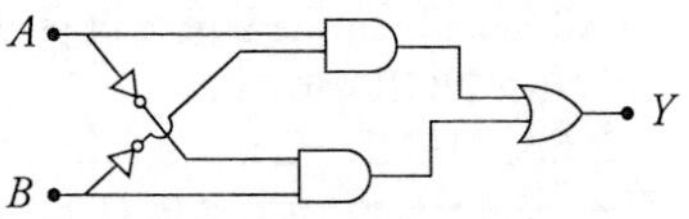

$Y = A\overline{B} + \overline{A}B$

A	B	$A\overline{B}$	$\overline{A}B$	Y
0	0	0	0	0
0	1	0	1	1
1	0	1	0	1
1	1	0	0	0

29. *(d)*

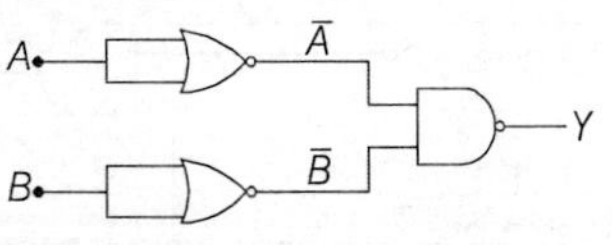

$Y=\overline{(\overline{A}\cdot\overline{B})} = \overline{\overline{A}} + \overline{\overline{B}}$

$= A + B =$ output of OR gate

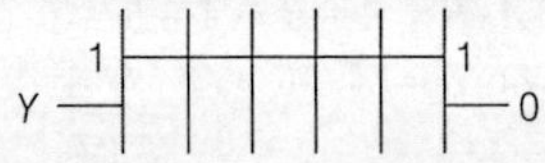

30. *(d)* Given,

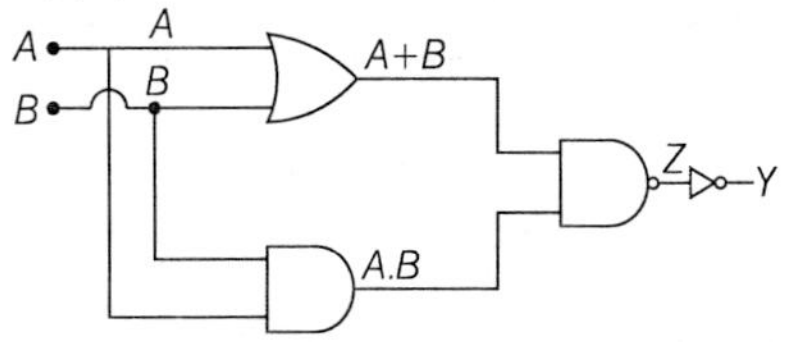

$Z = \overline{(A+B)(AB)},\ Y = \overline{Z} = (A+B)(AB)$

$= AAB + BAB = AB + ABB$

$= AB =$ output of AND gate

31. *(c)* Given, logic inputs are given as $(A, B) = (0, 0), (0,1), (1, 0)$ and $(1, 1)$ And the logic circuit is drawn as

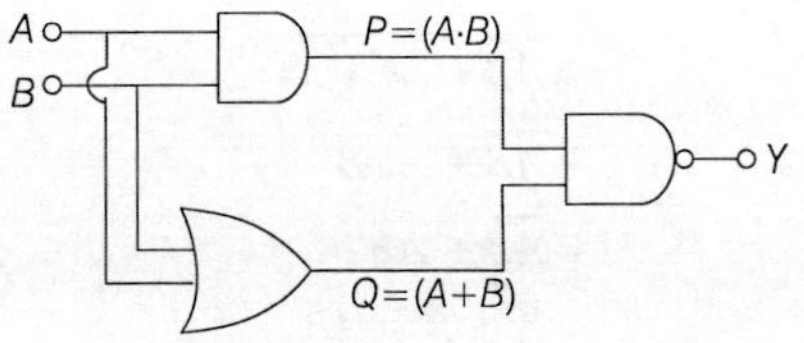

Output of the given logic circuit is given as

$$Y = \overline{AB\cdot(A+B)} = \overline{AB} + \overline{(A+B)}$$

[using de-Morgan's theorem]

$$= (\overline{A}+\overline{B}) + \overline{A}\cdot\overline{B} = \overline{A}+\overline{B}+\overline{A}\overline{B}$$

Now, by using truth table for above logic circuit

A	B	$\overline{A}$	$\overline{B}$	$\overline{A}\,\overline{B}$	P	Q	Y
0	0	1	1	1	0	0	1
0	1	1	0	0	0	1	1
1	0	0	1	0	0	1	1
1	1	0	0	0	1	1	0

32. *(c)* The given circuit can be drawn as

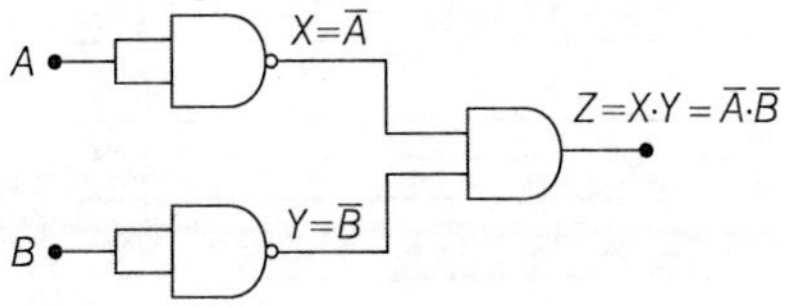

The output, $Z = \overline{A}\cdot\overline{B} = \overline{A+B}$

(Using de-Morgan theorem)

This is the expression of a NOR gate. So, the logic operation carried out by the given circuit is NOR.

Alternate method

The given question can be solved using truth table, which is as given below

A	B	X	Y	Z
0	0	1	1	1
1	1	0	0	0
1	0	0	1	0
0	1	1	0	0

This truth table is similar to that of a NOR gate.

33. *(b)* The given logic diagram can be shown as

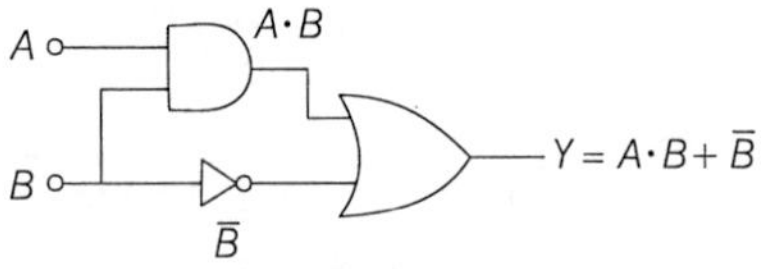

$\therefore$The Boolean expression for the above figure is $Y = A\cdot B + \overline{B}$

Now, let us draw the truth table for this expression as follows

A	B	$A\cdot B$	$\overline{B}$	$Y = A\cdot B + \overline{B}$
0	0	0	1	1
0	1	0	0	0
1	0	0	1	1
1	1	1	0	1

On comparing the above truth table with that given in option (b), we find that both are same.

34. *(b)* We can find the output of given combination of gates using Boolean algebra. Consider the given figure

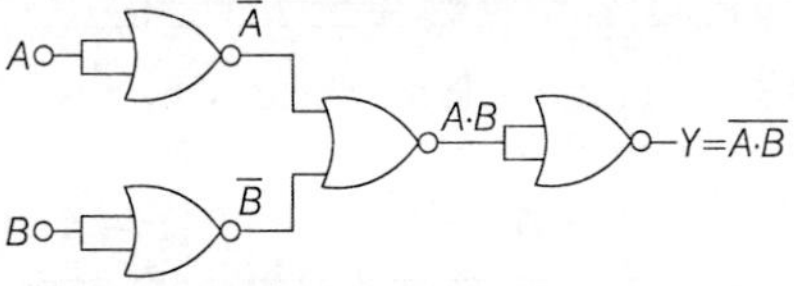

The output $Y = \overline{A\cdot B}$

which is the expresssion for NAND gate.

Alternate method

We can find the output of given combination of gates using truth table. Consider the given figure

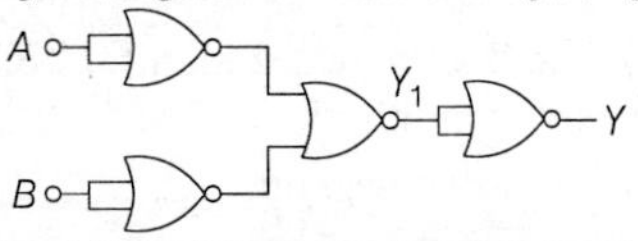

A	B	$Y_1=A\cdot B$	$Y=\overline{A\cdot B}$
0	0	0	1
0	1	0	1
1	0	0	1
1	1	1	0

The above truth table shows the output of NAND gate.

35. *(a)* Truth table for given circuit is

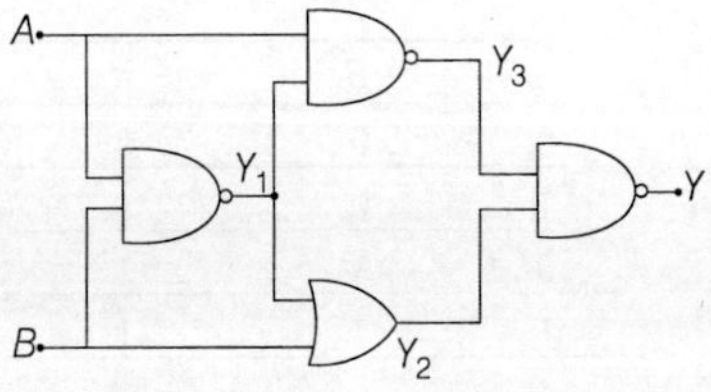

A	B	Y_1	Y_2	Y_3	Y
0	0	1	1	1	0
1	0	1	1	0	1
0	1	1	1	1	0
1	1	0	1	1	0

This is the same output produced by $A\cdot\overline{B}$ gate or

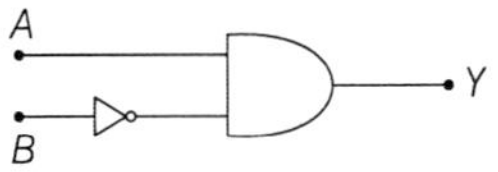

So, given circuit is equivalent to Boolean expression $A\cdot\overline{B}$.

36. *(a)*

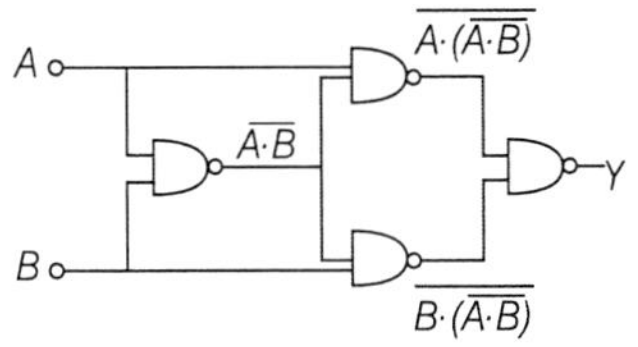

$$Y=\overline{\overline{\{A\cdot(\overline{A\cdot B})\}}\cdot\overline{[B\cdot(\overline{A\cdot B})]}}$$

[using de-Morgan's rules]

$$=\overline{(\overline{A}+A\cdot B)\cdot(\overline{B}+A\cdot B)}$$

[break the line, change the sign]

$$=\overline{(\overline{A}+A\cdot B)}+\overline{(\overline{B}+(A\cdot B))}$$
$$=A\cdot(\overline{A\cdot B})+B\cdot(\overline{A\cdot B})$$
$$=A\cdot(\overline{A}+\overline{B})+B\cdot(\overline{A}+\overline{B})$$
$$=A\cdot\overline{B}+B\cdot\overline{A}\;[\because A\cdot\overline{A}=B\cdot\overline{B}=0]$$

A	B	$\overline{A}$	$\overline{B}$	$A\cdot\overline{B}$	$B\cdot\overline{A}$	Y
0	0	1	1	0	0	0
0	1	1	0	0	1	1
1	0	0	1	1	0	1
1	1	0	0	0	0	0

It is the truth table of XOR gate.

37. *(d)* $Y=A\overline{B}C+\overline{A}\,\overline{C}$ (given)

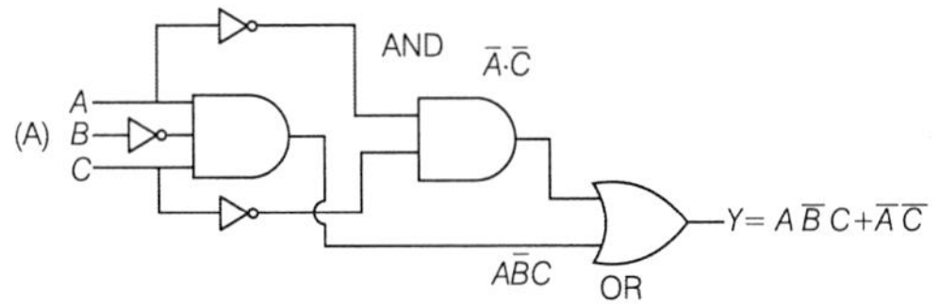

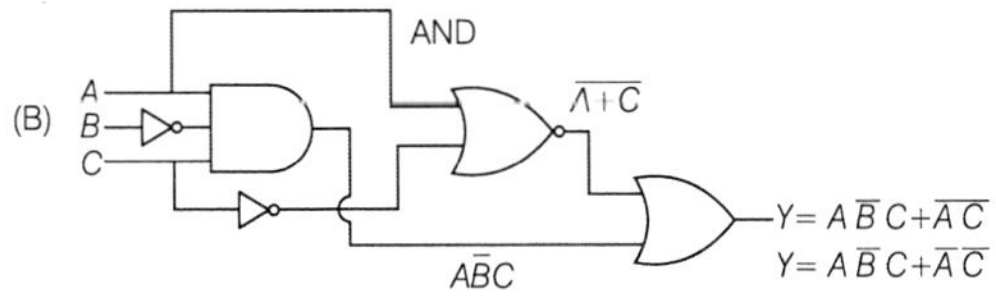

Only A and B are correct.

38. *(b)*

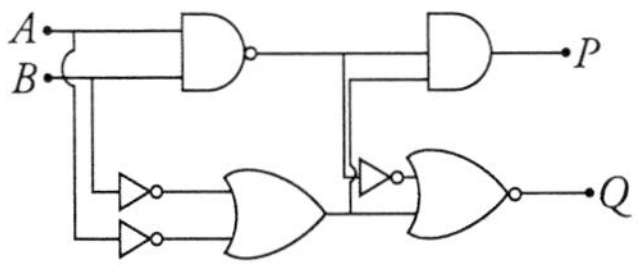

For the given logic circuit,

$$P=\overline{AB}\cdot(\overline{A}+\overline{B})$$
$$=(\overline{A}+\overline{B})\cdot(\overline{A}+\overline{B})=\overline{A}+\overline{B}$$

and $Q=\overline{\overline{\overline{AB}}+\overline{\overline{A}+\overline{B}}}=\overline{AB+(\overline{A}+\overline{B})}$

$$=\overline{AB}\cdot\overline{\overline{A}+\overline{B}}=\overline{AB}\,(\overline{\overline{A}}\cdot\overline{\overline{B}})$$
$$=\overline{AB}\cdot AB=0$$

When $A=1,\ B=1$

then $P=\overline{1}+\overline{1}=0+0=0$

and $Q=0$

39. *(d)*

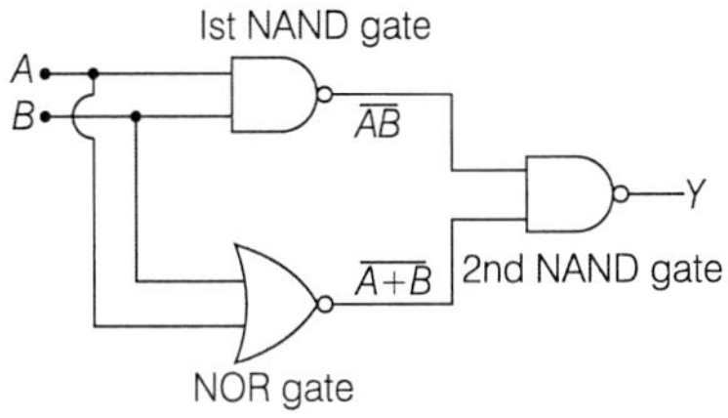

For 1st NAND gate, inputs are A and B while output is $\overline{AB}$.
For NOR gate, inputs are A and B,
while output is $\overline{A+B}$.

For IInd NAND gate, inputs are
$\overline{AB}$ and $\overline{A+B}$, so output

$$Y=\overline{\overline{AB}\cdot\overline{A+B}}$$

Applying de-Morgan's first law

$$Y=\overline{\overline{AB}}+\overline{\overline{A+B}}$$
$$Y=AB+(A+B)$$
$$=A[B+1]+B=A[1]+B$$
$$Y=A+B$$

OR gate

40. *(d)*

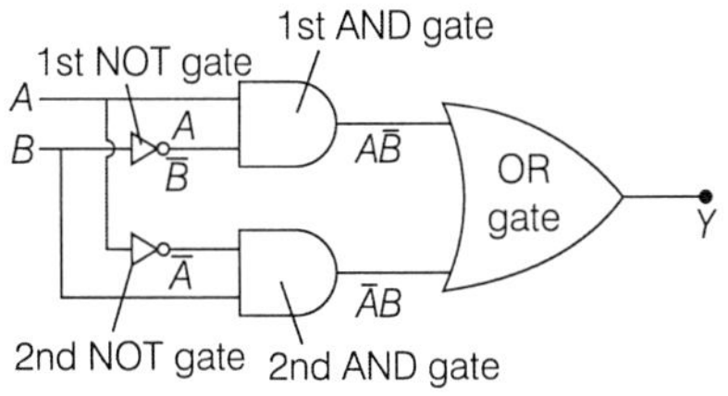

For 1st NOT gate, B is input,
So output is $\overline{B}$.
For 2nd NOT gate, A is input,
So output is $\overline{A}$.
For 1st AND gate, A and $\overline{B}$ are inputs.
So output is $A\overline{B}$.
For 2nd AND gate, $\overline{A}$ and B are input.
So, output is $\overline{A}B$.
Ultimately for OR gate, $A\overline{B}$ and $\overline{A}B$ are the inputs, so output is

$$Y=A\overline{B}+\overline{A}B$$

This is the XOR gate, whose truth table is

A	B	Y
0	0	0
0	1	1
1	0	1
1	1	0

Alternate method

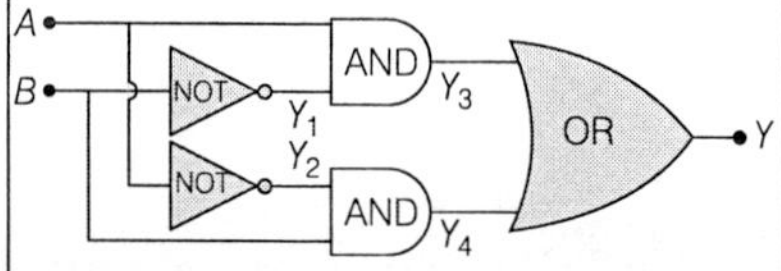

Truth table as per characterstics of different gates is as follows.

A	B	$y_1=\bar B$	$y_2=\bar A$	y_3	y_4	y
0	0	1	1	0	0	0
1	0	1	0	1	0	1
0	1	0	1	0	1	1
1	1	0	0	0	0	0

y_3 is output of $A+y_1$ by AND gate

$y_4=B$ and y_2

41. *(b)* Given combination of logic gate is as follows.

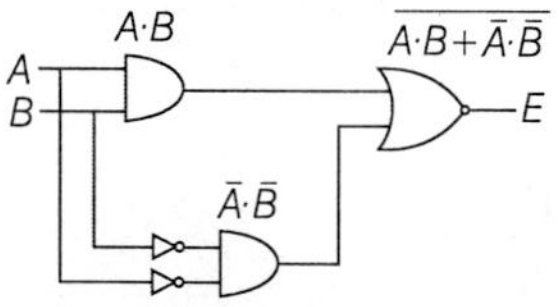

$$E=\overline{A\cdot B+\overline{A+B}}$$

$$=\overline{A\cdot B+\bar A\cdot\bar B}$$

(Using De-morgan's theorem)

$$\bar A\cdot\bar B=\overline{A+B}\;=\overline{\overline{A\cdot B}\cdot\overline{\overline{A+B}}}$$

$$=(\overline{A\cdot B})\,(A+B)$$

Now the truth table will be

A	B	E
0	0	0
0	1	1
1	0	1
1	1	0

So, the correct answer is 1, 1.

42. *(a)*

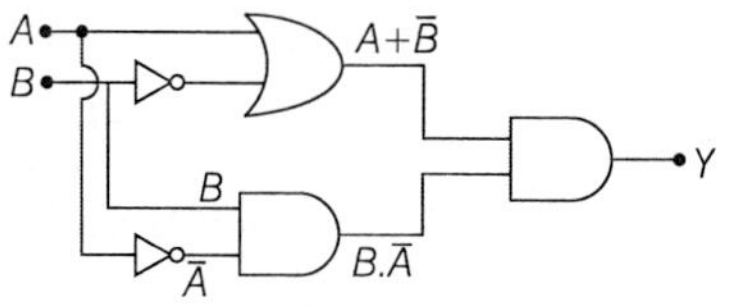

$$Y=(A+\bar B)\cdot(B\cdot\bar A)$$

$$=(AB+\bar B B)\cdot\bar A=(AB+0)\cdot\bar A$$

$$=AB\bar A=(A\bar A)B$$

$$=0B=0$$

43. *(b)*

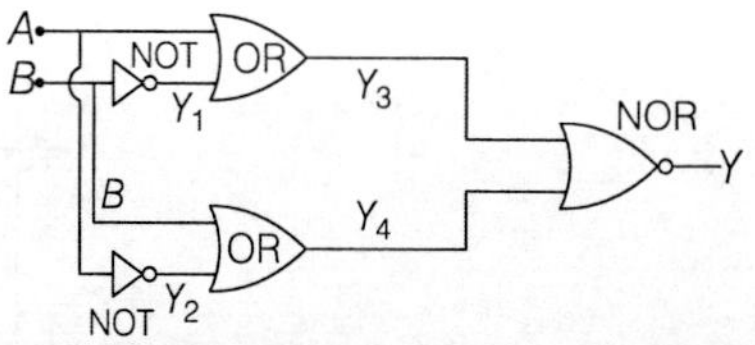

Truth table is

A	B	Y_1	Y_2	Y_3	Y_4	Y
0	0	1	1	1	1	0
1	0	1	0	1	0	0
0	1	0	1	0	1	0
1	1	0	0	1	1	0

Alternate Solution

$$Y=\overline{(A+\bar B)+(\bar A+B)}=\overline{A+\bar B}\cdot\overline{(\bar A+B)}$$

$$=(\bar A\cdot B)(A\cdot\bar B)$$

$$=(A\bar A)\cdot(B\bar B)=0.$$

44. *(a)* The logic circuit diagram is given as

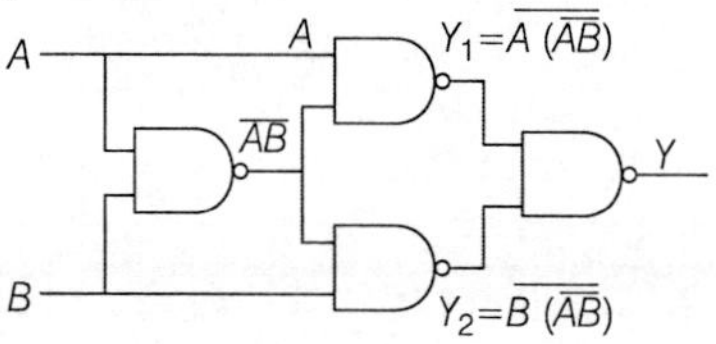

Output is given by, $Y=\overline{Y_1\cdot Y_2}$ ...(i)

where, $Y_1=\overline{A(\overline{AB})}$

$$=\bar A+\overline{\overline{AB}}=\bar A+AB$$

Similarly, $Y_2=\bar B+AB$

$\therefore$ From Eq. (i), we get

$$Y=\overline{(\bar A+AB)(\bar B+AB)}$$

$$=\overline{(\bar A+AB)}+\overline{(\bar B+AB)}$$

$$=(\bar{\bar A}\cdot\overline{AB})+(\bar{\bar B}\cdot\overline{AB})$$

$$=(A\cdot(\bar A+\bar B))+(B\cdot(\bar A+\bar B))$$

$$=A\bar A+A\bar B+B\bar A+B\bar B=A\bar B+B\bar A$$

$\therefore$ Truth table for the given logic gate is

A	B	$\bar A$	$\bar B$	$Y=A\bar B+B\bar A$
0	0	1	1	0
0	1	1	0	1
1	0	0	1	1
1	1	0	0	0

45. *(b)*

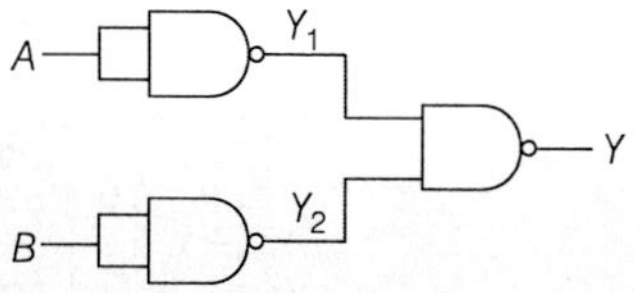

Truth table for given combination of gate is

A	B	Y_1	Y_2	$Y=\overline{Y_1\,Y_2}$
0	0	1	1	0
1	0	0	1	1
0	1	1	0	1
1	1	0	0	1

So, this logical circuit is of OR gate. Now, input and corresponding output waveforms are as shown,

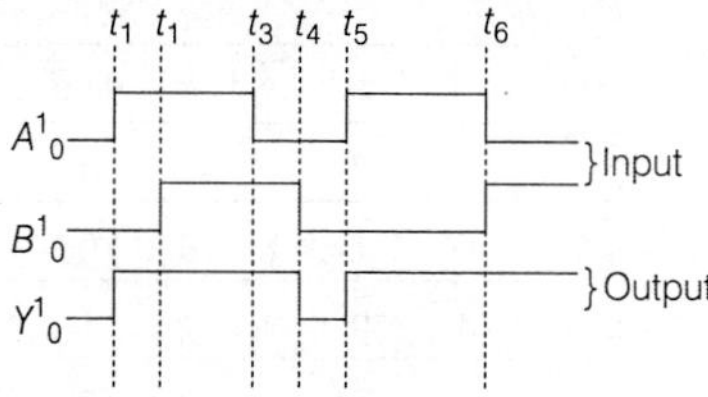

46. *(d)*

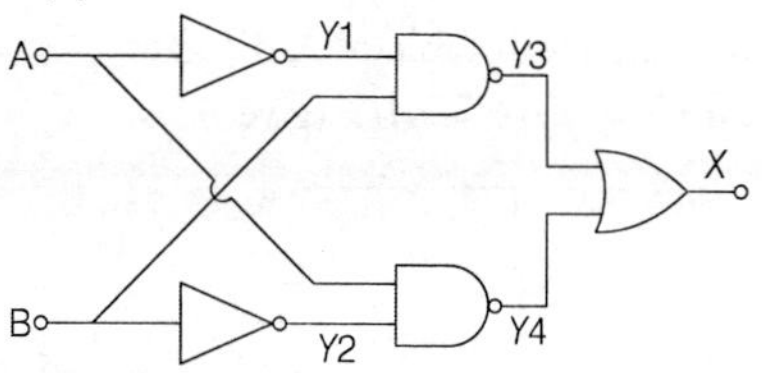

Truth table of above gate is

A	B	Y_1	Y_2	Y_3	Y_4	X
0	0	1	1	0	0	0
1	0	0	1	0	1	1
0	1	1	0	1	0	1
1	1	0	0	0	0	0

> ***Alternate Solution***
> The given circuit diagram is shown as
>
> 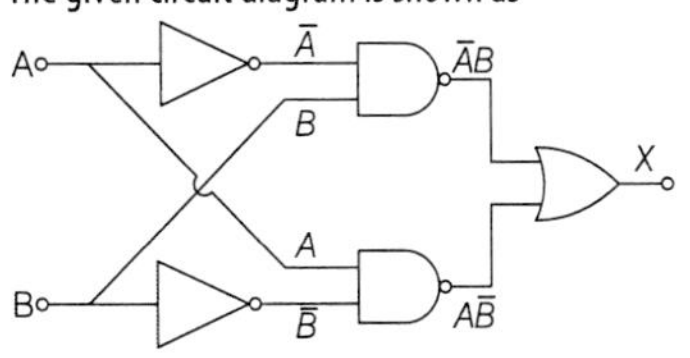
>
>
> Output of the gate is given as $X = \overline{A}B + A\overline{B}$
> Which is the output of XOR gate. Thus, truth table of XOR gate is given in option (d).

47. *(b)* In figure A, V_A and V_B are two input voltages (either 5 V or 0 V)
When, $V_A = 0, V_B = 0$
Both D_1 and D_2 are OFF, so $V_0 = 0$,
When, $V_A = 1$ i.e. 5 V, $V_B = 0$ i.e. 0 V
D_1 is ON but D_2 is OFF
So, $V_0 = 1$ i.e. 5 V
When, $V_A = 0$ i.e. 0 V, $V_B = 1$ i.e. 5 V
D_1 is OFF and D_2 is ON
So, $V_0 = 1$ i.e. 5 V.
Similarly, when $V_A = V_B = 1$
Both D_1 and D_2 are ON
So, $V_0 = 1$ i.e. 5 V.
We get the final truth table on the basis of inputs and their corresponding output.

A	B	Y
0	0	0
0	1	1
1	0	1
1	1	1

$\therefore$ The given circuit (A) is an OR gate.
In figure B, it is a transistor in common emitter configuration which gives high output at low input and low output at high input. Hence, this circuit behaves as NOT gate.

48. *(d)* The given circuit, with the output of the respective gates is as given below.

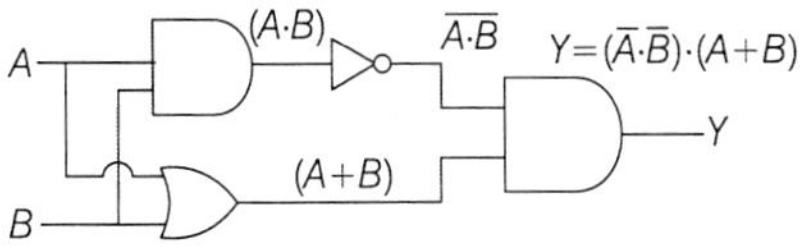

The Boolean expression of the output is

$$Y = (\overline{A \cdot B}) \cdot (A + B)$$

$$\Rightarrow \quad Y = (\overline{A} + \overline{B}) \cdot (A + B)$$

(using de-Morgan's theorem, $\overline{x \cdot y} = \overline{x} + \overline{y}$)

$$\Rightarrow \quad Y = \overline{A} \cdot B + \overline{B} \cdot A$$

This represents the Boolean expression for XOR gate.

> ***Alternate Solution***
> This question can also be verified from the following truth table.
>
> | A | B | $P = \overline{A \cdot B}$ | $Q = A + B$ | $Y = P \cdot Q$ |
> |---|---|---|---|---|
> | 0 | 0 | 1 | 0 | 0 |
> | 0 | 1 | 1 | 1 | 1 |
> | 1 | 0 | 1 | 1 | 1 |
> | 1 | 1 | 0 | 1 | 0 |

Hence, the output of this circuit represents the output of a XOR gate.

49. *(d)* According to the given figure signals are

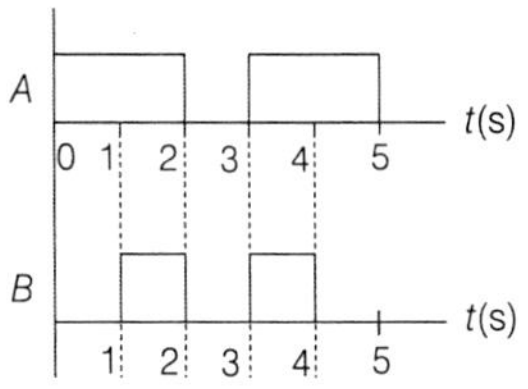

The output of given circuit diagram is

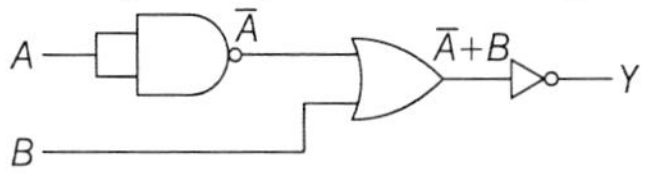

By using concept of de-Morgan's law,

$$Y = \overline{\overline{A} + B} = A \cdot \overline{B}$$

Truth table of given signals is

Time interval	A	B	$\overline{B}$	$A\overline{B}$
(0-1)s	1	0	1	1
(1-2)s	1	1	0	0
(2-3)s	0	0	1	0
(3-4)s	1	1	0	0
(4-5)s	1	0	1	1

Hence, output signal is,

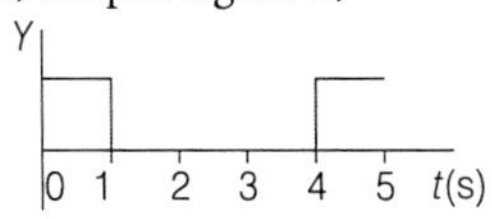

50. *(b)* Here A and B be the input and Y be the output.

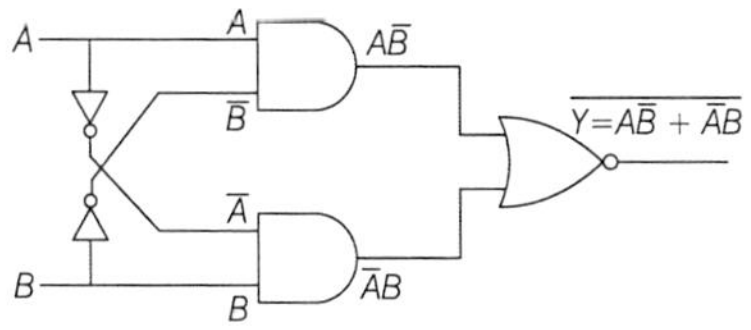

$$\because \quad Y = \overline{A\overline{B} + \overline{A}B} = \overline{A\overline{B}} \cdot \overline{\overline{A}B}$$

$$= (\overline{A} + \overline{\overline{B}}) \cdot (\overline{\overline{A}} + \overline{B})$$

$$= (\overline{A} + B)(A + \overline{B})$$

$$= \overline{A}A + \overline{A}\,\overline{B} + AB + B\overline{B}$$

$$= 0 + \overline{A}\,\overline{B} + AB + 0$$

$$= AB + \overline{A}\,\overline{B}$$

According to the truth table

A	B	$\overline{A}$	$\overline{B}$	AB	$\overline{A}\,\overline{B}$	$Y = AB + \overline{A}\,\overline{B}$
0	0	1	1	0	1	1
0	1	1	0	0	0	0
1	0	0	1	0	0	0
1	1	0	0	1	0	1

51. *(b)* The given circuit can be drawn as shown in the figure given below

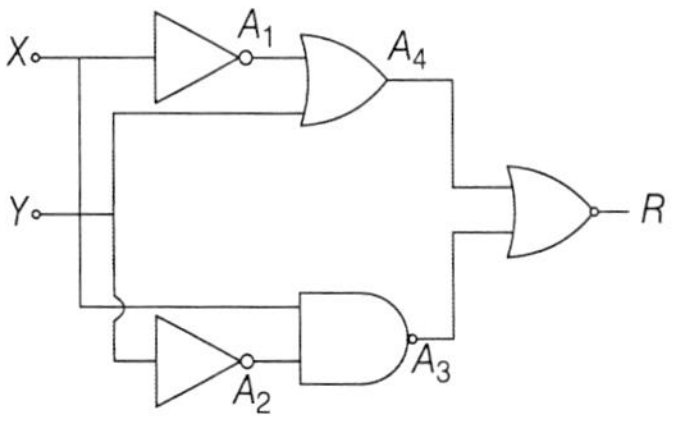

Truth table for this given logic gate is

Given inputs in the options		A_1	A_2	A_3	A_4	R
X	Y					
0	0	1	1	1	1	0
1	0	0	1	0	0	1
1	1	0	0	1	1	0
0	1	1	0	1	1	0

So to get output $R = 1$, inputs must be $X = 1$ and $Y = 0$.

52. *(0)* Consider the figure given in the question and solve it using Boolean identities.

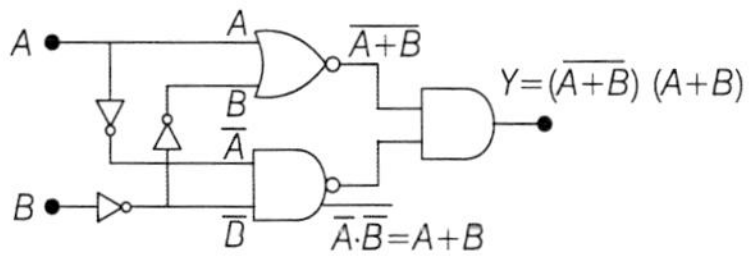

Now, put the value of A and B in the output, we get

$$Y = \overline{(0 + 1)}\,(0 + 1)$$

$$\Rightarrow \quad Y = \overline{1} \cdot 1 = 0$$

C-29

JEE Main Chapterwise Topicwise Physics

In JEE Main there is usually 0-1 question per paper is visible from this chapter. In past 3 years, 20 conceptual and 8 numerical questions were asked of moderate to difficult level from this chapter.

For detailed analysis scan QR code

EXPERIMENTAL PHYSICS

Topic 1 Experiments Related to Units and Measurements

Level 1

01 Given below are two statements.

Statement I In a vernier callipers, one vernier scale division is always smaller than one main scale division.

Statement II The vernier constant is given by one main scale division multiplied by the number of vernier scale divisions.

In the light of the above statements, choose the correct answer from the options given below.

[2025, 22 Jan Shift-I]

(a) Statement I is true but Statement II is false.
(b) Both Statement I and Statement II are true.
(c) Statement I is false but Statement II is true.
(d) Both Statement I and Statement II are false.

02 Given below are two statements, one is labelled as Assertion (A) and other is labelled as Reason (R).

Assertion (A) In vernier calliper if positive zero error exists, then while taking measurements, the reading taken will be more than the actual reading.

Reason (R) The zero error in vernier calliper might have happened due to manufacturing defect or due to rough handling.

In the light of the above statements, choose the correct answer from the options given below

[2024, 27 Jan Shift-II]

(a) Both A and R are correct but R is not the correct explanation of A.
(b) A is true but R is false.
(c) Both A and R are correct and R is the correct explanation of A.
(c) A is false but R is true.

03 In an experiment with vernier callipers of least count 0.1 mm, when two jaws are joined together the zero of vernier scale lies right to the zero of the main scale and 6th division of vernier scale coincides with the main scale division. While measuring the diameter of a spherical bob, the zero of vernier scale lies in between 3.2 cm and 3.3 cm marks, and 4th division of vernier scale coincides with the main scale division. The diameter of bob is measured as *[2023, 10 April Shift-II]*

(a) 3.26 cm (b) 3.22 cm
(c) 3.25 cm (d) 3.18 cm

04 **Assertion (A)** If in five complete rotations of the circular scale, the distance travelled on main scale of the screw gauge is 5 mm and there are 50 total divisions on circular scale, then least count is 0.001 cm.

Reason (R) Least count

$$= \frac{\text{Pitch}}{\text{Total divisions on circular scale}}$$

In the light of the above statements, choose the most appropriate answer from the options given below.

[2021, 27 July Shift-I]

(a) Both A and R are correct and R is the correct explanation of A.
(b) Both A and R are correct but R is not the correct explanation of A.
(c) A is correct but R is not correct.
(d) A is not correct but R is correct.

05 The vernier scale used for measurement has a positive zero error of 0.2 mm. If while taking a measurement, it was noted that 0 on the vernier scale lies between 8.5 cm and 8.6 cm vernier coincidence is 6, then the correct value of measurement is .... cm.

[2021, 17 March Shift-I]

(a) 8.56 cm (b) 8.36 cm (c) 8.54 cm (d) 8.58 cm

06 One main scale division of a vernier callipers is a cm and nth division of the vernier scale coincide with $(n-1)$th division of the main scale. The least count of the callipers in mm is *[2021, 16 March Shift-I]*

(a) $\left(\frac{n-1}{10n}\right)a$ (b) $\frac{10na}{(n-1)}$

(c) $\frac{10a}{n}$ (d) $\frac{10a}{(n-1)}$

07 A screw gauge has 50 divisions on its circular scale. The circular scale is 4 units ahead of the pitch scale marking, prior to use. Upon one complete rotation of the circular scale, a displacement of 0.5 mm is noticed on the pitch scale. The nature of zero error involved and the least count of the screw gauge are, respectively *[2020, 6 Sep Shift-I]*

(a) negative, 2 μm (b) positive, 10 μm

(c) positive, 0.1 mm (d) positive, 0.1 μm

08 A student measuring the diameter of a pencil of circular cross-section with the help of a vernier scale records the following four readings; 5.50 mm, 5.55 mm, 5.45 mm and 5.65 mm. The average of these four readings is 5.5375 mm and the standard deviation of the data is 0.07395 mm. The average diameter of the pencil should be therefore recorded as *[2020, 6 Sep Shift-II]*

(a) (5.5375 ± 0.0739) mm (b) (5.5375 ± 0.0740) mm

(c) (5.538 ± 0.074) mm (d) (5.54 ± 0.07) mm

09 Using screw gauge of pitch 0.1 cm and 50 divisions on its circular scale, the thickness of an object is measured. It should correctly be recorded as *[2020, 3 Sep Shift-I]*

(a) 2.121 cm (b) 2.124 cm

(c) 2.125 cm (d) 2.123 cm

10 If the screw on a screw gauge is given six rotations, it moves by 3 mm on the main scale. If there are 50 divisions on the circular scale, the least count of the screw gauge is *[2020, 9 Jan Shift-I]*

(a) 0.001 cm (b) 0.01 cm

(c) 0.02 cm (d) 0.001 cm

11 The least count of the main scale of a screw gauge is 1 mm. The minimum number of divisions on its circular scale required to measure 5 μm diameter of a wire is *[2019, 12 Jan Shift-I]*

(a) 50 (b) 100

(c) 200 (d) 500

Level 2

12 One main scale division of a vernier calliper is equal to m units. If nth division of main scale coincides with $(n+1)$th division of vernier scale, the least count of the vernier calliper is *[2024, 9 April Shift-I]*

(a) $\frac{m}{n(n+1)}$ (b) $\frac{n}{(n+1)}$ (c) $\frac{m}{(n+1)}$ (d) $\frac{1}{(n+1)}$

13 Least count of a vernier calliper is $\frac{1}{20N}$ cm. The value of one division on the main scale is 1 mm. Then, the number of divisions of main scale that coincide with N divisions of vernier scale is *[2024, 8 April Shift-II]*

(a) $\left(\frac{2N-1}{20N}\right)$ (b) $(2N-1)$

(c) $\left(\frac{2N-1}{2N}\right)$ (d) $\left(\frac{2N-1}{2}\right)$

14 In vernier calliper, when both jaws touch each other, zero of the vernier scale shifts towards left and its 4th division coincides exactly with a certain division on main scale. If 50 vernier scale divisions equal to 49 main scale divisions and zero error in the instrument is 0.04 mm, then how many main scale divisions are there in 1 cm? *[2024, 6 April Shift-II]*

(a) 10 (b) 20 (c) 40 (d) 5

15 While measuring diameter of wire using screw gauge the following readings were noted. Main scale reading is 1 mm and circular scale reading is equal to 42 divisions. Pitch of screw gauge is 1 mm and it has 100 divisions on circular scale. The diameter of the wire is $\frac{x}{50}$ mm. The value of x is .... *[2024, 6 April Shift-I]*

(a) 42 (b) 21 (c) 142 (d) 71

16 A vernier callipers has 20 divisions on the vernier scale, which coincides with 19th division on the main scale. The least count of the instrument is 0.1 mm. One main scale division is equal to *[2024, 5 April Shift-II]*

(a) 5 (b) 0.5 (c) 1 (d) 2

17 10 divisions on the main scale of a Vernier calliper coincide with 11 divisions on the Vernier scale. If each division on the main scale is of 5 units, the least count of the instrument is *[2024, 1 Feb Shift-I]*

(a) 10/11 (b) 1/2 (c) 5/11 (d) 50/11

18 In an experiment to find out the diameter of wire using screw gauge, the following observation were noted

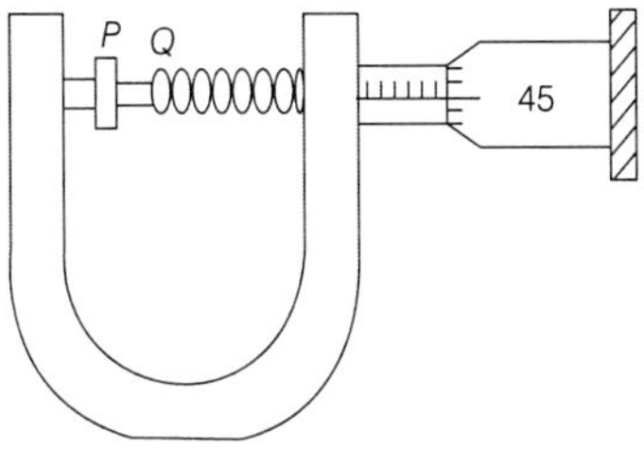

(a) Screw moves 0.5 mm on main scale in one complete rotation

(b) Total divisions on circular scale = 50

(c) Main scale reading is 2.5 mm

(d) 45th division of circular scale is in the pitch line

(e) Instrument has 0.03 mm negative error

Then, the diameter of wire is *[2022, 29 July Shift-I]*
(a) 2.92 mm (b) 2.54 mm
(c) 2.98 mm (d) 3.45 mm

19 In vernier calipers, 10 divisions of vernier scale is equal to 9 divisions of main scale. When both jaws of vernier callipers touch each other, the zero of the vernier scale is shifted to the left of zero of the main scale and 4th vernier scale division exactly coincides with the main scale reading. One main scale division is equal to 1 mm. While measuring diameter of a spherical body, the body is held between two jaws. It is now observed that zero of the vernier scale lies between 30 and 31 divisions of main scale reading and 6th vernier scale division exactly coincides with the main scale reading. The diameter of the spherical body will be *[2022, 26 July Shift-II]*
(a) 3.02 cm (b) 3.06 cm
(c) 3.10 cm (d) 3.20 cm

20 A screw gauge of pitch 0.5 mm is used to measure the diameter of uniform wire of length 6.8 cm, the main scale reading is 1.5 mm and circular scale reading is 7. The calculated curved surface area of wire to appropriate significant figures is (screw gauge has 50 divisions on its circular scale) *[2022, 26 July Shift-I]*
(a) 6.8 cm^2 (b) 3.4 cm^2
(c) 3.9 cm^2 (d) 2.4 cm^2

21 In a screw gauge, 5th division of the circular scale coincides with the reference line when the ratchet is closed. There are 50 divisions on the circular scale, and the main scale moves by 0.5 mm on a complete rotation. For a particular observation the reading on the main scale is 5 mm and the 20th division of the circular scale coincides with reference line. Calculate the true reading. *[2021, 26 Aug Shift-I]*
(a) 5.00 mm (b) 5.25 mm
(c) 5.15 mm (d) 5.20 mm

22 The pitch of the screw gauge is 1 mm and there are 100 divisions on the circular scale. When nothing is put in between the jaws, the zero of the circular scale lies 8 divisions below the reference line. When a wire is placed between the jaws, the first linear scale division is clearly visible while 72nd division on circular scale coincides with the reference line. The radius of the wire is *[2021, 25 Feb Shift-I]*
(a) 1.64 mm (b) 0.82 mm
(c) 1.80 mm (d) 0.90 mm

23 The pitch and the number of divisions on the circular scale, for a given screw gauge are 0.5 mm and 100, respectively. When the screw gauge is fully tightened without any object, the zero of its circular scale lies 3 divisions below the mean line. The readings of the main scale and the circular scale, for a thin sheet, are 5.5 mm and 48 respectively, the thickness of this sheet is *[2019, 9 Jan Shift-II]*
(a) 5.950 mm (b) 5.755 mm
(c) 5.725 mm (d) 5.740 mm

24 In a screw gauge, 5 complete rotations of the screw cause it to move a linear distance of 0.25 cm. There are 100 circular scale divisions. The thickness of a wire measured by this screw gauge gives a reading of 4 main scale divisions and 30 circular scale divisions. Assuming negligible zero error, the thickness of the wire is *[JEE Main 2018]*
(a) 0.3150 cm (b) 0.2150 cm (c) 0.4300 cm (d) 0.0430 cm

25 A spectrometer gives the following reading when used to measure the angle of a prism.

Main scale reading : 58.5 degree

Vernier scale reading : 09 divisions

Given that 1 division on main scale corresponds to 0.5 degree. Total divisions on the vernier scale is 30 and match with 29 divisions of the main scale. The angle of the prism from the above data is *[AIEEE 2012]*
(a) 58.59 degree (b) 58.77 degree
(c) 58.65 degree (d) 59 degree

Level 3

26 There are 100 divisions on the circular scale of a screw gauge of pitch 1 mm. With no measuring quantity in between the jaws, the zero of the circular scale lies 5 divisions below the reference line. The diameter of a wire is then measured using this screw gauge. It is found that 4 linear scale divisions are clearly visible while 60 divisions on circular scale coincide with the reference line. The diameter of the wire is *[2024, 8 April Shift-II]*
(a) 4.60 mm (b) 3.35 mm (c) 4.55 mm (d) 4.65 mm

27 The diameter of a sphere is measured using a vernier caliper whose 9 divisions of main scale are equal to 10 divisions of vernier scale. The shortest division on the main scale is equal to 1 mm. The main scale reading is 2 cm and second division of vernier scale coincides with a division on main scale. If mass of the sphere is 8.635 g, the density of the sphere is *[2024, 8 April Shift-I]*
(a) 2.2 g/cm^3 (b) 1.7 g/cm^3
(c) 2.5 g/cm^3 (d) 2.0 g/cm^3

28 The least count of the main scale of a vernier callipers is 1 mm. Its vernier scale is divided into 10th division and coincide with 9th division of the main scale. When jaws are touching each other, the 7th division of vernier scale coincides with a division of main scale and the zero of vernier scale is lying right side of the zero of main scale. When this vernier is used to measure length of a cylinder the zero of the vernier scale lies between 3.1 cm and 3.2 cm and 4th VSD coincides with a main scale division. The length of the cylinder is (VSD is vernier scale division) *[2020, 2 Sep Shift-I]*
(a) 3.2 cm (b) 2.99 cm (c) 3.07 cm (d) 3.21 cm

Numerical Value Type Questions

29 A tiny metallic rectangular sheet has length and breadth of 5 mm and 2.5 mm, respectively. Using a specially designed screw gauge which has pitch of 0.75 mm and 15 divisions in the circular scale, you are asked to find the area of the sheet. In this measurement, the maximum fractional error will be $\frac{x}{100}$, where x is .... *[2025, 28 Jan Shift-I]*

30 The least count of a screw guage is 0.01 mm. If the pitch is increased by 75% and number of divisions on the circular scale is reduced by 50%, the new least count will be ....$\times 10^{-3}$ mm. *[2025, 24 Jan Shift-I]*

31 In a screw gauge, there are 100 divisions on the circular scale and the main scale moves by 0.5 mm on a complete rotation of the circular scale. The zero of circular scale lies 6 divisions below the line of graduation when two studs are brought in contact with each other. When a wire is placed between the studs, 4 linear scale divisions are clearly visible while 46th division the circular scale coincide with the reference line. The diameter of the wire is .... $\times 10^{-2}$ mm. *[2023, 30 Jan Shift - I]*

32 The one division of main scale of vernier callipers reads 1mm and 10 divisions of vernier scale is equal to the 9 divisions on main scale. When the two jaws of the instrument touch each other, the zero of the Vernier lies to the right of zero of the main scale and its fourth division coincides with a main scale division. When a spherical bob is tightly placed between the two jaws, the zero of the Vernier scale lies in between 4.1 cm and 4.2 cm and 6th Vernier division coincides with a main scale division. The diameter of the bob will be .... $\times 10^{-2}$ cm. *[2022, 27 July Shift-I]*

33 The vernier constant of vernier callipers is 0.1 mm and it has zero error of (–0.05) cm. While measuring diameter of a sphere, the main scale reading is 1.7 cm and coinciding vernier division is 5. The corrected diameter will be... $\times 10^{-2}$ cm. *[2022, 29 June Shift-II]*

34 A student in the laboratory measures thickness of a wire using screw gauge. The readings are 1.22 mm, 1.23 mm, 1.19 mm and 1.20 mm. The percentage error is $\frac{x}{121}$%. The value of x is .... *[2022, 28 June Shift-II]*

35 In a vernier callipers, each cm on the main scale is divided into 20 equal parts. If tenth vernier scale division coincides with nineth main scale division. Then, the value of vernier constant will be .... $\times 10^{-2}$ mm. *[2022, 26 June Shift-I]*

36 The diameter of a spherical bob is measured using a vernier callipers. 9 divisions of the main scale, in the vernier callipers, are equal to 10 divisions of vernier scale. One main scale division is 1 mm. The main scale reading is 10 mm and 8th division of vernier scale was found to coincide exactly with one of the main scale division. If the given vernier callipers has positive zero error of 0.04 cm, then the radius of the bob is .... $\times 10^{-2}$cm. *[2021, 31 Aug Shift-II]*

37 Student A and Student B used two screw gauges of equal pitch and 100 equal circular divisions to measure the radius of a given wire. The actual value of the radius of the wire is 0.322 cm. The absolute value of the difference between the final circular scale reading observed by the students, A and B is .... [Figure shows position of reference O when jaws of screw gauge are closed] Given pitch = 0.1 cm *[20221, 25 July Shift-I]*

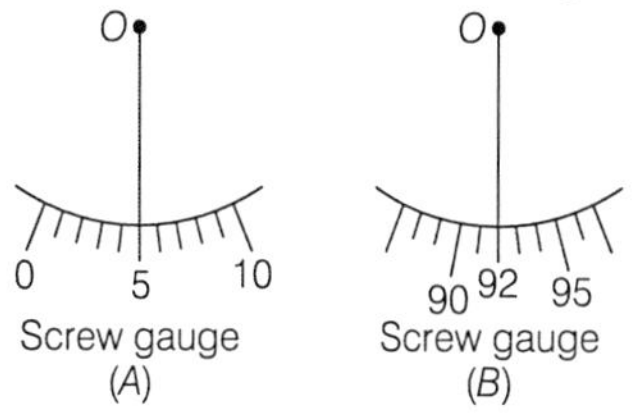

38 Suppose, you have taken a dilute solution of oleic acid in such a way that its concentration becomes 0.01 cm^3 of oleic acid per cm^3 of the solution. Then, you make a thin film of this solution (monomolecular thickness) of area 4 cm^2 by considering 100 spherical drops of radius $\left(\frac{3}{40\pi}\right)^{1/3} \times 10^{-3} cm^2$. Then, the thickness of oleic acid layer will be $x \times 10^{-14}$ m, where x is .... *[2021, 17 March Shift-II]*

Topic 2 Experiments Related to Oscillations and Waves

Level 1

01 To find the spring constant k of a spring experimentally, a student commits 2% positive error in the measurement of time and 1% negative error in measurement of mass. The percentage error in determining value of k is *[2024, 6 April Shift - I]*

(a) 3% (b) 4% (c) 5% (d) 1%

02 A student is performing the experiment of resonance column. The diameter of the column tube is 6 cm. The frequency of the tuning fork is 504 Hz. Speed of the sound at the given temperature is 336 m/s. The zero of the meter scale coincides with the top end of the resonance column tube. The reading of the water level in the column when the first resonance occurs is *[2021, 25 Feb Shift-I]*

(a) 13 cm (b) 16.6 cm (c) 18.4 cm (d) 14.8 cm

03 A simple pendulum is being used to determine the value of gravitational acceleration g at a certain place. The length of the pendulum is 25.0 cm and a stop watch with 1s resolution measures the time taken for 40 oscillations to be 50 s. The accuracy in g is *[2020, 8 Jan Shift-II]*
(a) 2.40% (b) 5.40% (c) 4.40% (d)3.40%

04 In an experiment to determine the period of a simple pendulum of length 1 m, it is attached to different spherical bobs of radii r_1 and r_2. The two spherical bobs have uniform mass distribution. If the relative difference in the periods, is found to be 5×10^{-4} s, the difference in radii, $|r_1 - r_2|$ is best given by *[JEE Main 2017]*
(a) 0.1 cm (b) 0.01 cm (c) 0.5 cm (d) 1 cm

Level 2

05 In the resonance experiment, two air columns (closed at one end) of 100 cm and 120 cm long, give 15 beats per second when each one is sounding in the respective fundamental modes. The velocity of sound in the air column is *[2025, 3 April Shift-II]*
(a) 335 m/s (b) 340 m/s (c) 360 m/s (d) 370 m/s

06 The period of oscillation of a simple pendulum is $T = 2\pi\sqrt{\frac{L}{g}}\cdot$ Measured value of L is 1.0 m from metre scale having a minimum division of 1 mm and time of one complete oscillation is 1.95 s measured from stopwatch of 0.01 s resolution.The percentage error in the determination of g will be *[2021, 24 Feb Shift-II]*
(a) 1.13% (b) 1.03% (c) 1.33% (d) 1.30%

07 In a simple pendulum, experiment for determination of acceleration due to gravity (g), time taken for 20 oscillations is measured by using a watch of 1 second least count. The mean value of time taken comes out to be 30 s. The length of pendulum is measured by using a meter scale of least count 1 mm and the value obtained 55.0 cm. The percentage error in the determination of g is close to *[2019, 8 April Shift-II]*
(a) 0.7% (b) 6.8% (c) 3.5% (d) 0.2%

Numerical Value Type Questions

08 In an experiment with sonometer when a mass of 180 g is attached to the string, it vibrates with fundamental frequency of 30 Hz. When a mass m is attached, the string vibrates with fundamental frequency of 50 Hz. The value of m is .... g. *[2023, 13 April Shift-II]*

09 In an experiment to find acceleration due to gravity g using simple pendulum, time period of 0.5 s is measured from time of 100 oscillations with a watch of 1 s resolution. If measured value of length is 10 cm known to 1 mm accuracy, the accuracy in the determination of g is found to be x%. The value of x is .... *[2022, 28 July Shift-II]*

10 Three students S_1, S_2 and S_3 perform an experiment for determining the acceleration due to gravity g using a simple pendulum. They use different lengths of pendulum and record time for different number of oscillations. The observations are as shown in the table

Student no.	Length of pendulum (cm)	No. of oscillations (n)	Total time for (n) oscillations	Time period (s)
1.	64.0	8	128.0	16.0
2.	64.0	4	64.0	16.0
3.	20.0	4	36.0	9.0

(Least count of length = 0.1 cm least count for time = 0.1 s)
If E_1, E_2 and E_3 are the percentage errors in g for students 1, 2 and 3 respectively, then the minimum percentage error is obtained by student no .... *[2021, 22 July Shift-II]*

Topic 3 Experiments Related to Properties of Solids and Liquids

Level 1

01 A student determined Young's modulus of elasticity using the formula $Y = \frac{MgL^3}{4bd^3\delta}$. The value of g is taken to be 9.8 m / s^2, without any significant error, his observation are as following.

Physical Quantity	Least count of the Equipment used for measurement	Observed Value
Mass (M)	1 g	2 kg
Length of bar (L)	1 mm	1 m
Breadth of bar (b)	0.1 mm	4 cm
Thickness of bar (d)	0.01 mm	0.4 cm
Depression (δ)	0.01 mm	5

Then the fractional error in the measurement of Y is *[2021, 1 Sep Shift-II]*
(a) 0.155 (b) 0.0083 (c) 0.083 (d) 0.0155

02 In order to determine the Young's modulus of a wire of radius 0.2 cm (measured using a scale of least count = 0.0010. cm) and length 1 m (measured using a scale of least count = 1 mm), a weight of mass 1 kg (measured using a scale of least count = 1 g) was hanged to get the elongation of 0.5 cm (measured using a scale of least count 0.001 cm). What will be the fractional error in the value of Young's modulus determined by this experiment ? *[2021, 16 March Shift -II]*
(a) 1.4 % (b) 0.14 % (c) 9 % (d) 0.9 %

03 The following observations were taken for determining surface tension T of water by capillary method

Diameter of capillary, $D = 1.25 \times 10^{-2}$ m

Rise of water, $h = 1.45 \times 10^{-2}$ m

Using $g = 9.80$ m s^{-2} and the simplified relation,

$T = \frac{rhg}{2} \times 10^3$ N m^{-1}, the possible error in surface tension is closest to *[JEE Main 2017]*

(a) 0.15 % (b) 1.5 % (c) 2.4 % (d) 10 %

Level 2

04 A load of mass M kg is suspended from a steel wire of length 2 m and radius 1.0 mm in Searle's apparatus experiment. The increase in length produced in the wire is 4.0 mm. Now, the load is fully immersed in a liquid of relative density 2. The relative density of the material of load is 8. The new value of increase in length of the steel wire is *[2019, 12 Jan Shift-II]*

(a) 4.0 mm (b) zero (c) 5.0 mm (d) 3.0 mm

05 In an experiment, brass and steel wires of length 1 m each with areas of cross-section 1 mm^2 are used. The wires are connected in series and one end of the combined wire is connected to a rigid support and other end is subjected to elongation. The stress required to produce a net elongation of 0.2 mm is, [Given, the Young's modulus for steel and brass are respectively, 120×10^9 N / m^2 and 60×10^9 N/m^2] *[2019, 10 April Shift-II]*

(a) 1.8×10^6 N/ m^2 (b) 0.2×10^6 N/ m^2

(c) 1.2×10^6 N/ m^2 (d) 4.0×10^6 N/ m^2

Numerical Value Type Questions

06 In an experiment to determine the Young's modulus, steel wires of five different lengths (1, 2, 3, 4 and 5 m) but of same cross-section (2mm^2) were taken and curves between extension and load were obtained. The slope (extension/load) of the curves were plotted with the wire length and the following graph is obtained. If the Young's modulus of given steel wires $x \times 10^{11}$ Nm^{-2}, then the value of x is .... *[2022, 27 July Shift-II]*

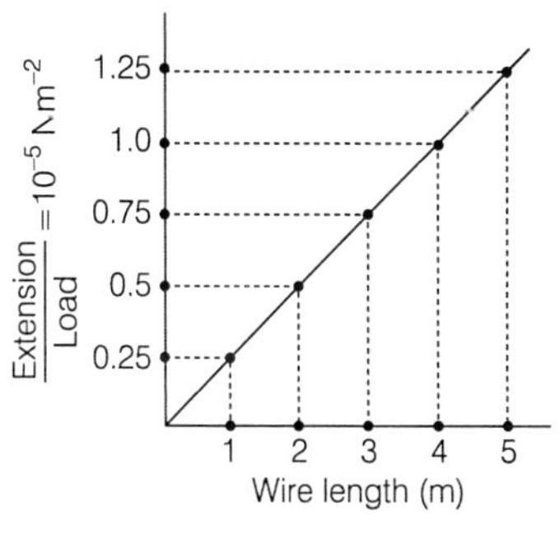

07 In an experiment to determine the Young's modulus of wire of a length exactly 1 m, the extension in the length of the wire is measured as 0.4 mm with an uncertainty of +0.02 mm when a load of 1 kg is applied. The diameter of the wire is measured as 0.4 mm with an uncertainty of ± 0.01 mm. The error in the measurement of Young's modulus (ΔY) is found be $x \times 10^{10}$ Nm^{-2}. The value of x is ....

(Take, $g = 10$ ms^{-2}) *[2022, 26 July Shift-I]*

Topic 4 Experiments Related to Current Electricity

Level 1

01 A student is provided with a variable voltage source V, a test resistor $R_T = 10\,\Omega$ two identical galvanometers G_1 and G_2 and two additional resistors, $R = 10\,M\Omega$ and $R_2 = 0.001\,\Omega$. For conducting an experiment to verify ohm's law, the most suitable circuit is *[2023, 6 April Shift-II]*

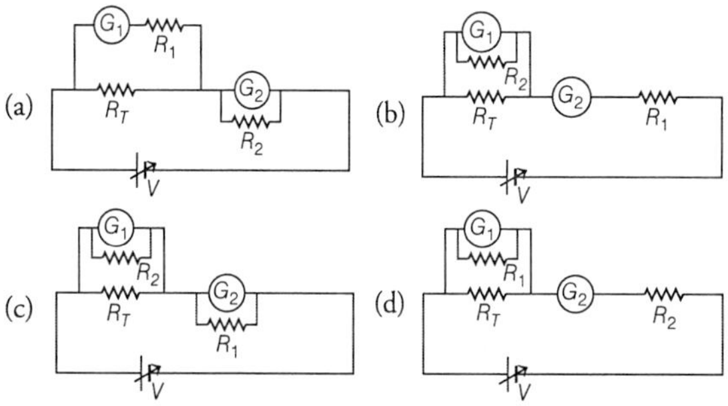

02 In the experiment of Ohm's law, a potential difference of 5.0 V is applied across the end of a conductor of length 10.0 cm and diameter of 5.00 mm. The measured current in the conductor is 2.00 A. The maximum permissible percentage error in the resistivity of the conductor is *[2021, 18 March Shift-I]*

(a) 3.9 (b) 8.4

(c) 7.5 (d) 3.0

03 In the potentiometer circuit arrangement, the balancing length AC is measured to be 250 cm. When the galvanometer connection is shifted from point (1) to point (2) in the given diagram, the balancing length becomes 400 cm. The ratio of emf of two cells, $\varepsilon_1 / \varepsilon_2$ is *[2021, 25 July Shift-II]*

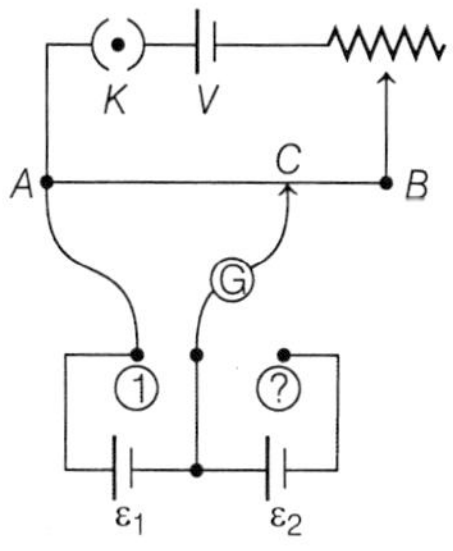

(a) 3/2 (b) 8/5

(c) 5/3 (d) 4/3

04 Which of the following will not be observed when a multimeter (operating in resistance measuring mode) probes connected across a component, are just reversed ? *[2020, 3 Sep Shift-II]*

(a) Multimeter shows no deflection in both cases, i.e. before and after reversing the probes if the chosen component is metal wire.

(b) Multimeter shows an equal deflection in both cases, i.e. before and after reversing the probes if the chosen component is resistor.

(c) Multimeter shows no deflection in both cases, i.e. before and after reversing the probes if the chosen component is capacitor.

(d) Multimeter shows a deflection, accompanied by a splash of light out of connected component in one direction and no deflection on reversing the probes if the chosen component is LED.

05 The length of a potentiometer wire is 1200 cm and it carries a current of 60 mA. For a cell of emf 5 V and internal resistance of 20 Ω, the null point on it is found to be at 1000 cm. The resistance of whole wire is *[2020, 9 Jan Shift-II]*

(a) 80 Ω (b) 100 Ω
(c) 60 Ω (d) 120 Ω

06 The actual value of resistance R, shown in the figure is 30 Ω. This is measured in an experiment as shown using the standard formula $R = V/I$, where V and I are the reading of the voltmeter and ammeter, respectively. If the measured value of R is 5% less, then the internal resistance of the voltmeter is *[2019, 10 Jan Shift-II]*

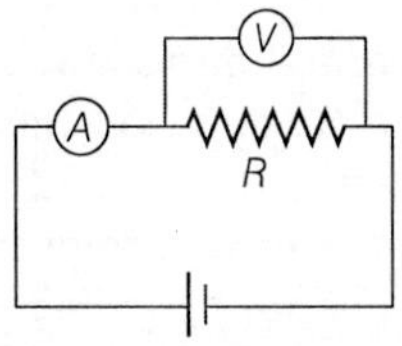

(a) 570 Ω (b) 600 Ω
(c) 35 Ω (d) 350 Ω

07 In a circuit for finding the resistance of a galvanometer by half deflection method, a 6 V battery and a high resistance of 11 kΩ are used. The figure of merit of the galvanometer is 60 μA per division. In the absence of shunt resistance, the galvanometer produces a deflection of $\theta = 9$ divisions when current flows in the circuit. The value of the shunt resistance that can cause the deflection of $\theta/2$ is closest to *[JEE Main 2018]*

(a) 55 Ω (b) 110 Ω
(c) 220 Ω (d) 550 Ω

08 In a potentiometer experiment, it is found that no current passes through the galvanometer when the terminals of the cell are connected across 52 cm of the potentiometer wire, If the cell is shunted by a resistance of 5 Ω, a balance is found when the cell is connected across 40 cm of the wire. Find the internal resistance of the cell. *[JEE Main 2018]*

(a) 1 Ω (b) 1.5 Ω (c) 2 Ω (d) 2.5 Ω

Level 2

09 A teacher in his physics laboratory allotted an experiment to determine the resistance G of a galvanometer. Students took the observations for 1/3 deflection in the galvanometer. Which of the below is true for measuring value of G ? *[2022, 25 June Shift-I]*

(a) 1/3 deflection method cannot be used for determining the resistance of the galvanometer.

(b) 1/3 deflection method can be used and in this case, the G equals to twice the value of shunt resistance (s).

(c) 1/3 deflection method can be used and in this case, the G equals to three times the value of shunt resistance (s).

(d) 1/3 deflection method can be used and in this case, the G value equals to the shunt resistance (s).

Numerical Value Type Questions

10 In the experiment to determine the galvanometer resistance by half-deflection method. the plat of $1/\theta$ *vs* the resistance R of the resistance box is shown in the figure. The figure of merit of the galvanometer is .... $\times 10^{-1}$ A/division. [The source has emf 2V] *[2024, 5 April Shift-I]*

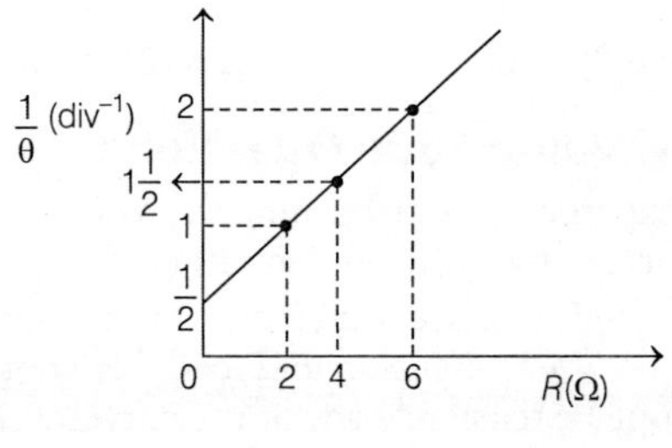

11 In an experiment to find emf of a cell using potentiometer, the length of null point for a cell of emf 1.5 V is found to be 60 cm. If this cell is replaced by another cell of emf E, the length of null point increases by 40 cm. The value of E is $x/10$ V. The value of x is .... *[2023, 1 Feb Shift-I]*

12 The balancing length for a cell is 560 cm in a potentiometer experiment. When an external resistance of 10Ω is connected in parallel to the cell, the balancing length changes by 60 cm. If the internal resistance of the cell is $N/10$ Ω, where N is an integer, then value of N is .... *[2020, 7 Jan Shift-II]*

Topic 5 Experiments Related to Optics

Level 1

01 For the determination of refractive index of glass slab, a travelling microscope is used whose main scale contains 300 equal divisions equals to 15 cm. The vernier scale attached to the microscope has 25 divisions equals to 24 divisions of main scale. The least count (LC) of the travelling microscope is (in cm) *[2025, 4 April Shift-II]*

(a) 0.002 (b) 0.001
(c) 0.0025 (d) 0.0005

02 Identify the physical quantity that cannot be measured using spherometer *[2024, 27 Jan Shift-I]*

(a) Radius of curvature of concave surface
(b) Thickness of thin plates
(c) Radius of curvature of concave surface
(d) Specific rotation of liquids

03 A travelling microscope has 20 divisions per cm on the main scale while its vernier scale has total 50 divisions and 25 vernier scale divisions are equal to 24 main scale divisions, what is the least count of the travelling microscope ? *[2022, 29 July Shift-I]*

(a) 0.001 cm (b) 0.002 cm
(c) 0.002 cm (d) 0.005 cm

04 An experiment is performed to find the refractive index of glass using a travelling microscope. In this experiment, distances are measured by *[AIEEE 2008]*

(a) a vernier scale provided on the microscope
(b) a standard laboratory scale
(c) a meter scale provided on the microscope
(d) a screw gauge provided on the microscope

Level 2

05 In an experiment to measure focal length f of convex lens, the least counts of the measuring scales for the position of object u and for the position of image v are Δu and Δv, respectively. The error in the measurement of the focal length of the convex lens will be *[2024, 4 April Shift-I]*

(a) $f^2\left[\frac{\Delta u}{u^2}+\frac{\Delta v}{v^2}\right]$ (b) $2f\left[\frac{\Delta u}{u}+\frac{\Delta v}{v}\right]$

(c) $f\left[\frac{\Delta u}{u}+\frac{\Delta v}{v}\right]$ (d) $\frac{\Delta u}{u}+\frac{\Delta v}{v}$

06 If 50 vernier divisions are equal to 49 main scale divisions of a travelling microscope and one smallest reading of main scale is 0.5 mm, the vernier constant of travelling microscope is *[2024, 30 Jan Shift-II]*

(a) 0.1 cm (b) 0.01 mm
(c) 0.1 mm (d) 0.01 cm

07 In an experiment a convex lens of focal length 15 cm is placed coaxially on an optical bench in front of a convex mirror at a distance of 5 cm from it. It is found that an object and its image coincide, if the object is placed at a distance of 20 cm from the lens. The focal length of the convex mirror is *[JEE Main 2017]*

(a) 27.5 cm (b) 20.0 cm (c) 25.0 cm (d) 30.5 cm

Level 3

08 In finding out refractive index of glass slab the following observations were made through travelling microscope 50 vernier scale division = 49 MSD; 20 divisions on main scale in each cm

For mark on paper
MSR = 8.45 cm, VC = 26
For mark on paper seen through slab
MSR = 7.12 cm, VC = 41
For powder particle on the top surface of the glass slab
MSR = 4.05 cm, VC = 1
(MSR = Main Scale Reading,
VC = Vernier Coincidence)

Refractive index of the glass slab is *[2024, 6 April Shift-II]*

(a) 1.24 (b) 1.42 (c) 1.35 (d) 1.52

09 A 2 metre long scale with least count of 0.2 cm is used to measure the locations of objects on an optical bench. While measuring the focal length of a convex lens, the object pin and the convex lens are placed at 80 cm mark and lm mark, respectively. The image of the object pin on the other side of lens coincides with image pin that is kept at 180 cm mark. The % error in the estimation of focal length is *[2023, 6 April Shift-II]*

(a) 0.51 (b) 1.70 (c) 0.85 (d) 1.02

Numerical Value Type Questions

10 In an experiment to measure the focal length (f) of a convex lens, the magnitude of object distance (x) and the image distance (y) are measured with reference to the focal point of the lens. The y - x plot is shown in figure. The focal length of the lens is .... cm.

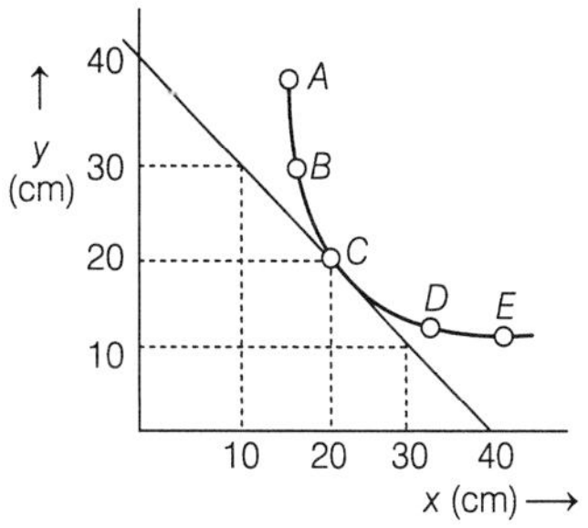

[2024, 30 Jan Shift-II]

11 In an experiment for estimating the value of focal length of converging mirror, image of an object placed at 40 cm from the pole of the mirror is formed at distance 120 cm from the pole of the mirror. These distances are measured with a modified scale in which there are 20 small divisions in 1 cm. The value of error in measurement of focal length of the mirror is $1/K$ cm. The value of K is ....
[2023, 30 Jan Shift -I]

12 A travelling microscope is used to determine the refractive index of a glass slab. If 40 divisions are there in 1 cm on main scale and 50 vernier scale divisions are equal to 49 main scale divisions, then least count of the travelling microscope is ...... $\times 10^{-6}$ m.
[2022, 26 June Shift-II]

Topic 6 Experiments Related to Electronics Devices

Level 1

01 An experiment is performed to determine the I-V characteristics of a zener diode, which has a protective resistance of $R = 100\,\Omega$ and a maximum power of dissipation rating of $1\,\Omega$. The minimum voltage range of the DC source in the circuit is *[JEE Main 2016]*
(a) 0 – 5 V (b) 0 – 24 V (c) 0 – 12 V (d) 0 – 8 V

Numerical Value Type Question

02 A common transistor radio set requires 12 V (DC) for its operation. The DC source is constructed by using a transformer and a rectifier circuit, which are operated at 220 V (AC) on standard domestic AC supply. The number of turns of secondary coil are 24, then the number of turns of primary are .... *[2021, 24 Feb Shift-I]*

Key Index

Topic 1

1 (a)	2 (a)	3 (d)	4 (d)	5 (c)	6 (c)	7 (b)	8 (d)	9 (b)	10 (a)
11 (c)	12 (c)	13 (d)	14 (b)	15 (d)	16 (d)	17 (c)	18 (c)	19 (c)	20 (b)
21 (c)	22 (b)	23 (c)	24 (b)	25 (c)	26 (c)	27 (d)	28 (c)	29 (3)	30 (35)
31 (220)	32 (412)	33 (180)	34 (150)	35 (5)	36 (52	37 (13)	38 (25)		

Topic 2

1 (c)	2 (d)	3 (c)	4 (a)	5 (c)	6 (a)	7 (b)	8 (500)	9 (5)	10 (1)

Topic 3

1 (d)	2 (a)	3 (b)	4 (d)	5 (*)	6 (2)	7 (2)

Topic 4

1 (a)	2 (a)	3 (c)	4 (c)	5 (b)	6 (a)	7 (b)	8 (b)	9 (b)	10 (5)
11 (25)	12 (12)								

Topic 5

1 (a)	2 (d)	3 (c)	4 (a)	5 (a)	6 (b)	7 (a)	8 (b)	9 (b)	10 (20)
11 (2)	12 (5)								

Topic 6

1 (b)	2 (440)

Detailed **Explanations**

Topic 1

01. *(a)* Statement I is true and Statement II is false.
In vernier scale, one VSD is smaller than one MSD. Also least count is given by one main scale division/number of vernier scale division for normal vernier calliper.

02. *(a)* Positive zero error subtracted from actual reading hence reading taken is more than actual reading.

03. *(d)* Given, Least count,

$$LC = 0.1 \text{ mm} = 10^{-4} \text{ m}$$

$$\text{Zero error} = 6\times LC = 0.6 \text{ mm} = 6\times10^{-4} \text{ m}$$

Reading

$$= MSR + VSR\times LC - \text{zero error}$$
$$= [32 \text{ mm} + (0.1)\times 4 \text{ mm}] - 0.6 \text{ mm}$$
$$= 31.8 \text{ mm} = 3.18 \text{ cm}$$

04. *(d)* According to question,
In 5 complete rotations of circular scale, distance travelled on the main scale of screw gauge is 5mm, so in one complete rotation of circular scale, distance travelled on the main scale of screw gauge will be $= \frac{5\text{mm}}{5} = 1$ mm
This is the required value of pitch.

$\because$ Least count

$$= \frac{\text{Pitch}}{\text{Total division on circular scale}}$$

$$\Rightarrow \text{Least count} = \frac{1}{50} = 0.02$$

It means, Assertion is not correct, but Reason is correct.

> ***Alternate Solution***
>
> $\therefore$ Least count $= \frac{\text{Pitch}}{\text{Total division of circular scale}}$
>
> Here, pitch $= \frac{\text{Distance travelled on main scale}}{\text{Number of rotations}}$
>
> $= \frac{5}{5} = 1\text{mm}$
>
> $\therefore \quad LC = \frac{1}{50} = 0.02 \text{ mm}$
>
> Hence, Assertion is not correct but Reason is correct.

05. *(c)* Given, zero error,

$$ze = 0.2 \text{ mm} = 0.02 \text{ cm}$$

Main scale reading, MSR $= 8.5$ cm
Vernier scale reading, VSR $= 6$
Least count, LC $= 0.1/10$ cm $= 0.01$ cm
We know that

$$\text{Reading} = MSR + (VSR\times LC) - \text{zero error}$$
$$= 8.5 + 6\times0.01 - 0.02$$
$$= 8.5 + 0.06 - 0.02$$
$$= 8.54 \text{ cm}$$

06. *(c)* Given, 1 MSD $= a$ cm
nth division of the vernier scale division are equal to $(n-1)$th division of main scale, we get

$$1 \text{ VSD} = \frac{(n-1)}{n} \text{ mm}$$

The least count is given by

$$= 1 \text{ MSD} - 1 \text{ VSD}$$
$$= \left(a - \frac{(n-1)}{(n)}\times a\right) \text{cm}$$
$$= \frac{na-(n-1)a}{n} \text{ cm} = \frac{a}{n} = 10\frac{a}{n} \text{ mm}$$

07. *(b)* Given that number of divisions on circular scale $= 50$
Pitch $= 0.5$ mm $= 5\times10^{-4}$ m
Least count of screw gauge

$$= \frac{\text{Pitch}}{\text{Number of divisions on circular scale}}$$
$$= \frac{5\times10^{-4}}{50} = 1\times10^{-5} \text{ m} = 10\ \mu\text{m}$$

Since, circular scale is 4 units ahead of the pitch scale marking which means screw gauge has positive zero error.

08. *(d)* There are three significant figures in the measured value (or readings) whereas there are 5 and 4 significant figures in average readings and standard deviation of the data, respectively. So, answer must be upto three significant figures. After rounding off the uncertain digits and considering least count of vernier scale, i.e. 0.01 mm, average diameter of pencil is recorded as (5.54 ± 0.07) mm.

> **• Concept Enhancer**
>
> If quantities have different number of significant figures, then after doing algebraic operations (Addition or subtraction), round off the result, so that it has the same number of significant figures as in least accurate quantity.

09. *(b)* Given, pitch $= 0.1$ cm
Number of divisions on circular scale $= 50$
Least count of screw gauge

$$= \frac{\text{Pitch}}{\text{Number of divisions on circular scale}}$$
$$= 0.1/50 = 0.002 \text{ cm}$$

Now, correctly recorded measurement is an integral multiple of LC.
So, amongst the given options, only 2.124 cm is the integral multiple of LC.
Thus, it is the correctly recorded value.

10. *(a)* Pitch of screw of a screw gauge,

$$\text{pitch} = \frac{\text{distance moved by screw}}{\text{number of rotations}}$$
$$= \frac{3 \text{ mm}}{6} = 0.5 \text{ mm}$$

and least count of screw gauge,
least count

$$= \frac{\text{pitch}}{\text{number of circular scale divisions}}$$
$$= \frac{0.5 \text{ mm}}{50} = 0.01 \text{ mm} = 0.001 \text{ cm}$$

11. *(c)* Given, Least count of main scale $= 1$ mm
N_d (Number of divisions)

$$= \frac{LC}{5\ \mu\text{m}} = \frac{1 \text{ mm}}{5\ \mu\text{m}} = \frac{10^3}{5} = 200$$

12. *(c)* Given, $1 \text{ VSD} = \frac{n}{(n+1)}$ MSD

$$LC = 1 \text{ MSD} - 1 \text{ VSD}$$
$$LC = 1\text{MSD} - \frac{n}{(n+1)}\text{MSD}$$
$$= \text{MSD}\left[1 - \frac{n}{n+1}\right] = \text{MSD}\left[\frac{n+1-n}{n+1}\right]$$
$$= \text{MSD}\frac{(1)}{(n+1)}$$
$$\therefore \quad LC = \frac{m}{n+1}$$

13. *(d)* Given,

$$\text{Least count} = \frac{1}{20N} \text{ cm} = \frac{10}{20N} \text{ mm}$$
$$1 \text{ MSD} = 1 \text{ mm}$$

Now, Least count $= 1$ MSD $- 1$ VSD
Let $\quad x$ MSD $= N$ VSD then,

$$\Rightarrow \frac{10}{20N} = 1 - \frac{x}{N} \Rightarrow \frac{1}{2N} = 1 - \frac{x}{N} = \frac{N-x}{N}$$
$$\Rightarrow N - x = \frac{1}{2} \quad \text{or} \quad N - \frac{1}{2} = x$$
$$\text{or} \quad x = \frac{2N-1}{2} \text{ divisions}$$

14. *(b)* Let 1 MSD $= x$

According to question,

50 VSD matches with 49 MSD $= 49x$

Least count, LC $= \frac{49}{50}x$

LC = 1 MSD − 1 VSD

$= x - \frac{49}{50}x = \frac{x}{50}$

Zero error = MSR + VSR (LC)

$0.04 \text{ mm} = 0 + 4\left(\frac{x}{50}\right)$

$\Rightarrow \quad x = 0.5 \text{ mm}$

Number of main scale division in 1 cm

$= \frac{1 \text{ cm}}{0.5 \text{ mm}} = 20$

15. *(d)* Given, Pitch = 1 mm

Main scale reading (MSR) = 1 mm

Least count,

$LC = \frac{\text{Pitch}}{\text{No. of circular scale division}}$

$= \frac{1}{100} = 0.01 \text{ mm}$

Circular scale reading (CSR) $= 42 \times LC$

$= 42 \times 0.01 = 0.42$ mm

∴Diameter = MSR + CSR

$= 1 + 0.42 = 1.42$ mm

$= \frac{71}{50} \text{ mm} = \frac{x}{50}$ (given)

$\therefore \quad x = 71$

16. *(d)* 20 VSD = 19 MSD

$\Rightarrow \quad 1 \text{ VSD} = \frac{19}{20} \text{ MSD}$

∴Least count = MSD − VSD

$\Rightarrow \quad 0.1 = \text{MSD} - \frac{19}{20}\text{MSD}$

$\Rightarrow \quad 0.1 = \text{MSD}\left(1 - \frac{19}{20}\right) \Rightarrow 0.1 = \frac{\text{MSD}}{20}$

$\Rightarrow \quad \text{MSD} = 0.1 \times 20 = 2 \text{ mm}$

17. *(c)* Given, 10 MSD = 11 VSD

So, $1 \text{ VSD} = \frac{10}{11}\text{ MSD} = \frac{10}{11} \times 5$ unit

(Given, 1 MSD = 5 units)

Now, Least count of instrument

$= 1 \text{ MSD} - 1 \text{ VSD} = 5 - \frac{10 \times 5}{11}$

$= \frac{55 - 50}{11} = \frac{5}{11}$ unit

18. *(c)*

> **key idea** To find the diameter of a wire, you can use following formula for screw gauge measurement,
>
> Diameter = MSR + (CSR × LC) − zero error
>
> (using the value of zero error with their sign)

MSR (Main scale reading) = 2.5 mm

In 50 circular scale division 0.5 mm of main scale is there.

∴ 1 CSR (Circular scale reading)

$= 0.5/50 = 0.01$ mm

∵ 45th division of circular scale is in the pitch line.

∴Total circular scale reading

$= (45 \times 0.01) \text{ mm} = 0.45$ mm

∴Diameter reading = MSR + Total circular scale reading + Negative zero error

$= 2.5 \text{ mm} + 0.45 \text{ mm} + 0.03 \text{ mm}$

$= 2.98$ mm

19. *(c)* Given, 1 MSD = 1 mm

As 10 VSD = 9 MSD

or $\quad 1 \text{ VSD} = \frac{9}{10}\text{ MSD}$

∴Least count = 1 MSD − 1 VSD

$= 1 \text{ MSD} - \frac{9}{10}\text{ MSD}$

$= (1 - 0.9) = 0.1 \text{ mm} = 0.01 \text{ cm}$

Zero error $= -(10 - 4) \times 0.1$ mm

$= -0.6 \text{ mm} = 0.06$ cm

Now, diameter of the spherical body

= MSR + Vernier scale reading × LC − Zero error

$= 3 + 6 \times 0.01 - (-0.06)$

$= 3 + 0.06 + 0.06 = 3.12 \text{ cm} \approx 3.10$ cm

20. *(b)* Given, pitch of screw gauge,

$\Delta x = 0.5 \text{ mm} = 5 \times 10^{-2}$ cm

∴Least count of screw gauge, LC

$= \frac{\text{Pitch}}{\text{Number of division on circular scale}}$

$= \frac{5 \times 10^{-2}}{50} = 10^{-3}$ cm

∴Diameter of wire, d = Main scale reading + Circular scale reading

$= 1.5 \text{ mm} + 7 \times LC$

$= 0.15 \text{ cm} + 7 \times 10^{-3} \text{ cm} = 0.157$ cm

∴Curved surface area of the wire

$= 2\pi r l = \pi l(2r) = \pi l d \quad (\because 2r = d)$

$= 3.14 \times 6.8 \times 0.157$

$(\because l = 6.8$ cm, given)

$= 3.354 \text{ cm}^2 = 3.4 \text{ cm}^2$

21. *(c)* Least count of a screw gauge

$= \frac{\text{Pitch}}{\text{Number of circular scale divisions}}$

$\Rightarrow LC = 0.5/50$ ...(i)

The actual reading (true reading)

= Main scale reading + Least count × Circular division − Zero error

$= 5 + \left(\frac{0.5}{50}\right) \times 20 - \frac{0.5}{50} \times 5$

[∵ Zero error = Least count × Coinciding division]

$= 5 + \frac{0.5}{50}(20 - 5) = 5 + \frac{15}{100} = \frac{515}{100}$ mm

$= 5.15$ mm

22. *(b)* Given, pitch of screw gauge,

$P = 1$ mm

Number of division, $n = 100$

∴ Least count, $LC = P/n = 1/100$

$= 0.01$ mm

As, zero of circular division lies 8 divisions below.

∴Zero error $= 8 \times LC = 8 \times 0.01$

$= 0.08$ mm

Since, 1st linear scale division coinside with 72nd circular scale division.

∴Radius (r)

$= \frac{[P + (72 \times LC) - \text{Zero error}]}{2}$

$= [1 + (72 \times 0.01) - 0.08]/2$

$= (1.72 - 0.08)/2 = 1.64/2 = 0.82$ mm

23. *(c)* Given, pitch = 0.5 mm, $N = 100$

Least count,

$LC = \frac{\text{Pitch}}{N} = \frac{0.5 \text{ mm}}{100} = 5 \times 10^{-3}$ mm

Thickness of sheet

= MSR + CSR × LC − Zero error

$= (5.5 + 48 \times 5 \times 10^{-3} - 3 \times 5 \times 10^{3})$ mm

$= (5.5 + 0.225) \text{ mm} = 5.725$ mm

24. *(b)* Given, Number of rotations = 5

Linear distance = 0.25 cm

∴Pitch $= \frac{0.25}{5} = 0.05$ cm

and least count $= \frac{\text{Pitch}}{100} = \frac{0.05}{100}$

$LC = 5 \times 10^{-4}$ cm

Thickness of wire

$= 4 \times \frac{0.25}{5} + 30 \times LC$

$= 4 \times 0.05 + 30 \times 5 \times 10^{-4}$

$= 0.20 + 0.0150 = 0.2150$ cm

25. *(c)* A vernier scale division = 29/30 main scale divison

$1 \text{ VSD} = \frac{29}{30} \times 0.5° = \left(\frac{29}{60}\right)°$

Thus, least count = 1 MSD − 1 VSD

$= \left(\frac{1}{2}\right)° - \left(\frac{29}{60}\right)° = \left(\frac{1}{60}\right)°$

∴Reading = Main scale reading + Vernier scale reading

$= \text{MSR} + n \times LC$

$= 58.5° + 9 \times (1/60°)$

$= 58.65°$

26. *(c)* Given,

Number of circular scale divisions $=100$

Pitch $=1$ mm

So, least count of screw gauge

$$=\frac{\text{pitch}}{\text{(Number of circular scale divisions)}}$$

$$=\frac{1}{100}=0.01 \text{ mm}$$

Also given zero error is, as shown,

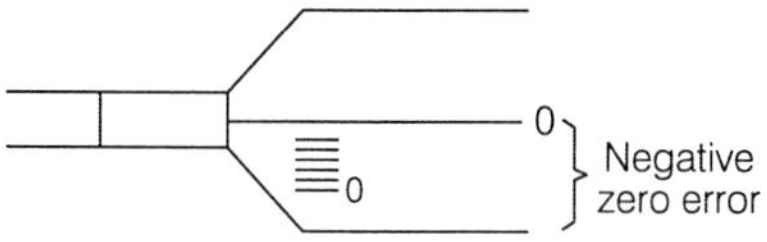

$\therefore$ Zero error $=-5\times \text{LC}=-5\times 0.01$

$=-0.05$ mm

Now while measuring diameter of wire,

Main scale reading $=4\times 1$ mm $=4$ mm

Vernier scale reading $=60$ division

$=60\times 0.01=0.6$ mm

$\therefore$ Measured diameter

$=\text{MSR}+(\text{VSR}\times \text{LC})-\text{zero error}$

$(4+0.6-0.05)$ mm $=4.55$ mm

27. *(d)* Given, $9\text{MSD}=10\text{ VSD}$

and $1\text{ MSD}=1$ mm

So, $1\text{ VSD}=\frac{9}{10}\text{ MSD}=\frac{9}{10}$ mm

$\therefore$Least count

$=1\text{MSD}-1\text{ VSD}$

$=1-\frac{9}{10}\text{mm}=\frac{1}{10}=0.1$ mm

Now, Diameter of sphere

$=\text{MSR}+(\text{VSR}\times \text{LC})$

$=2\text{ cm}+(2\times 0.1\text{ mm})$

$=(20+0.2)\text{ mm}=20.2$ mm

$\therefore$Radius of sphere $=10.1$ mm

Volume of sphere

$=\frac{4}{3}\pi(r)^3=\frac{4}{3}\pi\times(10.1)^3$

$=4315.71\text{ mm}^3=4.31571\text{ cm}^3$

$=4.3\text{ cm}^3$ (upto two significant figures)

Mass of sphere $=8.635$ g $=8.6$ g (given)

(correct upto two-significant figures)

So, density of sphere $=\frac{\text{Mass}}{\text{Volume}}$

$$=\frac{8.6}{4.3}=2\text{ g/cm}^3$$

28. *(c)* Given, least count of main scale $=1$ main scale division (MSD) $=1$ mm

Also, 10 vernier scale divisions $=9$ main scale divisions

$\Rightarrow$ 1 vernier scale division (VSD)

$=\frac{9}{10}\text{ MSD}=0.9\text{ MSD}=0.9$ mm

Vernier constant or least count of instrument is, $\text{LC}=1\text{ MSD}-1\text{ VSD}$

$=1\text{MSD}-\frac{9}{10}\text{MSD}$

$=\left(\frac{10-9}{10}\right)\text{MSD}=\frac{1}{10}\text{MSD}$

$=1\text{ mm}/10=0.1$ mm or 0.01 cm

The zero of vernier scale lies to the right of main scale. So, error is positive.

Zero error $=7\times \text{LC}=7\times 0.01$ cm

$=0.07$ cm

Given, main scale reading $=3.1$ cm

Vernier scale reading $=4$ divisions

$\therefore$Observed value

$=\text{MSR}+\text{VSR}\times \text{LC}-\text{Zero error}$

$=3.1+4\times 0.01-0.07$

$=3.14\text{ cm}-0.07\text{ cm}=3.07$ cm

29. *(3)* Least count of screw gauge

$=\frac{0.75}{15}=0.05$ mm

Now, area of rectangular sheet $=l\times b$

$$\frac{\Delta S}{S}=\frac{\Delta l}{l}+\frac{\Delta b}{b}$$

$$\frac{\Delta S}{S}=\frac{0.05}{5}+\frac{0.05}{2.5}=\frac{1}{100}+\frac{2}{100}=\frac{3}{100}$$

So, value of $x=3$

30. *(35)* Given, least count of screw gauge $=0.01$ mm

Least count is given by

$$\text{LC}=\frac{\text{Pitch}}{\text{Number of circular division}}$$

$$=\frac{P}{N}=0.01\text{ mm}$$

New pitch $=\frac{P(1+0.75)}{N(1-0.5)}$

$=\frac{P}{N}\left(\frac{1\cdot 75}{0\cdot 5}\right)$

$=(0\cdot 01)3\cdot 5$

$=35\times 10^{-3}$ mm

31. *(220)* Now, Least count

$$=\frac{\text{Pitch}}{\text{Number of circular scale division}}$$

$$=\frac{0.5}{100}=0.005\text{ mm}$$

Zero error, $Ze=6\times 0.005$ mm

$=0.03$ mm

Main scale reading $=4\times 0.5$ mm $=2$ mm

Circular scale reading

$=46\times 0.005$ mm $=0.23$ mm

So, diameter of wire

$=\text{MSR}+(\text{CSR}\times \text{LC})-Z_e$

$=4\times 0.5+46\times 0.005-0.03$

$=2.2\text{ mm}=220\times 10^{-2}$ mm

32. *(412)* Given, $10\text{ VSD}=9\text{ MSD}$

$\Rightarrow 1\text{ VSD}=\frac{9}{10}\text{MSD}$

Least count of Vernier scale

$\text{LC}=1\text{MSD}-1\text{VSD}$

$=1\text{ MSD}-\frac{9}{10}\text{MSD}=\frac{\text{MSD}}{10}=\frac{1}{10}$ mm

($\because$ 1 MSD = 1 mm)

According to given situation of Vernier callipers, Positive error = MSD + LC × Number of coinciding MSD

$=0+\frac{1}{10}\times 4=0.4$mm

$\therefore$Diameter of the bob

$=41+6\times\frac{1}{10}-0.4=41+0.6-0.4$

$=4.12\text{ cm}=412\times 10^{-2}$ cm

33. *(180)*

> **key idea** To find the diameter of a sphere using vernier callipers, you can use following formula.
>
> Diameter of sphere = MSR + (VSR × LC) zero error
>
> (Use the value of zero error with their sign)

Corrected reading of a vernier callipers i.e. observed value

$=\text{MSR}+(\text{LC}\times \text{VSR})-(\text{Zero error})$

$=1.7+\frac{0.1}{10}\times 5-(-0.05)$ cm

$=1.7+0.05+0.05=1.8$ cm

$=180\times 10^{-2}$ cm

34. *(150)* Mean thickness

$$=\frac{t_1+t_2+t_3+t_4}{4}$$

$$=\frac{1.22+1.23+1.19+1.20}{4}$$

$t_{\text{mean}}=\bar{t}=1.21$

Errors in measurement are $e=|\bar{t}-t_i|$

$\Rightarrow$ $e_1=|1.21-1.22|=0.01$

Similarly,

$e_2=0.02\ e_3=0.02$ and $e_4=0.01$

Mean error $=\bar{e}=\frac{e_1+e_2+e_3+e_4}{4}$

-0.015

Percentage error $=\frac{e_{\text{mean}}}{t_{\text{mean}}}\times 100$

$\frac{0.015}{1.21}\times 100=\frac{150}{121}\%=\frac{x}{121}\%$ [Given]

$\therefore$ $x=150$

35. *(5)* Given, the 1 cm in main scale is divided into 20 equal parts.

1 main scale division,

$\text{MSD}=1/20\text{ cm}=0.05$ cm

Given, 9th main scale coincide with 10th vernier scale,

$\therefore$ VC $= \left(1 - \frac{9}{10}\right) \times 0.05$

$= 5 \times 10^{-3}$ cm $= 5 \times 10^{-2}$ mm

36. *(52)* 9 MSD = 10 VSD

$\therefore 1$ VSD $= \frac{9}{10}$ MSD $= \frac{9}{10} \times 1$ mm

$= 9/10$ mm

MSR = 10 mm, $n = 8$

Zero error = 0.04 cm = 0.4 mm

The least count is given by

LC = 1 MSD − 1 VSD

$= 1 - \frac{9}{10} = \frac{1}{10}$ mm $= 0.1$ mm

Reading of diameter,

$d = 10 + n \times \text{LC} - \text{Zero error}$

$d = 10 + 8 \times 0.1 - 0.4 = 10.4$ mm

Radius, $r = \frac{d}{2} = 5.2$ mm $= 52 \times 10^{-2}$ cm

37. *(13)* **For student A** Pitch = 0.1 cm

$\text{LC} = \frac{0.1}{100} = 0.001$

Number of divisions on circular scale, $N = 100$

The reading is given by

Reading = MSR + CSR + Zero error

$0.322 = 0.3 + \text{CSR} + 5 \times 0.001$

$\text{CSR} = 0.017$ ...(i)

For student B $0.322 = 0.2 + \text{CSR} + 0.092$

$\text{CSR} = 0.030$

Difference $= 0.030 - 0.017 = 0.013$ cm

Division in circular scale $= \frac{0.013}{0.001} = 13$

38. *(25)* Given, the radius of the spherical drop

$r = \left(\frac{3}{40\pi}\right)^{1/3} \times 10^{-3}$ cm

The volume of 100 spherical drops,

$V = 100 \times \frac{4\pi r^3}{3}$

On substituting the value of the radius in the above equation, we get

$V = 100 \times \frac{4\pi}{3} \times \frac{3}{40\pi} \times 10^{-9}$

$\Rightarrow V = 10^{-8}$ cm^3

Now, Volume of the film = Area of the film × Thickness of the film

⇒ Volume of the film = $4 \times t$

The film is made up of 100 spherical drops.

∴ The volume of the film = Volume of 100 spherical drops

$4t = 10^{-8} \Rightarrow t = 25 \times 10^{-12}$ m

The thickness of the oleic layer,

$t_0 = 0.01 \times t = 25 \times 10^{-14}$ m

On comparing with $x \times 10^{-14}$, we get $x = 25$.

Topic 2

01. *(c)* We know that,

$$T = 2\pi\sqrt{\frac{m}{k}} \Rightarrow T^2 = 4\pi^2\left(\frac{m}{k}\right)$$

$$\therefore \frac{2\Delta T}{T}\% = \frac{\Delta m}{m}\% + \frac{\Delta k}{k}\%$$

$$2 \times 2\% = (-1\%) + \frac{\Delta k}{k}\% \Rightarrow \frac{\Delta k}{k}\% = 5\%$$

02. *(d)* Given, diameter of the column tube,

$d = 6$ cm $= 6 \times 10^{-2}$ m

Frequency of tuning fork, $f = 504$ Hz

Speed of sound at given temperature,

$v = 336$ ms^{-1}

As, this is a closed organ pipe.

We have,

$L + e = \lambda/4 = v/4f \quad (\because \lambda = v/f)$

and $e = 0.6 \times d/2$

$= \frac{6}{10} \times \frac{6 \times 10^{-2}}{2} = 0.018$

$\Rightarrow L + e = \frac{v}{4f} = \frac{336}{4 \times 504}$

$\Rightarrow L = \frac{336}{2016} - 0.018 = 0.1667 - 0.018$

$= 0.1487$ m $= 14.87$ cm $\simeq 14.8$ cm

03. *(c)* Given, length of pendulum, $l = 25.0$ cm.

So, there is an uncertainty of 0.1 cm in measurement of length.

Resolution of stopwatch is 1s. So, uncertainty in measurement of time is 1s.

Now using, $T = 2\pi\sqrt{\frac{l}{g}}$ or $g = \frac{4\pi^2 l}{T^2}$

We have, $\frac{\Delta g}{g} = \frac{\Delta l}{l} + \frac{2\Delta T}{T}$

$$\Rightarrow \frac{\Delta g}{g} \times 100\% = \left(\frac{\Delta l}{l} + \frac{2\Delta T}{T}\right) \times 100\%$$

Accuracy in measurement of g is

$$\frac{\Delta g}{g} \times 100\% = \left(\frac{0.1}{25} + \frac{2 \times 1}{50}\right) \times 100\%$$

$= (0.004 + 0.04) \times 100\%$

$= 4.40\%$

> **• Concept Enhancer**
>
> For an object placed at a height h or at depth d, the acceleration due to gravity is less as compared to that place on the surface of earth.

04. *(a)* Given, $\Delta T = 5 \times 10^{-4}$ s

We know that

Time period of simple pendulum,

$T = 2\pi\sqrt{\ell/g}$

$T \propto \sqrt{\ell}$

$\therefore \Delta T = \frac{1}{2}\Delta\ell \Rightarrow 5 \times 10^{-4} = \frac{|r_1 - r_2|}{2}$

$|r_1 - r_2| = 10 \times 10^{-4} = 10^{-3}$ m $= 0.1$ cm

05. *(c)* Fundamental frequency in closed organ pipe

$f = \frac{v}{4l}$

$\therefore f_1 = \frac{v}{4l_1}$ and $f_2 = \frac{v}{4l_2}$

$\because$ Beat $= f_1 - f_2$

$15 = \frac{v}{4}\left(\frac{1}{l_1} - \frac{1}{l_2}\right)$

$\Rightarrow 15 = \frac{v}{4}\left[\frac{1}{1} - \frac{1}{1.2}\right]$

$\Rightarrow v = 360$ m/s

06. *(a)* Given,

Time period, $T = 1.95$ s

Length of string, $l = 1$ m

Error in time period, $\Delta T = 0.01$ s $= 10^{-2}$ s

Error in length,

$\Delta L = 1$ mm $= 1 \times 10^{-3}$ m

$\because T = 2\pi\sqrt{L/g}$

Thus, we have,

$\Rightarrow \frac{\Delta g}{g} = \frac{\Delta L}{L} + \frac{2\Delta T}{T}$

$= \frac{10^{-3}}{1} + \frac{2 \times 10^{-2}}{1.95}$

$= 10^{-3} + 1.025 \times 10^{-2}$

$= 10^{-3} + 10.25 \times 10^{-3} = 11.25 \times 10^{-3}$

$\because \Delta g/g \times 100 = 11.25 \times 10^{-3} \times 10^2$

$= 1.125\% \simeq 1.13\%$

07. *(b)* Relation used for finding acceleration due to gravity by using a pendulum is given by

$g = 4\pi^2 l/T^2$

So, percentage error in value of g will be

$$100 \times \frac{\Delta g}{g} = \left[\frac{\Delta l}{l} + \frac{2\Delta T}{T}\right] \times 100 \quad \text{...(i)}$$

Given, $\Delta l = 0.1$ cm, $l = 55$ cm, $\Delta T = 1$ s and T for 20 oscillations = 30 s

$$100 \times \frac{\Delta g}{g} = \left[\frac{0.1}{55} + 2 \times \frac{1}{30}\right] \times 100$$

Hence, percentage error in g will be

$= \frac{10}{55} + \frac{20}{3} = 6.8\%$

08. *(500)* $f = \frac{1}{2l}\sqrt{\frac{T}{\mu}}$

$\therefore \quad f_1 = \frac{1}{2l}\sqrt{\frac{T_1}{\mu}}$ and $f_2 = \frac{1}{2l}\sqrt{\frac{T_2}{\mu}}$

Hence, $\frac{f_2}{f_1} = \frac{\frac{1}{2l}\sqrt{\frac{T_2}{\mu}}}{\frac{1}{2l}\sqrt{\frac{T_1}{\mu}}} \Rightarrow \frac{f_2}{f_1} = \sqrt{\frac{T_2}{T_1}}$

$\Rightarrow \left(\frac{50}{30}\right)^2 = \frac{mg}{180g}$

$\Rightarrow \frac{2500}{900} = \frac{m}{180}$

$\Rightarrow \quad m = \frac{25 \times 180}{9} = 500 \text{ g}$

$\Rightarrow \quad m = 500 \text{ g}$

09. *(5)*

> **key idea** To find out percentage error, you can use following formula.
> $\frac{\Delta g}{g} \times 100 = \frac{\Delta L}{L} \times 100 + \frac{2\Delta T}{nT} \times 100$

Given, time period of pendulum,
$t = 0.5$ s
Number of oscillations, $n = 100$
Resolution of watch, $\Delta T = 1$ s
Accuracy of length, $\frac{\Delta l}{l} = \frac{1 \text{ mm}}{10 \text{ cm}} = \frac{1}{100}$
Since, time period,
$T = 2\pi\sqrt{l/g}$
$\Rightarrow \quad g = 4\pi^2 l/T^2$
The accuracy in determination of g,
$\frac{\Delta g}{g} \times 100 = \frac{2\Delta T}{T} \times 100 + \frac{\Delta l}{l} \times 100$
$= \left(2 \times \frac{1}{(100 \times 0.5)} \times 100 + \frac{1}{100} \times 100\right)\%$
$= 4\% + 1\% = x\ \%$ [Given]
$\therefore \quad x = 5$

> **• Concept Enhancer**
> All bodies experience the same acceleration due to gravity irrespective of its mass. The value of acceleration due to gravity depend an mass of earth but not mass of object.

10. *(1)* The time period of simple pendulum is given by $T = 2\pi\sqrt{L/g}$

Also, $T^2 = 4\pi^2 \frac{L}{g}$, $g = 4\pi^2 \frac{L}{T^2}$

Percentage error,
$\frac{\Delta g}{g} \times 100 = \frac{\Delta L}{L} \times 100 + \frac{2\Delta T}{nT} \times 100$
ΔL and ΔT is same for all so, $\Delta g/g$ is minimum for highest value of L, n and T. So, the minimum percentage error is obtained by student 1.

Topic 3

01. *(d)* $Y = \frac{MgL^3}{4bd^3\delta}$, $g = 9.8 \text{ m/s}^2$ (constant)

Use error analysis, we get
$\frac{\Delta Y}{Y} = \frac{\Delta M}{M} + \frac{3\Delta L}{L} + \frac{\Delta b}{b} + \frac{3\Delta d}{d} + \frac{\Delta \delta}{\delta}$
$= \frac{1}{2000} + \frac{3}{1000} + \frac{0.1}{40} + \frac{0.03}{4} + \frac{0.01}{5}$
$= 0.0155$

02. *(a)* Given,
$r = 0.2$ cm, $\Delta r = 0.001$ cm
$L = 1$ m, $\Delta L = 0.001$ m
$m = 1$ kg
$\Delta m = 1 \text{ g} = 0.001$ kg
$l = 0.5$ cm, $\Delta l = 0.001$ cm
Young's modulus is given by
$Y = \frac{F/A}{l/L} = \frac{Mg/\pi r^2}{l/L} = \frac{MgL}{\pi r^2 L}$
Taking log and differentiating, we get
$\frac{\Delta Y}{Y} = \frac{\Delta m}{m} + \frac{\Delta L}{L} - \frac{2\Delta r}{r} - \frac{\Delta l}{l}$
[g is constant]
Error in Y, $\frac{\Delta Y}{Y} = \frac{\Delta m}{m} + \frac{2\Delta r}{r} + \frac{\Delta l}{l} + \frac{\Delta L}{L}$
$= \frac{0.001}{1} + \frac{2 \times 0.001}{0.2} + \frac{0.001}{0.5} + \frac{0.001}{1}$
$= 0.001 + 0.01 + 0.002 + 0.001 = 0.014$
Percentage error
$\frac{\Delta Y}{Y} \times 100 = 0.014 \times 100 = 1.4\ \%$

03. *(b)* Surface tension is given by
$T = \frac{rhg}{2} \times 10^3 \text{ N m}^{-1}$
$= \frac{Dhg}{4} \times 10^3 \text{ N m}^{-1}$
Possible error in the surface tension is
$\frac{\Delta T}{T} \times 100 = \frac{\Delta D}{D} \times 100 + \frac{\Delta h}{h} \times 100 + 0$
$= \left(\frac{0.01 \times 10^{-2}}{1.25 \times 10^{-2}} + \frac{0.01 \times 10^{-2}}{1.45 \times 10^{-2}}\right) \times 100$
$\frac{\Delta T}{T} \times 100 = \left(\frac{100}{125} + \frac{100}{145}\right)$
$\frac{\Delta T}{T} \times 100 = 0.8 + 0.689 = 1.489 \approx 1.5\ \%$

04. *(d)* In first case, the extension in the wire is
$x = Mg/k = V\rho g/k$ ...(i)
When the load is immersed in the liquid, then
$\Rightarrow V\sigma g + kx_1 = V\rho g$
$\Rightarrow \quad x_1 = Vg(\rho - \sigma)/k$ ...(ii)

Using Eqs. (i) and (ii), we get
$x_1 = \frac{Vgx}{Vg\rho}(\rho - \sigma) = x\left(1 - \frac{\sigma}{\rho}\right) = 4 \times \left(1 - \frac{2}{8}\right)$
$= 3$ mm

05. *(*)* The Young's modulus of two wires connected in series is
$Y = \frac{2Y_1Y_2}{Y_1 + Y_2} = \frac{2(120 \times 10^9)(60 \times 10^9)}{180 \times 10^9}$
$= 80 \times 10^9 \text{ N/m}^{-2}$
So, the stress required $= Y \times$ strain
$= 80 \times 10^9 \times \frac{0.2 \times 10^{-3}}{2}$
$= 8.0 \times 10^6 \text{ N/m}^2$
None of the given options is correct.

> ***Alternate Solution***
> Given, Total elongation $= 0.2 \text{ mm} = 0.2 \times 10^{-3}$ m
> We know that,
> Total elongation $= \Delta\ell_{\text{brass}} + \Delta\ell_{\text{steel}}$
> $0.2 \times 10^{-3} = \frac{F\ell_{\text{brass}}}{A\,Y_{\text{brass}}} + \frac{F\ell_{\text{steel}}}{AY_{\text{steel}}}$
> $\therefore \; 0.2 \times 10^{-3} = \frac{F}{A}\left(\frac{Y_{\text{brass}} + Y_{\text{steel}}}{Y_{\text{brass}}Y_{\text{steel}}}\right)$
> $\frac{F}{A} = \left(\frac{Y_{\text{brass}}Y_{\text{steel}}}{Y_{\text{brass}} + Y_{\text{steel}}}\right) \times 0.2 \times 10^{-3}$
> $= \frac{60 \times 10^9 \times 120 \times 10^9}{180 \times 10^9} \times 0.2 \times 10^{-3}$
> $= 8 \times 10^6 \text{ N/m}^2$.

06. *(2)* Slope $= \frac{\text{Extension/Load}}{\text{Length of wire}} = \frac{\Delta l/w}{L}$
Young's modulus,
$Y = \frac{\text{mg}/A}{\Delta l/L} = \frac{wL}{\Delta lA}$
$\therefore \quad Y = \frac{1}{(\text{slope})A}$
$\Rightarrow \quad Y = \frac{1}{2 \times 10^{-6}(0.25 \times 10^{-5})}$
$\Rightarrow \quad Y = 2 \times 10^{11} \text{ Nm}^{-2}$

07. *(2)* Given, $L = 1$ m
$l = 0.4$ mm
$= 0.4 \times 10^{-3}$ m
$d = 0.4$ mm
Extension of the length,

$\Delta l = \pm 0.02$ mm

Uncertainty in diameter of wire,

$\Delta d = \pm 0.01$ mm

Young's modulus, $Y = \dfrac{\text{stress}}{\text{strain}} = \dfrac{\frac{mg}{A}}{\frac{\Delta l}{L}}$

$$= \frac{mg \times L}{A \times \Delta l} = \frac{mgL}{\pi\left(\frac{d}{2}\right)^2 \Delta l} = \frac{4mgL}{\pi(d)^2 \Delta l} \quad \ldots\text{(i)}$$

After putting the given values in Eq. (i), we will get

$Y = 2 \times 10^{11}\ \text{Nm}^{-2}$

$$\frac{\Delta Y}{Y} = 2\frac{\Delta d}{d} + \frac{\Delta l}{l}$$

$$\frac{\Delta Y}{Y} = 2 \times \frac{0.01}{0.4} + \frac{0.02}{0.4}$$

$\Delta Y = 2 \times 10^{10}$

Topic 4

01. *(a)* To verify Ohm's law, one galvanometer is used as ammeter and other galvanometer as voltmeter. Hence, we should connect ammeter in series and voltmeter in parallel to measure current and potential difference, respectively. Also, when we use a galvanometer to create an ammeter, shunt resistance should be very small and should be in parallel. When we use as voltmeter series resistance should be large. All these criteria are satisfied in option (a).

02. *(a)* Given, $V = 5$ V

The length of the conductor, $L = 10$ cm

The measured value of the current in the conductor, $I = 2$ A

The diameter of the conductor,

$d = 5$ mm

As we know that

$R = \rho l / A$

Using Ohm's law,

$V = IR \Rightarrow R = V/I$

$$\frac{\rho l}{A} = \frac{V}{I} \Rightarrow \rho = \frac{V}{Il}\left(\frac{\pi d^2}{4}\right)$$

In error form,

$$\frac{\Delta\rho}{\rho} = \frac{\Delta V}{V} + \frac{\Delta I}{I} + \frac{\Delta l}{l} + 2\frac{\Delta d}{d}$$

$$\Rightarrow \frac{\Delta\rho}{\rho} = \frac{0.1}{5} + \frac{0.01}{2} + \frac{0.1}{10} + 2\frac{(0.01)}{(5)}$$

$$\Rightarrow \frac{\Delta\rho}{\rho} = 0.039 \Rightarrow \frac{\Delta\rho}{\rho} \times 100 = 3.9\%$$

03. *(c)* Length of $AC = 250$ cm

New balancing length $= 400$ cm

When, point (1) is connected

$\varepsilon_1 \propto \ell_1,$

where k is a constant.

When, point (2) is connected.

$\varepsilon_1 + \varepsilon_2 \propto \ell_2$

Using Eqs. (i) and (ii), we get

$$\frac{\varepsilon_1}{\varepsilon_1 + \varepsilon_2} = \frac{l_1}{l_2} = \frac{250}{400} = \frac{5}{8}$$

$8\varepsilon_1 = 5\varepsilon_1 + 5\varepsilon_2$

$$\therefore \quad \frac{\varepsilon_1}{\varepsilon_2} = \frac{5}{3}$$

04. *(c)* If a capacitor is put between probes of a multimeter, then during charging, a current flows in circuit, so there is a deflection which becomes soon zero.

$\therefore$Statement in option (c) is not correct.

> • Concept **Enhancer**
>
> A volt-ohm-milli ammeter (VOM) is a multimeter that can measure voltage, resistance and current. It is capable of measuring a wide range of electrical properties.

05. *(b)* The given situation is as shown below.

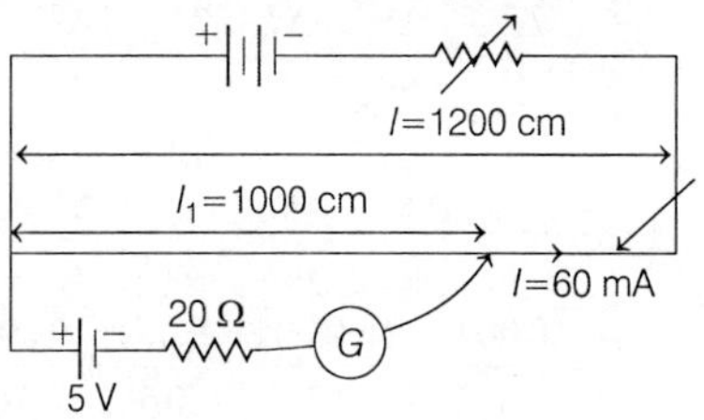

At null point, a length of 1000 cm balances emf of 5 V.

So, potential drop across 1000 cm of wire is equal to emf of cell,

$E = l_1 \cdot x \cdot I$

$\Rightarrow \quad 5 = 1000 \times x \times 60 \times 10^{-3}$

where, x = resistance per centimetre length of potentiometer wire.

$$\Rightarrow x = \frac{5}{1000 \times 60 \times 10^{-3}} \text{ or } x = \frac{1}{12}\,\Omega\ \text{cm}^{-1}$$

So, resistance of 1200 cm length of wire,

$$R = l \cdot x = 1200 \times \frac{1}{12} = 100\,\Omega$$

06. *(a)* Given,

$$R_a = 30\,\Omega,\ R_m = R_a - \frac{5}{100} \times R_a = 0.95\,R_a$$

As, $\quad 0.95R_a = \dfrac{R_a r_v}{(R_a + R_v)}$

$0.95R_a + 0.95R_v = r_v$

$0.95\,R_a = 0.05\,r_v \Rightarrow r_v = 19R_a = 19 \times 30$

$= 570\,\Omega$

07. *(b)* Initially, $I = \dfrac{\varepsilon}{R + G}$

$\therefore \quad G = \dfrac{1}{9}\,k\Omega$

Finally, $\dfrac{I}{2} = \dfrac{\varepsilon}{R + \dfrac{GS}{G+S}} \times \dfrac{S}{S+G}$

$$\Rightarrow \quad \frac{I}{2} = \frac{\varepsilon S}{R(S+G) + GS}$$

$$S = \frac{RG \times \frac{I}{2}}{\varepsilon - \frac{(R+G)I}{2}}$$

$$= \frac{11 \times 10^3 \times \frac{1}{9} \times 10^3 \times 270 \times 10^{-6}}{6 - \left(\frac{6}{2}\right)}$$

$\Rightarrow \quad S = 110\,\Omega$

08. *(b)* Without shunting condition On balancing, $\varepsilon_s = 52 \times k \quad \ldots$(i)

With shunting condition On balancing,

$$\varepsilon_s - \frac{\varepsilon_s}{(r+R)} r = 40 \times k$$

From Eqs. (i) and (ii), we get

$$\frac{1}{1 - \frac{r}{r+R}} = \frac{52}{40} \Rightarrow \frac{r+R}{R} = \frac{52}{40}$$

$$r = \left(\frac{52}{40} \times R\right) - R = \left(\frac{52}{40} \times 5\right) - 5$$

$\therefore \quad r = 1.5\,\Omega$

09. *(b)* A galvanometer as an ammeter is shown below

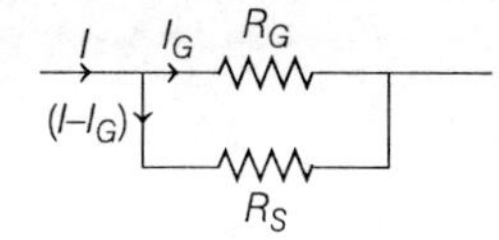

By KVL in closed loop,

$(I - I_G)R_S = I_G \cdot R_G$

$\Rightarrow \quad IR_S - I_G R_S = I_G R_G$

$\Rightarrow \quad IR_S = I_G R_S + I_G R_G$

$\Rightarrow \quad IR_S = I_G (R_S + R_G)$

$$\Rightarrow \quad \frac{I_G}{I} = \frac{R_S}{R_S + R_G}$$

According to given situation,

$$\frac{I_G}{I} = \frac{1}{3} \quad \therefore \quad \frac{1}{3} = \frac{R_S}{R_S + R_G}$$

$\Rightarrow \quad R_S + R_G = 3R_S \Rightarrow R_G = 2R_S$

10. *(5)*

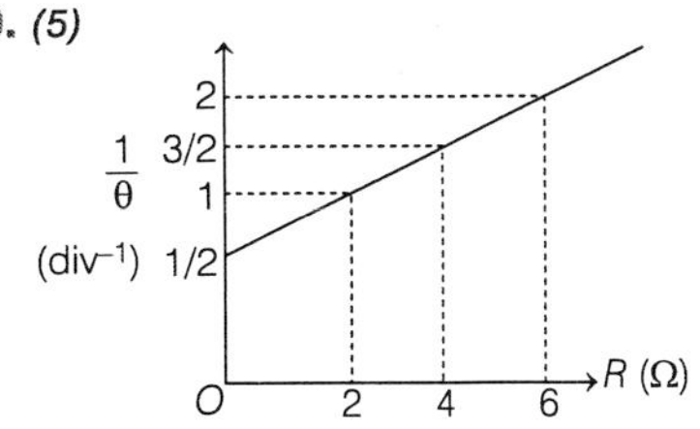

For galvanometer, $i = K\theta$

$$\frac{2}{G+R} = K\theta$$

[where i = current, $V = 2$ V, G = galvanometer resistance, θ = deflection]

$$\frac{1}{\theta} = \frac{(G+R)K}{2} = R\left(\frac{K}{2}\right) + \frac{KG}{2}$$

$$\frac{1}{\theta} = R\left[\frac{K}{2}\right] + \frac{KG}{2} \quad \ldots\text{(i)}$$

This is a straight line equation with intercept on Y-axis.

$$\text{Slope} = \frac{K}{2} \quad \ldots\text{(ii)}$$

$$\text{From graph slope} = \frac{\frac{3}{2}-1}{4-2} = \frac{1}{4} \quad \ldots\text{(iii)}$$

$$\frac{K}{2} = \frac{1}{4}$$

$$K = 0.5 = 5 \times 10^{-1} \text{ A}$$

Hence, figure of merit $= 5 \times 10^{-1}$ A.

11. *(25)* The emf E is given as

$$E = xl \quad \ldots\text{(i)}$$

In first situation,

$$E = 1.5 \text{ V}, l = 60 \text{ cm}$$

From Eq. (i),

$$1.5 = x \times 60 \Rightarrow x = \frac{1.5}{60}$$

In second situation,

$$E = x(60+40)$$

$$E = \frac{1.5}{60} \times 100 = \frac{25}{10} \text{ V}$$

On comparing with, $\frac{x}{10}$ volt, we get

$$x = 25$$

12. *(12)* In experiment of finding internal resistance,

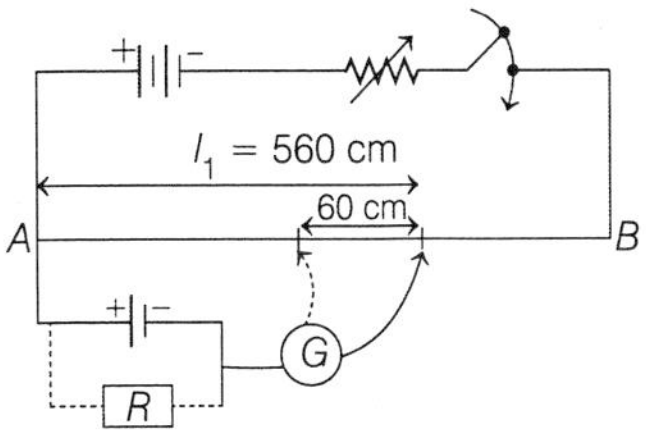

Balancing length with only cell,

$$l_1 = 560 \text{ cm}.$$

Balancing length with cell and resistance, $l_2 = 560 \text{ cm} - 60 \text{ cm}$

$$= 500 \text{ cm}$$

External resistance, $R = 10\,\Omega$

Now using

$$r = R\left(\frac{E}{V} - 1\right) = R\left(\frac{l_1}{l_2} - 1\right), \text{ we get}$$

$$r = 10\left(\frac{560}{500} - 1\right) = \frac{6}{5}\Omega$$

Now given, $r = \frac{N}{10} \Rightarrow \frac{6}{5} = \frac{N}{10}$

$$\Rightarrow \quad N = 60/5 = 12$$

Topic 5

01. *(a)* 300 MSD = 15 cm

$$1 \text{ MSD} = \frac{15}{300} \text{ cm}$$

$$= 0.05 \text{ cm}$$

25 VSD = 24 MSD

$$1 \text{ VSD} = \frac{24}{25} \text{ MSD}$$

L.C. = 1 MSD − 1 VSD

$$= \text{MSD} - \frac{24}{25} \text{MSD}$$

$$= \frac{1}{25} \text{MSD}$$

$$\text{L.C} = \frac{1}{25} \times 0.05 = 0.002 \text{ cm}$$

02. *(d)* Spherometer can't be used for measurement of specific rotation of liquid.

Spherometer is used to measure radius of curvature of any spherical surface and any small thickness.

03. *(c)* Travelling microscope has 20 division per cm on main scale.

$$1 \text{ MSD} = \frac{1}{20} \text{ cm}$$

Least count = 1 MSD − 1 VSD

$$= 1 \text{ MSD} - \frac{24}{25} \text{MSD}$$

$$\Rightarrow \frac{1}{25} \text{MSD} = \frac{1}{25} \times \frac{1}{20} \text{ cm}$$

LC = 0.002 cm

04. *(a)* If an experiment is performed to find the refractive index of glass using a travelling microscope, distances are measured by a vernier scale which is provided on the microscope.

05. *(a)* Using lens formula,

$$\frac{1}{f} = \frac{1}{v} - \frac{1}{u}$$

Taking 1st derivative

$$\frac{df}{f^2} = \frac{dv}{v^2} - \frac{du}{u^2}$$

For small change and maximum percentage error

$$df \to \Delta \cdot f$$

$$\frac{\Delta f}{f^2} = \frac{\Delta u}{u^2} + \frac{\Delta v}{v^2}$$

$$\Delta f = f^2\left[\frac{\Delta u}{u^2} + \frac{\Delta v}{v^2}\right]$$

06. *(b)* Given, 50 vernier scale divisions = 49 main scale divisions.

Also, 1 main scale division = 0.5 mm

Now, 1 vernier scale division $= \frac{49}{50}$ MSD

$$= \frac{49}{50} \times 0.5 \text{ mm} = 0.49 \text{ mm}$$

Least count of instrument is

LC = 1 MSD − 1 VSD

= 0.5 − 0.49 (mm)

= 0.01 mm

∴ Vernier constant = Least count

= 0.01 mm

07. *(a)* Here, $u = -20$ cm, $f_1 = 15$ cm

As we know,

$$\frac{1}{v} - \frac{1}{u} = \frac{1}{f} \Rightarrow \frac{1}{v} - \frac{1}{-20} = \frac{1}{15}$$

$$\Rightarrow \quad \frac{1}{v} = \frac{1}{15} - \frac{1}{20} = \frac{1}{60}$$

$$\therefore \quad v = 60 \text{ cm}$$

To obtain its real image on object's place. The center of curvature of the convex mirror must be at the position of virtual image, i.e, at 60 cm

Therefore, $5 + 2f_R = 60$

$\Rightarrow 2f_R = 55 \Rightarrow f_R = 27.5$ cm

08. *(b)* $1 \text{ MSD} = \frac{1 \text{ cm}}{20} = 0.05 \text{ cm}$

$$1 \text{ VSD} = \frac{49}{50} \text{MSD} = \frac{49}{50} \times 0.05 \text{ cm}$$

$$= 0.049 \text{ cm}$$

LC = 1 MSD − 1 VSD = 0.001 cm

For mark on paper,

$$L_1 = 8.45 + 26 \times 0.001 \text{ cm}$$

$$= 84.76 \text{ mm}$$

For mark on paper through slab,

$$L_2 = (7.12 + 41 \times 0.001) \text{ cm}$$

$$\Rightarrow \quad L_2 = 71.61 \text{ cm}$$

For the powder on the top of surface.

Zero error = 4.05 cm + 1 × 0.001 cm

= 40.51 mm

∴ Actual, $L_1 = 84.76 - 40.51$

= 44.25 mm

Actual, $L_2 = 71.61 - 40.51$

= 31.10 mm

$\therefore \quad L_2 = \dfrac{L_1}{\mu}$

$\Rightarrow \quad \mu = \dfrac{L_1}{L_2} = \dfrac{44.25}{31.10} = 1.42$

09. *(b)* Given, $u = 100 - 80 = 20$ cm

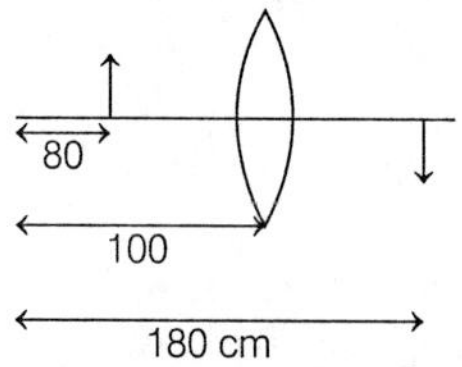

$v = 180 - 100 = 80$ cm

Using, $\dfrac{1}{f} = \dfrac{1}{v} + \dfrac{1}{u}$ or

$f = \dfrac{uv}{u+v} = \dfrac{20 \times 80}{20 + 80}$

$\Rightarrow \quad f = 16$ cm

For error analysis, $\dfrac{1}{f} = \dfrac{1}{v} + \dfrac{1}{u}$

differentiate both sides

$-\dfrac{df}{f^2} = -\dfrac{dv}{v^2} - \dfrac{\Delta u}{u^2}$

to calculate Δu and Δv

$u = (100 \pm 2) - (80 \pm 0.2) = 20 \pm 0.4$ cm

therefore $\Delta u = 0.4$ cm

$\Delta v = 0.4$ cm

Now, $\dfrac{\Delta f}{f} = f\left[\dfrac{\Delta v}{v^2} + \dfrac{\Delta u}{u^2}\right]$

$= 16\left[\dfrac{0.4}{(80)^2} + \dfrac{0.4}{(20)^2}\right]$

$\dfrac{\Delta f}{f} = \dfrac{16 \times 0.4}{(20)^2}\left[\dfrac{1}{16} + 1\right]$

$= \dfrac{16 \times 0.4}{(20)^2} \times \dfrac{17}{16}$

$= \dfrac{17 \times 0.4}{4\text{N}}$

Percentage error

$= \dfrac{\Delta f}{f} \times 100 = \dfrac{17 \times 0.4}{400} \times 100$

$= 1.70$

Option (b) is correct.

10. *(20)* If we draw $y = x$ line on graph, then it cuts the curve at point (20 cm, 20 cm).

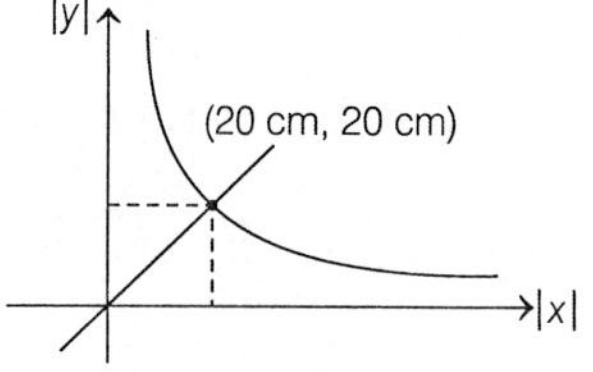

From the Newton's formula

$xy = f^2$

$\Rightarrow \quad 20 \times 20 = f^2 \Rightarrow f = 20$ cm

11. *(32)* As $\dfrac{1}{f} = \dfrac{1}{v} + \dfrac{1}{u}$

$\Rightarrow \quad \dfrac{-1}{f^2} \cdot \dfrac{df}{dx} = \dfrac{-1}{v^2} \cdot \dfrac{dv}{dx} - \dfrac{1}{u^2} \cdot \dfrac{du}{dx}$

$\Rightarrow \quad \dfrac{\Delta f}{f^2} = \dfrac{\Delta v}{v^2} + \dfrac{\Delta u}{u^2}$

$\Rightarrow \quad \dfrac{\Delta f}{f} = \text{Error in } f = f\left(\dfrac{\Delta v}{v^2} + \dfrac{\Delta x}{x^2}\right)$ ...(i)

Now, given,

$u = -40$ cm, $v = -120$ cm

$\therefore \quad \dfrac{1}{f} = \dfrac{1}{-120} + \dfrac{1}{-40} = -\left(\dfrac{1+3}{120}\right)$

$\Rightarrow \quad \dfrac{1}{f} = \dfrac{-1}{30}$

or $\quad f = -30$ cm

Least count of given scale is measure of 1 division $= \dfrac{1}{20}$ cm $= 0.05$ cm

So, error in measurement of focal length (from Eq. (i) is

$\Delta f = f^2\left(\dfrac{\Delta v}{v^2} + \dfrac{\Delta x}{x^2}\right)$

$\Rightarrow \quad \Delta f = 900\left(\dfrac{0.05}{(120)^2} + \dfrac{0.05}{(40)^2}\right)$

$= \dfrac{9 \times 5}{40 \times 40}\left(\dfrac{1}{3 \times 3} + 1\right)$

$= \dfrac{9 \times 5}{40 \times 40} \times \dfrac{10}{9} = \dfrac{1}{32}$

So, $\quad K = 32$

12. *(5)* Given, 50 VSD = 49 MSD

$\Rightarrow \quad 1 \text{ VSD} = \dfrac{49}{50}\text{MSD}$ ...(i)

$\therefore$ Least count = 1MSD − 1VSD

$= 1\text{MSD} - \dfrac{49}{50}\text{MSD}$

[From Eq. (i)]

$= \text{MSD}\left(1 - \dfrac{49}{50}\right)$

$\Rightarrow$ Least count $= \dfrac{1}{50}$MSD

But, 1 MSD $= \dfrac{1}{40}$ cm

$\therefore$ Least count $= \dfrac{1}{50} \times \dfrac{1}{40} = 5 \times 10^{-6}$ m

Topic 6

01. *(b)* Given power, $P = 1$ W

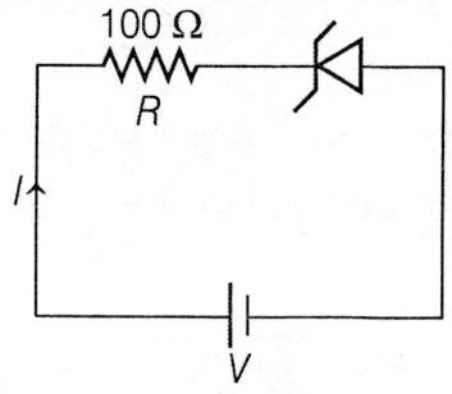

Potential drop across zener diode is,

$V_Z = V - IR = V - 100\,I$

Power, $\quad P = V_Z I_Z = (V - 100I)I$

$\therefore \quad (V - 100I)I = 1$

or $100I^2 - VI + 1 = 0$

For I to be real,

$V^2 - 4 \times 100 \times 1 \geq 0$

or $\quad V \geq 20$ V

02. *(440)* In a transformer,

$N_p / N_s = V_p / V_s$

where, N_p = number of turns in primary circuit,

N_s = number of turns in secondary circuit = 24,

V_p = potential of primary circuit = 220 V

and V_s = potential of secondary circuit = 12 V

$\Rightarrow \quad \dfrac{N_p}{24} = \dfrac{220}{12} \quad \Rightarrow \quad N_p = 440$